BIOLOGY
PRINCIPLES & EXPLORATIONS

Now with Holt BioSources!
There's no other program like it.

No other program is as easily adapted to your diverse classroom situations.
No other program meets such a wide range of students' needs.

Biology: Principles and Explorations offers you:

- complete, authoritative content coverage that reflects the National Science Education Standards
- a unique table of contents built for instructional flexibility
- a readable narrative style that highlights key principles
- a complete supplements package to meet the instructional needs of every student

- technology materials to match your daily instruction
- block-scheduling support materials
- a solid laboratory strand ranging from quick, simple activities to biotechnology experiments

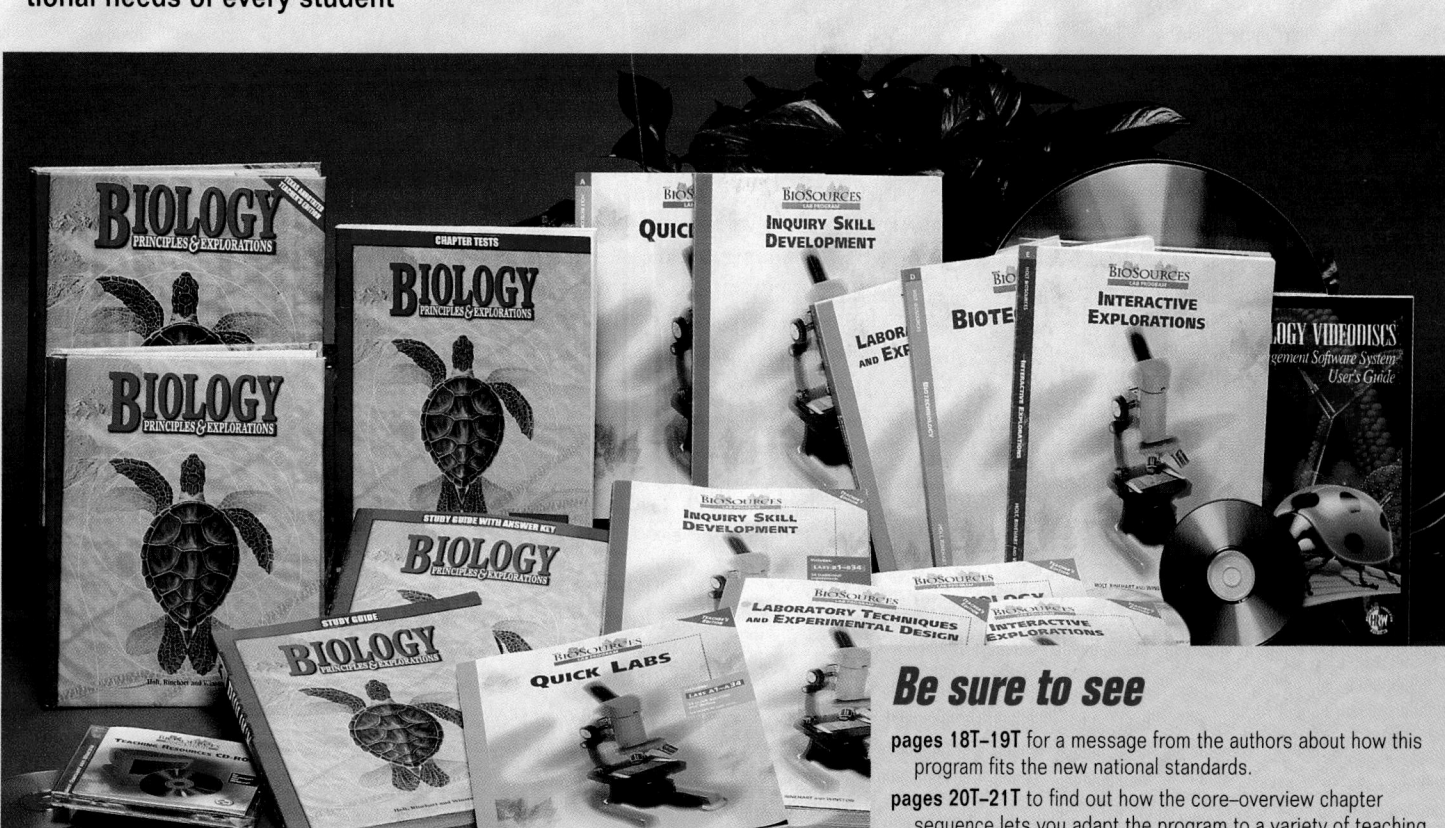

Be sure to see

pages 18T–19T for a message from the authors about how this program fits the new national standards.

pages 20T–21T to find out how the core–overview chapter sequence lets you adapt the program to a variety of teaching approaches and content emphases.

pages 26T–27T for a look at the most comprehensive supplements package available.

pages 48T–49T for an explanation of the thorough range of assessment instruments provided to help you evaluate student progress.

Acknowledgments

Authors

George B. Johnson

Professor of Biology
Washington University
St. Louis, Missouri

Peter H. Raven

Director, Missouri Botanical Gardens
Engelmann Professor of Botany
Washington University
St. Louis, Missouri

Contributing Writers

Tracey Cohen
Science Writer
Highland Park, NJ

Gary W. Goodnight
Biology Teacher
Denver South High School
Denver, CO

Deborah L. Jensen
Biology Teacher and Biology
Department Chairperson
Oak Ridge High School
Conroe, TX

Thomas R. Koballa, Jr.
Professor
Science Education Department
University of Georgia
Athens, GA

Kenneth Rainis
George Nassis
WARD'S Natural Science
Establishment, Inc.
Rochester, NY

Susan Green Talkmitt
Biology Teacher
Monterey High School
Lubbock, TX

Salvatore Tocci
Science Department
East Hampton High School
East Hampton, NY

Lab Reviewers

Alex Molinich
George Nassis
Laboratory Investigations
WARD'S Natural Science
Establishment, Inc.
Rochester, NY

Kenneth Rainis
Safety
WARD'S Natural Science
Establishment, Inc.
Rochester, NY

Reviewers

Hugh Clement Allen
Adjunct Professor of Natural Sciences
Miami-Dade Community College
Miami, FL

David Armstrong, Ph.D.
University of Colorado
Boulder, CO

Carol C. Baskin
Adjunct Professor
School of Biological Sciences
University of Kentucky
Lexington, KY

Jerry M. Baskin, Ph.D.
Professor
School of Biological Sciences
University of Kentucky
Lexington, KY

Barry Bogin, Ph.D.
Professor of Anthropology
Department of Behavioral Sciences
University of Michigan
Dearborn, MI

Linda Butler, Ph.D.
Lecturer
University of Texas at Austin
Austin, TX

Mark Coyne
Assistant Professor
Department of Agronomy
University of Kentucky
Lexington, KY

Joe Crim, Ph.D.
Professor of Zoology
University of Georgia
Athens, GA

Susan Chattan Dabb, Ed. D.
Chair, Science Department
King Senior High School
Tampa, FL

Mary Pitt Davis
Science Teacher
Glenelg High School
Glenwood, MD

Andrew Dewees, Ph.D.
Chair, Department of Biological Sciences
Sam Houston State University
Huntsville, TX

William J. Ehmann, Ph.D.
Chair, Department of Environmental
Science
Trinity College of Washington, DC
Washington, DC

William Forward
Biology Teacher/Science Chair
Rio Linda Senior High School
Rio Linda, CA
Science Methods Professor
California State University
Sacramento, CA

Bill Gasper
Science Department Chair
Clearwater Catholic Central High School
Clearwater, FL

Gary W. Goodnight
Biology Teacher
Denver South High School
Denver, CO

Jerry Halpern
Editor
Brookline, MA

Andrea L. Huvard, Ph.D.
Department of Biology
California Lutheran University
Thousand Oaks, CA

Wojciech Kedzierski, Ph.D.
Assistant Professor
University of Texas
Southwestern Medical Center
Dallas, TX

Jo Ann D. Lane
Science Department Chair
St. Ignatius High School
Cleveland, OH

Glenn E. Mitchell
Science Department Chair, Mentor
Teacher
Coalinga High School
Coalinga, CA

Martin Nickels, Ph.D.
Professor of Physical Anthropology
Illinois State University
Normal, IL

Nancy R. Parker
Associate Professor of Biological
Sciences
Southern Illinois University
Edwardsville, IL

Sharon Perlman
Biology Teacher
Coral Park Senior High School
Miami, FL

Celia T. Rainwater
Biology Teacher
Tom C. Clark High School NISD
San Antonio, TX

Irving Rashkover
Assistant Principal for Curriculum
Kinloch Park Middle School
Miami, FL

Marian Smith, Ph.D.
Associate Professor of Biology
Southern Illinois University
Edwardsville, IL

Gerald Summers
Associate Professor of Biological
Sciences
University of Missouri
Columbia, MO

Susan Green Talkmitt
Biology Teacher
Monterey High School
Lubbock, TX

William Thwaites, Ph.D
Biology Department
San Diego State University
San Diego, CA

Betty H. Tumminello
Assistant Principal/Curriculum
Coordinator
Pineville High School
Pineville, LA

Jerry Warren
Science Department Chair
Elk Grove High School
Elk Grove, CA

BIOLOGY
PRINCIPLES & EXPLORATIONS

The Flexibility You Want . . .
The Depth of Coverage You Need

Part 1: Biological Principles is devoted to Cell Biology, Genetics, Evolution, and Ecology. Biology: Principles and Explorations *covers fundamental content that lies at the heart of any biology curriculum in these 18 core chapters:*

Unit 1: Principles of Cell Biology teaches the basic principles of cell organization and cellular energetics.

Unit 2: Principles of Genetics expands the coverage of cell biology to cover genes—what they are and how they function.

Unit 3: Principles of Evolution then explains how natural selection causes particular changes in genes to become more common, enabling a species to evolve.

Unit 4: Principles of Ecology completes the core by explaining how coevolutionary accommodations among the many organisms living together within ecosystems create the web of relationships responsible for ecosystem stability.

Part 2: Biological Explorations

is devoted to Diversity. **Biology: Principles and Explorations** *divides the world of living organisms into five teaching units:*

OVERVIEW chapters cover the basic principles of each unit.

Supporting chapters discuss particular kinds of organisms or systems in more detail.

For example, this **OVERVIEW** chapter describes the animal body plan, following in order the development of evolutionary milestones such as tissues, body cavities, segmentation, and jointed appendages and showing how these innovations have led to the phyla we see today.

The three supporting chapters in this unit present a more detailed look at the major invertebrate phyla.

TEACHER'S GUIDE

UNIT 1 PRINCIPLES OF CELL BIOLOGY

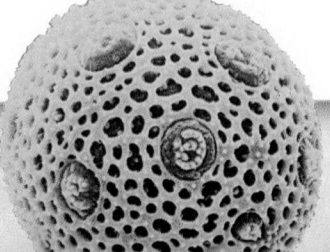
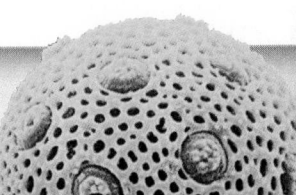

UNIT 2 PRINCIPLES OF GENETICS

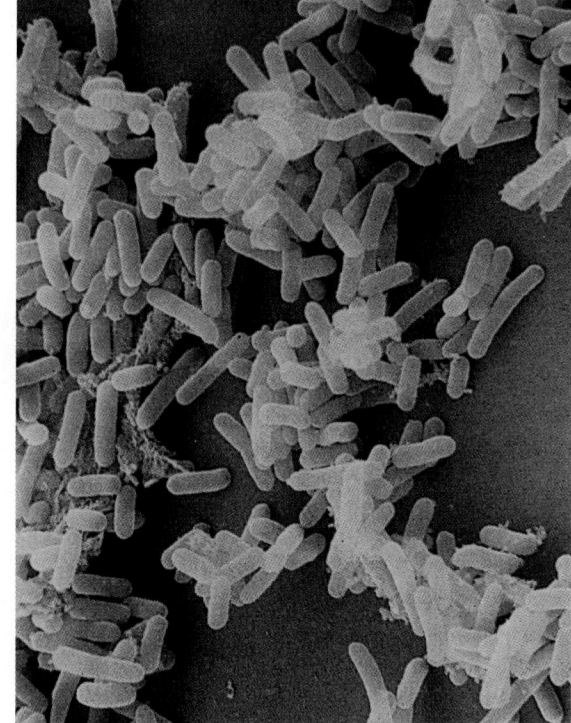

UNIT 3 PRINCIPLES OF EVOLUTION

UNIT 4 *PRINCIPLES OF ECOLOGY*

UNIT 5 EXPLORING DIVERSITY

UNIT 6 EXPLORING PLANTS

UNIT 7 EXPLORING INVERTEBRATES

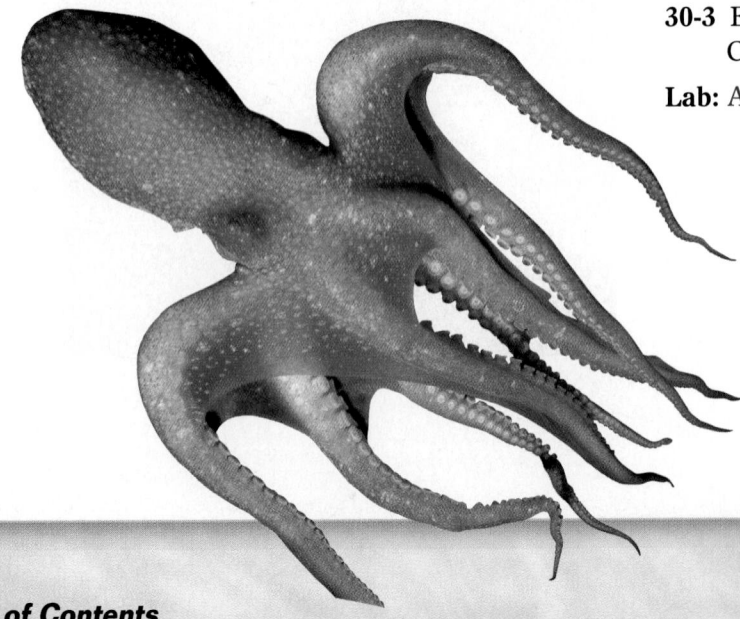

UNIT 8 EXPLORING VERTEBRATES

UNIT 9 EXPLORING HUMAN BIOLOGY

*F*EATURES

UP CLOSE

Dramatically illustrated Up Close features provide a detailed look at a variety of organisms that have been carefully selected for their biological significance. You will get a better understanding of how each organism has adapted to its environment by studying its internal and external structure.

SCIENCE➡TECHNOLOGY➡SOCIETY

Science, Technology, and Society articles present opinions on significant biological issues. You will analyze these persuasive essays to detect biased arguments and formulate your own informed opinions.

LABORATORY Investigations

Well-designed Laboratory Investigations will give you a firsthand look at the scientific processes and data-collection techniques used by biologists to answer questions and form conclusions.

Highlights

These features close each Overview chapter in the program. They provide quick graphic summaries of the key concepts and ideas of the unit and chapter.

A Message From the Authors

Peter H. Raven and George B. Johnson

*T*he last decade has seen a revolution in biology. The development of tools like monoclonal antibodies has allowed the detailed dissection of cell membranes, revealing a wealth of information about how proteins in the cell membrane control cell function. The cell surface receptors responsible for HIV infection, for drug addiction, and for normal growth and development have all been isolated and studied. Advances in genetic engineering are transforming medicine and agriculture so quickly that scarcely a week passes without some important announcement.

This revolution in biology has placed teachers in an awkward position. The textbooks are already long, and the curriculum is already packed with too much information to teach comfortably in the time available. In response to this concern, biology teachers began to demand a new approach, one that would allow them to teach the *ideas* of biology rather than reciting an ever-expanding body of information. A variety of studies were produced, which pointed to four recommendations for a successful biology curriculum.

Adopt a Higher Quality Standard for Texts

As active research scientists, we have taught together at Washington University in St. Louis for over 20 years, and have coauthored over 250 research papers and several successful college-level biology texts. We have worked hard to be accurate. We have also striven to make this text up-to-date and to reflect the most recent scientific philosophies. For example, we present biological diversity as comprising *six* kingdoms rather than five. The analysis of DNA data has produced almost universal agreement that the bacteria should be split into two kingdoms—and that our ancestors are the archaebacteria, not the traditionally taught eubacteria!

Coverage of the latest advances in cell biology makes teaching the rest of cell biology easier. A clear treatment of cell surface receptors in Chapter 3, for example, allows students to recognize that changes in receptors are the underlying cause of drug addiction.

Our text also stresses advances in cell biology and molecular biology—not because it makes us more fashionable, but rather because these advances make the teaching of the rest of biology easier. In every area of the text, from warm-blooded dinosaurs to DNA finger-printing, we have presented an account of the state of science that is as up-to-date as possible.

Focus on Principles Rather Than Information

In 1995—after more than a year of discussion with a wide array of teachers and teachers' organizations nationwide—the National Research Council published new National Science Education Standards for biology. These standards propose that biology curricula be organized around the teaching of a small number of key concepts and ideas in areas such as cell biology, genetics, evolution, ecology, and diversity. We believe strongly in what the new standards are trying to accomplish, which is in sum a return to the teaching of ideas. Both of us were involved in the actual drafting of the content portion of the standards. *Biology: Principles and Explorations* was designed by us from the outset to match the emerging standards.

Develop Flexible Curricula

There is great disparity around the country concerning mandated content. As a result, textbooks have tended to become ever more inclusive, covering any topic that anyone might be required to present. Today's texts simply contain too much material for any one course to teach. A new curriculum must provide **thorough coverage** of the basic prin-

ciples, and **flexibility** to allow teachers to explore the details of the topics they want their classes to concentrate on. *Biology: Principles and Explorations* does both.

Use of Technology

As every classroom teacher knows, hands-on exploration is the most effective way to learn biology. Students should be able to do experiments that explore the concepts and ideas they are attempting to understand.

Unfortunately, there is a limit to how much science can be taught hands on in a classroom or in a laboratory. Thus, it is with genuine excitement that teachers and students greet the new interactive CD-ROM technologies beginning to appear in today's classrooms. *Interactive Explorations in Biology: Human Biology* and *Interactive Explorations in Biology: Cell Biology and Genetics* contain 19 interactive CD-ROM labs for use with Macintosh® or Windows®. What does such a CD-ROM lab offer that slide collections and laser discs cannot? In a word, interactivity. Just as you cannot make a better painting simply by giving the artist more paint, you cannot make a concept more concrete by simply providing the student with more images to view. What is needed is true interaction, where inquiry and analysis can take place. This is what the CD-ROM interactive explorations provide, and it is why they play such an important part in the learning experience provided by this text.

Student interactive CD-ROMs are just one part of the wide range of technology offerings you can use with *Biology: Principles and Explorations*. To find out about other Holt technology products for biology, see pages 32T and 40T.

Turn the page to see how **Biology: Principles and Explorations** *allows you to tailor your teaching to meet your curriculum needs.*

Alternative Courses of Study and Pacing the Course

A content sequence designed to support your curricular needs

You can cover the five Overview chapters in any order, and then use as many supporting chapters as you wish to go into more depth. As long as the 18 core chapters and the five Overview chapters are covered, your students will be exposed to all the basic principles of biology, they will fulfill the requirements of the National Science Education Standards, and they will receive excellent preparation for further courses in biology.

The suggested pacing presumes the school year consists of *approximately* 180 blocks, each 45–50 minutes long. To implement one of the alternative courses of study shown, you will need to adjust the planning charts that precede each chapter to match these pacing suggestions.

If you do more lab work or add more chapters to a course sequence, you will also need to adjust the time spent on each chapter.

	BASIC COURSE 23 Chapters blocks	HUMAN BIOLOGY EMPHASIS 30 Chapters blocks
Unit 1 Principles of Cell Biology		
Chapter 1	6	5
Chapter 2	9	6
Chapter 3	8	6
Chapter 4	8	6
Chapter 5	8	6
Chapter 6	8	6
Unit 2 Principles of Genetics		
Chapter 7	8	6
Chapter 8	7	6
Chapter 9	8	6
Chapter 10	7	5
Unit 3 Principles of Evolution		
Chapter 11	6	4
Chapter 12	7	5
Chapter 13	7	5
Chapter 14	7	5
Unit 4 Principles of Ecology		
Chapter 15	7	6
Chapter 16	7	6
Chapter 17	9	6
Chapter 18	8	5
Unit 5 Exploring Diversity		
Chapter 19	11	8
Chapter 20	–	–
Chapter 21	–	–
Chapter 22	–	–
Unit 6 Exploring Plants		
Chapter 23	8	5
Chapter 24	–	–
Chapter 25	–	–
Chapter 26	–	–
Unit 7 Exploring Invertebrates		
Chapter 27	8	5
Chapter 28	–	–
Chapter 29	–	–
Chapter 30	–	–
Unit 8 Exploring Vertebrates		
Chapter 31	8	7
Chapter 32	–	–
Chapter 33	–	–
Chapter 34	–	–
Unit 9 Exploring Human Biology		
Chapter 35	6	7
Chapter 36	–	6
Chapter 37	–	8
Chapter 38	–	9
Chapter 39	–	7
Chapter 40	–	7
Chapter 41	–	6
Chapter 42	–	8

	Botany emphasis 26 Chapters blocks	Zoology emphasis 29 Chapters blocks	Zoology emphasis with Human Biology 36 Chapters blocks	Ecology emphasis 23 Chapters blocks	Cell Biology emphasis 32 Chapters blocks
Unit 1					
Chapter 1	5	5	3	5	5
Chapter 2	7	6	5	7	6
Chapter 3	7	6	5	7	6
Chapter 4	7	6	5	7	6
Chapter 5	7	6	5	7	6
Chapter 6	7	6	5	7	6
Unit 2					
Chapter 7	7	6	5	7	6
Chapter 8	7	6	5	7	6
Chapter 9	7	6	5	7	6
Chapter 10	6	5	4	6	5
Unit 3					
Chapter 11	6	5	4	6	5
Chapter 12	6	5	4	6	5
Chapter 13	6	5	4	6	5
Chapter 14	6	5	4	6	5
Unit 4					
Chapter 15	7	6	5	12	5
Chapter 16	7	6	5	12	5
Chapter 17	8	7	5	12	6
Chapter 18	6	5	4	12	4
Unit 5					
Chapter 19	8	9	6	9	7
Chapter 20	–	–	–	–	6
Chapter 21	–	–	–	–	5
Chapter 22	–	–	–	–	5
Unit 6					
Chapter 23	8	7	5	8	6
Chapter 24	7	–	–	–	5
Chapter 25	7	–	–	–	–
Chapter 26	7	–	–	–	–
Unit 7					
Chapter 27	7	7	6	7	5
Chapter 28	–	6	5	–	–
Chapter 29	–	5	4	–	–
Chapter 30	–	6	5	–	–
Unit 8					
Chapter 31	9	9	7	9	5
Chapter 32	–	7	6	–	–
Chapter 33	–	7	6	–	–
Chapter 34	–	7	6	–	–
Unit 9					
Chapter 35	8	8	6	8	6
Chapter 36	–	–	5	–	5
Chapter 37	–	–	5	–	–
Chapter 38	–	–	5	–	–
Chapter 39	–	–	5	–	6
Chapter 40	–	–	5	–	7
Chapter 41	–	–	5	–	5
Chapter 42	–	–	6	–	6

Strategically Organized Lessons

Biology Is Presented in a Way That Is Easily Taught and Readily Understood

Lessons

Every chapter in *Biology: Principles and Explorations* is divided into two to five lessons.

Each section begins with an interest-capturing introduction to the lesson.

Section Objectives alert students to the essential information covered in the lesson.

CHAPTER 5

PHOTOSYNTHESIS AND CELLULAR RESPIRATION

REVIEW
- proton pumps and chemiosmosis (Section 3-2)
- autotrophs and heterotrophs (Section 4-1)
- laws of thermodynamics (Section 4-1)
- oxidation-reduction reactions (Section 4-1)
- coenzymes and coupled reactions (Section 4-3)

Leaves reflecting sunlight.

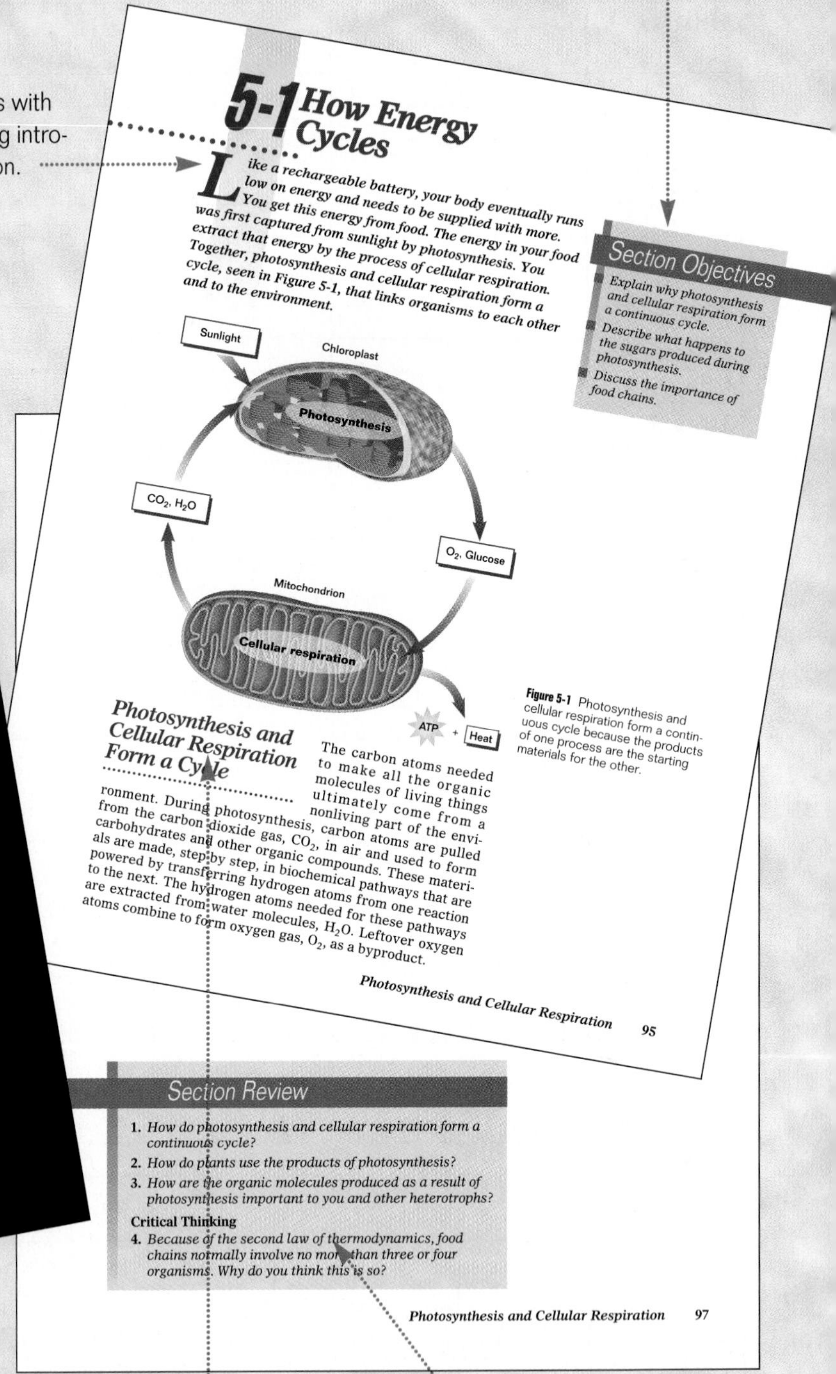

5-1 How Energy Cycles

Like a rechargeable battery, your body eventually runs low on energy and needs to be supplied with more. You get this energy from food. The energy in your food was first captured from sunlight by photosynthesis. You extract that energy by the process of cellular respiration. Together, photosynthesis and cellular respiration form a cycle, seen in Figure 5-1, that links organisms to each other and to the environment.

Section Objectives

- Explain why photosynthesis and cellular respiration form a continuous cycle.
- Describe what happens to the sugars produced during photosynthesis.
- Discuss the importance of food chains.

Sunlight

Chloroplast

Photosynthesis

CO_2, H_2O

O_2, Glucose

Mitochondrion

Cellular respiration

ATP + Heat

Figure 5-1 Photosynthesis and cellular respiration form a continuous cycle because the products of one process are the starting materials for the other.

Photosynthesis and Cellular Respiration Form a Cycle

ronment. During photosynthesis, carbon atoms from the carbon dioxide gas, CO_2, in air and used to form carbohydrates and other organic compounds. These materials are made, step by step, in biochemical pathways that are powered by transferring hydrogen atoms from one reaction to the next. The hydrogen atoms needed for these pathways are extracted from water molecules, H_2O. Leftover oxygen atoms combine to form oxygen gas, O_2, as a byproduct.

The carbon atoms needed to make all the organic molecules of living things ultimately come from a nonliving part of the environment.

Photosynthesis and Cellular Respiration **95**

Section Review

1. How do photosynthesis and cellular respiration form a continuous cycle?
2. How do plants use the products of photosynthesis?
3. How are the organic molecules produced as a result of photosynthesis important to you and other heterotrophs?

Critical Thinking
4. Because of the second law of thermodynamics, food chains normally involve no more than three or four organisms. Why do you think this is so?

Photosynthesis and Cellular Respiration **97**

REVIEW identifies topics from earlier chapters that students should understand before beginning the new chapter.

Subheads present a clear story line and identify the subtopics discussed within the section.

Section Review questions test students on content described by the Section Objectives. Each Section Review includes at least one Critical Thinking question, which requires students to apply the content in the section to new situations.

Features

Highlights features follow each of the five **OVERVIEW** chapters that begin the units in *Part 2: Exploring Biology*. These features provide four- to eight-page visual summaries of each **OVERVIEW** chapter's major content points.

Each supporting chapter in Units 5, 6, 7, and 8 contains at least one Up Close feature. Strategically placed near the accompanying discussion in the text, Up Close features use striking illustrations to examine the internal and external elements of representative organisms, showing students the relationship between structure and function.

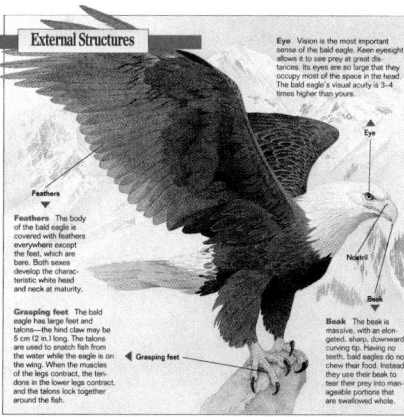

Science, Technology, and Society features consider the impact of technological developments on society from economic, philosophical, and ethical standpoints. Each feature closes with questions that lead students to recognize the writer's point of view.

Delivering Information Using Methods That Are Most Effective for the Learner

Engaging Story Line

The text's conversational but sophisticated tone presents comprehensive information in an interesting and easy-to-follow manner. Analogies allow students to apply complex scientific facts to real-world situations, and evolution is woven through the text as a unifying theme so that students understand how organisms have developed over time and why they are related to one another.

Capsule Summary short paragraphs offer concise on-site summaries of important lesson information.

reducing the surface area available for gas exchange. That is why a fish will suffocate on land, even though there is more oxygen in air than in water. Walking around on land requires a higher metabolism, which in turn uses greater amounts of oxygen. Thus, the heart also underwent change.

Some problems remained, however. For example, the eggs of amphibians are not watertight. Thus, amphibians generally seek out water or damp areas in which to repro-
in which their young can live as they grow and
ut evolution does not result in perfect solutions,
able ones. The adaptations that allowed the first
s to climb out onto land have enabled their
s to survive for over 350 million years. Today
approximately 4,200 species of amphibians,
familiar frogs, toads, and salamanders.

❏ CAPSULE SUMMARY

Amphibians were the first vertebrates to live on land. However, they lack watertight eggs and so cannot reproduce in environments in which their eggs would dry out.

10-1 What Is Genetic Engineering?

In 1973, biochemists Stanley Cohen and Herbert Boyer constructed a creature that was part bacterium and part frog. How did these scientists construct such an organism? Using organisms such as those shown in Figure 10-1, they first isolated the gene that codes for ribosomal RNA from the DNA of a frog. Then they inserted the frog gene into the DNA of the bacterium *Escherichia coli*. During transcription, the bacterium busily produced frog rRNA. Never before had such a genetically altered organism existed.

Section Objectives

- List the four steps involved in a genetic engineering experiment.
- Explain how recombinant DNA is produced.
- Describe how restriction enzymes are used in genetic engineering.

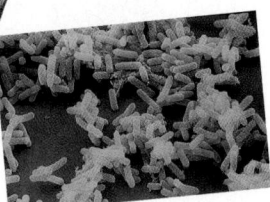

Figure 10-1 In their laboratories in San Francisco, Stanley Cohen and Herbert Boyer successfully put together the first genetically engineered organism. They used a gene from the frog *Xenopus laevis* and a chromosome from the bacterium *Escherichia coli*.

The Basics of Genetic Engineering

The Cohen and Boyer experiment ushered in a scientific revolution in biology. It yielded laboratory techniques that enable researchers to locate, isolate, and study small segments of DNA obtained from much larger chromosomes. These methods are changing basic research in agriculture, medicine, and many other fields. Today, certain human genes can be transferred into bacteria to produce enormous amounts of the protein encoded by the human gene.

The process used to isolate a gene from the DNA of one organism and transfer the gene into the DNA of another is called **genetic engineering**. Genetic engineering involves building **recombinant DNA**, a molecule made from pieces of DNA from separate organisms. Every genetic engineering

Gene Technology 203

Conquer

If you think of amphibians as the "first draft" of a manuscript about survival on land, then reptiles are
ok. Each of the adaptations that allowed
ead a terrestrial existence were refined in
gs, for example, were positioned to sup-
ore effectively. So reptiles not only were
could run. Changes in the lungs and heart
ns more efficient. The most significant
gh, were internal fertilization and watertight eggs made reptiles the first com-
vertebrates. Today there are about 7,000
s, mostly snakes and lizards, found in
abitat on Earth. The chameleon shown in
these species.

st evolved, about 320 million years ago,
a long, dry period. Early reptiles,
netrodon shown on page 733, were well
ns and quickly diversified. In particu-
conserve water allowed reptiles
dry conditions. This was something
. Within 50 million years, reptiles had
the large, terrestrial vertebrates. All
han a chicken were reptiles.
ctothermic. Their metabolism is too
heat to warm their bodies. They
he environment. **Endothermic** ani-
d birds, maintain a high, constant
se they produce heat internally
m. Therapsids, an order of extinct
ncestors of mammals, were prob-
re replaced by ectothermic the-
rs.

CONTENT LINK

More information on characteristics, classification, and anatomy of reptiles can be found in **Chapter 33**.

Figure 31-5 Unlike the moist skin of most present-day amphibians, reptilian skin is covered with scales that keep the body from drying out.

OVERVIEW *of Vertebrates* 707

Content Links identify topics that will be discussed at greater length in later chapters.

24T

dry land because they are watertight and contain their own supply of water. A reptilian egg contains a food source (the yolk) and a series of four membranes: the yolk sac, the amnion, the allantois (uh LAHN toh ihs), and the chorion. Each membrane plays a role in making the egg an independent life-support system. The outermost membrane of the egg is the chorion. This membrane allows oxygen to enter, but it retains water within the egg. The amnion encloses the developing embryo within a fluid-filled cavity. The yolk sac contains the yolk, which the embryo absorbs through blood vessels connected to its gut. The allantois surrounds a cavity into which waste products from the embryo are excreted. All modern reptiles show this pattern of membranes within the egg.

The majority of reptiles are oviparous. Their eggs are surrounded by a protective shell and are deposited in a suitable place in the environment. Heat from the environment incubates the eggs, which are usually left unprotected by the parents. However, several species of lizards and snakes, including rattlesnakes and some horned lizards, are ovoviviparous or viviparous and give birth to live, fully formed young.

Reptiles Are Ectothermic

The metabolism of present-day reptiles is too slow to generate enough heat to warm their bodies, so they must absorb heat from their surroundings. A reptile's body temperature is largely determined by the temperature of its environment. Many reptiles regulate their temperature by their behavior. They bask in the sun to warm up and seek shade to prevent overheating. Figure 33-5 shows that a lizard can maintain a fairly constant body temperature throughout the day by moving between sunlight and shade. You can also see why it is inaccurate to call ectotherms "cold blooded." The lizard's body temperature is higher than yours for part of the day.

Figure 33-5 Note the rapid increase in this lizard's body temperature as it basks in the morning sun. The data on the lizard's body temperature were transmitted by a temperature-sensitive instrument implanted in its abdomen.

...ng the body's cells. Sometimes cancer occurs ...cells that give rise to white blood cells. As a ...many white blood cells are produced, and a ...lled **leukemia** (loo KEE mee uh) develops. ...often fatal.

...rtain large cells in bone marrow ...aryocytes (mehg uh KAR ee oh ...rly pinch off bits of their ...ese unnucleated cell frag- ...**platelets** (PLAYT lihts), ...ole in blood clotting. ...humans, like other ver- ...closed circulatory sys- ...vessel gets a hole in its ...st be plugged quickly. If ...gged, all of the blood will ...tem, and death will occur. ...atelets start the clotting ...y encounter chemicals released ...vessel cells. The platelets then release a ...to the blood that initiates a series of ...brin begins to form a netlike covering ...site, as shown in Figure 37-6. Very ...esh of fibrin and platelets develops ...hat plugs the hole in the blood vessel. ...y gluey nature, the clot fits itself to ...re in the blood vessel and provides a ...ou learned in Chapter 7, the lack of ...eins causes hemophilia.

Before a blood transfusion can be ...erformed, it is important to know ...e blood type of the donor and ...cipient. If the blood of each is ...t compatible, a life-threatening ...uation can develop. Blood com- ...ibility is determined by pro- ...he outside of red blood cells. ...y human blood is called the ...cells have either A antigens, ...gens, or no antigens on their ...present on red blood cells ...danger of mixing bloods of ...n the presence of other pro- ...resent in the blood plasma. ...eign antigens and destroy ...ontains antibodies that will ...ith type B blood receives ...antibodies will react with

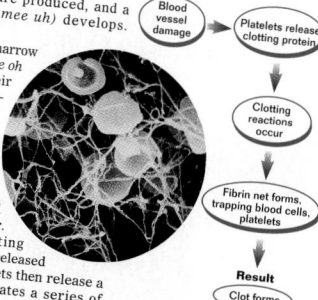

Stimulus

Blood vessel damage → Platelets release clotting protein → Clotting reactions occur → Fibrin net forms, trapping blood cells, platelets → **Result** Clot forms

Figure 37-6 The release of enzymes from platelets at the site of a damaged blood vessel initiates a "clotting cascade." During this cascade, a series of chemical reactions occurs, resulting in the formation of a fibrin net in which blood cells are trapped, *above left.* The net eventually forms into a clot in the hole of the damaged vessel.

Circ...

Functional Illustrations

State-of-the-art illustrations bring to life the concepts, structures, and processes of biology.

Thematic Development

Five key biological concepts —Evolution, Homeostasis, Flow of Energy, Levels of Organization, and Structure and Function—serve as unifying themes throughout the text.

THEMES OF BIOLOGY

Evolution

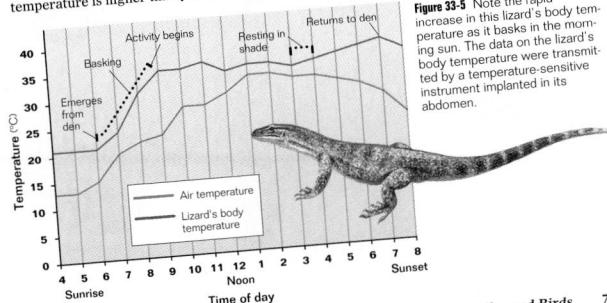

Evolution is the change of organisms over time. Although biologists had long considered that the diversity of life on Earth was the result of evolution, it was not until 1859 that a plausible mechanism for evolution was suggested. In that year, the English naturalist Charles Darwin published his book *On the Origin of Species,* in which he suggested that organisms change over time in response to changes in their environment. Darwin called this process natural selection. Simply stated, Darwin's theory of evolution by natural selection says: Those organisms that are better able to successfully respond to the challenges of their environment are more

likely to survive... thus, the chara... those organism... more common...

Today bio... nize that evo... unifying thre... throughout t... In this chap... you learned... things are ... more cells ... by a plasm... Additions... design— ... organelle... nucleus ... represe... embelli... simple ... Throug... biolog... ering ... spec... time ... envi... cep... stu... ful...

At a glance, the Euglena, Vorticella, and the in-line skater would seem to have little in common. Yet the cells of each contain energy-releasing mitochondria, an evolutionary inheritance found among all eukaryotes.

Section Review

1. A biologist wants to stud... exit a certain type of cell... disadvantages of using ... purpose.
2. Name five organelles. ... cell to display the prop...

Critical Thinking

3. Are lysosomes in anim... in plant cells? Explai...
4. What does the presence of mitochon... cells suggest about the evolution of eukaryotes?

27-1 Advent of Tissues and Symmetry

In this chapter you will discover how a series of key evolutionary innovations has led to today's animal phyla. These phyla are shown in Figure 27-1 and are described in Highlights: Invertebrate Evolution features that occur throughout the chapter. You will be provided with a broad overview of the body plan—the shape, symmetry, and internal organization—of different animals and learn how those body plans arose. An animal's body plan results from a pattern of development programmed into the animal's genes by natural selection.

Section Objectives

■ Describe the general characteristics of animals.

■ Define radial symmetry.

■ Compare and contrast sponges with cnidarians.

Mollusca (Clams, snails)
Arthropoda (Crustaceans, insects)
Chordata (Vertebrates)
Nematoda (Roundworms)
Annelida (Segmented worms)
Rotifera (Rotifers)
Echinodermata (Sea stars)
Platyhelminthes (Flatworms)
Cnidaria (Jellyfish)
Ctenophora (Comb jellies)
Porifera (Sponges)
Protist ancestors

Figure 27-1 This phylogenetic tree identifies the major animal phyla. Although many animals live in water, some inhabit land. Most terrestrial animals are mollusks, arthropods, or chordates. The circled numbers indicate important evolutionary stages, which are listed in the table below.

Evolutionary stages in the animal body	
Stage	**Milestone**
1	Multicellularity
2	Tissues
3	Bilateral symmetry
4	Body cavity
5	Coelom
6	Segmentation
7	Jointed appendages
8	Deuterostomes
9	Notochord

A Comprehensive Resource Package as Diverse as your Students

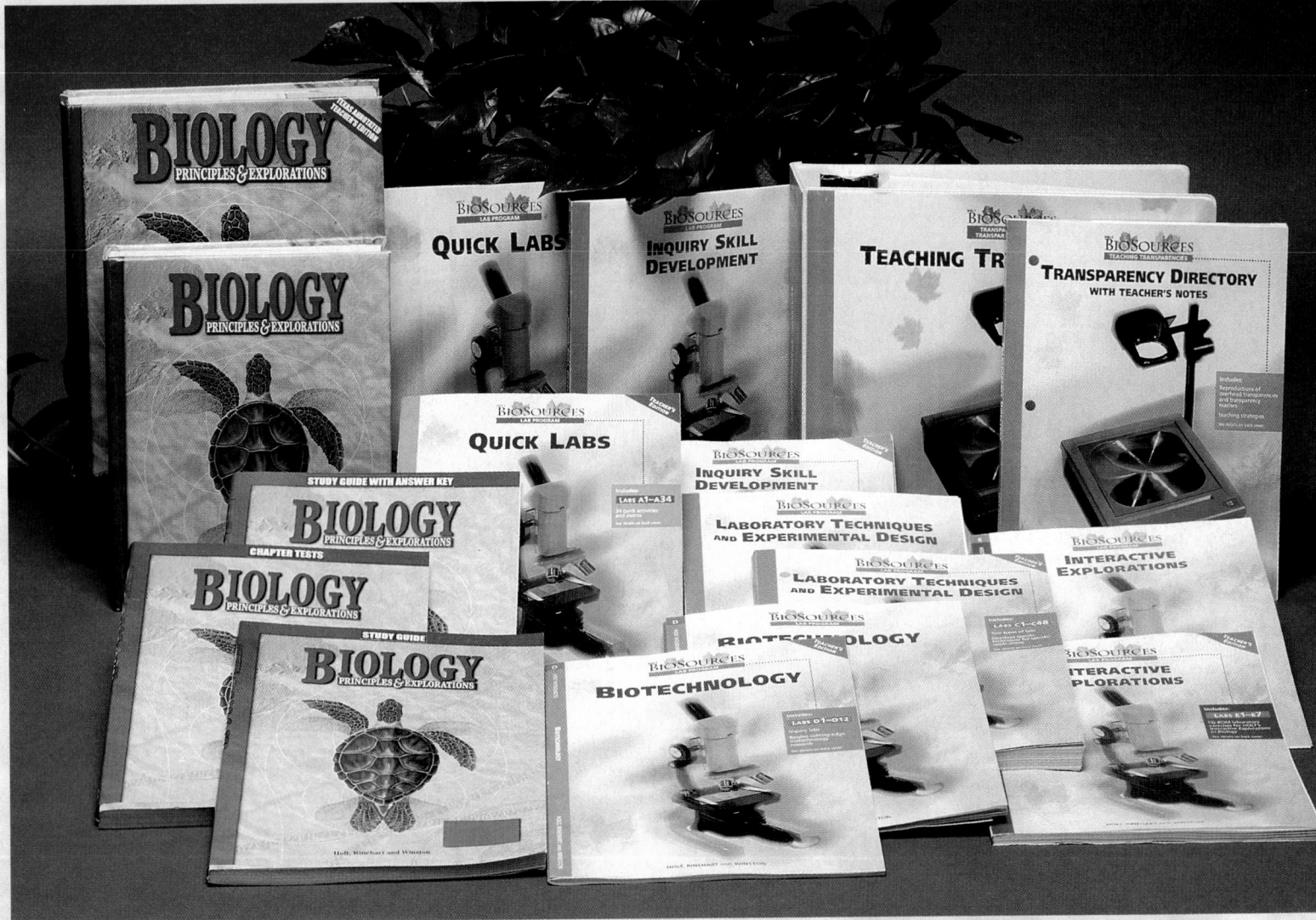

Biology: Principles and Explorations gives you the coordinated support you need to address various learning styles.

- **Study Guide**
  - Pretests
  - Concept Reviews
  - Science Skills

See page 34T for more information.

- **Study Guide Answer Key**

- **Chapter Tests with Answer Key**

- **Holt Biology Videodiscs Teacher's Correlation Guide to *Biology: Principles and Explorations***

Holt BioSources allows you to extend and customize your course to meet your students' needs.

- **Holt BioSources Lab Program**

Five laboratory manuals in blackline-master form provide all the options you need for hands-on learning. (See pages 29T–32T.)
  - **Quick Labs**
  - **Quick Labs, Teacher's Edition**
  - **Inquiry Skills Development**
  - **Inquiry Skills Development, Teacher's Edition**
  - **Laboratory Techniques and Experimental Design**
  - **Laboratory Techniques and Experimental Design, Teacher's Edition**

- **Biotechnology**
- **Biotechnology, Teacher's Edition**
- **Interactive Explorations in Biology Laboratory Manual**
- **Interactive Explorations in Biology Laboratory Manual, Teacher's Edition**

- **Holt BioSources Teaching Transparencies**

238 color and 71 blackline-master illustrations (See page 36T for examples.)

- **Holt BioSources Transparency Directory With Teacher's Notes**

Teaching strategies for each transparency and transparency master

Resources for all *your students* are at your fingertips in Holt's extensive collection of support materials for teaching biology.

• **Holt BioSources Teaching Resources CD-ROM for Macintosh and Windows**

Blackline-master classroom-management tools, and worksheets for practice, review, and enrichment (See page 35T for examples.)

• **Holt BioSources Test Generators for Macintosh and Windows**

CD-ROM software package with more than 3,000 items for building custom tests (See page 37T for more information.)

• **Holt BioSources Test Generator: Assessment Item Listing**

Complete listing of all items on the CD-ROM test generator

Holt Biology Technology **is your passport to the exciting world of curriculum-based multimedia instruction.**

• *Interactive Explorations in Biology: Human Biology* **for Macintosh and Windows**

Eleven interactive investigations for students to explore biological systems on CD-ROM (See page 32T for more information.)

• *Interactive Explorations in Biology: Cell Biology and Genetics* **for Macintosh and Windows**

Eight more interactive investigations on CD-ROM

• *Holt Biology Videodiscs*

Multimedia stills and video footage that are fully coordinated with *Biology: Principles and Explorations* (See page 40T for more information on media materials.)

• *Concepts of Biology*

Design-your-own multimedia presentations for *Biology: Principles and Explorations*

Laboratory Instruction

You've asked for options. You want the freedom to choose labs that best fit your curriculum, your students' abilities, your time frame, and your budget. Biology: Principles and Explorations provides these options and more in a new laboratory program that emphasizes inquiry, meeting the goals of the National Science Education Standards.

In the lab, students will
- work in groups to analyze data and defend conclusions
- acquire new skills and develop practical techniques and scientific habits
- relate newly acquired skills to past and present activities
- recognize that the process of science includes argumentation and explanation
- master biological techniques
- develop practical skills needed to achieve success in the real world
- manage and allocate resources
- find information and communicate effectively
- utilize appropriate technology

In-Text Laboratory Investigations

Each chapter contains a detailed two-page laboratory investigation.

Objectives and Process Skills
The *Objectives* and *Process Skills* used in each investigation are highlighted.

Inquiry and Analysis
Inquiry questions ask students about their actions and observations during the lab. Analysis questions require students to apply concepts they learned in the classroom to answer questions about the investigation and its implications.

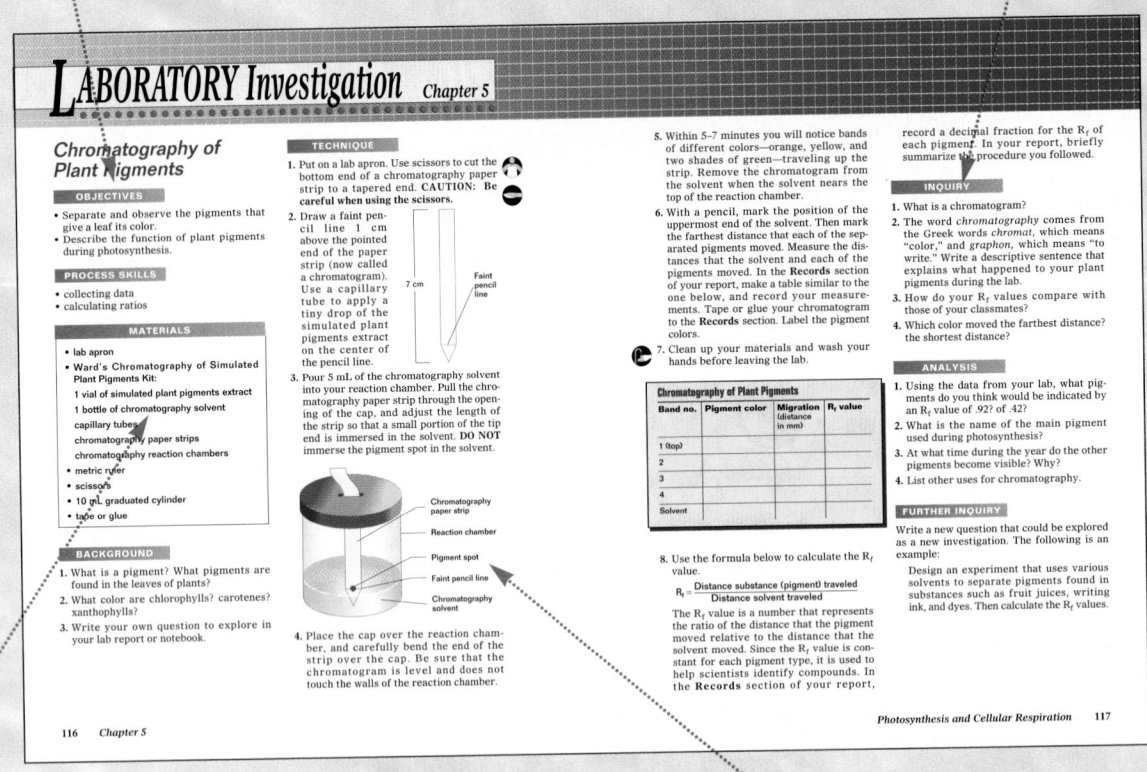

Materials
Whenever possible, required materials are kept to a minimum to reduce your costs and preparation time.

Illustrations
Photographs, diagrams, and tables guide students through the lab procedure.

Holt BioSources *Lab Program*

The Holt BioSources Lab Program gives you a collection of 135 labs covering a range of student ability levels and equipment requirements, from quick, simple activities to the latest in biotechnology.

Five lab manuals bring you unprecedented variety

A Quick Labs

Quick Labs are just that. Inexpensive, easy-to-obtain supplies are used in these short lab activities. Use the Quick Labs to demonstrate a concept or to help students build models for concepts.

A1	Imagining Solutions: Problem Solving	**A18**	Comparing Plant Adaptations
A2	Comparing Living and Nonliving Things	**A19**	Inferring Function From Structure
A3	Modeling Cells: Surface Area to Volume	**A20**	Relating Root Structure to Function
A4	Demonstrating Diffusion	**A21**	Recognizing Patterns of Symmetry
A5	Interpreting Labels: Stored Food Energy	**A22**	Comparing Animal Eggs
A6	Interpreting Information in a Pedigree	**A23**	Observing Some Major Animal Groups
A7	Making Models	**A24**	Observing Insect Behavior
A8	Making a Genetic Engineering Model	**A25**	Observing a Frog
A9	Comparing Observations of Body Parts	**A26**	Vertebrate Skeletons
A10	Analyzing Adaptations: Living on Land	**A27**	Comparing Skeletal Joints
A11	Comparing Primate Features	**A28**	Bias and Experimentation
A12	Making a Food Web	**A29**	Graphing Growth Rate Data
A13	Using Random Sampling	**A30**	Collecting Data Through a Survey
A14	Determining the Amount of Refuse	**A31**	Determining Lung Capacity
A15	Grouping Things You Use Daily	**A32**	Relating Cell Structure to Function
A16	Using Bacteria to Make Food	**A33**	Reading Labels: Nutritional Information
A17	Observing Protists	**A34**	Culturing Frog Embryos

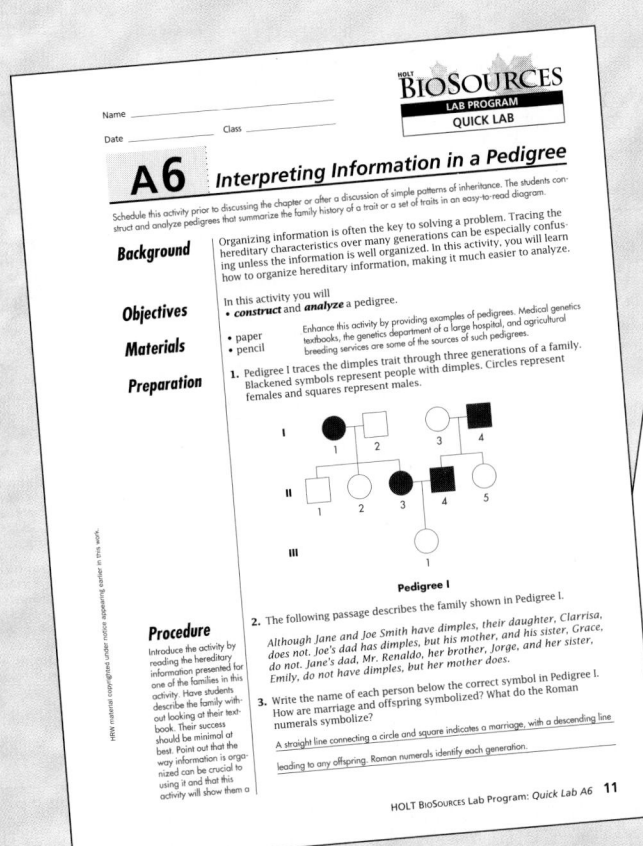

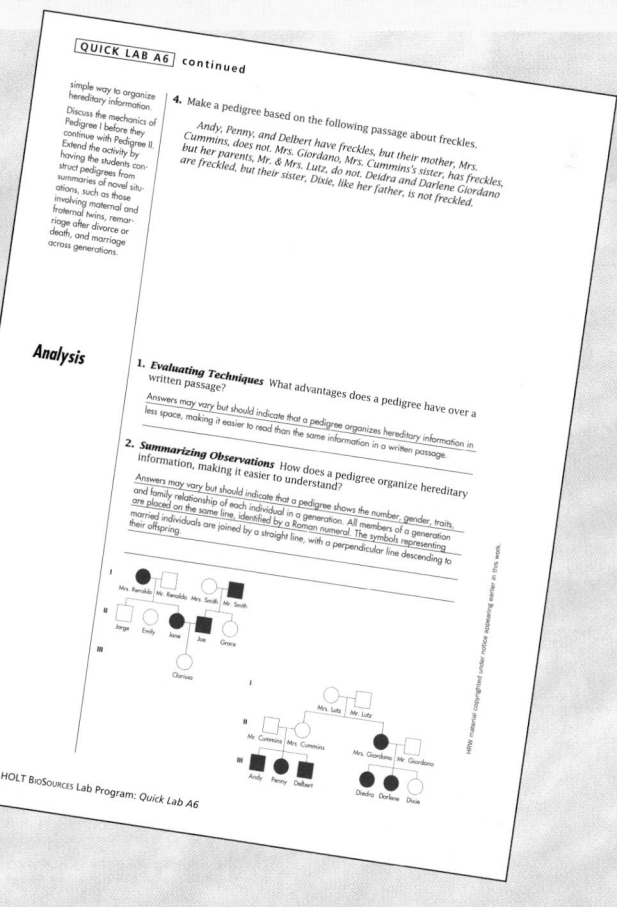

Name _____
Date _____ Class _____

HOLT BIOSOURCES
LAB PROGRAM
INQUIRY SKILLS

B8 Fossil Study

Skills • determining the evolutionary relationship of fossils based on age and morphology

Objectives
PREPARATION NOTES
Time Required: one 50-minute period
• *Analyze* characteristics of fossils to place them in similar lineage groupings.
• *Compare* the placement of fossils in rock strata to determine the relative age of fossils.
• *Develop* a model evolutionary tree based on the morphology and age of fossils.

Materials
• 11 in. × 17 in. paper (4 sheets)
• tape or glue stick
• scissors
• sample fossils

Purpose In this lab, you will categorize fossils by will then draw an evolutionary tree that

Background Fossils are traces of organisms that lived i carefully excavated and then analyzed. O age of the fossil. The absolute age of a fos dating. In **radiometric dating**, the amou pared to the amount of its decay element element (its half-life) is constant, the rati of a fossil. Analysis also includes a study istics, of the fossil so that the genus and

Preparation Tips
• Be sure that students are aware of the meanings of the following terms: *fossil, morphology, gradualism, radiometric dating, punctuated equilibrium, evolutionary tree.*
• Inform students that, unlike this simulated fossil record, the actual fossil record for most organisms on Earth is missing many of the pieces needed for a complete history. The "fossils" students are working with are unusual in that they have a complete history. The reasons for the gaps in the history of actual fossils include that most organisms are not preserved as fossils and that many fossils that do exist have not yet been found. However, the fact that there were

The age and morphologies of fossils in sequences that often show a pattern o This relationship is frequently depicte An **evolutionary tree** is a diagram tha among a group of organisms. The dia ary tree. Line A represents the origina represent two new lineages that evolv

Change in time

c

Change

There are two major theories on ho believe that organisms evolve throu called **gradualism**. Another theory

HOLT B

Name _____
Date _____ Class _____

HOLT BIOSOURCES
LAB PROGRAM
INQUIRY SKILLS

B23 Live Earthworms

Skills
• identifying anatomical structures of the earthworm
• organizing and analyzing data

Objectives
PREPARATION NOTES
• *Infer* how physical and behavioral characteristics of the earthworm reflect adaptations to its environment.
• *Describe* how an earthworm responds to light and moisture.

Materials
Time Required: two 50-minute periods
• paper towels (2)
• dissecting pan
• water, at room temperature
• large, live earthworm
• medicine dropper
• hand lens
• flashlight
• rubber band

Materials
Materials for this lab activity can be purchased from WARD'S. See the *Master Material List* for ordering instructions.
• red cellophane (2 pieces)
• blue cellophane (2 pieces)
• index card
• hole punch
• sandpaper, fine to medium grit
• 15 cm petri dish
• stereomicroscope
• clock with second hand

Purpose In this lab, you will look at live earthworms, observe their external structures, and test their responses to environmental stimuli.

Background Earthworms are classified as **annelids**, or segmented worms. They have digestive, circulatory, and nervous systems. Gas exchange is through the skin. They live in rich soil, which they eat, digesting the organic matter in it and passing the inorganic dirt particles out of the body. Earthworms are not very mobile animals. They spend their lives in one small area and as a result do not encounter many other earthworms. Their reproductive strategy is well suited to this type of life. Each earthworm is a **hermaphrodite**, that is, it has both male and female sex organs. Thus, any individual earthworm can cross fertilize with any other earthworm it encounters.

• You can also obtain live earthworms from bait shops or local soil. Buy or collect the largest worms available. They may be stored in a refrigerator if they are kept moist and covered.

Procedure

1.  Moisten a paper towel, and place it in a clean dissecting pan. Place the worm on the paper towel. **CAUTION: You are working with a live animal. Handle it gently, and follow all lab instructions carefully. Frequently during the lab, give your earthworm a "bath" with room temperature tap water from a medicine dropper, or the worm will dry up and become lethargic.** Watch the worm move, and notice which end leads. The worm's leading end is its **anterior** end. Identify the worm's **posterior** end, that is, the end away from the leading end.

Disposal
After the lab, earthworms may be released and returned to the soil.

2. To differentiate between the worm's **dorsal** (back) and **ventral** (stomach) sides, roll the worm over. Describe the worm's response to being put ventral side up.
It rolls over.

Name _____
Date _____ Class _____

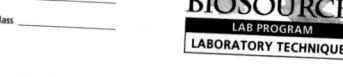
HOLT BIOSOURCES
LAB PROGRAM
LABORATORY TECHNIQUES

C46 Identifying Food Nutrients

Skills • qualitatively analyzing food for organic compounds

Objectives
PREPARATION NOTES
Time Required: 50 minutes
• *Perform* standard chemical identification tests for organic compounds.
• *Relate* indicator reactions to the presence of organic compounds.
• *Recognize* a standard.

Materials
Materials for this lab activity can be purchased from WARD'S. See the *Master Materials List* for ordering instructions.
• safety goggles
• lab apron
• test tubes (10)
• wax pencil
• test-tube rack
• dropping pipets (5)
• glucose solution
• starch solution
• water
• Benedict's solution
• hot water bath
• test-tube holder
• Lugol's iodine solution
• albumin (protein) solution
• biuret solution
• vegetable oil
• Sudan III

Purpose You have just started a job as a food-quality tester. Your task will be to develop a kit to test foods for sugar, starch, protein, and lipids. The test kit will be used by dietitians who will analyze foods to create nutritional menus for hospital patients. Your first assignment is to demonstrate proficiency using standard tests for nutrients in foods.

Preparation Tips
• Wear safety goggles and a lab apron if you prepare the solutions below.

Background Substances or compounds that supply your body with energy and the building blocks of macro-molecules are called **nutrients**. The food you eat contains nutrients important to your body. **Sugars** and **starches** make up a group of organic compounds called carbohydrates, which are important in supplying your body with energy. Some starches provide your body with indigestible fiber, or roughage, which aids digestion. Organic compounds called **proteins** are important for growth and repair. **Lipids** are organic compounds that can supply as much as four times the amount of energy as carbohydrates or proteins.

• Benedict's solution is available and should be purchased, not made. However, to make it yourself, dissolve 17.3 g of sodium citrate ($Na_2C_6H_5O_7 \cdot 2H_2O$) and 10 g of sodium carbonate (Na_2CO_3) in 80 mL of distilled water in a beaker. In a smaller beaker, dissolve 1.7 g of copper sulfate ($CuSO_4 \cdot 5H_2O$) in about 10 mL of distilled water. While stirring the first solution, slowly pour the copper sulfate solution into it. Pour the combined solutions into a 100 mL graduated cylinder and dilute to 100 mL.

You can perform **qualitative tests** to identify the presence of organic compounds in food using **indicators**, chemical substances that react in a certain way when a particular substance is present. **Benedict's solution** is used to identify the presence of reducing sugars, such as glucose. **Lugol's iodine solution** is used to indentify the presence of starch. **Biuret solution** is used to identify the presence of protein. **Sudan III** is used to identify the presence of lipids. A **standard** is a positive test for a known substance. Unknown substances can be tested and compared with the standard for positive identification of the substance.

Procedure

1. Put on safety goggles and a lab apron.

D Biotechnology

Biotechnology labs bring the latest in biotechnology research techniques into your classroom.

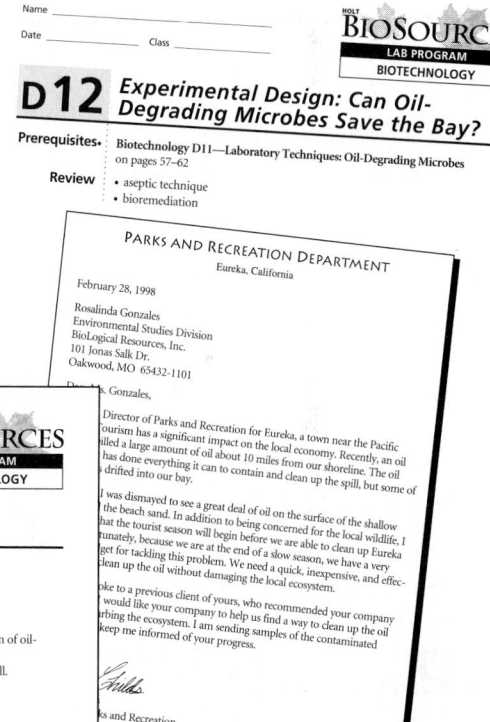

Interactive Explorations in Biology

Curriculum-based multimedia instruction

Interactive Explorations in Biology enables your students to observe phenomena that are difficult or impossible to study in the laboratory. Imagine students learning about photosynthesis. Until now, their learning experience consisted of viewing a text diagram of a chloroplast and listening to you describe how chlorophyll in the chloroplast captures light energy. Now, by using the CD-ROM

Interactive Exploration on *Photosynthesis*, the class can see photosynthesis in action. Students can witness the electrons energized by light traveling through the chloroplast membrane and driving the synthesis of ATP. Then—and this is the key to the hands-on learning experience—students can change the variables and see the effect of their manipulations by animation! Exploring the influences of variables such as wavelength and light intensity, helps students actually learn how photosynthesis works.

Interactive Explorations in Biology: Human Biology

1 Cystic Fibrosis
2 Active Transport
3 Life Span and Lifestyle
4 Evolution of the Heart
5 Diet and Weight Loss
6 Synaptic Transmission
7 Drug Addiction
8 Hormone Action
9 Immune Response
10 Heredity in Families
11 Pollution of a Freshwater Lake

Interactive Explorations in Biology: Cell Biology and Genetics

1 Hemoglobin
2 Cell Size
3 Active Transport
4 Thermodynamics
5 Oxidative Respiration
6 Photosynthesis
7 Meiosis: Down Syndrome
8 Gene Regulation

Each Interactive Exploration provides a simulation of a key biological concept. Each Exploration includes the following:

- a narrative overview of the biological concepts presented, followed by auditory instructions for using the meters and sliders in each simulation
- narration in both English and Spanish

- a navigation bar that makes it easy to access other parts of the program
- a User's Guide containing questions and answers for each Exploration
- additional videos and recommended readings

E Interactive Explorations in Biology Laboratory Manual

Interactive Explorations in Biology labs provide structured procedures and data-collection and analysis instructions for some of the explorations on the CD-ROM:

E1 Cell Size
E2 Oxidative Respiration
E3 Thermodynamics
E4 Gene Regulation
E5 Heredity in Families
E6 Hemoglobin
E7 Diet and Weight Loss

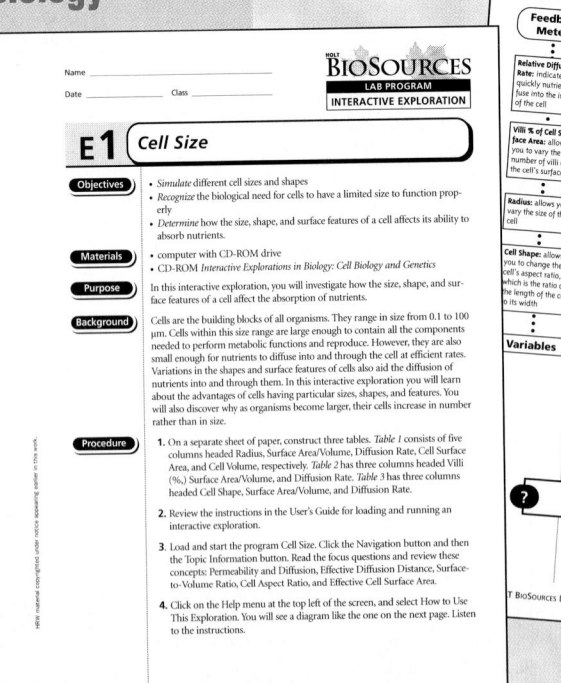

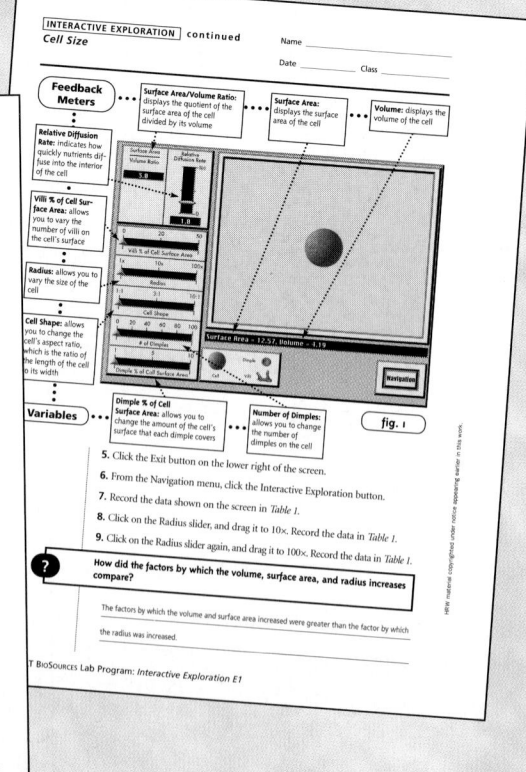

Process Skills Emphasized in Each In-Text Laboratory Investigation

Process Skill

Chapter	Observing	Analyzing Data	Relating	Inferring	Evaluating	Identifying	Designing Experiments	Demonstrating	Collecting Data	Measuring	Calculating	Comparing & Contrasting	Classifying	Modeling	Hypothesizing	Testing Hypotheses	Organizing Data	Predicting
1	✓	✓			✓			✓	✓	✓	✓		✓	✓			✓	✓
2	✓		✓		✓						✓	✓					✓	
3		✓	✓	✓				✓		✓	✓		✓				✓	
4		✓		✓				✓	✓		✓						✓	
5		✓						✓	✓	✓	✓						✓	
6			✓								✓			✓				
7		✓	✓	✓				✓			✓						✓	✓
8				✓								✓						
9				✓								✓					✓	✓
10											✓	✓						✓
11	✓					✓					✓						✓	✓
12		✓		✓					✓	✓							✓	✓
13	✓			✓							✓						✓	
14				✓					✓	✓		✓					✓	
15		✓						✓	✓	✓							✓	
16	✓	✓			✓			✓					✓	✓	✓		✓	
17		✓		✓				✓			✓						✓	
18	✓	✓		✓		✓							✓	✓			✓	
19	✓				✓						✓	✓						
20	✓				✓						✓						✓	
21	✓				✓						✓						✓	
22	✓	✓		✓		✓		✓									✓	
23	✓		✓		✓						✓						✓	
24	✓		✓		✓						✓						✓	
25	✓		✓					✓	✓		✓						✓	
26	✓			✓	✓			✓						✓	✓		✓	
27	✓		✓								✓						✓	
28	✓		✓	✓	✓												✓	
29	✓			✓	✓												✓	
30	✓	✓	✓	✓				✓									✓	
31	✓		✓									✓	✓					
32	✓			✓	✓												✓	
33	✓		✓		✓												✓	
34	✓	✓						✓		✓							✓	
35		✓	✓				✓	✓					✓	✓			✓	
36	✓	✓					✓	✓			✓						✓	
37	✓							✓	✓		✓						✓	
38	✓			✓		✓								✓	✓		✓	✓
39	✓			✓				✓									✓	
40		✓	✓	✓				✓	✓		✓		✓				✓	
41	✓	✓		✓						✓	✓						✓	
42	✓				✓						✓						✓	

Your total instructional package supports various teaching styles and learning modalities.

Biology: Principles and Explorations *supplements*
—for practice, review, and assessment

Study Guide

The Study Guide worksheets provide students with opportunities to practice their skills and to review concepts in preparation for the chapter test. Each chapter contains a **Pretest** *and a* **Concept Review.**

- Pretests follow multiple-choice, true-false, and short-answer formats and measure a students' mastery of the chapter materials.

- Concept Review worksheets strengthen a students' abstract-reasoning skills, requiring the student to recognize causal relationships using analogies between biological concepts.

- Science Skills worksheets require students to demonstrate proficiency in a particular science skill. They are found with chapters that emphasize that particular skill. Answers are included in a separate answer-key booklet.

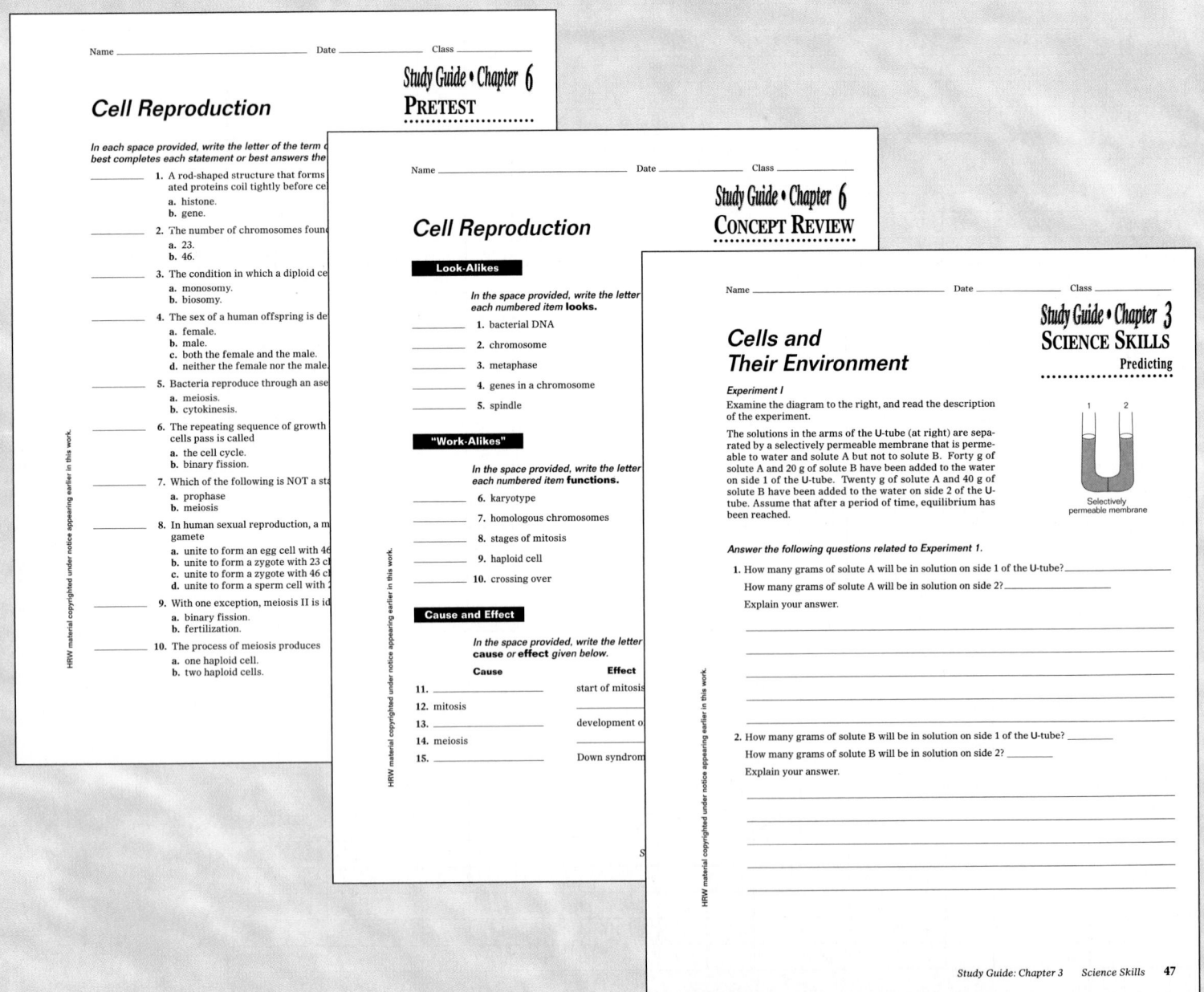

Chapter Tests

Each blackline-master Chapter Test consists of 25 items. Test items measure a student's mastery of section objectives. The following formats are used:

- true-false statements
- multiple-choice questions
- short-answer questions
- analogies
- essay questions

Answers are included in the back of the booklet.

Name _____ Date _____ Class _____

The following questions refer to the diagram that illustrates the life cycle of a mold.

_____ 12. Structure *A* is
　　a. a rhizoid.　　　　　　c. a hypha.
　　b. vascular tissue.　　　d. a stem.

_____ 13. Structure *B* is
　　a. commensal.　　　　　c. embryonic.
　　b. haploid.　　　　　　d. diploid.

_____ 14. The process that takes place at structure *C* is known as
　　a. meiosis.　　　　　　c. mitosis.
　　b. conjugation.　　　　d. fusion.

Complete each statement by writing the correct term or phrase in the space provided.

15. Fungi obtain food by _____ organic matter.

16. Mildews and yeasts are examples of _____ .

17. A fungal _____ is a haploid reproductive cell that is capable of developing into a new organism.

18. A(n) _____ is a saclike structure in which haploid spores are formed.

19. A lichen consists of a fungus and a(n) _____ living together in a symbiotic relationship.

Name _____ Date _____ Class _____

Chapter 22 • TEST

Fungi

Read each statement. If the statement is true, write T in the space provided. If the statement is false, write F in the space provided.

_____ 1. Most fungi are heterotrophic.

_____ 2. Sporangia are reproductive structures in which spores form asexually.

_____ 3. Ringworm is caused by a small wormlike animal.

_____ 4. Mitosis in the tips of ascomycetes hyphae gives rise to conidiophores.

In the space provided, write the letter of the term or phrase that best completes each statement or best answers each question.

_____ 5. Fungi play an important role in the biosphere because they
　　a. break down organic molecules.
　　b. help recycle nutrients.
　　c. are decomposers.
　　d. All of the above

_____ 6. The individual filaments that make up the body of a fungus are called
　　a. vascular tissue.　　　c. rhizoids.
　　b. hyphae.　　　　　　d. stems.

_____ 7. Fungi
　　a. do not contain chloroplasts.
　　b. have cell walls that contain chitin.
　　c. do not produce their own food.
　　d. All of the above

_____ 8. Reproductive structures in which spores form sexually are known as
　　a. gametophytes.
　　b. sporangia.
　　c. yeasts.
　　d. None of the above

_____ 9. Fungi digest food
　　a. through photosynthesis.
　　b. outside their bodies.
　　c. inside their bodies.
　　d. All of the above

_____ 10. An example of a mushroom that grows in moist, organic soils is
　　a. *Endothia parasitica.*　　c. *Rhizopus stolonifer.*
　　b. *Saccharomyces cerevisiae.*　d. *Amanita muscaria.*

_____ 11. Mycorrhizae
　　a. aid in the transfer of minerals from the soil to a plant.
　　b. cause a variety of plant diseases.
　　c. aid in the transfer of minerals to fungi.
　　d. are only found on aquatic fungi.

Holt BioSources *enrich and extend your base of resources.*

Teaching Transparencies

Contains 238 full-color images accompanied by 71 blackline masters

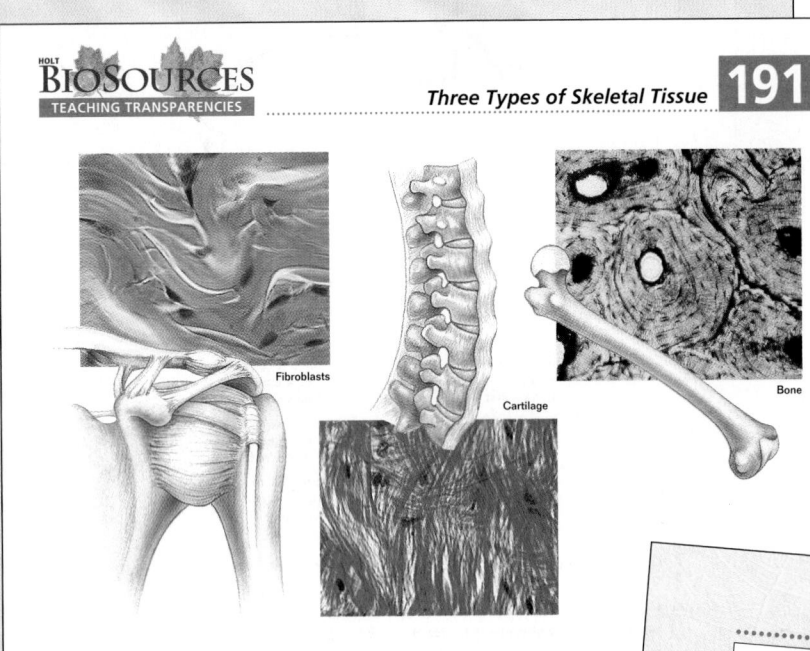

Transparency Directory with Teacher's Notes

Contains teaching strategies to guide your classroom use of each **Transparency** *or* **Transparency Master**

Customize tests using the *Holt BioSources Test Generator*

- available for Windows or Macintosh
- contains more than 3,000 multiple-choice, matching, completion, analogy, and essay items
- includes *Holt BioSources Assessment Item Listing for Biology: Principles and Explorations* listing all items
- core items keyed directly to text objectives organized by chapter and supplemental items organized by topic
- core and supplemental files that let you construct your own tests to assess student mastery of your selected objectives
- items coded to one of three levels of difficulty to challenge students to go beyond rote recall and to analyze and apply chapter content

- includes eight **Performance-Based Assessments,** laboratory-based activities that challenge students to design, perform, and analyze their own experiments to solve an assigned problem (To ensure that each student works independently, the Performance-Based Assessments are provided in pairs; as students use their skills to solve a scientific problem, others around them work simultaneously on a similar—but not identical—scientific problem.)

Name _____ Date _____

Read each statement. If the statement is true, write T in the space provided. If the statement is false, write F in the space provided.

1) ____ The circulatory system transports oxygen, carbon dioxide, food molecules, hormones, and other materials to and from the cells of the body.

2) ____ Blood consists entirely of plasma and hemoglobin.

3) ____ Blood vessels in the body include arteries, veins, and capillaries.

4) ____ The lymphatic system returns fluids from around cells back to the blood vessels.

5) ____ One of the functions of the human circulatory system is to distribute heat to all parts of the body.

6) ____ The innermost layer of an artery is a layer of elastic, smooth muscle tissue.

7) ____ New red blood cells are produced by root

8) ____ When a platelet encounters a damaged bl a clot to form.

9) ____ A person with type AB blood can donate

10) ____ Arteries contain valves that prevent the

11) ____ William Harvey discovered the circular century.

12) ____ Blood plasma is composed primarily o

13) ____ Type O blood contains neither A nor

1

14) ____ The upper chambers of the heart are called the ventricles.

15) ____ Systemic circulation carries blood from the heart to the rest of the body.

16) ____ When measuring blood pressure, the systolic pressure tells how much pressure is exerted when the heart relaxes.

17) ____ Hypertension is high blood pressure.

18) ____ The superior vena cava drains blood from the head and neck.

19) ____ Blood from the left ventricle flows into the aorta.

20) ____ Contractions of the heart muscle are initiated by the pacemaker (SA node).

Circle the letter of the term or phrase that best completes each statement or best answers each question.

21) The circulatory system is responsible for
   a) the distribution of oxygen throughout the body.
   b) carrying nutrients.
   c) transporting hormones throughout the body.
   d) All of the above.

22) Our understanding of the human circulatory system began with the work of
   a) Charles Darwin. b) William Harvey. c) Maurice Wilkins. d) Ernst Haekel.

23) The human circulatory system
   a) helps maintain a constant body temperature.
   b) carries cells that help protect the body from disease.
   c) helps the body maintain homeostasis.
   d) All of the above.

2

24) Red blood cells
   a) transport respiratory gases.
   b) combat bacterial infection.
   c) destroy viruses.
   d) transport cholesterol.

25) Defending the body against bacterial infection and invasion by foreign substances is a function of
   a) red blood cells. b) plasma. c) platelets. d) white blood cells.

26) An abnormality involving the platelets would probably affect the process of
   a) breathing. b) locomotion. c) fighting bacterial infections. d) blood clotting.

27) Blood cells called macrophages
   a) carry oxygen and hemoglobin.
   b) are smaller than red blood cells.
   c) scavenge worn-out or dead cells in the body.
   d) assist in clotting the blood.

e above diagram are
   b) platelets. c) red blood cells. d) white blood cells.

e above diagram
year.
the circulatory system.

3

Sample of student test printout

Teaching Resources CD-ROM

A collection of 89 worksheets that can be downloaded to a diskette or hard drive. Materials in the Scoring Rubrics and Classroom Management Checklists file are also fully editable and can be customized to match your evaluation needs.

The Teaching Resources Main Menu offers these options as *Adobe Acrobat files.*

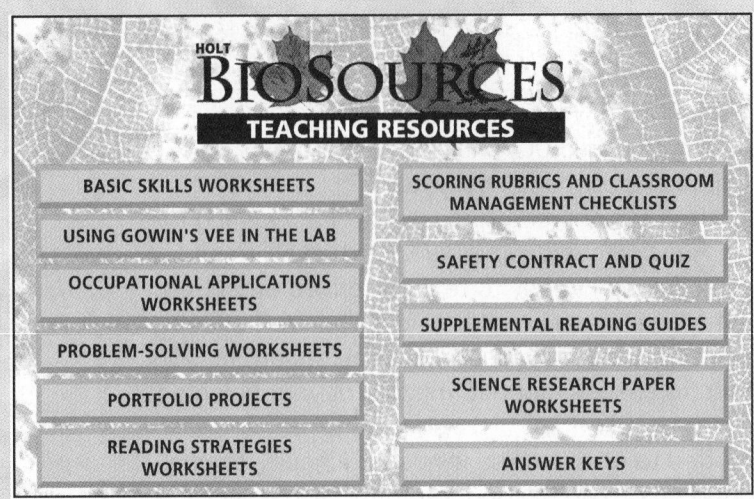

HOLT
BioSOURCES
TEACHING RESOURCES

BASIC SKILLS WORKSHEETS	SCORING RUBRICS AND CLASSROOM MANAGEMENT CHECKLISTS
USING GOWIN'S VEE IN THE LAB	SAFETY CONTRACT AND QUIZ
OCCUPATIONAL APPLICATIONS WORKSHEETS	SUPPLEMENTAL READING GUIDES
PROBLEM-SOLVING WORKSHEETS	SCIENCE RESEARCH PAPER WORKSHEETS
PORTFOLIO PROJECTS	
READING STRATEGIES WORKSHEETS	ANSWER KEYS

Basic Skills Worksheets

provide the critical additional practice that some students need in key skill areas such as graphing, using measurement units, and interpreting microscope magnifications.

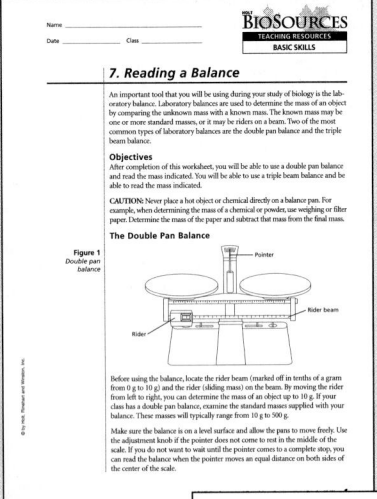

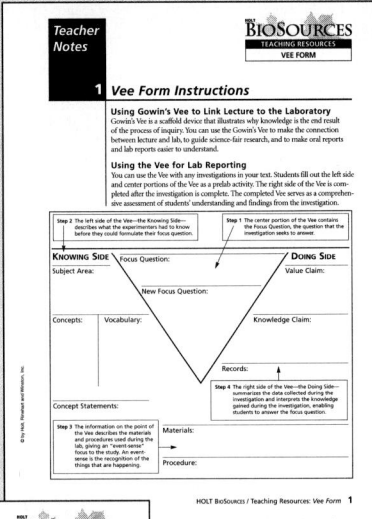

Gowin's Vee Templates

for the lab provide a lab-report format that helps students relate their lab work to what they learn in class.

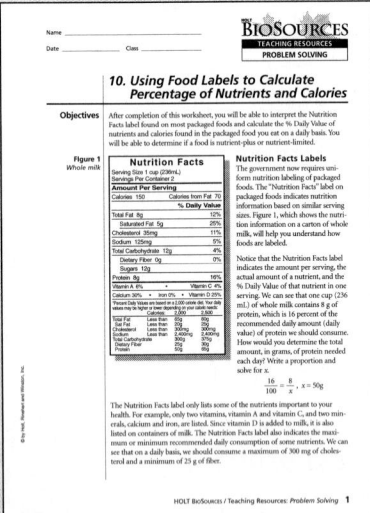

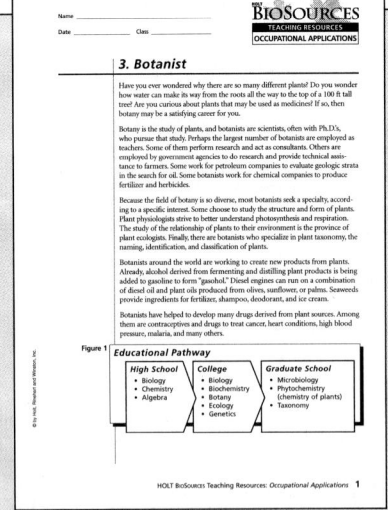

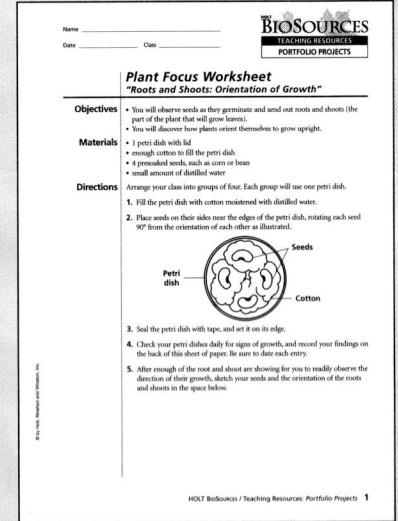

Problem-Solving Worksheets

provide practice with math skills related to biological concepts, such as genetics and probability, percentage composition, and methods for determining population size.

Occupational Applications Worksheets

show students the links between what they are learning in biology and the world of work. Students explore careers in respiratory therapy, botany, and veterinary technology, among others. These worksheets are especially useful for Tech Prep or vocational students.

Portfolio Projects

provide all of the information you need to assign two long-term research projects. Students can elect to explore topics related to plants or genetics. Scoring rubrics and a guide for oral presentations are included to help you evaluate students' research process.

Reading Strategies Worksheets provide helpful strategies and tips that can help poor readers achieve greater success in working with the textbook.

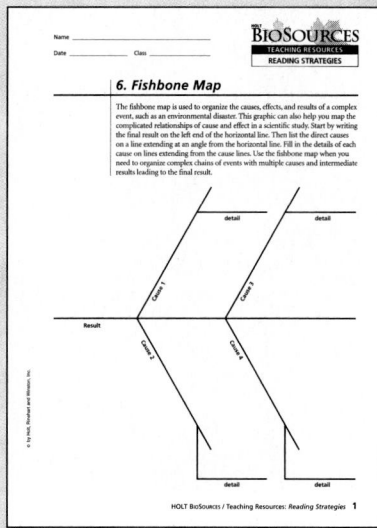

Supplemental Reading Guides help you manage the assignment of biology readings outside the textbook. The reading list includes both contemporary and classic works. Use the guided-reading worksheets to stimulate class discussions while students are reading the book. Students are given the opportunity to work on their language arts skills by providing written discussions about the readings. A test for each selection is also included.

Scoring Rubrics and Classroom Management Checklists help you successfully manage the new forms of assessment, such as performance-based and portfolio. *The Microsoft Word documents for the rubrics can be customized to match your evaluation needs. These documents are located in separate folders on the CD-ROM.*

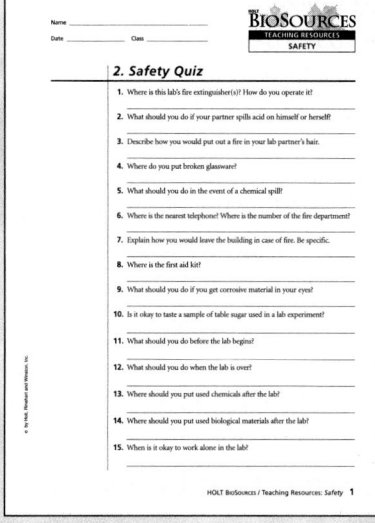

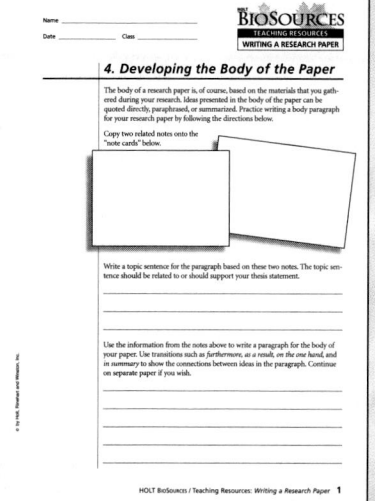

Science Research Paper Worksheets help you manage the development of a science research paper. The guided worksheets help students work through the phases of developing a research paper.

Lab Safety Contract and Safety Quiz ensures that your students have the appropriate safety knowledge before they begin working in the lab.

Answer Keys are provided for the Basic Skills Worksheets, Problem-Solving Worksheets, and Supplemental Reading Guides.

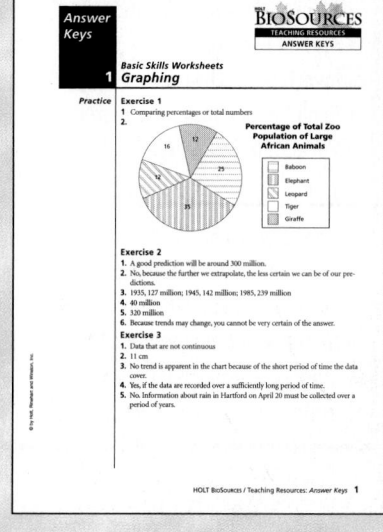

39T

Holt Biology Technology

Holt Biology Videodiscs offers you a comprehensive array of visuals tied directly to lessons found in *Biology: Principles and Explorations.*

Effective, easy to use, and well organized, *Holt Biology Videodiscs* and *Concepts of Biology* allow you to enhance your teaching with videodisc and multimedia teaching tools that are designed around the curriculum, not the technology.

Both *Holt Biology Videodiscs* and *Concepts of Biology* consist of 26 topics divided into four units.

UNIT 1: Cells, Energy, and Genetics

- Cell Structure and Function
- Cell Division
- The Chemistry of Life
- Energy and Enzymes
- Photosynthesis and Cellular Respiration
- Fundamentals of Genetics
- How Genes Work

UNIT 2: Evolution and Ecology

- Evolution and Natural Selection
- Evolution of Life on Earth
- Ecosystem Structure and Function
- Global Ecology
- Classification of Life
- Bacteria and Viruses
- Protists and Fungi

UNIT 3: Plants and Animals

- Plant Evolution
- Plant Structure and Function
- Animal Evolution
- Invertebrate Diversity
- Evolutionary Trends in Vertebrates
- Vertebrate Diversity

UNIT 4: Human Biology

- Body Systems
- Circulation and Respiration
- Digestion and Excretion
- Nervous and Endocrine Systems
- Drugs and the Human Body
- Reproduction and Development

Holt Biology Videodiscs

Four double-sided videodiscs provide thousands of images, including live-action and animated videos.

- More than 30 video segments offer visual examples of the ways in which complex processes work.

- More than 40 original animations allow students to explore sophisticated biological phenomena.

- More than 350 photographs reinforce material covered in *Biology: Principles and Explorations.*

- More than 120 step-frame sequences enable you to demonstrate and explain biological structures, processes, and conceptual relationships.

- More than 650 glossary terms—all of which are available on each videodisc—can be accessed as needed, giving your students valuable vocabulary support.

- Also included are concept maps, tables, graphic organizers, Gowin's Vees, Interactive Investigations, and suggestions for student portfolios.

Teacher's Correlation Guide to Biology: Principles and Explorations contains everything you need to begin using *Holt Biology Videodiscs* in your classroom right away, including barcodes and descriptors correlated specifically to *Biology: Principles and Explorations.*

Image Directory contains a topic index, a subject matter index, and a glossary index, enabling you to locate the images you want quickly and easily.

Also available: Management Software System, a Macintosh software program that enables you to create, save, and even print your own customized media play lists or to select preset media play lists included with the program.

Concepts of Biology CD-ROM adds interactive lesson plans and extensive tools for customization.

Concepts of Biology

Holt Biology Videodiscs

Four CD-ROM discs with:

- All media from *Holt Biology Videodiscs*
- Complete on-line lesson plans
- Media search and sort features
- Interactive glossary
- Customization features for multimedia presentations
- Fully functional word processor
- Digital labs

Teacher's Guide

Image Directory

User's Guide

Interactive Investigations and Laboratories, Teacher's Edition

Interactive Investigations and Laboratories, Pupil's Edition

1995 International GOLD CINDY Award winner

Students can use *Concepts of Biology* to prepare customized multimedia presentations.

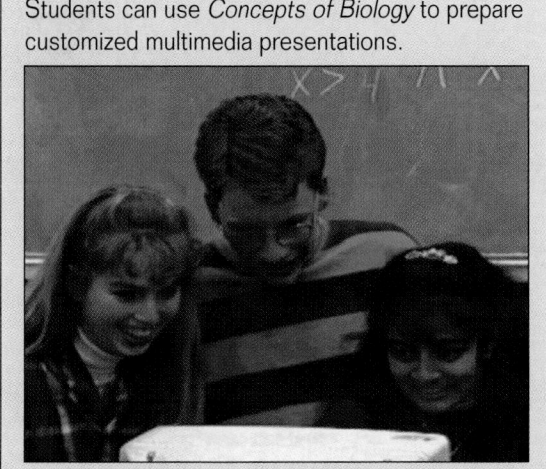

Annotated Teacher's Edition—
your guidebook to efficient instructional planning and effective classroom strategies

Instructional Planning

The Biology: Principles and Explorations Annotated Teacher's Edition is organized to help you plan on a yearly, weekly, and daily basis.

Yearly Planning
Alternative Courses of Study and Pacing are found on page 20T.

Weekly Planning
Teacher's Interleaf pages precede each chapter in the ATE.

Daily Planning
Teacher's Wrap surrounds the reduced pupil pages in the ATE.

Lesson Resources lists the resources available for each lesson categorized by the type of activity. Each listing shows a code that correlates with the Program Resource Key for easy reference. With these options you can create a lesson plan that supports various learning modalities and keeps students engaged for the longer periods required by block scheduling.

Lecture Resources are options *for you* to use during the instructional portion of the class.

Lecture Concepts summarizes the key points of each lesson.

Teacher's Interleaf— Weekly Planning

The Teacher's Interleaf is your master plan for integrating all the materials available for the program into your teaching plan.
Whether you teach on a traditional schedule or follow a block schedule, you know that an array of teaching strategies, classroom options, and assignment options is essential for sustaining student interest. The Teacher's Interleaf presents all of the options available from *Biology: Principles and Explorations* and *Holt BioSources* for use with each chapter.

UNIT 8
CHAPTER 31 OVERVIEW OF VERTEBRATES
Block Scheduling Guide

Each block represents about 45–50 minutes of instructional time. The resources cited in a block are coded to a specific program component using the Key on the next page.

Lecture Concepts	Lesson Resources

31-1 Story of Vertebrate Evolution pp. 702–710

BLOCK ❶

a. Jawless fishes were the first vertebrates. Reptiles were the first terrestrial vertebrates. Birds and mammals have reptile ancestors.
b. Lampreys and hagfishes are the only surviving jawless fishes. Sharks have a lightweight skeleton. Bony fishes have a heavier skeleton and use a swim bladder to provide buoyancy.
c. Amphibians were the first vertebrates to live on land. Because they lack watertight eggs, they must return to the water to reproduce and to keep their eggs from drying out.
d. The dry skin of a reptile prevents water loss. Reptile eggs protect the embryo from drying out on land.
e. Mammals are classified into three groups based on reproductive differences. Monotremes lay eggs. Marsupials bear underdeveloped young that mature in a pouch. Placental mammals nourish developing embryos by means of a placenta.

Lecture Resources
■ Opening Question, page 703
■ Demonstration: Exploring the Backbone
■ Visual Strategy: Figure 31-3
▤ Teaching Transparencies: 134, 164, 137A, 138A, 139A, 140A
◉ Holt Biology Videodiscs: Lesson 31-1

Classwork Options
■ Overcoming Misconceptions, page 708

■ Mammals and Milk, and Graphic Organizer, page 709
▲ Quick Labs: Lab A23 Observing Some Major Animal Groups
▲ Laboratory Techniques and Experimental Design C20: Observing Animal Behavior

Assignment Options
■ Section Review, page 710
■ Chapter Review, questions 1, 2, 8, 9, 17, 28, 30

31-2 Challenge of Obtaining Oxygen pp. 711–716

BLOCK ❷

a. In the gill of a bony fish, water and blood flow in opposite directions, maximizing the amount of oxygen that can be extracted from water.
b. The efficiency of gas exchange in the lung increases with increased surface area. The amphibian lung consists of folded sacs. Alveoli in reptiles provide more surface area than is provided by those of the amphibian lung. Mammals have more alveoli than reptiles.
c. Bird lungs consist of a series of air sacs connected to the lungs. This system, which provides air flow and blood flow similar to those of fish, provides maximum lung efficiency.

Lecture Resources
■ Opening Question, page 711
■ Demonstrations: Breathing Goldfish, Modeling the Increase in Lung Surface Area
▤ Teaching Transparencies: 166, 180, 181, 182

■ Visual Strategies: Figures 31-11, 31-13
◉ Holt Biology Videodiscs: Lesson 31-2

Assignment Options
■ Section Review, page 716
■ Chapter Review, questions 3, 11, 12, 19, 32

31-3 Evolution of a Better Heart pp. 717–722

BLOCKS ❸ and ❹

a. The fish heart is a tube of four connected chambers that collects deoxygenated blood from the body and pumps it to the gills.
b. The amphibian atrium is divided by a septum that sends deoxygenated blood to the lungs and oxygenated blood to the body, with some mixing of the two bloodstreams.
c. The reptile heart has a partial division of the ventricle, which reduces mixing of oxygenated and deoxygenated blood.
d. Crocodiles, birds, and mammals have completely separate ventricles, which prevents any mixing of blood.

Lecture Resources
■ Opening Question, page 717
■ Demonstrations: The Blood Pump, One Way Flow and Blood Flow in a Goldfish Tail
■ Visual Strategies: Figures 31-15, 31-16
▤ Teaching Transparencies: 171, 172

◉ Holt Biology Videodiscs: Lesson 31-3

Classwork Options
▲ Laboratory Techniques and Experimental Design C21: Observing Animal Behavior — Grant Application

Assignment Options
■ Section Review, page 722
■ Chapter Review, questions 4, 10, 16, 18

31-4 Challenge of Retaining Water pp. 723–726

BLOCK ❺

a. The excretory system of vertebrates increases in complexity from fish to mammals, enabling reptiles, birds, and mammals to handle the demands of living in a dry environment.
b. The evolution of protective coverings for eggs has enabled reptiles, birds, and mammals to reproduce in dry environments.

Lecture Resources
■ Opening Question, page 723
■ Demonstration: Fresh Kidneys
■ Visual Strategy: Figure 31-22

▤ Teaching Transparencies: 183
◉ Holt Biology Videodiscs: Lesson 31-4

Assignment Options
■ Section Review, page 726
■ Chapter Review, questions 5, 6, 13, 14, 17, 20, 29, 31

42T

701A   Chapter 31

Use the Block-Scheduling Guide to help you:

- **Plan lessons**

  Lesson Resources makes it easy for you to "switch gears" and keep your class engaged throughout the longer class periods required by block scheduling.

- **Review and assess**

  Review/Enrichment options help students prepare for the final chapter assessment and provide you with ideas and materials to expand course content. Select from the **Assessment Options** to evaluate content mastery.

The Program Resource Key lists the components of your comprehensive resource package.

Biology: Principles and Explorations resources are specifically correlated to each lesson and chapter of the text.

Holt BioSources can be used to extend and customize your course to meet your needs.

Holt Biology Technology gives you options for curriculum-based multimedia instruction.

- Select media to support and extend your lesson. *Holt Biology Videodiscs* provides you with thousands of images correlated lesson-by-lesson to *Biology: Principles and Explorations.* Use the topic from *Concepts of Biology* listed to enrich your lesson curriculum.

Classwork Options includes activities you can select *for students* to complete during class. These include labs and worksheets.

Assignment Options provide additional materials *for students* to complete outside of class.

PROGRAM RESOURCE KEY

- ■ Annotated Teacher's Edition
- ■ Pupil's Edition
- ■ Study Guide
- ■ Chapter Tests

Holt BioSources
- ■ Teaching Transparencies
- ■ Teaching Resources CD-ROM
- ◎ Test Generator/Assessment Item Listing

▲ Laboratory Program
- **A** Quick Labs
- **B** Inquiry Skills Development
- **C** Laboratory Techniques and Experimental Design
- **D** Biotechnology
- **E** Interactive Explorations in Biology Laboratory Manual

Holt Biology Technology
- ◎ Interactive Explorations in Biology CD-ROM
- ⊘ Holt Biology Videodiscs
- ◎ Concepts of Biology

For complete descriptions of all components, see the introductory material at the front of this book.

31-5 Reproduction and Development pp. 727–731

BLOCK ⑥
- a. Fertilization of eggs is external in most fishes and amphibians. After hatching, the larval amphibian feeds and grows until it reaches a certain size and transforms into an adult.
- b. Fertilization is internal in birds and reptiles. Most reptiles and all birds are oviparous.
- c. The placenta provides nourishment for a developing embryo while keeping the bloodstreams of the embryo and of the mother separate.

Lecture Resources
- ■ Opening Question, page 727
- ■ Demonstration: Stages of Development
- ■ Overcoming Misconceptions, page 728
- ■ Visual Strategy: Figure 31-25
- ⊘ Holt Biology Videodiscs: Lesson 31-5
- ▯ Teaching Transparencies: 228

Classwork Options
- ■ Laboratory Investigation: Comparing Vertebrate Characteristics, pages 742–743
- ■ Closure, page 730

Assignment Options
- ■ Section Review, page 730
- ■ Chapter Review, questions 7, 9, 15, 33

Review/Enrichment

BLOCK ⑦
- ■ Highlights: Vertebrate Evolution, pages 731–738
- ■ Highlights Review, questions 21–27
- ■ Study Guide: Concept Review and Pretest

Assignment Options
- ◎ Teaching Resources CD-ROM: Supplementary Reading Worksheets and Test, *The Dinosaur Heresies*
- ■ Activities and Projects, questions 34–36
- ■ Readings, questions 37–38

Assessment Options

BLOCK ⑧

Traditional Assessment
- ■ Chapter 31 Test
- ◎ Test Generator/Assessment Item Listing: Software item bank for preparing customized chapter tests.

Performance Assessment
- ◎ Test Generator : The Challenge of Retaining Water

Portfolio Assessment

Select student reports for one or more laboratory experiments from this chapter. The Direct Observation Checklist, on the Teaching Resources CD-ROM, should be completed during a laboratory or other cooperative-learning experience.

Concepts of Biology adds interactive lesson plans and extensive tools for customization using CD-ROM technology.

Answer to Concept Map
The following is one possible answer to the Concept Mapping exercise on page 739.

Holt Biology Videodiscs

Holt Biology Videodiscs gives you a powerful tool for teaching, review, and assessment. *Concepts of Biology* adds interactive lesson plans and extensive tools for customization using CD-ROM technology.

CONCEPTS OF BIOLOGY

Use the following topics from *Concepts of Biology* to help you teach this chapter:
- ■ Topic 19: Evolutionary Trends in Vertebrates
- ■ Topic 20: Vertebrate Diversity

For further information, see *Holt Biology Videodiscs Teacher's Correlation Guide to Biology: Principles and Explorations.*

Annotated Teacher's Edition, continued

The Lesson Wrap Provides Effective Strategies, Creative Reinforcement, and Thought-Provoking Extensions

Opening Question asks a question about material that has been covered previously.

Vocabulary Preview lists the boldfaced vocabulary terms found in each section.

Chapter Theme helps you integrate one or more of the biological themes important to the chapter into your lesson plan.

23-1 The Evolution of Plants

Plants (members of the kingdom Plantae) are complex multicellular organisms that are primarily terrestrial autotrophs—that is, they occur almost exclusively on land and produce their own organic molecules from inorganic materials by photosynthesis. Today, plants are one of the dominant groups of organisms on land, although some, like the tiny duckweeds seen in Figure 23-1, have returned to aquatic habitats. In this section, you will discover how plants began to adapt to life on land.

Section Objectives
- Identify several obstacles to living on land, and describe how plants overcame them.
- Distinguish nonvascular plants from vascular plants.
- Summarize alternation of generations in plants.
- Describe the moss life cycle.
- Describe the three basic features of vascular plants.

Figure 23-1 Duckweeds, the smallest flowering plants, live only in aquatic environments.

SECTION 23-1

Vocabulary Preview
cuticle
stomata
guard cell
vascular system
alternation of generations
archegonium
antheridium
vascular tissue
sporangium
phloem
xylem
meristem
shoot
root

Opening Question
Ask students to name three things that are important for the survival of an organism on land. Students should mention that survival requires a means of getting water and nutrients, a means of reproduction, and a means of maintaining homeostasis (i.e., self-regulation).

Demonstration
Vascular Plants vs. Nonvascular Plants
To emphasize the difference between living on land and living in water, as did the ancestors of plants, show students a potted plant and a beaker of aquarium water with algae. Ask: Which organism has more difficulty obtaining water? (*the potted plant because it lives in soil, not in water*) Which organism is more likely to dry out? (*the potted plant because it is surrounded by air, not water*) Which organism has more specialized structures? (*the potted plant*)

...e fossil record indicates that ...thing lived on the land sur-...es of our planet until about ... million years ago. For the ... 3 billion years of life's ... fined to the sea. Scientists ... for life to reach terrestrial ... tense solar radiation may ... d uninhabitable. With the ... h's oceans, oxygen gas ... ere. Some of this oxygen ... leading to the develop-... atmosphere. Then, as it ... d Earth's surface from

Did You Know?
...cientists estimate that the ...one concentration over the ...arctic drops by 50 percent ...he spring months. As ...ozone concentration ...eases, the amount of ultra-... radiation reaching Earth's ...e increases.

UNIT 6

CHAPTER 23

OVERVIEW

CHAPTER 23 OVERVIEW OF PLANTS

— A formal flower garden

REVIEW
- mitosis (Section 6-2)
- meiosis (Section 6-3)
- mycorrhizae (Sections 13-2 and 22-3)
- life cycles, gametophyte, and sporophyte (Section 19-2)
- characteristics of the kingdom Plantae (*Highlights: Kingdom Plantae* on page 447)

Chapter Themes

Evolution The development of structures that enabled plants to more efficiently survive on land can be traced by studying the 12 plant phyla. Note that the earliest plants to appear in the fossil record are also the simplest in structure and that the flowering plants, the most recent to appear in the fossil record, are also the most complex and highly specialized plants. Trace the evolution of plants in Table 23-1 and Highlights: Plant Evolution.

Structure and Function Plants do not have "behavior," and therefore the suitability of their structure to the functions they must perform is particularly critical. Look for ways to emphasize how structure relates to function in Figures 23-3, 23-4, 23-8, 23-10, 23-19, 23-20, 23-21, and 23-23.

▶ **Tapping Prior Knowledge**
- How do plants differ from other organisms?
- What makes plants green?
- What were the ancestors of plants?
- What role do plants play in ecosystems?

Opening Demonstration
Show students pictures of several different kinds of plants from different phyla. Ask students to rank the plants in the order in which they evolved based on appearance. Have students justify their sequencing of each plant. On the board or overhead projector, develop a list of the characteristics that students consider primitive and a list of the characteristics that they consider advanced.

Authors' Rationale
The plant kingdom's impact on our lives cannot be overstated. A broad understanding of plants as organisms—their anatomy, physiology, evolution, and diversity—is required for a complete understanding of biology. Life on Earth would not be possible without plants converting solar energy to chemical energy by photosynthesis. This chapter provides an overview of the plant kingdom, identifying the major plant groups and briefly discussing several topics that your students may take for granted, such as flowers, seeds, and fruit. Subsequent chapters build on the information presented in this chapter.

518 Chapter 23

OVERVIEW of Plants 519

Each **Demonstration** provides concrete examples for abstract ideas.

Tapping Prior Knowledge helps you assess how much your students know—and what misconceptions they may have—before you begin teaching.

Opening Demonstration or **Opening Question** prepares and motivates your students for the subjects covered in the chapter.

Each chapter begins with an **Authors' Rationale** concerning the intent of the chapter. Read this first.

Teaching Tips are strategies that help you teach the concepts presented in the lesson. A subhead identifies that concept.

What If . . . teaching tips describe cases in which "something goes wrong" (either temporarily or permanently) in a cell, an organism, or an ecosystem.

Application reveals the relationships between biology and subjects such as technology, sports, mathematics, health, medicine, and geography. This strand helps you show students how biology is relevant to their lives.

Demonstrations that appear on the lesson wrap pertain to content on that actual page.

Appearing once every chapter, **Research Updates** keep you up-to-date by summarizing recent biological research.

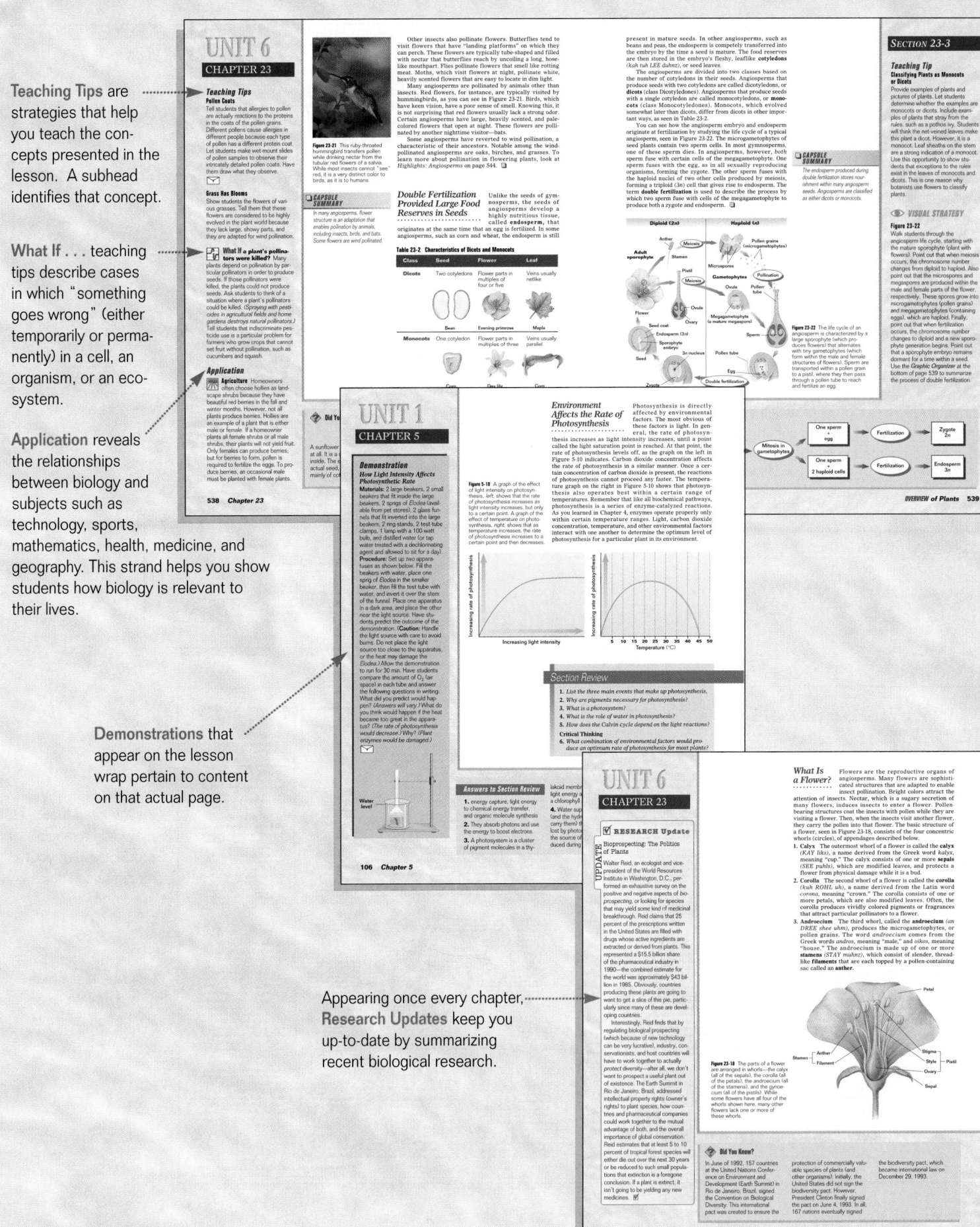

Lesson Wrap, continued

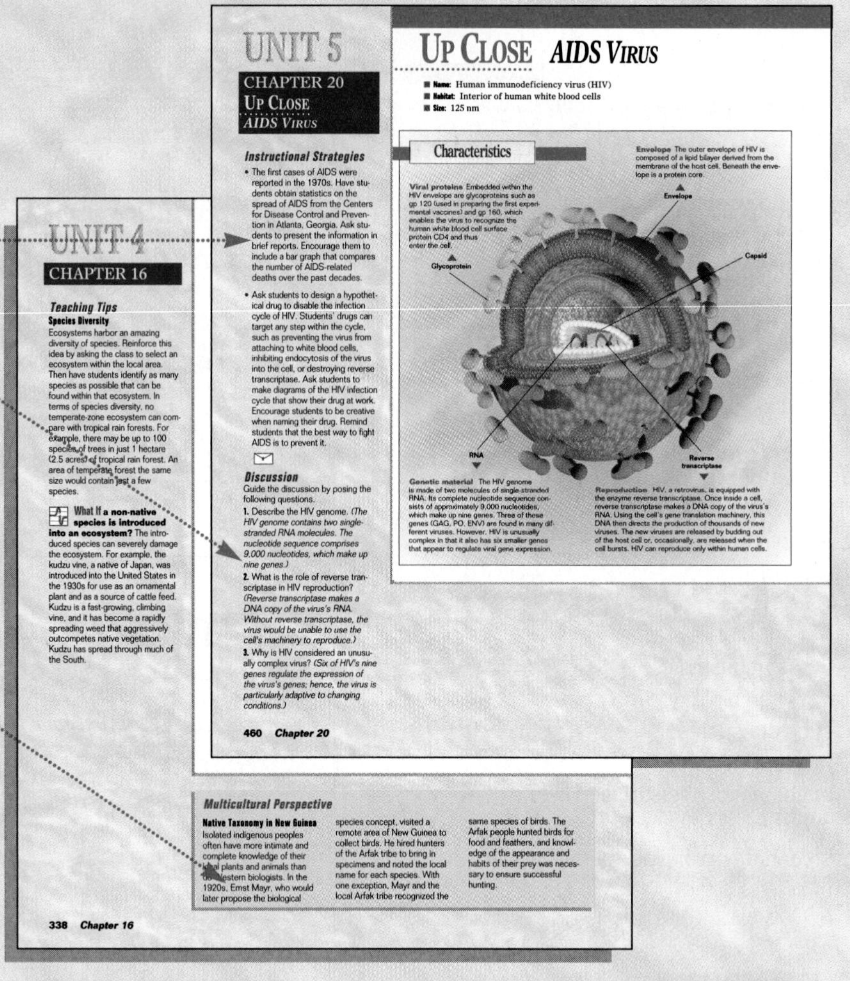

Instructional Strategies and **Discussion** provide background information and suggestions for using **Up Close** and **Highlights** features.

Multicultural Perspective provides information concerning people of various cultures associated with ideas presented in the text, or the influence of culture on a biological issue.

Portfolio icons point out elements in the lesson wrap that may yield a tangible product you or your students can place in a portfolio.

Visual Strategies are teaching ideas that directly relate to an illustration or photograph on the page.

Historical Note provides historical information you can integrate into your teaching of the subject material.

Overcoming Misconceptions identifies common student misconceptions and offers suggestions on how you can help them overcome those misconceptions.

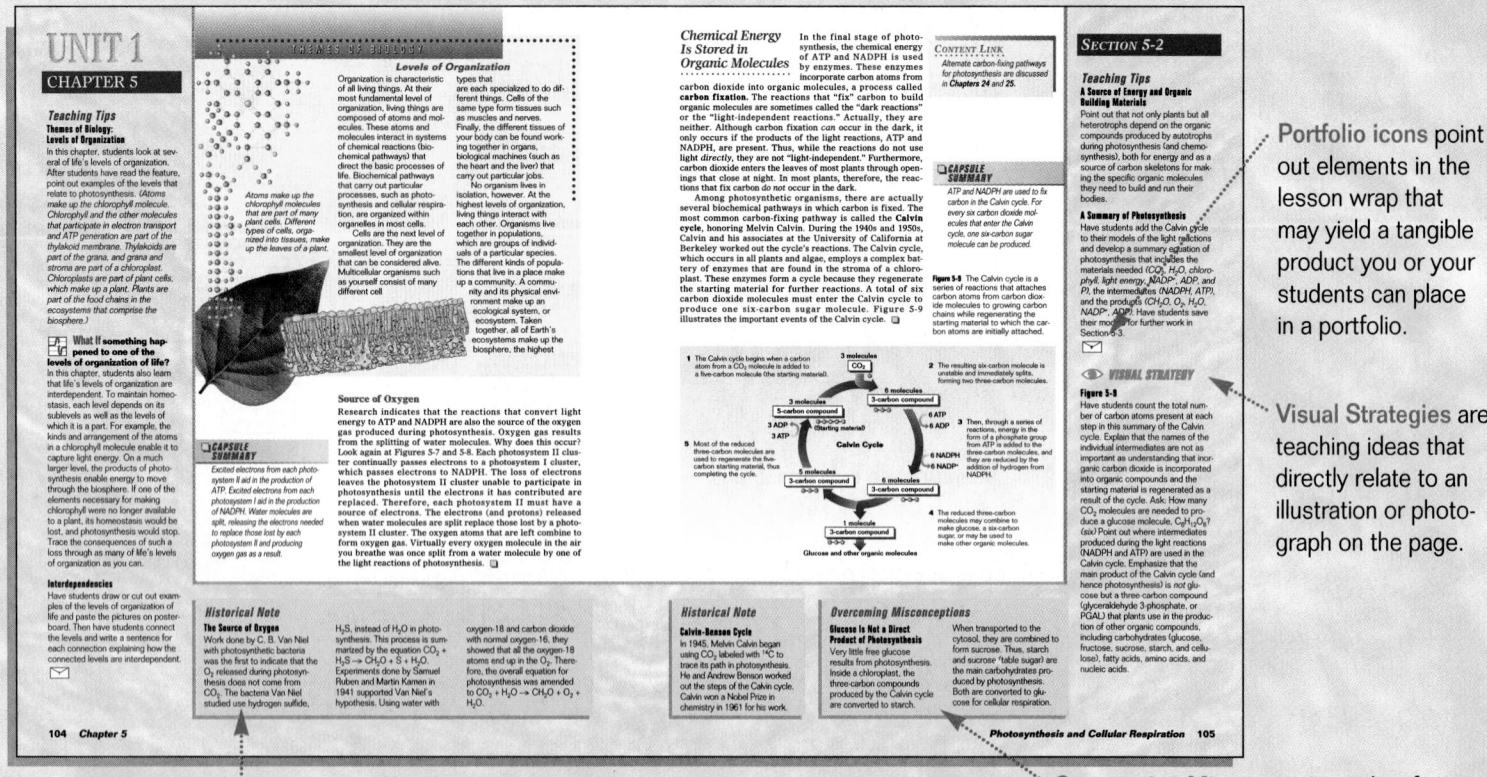

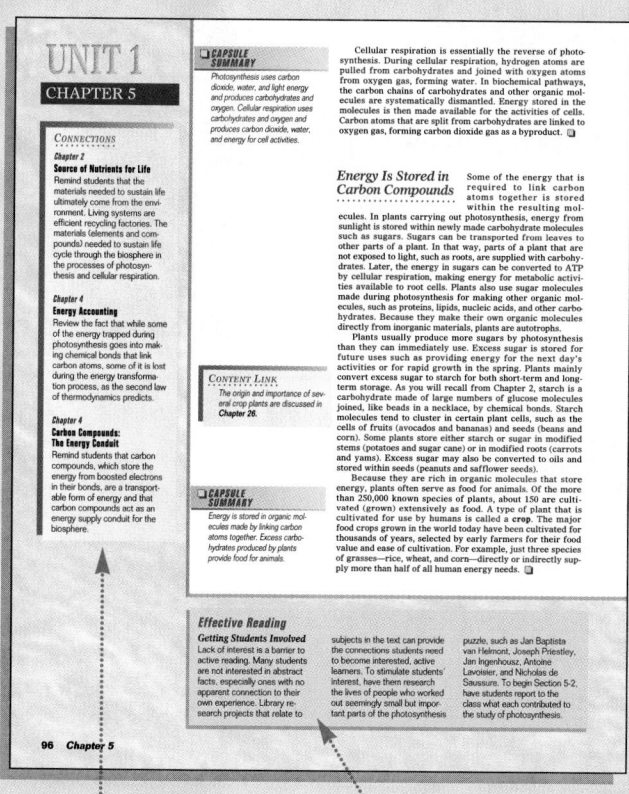

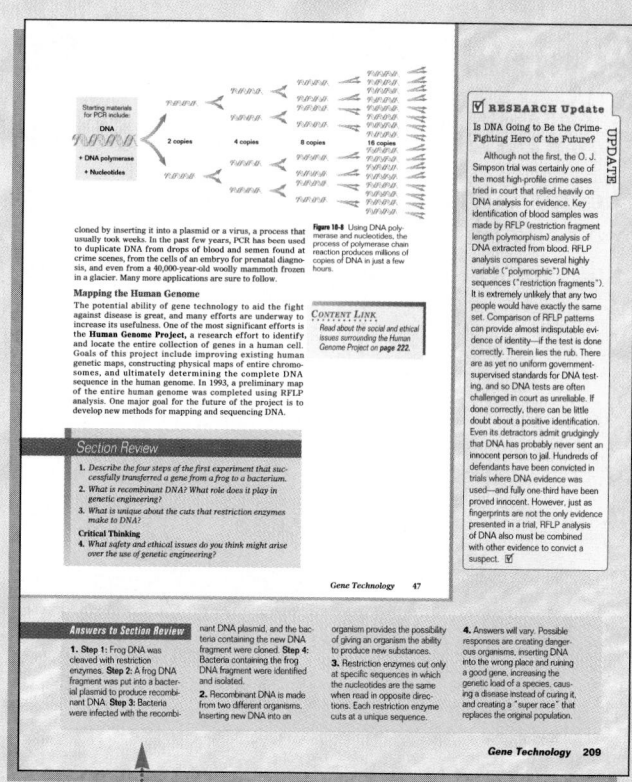

Connections discusses the ways in which a lesson topic relates to material the students have already learned in a previous chapter.

Effective Reading strategies offer tips for how students can increase their reading effectiveness.

Answers to Section Review appear on the same page as the Section Review questions and help you assess your students' mastery of section objectives.

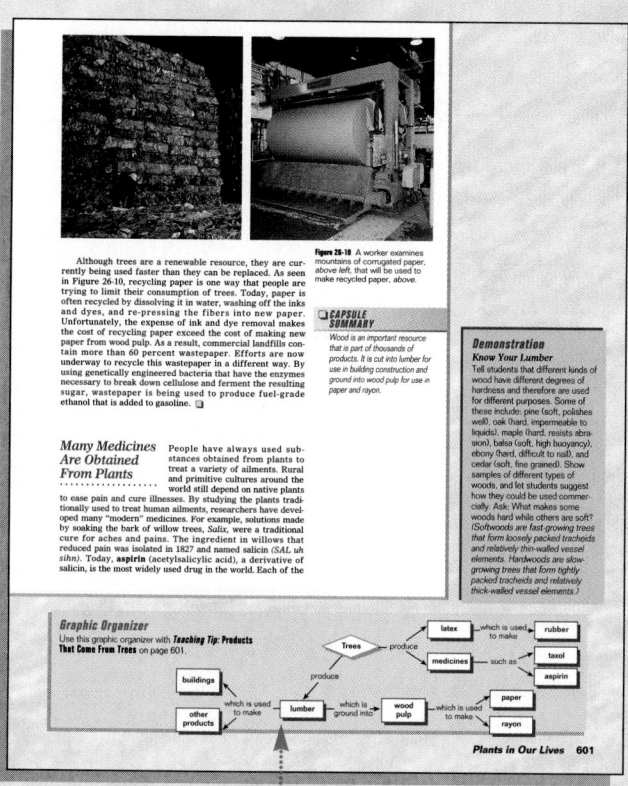

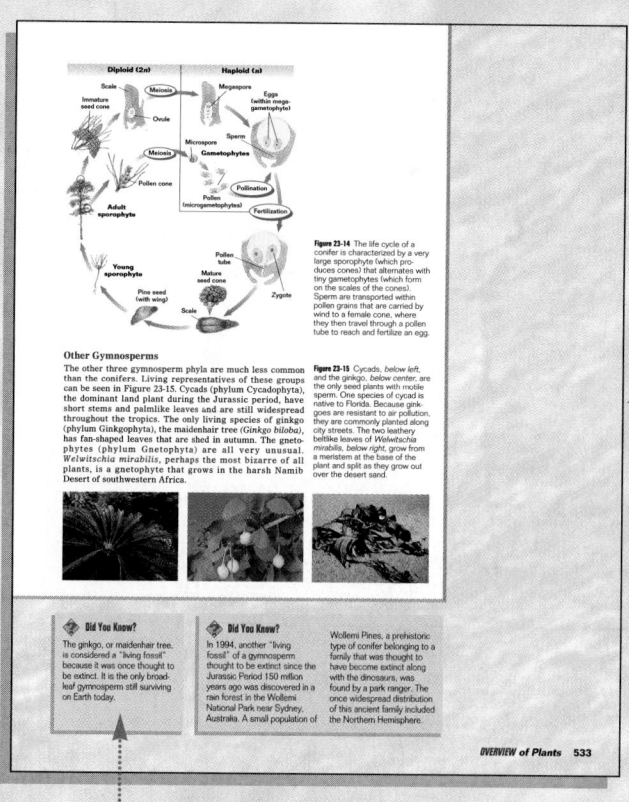

Graphic Organizers are visual tools that help you describe particularly complex information. A nearby Teaching Tip offers suggestions for using the graphic organizer.

Did You Know? offers interesting facts that extend and supplement the text.

Achieving Your Instructional Goals With a Wide Variety of Assessment Strategies

The objectives of a biology course begin with content mastery, but don't end there. Assessment activities throughout **Biology: Principles and Explorations** *help you evaluate students' progress.*

Section Reviews Help Students Check Their Content Mastery.

Section Objectives focus students on the section's most important content.

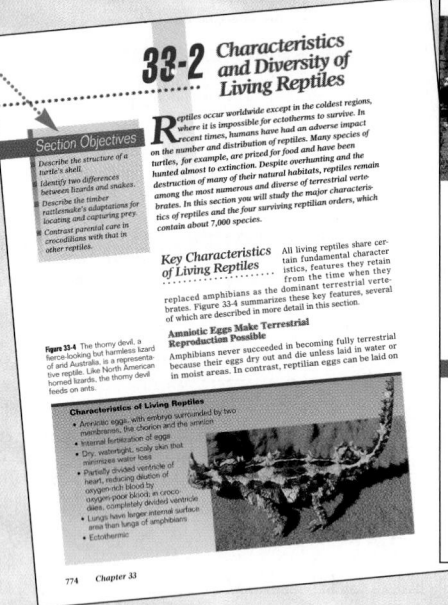

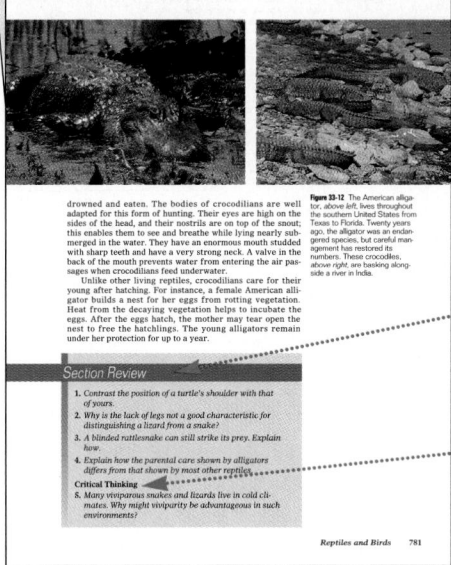

Section Review questions address the Section Objectives.

Critical Thinking questions require students to apply content knowledge to hypothetical situations.

Chapter Reviews Thoroughly Evaluate Content Mastery—and More.

Vocabulary lists the boldface terms from the chapter.

Highlights Review Chapters that include *Highlights* features also include questions about them in the Chapter Review.

Themes Review questions test students' conceptual understanding of chapter content.

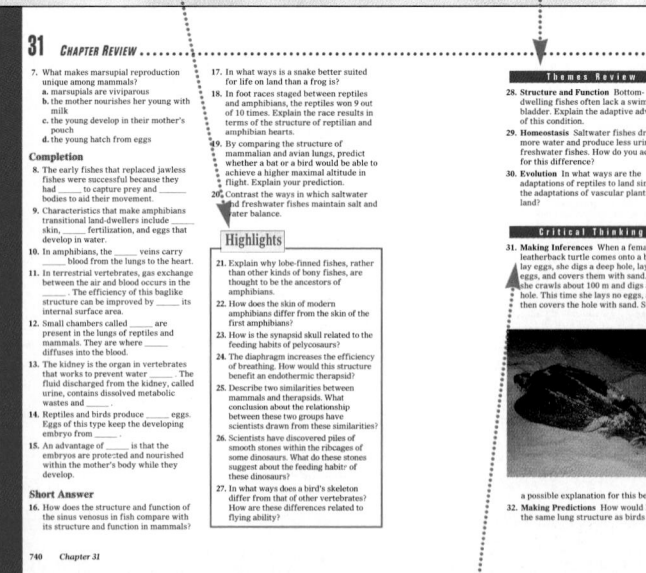

Concept Mapping Each chapter review includes a concept mapping activity that evaluates conceptual understanding by requiring students to link and relate chapter themes.

Review The Review section contains Multiple Choice, Short Answer, and Completion questions that evaluate students' mastery of factual knowledge.

Critical Thinking asks students to apply their knowledge to solving problems.

Holt BioSources *and* Biology: Principles and Explorations *give you the resources you need to make performance-based assessments a reality in your classroom.*

Portfolio Assessment

Students should select their best work and provide a rationale for their selections. You can use the *Portfolio Assignment Student Evaluation Form* on the Holt BioSources Teaching Resources CD-ROM. The portfolio icon denotes additional items that may be selected for inclusion in the portfolio. Use *the Assessment Rubric for Portfolio Items,* also on the CD-ROM, to evaluate portfolios.

Students can make selections in the following areas:

- *Content* One concept map from the chapter (See the rubric *Evaluating Concept Maps** for evaluation criteria.)

- *Reading Comprehension* Study Guide Pretest (Use the Answer Key to evaluate for accuracy.)
  Or: One *Reading Strategies Worksheet*.*

- *Writing* Vee form*, concept map, or report summarizing one of the following:
  - an activity or a reading in the Chapter Review
  - a newspaper or magazine article relating to the chapter topic

Or: A research paper on a subject related to the chapter topic using the *Science Research Paper Worksheets*.*

- *Performance Assessment* One lab report or Vee form* for a lab in *Biology: Principles and Explorations* or the *Holt BioSources Lab Program.*
  Or: A student-designed experiment (See the rubric *Evaluating Student Designed Experiments** for evaluation criteria.)

Teachers can make selections in the following areas:

- *Formal Assessment* Use the *Chapter Test* from *Chapter Tests* or *Holt BioSources Test Generator.* The teacher-scored test should be reviewed by the student and incorrect responses should be corrected by the student before the test becomes part of the portfolio.

- *Informal Assessment* Use the *Informal Assessment Direct Observations Checklist** during a laboratory or other cooperative learning experience.

- *Performance Assessment* Assign the appropriate *Performance-Based Assessment Activity* from *Holt BioSources Test Generator.*

* These materials are found on the *Holt BioSources Teaching Resources CD-ROM.*

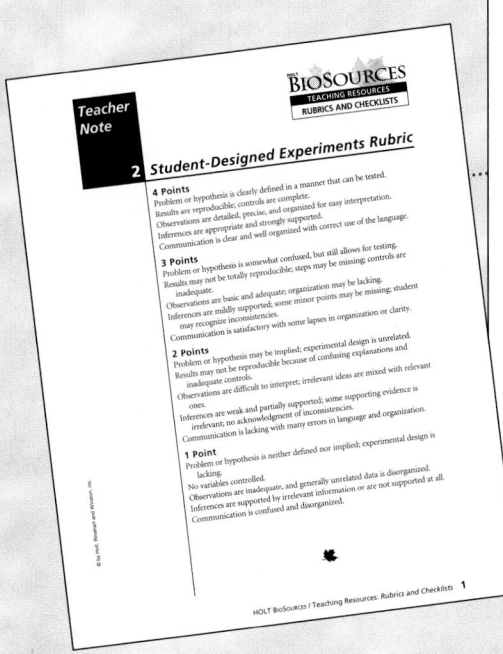

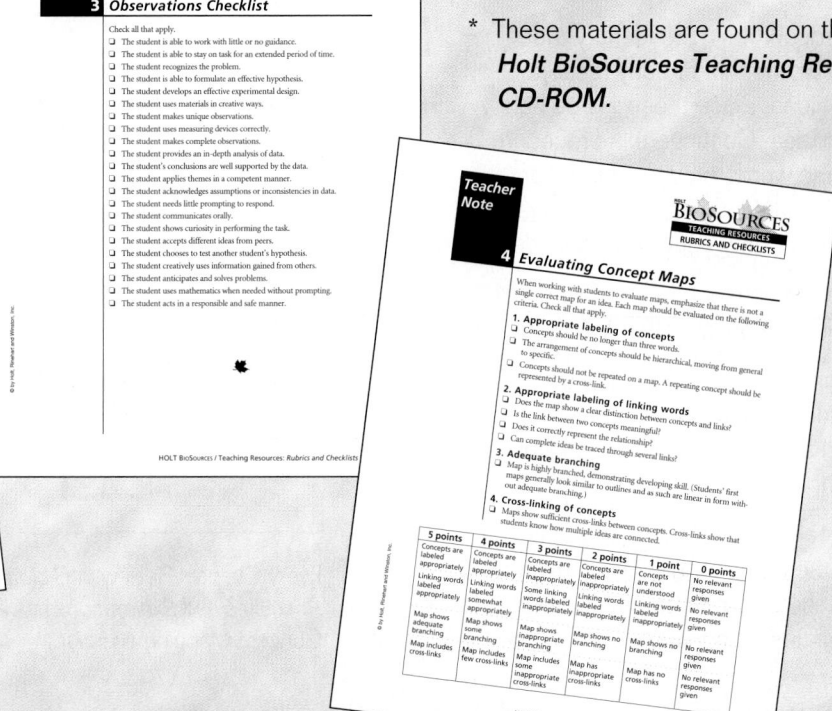

Concept Mapping Biology Concepts

Concept Mapping Helps Develop and Polish Conceptual Understanding

Introducing Concept Maps

Have students read Concept Mapping on page 1018 in their text. Let students know that you will expect them to learn and use this strategy. Be prepared for some initial resistance. If students understand that they will be graded on concept maps throughout the course, they will be more likely to learn the technique.

Concept maps help students move beyond rote memorization of facts by teaching them to organize and interrelate important concepts.

Making Concept Maps

Drawing concept maps involves the following steps:

1. Start with a list of concepts or ideas to be mapped. This list does not have to be complete, but it should be complete enough to allow you to choose the main idea of the map.

2. Look through the list to identify the concept words that directly relate to the main idea. Place these words below the main idea. Continue this procedure with all the words on your list and any supporting ideas until they are all placed in order of priority under the main idea.

3. Use lines to connect the concepts, based on relationships that link them.

4. Use connecting words to label the linking lines so that the relationship between any two concepts is a clear and complete thought.

5. Look for all possibilities to add cross-links to the map. Cross-links show how completely one understands the relationships among concepts.

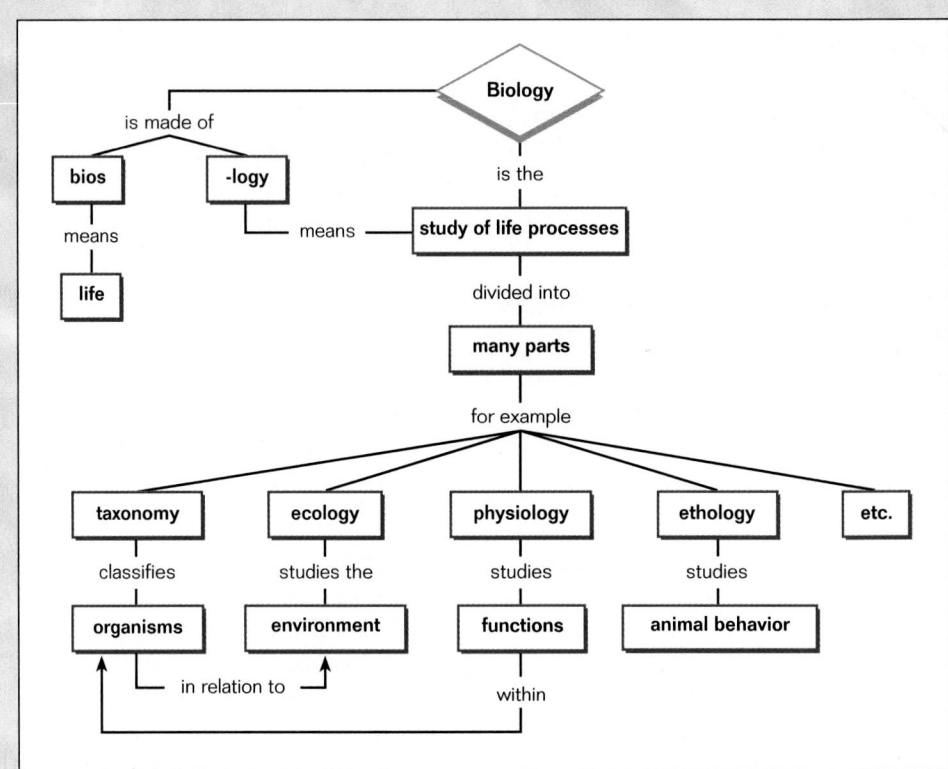

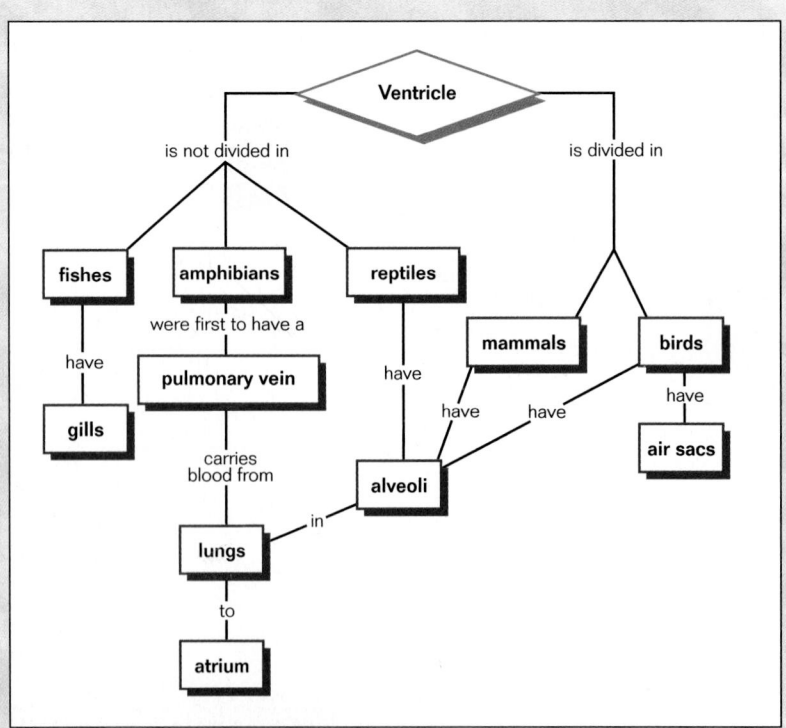

Every Chapter Review in *Biology: Principles and Explorations* has a Concept Mapping exercise.

Evaluating Concept Maps

When working with students to evaluate maps, emphasize that there is not a single correct map for an idea. Each map should be evaluated on the following criteria.

1. Appropriate labeling of concepts

Concepts should be succinctly represented—no longer than three words. The arrangement of concepts should be hierarchical from general to specific. Concepts should not be repeated on a map. A repeating concept should be represented by a cross-link.

2. Appropriate labeling of linking words

Does the map show a clear distinction between concepts and links? Is the link between two concepts meaningful? Does it correctly represent the relationship? Can complete ideas be traced through several links?

3. Adequate branching

Students' first maps generally look similar to outlines and as such are linear in form without adequate branching. Provide incentives for students to turn in highly branched maps so that they will work hard to develop this skill.

4. Cross-linking of concepts

The best maps show sufficient cross-links among concepts. Cross-links show that students know how multiple ideas are connected.

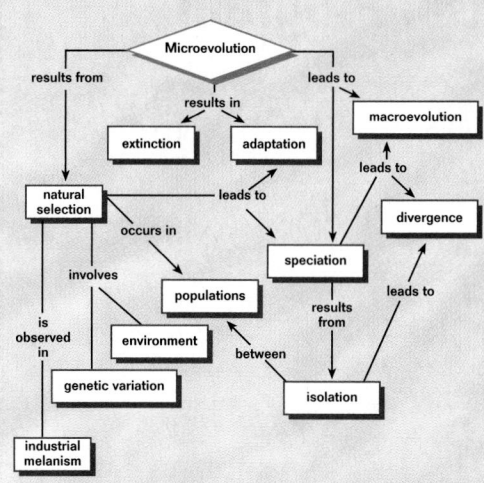

Additional opportunities for Concept Mapping appear in specialized Teaching Tips and associated Graphic Organizers throughout the lesson wrap.

Concept mapping helps students achieve a thorough understanding of the concepts presented in the chapter.

The model of translocation that most botanists favor was proposed in 1924 by the German botanist Ernst Münch. Münch's model was first tested by an experiment like the one seen in Figure 24-13. Using a concentrated sugar solution to represent the phloem near a source, water can be made to enter a tube and flow through it. The pressure created by water entering the tube pushes the water and some of the sugar in the solution to the other end of the tube. Therefore, Münch's model of translocation is often called the **pressure-flow model.** Once the tube is completely filled with water, sugar is also able to diffuse from one end to the other, as long as the sugar concentration remains higher at one end than the other. To see how Münch's model is thought to work

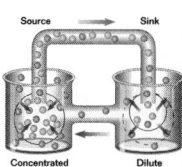

Source   Sink

Concentrated solution   Dilute solution

● Sugar ○ Water

Figure 24-13 In a simple osmometer, *above*, the diffusion of water [...] sugar solu[...]gar solution [...] to the "sink."

SECTION 24-2

◉ VISUAL STRATEGY

Figure 24-14
Walk students through this diagram. Ask volunteers to describe in their own words what happens to sugars and water during translocation. *(Sugars are actively transported into the phloem from a source cell, such as a leaf cell, and move toward a sink cell, such as a root cell. Water diffuses into the phloem because of its higher sugar concentration and flows toward a sink cell.)* Ask: Is it possible for a root cell to be a source? *(yes)* When? *(when upper parts of the plant need sugars and they are not being supplied by the leaves; for example, in the spring when new growth is beginning)* Are the new leaves developing in buds a source or a sink? *(a sink)* Why? *(because while they are in a bud, developing leaves cannot photosynthesize and therefore need sugars from elsewhere to make the energy needed for growth by cellular respiration)*

Chapter Closure

Have students make a concept map using as many of this chapter's vocabulary words as they can. Have them make their maps on large sheets of butcher paper and use markers to draw arrows and write connecting words.

UNIT 9
CHAPTER 37

Teaching Tip
Interconnected Systems
Remind students of how the organ systems are interrelated. Ask them to create a *Graphic Organizer* like the one shown at the bottom of page 892 that demonstrates the relationship between the circulatory, cardiovascular, immune, lymphatic, and respiratory systems. If students are having difficulty, you can put an empty skeleton of the graphic organizer on the chalkboard or overhead projector, and have them fill in the boxes with the correct organ systems.

◉ VISUAL STRATEGY

Figure 37-17
Tell students that the respiratory control center is located in the medulla oblongata region of the brain. Normally, the center is stimulated by changes in the pH of the cerebrospinal fluid that surrounds the brain. When carbon dioxide levels in the blood are high, carbon dioxide can diffuse into the cerebrospinal fluid. Since cerebrospinal fluid has no buffer in it, small changes in carbon dioxide concentration quickly affect its pH. The medulla responds to the low pH by increasing the rate of respiration.

□ **CAPSULE SUMMARY**
Oxygen is transported to tissues by combining with hemoglobin molecules inside red blood cells. Most carbon dioxide is transported to the lungs as bicarbonate ions inside the cytoplasm of red blood cells.

expelled during exhalation. The carbon dioxide carried in plasma diffuses into the alveoli as well. The carbon dioxide bound to hemoglobin is also released because hemoglobin has a greater affinity for oxygen than carbon dioxide. Therefore, the hemoglobin releases its bound carbon dioxide and takes up oxygen instead. The red blood cells, with their newly bound oxygen, then start their next journey back to the body's tissues.

The carbonic anhydrase reaction is critical to the removal of carbon dioxide from tissues because the difference in carbon dioxide concentrations of the blood and tissues is not large (only about 5 percent). We will see in the following section how the level of carbon dioxide in the blood regulates your breathing. □

The Brain Controls Breathing
....

You took your first breath within moments of being born. Since then, you have repeated the process over 200 million times. Every one of these breaths was initiated by the respiratory control center of the brain, as is shown in Figure 37-17.

When the body is at rest, the respiratory control center sends nerve signals to the diaphragm to initiate inhalations. During vigorous breathing, such as during exercise, the level of carbon dioxide in the blood increases, making the blood more acidic. The respiratory control center responds to this increased acidity by sending signals to the muscles between your ribs as well as to the diaphragm. The contraction of the rib muscles causes the chest cavity to expand even further, enabling you to take in more air. Right after inhalation occurs, other neurons from the respiratory center inhibit the stimulation of the diaphragm and the muscles between the ribs so that they relax, and the body exhales.

The respiratory control center also regulates breathing rate (how many breaths you take per minute). Again, blood pH is the controlling factor. When you are sleeping, your muscles are not being used very much, so they don't produce much carbon dioxide. As a result, blood pH remains relatively constant, and you breathe at a slow rate. When you run, intense muscle activity increases the level of carbon dioxide in the blood, and blood pH drops. In response, you breathe more rapidly.

The signals that travel from the breathing center of the brain are not subject to voluntary control. You cannot simply decide to stop breathing. You can hold your breath for a while, but ultimately your respiratory control center will take over and force your body to breathe.

Respiratory control center active — Inhalation (2 seconds) →
Breathing muscles contract →
Inhalation occurs →
Exhalation occurs →
Breathing muscles relax →
Expiration (3 seconds) — Respiratory control center inactive

Figure 37-17 The respiratory control center in the brain initiates each breath. When the center sends out a nerve impulse, inhalation occurs. When the center is inhibited, exhalation occurs.

[...]ding to Münch's [...]s are pushed [...] by the pres[...] the phloem by [...]ws indicate [...] and red arrows [...]vement.

[...]n is the move[...] from a source [...]rs are pushed [...] by the pressure [...]g into areas of [...]centration.

[...]tassium is nec[...]ate the move[...]nto and out of

guard cells; a lack of potassium would interfere with a plant's ability to open its guard cells. If the guard cells cannot open, gas exchange cannot occur, water cannot exit by transpiration, and water cannot be absorbed by the roots.

Plant Structure and Function 565

Graphic Organizer
Use this graphic organizer in *Teaching Tip: Interconnected Systems* on page 892.

Respiratory system
↕ exchanges gases with the
Cardiovascular system
Circulatory system — includes the
loses fluid to the / returns fluid to the
Lymphatic system
contain components of the
Immune system — protects the

Running a Safe Lab Program

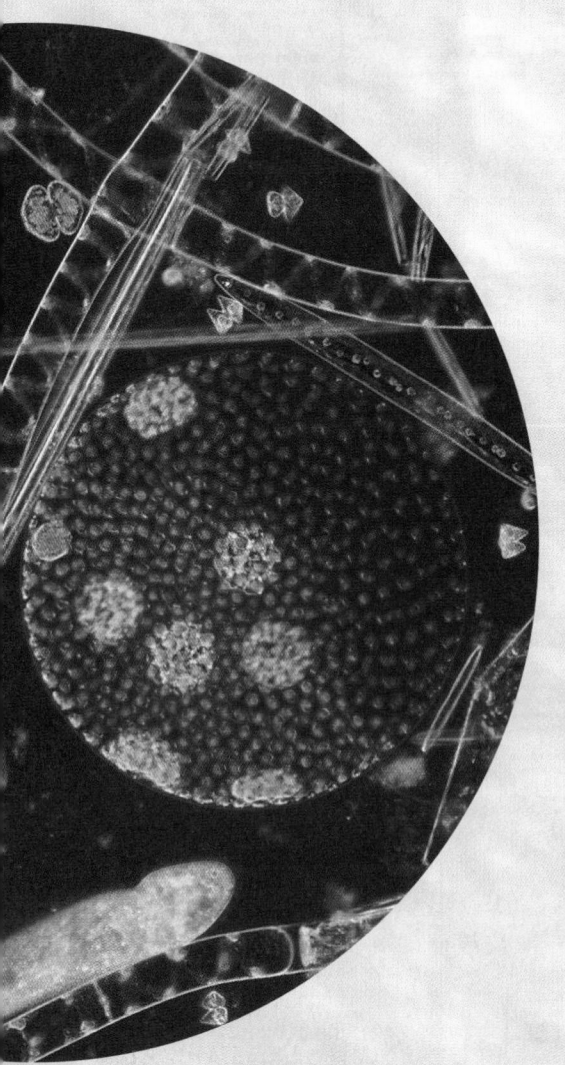

General Safety Guidelines

1. Post laboratory rules in a conspicuous place in the laboratory.

2. Before the class begins an experiment, review safety rules and demonstrate proper procedures.

3. Never permit students to work in your laboratory without your supervision. No unauthorized investigations should ever be conducted, nor should unauthorized materials be in the laboratory.

4. Lock your laboratory (and storeroom) when you are not present.

5. Mark locations of eyewash stations, safety shower, fire extinguishers (ABC tri-class), chemical spill kit, first-aid kit, and fire blanket in the laboratory and storeroom. Check this safety equipment prior to conducting each investigation.

6. Post an evacuation diagram and procedures by every entrance to the laboratory.

7. Provide labeled disposal containers for glass, sharps, and waste chemical reagents.

8. Allow no food or beverages in the laboratory. Caution students to keep their hands away from their face and to wash their hands with soap and water before leaving the laboratory.

9. Know the location for the master shut-off for laboratory circuits. Be sure that all outlets have correct polarity and have ground-fault interception. Polarity can be tested with an inexpensive (about $5) continuity tester available from most electronic hobby shops. All electrical equipment should have 3-prong plugs and 3-wire cords.

10. Follow prescribed procedures for any safety incident, including full documentation. Remind students that any safety incident, no matter how trivial, must be reported directly to you.

Personal Protective Equipment

Chemical goggles: (Meeting ANSI Standard Z87.1) These should be worn when working with any chemical or chemical solution other than water, when heating substances, using any mechanical device, or observing physical processes that could eject an object.

Face shield: (Meeting ANSI Standard Z87.1) Use in combination with eye goggles when working with corrosives.

Contact lenses: The wearing of contact lenses for cosmetic reasons should be prohibited in the laboratory. If a student must wear contact lenses prescribed by a physician, that student should be instructed to wear eye-cup safety goggles meeting ANSI Standard Z87.1 (similar to swimmer's cup goggles).

Eye-wash station: The device must be capable of delivering a copious, gentle flow of water to both eyes for at least 15 minutes. Portable liquid supply devices are not satisfactory and should not be used. A plumbed-in fixture or a perforated spray head on the end of a hose attached to a plumbed-in outlet is suitable if it is designed for use as an eye-wash fountain and meets ANSI Standard Z358.1. It must be within a 30-second walking distance from any spot in the room.

Safety shower: (Meeting ANSI Standard Z358.1) Location should be within a 30-second walking distance from any spot in the room. Students should be instructed in the use of the safety shower in the event of a fire or chemical splash on their body that cannot be simply washed off.

Gloves: Polyethylene, neoprene rubber, or disposable plastic may be used. Nitrile or butyl rubber gloves are recommended when handling corrosives.

Apron: Rubber-coated cloth or vinyl (nylon-coated) halter is recommended.

Emergency Preparedness

What would you do if a student dropped a liter bottle of concentrated sulfuric acid? RIGHT NOW? Plan now how to effectively react BEFORE you need to.

1. Post the phone numbers of your regional Poison Control Center, Fire Department, Police, Ambulance, and Hospital on your telephone.

2. Practice fire and evacuation drills. Also have drills on what students MUST do if they are on fire or have chemical contact.

3. Assure that all personal and other safety equipment is available and tested frequently.

4. Compile an MSDS file for all chemicals. This reference resource should be readily available in case of a spill or other accident.

5. Provide for spill control procedures. Handle only those incidents that you feel comfortable in handling. Situations of greater severity should be handled by trained hazardous material professionals.

6. Students should never fight fires or handle spills.

7. Be trained in first aid and basic life support (CPR) procedures. Have first-aid kits and spill kits readily available.

8. Fully document ANY INCIDENT that occurs.

Safety With Animals

It is recommended that teachers follow the "Guidelines for the Use of Live Animals" established by the National Association of Biology Teachers, which is reproduced in the Teacher's Editions of *Holt BioSources Lab Program Quick Labs* and *Inquiry Skills Development* lab manuals.

Safety in Handling Preserved Materials

The following practices are recommended when handling or dissecting any preserved specimen:

1. Wear protective gloves and splash-proof safety goggles at all times when handling preserving fluids, preserved specimens, and during dissection.

2. Wear lab aprons. Use of an old shirt or smock is recommended.

3. Conduct dissection activities in a well-ventilated area.

4. Do not allow preservation or body-cavity fluids to contact skin. Fixatives do not distinguish between living or dead tissue.

Biological supply firms use formalin-based fixatives, of varying concentrations, to initially fix zoological and botanical specimens. WARD'S Natural Science Establishment provides specimens that are freeze-dried, and rehydrated in a 10% isopropyl alcohol solution. In these specimens, no other hazardous chemical is present.

Many suppliers provide fixed botanical materials in 50% glycerin.

Reduction of Free Formaldehyde

Currently, federal regulations mandate a permissible exposure level of 0.75 ppm for formaldehyde. Contact your supplier for an MSDS that details the amount of formaldehyde present as well as gas-emitting characteristics for individual specimens.

Pre-washing specimens (in a loosely covered container) in running tap water for 1–4 hours will dilute the fixative. Formaldehyde may also be chemically bound (thereby greatly reducing danger) by immersing washed specimens in a 0.5–1.0% potassium bisulfate solution overnight or by placing them in 1% phenoxyethanol holding solutions.

Safety With Microbes

Pathogenic (disease-causing) microorganisms are not appropriate investigational tools in the high school laboratory and should never be used.

Safety With Chemicals

Label student reagent containers with the substance's name and hazard class(es) (flammable, reactive, etc.).

Dispose of hazardous waste chemicals according to federal, state, and local regulations. Refer to the Material Safety Data Sheet for recommended disposal procedures.

Remove all sources of flames, sparks, and heat from the laboratory when any flammable material is being used.

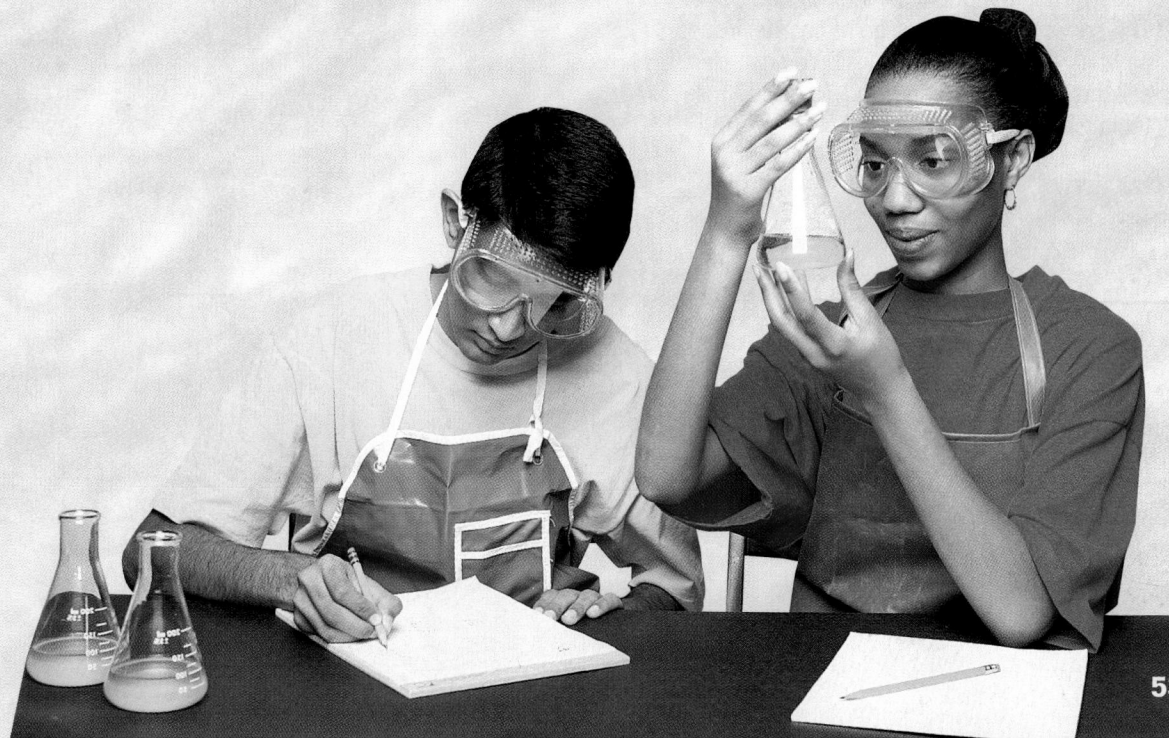

Material Safety Data Sheets

The purpose of a Material Safety Data Sheet (MSDS) is to provide readily accessible information on chemical substances commonly used in the science laboratory or in industry.

The MSDS should be kept on file and referred to BEFORE handling ANY chemical. The MSDS can also be used to instruct students on chemical hazards, to evaluate spill and disposal procedures, and to warn of incompatibility with other chemicals or mixtures.

Resources

American Chemical Society Health and Safety Service

This service will refer inquiries to appropriate resources for finding answers to questions about health and safety.

American Chemical Society (ACS)
1155 Sixteenth Street, N.W.
Washington, D.C. 20036
1-800-227-5558

Hazardous Materials Information Exchange (HMIX)

Sponsored by the Federal Emergency Management Agency and the U.S. Department of Transportation, HMIX can be accessed through an electronic bulletin board, and it provides information regarding instructional material and literature listings, hazardous materials, emergency procedures, and applicable laws and regulations.

HMIX can be accessed by a personal computer having a modem. Dial (630) 252-3275. The service is available free of charge. You pay only for the telephone call. You can also call the HMX hotline at 1-800-752-6367 for technical assistance.

Safety Information References

Gessner, G. H., ed. *Hawley's Condensed Chemical Dictionary.* 11th Ed. Van Nostrand Reinhold, 1987. (Revised by N. Irving Sax).

A Guide to Information Sources Related to the Safety and Management of Laboratory Wastes from Secondary Schools. New York State Environmental Facilities Corp., 1985.

Lefevre, M.J. *The First Aid Manual for Chemical Accidents.* Dowdwen, 1989. (Revised by Shirley A. Conibeau).

Pipitone, D., ed., *Safe Storage of Laboratory Chemicals.* John Wiley, 1984.

Prudent Practices for Disposal of Chemicals in Laboratories. Committee on Hazardous Substances in the Laboratory, National Research Council, National Academy Press, 1983.

Prudent Practices for Handling Hazardous Chemicals in Laboratories. Committee on Hazardous Substances in the Laboratory, National Research Council, National Academy Press, 1981.

Strauss, H. and M. Kaufman, ed. *Handbook for Chemical Technicians.* McGraw-Hill, 1981.

WARD'S MSDS *User's Guide.* WARD'S, 1989.

Storing Chemicals

Never store chemicals alphabetically, as this greatly increases the risk of promoting a violent reaction.

Storage Suggestions:

1. Always lock the storeroom and all its cabinets when not in use.

2. Students should not be allowed in the storeroom and preparation area.

3. Avoid storing chemicals on the floor of the storeroom.

4. Do not store chemicals above eye level or on the top shelf in the storeroom.

5. Be sure shelf assemblies are firmly secured to the walls.

6. Provide for anti-roll lips on all shelves.

7. Shelving should be constructed out of wood. Metal cabinets and shelves are easily corroded.

8. Avoid metal, adjustable shelf supports and clips. They can corrode, causing shelves to collapse.

9. Acids, flammables, poisons, and oxidizers should each be stored in their own locking storage cabinet.

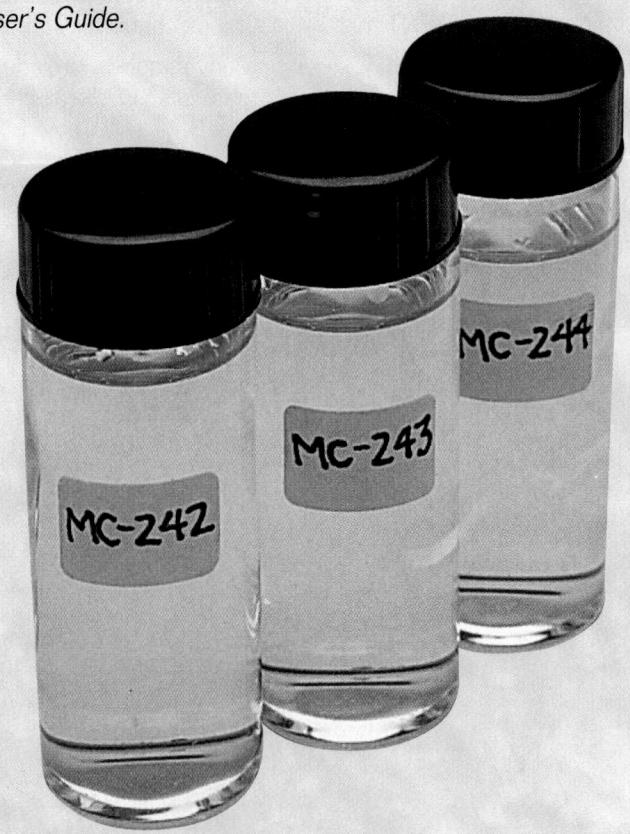

Bench-Tested and Teacher-Reviewed Lab Investigations

WARD'S Natural Science Establishment, Inc., the market leader in the production and distribution of science supplies, is the official materials and equipment supplier for *Biology: Principles and Explorations* and the *Holt BioSources Lab Program*. Holt, Rinehart and Winston's unique partnership with WARD'S guarantees the following:

- WARD'S lab materials and supplies are designed to specifically meet the requirements of the investigations in *Biology: Principles and Explorations* and *Holt BioSources* Lab Program.

- All investigations and lab activities in *Biology: Principles and Explorations* and the *Holt BioSources* Lab Program are bench tested by WARD'S technical staff to guarantee successful lab results and experiences.

- With exclusive WARD'S products like Simulated Blood™, Simulated Plant Pigments™ and others, WARD'S has eliminated the use of hazardous materials and dangerous chemicals in your classroom.

Ordering equipment and materials is convenient and easy with WARD'S software ordering system.

With WARD'S exclusive software-ordering system, specifically designed for use with *Biology: Principles and Explorations*, you can order your materials and supplies quickly and easily.

The software-ordering system lists all of the required and supplemental materials—per lab group or class size—needed for every lab investigation in *Biology: Principles and Explorations.*

- Click on the products you need, and the software automatically creates your "shopping list," keeping track of the materials you order and their costs.

- A Help window explains every function, making the program easy to use.
- The software-ordering system is available for both Macintosh and IBM-compatible computers.

Master Materials List

This list indicates all supplies needed for a class of 32 to perform the in-text investigations working in pairs or in groups of four. Order quantities for equipment and supplies may be obtained using the charts that follow. WARD'S is the exclusive supplier for *Biology: Principles and Explorations*. Catalog numbers for WARD'S and the quantities to order are provided.

WARD'S Natural Science Establishment, Inc.
5100 W. Henrietta Road
P.O. Box 92912
Rochester, New York 14692-9012

Phone: 1-800-962-2660
FAX: 1-800-635-8439

Biological Supplies

Item	WARD'S No.	Qty.	Inv.
Amoeba culture	87 M 0380	1	21
Angiosperm (*Tradescantia*)	86 M 7300	1	23
Antheridial head, microscope slide	91 M 4322	8	23
Archegonial head, microscope slide	91 M 4326	8	23
Bacillus megaterium culture	85 M 1154	1	20
Bacillus, microscope slide	90 M 0534	8	20
Bean seeds, 1lb.	86 M 8011	1	25
Beans, kidney,1lb.	86 M 8006	1	1
Black rams snail, pkg. 12	87 M 4161	1	16
Brine shrimp eggs	87 M 5102	1	17
Brown *Hydra* culture	87 M 2020	1	18, 28
Bryophyta (moss)	86 M 4360	1	23
Chlamydomonas culture	86 M 0102	1	16
Clam (mussels, freshwater), pkg. 10	87 M 4420	1	29
Clam shell	Local	8	29
Clover seeds	Local	pinch	16
Coccus, microscope slide	90 M 0554	8	20
Corn seeds, 1 lb.	86 M 8081	1	25
Crickets, pkg. 50	87 M 6102	1	16, 31, 32
Daphnia magna culture	87 M 5210	1	16, 28
Diatomaceous earth, jar/10	63 M 0240	1	21
Duckweed culture (*Lemna*)	86 M 7650	1	16, 26
Earthworm, live, pkg. 10	87 M 4660	3	16
Elodea, pkg. 10	86 M 7500	1	2, 16
Euglena culture	87 M 0100	1	21
Euplotes culture	87 M 1100	1	21
Feather set	69 M 2269	1	31
Ferns (Boston fern)	86 M 5550	1	23
Fish for aquarium set	86 M 7911	1	31

Item	WARD'S No.	Qty.	Inv.
Fontinallis strands	86 M 7456	1	16
Foxtail strands	86 M 7725	1	16
Frogs, live, pkg. 12	87 M 8217	1	30, 31
Fungal samples	Local		22
Grass seed, 1 oz.	86 M 8130	1	16
Guppies, pkg. 12	87 M 8110	1	16
Human blood tissue, microscope slide	93 M 6541	8	2
Human squamous tissue, microscope slide	93 M 6003	8	2
Invertebrate survey set	62 M 0044	1	27
Isopods, culture	87 M 5520	1	16, 30
Liverwort (*Conocephalum*)	86 M 4050	1	23
Mealworms, live, pkg. 100	87 M 6250	1	16
Micrasterias culture	86 M 0270	1	21
Micrococcus luteus culture	85 M 0966	1	20
Mung beans, 1 oz. pkg.	86 M 8007	1	12, 16
Paramecium culture	87 M 1300	1	18, 21
Peas, 2 oz. pkt.	86 M 8240	1	25
Physarum slime mold	85 M 4750	1	21
Pond snails, pkg. 6	87 M 4140	1	16
Ranunculus CS mature root, microscope slide	91 M 8142	8	24
Ranunculus CS stem, microscope slide	91 M 8144	8	24
Saccharomyces cerevisiae culture	85 M 5000	1	15
Sea star egg, unfertilized, microscope slide	92 M 8241	8	42
Sea star microscope slide, composite	92 M 8255	8	42
Sea star sperm, microscope slide	92 M 8238	8	42
Spirillum volutans culture	85 M 1140	1	20
Spirillum, microscope slide	90 M 0557	8	20
Spirogyra culture	86 M 0650	1	21
Stentor culture	87 M 1370	1	21
Tadpoles, live, pkg. 50	87 M 8211	2	41
Tree leaves	Local	1	12
Unknown slide 1 (plant mitosis)	93 M 2145	8	2
Unknown slide 2 (*Allium* mitosis)	91 M 7040	8	2, 24
Unknown slide 3 (animal cell slide)	93 M 2200	8	2
Vertebrate survey set	62 M 0045	1	31
Volvox culture	86 M 0800	1	21
Vorticella culture	87 M 1451	1	21
Zea mays stem, microscope slide	91 M 7448	8	24

Chemicals and Media

Item	WARD'S No.	Qty.	Inv.
Aspartic acid, 10 g	37 M 5542	1	11
Beef extract, 500 g	38 M 2123	1	28
Detain, 1/2 oz.	37 M 7950	8	17, 21
Disinfectant solution, 120 mL	37 M 7980	1	1
Gelatin, 100 g	38 M 3030	1	4
Glucose, 500 g	39 M 1457	1	38
Glutamic acid, 10 g	37 M 5543	1	11
Glycine, 25 g	39 M 1420	1	11
HCl 8% (in a 3 mL dropper bottle)	750 M 3915	8	29
Iodine solution, 100 mL	37 M 2379	1	15, 25
Knop's solution, 500 mL	38 M 3520	1	26
Methyl red, 10 g	38 M 8310	8	37
Methylene blue 1%, 500 mL	38 M 9522	1	20
Milk treatment product (liquid), 30 qt. supply	39 M 0140	1	38
Nutrient agar plates, pkg. 8	88 M 0906	3	22
Phenolphthalein agar, 800 mL jar	250 M 0828	1	3
Plant fertilizer, 2.7 g packet	20 M 6040	1	26
Pond water, 1 gal.	88 M 7010	3	16, 26, 41
Potassium chloride, 500 g	37 M 4634	1	40
Propionic acid, 500 mL	39 M 8957	1	22
Silicone culture gum, 1 oz.	37 M 9810	1	28
Sodium chloride, 500 g	37 M 5487	1	11, 40
Sodium hydroxide 0.4%, QS to 1L	37 M 9546	1	37
Sodium hydroxide, 500 g	39 M 8176	1	11
Thyroxin, 500 mg	38 M 5330	1	41
Vinegar, 500 mL	39 M 0138	1	1, 3
Washing soda (sodium carbonate), 500 g	37 M 5477	1	4
Water, distilled, 1 gal.	88 M 7005	1	26

Laboratory Equipment

Item	WARD'S No.	Qty.	Inv.
Alligator clips, 2"	15 M 9473	16	40
Aquarium, 10 gal.	21 M 5241	1	31, 32
Balance, triple-beam	15 M 6057	8	4, 11
Beaker, 250 mL	17 M 4040	9	1, 3
Beaker, 600 mL	17 M 4060	8	11, 29, 32, 41
Beaker, 50 mL	17 M 4010	8	4, 17, 25
Beaker, 150 mL	17 M 4030	8	4, 18, 20, 25, 40
Beaker, 400 mL	17 M 4050	16	18
Calculator	27 M 3055	8	12, 17
Clamp for ring stand	15 M 0699	8	11
Clear plastic tubing, 50' roll	18 M 5094	1	17
Clock	15 M 1492	8	11, 18
Copper wire, 1lb. spool	15 M 9235	1	40
Corks #2, pkg. 100	15 M 8362	1	17

Item	WARD'S No.	Qty.	Inv.
Corks #4, pkg. 100	15 M 8364	1	18
Coverslips, pkg. 100	14 M 3555	4	2, 11, 15, 21
DC millivoltmeter	21 M 0550	8	40
Depression slides, pkg. 72	14 M 3510	8	38
Desk lamp	15 M 5036	8	21
Dialysis tubing, 100 ft. roll	14 M 4512	1	40
Dissecting pan	14 M 7000	8	33
Erlenmeyer flask, 125 mL	17 M 2981	16	11
Filter paper, 90 mm, box 100	15 M 2805	1	28
Flashlight	15 M 3264	8	21
Flasks, 250 mL	17 M 2803	8	37
Food coloring, pkg. 4, 1/4 oz. bottles	15 M 0071	2	29
Forceps	14 M 1001	8	2, 21, 28
Funnel	18 M 1331	8	17
Glass jar, 1 gal.	17 M 2080	3	16
Glucose test strips, pkg. 100	14 M 4119	1	38
Graduated cylinder, 10 mL	17 M 0170	96	5, 37, 41
Graduated cylinder, 50 mL	17 M 0172	6	4, 11
Graduated cylinder, 100 mL	17 M 0173	8	18, 37
Graduated plastic pipets, pkg. 500, 1 mL	18 M 2971	1	15, 17, 21, 37
Hot plate	15 M 8055	8	11
Lab apron, vinyl	15 M 1005	32	1, 2, 4, 5, 11, 20, 22, 29, 39
Medicine dropper, 3", pkg.12	17 M 0230	2	11, 25, 28, 38
Meter stick	15 M 4065	8	17
Microscope-slide forceps	15 M 1645	8	20
Microscope slides, pkg. 72	14 M 3500	8	2, 11, 20, 21, 25, 28
Microscope, compound light	24 M 2310	8	2, 11, 15, 20, 21, 25, 28, 42, 31, 24
Nickel chromium alloy wire, 4 oz. spool	15 M 9360	1	40
Pasteur pipet, 5 3/4 in., pkg. 250	17 M 1145	1	4
Petri dish, 100 x 15 mm, pkg. 20	18 M 7101	8	17, 26, 29, 30, 41
Plant light	20 M 5100	8	17
Probe, blunt	14 M 0950	8	27, 30, 33
Ring stand, 4 x 6 in.	15 M 0719	8	11
Rubber stopper, 1lb., size 1	15 M 8461	1	17
Ruled microscope slides, pkg. 15	14 M 3120	8	15
Safety gloves, box 100	15 M 1071	5	20, 22, 29, 35, 3
Safety goggles	15 M 3052	32	37, 20, 1, 2, 3, 4, 11, 15, 22, 26, 9, 33, 39, 40

Item	WARD'S No.	Qty.	Inv.
Scalpel	14 M 0900	8	25, 29, 33
Scissors, 4 1/2" nickel-plated	14 M 0525	8	5, 8, 10, 21, 30, 40, 33
Screening, roll, 36" x 36"	15 M 0002	1	17
Screw clamp	15 M 3910	24	17
Screwdriver set, ten piece	12 M 0113	8	40
Spirometer, pkg. 4	14 M 5051	2	37
Stereomicroscope	24 M 4602	8	21, 22, 25, 26, 28, 29, 30
Stir rod, glass, pkg. 10	17 M 6010	1	8, 11, 29, 40
Stopwatch	15 M 0512	8	30, 34, 36
Swabs, sterile, pkg. 100	14 M 5502	1	20, 35
Tape measure, 5 ft.	15 M 2541	1	12, 36
Tape, roll	15 M 1959	2	4, 5
Terrarium, with lid	21 M 2100	8	32
Test-tube, Pyrex, 13 x 100 mm	17 M 0610	8	15
Test-tube racks, 6 well, LDPE	18 M 4231	8	17
Test tube rack, 40 wells, 20 mm	18 M 4214	8	20
Test tubes, Pyrex, 15 x 125 mm	17 M 0620	42	4
Test tubes, Pyrex, 16 x 125 mm	17 M 1403	24	17
Thermometer	15 M 1418	8	18
Tongs	15 M 0760	8	11
U-shaped glass tube, 150 mm	17 M 4991	8	18
WARD'S dual magnifier	24 M 1112	8	18, 23, 27
Wax pencil	15 M 1155	8	18, 41, 2, 4, 17, 25, 26

Miscellaneous

Item	WARD'S No.	Qty.	Inv.
5–7 lb. weight	Local	8	36
A Guide to Birds in N. America	32 M 0338	1	31
A Guide to Fishes, Whales and Dolphins	32 M 2115	1	31
Acetate sheets, 8.5 x 11 in.	Local		16
Aluminum foil, 12 x 25 in.	15 M 1009	1 roll	17
Antibacterial soap with iodine	15 M 9829	8	35
Beef or pork bones	Local		31
Chicken or turkey bones	Local		31
Chromosomal meiosis set	81 M 4501	1	6
Chromosomal mitosis set	81 M 4502	1	6
Colored pens for overhead, 4 colors	15 M 4635	1	16
Construction paper, black, pkg. 50	15 M 9825	1	21
Erasable bond paper	Local		35
Fabrics of different texture	Local	4 of each kind	30
Graph paper, pkg. 100	15 M 3835	3	1, 16, 41

Item	WARD'S No.	Qty.	Inv.
Gravel, rocks, and soil	Local		16
Green pop-it beads (peas), pkg. 300	36 M 1536	1	7
Guide to Mammals	32 M 0400	1	31
Guide to Reptiles and Amphibians	32 M 2123	1	31
Hot water bag, 2 qt.	14 M 8306	8	17
How to Know Protozoa, book	32 M 3331	1	21
Ice bag	Local	1	17
Lamp for heat source	36 M 4168	8	35
Marker, black	15 M 3083	8	8, 10, 19
Paper clips, pkg. 100	15 M 9804	1	10
Paper clips, pkg. 1,000	15 M 9815	1	8
Paper punch	15 M 9810	8	21
Paper towels	Local	5	1, 3, 20, 25, 35
Paper, pH, pkg. 100	15 M 2558	1	4
Pencil, colored, pkg. 12	15 M 4690	2	9, 41
Plastic knife	250 M 8128	8	3
Plastic soda straws, pkg. 500	15 M 9869	3	8, 37
Plastic spoon	250 M 8129	8	3
Plastic wrap	15 M 9858	2	4, 37
Pre-moistened alcohol pads, pkg. 200	36 M 5519	1	35
Protractor	15 M 4067	1	14
Pushpins, set of 5 colors	15 M 0505	1	8, 10
Rubber bands, 1/4 lb.	15 M 9824	1	25, 40
Ruler, 15 cm	14 M 0810	8	1, 3, 5, 8, 10, 12, 14, 25, 35
Strained spinach	Local		41
String, roll, 400 ft.	15 M 9863	1	40
Tape, masking, roll, pkg. 3	15 M 9828	1	17, 19, 22
Toothpicks, box 800	15 M 9840	2	21, 22, 38
Transparent tape	15 M 1957	1	30
Used tennis ball	Local	8	36
Various detergents	Local		4
WARD'S Simulated Plant Pigments Kit	36 M 0062	1	5
WARD'S Simulating Disease Transmission Kit	36 M 0070	1	39
White notebook paper	Local	8	21
Whole fresh chicken	Local	8	33
Whole milk	Local		38
Yellow pop-it beads (lentils), pkg. 300	36 M 1533	1	7
Zipper plastic bag, 6" x 9", pkg. 10	18 M 6922	2	1

Biology: Principles and Explorations

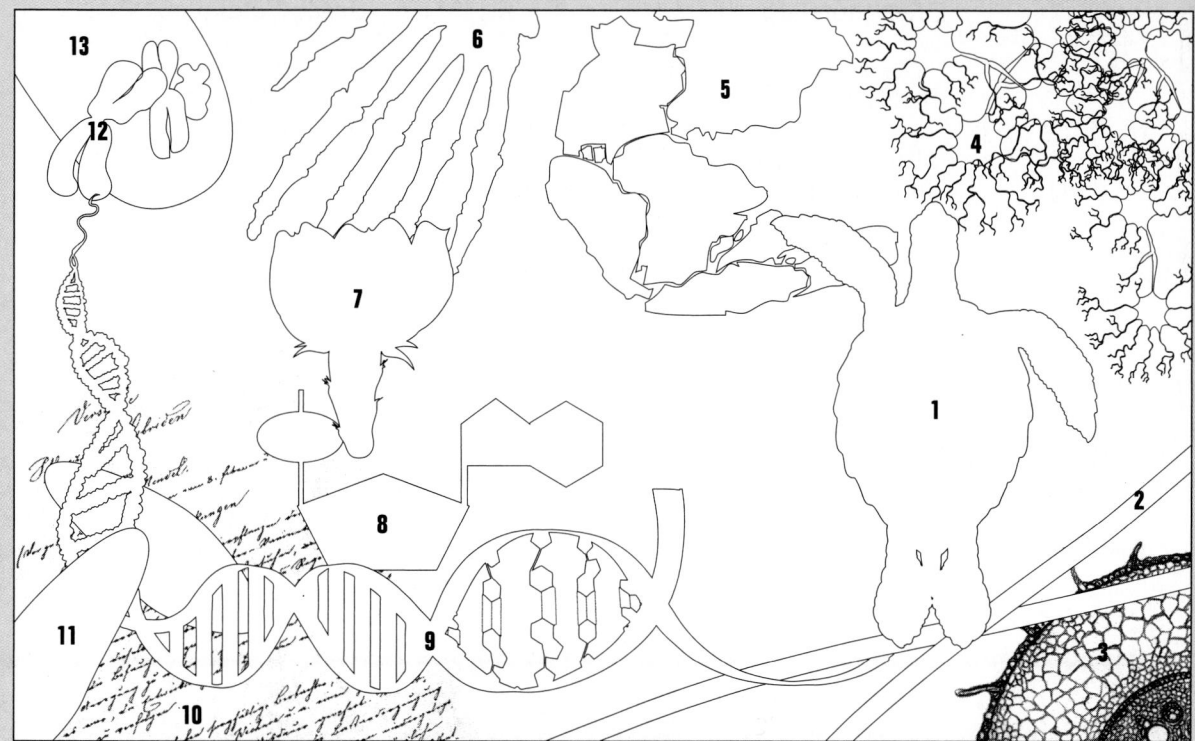

Cover illustration: Randy Gates, Morgan-Cain & Associates

1. A young green sea turtle (*Chelonia mydas*)
2. *Spirogyra*
3. Cross section of monocot root
4. Human neurons
5. Pangaea
6. Left hand of human skeleton
7. Prickly pear cactus flower
8. Nucleotide
9. DNA
10. Title page of Mendel's 1866 treatise on heredity
11. Paramecium
12. Human chromosomes
13. *Chilodonella*, a eukaryotic cell

Executive Editor
Ellen Standafer

Senior Editor
Susan Feldkamp

Project Editors
Carolyn Biegert
John Gallo
Mitchell Leslie
Jennifer Linn

Production
Beth Prevelige
Simira Davis
George Prevelige
Rose Degollado
Nancy Hargis
Susan Mussey

Editorial Staff
Jane Martin
Steve Oelenberger
Tanu'e White

Design and Production
Morgan-Cain & Associates

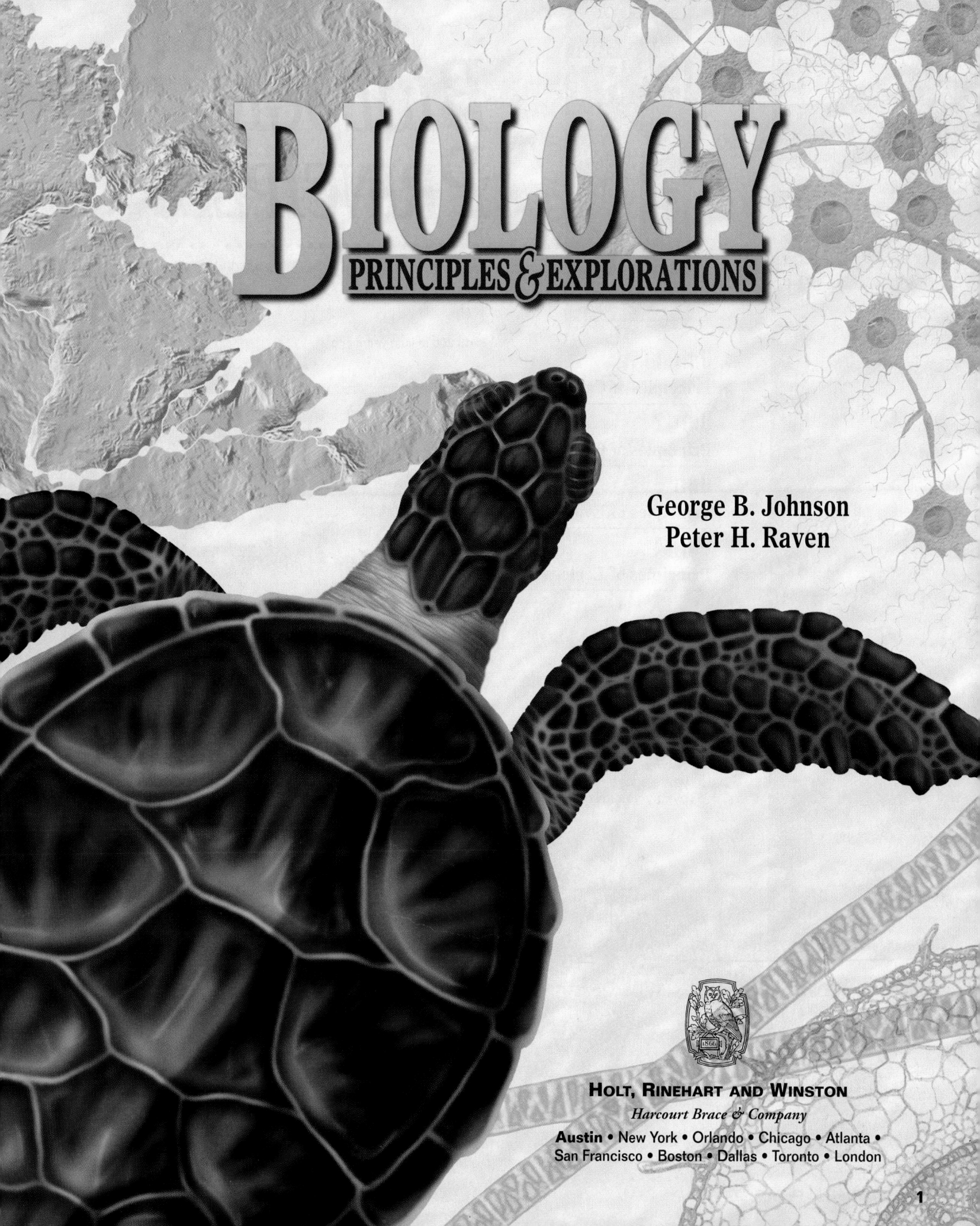

BIOLOGY
PRINCIPLES & EXPLORATIONS

George B. Johnson
Peter H. Raven

HOLT, RINEHART AND WINSTON

Harcourt Brace & Company

Austin • New York • Orlando • Chicago • Atlanta • San Francisco • Boston • Dallas • Toronto • London

PART 1

BIOLOGICAL PRINCIPLES

UNIT 1
Principles of Cell Biology

UNIT 2
Principles of Genetics

UNIT 3
Principles of Evolution

UNIT 4
Principles of Ecology

Earth 180 million years ago

Laurasia

Gondwana

Earth 200 million years ago

Pangaea

Chromosomes in cell nucleus

Eukaryotic cell

Glyptodont (armadillolike animal that lived in South America between 2 million and 15,000 years ago) an ancestor of the pygmy armadillo on the right.

> **❝** *During the voyage of the Beagle I had been deeply impressed by discovering great fossil animals covered with armour like that on the existing armadillos; secondly, by the manner in which closely allied animals replace one another in proceeding southward over the [South American] continent; and thirdly, by the South American character of most of the productions of the Galapagos archipelago, and more especially by the manner in which they differ slightly on each island of the group . . . It was evident that such facts as these, as well as many others, could be explained on the supposition that species gradually became modified; and the subject haunted me.* **❞**
> — Charles Darwin

Earth 60 million years ago

Africa

DNA in
Chromosomes

Pygmy armadillo, or pichis (armadillo found in South America)

Darwin's Observations

Visiting the Galapagos Islands off the coast of Ecuador in 1832, the young English biologist Charles Darwin recognized that each island was populated by different species of animals, yet the animals of each island resembled animals on the South American mainland. Darwin's observations led him to propose that organisms evolved, or changed, over time.

Cell Structure

Today we know more about the structure and function of living things than scientists in Darwin's day could have ever imagined. The findings of modern scientists reveal how species change over time, supporting Darwin's observations in a multitude of ways.

UNIT 1

Block Scheduling Guide

Each block represents about 45–50 minutes of instructional time. The resources cited in a block are coded to a specific program component using the Key on the next page.

Lecture Concepts | Lesson Resources

1-1 Biology at Work Today pp. 4–7

BLOCK 1

a. Scientists expand scientific knowledge by making observations and posing questions that can be examined through experimentation.

b. Sharing observations and studies is an important part of the scientific process.

Lecture Resources
- Opening Demonstration, page 4
- Opening Question, page 5
- Demonstrations: pH Scale, Acid Rain
- Visual Strategies: Figures 1-1, 1-2
- Effective Reading, page 6
- Teaching Transparencies: 1A
- Holt Biology Videodiscs: Lesson 1-1

Classwork Options
- Teaching Resources CD-ROM: Safety Contract and Quiz
- Laboratory Investigation: Observing Effects of Acid Rain, pages 22–23
- Quick Labs A1: Imagining Solutions: Problem Solving

Assignment Options
- Section Review, page 7
- Chapter Review, questions 1–3, 15, 16, 21, 22, 25

1-2 The Scientific Process pp. 8–11

BLOCK 2

a. Scientific investigation generally proceeds in six stages: collecting observations, forming hypotheses, making predictions, verifying predictions, performing control experiments, and forming a theory.

b. Control experiments validate conclusions by holding a single variable constant and sorting out other factors that could affect results.

c. Hypotheses not consistent with experimental results are rejected. Hypotheses repeatedly tested and supported by a body of evidence are accepted as theories.

d. Theories are always subject to revision or replacement by new theories that provide a better explanation.

Lecture Resources
- Opening Question, page 8
- Demonstrations: Using the Scientific Process; Predictions
- Visual Strategy: Figure 1-5
- Overcoming Misconceptions, page 9
- Demonstration: Setting Up the Proper Controls
- Holt Biology Videodiscs: Lesson 1-2

Classwork Options
- Forming a Theory and Graphic Organizer, pages 10–11
- Inquiry Skills Development B1: Introduction to Experimental Design and Data Presentation

Assignment Options
- Section Review, page 11
- Chapter Review, questions 4–6, 9, 12, 14, 19–21, 23, 24

1-3 Properties of Life pp. 12–14

BLOCK 3

a. Biologists recognize that living things share certain general properties: cellular organization, metabolism, homeostasis, reproduction, and heredity.

b. A cell is the smallest unit capable of all life functions.

c. Metabolism is the sum of all chemical reactions in an organism. Homeostasis is the process by which organisms maintain relatively stable internal conditions.

d. Reproduction and heredity are processes by which organisms produce more of their kind and pass on the information in cells needed to guide cell processes.

Lecture Resources
- Opening Question, page 12
- Demonstration: Are They Alive?
- Visual Strategy: Figure 1-11
- Effective Reading, page 12
- Historical Note, page 13
- Holt Biology Videodiscs: Lesson 1-3

Classwork Options
- Reproduction, page 14

Assignment Options
- Section Review, page 14
- Chapter Review, questions 8, 13, 17

1-4 Addressing Real-World Problems pp. 15–18

BLOCK 4

a. There are many areas in which biologists are actively working to solve many of today's problems.

b. Scientists are trying to develop more efficient crops without the intensive use of fertilizers and pesticides. They are also working on creating new crops through genetic engineering.

c. With research and new technologies, scientists may develop cures for a variety of diseases.

Lecture Resources
■ Opening Question, page 15
■ Demonstration: Making Environmentally Conscious Choices
■ Visual Strategies: Figures 1-13, 1-14
■ Overcoming Misconceptions, page 17
▧ Teaching Transparencies: 81
◿ Holt Biology Videodiscs: Lesson 1-4

Classwork Options
■ An Individual's Contributions, page 18
■ Closure, page 18

Assignment Options
■ Borrowing from Our Descendants, page 16
■ Section Review, page 18
■ Chapter Review, questions 7, 10, 11, 18

Review/Enrichment

BLOCK 5

■ Study Guide: Concept Review and Pretest

Assignment Options
◉ Teaching Resources CD-ROM: Supplemental Reading Worksheets and Test, *The Lives of a Cell*
■ Activities and Projects, questions 26–30
■ Readings, questions 31, 32

Assessment Options

BLOCK 6

Traditional Assessment
■ Chapter 1 Test
◉ Test Generator/Assessment Item Listing: Software item bank for preparing customized chapter tests.

Portfolio Assessment
Select student reports for one or more laboratory experiments from this chapter. *The Direct Observation Checklist* on the *BioSources Teaching Resources CD-ROM* should be completed during a laboratory or other cooperative learning experience.

Answer to Concept Map
The following is one possible answer to the Concept Mapping exercise on page 19.

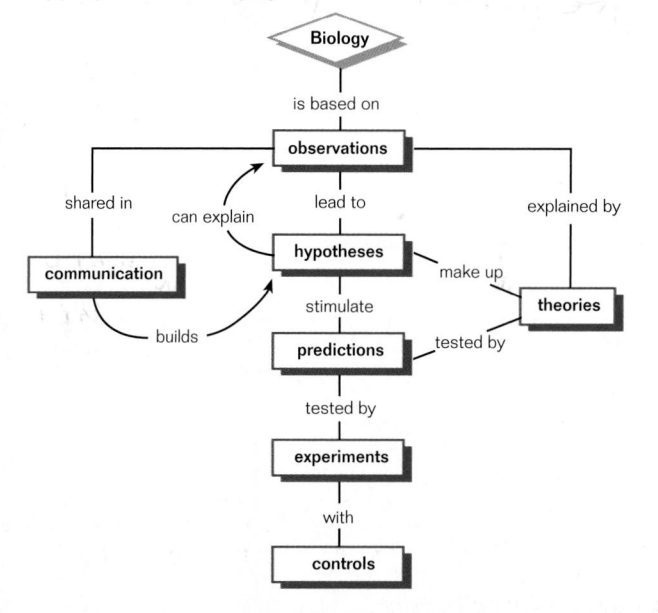

Holt Biology Videodiscs   CONCEPTS OF BIOLOGY

Holt Biology Videodiscs gives you a powerful tool for teaching, review, and assessment. *Concepts of Biology* adds interactive lesson plans and extensive tools for customization using CD-ROM technology.

Use the following videos from *Concepts of Biology* to help you teach this chapter:

■ Units 1–4: Scientists in Action

For further information regarding the media on the videodiscs, see *Holt Biology Videodiscs Teacher's Correlation Guide to Biology: Principles and Explorations.*

Chapter Theme

Students will encounter five themes that are consistently highlighted throughout the book. The first of these themes, Evolution, is introduced in Chapter 2. The remaining four— Homeostasis, Flow of Energy, Levels of Organization, and Structure and Function—will be introduced in Chapters 3–6. As students progress through their study of biology, they should become proficient at recognizing the underlying themes of each chapter.

Tapping Prior Knowledge

- What processes do biologists use in solving a problem or answering a question?
- How does a hypothesis differ from a theory?
- What properties do all life-forms share?
- Describe a current problem that biologists are attempting to solve.

Opening Demonstration

As a way of introducing the processes that biologists use, write the following five "trogs" on the board: kqrl, cwvb, qmlr, drse, and pxwo. Also list the following five "nontrogs": rsct, jump, wikx, akbl, and qust. Ask students to work in pairs to determine what makes a "trog." *(a four-letter combination whose first and last letters are adjacent in the alphabet, as are the second and third letters)* Have students describe how they arrived at their answers.

BIOLOGY AND YOU

This orphaned rabbit depends on our knowledge of its biology for its survival.

Authors' Rationale

An important tool in an abstract subject like biology is the scientific process. In Chapter 1, students are involved in the scientific process from the beginning by learning about John Harte's disappearing amphibians and the subsequent development of an explanation for their disappearance. This chapter also discusses the basic characteristics of life. Students can then apply what they know about life and how to study it to such seminal topics as protecting the Earth and combating disease.

1-1 Biology at Work Today

The science of biology developed from scientists sharing their observations and studies with each other. Their fascination with and investigation of the real world has led to a vast body of knowledge that is continuously growing. As you read this section, think about the changes that might be taking place in your surroundings and how they might eventually affect you. What type of changes, for example, have you observed in your neighborhood? What effects might those changes have on you, your family, and your neighbors?

Section Objectives

■ Assess the importance of observations in the world of science.

■ Explain why communicating is an important process in science.

■ Predict the effects of the release of airborne acids on the environment.

What Is Killing the Frogs?

Sometimes important things happen, right under our eyes, without anyone noticing. That thought occurred to David Bradford as he stood looking at a quiet lake high in the Sierras of California in the summer of 1988. Bradford, a biologist, had hiked all day to get to the lake, and when he got there his worst fears were confirmed. The lake was on a list of mountain lakes that Bradford had been visiting that summer in Sequoia–Kings Canyon National Parks while looking for a little frog with yellow legs, shown in Figure 1-1. The frog's scientific name was *Rana muscosa*, and it had lived in the lakes of the park for as long as anyone had kept records. But this silent summer evening the little frog was gone. The last major census of the frog's populations within the park had been taken in the mid-1970s, and *R. muscosa* had been everywhere, a common inhabitant of the many freshwater ponds and lakes within the park. Now, for some reason that Bradford did not understand, the frogs had disappeared from 98 percent of the ponds that had been their homes.

After Bradford reported this puzzling disappearance to other biologists, an alarming pattern soon became evident as other scientists reported similar findings. Throughout the American West, local populations of amphibians (frogs, toads, and salamanders) were becoming extinct. In Oregon, for example, 80 percent of the 30 populations of the Cascades frog, *Rana cascadae*, which had been studied in the mid-1970s, were now gone. Something was killing local populations of amphibians.

Amphibians have been around for 370 million years, since long before the dinosaurs. Their sudden disappearance from so many of their natural homes sounded an alarm among biologists. It was not just a concern for frogs.

Figure 1-1 Sequoia–Kings Canyon National Parks are outlined in the map of California, *top*. The many lakes and rivers of these parks are home to the mountain yellow-legged frog, *Rana muscosa, bottom.*

UNIT 1

CHAPTER 1

Teaching Tips

Identifying Factors

Students should recognize that an observable phenomenon, such as the salamander population's decline, may be the result of any one of a number of factors. A scientist's task in such a situation is to identify the factors responsible. Have each student list as many factors as possible that may have been responsible for the population's decline. Possibilities include increased predation, drastic temperature changes, habitat pollution, and decreased food availability.

Quantifying Observations

Students usually perceive biology as a qualitative science in which measurements are rarely made. Emphasize that the disappearance of the salamanders (a qualitative observation) was explained by experiments involving pH measurements and time periods (quantitative observations). Have students choose an article from a newspaper or magazine that discusses a current biological problem. Ask students to list both the qualitative and quantitative observations mentioned in the article.

👁 VISUAL STRATEGY

Figure 1-1 and Figure 1-2

Use these figures to point out the similarities and differences between frogs and salamanders. *(Both have thin, scaleless skins, and four legs; salamanders have tails throughout their lives, while frogs lose their tails as they mature from tadpoles into adults.)*

Figure 1-2 Biologist John Harte observed the decline in numbers of the tiger salamander, *Ambystoma tigrinum,* beginning in the early 1980s.

Scientists were afraid that the vanishing amphibians might be an example of "miner's canary" syndrome. In the nineteenth century, coal miners took canaries down into the mines as an early warning system to signal the presence of poisonous carbon monoxide gas. Canaries, which are very sensitive to the deadly gas, would die long before people would be affected. Amphibians are very sensitive to their environment, too, because their moist skins absorb chemicals from pond water. Biologists worried that the frogs' disappearance was an early warning that something very damaging was happening to the environment.

Searching for a Culprit

Important questions often do not have simple answers, and so it proved with this problem. Humans are putting stress on the environment in so many ways that many factors might be contributing to the decline of isolated amphibian populations. To understand what is going on, a scientist must focus on a particular situation, analyze it carefully, and try to learn the source of the problem for that particular amphibian population. Only after many populations are studied in this way can a general conclusion be drawn about the cause of the population decline.

One of the first amphibian populations to be analyzed in detail was that of a tiger salamander, *Ambystoma tigrinum,* shown in Figure 1-2. This salamander lives in a cluster of ponds located near Galena Lake, high on the western mountain slopes in west-central Colorado. John Harte, a biologist studying this population, had seen its numbers decline by 65 percent over a seven-year period.

There were many factors in the environment that might have been responsible for the salamander population's decline. However, Harte suspected acid rain. **Acid rain** results when sulfur in smoke produced by the burning of coal and oil reacts with water in the air to form sulfuric acid, which falls back to Earth in rain or snow. The acidity of a solution is described by **pH**—a number that represents the hydrogen ion concentration in a solution. Solutions with a low pH (between 0 and 7) are acidic. A solution with a pH of 2 is far more acidic than a solution with a pH of 6.

Colorado's Rocky Mountains, shown in Figure 1-3, get their moisture from winds that blow from the west, winds that carry acid from industry and urban smog. Over 90 percent of the annual moisture on the mountains falls as snow. Every spring when the snow melts, the acid accumulation of many months is released all at once into the high mountain ponds. Perhaps that pulse of acid was killing the salamanders.

 Did You Know?

A frog breathes through its lungs, but can also breathe through its skin. Thus, its skin is highly vascularized and must be kept moist to facilitate the diffusion of gases.

Effective Reading

Capsule Summaries

On page 7 is the first of many capsule summaries that students should take time to read. Emphasize the value of reading the capsule summaries as a way of reinforcing the major points of the material they have just read.

Testing an Idea

To test his idea, Harte first measured the amount of acid released into the ponds at spring snowmelt for the years 1984 through 1988. In all the ponds, the amount of acid shot up to very high levels when the snow melted in early to mid-June. After a few weeks, the added acid was neutralized by minerals in the rocks lining the ponds. Acid levels then fell back to normal. Harte concluded that a pulse of acid was indeed being released into the ponds at snowmelt.

Harte then asked himself whether the acid released by snowmelt could be killing the salamanders. To answer this question, he placed salamander eggs in pond water, added acid, and examined the eggs to see how many hatched. What happened depended on how much acid he added and when he added it. Many of the eggs were killed by the level of acid released during snowmelt—but only if the eggs were exposed to the acid between the fifth and tenth days of egg development. This is the time when organs develop within the growing embryo. Exposure to acid then is fatal.

So the question for Harte boiled down to this: Were the salamander eggs in the ponds at a critical stage of their development when snowmelt delivered the jolt of acid to the ponds? For every year but one, the answer proved to be "yes." The acid arrived from snowmelt just at the time the salamander eggs were sensitive to it. The one exception was 1988, a year of exceptionally early snowmelt. In that year egg-laying occurred after acid levels had fallen back to their normal values, and, therefore, the eggs in the pond survived. It was the only year in Harte's study in which the number of salamanders increased.

Thus, Harte concluded that the steady decline he observed in the tiger salamander population during the 1980s was probably caused by acid rain. In this one case, the most likely culprit responsible for a declining amphibian population was tentatively identified. ◻

Western United States

Galena Lake

Figure 1-3 Windblown acidic compounds from Arizona smelters and California smog may be responsible for the decline of the tiger salamander in the Colorado Rockies. The location of Harte's study is shown in the inset.

❑ CAPSULE SUMMARY

Scientists expand scientific knowledge by making observations and posing questions about what they observe. To test an idea, scientists gather and record specific information. It is very important in science to pose questions that can be examined through experimentation.

Section Review

1. *Identify one observation that made David Bradford aware that* Rana muscosa *had disappeared from the lakes in Sequoia–Kings Canyon National Park.*

2. *Why was communication so important among the biologists who studied amphibians?*

Critical Thinking

3. *Why do you think waterfowl in Harte's study area were not affected by acid rain?*

4. *Amphibians living in the lakes along the coast of California were not affected by acid rain. Offer an explanation for this observation.*

SECTION 1-1

Demonstration
pH Scale
Caution: Hydrochloric acid is a strong acid. Wear safety goggles, gloves, and a lab apron or chemically resistant lab coat when handling it. Prevent splattering by adding acid to water, not water to acid. Have students remain at least 10 ft. away from the demonstration area during the demonstration. If the acid spills on anyone, wash the area immediately with running water. Obtain a commercial preparation of 0.1 M hydrochloric acid (do not use concentrated HCl). Measure its pH using a pH meter or pH paper. *(pH = 1)* Next, pour 90 mL of distilled water into a graduated cylinder. Add 10 mL of the 0.1 M HCl to the water. Then, measure the pH of the diluted solution. *(pH = 2)* Students should see that a tenfold change in acid concentration results in only a onefold change in pH. To dispose of the HCl solution, neutralize it (test its pH) and pour it down the drain with a twentyfold excess of water.

Teaching Tip
Sources of Acid Rain
Have students locate both Arizona and California in Figure 1-3. Ask them to locate Wheeling, West Virginia, on a map and suggest the sources of acidic compounds that fall on this region of the country. Students should see that this area is an "ideal" site for acid rainfall. In fact, the pH of precipitation in Wheeling has been measured at 1.5. The acidic compounds probably originate from industries in the upper Midwest and coal mines in the more immediate area.

Answers to Section Review

1. The mid-1970s census showed that *R. muscosa* thrived everywhere in the parks. In the summer of 1988, Bradford observed that 98 percent of the ponds no longer served as homes to these frogs.

2. Such communication led to the realization that the frogs' disappearance was not limited to Sequoia–Kings Canyon National Parks.

3. Since birds' eggs develop on land, their embryological development is not affected by acid in the lakes.

4. Because the Pacific Ocean forms the western boundary of the coast of California, acid rain is not a problem in the local lakes.

Vocabulary Preview

observation
hypothesis
prediction
experiment
control
theory

Opening Question

Ask students to think of some current biological questions or problems. Then have students describe the steps they think might be involved in a scientific investigation of their questions or problems. Explain that the scientific process outlines a general method of investigating questions, but that many scientists have developed theories that grew out of personal instincts.

Demonstration

Using the Scientific Process

To show students how each stage of a scientific investigation leads logically to the next, have them perform the following exercise and identify each stage. Ask students to record the number of times they breathe during a one-minute period. (*stage: collecting observations*) Next have them suggest how exercise will affect this number. (*stage: forming hypotheses*) Then ask them to note the number of breaths they think they will take in the same period of time after they have jogged in place for one minute. (*stage: making predictions*) Instruct them to carry out the exercise and immediately record the number of breaths they take. (*stage: verifying predictions*) Ask the class how they know that exercise was the factor that affected the number of breaths they took. (*stage: performing control experiments*) Finally, have them suggest a connection between exercise and rate of breathing. (*stage: forming a theory*)

Section Objectives

■ Describe the six stages of a scientific investigation.

■ Distinguish between hypothesizing, predicting, and experimenting.

■ Define the elements of a control in an experiment.

■ Define theory and explain why theories form the framework of science.

Figure 1-4 Most scientists use variations of the scientific process when conducting their scientific investigations. Here, Asa Bradman, a student of John Harte, collects water samples from a Colorado pond.

1-2 The Scientific Process

John Harte is a biologist, and what he was doing in the Colorado mountains was science. Science is a particular way of investigating the world, of forming general rules about why things happen by observing particular situations. A scientist like Harte is an observer, someone who looks at the world in order to understand how it works. Stated briefly, a scientist determines principles from observation.

A Scientific Investigation Has Six Stages

It was once fashionable to claim that scientific investigations always progress by a rigid series of "either/or" steps called "the scientific method." In each of these steps, one of two incompatible alternatives is rejected. It is as if trial-and-error testing inevitably leads through the maze of uncertainty that always slows scientific progress. If this were true, a computer could be programmed to be a good scientist. But science is not done this way.

If you ask successful scientists like John Harte how they do their work, you will discover that without exception they design their experiments with a pretty good idea of the outcome. A scientist integrates all that he or she knows and allows his or her imagination full play in an attempt to get a sense of what might be true.

Although there is no single scientific method, all scientific investigations can be said to have six stages: collecting observations, forming hypotheses, making predictions, verifying predictions, performing control experiments, and forming a theory. Figure 1-4 summarizes these six stages.

Scientific Process
- Collecting observations
- Forming hypotheses
- Making predictions
- Verifying predictions
- Performing control experiments
- Forming a theory

Historical Note

Francis Bacon (1561–1626)

The English philosopher and statesman Francis Bacon is well known for his belief in experimental and scientific methods. In Book II of his *Novum Organum*, Bacon proposed a method of inductive reasoning to help people overcome their prejudices so that they could more objectively solve problems. He was a strong believer in the role of careful observations, analysis, and hypotheses in discovering truth. Moreover, in his book *The New Atlantis*, Bacon conceived of a utopic scientific institution that was based on collaboration and methodical research. Although his philosophy had weaknesses, his writings influenced the procedures of 19th century biologists.

As you will see, insight and imagination play a large role in scientific progress. It is for this reason that some scientists are better at science than others—in the same way that Beethoven and Mozart stand out above most other composers.

Collecting Observations

The heart of any scientific investigation is careful observation. **Observation** is the act of noting or perceiving objects or events by using one or more of the five senses. Harte had studied the Colorado salamander population for many years. He had noted a thousand details of how the salamanders lived and what their ponds, such as the one shown in Figure 1-5, were like. If he had not kept careful records of what he saw, he might not have noticed that the salamander population was slowly declining over the years.

Forming Hypotheses

Observing the decline, Harte made a guess about why the salamanders were dying—perhaps they were being killed by acid rain. We call such a guess a hypothesis. A **hypothesis** is a proposed explanation that might be true. Hypotheses must be able to be tested by additional observations or experimentation. It is important to note that a hypothesis is not just any guess. A scientist makes an educated or informed guess based on everything he or she already knows. Using what he knew, Harte guessed that acids created in the upper atmosphere by industry were falling onto the mountains in the winter snows. By making the ponds acidic, the melted snow was killing the salamander embryos.

Making Predictions

Harte knew that if his hypothesis was correct, he could reasonably expect several consequences. We call these expected consequences predictions. A **prediction** is what you expect to happen if a hypothesis is accurate. Harte predicted that if acid precipitation was killing the salamanders, then it should be possible to detect the acid entering the ponds when the snow melts. Moreover, if acid precipitation was the culprit, then the amount of acid entering the ponds should be enough to kill salamander embryos.

Verifying Predictions

Harte set out to test his hypothesis by trying to verify its predictions. We call the test of a hypothesis an **experiment**. Harte did two series of experiments. In the first, he sought to determine if a rise in acid levels in the ponds occurred at the time of snowmelt. He took water samples from both deep and shallow parts of several ponds at frequent intervals, starting before snowmelt began and continuing until well after snowmelt concluded. He found that a large amount of acid was indeed introduced into the ponds by the melting snow.

Salamanders lay their eggs only once a year, as soon as the pond ice melts. Any salamander eggs developing in the

Figure 1-5 Natural changes in a habitat can be mistaken as trends if detailed, accurate, long-term records are not kept. Biologists of the Rocky Mountain Biological Lab are performing long-term studies that carefully monitor conditions in lakes such as this one in Mexican Cut, Colorado.

Figure 1-6 Deformities like the curved spine and stunted gills of this salamander occur when animals develop in water of pH 5, common for lakes and ponds contaminated by acid rain.

❑ CAPSULE SUMMARY

Control experiments lend validity to conclusions by allowing a single variable to be held constant, sorting out other factors that could affect results.

ponds would thus be exposed to the acid from the melting snow. But was it enough to do real harm? To test his prediction that the melting snow would release enough acid to damage the eggs, Harte carried out a second series of experiments in the laboratory. He allowed captive salamanders to lay eggs in water from the ponds. Then he added acid to the water at different times, to correspond with the acid levels observed in the pond during snowmelt, and looked to see if the eggs developed properly into adult salamanders. Harte found that the amount of acid being delivered to the ponds by snowmelt was indeed enough to deform and kill developing salamander embryos, particularly when introduced from five to ten days after the eggs were laid. Figure 1-6 shows an example of a deformed salamander.

Performing Control Experiments

Harte was now able to conclude that salamander eggs were being exposed to lethal amounts of acid. He decided to test his hypothesis further by using a control. A **control** is that part of an experiment in which the key factor is not allowed to change. In this experiment, the key factor was the exposure of early embryos to lethal amounts of acid in the ponds. As a control, Harte used the data gathered in 1988, a year when the snowmelt was unusually early. Just as in other years of Harte's study, acid was released into the ponds. The only difference between 1988 and the other years was that the acid was released earlier in the year. It was released so early, in fact, that by the time the salamanders laid their eggs, the acid in the pond had already been neutralized by chemicals in the rocks. Had the salamanders continued to decrease in number in the absence of acid, then acid could not have been responsible for killing the salamanders, and Harte's hypothesis would have been rejected. Through careful monitoring of the number of salamanders in the population in that unusual year, however, Harte observed a dramatic increase in their numbers! The result of the control was thus consistent with Harte's prediction. ❑

Forming a Theory

The essence of science is to reject hypotheses that are not supported by observations. The predicted consequences of a hypothesis are tested by experiments. If the experimental results do not agree with the predictions, the hypothesis is rejected. If they do agree, the hypothesis is not rejected—yet.

A scientist works systematically by attempting to show that certain hypotheses are not valid, that they are not consistent with the results of experiments. A successful experiment is one in which one or more of the alternative hypotheses are shown to be inconsistent with observations and are thus rejected. Harte, for example, was able to show that enough acid was being introduced into the ponds to kill the salamander embryos; he could therefore retain the hypothesis that acid in the snow was killing the salamanders.

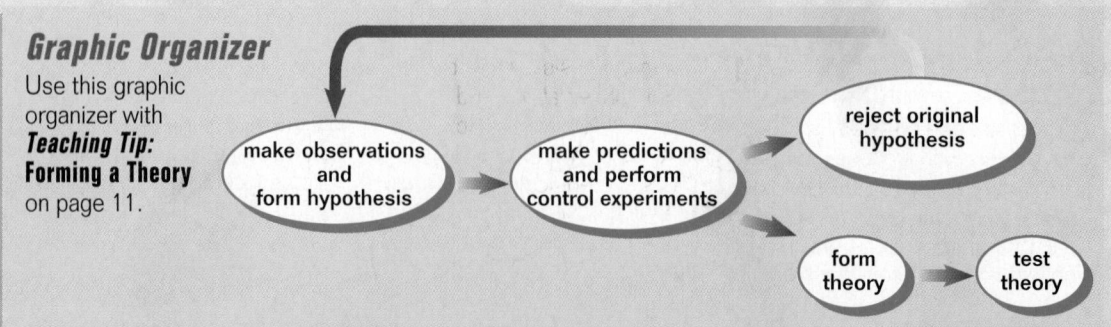

Scientific progress is made the same way a marble statue is, by chipping away the unwanted bits.

A collection of related hypotheses that have been tested many times is called a theory. A **theory** is a unifying explanation for a broad range of observations. The hypothesis that acid rain is contributing to the decline of isolated amphibian populations will require a great deal more evidence than Harte's before becoming generally accepted. Many other factors may play important roles, such as increased ultraviolet (UV) radiation due to ozone depletion, and it is important not to be misled by what happens in a single instance. Only after many studies like Harte's will scientists be able to assemble a picture that accurately reveals what is harming the amphibians.

Constructing a theory often involves contrasting ideas and conflicting hypotheses. For example, Harte's conclusions have been questioned by other scientists who suggest that the correlation he indicates may be a coincidence. Maybe in heavy snowpack years it isn't extra acid that kills the larvae, they suggest, but rather increased shading of the ponds (which retards photosynthesis, leading to oxygen depletion and larval death). Further experiments will tell who is right. Argument, disagreement, and unresolved questions are a healthy part of science, a true reflection of how science is done, and the speediest path to increased knowledge. ❑

The word *theory* is used very differently by scientists and by the general public. To a scientist a theory represents that of which he or she is most certain. To the general public, *theory* implies a lack of knowledge, a guess. How often have you heard someone say "It's only a theory" to imply lack of certainty? As you can imagine, confusion often results. In this text, the word *theory* will always be used in its scientific sense, as a generally accepted scientific principle.

Theories are the solid ground of science, that of which scientists are most certain. There is, however, no absolute certainty or scientific "truth" in a theory. The possibility always remains that future evidence will cause a theory to be revised or rejected. A scientist's acceptance of a theory is always provisional. ❑

Section Review

1. *Name the stages of a scientific investigation.*
2. *How do scientists use the following: hypotheses, predictions, and experiments?*
3. *What is the purpose of a control experiment?*

Critical Thinking
4. *Explain why there is no absolute certainty or scientific "truth" in a theory.*

SECTION 1-2

Teaching Tips
Forming a Theory
Solving a problem in science is much like finding a criminal in police work. While scientists reject hypotheses that are not supported by all the observations, police reject suspects that are not implicated by all the evidence. Both investigators will eventually arrive at a theory to explain what happened. Students will clearly see the common thread between science and detective work by reading one of the case studies found in *Medical Detectives* by Berton Roueche or in Sir Arthur Conan Doyle's chronicles of Sherlock Holmes. Have students select one such article and write a summary documenting the similarities between the investigations carried out by both scientists and detectives. Then, have them construct a *Graphic Organizer* similar to the one at the bottom of page 10.

Fact Versus Theory
Provide each student with a copy of a newspaper or magazine article describing a recent biological discovery. Have them write three headings on a piece of paper: Fact, Hypothesis, and Theory. Instruct the students to write down what they feel belongs under each heading as they read the article. Then divide the students into small groups to share and discuss their findings.

Answers to Section Review

1. The stages include collecting observations, forming hypotheses, making predictions, verifying predictions, performing control experiments, and forming a theory.

2. A hypothesis is used both to explain observations and to make predictions. Predictions are used to help design experiments. Experiments are used to test hypotheses.

3. An experiment designed with the proper controls ensures that only one factor is responsible for the results.

4. The possibility always exists that some future observation or experimental result will cause a theory to be revised or even rejected.

Vocabulary Preview

biology

cell

multicellular

metabolism

homeostasis

heredity

Opening Question

Ask the class to describe, using the scientific processes that were described in Section 1-2, how they would set about answering the question of whether life exists somewhere other than on Earth. Students may begin by proposing that observations scientists have collected on other planets be examined to see how those planets' environments compare to Earth's environment. Students might hypothesize that life is more likely to be found in an environment that is similar to that of Earth. They could also predict which planet might harbor life-forms. To verify their prediction, they would have to design an experiment, which could lead to a theory about the properties of life.

Demonstration

Are They Alive?

Prepare a slide of stained onion skin cells or human cheek cells. Either use a microprojector to display the slide to the class or have students use microscopes to examine their own slides. Ask students to suggest how they might go about determining if the cells are alive. They may suggest introducing an irritant in the cell's environment and watching for a response. Another suggestion may be to look for signs that a cell is dividing.

Section Objectives

■ Explain why life is difficult to define in terms of visually observable properties.

■ Identify the five properties of life.

Figure 1-7 Sensitivity to external stimuli is one of the key properties of life.

❏ CAPSULE SUMMARY

Some of the most obvious properties of life cannot be used alone to decide whether something is alive.

1-3 Properties of Life

What does the word alive mean? Most dictionaries will define alive as "living or having life." Stop for a moment and write a short definition of life. You will not find it an easy task. The problem is not a deficiency on your part but rather the loose way in which the word life is used in everyday speaking.

Living Things Share Certain Properties

Biology is defined as the study of life. The science of biology stems from knowledge acquired by scientists systematically studying living things, just as Harte did while investigating the effects of acid on salamander embryos. While most people are capable of distinguishing between living and nonliving, actually defining life can be quite difficult. When writing your definition of life, what sort of lifelike qualities did you mention? Perhaps you considered some of the most obvious examples: movement, sensitivity, the ability to respond to stimuli, change over time in the form of development, complexity, and even death. While unique to living things, these examples are terribly inadequate criteria of life. You can probably think of something that is not alive that has some of the qualities that you listed.

Clouds, for example, move when stimulated by the wind and develop from moisture that is suspended in the atmosphere. Clouds grow and change their shapes. Clouds are complex because of their chemical and structural makeup, and some might view their dissipation as being similar to death. However, death is not the same as disorder. All living things die, while inanimate objects do not. Clouds may break up and vanish, but they do not die.

So you see that movement, sensitivity, development, complexity, and death are properties you might expect to see in living things, but none of these properties individually is adequate to define life. While nonliving things may exhibit some of these properties, only living things exhibit all of them. Thus, you could argue that collectively these properties do define life. But is that enough?

Biologists recognize that all organisms with which we are familiar share certain general properties, as shown in Figure 1-8. It is by these properties that you can differentiate living things from nonliving things. ❏

Cellular Organization

All living things are composed of one or more cells—tiny chambers with thin coverings called membranes. A **cell** is

Effective Reading

Breaking Apart Words

One way of helping students master the vocabulary they will encounter in their study of biology is to look for prefixes and suffixes. For example, multicellular can be understood by recognizing the meaning of the prefix *multi*. Have students decode the words unicellular, asexual, binary fission, and spermatogenesis.

Properties of Life
- Cellular organization
- Metabolism
- Homeostasis
- Reproduction
- Heredity

Figure 1-8 Organisms like the soybean seedlings and the sand dollar have the same properties of life as you.

the smallest unit capable of all life functions. Some cells have more complex interiors than others, but all are able to grow and reproduce. The simplest organisms, such as the paramecium shown in Figure 1-9, have only a single cell. As a **multicellular** (composed of many cells) individual, your body contains more than 100 trillion cells.

Metabolism

All living things use energy to grow, to move, and to process information. That is why living things must eat to continue living, as shown in Figure 1-9. All the energy you use is captured from sunlight by plants and algae. To get your energy, you extract it from plants or plant-eating animals in a process called metabolism. **Metabolism** is the sum of all chemical reactions that an organism carries out.

Homeostasis

All living things maintain relatively stable internal conditions, often quite different from their surroundings, by a process called **homeostasis.** Your body, for example, attempts to maintain a temperature of about 37°C (98°F) regardless of how cold or warm the weather might be.

Figure 1-9 Living things like the paramecium, *below,* exhibit organization. Humans, *below left,* and other organisms extract energy from food sources to perform the many chemical activities that are essential to life.

SECTION 1-3

Application

Sports Metabolic rates increase sharply with exercise. For example, the oxygen consumption of running animals increases linearly with their speed. A person exercising consumes 15 to 20 times the amount of oxygen consumed while sitting or standing still.

Teaching Tip

What If an organism takes in more energy than it uses? Organisms store energy that they do not use. Humans store energy they don't use as fat.

Did You Know?

Most cells, like the paramecium shown in Figure 1-9, are far too small to be seen with the naked eye. However, some are quite large. For example, a neuron that runs down a giraffe's leg may be more than a meter long.

Historical Note

Charles Blagden's Experiment
To show how well homeostasis maintains body temperature, tell students the story of Dr. Charles Blagden, secretary of the Royal Society of London 200 years ago. Dr. Blagden took some friends, a dog, and a steak into a room that had been heated to 126°C. After 45 minutes, they emerged. Dr. Blagden, his friends, and the dog were fine. The steak, however, was cooked!

Figure 1-10 If they live to maturity, these hatchling snakes may someday produce their own offspring. Reproduction ensures the ongoing success of a species.

Figure 1-11 All living things pass on genetic information that makes their offspring similar yet unique.

Reproduction

The ability to reproduce from one generation to the next is characteristic of all species of living things. Rapidly growing bacteria divide into daughter cells every 15 minutes, and bristlecone pine trees that are 5,000 years old still produce seedlings. Since no organism lives forever, reproduction, as shown in Figure 1-10, is an essential part of living.

Heredity

All living things have DNA molecules inside their cells that encode information to direct growth and development—a set of blueprints, called genes, that determines what the organism will be like. In reproduction this set of instructions is passed on to the offspring in a process called **heredity**, resulting in family resemblances like those shown in Figure 1-11.

Section Review

1. *Why would it be difficult to use development as a single criterion to define life?*
2. *What are the properties of life?*

Critical Thinking

3. *Suppose you are a biologist who has found an object that looks like an organism. What steps might you take to determine if your discovery is indeed alive?*

1-4 Addressing Real-World Problems

No one can read a newspaper or magazine today without noticing, time and again, issues involving biology. We learn that tropical rain forests are being destroyed, that more Americans have died of AIDS than were killed in the Vietnam War, and that genetic engineering may offer a cure for cystic fibrosis. In this text you will encounter many areas in which biologists are actively working to solve many of today's problems.

Figure 1-12 The ever-increasing bulk of solid waste produced by modern lifestyles poses a troublesome challenge. Garbage in this dump site in Point Barrow, Alaska, pollutes ground water and despoils the tundra.

Protecting Our Earth

As the world's population approaches 6 billion, human needs are placing serious stress on the planet. Some of the problems described below may seem easy to solve, but in reality their solutions are anything but simple.

What to Do With the Garbage

What should we do with the garbage? Our growing population makes more and more of it, as shown in Figure 1-12, and disposal of this waste has become a global problem. Have you ever stopped to think about how much plastic you personally have used in your life? The National Solid Waste Management Association in Washington, D.C., estimated in 1990 that one person disposes of 113.4 g (0.4 lb.) of plastic per day. All that plastic is still around somewhere. Plastic does not break down in nature. Imagine if it were stacked beside you!

UNIT 1
CHAPTER 1

 VISUAL STRATEGY

Figure 1-13
Point out that the conditions over the Antarctic each fall are just right for an ozone hole to form. The chlorine from CFCs is not the only ingredient needed to make a hole in the ozone layer. Ice and sunlight are also essential. Ice crystals form in the cold air during the winter. These crystals serve as the site where chlorine reacts with ozone. The sunlight present during the summer provides the energy needed for the reaction to occur. The result is a large hole that appears each fall.

 VISUAL STRATEGY

Figure 1-14
Tell students that the destruction of tropical rain forests contributes to another problem—global warming. The burning of huge tracts of forests releases carbon dioxide into the atmosphere. Carbon dioxide is a "greenhouse gas" that traps solar energy. The more carbon dioxide in the atmosphere, the more solar energy trapped on Earth. Large amounts of carbon dioxide released from human activities, such as burning fossil fuels, results in too much solar energy trapped on Earth and thus too much heat.

Teaching Tip
Borrowing From Our Descendants
Ask each student to write a short essay that interprets the meaning of the saying "We have not inherited this land from our ancestors. We have merely borrowed it from our descendants."

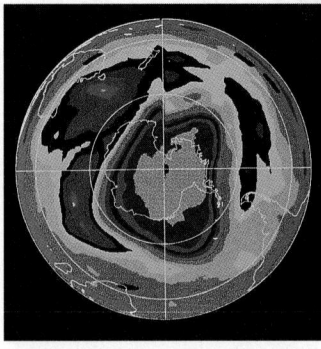

Figure 1-13 The pink area in the center of this satellite image represents the loss of ozone over the Antarctic. Industrial chemicals known as CFCs are responsible for this destruction of ozone.

Figure 1-14 The clearing of this rain forest will have disastrous environmental consequences for the many living organisms that depend on it for their survival.

Polluting the Atmosphere

Increasingly in the last century, humanity has taken to treating the atmosphere as if it were a garbage dump, one of limitless size into which chemicals could be "dumped" to be carried away by the wind. Only now are we beginning to realize that the Earth's atmosphere is far from limitless and that we cannot continue to treat it as a chemical sewer without paying a serious price. The tall smokestacks of power plants that burn high-sulfur coal release their smoke high in the sky, and it eventually falls back to Earth in rain or snow. This acid precipitation kills not only frogs, but also trees and many other creatures. The supposedly harmless aerosols and coolants made of chlorofluorocarbons (CFCs) have been released without concern into the atmosphere for decades. Only within the last few years have we learned that all those CFCs are still in the air, slowly destroying Earth's protective layer of ozone, as shown in Figure 1-13. Yet these compounds are still being produced and used. The ozone layer blocks harmful amounts of ultraviolet radiation from striking our skin. Its destruction would ultimately lead to ours.

Extinction

Earth's growing human population has also begun to seriously harm the other creatures with whom we share the planet. The world's tropical forests, home to one-fifth of the world's species of animals and plants, are being destroyed at the rate of more than an acre per second, as shown in Figure 1-14. At today's rate of destruction, all the rain forests will be gone in less than 30 years! With them will be lost over a million species, the greatest extinction event since the disappearance of the dinosaurs 65 million years ago. It is wise to remember that extinction is forever and that our children and their children and their children's children will never see any of these animals and plants alive again. Who knows what medicines and food plants we have forever discarded without knowing about them? Like burning a library without reading the books, extinction caused by humans is a tragedy beyond measure.

Feeding a Growing Population

Approximately 1.2 billion people live in a condition of extreme poverty, many of them malnourished. The demand for food is only going to increase.

To produce more food, we are going to have to develop new crops. Agricultural scientists are vigorously seeking new crops that will grow more efficiently in tropical soils, and crops that will grow without intensive use of fertilizers and pesticides. Also, genetic engineers are transplanting genes from one crop plant to another

Historical Note

Defeating Disease
Using the scientific method, medical researchers in the 1970s began to prove the validity of medicinal plants known to other cultures for thousands of years. The onslaught of scientific interest in traditional medicine has both positive and negative consequences. For example, a rare periwinkle flower found in the rain forest of Madagascar is now being used in America to treat leukemia and other forms of cancer, including Hodgkin's disease. However, in
their rush to obtain the cancer-fighting drug taxol from *Taxus brevifolia* (yew tree) roots, bark, and leaves, researchers are endangering both the Pacific yew of the northwest American coast and a rare yew species that grows in Nepal.

to create crop plants that are more resistant to insects. The success or failure of these efforts will have an enormous impact on the world's future. ◻

Combating Disease

Perhaps one of the most exciting ways that biology is affecting modern life is in the battle against disease. New technologies have enabled biological scientists to combat disease in ways scarcely imagined only a few years ago. Among the many diseases that you will encounter in this text, consider the following.

AIDS

It is estimated that by the end of this century 110 million people worldwide will be infected with HIV. **HIV** (human immunodeficiency virus), shown in Figure 1-15, is a virus that destroys the immune system, causing acquired immune deficiency syndrome, or AIDS. AIDS is fatal. All of the people infected with HIV are expected to develop AIDS. Over 300,000 Americans have already died, with 187,000 new cases reported in 1993–1994 alone. AIDS is now the leading cause of death among American males between the ages of 24 and 44. As the number of people with the HIV infection and AIDS grows, there is an increasing demand for more research to find a way to halt the spread of the disease and to help those already suffering from it.

Figure 1-15 HIV antibody testing and counseling are offered by public health clinics, *left,* to people who are at risk for infection by HIV, *above,* or who experience AIDS symptoms.

Cancer

At current rates, over one-third of the students who read this text will die of cancer someday. **Cancer** is a disorder of cells in which the normal controls on growth have been damaged and the cells divide unchecked within the body. Although the number of people with cancer is increasing, scientists studying this

◻ **CAPSULE SUMMARY**

Producing enough food has become a global problem. Scientists are looking for ways to develop more efficient crops without the intensive use of fertilizers and pesticides, and to develop new crops using genetic engineering.

SECTION 1-4

UPDATE

☑ **RESEARCH Update**

Smoking *Is* Bad for You

You aren't likely to greet this news with surprise. The point is that the study of biology tells us about important issues that can have a serious impact on our lives. Medical researchers, agricultural biologists, ecologists, epidemiologists, molecular biologists, botanists and others in related disciplines have contributed to our understanding of the link between smoking and cancer. According to Dr. M. Miles Braun of the National Cancer Institute, genetic inheritance has little or nothing to do with the development of lung cancer in smokers.

Braun studied 15,924 pairs of male twins who were born between 1917 and 1927. Some of these twins were identical, which means they share 100 percent of their genetic material, and some of these twins were fraternal, sharing only 50 percent of their genetic material. Braun tracked these men in their fifties to see who developed lung cancer and who didn't. By using fraternal and identical twins as control and experimental groups, Braun set out to see what percentage of lung cancer was genetic and what percentage was environmental (due to smoking). Previous estimates held that about 20 percent of smokers who developed lung cancer did so because they were genetically predisposed to the disease. Braun concluded instead that genes have virtually nothing to do with the development of lung cancer in smokers. If you smoke and get lung cancer, it's because of the smoke, plain and simple. ☑

Overcoming Misconceptions

Cancer

Many students think "cancer is cancer." Point out that cancer is actually a group of some 200 diseases in which a particular cell type escapes the factors that regulate normal cell growth and division. With few exceptions, any one of the approximately 200 cell types in the human body can become cancerous. The chances of a cure depend largely on the type of cell involved.

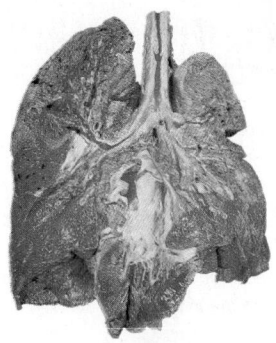

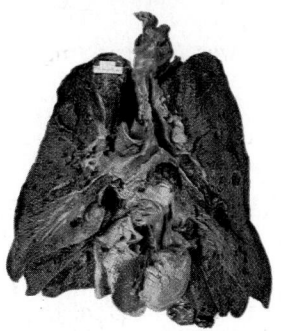

Figure 1-16 Cigarette smoking is the major cause of lung cancer. Its devastating effects on lungs are shown here, *bottom.*

disease have found that many cancer deaths are preventable. Almost all lung cancers and mouth cancers, for example, are due to the use of tobacco. Eliminating smoking and other uses of tobacco would prevent most lung and mouth cancers. Figure 1-16 shows how the use of tobacco can cause serious damage to your body. Your diet and the chemicals to which you expose yourself can affect the incidences of some major kinds of cancer. Research into everyday enemies, such as tobacco, could greatly lower the number of deaths caused by cancer in the future.

Cystic Fibrosis

Cystic fibrosis is an often fatal genetic disorder, passed from parent to child, that is caused by a defective gene encoding a cell membrane protein. Without a healthy copy of the gene, the child's cells cannot pump chloride ions in and out of cells correctly, causing thick mucus to build up in the lungs and other organs. In the 1990s biologists isolated a healthy version of the cystic fibrosis gene, and they are learning how to transfer it into cystic fibrosis patients. In 1993 the gene was first introduced into a human patient, carried into the cells of the lung aboard a cold virus. Researchers are hopeful that within a few years it will prove possible to cure humans by transferring into their lung cells, healthy copies of their defective gene. Other serious genetic disorders, such as muscular dystrophy, are also prime candi-dates for this gene-transfer therapy. The future for this approach looks very bright. ◻

◻ CAPSULE SUMMARY

Research and new technologies will help scientists battle such diseases as AIDS and cancer. Gene therapy could be a future cure for cystic fibrosis and muscular dystrophy in humans.

What You Can Contribute
....

It is clear that a scientific education has become necessary for everyone. The study of biology must play a major part in that education if you are to play any role in improving the standard of living for yourself and for future generations. Biological literacy is no longer a luxury for those who want to play a constructive role in improving the world; it is now a necessity.

Section Review

1. *Explain why saving the rain forests is such a pressing issue.*
2. *How is genetic engineering used in crop development?*

Critical Thinking

3. *Why is biological literacy essential in the battle against disease?*

Vocabulary

acid rain (6)
biology (12)
cancer (17)
cell (12)
control (10)
cystic fibrosis (18)
experiment (9)
heredity (14)
HIV (17)

homeostasis (13)
hypothesis (9)
metabolism (13)
multicellular (13)
observation (9)
pH (6)
prediction (9)
theory (11)

Concept Mapping

Construct a concept map that shows the relationships among scientific processes employed in biology. Use as many terms as needed from the vocabulary list. Try to include the following terms in your map: biology, observation, communication, hypotheses, predictions, experiments, and theories. Include additional terms in your map as needed.

Review

Multiple Choice

1. Scientific knowledge is ultimately based on
  a. beliefs.
  c. models.
  b. observations.
  d. experiments.

2. Biologists communicating about their observations of amphibian populations led to
  a. a decrease in the number of *Rana cascadae* in Oregon and California.
  b. the extinction of frogs in Kings Canyon National Park and other parks.
  c. a further decline in the number of amphibians in United States wetlands.
  d. the recognition of a widespread problem related to amphibian survival.

3. Harte suspected that acid rain was responsible for the tiger salamander population's decline because
  a. acid-laden winds provide the moisture received by the region.
  b. coal and oil are used to heat homes in nearby Colorado.
  c. the pulse of acid was detected long before the snow melted.
  d. local factories produce industrial acids which accumulate in the ponds inhabited by the tiger salamander.

4. A hypothesis is a testable statement that
  a. is considered unscientific because it is not a theory.
  b. can be supported in all circumstances.
  c. if false, can be demonstrated false through experimentation.
  d. is the final step in every research project.

5. Which of the following is *not* a stage of a scientific investigation, as described in the chapter?
  a. theorizing
  b. observing
  c. experimenting
  d. classifying

6. A girl wants to determine which of two foods produces the greatest weight gain in gerbils. She takes two gerbils of identical weight from the same litter and feeds one gerbil 20 g of food A each day and feeds the second gerbil 20 g of food B each day. After 30 days, she weighs the two gerbils and finds that the one fed food A has gained 50 g and the one fed food B has gained 80 g. A control in this experiment is
  a. gerbils of identical weight.
  b. the food fed the gerbils.
  c. the gerbils' weight gains.
  d. the color of the gerbils' fur.

CHAPTER REVIEW ANSWERS

Each item in the Chapter Review is correlated by text section in the assignment guide that follows.

ASSIGNMENT GUIDE

Section	Review	Themes Review	Critical Thinking
1-1	1–3, 15, 16, 21		22, 25
1-2	4–6, 9, 12, 14, 19–21		23, 24
1-3	8, 13, 17		
1-4	7, 10, 11, 18		

Concept Mapping

Map answer is shown on page 3B.

Review

Multiple Choice
Code in parentheses indicates section number and objective number.
1. b (1-1.1)
2. d (1-1.2)
3. a (1-1.3)
4. c (1-2.2)
5. d (1-2.1)
6. a (1-2.3)
7. b (1-4.1)
8. a (1-3.1)
9. a (1-2.3)

Completion
10. 30 (1-4.1)
11. fertilizers (1-4.2)
12. reject (1-2.2)
13. life (1-3.1)
14. observations, experiment (1-2.2)
15. acid rain (1-1.3)

CHAPTER REVIEW ANSWERS *continued*

Short Answer

16. The moist skin of frogs, like that of other amphibians, permits the absorption of chemicals from pond water that are not absorbed by reptiles or mammals. If the chemicals are hazardous, frogs are among the first organisms to be adversely affected. (1-1.1)

17. A virus is not alive. It is not composed of cells. (1-3.2)

18. News accounts are reported in newspapers and on radio and television. These media reach much of the population. Few people actually read scientific papers. (1-4.2)

19. The work of Harte led to the construction of related hypotheses and, thus, a theory that accounts for the decline of isolated amphibian populations. Harte's hypotheses and theory will stand as long as data tend to support them. If data are collected that do not support his hypotheses and theory, the hypotheses and theory will then be called into question. His hypotheses and theories could be modified or replaced if data indicate that it is necessary to do so. (1-2.2, 1-2.4)

20. Theory is that which a scientist is most certain of; however, theory implies a lack of knowledge to the general public. (1-2.4)

21. The more times an observation can be repeated, the greater the probability that it is valid and the greater the confidence that scientists have in it. (1-1.2, 1-2.3)

Critical Thinking

22. Such details enable scientists to evaluate the reliability of the experimental procedure and thus the conclusions, to understand the conditions under which the conclusions apply, and to try to repeat the experiment. (1-1.2)

23. The next step would be to repeat the controlled experiment with animals, such as rats, that have cancer. If successful, this series of experiments could be followed by tests of

---

1 CHAPTER REVIEW

7. Problems of waste management and air pollution are due to
 a. the spread of disease.
 b. growth of the human population.
 c. uncontrolled scientific experimentation.
 d. genetic engineering and global warming.

8. Using the criteria of movement and complexity to define life
 a. makes distinguishing living things from nonliving things a difficult task.
 b. is acceptable for children and adolescents.
 c. means that cars are alive but trucks and buses are not.
 d. means that single-celled organisms are alive whereas multicellular organisms are not.

9. To test the effectiveness of a new drug against human immunodeficiency virus (HIV), a biologist would
 a. take 100 cultures of HIV and put the new drug in 50 of them.
 b. take 100 cultures of HIV and put the new drug in all of them.
 c. put the new drug in none of the 100 cultures.
 d. avoid using cultures consisting of animal cells.

Completion

10. By one estimate, the tropical forests will be gone in less than _____ years.

11. Agricultural scientists seek new crops that will grow without intensive use of _____ and pesticides.

12. A scientist using a model to weigh the benefits and risks of a course of action is engaged in _____ .

13. Biology is defined as the study of _____ .

14. Before a hypothesis is proposed, _____ must be made. A test of a hypothesis is called a(n) _____ .

15. Harte hypothesized that _____ was the cause of the decline of the salamander population.

Short Answer

16. What characteristic of frogs makes them the "miner's canary" of wetlands?

17. A virus consists of DNA or RNA in a protein coat. It can reproduce and is capable of movement from one host to another. Is a virus alive? Explain.

18. Explain why news accounts are much more influential than scientific papers in educating the public about scientific issues.

19. "In science all hypotheses and theories are tentative." How does this statement apply to the work of John Harte?

20. Compare scientists' use of the term *theory* with that of the general public.

21. Scientists often try to repeat, or replicate, the results of other scientists' experiments. Why do you think this is an important part of scientific investigation?

Critical Thinking

22. Making Inferences One of the most important parts of any scientific paper is the part called "Methods and Materials," in which the scientist describes the procedures used in the experiment. Why do you think such details are so important?

23. Making Predictions An experiment was conducted to test the effect of a new drug on the growth of human cancer cells. In the experimental group, the drug was added to cancer cells growing in 100 test tubes. In the control group, genetically similar cancer cells were grown under exactly the same conditions as the experimental group but without the drug. The results of the experiment showed slower growth of cancer cells in the experimental group. What would be the next logical step in testing the effectiveness of the new drug to slow the growth of human cancer cells?

---

the drug on humans using a controlled experiment. (1-2.3)

24. For a hypothesis to be irrefutably proven, every case would have to be tested. Issues of concern to biologists make testing every case virtually impossible. (1-2.2)

25. Harte put together knowledge of acid rain, wind patterns in the Colorado Rockies, and the decline of amphibian groups in other areas. Using this knowledge, he developed hypotheses that were then tested. The findings of these tests allowed Harte to construct new hypothoses

about the decline in the tiger salamander population. (1-1.1)

Activities and Projects

26. Student answers will vary, but should identify amount of water, size of plants, sunlight, and temperature as controlled variables. The control and

24. **Making Inferences** Explain why it is inappropriate to say that a hypothesis has been proven.

25. **Making Inferences** Scientific knowledge is constructed like a tower built from blocks rather than discovered like a gold nugget in a stream. Explain how the report of John Harte's work with the tiger salamander provides support for this statement.

Activities and Projects

26. **Cooperative Group Project** Design an experiment to find out how the pH of the water used to water bean plants affects their growth. First, identify the variables that should be controlled and describe the control and experimental groups. Explain how the results from the control and experimental groups should be compared. Finally, write the components of your design on a poster and display it for other students to see.

27. **Cooperative Group Project** Create a television commercial intended to convince high school students to become biologically literate. To make your commercial as persuasive as possible, survey students about their opinions on the advantages and disadvantages of studying biology in high school. Address their responses in your commercial.

28. **Science-Technology-Society** Learn what you can about integrated pest management (IPM) by reading and talking with farmers and agricultural extension agents. What are some of its advantages and disadvantages? Why are agricultural companies that produce pesticides less than enthusiastic about IPM?

29. **Multicultural Perspective** Tropical rain forests are located mainly within the borders of impoverished nations in Central America, South America, and Africa and are home to at least one-third of all animal and plant species. It is predicted that in less than 100 years, all the world's rain forests will have been eliminated due to lumbering and clearing for farms. People in the United States and other wealthy nations argue that assisting the economic growth of the nations containing the rain forests is the best way to save the rain forests. Explore both sides of this argument. Report what you find to your class.

30. **Multicultural Perspective** Do library research to discover examples of ways that some cultures around the world speak of fire as a living being (many cultures, for example, refer to fire as "grandmother" or "grandfather"). Which of the five characteristics often associated with living things (movement, sensitivity, development, complexity, and death) do you think fire might possess? What about the five general properties of living things (cellular organization, metabolism, reproduction, homeostasis, and heredity)? Defend your answer.

Readings

31. Read the article "Silence of the Frogs," in *The New York Times Magazine*, December 13, 1992. Describe Cynthia Carey's research. What does her work seek to discover about the causes of red-leg disease in the western frog? Name three factors that scientists think could possibly be responsible for the disappearance of frog populations in various locations around the world.

32. Read the article "The Long Shot," in *Discover*, August 1993. Explain why scientific techniques using killed or weakened whole viruses have thus far failed to produce a successful vaccine for AIDS. Why is it difficult for scientists attempting to develop such a vaccine to perform experiments and thus verify their predictions regarding the success of any prospective vaccine?

are less than enthusiastic about IPM because it reduces their profits.

29. Student answers will vary, but should present reasons for giving the rain forest nations money to reduce poverty and reasons for not doing so.

30. Answers will vary. Students might point out that fire can move from one field to another; has sensitivity to water, air, and gases; shows development as it grows from a small spark into a large bonfire; demonstrates complexity in its various compositions of wood, paper, air, flint, and gases; and experiences death by a variety of fire extinguishers. Fire does not exhibit the five general properties of life, although some similarities exist.

Readings

31. Carey's research is seeking to discover if environmental factors such as ultraviolet radiation, acid rain, pesticides, or rapid temperature changes could damage the immune system of the western frog, resulting in fatal infections by the bacterium that causes red-leg. Acid rain, increases in ultraviolet radiation, and pesticides are three factors that scientists cite as possible causes of declining amphibian populations.

32. HIV is genetically unstable and changes frequently. Because it is a retrovirus, it can lie undetected for years within the infected individual's cells. When it emerges, it destroys white blood cells called T cells, which are essential for the normal functioning of the immune system. A vaccine made from a killed or weakened HIV would therefore fail to protect an individual from infection by a different form of HIV. Moreover, testing an AIDS vaccine on uninfected individuals would be extremely dangerous; if even one virus in the vaccine remained alive, it could cause the vaccinated individual to become sick with AIDS.

experimental groups should be identical except for the pH of the water; the pH of the control group should be 7 and the pH of the experimental group should be less than 7. Results may be compared by measuring the height of plants.

27. The message conveyed in the commercial should reinforce outcomes supportive of becoming biologically literate and downplay or discredit outcomes not supportive of the desired behavior.

28. IPM applies the knowledge of pests, crops, and the environment to keep the numbers of pests low. IPM programs cut the use of pesticides by 50 percent or more, saving farmers money and resulting in crops safer for human use and consumption. IPM programs are expensive to begin and require professional expertise. Chemical companies

LABORATORY Investigation Chapter 1

Preparation

Use unbroken bean, pea, or corn seeds since they germinate quickly and are usually mold resistant. Prepare solutions under a ventilated hood; wear goggles, face shield, impermeable gloves, and lab apron. Dilute acid spills with water; mop up with wet cloths designed for spill cleanup. Prepare 600 mL of mold inhibitor for 25 students (one part bleach to four parts water). Prepare a 0.01 M solution of sulfuric acid (H_2SO_4) with a pH of 2 by diluting 5 mL of 1.0 M H_2SO_4 to 1 L in a beaker. Dilute 50 mL of the 0.01 M solution to 1 L for pH 3. Repeat procedure using 5 mL and 0.5 mL of the 0.01 M H_2SO_4 to make pH 4 and pH 5 solutions, respectively. Test with pH paper or meter. Collect solutions in a single container and neutralize to pH 7 with 1.0 M base before discarding in sink.

Procedural Notes

- Take 40 minutes to complete Part A and 15 minutes on each day that detailed observations are made. Allow two days to elapse between Parts A and B. Lab will be completed 7–10 days after set up.

- At the end of Part A, collect left-over solutions in a single container and neutralize to a pH of 7 before discarding in the sink.

- You may want to review pH with your students. Discuss differences in germination among the different types of seeds (monocot and dicot) that the students will use.

- Put a large data chart on the chalkboard so that students can pool their data and get class results.

- Prepare a model graph on the chalkboard if your students have difficulty following the directions. Generally, plant growth is better in a pH range from 4.5 to 6.5.

Observing Effects of Acid Rain

OBJECTIVE

Observe the effects of acid rain on seeds from three different species of plants.

PROCESS SKILLS

- measuring pH
- predicting effects of acid rain
- organizing and graphing class data

MATERIALS

- safety goggles
- lab apron
- bean, pea, or corn seeds
- metric ruler
- 250 mL beakers
- mold inhibitor
- wax pencil or marker
- plastic bags (zipper type)
- paper towels
- 2 water solutions with different pH values
- graph paper

BACKGROUND

1. What threat does acid rain pose to the environment?

2. What are major sources of the pollutants that cause acid rain?

3. Write your own question to explore in your lab report or notebook.

TECHNIQUE

Part A: Measuring and Sowing Seeds

For this procedure, you will work with one water solution while your partner works with a different water solution. Before beginning the lab, put on safety goggles and a lab apron. You and your partner will use the same kind of seed.

1. Measure the length in millimeters of each of 20 seeds. Determine the average length. In the **Records** section of your report, make a table similar to the one shown on the following page. Record the seed type, your solution's pH, and the average seed length (Day 0) in the table.

2. Place the 20 seeds into a beaker, 10 seeds for you and 10 seeds for your partner. Slowly add mold inhibitor until the seeds are covered. Soak the seeds for 10 minutes. While the seeds soak, label a plastic bag with your name and the pH of the water solution assigned to you. Moisten three layers of paper towels with your assigned water solution. Your partner will also label a plastic bag and use a different assigned solution. Fold the moistened towels in half and put them inside the bag. Drain the mold inhibitor from the beaker and gently rinse the seeds with water. Place the seeds on clean paper towels.

3. Place your 10 seeds between the layers of the moistened paper towels. Slide the towels and seeds into your plastic bag. Close the bag. Your partner should do the same with the other 10 seeds.

Part B: Germination and Growth

4. After 2 days, examine your seeds. Note the changes in appearance, length, average seed size, and the number of germinated seeds. Record your observations in the **Records** Section of your report.

Background Answers

1. Acid rain can damage trees, plants, water-dwelling organisms, and human-made structures in the environment.

2. Air pollution from coal-burning factories, power plants, and automobile emissions causes acid rain.

3. What are the effects of acidic solutions on seeds from three diffferent species of plants?

Effects of Acid Rain on Seed Growth

	You	Your Partner
Seed Type		
pH of Solution		
Average Seed Length (Day 0)		
First Observation		
Second Observation		
Last Observation		
Number of Seeds That Germinated		
Overall Average Seed Growth		

5. Add your data to the class chart on the chalkboard. Compare your results with those of your classmates.

6. Moisten the paper towels with your assigned solution and return the towels and seeds to the plastic bag.

7. Observe your seeds again after 2–3 days. Add more of your solution when necessary. In the **Records** section of your report, record the average length and number of germinated seeds.

8. Make your final observations and measurements 7–10 days after you began this investigation. In the **Records** section, note any overall changes in appearance you observe. Check for changes in length. Note average seed size and the number of germinated seeds. Add your data to the class data.

9. Subtract the Day Zero average seed length from the Last Observation average seed length. This is the overall average seed growth. Record this in the **Records** section of your report and the class chart.

10. Make a graph to organize the class data on the effects of pH on the germination and growth of each type of seed. Label the vertical axis with the number of millimeters of overall seed growth. Along the horizontal axis, show the range of pH values for the entire class. Copy the graph in your report. In your report, briefly summarize the procedure you followed.

INQUIRY

1. Develop a hypothesis to explain the possible effects of acid on the germination of your seeds.

2. How does the pH value appear to affect seed germination?

ANALYSIS

1. In which solution did each type of seed germinate first?

2. What pH appears to be best suited for both successful germination and continued seed growth? Which is the least beneficial?

FURTHER INQUIRY

Write a new question that could be explored as a new investigation. The following is an example:

> Are certain stages in the plant life cycle more susceptible to acid rain than others?

Inquiry Answers

1. Answers will vary. Very acidic solutions will be injurious to seed germination.

2. Low pH values tend to prevent seed germination. pH values closer to neutral are better for seed germination.

Analysis Answers

1. Answers will vary.

2. Successful germination and continued growth should be better in the pH range from 4.5 to 6.5. The highly acidic pH of 1 should be the least beneficial.

Effects of Acid Rain on Seed Growth

	You	Your Partner
Seed Type		
pH of Solution		
Average Seed Length (Day 0)		
First Observation		
Second Observation		
Last Observation		
Number of Seeds That Germinated		
Overall Average Seed Growth		

Effects of pH on Seed Germination and Growth

Overall seed growth (mm) vs. pH value (0, 2, 4, 6, 8, 10, 12, 14)

> Each block represents about 45–50 minutes of instructional time. The resources cited in a block are coded to a specific program component using the Key on the next page.

Lecture Concepts   Lesson Resources

2-1 Cells: The Smallest Units of Life pp. 24–28

BLOCK ❶

a. Cells are the basic units of structure and function in living things.
b. All cells share certain structural characteristics: a cell membrane, cytoplasm, ribosomes, and DNA.
c. Except for bacteria, which are single-celled prokaryotes, organisms are made up of one or more eukaryotic cells. Prokaryotic cells lack internal membrane-bound compartments. Eukaryotic cells have a nucleus and other membrane-bound compartments.
d. The size of cells is limited by surface area to volume ratio, which determines the efficiency with which materials and information are transported between the cell's interior and exterior.

Lecture Resources
- Opening Question, page 25
- Demonstration: Where Does Cork Come From?
- Visual Strategy: Figure 2-2
- Overcoming Misconceptions, page 26
- Cell Membranes and Materials Transport, page 28
- Teaching Transparencies: 12A, 5A
- Holt Biology Videodiscs: Lesson 2-1

Classwork Options
- Teaching Resources CD-ROM: Basic Sills Worksheets; Length, Area, and Volume; Measurement and Scale Drawings
- Quick Labs A3: Modeling Cells: Surface Area to Volume
- Interactive Explorations in Biology Laboratory Manual E1: Cell Size

Assignment Options
- Section Review, page 28
- Chapter Review, questions 3, 4, 18, 20

2-2 The Chemistry of Living Cells pp. 29–33

BLOCKS ❷ and ❸

a. All matter is made of atoms, which consist of electron, protons, and neutrons.
b. Atoms share electrons in covalently bonded molecules. In ionically bonded substances, atoms transfer electrons. Hydrogen bonds form weak links between molecules.
c. Because it heats slowly and stores heat well, water is used by many organisms to eliminate excess heat.
d. Water molecules tend to form hydrogen bonds with each other (cohesion) and with other substances (adhesion). The polar nature of water molecules makes water a strong solvent.
e. Any compound that forms hydrogen ions when dissolved in water is called an acid. Any compound that forms hydroxide ions when dissolved in water is called a base. The pH scale relates the concentration of hydrogen ions from one solution to another.

Lecture Resources
- Opening Question, page 29
- Demonstrations: The Periodic Table, Surface Tension; The pH Scale
- Overcoming Misconceptions: pages 29, 31
- Probability and the Atom, page 30
- Covalent Versus Ionic Bonds: A Model, page 31
- Teaching Transparencies: 2, 3, 4

- Visual Strategies: Figures 2-7, 2-10
- Holt Biology Videodiscs: Lesson 2-2

Classwork Option
- Capillary Action in Celery, page 32

Assignment Options
- Section Review, page 33
- Chapter Review, questions 5, 8, 10, 11, 21

2-3 Chemical Building Blocks of Cells pp. 34–38

BLOCKS ❹ and ❺

a. Cells use carbohydrates to store energy and provide structural support.
b. Lipids store energy and are a component of cell membranes.
c. Proteins play an important role in structure and metabolic processes. The sequence of amino acids in a protein determines its shape, chemical properties, and function.
d. Nucleic acids contain the cell's hereditary information. DNA stores information in cells.
e. ATP is an energy-storage molecule necessary for cell processes.

Lecture Resources
- Opening Question, page 34
- Demonstrations: Carbohydrates, DNA: The Double Helix
- Overcoming Misconceptions, page 35
- Teaching Transparencies: 2A, 3A, 4A
- Health, page 36
- Visual Strategies: Figures 2-12, 2-13, 2-16
- Multicultural Perspective, page 36

- Holt Biology Videodiscs: Lesson 2-3

Classwork Options
- Evolution and Levels of Organization and Graphic Organizer, pages 34–35
- Visual Strategy: Figure 2-14
- ATP: The Cell's Energy Currency, page 38

Assignment Options
- Section Review, page 38
- Chapter Review, questions 7, 12, 13, 16, 17, 22

2-4 The Interior of the Cell pp. 39–48

BLOCKS ⑥ and ⑦

a. Light microscopes use a beam of light passing through one or more lenses to produce an enlarged image of a specimen being viewed. Electron microscopes use electrons to magnify the features of specimens.

b. Eukaryotic cells have the endoplasmic reticulum to transport materials within the cell.

c. The Golgi apparatus is the packaging and distribution center of the cell.

d. The nucleus directs all cell activities and stores the cell's DNA.

e. Mitochondria are the energy-releasing organelles. Chloroplasts enable algae and plants to capture the energy in sunlight and use it to make sugars.

f. Peroxisomes neutralize potentially dangerous molecules within the cell. Lysomes digest and recycle the cell's used components.

g. Cells have a cytoskeleton of protein fibers that supports the shape of the cell and anchors its organelles. Many eukaryotic cells also have flagella or cilia for movement.

Lecture Resources
- ■ Opening Question, page 39
- ■ Demonstration: The Light Microscope
- ■ Research Update, page 40
- ■ Visual Strategies: Figures 2-18, 2-23
- ■ Nuclear Envelope and Folds in Mitochondria, page 44
- ■ Cellular Recycling, page 45
- 🔒 Teaching Transparencies: 7, 8
- ✎ Holt Biology Videodiscs: Lesson 2-4

Classwork Options
- ■ Images Produced by Microscopes, page 41
- ⊙ Teaching Resources CD-ROM: Basic Skills Worksheets, Microscope Magnification
- 🧪 Quick Labs A32: Relating Cell Structure to Function
- ■ Laboratory Investigation: Animal and Plant Cells, pages 52–53
- ■ Closure, page 48

Assignment Options
- ■ Evolution of Eukaryotes, page 44
- ■ Section Review, page 48
- ■ Chapter Review, questions 1, 2, 6, 9, 14, 15, 19, 20, 23

Review/Enrichment

BLOCK ⑧

Review/Enrichment
- ■ Themes of Biology: Evolution, pages 48–49
- ■ Themes Review, question 20
- ■ Study Guide: Concept Review and Pretest

Assignment Options
- ■ Activities and Projects, questions 24–28
- ■ Readings, questions 29–30

Assessment Options

BLOCK ⑨

Traditional Assessment
- ■ Chapter 2 Test
- ⊙ Test Generator/Assessment Item Listing: Software item bank for preparing customized chapter tests.

Portfolio Assessment
Select student reports for one or more laboratory experiments from this chapter. *The Direct Observation Checklist* on the *BioSources Teaching Resources CD-ROM* should be completed during a laboratory or other cooperative learning experience.

Answer to Concept Map
The following is one possible answer to the Concept Mapping exercise on page 49.

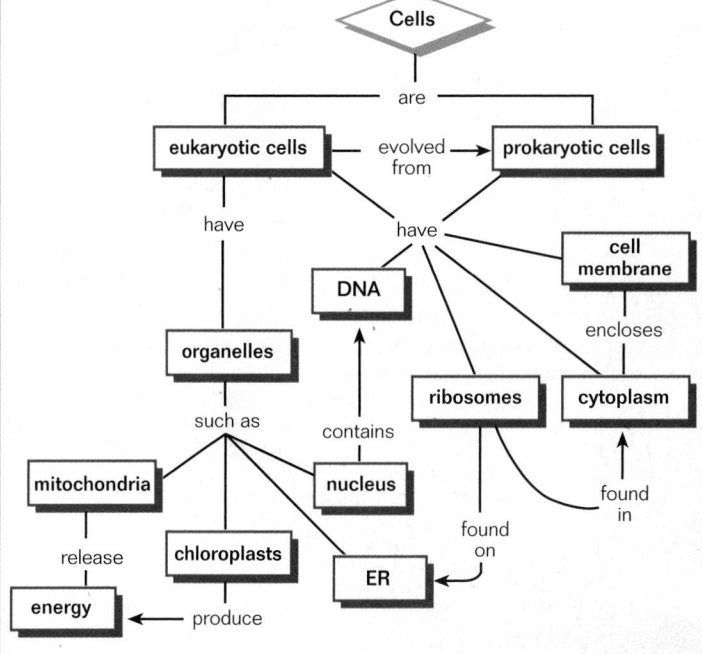

Chapter Theme

Evolution *Evolution—the theory that species change over time—is the first of five core themes that recur throughout this text. Organisms have changed dramatically since they first appeared on Earth. Evidence of this evolution includes fossils, homologous structures, similarities in biochemistry, and similarities in embryological development. Studying evolution helps scientists understand diversity and similarity among organisms. In this chapter, students will become acquainted with some precursor organisms that have become cell structures.*

Tapping Prior Knowledge

- Which is smaller, a cell or an atom?
- What are some characteristics that all cells share?
- Is there a relationship between the physical structures and the chemistry of cells?

Opening Demonstration

Put up pictures or diagrams of typical cells, both animal and plant, preferably poster sized. Ask students to list both the similarities and the differences they see among the cells displayed. Ask students to list one or two reasons why these cells might be different. Encourage students to think about some of the differences between animals and plants (e.g., animals move from place to place, but plants don't; some animals have an internal skeleton, but plants don't; etc.) and how these differences might be related to cell structure.

CHAPTER 2
NATURE OF CELLS

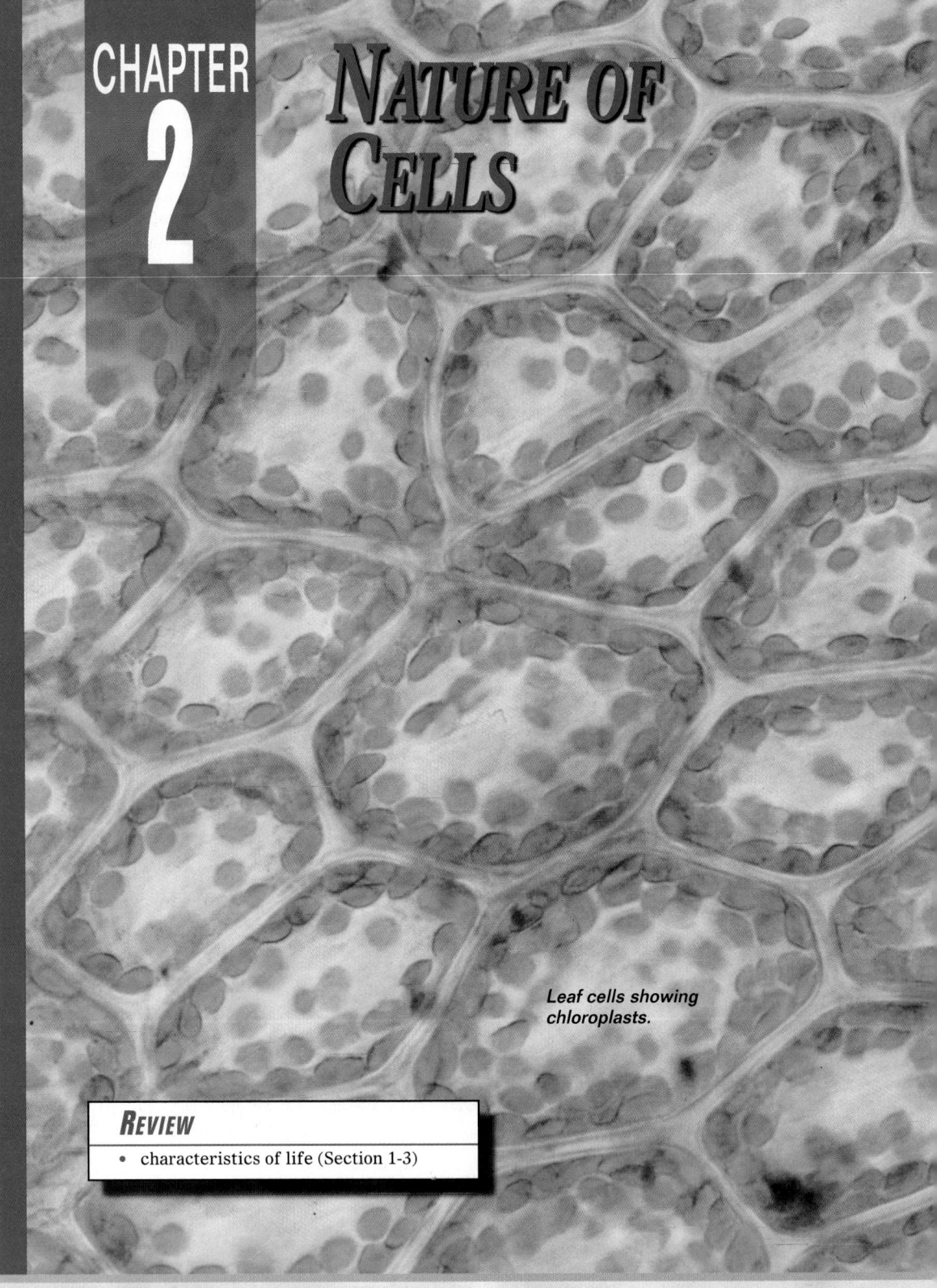

Leaf cells showing chloroplasts.

REVIEW

- characteristics of life (Section 1-3)

Authors' Rationale

Before tackling the broader concepts of "life," it is essential that students have a very good understanding of its smallest component parts—cells. Thorough understanding of this critical component of life is achieved through discussion of why cells exist, followed by a hierarchical description of the elements that actually combine to compose cells. This discussion includes enough chemistry for students to understand cell structure and function and a description of the individual organelles. The rest of biology builds on these concepts.

2-1 Cells: The Smallest Units of Life

Organisms are more than collections of chemicals. They are chemicals organized to carry out the functions of living. The organization of all living things begins with the cell, the smallest unit capable of carrying out the functions of life. All living things are composed of one or more cells.

Section Objectives

- State the parts of the cell theory.
- Identify the major differences between prokaryotes and eukaryotes, and explain the importance of each.
- Explain why cells must be small.

How Cells Were Discovered

Most cells are too small to see with the naked eye. A typical human body cell is many times smaller than a grain of sand! Scientists became aware of cells only after microscopes were invented in the 1600s. When the English scientist Robert Hooke used one of the first microscopes to observe a thin slice of cork in 1665, he saw "a lot of little boxes," shown in Figure 2-1. These little boxes reminded him of the small rooms in which monks live, so he called them cells. Later, Hooke observed the same pattern in the stems and roots of carrots and other plants. What Hooke still did not know, however, was that cells are the basic unit of living things.

Ten years later, the Dutch scientist Anton van Leeuwenhoek focused a microscope on what seemed to be clear pond water and discovered a wondrous world of living creatures! He named them "animalcules," or tiny animals. Today we know that they were not animals, but single-celled protists, among the most diverse of all living things.

Formation of the Cell Theory

It took scientists more than 150 years to fully appreciate the discoveries of Hooke and Leeuwenhoek. In 1838, the German botanist Matthias Schleiden concluded that cells compose not only the stems and roots but every part of a plant. A year later, the German zoologist Theodor Schwann made the same claim about animals. And in 1858, a German physician, Rudolph Virchow, observed that cells come only from other cells. The observations of Schleiden, Schwann, and Virchow form what is known today as the cell theory. The **cell theory** is usually stated in three parts:

1. All living things are composed of one or more cells.
2. In organisms, cells are the basic units of structure and function.
3. Cells are produced only from existing cells.

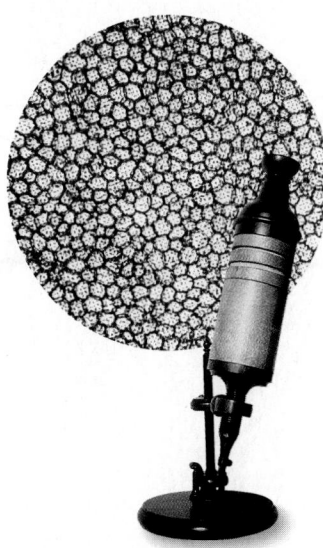

Figure 2-1 It was the English scientist Robert Hooke who first coined the term *cell.* Hooke used a simple microscope to view a slice of cork. The "little boxes" that he saw reminded him of the monastery cells that served as individual living quarters for monks, and a new biological term arose as a result. Although Hooke did not know it, all living things are made of one or more cells.

Historical Note

What Is Life's Lowest Common Denominator?
Early scientists from Aristotle to William Harvey, the discoverer of human circulation, and even Isaac Newton believed in spontaneous generation, a theory whose foundation rested on some intangible "vital principle" which supposedly produced life. It was not until the early to mid-1800s, the time of the invention of vital stains and the work of Spallanzani and Pasteur, that the theory of biogenesis finally prevailed. After about 1860, the cell theory—the synthesis of these lines of thought—became to biology what the atomic theory is to physics and chemistry.

Figure 2-2

Compare Figure 2-2 to Figure 2-3. Point out the obvious nucleus in Figure 2-3 on page 27 and the lack of a nucleus in both the fossilized and modern cyanobacteria. It is this lack of a nucleus, or any other membrane-bound organelle, that is the major characteristic of prokaryotes.

Teaching Tips

The Workings of Genetic Material

The chemical DNA provides the master code for all of a cell's activities. Compare DNA to the students' knowledge of computers. DNA can be thought of as a "hard disk" in the cell that contains information which is copied and transferred in small chunks, like transferring files to floppy disks, in order to direct activities all around the cell.

Sizes of Things

Give students a list of objects and their approximate metric sizes. Have students compare the sizes of the objects by converting the units to meter equivalents. For example, the width of a plastic pen cap is approximately 1 cm or 0.001 m. Other items to compare include a fish egg (0.1 mm), a human egg cell (0.1 mm), a human red blood cell (0.01 mm), a bacterium (1 mm), a virus (100 nm), a protein (10 nm), a water molecule (1 nm), and a hydrogen atom (0.01 nm). Have students make a table to compare the sizes of these objects. Encourage students to add their own objects to personalize their tables.

All Cells Share Certain Characteristics

All cells, even those that are very simple, share certain structural characteristics. All cells have a **cell membrane** or **plasma membrane** that separates the cell's contents from materials outside the cell. The cell membrane also regulates what moves in and out of a cell, helping it to maintain homeostasis. Without the cell membrane to contain the substances the cell needs for life, the cell would die. All cells have **cytoplasm** (*SYT uh plaz uhm*), which is everything inside the cell membrane except the cell's genetic material. The fluid portion of cytoplasm is called cytosol (*SYT uh sol*).

The cytosol is packed full of free-floating **ribosomes** (*RY buh sohmz*), the structures on which proteins are made. You will learn how ribosomes make proteins in Chapter 9.

All cells have the ability to reproduce themselves, and they all possess genetic material, which contains the instructions for making proteins and carrying out the cell's day-to-day activities.

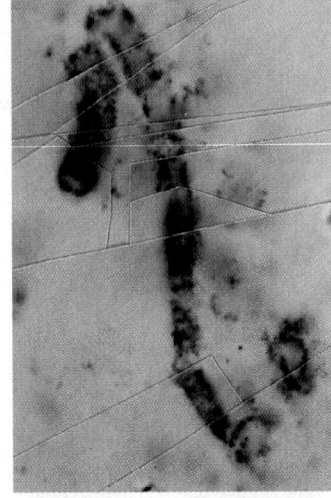

Figure 2-2 Fossils of prokaryotic microorganisms, *top,* have been found in Western Australia in rock formations over 3.4 billion years old. Modern cyanobacteria such as *Oscillatoria, bottom,* are little changed from their primitive ancestors. Like all prokaryotes, cyanobacteria lack nuclei and other membrane-bound organelles.

A Short History of Cells

The oldest fossils we have of cells are those of tiny cyanobacteria, shown in Figure 2-2. These **prokaryotic** (*pro KAR ee AHT ik*) cells lived at least 3.5 billion years ago. Prokaryotes are single-celled organisms that lack internal membrane-bound compartments. The term *prokaryote* is from the Greek *pro,* meaning "before," and *karyote,* meaning "kernel." Early cells were simple and small (1–2 µm in diameter). Like their fossil ancestors, modern prokaryotes are very small (1–15 µm) and do not have internal compartments. Without separate compartments that isolate materials, cells cannot carry out many specialized functions. In prokaryotes, the genetic material is a single, circular molecule that is not enclosed in a membrane-bound compartment. For nearly 2 billion years—half the age of Earth—prokaryotes were the only organisms that existed.

The First Cells With Internal Compartments Were Eukaryotes

The first cells with internal compartments evolved about 1.5 billion years ago. Much larger than any bacteria, these cells range from 2 to 1,000 µm in size. Such cells are **eukaryotic** (*yoo KAR ee AHT ik*). The term *eukaryotic* comes from the Greek words *eu,* meaning "true," and *karyote,* meaning "kernel" or "nucleus." Eukaryotes have a **nucleus**, a membrane-bound compartment that houses the cell's DNA. Eukaryotes possess other small, specific membrane-bound internal compartments called **organelles** that carry out specific functions. Such organization allows eukaryotic cells to function in more complex ways than do prokaryotic cells.

Overcoming Misconceptions

They All Contain Genetic Material

Point out that the genetic material not only is responsible for making proteins and carrying out the cell's daily activities, but also is intimately involved with cellular reproduction and repro-duction of the species. Otherwise, students may fail to make the connection between all of these processes. You may even choose to give a brief overview of how these processes are related.

How Eukaryotes Evolved

Most biologists who study eukaryotic cell structure think that eukaryotes evolved from prokaryotes. Many of the organelles of eukaryotes resemble bacteria, perhaps engulfed long ago by much larger cells. Scientists hypothesize that bacterial "trespassers" remained inside these cells, gradually losing their ability to live independently. These invading bacteria became organelles, and eukaryotic cells were the result. The fact that some organelles have their own distinctive DNA provides additional evidence for this hypothesis.

All living cells that are not bacteria are eukaryotes. Your cells are eukaryotic, as are tree cells and elephant cells. The "animalcules" seen by van Leeuwenhoek also were eukaryotic.

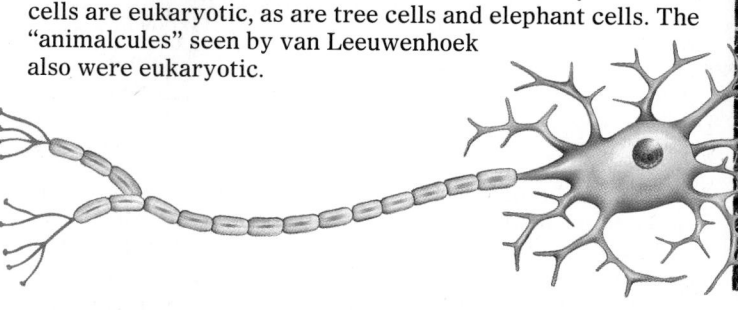

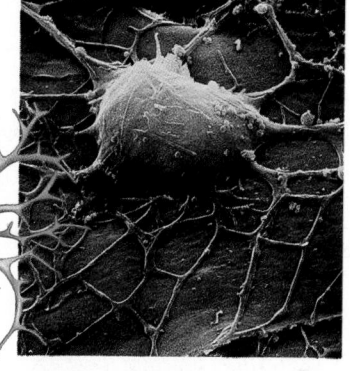

Figure 2-3 This nerve cell, or neuron, is one of the most highly specialized cell types in your body. The long cytoplasmic extensions extending outward from the cell body enable a neuron to receive and transmit electrical signals.

Multicellularity

Early eukaryotes were single-celled, but eventually many of them became aggregated (clustered) into multicellular organisms. Multicellular organisms are those that are composed of more than one cell. Being multicellular was a great evolutionary advance because it enabled particular cells to specialize in certain activities. For example, nerve cells, such as the one shown in Figure 2-3, are highly specialized cells that conduct messages in the form of nerve impulses from one part of the body to another. You are a multicellular individual. Your body is composed of trillions of cells whose specialized activities are coordinated with one another.

Not all eukaryotes are multicellular. In fact, if you were to survey all living organisms on Earth today, you would find that most living eukaryotes are unicellular protists, single-celled organisms. Whether single-celled or multicellular, the cells of all eukaryotes are similar in design, more similar to each other than to the prokaryotes. The basic design of eukaryotic cells will be explored in Section 2-4. ◻

Cells Must Be Small

There are some 100 trillion cells in the human body, typically ranging from 5 to 20 µm in diameter.

Why is your body made of many tiny cells instead of a few large cells? There are two limits that affect how efficiently cells work and that govern cell size. One limit is related to the exchange of materials between the

Teaching Tips

Symbiosis

The early invading organisms and their hosts established a symbiosis. Students may balk at the idea of one organism living within the body of another organism without harm to either. This type of symbiotic relationship called mutualism is not at all unusual. In one example of mutualism, a lichen (the organism growing on rocks that many people incorrectly call "moss") is a symbiotic relationship between a fungus and an alga. The fungus provides protection from water loss for the alga, while the alga accomplishes photosynthesis for itself and the fungus.

Homeostasis

The compartmentalization found in eukaryotes promotes homeostasis by segregating functional elements within protective membranes. Even though these membranes introduce some problems of their own (e.g., difficulty transporting materials), the advantage they confer is reflected in the complexity of eukaryotic function.

👁 VISUAL STRATEGY

Figure 2-3
Correlate the diagram of the neuron with the photomicrograph of the human nerve cell. Point out that while a neuron is microscopic, the long cytoplasmic extensions (axons) may be meters long.

◻ CAPSULE SUMMARY

Except for bacteria, organisms consist of one or more eukaryotic cells that contain membrane-bound organelles. Prokaryotes and eukaryotes share several characteristics: a cell membrane, cytoplasm, ribosomes, and DNA.

Effective Reading

Asking Questions and Anticipating Answers

Ask students to list the four main headings found in Section 2-1. *(The four main headings are How Cells Were Discovered, Formation of the Cell Theory, A Short History of Cells, and Cells Must Be Small.)* Tell students that these four headings can be formulated as questions. For example, the heading How Cells Were Discovered can be formulated as the question "How were cells discovered?" Ask students to think of one question that might be answered in the text following each heading.

Encourage students to write down their questions before reading Section 2-1. Ask them to keep their questions in mind as they read. At the end of the exercise, ask students to return to their questions and answer them.

Teaching Tip

Cell Membranes and Materials Transport

All of the methods of transporting materials, whether into the cell or out of it, are controlled by the cell membrane. These methods may be passive (i.e., they do not involve the expenditure of cellular energy) or energy dependent, but the larger the surface-area-to-volume ratio for transfer between the cell and its environment, the greater the efficiency of the transfer.

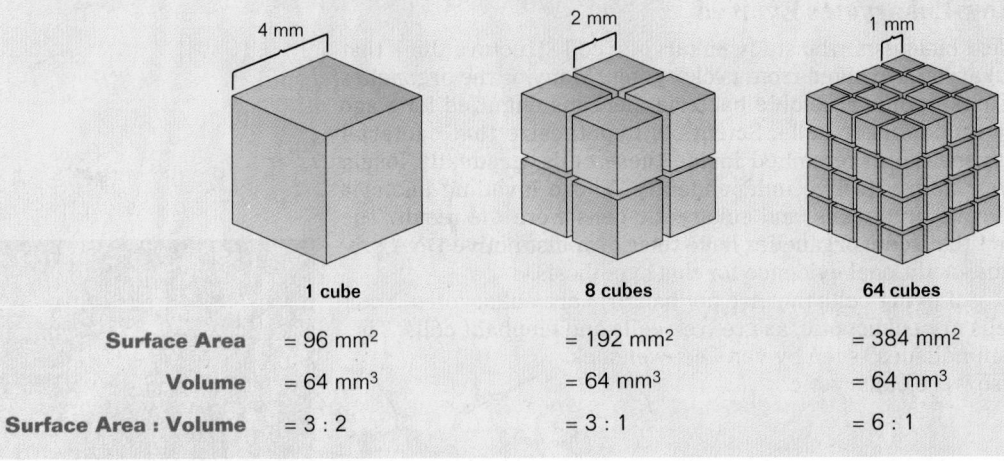

	1 cube	8 cubes	64 cubes
Surface Area	= 96 mm²	= 192 mm²	= 384 mm²
Volume	= 64 mm³	= 64 mm³	= 64 mm³
Surface Area : Volume	= 3 : 2	= 3 : 1	= 6 : 1

Figure 2-4 As a cell gets larger, its volume increases at a faster rate than its surface area. A cell's surface area must be large enough to meet the needs of its volume.

❑ CAPSULE SUMMARY

Smaller cells are more efficient than larger cells because it takes less time to transport information and materials across the surface to the inside where they are needed.

inside and the outside of the cell, and the other limit is related to the distribution of materials within the cell.

A cell's surface provides the only opportunity for interaction with its environment. All information and materials entering or leaving the cell, including wastes, must pass through "doors" in the cell's membrane. The efficiency of such an exchange depends on the ratio of the cell's surface area to its volume. The surface area of a cell is the measurement of the exterior of the cell. Large cells have far less membrane surface per unit of volume with which to supply materials to the cell's interior and to rid it of wastes. You can understand surface-area-to-volume ratios by looking at the imaginary cube-shaped cell in Figure 2-4. A cell with six times as much surface area per unit as volume per unit can move materials in and out of a cell more efficiently than a cell with a smaller surface-area-to-volume ratio.

A second limit governing the size of cells has to do with transport within the cell itself. As cell size increases, it takes longer for information and materials to reach their destination. Small cells, therefore, function more efficiently than larger cells. ❑

Section Review

1. *What observations were used to develop the cell theory?*
2. *What function do organelles serve?*
3. *Explain the importance of surface-area-to-volume ratio in a cell.*

Critical Thinking

4. *In prokaryotes, DNA exists as a single molecule. Why is it not considered an organelle?*

Answers to Section Review

1. Schleiden and Schwann observed that all parts of living things are made of cells. Virchow observed that all cells come from other cells.

2. Organelles compartmentalize the cell into functional areas.

3. If cells grow too large, they will outstrip their ability to supply needed materials to their interior.

4. The DNA in a prokaryote is not considered an organelle because it is not membrane bound.

2-2 The Chemistry of Living Cells

L iving cells carry on many complicated chemical processes. In order for you to understand living things, some knowledge of basic chemistry is necessary. The chemistry that you will encounter in this chapter will be useful in helping you understand biology, for all organisms are chemical machines.

Section Objectives

- Differentiate among atoms, ions, and molecules.
- Distinguish among covalent bonds, ionic bonds, and hydrogen bonds.
- Explain the solvent properties of water.
- Describe the differences between acids and bases.

Atoms Are the Cell's Smallest Components

All living and nonliving things are composed of **atoms**. Every atom consists of a cloud of tiny particles called electrons that spin in undefined paths around a small, very dense core called a nucleus, as shown in the model in Figure 2-5. The nucleus is a cluster of two kinds of particles, protons and neutrons. To understand atoms you must remember that **electrons** carry a negative charge and protons a positive charge (neutrons are not charged). It is the attraction between positive and negative charges that keeps the electrons spinning about the nucleus.

Kinds of Atoms

When the atoms in a sample of matter are all alike, the sample represents an element. An **element** is a substance that cannot be broken down to any other substance by ordinary chemical means. There are currently more than 100 known elements. Each is denoted with a one-, two-, or three-letter symbol. For example, carbon is represented by C, oxygen by O, and hydrogen by H.

Atoms as a whole have no electrical charge. Some atoms can react with other atoms to form particles with unequal numbers of electrons and protons. These kinds of atoms are called **ions.**

The Nature of Atoms

The chemical behavior of an atom is determined by the electron cloud that surrounds its nucleus. Almost all the volume of an atom is empty space. When two atoms meet, the electron clouds can overlap, but the nuclei never come into contact with each other.

Atoms Have Energy

Energy is the ability to do work. Electrons in atoms have energy; it takes energy to keep a negative electron from crashing into the positive nucleus of an atom.

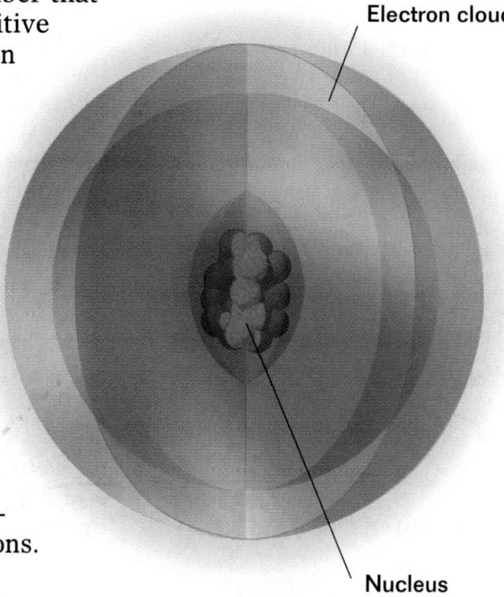

Electron cloud

Nucleus

Figure 2-5 The volume of an atom is mostly empty space. The cloud portions of the atom represent regions where moving electrons are most likely to be found.

Overcoming Misconceptions

Electrons Spin Around the Nucleus

Electrons do spin around the nucleus, but on their own axis rather than with the nucleus as the center of their path.

Electron spin is a model that allows us to conceptualize the electron's behavior.

SECTION 2-2

Vocabulary Preview

atom
electron
element
ion
compound
molecule
ionic bond
polar molecule
hydrogen bond
acid
pH scale
base

Opening Question

Ask students what the science of chemistry is, and have them speculate about how it relates to biology. Encourage students to conclude that chemistry is the study of the composition of substances and that chemistry relates to biology because all organisms are composed of chemical substances. Lead them to realize that the properties and reactions of these substances are the fundamental basis of all living things.

Demonstration

The Periodic Table

Give a brief introduction to the periodic table. Include the distinction between metals and nonmetals. *(Metals are found on the left side of the chart. Nonmetals appear on the right side of the chart.)* Point out common elements such as iron (Fe), copper (Cu), aluminum (Al), silver (Ag), carbon (C), sulfur (S), and the gases hydrogen (H) and oxygen (O). Stress that the general trend is increasing size of atoms from left to right and from top to bottom in the table. Point out the atomic mass numbers (atomic masses) and atomic numbers of a few representative elements. The five most important elements to living things are C, H, O, N, and S. Have students locate these on the table and note the names of these elements.

Teaching Tips

Probability and the Atom

Have students imagine that you have magnified an atom to the size of the room. Hang a marble or a lump of modeling clay in the center of the room to represent the nucleus. Point out that, in general, the farther away from the nucleus an electron is, the more energy it has. Similarly, the farther we stretch a rubber band, the more energy it has. Specify that regions exist in the atom that can contain certain set numbers of electrons. These regions are called energy levels. Point out that electrons "like" to possess as little energy as possible, but when regions of the atom near the nucleus are already occupied by other electrons, electrons must occupy regions farther from the nucleus.

Describe the movement of the electrons in the atom as more similar to the movements in a swarm of gnats than to the orbital motion of planets. Thus, we cannot describe the path, position, or energy level of any particular electron, but we can describe energy levels in terms of probability for groups of electrons. For example, electrons of a given energy are likely to be found in a certain energy level. Ask students why the solar system is not a very good model of an atom. (The planets have a predictable path, allowing us to always know their position. In an atom, the positions and paths of electrons cannot be predicted.)

The Shapes of Things

Mathematical modeling helps describe the shapes of the energy levels in atoms. The shapes of these energy levels influence the shapes of the molecules that are made by combinations of atoms. Discuss why scientists are concerned with the shapes of atoms. (The shapes of atoms help predict the shapes of molecules.)

❑ CAPSULE SUMMARY

All matter is composed of atoms. Atoms consist of electrons, protons, and neutrons. Electrons determine the nature of atoms and store considerable amounts of energy. Electrons that are close to the nucleus have lower energy levels than those that are farther away.

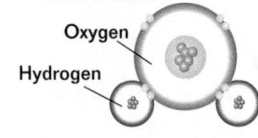

1. Electron energy level model

Oxygen

Hydrogen

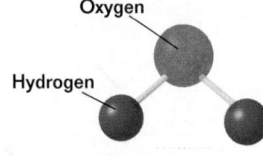

2. Ball and stick model

Oxygen

Hydrogen

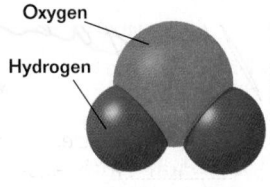

3. Space filling model

Oxygen

Hydrogen

Figure 2-6 Water is held together by the sharing of electrons (covalent bonds) between the hydrogen atoms and oxygen. Three different ways of representing the water molecule are shown.

Electrons are organized outside the nucleus by the amount of energy they possess. Electrons close to the nucleus are at low energy levels. Electrons farther from the nucleus are at higher energy levels. Because an electron is so small and moves so fast, we do not know its energy or position in space with a high degree of certainty. The locations of electrons, therefore, are described in terms of probability. We can mathematically determine how likely it is that an electron will be at a certain location. The pathways an electron takes in moving about the nucleus, however, are unknown. It was once thought that electrons orbited the nucleus like the planets orbit the sun. We now know that this model for the atom is inaccurate. ❑

Atoms Chemically React to Form Compounds

A **compound** is a group of atoms held together by chemical bonds. Compounds are represented by chemical formulas like NaCl (sodium chloride, or table salt) and H_2O (water). The formula identifies the elements in the compound as well as their proportions. The force that links the atoms of compounds is called a chemical bond. There are three kinds of bonds that are important to biological systems: covalent bonds, ionic bonds, and hydrogen bonds.

Covalent Bonds

Covalent bonds form when two atoms share electrons. Like the rivets and welds that link the steel girders in a skyscraper, covalent bonds are the strong links that hold together the atoms of most of the compounds in your body. The chemistry of living cells is based on the element carbon, which accounts for more than one-half the dry weight of cells. The ability of carbon atoms to form very stable carbon-carbon bonds by bonding covalently is of great significance in biology. A carbon atom has four outer electrons and can form four covalent bonds with another carbon atom or with a different kind of atom. A group of atoms held together by covalent bonds is called a **molecule.**

Stable atoms have filled outer energy levels. All atoms in living things (except hydrogen and helium) have outer levels that hold eight electrons. If an atom has only seven electrons in its outer energy level, it will react readily with an atom that has a single electron in its outer level. Water (H_2O) is a molecule made of oxygen (six outer electrons) that forms covalent bonds with two hydrogen atoms (one outer electron each).

There are two key properties of covalent bonds that make them ideal for their role in living systems. They are very strong and very directional, meaning they can form bonds in one or more specific directions. Thus, bonds can form between two or more specific atoms sharing electrons.

❓ Did You Know?

Even before prokaryotes evolved, "pre-living" complexes of chemicals probably existed. The complexes, consisting possibly of short strands of RNA surrounded by hydrophobic layers of lipid, created protective compartments that were resistant to degradation by the environment. The compartments themselves probably evolved from lipid micelles, globules of lipids forming a colloidal suspension with water.

Ionic Bonds

Ionic bonds form between two atoms of opposite charge. An actual exchange of an electron occurs to form the ions that form the ionic bond. The force of attraction between a positive and negative ion is an **ionic bond.** For example, sodium atoms become positively charged ions (Na^+) by losing their electrons to chlorine atoms to form chloride ions (Cl^-). Each chloride ion has an extra electron that attracts it electrically to surrounding sodium ions of opposite charge. Substances that form ionic bonds break apart when placed in water, producing free ions. Many such ions perform essential roles in biological activities. Sodium, for example, is essential for the functioning of nerve cells.

Hydrogen Bonds

Hydrogen bonds, which are weak bonds of a very special sort, play a key role in living systems. They differ from ionic and covalent bonds in that they link *molecules* together rather than atoms. Look at the water molecules in Figure 2-6. Oxygen forms covalent bonds with two hydrogen atoms. The shared electrons in water are more strongly attracted by the oxygen nucleus than by the hydrogen nuclei. Water molecules act like a molecular magnet, with positive and negative ends, or "poles." Molecules that have unequal areas of charge, like water, are **polar molecules**. A **hydrogen bond** is a weak chemical bond that forms between two polar molecules. The positive end of one polar molecule is attracted to the negative end of another, as shown in the model in Figure 2-7.

Hydrogen bonds are weak, so they do not form if there are long distances between molecules. Also, hydrogen bonds play critical roles in determining the shapes of many important biological molecules, such as DNA and proteins. ❑

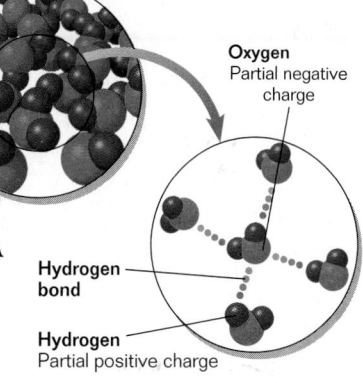

Figure 2-7 Ionic bonds in sodium chloride result from an electron transfer between a sodium atom and chlorine atom. The transfer produces ions that arrange in a cubic structure, *above.* Water molecules have slight positive and negative charges that cause the molecules to attract each other forming hydrogen bonds, *below.*

Sodium ion
Electron transfer
Positive (sodium) ion
Negative (chloride) ion
Chloride ion

Oxygen
Partial negative charge

Hydrogen bond

Hydrogen
Partial positive charge

Water Is a Major Component of Cells

When life on Earth was beginning between 3 and 4 billion years ago, water provided a medium in which other molecules could interact. Life as we know it could not have evolved without these interactions. Today, three-fourths of Earth's surface is covered by water. Every cell in your body contains water; in most cells there is an abundance of it. About two-thirds of the molecules in your body are water molecules.

Water Stores Heat

Water heats more slowly than most other substances. It also retains its temperature longer than other substances when

❑ CAPSULE SUMMARY

Atoms share electrons in covalently bonded substances, whereas atoms transfer electrons in ionically bonded substances. Hydrogen bonds form weak links between molecules.

SECTION 2-2

Teaching Tip
Covalent Versus Ionic Bonds: A Model
The following little fables might help in explaining the difference between ionic and covalent bonds.

Covalent bonds: Two classmates each have 25 cents. A soda costs 50 cents. They pool their resources, go to the soda machine, find a couple of straws, and obtain their desired refreshment. Neither of the two will break up their partnership until they have each consumed a fair share of the soda. This sharing of the soda has created a strong "bond" between the two. In a covalent bond, two atoms "need" an electron; by sharing, both of their requirements are fulfilled.

Ionic bonds: One student has 50 cents extra. Another student borrows that 50 cents to purchase a soda. The students do not share the soda, but, as it turns out, a bond exists between them because the lender has an attraction for the money lent to the other student. The lender will pursue the borrower until the money is repaid. Similarly, an atom might become a positive ion by giving up its electron to another atom, making the receiving atom a negative ion. The opposite charges attract, resulting in an ionic bond.

◉ VISUAL STRATEGY

Figure 2-7
Draw a water molecule on the board. Label the slight positive (hydrogens) and negative (oxygen) "poles" or ends of the molecule. Have students identify similar water molecules in the bottom of Figure 2-7. Emphasize the weak attraction between the positive and negative ends of each of the water molecules. It is this attraction between individual water molecules that results in many of the important properties of water. Emphasize that the charges are weaker than the charges in ionic bonds.

Demonstration

Capillary Action in Celery

Place a freshly cut stalk of celery in a beaker of water with a high concentration of food color. Red works particularly well. Allow the demonstration to proceed overnight. Have students observe the results during the next class meeting. Encourage them to pull one of the "strings" out of the celery stalk in order to examine it with a hand lens. Have students answer these questions: What evidence do you have that the "strings" in celery stalks are hollow? *(They fill up with colored water.)* How do you think the colored water was able to move into the hollow tubes of the celery stalk? *(At least partially by capillary action. Other forces may also be at work.)*

Demonstration

Surface Tension

Prior to class, coat a sewing needle with petroleum jelly. The grease keeps the water from wetting the needle; that is, it keeps the water molecules from being attracted more to the needle than they are to each other. Challenge students to predict whether you can float the needle on water in a beaker. Carefully lay the needle flat on the calm surface of the water using a pair of forceps. The needle will float. Have students observe the results without touching the beaker. Then, push the needle slightly beneath the surface of the water. It will sink. Have students answer these questions: Why did the needle float at first? *(It had not broken the surface tension among the molecules at the surface of the water.)* Why did the needle sink after it was disturbed? *(Touching the needle or the water breaks the surface tension of the water.)*

Water   Dye

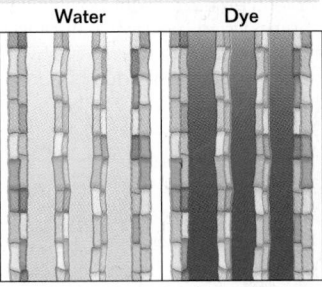

Liquid movement up a stem through capillary action

Figure 2-8 The capillary action of water is shown by the movement of dye up the stem to color the flower.

its surrounding environment cools—it stores heat well. Many organisms dispose of excess heat through water evaporation. For example, humans cool down by sweating.

Water Clings to Itself and Other Molecules

Water molecules readily form hydrogen bonds with one another, so water clings to itself in an attraction called cohesion *(koh HEE zhuhn)*. It is because of cohesion that water is a liquid and not a gas at room temperature. Hydrogen bonds link many individual water molecules together at the water's surface, like a crowd of people linked by holding hands. Surface tension forms across the surface of water because of the cohesive attraction between individual water molecules.

The attraction of water to a substance other than water is called adhesion *(ad HEE zhuhn)*. Water adheres to any substance that it can form hydrogen bonds with. That is why some things get "wet" and others, such as waxy substances, that are composed of nonpolar molecules, do not. The adhesion of water to substances with surface charges causes capillary action. Capillary action and cohesion are responsible for the upward movement of water as shown in Figure 2-8.

Water Ionizes

When the covalent bonds of water break, a hydrogen ion (H^+) and a hydroxide ion (OH^-) are produced.

$$H_2O \rightarrow H^+ + OH^-$$

This ionization process goes on continuously in water. As a result, pure water always has a low concentration of hydrogen and hydroxide ions; roughly 1 out of every 550 million water molecules exists as ions at any instant.

Water Is a Powerful Solvent

When a solvent, such as water, dissolves a solute, such as table sugar, a solution is formed. Many compounds dissolve in water because of water's polar nature. When covalent compounds dissolve in water, molecules are evenly dispersed in the solution. When ionic compounds are dissolved

Figure 2-9 Ionic substances dissolve in water to form charged particles (ions), *left*. Covalent substances dissolve in water to form molecules that may have very slight charges, *right*. Many substances found in the body are dissolved ions.

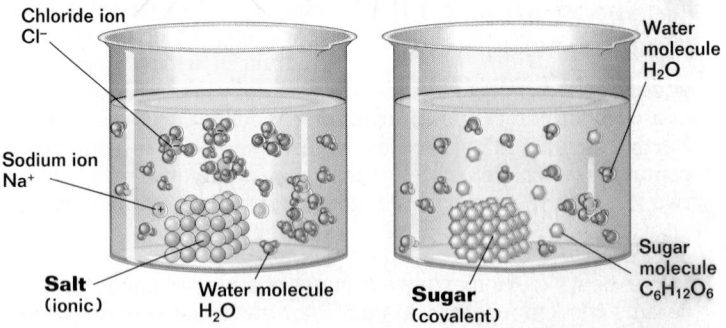

Chloride ion
Cl⁻

Sodium ion
Na⁺

Salt (ionic)

Water molecule
H_2O

Sugar (covalent)

Water molecule
H_2O

Sugar molecule
$C_6H_{12}O_6$

in water, ions are evenly dispersed in the solution. Elements in the body are present as compounds or dissolved ions in various fluids.

When nonpolar molecules (which do not form hydrogen bonds) are placed in water, the water molecules crowd the nonpolar molecules together. That is why oil and water do not mix. Nonpolar molecules, which are repelled by water, play many important roles in living things. They are responsible for fine-tuning the shapes of proteins and for maintaining the structure of cell membranes that surround every cell.

Acids and Bases Affect the Cell's Environment

Any compound that forms hydrogen ions when dissolved in water is called an **acid**. When an acid is added to water, the hydrogen ion concentration is increased. A convenient way of relating the amount of hydrogen ion from one solution to another is the **pH scale**. A simplified version of the pH scale is shown in Figure 2-10. The pH of most solutions falls within a numerical range from zero to fourteen. The pH of any solution can be determined using a pH meter or indicator papers.

Any substance that ionizes to form hydroxide ions when dissolved in water is called a **base**. Bases lower the hydrogen ion concentration of water because hydroxide ions react with hydrogen ions to form water molecules.

$$H^+ + OH^- \rightarrow H_2O$$

Bases have hydrogen ion concentrations that are lower than that of pure water. Bases thus have pH values above 7. Household ammonia has a pH of 11, and intestinal fluid has a pH of about 8. Your body's fluids are constantly monitored in a series of complex processes to keep the pH of these fluids within acceptable levels.

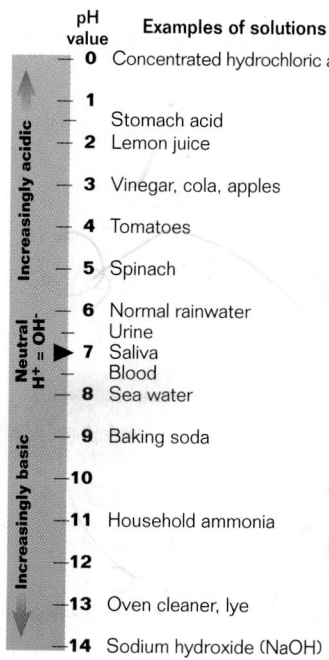

Figure 2-10 The pH scale enables you to compare the acidity of various materials. H⁺ concentration increases in the range from pH 7 to 1. OH⁻ concentration increases in the range from pH 7 to 14.

Section Review

1. *How does an element differ from a compound?*
2. *How do covalent bonds differ from ionic bonds?*
3. *Explain why water is such an excellent solvent.*

Critical Thinking

4. *Antacids are often taken to relieve symptoms of "heartburn." How might an antacid work to relieve such stomach conditions?*
5. *Explain why you think melted candle wax will not mix with water.*

Opening Question

Ask students what kind of a substance carbon is. Help them realize that carbon is an element as well as a major component of living things.

Demonstration

Carbohydrates

Draw the structure of glucose on the board. Have students count six carbon atoms on the diagram of one of the polysaccharides in Figure 2-11. Have them add the number of hydrogen atoms and the number of oxygen atoms attached to those six carbon atoms. The numbers will be 10 for hydrogen and 5 for oxygen. Explain that this is because one water molecule is removed from each glucose unit as the units are joined to each other.

2-3 Chemical Building Blocks of Cells

The basic chemical building blocks of your body are the same as those in all other organisms. Most of your body's molecules are organic compounds, which refers to a class of compounds containing carbon.

Macromolecules are built from small organic compounds the same way a railroad train is built, by linking a lot of units together into long chains. There are four principal kinds of macromolecules found in living organisms: carbohydrates, lipids, proteins, and nucleic acids.

Cells Use Carbohydrates to Store Energy and Provide Support

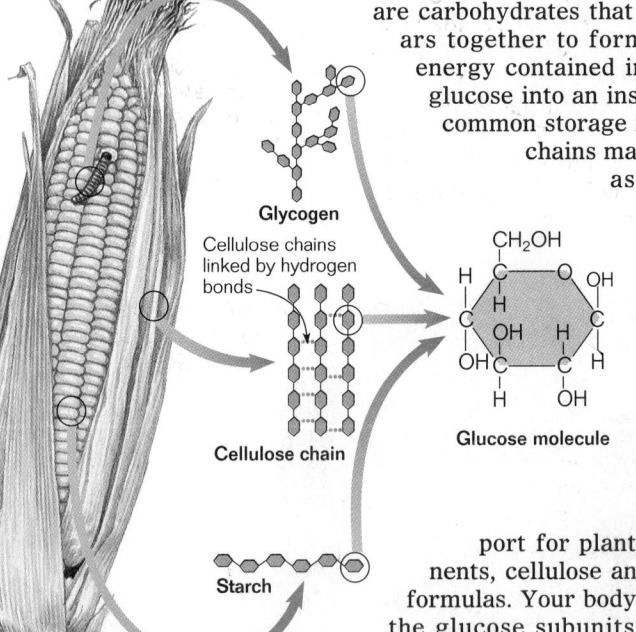

Figure 2-11 Starch, glycogen, and cellulose are made of glucose units organized in different structural arrangements.

Glycogen

Cellulose chains linked by hydrogen bonds

Cellulose chain

Starch

CH₂OH

Glucose molecule

Carbohydrates are composed of carbon, hydrogen, and oxygen atoms in the proportion of 1:2:1. A general formula for the carbohydrate class of compounds is $(CH_2O)_n$, where n is the number of carbon atoms. The sugar glucose is a small carbohydrate; its n equals 6. Its chemical formula is $C_6H_{12}O_6$.

Carbohydrates like glucose play a key role in the storing and transporting of energy in your body. **Polysaccharides** are carbohydrates that are made by linking individual sugars together to form long chains. Organisms store the energy contained in sugars like glucose by converting glucose into an insoluble form for future use. **Starch,** a common storage form of glucose, is composed of long chains made of hundreds of glucose molecules, as shown in Figure 2-11. When your body digests starch, the long starch chains are broken into short fragments. Your body then stores the glucose-containing fragments in longer chains called **glycogen**.

Many organisms use polysaccharides as structural material. In plants, for example, glucose molecules are joined together in long chains forming cellulose. **Cellulose,** a major component of the cell wall of plants, provides structural support for plants. Though they have similar components, cellulose and glycogen have different structural formulas. Your body is not able to break the links joining the glucose subunits in cellulose chains, so you cannot obtain energy from eating grass.

Graphic Organizer

Use this graphic organizer in **Teaching Tip: Evolution and Levels of Organization** on page 35.

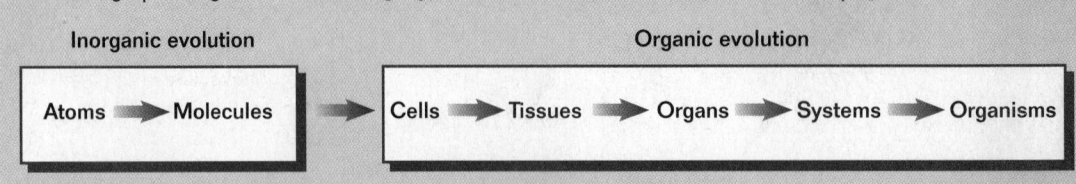

Inorganic evolution

Organic evolution

Atoms ➡ Molecules ➡ Cells ➡ Tissues ➡ Organs ➡ Systems ➡ Organisms

Lipids Store Energy and Are a Component of Cell Membranes

Lipids are a class of organic macromolecules that differ from other macromolecules in that they do not dissolve in water. Olive oil and vegetable oil are lipids, and so are waxes such as beeswax and earwax. Though lipids are a diverse class of compounds, most of their functions can be placed in one of three categories: energy storage, structural support in cell membranes, and specific reactants for metabolic reactions.

Fats are energy-storage lipid molecules that have more hydrogens bonded to their carbon chains than do carbohydrates. The structure of a fat molecule is shown in Figure 2-12. The fatty acid chains are usually 14 to 20 $-CH_2-$ units long. This structure enables fat to supply more energy than carbohydrates. A gram of fat provides nine calories; a gram of carbohydrate provides only four calories.

When all the carbon atoms on the fatty acid chains are bonded to hydrogen atoms ($-CH_2-$ units), these fats are called saturated fats. **Saturated fats** are called saturated because they contain the maximum number of C—H bonds possible. **Unsaturated fats** have carbon-carbon double bonds at various points along the fatty acid chain. They are called unsaturated because fewer hydrogen atoms can bond to the carbon chain when there are double bonds between carbon atoms.

Lipids also provide structural support in cell membranes. The membranes that surround the cells of your body are composed of lipids to which phosphorus molecules are attached. Such molecules are called phospholipids. Animal cell membranes also contain cholesterol, which is yet another kind of lipid called a **steroid.** Many of the hormones that your body uses to control its activities are steroids. There are many other kinds of lipids, including important pigments, light-absorbing substances such as the chlorophyll of green plants, and the pigment retinal found in your eyes. ❑

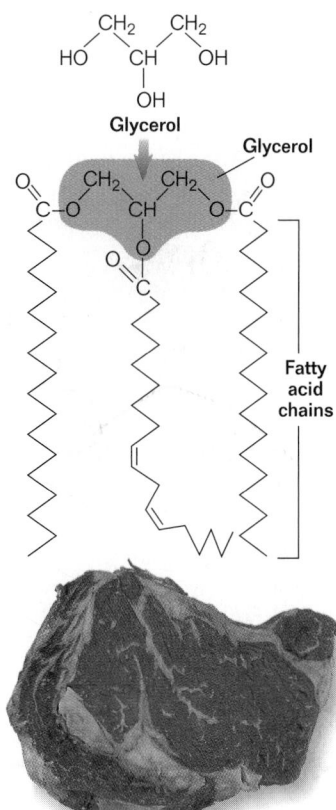

Figure 2-12 Fats, found in foods, consist of three long chains of fatty acids bonded to a glycerol backbone.

❑ **CAPSULE SUMMARY**

Lipids are not soluble in water. Fats store more energy per gram than carbohydrates because they have more carbon-hydrogen bonds.

Many Cellular Activities Involve Proteins

Proteins are the third major group of macromolecules. Proteins have many important structural functions. Your hair and muscles are made of protein, and so are a spider's web, a peacock feather, and the fibers of a blood clot. The most abundant protein in your body is **collagen,** a fibrous protein that forms the matrix of your skin, ligaments, tendons, and bones. Proteins also play a vital role in the metabolic (chemical and physical) activities of all living things. Proteins called **enzymes** assist the chemical reactions of metabolism. As you will learn in Chapter 4, few of the

SECTION 2-3

Teaching Tip
Evolution and Levels of Organization
Current scientific thought holds that the entire universe began with a "Big Bang." At some time after the Big Bang, the universe cooled down enough to allow atoms to coalesce. The *Graphic Organizer* at the bottom of page 34 is a very general summary of the evolution that followed.

◉ *VISUAL STRATEGY*

Figure 2-12
Point out the fatty acid chains in the figure. Indicate that the part of the molecule to which the fatty acid chains connect (the fat's "backbone") is actually a derivative of glucose called glycerol. Glycerol is formed, essentially, by cleaving glucose in half.

Overcoming Misconceptions

Saturated Fats
Students may be confused when they see that all of the carbon atoms in either saturated or unsaturated fats are not bonded to hydrogen atoms because they are bonded to other carbon atoms. Call their attention to the single, double, and triple bonds between carbon atoms. In fact, molecules have degrees of saturation.

Application

▲ **Health** High school students are often attracted to body-building schemes. One of the latest involves the use of amino acids taken in liquid form to promote "bulking up." Aside from the fact that these supplements have not been proven effective, the users of such preparations could get the same amino acids, although in an undigested form, simply by eating legumes and a variety of lean meats. The amino acid preparations are a very expensive source of these amino acids. Another disadvantage of the supplements is that, rather than supplying an actual need in the body, the amino acid mixtures increase the caloric intake of the athlete, undermining his or her efforts to obtain lean body weight.

👁 VISUAL STRATEGY

Figure 2-13
Correlate this figure to the *Application* feature above. Emphasize there are only 20 different amino acids. The same amino acids obtained from supplements are present in foods.

CONTENT LINK

*How the shape of an enzyme enables it to catalyze a particular chemical reaction is explained in more detail in **Chapter 4**.*

Figure 2-13 The formation of different proteins from the combinations of just 20 different amino acids is much like the formation of all the words in the English language from just 26 different letters.

☐ CAPSULE SUMMARY

Proteins are part of an organism's structure. As enzymes, proteins increase the rate of chemical reactions within cells. The sequence of amino acids in a particular protein determines its shape, chemical properties, and function.

chemical reactions that take place in your body can proceed quickly without enzymes.

Enzymes Control Chemical Reactions

An enzyme is a catalyst because it increases the rate of a chemical reaction without the enzyme itself being destroyed in the process. Organisms maintain internal balance because enzymes control chemical reactions. The long developmental process that turns you from a fertilized egg into an adult human is controlled by the proper starting and stopping of production of particular enzymes at the appropriate time.

Amino Acids

Amino acids are the building blocks of proteins. There are 20 different kinds of amino acids that humans use, as shown in Figure 2-13. Because amino acids differ in chemical character, it is not likely that any two proteins with different amino acids will be alike chemically. Long chains of amino acids are called **polypeptides.** A protein is composed of one or more polypeptides.

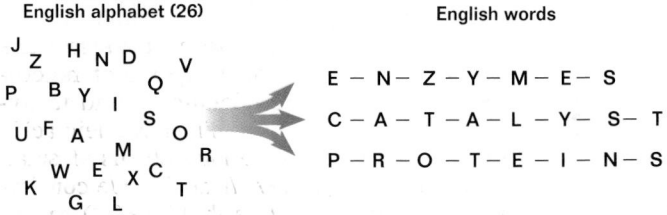

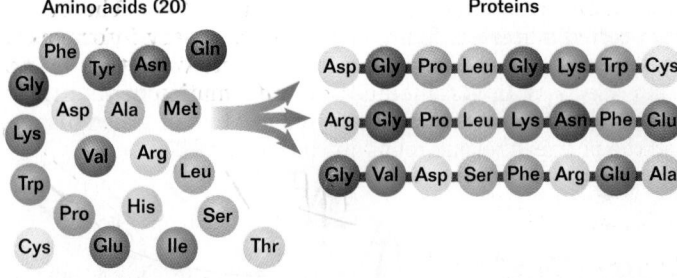

The amino acid chains of enzymes tend to fold into compact three-dimensional shapes, as shown in Figure 2-14. It is the precise shape of an enzyme that enables it to catalyze a particular chemical reaction.

Proteins also function as hormones and neurotransmitters. In these functions, proteins serve as signaling devices that are involved in regulating the activities of the cells of organisms. ☐

Multicultural Perspective

A Cultural Dependency on Milk?

Many mammals produce lactose in the milk intended for their offspring. Providing food energy in the form of lactose reserves the energy for the child because many adults lack lactase, the enzyme needed to digest the milk sugar lactose. For example, from 30 percent to 70 percent of African and Asian adults lack this enzyme and cannot metabolize lactose. Biologists believe that this difference may relate to a cultural dependency on milk products, since Asian and African diets often do not include the types of dairy products popular in Western diets.

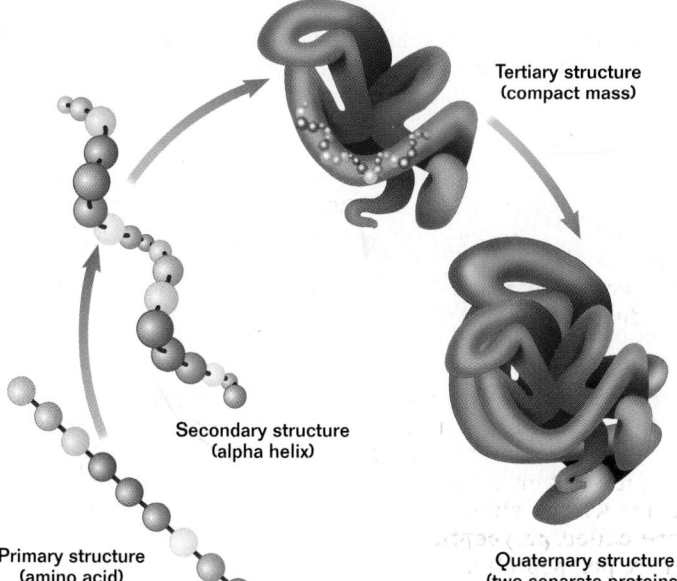

Tertiary structure
(compact mass)

Secondary structure
(alpha helix)

Quaternary structure
(two separate proteins)

Primary structure
(amino acid)

Figure 2-14
Have students make three-dimensional models of the primary protein structure in this figure, using clay to represent the amino acids and string to represent the peptide bonds between them. Rubber bands can be employed to represent the hydrogen bonds that hold the polypeptide in its secondary structure shape, and pipe cleaners can represent the chemical bonds that hold peptides together in tertiary proteins. Have students answer the following questions: At which level are hydrogen bonds important to the protein? (*Hydrogen bonds are important in holding a strand of amino acids into its polypeptide tertiary structure.*) Why are covalent bonds important to proteins? (*Covalent bonds join amino acids to each other and also join polypeptides to make proteins.*)

Figure 2-14 The biological activity of a protein depends on its structure. Four levels are used to describe protein structure. Primary structure shows amino acids linked by covalent bonds. Secondary structure can be a helix that shows recurring arrangements of amino acids. Tertiary structure shows the folding of that helix to make a polypeptide. Quaternary structure shows the arrangement of multiple tertiary proteins.

Nucleic Acids Contain the Cell's Hereditary Information

Nucleic acids are the fourth group of macromolecules found in living things. Nucleic acids are long chains of small repeating subunits, called **nucleotides**. A nucleotide consists of a five-carbon sugar with a phosphate (PO_4^-) group attached to one side of the sugar ring and an organic base attached to the other, as shown in Figure 2-15. The major function of nucleic acids is to store hereditary information that can be later translated to form new proteins. The nucleic acid that stores hereditary information in your cells is **DNA** (deoxyribonucleic acid).

CONTENT LINK

You will learn more about the structure and function of DNA in **Chapter 8**.

Figure 2-15 This portion of a DNA strand shows the arrangement of the sugar, base, and phosphate subunits in a nucleotide.

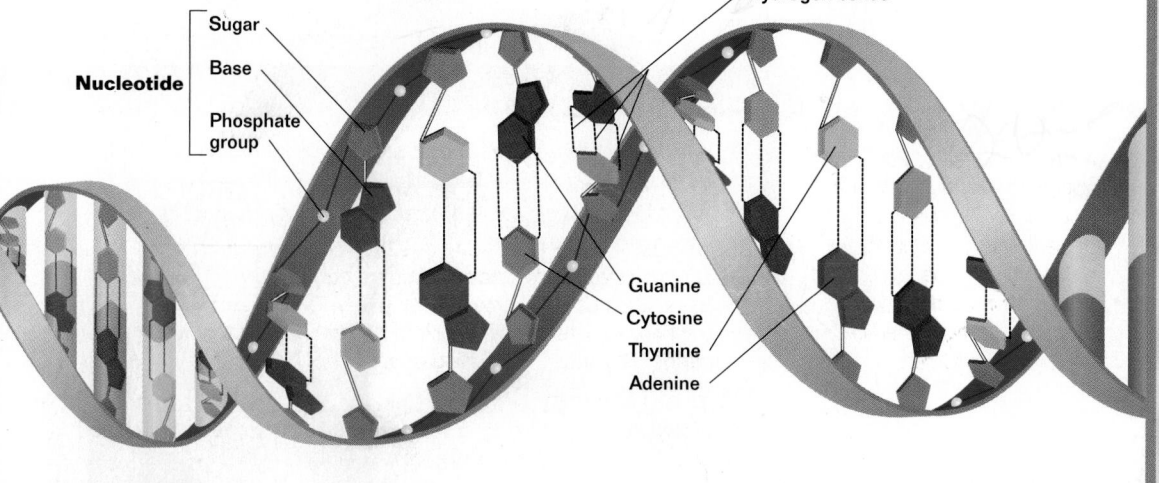

Nucleotide
Sugar
Base
Phosphate group

Hydrogen bonds

Guanine
Cytosine
Thymine
Adenine

Demonstration

DNA: The Double Helix

Take two 2–3 ft. long red or black licorice whips. With the two strands of licorice as sides, make a ladder by adding equally spaced rungs made of toothpicks. Finally, put a right-handed twist in the ladder. The sides represent alternating phosphate and sugar groups, while the rungs represent pairs of organic bases. The model may be kept until Chapter 8, where it can be used again. Have students answer the following questions: What are the three different types of molecules found in DNA? (*phosphate groups, five-carbon sugars, and organic bases*) Why is DNA important? (*It is the substance in which the hereditary information is stored.*)

? Did You Know?

Phosphate groups are very important chemical subunits throughout living systems. They not only help the cell store hereditary information, but also pass energy from place to place in the form of ATP, help in metabolizing glucose, form bone tissue, and perform other functions.

Teaching Tip

ATP: The Cell's Energy Currency

Construct a mental model of ATP with the students as follows. Imagine that the cell is a large amusement park. Imagine that glucose molecules are dollar bills. Most amusement parks will not allow a patron to simply purchase a ride on the roller coaster by giving the operator dollar bills. Patrons must purchase tickets in the amount of the ride's admission price. Only tickets are acceptable throughout the amusement park, while only the ticket booth can accept money for tickets. The action of ATP in the cell is similar to the tickets. ATP is usable throughout the cell in order to do the work of the living system. Even though glucose is the major source of energy within cells, it must first be exchanged for ATP molecules in order to transfer its energy to the cell. Have students answer the following: To what other biological molecule is ATP similar? *(nucleotides in DNA)* Why is ATP important? *(It is important because it transports energy in a usable form throughout the cell.)*

👁 VISUAL STRATEGY

Figure 2-16

Compare this figure with Figure 2-15 on page 37, and point out the similarities. Indicate that adenine is one of the four organic bases mentioned on the previous page as a component of DNA. DNA contains three other organic bases. RNA has three bases similar to DNA and one that is different.

CONTENT LINK

You will encounter RNA molecules in **Chapter 9** when you study how genes are expressed —how particular sequences of nucleotide bases are used by the cell to assemble particular proteins.

CONTENT LINK

You will learn more about ATP in **Chapter 4**.

Figure 2-16 ATP is structurally similar to the nucleotides of DNA, except it includes three phosphate groups instead of one.

A second nucleic acid found in organisms, called **RNA** (ribonucleic acid), has a slightly different structure. RNA plays a variety of roles in the process of making proteins.

ATP Is the Cell's Fuel

There is one additional biological molecule that should be mentioned because of its importance in living systems and its structural similarity to nucleic acids—ATP, adenosine triphosphate. ATP contains an organic base, a sugar, and three phosphate groups. The term *triphosphate* means the molecule has three phosphate groups.

ATP is the energy currency of the cell. Its phosphate groups store energy like a coiled spring, their negative charges repelling one another. It takes energy to put the terminal phosphate in place, and energy is released when it is removed. When living cells break down food molecules containing carbohydrates and fats, part of the energy from those reactions is stored temporarily in ATP. A steady supply of ATP is necessary to ensure that a cell can perform all the tasks essential for life.

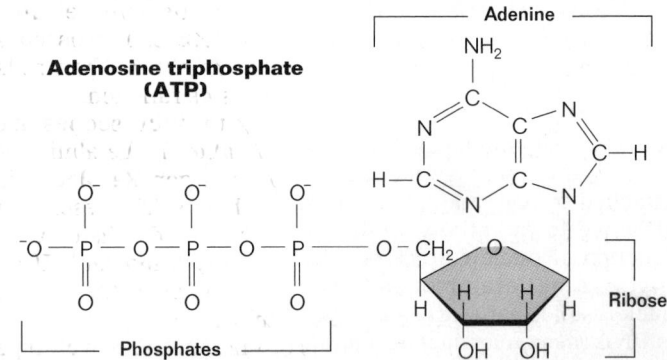

Section Review

1. *Name the elements found in all biological macromolecules.*

2. *Why do some athletes eat high-carbohydrate diets the day before competition?*

3. *List the subunit(s) that makes up each of the following:*
   a. *cellulose*   d. *fats*   g. *enzymes*
   b. *proteins*   e. *nucleic acids*   h. *starch*
   c. *polypeptides*   f. *glycogen*

Critical Thinking

4. *How is ATP similar to sugar? How is ATP similar to DNA?*

Answers to Section Review

1. Carbon, hydrogen, and oxygen are found in all biological macromolecules.

2. Athletes must store the high energy content of carbohydrates in order to meet their energy needs on the day of competition.

3. a. glucose; **b.** amino acids; **c.** amino acids; **d.** fatty acids and glycerol; **e.** phosphate groups, five-carbon sugars, and organic bases; **f.** glucose; **g.** amino acids (proteins); **h.** glucose

4. ATP is similar to sugar because it has a five-carbon sugar. It is similar to DNA because it has a five-carbon sugar, an organic base, and phosphate groups.

2-4 The Interior of the Cell

*I*magine yourself inside the cell. Scientists once thought that the cytoplasm was a structureless gel. Using microscopes, scientists have discovered that the cell's interior is filled with membranes and structures. If you journeyed into the cytoplasm, you would pass many structures, called organelles. In this section you will learn more about these principal structures of the eukaryotic cell.

Section Objectives

- Compare light microscopes with transmission and scanning electron microscopes in terms of magnification, resolution, and usage.
- Describe the structures and functions of the organelles found inside a eukaryotic cell.
- List the similarities and differences among the cells of plants, fungi, and animals.
- Recognize the role of evolution in biology.

Microscopes Reveal Cell Structure

Even before Robert Hooke first glimpsed cells of cork in 1665, scientists realized that they needed more than the human eye to study objects and living things. As microscopy has evolved, scientists have learned more about plant and animal life than Hooke and van Leeuwenhoek could have ever imagined. Modern microscopes serve as passports into the unseen world, enabling biologists to observe cellular processes and to see details of cell structure. Microscopes continue to provide scientists with new insight into how cells work—and ultimately how whole organisms function.

Two important concepts relating to microscopes are magnification and resolution. **Magnification** is the ability of a microscope to make an image appear larger. **Resolution** is the ability to distinguish small, close objects. The resolution of a microscope refers to its ability to show details clearly. Resolution and magnification are equally important. If an image appears larger but its details are unclear, the user of the microscope will see only a fuzzy blur.

There are several basic types of microscopes. Each type of microscope has its own strengths and limitations. Scientists have learned which microscope is the most appropriate for the organisms they wish to study.

Light Microscopes

Light microscopes use a beam of light passing through one or more lenses to produce an enlarged image of the object or specimen being viewed. Microscopes that use two sets of lenses are called compound microscopes. The ocular (*AHK yoo luhr*) lens is positioned near the viewer's eye. The objective lens set is positioned near the specimen. A typical compound microscope, such as the one shown in Figure 2-17, has a light bulb or mirror in the base that sends light upward through the specimen. Light rays pass through the objective and then through the ocular. The image you see is magnified by both sets of lenses. The total magnification is determined by multiplying the magnifications of the two lenses. If your

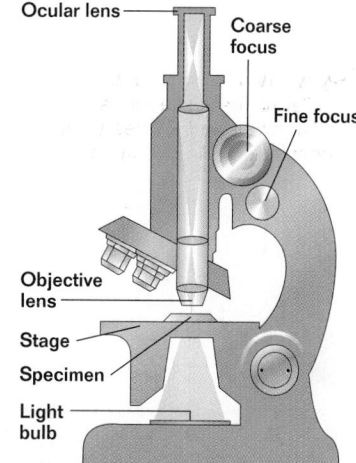

Figure 2-17 The compound light microscope illuminates a specimen with a beam of light. Two sets of lenses, the ocular (located in the eyepiece) and the objective (located just above the specimen), magnify the specimen, which is mounted on a glass slide.

SECTION 2-4

Vocabulary Preview

magnification
resolution
light microscope
transmission electron microscope (TEM)
scanning electron microscope (SEM)
endoplasmic reticulum (ER)
vesicle
Golgi apparatus
nuclear envelope
chromosome
mitochondrion
chloroplast
peroxisome
lysosome
cytoskeleton
microtubule
centriole
flagella
cilia
central vacuole

Opening Question

Ask students to name the principal organic compounds of living systems and some of their uses. Lead students to review the chemical building blocks discussed in Section 2-3. Make sure students include carbohydrates for storing energy, lipids for storing energy and forming cell membranes, proteins for structure and metabolism, nucleic acids for DNA, and ATP for energy.

Demonstration

The Light Microscope
Introduce the parts of the light microscope with a real microscope as a model in class. Have students compare the microscope you bring into class with the one pictured in Figure 2-17. Emphasize the functions of the parts in order to provide a background for students' lab work.

Historical Note

Biological Stains

Early microscopists were hampered in their explorations of the cell by the transparent nature of most of its components. Scientists had identified only a few cellular organelles by the mid-1800s. The relative prosperity brought to Europe by the Industrial Revolution fueled a demand for consumer goods, allowing German chemists to create brightly colored dyes for the textile industry. Soon, biologists such as Paul Ehrlich applied these dyes to biological preparations. This confluence of science and technology led to much of our current knowledge of cells.

☑ RESEARCH Update

Cell Speleology

With the advent of new technology, more is becoming known about the cell and its organelles. In the 1950s, electron microscopy first revealed dentike cavities (caveolae) in cell membranes, although their function has remained obscure. Michael Lisanti of the Whitehead Institute and his colleagues are beginning to develop some theories based on data they have been collecting. Lisanti's group, in an attempt to identify the composition of these cell membrane components, isolated organelles from endothelial tissues, concentrated them, and then began a detailed analysis of their components.

The researchers did not find anything new, but they did find the presence of some important and previously identified proteins: the CD36 receptor protein, the RAP-1 cancer-inhibiting protein, and porin, a transmembrane conduction protein. Lisanti speculates that the presence of caveolae on the surface of endothelial tissue lining the blood vessels may be what is responsible for arterial cholesterol buildup. He believes the caveolae may act to suck cholesterol into the endothelial cells, thereby providing a mechanism for the development of the plaque. Porin may be useful in the future as a system that enables drugs to enter cells. When? Not anytime soon. This research is just getting started. ☑

microscope has a 10× ocular, and the 40× objective is in place, the object you are looking at will appear 400 times larger than it actually is.

A biologist can use a compound microscope to study living cells. Cells appear to be essentially transparent, although there are small variations in thickness and density. As a result, the cell and some of its internal structures are visible, but the image is not very distinct. More details of the structures inside cells can be seen by slicing cells thinly and dyeing them with stains. Looking at a cell this way has obvious disadvantages—only one thin slice of cell is seen, and, of course, the cell is dead. However, sectioning and staining cells enables biologists to see many structures not visible in living cells.

Light microscopes are important tools, but they have one important shortcoming. As magnifications increase, the resolution decreases and the details of the object viewed appear fuzzy. The most powerful light microscopes can magnify an object 2,000 times. Practically speaking, bacteria with a diameter of 0.5 µm are about the smallest living things that can be distinguished using a good mass-produced light microscope.

Electron Microscopes

Microscopes using electrons instead of light to form images can magnify images at least 100 times as much as the light microscope. Because electrons would bounce off the gas molecules in air, the stream of electrons and the specimen to be viewed must be placed in a vacuum chamber. Therefore, living cells cannot be viewed with electron microscopes.

The **transmission electron microscope (TEM)** produces a stream of electrons that passes through a specimen and strikes a fluorescent screen. By replacing the fluorescent screen with a piece of photographic film, a photograph called a transmission electron micrograph can be made. Figure 2-18 shows a transmission electron micrograph of a *Giardia* cell. Sections of specimens that are to be viewed with a TEM are sliced much more thinly than sections prepared for a light microscope. These sections are treated with stains that block electrons, causing details to become visible.

The **scanning electron microscope (SEM)** enables biologists to see detailed three-dimensional images of cell surfaces, such as the image of the cells of *Giardia* shown in Figure 2-18. Specimens are not sliced but are placed on a small metal cylinder and coated with a very thin layer of metal. Like the picture on a television set, the image is formed one line at a time as the beam of electrons scans the specimen from side to side. The electrons that bounce off the specimen form an image that can be viewed on a video screen, or a scanning electron micrograph can be made.

The micrographs made with electron microscopes are always black and white—never in color. However, electron micrographs often have color added in the darkroom to make certain structures stand out in the micrograph.

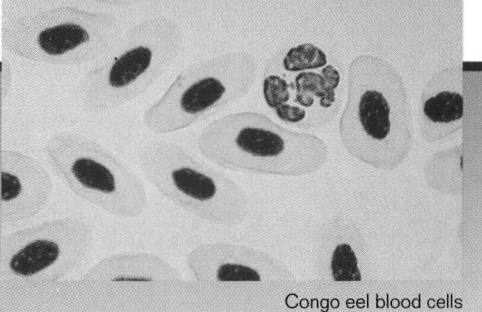

Light Microscopes
- Living specimens can be viewed.
- Selective stains enable specific organelles to be seen.
- Light microscopes are comparatively affordable, hence more available than electron microscopes.
- Light microscopes magnify up to 2,000×.

Congo eel blood cells

Electron Microscopes
- Specimens must be dead.
- Electron microscopes reveal details not visible with light microscopes.
- Electron microscopes are expensive, hence less available than light microscopes.

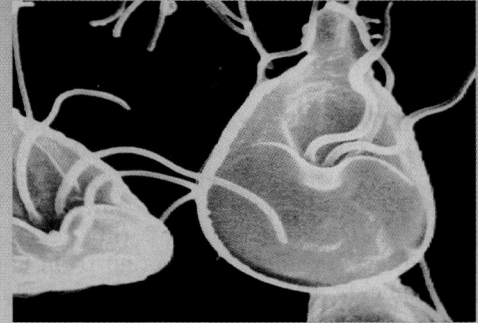

Giardia lamblia

Scanning (SEM)
- Specimens are coated with a thin layer of metal.
- A beam of electrons reveals surface details of specimens.
- The SEM magnifies up to 100,000×.

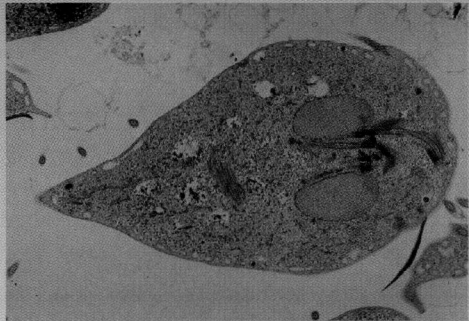

Giardia lamblia

Transmission (TEM)
- Specimens are thinly sliced, then stained.
- A beam of electrons passes through the specimen, revealing its internal structure.
- The TEM magnifies up to 200,000×.

The use of each type of microscope—light microscope, transmission electron microscope, and scanning electron microscope—has advantages and disadvantages. Figure 2-18 summarizes the differences among these three kinds of microscopes.

Figure 2-18 Light microscopes and electron microscopes each reveal different aspects of an organism's structure.

Scanning Tunneling Microscope
New video and computer techniques are extending the resolution and level of detail that can be detected by microscopes. The scanning tunneling electron microscope (STM) uses a needle-like probe to measure differences in voltage due to electrons that leak, or tunnel, from the surface of the object being viewed. A computer tracks the movement of the probe across the object, creating a three-dimensional image of the specimen's surface. The STM can be used to study living organisms.

Historical Note

Development of the Electron Microscope
The transmission electron microscope was independently invented in 1931 by two Germans, Ernst Ruska and Rheinhold Ruedenberg, but it was not until 1933 that Ruska built the first electron microscope that was more powerful (12,000×) than a light microscope. The first electron microscope did not become commercially available until 1935 in England. In 1945, Albert Claude, a Belgian American, first began electron microscope studies of the cell and discovered the endoplasmic reticulum and the details of the mitochondrion, for which he won the Nobel Prize in 1974.

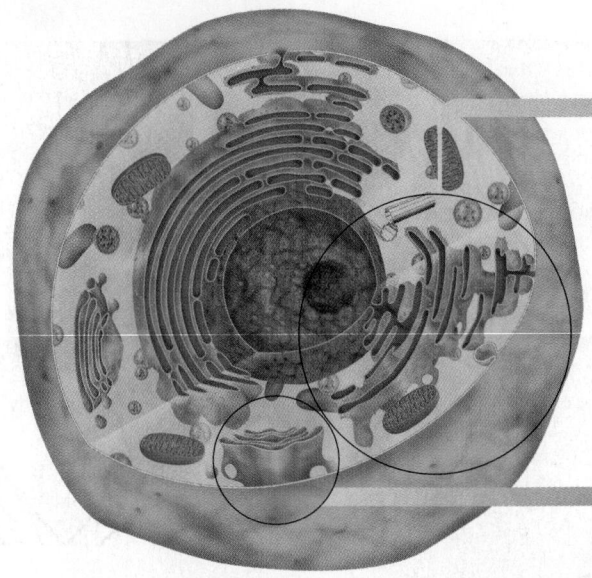

Teaching Tips
The Cell Factory

Making the connection between a factory, with which the students have some familiarity, and the cell makes these difficult and somewhat abstract concepts more accessible. Have students draw a cell modeled on a factory. Their diagrams should include the following types of analogies: head office (nucleus), store room (vacuole), powerhouse (mitochondrion), and fence (cell membrane). The metaphor can be carried further with the chloroplasts (solar panels), peroxisomes (part of the assembly line), lysosomes (janitors, security guards), cytoskeleton (the steel superstructure of the factory), and even ATP (the electricity supplying energy to the whole factory). Have students answer the following questions: Why is compartmentalization important in a factory? in a cell? *(It allows for specialization and protects the various machinery.)*

Compartmentalization and Homeostasis

Have students discuss difficulties presented by compartmentalization, as well as advantages. Have them list each in their notebooks. These points may be reviewed in the discussions of materials transport in Chapter 3.

Internal Membranes Transport Materials Within the Cell

In order for a eukaryotic cell to maintain homeostasis, it is necessary for supplies to be moved from one part of the cell to another. In prokaryotic cells, a molecule can go from one place to another fairly quickly. In eukaryotic cells, molecular traffic is directed more precisely by an extensive system of internal membranes called the **endoplasmic reticulum** *(ehn doh PLAZ mihk rih TIHK yuh luhm)*, or **ER**. Like the plasma membrane that surrounds the cell, the ER is composed of a lipid bilayer with embedded proteins. Weaving in sheets through the cell's interior, the ER creates a series of channels between the membranes that isolates some spaces as membrane-enclosed sacs called **vesicles**. This system of internal compartments is a fundamental distinction between eukaryotes and their prokaryote ancestors.

Manufacturing Centers

The cell manufactures many proteins and lipids on the ER's surface. Some proteins and lipids are used within the cell, for example, to replace damaged or worn parts of the plasma membrane. Other proteins and lipids, such as digestive enzymes or hormones, are exported from the cell.

Proteins that are exported from the cell are manufactured by ribosomes on the surface of portions of the ER. Ribosomes are complex molecules composed of dozens of different proteins and RNA. The endoplasmic reticulum that is densely studded with ribosomes is called rough ER, as shown in Figure 2-19. The endoplasmic reticulum of cells specialized in making lipids, such as many brain and intestinal cells, has relatively few or no ribosomes and is called smooth ER. Cells can have both kinds of ER.

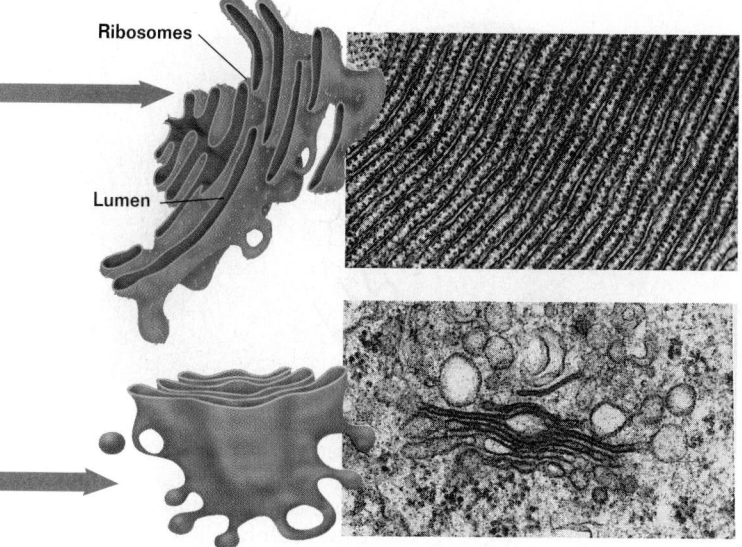

Ribosomes

Lumen

Figure 2-19 **Figure 2-19** The pebbly appearance of the cell's rough endoplasmic reticulum (rough ER) is due to the presence of ribosomes. It is here that the cell manufactures proteins intended for export.

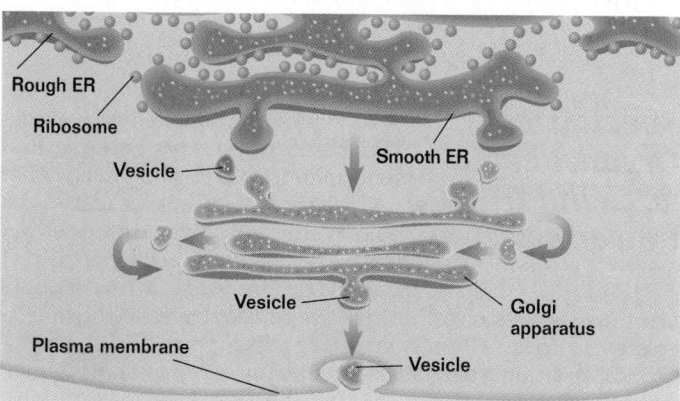

Figure 2-20 The Golgi apparatus packages and distributes proteins and lipids.

Packaging and Distribution Centers

Proteins and lipids destined for export are passed across the ER membrane as they are made. They then pass into an interior compartment of the ER called the lumen. These proteins and lipids move through the lumen to an area of smooth ER. The protein is then enclosed in a vesicle that buds off from the ER surface. The vesicle migrates across the cytoplasm to an adjacent organelle called a Golgi apparatus. The **Golgi** (*GOHL jee*) **apparatus,** shown in Figure 2-20, is the packaging and distribution center of the cell.

The Golgi apparatus contains a variety of enzymes that act on proteins and lipids and that serve as molecular address labels to determine where the protein or lipid will go. The newly made molecules are transported to different compartments of the cell or to the plasma membrane where they are exported from the cell. You can see this process in Figure 2-21.

Rough ER

Ribosome

Vesicle

Smooth ER

Vesicle

Golgi apparatus

Plasma membrane

Vesicle

Figure 2-21 Proteins made on the endoplasmic reticulum travel through the cytoplasm to the Golgi apparatus, where they are packaged in vesicles and sent to different cellular compartments or exported from the cell.

 VISUAL STRATEGY

Figures 2-19 and 2-20

Have students identify the channels of the endoplasmic reticulum and Golgi apparatus through which materials pass. Ask students to discuss why the ER is sometimes called "the highway of the cell."

Teaching Tip

What If a cell lost its ribosomes? The cell would die because the ribosomes produce the proteins necessary for life.

 VISUAL STRATEGY

Figure 2-21

Point out the organelles visible in this figure. Trace the movement of proteins through the cytoplasm from the ER to the Golgi apparatus and onto other destinations. Ask students to compare rough and smooth ER. *(Students should notice that rough ER has ribosomes.)* Be sure to point out the plasma membrane, which will be discussed in Chapter 3.

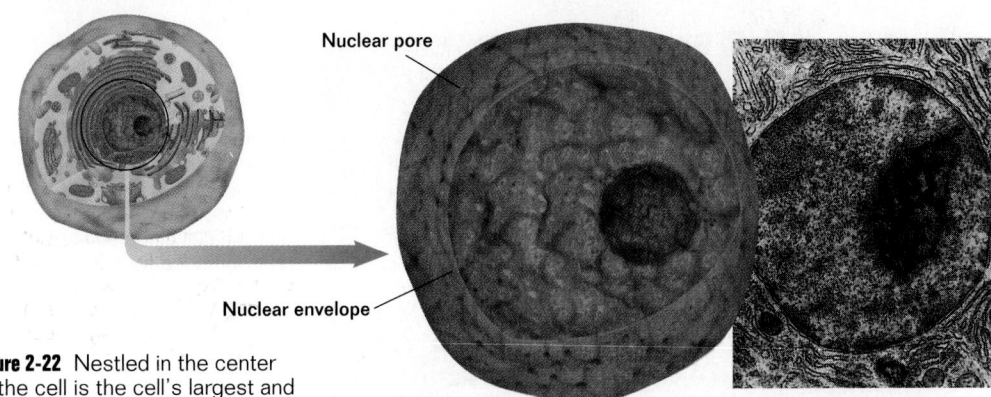

Figure 2-22 Nestled in the center of the cell is the cell's largest and most easily seen organelle, the nucleus. First described by the English botanist Robert Brown in 1831, the nucleus directs all cell activities. It also contains the cell's hereditary information, the DNA instructions for making the next generation.

☐ CAPSULE SUMMARY

The endoplasmic reticulum (ER) transports the cell's proteins and lipid vesicles. The Golgi apparatus packages and distributes proteins and lipids for export. The nucleus directs all of the cell's activities and houses the genetic material, DNA.

The Nucleus Directs Cell Activities and Stores DNA

Nestled in the center of the cell is the cell's largest and most easily seen organelle, the nucleus. The nucleus, shown in Figure 2-22, directs all cell activities and serves as the storage center for a eukaryotic cell's DNA.

Much like wearing two layers of clothing, the surface of the nucleus is bound by a complex double membrane (formed of a lipid bilayer) called the **nuclear envelope**. The cell's many activities are regulated by protein and RNA molecules that pass in and out of the nucleus across this nuclear envelope. How do they get across? Scattered over the surface of the nuclear envelope like craters on the moon are shallow depressions called nuclear pores that provide passageways. These pores contain many embedded proteins that act as molecular channels permitting certain molecules to pass into and out of the nucleus.

Inside the nucleus, DNA and proteins associated with DNA are organized into rod-shaped structures called **chromosomes.** All eukaryotic species have a characteristic number of chromosomes. The cells of your body each contain 46 chromosomes (except for egg or sperm cells, which have 23 chromosomes, and a few specialized tissues such as the liver, with 92, and red blood cells, with none). ☐

Specialized Organelles Act as Cellular Powerhouses

Two kinds of organelles play essential roles in energy release and food manufacture. The energy that drives the many activities of the eukaryotic cell is generated within organelles called **mitochondria** *(myt uh KAHN dree uh)*. Mitochondria are found in eukaryotic cells, where they release the stored energy in food. All of your energy is supplied by the mitochondria within your cells. Except for a few primitive protists, the cells of all eukaryotes have mitochondria.

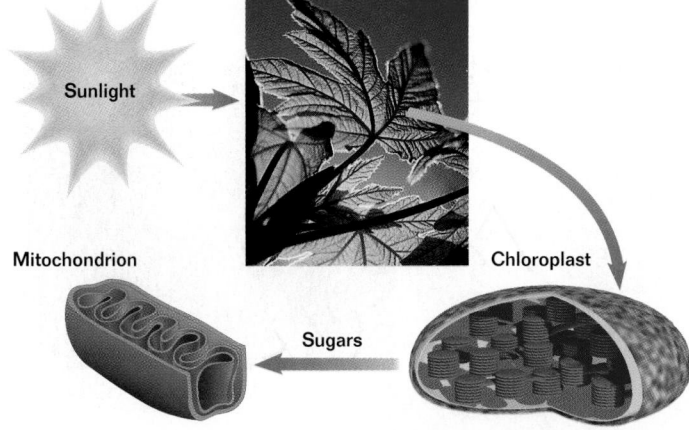

Sunlight

Mitochondrion

Chloroplast

Sugars

Figure 2-23 Chloroplasts in green plants capture sunlight, which enables them to make sugars. These sugars are the ultimate source of energy for all living things. A eukaryotic cell's mitochondria continuously release this energy, providing the cell with the energy necessary for life.

Chloroplasts are organelles that make food in the form of sugars, using water, carbon dioxide in the air, and energy from sunlight. This process is called photosynthesis. Chloroplasts are found only in algae, such as seaweed, and plants, and they are the only places within these organisms where photosynthesis occurs. Organelles like mitochondria and chloroplasts are thought to have evolved from separate organisms, such as bacteria, that were once ingested by a cell.

All the chemical energy that mitochondria extract from the food you eat originated from the photosynthetic process carried out by chloroplasts. Figure 2-23 shows how energy flows in connected pathways throughout the living world. ❑

Biochemical Factories of the Cell

Eukaryotic cells contain a variety of organelles that isolate specialized biochemical activities. For the same reason that a kitchen, bathroom, and bedroom of a house are separated, so a cell isolates certain activities.

Chemical Specialty Shops

Almost every eukaryotic cell contains small vesicles called peroxisomes *(puhr AHKS ih sohms)*, which are derived from the smooth ER. **Peroxisomes** contain several kinds of enzymes. Some peroxisome enzymes convert fats to carbohydrates. Others alter potentially harmful molecules within the cell by forming hydrogen peroxide, H_2O_2, which is converted to water.

Recycling Centers

Another group of spherical organelles the same size and appearance as peroxisomes are lysosomes *(LY seh sohms)*. **Lysosomes** are vesicles that contain the cell's digestive enzymes. The enzymes within lysosomes cause the rapid breakdown of proteins, nucleic acids, lipids, and carbohydrates. Lysosomes digest and recycle the cell's used components. ❑

CONTENT LINK

You will learn more about photosynthesis and cellular respiration in **Chapter 5**.

❑ **CAPSULE SUMMARY**

Mitochondria are the energy-releasing organelles of all eukaryotic cells. Chloroplasts enable algae and plants to capture energy during photosynthesis.

❑ **CAPSULE SUMMARY**

Peroxisomes neutralize potentially dangerous molecules within the cell. Lysosomes digest and recycle the cell's used components.

SECTION 2-4

👁 **VISUAL STRATEGY**

Figure 2-23

Use this figure to emphasize that the sun is the source of energy of most of life on Earth. The energy-requiring reactions of photosynthesis transform sunlight energy into chemical energy. The energy-releasing reactions of cellular respiration enable living things to use chemical energy to do work.

Teaching Tip
Cellular Recycling

Lysosomes can contain at least 40 different enzymes. If these enzymes were free in the cell, they could digest necessary structures and degrade newly synthesized macromolecules. Point out that the lysosome's membrane has the ability to resist destruction by the enzymes within it. The molecules broken down by lysosomes's can be recycled by the cell. An example of building-block recycling within the body is the reuse of components from old red blood cells (RBCs). After approximately 40 days in the circulation, the lipoprotein membrane of human RBCs wears out. The cells are then degraded in the spleen, and the basic building blocks are carried in the blood to the liver for resynthesis.

Teaching Tip
The Human Body as an Analogy for the Cell

Have students draw an analogy between the structure of a cell and the human body. On the board write the column head "Cell," and ask students to list the major organelles under the head. In a second column write "Human Body," and ask students to list an analogous part of the body. For example, nucleus = brain, endoplasmic reticulum = circulatory system, vacuoles = fat deposits, cell membrane = skin, microtubules = skeleton.

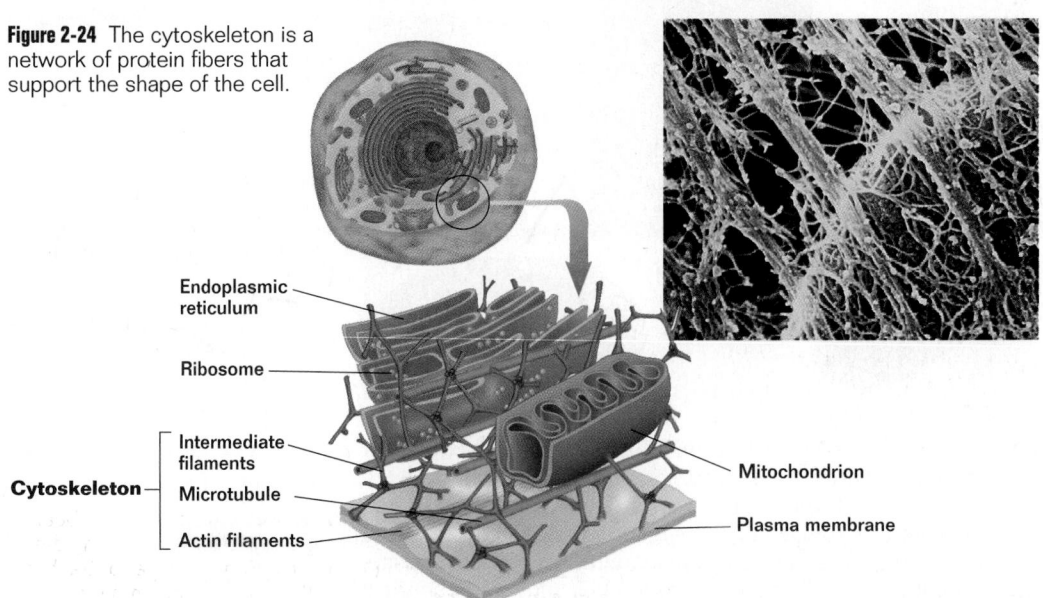

Figure 2-24 The cytoskeleton is a network of protein fibers that support the shape of the cell.

Endoplasmic reticulum

Ribosome

Cytoskeleton
— Intermediate filaments
— Microtubule
— Actin filaments

Mitochondrion

Plasma membrane

 CAPSULE SUMMARY

The cytoskeleton of cells is made of a network of protein fibers that supports the shape of the cell and anchors its organelles. Many eukaryotic cells possess flagella or cilia for movement.

Protein Fibers Provide an Internal Framework

The cytoplasm of eukaryotic cells is crisscrossed by a network of several kinds of protein fibers that supports the shape of the cell and anchors its organelles. This meshlike network is called the **cytoskeleton** and is shown in Figure 2-24. Within the cytoskeleton, hollow protein fibers called **microtubules** aid in moving chromosomes during cell division by forming cylindrical organelles called **centrioles**. The cells of plants and fungi lack centrioles. Two additional kinds of protein fibers, actin filaments and intermediate filaments, provide the cell with mechanical support and help to determine the shape of the cell.

Protruding from the surfaces of many eukaryotic cells are long threadlike organelles called **flagella,** which are used in locomotion. Flagella are actually complex cables of microtubules. Each flagellum consists of a circle of nine microtubules surrounding two central microtubules. This 9 + 2 arrangement is a fundamental feature of the flagella of eukaryotes, such as those on fast-swimming protists and those on the cells of sensory hairs inside the human ear. If flagella are numerous, shorter, and organized in tightly packed rows, they are called **cilia**. An example of cells that have cilia is shown in Figure 2-25. The 9 + 2 arrangement is also shown in Figure 2-25. ■

Interior Spaces Are Storage Areas

Plant cells, unlike animal and most fungal cells, store waste products in a large internal space called a **central vacuole**. Vacuoles store large amounts of water and nutrients, as well

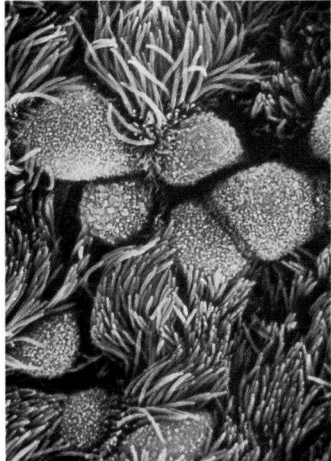 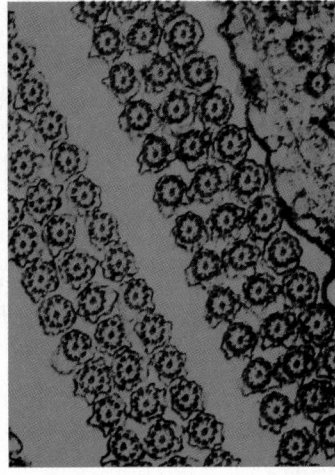 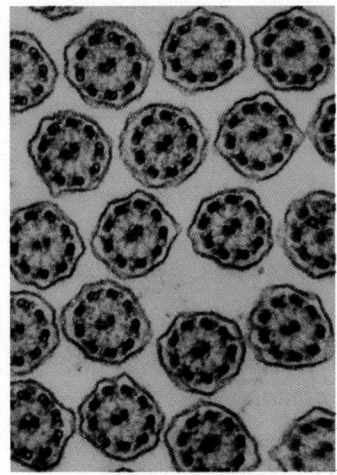

👁 **VISUAL STRATEGY**

Figure 2-25
Use this figure to emphasize that cells and tissues must look the way they do in order to perform their functions. Question students as to why parts of the human respiratory system might be lined with cilia. *(The cilia sweep foreign particles up and out of the delicate respiratory tissues.)*

as wastes. The pressure exerted by this stored water acts as a sort of "skeleton" for the plant, keeping it rigid. This rigidity allows a plant to stand upright; when a plant's vacuoles lack water, it will become limp and will wilt.

Figure 2-25 The continuous beating of the hairlike cilia lining the mammalian windpipe, or trachea, *left*, helps the body remove foreign particles. All cilia exhibit a 9 + 2 internal arrangement of microtubules. The cilia of a single-celled microorganism, *center*, appear strikingly similar to the cilia that line the human oviducts (the tubes that carry eggs from the ovary to the uterus), *right*.

Diversity and Themes in Biology

As you have learned in this chapter, eukaryotic cells share certain similarities. Yet they also differ in some characteristic ways. Plant cells, for example, have a rigid cell wall made of the polysaccharide cellulose, which lies just outside the plasma membrane. Plant cells also have central vacuoles and sometimes chloroplasts. Cells of fungi *(FUHN jeye)*, a group of organisms that includes mushrooms and molds, also have a rigid cell wall, but it is made of a different polysaccharide called chitin *(KEYE tihn)*. Animal cells lack a cell wall. An animal cell is surrounded only by its plasma membrane, making it possible for the cell—and ultimately the animal itself—to move. Within the three cellular "floor plans" seen in plants, fungi, and animals, an enormous variety exists.

Just as there are many kinds of cells, there is so much diversity among living things that it is easy to overlook the fact that all living things have much in common. As you study biology, you will discover that certain basic principles or themes tend to recur. When viewed together, these themes will make the underlying unity of life increasingly clear. The first of these themes—"Evolution"—is described on the following page. This theme will help you recognize how an understanding of cells fits into the "big picture" of biology. In the next four chapters, you will be introduced to four additional themes.

Teaching Tip

"Survival of the Fittest"

In spite of many students' misinformation, while Darwin did speak of the survival of the fittest, he did not mean to imply that the largest and most physically overpowering organism always survives to pass along its genes. Fitness for survival can be a subtle quality, often invisible except through biochemical studies.

Chapter Closure

Supply students with various construction materials, such as pipe cleaners, plastic foam, cardboard, etc., or have them bring these materials from home. Have the students construct a large model of the cell factory according to their own design. Encourage creativity. Each model must show all of the organelles listed in the chapter, their names and functions, and a clear correlation between form and function. Have students prepare questions based on their model to serve as quiz questions for the entire class. Use these questions as a review in preparation for assessment.

At a glance, the Euglena, Vorticella, and the in-line skater would seem to have little in common. Yet the cells of each contain energy-releasing mitochondria, an evolutionary inheritance found among all eukaryotes.

Evolution

Evolution is the change of organisms over time. Although biologists had long considered that the diversity of life on Earth was the result of evolution, it was not until 1859 that a plausible mechanism for evolution was suggested. In that year, the English naturalist Charles Darwin published his book *On the Origin of Species,* in which he suggested that organisms change over time in response to changes in their environment.

Darwin called this process natural selection.

Simply stated, Darwin's theory of evolution by natural selection says: Those organisms that are better able to successfully respond to the challenges of their environment are more likely to survive to reproduce; thus, the characteristics of those organisms become more common.

Today biologists recognize that evolution provides a unifying thread that runs throughout the living world. In this chapter, for example, you learned that all living things are made of one or more cells, each surrounded by a plasma membrane. Additions to this basic cell design—for example, organelles such as the nucleus and mitochondria—represent evolutionary embellishments upon this simple cellular "floor plan." Throughout your study of biology, you will see recurring examples of ways that species have changed over time in response to their environments. No other concept serves to unify the study of biology as powerfully as evolution.

Section Review

1. *A biologist wants to study how substances enter and exit a certain type of cell. Describe the benefits and disadvantages of using an electron microscope for this purpose.*

2. *Name five organelles. Describe how each enables the cell to display the properties of life.*

Critical Thinking

3. *Are lysosomes in animal cells similar to central vacuoles in plant cells? Explain your answer.*

4. *What does the presence of mitochondria in eukaryotic cells suggest about the evolution of eukoryotes?*

Answers to Section Review

1. Advantage: single cells and their organelles can be seen quite clearly under the electron microscope. Disadvantage: the cell must be killed before an electron microscope can be used.

2. Answers will vary. See pages 42–47 for possible answers.

3. Lysosomes are similar to vacuoles in that they contain liquid. They are dissimilar in that they contain digestive enzymes.

4. Answers might include that the presence of mitochondria points to the evolution of eukaryotic cells from separate organisms, such as bacteria, that were once ingested by the cell.

CHAPTER REVIEW

CHAPTER REVIEW ANSWERS

Each item in the Chapter Review is correlated by text section in the assignment guide that follows.

Vocabulary

acid (33)	enzyme (35)	peroxisome (45)
amino acid (36)	eukaryotic (26)	pH scale (33)
atom (29)	evolution (48)	plasma membrane (26)
base (33)	flagella (46)	polar molecule (31)
carbohydrate (34)	glycogen (34)	polypeptide (36)
cell membrane (26)	Golgi apparatus (43)	polysaccharide (34)
cell theory (25)	hydrogen bond (31)	prokaryotic (26)
cellulose (34)	ion (29)	protein (35)
central vacuole (46)	ionic bond (31)	resolution (39)
centriole (46)	light microscope (39)	ribosome (26)
chromosome (44)	lysosome (45)	RNA (38)
cilia (46)	magnification (39)	saturated fat (35)
collagen (35)	microtubule (46)	scanning electron microscope
compound (30)	mitochondrion (44)	(SEM) (40)
cytoplasm (26)	molecule (30)	starch (34)
cytoskeleton (46)	nuclear envelope (44)	steroid (35)
DNA (37)	nucleic acid (37)	transmission electron micro-
electron (29)	nucleotide (37)	scope (TEM) (40)
element (29)	nucleus (26)	unsaturated fat (35)
endoplasmic reticulum (ER) (42)	organelle (26)	vesicle (42)

ASSIGNMENT GUIDE

Section	Review	Themes Review	Critical Thinking
2-1	3, 4, 18	20	
2-2	5, 8, 10, 11		21
2-3	7, 12, 13, 16, 17		22
2-4	1, 2, 6, 9, 14, 15, 19	20	23

Concept Mapping

Construct a concept map that shows the similarities and differences between prokaryotic cells and eukaryotic cells. Use as many terms as needed from the vocabulary list. Try to include the following items in your map: organelles, DNA, cell membrane, cytoplasm, mitochondria, nucleus, endoplasmic reticulum, ribosomes, chloroplasts, and energy. Include additional terms in your map as needed.

Review

Multiple Choice

1. Organelles that are present in plant cells but absent from animal cells include the
   a. chloroplasts and central vacuole.
   b. flagellum and cell wall.
   c. mitochondria.
   d. endoplasmic reticulum, cell wall, and lysosomes.

2. What evidence supports the hypothesis that eukaryotic cells evolved from prokaryotic cells?
   a. Fossils have been found of eukaryotic cells that look like prokaryotic cells.
   b. Organelles of eukaryotic cells are similar to bacteria.
   c. Eukaryotic cells have DNA identical to that found in their organelles.
   d. Most living eukaryotes are unicellular protists.

3. The growth of a cell is limited by the ratio between its
   a. surface area and volume.
   b. organelles and surface area.
   c. organelles and cytoplasm.
   d. nucleus and cytoplasm.

4. Eukaryotic cells differ from prokaryotic cells in that eukaryotic cells
   a. lack organelles.
   b. have DNA, but not ribosomes.
   c. are single-celled.
   d. have a nuclear membrane.

Concept Mapping
Map answer is shown on page 23B.

Review
Multiple Choice
Code in parentheses indicates section number and objective number.
1. a (2-4.3)
2. b (2-4.4)
3. a (2-1.3)
4. d (2-1.2)
5. a (2-2.4)
6. a (2-4.2)
7. c (2-3.1)
8. c (2-2.3)
9. b (2-4.2)

CHAPTER REVIEW ANSWERS *continued*

Completion
10. ion (2-2.1)

11. covalently, hydrogen (2-2.2)

12. catalysts (2-3.2)

13. nucleic acid, hereditary (2-3.1)

14. rough ER, proteins (2-4.2)

15. cytoskeleton, support (2-4.2)

Short Answer
16. Proteins are made up of one or more polypeptides, and polypeptides are made up of long chains of amino acids. (2-3.2)

17. ATP and fat are both energy storage molecules. They are different in that the energy from ATP can be used immediately by the cell but the energy available in fat can be freed only through a series of chemical reactions. (2-3.3)

18. Multicellular organisms are specialized and usually larger than single-celled organisms. A large organism requires more food, oxygen, and living space. Specialization necessitates that each cell rely on others in order to survive. (21.2)

19. The detailed three-dimensional image indicates that a scanning electron microscope was used to produce the image. (2-4.1)

Themes Review
20. All cells have a plasma membrane, cytoplasm, ribosomes, and genetic material. These structures became common to all cells because they enable the cell to maintain homeostasis and to reproduce itself. (2-1.2, 2-4.4)

Critical Thinking
21. Surface tension forms across the surface of water because of the cohesive attraction between individual water molecules. Because the water molecules are more attracted to each other than to the molecules making up the water strider, the insect can stand on the water's surface. (2-2.1)

22. Answers will vary, but may include that unsaturated body fat

5. A substance that forms hydrogen ions when dissolved in water is called a(n)
   a. acid.
   b. hydrophilic solute.
   c. base.
   d. lipid.

6. An ionic bond forms when
   a. one atom transfers its electrons to another atom.
   b. two atoms share electrons.
   c. atoms act like molecular magnets.
   d. the neutrons of one atom fill the outer energy level of another atom.

7. Macromolecules that enable corn stalks to stand tall are
   a. lipids.
   b. proteins.
   c. carbohydrates.
   d. amino acids.

8. Water is a good solvent because it
   a. is a nonpolar molecule.
   b. has a pH of 7.
   c. has charged regions that interact with polar substances.
   d. gains electrons that produce oppositely charged ions.

9. Which of the following is the appropriate pairing of structure and function?
   a. smooth endoplasmic reticulum : protein manufacture
   b. Golgi apparatus : protein packaging and distribution
   c. mitochondrion : photosynthesis
   d. lysosomes : conversion of carbohydrates to fat

Completion
10. An atom with unequal numbers of electrons and protons is called a(n) _____ .

11. In water, one oxygen atom is _____ bonded to two hydrogen atoms. The oppositely charged regions of nearby water molecules are held together by weak _____ bonds.

12. Enzymes act as _____ that change the rate of a chemical reaction without being destroyed in the process.

13. DNA is an example of a(n) _____ . The role of DNA in the cell is to carry _____ information.

14. Endoplasmic reticulum that has ribosomes attached to its surface is called _____ . This type of endoplasmic reticulum is associated with the production of _____ .

15. The network of protein fibers that crisscrosses the cytoplasm of eukaryotic cells is called the _____ . This network provides _____ for the cell and holds organelles in place.

Short Answer
16. Describe how proteins, polypeptides, and amino acids are related to one another.

17. How are the functions performed by ATP and fat similar yet different?

18. Multicellular organisms are those composed of more than one cell. In Section 2-1, several advantages of multicellularity are described. What are two disadvantages of multicellularity?

19. What kind of microscope was used to produce the photograph of the insect shown below? State your reasons for your choice.

provides better protection against freezing by storing heat and supplying energy for maintaining body temperature than saturated body fat does. (2-3.2)

23. It can be, and is, used to fight human bacterial infections. Bacteria that are unable to synthesize protein will die. (2-4.2)

Themes Review

20. Evolution Name two structures found in all cells. How do these structures underscore the role of evolution in biology?

Critical Thinking

21. Making Inferences Look at the water strider in the photograph below. What property of water is responsible for the insect's ability to stand on the water's surface?

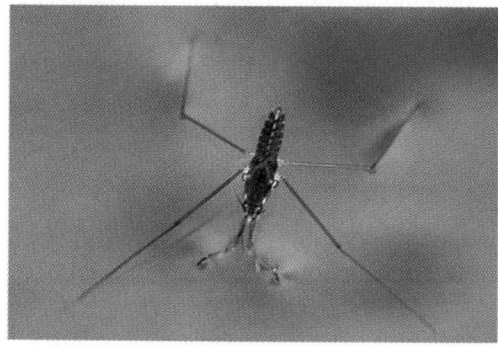

22. Making Inferences Animal fats usually exist as solid fats, and those in plants exist as oils. However, in many animals of the Arctic and Antarctic, animal fats are mostly oils. What adaptive advantage would the storage of body fat as oil instead of a solid be to animals that live in freezing climates?

23. Making Predictions The chemical erythromycin is particularly effective in inhibiting protein synthesis by the ribosomes in bacteria cells, but it does not inhibit protein synthesis in human cells. Suggest a possible use for erythromycin.

Activities and Projects

24. Cooperative Group Project Write and perform a series of skits to demonstrate the unique properties of water that make it essential for living things.

25. Multicultural Perspective Some of the world's earliest cultural symbols depict human relationships to the four elements of water, earth, fire, and air. Look for examples of these symbols in your library's art, anthropology, psychology, or science sections. What similarities do you find between cultures? What colors and shapes would you use to draw symbols for the four elements?

26. Researching and Writing Learn more about the work of Schleiden, Schwann, or Virchow. Write a brief report summarizing the processes that the researcher used to arrive at his conclusions about cells.

27. Cooperative Group Project Construct a cell organelle using modeling clay. After constructing your model, present it to the class. In your presentation, describe the organelle's structure, function, and relationship to other organelles.

28. Research and Writing Find out how the laboratory techniques of fractionation, centrifugation, and electrophoresis enable biologists to experiment with cells and analyze the substances that cells produce. In your report, describe the equipment used for each technique and what can be accomplished when the technique is employed.

Readings

29. Read the article "Revenge of the Killer Microbes" in *Time,* September 12, 1994. How do disease-causing bacteria become resistant to antibiotics?

30. Read the article "No-'Stick' Tips for Heart-Healthy Diets," in *Science News,* March 6, 1993, page 150. What percentage of calories in the experimental diet described in the article came from fat? What differences did the researchers observe between test subjects who used corn oil and those who used corn-oil margarine?

Activities and Projects

24. Skits should show water's effectiveness at dissolving polar solutes, capillary action, cohesion, and storing heat.

25. Answers will vary. Students will find interesting examples of symbols in books such as Elizabeth S. Helfman's *Signs and Symbols Around the World* (New York: Lothrop, Lee, and Shepard Co., 1967) or Henry Dreyfus's *Symbol Sourcebook* (New York: McGraw-Hill, 1972).

26. All three were Germans of the mid-1800s who made numerous microscopic observations of a variety of subjects. Also, the development of organic stains made it possible to study specific cell structures.

27. Encourage students to learn as much as they can about their organelle's structure, function, and relationship to other organelles.

28. Cell fractionation is used to release the components of a cell. Cells can be fractionated in a blender. By spinning fractionated cells at high speeds, centrifugation separates cell parts by type. A centrifuge is used for this technique. Electrophoresis is a process that separates fragments of large molecules, such as proteins and DNA, on the basis of their electrical charge and size. The molecules move through a porous material in an electrical field.

Readings

29. Because bacteria multiply so quickly, the evolutionary process is fast in bacteria. Bacteria are able to adapt to antibiotics by spontaneous mutations and by acquiring new DNA from other bacteria or viruses. Moreover, antibiotics promote new generations of resistant bacteria by wiping out the resistant bacteria's competitors for food and space.

30. The experimental diet derived 30 percent of its calories from fat. Subjects who used liquid corn oil had larger decreases in total blood cholesterol and in low-density lipoprotein ("bad") cholesterol than subjects consuming corn oil margarine.

Preparation

Obtain prepared slides of human epithelial and blood cells from a biological supply house. Purchase sprigs of *Elodea* at a pet store and keep plants under light for 12 hours before class to induce cytoplasmic streaming. For unknowns, make wet mount slides or order prepared slides from a biological supply house. Have students examine at least 3 unknowns.

Procedural Notes

• Students should carefully examine the known cells before examining the unknown cells.

• When viewing prepared slides of human blood, white blood cells appear blue, and platelets are violet or purple.

• When viewing the unknown slides, students should rotate from station to station, spending 2–5 minutes at each station. Remind students to record the identifying number from each slide.

Background Answers

1. Plant cells have a cell wall, chloroplasts, and a large central vacuole. Animal cells lack these structures. Plant cells often have a regular, rectangular shape, while animal cells are irregularly shaped.

2. What characteristics of animal and plant cells can you observe when you look at epithelial and *Elodea* cells under a microscope?

Animal and Plant Cells

OBJECTIVE

Differentiate between animal and plant cells by observing organelles and cellular structures of epithelial cells and *Elodea* cells.

PROCESS SKILLS

• observing cell organelles
• comparing and contrasting cell types

MATERIALS

• prepared slides of human epithelial cells from the skin lining the mouth
• compound light microscope
• sprigs of *Elodea*
• forceps
• microscope slides and coverslips
• safety goggles
• lab apron
• dropper bottle of Lugol's iodine solution
• prepared slides of three unknowns
• prepared slides of human blood

BACKGROUND

1. What characteristics distinguish plant cells from animal cells?

2. Write your own question to explore in your lab report or notebook.

TECHNIQUE

Part A: Epithelial Cells

1. Examine a prepared slide of epithelial cells under low power. Locate cells that are separate from each other and place them in the central field of view. Examine the cells under high power.

Adjust the diaphragm to reduce the light intensity for greater clarity.

2. In the **Records** section of your report, make a drawing of two or three cells as they appear under high power. Identify and label the cell membrane, the cytoplasm, the nuclear envelope, and the nucleus of at least one of the cells.

3. In the **Records** section of your report, make a drawing of these cells as you imagine they might look in the lining of your mouth.

Part B: Plant Cells

4. Carefully tear off a small leaf near the top of an *Elodea* sprig. Using forceps, place the whole leaf in a drop of water on a slide, and add a cover slip.

5. Observe the leaf under low power. The outermost part of the cell is the cell wall. The many small, green organelles in the cells are the chloroplasts. Look for a cell you can see clearly, and move the slide so that it is in the center of the field of view. Examine this cell under high power using the fine adjustment to bring it into focus.

6. Find an *Elodea* cell that is large enough to allow you to see the cell wall and the chloroplasts clearly. Draw this cell in the **Records** section of your report.

7. The chloroplasts may be moving in some of the cells. If no movement is observed, warm the slide in your hand or shine a bright lamp on it for a minute or two. Reexamine the slide under high power. Look for movement of the cell contents. Such movement is called cytoplasmic streaming.

8. Put on safety goggles and a lab apron. Because the cell membrane is pressed against the cell wall you may not see it. Due to the abundance of chloroplasts, other organelles might also be hidden. To see the vacuole, nucleus, and nucleolus,

Records

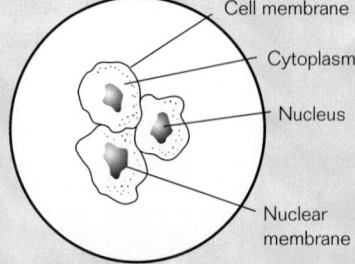

Epithelial cells

Cell membrane
Cytoplasm
Nucleus
Nuclear membrane

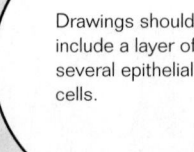

Drawings should include a layer of several epithelial cells.

Epithelial cells as they might appear in the lining of the mouth

prepare another wet-mount slide substituting Lugol's iodine solution for the water. Allow the iodine to diffuse throughout the cell. Then observe the stained cells under low and high power.

9. In the **Records** section of your report, draw a stained *Elodea* cell. Label the central vacuole, nucleus, nucleolus, chloroplasts, cell wall, and cell membrane if they are visible.

Part C: Blood Cells

10. Obtain a prepared slide of human blood and examine it under low and high power. The stain used on the red blood cells causes the endoplasmic reticulum to resemble a blue mesh. Other blood cells may be blue or purple.

11. Obtain prepared slides of three unknown specimens from your teacher, and examine each under low and high power. In the **Records** section of your report, make a table like the one below. Record the code number assigned to each unknown its classification (plant or animal), and your reasons for its classification.

Classification of Unknowns

Unknown (code number)	Classification (plant or animal)	Reasons for Classification

INQUIRY

1. Describe the shape of epithelial cells.
2. Describe the appearance of the cytoplasm.
3. What shape are the *Elodea* cells?
4. Describe the shape of the chloroplasts. Where are they located in the cell?

5. Describe the movement of the chloroplasts in the *Elodea* cells.
6. After staining with iodine, list the organelles that are visible in the *Elodea* cell.

ANALYSIS

1. In what observable ways are animal and plant cells structurally similar? In what ways are they different?
2. How does the structure of each of the three kinds of cells that you observed relate to the function of each type of cell? Based on what you know about the organism each cell is from, how might each structure help that organism survive?
3. If some of the epithelial cells were folded over on themselves but were still transparent, what could you conclude about their thickness?
4. Lugol's iodine solution causes the movement of chloroplasts to stop. Why?

FURTHER INQUIRY

Write a new question that could be explored as a new investigation. The following are examples:

Obtain other specimens from different plants and from different parts of plants, such as the roots or stems. Try to relate any differences you observe to special functions of each type of cell.

List cell structures that you were unable to see with light microscopy. Use library resources to locate electron micrographs of these structures.

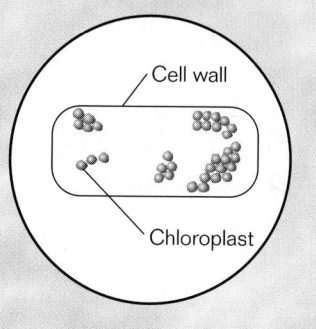

Elodea cell

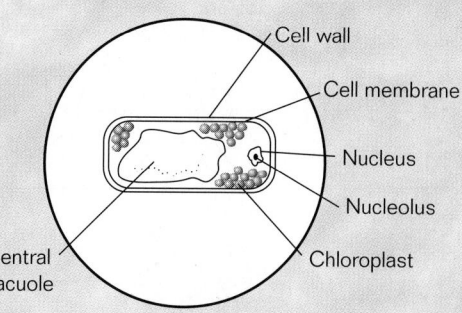

Elodea cell, stained

Block Scheduling Guide

Each block represents about 45–50 minutes of instructional time. The resources cited in a block are coded to a specific program component using the Key on the next page.

Lecture Concepts

Lesson Resources

3-1 Structure of the Plasma Membrane pp. 54–57

BLOCKS ① and ②

a. The envelope that makes up the outer surface of a cell is called the plasma membrane.
b. The membrane is made up of two main groups of organic molecules: phospholipids and proteins. The phospholipids form a lipid bilayer in which there are polar and nonpolar regions.
c. Various proteins penetrate the lipid bilayer and function as channels, receptors, and markers.
d. Channels serve as passageways, receptors serve as information receivers, and markers serve as cell identifiers.

Lecture Resources
- Opening Demonstration, page 54
- Opening Question, page 55
- Demonstrations: Diagramming the Membrane's Bilayer, Fatty Acids
- Multicultural Perspective, page 55
- Teaching Transparencies: 9
- Visual Strategy: Figure 3-2
- Overcoming Misconceptions, page 56
- Holt Biology Videodiscs: Lesson 3-1

Classwork Options
- Tapping Prior Knowledge, page 54
- Even Tiny Things Have Size and Shape, page 57
- Three-Dimensional Model, page 57

Assignment Options
- Section Review, page 57
- Chapter Review, questions 1, 2, 11, 18, 19, 23

3-2 Moving Materials Into and Out of Cells pp. 58–65

BLOCKS ③ and ④

a. Particles move into and out of a cell through its plasma membrane. Passive transport is the movement of a substance through a cell's membrane without the expenditure of cellular energy.
b. Diffusion is the movement of particles from an area of high concentration to an area of low concentration. Osmosis is the diffusion of water molecules through a membrane in the direction of higher solute concentration.
c. The plasma membrane is selectively permeable because it allows the passage of some solutes but not others. Diffusion through selective pores in the membrane is called facilitated diffusion.
d. Cells must expend energy to maintain their internal level of amino acids and sugars. Active transport is the use of cellular energy to transport particles through a membrane against a concentration gradient.
e. Proton pumps cause the production of ATP molecules in chloroplasts and mitochondria Sodium-potassium pumps use ATP molecules to create a high concentration of sodium ions outside the cell. Coupled channels carry the sodium ions along with food molecules back inside the cell.
f. Food particles too large to enter a cell through protein channels are engulfed and brought in by endocytosis. Waste materials and excretions are dumped outside the side by exocytosis.

Lecture Resources
- Opening Question, page 58
- Demonstration: Diffusion in a Single Medium
- Teaching Transparencies: 5, 6, 10, 11, 6A
- Visual Strategies: Figures 3-5, 3-6, 3-8, 3-9, 3-10
- What If Too Much Fertilizer Is Applied to a Lawn? page 60
- Concentration Gradients, page 62
- The Importance of ATP, page 62
- Facilitated Diffusion versus Pinocytosis, page 64
- Holt Biology Videodiscs: Lesson 3-2

Classwork Options
- Laboratory Investigation: Diffusion and Cell Size, pages 72–73
- Quick Labs A4: Demonstrating Diffusion
- Inquiry Skills Development B3: Diffusion and Cell Membranes
- Effective Reading, page 63
- Organizing the Information, page 64
- Climate Controlled Homeostasis and Graphic Organizer, pages 64–65

Assignment Options
- Section Review, page 65
- Chapter Review, questions 3–6, 10, 12, 15–17, 19–25

3-3 How Cells Communicate pp. 66–68

BLOCKS ⑤ and ⑥

a. The cell's communication machinery involves proteins on the cell surface known as receptors.
b. When the proper signal molecule locks on a receptor, a change will occur in the cell's cytoplasm. There may be chemical changes in cytoplasmic molecules, other messenger molecules may be produced, or channels in the plasma membrane may open.
c. Channels that open and close in response to the binding of a chemical to a receptor are said to be chemically gated.
d. Channels that open and close in response to electrical signals are said to be voltage gated.

Lecture Resources
- ■ Opening Question, page 66
- ■ Demonstration: Reviewing Membrane Structure
- ■ Visual Strategies: Figures 3-12, Figure 3-14
- 🔓 Teaching Transparency: 7A
- ■ Specialization of Communication Leads to Balance, page 67
- ■ What If Cellular Communication Is Disrupted? page 67

- ⊘ Holt Biology Videodiscs: Lesson 3-3

Classwork Options
- ■ Closure, page 68

Assignment Options
- ■ Section Review, page 68
- ■ Chapter Review, questions 7–9, 13, 14

Review/Enrichment

BLOCK ⑦

Review/Enrichment
- ■ Themes of Biology: Homeostasis, page 65
- ■ Themes Review, questions 22–23
- ■ Study Guide: Concept Review and Pretest

Assignment Options
- ⊙ Teaching Resources CD-ROM: Supplemental Reading Worksheets and Test, *The Lives of a Cell*
- ■ Activities and Projects, questions 26–28
- ■ Readings, questions 29–30

Assessment Options

BLOCK ⑧

Traditional Assessment
- ■ Chapter 3 Test
- ⊙ Test Generator/Assessment Item Listing: Software item bank for preparing customized chapter tests.

Portfolio Assessment
Select student reports for one or more laboratory experiments from this chapter. *The Direct Observation Checklist* on the *BioSources Teaching Resources CD-ROM* should be completed during a laboratory or other cooperative learning experience.

Answer to Concept Map

The following is one possible answer to the Concept Mapping exercise on page 69.

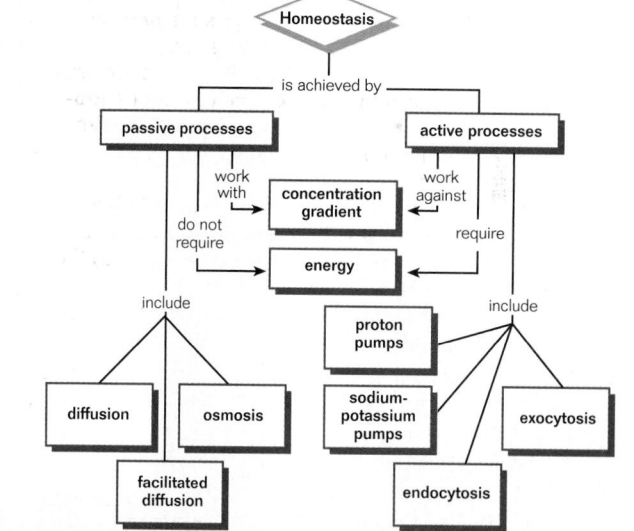

Holt Biology Videodiscs

Holt Biology Videodiscs gives you a powerful tool for teaching, review, and assessment. *Concepts of Biology* adds interactive lesson plans and extensive tools for customization using CD-ROM technology.

CONCEPTS OF BIOLOGY

Use the following topic from *Concepts of Biology* to help you teach this chapter:

- ■ Topic 1: Cell Structure and Function

For further information regarding the media on the videodiscs, see *Holt Biology Videodiscs Teacher's Correlation Guide to Biology: Principles and Explorations.*

Chapter Theme

Homeostasis *This chapter introduces homeostasis, one of the dominant themes of biology in this text. Emphasize that homeostasis is the result of the processes that living things use to maintain a constant internal environment even though external conditions change. Correlate homeostasis with the dominant theme of the previous chapter, evolution, by reminding students how compartmentalization helps maintain homeostasis.*

Tapping Prior Knowledge

- Name the chemicals of life. Identify which of these chemicals compose cell membranes.
- Draw a cell, showing the nucleus, the cytoplasm, and the cell membrane. List the job of each of the labeled parts in your drawing.
- Give three examples of hormones other than the sex hormones.

Opening Demonstration

Make a mask for the overhead projector that will allow light to pass through the bottom of a petri dish but will block out all other light. Prepare a solution of food coloring and water. Adjust the amount of food coloring so that a thin film of the solution in the petri dish will still allow light to pass through to the projection screen. Carefully add corn oil or another light cooking oil to the petri dish so that the oil stays contiguous. A dropping pipette will help. Point out and discuss the water-oil barrier that forms, comparing it to the cell membrane.

CELLS AND THEIR ENVIRONMENT

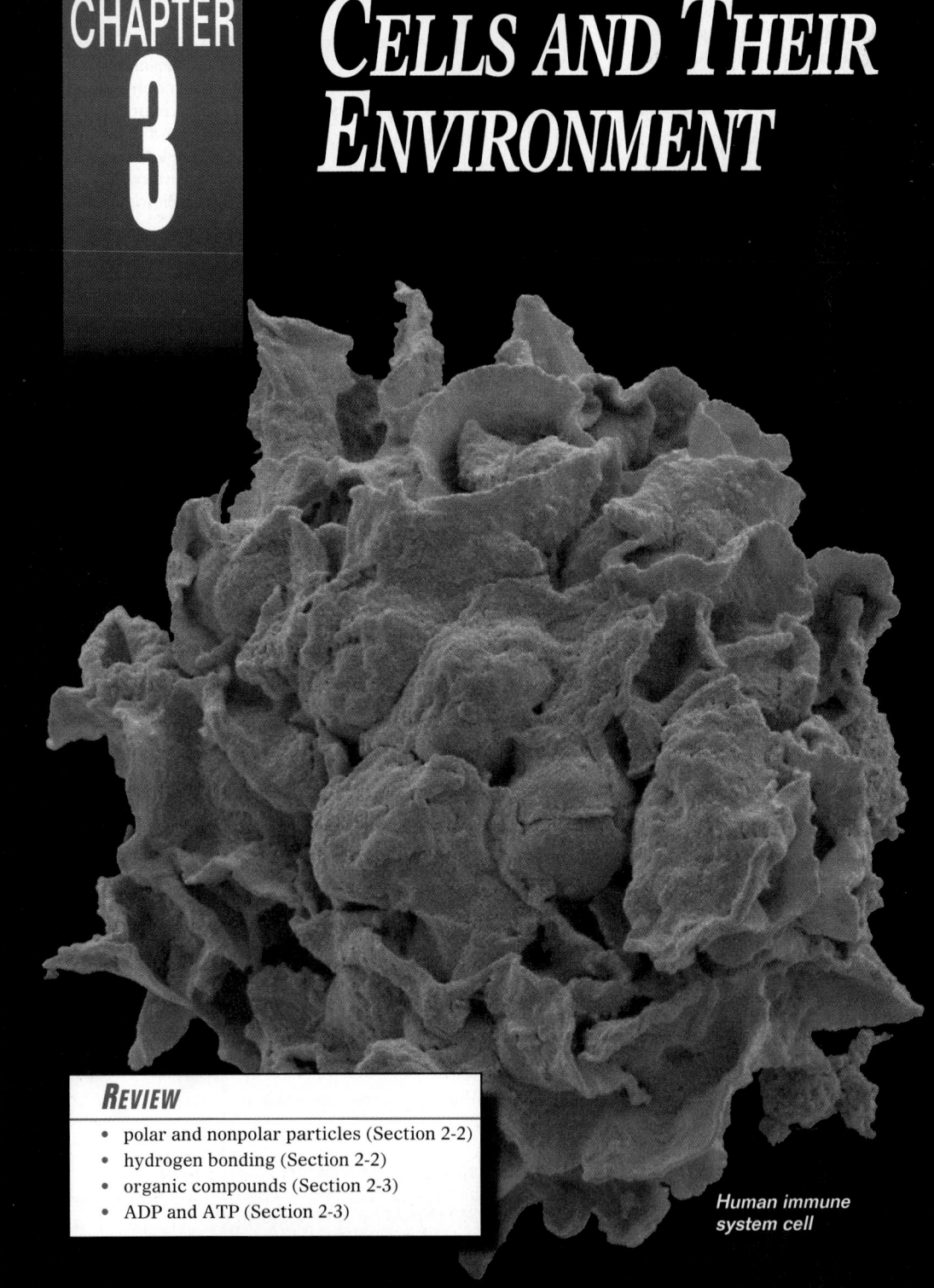

Human immune system cell

REVIEW

- polar and nonpolar particles (Section 2-2)
- hydrogen bonding (Section 2-2)
- organic compounds (Section 2-3)
- ADP and ATP (Section 2-3)

Authors' Rationale

The exploration of cell biology is continued and expanded in this chapter. As we work our way up the hierarchical ladder, cells interact with each other to produce larger structures. The interaction between cells and their environments is largely a consequence of their membranes. Students will learn how the cell membrane regulates the movement of materials into and out of the cell and how it allows communication with other cells. Regulation and communication at the cellular level are essential to the maintenance of homeostasis in the cell and in the organism.

3-1 Structure of the Plasma Membrane

The envelope that makes up the outer surface of a cell is called the plasma membrane. If you were small enough to stand on the outer surface of a cell, you would float, because most of the cell's plasma membrane is liquid. Floating alongside you would be proteins shaped like boulders, while towering overhead like huge trees would be a forest of other proteins. A plasma membrane is a complex and dynamic structure. It is the source of a cell's identity and the site of much of its activity.

Section Objectives

- Describe the arrangement of phospholipid molecules in a plasma membrane.
- Recognize the importance of cell surface proteins.
- Identify three main types of cell surface proteins.

A Lipid Bilayer Is the Foundation of the Membrane

The plasma membrane is composed of two main groups of organic molecules—phospholipids and proteins. A **phospholipid** is a molecule shaped like a head with two tails, as shown in Figure 3-1. The "head" is polar and the two "tails" are nonpolar. The polar head attracts water molecules, which are also polar. These attractions are known as hydrogen bonds. Since the polar water molecules cannot form hydrogen bonds with the nonpolar tails, the water tends to push them away. All this pushing and pulling of phospholipids by water results in the formation of a double layer of phospholipids called a **lipid bilayer**.

The lipid bilayer is not strong and firm like a hard shell, but is fluid like a soap bubble. The individual phospholipid molecules, arranged side by side, float within the bilayer. The most important feature of the lipid bilayer is its nonpolar interior zone. This zone is the true barrier that separates the cell from its surroundings. Most polar particles

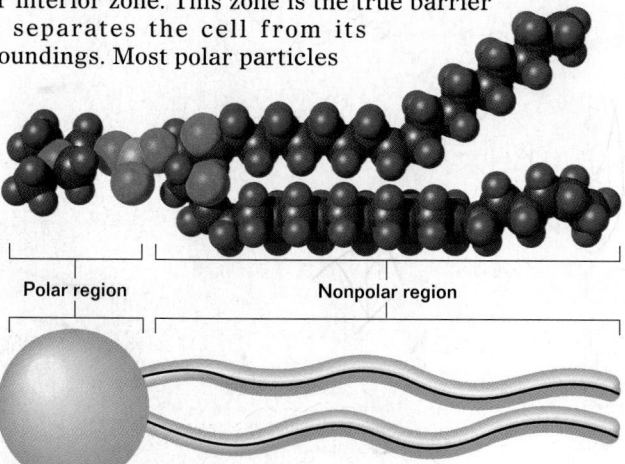

Polar region   Nonpolar region

Figure 3-1 The polar "head" of a phospholipid molecule contains a polar phosphorus group, and the two "tails" are long, nonpolar carbon chains.

SECTION 3-1

Vocabulary Preview

phospholipid
lipid bilayer
cell surface protein
channel
receptor protein
hormone
cell surface marker

Opening Question

Ask students to list the processes every living thing must maintain in order to stay alive. Encourage students to be thorough, listing procurement of materials, metabolism, excretion, growth, movement, response to stimuli, secretion, and reproduction.

Demonstration

Diagraming the Membrane's Bilayer

Draw the following diagram. Call attention to the labeled parts. Refer back to the drawing, calling attention to functions as they are mentioned in the text.

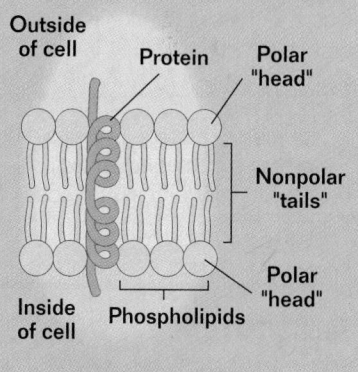

Outside of cell | Protein | Polar "head"
Nonpolar "tails"
Inside of cell | Phospholipids | Polar "head"

Multicultural Perspective

Fats

An Asian diet is typically low in fat, and because of the high amount of vegetables found in Asian meals, this fat is not as harmful as the fat found in Western diets. Why? Plant oils produce polyunsaturated fat, but animal fats are saturated fat, which can lead to arteriosclerosis. The typical North American meal, centered around a meat dish, consists of 35 percent fat, a higher percentage of lipids than is found in typical meals of any other country in the world.

Demonstration

Fatty Acids

Bring in samples of fatty acids, such as butter and household cooking oils like peanut oil, canola oil, olive oil, or corn oil, for students to observe. Demonstrate the oily properties of these compounds by rubbing them on brown grocery bags. *(The paper will become translucent.)* Have students predict whether the fatty acids will mix with water. *(They will not.)* Add the fatty acids to water. *(The fats will float on the water.)* Point out that the "tails" of the phospholipid in Figure 3-1 on page 55 are fatty acids.

👁 VISUAL STRATEGY

Figure 3-2

Use this figure to point out the various structures of the lipid bilayer. Begin by discussing the phospholipid structures, which make up the bilayer itself. Compare the phospholipids in Figure 3-2 to those in Figure 3-1. Point out the inner and outer polar surfaces that the phospholipids form, as well as the center, nonpolar region. Have students locate the cell surface proteins and the proteins' polar and nonpolar regions that correspond to those of the phospholipids.

Demonstration

The Cell Membrane's Bilayer

Obtain a black and white electron micrograph of a cell membrane. Make an overhead transparency of the micrograph. Point out the bilayer. Compare the micrograph's bilayer with the bilayer shown in Figure 3-2. Have students identify "heads" and "tails" of the individual lipids that make up the bilayer.

❑ CAPSULE SUMMARY

The foundation of a cell's plasma membrane is a fluid double layer of phospholipid molecules called the lipid bilayer.

cannot go across this zone, because they are repelled by the nonpolar interior of the bilayer. Sugars and many other food molecules are polar particles, as are proteins, ions, and most cell wastes. The nonpolar zone of the bilayer is such an excellent barrier that if the plasma membrane were made only of lipid bilayer, most food molecules and other polar substances would be unable to pass into and out of the cell.

The solution to this problem is to have passageways through the barrier that enable the cell to regulate precisely which substances go in and out. The passageways through plasma membranes are membrane proteins that traverse the lipid bilayer. ❑

Proteins Are Embedded Within the Bilayer

Every cell is isolated from its environment by its lipid bilayer, able to communicate with the outside world only by means of the proteins embedded within its lipid shell. The proteins within the plasma membrane of cells are often called **cell surface proteins**.

Figure 3-2 shows that a protein, which is typically polar, can extend all the way through a membrane's nonpolar bilayer zone. The protein is anchored into the membrane by its nonpolar middle region, unable to sink inside the cell or float up onto the surface. Notice that Figure 3-2 also indicates that there are different kinds of cell surface proteins. Each type plays a vital role in the life of a cell.

Figure 3-2 In a plasma membrane, phospholipid molecules align themselves in two layers to form a lipid bilayer. The polar heads of the phospholipids point toward water, intracellular water on one side and extracellular water on the other side. The lipid tails point toward each other. Various proteins penetrate the lipid bilayer. Indeed, from 50 to 70 percent of a typical body cell's membranes is protein.

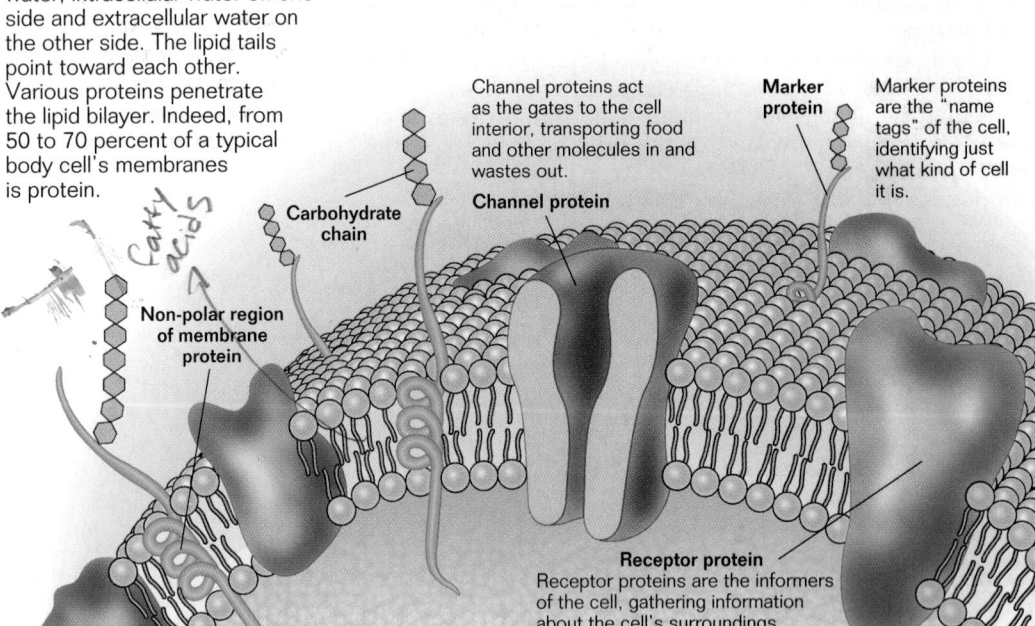

Channel proteins act as the gates to the cell interior, transporting food and other molecules in and wastes out.

Channel protein

Marker protein

Marker proteins are the "name tags" of the cell, identifying just what kind of cell it is.

Carbohydrate chain

Non-polar region of membrane protein

Fatty acids

Receptor protein
Receptor proteins are the informers of the cell, gathering information about the cell's surroundings.

Overcoming Misconceptions

Cell Surface Proteins

Students might be confused by the term *cell surface protein* since the protein extends through the whole bilayer. Explain that the part of the protein protruding through the layer to the outside is the only portion of the protein that a solute particle can "see" when it first encounters a cell, and it therefore "looks like" a surface protein. Likewise, a solute particle inside the cell sees only the part of the protein protruding on the inside. Explain that it is this surface recognition that is very important.

Channels Pass Through the Membrane

Cell surface proteins called **channels** have a series of non-polar amino acid sequences, causing them to loop back and forth through the membrane bilayer many times. This forms a doughnut-shaped channel through the bilayer. Polar sugars, amino acids, ions, and many other particles cross the membrane through these channels. However, particular channels fit only certain particles. The different cells of your body differ from one another largely by the kinds of channels they have. Nerve cells, for example, have a large number of channels specialized to transport sodium ions.

Receptors Transfer Information

Receptor proteins are specialized to transmit information from the world outside the cell to its interior. The end of the receptor that sticks out from the cell surface has a special shape that fits only a particular type of molecule. When a molecule of the right shape fits into the receptor—like a hand into a glove—it causes a change at the opposite end of the receptor protein, the end that protrudes into the cell interior. This change, in turn, triggers responses within the cell. Many of the hormones within your body act by binding to specific receptors on the surfaces of your cells. **Hormones** are chemicals that are secreted by tissues called glands.

Markers Identify Cells

Many cell surface proteins have long exterior arms, often with carbohydrates attached to them, called **cell surface markers**. Every cell of your body has markers on its surface saying that it is part of you and not of some other individual. Other markers say whether it is a liver cell or a heart cell or a brain cell. Still others convey additional information. Your body's immune defenses rely on cell surface markers to tell one cell from another. Cell surface markers will play an important role in Unit 9, where the immune system is discussed. ◻

◻ CAPSULE SUMMARY

Channels, receptors, and cell surface markers are proteins that are embedded within the lipid bilayer of plasma membranes. Channels serve as passageways, receptors serve as information receivers, and markers serve as cell identifiers.

Teaching Tips

Even Tiny Things Have Size and Shape

To account for channels fitting only certain particles, stress that different chemicals have different three-dimensional shapes and sizes. Have students create simple molecular models from toothpicks and small gumdrops to illustrate the three-dimensional shape of molecules. Two examples are shown below.

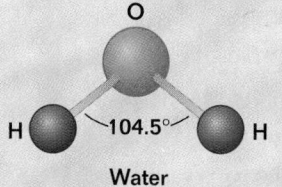

Water

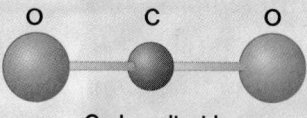

Carbon dioxide

Three-Dimensional Model

Have students make a three-dimensional model of the lipid bilayer membrane. Encourage their creativity by having them bring in their own construction materials. Have students label the parts of their models and list the functions of each part.

Hormones

Point out that hormones are secreted directly into the blood by special glands called endocrine glands. Emphasize that not all glands are endocrine glands; some, such as sweat glands and salivary glands, are exocrine glands. Ask students to link the prefixes *endo-* and *exo-* to the difference in the way endocrine and exocrine glands secrete.

Section Review

1. *Explain why the plasma membrane is an effective barrier to many molecules.*
2. *Draw a simplified diagram of a plasma membrane.*
3. *Describe the functions of the three types of cell surface proteins found in plasma membranes.*

Critical Thinking

4. *If the water molecules inside and outside a cell suddenly became nonpolar, would the transport of materials into and out of the cell be affected? Explain your answer.*

Answers to Section Review

1. Since most solutes in the cell's environment are polar in nature, the lipid bilayer's center, nonpolar section will not allow their passage.

2. See Figure 3-2 on page 56.

3. Channels regulate the passage of materials into and out of the cell. Receptors transfer information from outside the cell to inside the cell. Markers identify cells by distinguishing their function and determining whether they belong to the given organism.

4. If the water molecules inside and outside the cell suddenly became nonpolar, the entire organization of the lipid bilayer would be disrupted. The lipid tails would dissolve in the water, thus completely switching the orientation of the lipid molecules. The center regions of the proteins would attract water, while the polar protein ends would repel the water molecules.

Opening Questions

Have students recall that the cell membrane is made up of two types of chemicals—proteins and phospholipids. Ask students what problems the cell solves by having a membrane. Encourage students to understand that the cell membrane allows the cell to regulate the passage of materials into and out of the cell and to maintain homeostasis.

3-2 Moving Materials Into and Out of Cells

Section Objectives

■ *Differentiate between diffusion and osmosis.*

■ *Describe isotonic, hypotonic, and hypertonic solutions.*

■ *Predict the direction of water molecule movement through a membrane, given solute concentrations on each side.*

■ *Describe facilitated diffusion.*

■ *Define active transport.*

■ *Identify and describe the functions of three types of active transport channels.*

■ *Compare endocytosis with exocytosis.*

If a cell were simply a bag, with nothing but a phospholipid bilayer for a plasma membrane, its interior would be an inactive place. Practically nothing would be able to enter or leave the cell. Sitting inside such a cell would be like sitting inside a room with no windows or doors. No cell is like that, of course, because every cell has an assortment of proteins extending through its plasma membrane that act as its windows and doors.

Passive Processes Do Not Use Energy

Particles move into or out of a cell by passing through the cell's plasma membrane. **Passive transport** is the movement of a substance through a cell's membrane without the expenditure of cellular energy. Substances pass into or out of your cells by way of several different passive transport processes.

Diffusion

If you were submerged within a drop of water, with your body no bigger than a protein, what would you see? Water molecules as big as basketballs would be zipping around, many of them bashing into you! Molecules do not stand still; they are in constant motion. Water molecules and particles dissolved in water move randomly. Their path is not predictable.

This random movement of individual dissolved particles within water has an important consequence that *is* predictable, however. Since the movement is random, a particle is more likely to move from an area where there are a lot of them (an area of high concentration) to an area where there are fewer of them (an area of lower concentration). This net movement of particles from an area of high concentration to an area of lower concentration is called **diffusion** (*dif FYOO zhuhn*). In your lungs, oxygen diffuses into the bloodstream because there is a higher concentration of oxygen molecules in the lung's air sacs than there is in the blood. Eventually, dissolved particles diffuse within a liquid until they fill the volume uniformly, as shown in Figure 3-3. At this point the system can be described as being in **equilibrium**, the situation that exists when the concentration of a substance is the same throughout a space.

A substance that dissolves in another is called a **solute** (*SAHL yoot*). Sugars, amino acids, and ions are all solutes in cells. The more plentiful substance that dissolves the solute is called the **solvent** (*SAHL vuhnt*). In cells, the solvent is water. The mixture of solutes and solvent is called a **solution**.

Multicultural Perspective

Sweat Lodges

Different cultures throughout the world have used water and mineral salts to purify their bodies. Some Native American tribes used sweat lodges or community steam baths for physical and spiritual purification. The dome-shaped sweat lodge was made from a willow tree frame, with burning hot rocks placed in the center of the earthen floor. Small pieces of sage, sweetgrass, or juniper tossed on the rocks burned as incense to purify the air. Bathers drank tea made with medicinal plants and sprinkled it on the rocks. The tea increased the sweating process. Participants could lose more than half of a gallon of water through the sweating.

Lump of sugar

Figure 3-3 If you drop a lump of sugar into a beaker of water, the sugar particles will diffuse and become evenly distributed throughout the water.

Osmosis

Solute and solvent particles tend to diffuse from areas where their concentration is high to areas where their concentration is lower. Now imagine that a membrane separates two regions of a liquid. As long as solute particles and solvent (water) molecules can pass freely through the membrane, diffusion will soon equalize the amount of solute and solvent on the two sides. Equilibrium will be reached.

But what if a polar solute is added to one side and it cannot pass through the membrane? This situation arises in cells all the time. An amino acid cannot cross a lipid bilayer, and neither can an ion or a sugar molecule. What happens? Unable to cross the membrane, the polar solute particles form hydrogen bonds with the water molecules surrounding them, as shown in Figure 3-4. These "bound" water molecules are no longer free to diffuse through the membrane. In effect, the polar solute has reduced the number of free water molecules on that side of the membrane. This means the opposite side of the membrane (without solute) has more free water molecules than the side with the polar solute. As a result, water molecules move by diffusion from the opposite side toward the side with the polar solute.

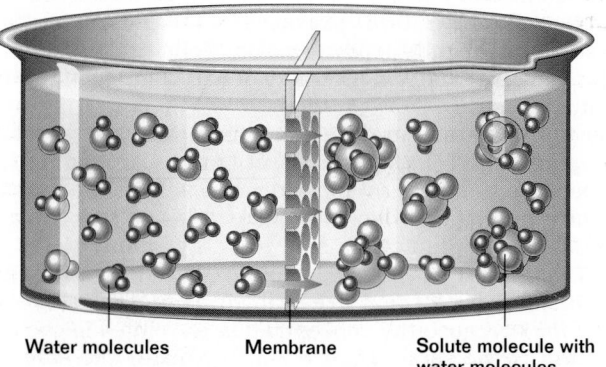

Water molecules   Membrane   Solute molecule with water molecules clustered around it

Figure 3-4 The addition of solutes to one side of a membrane reduces the number of water molecules that can move freely on that side. This is because the water molecules become bound to solute molecules. Water then moves by osmosis from the side where water molecule concentration is higher to the side where their concentration is lower.

Demonstration

Diffusion in a Single Medium

Obtain a large battery jar or small aquarium. Fill the container with water. Place a few drops of food coloring on the surface of the water near the edge. The color must be very intense to ensure that the color will be visible at the end of the demonstration. Allow the demonstration to proceed until the water becomes a uniform color. Have students record their observations and describe the process of diffusion in this demonstration.

Teaching Tip

Diffusion on a Macroscopic Level

Draw the following diagram.

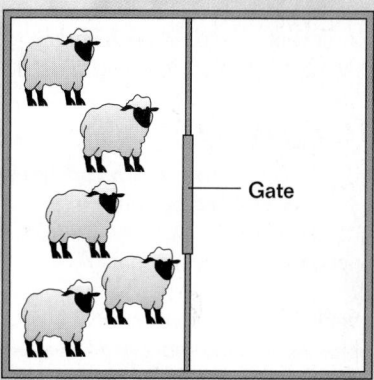

Gate

Ask students the following questions: What will happen immediately if the gate between the pens is opened? *(The sheep will move into the other pen, from a region of higher concentration of sheep to a region of lower concentration.)* Is the movement of any given sheep predictable? *(No, it is random.)* What will the pens and sheep look like in three hours? *(The sheep will be approximately evenly distributed throughout both pens.)* Did anyone have to supply energy to the system (herd the sheep) to get the results? *(No.)* Where did the energy that caused the change come from? *(The energy is intrinsic—it belongs to the sheep themselves.)*

VISUAL STRATEGY

Figure 3-5

Use this figure to correlate the effects of osmosis on animal cells subjected to hypotonic, isotonic, and hypertonic solutions with the movement of water. Stress the relative solute concentrations that affect the water movement. Point out that living plant cells are more resistant to these effects.

Application

Medicine A common medical treatment involves the use of intravenous fluids. Such fluids may deliver drugs or diagnostic chemicals, replace lost blood, or provide nourishment. The solutes in these solutions must be carefully adjusted to maintain isotonicity; otherwise, the patient's cells will be seriously disrupted, possibly resulting in death.

Teaching Tip

 What If too much fertilizer is applied to a lawn?

Grass wilts if too much fertilizer is applied to it. Have students relate the concept of osmosis to this event. (*High salt concentrations in water around the roots and leaves cause an osmotic flow of water out of cells. Wilting is the result of the water loss.*)

		Conditions	Environment Solution Is	Cell Solution Is	Water Will Move
Hypotonic solution		Solute concentration in the environment is lower than in the cell.	Hypotonic	Hypertonic	Into the cell, and cell will burst
Isotonic solution		Solute concentration in the environment is equal to that in the cell.	Isotonic	Isotonic	Equal amounts will move into and out of the cell, and cell volume is maintained
Hypertonic solution		Solute concentration in the environment is higher than that in the cell.	Hypertonic	Hypotonic	Out of the cell, and cell will shrivel

Figure 3-5 If almost any animal cell is suspended in a hypotonic solution, it will burst because of an increase in internal osmotic pressure. In an isotonic solution, a cell's volume will be maintained. In a hypertonic solution, a cell will shrivel.

Eventually, the concentration of free water molecules will equalize on both sides of the membrane. At this point, however, there are more water molecules (bound and unbound) on the side of the membrane with the polar solute. Net water movement through a membrane in response to the concentration of a solute is called **osmosis** (*ahz MOH sihs*). Stated another way, osmosis is the diffusion of water molecules through a membrane in the direction of higher solute concentration.

As a result of osmosis, extra water molecules accumulate on one side of the membrane. If these water molecules accumulate inside the cell, they will exert a pressure that can become very great—great enough to burst the cell! **Osmotic** (*ahz MAH tihk*) **pressure** is the increased water pressure that results from osmosis. Cells with strong cell walls (like plant and fungi cells) can withstand high internal osmotic pressures.

A cell immersed in pure water is said to be **hypertonic** (*heye puhr TAHN ihk*) with respect to the surrounding solution because the cell has a greater concentration of solutes. The surrounding solution, which has a lower concentration of solutes than the cell, is said to be **hypotonic** (*heye poh TAHN ihk*). Most of your body cells and the tissue fluid that circulates around them are said to be **isotonic** (*eye soh TAHN ihk*) because the concentration of solutes in the cells and fluid is the same. Figure 3-5 illustrates the effect of osmosis on animal cells.

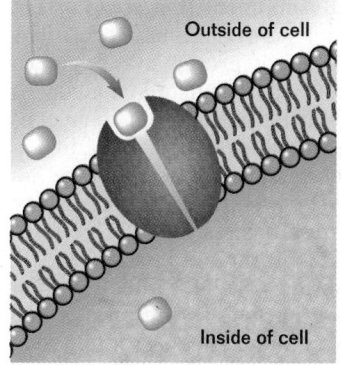
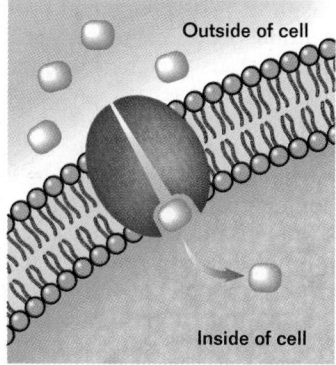

Figure 3-6 Facilitated diffusion is the transport of specific particles through a membrane by a channel protein. A molecule that is outside of the cell locks into the channel protein, *left*. The molecule is then transported through the channel to the inside of the cell, *right*.

 VISUAL STRATEGY

Figure 3-6
Use this figure to highlight the protein channels or pores of the cell membrane that are responsible for facilitated diffusion. Emphasize that the movement of particles is in two directions, although the net movement works to restore equilibrium.

Facilitated Diffusion

A cell is able to function because it is able to control what enters and leaves. Like a home, a cell has "doors with locks" that only certain particles can pass through. The doors of the eukaryotic cell are the protein channels in the plasma membrane. The structure of each protein channel is such that it can accept only particular particles, much as a lock is shaped to accept a particular key. The plasma membrane is said to be **selectively permeable** (*PUR mee uh buhl*) because it allows the passage of some solutes but not others. Selective permeability is one of the most important properties of the plasma membrane of every cell.

Many of the selective protein channels through the plasma membrane are "two-way" channels called pores. Pores transport molecules or ions through the membrane in either direction, much as an open door will allow you to pass into or out of a room. When a solute that fits a pore is more plentiful on one side of the membrane than the other, solute particles will pass by diffusion toward the side of lower concentration. Eventually, the concentration of the solute will become the same on both sides of the membrane. Different solute particles fit different pores. Their ability to fit depends on the size, polarity, and shape of the solute. Diffusion through selective pores is called **facilitated** (*fah SIHL uh tayt ehd*) **diffusion** and is shown in Figure 3-6. The particles move from higher to lower particle concentration. Like diffusion and osmosis, facilitated diffusion is a passive process, meaning that no energy is expended by the cell. Glucose is a sugar that moves into most cells by facilitated diffusion. ❏

CAPSULE SUMMARY

Diffusion, osmosis, and facilitated diffusion are particle transport processes that do not require an expenditure of energy by a cell. Of these three processes, only facilitated diffusion involves the movement of particles through membrane protein channels.

Active Processes Expend Energy

If facilitated diffusion were the only tool cells had to harvest particles from the environment, they would have serious problems. Many amino acids and sugars are even more scarce outside cells than inside. Facilitated diffusion of these particles would

Historical Note

The Cell's Internal Environment

Claude Bernard, a French physiologist, published his theories on the stability of the internal environment of living organisms in 1855. His concepts, now called homeostasis, have given rise to modern theories that explain the body's control mechanisms. American botanist Walter B. Cannon, building on Bernard's work, published six postulates governing homeostasis in the mid-1920s. Cannon actually coined the word *homeostasis*.

Concentration Gradients

Have students come up with analogies to explain what a concentration gradient is—for example, a skater moving through a crowded rink as opposed to an empty rink or a crowd of people gradually dispersing from a full movie theater.

The Importance of ATP

Emphasize the importance of energy at the cellular level: cellular processes such as chemical synthesis, nuclear division, and even movement require energy in the form of ATP. Emphasizing energy sets the stage for the energy studies in Chapter 4.

◉ VISUAL STRATEGY

Figure 3-8

Use this figure to point out the pathway of protons (H+) through the processes of active transport and diffusion. Emphasize that as the concentration of H+ builds up on one side of the cell, the diffusion of H+ back across the cell membrane produces ATP. At the same time, the active transport of protons requires energy derived from photosynthesis or food molecules.

◉ VISUAL STRATEGY

Figure 3-9

Use this figure to discuss the relationship between sodium ions, potassium ions, and glucose. Emphasize that three sodium ions leave the cell for every two potassium ions that are pumped into the cell, creating a higher concentration of sodium ions outside the cell than inside. This higher concentration of sodium ions outside the cell allows the diffusion of sodium ions, along with glucose molecules, back into the cell. Stress that one of the main purposes of the sodium-potassium pump is to create this higher concentration of sodium ions outside the cell.

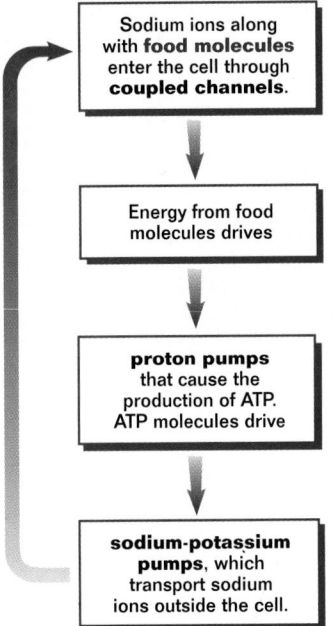

Figure 3-7 The interplay between coupled channels and two kinds of energy-driven pumps enables cells to maintain a steady inward flow of food molecules.

result in their pouring out of cells, an event that could affect cellular protein synthesis and energy production. Recall that amino acids are the building blocks of proteins, and sugars are a major energy source for cells. Cells must have a way to beat the diffusion game so that they can maintain their concentration of these important food molecules at a level different from the concentration level outside the cell. Figure 3-7 demonstrates how the interplay between several important transport mechanisms in cell membranes enables cells to maintain a high internal level of amino acids and sugars. **Proton pumps** cause the production of ATP molecules, the energy currency of the cell. **Sodium-potassium pumps** use some of this ATP to accumulate an abundance of sodium ions outside the cell. **Coupled channels** carry the sodium ions, along with food molecules, back inside the cell. Proton pumps and sodium-potassium pumps are highly specialized protein channels. As you will see, both use energy to transport ions (charged particles) against a concentration gradient (toward the side of higher particle concentration). Using energy to transport a particle through a membrane against a concentration gradient is called **active transport**.

Proton Pumps

A proton pump, illustrated in Figure 3-8, is one type of active transport channel. Proton pumps actively transport protons through the internal plasma membranes of mitochondria and chloroplasts. A proton is a hydrogen atom that is missing its electron. Proton pump channels are used to make ATP from ADP. ATP is the cell's key energy-storing molecule. This active transport of protons to make ATP is called **chemiosmosis** *(kehm ee ahz MOH sihs)*. All the energy harvested by plants in photosynthesis and practically all the energy you get from the food you eat is derived from chemiosmosis.

Figure 3-8 Inside a cell, proton pump channels actively transport protons through the internal membranes of chloroplasts and mitochondria. In a chloroplast, *right*, protons that build up on one side of the membrane diffuse back to the other side through protein channels by facilitated diffusion. These channels use the force of the proton pushing through to power the manufacture of ATP.

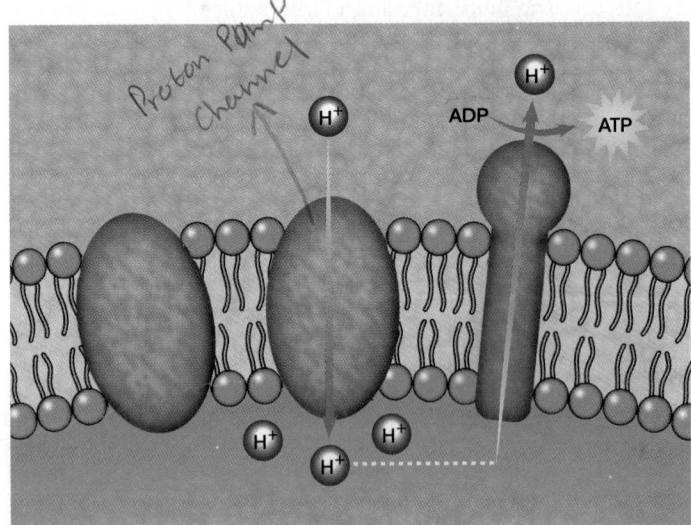

Sodium-Potassium Pumps and Coupled Channels

A second kind of active transport channel is called a sodium-potassium pump. A sodium-potassium pump uses energy stored in the form of ATP to power the active transport of sodium ions (Na^+) out through a cell's membrane. The action of the sodium-potassium pump is the most important energy-using process in your body. More than one-third of all the energy expended by a human cell that is not actively dividing is used to transport sodium ions in this way!

Why do your cells use so much of their energy pumping sodium and potassium ions? For one thing, nerve cells use the differences in sodium and potassium ion concentrations produced by sodium-potassium pumps to send signals throughout the body, like electrical signals passing over wires. The details of how nerve signals are created and transmitted will be discussed in Chapter 40. Sodium-potassium pumps also help to transport food particles into cells.

The transport of many food particles and a variety of other particles into your cells involves two different kinds of channels, as shown in Figure 3-9. One of the channels is the sodium-potassium pump. The other channels are coupled channels. In the first step, the active transport of sodium ions out of the cell by the sodium-potassium pump increases the sodium ion concentration outside the cell. Each channel is capable of transporting as many as 300 sodium ions per second when working full tilt. The fact that there are so many sodium ions outside the cell due to the action of the

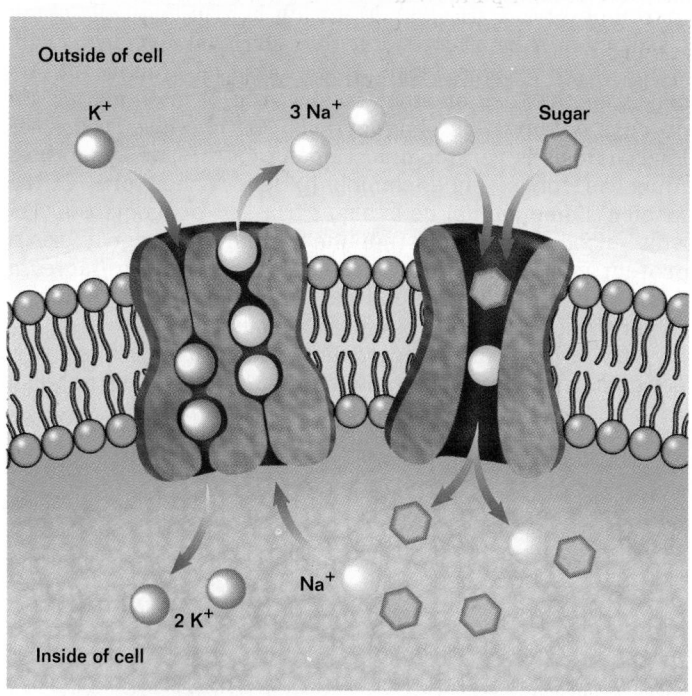

Figure 3-9 In a sodium-potassium pump, *left,* for every three sodium ions, Na^+, pumped out, the energy from a molecule of ATP is used, and two potassium ions, K^+, are pumped in. As a result of the pumping, most of your cells have a high concentration of sodium outside their membrane. As a sodium ion diffuses back into the cell through a coupled channel, *right,* it brings along with it a sugar molecule.

UPDATE

☑ RESEARCH Update

Life as a Consequence of Death

For an organism to survive, many of its cells have to die along the way during growth and development. This phenomenon is referred to as "programmed cell death." Researchers in several disciplines of biology have been aware of and are working on isolating the proteins that "turn on" or "turn off" cells. H. Robert Horvitz, a professor of biology at MIT, works on these small proteins in the soil nematode *Caenorhabditis elegans. C. elegans* is popular with genetic researchers because it has exactly 1,090 cells, so specific cells can be easily tracked. Horvitz has shown in his research that 131 of these cells die during development. Why? Horvitz and his colleagues have identified genes that create proteins which tell certain cells to die. This process is necessary for normal development, he contends.

By experimenting with worms, Horvitz has been able to show gross abnormalities when either an excess of the protein was produced or the protein was not produced at all. How does studying a microscopic roundworm help humans? Horvitz's group identified similar genes in the human genome after finding them in *C. elegans.* The ramifications of this are significant: drugs can now be found that will inhibit the uncontrolled growth of cells, which is cancer. Biotechnological and pharmaceutical companies are already hoping to make use of this information. ☑

Effective Reading

Using the Figures

Suggest that students take breaks from their reading to study the figures. As students read pages 61–64, they should use the corresponding diagrams to trace the events described in the text. Have students make a list of the mechanisms of transport featured in each figure by asking the following questions: Which figures in Section 3-2 show facilitated diffusion? (*Figures 3-6, 3-8, and 3-9*) Which figure shows a coupled channel? (*Figure 3-9*) Which figures show specialized pumps, and what are the names of the pumps? (*Figure 3-8, proton pump; Figure 3-9, sodium-potassium pump*) Which figure shows the use of membrane vesicles? (*Figure 3-10*) Grouping the diagrams will reinforce what the students have read.

UNIT 1
CHAPTER 3

Teaching Tips
Facilitated Diffusion Versus Pinocytosis

Ask students to explain the difference between facilitated diffusion and pinocytosis. *(Facilitated diffusion involves the use of surface membrane proteins in the cell membrane, while pinocytosis involves the formation of a pouch in the cell membrane.)*

Organizing the Information

Have students make a table of the transport mechanisms discussed in Section 3-2 and the materials that each mechanism transports. The table should include facilitated diffusion, transporting specific molecules such as glucose; active transport, transporting sodium and potassium ions; endocytosis, importing large amounts of fluids or nutrients; and exocytosis, exporting proteins or wastes.

👁 VISUAL STRATEGY

Figures 3-10

Use this figure to compare the directions of flow in endocytosis and exocytosis. Discuss the prefixes, connecting *exo-* to the word *exit* and *endo-* to the word *entry*.

□ *CAPSULE SUMMARY*

Proton pumps cause the production of ATP molecules in chloroplasts and mitochondria. Sodium-potassium pumps use ATP molecules to create a high concentration of sodium ions outside the cell. Many cells maintain a high internal concentration of food molecules by coupling the inward transport of food molecules with the inward migration of sodium ions.

Figure 3-10 Cells trap extracellular fluid and particles within membrane vesicles by endocytosis, *left.* Exocytosis is the dumping of excretions or waste materials outside a cell by discharging them from waste vacuoles that fuse with the plasma membrane, *right.*

sodium-potassium pump leads to the second step: sodium ions move back into the cell by means of coupled channels that also carry sugar molecules. The coupled channel actually contains two passageways through the membrane, both of which must be used for the channel to work. Imagine the two passageways as two turnstiles attached to the same gear. One passageway fits a particular molecule, such as a sugar molecule. The other fits sodium ions. Because there are so many sodium ions outside the cell due to the action of the sodium-potassium pump, sodium ions will diffuse back into the cell. The force of their entry is so great that it pulls sugar molecules into the cell too, even though the cell interior already has a generous supply of sugar. Just as it takes two hands to clap, so it takes two membrane channels to actively transport many food particles into the cells of your body. This two-step transport process is one of the most fundamental and important activities of any cell. □

Endocytosis and Exocytosis

Although, as you have read, individual food molecules are routinely imported through protein channels, other food particles are too large for that method of transport. In order to consume these larger meals, the cell literally engulfs the particle. This process, shown in Figure 3-10, is called **endocytosis** *(ehn doh seye TOH sihs)*. The cellular movements of endocytosis require an expenditure of energy by the cell. If the material brought into the cell is liquid and contains dissolved molecules, the endocytosis is referred to as **pinocytosis** *(peyen oh seye TOH sihs)*. Maturing human egg cells obtain dissolved nutrients secreted by the surrounding "nurse" cells of the ovary by pinocytosis. If the material brought into the cell is another cell or other fragment of organic matter, the process is known as **phagocytosis** *(fag oh seye TOH sihs)*. Phagocytosis is very common among unicellular eukaryotes. Amoebas, for example, commonly devour their prey in this fashion. Some animal cells also carry out phagocytosis. The white blood cells of your immune system use phagocytosis to protect you from infection, engulfing invading bacteria. Some types of white blood cells may take in as much as 25 percent of their cell volume each hour.

The reverse of endocytosis is exocytosis *(ek soh seye TOH sihs)*. **Exocytosis** is the dumping of excretions or waste

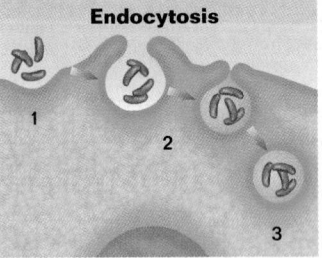
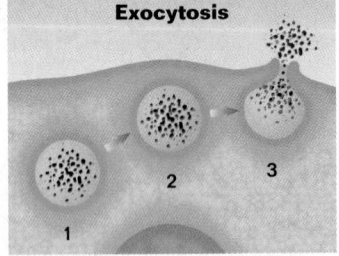

Graphic Organizer

Use the following graphic organizer with **Teaching Tip: Climate Controlled Homeostasis** on page 65.

(In warmer climates, "air conditioner" may be substituted for "furnace.")

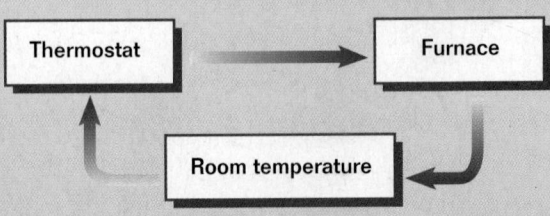

64 Chapter 3

materials outside a cell by discharging them from waste vacuoles that fuse with the plasma membrane. Exocytosis also occurs in cells that secrete important chemical products. Gland cells, for example, secrete hormones into the bloodstream by exocytosis.

THEMES OF BIOLOGY

Homeostasis

One of the major differences between a house and a tent is that a house does a better job of maintaining a constant environment by keeping out the cold, rain, and unwanted visitors. The same is true of your body and of every organism. On a cellular level, the plasma membrane of every cell determines what gets in and out. Cells use their membranes to maintain conditions inside that are far more constant than those on the outside. Without these constant internal conditions, many of the complex interactions that take place within cells would be impossible. Similarly, your body maintains consta... internal conditions. For example, it attempts to maintain a temperature of 37°C, no matter how cold or warm the weather might be. The levels of salt and glucose in your blood are maintained at nearly the same level at all times. The maintaining of constant internal conditions is called **homeostasis.** Control of internal conditions is an essential characteristic of all living things.

How do the emperor penguin and the athlete maintain constant internal body temperatures?

Section Review

1. *How does osmosis differ from diffusion?*
2. *If a living cell is placed in a hypertonic solution, in which direction will water molecules move?*
3. *Identify one similarity and two differences between facilitated diffusion and active transport.*
4. *Explain the role the sodium-potassium pump plays in the transport of glucose into some cells.*
5. *What is endocytosis?*

Critical Thinking

6. *A living cell is placed into a hypotonic solution. Compare the solute and solvent concentration inside the cell with the solute and solvent concentration in the surrounding solution.*

SECTION 3-2

Teaching Tips
Body Temperature
Have students explain the different body responses needed to maintain the internal temperatures of penguins and humans. Responses might be shivering, feather production, and a higher metabolic rate for the penguin in a cold climate, while humans might perspire, seek shade, and drink liquids on a warm day.

Climate Controlled Homeostasis
Explain that the heating or cooling system in the room is an example of a simple homeostasis system. The purpose of the system is to maintain the room temperature within reasonable limits in spite of changes in the external environment. The thermostat is an example of a "controller": when the temperature gets too low, the controller sends a signal to the furnace to begin heating the room. The heat brings the temperature up. The temperature acts on the thermostat, causing it to stop sending its signal to the furnace once the desired temperature is reached. Have students draw a flowchart like the *Graphic Organizer* on the bottom of page 64 to illustrate how the heating or cooling system maintains the temperature of the room.

Answers to Section Review

1. Osmosis is a special type of diffusion involving the passage of water through a selectively permeable membrane.

2. Water will move out of the cell because the amount of free water molecules inside will be greater than that on the outside.

3. Both involve the use of proteins to mediate movement. In facilitated diffusion, energy is not expended, and particles move toward the lower concentration of particles. Active transport requires energy and goes against a concentration gradient.

4. When sodium ions are pumped out of the cell, their concentration outside builds up so high that they diffuse back into the cell through coupled channels, which also carry glucose molecules.

5. Endocytosis is the engulfing of large particles by cells.

6. The solute is at a higher concentration and the solvent is at a lower concentration inside the cell than in the surrounding solution.

Vocabulary Preview

neurotransmitter

receptor

acetylcholine

gated channel

chemically gated

voltage-gated

Opening Questions

Ask students to name the three functions of cell surface proteins. Make sure students include channel proteins (serve as passageways), receptor proteins (receive information), and marker proteins (identify cells to other cells). Have students consider what keeps cell surface proteins from sinking into the cell. Students should recall that the middle region of the cell membrane is nonpolar. Lead students to conclude that this nonpolar region within the membrane keeps the (polar) proteins in place with molecular attractions.

Demonstration

Reviewing Membrane Structure

Display a diagram of a cell that shows the details of the cell membrane. Point out that the cell membrane separates two water environments. Ask students whether glucose and other molecules that the cell needs can cross the membrane passively. *(Yes, but the cell uses facilitated diffusion and active transport to transport these molecules into the cell quickly enough to meet its metabolic needs.)* Explain that this section will discuss gated channels, another means of passive transport, which the cell uses to communicate with other cells.

3-3 How Cells Communicate

Section Objectives

- Identify three ways cells communicate with each other.
- Describe how receptors can influence the cytoplasm of a cell.
- Distinguish between chemically gated and voltage-gated channels.

Bringing the "groceries" in and taking the "garbage" out are not the only things that cells do. Just as in your house you answer the phone and watch television, so the cells of your body communicate with the world outside. Your cells communicate with each other in order to coordinate the body's growth, development, and other activities.

Direct and Indirect Communication

Some of your cells are in physical contact with each other, as shown in Figure 3-11. However, most cells of your body communicate less directly. The two systems involved in indirect communication are the endocrine (ductless gland) system and the nervous system. The endocrine system communicates with body cells by using chemical signals called hormones. A nerve cell communicates with another nerve cell or a muscle cell with chemicals called **neurotransmitters** *(noo roh trans MIHT uhrs)*.

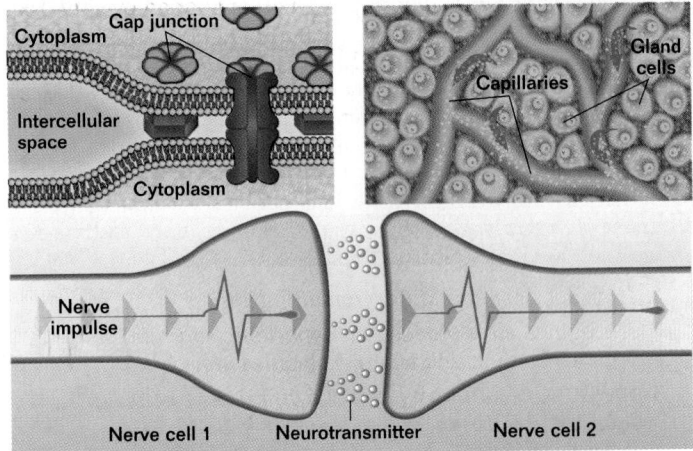

Figure 3-11 Cells communicate in several ways. Some cells have tiny openings called gap junctions between them that directly join their cytoplasms, enabling exchange of small particles, *top left*. Gland cells, *top right*, release hormones into the bloodstream. Hormones interact with specific cells elsewhere in the body. Communication between nerve cells, *bottom*, requires chemical communication across the short gap between adjacent cells. These chemicals are called neurotransmitters.

Receptors Receive Information

The key to understanding cell communication is to focus on the cell's communication "machinery," special proteins on the cell surface. Called **receptors**, these proteins are specialized antennae that signal the cytoplasm when a particular particle has just bumped into the cell surface. They are able to do this because each kind of receptor has a particular shape that fits only a particular signal particle. When that signal particle is encountered by the cell, it binds to the receptor specialized to fit it, like two pieces of a jigsaw puzzle.

The binding of a signal particle to its receptor can cause the receptor to influence the cytoplasm in one of three ways, as shown in Figure 3-12. First, the receptor can act as an enzyme, chemically changing molecules in the cytoplasm. Second, the receptor can cause the formation of another signal, called a second messenger. These second messengers will have an effect elsewhere in the cytoplasm. A modified form of ATP called cyclic AMP is a very common second messenger in eukaryotic cells. Third, the receptor can open a channel through the membrane. This is how neurotransmitters work.

Chemically Gated Channels

In order to better understand how receptors work, we will focus for a moment on a particular receptor, the receptor in muscle cells. The muscle receives the signal to contract from nerve cells. The chemical signal that the nerve releases into the gap between it and the muscle is a small neurotransmitter molecule called **acetylcholine** *(uh seet uhl KOH leen)*. This molecule binds to a receptor in the plasma membrane of the muscle cell, as shown in Figure 3-13. This acetylcholine (ACh) receptor contains a channel. Such a channel is called a **gated channel** because it may be opened or shut, like a gate in a fence.

The binding of acetylcholine to the receptor causes the pore to open up like the shutter of a camera. It doesn't stay open long, only a microsecond or so. That's how long the acetylcholine molecule stays bound to the receptor. During this brief period, sodium ions flood in, setting off the muscle contraction. The ACh receptor is said to be **chemically gated**, because its opening and closing depends on the binding of a chemical, the neurotransmitter acetylcholine. ❑

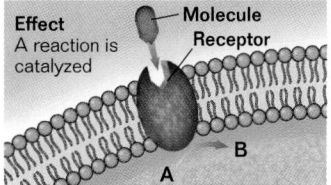
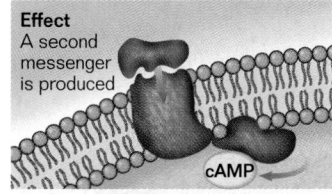
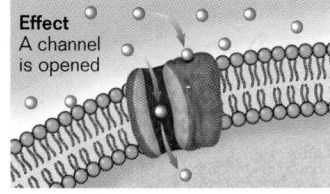

Figure 3-12 When the proper molecule locks into the membrane receptor of a cell, a change will occur within the membrane or cytoplasm of that cell. There may be chemical changes in cytoplasmic molecules, *top;* the production of cytoplasmic second messenger molecules, *center;* or the opening of channels within the plasma membrane, *bottom.*

❑ CAPSULE SUMMARY

The binding of an acetylcholine molecule to a receptor in the membrane of a muscle cell causes a sodium channel in the receptor to open. Sodium ions rush into the muscle cell, and the cell contracts.

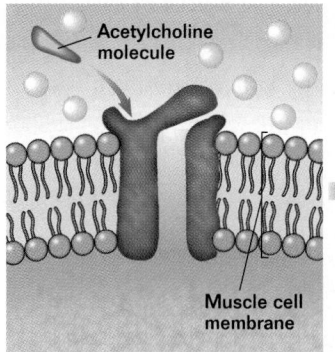

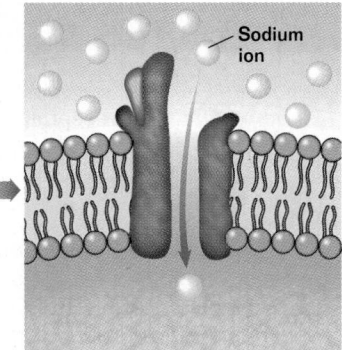

Closed channel   **Open channel**

Figure 3-13 An acetylcholine molecule released from a nerve cell binds to an acetylcholine receptor in the plasma membrane of the muscle cell. The receptor is a sodium channel that is closed; the binding of acetylcholine opens it.

SECTION 3-3

👁 VISUAL STRATEGY

Figure 3-12
Use this figure to help students identify membrane receptors, second messenger molecules, and channels in the membrane. Show how the messenger molecule interacts with the receptors. Discuss the relevance of the term *second messenger* in the center diagram. Help students to realize that the molecule that binds the membrane receptor is the first messenger, which initiates formation of the second messenger (cAMP, in this case).

Teaching Tips
Specialization of Communication Leads to Balance

Compare receptor-mediated communication among cells in a biological system to people living in a city. Just as each person's communication with others in the city is critical to his or her survival, receptor-mediated communication is critical to maintaining homeostasis. Have students come up with analogies. *(Answers might include the following: A person can use a credit card to gain access to money in an automated teller machine. Similarly, chemical messengers may open channels within the cell membrane, allowing important materials to enter or leave the cell. A letter put in a fax machine in one part of town can cause another fax machine across town to make a copy of the letter. Likewise, a messenger on the outside of the cell can cause the production of a second messenger inside the cell.)*

🔬 What If cellular communication is disrupted?

Disruption of cellular communication results in a breakdown of coordination between cells. For example, multiple sclerosis, a disability of young adults, results from a breakdown of myelin, an insulating sheath surrounding parts of nerve cells. Without myelin, nerve cells lose their ability to filter and respond to cellular messages.

UNIT 1

CHAPTER 3

 VISUAL STRATEGY

Figure 3-14
Use this figure to emphasize that as nerve cells communicate with one another, the voltage changes, opening and closing voltage-gated channels. In this figure, the signal to open the voltage-gated channel comes from within the cell. Compare this signal with the one shown in Figure 3-12. In Figure 3-12 the signal to open the channel comes from the binding of acetylcholine outside the cell. Why do the sodium ions flood into the cell? *(They diffuse into the cell because there is a larger concentration of sodium ions outside the cell than inside.)*

Chapter Closure
Have students draw an idealized cell, with its membrane drawn in detail. On the drawing, each student should illustrate three types of passive transport and one possible particle moved by each type. Additionally, three different sites on the drawing should be included to show different energy-dependent transport methods. Students should label the parts of the cell membrane and each type of transport illustrated.

 CAPSULE SUMMARY

Electrical charges along the membrane of a nerve cell cause a series of voltage-sensitive sodium channels in the membrane to open. The rush of sodium ions in through these channels creates the signal that is transmitted along the nerve cell.

Figure 3-14 When a voltage change occurs in the vicinity of a voltage-gated receptor in the plasma membrane of a nerve cell, the gate of this channel protein opens.

Voltage-Gated Channels

Not all gated channels are open and shut by chemicals. In nerve cells, another type of gated channel is found, as shown in Figure 3-14. Nerve cells are specialized to transmit electrical signals. The signals travel down long cytoplasmic extensions as electrical disturbances on the surface of the plasma membrane. What makes the signal move along the membrane? The nerve signal moves by opening a series of gated sodium channels one after another, like a line of falling dominoes. In this case, however, the "open" signal is electrical rather than chemical. These sodium channels are said to be **voltage-gated**. Like the chemically gated ACh receptor, voltage-gated sodium channels are composed of proteins that weave in and out of the membrane.

How does voltage act to open the channel? Some of the amino acids that make up the channel have ionized and become positively charged. This makes these charged areas of the channel very sensitive to voltage. When the immediate surroundings of the channel experience a change in voltage, the charged areas rotate out of the center of the channel, opening it to enable the passage of sodium ions. ☐

Normal state
(closed channel)

Nerve signal formation
(open channel)

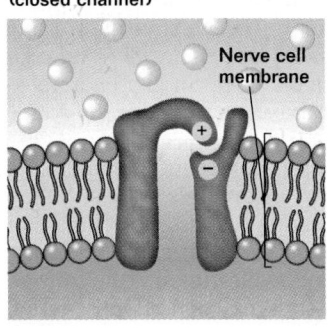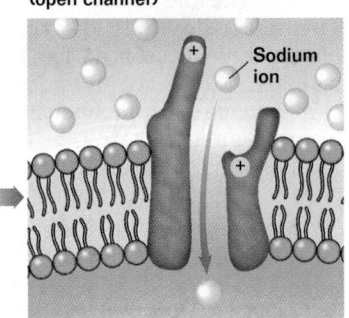

Section Review

1. *Compare and contrast hormones with neurotransmitters.*
2. *Define the term* receptor.
3. *What is acetylcholine, and what is its effect on the plasma membrane of a muscle cell?*

Critical Thinking

4. *Adrenaline is a hormone that binds with liver cells and causes them to release glucose by way of a cyclic AMP mechanism. Draw and label a cell diagram that demonstrates the action of an adrenaline molecule on a liver cell.*

Answers to Section Review

1. Hormones are chemicals that are produced in glands and whose sites of action are remote from the cells producing them. They are delivered to their targets by the bloodstream. Neurotransmitters are chemicals that are produced in nerve cells and whose sites of action are a muscle or another nerve just a short distance away. Neurotransmitters diffuse through intercellular spaces to reach their targets.

2. A receptor is a special protein on a cell's surface with a shape compatible to a specific "message" particle.

3. Acetylcholine is a neurotransmitter whose function is to open the sodium channels in the plasma memebrane of a muscle cell.

4. When adrenaline binds a liver molecule, cAMP is released, the channel opens, and glucose passes through the channel to the bloodstream.

3

CHAPTER REVIEW

CHAPTER REVIEW ANSWERS

Each item in the Chapter Review is correlated by text section in the assignment guide that follows.

Vocabulary

acetylcholine (67)
active transport (62)
cell surface marker (57)
cell surface protein (56)
channel (57)
chemically gated (67)
chemiosmosis (62)
coupled channel (62)
diffusion (58)
endocytosis (64)
equilibrium (58)
exocytosis (64)
facilitated diffusion (61)

gated channel (67)
hormone (57)
hypertonic (60)
hypotonic (60)
isotonic (60)
lipid bilayer (55)
neurotransmitter (66)
osmosis (60)
osmotic pressure (60)
passive transport (58)
phagocytosis (64)
phospholipid (55)
pinocytosis (64)

proton pumps (62)
receptor (66)
receptor protein (57)
selectively permeable (61)
sodium-potassium pump (62)
solute (58)
solution (58)
solvent (58)
voltage-gated (68)

ASSIGNMENT GUIDE

Section	Review	Themes Review	Critical Thinking
3-1	1, 2, 11, 18, 19	23	
3-2	3–6, 10, 12, 15, 16, 17, 19–21	22, 23	24, 25
3-3	7–9, 13, 14		

Concept Mapping

Construct a concept map that shows how cells maintain homeostasis. Use as many terms as needed from the vocabulary list. Try to include the following terms in your map: passive process, energy, concentration gradient, active process, diffusion, osmosis, facilitated diffusion, proton pump, sodium-potassium pump, endocytosis, and exocytosis. Include additional concepts in your map as needed.

3. If frog eggs are placed in a saltwater aquarium,
   a. water will move from inside the eggs to outside the eggs.
   b. water will move from outside the eggs to inside the eggs.
   c. the water balance will be maintained.
   d. the eggs will burst.

4. Facilitated diffusion in cells differs from ordinary diffusion in
   a. its use of energy.
   b. the direction of particle movement.
   c. the use of selective pores.
   d. the conversion of ATP to ADP.

5. The secretion of thyroid-stimulating hormone by cells of the pituitary gland is an example of
   a. pinocytosis.   **c.** chemiosmosis.
   b. diffusion.   **d.** exocytosis.

6. A two-step process involving the sodium-potassium pump and facilitated diffusion of sodium into the cell is required for
   a. gated channels to operate.
   b. glucose to enter cells.
   c. the functioning of cell surface markers.
   d. phagocytosis.

Concept Mapping

Map answer is shown on page 53B.

Review

Multiple Choice

1. Cell membrane functions such as serving as receptor sites and surface markers are accomplished by
   a. phospholipid molecules.
   b. ions.
   c. proteins.
   d. lipids.

2. The rate of diffusion of ink in a container of water can be increased by
   a. increasing the volume of the water.
   b. increasing the concentration of the ink.
   c. decreasing the permeability of the ink.
   d. decreasing the mass of the water.

Review
Multiple Choice
Code in parentheses indicates section number and objective number.
1. c (3-1.2)
2. b (3-1.2)
3. a (3-2.3)
4. c (3-2.4)
5. d (3-2.7)
6. b (3-2.6)
7. d (3-3.1)
8. a (3-3.2)
9. c (3-3.3)
10. b (3-2.7)

Completion
11. phospholipid (3-1.1)
12. produced, used (3-2.6)
13. hormone (3-3.1)
14. voltage, chemically (3-3.3)
15. active (3-2.5)
16. phagocytosis (3-2.7)
17. plasma membrane (3-2)

Short Answer

18. Both channel proteins and receptor proteins are cell surface proteins that enable the cell to communicate with the outside. Channel proteins form openings in the membrane bilayer through which certain particles pass. Receptor proteins have ends that protrude into and out of the cell. Molecules that attach to the outside end of the receptor protein bring about changes to the end that extends into the cell. These changes, in turn, cause changes in the cell. (3-1.2, 3-1.3)

19. The rubber balloon is analogous to the plasma membrane because the membrane is selectively permeable. Insects, dirt, etc., cannot enter the balloon, but air can escape. (3-1.1, 3-2.1)

20. Both require energy for the movement of water or particles. Additionally, the water or particles are moved against the concentration gradient, from an area of lesser concentration to an area of greater concentration. (3-2.5)

21. Putting salt on the pork chop will result in a hypertonic solution on the pork chop's surface. This will cause water and other liquids in the cells of the meat to be drawn to the surface of the pork chop and turned into vapors by the heat. (3-2.2)

Themes Review

22. The action of the sodium-potassium pump uses energy to transport sodium ions out of and potassium ions into most cells. The differences in sodium and potassium ion concentrations enable nerve cells to function. When the sodium-potassium pump is coupled with a second channel, it can affect the transport of sugars into cells. Nerve action and proper levels of sugar in the blood are necessary for maintaining constant conditions inside the human body. (3-2.6)

23. A cell with a nonpermeable membrane would be internally inactive because particles could not enter or leave the cell. Cells with selectively

---

7. Researchers observed that electrically charged ions and fluorescent-tagged tracer molecules of different sizes moved from cell to cell without appearing outside the cells. These observations support the existence of
a. receptors.
b. neurotransmitters.
c. ductless glands.
d. gap junctions.

8. Production of cytoplasmic second messengers and opening of plasma membrane channels are possible results of
a. signals received by membrane receptors.
b. the actions of amino acids.
c. the actions of sugar molecules.
d. the opening of gated channels.

9. Signals for muscle contraction are sent from nerve cells by way of
a. voltage-gated receptors.
b. peptide hormones.
c. chemically gated channels.
d. phospholipid molecules.

10. A white blood cell engulfs, digests, and destroys an invading bacterium through the process of
a. osmosis.   c. pinocytosis.
b. phagocytosis.   d. facilitated diffusion.

Completion

11. The _____ molecules in the plasma membrane arrange themselves so that the carbon chain ends of the molecules point toward each other and the phosphate ends point away from each other.

12. ATP molecules are _____ by proton pumps, and ATP molecules are _____ by sodium-potassium pumps.

13. A cheerleader experiences a burst of energy just before running onto the field. It is likely that the cheerleader's liver received a chemical signal called a _____ from the adrenal gland causing it to release glucose.

---

14. The sodium channels involved in the movement of nerve impulses are _____ gated channels, and the acetylcholine receptors associated with muscle contractions are _____ gated channels.

15. When a cell needs to accumulate particles in greater concentration than is found outside the plasma membrane, _____ transport is likely.

16. Particles too big to move through protein channels are able to enter the cell by _____.

17. The part of the cell that functions to maintain homeostasis relative to the cell's environment is the _____.

Short Answer

18. How are the functions of channel proteins and receptor proteins alike and different?

19. Inflated Mylar balloons and rubber balloons can be purchased at most any party store. Mylar balloons won't deflate unless punctured, but rubber balloons deflate within days without being punctured. Which balloon is analogous to a plasma membrane? Explain your answer.

20. Most boats of any size are equipped with a bilge pump to remove water that has leaked into the boat. Explain how the movement of water by a bilge pump is similar to the transport of particles through a membrane by active transport.

21. Use your understanding of osmosis to describe why putting salt on a pork chop before cooking it on a grill is likely to result in a dry and tough piece of meat.

22. Homeostasis A sodium-potassium pump is one kind of active transport channel. Describe how the function performed by this transport channel contributes to homeostasis in a human.

---

permeable membranes have an adaptive advantage over those with a nonpermeable membrane. (3-1.2, 3-2)

Critical Thinking

24. Ammonium hydroxide molecules are small and are passively diffused through the membrane. Since ammonium

hydroxide is a base, the reaction of ammonium hydroxide with phenolphthalein causes the pink color. (3-2.1)

25. A significant discrepancy can be observed between the distance that the solution traveled over time. Line A shows that the solution traveled only 10 mm over a 30-minute period. Line B shows that the average distance

traveled by the solution in 30 minutes is almost 45 mm. This discrepancy may be due to the first group's use of a sucrose solution that was different from the one used by the other groups (perhaps a less concentrated one) or use of a membrane that differs in permeability from that used by other groups. (3-2.1)

23. Evolution Explain why there are no cells with non-permeable membranes.

Critical Thinking

24. Making Inferences A gelatin block is prepared with some phenolphthalein solution added to it. (Phenolphthalein is an acid-base indicator that is clear in the presence of an acid, but turns pink in a base.) The block is enclosed in a membrane. The membrane-enclosed block is then suspended 25 cm above a beaker containing ammonium hydroxide. After half an hour, the block begins to turn pink. Account for the gelatin's pink color.

25. Evaluating Data An apparatus called an osmometer was used to determine the rate of osmosis during an experiment. Line A on the graph reflects the data collected by one student group. Line B reflects the class average, which was calculated using the data collected by the first student group and seven other groups. How do the data collected by the first group compare with the class average? What might account for any discrepancies identified between the two data sets?

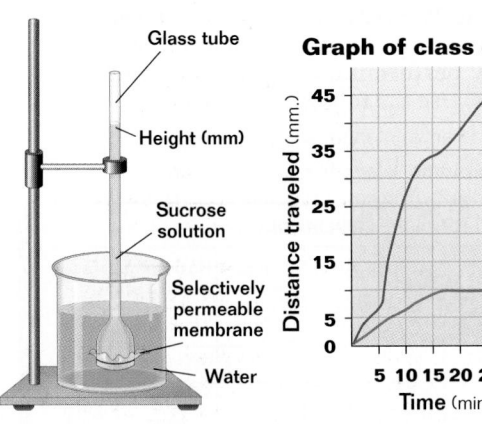

Glass tube
Height (mm)
Sucrose solution
Selectively permeable membrane
Water

Graph of class data

Activities & Projects

26. Multicultural Perspective Research the attitudes of Japanese people toward eating the meat of the Japanese puffer fish. Find out about the relationship between the tetrodotoxin sometimes present in the meat of the puffer fish and human fatalities.

27. History Find out about the model of the plasma membrane proposed by cell biologist J. D. Robertson in the 1950s. How does it compare with the "fluid mosaic model" described in this chapter and proposed by Singer and Nicholson in 1972?

28. Research and Writing Find out how the membrane of a bacterium compares with the membrane of a eukaryotic cell. In your report, include labeled cross-sectional sketches showing the similarities and differences between the two membrane types.

Readings

29. Read "Toward a Future With Memory: Researchers look high and low for the essence of Alzheimer's," in *Science News*, February 24, 1990, pages 120–123. Why are scientists studying the cell membranes of brain cells to discover the cause of Alzheimer's disease?

30. Read "Fiddling With Salt Intake," in *Bioscience,* June 1992, page 411. What is the scientific evidence that suggests diet may affect the development of sodium channels?

lipid metabolites (phosphomo-noesters, or PMEs) that accumulate in fetuses before extraneous nerve cells are destroyed. Other scientists think that a structural defect in cell membranes may enable a cell membrane protein (amyloid precursor protein, or APP) to break away and accumulate elsewhere in the bodies of Alzheimer's patients, thus causing the disease.

30. Scientists found that they could modify one kind of sodium channel by altering sodium intake in rat pups. When the salt intake of rat pups was restricted from the time of fetal development until 4 weeks of age, signals from sodium channels in the tongue were only 40 percent as responsive as those of rat pups fed a normal diet.

Activities and Projects

26. Many Japanese people consider the puffer fish a delicacy. Tetrodotoxin irreversibly binds with a receptor protein, which results in muscle paralysis and death.

27. Robertson's earlier "unit membrane model" is a three-layer structure, with two dense layers sandwiching a less-dense layer. It was thought that the dense layers were protein and the less dense layer was a double layer of lipids.

28. A bacterium has an inner cytoplasmic membrane, an intervening compartment called periplasm, and an outer membrane. Both the inner and outer membranes are phospholipid bilayers with proteins.

Readings

29. Some scientists think that Alzheimer's disease may result from membrane lipid abnormalities because Alzheimer's patients have higher than normal levels of the same

Preparation

Phenolphthalein agar may be obtained from a biological supply company or prepared in the laboratory. (Appropriate safety precautions must be taken.) Add drops of 0.1 M sodium hydroxide solution (prepared using 4 g NaOH diluted to 1 L) to turn the agar red. Pour the mixture into a flat pan to a depth of slightly more than 3 cm. After the agar hardens, cut it into 3 cm × 3 cm × 6 cm rectangles. Remind students to wear their safety goggles and lab aprons throughout the activity.

Procedural Notes

- Before students cut into their agar blocks, have them make a diagram showing how they plan to cut the block into cubes. Check students' diagrams before handing out the knives.
- Caution students to use their knife carefully to avoid injury.
- You may wish to use a timer to indicate when students should place their cubes in the vinegar and when they should remove them.

Background Answers

1. Factors that affect the rate of diffusion across a cell membrane include the concentrations of the substances on either side of the membrane.

2. The larger the ratio of surface area to volume, the more quickly diffusion occurs throughout a cell.

3. What are the relationships among rate of diffusion, cell size, and surface-area-to-volume ratio?

Records

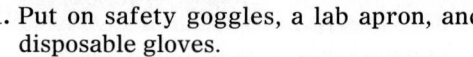

LABORATORY Investigation Chapter 3

Diffusion and Cell Size

OBJECTIVES

- Calculate ratios of surface area to volume.
- Recognize the relationship between rate of diffusion and cell size.

PROCESS SKILLS

- analyzing models
- predicting outcomes
- measuring rates of diffusion
- applying mathematical formulas

MATERIALS

- safety goggles
- lab apron
- disposable gloves
- block of phenolphthalein agar (3 cm × 3 cm × 6 cm)
- plastic knife
- metric ruler
- 250 mL beaker
- 150 mL of vinegar
- plastic spoon
- paper towel

BACKGROUND

1. What factors affect the rate of diffusion across a cell membrane?
2. What is the relationship between surface area, volume, and diffusion in cells?
3. Write your own question to explore in your lab report or notebook.

TECHNIQUE

1. Put on safety goggles, a lab apron, and disposable gloves.
2. **CAUTION: Use the knife carefully to avoid injury.** Trim the agar block with your knife to make three cubes: 3 cm, 2 cm, and 1 cm on a side, respectively. Each cube will represent a cell.

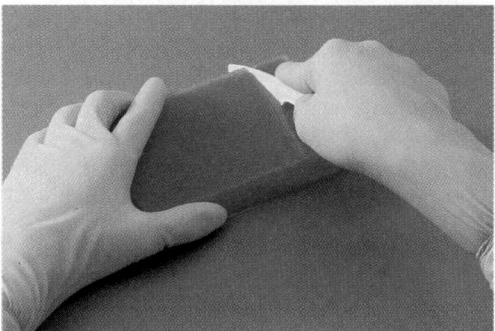

3. Place the three cubes in the beaker, and add vinegar until the cubes are submerged. Keep the cubes submerged 10 minutes, using the plastic spoon to turn the cubes frequently. Be careful not to scratch the surfaces of the cubes.
4. In the **Records** section of your report, make a table similar to Table 1. Complete it by performing the necessary calculations. Use the following formulas:

Surface area = length × width × number of sides

Volume = length × width × height

Table 1 Agar Cube Comparisons

Cube Dimension	Surface Area (cm²)	Volume (cm³)	Ratio of Surface Area/Volume*
3 cm			
2 cm			
1 cm			

*(reduce to simplest fraction)

Table 1 Agar Cube Comparisons

Cube Dimension (cm)	Surface Area (cm²)	Volume (cm³)	Ratio of Surface Area/Volume*
3	3 × 3 × 6 = 54	3 × 3 × 3 = 27	$\frac{54}{27}$ = 2:1
2	2 × 2 × 6 = 24	2 × 2 × 2 = 8	$\frac{24}{8}$ = 3:1
1	1 × 1 × 6 = 6	1 × 1 × 1 = 1	$\frac{6}{1}$ = 6:1

*(reduce to simplest fraction)

5. After 10 minutes, use the spoon to remove the agar cubes, and then blot them dry on a paper towel.

6. Cut the first cube in half. Measure the distance of diffusion of the vinegar in millimeters. Record the distance in a table similar to Table 2 below in the **Records** section of your report. A color change from red or pink to clear indicates diffusion. Be sure to thoroughly rinse and dry the knife before the next cut.

Table 2 Rate of Diffusion

Cube Dimension	Depth of Diffusion (mm)	Time (min.)	Rate (mm/min.)
3 cm			
2 cm			
1 cm			

7. Repeat step 6 with the other two cubes.

8. Record the total time in minutes in Table 2. Calculate the rate of diffusion as millimeters per minute, and record. In your report, briefly summarize the procedure you followed.

 9. Clean up your materials and wash your hands before leaving the lab.

INQUIRY

1. Why is it important not to scratch the surface of a cube when it is submerged in vinegar?

2. Propose a reason for rinsing and drying the knife thoroughly each time it is used to cut the cubes in half.

3. Cite evidence that vinegar diffuses into an agar cube.

4. List the agar cubes in order of size from largest to smallest. Then list them in order, from largest to smallest, according to their ratios of surface area to volume. How do the lists compare?

5. Compare the surface-area-to-volume ratio with the rate of diffusion for each cube.

6. Now compare the the surface-area-to-volume ratio with the extent of diffusion for each cube.

ANALYSIS

1. The size of some living cells is 0.01 cm. Using the formulas in the lab, calculate the surface area and volume of a cell that is 0.01 cm. Which cell has the greater surface area in proportion to its volume, the living cell or the largest agar block "cell"?

2. As cells increase in size, what happens to the surface-area-to-volume ratio?

3. Write a generalized statement about the relationship between the extent of diffusion and cell size.

4. In what ways do your agar models simplify or ignore the features of real cells?

FURTHER INQUIRY

Write a new question that could be explored as a new investigation. The following is an example:

How have the cell membranes of some cells of living organisms become modified to be more efficient?

Inquiry Answers

1. Scratching the surface of the agar changes the surface area of the cube and will provide misleading results.

2. Cleaning the knife after each cut prevents vinegar from the previous cube from contaminating the next cube.

3. The agar changes from red to clear where the vinegar diffuses into the cube.

4. Order of size—3 cm cube, 2 cm cube, 1 cm cube; order of surface-area-to-volume ratio—1 cm cube, 2 cm cube, 3 cm cube. The largest cube has the smallest surface-area-to-volume ratio, and the smallest cube has the largest surface-area-to-volume ratio.

5. As the size of the cube increases, the ratio of surface area to volume decreases. The rate of diffusion is constant.

6. As the size of the cube increases, the extent of the diffusion into the cube decreases.

Analysis Answers

1. Have students assume that the living cell has six sides in their calculations. The surface area of a 0.01 cm cell is 0.0006 cm^2. Its volume is 0.000001 cm^3, and its surface-area-to-volume ratio is 600:1. The surface area of the 3 cm cube is 54 cm^2. The volume of a 3 cm cube is 27 cm^3. The ratio of surface area to volume of a 3 cm cube is 2:1. A cell has a greater surface area in proportion to its volume than does a 3 cm cube.

2. As a cell increases in size, its surface-area-to-volume ratio decreases.

3. The extent of diffusion is greater in small cells.

4. Answers will vary but may include that the agar models ignore the selective permeability of cell membranes, the role of channels in facilitated diffusion and active transport, the internal structure of the cell, and the cell's mechanisms for exporting and importing substances based on its needs.

Table 2 Rate of Diffusion

Cube Dimension (cm)	Depth of Diffusion (mm)	Time (min.)	Rate (mm/min.)
3	should be same distance for each cube	10	should be the same rate for each cube
2		10	
1		10	

UNIT 1

ENERGY AND METABOLISM

Block Scheduling Guide

Each block represents about 45–50 minutes of instructional time. The resources cited in a block are coded to a specific program component using the Key on the next page.

Lecture Concepts	Lesson Resources

4-1 Energy and Living Things pp. 74–79

BLOCKS 1 and 2

a. Energy is the ability to cause matter to move or change. There are many forms of energy: mechanical, light, sound, thermal, and electric. Organisms use energy to carry out the many tasks of living.

b. Kinetic energy is the energy an object has because of its motion. Potential energy is the energy an object has because of its position.

c. All energy conversions are governed by the laws of thermodynamics. The first law of thermodynamics states that in a closed system, the total amount of energy is constant. The second law of thermodynamics says that in a closed system, energy tends to be converted to less organized, more stable forms.

d. All organisms need a constant supply of energy for their activities. Directly or indirectly, almost all of the energy for living systems comes from the sun.

e. The process that converts sunlight to chemical energy is called photosynthesis. Chemosynthesis is a process by which organisms convert inorganic chemical energy to organic chemical energy.

f. Organisms that harvest either light or chemical energy are called autotrophs. Heterotrophs obtain energy from other organisms.

g. Cellular respiration is a series of chemical reactions in which energy stored in food is converted to more useful form.

h. Chemical reactions that pass electrons from one atom or molecule to another are oxidation-reduction reactions.

Lecture Resources
- Opening Demonstration, page 74
- Opening Question, page 75
- Demonstration: Forms of Energy
- Potential versus Kinetic Energy, page 76
- Entropy, page 76
- Overcoming Misconceptions, page 77
- Visual Strategies: Figures 4-3, 4-4
- "Boosting" Electrons, page 77
- Connections, page 78

- Research Update, page 79
- Holt Biology Videodiscs: Lesson 4-1

Classwork Options
- Energy Transformations and Graphic Organizer, page 76
- Interactive Explorations in Biology Laboratory Manual E3: Thermodynamics

Assignment Options
- Teaching Resources CD-ROM: Basic Skills Worksheets, Reading a Thermometer and Temperature Conversions
- Section Review, page 79
- Chapter Review, questions 1–3, 11–14, 24, 25, 30, 31, 33, 34

4-2 Energy and Chemical Reactions pp. 80–84

BLOCKS 3 and 4

a. Chemical reactions produce new substances by breaking or forming chemical bonds between atoms.

b. The energy from chemical reactions that drives cell activities is called free energy. Exergonic reactions release free energy into their surroundings. Endergonic reactions absorb free energy from their surroundings.

c. Chemical reactions require an input of energy, called activation energy, to get started. Catalysts lower the amount of activation energy needed to start a chemical reaction.

d. Enzymes are catalysts that speed chemical reactions in cells. Enzymes have specific shapes that fit only certain substances.

Lecture Resources
- Opening Questions, page 80
- Demonstrations: Chemical and Physical Changes, Catalysis
- Teaching Transparencies: 8A, 9A
- Endergonic versus Exergonic, page 81
- ATP, page 81
- Visual Strategy: Figure 4-8
- Effective Reading, page 82
- Health, page 84
- Connections, page 84
- Holt Biology Videodiscs: Lesson 4-2

Classwork Options
- Laboratory Techniques and Experimental Design C5: Observing the Effect of Concentration on Enzyme Activity
- Laboratory Investigation: Observing Enzyme Activity, pages 92–93
- Laboratory Techniques and Experimental Design C6: Observing the Effect of Temperature on Enzyme Activity

Assignment Options
- Section Review, page 84
- Chapter Review, questions 4–6, 15–18, 26, 27, 32, 35

PROGRAM RESOURCE KEY

■ Annotated Teacher's Edition
■ Pupil's Edition
■ Study Guide
■ Chapter Tests

Holt BioSources
▣ Teaching Transparencies
◉ Teaching Resources CD-ROM
◉ Test Generator/Assessment Item Listing

▲ Laboratory Program
A Quick Labs
B Inquiry Skills Development
C Laboratory Techniques and Experimental Design
D Biotechnology
E Interactive Explorations in Biology Laboratory Manual

Holt Biology Technology
◉ Interactive Explorations in Biology CD-ROM
◉ Holt Biology Videodiscs
◉ Concepts of Biology

For complete descriptions of all components, see the introductory material at the front of this book.

4-3 Metabolism pp. 85–88

BLOCKS ⑤ and ⑥

a. Metabolism is the sum of chemical reactions within cells.
b. An ordered series of enzyme-catalyzed reactions that forms a product in a step-by-step manner is called a biochemical pathway. The product of each reaction in a pathway becomes the reactant of the next reaction in the pathway.
c. Enzymes that catalyze endergonic reactions also have an active site that splits ATP to release energy.
d. In many metabolic pathways, energy released from one set of reactions is transferred to another set of reactions by organic molecules called coenzymes.
e. Metabolism is regulated by controlling the kinds of enzymes present in a cell, their concentration, and their activity.

Lecture Resources
■ Opening Questions, page 85
■ Demonstration: Cells Release Energy Gradually
■ Connections, page 86
■ Overcoming Misconceptions, page 86
■ Visual Strategies: Figures 4-13, 4-14, 4-15
■ Coenzymes, page 87
■ Anabolism and Catabolism, page 87
◉ Holt Biology Videodiscs: Lesson 4-3

Classwork Options
▲ Laboratory Techniques and Experimental Design C7: Measuring the Release of Energy from Sucrose
▲ Laboratory Techniques and Experimental Design C8: Measuring the Release of Energy—Best Food for Yeast
■ Closure, page 88

Assignment Options
■ Section Review, page 88
■ Chapter Review, questions 7–10, 19–23, 29, 30, 33, 36

Review/Enrichment

BLOCK ⑦

■ Themes of Biology: Flow of Energy, page 79
■ Themes Review, questions 30–32
■ Study Guide: Concept Review and Pretest

Assignment Options
■ Activities and Projects, questions 37–40
■ Readings, questions 41, 42

Assessment Options

BLOCK ⑧

Traditional Assessment
■ Chapter 4 Test
◉ Test Generator/Assessment Item Listing: Software item bank for preparing customized chapter tests.

Portfolio Assessment
Select student reports for one or more laboratory experiments from this chapter. *The Direct Observation Checklist* on the *BioSources Teaching Resources CD-ROM* should be completed during a laboratory or other cooperative learning experience.

Answer to Concept Map

The following is one possible answer to the Concept Mapping exercise on page 89.

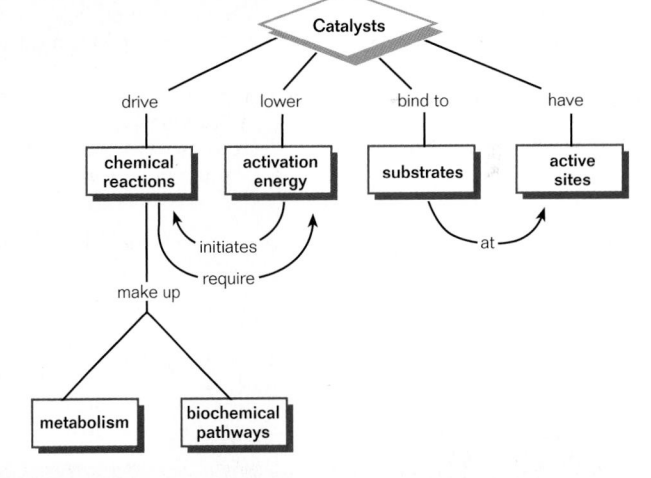

Holt Biology Videodiscs

Holt Biology Videodiscs gives you a powerful tool for teaching, review, and assessment. *Concepts of Biology* adds interactive lesson plans and extensive tools for customization using CD-ROM technology.

CONCEPTS OF BIOLOGY

Use the following topics from *Concepts of Biology* to help you teach this chapter:

■ Topic 4: Energy and Enzymes
■ Topic 10: Ecosystem Structure and Function

For further information regarding the media on the videodiscs, see *Holt Biology Videodiscs Teacher's Correlation Guide to Biology: Principles and Explorations.*

Chapter Theme

Flow of Energy *The total amount of energy in the universe is constant. Energy flows throughout the universe by being stored and released in different forms. Autotrophs absorb solar energy, which is then passed on to heterotrophs as chemical energy in the form of food. Thus, a food chain represents energy flow among organisms. In this chapter, students will become acquainted with some of the processes of storing and releasing energy within and among living systems.*

Tapping Prior Knowledge

- What is matter?
- What are the building blocks of matter?
- What is the major source of energy for all cells?
- What are some uses of energy within the cell?

Opening Demonstration

Set up a meter stick so that it can swing freely from a hole in one end. List the terms *kinetic energy* and *potential energy* on the board or overhead projector. Swing the meter stick, pointing out that this is an illustration of kinetic energy. Catch the stick in midswing, and explain that by catching the stick you have trapped energy in the system. This type of energy is called potential energy. As you release the stick again, point out that kinetic and potential energy are often interconvertible, i.e., they can be transformed into each other.

CHAPTER
4
ENERGY AND METABOLISM

Kestrel seeking food.

REVIEW

- chemical bonds (Section 2-2)
- carbohydrates, lipids, proteins, and nucleic acids (Section 2-3)
- structure of proteins (Section 2-3)
- structure of ATP (Section 2-3)

Authors' Rationale

No discussion of life would be complete without also discussing energy. Energy is required to drive all of life's processes, and energy is involved in how organisms both interpret and respond to their environments. This chapter first discusses the nature of energy—what it is, the forms it takes, and how it behaves. The concept of energy is critical to understanding chemical equations and metabolism. Students should understand that things require energy to work.

4-1 Energy and Living Things

Energy may sound like a topic that belongs in a physics course, but it is vital to living things. Organisms use energy to carry out the many tasks of living. The life processes of every cell—growth, reproduction, movement, and transport of molecules and ions across cell membranes—are all driven by energy. Plants make life on Earth possible by capturing energy from the sun. The capacity to use energy for life processes, as the insect in Figure 4-1 is doing, keeps us alive. In fact, biologists often view life as a constant flow of energy channeled by organisms to do the work of living.

Figure 4-1 This click beetle, *Pyrophorus noctilucus*, produces light by a chemical process that requires energy. Bioluminescence, an important form of communication that helps this animal locate a mate, is only one of the many ways in which organisms expend energy.

Energy Takes Many Forms

Energy is a quantity that is quite familiar to us all. You know that it is found in the motion of a falling boulder, in the sound of a guitar, in the blast of an explosion, and in the welcome warmth of a blazing fire. The food you eat also contains energy. But what, exactly, is energy? Although difficult to define, **energy** can be most easily understood as the ability to cause matter to move or change. Kick a football, for example, and the energy of your kick makes the football move from one place to another. Fry an egg, and the clear, liquid egg white changes color and solidifies as energy causes a change in the arrangement of its atoms and molecules. Because there are so many different ways that matter can move and change, energy appears to exist in many different forms. These forms include mechanical energy, light energy, sound energy, thermal energy (which is related to heat), and electric energy.

Teaching Tips

Potential Versus Kinetic Energy

Have students provide examples of potential and kinetic energy in their homes. *(potential—heating fuel, electrical outlets, food; kinetic—food processor blending, lights turned on, water heater heating water)*

Energy Transformations

Have students draw a graphic organizer of energy transformations. They should list at least two transformations in their diagrams. A sample *Graphic Organizer* is shown at the bottom of this page. Ask students to include a paragraph tracing the flow of energy they have charted.

Entropy

Explain that entropy is related to that portion of the total amount of energy in a system that cannot do useful work. The second law of thermodynamics implies that as the amount of usable (free) energy decreases in a system the amount of unusable energy (entropy) increases. Free energy will be discussed further in Section 4-2.

❑ CAPSULE SUMMARY

Energy has many different forms. Kinetic energy is work being done. Potential energy is the ability to do work.

Food	Calories per 100 g
Granola	482
Strawberries	31
Kidney beans	91
Cheddar cheese	375
Broccoli	28

Figure 4-2 Foods, *top*, contain stored energy. The amount of energy food contains is measured in calories, *bottom*. One food calorie is the amount of energy needed to raise the temperature of one kilogram of water one degree Celsius.

❑ CAPSULE SUMMARY

Energy can be changed from one of its forms to another. Energy cannot be created or destroyed.

The two most basic forms of energy are kinetic *(kih NEHT ihk)* energy and potential energy. **Kinetic energy** is the energy an object has because of its motion. All moving objects have kinetic energy. A cheetah racing across a savannah has a lot of kinetic energy, while a worm inching along the ground has much less. **Potential energy**, on the other hand, is the energy that is stored in an object because of its position. It is called potential energy because it is ready to make matter move or change when it is released. A boulder atop a hill, for example, has energy because of its position— *the energy that was required to lift the boulder against the force of gravity is stored as potential energy.* The higher the hill, the more potential energy the boulder has. ❑

Energy Can Be Transformed

A boulder can also be used to illustrate a fundamental characteristic of energy: energy can be transformed (changed from one form to another). When a boulder rolls downhill, energy stored by virtue of its position is released, becoming kinetic energy. In fact, most of the energy we use every day comes from converting one form of energy to another. An electric fan converts electric energy to mechanical energy. As gasoline burns in a car's engine, chemical energy (potential energy a substance has because of the position of the atoms in its molecules) is converted to mechanical energy and heat.

All energy conversions are governed by the laws of **thermodynamics** *(THUR moh deye NAM ihks)*, the study of energy transformations. One of these laws describes the observation that the total amount of energy in a closed system never changes. In science, a **system** is a collection of related objects (matter) that can be studied. For example, an organism is a system, as is the planet Earth and the universe. A **closed system** is one like the universe, into which no energy or matter can enter from the outside. A system that exchanges matter and energy with its surroundings (like Earth) is called an **open system**. The fact that the total amount of energy in a closed system remains constant is stated as the **first law of thermodynamics:** *Energy cannot be created or destroyed; it can only be converted from one form to another.*

Much of the work done by living things involves energy transformation. When you use food as a source of energy for movement, you are converting chemical energy into mechanical energy. When you use food to help maintain your body temperature, you are converting chemical energy into thermal energy (the random movement of all particles of matter). Figure 4-2 shows how much energy some foods contain. Just as potential energy is stored in a boulder's position at the top of a hill, chemical energy is stored in the arrangement of atoms in the molecules that make up food. ❑

Graphic Organizer

Use this graphic organizer with *Teaching Tip:* **Energy Transformations** on this page.

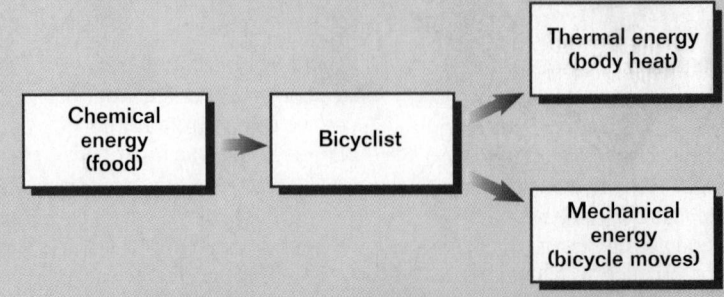

Energy Tends to Become Disorganized

Not all of the potential energy in food is recovered when it is transformed into kinetic energy. Almost half escapes into the environment as heat (thermal energy transferred by moving particles). Thermal energy is generally not a useful form of energy. It can do work only when it is concentrated and can flow from one object to another. Many systems, including living systems, cannot concentrate thermal energy enough to make it do work. A burning match can ignite wood, for example, but the thermal energy of air molecules cannot.

The reason some thermal energy is lost as heat when energy is transformed is stated in another law that also governs energy conversions. The basis of this law is that all systems tend to change in ways that make them more stable. Again, think of the boulder sitting on top of a hill. This is not a stable system, because the boulder may roll downhill. Once the boulder rolls to the bottom of the hill, it can go no farther and thus the system is more stable. To better understand this law, look at the soft-drink cans in Figure 4-3. Because thermal energy is the *random* motion of particles, it is a disorganized and stable form of energy. The amount of disorder in a system, or its amount of unavailable energy, is called **entropy**. The tendency of systems to become more stable by gaining entropy is stated in the **second law of thermodynamics**: *Disorder (entropy) in the universe constantly increases; in a closed system, energy tends to be converted to less organized (more stable) forms.* ❑

Energy Flows Through Living Systems

All organisms need a constant supply of energy to fuel the activities of life. Directly or indirectly, almost all of the energy for living systems comes from the sun. Energy from the sun enters living systems when sunlight is absorbed by molecules found in plants, algae, and certain bacteria. In the process, electrons in these molecules are boosted to higher energy states. Like a boulder on top of a hill, an electron that has been boosted to a higher energy state has additional energy because it is farther away from the nucleus of an atom. This additional energy is released by dropping the electron back to a lower energy state. The energy is then used to do work, or it is captured and stored.

Organisms that harvest energy by boosting electrons use it to produce energy-storing macromolecules. Most of these organisms boost electrons with energy from sunlight and convert light energy to chemical energy. The process that converts light energy to chemical energy is called **photosynthesis** (*foht oh SIHN thuh sihs*). Instead of using sunlight, certain bacteria obtain energy for boosting electrons from

Figure 4-3 If soft-drink cans are organized into a stack, *top*, the stack may tumble over. Therefore, the system is unstable. When the cans fall over, *bottom*, the system becomes less organized but more stable. To organize the system again (restack the cans), energy must be put into it.

❑ **CAPSULE SUMMARY**

In every energy transformation, some energy is converted to heat, increasing the disorder and the stability of a system.

👁 **VISUAL STRATEGY**

Figure 4-3
Use this figure to point out that, in general, the more organized a system is, the more energy it has. It took energy to stack the cans; thus, the stacked cans have stored energy. Cans can fall over, which takes energy. Cans that have fallen over do not have energy anymore. Like the stacked cans, living systems are quite organized. Maintaining the organization of living systems requires a constant influx of energy.

Teaching Tip
"Boosting" Electrons

Have students mentally construct a model of an apartment house that has several floors. This apartment house is populated by clumsy bowling enthusiasts who constantly change their residences within the building. The apartments on the upper floors cost more than those on the lower floors. When residents get raises, they move to a higher floor; when they lose their jobs or get pay cuts, they move to a lower floor. Each time they move to a new apartment, they take their bowling balls along. Clumsy as these bowlers are, they are always dropping their bowling balls on the steps as they move, and, of course, the balls fall to the bottom of the steps. Bowling balls coming from first-floor dwellers do not cause much damage, but those coming from top-floor dwellers have a lot of energy in them as they crash to the first floor.

Ask students the following questions. Energy added to atoms makes electrons move to different levels in the atom. How is this like your model? (*Pay raises make bowlers move up in the apartment house.*) Which has more energy, an electron in a lower level or an electron in a higher level? (*higher*)

Figure 4-4

Trace the flow of energy through the ecosystem shown in the figure. Emphasize the dissipation of heat at each level. Stress that cellular respiration even occurs in plants, so that they can use the energy they have stored during the day in carbohydrates (starches and sugars). Indicate that a living system uses the energy of sugars to make ATP within its cells, making the energy available to each cell's machinery.

CONNECTIONS

....

Chapter 2

Atoms Have Energy

Chemical energy is a form of potential energy due to the attraction between the nucleus and electrons. Electrons that are distant from the nucleus have more potential energy than those that are closer to the nucleus. Thus, the potential energy of an electron depends on its position in relation to the nucleus.

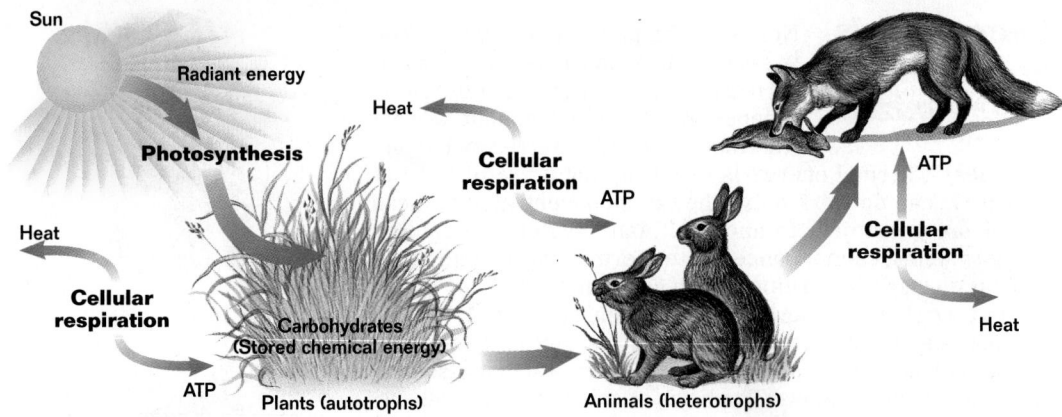

Figure 4-4 Energy flows through an ecosystem from sunlight or inorganic chemicals to autotrophs and then to heterotrophs. Because some heat escapes with every energy conversion, much of the energy captured by autotrophs eventually flows out of living systems.

📖 **CAPSULE SUMMARY**

Living systems are open systems. Energy from sunlight flows through living systems, from autotrophs to heterotrophs. Energy is captured during photosynthesis and made available for use by cells during cellular respiration.

inorganic molecules and convert it to organic chemical energy. The process that converts inorganic chemical energy to organic chemical energy is called **chemosynthesis** (*kee moh SIHN thuh sihs*). Organisms that harvest energy from either sunlight or chemicals are called **autotrophs** (*AWT oh trohfs*). The word *autotroph* comes from the Greek words *autos*, meaning "self," and *trophikos*, meaning "to feed."

Organisms that cannot harvest energy directly from sunlight or inorganic molecules but consume food instead are called **heterotrophs** (*HEHT uhr oh trohfs*). The word *heterotroph* comes from the Greek words *heteros*, meaning "other," and *trophikos*, "to feed." Heterotrophs obtain their energy from other organisms. Energy stored in food is released by a process that is similar to burning. This process, which is called **cellular respiration**, is a series of chemical reactions that converts energy stored in food to a more useful form. While burning converts almost all of the energy in a fuel to heat, much of the energy released during cellular respiration is used to make ATP, which fuels the activities of living. The flow of energy through the living world is illustrated in Figure 4-4. 🔲

Energy Is Carried by Electrons

....

Energy can be extracted from chemical bonds in food molecules. As you learned in Chapter 2, the chemical bonds that hold atoms together in food molecules are composed of electrons that are shared by two atoms. Like rubber bands, shared electrons keep two atoms from pulling apart. The energy in food is actually stored in the shared electrons of carbon-to-hydrogen bonds. The energy of these electrons is portable; it can be transferred to new chemical bonds by transferring the electrons. That energy is stored and released again in processes that involve chains of chemical reactions.

THEMES OF BIOLOGY

Flow of Energy

As you might imagine, there are many ways biological systems can be studied with an eye to understanding how they work. One of the best ways is to follow the energy needed for life. The energy-capturing process of photosynthesis and the energy-releasing process of cellular respiration are two of the fundamental pathways by which energy travels.

Taken together, the processes within cells that capture, transfer, and use energy are called metabolism.

All living things are united in a network in which each organism is linked to another by the common need for energy. Almost all the energy that powers life on Earth comes from the sun, captured by green plants, algae, and certain bacteria during photosynthesis. These organisms serve as sources of energy for other organisms, the energy passing from one organism to another. Ultimately, energy flow is a key factor in shaping ecosystems. ■

Energy that was stored as a result of photosynthesis is transferred to this broad-billed hummingbird when it drinks sugary nectar from the flowers of a thistle plant.

Chemical reactions that pass electrons from one atom or molecule to another are **oxidation-reduction reactions.** **Oxidation** *(AWKS ih DAY shun)* is the loss of electrons. **Reduction** is the gain of electrons. In organisms, however, electrons usually are not transferred alone. Instead, each electron moves with a proton as part of a hydrogen atom, H. Thus, as Figure 4-5 illustrates, the transfer of energy in organisms usually involves the removal of hydrogen atoms from one molecule and the addition of hydrogen atoms to another molecule. Oxidation-reduction reactions play key roles in organisms because they enable energy (carried by hydrogen atoms) to pass from one molecule to another.

Oxidation

$$C_6H_{12}O_6 + 6O_2 \rightarrow 6CO_2 + 6H_2O$$

Reduction

Figure 4-5 $C_6H_{12}O_6$ is oxidized when it loses hydrogen atoms. O_2 is reduced when it gains hydrogen atoms. Oxidation and reduction always take place together because every electron that is lost by one atom or molecule is gained by another.

Section Review

1. *Define kinetic energy and potential energy.*
2. *Summarize the first and second laws of thermodynamics.*
3. *What roles do photosynthesis and cellular respiration play in the flow of energy through living systems?*
4. *How are oxidation-reduction reactions important to cells?*

Critical Thinking

5. *Use the laws of thermodynamics to explain why a rock that tumbles off a cliff eventually comes to a stop.*

Answers to Section Review

1. Kinetic energy is the energy an object has because of its motion. Potential energy is the energy that is stored in an object because of its position.

2. First law: Energy can neither be created nor destroyed, only converted from one form to another. Second law: The entropy of the universe is increasing. In a closed system, energy tends to be converted to more stable forms.

3. Photosynthesis traps the energy of sunlight, storing it in sugars. All organisms use the sugars in cellular respiration to meet their energy needs.

4. Oxidation-reduction reactions enable energy to pass from one molecule (and even from one organism) to another.

5. Answers will vary. Encourage students to speculate, but be sure they use the laws of thermodynamics (energy can be converted from one form to another, and energy tends to be converted to less organized forms) in their explanations.

Vocabulary Preview

chemical reaction

reactant

product

free energy

exergonic reaction

endergonic reaction

activation energy

catalyst

catalysis

enzyme

substrate

active site

Opening Questions

Ask students why organisms need a constant supply of energy. Students should understand that organisms use energy to maintain their organization and, thus, to preserve homeostasis. Have students recall that oxidation-reduction reactions are the major type of chemical reaction used by organisms to pass energy from one molecule to another.

Demonstration

Chemical and Physical Changes

To demonstrate the difference between physical and chemical changes, take a wooden matchstick and break it in half. This is an example of a physical change. Then strike the match, producing a flame. This is a chemical change. Ask students to speculate about the difference between physical and chemical changes. Lead them to conclude that chemical changes change substances into other substances; hence, after burning, the match is no longer a wooden matchstick.

4-2 Energy and Chemical Reactions

*I*n a sense, organisms can be compared to race cars; both are complex "machines" powered by chemical energy. Chemical reactions power the movement of mechanical parts, creating the force that moves a race car. Chemical reactions also power most cellular activities. Just as a successful race car driver must "look under the hood" to learn how a race car's engine works, you must look at the chemical machinery of cells, such as the ones in Figure 4-6, to understand how organisms function. Understanding cell chemistry, however, requires some knowledge of chemical reactions.

Section Objectives

■ Distinguish between exergonic and endergonic chemical reactions.

■ Explain why activation energy and catalysts help chemical reactions occur.

■ Describe the role of enzymes in cell chemistry.

Figure 4-6 Chemical reactions power cellular activities that result in the movement, growth, response, and reproduction of an organism. For example, the contraction and relaxation of cells, like these human striated muscle cells, are translated into the movements of your body.

Chemical Reactions Absorb or Release Energy

A **chemical reaction** occurs when chemical bonds between atoms are broken or formed, resulting in the formation of one or more different substances. The substances that are combined or broken apart during chemical reactions are called **reactants**. The new substances that form are called **products**. You can think of the chemical reactions in cells as either gluing atoms or molecules together or tearing molecules apart. Extracting energy from sugar, for example, involves ripping apart the carbon backbones of sugar molecules. Making a protein is a matter of sticking amino acid molecules together to form long chains. Like most chemical reactions, the reactions that make proteins are reversible, meaning that they can also occur in the opposite direction. As you digest food, for instance, proteins that were assembled by other organisms are torn apart into individual amino acids.

The energy from chemical reactions that drives cell activities is called **free energy**. While most biochemical reactions release free energy into their surroundings, others absorb it. A reaction that releases free energy is called an **exergonic** *(ehks uhr GAHN ihk)* **reaction**. As the top graph in Figure 4-7 indicates, the products of exergonic reactions contain less energy than their reactants. Cellular respiration is an exergonic process. In contrast, a reaction that absorbs free energy is an **endergonic** *(ehn duhr GAHN ihk)* **reaction**. As the bottom graph in Figure 4-7 indicates, the products of endergonic reactions contain more energy than their reactants. Photosynthesis is an endergonic process. ◻

◻ CAPSULE SUMMARY

Chemical reactions produce new substances by breaking or forming chemical bonds between atoms. Exergonic reactions release free energy into their surroundings. Endergonic reactions absorb free energy from their surroundings.

Starting Chemical Reactions Requires Energy

As the second law of thermodynamics predicts, exergonic reactions readily occur because systems tend to become less organized and therefore more stable. If this is so, why haven't all exergonic reactions already occurred? For example, the burning of gasoline is an exergonic reaction. Therefore, you can think of a tank full of gasoline as crammed with chemical reactions waiting to happen. So why doesn't all the gasoline in the world burn up right now? It doesn't because gasoline needs a "kick in the pants" to burn. Energy from a spark or a match is needed to start the combustion reaction that burns gasoline. In order for the reaction's products (carbon dioxide and water) to form, existing chemical bonds in the reactants (gasoline and oxygen) must first be broken, and this takes energy.

The extra energy required to break existing chemical bonds and to initiate a chemical reaction is called **activation energy**. If you look at Figure 4-7 again, you will notice that the energy curve for the exergonic reaction initially rises. This "uphill" curve represents the reaction's activation energy. The heat from a spark or a match supplies the sizable amount of activation energy that is needed to start the chemical reactions waiting to happen in gasoline. How does heat become the activation energy for a chemical reaction? The heat from a spark or a match transfers free energy to the reactants. This extra energy causes the reactant particles to move and vibrate faster—they have more kinetic energy. In other words, the reactant particles are higher on the activation energy "hill." If the kinetic energy of reactant particles is the same as the required activation energy, it can be used to break chemical bonds in the reactants so that new bonds and products can form.

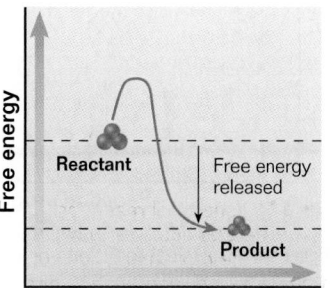

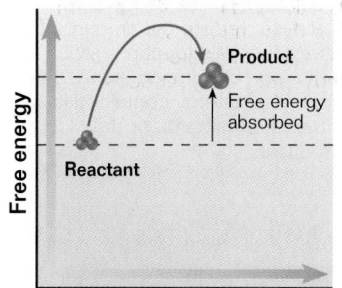

Figure 4-7 Exergonic reactions, *top,* release free energy, leaving their products with less total energy than their reactants. Endergonic reactions, *bottom,* absorb free energy, giving their products more total energy than their reactants.

SECTION 4-2

Teaching Tips
Endergonic Versus Exergonic

Ask students to come up with examples of endergonic and exergonic reactions. For example, many cellular reactions are endergonic because they use more energy than they release. An example of an exergonic reaction would be burning wood because the reaction gives off heat and light energy. Emphasize that whether a reaction is endergonic or exergonic depends on the relative amount of energy given to or taken from the environment.

ATP

Ask students to name an energy source for cellular endergonic reactions. Students should know that ATP briefly stores energy for use within the cell and that chemical reactions release the stored energy for the cell's use.

❏ CAPSULE SUMMARY

Chemical reactions require an input of energy to get started. The energy required to get a reaction started is called activation energy. A catalyst lowers the amount of activation energy needed to start a reaction.

A chemical reaction—exergonic or endergonic—will proceed at a much faster rate than it normally would if the necessary activation energy can be lowered. A substance that alters a chemical reaction so that it can proceed at a lower activation energy is called a **catalyst** (*KAT uh lihst*). The process of increasing the rate of a chemical reaction through the use of a catalyst is called **catalysis** (*kuh TAL uh sihs*). Substances that act as catalysts are not changed by the reactions they assist. Catalysts do not make chemical reactions begin spontaneously; that is, they do not reduce the activation energy to zero. The need for activation energy cannot be avoided, as you can see from the graphs in Figure 4-8. ❏

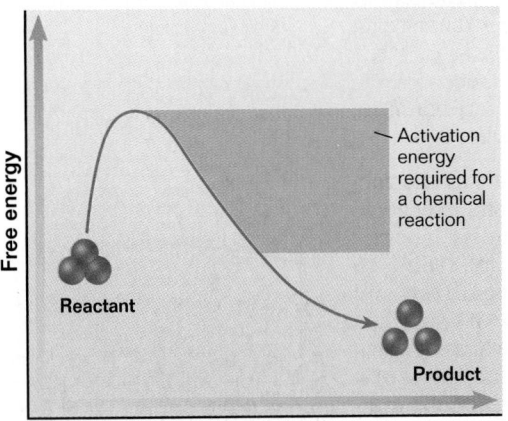
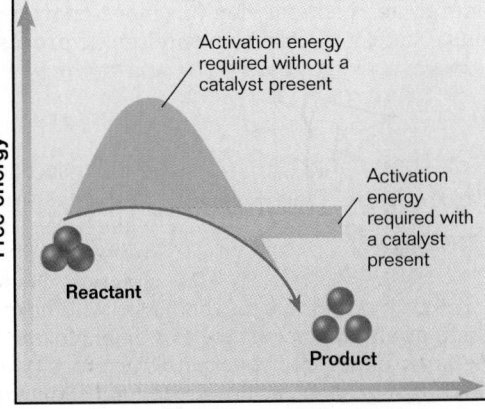

Figure 4-8 Chemical reactions are more likely to happen when the large amount of activation energy required, *left,* can be lowered. Catalysts reduce the amount of energy needed to start a reaction, *right,* but do not change the amount of energy contained in either the reactants or the products.

Enzymes Catalyze Chemical Reactions in Cells

Like a tank of gasoline, cells are crammed with chemical reactions waiting to happen. Each reaction waits for the nudge that will give it enough activation energy to start. Just as the heat from a flame supplies the activation energy needed for gasoline to burn, heat (molecular motion) can supply activation energy for chemical reactions that occur in cells. However, the amount of heat living things can tolerate is not enough to make these reactions occur as quickly as they must. As you might suspect, cells need catalysts to help start chemical reactions. And as you will learn later, cells control chemical reactions by controlling the activity of these catalysts.

Enzymes (*EHN zeyemz*) are the catalysts used by cells to trigger and control particular chemical reactions. An enzyme is a protein, often a rather large one, that works by binding to specific reactant molecules. Like other catalysts, *enzymes are not permanently changed by the reactions they catalyze.* The reactant molecules to which an enzyme binds are called **substrates.** Amylase (*AM uh lays*), for example, is the enzyme that catalyzes the breakdown of starch into glucose. Starch is this enzyme's substrate. Starch synthetase *(SIHN thuh tays)* is the enzyme that catalyzes the reverse reaction, the formation of starch from glucose. Glucose is this enzyme's substrate. Enzymes lower the activation energy required to start biochemical reactions, allowing them to occur more readily. *Without enzymes, chemical reactions necessary for life would not occur at a rate sufficient for sustaining life.*

Because of their shapes, enzymes bind to specific substrates. In Figure 4-9, you can see that an enzyme's surface provides a mold that fits the molecules of a certain substrate almost exactly. Other molecules that fit less perfectly simply don't adhere to the enzyme's surface. The site where a substrate binds to an enzyme is called the **active site**. At the active site, the enzyme and the substrate molecule interact in such a way that the activation energy for a particular reaction is lowered, making the substrate more likely to react.

Enzymes are very effective catalysts. As an example, consider a reaction that takes place in your bloodstream.

$$\text{carbon dioxide + water} \xrightarrow{\text{carbonic anhydrase}} \text{carbonic acid}$$

By itself, the reaction is very slow. In a cell, perhaps 200 molecules of carbonic acid form in an hour. However, in the presence of the right enzyme—carbonic anhydrase *(an HEYED rays)*—600,000 molecules of carbonic acid form every second! Thus, the enzyme speeds up the reaction rate about 10 million times. Carbonic anhydrase plays an important role in your blood. By allowing carbon dioxide and water in blood to combine and form carbonic acid so quickly, the enzyme prevents your blood from becoming saturated with carbon dioxide, a waste product. In your lungs, the reverse reaction occurs, and CO_2 is eliminated from your body when you exhale (breathe out). This enables your blood to continue removing carbon dioxide from your body efficiently.

Figure 4-9 One way that enzymes lower activation energy is by weakening chemical bonds, *left,* so that they will break more easily. For example, amylase speeds the breakdown of starch into glucose by weakening the bonds linking glucose subunits. Enzymes also bring molecules together in a certain order, *right,* by serving as templates (patterns). This is how starch synthetase works.

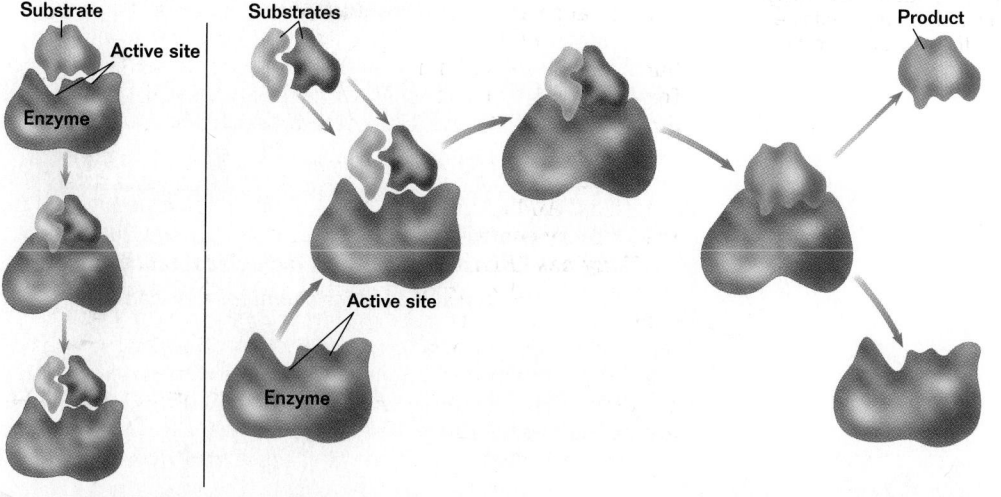

SECTION 4-2

Demonstration
Catalysis
Caution: Manganese dioxide is a strong oxidizer and should be kept from contact with organic materials. It is moderately toxic when ingested. Keep students 10 ft. away from demonstration area during the demonstration. The demonstrator should wear safety goggles. Dispose of H_2O_2, once bubbles cease, in the sink. Do not put MnO_2 in the trash—dry it out for use again. Write the following equation: $2H_2O_2 \rightarrow 2H_2O + O_2$. The reaction usually proceeds too slowly to detect. Ask students what the evidence for a reaction would be. *(Answers should include bubble formation.)* Show students a flask with a solution of ≤6% hydrogen peroxide (H_2O_2). (Never use 30% H_2O_2 because it is too strong.) Have students verify that no reaction is taking place. Add a tiny speck of manganese dioxide (MnO_2) to the hydrogen peroxide. Have students record their observations of the reaction. *(Bubbles appear in the solution. The amount of manganese dioxide does not diminish.)* Ask students the following questions: What is (are) the reactant(s) in the equation for this reaction? *(hydrogen peroxide)* What is (are) the product(s) in the equation for this reaction? *(water and oxygen)* What is the function of the manganese dioxide? *(It is a catalyst.)* What evidence is there that manganese dioxide performs this function? *(It speeds up the reaction without being used up.)*

Application

Health Fever is a response by the body to an infection. It is the body's attempt to kill the organisms responsible for the infection by making the organism's enzymes ineffective. The problem arises when the body temperature gets too high for too long, because then the body begins to inactivate its own enzymes. Temperatures in excess of 103°F, any fever accompanied by recurrent shaking or chills, or a fever that persists for several days should be investigated by a doctor.

CONNECTIONS

Chapter 3

Homeostasis

The action of the body's control mechanisms serves to maintain the proper pH and temperature in order to preserve homeostasis. Most of these control systems behave like a feedback system, similar to a thermostat that regulates the temperature indoors.

Figure 4-10 The darker fur of this Himalayan cat indicates cooler regions of its body. These regions have a high concentration of a dark pigment that is produced with the aid of an enzyme that becomes inactive at warmer temperatures.

❏ CAPSULE SUMMARY

Enzymes are catalysts that speed chemical reactions in cells. Enzymes have specific shapes that fit only certain substrates.

Figure 4-11 Pepsin and trypsin are digestive enzymes whose activities are affected by pH. Pepsin's activity peaks at a much lower pH than trypsin's. Pepsin works in the very acidic environment of the stomach, while trypsin works in the less acidic small intestine.

Factors that change the shape of an enzyme affect the enzyme's activity. One of these factors is temperature. For each enzyme, there is a range of temperatures within which the enzyme operates efficiently, as indicated by Figure 4-10. Temperatures outside of this range either cause some of the hydrogen bonds that determine an enzyme's shape to be broken or cause other hydrogen bonds to form, also changing the shape of the enzyme. Another factor that affects enzyme activity is pH (the relative acidity or alkalinity of a solution). Again, each enzyme has an optimal range of pH within which it can operate. A pH outside of this range can cause the breaking of certain bonds that determine an enzyme's shape. Figure 4-11 illustrates the relationship between the activity of two digestive enzymes and pH. ❏

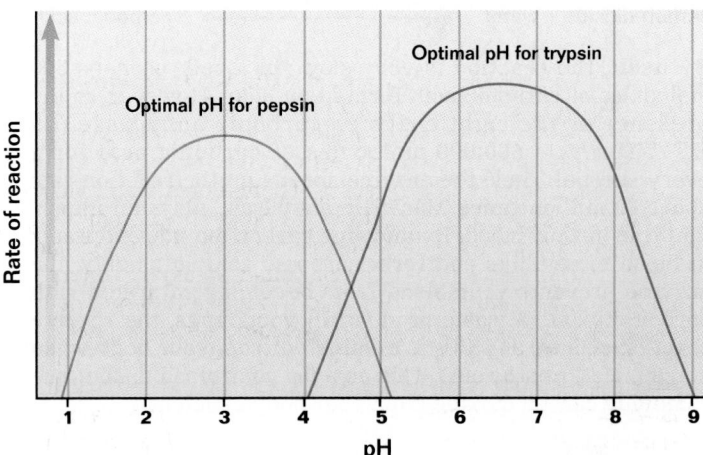

Section Review

1. *What is the difference between exergonic chemical reactions and endergonic chemical reactions?*
2. *What is activation energy?*
3. *How do catalysts help to start a chemical reaction?*
4. *Why are enzymes needed for chemical reactions in cells?*

Critical Thinking

5. *What effect might a molecule that changes the shape of carbonic anhydrase have on your body?*

Answers to Section Review

1. Exergonic reactions give off energy into their surroundings, while endergonic ones take energy from their surroundings.

2. Activation energy is the energy necessary to break existing bonds and to initiate a chemical reaction.

3. Catalysts lower the activation energy of a reaction and act as templates that bring reactants together.

4. Enzymes enable biochemical reactions to occur at lower temperatures and at the rate necessary for survival.

5. A molecule that changes the shape of carbonic anhydrase would keep the blood from carrying as much carbon dioxide as it does. Fatigue and even death would result since the body couldn't efficiently eliminate this metabolic waste.

4-3 Metabolism

Each of the significant properties we use to define life—growth, movement, reproduction, heredity—uses energy. To sustain life, energy must be continually supplied (like putting more logs on a fire) and converted into useful forms (like boiling water for cooking). But unlike burning wood in a fire, energy must be released slowly to prevent living cells from burning up. The energy must also be released in such a way that it can be put to use doing the work of cells. Thus, energy passes along a network of controlled chemical reactions that acts as a highway for energy transportation within a cell. The chemistry of living, this symphony of energy-transferring reactions within cells, is called *metabolism* (muh TAB uh lihz uhm).

Section Objectives

- Recognize several ways that cells use energy.
- Describe how ATP is used to power chemical reactions.
- Identify the energy carrier in biochemical pathways.
- Explain how enzymes are regulated to control the metabolic activities of cells.

Life Processes Are Driven by Energy

Cells use energy to do those things that require work. One of the most obvious of these is movement. Some cells swim through water, propelling themselves by rapidly spinning a long tail-like flagellum. During development, an embryo's cells use energy to crawl over one another and reach new positions in the embryo. Cells also expend energy to change their shape. When one of your white blood cells engulfs an invading bacterium, it accomplishes the necessary change in shape by extending its cytoskeleton. Movement also occurs within cells. Mitochondria and cellular materials are passed several feet along the narrow extensions of nerve cells that connect your feet to your spine. All of these movements use energy.

A second major way cells use energy is to build new molecules. Very few biological molecules are created by a single chemical reaction. Instead, most are manufactured by a long series of enzyme-catalyzed chemical reactions in which each reaction modifies the product of the previous reaction. An ordered series of enzyme-catalyzed chemical reactions that forms a product in a step-by-step manner, as illustrated in Figure 4-12, is called a **biochemical pathway**. Because building molecules takes energy, synthetic biochemical pathways (which make the specific molecules cells need) almost always involve at least one endergonic chemical reaction. Remember, these reactions consume energy because their products contain more energy than their reactants did. And nothing can happen until extra energy from an external source triggers the reaction.

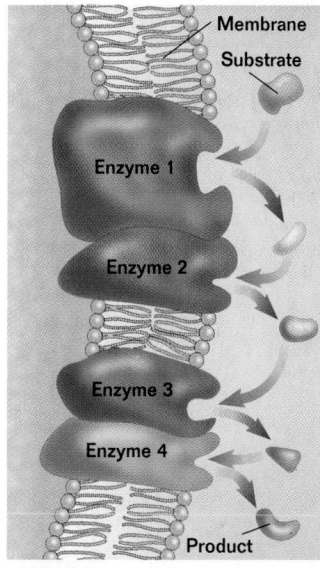

Figure 4-12 A biochemical pathway consists of a series of enzyme-catalyzed chemical reactions. The product of each reaction in a pathway becomes the reactant of the next reaction in the pathway.

SECTION 4-3

Vocabulary Preview

metabolism
biochemical pathway
coupled reaction
coenzyme
NAD$^+$
NADH
allosteric enzyme
allosteric site
feedback inhibition

Opening Questions

Ask students to name two factors that affect an enzyme's activity. Students should recall that temperature and pH influence enzyme activity. Ask how homeostasis and enzyme activity are related. Make sure students understand that enzyme activity helps maintain homeostasis.

Demonstration

Cells Release Energy Gradually

Use a staircase or step stool to demonstrate the advantage of releasing energy in a series of small steps. Climb several steps and then carefully jump to the floor. Exaggerate the impact on landing. Then climb to the same level on the staircase and walk down one step at a time. Tell students that cells release energy gradually in a series of steps that make up a biochemical pathway. Ask students what would happen if cells released large amounts of energy all at once. (*Students might suggest that cells would disrupt homeostasis, would waste energy, or would destroy themselves with the impact of so much energy being released at once.*)

Did You Know?

Cyanide is a deadly poison that disrupts the metabolic pathway by inactivating an enzyme important to metabolism and preventing the tissues from using oxygen. When energy and materials cannot be passed through the system, the cells die. Ingesting less than 3 g of cyanide can result in death in a short amount of time.

Figure 4-13
Use this figure to show one way that cells obtain energy from ATP. Point out that the enzyme that catalyzes the reaction of the reactant has two active sites, one for a reactant molecule and one for an ATP molecule. Also point out that the splitting of ATP into ADP and P (phosphate) is an exergonic reaction because ATP contains more energy than the products (ADP + P) and because energy is released into the environment. Emphasize that in a coupled reaction, the phosphate split from ATP and the energy released by the reaction are transferred to the reactant molecule, providing the energy necessary for the reaction that alters the reactant (in this case, the reactant appears to be splitting into two products).

CONNECTIONS

Chapter 3
Sodium-Potassium Pumps and Coupled Channels
Remind students of the action of the sodium-potassium pump and the coupled channels involved in glucose transport, an example of ATP-coupled reactions.

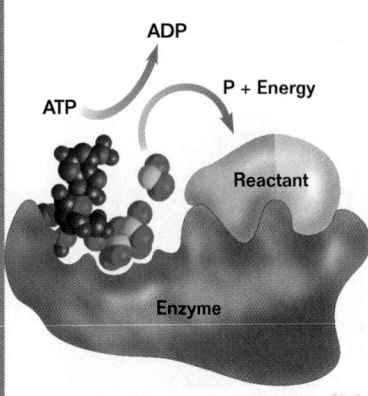

Figure 4-13 Coupling energy-requiring reactions with the splitting of ATP molecules is one of the key processes that help cells manage energy. The two parts of a coupled reaction (ATP-splitting and the endergonic reaction) take place in concert, separate parts of a single process.

📑 *CAPSULE SUMMARY*

Enzymes that catalyze endergonic reactions also have an active site that splits ATP to release energy.

ATP Supplies Energy for Metabolism

ATP supplies most of the energy that drives metabolism. In a sense, it is a cell's "energy currency"—the "money" it uses to "pay for" endergonic processes. As you learned in Chapter 2, each ATP molecule is made of three parts: ribose, a sugar; adenine, a nitrogen-containing base; and a chain of three phosphate groups. Energy that was temporarily stored in an ATP molecule is made available for use by a cell when the end phosphate group is transferred to another molecule. In this exergonic reaction, a sizable packet of energy is transferred along with the phosphate group. This energy then activates a reaction involving the molecule that received the phosphate group. Almost all of the endergonic reactions in cells require less activation energy than that transferred with a phosphate group from ATP. The breakdown of ATP is thus able to power many of a cell's endergonic activities.

An enzyme that catalyzes an endergonic reaction in cells has *two* active sites. As you can see in Figure 4-13, one is for the reaction's reactant and the other is for ATP. The ATP site splits the end phosphate group from an ATP molecule, releasing energy. The phosphate group (P) and some of the energy released are received by the reactant molecule attached to the second active site. This transfer of a phosphate group and its energy drives the reaction at the second active site. Since both reactions occur on the surface of the same enzyme, they are physically linked, or coupled. An endergonic reaction that is driven by the splitting of ATP molecules is therefore called a **coupled reaction**. In a similar way, you can make water in a swimming pool leap straight up in the air, despite the fact that gravity prevents water from rising spontaneously—just jump in the pool! The energy you add going in more than compensates for the force of gravity holding the water down. ◼

Coenzymes Help Transport Energy

In many metabolic pathways, energy released from one set of reactions is transferred to another set of reactions. Remember that in living systems, electrons carry energy from one atom or molecule to another in oxidation-reduction reactions. Often, high-energy electrons are passed from the active site of the enzyme that is catalyzing a reaction to an organic molecule called a **coenzyme** *(koh EHN zeyem)*. The coenzyme then carries the electrons to another enzyme that is catalyzing a different reaction, as you can see in Figure 4-14. Just as tanker trucks transport energy (in the form of gasoline) from storage tanks to gas stations, coenzymes shuttle energy (in the form of hydrogen atoms with high-energy electrons) from one place to another in a cell.

One of the most important coenzymes in cell metabolism is nicotinamide *(nihk uh TIHN uh meyed)* adenine

Overcoming Misconceptions

Transfer of Energy
Because ATP supplies most of the energy that drives metabolism, it is sometimes misleadingly called an energy-rich molecule and its phosphate bonds are sometimes misleadingly called high-energy bonds.

These terms are misleading because they imply that ATP contains an unusually high amount of energy. In fact, ATP serves as the cell's energy currency because its phosphate bonds are easy to break. If ATP's phosphate bonds con-

tained especially high amounts of energy, they would be difficult to break. ATP is the energy currency of cells because it is particularly suited to transferring energy efficiently in organisms, not because it is an energy-rich molecule.

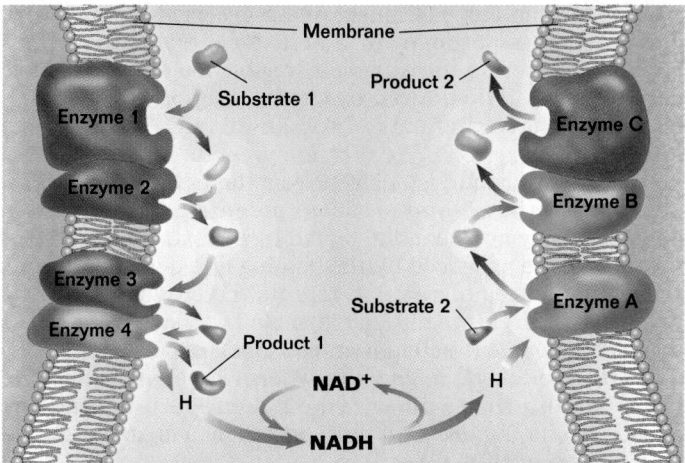

Figure 4-14 NAD⁺ accepts high-energy electrons (paired with protons as hydrogen atoms) and becomes NADH, which transfers the electrons from one set of enzyme-catalyzed reactions to another. When NADH reaches an enzyme in a second set of reactions, it releases the electrons (and their energy) to the reaction that this enzyme is catalyzing. NAD⁺ can then return to the enzyme in the first set of reactions and get more electrons.

dinucleotide, usually referred to by the abbreviation **NAD⁺**. When NAD⁺ acquires a hydrogen atom from the active site of an enzyme, it becomes **NADH** (and has been reduced). In cells, hydrogen atoms with high-energy electrons are stripped from food molecules and donated to NAD⁺, forming NADH. The NADH molecules then carry high-energy electrons. Like carrying money in a wallet, cells carry hydrogen atoms with high-energy electrons (one form of their energy "money") in molecules like NADH. And just as you can exchange one type of money for another, cells can exchange one type of energy carrier for another. The energy in NADH, as you will learn in Chapter 5, can be converted to the cell's main energy currency—ATP. ◻

☐ CAPSULE SUMMARY

NAD⁺ is a coenzyme that, when reduced to NADH, carries energy from one reaction to another in cells.

Enzymes Direct Metabolism in Cells

A cell may contain thousands of different kinds of enzymes, each specific to a particular substrate and each promoting a different chemical reaction. The enzymes that are active in a cell at any one time determine what happens in that cell, much as traffic lights determine the flow of traffic in a city. An increase in a particular enzyme's concentration accelerates the rate of the reaction it catalyzes. Conversely, a decrease in an enzyme's concentration slows the rate of the reaction it catalyzes. Furthermore, not all cells contain the same enzymes. The chemical reactions occurring in a nerve cell are very different from those occurring in a red blood cell because the two kinds of cells have a different array of enzymes. By controlling the concentration of enzymes and when they are active, a cell is able to control its chemical reactions, just as a conductor controls the music an orchestra produces by dictating the tempo and volume at which the music is played.

CONTENT LINK

More information about how cells regulate biochemical pathways is found in **Chapters 5, 35, and 41**.

👁 *VISUAL STRATEGY*

Figure 4-14

Point out the two biochemical pathways shown in this diagram. Remind students that high-energy electrons, which are carried by protons in the form of hydrogen atoms, are passed from the site of one enzyme-catalyzed reaction to the site of the next enzyme-catalyzed reaction in the pathway. Ask: What type of chemical reactions are represented here? *(oxidation-reduction reactions)* What happens to the electron-carrying hydrogen atom at the end of a pathway? *(It must be accepted by another molecule.)* Tell students that NAD⁺ is one type of molecule that accepts hydrogen atoms (and their electrons) and that the resulting molecule, NADH, is used by other biochemical pathways that require energy for their chemical reactions. Thus, energy released from ATP, which is used to power many reactions, may be used to produce NADH, another type of energy currency that can power biochemical reactions.

Teaching Tips
Coenzymes

Point out that vitamins are important in the diet because they serve as coenzymes that help body enzymes function properly. For example, niacin (vitamin B_3) is used by the body to produce NAD⁺. Without adequate amounts of niacin, the body does not synthesize enough NAD⁺ for its needs. Niacin deficiency can result in skin eruptions and irritations, fatigue, and digestive problems.

Anabolism and Catabolism

Ask students what kinds of reactions and energy transformations make up metabolism. Lead students to understand that metabolism includes destructive and constructive, i.e., energy-producing and energy-using, processes. Explain that anabolic reactions (constructive metabolism) synthesize macromolecules for use within the cell and that catabolic reactions (destructive metabolism) break larger macromolecules into smaller pieces.

VISUAL STRATEGY

Figure 4-15
Use this figure to point out that enzyme action can be fine-tuned through allosteric effects, either activating or repressing a given enzyme's activity. Students should understand that if the shape of the enzyme changes so that the substrate will not fit the active site, then the enzyme's activity is repressed. Ask students to explain how a signal molecule can activate an enzyme. (*A signal molecule activates an allosteric enzyme if the shape of the active site changes to accommodate the substrate when the signal molecule binds to the allosteric site of the enzyme.*)

Chapter Closure

Have students diagram the feedback loop described in question 5 of the **Section Review** at the bottom of this page. Students should label the initial source of glucose as glycogen in the liver. Their diagrams must show an inhibition loop, logical names describing the functions of the first and last enzymes in the pathway, and changes in the allosteric site of the initial enzyme in the pathway. Intermediate enzymes in the pathway can be represented by letters or nonsense names.

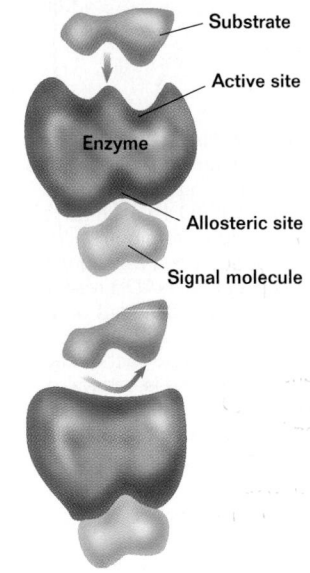

Figure 4-15 An allosteric enzyme has an active site and an allosteric site, *top.* When a signal molecule binds to the allosteric site, *bottom,* the enzyme and its active site change shape.

CAPSULE SUMMARY

Metabolism is regulated by controlling the kinds of enzymes present in a cell, their concentration, and their activity.

Because an enzyme must have a precise shape to work correctly, it is possible for a cell to control an enzyme's activity by altering the enzyme's shape. Many enzymes have shapes that can be altered by the binding of "signal" molecules to their surfaces. Such an enzyme is called an **allosteric** (al oh STEHR ihk) **enzyme**. *Allosteric*, which means "other shape," is derived from the Greek words *allos*, meaning "other," and *stereos*, meaning "solid." If an allosteric enzyme is unable to bind to a substrate because of the new shape produced by the binding of a signal molecule, the signal molecule is said to *repress* the enzyme's activity. If an allosteric enzyme is unable to bind to a substrate unless a signal molecule is bound to it, the signal molecule is said to *activate* the enzyme. The site where the signal molecule binds to an allosteric enzyme's surface is called the **allosteric site**. Figure 4-15 illustrates how a signal molecule may affect an allosteric enzyme.

When a cell already has an adequate amount of the chemical produced by a biochemical pathway, the pathway will often shut down. Thus, wasteful overproduction is avoided. How does a pathway "know" to shut itself down? The first enzyme in the pathway is an allosteric enzyme with a site that has the shape of the pathway's product. The binding of the product molecule to the allosteric enzyme *inhibits* the enzyme's activity so that when the concentration of the product is high, the first step in the pathway is effectively turned off. The shutting down of a biochemical pathway caused by a key enzyme's sensitivity to the level of the pathway's product is called **feedback inhibition**. When the product concentration drops, the pathway is reactivated. Feedback inhibition is another way that cells simply and effectively regulate their biochemical activities.

Section Review

1. *Name three types of processes that require a cell to expend energy.*
2. *What role does ATP play in coupled reactions?*
3. *Name two molecules that carry energy in biochemical pathways, and describe how each transfers its energy.*
4. *How is metabolism controlled in cells?*

Critical Thinking
5. *Excess glucose, the primary fuel for cellular respiration, is stored in your liver as the complex molecule glycogen. How do you suppose your body knows when to convert glucose to glycogen for storage and when to convert glycogen back to glucose for release from the liver?*

Answers to Section Review

1. Some examples are growth, reproduction, and movement of materials into and out of cells.

2. ATP supplies the energy that causes a reactant (substrate) to react in a coupled reaction.

3. ATP carries energy in the chemical bond holding its terminal phosphate to the rest of the molecule. NADH carries energy in the electron associated with the hydrogen atom it carries.

4. Metabolism is controlled through activation and inhibition of enzymes.

5. The body could use feedback to control glucose production and storage. Too much glucose could cause the pathways that release glucose into the blood to shut down by changing the shape of the enzyme that catalyzes its production. A shortage of glucose in the blood could cause activation of the metabolic pathway that produces glucose.

Vocabulary

activation energy (81)	coupled reaction (86)	NADH (87)
active site (83)	endergonic reaction (81)	open system (76)
allosteric enzyme (88)	energy (75)	oxidation (79)
allosteric site (88)	entropy (77)	oxidation-reduction reaction (79)
autotroph (78)	enzyme (82)	photosynthesis (77)
biochemical pathway (85)	exergonic reaction (81)	potential energy (76)
catalysis (82)	feedback inhibition (88)	product (80)
catalyst (82)	first law of thermodynamics (76)	reactant (80)
cellular respiration (78)	free energy (81)	reduction (79)
chemical reaction (80)	heterotroph (78)	second law of thermodynamics (77)
chemosynthesis (78)	kinetic energy (76)	substrate (82)
closed system (76)	metabolism (85)	system (76)
coenzyme (86)	NAD+ (87)	thermodynamics (76)

Concept Mapping

Construct a concept map that shows the function of enzymes in cells. Use as many terms as needed from the vocabulary list. Try to include the following terms in your map: metabolism, chemical reaction, activation energy, catalyst, substrate, active site, and biochemical pathway. Include additional terms in your map as needed.

Review

Multiple Choice

1. Which of the following is an example of potential energy?
   a. water flowing over a dam
   b. a rock rolling downhill
   c. a burning match
   d. a can full of gasoline

2. Energy flows through living systems from
   a. the sun, to heterotrophs, and then to autotrophs.
   b. autotrophs, to the environment, and then to heterotrophs.
   c. the sun, to autotrophs, and then to heterotrophs.
   d. the environment, to heterotrophs, and then to autotrophs.

3. When a molecule loses an electron, the molecule has been
   a. reduced.   c. activated.
   b. metabolized.   d. oxidized.

4. If the beaker in which two chemicals are mixed becomes hot, the reaction that occurred is probably a(n)
   a. energetically neutral reaction.
   b. exergonic reaction.
   c. endergonic reaction.
   d. coupled reaction.

5. Which statement is correct for the following chemical reaction?

   $$CO_2 + H_2O \xrightarrow{\text{carbonic anhydrase}} H_2CO_3$$

   a. H_2O is a product of the reaction.
   b. Carbonic anhydrase is the energy source for the reaction.
   c. H_2CO_3 is a catalyst for the reaction.
   d. CO_2 is a reactant of the reaction.

6. Enzymes speed up chemical reactions in cells by
   a. lowering the activation energy.
   b. remaining unaltered by the reaction.
   c. releasing the products of the reaction.
   d. taking the place of activation energy.

7. Energy is transferred from one place to another in a biochemical pathway by
   a. enzymes.   c. catalysts.
   b. electrons.   d. ATP.

CHAPTER REVIEW ANSWERS

Each item in the Chapter Review is correlated by text section in the assignment guide that follows.

ASSIGNMENT GUIDE

Section	Review	Themes Review	Critical Thinking
4-1	1–3, 11–14, 24, 25	30, 31	33, 34
4-2	4–6, 15–18, 26, 27	32	35
4-3	7–10, 19–23, 29, 30	33	36

Concept Mapping

Map answer is shown on page 73B.

Review

Multiple Choice

Code in parentheses indicates section number and objective number.

1. d (4-1.1)
2. c (4-1.3)
3. d (4-1.4)
4. b (4-2.1)
5. d (4-2.3)
6. a (4-2.3)
7. b (4-3.3)
8. a (4-3.2)
9. b (4-3.3)
10. c (4-3.4)

CHAPTER REVIEW ANSWERS *continued*

Completion

11. potential, kinetic (4-1.1)

12. closed systems, heat (4-1.2)

13. entropy (4-1.2)

14. photosynthesis, ATP (4-1.3)

15. activation energy, catalysis (4-2.2)

16. catalyst (4-2.2)

17. product, substrate (4-2.3)

18. shape (4-2.3)

19. energy (4-3.1)

20. biochemical pathway (4-3.1)

21. ATP, endergonic (4-3.2)

22. coenzyme, food (4-3.3)

23. feedback inhibition (4-3.3)

Short Answer

24. Organisms are open systems and are continuously incorporating energy that enables them to develop and maintain a high degree of organization. (4-1.2)

25. Oxidation-reduction reactions are reactions in which one molecule is oxidized (loses electrons) while another molecule is reduced (gains electrons). They are used to transport energy (carried by electrons) from one molecule to another. (4-1.4)

26. A catalyst increases the rate of a chemical reaction by lowering the reaction's activation energy. (4-2.2)

27. An enzyme's activity increases as substrate concentration increases because it is more likely to encounter a substrate molecule. With more substrate molecules filling active sites, more product is produced. The enzyme's activity will level off at the point when the enzyme's active sites are always filled with the substrate. (4-2.3)

28. Energy is stored in ATP by adding a third phosphate group to ADP. Energy is released from ATP by transferring the end phosphate to another molecule. (4-3.2)

29. Enzyme activity can be controlled by (1) a signal molecule that binds to an enzyme and changes its shape, activating or repressing the enzyme's activity; or (2) a pathway product that

4 CHAPTER REVIEW ········

8. Metabolism
   a. includes processes that capture, transfer, and use energy.
   b. is a series of exergonic reactions that provide energy to cells.
   c. functions only in closed systems.
   d. enables photosynthesis to occur without energy from the sun.

9. A coupled reaction
   a. increases the amount of ATP in a cell.
   b. includes the transfer of energy from ATP.
   c. has two parts that are endergonic.
   d. enables a cell to produce energy.

10. When a product molecule attaches to the allosteric site of an enzyme, the
   a. active site disappears.
   b. formation of product increases.
   c. enzyme's shape changes.
   d. substrate concentration decreases.

Completion

11. The energy of position is called _____ energy, while the energy of motion is called _____ energy.

12. Energy transformations tend to make _____ become more disorganized. _____ is a disorganized and less useful form of energy.

13. Simple organization and random molecular motion are characteristics of a system with a high degree of _____ .

14. Energy needed for life is captured during _____ and used by cells during cellular respiration. Much of the energy released during cellular respiration is used to make _____ .

15. All chemical reactions require _____ , which is lowered by the process of

_____ .

16. A substance that speeds the progress of a chemical reaction is called a(n) _____.

17. In a biochemical pathway, the _____ of one enzyme-catalyzed reaction becomes the _____ for the enzyme catalyzing the next reaction.

18. When an enzyme's _____ changes, its activity also changes.

19. Cells use _____ to move, to change their shape, and to build new molecules.

20. A series of enzyme-catalyzed reactions makes up a(n) _____ .

21. Coupled reactions transfer energy from _____ to power _____ reactions.

22. NAD$^+$ is a _____ that functions in cell metabolism. Your cells extract the electrons and protons needed to convert NAD$^+$ to NADH from _____ .

23. The binding of a product molecule to a key enzyme that shuts down a biochemical pathway is called _____ .

Short Answer

24. Living organisms, such as animals and plants, appear to violate the second law of thermodynamics because of their high degree of organization. Explain why these observations are misleading.

25. What are oxidation-reduction reactions? Describe how these reactions are used by cells.

26. Explain how the presence of a catalyst affects the rate of a chemical reaction.

27. An enzyme's efficiency increases with greater substrate concentration, but only to a point. Why?

28. Explain how energy is stored in and released from ATP molecules.

29. Describe two ways in which cells are able to control the activity of enzymes that are present in the cytoplasm.

Themes Review

30. Flow of Energy Describe the flow of energy that results in your being able to obtain energy for metabolism by eating a T-bone steak.

31. Homeostasis Explain why every reduction reaction is accompanied by an oxidation reaction.

binds to a key enzyme in a pathway and shuts down the pathway. (4-3.4)

Themes Review

30. Light energy from the sun is converted to chemical energy in grass by the process of photosynthesis. The grass is eaten by a cow that uses the energy

available in the grass to build body tissue. Steak that comes from the cow is eaten by humans and provides them with energy. (4-1.3)

31. A substance can lose an electron only when another substance accepts an electron. (4-1.4)

32. In cooler temperatures, the enzymes that catalyze the chemical reactions that produce pain-causing nerve impulses operate more slowly. (4-2.1, 4-2.3)

32. **Levels of Organization** A tennis trainer may treat a player's painful but minor elbow injury by spraying it with a surface anesthetic containing ethyl chloride. The anesthetic rapidly evaporates, cooling the skin and reducing the pain. Realizing that pain is caused by chemical reactions, explain how you think this treatment works.

Critical Thinking

33. **Using Analogies** Using an analogy, develop a model to describe the concept of entropy.

34. **Making Inferences** For every 100 units of energy available from alfalfa, cattle that eat the alfalfa capture only 10 units of that energy, and humans that eat the cattle as hamburger capture only 1 unit of that same energy. How do you account for this inefficient transfer of energy?

35. **Interpreting Data** How do the two reactions represented below compare in terms of energy?

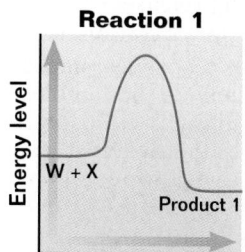
Reaction 1

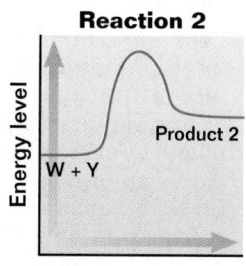
Reaction 2

36. **Making Predictions** The figure below depicts a metabolic pathway. Predict how a high concentration of Substrate C would affect the activity of Enzyme 1 if the shape of Enzyme 1 is changed when it binds with Substrate C.

Substrate A $\longrightarrow$ Substrate B $\longrightarrow$ Substrate C
$\quad\quad$ Enzyme 1 $\quad\quad\quad$ Enzyme 2

Activities and Projects

37. **Cooperative Group Project** Using modeling clay, build a model of an enzyme and its substrate. In your model, show the enzyme's active site, how the enzyme changes shape when it binds to its substrate, and how the enzyme returns to its original shape when the products of the reaction are released.

38. **Research and Writing** Find out about the relationship between vitamins and coenzymes. In a written report, identify several coenzymes and the vitamins with which they are associated.

39. **Research and Writing** Investigate the lethal effect of hydrogen cyanide on the human body. In a written report, discuss the chemical structure of hydrogen cyanide, and explain its effects in terms of its competitive inhibition of the enzyme cytochrome oxidase.

40. **Multicultural Perspective** Interview an expert on, or do research on, the following martial arts: tai chi, tae kwan do, and aikido. Find out what these forms of exercise have in common. How do they relate to the flow of energy? Where and how did each originate? Relate your findings in a report to your class.

Readings

41. Read Elizabeth Pennisi's article "Juiced-up fruit: Unbelievably flavorful," in *Science News*, September 14, 1991, page 173. How did Berger and his associates make use of natural biosynthetic pathways to enhance the flavor of fruits?

42. Read David Freedman's article "Life's Off-Switch," in *Discover*, July 1991, pages 61–67. Arthur Kornberg believes that "life is chemistry." What is his latest discovery? Why did Kornberg once receive a telephone call from President Lyndon Johnson?

Activities and Projects

37. Student models will vary, but they should parallel the actions shown in Figure 4-9 on page 83.

38. Vitamins are required for the synthesis of several coenzymes that affect the health of humans. Niacin is necessary for the presence of NAD^+; B_2, or riboflavin, is necessary for FAD; and B_1, or thiamine, is necessary for the coenzyme thiamine pyrophosphate.

39. Hydrogen cyanide irreversibly binds to the enzyme cytochrome oxidase, which is present in all cells and is important in cellular respiration. The irreversible binding of the poison to the enzyme blocks the enzyme's activity and accounts for the lethal effect.

40. Tai chi (China), tae kwon do (Korea), and aikido (Japan) are forms of philosophical, psychological, and artistic exercise based on a proper balance of mental concentration, physical energy, and harmony with the rhythms of nature.

Readings

41. Berger and his associates enhanced the flavors of apples, pears, and bananas by storing them in containers with alcohols that are the chemical precursors of fruity esters, which give fruits their flavors and odors.

42. Kornberg's most recent discovery is a protein (IciA) that prevents DNA from replicating—life's "off switch." In 1967, Kornberg received a call from President Johnson after news reports credited Kornberg with "creating life in a test tube."

Critical Thinking

33. Student analogies will vary, but they should show that a system to which energy is not added becomes more disorganized and more stable. An example is that students' bedrooms become disorderly if energy is not expended to keep them clean. (4-1.2)

34. As the second law of thermodynamics indicates, there is not a perfect exchange of energy between living systems. When energy is transferred from one organism to another, some energy is lost as heat. (4-1.2)

35. Reaction 1 is an energy-releasing (exergonic) reaction, while reaction 2 is an energy-absorbing (endergonic) reaction. (4-2.1)

36. A high concentration of C will shut off enzyme 1 by changing its shape, preventing it from converting substrate A to substrate B. (4-3.4)

Preparation
Students should bring in labeled samples (3–5 tablespoons) of each of five different detergents. Labels should include active ingredients listed on the container, as well as whether enzymes or enzyme action is mentioned. You may wish to have different groups use sugar-free and regular gelatin so that they can compare their results.

Procedural Notes

Day 1
- Students should be able to complete steps 1 through 5 in 30 minutes or less.
- Caution students to avoid burns by working carefully when heating and pouring boiling water.
- As students add the washing soda to the gelatin, the mixture will foam. During this reaction, CO_2 gas is released. The addition of washing soda to the gelatin raises the pH of the gelatin from 4 to 8, which is the optimum pH for protease activation in the detergent samples. Students can test and record the pH of the gelatin before and after they add the washing soda.

Day 2
- Students should be able to complete steps 6 through 10 in 30 minutes or less.
- To prepare a 10 percent solution of laundry detergent, students should dissolve 10 g of detergent in 90 mL of distilled water.

LABORATORY Investigation Chapter 4

Observing Enzyme Activity

OBJECTIVE

Recognize the presence and function of enzymes in commercial products.

PROCESS SKILLS

- organizing data
- predicting results

MATERIALS

- detergents (various brands)
- safety goggles
- lab apron
- 18 g of regular instant gelatin or 1.8 g of sugar-free instant gelatin
- 0.7 g washing soda (Na_2CO_3)
- 150 mL beaker
- glass stirring rod
- boiling water
- 6 test tubes
- test-tube rack
- graduated cylinder
- Pasteur pipette
- plastic wrap
- tape
- wax pencil
- 50 mL beakers
- pH paper
- balance

BACKGROUND

1. What is an enzyme? What is a substrate?
2. What environmental factors influence enzyme activity?
3. Many enzymes are named after their substrates. What kind of substrate might the enzymes known as *proteases* act upon?
4. Write your own question to explore in your lab report or notebook.

TECHNIQUE

Day 1

1. Bring a sample of laundry detergent from home in a plastic bag. Be sure to write down the ingredients of the product as they are listed on the label. Note any mention of the word *enzyme* either in the ingredients list or on the label.

2. Put on safety goggles and a lab apron.

3. Place 18 g of regular instant gelatin or 1.8 g of sugar-free instant gelatin in a beaker. Slowly add 50 mL of boiling water and mix, using a stirring rod. **CAUTION: Be careful handling boiling water.** To this hot gelatin solution, very slowly add 0.7 g of washing soda (Na_2CO_3) while stirring. Record your observations. What gas do you think is being released?

4. Pour 5 mL of the liquid gelatin–Na_2CO_3 mixture into each of 6 test tubes placed in a test-tube rack. Remove the bubbles on the top of each tube with a Pasteur pipette. Cover the tubes tightly with plastic wrap and secure with tape. Allow these to cool at room temperature or in a refrigerator until Day 2.

5. Clean up your materials and wash your hands before leaving the lab.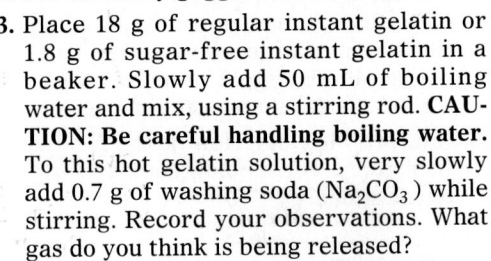

Day 2

6. Put on safety goggles and a lab apron.

7. Use a wax pencil to mark the test tubes at the upper level of the cooled gelatin in each tube. It is from this point that you will measure the breakdown, or hydrolysis, of the gelatin from day to day. Label the test tubes 1–6.

8. Prepare a 10 percent solution of each of five different detergents brought in by

Background Answers

1. An enzyme is a protein that triggers and controls particular chemical reactions. A substrate is the reactant molecule to which an enzyme binds while it is catalyzing a reaction.

2. Temperature, pH, and the presence of other substances influence enzyme activity.

3. Answers may vary. Proteases break down proteins. Examples include enzymes in laundry detergents that help break down food or other stains containing proteins, and enzymes in drain cleaners that help break down food, hair, and other drain-clogging substances made of protein.

4. What evidence is there that protease enzymes exist in laundry detergent?

you and your classmates. Test the pH of each with your pH paper. Record the pH for each numbered detergent sample in the **Records** section of your report.

Hydrolysis of Gelatin

Detergent	pH	Amount of Hydrolysis (mm)
1		
2		
3		
4		
5		

9. Add 15 drops (1 mL) of the first detergent solution to the gelatin surface of the first test tube. Repeat for each of the other samples. Reseal the tubes and place in a test-tube rack for each observation. To the sixth tube add only 15 drops (1 mL) of water to the gelatin surface. Put the tubes aside for 24 hours at room temperature.

 10. Clean up your materials and wash your hands before leaving the lab.

Day 3

11. After 24 hours at room temperature, draw another wax pencil line at the top of the gelatin layer. Then measure the distance in millimeters between the first line and the second line. This indicates the amount of hydrolysis of the protein in the liquid gelatin by the enzymes in the detergent. Record your data in the chart in the **Records** section of your report.

12. Clean up your materials and wash your hands before leaving the lab.

Inquiry Answers

1. The sixth test tube was a control to test whether water alone hydrolyzes protein in gelatin.
2. Enzymes in the detergents broke down protein in the gelatin, reducing the gelatin levels from Day 2 to Day 3.
3. The washing soda increases the pH of the gelatin from 4 to 8—the optimum pH for enzyme activity in this reaction.
4. Answers will vary.

Analysis Answers

1. Enzymes are added to detergents to help break down proteins from food and other substances that may stain clothes.
2. Answers will vary. Students should realize that enzymes tend to break down at high temperatures. However, students may consider that enzymes for laundry detergents can be genetically engineered to be more stable at high temperatures. Students may also conclude that such enzymes can be genetically engineered to be stable in the presence of bleach.
3. Answers will vary, but should focus on proteins such as those found in food stains.
4. Enzymes have an optimum temperature range for catalyzing reactions. Because the samples are kept for 24 hours at room temperature, the temperature can affect the activity of the enzymes if it becomes too cool or warm.

Records

Hydrolysis of Gelatin		
Detergent	pH	Amount of Hydrolysis (mm)
1		
2		
3		
4		
5		
6 Water		

UNIT 1

CHAPTER 5 — PHOTOSYNTHESIS AND CELLULAR RESPIRATION

Block Scheduling Guide

Each block represents about 45–50 minutes of instructional time. The resources cited in a block are coded to a specific program component using the Key on the next page.

Lecture Concepts	Lesson Resources

5-1 How Energy Cycles pp. 94–97

BLOCK 1

a. Photosynthesis and respiration form a continuous cycle because the products of one process are the starting materials for the other.

b. Photosynthesis uses carbon dioxide, water and light energy and produces carbohydrates and oxygen. Cellular respiration uses carbohydrates and oxygen and produces carbon dioxide, water, and energy for cell activities.

c. Energy is stored in organic molecules made by linking carbon atoms together. Excess carbohydrates produced by plants provide food for animals.

d. Energy stored in organic molecules is passed from producers to consumers through food chains.

Lecture Resources
- Opening Question, page 94
- Demonstration: Recycling in a Terrarium
- Overcoming Misconceptions, page 95
- Visual Strategies: Figures 5-1, 5-2
- Effective Reading, page 96
- What If Something Destroyed All of the Organisms in One of the Levels of a Food Chain? page 97
- Teaching Transparencies: 13

- Holt Biology Videodiscs: Lesson 5-1

Classwork Options
- Quick Labs A5: Interpreting Labels: Stored Food Energy
- Inquiry Skills Development B4: Plant and Animal Interrelationships

Assignment Options
- Section Review, page 97
- Chapter Review, questions 1, 11, 12, 18, 23

5-2 How Photosynthesis Works pp. 98–106

BLOCK 2

a. In plants and most other photosynthetic organisms, photosynthesis occurs inside chloroplasts. These membrane-bound organelles contain the pigment chlorophyll, a light-absorbing compound.

b. Light energy is packaged in photons. When a photon of light with the right amount of energy is absorbed by an atom, one of the atom's electrons is raised to a higher energy state.

c. Chlorophyll and other photosynthetic pigments absorb photons. These pigments are arranged in molecule clusters, each of which makes up a photosystem. Most photosynthetic organisms have two kinds of photosystems: photosystem I and photosystem II.

Lecture Resources
- Opening Questions, page 98
- Visual Strategies: Figures 5-3, 5-4, 5-5, 5-6
- Effective Reading, page 98
- Demonstration: Chocolate Mints as Thylakoids

- Research Update, page 100
- Connections, page 101
- Holt Biology Videodiscs: Lesson 5-2

BLOCKS 3 and 4

a. Light energy is converted to chemical energy by a series of reactions that is initiated when photons are absorbed by the photosystems. The two photosystems work in tandem.

b. Excited electrons from photosystem II aid in the production of ATP. Excited electrons from photosystem I aid in the production of NADPH. Water molecules are split, releasing the electrons needed to replace those lost by photosystem II. Oxygen gas is produced as a result.

c. ATP and NADPH are used to fix carbon in the Calvin cycle. For every six carbon dioxide molecules that enter the carbon cycle, one six-carbon sugar molecule can be produced.

d. Light, carbon dioxide concentration, temperature, and other environmental factors interact with one another to determine the optimum level of photosynthesis for a particular plant in its environment.

Lecture Resources
- Order of Photosystems, page 102
- Visual Strategies: Figures 5-7, 5-8, 5-9
- Teaching Transparencies: 12, 18, 19
- A Source of Energy and Organic Building Materials, page 105
- Overcoming Misconceptions, page 105
- Demonstration: How Light Intensity Affects Photosynthetic Rate
- Holt Biology Videodiscs: Lesson 5-2

Classwork Options
- Photosystem II and Graphic Organizer, page 102
- Photosystem I and Graphic Organizer, page 103
- Laboratory Investigation: Chromatography of Plant Pigments, pages 116–117
- A Summary of Photosynthesis, page 105

Assignment Options
- Section Review, page 106
- Chapter Review, questions 2–5, 13–15, 19, 20, 22, 24, 26–28

5-3 How Cellular Respiration Works pp.107–112

BLOCKS ⑤ and ⑥

a. Cellular respiration begins with glycolysis. A cell gains two pyruvate molecules, two ATP molecules, and two NADH molecules for every glucose molecule entering this anaerobic pathway.

b. When oxygen is present, oxidative respiration follows glycolysis and produces a large amount of ATP, as well as electron carriers and carbon dioxide.

c. Fermentation follows glycolysis when no oxygen is available for oxidative respiration.

d. Fats, proteins, and nucleic acids, in addition to glucose, can be oxidized by cellular respiration to make ATP.

e. A system of feedback inhibition enables cells to control the rate of cellular respiration.

Lecture Resources
- ■ Opening Questions, page 107
- ■ Demonstration: Burning an Organic Fuel
- ■ Visual Strategies: Figures 5-11, 5-12, 5-13, 5-14, 5-15, 5-16
- 🔖 Teaching Transparencies: 20, 21, 22, 23
- ■ Connections, page 109
- ■ Electron Transport Chain Analogy, page 110
- ■ Overcoming Misconceptions, page 110
- ■ The Making of ATP, page 111
- ■ Connections, page 112

⊘ Holt Biology Videodiscs: Lesson 5-3

Classwork Options
- 🧪 Laboratory Techniques and Experimental Design C8: Measuring the Release of Energy— Best Food for Yeast
- ■ A Summary of Cellular Respiration, page 111
- ■ Closure, page 112

Assignment Options
- ■ Section Review, page 112
- ■ Chapter Review, questions 6–10, 16, 17, 21, 22, 25, 28, 29

Review/Enrichment

BLOCK ⑦

Review/Enrichment
- ■ Themes of Biology: Levels of Organization, page 104
- ■ Themes Review, questions 23–25
- ■ Study Guide: Concept Review and Pretest

Assignment Options
- ■ Activities and Projects, questions 30–33
- ■ Readings, questions 34, 35

Assessment Options

BLOCK ⑧

Traditional Assessment
- ■ Chapter 5 Test
- ⊚ Test Generator/Assessment Item Listing: Software item bank for preparing customized chapter tests.

Portfolio Assessment
Select student reports for one or more laboratory experiments from this chapter. *The Direct Observation Checklist* on the *BioSources Teaching Resources CD-ROM* should be completed during a laboratory or other cooperative learning experience.

Answer to Concept Map
The following is one possible answer to the Concept Mapping exercise on page 113.

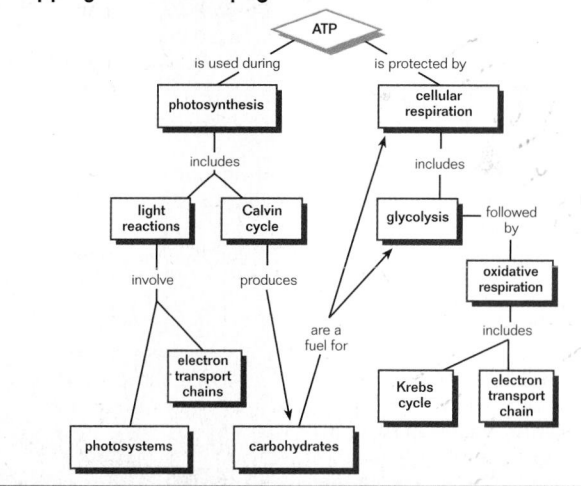

Holt Biology Videodiscs   CONCEPTS OF BIOLOGY

Holt Biology Videodiscs gives you a powerful tool for teaching, review, and assessment. *Concepts of Biology* adds interactive lesson plans and extensive tools for customization using CD-ROM technology.

Use the following topic from *Concepts of Biology* to help you teach this chapter:

- ■ Topic 5: Photosynthesis and Cellular Respiration

For further information regarding the media on the videodiscs, see *Holt Biology Videodiscs Teacher's Correlation Guide to Biology: Principles and Explorations.*

Chapter Themes

Flow of Energy *This chapter's most pervasive theme is apparent on both a global and cellular level. Figure 5-2 shows how energy flows through organisms. Figures 5-7, 5-8, and 5-14 show how energy flows within cells and between the living and nonliving world.*

Structure and Function *Though students will not be introduced to this theme until Chapter 6, use Figure 5-4 to help them infer that structure facilitates function.*

Levels of Organization *On Earth, the structure of living systems extends from the atomic level, where atoms combine to form molecules, to the biosphere, which includes all organisms. The feature on page 104 introduces this theme, and Figure 5-4 illustrates it.*

Tapping Prior Knowledge

- How is energy transported in living systems?
- What cell organelle makes food by trapping sunlight?
- What cell organelle releases energy stored in food?

Opening Question

To introduce the idea that certain chemicals are essential for life, ask students what raw materials living things need to sustain life. They should list water, oxygen, carbon dioxide (autotrophs only), and food (heterotrophs only). If students simply list "air" as the gas that living things need, stress that "air" is not a sufficient answer in this case because two of the gases that organisms get from air have different functions. Emphasize that air is made of several gases (N_2, 78.03%; O_2, 20.99%; CO_2, 0.023–0.050%; others, approximately 0.93%).

CHAPTER 5
PHOTOSYNTHESIS AND CELLULAR RESPIRATION

REVIEW

- proton pumps and chemiosmosis (Section 3-2)
- autotrophs and heterotrophs (Section 4-1)
- laws of thermodynamics (Section 4-1)
- oxidation-reduction reactions (Section 4-1)
- coenzymes and coupled reactions (Section 4-3)

Leaves reflecting sunlight.

Authors' Rationale

The discussion of energy continues here. Energy is stored in the chemical bonds of organic molecules when they are formed and released when these chemical bonds are rearranged. Essentially, the energy-storing bonds are formed during photosynthesis, and then their energy is released during the complementary process of cellular respiration. Through this chapter, students will see where energy comes from, what it does, and where it goes. Energy is examined as it is supplied by the sun, as it is trapped by photosynthesis, as it dwindles through a food chain, and as it is ultimately lost as heat.

5-1 How Energy Cycles

*L*ike a rechargeable battery, your body eventually runs low on energy and needs to be supplied with more. You get this energy from food. The energy in your food was first captured from sunlight by photosynthesis. You extract that energy by the process of cellular respiration. Together, photosynthesis and cellular respiration form a cycle, seen in Figure 5-1, that links organisms to each other and to the environment.

Section Objectives

- Explain why photosynthesis and cellular respiration form a continuous cycle.
- Describe what happens to the sugars produced during photosynthesis.
- Discuss the importance of food chains.

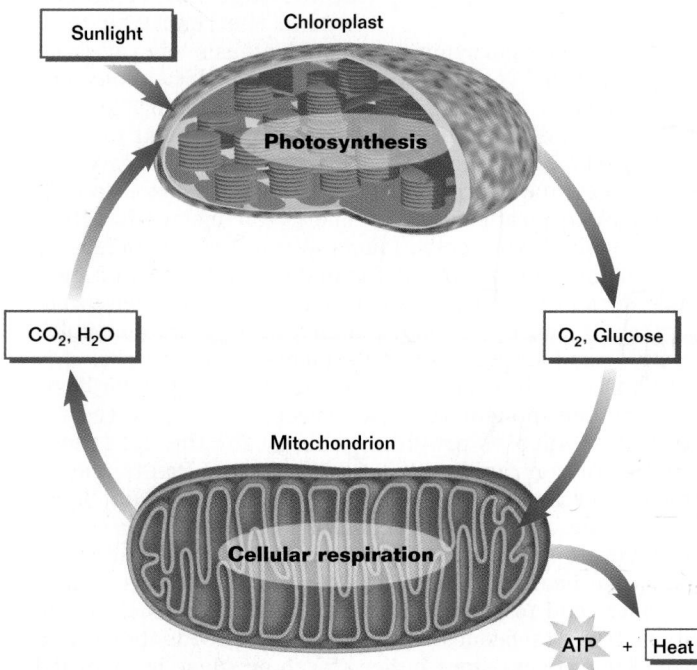

Figure 5-1 Photosynthesis and cellular respiration form a continuous cycle because the products of one process are the starting materials for the other.

Photosynthesis and Cellular Respiration Form a Cycle

The carbon atoms needed to make all the organic molecules of living things ultimately come from a nonliving part of the environment. During photosynthesis, carbon atoms are pulled from the carbon dioxide gas, CO_2, in air and used to form carbohydrates and other organic compounds. These materials are made, step by step, in biochemical pathways that are powered by transferring hydrogen atoms from one reaction to the next. The hydrogen atoms needed for these pathways are extracted from water molecules, H_2O. Leftover oxygen atoms combine to form oxygen gas, O_2, as a byproduct.

Overcoming Misconceptions

Plants Conduct Cellular Respiration

Students often think that because photosynthesis occurs only in autotrophs, cellular respiration occurs only in heterotrophs. Emphasize that cellular respiration occurs in *all* organisms, from the simplest bacteria to the most complex animals and plants.

SECTION 5-1

Vocabulary Preview

crop

producer

consumer

food chain

Opening Question

Ask students why energy is necessary to living systems. Students should recall that energy enables living systems to carry out the functions of life.

Demonstration

Recycling in a Terrarium

Materials: Aquarium or large glass jar with a tight-fitting lid, aquarium gravel, potting soil, 1 plant cutting (e.g., jade plant, geranium, impatiens), 1 small animal (e.g., pill bug, millipede, snail). **Procedure:** Place 3 cm of gravel in the bottom of the terrarium, cover with 6 cm of potting soil, and water until soil is moist. Plant the cutting in the soil, introduce the animal, seal the terrarium, and place it in *indirect* sunlight. Students can observe it for several days. Ask: Is the humidity in the terrarium high or low? *(high)* What is your evidence? *(water droplets on the sides)* Why put both a plant and an animal in the terrarium? *(They recycle raw materials for each other.)*

VISUAL STRATEGY

Figure 5-1

Emphasize that this figure summarizes two complex metabolic pathways, showing how they are linked. Point out the starting materials and products for each process and how the products of each process become the starting materials of the other. Also trace the flow of energy from the sun to the chloroplast to the mitochondrion to ATP. Ask: Is this an open system or a closed system? *(open)* What scientific law describes the production of heat? *(second law of thermodynamics)*

CONNECTIONS

Chapter 2

Source of Nutrients for Life

Remind students that the materials needed to sustain life ultimately come from the environment. Living systems are efficient recycling factories. The materials (elements and compounds) needed to sustain life cycle through the biosphere in the processes of photosynthesis and cellular respiration.

Chapter 4

Energy Accounting

Review the fact that while some of the energy trapped during photosynthesis goes into making chemical bonds that link carbon atoms, some of it is lost during the energy transformation process, as the second law of thermodynamics predicts.

Chapter 4

Carbon Compounds: The Energy Conduit

Remind students that carbon compounds, which store the energy from boosted electrons in their bonds, are a transportable form of energy and that carbon compounds act as an energy supply conduit for the biosphere.

☐ **CAPSULE SUMMARY**

Photosynthesis uses carbon dioxide, water, and light energy and produces carbohydrates and oxygen. Cellular respiration uses carbohydrates and oxygen and produces carbon dioxide, water, and energy for cell activities.

CONTENT LINK

The origin and importance of several crop plants are discussed in **Chapter 26.**

☐ **CAPSULE SUMMARY**

Energy is stored in organic molecules made by linking carbon atoms together. Excess carbohydrates produced by plants provide food for animals.

Cellular respiration is essentially the reverse of photosynthesis. During cellular respiration, hydrogen atoms are pulled from carbohydrates and joined with oxygen atoms from oxygen gas, forming water. In biochemical pathways, the carbon chains of carbohydrates and other organic molecules are systematically dismantled. Energy stored in the molecules is then made available for the activities of cells. Carbon atoms that are split from carbohydrates are linked to oxygen gas, forming carbon dioxide gas as a byproduct. ☐

Energy Is Stored in Carbon Compounds

Some of the energy that is required to link carbon atoms together is stored within the resulting molecules. In plants carrying out photosynthesis, energy from sunlight is stored within newly made carbohydrate molecules such as sugars. Sugars can be transported from leaves to other parts of a plant. In that way, parts of a plant that are not exposed to light, such as roots, are supplied with carbohydrates. Later, the energy in sugars can be converted to ATP by cellular respiration, making energy for metabolic activities available to root cells. Plants also use sugar molecules made during photosynthesis for making other organic molecules, such as proteins, lipids, nucleic acids, and other carbohydrates. Because they make their own organic molecules directly from inorganic materials, plants are autotrophs.

Plants usually produce more sugars by photosynthesis than they can immediately use. Excess sugar is stored for future uses such as providing energy for the next day's activities or for rapid growth in the spring. Plants mainly convert excess sugar to starch for both short-term and long-term storage. As you will recall from Chapter 2, starch is a carbohydrate made of large numbers of glucose molecules joined, like beads in a necklace, by chemical bonds. Starch molecules tend to cluster in certain plant cells, such as the cells of fruits (avocados and bananas) and seeds (beans and corn). Some plants store either starch or sugar in modified stems (potatoes and sugar cane) or in modified roots (carrots and yams). Excess sugar may also be converted to oils and stored within seeds (peanuts and safflower seeds).

Because they are rich in organic molecules that store energy, plants often serve as food for animals. Of the more than 250,000 known species of plants, about 150 are cultivated (grown) extensively as food. A type of plant that is cultivated for use by humans is called a **crop**. The major food crops grown in the world today have been cultivated for thousands of years, selected by early farmers for their food value and ease of cultivation. For example, just three species of grasses—rice, wheat, and corn—directly or indirectly supply more than half of all human energy needs. ☐

Effective Reading

Getting Students Involved

Lack of interest is a barrier to active reading. Many students are not interested in abstract facts, especially ones with no apparent connection to their own experience. Library research projects that relate to

subjects in the text can provide the connections students need to become interested, active learners. To stimulate students' interest, have them research the lives of people who worked out seemingly small but important parts of the photosynthesis

puzzle, such as Jan Baptista van Helmont, Joseph Priestley, Jan Ingenhousz, Antoine Lavoisier, and Nicholas de Saussure. To begin Section 5-2, have students report to the class what each contributed to the study of photosynthesis.

Energy Flows Through Food Chains

In a sense, energy moves through the living world on the "shoulders" of carbon atoms. The journey begins when carbon atoms captured from carbon dioxide gas are incorporated into organic molecules. In the process, energy is stored in organic molecules that can serve as food for other organisms. Autotrophs (plants, algae, and certain bacteria), which make food molecules by capturing energy and carbon atoms from their environment, are called **producers**. Once captured by producers, energy stored in organic molecules can be passed to heterotrophs, which are not able to capture energy and carbon directly from their environment. Because they must obtain energy and organic molecules from other organisms, heterotrophs are also called **consumers**. Some consumers obtain most of their food by eating producers. Some eat other consumers. When you eat beef, for example, you are a consumer eating another consumer (a cow) that ate a producer (grass). A series of organisms through which energy flows from a producer to one or more consumers, as illustrated in Figure 5-2, is called a **food chain**. ◻

◻ **CAPSULE SUMMARY**

Energy stored in organic molecules is passed from producers to consumers through food chains.

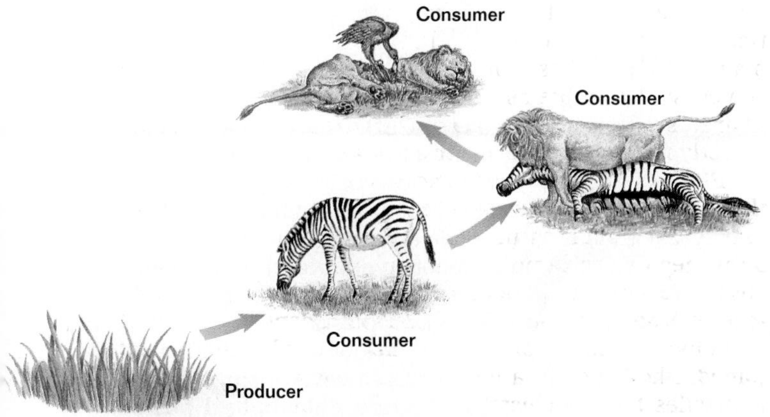

Consumer

Consumer

Consumer

Producer

Figure 5-2 A food chain consists of a series of organisms through which energy flows. Because some energy is lost every time it is transferred from one organism to another (second law of thermodynamics), most food chains consist of no more than three or four organisms.

Section Review

1. *How do photosynthesis and cellular respiration form a continuous cycle?*
2. *How do plants use the products of photosynthesis?*
3. *How are the organic molecules produced as a result of photosynthesis important to you and other heterotrophs?*

Critical Thinking

4. *Because of the second law of thermodynamics, food chains normally involve no more than three or four organisms. Why do you think this is so?*

SECTION 5-1

◉ **VISUAL STRATEGY**

Figure 5-2
Identify the producer and the consumers in this diagram. Emphasize that a loss of thermal energy (heat) occurs with each transfer. Point out that the vulture is a scavenger, which is a type of consumer that eats only dead organisms. Also point out that decomposers such as fungi and bacteria are also consumers and play an important role in cycling materials back to the producers.

Teaching Tip

What If something destroyed all of the organisms in one of the levels of a food chain? The balance (homeostasis) of an ecosystem depends on the presence of each of the types of organisms in its food chains. If a producer were destroyed, all of the consumers that traced the source of their food back to that producer would be affected (their numbers would decline or they might all die). The loss of a first-level consumer, such as the zebra, would cause the numbers of the next levels of consumers to decline. The loss of a second-level consumer, such as the lion, could initially cause the numbers of the first-level consumer (zebra) to increase, which could in turn cause the numbers of the producer to decrease and lead eventually to a decrease in the numbers of the first-level consumer. Even the decomposers return to the soil valuable nutrients needed by producers.

Answers to Section Review

1. The products of photosynthesis become the reactants of cellular respiration, and vice versa.

2. Plants use carbohydrates as an energy source to sustain themselves when no sunlight is available and for metabolic activities in their roots and other nonphotosynthetic tissues. They also use carbohydrates as raw materials for making the other carbohydrates, fats, proteins, and nucleic acids they need. Plants use oxygen in cellular respiration.

3. Heterotrophs depend on the organic molecules produced by photosynthesis as an energy source and as the raw material for making other organic compounds.

4. Energy is lost at each step of a food chain. Since no new energy can be created by each level, the total amount of usable energy in the system dwindles as it moves through a food chain. Ultimately, not enough energy is available to support another level.

Vocabulary Preview

thylakoid

granum

stroma

radiant energy

electromagnetic spectrum

photon

pigment

chlorophyll

carotenoids

photosystem

photosystem I

photosystem II

reaction center

electron transport chain

ATP synthetase

reducing power

NADPH

NADP⁺

carbon fixation

Calvin cycle

Opening Questions

Ask students how sunlight is important to living systems. Students should recall that it is the main source of the energy stored in carbon compounds. Ask what ATP is. Students should recall that it is the main energy currency of cells. Finally, ask how energy gets from sunlight to ATP. Make sure that students understand that energy is carried by electrons through a biochemical pathway that results in ATP production.

👁 VISUAL STRATEGY

Figure 5-3

Correlate this figure with the overview of photosynthesis. Point out the reactants and products. Emphasize that each of the three stages actually consists of many steps.

Effective Reading

Interpreting Formulas

Point out that the summary equation for photosynthesis contains the formula CH_2O. Explain that this is an empirical formula, representing the kinds of atoms present in a compound and their lowest

5-2 How Photosynthesis Works

Earth is bathed in energy streaming from the sun. Each day, our planet is bombarded with an amount of energy equal to the energy of about 1 million atomic bombs the size of the one dropped on Hiroshima, Japan, in 1945. Approximately 1 percent of this energy is captured by photosynthesis, providing the energy upon which almost all life on Earth depends.

Overview PHOTOSYNTHESIS

Photosynthesis takes place in three stages.

1. Energy is captured from sunlight.
2. Light energy is converted to chemical energy (ATP and NADPH).
3. ATP and NADPH power the synthesis of organic molecules, using carbon from carbon dioxide.

In plants, photosynthesis occurs primarily within leaves. Cells inside leaves contain organelles called chloroplasts, which house chlorophyll, the light-absorbing substance needed for photosynthesis. In all plants and photosynthetic protists, photosynthesis occurs within chloroplasts.

Photosynthesis also occurs in certain bacteria that contain chlorophyll but do not have chloroplasts. Some of the organic molecules produced during photosynthesis store energy for later needs. Others are used as building blocks for the macromolecules that make up a cell.

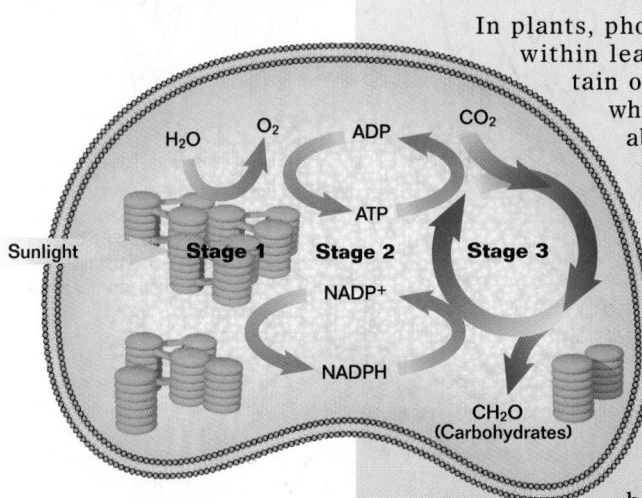

Figure 5-3 Photosynthesis occurs in three distinct phases: energy is captured from sunlight, converted to chemical energy, and stored in organic molecules.

The overall process of photosynthesis can be summarized by the following chemical equation:

$$CO_2 + H_2O + light \xrightarrow[enzymes]{chlorophyll} CH_2O + O_2$$

carbon dioxide   water   energy   carbohydrate   oxygen

whole-number ratio. Write this formula, the word *carbohydrate*, and the formula $C_6H_{12}O_6$ on the board or overhead projector. Have students make two lists, one of the similarities between the formulas and a second of the differences between them. Ask: What is the chemical formula for water?

(H_2O) What does the suffix -*hydrate* mean? (*water*) Why is it appropriate to call any compound with the empirical formula CH_2O a carbohydrate? (*Such compounds combine carbon with hydrogen and oxygen in the same ratio that would result if a carbon atom were added to a water molecule.*)

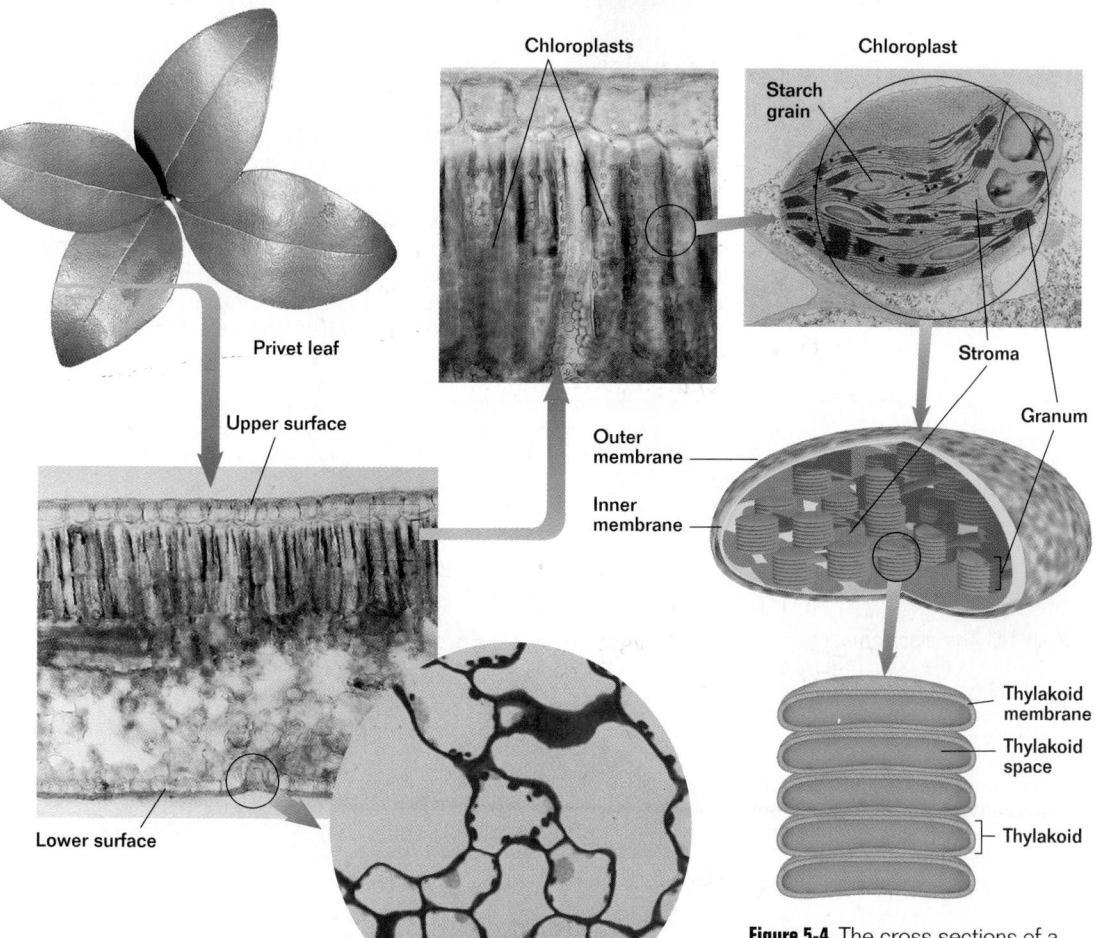

Chloroplasts

Chloroplast

Starch grain

Privet leaf

Upper surface

Lower surface

Opening

Outer membrane

Inner membrane

Stroma

Granum

Thylakoid membrane

Thylakoid space

Thylakoid

Figure 5-4 The cross sections of a privet leaf, *top left*, show that most of its chloroplasts are located in the upper part of the leaf. Tiny openings in the bottom surface of the leaf, *bottom center,* enable carbon dioxide to enter and oxygen and water vapor to exit. A photomicrograph of a chloroplast, *top right,* reveals its internal structure. The stroma surrounding the grana contains the enzymes needed for making carbohydrates. The cross section of a single granum, *bottom right,* shows that each thylakoid is a closed compartment. Chlorophyll molecules are embedded in the thylakoid membranes.

Plants Are Specialized for Photosynthesis

You may not think of plants as being exceptional. However, even the simplest leaf is a wonderfully complex and sophisticated photosynthetic machine, in which structure and function form a near-perfect union. Figure 5-4 shows the structure of one type of leaf and how it is adapted for photosynthesis. In plants and most other photosynthetic organisms, photosynthesis occurs inside chloroplasts, which are tiny green organelles found only in eukaryotic cells. If you could cut into a chloroplast, you would find that its interior is surprisingly complex. As Figure 5-4 also shows, the interior of a chloroplast is filled with many flattened membrane-bound sacs called **thylakoids** (*THEYE lah koydz*). Thylakoids are often stacked like coins, in columns called **grana** (*GRAN uh*). A chloroplast usually contains dozens of grana and individual thylakoids, which are all suspended in a fluid matrix called the **stroma** (*STROH muh*).

 VISUAL STRATEGY

Figure 5-4

Point out the photosynthetic and outer layers of the leaf. State that leaves are generally thin and their outer cells have no chloroplasts, which allows sunlight to penetrate to the photosynthetic cell layers inside. Openings in the leaf surface enable CO_2 to enter and O_2 to leave. Vascular cells (not shown) bring water from the roots and carry away the products of photosynthesis. Ask: Where in the leaf do you think most photosynthesis occurs? (*tightly packed upper layers*) Why? (*They have more chloroplasts and are closer to the light.*) How is having a loosely packed layer of photosynthetic cells on the lower side of the leaf an advantage? (*Spaces between cells enable CO_2 to reach inner layers of cells and O_2 to exit.*) Point out the starch grain (a carbohydrate) visible in the chloroplast.

Demonstration

Chocolate Mints as Thylakoids

Use chocolate mints approximately 5 cm in diameter with a center of a contrasting color. Make stacks of four or five mints. On one or two of the stacks, cut the top mint in half to expose the center before putting it on the stack. Point out that each stack of mints represents a column of thylakoids, or a granum. To make the model more realistic, connect the grana with strips of paper between each layer to represent the membranous connections between thylakoids. Point out that a fluid, the stroma, surrounds the grana. Ask: What color should this model really be? (*green*) Why is it better for a granum to be a stack of thylakoids instead of a single, larger membrane-bound sac? (*Stacks increase the surface area for light absorption in a chloroplast.*) Instruct students to make a sketch of one granum and label its parts. (*Student diagrams should label a thylakoid, the granum, and the stroma.*)

☑ **RESEARCH Update**

UPDATE

Will Photosynthesis Provide Job Security for Plants in the Future?

J. Barber at the Imperial College of Science in London and B. Andersson at Stockholm University contend that the light-harvesting mechanisms of photosynthesis have been all but worked out, down to the physical and chemical levels. The crux of energy conversion in photosynthesis is, according to plant biologists, the pigment-protein complexes (photosystems) found in the cell membrane of green plants (and other photosynthetic eukaryotes). These complexes confer a tremendous advantage to plants, allowing them to use several wavelengths of light in the visible-light spectrum as well as sporadic or unpredictable supplies of light. As a result of photosynthetic processes, autotrophs produce 100 times the amount of food required by humans and recycle the atmosphere's carbon dioxide every 300 years and its oxygen every 2,000 years.

Barber and Andersson suggest that with the right equipment, artificial photosynthesis, and therefore an increased exploitation of solar energy, may be possible in the future. Researchers in photosynthesis have already been able to construct crude artificial photochemical systems. Still, we cannot continue to destroy the phytoplankton and rain forests, our most important producers. ☑

Light Energy Is Packaged in Photons

What is there in sunlight that a plant can use to make sugars? Light is a form of **radiant energy**, which is energy that is transmitted in waves that can travel through a vacuum. The complete range of radiant energy forms is called the **electromagnetic spectrum**. As you can see in Figure 5-5, the many different forms of radiant energy differ both in wavelength and in the amount of energy they transmit. All forms of radiant energy actually consist of tiny packets of energy called **photons** (FOH tahnz).

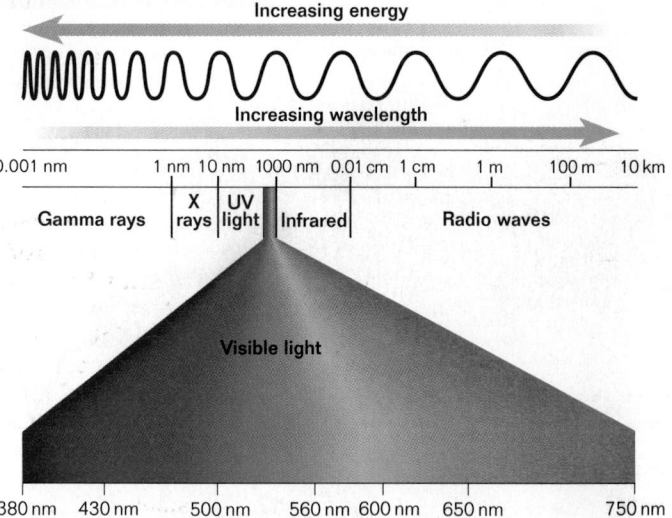

Figure 5-5 The electromagnetic spectrum is the complete range of all the types of radiant energy. The wavelengths of some electromagnetic waves are measured in millionths of a millimeter, or nanometers (nm). Photons of visible light (wavelengths of 380 nm to 750 nm), which includes all the colors of light in sunlight, have intermediate amounts of energy.

As sunlight shines on an object, the object's surface is bombarded by energetic photons. The photons striking an object are either reflected or absorbed by the object, or they are transmitted (passed) through it. An object's color results from the visible light photons it reflects. Human eyes see visible light because they have light receptors that can absorb those kinds of photons. Leaves absorb mostly photons of blue and red light and reflect mostly photons of green light, making them appear green to us. A plant uses some of the energy in the photons it absorbs to make sugars.

How do human eyes and plants "choose" which photons to absorb? The answer to this question is in an atom's structure. Recall that electrons spin in energy-specific regions near an atom's nucleus. When a photon strikes an atom, energy in the photon may boost one of the atom's electrons to a higher energy state. Just as you must lift your foot only so far to step to the next rung of a ladder, so boosting an electron to a particular energy state requires a precise amount of energy. Thus, a particular atom absorbs only photons of certain kinds of light—those with just the right amount of energy. ☐

☐ **CAPSULE SUMMARY**

When a photon of light with the right amount of energy is absorbed by an atom, one of the atom's electrons is raised to a higher energy state.

👁 **VISUAL STRATEGY**

Figure 5-5
Point out that the energy of a photon of electromagnetic radiation increases as the wavelength of the radiation decreases. Shorter wavelengths have more energy; longer wavelengths have less energy.

❓ **Did You Know?**

We are so accustomed to using visible light in picturing our world that we often forget that other wavelengths of the electromagnetic spectrum are also useful. Radar, which uses the longest waves in the spectrum, is useful for picturing large objects some distance away from the observer, while X rays, which are very short waves, are useful for picturing microscopic objects.

Photosynthetic Pigments Absorb Photons

A molecule containing atoms that enable it to absorb light is called a **pigment**. **Chlorophyll,** which is a green pigment, is the primary light-absorbing agent for photosynthesis. Most plants contain two types of chlorophyll, chlorophyll *a* and chlorophyll *b*, that both play important roles in photosynthesis. The yellow and orange plant pigments that produce fall colors, and the colors of many fruits, vegetables, and flowers, are **carotenoids** (*kuh RAH tuh noydz*). Carotenoids assist in photosynthesis by capturing energy from light of different wavelengths than those absorbed by chlorophyll. Figure 5-6 shows the wavelengths of light absorbed by chlorophylls and carotenoids as well as how these pigments affect photosynthesis.

The pigments used in photosynthesis are arranged in molecule clusters, each of which is called a **photosystem**. These clusters, which contain both chlorophylls and carotenoids, are embedded in the thylakoid membranes inside chloroplasts. In photosynthetic bacteria, which lack chloroplasts, photosystems are located in membranes inside the bacterial cell. Most photosynthetic organisms have two kinds of photosystems. **Photosystem I** clusters boost electrons to a higher energy state by absorbing light with a wavelength of 700 nm. **Photosystem II** clusters boost electrons by absorbing more energetic light with the slightly shorter wavelength of 680 nm. Energy from a photon strike anywhere in a photosystem is funneled to particular chlorophyll *a* molecules that are located in the **reaction center** of the photosystem. Upon receiving the right amount of energy from a photon strike, electrons from the reaction center's chlorophyll *a* are boosted to higher energy states.

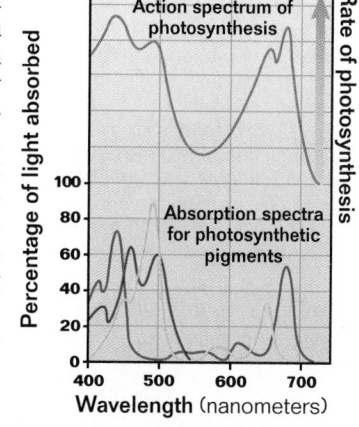

Figure 5-6 The absorption spectra, *bottom,* show the wavelengths of light absorbed by photosynthetic pigments. Chlorophylls absorb mostly red, blue, and violet light. Carotenoids absorb mostly blue and green light. The action spectrum, *top,* shows the combined effect of all photosynthetic pigments on the rate of photosynthesis.

Light Energy Is Converted Into Chemical Energy

In the second stage of photosynthesis, light energy is converted into chemical energy by a series of reactions that is initiated when light is absorbed by the photosystems. These reactions are commonly called the "light reactions" or the "light-dependent reactions." The two types of photosystems work in partnership, one passing electrons to the other like runners passing a baton in a relay race. The easiest way to understand how the two photosystems work together is to follow the "baton"—an electron.

Action of Photosystem II

In a sense, photosystem II clusters are the first to act during the light reactions. The absorption of light by photosystem II pigments initiates a series of events that results in the production of ATP. First, photons of light striking a chloroplast

SECTION 5-2

👁 VISUAL STRATEGY

Figure 5-6
Discuss the similarities and differences between these graphs. Have students record the wavelengths that seem to stimulate photosynthesis. Compare those wavelengths with the ones in Figure 5-5 on page 100, and have students identify the colors involved. Do these wavelengths correspond to the colors mentioned in the caption? (*yes*)

Teaching Tip
Photosystems I and II Are Light-Energy Antennas
Compare the structure and function of a photosystem to an antenna or a satellite dish. Antennas and satellite dishes gather and concentrate radio signals and pass them along to a receiver, such as a television or radio. Photosystems are clusters of light-absorbing pigments and associated proteins in the thylakoid membranes. Energy gathered by the light-absorbing pigments is passed to a central location, the reaction center (a special chlorophyll *a* molecule), which is able to relay the energy (by giving up an excited electron) to a receiver in the cell membrane.

CONNECTIONS

Chapter 4
Electron Emissions
Remind students that when electrons are boosted to higher energy states (levels) within an atom, they store energy. It follows, then, that when an electron drops back to a lower energy state, it emits energy. These emissions often take the form of electromagnetic waves that are always specific to the magnitude of the drop in energy. Emissions may be above the visible range of the electromagnetic spectrum, within the visible range, or below the visible range. Scientists use these specific emissions as a "fingerprint" to identify the atoms emitting the waves.

 Did You Know?

Orange and yellow vegetables are rich sources of carotenoids. One of the carotenoids is our most important dietary source of vitamin A, which is necessary for proper eyesight, for maintaining the health of membranes, and for tooth and bone development. The carotenoid beta-carotene breaks down into two molecules of vitamin A. Evidence also suggests that beta-carotene may provide some measure of protection against certain cancers.

Photosystem II

Have students draw an imaginative model of how ATP is generated as a result of the absorption of light energy by a photosystem II pigment cluster. Students can use the *Graphic Organizer* at the bottom of page 102 as a guide. Student models should also show that light is gathered by the photosystem, that a photon boosts an electron in the reaction center's chlorophyll, and that the electron is passed to an electron transport chain that siphons off energy each time the electron is passed to another molecule in the chain. Have students write a paragraph describing their model.

Order of Photosystems

Students may be confused by the fact that they are asked to look at photosystem II first. Tell them that photosystem I is so named because it evolved first.

VISUAL STRATEGY

Figure 5-7

Encourage students to follow the path of the electron coming from the photosystem II reaction center. Ask: How does the concentration of protons inside the thylakoid compare with that outside? *(It is greater inside.)* How is this difference in concentration used? *(to make ATP by chemiosmosis)* After students have read about photosystem I on page 103, return to this diagram and stress the relationship between photosystems II and I—electrons boosted in photosystem II are passed to photosystem I, not vice versa. Use the *Graphic Organizers* on pages 102 and 103 to summarize the processes shown in this figure. **Note:** The two types of photosystems do not really occur side by side in thylakoid membranes. They are shown side by side here to emphasize how they relate to one another.

Figure 5-7 Photosystems and electron transport chain molecules are embedded in the thylakoid membrane. Notice that the electron transport chain that accepts electrons from photosystem II contains a molecule that pumps protons into a thylakoid. As protons leave a thylakoid via ATP synthetase, ATP is generated.

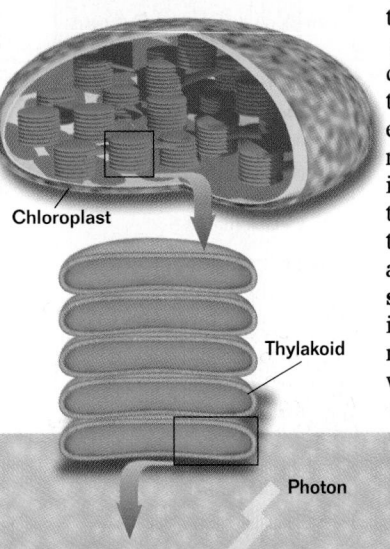

are absorbed by photosystem II pigment molecules, causing electrons from the reaction center's chlorophyll *a* to be boosted to a higher energy state. Traveling as part of a hydrogen atom, each excited electron leaves its chlorophyll molecule and jumps to a nearby membrane protein. Then, like the baton in a relay race, each excited electron is passed through a series of membrane-bound protein and pigment molecules called an **electron transport chain**. As excited electrons pass through this electron transport chain, some of their energy is used to power ATP production. Eventually, the electrons are accepted by photosystem I clusters.

To see how an electron's energy is used to generate ATP during photosynthesis, look at Figure 5-7. One of the electron transport chain molecules acts as a proton pump, using the energy of the excited electrons to "pump" protons (hydrogen nuclei) across the thylakoid membrane, into the thylakoid's interior. Because a thylakoid is a sealed compartment, protons cannot diffuse back out. Instead, they build up until there is enough pressure to drive them out through the only available exit—the ATP-generating protein channel **ATP synthetase**. Recall from Chapter 3 that making ATP by forcing protons through a membrane channel is called chemiosmosis. The ATP is released into the stroma of a chloroplast, where it is used in the production of organic molecules.

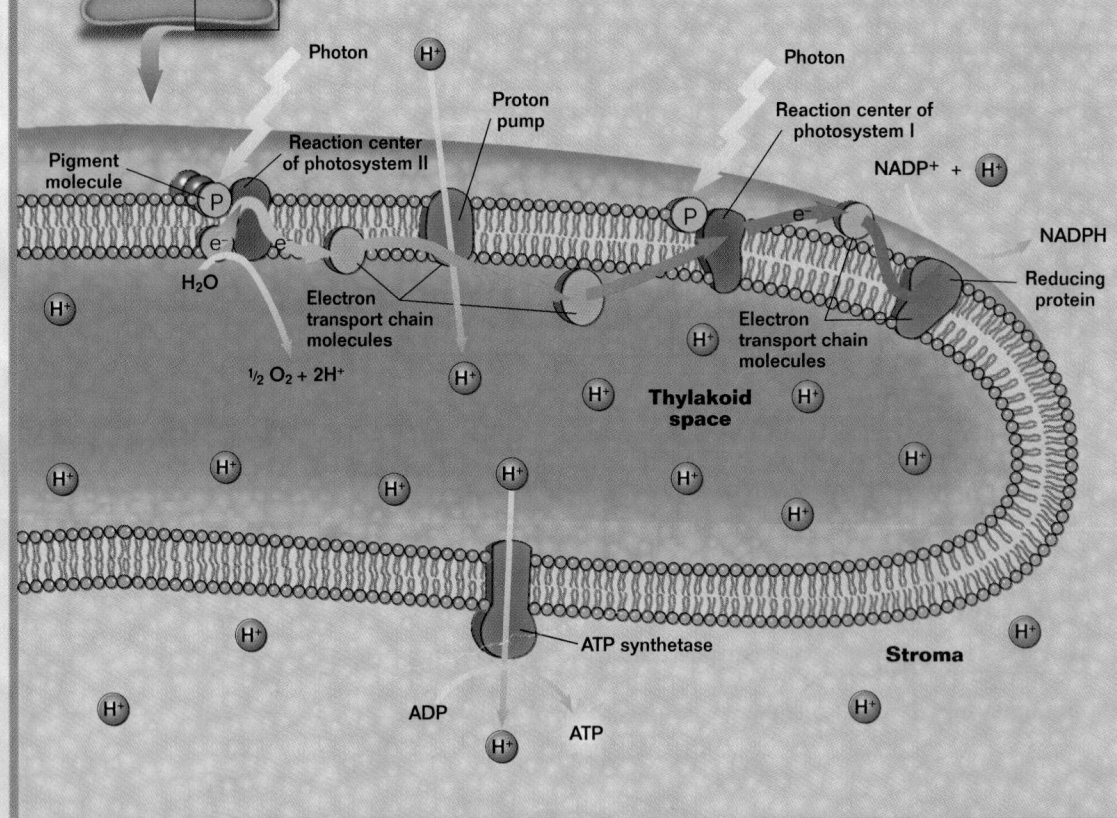

Graphic Organizer

Use this graphic organizer with *Teaching Tip:* **Photosystem II** and the *Visual Strategies* for **Figures 5-7** and **5-8.**

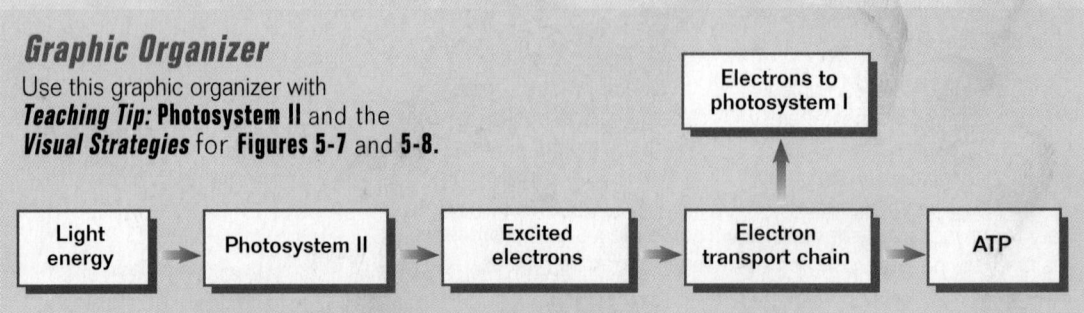

Action of Photosystem I

The point of photosynthesis, however, is not simply to make ATP. Rather, it is to extract carbon atoms from carbon dioxide and use them to make new organic molecules, such as sugars, lipids, proteins, and nucleic acids. Synthesizing such molecules takes a lot of energy. ATP provides this energy, but something else is needed as well. Organic molecules contain hydrogen as well as carbon and oxygen. Because carbon dioxide has no hydrogen atoms, a source of attachable hydrogen atoms is needed for synthesizing organic molecules. Recall that the addition of hydrogen atoms to a molecule is called reduction. Biologists therefore refer to a ready supply of attachable hydrogen atoms as **reducing power**.

The absorption of light by photosystem I pigments initiates a series of events that results in the generation of reducing power, which is in the form of NADPH molecules. **NADPH** is the reduced form of NADP+ (nicotinamide adenine dinucleotide phosphate). Like NAD+, **NADP+** is a coenzyme that carries hydrogen atoms (and energy) from one place to another in a cell. The NADPH is released into the stroma of a chloroplast, where both the hydrogen atoms and the energy NADPH carries are used in the production of organic molecules. Remember, NADPH does not result from the action of photosystem I alone. Figure 5-8 illustrates how the two types of photosystems work together to convert light energy into chemical energy.

Figure 5-8 The two types of photosystems work together simultaneously to convert light energy to chemical energy in the form of ATP and NADPH. Notice that electrons transferred to NADPH still contain excess energy.

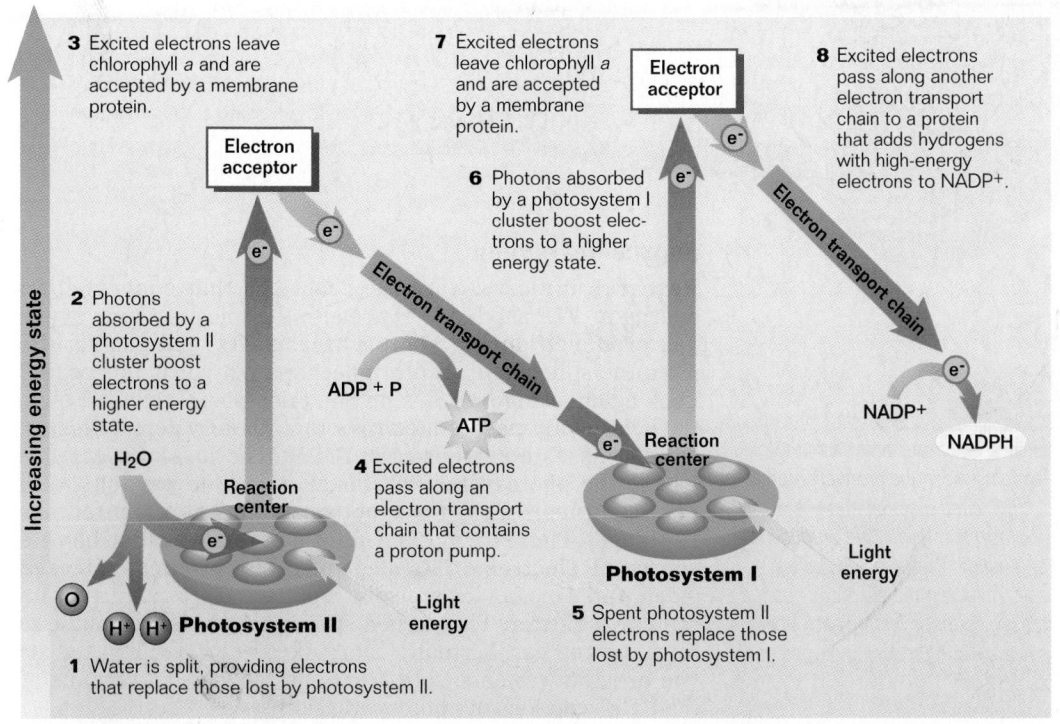

SECTION 5-2

Teaching Tips
Photosystem I
Have students add the action of a photosystem I pigment cluster to the model they drew for *Teaching Tip:* **Photosystem II** on page 102. Students can use the *Graphic Organizer* at the bottom of page 103 as a guide.

What If a plant had no available NADP+? The homeostasis of a photosynthesizing plant depends on a continual supply of materials needed for photosynthesis, including those supplied by the plant. If no NADP+ were available to accept electrons from photosystem I, the flow of electrons in the light reactions would back up like water in a clogged drain pipe. Photosystem I could not give up electrons and thus could not absorb light energy or accept electrons from photosystem II. With no acceptor for its electrons, photosystem II could not give up electrons and thus could not absorb additional light energy either.

👁 VISUAL STRATEGY

Figure 5-8
Use this figure, called a "Z diagram," to review the light reactions. Explain that electrons lost by photosystem II replace electrons lost by photosystem I. After students have read the subsection titled "Source of Oxygen" on page 104, return to this diagram and ask: What replaces the electrons lost by photosystem II? *(electrons from the splitting of water)* Also point out that the electrons passed to NADPH still contain some of the energy absorbed from light photons.

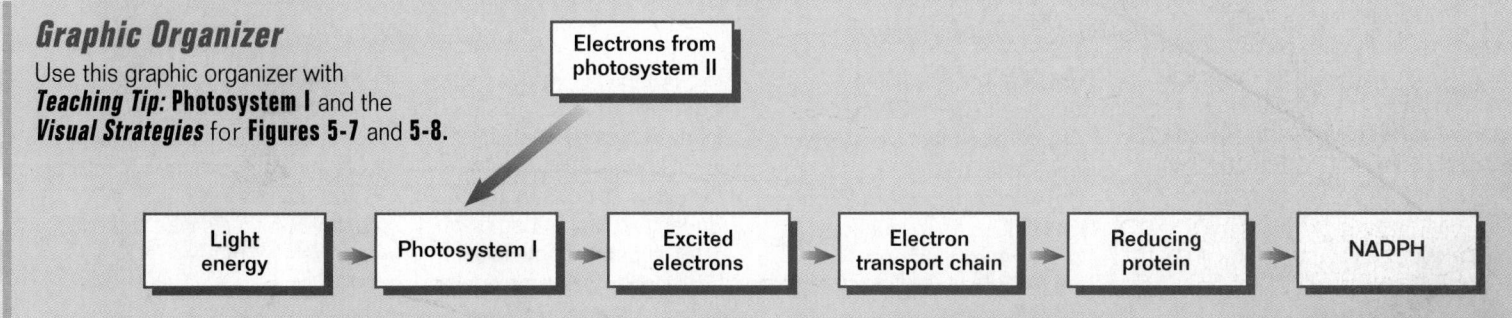

Graphic Organizer
Use this graphic organizer with *Teaching Tip:* **Photosystem I** and the *Visual Strategies* for **Figures 5-7** and **5-8**.

UNIT 1

CHAPTER 5

Teaching Tips

**Themes of Biology:
Levels of Organization**

In this chapter, students look at several of life's levels of organization. After students have read the feature, point out examples of the levels that relate to photosynthesis. (*Atoms make up the chlorophyll molecule. Chlorophyll and the other molecules that participate in electron transport and ATP generation are part of the thylakoid membrane. Thylakoids are part of the grana, and grana and stroma are part of a chloroplast. Chloroplasts are part of plant cells, which make up a plant. Plants are part of the food chains in the ecosystems that comprise the biosphere.*)

What If something happened to one of the levels of organization of life?

In this chapter, students also learn that life's levels of organization are interdependent. To maintain homeostasis, each level depends on its sublevels as well as the levels of which it is a part. For example, the kinds and arrangement of the atoms in a chlorophyll molecule enable it to capture light energy. On a much larger level, the products of photosynthesis enable energy to move through the biosphere. If one of the elements necessary for making chlorophyll were no longer available to a plant, its homeostasis would be lost, and photosynthesis would stop. Trace the consequences of such a loss through as many of life's levels of organization as you can.

Interdependencies

Have students draw or cut out examples of the levels of organization of life and paste the pictures on posterboard. Then have students connect the levels and write a sentence for each connection explaining how the connected levels are interdependent.

Atoms make up the chlorophyll molecules that are part of many plant cells. Different types of cells, organized into tissues, make up the leaves of a plant.

Levels of Organization

Organization is characteristic of all living things. At their most fundamental level of organization, living things are composed of atoms and molecules. These atoms and molecules interact in systems of chemical reactions (biochemical pathways) that direct the basic processes of life. Biochemical pathways that carry out particular processes, such as photosynthesis and cellular respiration, are organized within organelles in most cells.

Cells are the next level of organization. They are the smallest level of organization that can be considered alive. Multicellular organisms such as yourself consist of many different cell types that are each specialized to do different things. Cells of the same type form tissues such as muscles and nerves. Finally, the different tissues of your body can be found working together in organs, biological machines (such as the heart and the liver) that carry out particular jobs.

No organism lives in isolation, however. At the highest levels of organization, living things interact with each other. Organisms live together in populations, which are groups of individuals of a particular species. The different kinds of populations that live in a place make up a community. A community and its physical environment make up an ecological system, or ecosystem. Taken together, all of Earth's ecosystems make up the biosphere, the highest level of life's organization.

☐ CAPSULE SUMMARY

Excited electrons from each photosystem II aid in the production of ATP. Excited electrons from each photosystem I aid in the production of NADPH. Water molecules are split, releasing the electrons needed to replace those lost by each photosystem II and producing oxygen gas as a result.

Source of Oxygen

Research indicates that the reactions that convert light energy to ATP and NADPH are also the source of the oxygen gas produced during photosynthesis. Oxygen gas results from the splitting of water molecules. Why does this occur? Look again at Figures 5-7 and 5-8. Each photosystem II cluster continually passes electrons to a photosystem I cluster, which passes electrons to NADPH. The loss of electrons leaves the photosystem II cluster unable to participate in photosynthesis until the electrons it has contributed are replaced. Therefore, each photosystem II must have a source of electrons. The electrons (and protons) released when water molecules are split replace those lost by a photosystem II cluster. The oxygen atoms that are left combine to form oxygen gas. Virtually every oxygen molecule in the air you breathe was once split from a water molecule by one of the light reactions of photosynthesis. ☐

Historical Note

The Source of Oxygen
Work done by C. B. Van Niel with photosynthetic bacteria was the first to indicate that the O_2 released during photosynthesis does not come from CO_2. The bacteria Van Niel studied use hydrogen sulfide, H_2S, instead of H_2O in photosynthesis. This process is summarized by the equation $CO_2 + H_2S \rightarrow CH_2O + S + H_2O$. Experiments done by Samuel Ruben and Martin Kamen in 1941 supported Van Niel's hypothesis. Using water with oxygen-18 and carbon dioxide with normal oxygen-16, they showed that all the oxygen-18 atoms end up in the O_2. Therefore, the overall equation for photosynthesis was amended to $CO_2 + H_2O \rightarrow CH_2O + O_2 + H_2O$.

Chemical Energy Is Stored in Organic Molecules

In the final stage of photosynthesis, the chemical energy of ATP and NADPH is used by enzymes. These enzymes incorporate carbon atoms from carbon dioxide into organic molecules, a process called **carbon fixation.** The reactions that "fix" carbon to build organic molecules are sometimes called the "dark reactions" or the "light-independent reactions." Actually, they are neither. Although carbon fixation *can* occur in the dark, it only occurs if the products of the light reactions, ATP and NADPH, are present. Thus, while the reactions do not use light *directly*, they are not "light-independent." Furthermore, carbon dioxide enters the leaves of most plants through openings that close at night. In most plants, therefore, the reactions that fix carbon *do not* occur in the dark.

Among photosynthetic organisms, there are actually several biochemical pathways in which carbon is fixed. The most common carbon-fixing pathway is called the **Calvin cycle,** honoring Melvin Calvin. During the 1940s and 1950s, Calvin and his associates at the University of California at Berkeley worked out the cycle's reactions. The Calvin cycle, which occurs in all plants and algae, employs a complex battery of enzymes that are found in the stroma of a chloroplast. These enzymes form a cycle because they regenerate the starting material for further reactions. A total of six carbon dioxide molecules must enter the Calvin cycle to produce one six-carbon sugar molecule. Figure 5-9 illustrates the important events of the Calvin cycle. ▢

CONTENT LINK
....
*Alternate carbon-fixing pathways for photosynthesis are discussed in **Chapters 24 and 25.***

▢ **CAPSULE SUMMARY**

ATP and NADPH are used to fix carbon in the Calvin cycle. For every six carbon dioxide molecules that enter the Calvin cycle, one six-carbon sugar molecule can be produced.

Figure 5-9 The Calvin cycle is a series of reactions that attaches carbon atoms from carbon dioxide molecules to growing carbon chains while regenerating the starting material to which the carbon atoms are initially attached.

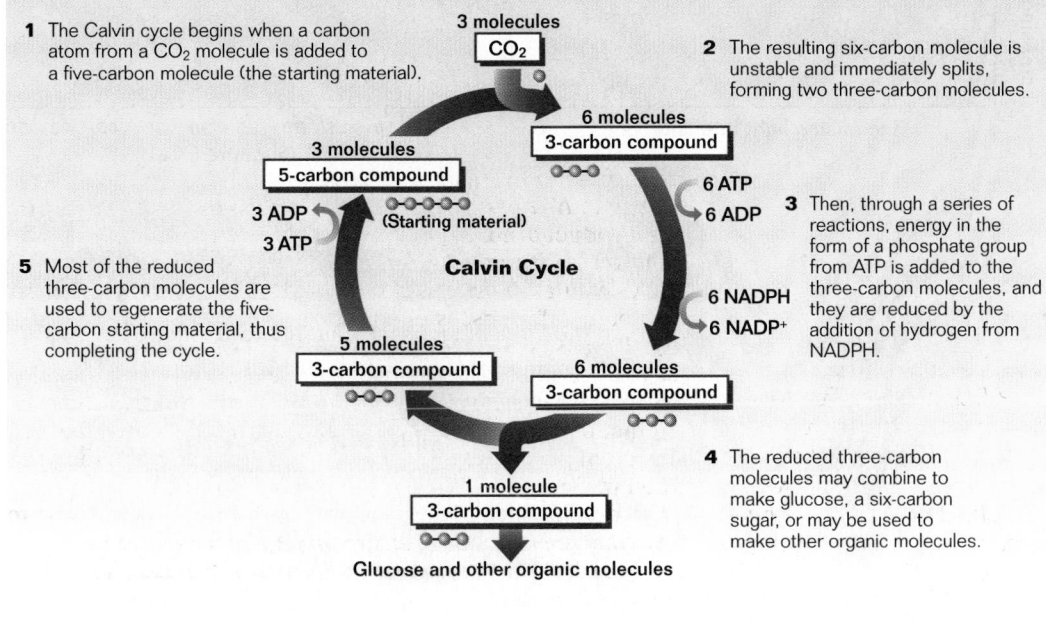

1 The Calvin cycle begins when a carbon atom from a CO_2 molecule is added to a five-carbon molecule (the starting material).

3 molecules
CO_2

2 The resulting six-carbon molecule is unstable and immediately splits, forming two three-carbon molecules.

6 molecules
3-carbon compound

3 molecules
5-carbon compound

3 ADP
3 ATP
(Starting material)

6 ATP
6 ADP

3 Then, through a series of reactions, energy in the form of a phosphate group from ATP is added to the three-carbon molecules, and they are reduced by the addition of hydrogen from NADPH.

Calvin Cycle

6 NADPH
6 NADP+

5 Most of the reduced three-carbon molecules are used to regenerate the five-carbon starting material, thus completing the cycle.

5 molecules
3-carbon compound

6 molecules
3-carbon compound

1 molecule
3-carbon compound

4 The reduced three-carbon molecules may combine to make glucose, a six-carbon sugar, or may be used to make other organic molecules.

Glucose and other organic molecules

SECTION 5-2

Teaching Tips

A Source of Energy and Organic Building Materials

Point out that not only plants but all heterotrophs depend on the organic compounds produced by autotrophs during photosynthesis (and chemosynthesis), both for energy and as a source of carbon skeletons for making the specific organic molecules they need to build and run their bodies.

A Summary of Photosynthesis

Have students add the Calvin cycle to their models of the light reactions and develop a summary equation of photosynthesis that includes the materials needed (CO_2, H_2O, chlorophyll, light energy, $NADP^+$, ADP, and P), the intermediates (NADPH, ATP), and the products (CH_2O, O_2, H_2O, $NADP^+$, ADP). Have students save their models for further work in Section 5-3.

👁 **VISUAL STRATEGY**

Figure 5-9

Have students count the total number of carbon atoms present at each step in this summary of the Calvin cycle. Explain that the names of the individual intermediates are not as important as understanding that inorganic carbon dioxide is incorporated into organic compounds and the starting material is regenerated as a result of the cycle. Ask: How many CO_2 molecules are needed to produce a glucose molecule, $C_6H_{12}O_6$? (*six*) Point out where intermediates produced during the light reactions (NADPH and ATP) are used in the Calvin cycle. Emphasize that the main product of the Calvin cycle (and hence photosynthesis) is *not* glucose but a three-carbon compound (glyceraldehyde 3-phosphate, or PGAL) that plants use in the production of other organic compounds, including carbohydrates (glucose, fructose, sucrose, starch, and cellulose), fatty acids, amino acids, and nucleic acids.

Historical Note

Calvin-Benson Cycle

In 1945, Melvin Calvin began using CO_2 labeled with ^{14}C to trace its path in photosynthesis. He and Andrew Benson worked out the steps of the Calvin cycle. Calvin won a Nobel Prize in chemistry in 1961 for his work.

Overcoming Misconceptions

Glucose Is Not a Direct Product of Photosynthesis

Very little free glucose results from photosynthesis. Inside a chloroplast, the three-carbon compounds produced by the Calvin cycle are converted to starch. When transported to the cytosol, they are combined to form sucrose. Thus, starch and sucrose (table sugar) are the main carbohydrates produced by photosynthesis. Both are converted to glucose for cellular respiration.

Demonstration
How Light Intensity Affects Photosynthetic Rate

Materials: 2 large beakers, 2 small beakers that fit inside the large beakers, 2 sprigs of *Elodea* (available from pet stores), 2 glass funnels that fit inverted into the large beakers, 2 ring stands, 2 test-tube clamps, 1 lamp with a 100-watt bulb, and distilled water (or tap water treated with a dechlorinating agent and allowed to sit for a day).

Procedure: Set up *two* apparatuses as shown below. Fill the beakers with water, place one sprig of *Elodea* in the smaller beaker, then fill the test tube with water, and invert it over the stem of the funnel. Place one apparatus in a dark area, and place the other near the light source. Have students predict the outcome of the demonstration. (**Caution:** Handle the light source with care to avoid burns. Do not place the light source too close to the apparatus, or the heat may damage the *Elodea*.) Allow the demonstration to run for 30 min. Have students compare the amount of O_2 (air space) in each tube and answer the following questions in writing: What did you predict would happen? (*Answers will vary.*) What do you think would happen if the heat became too great in the apparatus? (*The rate of photosynthesis would decrease.*) Why? (*Plant enzymes would be damaged.*)

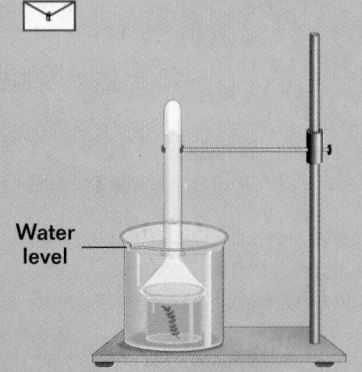

Water level

Figure 5-10 A graph of the effect of light intensity on photosynthesis, *left,* shows that the rate of photosynthesis increases as light intensity increases, but only to a certain point. A graph of the effect of temperature on photosynthesis, *right,* shows that as temperature increases, the rate of photosynthesis increases to a certain point and then decreases.

Environment Affects the Rate of Photosynthesis

Photosynthesis is directly affected by environmental factors. The most obvious of these factors is light. In general, the rate of photosynthesis increases as light intensity increases, until a point called the light saturation point is reached. At that point, the rate of photosynthesis levels off, as the graph on the left in Figure 5-10 indicates. Carbon dioxide concentration affects the rate of photosynthesis in a similar manner. Once a certain concentration of carbon dioxide is present, the reactions of photosynthesis cannot proceed any faster. The temperature graph on the right in Figure 5-10 shows that photosynthesis also operates best within a certain range of temperatures. Remember that like all biochemical pathways, photosynthesis is a series of enzyme-catalyzed reactions. As you learned in Chapter 4, enzymes operate properly only within certain temperature ranges. Light, carbon dioxide concentration, temperature, and other environmental factors interact with one another to determine the optimum level of photosynthesis for a particular plant in its environment.

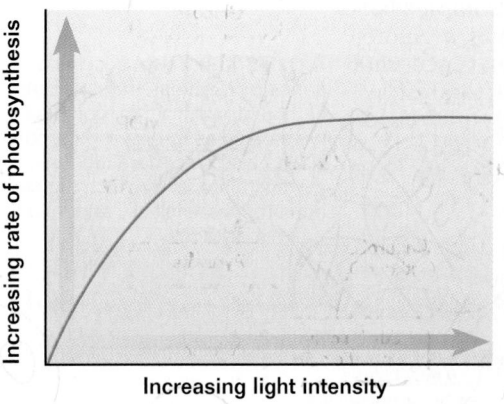

Increasing rate of photosynthesis

Increasing light intensity

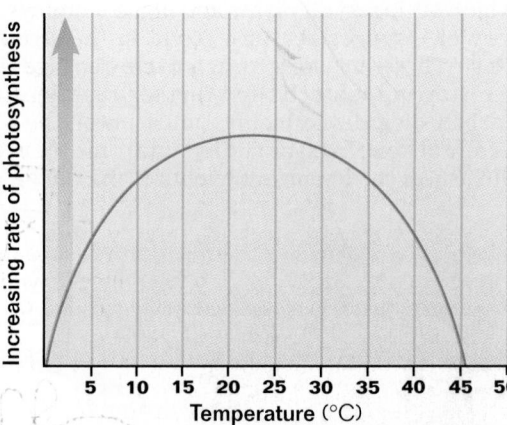

Increasing rate of photosynthesis

Temperature (°C)

Section Review

1. List the three main events that make up photosynthesis.
2. Why are pigments necessary for photosynthesis?
3. What is a photosystem?
4. What is the role of water in photosynthesis?
5. How does the Calvin cycle depend on the light reactions?

Critical Thinking

6. What combination of environmental factors would produce an optimum rate of photosynthesis for most plants?

5-3 How Cellular Respiration Works

Almost all organisms, autotrophs and heterotrophs, obtain energy for their activities from organic molecules assembled during photosynthesis. Energy that was invested in building these molecules is retrieved by stripping electrons from them and using these electrons to make ATP. This process, called cellular respiration, is possible because the electrons have extra energy obtained from a chlorophyll molecule's encounter with a photon of light.

Overview   CELLULAR RESPIRATION

Cellular respiration takes place in two stages.

1. Glucose is converted to **pyruvate** (*peye ROO vayt*), producing a small amount of ATP and NADH.
2. When oxygen is present, pyruvate and NADH are used to produce a large amount of ATP. When oxygen is absent, pyruvate is converted to lactic acid or ethyl alcohol.

The breakdown of glucose into pyruvate, which does not require oxygen, occurs in the cytosol of all cells. The conversion of pyruvate and NADH to ATP occurs in the mitochondria of all eukaryotic cells. In bacterial cells, which lack mitochondria, the conversion of pyruvate and NADH to ATP occurs in the cell membrane. Many of the intermediate products of cellular respiration are used to produce organic molecules needed for building and maintaining cells.

Metabolic processes that require oxygen are termed **aerobic** (*air OH bihk*), from the Greek word *aer*, meaning "air." Metabolic processes that do not require oxygen are termed **anaerobic** (*AN air oh bihk*), meaning "without air." Because far more ATP per food molecule results when oxygen is present, the aerobic pathways of cellular respiration are the primary source of energy for most cells.

The overall process of cellular respiration can be summarized by the following equation:

$$CH_2O + O_2 \xrightarrow{\text{enzymes}} CO_2 + H_2O + 36ATP$$

carbohydrate   oxygen   carbon dioxide   water   energy

Section Objectives

- Summarize the processes of cellular respiration.
- Identify the major events and outcomes of glycolysis, oxidative respiration, and fermentation.
- List the alternative fuels for cellular respiration.
- Describe how cellular respiration is controlled.

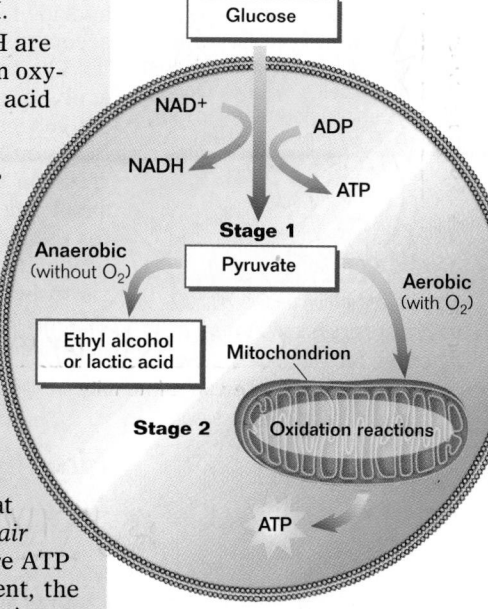

Figure 5-11 Cellular respiration occurs in two distinct stages. First, glucose is converted to pyruvate. The outcome of the second stage depends on the presence of oxygen.

SECTION 5-3

Vocabulary Preview

- pyruvate
- aerobic
- anaerobic
- glycolysis
- oxidative respiration
- acetyl-CoA
- Krebs cycle
- fermentation

Opening Questions

Ask students what relationship the products of photosynthesis have with cellular respiration. Students should recall that the products of photosynthesis are the reactants of cellular respiration. Ask students what types of organisms conduct cellular respiration. All organisms conduct cellular respiration.

Demonstration
Burning an Organic Fuel
Ignite the wick of an alcohol burner. Ask: What are the reactants of this reaction? (*alcohol, which is an organic compound, and O₂*) What are the products? (*CO₂, H₂O, and energy in the form of heat and light*) Tell students that the burning of a fuel is similar to the overall cellular respiration reaction.

VISUAL STRATEGY

Figure 5-11
Correlate this figure with the overview of cellular respiration. Point out that glucose is transported across a cell membrane and then converted to pyruvate in the cytosol. When oxygen is absent (anaerobic conditions), pyruvate is converted to ethyl alcohol or lactic acid in the cytosol, and no additional ATP is produced. When oxygen is present (aerobic conditions), further reactions that occur within the mitochondria produce additional ATP. Emphasize that the biochemical pathways of cellular respiration consist of many individual steps, just as with photosynthesis.

VISUAL STRATEGY

Figure 5-12

Guide students through this summary of glycolysis. Emphasize that glycolysis is another example of a metabolic (biochemical) pathway. Two coupled reactions, which were discussed in Chapter 4, use ATP and add two phosphate groups to a glucose molecule. The resulting six-carbon glucose phosphate is unstable and splits into two three-carbon sugar phosphates. The conversion of these sugar phosphates to pyruvate results in the production of ATP and NADH. Also, point out that cellular respiration uses NAD+ as an electron carrier instead of the NADP+ used by photosynthesis. Explain that the two molecules are very similar (NADP+ has a phosphate group and NAD+ does not). Have students count the total number of carbon atoms at each step of the process. *(six)* Ask students to calculate the net gain of ATP in glycolysis. *(−2 + 4 = 2)*

Teaching Tip
The Energy Conduit Revisited

Have students prepare a table of the organic compounds they have encountered so far in their study of photosynthesis (ATP, NADPH, and carbohydrates). In their tables, they should record the function of each compound, the source of the energy in the compound, and the immediate destination of the energy. A completed table is shown in the *Graphic Organizer* on page 108.

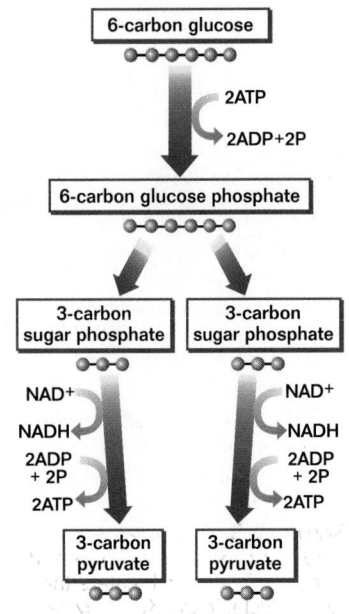

Figure 5-12 Glycolysis, which is the first stage of cellular respiration, is an anaerobic process that occurs in a cell's cytosol. During glycolysis, coupled reactions produce ATP by adding a phosphate group to ADP.

CAPSULE SUMMARY

Cellular respiration begins with glycolysis. A cell gains two pyruvate molecules, two ATP molecules, and two NADH molecules for every glucose molecule entering this anaerobic pathway.

Glucose Is Split During Glycolysis

Although other organic molecules can be used to produce ATP for a cell's energy needs, glucose is usually thought of as the starting material for cellular respiration. During the first stage of cellular respiration, glucose is split into smaller molecules in a biochemical pathway called **glycolysis** *(gleye KAHL uh sihs)*. The word *glycolysis* is derived from the Greek words *glykys*, meaning "sweet," and *lysis*, meaning "to dissolve." Glycolysis is a sequence of enzyme-catalyzed reactions that converts a six-carbon glucose molecule into two three-carbon molecules of pyruvate, as shown in Figure 5-12.

In addition to pyruvate, glycolysis produces a small amount of ATP and NADH. Coupled reactions that occur during the formation of each molecule of pyruvate produce a total of four ATPs. But because glycolysis also uses two ATPs to get started, there is a net gain of only two ATPs for each molecule of glucose entering the pathway. The formation of each pyruvate molecule also results in the removal of a hydride ion with high-energy electrons. Each hydride ion is donated to a molecule of NAD+, forming two NADH molecules for every glucose molecule that is split. Once a hydride ion (with its electrons) is passed to another hydrogen acceptor, NAD+ can return to glycolysis for more electrons.

The anaerobic reactions of glycolysis are thought to have evolved more than 3 billion years ago, when there was no oxygen gas in Earth's atmosphere. At that time, most life-forms on Earth were probably single-celled heterotrophs that "fed" on the ocean's rich supply of organic molecules. The small amount of ATP produced by glycolysis was sufficient for their energy needs. Although more efficient pathways for producing ATP evolved later, all organisms still use glycolysis to begin cellular respiration. In the absence of oxygen, this ancient energy-extracting pathway is the only way that a heterotrophic cell can harvest energy from food. ◻

Making More ATP Requires Oxygen

When oxygen is available, a series of reactions called **oxidative respiration** follows glycolysis. The reactions of oxidative respiration drive the production of large amounts of ATP. The pathways of oxidative respiration are thought to have appeared after photosynthetic bacteria began to fill the atmosphere with oxygen gas. The presence of this oxygen gas enabled the development of a far more efficient way of using glucose as an energy source.

In preparation for the ATP-producing part of oxidative respiration, three-carbon pyruvate molecules from glycolysis are first converted to a two-carbon fragment. As this takes place, a carbon dioxide molecule and a hydride ion with high-energy electrons are extracted from each pyruvate molecule.

Graphic Organizer

Use this graphic organizer with *Teaching Tip:* **The Energy Conduit Revisited** on page 108.

Compound	Function	Energy In	Energy Out
ATP	Cell's main energy currency; energy transport	Sunlight, via photosystem II electrons	Carbohydrates from Calvin cycle
NADPH	Energy transport; very short-term energy storage	Sunlight, via photosystem I electrons	Carbohydrates from Calvin cycle
Carbohydrate	Mid- to long-term energy storage and transport	ATP, NADPH	ATP for cell work, via cellular respiration

The carbon dioxide, a byproduct of cellular respiration, leaves the mitochondrion and then the cell. The hydride ions are donated to an NAD+ molecule, forming NADH that is used later in oxidative respiration. The remaining two-carbon fragment is called an acetyl *(uh SEET uhl)* group. As Figure 5-13 shows, this group is attached to a coenzyme (coenzyme A), forming a compound called **acetyl-CoA** *(uh SEET uhl-koh ay)*. If a cell already has a plentiful supply of ATP, acetyl-CoA is funneled into fat synthesis. Thus, high-energy electrons are stored for later needs. If the cell needs ATP immediately, acetyl-CoA is directed to the next stage of oxidative respiration.

The Krebs Cycle

The next phase of oxidative respiration is the **Krebs cycle,** which is a repeating series of reactions that produces ATP, electron carriers, and carbon dioxide. The cycle is named for the biochemist Hans Krebs, who first proposed the cycle in 1937. The cycle's reactions, however, were not worked out until the 1950s. The cycle, which is summarized in Figure 5-13, begins when the two-carbon fragment of acetyl-CoA is attached to a four-carbon molecule found in mitochondria. Once the Krebs cycle is completed, the same four-carbon molecule that began the cycle has been regenerated. High-energy electrons harvested by the cycle are taken away by hydride ions that are attached to NADH and FADH$_2$ (another electron carrier). These electron carriers now hold most of the chemical energy that was previously stored in glucose.

SECTION 5-3

CONNECTIONS

Chapter 4
Coenzymes
The suffix *-CoA* stands for "coenzyme A," an organic chemical that is necessary for the action of an enzyme. Remind students that some coenzymes, such as NAD+ and NADP+, aid certain enzymes by carrying energy in the form of hydrogen atoms with excited (high-energy) electrons. Coenzyme A attaches to the two-carbon fragment, called an acetyl (akin to acetic acid) group, that remains after a carbon dioxide molecule has been split from pyruvate. In this case, coenzyme A acts as a carrier and transports the acetyl group to the Krebs cycle.

Figure 5-13 During oxidative respiration, pyruvate is first converted to acetyl-CoA, which enters the Krebs cycle. In the Krebs cycle, the acetyl group is added to a four-carbon compound, the starting material. In an additional series of reactions, two carbon atoms are expelled in carbon dioxide, one ATP is generated, high-energy electrons are harvested, and the four-carbon starting material is regenerated.

👁 VISUAL STRATEGY

Figure 5-13
Walk students through this figure. Point out the materials that enter the Krebs cycle and where they come from. Also point out the products of each step in the figure. Have students account for the carbon atoms entering and leaving each step. *(For example, a two-carbon acetyl group is added to a four-carbon starting material to produce a six-carbon molecule.)* Explain that FAD is another type of coenzyme, similar to NAD+, that carries electrons during cellular respiration. Emphasize that the Krebs cycle must "turn" twice for every glucose molecule that is converted to pyruvate.

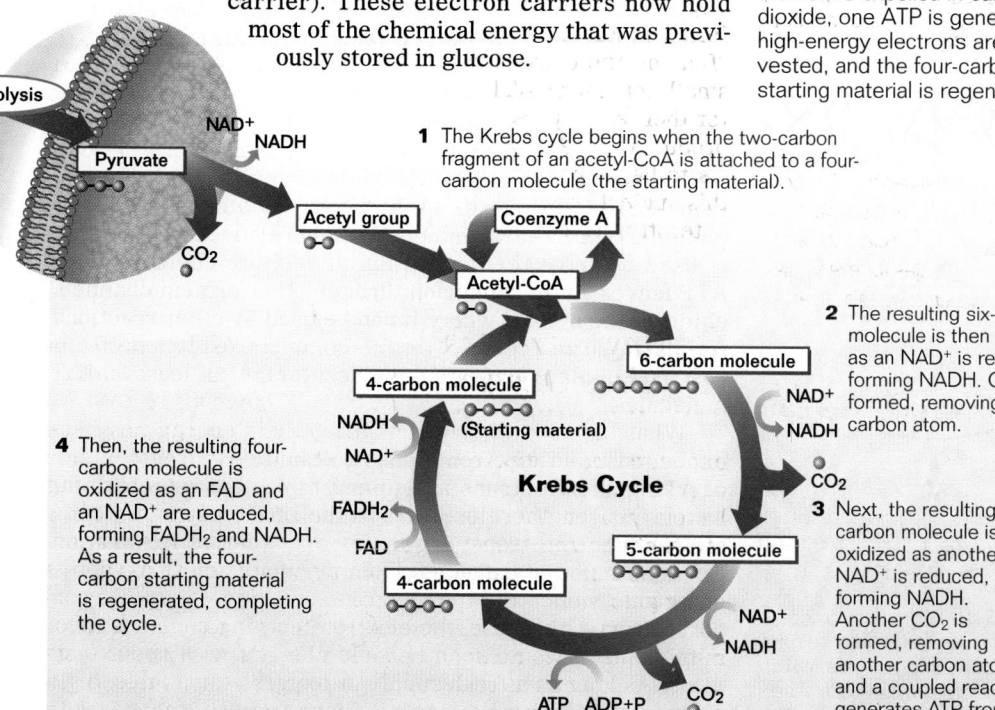

1 The Krebs cycle begins when the two-carbon fragment of an acetyl-CoA is attached to a four-carbon molecule (the starting material).

2 The resulting six-carbon molecule is then oxidized as an NAD+ is reduced, forming NADH. CO$_2$ is formed, removing a carbon atom.

3 Next, the resulting five-carbon molecule is oxidized as another NAD+ is reduced, forming NADH. Another CO$_2$ is formed, removing another carbon atom, and a coupled reaction generates ATP from ADP and P.

4 Then, the resulting four-carbon molecule is oxidized as an FAD and an NAD+ are reduced, forming FADH$_2$ and NADH. As a result, the four-carbon starting material is regenerated, completing the cycle.

Krebs Cycle

❓ Did You Know?

Glycolysis is one of the oldest biochemical pathways and is thought to have preceded photosynthesis in the evolution of life. According to the primordial soup model (described in Chapter 11), the earliest organisms would have had an abundance of organic molecules to metabolize for energy via glycolysis. Only when this supply was depleted did the ability to convert sunlight to energy-storing molecules become a selective advantage.

Historical Note

Hans Adolph Krebs
Krebs, a German biochemist, discovered the Krebs cycle in 1937. His work on the cycle, also known as the citric acid cycle, earned him a share of the Nobel Prize in physiology and medicine in 1953.

VISUAL STRATEGY

Figure 5-14

Have students follow the progress of an electron (the yellow to red arrow) through the electron transport chain. Point out that the electron transport chain molecules also act to pump protons out of the inner chamber of the mitochondrion. Ask: What happens when the concentration of protons builds up in the outer compartment? (*Protons are forced back into the inner chamber through ATP synthetase, resulting in the production of ATP.*) Point out that ATP leaves the inner chamber through a separate membrane channel.

Teaching Tip
Electron Transport Chain Analogy

Compare an electron transport chain to a line of people moving down an escalator to get in line for a cab. Ask students the following questions: What does the escalator represent in this analogy? (*the electron transport chain*) What does the cab represent? (*oxygen*) What would happen if no oxygen were available to pick up electrons and hydrogen ions? (*The electron transport system would back up, the inner membranes of the mitochondria would become clogged with electrons, and the process of oxidative respiration would stop.*)

 VISUAL STRATEGY

Figure 5-15

Point out that the rectangles in this graphic organizer represent substances and the ovals represent processes. Have students tally the total number of ATPs produced by cellular respiration in the presence of oxygen. (*36*) Some students may notice that the Krebs cycle on page 109 shows that only one ATP is produced. Remind them that converting each glucose molecule to pyruvate requires two turns of the Krebs cycle.

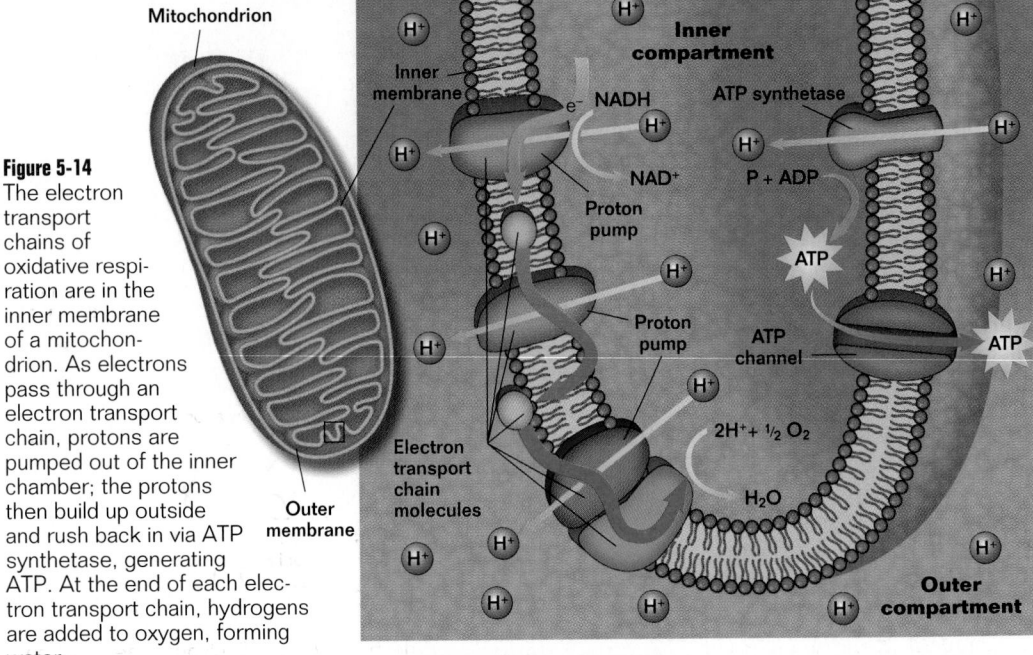

Figure 5-14
The electron transport chains of oxidative respiration are in the inner membrane of a mitochondrion. As electrons pass through an electron transport chain, protons are pumped out of the inner chamber; the protons then build up outside and rush back in via ATP synthetase, generating ATP. At the end of each electron transport chain, hydrogens are added to oxygen, forming water.

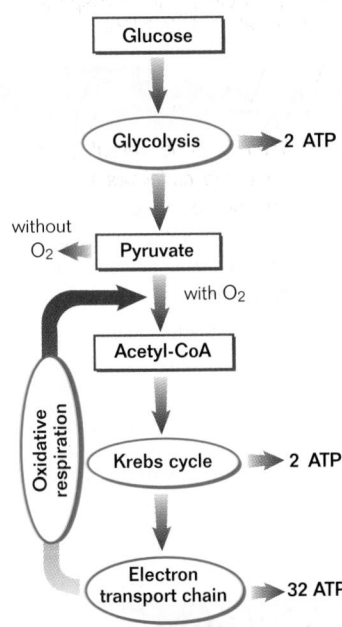

Figure 5-15 The complete oxidation of glucose by cellular respiration begins with glycolysis, which yields a net of only 2 ATPs. The pathways of oxidative respiration yield an additional 34 ATPs.

Electron Transport Chain

The NADH and FADH$_2$ made during the Krebs cycle, and the reactions that precede it, are used to generate the bulk of the ATP made by oxidative respiration. The electrons these molecules carry are passed through a series of membrane-bound proteins—an electron transport chain—by oxidation-reduction reactions. The electron transport chains for oxidative respiration, illustrated in Figure 5-14, are embedded in the inner membranes of mitochondria. In bacteria, they are located in the cell membrane. Each electron transport chain passes high-energy electrons to proton-pumping membrane channels. The energy of these electrons is used to drive the generation of ATP by chemiosmosis. ATP leaves a mitochondrion through other protein channels, entering the cytosol where it can be used by other reactions. As indicated in Figure 5-15, the complete oxidation of glucose by cellular respiration yields a total of at least 36 ATP molecules.

What happens to electrons after their energy has been expended in an electron transport chain? Hydrogen atoms carrying the electrons are joined to oxygen gas, forming water. Oxygen, therefore, acts as the final electron acceptor for the electron transport chains of oxidative respiration. Energy cannot be extracted from pyruvate unless oxygen is present to siphon off the electrons entering electron transport chains. Otherwise, the electron-ferrying components of mitochondria would soon become clogged with spent electrons. As long as a fresh supply of oxygen is available, high-energy electrons harvested from food molecules can continue to power the efficient production of ATP.

Overcoming Misconceptions

There Are Many Electron Transport Chains

Students often confuse the electron transport chain of oxidative respiration with the electron transport chains of photosynthesis. An electron transport chain is simply a series of membrane-bound molecules that transport energy in the form of electrons from one place to another. The two electron transport chains of photosynthesis are located in the thylakoid membranes, while the electron transport chain of oxidative respiration is located in the inner membranes of mitochondria.

Fermentation Occurs in the Absence of Oxygen

If there is no oxygen for oxidative respiration, the pyruvate produced by glycolysis has a different fate. As you read earlier, electrons extracted during glycolysis are carried away by NAD^+. In the absence of oxygen, these electrons do not enter electron transport chains but remain attached to their carriers. Soon a cell's NAD^+ becomes saturated with electrons. With no more NAD^+ available to carry away electrons, the pathways of oxidative respiration "back up," and glycolysis cannot proceed. To continue obtaining energy from food when oxygen is absent, another acceptor must be found for the electrons. Under anaerobic conditions, the electrons from glycolysis are added to organic molecules, a process that is called **fermentation**.

As Figure 5-16 indicates, there are two different products of fermentation among eukaryotes. In animals, such as yourself, lactic acid is produced simply by adding the electrons from glycolysis back to the pyruvate produced by glycolysis. If you lift a heavy weight up and down rapidly 100 times, for example, your muscle cells use up all the available oxygen. Consequently, the muscle cells have only the ATP made by glycolysis for further energy needs. This situation immediately produces a tired feeling in your muscles. Later, lactic acid in your muscle cells causes your muscles to feel sore. However, if oxygen becomes available, lactic acid is converted back to pyruvate, which then can enter oxidative respiration.

Fungi and plants have a different pathway for fermentation. They first convert pyruvate to two-carbon molecules by removing carbon dioxide. The electrons from glycolysis are added to these molecules, producing ethyl alcohol. For centuries, humans have used fungi (yeast) in the preparation of foods and beverages. Wine and beer contain ethyl alcohol made by yeast performing fermentation. Carbon dioxide made by yeast during fermentation causes bread to rise. ❑

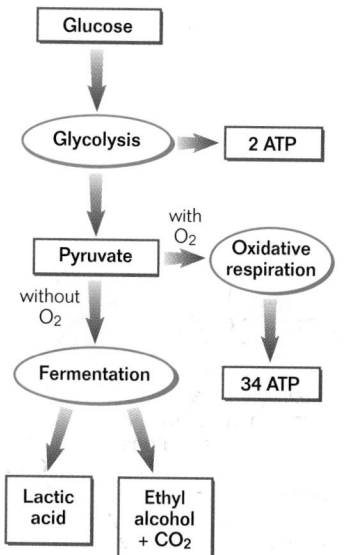

Figure 5-16 Under anaerobic conditions, the pyruvate generated by glycolysis undergoes fermentation, enabling cellular respiration to yield only 2 ATPs.

❑ CAPSULE SUMMARY

When oxygen is present, oxidative respiration follows glycolysis and produces a large amount of ATP. Fermentation follows glycolysis when no oxygen is available for oxidative respiration.

Many Fuels Are Used for Cellular Respiration

Thus far, only glucose has been discussed as a fuel for cellular respiration. The glucose for cellular respiration is obtained by eating carbohydrates such as starch and sugar. But if there are not enough carbohydrates in an organism's diet to supply its energy needs, other molecules must be oxidized to obtain energy. The long carbon chains of fatty acids have many hydrogen atoms, and thus fats provide a rich harvest of energy. In fact, humans obtain more energy from fats than they do from glucose (a gram of fat contains more than twice the energy of a gram of glucose). Although their subunits are normally used for building important cell components, proteins and nucleic acids also can be oxidized by cellular respiration to make ATP. ❑

❑ CAPSULE SUMMARY

Fats, proteins, and nucleic acids, in addition to glucose, can be oxidized by cellular respiration to make ATP.

SECTION 5-3

👁 VISUAL STRATEGY

Figure 5-16
Have students compare the total amount of ATP produced when a glucose molecule is metabolized in the presence of oxygen (36) with the amount produced when a glucose molecule is metabolized in the absence of oxygen (2).

Teaching Tips
The Making of ATP
Remind students that before the carbohydrates, fats, and proteins in food can be used in oxidative respiration, these molecules must be broken down into their components. Although any food molecule can be used as fuel to produce ATP, digested fats, proteins, and nucleic acids are normally used in more important ways. Ask: How could fat components be used? *(making phospholipids for cell membranes or other types of molecules that contain these components, such as hormones)* How could protein components be used? *(The amino acids could be used to make other proteins that the body needs.)* How could nucleic acid components be used? *(Nucleotides are used in cell division and protein synthesis.)* Lead students to conclude that a diet that does not provide enough carbohydrates to meet the body's energy needs forces the body to use fats, proteins, and nucleic acids to make ATP, rather than putting them to better use.

A Summary of Cellular Respiration
Have students create a model similar to the one they created summarizing photosynthesis. They could start with the summary equation. Their summary diagrams should include a title and the information found in the figures in Section 5-3. Have students highlight the materials needed, energy transformations, intermediates, and the organic compounds used and produced. Encourage students to be creative with their presentations.

CONNECTIONS

Chapter 3

Feedback and Homeostasis
Relate the concept of feedback to the discussions of homeostasis in Section 3-2, tying it to the *Graphic Organizer* on page 64.

Application

 Sports Trained athletes consume more oxygen than the average person. Vigorous exercise can lead to an oxygen debt in the muscles. For example, the body of an average person running a 100 yd. dash in 12 sec. would require 6 L (1.6 gal.) of air. However, that person's lungs and circulatory system could supply only about 1.2 L of air. As a result, an oxygen debt would be incurred, and the muscles would produce lactic acid. By contrast, most athletes take in at least 10 percent more oxygen, and trained marathon runners take in up to 45 percent more oxygen, a consequence of their more efficient respiratory and circulatory systems, and can exert greater effort without accruing an oxygen debt.

Chapter Closure

Have students summarize the energy conduit by combining the models summarizing photosynthesis (prepared for Section 5-2) and cellular respiration (prepared for Section 5-3). Require students to indicate the links between the two models with graphic representations of the transfer of energy and materials.

□ **CAPSULE SUMMARY**

A system of feedback inhibition enables cells to control the rate of cellular respiration.

Figure 5-17 Feedback inhibition controls the production of ATP in cells. Excess ATP shuts off a pathway when ATP binds to an allosteric site on a key enzyme, stopping ATP production. As ATP is used by a cell, the enzyme releases the ATP, and ATP production begins again.

Cells Control the Rate of Cellular Respiration

A cell's energy-producing machinery operates only when there is a need for energy. The rate of cellular respiration slows down when a cell already has ample supplies of ATP. This is very sensible, but how does a mitochondrion "know" whether to speed up or slow down cellular respiration? The control of cellular respiration works through the system of feedback inhibition illustrated in Figure 5-17. Key reactions early in glycolysis and the Krebs cycle are catalyzed by enzymes that have a regulatory allosteric site that is the same shape as ATP. When ATP levels in a cell are high, it is very likely that ATP molecules will bind to these allosteric sites, causing the enzymes to change shape. The new shape, however, deactivates the enzymes. Thus, high levels of ATP act to shut down the processes a cell uses to make ATP. □

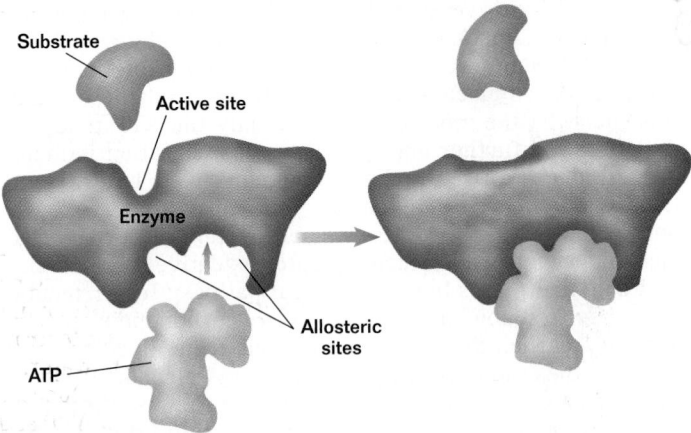

Substrate
Active site
Enzyme
Allosteric sites
ATP

Section Review

1. *How does oxygen affect the efficiency of cellular respiration?*
2. *What is the starting material for glycolysis? What is the product?*
3. *What are the three main events of oxidative respiration?*
4. *Using a concept map, indicate how ATP, lactic acid, and ethyl alcohol result from cellular respiration.*

Critical Thinking
5. *What happens when a cell's supply of ATP gets low? Why do you think this happens?*
6. *Considering what you have learned about cellular respiration, what are several things you could do to reduce your amount of body fat?*

Answers to Section Review

1. Oxygen increases the efficiency of the system by allowing it to get more energy out of each glucose molecule.

2. Glucose is the starting material; pyruvate is the product.

3. The three main events of oxidative respiration include the following: the conversion of pyruvate to acetyl-CoA; the production of NADH, FADH$_2$, and more ATP in the Krebs cycle; and the production of even more ATP in the electron transport chain.

4. Concept maps should indicate that glycolysis, the Krebs cycle, and electron transport chains produce ATP and that in the absence of oxygen, glycolysis can lead to fermentation, producing ethyl alcohol in fungi and plants and lactic acid in animals.

5. Release of ATP by allosteric sites on enzymes that catalyze glycolysis and the Krebs cycle activates the enzymes and initiates ATP production. The shape of the enzyme changes to enable substrates to bind.

6. Increase aerobic exercise and reduce caloric intake.

Vocabulary

acetyl-CoA (109)
aerobic (107)
anaerobic (107)
ATP synthetase (102)
Calvin cycle (105)
carbon fixation (105)
carotenoids (101)
chlorophyll (101)
consumer (97)
crop (96)
electromagnetic spectrum (100)

electron transport chain (102)
fermentation (111)
food chain (97)
glycolysis (108)
granum (99)
Krebs cycle (109)
NADP+ (103)
NADPH (103)
oxidative respiration (108)
photon (100)
photosystem (101)

photosystem I (101)
photosystem II (101)
pigment (101)
producer (97)
pyruvate (107)
radiant energy (100)
reaction center (101)
reducing power (103)
stroma (99)
thylakoid (99)

CHAPTER REVIEW ANSWERS

Each item in the Chapter Review is correlated by text section in the assignment guide that follows.

ASSIGNMENT GUIDE

Section	Review	Themes Review	Critical Thinking
5-1	1, 11, 12, 18	23	
5-2	2–5, 13–15, 19, 20, 22	24	26–28
5-3	6–10, 16, 17, 21, 22	25	28, 29

Concept Mapping

Construct a concept map that shows how photosynthesis and cellular respiration are related. Use as many terms as needed from the vocabulary list. Try to include the following terms in your map: ATP, photosynthesis, cellular respiration, light reactions, carbohydrates, glycolysis, Krebs cycle, electron transport chain, photosystems, and Calvin cycle. Include additional terms in your map as needed.

Review

Multiple Choice

1. The products that result from photosynthesis and serve as the starting materials for cellular respiration are
   a. carbohydrates and oxygen.
   b. carbon dioxide and water.
   c. NADP and hydrogen.
   d. ATP and water.

2. Carotenoids
   a. cause plants to look green.
   b. are found in the reaction centers of photosystems.
   c. trap light energy that chlorophyll cannot absorb.
   d. do not play a role in photosynthesis in most plants.

3. The thylakoid membranes of a plant cell are the sites where
   a. carbohydrates are formed.
   b. the light reactions occur.
   c. light energy is packaged into photons.
   d. ATP is used to produce NADPH.

4. The end product(s) of the electron transport chains of photosynthesis is (are)
   a. electrons.
   b. water.
   c. glucose.
   d. ATP and NADPH.

5. The oxygen that results from photosynthesis comes directly from the
   a. splitting of carbon dioxide molecules by the Calvin cycle.
   b. splitting of water molecules to provide electrons for photosystem II.
   c. action of proton pumps in the electron transport chains.
   d. absorption of photons by carotenoids in the photosystems.

6. Which of the following is the correct pairing of a metabolic process and its need for oxygen?
   a. conversion of glucose to pyruvate: no oxygen required
   b. fermentation: oxygen required
   c. conversion of pyruvate and NADH to ATP: no oxygen required
   d. synthesis of organic molecules from carbon dioxide: oxygen required

Concept Mapping

Map answer is shown on page 93B.

Review

Multiple Choice

Code in parentheses indicates section number and objective number.
   1. a (5-1.1)
   2. c (5-2.2)
   3. b (5-2.2)
   4. d (5-2.2)
   5. b (5-2.3)
   6. a (5-3.1)
   7. c (5-3.2)
   8. d (5-3.2)
   9. c (5-3.2)
   10. a (5-3.4)

Completion

11. cellular respiration, photosynthesis (5-1.1)

12. Crops, energy (5-1.2, 5-1.3)

13. pigments, chlorophyll (5-2.2)

14. ATP, NADPH, carbohydrates (5-2.4)

15. photons (5-2.2)

16. lactic acid, ethyl alcohol, carbon dioxide (5-3.2)

17. fats, proteins (5-3.3)

Short Answer

18. Through food chains, consumers receive the energy they need for life's activities. (5-1.3)

19. In the Calvin cycle, ATP and NADPH from the light reactions are used to provide the energy needed to incorporate carbon from carbon dioxide into organic molecules. The Calvin cycle cannot occur in the dark in most plants because their stomata close at night and they cannot take in CO_2. The Calvin cycle is not light independent because it requires products of the light reactions. (5-2.4)

20. When water is plentiful, the rate of photosynthesis will increase. When water is scarce, the rate of photosynthesis will slow. (5-2.5)

21. The anaerobic pathways of cellular respiration break down glucose without the use of oxygen, forming lactic acid or ethyl alcohol. The aerobic pathways of cellular respiration break down glucose using oxygen, forming carbon dioxide, water, and as many as 38 ATP molecules. (5-3.1)

22. They are alike in that both use enzyme-catalyzed biochemical pathways, electron transport, enzymes, and coenzymes. (5-2.1, 5-3.1)

Themes Review

23. Autotrophs fuel their lives by incorporating energy from the sun or an inorganic source. Heterotrophs obtain food energy that keeps them alive by eating autotrophs or other heterotrophs that have directly or indirectly received their food energy from autotrophs. (5-1.3)

---

5 CHAPTER REVIEW

7. Oxidative respiration involves all of the following *except*
   a. conversion of pyruvate to acetyl-CoA.
   b. the Krebs cycle.
   c. the Calvin cycle.
   d. electron transport.

8. The final electron acceptor in cellular respiration is
   a. carbon dioxide.   c. glucose.
   b. NADH.   d. oxygen.

9. In cellular respiration, the greatest number of ATPs is generated by
   a. glycolysis.
   b. the Krebs cycle.
   c. the electron transport chain.
   d. fermentation.

10. When a cell's supply of ATP is more than is needed,
   a. the rate of cellular respiration slows because ATP binds to an enzyme's allosteric site.
   b. the rate of cellular respiration increases in order to maintain homeostasis.
   c. the excess ATP is converted to ADP by the Krebs cycle.
   d. ATP enters glycolysis and binds with coenzyme A to make acetyl-CoA.

Completion

11. Both plants and animals use the biochemical pathways of _____ to break down carbohydrates formed during _____ .

12. _____ are plants that are grown for use by humans. As producers, these plants are the first step in many food chains, through which _____ from the sun is distributed to other organisms.

13. Photosynthesis begins when light is absorbed by _____ , which are arranged in clusters called photosystems. The principal light-absorbing compound in green plants is _____ .

14. In photosynthesis, the Calvin cycle uses _____ and _____ from the light reactions and produces _____ .

---

15. Light is a form of radiant energy that is packaged in units called _____ .

16. The product of fermentation in animals is _____ , and the product in plants is _____ and _____ .

17. Besides glucose, the possible energy sources for cellular respiration include _____ and _____ .

Short Answer

18. How are food chains important to the organisms that are consumers?

19. What happens in the Calvin cycle? Why are these reactions neither dark reactions nor light-independent reactions?

20. How could the availability of water affect the rate of photosynthesis?

21. Contrast the aerobic and anaerobic pathways of cellular respiration.

22. How are the processes of photosynthesis and cellular respiration alike?

Themes Review

23. Flow of Energy Why is it said that all life depends on autotrophs?

24. Levels of Organization Trace the movement of light energy through the structures of a leaf to the point where electrons are excited for photosynthesis.

25. Evolution The electron transport chains of oxidative respiration are located in the inner membranes of mitochondria in protists, plants, and animals. They are located in the cell membrane of bacteria. How might this difference indicate that mitochondria might have originated from bacteria?

Critical Thinking

26. Making Predictions In an experiment, several plants were placed in each of two growth chambers, one with red light and one with green light. Which plants should grow better? Why?

---

24. To excite electrons, light energy must pass through the upper layer of the leaf, through the cell membrane of a leaf cell with chloroplasts, through the outer and inner chloroplast membranes, through thylakoid membranes, to chlorophyll molecules in photosystems, and finally to electrons. (5-2.2)

25. If an ancient bacterium had been engulfed by a cell, it would have been inside a membrane-bound package (like a food vacuole), and thus its original cell membrane, the site of the electron transport chains of oxidative respiration, would be the inner membrane of the structure inside the larger cell. (5-3.4)

Critical Thinking

26. Students should predict that the plant in red light will grow better because plants reflect most green light, and without light absorption, photosynthesis could not proceed and the plant would not grow. (5-2.2)

27. **Interpreting Data** In an experiment conducted to test the effect of light intensity on the rate of photosynthesis, light intensity was varied by moving a light source to different distances from a water plant. The rate of photosynthesis was estimated by counting the number of oxygen bubbles generated per minute. Data collected during the experiment are presented in the graph below. What is the relationship between light intensity and the amount of oxygen produced? If light intensity were increased, would oxygen production continue to increase?

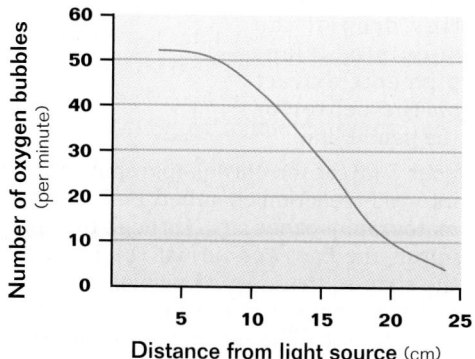

28. **Making Inferences** Evidence indicates that Earth's early atmosphere contained no oxygen. What types of metabolism would have been possible? What is the source of the oxygen present in Earth's atmosphere today?

29. **Making Inferences** Bread rises as a result of yeast fermentation. How might fermentation cause bread to rise? Why is sugar often added to bread dough?

Activities and Projects

30. **Cooperative Group Project** Design and produce a mural that shows how the processes of photosynthesis and cellular respiration link all organisms.

31. **Cooperative Group Project** Prepare a script for a three-act play that shows what takes place in the three stages of photosynthesis. With the other members of your group, perform the play for your class.

32. **Multicultural Perspective** For thousands of years, the seeds of the buffalo gourd, *Cucurbita foetidissima*, have been a source of oil for Native Americans. Research the characteristics of this plant, its uses in ancient Native American cultures, and its potential importance to modern society. In a written report or an oral report, relate your findings and explain why this plant is such a rich source of oil and other useful products.

33. **Research and Writing** Research several other ways, besides bread making, that fermentation is used in food preparation. Relate your findings in a written report.

Readings

34. Read Arthur W. Galston's article "Photosynthesis as a Basis for Life Support on Earth and in Space," in *BioScience*, July/August 1992, pages 490–493. What are the components of a CELSS system? How might the NASA-sponsored CELSS program help solve the problems humans face because of overpopulation?

35. Read James Utley's article "Chemistry that comes naturally," in *New Scientist*, July 31, 1993, pages 24–28. How do electrochemical reactions resemble the reactions that occur during cellular respiration and photosynthesis? How is electrochemistry being used to more safely make industrial chemicals and destroy toxic wastes?

Activities and Projects

30. Students' murals should include the products and reactants of both processes and illustrate how the products of one process become the reactants of the other. They should also show some examples of the types of organisms that use each process.

31. Students' plays should expand on the summary given in Figure 5-3 on page 98.

32. The buffalo gourd is a very large, vining cucurbit with an extensive root system. Many of its parts were used by Native Americans. In modern society, the buffalo gourd has been suggested as a possible source of starch and oil to use in the production of fuel alcohol. The plant has such great potential because of its size and because it is well adapted to hot sunny climates. It has a high rate of photosynthesis and thus can produce and store large quantities of organic compounds.

33. Students' reports should mention that fermentation is used in the production of certain cheeses and alcoholic beverages.

Readings

34. A CELSS (Controlled Ecological Life Support System) includes a food-producing subsystem, a waste-processing subsystem, a food-processing subsystem, and computer-driven controls to integrate the subsystems. CELSS research may ease overpopulation problems by providing more efficient food-production and recycling systems.

35. Electrochemical reactions resemble photosynthesis and cellular respiration in that all involve oxidation and reduction reactions. Electrochemistry, which works at lower temperatures and without harsh chemicals, is replacing other industrial processes in the manufacture of products such as fluorocarbons, nylon, and anthraquinone (a chemical needed for producing dyes) and in the destruction of toxic wastes such as benzene and nerve gas.

27. Students should note that as light intensity increases, the rate of photosynthesis also increases. Oxygen production would not continue to increase in this case because the light saturation point has already been reached and the rate of photosynthesis has leveled off. (5-2.5)

28. Students should infer that anaerobic processes (e.g., glycolysis or fermentation) would have been possible as would photosynthesis, which could have produced the oxygen present in Earth's atmosphere today. (5-2.1, 5-3.1)

29. Students should infer that yeast fermentation produces bubbles of carbon dioxide that become trapped in the bread dough, causing it to expand, or rise. Sugar is added to the dough to provide the glucose for fermentation. (5-3.2)

LABORATORY Investigation Chapter 5

Preparation
For every 30 students, order one Ward's "Chromatography of Simulated Plant Pigments" Kit, stock number 36T0062. Ward's refill kit (36T1256) contains all the consumable items.

Procedural Notes

- This investigation should take 30 minutes to complete.
- Review the procedure for creating a chromatogram, on which pigments in a mixture show up as colored streaks. Students' chromatograms will be almost identical to those of real plant pigments.
- Have students practice using a capillary tube to ensure getting a minute drop of extract on the chromatogram.
- Stress the importance of always using a pencil when marking the chromatogram.
- Stress the importance of placing only the tip of the chromatogram in the solvent. To prevent contamination of the solvent, the pigments should never make direct contact with the solvent.
- Tell students to mark the distances traveled by the pigments with a pencil as soon as the chromatogram is taken from the solvent and air dried. The simulated plant pigments fade over time.
- Allow students to use calculators to calculate R_f values.
- Reuse reaction chambers, uncontaminated solvent, and capillary tubes.
- **Safety:** Caution students to use care when working with the stock solutions so that they don't spill. Be sure students wash their hands before leaving the lab.
- **Safety:** Students should wear safety goggles and aprons.
- **Safety:** Caution students to take care when using the capillary tubes because they are breakable.

Chromatography of Plant Pigments

OBJECTIVES

- Separate and observe the pigments that give a leaf its color.
- Describe the function of plant pigments during photosynthesis.

PROCESS SKILLS

- collecting data
- calculating ratios

MATERIALS

- lab apron
- Ward's Chromatography of Simulated Plant Pigments Kit:
  - 1 vial of simulated plant pigments extract
  - 1 bottle of chromatography solvent
  - capillary tubes
  - chromatography paper strips
  - chromatography reaction chambers
- metric ruler
- scissors
- 10 mL graduated cylinder
- tape or glue

BACKGROUND

1. What is a pigment? What pigments are found in the leaves of plants?
2. What color are chlorophylls? carotenes? xanthophylls?
3. Write your own question to explore in your lab report or notebook.

Background Answers

1. Pigments are molecules that absorb some colors of light and reflect others. Plant pigments include chlorophylls, carotenoids (carotenes and xanthophylls), and anthocyanins.

2. Chlorophylls are green, carotenes are orange, and xanthophylls are yellow.

3. How can paper chromatography be used to separate plant pigments?

TECHNIQUE

1. Put on a lab apron. Use scissors to cut the bottom end of a chromatography paper strip to a tapered end. **CAUTION: Be careful when using the scissors.**

2. Draw a faint pencil line 1 cm above the pointed end of the paper strip (now called a chromatogram). Use a capillary tube to apply a tiny drop of the simulated plant pigments extract on the center of the pencil line.

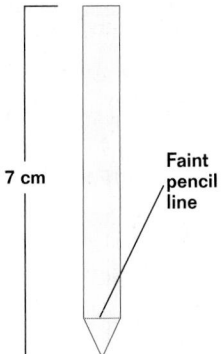

7 cm

Faint pencil line

3. Pour 5 mL of the chromatography solvent into your reaction chamber. Pull the chromatography paper strip through the opening of the cap, and adjust the length of the strip so that a small portion of the tip end is immersed in the solvent. **DO NOT** immerse the pigment spot in the solvent.

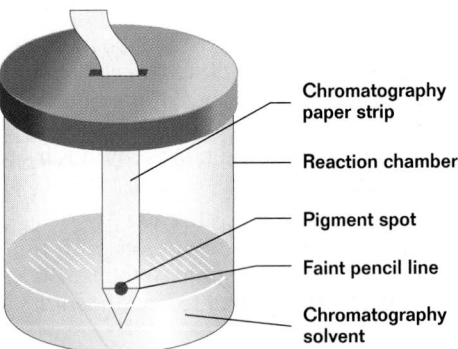

Chromatography paper strip

Reaction chamber

Pigment spot

Faint pencil line

Chromatography solvent

4. Place the cap over the reaction chamber, and carefully bend the end of the strip over the cap. Be sure that the chromatogram is level and does not touch the walls of the reaction chamber.

5. Within 5–7 minutes you will notice bands of different colors—orange, yellow, and two shades of green—traveling up the strip. Remove the chromatogram from the solvent when the solvent nears the top of the reaction chamber.

6. With a pencil, mark the position of the uppermost end of the solvent. Then mark the farthest distance that each of the separated pigments moved. Measure the distances that the solvent and each of the pigments moved. In the **Records** section of your report, make a table similar to the one below, and record your measurements. Tape or glue your chromatogram to the **Records** section. Label the pigment colors.

 7. Clean up your materials and wash your hands before leaving the lab.

Chromatography of Plant Pigments

Band no.	Pigment color	Migration (distance in mm)	R_f value
1 (top)			
2			
3			
4			
Solvent			

8. Use the formula below to calculate the R_f value.

$$R_f = \frac{\text{Distance substance (pigment) traveled}}{\text{Distance solvent traveled}}$$

The R_f value is a number that represents the ratio of the distance that the pigment moved relative to the distance that the solvent moved. Since the R_f value is constant for each pigment type, it is used to help scientists identify compounds. In the **Records** section of your report, record a decimal fraction for the R_f of each pigment. In your report, briefly summarize the procedure you followed.

INQUIRY

1. What is a chromatogram?

2. The word *chromatography* comes from the Greek words *chromat*, which means "color," and *graphon*, which means "to write." Write a descriptive sentence that explains what happened to your plant pigments during the lab.

3. How do your R_f values compare with those of your classmates?

4. Which color moved the farthest distance? the shortest distance?

ANALYSIS

1. Using the data from your lab, what pigments do you think would be indicated by an R_f value of .92? of .42?

2. What is the name of the main pigment used during photosynthesis?

3. At what time during the year do the other pigments become visible? Why?

4. List other uses for chromatography.

FURTHER INQUIRY

Write a new question that could be explored as a new investigation. The following is an example:

Design an experiment that uses various solvents to separate pigments found in substances such as fruit juices, writing ink, and dyes. Then calculate the R_f values.

Records

Chromatography of Plant Pigments

Band no.	Pigment color	Migration (distance in mm)	R_f value
1 (top)			
2			
3			
4			
Solvent			

UNIT 1

Block Scheduling Guide

> Each block represents about 45–50 minutes of instructional time. The resources cited in a block are coded to a specific program component using the Key on the next page.

Lecture Concepts

Lesson Resources

6-1 Chromosomes pp.118–125

BLOCKS ① and ②

a. DNA contains information that is encoded in segments called genes.

b. When a cell prepares to divide, the DNA coils into structures called chromosomes.

c. At the start of cell division, there are two copies of each chromosome, ensuring that each new cell will have the same genetic information as the old cell

d. The number of chromosomes in cells is constant within a species. Haploid cells contain one homologue of each chromosome. Diploid cells contain two homologues of each cell.

e. Deviations in chromosome number can cause abnormal development. In humans, Down syndrome is caused by an extra copy of chromosome 21, which is the result of nondisjunction.

f. A mutation in which a chromosome fragment is lost is called a deletion. Duplication occurs when a piece of a chromosome attaches to its homologue. Inversion occurs when a fragment reattaches to the original chromosome in a reverse order. A translocation is a mutation in which a chromosome fragment attaches to a nonhomologous chromosome.

g. Sex chromosomes carry information that determines an organism's sex. In humans and many other species, females are designated XX and males are designated XY.

Lecture Resources
- Opening Questions, page 119
- Demonstrations: Genetic Instructions, Preparing a Karyotype, Nondisjunction
- Effective Reading, page 120
- Overcoming Misconceptions, pages 122, 124
- Visual Strategies: Figures 6-3, 6-5, 6-6
- Medicine, page 125

- Holt Biology Videodiscs: Lesson 6-1

Classwork Options
- Incidence of Down Syndrome, page 123
- Prenatal Testing, page 124
- Abnormalities in Chromosome Number, page 125

Assignment Options
- Section Review, page 125
- Chapter Review, questions 1–3, 10–16, 18, 20, 28

6-2 Mitosis and Cell Division pp.126–131

BLOCKS ③ and ④

a. Bacteria divide by the process of binary fission, in which the DNA is copied and the cell splits in two.

b. In eukaryotic cells, cell division is preceded by the duplication of chromosomes and cell organelles and mitosis. During mitosis, the nucleus of the cell divides, forming two nuclei with identical genetic information. The cytoplasm divides during cytokinesis.

c. Mitosis can be divided into four stages: prophase, metaphase, anaphase, and telophase.

d. Cytokinesis is the division of the cell cytoplasm.

Lecture Resources
- Demonstrations: Chromosome Condensation, The Spindle
- Visual Strategies: Figures 6-8, 6-10
- Overcoming Misconceptions, page 130
- Teaching Transparencies: 24
- Holt Biology Videodiscs: Lesson 6-2

Classwork Options
- Laboratory Techniques and Experimental Design C9: Preparing a Root Tip Squash
- Laboratory Techniques and Experimental Design C10: Preparing a Root Tip Squash—Stopping Mitosis

Assignment Options
- Section Review, page 131
- Chapter Review, questions 5–8, 17, 21, 25–27, 29

6-3 Meiosis pp.132–134

BLOCKS ⑤ and ⑥

a. Meiosis is a form of cell division that halves the number of chromosomes in cells. Meiosis I separates homologous chromosomes. Meiosis II separates chromatids. Four haploid cells are produced from one diploid cell.

b. Meiosis II is identical to meiosis I except that the chromosomes do not replicate before they divide at their centromeres.

c. Crossing over is an efficient way to produce genetic recombination.

Lecture Resources
- Demonstration: The Importance of Meiosis
- Overcoming Misconceptions, page 132
- Visual Strategies: Figures 6-12, 6-13

Classwork Options
- Distinguishing Between Mitosis and Meiosis and Graphic Organizer, page 133

- Laboratory Investigation: Comparing Mitosis and Meiosis, pages 138–139

Assignment Options
- Section Review, page 134
- Chapter Review, questions 4, 9, 19, 22–24, 29

Review/Enrichment

BLOCK 7

■ Themes of Biology: Structure and Function, page 128
■ Themes Review, questions 23–27
■ Study Guide: Concept Review and Pretest

Assignment Options
⊙ Teaching Resources CD-ROM: Supplemental Reading Worksheets and Test, *The Lives of a Cell*
■ Activities and Projects, questions 30–32
■ Readings, questions 33, 34

Assessment Options

BLOCK 8

Traditional Assessment
■ Chapter 6 Test
⊙ Test Generator/Assessment Item Listing: Software item bank for preparing customized chapter tests.

Performance Assessment
Select the Performance Assessment items for this unit from the Test Generator/Test Item Listing

Portfolio Assessment
Select student reports for one or more laboratory experiments from this chapter. *The Direct Observation Checklist* on the *BioSources Teaching Resources CD-ROM* should be completed during a laboratory or other cooperative learning experience.

Answer to Concept Map
The following is one possible answer to the Concept Mapping exercise on page 135.

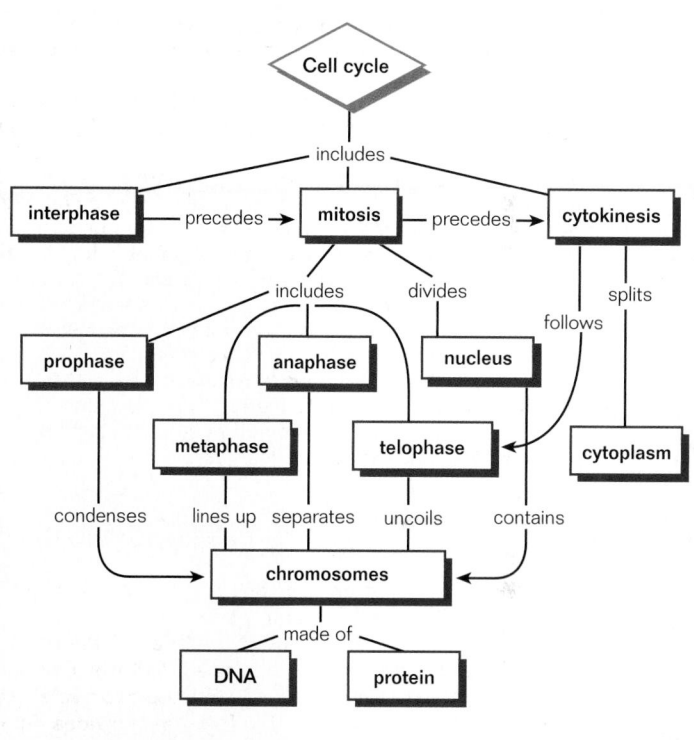

Holt Biology Videodiscs · CONCEPTS OF BIOLOGY

Holt Biology Videodiscs gives you a powerful tool for teaching, review, and assessment. *Concepts of Biology* adds interactive lesson plans and extensive tools for customization using CD-ROM technology.

Use the following topics from *Concepts of Biology* to help you teach this chapter:

■ Topic 2: Cell Division
■ Topic 6: Fundamentals of Genetics

For further information regarding the media on the videodiscs, see *Holt Biology Videodiscs Teacher's Correlation Guide to Biology: Principles and Explorations.*

Chapter Theme

Structure and Function

This chapter introduces the final of the five themes that form the framework of this book—structure and function. Students will learn that the responsibility of passing on genetic information to developing cells and to future generations lies with the chromosomes. Point out the difficulty of this task, considering the amount of information that is passed on, unaltered and intact. The major function of the chromosomes is carried out with relatively few errors largely because of their structure. The condensed nature of chromosomes allows DNA to be packed into a compact form that is easier for the cell to manipulate during mitosis and meiosis.

Tapping Prior Knowledge

- What role does the nucleus play in the cell?
- What role do the chromosomes play in the cell?
- What happens to the chromosomes when a cell divides to produce two new cells?
- What happens to the chromosomes when a sperm or egg cell matures?

Opening Demonstration

Hold up a meter stick. Tell the class that it represents a chromosome about to make a copy of itself. Then hold up another meter stick in the same hand, pointing out that each meter stick represents a chromatid, while your hand represents a centromere.
Compare the meter sticks to the chromosome shown in Figure 6-1 on page 119.

CHAPTER 6

CELL REPRODUCTION

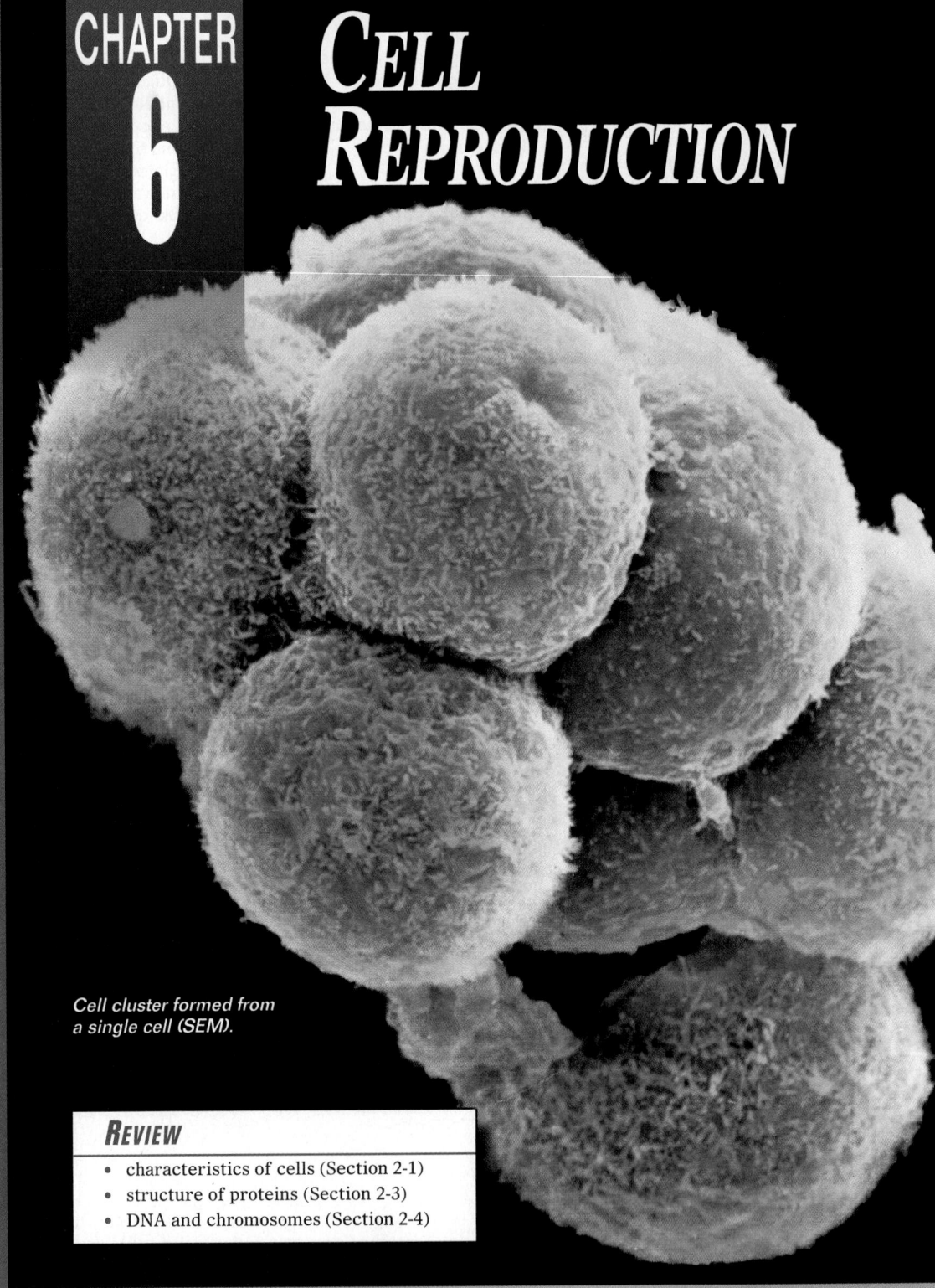

Cell cluster formed from a single cell (SEM).

REVIEW

- characteristics of cells (Section 2-1)
- structure of proteins (Section 2-3)
- DNA and chromosomes (Section 2-4)

Authors' Rationale

How does a multicellular organism develop from a single cell? To understand growth, students need to know how cells can divide to multiply. Cell division, along with all other cell functions, is regulated by the genes on the chromosomes in the nucleus. Discussion of chromosomes, which initiates this exploration, is followed by detailed discussion of cell division for growth (mitosis) and cell division for sex cell production (meiosis). It is important for students to understand the differences between these two processes and to understand why they occur.

6-1 Chromosomes

Your body is composed of approximately 100 trillion cells. The intricate organism that you are today is the result of millions of cell divisions. Each cell division is a carefully orchestrated performance carried out by many kinds of structures. In this section you will meet one of these structures, the chromosome, and learn how it is transmitted from one generation to the next.

Chromosome Structure

DNA (deoxyribonucleic acid) is a long, thin molecule that contains the information needed to direct a cell's activities and to determine a cell's characteristics. This vast amount of vital information encoded in DNA is organized into genes. A **gene** is a segment of DNA that transmits information from parent to offspring. A single molecule of DNA has thousands of genes, which are lined up like the railroad cars of a train. When genes are being used, the strand of DNA is extended, enabling other molecules to retrieve its information. However, when a cell prepares to divide, the DNA molecule coils and twists into a dense structure called a chromosome. A **chromosome** is a rod-shaped structure that forms when a single DNA molecule and its associated proteins coil tightly before cell division.

If DNA did not coil and form a chromosome, a single DNA strand would be about 5 cm (approximately 2 in.) long. This is too long to fit inside a cell. But when the thread of DNA is coiled around a protein scaffold, it can be compacted into a smaller, more manageable structure, as shown in Figure 6-1. Your chromosomes are approximately 40 percent DNA and 60 percent protein.

Section Objectives

- Describe the structure of a chromosome.
- Define the terms homologous chromosomes, haploid, diploid, and gamete.
- State the number of chromosomes found in the body cells of several different organisms.
- Explain how an abnormal number of chromosomes arises in a cell.
- Discuss the importance of the X and Y chromosomes in development.

Figure 6-1 A chromosome's scaffold is made of proteins called histones. Histones are rich in positively charged amino acids. Because DNA is a negatively charged molecule, it wraps tightly around the histones, causing the molecule to coil.

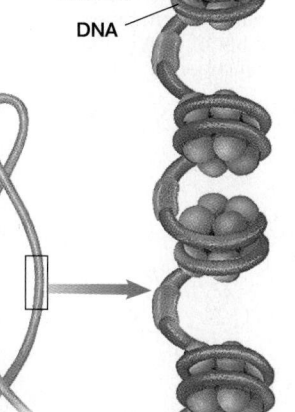

Histones
DNA

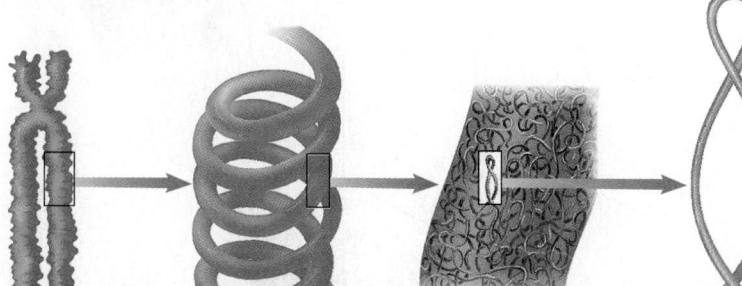

Chromosome.   Supercoil within chromosome   Further coiling within supercoil   One coil within supercoil   DNA and histones

SECTION 6-1

Vocabulary Preview

gene
chromosome
chromatid
centromere
homologous chromosome
diploid
gamete
haploid
zygote
trisomy
karyotype
Down syndrome
nondisjunction
amniocentesis
chorionic villi sampling
mutation
deletion
duplication
inversion
translocation
autosome
sex chromosome

Opening Questions

Have students list the things every new cell needs to function. Their answers should include the organelles common to all cells. Point out that the nucleus "disappears" prior to cell division; thus, dividing cells do not receive a nucleus. Ask students to explain how it is possible for each new cell to function without receiving an intact nucleus. Students should recognize that each cell receives what it needs to function by way of the chromosomes.

 Did You Know?

Although 23 (the number of chromosomes in a human) is not an impressive number, biologists estimate that 23 chromosomes contain 100,000 genes. Scientists throughout the world are now engaged in an effort to map all these genes. One group has traced genetic markers through three generations of 60 Mormon families living in Utah. Their efforts have resulted in the mapping of almost 500 genes.

Demonstration

Genetic Instructions

To help students visualize the behavior of chromosomes in mitosis and meiosis, provide each student with an identical set of a product's assembly instructions that consists of numbered steps. Have students take turns reading each step in the instructions until they recognize that each of them has a complete set of instructions to assemble the product. Tell students that chromosomes carry a complete set of the genetic instructions needed to form an adult organism. Correlate the complete lists of instructions with chromosomes during mitosis, in which each cell receives a complete set of genetic instructions. Next give each student another set of instructions. This time omit the even-numbered steps in the instructions given to female students and the odd-numbered steps in those given to male students. Have the class read and compare the instructions until they realize that each sex has only one-half of the information needed to produce a final product. Compare the partial lists of instructions to chromosomes in meiosis.

👁 VISUAL STRATEGY

Figure 6-3

Use this figure to point out how fertilization restores the diploid number. To introduce students to the terminology used for chromosome structure, each chromosome is shown here as two chromatids. In section 6-3, students will learn that chromatids separate during meiosis II. Ask students why sexually reproducing organisms possess an even number of chromosomes in the diploid state. If students don't realize that 2n must be an even number, then ask them to calculate the haploid number in each gamete if the diploid number in humans were 47 and not 46.

🔲 CAPSULE SUMMARY

DNA is a long molecule that directs cellular functions and heredity. Prior to cell division, DNA coils tightly around proteins and forms chromosomes. At this time, each chromosome consists of two copies of DNA.

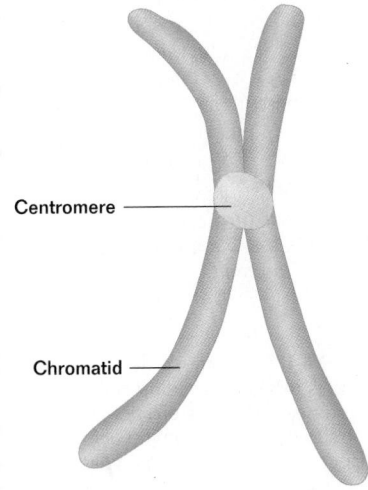

Figure 6-2 An easy way to determine the number of chromosomes in a cell is to count the number of centromeres. The number of centromeres will equal the number of chromosomes.

Labels: Centromere, Chromatid, **Chromosome**

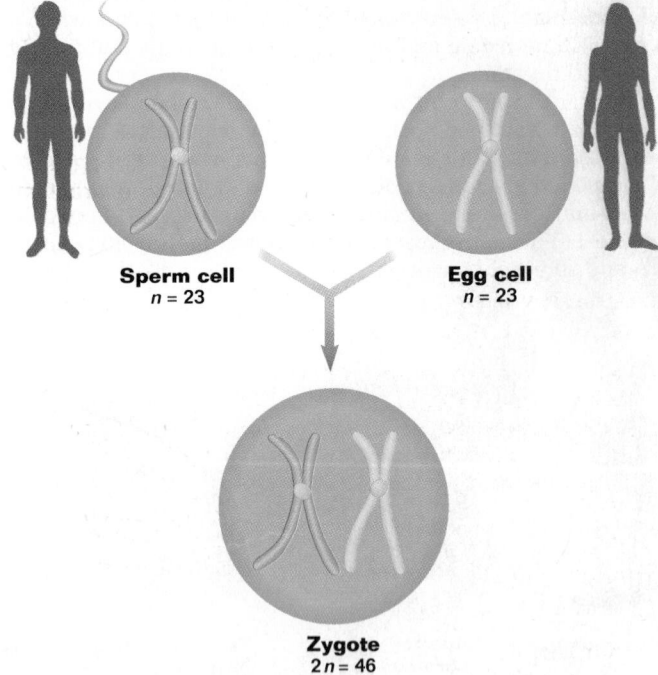

Figure 6-3 When haploid gametes fuse, they produce a diploid zygote.

Labels: **Sperm cell** n = 23, **Egg cell** n = 23, **Zygote** 2n = 46

Chromosomes become visible through a microscope only after they have condensed prior to cell division. By this time, each chromosome has formed a copy of itself. The two copies of each chromosome are called **chromatids** (*KROH muh tihdz*). Chromatids form prior to cell division when the DNA molecule duplicates itself, ensuring that each new cell will have the same genetic information as the old cell. The two chromatids are attached by a protein disk at a point called a **centromere** (*SEHN troh mihr*), shown in Figure 6-2. 🔲

Human cells have 23 different chromosomes. Your body cells (also called somatic cells) contain two copies of each chromosome, for a total of 46 chromosomes. The two copies of each chromosome are called **homologous** (*hoh MAHL uh gus*) **chromosomes**, or homologues. Homologous chromosomes are similar in shape and size and have similar genetic information. You received one homologue of each chromosome from your mother and the other homologue from your father. When a cell contains two homologues of each chromosome, it is termed **diploid** (*DIHP loyd*). Biologists use the symbol *2n* to represent the diploid number of chromosomes in a cell. For humans, *2n* = 46.

Not all cells are diploid. In the life cycle of animals, including humans, **gametes** (*GAH meets*)—egg cells and sperm cells—are haploid. A **haploid** (*HAP loyd*) cell contains only one homologue of each chromosome. The fusion of two haploid gametes forms a diploid zygote, as shown in Figure 6-3. A **zygote** (*ZY goht*) is a fertilized egg cell, the first cell of a new

Effective Reading

Prefixes

Have students use the prefixes in the words *haploid* and *diploid* to help them remember their meanings. It may be helpful to have students use the algebraic terms *n* and *2n* in their definitions. Remind students that haploid is *n*, or one set of chromosomes. Lead students to reason that since many organisms are diploid, having *2n* chromosomes, then *n*, or one set of chromosomes, is also one-half of the chromosomes their cells contain. Moreover, if haploid means a single set and diploid means a double set of chromosomes, have students predict what would be true of a polyploid cell.

Table 6-1 Chromosome Number of Various Organisms

Organism	Number
Fungi (haploid)	
Penicillium	1–4
Saccharomyces	18
Insect (diploid)	
Mosquito	6
Drosophila	8
Housefly	12
Plants (number of chromosome sets per cell varies)	
Garden pea	14
Corn	20
Sugarcane	80
Adder's tongue fern	1,262
Vertebrates (diploid)	
Frog	26
Mouse	40
Human	46
Chimpanzee	48
Orangutan	48
Gorilla	48
Horse	64
Dog	78
Duck	80

Saccharomyces (yeast)

Corn plant

Drosophila (fruit fly)

Orangutan

individual. Being haploid ensures that when an egg and a sperm fuse, the resulting zygote will contain the characteristic diploid number of chromosomes for that organism. Biologists use the symbol n to represent the haploid number of chromosomes. For humans, $n = 23$. Table 6-1 lists the characteristic chromosome number of a variety of organisms. ❑

❑ **CAPSULE SUMMARY**

The number of chromosomes in cells is constant within a species. Haploid cells contain one homologue of each chromosome. Diploid cells contain two homologues of each chromosome.

Chromosomes Affect Development

Each of the 46 human chromosomes has thousands of genes that play important roles in determining how a person's body develops and functions. All these genes must be present in an individual's cells for the same reason that a car must have an engine, a transmission, and wheels—to function properly. Therefore, a person must have the characteristic number of chromosomes in his or her cells. In most cases, humans who are missing even one chromosome do not

Teaching Tip

What If a human sperm or egg cell did not contain one member of each homologous pair of chromosomes? The resulting zygote might fail to develop. If the zygote did develop, the individual would not be normal because its cells would lack the important information contained on the missing genes.

👁 **VISUAL STRATEGY**

Table 6-1
Use the information provided in this table to have students determine the value of *n* for each of the organisms listed. (*Penicillium, 1–4; Saccharomyces, 18; mosquito, 3; Drosophila, 4; housefly, 6; garden pea, corn, sugarcane, Adder's tongue fern, depends on the number of sets of chromosomes per cell; frog, 13; mouse, 20; human, 23; chimpanzee, 24; orangutan, 24; gorilla, 24; horse, 32; dog, 39; duck, 40)*

Demonstration

The Impact of Chromosomes
Show the class two sets of pictures, one depicting a normal person, along with a corresponding karyotype, and one showing a person with either Klinefelter's or Turner's syndrome, along with the appropriate karyotype. Have students identify how the karyotypes differ. Relate the differences to the photographs of the individuals.

Historical Note

The Human Karyotype
Approximately 100 years ago, biologists first began counting the number of chromosomes in human cells. In the 1920s, a biologist reported the diploid number in human cells to be 48. For many years following

this report, textbooks stated that 48 was the normal diploid number of human chromosomes. However, it was not until 1956 that J. H. Tjio and A. Levan proved that 46 is the correct number of chromosomes in human cells. These two scientists worked out a procedure to culture white blood cells.

Their methods increased the number of cells undergoing mitosis and improved the spreading of the chromosomes, which allowed the scientists to count accurately the number of human chromosomes.

Demonstration
Preparing a Karyotype

To show students the procedure used to prepare a karyotype, hold up a test tube containing water, a drop of yellow food dye, and red sprinkles (like those used for decorating cakes). Tell students that the sprinkles represent red blood cells that can be removed from a small sample of blood. Pour the yellowish liquid, representing white blood cells in plasma, into another test tube to show how the "red blood cells" can be removed. To the "white blood cells," add a drop of liquid representing colchicine, the chemical reagent used to stop the cells from dividing. Tell students that although the cells will no longer divide, the chromosomes in the nuclei of white blood cells continue to make copies; thus, the number of chromosomes will increase within each white blood cell. Allow the chromosomes to make many copies. Next add some water to the test tube, informing students that the water will cause some of the "white blood cells" to burst. The tube is then centrifuged, causing the white blood cells and chromosomes to settle to the bottom. Place a drop of the bottom sample on a slide. The sample is then stained, observed under a microscope, and photographed. The photograph of the chromosomes is enlarged, and each chromosome is then cut out and arranged according to size and shape, producing a karyotype, such as the one shown in Figure 6-4. To reinforce the preparation of a karyotype, have students make a flowchart of the procedure.

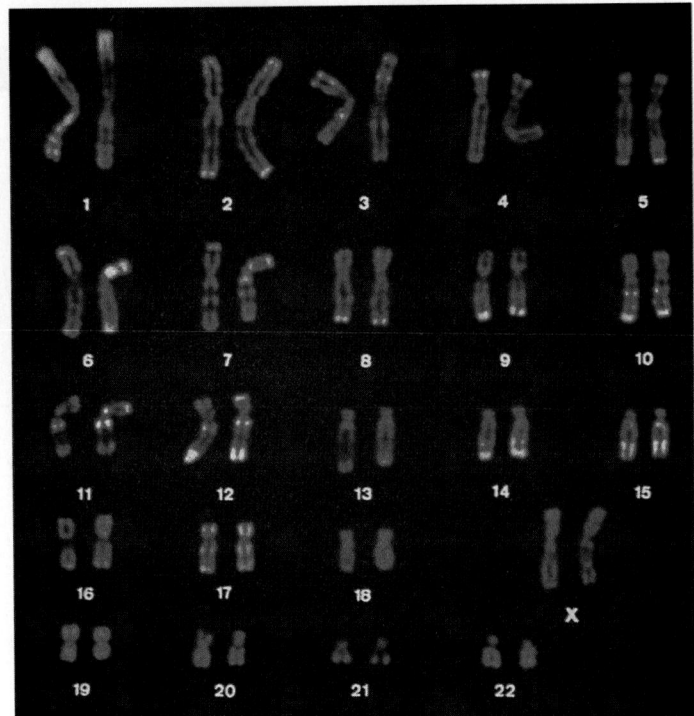

Figure 6-4 To examine an individual's karyotype, investigators chemically treat and stain the chromosomes in cells from a blood sample. Afterward, the chromosomes are usually photographed, cut out, arranged in pairs from largest to smallest, and numbered, *above right*. People with Down syndrome, *bottom left*, have three copies of chromosome 21 in their karyotypes, *top left*.

survive embryonic development. The condition in which a diploid cell is missing a chromosome is called monosomy (MAHN uh soh mee). And just as a car will not function correctly with two engines stuffed under the hood, a human embryo will not develop properly with more than two copies of most chromosomes. The condition in which a diploid cell has an extra chromosome is called **trisomy** (try SOH mee).

Deviations in chromosome number can be detected by analyzing a **karyotype** (KAR ee uh typ), the collection of chromosomes found in an individual's cells. Figure 6-4 shows a typical karyotype.

Figure 6-4 also shows a karyotype from an individual with an extra copy of chromosome 21. The traits produced by having an extra copy of chromosome 21 were first described in 1866 by the British physician J. Langdon Down and are collectively called **Down syndrome**, or trisomy 21 syndrome. The features that characterize Down syndrome include a short stature, a round face with upper eyelids that cover the inner corners of the eyes, and, most significantly, varying degrees of mental retardation.

Down syndrome occurs in all racial groups with the same frequency, approximately 1 in 1,000 children. It is much more common in children of older mothers. In mothers younger than 30 years old, the incidence is only about 1 in 1,500 births, while in mothers 30 to 35 years old, the incidence doubles to 1 in 750 births. In mothers older than 45,

Overcoming Misconceptions
Where Does That Chromosome Come From?

Most people believe that birth defects resulting from the presence of an extra chromosome, such as Down syndrome, are mainly attributable to maternal nondisjunction during the first meiotic division of chromosomes. However, recent studies have shown that in about 25 percent of Down syndrome cases, the extra chromosome comes from the father rather than the mother. Just as increasing age in females is associated with a higher incidence of Down syndrome, so is increasing age in males, with an increased risk of Down syndrome occurring after the father's age exceeds 55.

the risk is as high as 1 in 16 births. The reason that more babies with Down syndrome are born to older mothers is that all the eggs a female will ever produce are present in her ovaries when she is born. As the female ages, the eggs can accumulate an increasing amount of damage; males, in contrast, produce new sperm throughout adult life.

What events cause an individual to have an extra copy of a chromosome? When a cell divides normally, each chromosome and its homologue separate, an event called disjunction. When normal disjunction does not occur, one or more chromosomes may fail to separate properly. This accident in chromosome separation is called **nondisjunction** (*nahn dihs JUHNK shuhn*), which results in one new cell receiving both chromosomes and the other new cell receiving none. Trisomics arise as a result of nondisjunction. In the case of Down syndrome, nondisjunction occurs with chromosome 21.

Prenatal Testing

Because of the risk of Down syndrome, a pregnant woman over the age of 35 may be advised to have prenatal testing to check for an extra copy of chromosome 21 in the fetus. This can be done through procedures such as amniocentesis and chorionic villi sampling, which are illustrated in Figure 6-5.

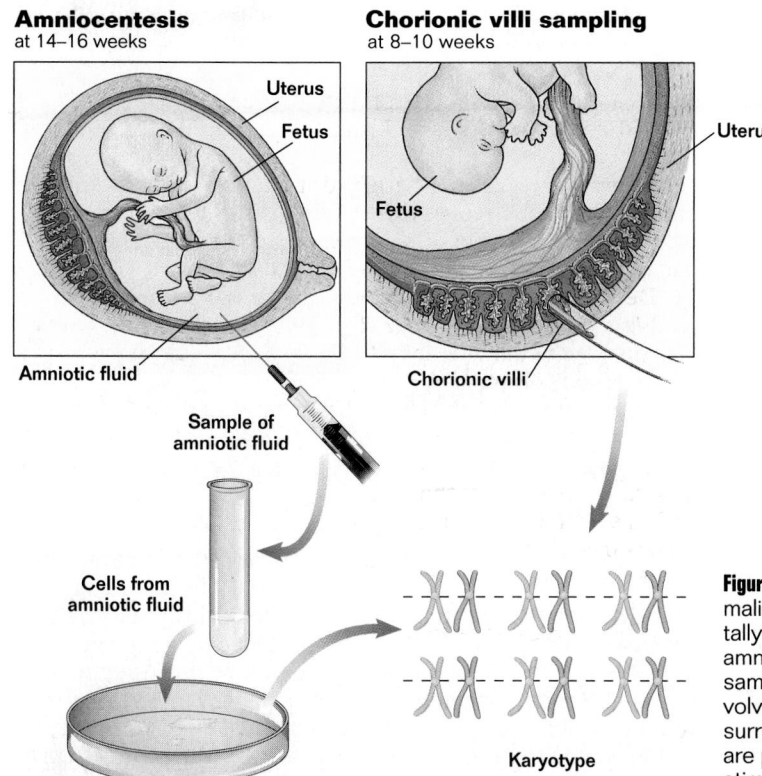

Figure 6-5 Chromosome abnormalities can be diagnosed prenatally with procedures such as amniocentesis and chorionic villi sampling. Both procedures involve taking cells from tissues surrounding the fetus. The cells are placed in a culture medium to stimulate growth, and then karyotypes are prepared.

CONNECTIONS

Chapter 3

Cells and Their Environment

Point out that cells placed in a culture, such as cells collected for use in prenatal testing, must be in an environment that mimics their normal one. Otherwise, the cells would not continue to grow and divide. Ask students to describe what they would add to a liquid that could be used for culturing human cells. (*Answers should include water with a concentration of nutrients, salts, vitamins, and minerals that is isotonic to human cells.*)

Teaching Tip

Prenatal Testing

Have students prepare a table that summarizes the similarities and differences between amniocentesis and chorionic villi sampling. Similarities should include that both procedures culture cells and prepare karyotypes. Differences between the procedures are in the tissues from which the cells are taken to prepare a karyotype.

VISUAL STRATEGY

Figure 6-6

Have students explain the difference between a duplication and a translocation. (*There are two copies of the duplicated gene[s] in duplication. In translocation there is only one copy of each gene.*) Point out that a translocation involving chromosomes 14 and 21 in humans accounts for a small percentage of Down syndrome cases.

CAPSULE SUMMARY

Deviations in a chromosome number can cause abnormal development. In humans, Down syndrome is caused by an extra copy of chromosome 21, which is a result of nondisjunction.

In **amniocentesis** (*am nee oh sehn TEE sihs*), a physician uses a needle and syringe to remove a small amount of the fluid from the amnion, the sac that surrounds the fetus. Fetal cells in the amniotic fluid are used to make a karyotype, which can then be analyzed. In **chorionic villi sampling**, a physician analyzes a karyotype made using cells grown from a sample of the chorionic villi, fingerlike extensions of the placenta that grow into the mother's uterus. Since the villi have the same genetic makeup as the fetus, the physician is able to detect abnormalities in the chromosome number. ❏

Alterations in Chromosome Structure

Although rare, changes in an organism's chromosome structure do occur. Some alterations cause **mutations**, changes in an organism's genetic material. Some of these mutations are illustrated in Figure 6-6. When a fragment of a chromosome breaks off, it can be lost when the cell divides, causing a mutation called a **deletion**. As a result of deletion, a new cell will lack a certain set of genes. In a mutation called a **duplication**, the chromosome fragment attaches to its homologous chromosome, which will then carry two copies of a certain set of genes. Sometimes the fragment reattaches to the original chromosome in the reverse orientation, producing a mutation called an **inversion**. Or, the fragment may join a nonhomologous chromosome, in an event called **translocation**.

Figure 6-6 When a chromosome breaks, its genes can become rearranged in different ways. Notice how each kind of mutation affects the shaded genes shown below.

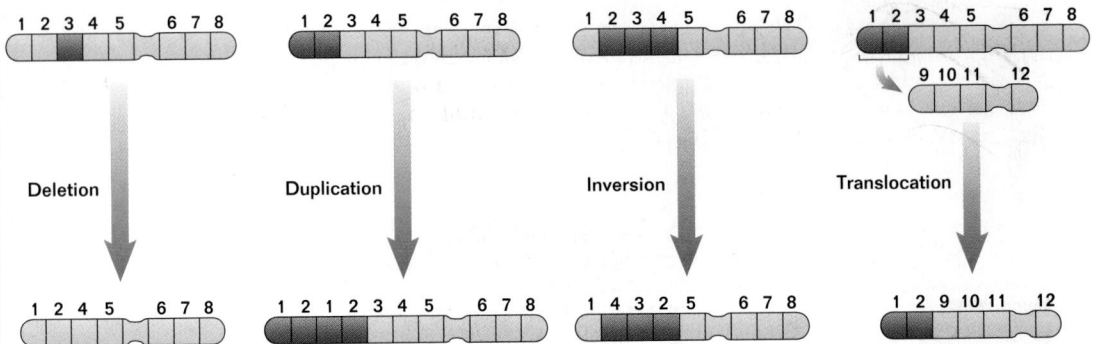

Chromosomes Determine Your Sex

Of the 23 pairs of chromosomes in human somatic cells, 22 pairs are the same in males and females. These chromosomes are called **autosomes** (*AWT uh sohmz*). The chromosomes that differ between males and females are called the **sex chromosomes** because they carry the genes that determine an individual's sex. Sex chromosomes exist in either of two forms—as an X chromosome or as a shorter Y chromosome. In humans and many other organisms, the genes that cause a fertilized egg to develop into a male are located on the Y chromosome. Thus, any individual with a Y chromosome is a male, and any individual without a Y chromosome is a female.

Overcoming Misconceptions

Sex Chromosomes

Students sometimes have the misconception that all traits associated with sex are controlled by genes on the sex chromosomes, while genes controlling traits that have nothing to do with sex are located on the autosomes. Point out that the genes for hemophilia and red-green colorblindness are located on the sex chromosomes, while those controlling the production of sex hormones are found on the autosomes.

Table 6-2 Sex Chromosome Abnormalities

Sex	Abnormality	Effects	Occurrence
Male	XXY	Klinefelter's syndrome: longer than average limbs, sparse body hair, underdeveloped genitalia, slight breast development; sterile	1 per 800–1,000 male births
	XYY	Some studies suggest increased risk of antisocial behavior; fertile	1 per 1,000 male births
Female	XO	Turner's syndrome: short stature. pronounced webbing of the neck, undeveloped secondary sexual characteristics; sterile	1 per 2,000 female births
	XXX	Triplo-X syndrome: most are of normal intelligence and are fertile	1 per 1,200 female births

In these cases, females are designated XX because they have two X chromosomes, and males are designated XY because they have one X chromosome and one Y chromosome. Because a female can donate only an X chromosome to her offspring, the sex of an offspring is determined by the male, who can donate either an X or a Y. Table 6-2 explains the abnormalities that arise when a person has an abnormal number of sex chromosomes.

In some insects, such as grasshoppers, there is no Y chromosome. In such cases, the females are characterized as XX and the males as XO (the O indicates the absence of a chromosome). In birds, moths, and butterflies, the male has two X chromosomes and the female only one. ❑

❑ CAPSULE SUMMARY

Sex chromosomes carry information that determines an organism's sex. In humans and many other organisms, females are designated XX, and males are designated XY.

SECTION 6-1

Teaching Tip
Abnormalities in Chromosome Number
Have students diagram events during nondisjunction that could account for each abnormality described in Table 6-2. For example, nondisjunction of the two X chromosomes in an egg would result in an abnormal gamete with an XX condition. If fertilized by a normal sperm carrying a Y chromosome, an XXY condition (Klinefelter's syndrome) would result.

Application
Rx **Medicine** While most nondisjunctions involving autosomes are lethal, those involving sex chromosomes usually are not. Although some very unusual combinations have been observed, a YO combination in humans is lethal, leading biologists to suspect that at least one X chromosome is necessary for development and survival.

Section Review

1. *When are chromosomes visible in a cell?*

2. *If a cell that is about to divide has 22 chromosomes, how many chromatids does it have? How many centromeres does it have?*

3. *Are homologous chromosomes normally found in gametes? Explain why or why not.*

4. *What is trisomy? How does trisomy arise in a cell?*

5. *Describe the karyotype of an individual with Down syndrome. What are the effects of this deviation?*

6. *What are sex chromosomes? In what combination do they exist in human females? in human males?*

Critical Thinking

7. *Why might two organisms have the same number of chromosomes but not the same traits?*

Answers to Section Review

1. The chromosomes become visible through a microscope after they have condensed.

2. A cell with 22 chromosomes has 44 chromatids and 22 centromeres.

3. Homologous chromosomes are found only in diploid cells, not in the gametes, which are haploid.

4. Trisomy is a condition in which a diploid cell has an extra chromosome as a result of nondisjunction.

5. The karyotype of a person with Down syndrome would show an extra chromosome 21. These individuals might exhibit short stature, a round face, upper eyelids that cover the inner corners of the eyes, and varying degrees of mental retardation.

6. The sex chromosomes carry the genes that determine an individual's sex. In humans, females are XX, while males are XY.

7. Traits are determined by the genes containing hereditary information on chromosomes, not by the number of chromosomes.

UNIT 2

Vocabulary Preview

binary fission

cell cycle

mitosis

cytokinesis

interphase

spindle fiber

kinetochore

Opening Questions

Have students develop a hypothesis to describe what happens to each chromosome when a somatic cell divides. Students' hypotheses should explain how the two new cells receive an exact replica of the hereditary information in the parent cell. Then have students explain why the new somatic cells produced by cell division are structurally and functionally similar both to the parent cell and to one another. Students should recognize that they are similar to each other because each new cell receives about half the parent cell's cytoplasm and organelles. Moreover, each new cell receives a complete set of genetic instructions that were present in the parent cell.

Demonstration

Chromosome Condensation

To show students one logical reason why chromosomes condense prior to cell division, place several long pieces of cord on a table top. Be sure that all the pieces are well entwined with one another. Tell the class that each piece represents a chromosome. Ask one student to distribute half of the pieces to one classmate and the other half to another classmate. Repeat this process after rolling each piece of cord into a small ball. Students should see that when chromosomes are condensed into tight bundles, they are easier to move around the cell than long, thin strands.

- Discuss the two stages of cell division in bacteria.
- Define cell cycle, and explain what occurs during each of its phases.
- Explain the events of mitosis, and describe the structures involved in them.

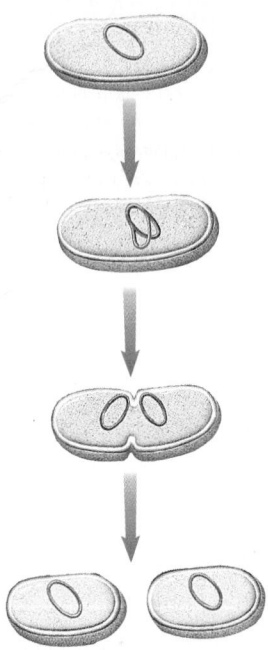

Figure 6-7 During binary fission, a bacterium makes a copy of its DNA and forms a new membrane and cell wall. Afterward, the cell splits into two cells.

6-2 Mitosis and Cell Division

*F*or cells, reproducing by cell division is essential for life. What triggers a cell to divide? Cell biologists have observed that the size of a cell seems to be the signal that stimulates it to divide. Recall that as a cell grows, its surface area becomes too small to enable the necessary amounts of nutrients to enter the cell and to allow wastes to pass out. Therefore, as the surface-area-to-volume ratio becomes relatively small, the cell may divide. In this section, you will explore cell division by first looking at how the simplest kinds of cells—bacteria—divide.

Bacteria Simply Split

A bacterium is a single, prokaryotic cell. It has a cell wall but lacks a nucleus and membrane-bound organelles. A bacterium's single DNA molecule is not coiled around proteins to form chromosomes. Instead, its DNA is a circular chromosome, attached to the inner surface of the plasma membrane like a rope attached to the inner wall of a tent. For these tiny organisms, cell division takes place in two stages: first the DNA is copied, and then the cell splits.

Before a bacterium can divide, it must have two copies of its DNA so that each of the two new cells will have a complete copy of the genetic information of that bacterium. To make a copy of itself, the DNA molecule begins to "unzip" lengthwise, exposing its two strands. As this happens, each strand is made into a complete DNA molecule. Shortly after the entire DNA circle has unzipped, the cell has two identical copies of its hereditary information, attached side by side to the interior plasma membrane, as illustrated in Figure 6-7.

Once the DNA has been copied and the cell has grown to an appropriate size, the bacterium splits into two equal halves through a process called binary fission. **Binary fission** is a form of asexual reproduction that produces identical offspring. First, a new plasma membrane is added at a point on the membrane between the two DNA copies. As new material is added in this zone, the growing plasma membrane pushes inward and the cell is constricted in two, like a long balloon being squeezed around its middle. A new cell wall forms around the new membrane. Eventually the dividing bacterium is pinched into two independent cells. Each cell contains one of the circles of DNA and is a complete, functioning bacterium.

Eukaryotic Cells Undergo Nuclear Division

The cells of eukaryotic organisms (protists, fungi, plants, and animals) have a far more complex internal structure than that of the simple cells of bacteria. Cell division in eukaryotic cells must take into account the nucleus with the chromosomes inside and the many other internal organelles, all of which must be strategically maneuvered before the cell can properly divide.

Cell Cycle

The life of a eukaryotic cell is traditionally diagrammed as a **cell cycle,** a repeating sequence of growth and division through which many kinds of eukaryotic cells pass. The five phases of the cell cycle are shown in Figure 6-8. The cell cycle may be summarized as follows:

$$G_1 \rightarrow S \rightarrow G_2 \rightarrow M \rightarrow C$$

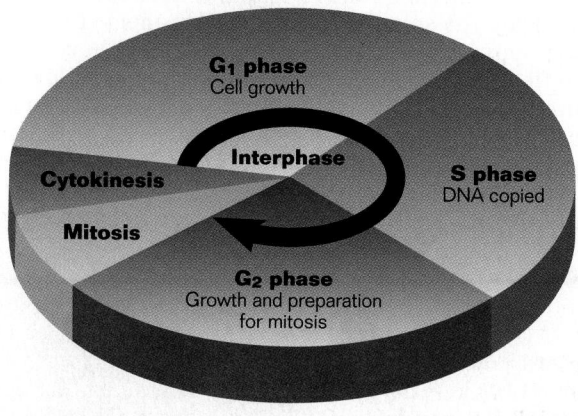

Figure 6-8 In the cell cycle, growth phases, *shaded blue,* alternate with a cell division phase, *shaded brown.*

1. The G_1 phase is the growth phase of a cell. During this phase, a cell grows rapidly and carries out its routine functions. For most organisms, this phase occupies the major portion of the cell's life between cell divisions.

2. The S phase is when the DNA is copied. At the end of this phase, an individual chromosome consists of two chromatids attached at the centromere.

3. In the G_2 phase, preparations are made for nuclear division. Mitochondria and other organelles replicate. Microtubules are reassembled; they will be used to form the spindle apparatus that moves the chromosomes.

4. The M phase is the phase in which mitosis occurs. **Mitosis** *(my TOH sihs)* is the process by which the nucleus of a cell is divided into two nuclei, each with the same number and kinds of chromosomes.

5. The C phase is when the cytoplasm divides during a process called **cytokinesis** *(syt oh kih NEE sihs).*

What Happens During Mitosis and Cell Division

A eukaryotic cell spends most of its life in the G_1, S, and G_2 phases, which are collectively called **interphase.** During interphase, a cell does a great deal of growing. Chromosomes are loosely wound, and genes are being used to make the enzymes that direct the activities of the cell.

Forming Spindle Fibers

As interphase ends and mitosis begins, the chromosomes begin to condense. Enzymes in the cell begin to break down the nuclear envelope. The cell starts building the equipment that will move copies of the chromosomes to opposite ends, or poles, of the cell. In the center of an animal cell, a pair of cylindrical structures called centrioles (*SEHN tree ohlz*) start to separate, each moving toward opposite poles of the cell. As the centrioles move apart, a network of protein cables, called the spindle, forms between them. The spindle will help move chromosomes apart. Each cable is called a **spindle fiber** and is made of microtubules, long hollow tubes of protein. Recall that plant cells do not have centrioles. However, plant cells form a spindle that is almost identical to that of an animal cell, as shown in Figure 6-9.

THEMES OF BIOLOGY

Structure and Function

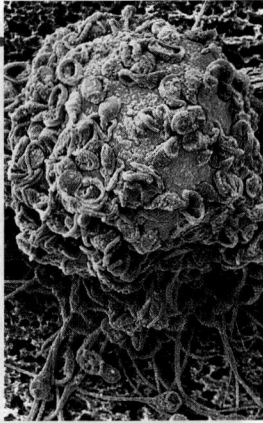

Because evolution has shaped life to meet the challenges of survival, it should come as no surprise that biological structures are very well suited to their functions. The excellent correspondence between structure and function will be seen time and again in your journey through biology. You have already seen how the shape of an enzyme is intimately related to the chemical reaction it carries out, and that the shapes of the proteins protruding from a cell's surface determine in large measure what goes on inside the cell. You have also seen how mechanisms within the cell precisely sort and distribute the cell's genetic materials at the time of cell division.

As you learn about the diversity of life on Earth, you will encounter a parade of structures, some bizarre and some commonplace. Each structure is precisely adapted to carry out its particular function, from the harpoon-like nematocysts that jellyfish use to capture prey, to the waterproof coverings that enable birds and reptiles to lay their eggs on dry land. Of all the lessons of biology, the relationship between structure and function is one of the most fundamental.

When a cell divides, top left, *or an egg and sperm fuse,* above right, *chromosomes,* bottom left, *ensure that an organism's genetic information will be safely transmitted to new cells.*

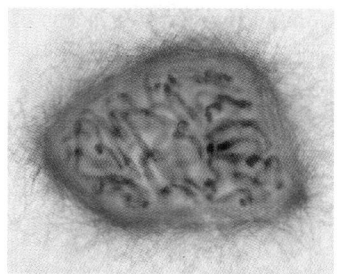

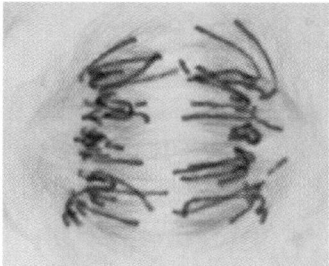
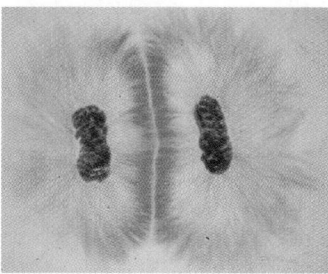

Attaching Spindle Fibers to Centromeres

As the chromosomes continue to condense, a second group of microtubules extends out from a region of the centromere of each chromosome called the **kinetochore** *(kuh NEHT uh kawr)*. The kinetochore is a disk of protein that serves as a platform for assembling the microtubules. The two sets of microtubules extend out toward opposite poles of the cell. Each set of microtubules continues to grow longer until it makes contact with the pole of the spindle. When the process is complete, as shown in Figure 6-9, one chromatid is attached by a set of microtubules to one pole, and the other chromatid is attached to the other pole.

Separating Chromatids

Once the microtubules are attached to the centromeres, the centromeres split, freeing the chromatids from each other. Mitosis is now simply a matter of reeling in the microtubules and dragging the chromatids, each now considered a chromosome, to the poles, as shown in Figure 6-9. At the poles, the ends of the spindle fibers are dismantled bit by bit. As the fibers become shorter and shorter, the attached chromosomes move closer and closer to the poles. When they finally arrive, each pole has one complete set of chromosomes.

Dividing the Cell

In the final step of cell division, a new nuclear envelope forms around each pole, forming two nuclei, and the chromosomes within uncoil. Cytokinesis then takes place; the cytoplasm of the cell divides in half, as shown in Figure 6-9. Each half includes one of the two new nuclei and an assortment of organelles, which replicated earlier in the cell cycle. ◻

Figure 6-9 In preparation for division, this cell from an African blood lily, *top left*, has formed a spindle. Then, spindle fibers attach to centromeres, *top center*. Next, the chromatids in this cell begin to separate, *bottom left*. Cell division produces two new cells with identical genetic information, *bottom center*.

❑ CAPSULE SUMMARY

The life of a eukaryotic cell includes periods of growth, replication, and division. During interphase, the DNA and the cellular organelles are copied, During mitosis, the nucleus of the cell divides, forming two nuclei with identical genetic information. The cytoplasm in the cell splits during cytokinesis.

RESEARCH Update

Deadly *Rascal*

Today people generally assume that anything is potentially carcinogenic. Although scientists have discovered more than 100 genes that cause cancer, a gene called *ras* is ultimately responsible in one-third of the cases. Researchers at the University of California at Berkeley and the California Institute of Technology in Pasadena have discovered that this *ras* gene makes "cells continuously replicate themselves under the misapprehension that the body's hormones are telling them to do so" in fruit flies and nematodes. Cell replication is directed from the nucleus, inside the cell. Hormone receptors are located on the outer membrane of the cell. The *ras* protein (a type of enzyme) behaves as an intermediary or a "molecular switch" that conveys the order to replicate from the external environment of the cell to the nucleus.

While several researchers have shown that *ras* is a DNA-activator, other scientists continue research on two other important culprits: the well-known p53 and newly discovered p16 tumor-suppressor genes. Myriad Genetics in Salt Lake City has worked on p16 recently. Scientists at Myriad have found that, like the p53 gene, p16 also inhibits rapid cell division (rapid cell division can lead to the production of a tumor). Experimentally, scientists have been able to manipulate these genes in the laboratory to show that when the genes are damaged, they lose their ability to suppress tumor growth. Scientists at Myriad and other biotechnology firms are excited about the results of these gene manipulation experiments because the cure for cancer is probably rooted in suppression of tumor growth or manipulation of the genes that are responsible for tumor growth. ☑

Historical Note

Chromosomal Mutations

Observations of alterations in chromosome structure first came from a 1993 study of the fruit fly, *Drosophila melanogaster*. The salivary glands of these flies, as in many other insects, are composed of cells that do not divide during the larval stage. However, the chromosomes continue to replicate, producing many chromosomes that are easy to study under a microscope.

VISUAL STRATEGY

Figure 6-10

Use this figure to trace the events of mitosis. Then ask students the following questions: What three phases of the cell cycle are incorporated in the interphase stage that is shown at the top of this figure? *(G₁, S, G₂ phases)* What occurs during these phases? *(G₁ phase—cell growth; S phase—DNA replication; G₂ phase—growth and preparation for mitosis)* How many chromosomes does the original cell have? *(The original cell has four chromosomes.)*

Stages of Mitosis

Biologists traditionally divide the process of mitosis into four general stages: prophase, metaphase, anaphase, and telophase. These four stages are illustrated in Figure 6-10, which shows mitosis and cytokinesis in a cell from a whitefish.

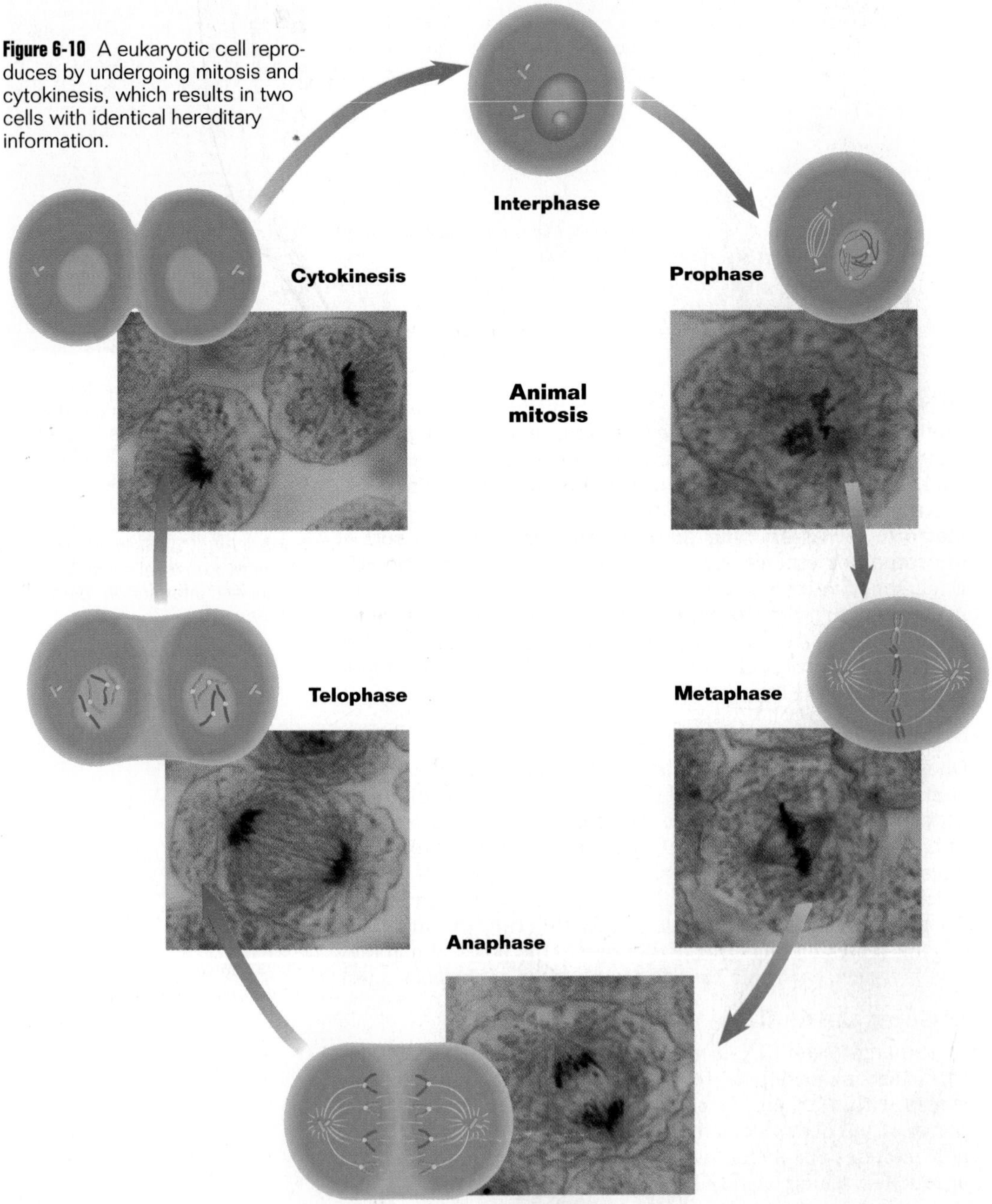

Figure 6-10 A eukaryotic cell reproduces by undergoing mitosis and cytokinesis, which results in two cells with identical hereditary information.

Interphase

Cytokinesis

Prophase

Animal mitosis

Telophase

Metaphase

Anaphase

Overcoming Misconceptions

Mitosis Versus Cytokinesis

Students often think that mitosis is the same as cell division. Be sure they understand that mitosis refers strictly to the division of the chromosomes, while cytokinesis refers to the division of the cytoplasm.

Remind students that cell division is just one of the four events that make up the cell cycle.

Prophase

During prophase, chromosomes begin condensing and become visible. The nuclear envelope surrounding the nucleus begins to break down, and the network of spindle fibers becomes visible.

Metaphase

In metaphase, the chromosomes move to the center of the cell and line up along the "equator." Once in place at the equator of the cell, each chromosome is held in place by the microtubules attached to the kinetochore.

Anaphase

In anaphase, the two chromatids physically separate when the centromere divides. The chromatids, each which now may be called a chromosome, move toward opposite poles of the cell as the fibers attached to them shorten.

Telophase

The chromosomes, now at opposite ends of the cell, uncoil. A new nuclear envelope forms. The spindle fibers break down and disappear, and mitosis is complete.

As mitosis ends, cytokinesis begins. During cytokinesis, the cytoplasm of the cell is cleaved in half, and the cell membrane grows to enclose both cells. Animal cells and other cells that lack cell walls are pinched in half by a belt of protein threads. Plant cells have a rigid cell wall and a different strategy of cell division. In plant cells, vesicles formed by the Golgi bodies fuse at the equator of the cell and form the cell plate, a membrane across the middle of the cell. A new cell wall then forms on both sides of the cell plate. Once the cell wall is complete, there are two new plant cells. ❏

Teaching Tip
Stages of Mitosis

Have students refer to Figure 6-10 on page 130 as they prepare a table that summarizes the events of each stage of mitosis and cytokinesis. Then ask them to draw and label the prophase, metaphase, anaphase, and telophase stages of a cell having a diploid chromosome number of 6. Encourage students to use colored pencils to distinguish between homologous chromosomes in their sketches.

❏ **CAPSULE SUMMARY**

Mitosis can be divided into four stages: prophase, metaphase, anaphase, and telophase. Cytokinesis is the division of the cytoplasm.

Section Review

1. *How does the structure of bacterial DNA differ from the structure of your DNA?*

2. *Describe the process in which bacterial DNA copies itself.*

3. *During which phase of the cell cycle does DNA replicate?*

4. *During which stage of mitosis do spindle fibers first appear? How do they function during metaphase? during anaphase?*

Critical Thinking

5. *What is cytokinesis? How does cytokinesis differ between plant cells and animal cells?*

6. *Why does it take more time for a cell to complete mitosis than to complete cytokinesis?*

Answers to Section Review

1. Bacterial DNA occurs in the form of circular chromosomes. Human DNA wraps tightly around proteins (histones) and forms linear chromosomes by means of supercoiling.

2. The DNA molecule "unzips" and each of the two strands is copied to make two identical copies of the DNA molecule.

3. DNA replicates during the S phase of the cell cycle.

4. Spindle fibers first appear during prophase. At metaphase, the chromosomes attach to the spindle fibers. At anaphase, the spindle fibers shorten, thus pulling the chromatids to opposite poles of the cell.

5. Cytokinesis is the division of the cell cytoplasm. In an animal cell, the cell membrane pinches inward to divide the cytoplasm. In a plant cell, a cell plate forms across the middle of the cell.

6. Many more steps are involved in mitosis, whereas cytokinesis only separates the cytoplasm into equivalent portions.

Vocabulary Preview

meiosis

spore

crossing-over

reduction division

genetic recombination

Opening Question

Ask students to explain how biologists were able to predict meiosis before it was actually observed. Explain that once biologists understood what was happening to the chromosomes during mitosis, they predicted that some other process must be involved when sperm and egg cells form. Lead students to understand that hereditary patterns in cells in which meiosis occurs differ from those in cells in which mitosis occurs (for example, in meiosis the new cells are not identical to the parent cells, but they are identical in mitosis). Help students to realize that scientists were able to predict events similar to mitosis to explain their observations.

Demonstration

The Importance of Meiosis

To help students understand the significance of the microscopic events of meiosis, have each student select one of the vertebrates listed in Table 6-1 on page 121. Ask students to list the cell types in which they would expect to find the diploid chromosome number in that organism. On the chalkboard, draw what would happen at fertilization if the sperm and egg cells of that organism also contained the diploid amount. If the gametes were diploid, what would the animal's diploid chromosome number be after five generations?

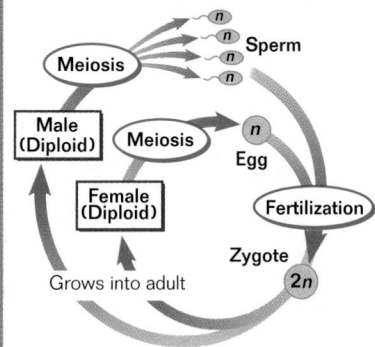

Figure 6-11 The figure above shows the pattern of alternation between the diploid and haploid chromosome number in a sexually reproducing organism such as yourself.

6-3 Meiosis

You have just learned that bacteria reproduce asexually by binary fission. Humans, like most animals and plants, reproduce sexually. In sexual reproduction, *gametes from the opposite sexes unite to form a zygote. Because the zygote is diploid, the number of chromosomes in the gametes must be halved. Otherwise, the number of chromosomes in a zygote would be twice the diploid number for the species. In this section, you will learn how the correct number of chromosomes is maintained from generation to generation.*

What Happens During Meiosis

The mechanism that halves the number of chromosomes in cells is a form of cell division called **meiosis** (my OH sihs). Meiosis consists of two successive nuclear divisions. Before the first division, the DNA is copied, just as it is before mitosis. In the first division, called meiosis I, homologous chromosomes separate into two cells. In the second division, called meiosis II, the two chromatids of each chromosome separate into two haploid cells. Thus, one diploid cell that undergoes meiosis produces four haploid cells. In animals, meiosis often results in haploid gametes, as shown in Figure 6-11. In plants, meiosis often leads to **spores**, haploid cells that later lead to the production of gametes.

During meiosis, two unique events occur:

1. **Crossing-over** In the beginning of meiosis I, homologous chromosomes pair up next to each other. While paired,

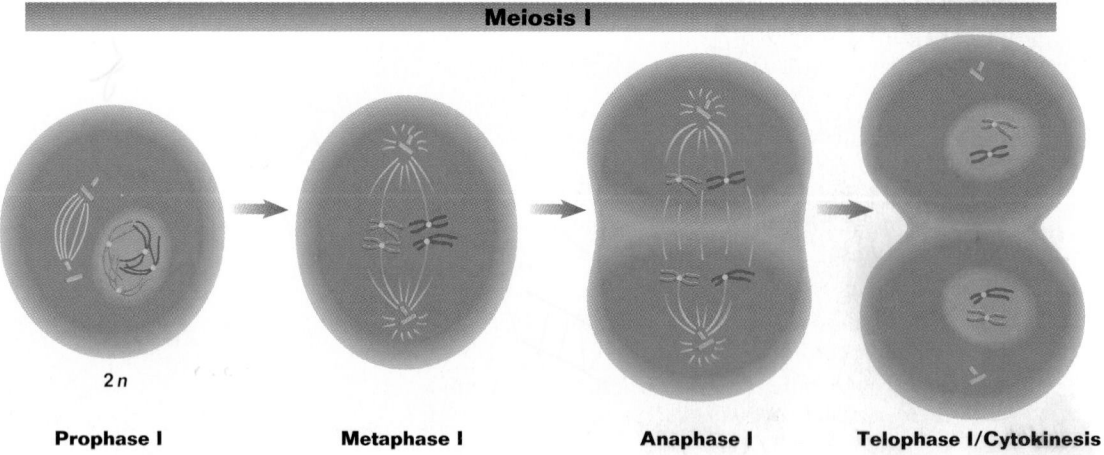

Meiosis I

| Prophase I | Metaphase I | Anaphase I | Telophase I/Cytokinesis |

2n

Overcoming Misconceptions

Reducing the Chromosome Number

Students may be confused when they learn that the cells produced by meiosis I as well as those produced by meiosis II are haploid. Point out that both sets of cells are haploid because they contain one-half of the number of chromosomes as the original cell. Ask students to determine the diploid chromosomal number for the cell that is in prophase I in Figure 6-12 on this page. (*The diploid number is four.*) Remind students that prior to prophase I, the original cell's chromosomes replicate. Since the cells produced by meiosis I and meiosis II each contain two chromosomes, they are haploid cells.

the arms of the chromosomes exchange reciprocal segments of DNA in a process called **crossing-over.**

2. **Skipping replication** Because there is only one replication of DNA but *two* divisions, meiosis halves the number of chromosomes.

Meiosis I Separates Homologues

The two nuclear divisions of meiosis are traditionally divided into eight stages, as illustrated in Figure 6-12. Although these stages have the same names as those of mitosis, the events differ in significant ways.

Prophase I

As in prophase of mitosis, the chromosomes condense, and the nuclear envelope breaks down. However, a unique event not seen in mitosis occurs: first the homologous chromosomes pair up, and then crossing-over occurs.

Metaphase I

In metaphase I, the pairs of homologous chromosomes are moved by spindle fibers to the equator of the cell. The homologues, each made up of two chromatids, remain together.

Anaphase I

In anaphase I, the homologues separate. As in mitosis, the chromosomes of each pair are pulled by action of the spindle fibers to opposite poles of the cell. The difference between anaphase of mitosis and anaphase of meiosis I is that in the latter, the chromatids do not separate at their centromeres. Each chromosome is still composed of two chromatids joined by a centromere. During this phase of meiosis, nondisjunction can occur.

Figure 6-12 Meiosis produces four cells, each with half as much genetic material as the original cell, *below across.*

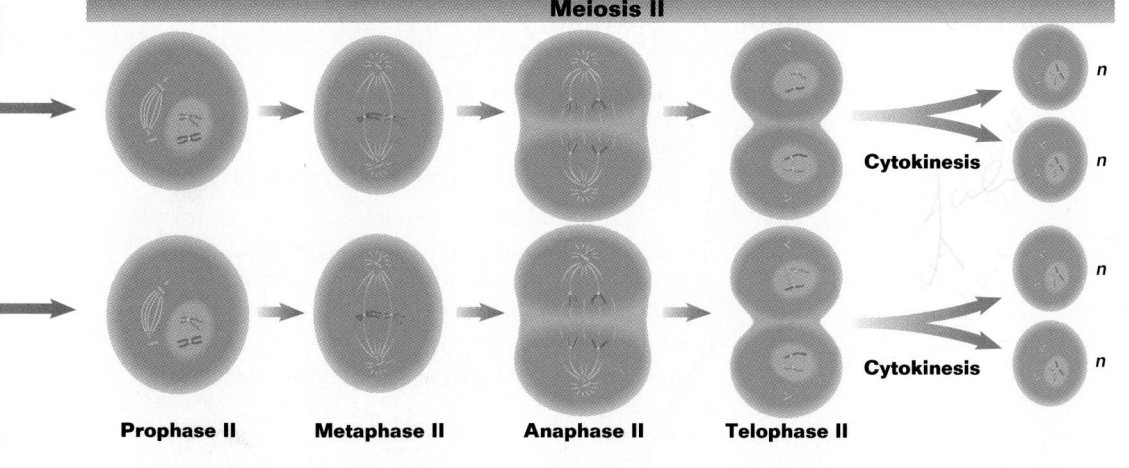

Meiosis II

Prophase II   Metaphase II   Anaphase II   Telophase II

Cytokinesis   *n*

Cytokinesis   *n*

Graphic Organizer

Use this graphic organizer with *Teaching Tip:* **Distinguishing Between Mitosis and Meiosis** on page 133.

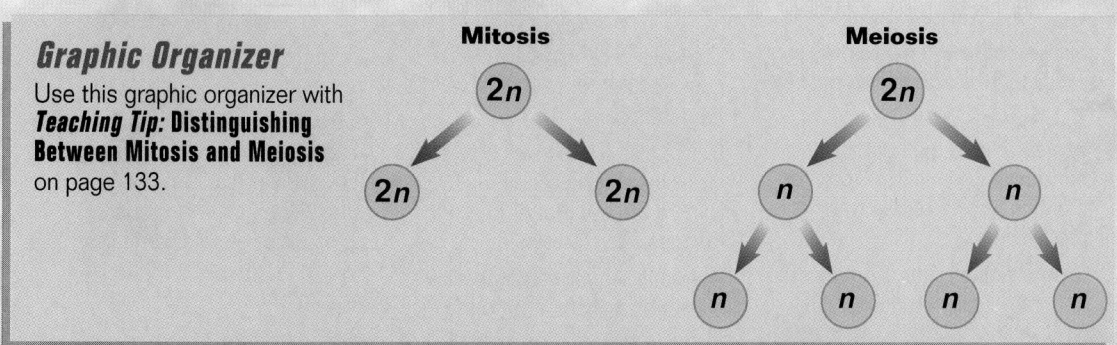

Mitosis

Meiosis

Teaching Tips

Meiosis II

Have students compare the stages of meiosis II as illustrated in Figure 6-12 to those of mitosis as drawn in Figure 6-10 so that they can see that they are identical.

On Stage

Tell the class to assume that each student is a chromosome. Then have them choreograph the stages of mitosis and meiosis.

 **VISUAL STRATEGY**

Figure 6-13

Use this figure to point out how crossing-over produces new genetic combinations. Because of the random orientation taken by chromosomes at metaphase I and because of crossing-over, students should see why they can never have children who will look exactly like them.

Chapter Closure

Have each student draw what anaphase I would look like in Figure 6-12 if nondisjunction were occurring. Tell students to complete their drawings to show what would happen to the four gametes that would be formed. Have them compare the four gametes they have drawn to those in Figure 6-12, which shows the four gametes after cytokinesis is complete.

Telophase I

In telophase I, individual chromosomes gather at each of the two poles. In most organisms, the cytoplasm divides, forming two new cells. Note that each of the cells produced now contains half the number of chromosomes of the original cell. For this reason, meiosis I is often called **"reduction division."**

□ **CAPSULE SUMMARY**

Meiosis is a form of cell division that halves the number of chromosomes in cells. Meiosis I separates homologous chromosomes. Meiosis II separates chromatids. As a result, four haploid cells are produced from one diploid cell.

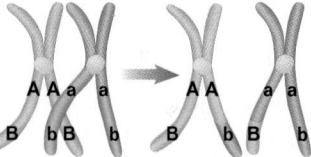

Figure 6-13 At sites of crossing-over, portions of a chromatid on one homologous chromosome are broken and exchanged with the corresponding portions of one of the chromatids on the other homologous chromosome.

Meiosis II Separates Chromatids

The stages of meiosis I are similar to those of mitosis, with some important differences. Meiosis II is identical to mitosis except that the chromosomes do not replicate before they divide at their centromeres. In anaphase II, the centromeres divide, and the chromatids, now called chromosomes, move to opposite poles of the cell. Meiosis II is followed by cytokinesis, in which new membranes are formed around the four products of meiosis to create four haploid cells. □

The Importance of Crossing-Over

Crossing-over is an efficient way to produce **genetic recombination**, the formation of new combinations of genes. As a result of crossing-over, the two chromatids of a chromosome no longer contain identical genetic material, as shown in Figure 6-13. Crossing-over thus provides a source of genetic variation. Since the speed at which a species can change is often limited by the amount of genetic variation available, crossing-over has an enormous impact on how rapidly organisms evolve.

Section Review

1. *Why is meiosis necessary in organisms that reproduce sexually?*
2. *What happens to homologous chromosomes in meiosis I? Why is meiosis I often called "reduction division"?*
3. *How is meiosis II similar to mitosis?*
4. *If a cell in a mosquito undergoes meiosis, how many chromosomes will each resulting cell contain? what if the cell is that of a dog?*

Critical Thinking

5. *What is crossing-over? How does it affect the rate of evolution?*

Answers to Section Review

1. Meiosis ensures that the gametes have the haploid number of chromosomes. Upon fertilization, the diploid number is restored.

2. In meiosis I, homologues are separated. Since each of the two cells receives only one homologue, meiosis I is the reduction division.

3. Meiosis II is similar to mitosis in that both processes involve the separation of chromatids.

4. In a mosquito, a gamete has 3 chromosomes. In a dog, a gamete has 39 chromosomes.

5. Crossing-over involves an exchange of portions of chromatids between homologous chromosomes. Crossing-over introduces an enormous amount of genetic variation, which increases the rate of evolution.

6

CHAPTER REVIEW

Vocabulary

amniocentesis (124)
anaphase (131)
autosome (124)
binary fission (126)
cell cycle (127)
centromere (120)
chorionic villi sampling (124)
chromatid (120)
chromosome (119)
crossing-over (133)
cytokinesis (127)
deletion (124)
diploid (120)

Down syndrome (122)
duplication (124)
gamete (120)
gene (119)
genetic recombination (134)
haploid (120)
homologous chromosome (120)
interphase (128)
inversion (124)
karyotype (122)
kinetochore (129)
meiosis (132)
metaphase (131)

mitosis (127)
mutation (124)
nondisjunction (123)
prophase (131)
reduction division (134)
sex chromosome (124)
sexual reproduction (132)
spindle fiber (128)
spore (132)
telophase (131)
translocation (124)
trisomy (122)
zygote (120)

Concept Mapping

Construct a concept map that shows the sequence of major events of the cell cycle. Use as many terms as needed from the vocabulary list. Try to include the following items in your map: cell cycle, mitosis, cytokinesis, interphase, prophase, metaphase, anaphase, telophase, chromosomes, DNA, and protein. Include additional concepts in your map as needed.

Review

Multiple Choice

1. In humans, a zygote will develop into a male if the
   a. sperm and egg both contribute an X chromosome.
   b. egg contributes an X chromosome and the sperm contributes a Y chromosome.
   c. sperm and egg both contribute a Y chromosome.
   d. egg contributes a Y chromosome and the sperm contributes an X chromosome.

2. The haploid number of chromosomes in a human gamete is
   a. 23.   c. 46.
   b. 22.   d. $2n$.

3. A human cell may have more or less than 46 chromosomes as a result of nondisjunction, a process in which
   a. chromosomes fail to separate during cell division.
   b. a fragment of a chromosome breaks off and is lost.
   c. homologous chromosomes pair up and exchange segments of DNA.
   d. genes are combined in new ways.

4. Meiosis is important for sexually reproducing organisms because it
   a. reduces the chromosomes in a cell from the diploid to the haploid number.
   b. ensures that each new cell will contain identical genetic information.
   c. prevents alterations in chromosome structure.
   d. ensures that the chromosome number is increased from one generation to the next.

5. Prokaryotes divide by binary fission, a form of asexual reproduction in which
   a. the nucleus divides into two nuclei.
   b. the number of chromosomes in the cell is reduced.
   c. a bacterium splits into two equal halves with identical genetic information.
   d. spindle fibers attach to the poles of the cell.

CHAPTER REVIEW ANSWERS

Each item in the Chapter Review is correlated by text section in the assignment guide that follows.

ASSIGNMENT GUIDE

Section	Review	Themes Review	Critical Thinking
6-1	1–3, 10–16, 18, 20		28
6-2	5–8, 17, 21	25–27	29
6-3	4, 9, 19, 22	23, 24	29

Concept Mapping

Map answer is shown on page 117B.

Review

Multiple Choice

Code in parentheses indicates section number and objective number.
1. b (6-1.5)
2. a (6-1.2)
3. a (6-1.4)
4. a (6-3.3)
5. c (6-2.1)
6. a (6-2.3)
7. b (6-2.2)
8. b (6-2.2)
9. b (6-3.4)

CHAPTER REVIEW ANSWERS *continued*

Completion

10. genes (6-1.1)

11. amniocentesis (6-1.5)

12. karyotype (6-1.5)

13. haploid, gametes (6-1.2, 6-3.3)

14. autosomes, sex (6-1.6)

15. homologous (6-1.2)

16. DNA, chromatids (6-1.1)

17. nucleus (6-2.1)

Short Answer

18. A doctor might recommend that a woman have an amniocentesis if she had a higher than normal risk of having genetically defective children. (6-1.5)

19. It results in daughter cells with half the number of chromosomes as the parental cells. (6-3.2)

20. Mutations are changes in an organism's chromosome structure. Deletion, duplication, inversion, and translocation are some ways a mutation can alter genetic material. (6-1.4)

21. Two sets of chromosomes must be present in order to be divided between two daughter cells and retain the diploid chromosome number. (6-2.2)

22. The male and female gametes unite to form a zygote. (6-3.1)

Themes Review

23. Meiosis separates the chromatids of a pair of chromosomes into single units that are combined independently during fertilization. When crossing-over occurs during prophase I, new combinations of genes are formed. (6-3.3)

24. Crossing-over is a process whereby portions of a chromatid on one homologous chromosome are broken and exchanged with corresponding portions of one of the chromatids on the other homologous chromosome during prophase I. Crossing-over produces genetic recombination and increases the rate of evolution, which improves a species' ability to adapt to its environment and survive. (6-3.4)

6. How does cell division differ between animal and plant cells?
   a. Plant cells do not have centrioles; animal cells do.
   b. Animal cells form a cell plate during cytokinesis.
   c. Animal cells are always haploid.
   d. Plant cells are pinched in half by a belt of protein threads.

7. During the metaphase stage of mitosis,
   a. the cell membrane begins to fold inward.
   b. chromosomes line up at the cell's equator.
   c. spindle fibers shorten, pulling chromosomes to the poles of the cell.
   d. chromosomes are at opposite ends of the cell.

8. During the C phase of the cell cycle the
   a. nucleus begins division.
   b. cytoplasm divides.
   c. chromosomes replicate.
   d. centromeres replicate.

9. Homologous chromosomes pair and undergo crossing-over during
   a. interphase.
   b. prophase of meiosis I.
   c. metaphase of mitosis.
   d. telophase of meiosis II.

Completion

10. The information that determines a cell's characteristics and directs its activities is organized into segments of DNA called _____.

11. Two methods of prenatal diagnosis are _____ and chorionic villi sampling.

12. Using a photograph of chromosomes from a cell, a _____ can be constructed by sorting and arranging the chromosomes by size and centromere position.

13. Meiosis results in new cells with the _____ number of chromosomes of the parent cell. When the _____ from a mother and a father fuse in the process of fertilization, the normal chromosome number is restored.

14. Humans have 23 pairs of chromosomes, 22 pairs of _____ , and one pair of _____ chromosomes.

15. Chromosomes that are similar in shape and size and carry similar genetic information are called _____ chromosomes.

16. A chromosome is composed of _____ wrapped around a protein scaffold. Before mitosis takes place, a chromosome is copied, resulting in two unseparated copies called _____ .

17. Mitosis is the process by which the _____ of a cell divides, while cytokinesis is the process by which the cytoplasm of a cell divides.

Short Answer

18. Under what conditions would a doctor recommend that a prospective mother have an amniocentesis?

19. Why is meiosis I often called the reduction division?

20. What is mutation? In what ways can a mutation alter the structure of a chromosome?

21. Why is DNA replication (forming a chromosome consisting of two chromatids) essential before mitosis begins?

22. What is the fundamental characteristic of sexual reproduction?

Themes Review

23. Evolution Explain how meiosis leads to genetic variation.

24. Evolution What is crossing-over? When does it occur? How does crossing-over help a species survive?

25. Homeostasis For a cell to function efficiently, the magnitude of its surface area must greatly exceed that of its volume. Explain how cell division functions to maintain this relationship between surface area and volume and in doing so maintains cell homeostasis.

25. As a cell grows, its surface area increases at a slower rate than its volume. If a cell becomes too large, nutrients cannot diffuse in, nor waste diffuse out, quickly enough for the cell to survive. Cell division results in a cell returning to a state where its surface area greatly exceeds its volume. (6-2.1)

26. Scientists think that centrioles do not function in mitosis since the plant spindle forms without them and mitosis continues according to schedule. (6-2.3)

27. Binary fission does not use a spindle and mitosis does. (6-2.1, 6-2.3)

Critical Thinking

28. The frequency of Down syndrome increases sharply with maternal age. (6-1.4)

29. Posters should illustrate the stages of mitosis (shown in Figure 6-10 on page 130) and meiosis (shown in Figure 6-12 on pages 132 and 133). Meiosis is likely to produce chromosomes that are genetically differ-

26. Structure and Function The events of mitosis in plants and animals are very similar with the exception of the absence of centrioles in plants. How has the absence of centrioles in plant cells influenced scientists' thinking about the function of centrioles in mitosis?

27. Structure and Function Binary fission and mitosis are successful evolutionary adaptations that ensure that DNA and cytoplasm are equally distributed between two daughter cells. How do binary fission and mitosis differ in the ways that this outcome is achieved?

Critical Thinking

28. Making Inferences Why is there a concern that more infants with Down syndrome will be born in the United States as more women delay having children?

29. Communicating Effectively Construct a poster comparing mitosis and meiosis in a specific organism. (Choose one with a small number of chromosomes.) First, depict a cell undergoing mitosis. Include diagrams of the cell and its chromosomes in all of the mitotic phases described in the text. Second, depict a cell undergoing meiosis, again including diagrams of the cell and its chromosomes in all of the meiotic phases described in the text. Indicate the number of chromosomes contained within each nucleus after the two processes are completed. Which process is likely to produce chromosomes genetically different from those of the original cell?

Activities and Projects

30. Research and Writing Do library research and talk with medical professionals to learn how cancer cells differ from normal cells in relation to the cell cycle. Share the results of your research with your class.

31. Research and Writing Find out what biologists observing nuclear division in primitive eukaryotes such as *Amoeba* have learned about the evolution of mitosis.

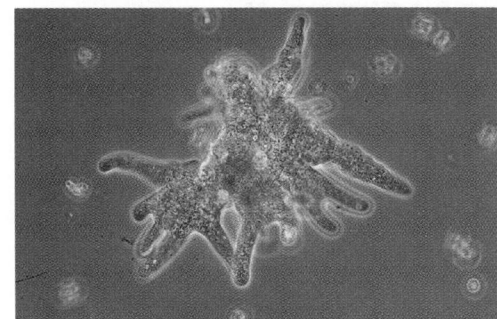

32. Multicultural Perspective Many cultures have stories and tales about dwarfs or giants who have special powers. Look for examples of cultural heroes and heroines who are chosen as leaders and protectors because of their physical differences. What attitudes do other characters in these tales demonstrate toward these individuals?

Readings

33. Read the article "An Introduction to DNA Fingerprinting," in *The American Biology Teacher*, April 1993, pages 216–221. What is DNA fingerprinting? What are possible sources for the DNA used in this technique? Describe some applications of this technique

34. Read the article "Cells That Reach Out For the Light," in *Discover*, June 1993, page 64. According to Albrecht-Buehler, how do cells know where to go as they differentiate? Explain the experiment he conducted that led to his hypothesis. What do other scientists think of this hypothesis?

may have resulted in the diminished role of the nuclear membrane in mitosis.

32. Answers will vary. Students may recall stories about Paul Bunyan, Jack and the Beanstalk, or Snow White. Often in cultural tales, individuals who are different will either have special powers or be victims of discrimination. Other characters in the stories may have feelings of disgust or admiration for these different individuals. One popular mythological hero in ancient Egypt was Bes, a robust, bearded dwarf who protected the queen's childbirth. Members of the Egyptian court were so fond of this cultural hero that many had his image carved on their dressing mirrors.

Readings

33. DNA fingerprinting is a forensic method that can be used to establish the identity of an individual or relationships among individuals. Possible sources of DNA that can be used in fingerprinting include blood, semen, saliva, hair, urine, bone, and muscle. DNA fingerprinting can be used in criminal and paternity cases.

34. According to Albrecht-Buehler, cells differentiate by watching each other. They see each other by homing in on infrared light that is emmitted by chemical bonds. Albrecht-Buehler grew hamster fibroblasts physically separated from each other by a thin glass film coated with different substances. The fibroblasts always lined up in the same pattern when light could pass through the glass. Few other scientists agree with Albrecht-Buehler's conclusions.

ent from those of the original cell because it halves the number of chromosomes of the original cell. Moreover, crossing-over can occur in prophase I of meiosis, forming new combinations of genes. (6-2.2, 6-3.3, 6-3.4)

Activities and Projects

30. Normally, the cell cycle is a highly regulated process that is controlled by chemical and physical signals that can inhibit or stimulate cell division. A cell that has become cancerous does not respond to the signals that stop cell division. Thus, cancer is the uncontrolled growth of cells.

31. Scientists believe that mitosis may have resembled binary fission, except that the chromosomes attached to the nuclear membrane rather than to the plasma membrane, as occurs in bacteria. Microtubules originally only supported the nuclear membrane and did not attach to the chromosomes. The change in the function of microtubules

Preparation

For each model, students need four pipe cleaners of one color and four pipe cleaners of another color, four wooden beads, 90 cm of yarn, and 16 small white labels.

Procedural Notes

• Review mitosis and meiosis before the lab.

Background Answers

1. Cells make new cells identical to themselves through the process of mitosis.

2. The process of meiosis ensures that gametes contain one-half as many chromosomes as body cells.

3. Meiosis occurs only in the ovaries and testes. Mitosis occurs in the rest of the human body, except in nerve cells and skeletal muscle cells.

4. Nondisjunction occurs when a chromosome and its homologue fail to separate properly. The condition in which a diploid cell has an extra chromosome is called trisomy.

5. How do mitosis and meiosis compare?

Technique Answers

4. During metaphase, the chromosomes are at the equator of the spindle fibers. The wooden bead represents the centromere, and it attaches to the spindle fiber.

5. During anaphase, the spindle fibers shorten, pulling the attached chromosomes to opposite ends in the cell. The centrioles divide and the chromatids separate. The chromatids are now called chromosomes.

6. During telophase, the cell membrane is pinched in half. In each new cell, the nuclear membrane forms around the genetic material, and the cytoplasm and organelles divide among the two cells.

9. During metaphase I, each lab student does not have the chromosome pairs in the same order. Combinations will vary.

12. Combinations will vary.

Comparing Mitosis and Meiosis

OBJECTIVES

• Understand the differences between mitosis and meiosis.
• Recognize the relationships between mitosis and genetic continuity, and between meiosis and genetic variation, among living things.
• Use models to understand such concepts as nondisjunction and trisomy.

PROCESS SKILLS

• comparing and contrasting structures
• relating structural features to functions
• comparing and contrasting processes

MATERIALS

• Colored pipe cleaners
• Yarn
• Wooden beads
• Small white labels for "genes"

BACKGROUND

1. How do cells make new cells identical to themselves?
2. What cellular procedures ensure that gametes contain one-half as many chromosomes as body cells?
3. Where in the human body does mitosis occur? Where does meiosis occur?
4. What is nondisjunction? What is trisomy?
5. Write your own question to explore in your lab report or notebook.

TECHNIQUE

Part A: Mitosis

As you go through the phases of mitosis and meiosis, record your drawings of the setup for each phase in the **Records** section of your report.

1. Select four pieces of equal-sized pipe cleaner in one color and four of equal size in a second color. Obtain four wooden beads and a piece of yarn approximately 90 cm long.

2. With the yarn, construct a "cell membrane" circle on your lab desk. When the cell is about to divide, the chromatin shortens and thickens into chromosomes inside the cell. Remember, the nuclear envelope has already disappeared.

3. The end of interphase and the beginning of prophase are indistinguishable and blend into one another, hence the use of the word *phase* instead of *stage*. Recall that nuclear division really begins with prophase, when chromosomes become distinct and visible under a microscope. Show your cell with two pairs of replicated chromosomes. Use the wooden bead as the centromere connecting the pipe cleaner chromosome and its double. Add labels for "genes."

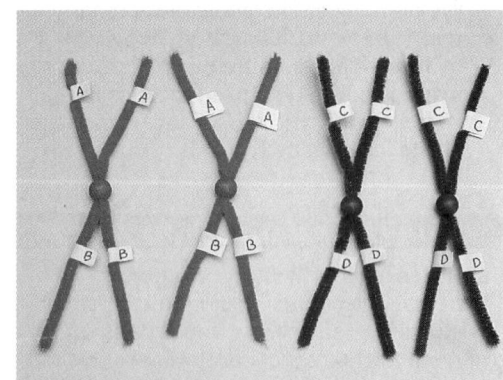

Part A   Mitosis

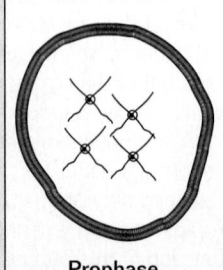
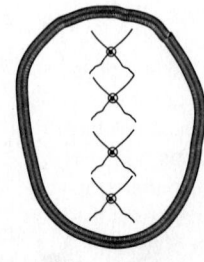
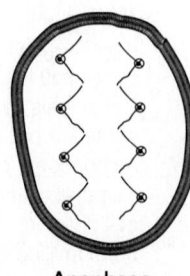
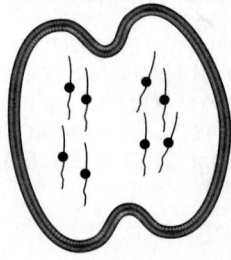

| Prophase | Metaphase | Anaphase | Telophase |

4. Now move on to metaphase. Where are the chromosomes at metaphase? What part of your chromosome model attaches to the spindle fibers that extend from the centrioles?

5. Now show your cell at anaphase. What is happening to the spindle fibers? What is happening to the chromatids? What are they now called?

6. Show your cell at telophase. What is happening to the cell membrane? Show that with your yarn. What other events occur?

Part B: Meiosis

7. Repeat steps 1–3, doing the same things you did for mitosis. At prophase I of meiosis, the homologous chromosomes pair up, forming tetrads. Show this with your materials.

8. In metaphase I, each chromosome attaches to a separate spindle fiber, rather than each chromatid, as during mitosis. Show this with your materials.

9. Does each lab student have the chromosome pairs in the same order (look at the "genes" or letters)? What are some of the gene combinations in your chromosome pairs?

10. During anaphase I, the spindle fibers contract. A pair of chromatids of each tetrad is pulled toward each pole. Telophase I is next.

11. Meiosis II follows the same procedure as mitosis, dividing the two cells resulting from meiosis I. The chromatids will separate. Four cells will result, so stretch out the yarn to make four cells. These cells represent gametes.

12. Take one of your gametes and combine it with the gamete belonging to another student. This is comparable to fertilization. What combinations of genes did you make?

13. Repeat step 10, only this time use your materials to show what would happen if a chromosome failed to separate from its homologue during anaphase.

INQUIRY

1. At metaphase during mitosis, do you and the students to either side of you have the chromosomes lined up in the same order? Does it matter?

2. After telophase of mitosis, how does the size of each new cell compare with that of the original cell?

3. What "genes" are in each of your gametes? When a parent has two different forms of a gene for a trait, how many forms of the gene for that trait are in each gamete produced by that parent?

ANALYSIS

1. Does a new cell forming as a result of mitosis have the same genes as the original cell? Explain why or why not.

2. How is mitosis similar to meiosis? How does it differ?

3. Give some examples of variety in a population. Why do you think variety is important to populations?

4. Describe some consequences that might arise if a gamete received an extra copy of a chromosome.

FURTHER INQUIRY

Write a new question that could be explored as a new investigation. The following are examples:

How often do different cells of the body undergo mitosis? Are there any body cells that do not divide?

How can certain chemicals disrupt the events of mitosis and meiosis?

Inquiry Answers

1. Students should not have the chromosomes lined up in the same order. The order of the chromosomes does not matter because one chromatid from each set of chromosomes will go to each of the new cells.

2. After mitosis, each new cell is smaller than the original cell. As the new cells grow, their surface-area-to-volume ratios will decrease, and the new cells will divide.

3. One set of genes is found in each gamete. Each gamete produced by a parent has one form of a gene for each trait.

Analysis Answers

1. A new cell forming as a result of mitosis has the same genes as the original cell. One copy of the genes goes to each new cell.

2. In both processes, DNA replicates once. In meiosis there are two separate stages of cell division, resulting in four haploid cells. In mitosis there is one stage of cell division, resulting in two cells identical to the original cell.

3. Answers will vary, but students might mention variations in height, weight, skin color, eye color, hair color, and facial features. Variations can be important as a pool of possible adaptations to different environments.

4. If a gamete receives an extra copy of a chromosome, trisomy can result when the gamete pairs with another gamete. Depending on which chromosome is the extra copy, Down syndrome or sterility can result. Often trisomy results in a nonviable fetus.

Part B   Meiosis

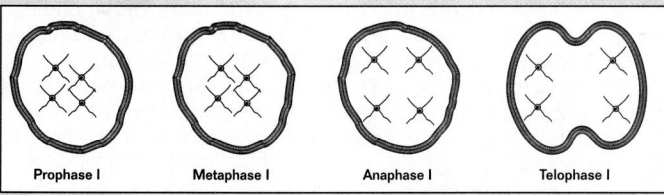

Prophase I   Metaphase I   Anaphase I   Telophase I

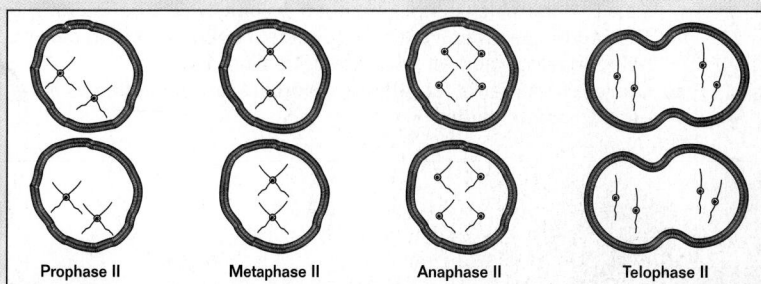

Prophase II   Metaphase II   Anaphase II   Telophase II

Block Scheduling Guide

Each block represents about 45–50 minutes of instructional time. The resources cited in a block are coded to a specific program component using the Key on the next page.

Lecture Concepts	Lesson Resources

7-1 Fundamentals of Genetics pp. 140–147

BLOCKS ① and ②

a. Mendel chose the garden pea because it has many varieties that grow quickly and are able to self-pollinate.
b. In Mendel's experiments, only the dominant traits were expressed in the F_1 generation. Recessive traits reappeared in the F_2 generation in a ratio of 3:1 dominant to recessive.
c. Mendel theorized that for each trait, an individual has two "factors," or genes, one from each parent. Each copy of a factor is called an allele.
d. If the two alleles contain the same information, the individual is said to be homozygous. If the information is different, the individual is said to be heterozygous. In a heterozygous individual, only the dominant allele is expressed.
e. The law of segregation states that the two alleles for a trait separate when gametes are formed. The law of independent assortment states that when two or more pairs of alleles are located on different chromosomes or far apart on the same chromosome, they separate independently of one another during gamete formation.

Lecture Resources
- Opening Demonstration, page 140
- Visual Strategies: Figures 7-3, 7-5, Table 7-1
- Teaching Transparencies: 25, 26, 27, 28, 29, 30
- Holt Biology Videodiscs: Lesson 7-1

Classwork Options
- Ad for Garden Peas, page 142

- Demonstration: Traits
- Independent Assortment and Graphic Organizer, pages 146–147

Assignment Options
- Section Review, page 147
- Chapter Review, questions 1–4, 7, 9, 10, 13, 14, 16, 18, 21–23

7-2 Analyzing Heredity pp. 148–153

BLOCKS ③ and ④

a. A monohybrid cross involves one pair of contrasting traits. A monohybrid cross between a homozygous dominant individual and a homozygous recessive individual yields heterozygous offspring expressing the dominant phenotype.
b. A monohybrid cross between two individuals who are heterozygous for a trait that has dominant and recessive forms yields offspring with a 1:2:1 genotypic ratio and a 3:1 phenotypic ratio.
c. A dihybrid cross between two homozygous individuals with contrasting traits yields heterozygous offspring expressing the dominant trait.
d. Patterns of inheritance are also affected by phenomena such as incomplete dominance, codominance, multiple alleles, continuous variation, and environmental influences.

Lecture Resources
- Demonstrations: Card Games, Blood Types
- Teaching Transparencies: 31, 32
- Visual Strategies: Figures 7-9, 7-10, 7-12
- Holt Biology Videodiscs: Lesson 7-2

Classwork Options
- Laboratory Investigation: Monohybrid Crosses, pages 162–163
- Laboratory Techniques and Experimental Design D12: Analyzing Corn Genetics

Assignment Options
- Section Review, page 153
- Chapter Review, questions 5, 6, 15, 17, 19

7-3 Human Genetics pp. 154–158

BLOCKS ⑤ and ⑥

a. Mutations in genetic material can cause genetic disorders such as sickle cell anemia and hemophilia. Both disorders are caused by faulty proteins. The alleles for both disorders are recessive.
b. By constructing and analyzing a pedigree, a scientist can determine the pattern of inheritance of a trait within a family. It may also be possible to determine whether a person is homozygous or heterozygous.
c. The techniques involved in genetic counseling can help parents at risk of having children with a genetic disorder.
d. Current treatments for genetic disorders are beginning to include gene transfer therapy.

Lecture Resources
- Opening Question, page 154
- Demonstration: Cultural Selection
- Hemophilia and HIV, page 155
- Pedigree Key, page 156
- Research Update, page 157
- Maternal PKU, page 158
- Holt Biology Videodiscs: Lesson 7-3

Classwork Options
- Laboratory Techniques and Experimental Design C13:

Preparing Tissue for Karyotyping
- Laboratory Techniques and Experimental Design C14: Karyotyping
- Laboratory Techniques and Experimental Design C15: Karyotyping— Genetic Disorders
- Interactive Explorations in Biology Laboratory Manual E5: Heredity in Families

Assignment Options
- Section Review, page 158
- Chapter Review, questions 8, 11, 12, 20, 24, 25

Review/Enrichment

BLOCK 7

- ■ Study Guide: Concept Review and Pretest

Assignment Options
- 💿 Teaching Resources CD-ROM: Supplementary Reading Study Guide and Test, *The Double Helix*
- ■ Activities and Projects, questions 26–28
- ■ Readings, questions 29, 30

Assessment Options

BLOCK 8

Traditional Assessment
- ■ Chapter 7 Test
- 💿 Test Generator/Assessment Item Listing: Software item bank for preparing customized chapter tests.

Portfolio Assessment
Select student reports for one or more laboratory experiments from this chapter. *The Direct Observation Checklist* on the *BioSources Teaching Resources CD-ROM* should be completed during a laboratory or other cooperative learning experience.

Answer to Concept Map

The following is one possible answer to the Concept Mapping exercise on page 159.

Holt Biology Videodiscs   CONCEPTS OF BIOLOGY

Holt Biology Videodiscs gives you a powerful tool for teaching, review, and assessment. *Concepts of Biology* adds interactive lesson plans and extensive tools for customization using CD-ROM technology.

Use the following topic from *Concepts of Biology* to help you teach this chapter:

- ■ Topic 6: Fundamentals of Genetics

For further information regarding the media on the videodiscs, see *Holt Biology Videodiscs Teacher's Correlation Guide to Biology: Principles and Explorations.*

UNIT 2

Chapter Theme

Evolution *The passing on of different genetic traits is a process fundamental to evolutionary change. The genetic variation provided by the possible combinations of parental genes and by crossing-over offers a means to change and an increase in the speed of evolution within a given species. Moreover, genetic mutations make evolution possible by introducing random changes into the inherited information that is passed on.*

Tapping Prior Knowledge

- How do parents transfer genetic information to their offspring?
- How could a chromosome transmit a genetic disorder?

Opening Demonstration

Have students bring photographs of parents or of other siblings to class. Ask them to match the names of fellow students with the photographs of their family members. Once the students have matched the photographs to their classmates, show the pictures individually and ask the matching student to stand. Ask students to identify how the two are similar and different. Once students have completed the activity, ask them to identify the strategies they used to match the family members. Ask students to identify other techniques they think scientists would use to find a family connection.

CHAPTER 7
MENDEL AND HEREDITY

> ### REVIEW
> - scientific method (Section 1-2)
> - gene (Section 6-1)
> - meiosis (Section 6-3)

Thoroughbred mares and colts

Authors' Rationale

In this chapter, students will become familiar with Mendel and his theories through active application of Mendel's observations. Why does someone look like her mother? How can someone's blood type be completely different from that of both his parents? Mendelian patterns of inheritance are explored through the construction of pedigrees, a powerful tool that can be used for tracking or even predicting the behavior of genetically inherited traits. Inherited mutations are studied, showing how Mendelian techniques can be used to identify individuals at risk of inheriting a genetic disease.

7-1 Fundamentals of Genetics

Many of your traits—the color and shape of your eyes, the texture of your hair, even your height and weight—resemble the characteristics of your parents. The transmission of traits from parent to offspring is called **heredity**. Humans have long been interested in heredity. From the beginning of recorded history, we have attempted to develop improved varieties of crop plants, like rice, and domestic animals, such as the horse. And almost certainly, parents of long ago wondered why a particular trait appeared in family members. Before DNA and chromosomes were discovered and understood, heredity was one of the greatest mysteries of science. In this section you will learn how this mystery was solved.

Section Objectives

- List four characteristics that make *Pisum sativum* a good subject for genetic studies.
- Outline the three major steps of Gregor Mendel's garden pea experiments.
- Explain how Mendel derived ratios from his observations.
- Define the terms dominant, recessive, heterozygous, homozygous, genotype, *and* phenotype.
- Compare and contrast Mendel's two laws of heredity.

Figure 7-1 Breeding the garden pea *Pisum sativum,* Gregor Mendel formulated the basis of modern genetics.

The Beginnings of Genetics

The key to understanding heredity was discovered over a century ago by an Austrian monk named Gregor Mendel. In the garden of a monastery, Mendel bred the garden pea *Pisum sativum,* shown in Figure 7-1. From these experiments, he developed a simple set of rules that accurately predicted patterns of heredity. The patterns he found form the basis of **genetics,** the branch of biology that studies heredity. When Mendel's rules became widely known, scientists all over the world set out to discover the physical mechanism behind these patterns. Eventually, their studies taught them that traits are

CONTENT LINK

The molecular basis of heredity is discussed in **Chapter 8**.

SECTION 7-1

Vocabulary Preview

genetics
self-pollination
cross-pollination
true-breeding
P generation
F_1 generation
F_2 generation
dominant
recessive
homozygous
heterozygous
allele
phenotype
genotype
law of segregation
law of independent assortment

Opening Question

A gardener notices that unlike the pea plants she has had in other years, some of the flowers are white. Previously, the pea flowers have been purple. Ask students what explanation would account for this difference in the garden. Have students suggest answers. Tell them this first section might give some insight. At the end of this chapter, come back to the question. (*Answer: Instead of buying the more expensive F_1 hybrid seeds as she usually did in the past, she decided to save money and plant pea seeds from the crop she harvested the previous year. Her plants were the F_2 generation, which shows a 3:1 ratio of purple to white flowers.*)

Demonstration

Traits

Display large pictures of a few flowering plants (or bring in real plants). Ask students to come up with a list of traits that could be inherited. Encourage students to think of many different traits, including flower shape, flower color, position of flowers on stems, leaf shape, leaf color, pattern of veins, pattern of stem growth, presence of hairs on stems, inner structure of flowers, etc.

Teaching Tips

Mendel's Journal

Have students write journal pages that represent comments Mendel would have written in a journal as he did his research. Students should put themselves in the place of Mendel and write as they think he would have written in his own journal.

Ad for Garden Peas

Have students design an ad for a newspaper that would have attracted someone like Mendel to purchase these peas for genetic research. Students should include all of the benefits of *Pisum sativum* that make it useful for genetic research. Ask students to use illustrations in their advertisements.

☐ CAPSULE SUMMARY

Genetics is the branch of biology that studies heredity, the transmission of traits from parents to offspring. Early investigators such as T. A. Knight and Gregor Mendel bred varieties of the garden pea Pisum sativum in attempts to understand heredity.

determined by genes, the instructions encoded in the DNA of the chromosomes an individual receives from each parent.

Mendel was not the first person to try to understand heredity by studying pea plants. Over 200 years ago, British farmers performed similar experiments and even obtained results comparable to Mendel's. In the 1790s, for example, the British farmer T. A. Knight bred a variety of the garden pea that had purple flowers with a variety that had white flowers. All the offspring had purple flowers. When two of the offspring were bred, however, some of their offspring had purple flowers and some had white. Knight's explanation of this phenomenon noted only that purple flowers had a "stronger tendency" to appear than white flowers. ☐

Mendel Studied the Garden Pea
....

Gregor Mendel was born in 1822 in a region of Austria that is now part of the Czech Republic. As a son of peasants, Mendel learned much about agriculture, knowledge that became invaluable later in his life. In 1843 Mendel entered an Augustinian monastery to study theology. Three years later, after he failed a teacher's examination, Mendel went to the University of Vienna to study science and mathematics. There, he learned how to study science through experimentation and how to use mathematics to explain natural phenomena.

After attending the university, Mendel returned to the monastery and joined a local science research society. Each member undertook a scientific investigation that could be discussed at meetings and published in the society's journal. Mendel repeated Knight's experiments with pea plants. However, his approach to the experiments differed substantially from that of his predecessor; Mendel attacked the problem in a mathematical fashion and *counted* the number of each kind of offspring. Quantitative approaches to science—those that include measuring and counting—were just becoming popular in Europe, so Mendel's method was on the cutting edge of research at the time. Now, most research in the natural sciences must be quantified.

Why Mendel Chose Peas

For his experiments, Mendel chose to study the garden pea. The garden pea is a good subject for genetic study for several reasons.

1. Many varieties of *P. sativum* exist. Mendel initially examined 32 varieties. From these he selected seven pairs of varieties that differed in easily distinguishable forms of various traits, such as flower color, seed color, and seed shape.

Figure 7-2 In 1857, Gregor Mendel began breeding garden peas in the garden of a monastery.

Historical Note

Mendel's Genetic Units

In 1902, Walter S. Sutton, a graduate student at Columbia University working in cell cytology, recognized what he considered to be a visible link between chromosome behavior and Mendel's genetic units. His work was instrumental in helping scientists learn that Mendel's genetic units were actually genes on chromsomes.

2. Mendel knew from earlier experiments that he could expect one of the two forms of each trait to disappear in one generation and then reappear in the next. This gave him something to count.

3. *P. sativum* is a small, easy-to-grow plant that matures quickly and produces a large number of offspring. Mendel would be able to conduct many experiments and obtain results quickly.

4. The male and female reproductive parts of *P. sativum* are enclosed within the same flower, as shown in Figure 7-3. When left undisturbed, the flower does not open fully; it simply fertilizes itself through a process called **self-pollination.** As a result, one individual plant can produce offspring. To cross two pea plants, Mendel first had to remove the anthers (the pollen-producing organs) from a flower of one plant. He could then dust the pistil (the egg-producing organ) with pollen from a flower of a different pea plant. Transferring the pollen from the flower of one plant to the flower of a different plant is called **cross-pollination.** Scientists use the term *cross* to refer to the breeding between two flowers from separate plants. ❑

❑ **CAPSULE SUMMARY**

The garden pea is a good subject for genetic study because it has many varieties that grow quickly and are able to self-pollinate, important characteristics for Mendel's experimental design.

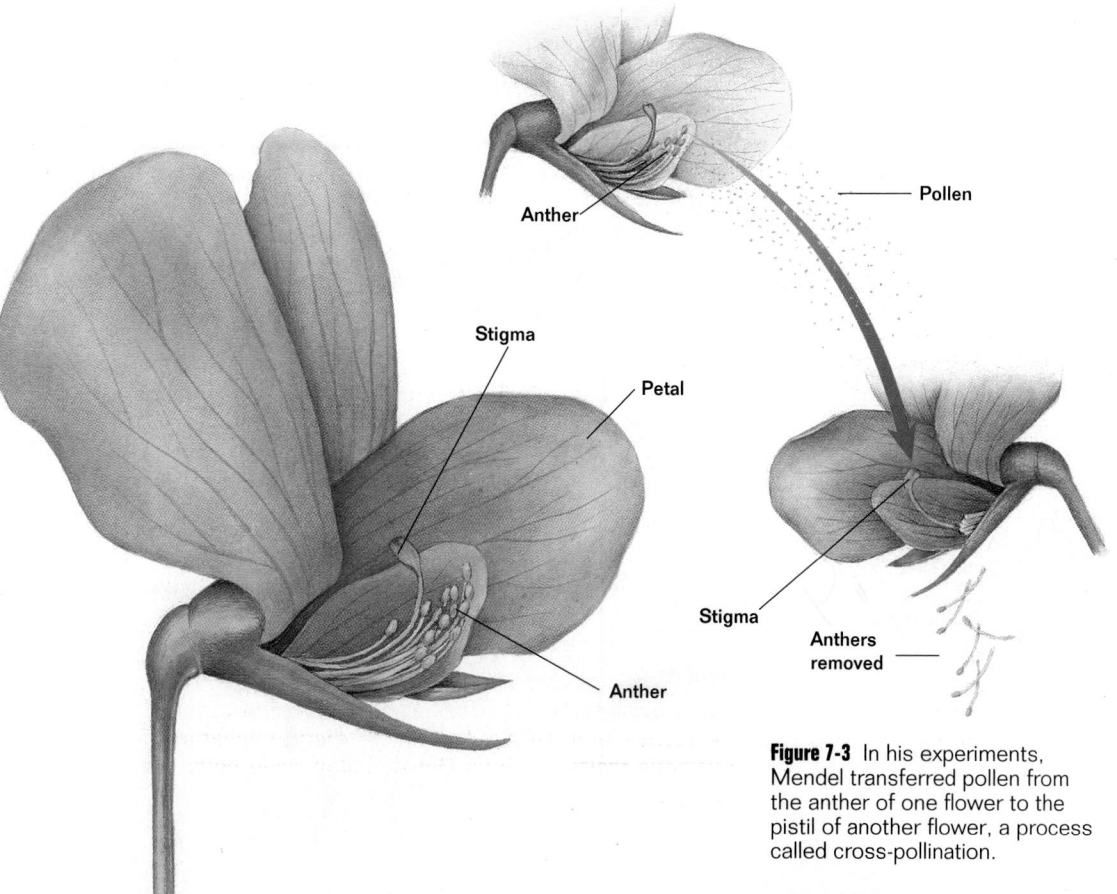

Anther

Pollen

Stigma

Petal

Anther

Stigma

Anthers removed

Figure 7-3 In his experiments, Mendel transferred pollen from the anther of one flower to the pistil of another flower, a process called cross-pollination.

Mendel's Experimental Design

Mendel carried out his experiments with garden peas in three steps.

1. **Step 1** Mendel began his experiments by allowing each variety of garden pea to self-pollinate for several generations. This method ensured that each variety was **true-breeding** for a particular trait, which means that all the offspring would display only one form of a particular trait. For example, a true-breeding, purple-flowering plant produced only plants with purple flowers in subsequent generations. Mendel called these plants the parental generation, or **P generation.**

2. **Step 2** Mendel then cross-pollinated two varieties from the P generation that exhibited contrasting traits, such as purple flowers and white flowers. He called the offspring of these plants the first filial generation, or **F_1 generation.**

3. **Step 3** Finally, Mendel allowed the F_1 generation to self-pollinate. He called the offspring of these plants the second filial generation, or **F_2 generation.** These were the plants that he counted. Figure 7-4 summarizes Mendel's experimental design.

Figure 7-4 Mendel studied traits in three generations of garden peas.

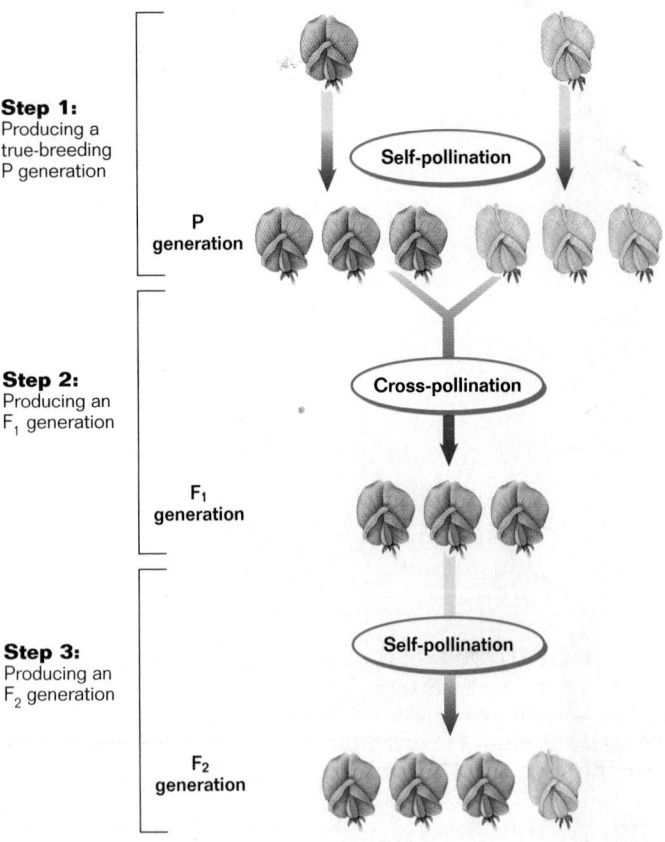

Step 1:
Producing a true-breeding P generation

Self-pollination

P generation

Step 2:
Producing an F_1 generation

Cross-pollination

F_1 generation

Step 3:
Producing an F_2 generation

Self-pollination

F_2 generation

Mendel Observed Two Ratios

For each cross, Mendel obtained F_1 generation plants that had only one form of the crossed traits. The contrasting trait had disappeared. Mendel described the remaining, or expressed trait, as **dominant**. The trait that was not expressed in the F_1 generation was described as **recessive**. Table 7-1 identifies the dominant and recessive forms of the seven traits that Mendel studied.

Table 7-1 Mendel's Crosses and Their Results

Trait	Dominant vs. Recessive		F_2 Generation Results		Ratio
			Dominant Form	Recessive Form	
Flower color	Purple ×	White	705	224	3.15:1
Seed color	Yellow ×	Green	6,022	2,001	3.01:1
Seed shape	Round ×	Wrinkled	5,474	1,850	2.96:1
Pod color	Green ×	Yellow	428	152	2.82:1
Pod shape	Round ×	Constricted	882	299	2.95:1
Flower position	Axial ×	Top	651	207	3.14:1
Plant height	Tall ×	Dwarf	787	277	2.84:1

When the F_1 generation was allowed to self-pollinate, the recessive trait reappeared in some of the plants in the F_2 generation. At this point Mendel *counted* each type of plant in the F_2 generation. For example, he counted 705 plants with purple flowers and 224 plants with white flowers. From these data, Mendel calculated a ratio of approximately 3 purple-flowering plants to every 1 white-flowering plant (3:1). For

Teaching Tip

F_1 Seeds

Obtain several seed packages that state that the seeds are F_1 to show students. Ask students why commercial growers would want to sell F_1 seeds. (*Answers would include larger flowers, vegetables, or fruits; better tasting products; more vigorous plants; etc.*)

 VISUAL STRATEGY

Table 7-1

Use this table to explain Mendel's research. Make sure students understand that each experiment begins with a cross between two parent plants having different traits and that the F_2 generation was produced by self-pollination of only one plant.

Teaching Tip
Recognizing Individuals

Using the letters of the alphabet, write different examples of individuals and gametes (including abnormalities) on separate index cards. For example, on one index card write AA. On another index card write A. Other examples could include Ee, ii, OO, Uu, AB, AaE, AABB, u, mm, AEU, mmc, sk, and DDEEr. Divide the class into groups, and give each group a stack of index cards. Ask students to devise a system for dividing the cards into groups. Have a spokesperson from each group explain their rules for sorting the cards. Many possible answers may be given. Help students arrange their cards into groups of individuals, gametes, and abnormalities. Have them come up with patterns from which they can form their own rules. (*Rules: Individuals have two of each letter. Gametes have only one of each different letter. Abnormalities have mixtures of the two.*) The cards can then be used to demonstrate the terms *allele, homozygous,* and *heterozygous.*

👁 VISUAL STRATEGY

Figure 7-5

Use this figure to emphasize that many genes are involved in giving an animal its overall appearance. Ask students to suggest one genotype for each animal shown. For example, students might suggest alleles for coat, color, ear length, or nose color. If they choose coat color, introduce the idea that more than one gene is involved in a trait that has variation as seen in the spotted rabbit.

☐ **CAPSULE SUMMARY**

In Mendel's experiments, only the dominant traits were expressed in the F₁ generation. The recessive traits reappeared in the F₂ generation, in which Mendel calculated a 3:1 ratio of dominant to recessive traits.

Figure 7-5 The physical appearance of an organism is its phenotype. These two rabbits have different phenotypes because their collection of genes, or genotypes, varies.

☐ **CAPSULE SUMMARY**

Mendel theorized that for each trait, an individual has two "factors," or genes, one from each parent. Each copy of a factor is called an allele. The two alleles may or may not contain the same information. When they differ, the allele with the dominant information will be expressed.

each cross, Mendel obtained the same 3:1 ratio of plants expressing the dominant trait to plants expressing the recessive trait. Table 7-1 lists the numbers of F₂ individuals and the ratios Mendel obtained for each cross.

Mendel's next question was, Will the 3:1 ratios continue in subsequent generations? He found that plants showing the recessive traits were true-breeding when they were allowed to self-pollinate. When plants with the dominant trait self-pollinated, Mendel found that only one-third of them were true-breeding, whereas two-thirds were not. For these plants, Mendel observed a 3:1 ratio of dominant to recessive traits. These results suggested that the 3:1 ratio in the F₂ generation was really a disguised 1:2:1 ratio: 1 true-breeding dominant plant to 2 not-true-breeding dominant plants to 1 true-breeding recessive plant. ☐

Mendel Proposed a Theory of Heredity

To explain his results, Mendel proposed a theory that has become the foundation of the science of genetics. His theory has five elements.

1. Parents do not transmit traits directly to their offspring. Rather, they pass on units of information that operate in the offspring to produce the trait. Mendel called these units of information "factors." In modern terminology, Mendel's factors are called genes. As you learned in Chapter 6, a gene is a segment of a DNA molecule that transmits hereditary information.

2. For each trait, an individual has two factors: one from its mother and one from its father. The two factors may or may not have the same information. If the factors have the same information (for example, if both factors have information for purple flowers), the individual is said to be **homozygous** (*hoh moh ZY guhs*). If the factors are different (for example, one factor has information for purple flowers and the other has information for white flowers), the individual is said to be **heterozygous** (*heht uhr oh ZY guhs*). Each copy of a factor, or gene, is called an **allele** (*uh LEEL*).

3. In modern terms, the physical appearance, or **phenotype** (*FEE noh typ*), of an individual is determined by the alleles that code for traits. The set of alleles that an individual has is called its **genotype** (*JEE noh typ*). See Figure 7-5.

4. An individual receives one allele from one parent and the other allele from the other parent. Each allele can be passed on when the individual matures and reproduces.

5. The presence of an allele does not guarantee that a trait will be expressed in the individual that carries it. In heterozygous individuals, only the dominant allele is expressed; the recessive allele is present but unexpressed. ☐

Graphic Organizer

Use this graphic organizer with *Teaching Tip:* **Independent Assortment** on page 147

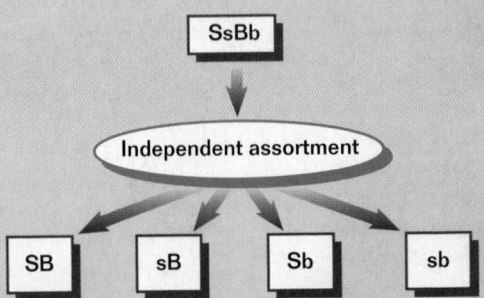

Mendel's Theory Became Laws of Heredity

Mendel's theory brilliantly predicts the results of his crosses and also accounts for the ratios he observed. Similar patterns of heredity have since been observed in countless other organisms. Because of its overwhelming importance, Mendel's theory is often referred to as the law of segregation. In modern terms, the **law of segregation** states that the members of each pair of alleles separate when gametes are formed.

Mendel went on to study how different pairs of genes are inherited, such as the genes for flower color and plant height. He found that for the pairs of traits he studied, the inheritance of one trait did not influence the inheritance of any other trait. This observation eventually became known as the law of independent assortment. The **law of independent assortment** states that pairs of alleles separate independently of one another during gamete formation. We now know that this principle applies only to genes located on different chromosomes or far apart on the same chromosome.

Mendel's paper describing his results was published in 1866. Unfortunately, it failed to arouse much interest and was forgotten. In 1900, sixteen years after Mendel's death, several scientists independently rediscovered the pioneering paper. They had been searching the literature in preparation for publishing their own findings, which were similar to those Mendel had quietly presented more than three decades earlier. ❑

❑ CAPSULE SUMMARY

The law of segregation states that the two alleles for a trait separate when gametes are formed. The law of independent assortment states that when two or more pairs of alleles are located on different chromosomes or far apart on the same chromosome, they separate independently of one another during gamete formation.

Section Review

1. *List four reasons why Mendel used* Pisum sativum *as a subject for genetic experiments.*

2. *Explain how Mendel designed the pea plant experiments that led to his theory of heredity.*

3. *At what point in his experiments did Mendel count the individual pea plants? How did he derive the 3:1 ratio?*

4. *At what point in his experiments did Mendel derive the 1:2:1 ratio? What hereditary characteristics in pea plants does this ratio represent?*

5. *What is the difference between the terms* dominant *and* recessive? heterozygous *and* homozygous? genotype *and* phenotype?

Critical Thinking

6. *How are Mendel's two laws explained in terms of meiosis?*

7. *Would Mendel's results have been different if he had experimented with squash plants, which usually do not self-pollinate? Why or why not?*

SECTION 7-1

Teaching Tips
Independent Assortment

Have students make a *Graphic Organizer* similar to the one at the bottom of page 146 that demonstrates the law of independent assortment. Ask students to provide at least one example to illustrate their charts and a brief explanation.

Mapping the Concepts

Have students use the boldfaced vocabulary words in Section 1-1 to draw a concept map. Students may choose to add a word other than one given on the list to begin the map, but the required vocabulary words should be highlighted in a single color. If students choose to add other words in developing the map, the new terms are highlighted in a different color. Additionally, students should add three cross-links to the map to show how the terms are related. Have students designate cross-links with a different color.

Answers to Section Review

1. See pages 142–143.

2. See Figure 7-4 on page 144.

3. In the F_2 generation, Mendel found three plants with purple flowers for every plant with white flowers.

4. Mendel discovered the 1:2:1 ratio by allowing the progeny of the F_2 generation to self-pollinate. The 1:2:1 ratio represents 1 true-breeding dominant plant to 2 not-true-breeding dominant plants to 1 true-breeding recessive plant.

5. See pages 145–146.

6. The law of segregation states that the members of each pair of alleles separate when gametes are formed. The law of independent assortment states that pairs of alleles separate independently during gamete formation.

7. Without self-pollination, Mendel would not have been able to ensure that his P generation was true-breeding because the phenotype for the heterozygous and homozygous dominant genotypes would be the same.

Opening Questions

Tell students that a man is heterozygous for the gene for cystic fibrosis. His wife does not carry the gene. What is the probability that their first child will get a gene for cystic fibrosis? Have students recall the process of gamete formation. Lead them to conclude that the child has a 50 percent probability of receiving the gene from the father. Will the child receiving the gene for cystic fibrosis exhibit the disease? Explain to students that cystic fibrosis is transmitted by an autosomal recessive allele, and that an individual must receive a copy of the cystic fibrosis allele from each parent in order to exhibit the disease. Students will learn more in Section 7-3 about human genetic disorders and how they are transmitted.

Demonstration

Card Games

Shuffle a deck of playing cards. Ask students to determine the probability of drawing an ace from the deck of cards. Students may suggest 4/52 or 1/13. Ask how they arrived at this conclusion. Deal 13 cards from the top of the deck. Compare the results with the students' predictions. If the numbers vary from the expected frequency, have students speculate about the reasons for the differences.

Section Objectives

- **Define** probability and explain how it is used to predict the results of genetic crosses.
- **Use** a Punnett square to predict the results of mono-hybrid and dihybrid genetic crosses.
- **Identify** five factors that influence patterns of heredity.

❑ CAPSULE SUMMARY

Geneticists use capital and lower-case letters to represent alleles. For each trait in an organism, three possible genotypes exist: TT, tt, or Tt.

7-2 Analyzing Heredity

Stripped of ratios and symbols, Mendel's work simply describes how units of information are passed from parents to offspring. The search for the physical nature of the units of information dominated the science of biology for more than half a century after Mendel's work was rediscovered in 1900. We now know that the units of heredity are genes, which are found on the chromosomes that an individual inherits from its parents.

Interpreting Mendel's Model

Geneticists still rely on Mendel's model to predict the likely outcome of genetic crosses. They use letters to represent the alleles of an organism. *Capital letters refer to dominant alleles, and lowercase letters refer to recessive alleles.* Note that capital and lowercase forms of the same letter must be used to designate the two forms of one gene. For example, the allele for the dominant trait of tallness in pea plants is represented by T, and the allele for the recessive trait of shortness by t. Since there are two alleles for each trait, the genotype of a pea plant that is homozygous dominant for tallness is TT. A pea plant that is homozygous recessive for shortness has the genotype tt. If these two plants are crossed with each other, the offspring will be heterozygous for the trait and will be designated Tt. ❑

Probability

Mendel's crosses can be interpreted according to rules of probability because these rules can predict how genes will be distributed among the offspring of two parents. **Probability** is the likelihood that a specific event will occur. For example, when you toss a coin, there's a chance that it will land with the "heads" side up. There's also a chance that it will land with the "tails" side up. The probability of either event happening can be determined by the following formula:

$$\text{Probability} = \frac{\text{number of one kind of possible outcome}}{\text{total number of all possible outcomes}}$$

Thus, when you toss the coin, the chance of its landing heads up is 1 out of 2 possibilities, or 1/2. The same formula can be used to predict the outcome of a genetic cross. For example, consider Mendel's crosses that studied seed color. Out of 8,023 F_2 pea plants, 6,022 had the dominant yellow seed color and 2,001 had the recessive green seed color.

Using the formula, the probability that the yellow seed color will appear in such a cross is 6,022/8,023, or 0.75 (75 percent). Expressed as a reduced fraction, the probability is 3/4. The probability that the green seed color will appear in the F_2 generation is 2,001/8,023, or 0.25 (25 percent). Expressed as a reduced fraction, the probability is 1/4. In other words, probability tells us that there are three chances in four that an offspring of two heterozygous individuals will have the dominant trait and one chance in four that it will have the recessive trait.

Monohybrid Crosses

A cross that provides data about one pair of contrasting traits is called a **monohybrid cross.** A cross between a pea plant that is true-breeding for tallness and one that is true-breeding for shortness is an example of a monohybrid cross. Biologists can also predict the probable outcome of a cross by using a diagram called a **Punnett square,** named for its inventor, Reginald Punnett. In the Punnett square in Figure 7-6, the genotype of the tall plant and the alleles (*TT*) it can contribute to its offspring are written on the top left side of the square. The genotype of the short plant and the alleles (*tt*) it can contribute to its offspring are written on the bottom left of the square. The interior of the square is a grid of boxes. Each box is filled with two letters—one letter from the left side of the square and one letter from the top of the square. These letters indicate the possible genotypes of the offspring.

In the case of the monohybrid cross in Figure 7-6, 100 percent of the offspring are expected to be heterozygous (*Tt*), expressing the dominant trait of tallness. Note that by convention, the dominant form of the trait is written first, followed by the lowercase letter for the recessive form of the trait. ▢

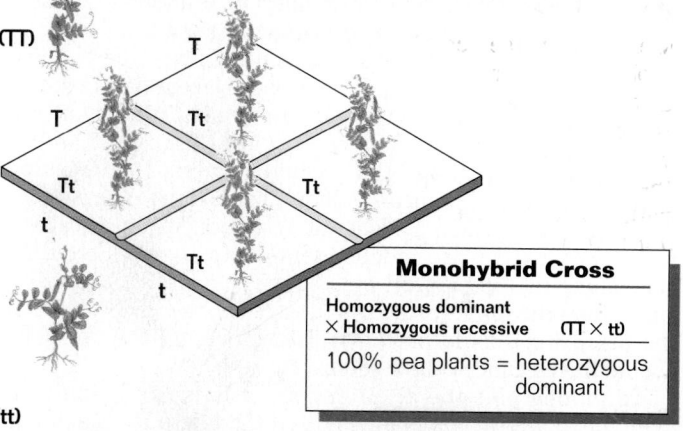

Monohybrid Cross

Homozygous dominant × Homozygous recessive	(*TT* × *tt*)
100% pea plants = heterozygous dominant	

Figure 7-6 A cross between a tall homozygous pea plant and a short homozygous pea plant will produce only tall heterozygous offspring.

Punnett squares can also be used to predict the outcome of a heterozygous cross. For example, in rabbits the allele for a black coat (*B*) is dominant over the allele for a brown

SECTION 7-2

Teaching Tip
Pick a Partner

Provide a list of traits that students recognize and that can be designated with a pair of alleles. Some examples include free ear lobes (ear lobes not attached to the skin of the face), ability to roll tongue, widow's peak, curved pinkie finger, hair on knuckles, and dimples. Students can also make up characteristics to go on the list.

Have students use the list of characteristics to describe a potential mate. For the purposes of this activity, assume that each characteristic is governed by a single gene. Students can design a person, or they can find a celebrity to represent that person. Students should then assign their potential mates a genotype for each trait. They should also make a genotype sheet for themselves.

Collect and shuffle the genotypic and phenotypic descriptions of the potential mates and redistribute them among the students. Have students use a coin toss to determine which alleles the offspring will receive. After the genotype is known, have students describe the phenotype of the "child."

Teaching Tip
The Runt

Give students the following scenario: A female labrador retriever has a litter of puppies. In the litter, she has nine black puppies with black noses, three black puppies with pink noses, three white puppies with black noses, and a white puppy with a pink nose that dies. Ask students what type of cross this example represents. *(dihybrid cross)* Ask students to suggest a possible explanation of the puppy's death. Lead students to conclude that a puppy with visible recessive genes may have other hidden recessive genes that can be lethal. If the puppy lives, often it will be the runt in the litter.

❑ **CAPSULE SUMMARY**

A monohybrid cross between two individuals that are heterozygous for a trait that has dominant and recessive forms is expected to yield offspring with a 1:2:1 genotypic ratio and a 3:1 phenotypic ratio.

Figure 7-7 Crossing two rabbits that are both heterozygous for black coats will produce two heterozygous black offspring, plus one true-breeding black and one true-breeding brown offspring.

❑ **CAPSULE SUMMARY**

A dihybrid cross between two homozygous individuals with contrasting traits yields heterozygous individuals expressing the dominant phenotype.

coat (*b*). Figure 7-7 shows a Punnett square that predicts the results of a monohybrid cross between two rabbits that are both heterozygous (*Bb*) for coat color. As you can see, one-fourth of the offspring would be expected to have the genotype *BB*, two-fourths (or one-half) would be expected to have the genotype *Bb*, and one-fourth would be expected to have the genotype *bb*. Since *B* is dominant over *b*, three-fourths of the offspring would have a black coat, and one-fourth would have a brown coat. Here you can see the two ratios that Mendel observed in his experiments— 1*BB* : 2*Bb* : 1*bb* (genotype) and 3 black : 1 brown (phenotype). ❑

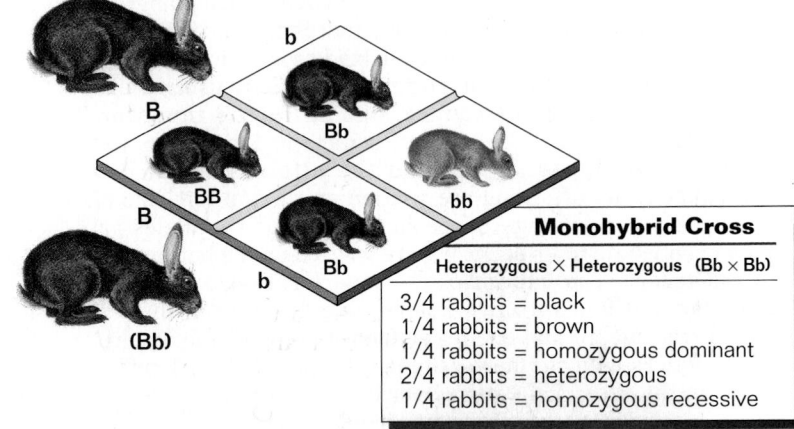

Monohybrid Cross

Heterozygous × Heterozygous (Bb × Bb)

3/4 rabbits = black
1/4 rabbits = brown
1/4 rabbits = homozygous dominant
2/4 rabbits = heterozygous
1/4 rabbits = homozygous recessive

Dihybrid Crosses

A **dihybrid cross** is a cross that involves two pairs of contrasting traits. Predicting the results of a dihybrid cross is more complicated than predicting the results of a monohybrid cross because you have to consider how the two alleles of each of the two traits from each parent can combine. For example, suppose you want to predict the results of crossing a pea plant that is homozygous for round, yellow seeds (*RRYY*) with one that is homozygous for wrinkled, green seeds (*rryy*). Figure 7-8 shows the Punnett square for a dihybrid cross, which consists of 16 boxes. When the alleles from each parent are independently sorted and listed, *RY* runs along the bottom left side of the Punnett square and *ry* runs along the top left side. As illustrated, the genotype of all the offspring should be *RrYy*. Therefore, all the offspring should have round, yellow seeds. ❑

In guinea pigs the allele for short hair (*S*) is dominant over the allele for long hair (*s*), and the allele for black hair (*B*) is dominant over the allele for brown hair (*b*). The Punnett square in Figure 7-9 predicts the probable offspring of a cross between two individuals heterozygous for both

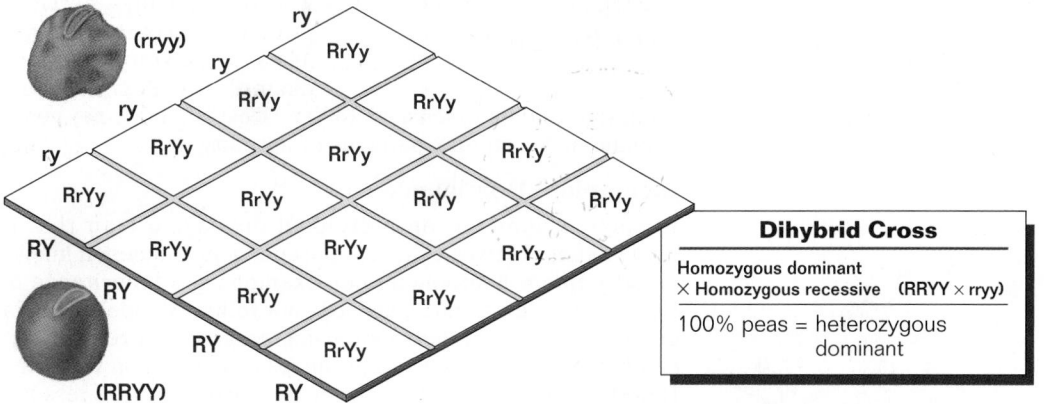

Dihybrid Cross

Homozygous dominant
× Homozygous recessive   (RRYY × rryy)

100% peas = heterozygous
dominant

Figure 7-8 This dihybrid cross produces only one type of offspring.

characteristics (*SsBb*). The offspring are likely to have nine different genotypes that will result in the following four phenotypes:

- Nine-sixteenths (9/16) of the guinea pigs will have short, black hair. These include individuals with the genotypes *SSBB, SsBB, SSBb,* and *SsBb.*

- Three-sixteenths (3/16) will have short, brown hair. These include individuals with genotypes *SSbb* and *Ssbb.*

- Three-sixteenths (3/16) will have long, black hair. These include individuals with the genotypes *ssBB* and *ssBb.*

- One-sixteenth (1/16) will have long, brown hair. These include individuals with the genotype *ssbb.*

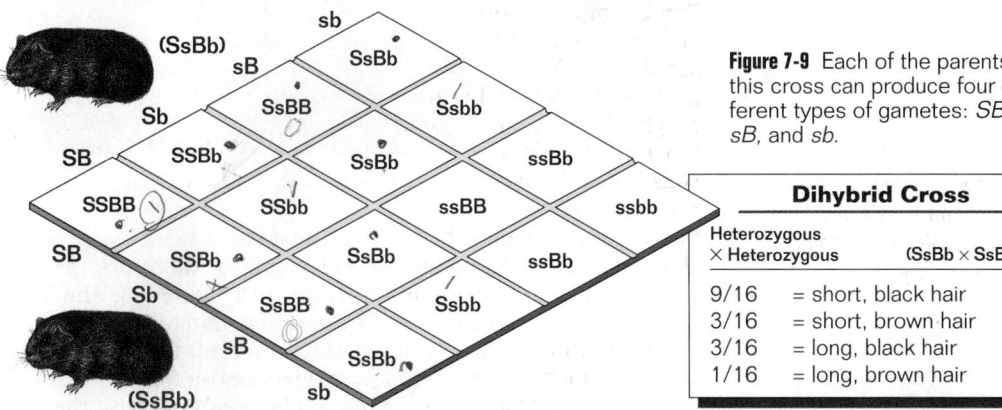

Figure 7-9 Each of the parents in this cross can produce four different types of gametes: *SB, Sb, sB,* and *sb.*

Dihybrid Cross

Heterozygous
× Heterozygous   (SsBb × SsBb)

9/16	= short, black hair
3/16	= short, brown hair
3/16	= long, black hair
1/16	= long, brown hair

Figure 7-10 These snapdragons appear pink because they have less red pigment than the red flowers.

Patterns of Heredity Can Be Complex

The relationships between genes and traits are not always as simple as the examples of dominant and recessive alleles discussed so far. Most of the time, genes display more complex patterns of heredity.

Incomplete Dominance

In some organisms, an individual displays a trait that is intermediate between the two parents, a phenomenon known as **incomplete dominance.** For example, the inheritance of flower color in snapdragons does not follow Mendel's idea of dominance. A cross between a snapdragon with red flowers and one with white flowers produces a snapdragon with pink flowers. The flowers appear pink because they have less red pigment than the red flowers. See Figure 7-10.

Codominance

In some cases, two dominant alleles are expressed at the same time, a phenomenon called **codominance.** Codominance is different from incomplete dominance because both traits are displayed. An example of codominance is the roan coat in horses. A cross between a homozygous red horse and a homozygous white horse results in heterozygous offspring with a roan coat, which consists of red hairs and white hairs, as seen in Figure 7-11.

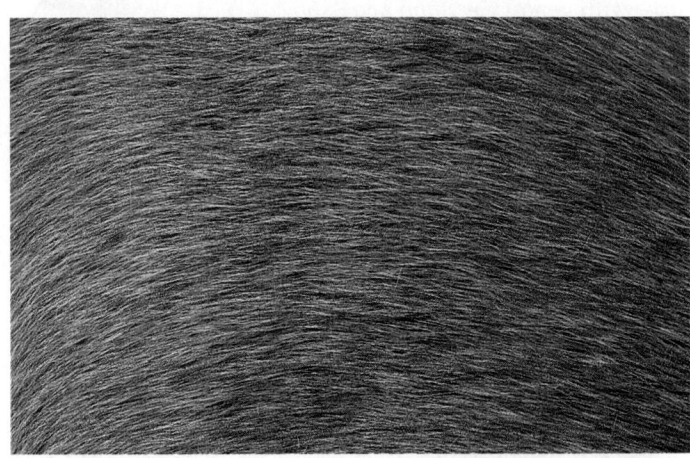

Figure 7-11 The roan coat of this horse consists of red hairs and white hairs.

Multiple Alleles

Some traits have genes with more than two alleles; these are referred to as **multiple alleles.** For example, there are three alleles that can determine human blood type—A, B, and O. The A and B alleles are both dominant over O, which is recessive, but neither is dominant over the other. When A and B are both present in the genotype, they are codominant. The existence of these multiple alleles explains why there are four different blood types—A, B, AB, and O.

Continuous Variation

When several genes influence a trait, such as height or weight, determining the effect of one of these genes is difficult, just as it is difficult to follow the flight of one bee within a swarm. Because the genes that determine a phenotype such as height or weight may segregate independently of one another, slight differences in phenotypes are expressed when many individuals are compared. These traits are said to be exhibiting **continuous variation** because you see a variety of phenotypes on a continuum from one extreme to the other.

Environmental Influences

An individual's phenotype often depends on conditions in the environment. For example, during the winter, the pigment-producing genes of the arctic fox shown in Figure 7-12 do not function due to the cold temperature. As a result, the coat of the fox is white, and the animal blends into the snowy background. In summer, the genes function to produce pigments and the coat darkens to a reddish brown, resembling the color of the tundra where the fox lives. ❑

❑ **CAPSULE SUMMARY**

Patterns of inheritance are more complex than those explained in Mendel's model. Heredity is affected by phenomena such as incomplete dominance, codominance, multiple alleles, continuous variation, and environmental influences.

Figure 7-12 Many arctic mammals develop white fur during the cold winter and dark fur during the warm summer.

Section Review

1. *Write the formula used to determine the likelihood that an event will occur. How is probability used in genetics?*

2. *Construct a Punnett square to predict the outcome of a monohybrid cross between two heterozygous tall pea plants. What are the expected phenotypic and genotypic ratios?*

3. *What is a dihybrid cross? What is the expected phenotypic ratio for a heterozygous dihybrid cross?*

4. *Describe three phenomena that can affect patterns of heredity. Give an example for each phenomenon.*

Critical Thinking

5. *Is it possible to have the blood type ABO? Why or why not?*

SECTION 7-2

Teaching Tips
Cooperative Learning

Place students into groups in which they are to become experts on one of the following topics: incomplete dominance, codominance, multiple alleles, or continuous variation. Ask students to derive a concrete example for teaching their assigned topics to their classmates. After their consultations, give each group five to ten minutes to explain their strategy to the class.

What If the population of an endangered species becomes so small that the animals begin inbreeding? If an endangered species becomes inbred due to its limited numbers, deleterious recessive genes can become prevalent. In many cases, traits do not cause disorders when the individual is heterozygous; however, inbreeding will increase the likelihood of individuals being homozygous for harmful disorders.

👁 VISUAL STRATEGY

Figure 7-12

Ask students to consider how the environment and genes affect these animals to make them change colors with changing seasons. Help students understand that the temperature triggers hormone responses that influence the genes.

Answers to Section Review

1. Probability = number of one kind of possible outcome/total number of all possible outcomes. Probability is used to predict how genes will be distributed among offspring.

2. The expected phenotypic ratio is 3:1. The genotypic ratio is 1:2:1.

3. A dihybrid cross is a cross that involves two pairs of contrasting traits. The expected phenotypic ratio is 9:3:3:1.

4. Three phenomena that can affect patterns of heredity include incomplete dominance in which intermediates can occur because the dominant gene is not totally expressed (pink-flowering snapdragons), codominance in which two traits are equally expressed (roan coat of horses), and multiple alleles in which more than two possible traits exist (ABO blood group).

5. It is not possible to have the ABO blood type because it would require that an individual have three alleles.

Opening Question

Ask students to name some possible genetic reasons for the high frequency of hemophilia among members of royal families throughout Europe. Explain that Queen Victoria was a carrier of sex-linked hemophilia. Because members of the European nobility usually married within their own social class, the hemophilia gene was passed via Queen Victoria's daughters to the Russian, German, and Spanish royal families, increasing the frequency of the recessive allele among European nobility.

Demonstration
Cultural Selection

To show students that many factors are involved in interpreting genetic data, have them consider the following scenario. A survey to determine the frequency of albinism in Native American communities in Arizona and New Mexico showed the numbers to range from very rare to nonexistent. However, in a Hopi tribe in Arizona, the frequency of albinos was determined to be 1 out of every 277 people. Why is the frequency so high? Lead students to consider the roles of albinos in this community. Emphasize that Hopi people have always had a high regard for albinos, and clan leaders have taken special care to protect them from the harsh desert sun. This type of selection would explain the increase in albinism in the community.

Section Objectives

- Explain how mutations can cause genetic disorders.
- List two genetic disorders, and describe their causes and symptoms.
- Explain what sex-linked traits are, and give an example.
- Describe how a pedigree is constructed and analyzed.
- Define genetic counseling, and list three techniques it employs.

CONTENT LINK

You will learn more about mutations in **Chapter 9**.

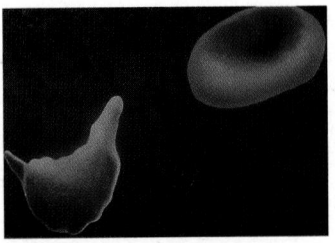

Figure 7-13 Sickle cell anemia is caused by a gene mutation that produces a defective form of hemoglobin. The cell bends, or sickles, in the absence of oxygen.

7-3 Human Genetics

*J*ust as flower color in Mendel's peas was determined by the particular chromosome the plant received, the degree to which you resemble your mother or father was established before your birth by the particular chromosomes you received. Many of the alleles present in human populations demand more serious consideration than the color of a flower. Alleles that code for defective forms of proteins can cause debilitating and even deadly disorders.

Mutations Cause Genetic Disorders

In order for a person to develop and function normally, the proteins encoded by his or her genes must function in a very precise manner. Unfortunately, sometimes genes become damaged or are copied incorrectly, resulting in faulty proteins. Changes in genetic material are called **mutations.** Most of the time, mutations are rare because cells have efficient systems for correcting errors. Still, some mutations occur, and they usually have harmful effects.

The harmful effects produced by mutated genes are called **genetic disorders.** Many mutations that cause genetic disorders are carried by recessive genes in heterozygous individuals. This means that two phenotypically normal people who are heterozygous carriers of a recessive mutation can produce children who are homozygous for the recessive gene. In such cases, the effects of the mutated genes cannot be avoided.

Sickle Cell Anemia

An example of a recessive genetic disorder is **sickle cell anemia,** a condition caused by a mutated allele that produces a defective form of the protein **hemoglobin** *(HEE moh gloh bihn)*. Hemoglobin is found within red blood cells, where it binds with oxygen and transports it through the body. In sickle cell anemia, the defective form of hemoglobin causes many red blood cells to bend into a sickle shape, as seen in the micrograph in Figure 7-13. The sickle-shaped cells rupture easily, impairing the oxygen-carrying capability of the blood. They also tend to get stuck in blood vessels, which can cut off blood supply to an organ altogether.

The recessive allele responsible for causing sickle-shaped red blood cells also helps protect the cells of heterozygous individuals from the effects of malaria, a disease caused by protozoans of the genus *Plasmodium*. (*Plasmodium* is a parasite that invades red blood cells.) People who are heterozygous have a mixture of sickled red blood cells and normal red blood cells. The sickled cells cannot

Multicultural Perspective

Understanding Sickle Cell Anemia

Working with limited laboratory facilities and a strong determination to fight the disease that was killing their friends and families, Dr. Angela Ferguson and Dr. Roland Scott published a paper on sickle cell anemia in the 1940s—25 years ahead of other researchers. Dr. Scott, known as the "father of sickle cell anemia research," is the founder and former director of Howard University's Center for Sickle Cell Anemia Research.

Dr. Ferguson was an associate professor of pediatrics at Howard University. She was listed in *Who's Who of American Women* in 1970.

readily be invaded by the parasite, and the normal cells support adequate oxygenation of tissues. Therefore, these people are protected from the effects of malaria that threaten the lives of individuals who are homozygous dominant for the gene. The defective hemoglobin allele apparently arose in populations that lived in malaria-prone areas of central Africa, where 1 person in 100 is homozygous for the defective allele and thus has the disease. In the United States, however, only 1 African American out of 500 has sickle cell anemia. This genetic disorder is almost unknown in other races.

Hemophilia

Another recessive genetic disorder is **hemophilia** *(hee moh FIHL ee uh)*, a condition that impairs the blood's ability to clot and can cause excessive and prolonged bleeding. A dozen genes code for the proteins involved in blood clotting, and mutations that cause hemophilia can occur in any of them. Two of these genes are found only on the X chromosome, a situation that has an important consequence for males. If a mutation appears in one of these genes in a male, he doesn't have a normal gene on the Y chromosome to compensate. Therefore, he will develop hemophilia. A trait that is determined by a gene found only on the X chromosome is said to be a **sex-linked trait.** Hemophilia and several other important genetic disorders are described in Table 7-2. ❑

CAPSULE SUMMARY

Mutations in genetic material can cause genetic disorders such as sickle cell anemia and hemophilia. Both disorders are caused by faulty proteins. The alleles for both disorders are recessive.

Table 7-2 Some Important Genetic Disorders

Disorder	Symptom	Defect	Dominant or Recessive	Frequency Among Human Births
Cystic fibrosis	Mucus clogs lungs, liver, and pancreas; usually don't survive to adulthood	Failure of chloride ion transport mechanism	Recessive	1/2,080 (whites)
Sickle cell anemia	Poor blood circulation	Abnormal hemoglobin molecules	Recessive	1/500 (African-Americans)
Tay-Sachs disease	Deterioration of central nervous system in infancy; usually don't survive to adulthood	Defective form of enzyme hexosamin-idase A	Recessive	1/1,600 (Jews)
Phenyl-ketonuria	Failure of brain to develop in infancy; usually don't survive to adulthood (if untreated)	Defective form of enzyme phenylalanine hydroxylase	Recessive	1/18,000
Hemophilia	Failure of blood to clot	Defective form of blood clotting factor VIII	Sex-linked recessive	1/7,000
Huntington's disease	Gradual deterioration of brain tissue in middle age; shortened life expectancy	Production of an inhibitor of brain cell metabolism	Dominant	1/10,000
Muscular dystrophy	Wasting away of muscles; shortened life expectancy	Muscle fibers degenerate and atrophy	Sex-linked recessive	1/10,000

Tracking Traits in Families

Imagine that you want to learn about an inherited trait present in your family. How would you find out whether the trait is dominant or recessive and what the chances are of transmitting it to your children? If you wanted to study a trait in organisms such as garden peas or fruit flies, you would conduct genetic crosses. However, studying human heredity requires a different approach.

To study human heredity, scientists look at family histories called **pedigrees.** By identifying which relatives exhibit a trait, scientists can determine whether the gene producing the trait is dominant or recessive and whether it is sex-linked or autosomal (located on one of the other 22 pairs of chromosomes). It is also possible to infer whether a family member is homozygous or heterozygous for a particular trait.

Analyzing a Pedigree for Albinism

How are pedigrees of a trait constructed? Consider the genetic disorder albinism, a condition in which the body is unable to synthesize pigment. As a result of albinism, the skin and hair appear white. In the pedigree of the family presented in Figure 7-14, each symbol represents one individual in the family history. The circles represent females and the squares represent males. Individuals who exhibit the trait being studied, in this case albinism, are indicated by yellow symbols. Marriages are represented by horizontal lines connecting a

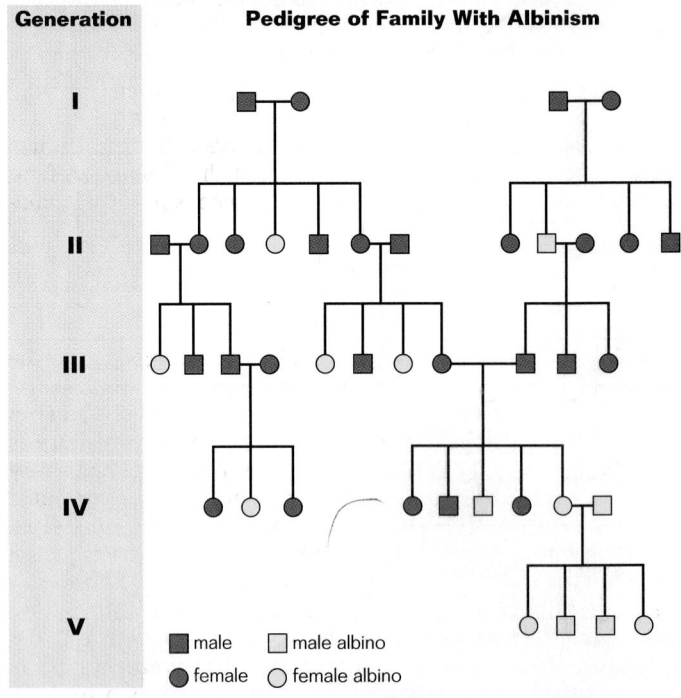

Figure 7-14 This pedigree shows how the genetic disorder albinism is distributed throughout five generations in a family.

circle and a square. Vertical lines indicate children, arranged from left to right in order of their birth.

How do you analyze the pedigree?

1. First, determine whether albinism is sex-linked or auto-somal. If the trait is sex-linked, it is usually seen only in males because they have only one X chromosome. If the trait is autosomal, it will appear in both sexes equally. In Figure 7-14, the proportion of males with albinism (5/13, or 39 percent) is similar to the proportion of females with albinism (8/21, or 37 percent). So you can conclude that the trait is probably carried on an autosomal chromosome rather than on a sex chromosome.

2. Second, determine whether albinism is inherited in a dominant or recessive manner. If the trait is dominant, every individual with albinism will have a parent with albinism. If the trait is recessive, an individual with albinism can have heterozygous parents who appear nor-mal. In Figure 7-14, most individuals with albinism have parents who do not exhibit the trait, which indicates that albinism is recessive.

3. Third, find out whether the trait is determined by a single gene or by several. If the trait is determined by a single recessive gene, then normal parents should produce affected children in a 3:1 ratio in the family. This means that approximately 25 percent (the 1 of 3:1) of the children should express the trait. If the trait were determined by several genes, the proportion of affected individuals would be much lower. In this case, 9/34, or 27 percent, of the individuals have albinism, strongly suggesting that only one gene determines the trait.

This example shows how a family pedigree can be ana-lyzed to understand the inheritance pattern of a particular trait. By looking at the pattern of inheritance of albinism in a single family, a scientist can predict that the albinism affect-ing these individuals is an autosomal recessive trait con-trolled by a single gene. ◻

Identifying and Treating People at Risk

Most genetic disorders cannot be cured, although progress is being made in many cases. If a person has a disorder for which there is no treatment, he or she may choose to avoid having children. How can a person find out if he or she is a potential carrier for a genetic disorder? People at risk of having children with a genetic disorder can often be identified through genetic counseling. In genetic counsel-ing, a trained genetic counselor helps an individual or family understand the nature of the hereditary disorder and the risks of passing it on to children. Genetic counseling can involve analyzing a family pedigree. It can also include

☑ **RESEARCH Update**

UPDATE

Knowing the Odds

Pedigrees are an important initial step in tracking and predicting inher-ited diseases. Dr. Ernest Beutler, at Scripps Institute in La Jolla, Califor-nia, tracks the perplexingly high num-ber (about a dozen) of autosomal recessive diseases in descendants of eastern European (Ashkenazi) Jews. Tay-Sachs, Gaucher's, and Niemann-Pick are three of these dis-eases, and all three involve a lipid storage problem due to faulty lyso-somal enzymes. Dr. Beutler and oth-ers have concentrated on tracking Gaucher's disease (which can be serious or fatal) in an attempt to find out why the Ashkenazi Jewish popu-lation in particular is afflicted by this disease. Molecular geneticists have gone on to conclude that there are *five individual mutations* responsible for the development of Gaucher's disease. Human geneticists, using the molecular and pedigree data, suggest that these mutations are evolutionarily recent and have, for some reason, been selected.

Dr. Jared Diamond of the Univer-sity of California Medical School in Los Angeles believes that there must be a compensatory advantage associated with the deleterious gene, probably in the heterozygous geno-type, as in the resistance to malaria conferred by sickle cell anemia. What could cause the selection for Gaucher's disease in eastern European Jews, and yet not affect neighboring non-Jewish communities? Religious segregation, sexual selec-tion, and protection from tuberculosis have all been suggested as selective factors, but researchers feel that the answer to this question is likely to remain highly speculative. ☑

◻ **CAPSULE SUMMARY**

By constructing and analyzing a pedigree, a scientist can deter-mine the pattern of inheritance of a trait within a family. It may also be possible to determine whether an individual is hetero-zygous or homozygous.

Did You Know?

The original diagnosis of neu-rofibromatosis in John Merrick, sometimes called the "elephant man," is considered a misdiag-nosis by many scientists who now think that his genetic disor-der was Proteus syndrome. Proteus syndrome, named after the Greek god Proteus, results in overgrowth of the head, enlarged hands and feet, and a distorted growth of part of the torso. All of the known cases to date have occurred from spo-radic mutations.

Teaching Tips
Maternal PKU

Women who have PKU often have babies with mental retardation, not because the baby has PKU, but because the mother's body chemistry is altered during pregnancy. These babies cannot be helped with a special diet; however, the mental retardation can be avoided if the mother follows a low phenylalanine diet before and during pregnancy.

Muscular Dystrophy

The gene that causes Duchenne muscular dystrophy, the most common type of muscular dystrophy, is an exceptionally large gene. A healthy gene that will code for dystrophin, the protein needed to eliminate the symptoms, can be introduced into cells using a virus as a vector, but the current problem with this gene is its size. Before scientists can incorporate it into human cells, they need to find a way to reduce the gene to a workable size that can be incorporated into a virus.

Chapter Closure

Ask students to make up examples of crosses to illustrate the following concepts: monohybrid cross, dihybrid cross, homozygous recessive trait, complete dominance, incomplete dominance, codominance, multiple alleles, and continuous variation. Encourage students to use attributes of hypothetical life-forms to help them visualize these concepts.

CONTENT LINK
*You will learn more about gene technology in **Chapter 10**.*

□ CAPSULE SUMMARY

The techniques involved in genetic counseling can help identify parents at risk of having children with a genetic disorder. Current treatments for genetic disorders are beginning to include gene transfer therapy.

examining the genetic makeup of an embryo using amniocentesis or chorionic villi sampling, two forms of prenatal testing that you learned about in Chapter 6.

In some cases, therapy is available to treat a genetic disorder if it is diagnosed early enough. For example, **phenylketonuria** (PKU) is a genetic disorder in which an individual lacks an enzyme that converts the amino acid phenylalanine into the amino acid tyrosine. As a result, phenylalanine builds up in the body, a condition that causes severe mental retardation. If PKU is diagnosed soon after birth, the newborn can be placed on a low-phenylalanine diet. Such a diet ensures that the baby gets enough phenylalanine to make proteins, but not so much that it causes any damage. The child maintains the low-phenylalanine diet until approximately 10 years of age, when the brain is fully developed and PKU is usually no longer a problem. Because this disease can be easily diagnosed after birth by inexpensive laboratory tests, many states require the testing of all newborns for PKU.

Gene technology is also making it possible for scientists to correct certain genetic disorders by replacing defective genes with copies of healthy ones. In 1990, this approach was tried for the first time on human patients. In one case a young girl was homozygous for a defective gene that normally encoded an enzyme necessary for a properly functioning immune system. Doctors were successful in transferring a healthy copy of the gene into the girl's cells, and she is considered to be cured of her genetic disorder. In another case, the transferred gene was a potent cancer-fighting gene. Currently, gene transfer therapy is being tested by scientists seeking cures for cystic fibrosis and muscular dystrophy. Scientists are hopeful that they will soon find cures for these genetic disorders. ❑

Section Review

1. *How does a mutation give rise to a genetic disorder?*
2. *Identify two genetic disorders. Explain the defect in the protein that causes each disorder.*
3. *Explain how males inherit hemophilia.*
4. *What is phenylketonuria? How is this disorder treated?*

Critical Thinking
5. *When analyzing a pedigree, how can you determine if an individual is heterozygous for the trait being studied?*
6. *What is genetic counseling? Why might a couple undergo genetic counseling?*

Answers to Section Review

1. A mutation leads to a genetic disorder by altering a gene and causing a harmful effect.

2. Refer to Table 7-2 on page 155 for possible answers.

3. A male inherits the recessive gene for hemophilia on the X chromosome he receives from his mother.

4. Phenylketonuria is a disease caused by lack of an enzyme that breaks down the amino acid phenylalanine. The disorder is treated by limiting phenylalanine in the diet.

5. An individual is heterozygous for a trait if the individual receives a dominant allele from one parent and if the individual's other parent is homozygous recessive.

6. Genetic counseling can include analyzing a family pedigree or examining the genetic makeup of an embryo. People at risk of having children with a genetic disorder go to genetic counselors to determine the probability of having a child with a genetic disorder, and to learn about treatments or support groups for the genetic disorder.

Vocabulary

allele (146)
codominance (152)
continuous variation (153)
cross-pollination (143)
dihybrid cross (150)
dominant (145)
F_1 generation (144)
F_2 generation (144)
genetic disorder (154)
genetics (141)
genotype (146)
hemoglobin (154)

hemophilia (155)
heredity (141)
heterozygous (146)
homozygous (146)
incomplete dominance (152)
law of independent
   assortment (147)
law of segregation (147)
monohybrid cross (149)
multiple alleles (152)
mutation (154)
pedigree (156)

P generation (144)
phenotype (146)
phenylketonuria (158)
probability (148)
Punnett square (149)
recessive (145)
self-pollination (143)
sex-linked trait (155)
sickle cell anemia (154)
true-breeding (144)

CHAPTER REVIEW ANSWERS

Each item in the Chapter Review is correlated by text section in the assignment guide that follows.

ASSIGNMENT GUIDE

Section	Review	Themes Review	Critical Thinking
7-1	1–4, 7, 9, 10, 13, 14, 16, 18	21, 22	23
7-2	5, 6, 14, 15, 17, 19		23
7-3	8, 11, 12	20	24, 25

Concept Mapping

Construct a concept map that shows the nature and results of Mendel's experiments and also the myths surrounding them. In constructing your map, use the following terms: *P. sativum,* varieties, segregation, independent assortment, P generation, F_1 generation, F_2 generation, genes, dominant trait, and recessive trait. Include additional concepts in your map as needed.

Concept Mapping

Map answer is shown on page 139B.

Review

Multiple Choice

1. Which of the following is not a good reason that *Pisum sativum* makes an excellent subject for genetic study?
   a. It has many varieties.
   b. It requires cross-pollination.
   c. It grows quickly.
   d. It demonstrates complete dominance.

2. Offspring that are the product of true-breeding parents are called the
   a. F_1 generation.
   b. F_2 generation.
   c. dominant offspring.
   d. phenotypic expression.

3. If two parents with dominant phenotypes have an offspring with a recessive phenotype, then
   a. the parents are heterozygous.
   b. other offspring must be homozygous dominant.
   c. the parents are homozygous.
   d. the mother carries a lethal gene.

4. The law of segregation
   a. is demonstrated by the phenotypic ratio of 9:3:3:1.
   b. states that pairs of alleles separate when gametes form.
   c. does not apply to the phenomenon of codominance.
   d. predicts the probability of a penny landing "tails" side up when tossed in the air 50 times.

5. Whenever a 9:3:3:1 ratio appears among the offspring of two parents who are both heterozygous for two different traits, it means that
   a. continuous variation is at work.
   b. the traits are expressions of incomplete dominance.
   c. the two sets of alleles probably sorted independently.
   d. mutations have likely occurred in both F_1 and F_2 generations.

Review

Multiple Choice

Code in parentheses indicates section number and objective number.

1. b (7-1.1)
2. a (7-1.2)
3. a (7-1.4)
4. b (7-1.5)
5. c (7-2.1, 7-2.2)
6. c (7-2.3)
7. c (7-1.5)
8. b (7-3.5)

Completion

9. 3:1, 1:2:1 (7-1.3, 7-1.4)
10. phenotype, genotype (7-1.4)
11. hemophilia, sex-linked (7-3.3)
12. genetic disorders, mutations (7-3.1)

CHAPTER REVIEW ANSWERS *continued*

Short Answer

13. Mendel was not the first person to investigate the transmission of traits from parent to offspring in pea plants. Mendel's work was preceded by the work of British farmers like T. A. Knight. (7-1.1)

14. If the allele for brown hair, H, were dominant over the allele for black hair, h, heterozygous brown parents, Hh, could produce black offspring. The genotype of the black offspring would be hh. Possible genotypes of the black offspring's siblings with brown hair would be HH and Hh. (7-1.4, 7-2.2)

15. The chance of the number being drawn is 1 in 100,000 (1/10 × 1/10 × 1/10 × 1/10 × 1/10 = 1/100,000). You should not count on winning the lottery. (7-2.1)

16. By allowing pea plants to self-pollinate for several generations, Mendel ensured that the plants were true-breeding for the trait he wished to study. (7-1.2)

17. The ratios are calculated by considering all of the possible outcomes and determining the likelihood that each outcome will occur. Thus, calculated ratios in a Punnett square predict the probable outcome of a genetic cross. (7-2.1, 7-2.2)

18. In cross-pollination, pollen is transferred from a flower of one plant to a flower of another plant. In self-pollination, a flower fertilizes itself. (7-1.1)

19. The genotypic ratios would be 1/16 YYAA, 2/16 YYAa, 2/16 YyAA, 1/16 YYaa, 4/16 YyAa, 2/16 Yyaa, 1/16 yyAA, 2/16 yyAa, and 1/16 yyaa. The phenotypic ratios for the pea plants would be 9/16 with yellow seeds and axial flowers, 3/16 with yellow seeds and top flowers, 3/16 with green seeds and axial flowers, and 1/16 with green seeds and top flowers.

6. An example of continuous variation is
   a. a pink snapdragon produced by crossing red and white snapdragons.
   b. phenylketonuria.
   c. the height of people.
   d. the changing color of the Arctic fox's coat.

7. Sickle cell anemia is an autosomal recessive disorder that occurs once in every 500 African American births. If the first child of two healthy African American parents is affected by the disorder, what is the probability that their second child will also have sickle cell anemia?
   a. 1 in 500
   b. 1 in 250
   c. 1 in 4
   d. 0

8. Genetic counseling involves
   a. separating pairs of alleles when gametes form.
   b. educating parents about hereditary disorders.
   c. replacing defective genes with healthy ones.
   d. supplementing an infant's diet with the amino acid tyrosine.

Completion

9. When Mendel crossed F_1 generation individuals, the phenotypic ratio he observed in their offspring was _____, while the genotypic ratio he inferred was _____ .

10. The color of a dog's coat can be described as the dog's _____ , which is determined by sets of alleles, or its _____ .

11. A recessive genetic disorder that impairs the blood's ability to clot is _____ . It is said to be a _____ trait because it is carried on the X chromosome.

12. Tay-Sachs disease and phenylketonuria are examples of _____ . They are caused by _____ in genetic material.

Short Answer

13. In science, it has often been said that one scientist's discovery is built on work of others that came before him. How does this saying apply to the work of Mendel?

14. Most of the members of one animal species are brown, but occasionally a black individual appears. Assuming that the trait for color shows complete dominance, how could two brown parents produce a black offspring? What is the genotype of the black offspring? What are the possible genotypes of the black offspring's siblings?

15. The law of independent assortment applies not only to genetics but also to many lottery games. Suppose that ten ping-pong balls are placed in a machine, each marked from 0 to 9. If they are mixed thoroughly and one is taken out, the chances are 1 in 10 that a particular number, such as 3, will be drawn. If two such machines are used, the chances of a 3 being drawn from both machines is 1 in 100.

$$1/10 \times 1/10 = 1/100$$

(1 in 10 for each jar, or a total of 1 in 100 for drawing 3 and 3)

Suppose you have chosen the number 35097 to win $1 million in a state lottery. What is the chance of your number being drawn? Should you count on winning the lottery?

16. Why did Mendel begin his experiments by allowing pea plants to self-pollinate for several generations?

17. Why are calculated ratios in a Punnett square probable rather than certain?

18. What is the difference between self-pollination and cross-pollination?

19. What would be the result of a cross between two pea plants that were heterozygous for both yellow seed color and axial flower position. Show the Punnett square, and state the genotypic and phenotypic ratios.

Themes Review

20. Squares represent males; circles represent females; lines show the relationships among family members; and black squares and circles represent individuals with the genetic disorder. Genetic counselors use pedigrees to determine whether a person is a potential carrier for a genetic disorder and to predict the risk of passing a disorder on to children. (7-3.4)

21. The alleles would not have sorted independently if the traits he studied were located close to each other on the same gene. Instead, the traits would have been linked during gamete formation. (7-1.5)

22. The genotype is the set of alleles that an individual has for a trait, which determines the phenotype, or the expression of the trait. (7-1.4)

Themes Review

20. Structure and Function Describe the organization of information in a pedigree chart. How is a pedigree chart used by a genetic counselor?

21. Structure and Function Many scientists claim that Mendel was very lucky because he studied traits determined by genes located on different chromosomes or located on the same chromosome, but some distance apart. How might Mendel's conclusions have differed if he had studied two traits determined by genes located close to one another on the same chromosome?

22. Levels of Organization How is the genotype for a particular trait related to the phenotype for that same trait?

Critical Thinking

23. Making Inferences Mendel based his conclusions about inheritance patterns on studies of large samples. Why do you think the use of large samples is advantageous when studying inheritance patterns?

24. Interpreting Graphics The partial pedigree below shows a family with a genetic disorder. Is the trait sex-linked? Is it inherited in a dominant or recessive manner?

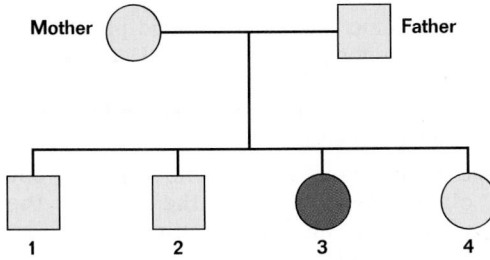

25. Communicating Effectively A 20-year-old man diagnosed with muscular dystrophy has a sister who is soon to be married. If you were the man, what would you tell your sister?

Activities and Projects

26. Research and Writing Interview health care workers or genetic counselors to learn why testing newborns for PKU is required by law and how the mental retardation associated with PKU can be avoided.

27. Cooperative Group Project Raise the fruit fly *Drosophila melanogaster* to investigate the results of different crosses.

28. Multicultural Perspective Hopi Native Americans have high numbers of albinos in their tribe. Research Dr. Frank C. Dukepoo's idea of cultural selection that explains this phenomenon.

Readings

29. Read the article "Cystic Fibrosis: Tests, Treatments Improve Survival," in *FDA Consumer*, June 1993. What are the symptoms associated with cystic fibrosis? How is CF inherited? How is the discovery of the CF gene enabling researchers to develop new diagnostic tests and treatments for the fatal disease?

30. Read the article "Origin of the Punnett Square," in *The American Biology Teacher*, April 1993, pages 204–208. Why did Punnett originally devise his "chessboard" approach to genetics?

Activities and Projects

26. PKU is treatable, and the mental retardation associated with PKU is preventable when the disorder is identified early. Mental retardation can be avoided by following a low phenylalanine diet during childhood.

27. Using *Drosophila* cross kits, students can study monohybrid, sex-linked, and dihybrid inheritance. Have students learn about the mating habits and developmental stages of fruit flies before beginning their crosses.

28. According to Dr. Frank C. Dukepoo, a Hopi Native American who grew up among albino Native Americans in the Southwest, the reason for the high number of albinos is cultural selection. Because Hopi people have a high regard for albinos, clan leaders have taken special care to protect them. Hopi albinos are not required to work out in the fields. Instead, they stay at home to weave, make pottery and other objects, and help prepare meals.

Readings

29. The symptoms associated with cystic fibrosis (CF) include mucus that builds up in the lungs, impairs breathing, and attracts infections; a pancreas that cannot release digestive enzymes; poor circulation resulting in stubbed fingers; infertility; and salty sweat. CF is inherited when both parents pass the gene on to their offspring. The discovery of the CF gene is enabling researchers to develop new tests for CF gene carriers, to treat patients with gene therapy, and possibly to find an early detection and treatment procedure for CF patients, as in patients with hypertension.

30. Punnett originally devised his "chessboard" approach to genetics as a compact method of writing out all the possible combinations of the gametes. He wanted an easy way to think about independent assortment of characters in gametes.

Critical Thinking

23. Patterns obtained from large samples are less likely to be distorted by rare events that can occur by chance. (7-1.3, 7-2.2)

24. The male children are not affected by the genetic disorder; thus, the trait is probably not sex-linked. Since neither of the parents is affected by the disorder, it is probably inherited in a recessive manner. (7-3.4)

25. Muscular dystrophy is a sex-linked recessive genetic disorder. The man's sister may be a carrier of the allele for muscular dystrophy and may wish to consult a genetics counselor before deciding to have children. (7-3.2, 7-3.3)

Preparation

Each group of students needs an even mix of lentils and green peas, and 2 petri dishes. You may choose to use pop-it beads, available from Ward's, instead of lentils and peas.

Procedural Notes

- You may wish to review monohybrid crosses and probability before beginning this investigation.
- Students can be grouped in pairs for this investigation.
- In step 1, make sure students place one green pea and one lentil into each petri dish.
- In step 3, remind students to return the seeds to their original petri dishes.

Background Answers

1. One trait and four alleles are involved in a monohybrid cross.

2. When a dominant allele combines with a recessive allele, the recessive allele will not be expressed. Sometimes the dominant allele is not completely dominant; in such cases the recessive allele may be partially expressed.

3. When gametes form, the alleles for each trait separate independently of one another.

4. What are the probable genotypic and phenotypic ratios of offspring that result from the random pairing of gametes from parents heterozygous for a given trait?

Monohybrid Crosses

OBJECTIVE

Using a Punnett square, predict the genotypic and phenotypic ratios of offspring resulting from the random pairing of gametes.

PROCESS SKILLS

- predicting results of monohybrid crosses
- organizing data using Punnett squares

MATERIALS

- lentils
- green peas
- petri dishes

BACKGROUND

1. How many traits are involved in a monohybrid cross? How many alleles?
2. What prevents the expression of a recessive allele?
3. When gametes form, what happens to the alleles for each trait?
4. Write your own question to explore in your lab report or notebook.

TECHNIQUE

You will model the random pairing of alleles by choosing lentils and peas from petri dishes. These dried seeds will represent the traits for seed color. A green pea will represent **G**, the dominant allele for green seeds; a lentil will represent **g**, the recessive allele for yellow seeds.

1. Place one lentil and one green pea in each of two petri dishes. Mark one petri dish "female gametes" and the other dish "male gametes."

2. Since each parent contributes one allele at random to each offspring, model a cross between these two parents by choosing 10 random pairings of the dried seeds from the two containers. Do this by simultaneously picking one seed from each container, without looking. Place the pair together on the lab table. The pair of seeds represents one offspring. Record the results in the **Records** section of your report in a table like the one below.

Table 1 Gamete Pairings

Trial	Offspring Genotype	Offspring Phenotype
1		
2		
3		
4		
5		
6		
7		
8		
9		
10		

3. Return the seeds to their original dishes and repeat step 2 nine more times.

4. Determine the genotypic and phenotypic ratios among the offspring. Using a data table like the one atop the next page, record your ratios in the **Records** section of your report.

5. Compare your ratios with those of your classmates. Now pool the data for the whole class and record it in the table on the next page. Compare the greater numbers with your small sample of 10. Calculate the genotypic and phenotypic

Records

Table 1 Gamete Pairings

Trial	Offspring Genotype	Offspring Phenotype
1		
2		
3		
4		

5		
6		
7		
8		
9		
10		

Table 2 Offspring Ratios

Genotypes:	Total	Genotypic Ratio
Homozygous dominant (GG)		___ : ___ : ___
Heterozygous (Gg)		
Homozygous recessive (gg)		
Phenotypes:		Phenotypic Ratio
Green fruit		___ : ___
Yellow fruit		

ratios for the class data, and record them in the table.

Table 3 Offspring Ratios (Entire Class)

Genotypes:	Total	Genotypic Ratio
Homozygous dominant (GG)		___ : ___ : ___
Heterozygous (Gg)		
Homozygous recessive (gg)		
Phenotypes:		Phenotypic Ratio
Green fruit		___ : ___
Yellow fruit		

6. Construct a Punnett square showing the parents and their offspring in the **Records** section of your report. Also, briefly summarize the procedure you followed.

INQUIRY

1. What are the genotypes and phenotypes of the parents?
2. What does each seed represent in the petri dish?

3. When the seeds were selected and paired, what did pairs represent?
4. Describe the genotypes of both parents using the terms *homozygous* or *heterozygous*, or both.
5. Did tables 2 and 3 reflect a classic monohybrid cross phenotypic ratio of 3:1?
6. What trait is being studied in this investigation?

ANALYSIS

1. When the class data were tabulated, did a classic monohybrid cross ratio of a phenotype of 3:1 result?
2. If a genotypic ratio of 1:2:1 is observed, what must the genotypes of both parents be?
3. Show what the genotypes of the parents would be if 50 percent of the offspring were green and 50 percent of the offspring were yellow.
4. Diagram the cross of a heterozygous black guinea pig and an unknown guinea pig whose offspring include a recessive white-furred individual. What are the possible genotypes of the unknown parent?

FURTHER INQUIRY

Write a new question that could be explored as a new investigation. The following is an example:

> How could you model a dihybrid cross of two parents that are heterozygous for two traits? Construct and complete a Punnett square for this cross.

Inquiry Answers

1. Both parents are phenotypically green. Both have the genotype Gg.
2. Each seed represents an allele.
3. The pairs represent the gametes that an offspring will receive.
4. Both parents are heterozygous green, Gg.
5. Answers will vary. In a sample size of 10 crosses, no clear ratio may be evident. When combining data from the entire class, the 3:1 ratio should be seen.
6. The trait being investigated is seed, or fruit, color.

Analysis Answers

1. When combining data from the entire class, the 3:1 ratio should be seen.
2. Both parents must be heterozygous, Gg.
3. One parent would be heterozygous, Gg, and the other parent would be homozygous recessive, gg.
4. Students should draw two Punnett squares. One should show a cross between two heterozygous individuals; the other should show a heterozygous individual crossed with a homozygous recessive individual. If B represents the dominant black fur color and b stands for white fur color, the unknown parent could be genotypically Bb or bb.

Table 2 Offspring Ratios

Genotypes:	Total	Genotypic Ratio
Homozygous dominant (*GG*)		___ : ___ : ___
Heterozygous (*Gg*)		
Homozygous recessive (*gg*)		
Phenotypes:		Phenotypic Ratio
Green fruit		___ : ___
Yellow fruit		

Table 3 Offspring Ratios (Entire Class)

Genotypes:	Total	Genotypic Ratio
Homozygous dominant (*GG*)		1 : 2 : 1
Heterozygous (*Gg*)		
Homozygous recessive (*gg*)		
Phenotypes:		Phenotypic Ratio
Green fruit		3 : 1
Yellow fruit		

UNIT 2

CHAPTER 8 DNA: THE GENETIC MATERIAL

Block Scheduling Guide

> Each block represents about 45–50 minutes of instructional time. The resources cited in a block are coded to a specific program component using the Key on the next page.

Lecture Concepts	Lesson Resources

8-1 Identifying the Genetic Material pp. 164–168

BLOCK 1

a. The transformation experiments of Griffith and Avery showed that DNA carried genetic information.
b. Alfred Hershey and Martha Chase used the bacteriophage T_2 and radioactive labels to identify DNA as the genetic material.

Lecture Resources
■ Opening Question, page 165
■ Demonstration: Structure of Protein and DNA
■ Visual Strategy: Figure 8-1
▤ Teaching Transparencies: 25A
◉ Holt Biology Videodiscs: Lesson 8-1

Classwork Option
■ Graphic Organizer, page 167
■ Role Playing, page 168

Assignment Options
■ Section Review, page 168
■ Chapter Review, questions 3, 5, 9, 16, 20, 21

8-2 The Structure of DNA pp. 169–173

BLOCK 2

a. The DNA molecule is a long chain of nucleotides. Each nucleotide is made of a phosphate group, the 5-carbon sugar deoxyribose, and one of the four different nitrogen bases.
b. The two strands of the double helix are complementary. Purines can form hydrogen bonds only with pyrimidines. Adenine can bond to thymine. Cytosine bonds with guanine.

Lecture Resources
■ Opening Question, page 169
■ Demonstration: The Parts of DNA
■ Visual Strategies: Figures 8-6, 8-8
▤ Teaching Transparencies: 33
■ Overcoming Misconceptions, page 170
◉ Holt Biology Videodiscs: Lesson 8-2

Classwork Options
■ Construct DNA, page 170
■ Study Guide: Science Skills, Interpreting Diagrams
▲ Biotechnology D2: Extracting DNA

BLOCKS 3 and 4

a. During replication, enzymes unwind and separate the double helix, and complementary bases are added to the exposed strands. Each new double helix consists of one old DNA strand linked to one new DNA strand.

Lecture Resources
■ Research Update, page 172
■ Visual Strategy: Figure 8-10
■ The Steps of Replication, page 173
◉ Holt Biology Videodiscs: Lesson 8-2

Classwork Option
▲ Biotechnology D3: Genetic Transformation of Bacteria

Assignment Options
■ Section Review, page 173
■ Chapter Review, questions 1, 2, 4, 6, 11–13, 17–19, 22, 25, 27–32

8-3 The Structure of a Gene pp. 174–176

BLOCK 5

a. Coding regions of DNA are called exons. Many eukaryotic genes are interrupted by segments of DNA that do not code for proteins. Noncoding regions are called introns.
b. Barbara McClintock's work involved the study of repeating genes called transposons, which can jump to new locations on a chromosome. Jumping genes can cause mutations.
c. It has been discovered that certain genes protect living things from damaging mutations. Such a gene is p53, which suppresses the development of cancerous tumors.

Lecture Resources
■ Opening Question, page 174
■ Demonstration: Exons
■ Finding Exons and Recognizing Introns, page 175
▤ Teaching Transparencies: 27A
◉ Holt Biology Videodiscs: Lesson 8-3

Classwork Options
■ Laboratory Investigation: DNA and Its Structure, pages 180–181
■ Closure, page 176

Assignment Options
■ Section Review, page 176
■ Chapter Review, questions 7, 8, 10, 14, 15, 23, 24, 26

Review/Enrichment

BLOCK ⑥

- ■ Study Guide: Concept Review and Pretest

Assignment Options
- ⊙ Teaching Resources CD-ROM: Supplementary Reading Worksheets and Test, *The Double Helix*
- ■ Activities and Projects, questions 33–35
- ■ Readings, questions 36, 37

Assessment Options

BLOCK ⑦

Traditional Assessment
- ■ Chapter 8 Test
- ⊙ Test Generator/Assessment Item Listing: Software item bank for preparing customized chapter tests

Portfolio Assessment
Select student reports for one or more laboratory experiments from this chapter. *The Direct Observation Checklist*, on the *BioSources Teaching Resources CD-ROM*, should be completed during a laboratory or other cooperative-learning experience.

Answer to Concept Map

The following is one possible answer to the Concept Mapping exercise on page 177.

Holt Biology Videodiscs

Holt Biology Videodiscs gives you a powerful tool for teaching, review, and assessment. *Concepts of Biology* adds interactive lesson plans and extensive tools for customization using CD-ROM technology.

CONCEPTS OF BIOLOGY

Use the following topic from *Concepts of Biology* to help you teach this chapter:
- ■ Topic 7: How Genes Work

For further information, see *Holt Biology Videodiscs Teacher's Correlation Guide to Biology: Principles and Explorations.*

Chapter Theme

Evolution *Patterns within DNA suggest that similar genes originated from ancestral genes that differentiated and became more specific with time. The 12 different hemoglobin genes seen in humans have enough similarity to suggest a common source of origin.*

Tapping Prior Knowledge

- Where is the genetic information in a eukaryotic cell?
- How does a newly formed cell get its genetic information?

Opening Question

Ask students to name some prestigious awards. Answers may include the Oscar or the Heisman Trophy. Ask students what they think would be a prestigious award for a leading scientist. Discuss the Nobel Prize. Alfred Nobel (1833–1896), the Swedish inventor of dynamite, provided a $9 million fund in his will that is to be used to award Nobel Prizes in the areas of chemistry, physics, physiology or medicine, and literature. Later, prizes in peace and economics were added. Tell students that a number of people discussed in this chapter have received a Nobel Prize for their work in medicine or physiology. Throughout the chapter point out individuals who have received a Nobel Prize. Encourage students to keep a list of these individuals and their prize-winning work. Answers will include Maurice Wilkins, James Watson, and Francis Crick (1962), Alfred Hershey, Max Delbrück, and Salvador Luria (1969), and Barbara McClintock (1983).

CHAPTER 8

DNA: THE GENETIC MATERIAL

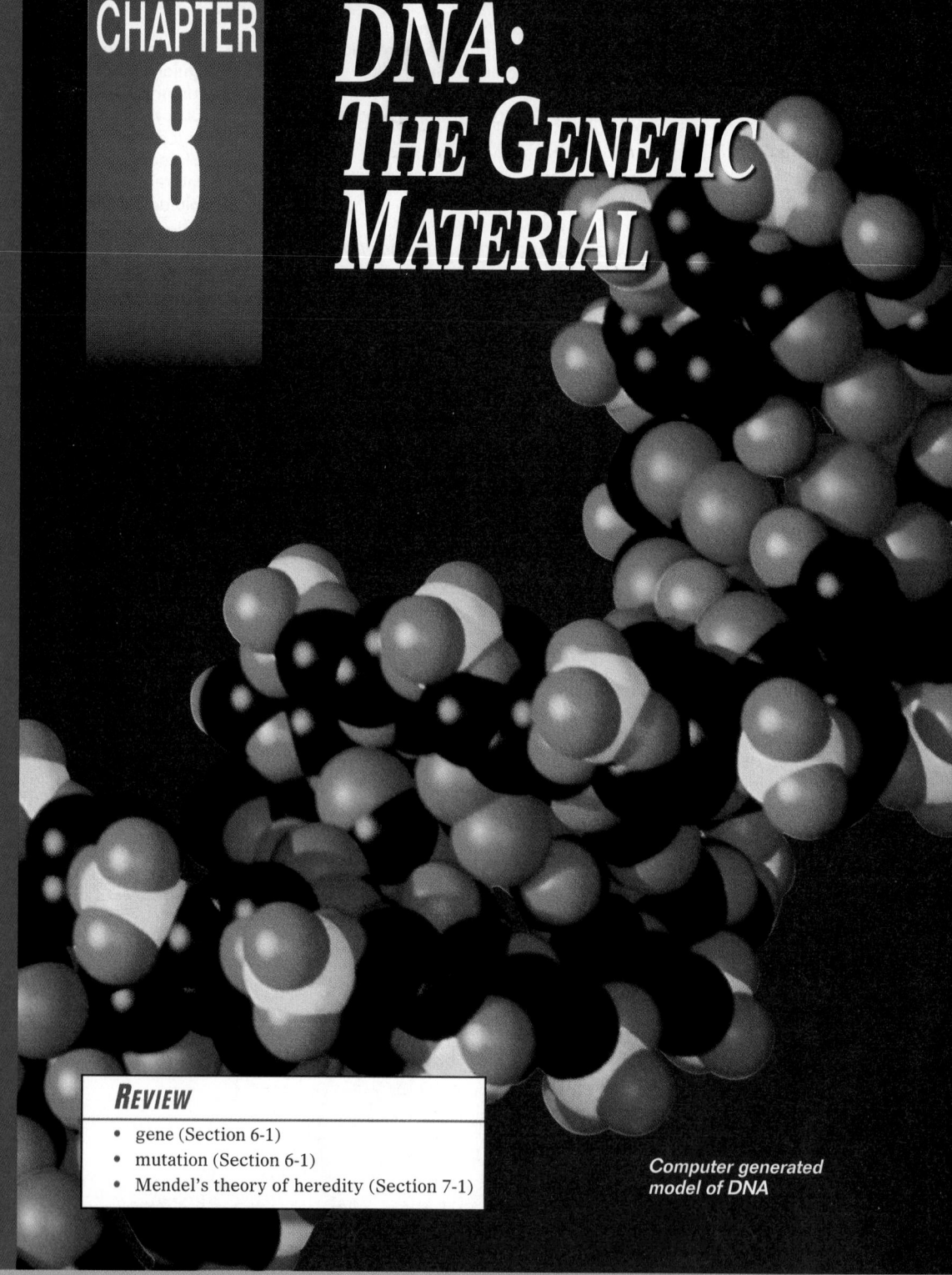

REVIEW

- gene (Section 6-1)
- mutation (Section 6-1)
- Mendel's theory of heredity (Section 7-1)

Computer generated model of DNA

Authors' Rationale

Clearly, a mastermind molecule must be at the helm of cellular activity, telling cells when to divide, what to specialize into, and what proteins to make. We now know that this molecule is DNA. The evidence for this comes from Griffith's and Hershey and Chase's historic experiments with DNA. After intro-ducing the experiments that confirmed DNA as the molecule of heredity, this chapter investigates the DNA molecule—what it is made of, how it is assembled, and how it interacts with other cellular structures. The chapter concludes with genes—what they are, where they are located, and how they work.

8-1 Identifying the Genetic Material

Mendel's experiments and results answered the question of why you resemble your parents. You resemble your parents because you have copies of their chromosomes, which contain sets of instructions called genes. Mendel's work, however, created more questions, such as, What are genes made of? Scientists believed that if they could answer this question they would understand how chromosomes function as the bearers of heredity. In this section, you will read about the events that led to the identification of the genetic material.

Section Objectives

- Describe what Frederick Griffith observed during his transformation experiments.
- Summarize the steps involved in Oswald Avery's transformation experiments, and state the results.
- Outline the experiment of Alfred Hershey and Martha Chase, and describe its significance in identifying the genetic material.

The Genetic Material: Protein or DNA?

By the 1940s, scientists knew that chromosomes consisted of deoxyribonucleic acid (DNA) and protein. While both were candidates for the role of the genetic material, proteins seemed the more likely choice because they were thought to be more chemically complex than DNA. However, little was known about DNA. It seemed far too simple a molecule to carry all the information needed for heredity. This view began to change when experiments yielded unexpected results.

Griffith Demonstrates Transformation

In 1928, an experiment completely unrelated to the field of genetics led to an astounding discovery about DNA. Frederick Griffith, a public health bacteriologist in England, was trying to prepare a vaccine against the pneumonia-causing bacterium *Streptococcus pneumoniae*. A **vaccine** *(vahk SEEN)* is a substance prepared from killed or weakened microorganisms that is introduced into the body to produce immunity. Griffith worked with two types, or strains, of *S. pneumoniae*. The first strain is enclosed in a capsule made of polysaccharides. The capsule protects the bacterium from the body's defense systems, which helps make the microorganism **virulent** *(VIHR yoo luhnt)*, or able to cause disease. The capsule also causes the bacteria to grow smooth-edged *(S)* colonies when grown in a Petri dish. The second strain of *S. pneumoniae* lacks the polysaccharide capsule and does not cause disease. When grown in a Petri dish, the second strain forms rough-edged *(R)* colonies.

SECTION 8-1

Vocabulary Preview

vaccine
virulent
transformation
bacteriophage

Opening Question

Ask students if the structure and composition of DNA is the same in almost all organisms, how does one person's or one organism's DNA differ from another's. Compare DNA to an owner's manual. Explain that DNA is an owner's manual that is written in the same language for almost all organisms. Students should understand that DNA encodes the instructions for each organism not with a different structure or language, but with a different sequence of nucleotides.

Demonstration
The Structures of Protein and DNA
Review the structures of both protein and DNA. Use photographs and figures to show that proteins are highly folded structures. Remind students that there are 20 different amino acids. Have students contrast the structure of proteins with that of DNA and its four bases. Have students discuss why scientists were reluctant to abandon the idea that protein, not DNA, is the genetic material.

Figure 8-1

Use this figure to go through each step of Griffith's experiment. Do not tell the students why Griffith obtained the results he did. Have them ask questions to which you can answer "yes" or "no." Encourage students to put the explanation into their own terms so that they will better understand the results and the implications of the experiment.

Teaching Tip
The Heat Factor

DNA survives temperatures that denature proteins. DNA can tolerate temperatures around 90°C without being altered, whereas proteins are denatured around 60°C. Ask students what effect temperature had on Griffith's work. *(The DNA of the heat-killed bacteria survived and transformed the DNA of the live bacteria.)*

Griffith knew that mice infected with the *S* bacteria grew sick and died. He also knew that mice infected with the *R* bacteria were not harmed and remained alive. To prepare a vaccine, Griffith weakened *S* bacteria by raising their temperature to a point at which the bacteria were "heat-killed," meaning that they could no longer divide. When Griffith injected the mice with heat-killed *S* bacteria, the mice lived. He then mixed the harmless live *R* bacteria and the harmless heat-killed *S* bacteria. Mice injected with this mixture of two harmless substances died. When Griffith examined the blood of the dead mice, he found that the live *R* bacteria had polysaccharide capsules. Somehow, the harmless *R* bacteria underwent a change, or **transformation** *(trans fuhr MAY shuhn)*, and became the virulent *S* bacteria. Griffith's transformation experiments are summarized in Figure 8-1.

Figure 8-1 Griffith discovered transformation when he showed that harmless bacteria could turn virulent when they were mixed with bacteria that caused disease.

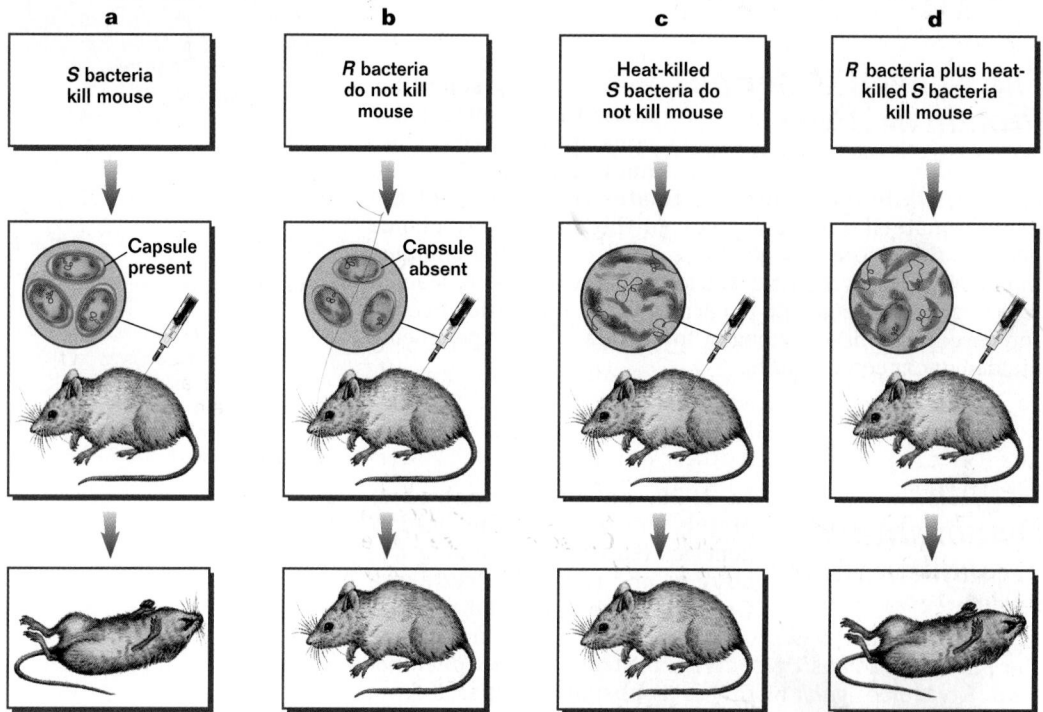

☐ CAPSULE SUMMARY

The transformation experiments of Griffith and of Avery yielded results that suggested DNA was the genetic material.

The search for the material responsible for transformation continued until 1944. In that year, Oswald Avery and his co-workers, biologists at the Rockefeller Institute in New York City, demonstrated that DNA was responsible for transformation. In an elegant series of experiments, they showed that the activity of the material responsible for transformation was not affected by protein-destroying enzymes, but was destroyed when a DNA-destroying enzyme was present. Avery and his colleagues made the announcement that the genetic material was DNA. ☐

Hershey and Chase Show That Genes Are Made of DNA

Even though Avery's experiments concluded that the genetic material was composed of DNA, many scientists remained skeptical. They thought that DNA was relevant only to certain kinds of bacteria and preferred to think that protein was the genetic material. In 1952, however, Alfred Hershey and Martha Chase, two scientists at Cold Spring Harbor Laboratory on Long Island, New York, performed an experiment that settled the controversy. In this experiment, Hershey and Chase used a type of bacteriophage (*bak TIR ee uh fayj*). A **bacteriophage** is a virus that infects bacteria. Hershey and Chase used the bacteriophage T_2, or just the T_2 phage for short, shown in Figure 8-2. The T_2 phage attaches to the surface of a bacterium and injects its hereditary information into the cell, functioning like a tiny hypodermic needle. The protein coat of the phage, including the tail, remains outside of the bacterium. Once inside the bacterium, this hereditary information replicates and directs the production of hundreds of new phages. When the new phages are mature, they burst out of the infected bacterium and attack new cells.

The only molecule in the phage that contains phosphorus is the DNA. Likewise, the only molecules that contain sulfur are the proteins in the phage's coat. The Hershey and Chase experiment, summarized in Figure 8-3, took advantage of this difference in chemical composition. Hershey and Chase grew one batch of phage in a nutrient medium that contained radioactive phosphorus (^{32}P). They grew another batch of phage in another nutrient medium that contained radioactive sulfur (^{35}S). The radioactive phosphorus became part of the phages' DNA, and the radioactive sulfur became part of the phages' protein coat. Radioactive elements were used because they can be followed, or traced, in a reaction or process. If these two kinds of labeled phages were used to infect a bacterial host, Hershey and Chase could use these labels to locate the genetic material of the phages after they infected bacteria.

In their experiments, Hershey and Chase first infected *Escherichia coli* bacteria with ^{35}S-labeled phages and allowed them to inject their genetic material. After a few minutes, the scientists mixed the infected bacteria in an ordinary kitchen blender. The violent agitation tore the phages off the surfaces of the bacteria. The investigators then used a rapidly spinning centrifuge to separate the bacteria and phages into two separate layers. When the layers were examined, Hershey and Chase found that most of the ^{35}S label was still part of the phage, meaning the protein was not injected into the bacteria. A very different result was observed in the second phase of the experiment when bacteria were infected with ^{32}P-labeled phages. When Hershey and Chase examined the layers, they found that most of the ^{32}P was now part of the bacteria, meaning the DNA had been

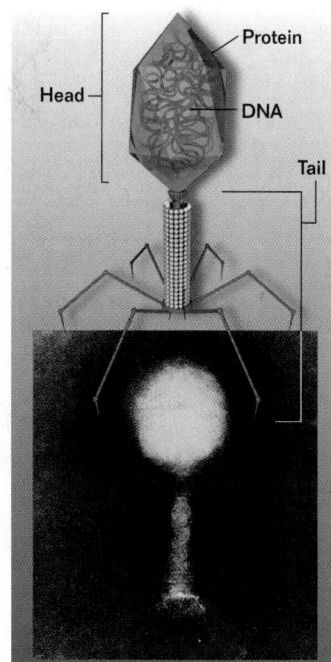

Figure 8-2 This T_2 phage, magnified 95,700X, is one of many phages that infect *E. coli*, a bacterium that normally lives in the intestines of mammals. These phages have a very simple structure—a core of DNA surrounded by a coat of protein.

Labels on figure: Protein, Head, DNA, Tail

SECTION 8-1

Teaching Tip

Hershey and Chase's Experiment

Have students design a *Graphic Organizer* like the one shown at the bottom of the page to summarize the work of Hershey and Chase. Ask students to include a paragraph that explains how the work of Hershey and Chase proved that DNA is the genetic material.

Graphic Organizer

Use this graphic organizer with ***Teaching Tip:*** **Hershey and Chase's Experiment** on this page.

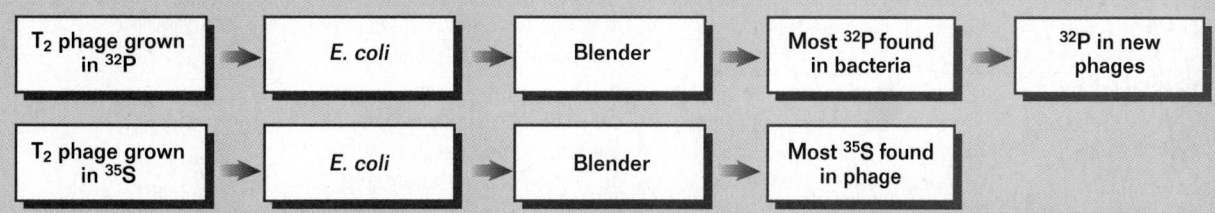

T_2 phage grown in ^{32}P → E. coli → Blender → Most ^{32}P found in bacteria → ^{32}P in new phages

T_2 phage grown in ^{35}S → E. coli → Blender → Most ^{35}S found in phage

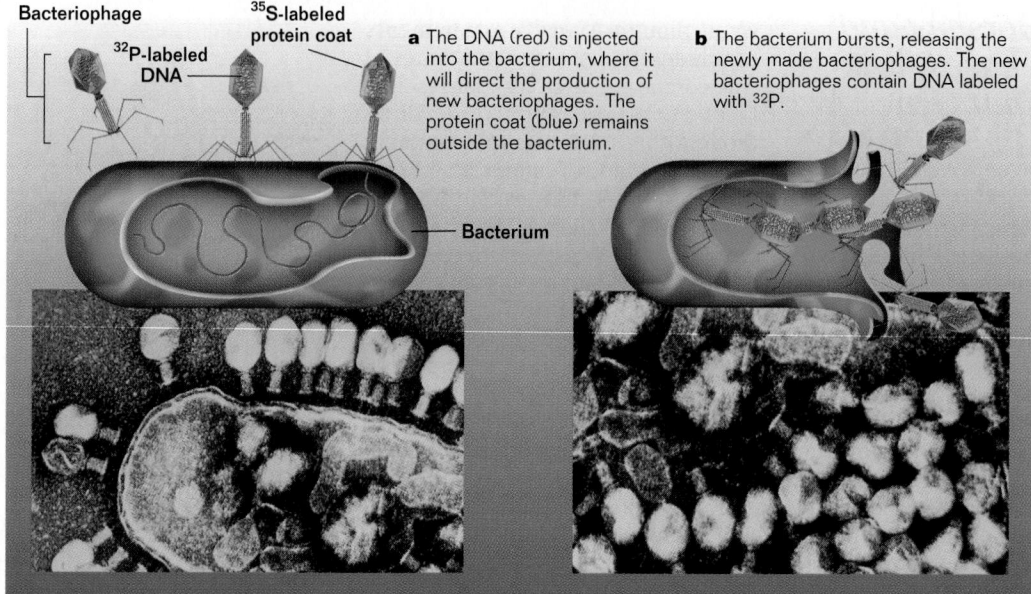

Bacteriophage · **^{32}P-labeled DNA** · **^{35}S-labeled protein coat**

a The DNA (red) is injected into the bacterium, where it will direct the production of new bacteriophages. The protein coat (blue) remains outside the bacterium.

b The bacterium bursts, releasing the newly made bacteriophages. The new bacteriophages contain DNA labeled with ^{32}P.

Bacterium

Figure 8-3 Hershey and Chase used bacteriophages to show that DNA, not protein, is the genetic material of viruses.

📘 CAPSULE SUMMARY

Alfred Hershey and Martha Chase used the bacteriophage T_2 and radioactive labels to identify DNA as the genetic material.

injected into the hosts. The new generation of phages that were produced by these bacteria also contained the radioactive DNA. The conclusion of the Hershey and Chase experiment was clear and indisputable—the genetic material is made of DNA and not protein.

These important experiments and many others have shown that DNA is the molecule that stores genetic information in living cells. As you will see in the next section, DNA is particularly well suited to this function. ◻

Section Review

1. *What did Frederick Griffith observe when he injected mice with living R bacteria and heat-killed S bacteria?*
2. *How did Oswald Avery's experiment supply evidence that DNA, and not protein, is the genetic material?*
3. *How did Alfred Hershey and Martha Chase alter the molecular structure of bacteriophage T_2 in their experiments? For what purpose did they use a kitchen blender? How did they conclude that DNA is the genetic material?*

Critical Thinking
4. *Why were many scientists reluctant to believe that DNA was the genetic material?*

8-2 The Structure of DNA

By the early 1950s, most scientists were convinced that genes were made of DNA. They began studying DNA in earnest, hoping that the mystery of heredity could be solved by understanding the structure of the molecule.

Section Objectives

- Describe the composition and structure of a DNA molecule.
- Explain the contributions of various scientists who helped determine the structure of the DNA molecule.
- Summarize the process of DNA replication.

Nucleotides Are the Building Blocks of DNA

DNA is an extraordinarily long, thin molecule made of subunits called **nucleotides** (*NOO klee uh tydz*) that are linked together like a chain. Each nucleotide is constructed of three parts: a phosphate group, a five-carbon sugar molecule, and a nitrogen base. Figure 8-4 shows how these three parts are arranged to form a nucleotide. The five-carbon sugar is called **deoxyribose** (*dee ahk see RY bohs*), from which DNA gets its full name, deoxyribonucleic acid.

While the sugar molecule and the phosphate group are the same for each nucleotide in a molecule of DNA, the nitrogen base may be any one of four different kinds. Figure 8-5 illustrates the molecular configurations of the four nitrogen bases: **adenine** (*AHD uh neen*), **guanine** (*GWAH neen*), **thymine** (*THY meen*), and **cytosine** (*SYT oh seen*). Adenine and guanine belong to a class of organic molecules called **purines** (*PYUR eenz*). Purines are large molecules, each with a double ring of carbon and nitrogen atoms. Thymine and cytosine are **pyrimidines** (*py RIHM uh deenz*). Pyrimidines have a single ring of carbon and nitrogen atoms.

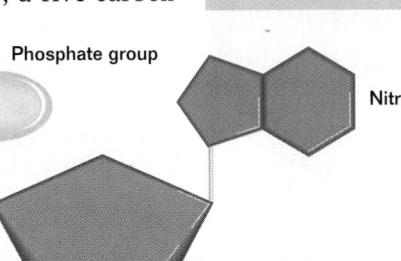

Phosphate group

Nitrogen base

Sugar (deoxyribose)

Figure 8-4 Each nucleotide is made of the sugar deoxyribose, a phosphate group, and a nitrogen base.

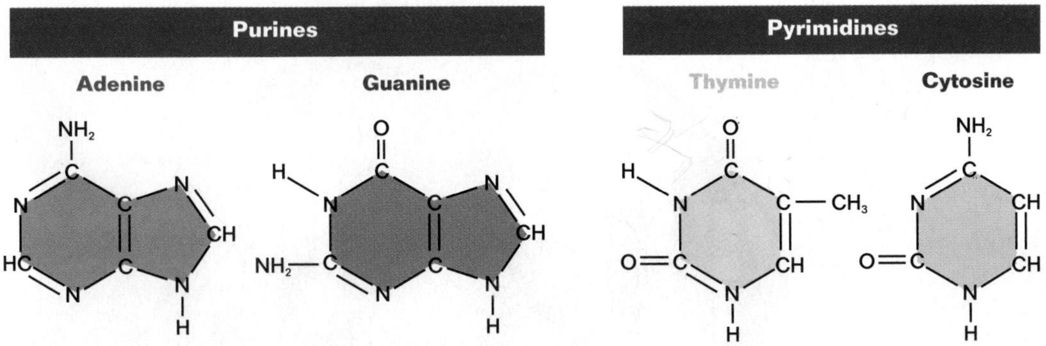

Figure 8-5 The nitrogen base in a nucleotide can be either a purine, which has a double ring of carbon and nitrogen atoms, or a pyrimidine, which has a single ring of carbon and nitrogen atoms.

Purines		Pyrimidines	
Adenine	Guanine	Thymine	Cytosine

SECTION 8-2

Vocabulary Preview

nucleotide
deoxyribose
adenine
guanine
thymine
cytosine
purine
pyrimidine
base-pairing rule
double helix
complementary
replication
helicase
replication fork
DNA polymerase

Opening Question

Ask students why DNA is often compared to a ladder. Make sure students understand how the structure of DNA is similar to that of a ladder: the rails of a ladder represent the sugar-phosphate backbone of DNA, and the rungs of a ladder represent the nitrogen bases. Ask students how a ladder is different from DNA. Point out that DNA is twisted, while a ladder is flat and that a "rung" in the DNA molecule is made of two bases, while a ladder's rung is a single unit.

Demonstration

The Parts of DNA

Draw a diagram of DNA on the chalkboard using different colors of chalk for the phosphate, deoxyribose sugar, and each nitrogen base (or use a model). Show students the structure of the molecule, and point out that certain bases always pair with each other. Ask students to determine the base-pairing rules from the diagram or model. Emphasize that the pairing allows DNA to make perfect copies of itself.

Effective Reading

Recalling the Text

Have students read about the structure of DNA in this section. Then give pairs of students index cards with the components of DNA written on them: sugar on one card, phosphate on one card, and each of the four nitrogen bases on four different cards. Each set of cards should include at least eight sugars, eight phosphates, and two of each of the nitrogen bases. Ask students to put the cards in a pattern to demonstrate the structure of DNA.

Students should build the molecule based on the information they obtained from reading this section. If they have difficulty, offer them another opportunity to read the material without the cards in front of them. Each student should be able to reconstruct a DNA molecule.

VISUAL STRATEGY

Figure 8-6

Use this figure to emphasize that the Watson-Crick DNA model was based on the work of many scientists. Point out the dark crescents at the top and bottom of the photograph. These crescents show that the purine and pyrimidine bases are stacked next to each other in a regular pattern.

Teaching Tips

How Is DNA Universal?

Point out that the structure and composition of DNA are the same for all organisms. Only the sequence of nucleotides is different.

Construct DNA

Have students work in teams to build their own DNA models. Let students choose their own construction materials to represent the sugars, phosphates, and different nitrogen bases and for connecting the components of the molecule. Once the molecules are finished, let students demonstrate how DNA is copied using their own models.

Application

Anthropology Scientists look for genes in mummy remains to obtain information about the past. They can find genes for diseases that give insight into a past population.

CAPSULE SUMMARY

The DNA molecule is a long chain of subunits called nucleotides. Each nucleotide is made of a phosphate group, the five-carbon sugar deoxyribose, and one of four different nitrogen bases.

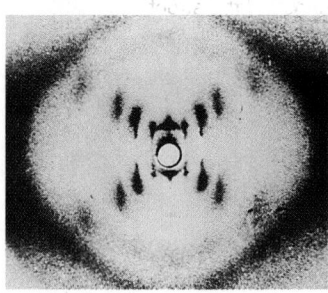
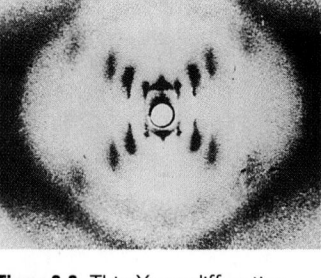

Figure 8-6 This X-ray diffraction photo of DNA fibers taken by Rosalind Franklin revealed the X pattern characteristic of a helix; the dimensions of the pattern indicated that the DNA helix had a diameter of about 2 nanometers and made a complete spiral turn every 3.4 nanometers.

In 1949, Erwin Chargaff, an American biochemist working at Columbia University in New York City, made an interesting observation about DNA. Chargaff's research data showed that for the DNA in each organism, the amount of adenine always equals the amount of thymine. Likewise, the amount of guanine always equals the amount of cytosine. However, the amount of adenine and thymine and of guanine and cytosine varied between different organisms. These findings, known as Chargaff's rules, or more commonly as the **base-pairing rules,** suggested that the precise arrangement of nucleotides within a DNA molecule specifies genes.

The DNA Molecule Is a Double Helix

The significance of Chargaff's rules became clear in the 1950s when scientists began using X-ray diffraction to study the structures of molecules. In X-ray diffraction, a beam of X rays is focused at an object. The X rays bounce off the object and are scattered in a pattern onto a piece of film. By analyzing the complex patterns on the film, scientists can determine the structure of the molecule. In the winter of 1952, Maurice Wilkins and Rosalind Franklin, two scientists working at King's College in London, developed some high-quality X-ray diffraction photographs of the DNA molecule. These photographs, such as the one in Figure 8-6, suggested that the DNA molecule resembled a tightly coiled helix and was composed of two or three chains of nucleotides.

The problem now was to discover the three-dimensional structure of the DNA molecule. The model had to take into account Chargaff's rules and the X-ray diffraction data. In 1953, James Watson and Francis Crick, two scientists at Cambridge University, used this information along with their knowledge of chemical bonding to come up with a solution. Using tin-and-wire models of molecules, they built a model of DNA with the configuration of a **double helix,** a "spiral staircase" of two strands of nucleotides twisting around a central axis. Figure 8-7 shows Watson and Crick next to their tin-and-wire model of DNA.

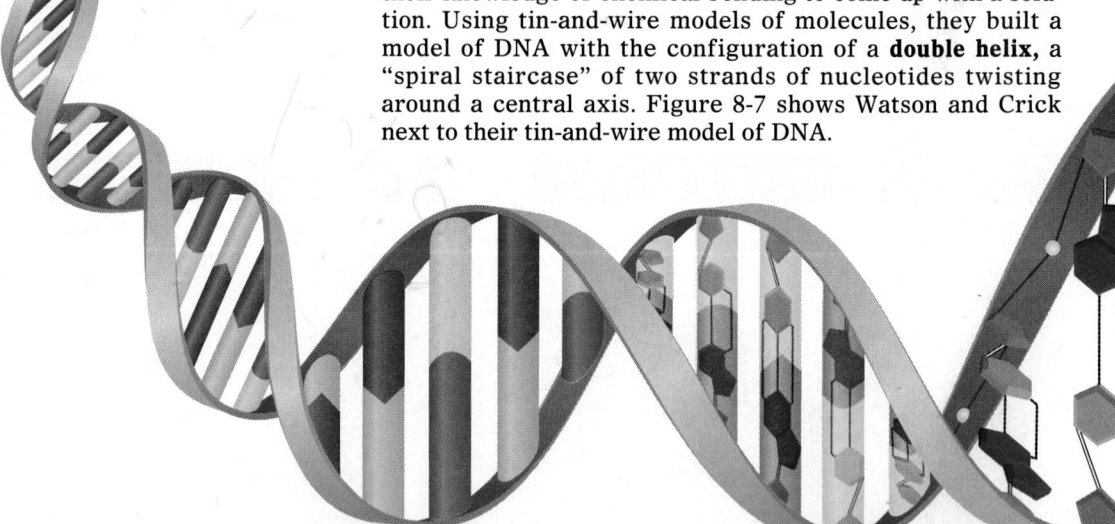

Overcoming Misconceptions

How Different Are We?

Even though people can seem very different, humans are very similar to each other biologically. The differences observed between two individuals are 2 million to 10 million nucleotide base pairs out of 3 billion—only 1 percent of the total DNA. Because these differences in base pairs are insignificant in terms of mapping the genome, the information sought by the Human Genome Project will be useful to all humans.

Figure 8-7 A breakthrough in genetics came in 1953 when James Watson and Francis Crick deduced the structure of DNA. They first built models of nucleotides and then assembled the nucleotides into a configuration that fit information from Chargaff's data and Franklin's photos.

As you can see in Figure 8-8, the double helix looks something like a twisted ladder. The sides of the ladder are constructed of alternating sugar and phosphate units, and each rung is a purine and a pyrimidine held together by hydrogen bonds. Why is a purine always paired with a pyrimidine? These base pairings are the only possible arrangement because adenine (A) can form hydrogen bonds only with thymine (T), and cytosine (C) can form hydrogen bonds only with guanine (G). Notice that this arrangement of the nitrogen bases explains Chargaff's observations. The strictness of base pairing results in two strands that are **complementary** to each

Figure 8-8 Watson and Crick's model of DNA is a double helix, a spiral staircase of two nucleotide chains that are hydrogen-bonded to each other and twisted around a central axis.

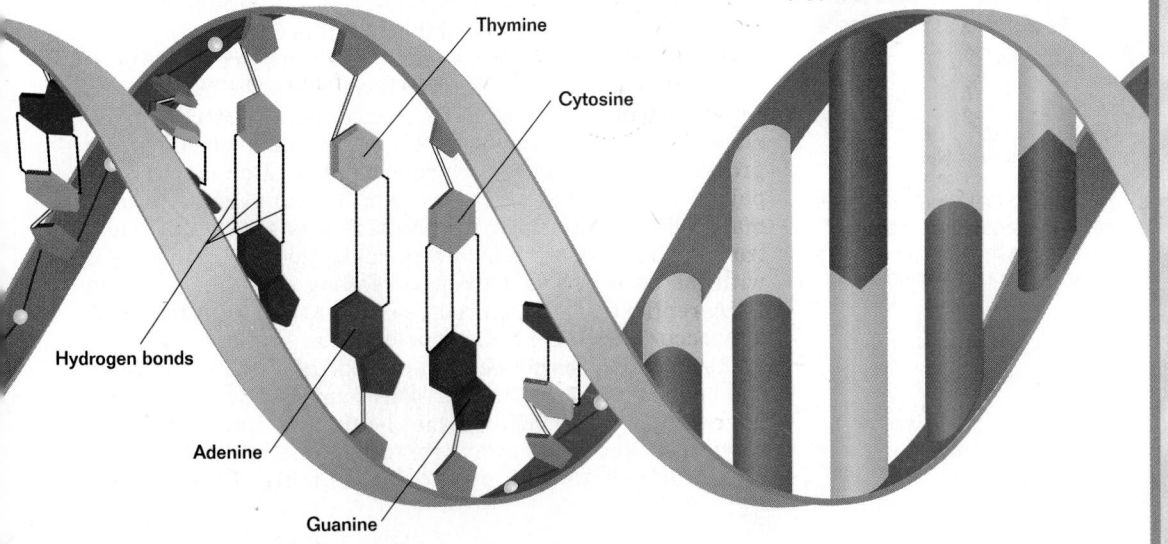

Thymine

Cytosine

Hydrogen bonds

Adenine

Guanine

SECTION 8-2

Teaching Tip

How Much Information Does DNA Contain?
Make a stack of books totaling about 10,000 pages. The stack of books represents only about one-fiftieth of the information contained in the DNA of every human cell. Correlate this with the great amount of information required to describe a human being.

CONNECTIONS
··················
Chapter 2
Chemical Bonds
The bonds along the backbone of the DNA molecule between the phosphate and deoxyribose sugar are covalent bonds. The bonds holding the two strands of DNA together in the double helix are hydrogen bonds between the purine and pyrimidine bases. Have students recall the differences between these two types of chemical bonds. Which type of bond is stronger? (*Covalent bonds are stronger than hydrogen bonds.*)

◉ VISUAL STRATEGY

Figure 8-8
Use this figure to point out how the structures of each component molecule of DNA come together in one long, helical DNA molecule. Have students verify that Chargaff's base-pairing rules are indeed followed in this DNA molecule. Ask students to identify the deoxyribose and the phosphate groups.

Historical Note

The Search for the Structure of DNA
James Watson was a biologist who studied bacteriophages, and Francis Crick was a physicist with a background in X-ray crystallography. Studying the work of their colleagues, these two scientists pooled knowledge acquired in their different backgrounds to determine the structure of DNA. However, Watson has since commented that Rosalind Franklin would have been able to determine the structure of DNA if she and Crick had simply talked with one another for an hour. Unfortunately, Franklin died of cancer at age 37 and did not live to receive public recognition for her share of the work that ultimately led to the Nobel Prize awarded to Maurice Wilkins, James Watson, and Francis Crick in 1962.

The Human Genome Project

The Human Genome Project (HGP) is an ambitious international attempt to map the entire human genome. One researcher in the limelight is Daniel Cohen of the Centre d'Étude du Polymorphisme Humain (CEPH) and Généthon in Paris. At the end of 1993, Cohen released a first draft map of almost 90 percent of the human genome. The map was pieced together with the genetic information of 10 individuals that had been cloned in mega-yeast artificial chromosomes (mega-YACs). Cohen arranged overlapping segments of the cloned DNA to make a rough map. This map involves almost 21,000 overlapping segments and shows about 2,000 specific landmarks on the genome that will contribute to the detailed mapping process.

While far from being considered the definitive work on the genome project, this map has been so eagerly sought after by other genome scientists that it has been put on the Internet for distribution. Cohen claims that the project is so immense that, if printed, it would form a pile several hundred meters high. With access to the rough map, other contributors to the Human Genome Project can fine-tune it, detect errors, and fill in gaps. The ultimate goal of the Human Genome Project is to determine the entire exact sequence of human DNA (about 100,000 genes) by 2005. Exact sequencing is going to be essential for understanding the organization and expression of genes, according to Francis Collins, director of the HGP at the National Institutes of Health. The biggest hurdle now is getting the funding for the technology that can make this possible. ☑

The two strands of the double helix are complementary because a purine can form hydrogen bonds only with a pyrimidine. Specifically, only adenine and thymine can form hydrogen bonds together, and only cytosine and guanine can form hydrogen bonds together.

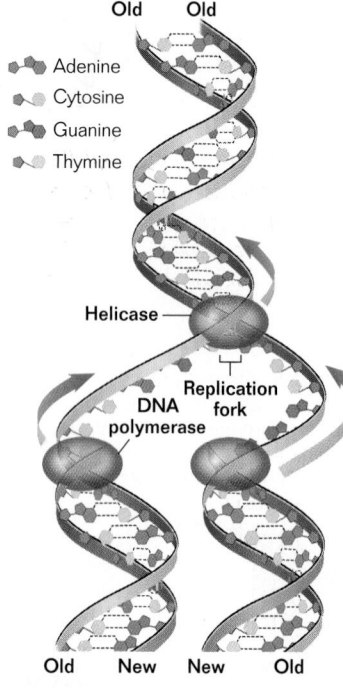

Old   Old

- Adenine
- Cytosine
- Guanine
- Thymine

Helicase

Replication fork

DNA polymerase

Old   New   New   Old

Figure 8-9 A DNA molecule replicates by separating into two strands. Each strand is used as a template to build a complementary strand. When the complementary strand is complete, it twists with the template strand to form a double helix.

❔ **Did You Know?**

Human DNA is composed of approximately 3 billion pairs of nucleotide bases, and it can copy all the information it contains (replication) in about six hours.

other; that is, the sequence of bases on one strand determines the sequence of bases on the other strand. For example, if the sequence of one strand of a DNA molecule is TCGAACT, the sequence on the other strand must be AGCTTGA. ❏

How DNA Is Copied

When the double helix was first discovered, scientists were very excited about the complementary relationship between the sequences of bases. They predicted that the complementary structure was used as a basis to make new DNA. Watson and Crick proposed that one strand could serve as a template, or surface, upon which the other strand is built. Within five years of the discovery of DNA's structure, scientists had firm evidence, from a number of different kinds of experiments, that the complementary strands of the double helix did indeed separate and serve as templates for building new DNA.

The process of synthesizing a new strand of DNA is called **replication.** Before replication can begin, the double helix must be unwound. This is accomplished by enzymes called **helicases,** which open up the double helix by breaking the hydrogen bonds that link the complementary bases. Once the two strands are separated, additional enzymes and proteins attach to the individual strands and hold them apart, preventing them from twisting.

The point at which the double helix separates is called the **replication fork** because of its Y shape. At the replication fork, enzymes known as **DNA polymerases** move along each of the DNA strands, adding nucleotides to the exposed bases according to the base-pairing rules. As the DNA polymerases move along, two new double helixes are formed, as shown in Figure 8-9. Once a DNA polymerase has begun adding nucleotides to a growing double helix, the enzyme remains attached to the strand until it reaches a signal that tells it to detach.

In the course of DNA synthesis, errors are sometimes made and the wrong nucleotide is added to the new strand. An important feature of DNA replication is that DNA polymerase has a "proofreading" role; it can add nucleotides to a growing strand only if the previous nucleotide is correctly paired to its complementary base. In the event of a mismatched nucleotide, DNA polymerase is capable of backtracking, removing the incorrect nucleotide, and replacing it with the correct one. This proofreading prevents errors in DNA replication. After proofreading, an error in the DNA may occur once per 1 billion nucleotides.

Replication does not begin at one end of the DNA molecule and end at the other. Circular DNA found in bacteria usually have two replication forks that begin at a single origin of replication and move away from each other until they meet on the opposite side of the DNA circle. Linear DNA

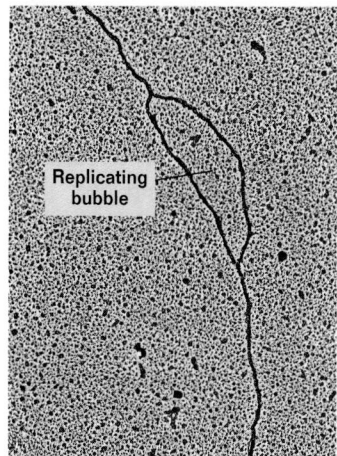

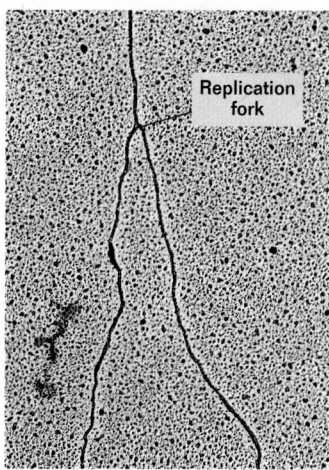

Figure 8-10 A replicating bubble enlarged 20,000X, *left*, has formed in the DNA of a mammalian cell. On the right, the bubble has enlarged into a replication fork with two double-stranded arms made of old and new DNA.

molecules found in eukaryotic organisms usually have many replication forks that begin in the middle and move in both directions, creating replicating "bubbles" along the molecule, as shown in Figure 8-10. If replication did not occur this way, it would take 16 days to copy just one DNA molecule of a fruit fly. But because approximately 6,000 replication forks exist simultaneously, replication of the fruit fly DNA takes only three minutes. Human DNA is also copied in segments, with a replication fork approximately every 100,000 nucleotides.

◻ CAPSULE SUMMARY

In DNA replication, enzymes work to unwind and separate the double helix and add complementary bases to the exposed strands. Each new double helix is composed of one old DNA strand and one new DNA strand.

Section Review

1. List the three parts of a nucleotide. Which part can be one of four molecules? Name these four different molecules.

2. Describe the structure of the DNA molecule. What kind of chemical bond holds its two strands together?

3. What two pieces of information enabled Watson and Crick to discover the structure of DNA?

4. Why are the two strands of the double helix described as "complementary"?

5. Define replication. Explain the two roles that DNA polymerases play in replication.

Critical Thinking

6. Suppose a strand of DNA has the base sequence CCA-GATTG. What is the base sequence of the complementary strand?

Vocabulary Preview

intron

exon

multigene family

transposon

Opening Question

Ask students why the discovery of DNA is sometimes called the greatest chemical discovery of modern time. Students should understand that DNA provides a set of instructions for the construction and maintenance of living things. The discovery of DNA provided a new frontier for understanding living things. Moreover, as researchers learn more about specific genes and genetic processes, there is the growing possibility that gene therapy can change the instructions.

Demonstration
Exons

Make a paper chain, mixing segments made of school colors with segments of other colors. Tell students that you want a chain made out of school colors to hang in the room. Ask them how they would go about making the chain only out of the school colors. *(Students should tell you to tear the chain apart at the sites where the segments need to be removed and attach the remaining segments together.)* Let the students follow through with the procedure. Compare their reconnecting the chain of school colors to the stitching of exons together.

Section Objectives

■ Describe how eukaryotic genes are organized, and explain why this organization is beneficial.

■ Explain the significance of a multigene family.

■ Define transposons, and identify the scientist responsible for discovering them.

■ Recognize the relationship between the p53 gene and cancer.

CONTENT LINK

You will learn how a gene is used to make a protein in **Chapter 9**.

8-3 The Structure of a Gene

Ever since it became clear that DNA is the hereditary material of the cell, scientists have peered ever more closely at DNA in an attempt to learn more about genes. New questions revolved around the nature of the information held in the DNA molecule. Scientists soon learned that genes hold information specifying how to build particular proteins. Remember that a protein is a string of amino acids. Each amino acid is coded for in the DNA. A gene affects the phenotype of an individual because of the activity of the protein that it specifies. If the protein is an enzyme that makes brown pigment, then the gene may affect hair color. Simply stated, genes are the DNA-encoded information that specifies particular proteins; each gene is made of a specific sequence of nucleotides.

Genes in Eukaryotes Are Often Interrupted

While it is tempting to think of a gene as an unbroken stretch of nucleotides that codes for a protein, this actually occurs only in the genes of bacteria. In all other organisms, much of the DNA does not code for protein. Most genes are frequently interrupted by long segments of nucleotides that have no coding information. These noncoding sequences are called intervening sequences, or **introns** (IHN trahns). The nucleotide segments that code for amino acids are called **exons** (EHK sahns) because they are expressed.

How does the cell make a protein from the fragments of genes? Inside the cell nucleus, both the introns and exons are used to make a very large molecule that is complementary to the DNA strand. Then, enzymes chop out all the introns from the molecule. The exons are "stitched" back together to form a smaller molecule that will be used to make a protein. These steps are summarized in Figure 8-11.

Why are so many genes of eukaryotic cells split up with noncoding sequences? Do these interruptions in genes fulfill some special purpose? Many biologists think that this organization of genes adds evolutionary flexibility. Each exon encodes a different part of a protein. One exon may influence which molecules an enzyme is able to recognize, while another may determine whether a protein will respond to particular signal molecules. By having introns and exons, cells can occasionally shuffle exons between genes and make new genes over time. Natural selection probably favored the intron-exon system of organization because it

Did You Know?

Family members have more in common than their appearance. They also have the same mitochondrial DNA. Mitochondrial DNA is the same in the offspring as it is in the mother because sperm do not donate mitochondria to the developing zygote. Mutations rarely occur in mitochondrial DNA. Families can be traced by this DNA because it remains essentially unchanged from generation to generation.

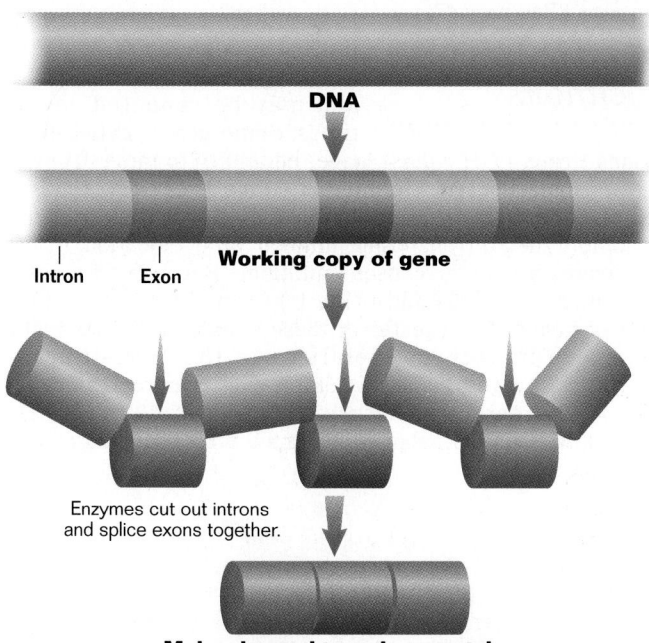

DNA

Intron | Exon

Working copy of gene

Enzymes cut out introns
and splice exons together.

Molecule used to make a protein

Figure 8-11 The genes of eukaryotic organisms contain noncoding information that must be cut out. The remaining fragments are then stitched together and used to make a protein.

enables cells to manufacture many different proteins by juggling exons between genes. The many thousands of proteins that occur in human cells appear to have arisen from only a few thousand exons. ☐

Some Genes Exist in Many Copies

Introns were not the only surprise that scientists found in eukaryotic genes. At first they assumed chromosomes carried one copy of each kind of gene—for example, one to make the enzyme that breaks down the sugar lactose, and another one to encode the protein hemoglobin that carries oxygen in your blood. They soon found out that this is not always true. Some genes in your cells exist in multiple copies, clusters of almost identical sequences called **multigene families.** Multigene families may contain as few as three or as many as several hundred versions of a gene. For example, your cells each contain 12 different hemoglobin genes in two families. Three genes are "silent," meaning they do not make proteins; five others are active only during embryonic or fetal development. Only four of the genes encode the two chains of human hemoglobins in adults. The nucleotide sequences of each of the 12 hemoglobin genes are clearly related to each other. Scientists theorize that the 12 different sequences arose from one ancestral gene, which duplicated and evolved into two gene families.

☐ CAPSULE SUMMARY

Many eukaryotic genes are interrupted by segments of DNA that do not code for proteins. Coding regions of DNA are called exons, and noncoding regions are called introns.

Teaching Tips
Finding Exons

Write a sentence on the board. Before showing it to students, insert extra letters between each word so that the sequence looks like nonsense. Ask students to remove the introns so that the exons make sense. Determine how many different ways the words (exons) can be juggled to make a sentence (a functional protein).

Recognizing Introns

Give students the following sequence of DNA: ACAATGGACAGTCAGCATTCAGGAGTCTGA. Ask them to build the strand from colored beads or paper clips using a designated color code for each nucleotide. Then, ask them to find the introns if the gene that codes for the protein is ACGACAGTCAGCAGGA. Students may find more than one set of introns to fit the data.

Multicultural Perspective

Genes and Language

Genetics professor Luigi Luca Cavalli-Sforza of Stanford University correlated family trees based on genetic migration and those based on language divergence. In one example, the linguistic classification of 400 African Bantu languages corresponds to the genetic classification of Bantu-speaking groups; thus, the linguistic Bantu category also applies to a genetically related group. Why does the correlation between genes and languages make sense?

Communities tend to speak the same language, just as they share the same gene pool. However, genes do not determine the language a person speaks; children learn to speak the language spoken by the people around them.

UNIT 2

CHAPTER 8

Teaching Tips

A Cause of Leukemia

One type of transposon, in which a piece of chromosome 22 breaks off and binds to chromosome 9, has been identified as a cause of leukemia.

p53 Suppresses Cell Growth

Up to 50 percent of all human cancers can be attributed to mutations in the p53 tumor suppressor gene. Ask students how a mutation in the p53 gene could cause cancer. *(Scientists have found that when working properly p53 can inhibit cell division by stimulating the production of a protein that blocks the cell cycle. When a mutation interrupts p53's normal behavior, cells can undergo unlimited growth, producing cancerous tumors.)*

Chapter Closure

Have students draw creative maps of what a portion of their own DNA might look like. Each map should reflect the correct structure of DNA and should contain a color-coded legend for identifying the component molecules. Ask students to prepare an "instructional manual" to go with their DNA. The manual should contain instructions for replication and should indicate which parts of the DNA are introns and exons.

Some Genes Can Jump to New Locations

In both bacteria and eukaryotes, individual genes are scattered randomly about on chromosomes and may be repeated several times. Some genes, called **transposons** *(trans POH zahns)*, have the ability to move from one chromosomal location to another. Once every few thousand cell divisions, a transposon jumps to a new location on a chromosome. When a transposon jumps to a new location, it often inactivates a gene or causes mutations. Because transposons can cause mutations and bring together different combinations of genes, the transfer of these mobile genes has had an enormous impact on evolution. In the 1950s Barbara McClintock, a geneticist at Cold Spring Harbor Laboratory, discovered transposons in the course of her studies of maize. In 1983, McClintock, shown in Figure 8-12, received a Nobel prize for her work.

Figure 8-12 The geneticist Barbara McClintock did research that revealed that transposons were the cause of the pigmented, spotted, and colorless kernels found in maize.

Some Genes Protect Cells From Mutation

Recently, scientists have discovered that some genes are able to prevent the ill effects caused by mutation. One such gene, which was the focus of intense research in 1993, is called p53 (short for protein 53). In normal cells, p53 acts as a tumor suppressor—its protein product helps coordinate a complex system of responses to the presence of damaged DNA that might otherwise lead to cancer. The p53 gene works by shutting off cell growth and division in damaged cells or, in some cases, by causing cells to self-destruct. However, when the p53 gene loses its activity by mutation, the cells of the body lose the protection it provides and, as a result, can begin to grow unchecked, signalling the development of cancer. Since 1989, researchers have found mutated forms of the p53 gene in more than 51 kinds of human tumors. As scientists learn more about the molecules that interact with p53, they will uncover new possibilities for drug development.

Section Review

1. *What is the difference between exons and introns? How are the noncoding segments of DNA extracted?*
2. *What is a multigene family? Give an example.*

Critical Thinking

3. *How do transposons and exons differ in the way they affect genes?*
4. *Why is p53 called a tumor suppressor gene?*

Answers to Section Review

1. Nucleotide segments that code for amino acids are exons; those that have no coding information are introns. Introns are removed by enzymes, and the exons are reattached.

2. A multigene family is a cluster of different versions of a gene. Gene families are thought to have evolved from a common gene. One multigene family is the hemoglobin family.

3. Transposons can inactivate genes or cause mutations.

Exons can be shuffled between genes to make new genes.

4. The p53 gene can shut off cell growth and division in damaged cells that might otherwise cause cancer.

176 *Chapter 8*

Vocabulary

adenine (169)
bacteriophage (167)
base-pairing rules (170)
complementary (171)
cytosine (169)
deoxyribose (169)
DNA polymerase (172)
double helix (170)

exon (174)
helicase (172)
intron (174)
guanine (169)
multigene family (175)
nucleotide (169)
purine (169)
pyrimidine (169)

replication (172)
replication fork (172)
thymine (169)
transformation (166)
transposon (176)
vaccine (165)
virulent (165)

Concept Mapping

Construct a concept map that shows the structure of DNA and how it is copied. In constructing your map, use the following terms: nucleotides, phosphate group, five-carbon sugar, nitrogen base, purine, pyrimidine, double helix, replication, DNA polymerases, and gene. Include additional concepts in your map as needed.

Review

Multiple Choice

1. A DNA nucleotide does *not* include a
   a. five-carbon sugar.
   b. phosphate group.
   c. double helix.
   d. nitrogen base.

2. James Watson and Francis Crick
   a. built a structural model of DNA.
   b. determined the matches between the four nitrogen bases.
   c. studied the structure of molecules using X-ray diffraction.
   d. discovered transposons.

3. The experiment of Hershey and Chase showed that
   a. bacteriophages can be injected into human cells.
   b. DNA controls heredity.
   c. bacteria undergo transformation.
   d. a vaccine for pneumonia could be produced.

4. If the sequence of bases in one strand of a DNA molecule is GCCATTG, what is its complementary sequence?
   a. CGGTAAC   c. ATTGCCA
   b. GCCATTG   d. TAAGCCG

5. In his experiments with *Streptococcus pneumoniae*, Griffith found that
   a. the DNA of the heat-killed smooth cells entered some of the rough cells.
   b. the S bacteria were transformed.
   c. the capsule does not protect the bacterium.
   d. mice infected with the R bacteria always died.

6. Multiple replication forks along the DNA molecule
   a. proofread and correct replication errors.
   b. reduce the time required for DNA replication.
   c. ensure that the new and old DNA strands are complementary.
   d. signals DNA polymerase to detach from the strand.

7. The results of Barbara McClintock's experiments with corn led her to conclude that
   a. transposons are inactive.
   b. some genes move about on a chromosome.
   c. chromosomes may contain many versions of a gene.
   d. genes are composed of sequences of nucleotides.

CHAPTER REVIEW ANSWERS

Each item in the Chapter Review is correlated by text section in the assignment guide that follows.

ASSIGNMENT GUIDE

Section	Review	Themes Review	Critical Thinking
8-1	3, 5, 9, 16, 20, 21		
8-2	1, 2, 4, 6, 11–13, 17–19, 22	25, 27, 28	29–32
8-3	7, 8, 10, 14, 15, 23, 24	26	

Concept Mapping

Map answer is shown on page 163B.

Review

Multiple Choice

Code in parentheses indicates section number and objective number.

1. c (8-2.1)
2. a (8-2.2)
3. b (8-1.3)
4. a (8-2.1)
5. a (8-1.1)
6. b (8-2.3)
7. b (8-3.3)
8. c (8-3.2)
9. a (8-1.3)
10. b (8-3.1)

Completion

11. Chargaff's (8-2.2)
12. pyrimidines, purines (8-2.1)
13. nucleotide, DNA (8-2.1)
14. exons, introns (8-3.3)
15. transposons (8-3.3)
16. Avery (8-1.2)
17. replication, new (8-2.3)

CHAPTER REVIEW ANSWERS *continued*

Short Answer

18. Chargaff's data showed that for the DNA in each organism, the amount of adenine always equals the amount of thymine and the amount of guanine always equals the amount of cytosine. He concluded that adenine pairs with thymine and guanine pairs with cytosine in the DNA molecule. (8-2.2)

19. They would not have been able to build the model. The photographs told them that the structure of DNA had to be a tightly coiled helix, composed of two or three nucleotides. (8-2.2)

20. Scientists thought that the results of Avery's work were specific to only certain kinds of bacteria and were not applicable to other kinds of organisms. In contrast, the findings reported by Hershey and Chase were considered to be generalizable to other organisms. (8-1.2, 8-2.2)

21. The chromosomes of eukaryotic cells consist of equal amounts of DNA and protein. Researchers were experimenting with bacteria, not eukaryotic cells. Additionally, they were unable to mark the DNA and protein to show which molecule was the transforming factor. (8-1.1, 8-1.2, 8-1.3)

22. Helicases open up the two strands of DNA. DNA polymerases link adjacent nucleotides on the newly synthesized strand. When the complementary strand is complete, it twists with the template strand to form a double helix. (8-2.3)

23. Introns are removed from the nucleotide sequence that will code for proteins, but the exons remain. Damage to exons can affect protein synthesis because exons are the coding regions of the DNA. (8-3.1)

24. The corn kernels are spotted because their cells contain transposons. (8-3.3)

8. Multigene families
 a. exist in prokaryotic cells.
 b. show that chromosomes carry one copy of a single gene.
 c. demonstrate convergent evolution.
 d. are nearly identical copies of the same gene.

9. What properties of the bacteriophage T_2 enabled Hershey and Chase to convincingly show that the genetic material is DNA?
 a. The protein shell of the T_2 phage attaches to a bacterium, and its DNA is injected inside.
 b. Phages containing radioactive sulfur and phosphorus are found in nature.
 c. The T_2 phage lacks protein.
 d. The T_2 phage is a eukaryote, whereas a bacterium is a prokaryote.

10. Eukaryotic genes generally differ from prokaryotic genes in that eukaryotic genes
 a. have no more than one copy of each kind of gene.
 b. are interrupted by noncoding segments of nucleotides.
 c. consist of nucleotides that code for proteins.
 d. never include exons.

Completion

11. According to _____ rules, the amount of thymine equals the amount of adenine, and the amounts of these nitrogen bases vary between different organisms.

12. Thymine and cytosine belong to a group of nitrogen bases known as _____, while guanine and adenine belong to a group of nitrogen bases known as _____.

13. A _____ consists of a phosphate group, a nitrogen base, and a five-carbon sugar. Many nucleotides joined together form a _____ molecule.

14. In eukaryotic cells, nucleotide sequences that code for amino acids are _____; those that are removed from the gene are _____.

15. In 1983, Barbara McClintock received a Nobel Prize for her discovery of _____, which are sometimes called "jumping genes."

16. Using protein- and DNA-destroying enzymes, _____ demonstrated that the transforming factor discovered by Griffith is DNA.

17. During the process of _____, the two strands of a double helix separate and are used as templates to build _____ strands of DNA.

Short Answer

18. What was the nature of the research done by Erwin Chargaff? What conclusions did Chargaff draw from his work?

19. Would Watson and Crick have been able to build their double helix model of DNA in 1953 without the photographs they obtained from Maurice Wilkins and Rosalind Franklin? Explain.

20. The majority of scientists were not convinced by the findings of Oswald Avery and his colleagues that DNA is the primary heredity mechanism in all cells. However, they were convinced by the findings reported by Hershey and Chase. Explain.

21. Why are the experiments of Griffith and Avery and his colleagues considered important but not conclusive steps in showing that DNA is the genetic material?

22. How is a DNA molecule copied in the process of replication?

23. Why is damage to exons very likely to affect the synthesis of a protein, while damage to introns is not?

24. A scientist observed that corn kernels that usually have no pigment were spotted with pigment. He knew that the spots were not due to the normal expression of the gene for kernel pigment. What genetic phenomenon could explain his observations?

Themes Review

25. A common evolutionary origin is suggested by the presence and function of DNA in all organisms. (8-2.1, 8-2.2)

26. New genes can be created by the shuffling of exons between genes. This situation enables cells to manufacture different proteins. (8-3.1)

27. The double-stranded structure of DNA allows it to replicate itself and to pass identical information to daughter cells. The sequences of four nitrogen bases allow DNA to form a code. (8-2.1)

28. A nucleotide consists of a five-carbon sugar molecule, a phosphate group, and a nitrogen base. (8-2.1)

Themes Review

25. Evolution What does the presence of DNA in all living organisms and its function as the genetic material suggest about how life evolved on Earth?

26. Evolution Why do scientists believe that evolutionary flexibility is increased by having exons separated by introns?

27. Structure and Function What are some ways in which the structure of a DNA molecule is related to its function?

28. Levels of Organization A molecule of DNA consists of two long strands of nucleotides bonded together. What molecules make up a nucleotide?

Critical Thinking

29. Making Inferences In what ways was the process of communication vital to the work of Watson and Crick?

30. Making Inferences Rosalind Franklin died of cancer at an early age. How might her work with X-ray diffraction have contributed to her death?

31. Interpreting Graphics The data in the table below were collected by using thin-layer chromatography. They show the amount of each type of base, by percentage, in several DNA samples. Do the data support Chargaff's rules? Explain.

	G	C	A	T
mold	15.2	34.3	14.9	35.1
plant	19.7	41.2	19.5	42.1
mollusk	17.4	32.3	17.9	34.7
reptile	12.9	35.6	13.2	36.7
mammal	14.6	39.5	13.8	37.6

32. Making Predictions A scientist extracted 4.6 picograms (or 10^{-12} grams) of DNA from mouse muscle cells. She also extracted 4.6 picograms of DNA from an equal amount of mouse liver cells. How much DNA could be extracted from the same amount of mouse kidney cells? how much from the same amount of mouse sperm? Explain.

Activities and Projects

33. Research and Writing Find out about DNA fingerprinting and how a DNA fingerprint can be used by police to solve a crime.

34. Research and Writing Shortly before Watson and Crick presented their model of DNA to the scientific community, Linus Pauling proposed a three-stranded model of DNA. Find out how the work of Linus Pauling influenced Watson and Crick's thinking about the structure of DNA. What led Watson and Crick to propose a double helix model of DNA rather than to accept Pauling's model?

35. Multicultural Perspective Discuss the kinds of obstacles that you think women scientists such as Rosalind Franklin and Barbara McClintock faced in the past. What kind of discrimination might they have had to overcome?

Readings

36. Read the article "DNA's Stroke of Genius," in *New Scientist*, volume 138, pages 21–23. What did reviewers of *The Double Helix* say about the uniqueness of Watson and Crick's contribution?

37. Read the article "Happy Birthday, Double Helix," in *Time*, March 15, 1993. Who gathered at Cold Spring Harbor Laboratory and what were they celebrating? Describe three ways their discovery has impacted the world.

Activities and Projects

33. DNA is extracted from white blood cells and snipped into fragments using enzymes. Electrophoresis is then used to align the DNA pieces by size. Next, the areas of DNA that can be used to establish identity are radioactively tagged. Using X-ray film, a pattern of bands is produced. The pattern of bands in a column is the DNA fingerprint, or profile. Police can use DNA fingerprinting to solve crimes by comparing DNA extracted from cells collected at a crime scene with DNA extracted from the white blood cells of a suspect.

34. Watson and Crick rejected Pauling's model because the product of mitosis is two daughter cells (not three) and the width of the helix is small.

35. Answers will vary. Women scientists have faced lower salaries; fewer opportunities for career advancement; longer work hours; less access to the best laboratory equipment, facilities, and grant funding; difficulties publishing and receiving credit for their work; and exclusion from universities, especially graduate schools.

Readings

36. Many reviewers, including Conrad Waddington, John Lear, and Erwin Chargaff, thought that Watson and Crick's contribution was overrated and that other scientists would have provided similar insights. Other reviewers, like Peter Medawar, wrote that the uniqueness of Watson and Crick's contribution was in its completeness—in the fact that they presented their contribution thoroughly in one historic paper.

37. James Watson, Francis Crick, and other scientists gathered to celebrate the 40th anniversary of the discovery of DNA's structure. Their discovery transformed biology, set the stage for the biotechnology industry, and affected business, industry, agriculture, food processing, and medicine.

Critical Thinking

29. Watson and Crick were able to use previously published information and concepts from many sources to build their DNA model. (8-2.2)

30. It is now known that radiation from X rays can cause cancer. It is suspected that Rosalind Franklin's cancer was a result of her work with X-ray diffraction. (8-2.2)

31. The data do not support Chargaff's rules because the amount of A should equal the amount of T and the amount of G should equal the amount of C. However, the data do show that the amounts of the bases vary between different organisms. (8-2.2)

32. The scientist can extract 4.6 picograms from the mouse liver cells. Only 2.3 picograms of DNA can be extracted from sperm cells because gametes will have one-half as much DNA as body cells. (8-2.1)

LABORATORY Investigation Chapter 8

Preparation

Obtain enough plastic soda straws to make 48 3-cm sections for each group. If students will be making their own straw sections, they will need a metric ruler and a pair of scissors. Students will also need 48 small, standard-sized paper clips and 48 colored pushpins (12 red, 12 blue, 12 yellow, and 12 green). The supplies for this lab can be saved and used again in Chapter 10.

Procedural Notes

- Caution students to be careful of the sharp tip on the push pins.
- This investigation should take one class period to complete.
- Make sure each group of students has enough materials to complete the activity before they begin.
- Account for all materials at the end of the lab, and have students store the nucleotides (push pins) by color.
- When completed, save the materials for use in Laboratory Investigation Chapter 10.

Background Answers

1. DNA provides the instructions that direct the activities in a cell.

2. DNA is a long, thin molecule shaped in a double helix. It is composed of nucleotide subunits.

3. Construct a DNA model and "unzip" it. Fill in unpaired base pairs with free nucleotides.

4. What is the relationship between the structure of DNA and the process of DNA replication?

DNA and Its Structure

OBJECTIVE

Construct and analyze a model of DNA.

PROCESS SKILLS

- constructing, identifying, and manipulating a model

MATERIALS

- plastic soda straws, 3 cm sections (48)
- metric ruler
- scissors
- permanent marker
- 48 pushpins (12 red, 12 blue, 12 yellow, and 12 green)
- 48 paper clips

BACKGROUND

1. What provides the instructions that direct the activities in a cell?

2. What is the structure of DNA?

3. How can can you use models of DNA to demonstrate the process of DNA replication?

4. Write your own question to explore in your lab report or notebook.

TECHNIQUE

1. Work with a classmate to complete the investigation.

2. **CAUTION: Pointed objects can cause injury if not properly used.** Cut the soda straws into 3-cm pieces to make 48 segments.

3. Insert a pushpin midway along the length of each straw segment. Push a paper clip into one end of each straw segment until it touches the pin.

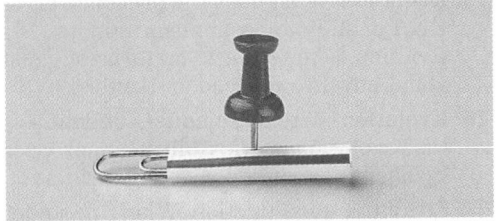

4. Keep the pins in a straight line, and insert the paper clip of a blue pushpin segment into the open end of a red pin segment. Add segments of straw to the red segment end in the following order: green, yellow, blue, yellow, blue, yellow, green, red, red, and green. Use the permanent marker to label the blue segment at the end "top." This strand of segments is one-half of your first model.

5. Construct the other half of your first model beginning with a yellow pin segment. Keep the pins in a straight line. Link segments together in the following order: green, red, blue, yellow, blue, yellow, blue, red, green, green, and red. Label the yellow segment at the end "top."

6. Place the strands parallel to each other on the table with the "top" blue pin of the

Records

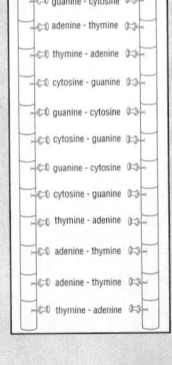

Original DNA model

Process of replication

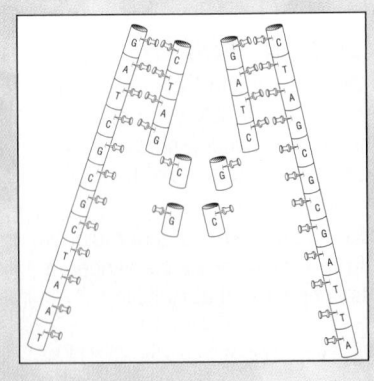

first strand facing the "top" yellow pin of the second strand.

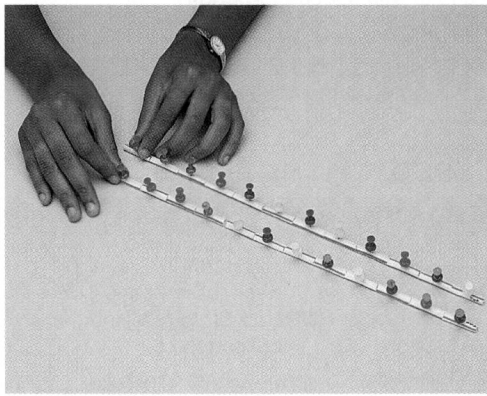

7. Demonstrate replication by simulating a replication fork at the top pair of pins. Add the remaining straw segments to complete a new DNA model. Be sure to use the base-pairing rules.

8. Assign the following bases to the pushpin colors: red = adenine, blue = guanine, yellow = cytosine, and green = thymine. Make a sketch of the original DNA model in the **Records** section of your report. Sketch the process of DNA replication in the **Records** section of your report. Also, briefly summarize the procedure you followed.

INQUIRY

1. How many nucleotides did the original DNA model contain?

2. What color pin is always across from a blue pin?

3. What color pin is always across from a green pin?

4. Write the base-pair order for the DNA molecule you created, using the following code: red = adenine, blue = guanine, yellow = cytosine, and green = thymine.

5. How does the replicated model of DNA compare to the original model of DNA?

ANALYSIS

1. What is the name given to the point at which replication starts on a DNA molecule?

2. What would the complementary bases be if one side of a DNA molecule had the bases adenine, cytosine, cytosine, thymine, thymine, and adenine?

3. Speculate what would happen if the base pairs in the replicated model were not in the same sequence as the original model.

4. Of what benefit is DNA replication to humans?

5. What are the advantages of having DNA remain in the nucleus?

FURTHER INQUIRY

Write a new question that could be explored as a new investigation. The following are examples:

How do DNA molecules differ among various species of animals and plants? How are they similar?

How might transposing genes produce cancer?

Inquiry Answers

1. The original DNA model contained 24 nucleotides, with 12 nucleotides on each side.

2. A yellow pin is always across from a blue pin.

3. A red pin is always across from a green pin.

4. The base-pair order for the DNA molecule is guanine-cytosine, adenine-thymine, thymine-adenine, cytosine-guanine, guanine-cytosine, cytosine-guanine, guanine-cytosine, cytosine-guanine, thymine-adenine, adenine-thymine, adenine-thymine, thymine-adenine.

5. The replicated model of DNA is identical to the original model.

Analysis Answers

1. The point at which replication starts on a DNA molecule is called the replication fork.

2. The complementary base pairs would be thymine, guanine, guanine, adenine, adenine, and thymine.

3. If the base pairs were out of sequence, the resulting organism may not be viable or could die as a result of a mutation.

4. DNA replication allows cells to pass genetic information on to other cells so that new cells have the information they need to survive and grow. Without replication, humans would not be able to survive, and there would be no inheritance.

5. The nucleus provides an isolated and protected area for storing genetic information. By having DNA in the nucleus, eukaryotes can house more DNA and undergo more complex cell division than prokaryotes.

Each block represents about 45–50 minutes of instructional time. The resources cited in a block are coded to a specific program component using the Key on the next page.

Lecture Concepts

Lesson Resources

9-1 From Genes to Proteins pp.182–189

BLOCKS ① and ②

a. Processing the information in DNA into proteins involves a sequence of events known as gene expression. Three types of RNA are essential for gene expression.

b. The first step in using DNA to direct the making of a protein is transcription, in which RNA polymerase assembles a molecule of RNA on a DNA template.

c. Transcription produces three types of RNA: messenger RNA (mRNA), transfer RNA (tRNA), and ribosomal RNA (rRNA).

d. All organisms have a genetic code made of three-nucleotide sequences called codons. Codons correspond to particular amino acids and to stop signals. The genetic code is nearly universal.

Lecture Resources
- Opening Question, page 182
- Demonstrations: Differentiating DNA and RNA, Comparing Structures, Triplet Spelling
- Visual Strategies : Figures 9-2, 9-3
- Teaching Transparencies: 34, 37

Classwork Options
- Biotechnology D1: Staining DNA and RNA

- Transcription versus Replication, page 185
- Holt Biology Videodiscs: Lesson 9-1

BLOCK ③

a. The equipment for translation is found in the cytoplasm, where a cell's supply of tRNA is kept. A cell's cytoplasm also contains thousands of ribosomes, the protein-making factories of the cell. Each ribosome is made of over 50 different proteins and several rRNA segments.

b. Translation begins when mRNA binds to the smaller subunit of a ribosome in the cytoplasm. After mRNA is bound, tRNA carries amino acids to the ribosome according to the three-base codons.

c. The amino acids are joined to form the chain that makes up the primary structure of the protein.

Lecture Resources
- Teaching Transparencies: 35, 36
- Demonstration: Protein Synthesis
- Holt Biology Videodiscs: Lesson 9-1

Classwork Options
- Translation, page 187
- Synthesizing Proteins, page 188

- Effective Reading, page 188
- Laboratory Investigation: DNA and Proteins, pages 200–201

Assignment Options
- Section Review, page 189
- Chapter Review, questions 1–5, 10–12, 18, 19, 23, 25–27, 30-32

9-2 Regulating Gene Expression pp.190–193

BLOCKS ④ and ⑤

a. The *lac* operon is a cluster of genes that enables a bacterium to build the proteins needed for lactose metabolism only when lactose is present.

b. Gene expression in eukaryotes involves mechanisms that must uncoil the appropriate regions of DNA in specific cells at specific times.

c. Enhancers are regions of DNA that stimulate transcription of certain genes in eukaryotic organisms.

Lecture Resources
- Demonstrations: Turning Genes On and Off, The Operon Model
- Overcoming Misconceptions, page 190
- Teaching Transparencies: 27A
- Holt Biology Videodiscs: Lesson 9-2

Classwork Options
- Inquiry Skills Development B6: Effect of Environment on Gene Expression

- Inquiry Skills Development B7: Gene Expression
- Gene Expression in Eukaryotes and Graphic Organizer, pages 192–193
- Interactive Explorations in Biology Laboratory Manual E4: Gene Regulation

Assignment Options
- Section Review, page 193
- Chapter Review, questions 6–8, 13, 14, 20–22, 28, 29

9-3 Genes, Mutation, and Cancer pp.194–196

BLOCK ⑥

a. Mutations are changes in the genetic message in an organism. A mutation may change the structure of a chromosome or may alter single nucleotides.

b. Cancer is the uncontrolled growth of cells that begins when the genes that control cell growth become mutated.

Lecture Resources
- Demonstration: Mutations
- Medicine, page 196
- Gene Therapy in the Brain, page 196
- Holt Biology Videodiscs: Lesson 9-3

Classwork Options
- Chernobyl, page 195
- Closure, page 196

Assignment Options
- Section Review, page 196
- Chapter Review, questions 9, 15–17, 24

Review/Enrichment

BLOCK ❼

- ■ Study Guide: Concept Review and Pretest

Assignment Options
- ⊙ Teaching Resources CD-ROM: Supplemental Reading Worksheets and Test, *The Double Helix*
- ■ Activities and Projects, questions 33–36
- ■ Readings, question 37

Assessment Options

BLOCK ❽

Traditional Assessment
- ■ Chapter 9 Test
- ⊙ Test Generator/Assessment Item Listing: Software item bank for preparing customized chapter tests.

Portfolio Assessment
Select student reports for one or more laboratory experiments from this chapter. *The Direct Observation Checklist* on the *BioSources Teaching Resources CD-ROM* should be completed during a laboratory or other cooperative learning experience.

Answer to Concept Map

The following is one possible answer to the Concept Mapping exercise on page 197.

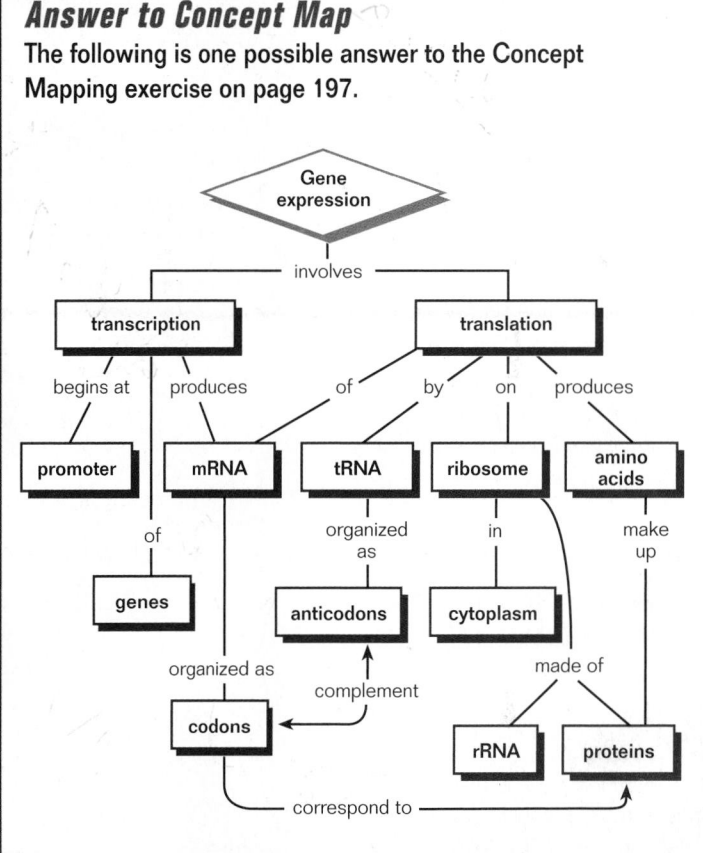

Holt Biology Videodiscs

Holt Biology Videodiscs gives you a powerful tool for teaching, review, and assessment. *Concepts of Biology* adds interactive lesson plans and extensive tools for customization using CD-ROM technology.

CONCEPTS OF BIOLOGY

Use the following topic from *Concepts of Biology* to help you teach this chapter:

- ■ Topic 7: How Genes Work

For further information regarding the media on the videodiscs, see *Holt Biology Videodiscs Teacher's Correlation Guide to Biology: Principles and Explorations.*

CHAPTER
9

GENE EXPRESSION

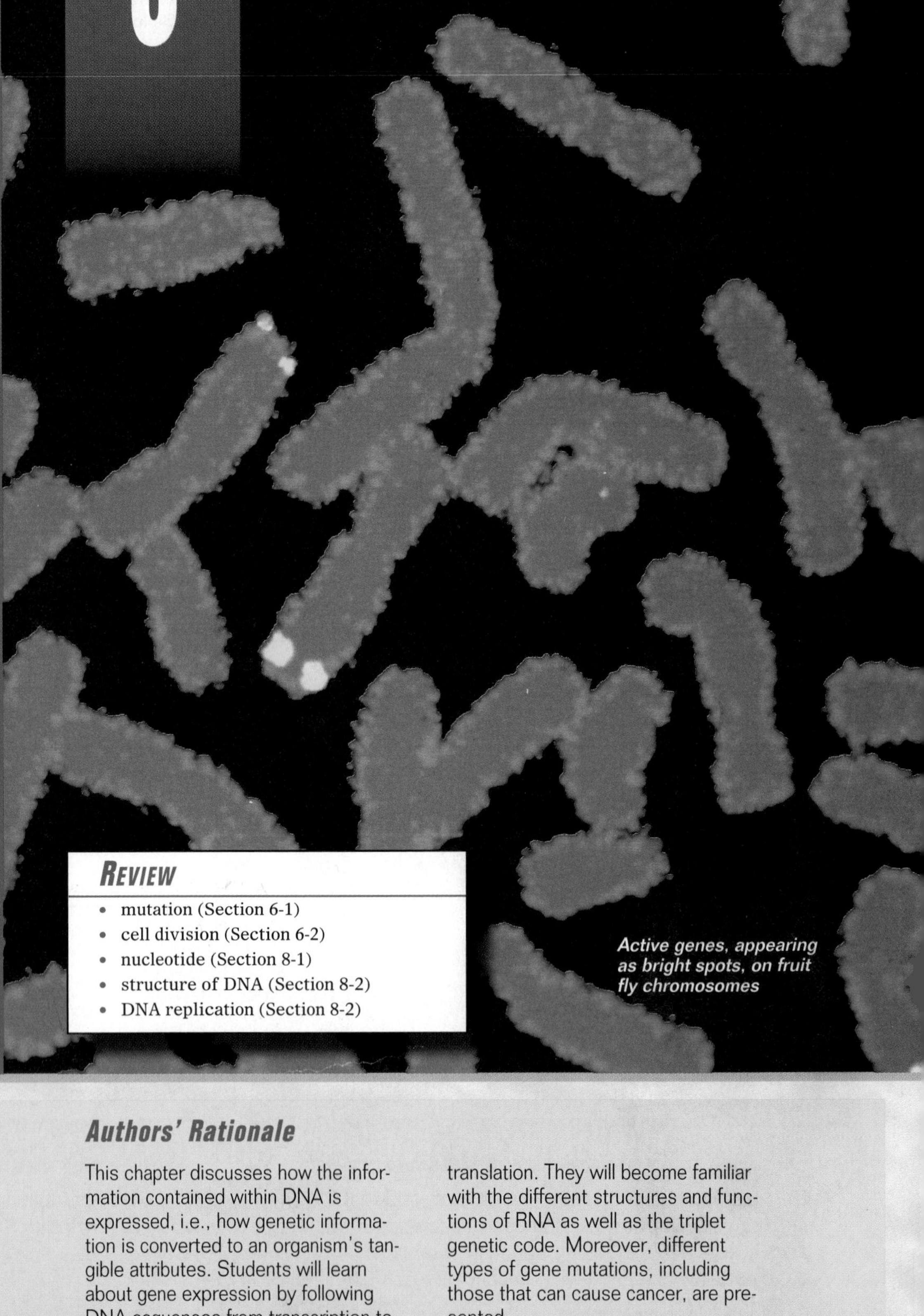

Chapter Themes

Homeostasis *Cells waste energy if they constantly transcribe every gene. Genes are often turned off in cells unless a protein the genes produce is needed. Feedback systems regulate which genes are turned on and which genes are turned off in a cell; thus, feedback systems prevent overproduction of unnecessary proteins in the cell. If viruses or environmental agents alter a feedback system, cancers can result.*

Structure and Function

Control of gene expression is required for cells to become specialized. Liver cells differ from brain cells because different genes are expressed within different cells. Gene regulation allows cells to differentiate in the early developmental stages of an organism.

Tapping Prior Knowledge

- How does DNA replicate itself?
- Why does DNA replication occur in the nucleus?

Opening Question

Have students divide a sheet of paper into five columns. They should label the five columns with the following headings: DNA, mRNA, amino acid, tRNA, and complementary DNA (the strand of DNA complementary to the original DNA strand). In the first column, assign the following sequence of nucleotides to be written down the page vertically: AACTACGGTCTCAGCACTCCC. In the fifth column, have students write the complementary sequence to the given DNA. (TTGATGCCAGAGTCGT-GAGGG) Have students complete the other columns as you go through this chapter.

REVIEW

- mutation (Section 6-1)
- cell division (Section 6-2)
- nucleotide (Section 8-1)
- structure of DNA (Section 8-2)
- DNA replication (Section 8-2)

Active genes, appearing as bright spots, on fruit fly chromosomes

Authors' Rationale

This chapter discusses how the information contained within DNA is expressed, i.e., how genetic information is converted to an organism's tangible attributes. Students will learn about gene expression by following DNA sequences from transcription to translation. They will become familiar with the different structures and functions of RNA as well as the triplet genetic code. Moreover, different types of gene mutations, including those that can cause cancer, are presented.

9-1 From Genes to Proteins

The discovery that genes are made of DNA led to more questions about heredity. How is the information in DNA used to determine an organism's characteristics? Investigators soon found a partial answer: an organism's traits are determined by proteins that are built according to the plans specified in its DNA. The next question was, How does DNA determine the nature of a protein?

The Path of Genetic Information

Proteins are not built directly from genes. Your cells preserve hereditary information by transferring the information in genes into sets of working instructions for use in building proteins. The working instructions of the genes are made of molecules of **ribonucleic** *(ry boh noo KLAY ihk)* **acid,** or RNA. RNA, like DNA, is a nucleic acid. However, it differs from DNA in three ways. First, RNA consists of a single strand of nucleotides instead of the two strands that form the DNA double helix. Second, RNA contains the five-carbon sugar ribose *(RY bohs)* rather than the sugar deoxyribose. And third, RNA has a nitrogen-containing base called **uracil** *(YUR uh sihl)*—abbreviated as U—instead of the base thymine found in DNA. Like thymine, uracil is complementary to adenine.

RNA is present in cells in three different forms, each of which has a different function: messenger RNA (mRNA), ribosomal RNA (rRNA), and transfer RNA (tRNA). All three types of RNA are essential for processing the information from DNA into proteins, a process called **gene expression**. Gene expression occurs in two stages—transcription and translation. In **transcription**, the information in DNA is transferred to mRNA. In **translation**, the information in mRNA is used to make a protein. The path of genetic information is summarized in Figure 9-1.

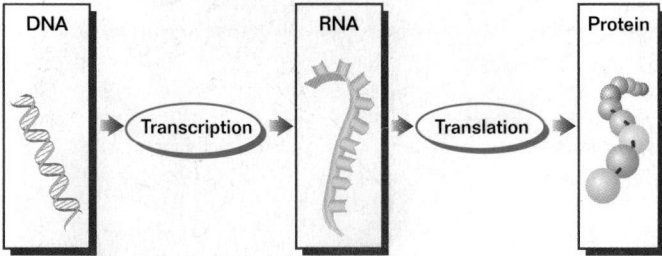

Figure 9-1 During gene expression, the information in DNA is first "rewritten" (transcribed) as a molecule of mRNA and then "deciphered" (translated) and used to build a protein.

SECTION 9-1

Vocabulary Preview

ribonucleic acid

gene expression

transcription

translation

RNA polymerase

promoter

terminator

messenger RNA

transfer RNA

ribosomal RNA

codon

genetic code

anticodon

A site

P site

Section Objectives

- Describe three roles of RNA in gene expression.
- Summarize the process of transcription.
- Recognize the evolutionary significance of the genetic code.
- Outline the major steps of translation.

Opening Question

Display pictures of two different people. Ask students to think of ways in which these two individuals differ. After they have listed several possibilities, ask students what role DNA played in the appearances of the two people. Help students understand that DNA codes for proteins that determine the attributes of each individual.

Demonstration

Differentiating DNA and RNA

Divide the chalkboard into two sides. Label one side "yes" and one side "no." Write the following terms on large sheets of paper: deoxyribose, thymine, adenine, ribose, uracil, anticodon, ribosome, nucleus, Watson and Crick, and replication. Terms that relate to DNA are "yes" terms, and those that relate to RNA are "no" terms. Mix the terms together. Hold up one term at a time, saying only "yes" or "no." Start with a "yes" term, followed by a "no" term. Say nothing other than "yes" or "no." Allow students to determine that the "yes" terms relate to DNA. Ask students to give a "yes" example when they know the concept.

Deoxyribose   **Ribose**

Thymine   **Uracil**

Transcription: Making RNA

The first step in using DNA to direct the making of a protein is transcription, the process that "rewrites" the information in a gene in DNA into a molecule of mRNA. In eukaryotic organisms, transcription occurs inside the nucleus; in prokaryotic organisms, it takes place in the cytoplasm. Transcription begins when an enzyme called **RNA polymerase** binds to the beginning of a gene on a region of DNA called a promoter. A **promoter** is a specific sequence of DNA that acts as a "start" signal for transcription. After RNA polymerase binds to a promoter, the enzyme starts to unwind and separate the double helix's two strands, exposing the DNA's nitrogen-containing bases. Like DNA replication, transcription uses DNA bases as a template for making a new molecule (RNA). In transcription, however, only one of the two strands of DNA serves as a template.

Once a portion of the DNA double helix has separated, RNA polymerase moves along the bases of the template strand like a train on a track, always in the same direction. The enzyme reads each nucleotide and pairs it with a complementary RNA nucleotide, as shown in Figure 9-2. In eukaryotic cells, the RNA nucleotides are found in the nucleus; in prokaryotic cells, they are in the cytoplasm. Transcription follows the same base-pairing rules as DNA replication except that uracil, rather than thymine, pairs with adenine. When the RNA nucleotides are added, they are linked together with sugar-to-phosphate covalent bonds. As RNA polymerase works its way down the strand, a single strand of RNA grows and

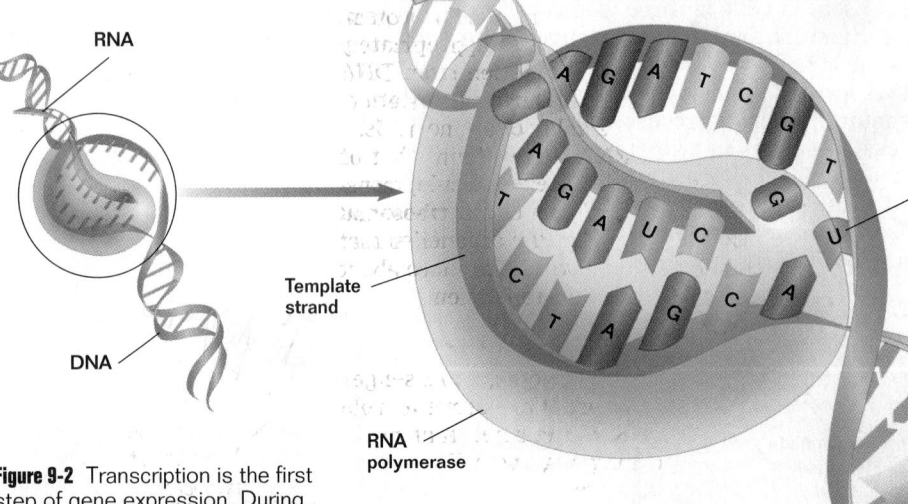

Figure 9-2 Transcription is the first step of gene expression. During transcription, the enzyme RNA polymerase rewrites the genetic information in DNA as a molecule of messenger RNA.

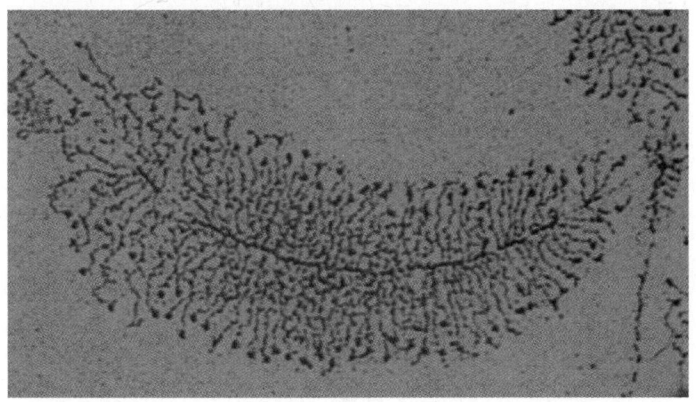

Figure 9-3 A cell can produce large amounts of a particular protein by transcribing a single gene with several molecules of RNA polymerase. The multiple copies of RNA being made in this electron micrograph give the DNA a feathery appearance.

dangles off the enzyme like a tail, as shown in Figure 9-3. Behind RNA polymerase, the two strands of DNA close up by forming hydrogen bonds between complementary bases, re-forming the double helix.

Transcription proceeds at a rate of about 60 nucleotides per second until the RNA polymerase reaches a stop signal on the DNA called a terminator. A **terminator** is a sequence of bases that tells the RNA polymerase to stop adding nucleotides. At this point the enzyme detaches from the DNA and releases the RNA molecule into the cell for the next stage of gene expression.

Three Types of RNA

Transcription manufactures three types of RNA: messenger RNA (mRNA), transfer RNA (tRNA), and ribosomal RNA (rRNA). **Messenger RNA** is an RNA copy of a gene used as a blueprint for a protein. When a cell needs a particular protein, a specific mRNA is made. Messenger RNA is appropriately named because it *carries* hereditary information from DNA and *delivers* it to the site of translation. During translation, mRNA serves as a template for the assembly of amino acids.

The functions of tRNA and rRNA differ from that of mRNA. **Transfer RNA** acts as an interpreter molecule, translating mRNA sequences into amino acid sequences; **ribosomal RNA** plays a structural role in ribosomes, the organelles that function as the sites of translation. You will learn more about these two molecules when you learn about translation.

Processing mRNA in Eukaryotic Cells

After transcription is completed in eukaryotes, a messenger RNA molecule must be processed before it can serve its role in building a protein. In Chapter 8 you learned that many eukaryotic genes are split into coding regions called exons, which are interrupted by noncoding regions called introns. Both exons and introns are transcribed into mRNA. Before mRNA leaves the nucleus, the introns are cut out. The exons are joined to form a single molecule of mRNA, which leaves the nucleus through a pore, and enters the cytoplasm. ❑

SECTION 9-1

 VISUAL STRATEGY

Figure 9-3
Ask students to suggest a protein that could ultimately be produced from the gene being transcribed. Once they have suggested a protein, ask them what type of cell they think requires the transcription of this gene. Possibilities could be insulin in a pancreas cell, growth hormone in a pituitary cell, or hemoglobin in a blood cell. Other possible answers include a transport protein for a membrane in any cell or a protein for photosynthesis in a plant cell.

Teaching Tip
Transcription Versus Replication
Ask students to contrast mRNA transcription with DNA replication. Students should include that uracil pairs with adenine instead of thymine. In transcription, only the portion of DNA designated between the promoter and termination sequences is transcribed, whereas in replication the entire DNA sequence is copied. The mRNA transcribes only one strand, called the "sense" strand, of DNA, but in replication both strands are copied.

❑ *CAPSULE SUMMARY*

During transcription, RNA polymerase assembles a molecule of RNA on a DNA template. Messenger RNA, one of three types of RNA, carries genetic information from the DNA of genes to the site of translation.

 Did You Know?

A fourth type of RNA, small nuclear RNA (snRNA), usually resides in the nucleus attached to nuclear proteins. A snRNA–nuclear protein complex is called a small nuclear ribonucleoprotein (snRNP), or a "snurp." Although scientists are not sure of the function of snRNPs, they think snRNPs are important because snRNA nucleotide sequences are highly conserved across mammalian, amphibian, and insect species. In fact, snRNPs may play a role in removing introns and splicing exons together.

CONNECTIONS

Chapter 8
The Structure of DNA

Although Watson and Crick published their structure of DNA in *Nature* in 1953, the procedure of protein synthesis had to be worked out to prove how DNA could control the cell. Putting together these final puzzle pieces won Watson, Crick, and Wilkins the Nobel Prize in medicine and physiology in 1962. Rosalind Franklin was not a recipient because she died of cancer in 1958. The rules of the Nobel Foundation do not allow the Nobel Prize to be granted posthumously.

 VISUAL STRATEGY

Table 9-1
Use this table to explain how mRNA codons code for specific amino acids. Ask students to identify the codons for phenylalanine (Phe) and alanine (Ala). (*phenylalanine—UUU, UUC; alanine—GCU, GCC, GCA, GCG*) Emphasize that the genetic code is degenerate, that is, more than one codon can specify a given amino acid. This fact was demonstrated by Francis Crick using experiments involving frameshift mutations. (Frameshift mutations are discussed in Section 9-3.)

☐ CAPSULE SUMMARY

All organisms have a genetic code made of three-nucleotide sequences called codons. Codons correspond to particular amino acids and to stop signals. The genetic code is nearly universal.

Table 9-1 In the genetic code, each amino acid is coded for by a three-nucleotide sequence called a codon. The first base in a codon is found along the left side of the chart, the second base is at the top, and the third base is found along the right side of the chart. Because any of four different nucleotides may be used at each of the three positions, there are 64 different possible codons (4 × 4 × 4 = 64) in the genetic code.

The Genetic Code

After transcription, the genetic message is ready to be translated from the language of RNA to the language of proteins. The instructions for building a protein are written as a series of three-nucleotide sequences called **codons** (*KOH dahnz*). Each codon along the mRNA strand either corresponds to an amino acid or signifies a stop signal. Recall from Chapter 2 that amino acids bond together to form a protein chain. You will learn more about the role of "stop" signals in translation later in this section.

From trial-and-error experiments, biologists worked out which codons correspond to which amino acids. In 1961, Marshall Nirenberg, an American biochemist, deciphered the first codon by making artificial mRNA that contained only the base uracil. When the mRNA was added to a test tube containing amino acids, it was translated into a polypeptide made of the amino acid phenylalanine. From this, Nirenberg learned that the codon UUU is the instruction for the amino acid phenylalanine. Soon, more elaborate techniques enabled scientists to decipher codons consisting of more than one kind of base. Table 9-1 presents the entire **genetic code**, the amino acids and stop signals that are coded for by each of the possible mRNA codons.

The genetic code is nearly universal. With few exceptions, it is the same in all organisms. For example, the codon GUC codes for the amino acid valine in bacteria, in eagles, in dogs, and in your own cells. Thus, it appears that all life forms had a common evolutionary ancestor with a single genetic code. The only exceptions biologists have found to this rule are in the ways cell organelles (such as mitochondria and chloroplasts) and a few microscopic protists (ciliates such as *Paramecium*) read stop codons. ☐

	U	C	A	G	
U	Phe	Ser	Tyr	Cys	**U**
	Phe	Ser	Tyr	Cys	**C**
	Leu	Ser	stop	stop	**A**
	Leu	Ser	stop	Trp	**G**
C	Leu	Pro	His	Arg	**U**
	Leu	Pro	His	Arg	**C**
	Leu	Pro	Gln	Arg	**A**
	Leu	Pro	Gln	Arg	**G**
A	Ile	Thr	Asn	Ser	**U**
	Ile	Thr	Asn	Ser	**C**
	Ile	Thr	Lys	Arg	**A**
	Met	Thr	Lys	Arg	**G**
G	Val	Ala	Asp	Gly	**U**
	Val	Ala	Asp	Gly	**C**
	Val	Ala	Glu	Gly	**A**
	Val	Ala	Glu	Gly	**G**

Translation: Making Proteins

The equipment for translation is located in the cytoplasm, where a cell keeps its supply of transfer RNA (tRNA). A tRNA molecule is a single strand of RNA folded into a compact shape with three loops, illustrated in Figure 9-4. One of the loops has a three-nucleotide sequence called an **anticodon** (*an tee KOH dahn*). It is called an anticodon because the three-nucleotide sequence is complementary to one of the 64 codons of the genetic code. This enables tRNA to bind to mRNA through hydrogen bonding. In most organisms, there is no tRNA molecule with an anticodon complementary to the codons UAG, UAA, or UGA, which is why these codons act as stop codons. Opposite the anticodon on a tRNA molecule is a site at which the molecule carries an amino acid. The amino acid that a tRNA molecule can carry corresponds to a particular codon.

A cell's cytoplasm also contains thousands of ribosomes, the protein-making factories of the cell. A ribosome, illustrated in Figure 9-5, is composed of two subunits, which are bound together only when they are involved in translation.

A ribosome has three binding sites that play important roles in translation. One binding site holds mRNA so that its codons are accessible to tRNA molecules. The other two binding sites recognize tRNA. The **A site** holds a tRNA molecule that is carrying its specific amino acid. The **P site** holds a tRNA molecule that is carrying its specific amino acid attached to the growing protein chain. As you can see in Figure 9-5, these two binding sites are next to each other on the ribosome.

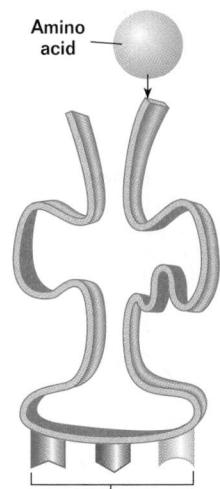

Figure 9-4 The unique shape of tRNA enables it to act as an interpreter molecule. Its anticodon complements a specific codon and corresponds to a specific amino acid.

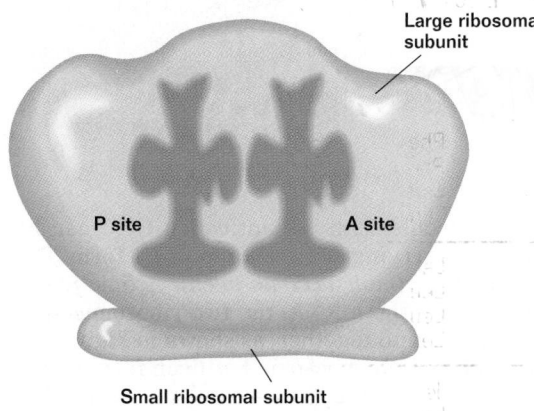

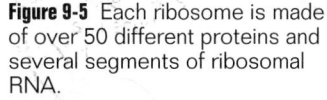

Figure 9-5 Each ribosome is made of over 50 different proteins and several segments of ribosomal RNA.

Assembling the Protein

Translation begins when mRNA binds to the smaller ribosomal subunit in the cytoplasm. The mRNA is now oriented so that the "start" codon, a codon that signals the beginning of a protein chain, is sitting in the P site. Research has shown that in most cases, the start codon has the sequence AUG.

SECTION 9-1

Teaching Tips
Translation
Have students complete the chart they started in the Opening Question on page 182. First, they should complete the mRNA and the tRNA columns. Have students justify the arrangement of the columns on the page. (*The complementary DNA sequence and tRNA are next to each other on the page. If the chart is completed correctly, they are the same, except for thymine in the complementary DNA and uracil in tRNA.*) Finally, have students complete the amino acid column using Table 9-1 on page 186. Save the chart for further discussion in Section 9-3.

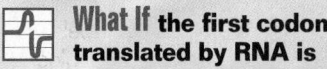

Translation Is Regulated by Structure
Ask students how translation is ultimately regulated by the structure of DNA. Students should realize that in translation, as in transcription, the tRNA anticodon nucleotides join only with complementary mRNA codons.

What If the first codon translated by RNA is UGA? Transcription will not begin because UGA is a termination, or stop, codon. This codon signals the synthesis to stop. Assembly of proteins will not begin unless the ribosome reads a start codon, such as AUG, to signal the beginning of a protein chain.

Did You Know?

A sequence of 100–200 adenines known as a poly-A tail is added to one end of the RNA before it leaves the nucleus. The poly-A tail may help the cell recognize mRNA and allow it to pass from the nucleus without being digested. Also, the tail could help with translation. Scientists identify mRNA in bioassays by radioactively labeling the poly-A tail.

Demonstration
Protein Synthesis

Make cutouts of DNA, mRNA, and tRNA out of bright-colored paper. Walk students through the steps of protein synthesis by manipulating the structures on the board. Use a strip of one color of paper for DNA, and write the nucleotide sequence given in the **Opening Question** on page 182 (AACTACG-GTCTCAGCACTCCC) on the strip of paper. Make a complementary DNA strip using the same color. Show how enzymes unzip the DNA and produce mRNA. Use a different color for mRNA. Move the RNA out of the nucleus to the cytoplasm. Show how the mRNA brings together the shapes representing ribosomal subunits that are in the cytoplasm. Use tRNA shapes with detachable anticodons and amino acid sequences to show the elongation of the newly made protein chain and its eventual termination. Show how the ribosomal units separate when a stop codon enters the A site, while the amino acid sequence remains intact.

Teaching Tip
Synthesizing Proteins

Have students use beads, squares of different colors of paper, or candy to sequence the DNA strand given in the **Opening Question** on page 182: AAC-TACGGTCTCAGCACTCCC. Provide each group with butcher paper to represent the cell. The sheets should be large enough to cover the table. Have students designate a corner of the paper as a part of the nucleus. From the DNA, students should make mRNA in the nucleus. Provide a template for ribosomal units and for tRNA, as well as construction paper and scissors for cutting out the structures. Have students move the mRNA into the cytoplasm through a nuclear pore and bind it to the ribosomal subunits with AUG in the P site. Have them act out translation and determine the amino acid sequence specified by the given DNA.

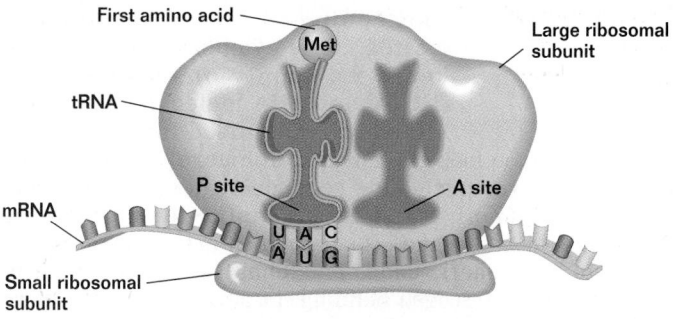

Figure 9-6 When translation begins, mRNA is bound to a complete ribosome so that the start codon is positioned in the P site, ready for the first amino acid of the protein chain.

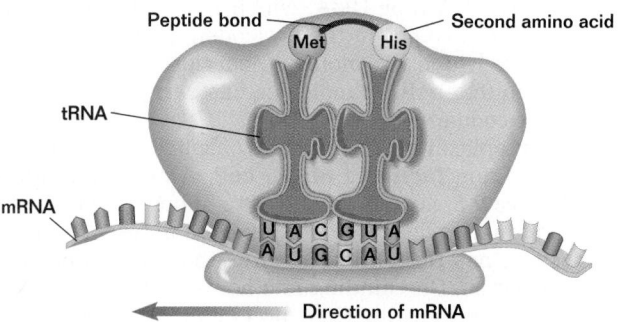

Figure 9-7 When both sites on the ribosome are filled, a peptide bond can form and link the amino acids.

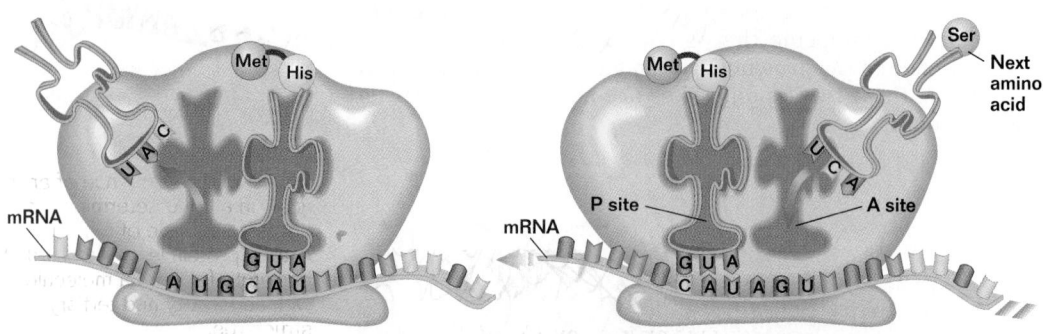

Figure 9-8 After a peptide bond is formed, mRNA shifts on the ribosome so that a new codon is present in the A site.

A tRNA molecule with the anticodon UAC can bind to the start codon, carrying with it a modified form of the amino acid methionine (*muh THY uh neen*). A functional ribosome is formed when the mRNA, the two ribosomal subunits, and the first tRNA bind together, as shown in Figure 9-6.

Once a complete ribosome has been formed, the codon in the vacant A site is ready to receive the next tRNA. A tRNA molecule with the complementary anticodon arrives and binds to the codon, carrying its specific amino acid with it. Now, both the A site and the P site are holding amino acids. Next, an enzyme helps form a peptide bond between the adjacent amino acids, forming the first link of the protein chain, shown in Figure 9-7. Afterward, the tRNA in the P site detaches and moves away from the ribosome, leaving behind its amino acid. The tRNA in the A site moves over to fill the

Effective Reading
Analogies

After students have read Section 9-1, have them devise analogies to represent the different molecules and the events necessary to carry out gene expression. For example, the Pentagon could represent the DNA molecule. Like DNA, it never leaves its site. Yet, it sends messages through generals, who take the messages to their troops. Like generals, molecules of mRNA carry messages from a "command center," the cell's nucleus. Other analogies could be the organization of the school, the message that runs a car, or the efficiency of a beehive. Students should explain their analogies with a chart or paragraph.

vacant P site, with the protein chain in tow. Because the anti-codon remains attached to the codon, the tRNA molecule and mRNA molecule move across the ribosome into the P site as a unit. As a result, a new codon is present in the A site, ready to receive the next tRNA and its amino acid, as shown in Figure 9-8. Amino acids are carried to the A site and bonded to the protein chain until the end of the mRNA sequence is reached. At this point, a stop codon is encountered for which, as you know, there is no anticodon. With nothing to fit into the empty A site in the ribosome, the ribosome complex falls apart and the newly made protein is released into the cell, as shown in Figure 9-9. ❑

❑ CAPSULE SUMMARY

After mRNA has bound to a ribosome, tRNA carries amino acids to the ribosome according to the three-base codons. The amino acids are joined to form a protein chain.

Teaching Tip
Codons and Anticodons
Give students the following list of anticodons: CCC, UGA, AGG, GUU, CAA. Ask students to determine the codons and amino acids specified by each anticodon. (GGG—glycine, ACU—threonine, UCC—serine, CAA—glutamine, GUU—valine) Make sure students understand that each anticodon corresponds to both a codon and an amino acid.

CONNECTIONS

Chapter 2
Protein Structure
Of the four levels used to describe protein structure, primary structure (the sequence of amino acids) is the structure specified by genes. When the proteins are correctly synthesized, they fold to form secondary and tertiary structures.

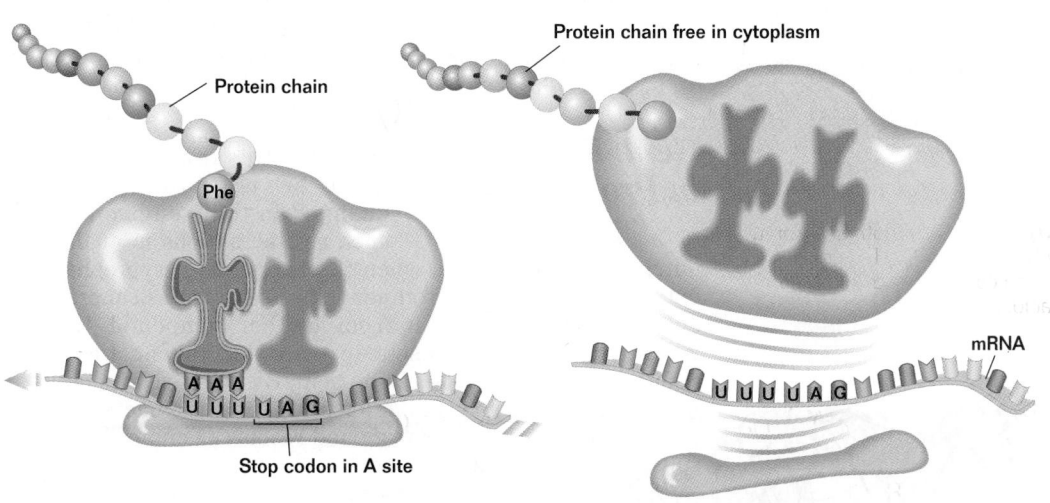

Figure 9-9 The sequence of amino acids in a gene determines the primary structure of a protein. After a protein is synthesized, it begins to fold into a molecule with secondary and tertiary structures.

Section Review

1. *Define the two stages of gene expression. Could translation take place if transcription did not occur? Explain why or why not.*

2. *What is the evolutionary significance in describing the genetic code as "nearly universal"?*

Critical Thinking

3. *Why is the term* transcription *appropriate for describing the process of making RNA?*

4. *Why is the term* translation *appropriate for describing the synthesis of proteins?*

5. *How does the structure of tRNA enable it to act as an interpreter molecule?*

6. *How do a ribosome's A site and P site enable it to function as a protein-making factory?*

Opening Questions

Ask students to recall differences and similarities between prokaryotes and eukaryotes. Lead students to recall that prokaryotes lack organelles, particularly nuclei, and that prokaryotic DNA is circular and found along the inner plasma membrane of the cell. Ask students how the differences between prokaryotes and eukaryotes relate to gene expression. Students should conclude that the simplicity of prokaryotes makes them easier for studying gene expression.

Demonstration

Turning Genes On and Off

To demonstrate the turning on and off of genes, plug in a lamp and turn it on and off. Have students compare the lamp to the *lac* operator. What corresponds to the inducer? *(the lamp switch in the "on" position)* What corresponds to the promoter? *(electricity)* What corresponds to the operator? *(the lamp socket)* What corresponds to the repressor? *(the lamp switch in the "off" position)*

9-2 Regulating Gene Expression

Section Objectives

- Discuss why the regulation of gene expression in living organisms is important.
- Explain how the lac operon is affected by the presence and absence of lactose.
- Explain how the steroid hormone estrogen acts to regulate gene expression.

Being able to translate a gene into a protein is only part of gene expression. Every cell must also be able to regulate when particular genes are used. Imagine if every instrument in an orchestra played at full volume constantly. All you would hear is noise! No orchestra plays that way, because music is not noise—it is the controlled expression of sound. Similarly, every function that a living organism carries out is the controlled expression of genes, each used at the proper moment to achieve precise effects.

Gene Regulation in Prokaryotes

In order to survive, bacteria must be able to adjust to changes in their environment, such as fluctuations in available nutrients. For example, when the amino acid tryptophan is not present in the environment, the bacterium *Escherichia coli* must manufacture it from another compound. Later, when tryptophan is present in the environment, the bacterium stops making the amino acid. By being able to adjust its metabolism to changes in its environment, a bacterium is saved from wasting its energy and resources on producing a substance that is readily available.

Scientists first studied gene expression in bacteria. An example of gene regulation that is well understood at the molecular level is found in *E. coli*. When you consume a milk product, the disaccharide lactose (milk sugar) is soon present in your intestinal tract and available to the *E. coli* living there. Before a bacterium can absorb the lactose, it must first make beta-galactosidase, an enzyme that cleaves lactose into glucose and galactose.

As with tryptophan, it is in a bacterium's best interest to focus its energy on using available nutrients. Therefore, *E. coli* should make beta-galactosidase only when lactose is present. To understand how this happens, you must first understand the basic mechanism that controls gene regulation in prokaryotes—the operon. An **operon** *(AHP uhr ahn)* is a cluster of genes that codes for proteins with related functions. The gene for beta-galactosidase is part of a group of genes called the **lac operon.** The *lac* operon is divided into several different regions, which are shown in Figure 9-10. One region is the promoter, the RNA polymerase binding site that signals the beginning of a gene. Another region is made of structural genes, the genes that code for polypeptides. Between the promoter and the structural genes is a region of DNA called the **operator**. Because of its position in

Overcoming Misconceptions

Efficient Transcription

Students often assume that all genes are transcribed at the same time in cells. Ask students to determine the significance of the mechanism of gene regulation found in the *lac* operon. Explain that genes are only transcribed as they are needed. Make sure students understand that a cell conserves energy and resources when it transcribes only the genes it needs.

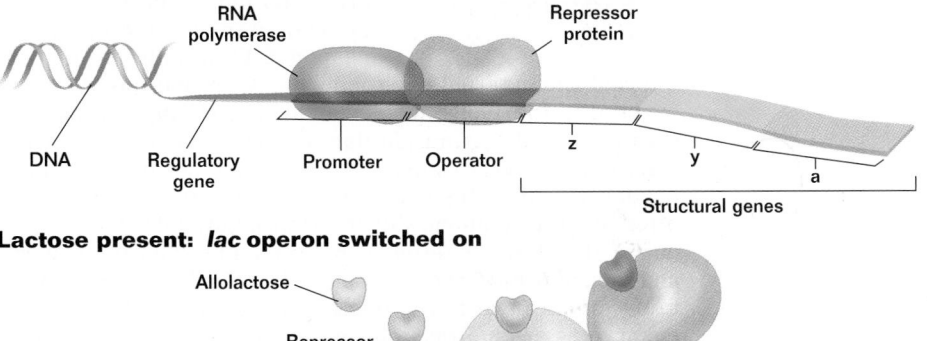

Lactose absent: *lac* operon switched off

RNA polymerase
Repressor protein
DNA
Regulatory gene
Promoter
Operator
z
y
a
Structural genes

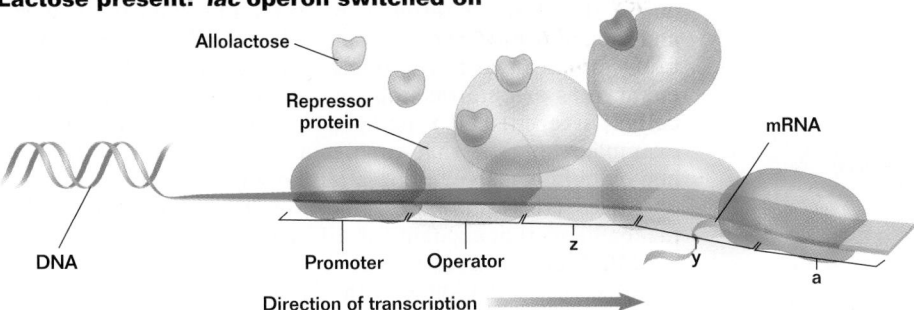

Lactose present: *lac* operon switched on

Allolactose
Repressor protein
mRNA
DNA
Promoter
Operator
z
y
a
Direction of transcription

Figure 9-10 The *lac* operon has three structural genes that code for proteins used to metabolize lactose. When lactose is absent from a bacterium's environment, the repressor protein is bound to the operator and the *lac* operon is switched off. When lactose is present, the repressor protein detaches from the operator and the *lac* operon is switched on.

Demonstration

The Operon Model

Cut shapes from colored paper to represent the components of the *lac* operon shown in Figure 9-10. These shapes should be large enough for students to see from their seats. Use the shapes to demonstrate the process by which genes are turned on and off in the *lac* operon. Emphasize the role of the feedback system in gene regulation.

Teaching Tip

A User-Friendly Operon

To help students better understand the *lac* operon, have them develop a strategy for teaching the mechanism for gene regulation to other members of the class. Their strategies could include cartoon characters, a story, or a simple flowchart that could make the concept easier to grasp. Have students share their ideas with the class.

Application

Health Lactose intolerance is characterized by a person's inability to digest lactose, the principal sugar found in cow's milk and dairy products derived from cow's milk. The lining of a person's intestine requires the enzyme lactase to break down lactose. Many people become less tolerant of lactose with age because the production of lactase in their intestines diminishes. Instead of avoiding dairy products, lactose-intolerant people can treat dairy products with a commercial preparation to break the lactose into disaccharides that can be easily digested.

the operon, the operator is able to control RNA polymerase's access to the structural genes; it acts like a switch, turning the operon on or off.

What determines whether the *lac* operon is in the "on" or "off" mode? When no molecules are bound to the operator, the *lac* operon is switched on; RNA polymerase can bind to the promoter, move across the operator, and transcribe the structural genes. The *lac* operon is switched off when a protein called a **repressor** is bound to the operator. A bound repressor creates a barrier, preventing RNA polymerase from transcribing the structural genes (although RNA polymerase can still bind to the promoter).

Transcription can resume when the repressor is removed by a molecule called an **inducer**. In the case of the *lac* operon, the inducer is allolactose, a molecule that is made from lactose when it enters the cell. When allolactose is present in the cell, it binds to the repressor and changes the shape of this protein. As a result, the repressor falls off the operator, as shown in Figure 9-10. Now, RNA polymerase has access to the structural genes. The bacterial cell can begin building the proteins it needs to metabolize lactose.

When lactose is not present in the intestinal tract, allolactose is not produced. Therefore, there is nothing to change the shape of the repressor, which remains bound to the operator. As a result, transcription of the *lac* operon is blocked and the structural genes remain unexpressed. By producing the enzymes only when the nutrient is available, the bacterium avoids wasting its energy making proteins it does not need. ❑

❑ CAPSULE SUMMARY

The lac operon is a well-understood mechanism of gene regulation in E. coli. It is a cluster of genes that enables a bacterium to build the proteins needed for lactose metabolism only when lactose is present.

Effective Reading

Definitions

Have students use dictionaries to determine the common meanings of the terms *operator*, *repressor*, and *inducer*. Then ask students to compare the dictionary definitions to the meanings given for these words in Section 9-2. Students can use their comparisons to develop ways of remembering how the operon works.

UPDATE

☑ **RESEARCH Update**

Sonic Hedgehog

In 1992, researchers discovered an important gene in the fruit fly *Drosophila melanogaster*. Mutant fruit fly embryos lacking the gene had a spiky, bristly surface; hence, the gene was given the name *hedgehog*. Cliff Tabin at Harvard Medical School, Andrew McMahon of Harvard University, and Philip Ingham of the Imperial Cancer Research Fund in Oxford, England, began independent investigations into the function of the *hedgehog* gene. *Hedgehog*, it turns out, produces proteins that tell cells to develop into specific tissues and organs. The *hedgehog* proteins were found to be important embryological signalers because they dictate which cell type will develop in which position.

Researchers recently found an analog of the *hedgehog* gene in vertebrates, which is responsible for the organization of the central nervous system, the orientation of limbs, and the location of digits in developing vertebrate embryos. Discoverers of the analog named it *Sonic hedgehog*, after a popular video game hero. *Sonic hedgehog* produces a protein that washes over developing cells, giving them developmental instructions. Apparently, the concentration of the protein determines which type of cells the newly developing cells will become, sculpting embryonic cells into limbs and organs during the first weeks of life. ☑

Gene Regulation in Eukaryotes

Like prokaryotic cells, eukaryotic cells must continually turn certain genes on and off in response to signals from their internal and external environments. After operons were discovered in the 1960s, molecular biologists studying eukaryotes expected to find similar mechanisms for regulating gene expression. However, operons have not been found in eukaryotic cells. Instead, genes with related functions are often scattered among different chromosomes. This is just one of the many differences between a prokaryotic genome and a eukaryotic genome. In addition, a eukaryotic cell contains much more DNA than a prokaryotic cell, and most eukaryotes are multicellular organisms made of specialized cells. Different cell types produce different proteins. For example, the gene that enables a red blood cell to produce hemoglobin is also present in all the other cell types, but it is unexpressed. Gene expression in eukaryotes must involve mechanisms that account for differences such as these.

Much of eukaryotic gene expression is still a mystery. What scientists do know is that gene expression is affected by the way the DNA in a chromosome is physically arranged. Transcription takes place in the regions of a chromosome where the DNA has uncoiled. Biologists are able to see evidence of this process in polytene (*PAHL ih teen*) chromosomes. Polytene chromosomes are giant chromosomes found in salivary gland cells of flies such as *Drosophila*. These giant chromosomes consist of a large number of partially replicated chromosomes stacked neatly side by side. Figure 9-11 shows a polytene chromosome from the fruit fly *Drosophila* in which DNA has uncoiled to form a chromosome puff. A **chromosome puff** is a region of intense transcription that forms when DNA loops out from the chromosome, perhaps making the genes in that region more accessible to RNA polymerase. Chromosome puffs along a chromosome change locations as the insect develops.

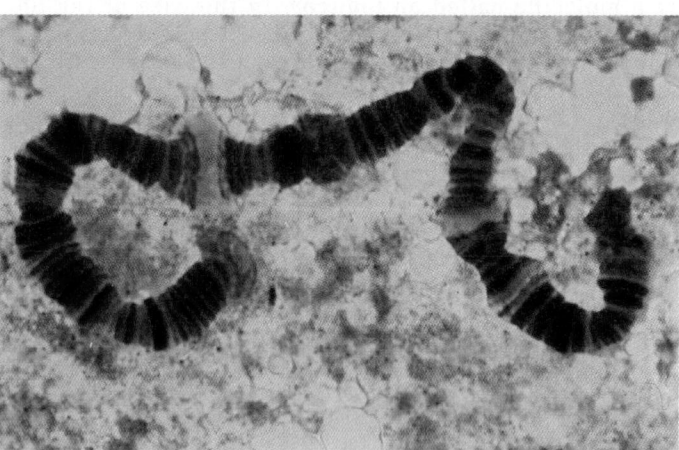

Figure 9-11 When the genes in a particular region of a polytene chromosome are activated, biologists are able to see obvious changes in the shape of the chromosome.

Graphic Organizer

Use this graphic organizer with *Teaching Tip: Gene Expression in Eukaryotes* on page 193.

| Inactive DNA coiled in chromosomes during mitosis and meiosis | → | DNA uncoils during transcription | → | mRNA moves from nucleus to cytoplasm | → | Translation in cytoplasm produces proteins |

Enhancer can expose RNA polymerase to binding sites of specific gene(s)

Once a eukaryotic gene is available to be expressed, what kinds of mechanisms regulate its expression? As with prokaryotes, gene expression can be regulated before, during, or after transcription. However, because a eukaryotic cell has a nuclear envelope that physically separates transcription from translation, there are more opportunities for regulating gene expression. Control mechanisms have been found to occur after the mRNA leaves the nucleus, before translation, and even after the protein is functional. ❏

Enhancer Control

Like prokaryotes, eukaryotes regulate gene expression primarily by controlling when RNA polymerase binds to the beginning of a gene. In eukaryotes, this binding cannot take place without the aid of a cluster of proteins called transcription factors. Sometimes one or more of the necessary transcription factors are located at a site called an enhancer, which is far distant from the gene being regulated. Because each such factor can be sensitive to different aspects of the cell, enhancers make eukaryotic gene regulation very flexible.

To understand how an enhancer stimulates transcription, look at the action of steroid hormones in the cells of vertebrates. A steroid hormone is a molecule made of lipids that acts as a chemical signal. Estrogen is a steroid hormone responsible for secondary sex characteristics in females. When estrogen passes through the cell membrane of specific cells it binds to a receptor protein in the nuclear envelope, forming a hormone-receptor complex. The hormone-receptor complex has the right shape to bind to a specific protein called an acceptor protein, which in turn binds to enhancer regions in the DNA. Binding of the acceptor protein to the enhancer region stimulates RNA polymerase to begin transcription. ❏

❏ **CAPSULE SUMMARY**

Gene expression in eukaryotes involves mechanisms that must uncoil the appropriate regions of DNA in specific cells at specific times.

❏ **CAPSULE SUMMARY**

Enhancers are regions of DNA that stimulate transcription of certain genes in eukaryotic organisms.

SECTION 9-2

Teaching Tips
Adult Versus Fetal Genes
A short time after birth, fetal hemoglobin is replaced by adult hemoglobin. The mechanism that stops the production of fetal hemoglobin and begins the production of adult hemoglobin is not known. Apparently, the genes for fetal hemoglobin are turned off and adult hemoglobin genes are turned on just after birth. A benefit of fetal hemoglobin is its higher affinity for oxygen. Thus, when the mother takes in oxygen, the oxygen goes to the baby's blood first.

Gene Expression in Eukaryotes
Ask students to construct a sequential chart like the *Graphic Organizer* at the bottom of page 192 to summarize the steps of gene expression in eukaryotes. Have students begin their charts with DNA coiled in chromosomes in the nucleus.

Section Review

1. *Why is gene regulation necessary in a living organism?*
2. *What effect does a repressor have on the* lac *operon when lactose is not present? when lactose is present?*
3. *How does the steroid hormone estrogen trigger transcription in a eukaryotic cell?*

Critical Thinking
4. *How would the* lac *operon in E. coli living in your intestinal tract be affected if you ate a bowl of ice cream?*
5. *Why are there more opportunities for regulating gene expression in eukaryotic cells than in prokaryotic cells?*

Answers to Section Review

1. Gene regulation enables cells to adjust to changes in their environment.

2. When lactose is not present, the repressor binds to the operon, switching the *lac* operon "off." When lactose is present, the repressor detaches from the operator, switching the *lac* operon "on."

3. It binds to a receptor protein on the nuclear envelope to form a hormone-receptor complex. The hormone-receptor complex binds to an acceptor protein for the enhancer region in the DNA, which stimulates RNA polymerase to begin transcription.

4. The lactose present in the ice cream would produce allolactase, an inducer that would bind to the repressor. The repressor would fall off the operator and give RNA polymerase access to the structural genes.

5. The physical separation of transcription and translation by the nuclear envelope provides more opportunities for regulating gene expression.

Opening Question

Ask students how a carcinogen such as asbestos or benzene might alter a cell. Lead students to conclude that carcinogens can cause mutations by changing the structure of a chromosome or by altering the original nucleotide sequence in the DNA. Mutations can upset gene regulation, causing cells to grow abnormally and uncontrollably.

Demonstration

Mutations

Have students refer to the DNA charts they completed in Section 9-1. Ask students what would happen to the amino acid sequence (AACTACGGT...) if the third nucleotide in the original DNA were changed to a T. *(Although the codon changes to UUA, the same amino acid, leucine, is specified.)* What if the fourth nucleotide is changed to a G? *(The change in nucleotide alters both the codon— AUG becomes CUG—and the amino acid—methionine becomes leucine.)* Explain that a change in the nucleotide would be a mutation. Emphasize that DNA has room for error because it can be altered without changing the amino acid sequence; however, DNA mutations can cause a change in amino acid sequence that would change the protein produced.

9-3 Genes, Mutation, and Cancer

Section Objectives

- Describe three ways point mutations alter genetic material.
- Discuss how the process of cell division can be affected by mutations.
- Give an example of an oncogene, and describe how it leads to cancer.

*I*f all the DNA in your cells could be pulled out, uncoiled, and connected end to end, it would be about 200 billion kilometers long. When all this DNA is being made, errors in replication could cause fatal mistakes. Fortunately, cells are equipped with mechanisms that work to preserve the integrity of your hereditary information. These mechanisms do not always perform perfectly, however, and errors do occur.

Mutations Are Changes in DNA

Although rare, changes in an organism's hereditary information can occur. As you learned in Chapter 6, a change in the DNA of a gene is called a mutation. The effects of a mutation vary, depending on whether it occurs in a gamete or in a body cell. Mutations in gametes can be passed on to offspring of the affected individual, but mutations in body cells affect only the individual in which they occur. Mutations are an important basis of evolution. Some mutations alter the structure of the chromosome itself. Other mutations called **point mutations,** illustrated in Figure 9-12, change one nucleotide or just a few

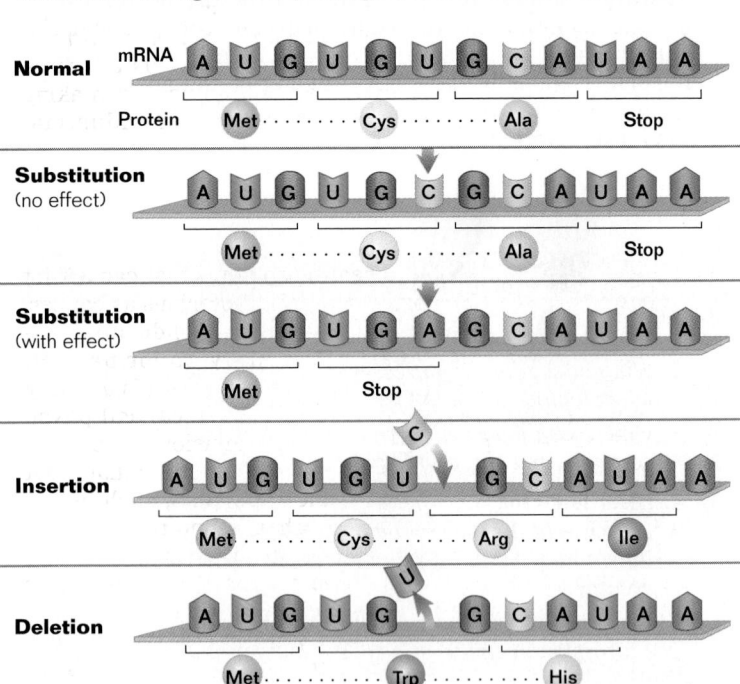

Figure 9-12 Point mutations arise when a chromosome breaks. The product of a normal gene is shown in the top row of this figure. Four effects of point mutations are shown below it.

Multicultural Perspective

Sickle Cell Anemia in Saudi Arabia

In 1972, physicians for an American-based oil company reported that many of the native Arabic people seen in their clinic had sickle cell anemia. However, the patients did not have the typical symptoms of the disease. This form of sickle cell anemia was labeled "benign." A surprising difference in these patients was that they had 25 percent fetal hemoglobin rather than the 5 percent fetal hemoglobin seen in other ethnic groups. Apparently, the fetal hemoglobin gene continued to produce fetal hemoglobin in these individuals, who were able to avoid the usual symptoms of sickle cell anemia.

nucleotides in a gene. There are two general types of point mutations. In point mutations called **substitutions,** one nucleotide in a gene is replaced with a different nucleotide. For instance, the codon UGU becomes UGC. Sometimes substitutions have little or no effect. In this case, the correct codon, UGU, and the mutated codon, UGC, both translate into the same amino acid—cysteine. However, if UGU changed into UGA, the codon would become a stop codon, and translation would end prematurely. As a result, the protein that the gene codes for would be shortened and incomplete.

In point mutations called **insertions** and **deletions,** one or more nucleotides are added to or deleted from a gene. Because the genetic message is read as a series of triplet codons, insertions and deletions can upset the triplet groupings. Imagine deleting the letter C from the sentence "THE CAT ATE THE RAT." Keeping the triplet groupings, the message would read "THE ATA TET HER AT," which is meaningless. A mutation that causes a gene to be read in the wrong three-nucleotide sequence is called a frameshift mutation because the reading pattern is displaced one or two positions. If a reading is displaced three positions, there may be no effect at all on the protein's function.

What Causes Mutations?

Some mutations are chemical mishaps that arise spontaneously. Other mutations, however, are induced by exposure to environmental agents called **mutagens** (*MYOOT uh jehnz*). Mutagens include ionizing radiation (such as X rays and gamma rays), ultraviolet light, and certain chemicals or irritants referred to as carcinogens (*kahr SIHN uh jihns*). **Carcinogens** are cancer-causing agents such as asbestos (fibers used in insulation), benzene (a liquid used in making detergents, plastics, and paints), and other industrial pollutants. ❏

Mutations Can Cause Cancer
....

Of all the health problems that can afflict humans, few are as mysterious as cancer. **Cancer** is a term used to indicate a disease characterized by abnormal cell growth. Normally, cell growth is a highly regulated process that is controlled by chemical and physical signals that inhibit or stimulate cell division. A cell that has become cancerous does not respond to the signals that stop cell division. As a result, cancerous cells, such as the ones shown in Figure 9-13, divide without stopping.

Health problems begin when a cell that has turned cancerous evades the body's immune system, which normally destroys cancerous cells. The cancerous cell then proliferates, forming a mass of cells called a **tumor.** Tumors are classified as either benign or malignant. A benign tumor does not

❏ **CAPSULE SUMMARY**

Mutations are changes in the genetic message in an organism. A mutation may change the structure of a chromosome or may just alter a single nucleotide.

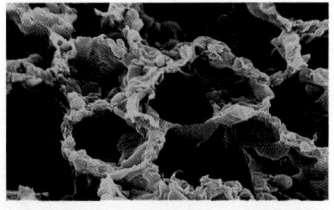

Figure 9-13 In the healthy lung tissue, *top,* cell division is highly regulated. When cells escape regulation and continue to grow, *bottom,* a tumor forms.

SECTION 9-3

Teaching Tips
Growing Genes?
A number of genetic disorders, including Huntington's disease, fragile X syndrome, and Kennedy disease (bulbar muscular atrophy) appear to become magnified or compounded as the genes are passed from one generation to the next. The mutations occur as numerous repeats of certain gene sequences. The gene is said to "stutter" as it repeats the same sequences. Often, symptoms of these disorders are mild, if noticeable, in parents. Frequently, in cases such as fragile X syndrome, parents do not know of their conditions until a child is born with the disorder.

Chernobyl
Have students read articles about the effects of the 1986 explosion of a nuclear reactor at the Chernobyl power plant in Ukraine. Scientists predicted that the residents of the surrounding towns would develop high rates of thyroid cancer and leukemia as a result of the radiation exposure they received from the accident. Have students discuss their findings in brief reports that link the elevated incidence of cancer to mutations caused by radiation damage to DNA.

Historical Note

Researching Cancer
Working at the National Cancer Institute of the National Institutes of Health for many years, Dr. Sarah Stewart (1923–1976) was among the first researchers to demonstrate that a virus causes many types of tumors in mice, rats, hamsters, rabbits, and guinea pigs and to prove that tumors could be spread from one animal to another. The daughter of a Mexican mother and an American father, Dr. Stewart was born in Mexico. She moved to the United States when she was five years old. She was the first woman to be featured in the "Medical Men of Georgetown" column of Georgetown University's *Medical Bulletin.*

Application

 Medicine A current treatment for cancer involves introducing reciprocal segments of the DNA strand that is not read during transcription, called "antisense" fragments, into cells. The antisense fragments target mRNA production and alter the gene expression of cancerous cells. These fragments are introduced into cells using viruses as vectors. What effect would antisense fragments have on a cancer cell? *(The fragments would alter the nucleotide sequence and hence the gene expression.)*

Teaching Tip
Gene Therapy in the Brain

A new gene therapy for inoperable brain tumors takes into the account the fact that normal brain cells do not divide. In this therapy, retroviruses containing suicide genes are delivered into the brain tissue. The retroviruses infect only the cells that are dividing. Since the tumor cells are the only cells dividing in the brain, they are infected by the retroviruses, which then destroy them.

Chapter Closure

Have students create their own DNA sequences of 24 nucleotides each. Shuffle and redistribute the DNA sequences among the class and ask students to determine the mRNA, tRNA, and amino acids specified by their DNA sequences. Have students introduce three different kinds of mutations into the original DNA sequence and determine the effects of their mutations on the amino acid sequence produced. Ask students to speculate about the possible causes of the mutations.

☐ CAPSULE SUMMARY

Cancer is the uncontrolled growth of cells. It begins when the genes that control cell growth become mutated. An example of such a mechanism is a class of genes called the ras *genes.*

invade surrounding tissues. They usually cause few problems and can be surgically removed. Tumors described as malignant, on the other hand, are very harmful. A malignant tumor spreads into other tissues and interferes with organ functions. The most devastating property of a malignant tumor is that its cells are able to break free of the tumor and enter blood and lymph vessels. These cancerous cells are then carried to new locations in the body and form new growths. The spread of malignant cells beyond their original site is called **metastasis** (*muh TAHS tuh sihs*).

Cancer Genes

Through many years of intense work, researchers have learned that a cell becomes cancerous when mutations occur in genes that regulate cell growth. An example of growth-regulating genes is a class of genes called the *ras* genes (because they were first discovered in viruses that cause tumors known as sarcomas in rats). The *ras* genes code for proteins that help prevent uncontrolled cell division. When there is a mutation in a *ras* gene, a faulty *ras* protein is built. As a result, the cell divides more rapidly than normal, a sign that it has become cancerous. The *ras* genes are examples of oncogenes (*ang KOH jeens*). An **oncogene** is a gene that, when mutated, can cause a cell to become cancerous.

Researchers have found that a mutated *ras* gene usually contains a point mutation. For example, in a form of human bladder cancer caused by a mutated *ras* gene, a single G nucleotide has been replaced with a T, transforming the normal amino acid glycine into a valine. However, the *ras* gene is only one of several controls that the cell normally exercises over unwanted growth. All of these controls must be inactivated before cancer results. This is why most cancers occur in people over 40 years old—it takes time for an individual cell to accumulate the necessary mutations. ☐

Section Review

1. *Which could have a more serious effect on a protein: a base-pair substitution or a frameshift mutation? Why?*
2. *What is cancer? How do the terms* tumor *and* metastasis *relate to this health problem?*
3. *What is the function of the protein encoded by the* ras *gene? How can a cell be affected when the* ras *gene is mutated?*

Critical Thinking

4. *Why is a malignant tumor considered more harmful than a benign tumor?*

Answers to Section Review

1. A frameshift mutation could have a more serious effect on a protein than a point mutation because it would change the entire sequence of the amino acids in the polypeptide chain.

2. Cancer is a group of diseases characterized by abnormal cell growth. A tumor is a mass of cancerous cells. Metastasis occurs when malignant cells spread beyond the original tumor to other parts of the body.

3. The protein produced by a *ras* gene keeps cell division under control. When a *ras* gene is mutated, a cell may divide more rapidly than normal.

4. A malignant tumor is considered more harmful because it invades other body tissues.

Each item in the Chapter Review is correlated by text section in the assignment guide that follows.

Vocabulary

anticodon (187)
A site (187)
cancer (195)
carcinogen (195)
chromosome puff (192)
codon (186)
deletion (195)
enhancer (193)
gene expression (183)
genetic code (186)

inducer (191)
insertion (195)
lac operon (190)
messenger RNA (185)
metastasis (196)
mutagen (195)
operator (190)
operon (190)
oncogene (196)
point mutation (194)

promoter (184)
P site (187)
repressor (191)
ribonucleic acid (183)
ribosomal RNA (185)
RNA polymerase (184)
substitution (195)
terminator (185)
transcription (183)
transfer RNA (185)
translation (183)

ASSIGNMENT GUIDE

Section	Review	Themes Review	Critical Thinking
9-1	1–5, 10–12, 18, 19, 23, 25–27	30, 31	32
9-2	6–8, 13, 14, 20–22	28, 29	
9-3	9, 15–17, 24		

Concept Mapping

Construct a concept map that shows the role of RNA in gene expression. In constructing your map, use the following terms: transcription, translation, mRNA, tRNA, rRNA, gene, promoter, codons, anticodons, proteins, amino acids, ribosome, and cytoplasm. Include additional concepts in your map as needed.

Concept Mapping

Map answer is shown on page 181B.

Review

Multiple Choice

1. RNA differs from DNA in that RNA
   a. includes thymine rather than uracil.
   b. consists of a single chain of nucleotides.
   c. contains the sugar deoxyribose.
   d. has four nitrogen bases.

2. A short chain of DNA has the nucleotide sequence ATA CCG GAC ATC. What is its complementary mRNA nucleotide sequence?
   a. TAT GCC CTG TAG
   b. AUA CCG GAC AUG
   c. UAU GGC CUG UAG
   d. ATA CCG GAC ATC

3. During translation, a codon pairs with a(n)
   a. complementary sequence of mRNA.
   b. three-nucleotide sequence of rRNA.
   c. complementary sequence of tRNA.
   d. specific amino acid.

4. What happens when RNA polymerase reaches a terminator on a strand of DNA?
   a. It kills the DNA.
   b. Transcription stops.
   c. Anticodons are no longer added to the growing strand of RNA.
   d. The promoter takes over, and translation starts.

5. In prokaryotes, transcription is regulated by
   a. the presence of enhancers.
   b. metastasis.
   c. the operon model.
   d. frameshift mutations.

6. Which of the following blocks transcription of lactose metabolizing genes?
   a. the absence of lactose
   b. the binding of lactose to the repressor
   c. the binding of lactose to the operator
   d. the presence of beta-galactosidase

7. What protein is able to stop RNA polymerase from transcribing structural genes in DNA?
   a. operator   c. promoter
   b. repressor   d. enhancer

Review

Multiple Choice

Code in parentheses indicates section number and objective number.

1. b (9-1.1)
2. c (9-1.2)
3. c (9-1.4)
4. c (9-1.2)
5. c (9-1.2)
6. a (9-2.2)
7. b (9-2.2)
8. a (9-2.1)
9. b (9-3.3)

Completion

10. transcription, translation (9-1.2, 9-1.4)
11. codons, 64 (9-1.2, 9-1.3)
12. peptide, tRNA (9-1.4)
13. operon, operator (9-2.2)
14. operons, physical arrangement (9-2.2, 9-2.3)
15. gametes, body cells (9-3.1)
16. tumor (9-3.2, 9-3.3)

Short Answer

17. A mutation in gametes is likely to be a source of genetic variation for evolution because gametes—egg and sperm—pass on the mutation to the next generation when forming a zygote. (9-3.1)

18. Messenger RNA (mRNA) carries hereditary information from DNA and delivers it to the site of translation. Transfer RNA (tRNA) translates mRNA sequences into amino acid sequences. Ribosomal RNA (rRNA) links with proteins to form the ribosome where the amino acid chains are assembled. (9-1.1)

19. RNA polymerase binds to the promoter on DNA causing the two DNA strands to unwind. It then moves along one DNA strand, pairing DNA and RNA nucleotides. Transcription ceases when RNA polymerase reaches a DNA terminator sequence. (9-1.2)

20. The operator is not bound by the repressor. Lactose concentration would be high. (9-2.2)

21. When allollactase is present in the cell, it binds to the repressor and changes the repressor's shape. The repressor then detaches from the operator, allowing RNA polymerase to reach the structural genes. This action enables the cell to begin building the proteins required to metabolize lactose. (9-2.2)

22. Strain A will multiply more rapidly and displace strain B. Strain A uses the energy saved by not carrying out transcription and translation to multiply. (9-2.1)

23. It is not possible for gene expression to occur without RNA. The three forms of RNA —mRNA, tRNA, and rRNA—are necessary for building proteins from DNA. (9-1.2)

24. Disagree. A deletion in the middle of a gene has a greater chance of being lethal because a larger number of codons follow the deletion and the amino acids for which they code would be affected. (9-3.1)

25. One site holds mRNA so that its codons are accessible to tRNA.

---

9 CHAPTER REVIEW ....

8. When polytene chromosomes puff, they are
a. undergoing transcription.
b. stimulating cancerous transformations.
c. initiating estrogen production.
d. keeping mRNA from leaving the cell nucleus.

9. Cancer is caused by
a. tumors.
b. mutations in growth regulating genes.
c. *ras* genes.
d. radiotherapy and chemotherapy.

Completion

10. The making of RNA from DNA is called _____ , while the construction of proteins from the information carried by mRNA is called _____ .

11. The genetic code is made of three-nucleotide sequences called _____ . The number of codons possible in the genetic code is _____ .

12. Amino acids of a protein chain are held together by _____ bonds. These bonds are formed as _____ is moved from the A site to the P site on the ribosome.

13. A(n) _____ is a group of structural genes that includes the promoter and operator. The _____ is located between the promoter and the structural genes.

14. Gene expression in prokaryotes is regulated by _____ . However, gene expression in eukaryotic cells is regulated by the _____ of DNA in a chromosome.

15. Mutations in _____ are likely to be genetically significant, while mutations in _____ affect only the organism in which they occur.

16. A cancerous cell can proliferate, forming a mass of cells called a _____ .

Short Answer

17. Mutations may occur in gametes or in body cells. In which cell type is a mutation likely to be a source of genetic variation for evolution? Why?

18. Three types of RNA are made from DNA. What are the three types of RNA? How does each kind function in protein synthesis?

19. What is the role of RNA polymerase in transcription?

20. What is the condition of the operator on the *lac* operon when transcription is occurring? Would this condition be a result of a high or a low concentration of lactose?

21. What is the function of allolactose in the operation of the *lac* operon?

22. Two strains of *E. coli* bacteria, A and B, are placed in a medium containing the amino acid tryptophan. Strain A is able to turn off the genes that code for the enzymes that produce tryptophan. Strain B is unable to do this. What is likely to happen to the two strains after several days in the medium? Why?

23. Could gene expression occur without RNA? Explain.

24. Students studying together for a test reached the conclusion that a deletion mutation that occurs at the end of a gene has a greater chance of being lethal than one that occurs near the middle of a gene. Do you agree with their conclusion? Explain.

25. A ribosome has three binding sites that function during translation. What is the function of each site?

26. During translation, a protein is made by linking amino acids. How are amino acids linked to form proteins?

27. How is translation initiated?

| Themes Review |

28. Structure and Function Chromosome puffs have been observed in the polytene chromosomes of *Drosophila* and other flies. How are chromosome puffs thought to promote transcription in eukaryotic cells?

Another site, called the A site, holds a tRNA molecule that is carrying a specific amino acid. The third site, called the P site, is positioned next to the A site and holds a tRNA molecule that is carrying its specific amino acid attached to the growing protein chain. (9-1.4)

26. A peptide bond is formed between the amino acid attached to the tRNAs at the P and A sites; the tRNA at the A site is shifted to the P site, leaving the A site open for the next tRNA. The cycle repeats until the "end" codon is reached. (9-1.5)

27. Translation begins when an mRNA molecule binds to the smaller ribosomal subunit and the "start" codon is sitting in the P site. (9-1.4)

29. Homeostasis How does gene regulation of the *lac* operon promote homeostasis in the *E. coli* bacteria that live in the human intestine?

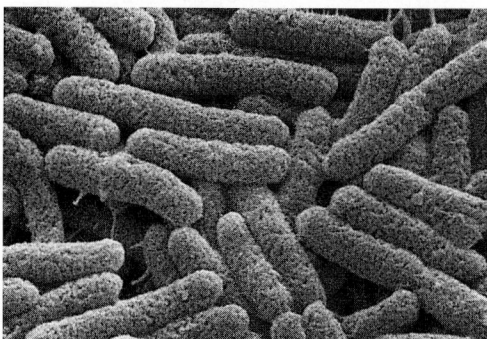

30. Evolution Genetic engineering involves inserting segments of DNA taken from one organism into the DNA of another organism. What would be the likely result of an experiment in which a scientist replaces a faulty stop codon in the DNA of mouse cells with the codon UAG taken from the DNA of a frog, a pine tree, or a clam? What do the results of this experiment suggest about the evolutionary ancestry of these organisms?

31. Levels of Organization Write the mRNA transcription of the DNA sequence presented below.

      CTG TTC ATA ATT

Next, write the tRNA anticodons that would pair with the mRNA transcription. Finally, write the abbreviations for the amino acids coded for by the mRNA transcription using Table 9-2.

Critical Thinking

32. Making Inferences In eukaryotic cells, there is a lag time between the beginning of transcription and the beginning of translation. This does not occur in prokaryotic cells. What reason can you give for this difference?

Activities and Projects

33. Research and Writing Do library research to find the latest information on oncogenes and their protein products. Present an oral report to the class in which you show how your findings support the theory that cancer arises from out-of-control cell division.

34. Cooperative Group Project Construct models showing gene expression in prokaryotes and eukaryotes. Your model of the prokaryote should include promoter, operator, and structural genes; repressor; RNA polymerase; inducer; mRNA; and ribosome. Your model of the eukaryote should show DNA, enhancer, receptor protein, acceptor protein, RNA polymerase, and ribosomes.

35. Research and Writing Tetracyclines are antibiotics that are commonly used as an antibacterial drug. Do library research and talk with health care professionals to learn how tetracyclines fight bacterial infection.

36. Multicultural Perspective Research the various types of cancer that seem to have a higher rate of incidence among particular cultures. For example, the Japanese have a very low rate of breast cancer compared with Americans, but they have a higher rate of other types of cancer. What possible reasons could explain these phenomena?

Readings

37. Read the article "How Does a Cell Become a Whole Body?" in *Discover*, November, 1992. What two discoveries helped scientists begin unraveling the mysteries of development? How does the homeobox differ from other gene clusters? What are the two sources for the signals that determine a cell's fate?

Critical Thinking

32. Transcription in eukaryotic cells occurs in the nucleus, while translation occurs in the cytoplasm. In prokaryotic cells that have no nucleus, both transcription and translation occur in the cytoplasm. The lag time is due to the time required for mRNA to pass through the nuclear membrane into the cytoplasm of eukaryotic cells. (9-1.2, 9-1.4)

Activities and Projects

33. Students should look for information about oncogenes and the mutations that affect them.

34. Students might choose to work in two dimensions with paper of different colors, or in three dimensions with materials such as clay, pipe cleaners, yarn, cotton balls, etc.

35. Tetracyclines block binding of tRNA to the A site on a prokaryotic ribosome.

36. Answers will vary. Students may want to contact the American Cancer Society for statistics. Possible reasons for different rates of cancer around the world include diet; pollution of air, water, and food; ozone layer depletion; public acceptance of social behaviors such as cigarette smoking; and the use of pesticides and herbicides on food and tobacco products.

Readings

37. The 1953 discovery of the structure of DNA and the 1984 discovery of the homeobox and homeotic genes that specify body-part formation have helped scientists unravel the mysteries of development. The homeobox differs from other gene clusters, in that its genes have stayed in the same order for one-half-billion years. The signals that determine a cell's fate can come initially from inside the egg or later from other nearby cells.

Themes Review

28. Genes in the regions of the chromosome puffs are more accessible to RNA polymerase because the DNA has uncoiled in these regions. (9-2.1)

29. Gene regulation promotes the production of proteins that metabolize lactose when lactose is present; it inhibits the production of the proteins when lactose is not present. This regulation enables *E. coli* to maintain homeostasis by conserving energy. (9-2.2)

30. The replacement codon would stop transcription. The results suggest that different organisms have a common evolutionary ancestor with a single genetic code. (9-1.3)

31. The mRNA transcription is GAC AAG UAU UAA. The tRNA anticodons are CUG UUC AUA AUU. The amino acids are aspartic acid, lysine, tyrosine, and stop. (9-1.2, 9-1.3)

LABORATORY Investigation Chapter 9

Preparation
Only paper and pencil are needed for this investigation.

Procedural Notes
- You may wish to discuss DNA, RNA, and protein synthesis prior to the lab period.
- Although this is a paper-and-pencil lab, students may work well in groups of two.

Background Answers

1. A DNA molecule is made of two strands of nucleotides twisted around a central axis. Each nucleotide is made of the sugar deoxyribose, a phosphate group, and a nitrogen base.

2. Messenger RNA, or mRNA, makes an RNA copy of DNA within the nucleus and carries the genetic information to the cytoplasm.

3. The building blocks of protein are amino acids.

4. A faulty protein can be coded for by the nucleotide sequence, if even one amino acid is coded incorrectly.

5. Enzymes, which control functions of the cell, are proteins.

6. A mutation is a change in the genetic material, such as changes in the nucleotide sequence. Mutations can change one nucleotide, several nucleotides, or displace the reading pattern of the DNA sequence.

7. Iodizing radiation, ultraviolet light, certain chemicals, and errors in coding can cause mutations.

8. Mutations can alter the nucleotide sequence so that it codes for different amino acids. The resulting protein will have a different primary structure and will not fold into the correct secondary or tertiary structures.

9. How does the cell's DNA determine the proteins made by the cell?

DNA and Proteins

OBJECTIVE

Demonstrate how the proteins made by a cell are determined by the cell's DNA.

PROCESS SKILLS

- analyzing models of protein synthesis
- recognizing the relationship between DNA structure and its function

MATERIALS

- paper and pencil

BACKGROUND

1. What is the basic structure of a DNA molecule?
2. How does DNA code for proteins, which are made in the cytoplasm, if it does not leave the nucleus?
3. What are the building blocks of protein?
4. What can result if only one amino acid is coded incorrectly?
5. Why are proteins important in cellular functions?
6. What is a mutation?
7. What conditions can cause mutations?
8. How can a mutation affect protein structure?
9. Write your own question to explore in your lab report or notebook.

Amino Acids	
Phenylalanine	UUU, UUC
Leucine	UUA, UUG, CUU, CUC, CUA, CUG
Isoleucine	AUU, AUC, AUA
Methionine or START	AUG
Valine	GUU, GUC, GUA, GUG
Cysteine	UGU, UGC
Tryptophan	UGG
Arginine	CGU, CGC, CGA, CGG, AGA, AGG
Glycine	GGU, GGC, GGA, GGG
Serine	UCU, UCC, UCA, UCG, AGU, AGC
Proline	CCC, CCA, CCU, CCG
Threonine	ACU, ACC, ACA, ACG
Alanine	GCU, GCC, GCA, GCG
Tyrosine	UAU, UAC
STOP	UAA, UAG, UGA
Histidine	CAU, CAC
Glutamine	CAA, CAG
Asparagine	AAU, AAC
Lysine	AAA, AAG
Aspartic acid	GAU, GAC
Glutamic acid	GAA, GAG

TECHNIQUE

1. Assume that the base sequence on one strand of a DNA molecule is:

C A C G C T T G G T G A C C G T A A

2. List the base sequences of the complementary DNA strand that would form during replication. Place the information in the **Records** section of your report.

3. List what the mRNA base sequences would be if the strand shown in step 1 were being read by mRNA.

4. Using the table on the preceding page, determine the amino acid sequence this mRNA codes for. List these amino acids in the **Records** section of your report.

5. If the fifth nucleotide in the DNA strand shown in step 1 were changed from cytosine to adenine, what would the resulting mRNA be? Record your answer in the **Records** section of the your report.

6. List the sequence of amino acids that would reflect the change described in step 5.

7. Here is the base sequence of a strand of DNA:

C T C C T C A G G A G T C A G C G T G C A A C A

   a. List the bases in the mRNA strand coded from this DNA strand in the **Records** section of your report.

   b. For which amino acids would this mRNA code?

   c. If the twelfth base in the original DNA strand were changed to cytosine, for what mRNA would the new DNA code?

   d. How are the proteins in b and c different?

8. In your report, briefly summarize the procedure you followed.

INQUIRY

1. How many mRNA bases code for an amino acid?

2. What is the significance of the start and stop codons?

3. If one base is changed, what happens to the protein that is formed?

4. Using the table on the preceding page as a guide, write a statement relating mRNA codons and amino acids.

ANALYSIS

1. Study the table shown on the preceding page. If the base G in the mRNA codon GAU (aspartic acid) were changed, what possible codons (and their corresponding amino acids) could result?

2. If one base is changed in the mRNA code, a mutation may result. Describe the effects of a mutation by contrasting normal red blood cells with "sickle-cell anemia" blood cells.

3. Write a generalized statement about mutations.

FURTHER INQUIRY

Write a new question that could be explored as a new investigation. The following is an example:

> How can changes in DNA that yield incorrect proteins cause genetic diseases such as sickle cell anemia, hemophilia, PKU, or cystic fibrosis?

Inquiry Answers

1. Three mRNA bases code for an amino acid.

2. They direct when to begin and end the process of making a protein.

3. One base change can change the amino acid that is coded for and thus change the protein formed.

4. Although any mRNA codon codes for only one amino acid, amino acids can be coded for by more than one codon.

Analysis Answers

1. The following amino acids would result if the G in the codon GAU changed to A, C, or U: AAU—asparagine, CAU—histidine, UAU—tyrosine.

2. If an amino acid is changed, the protein that forms can hinder the functioning of the cell and the individual. Such is the case when DNA codes for deformed red blood cells in sickle cell anemia. The sickle-shaped blood cells caused by this disease do not function properly and eventually lead to death.

3. Mutations are changes in the genetic message in an organism.

Records

2. The base sequence of the complementary DNA strand that would form during replication is GTGCGAACCACTGGCATT.

3. The mRNA strand is GUGCGA-ACCACUGGCAUU.

4. The amino acid sequence is valine, arginine, threonine, threonine, glycine, isoleucine.

5. The resulting mRNA would have uracil in the fifth position instead of guanine.

6. The resulting amino acid sequence is valine, leucine, threonine, threonine, glycine, isoleucine.

7a. The mRNA sequence is GAGGAGUCCUCAGUCGCACGUUGU.

b. The mRNA would code for glutamic acid, glutamic acid, serine, serine, valine, alanine, arginine, cysteine.

c. The twelfth base in the mRNA would be guanine instead of adenine.

d. The proteins in b and c are the same

UNIT 2

Block Scheduling Guide

Each block represents about 45–50 minutes of instructional time. The resources cited in a block are coded to a specific program component using the Key on the next page.

Lecture Concepts	Lesson Resources

10-1 What Is Genetic Engineering? pp. 202–209

BLOCK 1

a. Genetic engineering involves building recombinant DNA, a molecule made from pieces of DNA from separate organisms.

b. A genetic engineering experiment has four basic steps: cleaving DNA, producing recombinant DNA, cloning cells, and screening cells.

Lecture Resources
- Demonstration: Recombinant DNA
- Teaching Transparencies: 39, 40
- Holt Biology Videodiscs: Lesson 10-1

Classwork Options
- Effective Reading, page 203
- Laboratory Techniques and Experimental Design Lab C16: DNA Whodunit

BLOCKS 2 and 3

a. Restriction enzymes recognize specific nucleotide sequences and cleave DNA into fragments. DNA molecules cleaved by the same restriction enzyme will have complementary ends which allow DNA fragments from different organisms to join.

b. The Southern blot technique is used to screen clones to determine whether they contain specific recombinant genes.

c. RFLP analysis is based on the fact that restriction enzymes cut DNA into fragments that have specific lengths.

d. The polymerase chain reaction allows scientists to produce millions of copies of DNA in just a few hours.

e. The Human Genome Project is a research effort to identify and locate human genes.

Lecture Resources
- Defense Against Invaders, page 205
- Universality of DNA, page 205
- Visual Strategies: Figures 10-4, 10-7
- Overcoming Misconceptions, page 206
- Demonstration: Gel Electrophoresis
- Teaching Transparency: 31A
- Research Update, page 209
- Holt Biology Videodiscs: Lesson 10-1

Classwork Options
- Producing Recombinant DNA, page 206
- Biotechnology D4: Genetic Transformation—Antibiotic Resistance
- Biotechnology D5: Introduction to Agarose Gel Electrophoresis
- Biotechnology D6: DNA Fragment Analysis

Assignment Options
- Section Review, page 209
- Chapter Review, questions 1, 2, 8, 10–12, 16–18, 21, 24–26

10-2 The New Medicine pp. 210–213

BLOCK 4

a. Genetic engineering techniques are being used to manufacture proteins such as insulin and factor VIII, as well as vaccines.

b. Some human genetic disorders are being treated and "corrected" by inserting copies of the corresponding normal gene into individuals whose copy of the gene is defective.

Lecture Resources
- Demonstration: Vaccines
- Teaching Transparencies: 32A, 33A
- Curing Genetic Disorders, page 212
- Holt Biology Videodiscs: Lesson 10-2

Classwork Options
- Making Genetically Engineered Drugs and Graphic Organizer, page 211
- Biotechnology D7: DNA Ligation
- Biotechnology Lab D8: Comparing DNA Samples

Assignment Options
- Section Review, page 213
- Chapter Review, questions 3–5, 7, 9, 13, 15, 19, 22

10-3 The New Agriculture pp. 214–216

BLOCK 5

a. Genetic engineers have manipulated the genes of certain kinds of crop plants to make them resistant to herbicides and destructive pests.

b. Genetic engineers are looking for ways to transfer genes for nitrogen fixation from bacteria into crop plants.

c. The addition of genetically engineered growth hormone to the diets of livestock increases milk production in dairy cows and weight gain in cattle and hogs.

Lecture Resources
- Opening Question, page 214
- Demonstration: New Solutions to Old Problems
- Visual Strategy: Figure 10-12
- Getting Genes into Plants, page 215
- Teaching Transparency: 41
- Holt Biology Videodiscs: Lesson 10-3

Classwork Options
- Laboratory Investigation: Genetic Engineering Model, pages 220–221

Assignment Options
- Concept Mapping, page 217
- Section Review, page 216
- Chapter Review, questions 6, 14, 20, 23

Review/Enrichment

BLOCK ⑥

- ■ Study Guide: Concept Review and Pretest

Assignment Options
- ⊙ Teaching Resources CD-ROM: Supplemental Reading Worksheets and Test, *The Double Helix*
- ■ Activities and Projects, questions 27–29
- ■ Readings, questions 30, 31
- ■ Science, Technology, and Society, pages 222–223

Assessment Options

BLOCK ⑦

Traditional Assessment
- ■ Chapter 10 Test
- ⊙ Test Generator/Assessment Item Listing: Software item bank for preparing customized chapter tests.

Performance Assessment
- ⊙ Select the Performance Assessment Items for this unit from the Test Generator/Test Item Listing.

Portfolio Assessment
Select student reports for one or more laboratory experiments from this chapter. *The Direct Observation Checklist* on the *BioSources Teaching Resources CD-ROM* should be completed during a laboratory or other cooperative learning experience.

Answer to Concept Map
The following is one possible answer to the Concept Mapping exercise on page 217.

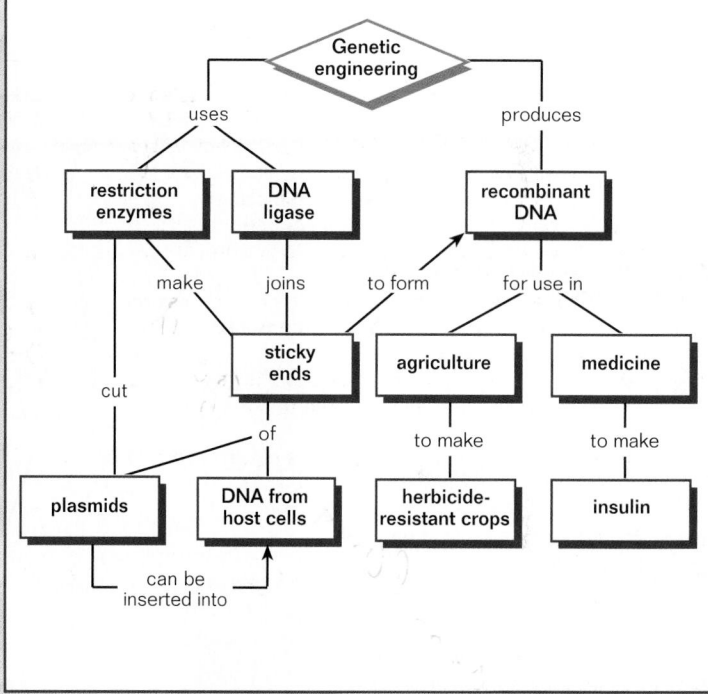

Holt Biology Videodiscs

Holt Biology Videodiscs gives you a powerful tool for teaching, review, and assessment. *Concepts of Biology* adds interactive lesson plans and extensive tools for customization using CD-ROM technology.

CONCEPTS OF BIOLOGY

Use the following topic from *Concepts of Biology* to help you teach this chapter:

- ■ Topic 7: How Genes Work

For further information regarding the media on the videodiscs, see *Holt Biology Videodiscs Teacher's Correlation Guide to Biology: Principles and Explorations.*

Chapter Theme

Homeostasis *Some genetic diseases are due to a failure to produce a particular substance, which in turn leads to the loss of homeostasis in the body. Diabetes leads to loss of control of blood sugar levels, PKU is a failure to metabolize an amino acid, leading to its buildup in the body, and hemophilia causes a breakdown in blood-clotting reactions, which leads to blood loss.*

Tapping Prior Knowledge

- What is the complementary strand for the DNA strand GAATTC?
- How does the DNA code determine which proteins an organism can make?
- Where do you find DNA in different kinds of cells, and is the DNA linear or circular?
- How do enzymes catalyze reactions?

Opening Question

Ask students to discuss why gene technology is controversial. Lead students to suggest that DNA technology allows researchers to produce new life-forms with specific characteristics or abilities. Students should understand that gene technology is a potentially powerful tool for fighting diseases and for understanding organisms. Explain that people are excited about the prospects of using this tool but concerned about its misuse.

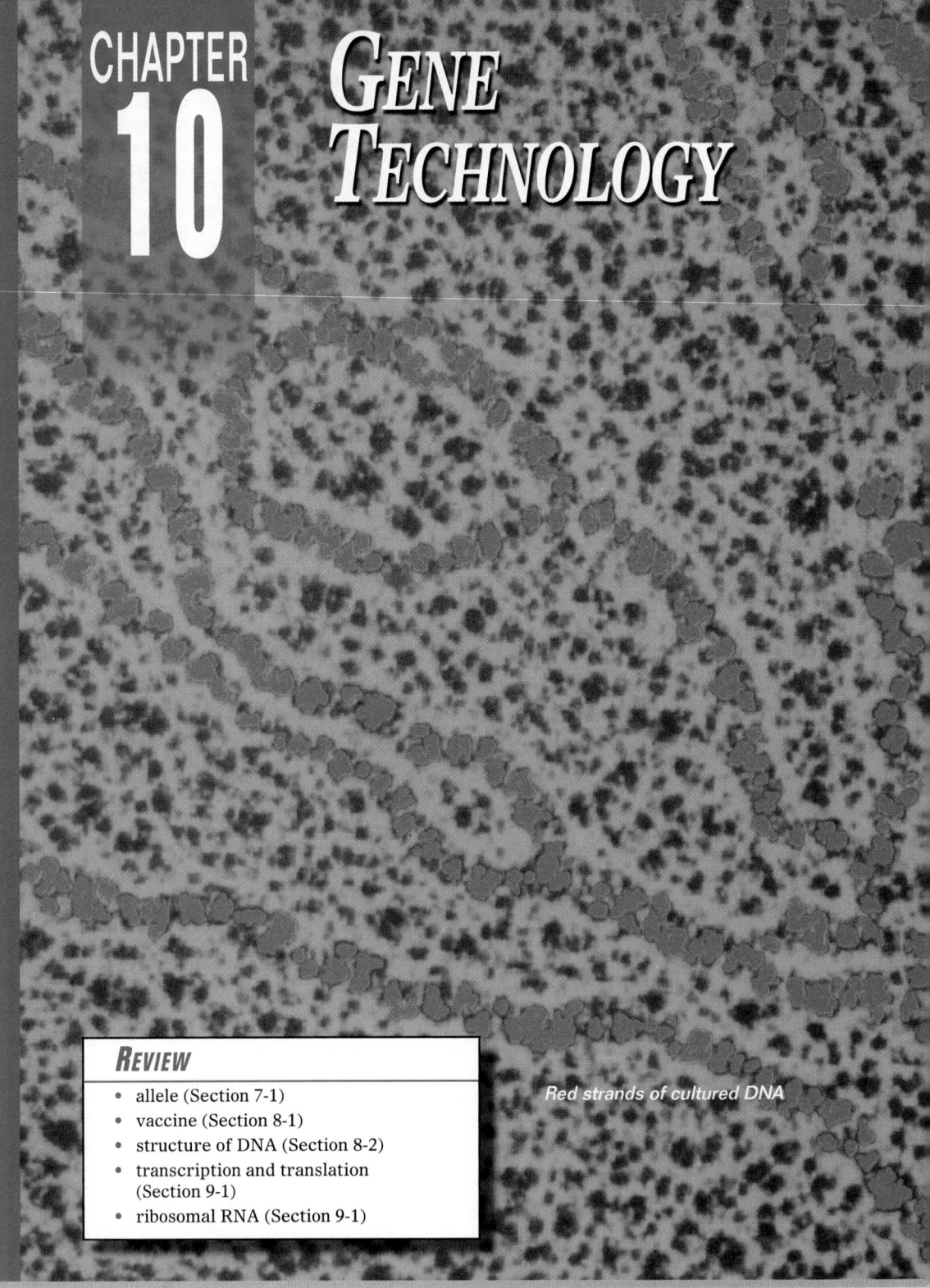

CHAPTER 10 GENE TECHNOLOGY

REVIEW

- allele (Section 7-1)
- vaccine (Section 8-1)
- structure of DNA (Section 8-2)
- transcription and translation (Section 9-1)
- ribosomal RNA (Section 9-1)

Red strands of cultured DNA

Authors' Rationale

Of all the fields in science, probably none is evolving as explosively as gene technology. Recent and ongoing developments enable researchers to manipulate DNA (and cells) in the laboratory as never before: to cut, rearrange, or amplify DNA; to take DNA from one organism and put it into a different one; and even to make synthetic DNA. This chapter introduces the concepts of genetic engineering and shows students some advantages of genetic manipulation in medicine and farming. Gene technology provides innovative approaches that have an impact on the entire world.

10-1 What Is Genetic Engineering?

*I*n *1973, biochemists Stanley Cohen and Herbert Boyer constructed a creature that was part bacterium and part frog. How did these scientists construct such an organism? Using organisms such as those shown in Figure 10-1, they first isolated the gene that codes for ribosomal RNA from the DNA of a frog. Then they inserted the frog gene into the DNA of the bacterium* Escherichia coli. *During transcription, the bacterium busily produced frog rRNA. Never before had such a genetically altered organism existed.*

Section Objectives

- List the four steps involved in a genetic engineering experiment.
- Explain how recombinant DNA is produced.
- Describe how restriction enzymes are used in genetic engineering.

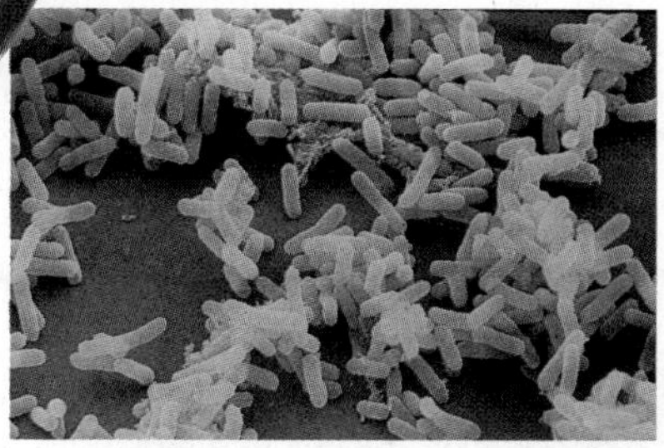

Figure 10-1 In their laboratories in San Francisco, Stanley Cohen and Herbert Boyer successfully put together the first genetically engineered organism. They used a gene from the frog *Xenopus laevis* and a chromosome from the bacterium *Escherichia coli.*

The Basics of Genetic Engineering

The Cohen and Boyer experiment ushered in a scientific revolution in biology. It yielded laboratory techniques that enable researchers to locate, isolate, and study small segments of DNA obtained from much larger chromosomes. These methods are changing basic research in agriculture, medicine, and many other fields. Today, certain human genes can be transferred into bacteria to produce enormous amounts of the protein encoded by the human gene.

The process used to isolate a gene from the DNA of one organism and transfer the gene into the DNA of another is called **genetic engineering**. Genetic engineering involves building **recombinant DNA**, a molecule made from pieces of DNA from separate organisms. Every genetic engineering

SECTION 10-1

Vocabulary Preview

genetic engineering
recombinant DNA
vector
plasmid
cloning
restriction enzyme
Southern blot
gel electrophoresis
restriction fragment length polymorphism analysis
DNA fingerprint
polymerase chain reaction
Human Genome Project

Opening Question

Ask students to write the DNA nucleotide sequence GAATTC and its complementary strand. Have students look at the two "strands" and list anything that they have in common. Students may mention that both strands contain A, C, T, G; both strands are DNA; both strands are six nucleotides long; and the strands are the same if read in opposite directions. Then have students make up their own six-nucleotide sequences that are the same if each strand is read in the opposite direction.

Demonstration

Recombinant DNA

Have students make a model of recombinant DNA that they can look at during their study of this chapter. Recombinant DNA technology can be effectively modeled with colored Velcro® loop and hook fasteners, colored paper, or colored transparencies on the overhead projector. Use one color for the plasmid DNA and assemble the model in a circle. Use a second color for the gene to be inserted and yet another color for the chromosome receiving the gene. When cutting the model at restriction sites, identify the scissors as a restriction enzyme. Reassemble the two sources of DNA.

Effective Reading

Heads Tell a Story

Have students make a list of the heads and subheads used on pages 203–207 of Section 10-1. Help students recognize that the head on the page above ("The Basics of Genetic Engineering") announces the main topic of Section 10-1. On page 204, the boldface numbered heads in the list call out the steps of genetic engineering. These boldface heads appear again as main heads on pages 204–207. Encourage students to recognize that the heads tell a story, and that they can be used to preview the chapter material before reading begins and again as a review after reading is completed.

Teaching Tip
Evolution at Work

Screening cells involves events similar to those that occur during evolution. The environment (a culture medium containing an antibiotic) selects the organisms to survive (the bacteria with the gene for antibiotic resistance). The surviving bacteria with the successful trait produce the future generations of bacteria.

CONNECTIONS

Chapter 8
The Genetic Code

Recombinant DNA technology is possible because of the universality of the genetic code among organisms as diverse as bacteria, elephants, and pine trees.

Chapter 9
Gene Expression

Environmental conditions that control gene expression can complicate genetic engineering studies. Sometimes the four steps in a gene transfer experiment can be flawless, but the gene isn't expressed because other controls haven't been "turned on."

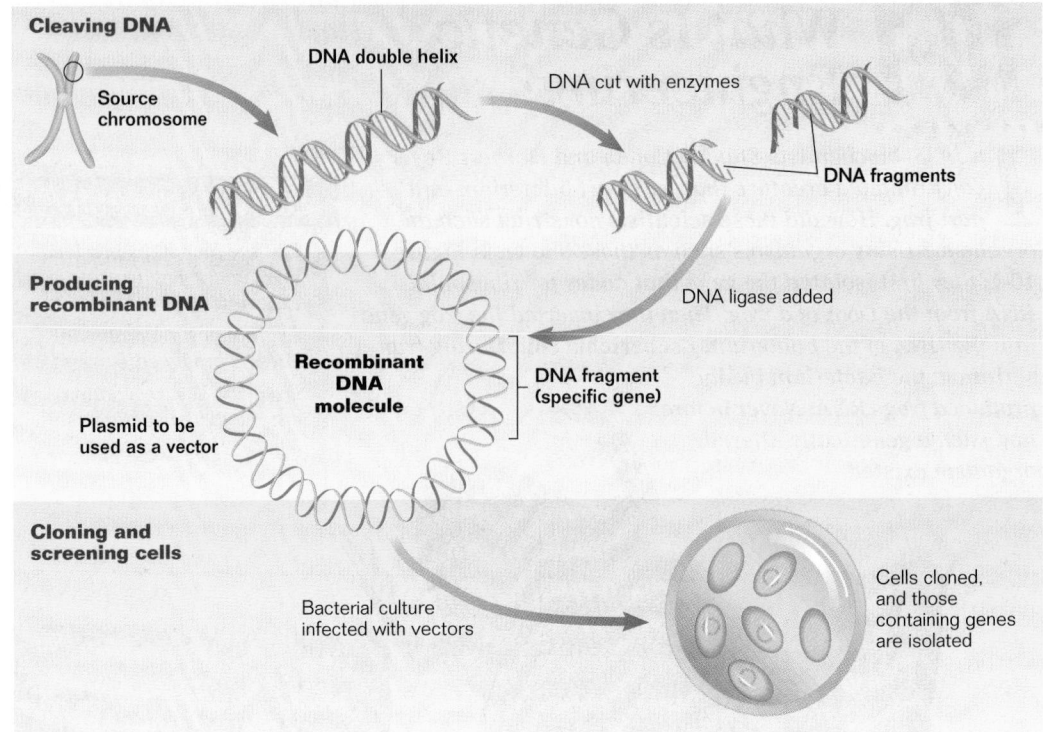

Figure 10-2 A genetic engineering experiment has four basic steps: cutting the gene of interest out of its source chromosome; inserting that gene into a vector that will carry the gene into bacterial cells; cloning the infected bacteria; and finding the cells that have taken up the vector with the gene.

experiment presents unique problems, but all share four distinct steps, which are illustrated in Figure 10-2.

1. **Cleaving DNA**   The DNA containing the gene of interest (the gene to be transferred) is cut into fragments using special enzymes that cleave, or separate, sequences of nucleotides.

2. **Producing recombinant DNA**   In genetic engineering, recombinant DNA is made when a DNA fragment is put into the DNA of a **vector,** an agent that is used to carry the fragment into another cell. Commonly used vectors include viruses and plasmids *(PLAZ mihds)*. A **plasmid** is a circular DNA molecule, usually found in bacteria, that can replicate independently from the main chromosome.

3. **Cloning cells**   A culture of bacteria is infected with the fragment-containing vectors. Some of the bacteria will take in the vectors. These cells are isolated and allowed to reproduce. Growing a large number of genetically identical cells from a single cell is called **cloning.**

4. **Screening cells**   Bacterial cells that have received the particular gene of interest are identified and isolated.

   Perhaps the best way to understand the steps in a genetic engineering experiment is to follow one from the beginning to the end. Let's go back and learn how Cohen and Boyer transferred the frog rRNA gene into the DNA of a bacterium.

 Did You Know?

Our chromosomes aren't pure "human" DNA. We probably all carry viruses that are now a part of our DNA and reproduce as our DNA replicates. In addition, our DNA contains evolutionary remains that link our DNA to other closely related organisms.

Cleaving DNA

In their experiment, Cohen and Boyer sought to transfer a gene that codes for ribosomal RNA in the African clawed frog, *Xenopus laevis*. To cut this gene out of the chromosome in which it was found, the investigators performed a type of molecular surgery using bacterial enzymes called restriction enzymes. **Restriction enzymes** cleave DNA at specific sequences, generating a set of small fragments of DNA. These sequences are unusual because they are made of two strands of DNA that have the same nucleotides running in opposite directions. For example, Cohen and Boyer used a restriction enzyme called *Eco*R1, which recognizes the sequence GAATTC. Try writing the sequence of the complementary strand; it is CTTAAG, the same as the original sequence written backward, as shown on the left side of Figure 10-3.

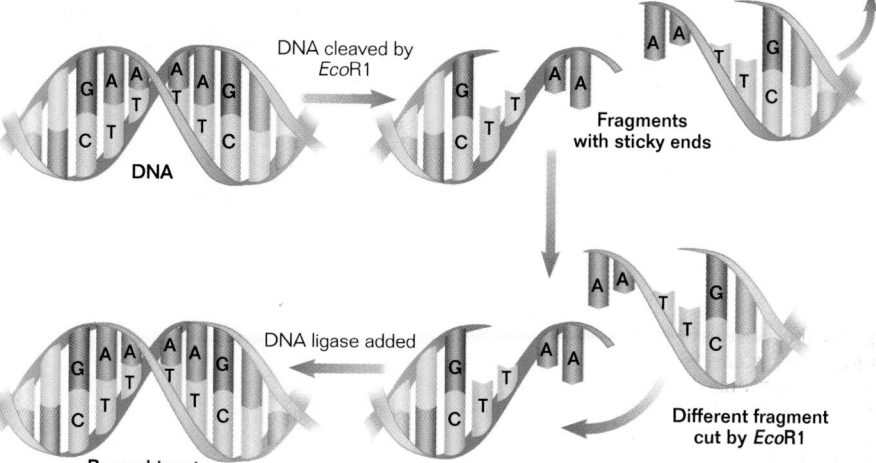

Figure 10-3 *Eco*R1 recognizes the nucleotide sequence GAATTC and makes its cut after the G. Any fragments cut by *Eco*R1 have the same sticky ends and therefore can be joined with DNA ligase to make recombinant DNA.

Many restriction enzymes do not make their incision in the center of the sequence; rather, the cut is staggered, made to one side of the sequence. For example, in the sequence GAATTC, *Eco*R1 makes its cut after the nucleotide G, as shown in Figure 10-3. The cuts produce fragments of DNA with short single strands dangling from each end. Because these dangling tails are complementary to each other, they are called cohesive ends, or "sticky ends." These ends can pair with each other, and the cuts can be sealed with the aid of an enzyme called DNA ligase. Or, the sticky ends can pair with *any other DNA fragment cut by the same restriction enzyme* because these would have the same complementary sticky ends. Any two fragments of DNA cut by the same restriction enzyme can be joined together, as shown in Figure 10-3. Fragments of elephant and ostrich DNA cleaved by the same enzyme can be joined just as readily as two fragments of bacterial DNA, because they have the same complementary sequences at their ends. ◻

◻ CAPSULE SUMMARY

Restriction enzymes recognize specific nucleotide sequences and cleave DNA into fragments with short sticky ends. The sticky ends enable DNA fragments from different organisms to join together.

SECTION 10-1

Teaching Tips
Defense Against Invaders
Restriction enzymes apparently evolved in bacteria as a defense against DNA invaders. The enzymes destroyed the foreign DNA before it could interfere with cell functions.

Universality of DNA
Restriction enzymes can leave single-stranded tails of DNA that will combine with any other complementary strand, whether the DNA is from the same organism or a different one. Emphasize to students that the nucleotide sequence is what is important, not the source of the DNA.

CONNECTIONS
Chapter 4
Energy and Metabolism
Restriction enzymes share the characteristics of enzymes found in energy metabolism. They are large proteins that speed reaction rates and are specific for certain reactants (substrates).

Historical Note

Guidelines for Genetic Research
A member of the Human Genome Project and an expert on nucleic acid chemistry and the biochemistry of animal viruses, Dr. Maxine Frank Singer was one of the first scientists to alert the National Academy of Sciences to the potential hazards of gene splicing. Due to the efforts of Dr. Singer and her colleagues in 1973, the National Institutes of Health (NIH) have developed specific guidelines for genetic research. The guidelines regulate the production and use of genetically engineered DNA as well as any organism that may carry recombinant DNA. All United States institutions receiving funding from the NIH are required to physically contain genetic research and to minimize the safety risks involved in their research projects.

Figure 10-4
Ask students if the DNA in the figure is linear or circular (*circular*), to identify the source of the DNA (the bacterium *E. coli*), and to describe how it got its name (*p = plasmid, SC = Stanley Cohen, 101 = the 101st plasmid they isolated*).

Teaching Tip
Producing Recombinant DNA

Give one-half of the students strips of colored paper and the other one-half strips of white paper. Students with colored paper should write these nucleotides on one side of their paper:

CCCTCGATTGAATTCGAACCC

and then write the complementary strand below it:

GGGAGCTAACTTAAGCTTGGG

Tape the colored paper in a circle. Students with white paper should write these nucleotides on one side of their paper:

AGAATTCCTCGTCAGAATTCG

and then write the complementary strand below it:

TCTTAAGGAGCAGTCTTAAGC

Both groups of students should cut the paper at any *Eco*R1 restriction site that they find; for example, find

GAATTC

and cut after the G.
Then each person with colored paper should pair up with a student with white paper and assemble their DNA fragments with tape. Discard DNA fragments that do not have restriction sites at both ends (which will be both ends of the white paper). Display the "recombinant DNA" around the room during the study of this chapter.

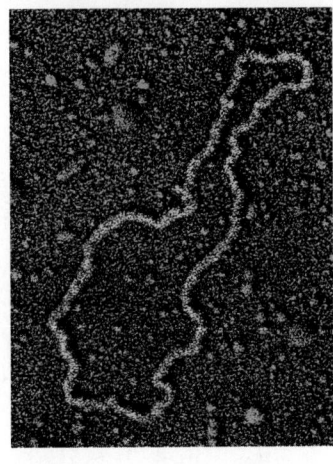

Figure 10-4 This circular molecule of DNA, pSC101, was the first plasmid to successfully carry a gene from a frog into a bacterium.

Producing Recombinant DNA

To make a molecule of recombinant DNA, Cohen and Boyer used the restriction enzyme *Eco*R1 to cut apart a large plasmid, which would be used as the vector to carry the ribosomal RNA frog gene. From this plasmid DNA they isolated a fragment that included two important genes: the gene for plasmid DNA replication and the gene that makes the cell carrying this plasmid resistant to the antibiotic tetracycline (*teh trah SY klihn*). These two genes would be important later in the experiment, when they would be used to identify cells that took in the frog gene.

Because both ends of this fragment from the large plasmid were cut by the same restriction enzyme, they could be joined together to form a circular molecule of DNA. This smaller plasmid, shown in Figure 10-4, was called pSC101 (because it was the 101st plasmid isolated by Stanley Cohen). Cohen and Boyer produced recombinant DNA by mixing the frog's DNA fragments and pSC101. This process is illustrated in Figure 10-5.

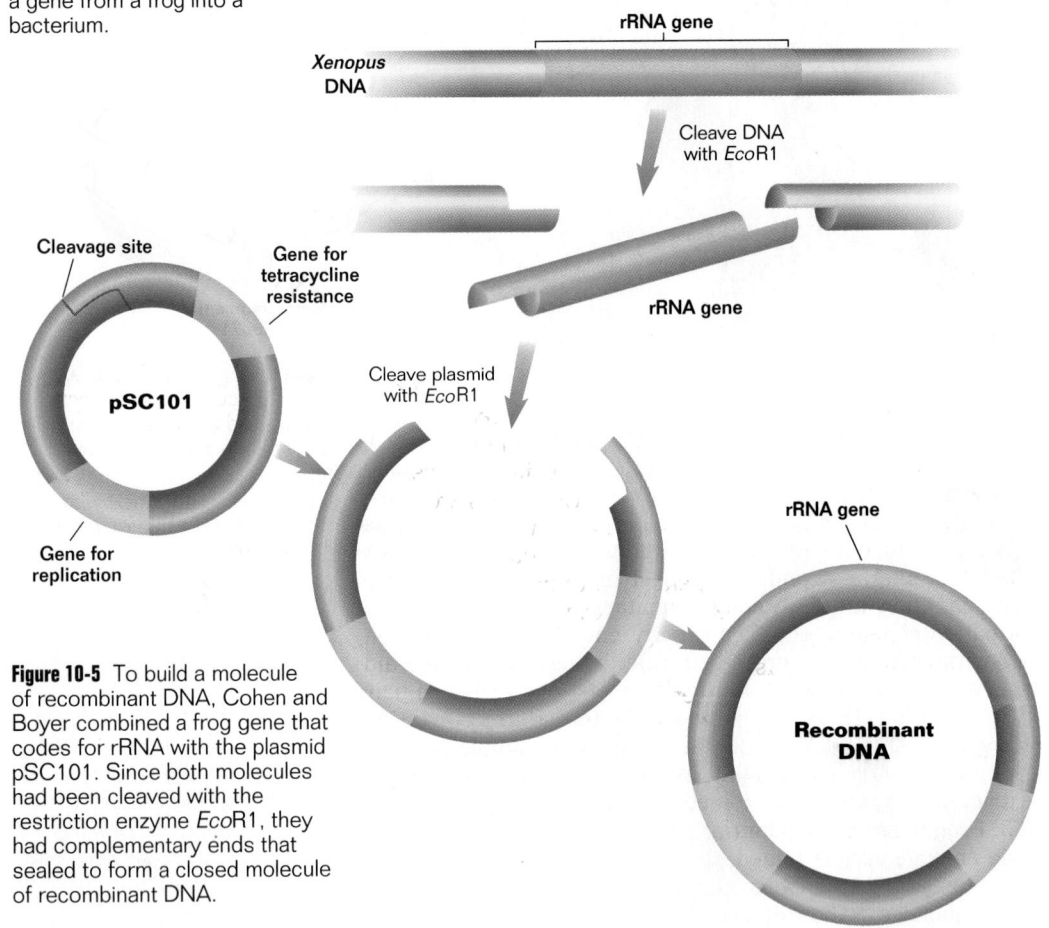

Figure 10-5 To build a molecule of recombinant DNA, Cohen and Boyer combined a frog gene that codes for rRNA with the plasmid pSC101. Since both molecules had been cleaved with the restriction enzyme *Eco*R1, they had complementary ends that sealed to form a closed molecule of recombinant DNA.

Overcoming Misconceptions

The Genetic Code Is Universal

Students often have trouble believing that a gene from one organism can be expressed in another organism, even one that is classified in a different kingdom. Remind students that because the genetic code is universal, the codons that specify an organism's hereditary information can be transcribed and translated into the same amino acids, and therefore, the same protein, in any other organism. You may wish to have students look again at Table 9-1 on page 186, which shows the genetic code. Ask students if genetic engineering could be possible without this universality.

Cloning Cells

Once the recombinant DNA that would serve as a vector was made, Cohen and Boyer were ready to introduce it into *E. coli* bacteria. They treated growing cultures of bacteria so that the cells would take up the recombinant DNA. Because only a few cells would take up the vectors, a method was needed to identify them. This is when the gene for tetracycline resistance became important. Cohen and Boyer added the antibiotic tetracycline to the bacterial cultures. As shown in Figure 10-6, the only cells that were not killed by the antibiotic were those that had become resistant to tetracycline because they had taken up the vector. All bacterial cells without vectors were eliminated. Every surviving cell was then allowed to reproduce, forming a colony of identical cells, or clones, all of which had the recombinant DNA. Cohen and Boyer obtained thousands of these clones.

□ *CAPSULE SUMMARY*

Cohen and Boyer made recombinant DNA consisting of a frog rRNA gene and part of a plasmid. The plasmid's gene for tetracycline resistance was used to identify bacteria that took up the recombinant DNA.

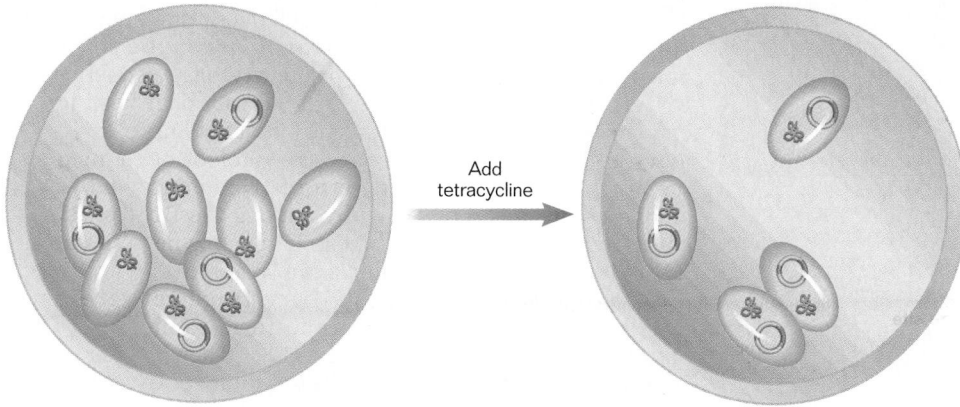

Add tetracycline

Screening Cells

Once Cohen and Boyer eliminated the bacteria that did not contain the recombinant DNA, they needed to test each clone to see if the frog rRNA gene was indeed present. A useful procedure for identifying a specific gene is known as the Southern blot. The **Southern blot** is a technique that uses radioactively labeled RNA or single-stranded DNA as a "probe" to identify a specific gene. The probes have specifically ordered nucleotide bases that are complementary to the sequence of bases in the gene being sought. In a Southern blot, the cloned DNA is first cleaved into fragments by restriction enzymes. The fragments are separated by **gel electrophoresis** (*ee lehk troh fuh REE sihs*), a technique that uses an electrical field within a gel to separate molecules in a mixture. Because DNA is negatively charged, the various fragments move through the gel according to their size, forming a pattern of bands. The DNA fragments are then split into single-stranded DNA, which is then blotted onto filter paper. Afterward, the filter paper is moistened with a solution

Figure 10-6 To identify bacterial cells that contained the vector, Cohen and Boyer added the antibiotic tetracycline to the bacterial cultures. Only the cells that had taken up the vectors were resistant to tetracycline and survived.

Teaching Tip
Basics of Genetic Engineering

Have students make a table that summarizes genetic engineering. Ask them to list the four steps in every genetic engineering experiment in the first column, describe the desired results of each step in the next column, and make sketches of each step (similar to those in Figure 10-2) in the last column.

Demonstration
Gel Electrophoresis

Divide your students into groups of one, two, or four individuals. Space the desks in your classroom randomly across the room. Have students line up at the back of the classroom and hold hands with members of their group. Signal the groups to walk to the front of the classroom all at the same time. They must find their way around the desks and not let go of their partners. The single people should arrive first and the groups of four last. Explain how the shorter strands of DNA also move through a gel faster than the long strands.

◉ VISUAL STRATEGY

Figure 10-7

Use this figure to explain how gel electrophoresis separates DNA by fragment length and how the radioactively labeled probes combine with specific DNA sequences. Have students look at Figure 10-7 and explain why some bands are closer to the bottom of the gel than others. *(The bands consisting of the largest DNA fragments move at the slowest rate and thus are found at the bottom of the gel.)* Then have students explain why some bands combined with the labeled probe and others did not. *(Bands that combined with the probe did so because the labeled probe was complementary to their base sequences, indicating that the bands represented the desired genes.)*

Application

Technology DNA fingerprinting is used not only to solve crimes, but also to reunite families separated by war and tragedy. Children whose parents have disappeared have been reunited with grandparents after years of separation. After her pregnant daughter was abducted in Argentina in 1977, Haydeé Lemos spent years searching for a missing grandchild she had never seen. Ms. Lemos's search led her to a woman who was raising an eight-year-old child as her own. In 1985 DNA fingerprints proved that the girl was actually Ms. Lemos's missing grandchild. Although Ms. Lemos still has no proof of what happened to her daughter and son-in-law, both of her granddaughters now live with her. Other parents and grandparents hope that DNA fingerprints will be able to provide some news about their own missing children and grandchildren.

◻ CAPSULE SUMMARY

Cohen and Boyer made radioactively labeled RNA that had a nucleotide sequence complementary to the frog rRNA gene. They used it to identify the gene in bacterial clones.

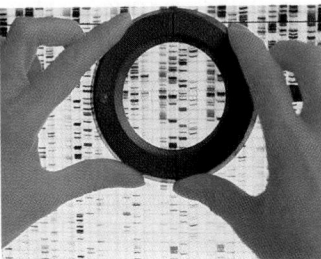

Figure 10-7 When fragments of DNA are separated by gel electrophoresis, they form a pattern of bands called a DNA fingerprint, which can be analyzed and used to establish identity.

containing the radioactively labeled probes. Among the thousands of DNA fragments, only the fragments that contain the gene of interest will bind with the probes, because of the complementary nucleotide sequence. Cohen and Boyer searched patiently and eventually uncovered cells containing the DNA that paired with the ribosomal RNA probe. These are the cells Cohen and Boyer sought—bacterial cells that contained the frog rRNA gene.

While the details vary from one experiment to another, all genetic engineering experiments employ the same basic strategy used in this first successful gene transfer. ◻

Identifying Sequences in DNA

The knowledge that restriction enzymes cut DNA at specific sites has produced powerful tools that have greatly affected the biological sciences. One such tool is **restriction fragment length polymorphism** (RFLP) **analysis**, used to identify base sequences in DNA. RFLP analysis is based on the fact that restriction enzymes cut DNA into fragments that have specific lengths. Because of differences in nucleotide sequences, the number of cutting sites for a restriction enzyme varies among alleles and, consequently, among the DNA of different individuals. Therefore, a restriction enzyme will cut the DNA from different individuals into fragments with different lengths. When samples of these fragments are separated by gel electrophoresis, each forms a characteristic pattern of bands, shown in Figure 10-7. These bands represent a unique array of RFLP sites, which serve as an individual's **DNA fingerprint.**

RFLP analysis can be used to establish the identity of a person. Because it can determine a sequence of nucleotides from a sample of DNA found in blood, semen, bone, and hair, RFLP analysis has been useful in forensics. RFLP analysis is also valuable for identifying the genes that cause genetic disorders, which sometimes have unique restriction sites. Alleles for Huntington's disease, sickle cell anemia, and a number of other genetic disorders are currently detected in this way.

In 1985, Kary B. Mullis, a biochemist working for a biotechnology corporation in California, discovered a simple way to make unlimited copies of a gene, a process now known as **polymerase chain reaction** (PCR). The starting materials for PCR include a single molecule of DNA (even a short fragment will do), DNA polymerase (the enzyme responsible for DNA replication), and a supply of all four nucleotides. When all these ingredients are incubated in a test tube, millions of copies of a DNA segment can be made in a few hours, as shown in Figure 10-8. PCR has revolutionized research in molecular biology, giving biologists as much of the particular DNA as they need. Prior to this technique, a piece of DNA was

Historical Note

A New Forensic Tool

The first time police used DNA fingerprinting on a murder suspect, the test cleared an innocent man of the charges. Although police in Narborough, England, were convinced that a suspect had committed two murders, and the man had even confessed to one of the murders, in December 1986 it was proven that he had not committed either murder. With the help of Alec Jeffreys, of the University of Leicester, and his DNA fingerprinting system, police collected and tested blood samples from every possible suspect (more than 4,500 men) in Narborough and a nearby town. DNA fingerprinting identified the real murderer, who was convicted of both murders.

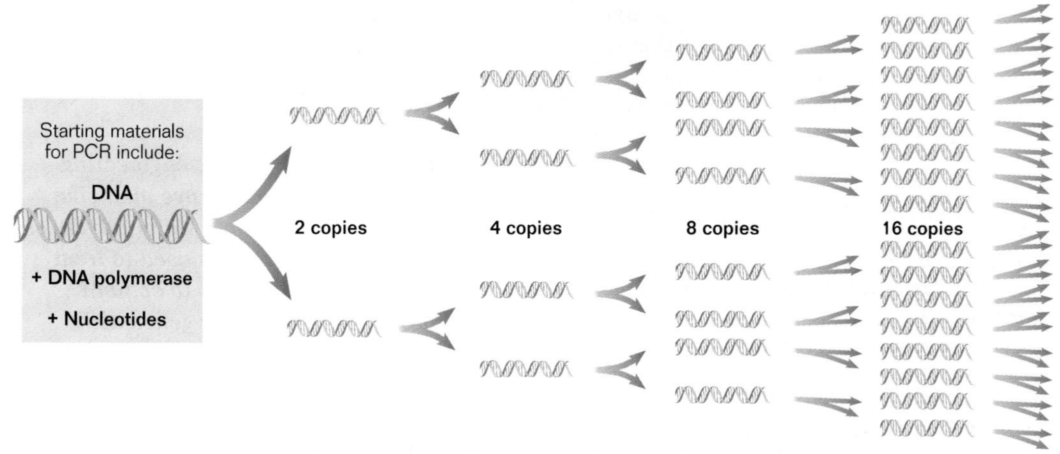

Starting materials
for PCR include:

DNA

+ DNA polymerase

+ Nucleotides

2 copies   4 copies   8 copies   16 copies

Figure 10-8 Using DNA polymerase and nucleotides, the process of polymerase chain reaction produces millions of copies of DNA in just a few hours.

cloned by inserting it into a plasmid or a virus, a process that usually took weeks. In the past few years, PCR has been used to duplicate DNA from drops of blood and semen found at crime scenes, from the cells of an embryo for prenatal diagnosis, and even from a 40,000-year-old woolly mammoth frozen in a glacier. Many more applications are sure to follow.

Mapping the Human Genome

The potential ability of gene technology to aid the fight against disease is great, and many efforts are underway to increase its usefulness. One of the most significant efforts is the **Human Genome Project,** a research effort to identify and locate the entire collection of genes in a human cell. Goals of this project include improving existing human genetic maps, constructing physical maps of entire chromosomes, and ultimately determining the complete DNA sequence in the human genome. In 1993, a preliminary map of the entire human genome was completed using RFLP analysis. One major goal for the future of the project is to develop new methods for mapping and sequencing DNA.

CONTENT LINK
....
Read about the social and ethical issues surrounding the Human Genome Project on **page 222.**

Section Review

1. *Describe the four steps of the first experiment that successfully transferred a gene from a frog to a bacterium.*

2. *What is recombinant DNA? What role does it play in genetic engineering?*

3. *What is unique about the cuts that restriction enzymes make to DNA?*

Critical Thinking

4. *What safety and ethical issues do you think might arise over the use of genetic engineering?*

Answers to Section Review

1. Step 1: Frog DNA was cleaved with restriction enzymes. **Step 2:** A frog DNA fragment was put into a bacterial plasmid to produce recombinant DNA. **Step 3:** Bacteria were infected with the recombinant DNA plasmid, and the bacteria containing the new DNA fragment were cloned. **Step 4:** Bacteria containing the frog DNA fragment were identified and isolated.

2. Recombinant DNA is made from two different organisms. Inserting new DNA into an organism provides the possibility of giving an organism the ability to produce new substances.

3. Restriction enzymes cut only at specific sequences in which the nucleotides are the same when read in opposite directions. Each restriction enzyme cuts at a unique sequence.

4. Answers will vary. Possible responses are creating dangerous organisms, inserting DNA into the wrong place and ruining a good gene, increasing the genetic load of a species, causing a disease instead of curing it, and creating a "super race" that replaces the original population.

Opening Question

Have students list all the ways they can think of that bacteria help humans live well. Answers may include food production (cheese, vinegar), recycling wastes, providing nitrogen for plants, and being part of the food chain. Then have students add the production of vaccines and drugs to their list.

Demonstration
Vaccines

Have students list the vaccines they have received. (If students are not sure which vaccines they have received, encourage them to find out by asking their parents or physicians.) Common vaccines include poliomyelitis, measles, rubella, and mumps. Why not be vaccinated against all diseases? (*Students' answers may include lack of vaccines, too many diseases, expense, slight risk of contracting the diseases, etc.*) Explain that many vaccines are attenuated-live preparations and that there is a danger that these vaccines can revert to virulence. However, genetic engineers can produce safer, noninfectious vaccines with recombinant DNA.

CONNECTIONS

Chapter 4
Energy and Metabolism

The use of human insulin produced through genetic engineering techniques allows people with diabetes mellitus to metabolize sugar normally.

Section Objectives

- Explain how genetic engineering can benefit human health.
- Describe two diseases that can be treated with products of genetic engineering.
- Explain how to genetically engineer a vaccine.

Figure 10-9 Large amounts of insulin can be obtained through genetic engineering. The gene for insulin is cut out of its chromosome and inserted into a bacterial plasmid. The bacteria containing the plasmids produce insulin, which can be collected to treat disorders such as diabetes.

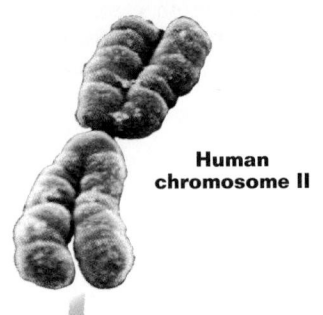

Human chromosome II

Insulin gene transferred

Bacterial plasmid

Gene for replication

Plasmid introduced into bacteria

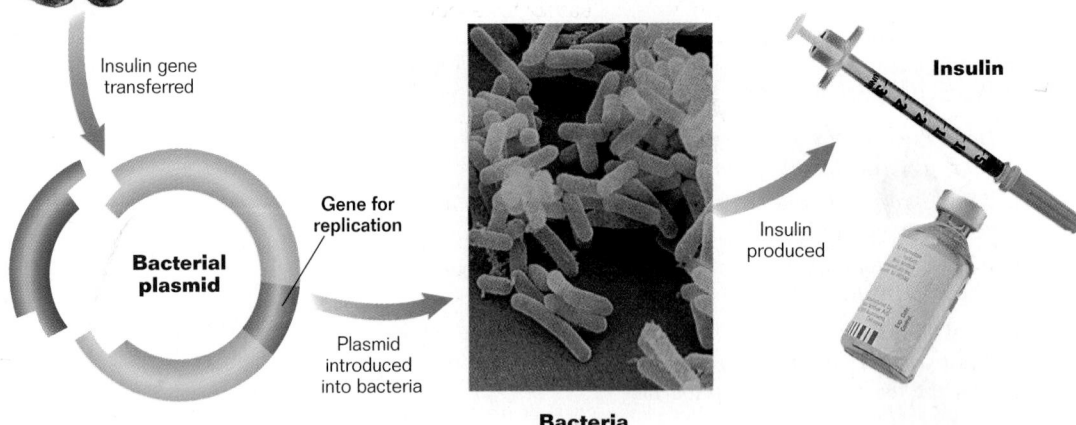

Insulin

Insulin produced

Bacteria culture

 Did You Know?

There has been an international debate over who owns a genome. Presently a lab that decodes a sample of DNA doesn't own the DNA, but genetically engineered organisms can be patented like any other invention.

10-2 The New Medicine

Much of the excitement about genetic engineering has focused on its potential to improve medicine by curing and preventing illnesses. Major advances have been made in the production of proteins used to treat illnesses, in the development of new vaccines to combat diseases, and in the replacement of defective genes with healthy ones.

Making Genetically Engineered Drugs

Many genetic disorders and other human illnesses occur when the body fails to make critical proteins that are essential for proper functioning. For example, **diabetes mellitus** (*muh LY tuhs*) **type I**, also called insulin-dependent diabetes, is an illness that occurs when the body cannot make sufficient amounts of the protein insulin. Diabetes mellitus type I can be treated by regular injections of insulin or by an insulin pump. However, insulin is typically present in the body in very low amounts, making the large quantities needed for pharmaceuticals difficult and expensive to obtain. With the availability of genetic engineering techniques, this problem has been largely overcome. The genes encoding the protein insulin can now be inserted into bacteria, which then produce insulin. Because the host bacteria can be grown cheaply in bulk, large amounts of insulin can be easily obtained, as shown in Figure 10-9.

In 1982, the U.S. Food and Drug Administration approved the use of the first commercial product of genetic engineering—human insulin. Today, hundreds of pharmaceutical companies around the world are busy producing other medically important proteins using the genetic engineering techniques you learned about in the last section. These products include **anticoagulants** (proteins involved in dissolving blood clots), which are effective in treating heart attack patients, and factor VIII, a protein that promotes blood clotting. A deficiency in factor VIII leads to hemophilia, an inherited disorder characterized by prolonged bleeding. For a long time, hemophiliacs received blood factors that had been isolated from donated blood. Unfortunately, some of the donated blood was infected with viruses such as HIV and the hepatitis B virus, which were then unknowingly transmitted to those people who received blood transfusions. Today, the use of genetically engineered factor VIII eliminates the risks associated with blood products obtained from other individuals. Other genetically engineered pharmaceutical products are listed in Table 10-1.

Table 10-1 Genetically Engineered Medicines

Product	Examples and Uses
Colony-stimulating factors	Growth factors that stimulate white blood cell production; used to treat immune system deficiencies and to fight infections
Erythropoetin	Stimulates red blood cell production; used to treat anemia in individuals with kidney diseases
Growth factors	Stimulate differentiation and growth of various cell types; used to promote wound healing
Human growth hormone	Used as a treatment for dwarfism
Interferons	Interfere with reproduction of viruses; also used to treat some cancers
Interleukins	Activate and stimulate different classes of white blood cells; can be used in treating wounds, HIV infections, cancer, immune deficiencies

Making Genetically Engineered Vaccines

A vaccine is a solution containing a harmless version of a pathogen (disease-causing microorganism) or its toxins. When a vaccine is injected, the recipient's immune system will recognize the pathogen's surface proteins. The immune system will then respond by making defensive proteins called **antibodies**, which will later combat the pathogen.

CONTENT LINK

You will learn more about the immune system in **Chapter 39**.

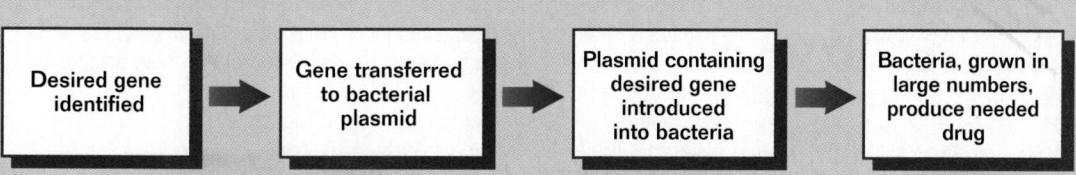

Teaching Tips

Making Vaccines

Have students fold a piece of paper into thirds and label each part this way:

No Vaccine, Catch the Disease

Vaccine From Killed or Weakened Pathogenic Microbes

Vaccine From Genetic Engineering

Students should then list or explain the pros and cons of each method of obtaining immunity.

What If a vaccine is improperly prepared?

There have been several recorded instances where vaccines have been improperly prepared, with the result that active disease-causing agents (instead of weakened, inactive forms of the pathogen) were accidentally introduced into individuals being vaccinated. In an incident several years ago, a number of children developed paralytic poliomyelitis after receiving polio vaccines containing active polio viruses.

Curing Genetic Disorders

Distinguish between a treatment for a disease and a cure for a disease. Genetic diseases have been treated for many years, but the prospect of cures has arrived just recently. Have students use information in the earlier part of this section to describe treatments for genetic diseases, then have them describe what a cure would require.

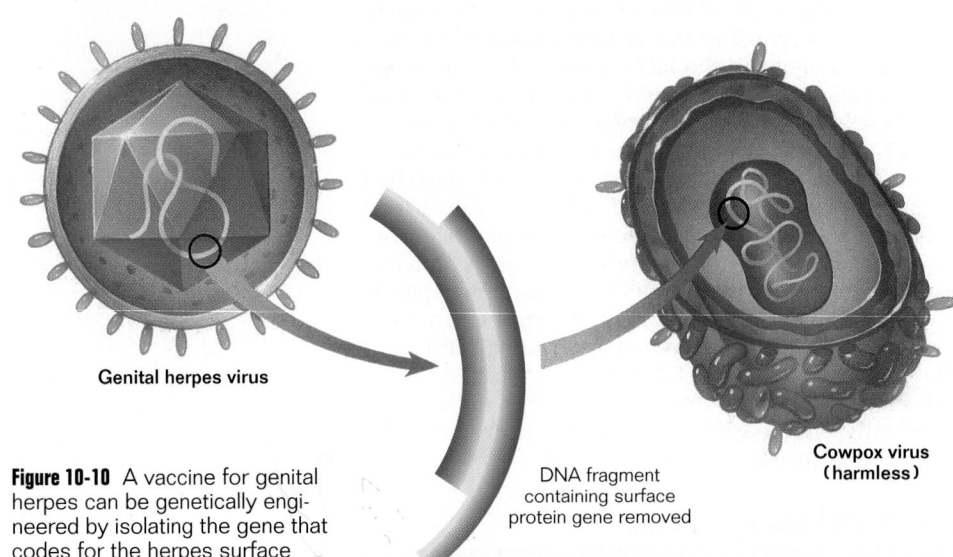

Genital herpes virus

DNA fragment containing surface protein gene removed

Cowpox virus (harmless)

Figure 10-10 A vaccine for genital herpes can be genetically engineered by isolating the gene that codes for the herpes surface protein and inserting it into a harmless cowpox virus. The virus will now manufacture the herpes surface proteins. A person vaccinated with the engineered virus will make the antibodies against the virus.

CAPSULE SUMMARY

Genetic engineering techniques are being used to manufacture proteins such as insulin and factor VIII, as well as vaccines.

Antibodies will stop the growth of the pathogen before the disease it causes can develop.

Traditionally, vaccines have been prepared either by killing a specific pathogenic microbe or by making it unable to grow. This ensures that the vaccine will not cause the disease. The problem with this approach is that any failure in the process to kill or weaken a pathogen will result in the transmission of the disease to the very patients seeking protection. While the majority of vaccines are safe, a fraction of a percentage of vaccines cause the treated individuals to contract the disease. This small, but real, danger is one of the reasons why rabies vaccines are administered only when a person has actually been bitten by an animal suspected of carrying rabies.

Today there is a new and much safer method of making vaccines, as illustrated in Figure 10-10. Using genetic engineering techniques, the genes that encode the pathogen's surface proteins can be inserted into the DNA of harmless bacteria (or viruses). The modified, but still quite harmless, bacteria become an effective and safe vaccine. These harmless bacteria can be used to stimulate the body to make the antibodies that will attack the pathogen. As a result, the body is protected against infection.

Among the vaccines now being manufactured in this way are ones directed against the herpes II virus (which produces small blisters on the genitals) and hepatitis B virus (which causes a sometimes fatal inflammation of the liver). A major effort is underway to produce a vaccine that will protect people against malaria, a disease caused by a protozoan for which there is currently no effective protection against infection. ◻

Multicultural Perspective

Genetic Diseases of Different Cultures

Scientists are currently using knowledge gained from genetic technology to find a prevention or cure for four serious genetic defects that affect people from particular environments: sickle cell anemia (affecting people from Africa, the Mediterranean, India, and Asia), thalassemias (affecting people in the malarial areas of the Middle East, Africa, the Mediterranean, and India), Tay-Sachs (affecting Ashkenazi Jews), and tyrosinemia (affecting French-Canadian children in the isolated Lac St. Jean-Chicoutimi region of Quebec).

Curing Genetic Disorders

In 1990 the first attempts were made to use genetic engineering to combat genetic disorders. Many genetic disorders arise when an individual lacks a normally functioning copy of a particular gene. One obvious way to cure such disorders is to give the person a working copy of the gene. Until recently this approach was not practical for three reasons. First, the defective gene was difficult to identify and isolate. Second, it was hard to transfer a healthy copy of such a gene into the cells of body tissues that use it. Finally, it was necessary to find a way to keep the altered cells or their offspring alive in the body for a long time. With genetic engineering, it is now possible to overcome these difficulties.

One of the first gene therapy attempts involved two young girls, shown in Figure 10-11, who suffered from an immune system disorder caused by a defective gene. Doctors extracted bone marrow cells from the girls and replaced the defective gene, which failed to produce an important immune-system enzyme, with a normal gene. These cells were returned to the girls' bones and began to produce the missing enzyme. Because this kind of bone marrow cell actively divides, researchers hope that offspring of the genetically engineered cells will continue to secrete the enzyme into their blood for a long time.

Genetic engineering is also providing a new and powerful weapon in the battle against cancer. All humans have white blood cells that secrete a protein called tumor necrosis factor (TNF). TNF attacks and kills cancer cells. Unfortunately this does not happen often. Genetic engineers recently developed a method of adding the TNF gene to a kind of white blood cell that is very effective at locating cancer cells but not very effective at harming them. Once armed with this TNF gene, however, these white blood cells will secrete TNF and kill cancer cells. Genetic engineering has enabled these cells to become like cruise missiles with a deadly payload zeroing in on cancer cells. ❑

Figure 10-11 Ashanthi DeSilva, *top,* and Cynthia Cutshall, *bottom,* were among the first patients to receive gene therapy. Three years after their treatment, they both appear to be thriving.

❑ CAPSULE SUMMARY

Some human genetic disorders are being treated and "corrected" by inserting copies of the corresponding normal gene into individuals whose copy of the gene is defective.

Section Review

1. *Discuss two ways human health has benefited from genetic engineering.*

2. *Explain how genetic engineering can be useful in the treatment of human illnesses such as diabetes mellitus type 1 and hemophilia.*

3. *Explain how vaccines can be genetically engineered.*

Critical Thinking

4. *What other illnesses or disorders can you think of whose treatment might benefit from genetic engineering?*

SECTION 10-2

Teaching Tip

Genetic Engineering Versus Evolution
Evolution favors efficiency in metabolism. Bacteria that are producing human genes take time and energy producing something that they don't need. Genetic engineers have to cope with the recurring problem that bacteria tend to get rid of genes they don't need, even if it took much time, trouble, and money to insert those genes. For example, bacteria that have received a gene for insulin do not need insulin; therefore, these bacteria may turn off the insulin gene to conserve energy.

Answers to Section Review

1. Answers will vary. Possible responses are that genetic engineering provides safe drugs to treat diseases, safe vaccines, cures instead of just treatments for genetic diseases, and new ways to fight disease.

2. Human insulin and human blood clotting factors can be produced by bacteria using genetic engineering techniques with no danger of infectious agents such as HIV or hepatitis virus.

3. The surface proteins of pathogenic bacteria are placed in harmless bacteria, then used for vaccines. The vaccinated subject's immune system produces antibodies to pathogenic bacteria without actually being exposed to them.

4. Answers will vary. Possible responses are cancer, allergies, mental illness, poisoning, treatment of stings and bites, and autoimmune diseases such as arthritis.

Opening Question

Ask students to design, draw, and name an organism that would produce "the perfect food." For instance, instead of a cow producing plain white milk, imagine one that produces chocolate milk. Then have the students list the biological changes that would be needed to produce that organism. (*Example: A chocolate milk cow would produce elevated milk sugar, and would need a gene to produce chocolate.*)

Demonstration
New Solutions to Old Problems

To illustrate how important pests, weeds, and fertilizer are to a gardener, bring in ads from nurseries, and distribute the ads to groups of two to three students. Have the students list the products that are sold to fight these three types of gardening problems in three columns. Students should then describe at the bottom of each column how the products could have a detrimental effect on the rest of the environment. Finally, have the students read Section 10-3, compare solutions offered by genetic engineering to traditional solutions, and add this information to each column.

◉ VISUAL STRATEGY

Figure 10-12
Have students compare this picture with Figure 10-4. If all plasmids are small circles of DNA, how can a genetic engineer tell one plasmid from another? (*Answers may include the DNA sequence, restriction sites, and use as a vector.*)

10-3 The New Agriculture

One of genetic engineering's greatest successes has been the manipulation of genes in crop plants and livestock. Gene transfers have resulted in crop plants that are more resistant to plant diseases and pests, as well as tomatoes that have a longer shelf life. Genetic engineering has also been used to boost milk production in dairy cows and increase the growth rate of certain livestock.

Transporting Genes Into Plants

The key to the great progress in genetic engineering of plants in recent years was the discovery of a suitable vector to transport a gene from one plant to another. For years, genetic engineering in plants was not possible because, unlike bacteria, plants have few viruses or plasmids that can perform this critical role. Said simply, genetic engineers lacked a suitable vector to carry the gene into plant cells.

The breakthrough came in the form of an unusual bacterial plasmid responsible for crown gall, a disease characterized by large bulbous tumors. This plasmid is called the Ti plasmid ("Ti" stands for tumor-inducing). The Ti plasmid easily infects broadleaf crop plants such as tomatoes, tobacco, and soybeans. When it has infected a plant cell, it proceeds to insert itself into the plant cell's chromosome. To make a genetic engineering vehicle, scientists removed the tumor-causing genes from the Ti plasmid. The vacant space in the now-harmless plasmid was then filled with DNA introduced by the scientists, as shown in Figure 10-12. This DNA could then be carried into the chromosomes of a target plant.

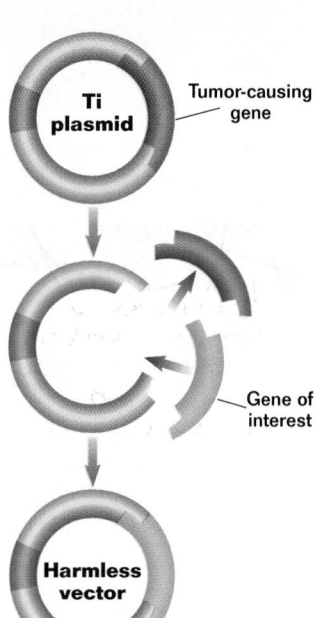

Figure 10-12 Genetic engineering can turn a tumor-causing Ti plasmid into a suitable vector for broadleaf crops. The tumor-causing gene is removed and replaced with a gene of interest, which can then be carried into plant cells by bacteria.

Making Crops Resistant to Herbicides and Insects

A recent improvement in agriculture is the development of crop plants that are resistant to the herbicide **glyphosate**, a powerful, biodegradable weedkiller. Glyphosate kills most actively growing plants by destroying an enzyme needed to make certain kinds of protein. Genetic engineers found a bacterial strain that could make the enzyme despite the presence of glyphosate. Then they isolated the gene encoding the enzyme. Using the Ti plasmid as a vector, genetic engineers then successfully introduced the resistance gene into non-cereal crop plants. More recently, genetic engineers successfully "shot" the Ti plasmid into cells of wheat plants using a gene gun.

Herbicide-resistant crops are advantageous; they lower the cost of producing a crop because a field of crops resistant to glyphosate does not need to be weeded. The farmer simply treats the field by dragging a rope soaked in glyphosate across the field. All growing plants die except the crop, which is resistant to the herbicide, as shown in Figure 10-13.

The development of glyphosate-resistant crops is also beneficial to the environment. Glyphosate is quickly broken down in the environment, which makes its use a great improvement over most herbicides. Perhaps even more important, the tragic loss of fertile topsoil to erosion, one of the greatest environmental challenges facing our country today, would be greatly reduced if cropland were not intensively cultivated to remove weeds.

Another important advance in agriculture that has resulted from genetic engineering is the development of crops resistant to insect pests. Such crops do not need to be sprayed with pesticides; this would be a great benefit to the environment.

Consider cotton. Its fibers are a major source of raw material for clothing throughout the world, yet the plant itself can hardly survive in a field because of the many insects that attack it. Over 40 percent of the chemical pesticides used today are employed to kill insects that harm cotton plants. The world's environment would be better off if these thousands of tons of pesticides were not needed. Biologists are now in the process of producing cotton plants that are resistant to insects, so that pesticides will not be needed.

One successful approach uses a kind of soil bacterium that produces enzymes that attack and kill the larvae of moths and butterflies, important pests of some crops. When the genes producing these enzymes were inserted into the chromosomes of tomato plants, the plants began to manufacture the new enzyme. The enzyme made the tomato plants highly toxic to tomato hornworms, a pest that seriously damages tomato crops. ■

Figure 10-13 The effects of genetic engineering can be seen in this field of soybeans. The plants that have been genetically engineered to resist glyphosate continue to thrive after treatment with the herbicide.

❏ *CAPSULE SUMMARY*

Genetic engineers have manipulated the genes of certain kinds of crop plants to make them resistant to herbicides and destructive pests.

Developing Crops That Need No Fertilizer

Nitrogen is an element that all plants must have in order to make proteins and DNA. The most abundant source of nitrogen in the environment is atmospheric nitrogen, N_2. However, plants cannot obtain nitrogen from the air. All of the nitrogen that plants need must be obtained from the soil. How does this nitrogen get in soil? Bacteria living within the roots of soybeans, peanuts, and clover "fix" nitrogen by converting N_2 gas from the atmosphere into nitrates, nitrites, and ammonia, which are forms of nitrogen that plants can use.

Teaching Tips
Nitrogen and Fertilizer

Have students use their textbook as a resource to explain why plants must have nitrogen to live, and to describe how the production and use of fertilizer can pollute the environment.

What If a genetically engineered organism escaped from the lab? Genetic engineers always must take steps to ensure that a transformed organism cannot survive in the environment. One way to do this is to use organisms that have exacting nutritional needs that can be supplied in a laboratory, but not outside in the "wild."

Chapter Closure

Have students imagine and design their own genetic engineering projects. Projects should have a clear goal, such as to prepare a vaccine or gene therapy for a specific disease. Students should provide a flowchart as well as an explanation of their projects. Remind students to include steps to prevent genetically engineered organisms from being able to survive outside the lab should they escape. Encourage students to name preparations, enzymes, and plasmids to suit their projects.

☐ CAPSULE SUMMARY

Genetic engineers are looking for ways to transfer genes for nitrogen fixation from bacteria into crop plants.

☐ CAPSULE SUMMARY

The addition of genetically engineered growth hormone to the diets of livestock increases milk production in dairy cows and weight gain in cattle and hogs.

Because crops rapidly consume nitrogen, farmers replenish the soil by adding high-nitrogen fertilizers. Farming would be much cheaper and far more productive if major crops such as wheat, rice, and corn could be grown without such massive applications of fertilizer—or without any fertilizer at all. The task of genetically engineering major crops to carry out nitrogen fixation has become the focus of many researchers. The problem has not been an inability to discover and isolate the necessary genes or to get them into crop plants. The problem is that when genes are introduced into plants, they do not seem to function properly in their new host. Researchers all over the world are working to find a way around this difficulty. ☐

Improving Livestock Production

A very interesting advance in agriculture has been the introduction of growth hormone into the diet of dairy cows, which greatly improves milk production. Instead of extracting this hormone at great expense from the brains of dead cows, the relevant gene has been introduced into bacteria. The bacteria then produce the hormone so cheaply that it is practical to add it as a supplement to the cows' diet.

Extra copies of the gene encoding the same growth hormone have been introduced directly into the chromosomes of both cattle and hogs to increase their weight. Though still underway, these attempts promise to create new breeds of very large and fast-growing cattle and hogs. The human version of this same growth hormone is now being tested as a potential treatment for dwarfism, a disorder in which the pituitary gland fails to make adequate amounts of growth hormone. ☐

Section Review

1. *How is the Ti plasmid used to insert genes into plant cells?*
2. *Explain how genetic engineering techniques can make crop plants resistant to weedkillers such as glyphosate.*
3. *How can genetic engineering reduce the amount of insecticides used in agriculture?*

Critical Thinking

4. *In spite of FDA approval, the use of genetically engineered growth hormone in dairy cows has been controversial. Would you have concerns about consuming milk from cows treated with growth hormone? Why or why not?*

Answers to Section Review

1. The Ti plasmid infects broadleaf plants and inserts itself in the host plant's DNA.

2. Glyphosate destroys an enzyme that plants need. If a bacterial gene for making that protein in the presence of glyphosate can be inserted into a plant, it will survive herbicide treatment.

3. Instead of spraying an entire field with insecticide, a gene can be inserted into plants so that they will make their own insecticide, poisoning only insects that try to feed on the plant.

4. Answers will vary but must be rationally supported. Students may have the opinion that as long as the United States government tested and approved the milk, it should be safe. Other students may cite concerns about the possible presence of hormones or other byproducts in the milk.

Vocabulary

antibody (211)
anticoagulant (211)
cloning (204)
diabetes mellitus type I (210)
DNA fingerprint (208)
Human Genome Project (209)

gel electrophoresis (207)
genetic engineering (203)
glyphosate (214)
plasmid (204)
polymerase chain reaction (208)
recombinant DNA (203)

restriction enzyme (205)
restriction fragment length
  polymorphism analysis (208)
Southern blot (207)
vector (204)

CHAPTER REVIEW ANSWERS

Each item in the Chapter Review is correlated by text section in the assignment guide that follows.

ASSIGNMENT GUIDE

Section	Review	Themes Review	Critical Thinking
10-1	1, 2, 8, 10–12, 16–18, 21	24	25, 26
10-2	3–5, 7, 9, 13, 15, 19	22	
10-3	6, 14, 20	23	

Concept Mapping

Construct a concept map that shows the process and outcomes associated with genetic engineering. In constructing your map, use the following terms: restriction enzymes, DNA ligase, sticky ends, plasmids, DNA from host cells, recombinant DNA, agriculture, and medicine. Include additional concepts in your map as needed.

Review

Multiple Choice

1. Broken pieces of DNA can be joined together by
   a. restriction enzymes.
   b. DNA ligase.
   c. recombinant DNA.
   d. RFLP analysis.

2. Which is a characteristic of the fragments of DNA generated by using restriction enzymes?
   a. They are between four to six base pairs long.
   b. Each fragment has short, single-stranded "sticky ends."
   c. They contain identical nucleotides running in the same direction.
   d. The fragments unzip completely, forming single-stranded DNA.

3. What can scientists engaged in genetic engineering experiments do to prevent potential hazards?
   a. Use organisms unable to survive outside the laboratory.
   b. Conduct only moderately dangerous experiments.
   c. Restrict their use of the Southern blot technique.
   d. Immediately call the local police when an accident occurs.

4. Which human illness can be treated using a product of genetic engineering?
   a. malaria   c. flu
   b. hemophilia   d. a sinus cold

5. The product of a technique in which the genes that code for a pathogen's surface proteins are inserted into the DNA of harmless bacteria is
   a. a genetically engineered vaccine.
   b. the human growth hormone.
   c. factor VIII.
   d. the Ti plasmid.

6. The genetic engineering of crop plants that are resistant to the herbicide glyphosate is significant because
   a. its use reduces the erosion of topsoil.
   b. glyphosate is stable in the environment for many years.
   c. it also kills moth and butterfly larvae.
   d. glyphosate use increases the need for insecticides.

Concept Mapping

Map answer is shown on page 201B.

Review

Multiple Choice

Code in parentheses indicates section number and objective number.

1. b (10-1.2)
2. b (10-1.3)
3. a (10-2.1)
4. b (10-2.2)
5. a (10-2.3)
6. a (10-3.3)
7. d (10-2.1)
8. d (10-1.1)

Completion

9. insulin (10-2.1)
10. restriction enzymes, sticky ends (10-1.1)
11. genetic engineering (10-1.1)
12. Cohen, Boyer, vector (10-1.1)
13. erythropoietin, interleukin (10-2.2)
14. Ti plasmid, gene gun (10-3.1)
15. tumor necrosis factor (10-2.1)

CHAPTER REVIEW ANSWERS *continued*

Short Answer

16. The steps are cleaving DNA, producing recombinant DNA, cloning cells, and screening cells. (10-1.1)

17. Restriction enzymes cleave DNA at specific sequences, generating a set of small fragments of DNA. Then, fragments of DNA from separate organisms are put together to make recombinant DNA. (10-1.3)

18. Answers will vary. The knowledge gained by mapping and sequencing DNA during the first 10 years of the project can be used to make the Human Genome Project more cost effective and to increase the usefulness of gene technology in fighting diseases. (10-1.2)

19. Genetically engineered vaccines tend to be less expensive to make and less likely to cause the disease in people injected with the vaccine. (10-2.3)

20. Tomato plants are less likely to be damaged by hornworms because genes producing enzymes that can kill hornworms can be inserted into the tomato plant. Additionally, the gene that makes plants resistant to the herbicide glyphosate can be introduced into tomato plants, allowing them to be grown with a minimum of cultivation. (10-3.2, 10-3.3)

21. RFLP analysis produces RFLP sites, or DNA fingerprints, that can be used to identify base sequences in DNA. PCR enables researchers to increase the amount of DNA they have to work with by generating unlimited copies of a gene within a few hours. (10-1.4)

Themes Review

22. By changing the genetic makeup of individuals, genetic engineering alters the course of natural selection. The fittest members of a naturally evolving population might not be able to compete with individuals containing newly engineered genes. Additionally, the newly engineered genes are passed on to offspring. (10-2.1)

23. The insect population will be destabilized because it will lose a

---

7. A legal issue likely to be raised by the Human Genome Project is
  a. how methods used in the project can be applied to map plant genomes.
  b. if doctors should use the project data to improve health care.
  c. whether the genomes of animals should be mapped.
  d. whether employers should know if a genetic disorder has been inherited.

8. The process used to make unlimited copies of a particular gene is called
  a. RFLP analysis.
  b. Southern blot.
  c. DNA fingerprinting.
  d. polymerase chain reaction.

Completion

9. Human _____ was the first product made by genetic engineering to be marketed.

10. Genetic engineering depends on the ability of _____ to cut DNA at specific sites along its length. When DNA from two different organisms is cut by the same restriction enzyme and mixed together, the _____ can undergo complementary base pairing.

11. The process of moving genetic material from the DNA of one organism to the DNA of another organism is called _____ .

12. The first experiment that successfully demonstrated genetic engineering was conducted by _____ and _____ . In the experiment, a plasmid from *E. coli* served as the _____ .

13. The genetically engineered pharmaceutical _____ is used to treat anemia in patients with kidney problems. Another product called _____ is used to treat HIV infection and cancer.

14. Genes have been introduced into soybean plants using _____ as a vector and "shot" into wheat plants using a _____ .

15. The gene for the protein _____ has been genetically engineered into white blood cells that locate cancer cells.

Short Answer

16. A scientist is conducting a genetic engineering experiment. What four steps will the scientist follow?

17. What role do restriction enzymes play in the production of recombinant DNA?

18. Why do you suppose a high priority has been given to developing new methods for mapping and sequencing DNA during the first 10 years of the Human Genome Project?

19. Vaccines for several human infectious diseases have traditionally been made from weakened or killed bacteria and viruses. The presence of specific surface proteins that trigger the production of antibodies is the key to a vaccine's success. Genetically engineered vaccines are now used in place of several traditionally made vaccines. Why are these vaccines preferred?

20. Describe two advances in genetic engineering that have made large scale tomato farming more profitable than ever before.

21. Restriction fragment length polymorphism analysis and polymerase chain reaction are processes used by genetic engineers. Explain how each process is used in gene technology.

Themes Review

22. Evolution Natural selection is a mechanism of evolution whereby the members of a population who are best able to adapt to their environment survive and produce offspring. How is natural selection affected by genetic engineering?

23. Homeostasis Homeostasis may be viewed as the tendency of organisms to remain relatively stable within a range of conditions. How does an insect-resistant crop that is the product of genetic engineering destabilize a population of insects that depends on the crop for food?

---

food source. The insects may then become extinct, or they may adapt by finding other habitats or food sources. In any case, destabilization of the insect population will affect other organisms. (10-3.2)

24. Sticky ends are cohesive tails that dangle from each end of a DNA fragment. The sticky

ends can pair with any other DNA fragment cut by the same restriction enzyme. Fragments can then be sealed together by DNA ligase to form recombinant DNA. (10-1.2, 10-1.3)

Critical Thinking

25. The regulations were prompted by concerns that

genetically engineered organisms might cause disease or have harmful effects on the environment. (10-1.1)

26. You would use the restriction enzymes *Bam*H1 and *Bal*1. The gene is 1,071 base pairs long. To determine whether the procedure was successful, you would reinsert the plasmid into

24. Structure and Function When restriction enzymes are used to cleave specific nucleotide sequences of DNA, the process results in two "sticky ends." What are sticky ends? How do they function in the preparation of recombinant DNA?

Critical Thinking

25. Making Inferences The United States government has stringent regulations requiring researchers to confine genetically engineered organisms that are considered high risk to the laboratory. What concerns might have led to the enactment of these regulations?

26. Interpreting Data Examine the restriction map of the *E. coli* plasmid pBR 322, shown below. This is a commonly used plasmid composed of 4,363 base pairs. The map shows sites where certain restriction enzymes cut the DNA of the plasmid. For example, *Sph* 1 cuts the plasmid at base pair 566. Suppose you want to isolate from the plasmid the gene that codes for resistance to the antibiotic tetracycline, which is indicated at Tcr. What restriction enzymes would you use? How many base pairs long is the Tcr gene? How might you check to be certain that your procedure was successful?

Restriction Map of pBR 322 DNA

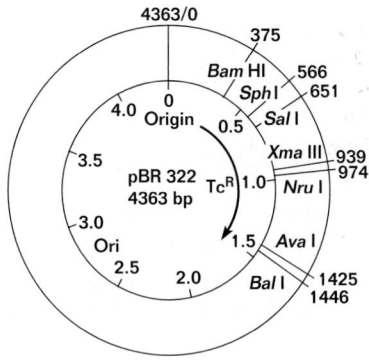

Activities & Projects

27. Research and Writing The question of awarding patents on genetically engineered organisms first arose when microbiologist Ananda Chakrabarty filed for a patent on a bacterium capable of digesting the components of crude oil. Find out about the bacterium that Chakrabarty engineered and about the court battle he waged for the right to obtain a patent for a genetically engineered organism.

28. Research and Writing The polymerase chain reaction (or PCR) is used by scientists to make multiple copies of single genes. Having multiple copies makes it easier to determine if the gene of interest is present in a sample. Find out how the PCR has been automated and how the automation has affected its use in detecting the virus responsible for AIDS.

29. Multicultural Perspective Find out which countries in Africa are suffering from drought. What kind of crops are grown in these countries? How could genetic engineering be used to help the people in these countries grow crops in spite of the recurrence of severe drought?

Readings

30. Read the article "The First Kids With New Genes," in *Time*, June 7, 1993. Describe the genetic disease that afflicts Ashanthi DeSilva and Cynthia Cutshall. What events were involved in developing and administering gene therapy to these two young girls?

31. Read the article "Whose Genome Is It, Anyway?" *Discover*, May 1992. How long is the Human Genome Project expected to take? How much is it expected to cost? Describe the characteristics of the hypothetical human whose genome will be the first to be sequenced. What are the reasons for this choice?

ence of HIV in a person's blood before a person has even begun to produce antibodies in response to the virus. Prior to the use of PCR, tests for AIDS detected the presence of HIV antibodies in blood.

29. The countries most likely to suffer from drought in Africa are those located in the north, within and near the Sahara Desert, including those in the Sahel region. Typical crops include peanuts, corn, millet, cacao, cocoa beans, cotton, etc. Genetic engineering could produce drought-resistant plants with thicker leaves, deeper roots, and a shorter growing season.

Readings

30. The two girls are afflicted with severe combined immunodeficiency (SCID), which results in an immune system that cannot protect the body. Before Cynthia could receive gene therapy, her illness had to be diagnosed, other treatments had to be found ineffective, and the doctors had to obtain permission to use gene therapy on humans. Then a gene capable of making the missing enzyme was inserted into a virus and added to bone marrow cells taken from Cynthia. The new gene entered the cells, and the altered cells were injected back into Cynthia.

31. The project is expected to take 15 years and to cost $3 billion. The hypothetical human whose genome the project will sequence will be a male (because males have both X and Y chromosomes). The composite genome will also be derived from people with genetic diseases because understanding and curing genetic diseases is one of the main goals of the project. It is not important whose genes are used to compile an average human genome since all humans are 99.9 percent alike in their genes.

the bacterial cell, clone the bacterium, and expose the new bacterial colony to tetracycline. If the bacteria are not resistant to tetracycline, the procedure has been successful. (10-1.3)

Activities and Projects

27. Chakrabarty identified enzymes that degrade different components of crude oil and combined them into a *Pseudomonas* bacterium. His patent request was brought before the U.S. Supreme Court, which ruled in 1980 that human-engineered

organisms are patentable under federal law.

28. The polymerase chain reaction can be automated by using a thermostable DNA polymerase, which does not have to be added to the mixture each time the strands of DNA are separated at high temperatures. Using PCR, scientists can detect the pres-

Procedural Notes

- Caution students to be careful of the sharp tip on the pushpins.
- This investigation should take one class period.
- You may want to review the structure of DNA and the steps of genetic engineering.
- Make sure each group of students has enough materials to complete the activity before they begin.
- Account for all materials at the end of the lab, and have students store the nucleotides (pushpins) by color.

Background Answers

1. The process is called genetic engineering.

2. You can build recombinant DNA by creating a model of bacterial DNA and inserting a model of a DNA fragment from a separate organism. The models are attached by connecting the nucleotide models according to base-pairing rules.

3. What steps are involved in transferring a gene from one organism to another one?

Records

LABORATORY Investigation Chapter 10

Genetic Engineering Model

OBJECTIVE

Construct and analyze a model representing genetic engineering.

PROCESS SKILLS

- manipulating models
- simulating a scientific process

MATERIALS

- plastic soda straws, 3 cm sections (56)
- metric ruler
- scissors
- permanent marker
- 56 pushpins (15 red, 15 green, 13 blue, and 13 yellow)
- 56 paper clips

BACKGROUND

1. What is the name of the process that isolates a gene from the DNA of one organism and transfers the gene into the DNA of another organism?

2. How can you demonstrate the manipulation of genetic material to produce new combinations of traits?

3. Write your own question to explore in your lab report or notebook.

TECHNIQUE

1. Work in cooperative groups of four students and divide into two-person teams.

2. **CAUTION: Pointed objects can cause injury if not properly used.** Cut the soda straws into 3 cm pieces to make 56 segments.

3. One two-person team should complete steps 4 and 5 while the other team completes steps 6–8. Work with your entire group to complete steps 9–12.

4. Make a model of a bacterial DNA molecule by arranging the nucleotides of the master strand in the following order: blue, red, green, yellow, red, red, blue, blue, green, red, blue, green, red, blue, blue, green, yellow, and red. Create a double-stranded DNA model by constructing a second strand starting with yellow and proceeding with the correct color pushpin to complement the original strand.

5. With your double-stranded DNA model lying on the table, form a circular molecule by carefully joining the opposite ends of each strand. In the **Records** section of your report, make a sketch of the molecule that shows the arrangement of the bases. Use the abbreviations B, Y, G, and R for the colors of pushpins.

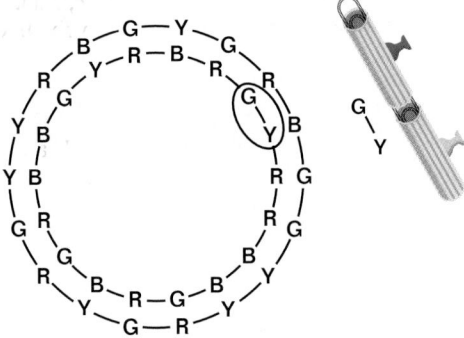

6. Make one strand of a donor, human DNA molecule with the following sequence: blue, blue, red, red, yellow, green, green, blue, red, and yellow.

7. Make a second strand of donor, human DNA having the following sequence:

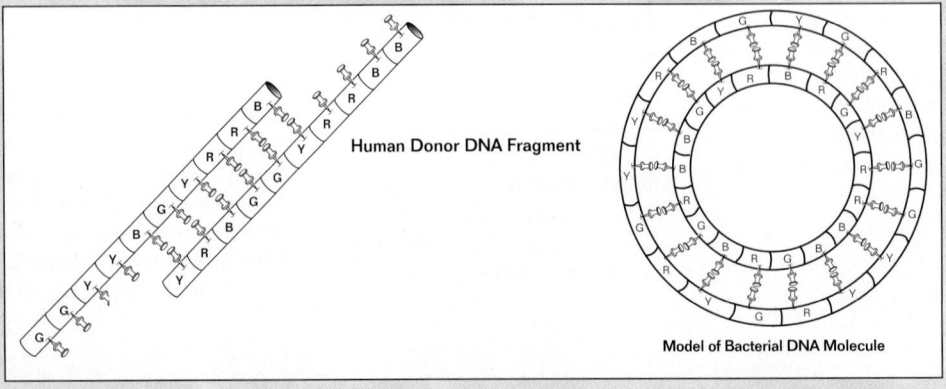

Human Donor DNA Fragment

Model of Bacterial DNA Molecule

blue, red, red, yellow, green, blue, yellow, yellow, green, and green.

8. Match the complementary portions of the two strands of your human DNA fragment. Make a sketch of the donor molecule in the **Records** section of your report.

9. Imagine that an enzyme moves around the circular molecule of bacterial DNA until it finds the sequence red-red-blue-blue and its complementary sequence, green-green-yellow-yellow. Find this sequence in the model of a bacterial DNA molecule that you drew in step 5.

10. Simulate the action of the enzyme by splitting the circular molecule at the sequence you identified in step 9. Separate the yellow nucleotide from the blue at one end of the sequence, and the green from the red at the opposite end of the sequence on the complementary strand. Make a sketch of the split molecule in the **Records** section of your report.

11. Move the double-stranded donor, human DNA fragment into the break in the bacterial DNA molecule.

12. Imagine that a second enzyme joins the ends of the donor and recipient DNA, creating a new DNA molecule. Make a sketch of the final bacterial DNA molecule in the **Records** section of your report. Also, briefly summarize the procedure you followed.

INQUIRY

1. How many nucleotides composed the bacterial DNA molecule?

2. What color pin is always across from a blue pin?

3. What color pin is always across from a red pin?

4. Use the following information to interpret the base-pair order for the DNA molecule you made in step 4, then write the code for the first DNA molecule you created: red = adenine, blue = guanine, yellow = cytosine, green = thymine. You may abbreviate the bases, using the letters A, G, C, and T.

5. Look at the model you created of donor, human DNA. What is unusual about the structure of this DNA fragment? (Hint: make sure to connect the bases to the correct complementary color.)

ANALYSIS

1. Compare and contrast the models of bacterial DNA and human DNA.

2. What knowledge did you have to possess in order to construct the donor, human DNA?

3. How does the original bacterial DNA molecule differ from the final DNA molecule?

4. Of what possible benefit could this process be to humans?

FURTHER INQUIRY

Write a new question that could be explored as a new investigation. The following is an example:

> Which bacteria are the best hosts for human gene insertion?

Inquiry Answers

1. The bacterial DNA molecule is composed of 36 nucleotides.

2. A yellow pin is always across from a blue pin.

3. A green pin is always across from a red pin.

4. The code for the first DNA molecule created in step 4 is G–C, A–T, T–A, C–G, A–T, A–T, G–C, G–C, T–A, A–T, G–C, T–A, A–T, G–C, G–C, T–A, C–G, A–T.

5. The donor DNA has a string of four unpaired nucleotides at each end.

Analysis Answers

1. Both the bacterial DNA model and the donor DNA model are double stranded. The bacterial DNA model is circular, and the donor DNA model is in the shape of a ladder.

2. Construction of the donor model was made possible by recognizing that yellow is always paired with blue and that green is always paired with red.

3. The new bacterial molecule has more nucleotides than the original molecule as a result of the addition of the human DNA fragment.

4. Answers may vary. Advances could be made in the production of proteins used to treat illnesses, in the development of new vaccines to combat diseases, and in the replacement of defective genes with healthy ones.

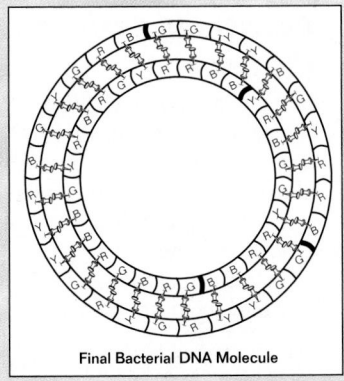

Final Bacterial DNA Molecule

Point of View: The Human Genome Project

Objectives

- Develop scientific literacy
- Identify an author's point of view
- Detect bias in a persuasive argument

Background

Each year scientific issues become more important in determining the nature and direction of our society, and yet most citizens have little or no training in evaluating the arguments that surround controversial issues. The purpose of this feature, Point of View, is to provide practice in identifying and evaluating a writer's point of view.

A critical step in this process is detecting bias. Much of the writing on critical issues is persuasive and not objective. A writer will include information that bolsters his or her point of view and suppress evidence that contradicts it.

Instructional Strategies

- As students read this article, remind them that it represents one person's point of view. They must read critically, looking for signs that would indicate biased or slanted writing. In particular, have students look closely for the following:

1. Does the article present both sides of the issue fairly and objectively?

2. Is the writer seeking to shape your opinion on the issues?

3. Does the writer use "loaded" language? Are some words chosen because they have an emotional rather than a logical appeal to the reader?

- Point out that because the genome project seeks to improve health, it is important to define sickness and disease. Have students list physical traits and behaviors that are

considered healthy. Then have them list traits and behaviors that are considered unhealthy. Ask students how the perception of a trait as positive, negative, or neutral can change with social conditions.

- Give students copies of a newspaper or magazine article that describes a new

scientific finding that involves genes. How are results presented? Are words like *suggests, may, hinted,* or *believed* used? What do such words imply about the certainty of the conclusions? What social attitudes are created or supported by the article?

POINT OF VIEW

THE HUMAN GENOME PROJECT

Does Biology's Biggest Venture Promise Too Much?

Chromosomes, *above,* provide the DNA that technicians, *left,* examine noting every band in the DNA "fingerprint".

BY TRACEY COHEN

The biggest research effort in the history of biology—the Human Genome Project has as its goal to map and sequence the tens of thousands of genes that make up the human genome. The maps will show where on the chromosomes each gene is located. Research is going on in national laboratories, scientific institutes, universities, and private companies around the world and will take about 15 years to finish. In the United States, the project is sponsored jointly by the National Institutes of Health and the Department of Energy and is expected to cost $3 billion.

Supporters of the Human Genome Project claim it will revolutionize medicine. They say that medical scientists will be able to diagnose and eventually cure large numbers of diseases. Many promises and predictions have been made to create support for the project. James Watson, former director for the project, has said that "never will a more important set of instruction books be made available to human beings." The Office of Technology Assessment wrote in a report that "sequencing the human genome will provide one of the most powerful tools humankind has ever had for deciphering the mysteries of its own existence."

Such statements make the project sound exciting and heroic. But are they true? Not everyone thinks so. In the scientific community, the project has caused many bitter arguments. Researchers critical of the project claim that a lot of money will be wasted on something that is poor science. They say that those scientists trying to get money for the project have made unrealistic claims to gain public support.

Will the Project Improve People's Health?

The Human Genome Project makes genetic diseases the focus of medical research. Much of the drive behind the Project is the claim that it will greatly improve the practice of medicine, and thus people's health. This statement might be true if diseases were mainly genetic in origin. However, even though researchers have identified more than 3,000 genetic disorders, most of these are rare. For example, the most

common lethal genetic disorder among Caucasians, cystic fibrosis, afflicts about 25,000 Americans. About 30,000 people of various races suffer from Huntington's disease, another deadly hereditary disease. Another 150,000 are thought to be at risk.

In contrast, over 900,000 people died of heart disease in 1990 alone according to the National Center for Health Statistics. Another 500,000 died of cancer. In the United States, heart disease, cancer, stroke, and accidental injury are the leading causes of sickness and death. For the last 20 years, nearly three-quarters of all deaths in this country have been the result of these four killers. These are not, strictly speaking, genetic diseases. Although people with rare disorders might benefit, the majority of society will not.

What Social and Ethical Problems Are Raised by the Project?

Genetic Discrimination Some evidence links genes with certain diseases. However, the actual role of these genes in disease is still unclear. They appear to make people susceptible to getting illnesses like cancer or heart disease. The

Answers to Analyzing the Issue

1. The writer opposes the Human Genome Project. Key facts might include the number of genetic diseases versus the leading causes of death; the distinction between risk of disease and actual cause of disease; and cases of genetic

Human Genome Project will make it easier to find out if a person has susceptibility genes.

But having a susceptibility gene does not mean the person will always get the disease. Nor does the presence of the gene give any clue about the severity of the disease. The gene does not show how the person will respond to treatment either. And, even without the gene, a person might still get that disease.

What is certain about susceptibility genes is that they present enormous opportunity for discrimination. A survey by researchers at Harvard Medical School has found about 30 cases of genetic discrimination in which people were denied jobs or insurance because they carried a gene linked to a particular disease. Although these people had no symptoms, employers and insurance companies saw them as either sick or likely to become sick.

Impact on Social Responsibilities When scientists talk about genes as "causes," they make it seem as if an individual's genes are responsible for many health conditions that are really social in origin. Poverty, for instance, is not a genetic condition. Yet poverty is clearly a crucial factor in health and disease.

For example, a study by the United States Department of Health and Human Services reported a clear relationship between family income and cancer rate. As income decreases, cancer rates increase. Survival rates are lower for low-income cancer patients as well. People with low incomes have less money for good nutrition and less access to good health care. Their diseases are discovered in later stages when the cure rates are lower.

Who Benefits From the Project?

There is money to be made in every step of the Human Genome Project. Companies that make automatic DNA sequencing equipment will benefit, as will companies that make computer equipment and software.

The creation of new genetic screening tests will also be very profitable. As more genes thought to be linked with diseases are identified, more people will be tested. New screening tests could become a billion-dollar industry. It is estimated that almost 3 million people a year could be tested to determine if they have genes for cystic fibrosis, sickle cell anemia, hemophilia, and muscular dystrophy. Companies that make screening tests, doctors, employers, and insurance companies will likely create social pressure for widespread use of these tests.

Scientists, too, have economic interests in genetic information. For instance, more than 30 leading genome researchers have made deals with investors and new companies to market the results of their work.

Unfortunately, the Human Genome Project is the wrong approach both to understanding human biology and to improving medicine and public health. Before any more work is done, there should be public debate about how this project is likely to affect everyone. Decisions about the Human Genome Project should not be left just in the hands of those who have their own narrow interests at stake.

Tracey Cohen is a freelance writer specializing in science and environmental issues.

Analyzing the Issue

1. **Detecting Bias** How does the writer feel about the Human Genome Project? List the key facts the writer gives to support that opinion. List instances where the writer uses words or phrases to sway your opinion without presenting the facts.

2. **Formulating an Opinion** Obtain at least three books and articles about the genome project. Several possible references are listed below. Write a persuasive essay supporting your own view of the genome project.

   Beckwith, Jon. "A Historical View of Social Responsibility in Genetics." *BioScience*, Vol. 43, No. 5, 1993, pp. 327–333.

   Hubbard, Ruth and Elijah Wald. *Exploding the Gene Myth.* Boston: Beacon Press, 1993.

   Kevles, Daniel J. and Leroy Hood, eds. *The Code of Codes: Scientific and Social Issues in the Human Genome Project.* Cambridge: Harvard University Press, 1992.

   Lewontin, R.C. "The Dream of the Human Genome." *The New York Review of Books*, Vol. 39, No. 10, May 28, 1992.

   Lee, Thomas F. *The Human Genome Project: Cracking the Code of Life.* New York and London: Plenum Press, 1991.

3. **Take Action** What legislation exists to protect people against genetic discrimination? Who is allowed to know the results of genetic screening tests? How can this information be used? Write to the following for information.

   The Council for Responsible Genetics
   19 Garden Street
   Cambridge, MA 02138

   National Center for Human Genome Research
   Cold Spring Harbor, NY 10098

4. **Examining Social Consequences** Are scientists responsible for the ways in which information from the Human Genome Project is used? Explain. Compare modern research in human genetics with the eugenics movement of the early 20th century. You may need to do library research. How is the comparison accurate? How is it false? Discuss the role of scientists in each case.

---

discrimination. Opinionated statements include assertions about who will profit and increased pressure for genetic tests.

2. Students' positions should be clear in their essays. Opinions should be supported by statistics and other facts.

3. Legislation varies from state to state. Students should go to the library to get information.

4. Answers will vary. Students should recognize that scientists can influence laws, social beliefs, and social practices, sometimes in dangerous and undesirable ways. A good article for students to read and discuss is "Eugenics: Past, Present, and Future," by Kenneth L. Garver and Bettylee Garver, *American Journal of Human Genetics* Vol. 49, 1991 pp. 1109–1118.

References

1. American Cancer Society. *Cancer and the Poor: A Report to the Nation.* Atlanta: American Cancer Society, 1989.

2. Draper, E. *Risky Business: Genetic Testing and Exclusionary Practices in the Hazardous Workplace.* Cambridge: Cambridge University Press, 1991.

3. Erickson, D. "Gene Rush: Companies Seek Profits in the Genome Project." *Scientific American,* January 1992, pp. 112–113.

4. Hood, L. "Biology and Medicine in the Twenty-First Century." In: *The Code of Codes.* D. J. Kevles and L. Hood, eds. Cambridge: Harvard University Press, 1992, pp. 136–163.

5. Hubbard, R., and E. Wald. *Exploding the Gene Myth.* Boston: Beacon Press, 1993.

6. Jerome, R. "Huntington's Cornered." *The Sciences,* May/June 1993, p. 7.

7. Kimbrell, A. *The Human Body Shop: The Engineering and Marketing of Life.* San Francisco: Harper San Francisco, 1993.

8. Lee, T. F. *The Human Genome Project: Cracking the Code of Life.* New York: Plenum Press, 1991.

9. Lewontin, R. C. *Biology as Ideology: The Doctrine of DNA.* New York: HarperPerennial, 1992.

10. Marx, J. "Genome Project Plans Described." *Science,* Vol. 260, April 9, 1993, pp. 152–153.

11. "Mortality Patterns—United States, 1989." *Journal of the American Medical Association,* Vol. 267, No. 11, March 18, 1992, pp. 1449–1450.

12. Nelkin, D. "The Social Power of Genetic Information." In: *The Code of Codes.* D. J. Kevles and L. Hood, eds. Cambridge: Harvard University Press, 1992, pp. 179–190.

13. Office of Technology Assessment. *Mapping Our Genes. Genome Projects: How Big, How Fast?* Baltimore: The Johns Hopkins University Press, 1988.

14. Rechsteiner, M. C. "The Human Genome Project: Misguided Science Policy." *Trends in Biological Sciences,* Vol. 16, December 1991, pp. 455–459.

UNIT 3
CHAPTER 11   THE ORIGIN OF LIFE

Block Scheduling Guide

> Each block represents about 45–50 minutes of instructional time. The resources cited in a block are coded to a specific program component using the Key on the next page.

Lecture Concepts	Lesson Resources

11-1 The Mystery of Life's Origin pp. 224–232

BLOCKS ① and ②

a. Many possible origins of life have been proposed. At present, only hypotheses based on the assumption that life arose naturally and spontaneously on Earth can be tested by scientific methods.

b. Miller demonstrated that basic organic molecules of life could have formed spontaneously from materials present on early Earth. The primordial soup model is being investigated because (1) life might have originated more quickly than previously assumed and (2) methane and ammonia might not have been present in Earth's early atmosphere, as was assumed in Miller's experiment.

c. Lerman's bubble model suggests that chemicals reacted within bubbles on the early ocean, forming simple organic molecules. More complex molecules formed by further reactions.

d. Short chains of RNA can be made to form spontaneously in water. Certain RNA molecules can act like enzymes. These discoveries have led scientists to hypothesize that RNA was the first self-replicating information storage molecule.

e. Lipids and proteins tend to aggregate in water, forming microspheres. These tiny spheres could have been the first step in the organization of cells.

Lecture Resources
- Opening Questions, page 225
- Demonstrations: Visualizing Ancient Earth, Growing Globs
- Teaching Transparencies: 43, 43A, 46, 47
- Visual Strategies: Figures 11-6, 11-7, 11-9
- Speeding up Reactions, page 230
- Overcoming Misconceptions, page 231
- Holt Biology Videodiscs: Lesson 11-1

Classwork Options
- Effective Reading, page 225
- Simple to Complex, page 226
- What Are Cells Made Of? page 227
- Summarizing with a Flow Chart and Graphic Organizer, pages 228–229
- Laboratory Investigation: Making Microspheres, pages 242–243

Assignment Options
- Section Review, page 232
- Chapter Review, questions 1–6, 11–13, 19, 20, 23–27

11-2 Evaluating the Spontaneous Origin Hypothesis pp. 233–235

BLOCK ③

a. Scientific estimates of Earth's age are derived from radiometric dating, which is based on the regular rate at which isotopes decay.

b. The diversity and complexity of organisms can be explained as the result of genetic variation and the power of natural selection.

c. Proteins cannot assemble spontaneously in water, but they can be assembled with the aid of a catalyst and an input of energy.

Lecture Resources
- Opening Questions, page 233
- Demonstration: Modeling Radioactive Decay
- Visual Strategy: Figure 11-10
- Teaching Transparencies: 45
- Direct and Indirect Evidence, page 234
- Home-Style Energy, page 234

- Seeds and the Second Law of Thermodynamics, page 235
- Holt Biology Videodiscs: Lesson 11-2

Assignment Options
- Concept Mapping, page 239
- Section Review, page 235
- Chapter Review, questions 7, 8, 14–16, 21, 25, 26, 28

11-3 Is There Life on Other Worlds? pp. 236–238

BLOCK ④

a. Earth has just the right mass and is just the right distance from the sun for life as we know it to exist.

b. Considering the number of stars in the universe, life might have developed many times across the universe.

Lecture Resources
- Opening Questions, page 236
- Is There Life on Mars? page 237
- Making Large Numbers More Meaningful, page 237
- Research Update, page 238
- Holt Biology Videodiscs: Lesson 11-3

Classwork Options
- Different Life for Different Worlds, page 236
- Closure, page 238

Assignment Options
- Section Review, page 238
- Chapter Review, questions 9, 10, 17, 18, 22, 29

Review/Enrichment

BLOCK ⑤

- ■ Study Guide: Concept Review and Pretest

Assignment Options
- ◉ Teaching Resources CD-ROM: Occupational Applications Worksheets, Biology Teacher
- ◉ Teaching Resources CD-ROM: Supplemental Reading Worksheets and Test, *Origin of Species*
- ■ Activities and Projects, questions 30–32
- ■ Readings, questions 33, 34

Assessment Options

BLOCK ⑥

Traditional Assessment
- ■ Chapter 11 Test
- ◉ Test Generator/Assessment Item Listing: Software item bank for preparing customized chapter tests.

Portfolio Assessment
Select student reports for one or more laboratory experiments from this chapter. *The Direct Observation Checklist* on the *BioSources Teaching Resources CD-ROM* should be completed during a laboratory or other cooperative learning experience.

Answer to Concept Map
The following is one possible answer to the Concept Mapping exercise on page 239.

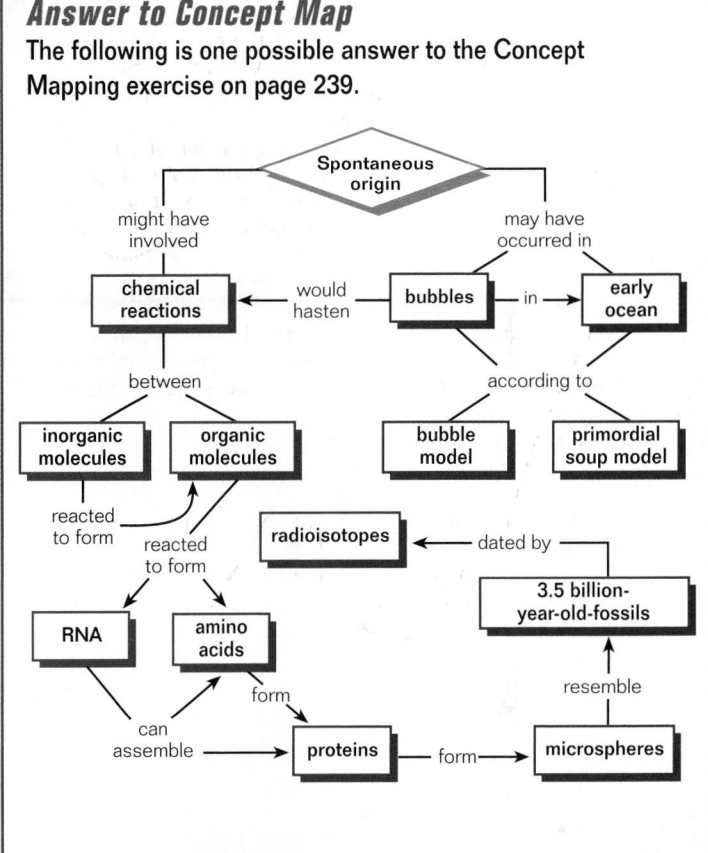

Holt Biology Videodiscs — CONCEPTS OF BIOLOGY

Holt Biology Videodiscs gives you a powerful tool for teaching, review, and assessment. *Concepts of Biology* adds interactive lesson plans and extensive tools for customization using CD-ROM technology.

Use the following topics from *Concepts of Biology* to help you teach this chapter:
- ■ Topic 8: Evolution and Natural Selection
- ■ Topic 9: Evolution of Life on Earth

For further information regarding the media on the videodiscs, see *Holt Biology Videodiscs Teacher's Correlation Guide to Biology: Principles and Explorations.*

CHAPTER 11
THE ORIGIN OF LIFE

Chapter Themes

Levels of Organization

In studying the scientific hypotheses about the origin of life, students learn how these levels may have developed. The steps in the spontaneous origin of life hypothesis mirror the increasing complexity of life, beginning at the atomic level and followed by the molecular, macromolecular, and biochemical pathway levels. Figures 11-3, 11-6, 11-8, and 11-9 illustrate this theme.

Structure and Function

A key to the spontaneous origin hypothesis is that the structures of molecules enable them to perform their functions. Review DNA replication and protein synthesis, and use Figures 11-2, 11-4, and 11-7 to illustrate this theme.

Tapping Prior Knowledge

- What are the components of a protein?
- What are the components of a nucleic acid?
- How do enzymes act as catalysts?
- How does the structure of a nucleic acid determine the structure of a protein?

Opening Question

Begin this unit by stating that diverse viewpoints can be respected while studying science. You may want to review the goal of science and the steps in a scientific method. Stress that science cannot answer every question and that this chapter explores scientists' hypotheses concerning a series of events that could have led to the formation of living matter from nonliving matter.

REVIEW

- polar nature of water (Section 2-2)
- structure of proteins and nucleic acids (Section 2-3)
- second law of thermodynamics (Section 4-1)
- role of enzymes in catalyzing chemical reactions (Section 4-2)
- natural selection (Section 5-3)
- roles of RNA and DNA in heredity (Sections 9-1 and 9-3)

Artist's view of the early Earth.

Authors' Rationale

After the discussion of life and its characteristics, you and your students can now address the universal philosophical question: Where did life come from? This chapter will help you look your students squarely in the eye and say, "I don't know." Remind your students that *not* knowing things is what gets the ball rolling in science. This chapter introduces students to major scientific hypotheses regarding the origin of life and discusses the evidence that supports these hypotheses, as well as their weaknesses. The chapter also explores the possibility that life exists on other planets.

11-1 The Mystery of Life's Origin

Everywhere you look there is life. It is found not only in fields, forests, and ponds but also in deserts, on polar icecaps, atop high mountains, and deep beneath the sea. Drops of water teem with life, mostly creatures too tiny to be seen without magnification as they dash about, bashing into each other. With life all around, it is hard to imagine there was ever a time when Earth was barren of life, but indeed there was. Studies of radioactive elements in rocks indicate that Earth is about 4.5 billion years old, or 1 billion years older than the fossil seen in Figure 11-1. Where did such life-forms come from? How did they arise? How did a barren mass of rock and water become a home for deer, butterflies, sea otters, and humans?

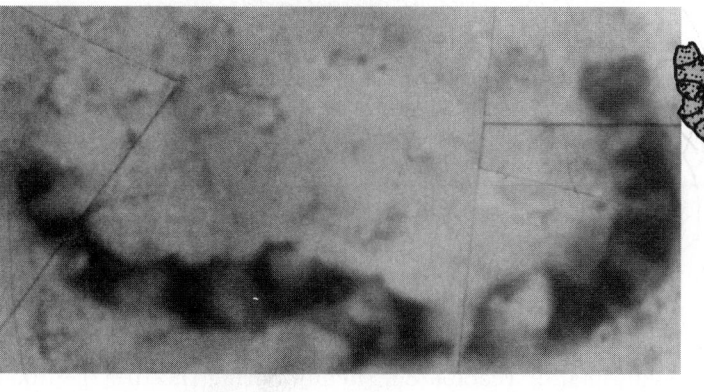

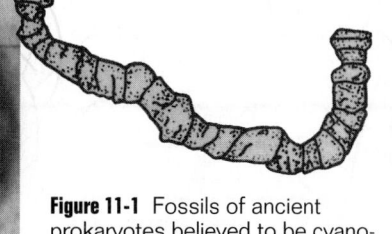

Figure 11-1 Fossils of ancient prokaryotes believed to be cyanobacteria, *left*, have been found in Australia. An artist's sketch of this fossil, *above*, shows that life had already become fairly complex 3.5 billion years ago.

There Are Several Ideas About the Origin of Life

Trying to explain how life might have originated on Earth is a difficult but fascinating quest and one that has been pursued by philosophers, theologians, and scientists alike. Still, we may never know *exactly* how life got here. No one was there to record what happened. What was it really like on the early Earth? Were the forces that caused life to begin ones that can be explained by science or ones that cannot be explained by science? A few scientists even question whether life originated on Earth. They have suggested instead that life on Earth had an **extraterrestrial** (EHKS truh tuh REHS tree uhl) **origin**—an origin outside of Earth. They hypothesize that life was carried here by an asteroid or by a meteorite like the one in Figure 11-2. Although there may always be questions about

Figure 11-2 Meteorites like this one might have brought life to Earth. Organic compounds make up about 2 percent of its weight.

Teaching Tips

Can Creation Stories Be Tested Scientifically?

Have students look up a creation story from a country or culture of their choice, then have them explain why the story can or cannot be confirmed by scientific methods. If you wish, have students write their own original stories about the origin of a plant or an animal. Have students either propose an experiment if they think their creation stories might fall within the realm of science, or give reasons why their stories cannot be duplicated in experiments.

Simple to Complex

Divide the class into groups of three to four students. Instruct each group to refer to Chapter 2 and draw two diagrams of each of the following kinds of molecules: simple sugars, polysaccharides, amino acids, proteins, nucleotides, and RNA. When their diagrams are done, have students take turns separating the diagrams into two piles—one of simple molecules (simple sugars, amino acids, and nucleotides) and one of complex molecules (polysaccharides, proteins, and RNA). When each student has successfully classified the molecules, ask: Which molecules would have to form before others could form? *(the simple ones)* Then have students mix up the diagrams and take turns pairing the simple molecules with the more complex ones they compose. *(e.g., amino acids and proteins)* Finally, ask students which one of the molecules might be able to make copies of itself. *(RNA)* Ask: What effect might this ability have on the kinds of molecules in the oceans of ancient Earth? *(There would be many more RNA molecules than the other kinds.)*

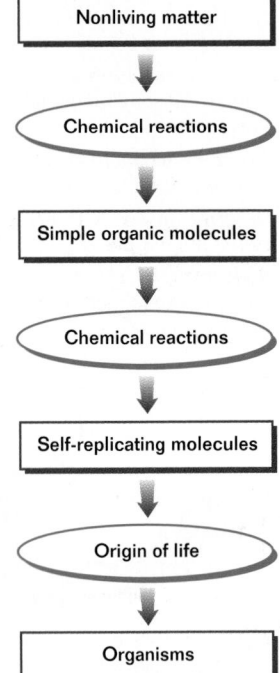

Figure 11-3 Most scientists think that life began spontaneously as the result of the process summarized here. A series of chemical reactions could have produced simple organic molecules, then more complex organic molecules, and eventually self-replicating, living matter.

❑ CAPSULE SUMMARY

Many possible origins of life have been proposed. Presently, only hypotheses based on the assumption that life arose naturally and spontaneously on Earth can be tested by scientific methods.

exactly how life arose on Earth, there are many possibilities. The two principal means by which life might have originated are discussed below.

Divine Creation

Traditionally, many cultures have believed that life was put on Earth by divine (relating to a god or gods) forces, as the act of a creator or creators. Belief in divine creation is common to many of the world's major religions, though the accounts of creation vary from one religion to another. By all accounts of divine creation, the process that gave rise to life on Earth was driven by forces that cannot be explained by science. Philosophers have debated the essence of these forces for centuries. It is important to understand, however, that a belief is not the same thing as a scientific hypothesis. The essence of any scientific hypothesis is that the proposed idea is subject to test—that the idea could, in principle at least, be proven false. As you learned in Chapter 1, science is a way of investigating the natural world (through observation and experimentation) and forming general rules about how things happen. A belief in divine creation, however, is not a scientific hypothesis that can be tested. Try to imagine an observation that would disprove divine creation. Whatever you propose, it is always possible to argue that a divine agent simply made things appear the way they do. Because the idea that life originated through divine creation cannot be tested by scientific methods, it falls outside the realm of science. This is not to say that the belief is wrong, but rather that science can never test it.

Spontaneous Origin

Most scientists think that life on Earth had a **spontaneous origin,** developing by itself through natural chemical and physical processes. They hypothesize that molecules of nonliving matter reacted chemically during the first 1 billion years of Earth's history, forming a variety of simple organic molecules. They further hypothesize that complex organic molecules, some of which were capable of replicating themselves and other molecules, formed associations that became increasingly complex. In this view, outlined in Figure 11-3, the process by which life arose was driven by the natural force of selection. Changes that increased the stability of certain molecules would have allowed (selected) those molecules to persist for a longer time. Molecules that could be replicated would have become more common than those that could not be replicated. In laboratory experiments, many of the organic building blocks of life have been made from molecules of nonliving matter. Thus, the assumption that life began spontaneously can be tested by scientific methods. Because it is the only assumption that can be tested by scientific methods, this chapter deals with hypotheses based on the idea that life on Earth originated through natural chemical and physical processes. ❑

Multicultural Perspective

Hindu and Hopi Creation Stories

The theme that Earth gave rise to life appears in the legends of many cultures. For example, according to a Hindu proverb, "The Earth is our Mother, and we are all her children." A Hopi proverb states, "Earth gives life and seeks the man who walks gently upon it."

Life's Basic Chemicals Can Form Spontaneously

For life to have formed naturally on Earth, the materials that make up living things must have been present. As you learned in Chapter 2, all organisms are built from the same chemicals, just as all cars are assembled from the same materials. Cars are made from steel, glass, plastic, and rubber. Cells are made from proteins, lipids, carbohydrates, and nucleic acids. If life originated spontaneously, then the first question that must be answered is, Where did these molecules—the building blocks of life—come from? In their attempts to answer this question, scientists have focused on Earth's early oceans and atmosphere. Most scientists hypothesize that the basic chemicals of life formed during chemical reactions that occurred there. To find out if these reactions could have taken place, you must take an experimental journey more than 4.5 billion years back in time. Scientists do this in their laboratories by analyzing data from newly forming stars and by re-creating conditions they suspect existed on the early Earth.

The Primordial Soup Model

Charles Darwin, who developed the theory that life evolves through natural selection, once speculated that life began in "a warm little pond." In the 1920s, the Russian scientist A. I. Oparin proposed a hypothesis that extended Darwin's idea. He suggested that Earth's oceans were once a vast primordial (preye MAWR dee uhl) soup containing large amounts of organic molecules. Oparin envisioned these molecules forming spontaneously in chemical reactions activated by energy from solar radiation, volcanic eruptions, and lightning. Oparin thought that over millions of years, these molecules had gradually come together to form living matter.

Oparin, Harold Urey of the University of Chicago, and other investigators of the solar system proposed that Earth's early atmosphere lacked oxygen. They hypothesized that the early atmosphere was instead rich in nitrogen (N_2) and hydrogen-containing gases such as hydrogen (H_2), water vapor (H_2O), methane (CH_4), and ammonia (NH_3). Electrons in these gases would have been frequently pushed to higher energy levels by photons crashing into them from the sun or by electrical energy in lightning. Today, high-energy electrons are quickly soaked up by the oxygen in Earth's atmosphere (air is 21 percent oxygen) because oxygen atoms have a great "thirst" for such electrons. But in the absence of oxygen, high-energy electrons would have been free to do other things.

In 1953, Oparin's hypothesis was tested by Stanley Miller, who was then a graduate student working with Urey. To find out what sort of chemical "mischief" these electrons might have done, Miller placed the proposed gases into an apparatus like the one seen in Figure 11-4. Then, to simulate lightning, he zapped the mixture with electrical sparks. After a

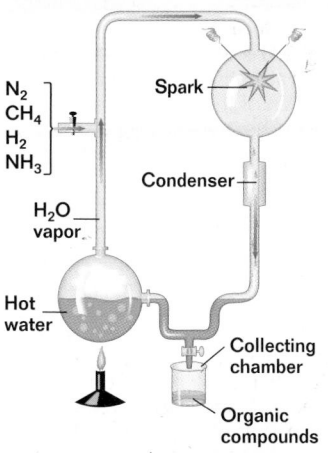

Figure 11-4 Lightning and heat from volcanic eruptions, *top*, might have been energy sources for the chemical reactions that led to the origin of life. In an apparatus similar to the one shown, *bottom*, Miller exposed a mixture of nitrogen and hydrogen-rich gases to electrical sparks simulating lightning. This produced high-energy electrons that jumped from molecule to molecule, causing the formation of chemical bonds and many new kinds of molecules.

SECTION 11-1

CONNECTIONS

Chapter 2
Elements of Life
Review the elements that make up the major macromolecules found in living things.

Chapter 5
Source of Oxygen
Remind students that the oxygen released during photosynthesis, which is thought to be responsible for most of the oxygen in Earth's atmosphere, comes from water molecules that are split during the light reactions.

Teaching Tips
Earth's Atmosphere Then and Now
Have students look up the types and amounts of the gases present in Earth's atmosphere today and compare this information to the atmosphere that Stanley Miller proposed for the early Earth. Ask: How are they alike? (*Both have much nitrogen and lesser amounts of the other gases.*) How are they different? (*The early Earth had almost no oxygen.*) Which component of the early atmosphere contained oxygen? (*water*) What is the source of oxygen in today's atmosphere? (*water molecules split during photosynthesis*)

What Are Cells Made Of?
Have students list the kinds of macromolecules found in cells (carbohydrates, lipids, proteins, and nucleic acids) across the top of a piece of paper. Then have students write in each column the chemicals in the ancient atmosphere and "primordial soup" of Earth that contain some of the same elements as the macromolecules. (*e.g., H_2O, CH_4, and H_2 would go under "carbohydrates," and H_2O, H_2, N_2, CH_4, and NH_3 would go under "proteins"*)

 Did You Know?

Ancient oceans were probably similar to freshwater lakes in salt content. The salinity of today's oceans is the result of millions of years of water running over rocks, picking up mineral salts, and flowing to the sea.

 Did You Know?

Experiments similar to Miller's that used UV light instead of electrical sparks as the energy source also produced amino acids.

Application

Health Remind students that the destructive force of ultraviolet (UV) light can be harmful to their health. For example, sunburn is caused by the UV rays in sunlight. Skin cancers can result from mutations caused by the UV rays in sunlight. Both sunburn and skin cancer can be deadly. Advise students that whenever they will have prolonged exposure to the sun, they should wear a sunscreen or protective clothing to reduce their risk of skin cancer and sunburn.

Teaching Tip

What If Earth's ozone layer were destroyed? Destruction of the protective ozone layer in Earth's atmosphere would enable far more ultraviolet light to reach Earth's surface. This increased amount of UV light not only would cause many more mutations and kill many organisms but also would make our planet's surface uninhabitable to most of its present life-forms. Thinning of Earth's ozone layer has already been reported and may be linked to a rise in skin cancers.

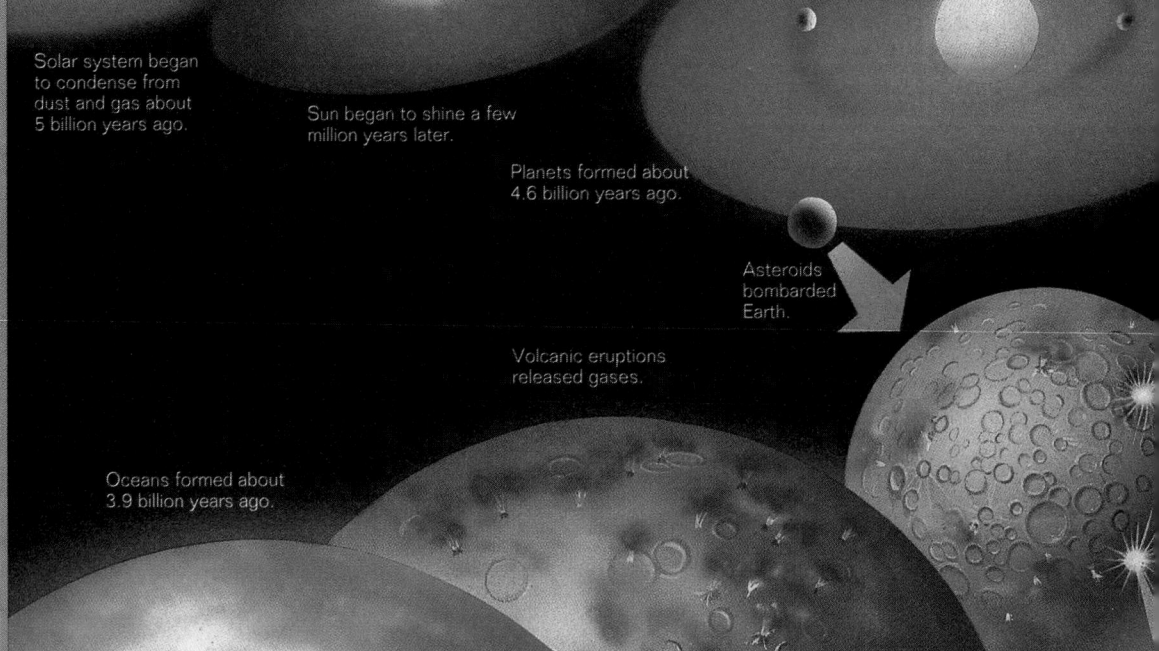

Solar system began to condense from dust and gas about 5 billion years ago.

Sun began to shine a few million years later.

Planets formed about 4.6 billion years ago.

Asteroids bombarded Earth.

Volcanic eruptions released gases.

Oceans formed about 3.9 billion years ago.

Figure 11-5 According to the current scientific model of the formation of the solar system, Earth would have been uninhabitable until about 3.8 billion years ago.

❑ CAPSULE SUMMARY

Miller demonstrated that basic organic molecules of life could have formed spontaneously from materials present on the early Earth. The primordial soup model is being reevaluated for two reasons: (1) life might have originated more quickly than previously assumed, and (2) methane and ammonia might not have been present in Earth's early atmosphere, as was assumed in Miller's experiment.

few days, Miller found a complex "chemical zoo" in the collecting chamber of his apparatus. Within this smelly mixture, he found some of life's basic building blocks: amino acids, fatty acids, and hydrocarbons (molecules made of carbon and hydrogen). These results demonstrated that some basic chemicals of life could have formed spontaneously on the early Earth under conditions like those in the experiment.

Recent discoveries have caused scientists to reevaluate Oparin's primordial soup model. At the time of Miller's experiment, scientists thought that life had taken more than a billion years to begin. However, today's model of Earth's formation, seen in Figure 11-5, and discoveries of 3.5-billion-year-old fossils indicate that the time available for life to begin was much shorter. Thus, it appears that life would have had to originate much faster than previously assumed. Another problem with the primordial soup model concerns the mixture of gases Miller used to simulate conditions on the early Earth. If the atmosphere had no oxygen 4 billion years ago, Earth would not have had a protective layer of ozone gas, O_3. Today, ozone shields our planet's surface from most of the sun's damaging ultraviolet light. Scientists think that without an ozone layer, ultraviolet light would have destroyed any ammonia and methane present in the atmosphere. When these gases are missing from experiments similar to Miller's, key biological molecules such as amino acids are not produced. This raises a very pointed question: If the necessary ammonia and methane were not in the atmosphere, where did they come from? ❑

Graphic Organizer

Use this graphic organizer with **Teaching Tip:** Summarizing With a Flow Chart on page 229.

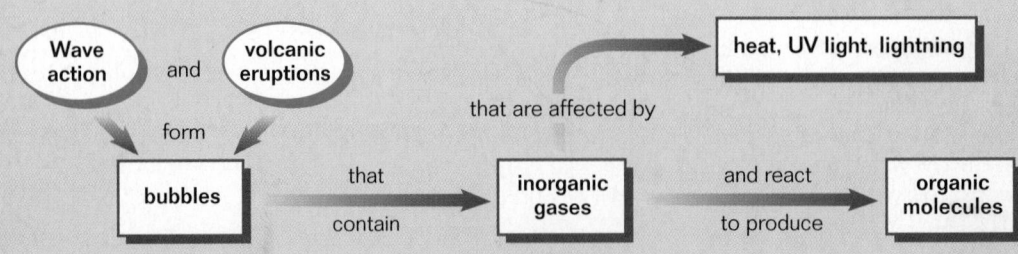

The Bubble Model

In 1986, the geophysicist Louis Lerman suggested that problems with Oparin's model could be solved if the model were "stirred up" a bit. Lerman suggested that the key chemical processes took place not in a primordial soup but within bubbles on the ocean's surface. Bubbles produced by wind, wave action, the impact of raindrops, and the eruption of volcanoes cover about 5 percent of the ocean's surface at any given time. As you learned in Chapter 2, water molecules are polar. Therefore, water bubbles tend to attract other polar molecules. Lerman's bubble model, illustrated in Figure 11-6, proposes that chemicals in the early ocean collected within bubbles and reacted to form the key biological molecules.

Lerman's bubble model solves two key problems with Oparin's primordial soup model. First, chemical reactions would proceed much faster in bubbles (where reactants would be concentrated) than in Oparin's stagnant primordial soup. Thus, life could have originated in a much shorter period of time than it could have according to Oparin's model. Second, inside the bubbles, the methane and ammonia required to produce amino acids would have been protected from destruction by ultraviolet light.

Figure 11-6 Louis Lerman's bubble model suggests that chemicals reacted within bubbles on the early ocean, forming simple organic molecules. After countless generations of bubbles within which constantly changing mixtures of these molecules reacted, very complex biological molecules eventually resulted.

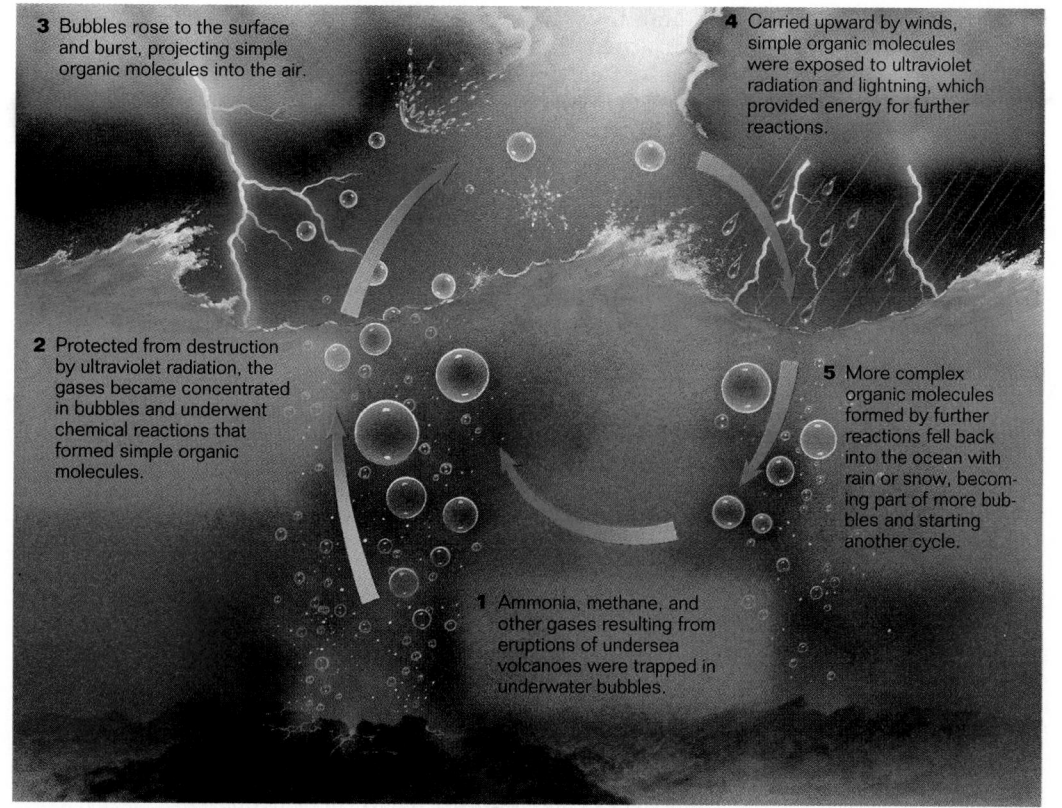

3 Bubbles rose to the surface and burst, projecting simple organic molecules into the air.

4 Carried upward by winds, simple organic molecules were exposed to ultraviolet radiation and lightning, which provided energy for further reactions.

2 Protected from destruction by ultraviolet radiation, the gases became concentrated in bubbles and underwent chemical reactions that formed simple organic molecules.

5 More complex organic molecules formed by further reactions fell back into the ocean with rain or snow, becoming part of more bubbles and starting another cycle.

1 Ammonia, methane, and other gases resulting from eruptions of undersea volcanoes were trapped in underwater bubbles.

Teaching Tip
Summarizing With a Flow Chart

Have students make a flow chart to summarize Lerman's bubble model using the following terms: bubbles, heat, inorganic gases, lightning, organic molecules, UV light, volcanic eruptions, and wave action. Have students add connecting words and phrases that link these terms. One possible flow chart is found in the *Graphic Organizer* on page 228.

CONNECTIONS
....

Chapter 2
Polarity of Water
Remind students that water molecules are polar and thus are able to stick to one another (cohere). These properties of water enable it to form bubbles.

 VISUAL STRATEGY

Figure 11-6
Walk students through the steps of Lerman's bubble model. Point out that bubbles could act as small reaction vessels by concentrating reactants. Tell students that simple organic molecules could have formed when inorganic reactants were exposed to energy sources such as lightning and UV light. Have students refer to Miller's experiment, described in Figure 11-4 on page 227. Ask: What are the similarities between Miller's experiment and Lerman's model? *(Both use the same starting materials and similar sources of energy.)* Tell students that the organic molecules resulting from chemical reactions could have been mixed in different combinations countless times through countless cycles over millions of years, providing many chances for simple organic molecules to combine into more complex organic molecules.

VISUAL STRATEGY

Figure 11-7
Ask students to describe how these iron pyrite crystals provide three-dimensional surfaces. *(The cubical crystals overlap, forming angles between their surfaces.)*

Teaching Tips
Speeding Up Reactions

Ask students to name sources of energy on the early Earth that could make molecules more reactive (*lightning, intense heat, UV light*) and to name things that could act as catalysts. (*RNA, iron pyrite, clay minerals*)

RNA Rummy

Have students make decks of cards for RNA rummy by writing the names *guanine, adenine, cytosine, uracil,* and *RNA enzyme* on index cards. Have each student make three sets of cards. Have groups of four players shuffle their cards, deal five cards to each person, and take turns drawing a card from the remaining face-down stack until someone wins by having all four nucleotides and an RNA enzyme. To add interest, place a few "natural disaster" cards like *lightning strikes* or *volcano erupts* in each deck. When one of these cards is played on would-be winners' runs, they must place their cards in the deck and draw five more cards.

CONNECTIONS

Chapter 8
Retroviruses and RNA

RNA-based heredity is observable today in retroviruses that contain RNA, not DNA. HIV and feline leukemia are two well-known retroviruses. Review the role of RNA in protein synthesis, and compare its structure to that of DNA. Point out that though it is smaller, RNA still contains the directions for making a protein.

Figure 11-7 Iron pyrite crystals have complex three-dimensional surfaces. Materials with such surfaces might have served as catalysts for the chemical reactions that produced the first building blocks of life.

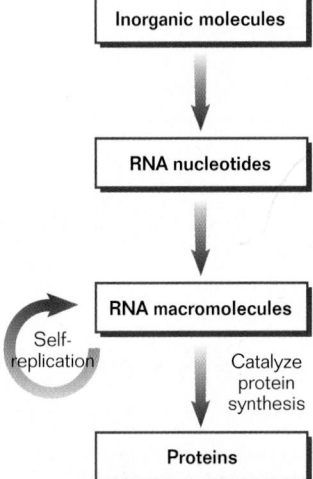

Figure 11-8 If RNA nucleotides from chemical reactions among inorganic molecules assembled into RNA molecules, these molecules might have been able to self-replicate and to catalyze the formation of proteins.

Proteins Can Be Assembled by RNA

Whatever the details of the process, most scientists accept that with the input of energy, the basic molecules of life could have formed spontaneously through simple chemistry. But knowing how the basic building blocks of life might have formed is like finding iron ore, sand, oil, and rubber trees—far from the steel, glass, plastic, and rubber needed to make a car. In other words, there is a long way to go from organic molecules to living cells. How did amino acids link together to form proteins? And how did nucleotides join to form long chains of DNA that store the instructions for assembling proteins? In the laboratory, scientists have not been able to make either of these macromolecules form spontaneously in water. Each link in an organic macromolecule is forged by a chemical reaction that also produces a water molecule. These reactions simply do not occur when their reactants are surrounded by a dense crowd of water molecules—that is, when they are dissolved in water. However, short chains of RNA, the nucleic acid that works with DNA to carry out DNA's instructions, can (with difficulty) be made to form spontaneously in water.

If the proteins and DNA that are necessary for life could not have formed spontaneously in water, how did they form? Some investigators speculate that early life could have developed on a solid surface rather than in water. The surfaces of clay minerals and iron pyrite (*PEYE reyet*) crystals have been suggested as possibilities. As you can see in Figure 11-7, materials such as these offer complex three-dimensional surfaces that might have acted as catalysts by serving as templates (patterns) for the formation of protein and DNA molecules. But while this might be possible in theory, researchers have not yet been able to make either protein or DNA molecules in this way, and most scientists are skeptical of the idea.

In the 1980s, Thomas Cech and his colleagues at the University of Colorado made a key discovery that may answer one question about life's mysterious origin. They found that certain RNA molecules can act like enzymes. RNA's three-dimensional structure provides surfaces with specific shapes for catalyzing reactions, much as protein shapes do. Like DNA, RNA acts as an information-storing molecule. Recall from Chapter 9 that messenger RNA molecules temporarily store the instructions for making proteins in their nucleotide sequences. As a result of Cech's work and experiments demonstrating that RNA molecules can form spontaneously in water, a very simple and attractive hypothesis has emerged: Perhaps RNA was the first self-replicating information-storage molecule. After it had formed, such a molecule could also have catalyzed the assembly of the first proteins, as suggested in Figure 11-8. But more important, such a molecule would have been capable of evolving through natural selection.

Did You Know?

Organic molecules with the same formula can have different shapes. Such molecules behave differently when they react with other molecules. Some molecules called stereoisomers are almost identical, but they are mirror images of each other. Some will work in living things, and others won't. For example, a right shoe will not fit well on a left foot. Molecules that are mirror images are distinguished with the letters D, meaning right-handed, and L, meaning left-handed. In a laboratory, proteins can be made with both D-amino acids and L-amino acids, but only L-amino acids are found in organisms. All sugars in organisms are D-sugars. The enzymes in living things work only on D-sugars and on proteins with L-amino acids. Therefore, if you ate only L-sugars and proteins with D-amino acids, you would starve!

Microspheres Might Have Led to Cells

When you consider how some of the basic molecules of life behave in water, it is not difficult to imagine how the first cells might have formed. Remember that every cell is surrounded by a plasma membrane. Observations show that the basic molecules of plasma membranes—proteins and lipids—tend to aggregate (gather together) in water. By shaking up a bottle of oil-and-vinegar salad dressing, you can see a similar thing happening; the small spherical globs of oil formed by the shaking action attract one another and grow in size by fusing with other globs. Phospholipids, which form the bilayer of a plasma membrane, do the same. Similarly, short chains of amino acids produced abiotically (without life) in a laboratory aggregate into tiny vesicles called **microspheres**.

Scientists think that microspheres, similar to those shown in Figure 11-9, might have been the first step toward cellular organization. Once the basic molecules were present, the early oceans would have contained untold numbers of microspheres—billions in each spoonful of sea water. At first, microspheres would have formed spontaneously, persisted for a while, and then dispersed. Over millions of years, those microspheres that could survive longer by more efficiently incorporating molecules and energy would have become more common than here-today-gone-tomorrow kinds. Still, microspheres could not be considered alive unless they had acquired the capacity to transfer their abilities to offspring. □

□ CAPSULE SUMMARY

Lipids and proteins tend to aggregate in water, forming microspheres. These tiny spheres could have been the first step in the organization of cells.

Demonstration
Growing Globs
Place a teaspoon of oil in a glass container (e.g., petri dish or beaker) of water, and set the container on an overhead projector. Mix up the oil to break it into many small globs, focus the projector on the globs, and watch as they join to form larger globs. A little food dye in the water, or Sudan IV dye for the oil, helps in the observation of the process. Ask students to explain why the oil drops are attracted to each other, (Because like attracts like, nonpolar molecules are attracted to other nonpolar molecules.) and what cell membranes have in common with oil drops. (Both contain nonpolar lipids.)

◉ VISUAL STRATEGY

Figure 11-9
Have students compare the shapes of the structures in these photos. Ask: What other similarity can you see? (Both occur in clusters.)

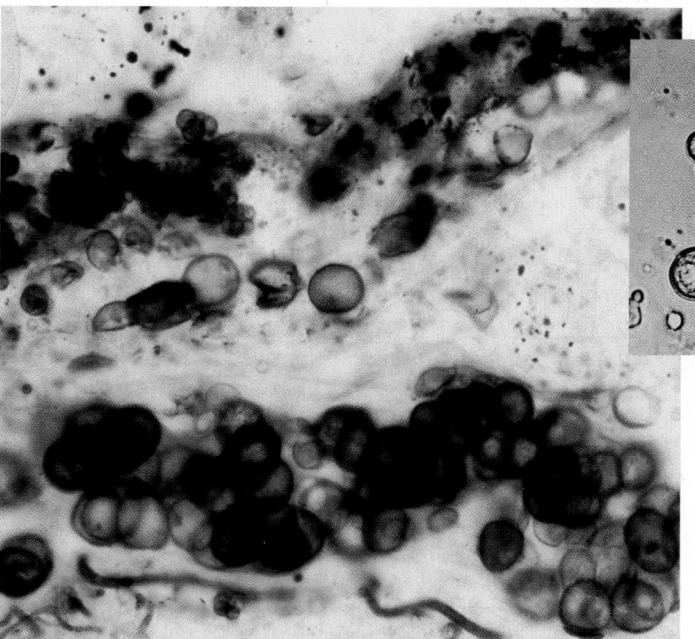

Figure 11-9 Ancient cells found in microfossils, *left*, resemble microspheres, *above*. Though they are not alive, microspheres share several characteristics with living cells. Both have a selectively permeable membrane, take in materials, grow in size, use energy to facilitate chemical reactions, and divide to form new individuals after reaching a certain size.

Overcoming Misconceptions

The Same Elements

Many students believe that living things are made of different "stuff" than nonliving things. Point out that although the molecules are different, we share the same elements with nonliving matter. Have students relate the elements found in living things, described in Chapter 1, to a nonliving part of Earth that contains that element. For example, carbon is found in diamonds and limestone, $CaCO_3$. Oxygen is a part of water and of many types of minerals in rocks. Hydrogen is found in water. Phosphorus is found in a mineral called apatite. Nitrogen is 78.03 percent of the atmosphere. Chlorine is in halite (rock salt). Calcium and sodium are in feldspar. Potassium, iron, and magnesium are in mica.

CONNECTIONS

Chapter 1

Characteristics of Life

Review the characteristics of life presented in Chapter 1. Remind students that the ability to reproduce is one of these characteristics. If life arose and could not reproduce itself, then it would not be able to persist.

Teaching Tip
Modeling Heredity

Students can model heredity by using interlocking plastic toy blocks to represent the building blocks of an organism. Begin with an original "organism" that consists of three blocks of specific size and color. Ask: What would happen if this organism were alone on the planet and could not reproduce itself? *(It would eventually die and decay.)* What could happen if this organism were able to reproduce itself? *(It could produce more organisms like itself.)* Have students model the reproduction of the "organism" by constructing additional ones from plastic blocks. Ask: How did you produce copies of the organism? *(recognized the pattern of blocks in the original organism and assembled the same colors of blocks in the same order)* Ask students what molecules perform this function in reproduction. *(RNA and DNA)*

❑ CAPSULE SUMMARY

Scientists now hypothesize that RNA was the first self-replicating information-storage molecule, but they cannot explain how heredity developed.

Origin of Heredity Remains a Mystery

There is considerable discussion among scientists about how hereditary mechanisms might have evolved. Most researchers now suspect that RNA was the first information-storing molecule to form and that RNA "enzymes" catalyzed the assembly of the earliest proteins. Scientists think that double-stranded DNA probably evolved later, as a way of ensuring the safety of hereditary information by storing it in a protected central location. However, scientists do not agree about whether RNA molecules first formed inside or outside of microspheres. Perhaps among microspheres that contained RNA, some might have developed a means of transferring their abilities to offspring (heredity). Once the mechanism of heredity developed, life as we know it began.

As you can see, the scientific vision of life's origin is at best a hazy outline viewed from a long distance through dark glasses. While scientists cannot disprove the hypothesis that life originated naturally and spontaneously, little is known about what actually happened. Many different scenarios—some of them quite imaginative—seem possible, and some have solid support from experiments. But because researchers do not yet understand how DNA, RNA, and hereditary mechanisms first developed, science is currently unable to resolve disputes concerning the origin of life. How life might have originated naturally and spontaneously remains a subject of intense interest, research, and discussion among scientists. ❑

Section Review

1. *List three ideas about how life could have originated on Earth.*
2. *Why can't hypotheses that assume life had a divine origin be tested scientifically?*
3. *How are the primordial soup model and the bubble model similar? How are they different?*
4. *What properties of RNA make it the most likely candidate for the first information-storage molecule?*
5. *What important process had to develop before life could begin?*

Critical Thinking
6. *Why are microspheres a logical first step toward cellular organization?*
7. *Besides DNA, what other important feature that is a characteristic of all cells had to develop before living cells could exist?*

Answers to Section Review

1. divine creation, spontaneous origin, extraterrestrial origin

2. Divine forces cannot be studied or re-created in an experiment nor explained by known laws of nature.

3. Similar: Both propose that the building blocks of life developed through chemical reactions, with products collecting in the ocean. Different: In Oparin's model, the chemical reactions occurred between molecules dissolved in a stagnant ocean and would take more time, while in Lerman's model the key chemical reactions occurred in bubbles on the ocean's surface.

4. RNA can act as a catalyst and can store the information needed to assemble proteins.

5. heredity

6. Answers will vary but may mention that microspheres are packages of proteins or lipids, are bound by a double-layered membrane, grow by taking in materials, and divide after reaching a certain size.

7. the plasma membrane

11-2 Evaluating the Spontaneous Origin Hypothesis

Most scientists accept spontaneous origin as the best available explanation of life's origin. However, like any scientific hypothesis, this explanation is subject to future modification or rejection based on new evidence. Indeed, strong objections to the spontaneous origin hypothesis have been voiced by people outside the scientific community. Evaluating these objections will provide you with a better understanding of the hypothesis.

Section Objectives

■ Explain how radioisotopes can be used in determining Earth's age.

■ Identify the biological processes that account for the diversity and complexity of life on Earth.

■ Explain why life processes do not violate the second law of thermodynamics.

■ Explain why proteins do not assemble spontaneously in water, and describe how they are assembled.

Is Earth Old Enough?

There is no doubt among scientists that our planet is very old. However, some people disagree, believing that Earth is no more than 10,000 to 20,000 years old—not nearly old enough for life to have arisen and then evolved as most scientists infer. The accepted scientific estimate of Earth's age is 4.5 billion years. This estimate is derived from **radiometric dating**, which involves the measurement of radioactive isotopes *(EYE soh tohps)* of certain elements found in rocks. Isotopes are forms of an element that differ in atomic mass. Radioactive isotopes, or **radioisotopes**, gradually change into other, more stable isotopes through a process called radioactive decay. For example, certain rocks contain minute traces of potassium-40, a radioisotope of the element potassium, K. As Figure 11-10 shows, it takes about 1.3 billion years for one-half of the potassium-40 in a rock to decay into other isotopes. The period of time it takes for one-half of a radioisotope to decay is called its **half-life**. By estimating how many half-lives have passed since a rock was formed, scientists can approximate the rock's age. Other long-lived radioisotopes are also used for radiometric dating.

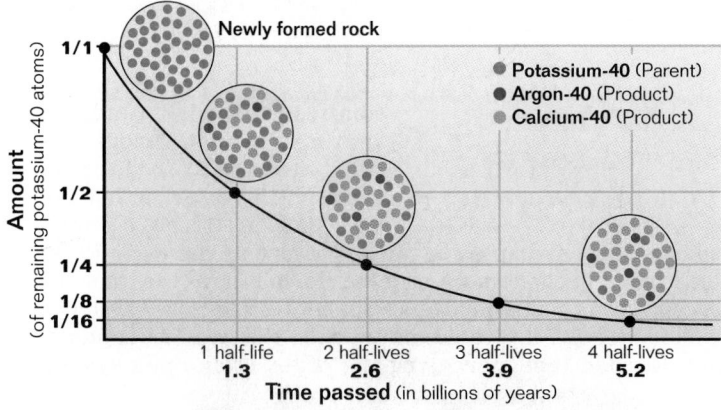

Figure 11-10 This graph shows the rate of decay for the radioisotope potassium-40. Rocks are dated by measuring the ratio of a radioisotope to its decay product. The radiometric clock starts "ticking" when rock minerals first crystallize from magma. After one *half-life* has passed, half of the original amount of a radioisotope remains, and the other half has been converted to decay product. The rest of the radioisotope continues to decay at a constant rate until all of it becomes the decay product.

SECTION 11-2

Vocabulary Preview

radiometric dating

radioisotopes

half-life

Opening Questions

Ask students to name several radioactive elements. Their responses might include uranium, plutonium, and radium. Ask students to write a definition of the term *radioactivity*. Students may recall that radioactivity is the emission of particles and energy by the nucleus of an unstable atom as it decays.

Demonstration

Modeling Radioactive Decay

Radioactive decay can be modeled by placing 100 pennies in a container (e.g., a shoe box), shaking it, and observing how they land. Have a volunteer remove and count the pennies that land tails up. State that these pennies have "decayed." Record this number on the board or overhead in a table with two columns headed *Heads-Up Pennies (not decayed)* and *Tails-Up Pennies (decayed)*. Repeat this procedure until no pennies remain in the container. Have students graph the number of pennies that landed heads up in each trial and compare their graphs to the one in Figure 11-10 on page 233.

👁 VISUAL STRATEGY

Figure 11-10

Point out that potassium-40 has *two* decay products. Still, 50 percent of the potassium-40 atoms decay to one or the other of these products after each half-life. Have students use this graph to estimate the age of rocks that contain potassium-40. For example, ask: If a rock had 1 g of potassium-40 when it formed and now has 0.5 g of potassium-40, how old is the rock? *(about 1.3 billion years old)*

Teaching Tips
Direct and Indirect Evidence

To introduce the idea that both direct and indirect evidence can be used in science, use either a news feature about a crime or a fictional account of someone being accused of a crime. Have students identify specific examples of evidence linking the suspect to the crime. Have students decide whether each example is direct or indirect evidence. Evidence such as an eyewitness account of the suspect committing the crime is *direct evidence*. Evidence such as fingerprints and hair samples is *indirect evidence*. Ask: What kind of evidence could convict the accused? *(both direct and indirect)* What kind of evidence merely shows that it was possible for the accused to have committed the crime? *(indirect)* What kind of evidence does radiometric dating provide? *(indirect)*

Home-Style Entropy

Entropy and the second law of thermodynamics are probably not things your students worry about, so give them humorous, everyday examples of how disorder tends to increase in the universe. Examples such as, What happens to a clean kitchen? or What happens when your little brother plays with your CD collection? will help students recognize their own experiences with entropy. Point out that an energy input can reverse the effects of entropy in a limited area. For instance, you can clean the kitchen by expending energy.

CONNECTIONS

Chapter 4
Entropy and the Second Law of Thermodynamics

Remind students that entropy is the amount of disorder in a system and that the second law of thermodynamics states that entropy tends to increase in a closed system, in which no energy can enter or leave.

☐ **CAPSULE SUMMARY**

Scientific estimates of Earth's age are derived from radiometric dating, which is based on the regular rate at which radioisotopes decay. For these rates to change, the fundamental nature of matter would have to change as well.

☐ **CAPSULE SUMMARY**

The diversity and complexity of organisms can be explained as the result of genetic variation and the power of natural selection.

◈ **Did You Know?**

It is important to use the right isotopes when dating fossils. Some isotopes, such as carbon-14 with a half-life of 5,730 years, decay so quickly that they are useless for dating materials that are more than 20,000 years old. The opposite is also true. Potassium-40, with a half-life of 1.3 billion years, and uranium-238, with a half-life of 4.5 billion years, are useless for dating a woolly mammoth that has been frozen in ice for only a few thousand years.

Is radiometric dating reliable? The regular and measurable rate at which radioisotopes decay is the basis for radiometric dating. The constancy of radioactive decay is an elementary principle of physics that springs from the fundamental forces that hold all atoms together. The behavior of these forces has been verified many times. Though the rate of radioactive decay does vary slightly for certain radioisotopes under specific conditions, any possible variations are taken into account during the dating process. These variations are not significant enough to invalidate dating techniques. For a radioisotope's decay rate to fluctuate significantly, the fundamental nature of matter would have to change. Still, every scientist knows that *no* technique is 100 percent accurate. Determining the age of a rock, therefore, involves multiple measurements, using several different radioisotopes if possible. When it is done properly, radiometric dating is one of the most reliable and reproducible measurements scientists can make. ☐

Is Life Too Complex to Have Arisen Naturally?

It may seem that organisms are too complex to have originated by random natural processes. To some, this implies that the complexity of life must be the result of an intelligent design. Most biologists, however, think that the complexity and diversity of life can result from the enormous amount of variation that can be encoded in DNA, coupled with the power of natural selection when applied over long periods of time. Natural selection—the very powerful agent of change that guides evolution—is neither random nor directionless. Instead, it produces very specific changes that are determined by the environment. Each favorable change is built upon a previous one, and, therefore, favorable changes are cumulative. With variation and natural selection working together over billions of years, very complex life-forms could have evolved naturally. ☐

Does Spontaneous Origin Violate the Second Law of Thermodynamics?

The second law of thermodynamics (disorder tends to increase in the universe) is often cited as a reason to reject the spontaneous origin hypothesis. For life to have begun spontaneously, simple chemicals present on the early Earth must have become more ordered. Recall, however, that the second law of thermodynamics applies only to closed systems, while Earth and its organisms are open systems. In Chapter 4, you learned that increasing the order in an open system simply requires an input of energy. And as you have also learned, radiant energy from the sun continually enters Earth's living systems through photosynthesis, fueling the processes that organize life from metabolism to evolution.

To better understand why the second law of thermodynamics does not apply to processes involving organisms, consider the changes that occurred as your body developed. Like every human being, you began life as a single fertilized egg cell, similar to the one in Figure 11-11. You are now a highly organized creature, far more complex than you were when you started life. You have not violated the second law of thermodynamics because you are an open system. Any open system is capable of increasing in complexity as it absorbs energy from its surroundings. Therefore, even life's first chemicals could have organized spontaneously if enough energy sources were present on the early Earth. As Miller demonstrated in his experiment, electrical energy from lightning could have been one of those energy sources.

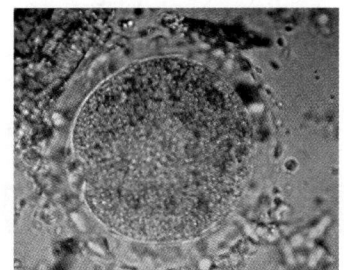

Figure 11-11 A fertilized human egg cell, like this one, grows and increases in complexity as it absorbs energy-containing nutrients from its surroundings.

Can Proteins Assemble Spontaneously?

Scientists have not been able to cause amino acids dissolved in water to join together to form proteins. The energy-requiring chemical reactions that join amino acids are freely reversible and do not occur spontaneously in water. However, most scientists no longer argue that the first proteins assembled spontaneously. Instead, they now propose that the initial macromolecules were composed of RNA, and that RNA later catalyzed the formation of proteins. The assembly of RNA from nucleotides is *not* reversible in water and could easily have occurred spontaneously. As scientists also point out, energy for such reactions could have come from many sources, including ultraviolet light, lightning, and the extreme heat of volcanic events. ▢

▢ **CAPSULE SUMMARY**

Proteins cannot assemble spontaneously in water, but they can be assembled with the aid of a catalyst and an input of energy.

Section Review

1. *How does radiometric dating indicate a rock's age?*
2. *What biological processes account for the diversity and complexity of life on Earth?*
3. *Does the second law of thermodynamics apply to the organization of life? Why or why not?*

Critical Thinking
4. *The cytosol of a cell is mostly water. Why, then, can proteins be assembled there?*

Teaching Tips
Seeds and the Second Law of Thermodynamics
Have students explain why the sprouting of a seed and the growth of a plant do not violate the second law of thermodynamics. Ask: What happens to a seedling that is kept in a dark cabinet after it sprouts? *(It will not grow.)* When students identify light as an energy source for plant growth, ask where the seed got the energy to sprout in the first place. *(chemical energy obtained from solar energy by the parent plant)* Finish by asking students to visualize how our planet would change if the sun stopped shining.

Where Do You Get Your Energy?
Have students apply the second law of thermodynamics to their own bodies. Ask: What is your energy source, *(food)* and what would happen if this energy source were to disappear? *(You would starve.)*

Demonstration
What a Difference an Enzyme Makes!
Appoint a student to be an "enzyme" to model how enzymes are important to chemical reactions. Place some pop beads or plastic blocks that can be assembled into a clear container labeled "random collisions," and give some of the beads or blocks to the student. Ask the student to assemble the beads while you shake the jar, causing random collisions. When the student is done, pour out the "molecules" you were shaking, and compare them to the structure assembled by the student. Ask: How long and how hard would you have to shake the container to make any of the pop beads connect? *(a very long time and very hard)* Ask a volunteer to summarize an enzyme's effect on a chemical reaction. *(The enzyme speeds the rate of the reaction by lowering the amount of activation energy and by bringing together the molecules to be joined.)*

Vocabulary Preview

(none)

Opening Questions

Ask students to pretend that they are space travelers visiting this solar system for the first time. Ask them to write down several things that they might see from outer space that would indicate that there is life on the third planet from the sun. These things include the color green, which indicates plant life, and features such as city lights, highways, dams, and other large structures.

Demonstration

Looking for Signs of Life

Show students photographs of Earth taken by satellites and by astronauts. Have students try to identify signs of life on our planet.

Teaching Tip

Different Life for Different Worlds

In order to creatively explore the idea of different types of life for different planets, have students design life-forms that could survive conditions radically different from those on Earth. This works best for small groups of two to three students who have the time to consider the possibilities of living where water is always solid ice, or where the gravitational force is half that of Earth. Have students draw or construct the life-forms, label their anatomical parts, and give each organism a scientific name.

Section Objectives

- *Relate Earth's size and distance from the sun to the life-forms it supports.*
- *Explain why life probably exists elsewhere in the universe.*
- *Describe how scientists are looking for intelligent life in other parts of our galaxy.*

*O*n a dark, clear night you can look up and see count-less stars in the sky. Have you ever wondered if we are alone in the universe? Does "someone" some-where out there look up and wonder if we exist? For cen-turies, people have speculated about whether life exists elsewhere in the universe. Jules Verne wrote novels about people on the Moon. Reports of canals on Mars, like those seen in Figure 11-12, led to stories about beings from Mars. Scientists now know there is no evidence of life on either the Moon or Mars. But what about all the places humans have not yet explored?

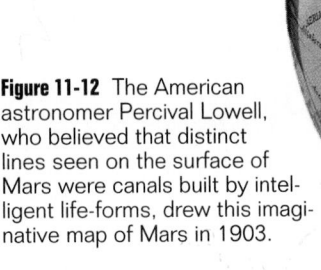

Figure 11-12 The American astronomer Percival Lowell, who believed that distinct lines seen on the surface of Mars were canals built by intelligent life-forms, drew this imaginative map of Mars in 1903.

Life on Earth Reflects the Nature of the Planet

Life as it has evolved on Earth closely reflects the nature of the planet and its history. For one thing, life as we recognize it exists mostly within a narrow range of temperatures—approximately –18° to 38°C (0° to 100°F). The spontaneous evolution of carbon-based life is probably possible only within the narrow range of temperatures that exists on Earth, a range directly related to its distance from the sun. If Earth were farther from the sun, it would be colder, and chemical processes would be greatly slowed down. Water, for example, would be a solid, and many carbon compounds would be brittle. If Earth were closer to the sun, it would be warmer, chemical bonds would be less stable, and few carbon compounds would be stable

 Did You Know?

The idea that Earth is just the right distance from the sun, so that it is not too hot and not too cold, has been called the Goldilocks hypothesis for obvious reasons. However, some scientists believe that life on Earth is essential to maintaining conditions favorable to life by fixing CO_2 when it becomes too plentiful in the atmosphere, and releasing it when there is not enough. When the CO_2 concentration rises, the planet warms. When the CO_2 concentration drops, the planet cools.

enough to persist. For life as we know it to exist on another planet, that planet would have to orbit its sun at just the right distance to have the necessary range of temperatures.

Earth's mass (which is a function of its size and density) is also just right for life to exist as we know it. Because of its mass, Earth has the right amount of gravitational pull to hold the gases found in our atmosphere. Earth's atmosphere helps insulate it from temperature extremes and radiation that is harmful to life, while allowing enough energy to reach its surface for fueling life's activities. If Earth were less massive (smaller or less dense), its gravitational pull would not be great enough to hold an atmosphere. If Earth were more massive (larger or more dense), it might hold such a dense atmosphere that all solar radiation would be absorbed before it reached the planet's surface. So for life as we know it to exist on another planet, that planet would have to have just the right mass (size and density) and gravitational pull to hold a suitable atmosphere. ❑

❑ **CAPSULE SUMMARY**

Earth has just the right mass and is just the right distance from the sun for life as we know it to exist.

Life Probably Exists Elsewhere
....

The universe as a whole is awash with places where life might have arisen. Within our solar system, the tiny moon of Jupiter seen in Figure 11-13 is the place most likely to support extraterrestrial life. In fact, conditions there would be far less hostile to life than the conditions that are thought to have existed in Earth's primordial oceans. Our own Milky Way galaxy and the nearby Andromeda galaxy each contain more than 100 billion stars. And the universe holds more than a billion galaxies. Astronomers estimate that the universe contains some 10^{20} (100,000,000,000,000,000,000) stars with physical characteristics that resemble those of our sun. At least 10 percent of these stars are thought to have planetary systems. New telescopes, such as the Hubble Space Telescope, reveal that several nearby stars seem to have planets orbiting them. If only 1 in 10,000 of the planets in the universe has the right combination of mass and distance from its sun to duplicate Earth's development, life could have arisen 10^{15} (a million billion) times. Undoubtedly, many other worlds have physical characteristics resembling those of Earth. Therefore, we might not be alone.

Life processes also might have arisen and evolved differently on other planets. A functional genetic system that is capable of accumulating and replicating changes is the basis of the evolution of life on Earth. But heredity does not require DNA—only a way to preserve and pass on information. Under different conditions, such a system theoretically could form from substances other than the carbon-based compounds and water that make up life on Earth. Silicon and ammonia are the most likely possibilities. Like

Figure 11-13 Europa, one of the moons of Jupiter, is the place in our solar system most likely to harbor extraterrestrial life. In 1995 scientists learned that its thin atmosphere contains oxygen gas. Beneath Europa's thick skin of ice, pressure due to gravity might create enough heat for water to exist in a liquid form. If that is so, life might have arisen and flourished there, hidden from view.

☑ **RESEARCH Update**

UPDATE

Cosmic Dust Wakes

Stanley Dermott and his collaborators at the University of Florida, in Gainesville, are tracking planets. Images relayed to Earth from the Infrared Astronomical Satellite (IRAS) show clouds of dust in planetary orbits (the bigger the planet, the more dust). This zodiacal cloud, as it is called, is the consequence of asteroidal collisions. The astronomers in Gainesville have used the IRAS images to create numerical simulations of orbital asteroidal dust particles. Their simulations show that Earth orbits the sun within a ring of dust and trails a dust cloud in its wake. Scientists hypothesize that due to the physical properties of the solar system, pieces of cosmic dust tend to spiral in toward the sun and become trapped in planetary orbits. Scientists find this idea exciting for two reasons: first, this cosmic dust may have been an early source of carbon, which at the right time and place in Earth's history may have contributed to the origin of life; and second, other scientists think that this dust-wake model can be used to predict the presence of presently unknown planets outside of our own solar system (that is, the wake may be "seen" even if the planet is not). Dermott and his colleagues also submit that close encounters with cosmic dust may have added organic matter to Earth's environment. ☑

☐ CAPSULE SUMMARY

Considering the number of stars in the universe, it is possible that life might have arisen numerous times across the universe.

Figure 11-14 The world's largest radio telescope, located in the hills above Arecibo, Puerto Rico, is a huge disk more than 305 m (1,000 ft.) in diameter.

carbon, silicon needs four electrons to fill its outer energy level. And ammonia is even more polar than water. Perhaps under radically different temperatures and pressures, these substances might have formed complex molecules as diverse and flexible as the carbon-based ones on Earth. ☐

Scientists Are Listening for Other Life-Forms

Serious attempts are now being made to look for messages from intelligent life-forms that might exist on planets circling distant stars. These efforts are referred to as the Search for Extraterrestrial Intelligence (SETI). The first SETI program was carried out by astronomer Frank Drake in 1960. He listened to two nearby stars for two weeks at one particular radio frequency but heard only sounds from military aircraft on Earth. Since then, other attempts (usually lasting several months or more) have been made to detect signals from intelligent life on other worlds. Although unusual signals have been detected, none have been found to repeat in a manner that could be detected again later.

In 1992, NASA stepped up the efforts to locate extraterrestrial life by aiming the world's largest radio telescope, seen in Figure 11-14, at approximately 1,000 of the stars nearest Earth. Using a new computer technology, NASA simultaneously monitored 8.4 million radio channels for signs of intelligent life. About 25 signals interesting enough to require further analysis were detected in the first year, although none were proved to be from alien life-forms. In 1993, however, Congress cut the program from the NASA budget, calling it "a great Martian chase" and "a waste of time." Astronomers are seeking ways to continue the search.

Section Review

1. *What evidence would you cite to support the hypothesis that life could exist elsewhere in the universe?*

Critical Thinking

2. *What do you think would happen to life on Earth if the planet's orbit suddenly shifted so that it was twice as far from the sun?*

3. *All stars give off radio waves, which result from the nuclear reactions that make stars shine. What, then, would you look for in extraterrestrial radio signals that might indicate they were produced by an intelligent life-form?*

Chapter Closure

Ask students to imagine traveling to another planet in our galaxy. Ask: What signs would you look for in trying to determine whether any life-forms exist there. (*Answers will vary but should include replication, metabolism, and responsiveness.*)

Certain meteorites contain organic compounds, which might indicate that carbon-based life exists elsewhere in the universe.

2. Answers will vary but should state that such a change would result in much colder temperatures on Earth and reduce available sunlight, slowing the rate of

photosynthesis. As a result, most of the life-forms currently living on Earth would probably die.

3. Answers will vary but should propose that signals created by intelligent life would have repeating patterns rather than the random patterns of "noise" produced by natural sources.

11
CHAPTER REVIEW

Vocabulary

extraterrestrial origin (225)
half-life (233)
microsphere (231)
radioisotope (233)
radiometric dating (233)
spontaneous origin (226)

Concept Mapping

Construct a concept map that shows how life might have originated by natural forces. Use as many terms as needed from the vocabulary list. Try to include the following items in your map: spontaneous origin, primordial soup model, bubble model, RNA, proteins, microspheres, and radioisotopes. Include additional terms in your map as needed.

Review

Multiple Choice

1. The divine creation explanation for the origin of life
   a. can be scientifically tested.
   b. is identical to the extraterrestrial origin explanation.
   c. is based on forces that cannot be explained by science.
   d. is virtually the same among the many religions of the world.
2. A reevaluation of the primordial soup model was prompted by
   a. the hypothesis that ammonia and methane may have been missing from Earth's early atmosphere.
   b. Miller's finding that amino acids can be formed from ammonia and other chemicals.
   c. evidence that 4.5 billion years were available for life to begin.
   d. the discovery that ozone gas was present in Earth's atmosphere 4 billion years ago.
3. Lerman's bubble model, which is a modification of the primordial soup model,
   a. accounts for the formation of bubbles in Earth's early oceans.
   b. describes how life could have formed in a short time in the presence of ultraviolet light.
   c. proposes how polar molecules and bubbles gave rise to cells.
   d. explains how ammonia, nitrogen, and oxygen combined to form amino acids.
4. RNA is thought to have been the first information-storage molecule because it
   a. can replicate on solid surfaces.
   b. can form spontaneously without water.
   c. can store information and catalyze reactions.
   d. can assist in the assembly of DNA.
5. Which of these sequences of chemical reactions is the most likely to have led up to the formation of proteins?
   a. inorganic molecules, RNA nucleotides, RNA macromolecules, proteins
   b. inorganic molecules, amino acids, microspheres, proteins
   c. silicon and ammonia, RNA nucleotides, microspheres, proteins
   d. methane and ammonia, amino acids, DNA, proteins
6. Cells are different from microspheres because cells
   a. contain amino acids.
   b. have an outer boundary made of two layers.
   c. grow by taking in molecules from their surroundings.
   d. transfer instructions for heredity.
7. The reliability of radiometric dating is based on the assumption that
   a. the rates of radioactive decay have not changed over time.
   b. radioisotopes gradually change to more stable isotopes.
   c. carbon isotopes do not decay.
   d. all rocks are less than 50,000 years old.

CHAPTER REVIEW ANSWERS

Each item in the Chapter Review is correlated by text section in the assignment guide that follows.

ASSIGNMENT GUIDE

Section	Review	Themes Review	Critical Thinking
11-1	1–6, 11–13, 19, 20	23–25, 26	27
11-2	7, 8, 14–16, 21	25, 26	28
11-3	9, 10, 17, 18, 22		29

Concept Mapping
Map answer is shown on page 223B.

Review
Multiple Choice
Code in parentheses indicates section number and objective number.
1. c (11-1.1)
2. a (11-1.2)
3. b (11-1.2)
4. c (11-1.3)
5. a (11-1.3)
6. d (11-1.4, 11-1.5)
7. a (11-2.1)
8. c (11-2.2)
9. b (11-3.1)
10. b (11-3.3)

11 *CHAPTER REVIEW* •••

Completion

11. extraterrestrial, asteroids or meteorites (11-1.1)
12. primordial soup, lightning (11-1.2)
13. microspheres, cellular (11-1.4)
14. radiometric dating, 4.5 billion (11-2.1)
15. open (11-2.3)
16. proteins, catalyst (11-2.4)
17. distance from the sun (11-3.1)
18. stars, likely (11-3.2)

Short Answer

19. According to the bubble model, the chemical reactions that may have led to the origin of life occurred within bubbles, where the chemical reactants would have been concentrated and more likely to react. In Oparin's model, the reactions occurred in the waters of the first oceans, where sea water mixed with organic chemicals to form a primordial soup. (11-1.2)
20. The development of heredity marks the point at which life actually began, according to the spontaneous origin model. (11-1.5)
21. Radiometric dating is not directly used to find the age of sedimentary rocks because the data obtained would indicate the age of the older pieces of rock or shell that make up a sedimentary rock and not the age of the sedimentary rock itself. (11-2.1)
22. Europa has a thick layer of ice, under which the liquid water necessary for life as we know it might exist. Mars, on the other hand, has lost most of its water and has no place for liquid water to exist. (11-3.1)

8. Scientists feel comfortable offering an explanation for the diversity and complexity of the biological world based on what is known about
   a. radioisotopes.
   b. supernatural forces.
   c. genetic variation and natural selection.
   d. the forces that hold all atoms together.

9. If Earth's mass were less than it is, life would not exist because
   a. Earth would be too close to the sun.
   b. the gases that form Earth's atmosphere would have dissipated into space.
   c. the atmosphere would be too dense for solar radiation to reach the surface.
   d. Earth's gravitational pull would make carbon compounds brittle.

10. What characteristics of radio waves might suggest that they were produced in another part of our galaxy by intelligent life-forms?
    a. They would blend with background noise.
    b. They would repeat and be louder and different from background noise.
    c. They would be transmitted at or above 2 million Hz.
    d. They would be transmitted at less than 2 million Hz.

Completion

11. Some scientists propose that life had a(n) _____ origin and was later transported to Earth by _____.
12. Stanley Miller's experiment tested the _____ model for the origin of life and showed that _____ might have served as an energy source for the formation of organic molecules from hydrogen-rich gases in Earth's early atmosphere.
13. Tiny vesicles called _____, which tend to form when lipids and proteins aggregate, are thought to be the first steps toward _____ organization.

14. Using a procedure called _____, scientists estimate that Earth's age is about _____ years.
15. The processes by which life is organized do not violate the second law of thermodynamics because Earth and each of its organisms are _____ systems.
16. Because the chemical reactions that form them are reversible, _____ cannot assemble spontaneously in water, but they can be assembled with the aid of a(n) _____.
17. Two critical factors that have affected the way in which life developed on Earth are the planet's mass and its _____.
18. Because of the tremendous number of _____ in the universe, it is _____ that life exists somewhere besides Earth.

Short Answer

19. How does Lerman's bubble model of the origin of life differ from Oparin's primordial soup model?
20. According to the spontaneous origin theory, how was the development of heredity important?
21. Sedimentary rocks are made up of the shells of dead organisms or pieces of older rocks. Why, then, is radiometric dating *not* used to determine the age of sedimentary rocks?
22. Why do scientists think that life as we know it is more likely to exist on Jupiter's moon Europa than on Mars?

Themes Review

23. **Evolution** The phrase *chemical evolution* is sometimes applied to the series of events that might have resulted in the spontaneous origin of life. How do the events that are thought to have produced living organic matter from nonliving inorganic matter represent a process of evolution?

Themes Review

23. The chemical reactions started with simple inorganic molecules and gradually produced complex organic molecules, which formed associations that gradually changed over time. (11-1.2, 11-1.3)

24. The organization of organisms begins at the atomic level. Atoms are organized into molecules that undergo chemical reactions, which control the activities of life. Organisms incorporate energy, inorganic materials, and organic molecules into their structures as they grow

and become more complex. (11-1.2, 11-1.3, 11-1.4)

24. **Levels of Organization** How do the steps in a spontaneous origin of life reflect the organization of organisms?

25. **Flow of Energy** How do the primordial soup model and the bubble model account for the input of energy needed to organize life? Where did the energy come from, how was it incorporated, and where did it go?

26. **Structure and Function** How does RNA's structure enable it to function as a catalyst, and how might RNA's structure have enabled it to evolve?

Critical Thinking

27. **Designing Experiments** Suppose that you are asked to design an experiment to test whether organic matter might have formed from inorganic matter on the early Earth. What variables would you test? What type of control would you use for each experiment?

28. **Interpreting Data** A sample of a certain type of igneous rock contains 25 g of a radioisotope. The same amount of this type of rock has 200 g of the radioisotope when it forms. Using the graph showing the radioisotope's decay rate, determine (a) about how long ago the rock sample was formed and (b) the radioisotope's approximate half-life.

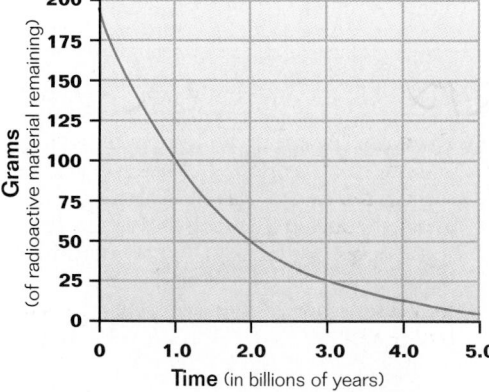

29. **Making Inferences** Suppose that while listening to recordings of radio waves coming from a solar system located in the Milky Way galaxy, you repeatedly hear strong radio signals that have patterns similar to music. What might you infer about the origin of such signals?

Activities and Projects

30. **Research and Writing** Thomas Cech and Sidney Altman shared a Nobel prize in 1989 for their work on RNA. Research their work and the rewards associated with winning a Nobel prize. Relate your findings in a written report.

31. **Cooperative Group Project** Collect information about the radio telescope at Arecibo, Puerto Rico, and about why the United States Congress decided in 1993 to stop funding NASA's use of the telescope to listen for signs of intelligent life. Then, debate the benefits and drawbacks of funding the listening project.

32. **Multicultural Perspective** Research stories from other cultures about visits by intelligent extraterrestrial beings. Relate your findings in an oral report.

Readings

33. Read R. Cowen's article "Taking a chemical look at the early Earth," in *Science News*, April 2, 1992, page 214. What radioisotope decayed to produce neodymium-142? What do Carlson and Jacobsen think that their findings may indicate about the formation of Earth's first crust and of the Moon?

34. Read Mitchell Waldrop's article "Finding RNA Makes Proteins Gives 'RNA' World a Big Boost," in *Science*, June 2, 1992, pages 1396–1397. How has Noller's research changed biologists' view of RNA's role in protein synthesis? How do Noller's findings support the hypothesis that RNA catalyzed the formation of the first proteins?

Critical Thinking

27. Answers will vary, but variables to test include energy source, mixture of gases, and presence of water. A control for each experiment might be, respectively: no outside energy source, mixture of gases in today's atmosphere, and absence of water. (11-1.2)

28. Answers will vary but should be (a) about 3 billion years and (b) about 1 billion years. (11-2.1)

29. One might infer that strong radio signals with repeating patterns that vary and resemble music may be produced by some type of intelligent life, as occurs on Earth. (11-3.3)

Activities and Projects

30. Cech and Altman shared a Nobel Prize for their discovery that during transcription, RNA functions as both a substrate and an enzyme. Nobel Prize winners receive a cash award and worldwide recognition.

31. The Arecibo telescope consists of a mesh that acts as an antenna and uses computers to collect radio signals. Arguments should be based on collected information rather than just on personal opinion.

32. Answers will vary.

Readings

33. Decay of sumarium-146 produced neodymium-142. Carlson and Jacobsen think the neodymium-142 in the rocks of Greenland may indicate that Earth melted and formed a solid crust within the first 100 million years it existed. Neodymium may provide evidence that the Moon was formed when a Mars-sized object struck Earth.

34. Noller's research shows that rRNA may be active in linking amino acids into proteins and in other events that occur in ribosomes. These findings support the "RNA world" hypothesis because they show that RNA can do things that people thought only proteins could do and that ribosomes may be descendants of the first RNA molecules that catalyzed the formation of functional proteins.

25. Both models contend that energy for organizing life came from sources such as electrical sparks, ultraviolet radiation, and intense heat. The energy was incorporated by endergonic chemical reactions and stored in organic molecules that resulted from the reactions. (11-1.2, 11-2.3)

26. RNA's three-dimensional structure provides surfaces with specific shapes for catalyzing reactions. RNA also stores information in its nucleotide sequence, which can be altered by mutation and can be self-replicating. Variation among such molecules could be acted upon by natural selection—the most "successful" molecules might increase in number. (11-1.3, 11-2.4)

LABORATORY Investigation Chapter 11

Making Microspheres

OBJECTIVE

Make microspheres from amino acids, and compare their structure with that of living cells.

PROCESS SKILLS

- observing
- comparing and contrasting

MATERIALS

- safety goggles
- lab apron
- 500 mL beaker
- hot plate
- two 125 mL Erlenmeyer flasks
- ring stand with clamp
- balance
- aspartic acid
- glutamic acid
- glycine
- glass stirring rod
- tongs
- clock or timer
- 1% NaCl solution
- 50 mL graduated cylinder
- dropper
- microscope slides
- coverslips
- compound light microscope
- 1% NaOH solution

BACKGROUND

1. What are microspheres?
2. Offer a scientific explanation for how life could have begun from something similar to simple microspheres of heated amino acids.
3. How would microspheres be similar to living cells? How would they differ?
4. Write your own question to explore in your lab report or notebook.

TECHNIQUE

1. Put on safety goggles and a lab apron.
2. Fill a 500 mL beaker half full with water and heat it on a hot plate. **CAUTION: Use care to avoid burns when working with the hot plate.** You will use the beaker as a hot-water bath. Leave space on the hot plate for a 125 mL Erlenmeyer flask to be added later.
3. While waiting for the water to boil, clamp a 125 mL Erlenmeyer flask to a ring stand. Add 1 g each of aspartic acid, glutamic acid, and glycine to the flask, and combine these dry powders with a stirring rod.
4. When the water in the beaker begins to boil, move the ring stand carefully so the flask of amino acids sits in the hot-water bath.

Preparation

Order enough amino acids to supply all lab stations with 1 g each of aspartic acid, glutamic acid, and glycine. Prepare a 1% NaCl solution by placing 5 g of NaCl in a 500-mL graduated cylinder and then dilute with distilled water to 500 mL. Prepare a 1% solution of NaOH by placing 1 g of NaOH in a 100-mL graduated cylinder and dilute with distilled water until the final volume is 100 mL.

Procedural Notes

- This investigation should take one 40-minute period to complete.
- You may want to review with students the structure of amino acids and how they combine to form protein chains.
- **Safety:** Caution students to exercise care when boiling liquids. For burns or scalds, have ice on hand for immediate application if needed.
- **Safety:** Caution students not to ingest the amino acid powders, and have students wash their hands after handling the amino acid powders.
- **Safety:** Caution students to be sure that electrical cords do not dangle from work stations; dangling cords are a tripping and shock hazard.
- **Safety:** Do not allow students to use hot plates with frayed or kinked cords.
- **Safety:** Caution students to be sure that the area under and around the hot plate is dry and to be sure that their hands are dry when using electrical equipment. Electrical cords should not lie in puddles of spilled liquid.
- **Safety:** Caution students to turn off and unplug hot plates before leaving the lab.

Background Answers

1. Microspheres are tiny vessels that are bounded by a double layer of amino acids.

2. Microspheres might have formed from amino acids that resulted from chemical reactions on the early Earth. Some of these microspheres might have incorporated RNA molecules

that could catalyze the production of proteins which could have been incorporated into the membranes of microspheres. Over time, those RNA molecules might have developed a way to replicate, enabling the microspheres to develop the ability to transfer copies of their RNA molecules to offspring and thus replicate themselves.

3. Microspheres are similar to living cells in that their contents differ from their surroundings, they shrink in a hypertonic solution and swell in a hypotonic solution, and they may be gram-negative or gram-positive like bacteria. Microspheres differ from living cells in that they do not have a plasma membrane that consists of phospholipids

5. When the amino acids have heated for 20 minutes, measure 10 mL of 1% NaCl solution in a graduated cylinder and pour the solution into a second Erlenmeyer flask. Place the second flask on the hot plate beside the hot-water bath.

6. When the NaCl solution begins to boil, use tongs to remove the flask containing the NaCl solution from the hot plate. Then, still holding the flask with tongs, slowly add the NaCl solution to the hot amino acids while stirring.

7. Let the NaCl–amino acid solution boil for 30 seconds.

8. Remove the NaCl–amino acid solution from the water bath, and allow it to cool for 10 minutes.

9. Use a dropper to place a drop of the solution on a microscope slide, and cover the drop with a coverslip.

10. Place the slide on the microscope stage. Examine the slide under low power for tiny spherical structures. Then examine the structures under high power. These tiny sphere-shaped objects are microspheres. In the **Records** section of your lab report, draw what you see.

11. Place a drop of a 1% NaOH solution at the edge of the coverslip to raise the pH as you observe the microspheres. What happens?

12. Make a chart similar to the one shown below in the **Records** section of your lab report.

Characteristics of Microspheres

Cell-like characteristics	Non-cell-like characteristics

13. Clean up your materials and wash your hands before leaving the lab.

INQUIRY

1. Why was the NaCl–amino acid solution heated in step 7?

2. Suggest how microspheres were formed.

3. What did you observe when the pH was raised in step 11?

ANALYSIS

1. Compare and contrast microspheres with living cells.

2. What characteristics would microspheres have to exhibit before they could be considered living?

3. How might the conditions you created in the lab be similar to those that are thought to have existed when life first evolved on land?

4. Predict what would happen to microspheres if they were placed in hypotonic or hypertonic solutions.

FURTHER INQUIRY

Write a new question that could be explored as a new investigation. The following are examples:

What do you think would happen if you added too much or too little heat? How can you test for the right amount of heat to use?

Do you think your microsphere experiment would have worked if you had substituted other amino acids? How can you test your hypothesis?

and proteins, and they lack heredity (are not able to replicate their contents).

4. What are some characteristics of microspheres, and how do they compare to living cells?

Inquiry Answers

1. Heating the solution speeds the process of microsphere formation so that microspheres can be produced within a reasonable time.

2. Answers will vary. Encourage speculation. Students might suggest that the amino acids were attracted to each other and formed clumps that later became spheres.

3. The microspheres split in two (underwent fission).

Analysis Answers

1. Like living cells, microspheres have a membrane-like boundary and undergo fission. Unlike living cells, microspheres have no organelles or nucleus.

2. Before they could be considered living, microspheres would have to have the ability to replicate themselves (pass their characteristics to offspring).

3. Answers will vary, but students should mention that the flask contained a mixture of amino acids like the early ocean and the temperatures were extremely hot as they were on the early Earth.

4. Students should predict that microspheres will swell in hypotonic solutions and shrink in hypertonic solutions.

Records

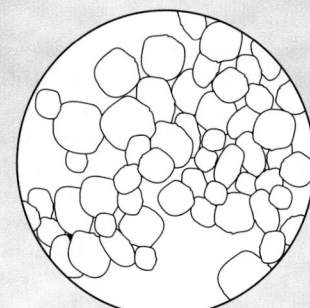

Characteristics of Microspheres

Cell-like characteristics	Non-cell-like characteristics
membranelike envelope	no nucleus or organelles
reproduction by fission	
take in material from exterior	

Block Scheduling Guide

Each block represents about 45–50 minutes of instructional time. The resources cited in a block are coded to a specific program component using the Key on the next page.

Lecture Concepts

Lesson Resources

12-1 The Work of Charles Darwin pp. 244–251

BLOCK 1

a. Lamarck believed that evolution occurs as structures develop through use, or disappear through disuse, and that these acquired characteristics are passed to offspring.

b. Darwin realized that individuals that have characteristics that give them an advantage in their environments are more likely to survive and reproduce. Thus, these characteristics will increase in a population and over time and the nature of the population will change.

c. Darwin's experience in breeding domestic animals, his observations during the voyage of the Beagle and Mathus's ideas on population led Darwin to develop the theory of evolution by natural selection.

d. Darwin did not publish his ideas about natural selection until similar ideas were presented to him by Alfred Russell Wallace. Today, Darwin's theory is widely accepted.

e. Microevolution is change within a species. Macroevolution is change among species.

Lecture Resources
- Opening Questions, page 244
- Demonstration: Organisms that Have Changed
- Overcoming Misconceptions, page 245
- Law of Use and Disuse, page 247
- Visual Strategies: Figures 12-3, 12-4, 12-7
- Overcoming Misconceptions, page 249
- What if an Organism's Environment Were to Change Suddenly?, page 249

- Teaching Transparencies: 52
- Darwin's Theory of Evolution, page 250
- Holt Biology Videodiscs: Lesson 12-1

Classwork Options
- Visual Strategy: Figure 12-5

Assignment Options
- Section Review, page 251
- Chapter Review, questions 1–4, 11, 12, 18, 19, 24, 26

12-2 Evidence of Macroevolution pp. 252–259

BLOCKS 2 and 3

a. Although incomplete, the fossil record contains striking evidence of evolution.

b. A family tree shows how organisms are related through evolution. Each branch point in the tree indicates a common ancestor.

c. DNA and other molecules contain a record of evolution.

d. Homologous structures provide evidence of common ancestry. Vertebrate embryonic development indicates that new genetic instructions have been layered on top of older ones.

e. Gradualism is the model of evolution in which change occurs gradually over time. Punctuated equilibrium is the model of evolution in which change occurs in spurts separated by long periods of equilibrium.

Lecture Resources
- Opening Questions, page 252
- Demonstrations: Examining Fossils, Making Mutations
- Visual Strategies: Figures 12-8, 12-9, 12-10, 12-11, 12-12, 12-14, 12-15
- Teaching Transparencies: 48, 49, 43A, 44A, 50, 51
- Effective Reading, page 253
- Connections, page 255
- Overcoming Misconceptions, page 256
- Gradualism vs. Punctuated Equilibrium, page 258

- Research Update, page 259
- Holt Biology Videodiscs: Lesson 12-2

Classwork Options
- Inquiry Skills Development B8: Fossil Study
- Laboratory Techniques and Experimental Design C19: Analyzing Amino-Acid Sequences to Determine Evolutionary Relationships

Assignment Options
- Section Review, page 259
- Chapter Review, questions 5–7, 13–15, 20, 21, 23, 25, 27, 28

12-3 Evidence of Microevolution pp. 260–266

BLOCKS 4 and 5

a. Experiments show that microevolution has occurred within populations of the European peppered moth.

b. Balancing selection maintains an allele in a population when the environment acts on the allele in opposing ways.

c. Directional selection causes an allele to become either more or less common.

Lecture Resources
- Opening Question, page 260
- Demonstration: Modeling Natural Selection
- Visual Strategies: Figures 12-17, 12-18, 12-19, 12-20
- Overcoming Misconceptions, page 261

- Connections, page 262
- How Natural Selection Affects Humans, page 263
- Teaching Transparencies: 63
- Modeling Microevolution, page 264
- Identifying Isolation, page 266
- Holt Biology Videodiscs: Lesson 12-3

12-3 Evidence of Microevolution (continued) pp.260–266

d. Speciation begins as a population adapts to its specific environment. Reproductive isolation keeps newly forming species from interbreeding. Over time, change within species caused by natural selection leads to the rise of new kinds of organisms.

Classwork Options
- ■ Sequential Diagram of Speciation and Graphic Organizer, page 265
- ■ Laboratory Investigation: Recognizing Patterns of Variation, pages 270–271
- 🧪 Inquiry Skills Development B9: Peppered Moth Survey
- ■ Closure, page 266

Assignment Options
- ■ Concept Mapping, page 267
- ■ Section Review, page 266
- ■ Chapter Review, questions 3, 8–10, 16, 17, 22, 24, 29

Review/Enrichment

BLOCK ⑥

- ■ Study Guide: Concept Review and Pretest

Assignment Options
- ⊙ Teaching Resources CD-ROM: Supplemental Reading Study Guide and Test, *Origin of Species*
- ■ Activities and Projects, questions 30–32
- ■ Readings, questions 33, 34

Assessment Options

BLOCK ⑦

Traditional Assessment
- ■ Chapter 12 Test
- ⊙ Test Generator/Assessment Item Listing: Software item bank for preparing customized chapter tests.

Portfolio Assessment
Select student reports for one or more laboratory experiments from this chapter. *The Direct Observation Checklist* on the *BioSources Teaching Resources CD-ROM* should be completed during a laboratory or other cooperative learning experience.

Answer to Concept Map
The following is one possible answer to the Concept Mapping exercise on page 267.

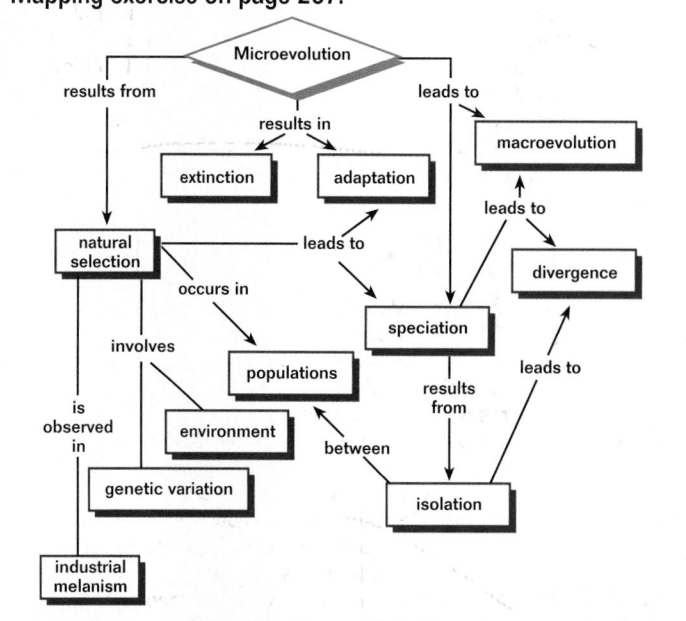

Holt Biology Videodiscs   CONCEPTS OF BIOLOGY

Holt Biology Videodiscs gives you a powerful tool for teaching, review, and assessment. *Concepts of Biology* adds interactive lesson plans and extensive tools for customization using CD-ROM technology.

Use the following topics from *Concepts of Biology* to help you teach this chapter:
- ■ Topic 8: Evolution and Natural Selection
- ■ Topic 9: Evolution of Life on Earth

For further information regarding the media on the videodiscs, see *Holt Biology Videodiscs Teacher's Correlation Guide to Biology: Principles and Explorations.*

Phyllium pulchrifolium
(beautiful moving leaf)

Chapter Themes

Evolution *Evolution, or change over time, not only is one of this program's themes but also has been called the central theme of life science.*

Levels of Organization *Evolution occurs at the population level of life but is linked to events, such as mutation, that occur at the molecular level of life. Briefly review the biological definition of the term* population.

Homeostasis *In this chapter, homeostasis is used in reference to populations rather than individual organisms. A threat to a population's homeostasis may be the primary cause of evolution.*

Structure and Function *This concept is important in understanding morphological adaptations, such as those seen in Figures 12-4, 12-9, 12-10, and 12-12.*

Tapping Prior Knowledge

- Who was Charles Darwin?
- What is meant by the term *natural selection*?
- What is a fossil?

Opening Questions

Have students study the photograph on page 244 of *Phyllium pulchrifolium*, which means "beautiful moving leaf." Ask students how they think this insect got its name. Students should recognize that it resembles a leaf. Ask students to name an advantage of looking like a leaf. Students should predict that insects that look like leaves may avoid being eaten by predators. Ask students to predict what would happen to this insect's chances of survival if it found itself in an environment that had only plants with red leaves. It would be less likely to be able to avoid predators and might be eaten.

THEORY OF EVOLUTION

REVIEW

- structure of proteins (Section 2-3)
- mutations (Section 7-3)
- structure of genes (Section 8-1)
- gene sequencing (Section 10-1)
- radiometric dating (Section 11-2)

Authors' Rationale

Evolution is the central unifying concept of biology—almost everything biological can be explained in terms of evolution. Evolution is introduced in this chapter with a discussion of Darwin's exquisite observations and theories, which have been expanded and improved upon since their release in 1859. Evidence of macroevolution—such as fossils and similarities in amino acid and nucleotide sequences, anatomy, and embryo development—follows. The mechanisms and rate of evolution are also discussed. Finally, the examples of microevolution presented show not only the process of evolution but also the role microevolution itself plays in macroevolution.

12-1 The Work of Charles Darwin

O f all the major themes of biology, evolution is perhaps the best known but least understood by the general public. Scientists have concluded that the great diversity of life on Earth is the result of more than 3.5 billion years of evolution, during which species were replaced over time by other species. The idea that life evolves was expressed by several eighteenth- and nineteenth-century scientists. Lacking a reasonable explanation for evolution, other scientists of the time adamantly opposed the idea. Then, in 1859, the English naturalist Charles Darwin provided convincing evidence that species do evolve and suggested a mechanism (means) by which evolution occurs.

Section Objectives

- Summarize the modern theory of evolution.
- Identify several observations that led Darwin to conclude that species evolve.
- Describe the process of natural selection and its outcome.
- Discuss each of the main points of Darwin's theory of evolution by natural selection as it is stated today.

Overview THEORY OF EVOLUTION

The theory of evolution consists of the following four major points.

1. Variation exists within the genes of every species (the result of random mutation).

2. In a particular environment, some individuals of a species are better suited for survival and so leave more offspring (natural selection).

3. Over time, change within species leads to the replacement of old species by new species as less successful species become extinct.

4. There is clear evidence from fossils and many other sources that the species now on Earth have evolved (descended) from ancestral forms that are extinct (evolution).

Scientific theories are unifying explanations for many related observations. Like all scientific theories, the theory of evolution was developed through decades of scientific observation and experimentation. The modern theory of evolution began to take shape as a result of the work of Charles Darwin, seen in Figure 12-1. Today, virtually all scientists recognize evolution as the basis for both the diversity and relatedness of life on Earth.

Figure 12-1 This portrait of Charles Darwin was painted shortly after he returned from an around-the-world voyage. Observations made during the voyage led Darwin to look for a mechanism by which evolution occurs.

SECTION 12-1

Vocabulary Preview

- **population**
- **natural selection**
- **adaptation**
- **microevolution**
- **isolation**
- **species**
- **extinct**
- **macroevolution**

Opening Questions

Ask students what the word *evolution* means. Students should recall that the word *evolution* means "change over time." Ask students to list examples of types of animals or plants that have changed over time. Domesticated plants and animals are the most likely examples to be listed.

Demonstration

Organisms That Have Changed

Show students pictures of several different varieties (breeds) of a commercially important plant (e.g., cotton, wheat, roses) or animal (e.g., cattle, horses, domesticated dogs and cats). Ask students how these varieties (breeds) originated. (*All originated through selective breeding by humans.*) If possible, show students pictures of the ancestors of these varieties (breeds). For example, quarter horses were developed by breeding American mustangs with English thoroughbreds. Ask: What has happened within this species? (*New varieties have developed over time.*) Tell students that observations of change in domesticated animals and plants helped people to recognize that species can change (and have changed) over time.

Overcoming Misconceptions

Like Gravity, Evolution Is Not "Just a Theory"

The word *theory*, as used here, refers to an accepted explanation for numerous observations that has stood the test of time and experiment. It does not imply that evolution (change over time) itself is hypothetical any more than the atomic theory implies that atoms are hypothetical or the theory of gravity implies that gravity is hypothetical. Evidence that organisms evolve is overwhelming. The mechanisms of change that result in evolution, and that are always being refined and updated, are the subject of modern hypotheses concerning evolution.

Darwin Became a Naturalist

Even though Charles Darwin became one of the greatest scientists of all time, he struggled in school. Darwin's father was a wealthy doctor who wanted Darwin to become either a doctor or a minister. Not inspired by the subjects his father urged him to study, Darwin frequently spent more time outdoors than in class. At the age of 16, Darwin was sent to Edinburgh, Scotland, to study medicine. Repelled by surgery, which at the time was done without anesthetics, Darwin repeatedly skipped lectures to collect biological specimens. In 1827, Darwin's father sent him to Cambridge University to prepare for the ministry. Although he did complete a degree in theology, Darwin again spent much of his time with friends who were interested in natural science.

In 1831, one of his professors at Cambridge recommended Darwin for a post as the unpaid naturalist on a naval voyage of the HMS *Beagle*. The ship and its route can be seen in Figure 12-2. Darwin was offered the position but regretfully declined when his father refused to let him go. Fortunately, Darwin's uncle interceded for him at the last moment. At the age of 22, Darwin was off on a journey that would change his life and forever change how we think of ourselves.

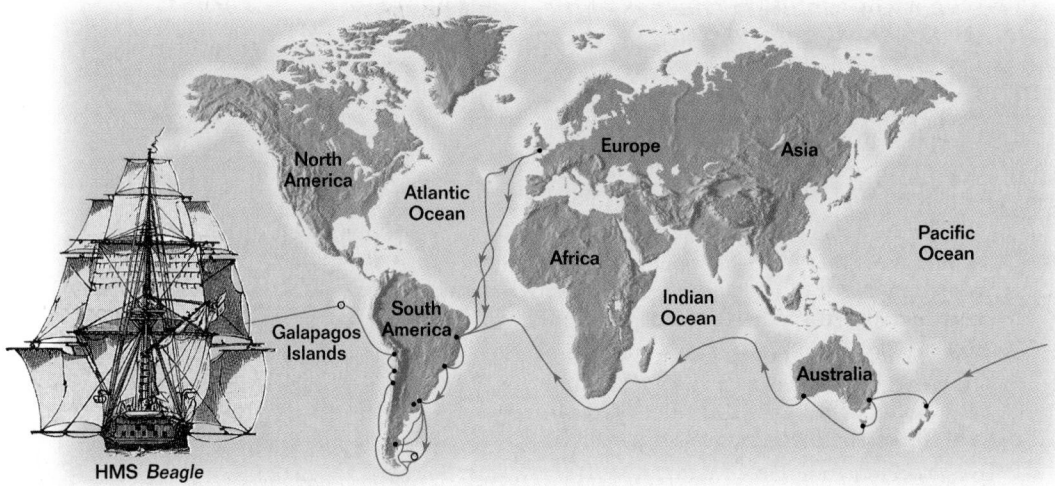

Figure 12-2 The HMS *Beagle* sailed around the world along the route shown on this map. The purpose of the ship's five-year voyage was to survey the coast of South America.

Darwin's Views Changed During the Beagle's Voyage

When the *Beagle* sailed on December 27, 1831, young Darwin accepted the prevailing view that each species was a divine creation, unchanging and existing as it was originally created. This view of divine creation was thought to explain why species are often uniquely adapted to their environments. However, scientists had begun to see that traditional views of divine creation could not explain the kinds and distribution of fossils they had found. Some scientists tried to explain their observations by modifying traditional accounts of creation, while others (including Darwin's own grandfather) proposed that the diversity of life resulted from evolution. In 1809, the French scientist Jean Baptiste Lamarck proposed a mechanism to explain how evolution occurs. Lamarck's hypothesis (now known to be incorrect) is illustrated in Figure 12-3.

While on the *Beagle,* Darwin's belief that species are unchanging began to weaken. During the voyage, Darwin read Charles Lyell's book *Principles of Geology,* which contains a detailed account of Lamarck's theory of evolution. As he visited different places, Darwin also saw things that he thought could only be attributed to a process of gradual change (evolution). For example, in South America, Darwin found fossils of extinct armadillos. These fossilized animals closely resembled, but were not identical to, the armadillos currently living in the area. Darwin wondered why living and fossilized species found in the same place would be similar and yet different. The most probable explanation, he reasoned, was that one species had given rise to the other.

On the Galapagos Islands, located about 1,000 km off the coast of Ecuador, Darwin found his most convincing evidence that species evolve. Darwin was struck by the fact that the plants and animals of the Galapagos Islands resemble those of the nearby coast of South America, as illustrated in Figure 12-4. If each species had been created independently and placed on the Galapagos Islands, why would they resemble the plants and animals of the adjacent South American coast? Why did they not instead resemble the plants and animals of similar islands, such as those near Africa, for instance? Darwin felt that the simplest explanation was that the ancestors of Galapagos species must have migrated there from South America long ago and changed after they arrived. Darwin referred to such change as "descent with modification"—evolution.

Figure 12-3 Lamarck believed that giraffes' necks became longer from stretching to reach higher for food and that the longer neck could be inherited by offspring. According to Lamarck, evolution occurs as structures develop through use, or disappear because of disuse, and as these "acquired characteristics" are passed to offspring.

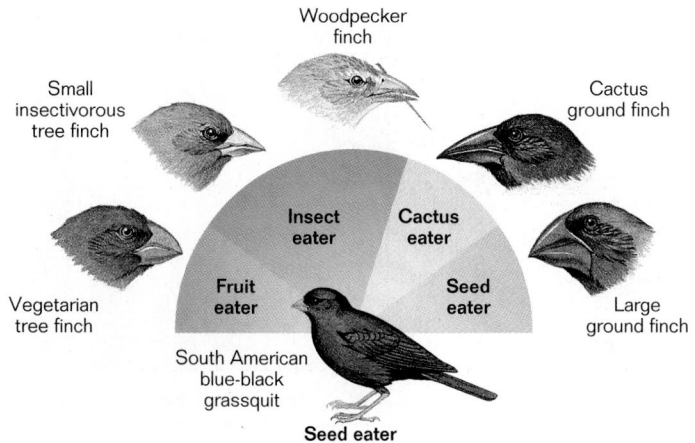

Woodpecker finch

Small insectivorous tree finch

Cactus ground finch

Insect eater

Cactus eater

Fruit eater

Seed eater

Vegetarian tree finch

Large ground finch

South American blue-black grassquit

Seed eater

Figure 12-4 On the Galapagos Islands, Darwin collected more than a dozen species of finches, each with a specialized diet and way of obtaining food. Close examination of these finches showed that all of the species closely resembled the blue-black grassquit, a South American finch species.

SECTION 12-1

Teaching Tip
Law of Use and Disuse
Tell students that Lamarck proposed the law of "use and disuse" to explain the evolution of organisms by the passing of acquired characteristics to offspring. Lamarck recognized that environment plays an important role in evolution. But according to Lamarck, when an organism's environment dictates that it use a certain part of its body, that part becomes more developed as a result of its use, and the improved part can be passed on to offspring. Tell students that if Lamarck were correct, the offspring of bodybuilders would be born with physiques as impressive as those of their parents, and they wouldn't need to work out and develop their muscles. Lamarck's notion that acquired characteristics are passed on was discredited when researchers discovered the role of genes in the inheritance of traits.

◉ VISUAL STRATEGY

Figure 12-3
Ask: How does Lamarck's law of use and disuse apply to the lengthening of the necks of the giraffes? *(Stretching to reach higher leaves caused the giraffes' necks to lengthen, and their offspring had longer necks as a result.)*

◉ VISUAL STRATEGY

Figure 12-4
Point out distinctive features of the finches' bills. Have students describe how each finch's bill is adapted for eating its food. *(For example, the large ground finch's bill is large and heavy for cracking seeds.)*

 Did You Know?

Despite the fact that every biology book cites "Darwin's finches" as a perfect example of species formation, Darwin never used the finches he collected on the Galapagos Islands as an example of evolution by natural selection and probably didn't realize the significance of his observations of these birds. It wasn't until the 1950s that the biologist David Lack did the classic study that revealed the Galapagos finches' unique place in evolutionary history.

Darwin Sought an Explanation for Evolution

When Darwin returned from his voyage at the age of 27, he continued his lifelong study of plants and animals. However, he did not publish his ideas about evolution until many years later. Shortly after his return, Darwin married and started a family. He and his wife, Emma, had 10 children. Darwin formulated his ideas on evolution over several years. During these full and active years, he painstakingly analyzed the specimens he had collected during his voyage and recorded his thoughts in several notebooks. As Darwin studied his data, his conviction that organisms had evolved grew ever stronger. However, he remained deeply puzzled about the most crucial question: *How* does evolution occur?

Malthus's Contribution

The key that unlocked Darwin's thinking about how evolution occurs was an essay written by the English economist Thomas Malthus. In his "Essay on the Principle of Population" (1798), Malthus stated that human populations have the potential to increase faster than the available food supply. As Malthus pointed out, a population grows in size by a geometric progression, such as the one illustrated in Figure 12-5. Food supply, however, increases at best by an arithmetic progression, which is also illustrated in Figure 12-5. According to Malthus, the human population would cover Earth's entire surface within a very short period of time if it could reproduce unchecked. This does not occur, Malthus pointed out, because death caused by disease, war, and famine intervenes.

The term *population*, as it is used today, does not refer only to the number of people living in a particular area. In biology, a **population** is a group of individuals that belong to the same species, live in a defined area, and breed with others in the group. For example, the alligators in the Everglades are a population because they live in a particular area and breed primarily with each other.

Figure 12-5 The "Food Supply Increase" graph, *left,* shows an arithmetic progression, in which the numbers increase by an added constant. The "Population Growth" graph, *right,* shows a geometric progression, in which the numbers increase by a multiplied constant.

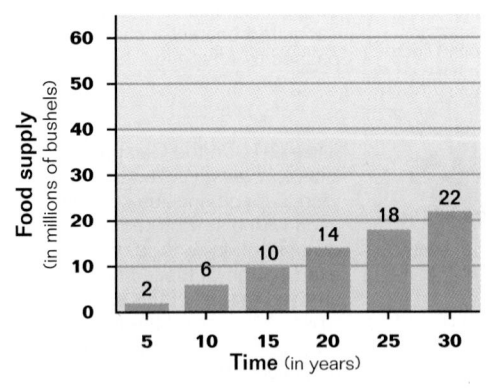

Food Supply Increase

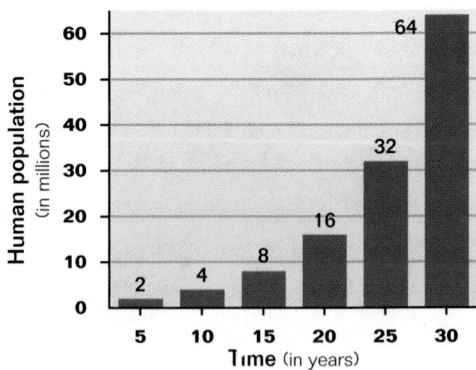

Population Growth

Natural Selection

Darwin realized that Malthus's principle of population applies to all species. Every organism has the potential to produce many offspring during its lifetime. In most cases, however, only a limited number actually survive to reproduce. Combining this observation with what he had seen on his voyage and with his own experiences in breeding domestic animals, Darwin made a key association: *Individuals that possess superior physical or behavioral attributes are more likely to survive than those that are not so well endowed.* Darwin saw that by surviving, individuals have the opportunity to reproduce and pass on their favorable characteristics to offspring. Thus, these characteristics will increase in a population, and the nature of the population will gradually change. Darwin called the process by which populations change in response to their environment **natural selection.**

Evolution by natural selection is a simple and logical explanation for the biological diversity and similarity Darwin saw on his voyage. As he saw it, organisms differed from place to place because their habitats presented different challenges to, and opportunities for, survival. Each species had evolved in response to its environment. The changing of a species that results in its being better suited to its environment is called **adaptation** (ad uhp TAY shun). Darwin also noticed that organisms more closely resemble those living in nearby geographic locations than those living in similar yet widely separated parts of the world. The idea that species were created individually does not explain this pattern of distribution. Thus, Darwin concluded that the species of a particular place evolved from species that previously lived in the same area or that migrated from areas nearby. ◻

◻ CAPSULE SUMMARY

Darwin developed the theory of evolution by natural selection by applying Malthus's ideas on population to the observations he had made during the voyage of the Beagle.

Darwin Published His Ideas in 1859

In 1844, Darwin finished writing down his ideas about evolution and natural selection in a preliminary manuscript that he showed only to a few scientists he knew and trusted. In that year, however, a burst of public criticism was directed against a book called *Vestiges of the Natural History of Creation.* The book claimed that evolution had occurred in Earth's past. Although it was a bestseller at the time, the book was criticized vehemently by both the government and the church. Lamarck's theory of evolution by the inheritance of acquired characteristics had also received severe criticism. Shrinking from such controversy, Darwin put aside his manuscript. For the next 14 years, he went about his life, enlarging and refining his notes on evolution but saying nothing about his ideas in public. While he published several books on other scientific subjects, his manuscript on evolution remained in a drawer in his study.

SECTION 12-1

Teaching Tip

⌁ **What If an organism's environment were to change suddenly?** When sudden changes occur in the environment, organisms that do not possess variations that enable them to survive in the new conditions often die. As a result, the characteristics of the individuals that die may be lost, because fewer offspring with those characteristics are produced. Students should realize that variations which enable organisms to survive environmental changes must already exist in the populations of each species, otherwise there would not be enough time to "develop" structures that enable survival, as Lamarck suggested. Emphasize that in natural selection, nature "selects" already existing variations for survival. Ask: How does nature "select" certain variations? *(by enabling better-adapted individuals to survive longer and produce more offspring)* How does this lead to change over time? *(In each generation, there will be more individuals like those that are best adapted to an environment.)*

Overcoming Misconceptions

Darwinism vs. Lamarckism

Both Darwinism and Lamarckism seem to dictate that, for adaptations, the rule is "use it or lose it." This can be very confusing to students. Tell them that a structure may be lost through natural selection because it provides no selective advantage. For this to occur, however, variations of the structure, from small and nonfunctional to large and functional, must exist in the genes of the species. According to Lamarck, parts that are not useful to individuals are small and nonfunctional because they are not used and developed.

Figure 12-6 This cartoon (from 1874) of Darwin with a monkey-like "ancestor" shows how some people ridiculed Darwin because of his work.

❑ CAPSULE SUMMARY

Darwin did not publish his ideas about natural selection until similar ideas were presented to him by Alfred Russel Wallace. Today, Darwin's theory is widely accepted.

The stimulus that finally brought Darwin's work into print was a letter he received in June of 1858. The letter and a short essay were sent from Malaysia by the young English naturalist Alfred Russel Wallace. Wallace's essay concisely described the idea of evolution by natural selection. In his letter, he asked if Darwin would help get the essay published. Darwin's scientific friends urged him to get his own work ready. They arranged for an abstract of Darwin's manuscript to be presented with Wallace's paper at a public scientific meeting. Neither Darwin nor Wallace actually attended the meeting on July 1, 1858. Wallace was still in Malaysia, and one of Darwin's children had died of scarlet fever two days prior to the meeting. At the time, little notice was taken of the two papers, but Darwin had finally begun to write what he considered to be a "short abstract of his work."

When Darwin's book *On the Origin of Species by Means of Natural Selection* appeared in November of 1859, it caused an immediate sensation. Many people were deeply disturbed by certain aspects of Darwin's theory, such as the implication that humans are related to apes. While recognizing that humans closely resemble apes, they found the possibility of a direct evolutionary link unacceptable, as Figure 12-6 suggests. Darwin's arguments and evidence were so compelling, however, that his views were soon widely accepted by biologists around the world. Today, after more than a century of scientific progress, the evidence supporting evolution and natural selection is even more convincing. Although some people still find Darwin's ideas unacceptable, virtually all scientists agree that they are correct. *On the Origin of Species* has had a strong and lasting impact for two reasons: it presented a vast body of evidence that evolution has occurred, and, more important, it presented a reasonable hypothesis explaining how evolution takes place. ◼

Darwin's Ideas Have Been Updated

Since Darwin's work was published, his hypothesis—that natural selection is the mechanism by which evolution occurs—has been carefully examined by biologists. New discoveries, particularly in the area of genetics, have given scientists new insight into how natural selection brings about the evolution of species. Darwin's ideas, restated in modern terms, are summarized below.

Natural Selection Causes Change *Within* Populations

Darwin's key inference was that within any population, those individuals best suited to survive and prosper in their environment will leave the most offspring. Thus, the traits of those individuals will become more common in successive generations. As a result, a population will adapt to its environment.

 Did You Know?

The most famous statement describing Darwin's theory, "survival of the fittest," did not appear in Darwin's original work. It was coined by another biologist, Herbert Spencer, upon learning about Darwin's theory. Darwin liked the phrase and used it to summarize what natural selection was all about. Unfortunately, the phrase came to be misunderstood. Soon the meaning of "fittest" was distorted to mean most powerful. Political leaders and industrialists used it to justify conquest, colonialism, and oppression under the guise of "natural law."

Scientists now know that genes are responsible for inherited traits. Therefore, certain traits become more common in a population because more individuals in the population carry the genes for those traits. In other words, natural selection causes the *frequency* of certain genes in a population to vary over time. Mutations and the recombination of genes that occurs during sexual reproduction are constant sources of new variations for natural selection to act upon. Today, biologists use the term **microevolution** to refer to change that occurs within a species over time. Figure 12-7 illustrates the extent to which microevolution can change a species.

Isolation Leads to Species Formation

The natural environment is not uniform; it differs from place to place. Therefore, populations of the same species living in different locations tend to evolve in different directions. The condition in which two populations of the same species are separated from one another is called **isolation**. As two isolated populations of the same species become increasingly different over time, the populations may no longer be able to interbreed (breed with one another). When the individuals of two populations can no longer interbreed, the two populations are considered to be different **species**.

Extinction Leads to Species Replacement

Over long periods of time, events such as climatic changes and natural disasters result in some species becoming **extinct**, which means that they disappear permanently. Species that are better suited to the new conditions may replace those that become extinct. Therefore, the organisms alive today are but the latest members of a long parade of life. Today, biologists use the term **macroevolution** to refer to change among species over time. The replacement of the dinosaurs by mammals is an example of macroevolution. ❑

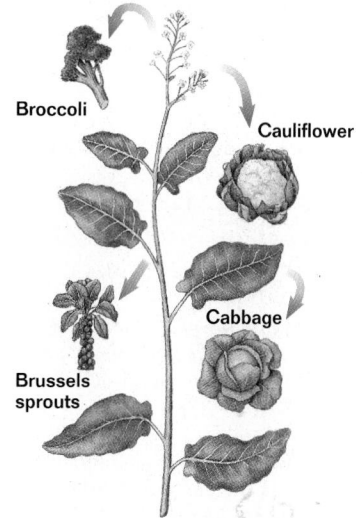

Figure 12-7 These familiar vegetables all belong to the same species, *Brassica oleracea.* Each was developed through selective breeding. Although the mechanism is *artificial* selection instead of *natural* selection, the breeding of crops is an example of microevolution.

❑ **CAPSULE SUMMARY**

Microevolution is change within a species. Macroevolution is change that leads to replacement of species.

👁 **VISUAL STRATEGY**

Figure 12-7
Point out the many different varieties of vegetables that have resulted from the same species, *Brassica oleracea.* Ask: How could these varieties of the same species look so different? *(There is much variation in the genes of this species.)* Is it possible for these varieties to become separate species? *(yes)* How? *(Through isolation, their genetic makeups may become so different that they can no longer interbreed.)*

Section Review

1. *List three observations made by Charles Darwin during his five-year voyage that led him to conclude that living species have evolved from species that are now extinct.*

2. *Briefly describe how natural selection occurs.*

3. *Using the terms* adaptation, microevolution, *and* macroevolution, *summarize the modern theory of evolution in as few words as you can.*

Critical Thinking

4. *Why are scientists' ideas about evolution stated as a theory? How is a scientific theory different from a scientific hypothesis?*

Answers to Section Review

1. Answers may vary but could mention that Galapagos species resemble South American species, living species resemble fossilized species, and that closely related species differ in appearance and diet.

2. Natural selection occurs as certain individuals with variations that make them more successful (more likely to survive and produce offspring) in their environments leave more offspring than less successful individuals, thus changing the genetic makeup of a population.

3. Answers will vary but should convey that through microevolution, adaptation occurs within populations, which over time leads to macroevolution, the replacement of species by other species.

4. Answers will vary but should indicate that evolution is a process that explains the patterns of variation seen in both living and extinct life-forms and has been observed repeatedly. Scientific hypotheses are ideas about how events occur that can be tested through observation and experimentation.

Opening Questions

Ask students if they think that "cave men" lived with dinosaurs. Many will think that they did. Ask students how we know that dinosaurs existed. Students should recognize that dinosaurs are known through fossils.

Demonstration
Examining Fossils

Have students study samples of fossils, or project pictures of several types of fossils. Ask: How are these organisms similar to modern organisms? (*Many of the same types of organisms living today are represented in the fossils. Some are more similar to living species than others, and some have no living relatives.*) Diatomaceous earth used in aquarium filters provides excellent examples of microfossils for examination.

👁 VISUAL STRATEGY

Figure 12-8

Have volunteers describe each of the organisms in these photos. Ask: How does the complexity of the fossil life-forms seen in these photos change as the rocks get younger? (*It increases.*)

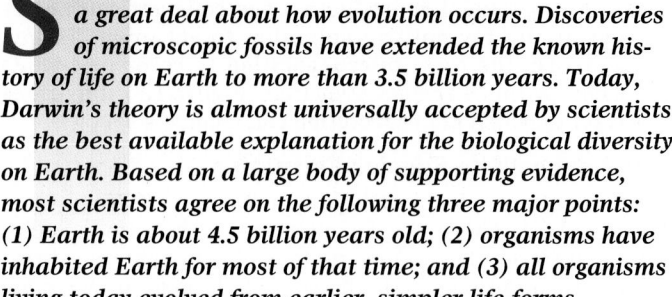

12-2 Evidence of Macroevolution

S *ince Darwin's death in 1882, scientists have learned a great deal about how evolution occurs. Discoveries of microscopic fossils have extended the known history of life on Earth to more than 3.5 billion years. Today, Darwin's theory is almost universally accepted by scientists as the best available explanation for the biological diversity on Earth. Based on a large body of supporting evidence, most scientists agree on the following three major points: (1) Earth is about 4.5 billion years old; (2) organisms have inhabited Earth for most of that time; and (3) all organisms living today evolved from earlier, simpler life-forms.*

Section Objectives

■ Describe how the fossil record supports evolution.

■ Explain how biological molecules such as proteins and DNA show evidence of macroevolution.

■ Explain how comparing the anatomy and development of living species provides evidence of macroevolution.

■ Differentiate between the gradualism and punctuated equilibrium models of evolution.

Figure 12-8 Fossils of simple unicellular life-forms, such as the coccoid cyanobacteria, *left*, characterize 700 million-year-old rocks found in western Canada. Fossils of more complex, multicellular life-forms, such as the crinoid, *center*, occur in 300 million-year-old rocks found in Indiana. Fossils of highly complex life-forms, such as the pterodactyl, *right*, occur in rocks formed during the Jurassic period 210 million to 140 million years ago.

Fossils Provide a Record of Macroevolution

The most direct evidence that macroevolution has occurred comes from fossils. A **fossil** is the preserved or mineralized remains (bone, tooth, or shell) or traces (footprint, burrow, or imprint) of an organism that lived long ago. Fossils, therefore, provide an actual record of Earth's past life-forms. Change over time (evolution) can be observed in this fossil record. For instance, fossilized species found in older rocks are different from those found in newer rocks, as you can see in Figure 12-8. After observing such differences, Darwin predicted that "missing links" (intermediate forms) between the great groups of organisms would eventually be found. Many of these links have been found. For example, fossil links have been found between fishes and amphibians, between reptiles and birds, and between reptiles

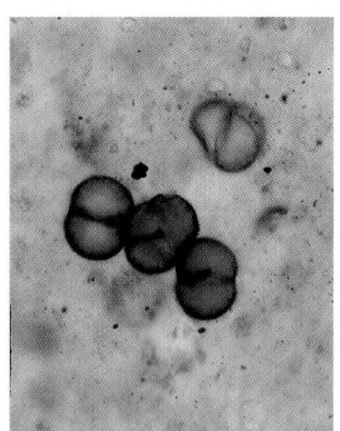

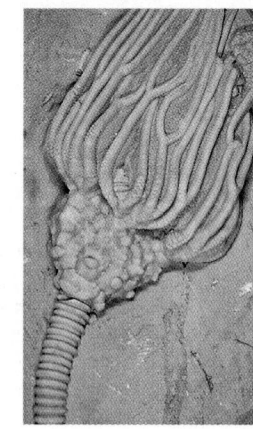

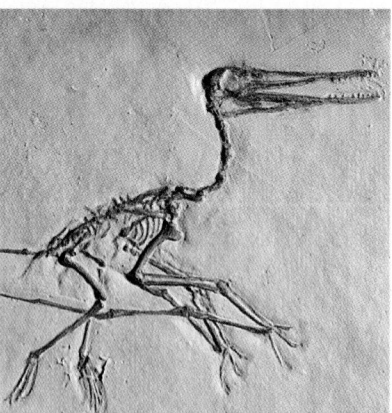

❓ Did You Know?

Geologists and paleontologists first speculated about, and provided convincing evidence of, evolution. The science of paleontology was in its infancy in Darwin's time. During the mid-1800s, most fossil hunting was done by a few amateurs.

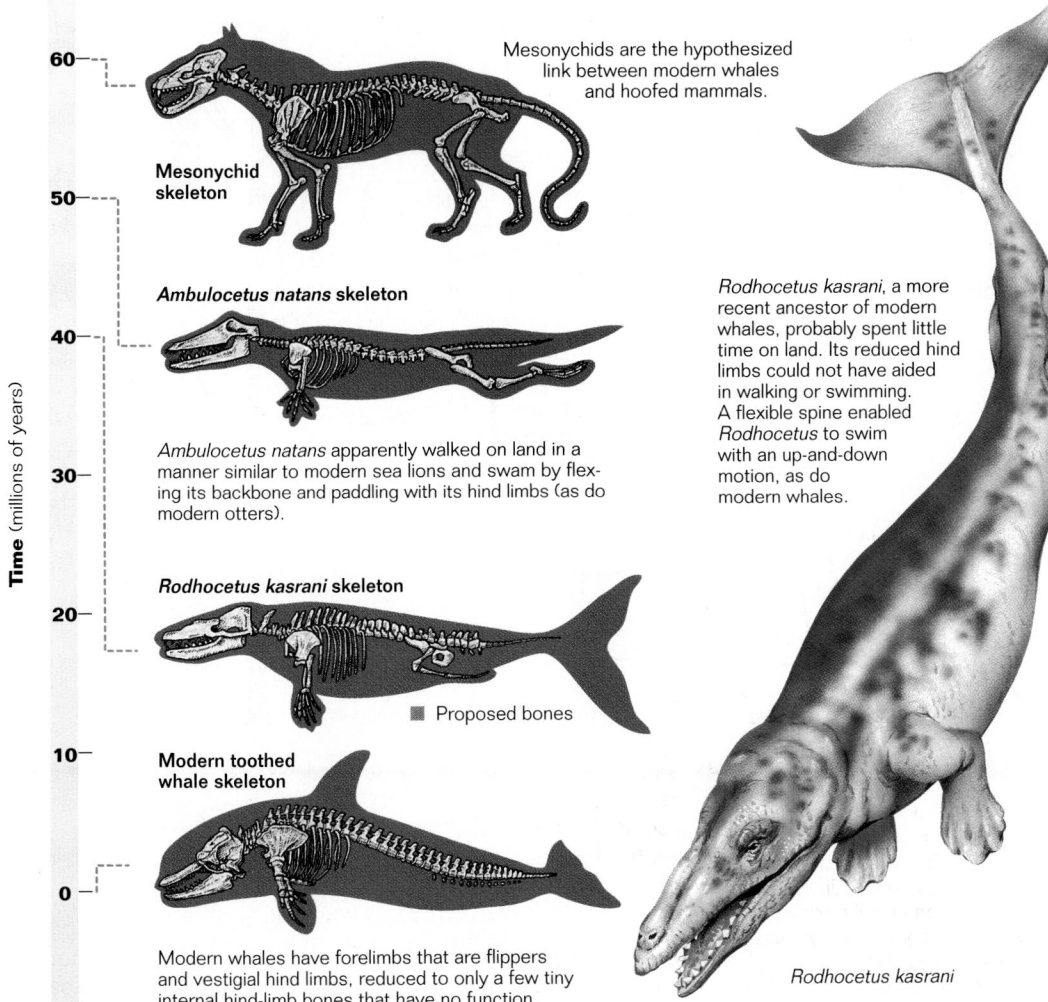

Mesonychids are the hypothesized link between modern whales and hoofed mammals.

Mesonychid skeleton

Ambulocetus natans skeleton

Ambulocetus natans apparently walked on land in a manner similar to modern sea lions and swam by flexing its backbone and paddling with its hind limbs (as do modern otters).

Rodhocetus kasrani skeleton

■ Proposed bones

Modern toothed whale skeleton

Modern whales have forelimbs that are flippers and vestigial hind limbs, reduced to only a few tiny internal hind-limb bones that have no function.

Rodhocetus kasrani, a more recent ancestor of modern whales, probably spent little time on land. Its reduced hind limbs could not have aided in walking or swimming. A flexible spine enabled *Rodhocetus* to swim with an up-and-down motion, as do modern whales.

Rodhocetus kasrani

Time (millions of years)

60 — 50 — 40 — 30 — 20 — 10 — 0 —

👁 VISUAL STRATEGY

Figure 12-9
Tell students that the top three skeletons are known only from their fossil remains. The bottom skeleton is that of a currently living whale species. Note that the backbone of the *Rodhocetus kasrani* skeleton is not complete, as the tail bones shown in gray indicate. Tell students that it is rare to find a complete skeleton in any one fossil. As more skeletons of this species are found, all the species' bones may eventually be found. Using their knowledge of anatomy, paleontologists project what these bones will look like. Have volunteers describe the similarities and differences in the bones of these animals. Ask: How did the backbone change in relation to the time these animals spent in water? *(The backbone became heavier.)* What is the advantage of this change? *(Whales use up-and-down motions of their bodies to swim, therefore a heavier backbone is needed to support the muscles used for this motion.)*

and mammals, making the fossil history of the vertebrates remarkably complete. In the early 1990s, the discoveries of the fossilized remains of two whale ancestors, illustrated in Figure 12-9, provided new links in the evolution of whales from four-legged land mammals.

However, the fossil record, and thus the record of the evolution of life on Earth, is not complete. Many species live in environments where fossils do not form. Most fossils form when organisms and traces of organisms are rapidly buried in fine sediments deposited by water, wind, or volcanic eruptions. Thus, the environments that are most conducive to fossil formation are wet lowlands, slow-moving streams, lakes, shallow seas, and areas near volcanoes that spew out volcanic ash. What are the chances that organisms living in upland forests, mountains, grasslands, or deserts

Figure 12-9 Whales are thought to have evolved from four-legged land mammals that are also the ancestors of modern hoofed mammals. *Ambulocetus natans* and *Rodhocetus kasrani* are recently discovered transitional (intermediate) forms, or "missing links," in whale evolution.

Effective Reading

Taking Notes as You Read
Encourage students to write summaries in their own words of each paragraph they read. For content-rich sections such as Section 12-2, taking notes as they read forces students to think about what they have just

read and will help them remember the material. The notes can be used later as a study guide. You might also encourage students to go over their individual notes with two or three other students and have them identify any differences that occur in

their understanding. Encourage each group to refer back to the text to verify information that has not been understood in the same way by all members of the group.

VISUAL STRATEGY

Figure 12-10

Have students make lists of the similarities and differences among the horses shown in the figure. Ask: What are the most noticeable changes that have occurred? *(Horses have become larger, their legs have lengthened in relation to their body size, and the number of toes on their feet has been reduced from four to one.)* How might these changes have been advantageous? *(Larger size makes the animal stronger and better able to survive attacks by predators. Longer legs in relation to body size enable the animal to run faster and escape predators. The one-toed hooves of the modern horse enable it to run over hard ground with less risk of injury.)* What do the side branches on the family tree represent? *(They represent other types of horses that have become extinct.)*

Teaching Tip

Where Might Fossils Form Today?

Have students identify specific places nearby where fossils might be likely to form. Then conduct a class discussion in which student volunteers can offer examples of places where fossils might be likely to form. Tell students they must justify their suggestions. Encourage other students to critique the suggestions and offer reasons why fossils might form, or why they might not form, in the places suggested. Develop a list of species that live in, or might visit, the places where the class decides fossils might be likely to form. Identify which of these species might form good fossils. Be sure students consider the likelihood that parts of an organism will remain intact long enough to be buried in sediment. Also develop a list of species that do not live in places where fossils might form. Ask: How many of this area's species will probably be found in the fossil record in the future? *(very few)*

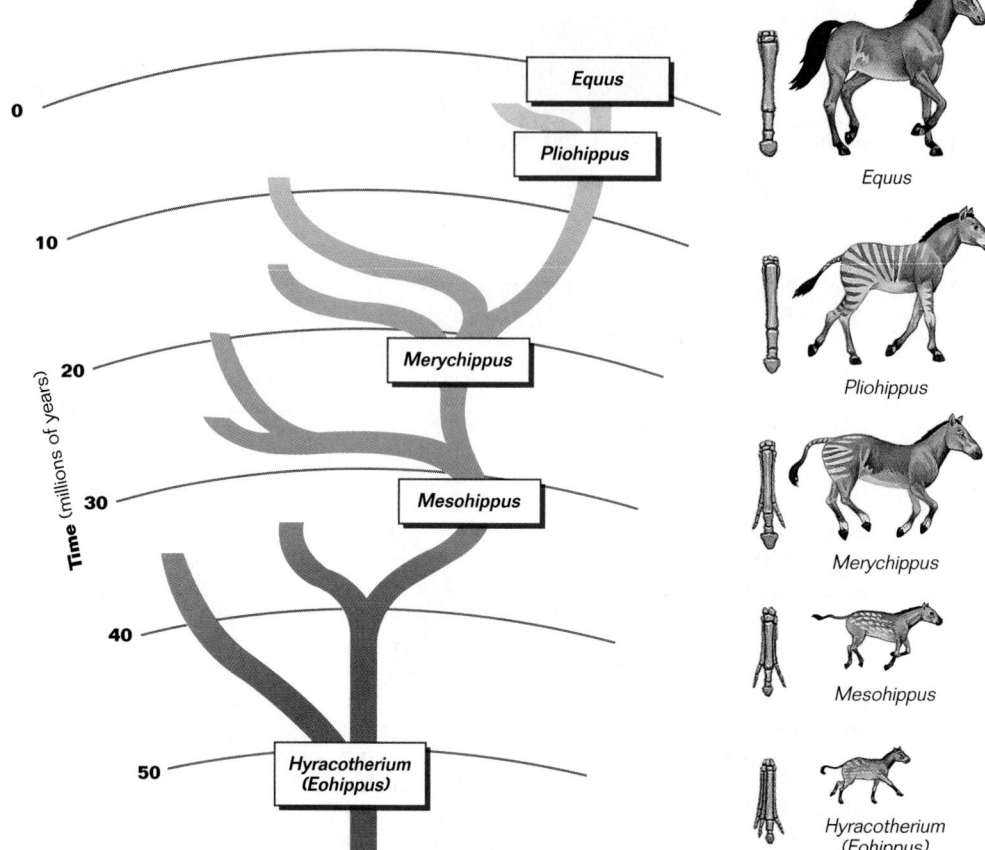

Figure 12-10 *Equus,* the modern horse, evolved from the dog-sized *Hyracotherium.* Notice the transition, *right,* from *Hyracotherium's* four-toed front foot to the one-toed front foot of the modern horse. By including more of the modern horse's ancestors and their descendants, a branching diagram called a family tree, *left,* can be constructed.

will die in just the right place to be buried in sediments? Even if an organism lives in an environment where fossils can form, the chances are slim that its dead body will be buried in sediment before it decays or is eaten and scattered by scavengers.

Although the fossil record will never be complete, it contains striking evidence that macroevolution has occurred. **Paleontologists,** scientists who study fossils, can determine the age of fossils fairly accurately by using radiometric dating, as you learned in Chapter 11. Radiometric dating enables paleontologists to arrange fossils in sequence from oldest to youngest. When this is done, patterns of successive change over time become evident, as you can see from the series of horses illustrated in Figure 12-10. The diagram on the left in Figure 12-10 is a **family tree,** which shows how organisms are related through evolution. Each branch point in a family tree indicates a **common ancestor,** which is a species from which *two or more* species diverged (separated).

 Did You Know?

Scientists estimate that 99 percent of all animal and plant species that ever existed are now extinct. Most people do not realize how many species have been found by paleontologists. In the last few years, many new species have been discovered. For example, in one small area in Wyoming, early Eocene rocks have yielded fossils of more than 50 species of animals, only a small portion of the animals that lived in the area at that time.

Molecules Contain a Record of Macroevolution

The picture of successive change seen in the fossil record enables scientists to make a prediction that can be tested. If species have changed over time, then the genes that determine their characteristics also should have changed. As species evolved, one change after another should have become part of their genetic instructions through mutation. Therefore, more and more changes in a gene's nucleotide sequence should accumulate over time.

This prediction was first tested by analyzing the amino acid sequences of proteins found in several species. Recall that the nucleotide sequence of the gene coding for a protein determines the amino acid sequence of that protein. If evolution has occurred, species that diverged from a common ancestor in the distant past should have more amino acid sequence differences between their proteins than do species that recently diverged. Comparing the beta chain of human hemoglobin (146 amino acids) with that of several species reveals these differences. The results of such a comparison are given in Table 12-1. Notice that species that recently shared a common ancestor (humans and gorillas) have few amino acid sequence differences, while species that shared a common ancestor in the distant past (humans and frogs) have many amino acid sequence differences.

Nucleotide substitutions (mutations) are responsible for changes in the amino acid sequence of a protein. Scientists estimate the number of nucleotide substitutions that have occurred in a gene since two species diverged by counting the resulting amino acid sequence differences. Today, gene sequencing also enables scientists to determine the exact nucleotide sequence of a gene. Using the data obtained from the study of this "molecular record," scientists produce family trees like the one in Figure 12-11. These family trees provide very strong evidence supporting evolution because

Table 12-1 Hemoglobin Comparison

Species	Amino Acid Differences
	Compared with human hemoglobin
Gorilla	1
Rhesus monkey	8
Mouse	27
Chicken	45
Frog	67
Lamprey	125

Hemoglobin Family Tree

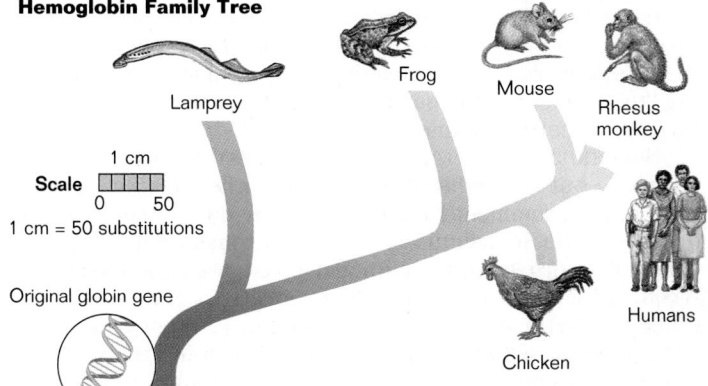

Lamprey

Frog

Mouse

Rhesus monkey

Scale
1 cm
0   50
1 cm = 50 substitutions

Original globin gene

Humans

Chicken

Figure 12-11 The length of the branches on this molecular family tree for the hemoglobin gene indicates the number of nucleotide substitutions that have occurred in the original "globin" gene's nucleotide sequence. The relationships indicated by this family tree are not a guess, but a direct observation of evolution, recorded in the DNA of many species.

UNIT 3
CHAPTER 12

 VISUAL STRATEGY

Figure 12-12
Have students describe the similarities in the structures of the forelimbs of these vertebrates. Be sure that they notice that each vertebrate has the same complement of bones. Ask: Why do the individual bones differ among these vertebrates? *(Each has been modified to perform a different function.)* Is the bone structure of a penguin's wing suited to flying? *(no)* Is the bone structure of a bat's wing suited to swimming? *(no)* Are the wings of birds and insects homologous structures? *(no)* Why? *(They have the same function but do not have the same structure.)*

Teaching Tip
Analyzing the Selective Advantage
Many cave-dwelling animals have vestigial eyes, as do worker ants. Ask: Is it a selective advantage for a cave dweller to invest energy in the development of structures such as eyes that have no use in its environment? *(no)* Might it be a selective advantage for the cave dweller to not need to invest energy in the development of structures that have no use? *(yes, because investing the energy elsewhere might extend the life of an organism and enable it to produce more offspring)* Which individuals, ones with large eyes or ones with vestigial eyes, will need to invest more energy in eye development as they grow? *(ones with large eyes)* In a cave, which individuals are likely to leave more offspring? *(ones with vestigial eyes)* How will a population of organisms change over time if it moves into a cave? *(There will be more and more individuals in the population with vestigial eyes.)* Tell students that if a characteristic does not contribute to the survival of the individuals with that characteristic, it may slowly be lost from a population over time.

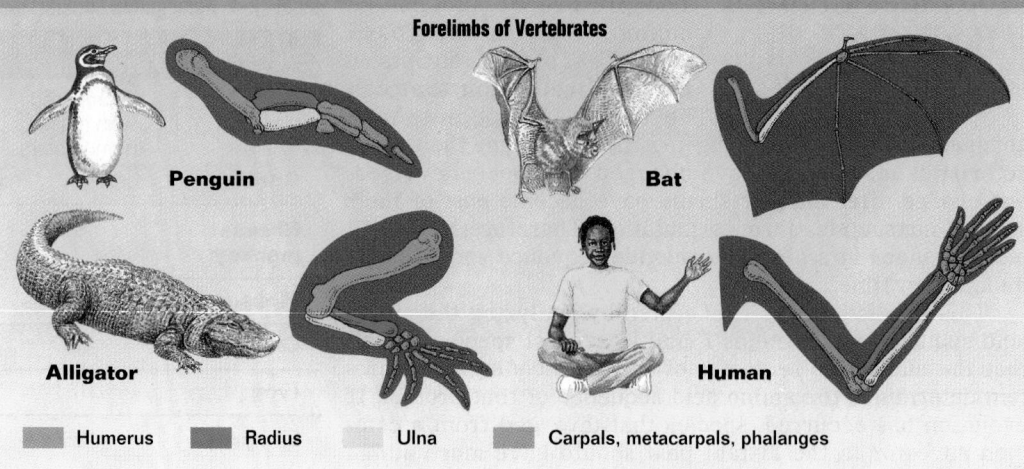

Forelimbs of Vertebrates

Penguin   Bat

Alligator   Human

Humerus   Radius   Ulna   Carpals, metacarpals, phalanges

Figure 12-12 Though they are modified for different functions, the forelimbs of vertebrates contain the same kinds of bones, which form in the same way during embryological development. Such structures are called homologous structures.

they show the same relationships indicated by the fossil record. Time after time, comparisons of amino acid and nucleotide sequences among organisms have provided a wealth of direct evidence of successive change over time.

Anatomy and Development Reflect Macroevolution

Comparisons of the anatomies (structures) of different types of organisms often reveal basic similarities in body structures, even though their functions may be very different. As vertebrates evolved, for example, particular sets of bones were sometimes put to different uses. And yet, the similarity in the structure of these bones can still be seen, suggesting that all vertebrates share a common ancestor. As you can see in Figure 12-12, the forelimbs of all vertebrates are constructed from the same basic array of bones. Such structures are said to be homologous *(hoh MAHL uh guhs)*, from the Greek words *homo*, meaning "same," and *legein*, meaning "to speak." **Homologous structures,** therefore, are structures that share a common ancestry.

Sometimes, bones (or other structures) are present in an organism but are reduced in size and either have no use or have a less prominent function than they do in other, related organisms. Such structures, which are considered to be evidence of an organism's evolutionary past, are called **vestigial** *(vehs TIJ ee uhl)* **structures.** The word *vestigial* is derived from the Latin word *vestigium*, meaning "footprint." An example of a vestigial structure can be seen in Figure 12-13. The human coccyx *(KAHK sihks)*, or tailbone, is another example of a vestigial structure. The fused vertebrae that make up the coccyx are homologous to those of other

Figure 12-13 The vestigial wings of this flightless cormorant, which lives in the Galapagos Islands, are too small to enable it to fly.

Overcoming Misconceptions

Vestigial Is Not Necessarily Useless
It is commonly thought that structures are labeled as vestigial because they are useless. Many such structures do, in fact, have functions. Structures are labeled vestigial if they are reduced in size and do not perform the same function, or do not perform it to the same degree, as homologous structures did in ancestral life-forms.

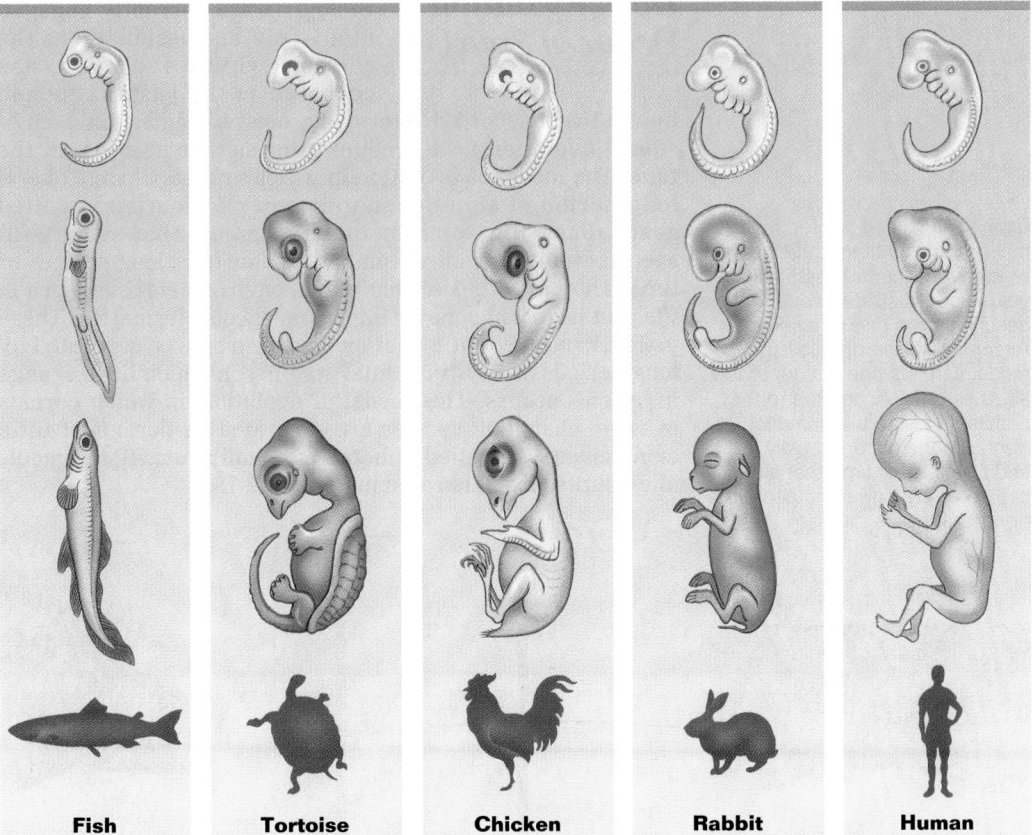

| Fish | Tortoise | Chicken | Rabbit | Human |

Figure 12-14

Have students locate the bony tail and pharyngeal pouches on each of the embryos. Ask: In which organism do the pharyngeal pouches persist the longest? *(fish)* In which organisms does the tail persist the longest? *(The tail persists into the adult stages of fish and tortoises. It also persists well into the embryological development of the chicken and the rabbit but is not obvious in the adult stage of either type of animal.)* Tell students that different sets of genes are activated as an animal's embryological development proceeds and that the most basic characteristics of an organism appear the earliest in its development. Ask: What is the most basic characteristic of all vertebrates? *(a backbone)* How might the differences among the vertebrates have been acquired? *(They might have been acquired through mutations that produced or activated new sets of genes.)*

vertebrate tails. Although the coccyx serves as a point of attachment for certain muscles, this small internal "tail" does not resemble the long external tail of most other vertebrates. Furthermore, the human coccyx has no function in locomotion, balance, or behavioral displays as the tails of other vertebrates do. Thus, the coccyx is a vestigial vertebrate tail.

The evolutionary history of organisms is also evident in their embryonic development. Compare the development of a human embryo with that of the other vertebrate embryos shown in Figure 12-14. Notice that in addition to their early similarity, each embryo develops a tail, buds that become limbs, and pharyngeal pouches (which house the gills of fish and amphibians). The tail remains in most adult vertebrates. Although vertebrate bones are homologous, limbs develop somewhat differently in each group of vertebrates. Only adult fish and immature amphibians retain pharyngeal pouches. In humans, the tail and pharyngeal pouches are relics (forms from the past) that disappear by the time of birth. These relics and the similarities among the embryos strongly suggest that the development of all vertebrates evolved as new genetic instructions were layered on top of older ones. ◻

Figure 12-14 Early in development, all vertebrate embryos are remarkably similar. As development continues, various structures are modified until they take on their characteristic adult forms.

◻ CAPSULE SUMMARY

Fossils and DNA molecules contain a record of evolution. Homologous structures and vestigial structures provide evidence of common ancestry. Vertebrate embryonic development indicates that new genetic instructions have been layered on top of older ones.

❓ Did You Know?

Although penguins are not able to fly in the air, the movements of their wings under water resemble the motion of the wings of birds that do fly. In other words, penguins literally fly underwater. The stouter bone structure of a penguin's wings is an advantage for moving through water, which is far denser than air. Also, heavier bones are not a disadvantage because of the buoyancy of water. The bones of birds and bats that fly in the air must be delicate to cut down on weight.

Teaching Tip

Gradualism vs. Punctuated Equilibrium

Encourage students to think of examples of environments where gradualism might occur *(stable environments that are not likely to experience catastrophic events)* and examples of environments where punctuated equilibrium might occur *(environments near active volcanoes, near an ocean shore line, in high altitudes or latitudes where glaciers might form, etc.)*. Ask: Can a catastrophic event occur in a stable environment? *(yes)* What is an example of such an event? *(meteorite strike, large volcanic eruption)*

 VISUAL STRATEGY

Figure 12-15

Walk students through this diagram. Ask: In which model do ancestral forms exist at the same time as new forms? *(punctuated equilibrium)* Have students write a paragraph describing in their own words how each model suggests that the evolution of titanotheres occurred.

Does Evolution Occur in Spurts?

Biologists are currently engaged in a lively discussion about the rate at which evolution has occurred in the past, as judged by the fossil record. Historically, most biologists have envisioned evolution as a gradual process that goes on all the time. The model of evolution in which gradual change over a long period of time leads to species formation is called **gradualism.** But some biologists suggest that successful species would remain virtually the same over long periods of time. They hypothesize that major environmental changes in the past have had a major impact on species formation. These biologists argue that evolution occurs in spurts, separated by long periods of environmental stability in which little change in species occurs. This model of evolution, in which periods of rapid change in species are separated by periods of little or no change, is called **punctuated equilibrium.** Both models of evolution are illustrated in Figure 12-15.

Figure 12-15 These hoofed mammals known as titanotheres, which lived from 50 to 35 million years ago, illustrate two models describing the rate of evolution. According to the gradualism model, *left,* change occurs gradually over time. According to the punctuated equilibrium model, *right,* change occurs rapidly in short periods of time that are separated by long periods of little or no change.

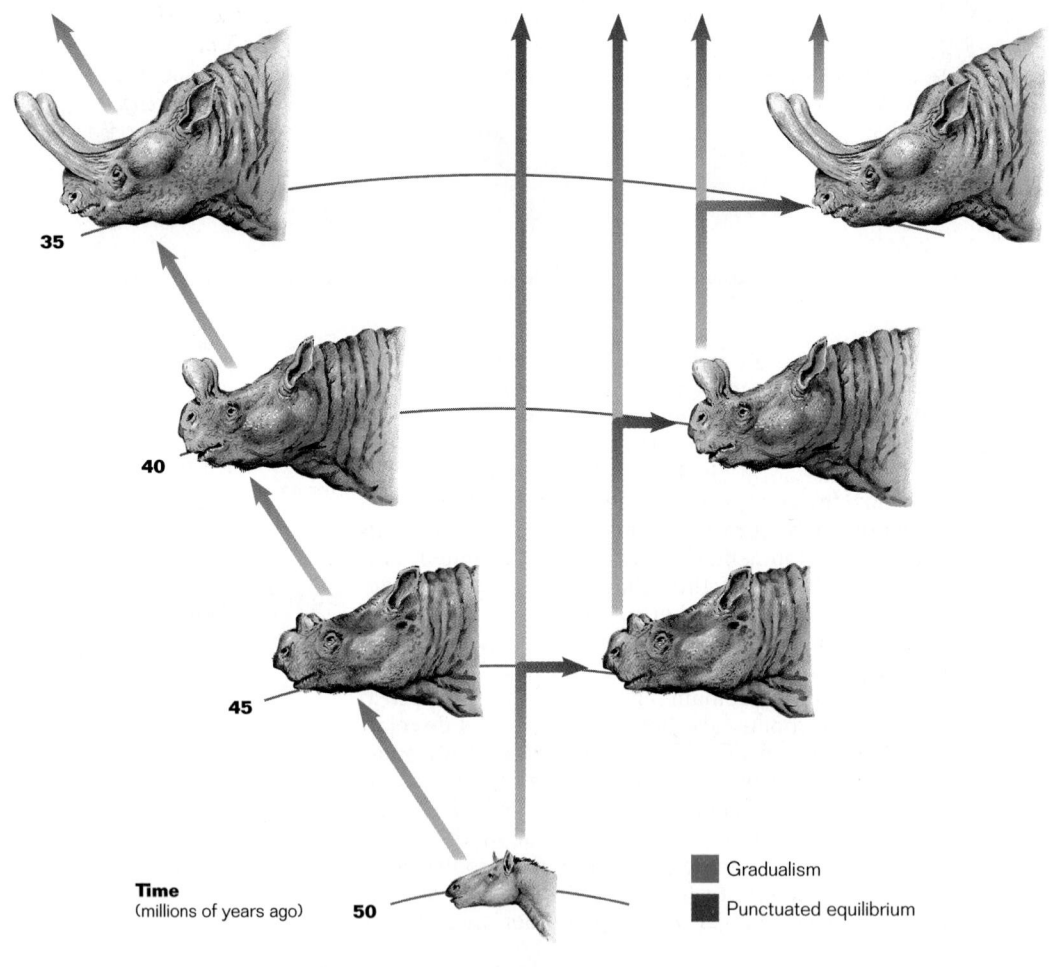

Time
(millions of years ago)

■ Gradualism
■ Punctuated equilibrium

What could cause major environmental changes, and why would they lead to spurts in evolution? The fossil record shows that drastic environmental changes have occurred very infrequently, separated by quiet periods that often last tens of millions of years. Events such as volcanic eruptions, asteroid impacts, and ice ages have been linked to sudden and drastic changes in climates, both locally and across the entire planet. Such changes have also been linked to the extinction of many groups of organisms. As a result, habitats that were once occupied became vacant and provided opportunities for colonization by species that could rapidly adapt to the new conditions through natural selection.

A prediction that can be made using the punctuated equilibrium model is that the fossil record should be very discontinuous. Is this prediction supported by evidence? There is considerable disagreement among biologists on this point, and the discussion continues. Of course, large gaps in the fossil record exist as a result of erosion and other destructive geologic processes. However, the fossil record seems to provide evidence of both types of evolution. Many groups of organisms appear suddenly in the fossil record. Some of these groups remain virtually unchanged for millions of years, while other groups disappear as suddenly as they appear. Still other groups of organisms change gradually through time, as predicted by the gradualism model. More careful study of the fossil record may reveal additional examples of both types of evolution. ◻

❏ CAPSULE
SUMMARY

Gradualism is the model of evolution in which change occurs gradually over time. Punctuated equilibrium is the model of evolution in which change occurs in spurts separated by long periods of equilibrium.

Section Review

1. *How does the fossil record indicate that macro-evolution has occurred?*

2. *What is a common ancestor?*

3. *What do the similarities in the development of vertebrate embryos indicate about vertebrates?*

4. *How does the punctuated equilibrium model of evolution differ from the gradualism model?*

Critical Thinking

5. *Why do fishes and mammals have more nucleotide sequence differences in their genes for a particular protein than do reptiles and mammals?*

6. *How are the wings of the flightless cormorant seen in Figure 12-13 homologous to the limbs of other vertebrates? Why are they considered vestigial structures?*

7. *Does acceptance of the punctuated equilibrium model mean that the gradualism model of evolution must be totally rejected? Why or why not?*

UPDATE

☑ **RESEARCH Update**

Evolution in Suburban Backyards?

Life is short in the backyard environment. It is the scene of repeated slaughter by both native predators and domestic pets, especially cats. Robert Garrot and P. J. White of the University of Washington and Callie Vanderbilt at the University of California at Davis contend that suburbs are, evolutionarily, new habitats, which tend to quickly become populated by opportunistic species. These ecologists point out that while conservationists bemoan the negative impact of invasions by foreign species on native species, they seem to ignore equally devastating invasions by indigenous species.

According to these concerned scientists, indigenous invaders are more than just pests. Indigenous species that adapt well are often responsible for forcing out species that do not adapt as well. While the construction of suburban areas destroys the habitats of many species, construction of houses, yards, and storm sewers creates new habitats in which adaptable populations thrive. As populations of "tough" species (raccoons, crows, etc.) increase, they prey on or otherwise force out "weaker" or less adaptable species, which may include rare plants or insects on which other rare species feed. This creates a chain reaction that may affect several species. Ecologists wonder what's going to happen as more natural habitats disappear. ☑

Answers to Section Review

1. Answers will vary but could state that many life-forms that lived in the past have been replaced by those living today. This record also shows a gradual progression in the complexity of life—from the oldest rocks to the youngest rocks—and many examples of transitions from one form to another.

2. A common ancestor is a species from which two or more species diverged.

3. These similarities indicate that all vertebrates had a common ancestor.

4. Punctuated equilibrium states that evolution occurs in spurts; gradualism states that evolution occurs at a slow, even rate.

5. Fishes are more distantly related to mammals than to reptiles, and thus fishes and mammals have more nucleotide sequence differences.

6. The wings are homologous to forelimbs of other vertebrates because they have the same types of bones, and they are vestigial because they are too small to enable the cormorant to fly.

7. No. Evidence suggests that both operate at the same time.

Opening Question

Ask students what environmental factor contributed to the evolution of polar bears and brown bears (seen in Figure 12-16) into separate species. Students should recognize that climate was the cause.

Demonstration

Modeling Natural Selection

Materials: (per group) one 30 cm (1 ft.) square of aluminum foil, 10 each 2.5 cm (1 in.) squares of aluminum foil and white paper. (Hint: Crinkle the foil and flatten it out before cutting the squares.)

Procedure: Divide the class into pairs or small groups. Have one student in each group spread out the 20 small squares on the larger aluminum square. Give another student in each group 10 seconds to pick up and remove as many squares as possible, one at a time, from the big sheet. Have each group count and report the number of aluminum squares and paper squares "captured." Keep a tally on the board or overhead. White paper squares should get "picked off" in greater numbers. Thus the camouflaged aluminum squares have an adaptive survival advantage. Ask: If the little squares could reproduce, which type would be more numerous in the next generation? (*small aluminum squares*) Why? (*More of them were left to reproduce.*)

12-3 Evidence of Microevolution

*T*he heart of Darwin's theory of evolution is its assertion that natural selection is the mechanism responsible for evolution. Darwin wrote: "Can we doubt . . . that individuals having any advantage, however slight, over others, would have the best chance of surviving and of procreating their kind? On the other hand, we may feel sure that any variation in the least degree injurious would be rigidly destroyed. This preservation of favorable variations, I call Natural Selection." In his writings, Darwin offered examples of how natural selection has shaped life on Earth. There are now many well-documented examples of how natural selection has acted to change the genetic makeup of species.

Section Objectives

■ Identify five steps in the process of natural selection.

■ Describe how natural selection has affected the European peppered moth.

■ Explain the role of natural selection in the distribution of sickle cell anemia.

■ Describe the process of species formation.

Overview NATURAL SELECTION

The process of natural selection depends on five main elements.

1. All species have genetic variation.

2. The environment presents many different challenges to an individual's survival.

3. Organisms tend to produce more offspring than their environment can support; thus, individuals of a species often compete with one another to survive (struggle for survival).

4. Individuals that are better able to cope with the challenges of their environment tend to leave more offspring than those less suited to the environment (survival of the fittest).

5. The characteristics of the individuals best suited to a particular environment tend to increase in a population over time.

The key lesson scientists have learned about evolution is that the environment dictates the direction and extent of change, as illustrated in Figure 12-16. Just as success determines which plays a football coach keeps in his team's game plan, so it determines which changes in a species are "kept" through natural selection.

Figure 12-16 Natural selection and the environment played important roles in the evolution of the polar bear (*Ursus maritimus*), *right*, and the grizzly bear (*Ursus horribilis*), *above*. The polar bear's white fur enables it to hunt more successfully in its snowy environment. The grizzly bear's brown fur enables it to hunt more successfully in its forested environment.

Industrial Melanism Is Natural Selection at Work

A particularly well studied example of natural selection in action is **industrial melanism,** the darkening of populations of organisms over time in response to industrial pollution. Although it occurs in many insect species, the best-known case of industrial melanism involves the European peppered moth, *Biston betularia.* Among the members of this species, there are two color variations, shown in Figure 12-17. The darker moths have genes for increased production of melanin (a black pigment). Once treasured by butterfly collectors, the dark variety of *Biston betularia* was extremely rare until the 1850s. Starting around 1850, however, dark peppered moths began to appear more often, usually in heavily industrialized areas. Every year, more dark moths were seen. After 100 years, almost all of the peppered-moth populations near industrial centers were composed of dark individuals.

The Concealment Hypothesis

Using Darwin's theory of evolution by natural selection, a hypothesis explaining the replacement of light moths by dark moths can be formed. Dark peppered moths are common in industrial regions where tree trunks are darkened by the soot of pollution. Perhaps dark moths are camouflaged against a background of soot-darkened bark and thereby escape being eaten by birds. Light moths, on the other hand, would stand out against a dark background and would be easy prey for hungry birds. A prediction can be made that the dark peppered moths would be favored in industrial areas because their dark color conceals them from birds that eat moths. This hypothesis is known as the concealment hypothesis.

Testing the Hypothesis

To see if natural selection could have caused the color change in the peppered-moth populations, the British ecologist H.B.D. Kettlewell performed an experiment during the late 1950s. Kettlewell raised populations of light and dark peppered moths in a laboratory. He then marked the underside of their wings with a dot of paint so they could be recognized later. Next, he released equal numbers of light and dark moths in two separate wooded areas of England. One of the wooded areas, near the city of Birmingham, was heavily polluted. The other wooded area, in the rural county of Dorset, was unpolluted. Finally, Kettlewell set rings of traps around the woods to recapture the moths and see which ones survived. As the graph in Figure 12-17 indicates, more of the moths matching the color of the tree trunks in each location survived. Many subsequent experiments confirmed these results. Hidden observers even saw birds passing by dark moths on polluted tree trunks and attacking the more conspicuous light moths. Kettlewell concluded that natural selection indeed causes industrial melanism in peppered-moth populations. ❑

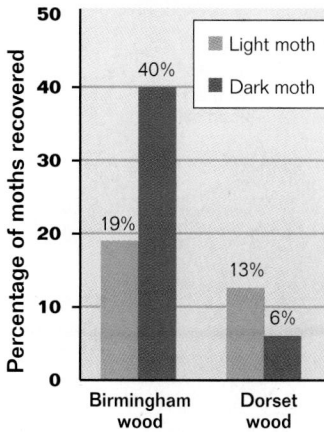

Figure 12-17 Two color variations, *top,* occur among European peppered moths, *Biston betularia.* A graph shows the results of Kettlewell's experiments, *bottom.* In the polluted woods near Birmingham, two-thirds of the surviving moths were dark. In rural Dorset, two-thirds of the surviving moths were light.

❑ **CAPSULE SUMMARY**

Experiments show that micro-evolution has occurred within populations of the European peppered moth.

SECTION 12-3

👁 *VISUAL STRATEGY*

Figure 12-17
Instruct students to examine the photograph of the two color variations of peppered moths. Ask: For which of these moths is color an advantage? *(The color of the light-colored moth is an advantage here because the moth blends in with the light-colored bark on which it is resting and is less likely to be detected by a predator.)* Under what circumstances would the dark-colored moth's color be an advantage? *(Dark color would be an advantage if the moths were resting on a dark-colored tree trunk.)*

Teaching Tip

What If there were no dark peppered moths?
When the Industrial Revolution began, air pollution upset the homeostasis of the peppered moth population of Great Britain. Ask: What would likely have happened to the peppered moth population during the Industrial Revolution if a dark variety had not existed in the gene pool of the moths? *(They likely would have become extinct in industrial areas.)*

Application

Agriculture Natural selection has worked to make insect pests harder to fight. When DDT was first introduced, for example, it was a highly effective insecticide. Over time, DDT became less and less effective; individuals that were resistant to the insecticide survived and produced the next generation. In fact, many populations of insects are now resistant to DDT. Although DDT is now banned in this country because of its persistent toxicity, farmers have repeatedly had to deal with insect populations that develop resistance to insecticides.

Overcoming Misconceptions

Environment Does Not Cause New Variations

The story of the peppered moth illustrates an important idea in the study of evolution that is frequently misunderstood: Individuals do not evolve, populations do. The darkened trees didn't cause the darker variant of moth to appear. That characteristic was already present in the natural genetic variation of the moth population. The "new" environment merely favored the survival of the darker moths.

CONNECTIONS

Chapter 7

Heterozygous and Homozygous Genotypes

Review the concepts of heterozygous and homozygous. Point out that these terms apply to an organism's genotype—the pair of alleles that diploid organisms possess for a particular gene locus on their chromosomes. Also remind students that the phenotype is the outward expression of the genotype.

Chapter 7

Punnett Squares

Review the construction of a Punnett square. Remind students that Punnett squares are used to predict the possible outcomes (genotypes) resulting from a cross.

👁 VISUAL STRATEGY

Figure 12-18

Have students compare the normal red blood cells to the sickled red blood cells. Ask: Which type of blood cell can more easily pass through blood vessels? *(normal red blood cells)* Tell students that the shape of sickled cells causes them to clog small arteries and capillaries, producing life-threatening blood clots. Also have students study the distribution of malaria and the sickle cell allele in Africa. Ask: How do these distributions compare? *(They are almost the same.)* What might this suggest about these two factors? *(The occurrence of one may affect the occurrence of the other.)*

Sickle Cell Anemia Reveals Evidence of Microevolution

One of the best-studied examples of the effect of natural selection involves hemoglobin proteins in humans. As you learned in Chapter 8, sickle cell anemia is a hereditary disease that affects hemoglobin molecules in human red blood cells. It arises from a single nucleotide change in the gene that codes for beta-hemoglobin and is one of the best understood of all human genetic disorders. In Figure 12-18, you can see how the sickle cell allele affects red blood cells. Because sickled red blood cells tend to clog tiny arteries, sickle cell anemia is usually lethal (deadly). Heterozygous individuals, who have both a defective hemoglobin gene and a normal hemoglobin gene, make enough functional hemoglobin to keep their red blood cells healthy.

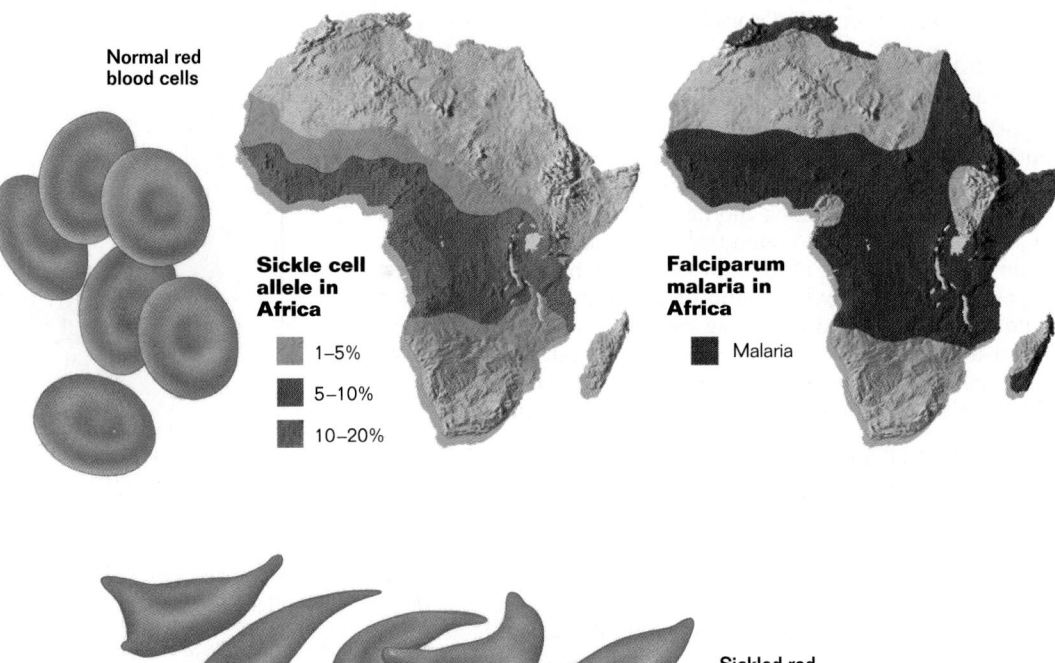

Normal red blood cells

Sickle cell allele in Africa

- 1–5%
- 5–10%
- 10–20%

Falciparum malaria in Africa

- Malaria

Sickled red blood cells

Figure 12-18 The red blood cells of people who are homozygous for the sickle cell allele collapse into sickled shapes, *bottom left,* when the oxygen level in the blood is low. The distribution of the sickle cell allele in Africa, *center,* coincides closely with that of falciparum malaria, *right.*

First detected in 1910 in Chicago, sickle cell anemia affects roughly 2 out of every 1,000 African Americans in the United States, but it is almost unknown within other ethnic groups. In central Africa where the sickle cell allele is thought to have originated, 1 in every 100 people is homozygous for the defective allele and develops the fatal disorder. Why has natural selection not eliminated the defective sickle cell allele from human populations? Why instead is this potentially fatal allele so common in Africa?

The sickle cell allele persists in human populations because having a single copy of the defective allele provides a definite advantage in certain environments. People who are heterozygous for the sickle cell allele are far more resistant to malaria (a leading cause of death in central Africa) than are people who are homozygous for normal hemoglobin. While one in a hundred individuals is homozygous for the sickle cell allele and dies of anemia, one in five individuals is heterozygous for the sickle cell allele and survives malaria. Thus, even though people who are homozygous for the sickle cell allele die, a population that lives where malaria is prevalent experiences far fewer deaths from sickle cell anemia than would occur from malaria if the allele were not present in the population. As Figure 12-18 also illustrates, the distribution of the sickle cell allele in Africa closely coincides with the occurrence of falciparum malaria, a particularly devastating type of malaria.

Balancing Selection

In central Africa, natural selection affects the sickle cell allele in opposing ways: (1) it tends to eliminate the sickle cell allele because it is lethal to homozygous individuals, and (2) it tends to preserve the sickle cell allele because heterozygous individuals are resistant to malaria. Selection that acts in opposite directions is called **balancing selection.** When malaria is present, *both* homozygous genotypes are selected against (neither homozygous genotype has a selective advantage). Thus, heterozygous individuals are the most "fit." Balancing selection, therefore, maintains both the normal hemoglobin allele and the sickle cell allele in the population of central Africa. Even defective alleles such as the sickle cell allele will remain in a population as long as the benefit balances the cost.

Directional Selection

While the sickle cell allele remains common in central Africa, it is gradually being eliminated from human populations in other parts of the world. In the United States, for example, the environment does not favor the sickle cell allele because its lethal potential is not balanced by any advantage. Because malaria no longer occurs in the United States, African Americans gain no advantage from being heterozygous for the sickle cell allele. As a result, the sickle cell allele is becoming less common among African Americans. When natural selection is unopposed, the frequency of a particular allele tends to move in one direction—in this case, toward elimination. Biologists call such unopposed selection **directional selection.** As it does with the peppered moth, the environment determines the direction and extent of evolutionary change. ▢

❏ CAPSULE SUMMARY

Balancing selection maintains an allele in a population when the environment acts on the allele in opposing ways. Directional selection causes an allele to become either more common or less common.

SECTION 12-3

Teaching Tip
How Natural Selection Affects Humans
Draw the three possible genotypes for the sickle cell trait as SS, Ss, and ss. Tell students that "S" is the allele for normal hemoglobin and "s" is the sickle cell allele. Ask: What is likely to happen to people in Africa with the ss genotype? With the SS genotype? With the Ss genotype? *(People with the ss genotype will die of sickle cell anemia, those with the SS genotype will be more likely to die of malaria, and those with the Ss genotype will be most likely to survive both sickle cell anemia and malaria.)* Place an "X" over the SS genotype to indicate those who die from malaria and an "X" over the ss genotype to indicate those who die from sickle cell anemia. Point out that natural selection tends to eliminate the sickle cell allele in one situation (ss) and favor it in another (Ss). Thus, balancing selection tends to maintain both alleles in the human population of Africa. Have students make a Punnett square for offspring that could result from matings between two Ss individuals. A completed Punnett square can be found below. You might also have students repeat this exercise to show what happens to the sickle cell allele in the United States, where directional selection tends to eliminate the sickle cell allele.

	S	s
S	SS	Ss
s	Ss	ss

Multicultural Perspective

Is All Variation Adaptive?

Based on his research on population genetics and molecular evolution, the Japanese geneticist Motoo Kimura proposed a neutral theory of evolution. Kimura argued that much of the variation in natural populations is selectively neutral, that is, the different forms of genetic traits neither help nor hinder survival and reproduction. According to Kimura, changes in the frequency of these traits are due to chance, not to natural selection.

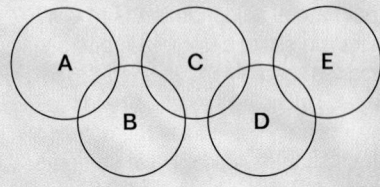
Species Formation Begins With Microevolution

Species formation occurs in a series of successive stages. Because natural selection favors changes that increase reproductive success, microevolution continually molds and shapes a species to improve the "fit" between a species and its environment. Over time, separate populations of the same species may become very different from one another as different kinds of changes accumulate in each isolated group. The accumulation of differences between groups (populations, species, genera, etc.) is called **divergence.** Within populations, divergence leads to the formation of new species. Biologists call the process by which new species form **speciation** *(spee see AY shun).*

Forming Ecological Races

Figure 12-19 illustrates what may be an early stage in the process of speciation. A species, such as the seaside sparrow, often occurs in several different kinds of environments. In each environment, natural selection acts to make the local population of a species better adapted to that environment. If their environments differ enough, local populations can become quite distinct. Over time, populations of the same

Figure 12-19 The formation of these ecological races of seaside sparrows may indicate that speciation has begun. If the populations of these ecological races remain isolated, they will become increasingly specialized as they adapt to their local environments.

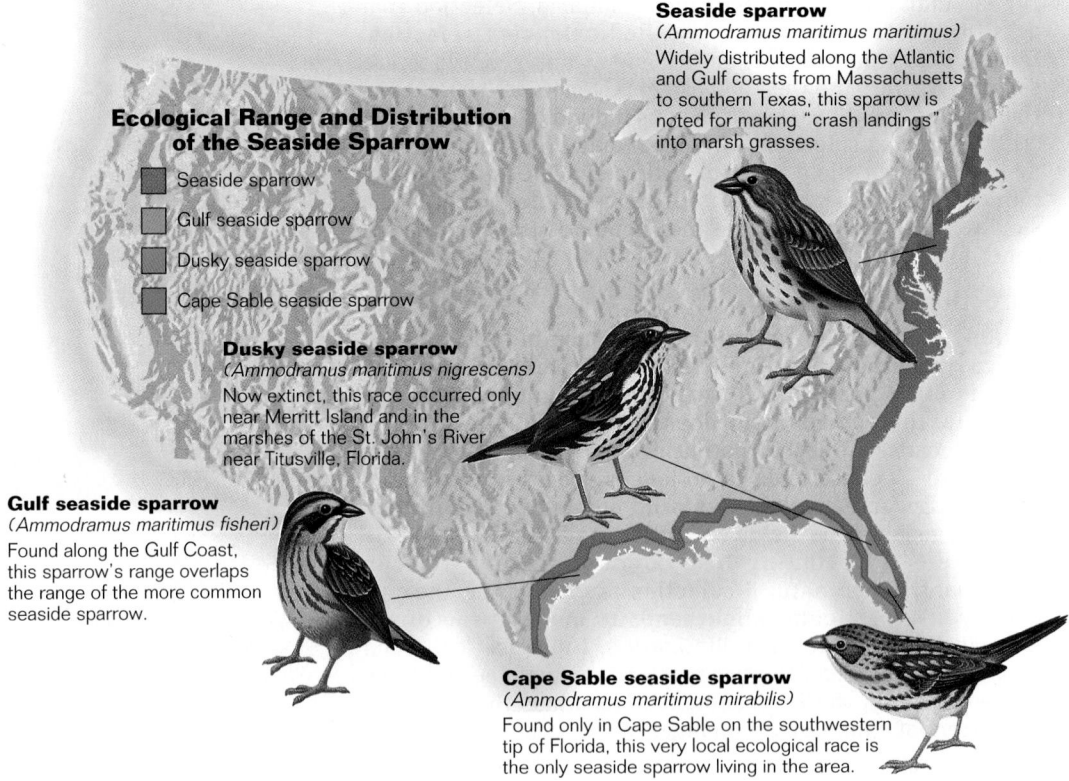

Ecological Range and Distribution of the Seaside Sparrow

- Seaside sparrow
- Gulf seaside sparrow
- Dusky seaside sparrow
- Cape Sable seaside sparrow

Seaside sparrow
(Ammodramus maritimus maritimus)
Widely distributed along the Atlantic and Gulf coasts from Massachusetts to southern Texas, this sparrow is noted for making "crash landings" into marsh grasses.

Dusky seaside sparrow
(Ammodramus maritimus nigrescens)
Now extinct, this race occurred only near Merritt Island and in the marshes of the St. John's River near Titusville, Florida.

Gulf seaside sparrow
(Ammodramus maritimus fisheri)
Found along the Gulf Coast, this sparrow's range overlaps the range of the more common seaside sparrow.

Cape Sable seaside sparrow
(Ammodramus maritimus mirabilis)
Found only in Cape Sable on the southwestern tip of Florida, this very local ecological race is the only seaside sparrow living in the area.

species that differ genetically because of adaptations to different living conditions become what biologists call **ecological races.** Although the members of different ecological races are not yet different enough for the groups to be considered different species, they have taken the first step on that road. Ecological races may continue to diverge and become more and more different from each other as natural selection favors different survival strategies in different environments. Eventually, the races may become so different that they can no longer interbreed successfully. Biologists then consider them separate species.

Maintaining New Species

What keeps new species that diverge from a common ancestor separate? Why are even closely related species usually unable to interbreed with one another? A variety of isolating mechanisms are responsible for preventing interbreeding between closely related species. Once ecological races become different enough, some type of barrier to reproduction, like the one illustrated in Figure 12-20, usually prevents different groups from breeding with each other. The prevention of mating between formerly interbreeding

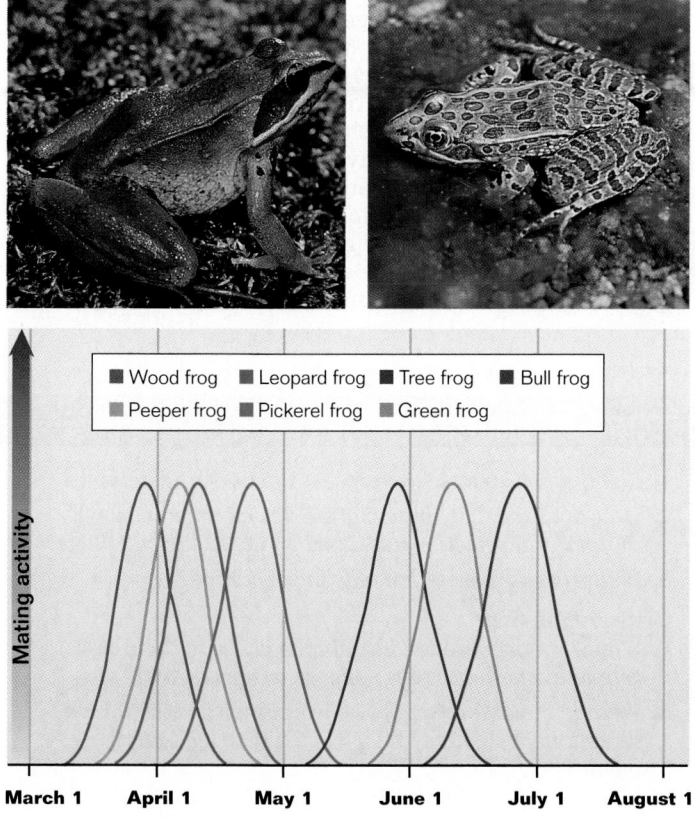

Figure 12-20 Though they appear to be very similar aside from their coloration, wood frogs *(Rana sylvatica), top left,* and leopard frogs *(Rana pipiens), top right,* are different species. The graph, *bottom,* shows that the time of peak mating activity varies among several species of frogs in the genus *Rana*. Such differences in mating activity represent one type of barrier to reproduction that promotes speciation.

SECTION 12-3

Teaching Tip
Sequential Diagram of Speciation
Have students make a sequential diagram for speciation using the following terms: divergence, isolation, natural selection, new species, variation. A completed sequential diagram is shown in the *Graphic Organizer* at the bottom of page 265.

 VISUAL STRATEGY

Figure 12-20
Have students compare the peak times of mating activity among these species of the genus *Rana*. Tell students that the differences in peak times of mating represent a type of reproductive isolation.

Graphic Organizer
Use this graphic organizer with
***Teaching Tip:* Sequential Diagram of Speciation** on page 265.

Teaching Tip
Identifying Isolation

Challenge students to think of as many examples of barriers to successful interbreeding as possible. Then, have them classify each example as one of the mechanisms for reproductive isolation listed in Table 12-2 on page 266. If students are confused by the similarity between the terms *reproductive isolation* and *reproductive failure*, tell them that reproductive failure is a type of reproductive isolation that results when two organisms are either unable to produce any offspring or unable to produce fertile offspring. For example, horses and donkeys are isolated by reproductive failure because their offspring (mules and hinnies) are sterile.

Chapter Closure

Tell students that extinction is not a freak event of nature but more likely the rule. Scientists estimate that more than 99 percent of the species that have ever lived are now extinct. Ask students to speculate about how mass extinctions, such as the one that occurred at the end of the Cretaceous period and included the extinction of the dinosaurs, might affect the rate of speciation and thus macroevolution. *(Rapid speciation might occur after mass extinctions, as many habitats would be open to colonization. There would also be many opportunities for directional selection to act on existing variation and enable isolated groups to adapt to their new environments.)*

Table 12-2 Mechanisms for Reproductive Isolation

Mechanism	Description
Geographical isolation	Groups are physically separated.
Ecological isolation	Groups occupy different habitats, and their hybrids are not suited to either environment.
Temporal isolation	Groups reproduce at different times of the day or year.
Behavioral isolation	Groups are not attracted to each other for mating.
Mechanical isolation	Structural differences prevent mating between individuals of different groups.
Reproductive failure	Matings between groups do not produce fertile offspring.

groups or the inability of these groups to produce fertile offspring is called **reproductive isolation.** Several types of barriers that may isolate two or more closely related groups are listed in Table 12-2.

□ CAPSULE SUMMARY

Speciation begins as a population adapts to its specific environment. Reproductive isolation keeps newly forming species from interbreeding. Over time, change within species caused by natural selection leads to the rise of new kinds of organisms.

Microevolution Leads to Macroevolution

Biologists have observed the stages of speciation in many different organisms. Thus, the way that natural selection leads to the formation of new species has been thoroughly documented. As changes continue to accumulate over time, living species may become very different from their ancestors and from other species that evolved from the same common ancestor. Biologists therefore agree that change within species caused by natural selection (microevolution) eventually leads to the appearance of new species and even new kingdoms of organisms (macroevolution). □

Section Review

1. *List the five steps in the process of natural selection.*
2. *How does the distribution of sickle cell anemia suggest that natural selection has acted on the sickle cell allele?*
3. *What role does isolation play in speciation?*

Critical Thinking
4. *Why were peppered moths able to adapt to their environment whether it was polluted or unpolluted?*
5. *What is the difference between balancing selection and directional selection? Which is likely to lead more quickly to evolution? Why?*

Answers to Section Review

1. The five steps are genetic variation, environmental stress, overproduction of offspring and a struggle for survival, survival of the fittest, and rise in number of individuals with characteristics suited to the environment.

2. The distribution of sickle cell anemia coincides with that of malaria, to which a person is resistant if he or she has one copy of the sickle cell allele.

3. Isolation keeps groups with different variations from interbreeding, which causes the differences between the groups to build up.

4. Peppered moths were able to adapt to their environments because they had variations that enabled some moths to be successful in an unpolluted environment and others to be successful in a polluted environment.

5. Balancing selection tends to maintain the variations of a gene in a population, while directional selection tends to eliminate a variation from a population. Directional selection more quickly leads to evolution, because it selects only certain variations.

Vocabulary

adaptation (249)	fossil (252)	paleontologist (254)
balancing selection (263)	gradualism (258)	population (248)
common ancestor (254)	homologous structures (256)	punctuated equilibrium (258)
directional selection (263)	industrial melanism (261)	reproductive isolation (266)
divergence (264)	isolation (251)	speciation (264)
ecological race (265)	macroevolution (251)	species (251)
extinct (251)	microevolution (251)	vestigial structure (256)
family tree (254)	natural selection (249)	

Concept Mapping

Construct a concept map that shows how microevolution leads to macroevolution. Use as many terms as needed from the vocabulary list. Try to include the following terms in your map: microevolution, macroevolution, natural selection, genetic variation, environment, speciation, divergence, and extinction. Include additional terms in your map as needed.

Review

Multiple Choice

1. According to the modern theory of evolution,
   a. extinct species are always replaced by gradualism.
   b. random gene mutation is a component of evolution.
   c. punctuated equilibrium has replaced natural selection.
   d. the diversity of life resulted from the inheritance of acquired characteristics.

2. Natural selection
   a. leads to a reduction in the diversity of life.
   b. involves the inheritance of acquired characteristics.
   c. allows organisms to live in environments for which they are best suited.
   d. causes species to become better suited to their environment.

3. Which of the following is *not* an element in the process of natural selection?
   a. Genetic variation exists within a species.
   b. Individuals struggle for survival.
   c. Organisms cause environmental change.
   d. Fit with the environment leads to a population increase.

4. According to the modern version of Darwin's theory of evolution by natural selection,
   a. the frequency of the genes in a population does not change over time.
   b. populations that are isolated from one another remain the same over time.
   c. microevolution results from species extinction.
   d. the individuals of two populations become distinct species when they can no longer interbreed.

5. Fossil evidence of all past life on Earth has not been found because
   a. fossils could not form in all environments in which organisms live.
   b. paleontologists need to perfect the techniques of radiometric dating.
   c. some organisms do not have the gene that codes for fossil formation.
   d. common ancestors first must be identified using family trees.

6. Anatomical structures that share a common ancestry are called
   a. vestigial structures.
   b. homologous structures.
   c. analogous structures.
   d. evolutionary structures.

CHAPTER REVIEW ANSWERS

Each item in the Chapter Review is correlated by text section in the assignment guide that follows.

ASSIGNMENT GUIDE

Section	Review	Themes Review	Critical Thinking
12-1	1–4, 11, 12, 18, 19	24	26
12-2	5–7, 13–15, 20	23, 25	27, 28
12-3	3, 8–10, 16, 17, 21, 22	24	29

Concept Mapping
Map answer is shown on page 243B.

Review
Multiple Choice
Code in parentheses indicates section number and objective number.
1. b (12-1.1)
2. d (12-1.2, 12-1.3)
3. c (12-1.3, 12-3.1)
4. d (12-1.4)
5. a (12-2.1)
6. b (12-2.3)
7. c (12-2.4)
8. a (12-3.2)
9. c (12-3.3)
10. a (12-3.4)

CHAPTER REVIEW ANSWERS *continued*

Completion

11. South America, islands (12-1.2)

12. microevolution, macroevolution (12-1.4)

13. Paleontologists, family trees (12-2.1)

14. amino acid, nucleotide (12-2.2)

15. punctuated equilibrium, gradualism (12-2.4)

16. balancing, directional (12-3.3)

17. Ecological races, species (12-3.4)

Short Answer

18. The theory of evolution is a set of principles that explains how organisms change over time and how the diversity of life on Earth today came to be. (12-1.1)

19. Darwin was the first to offer a logical explanation of how evolution occurs (natural selection), and he presented his ideas to a wide audience. (12-1.2)

20. The structures of proteins and DNA are used to construct family trees that show the evolution of particular genes within related groups of organisms. These family trees support fossil evidence because they are very similar to family trees constructed using the fossil record. (12-2.2)

21. Vestigial structures are evidence of common ancestry; they are homologous to functional structures possessed by other related organisms but are so reduced that they no longer have the same function. (12-2.3)

22. Environmental change resulted in conditions to which mammals were able to adapt but to which dinosaurs were not able to adapt. (12-3.1)

7. The punctuated equilibrium model differs from the gradualism model by stating that
  a. species formation occurs slowly and continuously.
  b. Darwin's theory of natural selection is inaccurate.
  c. periods of rapid species formation alternate with periods of little or no change.
  d. the fossil record should be continuous.

8. Kettlewell's experiments with peppered moths in polluted and unpolluted areas
  a. provided support for the concealment hypothesis.
  b. showed that light-colored moths are favored in industrial areas.
  c. suggested that birds prefer the taste of dark-colored moths.
  d. disproved the theory of evolution by natural selection.

9. The effect of the sickle cell allele on central Africa's population is an example of
  a. industrial melanism.
  b. punctuated equilibrium.
  c. balancing selection.
  d. directional selection.

Completion

11. Darwin observed that Galapagos species were more similar to the species of _____ than to the species of other similar but more distant _____ .

12. Evolution within populations of a species is called _____ , while the evolution of new species is called _____ .

13. _____ are scientists who study fossils. By arranging fossils from oldest to youngest, these scientists are able to construct _____ , which show how organisms are related through evolution.

10. The process by which isolated populations of the same species become new species is called
  a. speciation.
  b. microevolution.
  c. macroevolution.
  d. natural selection.

14. Evidence indicating that evolution has occurred is found in the _____ sequences of proteins and in the _____ sequences of DNA.

15. The _____ model of evolution proposes that new species form during brief spurts of rapid change, while the _____ model proposes that evolution occurs slowly over long periods of time.

16. The allele for sickle cell anemia exhibits _____ selection within the population of central Africa and _____ selection among African Americans.

17. _____ are populations of the same species that live in different environments and differ genetically. Over time, these groups may become new _____ .

Short Answer

18. What is the theory of evolution?

19. Charles Darwin was not the first person to suggest that organisms evolved. Yet his name is often the only one associated with the theory of evolution. Why?

20. How are the structures of proteins and DNA used to support fossil evidence that evolution has occurred?

21. How do structures such as the vestigial legs of whales and the human coccyx indicate that evolution has occurred?

22. At the time that dinosaurs became extinct, mammals became more numerous and diverse. Explain how such events might result from natural selection.

Themes Review

23. Structure and Function The wings of butterflies and the wings of bats serve similar functions but differ in structure. They are called analogous structures. Do analogous structures provide evidence of evolution? Explain.

24. Homeostasis How does natural selection help populations maintain homeostasis in changing environments?

Themes Review

23. Analogous structures may be considered to be evidence of evolution because structures that are similar in appearance and function tend to develop in similar environments. (12-2.3)

24. Natural selection helps populations maintain homeostasis (stability) by ensuring that a population is composed mainly of individuals that are well suited to survival in their environment. In other words, natural selection modifies the overall genetic makeup of a population. (12-1, 12-3)

25. All organisms must pass through a period of development as they become an adult. The fact that the larval forms of barnacles and lobsters are very similar indicates that they had a common ancestor but that different sets of genes, which developed over time, are activated in each group as its members become adults. (12-2.3)

25. Levels of Organization Look at the photos below. The larvae of barnacles and lobsters are virtually identical. What does this indicate about the evolutionary history of these organisms?

Critical Thinking

26. Making Inferences About 40 years after the publication of Darwin's book *On the Origin of Species*, genetics was recognized as a science. At this time, support for Darwin's theory of evolution by natural selection began to grow among scientists. How might these two events be related?

27. Using Analogies Darwin used the metaphor of an evolutionary tree to illustrate the diversity and relatedness of species. What do an evolutionary (family) tree's trunk, limbs, and twigs represent?

28. Interpreting Data Use the data in the table to answer the following questions. Which species is most closely related to humans? Which species is most distantly related to species B?

Species	Number of Amino Acid Differences Compared With Human Hemoglobin
Humans	0
Species A	8
Species B	17
Species C	39

29. Identifying Variables Why did H.B.D. Kettlewell mark the wings of the moths he released with a dot of paint? Name at least four variables that existed in Kettlewell's experiments.

Activities and Projects

30. Cooperative Group Project Locate and examine photographs and drawings of the tortoises Darwin observed on the Galapagos Islands. Plan and produce a mural showing the tortoises in their natural environments. Display the mural in your school.

31. Research and Writing Research Alfred Russel Wallace's contributions to the science of biology. Relate your findings in a written report.

32. Multicultural Perspective Most of Africa lies within the tropics, and more than one-half of the land is a desert. As a result, many African species have unique structural and behavioral adaptations for surviving in the desert. Use books, magazines, and films to research ways that African species, including humans, are adapted to life in the desert. Relate your findings to your class.

Readings

33. Read Jeremy Greenwood's article "Theory fits the bill in the Galápagos Islands," in *Nature*, April 22, 1993, page 699. How did *Geospiza fortis* respond to a change in its environment? How is the Grants' work important to the study of evolution and ecology?

34. Read David Raup's article "Extinction: bad genes or bad luck?" in *New Scientist*, September 14, 1991, pages 46–49. Mass extinctions affected species in many different habitats over wide geographic areas. What does Raup think this suggests about the process of extinction? How does he think most extinctions occur?

LABORATORY Investigation Chapter 12

Recognizing Patterns of Variation

Preparation

Each lab group needs 50–55 peanuts in the shell, 20 mung bean sprouts, and 10 tree leaves of the same species. Soak the mung beans overnight in cold water to prepare them for sprouting. Place the beans on a screen over moist paper towels or directly on moist paper towels in a baking dish. Rinse daily to prevent microbial infection. Grow the beans in a warm, dark place for 3–4 days until the sprouts are 1–3 cm long. Keep them moist.

Procedural Notes

- Two 40-minute periods are required to complete this investigation. Have students work in groups of two.
- Review the construction and labeling of a bar graph.
- During the first period, have students make the measurements for Parts A, B, and C and record their data. During the second period, have students graph their data and answer the questions.
- To compile class data, record each group's data for forearm length, leaf-blade length, and leaf-petiole length on the chalkboard. Pooling the data will increase the sizes of the populations studied, thus making students' statistical analyses more valid.
- **Safety:** Make sure that the leaves you collect are from a nonpoisonous species. Caution students to wash their hands before leaving the lab.

OBJECTIVES

- Observe genetic variation in particular traits among members of a population.
- Infer how individual variation can potentially contribute to the continuing evolution of a species.

PROCESS SKILLS

- measuring
- calculating means
- organizing data in tables and graphs

MATERIALS

- metric ruler
- mung bean sprouts
- graph paper (optional)
- peanuts in the shell
- tape measure
- calculator
- tree leaves

BACKGROUND

1. How much variation in genetic expression exists within a population?
2. How might variation within a species be useful? How might variation be harmful?
3. How can variation within plant and animal populations be measured?
4. How is variation within a species the basis for natural selection?
5. Write your own question to explore in your lab report or notebook.

TECHNIQUE

Part A: Bean Sprouts and Peanuts

You will need 50–55 peanuts in the shell and 20 mung bean sprouts.

1. Measure the lengths of 20 mung bean sprouts to the nearest millimeter from the tip to the site where the sprout first emerged from the seed.
2. Record your measurements in the **Records** section of your report. Organize your measurement values on a single line from the lowest to highest value. Beneath each value, write the number of sprouts of that length. On your report or on graph paper, make a graph showing the distribution curve.
3. Crack open the peanut shells and measure the lengths of 100 peanuts to the nearest millimeter. Record your results in a table in the **Records** section of your report, giving the lengths and the number of peanuts of each length.
4. Make a bar graph using your data.

Part B: Humans

5. Using the tape measure, measure your lab partner's left forearm from the wrist bone to the elbow. Record your measurements in the **Records** section of your report. Pool your data with those collected by others in your class. In the **Records** section of your report, make a bar graph listing the measurements on the horizontal axis and the number of forearms of each length on the vertical axis.
6. Calculate the mean forearm length. Record your results in the **Records** section of your report.

Part C: Tree Leaves

You will need 10 tree leaves collected from one tree species.

Background Answers

1. Most populations show a great deal of genetic variation.

2. Variation helps a population of a species adapt to changes in the environment. Individual variations that are harmful mutations may cause the death of many individuals but are not likely to destroy a population.

3. Variation within a population can be measured by comparing the traits of the individuals that make up the population.

4. Individuals with variations that make them better adapted to their environment are more likely to survive and reproduce.

5. How much variation exists in traits of genetic populations within a species?

Records

Bar graphs for parts A, B, and C should resemble the following sample graphs.

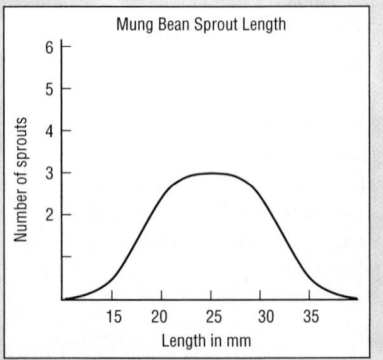

7. Measure the lengths of each of the 10 leaf blades. Record your results in the **Records** section of your report.

8. Measure the petiole (leaf stem) of each leaf and record the data in the **Records** section of your report.

9. Using the data from all class teams, record the lengths for the blades and petioles, beginning with the shortest length and ending with the longest length. Record your results in the **Records** section of your report. Briefly summarize the procedure you followed.

10. Clean up your materials and wash your hands before leaving the lab.

INQUIRY

1. What is the range of measurements for your mung bean sprouts?

2. The mode is the most frequently occurring value. What is the mode for your bean sprouts?

3. List the advantages and disadvantages for survival of the longest sprouts.

4. Analyze the bar graph showing the lengths of peanuts. Compare the numbers of peanuts of average length with the numbers of very large and very small peanuts.

5. Since the peanut supplies the seed embryo with energy, what is the relationship of peanut size to the amount of energy available to the peanut embryo?

6. For Part B, calculate the difference between the lowest measurement and the mode and the difference between the highest measurement and the mode. How does the mode differ from the mean?

ANALYSIS

1. What is variation?

2. Explain why it is advantageous for a species to show variation among individuals.

3. Describe the relationship between natural selection and variation.

4. If the environment changed so that a very large peanut was an advantageous variation, what would happen to the peanut population over time in response to such a change?

5. Describe an environmental change that would favor a much larger peanut. Can you think of a change that would favor a much smaller peanut?

6. How might leaf size be important to the success of a plant?

7. How might you determine whether a trait such as those measured in this lab investigation is a result of genetic or environmental factors?

FURTHER INQUIRY

Write a new question that could be explored as a new investigation. The following is an example:

What are some examples of variations found in other species? For example, measure adult human height, pine needle length, acorn weight, or the length of the hind legs of grasshoppers. For each characteristic measured, hypothesize about environmental changes that could lead to the selection of an extremely small or large size.

Inquiry Answers

1. Answers will vary. (Note: The range is the difference between the longest and the shortest mung bean sprout.)

2. Answers will vary.

3. Answers will vary. Longer sprouts can collect more sunlight but are more likely to be injured.

4. More peanuts will be of about average length.

5. The larger the peanut, the more energy there is available to the embryo.

6. Answers will vary. The mode is the most common measurement, while the mean is the average of all measurements.

Analysis Answers

1. Variation is the differences in the expressions of genetic traits among the individuals of a population.

2. Variation within a species is advantageous because it enables the populations of the species to adapt to changes in their environment.

3. Natural selection acts upon the variation within a population.

4. Over time, most of the peanuts produced by the population would be very large.

5. Answers will vary. Large peanuts might be favored if an environment suddenly had less sunlight for photosynthesis. Small peanuts might be favored if an environment suddenly became much drier.

6. Answers may vary. In environments where light is limited, larger leaves enable plants to gather more sunlight and grow larger. In hot, dry environments with plenty of sunlight, smaller leaves enable a plant to limit water use and still collect plenty of sunlight for growth.

7. You can determine whether a trait is genetic or not by observing whether it is passed on to offspring.

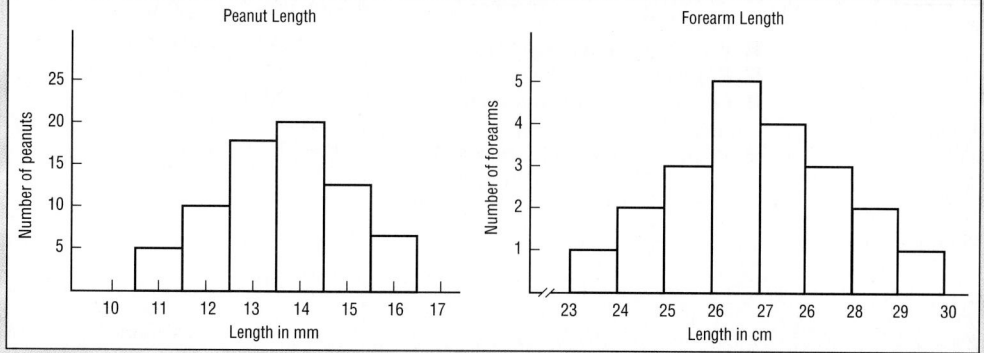

CHAPTER 13 HISTORY OF LIFE ON EARTH

Block Scheduling Guide

Each block represents about 45–50 minutes of instructional time. The resources cited in a block are coded to a specific program component using the Key on the next page.

Lecture Concepts	Lesson Resources

13-1 Life in the Ancient Seas pp. 272–279

BLOCKS ① and ②

a. The earliest fossil cells closely resemble modern-day bacteria.
b. Primitive cyanobacteria produced the first oxygen in Earth's atmosphere and are thought to be the ancestors of chloroplasts.
c. Archaebacteria probably gave rise to the first eukaryotes.
d. The evolution of multicellularity allowed for "division of labor" among cells and allowed cell specialization that led to increased complexity in organisms.
e. All major groups of organisms alive today, except plants, originated some time during the first hundred million years of multicellular life.
f. The have been five major mass extinction events on Earth. Human activity is contributing to what may become a sixth mass extinction.

Lecture Resources
■ Opening Question, page 273
■ Demonstration: Visible Fossils
■ Effective Reading, page 273
■ Visual Strategies: Figures 13-4, 13-7
■ Teaching Transparencies: 45A
■ Overcoming Misconceptions, page 275
■ Division of Labor, page 277
● Holt Biology Videodiscs: Lesson 13-1

Classwork Options
■ Remembering What Came from What and Graphic Organizer, page 274

■ Use a Chart to Organize Information, page 275
■ A Side-by-Side Comparison, page 276
■ Using the Timeline as a Way to Review, page 277

Assignment Options
■ Section Review, page 279
■ Chapter Review, questions 1–3, 12, 13, 21–24, 30, 32

13-2 Four Invasions of the Land pp. 280–283

BLOCK ③

a. For the majority of the time that life has existed on Earth, organisms have lived only in the water. Not until the formation of an ozone layer that blocks the sun's ultraviolet radiation was it possible for living things to survive on land. Ancient cyanobacteria produced the oxygen that was converted to ozone.
b. Plants and fungi, living symbiotically, were the first multicellular organisms to colonize the land. The first animals to venture onto land were the arthropods.
c. Vertebrates did not follow arthropods onto the land until 350 million years ago.

Lecture Resources
■ Opening Question, page 280
■ Demonstration: Personal Experience with UV Radiation
■ Big Bodies Need Big Support, page 281
■ Visual Strategy: Figures 13-11, 13-12
■ Research Update, page 283
● Holt Biology Videodiscs: Lesson 13-2

Classwork Options
■ Laboratory Investigation: Overview of Life on Earth, pages 292–293

Assignment Options
■ Section Review, page 283
■ Chapter Review, questions 4–6, 14, 15, 21, 25–27

13-3 Vertebrate Evolution pp. 284–288

BLOCKS ④ and ⑤

a. The first vertebrates were small, jawless fishes. Fishes with jaws appeared around 430 million years ago.
b. The first terrestrial vertebrates evolved from bony fishes. The evolution of lungs, a more efficient circulatory system, and legs allowed amphibians to move onto land for part of their lives.
c. A more complete solution to the problems of living on land is seen in the reptiles, which evolved from amphibians. Reptiles exhibit two key adaptations in vertebrate body design: a watertight skin that prevents dehydration in the atmosphere and a watertight, shelled egg that can be laid on land.
d. Birds and mammals, the dominant vertebrate groups on land today, evolved from reptiles.

Lecture Resources
■ Opening Question, page 284
■ Demonstrations: Inside and Out, Egg Watch
■ Clamping Down with Jaws, page 285
■ Watching Fish, page 285
■ Reptiles Aren't Slimy, page 286
■ Overcoming Misconceptions, page 286
■ Visual Strategy: Figure 13-19

● Holt Biology Videodiscs: Lesson 13-3

Classwork Options
■ Who Lives Where? Page 287
■ Closure, page 288

Assignment Options
■ Section Review, page 288
■ Concept Mapping, page 289
■ Chapter Review, questions 7–9, 16–20, 22, 28, 29, 31

Review/Enrichment

BLOCK 6

■ Study Guide: Concept Review and Pretest

Assignment Options
💿 Teaching Resources CD-ROM: Supplemental Reading Worksheets and Test, *Origin of Species*
■ Activities and Projects, questions 33–35
■ Readings, questions 36, 37

Assessment Options

BLOCK 7

Traditional Assessment
■ Chapter 13 Test
💿 Test Generator/Assessment Item Listing: Software item bank for preparing customized chapter tests.

Portfolio Assessment
Select student reports for one or more laboratory experiments from this chapter. *The Direct Observation Checklist* on the *BioSources Teaching Resources CD-ROM* should be completed during a laboratory or other cooperative learning experience.

Answer to Concept Map

The following is one possible answer to the Concept Mapping exercise on page 289.

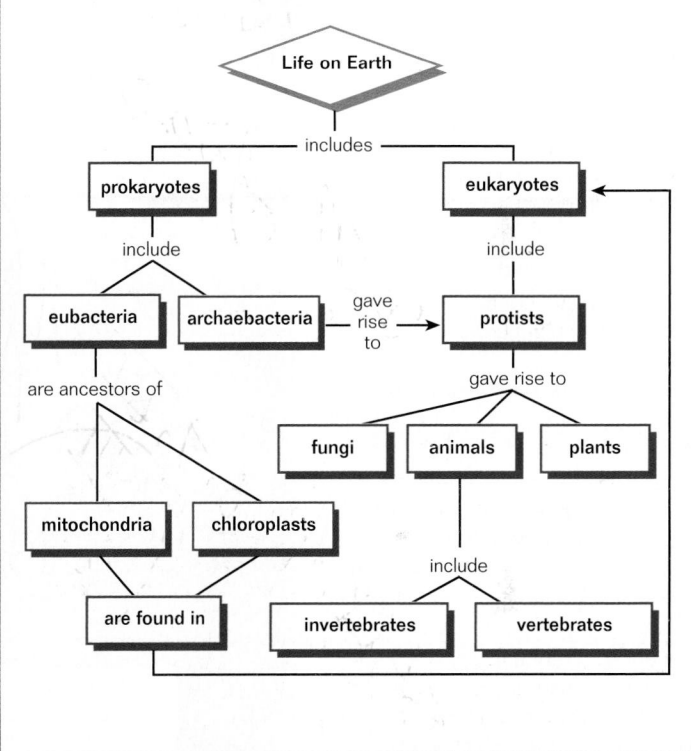

Holt Biology Videodiscs

Holt Biology Videodiscs gives you a powerful tool for teaching, review, and assessment. *Concepts of Biology* adds interactive lesson plans and extensive tools for customization using CD-ROM technology.

CONCEPTS OF BIOLOGY

Use the following topic from *Concepts of Biology* to help you teach this chapter:

■ Topic 9: Evolution of Life on Earth

For further information regarding the media on the videodiscs, see *Holt Biology Videodiscs Teacher's Correlation Guide to Biology: Principles and Explorations.*

Chapter Theme

Evolution *The history of life on Earth began when the first bacterial cells appeared, at least 3.5 billion years ago. Since that beginning, a rich diversity of organisms has inhabited Earth, each kind the result of evolution—the changing of organisms over time due to the process of natural selection.*

Tapping Prior Knowledge

- How can we tell that some fossils are older than others?
- What were the conditions on Earth before life began?
- Compare and contrast prokaryotic cells with eukaryotic cells.
- Compare and contrast chloroplasts with mitochondria.

Opening Demonstration

Have students look at examples of cyanobacteria that live on Earth today. These bacteria can be obtained commercially on prepared slides or collected from almost any polluted body of water. If you shine a bright light (not a flashlight) on an algae-rich beaker of water, you will be able to see oxygen bubbles form in the water. Ask students to tell you what happens to the bubbles when they rise to the surface of the water. *(They become part of the atmosphere.)* Relate this event to how ancient cyanobacteria released oxygen into Earth's early atmosphere.

CHAPTER 13
HISTORY OF LIFE ON EARTH

Eroded sandstone showing layers of sediment

REVIEW

- prokaryotes and eukaryotes (Section 2-1)
- chloroplasts and mitochondria (Section 2-4)
- photosynthesis (Section 4-1)
- origin of the Earth (Section 11-1)
- radiometric dating (Section 11-2)

Authors' Rationale

Now that students have a handle on *what* life is and *how* it is evolving, the actual evolution of the organisms on Earth (the six kingdoms) can be addressed. This is done in a sequential manner, starting with the appearance of the prokaryotes and moving through to the evolution of the vertebrates, with discussion of each major group in between. The appearances of the different groups are discussed along with the geologic periods in which they arose. Major events that affected evolution (such as development of ozone and invasion of land by plants, invertebrates, and vertebrates) are also discussed.

13-1 Life in the Ancient Seas

Several methods of radiometric dating have determined that Earth is approximately 4.5 billion years old. For most of the first several hundred million years, our planet was a fiery ball of molten rock, but eventually the surface cooled and formed a rocky crust over the hot molten interior. When Earth's surface cooled, water vapor in the air condensed to form great oceans. It was within these seas that life first evolved.

Section Objectives

- Contrast the Earth's age with the age of Earth's oldest fossil organisms.
- Describe the evolution of bacteria.
- Contrast bacteria with protists.
- Recognize an evolutionary advancement first seen in protists.
- Describe how mass extinction events have affected the evolution of life on Earth.

Bacteria Were the First Cells to Evolve

The earliest traces of life are found as tiny fossils in 3.5-billion-year-old rocks from the ancient seas. Earth's first cells were bacteria. Like bacteria that exist today, these ancient bacteria were prokaryotes. Unlike today's animal and plant cells, the first cells did not have internal compartments that could perform special functions. Instead, the interior of one of these early cells was like a warehouse, an open space within which all the cell's contents were free to move about.

Among the first of the bacteria to appear were the **cyanobacteria** (SY ahn oh bak TIHR ee ah). Cyanobacteria, shown in Figure 13-1, are photosynthetic. As ancient cyanobacteria carried out photosynthesis, they released oxygen gas into the oceans. This was something new,

Figure 13-1 The explosive growth of a particular kind of cyanobacteria in this lake, *left*, has caused what is known as a bloom. Cyanobacteria, *above*, were the first cells to use chlorophyll to capture light. Chlorophyll is the same pigment that green plants use today for photosynthesis.

SECTION 13-1

Vocabulary Preview

cyanobacteria

eubacteria

archaebacteria

protist

mass extinction

Opening Question

Scientists once thought that life mysteriously and suddenly appeared in the Cambrian period. Why did they miss the traces of the earliest life-forms fossilized in Precambrian rocks? (*The fossils were microscopic, and Precambrian fossil-bearing rocks are not common.*)

Demonstration

Visible Fossils

Make a model of rock layers with visible embedded fossils by mixing gelatin or agar, adding different colors to each layer, and putting "fossils" in each layer as you pour the solution into a large clear container. Good "fossils" include candy, coins, paper clips, or small plastic toys. Be sure to let one layer completely solidify before you add the next layer, and don't add liquid that is too hot. Choose "fossils" that look similar to each other but differ in complexity. Put simple ones in the lower layers, progressing to more complex forms toward the surface. If you cover and refrigerate the model, it should last for weeks.

You can use the model to see if students understand dating and the relationships of fossils. Can your students tell which fossils are older than the others? Do your students trace lines of descent by similar anatomy rather than location in the same column of rock? A good activity is to have your students create a phylogenetic tree for the "organisms" you placed in your model.

Effective Reading

Condensing Information

Have students write the subheadings for each section and summarize paragraphs under each subheading. For instance, under the subheading "Bacteria Were the First Cells to Evolve," students could write: (Para-graph 1) The first cells were prokaryotes. (Paragraph 2) Cyanobacteria are photosynthetic and give off oxygen. (Paragraph 3) Some early prokaryotes may have become mitochondria, and others chloroplasts.

CONNECTIONS

Chapter 11

The Origin of Cells

The precise manner by which the first cells appeared on Earth remains a mystery. But most scientists today agree that the first cells formed from organic compounds (proteins, lipids, carbohydrates, and nucleic acids) that were present in ancient seas.

Teaching Tip

Remembering What Came From What

The lines of descent of present-day bacteria and eukaryotes from early cells are difficult for students to understand and remember. Use the *Graphic Organizer* shown at the bottom of the page to help students with this. Put the diagram (minus all but one or two terms) on the chalkboard or overhead projector. Then list the remaining terms next to the diagram (mix them up first). Ask students to insert the terms into the correct rectangles. Then ask students to write a summary paragraph that explains the meaning of the graphic organizer. (*The first cells were prokaryotes that included two types of bacteria, archaebacteria and eubacteria. Archaebacteria are thought to have given rise to the first eukaryotic cells. Some eubacteria, called cyanobacteria, were photosynthetic, and it is thought that certain early cyanobacteria gave rise to the chloroplasts found inside some eukaryotic cells. Other eubacteria may have given rise to the mitochondria inside eukaryotic cells. Archaebacteria, cyanobacteria, other kinds of eubacteria, and eukaryotes all survive today.*)

Figure 13-2 Archaebacteria are rare today. They are found mainly in unfavorable environments where conditions resemble those of early Earth, such as in this hot spring in Yellowstone National Park. The densely growing bacteria stain the runoff channels orange.

CONTENT LINK

More information about the evolution of mitochondria and chloroplasts can be found in Chapter 20.

Figure 13-3 Some of the major events that have occurred during the evolution of life on Earth are shown in this continuous time line.

because oxygen gas was rare on Earth until then. After hundreds of millions of years, when the waters of the ancient oceans had become saturated with oxygen, the oxygen produced by cyanobacteria began to escape into the air. Over the billions of years that followed, more and more oxygen was added to the air. Today, oxygen gas is 21 percent of the Earth's atmosphere.

Research has shown that some of the early cyanobacteria may be ancestors of chloroplasts and that other kinds of early bacteria may have given rise to mitochondria. Remember that mitochondria and chloroplasts are organelles found only in eukaryotic cells. Scientists propose that bacteria may have first entered larger cells as parasites or undigested prey and then gradually evolved into the mitochondria and chloroplasts that characterize eukaryotic cells today.

Two Kingdoms of Bacteria

Fossils and other evidence indicate that ancient cyanobacteria are among the direct ancestors of the most common group of bacteria that exist today, the **eubacteria** *(yoo bak TIHR ee ah)*. Eubacteria include those bacteria that cause disease and decay. Other ancient bacteria were very different from eubacteria and formed a separate branch early during the evolution of the first cells. These bacteria are more closely related to a second group, or kingdom, of modern bacteria, the **archaebacteria** *(AHR kee bak TIHR ee ah)*, shown in Figure 13-2. The cell walls and membranes of archaebacteria are quite different from those of the eubacteria, and the way the archaebacteria produce protein from

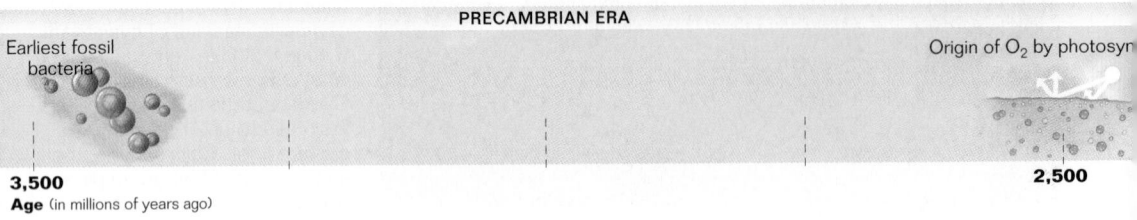

PRECAMBRIAN ERA

Earliest fossil bacteria

Origin of O₂ by photosyn

3,500
Age (in millions of years ago)

2,500

Graphic Organizer

Use this graphic organizer in *Teaching Tip*: **Remembering What Came From What** on page 274.

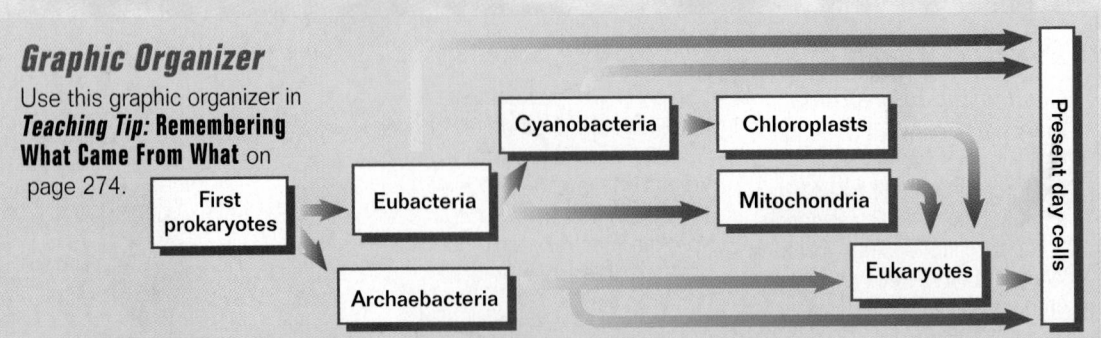

First prokaryotes → Eubacteria → Cyanobacteria → Chloroplasts; Eubacteria → Mitochondria; First prokaryotes → Archaebacteria → Eukaryotes; Present day cells

their DNA is also different. Since archaebacteria evolved before there was oxygen in the atmosphere, they had to develop ways of producing energy without using oxygen. Some modern archaebacteria get their energy not by photosynthesis but by combining hydrogen, H_2, with carbon dioxide, CO_2, to form methane, CH_4. Other kinds of archaebacteria produce energy using other chemical pathways. Chemical evidence suggests that the first eukaryotic cells, as shown in Figure 13-3, were more likely to have evolved from archaebacteria than eubacteria. Indeed, archaebacteria appear to be the ancestors of all eukaryotic cells, including yours! □

□ **CAPSULE SUMMARY**

The earliest fossil cells closely resemble modern-day bacteria. Primitive cyanobacteria produced the first oxygen in Earth's atmosphere and are believed to be the ancestors of chloroplasts. Archaebacteria probably gave rise to the first eukaryotic cells.

Dawn of the Eukaryotes

For over a billion years bacteria were the only living things on Earth. Then, starting about 1.5 billion years ago, a new kind of organism evolved from the archaebacteria—the **protist.** Protists were the first eukaryotes. They have DNA that is enclosed within a nucleus and complex systems of internal membranes.

Most biologists group all living things into six great categories called kingdoms, as shown in Figure 13-4. The two

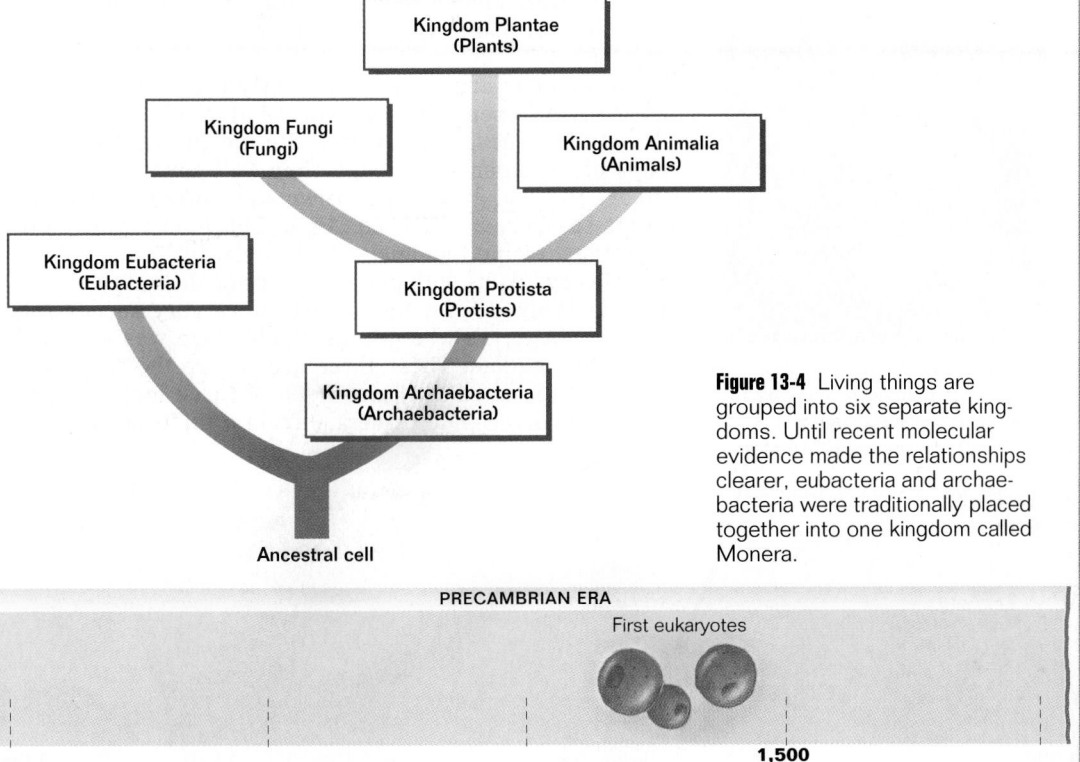

Figure 13-4 Living things are grouped into six separate kingdoms. Until recent molecular evidence made the relationships clearer, eubacteria and archaebacteria were traditionally placed together into one kingdom called Monera.

PRECAMBRIAN ERA

First eukaryotes

1,500

Teaching Tips

Use a Chart to Organize Information

To help students classify and distinguish the different kinds of life that are described in this chapter, have them begin a chart of organism characteristics that they will fill out. The headings for the chart may include **Eubacteria, Archaebacteria, Protists,** and **Multicellular organisms.** Students could draw diagrams of an example of each type of organism, locate its place on the phylogenetic tree (figure 13-4), and describe unique features of each.

Go to the Root

Many of the terms used in this chapter are derived from roots and prefixes that are almost self-explanatory. To help students learn the vocabulary in this chapter, have them make and decipher words based on the following roots and prefixes, and their meanings: cyano = blue, karyote = body, eu = true, pro = before, archae = ancient. An example of making a word is eu + karyote = eukaryote. Use terms in the Vocabulary Previews for additional items.

👁 *VISUAL STRATEGY*

Figure 13-4

Have your class reproduce the phylogenetic tree on the wall of your classroom using colored paper, pictures, and descriptions of the six kingdoms of living things. Pictures of representative organisms can be drawn, traced, or cut from magazines.

Overcoming Misconceptions

The Changing Atmosphere

Some students may think that the atmosphere is an infinite resource that can never "go away." Challenge this misconception by having students relate the changes in the atmosphere of early Earth to the changes that are happening now to our atmosphere. Begin by describing how the oxygen levels steadily rose in the ancient atmosphere due to the buildup of waste products of bacteria, then ask what kind of waste products we are producing today? *(carbon dioxide, methane, sulfur dioxide, nitrous oxide, and others)* Which gases are decreasing in our atmosphere, and which gases are increasing? *(Carbon dioxide is increasing and ozone is decreasing.)*

Figure 13-5 Protists, a very diverse group, are much larger than bacteria. Some protists, like these *Euglena*, have both plant and animal characteristics. What are these characteristics?

oldest kingdoms, the eubacteria and archaebacteria, are single-celled prokaryotes. The other four kingdoms (protists, fungi, plants, and animals) evolved later, and all are eukaryotes. Of these four kingdoms, the one that contains creatures of the greatest diversity is the kingdom Protista. Many fossil protists have elaborate shapes, and some display spines or highly branched filaments that extend out like antlers. Most protists, such as those shown in Figure 13-5, are single-celled, but there are some multicellular forms. Some protists are photosynthetic, while others hunt bacteria or other protists for food.

Multicellularity Evolves Many Times in Protists

The development of multicellular organisms marks a great step in the evolution of life on Earth. It allowed "division of labor" among cells and allowed cell specialization that led to organism complexity. Almost every creature large enough to see with the naked eye is multicellular. The first known fossils of multicellular organisms were found in 630-million-year-old rocks from southern Australia.

Multicellular organisms apparently evolved independently many different times among the protists. Some of the multicellular lines that resulted did not produce diverse groups of organisms, although they still survive. Among them are the red, green, and brown algae often seen swept onto the seashore as seaweed. Algae, shown in Figure 13-6, are protists, (some biologists used to classify these photosynthetic organisms with plants). Three of the multicellular lines that evolved from the protists were very successful, producing large, diverse groups. Each group was assigned a separate kingdom. These three groups are the fungi, the plants, and the animals. Each of these three multicellular kingdoms evolved independently from a different kind of protist.

Figure 13-6 Brown algae, called kelps, are multicellular protists that form vast underwater "forests" in some coastal waters.

PRECAMBRIAN ERA

Early eukaryotes

Diverse protists

1,500
Age (in millions of years ago)

1,000

  Did You Know?

Sponges are considered to be the simplest animals. Although they are multicellular, their cells are not organized into tissues. Tissues are found in every other kind of animal.

Today's Organisms Have Their Origins in the Cambrian Period

All major groups of organisms that survive today, except plants, originated sometime during the first hundred million years of multicellular life. This time period, called the Cambrian period, lasted from just less than 600 million years ago to about 500 million years ago. The Cambrian period was a time of great evolutionary experimentation. Many very unusual animals appeared at this time, animals for which there are no close living relatives. A particularly rich collection of Cambrian fossils, the Burgess Shale, has been uncovered on a rocky mountainside in Canada. The remains of bizarre creatures in the Burgess Shale, as shown in Figure 13-7, are unlike anything alive today. There appear to be more kinds of extinct animal phyla in the Burgess Shale than there are

Figure 13-7 By studying fossils from the Burgess Shale, artists were able to re-create a scene from the shallow seas of the Cambrian period. Animals of this environment included those with odd structural adaptations such as torpedo-shaped bodies, tripod-like tails, and elongated grasping organs.

Period of multicellular experimentation

Origin of all major animal phyla

CAMBRIAN PERIOD

ORDOVICIAN PERIOD

500

👁 VISUAL STRATEGY

Figure 13-7
Have students study the artist's rendition of life-forms in the Cambrian sea. Then have them choose an animal in the illustration and explain why they think that animal became extinct. Ideas might include a weather change, new predators, geological disasters, or disease.

Teaching Tips
Division of Labor

Have students relate division of labor to their own cell types. Have them name some cell types in a human while you list them on the board. As students name each cell type, have them explain what that cell can do that other cells cannot. For instance, a heart muscle cell contracts rhythmically. Ask students the following questions. What would happen if skin cells started growing in the heart or kidney cells started growing in the brain? (*They would interfere in the function of the organ.*) What would happen if a group of cells, such as lung cells, all died? (*The person would die.*)

Using the Timeline as a Way to Review

After each Chapter section, have students use the timeline to review major events discussed in that section. A table with two columns can be constructed. Students should describe the event in one column and record the approximate date of occurrence in the other column.

Historical Note

A Wonderful Life

Steven Jay Gould, professor of geology at Harvard University, wrote a book called *A Wonderful Life* (published in 1989), which is about the scientists who discovered the Burgess Shale and its fossils. Some of the creatures that produced the fossils are so different from the life on Earth today that it is difficult to interpret what kind of organism each one was. Gould's book describes what some of these organisms might have actually looked like. Although some of the descriptions appear to be accurate, others were refuted by scientists and were proven wrong.

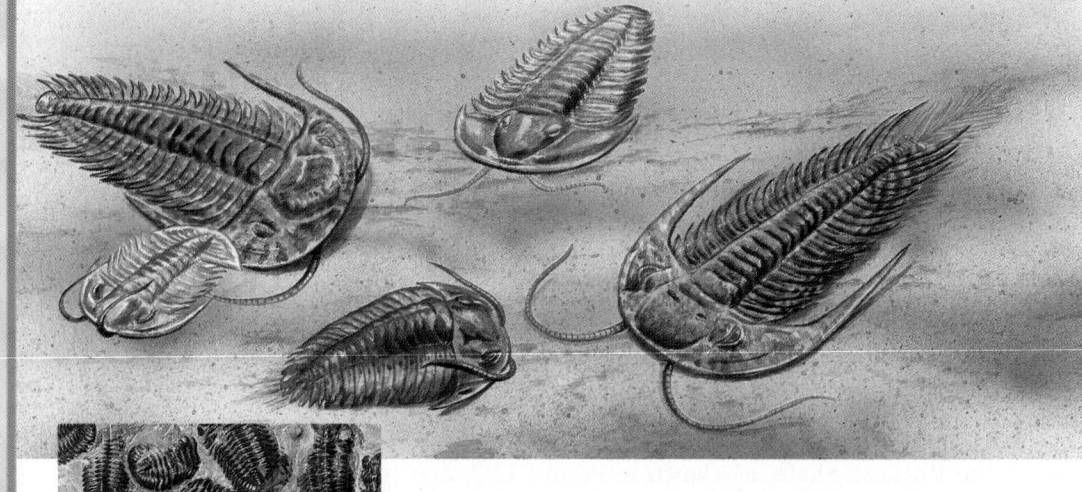

Figure 13-8 The fossil record contains nearly 4,000 trilobite species. These marine arthropods were extremely common 500–600 million years ago. By about 250 million years ago, they were all extinct.

kinds of living animal phyla today! Life was more diverse in the Cambrian seas than it has ever been since. In the period following the Cambrian, the Ordovician period, diverse animals continued to abound in the seas. Among them were the trilobites shown in Figure 13-8.

Mass Extinctions Have a Major Impact

The end of the Ordovician period is marked by a drastic change in the fossil record. A large portion of the organisms on Earth suddenly became extinct about 440 million years ago. This was the first of five major **mass extinctions** that have occurred during the history of life on Earth.

Another mass extinction of similar magnitude happened abruptly about 360 million years ago. Then, about a hundred million years later, the third and greatest of all mass extinctions literally devastated our planet. It happened at the end of what is called the Permian period, some 250 million years ago. About 96 percent of all species of animals living at the time became extinct. Approximately 35 million years later, a fourth less devastating mass extinction occurred. Although the specific causes for this and the other extinction events are not clear, evidence suggests that massive geological or climatic changes were likely contributing factors.

The last mass extinction event will be discussed in more detail in Chapter 33. It occurred 65 million years ago and

Animal diversity abounds, early vertebrates

ORDOVICIAN PERIOD

Jawless fishes

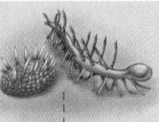

500

Age (in millions of years ago)

450

Did You Know?

The greatest number of fossils date from the Cambrian period, which lasted about 100 million years. This was when organisms first began developing skeletons and other hard body parts, such as shells. The hard tissues and shells easily fossilized in the clay on the ocean bottom.

Figure 13-9 Although tropical rain forests cover only 7 percent of the Earth's land surface, they contain more than one-half of all the world's animal and plant species. But rain forests are being destroyed at an alarming rate. This rain forest in Brazil is being burned for farmland.

brought about the extinction of about two-thirds of all land species, including the dinosaurs.

There is good reason to believe that another mass extinction may be occurring on Earth today. In contrast to previous mass extinctions, this extinction event has a known cause. It is occurring because the Earth's ecosystems, especially tropical rain forests, are being destroyed by human activity, as shown in Figure 13-9. ❑

❑ CAPSULE SUMMARY

There have been five major mass extinction events on Earth. Human activity is contributing to what may become a sixth mass extinction event. As humans destroy tropical rain forests, species are being eliminated at an alarming rate.

Section Review

1. *What is the age of the Earth's oldest fossil organisms, and what present-day organisms do they most closely resemble?*
2. *Explain why this statement is true: A eukaryotic cell is descended from both eubacteria and archaebacteria.*
3. *How is a protist different from a bacterium?*
4. *Define multicellularity, and identify the kingdom in which it first occurred.*

Critical Thinking

5. *Predict how human evolution would have been affected if mass extinctions had never occurred.*

Teaching Tips
Who Owns Earth?

Human-caused mass extinctions have been in progress throughout history. The auroch, Irish moose, moa, and dodo bird are just a few of the species now extinct due in part to human activities. Some feel that humans can use this planet for anything they wish, while others feel humans should protect the planet and its life. Have students discuss their views in small groups, then ask them to show their personal position by marking an *X* on a line that they draw on their paper. Have the students write "Humans can use Earth anyway they want" at one end of the line, and "Animals and plants have the same rights that humans do" at the other end of the line. Next to the *X*, have each student write a statement that summarizes his or her position.

What if we keep burning the rain forests? Besides causing the extinction of an enormous number of species, burning rain forests may also cause an increase in global temperature. Twenty percent of the carbon dioxide in the atmosphere comes directly from the burning of rain forests, leading to the greenhouse effect.

First mass extinction	SILURIAN PERIOD	DEVONIAN PERIOD
	Plants, arthropods, and fungi invade land; jawed fishes appear	

400

Answers to Section Review

1. They are 3.5 billion years old, and resemble bacteria.

2. Chemical evidence suggests that eukaryotic cells descended from cells like archaebacteria. Mitochondria and chloroplasts descended from eubacteria.

3. Protists are larger, have a complex system of internal membranes, and their chromosomal DNA is enclosed in a nucleus. Bacteria are smaller and contain no internal compartments.

4. Multicellularity describes organisms made of many cells that have specialized functions. Multicellularity first evolved in protists.

5. Mass extinctions created opportunities for new groups of animals to evolve. Without mass extinctions, the evolutionary history of life on Earth would have been much different. Since we humans are a part of this evolutionary history, our evolution probably would have been much different had there not been mass extinctions. In fact, it is possible that humans never would have had the opportunity to evolve at all.

Vocabulary Preview

ozone

mycorrhizae

symbiosis

arthropod

vertebrate

Opening Question

Before life invaded land, there were only rocks in terrestrial environments. What kind of organism can live on bare rocks? (*Students may mention lichens, which are a fungus and a cyanobacteria that live in a symbiotic relationship.*)

Demonstration

Personal Experience With UV Radiation

Display several types of sun-blocking agents, sunscreens, and suntan lotions. Ask students what UV radiation does to them personally. (*It can make them sunburned.*) Have they ever known anyone hurt by sunlight? (*They may know someone who has been sunburned or stricken with skin cancer.*) What would happen to them if they had to spend their entire lives outside, as other organisms do? (*They would be more likely to feel the effects of the sun—sunburn, skin cancer, dehydration.*) Explain to students how sun-blocking agents, sunscreens, and suntan lotions differ in their ability to protect them from the sun's ultraviolet radiation. (*Sun-blocking agents are opaque and reflect ultraviolet rays, and offer the greatest protection. Sunscreens absorb ultraviolet radiation and are rated according to the degree of protection that they give. Suntan lotions offer essentially no protection from the sun— some that contain oils even promote sunburn.*)

13-2 Four Invasions of the Land

Life *has existed on Earth for at least 3.5 billion years, and for 90 percent of that time living things remained in water. Only recently, in geologic terms, has life emerged from the seas to live on land.*

Section Objectives

- Describe the significance of the formation of ozone in Earth's primitive atmosphere.
- Identify the first multicellular organisms to populate land.
- Identify the first animals to live on land.
- Identify the first vertebrates to live on land.

The Importance of Ozone

During the Cambrian period and for millions of years afterward, while the seas teemed with life, there was no living thing on the dry, rocky surface of the land. Exposed to the sun's harsh rays, nothing could live there. During all this time, however, a subtle change was occurring. Photosynthesis by cyanobacteria was adding oxygen gas to Earth's atmosphere. As significant amounts of oxygen began to reach the upper atmosphere, the sun's rays caused the atoms of oxygen gas, O_2, to bond and form a new compound. This reaction of oxygen atoms resulted in the formation of a gas called **ozone,** O_3. In the upper atmosphere, the ozone acted like a great shield to block the harsh ultraviolet radiation of the sun from reaching the surface. For the first time, Earth's surface was a safe place to live.

Plants and Fungi Cooperate to Invade Land

The first multicellular creatures to populate the land were plants and fungi. Cooperating with each other, they solved a particularly difficult challenge—how to survive on bare rock. Each brought to the task a unique talent.

Plants, which evolved from photosynthetic protists, could carry out photosynthesis. Recall from Chapter 5 that in photosynthesis plants use the energy from sunlight to make their own food. They could not, however, extract needed minerals from bare rock. Fungi, which evidently evolved from a hunting kind of protist, could not produce food from sunlight but were able to absorb minerals, even from bare rock.

DEVONIAN PERIOD

Bony fishes become abundant

Second mass extinction

Early amphibians

400
Age (in millions of years ago)

350

Did You Know?

In addition to the ozone layer, there is a magnetic field around Earth that also protects us from harmful radiation. High-energy particles and rays are caught in the Van Allen belt that surrounds our planet and deflects the particles from our planet's surface. There is geological evidence that this magnetic field is disrupted occasionally, which results in very dangerous particles and rays hitting unprotected terrestrial organisms.

The solution to the challenge of living on bare rock was unique biological partnerships called mycorrhizae (MY koh REYE zee). **Mycorrhizae** are associations between fungi and the roots of plants, as shown in Figure 13-10. In mycorrhizae one creature lives inside the other, and each helps the other. The fungi provide minerals to the plant, and the plant provides the food to the fungi. This kind of "you-help-me-and-I-help-you" partnership is called **symbiosis** (sihm beye OH sihs). Both plants and fungi invaded the surface of the land at the same time, approximately 430 million years ago.

Figure 13-10 Mycorrhizae, the partnership between plants and fungi, have been very successful. Indeed, 80 percent of all living plants have mycorrhizae occurring within their roots. Fungi account for 15 percent of the total weight of the world's plant roots!

Arthropods Crawl out of the Sea

Within 100 million years of their initial invasion, plants covered the surface of the Earth and formed extensive forests. The landscape must have been eerily quiet when plants first invaded land, for no animals had yet left the sea. But the picture soon changed. Two groups of animals had been particularly successful in the oceans, and both groups were soon to populate the land. The first animals to venture forth were the **arthropods,** a kind of animal with a hard outer skeleton and jointed appendages. Crabs are arthropods, as are lobsters, insects, and spiders. As near as biologists are able to determine, the first arthropods to live on land were scorpions, like the one shown in Figure 13-11.

Figure 13-11 This fossil scorpion, a carnivorous relative of the spider, has two great pincers in front and a venomous stinger at the end of its tail.

Teaching Tips
Big Bodies Need Big Support
Help students relate what they know about sea life and terrestrial life to the challenges the first animals must have faced as they left the ocean to live on land. Fossils show that sea scorpions reached lengths of almost 10 ft. (3.1 m) long. Why aren't terrestrial scorpions that large? (They don't have the buoyancy of water to support a large body.)

Different Strokes for Different Folks
Have pairs of students find or sketch pictures of the structures that animals use for locomotion in the ocean. Pictures should include jointed legs of arthropods, fins of vertebrates, the foot and mantle of mollusks, or tube feet of echinoderms. Have the students list the pros and cons of the structures for movement on land. Finish by having the students discuss their conclusions. (It should be obvious from the discussion that the jointed legs of the arthropods were the best for terrestrial movement.)

VISUAL STRATEGY

Figure 13-11
Ask the students to identify some of the characteristics that may have enabled scorpions to be the first animals to successfully invade land. (Their jointed appendages made it easy for them to walk. Their strong outer skeleton (exoskeleton) protected them and supported their body weight. Their venomous sting and pincers made them successful predators.)

CARBONIFEROUS PERIOD

Amphibians dominate the land

Early reptiles

300

 Did You Know?

The common pill bug, or roly-poly, in the garden is a crustacean and thus has gills, not lungs. It must remain in damp leaf litter to keep its gills moist. The gills are easy to see under a low-power microscope or hand lens. Some terrestrial animals, such as earthworms, respire through their skin. Why do lungs allow animals to inhabit more varied habitats than do the other forms of respiration? (Lungs work in dry air, while the other methods require moist habitats.)

VISUAL STRATEGY

Figure 13-12

Have students use the timeline to name the period during which these swamp forests grew. *(Carboniferous period)* Tell students that the word *carboniferous* is defined as "producing or containing carbon or coal." Tell them that much of the coal that is mined today was formed from the organic matter (mostly wood) that composed these ancient forests.

Teaching Tip
Efficient Fertilization

Have students compare wind-pollinated plants such as pine trees and grasses with insect-pollinated plants. Ask how the insects and the plants they pollinate benefit. *(Pollination is more efficient for the plant, and the insect has a rich food supply.)*

Figure 13-12 Swamp forests 320 million years ago were dominated by tall, seedless canopy trees and shorter tree ferns. Dragonflies with wingspans of more than 1 m (3 ft.) hovered over these ancient swamps.

Figure 13-13 This blister beetle, an insect, is eating pollen produced by a flowering plant in India. Insects are important to plants because they help transfer pollen from one plant to another.

The arthropod invasion of the land is one of nature's great success stories. From the initial land invaders soon evolved a unique kind of terrestrial arthropod—the insect. Insects have become the most abundant and diverse group of animals ever. Today there are more than 200 million insects alive for each person on Earth! The special characteristic of insects that opened up the world to their invasion was the ability to fly. Insects were the first animals to evolve wings. Early forms, like the dragonfly shown in Figure 13-12, had two pairs of wings. Some insects that evolved later, like flies, had one pair.

Flying enabled individual insects to patrol the landscape in search of food, mates, and nesting sites. It also led directly to the great partnership between insects and flowering plants, as is shown in Figure 13-13. The oldest fossils of flowering plants are from about 127 million years ago, but the group may be much older than that. The association between flowering plants and insects has led to the current dominance of both these groups.

CARBONIFEROUS PERIOD	PERMIAN PERIOD		TRIASSIC PERIOD	
	Pelycosaurs dominate the land	Third mass extinction	Therapsids and thecodonts	Fourth mass extinction
300				The first dinosaurs and mammals

Age (in millions of years ago)   **200**

 Did You Know?

Most flowering plants are pollinated by bees. There are about 20,000 species of bees, 90 percent of which are solitary, meaning that they don't live in colonies with other bees.

Vertebrates Follow Onto Land

While plants and arthropods were dominant on land, vertebrates were widespread in the sea. **Vertebrates** are animals with backbones. An example is shown in Figure 13-14. The first vertebrates were fishes that evolved in the oceans 550 million years ago. Fishes soon came to dominate the seas, and for hundreds of millions of years that is where vertebrates stayed. The first vertebrates to inhabit the land did not venture out of the sea until 350 million years ago. Those first land vertebrates were amphibians, ancestors of today's frogs, toads, and salamanders. Because of their strong, flexible internal skeleton, vertebrates can be far larger than insects, and they soon dominated the landscape. □

□ CAPSULE SUMMARY

Ancient cyanobacteria produced the oxygen that was converted into ozone, which forms Earth's protective ozone layer. Once the Earth's surface was safe for habitation, fungi, plants, arthropods, and vertebrates were able to live on land.

Figure 13-14 This fish skeleton clearly shows the backbone, the structure that is characteristic of all vertebrate animals.

Section Review

1. *What are mycorrhizae, and what role did they play in the evolution of life on land?*
2. *What role did arthropods play in the evolution of life on land?*
3. *When did vertebrates first invade land?*
4. *List in order, from first to most recent, the four groups of organisms to invade land.*

Critical Thinking

5. *Identify and explain the relationship between the evolution of eubacteria and the evolution of the first land organisms.*

☑ **RESEARCH Update**

Guess Who's Coming to Dinner?

Get out the extra table leaf. Demographers are mostly in agreement about the population exploding from a 1990 statistic of 5.3 billion people to 10 billion people in 2050. What will all these people eat? Nothing, suggests the pessimist camp, championed by Dr. Paul Erlich and Dr. Anne Erlich, both professors of biology at Stanford University: "Human numbers are on a collision course with massive famines. If humanity fails to act, nature will end the population explosion for us—in very unpleasant ways—well before 10 billion is reached." The optimist enclave includes authorities such as University of Maryland professor Julian Simon, and University of Manitoba professor Vaclav Smil, among others. They acknowledge the fact that there are obvious problems but assert that food production will easily provide for 10 billion individuals.

Most scientists think that in countries that are economically developed, there isn't going to be a problem, but in the developing nations, the struggle to feed the population is probably going to continue due to economic limitations. ☑

JURASSIC PERIOD

Appearance of angiosperms

CRETACEOUS PERIOD

100

Answers to Section Review

1. They are a symbiotic association of fungi and plants roots. Mycorrhizae enabled plants and fungi to be the first organisms to successfully invade land 430 million years ago.

2. Arthropods were the first animals to invade land. Insects evolved symbiotic relationships with flowering plants; this helped both succeed.

3. Vertebrates first inhabited land 350 million years ago.

4. Plants and fungi invaded land at the same time, followed by arthropods, then vertebrates.

5. Cyanobacteria, a type of eubacteria, were among the first bacteria to evolve. Oxygen, a waste product of cyanobacteria metabolism, led to the formation of ozone in the atmosphere. Ozone protected Earth from ultraviolet radiation, which in turn made the invasion of the land by other organisms possible.

Opening Question

Ask students why it would be advantageous for vertebrates to begin living on land. Prompt your class with descriptions of the competition in the sea and with descriptions of the life already on land. *(Land offered vertebrates vast, untapped resources, such as space and food.)*

Demonstration

Cartesian Divers and Swim Bladders

Borrow a Cartesian diver from a physical science teacher and show your class the effect of pressure differences on the buoyancy of the "diver." As an in-class exercise, ask students to write a paragraph explaining how the Cartesian diver works. Furnish them with some physical science books to use as reference materials. *(Squeezing the bottle compresses the gas inside the Cartesian diver, which results in a decrease in the volume of the diver. A smaller volume means the diver is less buoyant, and so it sinks. The opposite scenario occurs when the bottle is released.)* The same experiment can be done using a goldfish instead of the Cartesian diver. The goldfish will sink or rise when the bottle is squeezed or released, just as the Cartesian diver does. *Be sure to leave the goldfish in the bottle for only short periods of time to prevent suffocation.* Ask students to explain why the goldfish sinks and rises. *(The goldfish has a swim bladder that contains gas. Squeezing and releasing the bottle causes the gas to compress and expand.)*

13-3 Vertebrate Evolution

O f all animals, vertebrates are the most familiar to us, both because we are vertebrates and because almost all other land animals bigger than our fist are vertebrates, too.

Section Objectives

- Recognize the advantages that sharks had over jawless fishes in the ancient seas.
- Describe the function of a swim bladder.
- Explain why reptiles are more completely adapted to life on land than amphibians.
- Explain how the Permian mass extinction turned out to be an opportunity for reptiles.
- Identify the dominant vertebrates on land today.

Figure 13-15 Early jawless fishes appear to have fed in a head-down position, their fins acting as stabilizers while their small mouths vacuumed organic particles from the bottom.

Fishes Are the First Vertebrates

The first vertebrates, shown in Figure 13-15, were small jawless fishes. Although many species of jawless fishes filled the seas 450 million years ago, only two types, hagfishes and lampreys, survive today.

Jaws Evolve

Fishes with jaws first appeared around 430 million years ago and rapidly replaced jawless fishes in the seas of the world. Jaws allowed fishes to bite and chew their food instead of sucking it up. As a result, jawed fishes became efficient predators. The earliest kinds of jawed fishes had bulky, armored bodies, but they were soon replaced by jawed fishes with flexible and agile bodies. These new species consisted of sharks and bony fishes.

Sharks

From the heavily armored jawed fishes evolved a very efficient predator—the shark. Sharks lack the bulky armor plating of their ancestors. Their skeletons are made of cartilage

CRETACEOUS PERIOD

Fifth mass extinction

Birds and mammals spread

| 100 | 80 | 60 |

Age (in millions of years ago)

rather than bone, making them lighter and more buoyant. They also have large, strong mobile fins, which allow them to swim fast and to quickly adjust their direction through the water. A Caribbean reef shark is shown in Figure 13-16.

Bony Fishes

Many hundreds of kinds of sharks evolved. About 250 species survive. Sharks were largely replaced by bony fishes, particularly versatile newcomers. The bony fishes have become an extremely diverse group. A unique way to regulate depth in the water evolved among bony fishes, shown in Figure 13-17. A gas-filled sac called a swim bladder enables bony fishes to change their depth with little effort. By letting gas into or out of the sac, bony fishes can sink, rise, or remain motionless at any depth without having to rapidly beat their fins.

Figure 13-16 Sleek and fast, sharks could swim much more skillfully than their armored relatives.

Amphibians Venture Onto Land

When you look at a frog, it is difficult to appreciate the major evolutionary advances that took place as the first amphibians evolved from bony fishes. After 200 million years of success in the sea, fishes were uniquely adapted for success in water. Major changes had to occur in a fish's body design to evolve into an animal capable of living on land.

Structural Innovations

Amphibians were able to adapt to land because of the development of several structural innovations. Early amphibians had moist breathing sacs called lungs, which they used to absorb oxygen from air. Although amphibians could also absorb oxygen through their skin, lungs eventually proved to be far more efficient. The lungs of amphibians required associated changes in the circulatory system, including the development of a new blood vessel and a modified heart. The development of limbs was another structural adaptation for life on land. The limbs of amphibians are derived from the bones of fish fins, but imagine trying to walk on the tips of flippers! What made walking possible was the evolution of a strong support system of bones in the region just behind the head. This platform of bone provided a rigid base for the limbs to work against.

Figure 13-17 Many bony fish species, like these herring, swim in large schools. About half of all species of vertebrates on Earth today are bony fishes.

TERTIARY PERIOD

Major mammal groups evolve

Mammals dominate the land

40   20   10

Demonstration
Inside and Out
Buy a long, soft candy such as "gummy worms," and challenge small groups of students to create an internal skeleton for one worm and an external skeleton for the other using toothpicks and tape. A successful skeleton is one that holds the worm straight when the worm is held upright at its base. When the groups are done, have them share their constructions with the class. Ask which skeleton requires less material and weighs less. (internal skeleton)

Teaching Tips
Clamping Down With Jaws
The lampreys that survive today are parasites. Help students appreciate the importance of jaws by having them share with the class any vertebrate predators that they can name. What does each predator do to catch and eat its prey? In every case, from eagles to frogs to sharks, direct the discussion to the role of the predator's jaws in its successful lifestyle.

Watching Fish
It is important for students to watch real fish moving around in water, either in an aquarium or on film. Give the students some direction for their observations with questions such as the following: How do the fish use just their fins for movement rather than flexing their whole body? How are fish that live on the bottom or top of the aquarium different from those that usually swim freely in the middle? Do any of the fish use their fins like legs? (catfish do) How do fish rise or sink in the water with no apparent body movement at all? (swim bladder)

Teaching Tip

Reptiles Aren't Slimy

Some students think that reptiles such as snakes and lizards are slimy. Make available real fish, amphibians, and nonvenomous reptiles for students to touch, or have students describe their experiences touching these types of animals. Ask students why fishes and amphibians have slimy skin but reptiles have dry skin. (The slime protects the underlying skin and helps keep it moist when exposed to air. The skin and scales of reptiles are waterproof, so they don't need a coating of slime.)

Demonstration

Egg Watch

Crack open an egg in a dish (setup A), and leave an intact egg in another dish (setup B). Weigh the egg and dish of each setup and write down the two weights. Watch the exposed egg for two days and make observations of the changes in it. Did bugs, mold, or bacteria feed on it? At the end of the two days, crack open the egg protected by its shell and compare it with the exposed egg. Also reweigh both setups. Which one, A or B, weighs less? Explain why.

Figure 13-18 An amphibian, like this salamander, has a body that stays close to the ground. In contrast, the body of a reptile, such as a lizard, is held higher off the ground, enabling its legs to function very efficiently.

❑ CAPSULE SUMMARY

Amphibians evolved from primitive bony fishes. The evolution of lungs, a more efficient circulatory system, and legs in amphibians enables them to spend part of their life on land. Skin and eggs that easily dry out require amphibians to remain near water.

Amphibians, such as the salamander shown in Figure 13-18, seem primitive compared with the reptiles that eventually replaced them as the dominant vertebrates on land. But amphibians are in fact a very successful group. Having survived for 350 million years, the amphibians evolved long before the dinosaurs and have thus far outlasted them by 65 million years.

Amphibians are an imperfect solution to the challenge of living on land. They must return to water to reproduce, and they must live in moist places because their bodies are in constant danger of losing too much water through their skin by evaporation. A more complete solution to the challenges of terrestrial living is seen in the reptiles. Reptiles evolved from amphibians while amphibians were still dominant on the land, some 300 million years ago. ❑

Reptiles Break the Ties to Water

Besides improving upon the land-adapted features of amphibians, reptiles exhibit two key adaptations in vertebrate body design. Reptiles evolved a watertight skin, which meant that they would not dehydrate by losing moisture to the atmosphere. They also developed a watertight egg. Unlike amphibians, reptiles can lay their eggs on dry land because the eggs are surrounded by a shell that prevents water loss.

The next 50 million years after the reptiles evolved was a period of widespread drought. Reptiles, better adapted for dry times, had an apparent advantage over amphibians. Gradually, reptiles became the dominant group on Earth. Then came the great Permian mass extinction.

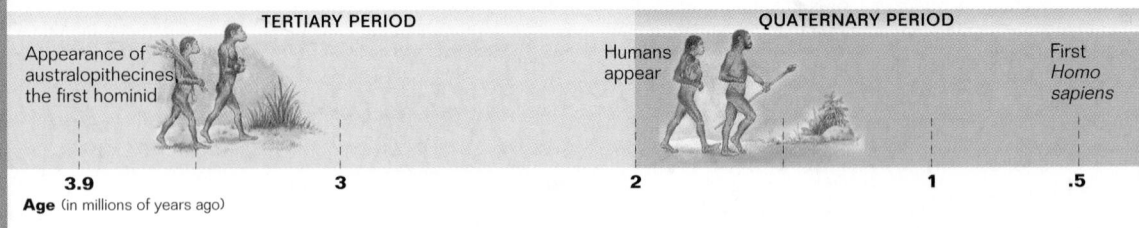

TERTIARY PERIOD

Appearance of australopithecines, the first hominid

QUATERNARY PERIOD

Humans appear

First *Homo sapiens*

3.9   3   2   1   .5

Age (in millions of years ago)

Overcoming Misconceptions

Remembering What Evolution Really Is

Some students view evolution as individuals changing into new forms, rather than populations changing over time periods longer than the lifetime of an individual organism. It may be necessary to review what students have already learned about how changes in DNA occur, and to relate those changes to adaptive traits present in a population.

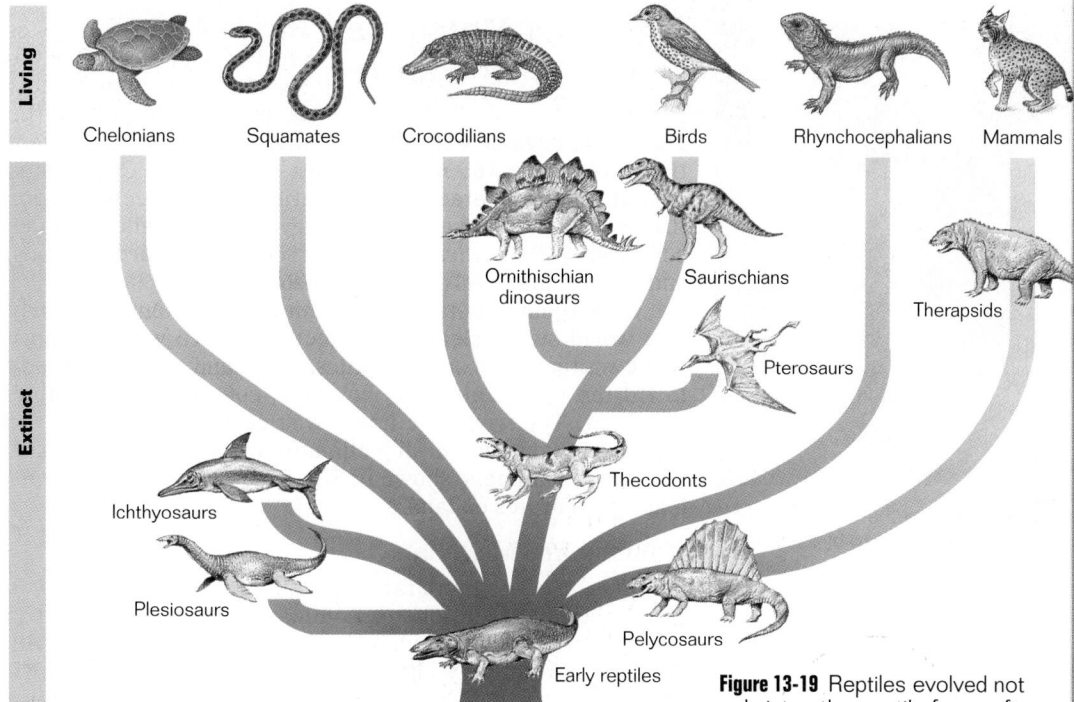

Living

Chelonians Squamates Crocodilians Birds Rhynchocephalians Mammals

Ornithischian dinosaurs

Saurischians

Therapsids

Pterosaurs

Extinct

Ichthyosaurs

Thecodonts

Plesiosaurs

Pelycosaurs

Early reptiles

Figure 13-19 Reptiles evolved not only into other reptile forms after the Permian extinction but also into two new vertebrate groups—the birds and mammals.

When the Permian extinction ended, over 96 percent of all species on Earth had disappeared, including many reptiles. But not all of them were gone.

Reptiles Branch Out

Following the Permian extinction, many new reptile groups evolved, as is shown by the evolutionary tree in Figure 13-19. One hundred million years after the Permian mass extinction, a vast assortment of reptile species dominated the land, water, and air. At about the same time, a geological event of massive proportion was just beginning. **Continental drift,** the movement of Earth's giant landmasses, has resulted in the present-day position of the continents. Understanding the movement of the continents provided scientists with an explanation for many formerly confusing geographical distributions. For example, there are a large number of marsupial (pouched) mammal species found in Australia and South America, continents that were formerly connected via Antarctica.

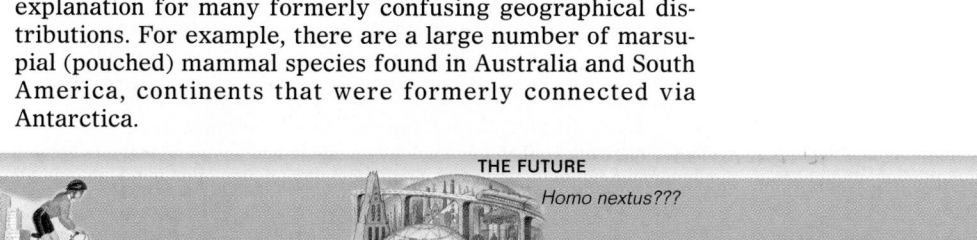

THE FUTURE

Homo nextus???

NOW

SECTION 13-3

👁 *VISUAL STRATEGY*

Figure 13-19
Place students in small groups and have them add the major groups of reptiles diagrammed in Figure 13-19 to the phylogenetic tree they started in the earlier part of this chapter (Figure 13-4). Be sure they add their reptile groups to the animal kingdom branch, rather than starting another branch.

Teaching Tip
Who Lives Where?
Divide the class into groups of two to three students and assign each group one of Earth's ecosystems described in Chapter 16. Each group should then use their textbook, library references, and/or personal knowledge to list at least 10 vertebrates that could be found in their ecosystem. Compare each group's lists for reptiles and mammals. Can reptiles live in climates where mammals cannot? *(no)* Can mammals live in places where reptiles cannot? *(yes)* Where? *(very cold climates)*

Did You Know?

Antarctica used to be home to more than just penguins and seals. There are fossils of tropical plants and animals in Antarctica, indicating that its climate was once much warmer than it is today. As the continents began to drift apart during the age of the reptiles, Antarctica started drifting toward the South Pole.

Chapter Closure

Divide the class into three groups, and assign one section from the chapter to each group. Ask each group to write at least 10 true/false questions that pertain to their assigned section. The members of one group can then take turns asking their questions to the other groups. Continue until all three groups have asked their questions. A tally can be kept to see which group finishes with the greatest number of correct answers.

Figure 13-20 Feathers and hair are structures unique to birds and mammals, respectively. These two vertebrate groups have successfully adapted to most of Earth's environments.

❏ CAPSULE SUMMARY

Birds and mammals are the dominant vertebrate groups on land today.

Birds and Mammals Dominate the Earth

Many evolutionary lines of reptiles appeared after the Permian extinction. Almost all of them, including the dinosaurs, became extinct in another mass extinction event that took place 185 million years later in the Cretaceous period. In the Cretaceous extinction, most species larger than a small dog disappeared forever. The smaller reptiles survived. We see them today as crocodiles, lizards, turtles, and snakes. Mammals and birds also survived.

Once again, the stage was set for a great evolutionary play in a world swept clean by extinction. There was an empty world waiting, with small reptiles, birds, and mammals ready to fill the void. This time, however, conditions were much different. The world's climate was no longer dry, and the reptiles' advantages in dry climates were not so important. Birds and mammals, shown in Figure 13-20, became the dominant vertebrates on land.

Humans Are Mammals

Humans are an important part of the mammal success story, the latest editions of dominant mammal forms. It is true that we humans have had many great accomplishments during our relatively short history, from creating magnificent art forms to discovering antibiotics. But it is also clear that our contributions have not always been positive ones. We have polluted the waters where fish still swim little changed from when vertebrates first ventured onto land. Humans are the first species to do major harm to the Earth, and it seems that the fate of our ancient home may ultimately lie in our hands. ❏

Section Review

1. *Identify the first vertebrates, and describe their lifestyle.*
2. *Explain why sharks have been a more successful group than lampreys.*
3. *Identify the fish group that has a swim bladder, and describe the function of this structure.*
4. *If a snake and a frog both lived in the same drought-stricken area, which animal would be more likely to survive? Why?*
5. *What effects did the Permian and Cretaceous mass extinctions have on the evolution of vertebrates?*

Critical Thinking
6. *Predict the impact humans will have on the Earth in the future.*

Answers to Section Review

1. The first vertebrates were jawless fishes. Most fed by vacuuming food particles from the ocean floor.

2. Lampreys are jawless fishes that lack paired fins. Sharks have jaws and have fins that help them maneuver in water, enabling them to be efficient predators.

3. Bony fishes have a swim bladder that allows them to rise or sink without having to use their fins.

4. The snake, because it has a watertight skin and lays shelled eggs that don't have to be laid in water for hatching.

5. The Permian extinction made available new space and food sources to reptiles, giving them the opportunity to diversify into many new forms. The Cretaceous extinction killed most of the reptiles, clearing the way for the diversification of mammals.

6. Answers will vary. Some may include descriptions of the results of ozone and rain forest destruction.

CHAPTER REVIEW ANSWERS

Vocabulary

archaebacteria (274) eubacteria (274) protist (275)
arthropod (281) mass extinction (278) symbiosis (281)
continental drift (287) mycorrhizae (281) vertebrate (283)
cyanobacteria (273) ozone (280)

Concept Mapping

Construct a concept map that shows the different types of life on Earth and the paths by which they evolved. Use as many terms as needed from the vocabulary list. Try to include the following items in your map: prokaryotes, eukaryotes, eubacteria, archaebacteria, protists, plants, fungi, animals, invertebrates, vertebrates, mitochondria, and chloroplasts. Include additional terms in your map as needed.

Review

Multiple Choice

1. It has been determined that the age of the Earth's oldest fossil organism is
   a. 4.5 billion years.
   b. 3.5 billion years.
   c. 2.0 billion years.
   d. 1.5 billion years.

2. Protists differ from bacteria in that protists
   a. contain DNA.
   b. are photosynthetic.
   c. have a nucleus.
   d. are the ancestors of mitochondria and chlororplasts.

3. Which of the following was not a characteristic of the first prokaryotic cells?
   a. The nucleus was absent.
   b. The flagella were absent.
   c. Special functions occurred in internal compartments.
   d. They were unicellular.

4. The formation of ozone made it less dangerous for organisms to invade the land because ozone
   a. blocks visible light.
   b. blocks ultraviolet radiation.
   c. blocks water loss from the land.
   d. encourages photosynthesis in cyanobacteria.

5. Which of the following is most likely to have been a problem for plants when invading the land?
   a. availability of light
   b. sources of water
   c. accessible minerals
   d. gravity

6. Which list reflects the correct order, from first to most recent, of the four groups of organisms to invade the land?
   a. fungi and plants, arthropods, amphibians
   b. amphibians, insects, fungi, plants
   c. plants, fungi, vertebrates and amphibians
   d. archaebacteria, fungi, arthropods, amphibians

7. Sharks were better adapted than jawless fishes to survive in the ancient seas because sharks
   a. had skeletons made of bone.
   b. were armor plated and had fins.
   c. used swim bladders.
   d. had jaws and agile, flexible bodies.

8. The gas-filled structure that allows bony fishes to remain motionless at any depth in the water is called a
   a. lung.   c. urinary bladder.
   b. swim bladder.   d. fin.

CHAPTER REVIEW ANSWERS

Each item in the Chapter Review is correlated by text section in the assignment guide that follows.

ASSIGNMENT GUIDE

Section	Review	Themes Review	Critical Thinking
13-1	1–3 12, 13 21–24		30, 32
13-2	4–6 14, 15 21 25–27		
13-3	9–11 16–20, 22	28, 29	31

Concept Mapping
Map answer is shown on page 271B.

Review
Multiple Choice
Code in parentheses indicates section number and objective number.
1. b (13-1.1)
2. c (13-1.3, 13-1.4)
3. c (13-1.2)
4. b (13-2.1)
5. c (13-2.2)
6. a (13-2.2, 13-2.3, 13-2.4)
7. d (13-3.1)
8. b (13-3.2)
9. c (13-3.3)
10. a (13-3.1, 13-3.3, 13-3.5)
11. d (13-3.4)

CHAPTER REVIEW ANSWERS *continued*

Completion

12. 5, 65 million (13-1.5)

13. eubacteria, archaebacteria (13-1.2)

14. mycorrhizae (13-2.2)

15. fungi, arthropods (13-2.2, 13-2.3)

16. jawless fishes, 550 (13-3.1)

17. sharks (13-3.1)

18. limbs, skin (13-3.3)

19. continental drift (13-3.4)

20. birds, mammals (13-3.5)

Short Answer

21. It came from cyanobacteria. Ozone had to form from the oxygen. (13-1.2, 13-2.1)

22. The destruction of the rain forests and other sensitive ecosystems must be stopped. (13-1.5, 13-3)

23. It is significant because it contains an extensive record of the diversity of organisms that lived during the Cambrian period. Almost all of the major groups of organisms that survive today, except plants, had their origin during the Cambrian period. (13-1.4)

24. Multicellularity allows for division of labor among cells and cell specialization leading to organism complexity. (13-1.5)

25. The partnership, called mycorrhizae, involves fungi living inside plant roots. The plant contributes food produced by photosynthesis to the fungus, and the fungus contributes minerals needed by the plant. (13-2.2)

26. It is a symbiotic partnership because both the flowering plants and bees benefit. (13.2-3)

27. Their strong, flexible internal skeleton enabled them to become larger than other animals. Their size allowed them to dominate other life-forms. (13-2.4)

9. What structural features of reptiles make them better suited for life on land than amphibians?
   a. lungs and walking legs
   b. the ability to absorb oxygen through their skin
   c. watertight skin and eggs
   d. an internal skeleton and sharp teeth

10. Which list reflects the correct sequence of evolution for groups of vertebrates?
   a. jawless fishes, bony fishes, amphibians, reptiles
   b. sharks, reptiles, amphibians, birds
   c. sharks, jawless fishes, reptiles, mammals
   d. bony fishes, jawless fishes, birds and mammals

11. Which of the following groups of vertebrates had a reptilian ancestor that existed prior to the Permian extinction?
   a. modern reptiles   **c.** mammals
   b. birds   **d.** all of the above

Completion

12. The fossil records suggest that _____ mass extinction events have occurred during the history of life on Earth. The most recent occurred about _____ years ago and brought about extinction of the dinosaurs.

13. One group of common bacteria that exists on Earth today, the _____ , are different in enough ways from another, less common group of bacteria, the _____ , that the two groups are considered to form two separate kingdoms of bacteria.

14. Most plants are able to absorb minerals from the soil because they have _____ in their roots.

15. The first multicellular organisms to live on land were _____ and plants. They were followed by animals—first by _____ and then by amphibians.

16. The first vertebrates, the _____ , evolved _____ million years ago.

17. Jawed fishes gave rise to _____ and bony fishes.

18. Structural innovations that enabled early amphibians to adapt to life on land included lungs and _____ . But their advances on land were hampered by _____ and eggs that easily dry out.

19. Understanding _____ has enabled scientists to explain the abundance of marsupial mammal species in Australia and South America.

20. Reptiles that survived the Permian extinction gave rise to many new species of reptiles as well as two new groups of vertebrates, the _____ and the _____ .

Short Answer

21. Oxygen has not always been a part of Earth's atmosphere. Where did Earth's first oxygen come from? Why was it necessary for oxygen to accumulate in Earth's atmosphere before life could inhabit land?

22. Scientists predict that a sixth mass extinction is imminent. What actions can be taken to prevent this mass extinction?

23. Why is the Burgess shale significant in terms of our understanding of the history of life on Earth?

24. Multicellularity first evolved in protists. What advantages do multicellular organisms have over single-celled organisms?

25. The invasion of the land by plants and fungi was possible because of a symbiotic partnership. What is the name given to this partnership? What are the contributions of plants and fungi to this partnership?

26. A partnership exists between flowering plants and bees. In the partnership, bees get nectar that they use as food from the flowers and the flowers are pollinated by the bees. Is this a symbiotic partnership? Explain.

27. What structural feature of vertebrates enabled them to become the most dominant life-forms on the land?

28. Structure and Function Amphibians evolved from bony fishes and were the first vertebrates to invade the land. What structural features not found in bony fishes did early amphibians have that enabled them to spend part of their life on land?

29. Evolution What are the dominant vertebrates on land today? How did a change in the Earth's environment favor the dominance of these vertebrates?

30. Making Inferences Cyanobacteria, a type of eubacteria, were once classified by scientists as members of the plant kingdom. Why do you think scientists did this?

31. Making Inferences Continental drift involves the movement of the Earth's great landmasses. How might the movement of the Earth's great landmasses have brought about mass extinction?

32. Interpreting Graphics The following diagram compresses the history of the Earth into a 12-hour clock to help you understand the relative time of different events. Note that the formation of the Earth occurred at midnight on the clock. The oceans formed at about 1:30 A.M., while the oldest human fossils date from just before noon. Based on fossil evidence, about what time on the clock did the first prokaryotes appear? About what time did the first eukaryotes appear?

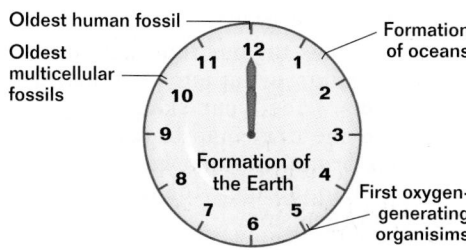

Oldest human fossil — Formation of oceans

Oldest multicellular fossils

Formation of the Earth

First oxygen-generating organisims

33. Research and Writing Lampreys are one of the two remaining kinds of jawless fishes. Examine biology books and science journals to learn more about the feeding habits of lampreys and how lampreys have affected commercial fishing in the Great Lakes.

34. Cooperative Group Project Demonstrate the operation of a swim bladder using 2-liter soft drink bottles. Fill several bottles with different amounts of water and place them in an almost full aquarium. Observe how the bottles float at different levels depending on the amount of water they contain. Relate the way in which you are able to regulate the depth at which the bottles float to the operation of the swim bladder in bony fishes. Have different group members gather materials, fill the aquarium with water, manipulate the soft-drink bottles, and share the findings with your class.

35. Cooperative Group Project Examine bacteria and protists using a microscope. Have each group member collect a different bacterium or protist and prepare a slide of the organism to be viewed by all group members. Exchange information about each organism viewed.

36. Read the article "Oldest Living Bacteria Tell All," in *Discover*, January 1992, pages 30–31. What two important discoveries did scientists make from the mastodon that was unearthed in an Ohio golf course?

37. Read the article "Life Beyond Boiling," in *Discover*, May 1993, pages 87–91. Why are scientists interested in studying the bacteria that grow in deep-sea thermal vents? What do scientists hypothesize about the structure of enzymes in ancient bacteria? Explain the scientific evidence that supports their hypothesis.

32. The first prokaryotes appeared at about 3:00 A.M. and the first eukaryotes at about 9:00 A.M. (13-1.1, 13-1.2, 13-1.3)

Activities and Projects

33. Lampreys are specialized predators. Their mouth is equipped with an oral disk and toothlike structures that are used to attach to fish and suck out their blood and tissues. Lampreys devastated commercial fishing in the Great Lakes, but treatments to kill lamprey larvae are now reducing their numbers significantly.

34. Students should observe that the more water in the bottle, the lower it will float in the water. Adding water to the bottle is analogous to a bony fish letting gas out of its swim bladder.

35. Organisms viewed by students will vary, as will the information available about each organism. Students should focus on the similarities and differences between the organisms viewed.

Readings

36. Scientists discovered 11,000-year-old living bacteria inside the mastodon's last meal, and also learned that mastodons probably ate wetland plants instead of spruce needles and twigs.

37. These bacteria are very similar to the first bacteria that lived on Earth. Ancient bacteria probably contained enzymes that were extremely stable. Scientists have isolated the enzyme rubredox from a hyperhot bacteria and have found that the enzyme has a unique chemical structure that gives it great stability in high temperatures.

Themes Review

28. The evolution of lungs, a more efficient circulatory system, and walking legs in amphibians enabled them to spend part of their life on land. (13-3.3)

29. The dominant vertebrates on land today are the birds and mammals. The development of a moister climate on Earth favored the dominance of these vertebrates. (13-3.5)

Critical Thinking

30. Cyanobacteria and plants are photosynthetic organisms. (13-1.2)

31. Movement at points in the Earth's history brought landmasses together and broke them apart. This movement of landmasses could have caused mass extinction by bringing about unfavorable climatic conditions as the landmasses drifted north and south of the equator. Also, the joining of landmasses brought species together that then competed for the same resources. (13-3.5)

LABORATORY Investigation Chapter 13

Procedural Notes

- This investigation should take two 40-minute periods to complete. During the first period, have students make their table and record data for as many stations as possible. During the second period, have students go through the remaining stations and complete the investigation.
- Include specimens of organisms from each kingdom, along with any supplementary resources you have available. These might include skeletons, photographs, photomicrographs, slides of entire organisms, or slides of particular tissue types or cross sections. If you have any samples that can be viewed under a stereoscope, display them at the appropriate stations. If you have a VCR in the classroom, provide short video segments of animals that students may be unfamiliar with, such as hydras.
- In the timeline that students prepare in step 3, each hour should be equivalent to approximately 438 million years.

Background Answers

1. Archaebacteria are relatively rare kinds of bacteria that have cell walls and membranes which are quite different from those of eubacteria. Archaebacteria are found in places such as hot springs and deep sea vents.

2. lungs for breathing and legs for walking

3. their ability to adapt to varied climates

4. How are the organisms on Earth similar, and how are they different? Approximately when did each kind of organism appear on Earth?

Overview of Life on Earth

OBJECTIVES

- Compare and contrast the distinguishing characteristics of representative organisms of the six kingdoms.
- Organize the appearance of life on Earth in a time line.

PROCESS SKILLS

- observing and inferring relationships
- organizing data

MATERIALS

You will need a variety of specimens, photographs, and other materials representing the six kingdoms. Some of these specimens will require viewing with a compound light microscope.

Examples could include:

- cyanobacteria
- archaebacteria
- eubacteria
- protists
- fungi
- plants
- arthropods
- lampreys
- sharks
- bony fishes
- amphibians
- reptiles
- birds
- mammals

BACKGROUND

1. What are some examples of archaebacteria?

2. What characteristics enabled amphibians to live on land?

3. What traits enabled birds and mammals to be successful on land?

TECHNIQUE

1. To observe organisms representative of the six kingdoms, first make a table similar to the one below in the **Records** section of your report. Then observe each specimen and record your data in the table.

Representative Organisms

Organism name	Kingdom	Characteristics/ adaptations for life on Earth	Sketch

2. Use care when handling the jars of preserved animals. The liquid preservative can leak if the jars are tilted. In your report, briefly summarize the procedure you followed.

3. Work in your lab group and use your book as a guide to create a time line of the history of the evolution of organisms. Base the time line on an 8-hour school day. For example, the 8-hour school day represents 3,500 million years. You will have to determine a scale for your time line based on how many millions of years each hour represents.

4. Add sketches to your time line to represent various organisms from the six kingdoms. Label your sketches. Be creative in the drawings to represent the various organisms that have existed on Earth. Also, account for the disappearance of organisms.

Records

Representative Organisms

Organism name	Kingdom	Characteristics/ adaptations for life on Earth	Sketch
amoeba	protist	pseudopods	
daisy	plant	flowers	
butterfly	animal	wings, antennae	

5. In your report, briefly summarize the procedure you followed.

INQUIRY

1. Cyanobacteria are green. What did these organisms contribute to the environment 2.5 billion years ago?

2. Look at your example of an archaebacterium.
  a. What type of environment does the archaebacterium live in?
  b. Why did early archaebacteria have to produce energy without oxygen?

3. Refer to your specimens from the kingdom Protista to answer the following.
  a. What structures are visible in the protists?
  b. If the protist is heterotrophic, what might it eat?
  c. Some protists that were green algae evolved into plants. What process enabled them to contribute oxygen to the environment?

4. Why is it an advantage for a plant to have a great amount of leaf surface area?

5. Look at the examples of fungi.
  a. How are fungi similar to plants? How are they different?
  b. What enabled fungi to live on land?

6. Examine your arthropod examples. What structures enabled them to live on land?

7. Compare the lamprey and the shark. What are several adaptations that allowed for the shark's success as a species?

8. What adaptation did bony fishes develop that enabled them to surpass the shark?

9. Compare the skins of the frog and the lizard. Describe the type of habitat each must live in because of their skin.

10. What enabled frogs to live on land?

11. What adaptations did reptiles develop to completely break their ties to water?

12. Birds and mammals are both endotherms. How did this adaptation allow for their success on land?

ANALYSIS

1. What organelle in your cells might be descended from early cyanobacteria?

2. Offer an explanation for the fact that multicellularity has evolved several times.

3. According to your time line, what period in the day did the mammals arise on Earth?

4. In terms of your time line, during what period of the 8-hour day did the first plants arise on Earth?

5. Write a sentence to describe, in terms of time, how long humans (mammals) have existed compared to all other organisms.

6. Speculate on what you think might have happened in the Earth's history if mass extinctions had not occurred.

FURTHER INQUIRY

Write a new question that could be explored as a new investigation. The following is an example:

> What characteristics might be used to determine which kinds of plants are the most successful? Which characteristics are valid indicators of success? Which characteristics are not?

CHAPTER 13

Inquiry Answers

1. oxygen

2. a. an environment that would be unfavorable to most organisms
   b. The atmosphere at the time probably contained very little oxygen.

3. a. nucleus, cilia, flagella
   b. bacteria, other protists
   c. photosynthesis

4. More area is exposed to the sun.

5. a. Their cells have cell walls, and most, like mushrooms, don't move around. Fungi are heterotrophs and plants are autotrophs.
   b. They formed mycorrhizal associations with plants.

6. exoskeleton, legs, wings

7. jaws, paired fins

8. swim bladder, more developed fins

9. Frog—wet habitat, lizard—dry habitat

10. development of lungs, legs

11. watertight skin and eggs

12. They could adapt to changing climates.

Analysis Answers

1. It is unlikely that any organelle in your cells descended from cyanobacteria.

2. Perhaps some of the first groups of multicellular organisms to evolve were eliminated by natural selection.

3. during the eighth hour

4. during the eighth hour

5. Humans have existed for only seconds.

6. Answers will vary. It's possible that many of the groups present today would never had evolved.

Each block represents about 45–50 minutes of instructional time. The resources cited in a block are coded to a specific program component using the Key on the next page.

Lecture Concepts

Lesson Resources

14-1 The Evolution of Primates pp. 294–295

BLOCKS ❶ and ❷

a. The first primates evolved about 60 million years ago from small, insect-eating mammals that lived in trees.
b. Primates have grasping fingers and toes, as well as binocular vision.
c. The first primates were prosimians.
d. Monkeys evolved from prosimian ancestors about 36 million years ago. The development of color vision and opposable thumbs in these diurnal primates is associated with a more developed brain.
e. Human ancestors diverged from the evolutionary line leading to gorillas and chimpanzees about 4 million years ago. DNA similarities suggest that humans have a closer evolutionary relationship to chimpanzees than to other primate species.

Lecture Resources
■ Opening Question, page 295
■ Demonstration: Colors Are Seen Better in the Light
■ Effective Reading, page 295
■ Look at Your Own Hands, page 296
■ Thumbs Are Important, page 297
■ Why Don't Apes Have Tails? page 297
■ What Are Primates Made Of? page 298
■ Visual Strategy: Figure 14-5
⊘ Holt Biology Videodiscs: Lesson 14-1

Classwork Options
■ Comparison of Primate Adaptations and Graphic Organizer, page 296
▮ Laboratory Techniques and Experimental Design C17: Analyzing Blood Serum to Determine Evolutionary Relationships

Assignment Options
■ Section Review, page 298
■ Chapter Review, questions 1–3, 11–13, 18–20

14-2 Early Hominids pp. 299–301

BLOCK ❸

a. The earliest known direct ancestors of humans belong to the genus *Australopithecus*. Australopithecines, which were hominids (belonging to the human line), exhibited two characteristics that were evolutionary milestones: they were bipedal and they had large brains relative to body weight.
b. The first australopithecine fossil, found in 1924, is thought to be 2.8 million years old. Since then, many australopithecine fossils have been found, some of which are estimated to be 3.9 million years old.
c. Because of an incomplete fossil record, scientists differ in their interpretations of how australopithecines evolved.

Lecture Resources
■ Opening Question, page 299
■ Demonstration: Comparing Brain Sizes
■ Who Am I?, page 299
■ How Do They Know That?, page 300
■ Visual Strategy: Figure 14-8
▯ Teaching Transparencies: 54, 55
⊘ Holt Biology Videodiscs: Lesson 14-2

Classwork Options
■ Lucy's Place, page 301
▮ Laboratory Techniques and Experimental Design C18: Analyzing Blood Serum— Evolution of Primates

Assignment Options
■ Section Review, page 301
■ Chapter Review, questions 4–6, 14, 24, 25

14-3 The Human Line pp. 302–308

BLOCKS ❹ and ❺

a. *Homo habilis* was the first hominid assigned to the human genus. *H. habilis* was known to make and use stone tools. It evolved from an australopithecine ancestor about 2 million years ago and survived in Africa for about 500,000 years.
b. *Homo erectus* was the second species of human to evolve. It arose in Africa 1.5 million years ago and by 500,000 years ago had migrated to Europe and Asia. *H. erectus* lived in groups, produced efficient stone tools, and was the first hominid to use fire.
c. Results from mitochondrial DNA studies suggest that *Homo sapiens* evolved from *Homo erectus* ancestors in Africa about 500,000 years ago. *Homo sapiens* then migrated throughout the world, evolving into the different human varieties.

Lecture Resources
■ Opening Question, page 302
■ Demonstration: Make a Hominid Lineup
■ Handy Man, page 303
▯ Teaching Transparencies: 53, 57, 64
■ The Wrong Ape, page 303
■ Overcoming Misconceptions, page 303
■ After Man, page 305
■ Mother's Mitochondria, page 305
■ Visual Strategies: Figures 13-19, 14-11

■ The Human Blip, page 308
⊘ Holt Biology Videodiscs: Lesson 14-3

Classwork Options
■ Laboratory Investigation: Human Evolution, pages 312–313
■ Closure, page 308

Assignment Options
■ Concept Mapping, page 309
■ Section Review, page 288
■ Chapter Review, questions 7–9, 16–20, 22, 28, 29, 31

14-3 The Human Line (continued) pp. 302–308

d. Neanderthals were powerfully built early Homo sapiens that lived in Europe and Asia 70,000 years ago. By 34,000 years ago they had been replaced by early modern humans.

Review/Enrichment

BLOCK 6

- ■ Study Guide: Concept Review and Pretest

Assignment Options
- Teaching Resources CD-ROM: Supplemental Reading Worksheets and Test, *Origin of Species*
- ■ Activities and Projects, questions 28–30
- ■ Readings, questions 31, 32

Assessment Options

BLOCK 7

Traditional Assessment
- ■ Chapter 14 Test
- Test Generator/Assessment Item Listing: Software item bank for preparing customized chapter tests.

Performance Assessment
- Select the Performance Assessment items for this unit from the Test Generator/Test Item Listing

Portfolio Assessment
Select student reports for one or more laboratory experiments from this chapter. *The Direct Observation Checklist* on the *BioSources Teaching Resources CD-ROM* should be completed during a laboratory or other cooperative learning experience.

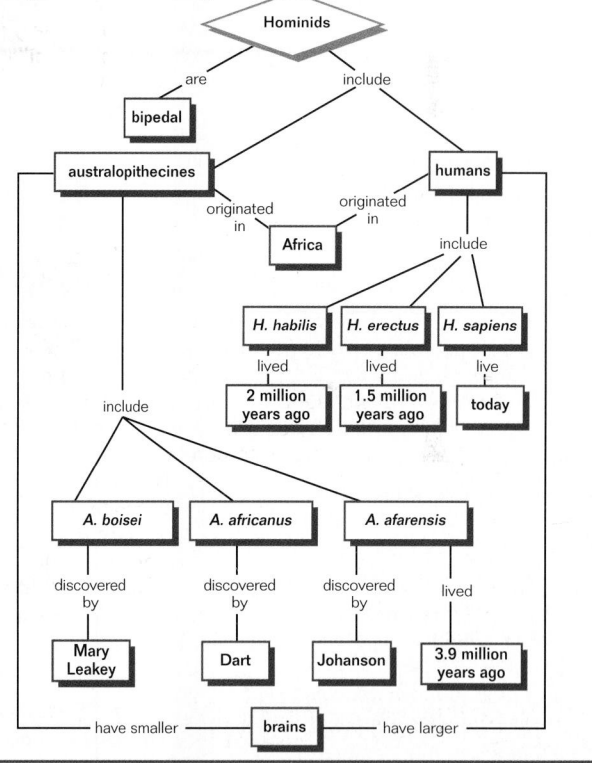

Answer to Concept Map

The following is one possible answer to the Concept Mapping exercise on page 309.

Holt Biology Videodiscs

Holt Biology Videodiscs gives you a powerful tool for teaching, review, and assessment. *Concepts of Biology* adds interactive lesson plans and extensive tools for customization using CD-ROM technology.

CONCEPTS OF BIOLOGY

Use the following topic from *Concepts of Biology* to help you teach this chapter:

- ■ Topic 9: Evolution of Life on Earth

For further information regarding the media on the videodiscs, see *Holt Biology Videodiscs Teacher's Correlation Guide to Biology: Principles and Explorations.*

Chapter Theme

Structure and Function *This chapter chronicles the successive structural changes in animals that eventually led to the development of modern humans. Changes in the hands and eyes of ancient mammals led to the success of the first primates living in trees. Changes in the hips and brains of later primates led to the success of the first hominids.*

Tapping Prior Knowledge

- What is a gene and what is a gene mutation?
- Define the term *natural selection*.
- From which group of vertebrates did mammals evolve?
- Use the timeline in Chapter 13 to determine approximately when the major groups of mammals evolved.

Opening Demonstration

Have your students demonstrate for themselves the importance of forward-facing eyes for depth perception by trying this: first hold one of your fingers about a foot in front of your nose and look through one eye, closing the other; then repeat for the other eye without moving your finger. Ask students how the two views look different. *(The finger will appear to move back and forth.)* Which position appears to change the most, that of close objects or of far away objects? *(close objects)* Each eye sends a different view of the world to the brain, which then uses this information to judge how close things are. Have students touch things at arm's length with only one eye open, then compare how easy it is to judge distance with both eyes open.

HUMAN EVOLUTION

Homo erectus, an early human

> ### REVIEW
>
> - DNA structure (Section 2-3)
> - protein structure (Section 2-3)
> - genes (Section 6-1)
> - gametes (Section 6-1)
> - natural selection (Section 12-1)

Authors' Rationale

In this chapter your students will become familiar with the characteristics of primates, many of which are unique among animals, and thus understand why humans are primates. This unique combination of characteristics steered human evolution to its present course. The fossil record, while far from complete, provides enough of the pivotal evidence upon which human evolutionary theory is based. The evidence collected to date is presented sequentially, along with the corresponding theory or theories that attempt to explain how we got where we are.

14-1 The Evolution of Primates

*I*n Charles Darwin's book **The Descent of Man,** *he proposed that humans, gorillas, and chimpanzees all evolved from a common ancestor. When the book was published in 1871, there was little fossil evidence to support Darwin's case. Although the fossil record of the origin of humans is still incomplete today, numerous fossil finds made since Darwin's death have added substantial validity to his hypothesis.*

Primates Have Unique Characteristics

The story of primate evolution begins about 80 million years ago, during the age of dinosaurs. Scurrying about in the trees at this time was an inconspicuous, insect-eating mammal the size of your fist. As is shown in Figure 14-1, this creature had big eyes and tiny, sharp teeth. Biologists think these ancient mammals were the ancestors of the first **primates,** the mammalian group that includes prosimians, monkeys, apes, and humans.

The first primates evolved about 60 million years ago. Evolution favored two distinct anatomical changes that made primates better than their ancestors at stalking and capturing insect prey in the branches of trees.

1. One change was the development of grasping fingers and toes. These fingers and toes are tipped with nails, not claws, as shown in Figure 14-2. Unlike the clawed, unbendable toes of their ancestors, primates have

Figure 14-1 The ancestor of primates closely resembled a modern-day mammal called a shrew, except that this early mammal lived in trees—a kind of "tree shrew."

Figure 14-2 This tarsier from the Philippines is a primate. Its bendable, clawed fingers and toes and forward-facing eyes are key adaptations for life in the trees.

Vocabulary Preview

primate

prosimian

diurnal

Opening Question

To stimulate student thinking about the characteristics that are unique to humans, ask your students to list the characteristics that make them different from all other animals. Have them save their list of answers and ask the same question at the end of the chapter to see if their ideas have changed. Student answers may at first be inaccurate, but by chapter's end, their answers should include: have large brains, walk upright, use fire, make and use complex tools, speak and write symbolic language that can be passed down from generation to generation.

Demonstration

Colors Are Seen Better in the Light

Turn off the lights and shade the windows enough so that objects in the classroom are barely distinguishable. Then display three brightly colored pieces of paper (one blue, one red, one yellow). Ask students to identify the color of each piece of paper. (If it is dark enough they will not be able to.) Then turn on the lights. Explain to students that without sufficient light, colors cannot be determined. The cone cells in the eyes of monkeys, apes, and humans contain photoreceptive pigments that respond to visible light. The photochemical responses of these cone cells to visible light are responsible for our ability to see the world in color.

Section Objectives

- Describe the mammal that gave rise to the first primates.
- List two distinctive characteristics of primates.
- Contrast prosimians with monkeys.
- Contrast monkeys with apes.
- Describe the evolutionary relationship between humans and apes.

Teaching Tips
Look at Your Own Hands

Have your students look at their own hands, the ways their fingers bend and the way the nails are attached to each finger. Ask them to name some of the uses for hands and some activities for which paws are used. (*hands: grabbing, climbing, pinching, holding, pushing, pointing, waving, slapping, tool-using; paws: running, hopping, digging, climbing, pushing*) Then ask why a dog can't climb a tree, and why a monkey can. (*A monkey has grasping hands. If students mention that cats climb well, point out that they have retractable claws that can grasp.*) Finish by having your students imagine how difficult it would be to do any of the things that they listed earlier for uses of hands if the hands had long claws or fingernails.

Comparison of Primate Adaptations

Remembering the special adaptations of the different kinds of primates can be difficult for students. Have them prepare a table similar to the one in the *Graphic Organizer* (minus the check marks) at the bottom of the page. After reading about each type of primate, have students check the appropriate boxes in the table.

❏ **CAPSULE SUMMARY**

Primates evolved about 60 million years ago from small, insect-eating mammals that lived in trees. Primates have grasping fingers and toes as well as binocular vision.

Figure 14-3 Like other prosimians, many lemur species are nocturnal. Nocturnal lemurs, like this sportive lemur, hunt for insects at night and sleep during the day. All 24 surviving species of lemurs live on Madagascar, an island about twice the size of the state of Arizona that is located off the southeast coast of Africa.

grasping hands and feet that let them grip limbs, hang from branches, seize food, cling to their mothers when they are young, and, significantly, use tools.

2. The second change was in the position of the eyes in primates. The eyes of their ancestors were located on the sides of the head so that their two fields of vision did not overlap. The eyes of primates are shifted forward to the front of the face. This forward placement of the eyes produces overlapping "binocular vision" that enables the primate brain to judge distance more precisely.

Judging distance accurately is a very important ability for an animal that leaps from branch to branch high above the ground. Other mammals have binocular vision, but only primates have both binocular vision and grasping hands. Three-dimensional sight and the ability to manipulate objects have played central roles in directing the evolution of increased intelligence in primates. ❏

The First Primates Were Prosimians

The term **prosimian** is from the Latin *pro*, meaning "before," and *simia*, meaning "ape." Fossils indicate that 38 million years ago prosimians were common in North America, Europe, Asia, and Africa. Only a few kinds of prosimians survive today, and their present range is severely limited. Modern prosimians include lorises, tarsiers, and lemurs. A lemur, shown in Figure 14-3, is about the size of a house cat and has a long tail used for balancing as it climbs through the trees. But the lemurs are in great danger today, and they may soon become extinct in the wild. The forest homes of these animals are being rapidly destroyed by the activities associated with an expanding human population. And as the lemurs' forest homes disappear, so does one of the oldest living links to our past.

Monkeys Adapt to Daylight Activity

About 36 million years ago, a revolution occurred in how primates lived. They became diurnal (*deye UR nuhl*). **Diurnal** primates are active during the day and sleep at night. This change had far-reaching effects. Because vision is far more important for daytime hunting, evolution favored many improvements in eye design. One of these was the development of specialized cells, called cone cells, in the sensory tissue that lines the back of the eye. Cone cells enable the eye to see in color, thus giving primates color vision. This improved sense of sight was accompanied by the

Graphic Organizer

Use this graphic organizer in *Teaching Tip:* **Comparison of Primate Adaptations** on page 296.

Type of primate	Grasping hands	Binocular vision	Color vision	No tail	Walks upright	Uses tools
Prosimian	✓	✓				
Monkey	✓	✓	✓			
Ape	✓	✓	✓	✓		✓
Hominid	✓	✓	✓	✓	✓	✓

development of a larger, more complex brain. These new day-active primates are called monkeys.

Monkeys rapidly replaced most of the diurnal prosimians; only nocturnal prosimians survive today in areas where monkeys or apes have lived. Feeding mainly on fruits and leaves rather than insects, monkeys were among the first primates with fully developed opposable thumbs. An **opposable thumb** stands out at an angle from the other fingers and can be bent toward them to grasp an object. This provides a hand with a much improved level of dexterity.

Monkeys, shown in Figure 14-4, appear to have evolved first in central Africa and quickly spread to Asia. Modern African and Asian species are commonly referred to as Old World monkeys. Some monkeys migrated to Central and South America, where they developed in isolation. These are the New World monkeys. ❑

❑ **CAPSULE SUMMARY**

Monkeys evolved from prosimian ancestors about 36 million years ago. The development of color vision and opposable thumbs is associated with a more developed brain in these diurnal primates.

Figure 14-4 Baboons, like this mandrill from Africa, *below*, are Old World monkeys that spend most of their time not in the trees but on the ground. Unlike their Old World relatives, many New World monkey species, like this wooly spider monkey from Brazil, *left*, grasp objects with their long, flexible tails.

The Path to Humans

Fossil evidence indicates that humans evolved from the evolutionary line that gave rise to apes. Apes evolved independently from Old World prosimian ancestors about 30 million years ago. Modern apes include gibbons, orangutans, gorillas, chimpanzees, and bonobos (pygmy chimpanzees). Apes have even larger, more developed brains than monkeys, and none of the apes have tails. Once common, apes are rare today. Modern apes are confined to relatively small areas in Africa and Asia. Apes never reached North or South America.

Teaching Tips
Thumbs Are Important

To help students appreciate how important a thumb is to survival, have them try to pick up a few items, such as a piece of paper, a paper clip, or a pencil, without using their thumbs. Have them think about how the lack of a thumb would affect a primate that is picking fruit from a tree, using a tool, or grabbing a branch.

Why Don't Apes Have Tails?

Ask students why it is an advantage for arboreal monkeys to have tails, but not for apes that spend most of their time on the ground. (*Monkeys that live in trees use their tails for balance. Some New World monkeys also use their tails for grabbing objects. Ground-dwelling apes have no real need for a tail.*) If they have trouble coming up with reasons, ask them how a long tail would interfere with their own activities.

❓ Did You Know?

Although some vertebrates, like birds and fishes, are able to distinguish colors, it is unusual for mammals to have color vision. Bulls don't see red and dogs can't see that the grass is green. Domestic cats have cones in the retinas of their eyes, but can't discriminate colors. Apparently, their brains can't make use of the information that their eyes send about the different wavelengths of light. Primates are unique among the mammals for their ability to sense so many colors so well.

Teaching Tip
What Are Primates Made Of?

The structural similarities of nonhuman primates to humans have been described for over a century, but comparison of biochemical similarities is relatively recent. Blood, enzymes, and DNA have been examined, and close similarities between humans and chimpanzees have been determined. The hemoglobin molecule of a chimpanzee and a human differs by a single amino acid, and there is a 1.6 percent difference in human and chimpanzee DNA nucleotide sequences. Ask students how both statements can be true. *(1.6% is an average of many DNA nucleotide sequences, not just those that encode hemoglobin.)*

◉ VISUAL STRATEGY

Figure 14-5

Some students think that evolution means that humans descended from gorillas or chimpanzees. Be sure to point out on Figure 14-5 that modern evolutionists theorize that gorillas and chimpanzees share a common ape-like ancestor with humans. Modern chimpanzees, gorillas, and other apes are as highly evolved as we are, they just followed a different evolutionary path.

☐ CAPSULE SUMMARY

Human ancestors diverged from the evolutionary line leading to gorillas and chimpanzees about 4 million years ago. DNA similarities between chimpanzees and humans suggest that humans have a closer evolutionary relationship to chimpanzees than to any other primate species.

Figure 14-5 This is a phylogenetic tree of apes and humans. The most primitive apes, those found in the evolutionary line leading to gibbons, diverged from other apes about 10 million years ago. Orangutans split off about 8 million years ago. The key split between the gorillas and chimpanzees and the line leading to humans occurred about 4 million years ago.

Studies of their DNA tell us a great deal about how apes evolved. The evolution of apes and humans is diagrammed in Figure 14-5. Because the split between the human line and the line leading to the gorilla and chimpanzees was so recent, the genes of humans and chimpanzees have not had time to evolve many differences. Human and chimpanzee DNA nucleotide sequences differ by only 1.6 percent. Your hemoglobin—a protein composed of 573 amino acids—and the hemoglobin of a chimpanzee differ in only a single amino acid. ☐

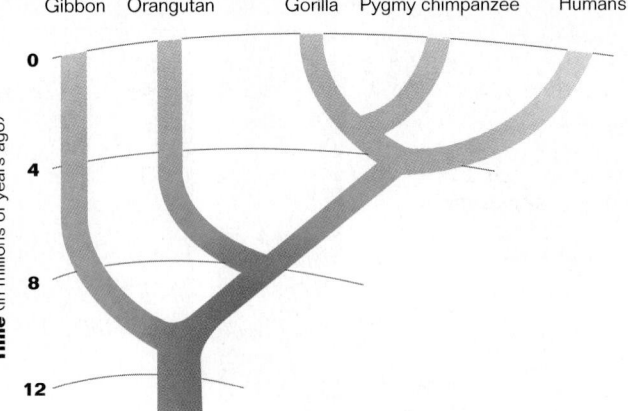

Gibbon   Orangutan   Gorilla   Pygmy chimpanzee   Humans

Section Review

1. Describe the ancestors of primates, and state when they lived.
2. List two primate characteristics that lemurs, chimpanzees, and humans have in common.
3. Describe the evolution of apes.
4. Identify and explain the evidence that closely links humans to chimpanzees.

Critical Thinking

5. Many trees produce brightly colored fruit on high branches. What advantages would monkeys have in finding and picking this fruit?

Answers to Section Review

1. Primates evolved about 60 million years ago from small, insect-eating mammals that lived in trees.

2. All primates share the two traits of grasping fingers tipped with nails and forward-facing eyes used for binocular vision.

3. Apes evolved independently from Old World prosimian ancestors about 30 million years ago.

4. Studies that compare the DNA and proteins (like hemoglobin) of humans to those of other primates have revealed that humans are more similar to chimpanzees than to any other kind of primate.

5. The binocular color vision of monkeys gives them an advantage in locating fruit by its distinguishing colors, and in picking the fruit at arm's length on high branches without having to climb out to the tip of the branches.

14-2 Early Hominids

***F**ifteen million years ago the world's climate began to get cooler, and the great forests of Africa were largely replaced with open savannas. In response to these environmental changes, natural selection resulted in the divergence of the human line from the gorilla and chimpanzee line about 4 million years ago.*

Section Objectives

- *State the genus and place of origin of the first hominids.*
- *Contrast apes with australopithecines.*
- *Describe several australopithecine species.*
- *Describe the evidence that indicates human ancestors walked upright before their brains enlarged.*

Australopithecines Were the First Hominids

Our earliest known direct ancestors belong to the genus *Australopithecus.* Australopithecines *(aw stray loh PIHTH uh seenz),* along with humans, are classified as **hominids** *(HAHM ih nihds),* belonging to the human line.

Australopithecines exhibited two characteristics that were early milestones on the path leading to the evolution of humans. First, they were **bipedal,** meaning they were able to walk upright on two legs. Table 14-1 compares the skeleton of a gorilla with that of an australopithecine. The structure of the ape skeleton makes it difficult for apes to stand and walk upright. In contrast, australopithecines had a skeletal structure that enabled them to be bipedal.

Second, most australopithecines had large brains—with a greater volume, relative to body weight, than apes had. Some australopithecines weighed about 18 kg (40 lb.) and were approximately 1.1 m (3.5 ft.) tall, about the size of a small

Table 14-1 Comparison of Gorilla and Australopithecine Skeletons

Gorilla

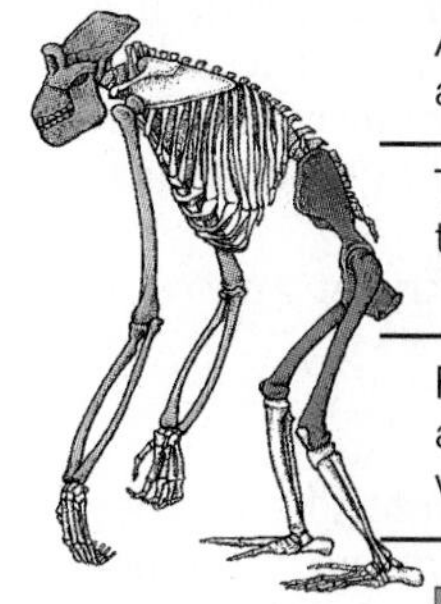

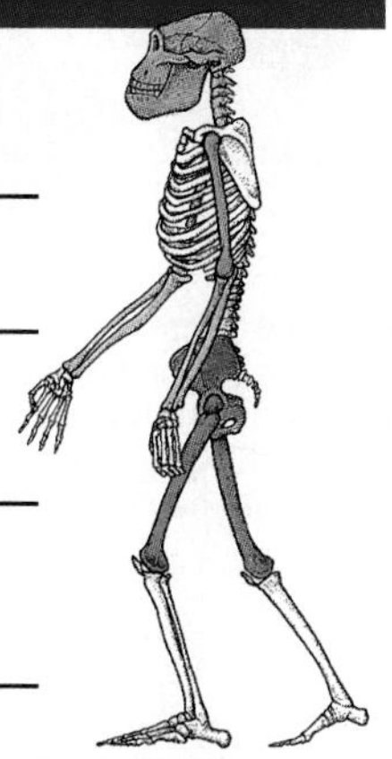

Gorilla	Australopithecine
Skull atop C-shaped spine; spinal cord exits near rear of skull	Skull atop S-shaped spine; spinal cord exits at bottom of skull
Arms longer than legs; arms and legs used for walking	Arms shorter than legs; only legs used for walking
Tall and narrow pelvis, allowing the body weight to shift forward	Bowl-shaped pelvis, centering the body weight over the legs
Femurs (thighbones) angled away from pelvis when walking upright	Femurs angled inward, directly below body to carry its weight

■ Skull ■ Spine ■ Arms ■ Pelvis ■ Femurs

Australopithecine

Section 14-2

Vocabulary Preview

hominid
bipedal

Opening Question

Ask students why it is highly unlikely that the first hominids evolved in North or South America. Students should remember that anatomical and biochemical evidence indicates that humans have evolved from the evolutionary line that gave rise to apes. Since apes never lived in North and South America, it is not logical to think that the hominids could have originated there.

Demonstration

Comparing Brain Sizes

Hominid brain sizes are mentioned throughout the last two sections of this chapter, and increasing brain size is an important part of human evolution. To help students appreciate how much hominid brain size has increased, get seven large, clear jars or bottles (all of the same size and shape), label each container for a hominid species, and fill each with the volume of water equal to the hominid's estimated brain size. Point out to students that one cubic centimeter equals one milliliter.

Teaching Tip

Who Am I?

Use a model or picture of a human skeleton to point out the parts of a skeleton that are different for gorillas and hominids. Finish your discussion by paraphrasing descriptions in Table 14-1 and challenging students to say whether the description applies to a gorilla or a hominid.

Did You Know?

Australopithecine teeth and bone fragments that are 4.4 million years old were found in Ethiopia in December 1993. The oldest hominid discovery to date, scientists have determined that the specimens are from a new species of australopithecine, and have named it *A. ramidus*. The species name "*ramidus*" is from a word in the local language that means "root," suggesting that *A. ramidus* may be the "root" species for the hominid line.

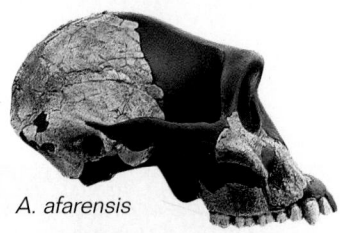

A. afarensis

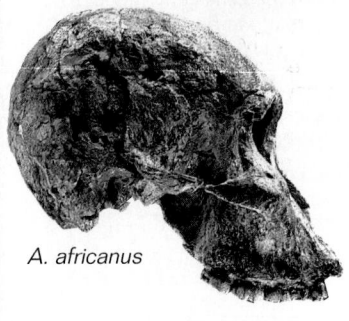

A. africanus

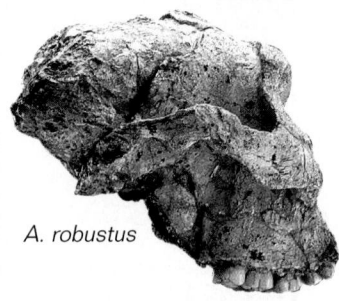

A. robustus

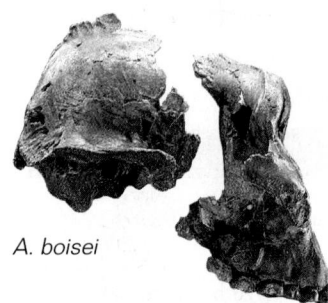

A. boisei

Figure 14-6 Most species of *Australopithecus* had brains quite a bit larger than those of apes, and their teeth were more like those of humans than of apes.

chimpanzee. Other australopithecine species were larger—more than 45 kg (100 lb.) and over 1.5 m (5 ft.) tall. Their brains were typically larger than a chimp's and in some species even larger than a gorilla's, occupying a volume of up to 550 cu. cm (34 cu. in.). While some australopithecine brains were bigger than a gorilla's brain, they were still much smaller than your brain, which is about 1,350 cu. cm (83 cu. in.).

The Discovery of Australopithecus

The first australopithecine fossil, a skull, was discovered in 1924 by Raymond Dart, an anatomy professor in Johannesburg, South Africa. Beautifully preserved, the skull was that of an individual who had died at about five years old. This skull has a rounded jaw, unlike the pointed jaw of apes, and the brain case is far larger than the brain case of an ape of similar size. What really attracted Dart's attention was the age of the skull. The rock in which the skull was embedded was from a geological formation that contained other kinds of fossils estimated to be several million years old. At that time, the oldest reported fossils of hominids were less than 500,000 years old, so the extreme age of this skull was unexpected and exciting. Scientists now believe the skull to be 2.8 million years old.

Other Australopithecine Finds

Dart named his discovery *Australopithecus africanus*. The name *Australopithecus africanus* is from the Latin *australis*, meaning "southern," the Greek *pithekos*, meaning "ape," and the Latinized *africanus*, meaning "African." Dart argued from the start that *Australopithecus* was a bipedal "manlike ape," the long-sought evolutionary link between humans and apes. At first, few thought he was right, but soon the evidence began to mount, and eventually the scientific community was persuaded. In 1938 a stockier species of *Australopithecus* was unearthed in South Africa. Called *A. robustus*, it had massive teeth and jaws. In 1959 in East Africa, Mary Leakey discovered a third species, *A. boisei*, which was even more stockily built. Nicknamed "nutcracker man," *A. boisei* had a great bony ridge on the crest of the head that anchored immense jaw muscles. Like the other australopithecines, *A. boisei* was very old—almost 2 million years old. In 1989 yet another species was discovered, a massively boned ancestor of *A. boisei*.

Figure 14-6 shows skulls of several different australopithecine species.

Lucy: The Oldest Hominid?

In 1974 the anthropologist Donald Johanson went to a remote desert region of northern Ethiopia in search of early human fossils, and he hit the jackpot. Johanson found the most complete and best-preserved skeleton of a prehuman hominid ever discovered. Nicknamed "Lucy," the skeleton was 40 percent complete. It was also nearly 3.2 million years old, the oldest australopithecine fossil then known. The skeleton was assigned to a new species and named *Australopithecus afarensis*; it is shown in Figure 14-7.

Since Johanson's discovery, many other specimens of *A. afarensis* have been unearthed, some of which are estimated to be 3.9 million years old. Most researchers agree that these smaller *A. afarensis* individuals represent the true base of the human family tree. They appear to be the first members of the genus *Australopithecus*. As shown in Figure 14-8, they are probably ancestors of all the other australopithecines as well as of the genus *Homo*, which includes our species.

Figure 14-7 The shape of Lucy's leg bones indicates that she must have walked upright. Her teeth are distinctly humanlike, but her head resembles that of an ape. Lucy's brain was no larger than an ape's, about 400 cu. cm.

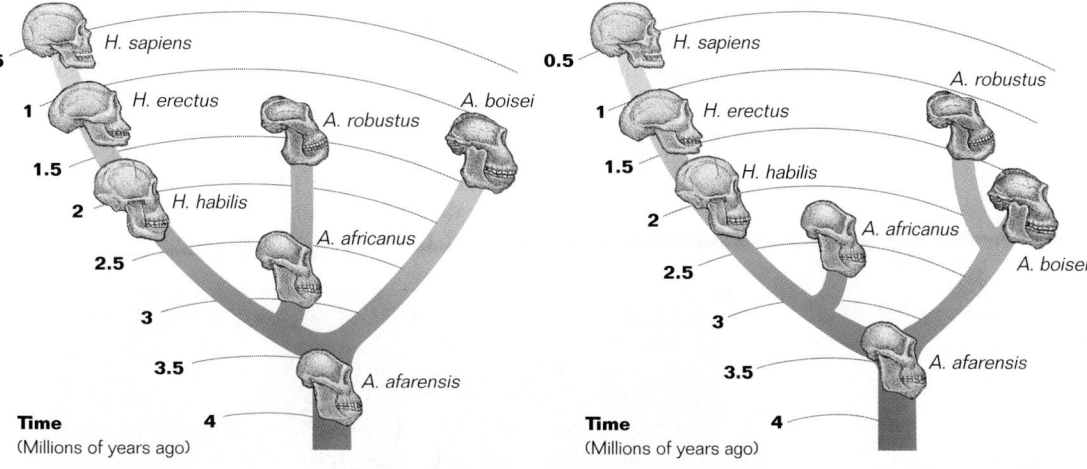

Figure 14-8 Because of an incomplete fossil record, scientists differ in their interpretations of how australopithecines evolved. Some researchers would put *A. robustus* on one branch, while others would put the same species on another branch.

SECTION 14-2

Teaching Tip
Lucy's Place

Have your students list the characteristics of "Lucy" that were ape-like and those that were human-like. Then have your students evaluate the two lists and make an argument *for or against* placing "Lucy" on the hominid family tree. Be sure they use data from their lists to support or reject her classification as a hominid.

👁 VISUAL STRATEGY

Figure 14-8

Homo erectus and *H. habilis* are thought to be ancestors of modern humans, while *A. robustus* and *A. boisei* are thought to be offshoots of the hominid family tree. Look at the skulls in Figure 14-8. What was human about *H. erectus* and *H. habilis*, and what was not human about *A. robustus* and *A. boisei*? (H. habilis *and* H. erectus *had large, rounded braincases, and human-like jaws.* A. robustus *and* A. bosei *had small braincases, and massive, ape-like jaws.)*

Section Review

1. *Where did the first hominids evolve, and what is their genus name?*

2. *How do the skeletal features of an ape and an australopithecine differ?*

3. *Explain why the discovery of "Lucy" was a significant scientific accomplishment.*

Critical Thinking

4. *All the australopithecine species were bipedal. Develop a scenario that explains why and how bipedalism evolved. Be sure to think in terms of selective advantages.*

Answers to Section Review

1. The first hominids evolved in Africa and their genus name is *Australopithecus*.

2. They differ by the way their skulls are attached to their spines, by the relative lengths of their arms and legs, by the shape and tilt of their hipbones, and by the shapes and angles of their femurs.

3. It was the most complete and best-preserved prehuman hominid skeleton found up until that time.

4. Answers will vary, but may include descriptions of situations in which it was an advantage to free the arms for tool use, hand signals, carrying food, or capturing prey. Other answers may focus on the ability to see over tall savanna grasses so that predators and prey could be more easily seen.

Opening Question

Have students look at the timeline at the bottom of the pages in Chapter 13. Ask them where the evolution of the genus *Homo* should be placed on the timeline. It belongs at 2 million years ago, which is found on page 286.

Demonstration
Make a Hominid Lineup

Split your class into five groups to make life-size hominid models. One group should make a grid that is marked off every 10 cm from the floor to 200 cm up the wall. The other four groups should pick *A. afarensis* (1.1 m tall), *H. habilis* (1.2 m tall), *H. erectus* (1.5 m tall), or *H. sapiens* to create a life-size paper profile of the hominid. Have students label each species and attach it to the grid on the wall.

Teaching Tip

What If stone tools from *H. habilis* and *H. erectus* were displayed, side by side? Could you tell who made which tools? *Homo habilis* made tools that were extremely limited in type and conformity. With *H. erectus,* the number of tool types increased to about a dozen and many of their tools had characteristic shapes. For example, teardrop-shaped hand axes are often found at sites where *H. erectus* lived.

14-3 The Human Line

We are the third and only surviving species of humans, members of the genus **Homo.** The first humans evolved from australopithecine ancestors about 2 million years ago. They were replaced in turn by a second species of human that moved out of Africa and spread across the Earth.

Section Objectives

- Compare and contrast Homo habilis with australopithecines.
- Describe the characteristics and evolution of Homo erectus.
- Contrast the two theories of the origin of Homo sapiens.
- Compare and contrast Neanderthals with modern humans.

Homo habilis Was the First Member of Our Genus

In the early 1960s more hominid bones were discovered close to the site where *Australopithecus boisei* had been unearthed. Scattered among the bones were stone tools. Although the fossils were badly crushed, painstaking reconstruction of the many pieces suggested a skull with a volume of about 640 cu. cm (39 cu. in.). This is much larger than the australopithecine brain volume of 400–550 cu. cm (24–34 cu. in.). There was much discussion at first about whether this fossil was human or australopithecine. Then, in 1972, Richard Leakey, shown in Figure 14-9, discovered a similar and virtually complete skull. Many of the critics who doubted that the

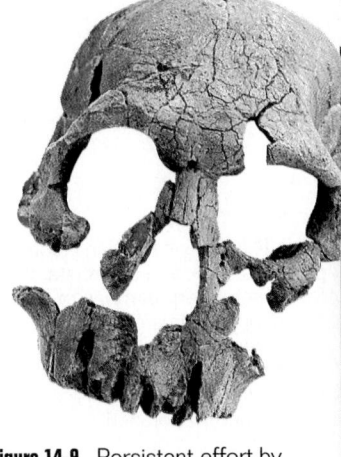

Figure 14-9 Persistent effort by Richard Leakey, far right, and his associates was rewarded. In 1972 Leakey's team found the most complete *Homo habilis* skull to date, *above.* About 1.6 million years old, it has a brain volume of 775 cu. cm (47 cu. in.) and many of the characteristics of modern human skulls.

Did You Know?

Richard Leakey is the son of Louis Leakey, the famous British anthropologist. Besides being a mentor to his son, the elder Leakey was also a mentor who encouraged Jane Goodall to work with chimpanzees and encouraged Dian Fossey to work with gorillas. Both women went on to conduct groundbreaking research on these two groups of primates.

first skull was human were silenced by Leakey's discovery. Because of its association with tools, this early human was named *Homo habilis*. The name *Homo habilis* is from the Latin *homo*, meaning "man," and *habilis*, meaning "handy." Skeletons discovered in 1987 indicate that *H. habilis* was short in stature, about 1.2 m (4 ft.) tall. *H. habilis* lived in Africa for 500,000 years and then became extinct, replaced by a new species of human with an even larger brain. ◼

Homo erectus Evolved Next

Our picture of *Homo habilis* lacks detail because it is based on only a few specimens. Because *H. habilis* had not evolved far from its australopithecine roots, some scientists still dispute the classification of *H. habilis* as a true human. There is no such doubt, however, about the species that replaced it, *Homo erectus*. Many specimens have been found, and they all indicate that *H. erectus* was distinctly human. The story of how this second species was discovered is a fascinating one.

The Discovery of Java Man

After the publication of Darwin's book *The Origin of Species* in 1859, there was much public discussion about the fossil ancestor common to both humans and apes. Puzzling over this, the Dutch physician and anatomist Eugene Dubois took a very simple approach to the problem—he went to the zoo. Looking at the apes there, he was most drawn to the orangutans. Dubois thought the orangutans resembled what a missing link should look like, so he decided to seek fossil evidence of the missing link in the area where orangutans live, in Indonesia on the island of Java.

Dubois set up a medical practice in a river village in eastern Java. Digging into a hill that villagers claimed had "dragon bones," he unearthed the top of a skull and a thighbone in 1891. He was very excited by his find for three reasons. First, the structure of the thighbone clearly indicated that the individual had long, straight legs and was an excellent, upright walker. Second, the size of the skull cap suggested a *very* large brain, about 900 cu. cm (54.9 cu. in.)—a great deal larger than any ape brain. Most surprising to Dubois, the bones seemed to be as much as 500,000 years old, judging by other fossils he unearthed with them.

The fossil hominid bones that Dubois had found were far older than any fossil hominid discovered up to that time, and at first few scientists were willing to accept that it was an ancient species of human. After years of trying to convince a doubting audience that the bones were human, Dubois became disgusted. He is said to have buried the "Java man" skull cap and thighbone under the floorboards of his dining room and for 30 years refused to let anyone see them.

◻ **CAPSULE SUMMARY**

Homo habilis, *the first hominid assigned to our own genus, was known to make and use stone tools. It evolved from an australopithecine ancestor about 2 million years ago and survived in Africa for about 500,000 years.*

Teaching Tips

Handy Man

Help students appreciate that the tools found with *H. habilis* would be easy for a layperson to miss. Flint knives may look like ordinary rocks, but closer examination reveals purposeful shaping of the rock by cleaving sections off one side to produce a sharp edge. Ask students to think of ways that rocks may be used as tools, and how they might show evidence of use. If they have trouble thinking of rocks as tools, suggest thinking of ways that Native Americans used rocks. (*arrowheads, axes, grinding stones, hammers*) Ask your students if they think that *H. habilis* used wooden tools, and have them cite reasons to support their opinion. (*Answers will vary.*)

The Wrong Ape

Have students discuss and evaluate Eugene Dubois's reasoning for seeking a human ancestor on the island of Java. Where might he go if he were alive today and looking for fossils of human ancestors? (*If he were looking for human ancestors today, the closest nonhuman primate to humans is the chimpanzee, which lives in Africa.*)

Overcoming Misconceptions

Cartoons About Early Humans

Cartoons perpetuate a myth of early humans wielding clubs and bashing each other over the head. Hominid fossil skulls found in the early twentieth century were crushed due to burial under heavy soil for long periods of time. An early paleontologist came to the erroneous conclusion that the hominids died from having their skulls crushed, and that erroneous conclusion has lived on in cartoons ever since.

UNIT 3

CHAPTER 14

Teaching Tips

A Successful Species

Homo erectus became numerous and widespread in Africa, then in Europe and Asia. Have pairs of students identify reasons for the success of this species, both by identifying successful adaptations mentioned in the text, and by making reasonable conjectures.

A Social Family

Members of the genus *Homo* often show evidence of living in social groups. Have students list all the reasons why living in a group is advantageous, then list all the reasons why living in a group could cause problems. If students have difficulty thinking of reasons, have them think about their own community and the problems and advantages they encounter being members of a group.

The Discovery of Peking Man

It was a generation before scientists were forced to admit that Dubois was right all along. In the 1920s a skull was discovered near Peking (now called Beijing), in China. This skull closely resembled that of Java man. Continued excavation at the site eventually uncovered 14 similar skulls. Many of the skulls, along with lower jaws and other bones, were excellently preserved. Crude tools were also found, as were the ashes of campfires. Casts made from the fossils were distributed to laboratories around the world for study. The originals were loaded onto a truck and evacuated from Peking at the beginning of World War II, only to disappear during the confusion of the time. No one knows what happened to the truck or its priceless cargo.

African Origins

Java man and Peking man are now recognized as belonging to the same species, *Homo erectus*. *Homo erectus*, depicted in Figure 14-10, was larger than *H. habilis*—about 1.5 m (5 ft.) tall. It had a large brain of about 1,000 cu. cm (60 cu. in.) and clearly walked erect. The shape of the skull interior suggests that *H. erectus* may have been capable of speech.

Where did *H. erectus* originate? It should come as no surprise to you that it came from Africa. In 1976 a complete *H. erectus* skull that is 1.5 million years old, a million years older than the Java and Peking finds, was discovered in East Africa. *Homo erectus* marked the beginning of a great human migration from Africa. Far more numerous than *H. habilis*, *H. erectus* quickly became widespread and

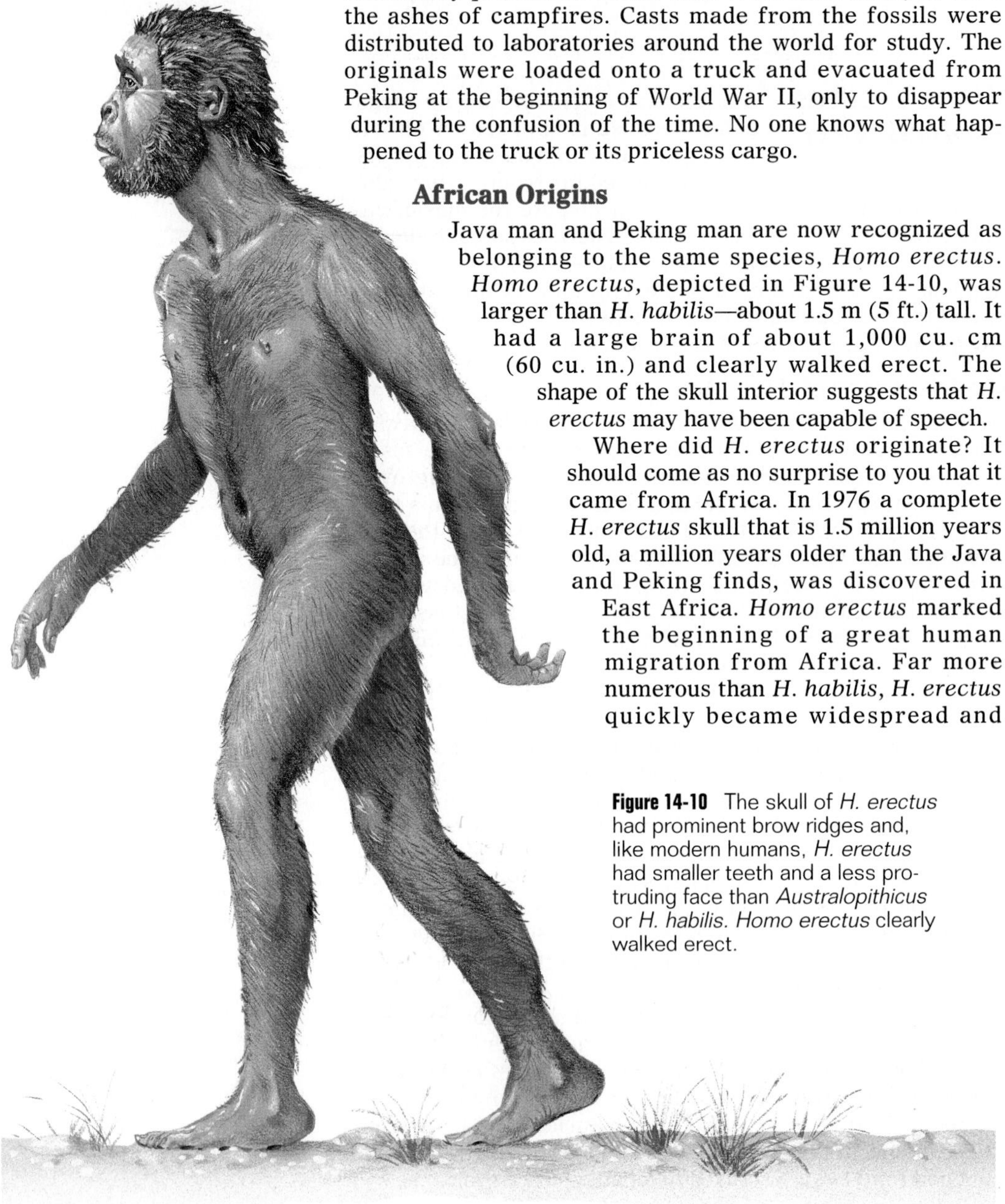

Figure 14-10 The skull of *H. erectus* had prominent brow ridges and, like modern humans, *H. erectus* had smaller teeth and a less protruding face than *Australopithicus* or *H. habilis*. *Homo erectus* clearly walked erect.

 Did You Know?

In 1886, at the age of only 28, Eugene Dubois became a lecturer on anatomy at the University of Amsterdam. The colleagues of Dubois predicted a long and happy academic career for the brilliant Dutch anatomist and physician. But Dubois had other plans for himself—he enlisted in the Dutch army. The physician joined the army as a surgeon, and volunteered to go to Java—if they would allow him time to dig for fossils.

abundant in Africa and eventually migrated into Asia and Europe. A social species, *H. erectus* lived in groups of 20 to 50 people. They often lived in caves, but there is evidence that they also built crude wooden shelters. They successfully hunted large animals, butchered them using flint and bone tools, and cooked the meat over fires. The Peking site contains the remains of horses, bears, elephants, deer, and rhinoceroses.

Homo erectus survived for over 1 million years, longer than any other species of our genus. These very adaptable humans disappeared in Africa and Europe only about 500,000 years ago, as early modern humans were emerging. Interestingly, they survived much longer in Asia, until about 250,000 years ago. *Homo erectus* was without serious doubt the direct ancestor of our species, *Homo sapiens.* ◻

❏ **CAPSULE SUMMARY**

Homo erectus *was the second species of human to live on Earth. It evolved in Africa 1.5 million years ago, and by 500,000 years ago it had migrated to Europe and Asia.* Homo erectus *lived in groups, produced efficient stone tools, and was the first hominid to use fire.*

Modern Humans Are Homo sapiens

The human evolutionary journey approaches present times with the appearance of *Homo sapiens*, our species, about 500,000 years ago. The name *H. sapiens* is from the Latin *homo*, meaning "man," and *sapiens*, meaning "wise." *Homo sapiens* is a newcomer to the human family. It has not been around nearly as long as *H. erectus* was. Early *H. sapiens* left behind many fossils and artifacts, including the first known paintings by humans.

African Origins Again?

The geographic origin of our species is a much-debated topic among researchers studying human evolution. Many scientists have argued that independent *H. erectus* lines living in Africa, Europe, and Asia interbred as they evolved and that *H. sapiens* thus arose as a new species more or less simultaneously all over the globe. Others think it unlikely that *H. erectus* evolved into *H. sapiens* more than once. They argue that *H. sapiens* appeared in one place, then spread over the world, replacing *H. erectus* as they went. Recently, scientists have added fuel to the fire of this controversy by studying the DNA within modern human mitochondria.

The reason these scientists looked at mitochondrial DNA to study evolution is that the DNA within mitochondria is transmitted only by females. It is possible to trace genetic variations within a mitochondrial gene back through a family tree, from mother to grandmother to great-grandmother. Since DNA accumulates mutations (changes in genes) over time, the oldest mitochondrial DNA should show the largest number of mutations.

It turns out that the greatest number of different mitochondrial DNA sequences occurs among modern Africans. The results from the DNA studies indicate that *H. sapiens* have been living in Africa longer than on any

Teaching Tips

After Man

The fictional book entitled *After Man: A Zoology of the Future* by Dougal Dixon describes hypothetical animal species that will evolve after the extinction of man. To build on this idea, have your students create and name a new *Homo* species that could replace *Homo sapiens*, and describe its special adaptations.

Mother's Mitochondria

We inherit our mitochondria from our mothers because these organelles are present in the cytoplasm of the ovum. Cytokinesis distributes the maternal mitochondria to every cell produced in our bodies. Visual aids will help students understand this process. Use diagrams of cell structure and cell division to review what mitochondria are, where they are found, and how they are inherited.

Competition

Introduce your students to the concept of competition by asking why the migration of *H. sapiens* around the world marked the end of *H. erectus*. Make the point that one species doesn't have to have face-to-face contact to eliminate another species. *H. sapiens* can eliminate other species by consuming and destroying their food supplies, or by spreading disease and pollution.

VISUAL STRATEGY

Figure 14-11

Use Figure 14-11 to correlate the different varieties of *H. sapiens* today with their geographical locations. Have students make a chart with seven columns to correspond with the color codes that indicate degree of genetic similarity. Start with the hypothetical cradle of mankind, Africa, where the nuclear genetic codes are <u>most</u> similar. Ask students to describe the physical features of the people who are endemic to this continent. Then in the next column, have them describe the features of European and Asian peoples, and so on until the features of the <u>least</u> genetically similar humans have been described. Ask students why there are different varieties of H. sapiens. *(The unique characteristics of each human variety represent adaptations to local conditions that evolved after our species migrated out of Africa.)*

Teaching Tip
Bones That Tell a Story

The evidence on which paleontologists based their conclusions about Neanderthal care of the injured and sick was from burial sites. Neanderthal skeletons have been found with healed broken bones. The fractures probably incapacitated the individuals, so the conclusion is made that someone had to be taking care of the individuals until their bones healed. Furthermore, an adult male skeleton was found with serious deformities that probably made the individual dependent on care from others his entire life. Describe the Neanderthal skeletons that were found to your students and see if they come to the same conclusions.

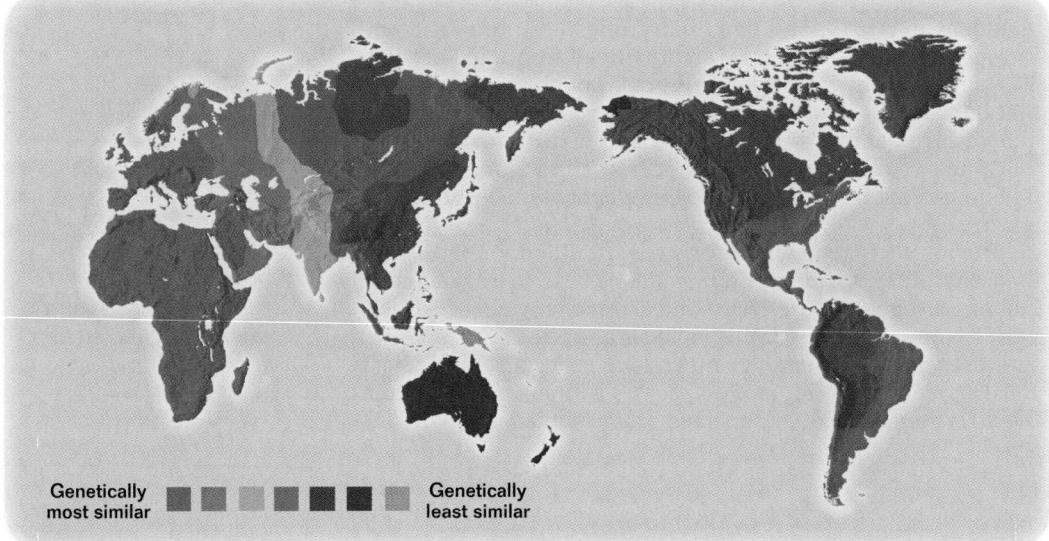

Genetically most similar ■ ■ ■ ■ ■ ■ Genetically least similar

Figure 14-11 By analyzing geographic patterns of variations in nuclear genes and their relative predominance, a map showing the migration of *Homo sapiens* from Africa throughout the world can be derived.

❏ CAPSULE SUMMARY

Results from mitochondrial DNA studies suggest that Homo sapiens, *our species, evolved from* Homo erectus *ancestors in Africa about 500,000 years ago.* Homo sapiens *then migrated throughout the world, evolving into the different human varieties.*

other continent. While there is not yet universal agreement among researchers, this evidence supports the hypothesis that *H. sapiens* evolved first in Africa. If this is true, the varieties of living humans evolved after that, and not independently from separate populations of *H. erectus*. This would mean our species was born in Africa and spread from there to all parts of the world. Results from nuclear gene research agree with the mitochondrial study findings, as indicated by the map in Figure 14-11. Indeed, it seems that *H. sapiens* retraced the path taken by *H. erectus* half a million years earlier. ❏

Homo sapiens Migrates Into Europe

Homo sapiens first appeared in Europe about 130,000 years ago. The first *H. sapiens* fossils were found in 1856 in the Neander Valley of Germany. These early European humans, called Neanderthals *(nee AN dur THALZ)*, are shown in Figure 14-12. The name *Neanderthal* is from the German *Neander*, the name of a river, and *thal*, meaning "valley." Compared with ourselves, the European Neanderthals were powerfully built, short, and stocky. Their skulls were massive, with protruding faces and heavy, bony ridges over the brows. Their brains were even larger than those of modern humans.

Rare at first outside of Africa, Neanderthals became progressively more abundant in Europe and Asia, and by 70,000 years ago they had become common. Neanderthals took care of their injured and sick and commonly buried their dead, often placing food, weapons, and even flowers with the bodies. Such attention to the dead suggests that they believed in a life after death. The Neanderthals were the first hominids to show evidence of the abstract thinking characteristic of modern humans.

Historical Note
Rose and Ralph Solecki

For some scientists, exploring the origins of humanity seems to be a family job. This was the case with archaeologists Ralph and Rose Solecki. In the 1950s, the Soleckis traveled to Iraq to look for ancient stone tools.

Although they worked together, Ralph Solecki specialized in Neanderthal fossils and Rose studied the archeology of newly emerging communities. In 1968, Rose's microscopic investigations revealed that the Neanderthal people used flowers to mourn the death of their family members. She found traces of pollen from eight different flowers. The plants that the pollen came from did not grow in the cave where a Neanderthal corpse had been found; they were definitely carried from the hillside and buried with the body.

RESEARCH Update

Ötzi Makes 5000 Year Comeback, Denies Rumors, Tells All

Two hikers in the Tyrolean Alps stumbled over a mummified corpse, which they thought was a murder victim, in September 1991. Experts radiocarbon-dated the mummy and determined that it was between 5100 and 5300 years old. Christened Ötzi, for the Ötz Valley glacier in which it was found, the mummy was not part of any archaeological site, and was therefore considered by some to be a fraud (perhaps a South American mummy planted in the ice).

The German scientists who did the initial and the authenticating research on the mummy used X rays, CT scans, and lab analysis of tissues—including DNA analysis. The DNA analysis showed that the mummy was genuine, not a "plant" from South America or elsewhere. Furthermore, the research evidence led Konrad Spindler, of the University of Innsbruck, to make some speculations about how Ötzi lived—and died. Spindler surmises that Ötzi was a herder who traveled away from his community, possibly became involved in an altercation with other individuals, and then, wounded, collapsed in the snow to die and become buried by the glacier. The glacier started to melt 5,000 years later, and Ötzi made his comeback. ☑

Modern *Homo sapiens*

About 34,000 years ago the European Neanderthals were abruptly replaced by *Homo sapiens* of essentially modern appearance. Early modern humans lived by hunting animals like the bison shown in Figure 14-13. It was the time of the

Figure 14-12 The Neanderthals lived in huts or caves and made diverse tools, including scrapers, spearheads, and hand axes.

Figure 14-13 Early modern humans used intricate tools and weapons made out of stone, horn, and bone. The animals they hunted can be seen in elaborate and often beautiful cave paintings throughout Europe. These paintings were discovered inside Altamira Cave, located near the north coast of Spain.

Teaching Tip
The Human Blip

The geometric growth of human populations in the last 10,000 years is due to the successful adaptation of humans, yet presents our species with increased problems of disease, war, pollution, and famine. When the number of individuals in a species suddenly increases in the geological record, the phenomenon is known as a "blip." A population cannot grow rapidly forever, so population growth is followed by change. Geologists have noticed that after populations exhibit a blip, they become extinct and are replaced by a new species, or they are reduced in numbers. Among the huge varieties of life on Earth, intelligence and cultural evolution is apparently a new adaptation, so challenge students to think about the long-term consequences of our population "blip." Will we go the way of other species, or will our future take a unique path? Have students present their thoughts in a one-page written report.

Chapter Closure

Ask students to draw a simplified phylogenetic tree that shows the evolution of primates, leading to modern humans. Include the following primate groups and species on the tree: first primates, prosimians, monkeys, gorillas, chimpanzees, australopithecines, *H. habilis*, *H. erectus*, and *H. sapiens*.

☐ CAPSULE SUMMARY

Neanderthals were powerfully built, early Homo sapiens *that lived in Europe and Asia 70,000 years ago. By 34,000 years ago they had been replaced by early modern humans, also* H. sapiens *whose physical features were very similar to ours.*

Figure 14-14 While not the only animal capable of conceptual thought, human beings have refined and extended this ability until it has become the hallmark of our species.

last great ice age, and Europe was covered with grasslands inhabited by large herds of grazing game. Early modern humans had complex patterns of social organization and are believed to have had sophisticated language capabilities. They eventually spread across Siberia and reached North America at least 13,000 years ago. They made this journey after the great fields of ice had begun to retreat and while a land bridge still connected Siberia and Alaska. There were no more than several million people living in the entire world 10,000 years ago, compared with more than 5.4 billion living today. ☐

Humans Are Unique

Like all living things, the humans shown in Figure 14-14 are the product of evolution. Our evolution has been marked by a regular increase in brain size. Our ability to make and use tools effectively is a capability that has, more than any other, been responsible for our dominant position in the animal kingdom. Humans use symbolic language and can shape concepts out of experience. Language has allowed us to transmit accumulated experience from one generation to another. Thus, humans have what no other animal has ever had—cultural evolution. Through culture, we have found ways to change and mold our environment to our needs, rather than changing ourselves in response to the demands of the environment. The human species controls its future in a way never before possible. This is both an exciting potential and an enormous responsibility.

Section Review

1. *State the genus and species name of the first humans, and describe when and how they lived.*
2. *Describe the evidence that identifies* Homo erectus *as the first human species to leave Africa.*
3. *Describe the appearance and lifestyle of* Homo erectus.
4. *Explain the scientific evidence that supports the hypothesis that* Homo sapiens *evolved in Africa.*
5. *When did H. sapiens with physical features like yours first appear?*

Critical Thinking
6. *The cause of the extinction of the Neanderthals is not known with certainty. Develop a hypothesis that explains the disappearance of these early* Homo sapiens.

Answers to Section Review

1. *Homo habilis* evolved about 2 million years ago, and survived in Africa for 500,000 years. They lived in groups and used tools.

2. Skeletons found on the island of Java and in Peking,
China, were identified as belonging to the species *H. erectus*, which originated in Africa. *Homo habilis* fossils have only been found in Africa.

3. *H. erectus* lived in groups of 20 to 50 people, inhabited caves, hunted, used tools, and cooked over fires.

4. Studies of the world's human populations indicate that the greatest number of mitochondrial DNA mutations occurs in African populations. Since populations with the most divergent mitochondrial DNA are the oldest, it is likely that *Homo sapiens* originated in Africa.

5. *Homo sapiens* like us appeared about 34,000 years ago.

6. Answers will vary, but may include violent encounters with the new modern humans and competition for food and shelter.

Vocabulary

bipedal (299)
diurnal (296)
hominid (299)
opposable thumb (297)
primate (295)
prosimian (296)

Concept Mapping

Construct a concept map that shows the hominids and their relationships to each other. Use as many terms as needed from the vocabulary list. Try to include the following items in your map: australopithecines, humans, Africa, *A. boisei, A. africanus, A. afarensis, H. habilis, H. erectus, H. sapiens,* bipedal, brain, Mary Leakey, Dart, Johanson, Lucy, Java man, 3.9 million years ago, 2 million years ago. Include additional terms in your map as needed.

Review

Multiple Choice

1. Which anatomical feature distinguishes primates from their ancestors?
   a. grasping hands
   b. canine teeth
   c. body hair
   d. monocular vision

2. Monkeys differ from most prosimians in that monkeys
   a. are color blind.
   b. have thumbs.
   c. sleep at night.
   d. live alone.

3. Among the apes, the most distant relatives to humans are the
   a. chimpanzees.
   b. orangutans.
   c. gorillas.
   d. gibbons.

4. The discovery of Lucy by Donald Johanson is most important because it shows that
   a. hominids inhabited the Earth 3 million years ago.
   b. our ancestors developed bigger brains after becoming bipedal.
   c. *A. africanus* is the base of the human family tree.
   d. Dart's theory about the evolution of man from apes was wrong.

5. Which pair reflects a correct match between hominid and its discoverer?
   a. nutcracker man : Mary Leakey
   b. *Australopithecus afarensis* : Raymond Dart
   c. *Homo erectus* : Donald Johanson
   d. *A. africanus* : Richard Leakey

6. Which of these is *not* a characteristic of an australopithecine skeleton?
   a. arms shorter than legs
   b. spinal cord that exits at bottom of skull
   c. tall and narrow pelvis
   d. skull atop S-shaped spine

7. Compared with australopithecines, *Homo habilis*
   a. used a more sophisticated language.
   b. developed better hand dexterity due to opposable thumbs.
   c. was more apelike.
   d. had greater brain volume.

8. The first hominid known to use fire and tools was
   a. *Homo erectus.*
   b. *A. africanus.*
   c. "handy man."
   d. *Homo sapiens.*

9. Studies of mitochondrial DNA from *Homo sapiens* suggest that
   a. egg cells carry mitochondria.
   b. modern man first evolved in Africa.
   c. populations of *Homo sapiens* evolved independently.
   d. Neanderthals lived in present-day Germany.

CHAPTER REVIEW ANSWERS

Each item in the Chapter Review is correlated by text section in the assignment guide that follows.

ASSIGNMENT GUIDE

Section	Review	Themes Review	Critical Thinking
14-1	1–3, 11, 14, 18–20	24	
14-2	5, 6, 14	24	25
14-3	7–10, 15–17, 21, 22	23	26-27

Concept Mapping

Map answer is shown on page 293B.

Review

Multiple Choice

Code in parentheses indicates section number and objective number.

1. a (14-1.2)
2. c (14-1.3)
3. d (14-1.5)
4. b (14-2.4)
5. a (14-2.3)
6. c (14-2.2)
7. d (14-3.1)
8. a (14-3.2)
9. b (14-3.3)
10. d (14-3.4)

CHAPTER REVIEW ANSWERS *continued*

Completion

11. shrew, 60 million, eyes (14-1.1)

12. prosimians, night (14-1.3)

13. apes, tail (14-1.4)

14. *Australopithecus,* Africa, 3.9 (14-2.1)

15. Homo, australopithecines (14-3.1)

16. Europe, fire, caves (14-3.2)

17. *sapiens,* extinct (14-3.4)

Short Answer

18. The eyes were positioned forward on the front of the face. The forward placement of the eyes produced binocular vision that enabled the first primates to better judge distances. Judging distances accurately is important to animals that leap from branch to branch high above the ground. (14-1.1)

19. Diurnal prosimians were unable to compete with monkeys that had color vision. Color vision is an advantage to daylight hunters. (14-1.3)

20. It is assumed that the nucleotide sequence has changed over time and that the fewer the differences between human DNA and the DNA to which it is compared, the closer the evolutionary relationship. (14-1.5)

21. Few fossils of *Homo habilis* have been found, making it difficult to classify them as true humans. But their use of tools and a greater brain volume support the hypothesis that *Homo habilis* is a true human. (14-3.2)

22. One theory is that *Homo sapiens* evolved simultaneously all over the globe, while a second theory is that *Homo sapiens* evolved in one location and then spread over the world. The evidence best supports the second theory described. (14-3.3)

---

10. Neanderthals differ most from modern humans in their
 a. tooth structure.
 b. walking gait.
 c. foot structure.
 d. forehead shape.

Completion

11. The mammal that gave rise to the first primates looked a lot like a modern-day _____. The first primates evolved about _____ years ago and had forward facing _____ and grasping fingers and toes.

12. Lorises and lemurs are _____. They live on the island of Madagascar, and many hunt for food at _____.

13. Orangutans, gorillas, gibbons, and chimpanzees are not monkeys, but _____. They are larger than monkeys, and none have a _____.

14. The first hominids were members of the genus _____. They evolved in _____ and lived on Earth between _____ and 1.5 million years ago.

15. Modern humans and "handy man" are included in the genus _____. The first members of this genus are thought to have evolved from _____ about 2 million years ago.

16. *Homo erectus* came from Africa and was the first human species to migrate into _____ and Asia. *Homo erectus* used _____ to cook meat and lived in _____ and wooden shelters.

17. Neanderthals and modern people are *Homo* _____. Neanderthals became _____ about 34,000 years ago, about the same time that modern people appeared.

Short Answer

18. Describe the positioning of the eyes in the first primates. How did this positioning of the eyes improve the primates' ability to survive?

---

19. In areas inhabited by monkeys today, very few diurnal prosimians are found. How can you explain this?

20. Differences in the nucleotide sequence of DNA from the hemoglobin gene provide evidence for the evolutionary relationship between human and ape species. What assumptions about the nucleotide sequence of DNA support this evidence?

21. Why do scientists dispute the classification of *Homo habilis* as a true human?

22. Two theories about the evolution of *Homo sapiens* have been proposed. What are they? Which theory is best supported by existing evidence?

Themes Review

23. Levels of Organization In trying to determine where and when modern humans evolved, scientists have made use of mitochondrial DNA. Why have they chosen to use mitochondrial DNA rather than DNA in the nuclei of cells?

24. Structure and Function Explain how you can tell that the skeleton shown below is from a primate? How do you know it is not from a hominid?

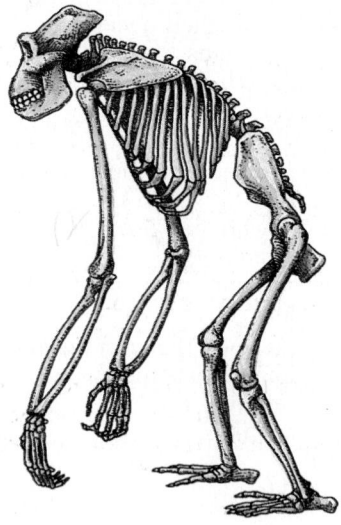

Themes Review

23. Mitochondrial DNA accumulates genetic mutations quickly over time, making it a faster "molecular clock" than nuclear DNA. Also, mitochondrial DNA is inherited through the maternal line, making it easier to reconstruct an evolutionary tree. (14-3.3)

24. The grasping fingers and toes and forward facing eyes tell you it is a primate's skeleton. It is not from a hominid because the arms are longer than the legs and the spine is C-shaped. (14-1.2, 14-2.2)

Critical Thinking

25. Making Predictions The reasons why our ancestors first stood on two feet are not known with certainty. But it has been proposed that bipedalism evolved as a mechanism to keep the brain cool on the open, equatorial savanna. Develop a hypothesis that explains how being bipedal and being hairless except for the top of the head function as interrelated adaptations to protect the brain from overheating.

26. Making Inferences Thermoluminescence and electron spin resonance are new dating techniques used by anthropologists. Using these new techniques, flint tools used by modern humans were determined to be about 92,000 years old. Flint tools used by Neanderthals were determined to be much younger—about 60,000 years old. What do these data suggest about the assumption that Neanderthals lived before modern humans?

27. Making Predictions The life expectancy of men and women today is about 70 years. By comparison, analyses of the bones of Neanderthals indicate that both males and females died at the end of the female reproductive cycle, at roughly 40 years of age. This suggests that grandparents did not exist in Neanderthal society. How do you think the absence of grandparents impacted the Neanderthal society as a whole?

Activities and Projects

28. Research and Writing In the past few years two new techniques, thermoluminescence (TL) and electron spin resonance (ESR), have been used by anthropologists to date fossils. Find out how these two techniques work and on what kinds of fossils the techniques can be used.

29. Cooperative Group Project Select a hominid described in this chapter. Then prepare a group report on the hominid. Have each member of your group choose one of the following tasks to complete. (1) Sketch a map of where the remains were found. (2) Write about who found the remains and when. (3) Write a description of the remains. (4) Draw an evolutionary tree that shows how the hominid is related to modern man. Consult scientific journals for the latest information on hominids. Finally, as a group, decide how the report will be presented to the class.

30. Multicultural Perspective Imagine that you are a biological anthropologist living in Africa 1.5 million years ago. You study human skeletons, both prehistoric and modern. In your imaginary visit to Africa, describe your encounter with a male and female *Homo erectus*. Do they behave like modern *Homo sapiens*? What similarities might you find? What differences? What technologies would *Homo erectus* have? Language? Fire-making skills? Can the *Homo erectus* couple tan animal hides?

Readings

31. Read the article "Climate and the Rise of Man," in *U.S. News and World Report,* June 8, 1992, pages 60–67. Explain how changes in climate might have caused bipedalism to evolve in the ancestors of humans. Why do some scientists regard the development of agriculture by humans as a last-ditch effort to survive?

32. Read the article "Women Leave Indelible Mark on Evolution," in *New Scientist,* July 24, 1993, page 15. Why do some scientists consider women rather than men to be the cornerstone of human evolution?

Activities and Projects

28. TL is used for dating inorganic materials like stone tools and rocks. It involves heating the object and measuring the light energy it emits. A formula is then used to calculate the age of the object. ESR is used to date organic materials. ESR involves exposing the ground-up material to a strong magnetic field. The magnetic field is affected by the number of trapped electrons in the sample. The more agitated the magnetic field, the older the fossil.

29. Information included in a group report will vary depending on the hominid chosen, but the report should reflect information contained in the chapter.

30. The *H. erectus* couple have faces that are similar to a modern human's face, except that *H. erectus* has prominent brow ridges. *H. erectus* is also shorter than a modern human. They make and use several kinds of stone tools, use fire, and communicate using a limited amount of spoken language. They butcher animals and tan their hides with the tools they have made.

Readings

31. A climate change about 7 million years ago may have caused the vegetation of Africa to change from thick jungles to a patchwork of grassy plains and forests. Prehominid ancestors crossed the vast stretches of grassland in search of forested areas. Walking upright would have made traveling across the grasslands easier. A drop in global temperature may have forced the sedentary Natufians to begin cultivating grains in order to survive.

32. It appears that imprinted genes from the mother, but not the father, play a critical role in the way a human embryo, and the subsequent baby, develops.

Critical Thinking

25. Standing on two feet reduces the body area exposed to solar radiation and makes the shielding provided by hair necessary only on the top of the head. The naked skin enables the loss of heat at a rate that exceeds that of other mammals. (14-2.2)

26. The data suggest that Neanderthals did not lead to early modern humans, but were their contemporaries. (14-3.4)

27. In modern hunter-gatherer societies, grandparents pass on knowledge of the environment and religion, and assist with child rearing. Without grandparents, child-rearing would have required more time from parents, and important aspects of the society's culture and history could have been lost. (14-3.4)

Procedural Notes

- This investigation should take 40 minutes to complete.
- Since the drawings of the human, ape, and fossil hominid skulls are approximately ⅙ to ½ of their actual size, the checklist of features has incorporated conversion factors of 40 for calculating area and 1,000 for calculating volume of skulls. If casts of actual skulls are available for students to measure, eliminate the factor of 40 in calculating area. Substitute 4.2 or ⅓ × pi for the factor of 1,000 in calculating volume.

Background Answers

1. Scientists compare the anatomical features of species to identify similar and dissimilar structures, which may help them to infer the evolutionary relationship between the two species.

2. What are some anatomical differences and similarities among apes, early hominids, and modern humans?

LABORATORY Investigation Chapter 14

Human Evolution

OBJECTIVES

- Identify anatomical differences and similarities between apes and humans.
- Categorize fossil forms of hominids by examining their anatomical features.

PROCESS SKILLS

- measuring and comparing anatomical features
- organizing and classifying data

MATERIALS

- metric ruler
- protractor

BACKGROUND

1. How do scientists use comparative anatomy to infer evolutionary relationships among species?
2. Write your own question to be explored in your lab report or notebook.

TECHNIQUE

Part A: Ape and Human Skulls

1. Refer to the drawings of the ape and human skulls and the checklist of anatomical features to identify similarities and differences between the brain areas, faces, teeth, and jaws of apes and humans. In the **Records** section of your report, make a table to record your observations and measurements for each of the features in the checklist.

Ape

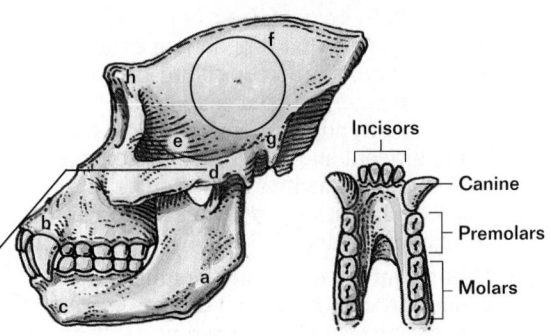

Human

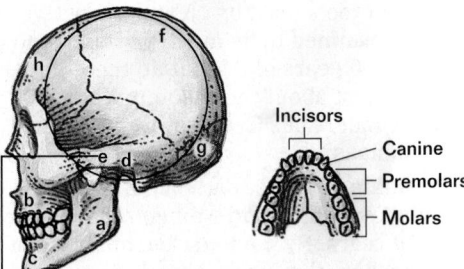

Checklist of Anatomical Features

Brain Capacity: The circle drawn on each skull represents the brain capacity. Measure the radius of each circle in centimeters, cube this number, and multiply by 1,000 to approximate the life-size brain capacity in cubic centimeters.

Lower Face Area: Measure *a* to *b* and *c* to *d* in centimeters for each skull. Multiply these two numbers together, and multiply the product by 40 to approximate the life-size lower face area in square centimeters.

Brain Area: Measure *e* to *f* and *g* to *h* in centimeters for each skull. Multiply these two numbers together, and multiply the product by 40 to approximate the life-size brain area in square centimeters.

Jaw Angle: Note the two lines in the nose area of each skull. Measure the inside angle with your protractor to determine how far outward the jaw projects.

Brow Ridge: Note the presence or absence of a bony ridge above the eye sockets.

Teeth: Note the number and kind of teeth in the lower jaw.

Records

Cranial Features of Modern Humans and Apes

	Brain capacity (cm³)	Lower face area (cm²)	Brain area (cm²)	Jaw angle (degrees)	Brow ridge	Teeth
Ape	422	420	230	130	yes	6 molars, 4 premolars, 2 canines, 4 incisors
Human	1953	161	264	80	no	same as above

Part B: Fossil Hominids

2. Refer again to the checklist of anatomical features. Measure any features possible for the four fossil hominid skulls. Record your measurements in a table in the **Records** section of your report

3. Classify each feature of the hominid skulls as being apelike, humanlike, or intermediate by writing an A, H, or I next to the feature. Note that a large brain capacity and the absence of a prominent brow ridge are characteristic of humans. Also, less lower face area and more brain area is typical of modern humans. Refer to your data for the ape and human skulls to help you classify the features of the hominid skulls. In the your report, briefly summarize the procedure you followed.

Fossil Hominids

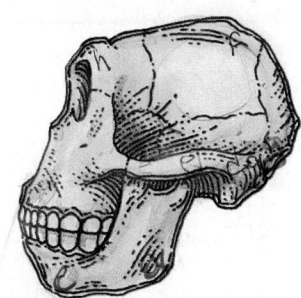

Australopithecus robustus

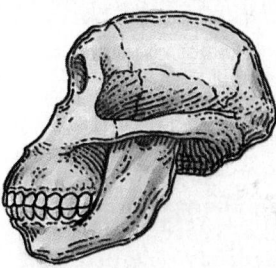
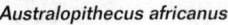

Australopithecus africanus

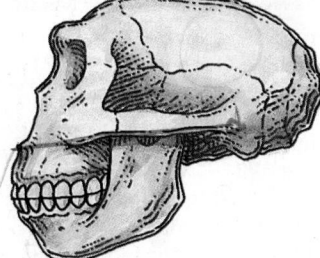

Homo erectus

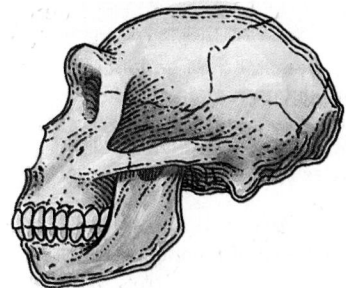

Neanderthal

INQUIRY

1. What is the relationship between the skull features and the brain size of humans compared with that of apes?
2. Based on your observations and measurements, which of the fossil hominid skulls is most apelike? Most humanlike?

ANALYSIS

1. What can you infer about the diets of apes and humans from the shapes of their teeth and jaws?
2. How can the hominid fossils be related to both humans and apes without actually being either?
3. Describe the overall changes in skull structure in the evolution of humans.
4. Most skeletons of fossil hominids are not complete. Why do you think this is so?

FURTHER INQUIRY

Write a new question that could be explored as a new investigation. The following is an example:

> What was the probable diet of fossil hominids, based on their facial features?

Inquiry Answers

1. The human skull has a high forehead that provides more space for the development of the front part of the brain. Apes have a low forehead, and thus show less development of the brain. Also, the ratio of brain area to face area is larger in humans than in apes.

2. Students should recognize that the Neanderthal skull is most humanlike because of its large brain area and nonprotruding jaw. Students may disagree about which hominid skull is most apelike, *Australopithecus robustus* or *Australopithecus africanus*. Both Australopithecine skulls have a small brain area, a brow ridge, and a protruding jaw, albeit to varying degrees.

Analysis Answers

1. The ape skull has sharp front teeth for biting into fruits, and large, flat back teeth used for grinding them. The spacing of the teeth enables apes to bite off leaves and pull branches through the spaces, retaining the leaves in the mouth for chewing. Human teeth are smaller and are adapted for a more varied, omnivorous diet.

2. The fossil hominids are intermediate species in the evolution of humans from an apelike ancestor.

3. The brow ridge disappeared, the brain area increased, and the jaw became less prominent.

4. Answers will vary. Students might suggest that early humans did not bury their dead; consequently, the bodies may have been dismembered and scattered by scavengers or by natural forces.

Cranial Features of Early Hominids

Hominid	Lower face area (cm²)	Brain area (cm²)	Jaw angle (degrees)	Brow ridge
A. africanus	252 (I)	190 (A)	125 (I)	yes
A. robustus	300 (I)	260 (I)	115 (I)	yes
H. erectus	312 (I)	252 (I)	111 (I)	yes
Neanderthal	300 (I)	308 (H)	115 (I)	yes

UNIT 4

CHAPTER 15 | POPULATIONS

Block Scheduling Guide

> Each block represents about 45–50 minutes of instructional time. The resources cited in a block are coded to a specific program component using the Key on the next page.

Lecture Concepts

Lesson Resources

15-1 How Populations Grow pp. 314–320

BLOCKS 1 and 2

a. A population consists of all the individuals of a species that live together in one place at one time. A population can be widely distributed or confined to a small area.

b. In the exponential model of population growth, the growth rate remains constant. Rapid, continuous growth is the result.

c. In the logistic model of population growth, the growth rate declines as population size rises. The population stabilizes at the carrying capacity, the maximal population size the environment can sustain.

d. Populations of species that are r-strategists are characterized by exponential growth. R-strategists tend to reproduce early in life and have many offspring that are small, mature rapidly, and receive little parental care.

e. Populations of species that are K-strategists are characterized by slow population growth. These species tend to reproduce late in life and have few offspring that are large, mature slowly, and often receive extensive parental care.

f. Human populations have evolved with many of the characteristics of K-strategists, however, the human population is now growing explosively.

Lecture Resources
- Opening Question, page 315
- Demonstrations: Age Structure, r- or K-Strategist?
- Population Density, page 316
- Visual Strategies: Figures 15-3, 15-5, 15-6, 15-7
- Effective Reading, page 316
- Dispersion, page 317
- Teaching Transparencies: 63A, 64A, 75
- Overcoming Misconceptions, page 319
- Holt Biology Videodiscs: Lesson 15-1

Classwork Options
- Comparing r-Strategists with K-Strategists, page 319
- Laboratory Investigation: A Population Study, pages 334–335
- Teaching Resources CD-ROM: Problem-Solving Worksheets, Population Size

Assignment Options
- Section Review, page 320
- Chapter Review, questions 1–3, 9-11, 17, 19, 22, 24, 25, 27

15-2 How Populations Evolve pp. 321–325

BLOCK 3

a. The Hardy-Weinberg principle states that the frequencies of alleles and genotypes in a population remain constant unless evolutionary forces act on them. This principle holds true only for large populations in which individuals mate randomly and in which forces that change the proportions of alleles are not acting.

b. The five forces that can cause allele frequencies to change are mutation, migration, nonrandom mating, genetic drift, and natural selection.

Lecture Resources
- Opening Question, page 321
- Mathematics, page 322
- Connections, page 322
- Multicultural Perspective, page 322
- Visual Strategy: Figure 15-10
- What If a Mutation Occurs in One of Your Cells? page 324
- Genetic Drift, page 324
- Research Update, page 325

- Holt Biology Videodiscs: Lesson 15-2

Classwork Options
- Calculating Frequencies, page 323

Assignment Options
- Mate Selection, page 324
- Section Review, page 325
- Chapter Review, questions 4–6, 12-14, 18, 21, 26

15-3 How Natural Selection Shapes Populations pp. 326–330

BLOCKS 4 and 5

a. Natural selection reduces the frequency of a harmful recessive allele slowly because very few individuals are homozygous recessive and express the allele.

b. There are three types of selection on polygenic traits. In directional selection, the range of phenotypes shifts toward one extreme. In stabilizing selection, the range of phenotypes narrows. In disruptive selection, the range of phenotypes "splits" and moves toward both extremes.

Lecture Resources
- Opening Questions, page 326
- Demonstrations: Selecting Phenotypes, Directional Selection
- A Recessive Condition, page 327
- Visual Strategy: Figure 15-15
- Holt Biology Videodiscs: Lesson 15-3

Classwork Options
- A Polygenic Trait, page 328

- The Three Types of Selection on Polygenic Traits and Graphic Organizer, page 329
- Inquiry Skills Development B9: Peppered Moth Survey
- Closure, page 330

Assignment Options
- Concept Mapping, page 331
- Section Review, page 330
- Chapter Review, questions 7–8, 15, 16, 20, 23

Review/Enrichment

BLOCK 6

■ Study Guide: Concept Review and Pretest

Assignment Options
◉ Teaching Resources CD-ROM: Supplemental Reading Worksheets and Test, *Silent Spring*
■ Activities and Projects, questions 28, 29
■ Readings, questions 30, 31

Assessment Options

BLOCK 7

Traditional Assessment
■ Chapter 15 Test
◉ Test Generator/Assessment Item Listing: Software item bank for preparing customized chapter tests.

Portfolio Assessment
Select student reports for one or more laboratory experiments from this chapter. *The Direct Observation Checklist* on the *BioSources Teaching Resources CD-ROM* should be completed during a laboratory or other cooperative learning experience.

Answer to Concept Map

The following is one possible answer to the Concept Mapping exercise on page 331.

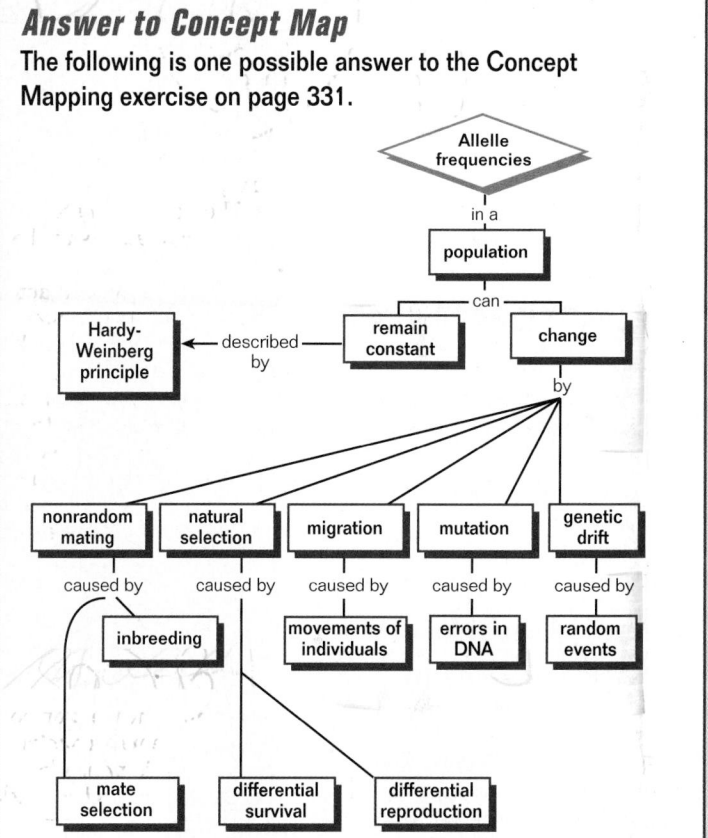

Holt Biology Videodiscs

Holt Biology Videodiscs gives you a powerful tool for teaching, review, and assessment. *Concepts of Biology* adds interactive lesson plans and extensive tools for customization using CD-ROM technology.

CONCEPTS OF BIOLOGY

Use the following topics from *Concepts of Biology* to help you teach this chapter:

■ Topic 8: Evolution and Natural Selection
■ Topic 11: Global Ecology

For further information regarding the media on the videodiscs, see *Holt Biology Videodiscs Teacher's Correlation Guide to Biology: Principles and Explorations.*

Chapter Theme

Evolution *Soon after Mendel's contribution to genetics was recognized, biologists began to extend the study of heredity beyond the analysis of the genetic makeup of individuals to the genetic makeup of populations. Population genetics allows scientists to determine whether evolution is occurring in populations. Scientists have identified five forces that cause populations to evolve—natural selection, mutation, migration, genetic drift, and nonrandom mating.*

Tapping Prior Knowledge

- What is a population?
- Under what conditions will a population grow, shrink, or remain stable?
- How might natural selection affect a population?

Opening Demonstration

Show the class a picture that depicts an abnormality caused by a dominant allele. An example might be brachydactylism (abnormally short fingers) or achondroplasia (hereditary dwarfism). Point out that scientists at first wondered why the dominant allele didn't simply "drive out" the recessive allele so that most, if not all, people would eventually inherit the condition. Inform students that in 1908 G. Hardy and W. Weinberg showed that the relative number of people with the normal condition should remain the same generation after generation—as long as certain conditions were satisfied.

CHAPTER 15 POPULATIONS

Atlantic Puffins

REVIEW

- Mendel's laws (Section 7-1)
- phenotype and genotype (Section 7-1)
- mutation (Section 9-3)
- natural selection (Section 12-1)

Authors' Rationale

In this chapter your students will become familiar with population dynamics. They will learn what populations are, how they evolve, and the extrinsic and intrinsic factors that control population size. The behavior of genes in a population can be studied in organisms with short generation times (such as fruit flies), but the overall concepts learned can be applied even to human populations. Students will be able to answer questions about population such as "Why is the human population the size it is?" and "What is going to happen if birth rates don't decline?"

15-1 How Populations Grow

*T**he commuters crowded into the New York subway shown in Figure 15-1 are members of a population of about 16 million people living in the greater New York City area. On a larger scale, they belong to the rapidly growing human population of the world. Since 1930, the world's human population has nearly tripled. What causes populations to grow? What determines how fast they grow? Is there anything that can slow their growth? You will be able to answer these questions after reading this section.*

Section Objectives

■ Distinguish between the three patterns of dispersion in a population.

■ Contrast exponential growth with logistic growth.

■ Identify the differences between r-strategists and K-strategists.

Figure 15-1 The world's human population has reached an unprecedented size of nearly 6 billion. Today, more people, including these rush-hour commuters, live in the New York City area than lived on Earth 10,000 years ago.

What Is a Population?

A **population** consists of all of the individuals of a species that live together in one place at one time. This is a flexible but useful definition; it allows scientists to use similar terms when speaking of the world's human population, the population of bacteria that live in your intestine, and the population of Devil's Hole pupfish that swims in the tiny pool shown in Figure 15-2.

Figure 15-2 A population can be widely distributed or confined to a small area. An extreme example is the Devil's Hole pupfish, *right.* This species consists of one population of several hundred individuals, which inhabit this small pool, *left,* in Death Valley, Nevada.

SECTION 15-1

Vocabulary Preview

population
demography
population size
population density
dispersion
model
exponential growth curve
carrying capacity
logistic model
r-strategist
K-strategist

Opening Question

Ask students to write a paragraph that shows a relationship between the following terms: meiosis, mutation, genetic variation, sexual reproduction, natural selection, and evolution.

Demonstration
Age Structure

One important attribute of any population is its age distribution. Age distributions differ between populations and depend partly on the rate of growth of the population. Draw a pyramid-shaped outline that depicts the age distribution for Mexico. Indicate that the base represents the youngest age group and that older groups are found higher in the pyramid. Traditionally the left side of the diagram shows the age distribution of males, and the right side shows the distribution of females.

Alongside this pyramid, draw a bottle-shaped outline that represents the age distribution for the United States. Ask students to describe how these two populations differ and what factors may account for these differences. For example, students should recognize that the youngest age group is significantly larger in the first population, likely signifying a much higher birth rate.

Every population tends to grow in size for the simple reason that individuals tend to have more than one offspring. The environment's capacity to support the population limits its size. The statistical study of populations is called **demography** (*duh MAW gruh fee*). This term is derived from two Greek words: *demos,* meaning "people" (the same root word as in "democracy"), and *graphos,* meaning "measurement." Demography helps to predict how the size of a population will change.

Three Key Features of Populations

To predict how a population will grow, it is first necessary to look at it closely. Every population has a set of key features that play a large role in determining its future. One of the most important features of any population is its current size. **Population size,** the number of individuals in a population, has an important effect on the ability of the population to survive. Many studies have shown that very small populations are most likely to become extinct. Random events or natural disturbances may threaten the continued survival of a small population containing only a few individuals more than they would endanger a large population. Inbreeding (breeding with relatives) may become common in small populations because only relatives are available as mates. Inbreeding produces a more genetically uniform population in which recessive traits, many of them unfavorable, are more likely to be homozygous and therefore expressed. Also, the reduced level of variability that results from inbreeding is likely to detract from the population's ability to adjust to changing conditions.

A second important feature of any population is its **population density.** Density refers to the number of individuals found in a given area, as shown in Figure 15-3. If the individuals of a population are widely spaced, they may rarely, if ever, encounter one another, making reproduction rare. This can happen even if the number of individuals over a wide area is relatively high.

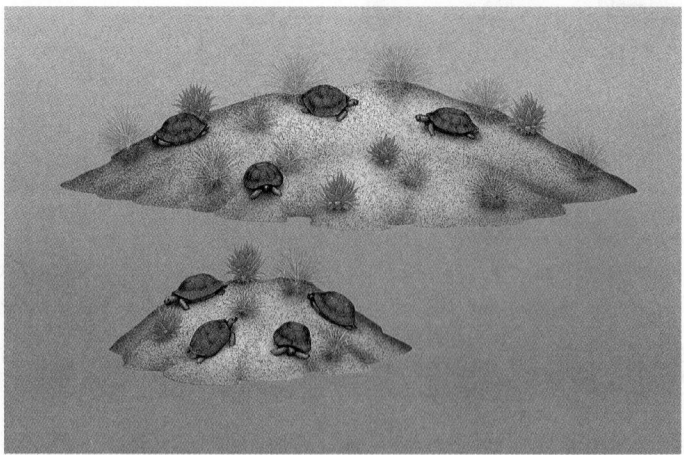

Figure 15-3 Both islands contain the same number of tortoises. However, because the upper island is twice as large as the lower island, it has half the population density of tortoises.

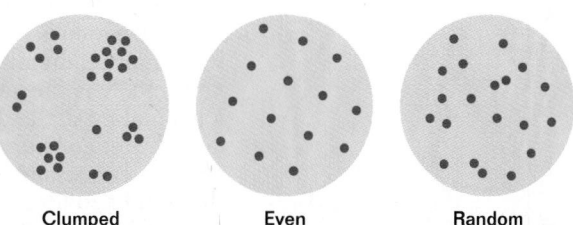

Clumped   Even   Random

Teaching Tip
Dispersion
Point out that the dispersion pattern for a particular population depends upon the scale of observation. For example, a population of birds may seem to be randomly distributed if an observer is looking only at one town. But the same population may seem to be clumped when viewed on a larger scale, such as the entire county.

A third important feature of any population is the way in which the individuals of the population are arranged, a measure called **dispersion.** The three main patterns of dispersion possible within a population are illustrated in Figure 15-4. If individuals are randomly spaced, the location of each individual is determined by chance. If individuals are evenly spaced, they are located at regular intervals. In a clumped distribution, individuals are bunched together in clusters. Each of these patterns reflects the interactions between the population and its environment. Clumped distributions are the most common type of distribution in nature. One reason they are common is that individual organisms tend to seek out particular sets of conditions (called microhabitats) that may occur only in certain spots—combinations of soil type, moisture, and host trees, for example.

Figure 15-4 This panel, *top right,* shows the three patterns of dispersion possible in populations. Turtles seeking the warmest basking sites form a clumped distribution, *bottom right.* Starlings arrange themselves evenly along telephone lines, *left.*

When Do Populations Grow?

When trying to predict how a population will grow, demographers construct a model of the population. A **model** of a population is a hypothetical population that has key characteristics of the real population being studied. By making a change in the model and observing what happens, it is possible to learn something about what might take place in nature if similar changes occur.

To learn how demographers study a population, you will construct a simple model of population growth. It is not difficult to do. You will develop your model in three stages by asking some basic questions.

VISUAL STRATEGY

Figure 15-5

Point out that exponential growth begins very slowly and then shoots up very rapidly as the number of reproducing individuals increases with each generation. This type of growth curve is seen mostly in microorganisms cultured under laboratory conditions.

Teaching Tips

Population Growth Rates and Interest Rates

The same formulas used to describe the exponential growth of populations can be used to describe the growth of bank accounts (or debts) with compound interest. Students have probably seen problems like this in math classes: If $100 is deposited in an account that pays 6 percent per year compounded annually, by how much will the account grow in the first year? The answer is simply the interest rate multiplied by the amount deposited, or $6. Each term in this equation corresponds to a term in the Stage II model: the interest rate corresponds to r, the principal to N, and the amount added to the account to ΔN. Students can use the Stage II model to calculate by how much a population will grow in a given period of time. For instance, the world's human population is about 5.6 billion and is growing by about 1.68 percent per year. The increase in world population in one year, ΔN, is the growth rate multiplied by the current population size, or 94 million.

What Values Can r Take?

Explain how r may have negative or positive values or may be 0. For a population in which the birth rate equals the death rate, r equals 0; for those in which the birth rate exceeds the death rate, r is positive; and for those in which the death rate exceeds the birth rate, r is negative.

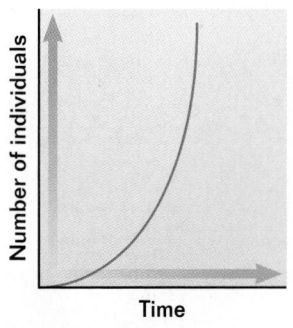

Figure 15-5 This J-shaped curve is characteristic of exponential growth.

Stage I Model

First, when does a population grow? The answer is obvious: when more individuals are born than die. So the first stage of your model is the simple statement that the rate of growth of a population is the difference between the birth rate and the death rate. For human populations, birth and death rates are usually expressed as the number of births and deaths per thousand people per year. The stage I model is usually written as a brief equation.

$$r \text{ (rate of growth)} = \text{birth rate} - \text{death rate}$$

Stage II Model

If you want to calculate the number of individuals that will be added to the population as it grows, symbolized by ΔN (read as "delta N"), multiply the size of the current population, symbolized by N, by the rate of growth.

$$\Delta N = r N$$

When population size is plotted against time, this stage II model produces what is called an **exponential growth curve.** In exponential growth, the rate of increase (r) remains constant, but the amount by which the population grows (ΔN) rises quickly as the size of the population increases. Note that a key assumption of this model is that the birth and death rates (which determine the rate of growth) do not change. Figure 15-5 illustrates exponential growth. In an exponentially growing population that initially contained just a single bacterial cell that divided every 30 minutes, there would be over a million bacteria after only 10 hours.

In fact, no population exhibits exponential growth for very long. Within a year, a population that began as a single bacterium and reproduced at its maximal rate could cover the Earth in a layer over a kilometer thick. The reason populations do not continue to grow unchecked is that they begin to run out of resources and to accumulate wastes. Eventually growth slows and the size of the population stabilizes. The population size that an environment can sustain is called the **carrying capacity,** symbolized by K.

Stage III Model

Your population growth model can be adjusted to compensate for diminishing resources by multiplying it by the fraction of resources still available. This fraction is $(K–N)/K$. As a population grows, smaller and smaller amounts of resources remain. The final model thus becomes:

$$\Delta N = r N [(K - N) / K]$$

Because this model accounts for the declining resources available to populations as they grow, it is called the **logistic model** of population growth. The word *logistics* means procuring, maintaining, and transporting materials—solving the day-to-day problems of living. The logistic model assumes that birth and death rates are not constant, but vary with population size. As the population grows in size, birth rates decline and death rates rise, and consequently the rate of growth falls. Eventually, as N approaches K, the population ceases to grow (that is, $\Delta N = 0$) because the birth rate equals the death rate. Logistic growth is illustrated in Figure 15-6.

The logistic model of population growth, though simple, provides an excellent description of how populations grow in nature. Competition for food, shelter, mating sites, and other resources, as well as the accumulation of toxic wastes, tends to increase as a population approaches its carrying capacity. ■

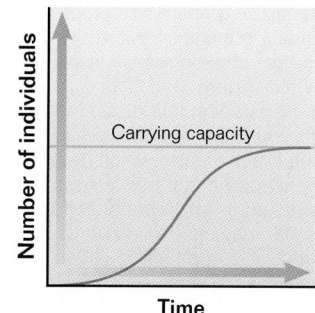

Figure 15-6 The curve of logistic growth looks like a stretched-out letter *s*.

Two Strategies of Population Growth

The population growth curve described by the logistic model illustrates two very different ways that a population can prosper in a competitive world. One approach to population growth depends largely on the r term (rate of growth) of the logistic equation, while the other is influenced predominantly by the K term (carrying capacity). In nature, most organisms employ one strategy or the other (not consciously, of course—natural selection simply favors those taking particular approaches). Some organisms, however, employ a strategy somewhere between the two extremes or change from one to the other as their environment changes.

Populations of *r*-strategists Can Grow Rapidly

Many species, including bacteria, some annual plants (those with only one growing season), and a number of species of insects, are r-strategists. Populations of species that are **r-strategists** are characterized by exponential growth—which results in temporarily large populations—followed by sudden crashes in population size. These species tend to live in unpredictable and rapidly changing environments, where it pays to be able to reproduce quickly when conditions are favorable. In general, r-strategists reproduce early in life and have many offspring each time they reproduce. Their offspring are small, mature rapidly, and receive little or no parental care. The parent thus spreads its reproductive investment among many offspring, investing little in each.

Populations of *K*-strategists Tend to Grow Slowly

Other species, such as redwood trees, whales, and rhinoceroses, have small population sizes and slow population growth. Populations of these species, called **K-strategists,** are characterized by a high degree of specialization. K-strategists tend to live in environments that are stable and predictable, where it pays to be able to compete effectively. K-strategists tend to reproduce late in life and to have few offspring each time they reproduce. These offspring are large, mature slowly, and often receive extensive parental care. Many of the plants and animals that are in danger of extinction today, such as tigers, rhinoceroses, and gorillas, are K-strategists.

Application

 **Mathematics** Have students calculate how many people are born each day, based on the estimate that 94 million people are added to the human population each year. (*Approximately 257,500 people are born every day.*)

👁 VISUAL STRATEGY

Figure 15-7
Inform students that it took about 150 years (1650–1800) for the human population to double from 500 million to 1 billion. The second doubling took only 130 years, the third took 45 years, and the fourth should take 45 years and be reached in about the year 2020.

Figure 15-7 Compare this graph of human population growth with the graph of exponential growth shown in Figure 15-5. The decrease in population size in the fourteenth century was caused by the Black Death, which devastated Europe and parts of Asia. This disease, which is transmitted by fleas, may have killed up to 50 percent of Europe's population between 1347 and 1352. Notice that World War I (1914–1918), in which almost 10 million died, and World War II (1939–1945), in which more than 50 million died, did not cause declines in the world's population.

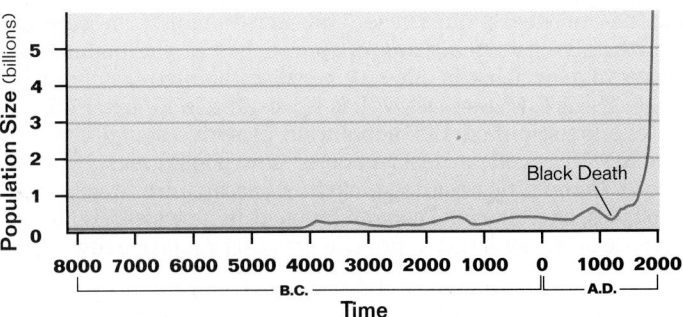

The Rapidly Growing Human Population

Human populations have evolved with many of the characteristics of K-strategists, including small families, extended parental care, and a large investment in each offspring. However, in recent times humans have greatly expanded the carrying capacity of their environment. The development of agriculture and technology has enabled humans to increase their food supply, combat pests, and cure diseases. This increase in the value of K has led to a rapid increase in ΔN. In other words, the human population is growing explosively, as shown in Figure 15-7. By 1995 there were more than 5.6 billion people on Earth, and the human population was growing at a rate of about 1.7 percent annually. Nearly 94 million people are added to the population every year. At this rate, the human population will double in just over 40 years. No one knows if the Earth can continue to support such a large population, although it seems clear that the human population cannot expand at this rate for long without seriously damaging the environment. Damage to the planet will eventually reduce the carrying capacity for humanity and therefore slow the growth of the human population.

Section Review

1. *Are fans attending a basketball game randomly dispersed, evenly dispersed, or clumped? Explain your answer.*

2. *The logistic model of population growth assumes that as population size increases, birth rates fall and death rates rise. Does this assumption reflect what is likely to happen in nature? Justify your answer.*

3. *Explain why an r-strategist might be better suited for an unpredictable environment than a K-strategist would be.*

Critical Thinking

4. *Suppose that a population is shrinking. What must be true of its birth and death rates?*

Answers to Section Review

1. The fans are clumped into different sections but are evenly dispersed within each section.

2. Yes, because as a population grows, the resources its members need are stretched thinner and thinner. Therefore, each individual gets less of these resources, making death more likely and reproduction less likely.

3. *r*-strategists reproduce and grow rapidly. Thus, if conditions become favorable even for a short time, they can still mature and have offspring. For a slow-growing *K*-strategist, long favorable periods are required to allow maturation and reproduction to be completed. Such favorable periods would be very rare in an unpredictable environment.

4. Its birth rate must not be keeping pace with its death rate, so *r* for this population is negative.

15-2 How Populations Evolve

Darwin proposed that biological diversity is the result of natural selection. In the modern statement of Darwin's theory, every natural population contains individuals with alternative versions of genes, known as alleles. Those individuals having alleles that improve the chances of survival and reproduction are favored, and so these alleles become more common. In Darwin's day, no one knew what caused genetic variation. Biologists now know it is produced by mutations, changes in DNA.

In the century since Darwin's death, the science of genetics has revolutionized how biologists think of heredity and has allowed them to construct detailed models of how natural selection alters the proportions of alleles within populations. Models of allele changes within populations have proven very useful, often allowing biologists to predict how a particular population will respond to a change in its habitat. Before you can understand how populations change in response to evolutionary forces, you need to understand how they behave in the absence of these forces.

Unchanging Populations

When Mendel's work was rediscovered in the early 1900s, biologists began to investigate how alleles might increase or decline in abundance. Specifically, they wondered if dominant alleles, which are usually more common than recessive alleles, would replace recessive alleles simply by sheer weight of numbers. In 1908, the English mathematician G. H. Hardy and the German physician Wilhelm Weinberg independently demonstrated that dominant alleles do not in fact replace recessive ones. With simple algebra and probability, they showed that the frequencies of alleles and proportions of heterozygous and homozygous individuals remain constant from generation to generation. Their demonstration, called the **Hardy-Weinberg principle,** states that populations do not change unless evolutionary forces act upon them. The Hardy-Weinberg principle holds true only for large populations in which individuals mate randomly and in which the forces that change the proportions of alleles are not acting.

The Hardy-Weinberg principle is usually stated as an equation. For a gene with two alternative alleles, *A* and *a,* the equation looks like the following.

Figure 15-8 In Chapter 12, you learned about the evolution of coloration in peppered moths like these. In this chapter you will learn how alleles responsible for traits such as coloration change in abundance under natural selection.

Vocabulary Preview

Hardy-Weinberg principle
allele frequency
gene flow
nonrandom mating
genetic drift

Opening Question

Darwin and Mendel were contemporaries. There is reason to believe that Mendel knew of Darwin's work. However, Darwin's writings about variation show that he was totally unaware of Mendel's work with garden peas. Ask the class to write a short script of what might have been said between Darwin and Mendel if they had stood together looking at a population of tortoises on the Galapagos Islands.

Demonstration

Prepare students for the Hardy-Weinberg equation by demonstrating that this equation can also be used to calculate the probabilities of some more common events. For example, suppose that a coin is flipped twice. The equation that describes the probabilities of the three possible outcomes—two heads, two tails, one head and one tail—is identical to the Hardy-Weinberg equation. Show students how to derive this equation for the case of coin flipping. Point out that although the probabilities of the two possible events—heads or tails—are equal in this example, the frequencies of two alleles in a population probably won't be.

Historical Note

An American Contribution

Five years before Hardy and Weinberg separately published their papers, an American scientist, William Castle, published a paper containing similar observations about population genetics. However, Castle's paper was generally neglected, and his contribution was recognized only after the publications of Hardy's and Weinberg's papers.

Application

Mathematics Remind students that if both sides of an equation are treated in the same way, the equation will remain unchanged. Thus if both sides of $p + q = 1$ are squared, the result is the equation $p^2 + 2pq + q^2 = 1$.

Teaching Tip

Inform students that the Hardy-Weinberg equation applies to any trait, no matter how many alleles are involved. For example, if three alleles, like those controlling blood types (A, B, and O), are present, the Hardy-Weinberg equation becomes $p + q + r = 1$, where p, q, and r represent the frequencies of the three alleles.

CONNECTIONS

Chapter 7

Patterns of Inheritance

Review how two individuals with a normal phenotype but a heterozygous genotype can produce offspring that are homozygous recessive.

Figure 15-9 White coloration in tigers is caused by a rare recessive allele. Do you think this allele is likely to become more abundant?

$$p^2 \quad + \quad 2pq \quad + \quad q^2 \quad = \quad 1$$

| frequency of individuals homozygous for allele A | frequency of heterozygous individuals, with alleles A and a | frequency of individuals homozygous for allele a | |

A frequency is the proportion of a group that is of one type. For example, the frequency of boys in your class is the number of boys divided by the total number of students in the class. In a population of 100 tigers containing 84 striped tigers and 16 white (albino) tigers, the frequency of striped tigers is 84/100, or 0.84; the frequency of white tigers is 16/100, or 0.16. The p and q symbols in the Hardy-Weinberg equation are **allele frequencies.** The frequency of allele A is the proportion of all alleles for this gene in the population that are A. Similarly, the frequency of allele a is the proportion of alleles that are a. By convention, the frequency of the more common of the two alleles is designated p, and the frequency of the rarer allele is designated q. Because there are only two alleles, $p + q$ must always equal 1.

Applying the Hardy-Weinberg Principle

You can use the Hardy-Weinberg principle to predict genotype frequencies (frequencies of homozygotes and heterozygotes) from allele frequencies. How can you do that? Remember that p represents the frequency of allele A, and q represents the frequency of allele a. Individuals homozygous for allele A have two copies of this allele, so they occur at a frequency of p times p, or p^2. Individuals homozygous for allele a have two copies of a and occur at the frequency of q times q, or q^2. Heterozygous individuals have one copy of A and one copy of a, but heterozygotes can occur in *two* ways—A from the father and a from the mother, or a from the father and A from the mother. Therefore, the frequency of heterozygotes is $2pq$. Every individual in the population must be either AA, Aa, or aa, so the sum of the three frequencies must equal 1.

$$p^2 + 2pq + q^2 = 1.$$

The importance of the Hardy-Weinberg equation is that it allows you to predict the frequency of each genotype in a population if you know the allele frequencies. The Hardy-Weinberg principle has proven very useful in assessing real-life situations. For example, as you learned in Chapter 7, the often-fatal human disorder cystic fibrosis is caused by a recessive allele. This disorder affects Caucasian North Americans at a frequency of about 1 in every 2,080 individuals, or 0.00048. What proportion of Caucasian North Americans are expected to be heterozygous for this allele? You can easily figure this out.

Multicultural Perspective

Allele Frequencies

Allele frequencies in humans often differ among populations. One well-studied example is the A, B, and O blood group alleles. This chart shows the A-B-O frequencies for several populations.

Group	Frequency of A	Frequency of B	Frequency of O
United States (Caucasians)	0.258	0.070	0.673
United States (African Americans)	0.168	0.131	0.701
China	0.181	0.158	0.662
Peru (Native Americans)	0.073	0.000	0.928

1. **Calculate the frequency of the recessive allele.** You know that q^2, the frequency of homozygous recessive individuals, is 0.00048. Therefore q, the frequency of the cystic fibrosis allele, is the square root of 0.00048, or 0.022.
2. **Calculate the frequency of the dominant allele.** You can calculate p, the frequency of the dominant allele, by subtraction. Because $p + q = 1$, $p = 1 - q$. So $p = 1 - 0.022$, or 0.978.
3. **Determine the frequency of heterozygotes.** The frequency of heterozygous individuals is expected to be $2pq$, or 2 times 0.978 times 0.022, or 0.043. This means that 43 of every 1,000 Caucasian North Americans are predicted to carry the cystic fibrosis allele unexpressed. Figure 15-10 summarizes the genotype frequencies for this example.

Genotype	Phenotype	Frequency
CC	Normal	$p^2 = 0.96$
Cc	Normal	$2pq = 0.043$
cc	Cystic fibrosis	$q^2 = 0.0048$

Figure 15-10 This chart summarizes the calculations of genotype and phenotype frequencies made in the text.

How valid are the estimates of genotype frequencies that are made using the Hardy-Weinberg equation? For all but small, isolated human populations, they prove to be very accurate. Most human populations are large and not inbred and therefore resemble the ideal population assumed by Hardy and Weinberg. Remember that the frequencies of genotypes deviate from those predicted by the Hardy-Weinberg equation only when there are evolutionary forces acting on the population that tend to favor one allele over another. *The Hardy-Weinberg equation is valid only when nature is blind to which allele an individual carries.* When the identity of the allele does matter, the equation no longer applies. ❑

❑ CAPSULE SUMMARY

The Hardy-Weinberg principle states that the frequencies of alleles and genotypes remain constant in populations in which evolutionary forces are not acting.

Five Forces Cause Populations to Evolve

In large, randomly mating (that is, not inbred) populations, such as most human populations, the frequencies of alleles and genotypes remain constant from generation to generation unless evolutionary forces act on the population. There are five evolutionary forces that can cause the proportions of homozygotes and heterozygotes in a population to differ significantly from those predicted by the Hardy-Weinberg equation.

Mutation

Although mutation from one allele to another can obviously alter allele frequencies, mutation rates in nature are so low

SECTION 15-2

Teaching Tip

Calculating Frequencies

Have students calculate the frequency of heterozygotes if a certain condition caused by a recessive allele appears in 1 in 10,000 infants. *($q^2 = 1/10{,}000$, or 0.0001. Thus, $q = \sqrt{0.0001}$, or 0.01, and $p = 1 - 0.01 = 0.99$. The frequency of heterozygotes, or $2pq = 2 \times (0.99)(0.01) = 0.0198$, or almost 0.02.)* Thus, about 2 percent of the population, or about 1 in every 50 people, is estimated to be a heterozygote for this condition.

VISUAL STRATEGY

Figure 15-10

Be sure that students understand the chart on how to calculate genotype and phenotype frequencies. Have them calculate the expected frequency of heterozygotes for albinism if 1 in every 20,000 people is an albino, a condition caused by a homozygous recessive condition *($2pq = 0.014$).* Be sure to emphasize that the calculations are based on the assumption that the trait is controlled by only one locus.

Did You Know?

Each person has about 100,000 genes. It is estimated that each human carries two new mutations. Point out that these are spontaneous mutations, and emphasize that different genes—and even different alleles—have different rates of mutation.

Figure 15-11 Normal fruit flies have two wings. This mutant has four. This rare mutation, like most mutations, is harmful. Beneficial mutations are the raw material for natural selection.

Figure 15-12 What characteristics does a female widowbird find attractive? As Malte Andersson of Oxford University discovered, she prefers long tails like the one on this male. Moreover, the longer the tail is, the more attractive it is. Andersson showed that females preferred to nest in areas controlled by males whose tails had been artificially extended. Darwin recognized that female preference for extreme male traits could explain the evolution of elaborate male ornamentation that seems to be detrimental to survival.

that proportions of common alleles are not measurably affected. Most genes mutate only about 1 to 10 times per 100,000 cell divisions, so mutation itself does not significantly change allele frequencies. It is, however, the ultimate source of all variation and thus makes evolution possible. Figure 15-11 illustrates an example of mutation.

Migration

Migration is the movement of individuals from one population to another. It can be a powerful force for genetic change. Migration of individuals to or from a population creates **gene flow,** the movement of alleles into or out of a population. Gene flow occurs because new individuals add alleles to the population, and departing individuals take alleles away.

Nonrandom Mating

Sometimes individuals prefer to mate with others of their own genotype, a situation called **nonrandom mating.** Mating with relatives, which is known as inbreeding, is a type of nonrandom mating that causes the frequency of heterozygotes to be much less than that predicted by the Hardy-Weinberg equation. Inbreeding does not change the frequencies of alleles; it increases the proportion of homozygotes in a population. For example, populations of self-fertilizing plants consist primarily of homozygous individuals. Nonrandom mating also results when organisms choose their mates. In animals, females often select males based on their size, coloration, ability to gather food, or other characteristics, as shown in Figure 15-12.

Genetic Drift

In small populations, the frequency of an allele can be changed drastically by a chance event, such as a fire, landslide, or lightning strike. When an allele is found in only a few individuals, the loss of even one individual from the population can have

major effects on its frequency. Since this sort of change in allele frequency appears to occur randomly, as if the frequencies were drifting, it is called **genetic drift.** Small populations that are isolated from each other can differ greatly as a result of genetic drift. Because we humans lived in small groups for much of our history, genetic drift must have been a particularly important factor in the evolution of our species. Cheetahs are another species whose evolution has been profoundly affected by genetic drift, as described in Figure 15-13.

Figure 15-13 Cheetahs are an endangered species, and efforts to save them are complicated by their history. Cheetahs have gone through at least two drastic declines in population size. As a result, the surviving cheetahs are descendants of only a few individuals, and the species is genetically uniform. One consequence of this genetic uniformity is reduced disease resistance—cheetah cubs are more likely to die from disease than are the cubs of lions or leopards. These chance reductions in the cheetah's population size, examples of genetic drift, may result in the extinction of this species.

Natural Selection

Natural selection causes deviations from the Hardy-Weinberg proportions by directly changing the frequencies of alleles. An allele can increase or decrease in frequency, depending on its effects on survival and reproduction. For example, the allele for sickle cell anemia is slowly declining in frequency in the United States because individuals homozygous for this allele rarely produce children. Selection is one of the most powerful agents of genetic change. You will learn more about how natural selection alters allele frequencies in the next section. ❏

❏ CAPSULE SUMMARY

The five forces that can cause allele frequencies in a population to change are mutation, migration, nonrandom mating, genetic drift, and natural selection.

Section Review

1. *The frequency of striped tigers in a population is 0.99. From this information, can you calculate the frequencies of alleles and genotypes? If not, what additional information is needed?*

2. *Distinguish between natural selection and genetic drift.*

Critical Thinking

3. *Recall from Chapter 12 that in Africa the frequency of the sickle cell allele is not changing. Is this an example of the Hardy-Weinberg principle? Explain your answer.*

SECTION 15-2

UPDATE

☑ RESEARCH Update

Hair-Raising Population Statistics

Rob Bonnichsen, an archaeologist at Oregon State University, was methodically sieving sediment at a 14,500-year-old site in Montana when it occurred to him that the small hairs often left behind on the sieving screens might just be an important kind of artifact. While the conventional bounty of the archaeologist tends to be bones and stone tools, contemporary scientists such as Bonnichsen are also turning to organic remains that could contain DNA clues about individuals, or even populations. In an attempt to collect these "microartifacts" to screen them for DNA, Bonnichsen collaborated with Marvin Beatty, a soil scientist from the University of Wisconsin. Together they were able to isolate fossil hairs, fish scales, feathers, plant materials, insects, and other tiny materials that normally would be discarded. The DNA contained in hairs, even fossil ones, can give scientists clues to the relationships of individuals in a population—if, of course, there is enough DNA to be isolated. From these miniature fossil artifacts, scientists have been able to get DNA "readings" from the hairs, but they have not been able to extract enough DNA to make positive statements about the artifacts' owners—such as how old they were or where they were from. If it does turn out to be early human DNA, Bonnichsen thinks important evidence will be contributed to the ongoing theories of human dispersion patterns. ☑

Answers to Section Review

1. Yes. From page 322, students know that white coloration is caused by a recessive allele. Therefore, the frequency of homozygous recessive tigers is 1–0.99, or 0.01. This value is q^2 and can be plugged into the Hardy-Weinberg equation.

2. Natural selection can operate on all populations, regardless of size. On the other hand, genetic drift occurs in small populations. Natural selection causes allele frequencies to change in a specific direction, while genetic drift leads to random changes in frequencies.

3. No. In Africa, the sickle cell allele remains in equilibrium because selection against homozygotes is balanced by selection for heterozygotes, so the Hardy-Weinberg principle does not apply.

Opening Questions

Have students imagine a rapidly expanding population that lives in an ideal environment where every individual that is born lives and reaches reproductive age. Is natural selection acting on this population? Have students explain the reasons for their answers. Now have them imagine another rapidly expanding population that lives in an environment where large individuals are more likely to die. Is natural selection acting on this population? Which individuals will make the greatest genetic contribution to future generations? Predict what this population will look like after 10 generations.

Demonstration

Selecting Phenotypes

Students must recognize that natural selection operates on phenotype and not genotype. Show the class two photographs, one of a human population where differences among phenotypes are obvious, and a second population where phenotypic differences are not very apparent. The latter could be one of penguins, which, like humans, form large populations of sexually reproducing individuals. Point out that no two penguins are exactly alike, something that is not obvious to humans but is to penguins. Every penguin knows its mate, its offspring, and its nesting-ground neighbors. Some of these variations could be favored by natural selection. For example, some penguins are slightly bigger, which may reduce their likelihood of being eaten by predators.

15-3 How Natural Selection Shapes Populations

Darwin argued, and more than a century of research has demonstrated, that natural selection is a powerful force in nature, constantly adjusting populations to accommodate an ever-changing environment. However, although selection is a powerful agent of genetic change, there are limits to what it can accomplish. These limits arise because selection does not act directly on genes. It does not favor certain individuals just because they have particular genes, but because of the consequences of possessing those genes—the individuals are bigger, or smarter, or have other favorable characteristics. Stated differently, natural selection acts on phenotype, not genotype.

Selection Against Unfavorable Recessive Alleles Is Slow

Figure 15-14 Alexis, *far right*, the son of Nicholas II, last Tsar of Russia, suffered from severe hemophilia, a disease caused by a recessive allele. For Alexis, even a fall or cut could lead to uncontrollable and potenially dangerous bleeding. Why hasn't natural selection eliminated such a harmful allele? You will learn the answer in this section.

When a mutation creates a new allele by altering the DNA sequence of a gene, the effect of the new allele on the phenotype is almost always recessive. Remember that genes code for proteins that influence the phenotype (what the individual is like) through their activity as enzymes, regulators, or structural elements. Most random changes in a gene are destructive, resulting in a protein whose shape is no longer well suited to its function. In a heterozygous individual, both altered and normal forms of the protein are made. As long as enough of the normal form is produced to perform its function, the phenotype of the heterozygous individual is normal.

Now think carefully about how natural selection might operate on a mutant allele. Only characteristics that are expressed can be targets of natural selection. It follows that selection cannot operate efficiently against rare recessive alleles, even if they are unfavorable. Only when the allele becomes common enough that individuals carrying it come together and produce homozygous offspring does natural selection have an opportunity to act.

To better understand this limitation on natural selection, go back for a moment to the Hardy-Weinberg equation. When a recessive allele (a) is present at a frequency equal to 0.1, only 1 out of 100 individuals will be homozygous recessive (aa) and display the phenotype associated with this allele. But 18 out of 100 individuals will be heterozygotes (Aa) and carry the allele unexpressed. So natural selection can act on only 1 out of every 19 individuals that carry the allele. If the frequency in the population of the recessive allele is 0.01, the frequency of homozygous recessive individuals in that population will be only 1 in 10,000. For a slightly higher frequency of 0.02, the frequency of homozygous recessive individuals is still only 1 in 2,500. Many human diseases caused by recessive alleles have frequencies similar to this. These genetic disorders are not eliminated by natural selection, because very few of the individuals bearing the alleles express them. ◻

◻ **CAPSULE SUMMARY**

Natural selection reduces the frequency of a harmful recessive allele slowly because very few individuals are homozygous recessive and express the allele.

Many Genes Have More Than One Common Allele

....

Selection is a very powerful force for genetic change in biological communities, for the simple reason that not all alleles are rare. When biologists first learned how to measure the amount of genetic variation in natural populations, they found that a minimum of 15 percent of the genes of an average insect are heterozygous. Most of the variation is due to a small number of common alleles. When a gene in a population has more than one allele appearing at a significant frequency, the population is said to be polymorphic *(pah lee MAWR fihk)* for that gene. The genetic variation that results is called **genetic polymorphism**. Vertebrates are a little less polymorphic than insects. Humans, for example, are heterozygous for a minimum of 5 percent of their genes. This is an enormous pool of variation on which natural selection can act. Figure 15-15 illustrates one example of polymorphism in vertebrates.

Figure 15-15 All of the geese in this photo are lesser white snow geese. However, the two forms were once considered separate species. Now scientists know that one polymorphic gene with two alleles is responsible for the color difference. White coloration is recessive. Because geese prefer mates of their own color (an instance of nonrandom mating), both colors are maintained in the species.

Teaching Tip
A Recessive Condition

Point out that the disease phenylketonuria is caused by a recessive allele. Affected individuals lack the enzyme to break down the amino acid phenylalanine, which accumulates in the body and causes neurological damage. Treatment is a diet low in phenylalanine. About 1 in every 15,000 infants born in the United States is homozygous recessive for phenylketonuria. Ask students to calculate the frequency of the recessive allele. *($q^2 = 1/15,000$; $q = 0.0082$)*

 VISUAL STRATEGY

Figure 15-15

Ask students which genotype (in terms of p and q) will be least frequently found in these geese *(2pq, which represents the heterozygotes that would be produced by a mating between a white and a blue goose)*, and which genotypes would be most commonly found *(p^2 and q^2, which represent the homozygotes).*

 Did You Know?

The blood types A, B, AB, and O represent the most thoroughly studied polymorphic trait in the human population. Yet little is known about how this polymorphism is maintained. For example, the B allele is totally absent in the Native American population. Most Native Americans are type O. However, the Blackfoot tribe has the highest frequency of type A blood found anywhere in the world.

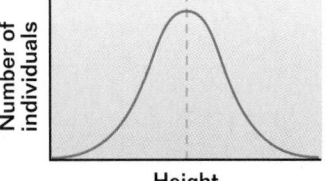

Height

Selection on Traits Controlled by More Than One Gene
·······························

A second reason that natural selection acts powerfully to shape populations is that many important characteristics of plants, animals, and other organisms are affected by a large number of genes. A characteristic influenced by several genes is called a **polygenic trait** *(pah lee JEHN ihk)*. Human height, for example, is determined by dozens of genes, each adding a little bit to the overall effect. Natural selection thus can alter the allele frequencies of many different genes governing a trait, influencing most strongly those genes that make the greatest contribution to the phenotype. It is therefore difficult to keep track of any one particular gene—it is like trying to follow one duck in a flock. Instead, what biologists do is more like keeping track of the entire flock. They measure the polygenic trait in each individual in the population and calculate the average. If selection is favoring increased height, for instance, then over time the average height within the population should increase.

Because of genetic polymorphism, polygenic traits tend to exhibit a range of phenotypes clustered around an average value. If you were to plot the height of everyone in your class on a graph, the values probably would form a hill-shaped curve called a **normal distribution,** with the average value at the summit, as illustrated in Figure 15-16.

Figure 15-16 This hill-shaped curve, *top,* is a normal distribution. The dotted line indicates the average height. In this photo taken around the turn of the century, cadets of the Connecticut Agricultural College, *bottom,* showed that height was distributed normally in their class.

Directional Selection

When selection acts to eliminate one extreme from a range of phenotypes, the genes promoting this extreme become less common in the population. This can be demonstrated in a simple experiment. If fruit flies are raised in the dark and then given a chance to fly toward light, some will fly toward light and some will not. In generation after generation, investigators selected those flies that had the strongest tendency to fly toward light as parents for the next generation (thus selecting *against* the light-avoiders).

After 20 generations, the average tendency to fly toward light had increased 50 percent within the population. Elimination of flies that failed to move toward light caused the population to contain fewer individuals with alleles that resulted in light-avoidance.

As a consequence, a fly picked at random from the new fly population would be more likely to move spontaneously toward light than a fly selected from the original population. The population has been changed by selection in the direction of greater attraction to light.

In selection against one extreme form of a polygenic trait, the average value is shifted toward the other extreme. For this reason, this form of selection is called **directional selection.** It is illustrated in the upper panel of Figure 15-17.

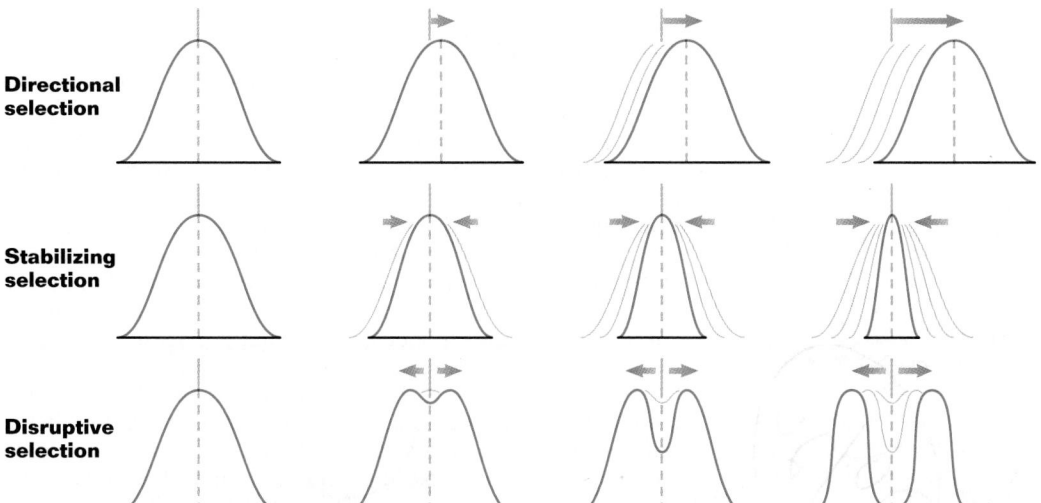

Directional selection

Stabilizing selection

Disruptive selection

Stabilizing Selection

When selection acts to eliminate extremes at both ends of a range of phenotypes, the frequencies of the intermediate phenotypes increase. When selection acts in this way, the population will contain fewer individuals with alleles promoting extreme types. This type of selection is very common in nature. In humans, for example, babies with intermediate weight at birth have the highest survival rate. In ducks and chickens, eggs of intermediate weight have the highest hatching success.

In selection against both extremes of a polygenic trait, the average value (the peak of the normal distribution) does not change. Instead, as you can see in the middle panel of Figure 15-17, the distribution becomes narrower, tending to "stabilize" the average by increasing the proportion of individuals that are similar. Therefore, this form of selection on polygenic traits is called **stabilizing selection.**

Figure 15-17 This figure shows the three kinds of selection on polygenic traits. As in Figure 15-16, the dotted line marks the average value of the trait.

SECTION 15-3

Demonstration
Directional Selection
Display photographs of various types of professional sports teams. Ask students how the selection of players by a team is similar to how directional selection in nature operates. For example, students might point out that only the tallest players would be considered by a basketball team looking for a center.

Application
Agriculture Directional selection is often used to improve domestic plants and animals. For example, farmers will select the varieties of corn plants that produce the highest yields and cross them to produce even higher-yielding varieties.

Teaching Tips
Directional vs. Stabilizing Selection
Have students describe how the weights of human babies and of duck and chicken eggs would be affected if directional selection, rather than stabilizing selection, were operating. (*Students should realize that under such circumstances either a low or high weight, rather than an intermediate weight, would be favored.*)

The Three Types of Selection on Polygenic Traits
Have students create a *Graphic Organizer* that describes the differences between directional, stabilizing, and disruptive selection. A completed *Graphic Organizer* is shown at the bottom of the page.

Graphic Organizer
Use the following graphic organizer in *Teaching Tip:* **The Three Types of Selection on Polygenic Traits** on page 329.

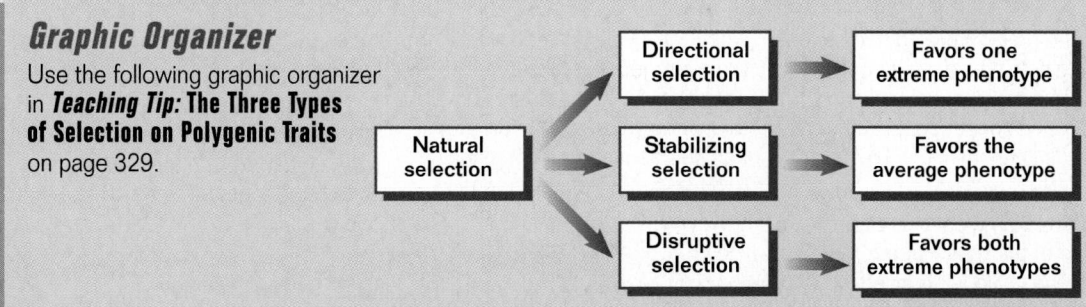

CONNECTIONS

Chapter 12
Natural Selection
Point out that disruptive selection is more likely to result in the formation of two new species than either directional or stabilizing selection. Both directional and stabilizing selection favor only one phenotype, whereas disruptive selection favors at least two phenotypes, each of which may evolve into a new species.

Chapter Closure

Have each student write a short essay that answers the question, "Is the human population, which is obviously growing, also evolving?" Inform students that they must support their position based on what they have learned about evolution and populations.

Figure 15-18 In Africa, the swallowtail butterfly mimics several species of butterflies. The left column in the photo shows two individual swallowtail butterflies, and the right column shows the species each individual mimics.

□ **CAPSULE SUMMARY**

There are three types of selection on polygenic traits. In directional selection, the range of phenotypes shifts toward one extreme. In stabilizing selection, the range of phenotypes narrows. In disruptive selection, the range of phenotypes "splits" and moves toward both extremes.

Disruptive Selection

Sometimes selection acts to eliminate rather than favor the intermediate type. When selection acts in this way, the population will contain more individuals exhibiting the two extremes. For example, some butterflies that birds find palatable mimic the coloration and patterns of species that birds find distasteful and thus gain protection from being eaten. In different localities, a palatable butterfly species may closely resemble different distasteful species, as illustrated in Figure 15-18. However, any individuals that do not closely resemble one of the distasteful species are readily detected and eaten by birds, so intermediate patterns are selected against.

In selection against the intermediate types of a polygenic trait, the average value does not change. Instead, the distribution becomes separated or "disrupted" into extreme classes, as you can see in the lower panel of Figure 15-17. This form of selection on polygenic traits is called **disruptive selection.** □

Section Review

1. *Explain why you would expect natural selection against a harmful dominant trait to proceed rapidly.*

2. *In Mendel's peas, height was controlled by a single gene with two alleles. How would a graph of height versus frequency for peas differ from the graph for humans shown in Figure 15-16?*

3. *In disruptive and stabilizing selection, the average value of the trait does not change. Explain how these two forms of natural selection differ.*

Critical Thinking

4. *What role might disruptive selection play in speciation?*

Answers to Section Review

1. Any harmful dominant trait would be immediately selected against because the trait would appear in both homozygous dominant and heterozygous individuals.

2. The graph showing height in pea plants would depict only two phenotypes, tall and short, and not a continuous distribution as seen in Figure 15-16.

3. In disruptive selection, few individuals have the average value of the trait. Rather, the average value is obtained only when the two extreme values are averaged. In stabilizing selection, most individuals have the average value. This distinction can be made by comparing two classes of students, one in which students scored in the 70s or 90s, and the other in which everyone scored in the 80s.

4. Disruptive selection would favor speciation by promoting two very different phenotypes among the members of a population.

Vocabulary

allele frequency (322)
carrying capacity (318)
demography (316)
directional selection (329)
dispersion (317)
disruptive selection (330)
exponential growth curve (318)
gene flow (324)

genetic drift (325)
genetic polymorphism (327)
Hardy-Weinberg principle (321)
K-strategist (319)
logistic model (318)
model (317)
nonrandom mating (324)

normal distribution (328)
polygenic trait (328)
population (315)
population density (316)
population size (316)
r-strategist (319)
stabilizing selection (329)

Concept Mapping

Construct a concept map that shows how the forces of genetic change cause evolution. Use as many terms as needed from the vocabulary list. Try to include the following items in your map: Hardy-Weinberg principle, genetic drift, nonrandom mating, natural selection, mutation, migration. Include additional terms in your map as needed.

Review

Multiple Choice

1. Fish commonly called king mackerel follow schools of menhaden, their primary food source. The distribution of king mackerel in the ocean is most likely
a. randomly spaced.
b. evenly spaced.
c. clumped.
d. normal.

2. The growth exhibited by a colony of bacteria that has a limited food supply will most likely be
a. exponential.   c. natural.
b. logistic.   d. random.

3. Which of the following reflects a *correct* match between a species and the population growth strategy it employs?
a. rhinoceros: *r*-strategist
b. human: *K*-strategist
c. bacterium: *K*-strategist
d. cockroach: *K*-strategist

4. According to the Hardy-Weinberg principle, allele frequencies in randomly mating populations
a. change when birth rate exceeds death rate.
b. increase and then decrease.
c. achieve disequilibrium.
d. do not change.

5. The frequency of homozygous recessive albino rats in a population is 0.01. What is the expected frequency of the dominant allele in this population?
a. 0.9   b. 0.09   c. 0.18   d. 1.8

6. Which of the following is *not* a cause of genetic change?
a. genetic drift   c. natural selection
b. random mating   d. mutation

7. Why is it unlikely that the frequency of muscular dystrophy in the human population will be reduced quickly by means of natural selection?
a. Natural selection acts only on homozygous recessive individuals.
b. Muscular dystrophy is not a genetic disorder.
c. The frequency of homozygous recessive individuals is too great.
d. Homozygous dominant individuals can have affected children.

8. In an effort to develop the best-tasting apple, a grower eliminated all trees that produced apples considered too sweet or too sour. What type of selection is being carried out by the grower?
a. directional   c. disruptive
b. stabilizing   d. choice

CHAPTER REVIEW ANSWERS

Each item in the Chapter Review is correlated by text section in the assignment guide that follows.

ASSIGNMENT GUIDE

Section	Review	Themes Review	Critical Thinking
15-1	1–3, 9–11, 17, 19	22, 24	25, 27
15-2	4–6, 12–14, 18	21	26
15-3	7, 8, 15, 16, 20	23	

Concept Mapping
Map answer is shown on page 313B.

Review
Multiple Choice
Code in parentheses indicates section number and objective number.
1. c (15-1.1)
2. b (15-1.2)
3. b (15-1.3)
4. d (15-2.1)
5. a (15-2.1)
6. b (15-2.2)
7. a (15-3.1)
8. b (15-3.3)

Completion
9. density (15-1.1)
10. shrink, grow (15-1.2)
11. temporarily large, rapidly, small, slowly (15-1.3)
12. constant (15-2.1)
13. recessive, dominant (15-2.1)
14. nonrandom mating, genetic drift (15-2.2)
15. homozygous recessive (15-3.1)
16. polygenic, normal (15-3.2)

CHAPTER REVIEW ANSWERS *continued*

Short Answer

17. Population size is the number of individuals the population contains. Population density is the number of individuals per unit of area. (15-1.1)

18. In a small population, some alleles are likely to be found in only a few individuals. Chance events that affect these individuals can dramatically change the frequencies of the alleles they carry. (15-2.2)

19. *r*-strategists usually grow fast, reproduce early in life, and have many offspring. Thus, their populations are able to quickly rebound from control efforts. (15-1.3)

20. It was a case of directional selection. The population shifted toward one phenotype—from being predominantly pale to being predominantly dark. (15-3.3)

Themes Review

21. mutation, migration, nonrandom mating, genetic drift, and natural selection (15-2.2)

22. A population is all the individuals of a species that occur together at one time and in one place. All the individuals in a population belong to the same species, and a species is composed of one or more populations. (15-1.1)

23. The graph should show that the intermediate type of the population is eliminated, leaving the extreme phenotypes of the original population. Disruptive selection splits the population into two extreme phenotypes. (15-3.3)

24. They are more likely to be *r*-strategists. *r*-strategists reproduce and grow quickly, and so they are well suited to invading and occupying the vacant habitat. (15-1.3)

Critical Thinking

25. The carrying capacity is around 1,600–1,700. The density is about 1,700/25 sq. km, or about 68 birds per sq. km. (15-1.2)

26. She should redo her calculations. They should read: (1) 0.045; (2)

---

15 CHAPTER REVIEW ....

Completion

9. Population growth can be predicted by knowing something about a population's size, its _____ , and its dispersion.

10. A population will _____ if the number of deaths exceeds the number of births. But, if the number of births exceeds the number of deaths, the population will _____ .

11. Populations of *r*-strategists tend to be _____ in size and grow _____ , while populations of *K*-strategists tend to _____ in size and grow _____ .

12. According to the Hardy-Weinberg principle, the frequency of dominant and recessive traits within a population will remain _____ unless the population is evolving.

13. In the Hardy-Weinberg equation, q^2 represents homozygous _____ individuals, p^2 represents homozygous _____ individuals, and $2pq$ represents heterozygous individuals.

14. One of the evolutionary forces that disrupts the constant state of allele frequency as predicted by the Hardy-Weinberg principle is _____ , which may involve inbreeding. Another is _____ , which results from random environmental events such as landslides.

15. Selection against an unfavorable recessive allele is slow because natural selection can act only on _____ individuals.

16. Human characteristics such as height are _____ traits because they are influenced by several genes. A graph of the distribution of these types of traits within a population usually shows a _____ distribution.

Short Answer

17. Distinguish between the size of a population and its density.

18. Explain why genetic drift is more likely to occur in a small population than in a large one.

19. Many of our most serious pests, including rats, cockroaches, and mosquitoes, are *r*-strategists. What features of *r*-strategists make them particularly difficult to control?

20. Is the evolution of the peppered moth discussed in Chapter 12 a case of directional, disruptive, or stabilizing selection? Explain your answer.

Themes Review

21. **Homeostasis** Homeostasis may be viewed as a state of biological equilibrium. According to the Hardy-Weinberg principle, allele frequencies remain in a state of equilibrium from one generation to the next unless certain forces act to upset the equilibrium. What forces can upset the equilibrium and bring about evolution?

22. **Levels of Organization** What is a population? What is the relationship between the biological concepts of population and species?

23. **Structure and Function** Draw a graph that depicts the distribution of a population that has undergone disruptive selection. How does disruptive selection affect the population?

24. **Evolution** After a forest fire, plants quickly recolonize the burned area. Are these "pioneer" plants more likely to be *r*-strategists or *K*-strategists? Explain your answer.

0.002; (3) 0.955; (4) 0.913; and (5) 0.086. (15-2.1)

27. Under the first condition, you would receive $30 million, but under the second condition you would receive $53.6 million on day 29 and over $107 million on day 30. Exponential growth is very deceiving. (15-1.2)

Critical Thinking

25. Interpreting Data Biologists introduced pheasants onto an island in Washington State in the 1930s. Using the data shown on the graph below, answer the following questions. What is the island's carrying capacity? Describe how you reached your answer. If the island's area is 25 sq. km, what was the population density at the beginning of the fourth year?

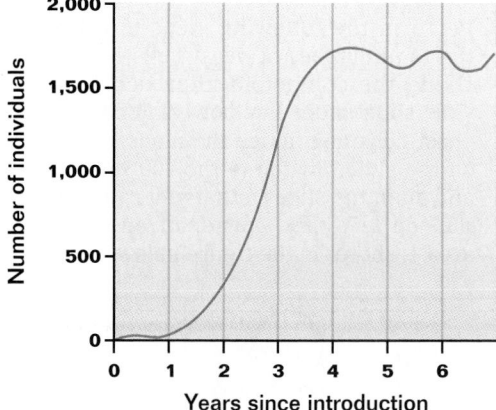

26. Responding Critically The frequency of sickle cell anemia among the African Americans is 1 in 500 individuals. A reporter writing about the disease has asked you to check the accuracy of the data, which show (1) frequency of the sickle cell allele as 0.25; (2) frequency of homozygous recessive individuals as 0.065; (3) frequency of the dominant allele as 0.75; (4) frequency of homozygous dominant individuals as 0.563; and (5) frequency of heterozygotes as 0.375. What advice would you give the reporter?

27. Evaluating Data Would you rather get $1 million a day for 30 days or receive a nickel the first day, two nickels the second day, four nickels the third day, eight nickels the fourth day, and so on, for the same 30-day period? Carry out the calculations to check your decision.

Activities and Projects

28. Research and Writing China and Thailand greatly reduced their birth rates through government-sponsored population control programs. Research the population control measures taken in both countries, and write a report that summarizes what you have learned. In your report you should evaluate each country's program and explain whether it could be applied to other countries.

29. Cooperative Group Project Research human population growth in selected countries on the continents of North America, South America, Asia, Europe, and Africa. Make a presentation to your class describing population trends around the world. In your presentation, be sure to indicate where population growth is most rapid and if population growth is leveling off in any countries. Different group members can take responsibility for researching the population growth in different countries, preparing graphics for the presentation, and presenting the group's findings to the class.

Readings

30. Read the article "Sweet Death," in *Natural History*, February 1992, pages 2–6. According to Jared Diamond, what advantage might the genes that cause diabetes have once provided? What lifestyles changes have made these genes harmful?

31. Read the article "The Allure of Symmetry," in *Natural History*, September, 1993, pages 30–36. How did Dr. Thornhill demonstrate that female scorpion flies prefer the most symmetrical males? Suggest a hypothesis that explains why the females have this preference.

Activities and Projects

28. Thailand's program involves distribution of free contraceptives and a public-relations campaign to encourage people to have fewer children and to use contraceptives. China's program also includes free contraception and a public-relations campaign, but it is more intrusive and comprehensive. Couples who have only one child receive preferential treatment in housing, education, and employment, and they are paid more, receive larger pensions, and get free medical care. Couples who promise to have only one child and then break their promise lose all their benefits. Communist party officials pressure couples to have small families. One parent must be sterilized in families that have two children. Thailand's program is voluntary and would be applicable to all countries. China's program is intrusive and coercive, and it wouldn't be acceptable for most countries.

29. Population growth is most rapid in Africa, Asia, and South America. In Europe, Australia, New Zealand, Japan, and North America, population growth is leveling off or has ceased.

Readings

30. In the past, food supplies were irregular and uncertain. The genes that cause diabetes may have allowed food to be stored as fat during times of plenty. Today, most people eat three fat- and sugar-rich meals each day and get little exercise.

31. He placed female scorpion flies in a wire cylinder and allowed them to choose between two males that differed in degree of symmetry. The females may be using symmetry as an indication that the male had not been infected by parasites or pathogens during his development.

Preparation

Prepare the yeast population by dissolving a fresh cake of yeast in 80 mL of warm water. Add 20 g of sugar and stir. Fresh yeast may be prepared up to 4 hours before the investigation. If substituting dried yeast or freeze-dried yeast (available from Ward's), prepare about a week in advance. Keep the yeast in a warm, dark area for the duration of the investigation.

Procedural Note
• Have students work in teams of two.

Background Answers

1. A population is a group of individuals of the same species living in the same place.

2. Carrying capacity is the maximum number of individuals the environment can support for an indefinite period of time.

3. Common limiting factors include food, water, space, presence of predators, and extremes of temperature.

4. *r*-strategists are small, grow and mature rapidly, reproduce early in life, and have many offspring that receive little parental care. They usually show exponential population growth, leading to temporarily large populations, followed by a population crash. *K*-strategists are large, grow and mature slowly, reproduce later in life, and have few offspring, which receive large amounts of parental care. They usually show slow population growth and have small population sizes.

5. To be accurately sampled, a population can be divided into several smaller units. Several of these units are randomly chosen, then counted and averaged. The average is multiplied by the number of units to give an estimate for the entire population.

6. How does the size of a population of yeast cells that has limited resources change over time?

A Population Study

OBJECTIVE

Observe the growth and decline in a population of yeast cells and apply the underlying principles to changes in human populations.

PROCESS SKILLS

• sampling population density
• graphing changes in populations
• interpreting data

MATERIALS

• yeast culture
• 1 mL pipette
• test tube
• iodine in dropper bottle
• ruled microscope slide (2 mm × 2 mm)
• safety goggles
• coverslip
• compound microscope

BACKGROUND

1. What is a population?
2. What is carrying capacity?
3. What are some common limiting factors that prevent populations from exceeding their carrying capacity?
4. How do populations of *r*-strategists differ from populations of *K*-strategists?
5. How can populations be sampled to achieve an accurate count?
6. Write your own question to explore in your lab report or notebook.

TECHNIQUE

1. Your teacher will transfer approximately 1 mL of the yeast culture to a test tube, and then add 2 drops of iodine to the test tube.

2. Put on safety goggles. Use the 1 mL pipette to transfer 0.1 mL (one drop) from the test tube to a ruled microscope slide. **CAUTION: Work with care—iodine is a poison and an eye irritant.** If you get iodine in your eyes, flush with water and alert your teacher. Carefully lower a coverslip over the drop.

3. Perform the following steps to estimate the total number of yeast cells in 0.1 mL. Using the compound microscope, view your slide under low power. Focus on the yeast cells and notice the black grid lines on the slide. Switch to the 400× objective and align the slide so that you can just see the top left-hand corner of one square, area 1, shown in the figure below.

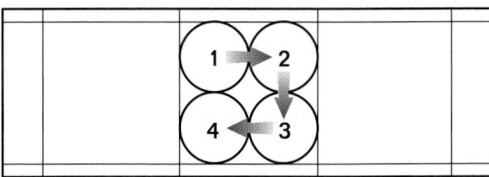

4. Count all the yeast cells in area 1 and record the number. Then move the slide to area 2. Continue counting cells and recording data until you have counted the cells in each of the four areas that make up one square. Add the total number of cells in the square and record this number in the **Records** section of your report.

5. Switch the microscope to low power and move the slide one square to the left. Under high power, count the cells and record the number.

6. Repeat steps 4 and 5 to count the cells in a total of six squares. Add the total number of cells you counted in the six squares. Calculate the average number

Records

Time (hours)	Estimated Number of Yeast Cells in 0.1 mL of Culture — Answers will vary.							
	Number of cells per 2mm square						Average	Population size (cells /0.1 mL)
	1	2	3	4	5	6		
0								
24								
48								
72								
96								

of yeast cells in a 2 mm square by dividing the total by 6. Record this initial number in the **Records** section of your report

7. Clean up your materials and wash your hands before leaving the lab.

8. Repeat steps 1 through 6 each day for four more days. Record your data for 24, 48, 72, and 96 hours. If possible, repeat steps 1 through 6 again after the weekend.

9. To find the total population of yeast cells in 1 mL (the amount in the test tube), multiply the average number of cells counted in a 2 mm square by 2,500.

10. Graph your data using the values you calculated for the entire test tube population over the five-day period.

11. In your report, briefly summarize the procedure you followed.

INQUIRY

1. What effect does iodine have on yeast cells?

2. Why were several areas counted and then averaged each day?

3. From your graphed data, what was the change in population size each day?

4. When was the most rapid growth? the slowest growth?

5. When was the growth peak reached?

6. To find the total number of cells in 1 mL, you multiplied the number of cells you counted in a 2 mm square by 2,500.

Show the equation that gives the ratio of the number of cells counted in one square to the total population in 1 mL. (Hint: The cells and liquid formed a layer about 0.1 mm thick on the microscope

slide. As a result, each small square held about 0.4 mm³ of liquid. To find the number of cells in 0.1 mL, you must convert mm³ to mL. Remember that 1 mL = 1 cm³ and that 1 cm³ = 1,000 mm³.)

ANALYSIS

1. How did the yeast population change over time?

2. Why did the population grow?

3. What limiting factors probably caused the yeast population to decline?

4. If you kept counting for several more days, what do you think would happen to the yeast population?

5. What limiting factors that affect yeast also affect human populations?

6. Is yeast an *r*-strategist or a *K*-strategist? Defend your answer.

7. In what ways do population growth and decline in a yeast population resemble growth and decline in a human population? In what ways do they differ?

FURTHER INQUIRY

Write a new question that could be explored as a new investigation. The following is an example:

Is it possible to set up a population of yeast that continues to grow, without declining, for a week? Would it be possible to keep the population size growing indefinitely?

Inquiry Answers

1. Iodine kills the cells. Dead cells don't move around and are therefore easier to count. Iodine also stains the cells, making them easier to see.

2. An average was taken to allow for variation within the population.

3. Answers will vary.

4. The most rapid growth was between 24 and 48 hours. Growth was slowest from 0 to 24 hours. Some students may interpret this question about growth with an answer about decline. Decline is not growth.

5. The growth peak was reached at 72 hours.

6. (# cells/0.4 mm³) × (1,000 mm³/1 cm³) × (1 cm³ /1 mL) = # cells × 2,500/1 mL

Analysis Answers

1. The yeast population grew rapidly and then declined.

2. The yeast cells initially had a wealth of food and plenty of space, so they reproduced rapidly.

3. The yeast cells were limited by a lack of food and lack of space. They could also be poisoned by their own waste products.

4. The yeast population would die out when the food supply was exhausted.

5. The limiting factors of yeast—lack of food and space and an excess of waste products—are some of the same limiting factors that affect human populations.

6. Yeast is an *r*-strategist. Its pattern of population growth—swift, exponential growth followed by a population crash—is characteristic of *r*-strategists.

7. Human population growth is exponential and rapid and resembles the early, expanding phase of the yeast population. So far, the human population has not crashed like the yeast population did.

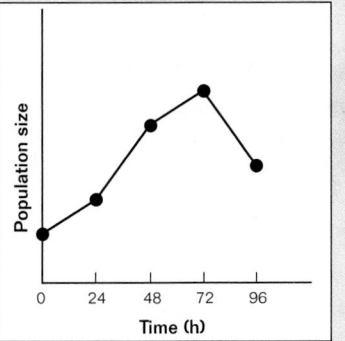

Estimations of the yeast population size will vary. A graph of population size should look something like the following.

Block Scheduling Guide

Each block represents about 45–50 minutes of instructional time. The resources cited in a block are coded to a specific program component using the Key on the next page.

Lecture Concepts | Lesson Resources

16-1 An Introduction to Ecosystems pp. 336–342

BLOCK ①

a. Ecology is the study of the interactions of organisms with the living and nonliving parts of their environment. An interacting group of organisms and their environment constitute an ecosystem.

b. Ecosystems change through the process of succession. Succession on newly formed habitat is called primary succession. Succession in a previously inhabited area is called secondary succession.

Lecture Resources
- Opening Questions, page 337
- Demonstrations: A Small Ecosystem, Pioneers
- Species Diversity, page 338
- Visual Strategies: Figures 16-4, 16-5, 16-6
- Teaching Transparencies: 70
- Holt Biology Videodiscs: Lesson 16-1

Classwork Options
- Succession, page 341
- The Role of Each Species in Succession, page 342
- Inquiry Skills Development B10: Ecology Scavenger Hunt

Assignment Options
- Section Review, page 342
- Chapter Review, questions 1, 2, 5, 10, 11, 17, 18

16-2 Energy Flows Through Ecosystems pp. 343–348

BLOCKS ② and ③

a. Ecologists assign every organism in an ecosystem to a trophic level, which is determined by the organism's source of energy. The lowest trophic level of any ecosystem is occupied by the producers, organisms that make energy-storing molecules. Consumers, organisms that get energy by eating producers, occupy the higher trophic levels of the ecosystem.

b. The path of energy through the trophic levels of an ecosystem is called a food chain. Food chains interconnect to form food webs.

c. Energy transfers between trophic levels are very inefficient. On average, only 10 percent of the energy in any trophic level will be incorporated into the next level.

Lecture Resources
- Opening Question, page 343
- Demonstration: A Simple Food Web
- Overcoming Misconceptions, page 343
- Visual Strategies: Figures 16-8, 16-9, 16-10
- Teaching Transparencies: 68, 69, 84, 85
- Pyramids of Energy and Numbers, page 348
- An Inverted Pyramid of Numbers, page 348
- Holt Biology Videodiscs: Lesson 16-2

Classwork Options
- Understanding the Flow of Energy and Graphic Organizer, page 345
- Pyramid of Energy, page 347
- Concept Mapping, page 355
- Quick Labs A12: Making a Food Web
- Laboratory Techniques and Experimental Design C22: Owl Pellets

Assignment Options
- Section Review, page 348
- Chapter Review, questions 3, 4, 6, 12–14, 19, 20, 23, 24

16-3 Materials Cycle Within Ecosystems pp. 349–354

BLOCKS ④ and ⑤

a. In the water cycle, water falls to the surface as precipitation. Some water returns to the atmosphere through evaporation. Water also collects in lakes, rivers, and oceans. Some water seeps into the soil and joins the ground water.

b. Carbon enters the living portion of the carbon cycle through photosynthesis. Cellular respiration releases carbon dioxide from organisms.

c. Carbon trapped in rock and fossil fuels is freed by erosion and burning.

d. Nitrogen-fixing bacteria living in the soil or plant roots transform nitrogen gas into ammonia. Plants lacking these bacteria take up ammonia and nitrate from the soil. Animals get nitrogen by eating plants or other animals.

Lecture Resources
- Opening Question, page 349
- Demonstration: Recycling
- Visual Strategies: Figures 16-14, 16-16, 16-17
- Teaching Transparencies: 71, 72, 73
- Transpiration, page 350
- Research Update, page 351
- Effective Reading, page 353
- Connections, page 354
- Holt Biology Videodiscs: Lesson 16-3

Classwork Options
- Atmospheric Carbon Dioxide, page 352
- Closure, page 354
- Laboratory Investigation: Ecosystem in a Jar, pages 358–359
- Laboratory Techniques and Experimental Design C23: Owl Pellets—NW vs. SE

Assignment Options
- Section Review, page 354
- Chapter Review, questions 7–9, 15, 16, 21, 22, 25–28

16-3 Materials Cycle Within Ecosystems pp. 349–354 (continued)

e. Phosphorus, an essential element of all living organisms, is usually present in soil and rock as calcium phosphate. Phosphate absorbed by the roots of plants is used to build organic molecules. When animals eat plants, organic phosphorus is reused.

Review/Enrichment

BLOCK 6

■ Study Guide: Concept Review and Pretest

Assignment Options
◉ Teaching Resources CD-ROM: Supplemental Reading Worksheets and Test, *Silent Spring*
■ Activities and Projects, questions 29, 30
■ Readings, questions 31–33

Assessment Options

BLOCK 7

Traditional Assessment
■ Chapter 16 Test
◉ Test Generator/Assessment Item Listing: Software item bank for preparing customized chapter tests.

Portfolio Assessment
Select student reports for one or more laboratory experiments from this chapter. *The Direct Observation Checklist* on the *BioSources Teaching Resources CD-ROM* should be completed during a laboratory or other cooperative learning experience.

Answer to Concept Map
The following is one possible answer to the Concept Mapping exercise on page 355.

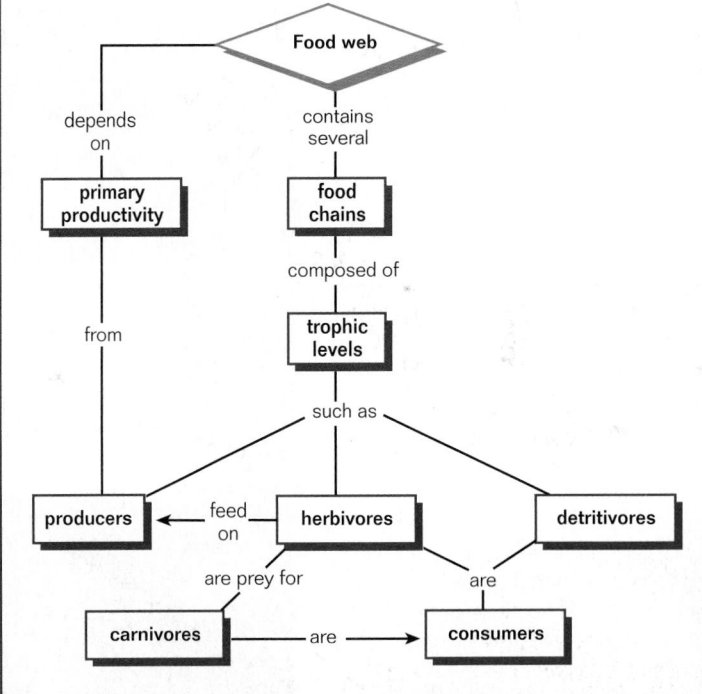

Holt Biology Videodiscs

Holt Biology Videodiscs gives you a powerful tool for teaching, review, and assessment. *Concepts of Biology* adds interactive lesson plans and extensive tools for customization using CD-ROM technology.

CONCEPTS OF BIOLOGY

Use the following topic from *Concepts of Biology* to help you teach this chapter:

■ Topic 10: Ecosystem Structure and Function

For further information regarding the media on the videodiscs, see *Holt Biology Videodiscs Teacher's Correlation Guide to Biology: Principles and Explorations.*

Chapter Theme

Flow of Energy *An ecosystem is self-sustaining only if there is a continuous flow of energy into it. The energy flow is always one-way, originating as radiant energy emitted by the sun that is slowly diverted into living systems. Of the solar energy that reaches Earth, only an estimated 0.1 percent is trapped and made useful to living things. Yet that 0.1 percent accounts for about 120 billion metric tons (132 billion tons) of organic matter that is produced by photosynthetic organisms each year.*

Tapping Prior Knowledge

- What is an ecosystem?
- Explain the differences between the terms *ecosystem, habitat,* and *community.*
- What crucial role do photosynthetic organisms play in most ecosystems?

Opening Demonstration

Students should appreciate the fragility of ecosystems. A case in point is the fertile farmlands of Ireland. Show the class a potato. Point out that although the Irish farmlands could grow any number of crops, the farmers concentrated on just one, the potato. An ecosystem dominated by one species can be extremely fragile and susceptible to destruction by opportunistic invaders. In Ireland, a protist *(Phytophthora)* invaded and destroyed the potatoes—virtually the entire potato crop was wiped out in a single week in the summer of 1846. As a result, more than 1 million people in Ireland died of starvation. An additional 3 million people left Ireland, reducing its population by one-half.

CHAPTER 16
ECOSYSTEMS

Coral reef ecosystem

REVIEW

- first and second laws of thermodynamics (Section 4-1)
- photosynthesis (Section 5-2)
- cellular respiration (Section 5-3)
- six kingdoms of life (Section 13-1)

Authors' Rationale

Recognizing that species are continually interacting with other species *and* their nonliving environment is critical to understanding ecology. Ecosystem dynamics are addressed in this chapter, showing your students the interdependence of organisms and how changing one component in an ecosystem, or upsetting the "balance of nature," can have disastrous results. Students will gain an understanding of ecosystem dynamics by studying energy flow between trophic levels, pyramids of energy, and nutrient cycles.

16-1 An Introduction to Ecosystems

The coasts of the United States are dotted with salt marshes similar to the one shown in Figure 16-1. Several years ago in one such marsh, a team of biologists captured a little sparrow like the one shown on the opposite page. This bird was unusual because he was a dusky seaside sparrow, the last member of this subspecies alive in nature. He was a ragged bird, no longer in his prime; the biologists who caught him estimated his age as over 10 years. They had caught him, and earlier two other elderly males, in an attempt to preserve some of the genes of this almost extinct subspecies through crossbreeding with another subspecies of seaside sparrow. The other two males died in 1986, and so for a year this sparrow was alone, the last of his kind. With his death on July 14, 1987, the dusky seaside sparrow disappeared forever.

Living Things Are Interdependent

Why be concerned about the loss of a sparrow? There are, after all, lots of kinds of sparrows. Why mourn this one? Few of us would have been able to recognize a dusky seaside sparrow if we had been lucky enough to see one. And what does an extinct sparrow have to do with biology anyway?

The answers to these questions lie at the heart of biology. Although it is easy to think of the environment as external to us—something we always use, sometimes enjoy, and occasionally pollute—this viewpoint ignores all that biologists have learned about life on Earth. In fact, we share the world with all of Earth's other organisms. Together, we and the Earth's other inhabitants weave an intricate tapestry of interactions that defines what our environment is like. Just as removing one resistor from an enormous mainframe computer can alter the interactions of its many components in ways that influence the computer's operation, so removing something as simple as a sparrow from our environment can have many diverse consequences, not all of them easily predictable.

In 1866, the German biologist Ernst Haeckel gave a name to the study of how organisms fit into their environment. He called it *ecology*, from the Greek words *oikos*, meaning "house, or place where one lives," and *logos*, meaning "study of." **Ecology** is the study of the interactions of living organisms with one another and with their physical environment (soil, water, weather, and so on). Our study of

Figure 16-1 A salt marsh in Florida was the home of the last dusky seaside sparrow until his capture. This subspecies of sparrow was driven to extinction by pesticides used to control mosquitoes.

Vocabulary Preview

ecology
habitat
community
ecosystem
species diversity
diversity
succession
primary succession
secondary succession

Opening Questions

To get students thinking about interdependence in ecosystems, ask them these questions about their interactions with other organisms. Your cells need oxygen to carry out cellular respiration. What is the ultimate source of this oxygen? Your cells produce carbon dioxide as a waste product of cellular respiration. What becomes of this carbon dioxide after you exhale it?

Demonstration

A Small Ecosystem
Students should recognize that both abiotic (nonliving) and biotic (living) factors influence the nature of an ecosystem. If an aquarium or terrarium is set up in the classroom, use it as an example of an ecosystem. If not, simply display a photograph of an ecosystem. Prepare a table on the board with two headings: "Living Factors" and "Nonliving Factors." Ask students to identify as many components in the ecosystem as possible, placing each one under its appropriate heading. Then have students describe an example of interdependence in the ecosystem. For example, oxygen supplied by a plant through photosynthesis is used by a fish that, in turn, supplies carbon dioxide needed by the plant.

ecology, then, is the study of the house in which we live. The place where a particular population of a species lives is its **habitat.** The many different species that live together in a habitat are called a **community.** A community and all the physical aspects of its habitat—the soil, water, and weather—are called an **ecosystem,** or ecological system.

The Inhabitants of an Ecosystem

Most of us know that tropical rain forests teem with life; an area the size of a football field may contain hundreds of species of plants and thousands of species of insects. But almost any ecosystem is rich with species. **Species diversity,** or **diversity,** is the number of species living within an ecosystem.

To gain some idea of the diversity of ecosystems, consider for a moment a pine forest in the southeastern United States, such as the one illustrated in Figure 16-2. If you could fence in a square kilometer of this forest and then go in and collect every animal, what would you expect to get? Starting with the largest animals, you might find a black bear and almost certainly would find a white-tailed deer. In times past, before human activities drove them away, you might have found a cougar or a red wolf as well. The woods also contain smaller mammals—raccoons, foxes, gray squirrels, rabbits, and chipmunks. Lizards dart among the leaves, while snakes and toads often remain hidden. Beneath the ground, moles tunnel through the soil. A host of large birds can be found, including red-tailed hawks, turkey vultures, turkeys, and quail. The trees are alive with smaller birds: warblers, sparrows, cardinals, and wrens. If the square kilometer included a lake, you might find catfish, bass, perch, a variety of turtles, and perhaps an alligator.

This collection includes only a few of the members of the animal kingdom to be found in this forest ecosystem. The soil contains immense numbers of earthworms and flatworms.

Figure 16-2 Pine forests like this one are common in the southeastern United States.

Figure 16-3 Unlike most other spiders, jumping spiders, *left,* do not build webs. Instead, they lie in wait and pounce on passing insects. Male stag beetles, *right,* use their large jaws to compete for mates.

Hidden under the bark of trees and beneath the pine needles and leaves covering the ground are many different species of arthropods, two of which are shown in Figure 16-3.

Now stop and consider this: the animals you have been collecting belong to only one of the six kingdoms of life. Among the members of the plant kingdom, the pine trees are obvious, but they are by no means the only kind of plant present. There are also a variety of smaller trees and shrubs, and beneath them are vines, ferns, and mosses. Trees in part of the forest may be sheathed in kudzu vine, an introduced species from Asia that covers shrubs and trees with a dense, leafy blanket. Grasses and many kinds of flowers grow on the forest floor.

The fungi constitute a separate kingdom. Most fungi are multicellular organisms. They obtain their food by secreting materials that digest organic material and then absorbing the nutrients. You would find many kinds of fungi growing on fallen trees and spreading as fine threads through the decaying material on the forest floor. Figure 16-4 shows two examples of fungi. The roots of nearly all the kinds of trees will be surrounded by fungi growing in intimate associations called mycorrhizae. Other fungi might be present on the surface of trees or rocks as lichens, which are associations between fungi and algae or cyanobacteria.

A fourth kingdom of life contains the protists, most of which are single-celled and too small to see without a microscope. Protists in a pine forest include the algae and related organisms that grow in fresh water, and the many different kinds of microscopic creatures that inhabit any drop of water.

The fifth and sixth kingdoms of organisms contain bacteria—simple, single-celled organisms that occur everywhere in the forest, although mainly in the soil. A single teaspoon of soil is home to billions of individual bacteria. They play many critical roles in the life of the forest, including changing atmospheric nitrogen gas into a form that organisms can use to build their tissues (you will learn more about this function later in the chapter). In addition, bacteria and fungi are responsible for the decomposition of dead organisms. Without bacteria and fungi, organisms would not decompose, and the

Figure 16-4 Among the fungal residents of pine forests are these mushrooms, *top,* and these shelf fungi, *bottom,* which are digesting the tree on which they are growing.

CONNECTIONS
....

Chapter 13
The Six Kingdoms
Review the six-kingdom classification system. Members of five kingdoms—Eubacteria, Fungi, Protista, Plantae, and Animalia—are likely residents of the ecosystems that students are familiar with. The sixth kingdom, Archaebacteria, contains prokaryotes that live predominantly in hostile environments such as hot springs and extremely salty lakes—although some methane-generating archaebacteria live in the mud of swamps and lakes and in the guts of cows and other grazing animals.

◄● VISUAL STRATEGY

Figure 16-4
Tell students that in a forest community, more than 90 percent of the energy made available by producers is eventually consumed by organisms such as these fungi rather than by herbivores such as deer. The organisms that consume most of the net primary production are known as detritivores. Detritivores are organisms that live on the detritus of a community, which includes dead leaves, branches, logs, and animal carcasses. Detritivores include both decomposers, such as fungi and bacteria, and scavengers, such as vultures and hyenas.

CONTENT LINK

In **Chapter 19**, you will learn more about the characteristics of the six kingdoms.

nutrients contained within their bodies would be unavailable to other organisms. No ecosystem can persist without bacteria and fungi to break down the bodies of its dead members.

If you were to collect and remove every single organism of each of the six kingdoms from your square kilometer, it would be bare, stripped of life. Only the nonliving surroundings would remain—the soil, rocks, and water; the wind that blows over it; and the rain and sunlight that fall upon it. This portion of the forest ecosystem is the physical habitat. The minerals and water that a physical habitat contains and the weather to which it is exposed determine in large measure what kinds of organisms are able to live there.

Figure 16-5 No ecosystem is completely isolated. This coconut, which may have drifted thousands of kilometers, is sprouting on the beach of a small Pacific island.

What Determines the Boundaries of an Ecosystem?

The physical boundaries of an ecosystem are not always obvious and depend in large measure on how the ecosystem is being studied. For example, a scientist might consider a single rotting log on the forest floor to be an ecosystem if he or she is interested in the fungi and insects living there. Quite often, individual fields, forests, or lakes are studied as isolated ecosystems. In truth, of course, no one location is ever totally isolated from other places. Even distant oceanic islands get occasional migrant visitors—birds blown off course, seeds carried by the wind or waves, and lizards that arrive on floating logs or mats of vegetation.

Natural Changes in Ecosystems

Natural changes in the physical environment of ecosystems happen all the time. When a volcano forms a new island, a glacier recedes and exposes bare soil, or a fire burns all of the vegetation in an area, new habitat is created. This change sets off a process of colonization and ecosystem development. Small, fast-growing plants quickly invade this new habitat. They do not persist, however, because their pioneering efforts make the ground more hospitable for other species. As a result, later waves of plant immigrants soon outcompete and replace the original inhabitants, only to be replaced in turn by still other species that are better able to compete in the new environment.

This regular progression of species replacement is called **succession.** When succession occurs on land where nothing has grown before, it is called **primary succession.** When succession occurs in areas where there has been previous growth, such as in abandoned fields or forest clearings, it is called **secondary succession.** It used to be thought that the stages of succession were predictable and that they always led to the same final community of organisms within any particular ecosystem, a group called a climax community. Ecologists now recognize that chance plays a major role in the competitive interactions that govern each successional change. For this reason, no two successions are exactly alike, although the progression of changes will tend toward similar communities in similar physical conditions. That is why characteristic communities such as tropical rain forest, desert, and tundra occur over large areas of the planet. ◻

An Example of Ecological Succession: Glacier Bay

You can better understand succession by looking at a real example. A good choice is a receding glacier, because land is continually being exposed as the face of the glacier moves back. So imagine that you have traveled to the glacier that dominates the head of Glacier Bay, Alaska. The face of the glacier has been melting and receding for the last 200 years, and has moved back some 100 km (62 mi.). A walk across the ground exposed by the receding glacier is in a sense a walk through time—the farther one walks away from the glacier face, the longer the land has been exposed. In Figure 16-6, you can follow the changes that take place as time passes.

The most recently exposed areas, at the face of the receding wall of ice, are piles of bare rock and gravel that lack the usable nitrogen essential to plant and animal life. For about 10 years the piles remain devoid of life. The seeds and spores of the first pioneering plants to colonize this barren landscape—mosses, fireweed, willows, cottonwood, and *Dryas*, a sturdy plant with clumps about 30 cm (1 ft.) across —are carried in by the wind. At first all of these plants grow close to the ground, severely stunted by mineral deficiency. But *Dryas* has mycorrhizae on its roots that supply the nutrients it needs. After a few years, *Dryas* uses this competitive advantage to crowd out the other plants. What remains is a dense mat of *Dryas*.

After about 10 years, seeds of alder arrive, blown in from distant sites. It is a matter of chance when they come, but inevitably some finally make their way to the site, landing on the mat of *Dryas*. Alder roots have nitrogen-fixing nodules, so they are able to grow even more rapidly than *Dryas*. Dead leaves and fallen branches from the alder trees add even more usable nitrogen to the soil, in time allowing the willows and cottonwoods to invade and grow with vigor. After about 30 years, there are dense thickets of alder, willow, and cottonwood that shade and eventually kill the pioneering *Dryas*.

The alder thicket matures until, some 80 years after the glacier first exposed the land, Sitka spruce invades the

◻ **CAPSULE SUMMARY**

Ecology is the study of the interactions of organisms with their living and nonliving environment. An interacting group of organisms and their nonliving environment constitute an ecosystem. Ecosystems change through the process of succession. Succession occurring on newly formed habitat is primary succession. Secondary succession occurs on habitat that has previously supported growth.

Teaching Tip
Succession

Have each student draw a time line representing succession that occurs after a forest fire has burned all of the vegetation in an area. Starting at time zero, the land should be barren, followed by the appearance of grasses and weeds, then small bushes, and finally by trees that represent the climax community.

CONNECTIONS

Chapter 2
Nitrogen
Remind students that nitrogen is required by organisms to synthesize proteins. Nitrogen is also a component of the nitrogenous bases in RNA and DNA.

 Did You Know?

Primary succession is a slow process. Scientists estimate that the primary succession from sand dunes to the climax beech-maple forest that occurred along the shores of Lake Michigan took about 1,000 years. On the other hand, secondary succession, such as that which occurs when a forest is cleared or burned, may take less than 100 years.

Teaching Tip

The Role of Each Species in Succession

Have students prepare a table with the following headings: Organism, Adaptations, and Impact on Environment. Then have the class fill in the table as they reread the section describing succession on Glacier Bay. Help them begin by listing the following information under the appropriate headings: Organism: *Dryas;* Adaptations: mycorrhizae; Impact on Environment: pioneer organism.

 VISUAL STRATEGY

Figure 16-6

Ask students to describe what would happen if the spruce-hemlock climax community were destroyed by fire. Students should recognize that succession would occur, beginning with pioneer plants similar to those that first appeared at Glacier Bay.

Figure 16-6 At first, land exposed by the receding glacier is lifeless because it lacks nutrients. An early "pioneer" of this land is the rockrose *Dryas, above left.* After several decades trees such as alder and shrubs grow large enough to shade and kill off the low-growing mat of *Dryas, above center.* After several more decades, these trees and shrubs are replaced by spruce and hemlock, *above right.*

thickets. Spruce trees cannot fix nitrogen, but the nitrogen released by the alders enables them to form a dense forest that shades out the alders and willows. In the competition for light needed to carry out photosynthesis, the alders lose, just as *Dryas* did before them.

After the spruce forest is established, hemlock trees arrive at the site. They are very shade-tolerant and have a root system that competes well against spruce for soil nitrogen. Hemlock trees soon become dominant in the forest. This last community of spruce and hemlock proves to be a very stable ecosystem—a climax community. Similar communities are found over broad areas of Alaska, Canada, and Russia.

Climax communities are not permanent creations of nature, fixed and unchangeable. As local climates change, the distribution of particular species within the forest ecosystem may change too. Whatever these fluctuations, it remains true that climax communities are among the most stable of ecosystems, remaining relatively unchanged for long periods of time.

Section Review

1. *What components of an ecosystem are not part of a community?*

2. *What equipment would help an ecologist to obtain a more accurate count of the number of species in the pine forest ecosystem described previously?*

3. *A lawn is not a climax community. Explain why it usually does not go through succession.*

Critical Thinking

4. *Consider a pond ecosystem. In what ways does the surface of the water constitute a boundary for this ecosystem? Describe some ways in which the surface is not a boundary.*

Answers to Section Review

1. All the nonliving components of an ecosystem (water, soil, and rocks) are not part of a community.

2. Since many of the organisms in an ecosystem are inconspicuous or small, equip-ment that would help locate these organisms would be most helpful. For instance, a shovel could be used to dig up soil-dwelling organisms

3. Lawns would proceed through ecological succession if they were not regularly mowed.

4. The surface of the water is a boundary in that it limits the movement of some organisms, such as fish, However, the surface is not a boundary to organisms such as frogs and turtles that can move into and out of the water. In addition, some abiotic factors such as sunlight and oxygen can pene-trate the surface.

16-2 Energy Flows Through Ecosystems

*E*verything that organisms do in ecosystems—running, breathing, burrowing, growing—requires energy. Of all the factors that organize an ecosystem, determining what organisms and how many of each it will contain, none is more important than the flow of energy. In this section you will learn where organisms get their energy and how this energy moves within an ecosystem.

Section Objectives

- Distinguish between producers and consumers.
- Contrast food webs with food chains.
- Explain why food chains are rarely longer than three or four links.

The Path of Energy: Who Eats Whom in Ecosystems

Energy flows into the biological world from the sun. Life exists on Earth because photosynthesis makes it possible to capture some of the light energy from the sun and transform it into the chemical energy of organic molecules. These organic compounds compose what we call food. The amount of organic material that the photosynthetic organisms of an ecosystem produce is called **primary productivity**.

Primary productivity determines the energy budget of an ecosystem. All of the organisms in an ecosystem are chemical machines driven by the energy captured in photosynthesis. The organisms that first capture energy, the **producers,** include plants, some kinds of bacteria, and algae. Producers make energy-storing molecules. All other organisms in an ecosystem are **consumers,** which obtain the energy to build their molecules by consuming plants or other organisms.

Ecologists assign every organism in an ecosystem to a **trophic level,** which is determined by the organism's source of energy. Energy moves from one trophic level to another, as illustrated in Figure 16-7. The lowest trophic level of any ecosystem is occupied by the producers: plants in most terrestrial ecosystems and algae and bacteria in aquatic ones. Producers not only use the energy of the sun to build energy-rich sugar molecules, but also absorb nitrogen and other key substances from the environment and incorporate them into biological molecules. It is important to realize that plants respire as well as produce. The roots of a plant, for example, do not carry out photosynthesis, because there is no light underground. Roots obtain their energy the same way that you do—by using energy-storing molecules produced elsewhere (in this case, in the leaves of the plant).

At the second trophic level are **herbivores,** animals that eat plants. They are the primary consumers in ecosystems. Cows and horses are herbivores, as are caterpillars and ducks. A herbivore must be able to break down the plant's molecules. Simple sugars and starches present no problem,

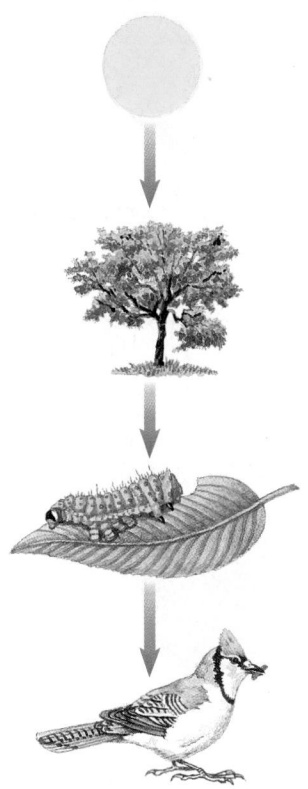

Figure 16-7 The sun is the ultimate source of energy for organisms. Producers, the first trophic level, trap solar energy in organic molecules. Animals at the second trophic level get energy by eating producers, and animals at the third level get energy by feeding on other animals.

Vocabulary Preview

primary productivity
producer
consumer
trophic level
herbivore
carnivore
omnivore
detritivore
decomposer
food chain
food web
biomass

Opening Question

Ask students to explain how the first and second laws of thermodynamics are involved when a person eats a piece of beef from a cow that has grazed on grass. The first law applies to conversions of energy. For example, grass carries out this reaction: light $\longrightarrow$ chemical energy (stored in carbohydrates). Both the cow eating grass and the person eating beef carry out this reaction: chemical energy (stored in carbohydrates and other macromolecules) $\longrightarrow$ chemical energy and heat. According to the first law, each of these conversions can be, at most, 100 percent efficient. The second law is exemplified by the heat energy that is "lost" whenever a primary or secondary consumer feeds. In other words, none of these transformations can be 100 percent efficient.

Demonstration

A Simple Food Web

List the following organisms that can be found in an open field: robin, hawk, snake, frog, grasshopper, mouse, and rabbit. Have students draw arrows to show who eats whom in this field ecosystem. Students should see the complexity of even this simple food web, in which each predator can take more than one type of prey and each type of prey can be exploited by several different species of predators.

Overcoming Misconceptions

Plants Respire Too

Emphasize that plants respire 24 hours a day, but they can carry out photosynthesis only during the daylight. Healthy plants capture more energy than they consume, using the excess for building tissue or putting it into storage molecules such as starch. Also point out that the carbon dioxide produced when plants respire is released into the atmosphere.

CONNECTIONS

Chapter 4
Enzymes
Remind students that chemical reactions in all organisms are carried out by enzymes. Humans do not possess the enzyme that can hydrolyze cellulose. The bacteria that live in the gut of a cow do, and therefore the cow can digest the cellulose in the grass it eats.

Teaching Tip
Trophic Levels

Ask a volunteer from the class to describe his or her breakfast. Then have students describe what trophic level the volunteer functioned at when eating each different food in the meal.

VISUAL STRATEGY

Figure 16-8
Have students identify the producers, herbivores, and carnivores in this diagram. Students should recognize that algae are producers, krill are herbivores, and cod, leopard seals, and killer whales are carnivores. Then ask students which organisms, if any, are omnivores or detritivores. Students should recognize that there are no omnivores or detritivores in this food chain.

but the digestion of cellulose, a molecule made of sugar units linked together, is a chemical feat that only a few organisms have evolved the ability to perform. Most herbivores rely on helpers. A cow, for instance, has a thriving colony of bacteria in its gut that digests cellulose. Humans cannot digest cellulose because we lack these bacteria. This is why a cow can live on a diet of grass but you cannot.

At the third trophic level are secondary consumers, animals that eat herbivores. Such flesh-eating animals are called **carnivores.** Tigers, wolves, and snakes are carnivores. Some animals, such as bears, are both herbivores and carnivores, eating both plants and animals. They use the simple sugars and starches stored in plants as food, but they cannot digest cellulose. Such animals are called **omnivores.** Humans are omnivores. Many ecosystems contain a fourth trophic level made up of carnivores that consume other carnivores. They are called tertiary consumers, or top carnivores. A hawk that eats a snake is a tertiary consumer. Very rarely do ecosystems contain more than four trophic levels, for reasons that will become clear in a moment.

In every ecosystem there is a special class of consumers called **detritivores,** which include fungal and bacterial decomposers, vultures, and worms. Detritivores obtain their energy from the organic wastes and dead bodies that are produced at all trophic levels. Bacteria and fungi are known as **decomposers** because they cause decay. Decomposition of bodies and wastes releases nutrients back into the environment to be used again by other organisms.

A path of energy through the trophic levels of an ecosystem is called a **food chain,** shown in Figure 16-8. In most ecosystems, energy does not follow simple linear paths because animals feed at several trophic levels. This creates a complicated, interconnected path of energy called a **food web,** illustrated in Figure 16-9.

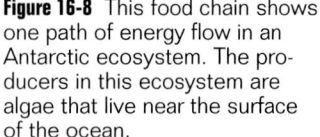

Figure 16-8 This food chain shows one path of energy flow in an Antarctic ecosystem. The producers in this ecosystem are algae that live near the surface of the ocean.

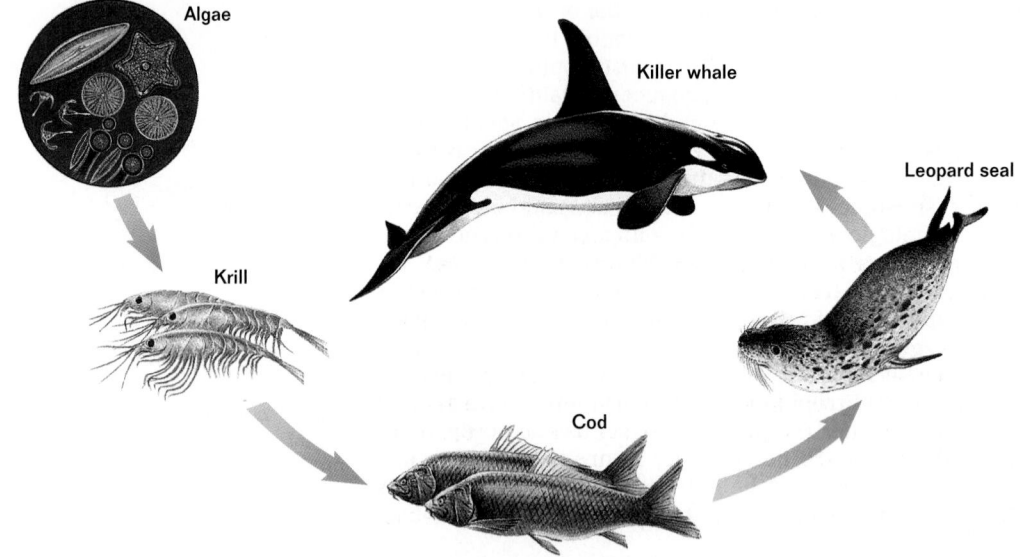

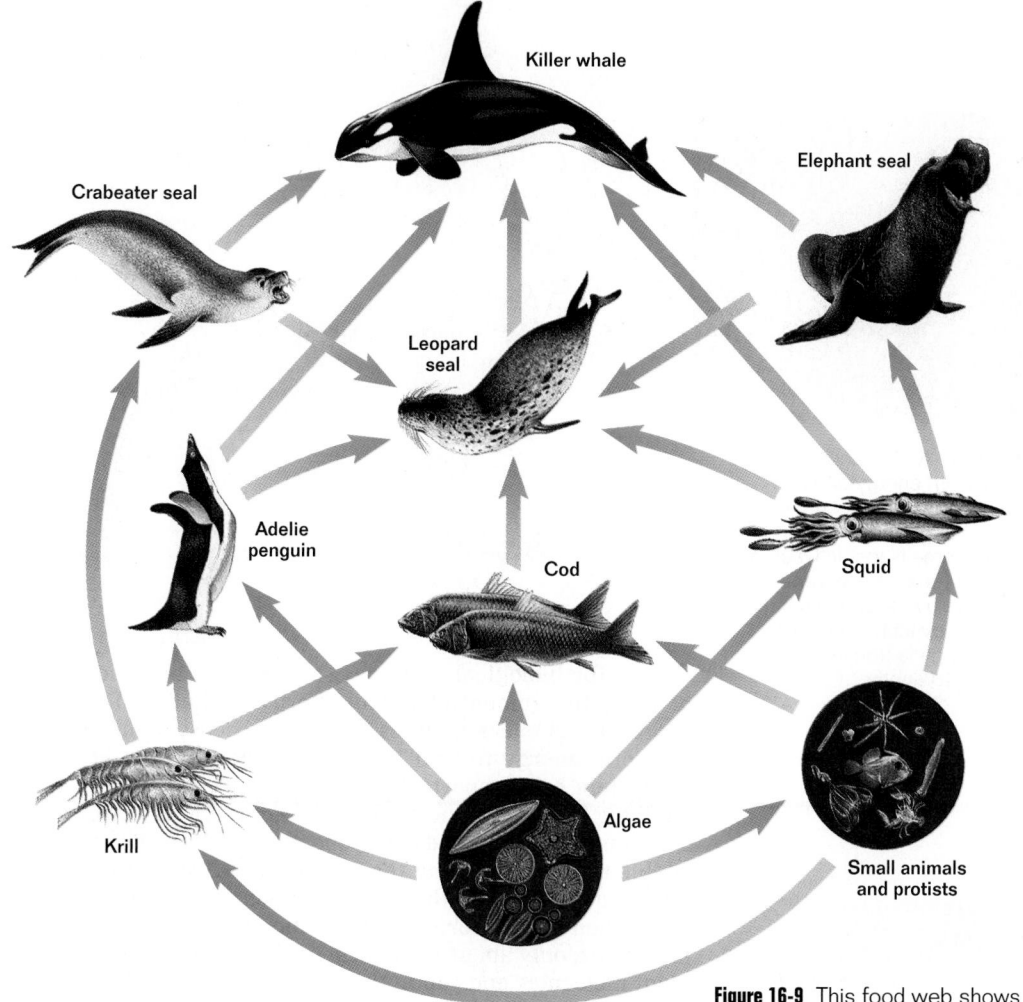

Figure 16-9 This food web shows a more complete picture of the feeding relationships in an Antarctic ecosystem. Notice that the food chain shown in Figure 16-8 is just one strand of this complex food web.

SECTION 16-2

 VISUAL STRATEGY

Figure 16-9
Have students describe how this ecosystem would be affected if algae were eliminated. Students should recognize that eliminating algae from this ecosystem is analogous to destroying all of the plants in a terrestrial ecosystem. Like plants, algae are producers that capture the energy needed by the other members of the ecosystem. Eliminating algae would remove the ecosystem's energy source, and the other organisms would eventually starve.

Teaching Tip
Understanding the Flow of Energy
Have students draw a *Graphic Organizer* that summarizes the flow of energy from producers to herbivores, omnivores, carnivores, and detritivores. A completed *Graphic Organizer* is shown at the bottom of the page.

Energy Transfers Between Trophic Levels Are Inefficient

The deer that you see browsing on leaves in Figure 16-10 is busy acquiring energy. There is potential energy in the leaves, stored in the chemical bonds within their molecules. What happens to a leaf's energy after the deer consumes it? You can see what happens to the energy in Figure 16-10. Some of the energy is transformed to other forms of potential energy, such as fat. Another portion accomplishes mechanical work such as running, breathing, and eating more leaves. However, almost half is dissipated to the environment as heat.

Every transfer of energy within an ecosystem dissipates energy as heat. Although heat can be harnessed to do work (as in a steam engine), it is generally not a useful source of

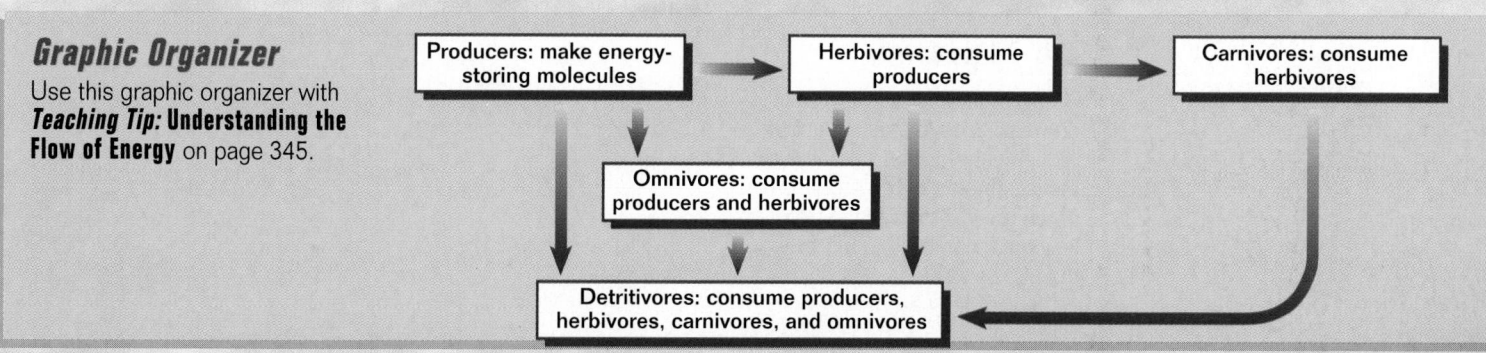

Graphic Organizer
Use this graphic organizer with *Teaching Tip:* **Understanding the Flow of Energy** on page 345.

Producers: make energy-storing molecules → Herbivores: consume producers → Carnivores: consume herbivores

Omnivores: consume producers and herbivores

Detritivores: consume producers, herbivores, carnivores, and omnivores

CONNECTIONS

Chapter 4

The Laws of Thermodynamics
Remind students that the first law of thermodynamics states that energy cannot be created or destroyed but only converted between different forms. The second law states that energy conversion between different forms is never 100 percent efficient.

 VISUAL STRATEGY

Figure 16-10
Point out that the energy dissipated to the environment as heat is considered "lost" because it is no longer available to do useful work. By the first law of thermodynamics, the energy cannot actually be lost; it is just transformed into an unusable form—heat.

Application

Mathematics If an average of 1,500 kilocalories of light energy per day falls on a square meter of land surface covered by plants, approximately 15 kilocalories are incorporated into chemical compounds synthesized through photosynthesis. Have students calculate the approximate number of kilocalories of energy that wind up in a person who has eaten a steak from a cow that fed on the plants growing on this square meter. *(about 0.15 kilocalories)*

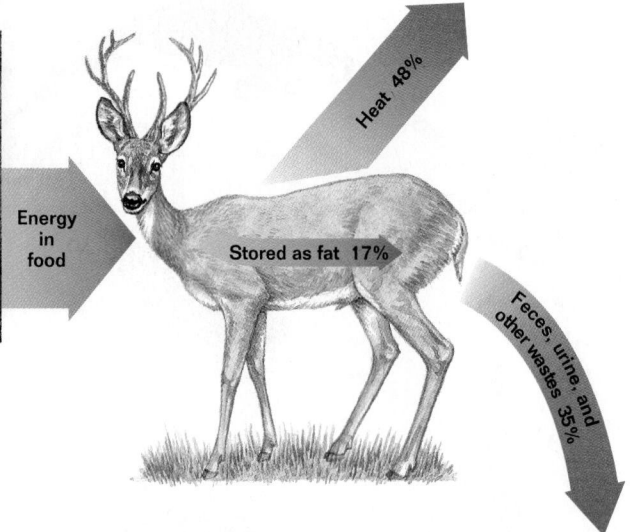

Figure 16-10 A deer acquires energy from energy-rich molecules in the vegetation it eats. Some of that energy is stored as fat, and some is lost in wastes—urine and feces. However, most of the energy escapes as heat, which is inevitably produced when energy is transformed from one form to another.

 **CAPSULE SUMMARY**

Energy transfers between trophic levels are very inefficient. On average, only 10 percent of the energy in any trophic level will be incorporated into the next level.

energy for biological systems. Thus, from a biological point of view, the amount of useful energy available to do work decreases as energy passes through an ecosystem. The loss of useful energy to heat limits how many trophic levels an ecosystem can support. When a plant harvests energy from sunlight to make structural molecules such as cellulose, it stores in chemical bonds only about one-half of the energy it is able to capture, losing the rest as heat. This is the first of many such losses as the energy passes through the ecosystem. When a herbivore uses plant molecules to make its own molecules, only about 10 percent of the energy present in the plant molecules ends up in the herbivore's molecules; 90 percent of the energy is lost. And when a carnivore eats the herbivore, 90 percent of what little energy remains is then lost in making carnivore molecules. At each trophic level, the energy stored by the organisms is about one-tenth of that stored by the organisms in the level below. ◻

Ecologists often portray the flow of energy through ecosystems by means of a block diagram called a pyramid of energy. In this diagram, each trophic level is represented by a block, and the blocks are stacked atop one another, with the lowest trophic level on the bottom. The width of each block is determined by the amount of energy stored in the organisms at that trophic level. Because the energy stored by the organisms at each trophic level is about one-tenth of that stored by the organisms in the level below, the diagram is a pyramid. In the river ecosystem shown in Figure 16-11, eelgrass and algae are producers; turtles, snails, and caddis flies are herbivores; and beetles, bass, and sunfish are carnivores. The pyramid of energy for this ecosystem is shown in Figure 16-11.

 Did You Know?

On average, energy transfers between ecosystems are about 10 percent efficient. Actual measurements of the efficiency of energy transfers range from less than 1 percent to more than 20 percent.

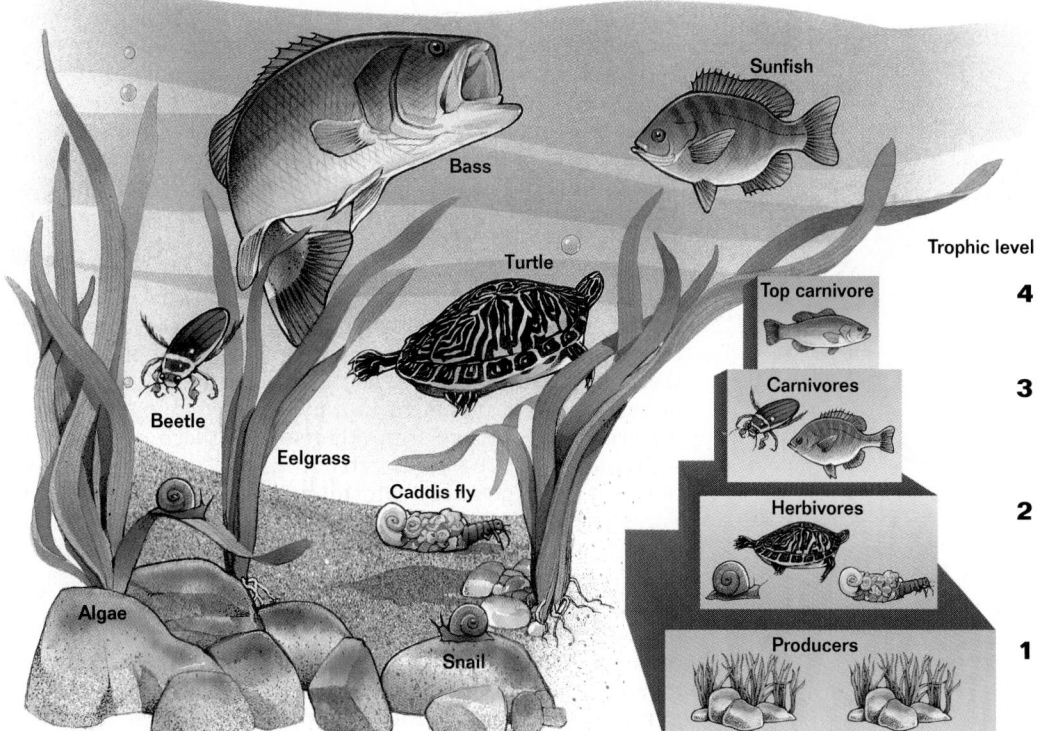

Trophic level

4 Top carnivore

3 Carnivores

2 Herbivores

1 Producers

Teaching Tip
Pyramid of Energy
Have students draw a pyramid of energy that accurately reflects the following energy transfer: 1,000 calories (algae) ⟶ 150 calories (small aquatic animals) ⟶ 30 calories (smelt) ⟶ 6 calories (trout) ⟶ 1.2 calories (human).

Figure 16-11 This simple aquatic ecosystem contains four trophic levels. Each trophic level contains about 90 percent less energy than the level below it. Because of space considerations, the trophic levels are not shown to scale here.

Energy Loss Limits the Number of Trophic Levels in an Ecosystem

From what you have learned, you should not be surprised that there is a limit to the number of trophic levels in an ecosystem. Most terrestrial ecosystems involve only three or, rarely, four levels. Too much energy is lost at each level to allow more levels. For example, a large human population could not survive by eating lions captured on the Serengeti Plain of Africa; there are simply too few lions to make this possible. The amount of grass in that ecosystem cannot support enough zebras to maintain a large enough population of lions to feed lion-eating humans. The ecological complexity of the world is thus fixed in a fundamental way by the loss of potential energy that occurs at each trophic level. This loss is a consequence of the second law of thermodynamics, which you studied in Chapter 4.

The loss of useful energy to heat as energy passes from one trophic level to another is a fact of nature that we cannot change. There is an important lesson in the fact that it takes a very large population of zebras and wildebeest to support a small population of lions on the Serengeti Plain (lions are outnumbered by their prey by about 1,000 to 1). As we seek

 Did You Know?

A study of 100 top predators (animals that are not preyed on) revealed that there are usually only four trophic levels in a food chain. Only one of these predators was part of a food chain consisting of six levels.

Teaching Tips

Pyramids of Energy and Numbers

Point out the differences between pyramids of energy and numbers. Unlike a pyramid of numbers, which indicates only how many individuals are present at any one time, a pyramid of energy provides information about the rate at which energy is stored over a period of time (often expressed as kilocalories/sq.m/ year). Ask students if pyramids of energy can be inverted. *(No. Except for the first trophic level, each level depends on the level below for energy. By the second law of thermodynamics, all of the energy in one level cannot be transferred to and stored in the level above, so no trophic level can contain more energy than the level below.)*

An Inverted Pyramid of Numbers

Ask the class what the pyramid of numbers would look like for a forest community in which caterpillars and birds greatly outnumber trees. *(The second trophic level, containing caterpillars, and the third trophic level, containing birds, would be larger than the first, containing trees.)*

Figure 16-12 Because energy transfer between trophic levels is inefficient, about 10 times more grain is required to feed carnivorous humans like the boy eating the steak than to feed herbivorous humans like the girl eating a slice of bread.

ways to maintain ever-larger human populations, we should remember the lion's situation. Humans are omnivores and, unlike the lion, can choose to eat either meat or plants. The choice makes a difference for our future because we eat steak and hamburger only at a great cost in energy. As explained in Figure 16-12, 10 kg of grain are needed to build 1 kg of human tissue if we eat the grain, but 100 kg of grain are needed to build 1 kg of human tissue if a cow eats the grain and we eat the cow.

There are more zebras and wildebeest than lions on the Serengeti Plain because the amount of available energy determines the numbers of animals living in an ecosystem at any one time. However, because some organisms are much bigger than others and therefore use more energy, the numbers of organisms often do not form a pyramid when one compares different trophic levels. In other words, there may be more individuals in a particular trophic level than in the level below it. For instance, caterpillars and other insect herbivores greatly outnumber the trees they feed on. The number of individuals in a trophic level can be an inaccurate indicator of the amount of energy in that level. To better assess the amount of energy present in trophic levels, ecologists usually measure **biomass,** the dry weight of tissue and other organic matter. By collecting, drying, and weighing all of the organisms in each trophic level of an ecosystem, ecologists obtain a pyramid of biomass, with each trophic level containing only about 10 percent of the biomass of the level below it.

Section Review

1. *Suppose you want to create an ecosystem in an aquarium. Explain why your ecosystem would require producers.*
2. *Draw a food web that reflects the feeding relationships in the ecosystem illustrated in Figure 16-11. What information not included in Figure 16-11 would help you to draw a more accurate food web?*
3. *Explain why a given area of land could support more vegetarian humans than omnivorous humans.*

Critical Thinking

4. *How would you modify the food web in Figure 16-9 to include decomposers? Explain your answer.*
5. *In which trophic level would you place humans? Justify your answer.*
6. *Nearly all the mammals that humans eat, including cows, sheep, and goats, are herbivores, not carnivores or omnivores. Explain why a herbivore is a more efficient meat producer from an energetic point of view.*

Answers to Section Review

1. Only producers can capture energy and convert it into a usable form.

2. The web should have links between organisms at one trophic level and the level below. The energy pyramid does not indicate whether each species at one trophic level feeds on *all* of the organisms in the level below.

3. If plants are fed to a herbivore to produce meat, 90 percent of the energy in those plants is lost. Vegetarians skip this energy-losing step, and more energy is available to support more people.

4. Since decomposers feed on the wastes and bodies produced at all trophic levels, an arrow should be drawn from each species in the food web to the decomposers.

5. Humans can be placed in a number of different trophic levels, depending on what they consume. A person who eats cereal would be in the second trophic level, and a person who eats a steak would be in the third level.

6. The logic here is the same as that in question 3.

16-3 Materials Cycle Within Ecosystems

Unlike energy, which flows through the Earth's ecosystems in one direction (from sun to producers to consumers), the physical components of ecosystems often cycle constantly. These components include all of the inorganic (noncarbon) substances that make up the soil, water, and air. All materials that cycle through living organisms are important in maintaining the health of ecosystems, but four substances are particularly important: water, carbon, nitrogen, and phosphorus.

Materials Cycle Between Organisms and the Nonliving Environment

The paths of water, carbon, nitrogen, and phosphorus, as they pass from the nonliving environment to living organisms and then back to the environment, form closed circles, or cycles, called biogeochemical cycles. In each biogeochemical cycle, a substance enters living organisms from the atmosphere, water, or soil, resides for a time in the organisms, then returns to the nonliving environment. Ecologists often speak of such substances as cycling within an ecosystem between a living reservoir (organisms that live in the ecosystem) and a nonliving reservoir. In almost all cases, there is much less of the substance in the living reservoir than in the nonliving reservoir.

Figure 16-13 By taking in carbon dioxide during photosynthesis, these trees are participating in the carbon cycle.

The Water Cycle Is Driven by the Sun

Of all the nonliving components of an ecosystem, water has the greatest influence on the ecosystem's inhabitants. To a great degree, the availability of water determines the diversity of ecosystems, as you will see repeatedly in the detailed descriptions of the world's principal ecosystems presented in the next chapter.

Water cycles within ecosystems in two ways, each of which is driven by the sun. In the nonliving portion of the water cycle, water vapor in the atmosphere condenses and falls to the Earth's surface as rain or snow. A portion of this water seeps into the soil and becomes, for a time, part of the **ground water,** which is water retained beneath the surface of the Earth. Most of the remaining water that has fallen to the Earth does not stay long at the surface. Instead, heated

 Did You Know?

In the water cycle, most of the water is in bodies of water. Almost 98 percent of the water on Earth is in liquid form—in oceans, lakes, or other bodies of water. The remaining 2 percent is found as ice in the polar regions, as water vapor in the atmosphere, and as a liquid both in the soil and in the bodies of living organisms.

VISUAL STRATEGY

Figure 16-14
Have students imagine a molecule of water that has just fallen to the ground in a raindrop. Have students trace the paths this molecule could follow through the water cycle.

Teaching Tip
Transpiration
Have each student make a list of as many factors as possible (in addition to sunlight) that affect transpiration. Answers should include amount of water in soil *(low water content reduces transpiration)*, wind *(high winds increase transpiration)*, humidity *(low humidity increases transpiration)*, and the nature of the leaf surface *(for example, the waxy cuticle on plants that grow in arid environments reduces evaporation of water from leaves)*.

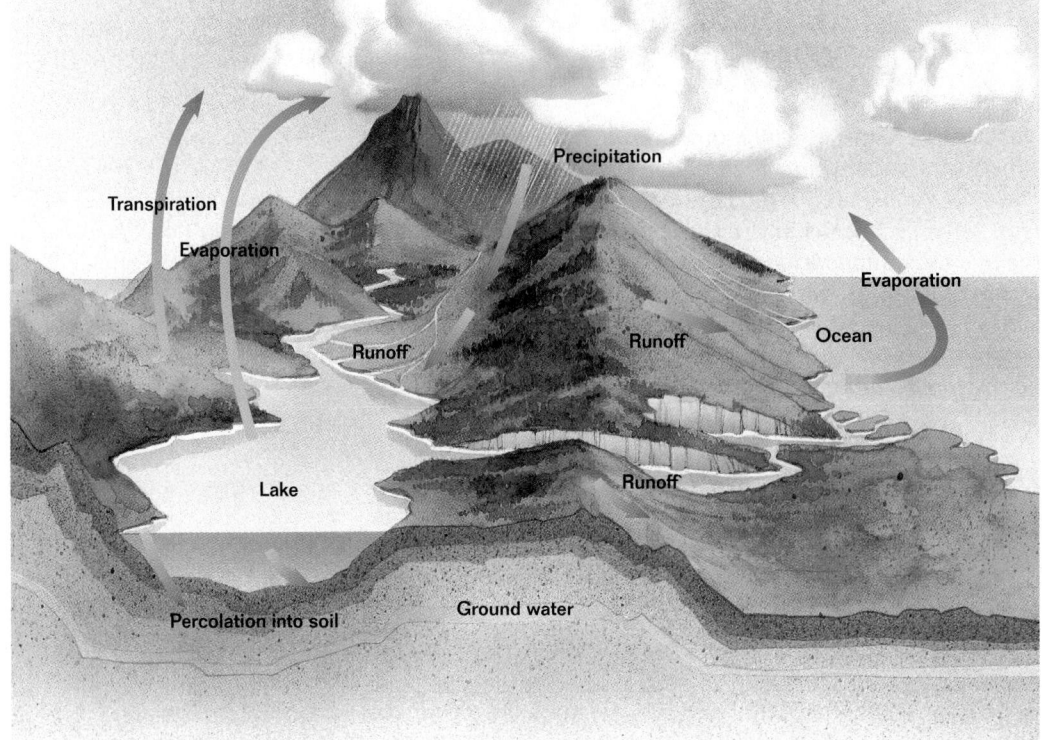

Figure 16-14 This diagram shows the major steps in the water cycle.

☐ CAPSULE SUMMARY

In the water cycle, water falls to the surface as precipitation. Some water returns to the atmosphere through evaporation. Water also collects in lakes, rivers, and oceans. Some water seeps into the soil and joins the ground water.

by the sun, it reenters the atmosphere by evaporation from lakes, rivers, and oceans. The path of water within an ecosystem is illustrated in Figure 16-14.

In the living portion of the water cycle, water is taken up by the roots of plants. After passing through a plant, the water moves into the atmosphere by evaporating from the leaves, a process called transpiration. Transpiration is also a sun-driven process: the sun heats the Earth's atmosphere, creating wind currents that draw moisture from the tiny openings in the leaves of plants.

In aquatic ecosystems (lakes, rivers, and oceans), the nonliving portion of the water cycle predominates; almost all of the water that falls to the Earth returns to the atmosphere by evaporation. In terrestrial ecosystems, the nonliving and living parts of the water cycle play important roles. In thickly vegetated ecosystems, such as tropical rain forests, more than 90 percent of the moisture in the ecosystem passes through plants and is transpired from their leaves. In a very real sense, these plants create their own rain. Moisture travels from plants to the atmosphere and falls back to the Earth as rain. ☐

When forests are cut down, the water cycle is disrupted and less moisture is returned to the atmosphere. Some of the water drains into rivers and ultimately into the sea instead of

Historical Note

How Vegetation Affects the Runoff of Water and Nutrients
During the winter of 1965–1966, scientists at the Hubbard Brook Experimental Forest in New Hampshire cut down all the trees, saplings, and shrubs in a forested area. They wanted to determine how cutting down a forest would affect the cycling of water and nutrients. The runoff of water in the denuded forest was four times higher than in the previous year. This increase in water runoff severely hindered reforestation. In addition, greater amounts of mineral nutrients such as calcium and potassium were lost in runoff after deforestation. From these data, the scientists concluded that vegetation plays an important role in the cycling of water and minerals within an ecosystem.

rising to the clouds and falling again on the land. Because less water cycles within an ecosystem after the trees are cut down, extensive cutting can convert lush forests into wasteland with too little rain to support the return of the forest, as shown in Figure 16-15. It is a tragedy of our time that just such a transformation is occurring in many tropical areas.

The Carbon Cycle Is Linked to the Flow of Energy

Carbon also cycles between the nonliving environment and living organisms. You can follow the carbon cycle in Figure 16-16. The Earth's atmosphere is about 0.035 percent carbon dioxide. Carbon dioxide in the air or dissolved in water is used by photosynthesizing plants, algae, and bacteria as a raw material to build organic molecules. In effect, they trap the carbon atoms of carbon dioxide within the living world. Carbon atoms return to the pool of carbon dioxide in the air and water in three ways.

One way is through cellular respiration. Nearly all living organisms, including plants, perform cellular respiration. They use oxygen to oxidize organic molecules during cellular respiration, and carbon dioxide is a byproduct of this reaction.

Carbon also returns to the atmosphere through combustion, or burning. Much carbon is contained in wood and may stay there for many years, returning to the atmosphere only when the wood is burned. Sometimes carbon can be locked away beneath the Earth for a very long time—thousands or

Figure 16-15 Removal of the trees from this area in Brazil disrupted the water cycle, reduced rainfall, and left the soil without the protection normally provided by roots. The rain that did fall carried off the thin layer of topsoil, further diminishing the chances that vegetation could return.

Figure 16-16 This diagram shows the major steps of the carbon cycle. Humans are disrupting the carbon cycle by burning fossil fuels and by destroying vegetation that would have absorbed carbon dioxide.

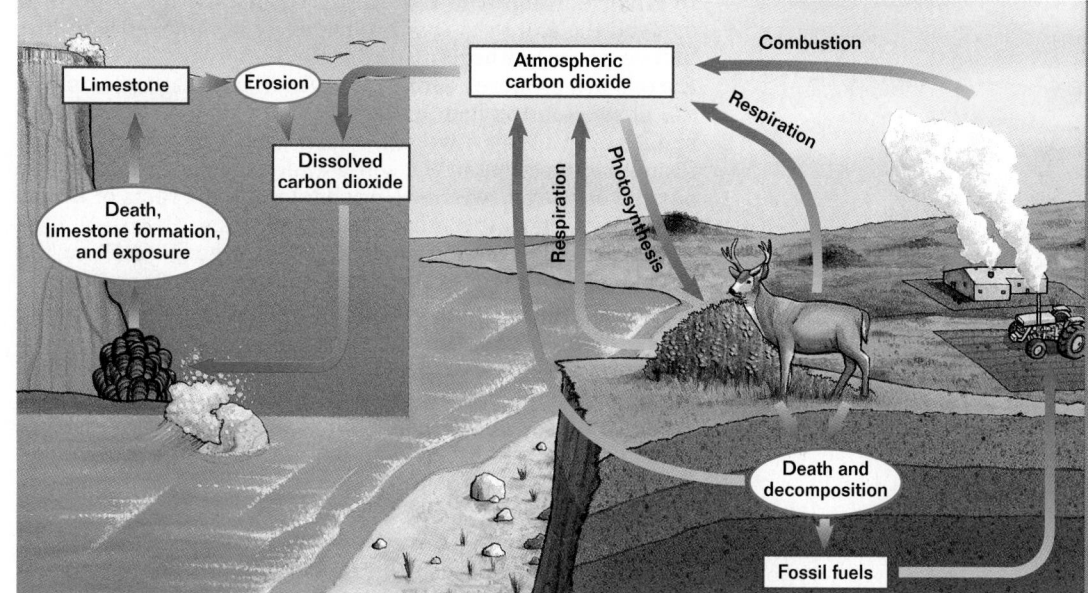

SECTION 16-3

☑ RESEARCH Update

How Much Will It Cost to Shop Earth's Energy Stores?

The World Energy Council (WEC) is predicting that by 2020 the global demand for energy could be twice what it is today. The WEC and the International Energy Agency (IEA) have compiled data that suggest that the bulk of these energy demands will come from developing countries.

How are these increased demands for energy going to be met? The WEC estimates that there is a 60-year supply of oil in the Earth's crust (and this is reported as *good* news). Various international energy agencies point to untapped oil reserves in China and Russia, but they also point to the unrealistic amount of capital that would be required to deliver these resources and the political instability associated with ongoing change in these countries. Natural gas consumption will increase, which is good since it is typically cleaner and cheaper than oil. Renewable energy sources will succeed fossil fuels in the distant future. Wind, solar, and biomass energies, while already in use in some areas, will require decades to become significant sources of energy. ☑

👁 VISUAL STRATEGY

Figure 16-16

Have students trace the steps in the carbon cycle. Point out that the burning of fossil fuels has released much more carbon dioxide into the atmosphere than can be recycled through photosynthesis or absorbed by the oceans. Because carbon dioxide plays a major role in the greenhouse effect that gives the Earth a moderate climate, scientists are concerned about the possibility of global warming caused by the excess carbon dioxide.

Teaching Tip
Atmospheric Carbon Dioxide
Have each student make a list indicating the combustion reactions that they depend on that contribute to rising levels of atmospheric carbon dioxide. Answers could include burning gasoline in a car, propane in a stove, charcoal in a barbecue, and fuel oil in a furnace.

Application

 Foreign Language Beans are leguminous plants, which can carry out nitrogen fixation. Other such plants include clover, peas, vetch, and alfalfa. The French, recognizing the importance of leguminous plants, use the word *legume* for "vegetable."

❏ CAPSULE SUMMARY

Carbon enters the living portion of the carbon cycle through photosynthesis. Organisms release carbon through cellular respiration. Carbon long trapped in rock and fossil fuels is freed by erosion and burning, respectively.

CONTENT LINK

Over the last 150 years, the burning of fossil fuels has led to a 30 percent increase in the concentration of carbon dioxide in the atmosphere. You will find out about the possible consequences of this atmospheric change in Chapter 18.

even millions of years. The remains of organisms that become buried in sediments may be gradually transformed by heat and pressure into fossil fuels—coal, oil, and natural gas. The carbon contained in these remains is released back into the atmosphere only when the fossil fuels are burned.

Erosion releases carbon as well. Marine organisms extract a substantial amount of the carbon dioxide dissolved in sea water and use it to build their calcium carbonate shells. When these marine organisms die, their shells sink to the ocean floor, become covered with sediments, and form limestone. Eventually, as the limestone becomes exposed and erodes, the carbon in it is returned to the pool of available carbon. This process takes millions of years. ❏

The Nitrogen and Phosphorus Cycles

Organisms contain large amounts of nitrogen because proteins and nucleic acids are both nitrogen-rich. The atmosphere is 79 percent nitrogen gas, N_2. However, organisms obtain that nitrogen only with great difficulty; most organisms are unable to use the nitrogen gas that is so plentiful in the atmosphere. The two nitrogen atoms in a molecule of nitrogen gas are connected by a particularly strong triple covalent bond that is very difficult to break. The profusion of life that blankets our globe is possible only because a few bacteria have enzymes that can break this triple bond and bind the nitrogen atoms to hydrogen, forming ammonia, NH_3. The process of combining nitrogen gas with hydrogen to form ammonia is called **nitrogen fixation.** Bacteria evolved the ability to fix nitrogen early in the history of life, before photosynthesis had introduced oxygen gas into the Earth's atmosphere. Today, nitrogen fixation occurs only in the absence of oxygen; even a trace of oxygen poisons the process. In today's world, awash with oxygen, nitrogen-fixing bacteria live in the soil within capsules that admit no oxygen or within swellings, or nodules, on the roots of beans, alder trees, and a few other kinds of plants.

The nitrogen cycle, diagrammed in Figure 16-17 on the next page, is a complex process with four important stages.

1. **Assimilation**   The ammonia produced by nitrogen-fixing bacteria spreads through the soil and is picked up by plants, which use the nitrogen atoms to build proteins, nucleic acids, and a variety of other nitrogen-containing molecules. When animals eat these plants, they use the nitrogen to build their own molecules. Nitrogen assimilation is the absorption and incorporation of nitrogen into plant and animal compounds.

2. **Ammonification**   Many animals excrete excess nitrogen in their urine as urea or uric acid, which a second kind of soil bacterium converts back into ammonia. When an

◇ Did You Know?

Farmers often rotate a nonleguminous crop, such as corn, with a leguminous one, such as alfalfa. The alfalfa will fix nitrogen and release some of it into the soil. Moreover, if a crop of alfalfa is plowed back into the soil, it may add as much as 350 kg (770 lb.) of nitrogen per hectare (2.5 acres) of soil, enough to grow a crop of non-leguminous plants without the need for any additional fertilizer.

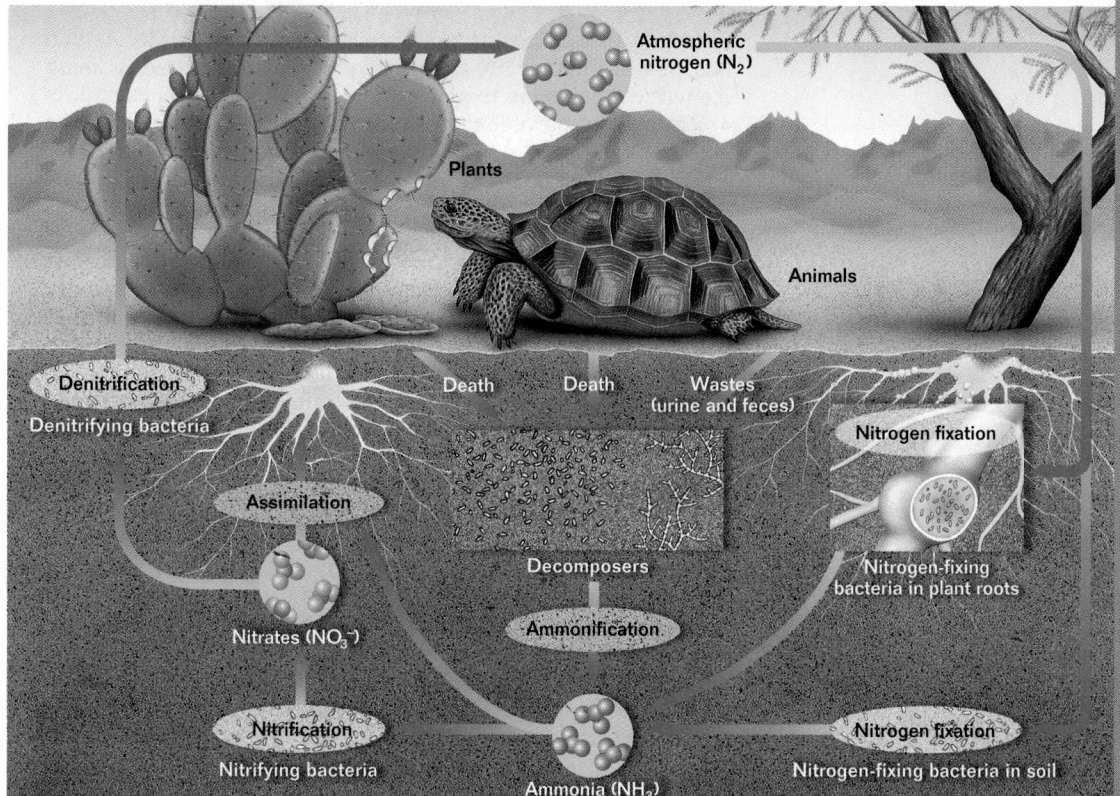

Figure 16-17 Bacteria carry out many of the important steps in the nitrogen cycle, including the conversion of atmospheric nitrogen into a usable form, ammonia.

organism dies, these bacteria also convert the nitrogen compounds in the dead tissue to ammonia. Some of this ammonia is then reabsorbed by plants. Ammonification is the production of ammonia by bacteria during the decay of nitrogen-containing organic matter.

3. **Nitrification** Some of the ammonia in the soil is converted by several kinds of bacteria to nitrate, NO_3^-. Some nitrate is then absorbed by plants, which can assimilate nitrate as well as ammonia. Nitrification is the production of nitrate from ammonia.

4. **Denitrification** Another kind of bacterium acts on the remaining nitrate, converting it back into nitrogen gas. The nitrogen gas is then released into the atmosphere, completing the nitrogen cycle. Denitrification is the conversion of nitrate to nitrogen gas. ■

The growth of plants in ecosystems is often limited by the availability of nitrate and ammonia in the soil, which is one of the reasons farmers fertilize fields. Today most of the ammonia and nitrate added to the soil by farmers is not organic. It is produced in factories by industrial, rather than bacterial, nitrogen fixation. The amount of nitrogen gas converted to ammonia or nitrates in chemical factories is immense, amounting to perhaps 30 percent of the total

SECTION 16-3

◆ **VISUAL STRATEGY**

Figure 16-17
Point out the difference between nitrogen fixation (nitrogen gas ⟶ ammonia) and nitrification (ammonia ⟶ nitrates). Tell students that lightning also results in nitrogen fixation, although such atmospheric action amounts to less than 10 percent of that carried out by organisms through nitrogen fixation. Finally, have students recognize that nitrogen fixation and denitrification have opposite results.

Effective Reading

Connecting Text With Visuals

Help students to see that the four stages of the nitrogen cycle, introduced in boldface type on pages 352–353, can be visualized by studying Figure 16-17. Tracing the path nitrogen follows will help students see that nitrogen does move through a cycle. Point out that studying a visual while reading the text can be an effective way to understand the material.

CONNECTIONS
....

Chapter 2
ATP

Draw the structure of ATP so that students can see that phosphorus is a component of this molecule.

Chapter 8
Nucleic Acids

Review the structure of DNA and RNA, and point out that phosphorus is part of the "backbone" of these molecules but that it has no role in information coding.

Chapter Closure

Have students make a list of all the things they throw away during one week. Then, have them divide their list into recyclable and nonrecyclable items. Recyclable items might include aluminum cans, plastic bottles, newspaper, office paper, and yard waste (which can be used as compost), while nonrecyclable items might include plastic packaging and ballpoint pens. Then, have them write a short essay in which they address this topic: Suppose that no cycling of materials occurred in an ecosystem. Could the ecosystem persist?

nitrogen that cycles in biological systems worldwide. As you learned in Chapter 10, scientists are very interested in using genetic engineering to place the nitrogen-fixing genes of bacteria into the chromosomes of crop plants. If these attempts succeed, the plants themselves will be able to fix nitrogen, eliminating the need for nitrogen-supplying fertilizers.

The Phosphorus Cycle

Phosphorus is an essential element of all living organisms. It is a key part of both ATP and DNA. Phosphorus is usually present in soil and rock as calcium phosphate, which dissolves in water to form phosphate ions, PO_4^{3-}. This phosphate is absorbed by the roots of plants and used to build organic molecules such as ATP and DNA. Animals that eat the plants reuse the organic phosphorus. When the plants and animals die and decay, bacteria in the soil convert the phosphorus in organic molecules back into PO_4^{3-}.

Phosphorus is sometimes transferred from one ecosystem to another. For example, phosphorus from a forest ecosystem might be carried down a river to the sea, where it would be trapped in sediments that would eventually turn into rock. Millions of years later the sea floor might rise and expose the rock, and the phosphorus would be released by weathering and made available for living organisms once again.

The phosphorus level in freshwater lake ecosystems is often low, which prevents much growth of photosynthetic algae in these ecosystems. Phosphorus added to lakes as a result of human activities (some detergents and fertilizers are rich in phosphorus) can have disastrous consequences. Fertilization of a lake by the continual, inadvertent addition of phosphorus to its waters produces a green scum of algal growth on the surface of the lake that will eventually kill most of the organisms in the lake. After an initial "bloom" of rapid algal growth, the algae begin to die. Bacteria that feed on the dead algal cells use up so much of the lake's dissolved oxygen that fishes and invertebrate animals suffocate, as shown in Figure 16-18.

Figure 16-18 These fish suffocated after fertilizer, washed into the lake by runoff from farmland, caused a population explosion of algae. Bacteria that decomposed the algae used up the oxygen in the lake.

Section Review

1. *Explain how cutting down a rain forest disrupts the water cycle.*
2. *The burning of rain forests leads to increased atmospheric carbon dioxide levels in two ways. Explain how.*
3. *Describe three functions performed by bacteria in the nitrogen cycle.*

Critical Thinking
4. *Nutrients can be reused, but energy cannot. Explain why.*

Vocabulary

Vocabulary

biomass (348)
carnivore (344)
community (338)
consumer (343)
decomposer (344)
detritivore (344)
diversity (338)
ecology (337)

ecosystem (338)
food chain (344)
food web (344)
ground water (349)
habitat (338)
herbivore (343)
nitrogen fixation (352)
omnivore (344)

primary productivity (343)
primary succession (340)
producer (343)
secondary succession (340)
species diversity (338)
succession (340)
trophic level (343)

Concept Mapping

Construct a concept map that describes the role of organisms in the flow of energy through an ecosystem. Use as many terms as needed from the vocabulary list. Try to include the following items in your map: trophic level, food web, food chain, producer, consumer, primary productivity, carnivore, detritivore, herbivore. Include additional terms in your map as needed.

Review

Multiple Choice

1. Extinction of the dusky seaside sparrow is noteworthy because
   a. it was the last remaining subspecies of seaside sparrow.
   b. of its possible effects on other organisms in the same ecosystem.
   c. bird-watchers will never again see the dusky seaside sparrow.
   d. other organisms will fill the habitat that it left vacant.

2. What critical role is played by fungi and bacteria in any ecosystem?
   a. primary production
   b. decomposition
   c. boundary setting
   d. physical weathering

3. A mountain lion is a(n)
   a. omnivore.   c. detritivore.
   b. herbivore.   d. carnivore.

4. Plants are called producers, rather than consumers, because they
   a. produce carbon dioxide when they respire.
   b. capture energy from the sun and build carbohydrates.
   c. produce leaves and roots in the spring.
   d. use the energy they produce in photosynthesis.

5. Which sequence shows the correct order of succession at Glacier Bay, Alaska?
   a. *Dryas,* alder, Sitka spruce
   b. mosses, hemlock, Sitka spruce
   c. *Dryas,* hemlock, alder
   d. alder, *Dryas,* hemlock

6. Assume that the energy stored by organisms at one trophic level is about 10 percent of that stored by organisms in the level below. How much energy would be available to organisms at the third trophic level of an energy pyramid if the energy stored by organisms at the first trophic level were 1,000 kcal?
   a. 1,000 kcal   c. 10 kcal
   b. 100 kcal   d. 1 kcal

7. In terrestrial ecosystems, plants return water to the atmosphere by
   a. condensation.   c. assimilation.
   b. transpiration.   d. desiccation.

8. Humans are adversely affecting the carbon cycle by
   a. planting trees.
   b. burning fossil fuels.
   c. mining limestone.
   d. breathing.

CHAPTER REVIEW ANSWERS

Each item in the Chapter Review is correlated by text section in the assignment guide that follows.

ASSIGNMENT GUIDE

Section	Review	Themes Review	Critical Thinking
16-1	1, 2, 5, 10, 11, 17, 18		
16-2	3, 4, 6, 12–14, 19, 20	23, 24	
16-3	7–9, 15, 16, 21, 22	25	26–28

Concept Mapping

Map answer is shown on page 335B.

Review

Multiple Choice

Code in parentheses indicates section number and objective number.

1. b (16-1.2)	**6.** c (16-2.3)
2. b (16-1.2)	**7.** b (16-3.1)
3. d (16-2.1)	**8.** b (16-3.2)
4. b (16-2.1)	**9.** d (16-3.3)
5. a (16-1.3)	

Completion

10. ecosystem, ecology (16-1.1)
11. secondary succession, primary succession (16-1.3)
12. herbivores, omnivores (16-2.1)
13. energy, trophic (16-2.2)
14. chain, web (16-2.2)
15. carbon dioxide, fossil fuels or wood (16-3.2)
16. ATP or RNA (16-3.3)

Short Answer

17. Removing bacteria and fungi would eliminate decomposers from the ecosystem. Without decomposers, dead organisms and wastes would not decay and essential elements like carbon and nitrogen would not be recycled. Eventually, the ecosystem could self-destruct from lack of nutrients. (16-1.2)

CHAPTER REVIEW ANSWERS continued

Short Answer continued

18. *Dryas* is outcompeted for light by alders, which grow faster than *Dryas* because they produce their own nitrogen with nitrogen-fixing nodules on their roots, and by cottonwoods and willows, which grow larger than *Dryas*. (16-1.3)

19. The number of trophic levels in a food chain is limited because so little of the energy stored by the organisms at one trophic level is transferred to organisms at the level above. As a result, there is not enough energy at the highest level to support an additional level. (16-2.3)

20. The number of producers may be lower than the number of consumers when consumers are much smaller than producers, as is the case when several consumers feed on one large tree. Also, producers such as aquatic phytoplankton that have short life spans and are eaten as fast as they develop may be difficult to count at a given time. (16-2.3)

21. Respiration, erosion, combustion, fossil fuel formation, and photosynthesis are the possibilities. (16-3.2)

22. In assimilation, ammonia produced by nitrogen-fixing bacteria is absorbed by plants; in ammonification, the nitrogen compounds in animal wastes and dead tissue are converted to ammonia; in nitrification, nitrate is produced from ammonia; and in denitrification, nitrate is converted to nitrogen gas. (16-3.3)

Themes Review

23. Because one Komodo dragon requires less energy than one lion, the Serengeti Plain could support more Komodo dragons than lions, and the predator-to-prey ratio would be higher. (16-2.3)

24. An omnivore can choose from many types of food. Thus, if the supply of one type of food dwindles, an omnivore can switch to another type to avoid starving. (16-2.1)

25. Oxygen interferes with the nitrogen-fixing reactions, so they must occur in the absence of oxy-

9. Which role is not performed by bacteria in the nitrogen cycle?
a. converting nitrogen gas to ammonia
b. changing urea to ammonia
c. turning nitrates into nitrogen gas
d. converting nitrates to uric acid

Completion

10. A(n) _____ includes the interactions among living organisms and everything that affects their lives. The study of these interactions was named _____ by Ernst Haeckel.

11. Weeds growing in a recently burned patch of forest represent the first stage in _____ , while lichens that colonize a newly formed volcanic island are the first stage of _____ .

12. Cows, which eat only plants, are called _____ , but humans, who eat both plants and animals, are called _____ .

13. A food chain is used to describe the passing of _____ from one organism to another. An organism's position in the food chain is called its _____ level.

14. A linear pathway that describes what organisms eat is called a food _____ . When all of the linear pathways in an ecosystem are linked together, they form a food _____ .

15. Carbon enters living things as _____ during photosynthesis. Carbon is returned to the nonliving environment by the burning of _____ .

16. Phosphorus is an essential element in many biologically important chemicals, such as _____ and DNA.

Short Answer

17. Picture a southeastern pine forest. How would it change if all bacteria and fungi were removed?

18. Explain why alders, willows, and cottonwoods replace *Dryas* at Glacier Bay, Alaska.

19. Food chains usually consist of no more than three or four trophic levels. Explain why.

20. Ecological pyramids can be constructed using biomass, numbers of individuals, and energy. Typically, the greatest amount of biomass, number of individuals, or amount of energy is found in the lowest trophic level, that of the producers. However, there are occasionally fewer individuals in the lowest trophic level than in the level above. Explain how this can occur.

21. Name two major processes involved in the carbon cycle.

22. Name the four main stages of the nitrogen cycle, and describe what occurs at each stage.

Themes Review

23. Flow of Energy Reptiles, such as the Komodo dragon shown below, require about one-tenth as much food as mammals, which have much faster metabolic rates. On the Serengeti Plain of Africa, the ratio of lions to their prey is about 1 to 1,000. If Komodo dragons instead of lions were the top predators on the Serengeti Plain, would the ratio of predators to prey be higher, lower, or the same? Explain your answer.

gen. That is why nitrogen-fixing bacteria live within airtight nodules on the plants' roots. (16-3.3)

Critical Thinking

26. Carbon dioxide levels fall in the spring, when new plant growth is removing large amounts of carbon dioxide from the atmosphere. In the fall, plants die and trees lose their leaves. Carbon dioxide levels rise because of the decomposition of leaves and other plant parts and because photosynthesis stops due to low temperatures. (16-3.2)

27. The water cycle would involve the processes of evaporation, precipitation, runoff, and percolation only. There would be no transpiration. (16-3.1)

28. Producers, such as plants, use carbon dioxide to build carbohydrates. They also release carbon dioxide when carbohydrates are broken apart during respiration. Consumers take in carbon as carbohydrates when

24. **Evolution** Humans, raccoons, and bears are omnivores. What adaptive advantage might this feeding strategy provide?

25. **Structure and Function** Explain why the nitrogen-fixing bacteria associated with plants live in airtight nodules on the plants' roots.

Critical Thinking

26. **Explaining Observations** A scientist measuring atmospheric carbon dioxide levels in a Michigan deciduous forest (deciduous trees, such as elms and maples, shed their leaves in the fall) finds that the carbon dioxide levels fluctuate during the year, as shown in the graph below. Using your knowledge of the carbon cycle, explain the cause of these fluctuations.

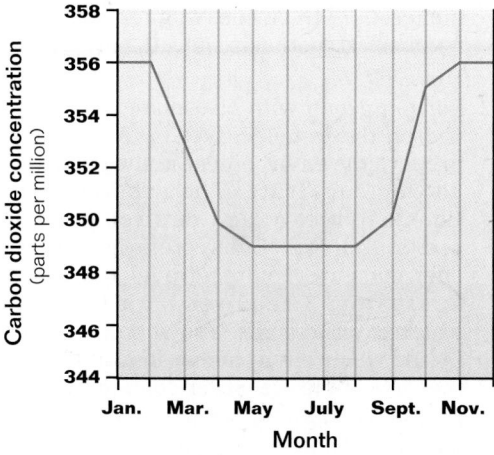

27. **Making Predictions** How would the water cycle change if all living organisms became extinct?

28. **Making Inferences** Energy flow in an ecosystem is described in terms of the activities of producers, consumers, and decomposers. What is the role of each of these in the carbon cycle?

Activities and Projects

29. **Research and Writing** Write a report that describes how ecologists are using stable-isotope tracing to construct food webs without observing feeding behavior. In your report, discuss the nature of the sample needed for analysis, and discuss the long-standing assumption that was questioned by ecologists as a result of data collected using stable-isotope tracing.

30. **Cooperative Group Project** Build your own ecosystem in an aquarium or terrarium. Decide as a group what kind of ecosystem you will build. Then assign each group member the task of collecting representative producers, consumers, or decomposers for the ecosystem. Finally, as a group, construct a food web for the ecosystem.

Readings

31. Read the article "The Recovery of Spirit Lake," in *American Scientist*, March–April 1993, pages 166–177. What was the condition of Spirit Lake before and soon after the eruption of Mount St. Helens in 1980? How has Spirit Lake changed since the eruption?

32. Read the article "Ecosystems For Industry," in *New Scientist*, February 5, 1994, pages 21–22. How are current industries and ecosystems different? How would we all benefit if industries functioned more like ecosystems?

33. Read the article "Why American Songbirds Are Vanishing," in *Scientific American*, May 1992, pages 98–104. What types of ecological damage explain the population decline of many bird species? What can be done to increase the numbers of songbirds in North America?

30. Students should use creativity in assembling their ecosystem, but it should exhibit diversity. Recommend that students construct an ecosystem for which organisms and nonliving parts can be obtained with relative ease. Discourage students from collecting large numbers of organisms, which could disrupt local populations.

Readings

31. Before the eruption, the lake was clear, contained low levels of dissolved minerals, and supported few fish. After the eruption, the lake was shallower, contained very high concentrations of inorganic chemicals and organic matter, and had an active and prolific microbial population. The populations of photosynthetic organisms—including Eurasian water milfoil—have risen since the eruption, accelerating the lake's aging.

32. Industries are usually one-way systems; raw materials enter, and wastes and finished products exit. The movement of materials within an ecosystem is cyclical; the wastes of one organism serve as a raw materials for another organism. Industries operating more like ecosystems would produce less pollution since their wastes would be used by another industry. Also, the reuse of materials means that less raw material needs to be obtained, thus helping to conserve supplies of resources.

33. Growth of suburbs has drawn in more nest parasites, such as bluejays, raccoons, and opossums. Deforestation in the tropics is eliminating the birds' overwintering grounds. Consolidating and expanding the largest tracts of forest, preserving wetlands, and slowing tropical deforestation can increase the number of songbirds in North America.

they eat foods and use the carbohydrates to build their own bodies. Like producers, consumers release carbon to the air as carbon dioxide during respiration. Decomposers use carbon in much the same way as do consumers. They take in carbon as carbohydrates when they break down the bodies of pro-

ducers and consumers. Decomposers also return carbon to the air as carbon dioxide. (16-2.1)

Activities and Projects

29. Stable isotope tracing determines the ratio of carbon-13 to carbon-12 in a tissue sample. The ratio is compared with a

known "carbon signature" for different primary producers, allowing the identification of the primary producer of a food web. The tissue sample required for the procedure can be as small as a nail paring. Use of stable isotope tracing has revealed that kelp, not phytoplankton, are the primary producers in a number of coastal ecosystems.

Preparation

Organisms may be obtained from a biological supply house. You may also be able to find earthworms and crickets in the local environment or at pet stores or bait shops. If you use tap water, allow it to stand in an open jar at least 24 hours so that the chlorine will escape.

Procedural Notes

• Allow one week between Parts A and B to allow plants and algae to grow. For Part B, students will need a few minutes each day for a period of a few weeks to make observations.

• Encourage students to think about the ecological roles that organisms have, such as primary producers, herbivores, and carnivores.

• Recording the number of organisms may be tricky in some cases, so estimates may be required. Discuss the sources of error in estimates versus direct counts.

• The ecosystems may be kept for the rest of the school year and then given to students to take home over the vacation.

Background Answers

1. Organisms are limited by such abiotic factors as the type of soil, rock, or water that they live on or in, the extremes of temperature, and the amount and frequency of precipitation. Organisms can also be killed by extreme weather conditions, such as storms, or by major changes in climate.

2. Predation, competition, mutualism (in which both organisms benefit), and commensalism (in which one organism benefits, the other is unaffected) are the four main ways in which organisms interact.

3. How do species interact in a closed ecosystem?

LABORATORY Investigation Chapter 16

Ecosystem in a Jar

OBJECTIVE

Observe the interaction of organisms in a closed ecosystem, and compare this ecosystem with others observed in nature.

PROCESS SKILLS

• observing a model ecosystem
• recognizing relationships among components of an ecosystem

MATERIALS

• large glass jar with lid
• pond water or dechlorinated tap water
• gravel, rocks, and soil

Ecosystem 1
  • pinch of grass seeds
  • pinch of clover seeds
  • 10 mung bean seeds
  • 3 earthworms
  • 4–6 isopods
  • 6 mealworms
  • 6 crickets

Ecosystem 2
  • Strands of *Anacharis*, *Fontinallis*, and foxtail
  • duckweed
  • *Chlamydomonas* culture
  • black ram's horn snail
  • 4 guppies or platys

Ecosystem 3
  • pinch of grass seeds
  • pinch of clover seeds
  • pond snails
  • 2 *Anacharis* strands
  • 10 *Daphnia*
  • 3 *Fontinallis* strands
  • *Chlamydomonas*
  • 4 guppies or platys

• graph paper
• 8½ × 11 in. acetate sheets
• several colors of pens for overhead transparencies

BACKGROUND

1. How are living things affected by nonliving things in the environment?

2. How do different types of organisms interact with one another?

3. Write your own question to explore in your lab report or notebook.

TECHNIQUE

Part A: Experimental Setup

1. In this investigation, you will be observing organisms in one of three different ecosystems. Look at the organisms listed in the materials list for each of the three systems, and choose the system you would like to observe. Think about the organisms that will live in your ecosystem. Which ones might be the most numerous? Which ones might decrease in number? Hypothesize how the organisms will interact.

2. Form a group with classmates who have chosen the same ecosystem. As a group, prepare the environment in the jar with a chosen substrate. The substrate for Ecosystem 1, a land environment, is rocks and soil. Ecosystem 2, a water environment, requires gravel and water. Ecosystem 3 requires both land and water environments. You will need to set up the water environment in a small dish that fits inside the larger container.

3. Plant the seeds and/or add the plants and/or algae to your ecosystem. Put a lid on the container and let the ecosystem remain undisturbed in indirect sunlight for a week.

4. Clean up your materials and wash your hands before leaving the lab.

Records

Graphs for each species in the chart will vary.

Population Change in an Ecosystem Over Time

Species	Number of individuals						
	Day 1	Day 2	Day 3	Day 4	Day 5	Day 6	Day 7

Part B: Observing Your Ecosystem

5. Place the chosen animals into the jar and lightly close it.

6. Observe the jar for a few minutes daily.

7. Make a chart in the **Records** section of your report to record the number of each species in your original ecosystem. Record daily any changes you observe.

8. In the **Records** section of your report, make a graph for each species in your chart, plotting the number of organisms as a function of time. Place a clear acetate sheet over each graph. Using a pen for overhead transparencies, trace each graph onto an acetate sheet. Use a different color and a different acetate sheet for each organism.

9. Compare the acetate sheets of two organisms that you hypothesized would interact—a predator and its prey, for example. Hold one sheet on top of the other and analyze both graphs. Record the results in the **Records** section of your report. Also, briefly summarize the procedure you followed.

10. Clean up your materials and wash your hands before leaving the lab.

INQUIRY

1. What happened to the organisms in your ecosystem?

2. What are some possible causes of the changes in the populations you observed?

3. Construct a food chain for the ecosystem you observed.

4. What could be learned if more than one jar was set up in an identical manner?

5. How does your ecosystem resemble a natural ecosystem? How does it differ?

ANALYSIS

1. How did your observation compare with your hypothesis? If the results differed from what you expected, explain what might have caused the difference.

2. Looking at your graphs, what kind of relationship can you find between predator and prey populations?

3. How would you modify the ecosystem if you were to repeat this investigation?

FURTHER INQUIRY

Write a new question that could be explored as a new investigation. The following are examples:

What are the effects of certain abiotic factors—including temperature, light, and moisture—on the organisms in an ecosystem?

How could a scientist set up an experiment to find out how certain pesticides or fertilizers might affect an ecosystem?

Inquiry Answers

1. Answers will vary.

2. Answers will vary but may include that plant and algae populations decreased after animals started eating them.

3. Answers will vary.

4. It would help to determine whether the observed changes were random or the result of the interactions among the ecosystem's inhabitants.

5. Real ecosystems and the laboratory ecosystem contain a diversity of organisms from several trophic levels, have both living and nonliving components, and depend on the sun for energy. The laboratory ecosystem is less diverse and much younger and has more definite boundaries than a real ecosystem.

Analysis Answers

1. Answers will vary.

2. Students should note that as the number of predators increases, the number of prey decreases; the trend reverses as the number of prey reaches a minimum.

3. Answers will vary.

Population Change in an Ecosystem Over Time

Species	Number of individuals						
	Day 8	Day 9	Day 10	Day 11	Day 12	Day 13	Day 14

UNIT 4

Block Scheduling Guide

Each block represents about 45–50 minutes of instructional time. The resources cited in a block are coded to a specific program component using the Key on the next page.

Lecture Concepts	Lesson Resources

17-1 How Organisms Interact in Communities pp. 360–364

BLOCKS 1 and 2

a. Present-day interactions between organisms are the result of a long evolutionary history in which the participants have adjusted to one another. There are a number of interactions: predation, parasitism, competition, and symbiosis.

Lecture Resources
- Opening Question, page 361
- Demonstrations: Adaptations of a Parasite, Overcoming Defenses
- Visual Strategy: Figure 17-5
- Holt Biology Videodiscs: Lesson 17-1

Classwork Options
- Who Benefits and Who Loses in Species Interactions? and Graphic Organizer, page 363

- Laboratory Techniques and Experimental Design C24: Mapping Biotic Factors in the Environment

Assignment Options
- Secondary Compounds, page 362
- Section Review, page 364
- Chapter Review, questions 1, 2, 12, 32

17-2 How Competition Shapes Communities pp. 365–371

BLOCKS 3 and 4

a. The functional role of a particular species in an ecosystem is called its niche. The entire range of conditions a species can tolerate is its fundamental niche. The part of its fundamental niche that a species actually occupies is called its realized niche.

b. Competition can cause changes in a community. In the short term, the species that is the superior competitor may drive inferior competitors to extinction. Over the long term, competing species may evolve differences that reduce their use of common resources.

Lecture Resources
- Opening Question, page 365
- Demonstration: Winner Takes All
- Habitat vs. Niche, page 365
- Visual Strategies: Figures 17-7, 17-8, 17-9, 17-10
- Teaching Transparencies: 78, 87
- Character Displacement, page 368
- Competitive Exclusion, page 369
- Niche Overlap, page 370
- What If A Predator Is Removed from an Ecosystem? page 371

- Ecosystem Complexity and Fragility, page 371
- Holt Biology Videodiscs: Lesson 17-2

Classwork Options
- Laboratory Investigation: Habitat Selection, pages 386–387

Assignment Options
- Section Review, page 371
- Chapter Review, questions 3–5, 13, 14, 18–20, 25–28, 30, 31

17-3 The Influence of the Physical Environment pp. 372–375

BLOCK 5

a. The great diversity of ecosystems is due in part to variations in climate from place to place. Latitude is the most important factor for determining climate in an area. Other important factors are distance from the ocean, elevation, and position relative to mountain ranges that can intercept moisture-carrying winds.

Lecture Resources
- Opening Question, page 372
- Demonstration: The Sun and the Moon
- Effective Reading, page 372
- Visual Strategy: Figure 17-13
- Overcoming Misconceptions, page 373
- The Rain-Shadow Effect, page 374
- Ocean Currents, page 375
- Holt Biology Videodiscs: Lesson 17-3

Classwork Options
- Laboratory Techniques and Experimental Design C25: Assessing Abiotic Factors in the Environment

Assignment Options
- Rainfall, page 374
- Section Review, page 375
- Chapter Review, questions 6, 7, 10, 15, 21, 29

17-4 Major Biological Communities pp. 376–382

BLOCKS 6 and 7

a. The marine environment consists of three major kinds of habitats: shallow ocean waters, open sea surface, and deep sea waters.

b. Freshwater habitats are distinct from both marine and terrestrial habitats and are very limited in area.

Lecture Resources
- Opening Question, page 376
- Demonstration: Biological Communities
- Connections, page 376
- Teaching Transparencies: 65, 79
- Research Update, page 377

- Connections, page 378
- Visual Strategies: Figures 17-22, 17-23
- Agriculture in the Rain Forest, page 379
- Overcoming Misconceptions, page 379
- Numbers vs. Diversity, page 381

17-4 Major Biological Communities pp.376–382 (continued)

c. A major terrestrial community that is found in different areas with similar climates is called a biome. A biome's structure and appearance are similar throughout its distribution.

d. The world's biomes are tropical rain forest, savanna, desert, temperate grassland, temperate deciduous forest, taiga, and tundra.

- ■ Sunlight in the Tundra, page 382
- Holt Biology Videodiscs: Lesson 17-4

Classwork Options
- ■ Concept Mapping, page 383
- Laboratory Techniques and Experimental Design C26: Assessing and Mapping Factors in the Environment

Assignment Options
- ■ Deserts in the United States, page 380
- ■ Closure, page 382
- ■ Section Review, page 382
- ■ Chapter Review, questions 8–11, 16, 17, 22–24

Review/Enrichment

BLOCK 8

- ■ Study Guide: Concept Review and Pretest

Assignment Options
- ■ Activities and Projects, questions 33, 34
- ■ Readings, questions 35, 36

Assessment Options

BLOCK 9

Traditional Assessment
- ■ Chapter 17 Test
- Test Generator/Assessment Item Listing: Software item bank for preparing customized chapter tests.

Portfolio Assessment
Select student reports for one or more laboratory experiments from this chapter. *The Direct Observation Checklist* on the *BioSources Teaching Resources CD-ROM* should be completed during a laboratory or other cooperative learning experience.

Answer to Concept Map
The following is one possible answer to the Concept Mapping exercise on page 383.

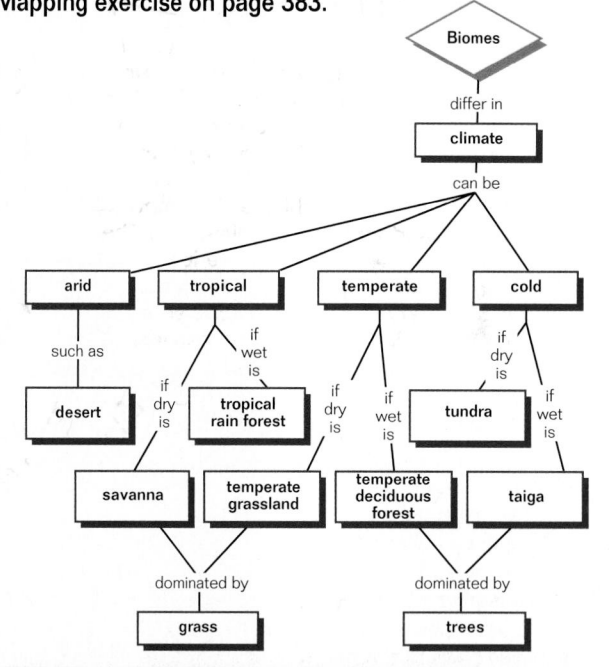

Holt Biology Videodiscs

Holt Biology Videodiscs gives you a powerful tool for teaching, review, and assessment. *Concepts of Biology* adds interactive lesson plans and extensive tools for customization using CD-ROM technology.

CONCEPTS OF BIOLOGY

Use the following topic from *Concepts of Biology* to help you teach this chapter:

- ■ Topic 10: Ecosystem Structure and Function

For further information regarding the media on the videodiscs, see *Holt Biology Videodiscs Teacher's Correlation Guide to Biology: Principles and Explorations.*

Chapter Theme

Evolution *Present-day interactions between organisms have been shaped by natural selection. Natural selection favors those organisms best suited to their environment, which includes living as well as nonliving components. Therefore, living things evolve in response to other organisms in their environment, not just in response to their physical environment.*

Tapping Prior Knowledge

- What is a community?
- What are the types of species interactions in communities?
- How does the shape of Earth and its angle with respect to its orbit affect climate?
- How does the ocean influence climates on land?

Opening Demonstration

Coevolution takes place when species interact so closely that they serve as a strong selective force on each other. The intimate relationship between coevolving species often stimulates the evolution of complementary adaptations. Show the class a photo of a hummingbird feeding at a flower. The deep, bell-shaped flower conforms to the long beak of the bird. When the hummingbird inserts its beak into the flower, it brushes against the female parts of the flower, leaving any pollen it already carries on their surface. It also contacts the male pollen-bearing structures and is dabbed with pollen, which it will transport to another flower. Emphasize that frequently each member of a plant-pollinator pair evolved so that it primarily served the other.

CHAPTER 17 BIOLOGICAL COMMUNITIES

REVIEW

- community (Section 16-1)
- habitat (Section 16-1)
- ecosystems (Section 16-1)
- primary productivity (Section 16-2)
- energy flow in ecosystems (Section 16-2)

Desert biome in winter

Authors' Rationale

This close look at the biology of communities shows the nature of species interactions, how they evolve, and what purposes they serve. Your students will find that predation and competition play very important roles in maintaining the stability of ecosystems. Competition is an important resource-allocating as well as selective force in communities. Species are affected not only by other organisms but also by climate. These important interactions are investigated in marine, freshwater, and terrestrial communities.

17-1 How Organisms Interact in Communities

Y ou can view the community of organisms that inhabits an ecosystem as a web of interactions. Each species competes with other species for resources. Many species cooperate with other species, and every animal species feeds on some species and is fed on by others. As you will read in this chapter, these interactions between species play major roles in shaping ecosystems.

Vocabulary Preview

coevolution
predation
parasitism
secondary compound
competition
symbiosis
mutualism
commensalism

Section Objectives

- Describe coevolution of predators and prey.
- Identify one example of competition.
- Contrast mutualism with commensalism.

Opening Question

Ask each student to choose two animals that can be found in the same community. Have each student describe the various ways the two animals interact. Prepare a list on the board with three headings: Benefits/Benefits, Benefits/Not Affected, and Benefits/Harmed. After students provide their descriptions, ask under which category they would place the interaction they have described.

Interacting Species Evolve in Response to One Another

The interactions between species in ecosystems today are the result of a long evolutionary history in which the participants have adjusted to one another. Thus, plants evolved flowers that promoted efficient dispersal of their pollen by insects and other animals, and pollinators evolved a number of traits that enabled them to obtain food or other resources efficiently from the flowers of the plants they pollinate. Natural selection has often led to a close match between the characteristics of flowers and their pollinators, as you can see in Figure 17-1. Back-and-forth evolutionary adjustments between interacting members of an ecosystem are called **coevolution**.

Figure 17-1 With its long tongue, the hawk moth is able to reach the nectar deep within these flowers.

Coevolving in Opposition: Predators and Prey

Predation occurs when one organism feeds on another, as the shark shown in Figure 17-2 is doing. Examples of predation include familiar situations such as lions eating zebras and snakes eating mice. A special case of predation is **parasitism,** in which one organism feeds on and usually lives on or in another, typically larger, organism. Parasites often do not kill their prey (known as the "host"), since they depend on it for food and a place to live, and as a means to transmit their offspring to new hosts. Among the parasites that may have fed on you at some time are ticks, mosquitoes, and fleas.

How would you expect predators and prey to coevolve? To answer this question, you must consider the effect of predation on both participants. The predator benefits from

Figure 17-2 This blacktip shark is devouring a tuna.

Demonstration

Adaptations of a Parasite
Show the class a photograph of a tapeworm. Point out the various adaptations that enable it to be a successful parasite. Hooks located on the head enable the tapeworm to attach to the intestinal wall of its host. The tapeworm's body wall is permeable so that nutrients can be easily absorbed from the host. Most of the body consists of segments that are specialized for just one purpose—reproduction. Tapeworms in humans may grow as long as 6 m (20 ft.). A parasite of this size can cause illness not only by stealing nutrients but also by causing an obstruction in the intestinal tract.

Historical Note

Coevolution and the Conquistadores

When Europeans conquered the New World beginning in the late 1400s, they introduced their animals, plants, culture, and epidemic diseases. By some estimates, 95 percent of the population of the New World perished from diseases such as smallpox, measles, and influenza brought from the Old World. Why were the residents of the New World so much more susceptible than the Europeans? Coevolution. These diseases had existed in the Old World for thousands of years, and the Europeans had over many centuries evolved a natural resistance. The residents of the New World, by contrast, had never experienced smallpox, measles, or influenza, and few people had any resistance.

feeding on its prey. The prey, in contrast, is always harmed. At best, it only loses some tissue, which requires energy to replace. At worst, it is killed. Therefore, you would expect the prey to evolve ways to elude, avoid, or fight off predators. In response, the predator should evolve countermeasures to the defenses of its prey. Next, you will follow a real example of predator-prey evolution, the coevolution of plants and herbivores (plant-eating animals).

How Plants Defend Themselves From Herbivores

The most obvious ways that plants protect themselves from herbivores are with thorns, spines, and prickles. But many chemical compounds that occur in plants are much more widely distributed and even more important in defending the plants in which they occur. Virtually all plants contain defensive chemicals called **secondary compounds.** Estimates suggest that perhaps 50,000 to 100,000 different secondary compounds may exist; the chemical structures of some 15,000 of these compounds already have been characterized.

As a rule, each group of plants makes its own special kind of defensive chemical. For example, the mustard plant family produces a characteristic group of chemicals known as mustard oils. These oils give pungent aromas and tastes to such plants as mustard, cabbage, radish, and horseradish. The same tastes that we enjoy signal the presence of chemicals that are toxic to many groups of insects.

Figure 17-3 Defensive chemicals of the poison ivy plant, *right,* cause an itchy rash, *above.*

Some of the best-known plants with defensive chemicals are the species of the genus *Toxicodendron,* which includes poison ivy and poison oak. These plants produce a gummy oil called urushiol *(OO roo shee awl)* that causes a severe, itchy rash in susceptible people, as shown in Figure 17-3. This irritating substance almost certainly functions to protect the plants that produce it.

How Herbivores Overcome Plant Defenses

For each group of plants that is protected by a particular kind of defensive chemical, there are certain groups of herbivores that are able to feed on these plants, often as their exclusive food source. For example, the larvae of cabbage butterflies feed almost exclusively on plants of the mustard and caper families, which are defended from other herbivores by mustard oils. How do these animals manage to avoid the chemical defenses of the plants? Cabbage butterflies have evolved the ability to break down mustard oils and thus feed on mustards and capers without harm. This is one example of predator-prey coevolution. ◻

Competition: Common Use of Scarce Resources

When two species use the same resource, they are said to compete, and their interaction is called **competition.** Resources for which species compete include food, nesting sites, living space, light, mineral nutrients, and water. For competition to occur, however, the resource must be in short supply. In Africa, for example, lions and hyenas compete for prey. Fierce rivalry between these species can lead to battles that cause injuries to both sides. Though the term *competition* brings such battles to mind, most competitive interactions do not involve fighting. In fact, some competing species never encounter one another. They interact only by means of their effects on the abundance of resources. In the next section, you will learn how competition affects the nature of communities.

Coevolving in Cooperation

In **symbiosis** (sihm beye OH sihs), two or more species live together in a close, long-term association. The two types of symbiotic relationships are **mutualism,** in which both participating species benefit, and **commensalism,** in which one species benefits and the other is neither harmed nor helped. Cooperation is the hallmark of symbiosis. Examples of symbiosis include lichens, which are associations of certain fungi with green algae or cyanobacteria, and mycorrhizae, which are associations of fungi and the roots of plants.

One well-known instance of mutualism involves ants and aphids, as illustrated in Figure 17-4. Aphids are small insects that use their piercing mouthparts to suck fluids from the sugar-conducting vessels of plants. They extract a certain amount of the sucrose and other nutrients from this fluid. However, much of the fluid—so-called honeydew—runs out

Figure 17-4 The small green insects on this plant stem are aphids. They are pampered and protected by their ant guards, which feed on the sugary fluid the aphids secrete.

❑ CAPSULE SUMMARY

In a predator-prey interaction, the prey is harmed while the predator benefits. Thus, prey often evolve ways to escape being eaten, and predators evolve ways to overcome the defenses of the prey. For instance, many species of plants produce chemicals to protect them from herbivores. Many species of herbivores have evolved ways to break down these chemicals.

Demonstration
Overcoming Defenses
Show the class a specimen or photograph of a monarch butterfly, which has evolved enzymes that enable it to feed on milkweeds without being poisoned (the caterpillars, not the adults, eat the plants). The bitter, white sap of milkweed plants contains a compound that acts as a heart poison in vertebrates, which are potential predators of this plant.

Teaching Tip
Who Benefits and Who Loses in Species Interactions?
Have students create a *Graphic Organizer* in the form of a table that summarizes the effects of each type of species interaction on the participants. Students should indicate benefit with a "+," harm with a "−," and no effect with a "0." A completed *Graphic Organizer* is shown at the bottom of the page.

Graphic Organizer
Use this graphic organizer with *Teaching Tip:* **Who Benefits and Who Loses in Species Interactions?** on page 363.

Interaction	Effect on One Species	Effect on Other Species
Predation	+	−
Competition	−	−
Mutualism	+	+
Commensalism	+	0

Teaching Tip

 What If one species takes advantage of a mutualistic relationship among other species? In some mutualistic relationships, small fish can approach larger fish without fear because they feed off parasites that grow on the larger fish's body. The larger fish recognize these "cleaners" by their distinctive markings. Other species of fish, known as mimics, resemble the cleaners. However, instead of removing parasites, mimics take large bites of flesh from the larger fish. Ask students to predict what would happen if the number of mimics began to approach the number of cleaners. (*Students should recognize that the larger fish might start to attack both cleaners and mimics, thus ending the mutualistic relationship.*)

👁 VISUAL STRATEGY

Figure 17-5

Inform students that sea anemones are carnivorous animals whose tentacles are equipped with specialized stinging cells that paralyze their prey. Sea anemones are related to jellyfish and corals.

Figure 17-5 The clown fish can survive the stings of the sea anemone. Because it can live among the anemone's tentacles, the clown fish is protected from predators and can feed on the leftovers of its host's meal. The anemone is apparently unaffected by the fish.

in an altered form through their anus. Certain ants have taken advantage of this habit, in effect domesticating the aphids. The ants move the aphids from plant to plant, like a herd of dairy cows, "milking" them for the honeydew, which the ants use as food.

Among the best-known examples of commensalism are the relationships between certain small tropical fishes and sea anemones, marine animals that have stinging tentacles. These fishes, such as the clown fish shown in Figure 17-5, have evolved the ability to live among the tentacles of the sea anemones, even though these tentacles would quickly paralyze other fishes that touched them. The clown fishes feed on the leftovers from the meals of the host anemone.

Section Review

1. *Predator-prey coevolution has been described as an "arms race." Explain why this is an apt description.*
2. *Describe some ways that humans compete with other organisms.*
3. *Would you expect the unaffected member of a commensal relationship to evolve in response to the other member? Explain your answer.*

Critical Thinking
4. *The relationship between a hawk and the tree in which it nests may be a case of commensalism. However, the tree may benefit from the hawk's presence, perhaps gaining protection from herbivores. Describe an experiment to determine whether the hawk-tree relationship is mutualism or commensalism.*

Answers to Section Review

1. Each is in competition with the other to evolve features that will give it an advantage over its opponent.

2. Humans compete by using up resources that would otherwise be available to other organisms. For example, by constructing roads and buildings, humans take over habitat in which other organisms might live.

3. The unaffected member would not be expected to evolve. As it neither gains nor loses from the relationship, there is no stimulus for evolution.

4. The easiest way to characterize the interaction is by comparing trees that have nests with trees lacking nests. If both types of trees are in the same physiological state, then the relationship between hawk and tree is commensalism. If the trees with nests are in better shape (larger leaves, faster growth rate, etc.), then the relationship is mutualism.

17-2 How Competition Shapes Communities

*E*very biological community is part of an ecosystem, which encompasses not only living organisms but also the physical environment in which they live. In the previous section, you learned that a biological community is composed of organisms that interact in ways shaped by a long history of coevolution. Interactions among species have molded ecosystems over long periods of time, creating the world you see today. What determines which species live where? Why, for example, does the jaguar shown in Figure 17-6 live in Central America but not in Canada? To some degree, accident and history play a role. To a greater degree, however, biological interactions and the characteristics of the physical environment determine what animals and plants are able to live in a particular ecosystem. This section examines the role of biological interactions in shaping the nature of communities; the role of the physical environment is investigated in the next section.

Section Objectives

- Understand what a niche is.
- Distinguish between fundamental and realized niches.
- Contrast character displacement with competitive exclusion.
- Explain why competitive exclusion is not inevitable.

SECTION 17-2

Vocabulary Preview

niche

fundamental niche

realized niche

character displacement

competitive exclusion

principle of competitive exclusion

Opening Question

Ask students why competition is usually most intense between closely related organisms. Students should recognize that closely related organisms are likely to be very similar and therefore likely to use resources in similar ways.

Demonstration

Winner Takes All

Competition between two species occasionally results in one being eliminated from the community. Show the class photographs of a starling and a bluebird. Starlings were first introduced into Central Park in New York City in 1891. Today starlings are found throughout the United States, and in many areas they have outcompeted bluebirds for nesting sites, causing a drastic decline in bluebird numbers.

Teaching Tip

Habitat vs. Niche

An analogy may help students understand the difference between habitat and niche. Point out that their house or apartment is their habitat. What they do in their habitat is their niche.

Figure 17-6 The jaguar is the largest cat native to the Americas. It once ranged from Argentina through Central America to central Texas and southern Arizona. Humans have eliminated the jaguar from much of its former range through overhunting and habitat destruction. It is no longer found in the United States and is rare in Central America.

Each Species in a Community Plays a Role

To understand how biological interactions influence the makeup of communities, you must focus on the day-to-day events within the community—how organisms get food, where they live, and so on. The functional role of a particular species in an ecosystem is called its **niche.** A niche is how an organism lives, the "job" it performs within the ecosystem. A niche is often described in terms of how the organism affects energy flow within the ecosystem. Viewed in this way, the niche of a shrub growing in a forest

UNIT 4

CHAPTER 17

VISUAL STRATEGY

Figure 17-7

Point out that a niche is very complex because it comprises all the ways an organism affects and is affected by its environment. This figure shows only four aspects of the jaguar's niche. Have the students make a list of any additional aspects of the jaguar's niche they can think of. Then, divide the list into abiotic (nonliving) and biotic (living) aspects. *(Abiotic aspects will include temperature tolerance and water and nutrient requirements. Biotic aspects will include interactions with other organisms, such as potential competitors, mutualists, and parasites.)*

CONNECTIONS

Chapter 16

Top Predators

Jaguars are top predators—carnivores that have no predators—and belong to the highest trophic level of their ecosystem. Thus, they would be rare even if humans had not persecuted them so relentlessly.

Diet

Jaguars feed on mammals, fish, and turtles.

Time of Activity

Jaguars hunt by day and by night.

Reproduction

Jaguars give birth from June to August, during the rainy season.

Figure 17-7 All the ways in which the jaguar interacts with its environment—including where it lives, what it eats, when it is active, and when it reproduces—constitute its niche.

meadow is that of a producer, while the niche of a deer that eats the shrub is that of a herbivore, and the niche of a cougar that eats the deer is that of a carnivore. Niches can also be described in terms of how organisms use space. For instance, arthropod carnivores feed at different heights from the ground: wolf spiders hunt their insect prey within leaf litter on the forest floor, ladybird beetles hunt on the stems of plants, and dragonflies hunt in the air. Within every ecosystem, each organism has its own niche, its particular way of making a living and influencing its surroundings. Figure 17-7 summarizes some aspects of the jaguar's niche in the Central American rain forest.

A Closer Look at Niches

To gain a better understanding of what a niche is, you must look more closely at a particular species, say a Cape May warbler (a small, insect-eating songbird) flying in a forest and landing to search for dinner in a spruce tree. In a broad sense, the niche of this bird is the temperature and humidity it prefers (this species summers almost exclusively in the northeastern United States and Canada), the time of year it nests (mid-summer), what it likes to eat (small insects), and where on the spruce tree it finds its food (high on the tree at the tips of the branches). The total niche an organism is potentially able to occupy within an ecosystem, the entire range of conditions it can tolerate, is its **fundamental niche.**

Now reconsider what the warbler is doing. It feeds mainly at the very top of the spruce tree. Why is this? Insects that the warbler could eat are located all over the tree, but this bird, like most other Cape May warblers, centers its attention on insects in the upper branches. Seeming to defy good sense, Cape May warblers occupy only a portion of their fundamental niche.

Closer study of the Cape May warbler and its potential competitors reveals that this surprising behavior is part of a larger pattern of niche restriction. In the late 1950s, Robert MacArthur, an ecologist from Princeton University, carried out a classic investigation of the feeding habits of five warbler species. MacArthur found that all five species fed on insects in the same spruce trees at the same time but that each species concentrated on a different part of the tree, as shown in Figure 17-8. Some of the species of warblers fed on insects near the ends of the branches, while others regularly penetrated well into the foliage. Some, such as the Cape May warbler, stayed high on the trees, while others, such as the myrtle warbler, hunted on the lower branches. Thus, although

Figure 17-8 In the forests of the northeastern United States, these five warblers are potential competitors because they feed on insects in the same trees. However, each species hunts only in a portion of the tree. Ecologist Robert MacArthur proposed that this feeding strategy reduced competition among these species of warblers.

Cape May warbler

Blackburnian warbler

Black-throated green warbler

Bay-breasted warbler

Myrtle warbler

👁️ **VISUAL STRATEGY**

Figure 17-8
Ecologists often use the phrase "resource partitioning" to describe the patterns of resource use in a community. Ask students why resource partitioning is an appropriate description of the feeding behavior of the five warbler species that inhabit a spruce tree. (*The warblers partition, or divide, the insects on which they feed by foraging in different parts of the tree.*)

Teaching Tip

Tell students that all five warbler species that MacArthur studied belong to the same genus (*Dendroica*). Because closely related species are often competitors, MacArthur was interested in how these very similar species could coexist.

CONNECTIONS
....

Chapter 1
The Scientific Process
Be sure students recognize that MacArthur's conclusions about the foraging habits of warblers stemmed solely from careful observations of the birds in nature. Remind students that observations are an important part of scientific investigation.

Teaching Tips

Realized Niches

Point out that an organism's realized niche is a subset of its fundamental niche. Thus, the realized niche can be smaller than, or the same size as, the fundamental niche, but never larger.

Character Displacement

Emphasize that character displacement is an evolutionary change in some feature of the participating species in response to competition. The change may involve physical characteristics such as bill or jaw size, or it may involve behavioral characteristics such as time of foraging or food preference. For instance, the graph below shows character displacement in two hypothetical species of birds with similar feeding habits and overlapping ranges.

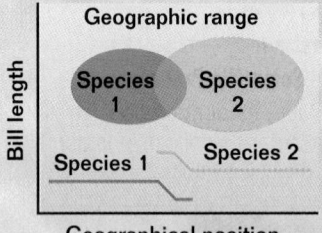

☐ CAPSULE
SUMMARY

An organism's niche is its "profession," its way of life. The entire range of conditions it can tolerate is its fundamental niche. In the face of competition, an organism may occupy only part of its fundamental niche. That part is called its realized niche.

all five species of warbler had very similar fundamental niches, they did not actually use the same resources. In effect, they divided the range of resources among them, each taking a different portion.

The part of its fundamental niche that a species actually occupies is called its **realized niche.** Stated in these terms, the realized niche of the Cape May warbler is only a small portion of its fundamental niche. What is the advantage in not foraging throughout the tree? MacArthur suggested that this feeding pattern reduces competition. Because each of the five warbler species uses a different set of resources by occupying a different realized niche, the species are not in competition with one another. In general, ecologists agree with MacArthur's conclusion that natural selection has favored a range of preferences and behaviors among the five species that "carves up" the available resources. In nature, similar, potentially competing species often differ more where their ranges overlap than where each occurs alone. This increased difference when living together is called **character displacement.** Many ecologists think character displacement is the result of evolution to reduce competition, although this conclusion remains controversial. ☐

Demonstrating Competition in Nature
....

It is one thing to propose that one species can limit the realized niche of another, as MacArthur did based only on observed patterns of resource use by warblers. But it is quite a different thing to demonstrate such an effect experimentally. A very clear case of competition was provided by experiments carried out in the early 1960s by Joseph Connell of the University of California. Connell worked with two species of barnacles that grow on the same rocks along the coast of Scotland. Barnacles are marine animals that are related to crabs, lobsters, and shrimp. They have free-swimming, microscopic larvae that cement themselves to rocks and remain attached there for the rest of their lives. As you can see in Figure 17-9, one species, *Chthamalus stellatus*, lives in shallow water, where it is often exposed to air by receding tides. A second species, *Balanus balanoides*, lives lower down on the rocks, where it is rarely exposed to the atmosphere.

When Connell removed *Balanus* from the deeper zone, *Chthamalus* was easily able to occupy the vacant surfaces, indicating that intolerance of the physical environment did not prevent *Chthamalus* from becoming established there; its fundamental niche clearly includes the deeper zone. However, when *Balanus* was reintroduced, it could always outcompete *Chthamalus* by crowding it off the rocks. In contrast, *Balanus* could not survive when placed in the

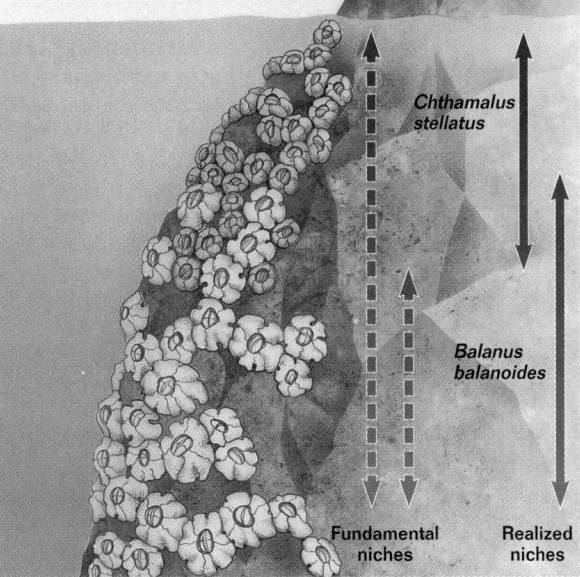

Figure 17-9 On the rocky coast of Scotland, *Balanus*, the larger species, and *Chthamalus* live on the same rocks, *above left*. *Balanus*'s superiority in the competition for space on the rocks stems from its faster growth rate: a *Balanus* individual can actually force a *Chthamalus* individual off a rock. Though their fundamental niches overlap considerably, *above right*, the realized niches of these two barnacles overlap little because of competition.

shallow-water habitats where *Chthamalus* normally occurs. *Balanus* apparently lacks the adaptations that permit *Chthamalus* to survive long periods of exposure to air. Connell's experiments show that *Chthamalus* occupies only a small portion of its large fundamental niche; the rest is unavailable because of competition with *Balanus*. As MacArthur suggested, competition can limit how species use resources.

The Outcome of Competition

Competition among species does not occur just because they use the same resource. They compete only when that resource is in limited supply. In nature, shortage is the rule, and species that use the same resource are almost sure to compete with each other. Such competitive interactions can have a profound effect on the structure of an ecosystem by stimulating character displacement. Alternatively, the species that is the better competitor may drive out the other, as occurred with Connell's barnacles (*Balanus* excluded *Chthamalus* from the rock surfaces). Local elimination of one competing species is known as **competitive exclusion.**

Darwin noted that competition should be most acute between very similar kinds of organisms because they tend to use the same resources in the same way. Does it follow, then, that when very similar species compete, one will always become extinct locally? In a series of carefully controlled laboratory experiments performed in the 1930s, the Russian biologist G. F. Gause looked into this question. In his initial

◉ VISUAL STRATEGY

Figure 17-9
Point out that the fundamental niche and realized niche for *Balanus* are equivalent, while the realized niche for *Chthamalus* is considerably smaller than its fundamental niche. Ask students whether the reduction in *Chthamalus*'s realized niche is an example of character displacement. (*No, because there has been no change in* Chthamalus's *habitat preference. It will live lower on the rocks if* Balanus *isn't there.*)

Teaching Tip
Competitive Exclusion
Inform students that the elimination of one competing species may be the result of aggressive interactions—including fighting—between the two species (interference competition) or the result of the use of a resource, leaving less for the other species (exploitative competition).

CONNECTIONS

Chapter 1
The Importance of Controls
Field studies are often hampered by a lack of controls that would otherwise permit ecologists to arrive at firm conclusions, especially regarding the outcome of competition. An example can be seen with the case of the California vole, which was accidentally introduced onto a small island in San Francisco Bay in the late 1950s. The vole population grew rapidly, while the resident house mouse population quickly declined. Eventually, the house mouse became extinct on the island. Whether its extinction was due to competition from the voles or some other factor remains in question because the process was not thoroughly investigated. Thus, other possible causes cannot be eliminated.

VISUAL STRATEGY

Figure 17-10
Be sure students do not develop the misconception that the same species will always be the one to eliminate the other. Point out that it is possible to alter the environmental conditions so that the outcome is reversed. For example, Thomas Park and colleagues at the University of Chicago conducted competition experiments on two species of flour beetles *(Tribolium)*. The temperature and humidity at which the beetles were raised determined which species was the superior competitor.

Teaching Tip
Niche Overlap

Ecologists use the term *niche overlap* to describe the use of the same resources by two or more species. Be sure students recognize that *Paramecium caudatum* and *P. bursaria* both survived because they used different resources—their niches did not overlap. However, *P. aurelia* and *P. caudatum* fed on the same food, and so their niches overlapped to a great degree. Hence, these two species could not coexist when food was scarce.

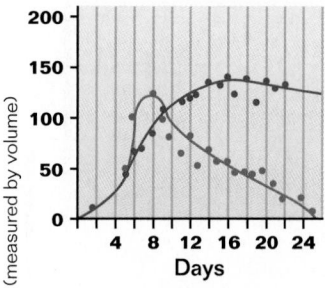

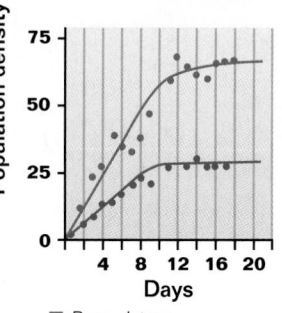

■ *P. caudatum*
■ *P. aurelia*
■ *P. bursaria*

Figure 17-10 These graphs summarize Gause's experiments on competition. The lower graph also demonstrates the negative effect of competition on the participants. Both species reach about twice the density when grown without a competitor as when grown with one.

☐ CAPSULE SUMMARY

Competition can cause changes in a community. In the short term, the species that is the superior competitor may drive inferior competitors to extinction. Over the long term, competing species may evolve differences that reduce their use of common resources.

experiments, Gause grew two species of the protist *Paramecium* in the same culture tubes, where they had to compete for the same food (bacteria). Invariably, the smaller of the two species, which was more resistant to bacterial waste products, drove the larger one to extinction. From these results Gause formulated what is now called the **principle of competitive exclusion.** This principle states that if two species are competing, the species that uses the resource more efficiently will eventually eliminate the other locally—no two species can have the same niche.

When Can Competitors Coexist?

Is competitive exclusion the inevitable outcome of competition for limited resources, as Gause's principle states? No. The outcome depends on the fierceness of the competition and on the degree of similarity between the fundamental niches of the competing species. If it is possible for the species to avoid competing, they may coexist.

In a revealing experiment, Gause challenged *Paramecium caudatum*—the defeated species in his earlier experiments—with a third species, *P. bursaria*. Since these two species were also expected to compete for the limited bacterial food supply, Gause thought one would win out, as had happened in his previous experiments. But that's not what happened. Both species survived in the culture tubes. Like MacArthur's warblers, the paramecia found a way to divide the food resources. How did they do it? In the upper part of the culture tubes, where oxygen concentration and bacterial density were high, *P. caudatum* was dominant because it was better able to feed on bacteria than was *P. bursaria*. However, in the lower part of the tubes, the lower oxygen concentration favored the growth of a different potential food, yeast, which *P. bursaria* was better able to eat. The fundamental niche of each species was the whole culture tube, but the realized niche of each species was only a portion of the tube. Because the niches of the two species did not overlap too much, both species were able to survive. Figure 17-10 summarizes Gause's experiments. ☐

Predation Can Lessen Competition Among Prey
....

Many studies of natural ecosystems have shown that predation lessens the effects of competition. A very clear example is provided by the studies of Robert Paine of the University of Washington. Paine examined how sea stars affect the numbers and types of species within marine intertidal communities. Sea stars are fierce predators of marine animals such as clams and mussels. When sea stars were kept out of experimental plots in a rocky intertidal zone, the number of prey species fell from 15 to 8. As you can see in Figure 17-11, the eliminated

species were crowded out by mussels, the chief prey of sea stars. By preying on mussels, sea stars keep the mussel populations low, neutralizing the mussels' ability to outcompete the other animals for space on the rocks. When the sea stars were removed, the mussels were free to assert their competitive dominance. As a direct result, the number of species in the ecosystem declined. Thus, predation can have a very healthy effect on ecosystems: it can promote diversity by minimizing competition.

Figure 17-11 The sea star *Pisaster, above left,* altered the outcome of competition for space in the coastal ecosystem studied by Robert Paine. When freed of the control exerted by sea stars, mussels—which are the superior competitors—crowded seven other species out of the ecosystem and overgrew the habitat, *above right.*

Teaching Tips

What If a predator is removed from an ecosystem? The diversity of the ecosystem can suffer. For example, when a viral epidemic spread among the rabbit population in England, the grasses, once held in check by the rabbits, grew out of control. On the other hand, the many species of wildflowers that once thrived in the chalky soils disappeared.

Ecosystem Complexity and Fragility
Paine's experiments show that even a low-diversity ecosystem (only 15 species) is a complex and highly interconnected system and that disturbing such a system can have unexpected and harmful effects. To predict that the removal of the sea star would result in the disappearance of seven other species, Paine would have had to understand the relationships between *Pisaster* and its prey, as well as the relationships among the prey species. Humans regularly make far greater changes to ecosystems that are less well understood. Students will see some of the consequences of these changes in the next chapter.

Section Review

1. *Explain the difference between niche and habitat.*

2. *Can an organism's realized niche be larger than its fundamental niche? Justify your answer.*

3. *Why can character displacement occur only if competitive exclusion has not occurred?*

4. *How did diversity of habitat promote coexistence of the two species of barnacles studied by Connell?*

Critical Thinking

5. *A scientist finds no evidence that species in a community are competing and concludes that competition never played a role in the development of this community. Is this conclusion valid? Justify your answer.*

Answers to Section Review

1. Habitat is where an organism lives. Its niche is the role it plays in that habitat.

2. The realized niche can be smaller than, or the same size as, the fundamental niche, but never larger. The fundamental niche describes the range of conditions an organism can tolerate, whereas the realized niche describes exactly what portion of the fundamental niche the organism utilizes.

3. If competitive exclusion had occurred, then one species would have been eliminated and could not have undergone character displacement.

4. The habitat contained two distinct zones. *Chthamalus,* the inferior competitor, had a refuge in which *Balanus* could not live.

5. The conclusion is not valid. Competition in the past may have led to competitive exclusion, which could eliminate the evidence of competition from the present-day community. Alternatively, today's species could differ in resource use because past competition stimulated character displacement.

Vocabulary Preview

rain-shadow effect

Opening Question

Ask students to recall their study of the early Earth in Chapter 13 and then contrast Earth's climate when it was first formed some 4.5 billion years ago with its climate today. *(Then: extremely hot, very unpredictable, frequent thunderstorms, turbulent winds. Today: seasonally mild, somewhat predictable, periodic thunderstorms, and milder wind conditions)* In this section students will study the basis for today's climatic patterns.

Demonstration

The Sun and the Earth

Help students visualize how the sun is responsible for Earth's climate by using a globe and a flashlight. First, demonstrate that areas near the equator receive concentrated, direct sunlight. Hold the flashlight perpendicular to the surface a few inches above the equator and turn the light on. The beam illuminates a small area. Then, show how areas far from the equator receive more diffuse sunlight. Tilt the flashlight upward so that it shines on the central United States. The beam clearly illuminates a much larger area, and its energy is spread over more of the surface.

Teaching Tip

Latitude and Longitude

Using a globe, remind students that longitude indicates east-west position and that latitude indicates north-south position. Latitude profoundly affects climate, but longitude is essentially irrelevant to climate.

17-3 The Influence of the Physical Environment

Section Objectives

- Explain the influence of the Earth's shape on climate.
- Describe the cause of the rain-shadow effect.
- Explain how ocean currents affect terrestrial climates.

If you traveled across the country by car you would notice dramatic changes in the plants and animals outside your window. For example, the drought-tolerant cactuses of Arizona deserts do not live in the forests of Maine or the swamps of Florida. Although it is not surprising that different environments harbor different sets of organisms, it is surprising that there are relatively few sets. Certain characteristic groups of animals and plants occur together over very broad areas. Why is this? The physical environment, together with biological interactions, determines which animals and plants are able to live in a particular ecosystem. Climate has a particularly important influence on the character of ecosystems.

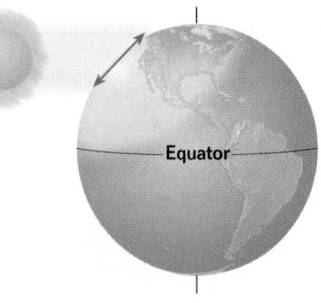

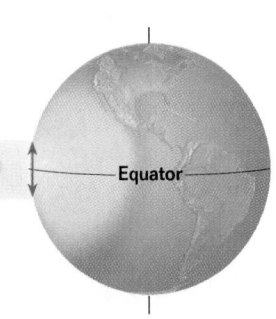

Figure 17-12 Near the equator, direct sunlight falls on the Earth's surface, *bottom.* Therefore, solar energy is concentrated and produces greater warming than at higher latitudes, *top.* There, the curvature of the Earth spreads the solar energy over a greater area.

What Causes Climate?

The distribution of the Earth's ecosystems results from the interaction of the physical features of the Earth (such as the different soil types and the occurrence of mountains, valleys, lakes, and oceans) with two key factors: (1) the amount of energy from the sun that reaches different parts of the Earth and the seasonal variation in that energy, and (2) the global pattern of atmospheric and oceanic circulation created by the unequal global distribution of solar energy. Together, these factors determine local climates, particularly the amount and distribution of precipitation.

The great diversity of ecosystems on Earth is due in part to variations in climate from place to place. Miami, Florida, and Bangor, Maine, for example, are likely to have very different weather on every day of the year. There is no mystery about the causes of climatic differences. Because the Earth is nearly spherical, different places on the surface receive different amounts of solar energy, as illustrated in Figure 17-12. These differences are directly responsible for many of the major climatic differences that occur over the Earth's surface and indirectly responsible for much of the variety of ecosystems. At regions near the equator, the sun's rays arrive almost perpendicularly, making the tropics warmer than the temperate regions. Nearer the poles, the angle at which the sun's rays hit the Earth spreads them out over a much greater area, providing less energy per unit of area.

The Earth's daily rotation on its axis and its revolution around the sun are also important in determining world

Effective Reading

Making Connections

Students too often read a text as a collection of isolated and unrelated sections. Emphasize that what they will read in this section is just as important in shaping an ecosystem as the living environment covered in the previous section.

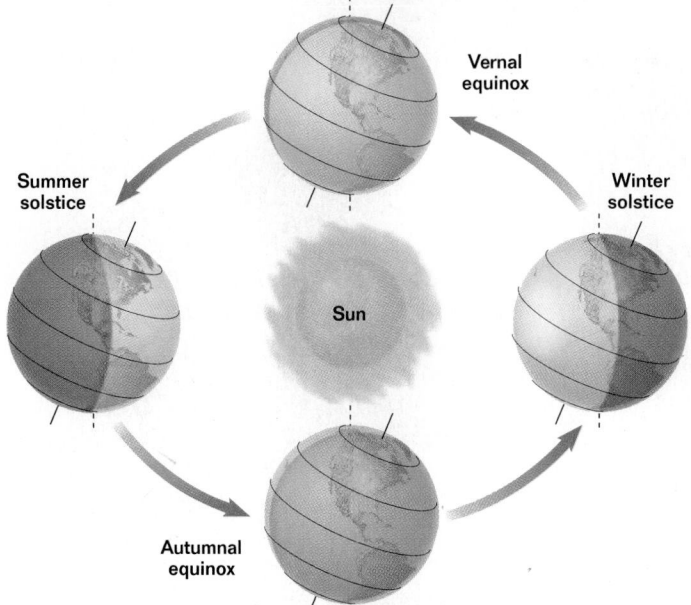

Vernal equinox

Summer solstice

Winter solstice

Sun

Autumnal equinox

Figure 17-13 Because the Earth is tilted 23.5°, the relationship shown in Figure 17-12 changes as the Earth orbits the sun. Between the vernal equinox (March 21) and the autumnal equinox (September 23), the Northern Hemisphere tilts toward the sun and therefore receives more direct sunlight than does the Southern Hemisphere, which tilts away from the sun. The result is spring and summer in the Northern Hemisphere and fall and winter in the Southern Hemisphere. After the autumnal equinox, the pattern reverses. For the next six months, the Southern Hemisphere tilts toward the sun and experiences spring and summer, while the Northern Hemisphere tilts away and has fall and winter.

climate. Because of the Earth's daily rotation, periods of cooling (night) and warming (day) alternate, preventing the buildup of extreme temperatures. The axis of the Earth is tilted 23.5° off vertical, so the revolution of the Earth around the sun produces a progression of seasons at higher latitudes (areas far from the equator), as diagramed in Figure 17-13.

Solar Heating Drives the Circulation of the Earth's Atmosphere

Since hot air rises, air near the equator, which is warmed by the sun, also rises. This air cools and loses most of its moisture, producing heavy rains in the tropics. Because this region of rising air is one of low pressure, it draws air from both north and south of the equator. As the rising air cools, it moves toward the poles. When the rising air masses reach about 20° to 30° north and south latitude, the air, now cooler and drier, sinks and heats, producing a zone of low precipitation; the sinking air acts like a sponge, because warm air holds more moisture than cooler air. Since air at these latitudes is still warmer than air at the polar regions, it continues to flow toward the poles, rising again at about 60° north and 60° south latitude. When this air mass descends near the poles, it produces a zone of very low temperature and precipitation. These movements of air result in the major pattern of atmospheric circulation that is illustrated in Figure 17-14.

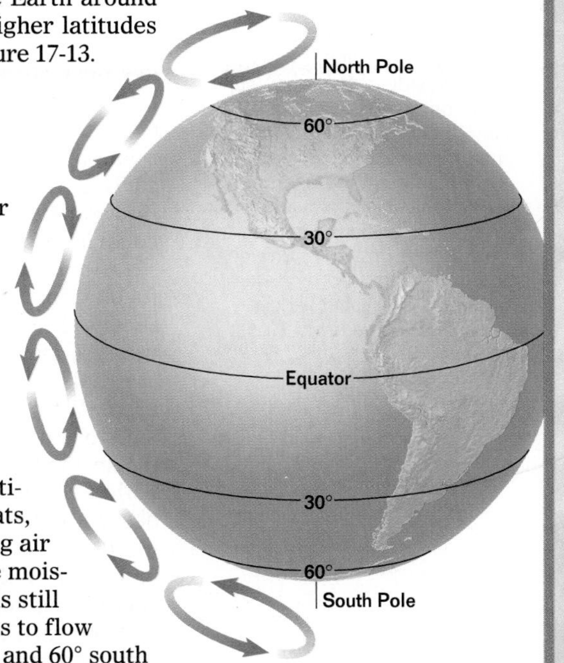

North Pole
60°
30°
Equator
30°
60°
South Pole

Figure 17-14 This diagram shows the global pattern of atmospheric circulation caused by the sun's heating of the Earth.

👁 VISUAL STRATEGY

Figure 17-13
The tilt of the Earth with respect to its axis is the basis for seasons. Because of this tilt, the latitude receiving direct solar radiation (known as the solar equator) changes as the Earth revolves around the sun. On March 21, the vernal equinox, the solar equator and the equator coincide, and direct sunlight is falling on the equator, 0°. Then the solar equator moves into the Northern Hemisphere, reaching its most northerly position on June 22, the summer solstice, when the solar equator is at 23.5° north latitude. After the summer solstice, the solar equator moves south until, on September 23 (the autumnal equinox), it is again over the equator. The solar equator then moves into the Southern Hemisphere, reaching its maximum southerly position at 23.5° south on December 22, the winter solstice. After December 22, the solar equator moves north again.

Teaching Tip
Diametrically Opposite Seasons
Point out that the seasons are the same in the Northern and Southern Hemispheres, but that they occur six months out of phase. Summer in the Northern Hemisphere is winter in the Southern Hemisphere, and so on. Students should see the cause for this difference in the movement of the solar equator. They may notice that the key dates in the year—summer and winter solstices, vernal and autumnal equinoxes— refer to the position of the sun relative to the Northern Hemisphere. This terminology is a matter of historical convention.

Overcoming Misconceptions

The Cause of the Seasons
Students sometimes think that the seasons are a consequence of the Earth's changing distance from the sun. During the Earth's elliptical orbit, the distance between the Earth and the sun does vary, but this variation is not the cause of the seasons. In fact, the Earth is closest to the sun during winter in the Northern Hemisphere.

Have students check library resources to identify areas in the world that receive the highest amounts of rainfall. Have them write a report that explains the physical factors responsible for the heavy rainfalls in these areas.

The Rain-Shadow Effect

Students will better understand the rain-shadow effect if they remember three facts about the behavior of air. First, air rises when heated and sinks when cooled. Second, because atmospheric pressure varies inversely with altitude, rising air expands and cools, and sinking air contracts and warms. Third, the amount of moisture air can hold depends on its temperature—the warmer the air, the greater its moisture capacity. When air cools, therefore, it tends to lose some of its water as precipitation.

Coast ranges

Sierra Nevada range

Figure 17-15 Winds blowing in from the Pacific Ocean must ascend to cross the Sierra Nevada range, causing the air to cool and release its moisture. Abundant precipitation on the western side of the range supports dense forests, *top left.* Having already lost most of their moisture, the winds drop little precipitation on the eastern side of the Sierra Nevada range. The arid conditions there permit only sparse growth of drought-tolerant plants, *top right.* A similar but smaller rain shadow is caused by the lower Coast ranges.

Why It Rains Where It Does

The moisture-holding capacity of air increases when it is warmed and decreases when it is cooled. Consequently, precipitation is generally low near 30° north and 30° south latitude, where cool air is falling and warming, and relatively high near 60° north and 60° south latitude, where air is rising and cooling. Partly as a result of these factors, most of the great deserts of the world, such as the Sahara, lie near 30° north or 30° south latitude. And some of the great temperate forests are located near 60° north latitude (there is virtually no land at 60° south latitude).

Other major deserts, such as the Gobi Desert of Asia, are formed in the interiors of continents. Continental interiors receive limited precipitation because of their distance from the sea, and sometimes because mountain ranges intercept the moisture-laden winds from the sea, as shown in Figure 17-15. When mountains force incoming air upward, the air's moisture-holding capacity decreases as the air cools, resulting in increased precipitation on the windward side (from which the wind is blowing) of the mountains. As the air descends the other side of the mountains, known as the leeward side, it is warmed, and its moisture-holding capacity increases. Therefore, it tends to draw up moisture from the surface, rather than releasing the little moisture it contains. The leeward sides of mountains are often much drier than their windward sides, and the vegetation is often very different. This phenomenon is called the **rain-shadow effect.** In the United States, the Mojave and Great Basin Deserts lie in the rain shadow of the Sierra Nevada range, and the Great Plains lie in the rain shadow of the Rocky Mountains.

Patterns of Circulation in the Ocean

The patterns of circulation in the ocean, shown in Figure 17-16, are determined by the major patterns of atmospheric circulation discussed previously, but they are modified by the location of the landmasses around and against which the ocean currents must flow. Oceanic circulation is dominated by huge, circling surface currents that move around the subtropical oceans at about 30° north and 30° south latitude. These currents move clockwise in the Northern Hemisphere and counterclockwise in the Southern Hemisphere. By redistributing heat, oceanic currents profoundly affect life not only in the oceans but also on coastal lands. For example, in the Atlantic Ocean the warm Gulf Stream swings away from North America near Cape Hatteras, North Carolina, and reaches Europe near the southern British Isles. Because of the Gulf Stream, western Europe is much warmer than eastern North America at the same latitudes. ☐

☐ **CAPSULE SUMMARY**

Latitude, or distance from the equator, is the most important factor determining the climate of an area. Other important factors are distance from the ocean, elevation, and position relative to mountain ranges that can intercept moisture-carrying winds.

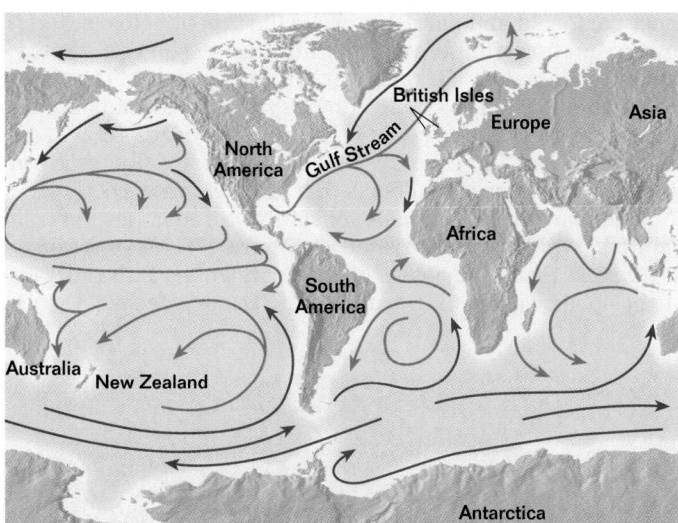

Figure 17-16 This diagram shows the major oceanic currents. Warm currents, which begin in tropical waters, are indicated in red. Cold currents are shown in blue.

Section Review

1. *Why are the tropics warmer than the temperate regions?*
2. *Describe two geographical factors that promote the development of deserts.*
3. *How does the Gulf Stream moderate the climate of western Europe?*

Critical Thinking

4. *Would there still be seasons if the Earth were not tilted in its orbit? Explain your answer.*

Answers to Section Review

1. Because of the Earth's curvature, the tropics receive more direct sunlight than temperate regions.

2. One factor is latitude; many of the world's deserts are located near 30°, where cool, dry air is descending and warming. Another factor is continental position. Many deserts are located in the interior of continents, far from moisture-laden winds blowing from the ocean. These winds may also be intercepted by mountain ranges before they reach the interior of a continent.

3. The Gulf Stream brings warm tropical water to western Europe, raising the air temperature.

4. If the Earth were not tilted, direct sunlight would fall on the equator all year, and higher latitudes would always receive less-concentrated sunlight. Because the latitude receiving direct sunlight would not change, there would be no seasons.

Opening Question

The kangaroo rat can survive the hot climate typical of the deserts of the southwestern states and Mexico without drinking water. Ask students to hypothesize how the kangaroo rat obtains the water it needs to survive. Tell students that it survives on the water in its food and supplements this with water it produces through respiration. The kangaroo rat also conserves water by excreting a very concentrated urine and spending the day in a cool, humid burrow.

Demonstration

Biological Communities

Show the class photographs of the five major biomes that can be found in the lower 48 states or Alaska—desert, temperate grassland, temperate deciduous forest, tundra, and taiga. Ask students to describe what they can deduce about the physical and biological characteristics of these biomes from examining the photos.

CONNECTIONS

Chapter 5

Photosynthetic Organisms

When sunlight strikes the ocean surface, red, orange, and yellow wavelengths of light are absorbed first. Only blue and green can penetrate deeply. Thus, below a depth of a few meters, only those photosynthetic organisms capable of using these shorter wavelengths of light can survive.

17-4 Major Biological Communities

I *f you were to tour the world and look at biological communities in the oceans and on land, you would soon learn one of the great generalizations of ecology: very similar communities occur in many different places that have similar climates and geographies. In this section you will take just such a tour and visit the major types of biological communities found on Earth.*

Marine Communities

Nearly three-fourths of the Earth's surface is covered by ocean. The seas have an average depth of more than 3 km (1.9 mi.), and they are, for the most part, cold and dark. Heterotrophic (nonphotosynthetic) organisms are found even at the greatest ocean depths, which reach nearly 11 km (6 mi.) in the Marianas Trench of the western Pacific Ocean. Photosynthetic organisms are confined to the upper few hundred meters of water. Organisms that live below this level obtain almost all of their food from organic debris that drifts downward.

The marine environment consists of three major kinds of habitats, which are illustrated in Figure 17-17: (1) the shallow waters along the coasts of the continents, (2) the surface of the open sea, and (3) the depths of the open ocean.

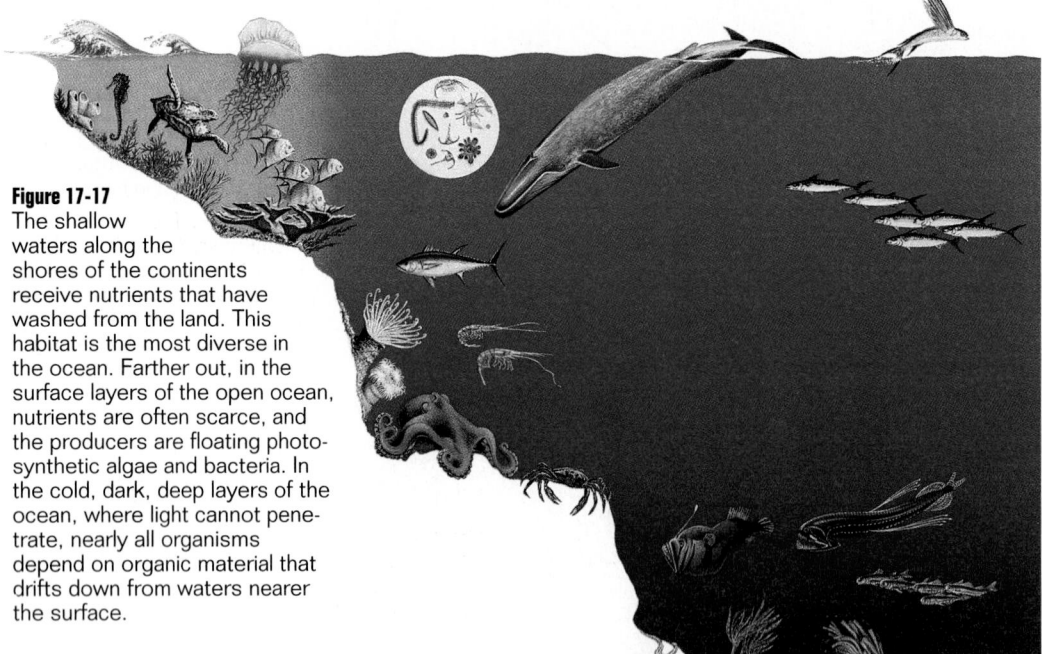

Figure 17-17
The shallow waters along the shores of the continents receive nutrients that have washed from the land. This habitat is the most diverse in the ocean. Farther out, in the surface layers of the open ocean, nutrients are often scarce, and the producers are floating photosynthetic algae and bacteria. In the cold, dark, deep layers of the ocean, where light cannot penetrate, nearly all organisms depend on organic material that drifts down from waters nearer the surface.

Figure 17-18 Coral reefs occur in shallow ocean waters in tropical regions. In numbers of species, coral reefs rival tropical rain forests.

Shallow Ocean Waters

The zone of shallow water is small in area, but compared with other parts of the ocean, it is inhabited by large numbers of species. The world's great fisheries are located on banks in the coastal zones, where nutrients, washed from the land, are often more abundant than in the open ocean.

Open Sea Surface

Drifting freely in the upper waters of the ocean is a diverse biological community called **plankton,** which consists mostly of microscopic organisms. Plankton includes bacteria, algae, fish larvae, and other small animals. Some of the members of the plankton, including algae and some bacteria, are photosynthetic. Collectively these organisms account for about 40 percent of all the photosynthesis that takes place on Earth.

Deep Sea Waters

In the deep waters of the sea, below the depth of 300 m (1,000 ft.), live some of the most bizarre organisms found on Earth. Many of these animals can produce their own light, which they use to communicate with one another or to attract prey.

 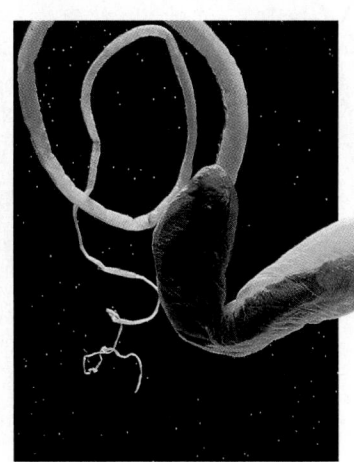

Figure 17-19 These schooling mackerel are speedy inhabitants of the open ocean surface.

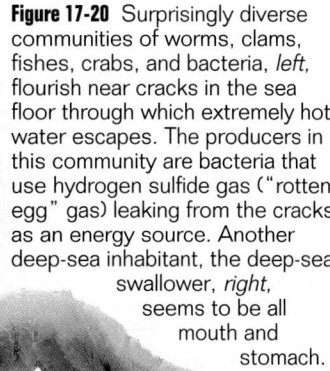

Figure 17-20 Surprisingly diverse communities of worms, clams, fishes, crabs, and bacteria, *left,* flourish near cracks in the sea floor through which extremely hot water escapes. The producers in this community are bacteria that use hydrogen sulfide gas ("rotten egg" gas) leaking from the cracks as an energy source. Another deep-sea inhabitant, the deep-sea swallower, *right,* seems to be all mouth and stomach.

Historical Note

Deep-Sea Waters

Humans first reached the floor of the Marianas Trench and observed some of these deep-water organisms in 1960. Two scientists used the bathyscaphe *Trieste* to descend nearly 11 km (7 mi.) below the ocean's surface. The craft carried gasoline for buoyancy and buckshot for ballast. It descended when water was pumped into air tanks at either end, and it ascended when water was pumped out and buckshot released. The *Trieste* obtained much data about the conditions at such depths.

☑ RESEARCH Update

What Is Killing the Coral?

"Coral bleaching" is a term that applies to the loss of the vibrant color most corals have. They slowly turn white, and then they die. Why? Many investigators are working on the phenomenon of coral bleaching. Recently two geoscientists from the U.S. Naval Academy, Alan Strong and R. E. Montgomery, have compiled the evidence and announced that if communities as productive as coral reefs are endangered, life itself can be endangered. The corals, these researchers affirm, are threatened by the loss of their zooxanthellae—symbiotic algae that live within the coral tissues. The relationship is a true mutualism, for the coral cannot live without its tenants. The zooxanthellae are very sensitive to temperature, UV light, and nutrient composition of the water. A mild rise in temperature (such as one that may accompany global warming) or increase in incidence of UV light can be fatal to the algae.

Researchers working on the Great Barrier Reef in Australia have been investigating the effects of human pollution that ultimately ends up in the ocean. These pollutants—usually agricultural fertilizers—can disrupt the nutrient balance of sea water. In fact, after repeated experiments on the Great Barrier Reef, Australian scientists have determined that phosphorus appears to be the major culprit in coral bleaching on this reef. The runoff from farming over-fertilizes the coral and impedes development of the hard skeleton. With skeletal development obstructed, corals grow in size but are exceedingly brittle, which also contributes to the early demise of some corals. Corals play important roles in oceanic food webs, so what can humans do to check these impacts? The Australian scientists suggest regulating phosphorus runoff. ☑

VISUAL STRATEGY

Figure 17-22

Emphasize that a biome is a category and not a place. A tropical rain forest does not refer to any specific geographical location on Earth, but rather to all regions of the planet where such a biome can be found. Point out that biomes are characterized by their dominant vegetation—drought-tolerant plants in deserts, coniferous trees in the taiga, and so on. Also emphasize that the boundaries of biomes are not as well defined as they are shown in this figure.

Figure 17-21 This pond in Pennsylvania is a typical freshwater ecosystem.

Figure 17-22 The seven biomes discussed in this chapter cover most of the Earth's land surface. Areas that cannot be assigned to one of these biomes are shown in gray. Because climate changes with elevation, mountains contain a variety of communities and so do not belong to any one biome. Mountainous areas are indicated on the map. Also, Antarctica is not shown because it has no biomes.

Freshwater Communities

Freshwater habitats (lakes, ponds, streams, and rivers) are distinct from both marine and terrestrial habitats, and they are very limited in area. Lakes cover about 1.8 percent of the Earth's surface, and rivers and streams cover about 0.3 percent. All freshwater habitats are strongly connected to terrestrial ones, with freshwater marshes and swamps constituting intermediate habitats. In addition, a large amount of organic and inorganic material continuously enters bodies of fresh water from communities on the land.

Like the oceans, large lakes have three zones in which organisms live: a shallow zone near the shore, an open-surface zone inhabited by plankton, and a deep-water zone below the limits of effective light penetration.

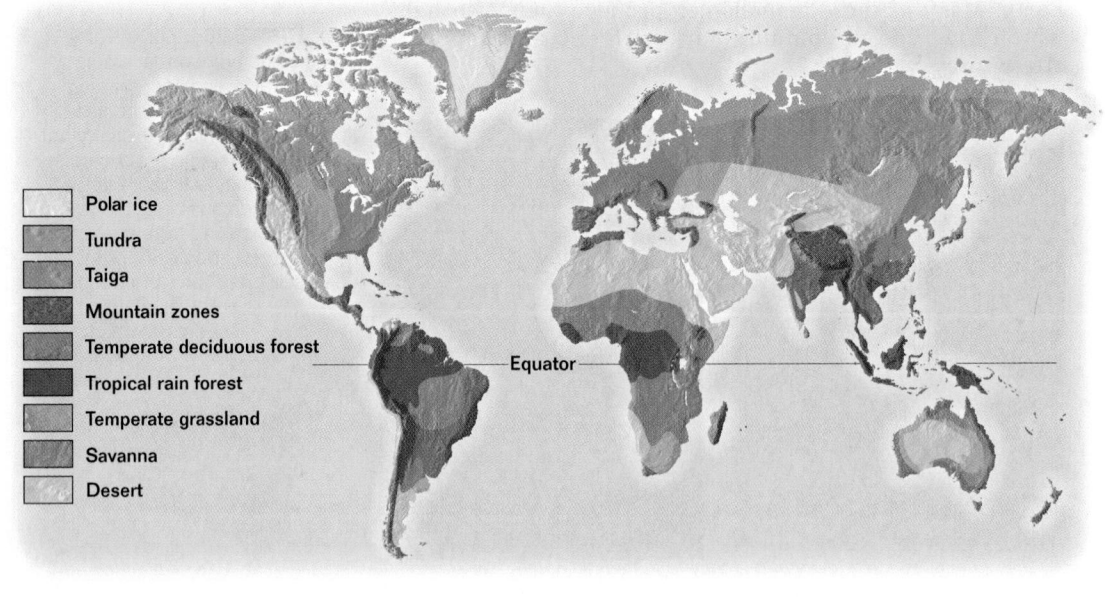

Polar ice
Tundra
Taiga
Mountain zones
Temperate deciduous forest
Tropical rain forest
Temperate grassland
Savanna
Desert

Equator

Terrestrial Communities

A major terrestrial community that is found in different areas with similar climates is called a **biome**. A biome's structure and appearance are similar throughout its distribution. Biomes can be classified in a number of ways; the classification used here is chosen merely as a convenient means of discussing the properties of organisms from an ecological perspective. This classification recognizes seven biomes: (1) tropical rain forest, (2) savanna, (3) desert, (4) temperate grassland, (5) temperate deciduous forest, (6) taiga, and (7) tundra. These biomes differ greatly from one another because they have developed in regions with very different climates. The global distribution of these biomes is shown in Figure 17-22.

Tropical Rain Forests: Lush Equatorial Forests

The rainfall in tropical rain forests is generally 200–450 cm (80–180 in.) per year, with little difference in distribution from season to season. The tropical rain forest is the richest biome in terms of number of species, probably containing at least half of the Earth's species of terrestrial organisms—more than 2 million species. In a single square mile (just over 2.5 sq. km) of tropical forest in Peru or Brazil, there may be 1,200 or more species of butterflies, twice the total number of species found in the United States and Canada combined. However, although the communities that make up tropical forests are rich in species, each kind of animal, plant, or microorganism in a given area is often represented by very few individuals.

Tropical rain forests have a high primary productivity, even though they exist mainly on quite infertile soils. Most of the nutrients are held within the plants themselves and are rapidly recycled when the plants die or when parts, such as leaves, fall off and decompose. Most of the roots of the tall trees spread through the top 1 or 2 cm (less than 1 in.) of the soil and extract the nutrients from decomposing leaves and other plant parts.

Figure 17-23 Tropical rain forests, such as this one in Costa Rica, have very high species diversity.

SECTION 17-4

◉ *VISUAL STRATEGY*

Figure 17-23

Point out that the critical competition among trees in a rain forest is for light. About 70 percent of all the species of plants are trees. Their trunks are usually slender and tall. Their leaves are usually large and dark green. All these are adaptations to obtain and utilize the most sunlight for photosynthesis.

Teaching Tip

Agriculture in the Rain Forest

Emphasize the surprising fact that the soil in tropical rain forests is typically thin and poor in nutrients. Consequently, farms created from cleared areas of tropical rain forest are usually productive for only a few years before the soil nutrients are exhausted. The farmers must move on and clear another patch of forest. This cycle of clearing and abandoning contributes to the swift destruction of tropical rain forests occurring today.

Desert covers less than 5 percent of North America. The North American deserts are the Mojave, Sonoran, Great Basin, and Chihuahuan. Have students write a report that contrasts the animals, plants, geography, and climate of these four deserts.

Application

Geography The Sahara stretches across Africa and covers a land area almost the size of the United States.

Application

Agriculture Several of today's domesticated plants and animals, including cattle, sheep, horses, wheat, barley, and oats, are descendants of inhabitants of the grasslands of Eurasia.

Figure 17-24 Massive herds of wildebeest and zebras migrate across the Serengeti Plain, part of East Africa's savanna. Carnivores such as lions, leopards, African wild dogs, and hyenas attack the herds.

Figure 17-25 In deserts, conservation of water is essential. Spines on these cactuses help protect their moist tissues from thirsty herbivores.

Figure 17-26 Before European settlers arrived in the New World, much of the central United States and southern Canada was covered by prairie, or temperate grassland. As many as 60 million bison grazed on these prairies. Today, only fragments of undisturbed prairie and fewer than 100,000 bison remain.

Savannas: Dry Tropical Grasslands

Dry climates often favor the development of grassland. The world's great dry grasslands, called savannas, are found in tropical areas that have relatively low annual precipitation or prolonged annual dry seasons. On a global scale, the savanna biome is found in a transitional zone located between tropical rain forest and desert. Annual rainfall is generally 90–150 cm (35–60 in.) in savannas. There is a wider fluctuation in temperature during the year than in the tropical rain forests, and there is seasonal drought. These factors have led to the evolution of an open landscape with widely spaced trees. Many of the animals are active only during the rainy season. The huge herds of grazing mammals and their associated predators that inhabit the savannas of Africa are well known and spectacular.

Deserts: Arid Lands

Typically, fewer than 25 cm (10 in.) of precipitation falls annually in the world's desert areas. The scarcity of water is the overriding factor influencing most biological processes in the desert. In desert regions, the vegetation is characteristically sparse. Deserts are most extensive in the interiors of continents, especially in Africa (the Sahara), Asia, and Australia. Less than 5 percent of North America is desert. Lands in which water is not so scarce are called semideserts. These two dry biomes are combined in Figure 17-22 on page 378. In deserts and semideserts, the amount of water that actually falls at a particular place can vary greatly, both during a given year and between years.

Temperate Grasslands: Seas of Grass

Moderate climates promote the growth of rich temperate grasslands. Temperate grasslands once covered much of the interior of North America and were widespread in Eurasia and South America. Such grasslands are often highly productive when converted to agriculture, and much of the rich agricultural land in the United States and southern Canada was originally occupied by prairie, another name for

Figure 17-27 Temperate deciduous forests, such as this one in Pennsylvania, are dominated by hardwood trees that lose their leaves in autumn. In North America, the area occupied by temperate deciduous forests has shrunk considerably since the arrival of European settlers.

temperate grassland. The roots of grasses characteristically penetrate far into the soil, which tends to be deep and fertile.

Temperate Deciduous Forests: Rich Hardwood Forests

Relatively mild climates and plentiful rain promote the growth of forests. Temperate deciduous forests (deciduous trees shed their leaves all at once in the fall) grow in areas with relatively warm summers, cold winters, and sufficient precipitation. The annual precipitation generally ranges from 75 to 250 cm (30 to 100 in.) and is well distributed throughout the year. Moisture is generally unavailable to animals and plants in the winter, because it is usually frozen.

This biome covers very large areas, including much of the eastern United States and southeastern Canada and extensive areas of Europe and Asia. In North America, deciduous forests are home to deer, bears, beavers, raccoons, and the other familiar animals of the temperate regions. The trees are hardwoods (oak, hickory, and beech), and shrubs and herbs grow on the forest floor.

Figure 17-28 Coniferous trees predominate on the taiga. This photo of taiga was taken near Churchill, Manitoba, in north-central Canada.

Taiga: Great Coniferous Forests of the North

Cold, wet climates promote the growth of coniferous forests. A great ring of northern forests of coniferous trees, primarily spruce and fir, extends across vast areas of Eurasia and North America. This biome, one of the largest on Earth, is called by its Russian name, taiga (*TEYE guh*). Winters in the taiga are long and cold, and most of the precipitation falls in the summer. At the northern

 Did You Know?

The northernmost areas of taiga receive only six to eight hours of sunlight during the winter, but nearly 19 hours during the summer.

Figure 17-29 The scarcity of trees on the tundra is apparent in this view of Denali National Park in Alaska.

latitude where taiga occurs, the days are short in winter (as little as six hours) and long in summer. During the summer, plants may grow rapidly. Marshes, lakes, and ponds are common and are often fringed by willows or birches. Most of the trees in the taiga tend to occur in dense stands of one or only a few species. Many large mammals live in the taiga, including herbivores such as elk, moose, and deer, and carnivores such as wolves, bears, lynxes, and wolverines.

Tundra: Cold Plains of the Far North

In the Far North, there are few trees. Between the taiga and the permanent ice surrounding the North Pole is the open, sometimes boggy, biome known as the tundra. This enormous biome covers one-fifth of the Earth's land surface. Annual precipitation in the tundra is very low, usually less than 25 cm (10 in.), and water is unavailable for most of the year because it is frozen. During the brief Arctic summers, when water is not frozen, the surface of the tundra is often extremely boggy because permafrost, or permanent ice, usually exists within 1 m (about 3 ft.) of the surface.

A land of grasses, sedges, dwarf willows, and mosses, the tundra is open and windswept. Foxes, lemmings, owls, and caribou are among the vertebrate inhabitants. Trees are small and usually confined to the edges of streams and lakes. The tundra's appearance is extremely uniform.

Section Review

1. *Why can't photosynthesis occur in the deepest parts of the ocean?*
2. *What common environmental hardship confronts organisms in deserts and tundra?*

Critical Thinking
3. *Explain why there are no biomes in Antarctica.*

Vocabulary

biome (379)
character displacement (368)
coevolution (361)
commensalism (363)
competition (363)
competitive exclusion (369)

fundamental niche (367)
mutualism (363)
niche (365)
parasitism (361)
plankton (377)
predation (361)

principle of competitive
   exclusion (370)
rain-shadow effect (374)
realized niche (368)
secondary compound (362)
symbiosis (363)

Concept Mapping

Construct a concept map that shows how the biomes can be classified based on precipitation, temperature, and geographical location. Use as many terms as needed from the vocabulary list. Try to include the following terms in your map: tropical rain forest, savanna, desert, temperate deciduous forest, temperate grassland, taiga, and tundra. Include additional terms in your map as needed.

Review

Multiple Choice

1. In predator-prey coevolution, if the prey evolves a defense to stop predation then the predator likely will evolve
   a. in a way that enables it to overcome the prey's defense.
   b. so that it can parasitize the prey.
   c. countermeasures to secondary compounds.
   d. into the prey.

2. The interaction between a spruce tree and a hemlock tree, both of which are trying to get nitrogen from the soil, is an example of
   a. mutualism.
   b. commensalism.
   c. competition.
   d. succession.

3. The ways in which an organism interacts with its environment make up its
   a. niche.   c. habitat.
   b. space.   d. ecosystem.

4. Connell's experiment with barnacles demonstrated
   a. that barnacles can be grown in captivity.
   b. how barnacles extend their realized niche.
   c. character displacement in the ocean.
   d. competitive exclusion in nature.

5. The principle of competitive exclusion indicates that
   a. a niche can be shared by two species.
   b. niche subdivision may occur.
   c. one species will give way to another if their niches are very similar.
   d. competition ends in worldwide elimination of a species.

6. Compared with the sun's rays falling on a location at the equator, the sun's rays striking a location at 30 degrees north latitude would be
   a. spread out over a greater area.
   b. focused on a smaller area.
   c. almost perpendicular.
   d. tilted at 23.5 degrees off vertical.

7. The Great Plains of the United States get little rain because
   a. of the rain-shadow effect of the Sierra Nevada range.
   b. of the dry winds that blow across the western deserts.
   c. they lie on the leeward side of the Rocky Mountains.
   d. they lie on the windward side of the Cascade range.

8. The highest species diversity is found in which biome?
   a. taiga   c. deciduous forest
   b. grassland   d. tropical rain forest

CHAPTER REVIEW ANSWERS

Each item in the chapter review is correlated by text section in the assignment guide that follows.

ASSIGNMENT GUIDE

Section	Review	Themes Review	Critical Thinking
17-1	1, 2, 12		32
17-2	3–5, 13, 14, 18–20	25, 26	27, 28, 30, 31
17-3	6, 7, 10, 15, 21		29
17-4	8, 9, 11, 16, 17, 22, 23	24	

Concept Mapping

Map answer is shown on page 359B.

Review
Multiple Choice

1. a (17-1.1)   7. c (17-3.2)
2. c (17-1.2)   8. d (17-4.2)
3. a (17-2.1)   9. c (17-4.1)
4. d (17-2.2)   10. b (17-3.3)
5. c (17-2.3)   11. a (17-4.2)
6. a (17-3.1)

Completion

12. commensalism, mutualism, ants (17-1.3)
13. functions or lives, energy (17-2.1)
14. warblers, realized (17-2.2)
15. leeward, windward (17-3.2)
16. shallow ocean waters, deep-sea waters (17-4.1)
17. tundra, tropical rain forest (17-4.2)

Short Answer

18. Connell sought to determine how the removal and reintroduction of *Balanus*, a deep-zone barnacle, affected the realized niche of *Chthamalus*, a shallow-zone barnacle. He concluded that *Balanus* limits the realized niche of *Chthamalus* by out-competing it in deeper water. (17-2.3)

CHAPTER REVIEW ANSWERS *continued*

Short Answer *continued*

19. One possible outcome is that the two species will limit their feeding to particular nuts. For example, one species will feed on nuts that have fallen to the ground, while the other species only feeds on nuts that remain on the branches. A second possible outcome is that one species will be eliminated as a result of the competition. A third outcome is the evolution of reduced use of this type of nut. (17-2.4)

20. Gause observed that when two species of *Paramecium* competed for the same resources, the one better able to compete eventually eliminated the other locally. (17-2.3)

21. Two factors are latitude and proximity to the ocean. More solar energy is received near the equator than at higher latitudes. Ocean currents, such as the Gulf Stream, can warm or cool adjacent areas on land. (17-3.1)

22. Shallow ocean waters are nearest to the coast and are inhabited by a large number of species. The open sea surface is the top 300 m (1,000 ft.) of open ocean. It is often rich in plankton, but may be short of nutrients. The deep-sea waters are below 300 m and contain unusual carnivores and no photosynthetic organisms. (17-4.1)

23. A biome is a major type of terrestrial biological community that is found over a large area. Factors that determine a biome's characteristics include temperature, precipitation, and seasonal differences in climate. The characteristics of the seven major biomes are listed on pages 379–382. (17-4.2)

Themes Review

24. Blood circulating through the ears helps dissipate heat from the interior of the body to the environment. Resting in the shade reduces heat production through physical exertion and keeps the rabbit out of the direct sun. (17-4.2)

25. Grass is the producer, capturing solar energy and storing some of it in complex molecules. The zebra feeds

9. Marine life is most diverse in the
  a. deep sea waters.
  b. open sea surface.
  c. shallow ocean waters.
  d. polar oceans.

10. The Gulf Stream affects the climate of coastal lands by
  a. lowering air temperature.
  b. raising air temperature.
  c. lowering the freezing point.
  d. causing beach erosion.

11. Cold and long winters, very few trees, and little precipitation describe the
  a. tundra.   c. deciduous forest.
  b. taiga.   d. grasslands.

Completion

12. An interaction in which one species benefits and the other is not affected is _____ . The other type of symbiosis is _____ , in which both species benefit. An example of this interaction is the relationship between _____ and aphids.

13. A niche is how an organism _____ in an ecosystem. It may be described in terms of _____ flow or how organisms use space.

14. Robert MacArthur studied niche restriction among five species of _____ . His work showed that a species' _____ niche may be a small portion of its fundamental niche.

15. The rain-shadow effect predicts that the _____ side of mountains will get less rain than the _____ side.

16. In marine environments, species diversity is highest in the _____ , and photosynthetic organisms are least abundant in the _____ .

17. The two terrestrial biomes that receive less than 25 cm of precipitation annually are desert and _____ . The biome that receives the most precipitation is _____ .

on the grass, transforming some of the energy in the grass into its own molecules. The lion acquires some of this energy when it feeds on the zebra. (17-2.1)

26. Because competition has a negative effect on the individuals of both participating species, natural selection favors reduced use of common resources. Thus,

Short Answer

18. Briefly summarize Joseph Connell's barnacle experiment. What did he conclude from his work?

19. What are two possible outcomes of competition between two species of squirrel for the same type of nut?

20. What observations led to the formulation of the competitive exclusion principle by the Russian biologist G. F. Gause?

21. Identify two factors that affect the temperature patterns in a region. How are the temperature patterns affected by each factor?

22. List the major kinds of marine habitats and describe the characteristics of each.

23. What is a biome? What climatic factors determine a biome's characteristics? List the seven major biomes and describe the characteristics of each.

Themes Review

24. Homeostasis The blacktail jackrabbit, *Lepus californicus*, shown below, lives in the deserts of the southwestern United States. How do its large ears, which are richly supplied with blood vessels, and its tendency to rest in deep shade on hot days help the rabbit maintain a relatively stable body temperature?

competing species may evolve differences that reduce their competition and promote coexistence. (17-2.3)

Critical Thinking

27. No. To establish that character displacement had occurred, he would have had to show that the differences between the war-

blers were greatest in areas in which all five species were present. The case for character displacement could be strengthened by observations of each species in areas in which some or all of the other species didn't live. If the differences were due to character displacement, each species should have foraged in more of

25. **Energy Flow** Describe the niches of a lion, a zebra, and the grass that grows on the African plain in terms of how each species affects energy flow in the ecosystem.

26. **Evolution** Explain the role of coevolution in the coexistence of potential competitors.

Critical Thinking

27. **Evaluating Data** Do MacArthur's observations of warblers furnish enough evidence to conclude that character displacement had occurred in these birds? If not, specify the observations required to make a stronger case for character displacement.

28. **Making Predictions** In Gause's experiments, *Paramecium caudatum* could coexist with *P. bursaria* but not with *P. aurelia*. Predict what would happen if *P. aurelia* and *P. bursaria* were grown together.

29. **Making Predictions** Tillamook, Oregon, is on the windward side of the Pacific coastal mountains, and Portland, Oregon, is on the leeward side. In which city would you expect the least precipitation? Explain why.

30. **Designing an Experiment** Suggest an experiment to test the hypothesis that the extinction of the bluebirds native to New York City's Central Park was due to their being outcompeted by starlings released in the park by humans.

31. **Making Inferences** A scientist captures two species of *Paramecium* in a pond. When the two species are grown together in cultures in the laboratory, one species always outcompetes and eliminates the other. In the pond, however, the two species coexist. Suggest a hypothesis to account for this difference in outcomes. How would you test your hypothesis?

32. **Making Inferences** Early human settlers of Hawaii introduced predators— including cats, dogs, rats, and pigs— that the native animals had never encountered. These introduced predators proved devastating to the native animals, many of which became extinct. Using your knowledge of coevolution, explain why prey should be more vulnerable to introduced predators than to native predators.

Activities and Projects

33. **Research and Writing** What is dendrochronology? Find out how the work of dendrochronologists provides insights into the past physical environment of an area.

34. **Multicultural Perspective** The Australian Aborigines inhabited all of Australia, even the arid central region. Research the Aborigines' way of life to learn how they were able to live in such a hostile environment with such simple technology.

Readings

35. Read the article "Singing Caterpillars, Ants and Symbiosis," in *Scientific American*, October 1992, pages 76–82. In the mutualistic relationship between ants and caterpillars, what does each species gain? How do the sounds produced by the caterpillars affect the ants?

36. Read the article "Living Together," in *Scientific American*, January 1992, pages 122–133. Describe the traditional view of how coevolution between host and parasite should proceed. According to Dr. Paul Ewald, some parasites should not evolve in this way. Explain his reasoning.

pond might differ in abundance or distribution from those in the laboratory. Or, each species might have a refuge in the habitat, which is more complex than that provided in the laboratory. For the second part of the question, accept any feasible experiment or set of observations that would test the possibility chosen. (17-2.4)

32. Prey species adjust to the characteristics of their predators. Thus, prey defenses tend to be directed at particular hunting adaptations of their predators. These defenses may not work against an introduced predator, which may have different ways of finding and capturing prey. (17-1.1)

Activities and Projects

33. A dendrochronologist is a person who can make precise inferences about past physical environments by examining cross sections of trees. Of particular interest to a dendrochronologist are tree rings. The rings reveal information about past temperature, rainfall, light availability, altitude, and length of growing season.

34. The Australian Aborigines had a deep and intimate understanding of the climate, land, plants, and animals. They knew where to find underground water, what plants could provide moisture, and even how to get water from animals—they would dig up hibernating frogs that store water. They also used water sparingly.

Readings

35. The caterpillars are protected from predators, and the ants receive a nutritious fluid from the caterpillars. The "songs" of the caterpillars mimic sounds made by ants and help keep the ants near the caterpillars.

36. The standard view of parasite-host coevolution is that the parasite should evolve to become less damaging to its host. A parasite that allows its host to live longer will be transmitted to more new hosts than one that kills its host swiftly. According to Dr. Ewald, the standard view should not apply to parasites that do not require that their host be able to move and come into contact with new potential hosts in order to be transmitted.

the tree when the other species weren't present. (17-2.3)

28. They should coexist. *P. aurelia* and *P. caudatum* both fed on bacteria, but *P. bursaria* fed on yeast. *P. bursaria* and *P. aurelia* should coexist because they feed on different foods. (17-2.4)

29. Portland would receive the least precipitation because the

air will have lost its moisture as it climbs the windward side of the mountains. (17-3.2)

30. Populations of bluebirds and starlings could be maintained in separate netted areas that are identical. Observations over time would indicate if the two bird populations are affected by species native to the area. If the bluebirds are not affected by

native species, then the populations could be allowed to mix. Continual observation of the mixed populations should indicate if the starlings are outcompeting the bluebirds. (17-2.3)

31. There are several possibilities for this discrepancy. A predator in the pond might be keeping the population of the superior competitor low. The resources in the

LABORATORY Investigation Chapter 17

Habitat Selection

Preparation

Order brine shrimp in advance from a biological supply company. Allow extra time to grow large cultures. Use 0.5 inch internal diameter plastic tubing cut into 44-cm lengths. Detain™ or methyl cellulose may be used to slow shrimp without killing them. For the reaction to light, avoid incandescent or other bulbs that generate much heat. Fiberglass screening is available from hardware stores or can be ordered from Ward's.

Procedural Notes

- Divide the class into teams of six, with two students in each team working on each part (A, B, or C).
- When tubing is divided as directed, it allows for space taken up by stoppers and forms equal quarters.
- Students may count shrimp by viewing them in the petri dish or by holding the pipette up to the light.

Background Answers

1. Organisms select habitats in which conditions such as temperature, light levels, salinity, and pH are within their tolerance limits. Brine shrimp, for example, select for temperature, light, and pH. The niche is all an organism's interactions with its living and nonliving environment, including its habitat preferences.

2. What are the habitat preferences of brine shrimp with respect to temperature and light?

Records

Answers will vary depending on species of brine shrimp used. Students should draw histograms that summarize their data.

OBJECTIVE

Assess the effect of environmental variables, including temperature and light, on habitat selection by brine shrimp.

PROCESS SKILLS

- identifying and controlling variables
- graphing and interpreting data

MATERIALS

- marking pen
- clear, flexible plastic tubing
- meter stick
- 4 test tubes with stoppers and test tube rack
- 2 corks to fit tubing
- graduated cylinder or beaker
- funnel
- brine shrimp culture
- aluminum foil
- 3 screw clamps
- 1 pipette
- petri dish
- Detain™ or methyl cellulose
- tape
- ice bag
- hot water bag
- fluorescent lamp or grow light
- 14 pieces of screening
- calculator

BACKGROUND

1. What variables are involved for habitat selection? What is a niche?

2. Write your own question to explore in your lab report or notebook.

TECHNIQUE

You and your partner will complete Part A, B, or C and then share your results with the other students on your team.

Part A: Control Setup

1. Use a marking pen to mark the plastic tubing at 12 cm, 22 cm, and 32 cm from one end. You have now divided the tube into 4 sections. Start at one end and label the sections *1*, *2*, *3*, and *4*. Label 4 test tubes *1*, *2*, *3*, and *4*.

2. Place a cork in one end of the tubing. Use a graduated cylinder or beaker and a funnel to transfer 50 mL of brine shrimp culture to the tubing. Cork the open end and lay the tubing on a desk top.

3. Cover the tubing with aluminum foil and let it remain undisturbed for 30 minutes.

4. After the time has passed, attach screw clamps to each spot that you marked on the tubing. While your partner holds the corks firmly in place, tighten the middle clamp first and then the outer clamps.

5. Immediately pour the contents of each section of tubing into the test tube labeled with the same number.

6. Stopper test tube 1 and invert it gently to distribute the shrimp. Use a pipette to draw a 1 mL sample of shrimp culture, and transfer it to the petri dish. Add a few drops of Detain™ to the petri dish to slow down the shrimp, and then count the live shrimp. Record the count in the **Records** section of your report. Dispose of the shrimp as your teacher directs, and repeat this procedure three more times for a total of four counts.

   Calculate the average number of shrimp in test tube 1 and record the result in the **Records** section of your report.

7. Repeat step 6 for the contents of each of the remaining three test tubes.

Environmental Conditions Preference of Brine Shrimp

	Average number of Brine Shrimp			
Section	1	2	3	4
Control				
Temperature gradient				
Light gradient				

8. Clean up your materials and wash your hands before leaving the lab.

9. Make a histogram showing the number of brine shrimp in each section of tubing.

10. In your report, briefly summarize the procedure you followed.

Part B: Temperature Gradient

1. Repeat steps 1 and 2 of Part A.

2. Tape the tubing to the desk top and cover it with aluminum foil. Mark the aluminum foil to show the approximate positions of sections 1 and 4.

3. Place a bag of ice over section 1 of the tubing. Place a hot-water bag over section 4 of the tubing. **CAUTION: Handle the hot-water bag carefully.** Do *not* use water over 70°C, which can burn you.

4. After 10 minutes, replace the hot water bag with a fresh bag. Replace the bag again after 20 minutes. After 30 minutes, quickly complete Steps 4 and 5 of Part A.

5. *Immediately* read and record the temperature of the test tube contents in the **Records** section of your report.

6. Make a histogram showing the number of shrimp in each section of tubing. Identify each section with its temperature.

7. Complete steps 6 through 10 of Part A.

Part C: Light Gradient

1. Repeat steps 1 and 2 of Part A.

2. Set a light source about 20 cm away from the tubing. Use a low-wattage light bulb or soft white fluorescent tubes to give light without much heat. **CAUTION: Light bulbs get very hot and can burn your skin. Do not touch the bulb.**

3. Cover section 1 of the tubing with eight layers of screening. Place four layers on section 2, two layers on section 3, and leave section 4 uncovered. Wait 30 minutes.

4. Make a histogram showing the number of shrimp in each section. Identify each section with the amount of screening.

5. Complete steps 4 through 10 of Part A.

INQUIRY

1. How did the brine shrimp react to changes in temperature and light?

2. Why were five counts taken in each test tube?

3. Why was a control (Part A) necessary?

ANALYSIS

1. After examining the histograms made by other teams, describe the niche of brine shrimp.

2. Brine shrimp cannot migrate, or move, from one body of water to another. How is it helpful for shrimp to react to changes in their environment?

3. How might a positive phototactic response (moving toward light) be advantageous for brine shrimp? How could a negative response be helpful?

FURTHER INQUIRY

Write a new question that could be explored as a new investigation. The following is an example:

How do brine shrimp react to a gradient of motion? How would this relate to turbidity?

Inquiry Answers

1. Answers will vary depending on the species of *Artemia* used.

2. Five counts were taken to make allowances for variations in populations and to allow an average to be calculated.

3. The control was necessary to show the brine shrimp did not inherently prefer one part of the tube—the ends or the middle, for example.

Analysis Answers

1. Answers will vary depending on the species of *Artemia* used. Generally they prefer temperatures of 25°–30°C.

2. They are able to move among different parts of their habitat that may have different conditions. For example, a shallow area receiving direct sunlight may stay warmer than deeper and shaded areas.

3. A positive response might help them find food—brine shrimp are filter feeders. A negative reaction could protect them from drying out in shallow water.

UNIT 4

Block Scheduling Guide

Each block represents about 45–50 minutes of instructional time. The resources cited in a block are coded to a specific program component using the Key on the next page.

Lecture Concepts

Lesson Resources

18-1 Global Change pp. 388–394

BLOCKS ① and ②

a. Human-induced environmental changes that affect ecosystems worldwide are referred to as "global change."

b. Acid rain is any precipitation with higher than normal acidity. Acid rain is caused by the interaction between pollutants and water in the atmosphere.

c. Destruction of the ozone layer is caused by chlorofluorocarbons and several other kinds of chemicals.

d. Humans are increasing the concentrations of greenhouse gases such as carbon dioxide. Because these gases trap heat in the atmosphere, global temperatures may increase.

Lecture Resources
- Opening Questions, page 389
- Acid Rain, page 390
- Teaching Transparencies: 74, 81
- "Good" Ozone and "Bad" Ozone, Ozone Levels, and The Ozone Hole, page 391
- Visual Strategy: Table 18-1
- Effective Reading, page 393
- What If the Global Warming Trend Continues?, page 394
- Connections, page 394

- Holt Biology Videodiscs: Lesson 18-1

Classwork Options
- Effects of Ozone Depletion and Graphic Organizer, page 392

Assignment Options
- Section Review, page 394
- Chapter Review, questions 1–3, 10–12, 16–18, 20, 22, 24, 25

18-2 Ecosystem Damage pp. 395–400

BLOCKS ③ and ④

a. Two serious environmental problems are pollution and destruction of nonreplaceable resources. Three such resources in particular danger are topsoil, ground water, and species.

b. In addition to the problems of pollution and consumption, a more fundamental problem is the rapid growth in the human population. The number of people that Earth can support is unknown.

Lecture Resources
- Opening Question, page 395
- Demonstration: Environmental Pollution
- Topsoil, page 396
- Research Update, page 396
- Overcoming Misconceptions, page 396
- Sustainability, page 397
- Balancing Development and Species Conservation, page 398
- How Many Species Are Becoming Extinct?, page 398
- Visual Strategies: Figures 18-10, 18-11
- Teaching Transparencies: 76
- Connections, page 399
- Population Growth and Environmental Damage, page 399
- Holt Biology Videodiscs: Lesson 18-2

Classwork Options
- Laboratory Investigation: Effects of Thermal Pollution, pages 410–411
- Laboratory Techniques and Experimental Design Lab C27: Studying an Algal Bloom
- Laboratory Techniques and Experimental Design Lab C28: Studying an Algal Bloom—Phosphate Pollution

Assignment Options
- The Endangered Species Act, page 397
- The Cairo Population Conference, page 400
- Teaching Resources CD-ROM: Wildlife Biologist
- Section Review, page 400
- Chapter Review, questions 4–7, 13, 14, 19, 21, 23, 26

18-3 Solving Environmental Problems pp. 401–406

BLOCKS ⑤ and ⑥

a. In our economy, pollution can be profitable because the environmental damage caused by the pollution is usually not factored into the price of products.

b. There are two ways to factor the costs of environmental damage into the prices of goods and services. One way is to require pollution control devices. The other is to tax products or services that create pollution.

c. Environmental problems must be documented and understood before they can be solved. There are five components to solving environmental problems: assessment, risk analysis, public education, political action, and follow-through.

Lecture Resources
- Opening Question, page 401
- Demonstrations: Air Pollution, Catalytic Converter
- Visual Strategy: Figure 18-13
- Paying for the Damage, page 403
- Connections, page 405
- Student Involvement, page 406
- Holt Biology Videodiscs: Lesson 18-3

Classwork Options
- Inquiry Skills Development B11: Composting
- Biotechnology D11: Oil-Degrading Microbes
- Biotechnology D12: Can Oil-Degrading Microbes Save the Bay?

Assignment Options
- Solving Environmental Problems, page 404
- Closure, page 406
- Section Review, page 406
- Chapter Review, questions 8, 9, 15

Review/Enrichment

BLOCK 7

■ Study Guide: Concept Review and Pretest
Assignment Options
■ Science, Technology, and Society, pages 412–413
⊙ Teaching Resources CD-ROM: Supplemental Reading Worksheets and Test, *Silent Spring*
■ Activities and Projects, questions 28, 29
■ Readings, questions 30–32

Assessment Options

BLOCK 8

Traditional Assessment
■ Chapter 18 Test
⊙ Test Generator/Assessment Item Listing: Software item bank for preparing customized chapter tests.
Performance Assessment
⊙ Select the Performance Assessment items for this unit from the Test Generator/Test Item Listing
Portfolio Assessment
Select student reports for one or more laboratory experiments from this chapter. *The Direct Observation Checklist* on the *BioSources Teaching Resources CD-ROM* should be completed during a laboratory or other cooperative learning experience.

Answer to Concept Map

The following is one possible answer to the Concept Mapping exercise on page 407.

Holt Biology Videodiscs

Holt Biology Videodiscs gives you a powerful tool for teaching, review, and assessment. *Concepts of Biology* adds interactive lesson plans and extensive tools for customization using CD-ROM technology.

CONCEPTS OF BIOLOGY

Use the following topic from *Concepts of Biology* to help you teach this chapter:

■ Topic 11: Global Ecology

For further information regarding the media on the videodiscs, see *Holt Biology Videodiscs Teacher's Correlation Guide to Biology: Principles and Explorations.*

Human Impact on the Environment   **387B**

Chapter Theme

Homeostasis *One feature of ecosystems is stability, the capacity to absorb perturbations without wholesale change. In terrestrial ecosystems, for example, increased carbon dioxide levels stimulate increased photosynthesis by plants, causing a decline in carbon dioxide levels. In this chapter, students will see that human activities often overwhelm the capacity of ecosystems to respond, resulting in permanent change.*

Tapping Prior Knowledge

- What are mass extinctions, and how have they shaped the history of life?
- How did the ozone layer form? What is its importance to terrestrial living things?
- Describe the major steps of the carbon cycle.

Opening Demonstration

So much has been written and said about the negative impact humans have had on the environment that students are often "turned off" as soon as the issue is raised. To stimulate their interest, provide a challenge. Ask each student to locate a picture that graphically depicts how humans have affected our environment, and ask them to uncover a statistic that reflects the magnitude of the problem. For example, show them a picture of the Fresh Kills landfill in New York State, and point out that it is the largest "structure" ever assembled by humans. Begun in 1948 atop a swamp, the landfill covers nearly 20 sq. km (8 sq. mi.). The trash this landfill holds could cover 1,000 football fields to a height of 21 stories.

CHAPTER 18

HUMAN IMPACT ON THE ENVIRONMENT

Bald eagle, an endangered species

REVIEW

- pH (Section 1-1)
- ozone layer (Section 13-2)
- population growth (Section 15-1)
- carrying capacity (Section 15-1)
- diversity (Section 16-1)
- ground water (Section 16-3)
- carbon cycle (Section 16-3)

Authors' Rationale

Now that your students are conversant with the workings of the Earth, they are in the best position to appreciate what our species is doing to destroy it. They can understand why global changes such as global warming, loss of ozone, and acid rain threaten even "tough" ecosystems. It is very important to raise their consciousness (especially at the high school level) about the damage that is being done to the planet under the guise of "progress." Among the topics your students will learn about are industrial pollution, the damaging effects of burning fossil fuels, the expanding human population, and what can be done to remedy the situation—including at the individual level.

18-1 Global Change

As you read this, the Earth faces unprecedented environmental problems, ranging from global changes in the atmosphere to the loss of topsoil. This section will focus on a collection of environmental issues that affect the entire world—changes in the atmosphere that influence every ecosystem on Earth. Human-induced environmental changes that affect ecosystems worldwide are referred to as "global change." This section focuses on three such human-induced environmental changes: acid rain, ozone destruction, and global warming. Although other global environmental changes are also occurring, these three serve to clearly illustrate the general nature of the changes.

Section Objectives

- Recognize the causes and effects of acid rain.
- Describe the cause and effects of ozone depletion.
- Explain how the burning of fossil fuels has changed the atmosphere, and describe the controversy over global warming.

Burning High-Sulfur Coal Creates Acidic Precipitation

The power plant you see in Figure 18-1 burns coal and sends the smoke high into the atmosphere through stacks more than 65 m (210 ft.) tall. This smoke contains high concentrations of sulfur because the coal that the plant burns is rich in sulfur. Burning high-sulfur coal causes several unpleasant effects. The sulfur-rich smoke smells like rotten eggs, and it combines chemically with the limestone in many public buildings, producing unsightly black surfaces. The intent of those who designed the power plant was to release the sulfur-rich smoke high into the atmosphere where winds would disperse and dilute it. Tall smokestacks were first introduced in Britain in the mid-1950s, and they rapidly became popular in

Figure 18-1 The Four Corners Power Plant in New Mexico burns high-sulfur coal and releases sulfur dioxide, which can combine with water in the atmosphere to produce acid rain.

SECTION 18-1

Vocabulary Preview

acid rain
chlorofluorocarbon
greenhouse effect
global warming

Opening Questions

Ask students to explain the major steps of the water cycle. Answers are shown on page 350 in Chapter 16. Point out that precipitation can transport substances such as sulfuric acid from the atmosphere to the surface. Then ask the class to list the steps of the carbon cycle that remove carbon dioxide from the atmosphere (photosynthesis, absorption by ocean) and those steps that add carbon dioxide to the atmosphere (respiration, combustion, decomposition).

Demonstration
The Greenhouse Effect
Model the greenhouse effect by placing a small thermometer under an inverted glass container and exposing it to sunlight. Place a second thermometer outside the container and note both temperatures after some time has elapsed.

UNIT 4
CHAPTER 18

CONNECTIONS

Chapter 1
Acid Rain

Remind students that acid rain may be responsible for the disappearance of frogs from Sequoia–Kings Canyon National Parks.

Teaching Tip
Acid Rain

Students have had some exposure to chemical equations. Writing the equations on the chalkboard for the formation of acid rain will help them visualize how acid rain forms. Sulfur released by smokestacks reacts with oxygen in the atmosphere to form sulfur trioxide.

$$2S + 3O_2 \longrightarrow 2SO_3$$

In turn, the sulfur trioxide reacts with water vapor in the atmosphere to form sulfuric acid.

$$SO_3 + H_2O \longrightarrow H_2SO_4$$

Another reaction leading to the production of acid rain occurs when sulfur dioxide (also formed by burning coal) reacts with water vapor to produce sulfurous acid.

$$SO_2 + H_2O \longrightarrow H_2SO_3$$

Figure 18-2 The bare trees in this North Carolina forest died as a result of acid rain. Acid rain can change the acidity of the soil and interfere with nutrient absorption by trees and other plants.

the United States and Europe—there are now over 800 of them in the United States alone.

In the 1970s, ecologists began to report evidence that the tall stacks were not eliminating the problems of high-sulfur coal, just exporting the ill effects elsewhere. Throughout northern Europe, lakes were reported to have suffered drastic drops in species diversity; some even became lifeless. The trees of the great Black Forest of southern Germany seemed to be dying as well. But the ill effects were not limited to Europe. Forests in the eastern United States and southern Canada were also being damaged, as you can see in Figure 18-2. It is now estimated that at least 570,000 hectares (1.4 million acres) of forest in the Northern Hemisphere have been adversely affected. Scientists discovered that the sulfur introduced into the atmosphere by smokestacks combines with water vapor to produce sulfuric acid. Rain and snow carry the sulfuric acid back to the surface. This acidified precipitation is called **acid rain.** In North America, acid rain is most severe in the northeastern United States and southeastern Canada, downwind from the coal-burning plants in the Midwest. In 1989, when schoolchildren in a nationwide project measured the pH of rainwater, locations around the United States rarely had a pH lower than 5.6—except in the northeastern United States, where rain and snow had a pH of about 3.8, almost 100 times as acidic as the typical values for the rest of the country.

Acid rain destroys many forms of life. Thousands of lakes in Sweden and Norway no longer support fish, and in the United States and Canada, tens of thousands of lakes are dying as their pH levels fall below 5.0. Throughout the United States and Europe, local populations of frogs and other amphibians are becoming extinct, at least partly because their young cannot develop properly in acidic water. Destruction of the ozone layer, which will be discussed next, may also be a factor in these extinctions.

Historical Note
Acid Rain

A study performed in the Adirondack Mountains of New York between 1929 and 1937 found that only 4 percent of the lakes were acidic and had no fish life. A similar study of the same region in 1977 revealed that 51 percent of the lakes had a pH below 5.0 and that 90 percent of these acidic lakes had no signs of any fish life.

The solution to acid rain at first seems obvious: capture the sulfur-rich emissions instead of releasing them into the atmosphere. But there have been serious problems with implementing this solution. For one thing, it is expensive. In the United States, it is estimated that installation of the necessary smokestack "scrubbers," which would remove most of the sulfur from the emissions, would cost about $5 billion. An additional difficulty is that the polluter and the recipient of the pollution often are far from one another, and neither wants to pay for what is viewed as the other's problem. Revisions of the Clean Air Act passed by Congress in 1990 have begun to address this problem by requiring some cleaning of emissions in the United States. Much still remains to be done worldwide, however. ❑

❑ CAPSULE SUMMARY

Tell rain is any precipitation with higher than normal acidity. Acid rain is caused by the interaction between pollutants and water in the atmosphere.

Destruction of the Ozone Layer

As you learned in Chapter 13, life was trapped in the oceans for 3 billion years because ultraviolet radiation from the sun reached the Earth's surface unchecked, and nothing could survive that bombardment of destructive energy. Living things were able to leave the oceans and colonize the surface of the Earth only after a protective shield of ozone, O_3, had developed in the upper atmosphere. Imagine if that shield were taken away. Alarmingly, it appears that the ozone layer is disappearing and that we are destroying it ourselves.

In 1985, a British researcher in Antarctica noticed that ozone levels in the atmosphere seemed to be as much as 30 percent lower than they had been 10 years earlier. Satellite images taken over the South Pole revealed that the ozone concentration was unexpectedly lower over Antarctica than elsewhere in the Earth's atmosphere, as if some "ozone-eater" were chewing up the ozone and leaving a mysterious zone of below-normal concentration, an ozone "hole." Alarmed, scientists examined satellite images taken in previous years. They found that the disintegration of the Earth's ozone shield was evident as far back as 1978. Every year since then, more ozone has disappeared, and the ozone hole has grown larger. Moreover, a smaller hole has appeared over the Arctic, as you can see in Figure 18-3. Currently, the global concentration of ozone in the upper atmosphere has fallen by more than 3 percent. Because more ultraviolet radiation is reaching the Earth's surface, scientists expect more cases of diseases caused by exposure to ultraviolet radiation: skin cancer, cataracts (a disorder in which the lens of the eye becomes cloudy), and cancer of the retina, the light-sensitive part of the eye. For example, in the United States, the number of cases of malignant melanoma, a potentially lethal form of skin cancer, has almost doubled since 1980.

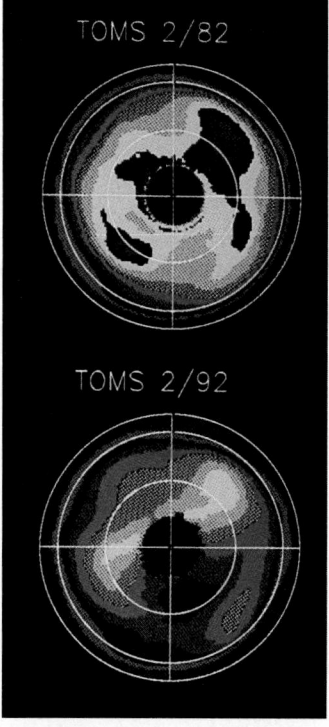

Figure 18-3 Ozone is invisible, but it can be detected by instruments on the ground or carried in satellites. In these computer-enhanced satellite images of the Northern Hemisphere, dark colors indicate zones with the lowest ozone concentration.

SECTION 18-1

Teaching Tips

"Good" Ozone and "Bad" Ozone
Tell students that the ozone layer lies between 17 and 26 km (11–16 mi.) above sea level. The ozone layer absorbs about 99 percent of the ultraviolet radiation that reaches the Earth. Ozone also forms lower in the atmosphere, typically when emissions from automobiles or factories react with sunlight. Ozone is a corrosive, reactive substance that is a serious pollutant in the lower atmosphere. In humans, ozone irritates and damages the lungs, air passages, and eyes, suppresses the immune system, and aggravates respiratory and heart diseases. It also damages many kinds of plants.

Ozone Levels
Before humans began destroying the ozone layer, ozone levels in the upper atmosphere were in dynamic equilibrium. When sunlight strikes ozone, highly reactive oxygen atoms are formed. These recombine almost immediately to re-form ozone, a process that had kept the quantity of ozone nearly constant for hundreds of millions of years. However, the amount of ozone will be reduced by an estimated 5–10 percent by the year 2000 as a result of human activities.

The Ozone Hole
The ozone hole over Antarctica appears for about six weeks each year during late summer and early fall, when conditions for its formation (temperature and amount of sunlight) are just right. The ozone layer over the Antarctic is, on average, 50 percent thinner than it was 15 years ago. By 1993, the region of ozone-depleted air had grown to over 23 million sq. km (9 million sq. mi.), almost the size of the entire North American continent.

Historical Note

Ancient Pollution
Widespread pollution is usually associated with industrialization, which began with the Industrial Revolution in the late 1700s. Recent studies have extended the history of large-scale pollution back several thousand years. Sediment samples from the beds of Swedish lakes show that lead concentrations increased well above natural levels beginning about 2,600 years ago. Ice cores removed from Greenland's icecap also have higher-than-normal lead levels starting at approximately the same time. What caused this ancient pollution? It is most likely the fallout from lead smelting conducted by the ancient Greek and Roman civilizations.

Teaching Tips

Effects of Ozone Depletion

It is estimated that every 1 percent drop in the concentration of ozone in the upper atmosphere will lead to a 6 percent increase in skin cancer. However, ozone depletion's most serious effects may be felt by organisms that cannot escape the increased incidence of UV radiation. For instance, corn, rice, soybeans, cotton, peas, and wheat are easily damaged by UV radiation, and the yields of these crops are expected to decline as ozone levels fall. Perhaps more serious, some kinds of photosynthetic marine algae are very sensitive to UV levels. Since photosynthetic algae are the producers in many aquatic ecosystems, reductions in their populations could have dramatic and widespread effects on many marine organisms. Have students draw a *Graphic Organizer* that shows how release of CFCs into the atmosphere can affect their lives. A completed *Graphic Organizer* is shown at the bottom of the page.

CFCs

On the chalkboard, write the formula for a Freon refrigerant (CCl_3F), some of the most widely used CFCs. Ask students how the name CFC was derived. (*Each of the letters designates one of the elements in these compounds—chlorine* [C], *fluorine* [F], *and carbon* [C].)

CONNECTIONS

Chapter 8

UV Radiation and Mutation

Ultraviolet radiation is a mutagen because it has enough energy to break the covalent bonds in DNA and thus cause rearrangement of the genetic material. Such mutations can lead to cancer.

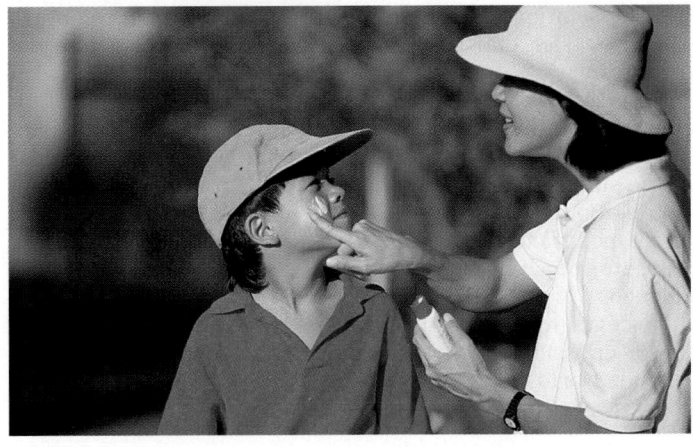

Figure 18-4 Because of the thinning ozone layer, it is wise to minimize your exposure to the sun. When out in the sun, you should cover as much of your skin as possible and apply sunscreen to any exposed areas.

What Is Destroying the Ozone?

The major culprit of ozone destruction is a class of chemicals called **chlorofluorocarbons** (CFCs). High over the South and North Poles, where it is very cold, CFCs stick to frozen water vapor and catalyze the conversion of ozone, O_3, into molecular oxygen, O_2. Invented in the 1920s, CFCs were considered miracle chemicals because they were stable, supposedly harmless, and a nearly ideal heat exchanger. Throughout the world, CFCs were used in large amounts as the coolant in refrigerators and air conditioners, as the propellant in aerosol dispensers, and as the foaming agent in the production of plastic foam cups and containers. Though CFCs were escaping into the atmosphere, at first no one worried because CFCs were thought to be chemically inert and because the atmosphere was thought of as limitless. CFCs are very stable chemicals, so over many years they accumulated in the atmosphere. In the early 1970s, however, chemists Sherwood Rowland and Mario Molina warned that CFCs reacted with ozone and therefore could damage the ozone layer. As a result, CFCs were banned as aerosol propellants in spray cans in the United States. Discovery of the ozone hole in 1985 substantiated Rowland and Molina's warning and stimulated the international community to control CFCs.

International agreements to end CFC production by 1996 were signed by the United States and 92 other countries in the early 1990s. Although this development is important and encouraging, the vast majority of the CFCs that have already been manufactured have not yet reached the upper atmosphere. Also, because of their great stability, CFCs currently in the atmosphere will remain there and continue to destroy ozone for more than a century. Moreover, several other kinds of chemicals destroy ozone. Agreements to ban some of these chemicals have been signed, but other ozone-destroying chemicals remain unregulated. Table 18-1 summarizes the uses and regulatory status of some of the major ozone-destroying chemicals. ❑

❑ CAPSULE SUMMARY

Destruction of the ozone layer is caused by chlorofluorocarbons (CFCs) and several other kinds of chemicals. Many, but not all, of these ozone-destroying chemicals are controlled by international treaties.

Graphic Organizer

Use this graphic organizer in *Teaching Tip:* **Effects of Ozone Depletion** on page 392.

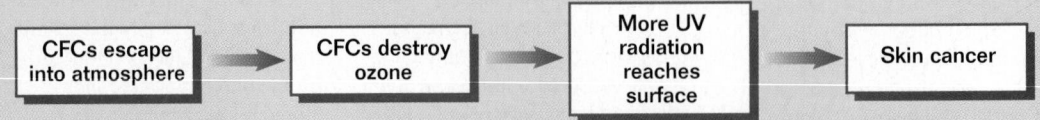

| CFCs escape into atmosphere | → | CFCs destroy ozone | → | More UV radiation reaches surface | → | Skin cancer |

Table 18-1   Uses and Status of Some Known Ozone-Depleting Chemicals

Chemical	Use	Status
CFCs (Chlorofluorocarbons)	Air conditioners, refrigerators, several industrial processes	Production to cease in 1996
Halon	Fire extinguishers	Production ceased in 1994
Carbon tetrachloride	Industrial solvent	Production to cease in 1996
HCFCs (Hydrochlorofluorocarbons)	Temporary replacements for CFCs	Production to cease in 2030
Methyl bromide	Pesticide	Unregulated globally

Is the Climate Warming?

For over 150 years the growth of our industrial society has been fueled by cheap energy, most of it obtained by burning fossil fuels—coal, oil, and natural gas. Fossil fuels are the remains of ancient organisms that have been transformed by pressure and heat into carbon-rich substances. When fossil fuels are burned, carbon atoms combine with oxygen atoms, yielding carbon dioxide, CO_2. Industrial society's burning of fossil fuels has released huge amounts of carbon dioxide into the atmosphere. Additional carbon dioxide has been added to the atmosphere by the burning of vegetation to clear land for agriculture. As with CFCs, no one worried because carbon dioxide was thought to be harmless, and the atmosphere was thought to be a limitless reservoir that was able to absorb and disperse any amount of carbon dioxide. That, as it turns out, is not true.

Carbon dioxide in the atmosphere influences global temperatures. The chemical bonds in carbon dioxide molecules absorb solar energy, trapping heat within the atmosphere like glass traps heat within a greenhouse. The warming of the atmosphere that results from the heat-trapping ability of carbon dioxide and several other gases (including methane, CH_4, and nitrous oxide, N_2O) is known as the **greenhouse effect.** Earth's moderate climate results, at least in part, from the insulating effect of carbon dioxide and these other gases.

However, human activities are greatly increasing the concentrations of the greenhouse gases in the atmosphere. Since studies show that the world's temperatures are directly correlated with the concentration of greenhouse gases in the atmosphere, many scientists fear that the rising

Application

 Technology CFCs are organic compounds containing chlorine and fluorine. In the stratosphere, solar energy can break the covalent bond holding the Cl atom to the rest of the molecule. Once free, Cl atoms can begin their destruction of the ozone layer by converting O_3 into O_2.

VISUAL STRATEGY

Table 18-1
Point out that HCFCs also contain chlorine, but these compounds break down more rapidly than CFCs in the lower atmosphere and thus do not pose as great a threat to the ozone layer, which is higher in the atmosphere.

Effective Reading

Punctuation Signals
Ask students why the head on page 393 is followed by a question mark. They should realize that questions still remain and that scientists have not identified all the variables involved in this issue.

 Did You Know?

Earth's atmosphere is about 0.035 percent carbon dioxide. Carbon dioxide and other greenhouse gases help to keep the average temperature of the Earth's surface at 17°C (63°F). Without these molecules, scientists estimate the average temperature would be −25°C (−13°F), about the same as on Mars.

UNIT 4
CHAPTER 18

Teaching Tip

What If the global warming trend continues?

Average global temperature might increase by as much as 1.5–3.5°C (2.7–9.9°F). This increase would be enough to trigger massive droughts in some inland regions as rainfall patterns changed and lakes and rivers dried up, and it would cause catastrophic floods in coastal regions as the polar icecaps melted.

CONNECTIONS

Chapter 1

Models

Point out that scientists base their predictions of global warming on computer models. Because so many variables involving climate need to be considered—some of which have yet to be identified—such models may give inaccurate predictions.

Figure 18-5 This graph shows the average atmospheric CO_2 concentration and the average change in global temperature over the last 130 years. Currently, average global temperatures are 0.3°C–0.6°C (0.5°F–1.1°F) higher than in 1860. The concentrations of other greenhouse gases, such as methane, nitrous oxide, and CFCs (which also trap heat in the atmosphere) have been increasing as well.

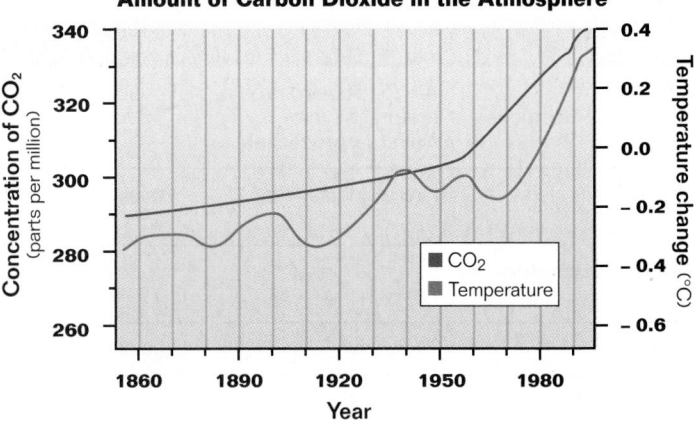

Amount of Carbon Dioxide in the Atmosphere

☐ CAPSULE SUMMARY

Humans are increasing the concentrations of greenhouse gases such as carbon dioxide in the atmosphere. Since these gases trap heat in the atmosphere, scientists worry that global temperatures may increase.

levels of greenhouse gases are causing **global warming,** an increase in global temperatures. Temperature records of the last 50 years seem to support this concern; global levels of greenhouse gases are rising, and so are global temperatures, as shown in Figure 18-5. In fact, eight of the warmest years on record occurred between 1980 and 1992, and 1990 was the hottest ever measured. Some scientists do not agree that these high temperatures can be blamed on rising carbon dioxide levels. They argue that these years of warm weather are nothing unusual and that such fluctuations have often occurred in the past. As you might expect, there is considerable debate among governments about what ought to be done about global warming. Some countries, such as Denmark and Netherlands, are already reducing their production of carbon dioxide. In late 1993, the United States announced a plan to hold production of greenhouse gases at 1990 levels—although compliance by industry is largely voluntary. ☐

Section Review

1. *Describe two harmful effects of acid rain.*
2. *How will the depletion of the ozone layer affect human health?*
3. *Seven of the eight warmest years on record occurred in the 1980s. Explain why this is not conclusive evidence of global warming.*

Critical Thinking

4. *Some of the pollution from power plants in the United States causes acid rain in Canada. Should the United States compensate Canada for damage to its ecosystems? Justify your answer.*

Answers to Section Review

1. Acid rain destroys trees and reduces the pH of bodies of surface water, sometimes to levels that organisms cannot tolerate.

2. Ozone depletion will cause an increase in the number of people who develop cataracts and skin and retinal cancers.

3. Eight years is not nearly a long enough time period on which to base any firm conclusions concerning changes in climate.

4. A student could argue that the United States should compensate Canada since it was responsible for introducing the pollution into the atmosphere. The opposite position could be defended on the grounds that such action was not a deliberate attempt to cause damage because no one could have foreseen the problems that arose from burning high-sulfur coal.

18-2 Ecosystem Damage

Many of the most serious aspects of the environmental crisis concern changes that do not occur on the same scale as ozone depletion or global warming. This section focuses on the two most important examples of local environmental problems—pollution and consumption of nonreplaceable resources. You will also learn about the core environmental problem that drives local and global pollution and consumption: human population growth.

Section Objectives

- Identify three nonreplaceable resources.
- Describe the history of human population growth.
- Contrast population growth in developing and industrialized countries.

Pollution

Since the industrial revolution, society has had a tendency to assume that the environment can absorb any amount of pollution. Lake Erie, shown in Figure 18-6, and other large lakes became polluted because of the unthinking assumption that they could absorb unlimited amounts of industrial chemicals. Because of overly casual attitudes and poor regulation in industrialized and developing countries, the problem of pollution has grown very serious in recent years.

Many of the most disastrous incidents of pollution involve industrial chemicals that are toxic or carcinogenic (cancer-causing). Until recently, there has been relatively little regulation of the manufacture, transportation, storage, and destruction of such chemicals. A particularly clear example of the casual attitude that has prevailed until recently occurred in Basel, Switzerland, in 1986. Firefighters putting out a warehouse fire inadvertently washed 27,000 kg (30 tons) of mercury and pesticides that were stored in the warehouse into the Rhine River. These poisons flowed down the Rhine, through Germany and Netherlands, and into the North Sea, killing fish and other aquatic animals and plants. Today, the river is recovering, but its species diversity remains far lower than before the disaster. In response, European governments have recently passed laws prohibiting the location of dangerous chemicals so close to major rivers. In the United States, few such laws exist.

In an example of pollution closer to home, an oil tanker named the *Exxon Valdez* ran aground on the coast of Alaska in 1989, as shown in Figure 18-7. Oil that gushed from a tear in the tanker's hull heavily polluted many kilometers of coastline and killed thousands of marine animals. If the tanker had been loaded no higher than the waterline, little oil would have been lost. But it was loaded far higher than that, and the weight of the oil above the waterline forced 42 million L (11 million gal.) of oil out the hole in the ship's hull. Today, years after the spill, cleanup efforts continue, and the evidence of damage to local wildlife continues to mount.

Figure 18-6 In the early 1970s, Lake Erie was so polluted that few fish could survive there. This aerial photo shows a slick of raw sewage spreading into the lake from a damaged sewage treatment plant in Cleveland, Ohio.

SECTION 18-2

Vocabulary Preview

aquifer

Opening Question

Ask students why the human population size has not been limited by the carrying capacity of its environment to the same extent as all other animal populations. Students should be aware that for animal species the carrying capacity is mainly determined by food supply and availability of suitable habitat. Humans have control over these two factors to a much greater extent than any other animal species and thus have increased the carrying capacity of their environment.

Demonstration

Environmental Pollution
Display pictures of different forms of environmental pollution, such as air pollution, water pollution, soil pollution, noise pollution, nuclear radiation, solid wastes, and acid rain. Encourage students to add their own examples of environmental pollution to the display. Ask students to consider some causes and effects of each kind of pollution. For example, the emission of gases and particles into the atmosphere causes air pollution, which in turn affects the health of humans, animals, and plants. Sewage, agricultural drainage, and industrial wastes pollute water, which can harm plants and animals that depend on the water.

Teaching Tip

Topsoil

Point out that soil typically has three layers. The top layer (topsoil) contains the most organic matter, also known as humus. The second layer (subsoil) consists of inorganic particles and minerals that have washed down from the topsoil. The lowest layer consists of loose rock that extends down to the bedrock.

☑ RESEARCH Update

UPDATE

Bioremediation 101: The *Exxon Valdez* Experiment

Oil spills are not particularly uncommon, but the 1989 *Exxon Valdez* spill, which dumped 146 million kg (33,000 tons) of crude oil into Prince William Sound in Alaska, ranks as one of the best-known oil spills. Immediately after the spill, Exxon undertook one of the biggest cleanup jobs ever: removing all the crude oil from Alaska's previously scenic coastline. Much of the cleanup involved high-pressure water sprays, but the technique that showed the most promise was the use of biodegradation, the addition of chemicals or microorganisms that catalyze the breakdown of crude oil.

The use of oil-eating microbes was enhanced with the use of an oleophilic fertilizer, and improvements were actually seen on Alaska's beaches. Scientists also noticed a positive association with levels of nitrogen in the water, suggesting that bioremediation requires constant monitoring of water nutrient levels for maximum efficacy. Biostatisticians and geochemists have recently come up with a technique that will allow them to quantify the cleanup. Previously, the only means of assessment was "eyeballing," an imprecise method. ☑

Figure 18-7 The *Exxon Valdez* ran aground on the Alaskan coast in 1989. The spilled oil, *above right,* polluted 1,600 km (1,000 mi.) of coastline. About 6,000 sea otters, *above left,* were killed by the spill. By 1994, Exxon, owner of the tanker, had spent over $3 billion in cleanup costs and fines.

Obviously, a "back to nature" approach—one that ignores the benefits of using chemicals intelligently—will not allow us to meet the needs of a very crowded world or to feed the additional billions of people who will join us during the next few decades. But it is essential that we use our technology as intelligently as possible, with due attention to the protection of the productive capacity of all parts of the Earth, on which we all depend.

Consuming Nonreplaceable Resources

Among the many ways that ecosystems are being damaged, one class of problems stands out as potentially more serious than the rest: the consumption or destruction of resources that we cannot replace. Though a polluted stream can be cleaned up, no one can restore an extinct species. Three kinds of nonreplaceable resources are being consumed or destroyed at alarming rates: topsoil, ground water, and species.

Loss of Topsoil

The United States is one of the most productive agricultural countries on Earth, largely because of its particularly fertile soils. These soils have accumulated over tens of thousands of years. The Midwestern farm belt sits astride what was once a great prairie. The topsoil of that ecosystem accumulated bit by bit as the remains of countless animals and plants decayed. By the time humans came to plow the prairie, the topsoil extended down more than a meter.

We can never replace this rich topsoil, and yet we are allowing it to be lost at a rate of several centimeters each decade. By repeatedly turning over the soil to eliminate weeds, by allowing animals to overgraze ranges and pastures, and by practicing poor land management, we permit wind and rain to remove more and more of the topsoil, as shown in Figure 18-8. Our country has lost one-fourth of its topsoil since 1950.

Overcoming Misconceptions

Our Biggest Polluter

Most people think industry is primarily responsible for water pollution. Actually, agriculture is. Agriculture has polluted over 160,000 km (100,000 mi.) of river, while industry has been responsible for polluting nearly 25,600 km (16,000 mi.). The runoff of silt, pesticides, and fertilizers from agricultural land constitutes the largest single source of water pollution.

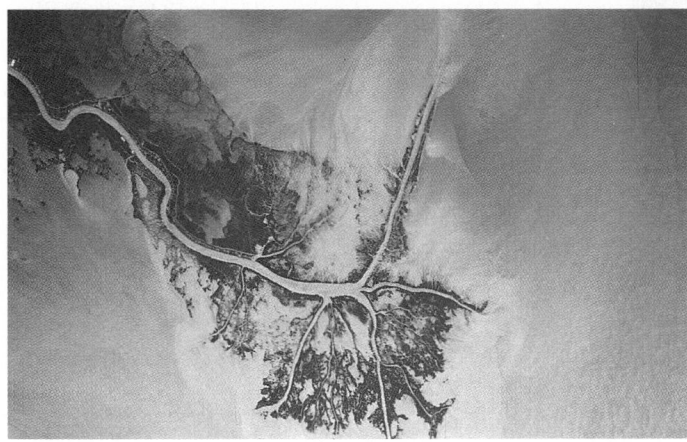

Pollution and Depletion of Ground Water

A second resource that we cannot replace is ground water—water trapped beneath the soil, much of it within porous rock reservoirs called **aquifers** *(AWK wuh furz)*. Water seeps into aquifers at too slow a rate to replace the large amount of water now being withdrawn. In most areas of the United States, there is relatively little control over the use of ground water, with the unfortunate consequence that a very large portion is wasted watering lawns, washing cars, and running fountains. A great deal more is being polluted by poor disposal of chemical wastes. Once pollution enters the ground water, there is no effective means of removing it.

Extinction of Species

Over the last 50 years, about half of the world's tropical rain forests have been destroyed, burned to make pasture and farmland or cut for timber. About 150,000 sq. km (58,000 sq. mi., about the area of the state of Georgia) will be destroyed this year. At this rate, all of the rain forests of the world will be gone within your lifetime. As the rain forests disappear, so do their inhabitants, and tropical rain forests have the highest species diversity of any biome. It is estimated that at least one-fifth of the world's species of animals and plants—about 1 million species—will become extinct during the next 50 years. This is an extinction event unparalleled for at least 65 million years, since the end of the age of the dinosaurs. And the number of species in danger of extinction during your lifetime is far greater than the number that became extinct at the end of the Cretaceous period. ☐

This disastrous loss of species is important to every one of us, because as species disappear, so do our chances to learn about them and their possible benefits. The fact that much of our supply of food is based on only 100 species of plants, out of the tens of thousands of species that have been used at least occasionally as food, should give us pause. Like burning a library without reading the books, we don't know what it is we

Figure 18-8 The Mississippi River, shown here where it flows into the Gulf of Mexico, *above left*, dumps the fertile soil of the Midwest into the ocean. Runoff carries this topsoil into the river from farmland. Loss of topsoil is a worldwide problem. For example, this photo shows Melbourne, Australia, *above right*, during a choking dust storm. Winds pick up topsoil from overgrazed rangeland and abandoned farms in the interior of Australia and carry it to the coast.

☐ CAPSULE SUMMARY

Two serious environmental problems are pollution and destruction of nonreplaceable resources. Three nonreplaceable resources in particular danger are topsoil, ground water, and species.

Teaching Tips

Balancing Development and Species Conservation

One way to preserve a species is to set aside habitat as a park or reserve. However, "locking away" large tracts of land may not be practical, especially in developing countries with large, rapidly growing populations that depend on the land for subsistence. One way to reconcile species conservation with the needs of rural populations is through extractive reserves, areas in which the habitat is preserved but the sustainable harvesting of natural products such as fruits, seeds, and rubber is permitted. Extractive reserves benefit local populations by providing a long-lasting source of income. Also, because continued production depends on keeping the natural habitat intact, species are protected. Ecologists Charles Peters, Alwyn Gentry, and Robert Mendelsohn calculated that 1 hectare of forest in Peru, if managed as an extractive reserve, could yield products with a net worth of about $400 each year. If the same area were logged, the value of the timber would be higher—about $1,000—but the land would not produce any additional income for at least several decades (if ever again) until the forest regenerated.

How Many Species Are Becoming Extinct?

Inform students that the precise rate of extinction today is not known because no one knows how many species exist. About 1.5 million species have been described and named, but millions more certainly await discovery. Estimates for the total number of species on Earth range from 5 million to 100 million.

Figure 18-9 Why should we save species? One reason is that they may be useful, as are these four species. The rosy periwinkle, *top left*, grows in Madagascar, a country already devastated by deforestation. Vinblastine and vincristine, two potent anti-cancer drugs, have been isolated from its leaves. In tropical areas, cattle provide meat but cause great destruction to the land. A potential alternative source of meat is the green iguana, *bottom left*. Studies have shown that, compared with cattle, green iguanas can provide 10 times more meat on the same amount of land while causing far less environmental damage. Corn is an annual plant—it only lives for one growing season. One of its wild relatives from Mexico, *above center*, lives for more than one year and is resistant to many pests. Scientists are trying to transfer the genes for these desirable traits from wild corn to domestic corn. Grain amaranth, *above right*, was a common food in the New World before the European conquest. Its grain and leaves are edible, and it is drought-resistant and fast-growing.

waste. All we can be sure of is that we cannot retrieve the lost knowledge. Figure 18-9 shows four tropical species with actual or potential benefits.

You should not be lulled into thinking that extinction is a problem limited to the tropics. The ancient forests of the northwestern United States are being cut swiftly, largely to supply jobs, and much of the cost of logging is subsidized by our government (the U.S. Forest Service builds the necessary access roads, for example). At the current rate, very little of these ancient forests will remain in a decade. It is hypocritical of us to scold tropical nations for destroying their rain forests when we do such a poor job of preserving our own country's species.

The Core Problem: Population Growth

If we were to solve all of the problems of pollution and consumption mentioned in this section, all we would gain is more time to address a more fundamental problem. The human population is growing extremely rapidly, as illustrated in Figure 18-10.

Human beings first reached North America at least 12,000 years ago, by crossing the narrow strait between Siberia and Alaska, and then spread throughout North America and South America. Ten thousand years ago the continental ice sheets that covered northern Europe and North America withdrew, and agriculture first developed. There were only about 5 million people on Earth then, distributed over all of the continents except Antarctica. With the new and much more dependable sources of food that became available as a result of agriculture, the human population began to grow more rapidly. By about 2,000 years ago, there were an estimated 130 million people on Earth. By 1650, the world's population had nearly quadrupled, reaching 500 million.

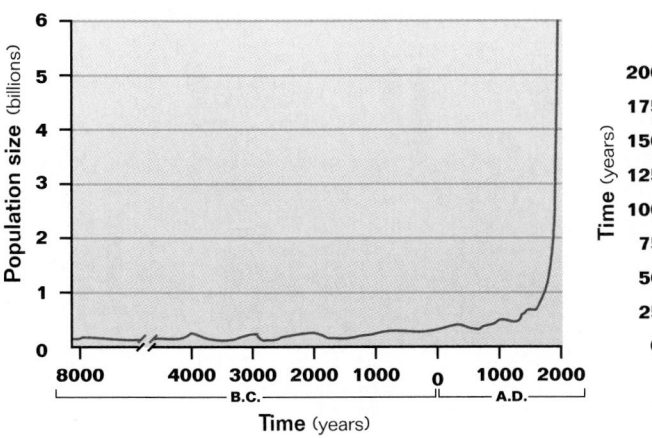

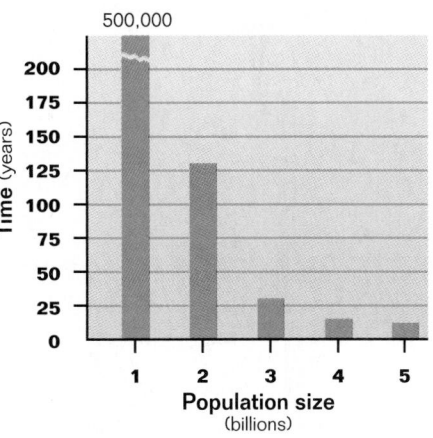

Figure 18-10 (years)

Since at least 1650, and probably for much longer, the human birth rate (as a global average) has remained fairly constant, near 30 births per 1,000 people per year. However, with the spread of better sanitation and improved medical techniques, the death rate has fallen steadily, to an estimated 1994 level of about 9 deaths per 1,000 people per year. The difference between the current annual birth rate (now estimated to be 26 births per 1,000 people) and death rate amounts to an annual worldwide increase in the human population of approximately 1.7 percent. This number may seem small, but don't be deceived. The world's population will double in just over 40 years *if it continues to grow at this rate* (but most scientists think it will not).

The world's population exceeded 5 billion in early 1987, and the annual increase now amounts to about 94 million people; about 260,000 people are added to the world population each day, or about 180 every minute. The world's population is expected to exceed 6 billion by the year 2000. Population growth is fastest in the developing countries of Asia, Africa, and Latin America and slowest in the industrialized countries of North America, Europe, Japan, New Zealand, and Australia. For example, the growth rate of the American population is only 0.8 percent, less than half of the global rate. Most European countries are growing even more slowly, and the populations of Germany and Russia are actually shrinking. By contrast, Kenya's population is increasing by about 3.7 percent per year.

Many countries are devoting considerable attention to slowing the growth rate of their populations, and there are genuine signs of progress. For example, by encouraging families to have only two children, Thailand reduced its growth rate from 3.2 percent to 2.4 percent between 1960 and 1994. In Mexico over the last 30 years, the average number of children per family has decreased from five to less than three. Though the global rate of population growth has been declining, experts project that the world's population will increase

Figure 18-10 The graph on the left shows the soaring human population. Another way to visualize the speed of human population growth is to plot the time required for the population to increase by 1 billion, as done in the graph on the right. It took over 500,000 years—from the origin of our species until about the year 1800—for the human population to reach 1 billion. The population grew from 1 billion to 2 billion in only 130 years, and from 4 billion to 5 billion in only 12 years.

UNIT 4

CHAPTER 18

 VISUAL STRATEGY

Figure 18-11

Point out that the left hand figures in this diagram are the 1994 populations for these countries. Have students calculate the average growth rate per year for the United States, Mexico, Russia, China, and Kenya. Students should assume that the 1994 value represents the population at the beginning of that year and that the 2025 figure is the population on January 1, 2025—the time span for their calculation is 31 years. (*United States: 2.5 million per year; Mexico: 1.5 million per year; Russia: –190,000 per year; China: 10 million per year; Kenya: 1.2 million per year*)

Teaching Tip

The Cairo Population Conference

Have students use the library to research the decisions of the Cairo Population Conference, which took place in late 1994. Students should write a report that describes the nature of the debate at the conference and that summarizes the proposals finally agreed to.

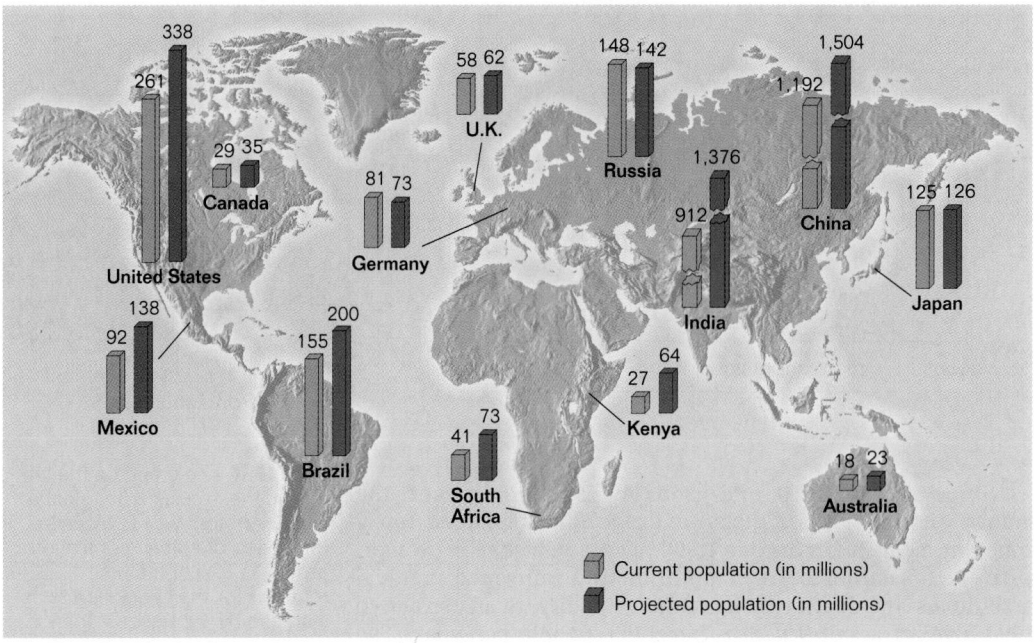

Figure 18-11 This map shows the current populations of several countries (figure on the left in each entry) and their projected populations in 2025 (figure on the right in each entry). Data were obtained from the Population Reference Bureau. These estimates assume that growth rates will continue to slowly decline.

to 8.5 billion by the year 2025. Figure 18-11 shows the current populations of several countries and their projected populations for the year 2025. If the world's growth rate continues to decrease, it is estimated by the United Nations that the population may stabilize by the end of the next century at about 10 billion to 15 billion people. No one knows whether the world can support so many people indefinitely. Finding a way for it to do so is the greatest task facing humanity. The quality of life that will be available to your children will depend to a large extent on our success.

Section Review

1. *Explain why supplies of ground water and soil are dwindling even though these resources are replenished by natural processes.*
2. *How does the growth of the human population affect the populations of other species?*
3. *Describe two instances in which technology has caused the growth rate of the human population to increase.*

Critical Thinking

4. *Unlike most other organisms, humans often use more resources than are needed for survival. How does our standard of living affect the Earth's human carrying capacity?*

Answers to Section Review

1. These resources are being used faster than they are replenished. Hence, their supplies are dwindling.

2. As the human population grows, humans take a larger share of the world's resources,

leaving a smaller share available for other organisms, whose populations consequently shrink. Populations of the few species that flourish in the artificial environments created by humans—such as rats and cockroaches—usually increase.

3. Two possible answers

include advances in farming techniques that have dramatically increased food resources, and methods to eliminate diseases that would otherwise result in a population decline.

4. The higher the standard of living, the lower the carrying capacity becomes. Since the

amount of resources available to support the human population is finite, the more resources each individual uses, the smaller the number of people who can survive on those resources.

18-3 Solving Environmental Problems

*T*he pattern of global change and ecosystem damage that is overtaking our world is very disturbing. Human activities are placing severe stresses on ecosystems worldwide, and we must quickly find ways to reduce their harmful impact. Worldwide attention is focused on solving these problems, and a great deal of progress is being made. In this section you will learn how environmental problems can be solved.

Section Objectives

- Explain how our economic system can make it profitable to pollute.
- Identify the five major steps necessary to solve environmental problems.

Reducing Pollution

One of the most encouraging developments of the early 1990s was the worldwide increase in efforts to reduce pollution. International agreements to stop CFC production are but one example. The release of many dangerous industrial and agricultural chemicals—notably the insecticide DDT and the carcinogens asbestos and dioxin—has been restricted in the United States. A great deal of progress has also been made in reducing air pollution. Emissions of sulfur dioxide, carbon monoxide, and soot—three pollutants produced by the burning of coal—have been cut by over 30 percent in 10 years. The number of secondary sewage treatment facilities, which remove chemicals as well as bacteria from sewage, has increased 72 percent in the last 10 years. The Environmental Protection Agency estimates that businesses and private agencies are spending about $100 billion a year on pollution control,

Figure 18-12 This power plant is equipped with scrubbers that capture pollutants before they are released into the environment.

SECTION 18-3

Vocabulary Preview

(none)

Opening Question

Have students consider this question: Could an organism that does not interfere with its environment in any way survive? The answer is no. All organisms require substances from their environment and change the abundance of these substances by using them. Likewise, all organisms produce wastes that must be released into the environment. Students should see that humans are not alone in exploiting and polluting the environment. However, no organism can disrupt the environment to such a degree or on such a scale as can humans.

Demonstration

Air Pollution

Show the class a photograph of Los Angeles taken on a smoggy day. Point out that in the 1950s, the residents of Los Angeles were suffering from the consequences of the rapid industrialization that had taken place during the preceding decade. At times, residents would be plagued by sudden "gas attacks" that irritated the eyes, diminished visibility, and produced an unpleasant odor. During the 1950s, the situation sometimes got so bad that messengers working for the Rapid Blueprint Company had to wear gas masks when driving their motorcycles through the streets of Los Angeles to deliver packages.

VISUAL STRATEGY

Figure 18-13

Use this figure to discuss some causes and effects of air pollution. Point out that a primary cause of air pollution is combustion. Ask students to think of some common burning processes that pollute air, such as combustion in furnaces and motor vehicles. Then have students list as many harmful effects of air pollution as they can, including its effects on human health, plant life, outdoor structures, climate, the Earth's ozone layer, and the overall environment.

Teaching Tip
Pollution Permits as a Method of Pollution Control

Have students research the practice of selling pollution permits and write a report about what they have learned. In their reports, students should present the arguments for and against this practice and evaluate its effectiveness as a method of pollution control.

Demonstration
Catalytic Converter

Show the class a catalytic converter. Tell students that a few grams of platinum, palladium, or rhodium are embedded as pellets in the catalytic converter, where they act as catalysts in reactions that break down toxic emissions that would otherwise pass through the exhaust system and into the atmosphere.

Figure 18-13 Pollution sometimes kills. In 1952, unusual weather conditions trapped London's pollution, generated mainly by the burning of coal, over the city for five days. This photo, *right,* was taken at noon during the "killer smog." In this case, at least some of the costs of pollution are obvious: 4,200 people died, most from respiratory illnesses aggravated by the choking smog. Pollution also erodes buildings, cars, and statues like this one in England, *above.*

twice the amount spent 10 years ago and five times the amount spent in 1970. However encouraging, this progress represents only a beginning. A serious attempt to address the overall problem of pollution requires more fundamental changes in how the economy of our society operates.

Why Pollution Is Profitable

To learn what needs to be done, it is first necessary to understand the ultimate cause of pollution. In essence, it is a failure by our economy to set an appropriate value on environmental health. To understand how this happens, you must think for a moment about money. The economy of the United States (and much of the rest of the industrialized world) is based on a simple feedback system of supply and demand. As something gets scarce, its price increases. This added profit acts as an incentive for the production of more of the item. If too much of the item is available, the price falls, and because it is no longer so profitable to produce the item, less of it is made.

This system works very well and is responsible for the economic strength of our nation. But it has one great weakness: if demand is set by price, then it is very important that all of the costs be included in the price. Imagine if the person selling the item is able to pass off part of the production cost to a third person; the seller would be able to set a lower price and sell more of the item. And, stimulated by the lower price, the buyer would purchase more than if all the costs had been added into the price.

That sort of pricing error is what has driven industry's pollution of the environment over the last century. The true costs of energy and of the many things made by industry are composed of direct production costs, such as materials and wages, and indirect costs, such as pollution and the risk of unanticipated ill effects to the environment. Imagine if all the medical costs associated with a 20 percent worldwide increase in skin cancer were factored into the price of a refrigerator or air conditioner (because of the ozone-destroying CFCs they contain). Indirect costs from the use of fossil fuels include reduced harvests of fish and shellfish due to oil spills, and crop and timber losses caused by air pollution. As shown in Figure 18-13, pollution from fossil fuels also damages buildings and causes illness and death. Since the indirect costs are not included in the price that the consumer pays, far more is consumed than if they had been included. By not adding the indirect costs to the price of energy and manufactured goods, our society has made it profitable to pollute. The indirect costs do not disappear because we ignore them; they are simply passed on to future generations, who must pay the bill in terms of damage to their own health and to the ecosystems on which they depend. ❑

Paying for Environmental Damage

Two effective approaches have been taken to reduce pollution in this country. The first approach has been to pass laws forbidding it. In the last 20 years, laws have begun to significantly slow the spread of pollution by imposing stiff standards for what can be released into the environment. All cars are required to have effective catalytic converters to reduce emissions, for example. Similarly, the Clean Air Act of 1990 requires that power plants install "scrubbers" on their smokestacks to reduce sulfur emissions. Catalytic converters make cars more expensive, and the installation and maintenance of smokestack scrubbers increase the price of energy. The overall effect is that the consumer pays to avoid polluting the environment. The new higher prices are closer to the true costs, and they lower consumption to more appropriate levels.

A second approach to reducing pollution has been to increase the consumer costs directly by placing a tax on pollution, such as the sulfur emitted by smokestacks. The problem with this approach is that the taxes have never been high enough to reflect the true costs of the pollution. If they were, the consumer costs might be so high as to inhibit industrial growth and development, thus eliminating jobs and depressing economic growth. In recent years, some economists have instead advocated an artificial price hike imposed by the government in the form of a tax added to the cost of goods and energy. This added cost would lower consumption, too, but by adjusting the tax, the government could attempt to balance the conflicting demands of environmental safety and economic growth. Such taxes, often imposed as "pollution permits," are becoming an increasingly important part of laws regulating pollution. They are a key part of the Clean Air Act of 1990. An example of taxes that help control pollution is shown in Figure 18-14.

Teaching Tip
Paying for the Damage
Tell students that consumers must pay more for cleaner gasoline. In major cities throughout the country, motorists must use a cleaner-burning fuel—a blend of hydrocarbons and ethanol or a methanol-based chemical. In most places, this fuel costs only a few pennies more per gallon. However, in some regions of Alaska, this fuel may cost 14 cents more per gallon.

❑ **CAPSULE SUMMARY**

In our economy, pollution can be profitable because the environmental damage caused by the pollution is usually not factored into the price of products.

Historical Note

Auctioning Pollution Permits
The first auction of pollution permits in the United States was held in 1993. At this auction, some 150,000 permits were sold for over $21 million. The top buyer was California Power and Light, which spent $11.5 million for permits. Some public interest groups also bid in an effort to keep the pollution permits out of the hands of industry. Their efforts amounted to less than 1 percent of the bidding.

Figure 18-14 One way to reduce pollution is to tax products whose use or production creates pollution. For example, though the costs of producing gasoline are about the same in the United States and Great Britain, the British pay more than twice as much for gasoline because of higher taxes. At mid-1994 exchange rates, the price of gasoline at this British station is equivalent to $3.20 per gallon. Such high taxes, though primarily designed to increase government revenues, also reduce the use of gasoline and therefore decrease the amount of pollution.

❑ CAPSULE SUMMARY

There are two ways to factor the costs of environmental damage into the prices of goods and services. One way is to require pollution control devices. The other way is to tax products or services that create pollution.

Environmentalists have objected to this added-cost approach, pointing out that true environmental costs are very hard to calculate. How do you put a price on the possibility that sulfur emissions might drive a distant frog population to extinction? More important, in most cases we do not know enough about ecosystems to predict the effects of pollution with any certainty. What if the long-range consequences of pollution turn out to be disastrous? Pollution permits do not take into account these very real biological risks. ❑

How to Solve Environmental Problems

It is easy to get discouraged when considering the world's many serious environmental problems. But do not lose track of the single most important conclusion that emerges from our examination of these environmental problems—each of the world's many problems is solvable. A polluted lake can be cleaned up, a dirty smokestack can be altered to remove noxious gases, and the waste of key resources can be stopped. What is required is a clear understanding of the problem and a commitment to do something about it. The extent to which American families recycle aluminum cans and newspapers is evidence of the degree to which people want to become part of the solution, rather than part of the problem.

If one looks at how success was achieved in cases where environmental problems have been overcome, a clear pattern emerges. Viewed simply, there are five components to successfully solving any environmental problem.

1. **Assessment** The first stage of addressing any environmental problem is scientific analysis, the gathering of information about what is happening, as the scientist shown in

Figure 18-15 An environmental problem must be documented and understood before it can be solved. Here, biologist David Mech is fitting a radio-tracking collar to a rare gray wolf so that its movements can be tracked. Such information is necessary before a plan to conserve gray wolves can be created.

SECTION 18-3

CONNECTIONS
....

Chapter 1
The Scientific Process
Have students compare what is involved in the five components of solving an environmental problem with the processes scientists use in answering a question or solving a problem. For example, assessment involves making observations.

Figure 18-15 is doing. Data must be collected and experiments must be performed to construct a model of the ecosystem that describes how the ecosystem is responding to the situation. Such a model can then be used to make predictions about the future course of events in the ecosystem.

2. **Risk analysis** Using the information obtained by scientific analysis, it is possible to predict the consequences of environmental intervention—what could be expected to happen if a particular course of action were followed. It is necessary to evaluate not only the potential for solving the environmental problem, but also any adverse effects that a plan of action might create.

3. **Public education** When a clear choice can be made among alternative courses of action, the public must be informed. This involves explaining the problem in understandable terms, presenting the alternative actions available, and explaining the probable costs and results of the different choices.

4. **Political action** The public, through its elected officials, selects and implements a course of action. Individuals can have a major impact at this stage by exercising their right to vote and by contacting their elected officials. Many voters do not realize how much they can achieve by writing letters and supporting special interest groups.

5. **Follow-through** The results of any action should be carefully monitored to see if the environmental problem is being solved. The results can also be used to evaluate and improve the initial assessment and model of the problem. We learn by doing.

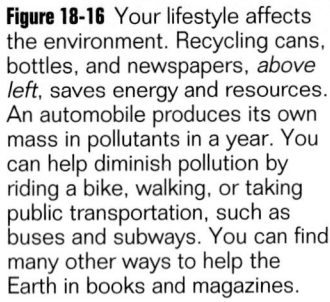

Figure 18-16 Your lifestyle affects the environment. Recycling cans, bottles, and newspapers, *above left*, saves energy and resources. An automobile produces its own mass in pollutants in a year. You can help diminish pollution by riding a bike, walking, or taking public transportation, such as buses and subways. You can find many other ways to help the Earth in books and magazines.

What You Can Contribute

You cannot hope to preserve what you do not understand. Although solving the world's environmental problems will take the efforts of many people, including politicians, economists, and engineers, the issues are largely biological. When all is said and done, your knowledge of ecology is the essential tool that you will need to contribute to the effort. If economists and politicians understood as much ecology as you have learned in these chapters, the world might not be as polluted as it is today. Figure 18-16 shows two simple ways you can participate in solving the problems described in this chapter. Humans rely on the Earth's ecosystems for food and for all of the other materials on which our civilization depends. It has been said that we do not inherit the Earth from our parents but borrow it from our children. We must preserve for them a world in which they can live.

Section Review

1. *Explain two ways to factor the costs of environmental damage into the price of products.*
2. *At which step in the solution of an environmental problem could you have the greatest influence? Explain your answer.*

Critical Thinking
3. *Of the five steps for solving environmental problems, which do you think might be the most likely to fail? Explain your answer.*

Vocabulary

acid rain (390)
aquifer (397)
chlorofluorocarbon (392)
global warming (394)
greenhouse effect (393)

Concept Mapping

Construct a concept map that shows how human activities are disrupting the atmosphere and that describes the effects of these disruptions. Use as many terms as needed from the vocabulary list. Try to include the following terms in your map: greenhouse effect, carbon dioxide, greenhouse gases, global warming, CFCs, ozone layer, acid rain, and high-sulfur coal. Include additional terms in your map as needed.

Review

Multiple Choice

1. Which of the following causes acid rain?
   a. releasing chlorofluorocarbons
   b. burning high-sulfur coal
   c. polluting ground water
   d. scrubbing smokestack emissions

2. An expected effect of ozone depletion is
   a. increased incidence of skin cancer.
   b. depletion of water stored in aquifers.
   c. increased availability of CFCs.
   d. a rise in pH levels in lakes and streams.

3. Burning of fossil fuels has changed the atmosphere by
   a. increasing the global concentration of ozone in the upper atmosphere.
   b. reducing the amount of CFCs.
   c. producing an ozone hole.
   d. increasing the concentration of carbon dioxide.

4. Which of the following is *not* true of the drastic extinction of species that is now occurring?
   a. It is the largest extinction event since the dinosaurs disappeared.
   b. One of its causes is the destruction of tropical rain forests.
   c. It is confined to tropical countries.
   d. Potentially useful species are becoming extinct.

5. In the last 100 years the human population has grown more rapidly than ever before. What factor has contributed the most to this growth?
   a. destruction of the rain forest
   b. the end of global wars
   c. improved sanitation
   d. discovery of new species

6. During the last 200 years, the human death rate has
   a. increased.
   b. decreased.
   c. remained stable.
   d. decreased then increased.

7. Which of these countries has the largest population growth rate?
   a. Germany   c. Kenya
   b. United States   d. Russia

8. The fact that the pollutants produced by industries can be dumped into the environment at no cost has contributed to
   a. goods being sold at prices below their true cost.
   b. bargain prices that weaken consumer confidence.
   c. an increase in the costs of cars since 1970.
   d. government reluctance to tax industries.

9. Of the five major steps to solving environmental problems, which involves determining the potential outcomes of an environmental plan before it is tried?
   a. assessment   c. follow-through
   b. risk analysis   d. political action

CHAPTER REVIEW ANSWERS

Each item in the Chapter Review is correlated by text section in the assignment guide that follows.

ASSIGNMENT GUIDE

Section	Review	Themes Review	Critical Thinking
18-1	1–3, 10–12, 16–18	20	22, 24, 25
18-2	4–7, 13, 14, 19	21	23, 26
18-3	8, 9, 15		

Concept Mapping
Map answer is shown on page 387B.

Review
Multiple Choice
Code in parentheses indicates section number and objective number.
1. b (18-1.1)
2. a (18-1.2)
3. d (18-1.3)
4. c (18-2.1)
5. c (18-2.2)
6. b (18-2.2)
7. c (18-2.3)
8. a (18-3.1)
9. b (18-3.2)

Completion
10. water vapor, lowered (18-1.1)
11. CFCs, ultraviolet, increase (18-1.2)
12. greenhouse effect (18-1.3)
13. aquifers, conserve (18-2.1)
14. 5 million, 130 million, 6 billion (18-2.2)
15. below, pollution (18-3.1)

Short Answer

16. The tall stacks function to release the sulfur-rich smoke high into the atmosphere so that the bad-smelling smoke is blown away from the power plant and diluted. (18-1.1)

17. Because of their stability, CFCs that reach the ozone layer can attack ozone for more than 100 years, destroying many ozone molecules, before they finally degrade. (18-1.2)

18. The greenhouse effect is the moderation of Earth's climate by gases in the atmosphere that trap heat. Global warming is an increase in global temperatures that could result from excessive levels of greenhouse gases. (18-1.3)

19. This increase in growth rate was primarily the result of better sanitation and hygiene, which greatly reduced death rates from disease. (18-2.2)

Themes Review

20. The nonliving material functions as a physical barrier that prevents ultraviolet radiation from reaching the animal's living tissue and doing damage. (18-1.2)

21. Evolution is slow. New species may indeed evolve to take the place of those that have died out, but that process may take millions of years. Extinct species certainly will not be replaced within our lifetime, and therefore their functions in ecosystems and potential uses for humans are effectively lost forever. (18-2.1)

Critical Thinking

22. Due to the thinning of the ozone layer, more harmful ultraviolet radiation is reaching Earth's surface. A hat and sunscreen will protect the skin and reduce the likelihood of skin cancer. (18-1.2)

23. Student responses will vary but should center on factors that they suspect are more closely linked to birth rate than is per-capita income. Factors suggested by Paul and Anne Ehrlich, authors of *The Population Explosion,* include adequate nutrition, basic health care, proper sani-

Completion

10. Acid rain is created when sulfur combines with _____ , forming sulfuric acid. The accumulation of acid rain has _____ the pH in many lakes in the United States and Canada, often killing the organisms that live there.

11. Scientists suspect that _____ are primarily responsible for the thinning of the Earth's ozone layer. A decreased ozone concentration will allow higher levels of _____ radiation to reach the Earth's surface and will _____ the number of cases of skin cancer.

12. The _____ is due to the accumulation of carbon dioxide and other gases in the atmosphere.

13. Ground water, which collected in _____ , is one nonreplaceable resource. Ways to _____ ground water include controlling its use and disposing of chemical wastes properly.

14. Ten thousand years ago, about _____ people inhabited the Earth. About 2,000 years ago, the human population was about _____ . In the year 2000, the world's population will exceed _____ people.

15. The price paid for most goods is _____ the actual cost because the cost of _____ and the unexpected harmful effects on the environment are not included.

Short Answer

16. Tall smokestacks are a part of many coal-burning power plants. What is the purpose of these tall smokestacks?

17. The stability of CFCs, long considered one of their virtues, is now recognized to be a serious drawback. Explain how the stability of these chemicals multiplies their destructive effects on the ozone layer.

18. What is the difference between the greenhouse effect and global warming?

19. The human population began to grow rapidly about 300 years ago. What caused this increase in the population growth rate?

Themes Review

20. Structure and Function A number of land animals have one or more layers of nonliving material covering their living tissue. The exoskeleton (external skeleton) of insects and the feathers of birds are two examples. How might such an adaptation improve an animal's chances for survival and reproduction in an environment with high levels of ultraviolet radiation?

21. Evolution The Earth's species diversity has declined drastically several times in the past, during so-called mass extinction events. Each time, however, new species evolved, and species diversity rebounded. Some critics of efforts to save species have argued that we need not worry about the loss of species, since new ones will evolve to take their place. Using your knowledge of evolution, identify a flaw in this argument.

Critical Thinking

22. Making Inferences Why is it wise to wear a hat and apply sunscreen to exposed areas of your body when you plan to be outdoors for an extended period of time?

23. Interpreting Data It is a widely held belief that the birth rate will decline when countries make progress toward industrialization. However, data show that birth rates have declined in China, Sri Lanka, and Costa Rica, countries with minimal industrial development, and have remained constant in Brazil, which has made progress toward industrialization. How might these data be explained?

tation, and education of women. (18-2.3)

24. Scrubbers on smokestacks reduce the amount of sulfur dioxide and nitrogen oxide released into the air; less acid rain is formed. Another possibility is to reduce the amount of fuel the plant uses, perhaps through greater efficiency. (18-1.1)

25. The thinning ozone layer lets more harmful ultraviolet light reach Earth; this could kill large numbers of plankton, thus increasing the amount of carbon dioxide in the atmosphere. (18-1.2)

26. Both problems will hamper efforts to increase food production. Loss of topsoil diminishes productivity of agricultural land.

Wild species can provide new foods. Also, they can be used to improve existing varieties of crops and domestic animals through cross-breeding or transfer of genes by genetic engineering. (18-2.1)

24. **Making Predictions** Acid rain can damage or even kill many plants and animals. The photo shows a situation that contributes to the production of acid rain. What methods could be used to reduce the pollutants that cause acid rain?

25. **Making Predictions** Harmful ultraviolet radiation kills photosynthetic plankton, which absorb large amounts of carbon dioxide from the air. What effect would ozone depletion have on the levels of atmospheric gases such as carbon dioxide?

26. **Making Predictions** To prevent starvation, food production must increase at least as fast as population size. How will loss of topsoil and extinction of species hamper efforts to increase the production of food?

Activities and Projects

27. **Research and Writing** Contact local or state wildlife officials to obtain a list of endangered species that live in your state or area. Select one of these species and research the efforts being made to protect it. Write a report that explains and evaluates these efforts. In your report you should also describe what environmental changes caused the species to become endangered. Share your findings with your class.

28. **Research and Writing** CFCs deplete ozone in the upper atmosphere and contribute to the greenhouse effect. For these reasons, the chemical industry is now developing a range of alternatives to CFCs. Write a report that describes the requirements established for CFC replacements and that identifies several alternatives to CFCs.

29. **Multicultural Perspective** Many medicines are derived from chemicals produced by plants. One such plant is the neem tree of Asia. Research the medicines and other products obtained from this tree in India, where it is known as the "village pharmacy." Share your findings with your class.

Readings

30. Read *The Population Explosion* by Paul and Anne Ehrlich. Explain the logic behind the Ehrlichs' assertion that the United States is overpopulated. What does each term in the equation I=PAT represent? How does the magnitude of each term in the equation differ between developed and developing countries?

31. Read the article "The Scandal of Siberia," in *New Scientist*, November 27, 1993, pages 28–33. What kinds of damage have been done to the forest and swamp ecosystems of Siberia as a result of oil exploration?

32. Read the article "Plight of the Plover," in *Science News*, December 7, 1991, pages 382–383. How has the presence of humans on the beaches in the eastern United States affected the natural habitat of piping plovers? What actions have been taken to ensure that the plovers have peaceful nesting sites? What can the average citizen do to help save these birds from extinction?

Readings

30. Carrying capacity is the population an area can sustain over the long term. A requirement for long-term survival is that the population use its resources sustainably, which the United States is not doing. *I* is environmental impact. *P* is the population size. *A* is the per-capita resource use, a measure of affluence. *T* is the environmental destructiveness of the technology used by the population. In developed countries, *P* (population) is low but *A* and *T* are high. In the developing countries, *P* is high, but *A* and *T* are low. The result in both cases, however, is environmental destruction.

31. Fires caused by sparks from machinery have burned 200,000 hectares (500,000 acres) of forest, road-building has upset the natural drainage systems, and leaks from pipelines have polluted swamps and contaminated underground water.

32. Humans disturb the birds, keeping them from foraging for food. Government-owned beaches were closed during plover breeding season. The average citizen can stay away from the birds when on the beach and keep the beach clean, so as not to attract unwanted predators.

Activities and Projects

27. Answers will depend on species chosen.

28. Requirements for CFC replacements include **(1)** similar boiling points to CFCs, **(2)** non-flammability and low toxicity, **(3)** technically and economically viable, **(4)** chemically stable, and **(5)** low potential for ozone depletion and global warming. Alternatives to CFCs include hydrochlorofluorocarbons (HCFCs) and hydrofluorocarbons (HFCs). Their primary uses will be as refrigerants and propellants, the same as CFCs.

29. The neem tree yields drugs for relieving pain and treating skin diseases, a toothpaste-like material, an insect repellent, and a general tonic.

*L*ABORATORY Investigation *Chapter 18*

Effects of Thermal Pollution

Preparation

Paramecium cultures may be purchased from a biological supply company. The cultures may be mixed with bottled spring water or tap water that has been left in an open container at least 24 hours to eliminate the chlorine. Do not let students mix the cultures with pond water because it might contain other organisms.

Procedural Notes

- Students may work in groups of two to four.
- The U-shaped tubes must be completely filled to distribute the paramecia.
- If large air bubbles form in the tubing, the bubbles will move to the bend of the tube when it is inverted into the beakers, and the paramecia will not be able to move freely.
- Check that glass tubing has no cracks or fractures before students begin investigation.

Background Answers

1. A pollutant is anything introduced into the environment that can be harmful.

2. If heat, noise, or anything else causes harm to living things, it is considered a pollutant.

3. Some power plants use water from rivers as a coolant. After the water has been used, it is much hotter than normal river water.

4. Power plants can cool the water before returning it to the environment.

5. What are the effects of thermal pollution on *Paramecium*?

Records

Students' data may vary, but should include their observations, the number of *Paramecium* involved, and the time involved.

OBJECTIVE

Model the effects of thermal pollution on organisms in the laboratory, and apply the underlying scientific principles to current environmental issues.

PROCESS SKILLS

- forming a hypothesis
- designing an experiment
- observing
- analyzing data

MATERIALS

- two 400 mL beakers
- ice
- hot water
- thermometer
- U-shaped glass tube, 30 cm long
- two corks to fit ends of tubing
- 125 mL beaker
- graduated cylinder
- water
- *Paramecium* culture
- hand lens
- clock or watch
- glass-marking pen or wax pencil

BACKGROUND

1. What is a pollutant?
2. How can heat be considered a pollutant?
3. How can power plants pollute water, even when they have added no chemicals or waste products to the water?
4. How can power plants release nonharmful water?
5. Write your own question to explore in your lab report or notebook.

TECHNIQUE

1. Discuss the objective of this investigation with your partners and develop a hypothesis concerning the effect of temperature on *Paramecium*.

2. Design an experiment to test your hypothesis. In your experiment, *Paramecium* will be contained in the U-shaped tube. One large beaker will be filled with ice, and the other large beaker will be filled with hot water. Other available supplies are listed in your materials list.

3. In designing your experiment, decide which factor will be an independent variable. Plan how you will vary the independent variable.

4. Decide which factor will be the dependent variable. Plan how you will measure changes in the dependent variable.

5. Remember that in most experiments a control is necessary. Plan a control for this investigation.

6. Discuss your planned experiment with your teacher. Proceed with the experiment once you have received approval from your teacher.

7. Once you begin the investigation, proceed until it is completed. Remember to record your results, including the numbers of individual *Paramecium* and the time involved.

8. Fill a 400 mL beaker with ice and water. Make sure ice remains in the beaker for the entire experiment. Fill another 400 mL beaker with 60°C tap water. **CAUTION: Be careful handling hot water.** Water more than 60°C can scald.

9. In a 125 mL beaker, gently swirl 20 mL of water and 20 mL of *Paramecium* culture. **Note:** Your teacher will supply you with aged tap water or spring water. (Chlorinated water would kill *Paramecium*.)

10. While your partner holds the tube steady, carefully pour the *Paramecium* and water mixture into the U-shaped tube. Fill the tube completely, leaving just enough room for a cork. Make sure there are no large bubbles of air in the tube. Place a cork in each end of the tube.

11. Proceed with your experiment using the tubing, ice water, hot water, and hand lens to observe any response of *Paramecium* to their environment.

12. Record the data you collect in the **Records** section of your report. Also, briefly summarize the procedure you followed.

13. Clean up your materials and wash your hands before leaving the lab.

INQUIRY

1. What effect did heat have on *Paramecium* in your experiment? What was the effect of cold?

2. What evidence did you have that *Paramecium* preferred one temperature range to another?

3. What was the independent variable? How did you vary it?

4. What was the dependent variable? How did you measure changes in it?

5. What controls did you use?

6. Why are the length of time and the temperature range important factors?

ANALYSIS

1. Did the results of your experiment support your hypothesis? Explain.

2. What are some possible sources of error in your experiment?

3. How could a pollutant cause an increase in the number of organisms? Explain.

4. Judging from your experiment, how do you think other organisms might react to a change in water temperature?

5. How could a power plant change the type of organisms that live in the water where it releases its cooling water?

FURTHER INQUIRY

Write a new question that could be explored as a new investigation. The following is an example:

How can an experiment be designed to test how acid rain affects *Paramecium*?

Inquiry Answers

1. The paramecia avoided both hot and cold, moving to the center of the U-shaped tube.

2. They move away from sections of the tube with extreme temperatures.

3. The independent variable was temperature, and it was varied with ice and hot water.

4. The dependent variable was concentration of paramecia. It was counted.

5. Answers may vary, but a duplicate setup placed in beakers of room-temperature water would serve as a control.

6. If too short a time is allotted, the organisms will not have moved to their new area. If the temperature is within the organisms' tolerance range, that too will result in no movement. If the temperatures are too high or low, the paramecia may be killed before they can move.

Analysis Answers

1. Answers will vary.

2. Answers will vary. Errors could include jarring the tube, unhealthy paramecia, using temperatures that do not affect the organisms, inaccurate counting, or giving the paramecia insufficient time to respond.

3. A pollutant may be harmful to some organisms but beneficial to others. For instance, those organisms suited to higher temperatures will increase when thermal pollution raises water temperatures.

4. Organisms might move out of the area, die off, or increase in number, depending on the type of organism and the exact temperature.

5. Only organisms that can tolerate the higher temperatures are able to live in thermally polluted water. Some fish may die out, while certain algae and other organisms may flourish and successfully compete with organisms that had previously lived in the river.

Point of View: Biological Diversity

Objectives
- Develop scientific literacy
- Identify an author's point of view
- Detect bias in a persuasive argument

Background

Each year scientific issues become more important in determining the nature and direction of our society, and yet most citizens have little or no training in evaluating the arguments that surround controversial issues. The purpose of this feature, Point of View, is to provide practice in identifying and evaluating a writer's point of view.

A critical step in this process is detecting bias. Much of the writing on critical issues is persuasive and not objective. A writer will include information that bolsters his or her point of view and suppress evidence that contradicts it.

Instructional Strategies

- As students read this article, remind them that it represents one person's point of view. They must read critically, looking for signs that would indicate biased or slanted writing. In particular, have students look closely for the following:

1. Does the article present both sides of the issue fairly and objectively?
2. Is the writer seeking to shape your opinion on the issues?
3. Does the writer use "loaded" language? Are some words chosen because they have an emotional rather than a logical appeal to the reader?

- Ask students to define the function of a keystone in an archway. Then tell them that many ecosystems have keystone species. Have them predict what would happen to an ecosystem in which

BIOLOGICAL DIVERSITY

POINT OF VIEW

How Are We Changing Life on Earth?

The extinction of the tiger means the elimination of a single species. The destruction of rain forests results in the loss of hundreds of species.

BY TRACEY COHEN

Between 1989 and 1992, 18 tigers were killed by poachers in India's Ranthambhore National Park. Although the country is dedicated to protecting the last big carnivores in its parks, India's tiger populations have fallen about 35 percent over the last five years. Elsewhere in Asia, tiger populations are also endangered. Probably no more than 5,000 to 7,000 of the great striped cats are left in the wild. The situation is so dire that many conservationists fear that wild tigers will be extinct in only a few years. Sadly, if the tiger becomes extinct, it will be just one of thousands of species driven to extinction by human activities.

Extinction is actually a natural process. Scientists estimate that every million years for the last 200 million years about 900,000 vertebrate species have become extinct. For higher plants, the extinction rate has been about 37,000 species every million years for the past 400 million years.

If extinction is part of the course of natural events, why does it matter if tigers and other organisms die out?

Why Are Different Species Important?

From a human perspective, natural ecosystems and the organisms in them are important for a number of reasons. The most obvious is that plants and animals provide us with food. For example, throughout history people have eaten about 7,000 different kinds of plants. At least 75,000, however, are known to be edible. Plants and animals also provide raw materials for clothing, fuels, oils, dyes, industrial chemicals, and building materials. Over 40 percent of the medicines used in the industrialized world come directly from organisms or were synthesized based on chemical compounds taken from them: 25 percent from plants, 13 percent from microorganisms, and 3 percent from animals. Aspirin, antibiotics, and the anticancer drugs vincristine and vinblastine are just a few examples. The possibilities for new foods, drugs, and other substances from the world's genetic library are enormous.

Biological diversity is crucial for other less visible, but no less important, reasons. The vast web of living things provides a variety of "life-support" services on a scale that would be impossible to replace. Plants and algae, for instance, capture the sun's energy and make it available to other organisms through photosynthesis. They also purify the planet's air by taking in carbon dioxide and releasing oxygen. The roots of plants break apart rocks and, with the help of specific microbes, form soil and keep it fertile. The global cycling of nutrients and the decomposition of wastes depend on countless organisms in a variety of ecosystems, as do climate patterns and water circulation and purification.

What Threatens Biological Resources?

According to some estimates, human activities have increased the natural rate of extinction by as much as 10,000 times. For example, 11 percent of the world's bird species are endangered. About 20 percent of the world's freshwater fish species are extinct or nearly so. In the United States, over 200 plant species are extinct and another 680 species may be gone by the end of the decade.

The single greatest threat to biodiversity is the destruction of

a keystone species is removed. Point out that in many cases, the function of a keystone species becomes apparent only after it is extinct.

- Ask students if they can imagine ants, bees, or cockroaches becoming endangered. Point out that invertebrate species are not immune to habitat

loss, pollution, and other factors that threaten vertebrates and plants. In Austria, for example, 22 percent of invertebrate species are threatened or endangered, as are 17 percent of insect species in England.

- Have students identify the ways that their lifestyle choices affect biological diver-

sity. Ask them to list the ways that they use resources, generate pollution, destroy habitats, etc. Remind them to consider the impact of their activities magnified by those of millions of people. Then ask them what they would be willing to change or give up to protect biodiversity.

habitat. According to the International Union for the Conservation of Nature and Natural Resources, 73 percent of those species with declining populations are suffering from habitat destruction. Loss of habitat occurs when people convert ecosystems from one type to another, say from forest to farmland or from wetlands to suburban housing developments or shopping malls. Habitat destruction also happens as a result of pollution. Such changes destroy hundreds of thousands of square kilometers of habitat. According to biologist E. O. Wilson, of the many habitats in the world that cover at least 1 sq. km, not many have fewer than a thousand species of plants and animals. Patches of rain forest and coral reefs—the most productive habitats—have tens of thousands of species.

Introduction of foreign species also seriously threatens biodiversity. Plants, animals, and other organisms have been spread by humans to new locations both accidentally and on purpose. Foreign species displace or destroy native ones. For example, in 1959 British colonists released Nile perch into Lake Victoria, in Africa. This predator fish has greatly reduced the native fish population and caused the extinction of some species in the lake. In the United States, the most serious threat to ecosystems in the national park system are new species of plants introduced into a park.

What Can Be Done to Preserve Biological Diversity?

Scientists agree that the first step is to collect as much information as possible. One approach, pioneered by Conservation International, is the Rapid Assessment Program (RAP). RAP is designed to quickly survey possible ecosystems that are "hot spots"—areas with the largest number of

endangered species. Emergency recommendations can then be made about conservation strategies for these areas.

At the same time, there are some immediate actions that can help. Zoos, for instance, are breeding endangered and threatened animals so that they can provide a reservoir for restoring wild populations—assuming, of course, that natural habitats remain.

Botanical gardens and seed banks are helping to preserve plant species. Employees of botanical gardens grow plants and carry out long-term research of their biology, reproduction, and roles in ecosystems. Seed banks conserve plant genetic material through seed storage. There are currently more than 50 seed banks worldwide. Many are devoted to preserving the numerous local varieties of food crops like corn, wheat, and potatoes.

Efforts to restore destroyed or degraded ecosystems are also underway. Projects range in size from small local streams to the Florida Everglades.

The Limits of Science and Technology: Can Biological Diversity Really Be Saved?

People tend to think that the crisis of biodiversity, as well as other environmental problems

created by human activities, can be corrected with things like better technology. Science and technology can provide crucial information and strategies, but there are limits to their usefulness. Captive breeding programs, for instance, can help save no more than a few hundred of the 1 million known animal species. On a larger scale, too many interdependent species are being lost, often before their ecological roles are fully known. And it would be impossible to create artificially all the conditions necessary for complex, healthy ecosystems.

The true problem of preserving life on Earth is ethical. It means recognizing that other species have a right to continue living. This recognition involves drastic changes in the ways we see and use natural resources. It also means changing the way we see ourselves in the larger community of living things. We need what environmental science professor David Orr called a "biophilia revolution." Biophilia is, in essence, a love of life and things that are alive. A biophilia revolution would combine a reverence for life with efficient use of resources and simpler lifestyles. Until that happens, the plight of biodiversity will not be solved.

Tracey Cohen is a freelance writer specializing in science and environmental issues.

attacked by loggers, ranchers, land developers, various industries, and other groups for the ways that it limits their economic activities.

References

1. Conway, W. "Can Technology Aid Species Preservation?" In: *Biodiversity*. E. O. Wilson, ed. National Academy Press, Washington, D.C., 1988, pp. 263–268.

2. Devine, R. "Botanical Barbarians." *Sierra*, Vol. 79, No. 1, Jan./Feb. 1994, pp. 50–57, 71.

3. Ehrenfeld, D. *Beginning Again, People and Nature in the New Millenium.* Oxford University Press, New York, 1993.

4. Ehrenfeld, D. "Why Put a Value on Biodiversity?" In: *Biodiversity*. E. O. Wilson, ed. National Academy Press. Washington, D.C., 1988, pp. 212–216.

5. Ehrlich, P. R. and A. H. Ehrlich. "The Value of Biodiversity." *Ambio*, Vol. 21, No. 3, 1992, pp. 219–226.

6. Linden, E. "Tigers on the Brink." *Time*, March 28, 1994, pp. 44–51.

7. McNeely, J. A., et al. *Conserving the World's Biological Diversity.* International Union for Conservation of Nature and Natural Resources, World Resources Institute, Conservation International, World Wildlife Fund-U.S., and World Bank. Washington, D.C., 1990.

8. Orr, D. "Love It or Lose It: The Coming Biophilia Revolution." *Orion*, Winter 1994, pp. 8–15.

9. Orr, D. "The Challenge of Sustainability." *Phytopathology*, Vol. 83, No. 1, 1993, pp. 38–40.

10. Ryan, J. C. "Conserving Biological Diversity." In: *State of the World. A Worldwatch Institute Report on Progress Toward a Sustainable Society.* L. R. Brown et al., eds. W. W. Norton & Co., New York, 1992, pp. 9–26.

11. Spellerberg, I. F. and S. R. Hardes. *Biological Conservation.* Cambridge University Press, Cambridge, 1992.

12. Wilson, E. O. *The Diversity of Life.* W. W. Norton & Co., New York, 1992.

Analyzing the Issue

1. **Reading Critically** What is the thesis, or main point, of the article? List the facts that the writer uses to support the thesis.

2. **Formulating an Opinion** Many scientists believe that humans have an ethical responsibility to preserve biodiversity. Yet, according to biologists Paul and Ann Ehrlich, "One cannot assert this ethical responsibility on scientific grounds."
   a. What do you think the Ehrlichs mean by this statement? Do you agree? Explain your answer.
   b. What role do scientists have in helping societies make ethical or moral decisions related to nature?

3. **Take Action** Find out what legislation exists at the local, state, and federal level to protect species and ecosystems. How does this legislation work? Has there been criticism of it? Why? By whom?

Answers to Analyzing the Issue

1. The thesis of the article is that biological diversity is seriously threatened by human activities. Facts may include the number of endangered freshwater fish species, plant species in the United States, and bird species worldwide.

2. (a) Answers will vary. Science supports preserving biological diversity on practical grounds; however, the ethical basis for species preservation arises mainly from the view that species have intrinsic value and a right to continue to exist.
(b) Scientists can provide information on the likely consequences of specific actions for different species. The evaluation of consequences is an essential part of ethical decision making.

3. Local and state legislation varies. At the federal level, the Endangered Species Act is designed to protect particular species. The act, which is up for reauthorization, has been

PART 2

BIOLOGICAL EXPLORATIONS

Protist

South African bullfrog

Nile crocodile

Catfish

> *Since our genes stretch back in time in an unbroken line to the first lifeforms—just as the genes of mushrooms, bacteria, and rabbits do—it is questionable to assume that we and our domesticated mammals are higher forms of life. Certainly more recent, in some ways more complex, our intelligence is, nonetheless, simply another successful strategy for survival. . . . We simply exist today because our particular collections of genes, like stacks of chips growing on a roulette table, have not yet exhausted their winning streaks.*

— Dorion Sagan and Lynn Margulis
Garden of Microbial Delights

Lappet-faced
vulture

African
elephant

Mountain biker

Diversity of Life

Although they appear dissimilar, the African elephants, catfish, and underwater plants (and the unseen bacteria and fungi also present in this scene) have much in common. Living things have evolved surprisingly similar strategies for obtaining energy, retaining water, reproducing, and a host of other tasks essential for life.

Each block represents about 45–50 minutes of instructional time. The resources cited in a block are coded to a specific program component using the Key on the next page.

Lecture Concepts

Lesson Resources

19-1 Origin of Eukaryotes pp. 416–421

BLOCKS ① and ②

a. Bacteria, which are prokaryotes, are the oldest and most abundant form of life. These organisms evolved into two branches: the archaebacteria and the eubacteria.

b. The evolution of organelles such as mitochondria and chloroplasts is explained by the theory of endosymbiosis, which proposes that eukaryotic cells evolved when certain prokaryotes established symbiotic relationships within larger cells.

Lecture Resources
- Opening Question, page 417
- Demonstration: Benefits of Bacteria
- Visual Strategies: Figures 19-1, 19-2, 19-3, 19-4, 19-5
- Teaching Transparencies: 61, 88, 88A
- Methane Sources, page 418
- Another Look at Mitosis, page 421
- Holt Biology Videodiscs: Lesson 19-1

Classwork Options
- Visualizing the Difference, page 418
- Enhancing Listening Skills, page 419

Assignment Options
- Write a Story, page 420
- Concept Mapping, page 449
- Section Review, page 421
- Chapter Review, questions 1, 2, 9, 13, 27, 29

19-2 Evolution of Sexual Reproduction pp. 422–425

BLOCKS ③ and ④

a. Asexual reproduction is the simplest and most primitive form of reproduction.

b. One of the most important evolutionary innovations of eukaryotes was the ability to reproduce sexually. Sexual reproduction may have evolved as a mechanism to repair damaged DNA.

c. Eukaryotic organisms can have one of three kinds of sexual life cycles: zygotic meiosis, gametic meiosis, or sporic meiosis.

Lecture Resources
- Opening Question, page 422
- Demonstration: Modes of Reproduction
- Visual Strategies: Figures 19-6, 19-7, 19-8, 19-9, 19-10
- Holt Biology Videodiscs: Lesson 19-2

Classwork Options
- Effective Reading, page 422
- Concept Map of Life Cycles, page 425

Assignment Options
- Section Review, page 425
- Chapter Review, questions 3–6, 11, 14

19-3 Evolution of Multicellularity pp. 426–430

BLOCKS ⑤ and ⑥

a. The evolution of multicellularity enabled organisms to grow more complex.

b. Colonial organisms and aggregations are not considered to be multicellular. They are a permanent or temporary collection of cells.

c. Cell specialization and coordination enable complex multicellular organisms to have different cells and different functions.

Lecture Resources
- Opening Question, page 426
- Demonstration, page 426
- Visual Strategies: Figures 19-11, 19-15
- Cell Specialization, page 430
- Holt Biology Videodiscs: Lesson 19-3

Classwork Options
- Distinguishing Organisms, page 427
- Family Tree, page 429
- Grouping by Complexity, page 429

Assignment Options
- Section Review, page 430
- Chapter Review, questions 7, 12, 25, 28

19-4 Classification of Living Things pp. 431–442

BLOCKS ⑦ and ⑧

a. To ensure accurate communication about organisms, biologists use binomial nomenclature to name species. Binomial nomenclature is a system of scientific names developed by the eighteenth-century Swedish biologist Carl Linnaeus.

b. Scientists classify organisms into a hierarchical system of groups. Today, there are seven basic levels in the hierarchy: kingdom, phylum, class, order, family, genus, and species.

c. A species is defined as a group of organisms that share many characteristics and, in nature, interbreed with each other and not with members of other species.

Lecture Resources
- Opening Question, page 431
- Demonstrations: Naming Organisms, Deforestation of Rain Forest
- Identify the Organism, page 432
- Visual Strategy: Figure 19-1
- Overcoming Misconceptions, page 433
- Research Update, page 434
- Teaching Transparencies: 56
- Connections, page 436
- Holt Biology Videodiscs: Lesson 19-4

Classwork Options
- Mnemonic Device, page 435

- Comparing Classifications and Graphic Organizer, page 435
- Concept Map for Species, page 436
- Loss of Biodiversity, page 437
- Quick Labs A15: Grouping the Things You Use Daily
- Laboratory Techniques and Experimental Design C29: Classifying Mysterious Organisms

Assignment Options
- These Are the Names of My Favorite Things, page 435

19-4 Classification of Living Things pp. 431–442 (continued)

BLOCK 9

a. An organism's evolutionary history is its phylogeny. Cladistics is a method of taxonomy that uses information about the presence of traits among a group of organisms to provide information about the relative relationships among organisms.

b. Evolutionary systematics is a subjective method of taxonomy in which a biologist uses judgement to consider the importance of the characters among a group of organisms.

Lecture Resources
- ■ Demonstrations: Looks Can Be Deceiving, Constructing a Cladogram
- ■ Visual Strategy: Figure 19-24
- ■ Analogous Structures, page 438
- ■ Cladogram vs. DNA, page 440
- 🔒 Teaching Transparencies: 59

Classwork Options
- ■ Visual Strategies: Tables 19-2, 19-3

- ■ Laboratory Investigation: Dichotomous Keys, pages 452–453
- ▲ Inquiry Skills Development B12: Classification
- ■ Closure, page 442

Assignment Options
- ■ Section Review, page 442
- ■ Chapter Review, questions 8, 10, 24

Review/Enrichment

BLOCK 10

- ■ Highlights: *Six Kingdoms of Life*, pages 443–448
- ■ Highlights Review, questions 19–24
- ■ Study Guide: Concept Review and Pretest

Assignment Options
- ◎ Teaching Resources CD-ROM: Supplemental Reading Worksheets and Test, *Microbe Hunters*
- ■ Activities and Projects, questions 30, 31
- ■ Readings, question 32

Assessment Options

BLOCK 11

Traditional Assessment
- ■ Chapter 19 Test
- ◎ Test Generator/Assessment Item Listing: Software item bank for preparing customized chapter tests.

Portfolio Assessment
Select student reports for one or more laboratory experiments from this chapter. *The Direct Observation Checklist* on the *BioSources Teaching Resources CD-ROM* should be completed during a laboratory or other cooperative learning experience.

Answer to Concept Map

The following is one possible answer to the Concept Mapping exercise on page 449.

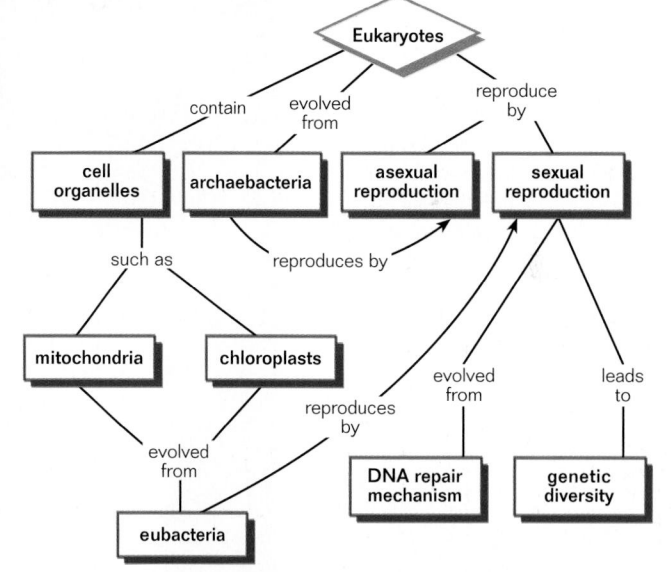

Holt Biology Videodiscs   CONCEPTS OF BIOLOGY

Holt Biology Videodiscs gives you a powerful tool for teaching, review, and assessment. *Concepts of Biology* adds interactive lesson plans and extensive tools for customization using CD-ROM technology.

Use the following topics from *Concepts of Biology* to help you teach this chapter:
- ■ Topic 9: Evolution of Life on Earth
- ■ Topic 12: Classification of Life

For further information regarding the media on the videodiscs, see *Holt Biology Videodiscs Teacher's Correlation Guide to Biology: Principles and Explorations.*

Chapter Theme

Evolution *Populations of organisms must be able to adapt to changing environments in order to survive. As the Earth has changed over time, a fantastic array of organisms has evolved to survive in practically every kind of environment found on Earth.*

Tapping Prior Knowledge

- What are the major differences between prokaryotes and eukaryotes?
- What is the theory of evolution?

Opening Demonstration

Collect photographs of places such as Antarctica, Old Faithful, tundra, sea vents in the ocean floor, and the desert. And collect microphotographs of things such as skin, teeth and gums, and intestinal linings. Present the photos to the class, and ask students to determine what all the places and things have in common. After discussing possibilities, explain that they all house bacteria. Emphasize that bacteria have evolved to survive in drastically different environments.

OVERVIEW

CHAPTER 19 OVERVIEW OF DIVERSITY

REVIEW

- characteristics of prokaryotes and eukaryotes (Section 2-1)
- binary fission and mitosis (Section 6-2)
- crossing-over and meiosis (Section 6-3)
- introns (Section 8-3)
- homologous characters (Section 12-2)
- reproductive isolation (Section 12-3)
- six kingdoms of life (Section 13-1)
- symbiotic relationships (Section 13-2)

Diatom surface (SEM), a eukaryotic single-celled organism

Authors' Rationale

An understanding of diversity is necessary to fully comprehend the mechanics of biological systems. A survey of organic evolution is presented in a sequential manner, beginning with prokaryotes—the most primitive life-forms. The more "advanced" forms of life, the eukaryotes, are then discussed. The significance of the events that allowed for advances in evolution, such as sexual reproduction and multicellularity, are examined in detail, and their impacts are assessed. Finally, students are coached in how to categorize all this diversity with a discussion of biological classification.

19-1 Origin of Eukaryotes

L ife on Earth is over 3.6 billion years old. For the first 2 billion years, the biggest organisms were prokaryotes no larger than 6 μm thick. Evidence of different life-forms is found in 1.5-billion-year-old fossils of larger cells (some as large as 60 μm in diameter) that appear to have had internal membranes and small, membrane-bound structures. These larger organisms mark one of the most important events in the evolution of life—the appearance of the eukaryotic cell. Highly adaptable, eukaryotes evolved rapidly, populating the Earth with a diversity of larger, more complex organisms such as yourself.

Section Objectives

- Describe the characteristics of the two kingdoms of bacteria.
- Identify the characteristics of *Pelomyxa palustris*.
- Discuss the evolution of mitochondria and of mitosis.

Prokaryotes: The Oldest Organisms

Prokaryotes are the planet's most abundant inhabitants; one teaspoon of soil can be home for over 1 billion bacteria. Although microscopic and structurally simple, bacteria are the most diverse of all living organisms. They are found in the widest possible range of habitats, playing important roles in almost all of them. The reason that bacteria are able to play such varied roles is their metabolic diversity. For example, some bacteria obtain energy from inorganic sources such as sulfur or ammonia; others can metabolize petroleum. Such metabolic diversity is what has enabled bacteria to adapt to environmental niches too harsh to support other forms of life.

Early in the history of life, two structurally and metabolically different groups of bacteria evolved—archaebacteria and eubacteria. Figure 19-1 illustrates their evolution from an ancestral prokaryote.

Even though both archaebacteria and eubacteria are prokaryotes, they are so different from each other that biologists assign them to two separate kingdoms—kingdom Archaebacteria and kingdom Eubacteria, which are the subjects of the *Highlights* features on pages 443 and 444. The characteristics that

Figure 19-1 An ancestral prokaryote gave rise to two different kinds of bacteria—archaebacteria and eubacteria. *Sulfolobus* is an archaebacterium that thrives in hot, sulfur springs. *E. coli* is a eubacterium that lives inside your intestines.

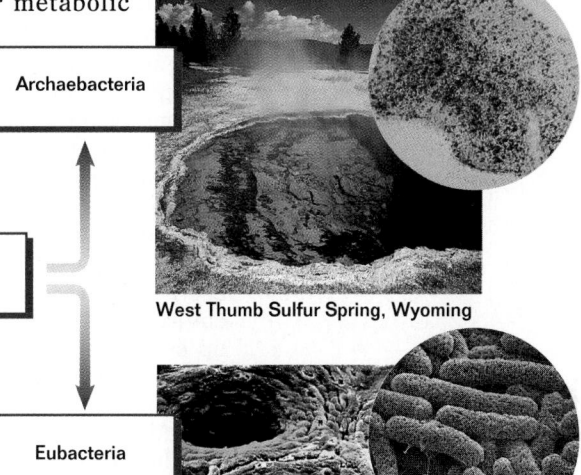

Archaebacteria

Ancestral prokaryote

Eubacteria

West Thumb Sulfur Spring, Wyoming

Lining of human intestine

VISUAL STRATEGY

Figure 19-2

Use this figure to review the differences between eubacteria and archaebacteria. Have students write a short paragraph that includes all the information summarized in this concept map.

Teaching Tips
Visualizing the Difference

Have students draw pictures to represent the differences between archaebacteria and eubacteria. Encourage them to be creative and to illustrate their work using characters or other helpful props. This can be a brainstorming activity that allows students to work in groups.

Methane Sources

Ask students to think of possible sources of methane gas in the atmosphere. After they read this page, they will know that archaebacteria called methanogens are one source. Tell them that human activities such as coal mining, rice farming, and landfills also contribute to the levels of methane gas in the atmosphere.

Application

Technology In biomining for copper, bacteria are used to break down ore and release the copper. This strategy is used with low quality ores; it improves the recovery rate and reduces the operational costs. The newest advances in biomining include extracting gold from ores originally considered worthless. Pilot programs in Africa show that the recovery rate from an ore can increase from the original 70 percent to a more efficient 95 percent. A current setback in this industry is the high temperatures produced when ore is oxidized. These temperatures can hinder bacterial growth or kill them completely. Currently, scientists are experimenting with a thermophilic archaebacterium, which thrives in temperatures of 100°C or higher.

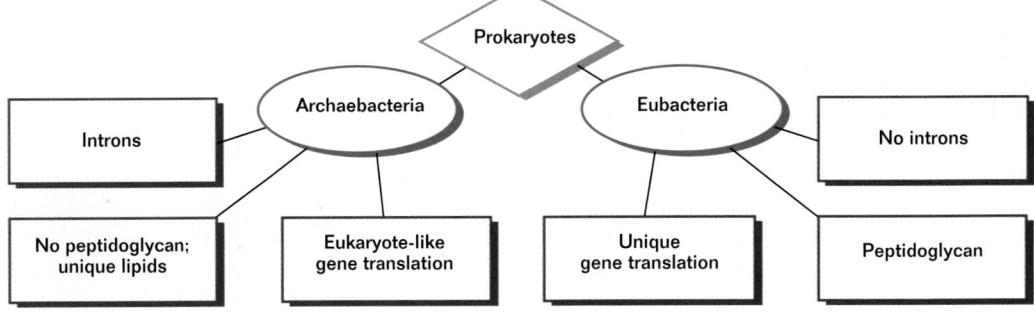

Figure 19-2 Archaebacteria and eubacteria are so different from each other that biologists assign them to separate kingdoms.

CONTENT LINK

The kingdom Eubacteria will be discussed in more detail in **Chapter 20.**

CAPSULE SUMMARY

Bacteria, the oldest and most abundant form of life, evolved into two branches: the archaebacteria and the eubacteria.

differentiate these two kingdoms of bacteria are summarized in Figure 19-2.

1. **Cell wall** All bacteria have a plasma membrane surrounded by a cell wall. The cell walls of eubacteria are made of **peptidoglycan** (pep tih doh GLY kan), a carbohydrate-protein compound that forms a strong mesh that strengthens the cell wall. Peptidoglycan is not found in the cell walls of archaebacteria.

2. **Plasma membrane** In all bacteria, the plasma membrane is made of a phospholipid bilayer. The plasma membranes of archaebacteria, however, contain unique lipids that are not found in any other organism.

3. **Gene translation machinery** Archaebacteria have a ribosomal protein and an RNA polymerase that are very similar to those found in eukaryotic cells. The gene translation machinery of eubacteria, however, is distinctly different from that of both archaebacteria and eukaryotes.

4. **Gene architecture** The genes of eubacteria are not interrupted by introns as are those of eukaryotes. However, some of the genes of archaebacteria have been found to have introns.

Most of the archaebacteria that survive today are methanogens (muh THAN uh jehnz), bacteria that use hydrogen gas (H_2) to reduce carbon dioxide (CO_2) to methane (CH_4). Methanogens are **obligate anaerobes,** organisms that are poisoned by oxygen. Swamps and marshes are perfect environments for them; the methane they produce bubbles up as "marsh gas." Methanogens also live in the digestive tracts of cows and other herbivores that consume a diet rich in cellulose, converting carbon dioxide to methane gas. An essential part of the sewage treatment process is encouraging the growth of these organisms in anaerobic digestion tanks, where sewage sludge is converted into methane gas.

Two other types of archaebacteria live in unusually harsh environments. The archaebacteria called extreme halophiles live in very salty places such as the Dead Sea (an inland body of salt water on the Israel-Jordan border) and the Great Salt Lake (a shallow saltwater lake in Utah). The thermoacidophiles grow in hot, acidic environments such as the sulfur springs of Yellowstone National Park.

Did You Know?

The classification of the single-celled organism *Epulopiscium fishelsoni*, which lives in the intestines of surgeonfish, puzzled scientists. The organism looks like a bacterium on the inside, yet it is over a million times larger than *E. coli*. Its massive size led scientists to think the organism was a protist. However, they now classify the organism as a bacterium, instead. A comparison of the organism's genetic material with that of other organisms shows that a gram-positive bacterium is the closest relative. If this organism is a bacterium, it discredits the rule that bacteria do not grow large in size.

The First Eukaryotes: A New Kind of Cell

One and a half billion years ago the first eukaryotes appeared. Recall that a eukaryotic cell has a complex system of internal membranes and DNA that is enclosed within a nucleus. What were the first eukaryotes like? Although biologists cannot be sure, an organism that may resemble a very early stage in the evolution of the eukaryotic cell is *Pelomyxa palustris*, shown in Figure 19-3.

Pelomyxa is a single-celled, nonphotosynthetic organism found on the bottom of freshwater ponds. Like some of the early eukaryotes, *Pelomyxa* is much larger than a prokaryotic cell and contains a complex system of internal membranes. It also resembles some of the fossils of early eukaryotes. However, *Pelomyxa* still resembles bacteria in two ways. First, it does not have mitochondria, the sites of cellular respiration in eukaryotic cells. However, it does contain two kinds of bacteria that may play the same role that mitochondria do in all other eukaryotes. Second, *Pelomyxa* does not undergo mitosis like eukaryotic cells. Instead, its nuclei divide in a manner similar to binary fission in bacteria: they divide by pinching apart into two new nuclei around which membranes form. Many of the fundamental characteristics of *Pelomyxa* resemble those of the archaebacteria far more than those of the eubacteria.

The Evolution of Mitochondria

If one of the main differences between prokaryotes and eukaryotes is the presence of mitochondria, how did these organelles arise in eukaryotic cells? To explain the origins of mitochondria, most biologists accept the theory of endosymbiosis. The **theory of endosymbiosis,** presented in the 1970s by Lynn Margulis, a biologist at the University of Massachusetts at Amherst, proposes that mitochondria are the descendants of symbiotic, aerobic (oxygen-requiring) eubacteria. The symbiotic bacteria may have first entered large cells (similar to *Pelomyxa*) as parasites or undigested prey and eventually taken up residence there, as illustrated in Figure 19-4. The eubacteria most similar to mitochondria are the nonsulfur purple bacteria, which are able to carry out the key metabolic process of oxidative respiration. Before they had acquired these bacteria, the larger host cells were unable to perform oxidative respiration, a process essential for living in an atmosphere that contained increasing amounts of oxygen.

Transitional Prokaryote

Resembles prokaryotes
- Lacks mitochondria
- Reproduces in a manner similar to binary fission

Resembles eukaryotes
- Larger than prokaryotic cells
- Contains complete system of internal membranes

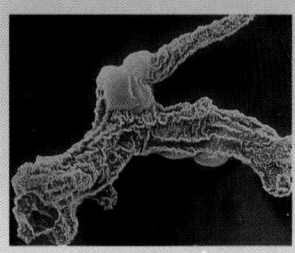

Figure 19-3 The protist *Pelomyxa palustris* is a eukaryotic cell with prokaryotic features. It may resemble some of the first eukaryotic cells that appeared on Earth.

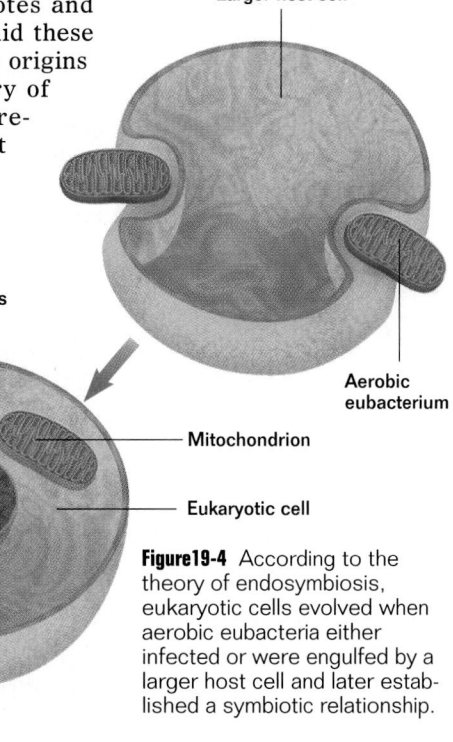

Larger host cell

Nucleus

Aerobic eubacterium

Mitochondrion

Eukaryotic cell

Figure 19-4 According to the theory of endosymbiosis, eukaryotic cells evolved when aerobic eubacteria either infected or were engulfed by a larger host cell and later established a symbiotic relationship.

Multicultural Perspective

Chuck Haines, Ph.D.

Dr. Chuck Haines is a microbiologist who investigated the mutagenesis of eukaryotes at the University of Kansas. He is also an ethnobotanist. Combining his love of science with his fascination of the medicinal plants of indigenous people, Dr. Haines wrote his doctoral dissertation on the medicinal and microbiological properties of cinnamon. He teaches biology, ethnobotany, and microbiology at Haskell Indian Nations University in Lawrence, Kansas— the only school in the United States that brings together Native American students from over 160 tribes. The grandson of a Cree Indian, Dr. Haines is the first in his family to obtain a college degree.

Figure 19-5

Use this micrograph to show students the plasma membrane infoldings in *Nitrobacter*. Have students compare the structure of the plasma membrane in the bacterium with a plasma membrane seen in a typical eukaryotic cell. Ask them to determine the significance of this difference. (*It has an increased surface area, which indicates one or more important reactions are occurring at this site.*) If this bacterium developed a symbiotic relationship with another cell and maintained its membrane through the evolutionary process, what could you derive about the membrane? (*An important process continued to occur at this site.*)

Teaching Tip
Write a Story

Tell students to imagine themselves as a mitochondrion or a chloroplast and to write an autobiography. They should explain how and why they entered the larger cell and why they decided to stay. Have them include an explanation about their role in the symbiotic relationship.

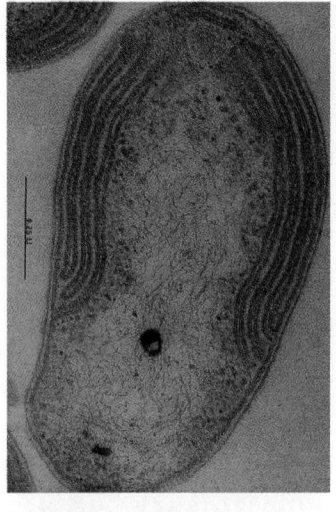

Figure 19-5 The theory of endosymbiosis is supported by the observation that the plasma membrane infoldings of aerobic eubacteria, such as *Nitrobacter*, function in cellular respiration and resemble the membranous layer found in mitochondria.

What observations support the idea that mitochondria are descended from bacteria? There are several key lines of evidence.

1. **Size and structure**   Mitochondria are sausage-shaped organelles 1–3 μm long, about the same size as most eubacteria. They are bound by two membranes. The smooth outer membrane is thought to be derived from the endoplasmic reticulum of the larger host cell. The inner membrane is folded into numerous layers, resembling the plasma membranes of aerobic eubacteria, shown in Figure 19-5. Embedded within this membrane are proteins that carry out oxidative respiration.

2. **Genome**   Mitochondria have DNA in the form of a circular, closed molecule similar to the chromosomes found in bacteria. This DNA molecule contains the genes encoding the essential proteins of oxidative respiration. However, during the billion and a half years in which mitochondria have existed within eukaryotic cells, most of their genes have been transferred to the chromosomes of the host cells.

3. **Gene translation machinery**   Genes within mitochondrial DNA are expressed within mitochondria. Mitochondrial ribosomes resemble bacterial ribosomes in size and structure.

4. **Reproduction**   Mitochondria, like bacteria, reproduce by simple fission. However, genes in the cell nucleus direct mitochondrial reproduction.

Other Examples of Endosymbiosis

In addition to mitochondria, many eukaryotic cells contain other organelles thought to be descended from bacteria. Plants and photosynthetic protists like algae contain chloroplasts. According to the theory of endosymbiosis, chloroplasts in the cells of plants and algae are descendants of photosynthetic prokaryotes that were incorporated into a larger host cell. The chloroplast is bound by two membranes that resemble those of mitochondria and were apparently derived in the same fashion. As in mitochondria, the gene translation machinery of the chloroplast closely resembles that of bacteria.

While all mitochondria almost certainly arose from a single symbiotic event, it has only recently become clear that chloroplasts also have a single origin. Three biochemically distinct classes of chloroplasts exist, each resembling a different bacterial ancestor. Red algae have chloroplasts containing pigments similar to those of cyanobacteria. The chloroplasts of plants and green algae, however, contain pigments that more closely resemble those of the photosynthetic bacterium *Prochloron*. Brown algae contain pigments resembling yet a third group of bacteria. This diversity of chloroplasts suggested that eukaryotic cells acquired chloroplasts by symbiosis at least three different times. However, recent comparisons of chloroplast DNA sequences indicate that there was a single origin of chloroplasts, followed by very different evolutionary histories. In each of the three main lines, different genes became relocated to the nucleus, lost, or modified. Chloroplasts seem to have had a rich evolutionary history after their entry into eukaryotic cells.

Centrioles, organelles associated with the assembly of microtubules, also appear to have had an endosymbiotic origin. They resemble spiral-shaped bacteria (spirochetes) and contain bacteria-like DNA involved in the production of their structural proteins. The body of evidence supporting endosymbiotic spirochetes as the evolutionary forerunner of centrioles is still incomplete and is an area of active research. ▢

The Origin of Mitosis Is Unknown

Another characteristic that distinguishes eukaryotes from prokaryotes is mitosis, the process of nuclear division. How did this intricate mechanism in a eukaryotic cell arise? Mitosis did not evolve all at once. There are traces of very different, and possibly intermediate, mechanisms in some of the eukaryotes surviving today. In fungi and some groups of protists, for example, the nuclear envelope does not disintegrate. Only after the chromosomes have been replicated and sorted is the nuclear envelope constricted to form two new nuclei. This is followed by cytokinesis. This separate nuclear division phase of mitosis does not occur in most protists or in plants or animals. Biologists are not certain whether it represents an intermediate step on the evolutionary journey to the form of mitosis that is characteristic of most eukaryotes today, or whether it is simply a different way of preserving the number of chromosomes in a cell. Scientists have yet to find fossils in which the interiors of dividing cells can be seen well enough to trace the history of mitosis.

Section Review

1. *List four differences between* Sulfolobus *and* E. coli. *Do you agree or disagree with biologists who place the two bacteria in separate kingdoms? Explain.*

2. *Why is* Pelomyxa palustris *thought to resemble the first eukaryotes?*

3. *Explain the theory of endosymbiosis. What evidence would you present to support this theory?*

Critical Thinking

4. *How does mitosis in a mushroom differ from mitosis in your body? Does the form of mitosis in the mushroom have any evolutionary significance? Why or why not?*

▢ **CAPSULE SUMMARY**

The evolution of organelles such as mitochondria and chloroplasts is explained by the theory of endosymbiosis. This theory proposes that eukaryotic cells evolved when certain prokaryotes established symbiotic relationships within larger cells.

Teaching Tip
Another Look at Mitosis

Review the process of mitosis as explained in Chapter 6. Explain to students that these examples illustrate mitosis in a plant cell and in an animal cell. Have them determine the stage that differs in mitosis in a fungus cell. *(prophase because the nuclear envelope does not break down)*

Answers to Section Review

1. See Figure 19-1 on page 417 and Figure 19-2 on page 418. Answers will vary as to whether students think the bacteria should be placed into different kingdoms.

2. Although it has similarities to prokaryotes, *Pelomyxa palustris* is considered to be similar to the first eukaryotes because of its large size, complex internal membrane system, and resemblance to fossils of earlier eukaryotes.

3. See pages 419 and 420.

4. In a mushroom cell, mitosis occurs within the nuclear envelope, which eventually pinches in two. Afterward, cytokinesis occurs. In a human cell, the nuclear envelope disappears in the early phases of mitosis. It reappears during cytokinesis. It is not known whether mushroom cell mitosis is an intermediate step to human cell mitosis, or just another way of preserving chromosome numbers.

Opening Question

Inform students that most animals reproduce sexually (a process that requires two parents) but that bacteria reproduce asexually (a process that requires only one individual). Ask students to predict how the offspring of each kind of reproduction would compare to the parents. Have them list possible benefits of each kind of reproduction, and discuss the points in class.

Demonstration

Modes of Reproduction

Display pictures of different plants, animals, protists, and bacteria. Ask students to suggest a means of reproduction for each. Ask them to formulate a general rule about eukaryotes and prokaryotes in terms of reproduction.

👁 VISUAL STRATEGY

Figure 19-6

Use this photograph to help explain budding, the form of asexual reproduction used by hydra. Ask students why budding is considered asexual reproduction? (*because only one parent is involved*) Why is it beneficial for a hydra to reproduce asexually? (*It can produce a large number of well-adapted offspring quickly.*)

Section Objectives

- Differentiate between asexual and sexual reproduction. Explain the benefits of sexual reproduction.
- Discuss a current theory about the evolution of sexual reproduction.
- Describe three major sexual life cycles found in eukaryotes.

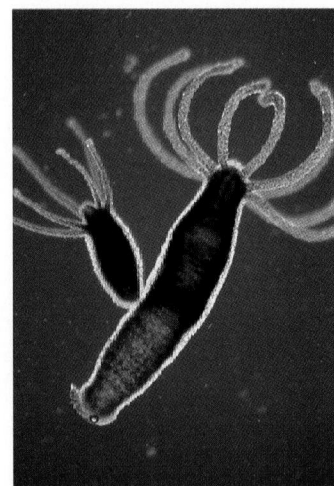

Figure 19-6 Some organisms, such as this *Hydra,* are able to reproduce by budding. In this form of asexual reproduction, a new individual grows out of the body of the original.

19-2 Evolution of Sexual Reproduction

O f all the differences between prokaryotes and eukaryotes, the most essential is found in their mode of reproduction. Prokaryotes reproduce by binary fission, a type of asexual reproduction involving one parent that produces genetically identical offspring without forming gametes. Eukaryotes are able to undergo sexual reproduction. In sexual reproduction two parents contribute genetic material that recombines during the formation of offspring.

Asexual Reproduction Is Primitive and Prevalent

Asexual reproduction is the simplest and undoubtedly the most primitive method of reproduction. It is advantageous for organisms in a stable environment, enabling them to produce many well-adapted offspring in a short period of time. Also, an organism that reproduces asexually does not need to expend energy producing gametes or finding a mate.

The oldest groups of eukaryotes are all protists that are able to reproduce asexually. Some protists are also able to reproduce sexually. For example, the flagellates, a group of single-celled eukaryotic organisms that propel themselves with whiplike flagella, are predominantly asexual. Other protists, such as the green algae, exhibit a true sexual cycle, but only occasionally. Indeed, many protists reproduce asexually almost all the time. The fusion of two haploid cells to create a diploid zygote, the essential act of sexual reproduction, usually occurs in the presence of stressful conditions in the environment, such as a shortage of nutrients.

In some forms of asexual reproduction a single parent fragments or forms buds to give rise to offspring. These methods are found in multicellular eukaryotes. For example, consider a sponge, a representative of a group of animals living in the sea. Frequently, a sponge reproduces by fragmentation, the breaking of the body into several small pieces. Each small portion is able to grow into a whole new individual. On the other hand, some organisms, like the tiny, cylindrical *Hydra* shown in Figure 19-6, reproduce by budding. In budding, a small part of the parent's body grows into a new individual. The bud may break from the parent and become an independent organism, or it may remain attached to the parent and eventually give rise to a colony composed of many other individuals.

Effective Reading

Annotating Text

Photocopy pages 422 and 423 for students and have them annotate their reading as follows. First number the paragraphs on both pages beginning with #1 and ending with #6. Next note important facts in the margins using numbers, fast sketches, or words, which will serve as reminders about the material. Once students have completed the reading, ask the following questions:

1. How does a hydra reproduce asexually? (*budding*)

2. What causes most protists to reproduce sexually? (*stressful environmental conditions*)

3. One theory explains that meiosis originally occurred to _____. (*repair DNA*)

4. How do the earliest eukaryotes reproduce? (*sexually and asexually*)

Sexual Reproduction Has Advantages

One of the most important evolutionary innovations of eukaryotes was the ability to reproduce sexually. Sexual reproduction provides a powerful means of shuffling genetic material, quickly generating different combinations of genes among individuals. The rapid generation of genetic diversity produced by sexual reproduction has been a principal factor in the evolutionary success of eukaryotes. Genetic diversity is the raw material for evolution. In many cases, the pace of evolution appears to be related to the level of genetic diversity available for selection to act upon: the greater the genetic diversity, the more rapid the evolutionary pace. The processes that occur during sexual reproduction and give rise to genetic diversity are shown in Figure 19-7.

However, as you learned in Chapter 12, evolution is the result of changes that occur at the level of *individual* survival and reproduction rather than at a species level. It is not immediately obvious what advantage is gained by the offspring of an individual that reproduces sexually. The segregation of chromosomes that occurs in meiosis tends to disrupt advantageous combinations of genes more often than it assembles new, better-adapted ones. As a result, some of the offspring produced by sexual reproduction are not as well adapted as their parents. Because all of the offspring could maintain a parent's successful gene combinations if the parent reproduced asexually, the appearance of sexual reproduction among eukaryotes raises a question: What are the benefits that promoted its evolution?

How Sexual Reproduction Evolved

In attempting to answer this question, biologists have looked more carefully at the origins of sexual reproduction among the protists. Many species of protists form a diploid cell in response to stress in the environment. Why? Biologists think this occurs because only in a diploid cell can certain kinds of chromosome damage be effectively repaired, particularly breaks in both strands of DNA. As organisms became larger and lived longer, it must have become increasingly important for them to be able to repair such damage. The early stages of meiosis, in which the two copies of each chromosome line up and pair with each other, may have originally been a mechanism for repairing damage to DNA. Perhaps the undamaged version of the chromosome was used as a template to fix the damaged one. Indeed, the molecular events that occur during crossing-over in meiosis involve many of the enzymes that repair DNA damage. In yeasts, mutations that inactivate the system that repairs double-stranded breaks in chromosomes

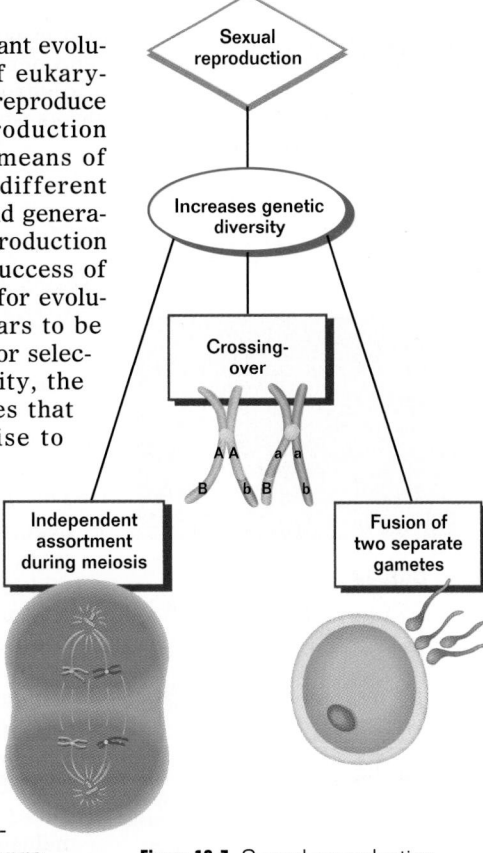

Figure 19-7 Sexual reproduction represents an advance in the ability of an organism to generate genetic variability.

👁 *VISUAL STRATEGY*

Figure 19-7
Use this figure to remind students that sexual reproduction is advantageous because it increases genetic diversity. Ask students to list three ways this is accomplished. *(independent assortment during meiosis, crossing-over, fusion of two separate gametes)*

VISUAL STRATEGY

Figure 19-8
Use this figure to discuss the steps involved in a life cycle characterized by zygotic meiosis. Be sure students know the difference between a diploid cell and a haploid cell. Ask students to point out the diploid stage in the life cycle. *(the zygote)* Ask them what the result of meiosis is. *(haploid individuals)* Have students explain the term *zygotic meiosis*. *(The zygote undergoes meiosis.)*

VISUAL STRATEGY

Figure 19-9
Use this figure to discuss the steps involved in a life cycle characterized by gametic meiosis. Ask students to point out the diploid stage in the life cycle. *(the zygote, which grows into an individual)* Ask students what the result of meiosis is. *(haploid gametes)* Have students explain the term *gametic meiosis*. *(Meiosis produces gametes.)*

□ CAPSULE SUMMARY

Sexual reproduction may have evolved as a mechanism to repair damaged DNA.

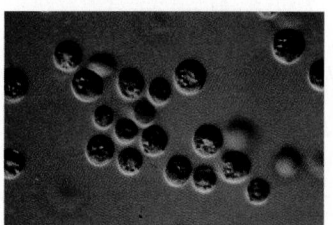

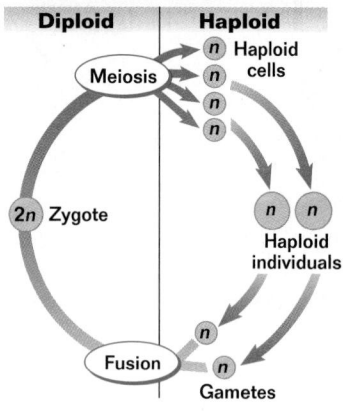

Figure 19-8 Some organisms, such as the green alga *Chlamydomonas,* have a life cycle characterized by zygotic meiosis. In zygotic meiosis a diploid zygote undergoes meiosis and develops into haploid individuals.

Figure 19-9 Some organisms, such as humans, have a life cycle characterized by gametic meiosis. In gametic meiosis a diploid reproductive cell undergoes meiosis and gives rise to haploid gametes.

also prevent crossing-over. Thus it seems likely that sexual reproduction and the close association between pairs of meiotic chromosomes first evolved as mechanisms to repair chromosomal damage. □

Eukaryotes Have Life Cycles

Many protists are haploid all their lives. On the other hand, animals and most plants are diploid for most of their life cycle. When organisms are diploid, their cells contain two sets of chromosomes, one set derived from the male parent and one set derived from the female parent. The production of haploid gametes by meiosis, followed by the union of two gametes in sexual reproduction, is a **sexual life cycle.** Eukaryotes are characterized by three types of sexual life cycles.

1. **Zygotic meiosis**   In the simplest of sexual life cycles, the zygote is the only diploid cell. This sort of life cycle is said to exhibit **zygotic meiosis** because the zygote undergoes meiosis immediately after it is formed. Thus, haploid cells occupy the major portion of the life cycle. Zygotic meiosis is found in many algae, such as the unicellular *Chlamydomonas* shown in Figure 19-8.

2. **Gametic meiosis**   In almost all kinds of animals, including humans, the gametes are the only haploid cells; all of the other cells of the individuals in the life cycle are diploid. This sort of sexual life cycle is said to exhibit **gametic meiosis** because meiosis produces gametes. In this type of life cycle, illustrated in Figure 19-9, the diploid zygote occupies the major portion of the life cycle.

3. **Sporic meiosis**   Plants have a life cycle that regularly alternates between a haploid phase and a diploid phase.

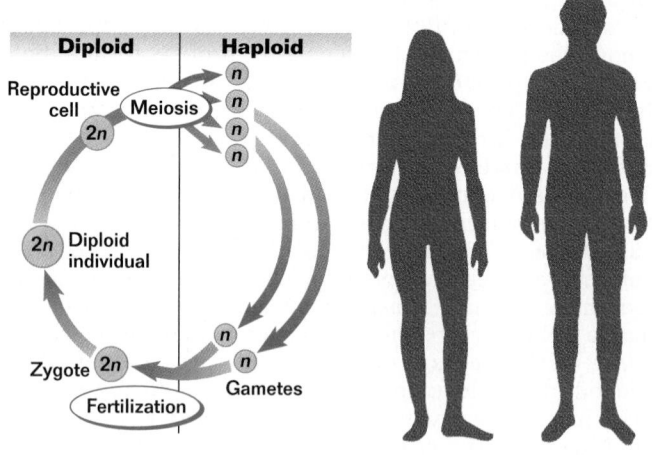

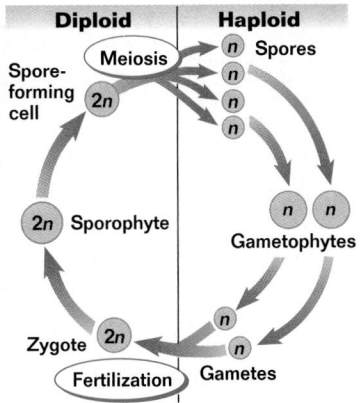

Diploid | **Haploid**

Meiosis

Spore-forming cell — 2n

n Spores
n
n
n

2n Sporophyte

n n Gametophytes

Zygote — 2n

Fertilization

n
n Gametes

Figure 19-10 Some organisms, such as tree ferns, have a life cycle characterized by sporic meiosis. In sporic meiosis a diploid spore-forming cell undergoes meiosis and gives rise to haploid spores.

The diploid phase, called a **sporophyte** *(SPOHR uh fyt)*, produces spores. The spores give rise to the haploid phase, called a **gametophyte** *(guh MEET uh fyt)*, which produces gametes that fuse to give rise to the diploid phase. Plants are said to exhibit **sporic meiosis.** In this form of meiosis, illustrated in Figure 19-10, certain diploid cells of the sporophyte undergo meiosis to form haploid spores. ◻

❏ **CAPSULE SUMMARY**

Eukaryotic organisms can have one of three kinds of sexual life cycles—zygotic meiosis, gametic meiosis, or sporic meiosis—depending on when meiosis occurs and what kind of cell it produces.

Section Review

1. *What advantage does an asexually reproducing organism have over one that reproduces sexually?*

2. *How have eukaryotes benefited from sexual reproduction?*

3. *What kind of sexual life cycle is found in the green alga Chlamydomonas? When does meiosis take place in these organisms?*

4. *What kind of organisms exhibit gametic meiosis? What is the function of meiosis in organisms that have gametic meiosis?*

5. *What kind of organisms display sporic meiosis? What are the names of each of the phases in this kind of life cycle? What kind of cells are produced during each phase?*

Critical Thinking

6. *What is the mechanism in ancestral cells that may have served as the basis for the evolution of sexual reproduction? In what form is this mechanism still present in cells involved in sexual reproduction?*

Answers to Section Review

1. An asexually reproducing organism can produce many offspring in a short period of time without utilizing energy to find a mate.

2. Eukaryotes have benefited from sexual reproduction by having increased genetic diversity.

3. *Chlamydomonas* exhibits zygotic meiosis. Meiosis takes place after the zygote is formed.

4. Most animals exhibit gametic meiosis. Meiosis functions to produce haploid eggs and sperm, which fuse to form a diploid individual.

5. Plants exhibit sporic meiosis. The two phases are the diploid sporophyte and the haploid gametophyte. The sporophyte produces spores, and the gametophyte produces gametes.

6. One theory is that ancestral cells underwent sexual reproduction to repair damaged DNA. The mechanism of crossing-over allows a means for genetic information to be exchanged and repaired between chromosomes.

Vocabulary Preview

multicellular organism

tissue

organ

colonial organism

aggregation

cell specialization

intercellular coordination

Opening Question

Ask students to list the advantages of being a multicellular organism and the advantages of being a unicellular organism. Discuss the advantages of each in class.

Demonstration

Ask five students to volunteer for a class demonstration. Place four students in a group, and have them hold on to a meter stick. Tell the class that the students represent two different protists. The single student is a unicellular protist, and the group of four students is a multicellular protist. Ask students what the group of students would have to do to represent a multicellular organism. Let them list the benefits of both types of organism on the chalkboard.

◉ VISUAL STRATEGY

Figure 19-11

Use these photographs to illustrate the similarities and differences between unicellular and multicellular protists. Ask students to list the similarities and differences between the two organisms. Review the lists in class. Students may be surprised that there are more differences than similarities. Inform them that protists are a collection of extremely diverse organisms.

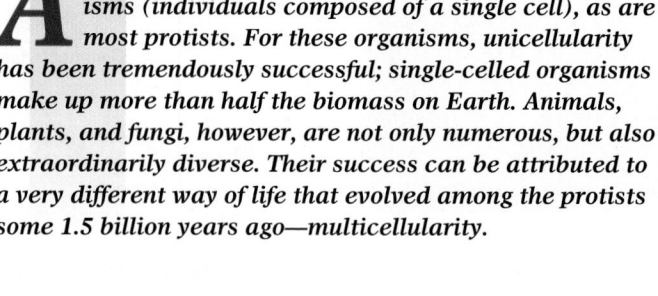

19-3 Evolution of Multicellularity

*A*rchaebacteria and eubacteria are unicellular organisms (individuals composed of a single cell), as are most protists. For these organisms, unicellularity has been tremendously successful; single-celled organisms make up more than half the biomass on Earth. Animals, plants, and fungi, however, are not only numerous, but also extraordinarily diverse. Their success can be attributed to a very different way of life that evolved among the protists some 1.5 billion years ago—multicellularity.

Section Objectives

■ Explain the advantages of multicellularity.

■ Compare and contrast colonial organisms and aggregates, and give examples of each.

■ Describe the protist ancestors of plants, animals, and fungi.

■ Explain two characteristics that distinguish complex multicellular organisms.

Multicellularity Allows for Specialization

An organism that is unicellular, such as *Vorticella* shown in Figure 19-11, has size limits. It can only grow to a certain size before encountering serious surface-area-to-volume problems; as cell size increases, there is soon too little surface area to meet the needs of the cell volume. The evolution of multicellularity solved this dilemma. A **multicellular organism** is one composed of many cells that are permanently associated with one another and that integrate their activities, such as the green algae shown in Figure 19-11. The main advantage of multicellularity is that it allows for specialization. Distinct types of cells form **tissues,** groups of cells with a common structure and function. Different tissues are organized into **organs,** specialized structures with specific functions. With such functional "division of labor" within its body, a multicellular organism can have cells devoted specifically to protecting the body, others devoted to moving it about, and still others devoted to seeking mates and pursuing prey. In short, multicellularity enables an organism to carry on a host of activities of a complexity that would have been impossible

Figure 19-11 Multicellularity evolved among the protists. Unicellular protists such as *Vorticella, left,* are more complex than prokaryotes, but multicellular protists such as green algae, *right,* have specialized cells with distinct functions.

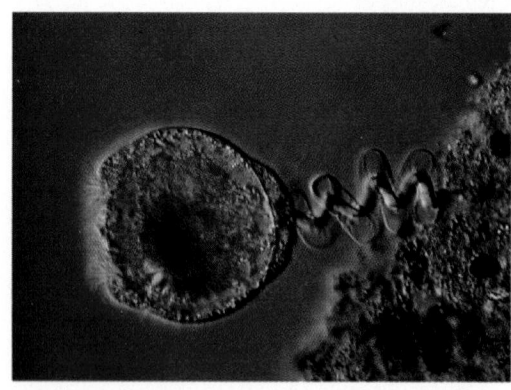

for its unicellular ancestors. In just this way, a small city of 50,000 people is vastly more complex and organized than a crowd of 50,000 people in a football stadium; each inhabitant plays a role that keeps the city functioning, rather than just being another body in a mob. ◻

Colonies and Aggregates Are Not Truly Multicellular
....

True multicellularity, in which the activities of the individual cells are coordinated and the cells themselves are in contact, occurs only in eukaryotes; it is one of their major characteristics. Occasionally, the cell walls of bacteria adhere to one another, and bacterial cells may even be held together within a common sheath. In fact, some bacteria, such as the cyanobacteria in Figure 19-12, form filaments, sheets, or three-dimensional formations of cells. However, these formations cannot be considered truly multicellular because little integration of cell activities occurs. Such bacteria may properly be considered colonial (living together). A **colonial organism** is a collection of cells that are permanently associated, but in which little or no integration of cell activities occurs. Many protists form colonies, consisting of many cells with little differentiation or integration.

An **aggregation** is a more temporary collection of cells that come together for a period of time and then separate. For example, a plasmodial slime mold (a member of the kingdom Protista) is a unicellular organism that spends most of its life moving about and feeding as a single-celled amoeba. When starved, however, these cells aggregate into a large colony, shown in Figure 19-13. This weblike mass produces spores, which can travel to distant locations where there may be more food.

In some protists, the distinction between being a colonial organism and one that is multicellular is blurry. For example,

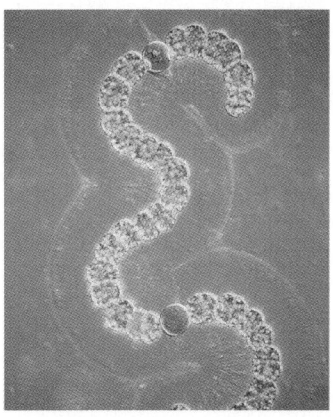

Figure 19-12 Although the cyanobacterium *Anabaena* can appear as a filament made of many cells, it is not considered a multicellular organism.

Figure 19-13 Plasmodial slime molds are brightly colored protists that feed as a large mass, which may grow to a diameter of several centimeters. These masses are considered to be aggregations, not multicellular.

VISUAL STRATEGY

Figure 19-15
Use this figure to illustrate the differences among the three types of multicellular algae. Ask students which type of algae is the possible ancestor of plants. *(green algae)*

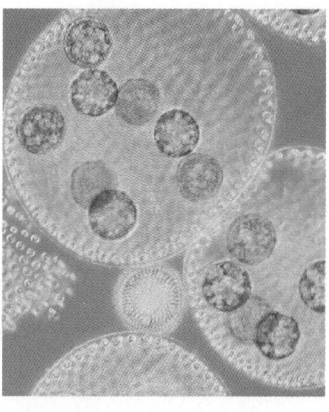

Figure 19-14 *Volvox* is a colonial organism shaped like a hollow ball. Its wall is composed of hundreds or thousands of flagellated cells embedded in a jellylike layer.

☐ CAPSULE SUMMARY

Colonial organisms and aggregations are not considered to be multicellular, but rather a permanent or temporary collection of cells.

Figure 19-15 Red algae, brown algae, and green algae are multicellular protists that can grow to be many meters in length.

the green alga *Volvox*, shown in Figure 19-14, is a colonial protist composed of hundreds or thousands of individual cells aggregated into a hollow ball. *Volvox* moves by a coordinated beating of the flagella of individual cells—like scores of rowers all pulling their oars in concert. A few cells near the rear of the moving colony are reproductive cells, but most are relatively undifferentiated. Some species have cytoplasmic connections between the cells that might permit coordination of some activities. Does this make *Volvox* a multicellular organism? Is coordination of flagella evidence of integration? To return to the analogy comparing the inhabitants of a small city with a crowd in a football stadium, imagine that the stadium crowd is carrying out a "wave," in which sections of the crowd stand and raise their arms one after another like falling dominoes. Does this indicate that the football watchers are truly integrated? No. In this sense, *Volvox* is more properly considered a colonial organism, although the distinction is a difficult one. ☐

Multicellularity Appears in Four Kingdoms

Multicellularity has evolved many times among the protists. Three groups of photosynthetic protists that contain species with independently attained true multicellularity are red algae (phylum Rhodophyta), brown algae (phylum Phaeophyta), and green algae (phylum Chlorophyta), all shown in Figure 19-15. In these algae, individuals are composed of many cells that interact with one another and coordinate their activities, and individual cell specialization is fairly complex. Multicellularity does not imply small size or limited adaptability. Some marine algae grow to be enormous. An individual kelp, a type of brown algae, may grow to dozens of meters in length—some taller than a redwood tree! Red algae grow at great depths in the sea, far below where kelp or other algae are found. Nevertheless, not all algae are multicellular. Green algae, for example, include many kinds of multicellular organisms, but an even larger number of unicellular ones.

Plants, animals, and fungi are composed of many kinds of highly specialized cells that coordinate their activities. Protists were the ancestors of the three complex multicellular kingdoms whose representatives are shown in Figure 19-16.

Plantae Plants, such as the cactus shown in Figure 19-16, are descendants of multicellular green algae. Most green algae are aquatic and have simpler reproductive and vegetative structures than plants.

Animalia Animals, such as the meerkat in Figure 19-16, arose from a unicellular protist ancestor. Formerly, several phyla of heterotrophic protists were grouped with the animals, but today, scientists include only multicellular organisms in this kingdom. The simplest and seemingly most primitive animals today, the sponges, seem to have evolved from a kind of flagellated protist.

Fungi Fungi, such as the mushrooms shown in Figure 19-16, also arose from a different heterotrophic unicellular protist than the one that gave rise to animals. Certain protists, including slime molds and water molds, have been considered to be fungi, although they are in fact protists.

Figure 19-16 The three kingdoms of multicellular eukaryotes arose from protists. Cactuses, *above left,* and all other plants are the descendants of a multicellular green alga. Mushrooms, *above center,* and other fungi evolved from a unicellular protist. A different type of unicellular protist gave rise to animals, such as meerkats, *above right.*

Multicellularity Requires Cell Specialization and Coordination

The main difference between the more complex multicellular organisms and multicellular protists like marine algae is **cell specialization.** In an animal, plant, or fungus, the body of an individual has different kinds of cells with very different structures and functions, which you can learn about in *Highlights: Six Kingdoms of Life* on pages 446–448. Even the simple sponges, for example, which lack specialized tissues and organs, have a variety of different kinds of cells. The outer body of the sponge is lined with flattened cells that function as a protective coat of skin. The inner layer, or body cavity, is lined by special collar cells that catch food particles in the water and move water through the sponge by beating the flagella back and forth. Between the

❏ CAPSULE SUMMARY

Cell specialization and coordination enable complex multicellular organisms to have different cells and different functions.

two layers of cells, several kinds of amoeboid cells circulate, some of which produce a tough protein skeleton of spongin (familiar as the bathtub sponge). Each of these kinds of cells has a different structure and a different job to do.

Multicellularity implies something very important about the genes of an individual: *different cells are using different genes!* For example, the genes encoding the flagellar protein of collar cells are not being expressed in amoeboid cells—they are "turned off," which is why the amoeboid cells do not have flagella. The process whereby a single cell (in humans a fertilized egg) becomes a multicellular individual with many different kinds of cells is called **differentiation.** The cell specialization that is the hallmark of complex multicellular life is the result of differential development—different cells developing in different ways by activating different genes.

A second characteristic of complex multicellular organisms is **intercellular coordination,** the adjustment of a cell's activity in response to what other cells are doing. Consider a fir tree. The cells growing at the very top of the tree secrete a chemical called auxin, which diffuses downward through the plant and suppresses the growth of side branches. Farther down from the top of the tree, less auxin is present so the side branches grow longer. The triangular appearance of the fir tree is thus a direct consequence of the chemical signal passed down from the topmost cells. Auxin is an example of a hormone, a chemical signal used to communicate between the cells of complex multicellular organisms. The cells of all complex multicellular organisms communicate with one another with hormones. In some organisms, like sponges, there is relatively little coordination between the cells; in other organisms, like humans, almost every cell is under complex coordination. ❏

Section Review

1. *What advantage does a multicellular organism have over one that is unicellular?*

2. *Describe* Anabaena, *plasmodial slime molds, and* Volvox. *Is each considered to be a multicellular organism? Explain.*

3. *What three kingdoms arose from protists? Did they all arise from the same kind of protist? Explain.*

Critical Thinking

4. *In terms of multicellularity, why can a fir tree grow on land, whereas a sponge must live in the sea?*

5. *Name three species that are complex multicellular organisms. What can you infer about the genes of these organisms?*

19-4 Classification of Living Things

O ur world is populated by at least 10 million different kinds of organisms. In order to study and discuss these organisms, scientists give each kind they discover a name. It would be impossible to remember the name of every kind of organism, just as it would be impossible for a postal worker to sort mail bearing only the addressee's first name. To make the process easier, the postal worker sorts mail first by zip code, then by street name and house number. By categorizing individuals according to their addresses, the post office can accurately locate a specific individual. Similarly, to help identify an organism, a biologist groups specific kinds into categories, which are assigned to larger and more inclusive categories.

Section Objectives

- Describe Linnaeus's role in developing the modern system of naming organisms.
- Explain the scientific system for naming a species.
- Distinguish scientific naming from biological classification.
- Define the term species.
- Describe how classification reflects evolutionary history.

Linnaeus Assigned Organisms Two-Word Names

Humans have been naming and describing organisms since the beginning of language. Over 2,000 years ago, the Greek philosopher and naturalist Aristotle (384–322 B.C.) grouped plants and animals on the basis of their structural similarities. This simple classification system was expanded by later Greeks and Romans, who grouped plants and animals into basic categories such as oaks, dogs, and horses. Eventually each unit of classification was called a **genus** (*JEE nuhs*), the Latin word for "group." The plural of genus is genera (*JEHN uhr uh*). Starting in the Middle Ages, a genus was given a name in Latin, the language of scholars. Thus, cats were assigned to the genus *Felis*, dogs to *Canis*, and horses to *Equus*. The science of naming and classifying organisms is called **taxonomy.**

Until the mid-1700s, biologists referred to a particular species by adding a series of descriptive terms to the name of the genus. These phrases, sometimes consisting of 12 or more Latin words, were called polynomials (from *poly*, meaning "many," and *nomen*, meaning "name"). As you can see in Figure 19-17, polynomials became quite unwieldy and awkward. Polynomials were sometimes altered by various biologists so that a given organism rarely had a universal name. A simplified system for naming organisms came from the work of the Swedish biologist Carl Linnaeus (1707–1778), whose ambition was to catalog all the known kinds of organisms. In the 1750s he published several books that employed the well-established polynomial system. But as a kind of shorthand, Linnaeus included a two-word Latin name for each species.

Figure 19-17 The polynomial system of naming organisms was cumbersome. For example, under this system the European honeybee had a 12-part name: *Apis pubescens, thorace subgriseo, abdomine fusco, pedibus posticis glabis, untrinque margine ciliatus.*

Historical Note

Linnaeus Preceded Darwin

Carolus Linnaeus arranged organisms into groups based on similarities. He did not associate the similarities with evolution. His work was used a century later as argument supporting evolution.

Vocabulary Preview

genus
taxonomy
binomial nomenclature
scientific name
family
order
class
phylum
kingdom
division
biological species concept
hybrid
convergent evolution
analogous character
phylogeny
cladistics
derived trait
cladogram
out-group
in-group
evolutionary systematics

Opening Question

Ask students if they have ever met someone or heard of someone who had their same name. Have them consider what kinds of problems could arise from this situation. Ask students what other information they would use to identify themselves from an individual with the same name.

Demonstration

Naming Organisms

Show students two plants that have obvious differences, such as an ivy and a geranium in bloom. Ask the students to devise a single-word name that is descriptive of these two plants. Next have them distinguish the plants from each other by adding a second word to the name for each plant. After students share their names, point out that they did not all use the same names.

For example, the European honeybee became *Apis mellifera.* Linnaeus's system for naming organisms is called **binomial nomenclature** because each name is composed of two words.

Scientific Names Are Universal

For nearly 250 years, Linnaeus's binomial nomenclature has remained the standard way of identifying a species. The unique two-word name for a species is its **scientific name.** The first word in a scientific name is the genus to which the organism belongs. An organism is assigned to a genus based on its major characteristics. For example, oak trees, all of which produce acorns, are placed in the genus *Quercus.* The second word in a scientific name identifies one particular kind of organism within the genus. Table 19-1 lists and describes two species of oak in the genus *Quercus.*

As you can see, when you write a scientific name, the first letter of the genus name is always capitalized and the first letter of the second word is always lowercased. As with all foreign words, scientific names are italicized or underlined.

Table 19-1 Two Species of Oaks

	Common Name	Genus	Scientific Name	Traits
	Red oak	*Quercus*	*Quercus rubra*	Lobed leaves; produces acorns approximately 25 mm (1 in.) long
	Willow oak	*Quercus*	*Quercus phellos*	Unlobed leaves; produces acorns approximately 15 mm (0.6 in.) long

Figure 19-18 Common names often differ from place to place. A "robin" in Great Britain is *Erithacus rubicula, left,* but in North America it is *Turdus migratorius, right.*

The scientific name of an organism is the same throughout the world, providing a standard for communication among biologists, regardless of their native language. This system of classification is a great improvement over the use of common names, which often vary from place to place. For example, the birds shown in Figure 19-18 are both commonly called robins, one in Great Britain, the other in North America. However, they belong to different species, genera, and families. In order to avoid ambiguity in published research, scientists refer to them by their scientific names, *Erithacus rubicula* and *Turdus migratorius.*

The name given to a species must conform to rules formulated by an international commission of scientists. All scientific names must be Latin words or terms constructed according to the rules of Latin grammar. Two different organisms cannot be assigned the same scientific name. Since all the members of a genus will share their genus name, the second word in the name of each member of that genus must be different. For example, only one species of the genus *Homo* can be given the name *sapiens.* Organisms in different genera cannot have the same genus name but can share the second word of their scientific names. For example, the green anole lizard, *Anolis carolinensis,* and the Carolina chickadee, *Parus carolinensis,* share the name *carolinensis,* meaning they were described from specimens collected in North Carolina and South Carolina. ❑

CAPSULE SUMMARY

To ensure accurate communication about organisms, biologists use binomial nomenclature to name species. Binomial nomenclature is a system of scientific names developed by the eighteenth-century Swedish biologist Carl Linnaeus.

All Organisms Are Classified in a Hierarchy

Linnaeus worked out a fairly extensive system of classification for both plants and animals, emphasizing an organism's form and structure as the basis for arranging specimens in a collection. Subsequently, the genera and species that he described were organized into a hierarchical system of groups that increase in inclusiveness. The different groups into which organisms are classified have

Ecologists have realized that in order to protect plant and animal species, they would have to devise a way to assess the impact of maintaining species diversity. In March of 1994, fifty scientists formed the Scientific Committee on Problems of the Environment (SCOPE). SCOPE scientists reviewed evidence from observations of many ecosystems in an attempt to find out "how species richness affects the workings of an ecosystem." Researchers could find no reliable way to quantify nutrient values in an ecosystem, but they did come to one important realization: greater diversity does make a difference when it comes to increased production of plant biomass. If total production of biomass represents ecosystem success, scientists were in a position to set up experiments to prove it.

Ecologists propose different views on the importance of individual species. Paul and Anne Ehrlich of Stanford University suggest that each species contributes to the overall strength of the ecosystem. Removing a species at a time will weaken the system until it can no longer support itself. Brian Walker, from the Australia Commonwealth Scientific and Industrial Research Organization, proposes alternatively that some species are "key" and others are superfluous, and the impact of species loss on the ecosystem will vary depending on which species in particular are lost. Experiments conducted by Shahid Naeem and John Lawton in London tend to support the Ehrlichs' hypothesis. These researchers set up several controlled artificial environments and showed that those environments that had more species in them under identical conditions produced more biomass than those communities which had fewer species in them. ☑

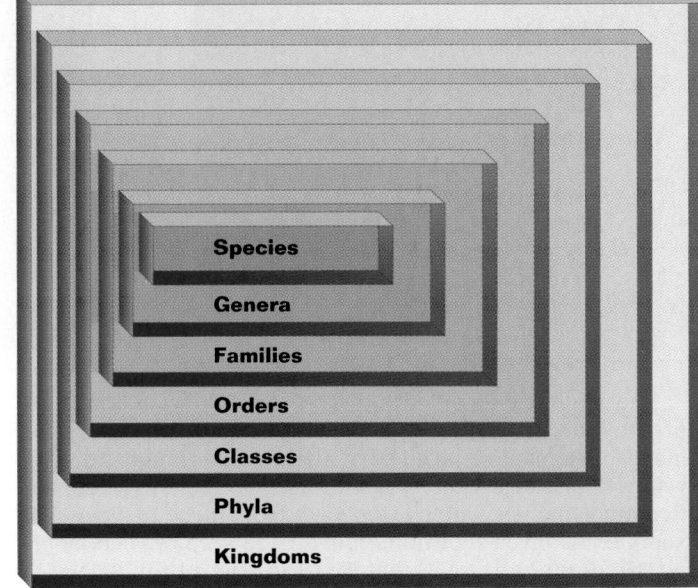

Figure 19-19 Each living thing is assigned to a series of groups that increase in inclusiveness, beginning with species (least inclusive) and ending with kingdom (most inclusive).

☐ **CAPSULE SUMMARY**

Scientists classify organisms into a hierarchical system of groups. Today, there are seven basic levels in the hierarchy: kingdom, phylum, class, order, family, genus, and species.

expanded since Linnaeus's time and now consist of seven levels, as illustrated in Figure 19-19. Genera with similar properties are clustered into a **family.** Similar families are combined into an **order.** Orders with common properties are united in a **class.** Classes with similar characteristics are assigned to a **phylum** *(FY luhm).* Finally, similar phyla are collected into a **kingdom.** The term **division** is an alternative term for phylum in the classification of bacteria, fungi, and plants. As you learned in Chapter 13, living things are grouped into six kingdoms—Archaebacteria, Eubacteria, Protista, Fungi, Plantae, and Animalia. The seven-level hierarchy can be subdivided into more specific categories, such as superclass, subclass, superorder, and suborder. In all, more than 30 taxonomic levels are recognized.

Each category at every level of classification is based on characteristics shared by all the organisms it contains. Consider again the classification of the honeybee, illustrated in Figure 19-20. Its scientific name, *Apis mellifera,* indicates that it belongs to the genus *Apis,* which is classified in the family Apidae. All members of the family Apidae are bees that either live alone or in hives, as does *A. mellifera.* The order to which the honeybee belongs, Hymenoptera, includes ants, bees, and wasps, which usually have two pairs of wings and are likely to be able to sting. At the next level of classification, *A. mellifera* belongs to the class Insecta, meaning it is an insect with three major body segments and three pairs of legs. Its phylum, Arthropoda, indicates that the honeybee is an arthropod, an organism with a hard cuticle of chitin and jointed appendages. Its kingdom, Animalia, tells you that *A. mellifera* is a multicellular heterotroph whose cells lack walls. ☐

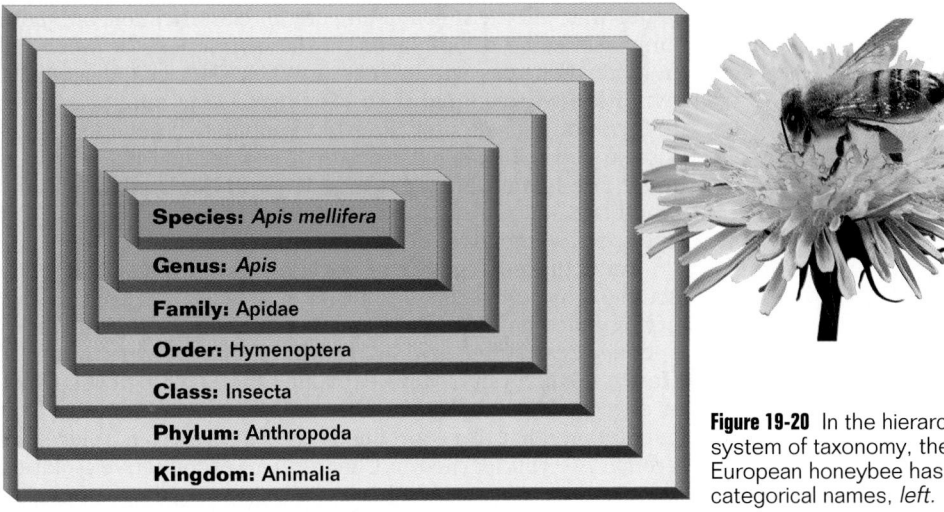

Species: *Apis mellifera*

Genus: *Apis*

Family: Apidae

Order: Hymenoptera

Class: Insecta

Phylum: Anthropoda

Kingdom: Animalia

Figure 19-20 In the hierarchical system of taxonomy, the European honeybee has seven categorical names, *left.*

Teaching Tips
Mnemonic Device
Have students devise a mnemonic to help them learn the sequence of the classification scheme showing the seven levels. One example is: King Philip came over for great spaghetti.

Comparing Classifications
Ask students to devise a graphic organizer that compares humans, chimpanzees, and gorillas in a hierarchy that begins with the kingdom. Provide books that will help students acquire needed information on phylum, class, order, family, genus, and species for each organism.

These Are the Names of My Favorite Things
Have students write down the common names of their favorite plants and animals. Then have them research the classification of these organisms and report to the class the seven names for each organism. Make sure most students research different organisms. Remind them of insects and reptiles.

Recognizing a Species

The basic biological unit in the Linnaean system of classification is the species. A definition of species was proposed by John Ray (1627–1705), an English clergyman and scientist. A species, Ray suggested, is a group of individuals that can breed with one another and produce fertile offspring. Even offspring that looked very different were considered the same species. By Ray's definition, all dogs are one species, all pigeons are one species, and so on. However, carp and goldfish are not the same species because they cannot interbreed. And although a horse and a donkey can mate, they are not the same species because the offspring, a mule, shown in Figure 19-21, is sterile.

With Ray's definition, the species became regarded as an important biological unit that could be cataloged—the task that Linnaeus undertook a generation later. The

Figure 19-21 According to Ray's definition of species, a horse, *below left,* and a donkey, *below right,* belong to different species because their offspring, a mule, *below,* is sterile.

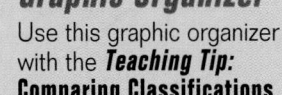

Graphic Organizer

Use this graphic organizer with the *Teaching Tip:* **Comparing Classifications** on page 435.

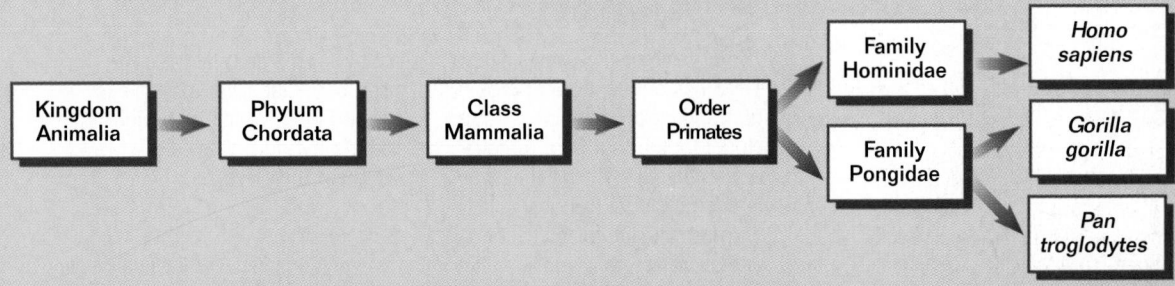

Kingdom Animalia → Phylum Chordata → Class Mammalia → Order Primates → Family Hominidae → *Homo sapiens*

Order Primates → Family Pongidae → *Gorilla gorilla*

Family Pongidae → *Pan troglodytes*

Teaching Tip

Concept Map for Species

Have the class devise a concept map for "species" using the explanations of Ray and of Mayr. Use arrows and propositions to link the concepts. Some terms to include are species, Ray, Mayr, fertile, sterile, reproduction, organisms, population, isolated, reproductive barriers, interbreed, biological species concept.

CONNECTIONS

Chapter 10

Hybrids are determined in the wild by comparing DNA from different organisms. Analysis of the DNA from the red wolf shows that it has DNA fingerprints from both the coyote and the gray wolf.

genus was still the fundamental unit of classification, but more emphasis came to be placed on the distinctions between the individual species within genera. When Darwin's ideas about evolution were joined with Mendel's ideas about genetics to form the area of study known as population genetics, biologists found that Ray's definition of species had its shortcomings. For example, sometimes it is difficult to determine whether a set of organisms represents several species or just one species composed of dissimilar individuals.

With the emergence of population genetics, biologists sought a more precise definition of species. A result of this effort was the **biological species concept,** a definition of species first stated in 1942 by the biologist Ernst Mayr of Harvard University. Mayr explained that a biological species is a group of actually or potentially interbreeding natural populations that are reproductively isolated from other such groups. As you learned in Chapter 12, reproductive isolation arises when a barrier isolates two or more groups of organisms and prevents them from interbreeding. The biological species concept emphasizes interfertility between organisms. However, reproductive barriers between sexually reproducing species are not always effective. Sometimes, individuals of different species interbreed and produce offspring called **hybrids.** For example, wolves and dogs are members of separate species in the genus *Canis*. However, interbreeding between wolves and dogs produces fertile offspring, such as the hybrid shown in Figure 19-22. Coyotes, members of another separate species in the genus *Canis*, also can interbreed with dogs to produce fertile hybrids.

The biological species concept works fairly well for members of the kingdom Animalia, in which strong barriers to hybridization usually exist. It accurately defines many species in which members reproduce sexually. However, the biological species concept fails to describe species that are predominantly asexual in

Figure 19-22 Even though a wolf, *below left,* and a dog, *below right,* belong to separate species, they can interbreed and produce fertile offspring such as this dog-wolf hybrid, *below.*

 Did You Know?

The Endangered Species Act of 1973 has a clear goal to protect endangered species, but a closer look at some of the protected animals shows that they may be hybrids. Hybrids do not currently qualify for protection under the act, and some animals risk removal from the protection list. One example is the red wolf, which is thought to be a hybrid between the gray wolf and the coyote.

their reproduction, including all bacteria, as well as some protists, fungi, plants, and even a few animals. Most of the time, such asexually reproducing organisms give rise to clones that are genetically identical to a single parent. These species clearly cannot be defined in the same way as sexually reproducing animals and plants that derive genetic information from two parents.

Essentially, there are no barriers to interbreeding between the species within many groups of organisms. Most species of plants, some mammals, and many fishes are able to form fertile hybrids with one another, even though they may not do so in nature. In practice, biologists today recognize species much the way their predecessors did—by studying an organism's visible features. ▢

How Many Species Exist?

Since the time of Linnaeus, approximately 1.5 million species have been named, a far greater number of organisms than Linnaeus ever thought existed. But the actual number of species in the world is undoubtedly much greater, judging from the large numbers still being discovered, such as *Lecythis prancei* shown in Figure 19-23. Scientists estimate that there are at least 10 million different kinds of organisms on Earth and that many of them (more than 5 million species) live in the tropics. Considering that no more than half a million tropical species have been named, our knowledge of these organisms is obviously very limited.

After naming and classifying approximately 1.5 million species of organisms, what have biologists learned? Identifying particular species has enabled biologists to understand which species are unique sources of food and medicine. For example, being able to distinguish the fungus *Penicillium* from the fungus *Aspergillus* is necessary for the production of the antibiotic penicillin. Taxonomy has also given biologists the ability to catch a glimpse of the evolutionary history of life on Earth.

Figure 19-23 *Lecythis prancei* is a species of a nut tree found in rain forests along the Amazon River in Brazil. First described in 1990, virtually nothing is known about the biology of this spectacular new species.

Demonstration
Deforestation of Rain Forest
Show students a map of Belgium. Tell them that three times this much rain forest is destroyed each year in the Amazon. This is roughly 50 acres per minute.

Teaching Tip
Loss of Biodiversity
Give students some of the following data regarding the diversity of life in the rain forest: in 19 trees, 1,200 species of beetles were found; on 2 acres, 700 different species (as many as found in the U.S.) can exist; Brazil has more primates and terrestrial vertebrates than any other nation; and the rain forests are thought to house 30 million species of insects.

Ask students to write an essay explaining why they feel the rain forest in Brazil is so rich in diversity. Once they have written their essay, let them discuss the impact deforestation will have on this ecosystem.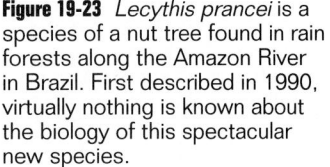

Demonstration

Looks Can Be Deceiving

Show students a pencil cactus, and ask them what kind of plant they think it is. They will probably answer "cactus." Inform them that the pencil cactus is a kind of plant called a euphorb and is not really a cactus. Use this opportunity to introduce analogous structures.

👁 VISUAL STRATEGY

Figure 19-24

Use this figure to illustrate an example of analogous characters and convergent evolution. Have students find the analogous characters in the two photographs and reason why both plants developed such similar traits.

Teaching Tip

Analogous Structures

Let students brainstorm and derive a list of analogous structures that are found in organisms. Once they have determined the list, have them justify why each organism developed the structure.

Example:

Bat wing	Bird wing
Platypus bill	Duck bill

Reconstructing Evolutionary Histories

Linnaeus's classification system was based on the fact that organisms exhibit different degrees of similarity. For instance, tigers resemble gorillas more closely than they resemble lampreys. According to Darwin's views, organisms that are similar descended from a common ancestor; therefore, classification provided strong evidence supporting evolution. However, making evolutionary connections based on similar traits can be misleading because not all traits are inherited from a common ancestor. Consider the wings of a bird and the wings of an insect. Both equip the respective organism for flight, but the two kinds of wings are built differently and evolved independently of each other. Similar traits such as wings are the result of convergent evolution. In **convergent evolution,** organisms evolve similar features independently, often because they live in similar habitats. Similar features that evolved through convergent evolution are called **analogous characters.** Figure 19-24 illustrates an example of convergent evolution found in two different plant families—the cactus family and the spurge family.

Figure 19-24 Thick water-storing stems are not only found in the Cactaceae (cactus family) of North and South American deserts, *above left,* but also in members of the Euphorbiaceae (spurge family) native to the African deserts, *above right.*

Biologists must be able to distinguish homologous traits from analogous ones in order to reconstruct evolutionary history. The evolutionary history of a species is called its **phylogeny** (fy LAHJ uh nee). How do taxonomists determine the phylogeny of a species? In general, this is determined by the overall similarity between the characteristics of different kinds of organisms. In one approach, called **cladistics** (kluh DIHS tihks), biologists reconstruct a phylogeny in which relationships are inferred based on similarities derived from a common ancestor. Cladistics is used to determine the sequence in which different groups of organisms evolved. To do this, cladistics focuses on a set of unique characteristics found in a particular group of organisms. These unique characteristics are called **derived traits.** Using patterns of shared derived traits, a biologist using cladistics constructs a branching diagram called a **cladogram,** which shows the evolutionary relationships among groups of organisms. The key to cladistics is identifying morphological, physiological, or behavioral traits that differ among the organisms being studied and that can be attributed to a common ancestor.

In practice, a biologist constructing a cladogram is interested in studying the evolutionary relationships of certain groups of organisms, such as species within a genus or genera within a family. Cladograms do not convey direct information about ancestors and descendants, showing who came from whom. Instead, cladograms convey comparative information about relationships. Organisms that are grouped more closely on a cladogram share a more recent common ancestor than those farther apart. Because the analysis is comparative, a cladogram deliberately includes an organism that is only distantly related to the other organisms. This distantly related organism is called an **out-group.** The out-group serves as a baseline for comparisons among the other organisms being evaluated, the **in-group.**

How to Construct a Cladogram

To see how a biologist might go about reconstructing a phylogeny using a cladogram, consider a collection of seven vertebrates: a lamprey, a lizard, a tiger, a salamander, a shark, a gorilla, and a human. For each of these animals, seven different traits have been recorded in Table 19-2; a plus sign (+) indicates the presence of the trait, and a minus sign (–) indicates its absence.

Table 19-2 Traits Identified in Collection of Vertebrates

Organism Traits:	Jaws	Dry Skin	Hair	Lungs	Tail	Pectoral Fins	Bipedal
Lamprey	–	–	–	–	+	–	–
Lizard	+	+	–	+	+	–	–
Tiger	+	+	+	+	+	–	–
Salamander	+	–	–	+	+	–	–
Shark	+	–	–	–	+	+	–
Gorilla	+	+	+	+	–	–	–
Human	+	+	+	+	–	–	+

As you learned in Chapter 13, the lamprey is only distantly related to the other six vertebrates. It is an agnathan (jawless fish) and was among the first vertebrates to appear before jaws or paired appendages evolved. Compared with the other groups being considered, the lamprey is the out-group, the one with the fewest traits in common with the other vertebrates. Any deviation from the basic characteristics that are

VISUAL STRATEGY

Table 19-2
Have students review the table to see how it is organized. Let them work in groups of two and devise a table of their own for the following animals: eagle, frog, goldfish, dog, mouse. They need at least five different traits to compare. Let students rank the animals as they choose.

VISUAL STRATEGY

Table 19-3
Once again, have students review the table to see how it is organized. Have them take their data and insert a one or a zero for the trait. Have them calculate the values as shown in the table.

Teaching Tip
Cladogram vs. DNA
Review DNA's role as the carrier of genetic information. Emphasize that much of what taxonomists use to classify organisms—basic structures, reproductive patterns, life cycles, development from fertilization to adulthood—is controlled by DNA. Ask students to discuss which method of relating organisms they think is more accurate—comparing two organisms' DNA or constructing a cladogram. Do they feel one is better than the other? If so, they should justify their answer.

Table 19-3 Identification of Derived Traits

Organism Traits:	Jaws	Dry Skin	Hair	Lungs	Tail	Pectoral Fins	Bipedal
Lamprey	0	0	0	0	0	0	0
Lizard	1	1	0	1	0	0	0
Tiger	1	1	1	1	0	0	0
Salamander	1	0	0	1	0	0	0
Shark	1	0	0	0	0	1	0
Gorilla	1	1	1	1	1	0	0
Human	1	1	1	1	1	0	1
Total Number of Taxa	6	4	3	5	2	1	1

present in the out-group is considered an evolutionary change, or a derived trait.

The first step in constructing a cladogram is identifying the derived traits. In Table 19-3, the traits in the row for the out-group are marked with a zero (0). When a vertebrate has a trait not found in the out-group, the trait is considered a derived trait and is marked with a one (1). Next, the numbers of shared derived traits are calculated. This is done by adding the numbers of derived traits in each column.

It is obvious that two traits, pectoral fins and bipedalism, are each present in only one group of organisms; they are not shared with any of the other animals under consideration. Such traits are not useful in reconstructing the phylogeny because they do not allow the elimination of any potential relationships. Now you can begin sketching the cladogram.

1. Starting with a diagonal line, the out-group is drawn as the first branch of the cladogram and is labeled *Lamprey*. Just past this first branch, the most common derived trait present in the organisms is listed. In this example, jaws are present in six organisms (all but the out-group lampreys), so the derived trait *Jaws* is written on the diagonal line just past the branch representing lampreys, as shown in the cladogram in Figure 19-25.

2. Next, the second most common derived trait is determined, which in this case is lungs, found in five of the seven organisms. The two organisms that lack lungs are lampreys and sharks. Since lampreys are already in the cladogram, a second branch labeled *Shark* is now drawn above the first branch on the cladogram. The derived trait *Lungs* is listed on the main line past this second branch point.

3. The third most common derived trait is dry skin, present in four organisms. All the organisms except lampreys, sharks, and salamanders have dry skin. A third branch labeled *Salamander* is drawn above the second branch of

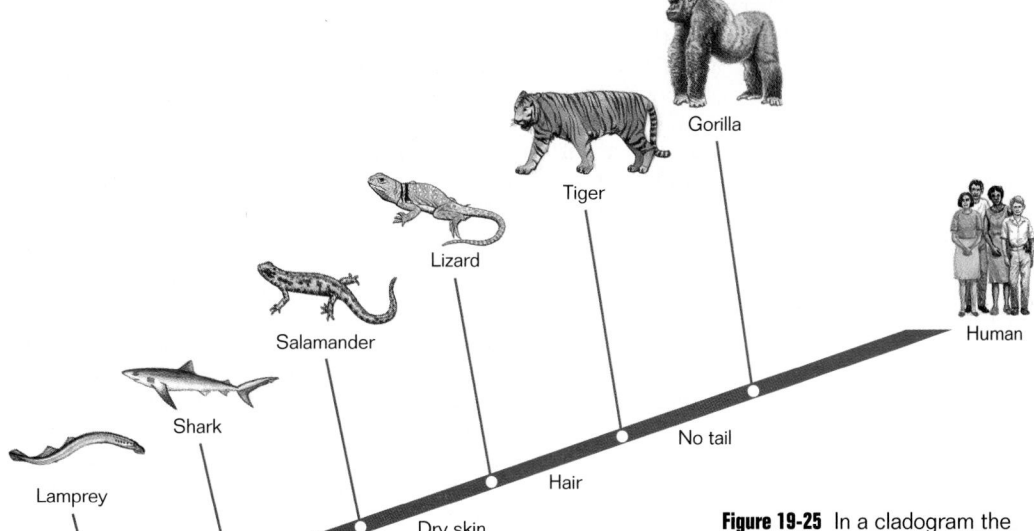

Gorilla

Tiger

Lizard

Salamander

Shark

Lamprey

Human

No tail

Hair

Dry skin

Lungs

Jaws

Figure 19-25 In a cladogram the derived traits listed along the horizontal axis are shared by all the organisms to the right of the branch point (white dot) and are not present in any organisms to the left.

the cladogram with the derived trait *Dry skin* indicated past the branch.

4. The fourth most common derived trait is hair. Of the four kinds of organisms remaining, all have hair except lizards, which are labeled on a fourth branch of the cladogram with the derived trait *Hair* indicated past the branch.

5. The fifth most common derived trait is a tail. Of the three kinds of organisms remaining, all lack a tail except tigers, which are labeled on the fifth branch of the cladogram with the trait *No tail* indicated just past it.

6. The other two traits, each present in only one taxon, are of no use in constructing the cladogram. While it is true that gorillas do not walk upright and humans do, a branch off the main axis with gorilla on it and humans as the remaining group exhibits exactly the same information as a branch off the main axis with human on it and gorilla as the remaining group. ▢

❏ Capsule Summary

An organism's evolutionary history is its phylogeny. Cladistics is a method of taxonomy that uses information about the presence of traits among a group of organisms to provide information about the relative relationships among organisms.

Considering the Weight of a Character

The great strength of a cladogram is its objectivity. A computer fed the data will generate exactly the same cladogram time and again. The great disadvantage of a cladogram is that it ignores too much information. It simply indicates that a character does or does not exist. A cladogram cannot take into account variations in the "strength" of a character, such

SECTION 19-4

Demonstration
Constructing a Cladogram
Show the class photographs of a bird, a crocodile, a turtle, a lizard, and a snake. Point out that in a traditional classification, the last four are grouped in the class *Reptilia*, while the bird belongs to the class *Aves*. Have students discuss what observable characteristics the four have in common to belong to the class *Reptilia*. Cladists focus on the order in which evolutionary lines diverged, or branched, as revealed by derived characters. Consequently, a cladist would group crocodiles with birds since they share several derived characters. Traditional taxonomists argue that putting crocodiles and birds in the same group with their obvious physical differences makes no sense.

✉

Chapter Closure

Lynn Rothschild, a NASA biologist, found bacteria living in a most unexpected place, a clump of salt from a salt flat in Baja. The broken chunks contained red and green stripes, which she verified were products of bacteria. These bacteria were efficiently removing nitrogen and carbon dioxide gas from the atmosphere for biological processes. If bacteria can survive in a place as harsh as a chunk of salt, they might exist in other places that we would not expect to find them. The theory is that these bacteria lived in water, but as the water evaporated, the bacteria migrated into the salt crystals. The same scenario could exist on Mars. Scientists speculate that Mars could have had seas approximately 3.5 billion years ago. If the seas dried up, the bacteria could have moved into the remaining salt crystals. The Martian atmosphere could provide carbon dioxide, nitrogen, and water vapor.

Knowing the theory that life evolved from prokaryotes, which yielded eukaryotes through endosymbiosis, have students speculate what life-forms could be living in the Martian soil. Have them design an organism specially adapted to the Martian environment. This can be a drawing or a model. Then have them write a description as to how their organism has adapted to life on the planet.

as the size or location of a fin, the effectiveness of a lung, and so on. Each character is treated the same. Because evolutionary success depends so much on high-impact events, such as the evolution of feathers, cladograms sometimes fail to look at information of great potential importance. Thus, a cladogram of vertebrate evolution will group birds among the reptiles with crocodiles, accurately reflecting their true ancestry but ignoring the immense evolutionary impact of a derived character like feathers.

In order to avoid this pitfall, most practicing taxonomists attempt to weigh the evolutionary significance of the characters they study and to produce a more subjective analysis of evolutionary relationships. This approach, called **evolutionary systematics,** places birds in an entirely separate class from reptiles, giving extra weight to the characters like feathers that made powered flight possible. In evolutionary systematics, the full observational power and judgment of the biologist is brought to bear—along with any biases he or she may have.

In practice, evolutionary systematics is the approach of choice when a great deal of information about the organisms is available to consider. You cannot give a character due evolutionary weight without having enough information to make an accurate judgment. When little information is available about how the character affects the life of the organism, cladistics is the approach of choice. ◻

◻ CAPSULE SUMMARY

Evolutionary systematics is a subjective method of taxonomy in which a biologist uses judgment to consider the importance of the characters among a group of organisms.

Section Review

1. *Why is Linnaeus sometimes called "the father of modern taxonomy"?*
2. *Why do biologists use both words of a scientific name to correctly identify an organism?*
3. *From least to most inclusive, what are the names of the seven levels of the hierarchical system of classification?*
4. *What is cladistics? What kind of information does it reveal about evolutionary lines?*
5. *What is the biological species concept? Why does it fail to define many species that reproduce asexually?*

Critical Thinking

6. *What is the relationship between an organism's evolutionary history and its classification?*
7. *Why is evolutionary systematics considered more subjective than cladistics?*

Answers to Section Review

1. Linnaeus is thought of as "the father of modern taxonomy" because he developed a method of classification that reduced the lengthy, cumbersome scientific name of an organism to two words.

2. See page 432.

3. From least to most inclusive, the seven levels of the hierarchical system of classification are species, genus, family, order, class, phylum, kingdom.

4. Cladistics is a method of relating organisms based on given traits. It is used to show a time line as to when traits developed.

5. The biological species concept identifies a species as an interbreeding, natural population that is isolated. This concept does not work for asexual species because it describes a species as an individual that breeds with others in its population.

6. A similarity is seen in an organism's evolutionary history and its classification. Organisms that evolved from common ancestors are classified more closely together.

7. See page 442.

Kingdom Archaebacteria

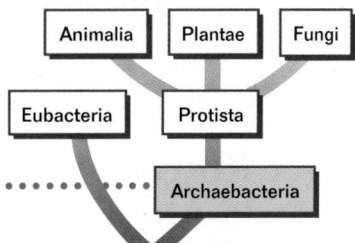

Animalia | Plantae | Fungi
Eubacteria | Protista
Archaebacteria

Characteristics of Archaebacteria

- Unicellular prokaryotes
- Cell walls lack peptidoglycan
- Genes have introns
- Unique lipids in plasma membranes
- Either heterotrophic or autotrophic
- May be ancestors of eukaryotic cells
- Three groups that live in extreme environments—methanogens, extreme halophiles, and thermoacidophiles

Morning glory pool

Methanogens

Methanogens make up the largest group of archaebacteria identified so far. They are among the most strictly **anaerobic** of all organisms, poisoned by even traces of oxygen. Methanogens convert carbon dioxide into methane. Their metabolism is ideally suited to the kind of atmosphere thought to have existed on the primitive Earth. This methanogen, *Methanobacterium formicum,* has a cell wall made mostly of protein.

In ancient times methanogens could have lived anywhere, but today they live only where oxygen has been excluded and hydrogen and carbon dioxide are available. Methanogens are found in stagnant water, in sewage treatment plants, and in the intestinal tracts of animals. They can be found living on the ocean bottom and in hot springs. In spite of their intolerance to oxygen, they are obviously distributed throughout the world.

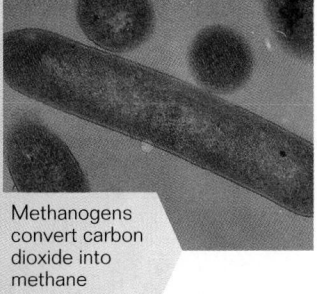

Methanogens convert carbon dioxide into methane

Thermoacidophiles

As their name implies, thermoacidophiles are archaebacteria that thrive in environments that are both **acidic** and **hot.** These bacteria flourish in temperatures between 60°C and 80°C (140°F and 176°F) and in pH between 2 and 4. The hot sulfur springs in Yellowstone National Park are inhabited by the thermoacidophile *Sulfolobus,* which obtains its energy by oxidizing sulfur.

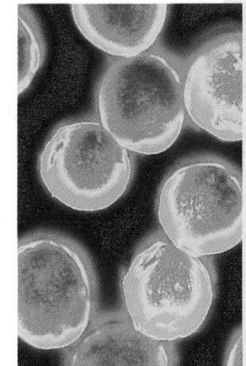

Extreme Halophiles

The extreme halophiles are archaebacteria that require high concentrations of salt in order to survive. They grow in salty habitats along shorelines and in inland waters such as the Great Salt Lake and the Dead Sea. Some species require an environment 10 times saltier than sea water to grow. Colonies of halophiles often form in seawater evaporating ponds used in commercial salt production; the halophilic bacteria are harmless.

Halobacterium

Sewage treatment plant

Instructional Strategy

Instruct the class to break up into groups. Inform them that each group will be responsible for learning about the environments in which one of the three kinds of archaebacteria thrive. For example, one group will research the environments that are home to methanogens, another group will study the hot, acidic surroundings of the thermoacidophiles, and a third group will collect information regarding the habitat of the extreme halophiles. Have each group write a two-page paper that describes the environment and that explains why the environment is hostile to other forms of life. Make sure students include photos and maps in their papers.

Discussion

Guide a discussion by posing the following questions.

1. Why are archaebacteria confined to such extreme habitats that cannot support other life-forms? *(Archaebacteria evolved at a time when the Earth was covered in harsh environments. As the Earth's surface changed, these organisms died out in many areas. Those surviving today are confined to areas similar to those in which they originally evolved.)*

2. How does the existence of archaebacteria provide the possibility that life exists on other planets? *(They live in extreme conditions, such as oxygen-free, high-temperature, salty, and low-pH environments. If life can exist in such harsh conditions, perhaps it can exist on another planet with similar environments.)*

3. What roles do you think archaebacteria play in their ecosystems? *(Student answers will vary; they should include that the bacteria act as decomposers and are a food source.)*

4. Why are archaebacteria thought to be ancestors of eukaryotes? *(They are more similar in genetic structure to eukaryotes than are the eubacteria. Carl Woese of the University of Illinois in Urbana found the RNA of archaebacteria to be more similar to eukaryotes than that of eubacteria.)*

UNIT 5

CHAPTER 19
Highlights
Six Kingdoms

Instructional Strategies

• Have students research information about diseases caused by *Clostridium,* such as tetanus (also called lockjaw), botulism, and gas gangrene. In a short report, they should describe the symptoms, effects, and treatments of the disease, the name of the organism that causes it, and how the disease can be avoided.

• The genus *Clostridium* contains rod-shaped bacteria that form endospores. Have students research other major groups of eubacteria and report on their major characteristics.

• Bring a copy of *Bergey's Manual of Systematic Bacteriology* to class and have students browse through it. Give students an idea of the complex nature of bacterial taxonomy by telling them that this book classifies bacteria according to information from the analyses of the compositions of nucleic acids, from DNA hybridizations, and from chemical analyses of cellular components. Have students research the meaning of the term *bacterial species.* (*A bacterial species is defined as a population of cells with similar characteristics— possession of enzymes, reactions to serological tests, phage typing, amino acid sequences, etc.*)

Discussion

Guide a discussion by posing the following questions.

1. Why should individuals maintain their tetanus vaccines? (*Tetanus vaccines must be repeated because the body does not build a permanent immunity against the bacterium.*)

2. Why should you be concerned about a can of food that has been stored for an undetermined amount of time? (*Clostridium grows in an oxygen-free environment. If they are not destroyed in a sterilization*

Highlights
Six Kingdoms of Life

Kingdom Eubacteria

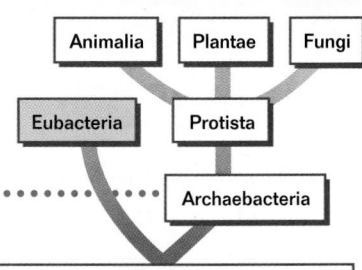

Characteristics of Eubacteria

■ Unicellular prokaryotes
■ Cell walls contain peptidoglycan
■ Genes lack introns
■ Reproduce asexually by binary fission
■ Either autotrophic or heterotrophic
■ Extremely diversified groups
■ Certain types may be ancestors of mitochondria

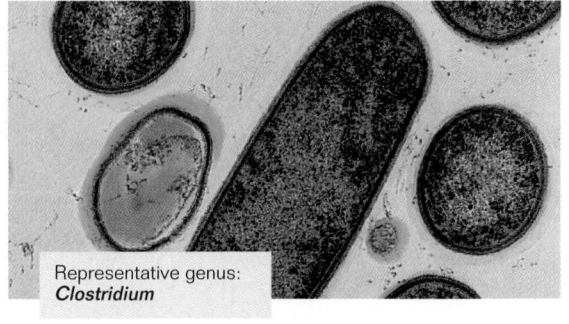

Representative genus: *Clostridium*

Clostridium is a typical eubacterium that thrives in the soil. It is an **obligate anaerobe** (poisoned by oxygen) because it lacks the enzymes that break down the toxic peroxides produced by aerobic metabolism. *Clostridium* lives in tiny oxygen-free pockets in the soil, where it acts to decompose cellulose plant matter.

Forming Endospores

Clostridium is one of the few eubacteria able to form **endospores,** heat-resistant packages containing a copy of the DNA. The endospores can persist in the environment for as long as 50 years and then reinitiate growth.

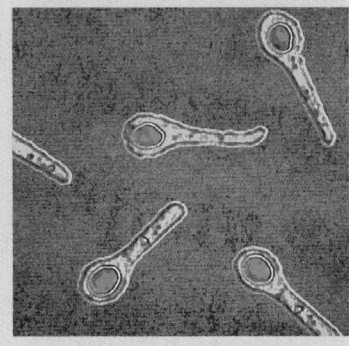

Causing Diseases

Several species of *Clostridium* can cause disease in humans. For example, *Clostridium botulinum* causes a form of food poisoning called **botulism.** *C. botulinum* produces an extremely potent toxin that affects nerve activity. Additional diseases caused by other species of *Clostridium* include tetanus and gas gangrene.

Botulism-tainted peppers

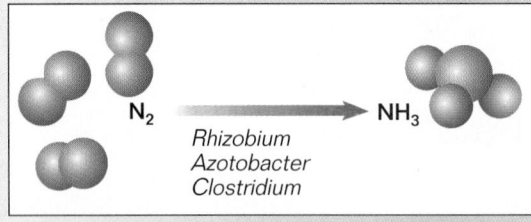

N_2 → NH_3

Rhizobium
Azotobacter
Clostridium

Fixing Nitrogen

A large portion of the Earth's **nitrogen fixation** is carried out by *Clostridium pasteurianum,* along with other eubacteria such as *Azotobacter* and *Rhizobium.* These eubacteria are able to convert nitrogen gas (N_2) into ammonia (NH_3).

process, they can survive because they form heat-resistant endospores.)

3. How could the toxin from *Clostridium botulinum* prove beneficial in medical treatment? (*Type A toxin, one of the seven neurotoxins produced by* C. *botulinum, causes temporary paralysis of muscles in low doses. It is useful in the treatment of*

muscle spasms in the eyes, face, arms, and legs. This paralysis could be beneficial in diseases that result from excessive muscle contraction.*)

4. Why is the nitrogen fixation of *C. pasteurianum* and other eubacteria necessary for the homeostasis of the ecosystem? (*Nitrogen must be converted to a form that plants can use.*

Plants require nitrogen for their growth and metabolism, but they can not use it in the nitrogen gas form. Eubacteria convert the nitrogen gas to ammonia, a usable form of nitrogen for plants.*)

Kingdom Protista

Animalia   Plantae   Fungi

Eubacteria   Protista

Archaebacteria

Characteristics of Protists

- Eukaryotic organisms
- Mostly unicellular; some multi-cellular
- Autotrophic, heterotrophic, or both
- Reproduce asexually; some also sexually

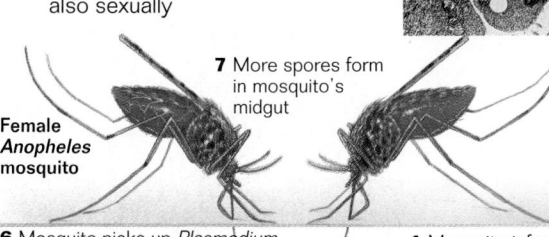

Representative genus:
Plasmodium

The protist *Plasmodium* is a **spore-forming parasite** with a complex life cycle that alternates between two hosts. *Plasmodium* causes malaria, a serious disease characterized by severe chills, fever, sweating, an enlarged spleen, confusion, and great thirst.

Female Anopheles mosquito

7 More spores form in mosquito's midgut

6 Mosquito picks up *Plasmodium* from infected individual

5 Spores released as red blood cells burst

4 Spores reproduce in host's red blood cells

1 Mosquito infects host with *Plasmodium*

2 Asexual spores form in host's liver cells

3 Spores burst out of liver cells

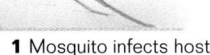

Life Cycle of *Plasmodium*

Plasmodium spends part of its life cycle growing in mosquitoes of the genus *Anopheles,* and another part of its life cycle in humans. A female mosquito acquires *Plasmodium* when she bites an infected human. Later, when she feeds off a different human, she injects saliva into the wound to prevent her victim's blood from clotting. Her saliva transfers *Plasmodium* into the bloodstream.

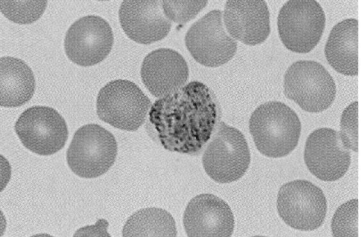

In humans, *Plasmodium* parasites are carried to the liver, where they reproduce. They reenter the bloodstream and invade blood cells. *Plasmodium* divides so rapidly that by 48 hours the cells rupture, releasing toxic substances into the bloodstream and causing a high fever. The cycle repeats itself every 48 hours until the infection is brought under control by the person's immune system, or until the person dies.

Affected Areas

Over 100 million people are infected with *Plasmodium* at any one time, and every year about 1 million of them (mostly children under the age of 5) die of malaria. Since *Anopheles* mosquitoes are abundant in the tropics, nearly all of malaria's victims live in Africa, Asia, and South America.

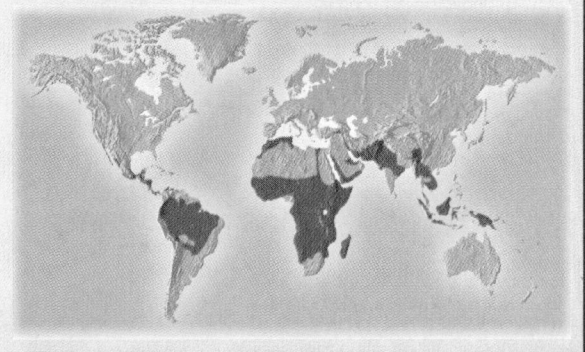

Instructional Strategies

- Malaria was once common in the United States. Have students research the methods used in efforts to eliminate the disease from this country. Then have them research the efforts taken by another malaria-prone country. Students can take notes and present a brief, informal report to the class.

- Challenge students to devise a worldwide plan to eliminate malaria, a disease that kills more children than adults. Let them work in groups to brainstorm ideas to present to the class for consideration. Let the class select the best plan to follow by popular vote.

- Have students research other human diseases caused by protists, such as toxoplasmosis, Chagas' disease, giardia, or amoebic dysentery. Students should describe the disease, identify its causative agent, explain its mode of transmission, and list methods of treatment and prevention.

- Have students compare the map in the Highlight to a large, more detailed map of the world. Locate the countries shaded and write their names on the chalkboard.

Discussion

Guide a discussion by posing the following questions.

1. Why is malaria still a major killer in the world today? (*Malaria is controlled by eliminating the* Anopheles *mosquito, which has been building resistance to pesticides. People who live in environments favorable for the mosquito are still at risk of contracting the disease.*)

2. The main symptoms of malaria are chills and fever. Which part of the *Plasmodium* life cycle do you suppose causes these symptoms in malaria victims? (*When blood cells explode, toxic substances are released into the bloodstream and the body temperature increases in an attempt to destroy the invaders.*)

Instructional Strategies

- Have students describe and illustrate how a mushroom would develop in their own yard or in an area outside of the school. They should describe the optimal conditions that would allow that mushroom to develop.

- Assign students the responsibility of setting up a mushroom farm to make a profit. Let them do some research on the types of mushrooms that would be most prosperous, and provide them a set amount of money that they must budget for the project. Let them present their proposed project to the class.

Discussion

Guide a discussion by posing the following questions.

1. What role do the fungi serve in an ecosystem? *(They break down dead and decaying matter by digesting the organic matter with enzymes released from the hyphae. They provide a means to recycle organic material in the environment.)*

2. Why are spores a good way for mushrooms to reproduce? *(Spores can travel to new locations by wind, water, or animal, where they germinate and produce hyphae that will repeat the cycle.)*

Highlights
Six Kingdoms of Life

Kingdom Fungi

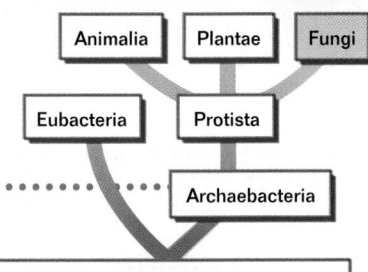

Animalia   Plantae   Fungi

Eubacteria   Protista

Archaebacteria

Characteristics of Fungi

- Eukaryotic organisms
- Mostly multicellular; some unicellular
- Cell walls made of chitin
- Filamentous bodies
- External heterotrophs; principal decomposers
- Reproduce sexually or asexually

Representative organism: **Mushroom**

Fungi, such as this mushroom *Amanita muscaria*, are external heterotrophs that acquire their nutrition by **absorption.** They digest food outside their bodies by secreting an enzyme that breaks organic materials into molecules fungi can absorb.

Mushroom

Hyphae

Mycelium

Filamentous Bodies

Most fungi are composed of filaments called **hyphae.** Hyphae are woven together to form a dense mat known as a **mycelium.** The mycelium of a fungus is usually hidden within the tissues of its food source.

Forming Spores

Mushrooms are reproductive structures. The underside of the cap is lined with clublike structures that give rise to haploid spores. One mushroom with a cap 3 in. across can produce as many as 40 million sexual spores per hour! Carried by wind or water, a spore can germinate if it lands in a moist environment.

Spores are haploid. When they germinate, they give rise to haploid hyphae. When two such haploid hyphae touch, they fuse together, forming secondary hyphae with paired but separate nuclei. It is from these filaments that mushroom caps form.

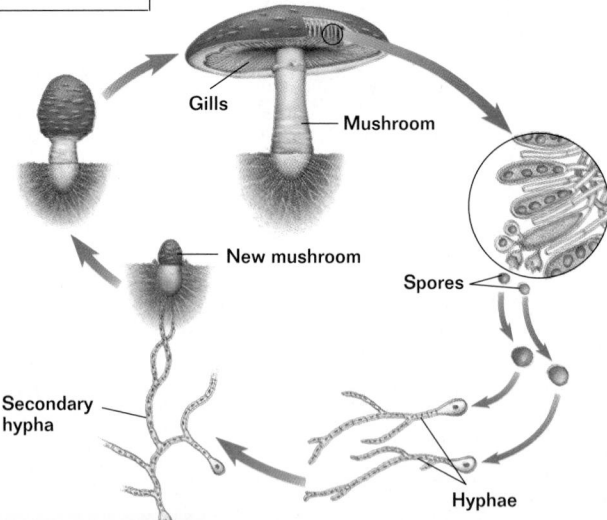

Gills

Mushroom

New mushroom

Spores

Secondary hypha

Hyphae

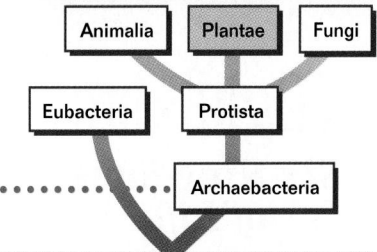

Animalia | Plantae | Fungi
Eubacteria | Protista
Archaebacteria

Kingdom Plantae

Characteristics of Plants

- Eukaryotic organisms
- All multicellular
- Cell walls made of cellulose
- Most consist of roots and shoots
- Mostly autotrophic and terrestrial
- Highly specialized structures for reproduction and survival on land
- Reproduce sexually; some capable of reproducing asexually

Representative organism:
Cherry tree

Flowering plants such as a cherry tree are called **angiosperms,** a name derived from the Greek words *angion,* meaning "vessel" and *sperma,* meaning "seed." A flower is a reproductive structure that encourages cross-pollination by means of animals. Cherry flowers have nectar, which attracts bees. In seeking nectar within a flower, a bee is covered with pollen, which will produce male gametes; it carries this pollen to ovules, which will produce female gametes in other flowers as it seeks more nectar.

Forming Seeds

Fertilization in angiosperms is a unique process called **double fertilization,** in which sperm possess two nuclei. One sperm nucleus fertilizes the egg to form the zygote, which produces the embryo. The other sperm nucleus fuses with other nuclei to form nutritive tissue called **endosperm.** The embryo and the endosperm are enclosed within the seed coat. **Seeds** are usually contained within fruits, structures that aid in their dispersal.

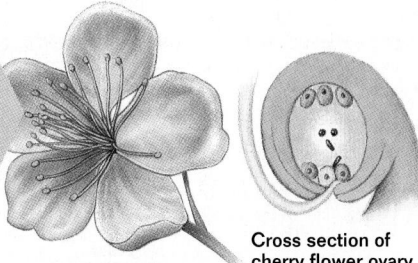

Cross section of cherry flower ovary

Transporting Substances

Like most plants, the cherry tree is a vascular plant. Vascular plants have tissues specialized for conducting water **(xylem)** and transporting carbohydrates **(phloem)** throughout the plant. The veins seen in leaves are bundles of vascular tissues.

Fruits with fleshy coverings, such as cherries, are eaten by birds and other vertebrates. The fruit's bright red color signals an abundant food supply. By feeding on these fruits, the animals may carry seeds to suitable habitats where new trees can grow.

Instructional Strategies

- Have students make a collage about a kind of angiosperm. Let them select one from a designated list. Each collage can include pictures of anything related to the particular angiosperm, particularly economically important products.

- Ask students to make journal entries on plants they find near the school or near their home. They should illustrate the plant and describe an interesting feature about the plant.

- Double fertilization can be a difficult topic for students. Have students work in groups to illustrate the process once you have explained it. Then, let students work in teams to master the topic. Select a member from each group to answer questions for the group, but do not tell the students which person will be selected until they have worked in their groups to understand double fertilization.

Discussion

Guide a discussion by posing the following questions.
1. What is the benefit of double fertilization? *(One sperm combines with the egg to produce the zygote while the other sperm combines with polar bodies to produce a food source to feed the developing embryo.)*
2. Why have plants evolved to have fruits? *(Fruits contain seeds, which protect the tiny, delicate embryo within. Fruits also offer a means of seed dispersal.)*
3. What characteristics have enabled plants to evolve from small, primitive plants such as liverworts to large, advanced plants like cherry trees? *(Plants developed cuticles to maintain their moisture, vascular tissue to transport water and nutrients long distances, and seeds to protect and disperse their offspring.)*

Instructional Strategies

- Purchase the larvae of ladybird beetles and watch them develop in the classroom. Have students research the conditions needed for the survival of the beetles. Have students illustrate the life cycle of the ladybird beetle and identify the different stages of metamorphosis.

- For interest, have students develop T-shirt designs to emphasize the importance of the ladybird beetle in the environment.

- Have students research the importance of ladybird beetles in the agriculture industry. Invite an entomologist to class to discuss the importance of these beetles in controlling harmful insects.

Discussion

Guide a discussion by posing the following questions.

1. Have students speculate as to why they think beetles make up such a large number of the animal world. *(They are well adapted to their environments.)*

2. Although most people think of insects as pests, many are helpful. List some ways in which insects are useful. *(Answers may vary. Students can cite examples such as honey and beeswax from bees and silk from silkworms. More important, insects are necessary for the cross-fertilization of many fruits and other crops.)*

Highlights
Six Kingdoms of Life

Kingdom Animalia

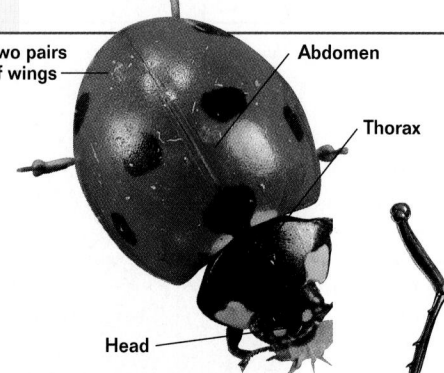

| Animalia | Plantae | Fungi |
| Eubacteria | Protista |
| Archaebacteria |

Characteristics of Animals

- Eukaryotic organisms
- Multicellular; no cell walls
- Most reproduce sexually
- Interior heterotrophs
- Specialized tissues for impulses and movement
- Inhabit nearly all environments in biosphere

Two pairs of wings

Abdomen

Thorax

Head

Jointed appendage

Representative organism: *Ladybird beetle*

The ladybird beetle, like all arthropods, has a rigid **exoskeleton,** in which chitin is an important element. The exoskeleton provides places for muscle attachment, protects the animal from injury, and prevents water loss. Insects have three body sections —the head, thorax, and abdomen—and are noted for possessing jointed appendages, such as legs and wings.

Fungi

Flowering plants

Beetles

Protists

Other

Bees, wasps

Mollusks

Vertebrates

Butterflies, moths

Spiders

Other insects

Flies, mosquitoes

Millipedes, centipedes

Arthropod Diversity

Insects such as the ladybird beetle are **arthropods,** the most successful group of animals that has ever lived. Two-thirds of all named species of organisms on Earth are arthropods. Of the 770,000 named species of insects, 350,000 are beetles.

Increasing in Size

As an individual arthropod grows larger, its exoskeleton splits and is shed, allowing the animal to increase in size. The eggs of arthropods develop into immature forms that may bear little or no resemblance to the adult. Most arthropods change their characteristics as they develop, a process called **metamorphosis.**

Although many of the millions of insects are harmful to humans, many others are helpful. Ladybird beetles are fierce predators, feeding on plant-eating insects.

A hallmark of insects such as the ladybird beetle is the diversity of the specialized appendages attached to their heads. Insects have **antennae** specialized for sensing the environment. Insects also have jaws, or **mandibles,** and other mouthparts that have evolved into different shapes for grinding, scraping, piercing, and sucking.

Vocabulary

aggregation (427)
analogous character (438)
binomial nomenclature (432)
biological species concept (436)
cell specialization (429)
cladistics (438)
cladogram (438)
class (434)
colonial organism (427)
convergent evolution (438)
derived traits (438)
differentiation (430)
division (434)

evolutionary systematics (442)
family (434)
gametic meiosis (424)
gametophyte (425)
genus (431)
hybrid (436)
in-group (439)
intercellular coordination (430)
kingdom (434)
multicellular organism (426)
obligate anaerobe (418)
order (434)

organ (426)
out-group (439)
peptidoglycan (418)
phylogeny (438)
phylum (434)
scientific name (432)
sexual life cycle (424)
sporic meiosis (425)
sporophyte (425)
taxonomy (431)
theory of endosymbiosis (419)
tissue (426)

CHAPTER REVIEW ANSWERS

Each item in the Chapter Review is correlated by text section in the assignment guide that follows.

ASSIGNMENT GUIDE

Section	Review	Themes Review	Critical Thinking
19-1	1, 2, 9, 13	27	29
19-2	3–6, 11, 14		
19-3	7, 12	25	28
19-4	8, 10	24	

Concept Mapping

Construct a concept map that shows the evolutionary events leading to the origin and development of eukaryotes. In constructing your map, use the following terms: archaebacteria, cell organelles, mitochondria, chloroplasts, eubacteria, sexual reproduction, asexual reproduction, DNA repair mechanism, and genetic diversity. Include additional concepts in your map as needed.

Concept Mapping

Map answer is shown on page 415B.

Review

Review

Multiple Choice

Multiple Choice

1. Archaebacteria differ from eubacteria because archaebacteria
   a. have cell walls made of peptidoglycan.
   b. have plasma membranes made of phospholipids.
   c. contain genes interrupted by introns.
   d. cannot live in harsh environments.

2. According to the theory of endosymbiosis, mitochondria evolved from
   a. eubacteria.
   b. archaebacteria.
   c. chloroplasts.
   d. protists.

3. Which of the following does *not* relate to the subject of asexual reproduction?
   a. binary fission   **c.** budding
   b. fragmentation   **d.** zygotic meiosis

4. Sexual reproduction is believed to have evolved originally as a way for cells to
   a. shuffle genetic material.
   b. repair damaged DNA.
   c. produce diploid individuals.
   d. increase their population growth at a maximum rate.

5. Sporic meiosis differs from zygotic meiosis and gametic meiosis because
   a. sporic meiosis produces a gametophyte stage.
   b. sporic meiosis produces a sporophyte stage.
   c. gametes do not fuse in sporic meiosis life cycles.
   d. meiosis in the sporic meiosis life cycle occurs immediately after the formation of the zygote.

6. The main advantage of multicellularity is that
   a. it allows the most rapid population growth.
   b. it tends to regulate population growth more efficiently.
   c. it allows division of labor.
   d. it prevents cells from specializing in only one or a few functions.

Multiple Choice
1. c (19-1.1)
2. a (19-1.3)
3. d (19-2.1)
4. b (19-2.2)
5. a (19-2.3)
6. c (19-2.1)
7. b (19-3.2)
8. c (19-4.2)

Completion
9. Archaebacteria, gene translation machinery (19-1.1)
10. seven, generic, specific, binomial (19-4.3)
11. eukaryotes, increase (19-2.1, 19-2.2)
12. colonial, aggregation (19-3.2)
13. eukaryotes, mitochondria (19-1.2)
14. sporic meiosis, haploid (19-2.3)

Short Answer
15. The cell walls of eubacteria contain peptidoglycan; the cell walls of archaebacteria do not contain peptidoglycan. The plasma membranes of organisms in both kingdoms are made of phospholipids, but those of archaebacteria also contain unique lipids. The genes of archaebacteria have introns, whereas those of

CHAPTER REVIEW ANSWERS continued

Short Answer continued

eubacteria do not. Archaebacteria contain ribosomal protein and an RNA polymerase similar to that found in eukaryotes; these are not present in the genes of eubacteria, which have unique gene translation machinery. (19-1.1)

16. Three organelles with possible endosymbiotic origins are mitochondria, chloroplasts, and centrioles. Mitochondria evolved from eubacteria similar to nonsulfur purple bacteria that are able to perform oxidative respiration; chloroplasts evolved from photosynthetic bacteria similar to *Prochloron;* and centrioles are believed to have evolved from bacteria similar to spiral-shaped bacteria called spirochetes. (19-1.3)

17. Bacteria reproduce by binary fission, a process in which one bacterium splits into two identical cells. *Hydra* reproduce by budding, a form of asexual reproduction in which a new individual grows from the parent's body and eventually breaks off. (19-2.1)

18. Both life cycles include haploid gametes that unite to form a diploid zygote. In the life cycle of *Chlamydomonas,* the zygote undergoes meiosis to produce haploid individuals, but in the human life cycle gametes are the products of meiosis. In addition, the life cycle of *Chlamydomonas* includes both sexual and asexual reproductive phases, while the human life cycle has only a sexual reproductive stage. (19-2.3)

Highlights Review

19. Representatives of the two kingdoms of bacteria differ in terms of cell wall, plasma membrane, composition, gene architecture, and gene translation machinery.

20. *Clostridium tetani* causes tetanus, a harmful infection that can occur in deep puncture wounds. *C. botulinum* poisons food with a toxin that affects the human nervous

7. A colonial organism differs from an aggregation because a colonial organism
   a. is a temporary collection of cells.
   b. is a permanent collection of cells.
   c. is truly multicellular.
   d. forms only upon starvation or stress of its component cells.

8. A difference between the scientific name of an organism and the classification of that organism is that
   a. the scientific name includes the family and class of the organism.
   b. the scientific name always contains three words (trinomial nomenclature).
   c. the classification includes more categories than the scientific name.
   d. classification can vary from place to place.

Completion

9. Prokaryotes belong to kingdom _____ or kingdom Eubacteria. Members of these two kingdoms have different cell walls, plasma membranes, gene architectures, and _____ .

10. According to the text, biological classification now uses a total of _____ groups or levels. The names of the _____ group and the _____ group of an organism compose its scientific name according to the modern system of _____ nomenclature.

11. Prokaryotes reproduce without producing gametes, while many _____ are able to undergo sexual reproduction. An advantage of sexual reproduction is that it can _____ genetic diversity.

12. *Volvox* is a(n) _____ organism; its cells are permanently associated. However, individual cells of *Dictyostelium discoideum* come together as a(n) _____ only when food is scarce.

13. *Pelomyxa palustris* represents an early stage in the evolution of _____ . Like archaebacteria, *Pelomyxa palustris* lacks _____ and does not undergo mitosis.

14. Sexually reproducing plants undergo _____ meiosis. During the sexual life

cycle, the diploid sporophyte generation produces _____ spores through meiosis.

Short Answer

15. How do representatives of the two kingdoms of bacteria differ in cell wall and plasma membrane composition? How do they differ in gene architecture and translation machinery?

16. Name three organelles found in eukaryotic cells that have endosymbiotic origins. Briefly describe the likely prokaryotic ancestors of these organelles.

17. Name and describe two forms of asexual reproduction, and identify an organism that uses each form.

18. Examine the life cycles of *Clamydomonas* and humans in Figures 19-8 and 19-9. How are these two life cycles similar? How are they different?

Highlights Review

19. Until recently, all bacteria were classified in the kingdom Monera. What observations led scientists to classify bacteria into the two kingdoms Archaebacteria and Eubacteria?

20. Name two species of *Clostridium,* and describe their harmful and beneficial effects on humans.

21. What criteria are used to classify organisms as members of the kingdom Protista?

22. Name several external heterotrophs and internal heterotrophs. In what kingdom are they found? How do such organisms differ in the way they obtain nutrients?

23. Describe the process of fertilization found in angiosperms. How does this process differ from that found in animals?

24. Complete the chart at the top of next page.

system. *C. pasteurianum* converts nitrogen gas to the ammonia required by crops.

21. Protists are mostly unicellular organisms that are autotrophic, heterotrophic, or both. They reproduce asexually, but some can reproduce sexually as well.

22. Mushrooms and other members of the kingdom Fungi are

external heterotrophs; they obtain food by absorption. Ladybird beetles and all other members of the kingdom Animalia are internal heterotrophs; they acquire food by eating.

23. Double fertilization involves sperm with two nuclei. One sperm fertilizes the egg, forming the zygote, and the second sperm fuses with other nuclei,

producing the endosperm. Both the zygote and endosperm are enclosed in a seed coat. Fertilization in animals involves the union of sperm and egg to form a zygote.

24. (Answers for table are listed by rows) Prokaryote; Eubacteria; Unicellular; *Clostridium;* Protista; Heterotroph, autotroph, or both; Eukaryote; External heterotroph;

Kingdom	Prokaryote or Eukaryote	Cellular Organization	Mode of Nutrition	Representative Organism
Archaebacteria		Unicellular	Heterotroph	*Sulfolobus*
	Prokaryote		Heterotroph or autotroph	
	Eukaryote	Unicellular; multicellular		*Plasmodium*
Fungi		Mostly multicellular		*Amanita*
Plantae	Eukaryote			
Animalia			Heterotroph	

Themes Review

25. Evolution Compare a unicellular alga, such as *Vorticella*, with a green plant. What distinguishing characteristics of multicellular organisms are present in the green plant, but not in the alga?

26. Levels of Organization Rank the seven major taxonomic groups from the group at the top of the hierarchy to the group at the bottom of the hierarchy. Which two taxonomic groups provide the scientific name of an organism? Where are they located in the hierarchy?

27. Evolution Describe the evidence that supports the theory that mitochondria evolved from eubacteria.

Critical Thinking

28. Making Inferences How does the cladistic approach to the classification of birds differ from the method used in evolutionary systematics? Which method best reflects the ancestry of the birds? Defend your answer.

29. Making Inferences Many organisms are repulsed by the odor of decaying organic matter, which often contains infectious bacteria. What adaptive advantage might this reaction give an organism?

Activities and Projects

30. Research and Writing The endosymbiotic hypothesis is one of two hypotheses that explain the origin of eukaryotic cells. The second hypothesis is the autogenous hypothesis. Research the autogenous hypothesis. In your report, present the highlights of both hypotheses and explain why most scientists favor the endosymbiotic hypothesis.

31. Research and Writing The bacterium *Escherichia coli* usually reproduces by binary fission, but it can also transfer genetic material through a sex pilus. Research how this transfer of genetic material takes place. Summarize this process and speculate how the transfer of genetic material might impact the use of antibiotics.

Readings

32. Read the article "Marriage of Convenience" in *Sciences*, September–October 1990. To whom did Lynn Margulis and Mark McMenamin give credit for first proposing endosymbiotic origins for chloroplasts, mitochondria, and centrioles? What 1960s technological advancement enabled Margulis to provide evidence for the endosymbiotic origins of eukaryotic organelles?

Critical Thinking

28. The cladistic method places birds among the reptiles (class Reptilia), whereas the evolutionary systematics approach erects a special class for the birds alone (Aves). The cladistic approach to classification best reflects the ancestry of the birds. The evolutionary systematic approach emphasizes particular character differences. (19-4.5)

29. Many infectious bacteria can be found in decaying matter. An organism that is repulsed by such substances would likely avoid any contact and thus be less likely to become infected. (19-1.1)

Activities and Projects

30. The autogenous hypothesis states that a gradual increase in complexity of prokaryotic cells led to the evolution of eukaryotic cells. The endosymbiotic hypothesis indicates that the union of prokaryotic cells played an important part in the formation of eukaryotic cells. The similarities between certain cell organelles and prokaryotes cause scientists to favor the endosymbiotic hypothesis, but neither can be discounted because intermediate stages of evolution between prokaryotes and eukaryotes have not been found.

31. The transfer of genetic material has the same effect as sexual reproduction in terms of genetic diversity. The evolution resulting from the genetic diversity has led to strains of bacteria that are resistant to antibiotics that have killed them in the past.

Readings

32. Margulis and McMenamin credit Konstantin Sergeivich Mereschkovskii for first proposing endosymbiotic origins for chloroplasts, Ivan Wallin for mitochondria, and Boris Kozo-Polyanski for centrioles. Margulis discovered evidence for the theory using the electron microscope.

Multicellular; Cherry tree; Eukaryote; Multicellular; Ladybird beetle

Themes Review

25. The green plant has tissues that are made up of many different kinds of cells and that are devoted to specific functions; the alga does not. The cells in the green plant are able to communicate with each other and coordinate their activities. (19-3.4)

26. The seven major taxonomic groups are kingdom, phylum (or division), class, order, family, genus, and species. An organism's scientific name is composed of its genus and species names, which are the two most exclusive groups. (19-4.2, 19-4.3)

27. Mitochondria are about the same size as most eubacteria, and they contain a folded membrane that resembles the plasma membrane of eubacteria. Like eubacteria, mitochondria have a circular molecule of DNA and ribosomes. Mitochondria reproduce by fission. (19-1.1)

LABORATORY Investigation Chapter 19

Procedural Notes

- You may want to have students complete Parts A and B on two different days. Part A will take about 10 minutes to complete. Part B will require about 40 minutes.

- Having students make dichotomous keys allows them to practice observing, recording, and organizing data. Lead students through a discussion on how to use the process of elimination when working with a dichotomous key.

- Encourage students to select characteristics that are useful in distinguishing one shoe from another. Characteristics might include type of shoe, style, color, size, or function.

- Emphasize that a dichotomous key includes pairs of descriptions that lead to the identification of an object.

Background Answers

1. Carolus Linnaeus developed the classification system used by scientists.

2. Scientists classify organisms by their characteristics, such as structural features and chemical makeup. Classification enables scientists to identify a specific organism and to communicate accurate information efficiently.

3. A dichotomous key uses pairs of contrasting descriptive statements to lead to the identification of an organism (or some other object).

4. How can a dichotomous key be used to identify leaves?

Records

A chestnut oak
B red bud
C shingle oak
D lombardy poplar
E live oak
F English oak
Dichotomous keys of students' shoes will vary.

Dichotomous Keys

OBJECTIVES

- Use a dichotomous key to identify leaves.
- Construct a dichotomous identification key.

PROCESS SKILLS

- identifying and comparing characteristics of objects
- organizing data

MATERIALS

- shoes
- masking tape
- marker

BACKGROUND

1. Who first developed the classification system used by scientists today? What is taxonomy?
2. How do scientists classify organisms? Why is classification an essential tool of biology?
3. What is a dichotomous key?
4. Write your own question to explore in your report.

TECHNIQUE

Part A: Using a Dichotomous Key

1. Field guides often use dichotomous keys to identify organisms. Use the dichotomous key shown here to identify the tree leaves below. Begin with descriptions 1a and 1b, and follow the directions. Proceed through the list of paired descriptions until you identify the leaf in question. In the **Records** section of your report, write the names of the leaves.

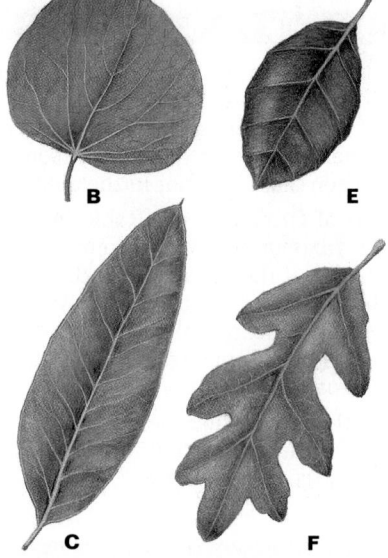

1a. If the edge of the leaf has no teeth, waves, or lobes, go to 2 in the key.
1b. If the edge of the leaf has teeth, waves, or lobes, go to 3 in the key.
2a. If the leaf has a single bristle at its tip, it is a shingle oak.
2b. If the leaf has no single bristle at its tip, go to 4 in the key.
3a. If the leaf edge is toothed, it is a lombardy poplar.

3b. If the leaf edge has waves or lobes, go to 5 in the key.

4a. If the leaf is heart-shaped with veins branching from the base, it is a redbud.

4b. If the leaf is not heart-shaped, it is a live oak.

5a. If the leaf edge has lobes, it is an English oak.

5b. If the leaf edge has waves, it is a chestnut oak.

Part B: Making a Dichotomous Key

2. Ask 10 or more student volunteers to remove their shoes and use masking tape and a marker to label the sole of one of their shoes with their name. The labeled shoes should then be placed on a single table in the classroom.

3. Form into small groups. Discuss the appearance of the shoes. In the **Records** section of your report, make a table that lists some general characteristics of the shoes, such as the type and size. Also list the names of the students who own the shoes. Complete the chart by describing the characteristics of each person's shoe.

4. Use the information in your table to make a dichotomous key that can be used to identify the owner of each shoe. Remember that a dichotomous key includes pairs of opposing descriptions. At the end of each description, the key should either identify an object or give directions to go to another specific pair of descriptions. Write your dichotomous key in the **Records** section of your report.

5. After all groups have completed their key, exchange keys with a member of another group. Use the key to identify the owner of each shoe, and then verify the accuracy of your identification by reading the label on the shoe. If the key has led you to an inaccurate identification, return the key so that corrections can be made.

6. In your report, briefly summarize the procedure you followed.

INQUIRY

1. What other characteristics might be used to identify leaves with a dichotomous key?

2. How was the shoe identification key that your group designed dichotomous?

3. Were you able to identify the shoes using another group's key? If not, describe the problems you encountered.

ANALYSIS

1. How was it helpful to list the characteristics of the shoes before making the key?

2. Does a dichotomous key begin with general descriptions and then proceed to more specific descriptions, or vice versa? Explain your answer, giving an example from the key you made.

FURTHER INQUIRY

Write a new question that could be explored as a new investigation. The following is an example:

What characteristics might be used to identify birds or other animals using a dichotomous key?

1. Other leaf characteristics might include whether the leaf is compound or simple, whether or not the leaf is needlelike, the arrangement of leaves on the branches, and the leaf's vein pattern.

2. The key was dichotomous because the shoe's descriptive statements were written as pairs of contrasting statements.

3. Answers will vary. Students might have discovered that some of the opposing statements were not contrasting, that some of the shoes were not correctly described, or that the key did not include a sufficient number of descriptions to distinguish all the shoes from one another.

Analysis Answers

1. Listing the shoes' characteristics beforehand makes it easier to identify those characteristics that distinguish one shoe from another.

2. Dichotomous keys proceed from general characteristics to specific characteristics. Examples from keys will vary but should reflect this gradation.

UNIT 5

CHAPTER 20 — VIRUSES AND BACTERIA

Block Scheduling Guide

Each block represents about 45–50 minutes of instructional time. The resources cited in a block are coded to a specific program component using the Key on the next page.

Lecture Concepts | Lesson Resources

20-1 Viruses pp. 454–461

BLOCKS 1 and 2

a. Viruses are particles that consist of segments of a nucleic acid contained in a protein sheath. Because viruses require living cells for reproduction, biologists do not consider them to be living organisms.

b. Viruses reproduce inside living cells. They can enter a cell by injecting their genetic material into the cell, slipping through tears in the cell wall, or binding to molecules on the cell surface and triggering endocytosis.

c. Viruses are able to reproduce by taking over a host cell's machinery. Retroviruses, such as HIV, are equipped with reverse transcriptase, a unique enzyme that can transcribe DNA from an RNA template.

Lecture Resources
- Opening Question, page 455
- Demonstration: TMV Resistance
- Visual Strategies: Figures 20-3, 20-7
- Teaching Transparencies: 93, 94, 95, 96
- Viruses Are Not Living Organisms, page 457
- Headlines, page 459
- New Viruses, page 461
- Up Close, page 460
- Holt Biology Videodiscs: Lesson 20-1

Classwork Options
- Design a Virus, page 456
- Laboratory Techniques and Experimental Design C30: Screening for Resistance to Tobacco Mosaic Virus

Assignment Options
- Teaching Resources CD-ROM: Occupational Applications Worksheets, Forensic Toxicologist
- Section Review, page 461
- Chapter Review, questions 1–3, 8, 9, 14, 17

20-2 Bacteria pp. 462–468

BLOCKS 3 and 4

a. Bacteria can be classified into two groups according to cell wall structure. Gram staining can be used to identify bacteria because it distinguishes between two different kinds of bacterial cell walls.

b. Bacteria and eukaryotes differ in their cellular organization, cell structures, and in metabolic activities.

c. Bacteria can be classified according to the ways in which they get energy. Photosynthetic bacteria are autotrophs. Chemoautotrophic bacteria use inorganic molecules as a source of energy. Heterotrophic bacteria obtain energy by feeding on organic matter.

Lecture Resources
- Opening Question, page 462
- Demonstration: Bacteria Are Useful
- Visual Strategy: Figure 20-9
- Teaching Transparencies: 90, 91
- Bacteria in Food, page 463
- Overcoming Misconceptions, page 463
- Up Close, page 465
- Holt Biology Videodiscs: Lesson 20-2

Classwork Options
- Breakthroughs in Science, page 466
- Classifying Bacteria and Graphic Organizer, page 467

- Quick Labs A16: Using Bacteria to Make Food
- Laboratory Techniques and Experimental Design C31: Aseptic Technique
- Laboratory Investigation: Identifying and Staining Bacteria, pages 474–475
- Laboratory Techniques and Experimental Design C32: Gram Staining of Bacteria

Assignment Options
- Section Review, page 468
- Chapter Review, questions 4–6, 10, 11, 15

20-3 Viruses and Bacteria as Pathogens pp. 469–471

BLOCKS 5 and 6

a. Pathogenic bacteria are harmful because they attack cell or secrete toxins. Antibiotics are used to fight pathogenic microorganisms and work by interfering with the microbe's cellular processes.

b. Viruses damage cells and cause illness.

Lecture Resources
- Opening Question, page 469
- Demonstration: Talking about Diseases
- Teaching Transparencies: 92, 97
- "Rational" Drug Design, page 470
- Preventing Drug Resistance, page 471
- Holt Biology Videodiscs: Lesson 20-3

Classwork Options
- Design a Cartoon, page 470
- Laboratory Techniques and Experimental Design C33: Gram Staining of Bacteria—Treatment Options
- Closure, page 472

Assignment Options
- Section Review, page 471
- Chapter Review, questions 7, 12, 13, 16, 18, 19

Review/Enrichment

BLOCK 7

■ Study Guide: Concept Review and Pretest

Assignment Options
- ◉ Teaching Resources CD-ROM: Supplemental Reading Worksheets and Test, *Microbe Hunters*
- ■ Activities and Projects, questions 20, 21
- ■ Science, Technology, and Society pages 476–477

Assessment Options

BLOCK 8

Traditional Assessment
- ■ Chapter 20 Test
- ◉ Test Generator/Assessment Item Listing: Software item bank for preparing customized chapter tests.

Portfolio Assessment
Select student reports for one or more laboratory experiments from this chapter. *The Direct Observation Checklist* on the *BioSources Teaching Resources CD-ROM* should be completed during a laboratory or other cooperative learning experience.

Holt Biology Videodiscs

Holt Biology Videodiscs gives you a powerful tool for teaching, review, and assessment. *Concepts of Biology* adds interactive lesson plans and extensive tools for customization using CD-ROM technology.

CONCEPTS OF BIOLOGY

Use the following topic from *Concepts of Biology* to help you teach this chapter:

■ Topic 13: Bacteria and Viruses

For further information regarding the media on the videodiscs, see *Holt Biology Videodiscs Teacher's Correlation Guide to Biology: Principles and Explorations.*

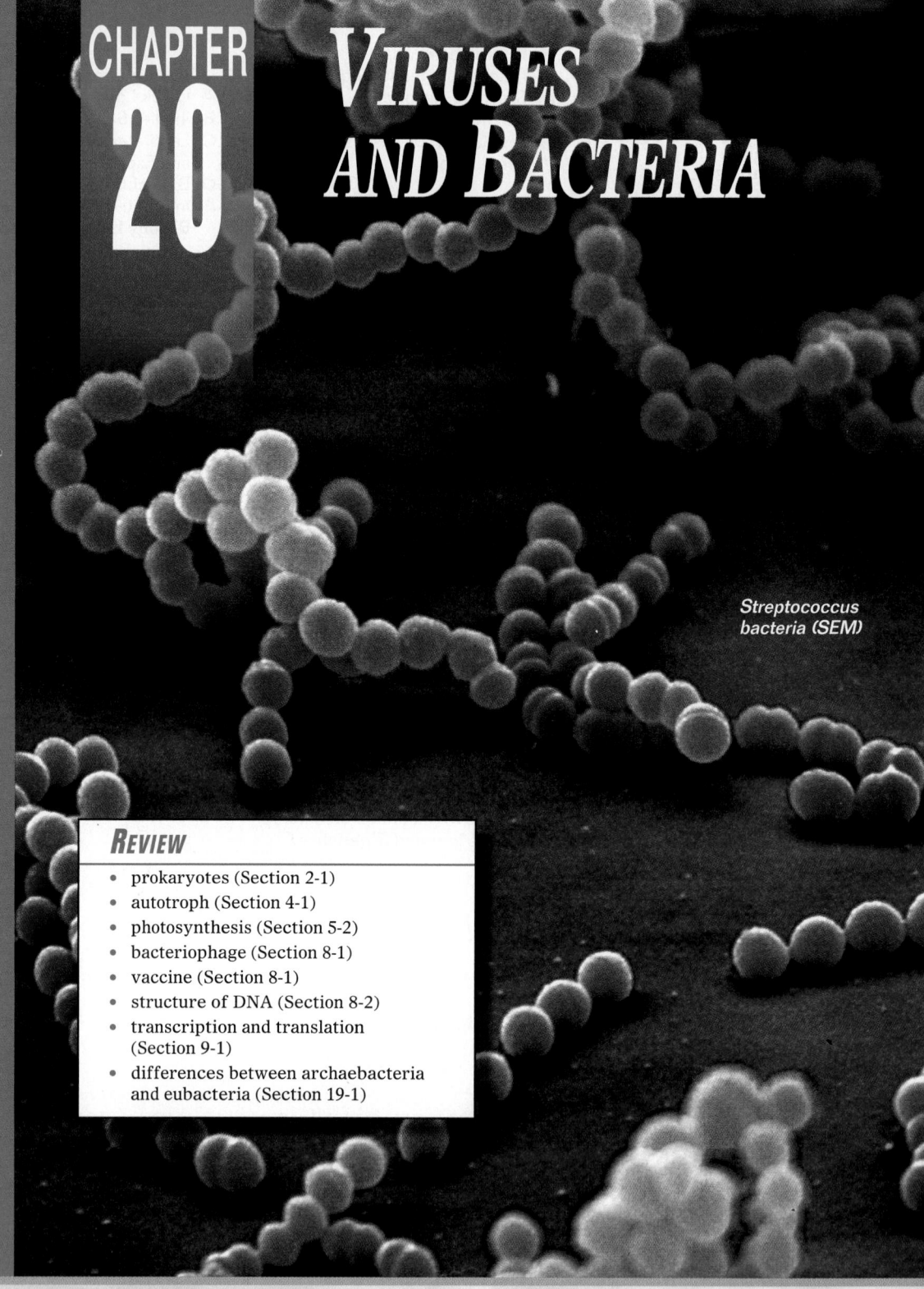

Chapter Theme

Structure and Function

One reason why viruses and bacteria are successful is that they develop modified structures, which allow them to adapt and reproduce in environments that would otherwise be unfavorable.

Tapping Prior Knowledge

- What components are necessary to classify a structure as a living cell?
- How is the structure of prokaryotes different from that of eukaryotes?

Opening Questions

Ask students which of the following diseases are caused by viruses: measles, AIDS, tuberculosis, syphilis, influenza, chickenpox, botulism, bubonic plague, polio, mumps, Lyme disease, strep throat, the common cold. Measles, AIDS, influenza, chickenpox, polio, mumps, and colds are caused by viral infections; the other diseases listed above are caused by bacterial infections. Explain that although antibiotics can cure many bacterial infections, drugs have not been effective at curing viral infections. Ask students how a viral illness is cured. Tell them that the immune system produces antibodies that help destroy the virus.

CHAPTER 20

VIRUSES AND BACTERIA

Streptococcus bacteria (SEM)

REVIEW

- prokaryotes (Section 2-1)
- autotroph (Section 4-1)
- photosynthesis (Section 5-2)
- bacteriophage (Section 8-1)
- vaccine (Section 8-1)
- structure of DNA (Section 8-2)
- transcription and translation (Section 9-1)
- differences between archaebacteria and eubacteria (Section 19-1)

Authors' Rationale

This chapter's overview of viral and bacterial biology is particularly pertinent in light of current diseases and antibiotic failures. Students will learn how to classify the major groups of viruses and bacteria and about their origins, structures, and functions. In addition, students will become familiar with how viruses and bacteria can be useful. This chapter also summarizes common diseases caused by viruses and bacteria, routes of infection, and treatments.

20-1 Viruses

*I*n Chapter 1 you learned about the properties of life. All living things are cellular, able to grow and reproduce, and are guided by information encoded in the nucleic acid DNA. The smallest organisms that have these properties are bacteria, single cells that were the earliest forms of life. Even smaller than bacteria are viruses, mere segments of nucleic acids wrapped in a protein coat. Because viruses depend on the cells of other living organisms in order to reproduce, biologists do not consider viruses to be alive. However, when they are able to reproduce, viruses often cause damage to the host organism. For this reason, viruses have had, and continue to have, a major impact on the living world, as you can see in Figure 20-1.

Is a Virus a Living Organism?

Biologists first suspected the existence of viruses near the end of the nineteenth century. At that time, European scientists were seeking to identify the infectious agent responsible for tobacco mosaic disease, a disease that stunts the growth of tobacco plants and makes their leaves blotchy. One experiment used fine-pored porcelain filters to strain bacteria from sap extracted from infected tobacco plants. The pores in these filters were so small that bacteria could not pass through them. However, the infectious agent passed through the filters without difficulty. Scientists concluded that the infectious agent must be smaller than a typical bacterial cell. After further investigation, they found that the agent could reproduce only inside the living cells it infected. They called the agent a *virus*, a Latin word meaning "poison."

For many years after this discovery, viruses were erroneously regarded as primitive forms of life, tiny cells that were perhaps the ancestors of bacteria. The true nature of viruses was discovered in 1933 when the biologist Wendell Stanley of the Rockefeller Institute tried to purify an extract of the tobacco mosaic virus (TMV). To Stanley's great surprise, the purified TMV extract formed crystals, a property of chemicals. The crystals retained the ability to infect healthy tobacco plants and were therefore the virus itself. Stanley concluded that TMV is chemical matter rather than a living organism.

Within a few years of Stanley's findings, scientists were able to disassemble TMV and confirm Stanley's conclusion. TMV is a chemical, not a cell. In fact, each particle of TMV is made of only two kinds of molecules: the nucleic acid RNA

SECTION 20-1

Section Objectives

■ Discuss the events that led to the discovery and understanding of tobacco mosaic virus.

■ Define virus, and explain why a virus is not considered a living organism.

■ Describe the basic structure of a virus.

■ Describe the structure of HIV and explain how it reproduces.

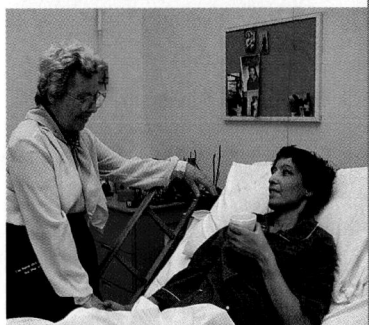

Figure 20-1 This AIDS patient is receiving treatment in a clinic at San Francisco General Hospital. In 1994, nearly 81,000 new AIDS cases were reported.

Vocabulary Preview

virus

capsid

envelope

glycoprotein

icosahedron

bacteriophage

pathogen

antibody

reverse transcriptase

retrovirus

Opening Question

Ask students to discuss what tomatoes, white potatoes, and garden peppers have in common. Explain that all three are from the nightshade family, Solanaceae, and that they can catch the same diseases. Another familiar member of the nightshade family is tobacco. A virus that plagues this family worldwide was first recognized on tobacco. The virus is tobacco mosaic virus (TMV).

Demonstration

TMV Resistance

Show students examples of labels found on tomato plants at a nursery or ads in nursery catalogs that state that the plants are resistant to TMV. Explain that nursery workers, farmers, and gardeners who handle tobacco products can carry the virus on their hands. Simply by handling a plant, they can infect it with the virus, which can be fatal to the plant. An infected plant will develop a pattern of light and dark patches on its leaves that are symptoms of the disease.

 Did You Know?

The oil-eating bacteria that are used to clean up oil spills were the first organisms protected under a United States patent. Ananda Chakrabarty engineered the bacterium in 1972.

UNIT 5
CHAPTER 20

Teaching Tip
Design a Virus

Have students create a fictitious virus. Students should produce a drawing or a model of their virus. They should also provide historical information about the origin of the virus, its components, the host it infects, and the effects of the virus on its host. Display the students' work after individuals share their creations with the class.

Figure 20-3

Use this figure to point out the standard parts of a virus: the capsid (proteins) and nucleic acid (RNA, in this case). Note that most viruses contain DNA, and some contain RNA. Ask students what the membranous envelope surrounding the capsid of the influenza virus is made of. *(proteins, lipids, and glycoproteins from the host cell)* Emphasize that not all viruses look alike; some are larger and more complex than others.

and protein. As you can see in Figure 20-2, TMV consists of a core of RNA surrounded by a coat of protein. Later scientists were able to separate the RNA from the protein. When they reassembled the two components, the reconstructed TMV particles were fully able to infect healthy tobacco plants. Clearly, the chemicals were the virus itself, not merely derivatives of it. From these experiments, and a host of others carried out using other viruses, a general picture of virus structure and function has emerged. A **virus** is a strand of nucleic acid encased in a protein coat that can infect cells and replicate within them. Biologists do not consider viruses to be living organisms.

Figure 20-2 Tobacco mosaic virus is rod-shaped and has a coat of proteins spiraling around a single strand of RNA. TMV causes a disease that stunts the growth of tobacco plants and discolors their leaves.

Figure 20-3 Influenza virus is made of a coil of RNA surrounded by a lipid-rich envelope studded with protein spikes.

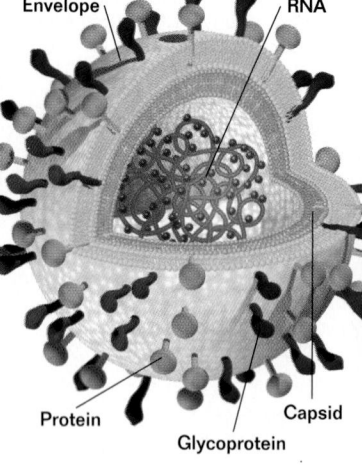

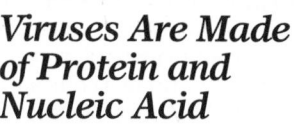

Envelope — RNA

Protein — Capsid

Glycoprotein

Viruses Are Made of Protein and Nucleic Acid

Like TMV, most viruses have a protein sheath, or **capsid,** surrounding a core of nucleic acid. Many plant viruses, as well as some animal viruses such as the human immunodeficiency virus (HIV) that causes AIDS, contain RNA. However, the nucleic acid found in most viruses is DNA. Many viruses found in animals, such as the influenza virus shown in Figure 20-3, have a membranous **envelope** surrounding the capsid. The envelope helps the virus gain entry into cells. It contains proteins, lipids, and **glycoproteins** (proteins with carbohydrate molecules attached) derived from the host cell.

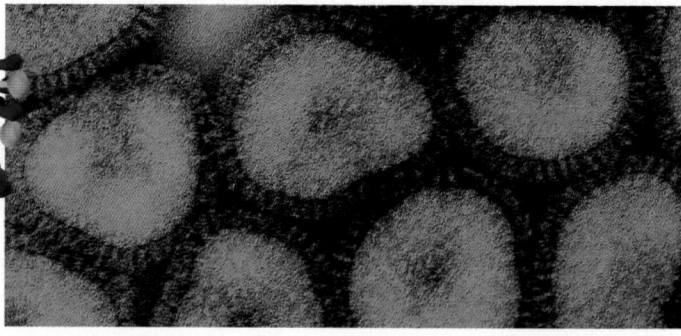

Viruses differ greatly in appearance. The simplest viruses consist of a single molecule of a nucleic acid and a capsid made of a single protein or a few different protein molecules repeated several times. The more complex viruses may consist of several different segments of DNA or RNA contained within a capsid made of several different kinds of protein. Most viruses have an overall structure that is either helical or polyhedral. A helical virus, like the tobacco mosaic virus, is rodlike in appearance, with capsid proteins winding around the core in a helix. A polyhedral virus has many sides and is roughly spherical. The capsid of most polyhedral viruses is in the shape of an **icosahedron** (*eye koh suh HEE druhn*), a shape with 20 triangular faces and 12 corners. By arranging themselves as an icosahedron, the proteins form an external shell that has more volume than would be possible with other shapes. Figure 20-4 shows the polyhedral shape of an adenovirus, which can cause upper respiratory infections in humans.

Some viruses, particularly bacterial viruses called **bacteriophages,** have very complicated structures. By taking a close look at Figure 20-5, you can see that the capsid is a polyhedral head attached to a helical tail. A long DNA molecule is coiled within the head.

Figure 20-4 Adenovirus, *right*, is in the shape of an icosahedron, *left*. It has a capsid with a protein spike at each of its 12 corners.

Figure 20-5 Many bacteriophages (viruses that infect bacteria) have a complex capsid made of a polyhedral head attached to a tail. Such viruses usually inject their genetic material into a cell like a hypodermic needle.

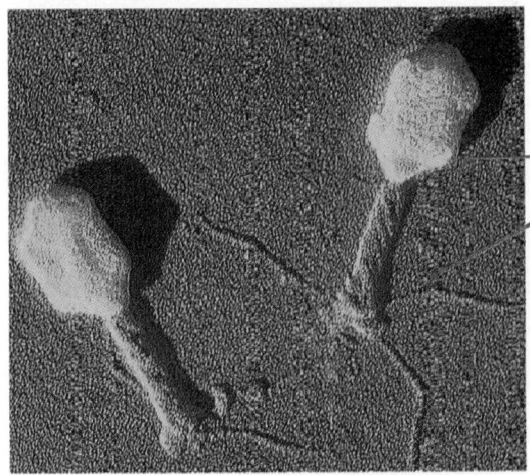

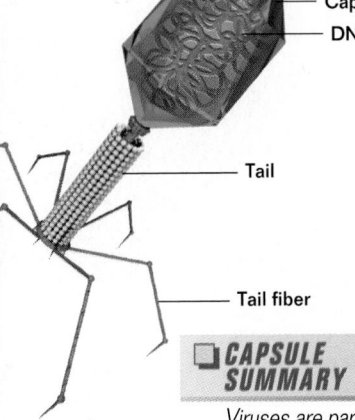

— Capsid
— DNA

— Tail

— Tail fiber

The smallest viruses are only about 17 nm in diameter. The largest ones may be up to 100 nm, barely big enough to be visible with a light microscope. Most viruses can be detected only by using the higher resolution of an electron microscope. ◻

◻ CAPSULE SUMMARY

Viruses are particles that consist of segments of a nucleic acid contained in a protein sheath. Because viruses require living cells for reproduction, biologists do not consider them to be living organisms.

Historical Note

Hantavirus
On the Navajo reservation in the "four corners" region of Colorado, New Mexico, Utah, and Arizona, a viral infection previously unknown to Americans killed 16 people within six months in 1993. This pulmonary syndrome, which is spread by deer mice, is caused by a hantavirus, named after the Hantaan River in Korea, where U.S. military researchers first encountered it. The American hantavirus variant attacks primarily the lungs, unlike Asian and European strains, which cause hemorrhagic fever and kidney disease. Health workers still don't know whether the American hantavirus variant is really new or whether it has simply gone unnoticed until recently.

Application

 Medicine Although viruses are species-specific, some scientists fear that vaccines made from live viruses may help viruses cross from one species to another. One possible example is the canine parvovirus (CPV), which first appeared in dogs in 1977. This virus, which causes diseases of the heart and the intestines in dogs, is very similar to the virus found in a vaccine used to prevent feline panleukopenia virus (FPLV) in cats. One theory regarding the origin of CPV is that the FPLV vaccine could have been injected into a dog where the weakened virus adapted to the new host and gave rise to CPV.

Viruses Reproduce Inside Living Cells

Viruses lack the enzymes for metabolism and have no ribosomes or other equipment for protein synthesis. Therefore they must rely on living cells for reproduction. Before a virus can reproduce, it must first infect a living cell. How does a virus get into a cell? A bacterial virus, like bacteriophage T4, punches a hole in the bacterial cell wall and injects its DNA into the cell like a hypodermic needle. A plant virus, like TMV, enters a plant cell through tiny rips in the cell wall at points of injury. An animal virus enters its host cell by endocytosis, the process by which the cell engulfs materials that are too large to enter through channels in the cell membrane. Once they are inside a cell, many viruses are **pathogens** (*PATH uh jehnz*), agents that cause disease.

Before any virus can be engulfed by a cell, the virus must first bind to the cell membrane. What does the virus bind to? The envelope of an animal virus has spikes of glycoproteins and lipids that are able to bind to specific receptor molecules on the cell membrane. An animal virus is able to infect only cells with surface receptor proteins to which the virus's envelope molecules can attach. This is why human immunodeficiency virus (HIV), shown in Figure 20-6, infects only human white blood cells. It is also why the polio virus enters only certain spinal nerve cells and the hepatitis virus enters only liver cells.

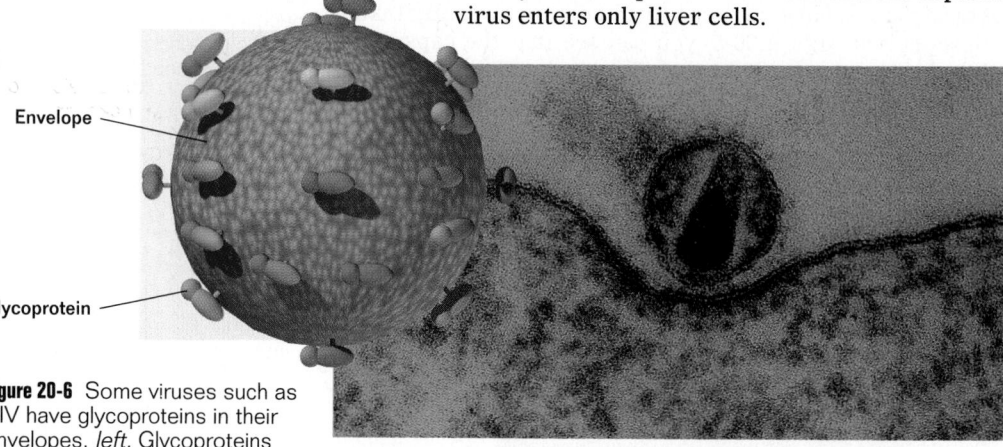

Envelope

Glycoprotein

Figure 20-6 Some viruses such as HIV have glycoproteins in their envelopes, *left*. Glycoproteins help these viruses enter an animal cell by binding to specific receptors on the cell surface, *right*.

Mammals protect themselves from viral infections by producing antibodies to the virus envelope's glycoproteins. An **antibody** is a protein secreted by cells in the immune system in response to a foreign substance in the body. The antibodies enable mammalian immune systems to identify and destroy viruses. However, mutations in viruses often change their glycoproteins and therefore make it difficult for the antibodies to recognize the virus. That is why new strains of influenza (flu) to which people are not resistant continually arise. As time passes, the envelope of the flu virus keeps changing. Like a fugitive changing disguises, the altered virus avoids detection.

❔ Did You Know?

By studying the structure and function of viruses, Dr. Nir Kossovsky and his colleagues at the University of California have come up with a way to produce decoy viruses that could be used in vaccines. Dr. Kossovsky coats a ceramic core with sugars and viral proteins. Since the fake virus does not contain DNA or RNA, it would be safe to use in a vaccine. The immune system responds to the outer shell of a virus to protect the body; thus, the body would produce antibodies against the fake virus. Although his vaccines will not be tested on humans for a long time, Dr. Kossovsky has already developed vaccines with decoys for several viruses, including HIV, for use in lab animals.

Mutations in the virus genes that encode the structure of its glycoproteins may also enable the virus to bind to a receptor protein that it failed to recognize earlier. Many scientists think that HIV became a human pathogen when a mutation occurred in the HIV gene that encodes a glycoprotein (gp 160) in the envelope. The mutation altered the glycoprotein so that it was able to recognize and bind to the human white blood cell protein called CD4. The cells with the CD4 protein play a key role in the production of antibodies. Once able to bind to these cells, HIV enters and destroys them, disrupting the human immune system and eventually causing the disease AIDS. ❑

◻ **CAPSULE SUMMARY**

Viruses reproduce inside living cells. They can enter a cell by injecting their genetic material into the cell, slipping through tears in the cell wall, or binding to molecules on the cell surface and triggering endocytosis.

Teaching Tip
Headlines

Ask students to write headlines they expect to see in the year 2010 regarding HIV. Discuss the headlines and the predictions behind them. Then, have students write short essays to explain their predictions about AIDS in the future. Post the headlines and essays around the room.

Viruses Take Over a Cell's Machinery

For an example of how a virus is able to reproduce within a host cell, study Figure 20-7, which illustrates how HIV infects human white blood cells. HIV gains access to a white blood cell first by binding to the cell membrane. The binding triggers endocytosis, and the cell membrane folds inward and surrounds the virus. The virus enters the cell within a membrane-bound vesicle, which soon releases the virus into the cell cytoplasm.

Once within the host cell, HIV sheds its envelope and capsid, leaving two strands of the virus's RNA floating in the cytoplasm. HIV also contains an enzyme called **reverse transcriptase,** which manufactures DNA that is complementary to the virus's RNA. A virus that transcribes DNA from an RNA template is called a **retrovirus.** Reverse transcriptase and other structural features of HIV are shown in *Up*

CONTENT LINK

HIV's devastating effect on the human body's immune system will be discussed in **Chapter 39**.

Figure 20-7 After HIV enters a cell by endocytosis, its nucleic acid (RNA) is used to make viral proteins and RNA.

👁 **VISUAL STRATEGY**

Figure 20-7
Use this figure to trace the steps of HIV reproduction within a white blood cell. After HIV builds an outer envelope that binds to a white blood cell's cell membrane, the virus is engulfed into the cytoplasm. Within the cell, reverse transcriptase generates DNA using the virus's RNA template, and the genes are translated into HIV proteins. The cell's protein synthesis machinery continues to manufacture HIV until the cell bursts, releasing many viruses to infect other white blood cells.

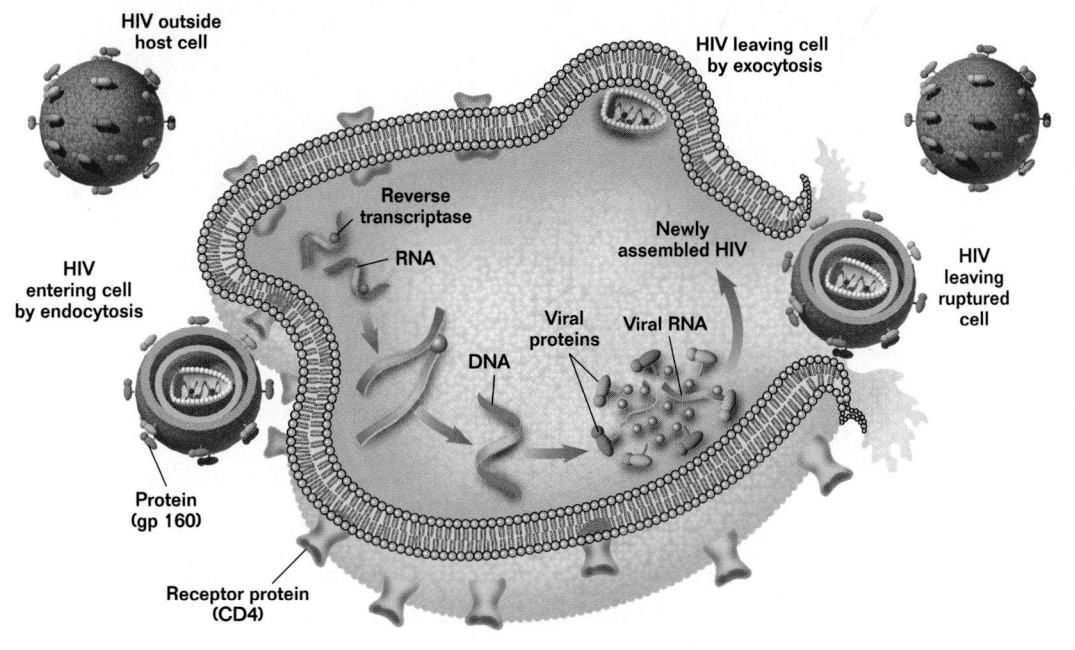

HIV outside host cell

HIV leaving cell by exocytosis

HIV entering cell by endocytosis

Reverse transcriptase

RNA

Newly assembled HIV

HIV leaving ruptured cell

Viral proteins

Viral RNA

DNA

Protein (gp 160)

Receptor protein (CD4)

Historical Note

Eradicating Smallpox

Smallpox is the only disease that has ever been eradicated. By producing a highly effective vaccine against the smallpox virus and using it to vaccinate populations worldwide, the World Health Organization was able to wipe out the disease. The last known smallpox case was reported in 1977 in Somalia. Since the virus was officially declared to be eradicated in 1980, samples of the smallpox virus have been stored in Atlanta, Georgia, and Moscow, Russia. Although the samples are scheduled to be destroyed, there is a wave of controversy surrounding them. Many scientists would like to maintain the samples for future study, while others would like to destroy the virus to avoid releasing it through misuse or accident.

Instructional Strategies

- The first cases of AIDS were reported in the 1970s. Have students obtain statistics on the spread of AIDS from the Centers for Disease Control and Prevention in Atlanta, Georgia. Ask students to present the information in brief reports. Encourage them to include a bar graph that compares the number of AIDS-related deaths over the past decades.

- Ask students to design a hypothetical drug to disable the infection cycle of HIV. Students' drugs can target any step within the cycle, such as preventing the virus from attaching to white blood cells, inhibiting endocytosis of the virus into the cell, or destroying reverse transcriptase. Ask students to make diagrams of the HIV infection cycle that show their drug at work. Encourage students to be creative when naming their drug. Remind students that the best way to fight AIDS is to prevent it.

Discussion

Guide the discussion by posing the following questions.

1. Describe the HIV genome. (*The HIV genome contains two single-stranded RNA molecules. The nucleotide sequence comprises 9,000 nucleotides, which make up nine genes.*)

2. What is the role of reverse transcriptase in HIV reproduction? (*Reverse transcriptase makes a DNA copy of the virus's RNA. Without reverse transcriptase, the virus would be unable to use the cell's machinery to reproduce.*)

3. Why is HIV considered an unusually complex virus? (*Six of HIV's nine genes regulate the expression of the virus's genes; hence, the virus is particularly adaptive to changing conditions.*)

UP CLOSE   *AIDS VIRUS*

- **Name:** Human immunodeficiency virus (HIV)
- **Habitat:** Interior of human white blood cells
- **Size:** 125 nm

Characteristics

Viral proteins Embedded within the HIV envelope are glycoproteins such as gp 120 (used in preparing the first experimental vaccines) and gp 160, which enables the virus to recognize the human white blood cell surface protein CD4 and thus enter the cell.

▲ Glycoprotein

Envelope The outer envelope of HIV is composed of a lipid bilayer derived from the membrane of the host cell. Beneath the envelope is a protein core.

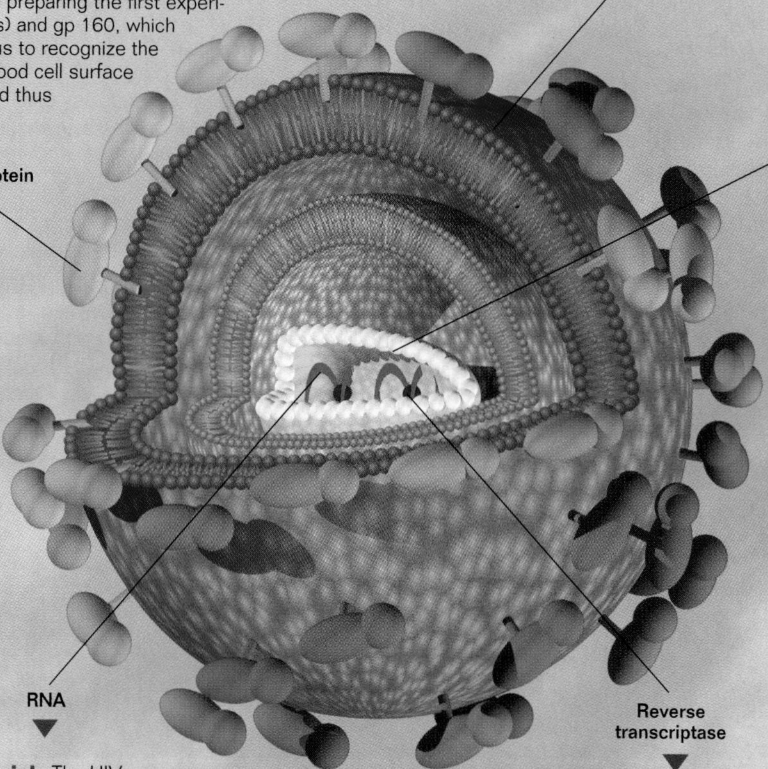

▲ Envelope

Capsid

RNA ▼

Reverse transcriptase ▼

Genetic material The HIV genome is made of two molecules of single-stranded RNA. Its complete nucleotide sequence consists of approximately 9,000 nucleotides, which make up nine genes. Three of these genes (GAG, PO, ENV) are found in many different viruses. However, HIV is unusually complex in that it also has six smaller genes that appear to regulate viral gene expression.

Reproduction HIV, a retrovirus, is equipped with the enzyme reverse transcriptase. Once inside a cell, reverse transcriptase makes a DNA copy of the virus's RNA. Using the cell's gene translation machinery, this DNA then directs the production of thousands of new viruses. The new viruses are released by budding out of the host cell or, occasionally, are released when the cell bursts. HIV can reproduce only within human cells.

Close: AIDS Virus on page 460. In some viral infections, such as herpes, the DNA then inserts itself into the host cell's chromosome, where it remains inactive until some future event causes it to be removed from the chromosome and to resume activity. In other viral infections, the DNA does not insert itself into the host chromosome. Whether HIV inserts itself into the chromosomes of the white blood cells it infects is currently being disputed. Individuals infected with HIV typically do not exhibit high levels of the virus for an average of eight years. There is some evidence that the immune system succeeds in suppressing an ongoing infection over this long period, until its resources eventually become depleted and the infection escapes control.

After the viral RNA is transcribed into DNA, the genes are translated into HIV proteins. The host cell's machinery is then used to produce and assemble many copies of the virus. Some of the newly assembled virus particles leave the cell by exocytosis (budding out). Eventually the host cell ruptures, releasing thousands of additional virus particles. These newly released virus particles are then free to infect other white blood cells and continue the cycle of infection. Most other animal viruses follow a similar course of infection, although details may differ in individual cases. ❑

❑ **CAPSULE SUMMARY**

Viruses are able to reproduce by taking over a host cell's machinery. Retroviruses, such as HIV, are equipped with reverse transcriptase, a unique enzyme that can transcribe DNA from an RNA template.

Origins of Viruses

Viruses are considered to be escaped fragments of host genomes. This is why viruses are almost always highly specific to the hosts they infect. Because viruses originated as fragments of bacterial and eukaryotic genomes, their great diversity is not surprising. Biologists think there are at least as many kinds of viruses as there are kinds of organisms. Because a given organism often is susceptible to many kinds of viruses, the actual number of kinds of viruses may be much greater than the number of organisms.

Section Review

1. How did Stanley's experiment with tobacco mosaic virus help reveal the nature of viruses?

2. What is a virus? Why don't biologists consider viruses to be living organisms?

3. Why do viruses rely on living cells for reproduction?

4. Why are some viruses able to avoid detection by the immune system?

Critical Thinking

5. Based on your knowledge of HIV structure and reproduction, describe one way to interrupt its life cycle.

Answers to Section Review

1. When Stanley purified an extract of TMV, the extract formed crystals.

2. A virus is a nucleic acid surrounded by a protein coat that can infect cells and replicate within them. Biologists do not consider viruses to be living organisms because they lack the components needed to live and replicate on their own.

3. Viruses require living cells to reproduce because the host cells provide the machinery needed for reproduction.

4. Some viruses are able to avoid detection by the immune system by constantly changing their glycoproteins.

5. Answers might include preventing the virus from binding to white blood cells, degrading the retrovirus's reverse transcriptase, or otherwise interrupting the replication of HIV.

Opening Question

Ask students to name some places where they might find bacteria. Write their responses on the chalkboard for comparison. Tell students that they are all correct because bacteria are basically everywhere—in the human body and on everything we touch and eat. Emphasize that bacteria are an important part of the web of life.

Demonstration

Bacteria Are Useful

Show students pictures of an oil spill, a ski resort, and a field with a growing crop. Ask students what useful roles bacteria might play in these pictures. Explain that bacteria are being "domesticated" by biotechnologists to perform a variety of useful tasks, including cleaning up spilled petroleum products and making snow for ski resorts (See **Did You Know?** on page 466.). In agriculture, bacteria can take the place of some fertilizers and pesticides and can even prevent frost damage.

20-2 Bacteria

Bacteria are the oldest, simplest, and most abundant form of life on Earth. In a single gram of soil, there may be 2.5 billion bacteria. It is not surprising, then, that bacteria play a very important part in the web of life on Earth. They play a key role in recycling minerals within the Earth's ecosystems. In fact, photosynthetic bacteria were responsible in large measure for the introduction of oxygen into the Earth's atmosphere. Bacteria cause some of the most deadly diseases that cause injury to plants and animals, including humans. Our constant companions, bacteria are present in everything we eat and on everything we touch.

Section Objectives

■ Differentiate between gram-positive and gram-negative bacteria.

■ List seven differences between bacteria and eukaryotic cells.

■ Describe the external and internal structure of Escherichia coli.

■ Describe three different ways bacteria can obtain energy.

Bacteria Differ in Cell Wall Structure

Bacteria are small, single cells and are the only ones characterized by prokaryotic organization. Early in the evolution of life, bacteria split into the two branches archaebacteria and eubacteria, which are now classified in separate kingdoms. Too tiny to see with the naked eye, a bacterial cell is usually one of three basic shapes, as shown in Figure 20-8: **bacillus** *(buh SIHL uhs)*, a rod-shaped cell; **coccus** *(KAHK us)*, a spherical cell; or **spirillum** *(spy RIHL uhm)*, a spiral cell. A few kinds of bacteria aggregate into stalked structures or filaments.

A bacterium's plasma membrane is encased within a cell wall. Members of the kingdom Eubacteria have a cell wall made of peptidoglycan, a network of polysaccharide molecules linked together with chains of amino acids. Some eubacteria have a cell wall covered with an outer membrane layer made of large molecules called lipopolysaccharides.

Figure 20-8 Approximately 4,800 different kinds of bacteria are currently recognized, but undoubtedly thousands more await discovery and description.

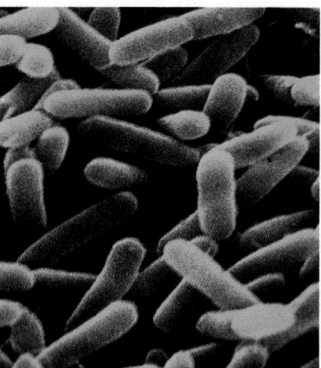
Pseudomonas

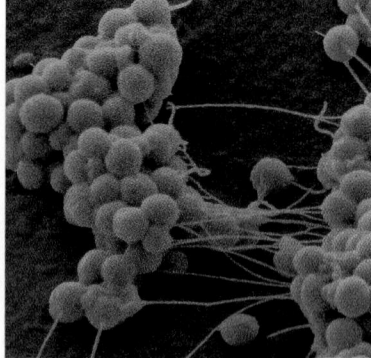
Staphylococcus

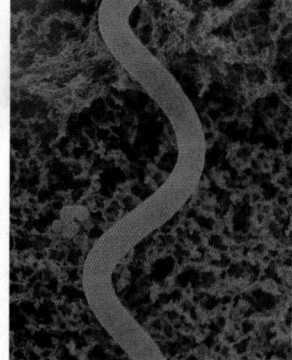
Spirillum

Multicultural Perspective

Making Yogurt

Originally, plain yogurt was made by nomadic Middle-Eastern tribes. The hot desert sun, aided by the gentle rocking of camel transportation, provided the perfect environment for making yogurt. Yogurt is created from milk that has been curdled by two types of bacteria: *Lactobacillus bulgaricus* and *Streptococcus thermophilus*. Because yogurt is cultured, it is easier to digest than regular milk. In addition, the bacteria in yogurt can produce B vitamins in the intestines. Today, many Americans prefer to eat yogurt that has been sweetened and flavored by adding fruit.

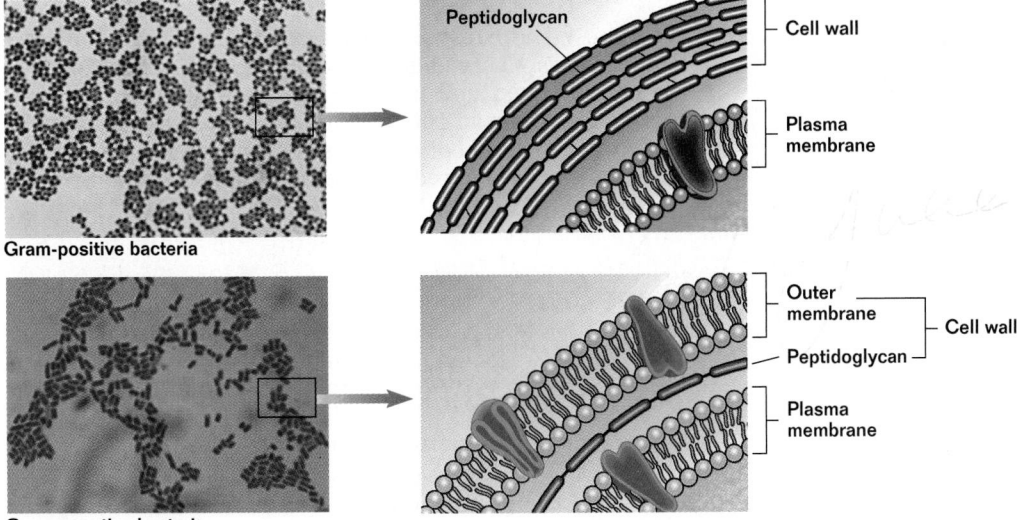

Gram-positive bacteria

Gram-negative bacteria

Peptidoglycan — Cell wall

Plasma membrane

Outer membrane — Cell wall

Peptidoglycan

Plasma membrane

Figure 20-9 The peptidoglycan layer is much thicker in gram-positive bacteria than in gram-negative bacteria, causing them to retain violet dye and appear purple.

(A lipopolysaccharide is a chain of sugar molecules with a lipid attached to one end.) Outside of the cell wall and membrane, many bacteria have a gelatinous layer called a **capsule.**

Eubacteria are commonly classified by differences in their cell walls. A bacterium with a cell wall containing a large amount of peptidoglycan is classified as **gram-positive.** A bacterium with a cell wall containing a thin layer of peptidoglycan covered by an outer membrane is classified as **gram-negative.** These terms refer to a bacterium's reaction to a staining procedure developed by the Danish microbiologist Hans Gram. These reactions are illustrated in Figure 20-9. In this procedure, a sample of bacteria is covered in a series of chemicals, beginning with a purple dye that stains the cell wall and ending with an alcohol rinse that breaks down cell membranes. Gram-positive bacteria retain the purple dye because their thick peptidoglycan layer remains intact and prevents the dye from leaving the cell. Gram-negative bacteria are not stained purple because the outer membrane is removed by the alcohol rinse and their thin peptidoglycan layer allows the dye to escape the cell. Gram-negative bacteria are usually identified by a pink stain that is applied after the alcohol wash.

Gram staining is an important technique in medicine. In many cases, the reaction to a Gram stain provides valuable information for the treatment of a bacterial disease. For example, gram-positive bacteria tend to be killed by penicillin, an antibiotic that prevents the proper formation of peptidoglycan in cell walls. Because of their outer membranes, gram-negative bacteria tend to be resistant to penicillin, but much more susceptible to the antibiotic tetracycline. Thus Gram stain identification of a bacterium can help determine which drug will be most effective against a disease.

Some bacteria form thick-walled **endospores** around their chromosomes and a small bit of cytoplasm when they

SECTION 20-2

👁 VISUAL STRATEGY

Figure 20-9
Use this figure to point out the differences in the cell walls of gram-negative and gram-positive bacteria. Ask students how the cell wall determines a bacterium's reaction to Gram stain. (*Because the cell wall regulates what enters and leaves the cell, it determines whether the stain will enter and color the cell.*) Which type of bacteria are antibiotics more effective at fighting? (*gram-positive bacteria*)

Teaching Tip
Bacteria in Food

Ask students to list some foods that bacteria help produce. (*Answers should include sauerkraut, cheese, yogurt, vinegar, and some alcoholic beverages.*) Explain that bacteria convert sugars to these food products through fermentation. Encourage students to find out more about how these foods are produced using bacteria.

CONNECTIONS

Chapter 10
Genetic Engineering

Cotton has been genetically engineered to contain a gene from a soil bacterium, *Bacillus thuringiensus.* The bacterium makes a protein in the leaves of cotton that controls an economically devastating pest, the bollworm. This biological control is harmless to humans and other animals, including beneficial insects.

Overcoming Misconceptions

"Bad" Bacteria

Students tend to think of bacteria as "bad" organisms because they are often described as sources of disease. However, only one bacterium out of one thousand is harmful. Bacteria are nature's recyclers, and scientists are finding ways to get bacteria to produce what people want and to degrade substances that people do not want. Biotechnologists insert genes into bacterial cells that allow the cells to make plastics, pharmaceuticals, pesticides, and foods. Bacteria can even be used to mine metals such as copper and gold and to clean up industrial wastes.

Teaching Tip
Movement and Locomotion

Explain that bacteria have several mechanisms of locomotion including sliding over slimy surfaces, twisting through fluids, and propelling themselves with flagella. However, not all bacteria move. Point out that bacteria that are capable of movement are called motile bacteria and bacteria that cannot move are called non-motile bacteria.

□ **CAPSULE SUMMARY**

Bacteria can be classified into two groups according to cell wall structure. Gram staining can be used to identify bacteria because it distinguishes between two different kinds of bacterial cell walls.

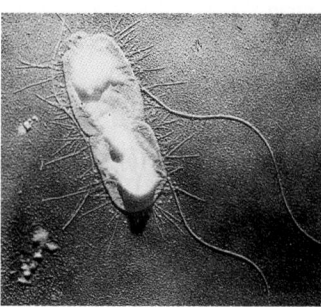

Figure 20-10 The short, hairlike fibers covering the *E. coli* above are pili, structures that help the bacterium adhere to surfaces. The two long strands are flagella, structures used for locomotion.

□ **CAPSULE SUMMARY**

Bacteria and eukaryotes differ in their cellular organization, cell structures, and in metabolic diversity.

are exposed to harsh conditions such as drought or high temperatures. These endospores are highly resistant to environmental stress and may germinate after years to form new, active bacteria. The formation of endospores in the bacterium *Clostridium botulinum* is responsible for botulism, a disease often considered to be a form of food poisoning. □

Comparing Bacteria With Eukaryotic Cells

Bacteria, which outnumber all eukaryotes combined, differ from eukaryotes in at least seven important respects. You can study the characteristics of *Escherichia coli*, a eubacterium that resides in your intestinal tract, in *Up Close: Escherichia coli* on page 465.

1. **Internal compartmentalization** Bacteria are prokaryotes and lack a cell nucleus, unlike eukaryotes. The cytoplasm of bacteria has very little internal organization. It contains no internal compartments or membrane systems.

2. **Cell size** Most bacterial cells are about 1 μm in diameter; most eukaryotic cells are well over 10 times that size.

3. **Multicellularity** All bacteria are single cells. Some bacteria may stick together in a matrix or may form filaments. However, these formations cannot truly be considered multicellular because the cytoplasm in the cells does not directly interconnect as is the case with many multicellular eukaryotes. Also, the activities of the cells are not specialized.

4. **Chromosomes** Bacterial chromosomes consist of a single circular piece of DNA. Eukaryotic chromosomes are linear pieces of DNA that are completely integrated with proteins.

5. **Cell division** Cell division in bacteria typically takes place by binary fission, a process in which one cell simply pinches into two cells. In eukaryotes, however, microtubules pull chromosomes to opposite poles of the cell during the nuclear division process called mitosis. Afterward the cytoplasm of the eukaryotic cell divides in half, forming two cells.

6. **Flagella** Bacterial flagella are simple structures composed of a single fiber of protein that spins like a corkscrew to move the cell. Eukaryotic flagella are more complex structures made of microtubules that whip back and forth rather than spin. Some bacteria also have shorter, thicker outgrowths called **pili** that act as docking cables, helping the cell to attach to surfaces or to other cells. Both of these bacterial appendages can be seen in Figure 20-10.

7. **Metabolic diversity** Bacteria have many metabolic abilities that eukaryotes lack. For example, bacteria perform several different kinds of anaerobic and aerobic respiration while eukaryotes are aerobic organisms. Unlike eukaryotes, certain bacteria can obtain their energy by oxidizing inorganic compounds such as sulfur. Other bacteria have the ability to fix atmospheric nitrogen. □

Effective Reading

Keeping Track of the Facts

Have students read "Comparing Bacteria With Eukaryotic Cells." Ask students to make two columns, labeled Bacteria and Eukaryotes, on a sheet of paper. Students should fill in the columns with comparisons of internal compartmentalization, cell division, flagella, and metabolic diversity. If students cannot recall information for each category, encourage them to reread the section. After they have completed their charts, have students check them against the text for accuracy and thoroughness.

UP CLOSE ESCHERICHIA COLI

- **Scientific Name:** *Escherichia coli*
- **Habitat:** Inhabits the intestines of many mammals
- **Size:** Up to 1 μm
- **Mode of nutrition:** Heterotrophic

Characteristics

Cell structure *E. coli* is a prokaryotic cell; its genetic material and organelles are not enclosed within membranes. It has a rigid cell wall composed of a strong network of polysaccharides cross-linked by polypeptide chains. *E. coli* are gram-negative eubacteria—a lipopolysaccharide layer is deposited over the cell wall, forming an outer membrane.

Locomotion *E. coli* is motile, that is, it has the ability to move on its own. By rotating its slender whiplike flagella, *E. coli* propels itself through its environment.

Adherence Like many gram-negative bacteria, *E. coli* has pili—short, thin, hairlike appendages attached to the bacterial cell. Pili have two main functions. The first is to adhere to surfaces, including the surfaces of intestinal-lining cells. The second function of pili is to join bacterial cells prior to the transfer of DNA from one cell to another.

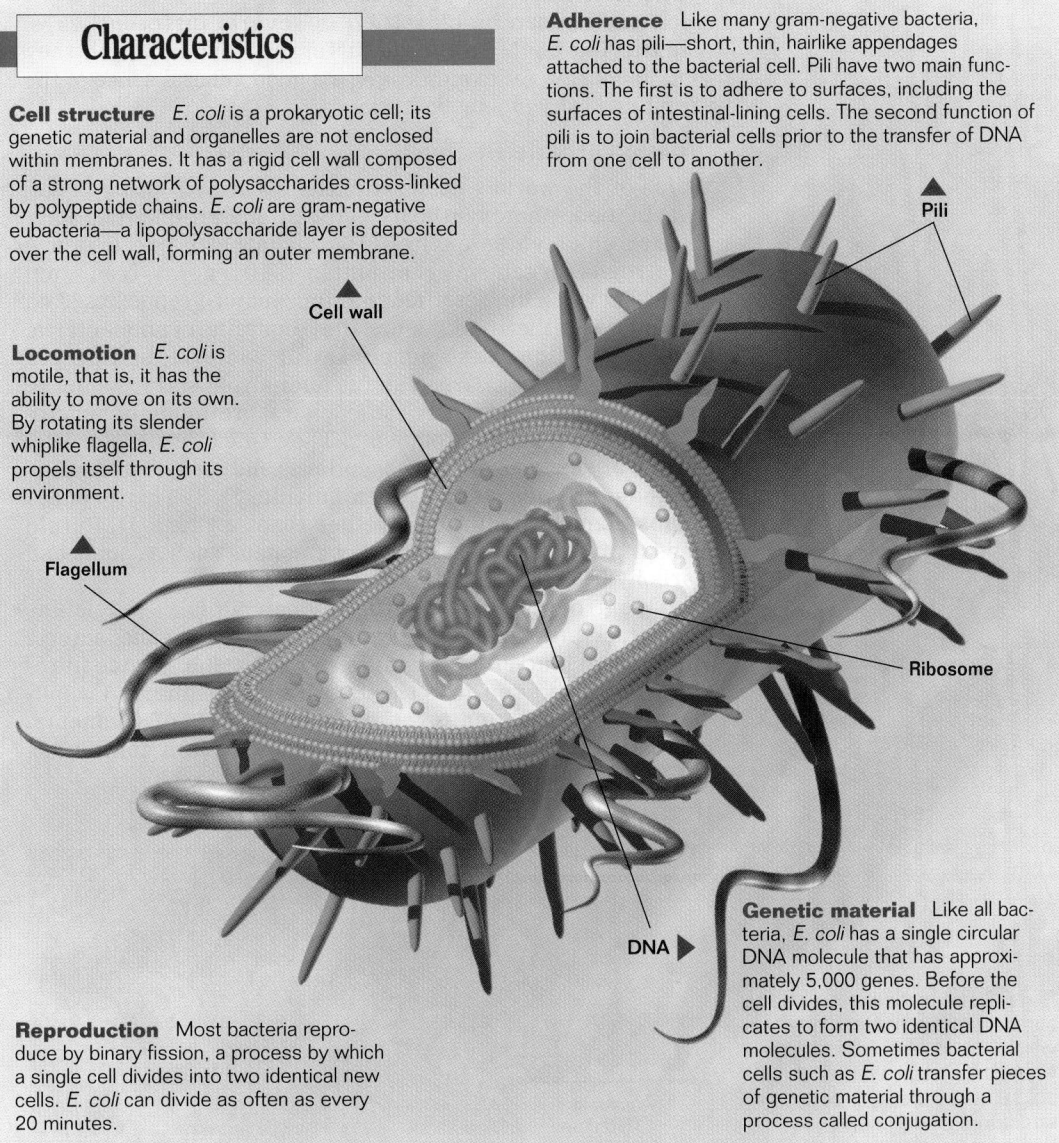

Pili

Cell wall

Flagellum

Ribosome

DNA ▶

Reproduction Most bacteria reproduce by binary fission, a process by which a single cell divides into two identical new cells. *E. coli* can divide as often as every 20 minutes.

Genetic material Like all bacteria, *E. coli* has a single circular DNA molecule that has approximately 5,000 genes. Before the cell divides, this molecule replicates to form two identical DNA molecules. Sometimes bacterial cells such as *E. coli* transfer pieces of genetic material through a process called conjugation.

SECTION 20-2
UP CLOSE
ESCHERICHIA COLI

Instructional Strategy

In 1993, the deaths of four children were attributed to *E. coli* found in undercooked hamburger meat, and more than 600 cases of food poisoning were reported in northwestern states. The symptoms of *E. coli* food poisoning include bloody diarrhea and kidney failure. In severe cases, *E. coli* food poisoning can cause permanent damage to organs or even death. The Centers for Disease Control and Prevention estimate that approximately 20,000 individuals per year suffer symptoms from *E. coli* found in undercooked food. Ask students how they can avoid getting food poisoning. (*Answers include choosing clean restaurants, making sure that foods are served and stored at their proper temperatures, checking that meats are thoroughly cooked, etc.*)

Discussion

Guide the discussion by posing the following questions.

1. Are *E. coli* gram-positive or gram-negative bacteria? (*gram-negative*)

2. How do *E. coli* reproduce? (*binary fission*) How fast can they divide? (*as often as every 20 minutes*)

3. Describe the genetic material of an *E. coli* cell. (*It has one circular DNA molecule containing about 5,000 genes.*)

4. What is the function of pili? (*Pili serve to attach* E. coli *to surfaces and to join bacterial cells during conjugation.*)

Teaching Tips

Breakthroughs in Science

Organize students into small groups to discuss the technological use of bacteria. Have students propose future uses of bacteria and ask them to imagine the potential impact of their breakthroughs. Encourage students to consider potentially harmful, beneficial, and benign effects of their ideas on quality of life, economy, health, and environment. Then have students produce a diagram of their discussion to share with the class. Diagrams might feature the technological development in the center with its effects on society represented by spokes coming out from the center.

Bacteria in Cheese

Encourage students to research the roles of bacteria and heat in making cheese. The production of cheese involves bacteria that break down lactose in milk, producing lactic acid as a waste product. The acid causes the milk to separate into curds, solid components from which cheese is made, and whey, a liquid product that is removed. Cheeses require a heating process that can destroy the bacteria if the conditions are excessive, hence different types of cheeses are made with different kinds of bacteria. Cheddar cheese is made with bacteria that require moderate temperatures, while Swiss cheese is made with bacteria that can tolerate higher temperatures.

Grouping Bacteria According to How They Obtain Energy

Over 4,000 species of bacteria have been named, and undoubtedly many thousands more exist. Bacteria occur in the widest possible range of habitats and play key ecological roles in virtually all of them. They thrive in hot springs, where the usual temperature may range as high as 100°C (210°F); they have been found living beneath 430 m of ice in Antarctica. Bacteria are abundant in ground water and are even present at high pressures in the deep ocean, growing around deep sea vents where the water temperature is as high as 360°C (680 °F).

Bacteria can be classified in several different ways. Classifying bacteria by the different ways in which they obtain energy, for example, gives a good general sense of the great diversity among bacteria.

Photosynthetic Bacteria

Much of the world's photosynthesis is carried out by bacteria. Indeed, photosynthesis evolved among bacteria long before eukaryotes existed. Photosynthetic bacteria are autotrophs, organisms that obtain their energy from sunlight. They can be classified into four major groups based on the photosynthetic pigments they contain: cyanobacteria, green sulfur bacteria, purple sulfur bacteria, and purple nonsulfur bacteria. Green sulfur and purple sulfur bacteria, which are not necessarily these colors, are found in anaerobic (oxygen-free) environments. They cannot use water as a source of electrons for photosynthesis and instead use sulfur compounds such as hydrogen sulfide, H_2S. This process forms elemental sulfur, S, rather than oxygen, O_2. Purple nonsulfur bacteria use organic compounds such as acids and carbohydrates for photosynthesis.

One group of photosynthetic bacteria is of particular importance. In Chapter 13 you learned that cyanobacteria (also called blue-green algae) were responsible for the introduction of oxygen into the Earth's atmosphere. Cyanobacteria often clump together in large mats of filaments. Each filament is a chain of cells encased in a continuous

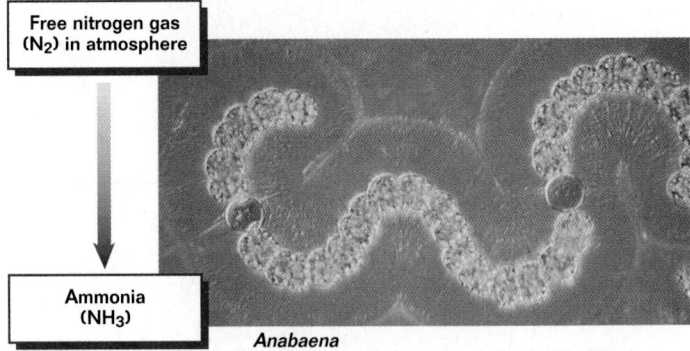

Figure 20-11 *Anabaena* is a photosynthetic cyanobacterium in which individual cells adhere in filaments. The two large orange-colored cells are heterocysts, structures in which nitrogen fixation occurs.

Free nitrogen gas (N_2) in atmosphere

Ammonia (NH_3)

Anabaena

 Did You Know?

Pseudomonas syringae is a bacterium used to make artificial snow for ski slopes. This microbe has the ability to produce ice crystals because of the way that water molecules attach to proteins in the cell wall. The artificial snow is made from a powder containing dead *P. syringae*. When the powder is mixed with water and sprayed on slopes, a snow substitute forms.

 Did You Know?

Although born with no bacteria, within hours of birth the human body is host to about a quarter of a pound of bacteria—literally billions of bacteria.

jellylike capsule. A considerable number of cyanobacteria, such as *Anabaena*, shown in Figure 20-11, are capable of fixing nitrogen. Structures called **heterocysts** contain enzymes that fix nitrogen gas, N_2, into ammonia, NH_3, for use by the growing cell. Cyanobacteria such as *Anabaena* come the closest to multicellularity among the bacteria.

Chemoautotrophic Bacteria

Not all autotrophs obtain energy from sunlight. Bacteria called chemoautotrophs obtain energy by removing electrons from inorganic molecules such as ammonia, NH_3, methane, CH_4, or hydrogen sulfide, H_2S. In the presence of one of these hydrogen-rich chemicals, chemoautotrophic bacteria can manufacture all their own amino acids and proteins. Chemoautotrophic bacteria that live in the soil, such as *Nitrosomonas* and *Nitrobacter*, are of great importance to the environment and to agriculture. As depicted in Figure 20-12, they have a crucial role in a sequence of reactions in the nitrogen cycle called **nitrification,** a process that involves the oxidation of ammonia into nitrate. Nitrate is the form of nitrogen most commonly used by plants.

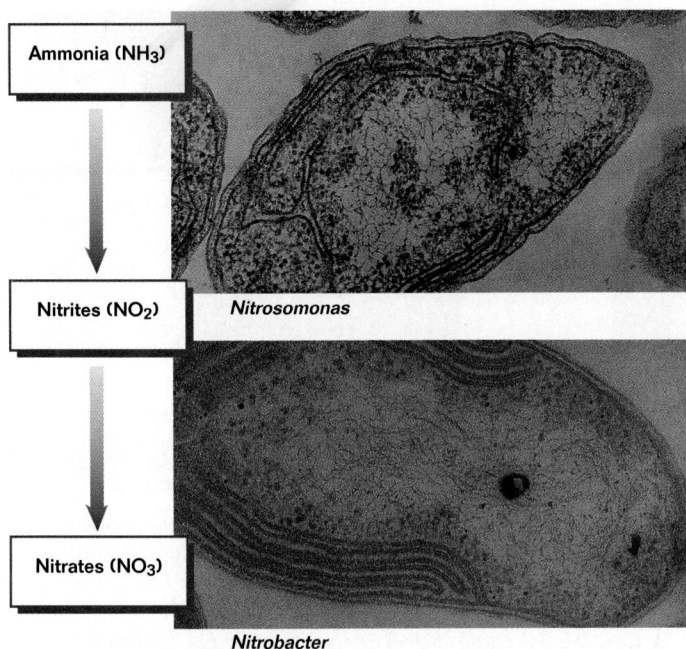

Ammonia (NH_3)

Nitrites (NO_2)

Nitrosomonas

Nitrates (NO_3)

Nitrobacter

Figure 20-12 *Nitrosomonas, above,* and *Nitrobacter, below,* are chemoautrophic bacteria that obtain their energy from inorganic molecules. They both play an important role in the nitrogen cycle. *Nitrosomonas* oxidizes ammonia into nitrites. *Nitrobacter* oxidizes nitrites into nitrates.

Heterotrophic Bacteria

Most bacteria are heterotrophs, feeding on organic material formed by other organisms. Together with fungi, heterotrophic bacteria are the principal decomposers of the living world; they break down the bodies of dead organisms and make the nutrients in these molecules available for recycling. Most of the odors associated with soil come from

Graphic Organizer
Use this graphic organizer with *Teaching Tip:* **Classifying Bacteria** on this page.

Type of bacteria	Energy source	Examples
Photosynthetic	Sunlight	Cyanobacteria, green sulfur bacteria, purple sulfur bacteria, purple nonsulfur bacteria
Chemoautotrophic	Electrons in inorganic molecules	*Nitrosomonas, Nitrobacter*
Heterotrophic	Organic material formed by other organisms	*Streptomyces, Rhizobium*

Teaching Tips
Food Poisoning

Tell students that the most toxic form of food poisoning is caused by *Clostridium botulinum*, bacteria that can secrete a toxin in food. Less than one-millionth of a gram of the toxin can be lethal when ingested. *Clostridium botulinum* grows in soil and in improperly sterilized canned goods, but cannot grow in fresh or frozen foods. Ask students to speculate why *Clostridium botulinum* cannot grow in fresh or frozen foods. (*Clostridium botulinum* cannot survive in the presence of oxygen.)

What If nitrogen-fixing bacteria stopped providing nitrogen to plants and soil?
The amount of available nitrogen would fall, plant growth would be reduced, and the amount of energy available to higher trophic levels would decline.

substances produced by heterotrophic bacteria. In recent decades, mutant strains of bacteria that break down synthetic products such as nylon and pesticides have been discovered.

Other activities of heterotrophic bacteria may be helpful or harmful to humans. For example, more than half of our antibiotics are produced by species of *Streptomyces*, a filamentous bacterium abundant in soil. On the other hand, one species of *Staphylococcus* secretes a toxin into food, causing severe reactions, including nausea, diarrhea, and vomiting, within a few hours after the food is eaten.

Species of the symbiotic bacteria *Rhizobium* are by far the most important of all nitrogen-fixing organisms. *Rhizobium* species are heterotrophic bacteria that usually live within nodules (lumps) on the roots of legumes (plants such as soybeans, beans, peas, peanuts, alfalfa, and clover), as shown in Figure 20-13. The plant furnishes anaerobic conditions and growth nutrients for the bacteria, and the bacteria fix nitrogen, which is incorporated into plant protein. Farmers take advantage of *Rhizobium*'s nitrogen-fixing abilities when they "rotate" their crops every few years and grow leguminous plants to replenish the soil with nitrogen.

Figure 20-13 Nitrogen fixation is carried out primarily by bacteria of the genus *Rhizobium*. This bacterium forms nodules in the roots of legumes.

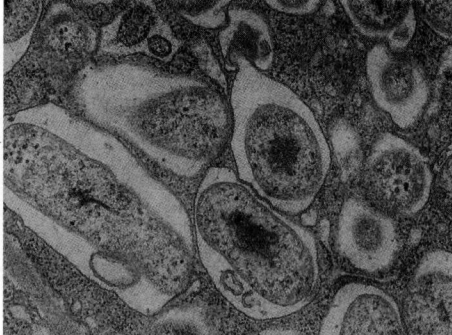

Rhizobium

Section Review

1. Draw a cross section of the cell wall of a gram-positive bacterium and a gram-negative bacterium and explain the Gram staining procedure. How can the information obtained from a Gram stain be useful?

2. Construct a table that lists the seven ways bacteria differ from eukaryotic cells.

3. What two purposes do pili serve in E. coli?

4. In what three ways can bacteria obtain energy?

Critical Thinking

5. How would you classify bacteria used in cleaning up oil spills?

Answers to Section Review

1. Diagrams should show that the gram-positive bacterium has a thicker peptidoglycan layer than the gram-negative bacterium. The color of a bacterium after Gram stain has been applied suggests which type of cell wall a bacterium has. Antibiotics are often effective against gram-positive bacteria.

2. See "Comparing Bacteria With Eukaryotic Cells" on page 464.

3. Pili help bacteria adhere to surfaces and join bacterial cells before they transfer DNA from one cell to another.

4. Three ways that bacteria obtain energy are by photosynthesizing, by removing electrons from inorganic molecules, and by feeding on organic compounds.

5. Bacteria used to clean up oil spills might be classified as heterotrophic because they feed on organic materials formed by other organisms.

20-3 Bacteria and Viruses as Pathogens

• • • • • • • • • • • • •

Bacteria are beneficial to humans in many ways. They recycle nutrients in ecosystems and play key roles in the manufacture of foods and drugs. However, bacteria have their most noticeable impact on humans as pathogens, agents of disease. At times, you may have blamed an illness on bacteria when in fact it was caused by a virus.

Diseases Caused By Bacteria

Occasionally your body becomes a host for parasitic bacteria that may cause infection and disease. Pathogenic bacteria are harmful because they attack cells or secrete toxins. For example, tuberculosis is a bacterial disease that once ranked among the most common causes of death. **Tuberculosis** is a disease of the respiratory tract caused by the bacterium *Mycobacterium tuberculosis*, shown in Figure 20-14. In most instances, a person becomes infected with *M. tuberculosis* by inhaling tiny droplets of moisture that contain the bacteria. Some bacteria settle in the lungs. The body responds by forming small nodules called tubercles around the organisms. Bacteria can survive in tubercles for an indefinite amount of time. Eventually the tubercles may become scar tissue and trap the bacteria, rendering them harmless. Or, if the body's immune system becomes weakened, the bacteria may break out of the tubercles and enter a blood vessel. They then may spread to new sites. This condition leads to a progressive disease characterized by coughing up sputum and blood, chest pain, fever, fatigue, weight loss, and loss of appetite. Tuberculosis commonly occurs as a long-term, progressively worsening disease, although it can lead to a rapid death. Other bacterial diseases that have had notable impacts on human populations are described in Table 20-1.

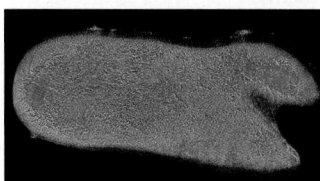

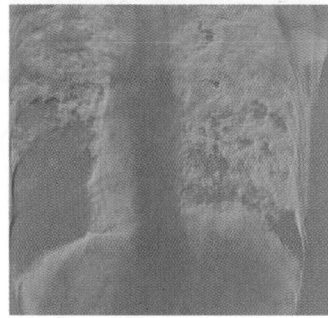

Figure 20-14 When inhaled, the bacterium *Mycobacterium tuberculosis, above,* forms small nodules called tubercles in lungs, *below.*

Treating Bacterial Diseases

In 1928, the British bacteriologist Alexander Fleming observed a mold of the genus *Penicillium* growing on a culture of bacteria. He noticed that bacteria did not grow near the mold. Apparently the mold was secreting a substance that killed the bacteria. Fleming isolated the substance and named it penicillin. In the early 1940s, scientists found that penicillin was very effective in treating many bacterial diseases, and its various forms have since been used in treating pneumonia, scarlet fever, rheumatic fever, and many other diseases.

Penicillin is an example of an **antibiotic**, a substance obtained from bacteria or fungi that is used as a drug to

Application

Medicine A new technique can help doctors find out whether a patient is infected with a drug-resistant strain of tuberculosis (TB). Strains that respond to TB treatment are weakened and produce less ATP. Resistant strains are healthy and make higher quantities of ATP. Luciferase and luciferin, the chemicals that make fireflies glow, use ATP in the light-producing reaction. By adding these substances to the strain of TB infecting a patient, doctors can identify whether the treatment will be successful because resistant microbes will have a faint glow, while susceptible microbes will not glow.

Teaching Tips

Design a Cartoon

Antibiotics are losing their effectiveness against bacterial infections that they once treated. Four strategies have been found in bacteria that help them develop resistance: (1) developing pumps to remove the antibiotics; (2) altering cell walls to keep out the antibiotics; (3) making enzymes to attack the antibiotics; and (4) making new proteins that are resistant to the antibiotics. Have students illustrate one of these survival tactics in a cartoon that explains how the strategy provides resistance to antibiotics.

"Rational" Drug Design

One strategy that researchers are using to control resistant bacterial strains is called "rational" drug design. By looking at the genes and enzymes involved in drug resistance, scientists are trying to create drugs that bind to the active sites of the enzymes and render them inactive. Ask students how binding the active sites of enzymes in bacterial cells can control resistant strains of bacteria. *(Without the work of the enzymes, the bacteria will be susceptible to antibiotics again.)*

Table 20-1 Important Bacterial Diseases

Disease	Symptoms	Infectious Agent	Mode of Transmission
Bubonic plague	Fever, bleeding lymph nodes form swellings called buboes; often fatal	*Yersinia pestis*	Carried from infected rodents by fleas
Cholera	Severe diarrhea and vomiting; often fatal	*Vibrio cholerae*	Contaminated water
Dental caries	Destruction of minerals in tooth enamel	species of *Streptococcus*	Dense collections of bacteria in mouth
Dysentery	Fever, diarrhea, vomiting	*Shigella dysenteriae*	Contaminated food or water
Lyme disease	Rash, pain, swelling in joints	*Borrelia burgdorferi*	Carried from infected animals by ticks
Typhoid fever	Headache, fever, diarrhea, rash; often fatal	*Salmonella typhi*	Contaminated water or food

fight pathogenic microorganisms. Antibiotics work by interfering with the microorganism's cellular processes. In most cases, this includes preventing cell wall formation, breaking up cell membranes, or disrupting chemical processes. Because these processes do not occur in viruses, antibiotics cannot be used for fighting viral diseases.

Since penicillin's discovery, other antibiotics such as tetracycline, streptomycin, and ampicillin have been discovered in nature or produced chemically. In the past, these drugs helped conquer diseases such as tuberculosis, bacterial pneumonia, syphilis, and gonorrhea. However, in recent years, strains of bacteria have become resistant to the antibiotics that once killed them. Antibiotic resistance arises when a few bacteria harbor mutant genes that make them immune to the drug. These bacteria survive and pass on their resistance genes to their offspring. Doctors have recently seen an increase in antibiotic-resistant strains of bacteria that cause diseases such as tuberculosis, pneumonia, meningitis and gonorrhea. New drugs, as well as new strategies, will be needed to tackle these invulnerable bacteria.

Diseases Caused By Viruses

Like pathogenic bacteria, viruses damage a body's cells and cause illness. For example, the common cold, which is usually spread by inhalation, is familiar to all of us. More than 200 kinds of viruses can cause a cold. Influenza, one of the most devastating diseases of all times, is caused by a virus found in wild ducks living in central and northern Asia.

Multicultural Perspective

Accidental Spread of Viral Diseases

Isolated communities are at particular risk of epidemics when outsiders visit. Diseases spread by travelers can wipe out a small, isolated community. In the northern Quebec community of Ungava Bay, 99 percent of the native people were afflicted with measles in 1952. Even with the help of modern medicine, almost 7 percent of the population died. Measles also killed a high number of native Brazilian people in Xingu National Park in 1952. Today, the Yanomamo tribe of Brazil and Venezuela is dying rapidly because of the onslaught of malaria, influenza, measles, and chickenpox brought by miners in search of gold.

Table 20-2 Important Viral Diseases

Disease	Symptoms	Mode of Transmission
Chickenpox	Blisters, painful rash, fever	Air currents
Measles	Blotchy rash, respiratory congestion, high fever	Air currents
Rubella	Rash, swollen glands	Air currents
Mumps	Painful swelling in salivary glands	Air currents
Smallpox	Blisters, lesions, fever, malaise; often fatal	Air currents
Infectious hepatitis	Fever, chills, nausea, swollen liver, jaundice, painful joints	Contaminated food or water
Polio	Headache, stiff neck, possible paralysis	Contaminated food or water
AIDS	Immune system failure; fatal	Sexual contact, contaminated-blood products or hypodermic needles,

Commonly known as the flu, influenza is a viral disease of the respiratory tract characterized by chills, fever, and muscular aches. When the virus is inhaled, it comes in contact with the cells of the upper respiratory tract and penetrates them. There, the viruses reproduce and spread to new cells. People usually develop antibodies, substances that attach to the influenza viruses and prevent them from infecting new cells. However, the influenza virus mutates readily, constantly changing its surface proteins. Occasionally a new influenza variant is so different that the human immune system fails to recognize it. When this happens, the body has little defense against the infection, and influenza can be fatal. Other viral diseases that have had a serious effect on society are described in Table 20-2.

Section Review

1. *What causes tuberculosis? What are the symptoms of this disease?*

2. *How does antibiotic resistance in bacteria arise?*

3. *Describe five diseases caused by bacteria and five caused by viruses.*

Critical Thinking

4. *Why wouldn't a doctor prescribe an antibiotic to treat influenza?*

SECTION 20-3

Teaching Tip
Preventing Drug Resistance

Ask students how drug resistance can be prevented. *(One way to prevent drug-resistant strains of deadly bacteria is to use fewer antibiotics. Taking antibiotics for viral diseases and injecting healthy livestock with antibiotics to help them grow fatter are uses that contribute to the rapid development of drug-resistant bacterial strains.)*

Application

Health Because influenza is an airborne virus, it is highly contagious. A common flu virus can be lethal to anyone with a weak immune system, such as elderly people and young children. Sometimes a new flu virus can be very lethal. For example, in 1918 a strain of influenza killed 20 million people—more people than were killed in combat in World War I.

Chapter Closure

Have students choose two diseases, one caused by a bacterium and one caused by a virus. Students can use information from the text, do outside research, or create fictitious diseases. Ask students to draw plausible structures for the pathogens and to describe methods of infection, prevention, and treatment. Students' drawings must depict the characteristics of bacteria and viruses described in this chapter.

Answers to Section Review

1. *Mycobacterium tuberculosis* causes tuberculosis, a respiratory disease characterized by nodules in the lungs. Symptoms include coughing up sputum and blood, chest pain, fever, fatigue, weight loss, and loss of appetite.

2. Mutations sometimes render a bacterium immune to an antibiotic. The mutant bacterium passes on the resistance to offspring, creating a new, resistant strain.

3. See Table 20-1 and Table 20-2 on pages 470–471.

4. Antibiotics work by interfering with a microorganism's cellular processes. Because viruses are chemicals and not living organisms, antibiotics are not effective against viruses.

CHAPTER REVIEW ANSWERS

Each item in the Chapter Review is correlated by text section in the assignment guide that follows.

ASSIGNMENT GUIDE

Section	Review	Themes Review	Critical Thinking
20-1	1–3, 8, 9	14	17
20-2	4–6, 10, 11	15	
20-3	7, 12, 13	16	18, 19

Review

Multiple Choice

Code in parentheses indicates section number and objective number.

1. a (20-1.1, 20-1.2)

2. c (20-1.3)

3. a (20-1.4)

4. c (20-2.2)

5. d (20-2.4)

6. b (20-2.2)

7. a (20-3.4)

Completion

8. helical, icosahedron, 100 (20-1.3)

9. retrovirus, white blood, AIDS (20-1.4)

10. gram-positive, gram-negative (20-2.1)

11. oxygen, nitrogen (20-2.4)

12. bacteria, viruses (20-3.3, 20-3.4)

13. tubercles (20-3.1)

Vocabulary

antibiotic (469)	envelope (456)	pili (464)
antibody (458)	glycoprotein (456)	retrovirus (461)
bacillus (462)	gram-negative (463)	reverse transcriptase (459)
bacteriophage (457)	gram-positive (463)	spirillum (462)
capsid (456)	heterocyst (467)	tuberculosis (469)
capsule (463)	icosahedron (457)	virus (456)
coccus (462)	nitrification (467)	
endospore (463)	pathogen (458)	

Review

Multiple Choice

1. What evidence led Stanley to conclude that tobacco mosaic virus (TMV) is not a living organism?
   a. The extract of TMV crystallized.
   b. TMV is made of RNA and protein.
   c. TMV reproduces only in cells.
   d. The virus poisons tobacco plants.

2. The basic components of all viruses are a nucleic acid and a(n)
   a. endospore.   **c.** protein coat.
   b. glycoprotein.   **d.** icosahedron.

3. What triggers the entry of HIV into human white blood cells?
   a. HIV glycoproteins bind to receptor proteins.
   b. The cells absorb genetic material.
   c. HIV injects its genetic material.
   d. The cell begins to divide.

4. Bacteria are different from eukaryotic cells in that
   a. only bacteria have flagella.
   b. eukaryotic cells are smaller than bacteria.
   c. bacteria perform aerobic and anaerobic respiration.
   d. eukaryotic cells lack a nucleus.

5. Bacteria that do not require sunlight but that obtain energy by removing electrons from hydrogen-rich chemicals are called
   a. heterotrophs.
   b. photosynthetic bacteria.
   c. cyanobacteria.
   d. chemoautotrophs.

6. Which of the following is *not* a mechanism by which antibiotics kill bacteria?
   a. preventing cell wall construction
   b. inhibiting mitosis
   c. breaking up cell membranes
   d. disrupting chemical processes

7. Which is the correct match of a viral disease and its effect on the human body?
   a. polio: paralysis
   b. influenza: skin blisters and lesions
   c. mumps: immune system failure
   d. dental caries: tooth decay

Completion

8. The shape of a virus is typically either _____ or polyhedral. The capsid of most polyhedrals is in the shape of a(n) _____ , which provides more volume than other shapes. Viruses are very small, ranging in size from about 17 nm in diameter to about _____ nm.

9. HIV is a _____ , which means that it transcribes DNA from RNA. It attacks human _____ cells and apparently causes the disease _____ , for which there is no known cure.

10. Bacteria with much peptidoglycan in their cell walls (and typically susceptible to penicillin) are called _____ bacteria, while those with less peptidoglycan (and typically susceptible to tetracycline) are called _____ bacteria.

11. Cyanobacteria introduced _____ into the Earth's atmosphere and are capable of fixing _____ .

12. Antibiotics are used to treat diseases caused by _____ , although they are usually ineffective against diseases caused by _____ .

13. A symptom of tuberculosis is the formation in the lungs of nodules called _____ .

14. **Structure and Function** The typical virus consists of either DNA or RNA encased in a protein coat and ranges in diameter from about 17 to 100 nm. What advantages do viruses derive from their relatively simple composition and extremely small size?

15. **Flow of Energy** Heterotrophic bacteria function as decomposers. Describe how these bacteria contribute to the flow of energy in an ecosystem.

16. **Evolution** Suppose cold viruses invade your body. Your body's immune system may destroy most but not all of these viruses. How does the response of your body's immune system affect the evolution of the cold viruses?

17. **Making Inferences** Some medical experts suggest that the drug AZT (azidothymidine) can help patients with AIDS. This drug blocks the action of the enzyme reverse transcriptase. Explain how AZT might help these patients.

18. **Making Inferences** In the 1520s, the Spanish explorer Cortes and his armies introduced smallpox to the Americas. The death rate among the native Aztecs ranged between 50 to 90 percent compared with a death rate of about 10 percent among people in Europe. What accounts for the difference in death rates?

19. **Making Predictions** Over the last 20 years, the number of antibiotic-resistant bacterial pathogens has steadily increased. This is thought be a result of antibiotic abuse by patients and doctors. Doctors tend to overprescribe antibiotics for patients who demand a quick fix for their illness. Just recently, the World Health Organization has established a global computer database so that doctors can report outbreaks of antibiotic resistance. What are the potential benefits of this database?

20. **Multicultural Perspective** AIDS, the disease caused by HIV, is a major health concern worldwide. Locate statistics on AIDS cases for as many countries as possible. Then, draw a world map on poster-size paper and devise a color legend for the map that shows the number of AIDS cases in countries for which data were located. Color the map to match the legend, and give it a title. Next, write a set of questions that can be answered using the map. Display the map and the questions where they can be seen by other students. Different students can take responsibility for gathering the AIDS data, drawing and coloring the map, creating the legend, and writing the questions.

21. **Research and Writing** Edward Jenner developed the first successful vaccine for smallpox in 1796. His vaccine was based on the fact that people who contracted the milder disease cowpox did not contract smallpox. Jenner inoculated people with the cowpox virus. Research the concerns people of Jenner's time had about his vaccine. Then, write a message to persuade them to get vaccinated.

Themes Review

14. Because of their small size and simplicity, viruses can infect a variety of host cells, and they can be assembled rapidly and in large numbers (20-1.3).

15. Heterotrophic bacteria break down the bodies of dead organisms and make the nutrients in these molecules available for recycling. (20-2.4)

16. The few cold viruses that are not destroyed by the body's immune system will have been naturally selected and will be resistant to immediate attack by your immune system. Natural selection favors the evolution of those cold viruses best able to resist attacks by your immune system. (20-3.4)

Critical Thinking

17. AIDS is caused by the retrovirus HIV, which must use the enzyme reverse transcriptase to reproduce. By blocking reverse transcriptase, AZT can prevent HIV from reproducing and infecting more cells. (20-1.4)

18. The people native to the Americas had no natural immunity to smallpox because they had not been previously exposed to the disease. However, the Europeans had been exposed to smallpox, and their immune systems were prepared to combat the virus. (20-3.4)

19. Doctors will have almost immediate access to information about bacterial pathogens that are resistant to various antibiotics. Over time, the database will help reduce the number of unnecessary prescriptions written for antibiotics. This, in turn, should help preserve the effectiveness of antibiotics. (20-3.2)

Activities and Projects

20. Statistics on AIDS cases can be obtained from several sources, including the Centers for Disease Control and Prevention in Atlanta, Georgia.

21. The message should highlight the fact that people who received the vaccine did not contract the deadly smallpox disease, and it should discourage belief in the myth that cow parts grow out of vaccinated people.

LABORATORY Investigation Chapter 20

Preparation
Unknown bacterial cultures of cocci, spirilli, and bacilli may be made from live or freeze-dried bacteria available from a biological supply company. *Micrococcus luteus, Spirillum volutans,* and *Bacillus megaterium* are acceptable choices. Each lab group should receive three test tubes, each containing a different bacterial culture.

Procedural Notes
- This investigation can be completed during one 40-minute period.
- **Safety:** Make sure students wear safety goggles, lab aprons, and disposable gloves before obtaining their bacterial cultures.
- Each lab group should be given a set of bacterial cultures. If you allow groups to obtain samples directly from stock cultures, the stocks are likely to become contaminated.
- Have students view the bacteria through oil immersion objectives, if possible.
- In step 3, remind students to use methylene blue carefully and to keep it away from their faces, skin, and clothing.
- **Safety:** Make sure that students dispose of their materials properly and wash their hands before leaving the lab.

Background Answers
1. Bacteria are prokaryotes.

2. Prokaryotes lack organelles. Eukaryotes have organelles, such as mitochondria and nuclei.

3. Bacteria reproduce by binary fission.

4. What physical differences of coccus, bacillus, and spirillum bacteria can be seen under a compound light microscope?

Identifying and Staining Bacteria

OBJECTIVE

Identify types of bacteria by their shape.

PROCESS SKILLS

- comparing and contrasting different types of bacteria

MATERIALS

- prepared slides of coccus, bacillus, and spirillum bacteria
- compound light microscope
- safety goggles
- lab apron
- disposable gloves
- 3 culture tubes of bacteria, marked A, B, and C
- test tube rack
- sterile cotton swabs
- 3 microscope slides
- microscope slide forceps or wooden alligator-type clothespin
- 150 mL beaker
- methylene blue stain in dropper bottle
- paper towels

BACKGROUND

1. Are bacteria prokaryotes or eukaryotes?
2. How do prokaryotes and eukaryotes differ?
3. How do bacteria reproduce?
4. Write your own question to explore in your lab report or notebook.

TECHNIQUE

Part A: Prepared Slides

1. Observe each of the three prepared slides of bacteria under low power and high power. Draw and label each type of bacterium in the **Records** section of your report. Compare your slides to the photographs shown below.

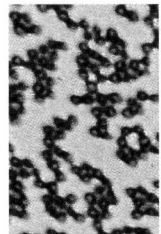 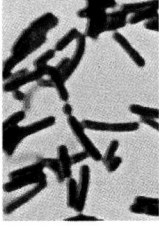 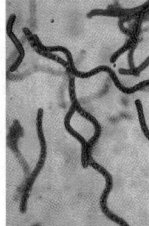

Part B: Live Bacteria

In this part of the investigation, you will transfer bacteria from each culture tube (A, B, and C) to three microscope slides for staining and observation. **CAUTION: Put on safety goggles, a lab apron, and disposable gloves.**

2. Have your partner remove the cap from the culture tube marked "A" with the tips of his or her fingers. The cap should not be placed on the table. Insert a sterile cotton swab into the test tube and transfer a very small amount of bacterial culture to a microscope slide. Spread the culture on the slide with the swab and allow it to dry. Dispose of the swab properly.

3. Use the microscope slide forceps or the wooden alligator-type clothespins to hold the slide. Place the slide across the mouth of a beaker half filled with water. Apply drops of methylene blue stain with a dropper. **CAUTION: Methylene blue is an eye irritant and stains skin and clothing. Avoid contact with your eyes and**

Records

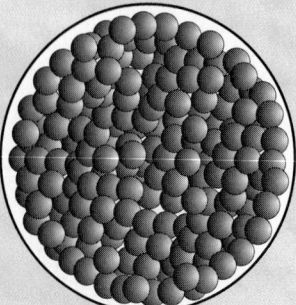

Coccus bacteria

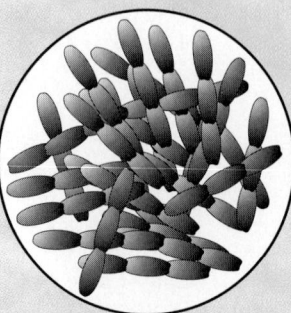

Bacillus bacteria

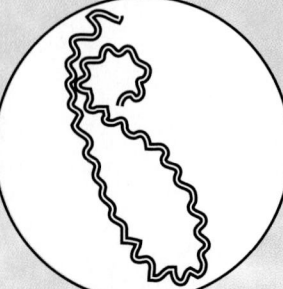

Spirillum bacteria

skin. **Do not ingest.** Flood the slide with the stain, but do not allow the stain to spill into the beaker. Allow the stain to remain on the slide for two minutes.

4. Rinse the slide by dipping it into the water in the beaker several times. Blot the slide dry with a paper towel, being careful not to rub the slide. Allow the slide to dry.

5. When the slide is dry, observe it under the microscope and determine the type of bacteria on the slide—coccus, bacillus, or spirillum—based on your observations of the prepared slides. In a table similar to the one shown below, record the identity of the bacteria in the **Records** section of your report.

Types of Bacteria

Culture Tube	Type
Tube A	
Tube B	
Tube C	

6. Repeat steps 2 through 5 using the culture tubes marked "B" and "C." In your report, briefly summarize the procedure you followed.

7. Clean up your materials and wash your hands before leaving the lab. Dispose of all materials according to instructions provided by your teacher.

INQUIRY

1. When observing the prepared slides, which power provided the clearest view of the bacteria?

2. Describe the shapes of coccus, bacillus, and spirillum bacteria.

3. Why should the test tube cap in Part B not be placed on the table?

ANALYSIS

1. What were some possible sources of contamination as you transferred the bacteria from the test tubes to the slides?

2. What is the advantage of staining bacteria before observing them under a microscope?

3. Why is caution necessary in handling bacteria, even if you are working with a bacterial species known to be harmless?

FURTHER INQUIRY

Write a new question that could be explored as a new investigation. The following is an example:

Are antiseptics equally effective against the three types of bacteria?

Inquiry Answers

1. High power should provide the clearest view of the bacteria.

2. The shapes of the bacteria are as follows: coccus—spherical; bacillus—cylindrical or rodshaped; spirillum—corkscrewshaped. Coccus bacteria show cells that are attached in pairs, groups of four, chains, or clusters. Bacillus bacteria usually appear as single cells, although sometimes they are arranged in pairs or in chains. Spirillum bacteria almost always appear singly, and they differ in length and in the number and size of their spirals.

3. Holding the cap helps prevent contamination of the bacterial cultures.

Analysis Answers

1. Answers will vary. Possible sources of contamination include the air and contact with an improperly sterilized cotton swab or a contaminated test-tube cap.

2. Most bacteria are colorless and cannot be seen under a microscope unless they are stained.

3. Even bacteria that are generally considered harmless might be harmful under certain conditions; thus, it is best to handle all bacteria as if they were harmful. Also, pathogens might have been accidentally introduced into the culture tubes.

UNIT 5

SCIENCE, TECHNOLOGY, AND SOCIETY

Point of View: Biological Weapons

Objectives

- Develop scientific literacy
- Identify an author's point of view
- Detect bias in a persuasive argument

Background

Each year scientific issues become more important in determining the nature and direction of our society, and yet most citizens have little or no training in evaluating the arguments that surround controversial issues. The purpose of this feature, Point of View, is to provide practice in identifying and evaluating a writer's point of view.

A critical step in this process is detecting bias. Much of the writing on critical issues is persuasive and not objective. A writer will include information that bolsters his or her point of view and suppress evidence that contradicts it.

Instructional Strategies

- As students read this article, remind them that it represents one person's point of view. They must read critically, looking for signs that would indicate biased or slanted writing. In particular, have students look closely for the following:

  1. Does the article present both sides of the issue fairly and objectively?

  2. Is the writer seeking to shape your opinion on the issues?

  3. Does the writer use "loaded" language? Are some words chosen because they have an emotional rather than a logical appeal to the reader?

- Point out that both offensive and defensive research on biological weapons involve similar protocols. Vaccine development is a good example. Both sides in a war are at risk of infection, so a country

BIOLOGICAL WEAPONS

Should the Research Continue?

Can soldiers defend themselves from deadly biological weapons, like *Bacillus anthracis* bacteria?

BY TRACEY COHEN

The following article is an editorial that presents one person's viewpoint on the uses of scientific research and the social responsibility of researchers. Read the article carefully and see if you can determine the author's point of view and whether or not the evidence presented is balanced and fair.

*I*magine that you turn on the radio one morning and hear the following news.

" *The number of civilian casualties in the eastern European conflict continues to rise following the deliberate poisoning of a reservoir with a biological warfare agent last week. Another 32 people have died, bringing the death toll to 465, health officials reported. Hundreds more people have become ill, and hospitals are overrun with sick, frightened people. Medical experts have identified the pathogen as a new strain of intestinal bacteria that has been genetically altered to make it resistant to the three most widely used antibiotics.* **"**

Biological weapons use bacteria, viruses, fungi, and other organisms, or substances derived from them, to cause illness and death in humans, animals, or plants. Two well known agents developed for military use are anthrax and botulin toxin. Anthrax, a disease caused by the bacterium *Bacillus anthracis*, quickly kills between 95 and 100 percent of untreated infected victims. Botulin toxin, a bacterial toxin derived from *Clostridium botulinum*, is one of the most potent natural poisons known.

Warfare by contamination is actually not new. There are historical accounts dating back 2,000 years. But the development of biological weapons did not begin in earnest until the twentieth century. Though the full details were not known at the time, Japan started a large-scale biological weapons program in 1935. Fear that Nazi Germany was conducting a similar program led the United States to begin its own research on biological weapons in 1941. Today, at least 25 nations are developing biological weapons or already have them, according to the Pentagon.

Biological weapons are so terrible and uncontrollable that

most nations have decided to outlaw their use. In support of this view, 126 nations signed the Biological Weapons Convention (BWC) of 1972. This treaty bans the development, production, and stockpiling of biological weapons (BWs). Although the treaty prohibits research leading to the development of offensive weapons, it does permit countries to continue research aimed at defending themselves against attack by biological weapons.

Can We Really Defend Against BWs?

Many bacteria, viruses, fungi, and other agents known to cause disease can be adapted to serve as weapons. Genetic engineering and other biotechnologies can increase the diversity of potential pathogens further. It would be technically and economically impossible to vaccinate entire populations of people or livestock against all potential viruses or to develop antibiotics for all the pathogenic bacteria. Protecting plants would be no easier. Disease-resistant crops would have to be planted months ahead or enough of the right fungicide would have to be available at the right time. But, as in the case of

planning to use biological weapons must first vaccinate its own troops.

Answers to Analyzing the Issue

1. (a) This article is a point of view because the writer lists opinions and draws a conclusion. **(b)** The writer's thesis

is that scientists should not work on biological weapons. **(c)** Opinions include the following statements: biological weapons are terrible and uncontrollable; research efforts should focus on work that does not involve producing biological weapons. **(d)** Facts include the following statements: 126 nations signed a treaty in 1972

that bans the development of biological weapons; genetic engineering can increase the diversity of potential pathogens. **(e)** Answers will vary. Without biological weapons research a country may be more susceptible to attack by biological weapons and less prepared to defend itself against them.

human and animal pathogens, advance knowledge of exactly what weapons the enemy might use would be needed to enact defensive measures. This kind of information is not likely to be widely available.

Can Research Be Limited?

Given the horrible consequences of biological warfare, there is a growing concern that some defense research may apply to creating offensive weapons. In 1993, a study was issued by the Center for Public Integrity, an independent group that examines public services and ethic-related issues. They reported that approximately one-quarter of the current research could be converted to the development of offensive weapons. In their study, they cited several cases. In one instance, a biologist had created a new, highly infectious and antibiotic-resistant strain of anthrax. In another case, researchers had altered the deadly botulin neurotoxin so that it could not be treated by the usual antidote.

Who Will Decide the Future of Biological Weapons Research?

Creating a defense against biological weapons requires the expertise of many scientists from a variety of backgrounds. Specialists in public health, medicine, and biology all have a role to play in deciding the fate of future research. In a democracy, however, every citizen also has a view in these decisions–not just the scientists. It is in the public's interest that these issues be debated in the national forum where an informed citizen can be a part of the decision-making process.

What Steps Are Possible?

The United States has been a leader in the effort to eliminate

the use of biological weapons. It may be possible, however, for the United States to take an even more active role in eliminating this threat. One possibility is refocusing its research efforts on work that does not involve producing pathogens or toxins. For instance, more research funds could be diverted to developing better protective clothing and equipment for military personnel who may be exposed to biological weapons in the field.

Another possibility is to transfer research on pathogens and toxins to civilian agencies such as the National Institutes of Health. This would increase public access to information and improve accountability. It would also help to counter another criticism made in the Center for Public Integrity's 1993 report: The study found that in 1989-90 (the most recent year for which complete data is available), just 3 percent of the $84.8 million spent on research went to institutions well known for biomedical research.

What Is the Role of Scientists in Shaping Research?

Through their work, scientists are able to cause broad and lasting changes in the world. This power carries with it a special obligation for scientists to consider the consequences of their work. It is not hard for example, for biomedical researchers to anticipate the impact of creating a deadly new virus.

Recently, several hundred scientists at universities and in industry signed a nationwide pledge not to accept research grants to develop biological warfare agents. These scientists have made a personal choice about the nature and extent of their involvement in research. Their actions, however, serve to remind all of us that every citizen has a stake in determining the future of biological weapons research.

Tracey Cohen is a freelance writer specializing in science and environmental issues.

Analyzing the Issue

1. **Detecting Bias** There are many different kinds of writing. For example, some articles attempt to describe events objectively, leaving readers to draw their own conclusions. Other articles present an analysis with an obvious point of view.
a. What kind of article is this? How do you know?
b. What is the writer's thesis, or main point?
c. Read through the article carefully. List all the opinions or unsupported judgments made in the article.
d. List the facts the writer uses to support the thesis.
e. Writers use facts selectively to support their arguments. What

information not used in the article could challenge the writer's viewpoint?

2. **Researching the Issue** Research the history of the Manhattan Project. Find any accounts of how work on the project adversely affected the scientists who worked on it.

3. **Take Action** Read at least three articles concerning this issue. Your teacher will have a list of references that you can use. Based on your reading, list four reasons why the United States should continue defensive biological weapons research and four reasons why all such research should be abandoned.

---

2. The life of J. Robert Oppenheimer is an example of the far reaching effects of the Manhattan Project. Oppenheimer resigned as director of the Los Alamos Laboratory the same year the first atomic bomb was dropped. He was later part of a group of scientists that opposed the development of the hydrogen

bomb. His views and ideas were such that he was thought to be a security risk, which resulted in his being brought to trial on the charge of treason. Though he was found not guilty, he lost his security clearance and his job as an advisor to the Atomic Energy Commission.

3. Answers will vary. Look for rational arguments to support each side of the issue.

SCIENCE, TECHNOLOGY, AND SOCIETY

References

1. Bartfai, T., et al. "Benefits and Threats of Developments in Biotechnology and Genetic Engineering." *SIPRI Yearbook* 1993: World Armaments and Disarmament.

2. Horgan, J. "Biowarfare Wars: Critics Ask Whether the Army Can Manage the Program." *Scientific American,* 1994, p. 22.

3. King, J. "Biology Goes to War: Marching Toward a Biological Arms Race." *Science for the People,* Jan/Feb 1988, pp. 17–20.

4. King, J. and H. Strauss. "The Hazards of Defensive Biological Warfare Programs." In: *Preventing a Biological Arms Race.* Susan Wright, ed. MIT Press, Cambridge, Mass., 1990, pp. 120–312.

5. Lappe, M. "Ethics in Biological Warfare Research." In: *Preventing a Biological Arms Race.* Susan Wright, ed. MIT Press, Cambridge, Mass., 1990, pp. 78–99.

6. Piller, C. and K. R. Yamamoto. *Gene Wars: Military Control Over the New Genetic Technologies.* Beech Tree Books, William Morrow, New York, 1988.

7. Shulman, S. *Biohazard: How the Pentagon's Biological Warfare Research Program Defeats Its Own Goals.* The Center for Public Integrity, Washington, D. C., 1993.

8. Shulman, S. "Funding for Biological Weapons Research Grows Amidst Controversy." Bioscience Vol. 37, June 1987, pp. 372–375.

9. Shulman, S. "Resisting Biological Warfare." *Science for the People,* Jan/Feb. 1988, p. 18.

10. Stock, T. "Chemical and Biological Weapons: Developments and Proliferation." *SIPRI Yearbook* 1993: World Armaments and Disarmament.

11. Wright, S. and S. Ketcham. "The Problem of Interpreting the U. S. Biological Defense Research Program." In: *Preventing a Biological Arms Race.* Susan Wright, ed. MIT Press, Cambridge, Mass., 1990, pp. 169–196

Block Scheduling Guide

Each block represents about 45–50 minutes of instructional time. The resources cited in a block are coded to a specific program component using the Key on the next page.

Lecture Concepts

Lesson Resources

21-1 Characteristics of Protists pp. 478–483

BLOCKS ① and ②

a. The kingdom Protista contains the most diverse groups of eukaryotic organisms. While protists have some characteristics found in the other three eukaryotic kingdoms, they differ enough to be assigned their own kingdom.

b. Protists live in moist environments and can be either free-living or parasitic.

c. Some protists are able to reproduce sexually in times of stressful environmental conditions.

Lecture Resources
- Opening Question, page 479
- Demonstration: Classifying Protists into Groups
- Finding Protists, page 480
- Teaching Transparencies: 89A, 103, 104
- Visual Strategy: Figure 21-2
- Adaptive Advantage, page 482
- Pond Scum, page 482
- Holt Biology Videodiscs: Lesson 21-1

Classwork Options
- Effective Reading, page 479

- Comparing Environments, page 480
- Reproduction in Multicellular Protists and Graphic Organizer, page 482
- Mapping the Concepts, page 483
- Quick Labs A17: Observing Protists
- Inquiry Skills Development B14: Protists—A Comparison

Assignment Options
- Section Review, page 483
- Chapter Review, questions 1–3, 10, 11, 16

21-2 Protist Diversity pp. 484–492

BLOCKS ③ and ④

a. Amoebas and forams are protists that move using cytoplasmic extensions. Diatoms are protists with glasslike shells.

b. The three kinds of algae—red, green, and brown—are distinguished by the types of chlorophyll they contain. Some algae are multicellular.

c. Dinoflagellates are protists that move with flagella. Protists that use cilia to swim are called ciliates.

d. Protistan molds are not fungi. Cellular slime molds resemble amoebas. Plasmodial slime molds are masses of cytoplasm containing many nuclei.

e. Sporozoans are nonmotile, spore-forming, unicellular parasites.

Lecture Resources
- Opening Question, page 484
- Demonstration: A Giant Amoeba
- Teaching Transparencies: 90A, 99, 100
- Overcoming Misconceptions, page 486
- Red Tides, page 488
- Visual Strategies: Figures 21-6, 21-13, 21-15
- Locomotive Features in Human Cells, page 489
- Up Close, Paramecium page 490
- Holt Biology Videodiscs: Lesson 21-2

Classwork Options
- Pseudopodia, page 485
- Comparing Algae and Graphic Organizer, page 487
- Growing Slime Mold, page 491

Assignment Options
- Have You Had Your Algae Today?, page 487
- Comparing Protistan Molds, page 491
- Section Review, page 492
- Chapter Review, questions 4–6, 12–14, 17–19

21-3 Protists and Human Health pp. 493–495

BLOCK ⑤

a. Protists can cause serious diseases such as malaria.

Lecture Resources
- Opening Question, page 493
- Demonstration: Combating Mosquitos
- Teaching Transparencies: 101, 102
- Research Update, page 494
- Connections, page 495
- Holt Biology Videodiscs: Lesson 21-3

Classwork Options
- Closure, page 495
- Laboratory Investigation: Comparative Protists, pages 498–499

Assignment Options
- Section Review, page 495
- Chapter Review, questions 7–9, 15, 20

Review/Enrichment

BLOCK ⑥

■ Study Guide: Concept Review and Pretest

Assignment Options
◉ Teaching Resources CD-ROM: Supplemental Reading
   Worksheets and Test, *Microbe Hunters*
◉ Teaching Resources CD-ROM: Occupational Applications
   Worksheets, Sanitarian
■ Activities and Projects, questions 21, 22

Assessment Options

BLOCK ⑦

Traditional Assessment
■ Chapter 21 Test
◉ Test Generator/Assessment Item Listing: Software item bank
   for preparing customized chapter tests.

Portfolio Assessment
Select student reports for one or more laboratory experiments
from this chapter. *The Direct Observation Checklist* on the
BioSources Teaching Resources CD-ROM should be completed
during a laboratory or other cooperative learning experience.

Holt Biology Videodiscs   CONCEPTS OF BIOLOGY

Holt Biology Videodiscs gives you a powerful tool for
teaching, review, and assessment. *Concepts of Biology*
adds interactive lesson plans and extensive tools for
customization using CD-ROM technology.

Use the following topic from *Concepts of Biology*
to help you teach this chapter:

■ Topic 14: Protists and Fungi

For further information regarding the media on the videodiscs,
see *Holt Biology Videodiscs Teacher's Correlation Guide to
Biology: Principles and Explorations.*

Chapter Theme

Evolution *Modern biological classification is based on evolutionary relationships, though on the surface, classification seems to be based on morphology, mode of locomotion, or some other characteristic. Classification of protists reflects what scientists know about evolutionary divergence among organisms.*

Tapping Prior Knowledge

• How does an individual produced by sexual reproduction differ from its parents? How is it similar?

• Name two ways in which unicellular and multicellular organisms differ.

Opening Demonstration

Show students prepared slides or pictures of protists (examples can be found in Section 21-2). Have students derive some general rules for classifying protists, such as being microscopic, unicellular, eukaryotic, etc. Write students' suggestions on cards, and post them where students can view them during their study of protists. Point out that protists do not have a lot in common with each other but they are classified in one kingdom because they lack the specialized features of members of other kingdoms.

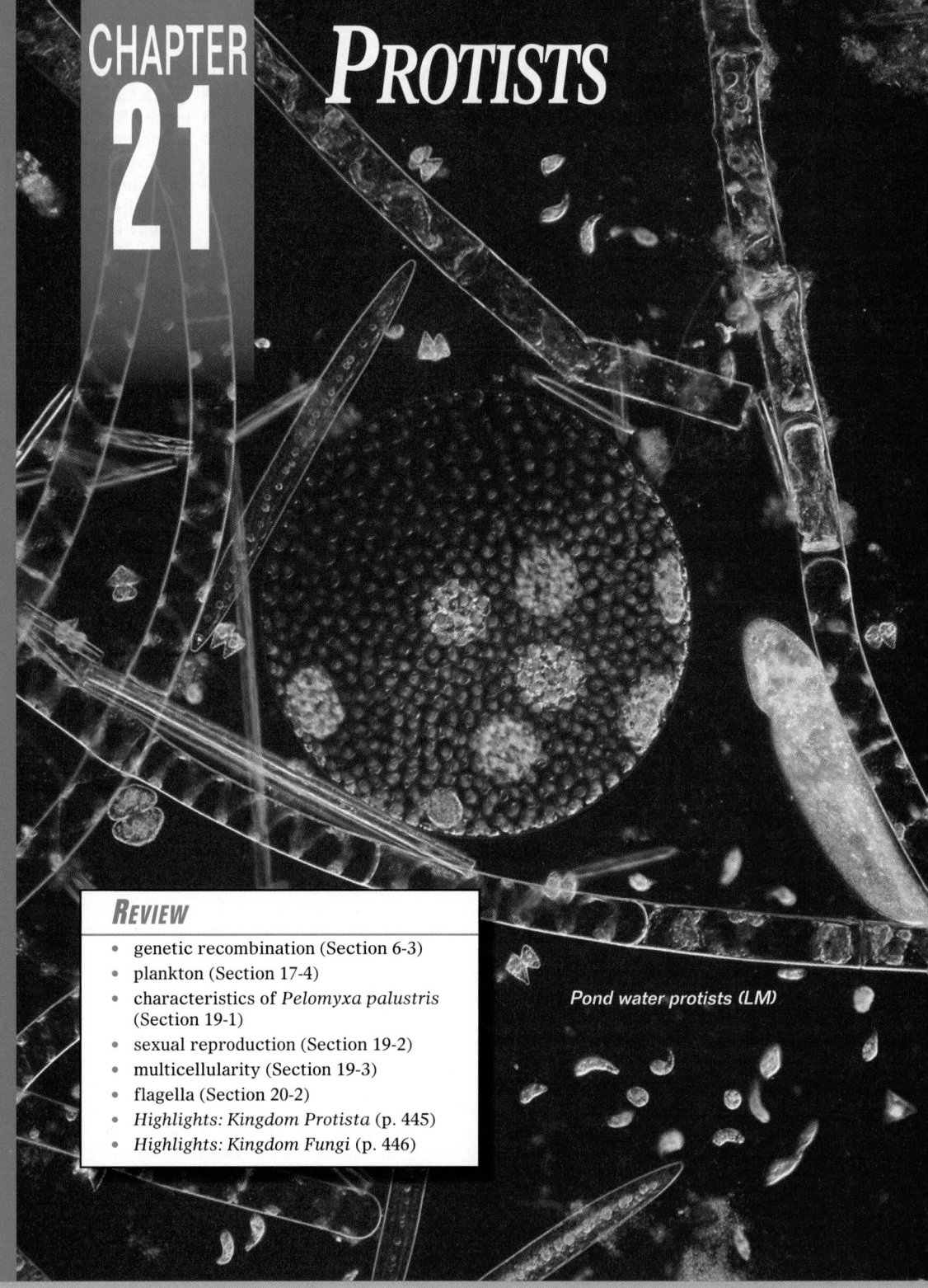

CHAPTER 21 PROTISTS

Pond water protists (LM)

REVIEW

• genetic recombination (Section 6-3)
• plankton (Section 17-4)
• characteristics of *Pelomyxa palustris* (Section 19-1)
• sexual reproduction (Section 19-2)
• multicellularity (Section 19-3)
• flagella (Section 20-2)
• *Highlights: Kingdom Protista* (p. 445)
• *Highlights: Kingdom Fungi* (p. 446)

Authors' Rationale

Protists are the earliest eukaryotic group. Understanding protists is crucial to developing a good understanding of the subsequent multicellular eukaryotic kingdoms because protists probably represent the ancestral group from which the other eukaryotes evolved. There is a tremendous amount of diversity among protists—students should notice that some protists seem to be "protoplants," "protoanimals," or "protofungi." Students will also learn about the pathogenic implications of some protist species, such as *Plasmodium*.

21-1 Characteristics of Protists

The most structurally diverse collection of organisms is the kingdom Protista. Most protists are unicellular, microscopic organisms, but a few are complex and multicellular. Protists are also the most diverse in terms of their life cycles, which can involve asexual or sexual reproduction. Composed of metabolically diverse organisms, the kingdom Protista contains the ancestral life-forms that gave rise to the three kingdoms of multicellular organisms—the fungi, the plants, and the animals.

Section Objectives

- Describe the eukaryotic features that first evolved in protists.
- Describe the unifying features of protists.
- List three environments where protists can be found.
- Discuss asexual and sexual reproduction of the protist Chlamydomonas.
- Describe three ways multicellular protists reproduce.

Protists Were the First Eukaryotes

The first eukaryotic cells are thought to have evolved approximately 1.5 billion years ago, but there is a gap in the evolutionary record between prokaryotes and the first eukaryotes. And because eukaryotes differ from bacteria in so many ways, they could not have evolved quickly. Although no living organism has the characteristics associated with the first transitional eukaryotes, the amoeba *Pelomyxa palustris* exhibits many of them. *Pelomyxa* is a unique protist that lacks mitochondria and does not undergo mitosis; it may represent a very early stage in the evolution of eukaryotic cells.

Two important eukaryotic features that evolved among the protists are sexual reproduction and multicellularity. Many protists reproduce only asexually, by mitosis; some employ meiosis and sexual reproduction in times of environmental stress, and others reproduce sexually most of the time. Multicellularity, involving a small degree of coordination and interaction among specialized cells, also evolved independently in different groups of protists at different times. And, early during the evolution of protists, complex flagella and cilia appeared. These organelles are composed of nine pairs of microtubules surrounding two single microtubules in the center. Biologists refer to this as the "9 + 2" structure, which is present in all eukaryotes that have these organelles. Figure 21-1 lists the eukaryotic features that evolved in protists.

What Unites Members of the Kingdom Protista?

The kingdom Protista contains all eukaryotes that cannot be classified as animals, plants, or fungi. Protists are distinguished from these other organisms because they lack specialized features that characterize the three other multicellular kingdoms. For example, unlike plants and animals, protists do not reproduce by forming embryos, nor do they develop complex multicellular reproductive structures.

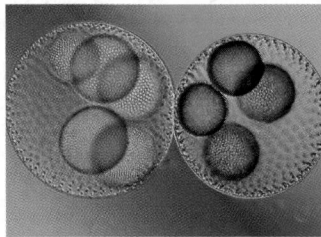

Eukaryotic Features That Evolved in Protists

Sexual reproduction

Meiosis

Mitosis

Multicellularity

Flagella and cilia with "9 + 2" structure

Figure 21-1 The protist *Volvox* is a colony of hundreds or thousands of flagellated cells. Like all other eukaryotes, *Volvox* exhibits key eukaryotic characteristics that first evolved among protists.

SECTION 21-1

Vocabulary Preview

- eyespot
- zygospore
- zoospore
- conjugation
- sporangia

Opening Question

Ask students how an organism can ensure the survival of its genes in a future generation. Encourage students to propose strategies that could help organisms perpetuate their genes. Students' suggestions might include improved ways of storing and protecting the genetic information or alternative forms of reproducing under different conditions. Lead them to understand that there are adaptive advantages to different ways of passing on genetic information.

Demonstration
Classifying Protists Into Groups

Give small groups of students folders that contain pictures of protists (try to include some representatives of each group shown in Table 21-1 on page 480). These pictures can be from magazines, catalogs, or textbooks. Ask students to group the protists into subgroups. Have representatives from each group of students present their classification of the protists. This activity will give students an idea of how difficult classifying protists can be for scientists.

Effective Reading

Organizing the Information

Give each student a photocopy of page 479 of the textbook. Remove Figure 21-1, which summarizes the eukaryotic features that evolved in protists, from the copy. In the margins of their photocopies, have students write key words and draw pictures and symbols to help translate the reading into their own terms. Once students have finished annotating their reading, ask them to summarize in a chart the features that first evolved in protists. Let them compare their charts to the one shown in Figure 21-1.

Teaching Tip
Finding Protists

Ask students where they might be able to find protists. *(Students' answers will vary, but should include wet areas, such as marshes, lakes, ponds, rivers, oceans, moist soil or sand, and the bloodstreams and tissues of animals.)*

Demonstration
Comparing Environments

Collect water from local ponds, and make slides of samples from each pond. Try to find samples of pond scum, seaweed, and plankton. Have students view the slides and document their findings with drawings. Why are ponds a good place to look for protists? *(Protists live almost anywhere there is water.)*

□ CAPSULE SUMMARY

The kingdom Protista contains the most diverse groups of eukaryotic organisms. While protists have some characteristics found in the other three eukaryotic kingdoms, they differ enough to be assigned their own kingdom.

Instead of containing a well-defined group of similar organisms, the kingdom Protista consists of an unusual assortment of organisms that have different ways of meeting life's needs. For example, some protists have chloroplasts and manufacture their own food (like plants), some ingest their food (like animals), and some absorb their food (like fungi). The major phyla of protists are strikingly different from one another and, with a few exceptions, are only distantly related.

In older systems of classification, photosynthetic protists (algae) were classified in the plant kingdom, as were protists that absorbed their food (water molds and slime molds). Heterotrophic protists were thought to consume food "like animals" and were considered small, simple members of the animal kingdom. Today all groups of eukaryotic organisms that do not have the specialized characteristics of either animals, plants, or fungi are classified in the kingdom Protista, as shown in Table 21-1. □

Table 21-1 Members of Kingdom Protista

Common Name	Approximate Number of Species
Amoebas	300
Brown algae	1,500
Cellular slime molds	70
Chytrids	575
Ciliates	8,000
Diatoms	more than 11,500
Dinoflagellates	2,100
Euglenoids	1,000
Foraminiferans (Forams)	300
Green algae	more than 7,000
Plasmodial slime molds	500
Red algae	4,000
Sporozoans	3,900
Unicellular flagellates	3,000
Water molds	580

Protists Are Confined to Moist Environments

Protists are found almost anywhere there is water. Many live in lakes and oceans, floating as plankton or anchored to rocks. They are also common inhabitants of damp soil and sand, and thrive in other moist terrestrial environments such

Historical Note
Anton van Leeuwenhoek (1632–1723)

Leeuwenhoek, a Dutchman sometimes described as an amateur scientist, is credited with the first descriptions of microorganisms such as bacteria and protists. Leeuwenhoek was a cloth merchant. He developed some of the first powerful (270×) microscopes in order to inspect cloth. Leeuwenhoek called the tiny organisms he viewed through his microscopes *animalcules* because he assumed that they were small animals.

as leaf litter. Some species of protists are parasites. They live in the tissues and bloodstream of humans and other animals, where they can cause potentially fatal diseases. Other protists parasitize plants.

Many protists have mechanisms for monitoring and responding to stimuli in their environment. For example, some protists have **eyespots**, small organelles containing light-sensitive pigments that detect changes in the quality and intensity of light. Protists are also sensitive to touch and chemical changes in their environment. When certain protists encounter a noxious chemical, for example, they will back up and try to bypass it. ▪

CAPSULE SUMMARY

Protists live in moist environments and can be either free-living or parasitic.

Some Protists Can Reproduce Sexually

Some kinds of protists reproduce only asexually, giving rise to new genetically identical individuals. Others can reproduce sexually or asexually. In these protists, sexual reproduction is often triggered by periods of environmental stress.

Reproduction Among Unicellular Protists

The unicellular green alga *Chlamydomonas* reproduces in a manner typical of unicellular protists. *Chlamydomonas* exhibits sexual as well as asexual reproduction, as outlined in Figure 21-2. As a mature organism, the single-cell protist is haploid. *Chlamydomonas* reproduces sexually in times of environmental stress such as a shortage of nutrients. Under such circumstances, mitosis produces haploid gametes. After they are released, gametes of opposite mating types fuse to form pairs. There are two different mating types (designated + or –) in *Chlamydomonas*. Each mating type has distinctive

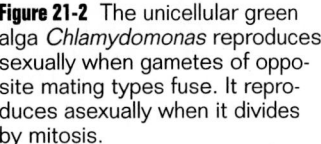

Figure 21-2 The unicellular green alga *Chlamydomonas* reproduces sexually when gametes of opposite mating types fuse. It reproduces asexually when it divides by mitosis.

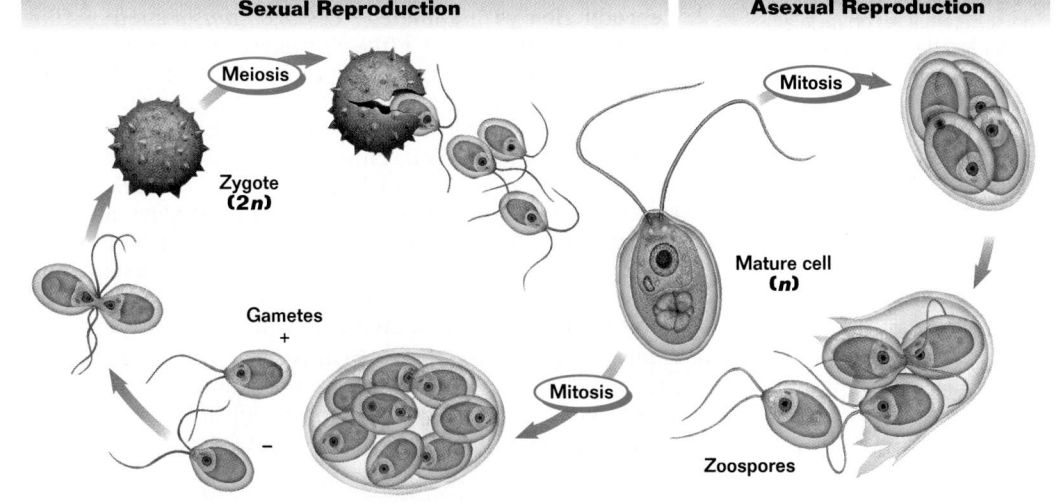

SECTION 21-1

Teaching Tip
Eyespots

Ask students what the function of an eyespot is. *(Eyespots detect the quality and intensity of light.)* Why might a protist want to monitor the characteristics of light? Lead students to predict that some protists are photosynthetic; thus, these protists use information about light to maintain a homeostatic metabolism. This discussion will set the stage for Section 21-2.

 VISUAL STRATEGY

Figure 21-2

Use this figure to trace sexual and asexual reproduction in *Chlamydomonas*. Which form of reproduction does *Chlamydomonas* undergo in times of environmental stress? *(sexual reproduction)*

cell-surface proteins. These mating types ensure that only individuals with different genetic lineages will participate in sexual reproduction. When gametes of opposite mating types pair, they shed their cell walls and fuse into a diploid zygote with a thick protective wall called a **zygospore.** A zygospore can withstand unfavorable environmental conditions for long periods. When conditions become favorable again, meiosis within the zygospore produces four haploid individuals, which break out of the zygospore wall. These haploid cells, two of each mating type, grow into mature cells and complete the sexual life cycle.

When it reproduces asexually, *Chlamydomonas* first absorbs its flagella. The haploid cell then divides mitotically one to three times, producing two to eight haploid cells called **zoospores,** which remain within the wall of the parent cell. After developing flagella, mature zoospores break out of the parent cell and complete the asexual life cycle.

Sexual Reproduction Among Multicellular Protists

Sexual reproduction among multicellular protists occurs in many different ways, some of which are quite complex. Three modes of reproduction found among the green algae demonstrate the reproductive variation that exists among multicellular protists.

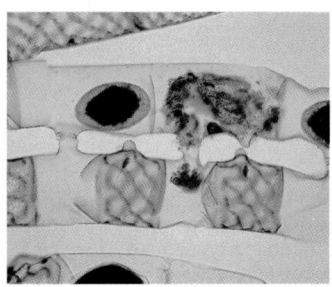

Figure 21-3 *Spirogyra* is a filamentous green alga that contains large spiral-shaped chloroplasts. During sexual reproduction, adjacent filaments join by conjugation tubes. Genetic material from the cell of one filament passes into the adjacent cell of the other filament.

1. **Conjugation** *Spirogyra*, a filamentous green alga, reproduces sexually by a process called **conjugation.** Conjugation begins when two filaments align side by side. Portions of the walls between adjacent cells then dissolve and form a cytoplasmic bridge between the cells called a conjugation tube. As seen in Figure 21-3, the contents of one cell then pass through the tube into the cell of the adjacent filament, where the two nuclei fuse to form a diploid nucleus. The resulting zygote then develops a thick wall, falls from the parent filament, and becomes a resting spore. When conditions are favorable, the resting spore undergoes meiosis and produces a new haploid filament.

2. **Reproduction with gametes** *Oedogonium* is another filamentous green alga. Unlike *Spirogyra*, *Oedogonium* has specialized cells for producing and holding gametes. One type of cell produces flagellated sperm; the other type produces an egg. When the sperm are released, they swim to the egg-carrying cell, enter it, and fuse with the egg. The resulting zygote is released and forms a thick-walled resting spore. This diploid spore then undergoes meiosis, forming four haploid zoospores that are released into the water. Each zoospore settles and divides by mitosis. When a zoospore divides, one of the new cells becomes an anchor, while the others divide to form the new filament.

3. **Alternation of generations** *Ulva* is a very common marine green alga. Figure 21-4 shows that the reproductive cycle of *Ulva* is characterized by two distinct multicellular phases: a haploid, gamete-producing phase called the gametophyte

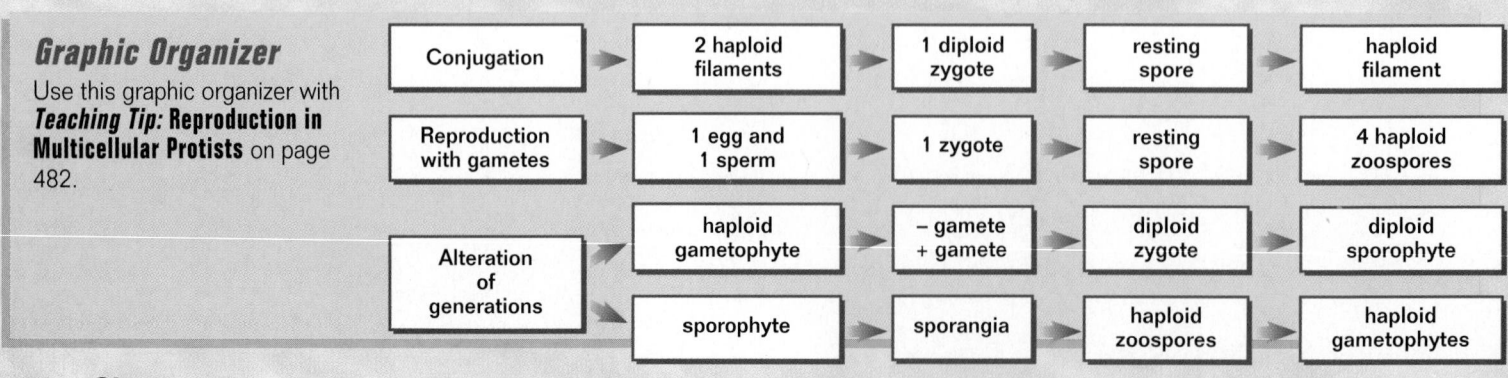

Conjugation	2 haploid filaments	1 diploid zygote	resting spore	haploid filament
Reproduction with gametes	1 egg and 1 sperm	1 zygote	resting spore	4 haploid zoospores
Alteration of generations	haploid gametophyte	– gamete + gamete	diploid zygote	diploid sporophyte
	sporophyte	sporangia	haploid zoospores	haploid gametophytes

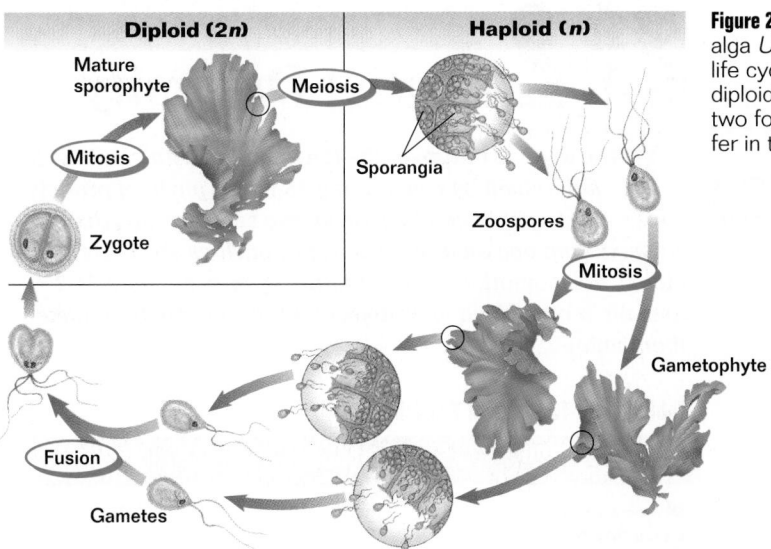

Diplod (2*n*)

Mature
sporophyte
Meiosis
Mitosis
Zygote
Sporangia
Fusion
Gametes

Haploid (*n*)

Zoospores
Mitosis
Gametophyte

Figure 21-4 The multicellular green alga *Ulva*, or sea lettuce, has a life cycle in which haploid and diploid individuals alternate. The two forms look the same, but differ in their chromosome number.

SECTION 21-1

Teaching Tip
Mapping the Concepts
Have students work individually or in small teams to make a concept map of the following terms: zoospores, haploid, diploid, unicellular, multicellular, zygospore, conjugation, gametes, alternation of generations, *Chlamydomonas, Spirogyra, Ulva, Oedogonium*, zygote, sporangia. Ask students to highlight these words in one color. In another color, students should draw three cross-links between the words.

generation, and a diploid, spore-producing phase called the sporophyte generation. The adult sporophyte alga, called sea lettuce, is large and distinctive in appearance. It has reproductive cells called **sporangia** (*spoh RAN jee uh*), which produce haploid zoospores by meiosis. The spores land on rocks and grow into multicellular haploid gametophytes. When mature, the gametophyte superficially resembles the sporophyte. The mature gametophyte eventually produces haploid gametes that unite and form zygotes. These diploid zygotes then complete the life cycle by dividing mitotically to form a new diploid sporophyte. ❑

CAPSULE SUMMARY

Some protists are able to reproduce sexually in times of stressful environmental conditions.

Section Review

1. *What are eukaryotic features that first evolved in protists?*
2. *Why aren't protists classified in the other three multicellular kingdoms?*
3. *Describe three environments in which protists thrive.*
4. *Draw the asexual and sexual life cycles of the protist* Chlamydomonas. *What causes the protists to reproduce sexually?*
5. *Describe three ways sexual reproduction can occur in multicellular protists.*

Critical Thinking
6. *Why is kingdom Protista casually referred to as a "catchall" kingdom?*

Answers to Section Review

1. The eukaryotic features that first evolved in protists include sexual reproduction, meiosis, mitosis, multicellularity, and flagella and cilia with the "9+2" structure.

2. Protists are not classified in the other three multicellular kingdoms because they lack the specialized features observed in members of those kingdoms.

3. Answers should include moist environments such as bodies of water, damp soil or sand, and bloodstreams and tissues of animals.

4. See Figure 21-2 on page 481 for the sexual and asexual life cycles of *Chlamydomonas*. Environmental stress causes the protists to reproduce sexually.

5. Multicellular protists reproduce by conjugation, by the production of gametes, and by alternation of generations. See pages 482–483 for further information.

6. The kingdom Protista consists of an assortment of organisms. It includes eukaryotes that lack the specialized characteristics of animals, plants, and fungi.

Vocabulary Preview

amoeba

pseudopodia

foram

diatom

dinoflagellate

zoomastigote

euglenoid

pellicle

plasmodium

oocyte

Opening Question

Ask students to recall the characteristics that unite members of the kingdom Protista. Have them review the suggestions that you posted around the room in Section 21-1. Are they all true? Modify their original statements using the information students learned in Section 21-1. Students should know that protists are eukaryotes that are not classified in other kingdoms. Help students understand that membership in the kingdom Protista is diverse and is defined by a lack of features.

Demonstration

A Giant Amoeba

Obtain a culture of live amoebas. Using a microprojector, project an amoeba on a screen for students to view. Discuss the general characteristics of the amoeba and point out its nucleus. Show students how amoebas move by means of pseudopodia.

Section Objectives

- Explain how amoebas and forams move.
- Describe the structure of diatoms.
- Compare and contrast the three kinds of algae.
- Name three different kinds of flagellates.
- Describe the general characteristics of Euglena and Paramecium.
- Discuss the unique features that distinguish protistan molds and sporozoans from other protists.

21-2 Protist Diversity

The diverse nature of the kingdom Protista can best be understood by considering 15 major phyla of protists. These phyla can be placed into seven groups distinguished from one another by features such as structure, means of locomotion, and formation of spores. Table 21-2 lists the major phyla of protists and the features that make them unique.

Table 21-2   15 Phyla of Protists

Distinguishing Features	Phylum	Mode of Nutrition
Move using pseudopodia	Rhizopoda (amoebas)	Heterotrophic
	Foraminifera (forams)	
Have double shells made of silica	Bacillariophyta (diatoms)	Photosynthetic
Photosynthetic protists; can be multicellular	Chlorophyta (green algae)	Photosynthetic
	Rhodophyta (red algae)	
	Phaeophyta (brown algae)	
Move using flagella	Dinoflagellata (dinoflagellates)	Photosynthetic
	Zoomastigina (unicellular flagellates)	Heterotrophic
	Euglenophyta (euglenoids)	Most are heterotrophic; some are photosynthetic
Move using cilia	Ciliophora (ciliates)	Heterotrophic
Funguslike protists	Acrasiomycota (cellular slime molds)	Heterotrophic
	Myxomycota (plasmodial slime molds)	
	Oomycota (oomycetes)	
	Chytridiomycota (chytrids)	
Form resistant spores	Sporozoa (sporozoans)	Heterotrophic

 Did You Know?

Green algae have been tested in the space program as a means of eliminating excessive amounts of cargo in traveling great distances. The algae would feed shrimp, a food source for the astronauts, and would also create oxygen for the astronauts. By taking an oxygen source, the supplies needed for the trip could be reduced.

Protists That Move Using Cytoplasmic Extensions

An interesting group of protists consists of amoebas (uh MEE buhs) and forams. These protists are distinguished by their unique form of locomotion, in which they use extensions of cytoplasm. Amoebas and forams are unicellular heterotrophs.

Amoebas, members of the phylum Rhizopoda, are protists that live in fresh and salt waters and are especially abundant in soil. Because an amoeba has no cell walls or flagella, it is extremely flexible. It moves through its environment using extensions of cytoplasm called **pseudopodia** (soo doh POH dee uh), from the Greek words pseudo, meaning "false," and podium, meaning "foot." A pseudopodium bulges from the cell surface, stretches outward, and anchors itself to a nearby surface. The cytoplasm from the rest of the amoeba then flows into the pseudopodium. Pseudopodia are also used to surround and engulf food particles in the process of endocytosis, as you can see in Figure 21-5. Meiosis and sexual reproduction do not occur in amoebas. They reproduce by fission, simply dividing into two new cells. The majority of amoebas are free-living, but some species are parasites, such as Entamoeba histolytica, the protist that causes amoebic dysentery in humans. These organisms are transmitted by contaminated food or water.

Forams, members of the phylum Foraminifera, are marine protists that live in sand or attached to other organisms or rocks. Forams are characterized by their porous shells, called tests. Tests usually have many chambers arrayed in a spiral shape resembling a tiny snail and consisting of organic material that contains grains of calcium carbonate. Long, thin projections of cytoplasm extend through the pores in the tests to aid in swimming and in capturing prey. For 200 million years, tests have accumulated in limestone deposits and are important components of land formations such as the white cliffs of Dover, the famous landmark in southern England, shown in Figure 21-6. The life cycle of forams is complex and involves alternation between haploid and diploid generations.

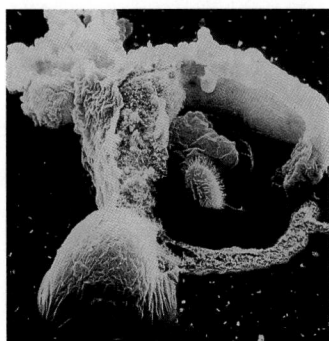

Figure 21-5 Members of the phylum Rhizopoda, such as this amoeba, ingest prey by endocytosis, capturing and engulfing prey with pseudopodia.

Figure 21-6 Forams live in snail-like shells made of calcium carbonate, *below right*. Fossils of these shells make up the sedimentary deposits in the white cliffs of Dover in England, *below left*.

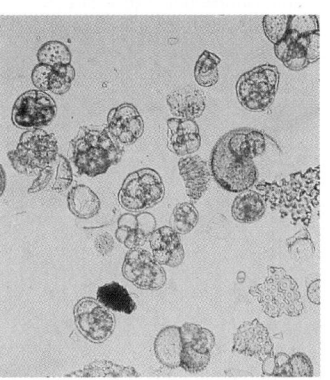

SECTION 21-2

Teaching Tip
Pseudopodia
Ask students to depict the movement of an amoeba using a pseudopodium. Drawings should demonstrate a pseudopodium forming and extending from the edge of the amoeba. They should also show the cytoplasm of an amoeba flowing into the pseudopodium. Ask students to add a brief paragraph to explain their drawings.

 VISUAL STRATEGY

Figure 21-6
Explain how the white cliffs of Dover formed. Millions of years ago these cliffs were covered by a sea that was rich in forams. As the forams died, their remains dropped to the bottom of the ocean and formed sedimentary deposits. As the water receded, the cliffs that were once at the bottom of the ocean were exposed.

Teaching Tips

Diatom Deposits

Diatoms make up a major portion of sediments on the ocean floor. One of the thickest deposits measured is about 1 km deep and is located off the coast of Lompoc, California. Diatom deposits are mined in Nevada, Washington, Kansas, and Oregon. Ask students to find out some uses of diatoms. *(Fossil diatom deposits can be used to make abrasive powders for cleaning and polishing products and for use in filtration processes.)*

Diatometer

Dr. Ruth Patrick, a limnologist who has spent many hours wading through lakes, streams, and rivers throughout North and South America, invented the diatometer—a device used by scientists throughout the world to identify particular types of diatoms. Encourage students to find out more about the uses of a diatometer.

Demonstration

Observing Algae

If possible, obtain live or preserved specimens of various algae from a biological supply house, or display pictures of algae from scientific reference books. Have students examine the algae and compare their characteristics.

☐ CAPSULE SUMMARY

Amoebas and forams are protists that move using cytoplasmic extensions. Diatoms are protists with glasslike shells.

Figure 21-7 Members of the phylum Bacillariophyta, such as these diatoms, consist of two shells that fit together like a tiny box. Small pores in the shells enable gases and other substances to enter and leave the cell.

Diatoms Have Double Shells
....

Diatoms, members of the phylum Bacillariophyta, are photosynthetic, unicellular protists with unique double shells made of silica, which are often strikingly and characteristically marked, as shown in Figure 21-7. Their shells are like small boxes with lids, one half fitting inside the other. Abundant in oceans and lakes, diatoms are important food producers. Diatoms can have one of two types of symmetry: radial (like a wheel) or bilateral (two-sided). The empty shells of diatoms form thick deposits that are mined commercially as "diatomaceous earth," which is often used as an abrasive or to add the sparkling quality to paint used on roads. Diatoms are capable of a gliding movement made possible by chemicals secreted out of holes in their shells. Individuals are diploid and usually reproduce asexually. The two halves of the shell separate, and each half regenerates another matching half. As a consequence of this mode of reproduction, diatoms tend to get smaller and smaller with each generation. When an individual gets too small because of repeated division, it slips out of its shell, grows to full size, then regenerates a new shell. Sexual reproduction in diatoms is rare. ☐

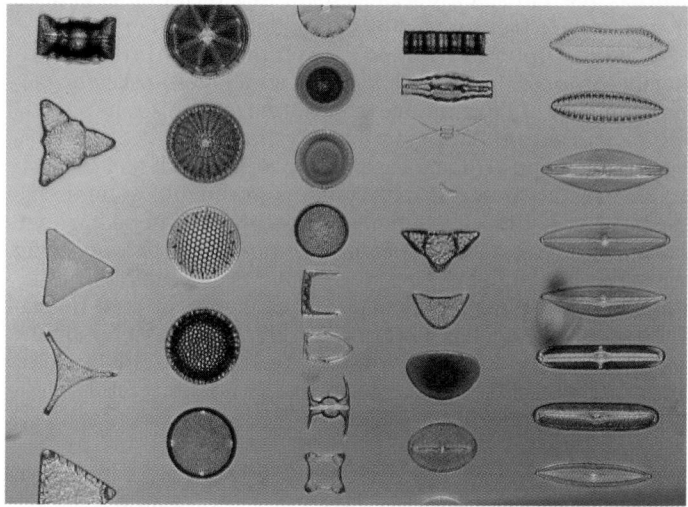

Some Algae Are Multicellular
....

The kingdom Protista also contains some of the fastest growing and most photosynthetically productive organisms—the algae. The three kinds of algae—green algae, red algae, and brown algae—are distinguished by the types of chlorophyll they contain.

Green algae, members of the phylum Chlorophyta, are an extremely varied group of protists. Most green algae are freshwater, unicellular organisms such as *Chlamydomonas,* but some are large multicellular marine organisms like *Ulva,*

 Did You Know?

Diatomaceous earth has been used in the manufacture of dynamite. When nitroglycerin is absorbed by diatomaceous earth, it is stable and safe to handle.

Overcoming Misconceptions

Red and Brown Algae Also Have Green Pigments

Both brown and red algae contain chlorophyll in addition to the pigments that make them look brown and red. A misconception is that they lack chlorophyll, but the green pigment is hidden by the other colors.

shown in Figure 21-8, a species found in marine intertidal zones. Other species are part of the marine plankton, inhabit damp soil, or even thrive within the cells of other organisms as photosynthetic symbionts. Green algae have chloroplasts that contain chlorophylls *a* and *b*, the same pigments found in the chloroplasts of plants. This is one reason why green algae are considered to be the ancestors of the plant kingdom. Most green algae have complex life cycles with sexual and asexual reproductive stages.

Red algae, members of the phylum Rhodophyta, are multicellular organisms found in warm ocean waters. Their color results from red pigments called phycobilins (*fy koh BYLIHNZ*), which are especially efficient at absorbing the green, violet, and blue light that penetrates into deep waters. For this reason, red algae can thrive at greater depths than other photosynthetic organisms. Red algae have complex bodies made up of interwoven filaments of cells. Some, such as the coralline algae shown in Figure 21-9, have calcium carbonate in their cell walls. Others have cell walls with a slippery outer layer that is used to make commercial products such as agar and carrageenan. Red algae have a complex life cycle, usually involving alternation of generations. None have flagella or centrioles, suggesting that red algae may be one of the most ancient groups of eukaryotes.

Brown algae, members of the phylum Phaeophyta, are all multicellular and almost exclusively marine. They are the most abundant seaweeds in many northern regions along rocky shores. The larger brown algae known as kelp grow in massive groves in relatively shallow water along coasts and provide food and shelter for many different kinds of organisms. Many have flattened blades, stalks, and anchoring bases and often contain complex internal conducting tissues like those of plants. Among the larger brown algae are genera such as *Macrocystis*, shown in Figure 21-10, whose blades float on the surface of the water while the base is anchored many meters below. The chloroplasts of brown algae resemble those of diatoms and dinoflagellates. Their life cycle involves an alternation of generations, with the sporophyte (diploid) generation consisting of the largest individuals.

Figure 21-8 *Ulva* is a green alga composed of a sheet of cells that is two cells thick. It grows attached to rocks in intertidal zones.

Figure 21-9 This species of coralline algae contributes to the great coral reefs.

Figure 21-10 Massive groves of the brown alga *Macrocystis* grow in temperate coastal waters throughout the world. These great kelp beds contain individuals that grow to a length of more than 60 m in a single season.

Teaching Tips

Red Tides

The toxic substance produced by the dinoflagellate that is responsible for "red tides" can paralyze the nervous system of humans. Ask students how red tides might affect humans. (*Humans who ingest mussels that feed on the toxic dinoflagellates become ill and may even die.*)

Glowing Protists

The dinoflagellate *Noctiluca* glows because of a chemical reaction that produces light. The waters that contain these organisms create luminescent, or sparkling, waves. Encourage students to find out more about *Noctiluca* and its luminescence.

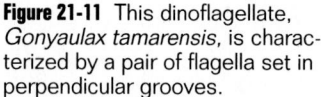

Some Protists Move With Flagella

Flagellates are protists that move using flagella. The three major phyla of flagellates are the dinoflagellates, the zoomastigotes, and the euglenoids.

Dinoflagellates (*dy noh FLAJ uh layts*), members of the phylum Dinoflagellata, are unicellular photosynthetic protists, most of which have two flagella. A few kinds of dinoflagellates are found in fresh waters, but the majority are marine, and they are often a component of plankton. Most dinoflagellates have a protective coat made of cellulose that is often encrusted with silica, giving them unusual shapes, as shown in Figure 21-11. Their flagella beat in two grooves—one encircling the body like a belt, the other perpendicular to it. As a result, dinoflagellates spin through the water like a top. Some species are luminous and produce a twinkling light that can be seen at night in tropical seas. A few dinoflagellates produce powerful toxins. The poisonous "red tides" that occur frequently in coastal areas are often associated with population explosions of dinoflagellates. Dinoflagellates reproduce asexually by mitosis.

Figure 21-11 This dinoflagellate, *Gonyaulax tamarensis*, is characterized by a pair of flagella set in perpendicular grooves.

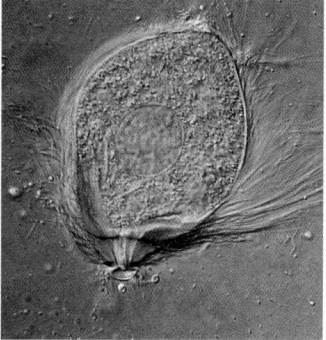

Figure 21-12 *Trichonympha* is a protist that inhabits the digestive tract of termites, where it aids in digesting cellulose. Rows of flagella protrude from one end of the cell.

Zoomastigotes (*zoh oh MAS tih gohts*), members of the phylum Zoomastigina, are unicellular, heterotrophic organisms that vary greatly in form. Each has at least one flagellum; some species have thousands. Most zoomastigotes reproduce only asexually, but some are known to produce gametes and reproduce sexually. Some zoomastigotes such as *Trichonympha*, shown in Figure 21-12, live symbiotically in the guts of termites, where they provide the enzymes that digest wood. Others, such as the trypanosomes, are dangerous pathogens in humans and domestic animals. The choanoflagellates (*koh an o FLAG uh layts*) are zoomastigotes that closely resemble collar cells in sponges and are thought to have given rise to the sponges and thus all other animals.

 Did You Know?

Dinoflagellates are essential in the formation of coral reefs, coral islands, and atolls. By inhabiting the living tissues of reef corals, mollusks, and other marine organisms, dinoflagellates are able to use the intense sunlight that reef waters receive. In turn, dinoflagellates contribute to the accumulation of reefs by speeding up the formation of calcium carbonate in coral skeleton.

Euglenoids *(yoo GLEE noyds),* members of the phylum Euglenophyta, are freshwater protists with two flagella. They clearly illustrate the impossibility of classifying protists as animals or plants. About one-third of the 1,000 known species of euglenoids have chloroplasts and are photosynthetic; other species lack chloroplasts, ingest their food, and are heterotrophic. Some photosynthetic euglenoids may reduce the size of their chloroplasts and become heterotrophic if they are kept in a dark environment. If they are put back in the light, their chloroplasts return to normal size within a few hours, and photosynthesis resumes. Euglenoids are clearly related to zoomastigotes, and many taxonomists merge the two phyla and consider them to be one. *Euglena,* shown in Figure 21-13, has a protein scaffold called a **pellicle** *(PEHL ih kuhl)* inside the cell membrane. Since the pellicle is flexible, the euglenoid can change shapes. A light-sensitive organ called the eyespot helps orient the movements of these organisms toward light. Reproduction in this phylum occurs by mitosis; sexual reproduction has not been observed in this group.

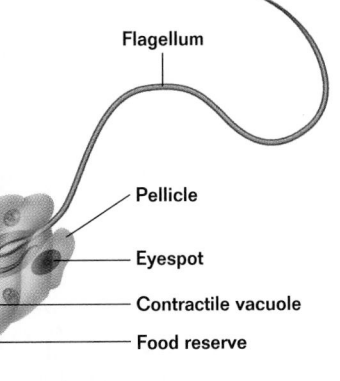

Flagellum

Nucleus

Pellicle

Eyespot

Contractile vacuole

Food reserve

Chloroplast

Figure 21-13 *Euglena* is a versatile protist. It contains chloroplasts and is photosynthetic, but it can also absorb organic nutrients and can live without light.

Protists That Use Cilia to Swim

As the name *ciliates* indicates, all members of the phylum Ciliophora have large numbers of cilia, usually arranged in long rows down the body or in spirals around it, as shown in Figure 21-14. Ciliates are complex unicellular heterotrophs. The body wall of ciliates is a tough but flexible outer pellicle that enables the organism to squeeze through or move around many obstacles. The pellicle consists of an outer membrane with numerous fluid-filled cavities beneath it. Ciliates, such as *Paramecium* shown in *Up Close: Paramecium* on page 490, form vacuoles for ingesting food and regulating their water balance. In addition to their characteristic cilia, most ciliates have two types of nuclei within their cells: small micronuclei and larger macronuclei. The micronuclei contain normal chromosomes that divide by mitosis. Macronuclei contain small pieces of DNA derived from micronuclei. Reproduction is usually by mitosis, with the body splitting in half. Cells divide asexually for about 700 generations and then die if sexual reproduction has not occurred. Most ciliates engage in the sexual process of conjugation, in which two cells unite and exchange genetic material.

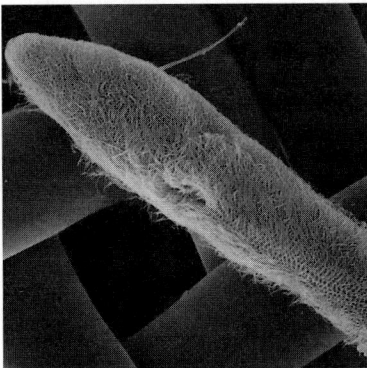

Figure 21-14 *Paramecium* is a fresh water protist that uses its thousands of cilia to move and feed.

UP CLOSE PARAMECIUM

- **Scientific name:** *Paramecium caudatum*
- **Habitat:** Lives in freshwater streams and ponds
- **Size:** Microscopic; up to 1 mm long
- **Diet:** Bacteria, small protists, organic debris

Instructional Strategies

- Ask students to draw a flowchart to trace the flow of food through a paramecium. Charts should begin with food entering the paramecium via the oral groove. At the cell mouth, food is engulfed by a food vacuole containing digestive enzymes. Undigested food is excreted by exocytosis.

- Ask students to compare the methods used by amoebas and paramecia for capturing food. Remind students that amoebas use their pseudopodia to surround and engulf food by endocytosis. Paramecia use cilia to create a whirlpool to pull food into the oral groove, where the food is engulfed by a food vacuole.

Discussion

Guide the discussion by posing the following questions.

1. How would a paramecium living in a clean environment differ from one living in a contaminated environment? (*Answers might include that paramecia in polluted water may be less efficient at maintaining water balance, acquiring nutrients, and moving due to the pollutants and subsequent damage to cilia.*)

2. Describe an ideal environment for *Paramecium.* (*Answers will vary. Answers should include a food source and a freshwater environment that will allow a paramecium to maintain homeostasis.*)

Characteristics

Surface *Paramecium*, a ciliate, is covered with thousands of cilia arranged in rows along the cell. Cilia beat in waves that move diagonally across the cell, causing the protist to spin through the water. *Paramecium* is surrounded by a rigid protein covering called a pellicle.

▲ **Surface**

Maintaining Water Concentration Like other freshwater protists, *Paramecium* is constantly absorbing water by osmosis. Since these organisms need to maintain a relatively low concentration of water inside the cell in order to function normally, they must get rid of excess water. *Paramecium* does this with contractile vacuoles, saclike organelles that expand, collecting excess water, and then contract, squeezing water out of the cell.

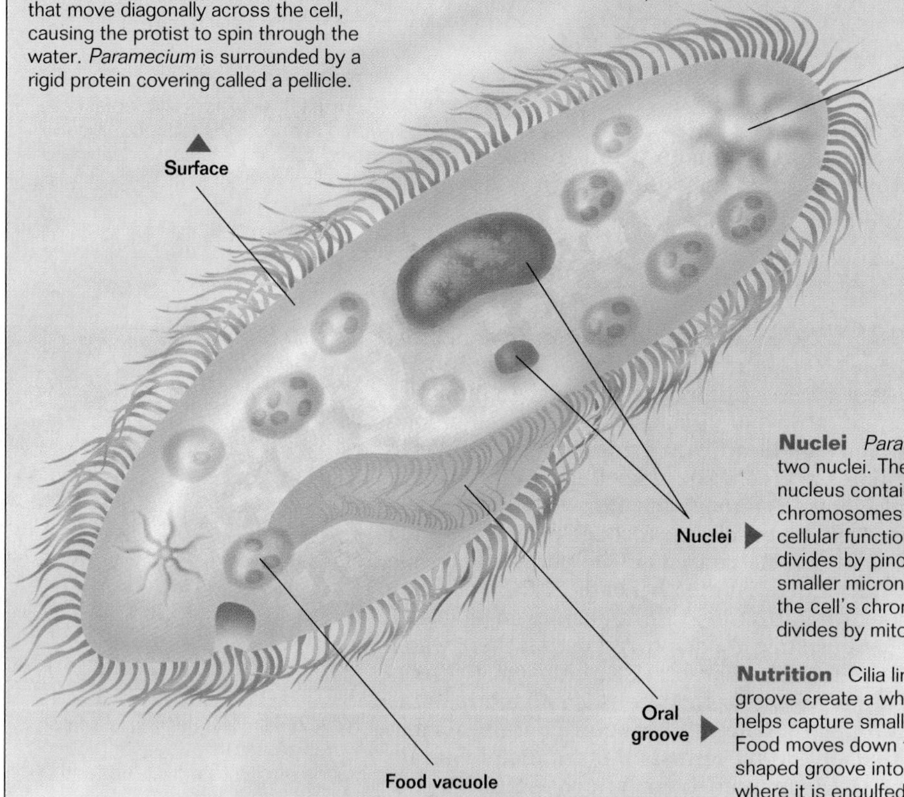

▲ **Contractile vacuole**

Nuclei ▶

Nuclei *Paramecium* has two nuclei. The larger macronucleus contains fragmented chromosomes used in routine cellular functions, and it divides by pinching in two. The smaller micronucleus contains the cell's chromosomes and divides by mitosis.

Nutrition Cilia lining the oral groove create a whirlpool effect that helps capture small bits of food. Food moves down the funnel-shaped groove into the cell mouth, where it is engulfed in a food vacuole by endocytosis. As a food vacuole moves throughout the cell, it combines with digestive enzymes. Undigested food is released from the cell by exocytosis.

Oral groove ▶

Food vacuole

Genetic Variation *Paramecium* generally reproduces asexually by binary fission. Genes are shuffled during a sexual process called conjugation. During conjugation, two individuals exchange haploid micronuclei, which fuse to other micronuclei, forming a diploid nucleus that contains nuclei from the two individual paramecia.

Protistan Molds Are Not Fungi

Protistan molds are heterotrophs with restricted mobility. They were once thought to be related to fungi because they have a similar appearance and lifestyle. They are not fungi however; protistan molds have cell walls made of carbohydrates, like those found in other protists, whereas fungi have cell walls made of chitin. Also, protists carry out normal mitosis, whereas mitosis in fungi is unusual, as you will learn in Chapter 22. The four major phyla of protistan molds, each with different structures and life cycles, are not related to each other.

Cellular slime molds, members of the phylum Acrasiomycota, resemble amoebas in the phylum Rhizopoda but have many distinct features. The individual organisms behave as separate amoebas, moving through the soil and ingesting bacteria. In times of environmental stress, the individual amoebas aggregate and move toward a fixed center, as shown in Figure 21-15. There, they form multicellular colonies called slugs. Each slug develops a base, a stalk, and a swollen tip that develops spores. Each of these spores, when released, becomes a new amoeba, which begins to feed and repeat the life cycle. There are 70 known species of cellular slime molds, the best known of which is *Dictyostelium discoideum*.

Plasmodial slime molds, members of the phylum Myxomycota, are a group of bizarre organisms that stream along as a **plasmodium,** a mass of cytoplasm that looks like an oozing slime, as shown in Figure 21-16. As they move, they engulf and digest bacteria and other organic material. A plasmodial slime mold contains many nuclei, but these are not separated by cell walls. All nuclei undergo mitosis at the same time, in coordinated fashion. If the plasmodium begins to dry out or starve, it moves away rapidly, then stops, and often divides into many small mounds. Each mound produces a stalk tipped with a capsule in which haploid spores develop. The spores are highly resistant to unfavorable environmental conditions and can survive for years. When conditions are favorable, the spores germinate and become haploid cells that are either amoeboid or flagellated. These haploid cells are able to fuse into diploid zygotes, which undergo mitosis and form a new plasmodium.

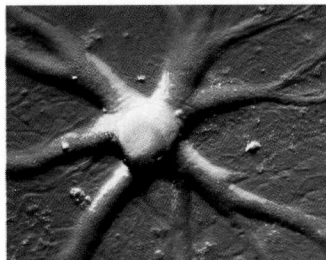

Figure 21-15 When deprived of food, the individual amoebas of the cellular slime mold *Dictyostelium discoideum* aggregate, *top,* and form a moving mass called a slug, *center.* The slug will move to a new habitat and transform into a stalked structure that contains spores, *bottom.*

Figure 21-16 A plasmodial slime mold is a mass of cytoplasm containing many nuclei. Plasmodia can flow around objects and even pass through cloth.

Application

Agriculture A new, more virulent form of the blight that was responsible for the Irish potato famine of 1845–1847 threatens current potato crops. It is resistant to metalaxyl, the fungicide used against potato blight in the United States. The disease threatens a major food source because potatoes are so widely grown and consumed throughout the world. Currently, scientists are working with wild potatoes to produce new disease-resistant varieties of potatoes.

Figure 21-17 Water molds usually help decompose dead animals in the water.

❏ **CAPSULE SUMMARY**

Protist diversity consists of organisms that vary in cellularity, in methods of locomotion and of reproduction, and in structure.

Oomycetes (*oh oh MY seets*), members of the phylum Oomycota, are the water molds, white rusts, and downy mildews that often grow on dead algae and dead animals in fresh water, as seen in Figure 21-17. All members of the group are either parasites or feed on dead organic matter. Oomycetes are unusual in that their spores have two flagella: one pointed forward, the other backward. Many oomycetes are plant pathogens, including *Phytophthora infestans*, which causes late blight in potatoes. This protist was responsible for the Irish potato famine of 1845–1847, during which about 400,000 people starved to death.

Finally, the Chytridiomycota, or chytrids, are a group of protists distinguished by their motile cells, which have a single, whiplike flagellum. There is some evidence that there may be an evolutionary relationship between chytrids and fungi.

Some Protists Form Resistant Spores

Sporozoans, members of the phylum Sporozoa, are nonmotile, spore-forming, unicellular parasites. They infect animals with small spores that are transmitted from host to host. All sporozoans have a unique arrangement of microtubules and other organelles clustered at one end of the cell. Sporozoans have complex life cycles that involve both asexual and sexual reproduction. Sexual reproduction involves the fertilization of a large female gamete by a small, flagellated male gamete. The zygote that results soon becomes a thick-walled cyst called an **oocyte** (*OH oh syt*), which is highly resistant to drought and other unfavorable environmental conditions. The best known sporozoan is the malaria-causing parasite *Plasmodium*, which is shown in Figure 21-18. ❏

Section Review

1. *What are pseudopodia? How do amoebas and forams use them to move?*
2. *What are the shells of diatoms made of? What commercial uses are there for diatoms?*
3. *Construct a table that compares three kinds of multicellular algae.*
4. *How does* Euglena *differ from the other two phyla of flagellates?*
5. *Describe five general characteristics of* Paramecium.

Critical Thinking
6. *Why were protistan molds once classified as fungi?*

Answers to Section Review

1. Pseudopodia are cytoplasmic extensions that some protists use for movement. To move, these organisms stretch their cytoplasm in the desired direction, anchor its mass, and then adjust the cell cytoplasm around the attachment site.

2. The shells of diatoms are made of silica. Commercially, they are used in abrasive cleaners and sparkle chips in paint on highways.

3. Answers will vary. See the *Graphic Organizer* on page 487 for one example.

4. *Euglena* can be either autotrophic or heterotrophic. It has a pellicle that allows it to change shape and an eyespot that helps it find light.

5. See *Up Close: Paramecium* on page 490.

6. Protistan molds are similar to fungi in appearance and lifestyle.

21-3 Protists and Human Health

*O*ne of the greatest impacts protists have on humans is as pathogens, agents of disease. In this section you will consider malaria, a significant human disease caused by a protist. Malaria is by no means the only disease caused by protists, but it will serve to acquaint you with the complex nature of the life cycles of pathogenic protists. Table 21-3 lists other diseases caused by protists.

Section Objectives

- Explain how malaria is transmitted.
- Describe the methods used to control malaria.
- Discuss three other human diseases caused by protists.

Plasmodium Causes Malaria

Malaria, caused by a sporozoan, is one of the most serious infectious diseases of recent history. Over 100 million people have malaria at any one time, and over a million, mostly children, die from it every year. The symptoms include severe chills, fever, sweating, confusion, and great thirst. Victims die of anemia, kidney failure, or brain damage unless the disease is brought under control by the person's immune system or by medical treatment.

The malarial sporozoan parasite is *Plasmodium*, shown in Figure 21-18. *Plasmodium* is spread from person to person by mosquitoes of the genus *Anopheles*. When an *Anopheles* mosquito "bites" a human to obtain blood, it injects saliva mixed with a substance that prevents the blood from clotting. If the mosquito is infected with *Plasmodium*, it will also inject about 1,000 elongated cells of this protist into the bloodstream of its victim. There are three stages in the *Plasmodium* life cycle, illustrated in Figure 21-18. The stage of *Plasmodium* while it

Figure 21-18 *Plasmodium* is a sporozoan that causes the disease malaria. *Plasmodium* has a complex life cycle that involves the mosquito *Anopheles* and human blood and liver cells.

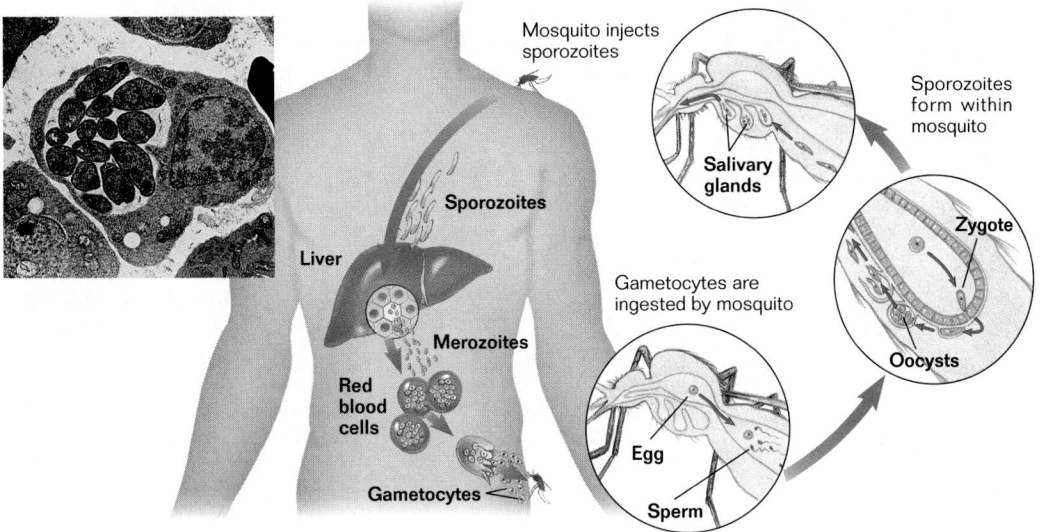

Mosquito injects sporozoites

Sporozoites form within mosquito

Salivary glands

Sporozoites

Liver

Zygote

Merozoites

Gametocytes are ingested by mosquito

Red blood cells

Oocysts

Gametocytes

Egg

Sperm

SECTION 21-3

Vocabulary Preview

sporozoite

merozoite

gametocyte

Opening Question

Ask students to list some characteristics of an ideal habitat for a microbe. Students may suggest a moderate or warm climate in which the microbe can be active year round, poor sanitation conditions in which the microbe is spread easily, organisms with weakened immune systems that once infected cannot fight back, etc. Lead students to conclude that these conditions lower a population's ability to combat diseases.

Demonstration

Combating Mosquitoes

Show students pictures of mosquito-infested areas, such as forests, tropical areas, swamps, etc. Ask students to think of ways to prevent a deadly disease that is spread by mosquitoes. Students may suggest killing mosquitoes with pesticides, keeping mosquitoes away from people through the use of repellents or mosquito netting, or taking drugs to kill the disease-causing microbe carried by mosquitoes once a person is infected. Other answers could include vaccines or genetic manipulation. Tell students that scientists have worked on all of these ideas and others; however, malaria, a disease spread by mosquitoes, is still not under control in many areas of the world.

 Did You Know?

Plasmodium falciparum is 95 percent more likely to cause death than other strains of malaria. *P. falciparum* has been observed in man for only the last 10,000 years, so it is considered a new strain of malaria.

☑ **RESEARCH Update**

A Vaccine for Malaria?

Malaria is one of the most devastating communicable diseases in existence. Caused by a sporozoite protist, *Plasmodium,* the disease is difficult to treat because of the complex life cycle of the protist and the mode of transmission of the disease by mosquitoes. Worldwide, researchers in tropical medicine have been working for years on malaria vaccines—with little success. However, recent controversial trials in Colombia are giving researchers hope. Dr. Manuel Patarroyo, a biochemist who took a radically different approach from that of his colleagues, developed a synthetic polymeric vaccine. He proceeded to test this vaccine on owl monkeys, the only nonhuman species known to carry the malarial protozoan. When he obtained encouraging results on that test, he proceeded to test the vaccine on humans.

Initial tests were met with resistance from a number of groups because Patarroyo was essentially experimenting on humans, and he was not using any controls in his experiment. However, on recent subsequent trials, he obtained positive results (with a control this time) and got protective efficacy results in 22 percent of 15- to 44-year-olds, and in 77 percent of 1- to 4-year-olds—the group most at risk. Patarroyo's vaccine, SPf66, is being tested all over the world now. The biggest test the vaccine will face is protecting children in Africa, where the disease is rampant. "From the point of view of public health," comments a researcher in England, "fewer days would be lost by sickness if everyone were immunized with a vaccine, even only a partially effective one." ☑

lives in mosquitoes and is injected into humans is called the **sporozoite.** Sporozoites make their way through the bloodstream to the human liver in about three minutes. In the liver, they rapidly divide and produce millions of cells of the second stage of the life cycle, called the **merozoite.** Merozoites re-enter the host's bloodstream, invade red blood cells, and divide rapidly. In about 48 hours the blood cells rupture, releasing merozoites and toxic substances throughout the host's body, initiating a cycle of fever and chills that characterizes malaria. The cycle repeats itself regularly every 48 hours as new waves of blood cells are infected.

Some of the merozoites in the human bloodstream undergo a sexual phase and develop into the third stage of the *Plasmodium* life cycle, called the **gametocyte.** In the human bloodstream, gametocytes are incapable of undergoing meiosis to form haploid gametes. However, if they are extracted from an infected person by a mosquito, they form sperm and egg cells within the gut of the mosquito. Gametes fuse to form a zygote, which develops in the wall of the mosquito's gut and produces large numbers of sporozoites. Sporozoites migrate to the salivary glands of the mosquito, where they may be injected by the mosquito into the bloodstream of a human, completing the life cycle. ❑

Treating and Preventing Malaria

Chemical treatments for malaria are hundreds of years old. In the middle of the seventeenth century, quinine, a chemical derived from the bark of the cinchona tree (*Cinchona officialis,* found in South America), was discovered to be a remedy for the disease. The native name for the tree, shown in Figure 21-19, was *quina,* hence quinine. Today, derivatives of quinine, such as chloroquine and primaquine, are used to treat infected individuals and prevent malaria in healthy individuals.

One way to reduce the number of cases of malaria is to reduce the size of mosquito populations. Efforts to eradicate malaria have focused on breaking the life cycle by eliminating *Anopheles* mosquitoes. Widespread application of the powerful insecticide DDT 30 to 50 years ago eliminated these mosquitoes from the United States, but DDT-resistant strains evolved in many regions. Long banned in the United States because of its devastating effects on natural ecosystems, DDT is still used in many developing countries to control mosquitoes.

Biologists continue to search for drugs that poison the parasites once they have entered the human body, and several effective agents are known. However, strains of *Plasmodium* have appeared that are resistant to all known antimalarial drugs. As a result of drug resistance, the number of new malaria cases has tripled since the mid-1970s. Recently, a potent antimalarial drug called artemisim has been isolated from the wormwood, *Artemisia annua.* This species of plant has been used in China for various medicinal

❑ *CAPSULE SUMMARY*

Protists can cause serious human diseases such as malaria.

Figure 21-19 The bark of *Cinchona officinalis* yields quinine, a remedy for malaria.

Multicultural Perspective

Controlling Malaria in Africa

Ten percent of the world's population lives in tropical Africa, where 90 percent of the world's malarial infections occur. Three species of African mosquitoes carry and spread the pathogen to humans while feeding on their blood. In tropical Africa, one person can receive more than 300 bites by infected mosquitoes per year. International researchers hope to introduce a pathogen-inhibiting gene into the genomes of these mosquitoes. Releasing transgenic mosquitoes to spread the new gene throughout the *Plasmodium*-carrying mosquito populations could spare other continents, such as South America, from future malaria epidemics.

purposes for over 2,000 years. Attempts to produce a vaccine using the techniques of genetic engineering are starting to produce promising results, and human trials are underway.

Table 21-3 Diseases Caused by Protists

Disease	Symptoms	Protist	Mode of Transmission
Amoebic dysentery	Amoebas feed on intestinal lining, causing bloody diarrhea	*Entamoeba histolytica*	Contaminated food or water
Giardiasis	Cramps, nausea, diarrhea, and vomiting	*Giardia lamblia*	Contaminated water
Leishmaniasis	Skin sores and deep, eroding lesions	*Trypanosoma*	Bites from sand flies
Sleeping sickness	Fever, weakness, lethargy	*Trypanosoma gambiense, Trypanosoma rhodesiense*	Bite from infected insects such as tsetse fly
Chagas' disease	Fever, severe heart damage	*Trypanosoma cruzi*	Bite from infected kissing bug
Toxoplasmosis	Primary danger is fetal infection; can cause convulsions, brain damage, blindness, and death in fetuses	*Toxoplasma gondii*	Contact with infected cats or improperly cooked meat
Late blight	Parasitizes and destroys potato plants	*Phytophthora*	Spores spread from diseased plants to healthy ones

Section Review

1. What are the three stages in the life cycle of Plasmodium? Which one causes the characteristics of malaria?
2. Describe three methods used to treat and control malaria.
3. What are three additional human diseases caused by protists?

Critical Thinking

4. What might be a way to control sleeping sickness, a disease caused by a trypanosome and spread by the bite of the tsetse fly?

Answers to Section Review

1. The three stages in the life cycle of *Plasmodium* are the sporozoite, the merozoite, and the gametocyte. Malaria symptoms develop in the merozoite stage.

2. The methods used to control malaria include eliminating *Anopheles* mosquitoes and treating infected patients with antimalarial drugs. New research involves a vaccine produced by genetic engineering.

3. See Table 21-3 on page 495 for a list of other diseases caused by protists.

4. Answers might include eliminating tsetse flies or developing a vaccine for sleeping sickness.

CONNECTIONS

Chapter 7
Human Genetics
Sickle cell anemia offers an evolutionary advantage to individuals with the gene. The malaria-causing protists become trapped inside blood cells that become sickled. Once cells are altered, the protists die, and the person does not become ill with malaria.

Chapter 18
Global Warming
Until efforts to control malaria began to restrict its geographical boundaries, it was found in areas where summer temperatures exceeded 16°C (61°F). Global warming has extended the areas of the world where conditions are favorable for the spread of malaria.

Teaching Tip
Precautions Against Pathogens
Ask students to suggest how they would protect themselves from the diseases discussed in Section 21-3 if they were exposed to them. *(Precautions might include getting the appropriate vaccinations or medicines, watching for possible food or water contamination, and protecting themselves from pests that can carry pathogens.)*

Chapter Closure
Ask students to write a short essay exploring the rise of drug-resistant protistan pathogens. Have students consider the phrase "survival of the fittest." Does evolution occur faster in microbes than in other organisms? *(Yes, they multiply more quickly and thus have more opportunity for mutation. By killing the nonresistant microbes with drugs, more resources are available for the resistant strains to multiply.)*

CHAPTER REVIEW ANSWERS

Each item in the Chapter Review is correlated by text section in the assignment guide that follows.

ASSIGNMENT GUIDE

Section	Review	Themes Review	Critical Thinking
21-1	1–3, 10, 11	16	
21-2	4–6, 12–14	17, 18	19
21-3	7–9, 15		20

Review

Multiple Choice

Code in parentheses indicates section number and objective number.

1. c (21-1.1)
2. c (21-1.3)
3. c (21-1.5)
4. d (21-2.2, 21-2.5)
5. a (21-2.2)
6. b (21-2.3)
7. a (21-3.2)
8. a (21-3.3)
9. d (21-3.3)

Completion

10. protists (21-1.2)
11. zoospores, zygospore (21-1.4)
12. pseudopodia, tests (21-2.2)
13. *Euglena*, pellicle (21-2.5)
14. plasmodium, amoebas (21-2.6)
15. *Plasmodium*, sporozoite, merozoite, gametocyte (21-3.1)

---

21 CHAPTER REVIEW ....

Vocabulary

amoeba (485)	foram (485)	pseudopodia (485)
conjugation (482)	gametocyte (494)	sporangia (483)
diatom (486)	merozoite (494)	sporozoite (494)
dinoflagellate (488)	oocyte (492)	zoomastigote (488)
euglenoid (489)	pellicle (489)	zoospore (482)
eyespot (481)	plasmodium (491)	zygospore (482)

Review

Multiple Choice

1. Two of the most important eukaryotic features that evolved in protists are
  a. photosynthesis and silica shells.
  b. forams and pseudopodia.
  c. multicellularity and sexual reproduction.
  d. spores and microtubules.

2. Which habitat is least likely to harbor any species of Protista?
  a. ocean waters   **c.** a desert
  b. the human liver   **d.** leaf litter

3. Which pair shows a correct match between a protist and its manner of reproduction?
  a. *Spirogyra*: sporangia
  b. *Ulva*: conjugation
  c. *Oedogonium*: gametes
  d. *Spirogyra*: alternation of generations

4. A photosynthetic single-celled protist that moves using flagella would likely be classified as a member of what phylum?
  a. Apicomplexa   **c.** Bacillariophyta
  b. Oomycota   **d.** Euglenophyta

5. Photosynthetic protists with boxlike shells are
  a. diatoms.   **c.** zoomastigotes.
  b. plankton.   **d.** euglenoids.

6. Red algae are different from green and brown algae because red algae
  a. are multicellular.
  b. have the pigment phycobilin.
  c. inhabit marine environments.
  d. display alternation of generations.

7. Why have efforts to control the spread of malaria not eradicated the disease?
  a. Resistant strains of mosquitoes and *Plasmodium* have evolved.
  b. The use of DDT has been erratic around the world.
  c. The symptoms of the disease are confused with those of the flu.
  d. The life cycle of *Plasmodium* is not well understood.

8. How is amoebic dysentery spread from person to person?
  a. by drinking unsanitary water
  b. from the bite of the conenose bug
  c. from the bite of the tsetse fly
  d. by eating overcooked pork

9. Which of the following identifies the correct mode of transmission of a human disease caused by a protist?
  a. Giardiasis is transmitted through the sting of a small wasp.
  b. Leishmaniasis is transmitted through the bite of a small bee.
  c. Giardiasis is transmitted through the bite of a sand fly.
  d. Leishmaniasis is transmitted through the bite of a sand fly.

Completion

10. Eukaryotes that do not belong to the fungus, plant, or animal kingdoms are classified as _____ .

11. *Chlamydomonas* produces haploid cells called _____ during the asexual phase of its life cycle and a protected diploid zygote called a _____ during the sexual phase of the life cycle.

12. Amoebas and forams are alike because they both use _____ to move around. Amoebas and forams are different because forams secrete external shells called _____ that are made of calcium carbonate.

13. The protist _____ is found in fresh water and exhibits both plantlike and animal-like characteristics. Inside the cell membrane of this protist is the flexible _____ , which allows the organism to change shape and move around obstacles.

14. The feeding stage of a plasmodial slime mold is a large mass of oozing cytoplasm called a _____ , whereas the feeding stage of a cellular slime mold is made up of separate _____ .

15. The _____ life cycle has three stages. The stage that lives in mosquitoes and is injected into humans is called the _____; the stage that invades human red blood cells is called the _____ ; and the stage that is extracted from a human before the formation of eggs and sperm in the gut of a mosquito is called _____ .

Themes Review

16. **Structure and Function** A zygospore is formed during the sexual phase of the life cycle of *Chlamydomonas*. Describe the structure of the zygospore and its function in the life cycle.

17. **Evolution** While red, brown, and green algae are alike in many ways, scientists hypothesize that green algae are the evolutionary ancestors of plants. What evidence supports this hypothesis?

18. **Homeostasis** Many freshwater protists that lack a cell wall have a contractile vacuole. How does the contractile vacuole maintain conditions inside the protist within the limits required by living cells?

Critical Thinking

19. **Interpreting Data** As you peer through your microscope, you see an organism that is single-celled, flexible, and that moves about with the aid of "false feet." You reason that it must be a protist. In which of the major protist phyla would you classify the organism? Explain your choice.

20. **Making Inferences** *Euglena* is often used in experiments in high school laboratories. In view of the fact that *Euglena* can be autotrophic, why is it a good choice for use in the lab?

Activities and Projects

21. **Multicultural Perspective** The Irish potato famine of 1845–1847 resulted in the deaths of many people and also the emigration of Irish people to other countries including the United States. Research this time period. Construct a graph that shows the number of Irish people who came to the United States between the years 1835 and 1857. Using the graph, write a paragraph that describes the impact of the Irish potato famine on the immigration of Irish people to the United States.

22. **Cooperative Group Project** Carrageenan and agar are extracts of red algae. Algin is an extract of brown algae that grow in ocean waters. Carrageenan, agar, and algin are added to many foods to make them thicker, smoother, and better-tasting. Examine food packages at home and in a grocery store to identify products that contain extracts of these algae. Make a poster-size chart that contains (1) the product names, (2) the algal extract added, and (3) how the product is improved by the addition of the extracts.

Themes Review

16. A zygospore is a thick-walled diploid zygote. It functions to protect the zygote during times of unfavorable environmental conditions. (21-1.4)

17. Green algae have chloroplasts that contain chlorophylls a and b, the same pigments found in the chloroplasts of green plants. (21-2.3)

18. It regulates the amount of water in the cell and thus prevents the cell from bursting as a result of osmotic water intake from the environment. (21-2.5)

Critical Thinking

19. Rhizopoda; the description seems to be that of an amoeba. (21-2.1)

20. Since *Euglena* can be autotrophic, they do not need to be fed. (21-2.5)

Activities and Projects

21. The number of Irish people who came to the United States increased during the years immediately following the Irish potato famine.

22. Carrageenan is used to thicken chocolate milk and milkshakes. Some soups and cake frostings are thickened with agar. Ice creams are made smoother by adding algin.

Preparation

Set up a stereomicroscope station for observation of *Dictyostelium*. Make a mixture of the following protists, which can be obtained from a biological supply house: *Amoeba, Paramecium, Volvox, Spirogyra, Euplotes, Micrasterias, Vorticella,* and *Stentor*.

Procedural Notes

- This investigation may take more than one class period. To save time, you may wish to make the "sun shades" in advance.
- Establish an area where students can obtain culture material and make wet mounts.
- Point out the stereomicroscope station that students can use to observe *Dictyostelium*.
- For measuring the size of various protists, use of a calibrated ocular micrometer is preferred. If unavailable, tell students that a compound microscope has a diameter of approximately 1500 μm at low power (100×) and a diameter of 375 μm at high power (430×). Using these figures, students can estimate the sizes of protists; for example, under low-power magnification, a paramecium is one-fifth the field diameter, or 300 μm.

Background Answers

1. All members of the kingdom Protista are eukaryotes that cannot be classified as animals, plants, or fungi. They may be single-celled or multicellular, producers, consumers, or decomposers.

2. Protists have simpler structures than plants or animals, but all cells are capable of carrying out necessary life functions.

3. Some protists, like plants, have chloroplasts and make their own food; others, like animals, are heterotrophs that ingest food and are capable of movement; and others, like fungi, absorb their food.

*L*ABORATORY *Investigation* Chapter 21

Comparative Protists

OBJECTIVES

- Observe live protists under the microscope.
- Compare structural characteristics, methods of locomotion and feeding, and behavioral strategies among protists.
- Determine why protists are placed in a separate kingdom, based upon your observations.

PROCESS SKILLS

- observing living organisms using a compound microscope
- comparing and contrasting organism traits and characteristics

MATERIALS

- Detain™ (protist-slowing agent)
- microscope slides
- mixture of various protists, including *Amoeba, Paramecium, Euglena, Volvox, Spirogyra, Euplotes, Micrasterias, Vorticella, Stentor*
- toothpicks
- coverslips (22 × 22 mm)
- compound microscope
- references for identifying protists
- black construction paper
- paper punch
- scissors
- desk lamp or sunlit window
- white paper
- forceps
- plastic pipettes with bulbs
- vial of diatomaceous earth
- flashlight
- stereomicroscope
- culture tube of slime mold

4. How do some sample protists compare?

Technique Answers

5. Students should find that euglenoids are evenly distributed throughout the microscopic field.

7. Students should observe that euglenoids congregate in the center area that is exposed to light.

10. Students should note that tiny facets of the diatoms reflect light rays.

BACKGROUND

1. What are some of the general characteristics of the kingdom Protista?
2. Are protists simple or complex organisms?
3. How are protists similar to plants? to animals? to fungi?
4. Write your own question to explore in your lab report or notebook.

TECHNIQUE

1. Create a table in the **Records** section of your report that will allow you to name each protist and to draw or describe its size, shape, and color; its type of movement, if any; how food is obtained; and its behavioral responses.

Part A: Protist Mixture

2. Place a drop of protist-slowing agent on the center of a clean microscope slide. Then add another drop from the bottom of the protist culture mixture. Mix the drops using a toothpick. Add a coverslip.

3. Observe your wet mount preparation under both low (100×) and high power (430×) of a compound microscope. Use references to identify at least five different organisms.

4. Using the drawing on the next page as a guide, prepare a protist "sun shade" by punching a hole in the center of a 40 × 20 mm piece of black construction paper that has a slight "curl" to it.

5. Prepare another wet mount without using any protist-slowing agent. View this under low power (100×) of a compound microscope, making note of the kind and distribution of all protists you encounter. Record your findings in the **Records** section of your report.

6. Place the wet mount on top of a piece of white paper on a sunlit windowsill or under a table lamp if it is cloudy. Position the curled protist "sun shade" so that the hole is in the center of the coverslip.

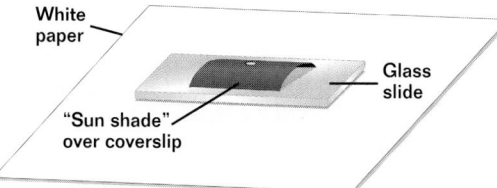

White paper

Glass slide

"Sun shade" over coverslip

7. After 10 minutes, gently pick up your slide, being careful not to disturb your protist "sun shade." Place the slide on the stage of your compound microscope. Use the low-power objective to focus in the center of the exposed opening. Record your observations. Have your partner use forceps to carefully remove the "sun shade" while you continue observing the protists. Record your observations in the **Records** section of your report.

Part B: Diatomaceous Earth

8. Using a pipette, place a drop of water in the center of a clean microscope slide.

9. Using a toothpick, mix a small amount of diatomaceous earth with the water drop. Add a coverslip and observe under both low and high power of a compound microscope. Use references to identify various diatom types—elongate (pennate) or circular (centric).

10. Observe the wet mount under low power as your partner shines a flashlight (at a 45° angle) onto the slide as it rests on the stage. Turn off your light source so that only the flashlight illuminates the diatomaceous earth. Record your observations in **Records** section of your report.

Part C: Slime Mold Culture

11. Use a stereomicroscope to observe a sealed tube containing a slime mold culture, as your teacher directs. Identify the two primary stages of its life cycle.

12. In your report, briefly summarize the procedures you followed in this investigation.

13. Clean up your materials and wash your hands before leaving the lab.

INQUIRY

1. Describe the different ways protists move. Give examples.

2. How do protists obtain food?

3. Which protists were affected by light? How? What structures were responsible?

4. Were all protists single cells?

5. Which observed protists had animal characteristics? plant characteristics? Which protist resembled a fungus?

ANALYSIS

1. Based on your observations in step 10, to what commercial use might the protists in diatomaceous earth be put?

2. How do the motions of *Vorticella* resemble the movement of muscle fibers?

FURTHER INQUIRY

Write a new question that could be explored as a new investigation. The following is an example:

Do samples from the middle and upper areas of the culture jar have different protist populations than those taken from the bottom?

Records

Students should create a table that includes the following information for each protist they view: protist name; sketch of the protist; general description of the protist's size, shape, and color; movement type; feeding mechanism; and behavioral responses.

Inquiry Answers

1. Protists move by the movement of cilia, flagella, or pseudopodia. For example, *Amoeba* uses pseudopodia, *Paramecium* uses cilia, *Euglena* uses a flagellum, and *Volvox* uses flagella.

2. Some protists are autotrophs, which make and store food using sunlight. Others are heterotrophs, which obtain metabolic energy from organic molecules previously assembled by autotrophs. Certain heterotrophs (decomposers) live on decaying organic material.

3. Euglenoids congregated within the area exposed to light (passing through the punched hole). Once the shade was removed, they scattered, dispersing themselves evenly throughout. Euglenoids have an eyespot used to detect light needed for photosynthesis.

4. No, *Volvox* and *Spirogyra* are multicellular.

5. Answers will vary. Animal characteristics include locomotion using cilia (*Paramecium, Euplotes, Vorticella, Stentor*), flagella (*Volvox, Euglena*), or pseudopodia (*Amoeba, Dictyostelium*) and ingestion of food (*Vorticella, Stentor, Euplotes, Amoeba, Paramecium*). Plant characteristics include photosynthetic behaviors and structures (*Volvox, Micrasterias, Euglena, Spirogyra,* diatoms). The slime mold resembles a fungus.

Analysis Answers

1. Silica facets in diatoms reflect light. Mixing them with paint makes a reflective paint that can be used to paint the center line of a highway or used on other materials that need to be seen at night.

2. Like muscle, *Vorticella* can contract because of the presence of bundles of contractile fibrils in its stalk.

UNIT 5

Block Scheduling Guide

Each block represents about 45–50 minutes of instructional time. The resources cited in a block are coded to a specific program component using the Key on the next page.

Lecture Concepts	Lesson Resources

22-1 Characteristics of Fungi pp. 500–504

BLOCK 1

a. Fungi are eukaryotic and heterotrophic. Their bodies are made up of slender woven filaments. Fungal cells contain chitin and undergo nuclear mitosis.

b. Fungi reproduce by releasing spores that are produced sexually or asexually.

Lecture Resources
- Opening Question, page 501
- Demonstration: Observing Fungal Structure
- Comparing Fungi and Plants, page 502
- Teaching Transparencies: 107
- Visual Strategy: Figure 22-3
- Distinguishing Terms, page 504
- Holt Biology Videodiscs: Lesson 22-1

Classwork Options
- Moldy Oranges and Graphic Organizer, page 502
- Spore Prints, page 504
- Laboratory Techniques and Experimental Design C34: Fungal Growth

Assignment Options
- Section Review, page 504
- Chapter Review, questions 1–3, 9–11, 13, 17

22-2 Fungal Diversity pp. 505–510

BLOCK 2

a. Fungi are classified into phyla based on their sexual reproductive structures. Zygomycota form zygosporangia. Ascomycota form sacs of spores.

b. Basidiomycota are fungi that form clublike structures. Fungi in which sexual reproduction has not been observed are placed in the phylum Deuteromycota.

Lecture Resources
- Opening Question, page 505
- Teaching Transparencies: 106, 108, 109, 110, 111
- Effective Reading, page 505
- Visual Strategy: Figure 22-7
- Research Update, page 508
- Up Close, Mushroom page 509
- Holt Biology Videodiscs: Lesson 22-2

Classwork Options
- Growing Bread Mold, page 506

Assignment Options
- Section Review, page 510
- Chapter Review, questions 4, 5, 12, 14, 16, 18–20

22-3 Fungal Associations pp. 511–513

BLOCKS 3 and 4

a. The fungal partner in a lichen protects the photosynthetic partner and provides it with minerals. The photosynthetic partner provides the fungus with carbohydrates.

b. Mycorrhizae are symbiotic associations in which a fungus transfers minerals to a plant's roots, which in turn supply carbohydrates to the fungus. Endomycorrhizae penetrate root cells; ectomycorrhizae wrap around roots.

Lecture Resources
- Opening Question, page 511
- Demonstrations: A Natural Indicator, Describing Lichens
- Visual Strategies: Figures 22-14, 22-15
- Holt Biology Videodiscs: Lesson 22-3

Classwork Options
- Closure, page 513
- Laboratory Investigation: Fungi on Food, pages 516–517

Assignment Options
- Section Review, page 513
- Chapter Review, questions 6–8, 15

Review/Enrichment

BLOCK ⑤

■ Study Guide: Concept Review and Pretest

Assignment Options
◉ Teaching Resources CD-ROM: Supplemental ReadingWorksheets and Test, *Microbe Hunters*
■ Activities and Projects, questions 21, 22

Assessment Options

BLOCK ⑥

Traditional Assessment
■ Chapter 22 Test
◉ Test Generator/Assessment Item Listing: Software item bank for preparing customized chapter tests.

Performance Assessment
◉ Select the Performance Assessment items for this unit from the Test Generator/Test Item Listing

Portfolio Assessment
Select student reports for one or more laboratory experiments from this chapter. *The Direct Observation Checklist* on the *BioSources Teaching Resources CD-ROM* should be completed during a laboratory or other cooperative learning experience.

Holt Biology Videodiscs

Holt Biology Videodiscs gives you a powerful tool for teaching, review, and assessment. *Concepts of Biology* adds interactive lesson plans and extensive tools for customization using CD-ROM technology.

CONCEPTS OF BIOLOGY

Use the following topic from *Concepts of Biology* to help you teach this chapter:

■ Topic 14: Protists and Fungi

For further information regarding the media on the videodiscs, see *Holt Biology Videodiscs Teacher's Correlation Guide to Biology: Principles and Explorations.*

CHAPTER 22 FUNGI

Chapter Themes

Structure and Function

Fungi develop structures that facilitate their survival. For instance, their filamentous hyphae spread and penetrate a host to obtain nutrients. Once the food source is limited or the environment changes, they develop reproductive bodies to disperse their spores.

Levels of Organization

Fungi are important decomposers in the environment. They provide the basic components that other living things need for survival.

Tapping Prior Knowledge

- How does fermentation yield energy?
- How do the products of mitosis and meiosis differ?
- What happens to the surface-area-to-volume ratio as an object increases in size?
- What is a spore?

Opening Demonstration

Prior to class, put a mushroom in a paper bag. Have students determine the contents of the bag by asking "yes and no" questions. (This strategy will help you determine student misconceptions regarding fungi.) Once the class has determined that a mushroom is in the bag, ask them which questions hindered their conclusion. Analyze the responses, and point out misconceptions students have about fungi (for example, they are plants, or they make their own food).

Scarlet waxy cap

REVIEW

- surface-area-to-volume ratio (Section 2-1)
- stages of mitosis (Section 6-2)
- spore (Section 6-3)
- mycorrhizae (Section 13-2)
- lichen (Section 16-1)
- mating types (Section 21-1)

Authors' Rationale

The evolutionary story of life on land must include a study of fungi. The first organisms were able to invade land through symbiosis with fungi. This relationship involved plants as the providers of food through photosynthesis and fungi as the providers of inorganic nutrients extracted from the ground.

The multicellular fungi are placed in their own kingdom because they are so dissimilar to the other eukaryotes. The uniqueness of fungi is emphasized through the discussion of fungal nutrition, fungal reproduction, and fungal associations with other organisms.

22-1 Characteristics of Fungi

Some of the most unusual and peculiar organisms on Earth are members of the kingdom Fungi. For example, mushrooms and molds are familiar fungi that grow so rapidly they sometimes appear overnight. Together with heterotrophic bacteria, fungi are the major decomposers of the biosphere, breaking down organic molecules and making them available for recycling.

Section Objectives

- Explain the role fungi play in ecosystems.
- List the characteristics of the kingdom Fungi.
- Describe the structure of a typical fungus body.
- Explain how fungi obtain nutrients.
- Describe how fungi reproduce.

Vocabulary Preview

chitin

hypha

mycelium

septa

dikaryotic

dikaryon

Opening Question

Ask students to describe an ideal environment in which fungi would thrive. Have them make a list of specific characteristics of this environment. Discuss the lists in class.

Demonstration

Observing Fungal Structure
Give pairs of students a piece of moldy fruit or a mushroom, a plastic knife, and a hand lens. Have them cut the fruit or mushroom in half longitudinally and examine the structures with a hand lens. Provide slides, coverslips, microscopes, and thin slices of tissues. Have students make wet mounts. Have them observe the cell structure of the hyphae and make a graphic organizer about the structure and function of hyphae.

Fungi Are Classified in Their Own Kingdom

The kingdom Fungi consists of eukaryotic, mostly multi-cellular organisms. They are an ancient group of organisms at least 400 million years old. Traditionally, biologists grouped fungi with plants, probably because fungi are immobile and appear "rooted" in the soil. However, fungi are as different from plants as they are from animals. Compare a familiar fungus, such as the mushrooms shown in Figure 22-1, with a plant you've recently studied, such as a pea plant. Upon careful examination, the following differences between fungi and plants become evident.

1. **Fungi are heterotrophic** Perhaps the most obvious difference between fungi and plants is their color: the stalk and cap of the mushroom are not green like the stem and leaves of the pea plant. Plants appear green because they contain chlorophyll, the pigment that enables them to make their own food using energy from the sun. Fungi do not contain chlorophyll; they are heterotrophs. They obtain energy by absorbing organic molecules from their surroundings.

2. **Fungi have filamentous bodies** Unlike the pea plant, which consists of a variety of cell and tissue types, the mushroom consists of long slender filaments. These filaments weave tightly together to form the fungus body and reproductive structures such as a mushroom.

3. **Fungal cells contain chitin** The cells of the mushroom, like the cells of all fungi, have walls made of **chitin** (KY tihn), the tough material found in the exoskeleton of insects and other arthropods. Cells of the pea plant, on the other hand, have walls made of cellulose, a different polysaccharide. Chitin is more resistant to bacterial decomposition than is cellulose.

4. **Fungi have nuclear mitosis** Mitosis in the mushroom is different from that in the pea plant and most other eukaryotic organisms, in which the nuclear envelope

Figure 22-1 These mushrooms are actually the reproductive structures of an extensive network of filaments that make up the body of a fungus.

 Did You Know?

A fungus in Washington State covers an area of 1,500 acres. This giant fungus feeds on dying trees. A healthy, living tree secretes a toxic substance that limits the size of the fungus. However, when a tree dies, it can no longer secrete the toxin and the fungus grows out of control.

Teaching Tip
Comparing Fungi and Plants

Have groups of students write the following words on index cards or strips of paper (one word per card): autotroph, heterotroph, hyphae, roots, stem, leaves, spore, chitin, cellulose, mushroom, pea. Have students sort the terms into two groups based on differences between plants and fungi. Once they have organized the groups, discuss the differences between fungi and plants. Have students change any terms that are grouped incorrectly.

Teaching Tip
Moldy Oranges

Bring an orange covered with mold to class. Cut the orange in half and compare one of the halves with the orange in the figure. Use this figure to emphasize the structure of a fungus. Have students locate in the orange the fungus's mycelium and reproductive structures.

☐ CAPSULE SUMMARY

Fungi are eukaryotic and heterotrophic. Their bodies are made up of slender woven filaments. Fungal cells contain chitin and have nuclear mitosis.

Figure 22-2 This orange is covered with the fungus *Penicillium*. The green and white fuzz growing on the orange's surface is the fungus's reproductive structures. Throughout the rest of the orange, the fungus grows as a mycelium, a mass of tangled filaments called hyphae, *below right.*

disintegrates in prophase and re-forms in telophase. In dividing mushroom cells, by contrast, the nuclear envelope remains intact from prophase to anaphase. Consequently, spindle fibers form within the nucleus, dragging chromosomes to opposite poles of the nucleus, not opposite poles of the cell. Mitosis is completed when the nuclear envelope pinches in two.

The unique features of fungi strongly suggest that they are not closely related to any other group of organisms. The first fungi were probably unicellular eukaryotic organisms. The oldest fossils that have been identified as fungi are 450–500 million years old. ☐

Fungi Are Well Suited for Absorbing Nutrients

In Figure 22-2, the fungus *Penicillium* is growing on an orange. The green and white fuzz you recognize as mold is actually the reproductive structures of the fungus. The body of the fungus lies woven within the tissues of the orange. All fungi except yeasts have bodies composed of slender filaments called **hyphae** *(HY fee)*. When hyphae grow, they branch and form a tangled mass called a **mycelium** *(my SEE lee uhm)*, shown in Figure 22-2. A mycelium can be made of many meters of individual hyphae. This body organization creates a high surface-area-to-volume ratio, which makes a fungus well suited for absorbing food from the environment.

Each hypha is a long string of cells divided by walls called **septa**. In most kinds of fungi, septa do not form a complete barrier between cells (another characteristic that makes fungi different from other eukaryotes). From one cell to the next, cytoplasm flows freely throughout

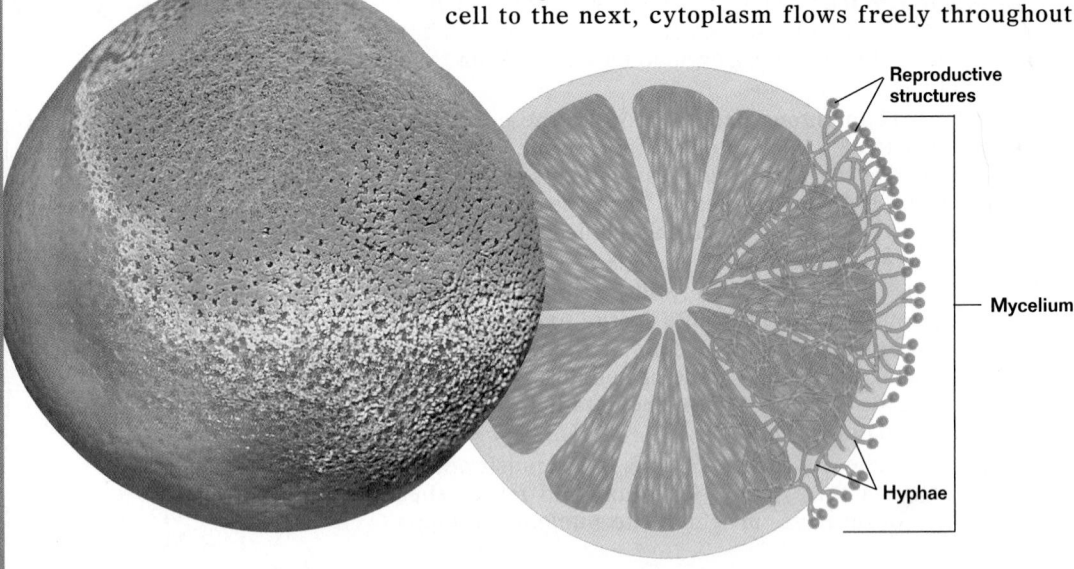

Graphic Organizer

Use this graphic organizer with the *Teaching Tip:* **Moldy Oranges** on page 502.

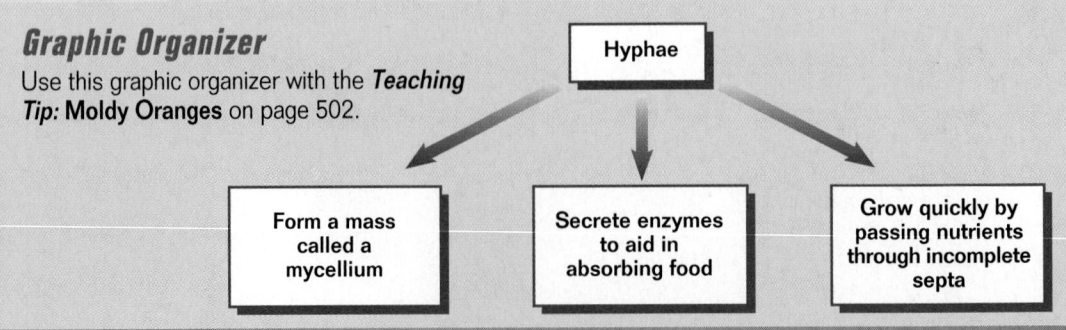

the hypha through perforations in the septa, as shown in Figure 22-3. Other organelles, such as ribosomes, mitochondria, and nuclei, also pass through these perforations. A typical fungal cell usually has many nuclei streaming in the cytoplasm.

How Fungi Obtain Food

All fungi digest food outside their bodies. Through the tips of their hyphae, they secrete powerful digestive enzymes that break down organic matter into small molecules. Fungi absorb these molecules and use them for energy. In their search for food sources, many fungi attack nonliving organic matter—leaves, branches, animal corpses, and waste—and decompose these materials. Fungi, such as the carbon fungus in Figure 22-4, are often found growing on dead trees, where they are the only major group of organisms capable of breaking down lignin, a major component of wood. Other fungi absorb nutrients from living hosts, which sometimes become weakened and succumb to infection or disease.

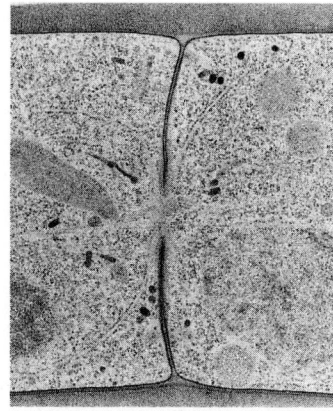

Figure 22-3 Fungal hyphae may be divided into cells by walls called septa. Septa rarely form complete barriers, so cytoplasm streams freely along the hypha, carrying proteins and nutrients to the rapidly growing tips.

Figure 22-4 The carbon fungus *Hypoxylon fragiforme* appears in clusters on the wood it decomposes.

As decomposers, fungi often come into conflict with human interests. A fungus makes no distinction between a fallen log in the forest and bread, fruit, vegetables, meat or other items stored in a refrigerator. In fact, fungi have been known to attack materials humans consider inedible—paper, cardboard, cloth, paint, leather, waxes, fuel, and petroleum.

The success of fungi as commercial pests is due to their ability to grow under a wide range of conditions, a quality that also makes them commercially valuable. Fungi called yeasts produce substances such as carbon dioxide and ethanol, which are useful in baking, brewing, and wine-making. Other fungi provide the pungent flavors and aromas of specific kinds of cheese. Many kinds of antibiotics, such as penicillin, are produced by fungi. An extraordinary example

SECTION 22-1

 VISUAL STRATEGY

Figure 22-3
Use this photograph of fungal septa to remind students of the unique characteristics of the kingdom Fungi. Tell students to note that the septa do not form a complete wall. Ask them if incomplete septa occur in other eukaryotic organisms.

Teaching Tip
Design a Bumper Sticker
Have groups of students design bumper stickers that promote fungi as weapons that can be used to clean up environmental waste. Students may want to research this topic first.

Application

Technology Fungi are used commercially to clean pollutants, including pesticides and chemicals produced in industrial waste. *Penicillium* is used to clean contaminants from polluted waters, such as selenium produced by industry. Fungi such as white rot fungus are cost effective at the commercial level because they can be grown on newspaper, corncobs, or wood chips. Additionally, they grow at a rapid rate, and they are produced from sources that are readily available. Fungi offer great potential as environmental weapons because they are unique in their ability to break down more than one pollutant at a time.

Teaching Tips

Spore Prints

Have students place mushroom caps on paper with the gill side down and cover the mushrooms to keep them from drying. After one to two days, they will have spore prints. (You may want to have students place the cap on paper that has been constructed so that one side is black and one side is white. Spores may be light or dark, and this will ensure that students see the spores.) Spore prints are actually used by mycologists in the process of identifying and classifying basidiomycetes.

Distinguishing Terms

Write the prefix *karyo-* on the board. Ask students to recall two words in which this prefix is used and explain what the words mean. (Prokaryote—*a cell without a nucleus*; eukaryote—*a cell that contains a nucleus*) Ask students to infer what *karyo-* means. (It means "the nucleus of a cell.") Then challenge students to infer the meanings of *dikaryotic* and *dikaryon*.

Figure 22-5 *Lycoperdon perlatum* is a kind of fungus called a puffball. Such fungi release hundreds of thousands of spores through a small opening in the top of the puffball.

❏ CAPSULE SUMMARY

Fungi reproduce by releasing spores that are produced asexually or sexually.

of the use of compounds derived from fungi is cyclosporine, a drug derived from a fungus that dwells in the soil. Cyclosporine suppresses the reactions of the immune system that cause rejection of transplanted organs. This reduces the possibility of transplant rejection without the undesirable side effects caused by other drugs used for this purpose.

How Fungi Reproduce

Fungi reproduce by releasing spores. Spores form sexually or asexually in reproductive structures at the tips of hyphae. When these reproductive structures form, they are cut off from the rest of the fungus by complete septa. Reproductive structures extend high above the food source so that air currents can carry the spores to a new habitat. Spores are well suited to the needs of an organism that is anchored to one place. As you can see in Figure 22-5, spores are so small and light that they remain suspended in the air for long periods of time and are carried great distances. When a spore lands in a suitable place, it begins to divide and soon gives rise to a new fungal hypha.

Fungal spores are haploid. Most spores are formed by mitosis during asexual reproduction. Like protists, fungi usually resort to sexual reproduction only in times of environmental stress. In sexual reproduction, hyphae from two mating types undergo cytoplasmic fusion. However, unlike sexual reproduction in plants and animals, in which nuclei fuse to form a diploid zygote, most fungal nuclei pair up but do not immediately fuse. Instead, the nuclei remain paired but separate in the same cytoplasm for most of the life of the fungus. A fungal hypha that has nuclei derived from two genetically different individuals is called a **dikaryotic** hypha. A mycelium with paired nuclei is called a **dikaryon**. ❏

Section Review

1. *What role do fungi play in the environment?*
2. *How does mitosis differ between fungi and plants?*
3. *What is a hypha? How does its structure enable fungi to obtain nutrients?*
4. *How do fungi reproduce? How does sexual reproduction in fungi differ from that in other eukaryotes?*

Critical Thinking

5. *Given fungi's ability to break down many kinds of substances, how might they be used by industries that produce waste?*

Answers to Section Review

1. Fungi are decomposers, breaking down organic matter in the environment.

2. Unlike a plant cell, mitosis in a fungal cell occurs with the nuclear envelope in tact. The chromosomes are pulled to opposite poles inside of a nucleus, and the nucleus then splits in half. In plants, the nuclear envelope disappears, the chromosomes separate, the nuclear envelope reforms as the two cells separate.

3. A hypha is a filament made of long strings of cells. It enables fungi to grow rapidly because the cells are not totally separated by walls or septa. As a result, nutrients can pass quickly from one cell to the next, and growth can occur at a faster rate.

4. Fungi reproduce asexually when hyphae break off and develop into new fungi and when they produce spores, which develop into new individuals.

5. They might be used to break down wastes that pollute the environment.

22-2 Fungal Diversity

*T*he one characteristic of fungi that distinguishes them from all other multicellular organisms is that their cells share nuclei. Three phyla of fungi—Zygomycota, Ascomycota, and Basidiomycota—are classified according to the way nuclei from different individuals are sorted in fused hyphae during sexual reproduction. These differences are summarized in Table 22-1. A fourth phylum, Deuteromycota, contains fungi in which sexual reproduction is unusual, rare, or has not been observed.

Table 22-1 Three Sexually Reproducing Phyla of Fungi

Phylum	Distinctive Characteristics	Number of Species and Examples
Zygomycota	Haploid nuclei from different mating types fuse to form diploid zygotes, which develop into zygosporangia	665 species; black bread molds
Ascomycota	Formation of fine asexual spores; sexual spores in asci; hyphae divided by perforated septa; dikaryons	30,000 species; morels, truffles, yeasts, cup fungi
Basidiomycota	Formation of sexual spores in basidia; hyphae divided by perforated septa; dikaryons	16,000 species; mushrooms, puffballs rusts, smuts

Fungi That Form Zygosporangia

If you place an uncovered loaf of bread near a windowsill, a cottony mold will soon cover its surface. This common black bread mold, shown in Figure 22-6, is *Rhizopus stolonifer,* a member of the phylum Zygomycota *(zy goh my COHT uh).* The term *Zygomycota* refers to the thick-walled sexual structures called **zygosporangia** that characterize the members of this phylum. *Rhizopus* and other zygomycetes are fungi that live in the soil and feed on decaying plant and animal matter.

Asexual reproduction in zygomycetes occurs much more frequently than sexual reproduction. During asexual reproduction, haploid spores are produced in specialized hyphae that terminate with spore-producing structures called sporangia. When mature, spores

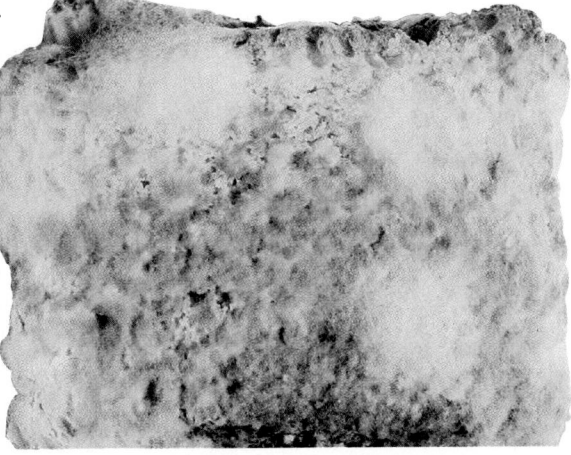

Figure 22-6 *Rhizopus stolonifer* is a familiar member of the phylum Zygomycota that is often found growing on bread.

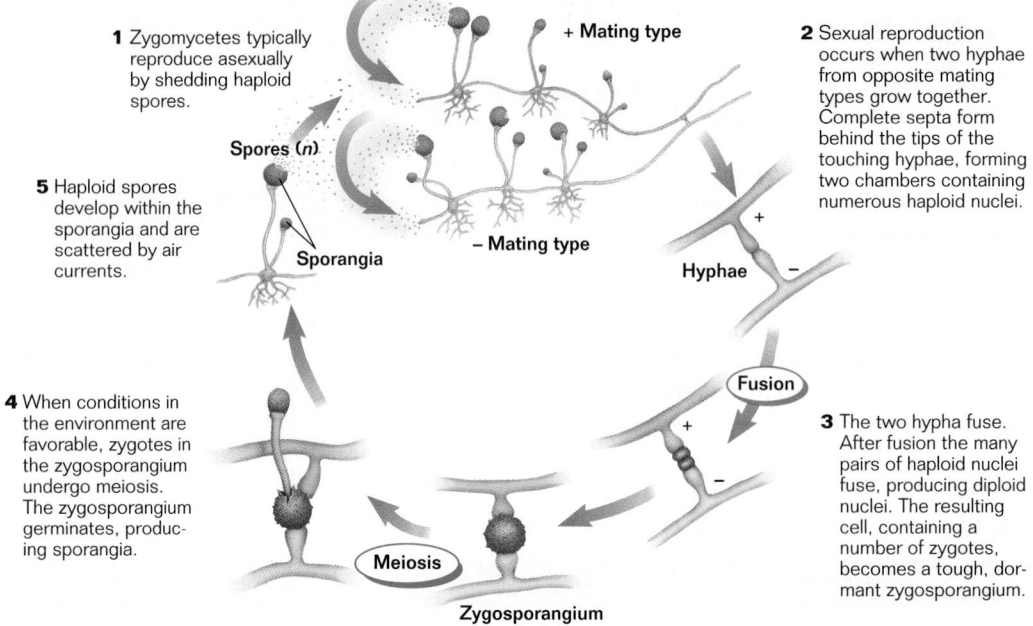

1 Zygomycetes typically reproduce asexually by shedding haploid spores.

2 Sexual reproduction occurs when two hyphae from opposite mating types grow together. Complete septa form behind the tips of the touching hyphae, forming two chambers containing numerous haploid nuclei.

5 Haploid spores develop within the sporangia and are scattered by air currents.

Spores (*n*)

Sporangia

+ Mating type

− Mating type

Hyphae

Fusion

4 When conditions in the environment are favorable, zygotes in the zygosporangium undergo meiosis. The zygosporangium germinates, producing sporangia.

3 The two hypha fuse. After fusion the many pairs of haploid nuclei fuse, producing diploid nuclei. The resulting cell, containing a number of zygotes, becomes a tough, dormant zygosporangium.

Meiosis

Zygosporangium (2*n*)

Figure 22-7 In the sexual reproduction of a zygomycete, the fusion of hyphae from opposite mating strains produces a dormant zygosporangium. Meiosis and germination produce a sporangium, which bears haploid spores.

are shed and carried by the wind to new locations where they germinate and start new mycelia. Reproduction of *Rhizopus* is illustrated in Figure 22-7.

Fungi That Form Sacs of Spores

Prior to the twentieth century, the chestnut tree *Castanea dentata* was one of the dominant trees in forests in the eastern United States. Around 1890, a disease called chestnut blight wiped out virtually all the chestnut trees within a few years. Chestnut blight is caused by the fungus *Endothia parasitica,* a member of the phylum Ascomycota. The fungus that causes Dutch elm disease, another devastating plant disease, is also caused by a member of Ascomycota, *Ceratocystis ulmi.* Other ascomycetes are more familiar and economically beneficial fungi, such as yeasts used in baking and brewing, flavorful morels and truffles prized by gourmet chefs, and the salmon-colored bread mold *Neurospora,* which has played an important role in the development of modern genetics. The ascomycetes are named for their characteristic reproductive structure, the microscopic, club-shaped **ascus** *(AS kuhs),* a saclike structure in which haploid spores are formed. Asci usually form within the tightly interwoven hyphae of a complex structure called an **ascocarp.** In cup fungi and morels, the ascocarps are open, and the asci line the open cups. Other fungi have ascocarps that are closed or have a small opening at the top.

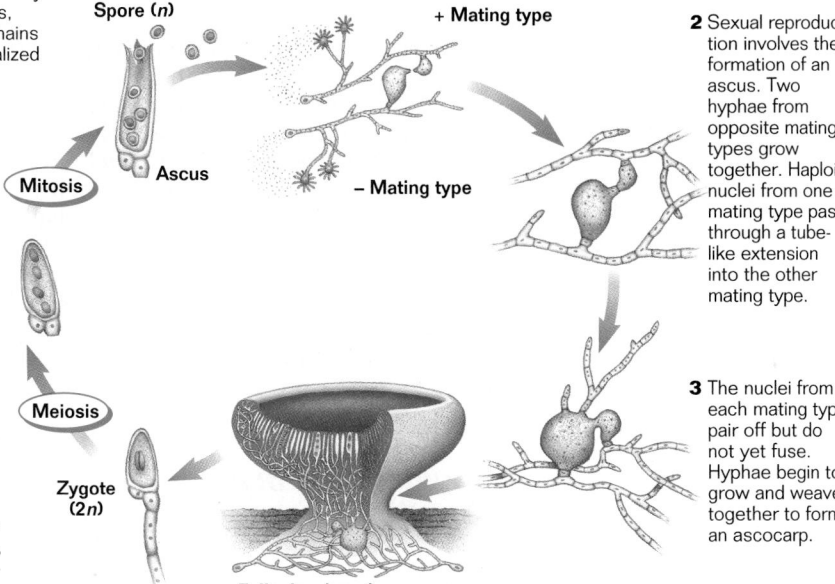

1 Ascomycetes commonly form asexual spores, either singly or in chains at the tips of specialized hyphae.

Spore (*n*)

+ Mating type

Mitosis

Ascus

– Mating type

5 These four nuclei divide mitotically, producing eight haploid nuclei. Each nuclei develops into a spore. They are contained in an ascus, which releases the spores when they are mature.

Meiosis

4 Eventually some of the paired nuclei fuse and form a diploid zygote. The zygote immediately undergoes meiosis, producing four haploid nuclei.

Zygote (2*n*)

Fully developed ascocarp

2 Sexual reproduction involves the formation of an ascus. Two hyphae from opposite mating types grow together. Haploid nuclei from one mating type pass through a tube-like extension into the other mating type.

3 The nuclei from each mating type pair off but do not yet fuse. Hyphae begin to grow and weave together to form an ascocarp.

Like zygomycetes, ascomycetes usually reproduce asexually. Asexual reproduction occurs when complete septa separate the tips of hyphae from the rest of the fungus. Mitosis in these tips gives rise to reproductive structures in which specialized spores called conidia form. When conidia are released, air currents carry them to other places, where they may germinate and form new mycelia. Reproduction in a typical ascomycete is illustrated in Figure 22-8.

Figure 22-8 In the sexual reproduction of an ascomycete, the fusion of hyphae from opposite mating strains produces structures called asci. Meiosis and mitosis within an ascus form haploid spores.

Yeasts

In general, **yeast** is the common name given to unicellular ascomycetes. There are about 350 named species of yeasts, including *Saccharomyces cerevisiae*, or baker's yeast, used for thousands of years in the production of baked goods, such as the bread shown in Figure 22-9, and many kinds of alcoholic beverages. Other yeasts, such as *Candida*, are human pathogens. *Candida* is a common source of thrush, a

Figure 22-9 The ability of yeast such as *Saccharomyces cerevisiae, below left,* to ferment carbohydrates by breaking down glucose to produce ethanol and carbon dioxide is fundamental to the production of breads and other baked goods.

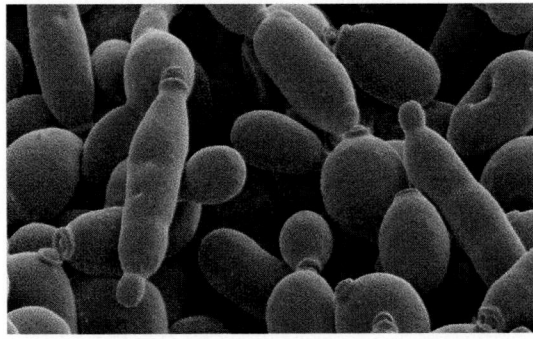

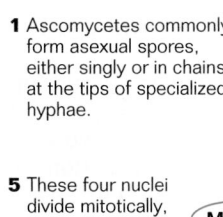
CONNECTIONS

Chapter 10
Genetic Engineering
Since scientists synthesized a functional artificial chromosome in *Saccharomyces cerevisiae*, yeast have become a preferred organism in genetic research. Explain to students that while yeasts are simple eukaryotes with true cellular organelles, their biochemistry closely resembles that of plant and animal genetics. Yeast are used as models for identifying human genes that influence cancer, Down syndrome, and other genetic disorders.

Chapter 20
Viruses
Chestnut blight, a fungal disease that has eradicated the chestnut tree from forests in the eastern United States, has a new enemy. Scientists have found a milder version of the blight that can be inactivated by a virus. By transferring this virus to the more virulent fungus, they hope to tame the blight and hinder its ability to fatally infect and eliminate trees.

Demonstration
Comparing Structures
Provide students with pictures of several types of ascomycetes, such as the morel, cup fungus, and yeast. For each picture, help students identify the ascocarp. Have them compare the different forms, emphasizing that despite their different appearances, they perform similar functions. Ask them to speculate on the adaptive functions of the different forms.

UPDATE

☑ RESEARCH Update

Killer Spores: An Alternative To Pesticides?

Agricultural researchers are working to find an acceptable method of controlling pests such as grasshoppers and locusts. Conventional chemical pesticides such as DDT are proving to be increasingly harmful to the ecosystem. Scientists have been working on an alternative pesticide—fungi. Dr. Hans Herren of the International Institute for Tropical Agriculture in Nigeria is preparing a strain of *Metarhizum flavoviride* for testing against locusts and grasshoppers. Dr. John Henry, an entomologist at the University of Montana researching the potential use of *Beauveria bassiana*. Both of these fungi operate the same way: when sprayed on the offending locusts or grasshoppers, the fungus penetrates the exoskeleton of the insect and grows inside it. As a result, the insect sickens and dies, usually in about four days.

The benefits outweigh the potential risks. Agricultural researchers say that the fungus can be sprayed over the invading insects and that it will kill them before dying off due to desiccation. This leaves no residual trace of pesticides in the crops or soil. However, Dr. Jeffrey Lockwood at the University of Wyoming is concerned that these fungi could wipe out other closely related species as well. "Fewer than 15 of 300 grasshopper species in the United States endanger crops, and the others appear to be harmless and may even be beneficial." ☑

disease characterized by the formation of milk-white lesions on the mouth, lips, and throat.

Most yeasts reproduce asexually by fission or budding (the formation of a small cell from a portion of a larger one). Sexual reproduction among yeasts occurs when two cells fuse. One of these cells, containing two nuclei, functions as an ascus. Meiosis of the fused nuclei produces four spores that develop directly into new yeast cells.

Over the past few decades, yeasts have played an important role in genetic research. They were the first eukaryotic cells to be the subject of genetic engineering techniques. Because they reproduce rapidly, yeasts have become the eukaryotic cells of choice for many types of experiments in molecular and cellular biology.

Fungi That Form Clublike Structures

The fungi with which you are probably most familiar—mushrooms—are members of the third phylum of fungi, Basidiomycota. Other basidiomycetes include toadstools, puffballs, jelly fungi, and shelf fungi. Basidiomycetes are named for their characteristic club-shaped sexual reproductive structure, the **basidium** (*buh SIHD ee um*), which you can see in the *Up Close: Mushroom* feature on page 509. Unlike the other fungal phyla, asexual reproduction is rare among the basidiomycetes, except in some of the rusts and smuts, two important groups of plant pathogens that affect many crop plants. Sexual reproduction of a typical basidiomycete is illustrated in Figure 22-10.

Figure 22-10 In the sexual reproduction of a basidiomycete, structures called basidia form on the gills that line the mushroom cap. Meiosis within basidia produces spores.

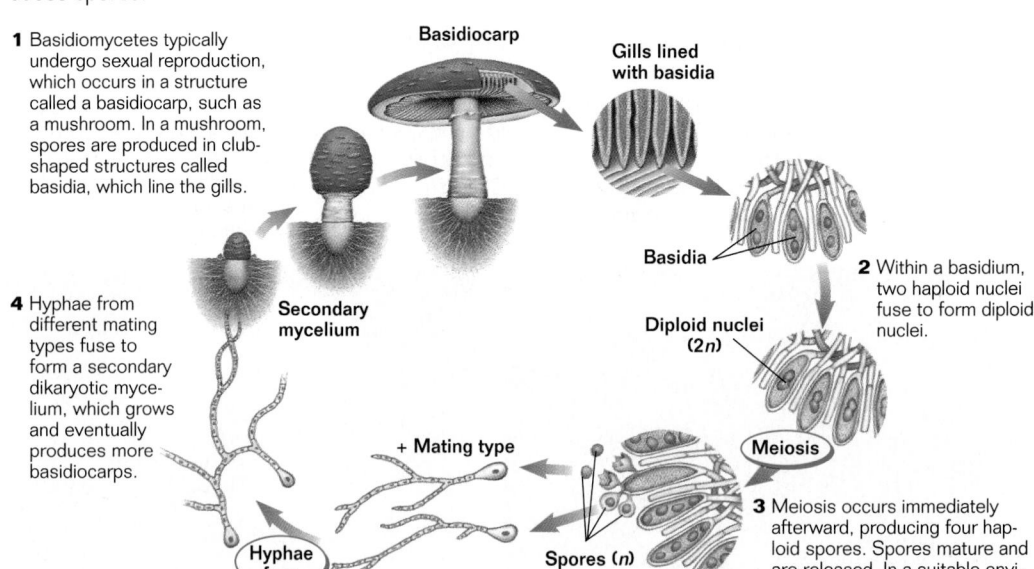

1 Basidiomycetes typically undergo sexual reproduction, which occurs in a structure called a basidiocarp, such as a mushroom. In a mushroom, spores are produced in club-shaped structures called basidia, which line the gills.

4 Hyphae from different mating types fuse to form a secondary dikaryotic mycelium, which grows and eventually produces more basidiocarps.

Basidiocarp

Gills lined with basidia

Basidia

2 Within a basidium, two haploid nuclei fuse to form diploid nuclei.

Diploid nuclei (2*n*)

Secondary mycelium

Meiosis

+ Mating type

Spores (*n*)

3 Meiosis occurs immediately afterward, producing four haploid spores. Spores mature and are released. In a suitable environment, a spore germinates into a haploid hypha.

Hyphae fuse

– Mating type

 Did You Know?

Although yeast do not have a body composed of filaments, they have been observed to grow filaments in the lab when deprived of nutrients. Apparently, they have genes from ancient ancestors, common to other fungi, which are no longer expressed; however, stressful conditions in the lab can activate the expression of these genes.

UP CLOSE MUSHROOM

- **Scientific Name:** *Amanita muscaria*
- **Habitat:** Moist organic soils
- **Size:** 10–15 cm
- **Nutrition:** Absorbs organic material in soil

Characteristics

Body structure   The multicellular body of a fungus is basically filamentous, consisting of long strings of cells called hyphae. Hyphae are woven together to form a dense mat called a mycelium. Usually, the majority of a mycelium is hidden within a substrate such as soil.

Cell structure   *Amanita* and other fungi have cell walls made of chitin, a complex polysaccharide also found in the external skeleton of insects and other arthropods. Some fungi have hyphae that are not divided into separate cells and have many nuclei in the same cytoplasm. Other fungi have hyphae that are divided into cells by perforated walls called septa.

Reproduction   Under proper conditions, underground hyphae grow upward and weave together to produce a mushroom, the reproductive structure of fungi such as *Amanita*. A mushroom has a flattened cap attached to a stem called a stalk. The underside of the mushroom cap is lined with rows of gills. Thousands of club-shaped reproductive cells called basidia form on the gills. Through fusion and meiosis, each basidium produces spores that are released and germinate into new hyphae.

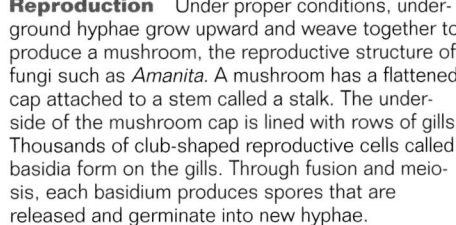

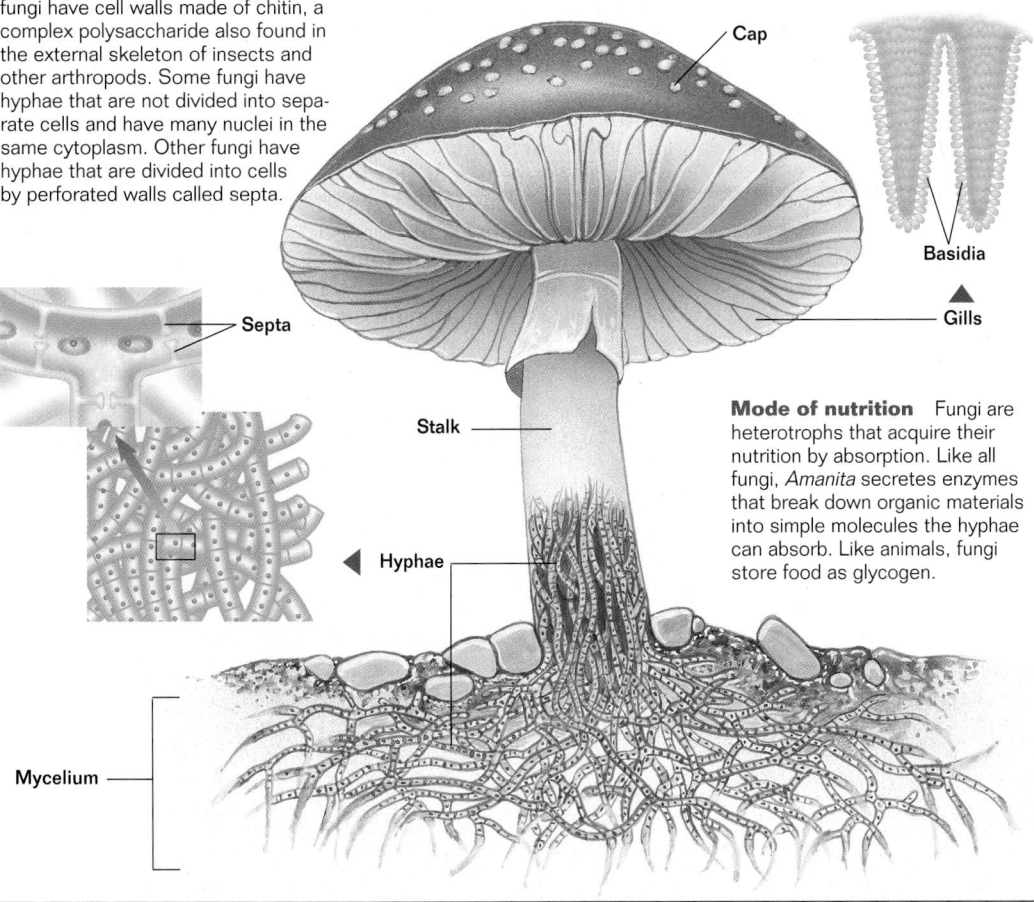

Cap

Basidia

Gills

Septa

Stalk

Hyphae

Mycelium

Mode of nutrition   Fungi are heterotrophs that acquire their nutrition by absorption. Like all fungi, *Amanita* secretes enzymes that break down organic materials into simple molecules the hyphae can absorb. Like animals, fungi store food as glycogen.

Instructional Strategies

1. Bring a field guide of mushrooms and other fungi to class and have students look through it to obtain an accurate idea of fungal diversity. Have each student select a particular mushroom or fungus and draw a large, colored picture to display in the class. Each illustration should include the scientific name of the fungus.

2. Write a journal entry about mushrooms. Include a spore print, a labeled drawing, a recipe, and an abstract from an article regarding mushrooms.

Discussion

Guide a discussion by posing the following questions.
1. Why is it incorrect to classify a mushroom as a plant? *(As discussed in the text, fungi are classified in their own kingdom because they are heterotrophic, have filamentous bodies, have cells that contain chitin, and have nuclear mitosis. None of these characteristics is found in the plant kingdom.)*
2. What are some characteristics of mushrooms that enable them to survive? *(Some of the characteristics include: hyphae, which penetrate decaying materials and secrete enzymes to absorb nutrients; chitin, which protects fungal cells from bacterial destruction; and rapidly-growing reproductive bodies, which produce large quantities of spores.)*
3. In what kinds of environments are you most likely to find mushrooms growing? *(They grow best in moist environments with an available food source, so they can most likely be found in damp forests, in swampy regions, on lawns after a rain.)*

Application

Medicine Alexander Fleming's discovery of penicillin was not fully appreciated until World War II, when soldiers needed the drug to fight off infections caused by wounds. At that time, scientists in a lab in Peoria, Illinois, asked Americans to donate mold in order to get adequate supplies of the drug. People responded eagerly by sending their moldy garbage, among which was a specific, moldy cantaloupe. This cantaloupe had a mutant, a mold that yielded higher amounts of penicillin. Today, most of the penicillin derived commercially can be traced back to the fungus on that original, moldy cantaloupe.

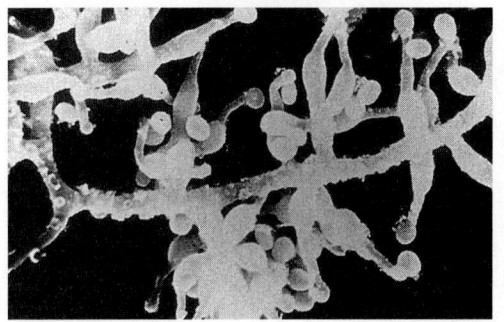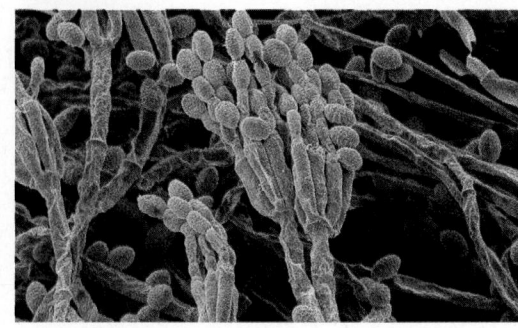

Figure 22-11 *Tolypocladium inflatum, above left,* a soil fungus, is one of the sources of cyclosporine, a drug that suppresses the immune system. Some species of *Penicillium, above right,* are the sources of the well known antibiotic penicillin.

CAPSULE SUMMARY

Based on their sexual reproductive structures, fungi are classified into three phyla: Zygomycota, Ascomycota, or Basidiomycota. Fungi in which sexual reproduction has not been observed are placed in the phylum Deuteromycota.

Fungi in Which Sexual Reproduction Has Not Been Observed

As you have seen, the three phyla of fungi that have been discussed differ primarily in their mode of sexual reproduction. Not all fungi fit into one of these three phyla. In some 17,000 described species of fungi, sexual reproduction has not been observed. These fungi are placed in the phylum Deuteromycota, or **Fungi Imperfecti.** The Fungi Imperfecti are classified on the basis of their asexual reproductive structures.

Many of the Fungi Imperfecti, such as the two species in Figure 22-11, have great commercial importance. Some species of *Penicillium* are sources of the antibiotic penicillin, while other species of this genus give the characteristic flavors and aromas to cheeses such as Roquefort and Camembert. Species of *Aspergillus* are used for fermenting soy sauce and for the commercial production of citric acid. Most of the fungi that cause skin diseases in humans, including athlete's foot and ringworm, are also Fungi Imperfecti. ◻

Section Review

1. *On what basis are fungi classified into phyla? List the four phyla and give an example of each.*

2. *When does meiosis occur in the life cycle of a zygomycete? What is a zygospore? When do spores form?*

3. *What is an ascus? When does an ascus form in the life cycle of an ascomycete? What is contained within an ascus? When do spores form?*

4. *What is a basidium? Describe the activity within a basidium that leads to the formation of spores.*

Critical Thinking

5. *Why are yeasts gaining popularity in scientific research?*

Answers to Section Review

1. Fungi are classified into phyla by their reproductive structures. The four phyla and examples of each include Zygomycota (bread mold), Ascomycota (yeast), Basidiomycota (mushroom), and Deuteromycota *(Penicillium).*

2. In zygomycetes, meiosis occurs before the spores are released, and once they are released, they develop by growing hyphae. A zygospore is a structure that forms when two different hyphae combine. Once the zygospore is produced, spores form within the structure.

3. An ascus is a sac that is formed by sexual reproduction in ascomycetes. It forms when two different strains of hyphae combine. Inside, it has spores. The spores form after the haploid hyphae combine. The resulting diploid cell undergoes meiosis to produce the spores.

4. A basidium is the reproductive structure of a basidiomycete. The nuclei fuse in the basidium and form a zygote. The zygote then undergoes meiosis to form spores.

5. Yeasts are single-celled eukaryotes that reproduce rapidly.

22-3 Fungal Associations

Fungi are involved in a variety of intimate symbiotic associations with algae and plants that play very important roles in the biological world. Recall from Chapter 17 that mutualism is a form of symbiosis in which each partner benefits. These symbiotic associations typically involve a sharing of abilities between a heterotroph (a fungus) and a photosynthesizer (an alga or a plant). The fungus contributes the ability to absorb minerals and other nutrients efficiently from the environment; the photosynthesizer contributes the ability to use sunlight to power the building of organic molecules.

Section Objectives

- *Discuss two symbiotic relationships that involve fungi.*
- *Describe lichens, and explain their sensitivity to pollution.*
- *Describe two types of mycorrhizae.*

Lichens: Fungi and Photosynthetic Partners

A **lichen** is a symbiotic association between a fungus and a photosynthetic partner such as a green alga, a cyanobacterium, or both. The photosynthetic partner provides energy-rich compounds for both partners. It is protected from the environment by the fungal partner, which provides it with mineral nutrients. In most of the 15,000 described species of lichens, the fungal partners are ascomycetes. When you look at a lichen, such as the ones in Figure 22-12, you are seeing the fungus. The photosynthetic partner is hidden between the layers of hyphae, as shown in Figure 22-13. Sunlight penetrates the translucent layers of hyphae to fuel photosynthesis. Sometimes specialized fungal hypha penetrate the photosynthetic cells and serve as highways for transferring sugars and other organic molecules to the fungus. The fungus transmits biochemical signals that direct the cyanobacterium or the green alga to produce metabolic substances that it would not make if growing independently of the fungus. Many biologists characterize this particular symbiotic relationship

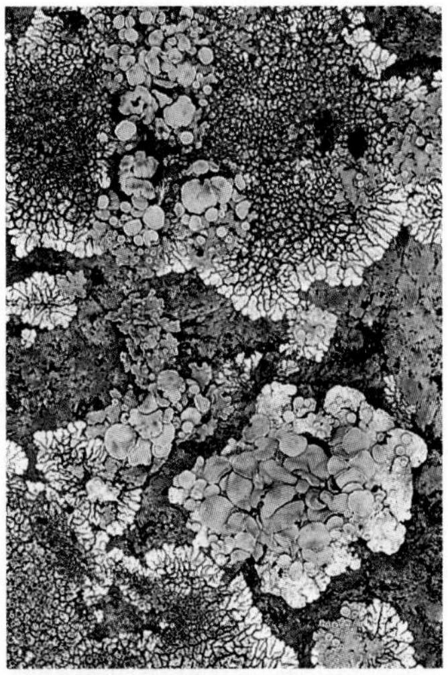

Figure 22-12 Lichens absorb some minerals from their substrate, but most of their minerals are derived from air and from rainfall.

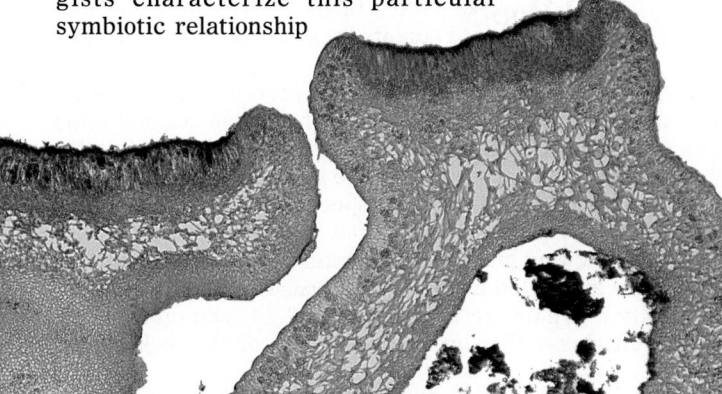

Figure 22-13 This cross section of a species of lichen called *Physica* shows algal cells, *colored magenta,* penetrating fungal cells, *colored green.*

Section 22-3

Vocabulary Preview

lichen
endomycorrhizae
ectomycorrhizae

Opening Question

Show students photos of colorful lichens, or bring in live lichens growing on small rocks. Ask students what kind of organism they think a lichen is and in which kingdom it is placed.

Demonstration

A Natural Indicator

Orchil is a commercial dye made from lichens that is used in making litmus paper. Using vinegar and ammonia, show students how litmus can determine whether a substance is an acid or a base. Ask them why the lichens, from which litmus is derived, are good indicators of pollution in the environment.

Did You Know

Finland leads the world in mapping pollution with lichens. They are able to map sources of pollution from hundreds of miles away by tracing wind patterns that account for changes in lichens.

□ *CAPSULE SUMMARY*

The fungal partner in a lichen protects the photosynthetic partner and provides it with minerals. The photosynthetic partner provides the fungus with carbohydrates.

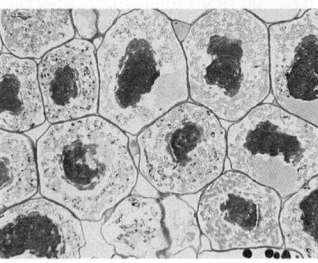

Figure 22-14 The roots of leek plants, *top*, have cells that contain endomycorrhizae with a zygomycete component, *bottom*.

as one of slavery rather than cooperation, a controlled parasitism of the photosynthetic organism by the fungal parasite.

The tough construction of the fungus, combined with the photosynthetic abilities of the alga, has enabled lichens to invade the harshest of habitats. They are extremely widespread in nature. Lichens have been found in arid desert regions, in the Arctic, growing on bare soil, on tree trunks, and on sunbaked rocks. In harsh, exposed areas, lichens are often the first colonists. They break down rocks and prepare the environment for other organisms. Lichens are a key component of primary succession because they are able to carry out nitrogen fixation and introduce useful forms of nitrogen into the environment.

Lichens are able to survive drought and freezing by becoming dormant. When moisture and warmth return, the lichen recovers quickly and resumes its normal activities, such as photosynthesis. In harsh environments, lichens may grow extremely slowly. Some lichens that grow high in the mountains appear to be thousands of years old and cover an area no larger than a fist. These lichens are among the oldest living organisms on Earth. □

Sensitivity to Pollution
Lichens are extremely sensitive to pollutants in the atmosphere because they readily absorb substances dissolved in rain and dew. Pollutants such as sulfur dioxide, a byproduct of automobile engine exhaust and industrial activity, quickly destroy a lichen's chlorophyll, decreasing its rate of photosynthesis. As a result, the physiological balance of the symbiotic relationship is upset. For this reason, lichens do not grow in or around cities. Biologists use the relative health of lichens and their chemical compositions as indicators of an environment's health. Recently, biologists have discovered that lichens are disappearing from national parks and other remote areas, which, despite their distance from the sources of pollution, are clearly being affected by the quality of air that reaches them.

Mycorrhizae: Fungi and Roots
Certain fungi play important roles in the nutrition of vascular plants by forming symbiotic associations with their roots. Associations of this kind are called mycorrhizae. The fungal filaments aid in the direct transfer of phosphorus and other minerals from the soil into the roots of the plant, while the plant supplies carbohydrates to the symbiotic fungus.

In the mycorrhizae of most species of plants, the fungal hyphae penetrate the outer cells of the root and form coils, swellings, and tiny branches that extend into the surrounding soil. These are called **endomycorrhizae.** The fungus involved is usually a zygomycete. In Figure 22-14, you can see the endomycorrhizae that grow in the roots of the leek *Allium* *porrum.* Fossils reveal that the rootlike appendages of the earliest plants often had endomycorrhizae, which may have played an important role in the invasion of land by plants. The soil of that time was completely lacking in organic matter, and mycorrhizal plants are particularly successful in infertile soil. Some archaic primitive vascular plants surviving today continue to depend strongly on endomycorrhizae.

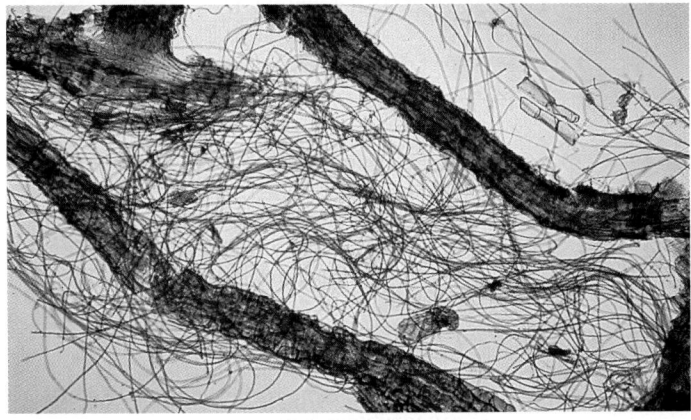

Figure 22-15 The hyphae of the fungus in ectomycorrhizae sometimes appear as a tangled mass around the root of the plant.

In at least 10,000 species of plants, the mycorrhizae do not physically penetrate the plant root but instead wrap around it. These are called **ectomycorrhizae.** In contrast to endomycorrhizae, the nonpenetrating ectomycorrhizae represent highly specialized relationships in which a particular species of plant has become associated with a particular fungus (usually a basidiomycete). These kinds of ectomycorrhizae are important because they involve many commercially significant trees that grow in temperate regions, such as pines, oaks, beeches, and willows. Figure 22-15 shows the roots of a pine tree surrounded by a tangled mass of ectomycorrhizae. At least 5,000 species of fungi have been identified in different ectomycorrhizae. They are mostly formed with basidiomycetes, but some involve associations with ascomycetes.

☐ **CAPSULE SUMMARY**

Mycorrhizae are symbiotic associations in which a fungus transfers minerals to a plant's roots, which in turn supply carbohydrates to the fungus. Endomycorrhizae penetrate root cells; ectomycorrhizae wrap around roots.

Section Review

1. *What is a lichen? What are the benefits for each partner in this symbiotic association?*
2. *Why are lichens so sensitive to pollution?*
3. *How do plants benefit from having mycorrhizae?*

Critical Thinking

4. *How might the Earth's landscape appear today if the earliest plants had not had mycorrhizae?*

 SECTION 22-3

 VISUAL STRATEGY

Figure 22-15

Use this diagram to explain the structure of ectomycorrhizae. Have them point out the roots of the pine tree. Then have them find the fungal component of the ectomycorrhizae that wraps around the root of the tree.

Chapter Closure

The purpose of this activity is to show students the impact fungi have on humans. Have students design a graphic organizer with three tiers. On the top tier place the word "Fungus." The second tier should contain the following terms: medicine, agriculture, industry, ecology, economics. The third tier will be derived by the students. They should give as many examples as possible for the terms in the second tier. Once they have completed their graphic organizers, have them go back and designate each example with a plus or a minus symbol to show how it affects humans. Afterward, have students write a two-page paper that answers the following questions: When the negatives and the positives are added together, which one prevails? Can this measure indicate whether a fungus is good or bad in human terms? Justify your response.

✉

CHAPTER REVIEW ANSWERS

Each item in the Chapter Review is correlated by text section in the assignment guide that follows.

ASSIGNMENT GUIDE

Section	Review	Themes Review	Critical Thinking
22-1	1, 2, 3, 9, 10, 11,13	17	
22-2	4, 5, 12, 14	16, 18	19, 20
22-3	6, 7, 8, 15		

Review

Multiple Choice

1. d (22-1.2)
2. d (22-1.3)
3. a (22-1.4)
4. d (22-2.1)
5. b (22-2.2)
6. c (22-3.1, 22-3.3)
7. d (22-3.2)
8. b (22-3.3)

Completion

9. Fungi (22-1.2)
10. mycelia, chitin (22-1.2)
11. enzymes, absorbing (22-1.4)
12. fusion, spores (22-2.3)
13. paired, dikaryon (22-1.5)
14. sexual, zygospore (22-2.3)
15. endomycorrhizae, ectomycorrhizae (22-3.3)

Vocabulary

ascocarp (506)	dikaryotic (504)	lichen (511)
ascus (506)	ectomycorrhizae (513)	mycelium (502)
basidium (508)	endomycorrhizae (512)	septa (502)
chitin (501)	hypha (502)	yeast (507)
dikaryon (504)	Fungi Imperfecti (510)	zygosporangia (505)

Review

Multiple Choice

1. Fungi differ from plants in that fungi
   a. are multicellular.
   b. are immobile.
   c. have cell walls.
   d. are heterotrophic.

2. The cross walls that separate cells in hyphae are known as
   a. rhizoids.　　　　c. asci.
   b. gills.　　　　　d. septa.

3. Which of the following characteristics is shared by all fungi and helps them obtain food?
   a. external digestion
   b. phagocytosis
   c. feed on nonliving matter
   d. anesthetize prey

4. Members of Deuteromycota are more difficult to classify than other fungi because
   a. they develop from zygosporangia.
   b. they are sexual and parasitic.
   c. they undergo meiosis.
   d. if sexual reproduction occurs, it is unknown.

5. The common edible mushroom is classified in the phylum
   a. Zygomycota.　　c. Basidiomycota.
   b. Ascomycota.　　d. Deuteromycota.

6. Mycorrhizae are symbiotic relationships of fungi and
   a. algae.　　　　c. roots.
   b. lichens.　　　d. chloroplasts.

7. Some fungal associations no larger than a fist appear to be thousands of years old. These have been found
   a. in temperate forests.
   b. on well-irrigated alluvial plains.
   c. in fields of corn.
   d. in harsh environments, high in the mountains.

8. One might expect that plants without mycorrhizae are
   a. more likely to get fungal diseases.
   b. less successful in the transfer of minerals from the soil into the roots.
   c. best suited to poor soil conditions.
   d. primitive and might soon become extinct.

Completion

9. The kingdom _____ contains eukaryotic organisms that are immobile, lack chlorophyll, and produce chitin, hyphae, and spores.

10. Fungal _____ consists of a tangled network of hyphae. Each hypha is covered by a cell wall made of _____ .

11. Most fungi feed by secreting _____ into the organic matter that surrounds them and then _____ the digested food.

12. A mushroom develops from the _____ of different mating strains. On the hyphae lining the gills of mushrooms, haploid _____ are produced.

13. A mycelium with _____ but separate nuclei is called a _____ .

14. During the _____ phase of the life cycle of *Rhizopus*, hyphae from two mating types fuse and produce a cell that develops into a dormant _____ .

15. In _____ the fungal hyphae penetrate the outer cells of the plant root, but in _____ the fungal hyphae wrap around the plant root.

Themes Review

16. **Structure and Function** Describe the structure of the zygosporangium of *Rhizopus stolonifer*. How does the zygosporangium function to ensure survival of the species?

17. **Levels of Organization** Organisms are identified as producers, consumers, or decomposers based on the role they play in an ecosystem. What role do fungi play in most ecosystems? Give an example to support your answer.

18. **Levels of Organization** When observed, a member of the phylum Deuteromycota exhibits sexual reproduction involving the development of zygospores. Given this observation, should the fungus be reclassified? If so, in which phylum should it be placed?

Critical Thinking

19. **Making Inferences** The bacterium *E. coli* has long been a popular organism for use in studying the molecular biology of cells. But in recent years, the baker's yeast *Saccharomyces cerevisiae* has been used more often, especially when research questions relate to the molecular biology of human cells. Why do you suppose *Saccharomyces cerevisiae* is the preferred research organism for questions concerning the molecular biology of human cells?

20. **Interpreting Data** In the last 50 years, the truffle harvest in a certain area in Europe has drastically declined. During the same period, oak forests have been cut and the land cleared for farming and housing. Some truffle hunters in the area suspect that the decline in the truffle harvest is related to forest clearing. Based on the data presented in the graphs below, what is the relationship between the truffle harvest and forest clearing? What would you advise land-owners in the area to do if they would like to maintain the truffle harvest at its current level?

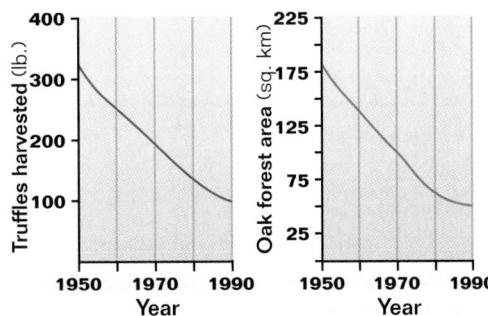

Activities and Projects

21. **History** Downy mildew caused by the fungus *Plasmopara viticola* devastated the crops of French grape growers during the 1870s and early 1880s. Then the Bordeaux mixture was invented by Pierre Millardet of the University of Bordeaux. Find out what chemicals are in the Bordeaux mixture and how the mixture prevents downy mildew. Also, learn how Millardet got the idea to develop the mixture. Write a paper that details what you learn, and share your results with the class.

22. **Research and Writing** Use the library to research ways in which medical science has been influenced by fungi. Investigate the discovery of penicillin and other drugs derived from fungi. Investigate the role played by fungi in various diseases. Present the findings of your research in a report.

Themes Review

16. The zygosporangium is a thick-coated dormant structure. It can withstand drought and freezing and undergoes meiosis and germinates when favorable conditions return. (22-2.2, 22-2.3)

17. Fungi function as decomposers. Bracket fungi in a forest ecosystem break down dead tissue in tree trunks into nitrogen, carbon and other elements. These elements are extracted from the soil by autotrophs. (22-1.5)

18. The fungus should be reclassified in phylum Zygomycota because it develops zygospores. (22-2.2)

Critical Thinking

19. Both yeast and humans are eukaryotes; their chromosomes are structurally similar. Yeast can be infected by viruses similar to the one that causes AIDS and make a number of proteins similar to those made by human cells. (22-2.2)

20. It appears that the destruction of oak forests does affect the truffle harvest. To maintain the truffle harvest at its current level, oak forest destruction must stop. (22-2.2)

Activities and Projects

21. It is a mixture of copper hydroxide and lime. The very insoluble copper hydroxide is dissolved by the acid produced by germinating fungal spores. The dissolved copper kills the spores, but the remaining undissolved copper does not harm the grape plant. Millardet observed that grapes sprayed with copper sulfate and lime to prevent people from eating the grapes did not have downy mildew. (22-2.2)

22. Some drugs that students might research include streptomycin, Chloromycetin, and Tetramycin. Some fungal diseases that students can find out about include cryptococcosis, thrush, histoplasmosis, and dermatomycosis. (22-2.2)

LABORATORY Investigation Chapter 22

Fungi on Food

Preparation

Fungal samples may be grown on food items (e.g., unpreserved bread). Place a piece of damp filter paper under the food item in a bowl. Store the bowls in an incubator at room temperature for a week. Keep the filter paper moist. A dish of water in the incubator will provide additional moisture. To prevent contamination when putting the cultures out in the lab, cover each bowl with plastic wrap.

Prepare media from Sabouraud dextrose agar (500 g bottle) and 0.3% propionic acid mold inhibitor. Melt the agar in a microwave or hot-water bath and divide the contents evenly into two flasks. Add 0.75 mL of the propionic acid to one flask. Sterilize both flasks and 48 petri dishes at 15 psi for 20 minutes. Label the petri dishes for the presence or absence of propionic acid (24 of each). Pour the contents of each flask into the appropriate dishes. Refrigerate dishes until lab time.

Procedural Notes

- Part A can be completed in 20 minutes. Inoculated petri dishes should be incubated for one week. Allow 30 minutes for students to complete Part B.
- Have students develop a list of possible environmental conditions that affect fungal growth, such as nutrients, moisture, warmth, light, absence of inhibitory or toxic chemicals.
- You might have students sketch the fungal colonies as they appear in the Petri dish.

OBJECTIVES

- Recognize fungal growth on food.
- Identify environmental conditions that favor the growth of fungi on food and those that inhibit it.

PROCESS SKILLS

- designing an experiment
- analyzing results

MATERIALS

- safety goggles
- lab apron
- disposable gloves
- 2 sterile petri dishes with nutrient medium
- 2 sterile petri dishes with nutrient medium and propionic acid
- fungal samples
- stereomicroscope
- toothpicks
- wax pencil
- masking tape

BACKGROUND

1. How do multicellular fungi, such as molds, obtain nutrients?
2. How do multicellular fungi reproduce and grow?
3. Write your own question to explore in your lab report or notebook.

TECHNIQUE

Part A: Experimental Setup

1. **CAUTION:** Put on safety goggles, a lab apron, and disposable gloves. Obtain four sterile petri dishes, two with nutrient medium and two with nutrient medium plus propionic acid. Be sure the dishes are labeled for the presence or absence of propionic acid.

2. Examine the fungal samples through a dissecting microscope. Select a dense growth of a fungus, from which you will take samples.

3. Use a toothpick to scoop up a small sample of the fungus you selected. Gently touch the sample to the medium in four places in the two petri dishes without propionic acid. Raise the lids of the dishes as little as possible to do so. Do the same for the two petri dishes with propionic acid, using a clean toothpick and another small sample of the same fungus. Properly dispose of the toothpicks.

4. Place a piece of masking tape on opposite sides of each dish to hold the lid and bottom together. Label each petri dish with your name and the food source of the fungal sample you selected.

5. Design an experiment to determine which of two opposite environmental conditions is best for fungal growth. Some possible combinations are warm/cold, light/dark, and moist/dry. Label one of the two environmental conditions you selected on a dish with propionic acid and on a dish without propionic acid. Label the other environmental condition on the other two petri dishes (one with propionic acid and one without). Then incubate all four dishes under the appropriate conditions. Record the food source

Background Answers

1. Fungi obtain nutrients by releasing enzymes that break down living or dead matter. The smaller particles are then absorbed by the fungi.

2. Fungi reproduce by producing spores. Slender filaments called hyphae grow from a spore and tangle into a thick mat called a mycelium that grows within a food source.

3. Under what kinds of environmental conditions do fungi grow best?

Records

Students should use a table similar to the one on page 517 to record their data.

of the fungal sample and the two environmental conditions that you are testing; use a table like the one below in the **Records** section of your report.

Fungal Growth in Different Environments

Dish	Environmental Condition	Propionic Acid?	Source of Fungus	Growth
1				
2				
3				
4				

Part B: Comparing Amounts of Growth

6. After one week, examine each dish under the stereomicroscope without opening the dish.

7. Record your observations in the data table in the **Records** section of your report.

8. Examine dishes belonging to other groups, especially those grown under different environmental conditions.

9. Add your observations of those dishes to the **Records** section of your report. Also, briefly summarize the procedure you followed.

10. Clean up your materials and wash your hands before leaving the lab. Dispose of all materials according to instructions provided by your teacher

INQUIRY

1. Besides the environmental conditions you chose, what additional factor was tested in your experiment?

2. What steps were taken in your experiment to avoid contamination of the plates?

3. How would contamination of the plates affect the results of your experiment?

4. What does extensive fungal growth on a plate indicate?

5. What effect does propionic acid have on fungal growth? How do you know?

ANALYSIS

1. Why do you think propionic acid is added to foods?

2. Which environmental conditions favor fungal growth? Which inhibit it?

3. Compare the results for the different kinds of fungi grown. Did fungi from certain food sources grow more rapidly than others?

4. Based on your conclusions, under what conditions would you keep a nonsterile food product if you wanted to prevent it from becoming moldy?

FURTHER INQUIRY

Write a new question that could be explored as a new investigation. The following is an example:

On what kinds of food do fungi grow best?

a. additive-free bread or bread containing chemical additives

b. natural cheese or processed cheese

c. regular strawberry preserves or low-sugar strawberry spread

Inquiry Answers

1. The effect of propionic acid on fungal growth was also tested.

2. To avoid contamination, the lids of the petri plates were raised only slightly.

3. Contamination may have resulted in the growth of bacterial colonies on the plates, which may have interfered with fungal growth.

4. Extensive fungal growth indicates that the environmental condition under which it was grown promotes the growth of fungi.

5. Propionic acid inhibits the fungal growth. Less fungi grew on the petri dishes that were treated with the acid.

Analysis Answers

1. Propionic acid is added to foods to help keep them fresh by retarding fungal growth.

2. Warmth and moisture favor fungal growth, while cold and dryness inhibit it. Light and darkness have little effect on fungal growth.

3. Answers will vary, depending on the food sources used.

4. To keep food products fresh, they should be wrapped or placed in a sealed container in a refrigerator.

Block Scheduling Guide

Each block represents about 45–50 minutes of instructional time. The resources cited in a block are coded to a specific program component using the Key on the next page.

Lecture Concepts

Lesson Resources

23-1 The Evolution of Plants pp. 519–528

BLOCKS 1 and 2

a. Before plants were able to live in terrestrial habitats, they needed ways to absorb minerals, conserve water, and reproduce on land.
b. Many plants have vascular tissues that act as pipelines for carrying water from roots to leaves and for moving carbohydrates from leaves to roots.
c. Plants exhibit a life cycle in which a haploid individual that produces gametes alternates with a diploid individual that produces spores.
d. The first plants to invade land lacked roots, stems, and leaves. They took in water and other substances by osmosis and diffusion.
e. Nonvascular plants, such as mosses, liverworts, and hornworts, have vascular tissue that is very simple in design.
f. Modern vascular plants are distinguished by the following features: a dominant sporophyte, specialized conducting tissues, and a distinctive body form.

Lecture Resources
- Opening Question, page 519
- Demonstrations: Vascular Plants vs. Nonvascular Plants, Identifying Plant Body Structures
- Visual Strategies: Figures 23-2, 23-3, 23-4, 23-5, 23-6, 23-7, 23-8, 23-9
- Analyzing Table 23-1, page 523
- What's in a Name? page 523
- Life Cycle Review, page 524
- Teaching Transparencies: 113A, 114A, 115A, 118, 121
- Identifying Vascular Plants, page 527

- Holt Biology Videodiscs: Lesson 23-1

Classwork Options
- Growing Tall, page 522
- Making Greenhouses, page 522
- Effective Reading, page 523
- Charting Information, page 524
- Liverworts, page 525
- Mosses, page 526

Assignment Options
- Section Review, page 528
- Chapter Review, questions 1–3, 11, 12, 16, 28, 31

23-2 The Evolution of Seeds pp. 529–534

BLOCK 3

a. The first vascular plants, such as ferns, were seedless and required a film of water for fertilization.
b. The first seed-bearing plants were gymnosperms, which produce seeds that develop in cones.
c. Angiosperms are flowering plants, which produce seeds that develop within fruits.
d. A seed is a sporophyte plant embryo surrounded by a protective coat.

Lecture Resources
- Opening Questions, page 529
- Demonstration: Can Seeds Fly?
- Overcoming Misconceptions, page 530
- Visual Strategies: Figures 23-11, 23-14
- How Are Seeds Dispersed? page 534
- Teaching Transparencies: 117, 132, 133

- Holt Biology Videodiscs: Lesson 8-2

Classwork Options
- Growing a Fern, page 530
- Observing Pollen, page 531
- Prepared Slides, page 531

Assignment Options
- Section Review, page 534
- Chapter Review, questions 4–7, 13, 17, 29, 32, 33

23-3 The Evolution of Flowers pp. 535–540

BLOCKS 4 and 5

a. Angiosperms are the most successful plants, comprising 90 percent of all living plant species.
b. Flowers are reproductive structures that generally consist of four whorls of appendages.
c. In many angiosperms, flower structure is suited for a particular type of pollination.

Lecture Resources
- Opening Questions, page 535
- Demonstrations: Comparing Cones and Flowers, Classifying Flowering Plants
- Overcoming Misconceptions, page 535
- Research Update, page 536
- Grass Has Blooms, page 538
- Teaching Transparencies: 131, 132

- Holt Biology Videodiscs: Lesson 23-3

Classwork Options
- Recognizing Flower Parts, page 537
- Inquiry Skills Development B15: Flower Structures

Assignment Option
- Pollen Coats, page 538

23-3 The Evolution of Flowers (continued) pp. 535–540

BLOCK 6

a. The typical angiosperm undergoes double fertilization, a process in which two sperm fuse with cells in the megagametophyte to produce both a zygote and endosperm.
b. Angiosperms are classified as either monocots or dicots, depending on the number of cotyledons in their seeds.
c. Fruits are specialized for seed dispersal by agents such as animals, wind, and water.

Lecture Resources
- ■ Classifying Plants as Monocots or Dicots, page 539
- ■ Visual Strategy: Figure 23-22
- ■ Comparing Fruits, page 540
- Teaching Transparencies: 115, 116, 132
- Holt Biology Videodiscs: Lesson 23-3

Classwork Options
- ■ Closure, page 540

- ■ Laboratory Investigation: Plant Diversity, pages 548–549
- Inquiry Skills Development B16: Fruits and Seeds

Assignment Options
- ■ Section Review, page 540
- ■ Chapter Review, questions 8–10, 14, 15, 18, 19, 30, 34

Review/Enrichment

BLOCK 7

- ■ Highlights: Plant Evolution, pages 541–544
- ■ Highlights Review, questions 20–27
- ■ Study Guide: Concept Review and Pretest

Assignment Options
- ■ Activities and Projects, questions 35–37
- ■ Readings, questions 38, 39
- Teaching Resources CD-ROM: Occupational Applications Worksheet, Botanist

Assessment Options

BLOCK 8

Traditional Assessment
- ■ Chapter 23 Test
- Test Generator/Assessment Item Listing: Software item bank for preparing customized chapter tests.

Portfolio Assessment
Select student reports for one or more laboratory experiments from this chapter. *The Direct Observation Checklist* on the *BioSources Teaching Resources CD-ROM* should be completed during a laboratory or other cooperative learning experience.

Answer to Concept Map
The following is one possible answer to the Concept Mapping exercise on page 545.

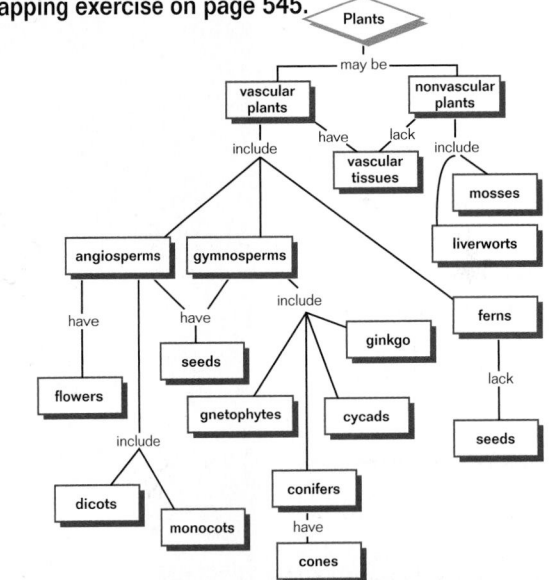

Holt Biology Videodiscs

Holt Biology Videodiscs gives you a powerful tool for teaching, review, and assessment. *Concepts of Biology* adds interactive lesson plans and extensive tools for customization using CD-ROM technology.

CONCEPTS OF BIOLOGY

Use the following topics from *Concepts of Biology* to help you teach this chapter:
- ■ Topic 15: Plant Evolution
- ■ Topic 16: Plant Structure and Function

For further information regarding the media on the videodiscs, see *Holt Biology Videodiscs Teacher's Correlation Guide to Biology: Principles and Explorations.*

Chapter Themes

Evolution *The development of structures that enabled plants to more efficiently survive on land can be traced by studying the 12 plant phyla. Note that the earliest plants to appear in the fossil record are also the simplest in structure and that the flowering plants, the most recent to appear in the fossil record, are also the most complex and highly specialized plants. Trace the evolution of plants in Table 23-1 and* Highlights: Plant Evolution.

Structure and Function *Plants do not have "behavior," and therefore the suitability of their structure to the functions they must perform is particularly critical. Look for ways to emphasize how structure relates to function in Figures 23-3, 23-4, 23-8, 23-10, 23-19, 23-20, 23-21, and 23-23.*

Tapping Prior Knowledge

- How do plants differ from other organisms?
- What makes plants green?
- What were the ancestors of the plants?
- What role do plants play in ecosystems?

Opening Demonstration

Show students pictures of several different kinds of plants from different phyla. Ask students to rank the plants in the order in which they evolved based on appearance. Have students justify their sequencing of each plant. On the board or overhead projector, develop a list of the characteristics that students consider primitive and a list of the characteristics that they consider advanced.

OVERVIEW

CHAPTER 23

OVERVIEW OF PLANTS

A formal flower garden

REVIEW

- mitosis (Section 6-2)
- meiosis (Section 6-3)
- mycorrhizae (Sections 13-2 and 22-3)
- life cycles, gametophyte, and sporophyte (Section 19-2)
- characteristics of the kingdom Plantae (*Highlights: Kingdom Plantae* on page 447)

Authors' Rationale

The plant kingdom's impact on our lives cannot be overstated. A broad understanding of plants as organisms—their anatomy, physiology, evolution, and diversity—is required for a complete understanding of biology. Life on Earth would not be possible without plants converting solar energy to chemical energy by photosynthesis. This chapter provides an overview of the plant kingdom, identifying the major plant groups and briefly discussing several topics that your students may take for granted, such as flowers, seeds, and fruit. Subsequent chapters build on the information presented in this chapter.

23-1 The Evolution of Plants

Plants (members of the kingdom Plantae) are complex multicellular organisms that are primarily terrestrial autotrophs—that is, they occur almost exclusively on land and produce their own organic molecules from inorganic materials by photosynthesis. Today, plants are one of the dominant groups of organisms on land, although some, like the tiny duckweeds seen in Figure 23-1, have returned to aquatic habitats. In this section, you will discover how plants began to adapt to life on land.

Section Objectives

- Identify several obstacles to living on land, and describe how plants overcame them.
- Distinguish nonvascular plants from vascular plants.
- Summarize alternation of generations in plants.
- Describe the moss life cycle.
- Describe the three basic features of vascular plants.

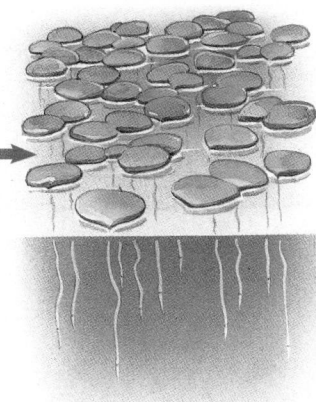

Figure 23-1 Duckweeds, the smallest flowering plants, live only in aquatic environments.

Plants Overcame Obstacles to Living on Land

The fossil record indicates that nothing lived on the land surfaces of our planet until about 440 million years ago. For the first 3 billion years of life's existence on Earth, life was confined to the sea. Scientists are not sure why it took so long for life to reach terrestrial habitats, but they suspect that intense solar radiation may have made the surface of the land uninhabitable. With the advent of photosynthesis in Earth's oceans, oxygen gas began to accumulate in the atmosphere. Some of this oxygen gas, O_2, was converted to ozone, O_3, leading to the development of an ozone layer high in the atmosphere. Then, as it does today, the ozone layer shielded Earth's surface from much of the harmful solar radiation.

Did You Know?

The duckweeds seen in Figure 23-1 have a big job in an ecosystem. These small plants remove pollutants from the environment in which they live. Their root hairs absorb metals and other chemicals that devastate the wildlife in an aquatic habitat.

Did You Know?

Scientists estimate that the ozone concentration over the Antarctic drops by 50 percent in the spring months. As the ozone concentration decreases, the amount of ultraviolet radiation reaching Earth's surface increases.

SECTION 23-1

Vocabulary Preview

- cuticle
- stomata
- guard cell
- vascular system
- alternation of generations
- archegonium
- antheridium
- vascular tissue
- sporangium
- phloem
- xylem
- meristem
- shoot
- root

Opening Question

Ask students to name three things that are important for the survival of an organism on land. Students should mention that survival requires a means of getting water and nutrients, a means of reproduction, and a means of maintaining homeostasis (i.e., self-regulation).

Demonstration

Vascular Plants vs. Nonvascular Plants

To emphasize the difference between living on land and living in water, as did the ancestors of plants, show students a potted plant and a beaker of aquarium water with algae. Ask: Which organism has more difficulty obtaining water? (*the potted plant because it lives in soil, not in water*) Which organism is more likely to dry out? (*the potted plant because it is surrounded by air, not water*) Which organism has more specialized structures? (*the potted plant*)

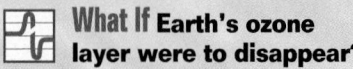

Teaching Tip

 What If Earth's ozone layer were to disappear?
Some scientists are concerned that exposure to greater amounts of ultraviolet radiation will destroy phytoplankton, the most important producers in Earth's food chains. Ask students to describe how life on this planet would differ if ozone became severely depleted.

◉ VISUAL STRATEGY

Figure 23-2
Have students compare the sizes of the pine seedlings, and ask them how the two seedlings could be the same age but differ so much in size. *(The larger seedling must have greater access to nutrients that enable it to grow.)* Point out that in an external mycorrhiza the fungal mycelium may penetrate the root but that it grows only on the *outside* of the root cells, while in an internal mycorrhiza the fungal mycelium penetrates individual root cells.

Soon after the appearance of significant amounts of oxygen in Earth's atmosphere, plants and fungi invaded the land. Both plants and fungi probably evolved from multicellular protists. Multicellularity enabled plants to develop the complex structures and associations that have contributed to their success on land. However, the multicellular, aquatic green algae that were the ancestors of modern plants could not survive on land. Before the descendants of the earliest plants could thrive in terrestrial habitats, they had to overcome three obstacles: they had to be able to absorb minerals from the rocky surface; they had to be able to conserve water; and they had to have a way to reproduce on land.

Absorbing Minerals

Mutualistic associations similar to mycorrhizae may have played a key role in the initial occupation of land. As you learned in Chapter 13, mycorrhizae are symbiotic relationships between fungi and the roots of plants. The plants provide the fungi with carbohydrates and other organic molecules made during photosynthesis. The fungi absorb from the soil the phosphorus and other minerals that are needed by plants. Figure 23-2 shows how a plant benefits from mycorrhizae. Although the first plants lacked roots, fungi have been seen within and among the root cells of many fossilized early plants. Thus, some botanists (scientists who study plants) think that mutualistic associations similar to mycorrhizae may have enabled the first plants to absorb minerals from Earth's rocky surface. Today, approximately 80 percent of all living plant species form mycorrhizae.

Conserving Water

One of the key challenges to living on land is to avoid drying out. The first plants were very small and lived at the edges of oceans, where water was abundant. However, to occupy drier habitats, a means of conserving water was needed. For plants, the solution to this problem was the development of a watertight outer covering called a **cuticle.** Made of a waxy

Figure 23-2 The two pine seedlings, *below,* are the same age. The roots of the small pine seedling, *left,* have no mycorrhizal fungi. Mycorrhizal fungi growing within and between the root cells of the large pine seedling, *right,* have enhanced the seedling's growth.

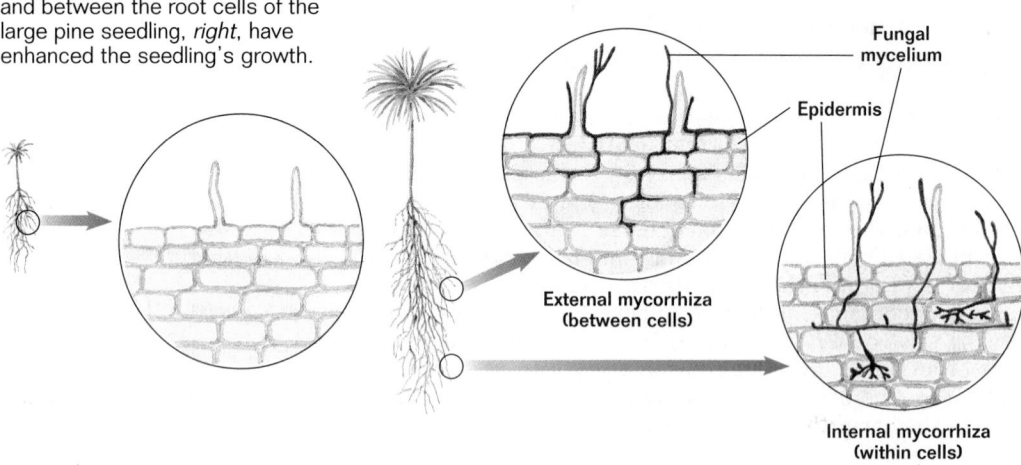

External mycorrhiza (between cells)

Fungal mycelium

Epidermis

Internal mycorrhiza (within cells)

 Did You Know?

Plants can be used to mine minerals because they have the ability to remove heavy metals from soil. For example, nickel is a metal that can be obtained by using plants in areas where typical mining procedures would be impractical because of the low quantities. Certain weeds, such as mustards, can survive and absorb nickel from soils that are not suitable for growing crops. When incinerated, the ash of the plants yields 15 to 20 percent nickel, which sells for $3 per pound.

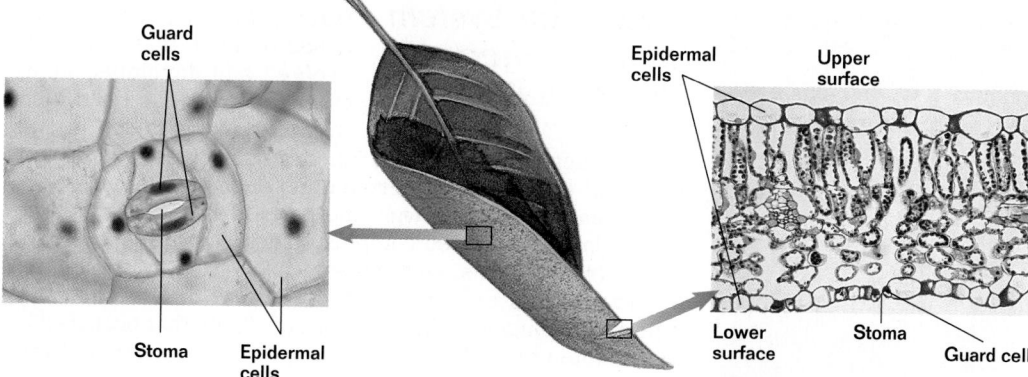

Figure 23-3 A leaf, *center*, has numerous stomata. A pair of guard cells, *left*, surrounds each stoma. The cross section of the leaf, *right*, shows that a stoma is an opening through which air and water vapor can enter and exit the leaf.

👁 VISUAL STRATEGY

Figure 23-3
Point out the guard cells and the stoma in the plane view and cross section of a leaf. Be sure students understand that the guard cells form the opening that is called a stoma. Ask: On what surface of the leaf do these stomata appear? *(lower)* Tell students that in many (but not all) leaves, most stomata are on the lower surface.

CONNECTIONS

Chapter 5
Role of Gases in Plants
Review the role that the gases exchanged through the stomata play in photosynthesis and cellular respiration.

substance, the cuticle covers the aboveground parts of a plant and prevents these tissues from losing water to the air. Like the wax on a shiny car, the cuticle is impermeable to water, but it is also impermeable to the gases required by plants for photosynthesis and cellular respiration.

Passages through the cuticle, in the form of specialized pores called **stomata** *(STOH muh tuh)*, developed and enabled plants to exchange gases. The word *stoma* means "mouth" in Greek. Two views of a leaf's stomata are seen in Figure 23-3. Occurring on at least some portions of all plants except liverworts, stomata permit carbon dioxide gas to enter a plant body and permit water vapor and oxygen gas to exit. Because of the cuticle and stomata, water enters most plants primarily through their roots (which do not have a cuticle) and exits primarily through stomata. A pair of specialized cells called **guard cells** borders each stoma and controls its size by expanding and contracting. The timing of the opening and closing of stomata is critical to preventing excessive water loss while admitting the carbon dioxide required for photosynthesis.

Reproducing on Land

To reproduce sexually, an organism's male and female gametes must be able to reach one another. The male gametes of aquatic algae are able to swim through water to fertilize the female gametes. The gametes of plants that live on land, however, must be able to move in an environment where water is not abundant. Also, they must be protected from drying out while they are being transferred. The eggs of the first plants were surrounded by jackets of cells, and a film of water was required for a sperm to swim to an egg and fertilize it. Today, mosses, ferns, and several other groups of primitive plants still reproduce in this way. In more advanced plants, the sperm are enclosed in multicellular structures (pollen grains) that keep them from drying out. Such structures enable the male gametes of more advanced plants to be transmitted to female gametes by wind or animals rather than by water. ∎

🖥 CONTENT LINK

The location, structure, and function of stomata and guard cells are covered in more detail in **Chapter 24.**

🖵 CAPSULE SUMMARY

Adaptations that have enabled plants to survive on land include mycorrhizae, a waxy cuticle, and structures that protect gametes.

⬥ Did You Know?

Pine trees often have sunken stomata. This adaptation helps them conserve water, an ability that is important for plants that grow in dry or cold climates. (Water is limited where soil is frozen, so water conservation is essential.)

UNIT 6

CHAPTER 23

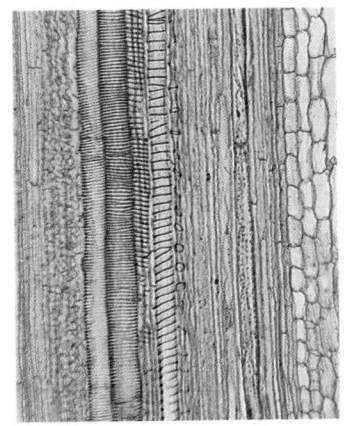

Figure 23-4 Thick-walled, tubular cells like these carry water from the tips of roots to the tips of leaves. Stacked end to end, these cells form tiny pipes called vessels.

A Vascular System Enabled Plants to Thrive on Land

Once plants became established on land, they gradually evolved and developed many other features that aided them in achieving success in this new and demanding habitat. For example, there was no fundamental structural difference between the aboveground and, when present, belowground parts of the earliest plants. Later, plants developed complex structures with specialized tissues (roots, stems, and leaves), each suited to its function and immediate environment. However, one of the most important changes in the structure of plants that occurred as they adapted to land was the development of an efficient way to move water and other materials through the plant body.

In order to survive in environments that have a limited water supply, most plants need an efficient "plumbing" system to carry water from their roots up to their leaves and to carry carbohydrates from their leaves down to their roots. These plumbing systems consist of specialized strands of hollow cells connected end to end like a pipeline, as you can see in Figure 23-4. The tissues that transport water and other materials within a plant make up the **vascular system.** The word *vascular* is derived from the Latin word *vasculum*, meaning "vessel" or "duct." In the dominant group of plants today, the cells of the vascular system run from near the tips of the roots to the tips of the stems and into the leaves. Not all plants have efficient vascular systems, however. Of the twelve phyla of living plants listed in Table 23-1, the three phyla that are referred to as nonvascular plants either have no vascular system or have only very simple vascular tissue. The nine remaining plant phyla have well-developed vascular systems and are referred to as vascular plants.

Table 23-1 Phyla of Living Plants

Phylum	Number of Species	Main Characteristics
Nonvascular Plants		
Hepatophyta Liverworts	6,000	Simplest plants; small, having a dominant gametophyte with a flattened or "leafy" body that lacks vascular tissue, a cuticle, stomata, roots, stems, and leaves
Anthocerophyta Hornworts	100	Small, with a flattened, dominant gametophyte that has stomata but lacks vascular tissue, roots, stems, and leaves
Bryophyta Mosses	10,000	Small; most have simple vascular tissue, a sporophyte consisting of a bare stalk and a spore capsule, and a dominant, "leafy" green gametophyte that lacks roots, stems, and leaves

 Did You Know?

The 1993 International Botanical Congress in Yokohama, Japan, ruled that the terms *division* and *phylum* are equivalent. Therefore, to eliminate the confusion connected with using different terms to represent the same phylogenetic relationships, this book uses the term *phylum* to identify the major subdivisions of the plant kingdom instead of the historically traditional term *division.*

Phylum		Number of Species	Main Characteristics
Vascular Plants			
Psilotophyta Whisk ferns		21	Seedless, with a small, independent gametophyte and a dominant sporophyte that is highly branched and has tiny leaves but is not differentiated into roots and stems
Sphenophyta Horsetails		15	Seedless, with a small, independent gametophyte and a dominant sporophyte consisting of roots and ribbed and jointed stems with soft needlelike leaves at the joints
Lycophyta Club mosses		1,000	Seedless, with a small, independent gametophyte and a dominant, mosslike sporophyte with roots, stems, and leaves
Pterophyta Ferns		12,000	Seedless, with a small, independent gametophyte and a dominant sporophyte consisting of roots, horizontal stems, and leaves called fronds; spores are produced in clusters of sporangia on lower surfaces of leaves
Coniferophyta Conifers		550	Gymnosperms (seed plants with tiny gametophytes, a large sporophyte, and ovules not enclosed by an ovary) that produce cones; mostly evergreen trees and shrubs with leaves modified as needles or scales
Cycadophyta Cycads		100	Gymnosperms with palmlike leaves; produce male and female cones on separate plants
Ginkgophyta Ginkgo		1	Gymnosperm; deciduous tree with fanlike leaves; produces conelike male reproductive structures and uncovered seeds on separate individuals
Gnetophyta Gnetophytes		70	Gymnosperms; diverse group of shrubs and vines
Anthophyta Flowering plants		250,000	Angiosperms (seed plants with tiny gametophytes, a large sporophyte, and ovules enclosed by an ovary); a very diverse group including trees, shrubs, vines, and herbs that produce flowers and fruits

Teaching Tips
Analyzing Table 23-1
Have students answer the following questions with examples from the table. Ask: Which plants would require a wet environment to reproduce? Why? (*All plants lacking seeds would need a moist environment to enable the sperm to swim to the egg.*) Which plants could grow tall? Why? (*plants with a vascular system, because they have the ability to distribute water and nutrients over long distances*)

What's in a Name?
Point out that whisk ferns are not placed in the same phylum as ferns and that club mosses are not placed in the same phylum as mosses. Ask: From the information given, why are whisk ferns not classified in the same phylum as the ferns? (*Whisk ferns do not have roots or leaves that resemble fronds.*) Why are club mosses not classified in the same phylum as mosses? (*Club mosses have a vascular system, roots, stems, and leaves, while mosses do not.*) Tell students that common names are often misleading and often indicate that similarly named organisms resemble each other outwardly. Tell students that plants known as ball moss and Spanish moss are actually flowering plants in the phylum Anthophyta, and that reindeer moss is not a plant at all, but a lichen. Ask: What similarities do these organisms have? (*All are small, grow in clumps, and are often found growing in moist areas.*) Tell students that the word *moss* comes from the Old English word *mos*, which meant "a swamp."

Effective Reading
Concept Mapping
Have students construct a concept map illustrating the relationships and distinguishing characteristics of the plant phyla listed in Table 23-1.

Plants Evolved With Alternation of Generations

Figure 23-5 The life cycles of all plants exhibit sporic meiosis and alternation of generations. The life cycle of a nonvascular plant, *left*, is characterized by a gametophyte that is larger than the sporophyte. The life cycle of a vascular plant, *right*, is characterized by a large sporophyte and a very small gametophyte.

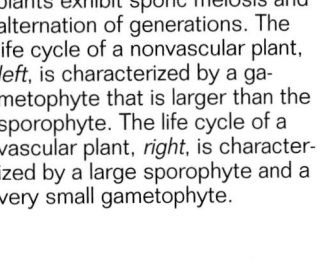

Plants evolved with a distinctive pattern of development in their life cycles. Among many algae, the zygote is often the only diploid (2*n*) cell, and it undergoes meiosis immediately after fertilization (zygotic meiosis) to form haploid cells. In early plants, however, meiosis was delayed. Recall from Chapter 19 that plant life cycles exhibit sporic meiosis, in which spores form by meiosis and gametes form by mitosis. The zygote divides by mitosis to produce many diploid cells that persist for a long portion of the life cycle. As a result, plants developed a life cycle in which a multicelled haploid individual that produces gametes, the gametophyte, alternates with a multicelled diploid individual that produces spores, the sporophyte. The pattern among life cycles in which a haploid individual alternates with a diploid individual is called **alternation of generations.**

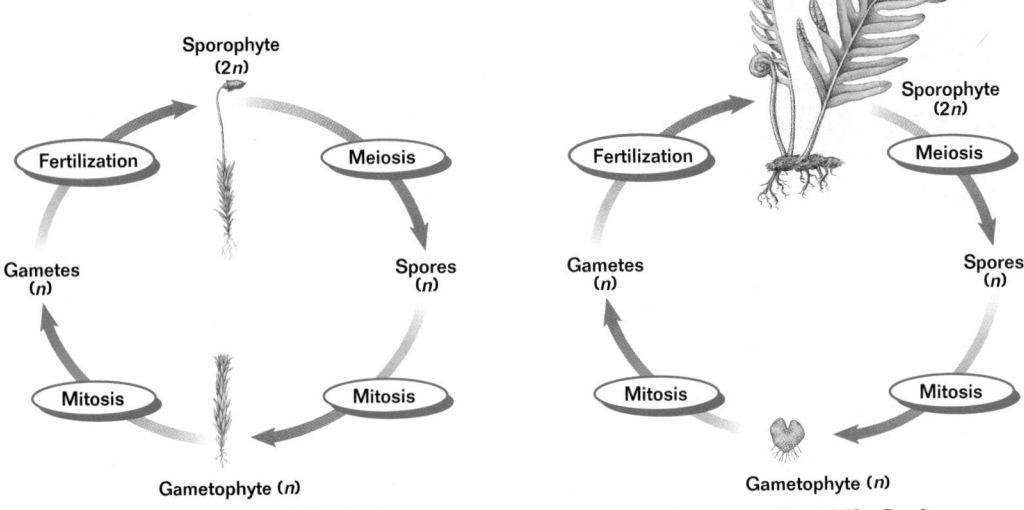

Nonvascular Plant Life Cycle

Vascular Plant Life Cycle

CAPSULE SUMMARY

Plant life cycles are characterized by alternation of generations in which a diploid sporophyte generation alternates with a haploid gametophyte generation. Over time, the plant life cycle evolved from one in which the gametophyte generation is dominant to one in which the sporophyte generation is dominant.

As plants evolved, however, a fundamental difference arose between the life cycles of the simpler, nonvascular plants and those of the more complex, vascular plants. As Figure 23-5 illustrates, the relative sizes of the gametophytes and sporophytes of nonvascular and vascular plants differ. In the nonvascular plants (mosses and liverworts), the gametophyte generation is the dominant (most noticeable) generation. In the vascular plants (ferns, gymnosperms, and angiosperms), the sporophyte generation is dominant. In fact, the very tiny gametophytes of most vascular plants grow *within* tissues of the sporophytes. ▫

Graphic Organizer

Use this graphic organizer with *Teaching Tip:* **Charting Information** on page 524.

Phylum	Vascular Tissue	Roots	Seeds	Flowers	Dominant Gametophyte	Dominant Sporophyte
Liverworts					✔	
Mosses	✔				✔	
Ferns	✔	✔				✔
Conifers	✔	✔	✔			✔
Flowering Plants	✔	✔	✔	✔		✔

Marchantia, a common liverwort

Male gametophyte stalk

Antheridia

Female gametophyte stalk

Archegonia

Sporophytes

Sporophytes

Figure 23-6 *Marchantia* is a common liverwort found living in moist, shady areas. The flattened, green gametophytes produce male and female gametes on separate stalks. Antheridia formed at the tip of a male stalk, *top right,* produce sperm. Archegonia formed at the tip of a female stalk, *bottom left,* produce eggs. The very tiny liverwort sporophytes, *bottom right,* grow from the archegonia under the caps of the female stalks.

The First Plants Lacked a Vascular System

The first plants to successfully make the transition to living on land probably had no vascular system for transporting materials throughout their bodies. All materials had to be transported by osmosis and diffusion, which greatly limited the maximum size of the plant body. Only two phyla of living plants, the liverworts (phylum Hepatophyta) and the hornworts (phylum Anthocerophyta), completely lack a vascular system. The members of these two groups have historically been grouped with the mosses and are still frequently referred to as "bryophytes." However, botanists no longer think that these three groups of relatively simple plants are directly related to one another. Liverworts and hornworts are usually inconspicuous and found growing in moist, shady places.

A common liverwort can be seen in Figure 23-6. The name *liverwort* dates back to the Middle Ages, when it was thought that plants resembling certain body parts might contain substances that could cure diseases of those body parts. The word *wort* meant "herb" in Old English. While the shape of some liverworts resembles a liver, the dominant gametophyte of most liverworts consists of simple leaflike and stemlike structures. Projections called rhizoids help anchor liverworts to the surfaces on which they grow. Gametes (eggs and sperm) are formed by mitosis in separate multicellular structures. **Archegonia** (*ark uh GOHN ee uh*) produce eggs, and **antheridia** (*an thuhr IHD ee uh*) produce sperm. When water is available, the sperm swim to a nearby archegonium and fertilize the egg within it. The resulting zygote grows into a very tiny diploid sporophyte. ☐

☐ CAPSULE SUMMARY

Liverworts and hornworts are simple, nonvascular plants that lack roots, stems, and leaves. Water and other materials are distributed throughout their bodies by osmosis and diffusion.

👁 **VISUAL STRATEGY**

Figure 23-6
Point out that the male and female gamete-bearing stalks of *Marchantia* rise from the flattened, green portion of this liverwort's gametophyte. Then ask students to locate the sporophytes of *Marchantia*. Ask: How do the liverwort gametophyte and sporophyte compare in size? (*The sporophyte is much smaller than the gametophyte.*) Ask students whether they think liverworts or mosses are more primitive, based on the relative sizes of their gametophytes and sporophytes. (*Liverworts appear to be more primitive than mosses because the trend in more advanced plants is for the sporophyte to be much larger than the gametophyte.*)

Teaching Tip
Liverworts

Provide examples of liverworts in the classroom. Show students how the gametophytes look like islands. The female gamete-bearing structures (gametophores) resemble palm trees on their islands, and the male gamete-bearing structures resemble mushrooms on their islands. (Note that the male and female gametophytes are separate.) Fertilization of the female gamete in an archegonium produces a sporophyte that resembles a coconut on a palm tree. Tell students that the archegonia of liverworts and mosses release substances that attract sperm, guiding sperm to the eggs. Once the students have been introduced to liverworts, give them a small sample of a liverwort to add to their greenhouse (described in *Teaching Tip:* **Making Greenhouses** on page 522).

☐ **CAPSULE SUMMARY**

Mosses are considered nonvascular plants and are often grouped with liverworts and hornworts because their vascular tissue is very simple and because all three groups have similar life cycles.

Mosses Have Simple Vascular Tissue

The mosses (phylum Bryophyta) include many species in which a central strand of specialized conducting cells distributes water and carbohydrates throughout the plant. These conducting strands make up what is called **vascular tissue.** Simple in design, the vascular tissue in mosses is composed of conducting cells that lack thickened walls. To learn more about mosses and the other bryophytes, look at *Highlights: Bryophytes* on page 541.

Because their vascular tissue is so simple, mosses are still grouped with the liverworts and hornworts, and all three groups are considered to be nonvascular plants. The three phyla of nonvascular plants share an important similarity. Their life cycles, represented by the moss life cycle seen in Figure 23-7, are dominated by the gametophyte generation. The archegonia and antheridia of mosses are produced on separate gametophytes. The moss sporophyte consists of a bare stalk that supports a spore capsule, or **sporangium,** in which haploid spores are produced by meiosis. ☐

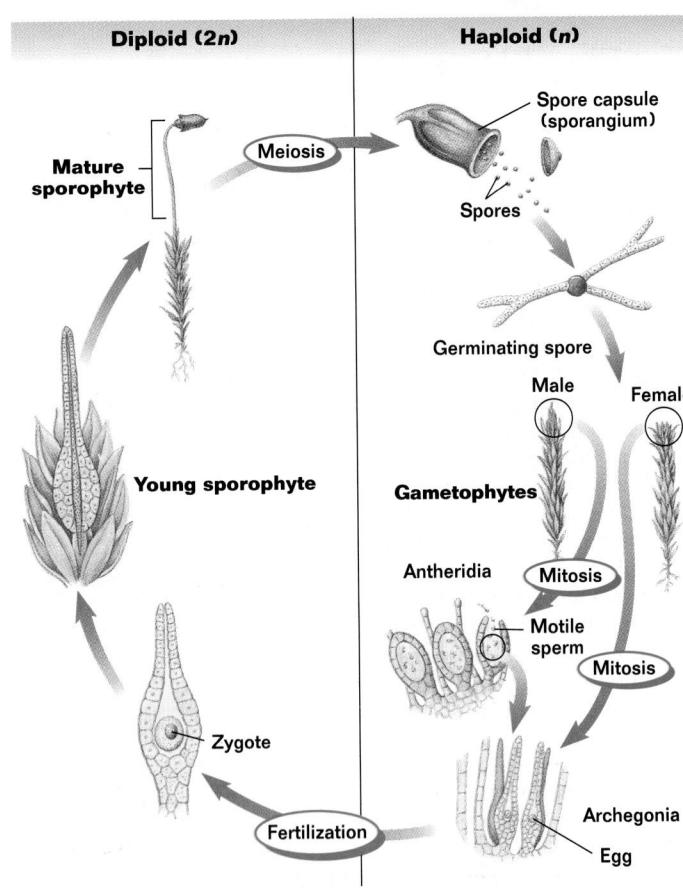

Figure 23-7 The life cycle of a moss is characterized by a "leafy," green gametophyte that alternates with a smaller sporophyte, which consists of a bare stalk and a spore capsule. Motile sperm must swim through a film of water to fertilize an egg. The sporophyte grows on top of the gametophyte.

Vascular Plants Are Characterized by Several Features

The first vascular plants appeared approximately 430 million years ago, but only incomplete fossils of these plants have been found. The earliest known vascular plants for which there are relatively complete fossils, *Rhynia* and *Cooksonia,* are illustrated in Figure 23-8. Today, vascular plants occupy almost all terrestrial habitats except those perpetually covered by ice and snow. Unlike the nonvascular plants and the earliest vascular plants, many modern vascular plants grow very tall. For example, some trees are 50 m (163 ft.) or more in height and weigh many tons.

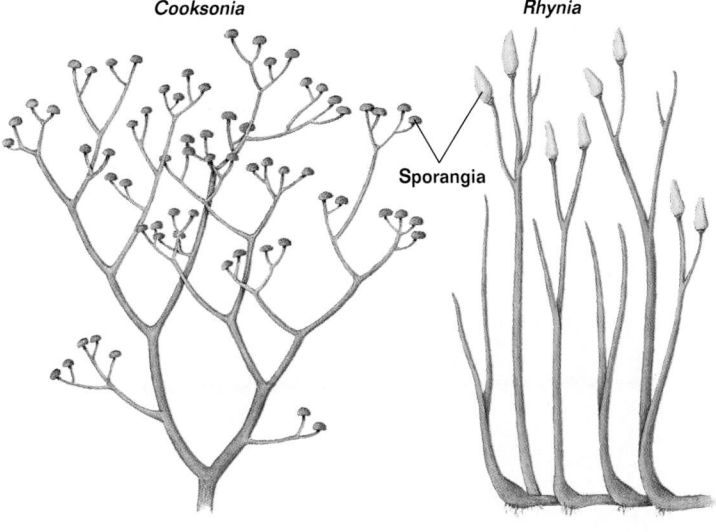

Cooksonia

Rhynia

Sporangia

Figure 23-8 Members of the extinct phylum Rhyniophyta, which appeared about 410 million years ago, included *Cooksonia, left,* the oldest known vascular plant, and *Rhynia, right.* These ancient plants had branched, leafless stems that were only a few centimeters long. Spore-forming sporangia were located at the tips of these stems. *Rhynia* also had horizontal underground stems, or rhizomes.

All vascular plants are distinguished by the following features: a dominant sporophyte, specialized conducting tissue, and a distinctive body form.

1. **Dominant sporophyte**   In contrast to nonvascular plants, the life cycles of vascular plants are dominated by a diploid sporophyte that is much larger than the gametophyte.

2. **Specialized conducting tissue**   The earliest vascular plants had two types of conducting tissue, still found largely unchanged in today's vascular plants. Both types of conducting tissue consist of strands of elongated cells that occur end to end like sections of pipe. Relatively soft-walled cells that conduct carbohydrates away from the areas where they are made, such as in leaves and stems, make up a kind of tissue called **phloem** *(FLOH uhm).* Hard-walled cells that transport water and dissolved minerals up from the roots make up a kind of tissue called **xylem** *(ZEYE luhm).* The reinforced walls of water-conducting cells can withstand considerable water pressure. Thus, water is able to rise to great heights in vascular plants.

> **CONTENT LINK**
>
> *A more detailed discussion of the structure and function of xylem and phloem is contained in* **Chapter 24.**

Teaching Tip
Identifying Vascular Plants
Show students a familiar plant, such as a houseplant or a bedding plant. Ask them if it is a vascular plant. Then have them read the information on pages 527 and 528. Ask students if they still agree with their answer. After reading these pages, all students should agree the plant is a vascular plant. Have students give three reasons to support their answer. *(The plant has a dominant sporophyte, specialized conducting tissue, and a body consisting of roots, stems, and leaves.)*

 VISUAL STRATEGY

Figure 23-8
Ask students how *Cooksonia* and *Rhynia* resemble plants that are alive today. *(The upright growth habit of these plants and the location of reproductive structures at the tips of stems resemble modern plants.)* Ask students how they differ from the plants they are most familiar with. *(These primitive plants have no leaves, roots, flowers, seeds, or fruits.)*

Demonstration
Identifying Plant Body Structures
Display various plants or pictures of plants in the classroom. Have students identify the leaves, stems, and roots of these plants. Use typical examples as well as some less typical examples such as banana trees, which grow from underground stems. (Note: The structures that appear as banana tree trunks are modified leaf sheaths, which are wrapped around each other to make a stemlike structure.)

Figure 23-9
Point out that the aboveground portion of a plant, consisting of stems and leaves, is called the shoot. Roots are the underground, water-absorbing portion of a plant's body. Ask students if they think it is possible for a part of the shoot to grow underground. *(Yes.* Rhynia *is an example of a plant that had underground stems.)* Ask students to speculate how biologists distinguish underground stems from roots. *(Their structures are different.)*

❏ **CAPSULE SUMMARY**

Vascular plants are characterized by a dominant sporophyte generation, a vascular system with specialized conducting tissue, and a distinctive body form.

CONTENT LINK
....
Meristems and their role in plant growth are discussed in more detail in **Chapter 25**.

Figure 23-9 A typical vascular plant has an aboveground shoot with stems and leaves and a belowground root. Plants grow from their tips as the cells of the meristems produce new cells by cell division.

3. Distinctive body form All vascular plants have a body that consists of a vertical shaft from which specialized structures branch. This body form, which developed in the earliest plants, results as plants grow by adding new cells to the tips of their bodies. Zones of actively dividing cells, called **meristems** *(MEHR uh stehmz)*, produce plant growth. In the first vascular plants, there was no differentiation of the plant body into the aboveground structures, or **shoots**, and belowground structures, or **roots**, that characterize most plants. As the vascular plants became better adapted to living on land, however, they developed the familiar plant architecture—roots and a shoot consisting of stems and leaves—seen in Figure 23-9. ❏

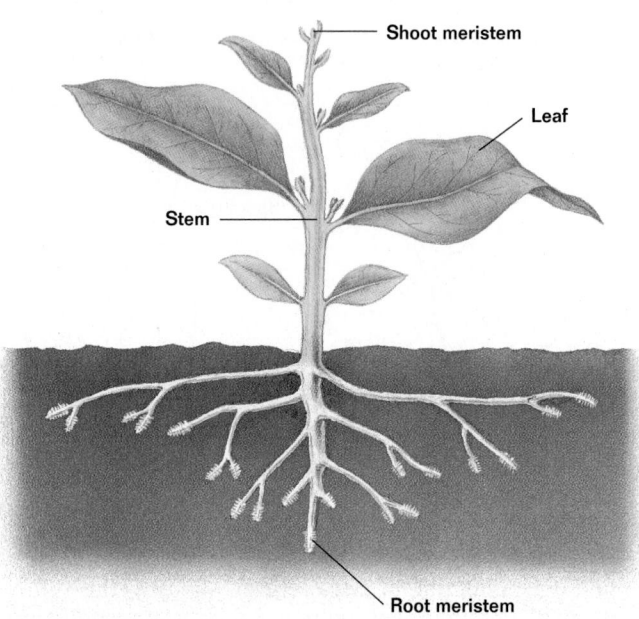

Shoot meristem
Leaf
Stem
Root meristem

Section Review

1. *What were three major obstacles that plants had to overcome to live successfully on land?*
2. *Describe three stages in the evolution of the plant vascular system. Use modern plants as examples of each stage.*
3. *How does alternation of generations occur in plants?*
4. *Summarize the events in the moss life cycle.*
5. *What are the three main characteristics of all vascular plants?*

Critical Thinking
6. *Why are vascular plants the most successful land plants?*

1. Major obstacles to life on land include having to absorb minerals, conserve water, and reproduce with limited water.

2. The possible stages in the evolution of the plant vascular system include: no conducting tissue (liverworts); simple conducting tissue (mosses); and an efficient vascular system (ferns, gymnosperms, angiosperms).

3. In plants, a haploid gamete-producing generation (gametophyte) alternates with a diploid spore-producing generation (sporophyte). Gametes result from mitosis, while spores result from meiosis.

4. Answers will vary but should mention that gametes are produced by mitosis at the tips of "leafy" green, haploid gametophytes, sperm fuse with eggs to produce diploid sporophytes, and spores form by meiosis in spore capsules at the tips of the bare stalks of sporophytes.

5. All vascular plants have a dominant sporophyte, specialized conducting tissue, and a body with shoots and roots.

6. They have the most efficient system for gathering and distributing water and minerals.

23-2 The Evolution of Seeds

Following the development of a vascular system, the next great evolutionary advance in plants was the seed. Most plants living today produce seeds, which consist of a sporophyte plant embryo surrounded by a protective coat. A seed has several important functions that were crucial in the successful adaptation of plants to life on land. One of these functions is to protect a new plant during its most vulnerable stage—the embryo. Another function is to disperse (distribute) offspring to new locations away from their parents. Many kinds of seeds, such as the spruce seeds shown in Figure 23-10, have unique devices that are adaptations for carrying them farther away from their parent plants. Plants that produce seeds are called seed plants.

Section Objectives

- Summarize the characteristics and the life cycle of ferns.
- Explain how gymnosperms differ from other seed plants.
- Compare and contrast the life cycles of the gymnosperms and the ferns.
- Describe a seed, and state its importance to the adaptation of plants to life on land.

Figure 23-10 The seeds of a white spruce (Picea glauca) have a thin, papery "wing" that causes them to spin like a helicopter blade and keeps them airborne long enough to float some distance from their parent plant.

The First Vascular Plants Lacked Seeds

Although today's great forests are dominated by trees that produce seeds, the first forests were composed of vascular plants that did not produce seeds. Like the nonvascular plants, ferns and other seedless vascular plants have swimming sperm and require a film of water for fertilization. Forests of these plants flourished in the warm, humid climate of the late Paleozoic era. In these forests, plenty of water was available for successful reproduction. Ferns (phylum Pterophyta) are the most abundant and most familiar group of seedless vascular plants today. Though they are found throughout the world, ferns are most abundant in the tropics. Many ferns are small, measuring only a few centimeters in diameter. However, some of the largest living plants are tree ferns that can have trunks more than 24 m (79 ft.) tall and leaves up to 5 m (16 ft.) long. To learn more about ferns and other seedless vascular plants, look at *Highlights: Seedless Vascular Plants* on page 542.

SECTION 23-2

Vocabulary Preview

frond

gymnosperm

angiosperm

microgametophyte

megagametophyte

microspore

megaspore

pollen grain

ovule

pollination

pollen tube

conifer

seed

seed coat

Opening Questions

Ask students to name the reproductive structure that most plants have in common. Students should recognize that most plants reproduce by producing seeds. Ask students to suggest some advantages to reproducing with seeds. Students should recognize that seeds protect developing plant embryos from the drying conditions on land and help to disperse plant offspring.

Demonstration
Can Seeds Fly?
Bring mature cones of a conifer (pine, spruce, fir) to class. Pass the cones around, and ask students to look for the seeds. (They lie on top of the scales and have thin papery wings.) Extract a few seeds, and have volunteers take turns tossing the seeds into the air. Have students describe the motion of the seeds.

Historical Note
Hawaii's Pig War
Pigs were introduced on the island of Hawaii by settlers as a source of food, but these animals wiped out native vegetation in the area now called Hawaii Volcanoes National Park. Additionally, the settlers brought their own plants, which eventually took over areas of island plants. The people of Hawaii are now trying to reclaim their native plants. Tree ferns are one species that is now reestablishing in the area.

◆ *VISUAL STRATEGY*

Figure 23-11

Walk students through the fern life cycle, starting with the mature sporophyte. Point out that when meiosis occurs, the chromosome number changes from diploid to haploid. Also point out that the spores grow into the gametophytes, which are haploid. Finally, point out that when fertilization occurs, the chromosome number changes to diploid and a new sporophyte generation begins. Point out that a young sporophyte grows and feeds on the gametophyte because it initially lacks roots to absorb nutrients.

Teaching Tip
Growing a Fern

Show students the clusters of sporangia (sori) on the lower surface of fern fronds. Let them remove a sorus and crush it using the eraser end of a pencil on white paper. Next, have them sprinkle some spores on the peat pellets in their greenhouses. The first structures they will observe will be the little green heart-shaped gametophytes.

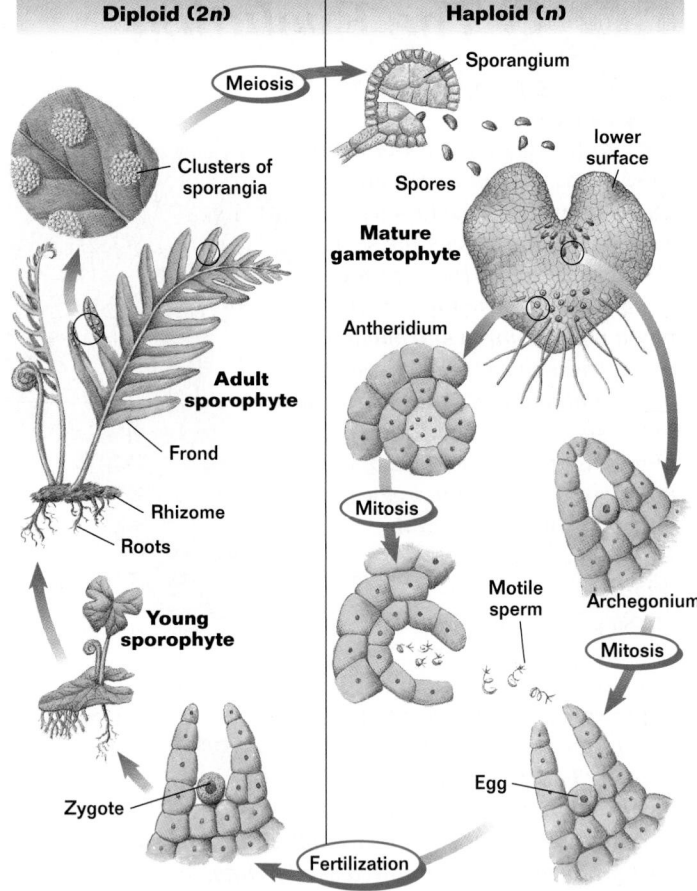

Figure 23-11 The life cycle of a fern is characterized by a large sporophyte (with leaves called fronds) that alternates with a small, heart-shaped gametophyte. Motile sperm must swim through a film of water to fertilize an egg.

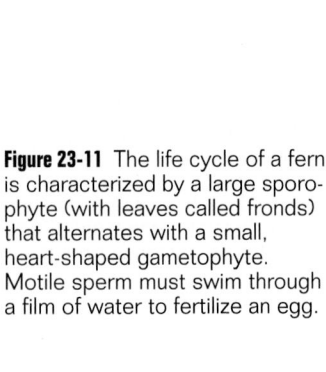

■ *CAPSULE SUMMARY*

Ferns are the most familiar seedless vascular plants. The fern life cycle has a dominant sporophyte generation and a smaller, but independent, gametophyte generation.

The fern life cycle, illustrated in Figure 23-11, represents an intermediate stage in the revolutionary change that took place in plant life cycles. Remember, the life cycles of non-vascular plants are dominated by a gametophyte that supports a smaller, dependent sporophyte. In ferns and other seedless vascular plants, however, the sporophyte is dominant, and the gametophyte is smaller, independent, and self-sufficient. The fern gametophyte is a thin, heart-shaped photosynthetic plant that lives in moist places and is usually no more than 1 cm in diameter. Fern gametophytes produce eggs in archegonia and sperm in antheridia, both of which are located on the lower surface of the plant. In ferns, the archegonia and antheridia are produced by the same individual. In other seedless vascular plants, the male and female structures are produced by separate gametophytes. When a film of water is available, sperm are able to swim to eggs and fertilize them. Fern sporophytes consist of roots, horizontal underground stems called rhizomes, and long, often highly divided leaves called **fronds.** Clusters of spore-producing sporangia form on the lower surfaces of fronds. ■

Overcoming Misconceptions

Diseased or Not Diseased?

Florists often get calls about flower arrangements that include fern leaves. When the fern leaves have sporangia on them, people think they are getting insects with their flower arrangement. A close look at these spore-producing structures reveals that they are lined up in rows across from each other.

The First Seed Plants Were Gymnosperms

Seeds apparently arose only once among the vascular plants, as the plant life cycle continued to shift toward a more dominant sporophyte generation and a more reduced gametophyte generation. Of the five phyla of living seed plants, four are collectively called **gymnosperms** (*JIHM noh spurmz*). The word *gymnosperm* comes from the Greek words *gymnos*, meaning "naked," and *sperma*, meaning "seed," and refers to the fact that gymnosperm seeds do not develop within a fruit (a mature ovary). First appearing about 380 million years ago, gymnosperms were the first seed plants. The flowering plants, or **angiosperms** (*AN jee oh spurmz*), evolved from gymnosperms and make up the fifth phylum of seed plants. The word *angiosperm* comes from the Greek words *angeion*, meaning "case," and *sperma*, meaning "seed," and refers to the fact that angiosperm seeds develop within a fruit. First appearing between 150 million and 200 million years ago, angiosperms are the most recently evolved of all plant phyla.

As Figure 23-12 illustrates, the gametophytes of seed plants have become highly reduced during the course of evolution. Developing from spores that are produced *within* the tissue of the sporophyte individuals, the gametophytes of seed plants are entirely dependent upon those sporophyte individuals for nutrients and water. Seed plants produce two kinds of gametophytes: a very tiny male gametophyte, or **microgametophyte,** that produces sperm, and a relatively large female gametophyte, or **megagametophyte,** that produces eggs. Thus, the spores that produce the microgametophytes are called **microspores,** and those that produce the megagametophytes are called **megaspores.** A **pollen grain,** which consists of only a few haploid cells surrounded by a thick protective wall, is a mature microspore that contains a microgametophyte. Each megagametophyte develops from a megaspore within an **ovule** (*AHV yool*), a multicellular structure that is part of the sporophyte. If the egg inside of an ovule is fertilized, the ovule and its contents become a seed.

Wind, insects, or other animals transport pollen grains to the female reproductive structures that contain ovules. The transportation of pollen grains from a plant's male reproductive structures to a female reproductive structure of a plant of the same species is called **pollination.** When a pollen grain reaches a female reproductive structure, the pollen grain cracks open. A **pollen tube** then grows from the pollen grain to an ovule and enables a sperm to pass directly to an egg. Thus, in seed plants there is no need for a film of water during the fertilization process. ◻

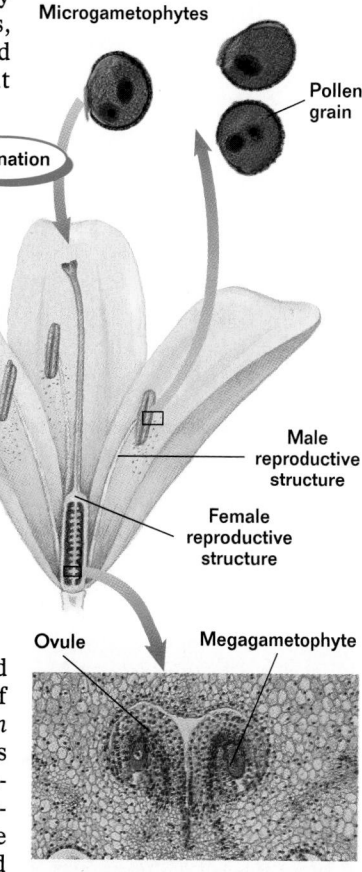

Microgametophytes

Pollen grain

Pollination

Male reproductive structure

Female reproductive structure

Ovule

Megagametophyte

Figure 23-12 Reproductive structures found in cones or flowers produce the tiny gametophytes of seed plants. The transfer of pollen from a male reproductive structure to a female reproductive structure of the same plant species is called pollination.

◻ CAPSULE SUMMARY

The seed plants include gymnosperms, which produce seeds that do not develop within fruits, and angiosperms, which produce seeds that develop within fruits.

CONTENT LINK

Plant products that are important to humans are discussed in **Chapter 26**.

Figure 23-13 This branch of an Austrian pine *(Pinus nigra)* has an immature seed cone and many pollen cones.

□ CAPSULE SUMMARY

Conifers produce massive amounts of pollen grains, which are carried to seed cones by wind.

Most Living Gymnosperms Are Conifers

Members of the most familiar phylum of gymnosperms are trees that produce seeds in cones and thus are called **conifers** *(KAHN uh fuhrz)*. Conifers (phylum Coniferophyta) include cedar, cypress, fir, hemlock, pine, redwood, spruce, and yew. The tallest living vascular plants, the giant redwoods of coastal California and Oregon, are conifers. One of the biggest redwoods, a giant sequoia *(Sequoia gigantea)* named after General Sherman of the Civil War, stands more than 80 m (262 ft.) tall and measures 20 m (66 ft.) around its base. Some individuals of another, much smaller species of conifer, the bristlecone pine *(Pinus longaeva)* that lives in the Rocky Mountains, are more than 5,000 years old—the oldest trees in the world. Most conifers have needle-like leaves that are an adaptation for limiting water loss. Conifers are often found growing in seasonally dry regions of the world, including the vast taiga forests of the northern latitudes. Many species of conifers are very important sources of timber and pulp. To learn more about conifers, look at *Highlights: Gymnosperms* on page 543.

Life Cycle of a Conifer

Conifers form two kinds of cones, which can be seen in Figure 23-13. Seed cones produce ovules on the surface of their scales. At the time of pollination, the scales of a seed cone are open, exposing the ovules. Pollen cones produce pollen grains within sacs that develop on the surface of their scales. The pollen grains of conifers are small and light, and they are carried by wind to seed cones. In pines and some other conifers, each pollen grain has a pair of air sacs that help to carry it in the wind. Because it is very unlikely that any particular pollen grain will be carried to a seed cone of the same species (the wind can take it anywhere), a great many pollen grains are needed to ensure that at least a few succeed in pollinating the species' seed cones. For this reason, pollen cones produce huge quantities of pollen grains. When pollen cones shed, the pollen grains often form a yellow layer on the surfaces of ponds, lakes, pavement, and car windshields.

When a pollen grain lands near the ovule on a scale of a female cone, a slender pollen tube grows out of the pollen grain and into the ovule. Thus, the pollen tube delivers a sperm to the egg inside the ovule. Fertilization occurs when the sperm fuses with the egg, forming a zygote that is the beginning of a new sporophyte generation. Instead of growing directly into an adult sporophyte (a tree)—just as you grow directly into an adult from a zygote—the zygote first develops into a small embryo and then becomes dormant. While its further growth is postponed, the zygote and the sporophyte tissues that surround and protect it form a seed. Trace the stages in a conifer's life cycle in Figure 23-14. □

Diploid (2n)   **Haploid (n)**

Immature seed cone — Scale — (Meiosis) → Megaspore — Eggs (within mega-gametophyte)

Ovule

Adult sporophyte

(Meiosis) — Microspore — Sperm — **Gametophytes**

Pollen cone

Pollen (microgametophytes) — (Pollination)

(Fertilization)

Young sporophyte

Pollen tube

Mature seed cone

Pine seed (with wing)

Scale — Zygote

Figure 23-14 The life cycle of a conifer is characterized by a very large sporophyte (which produces cones) that alternates with tiny gametophytes (which form on the scales of the cones). Sperm are transported within pollen grains that are carried by wind to a female cone, where they then travel through a pollen tube to reach and fertilize an egg.

Other Gymnosperms

The other three gymnosperm phyla are much less common than the conifers. Living representatives of these groups can be seen in Figure 23-15. Cycads (phylum Cycadophyta), the dominant land plant during the Jurassic period, have short stems and palmlike leaves and are still widespread throughout the tropics. The only living species of ginkgo (phylum Ginkgophyta), the maidenhair tree *(Ginkgo biloba)*, has fan-shaped leaves that are shed in autumn. The gnetophytes (phylum Gnetophyta) are all very unusual. *Welwitschia mirabilis*, perhaps the most bizarre of all plants, is a gnetophyte that grows in the harsh Namib Desert of southwestern Africa.

Figure 23-15 Cycads, *below left,* and the ginkgo, *below center,* are the only seed plants with motile sperm. One species of cycad is native to Florida. Because ginkgoes are resistant to air pollution, they are commonly planted along city streets. The two leathery beltlike leaves of *Welwitschia mirabilis, below right,* grow from a meristem at the base of the plant and split as they grow out over the desert sand.

SECTION 23-2

 VISUAL STRATEGY

Figure 23-14

Walk students through the gymnosperm life cycle, starting with the mature sporophyte (pine tree). Point out that when meiosis occurs, the chromosome number changes from diploid to haploid. Also point out that the microspores and megaspores are produced within the male and female cones, respectively. These spores grow into microgametophytes (pollen grains) and megagametophytes (containing eggs), which are haploid. Finally, point out that when fertilization occurs, the chromosome number changes to diploid and a new sporophyte generation begins. Point out that a sporophyte embryo remains dormant for a time within a seed.

Application

Technology Numerous products are made from gymnosperms. In school, some of these products include paper, pencils, lumber, and linoleum.

Did You Know?

The ginkgo, or maidenhair tree, is considered a "living fossil" because it was once thought to be extinct. It is the only broadleaf gymnosperm still surviving on Earth today.

Did You Know?

In 1994, another "living fossil" of a gymnosperm thought to be extinct since the Jurassic Period 150 million years ago was discovered in a rain forest in the Wollemi National Park near Sydney, Australia. A small population of Wollemi Pines, a prehistoric type of conifer belonging to a family that was thought to have become extinct along with the dinosaurs, was found by a park ranger. The once widespread distribution of this ancient family included the Northern Hemisphere.

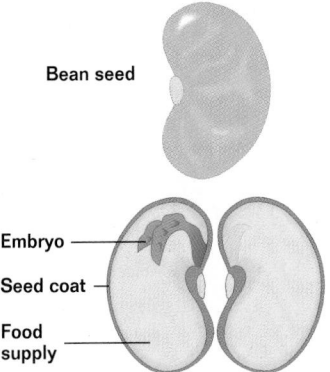

Bean seed

Embryo

Seed coat

Food supply

Figure 23-16 A bean, *top*, is a relatively large seed. It consists of an embryo and a stored food supply, both of which are surrounded by a seed coat, *bottom*.

CAPSULE SUMMARY

A seed is a sporophyte plant embryo surrounded by a protective coat. Seeds disperse a plant's offspring, nourish the embryo, and help control the time at which a new plant begins to grow.

What Is a Seed?

By providing the offspring of plants with several survival advantages, seeds have had an enormous influence on the evolution of plants on land. As you can see in Figure 23-16, a **seed** is a sporophyte plant embryo surrounded by a protective coat. The hard cover of a seed is called the **seed coat.** Formed from the sporophyte tissue of the parent plant, the seed coat protects the embryo and other tissues in the seed from drying out. In addition to their role in protecting a plant embryo, seeds have enabled plants to become better adapted to living on land in at least three other respects.

1. **Dispersal**   Seeds enable the offspring of plants, which are anchored in one place by their roots, to be dispersed to new locations. Many seeds have appendages, such as wings, that help wind, water, or animals carry them away from their parent plant. The dispersal of a plant's offspring prevents the parent and offspring from competing with each other for water, nutrients, light, and living space. Seed dispersal also facilitates the migration of a plant species to new habitats.

2. **Nourishment**   Most kinds of seeds have abundant food stored in them. Playing a role similar to that of the yolk in an egg, this food supply is a ready source of energy for a plant embryo as it starts its growth. Thus, seeds offer a young plant nourishment during the critical period just after germination when the seedling must establish itself.

3. **Dormancy**   Once a seed has fallen to the ground, it may lie dormant for many years. When conditions are favorable, particularly when moisture is present, the seed will begin to grow into a young plant. By remaining dormant until conditions improve, seeds enable plants to postpone development during unfavorable conditions such as a drought or a cold period. Thus, seeds aid in synchronizing the growth of a new plant with the season of the year. ❑

Section Review

1. *Briefly describe the fern life cycle.*
2. *What is a fundamental difference between gymnosperms and angiosperms?*
3. *How are the life cycles of ferns and gymnosperms similar? How are they different?*
4. *What is a seed? How are seeds an important adaptation to life on land?*

Critical Thinking

5. *Why is wind pollination more advantageous than the transfer of sperm to eggs through a film of water?*

23-3 The Evolution of Flowers

*T*he last difficult problem posed to plants by terrestrial living was resolved in angiosperms. Plants had previously been limited by a conflict between the need to obtain water and nutrients (solved by roots) and the need to find mates (solved by male gametes that can be carried to other plants). This problem was never completely solved in gymnosperms, whose lightweight pollen grains are carried passively by wind. Large numbers of pollen grains are therefore needed to improve the chances of a lucky encounter with a female cone—a very inefficient system. The pollen of many angiosperms, however, is delivered directly from one individual of a species to another. How? As Figure 23-17 indicates, animals carry the pollen for them! The innovation that made this great advantage possible is the flower.

Section Objectives

- Identify the parts of a flower.
- Explain why there are different kinds of flowers.
- Summarize the angiosperm life cycle.
- Describe the process of double fertilization, and state its importance.
- Relate the characteristics of fruits to their role in seed dispersal.

Figure 23-17 Pollen covers the faces of four lesser long-nosed bats (*Leptonycteris curasoae*), *above*, after a nighttime visit to organ pipe cactus flowers in the southwestern United States, *above left*. Worldwide, bats are important pollinators of cactuses and tropical fruit trees.

Angiosperms Achieved Evolutionary Success on Land

The most successful of all plants are angiosperms, seed plants that produce flowers. The remarkable evolutionary success of the angiosperms is the culmination of the plant kingdom's adaptation to life on land. Ninety percent of all living plants—more than 250,000 species of trees, shrubs, herbs, fruits, vegetables, and grains—are angiosperms. In short, nearly all of the plants that you see every day are angiosperms. Virtually all of your food comes directly or indirectly from angiosperms. In fact, more than half of the calories that humans consume come from just three species of angiosperms: rice, corn, and wheat.

SECTION 23-3

Vocabulary Preview

- calyx
- sepal
- corolla
- androecium
- stamen
- filament
- anther
- gynoecium
- pistil
- ovary
- style
- stigma
- complete flower
- perfect flower
- imperfect flower
- incomplete flower
- endosperm
- cotyledon
- dicot
- monocot
- double fertilization
- fruit

Opening Questions

Ask students which phylum most of the plants they are familiar with belong to. Students should recognize that most of the plants they encounter are flowering plants and therefore belong to the phylum Anthophyta. Ask students why they think flowering plants are so numerous. Students should recall that flowers increase the efficiency of plant reproduction.

Demonstration

Comparing Cones and Flowers
Bring several examples of cones (pine, spruce, fir, yew) and flowers (iris, daylily, gladiolus) to class, and have students examine each of these types of reproductive structures. Ask: How are these structures similar? (*Both produce seeds.*) How are they different? (*Flowers have petals, stamens, and pistils; cones do not.*)

Overcoming Misconceptions

Bats Are Beneficial

There are several misconceptions about bats. Some people think that all bats drink blood, which is far from the truth. Only a small number of bats (vampire bats) feed on the blood of animals such as cattle and horses.

Many bats eat plant products such as nectar and fruits. Important crops, such as bananas, are pollinated by bats. Bats are also important dispersal agents for many tropical fruit plants.

Bioprospecting: The Politics of Plants

Walter Reid, an ecologist and vice-president of the World Resources Institute in Washington, D.C., performed an exhaustive survey on the positive and negative aspects of *bioprospecting,* or looking for species that may yield some kind of medicinal breakthrough. Reid claims that 25 percent of the prescriptions written in the United States are filled with drugs whose active ingredients are extracted or derived from plants. This represented a $15.5 billion share of the pharmaceutical industry in 1990—the combined estimate for the world was approximately $43 billion in 1985. Obviously, countries producing these plants are going to want to get a slice of this pie, particularly since many of these are developing countries.

Interestingly, Reid finds that by regulating biological prospecting (which because of new technology can be very lucrative), industry, conservationists, and host countries will have to work together to actually *protect* diversity—after all, we don't want to prospect a useful plant out of existence. The Earth Summit in Rio de Janeiro, Brazil, addressed intellectual property rights (owner's rights) to plant species, how countries and pharmaceutical companies could work together to the mutual advantage of both, and the overall importance of global conservation. Reid estimates that at least 5 to 10 percent of tropical forest species will either die out over the next 30 years or be reduced to such small populations that extinction is a foregone conclusion. If a plant is extinct, it isn't going to be yielding any new medicines. ☑

What Is a Flower?

Flowers are the reproductive organs of angiosperms. Many flowers are sophisticated structures that are adapted to enable insect pollination. Bright colors attract the attention of insects. Nectar, which is a sugary secretion of many flowers, induces insects to enter a flower. Pollen-bearing structures coat the insects with pollen while they are visiting a flower. Then, when the insects visit another flower, they carry the pollen into that flower. The basic structure of a flower, seen in Figure 23-18, consists of the four concentric whorls (circles), of appendages described below.

1. **Calyx** The outermost whorl of a flower is called the **calyx** *(KAY liks),* a name derived from the Greek word *kalyx,* meaning "cup." The calyx consists of one or more **sepals** *(SEE puhls),* which are modified leaves, and protects a flower from physical damage while it is a bud.

2. **Corolla** The second whorl of a flower is called the **corolla** *(kuh ROHL uh),* a name derived from the Latin word *corona,* meaning "crown." The corolla consists of one or more petals, which are also modified leaves. Often, the corolla produces vividly colored pigments or fragrances that attract particular pollinators to a flower.

3. **Androecium** The third whorl, called the **androecium** *(an DREE shee uhm),* produces the microgametophytes, or pollen grains. The word *androecium* comes from the Greek words *andros,* meaning "male," and *oikos,* meaning "house." The androecium is made up of one or more **stamens** *(STAY muhnz),* which consist of slender, thread-like **filaments** that are each topped by a pollen-containing sac called an **anther.**

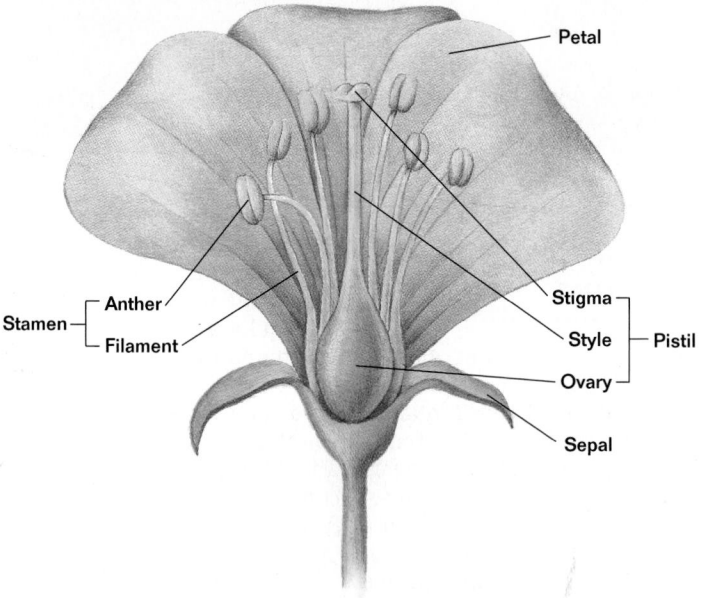

Figure 23-18 The parts of a flower are arranged in whorls—the calyx (all of the sepals), the corolla (all of the petals), the androecium (all of the stamens), and the gynoecium (all of the pistils). While some flowers have all four of the whorls shown here, many other flowers lack one or more of these whorls.

In June of 1992, 157 countries at the United Nations Conference on Environment and Development (Earth Summit) in Rio de Janeiro, Brazil, signed the Convention on Biological Diversity. This international pact was created to ensure the protection of commercially valuable species of plants (and other organisms). Initially, the United States did not sign the biodiversity pact. However, President Clinton finally signed the pact on June 4, 1993. In all, 167 nations eventually signed the biodiversity pact, which became international law on December 29, 1993.

4. Gynoecium The fourth and innermost whorl, called the **gynoecium** *(jeye NEE shee uhm),* houses the ovules, in which the megagametophytes develop. The word *gynoecium* comes from the Greek words *gyne,* meaning "female," and *oikos,* meaning "house." The gynoecium consists of one or more **pistils** that are found in the center of a flower. Ovules develop in a pistil's swollen lower portion, which is called the **ovary.** An ovary may have one or more chambers. Usually, a slender stalk, called the **style,** rises from an ovary. The style has a swollen, sticky tip called the **stigma,** on which pollen lands and adheres. When a flower is pollinated (by pollen from the same species), a pollen tube emerges from each pollen grain and grows through the style and into the ovary.

Flowers may or may not have all four of these parts. A flower that has all four whorls of appendages is called a **complete flower.** If a flower has both a gynoecium and androecium, it is a **perfect flower.** Many flowers lack either a gynoecium or an androecium and are **imperfect flowers. Incomplete flowers** are those that lack any one of the four whorls. Therefore, flowers that lack *either* stamens or pistils, such as the squash flowers seen in Figure 23-19, are *both* incomplete and imperfect. ❑

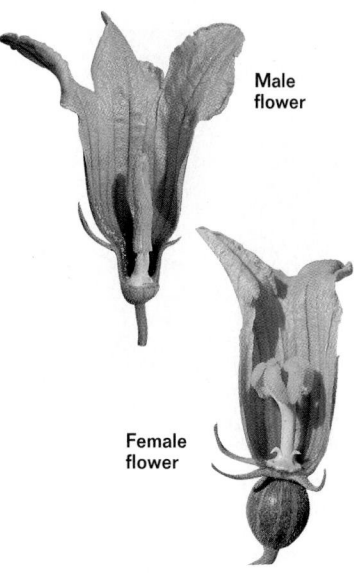

Figure 23-19 Squash and other members of the Cucurbitaceae family produce imperfect (and incomplete) flowers.

❑ **CAPSULE SUMMARY**

Flowers are the reproductive structures of angiosperms. Most flowers consist of four whorls of appendages.

Flowering Plants Coevolved With Animals

If you were to watch insects visiting flowers, you would quickly discover that the visits are not random. Instead, certain insects are attracted by particular flowers. As you learned in Chapter 17, insects and plants have coevolved so that certain insects specialize in visiting particular kinds of flowers. An insect recognizes a particular color pattern and searches for flowers with that pattern. As a result, a particular insect carries pollen from the flowers of one individual to the flowers of another individual *of the same species.* This specificity is the key to successful insect pollination, making it much more effective than wind pollination.

Of all insect pollinators, the most numerous are bees. Bees evolved approximately 100 million years ago, about the time that flowering plants began to diversify greatly. Today, there are over 20,000 species of bees. Bees locate sources of nectar by odor at first, then by homing in on a flower's color and shape. Bee-pollinated flowers are usually yellow or blue and, as you can see in Figure 23-20, frequently have lines of dots, or "guides," to indicate the location of the nectar (usually in a flower's throat). Some of these markings are visible only to insects. While inside a flower, bees become coated with pollen. This coating is far from accidental. Most of the bees visiting flowers actively seek the pollen, which is a rich source of protein that they feed to their larvae.

Figure 23-20 A yellow color, nectar guides, and a landing platform attract bees to the flowers of the unicorn plant, *Proboscidea altheifolia.*

Teaching Tip
Recognizing Flower Parts
Have students bring a flower to class. After going over the general parts of a flower using a model or a diagram, let students look for these parts on their flowers. Many of their flowers will have modifications. Have them determine if the flowers are complete or incomplete and perfect or imperfect. After they have observed the flowers, let them trade with a partner to identify the parts of a different flower. Have students sketch and label the parts of the flower that appeals to them most.

Application

Agriculture Production of hybrids in some commercially important plants often requires a boost. An egg and a sperm can be joined in about an hour using a jolt of electricity. Sperm are easily removed from a flower, but eggs are more difficult. When both types of gametes are put into a salt solution, electricity forces the sperm into the egg. This technique is especially useful in forming hybrids between plants that are difficult to cross-pollinate.

Demonstration
Classifying Flowering Plants
Plants are classified into families by their flowers, and members of the same family have similar flowers. Show students photographs of the flowers of tomatoes, potatoes, eggplant, peppers, and a weed called silverleaf nightshade, which all belong to the Solanaceae family. Ask students what similarities they see in the flowers. *(The flowers of each plant have the same structure; all but one are the same color—yellow.)*

Teaching Tips
Pollen Coats

Tell students that allergies to pollen are actually reactions to the proteins in the coats of the pollen grains. Different pollens cause allergies in different people because each type of pollen has a different protein coat. Let students make wet-mount slides of pollen samples to observe their intricately detailed pollen coats. Have them draw what they observe.

Grass Has Blooms

Show students the flowers of various grasses. Tell them that these flowers are considered to be highly evolved in the plant world because they lack large, showy parts, and they are adapted for wind pollination.

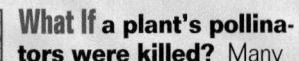

What If a plant's pollinators were killed? Many plants depend on pollination by particular pollinators in order to produce seeds. If those pollinators were killed, the plants could not produce seeds. Ask students to think of a situation where a plant's pollinators could be killed. *(Spraying with pesticides in agricultural fields and home gardens destroys natural pollinators.)* Tell students that indiscriminate pesticide use is a particular problem for farmers who grow crops that cannot set fruit without pollination, such as cucumbers and squash.

Application

Agriculture Homeowners often choose hollies as landscape shrubs because they have beautiful red berries in the fall and winter months. However, not all plants produce berries. Hollies are an example of a plant that is either male or female. If a homeowner plants all female shrubs or all male shrubs, their plants will not yield fruit. Only females can produce berries; but for berries to form, pollen is required to fertilize the eggs. To produce berries, an occasional male must be planted with female plants.

Figure 23-21 This ruby-throated hummingbird transfers pollen while drinking nectar from the tubular red flowers of a salvia. While most insects cannot "see" red, it is a very distinct color to birds, as it is to humans.

□ **CAPSULE SUMMARY**

In many angiosperms, flower structure is an adaptation that enables pollination by animals, including insects, birds, and bats. Some flowers are wind pollinated.

Other insects also pollinate flowers. Butterflies tend to visit flowers that have "landing platforms" on which they can perch. These flowers are typically tube-shaped and filled with nectar that butterflies reach by uncoiling a long, hose-like mouthpart. Flies pollinate flowers that smell like rotting meat. Moths, which visit flowers at night, pollinate white, heavily scented flowers that are easy to locate in dim light.

Many angiosperms are pollinated by animals other than insects. Red flowers, for instance, are typically visited by hummingbirds, as you can see in Figure 23-21. Birds, which have keen vision, have a poor sense of smell. Knowing this, it is not surprising that red flowers usually lack a strong odor. Certain angiosperms have large, heavily scented, and pale-colored flowers that open at night. These flowers are pollinated by another nighttime visitor—bats.

Some angiosperms have reverted to wind pollination, a characteristic of their ancestors. Notable among the wind-pollinated angiosperms are oaks, birches, and grasses. To learn more about pollination in flowering plants, look at *Highlights: Angiosperms* on page 544. □

Double Fertilization Provided Large Food Reserves in Seeds

Unlike the seeds of gymnosperms, the seeds of angiosperms develop a highly nutritious tissue, called **endosperm,** that originates at the same time that an egg is fertilized. In some angiosperms, such as corn and wheat, the endosperm is still

Table 23-2 Characteristics of Dicots and Monocots

Class	Seed	Flower	Leaf
Dicots	Two cotyledons Bean	Flower parts in multiples of four or five Evening primrose	Veins usually netlike Maple
Monocots	One cotyledon Corn	Flower parts in multiples of three Day lily	Veins usually parallel Corn

 Did You Know?

A sunflower seed is not a seed at all. It is a dry fruit with a seed inside. The edible part is the actual seed, which consists mainly of cotyledons.

 Did You Know?

Vitamin C is produced commercially from rose hips, which are the fruits of roses.

present in mature seeds. In other angiosperms, such as beans and peas, the endosperm is competely transferred into the embryo by the time a seed is mature. The food reserves are then stored in the embryo's fleshy, leaflike **cotyledons** (kah tuh LEE duhnz), or seed leaves.

The angiosperms are divided into two classes based on the number of cotyledons in their seeds. Angiosperms that produce seeds with two cotyledons are called dicotyledons, or **dicots** (class Dicotyledones). Angiosperms that produce seeds with a single cotyledon are called monocotyledons, or **monocots** (class Monocotyledones). Monocots, which evolved somewhat later than dicots, differ from dicots in other important ways, as seen in Table 23-2.

You can see how the angiosperm embryo and endosperm originate at fertilization by studying the life cycle of a typical angiosperm, seen in Figure 23-22. The microgametophytes of seed plants contain two sperm cells. In most gymnosperms, one of these sperm dies. In angiosperms, however, *both* sperm fuse with certain cells of the megagametophyte. One sperm fuses with the egg, as in all sexually reproducing organisms, forming the zygote. The other sperm fuses with the haploid nuclei of two other cells produced by meiosis, forming a triploid ($3n$) cell that gives rise to endosperm. The term **double fertilization** is used to describe the process by which two sperm fuse with cells of the megagametophyte to produce both a zygote and endosperm. ❏

❏ **CAPSULE SUMMARY**

The endosperm produced during double fertilization stores nourishment within many angiosperm seeds. Angiosperms are classified as either dicots or monocots.

Figure 23-22

Diploid (2n) — **Haploid (n)**

Anther
Meiosis
Pollen grains (microgametophytes)
Adult sporophyte
Stamen
Microspores
Pistil
Meiosis
Gametophytes
Pollination
Ovule
Pollen tube
Flower
Ovule
Megagametophyte (a mature megaspore)
Ovary
Seed coat
Endosperm (3n)
Sperm
Sporophyte embryo
3n nucleus
Pollen tube
Seed
Egg
Zygote
Double fertilization

Figure 23-22 The life cycle of an angiosperm is characterized by a large sporophyte (which produces flowers) that alternates with tiny gametophytes (which form within the male and female structures of flowers). Sperm are transported within a pollen grain to a pistil, where they then pass through a pollen tube to reach and fertilize an egg.

SECTION 23-3

Teaching Tip
Classifying Plants as Monocots or Dicots
Provide examples of plants and pictures of plants. Let students determine whether the examples are monocots or dicots. Include examples of plants that stray from the rules, such as a pothos ivy. Students will think the net-veined leaves make this plant a dicot. However, it is a monocot. Leaf sheaths on the stem are a strong indication of a monocot. Use this opportunity to show students that exceptions to the rules exist in the leaves of monocots and dicots. This is one reason why botanists use flowers to classify plants.

◆ VISUAL STRATEGY

Figure 23-22
Walk students through the angiosperm life cycle, starting with the mature sporophyte (plant with flowers). Point out that when meiosis occurs, the chromosome number changes from diploid to haploid. Also point out that the microspores and megaspores are produced within the male and female parts of the flower, respectively. These spores grow into microgametophytes (pollen grains) and megagametophytes (containing eggs), which are haploid. Finally, point out that when fertilization occurs, the chromosome number changes to diploid and a new sporophyte generation begins. Point out that a sporophyte embryo remains dormant for a time within a seed. Use the *Graphic Organizer* at the bottom of page 539 to summarize the process of double fertilization.

Graphic Organizer

Use this graphic organizer with *Visual Strategy:* **Figure 23-22** on page 539.

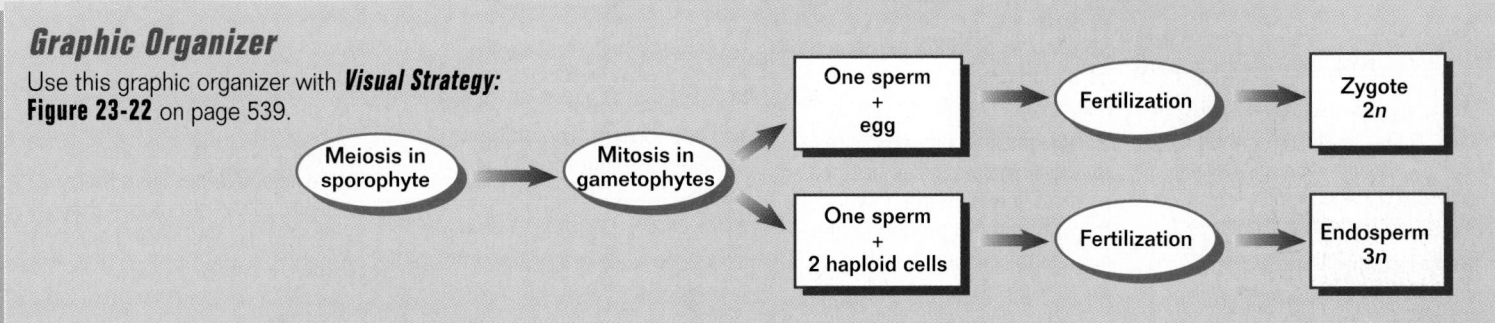

Meiosis in sporophyte → Mitosis in gametophytes →

One sperm + egg → Fertilization → Zygote 2n

One sperm + 2 haploid cells → Fertilization → Endosperm 3n

Fruits Enabled Efficient Seed Dispersal

Among angiosperms, fruits evolved and enabled more efficient dispersal of seeds. A **fruit** consists of a mature ovary that contains one or more seeds and often includes other flower parts. Animals, which aid in pollination because they are attracted to and obtain nourishment from flowers, also aid in seed dispersal. Many mammals and birds are attracted to and eat fruits that are fleshy and tasty. You might have noticed that as fleshy fruits ripen, they often change from green and odorless to brightly colored and sweet smelling. The mature seeds within such ripe fruits are often resistant to chewing and digestion. Sometimes, virtually undamaged seeds pass out of an animal with its feces and are ready to germinate at a new location far from the parent plant. Most fruits, however, are not eaten by animals. Some are specialized for sticking to an animal's fur. Others are adapted for floating on wind currents or water. Figure 23-23 shows several types of angiosperm fruits that are specialized for different methods of seed dispersal. ❑

❑ **CAPSULE SUMMARY**

The fruits of angiosperms are specialized for seed dispersal by agents such as animals, wind, and water.

Figure 23-23 These fruits disperse seeds by different means. The fleshy fruits of raspberries, *left*, are eaten by birds and other animals. Cockleburs, *center left*, stick to animal fur. Wings enable silver maple fruits, *center right*, to fly. Coconuts, *right*, float on water.

Section Review

1. *What is the function of each whorl in a typical flower?*
2. *What are four flower adaptations that attract insects and other animal pollinators?*
3. *Briefly summarize the angiosperm life cycle.*
4. *What happens during double fertilization?*
5. *What function do the fruits of angiosperms serve?*

Critical Thinking

6. *Why is the angiosperm method of seed dispersal more effective than the gymnosperm method?*

Highlights
Plant Evolution

Bryophytes

Becoming a Plant

The **nonvascular plants,** commonly referred to as **bryophytes,** represent the earliest stages in plant evolution. Arising from multicellular aquatic green algae, the first plants were small and were limited to habitats with abundant water, such as the edges of oceans or ponds. Like all bryophytes, they lacked roots, stems, and leaves. Included among the bryophytes are the two simplest groups of plants—liverworts (phylum Hepatophyta) and hornworts (phylum Anthocerophyta). Liverworts, such as this *Marchantia,* lack vessels that conduct water and other materials.

Hornworts also lack conducting tissue.

Marchantia

Developing Vascular Tissue

Mosses (phylum Bryophyta) are the most familiar bryophytes. Like this *Sphagnum* moss, most mosses have simple conducting tissue and thus may represent the next phase in plant evolution—the beginning of **vascular tissue.** As in all bryophytes, the moss body is not differentiated into roots, stems, and leaves, and the most conspicuous phase of the moss life cycle is the haploid gametophyte generation. Moss sporophytes consist of a bare stalk topped by a spore capsule and grow from the tips of the "leafy," green gametophytes.

Sphagnum

Small Size

The sides of a moss's water-conducting cells are not thickened for added strength, as they are in true vascular plants. Individual mosses never become very large because they are able to conduct water only short distances.

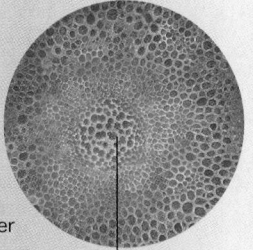

Conducting strand

An Economically Important Moss

Sphagnum, or peat moss, is the most economically important moss. More than 290 species of *Sphagnum* mosses form extensive bogs in areas with cool climates and sufficient rainfall. The dense and deep mats formed by peat mosses are often cut into blocks, dried, and burned as fuel. Because of its moisture-holding capacity and its acidity, peat is also used extensively in gardening.

Tied to Water

Mosses almost always grow in mats in very moist places. When the mats are covered by dew or rain, motile moss sperm can easily swim to neighboring female gametophyte plants.

Instructional Strategies

- Show students examples of the gametophyte stages of mosses and liverworts. Have them make a concept map using pictures of the type of environment in which these plants would live based on their appearances.
- Divide students into teams. Give each team a *Sphagnum* peat pellet, and have them determine the mass and volume of the pellet. Students should then place their pellet into a petri dish and add water to the dish until the pellet can absorb no more water. Once again, students should find the mass and volume of their peat pellet. Have them calculate the percentage of water the peat pellet holds.
- Challenge students to predict how the world would be different without bryophytes. They can determine the role of bryophytes by reading the text or other sources.
- To stimulate interest in bryophytes, have students establish and maintain a terrarium containing several different types of bryophytes.

Discussion

Guide the discussion by posing the following questions.

1. Why do nonvascular plants thrive in moist, humid environments? *(Nonvascular plants lack conducting tissue and must obtain water from their surroundings by diffusion. They also require water for fertilization.)*

2. What characteristic of bryophytes limits their size? *(Bryophytes lack a vascular system that would help transport water and nutrients long distances. Therefore, they must stay small so that water and nutrients can reach all of their cells by diffusion.)*

3. What are the benefits of adding peat moss to a potting soil mix? *(Peat moss holds water, aerates the soil, supplies organic nutrients, and lowers soil pH.)*

Instructional Strategies

- Ask students to write three "fat" questions and three "skinny" questions about the reading material. A "fat" question requires a yes or no answer. A "skinny" question requires an explanation. Once students have written their questions, have them exchange papers with other students and answer the questions they receive.

- Have students work in small groups to make murals describing one of the four phyla of seedless vascular plants. Tell students to use the text and other sources to get ideas for the topics to include in their murals.

- Challenge students to determine a suitable environment for growing ferns in the classroom. Allow them to create different environments for growing ferns, and have them compare the growth observed in each of the environments.

Discussion

Guide the discussion by posing the following questions.

1. How are ferns more complex than bryophytes? *(Ferns have a larger sporophyte with a vascular system and roots, stems, and leaves. Their fronds are also protected by a cuticle.)*

2. Why would a fern have a difficult time surviving in a desert? *(Although ferns have a vascular system, a cuticle, and true roots that absorb water, they still require a film of water so that sperm can swim to eggs for fertilization.)*

3. How could a fern be improved to enable it to survive in a drier environment? *(A mechanism to facilitate fertilization on dry land would make a fern more successful in a dry environment.)*

Highlights
Plant Evolution

Seedless Vascular Plants

Developing Shoots and Roots

Seedless vascular plants, such as this tree fern, *Cyathea,* were the first plants to develop a **vascular system.** A vascular system enabled the development of a body differentiated into **shoots** and **roots.** Because they had conducting cells with reinforced walls, vascular plants were able to grow much larger than nonvascular plants. The development of a waxy covering called a **cuticle** and spores that were resistant to drying aided in the survival of the seedless vascular plants on land. A life cycle dominated by the diploid sporophyte generation evolved with the seedless vascular plants. Today, ferns (phylum Pterophyta) are the only widely successful seedless vascular plants.

Cyathea sp.

Reduced Gametophyte

Less than 1 cm (0.5 in.) across, this fern gametophyte is much smaller than the sporophyte but is independent of it.

Roots, Stems, and Leaves

Most fern sporophytes have horizontal stems, called **rhizomes,** that creep along below the ground and are anchored by **roots** that absorb water and minerals. The leaves, called **fronds,** stand vertically. In most ferns, the young leaves are called **fiddleheads.**

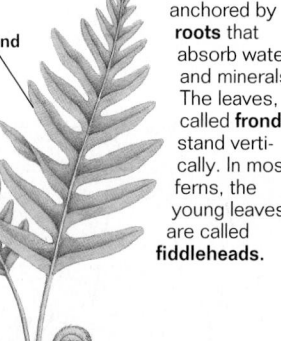

Frond

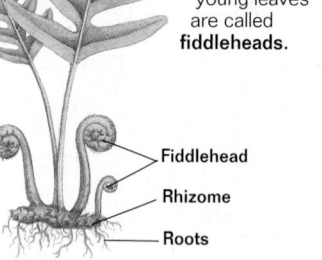

Fiddlehead

Rhizome

Roots

Drought-Resistant Spores . . .

Fern spores, which are highly resistant to drying, are produced in sporangia that grow in clumps on the lower side of the fronds.

. . . But Still Tied to Water

Ferns are most abundant in the tropics and prefer moist habitats. A film of water is required for the motile sperm to swim to eggs.

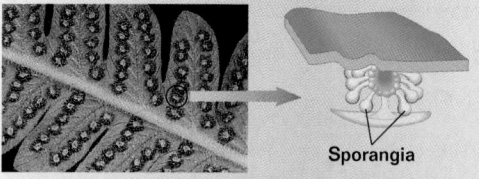

Sporangia

Other Seedless Vascular Plants

Three other phyla of seedless vascular plants that are well represented in the fossil record have only a few surviving species.

Phylum Sphenophyta (the horsetails)

Phylum Psilotophyta (the whisk ferns)

Phylum Lycophyta (the club mosses)

Gymnosperms

Reproducing With Seeds

Gymnosperms, represented by this lodgepole pine, *Pinus contorta*, were the first **seed plants**. A **seed** (which is a mature ovule) consists of a diploid sporophyte embryo surrounded by a protective coat. By postponing development until conditions are suitable for growth, seeds permit a plant embryo to survive long periods of unfavorable conditions. Seeds also serve to disperse an embryo far from its parent plant. With the evolution of gymnosperms, the sporophyte generation of the plant life cycle became completely dominant.

Pinus contorta

Greatly Reduced Gametophytes

The greatly reduced gametophytes of seed plants consist of the **microgametophytes**, encased in **pollen grains**, and the **megagametophytes**, housed in **ovules**. In most gymnosperms, the gametophytes are produced in cones, shown here on a lodgepole pine.

Seed cone   Pollen cones

Conserving Water

Like the lodgepole pine, most conifers have water-conserving needle-shaped leaves, an adaptation to dry environments. Some conifers, such as this juniper, have leaves that are reduced to tiny scales.

Scale-like leaves   Seed cone

Wind Pollination

The sperm of conifers do not swim but are carried within pollen grains that drift with the wind. Conifers and most other gymnosperms are wind pollinated. For this reason, they tend to occur in dense stands of one species. Extensive conifer forests are found in northern latitudes.

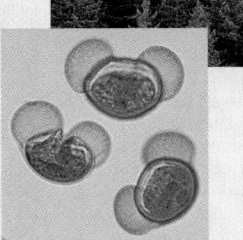

Pine pollen is equipped with two bladderlike wings that help to carry it on the wind.

Soft Wood

Secondary growth produces the woody stems of conifers. Many conifers grow very fast, so their wood is not as dense as other trees and is often referred to as "soft wood."

Secondary growth

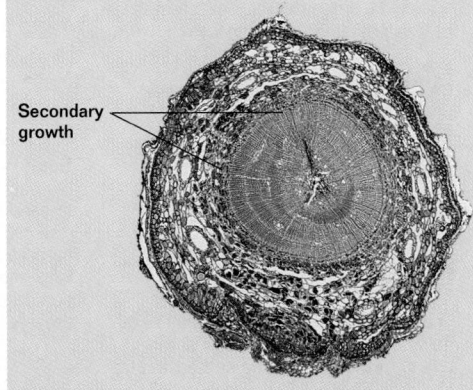

Highlights
Plant Evolution

Instructional Strategies

- Display examples or photographs of seeds, cones, wood, and other parts of a variety of gymnosperms around the classroom.
- Have students research the discoveries of the ginkgo, dawn redwood, and Wollemi pine. Then conduct a class discussion about these "living fossils" of gymnosperms that have survived since the age of the dinosaurs.

Discussion

Guide the discussion by posing the following questions.

1. What adaptive advantage do the gymnosperms have over the bryophytes and the seedless vascular plants? (*Gymnosperms have developed structures that make them more tolerant of harsh conditions, including needle-shaped leaves with thick cuticles, seeds that protect the embryo from drying out and provide a means of dispersal, and pollen grains that can carry sperm long distances over dry land on the wind.*)

2. In what types of environments would you expect to find gymnosperms? (*Answers will vary but should suggest that gymnosperms might be found in many different types of environments.*)

3. What factors could have enabled gymnosperms to survive the environmental changes that the dinosaurs could not survive? (*Answers will vary but should suggest that plant seeds can remain dormant for many years until conditions are favorable for growth and survival. Also, plants can self-fertilize and are more likely to survive in small, isolated populations than are animals.*)

Highlights
Plant Evolution

Angiosperms

Instructional Strategies

- Provide students with examples of different types of flowers. (If flowers are needed, check with a local florist for older flowers that they may want to discard.) Give students sheets of construction paper, and have them carefully separate a flower into its components. Have students group the parts of their flower, attach them to the paper with tape, and label the parts. Then have them determine if the flower is a perfect flower, a male flower, or a female flower. Also have them suggest how their flower is pollinated.

- Cut pictures of flowers from magazines and seed catalogs, and glue them to cards (one picture per card). Place the cards around the classroom, and have students rotate from station to station and predict how each flower is pollinated.

- To add interest, display examples or photographs of some of the more unusual mechanisms flowers have for attracting pollinators. For example, many flowers have patterns that are visible to insects but can be detected by the human eye only in ultraviolet light. Other flowers mimic shapes that attract insects or have certain smells that attract specific insects.

Discussion

Guide the discussion by posing the following questions.

1. Why are angiosperms the most successful land plants? *(Answers will vary but should mention flowers that attract pollinators, seeds that protect embryos, pollen that transports sperm without water, and fruits that protect and disperse seeds.)*

2. Which type of pollination mechanism in angiosperms is the most advantageous? *(Answers will vary.)*

Reproducing With Flowers

Angiosperms, such as the wild rose (*Rosa* sp.), represent the culmination of the adaptation of plants to land. The great evolutionary advance of angiosperms is the **flower**, a reproductive structure that promotes pollination. The ovules of angiosperms are completely enclosed by the **ovary** of a flower. Mature ovaries form **fruits**, which greatly enhance seed dispersal among angiosperms. The evolution of the flower enabled the development of mechanisms for promoting **cross-pollination**, which tends to increase the genetic variation of a population and makes it better able to adapt via natural selection. For example, some flowers attract insects and other animals that carry pollen and thus increase the likelihood of cross-pollination.

Rosa sp.

Early angiosperms may have been pollinated by beetles, which were already abundant when angiosperms first evolved.

Self-Pollination

Many angiosperms with **perfect flowers** (both male and female parts), such as this garden pea, are adapted for **self-pollination**, which ensures pollination and fertilization (fusing of egg and sperm). Self-pollination occurs before a flower opens.

Preventing Self-Fertilization

In some angiosperms, such as oaks, **self-fertilization** (the fertilization of an egg by a sperm from the same individual) is unlikely because the male and female gametophytes form in separate flowers.

Female flower

Male flower

The male and female gametophytes of many angiosperms mature at different times, further preventing self-fertilization.

Ensuring Cross-Pollination

Some angiosperms, such as poplars, have two kinds of individuals: males (with flowers that produce only pollen) and females (with flowers that produce only ovules). The offspring of such plants result from **cross-fertilization** (the process in which a sperm from one individual fertilizes an egg of a different individual).

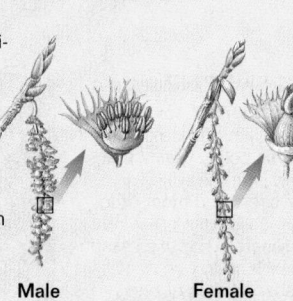

Male flower

Female flower

Wind Pollination

Some angiosperms, such as oaks, birches, and grasses, have reverted to wind pollination. Their small, greenish flowers do not attract pollinators. These plants grow in very large masses, making wind pollination particularly efficient and making cross-pollination more likely.

3. What type of flower is the most highly evolved of all flowers? *(Answers will vary but should suggest that energy conservation and efficiency are desirable traits. An example of a highly evolved flower is the grass flower, which is a small, efficient seed producer that is wind pollinated. Primitive flowers, such as the magnolia flower, tend to be very large and showy. Growing such flowers requires the plant to expend a great deal of energy in order to achieve pollination.)*

Vocabulary

alternation of generations (524)
androecium (536)
angiosperm (531)
anther (536)
antheridium (525)
archegonium (525)
calyx (536)
complete flower (537)
conifer (532)
corolla (536)
cotyledon (539)
cuticle (520)
dicot (539)
double fertilization (539)
endosperm (538)
filament (536)

frond (530)
fruit (540)
guard cell (521)
gymnosperm (531)
gynoecium (537)
imperfect flower (537)
incomplete flower (537)
megagametophyte (531)
megaspore (531)
meristem (528)
microgametophyte (531)
microspore (531)
monocot (539)
ovary (537)
ovule (531)
perfect flower (537)
phloem (527)

pistil (537)
pollen grain (531)
pollen tube (531)
pollination (531)
root (528)
seed (534)
seed coat (534)
sepal (536)
shoot (528)
sporangium (526)
stamen (536)
stigma (537)
stomata (521)
style (537)
vascular system (522)
vascular tissue (526)
xylem (527)

CHAPTER REVIEW ANSWERS

Each item in the Chapter Review is correlated by text section in the assignment guide that follows.

ASSIGNMENT GUIDE			
Section	Review	Themes Review	Critical Thinking
23-1	1–3, 11, 12, 16	28	31
23-2	4–7, 13, 17	29	32, 33
23-3	8–10, 14, 15, 18, 19	30	34

Concept Mapping

Construct a concept map that shows how plants are classified. Use as many terms as needed from the vocabulary list. Try to include the following terms in your map: vascular plants, nonvascular plants, ferns, angiosperms, gymnosperms, mosses, cones, vascular tissue, seeds, and flowers. Include additional terms in your map as needed.

Review

Multiple Choice

1. What evolutionary development enabled plants to conserve water and, thus, helped them to adapt to life on land?
   a. mycorrhizae   c. pollen
   b. cuticle   d. seed

2. Because plants have an alternation of generations, their life cycles include a diploid individual that is called a(n)
   a. sporophyte.   c. zygophyte.
   b. gametophyte.   d. epiphyte.

3. Which of the following is *not* a characteristic of vascular plants?
   a. xylem and phloem
   b. a dominant haploid generation
   c. stems and leaves
   d. specialized underground structures

4. The first true seed plants were
   a. angiosperms.   c. ferns.
   b. gymnosperms.   d. mosses.

5. Gymnosperms and angiosperms are different in that gymnosperms
   a. are pollinated by wind.
   b. lack megagametophytes.
   c. have a diploid sporophyte generation.
   d. do not bear fruit.

6. The life cycle of a conifer differs from the life cycle of a fern in that
   a. alternation of generations does not occur in ferns.
   b. fertilization in conifers produces a haploid zygote.
   c. water is not required for fertilization in conifers.
   d. the sporophyte of a fern is larger than the gametophyte.

Concept Mapping

Map answer is shown on page 517B.

Review

Multiple Choice

Code in parentheses indicates section number and objective number.

1. b (23-1.1)
2. a (23-1.2)
3. b (23-1.4)
4. b (23-2)
5. d (23-2.1, 23-2.3)
6. c (23-2.4)
7. a (23-2.2)
8. d (23-3.2)
9. c (23-3.3)
10. a (23-3.4)

Completion

11. mycorrhizae (23-1.1)
12. sporophyte, gametophyte, archegonia, antheridia (23-1.2)
13. gametophyte, sporophyte (23-2.1)
14. stamen, pistil, ovary (23-3.1)
15. zygote, endosperm (23-3.3)

CHAPTER REVIEW ANSWERS *continued*

Short Answer

16. Mosses do not have specialized conducting tissue, and they lack other characteristics associated with vascular plants, such as a dominant sporophyte, and specialized body structures, such as roots, stems, and leaves. (23-1.3)

17. The survival of the species would be in jeopardy because conditions suitable for plant development might not exist at the time of germination. (23-2.4)

18. Imperfect flowers are an advantage because they ensure cross-pollination and thus enhance the variation within populations. (23-3.1)

19. Answers will vary. Some possible responses include the following. Seeds within ripe fruit are eaten by animals and deposited in a different location with the animal's feces. The bright color and sweet smell of the fruit facilitate this method of seed dispersal. Seeds may also be dispersed by water. The enclosure of seeds in a fruit that floats on water, such as a coconut, facilitates this method of dispersal. (23-3.4)

Highlights *Review*

20. Bryophytes include the mosses as well as the other two groups of nonvascular plants—the liverworts and the hornworts.

21. The conducting strands of mosses do not have reinforced walls and therefore cannot conduct water to the heights reached by oak trees.

22. The seedless vascular plants are the ferns (Pterophyta), club mosses (Lycophyta), whisk ferns (Psilotophyta), and horsetails (Sphenophyta).

23. *Cyathea* produces spores, which grow into a gametophyte that produces eggs in archegonia and sperm in antheridia. When sperm swim to eggs and fertilize them, the resulting zygote grows into the sporophyte.

24. Plant reproduction changed from a process that requires water and involves few specialized structures

7. Seeds have helped plants adapt to life on land by
  a. providing nourishment for embryos.
  b. protecting embryos from air pollution.
  c. allowing development to occur during unfavorable conditions.
  d. limiting the dispersal of plant progeny.

8. Which of the following describes an imperfect flower?
  a. petals, pistil, and stamens
  b. anthers, ovary, and sepals
  c. pistil, stamen, and corolla
  d. calyx, corolla, and stamens

9. Which of the following flower characteristics would attract hummingbirds?
  a. a large white corolla
  b. a strong odor
  c. a bright red, tubular corolla
  d. heavily scented, blue petals

10. What is the function of the endosperm found in the seeds of an angiosperm?
  a. to provide nourishment
  b. to stimulate root growth
  c. to inhibit embryonic development
  d. to bring on a period of dormancy

Completion

11. Associations similar to _____ probably helped the first plants absorb water and minerals.

12. In the moss life cycle, the _____ grows on top of the leafy, green _____ . In mosses, eggs form in _____ , while sperm form in _____ .

13. The fern _____ is a heart-shaped plant that produces eggs and sperm. Fronds, rhizomes, and roots characterize the adult fern _____.

14. The male reproductive part of a flower is the _____ ; it includes a filament and an anther. The female reproductive part of a flower is the _____ ; its enlarged base is called the _____ .

15. During double fertilization, one sperm fuses with the egg to form the _____ , while the other sperm fuses with the nuclei of other cells to form the _____ .

Short Answer

16. Why are mosses considered to be nonvascular plants even though they contain simple vascular tissue?

17. Suppose that the seeds of a particular plant species always germinate as soon as they land on soil. How might this affect the survival of the species?

18. How might imperfect flowers be an advantage to a plant species?

19. Describe two methods of seed dispersal in angiosperms. What plant adaptations facilitate the process of seed dispersal?

Highlights *Review*

20. What are the bryophytes?

21. Why can mosses such as *Sphagnum* never grow to be as large as oak trees?

22. Name four types of seedless vascular plants.

23. Explain how *Cyathea* is able to reproduce without seeds.

24. How did the process of plant reproduction change as plants evolved?

25. How does growth in stands of one species facilitate pollination in conifers?

26. Contrast male and female oak flowers.

27. Describe two ways that cross-pollination is promoted among angiosperms.

Themes Review

28. **Evolution** Alternation of generations occurs in all plant life cycles. Explain how this fact plus comparisons of plant structure and plant life cycles indicate that evolution has occurred in plants.

29. **Levels of Organization** Name the four phyla of living seed plants that are collectively called gymnosperms. What do these groups have in common? In addition, give an example of and a distinguishing characteristic of each group.

(sporangia, archegonia, and antheridia) to a process that involves specialized structures that transport sperm (pollen), enhance pollination (flowers), and enclose, nourish, and disperse embryos (seeds and fruits).

25. Dense stands of one species increase the likelihood of successful pollination of conifers, which are wind pollinated.

26. In oaks, female flowers have pistils and are solitary, and male flowers have stamens and occur in long clusters.

27. Cross-pollination is promoted in angiosperms by imperfect flowers and by separate male and female plants.

Themes Review

28. Alternation of generations in all plants indicates a common ancestor. Successive change toward a larger sporophyte that is more specialized for reproducing on land shows that evolution occurred in plants. (23-1.2, 23-1.3)

30. **Structure and Function** Name the four whorls of appendages that make up the basic structure of a flower. What is the function of each whorl?

Critical Thinking

31. **Making Inferences** Some liverworts have leaflike structures that absorb water rapidly. How might this feature be important to a nonvascular plant?

32. **Making Predictions** Suppose that the sporophyte of a type of fern has a diploid chromosome number of 14. What will be the chromosome number of the fern's gametophyte? Explain.

33. **Communicating Effectively** Write a paragraph that you could read to a classmate who cannot see Figure 23-22. Your paragraph should describe the important aspects of the angiosperm life cycle.

34. **Making Inferences** Look at the photograph of a cocklebur plant's seed. How might this seed be dispersed? Explain.

Activities and Projects

35. **Research and Writing** Identify a plant that grows well in your area. Make careful observations of the plant and do library research to determine what adaptations enable the plant to thrive in its environment. Describe what you learn in a written report.

36. **Multicultural Perspective** In Japan, gardening has long been a fine art. A well-planned Japanese garden combines both visual and musical art with botany, mythology, and the psychology of relaxation. Research the building of Japanese gardens and their importance to the Japanese people. Relate your findings in a written or an oral report that includes visuals such as slides or photographs from magazines and books.

37. **Cooperative Group Project** Investigate the plants in an area near your school. Then, draw a map of the area on poster-sized paper, and mark on the map the location of the different types of plants (liverworts, mosses, ferns, angiosperms, and gymnosperms) that you find. To identify some plants, you may need to use an identification key. Include on your map a legend that lists the groups of plants found and the symbol used to identify the plants on the map. Display your map in your classroom or school.

Readings

38. Read the article "Bogs: Thick Mats of Debris," in *American Horticulturist*, May 12, 1994, page 12. The author describes how *Sphagnum* mosses affect succession in wetlands. Why is a peat bog *not* an inviting place for most types of plants? What adaptations enable certain plants to live successfully in peat bogs?

39. Read the article "Angiosperm Endozoochory: Were Pterosaurs Cretaceous Seed Dispersers?" by Theodore Fleming and Karen Lips, in *The American Naturalist*, October 1991, pages 1058–1063. The authors hypothesize that flying reptiles of the Cretaceous period ate the fruit of angiosperms and contributed to the adaptive radiation of the angiosperms. What evidence supports their hypothesis?

Critical Thinking

31. Because the plant has no vascular system, the liverwort must absorb as much water as possible over as much of its surface as possible. (23-1.1)

32. The chromosome number would be 7 because the haploid number is one-half of the diploid number. (23-2.1)

33. Answers will vary but should discuss how gametophytes develop within sporophytes, pollination, double fertilization, and how a new sporophyte develops within a seed. (23-2.1)

34. Cocklebur seeds are dispersed by animals. The tiny hooks catch on an animal's fur or a person's clothing, enabling the seed to be carried to another location. (23-3.4)

Activities and Projects

35. Answers will vary depending on the plant selected. Some adaptations that may enable the plant to thrive include a thick cuticle or fleshy leaves and stems that prevent the plant from drying out, a root system that gathers available water, a method of promoting effective pollination, and a method of seed dispersal that places seeds where favorable conditions exist.

36. Student reports should include information on how gardens are built, their size, how plants are selected, and how the gardens are cared for.

37. Encourage students to find as many examples of each of the major groups of plants as possible.

Readings

38. Peat bogs are highly acidic, which interferes with water uptake, and tend to have little available oxygen and nutrients. Tough, thick leaves enable plants to conserve water, air roots enhance oxygen uptake, being carnivorous enables plants to obtain nutrients that are not available in the soil, and symbiotic relationships with fungi enable plants to obtain needed nutrients.

39. Evidence supporting the hypothesis includes: the reptiles resembled birds in several ways, their ecological radiation was extensive, and fruit is eaten by some modern reptiles.

29. Coniferophyta—has needle-like or scalelike leaves and cones and includes pines and junipers; Ginkgophyta—has fan-shaped leaves and fleshy seeds on separate female plants and consists only of the Ginkgo; Gnetophyta—are vines and shrubs such as welwitchia; Cycadophyta—has palmlike leaves and includes the cycads. All gymnosperms produce seeds that do not develop within a fruit (ripened ovary). (23-2.2)

30. The calyx, or outermost whorl, protects a flower bud from damage. The corolla, or second whorl, has colors and smells that attract pollinators. The androecium, or third whorl, produces microgametophytes. The gynoecium, or innermost whorl, houses ovules within which megagametophytes develop. (23-3.1)

LABORATORY Investigation *Chapter 23*

Plant Diversity

OBJECTIVES

- Compare similarities and differences among phyla of living plants.
- Relate structural adaptations to the evolution of plants.

PROCESS SKILLS

- classifying plants
- relating the structure of plant parts to their function

MATERIALS

- live and preserved specimens representing four plant phyla
- stereomicroscope or hand lens
- compound microscope
- prepared slides of archegonia and antheridia of mosses and ferns

BACKGROUND

1. How do plants you commonly see compare with their ancestors, the green algae?
2. What is alternation of generations? Is it found in all plants?
3. What do you think was the evolutionary pressure for flowers to become colorful?
4. Write your own question to explore in your lab report or notebook.

TECHNIQUE

You will travel to four stations to observe plants that are representatives of four phyla of plants. Record the answers to the following questions in the **Records** section of your report.

Station 1: Mosses

1. Use a stereomicroscope or a hand lens to examine the samples of mosses. Which part of the moss is the gametophyte?

Which part of the moss is the sporophyte? Make a sketch of your observations in the **Records** section of your report. In your drawing, label the gametophyte and sporophyte portions of the moss plant and indicate whether each is haploid or diploid.

2. Use the compound microscope to look at the prepared slides of moss archegonia and antheridia. What kinds of reproductive cells are produced in each of these structures? Draw the archegonia and antheridia in the **Records** section of your report.

3. Do mosses have roots? How do mosses obtain water and nutrients from the soil?

Station 2: Ferns

4. Look at the examples of ferns at this station. The leafy green fern is called a frond.
   a. How does water travel throughout a fern? List observations supporting your answer.
   b. In the **Records** section of your report, make a drawing of the fern plant. Indicate whether the leafy, green frond in your drawing is haploid or diploid.
   c. Search the underside of the fern fronds for evidence of reproductive structures. In the **Records** section of your report, make a drawing of your findings. What kind of reproductive cells are produced by these structures?

5. Examine the examples of fern gametophytes.
   a. Locate and identify the reproductive organs found on the gametophytes. In the **Records** section of your report, sketch and label these organs and identify the reproductive cells produced by each.

Preparation

Set up eight display stations, two stations each for mosses, ferns, gymnosperms, and angiosperms. Use a variety of these types of plants and plant parts that are available in your local area or can be ordered from a biological supply house. Specimens that can be ordered from Ward's include—mosses: *Polytrichum, Sphagnum;* ferns: Boston fern, woodland fern; gymnosperms: Norfolk Island pine, *Podocarpus;* and angiosperms: *Tradescantia,* Rex begonia. Stations with mosses and ferns require a stereomicroscope or hand lens, a compound microscope, and prepared slides of moss and fern archegonia and antheridia.

Procedural Notes

- Direct students to spread out among the stations. They do not need to begin the investigation at Station 1.

Background Answers

1. Flowering plants are much more complex than the green algae but have the same types of chlorophyll and cell walls.

2. Alternation of generations is the alternating of a haploid gametophyte with a diploid sporophyte in the life cycle. It occurs in all plants.

3. Answers will vary but should suggest that flowers that were brightly colored attracted animal pollinators.

4. How do the internal structures of roots, stems, and leaves differ?

Records

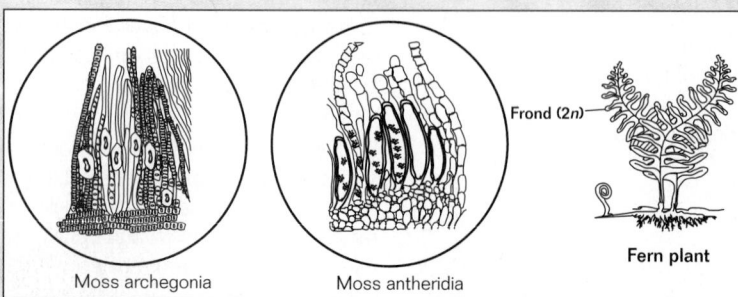

Moss archegonia   Moss antheridia   Frond (2*n*) → Fern plant

Technique Answers

1. The leafy, green part is the haploid moss gametophyte. The bare stalk with a spore capsule is the diploid moss sporophyte.

2. Eggs are produced in archegonia, sperm are produced in antheridia.

3. No. Mosses obtain water and nutrients from their surroundings through osmosis and diffusion.

4. (a) Water travels through ferns in vascular tissue. Ferns have roots and veins in their leaves. **(b)** The leafy green frond is diploid. **(c)** spores

5. (a) Antheridia produce sperm, and archegonia produce eggs. **(b)** haploid **(c)** The zygote immediately follows fertilization. The fern sporophyte appears after fertilization.

b. Are the gametophytes haploid or diploid?

c. What portion of the fern life cycle immediately follows fertilization? What part of the fern appears after fertilization?

6. In what ways are ferns like bryophytes? In what ways are they different?

Station 3: Conifers

7. The gymnosperms most familiar to us are conifers. Look at the samples of conifers at this station.

a. When you look at the limb of a pine tree, which portion (gametophyte or sporophyte) of the plant life cycle are you seeing?

b. In what part of the conifer would you find reproductive structures?

8. Name an evolutionary advancement found in gymnosperms but lacking in ferns.

Station 4: Angiosperms

9. Draw one of the representative angiosperms at this station in the **Records** section of your report. In your drawing, label the sporophyte portion of the angiosperm's life cycle. Where are the sperm and eggs produced in an angiosperm?

10. What is an evolutionary development that is present in both gymnosperms and angiosperms but absent in bryophytes and ferns?

11. How do the seeds of angiosperms differ from those of gymnosperms?

12. Examine the fruits found at this station. How have fruits benefited angiosperms?

13. In your report, briefly summarize the procedure you followed in this investigation.

INQUIRY

1. How do the sperm travel from the bryophyte antheridium to the archegonium?

2. In an angiosperm, how does the sperm get to the part of the flower containing the egg?

3. Which portion of the plant life cycle is dominant in bryophytes? Which portion is dominant in ferns, gymnosperms, and angiosperms?

ANALYSIS

1. What is a seed? Why is the seed a helpful adaptation to terrestrial life?

2. Why are gymnosperms referred to as "naked seed plants"?

3. Which type of plants are the most successful and diverse today? What are some adaptations found among members of this group?

FURTHER INQUIRY

Write a new question that could be explored as a new investigation. The following is an example:

> How are the geographic distributions of the phyla of living plants related to their structures?

6. Ferns are like bryophytes because their sporophytes grow directly from their gametophytes. They are different because the fern sporophyte is the predominant generation, while the gametophyte is the predominant bryophyte generation.

7. (a) sporophyte (b) cones

8. seeds, pollen

9. The leaves, stems, roots, and flowers make up an angiosperm sporophyte. Sperm are produced in the anthers of stamens. Eggs are produced in ovules in the ovaries of pistils.

10. seeds, pollen

11. The seeds of angiosperms have cotyledons and develop within fruits; the seeds of gymnosperms do not.

12. Fruits protect seeds and promote seed dispersal.

Inquiry Answers

1. Sperm swim through water from the antheridium to the archegonium.

2. After pollination occurs, a pollen tube grows from each pollen grain to the ovary, and the sperm pass through the pollen tube to the ovule.

3. The gametophyte is the dominant portion of the bryophyte life cycle, and the sporophyte is the dominant portion of the life cycle in ferns, gymnosperms, and angiosperms.

Analysis Answers

1. A seed is a dormant sporophyte embryo surrounded by a protective coat. Inside the seed, the embryo is protected from drying out, as well as from other stresses.

2. The seeds of a gymnosperm are not covered by a fruit.

3. Flowering plants are today's most successful and diverse plants. Flowers are adaptions for pollination and fertilization without water. Fruits are adaptions for efficient seed dispersal. Seeds with endosperm are adaptations for nourishing young plants through early growth.

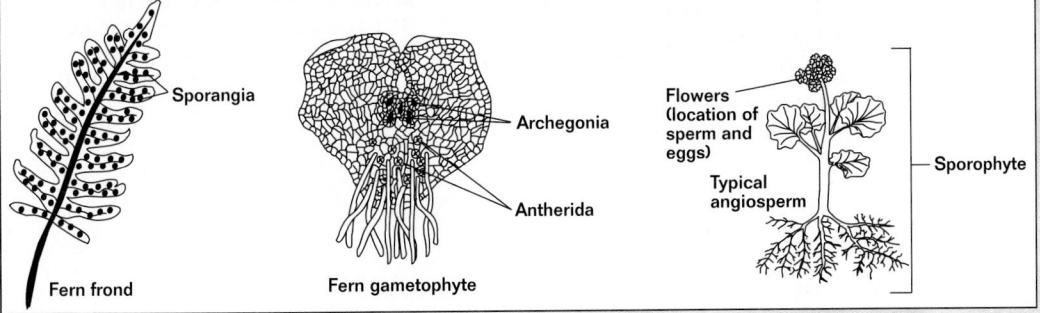

Sporangia

Fern frond

Archegonia

Antherida

Fern gametophyte

Flowers (location of sperm and eggs)

Typical angiosperm

Sporophyte

Each block represents about 45–50 minutes of instructional time. The resources cited in a block are coded to a specific program component using the Key on the next page.

Lecture Concepts

Lesson Resources

24-1 The Vascular Plant Body   pp. 550–559

BLOCK 1

a. Vascular plants are composed of three basic types of tissue: dermal tissue, ground tissue, and vascular tissue.
b. Dermal tissue acts to cover and protect other plant tissues. Ground tissue includes a variety of cells that function in photosynthesis, storage, transport, and support. It makes up the majority of the plant body.
c. Vascular tissue consists of two types of conducting cells (xylem and phloem) that form a system of pipes, which transports materials throughout the plant body.

Lecture Resources
- Opening Question, page 552
- Demonstrations: Three Tissue Systems, Wetting a Waxy Surface
- Visual Strategies: Figures 24-3, 24-4
- How Cells Become Specialized, page 554

- Teaching Transparencies: 120, 121, 125, 126
- Holt Biology Videodiscs: Lesson 24-1

Classwork Option
- Effective Reading, page 551

BLOCKS 2 and 3

a. The typical vascular plant body has an aboveground portion, the shoot, that is anchored by an underground portion, the root.
b. Leaves, the primary photosynthetic organs of a plant, can be specialized for particular purposes, such as protection, water conservation, and climbing.
c. Stems provide a supporting framework and house a plant's vascular tissues. They can be modified for other functions, such as asexual reproduction and storage for food or water.
d. In addition to anchoring a plant, roots absorb water and nutrients from the soil.

Lecture Resources
- Visual Strategies: Figures 24-5; 24-6 and 24-7; 24-8; 24-9; Table 24-1
- Sapwood and Heartwood, page 557
- Research Update, page 558
- Overcoming Misconceptions, page 558
- Teaching Transparencies: 119, 122, 123, 124, 116A, 117A
- Holt Biology Videodiscs: Lesson 24-1

Classwork Options
- Compound Leaves, page 555

- Summary of Plant Cells and Tissues, page 557
- Up Close: Sugar Maple, pages 560–561
- Quick Labs A19: Inferring Function From Structure
- Quick Labs A20: Relating Root Structure to Function

Assignment Options
- Section Review, page 559
- Chapter Review, questions 1–5, 7, 11–13, 18–20

24-2 Transport in Vascular Plants   pp. 562–565

BLOCKS 4 and 5

a. Transpiration is the loss of water vapor from a plant through its stomata. The properties of water molecules help pull water up a plant body.
b. Guard cells regulate the rate of transpiration by opening or closing stomata.
c. Translocation is the movement of sugars through a plant from a source to a sink.

Lecture Resources
- Opening Questions, page 562
- Demonstrations: Transport in Xylem, Adhesion and Cohesion of Water, Modeling Guard Cells
- Visual Strategies: Figures 24-10, 24-14
- Overcoming Misconceptions, page 562
- Teaching Transparencies: 127
- Holt Biology Videodiscs: Lesson 24-2

Classwork Options
- Closure, page 565

- Inquiry Skills Development B17: Transpiration and Stem Structure
- Laboratory Techniques and Experimental Design C35: Staining and Mounting Stem Cross Sections
- Laboratory Investigation: Roots, Stems, and Leaves, pages 568–569

Assignment Options
- Solve the Puzzle, page 563
- Locating Guard Cells, page 564
- Section Review, page 565
- Chapter Review, questions 6, 8–10, 14–19, 21, 22

Review/Enrichment

BLOCK ⑥

■ Study Guide: Concept Review and Pretest

Assignment Options
■ Activities and Projects, questions 23–25

Assessment Options

BLOCK ⑦

Traditional Assessment
■ Chapter 24 Test
◉ Test Generator/Assessment Item Listing: Software item bank for preparing customized chapter tests.

Portfolio Assessment
Select student reports for one or more laboratory experiments from this chapter. *The Direct Observation Checklist* on the *BioSources Teaching Resources CD-ROM* should be completed during a laboratory or other cooperative learning experience.

Holt Biology Videodiscs   CONCEPTS OF BIOLOGY

Holt Biology Videodiscs gives you a powerful tool for teaching, review, and assessment. *Concepts of Biology* adds interactive lesson plans and extensive tools for customization using CD-ROM technology.

Use the following topic from *Concepts of Biology* to help you teach this chapter:

■ Topic 16: Plant Structure and Function

For further information regarding the media on the videodiscs, see *Holt Biology Videodiscs Teacher's Correlation Guide to Biology: Principles and Explorations.*

Chapter Themes

Structure and Function

Plants are composed of structures (leaves, stems, and roots) that are modified to meet the needs of the plant. The structures of a plant enable it to survive in the environment in which it lives. Figures 24-3 and 24-4, as well as Tables 24-1 and 24-2, illustrate how structure is linked to function.

Levels of Organization

The cells, tissues, and organs of a plant are described in this chapter. Figures 24-1, 24-6, 24-8, and 24-9 describe levels of organization in plants.

Tapping Prior Knowledge

- What are vascular plants?
- What are xylem and phloem?
- What are adhesion and cohesion?
- What is osmosis, and how is it triggered?
- How does gas exchange occur in plants?

Opening Demonstration

Show students three very different plants, such as a cactus, a tomato, and a Venus' flytrap. Have the class compose a list of things that the three plants have in common. They should derive that all plants have the same needs—water, oxygen, light energy, and mineral nutrients for making organic molecules. Ask: Why do these three plants look so different? *(They have specialized tissues to accommodate their needs.)*

CHAPTER 24

PLANT STRUCTURE AND FUNCTION

Flower of a widow's tears, Commelinantia anomala

REVIEW

- adhesion and cohesion of water (Section 2-2)
- cell structure and organelles (Section 2-4)
- osmosis and active transport (Section 3-2)
- carbon fixation and the Calvin cycle (Section 5-2)
- stomata and guard cells (Section 23-1)
- roots, shoots, and meristems (Section 23-1)
- xylem and phloem (Section 23-1)

Authors' Rationale

Now that students have a basic understanding of plants, their characteristics, major groups, and evolution, they will study the structures and functions of the largest group of plants—the vascular plants. Your students should know why these plants are referred to as "vascular." In this chapter, the tissues of the plant body are described, as are the physiological mechanisms plants use to stay alive. Which tissues make up wood? Which tissues take up water? Which tissues store food? How do plants raise water to great heights? These are some of the questions this chapter will answer.

24-1 The Vascular Plant Body

Vascular plants are among the most successful groups of organisms on Earth, literally covering its surface with a carpet of life. Some vascular plants are smaller than your fingernail, while others are larger than a house. Yet, all share the same basic body plan. Like your body, a vascular plant's body is composed of tissues that form its organs—roots, stems, and leaves. In this section, you will look first at the cells and tissues that form a vascular plant's body and then at the structure of leaves, stems, and roots.

Section Objectives

- Describe the three major tissue systems that compose the vascular plant body, and state the function of each.
- Explain how leaves, stems, and roots are adapted for the functions they perform.
- Describe several distinguishing features of sugar maple trees.

Three Tissue Systems Make Up the Vascular Plant Body

Like other complex multicellular organisms, the body of a vascular plant is made of different types of tissues. The fundamental plant tissues are organized into three basic functional units, or systems.

1. **Dermal tissue system** The **dermal tissue system** serves as a protective outer layer.

2. **Ground tissue system** The **ground tissue system** performs photosynthesis, stores water and carbohydrates, assists in transport, and surrounds and supports the conducting tissues.

3. **Vascular tissue system** The **vascular tissue system** conducts water, mineral nutrients, and carbohydrates made by photosynthesis.

The basic body plan of a vascular plant is seen in Figure 24-1, which shows how plant tissues are arranged.

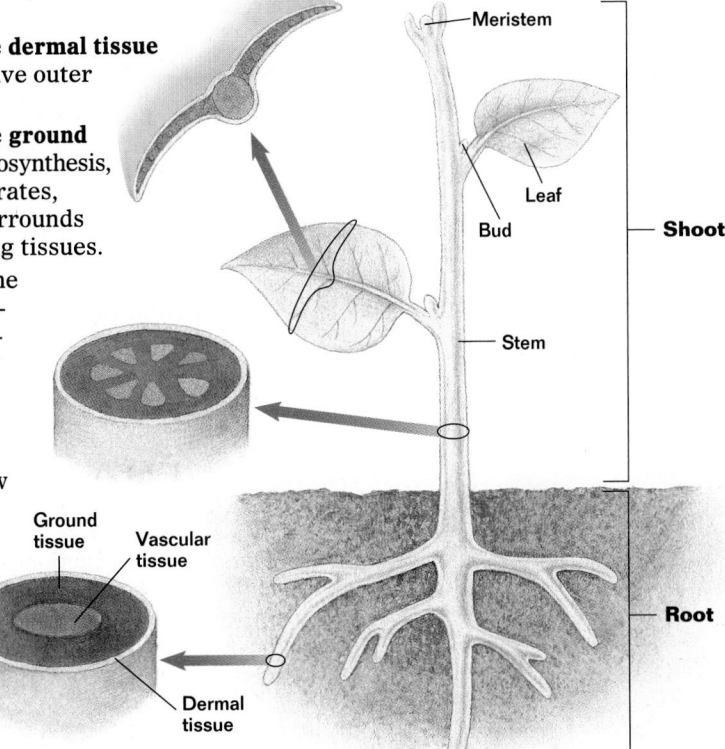

Figure 24-1 Most vascular plants consist of an aboveground shoot and an underground root. Cross sections of various points in the plant body show that the tissue systems are found throughout the plant.

Labels: Meristem, Leaf, Bud, Shoot, Stem, Ground tissue, Vascular tissue, Dermal tissue, Root, Meristem

Effective Reading

Defining Terms

There are many unfamiliar terms in this chapter. Ask students to write a definition of each boldface term they encounter as they read. Tell students that to make sense of paragraphs with two or more boldface terms, they should reread those paragraphs *after* they have written their definitions.

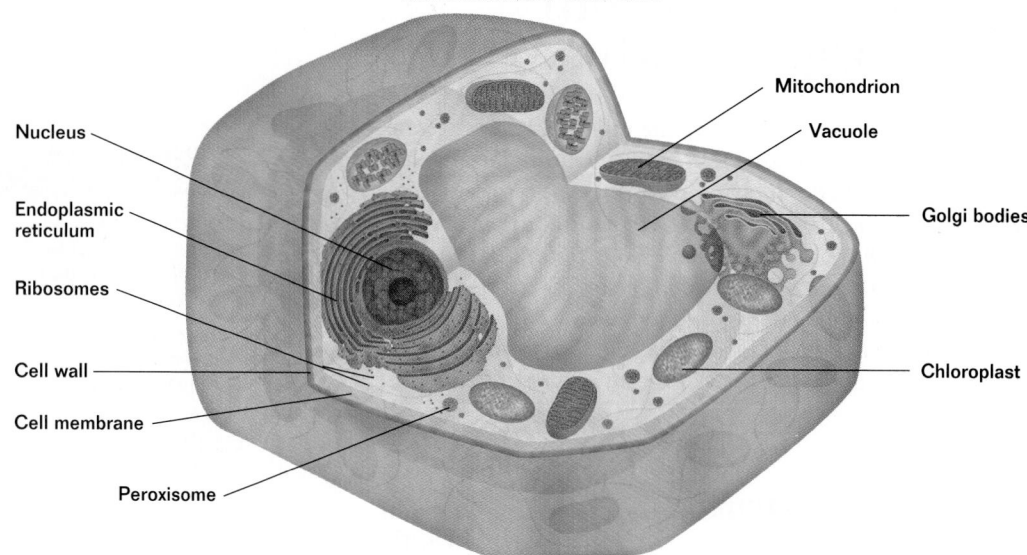

Generalized Plant Cell

Nucleus · Endoplasmic reticulum · Ribosomes · Cell wall · Cell membrane · Peroxisome · Mitochondrion · Vacuole · Golgi bodies · Chloroplast

Figure 24-2 This diagram of a plant cell shows structures that occur in plant cells. However, each individual type of plant cell typically lacks one or more of these structures.

Each of the tissue systems is composed of one or more distinctive kinds of cells that are specialized for particular functions. The generalized plant cell in Figure 24-2 has the components typically found in plant cells. However, many specialized plant cells lack one or more of these components. Some plant cells perform their functions only after they have lost all of their contents. All plant tissues arise from the masses of actively dividing cells called meristems.

Dermal Tissue

Dermal tissue covers all parts of a plant's body. The outer covering, or "skin," of a plant is called the **epidermis** and is composed primarily of relatively flat cells. The word *epidermis* is derived from the Greek words *epi*, meaning "upon," and *derma*, meaning "skin." The epidermis of most aboveground plant parts is only one cell thick and is usually coated with a waxy cuticle that retards water loss.

Ground Tissue

Much of the body of a plant—including its roots, stems, and leaves—is made up of ground tissue. Ground tissue consists primarily of masses of thin-walled cells that are alive at maturity and have a functional nucleus. A variety of thick-walled cells that provide support are also found in ground tissue. In leaves, most of the ground tissue is packed with chloroplasts and is specialized for photosynthesis. In stems and roots, the ground tissue functions mainly to support the plant body and to store water and carbohydrates. Ground tissue also assists in the movement of water, minerals, and carbohydrates throughout the plant body by osmosis and active transport. ❏

Historical Note
The Cell Theory

The study of plant tissue led the way in the development of the cell theory. In 1665, the English physicist Robert Hooke was the first to observe "little boxes" in sections of cork viewed under a microscope. Hooke used the term *cell* to refer to the chambers in cork, which are formed by the cell walls of plant cells. In 1838, the German botanist Matthias Schleiden proposed that all plants are made of cells. The German zoologist Theodore Schwann concluded in 1839 that all animals are made of cells. The ideas of Schleiden and Schwann were combined with the conclusion that new cells come from existing cells, made by the German physician Rudolph Virchow in 1855, to form the cell theory.

Vascular Tissue

As you learned in Chapter 23, plants have two kinds of vascular tissue. Both xylem and phloem form tiny pipes that act like a plumbing system, carrying fluids up and down the plant body. However, the structure and function of the cells in each type of vascular tissue are very different. One major difference between the two types of vascular tissues is that mature xylem tissue consists of dead cells, while mature phloem tissue consists of living cells.

Xylem contains two principal types of conducting elements (cells), which can be seen in Figure 24-3. Both types of xylem cells lose their contents (plasma membrane, nucleus, and cytoplasm) at maturity and consist only of their cell walls. **Tracheids** (*TRAY kee ihdz*) are narrow, elongated, thick-walled cells that are tapered at each end. Water flows from one hollow tracheid to the next through pits in their cell walls. **Vessel elements,** the second type of xylem cells, are often wider than tracheids and develop large perforations in their ends. These perforations enable water to flow more quickly between the elements. Stacked end to end, a series of vessel elements forms a tube called a **vessel.**

Figure 24-3 Two types of conducting cells are found in xylem tissue. Tracheids are thin, elongated cells with tapered ends. Vessel elements are wider, and their ends have large perforations that enable water to flow more readily than it does through tracheids.

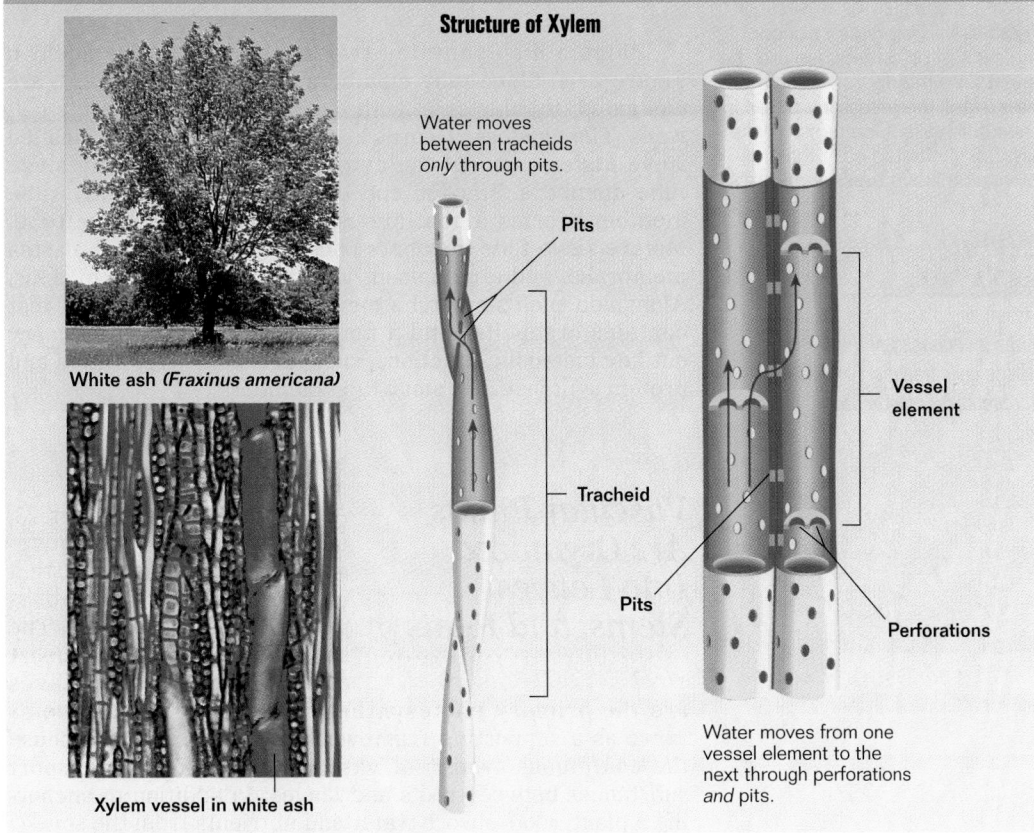

Structure of Xylem

White ash (*Fraxinus americana*)

Xylem vessel in white ash

Water moves between tracheids *only* through pits.

Pits

Tracheid

Pits

Vessel element

Perforations

Water moves from one vessel element to the next through perforations *and* pits.

? Did You Know?

Gymnosperm xylem contains only tracheids, which are considered a primitive type of xylem cell. The vessel elements in angiosperms evolved from tracheids, which are also found in angiosperms.

Demonstration

Wetting a Waxy Surface

Materials: (per group) wax paper, water, liquid detergent, 2 beakers.
Procedure: Have students place two or three drops of water on a piece of wax paper and draw what they see. Ask: What happens when the wax paper is tilted? (*The drops of water roll off the paper.*) Then have students place two or three drops of soapy water on the wax paper and draw what they see. (*The drops of soapy water spread out over the surface of the wax paper.*) Tell students that the detergent acts as a "wetting agent" by interfering with the attraction water molecules have for each other and enabling the water drops to spread out over the waxy surface. Have students write a short paragraph about how detergent alters the behavior of water on wax paper.

Application

Agriculture The waxy cuticle on leaves slows the absorption of many fertilizer and pesticide sprays applied by farmers and home gardeners. These chemicals are dissolved in water, which beads up and tends to drop off before the chemical can penetrate the leaf. The addition of a wetting agent, a special detergent that breaks down the surface tension of water, allows the chemical to remain on the leaves long enough to penetrate the waxy cuticle.

VISUAL STRATEGY

Figure 24-3
Have students compare the structure of a tracheid with that of a vessel element.

VISUAL STRATEGY

Figure 24-4
Have students compare and contrast the structure of the phloem sieve tube member with the xylem vessel element seen in Figure 24-3 on page 553.

Teaching Tips
How Cells Become Specialized

Remind students that specialization begins at the genetic level. Ask: How do dermal cells differ from ground cells? *(Dermal cells are flattened and have no plastids, while ground tissue cells are rounded or columnar and have plastids.)* How do these differences develop? *(When different sets of genes are transcribed, cells develop different structures and thus become specialized for particular activities.)*

What If the conducting cells of a plant became clogged? In fact, the conducting cells of plant vascular tissue frequently become clogged by materials such as air bubbles and fungal mycelia. When this happens, materials either move laterally into adjacent vascular cells or transport in that region stops. If too many cells are blocked, the plant will die.

Structure of Phloem

American basswood (*Tilia americana*)

Both sieve tube members and companion cells are living at maturity, but sieve tube members lose their nucleus and organelles.

The cytoplasms of adjacent phloem cells are connected through the tiny pores in their cell walls.

Pores

Nucleus

Companion cell

Sieve plate

Sieve tube member

Phloem sieve tube in American basswood

Figure 24-4 The primary conducting cells of phloem tissue are the sieve tube members, which are named for the many tiny pores in their cell walls. Companion cells are the nucleated cells that lie alongside sieve tube members.

☐ CAPSULE SUMMARY

Xylem and phloem are vascular tissues that consist of tubular cells. These cells transport materials throughout a plant's body.

Phloem also contains two kinds of cells, as seen in Figure 24-4. **Sieve tube members,** the conducting cells, are elongated, tubular cells with clusters of pores in their cell walls. The clusters of pores at the ends of these cells, called **sieve plates,** connect the cytoplasms of neighboring sieve tube members. Stacked end to end, a series of sieve tube members forms a continuous strand called a **sieve tube.** Mature sieve tube members consist of a cell wall, a plasma membrane, and a cytoplasm with no organelles or nucleus. Alongside each sieve tube member is a **companion cell** that contains organelles and a nucleus. Companion cells carry out key metabolic functions, such as cellular respiration and protein synthesis, for sieve tube members. ☐

Vascular Plants Are Organized Into Leaves, Stems, and Roots

As you learned in Chapter 23, most plants have an aboveground portion, the shoot, that is anchored by an underground portion, the root. The shoots of most plants consist of leaves and stems. Leaves are the primary photosynthetic organs of plants. Stems serve as a supporting framework for the leaves and house the continuous system of vascular strands that transport substances between roots and leaves. In addition to anchoring a plant, roots absorb water and nutrients from the soil.

 Did You Know?

Sieve cells, which are conducting cells found in the phloem tissue of gymnosperms and ferns, are thought to be the precursors of the sieve tube members in angiosperms. Sieve cells, which also lack a nucleus, are not stacked end to end; instead, their ends overlap, as do the ends of tracheids. Also, sieve cells lack companion cells, but nearby ground cells provide the same services to sieve cells that companion cells provide for sieve tube members.

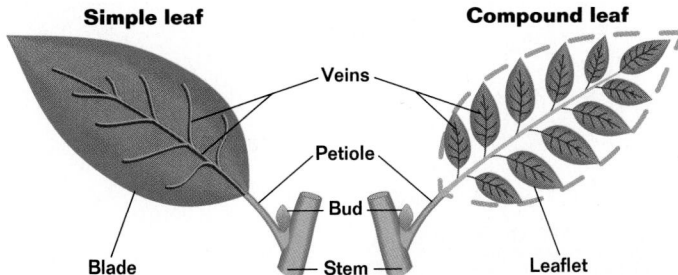

Simple leaf   **Compound leaf**

Veins

Petiole

Bud

Blade

Stem

Leaflet

Figure 24-5 A simple leaf, *left,* consists of a continuous, flattened blade with many veins. The blade of a compound leaf, *right,* is divided into two or more leaflets. Most leaves are attached to a stem by a petiole.

Leaves

Most leaves have the basic structure seen in Figure 24-5. They consist of a flattened surface, the blade, that is often attached to the stem by a slender stalk, the **petiole** *(PEHT ee ohl).* Leaves with an undivided blade are called simple leaves. Those with a blade divided into two or more sections, or **leaflets,** are called compound leaves. Veins, which are bundles containing strands of both xylem and phloem tissue, are the plumbing system of a leaf. These veins are extensions of vascular bundles that run from the tips of roots to the edges of leaves. Recall from Chapter 23 that the veins in the leaves of most monocots run parallel to one another, while the veins in the leaves of most dicots branch and form a network. Many plants have highly modified leaves that are specialized for particular purposes, such as protection, water conservation, and climbing. Table 24-1 compares the structure and function of three types of modified leaves.

Table 24-1 Modified Leaves

Leaf	Name	Function
	Cactus spines	Protection, water conservation
	Garden pea tendrils	Climbing
	Venus' flytrap leaves	Photosynthesis, trapping insects to obtain nitrogen

 Did You Know?

A "fuzzy" or "woolly" leaf is covered with hairs called trichomes, which are appendages of certain epidermal cells. Some trichomes are stiff and serve a protective function, like horns. Others branch, sometimes elaborately, and aid in the establishment of a buffer zone between the surrounding air and the leaf surface. Such buffer zones protect leaves from temperature extremes and from drying conditions.

SECTION 24-1

 VISUAL STRATEGY

Figure 24-5

Have students describe the differences between simple and compound leaves. Point out that one way to tell a leaf from a leaflet is to look for the bud in the axil of the leaf (the angle between a leaf and the stalk or stem to which it is attached). Ask: Are there buds in the axils of leaves? *(yes)* Are there buds in the axils of leaflets? *(no)* Also point out that compound leaves help plants maintain homeostasis in arid environments by reducing the leaf surface area and water loss.

Teaching Tip
Compound Leaves

Just as the veins in a leaf are arranged either pinnately (like a feather) or palmately (like the palm of a hand), the leaflets of a compound leaf are arranged in the same manner. Tell students that the compound leaf seen in Figure 24-5 is pinnately compound. Some common plants with pinnately compound leaves include pecans, walnuts, hickories, ashes, acacias, mimosas, and mesquites. Have students sketch what they think is a palmately compound leaf. *(Leaflets should branch from a central point.)* Aralias and scheffleras are common houseplants with palmately compound leaves. Common weeds with palmately compound leaves include lupines, clovers, and wood sorrels. Next, tell students that leaves are often doubly or even triply compound. Challenge students to draw what they think are twice pinnately compound and thrice pinnately compound leaves. For extra credit, have students do library research to find the names of some plants with compound leaves like the ones they drew.

 VISUAL STRATEGY

Table 24-1

Have students describe how the structure of each of these modified leaves helps them perform their specialized functions.

Figure 24-6 This cross section of a leaf illustrates one type of structure that is characteristic of many leaves. Other types of leaves may have different internal arrangements.

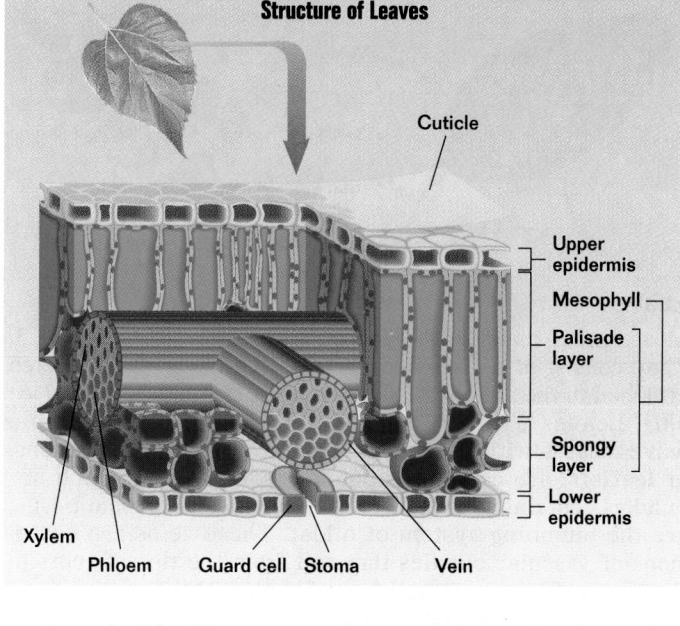

Structure of Leaves

- Cuticle
- Upper epidermis
- Mesophyll
- Palisade layer
- Spongy layer
- Lower epidermis
- Xylem
- Phloem
- Guard cell
- Stoma
- Vein

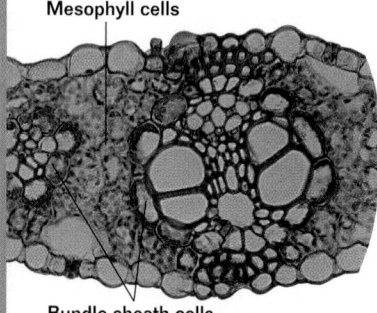

Mesophyll cells

Bundle sheath cells

Figure 24-7 The leaf of a corn plant, a C_4 plant, has a different internal structure than the leaf illustrated in Figure 24-6. The cells of the mesophyll are not differentiated into palisade and spongy layers. The mesophyll cells fix carbon by making a four-carbon compound. This compound is later transferred to the bundle sheath cells, where carbohydrates are made in the Calvin cycle.

A typical leaf is a mass of ground tissue that has veins running through it and that is encased in an envelope of epidermis. Figure 24-6 shows the internal structure of a typical leaf. The ground tissue in a leaf is called the **mesophyll** (MEHS oh fihl), which comes from the Greek words *mesos*, meaning "middle," and *phyllon*, meaning "leaf." Two kinds of mesophyll are found in most plants. Just beneath the upper epidermis of many kinds of leaves is the **palisade layer**, which consists of one or more rows of closely packed, columnar cells. The lower portion of the mesophyll usually consists of loosely packed, spherical cells and is called the **spongy layer**. The cells of the palisade and spongy layers are packed with chloroplasts, in which photosynthesis takes place. Scattered throughout the spongy layer are large air spaces, through which gases and water vapor travel. Stomata, the tiny holes that dot leaf surfaces, connect the air spaces of the mesophyll to the outside air.

The leaves of many plants have a modified internal structure that is an adaptation for a type of photosynthesis that operates very efficiently in hot climates. Recall from Chapter 5 that plants fix carbon (use carbon from carbon dioxide to make sugars) during photosynthesis by means of the Calvin cycle. Because the first detectable product of the Calvin cycle is a three-carbon compound, plants that fix carbon only with the Calvin cycle are called C_3 **plants.** In some plants, such as the plant shown in Figure 24-7, carbon is also fixed by an alternative pathway in which the first detectable product is a four-carbon compound. These plants are called C_4 **plants.** Because they fix carbon efficiently in high temperatures and intense light, C_4 plants are plentiful in the tropics.

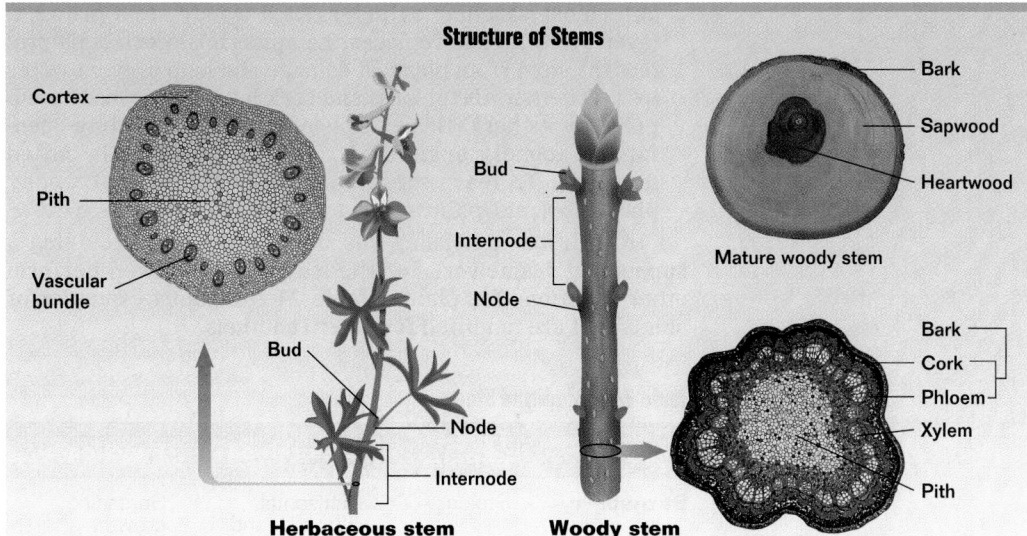

Structure of Stems

Cortex

Pith

Vascular bundle

Bud

Internode

Node

Bud

Node

Internode

Herbaceous stem

Woody stem

Bark

Sapwood

Heartwood

Mature woody stem

Bark

Cork

Phloem

Xylem

Pith

Stems

Figure 24-8 shows the external and internal structures of typical plant stems. The places where leaves attach to a stem are called **nodes. Internodes** are the areas of a stem between nodes. Also located at the nodes are **lateral buds** that grow into the branches of a stem. The bud at the tip of a stem is called the **terminal bud.** A typical stem consists of bundles of vascular tissues embedded in ground tissue. The outer layers of ground tissue in a stem are called the **cortex,** and the inner layers are called the **pith.** Other features of a stem vary, depending on whether the stem is woody or non-woody.

1. **Herbaceous stems**   Flexible, relatively soft, and usually green stems, like those of violets and petunias, are called **herbaceous** *(huhr BAY shuhs)* **stems.** An epidermis forms the outermost layer of a herbaceous stem. Stomata in this epidermis enable the stem to exchange gases. The vascular tissues of a herbaceous (non-woody) stem are distributed within the ground tissue and arranged in **vascular bundles** that contain both xylem and phloem.

2. **Woody stems**   Stiff, usually nongreen stems that contain layers of wood, like the trunks of trees, are called **woody stems.** In woody stems, the vascular tissues are arranged in solid cylinders. Xylem cells form the innermost cylinder. A new belt of xylem cells forms each year, adding to the width of the stem. Wood consists primarily of these xylem cells. The darker wood in the center of a tree trunk is called **heartwood.** Xylem cells in heartwood no longer conduct water because they have been filled with substances that help to strengthen the stem. The lighter wood in a tree trunk, which contains xylem cells that still conduct water, is called **sapwood.** Phloem cells form the

Figure 24-8 Compare two basic types of stems, *above.* Herbaceous stems, *left,* are typically soft and green. Woody stems, *right,* are typically stiff and nongreen with several layers of xylem cells. In herbaceous stems, the vascular bundles remain distinct. In woody stems, vascular bundles occur in young stems but fuse into solid cylinders in mature stems.

CONTENT LINK

The growth of woody and non-woody plants is discussed in **Chapter 25.**

SECTION 24-1

👁 *VISUAL STRATEGY*

Figure 24-8

Have students compare the external structures of the herbaceous and woody stems. Ask: What structures are seen on the woody stem and not on the herbaceous stem? *(lenticles)* Have students compare the internal structures of the herbaceous and woody stems. Ask: What structures are seen in the woody stem and not in the herbaceous stem? *(Herbaceous stems do not have cork or bark.)*

Teaching Tips
Sapwood and Heartwood

Show students cross sections of logs, and let them find the sapwood and the heartwood. Ask students how the redwoods of California that had roads tunneled through them were able to survive. Point out that heartwood does not conduct the water important to the trees' survival, but simply provides support and strength.

Summary of Plant Cells and Tissues

Have students construct a table summarizing the types of plant cells and tissues they have studied so far. In the table, students should classify each type of cell encountered as part of the dermal tissue system, ground tissue system, or vascular tissue system. A completed table is found in the *Graphic Organizer* on page 556.

 Did You Know?

The eyes of potatoes are actually buds on a stem.

Multicultural Perspective
Cattails

Native Americans once ate the young stems of cattails, which taste very much like the asparagus we eat today.

outermost vascular cylinder. As a woody stem grows, a layer of **cork cells** replaces the epidermis. Cork cells protect the stem from physical damage and help prevent water loss. Together, the phloem and cork layers of a woody stem make up its bark. Gas exchange occurs through tiny openings in loosely organized groups of cork cells called **lenticels.** To learn more about the structure of woody plants, look at *Up Close: Sugar Maple* on pages 560–561.

In addition to housing vascular tissues and providing a supporting framework for the leaves, stems often perform other functions for plants. Table 24-2 contains examples of stems that are modified for other functions.

Table 24-2 Modified Stems

Name	Type of Stem	Description	Function
Strawberry	Stolon	Horizontal, aboveground stem	Spreading growth, asexual reproduction
Potato	Tuber	Enlarged underground stem	Food storage
Cactus	Succulent	Fleshy, often leafless stem	Water storage

Roots

Roots have a simpler structure than stems. As Figure 24-9 shows, roots lack external features such as nodes, leaves, and buds. They also lack pith. Instead, the vascular bundles are at the center of a root. The vascular tissue of a root is surrounded by a thick layer of cortex that is covered by an outer sheath of dermal tissue. Epidermal cells cover the end of a root and are replaced by layers of cork cells as the root grows. The epidermal cells near the root tip produce slender projections called **root hairs.** By extending the cell membranes of epidermal cells, root hairs greatly increase the surface area of a root and play a critical role in the absorption of water and minerals. The actively growing tip of a root is covered by a protective layer of cells called the **root cap.**

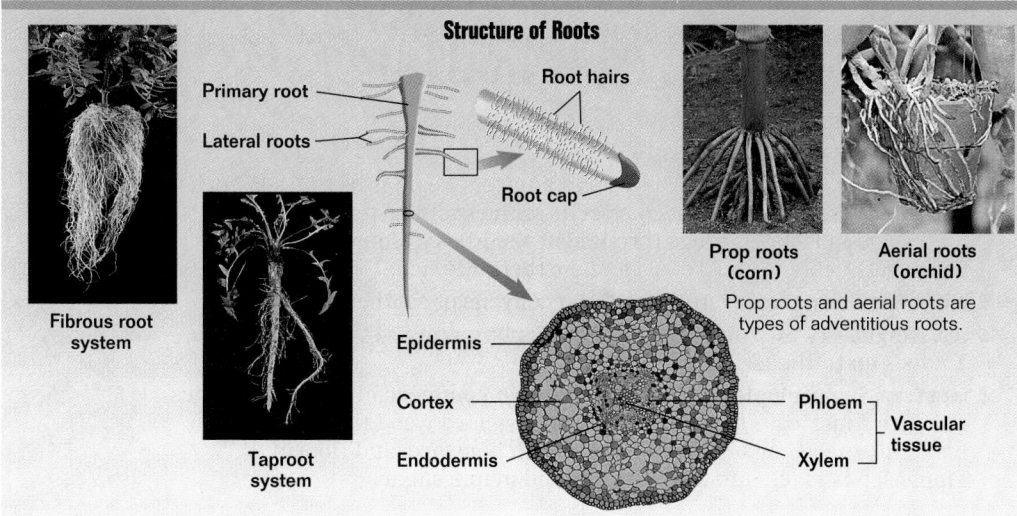

Structure of Roots

Primary root
Lateral roots
Root hairs
Root cap
Prop roots (corn)
Aerial roots (orchid)
Prop roots and aerial roots are types of adventitious roots.

Fibrous root system
Taproot system

Epidermis
Cortex
Endodermis
Phloem
Xylem
Vascular tissue

Based on branching patterns, there are two different types of root systems—taproot systems and fibrous root systems. Plants with a **taproot system,** such as carrots and radishes, have a large central root called a taproot. The lateral roots that branch from the taproot are usually much smaller than the taproot. Most dicots have a taproot system. The roots of a highly branched **fibrous root system** are all about the same size. Most monocots, such as grasses, have a fibrous root system. Many plants, such as orchids, also have **adventitious roots,** which grow from aboveground parts such as stems and leaves. Just like underground roots, adventitious roots provide support and absorb water and minerals. ◻

Figure 24-9 Most plants have either a taproot or fibrous root system. Some types of roots, such as prop roots and aerial roots, grow above ground and provide support as well as absorb water. Near a root tip, there are many root hairs, which enhance water absorption.

CAPSULE SUMMARY

The leaves, stems, and roots of plants are made of the three types of plant tissues and are adapted to the particular functions they perform for the plant.

VISUAL STRATEGY

Figure 24-9
Walk students through this figure. Be sure they notice that the internal structure of a root differs from that of a stem. Ask: How could you tell whether you were looking at a cross section of a stem or a cross section of a root? *(In stems, the vascular tissue is arranged in many bundles. In roots, all of the vascular tissue is found in the center of the root.)* Tell students that this central cylinder of vascular tissue is called the stele.

Teaching Tip
Pick a Pot
Tell students that you need to buy a pot for a houseplant that has outgrown its container. Tell them that the plant is a pothos ivy, which is a monocot. Ask students to sketch the shape of the pot that would be most suited to the pothos ivy, and have them write a paragraph summarizing why you should buy that shape of pot. *(The pot must accommodate a fibrous root system, which is characteristic of monocots, and thus should be wider than it is tall.)*

Section Review

1. *What are three functions of the ground tissue system?*
2. *Name and describe the two main types of conducting cells in xylem and phloem.*
3. *How does the structure of a leaf help it perform photosynthesis efficiently?*
4. *How does the structure of a root enable it to anchor a plant and to absorb water?*
5. *What structural features of the sugar maple make it economically important?*

Critical Thinking
6. *Why might a taproot system be an advantage to some plants, while a fibrous root system is an advantage to others?*

Answers to Section Review

1. photosynthesis, store water and food, support vascular tissue

2. The vessel elements of xylem are thick-walled, tubular cells with pits in their cell walls. The sieve tube members of phloem are thin-walled, tubular cells with clusters of tiny pores in their cell walls.

3. Colorless epidermal cells admit the maximum amount of light, while the chlorophyll-containing cells are usually clustered near the upper leaf surface for maximum collection of light. Air spaces in the spongy layer connect inner cells to the outside air for gas exchange.

4. Lateral roots help anchor plants. Root hairs increase surface area for water absorption.

5. The hard wood of the sugar maple is very durable and has desirable grains. The phloem conducts a rich supply of sap that is collected and used for making syrup and sugar.

6. Taproots offer large plants support and enable plants to reach water deep in the ground. Fibrous roots offer support in shallow soils and maximize water collection while they hold soil and prevent erosion.

Instructional Strategies

- Have students do library research on the sugar maple. Then have each student write an interesting fact about sugar maples on a piece of paper cut in the shape of a maple leaf. Draw the outline of a maple tree on a large sheet of butcher paper, and have students attach their leaves to the branches of the tree.

- If any of the many species of maple trees grow in your area, have students bring leaves and samaras from those trees and compare them with the leaves and samaras of the sugar maple. Remind students to seek permission to collect leaves and samaras if they are collecting from someone else's trees.

- Display samples of maple wood and photographs of various types of maple trees at different times of the year.

- For added interest, provide examples of foods that contain maple flavoring or maple sugar.

UP CLOSE   SUGAR MAPLE

- **Scientific Name:** *Acer saccharum*
- **Range:** Northeastern United States and adjacent regions of southeastern Canada. Closely related species occur in the southeastern United States and in isolated areas throughout the Rocky Mountains, southwestern Oklahoma and Texas, and northern Mexico.
- **Habitat:** Northern temperate forests and canyons of southern mountains
- **Size:** Height—12 to 37 m (40 to 120 ft.), Canopy—up to 14 m (45 ft.) wide, Trunk—up to 1 m (3 ft.) in diameter
- **Importance:** Sugar maples are among the most commercially valuable trees in North America. They have high-quality, hard wood that is used to make furniture, musical instruments, kitchenware, and flooring. Maple sap is made into maple syrup and maple sugar.

External Structures

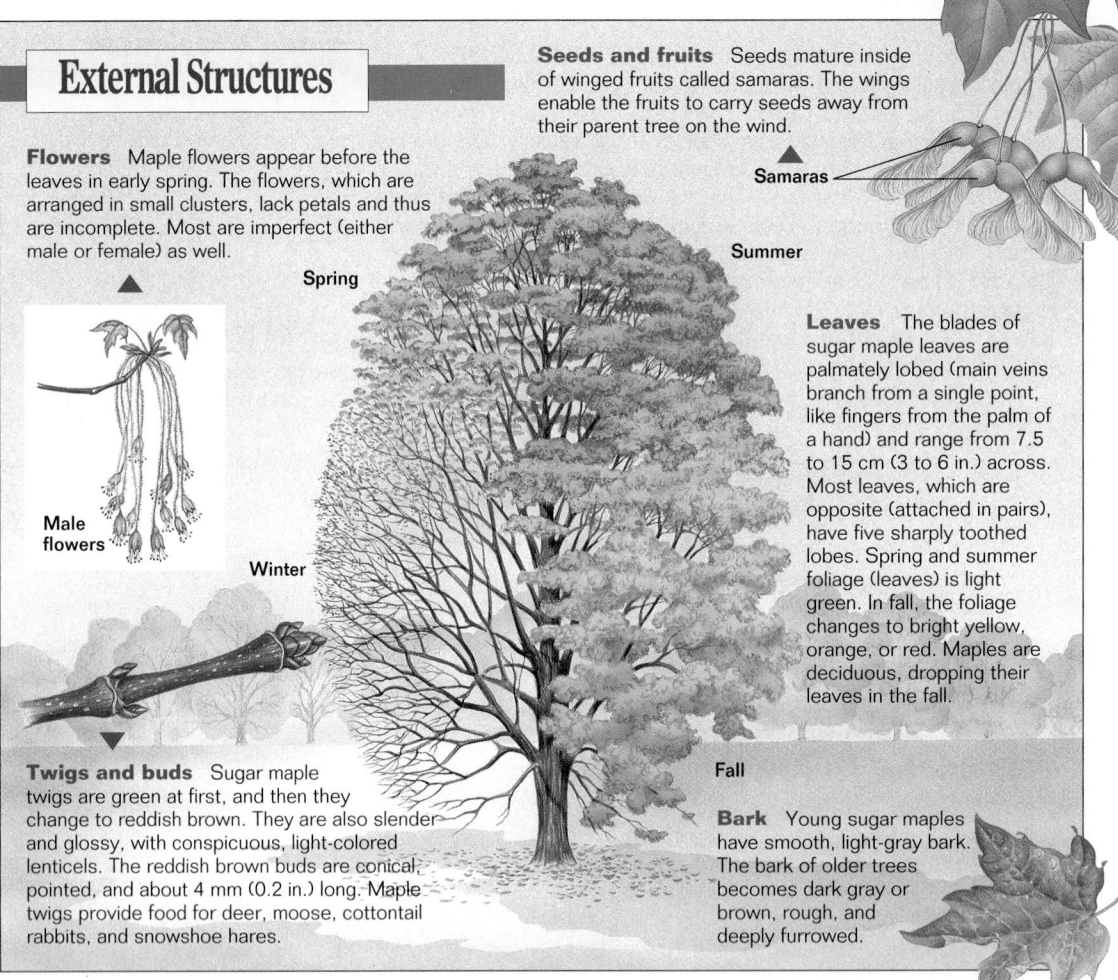

Seeds and fruits Seeds mature inside of winged fruits called samaras. The wings enable the fruits to carry seeds away from their parent tree on the wind.

Samaras

Summer

Flowers Maple flowers appear before the leaves in early spring. The flowers, which are arranged in small clusters, lack petals and thus are incomplete. Most are imperfect (either male or female) as well.

Spring

Male flowers

Winter

Leaves The blades of sugar maple leaves are palmately lobed (main veins branch from a single point, like fingers from the palm of a hand) and range from 7.5 to 15 cm (3 to 6 in.) across. Most leaves, which are opposite (attached in pairs), have five sharply toothed lobes. Spring and summer foliage (leaves) is light green. In fall, the foliage changes to bright yellow, orange, or red. Maples are deciduous, dropping their leaves in the fall.

Fall

Twigs and buds Sugar maple twigs are green at first, and then they change to reddish brown. They are also slender and glossy, with conspicuous, light-colored lenticels. The reddish brown buds are conical, pointed, and about 4 mm (0.2 in.) long. Maple twigs provide food for deer, moose, cottontail rabbits, and snowshoe hares.

Bark Young sugar maples have smooth, light-gray bark. The bark of older trees becomes dark gray or brown, rough, and deeply furrowed.

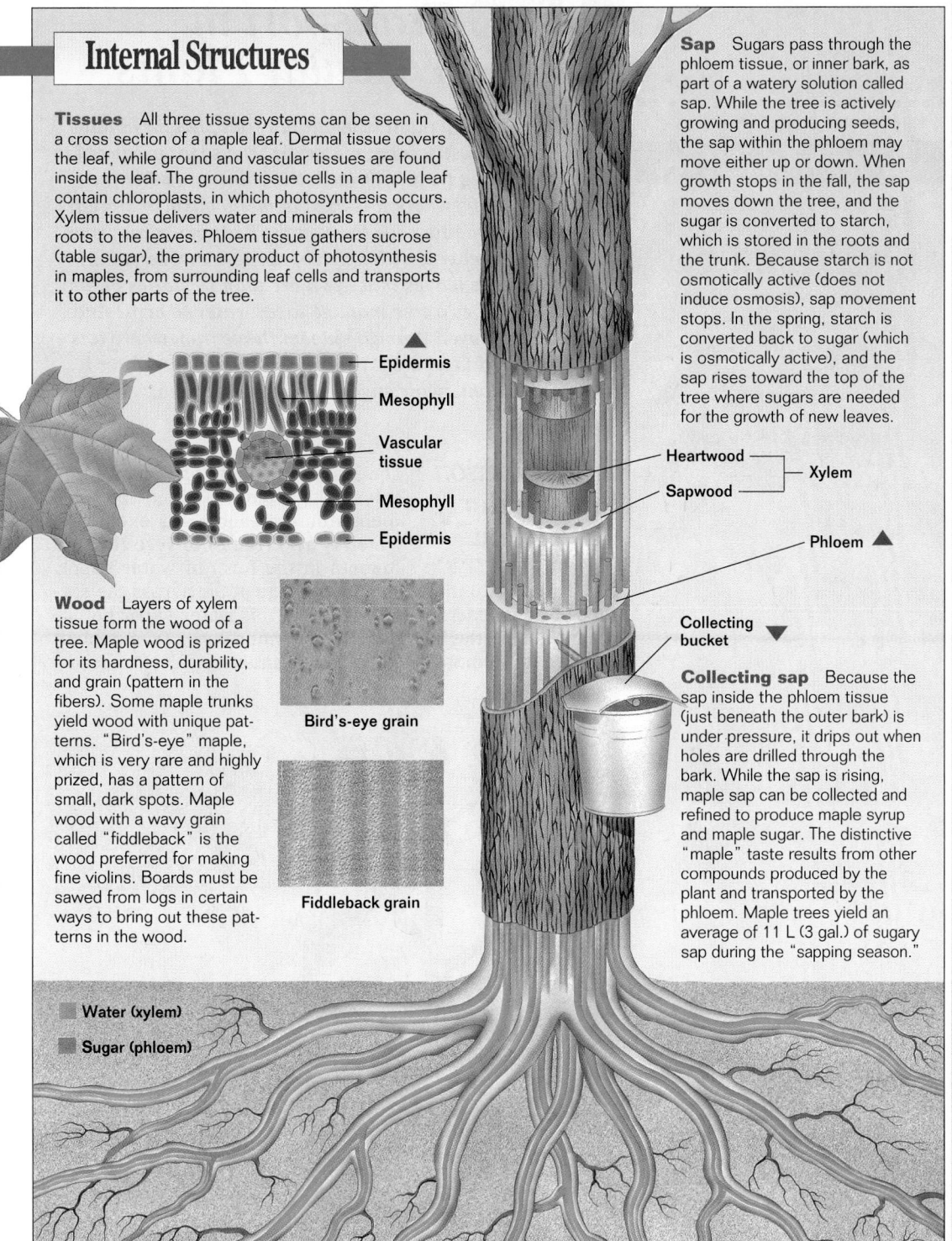

Internal Structures

Tissues All three tissue systems can be seen in a cross section of a maple leaf. Dermal tissue covers the leaf, while ground and vascular tissues are found inside the leaf. The ground tissue cells in a maple leaf contain chloroplasts, in which photosynthesis occurs. Xylem tissue delivers water and minerals from the roots to the leaves. Phloem tissue gathers sucrose (table sugar), the primary product of photosynthesis in maples, from surrounding leaf cells and transports it to other parts of the tree.

Epidermis

Mesophyll

Vascular tissue

Mesophyll

Epidermis

Wood Layers of xylem tissue form the wood of a tree. Maple wood is prized for its hardness, durability, and grain (pattern in the fibers). Some maple trunks yield wood with unique patterns. "Bird's-eye" maple, which is very rare and highly prized, has a pattern of small, dark spots. Maple wood with a wavy grain called "fiddleback" is the wood preferred for making fine violins. Boards must be sawed from logs in certain ways to bring out these patterns in the wood.

Bird's-eye grain

Fiddleback grain

■ Water (xylem)
■ Sugar (phloem)

Sap Sugars pass through the phloem tissue, or inner bark, as part of a watery solution called sap. While the tree is actively growing and producing seeds, the sap within the phloem may move either up or down. When growth stops in the fall, the sap moves down the tree, and the sugar is converted to starch, which is stored in the roots and the trunk. Because starch is not osmotically active (does not induce osmosis), sap movement stops. In the spring, starch is converted back to sugar (which is osmotically active), and the sap rises toward the top of the tree where sugars are needed for the growth of new leaves.

Heartwood
Sapwood
Xylem

Phloem ▲

Collecting bucket ▼

Collecting sap Because the sap inside the phloem tissue (just beneath the outer bark) is under pressure, it drips out when holes are drilled through the bark. While the sap is rising, maple sap can be collected and refined to produce maple syrup and maple sugar. The distinctive "maple" taste results from other compounds produced by the plant and transported by the phloem. Maple trees yield an average of 11 L (3 gal.) of sugary sap during the "sapping season."

Discussion

Guide the discussion by posing the following questions.

1. What is the economic importance of sugar maple trees in the United States? *(They provide raw materials for making many important products such as maple syrup, maple sugar, furniture, musical instruments, and wood flooring.)*

2. How might a sugar maple tree's production of maple sap be affected if it were growing in an environment where the temperature stayed the same year round, such as the tropics? *(In an environment with constant temperature, there might not be as much sugar converted to starch for storage, and some sap movement would occur within the tree all the time. As a result, there would not be a large amount of sap rising at any one time, and collecting the sap would be difficult.)*

3. Why is maple a more desirable wood for making furniture than oak or pine? *(Maple is a hardwood, which is more durable than a softwood such as pine, but it is not as hard as oak and therefore easier to work with than oak. Maple also takes a stain better than harder woods and does not split as easily as oak.)*

Vocabulary Preview

transpiration

tension-cohesion theory

source

sink

translocation

pressure-flow model

Opening Questions

Ask students how phloem and xylem cells are similar. Students should recall that the cells of both types of vascular tissue are elongated and tubular. Ask how the two cell types differ. Students should recall that xylem cells are thick-walled and phloem cells are thin-walled.

Demonstration

Transport in Xylem

Slice a stalk of celery lengthwise to just below the leaves. Place the two halves in separate beakers, each containing a different color of water. (Hint: Red and blue food coloring work best.) Students should be able to see the veins in the leaves turn color. Have students make cross sections of the celery stalks and locate the xylem.

👁 VISUAL STRATEGY

Figure 24-10

Trace the path that water takes through a plant. Ask: How could a plant prevent water loss? *(by closing its stomata)* How would the transpiration rate differ between a humid environment and a dry environment? *(The transpiration rate would be greater in a dry environment because water would evaporate from a leaf more quickly.)*

The body of a plant is beautifully adapted for obtaining and distributing materials. The sugar produced in the leaves travels through the phloem to other parts of the plant body, where it is used in metabolism or stored. Water and nutrients move through the body of a plant, from its roots to its leaves, within xylem tissue. However, many large trees have leaves that are more than 10 stories off the ground. How does a tree manage to lift water so high? And how is sugar moved through phloem tissue from where it is made or stored to where it is needed? Simply put, water is pulled up a plant, while sugar is pushed through it.

Section Objectives

■ Explain how transpiration helps move water up a plant.
■ Relate the cohesive and adhesive properties of water to its movement in a plant.
■ Explain how guard cells regulate the rate of transpiration.
■ Describe the process by which sugars are translocated throughout the body of a plant.

Transpiration Pulls Water Up a Plant

The leaves of plants have many tiny holes—the stomata. When they are open, stomata enable gas exchange. Water is also free to diffuse through stomata in the form of water vapor. As Figure 24-10 indicates, the passage of air across the surface of a leaf carries away much of this water vapor before it can reenter the leaf. The loss of water vapor from a plant through its stomata is called **transpiration**.

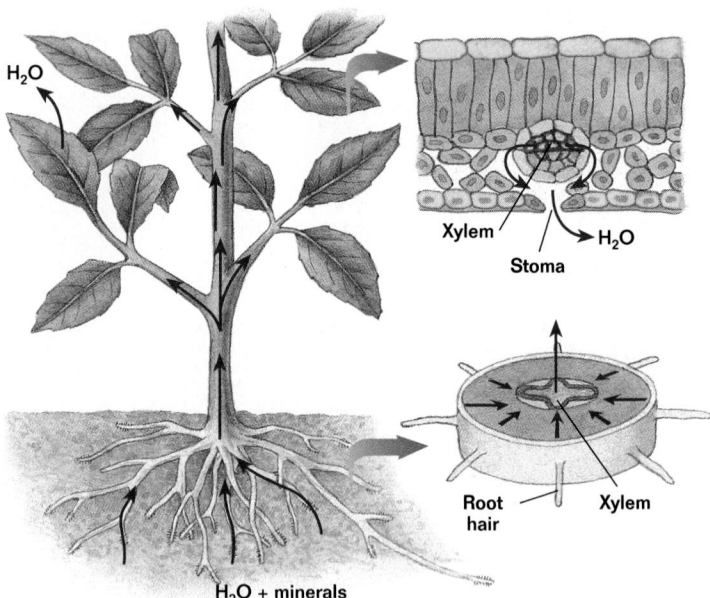

Figure 24-10 Transpiration is the loss of water from a plant. Most of this water loss is from the leaves. The water vapor that exits a leaf by transpiration first entered the plant through its roots and traveled to the leaves through the xylem.

Overcoming Misconceptions

Movement in Xylem and Phloem

Transpiration is *not*, as it is often described, the movement of water within xylem. Water flows through xylem, aided by capillary action. Transpiration is merely the loss of water from leaves through evaporation. It drives the upward movement of water by creating a suction. Also, stating simply that xylem carries water and dissolved nutrients *up* a plant and that phloem carries food *down* a plant is an oversimplification. Water does move upward from roots, but it also moves laterally. Food is carried downward, but it also moves laterally and upward from storage tissues in the roots and stems.

Water is pulled up a plant because the loss of water by transpiration creates a suction that draws water out of the tracheids and vessels of the plant's xylem. This water extends in an unbroken column down through the stems and into the roots, where water is absorbed from the soil. As long as the column of water in the xylem remains unbroken, water will continue to move upward because of the pull of transpiration. More than 90 percent of the water taken in by the roots of a typical plant is ultimately lost through the plant's leaves in this way.

According to the **tension-cohesion theory,** two properties of water itself and a simple element of plant structure assist the pull of transpiration in moving water up a plant. Recall from Chapter 2 that water is a polar substance. As a result, water molecules readily form hydrogen bonds that enable them to stick to each other (cohere) and stick to other polar substances (adhere). The cohesion of water molecules, illustrated in Figure 24-11, gives a column of water great tensile strength. In other words, it can withstand a lot of tension (pull) without breaking. Thus, cohesion of water molecules helps to maintain an unbroken column of water in xylem tissue. Because of adhesion, water is able to move up the sides of a narrow tube by capillary action. Water absorbed by the roots of a plant moves through xylem cells that are elongated and very narrow, like straws. Thus, the adhesion of water molecules to the walls of the very narrow xylem cells helps to draw water to the top of a plant. ◻

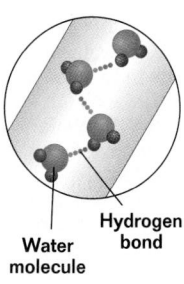

Figure 24-11 Water molecules are held together so strongly by hydrogen bonds that a column of water can withstand great tension without breaking.

◻ **CAPSULE SUMMARY**

Transpiration is the loss of water from a plant through its stomata. The cohesion and adhesion of water molecules help water to move up a plant in an unbroken column.

Guard Cells Regulate the Rate of Transpiration

The rate of transpiration must be regulated so that a plant does not lose too much water. Water loss by transpiration can be prevented only by the closing of a plant's stomata. However, stomata must be open at least part of the time so that the carbon dioxide needed for photosynthesis can enter the plant. Therefore, every plant must strike a balance between the conflicting demands of water conservation and photosynthesis.

A stoma is surrounded by a pair of guard cells that are shaped like two cupped hands. The stoma opens and closes because of changes in the water pressure within the guard cells. When guard cells take in water by osmosis, they become turgid (plump and swollen) and bowed in shape. The bowed shape results because the inner wall of a guard cell is thicker than the rest of the cell wall and cannot stretch when the guard cell swells. If you were to use a piece of tape to thicken one side of a long balloon, it would also bend when inflated. As the inner walls of a pair of guard cells separate, the stoma opens. When water leaves the guard cells, they lose turgor, their inner walls come back together, and the

Teaching Tip
Solve the Puzzle

Use the demonstration below to teach students about the properties of water that enable it to move up a tree. After students have seen the demonstration, pose the following question: How does water move up the untreated paper towel? *(The water first begins to move up the paper towel because it adheres to the towel's surface. Other water molecules follow because of cohesion.)* Why doesn't water move up the treated paper towel? *(Water does not move up the treated paper towel because Scotchgard® is a nonpolar substance to which water will not attach.)* Let students take turns asking questions to which you can respond "yes" or "no." From your responses, students should form hypotheses to explain the problems.

Demonstration
Adhesion and Cohesion of Water

Materials: 2 paper towels from the same roll (per class), 1 can of Scotchgard®, 2 large beakers of water, blue and green food color.
Procedure: Prior to the lesson, spray one of the paper towels with Scotchgard® and let it dry. Fill both of the beakers half full of water, and add blue food color to one beaker and green food color to the other. (Note: The purpose of the food color is to make it easier for students to see any movement of water. Students should determine that color makes no difference to the outcome of this demonstration.) Place the beakers in front of the class. Take the two paper towels and roll them lengthwise so that they can be dipped into the beakers. Slowly lower one paper towel into each beaker at the same time. Do not say anything. Just let students observe what happens, and ask the questions posed above in *Teaching Tip: Solve the Puzzle.*

Teaching Tips
Locating Guard Cells

Have students paint a small area on the lower side of a leaf with clear fingernail polish. Once the polish is dry, have students use forceps to carefully remove the dried polish in strips that are large enough to make wet-mount slides. Let students use a microscope to observe the images of the guard cells in the polish.

What If the epidermis of a plant became covered with an impermeable substance?

Covering the epidermis of a plant blocks gas exchange through the stomata. In the case of leaves, the inability to obtain carbon dioxide would cause photosynthesis to slow or stop, and the inability to transpire would cause water movement up the plant to stop. Since it doesn't rain inside, occasionally washing the dust from leaves of houseplants prevents these problems from occurring.

Demonstration
Modeling Guard Cells

Materials: 2 long, straight balloons; roll of electrical tape.
Procedure: To demonstrate how guard cells work, partially inflate two long, straight balloons. Place a strip of tape lengthwise on one side of each balloon. Then continue inflating the balloons until they are firm. When the balloons are fully inflated, they curve toward the side with the tape. By joining the ends of the balloons together so that the strips of tape face each other, you can demonstrate how a stoma is opened by the curving of turgid guard cells. Finally, allow the balloons to slowly deflate. Ask: How does this resemble what happens to the stomata of plants? (*As the guard cells deflate, they relax and close the stomata.*) What do guard cells lose as they deflate? (*water*)

Figure 24-12 When potassium ions enter a pair of guard cells, the cells take in water by osmosis. The thickened inner walls of the guard cells enable a stoma to open when this water causes the cells to swell, *left*. When potassium ions leave the guard cells, water also leaves by osmosis, and the guard cells relax, closing the stoma, *right*. Blue arrows indicate water movement.

CAPSULE SUMMARY

Thickened inner walls of guard cells bend when the cells are turgid, thus opening a stoma. Guard cells regulate the rate of transpiration in response to a mechanism that causes potassium ions to move into and out of the cells.

CAPSULE SUMMARY

Translocation is the movement of sugars through a plant from a source to a sink.

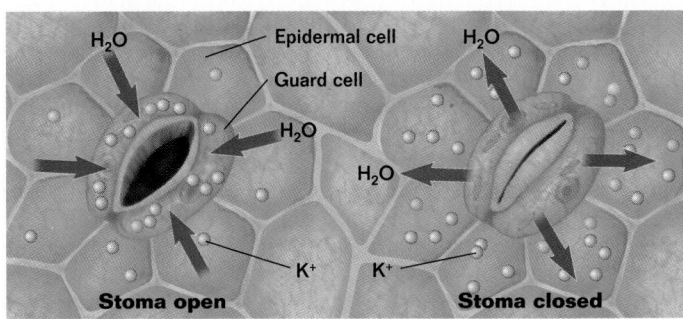

Stoma open   Stoma closed

stoma closes, as illustrated in Figure 24-12. Thus, loss of water from the guard cells for any reason causes stomata to close and stops further water loss. This is homeostasis in action.

Although the exact mechanism is not well understood, potassium ions, K^+, play a critical role in opening and closing stomata. An active transport process that is triggered by light causes potassium ions to move into guard cells. The increased concentration of potassium ions inside the cells causes water to enter them by osmosis. When the movement of potassium ions is reversed and the potassium ion concentration becomes high in the surrounding cells, water diffuses out of the guard cells. The resulting loss of turgor by the guard cells causes the stoma to close. ◻

Sugar Is Pushed Through a Plant

In a plant, sugar moves from where it is made or stored to where it is needed through the sieve tubes of phloem. Botanists use the term **source** to refer to a part of a plant that provides sugar for other parts of the plant. For example, a leaf is a source because it makes sugar during photosynthesis. A root is also a source when sugar stored there is moved to other parts of the plant. Botanists use the term **sink** to refer to a part of a plant to which sugar is delivered. Areas of active growth where sugar is needed for metabolism, such as root tips and developing fruits, are examples of sinks. The movement of sugar within a plant from a source to a sink is called **translocation**. Look back at *Up Close: Sugar Maple* on pages 560–561 to learn more about how sugars move in a plant.

The movement of sugar in a plant is more complex than the movement of water. First, water flows freely through empty xylem elements, but sugar must pass through the cytoplasm of living cells. Second, water only moves upward within the xylem, while sugar moves both upward and downward in the same sieve tube, but at different times. Last, water diffuses freely through a plasma membrane, but sugar cannot. How, then, is sugar distributed throughout a plant? Many attempts have been made to answer this question. ◻

Did You Know?

Scientists cannot tap into a phloem sieve tube without causing the sieve plates to immediately seal and disrupt the flow of sugars. Yet a tiny insect that feeds on sugar from plants, the aphid, has the ability to tap into the phloem without disrupting the flow of sugars.

The model of translocation that most botanists favor was proposed in 1924 by the German botanist Ernst Münch. Münch's model was first tested by an experiment like the one seen in Figure 24-13. Using a concentrated sugar solution to represent the phloem near a source, water can be made to enter a tube and flow through it. The pressure created by water entering the tube pushes the water and some of the sugar in the solution to the other end of the tube. Therefore, Münch's model of translocation is often called the **pressure-flow model.** Once the tube is completely filled with water, sugar is also able to diffuse from one end to the other, as long as the sugar concentration remains higher at one end than the other. To see how Münch's model is thought to work in a plant, look at Figure 24-14.

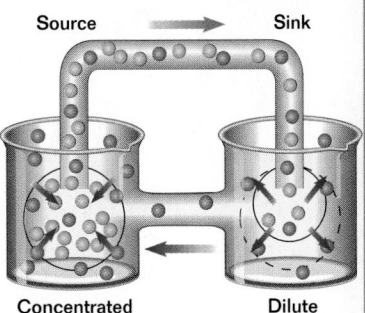

Source　　　　Sink

Concentrated solution　　Dilute solution

● Sugar　● Water

Figure 24-13 In a simple osmometer, *above,* the diffusion of water into a concentrated sugar solution pushes the sugar solution from the "source" to the "sink."

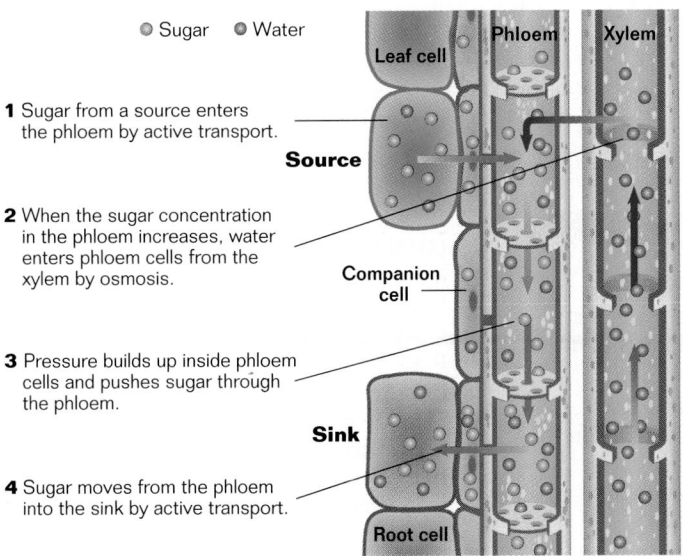

● Sugar　● Water

Leaf cell　Phloem　Xylem

1 Sugar from a source enters the phloem by active transport.

Source

2 When the sugar concentration in the phloem increases, water enters phloem cells from the xylem by osmosis.

Companion cell

3 Pressure builds up inside phloem cells and pushes sugar through the phloem.

Sink

4 Sugar moves from the phloem into the sink by active transport.

Root cell

Figure 24-14 According to Münch's hypothesis, sugars are pushed through the phloem by the pressure that results from the movement of water into the phloem by osmosis. Blue arrows indicate water movement, and red arrows indicate sugar movement.

Section Review

1. What is transpiration, and how does it pull water up a plant?

2. What properties of water assist in its rise up a plant?

3. How do guard cells regulate the rate of transpiration?

4. What is translocation, and how does it occur?

Critical Thinking

5. Why might the lack of potassium, K, in the soil around a plant's roots interfere with the plant's ability to exchange gases and absorb water?

SECTION 24-2

◉ VISUAL STRATEGY

Figure 24-14
Walk students through this diagram. Ask volunteers to describe in their own words what happens to sugars and water during translocation. *(Sugars are actively transported into the phloem from a source cell, such as a leaf cell, and move toward a sink cell, such as a root cell. Water diffuses into the phloem because of its higher sugar concentration and flows toward a sink cell.)* Ask: Is it possible for a root cell to be a source? *(yes)* When? *(when upper parts of the plant need sugars and they are not being supplied by the leaves; for example, in the spring when new growth is beginning)* Are the new leaves developing in buds a source or a sink? *(a sink)* Why? *(because while they are in a bud, developing leaves cannot photosynthesize and therefore need sugars from elsewhere to make the energy needed for growth by cellular respiration)*

Chapter Closure

Have students make a concept map using as many of this chapter's vocabulary words as they can. Have them make their maps on large sheets of butcher paper and use markers to draw arrows and write connecting words.

Answers to Section Review

1. Transpiration is the loss of water from a plant. It causes water to move up a plant because each cohesively bound water molecule that is lost at the top of a column of water pulls another molecule into its place.

2. Because of its polarity, water binds to itself (cohesion) and to other objects (adhesion).

3. Guard cells slow the rate of transpiration by closing and speed the rate of transpiration by opening.

4. Translocation is the movement of sugar from a source to a sink. Sugars are pushed through phloem by the pressure of water moving into areas of high sugar concentration.

5. Because potassium is necessary to regulate the movement of water into and out of guard cells, a lack of potassium would interfere with a plant's ability to open its guard cells. If the guard cells cannot open, gas exchange cannot occur, water cannot exit by transpiration, and water cannot be absorbed by the roots.

CHAPTER REVIEW ANSWERS

Each item in the Chapter Review is correlated by text section in the assignment guide that follows.

ASSIGNMENT GUIDE

Section	Review	Themes Review	Critical Thinking
24-1	1–5, 7, 11–14	18	20
24-2	6, 8–10, 14–17	18, 19	21, 22

Review

Multiple Choice

Code in parentheses indicates section number and objective number.

1. c (24-1.1)
2. d (24-1.1)
3. c (24-1.1)
4. d (24-1.2)
5. b (24-1.2)
6. a (24-2.1)
7. a (24-1.3)
8. b (24-2.2)
9. c (24-2.3)
10. b (24-2.4)

Completion

11. xylem, phloem (24-1.1)
12. sieve tube, companion cells (24-1.1)
13. mesophyll, palisade, spongy (24-1.2)
14. translocation, transpiration (24-2.1, 24-2.4)
15. polar, cohesion (24-2.2)
16. water, open, closed (24-2.3)
17. source, sink (24-2.4)

Vocabulary

adventitious root (559)	leaflet (555)	sink (564)
C_3 plant (556)	lenticel (558)	source (564)
C_4 plant (556)	mesophyll (556)	spongy layer (556)
companion cell (554)	node (557)	taproot system (559)
cork cell (558)	palisade layer (556)	tension-cohesion theory (563)
cortex (557)	petiole (555)	terminal bud (557)
dermal tissue system (551)	pith (557)	tracheid (553)
epidermis (552)	pressure-flow model (565)	translocation (564)
fibrous root system (559)	root cap (558)	transpiration (562)
ground tissue system (551)	root hair (558)	vascular bundle (557)
heartwood (557)	sapwood (557)	vascular tissue system (551)
herbaceous stem (557)	sieve plate (554)	vessel (553)
lateral bud (557)	sieve tube (554)	vessel element (553)
internode (557)	sieve tube member (554)	woody stem (557)

Review

Multiple Choice

1. Which of the following tissues transports water and carbohydrates through a plant?
   a. dermal
   b. cork
   c. vascular
   d. mesophyll

2. Which of the following is *not* a function of ground tissue in plants?
   a. storage
   b. support
   c. photosynthesis
   d. protection

3. The dermal tissue system functions in
   a. transport of water.
   b. transport of carbohydrates.
   c. protection.
   d. sexual reproduction.

4. Xylem cells that no longer conduct water make up the
   a. pith.
   b. sapwood.
   c. cortex.
   d. heartwood.

5. Which of the following is a modified stem?
   a. carrot
   b. Irish potato
   c. pine needle
   d. cactus spine

6. Transpiration rate is regulated by
   a. stomata.
   b. tracheids.
   c. pressure-flow.
   d. movement from source to sink.

7. Which of the following is *not* a characteristic of sugar maple trees?
   a. large, brightly colored flowers
   b. winged fruits called samaras
   c. five-pointed, palmately lobed leaves
   d. glossy, reddish brown twigs

8. Capillary action in the xylem of plants is made possible by
   a. pressure-flow.
   b. hydrogen bonding.
   c. low surface tension.
   d. high vapor pressure.

9. Guard cells become turgid and bow in shape when
   a. water moves out of the cells.
   b. potassium moves out of the cells.
   c. water moves into the cells.
   d. water levels are low in the cells.

10. According to the pressure-flow model, sugar moves through the phloem by
    a. osmosis, from areas of low sugar concentration to areas of high sugar concentration.
    b. pressure created when water moves by osmosis into areas of high sugar concentration.
    c. the diffusion of sugar molecules from sink to source.
    d. active transport of water from companion cells in the leaves.

Completion

11. In vascular tissue, _____ cells are dead at maturity, but _____ cells are living.

12. Phloem contains _____ members through which carbohydrates flow and _____ that control their activities.

13. In some leaves, the _____ includes two distinct layers of photosynthetic cells. The cells of the _____ layer are tightly packed and columnar. The _____ layer consists of loosely packed, spherical cells.

14. The movement of the products of photosynthesis within a plant is called _____. Water moves up a plant because of _____, the loss of water from leaves.

15. Because water is a _____ molecule, its movement in xylem is aided by adhesion and _____.

16. The guard cells regulate _____ loss in plants. The rate of transpiration increases when stomata are _____ and decreases when stomata are _____.

17. According to the pressure-flow model, sugars travel from a _____ to a _____.

Themes Review

18. **Structure and Function** Suppose that you are a water molecule passing through a plant. Trace your path from your entry into a root hair to your exit between two guard cells. Identify each structure that you encounter, and describe its function.

19. **Homeostasis** When a plant wilts, its stomata close. How, then, does wilting help a plant maintain homeostasis?

Critical Thinking

20. **Making Inferences** Compare taproot and fibrous root systems in Figure 24-9 on page 559. Would plants with taproot systems or plants with fibrous root systems be more likely to prevent erosion on a steep hillside? Explain.

21. **Interpreting Data** The rate of water movement in a plant is assumed to indicate the rate of transpiration. After measuring the rate of water movement in a plant during high and low humidity, the data was graphed, below. Which line, A or B, indicates a lower transpiration rate? Justify your answer.

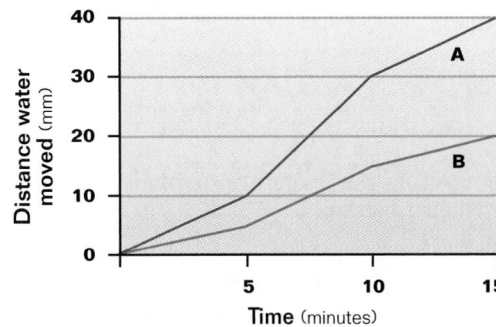

22. **Making Inferences** In the graph above, which line, A or B, indicates the rate of transpiration during high humidity? Which indicates the rate of transpiration during low humidity? Explain.

Activities and Projects

23. **Cooperative Group Project** Obtain a Venus' flytrap, a pitcher plant, and a sundew plant from a nursery or a biological supply company. Observe each plant to find out how its leaves trap insects. Make a videotape of the plants capturing insects. Report your findings to your class, using the videotape to enhance your report.

24. **Cooperative Group Project** Build clay models of cross sections of a C_3 and a C_4 leaf. Show the epidermis, mesophyll, veins, and guard cells of each. Using the models, explain to your class each leaf's structure and the function of its parts.

25. **Research and Writing** Research how sugar is obtained from plants. Relate your findings in a written report.

Critical Thinking

20. Plants with fibrous root systems branch throughout a large volume of soil and therefore would do a better job of controlling soil erosion than plants with taproot systems. (24-1.2)

21. The rate of transpiration is lower in "B." Water in "B" did not move as far as it did in "A" in the same amount of time. Water does not move through a plant as quickly when the transpiration rate is low. (24-2.1)

22. "B" shows the rate of transpiration during high humidity, and "A" shows the rate during low humidity. When the humidity is high, water would not evaporate from a leaf as quickly as it would in low humidity; and when the evaporation rate is low, the transpiration rate is low. (24-2.1)

Activities and Projects

23. By experimenting, students should learn the following: the leaves of the Venus' flytrap close when one of the trigger hairs is touched three times in succession; when an insect falls into a pitcher plant, the plant's downward-pointing hairs prevent the insect from crawling out; and when an insect gets stuck in the sticky tendrils of the sundew plant, the sundew plant begins to curl over it.

24. The models should resemble the internal structure of a leaf shown in Figure 24-6.

25. Sugar is obtained from sugar cane, sugar beets, and sugar maple trees.

Themes Review

18. Answers will vary but should mention moving into the root hair, into a xylem tracheid or vessel element, up the plant, into its branches, through a petiole, and into a leaf within a xylem vessel. Finally, the water molecule would move into an air space between the cells of the spongy layer and out through a stoma. (24-1.1, 24-1.2, 24-2.1)

19. When stomata are closed and transpiration decreases, water inside the plant is conserved. Thus, wilting reduces a plant's need for water from the soil. When water is available in the soil, a plant will respond by absorbing water and becoming turgid again. (24-2.3)

LABORATORY Investigation Chapter 24

Preparation

Set up stations for observing slides of roots, stems, and leaves. Each station will need a compound light microscope and one of the types of prepared slides listed below. Prepared slides may be ordered from biological supply companies. Suitable products that can be ordered from Ward's include the following: *Allium* root tip, *Ranunculus* root cross section, *Zea mays* stem cross section, *Ranunculus* stem cross section, and lilac leaf cross section.

Procedural Notes

- Emphasize that it is best to focus at low power and then switch to high power.

Background Answers

1. Root hairs absorb water and minerals.

2. Translocation within phloem tissue transports food throughout a plant.

3. Herbaceous stems are soft and usually green. Woody stems are rigid and nongreen.

4. The vascular tissues, xylem and phloem, are continuous in the root, stem, and leaf.

5. The closing of stomata by guard cells and the cuticle covering the epidermis are two ways that a leaf conserves water.

6. What tissues make up roots, stems, and leaves, and what are the functions of these tissues?

Roots, Stems, and Leaves

OBJECTIVE

Observe the tissues that make up roots, stems, and leaves, and examine their structure.

PROCESS SKILLS

- identifying tissues found in roots, stems, and leaves
- relating the structure of plant tissues to their function

MATERIALS

- Prepared slides of the following tissues:
  - *Allium* root tip
  - *Ranunculus* root cross section
  - *Zea mays* stem cross section
  - *Ranunculus* stem cross section
  - Lilac leaf cross section
- Compound light microscope

BACKGROUND

1. Which plant tissues are responsible for the absorption of water and dissolved minerals?

2. How is food, produced in the leaf, moved to other parts of the plant?

3. How do woody and herbaceous stems compare?

4. What tissues are continuous in the root, stem, and leaf?

5. How does the leaf conserve water?

6. Write your own question to explore in your lab report or notebook.

TECHNIQUE

Part A: Roots

1. Study the cross section of the root tip shown in the photograph in Figure 24-9 on page 559.

2. Observe the prepared slide of the *Allium* root tip under low power. Locate the root cap and the root tip meristematic cells. Note the long root hairs in the area above the root tip.

3. Change slides to the *Ranunculus* cross section. The inner core is the vascular tissue, which is surrounded by the endodermis. This area is the tissue for transport of water, minerals, and food. Look for the star-shaped xylem and the smaller phloem cells surrounding the xylem. Draw what you see, and identify the tissues in the **Records** section of your report.

4. Locate the cortex, where starch is stored, surrounding the vascular cylinder. Outside the cortex you will find the epidermal cells and their root hairs. Draw a one-fourth section of the root tissues, as if you were cutting a pizza slice out of the cross section. Label all the tissues in the **Records** section of your report.

Part B: Stems

5. Observe a prepared slide of the stem of *Zea mays*, a monocot, and find the epidermis and the photosynthetic layer. In the center, look for the vascular bundles made up of xylem and phloem. Draw a diagram showing the location of the vascular bundles and the epidermis layer as they appear when viewed under low power.

6. Switch to high power and observe a vascular bundle. Draw the vascular bundle and label the tissues.

7. Observe a cross section of a herbaceous dicot stem such as that of *Ranunculus*.

Records

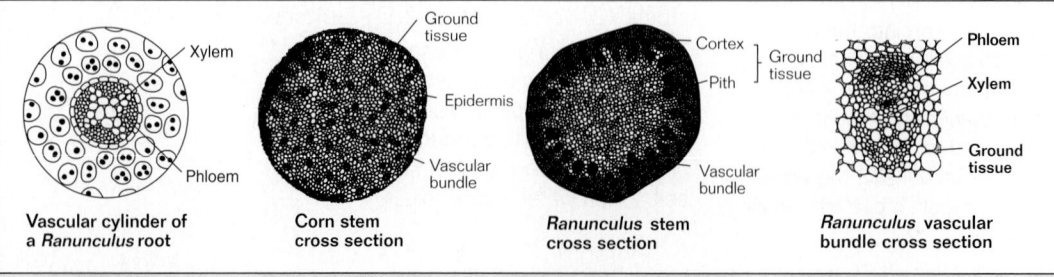

Vascular cylinder of a *Ranunculus* root — Xylem, Phloem

Corn stem cross section — Ground tissue, Epidermis, Vascular bundle

Ranunculus stem cross section — Cortex, Pith } Ground tissue, Vascular bundle

Ranunculus vascular bundle cross section — Phloem, Xylem, Ground tissue

Look for the epidermis and cortex layers. Notice that the stem is more complex than a dicot root. Note the arrangement of the vascular bundles. In the **Records** section of your report, draw what you observed.

8. Now focus on a vascular bundle under high power. Draw and label a diagram in the **Records** section of your report.

Part C: Leaves

9. Observe a prepared slide of a lilac leaf cross section under low power, and find the lower epidermis.

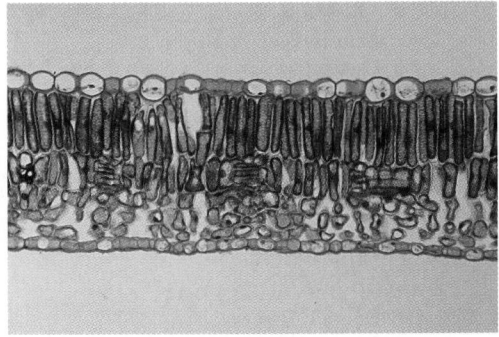

10. Identify the stomata on the lower epidermis. Find the guard cells that open and close a particular stoma. Locate an open stoma and a closed stoma. Draw and label diagrams of the stomata and guard cells in the **Records** section of your report.

11. Note the spongy texture of mesophyll. Locate a vein containing xylem and phloem. Continue toward the palisade layer into the upper epidermis until you reach the clear continuous noncellular layer on top. This layer is called cutin. Draw and label a diagram of your observations in the **Records** section of your report. Also, briefly summarize the procedure you followed.

INQUIRY

1. In the root, where are phloem and xylem located?
2. Where are the xylem and phloem found in the herbaceous stem?
3. How are the vascular bundles different in monocot and dicot stems?

ANALYSIS

1. How are the root cap cells different from the root tip meristematic cells?
2. What is the function of the root hairs?
3. How different is the arrangement of xylem and phloem in roots, stems, and leaves?
4. What is the function of a stoma?
5. What is the function of the air space in the mesophyll of the leaf?
6. Which leaf structures help to conserve water?
7. Which tissues of the leaf are continuous with the stem and root tissues? How is this functional?

FURTHER INQUIRY

Write a new question that could be explored as a new investigation. The following is an example:

> How have the stems and roots of some plants become modified for storing sugars and starches?

4. A stoma allows water vapor, carbon dioxide, and oxygen to pass into and out of a leaf.

5. Air spaces enable gases to diffuse between the outside of a leaf and the cells inside of the leaf.

6. The guard cells and the cuticle help conserve water.

7. Xylem and phloem are continuous from root to stem to veins of leaves, an arrangement that facilitates the movement of water from the roots to the rest of the plant and the movement of sugars from where they are made or stored to where they are needed.

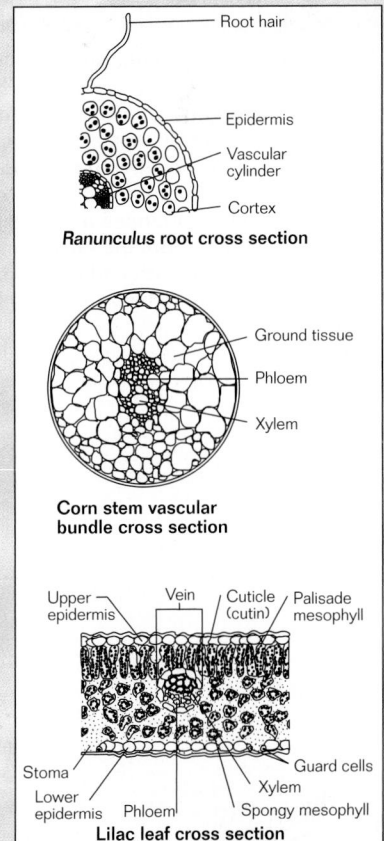

Ranunculus root cross section

Corn stem vascular bundle cross section

Lilac leaf cross section

Inquiry Answers

1. Xylem and phloem are located in a cylinder at the center of a root.

2. In herbaceous stems, xylem and phloem are found in vascular bundles, with phloem toward the outside of the stem and xylem toward the center of the stem.

3. In monocots, vascular bundles are scattered throughout the stem. In dicots, vascular bundles are arranged in an orderly ring near the outside of the stem.

Analysis Answers

1. Root cap cells are large and loosely packed. Meristematic cells are very small and tightly packed.

2. Root hairs absorb water and minerals from the soil.

3. Xylem and phloem are found in the center of roots, in vascular bundles either scattered throughout or arranged in rings at the outside of stems, and in the veins of leaves.

Block Scheduling Guide

> Each block represents about 45–50 minutes of instructional time. The resources cited in a block are coded to a specific program component using the Key on the next page.

Lecture Concepts

Lesson Resources

25-1 How Plants Grow and Develop pp. 571–577

BLOCKS ① and ②

a. A seed germinates when the dormant embryo within the seed resumes growing.
b. Primary growth is produced at apical meristems located at the tips of shoots and roots.
c. Secondary growth is produced at lateral meristems that form from primary stem tissues.
d. Plants develop continuously as new cells form and differentiate.
e. Plant development is considered reversible because cells of certain tissues can produce unspecialized cells that then differentiate into all of the mature plant's tissues.
f. Depending on how long they live, plants can be classified as one of three types: annuals, biennials, or perennials.

Lecture Resources
- Opening Question, page 571
- Demonstration: Corn and Bean Seeds
- Visual Strategies: Figures 25-2, 25-4, 25-6
- Teaching Transparencies: 128, 129, 120A
- Finding the Vascular Cambium, page 575
- How is Plant Development "Reversible"? page 575
- Overcoming Misconceptions, page 575
- Holt Biology Videodiscs: Lesson 25-1

Classwork Options
- Effective Reading, page 572

- Solve the Problem, page 572
- Prepared Slides, page 573
- Summarizing Primary and Secondary Growth, page 574
- Nature Walk, page 577
- Design a Flower Bed, page 577
- Laboratory Techniques and Experimental Design Manual C37: Growing Plants in the Laboratory
- Laboratory Techniques and Experimental Design Manual C38: Growing Plants in the Laboratory—Fertilizer Problem

Assignment Options
- Section Review, page 577
- Chapter Review, questions 1–6, 11–14, 18, 19, 21

25-2 Regulating Growth and Development pp. 580–585

BLOCKS ③ ④ and ⑤

a. Plants use photosynthesis to produce all of their organic molecules, but require several inorganic nutrients to convert carbohydrates to other important organic molecules.
b. Plant growth and development is regulated by hormones.
c. Plant hormones are produced in response to stimuli from the environment. Therefore, plant growth and development respond to the environment in many ways.

Lecture Resources
- Opening Question, page 580
- Demonstrations: Dormant Apple Seeds, Ethylene and Leaf Drop, Plant Responses
- Visual Strategies: Figures 25-8, 25-12, 25-15
- What Do Apple Seeds Need to Grow? page 581
- Research Update, page 582
- Teaching Transparencies: 130, 119A
- Holt Biology Videodiscs: Lesson 25-2

Classwork Options
- Effective Reading, page 580

- Closure, page 585
- Laboratory Investigation: Seed Structure and Seedling Development, pages 588–589
- Inquiry Skills Development B18: Mineral Deficiencies in Plants
- Inquiry Skills Development B19: Gravitropism and Phototropism in Plants

Assignment Options
- Identify the Photoperiod, page 584
- Section Review, page 585
- Chapter Review, questions 7–10, 15–17, 20, 22, 23

Review/Enrichment

BLOCK ⑥

- ■ Up Close: Kalanchoë, pages 578–579
- ■ Study Guide: Concept Review and Pretest

Assignment Options
- ■ Activities and Projects, questions 24, 25

Assessment Options

BLOCK ⑦

Traditional Assessment
- ■ Chapter 25 Test
- ⊙ Test Generator/Assessment Item Listing: Software item bank for preparing customized chapter tests.

Portfolio Assessment
Select student reports for one or more laboratory experiments from this chapter. *The Direct Observation Checklist* on the *BioSources Teaching Resources CD-ROM* should be completed during a laboratory or other cooperative learning experience.

Holt Biology Videodiscs

Holt Biology Videodiscs gives you a powerful tool for teaching, review, and assessment. *Concepts of Biology* adds interactive lesson plans and extensive tools for customization using CD-ROM technology.

CONCEPTS OF BIOLOGY

Use the following topic from *Concepts of Biology* to help you teach this chapter:

- ■ Topic 16: Plant Structure and Function

For further information regarding the media on the videodiscs, see *Holt Biology Videodiscs Teacher's Correlation Guide to Biology: Principles and Explorations.*

Chapter Themes

Levels of Organization As plants grow, they develop the specialized cells, tissues, and structures of their bodies. Use Figures 25-3, 25-4, and 25-5 to illustrate this theme.

Structure and Function
Plants develop specialized structures that enable them to survive in their environments. Some of these structures are seen in Figures 25-1, 25-2, 25-9, and 25-15.

Homeostasis The many changes in a plant, which begin with the germination of a seed and result in the production of new seeds, require a balance of nutrients obtained from the environment and a balance of growth-regulating hormones produced by the plant. Use Figures 25-10, 25-11, 25-12, 25-13, and 25-14 to illustrate this theme.

Tapping Prior Knowledge

- What conditions do plants need to grow?
- What are the three tissue systems of a plant?
- How do monocot seeds differ from dicot seeds?

Opening Question

Ask students how they think the bristlecone pine in the chapter opening photograph on page 570 developed its shape and how this shape helps the tree survive in its environment. Help students to derive that this tree grows at a high altitude, where, with few other trees around, it is greatly affected by strong winds that blow mainly from one direction and cause the tree to bend away from the direction of the wind as it grows. The "flagged" (like a flag in the wind) shape reduces the tree's wind resistance and helps prevent broken limbs and uprooting.

Bristlecone pine "flagged" by the wind.

PLANT GROWTH AND DEVELOPMENT

REVIEW

- requirements for photosynthesis and cellular respiration (Section 5-1)
- meristem, vascular tissue, xylem, and phloem (Section 23-1)
- seeds (Section 23-2)
- endosperm and double fertilization (Section 23-3)
- cotyledons, monocots, and dicots (Section 23-3)
- stem and root structure (Section 24-1)

Authors' Rationale

Life is not possible without plants, so it stands to reason that an understanding of how plants grow is essential. In this chapter, plant development is traced from the germination of a seed through all stages of growth. Your students may be surprised to learn that many different factors play a role in plant growth and that plants often require specific light cycles or specific temperature cycles for proper growth and development.

25-1 How Plants Grow and Develop

A seed sprouts with a burst of growth in response to certain changes in the environment. These changes, such as warming temperatures and increasing soil moisture, signal the start of a season of favorable growing conditions. As a new plant grows (increases in size), the various tissues and organs of its body develop (differentiate and take form). All seed plants share certain fundamental patterns of growth and development.

Growth and Development Begin as a Seed Forms

You learned in Chapter 23 that a new flowering plant begins its life after double fertilization occurs within the ovary of a flower. The resulting zygote and endosperm cell divide by mitosis, forming an embryo and endosperm. Then, a significant event occurs—the layers of protective tissue surrounding the embryo and endosperm toughen and become impermeable to both water and oxygen. Denied water and oxygen, the embryo stops growing, and a mature seed forms.

Compare the structures of two familiar types of seeds—a bean seed and a corn seed—illustrated in Figure 25-1. As you can see, most of the interior of a bean seed is occupied by two large, fleshy cotyledons. Before a bean seed fully matures, the endosperm is consumed by the embryo. The embryo's energy reserves are now stored in its cotyledons. Much of the interior of a corn seed, which has only one cotyledon, is occupied by endosperm. In both bean and corn seeds, the cotyledons are attached to the embryo, which has begun to develop specialized tissues. The embryo's shoot develops above the cotyledons and consists of an embryonic stem and tiny embryonic leaves. The embryo's root develops below the cotyledons.

Figure 25-1 A bean and a kernel of corn are two representative types of seeds that vary somewhat in structure.

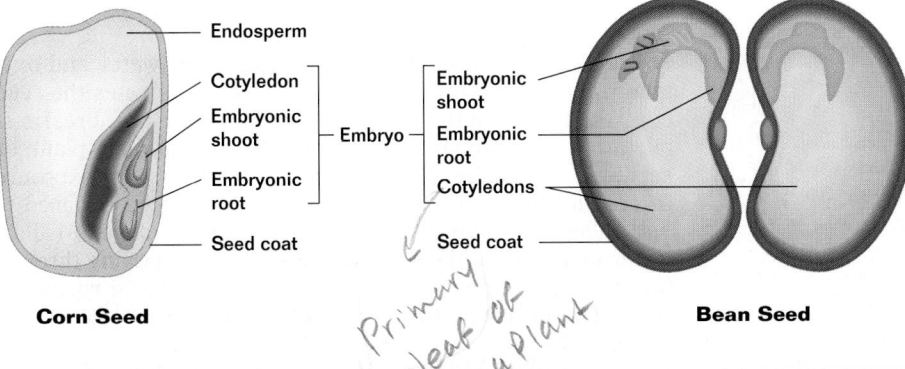

Corn Seed

- Endosperm
- Cotyledon
- Embryonic shoot — Embryo
- Embryonic root
- Seed coat

Primary leaf of a plant

Bean Seed

- Embryonic shoot
- Embryonic root
- Cotyledons
- Seed coat

VISUAL STRATEGY

Figure 25-2

Have students compare the structures of the germinating corn seed and bean seed. Ask: How do monocot and dicot seedlings differ in the way they emerge from the soil? *(Dicot seedlings have a stem that hooks as it emerges to protect the delicate tip, whereas monocots grow straight out of the soil. Monocot seedlings have a protective sheath that guards the tip.)* What other differences do you notice? *(The cotyledon and seed coat of the corn seed remain underground, while the cotyledons and seed coat of the bean emerge from the soil.)* Tell students that the seed coats of beans do not always emerge with the cotyledons and that the cotyledons of some dicot seeds also remain underground.

Teaching Tip
Solve the Problem

To solve the following problem, have students ask questions to which you can answer "yes" or "no." Tell them to form a hypothesis based on your answers. **Problem:** A highway was seeded with Indian paintbrush, a wildflower. Two different loads of seeds were used for the project, and they were planted on opposite sides of the highway. In the spring, the seeds on only one side of the highway germinated. Why did the seeds on the other side fail to germinate? **Answer:** Indian paintbrush seeds require both a period of cool temperature and moisture for the embryos in the seeds to develop. The seeds that failed to germinate did not receive one of the required treatments. After students have developed their hypothesis *(either the cool temperature or moisture treatment was missing)*, have each student design an experiment that would test their hypothesis.

Figure 25-2 The events in the germination of a corn seed, *below left,* differ from those of a bean seed, *below right.* As the corn shoot elongates and emerges from the soil, its leaves push through their protective cover, unfurl, and start to photosynthesize. Once the hook of a bean seedling breaks through the soil, it straightens out and pulls the cotyledons and leaves up into the air. As the food stored in a bean's cotyledons is used up, the cotyledons shrivel and eventually fall off.

Growth Continues When a Seed Germinates
....

The **germination** of a seed is the resumption of growth by a plant embryo. The first visible evidence that a seed is germinating is the emergence of the embryo's root. What happens next varies somewhat from one type of plant to another, as you can see in Figure 25-2. For example, the cotyledons of some plants emerge with the shoot, while the cotyledons of other plants remain underground. The embryonic shoots of many plants, such as the common bean (*Phaseolus vulgaris*), curve until they are shaped like a hook. This hook enables the shoots to push through the soil and prevents the tips of the shoots from being damaged. The embryonic shoots of other plants, such as corn (*Zea mays*), do not curve but are surrounded by a protective sheath.

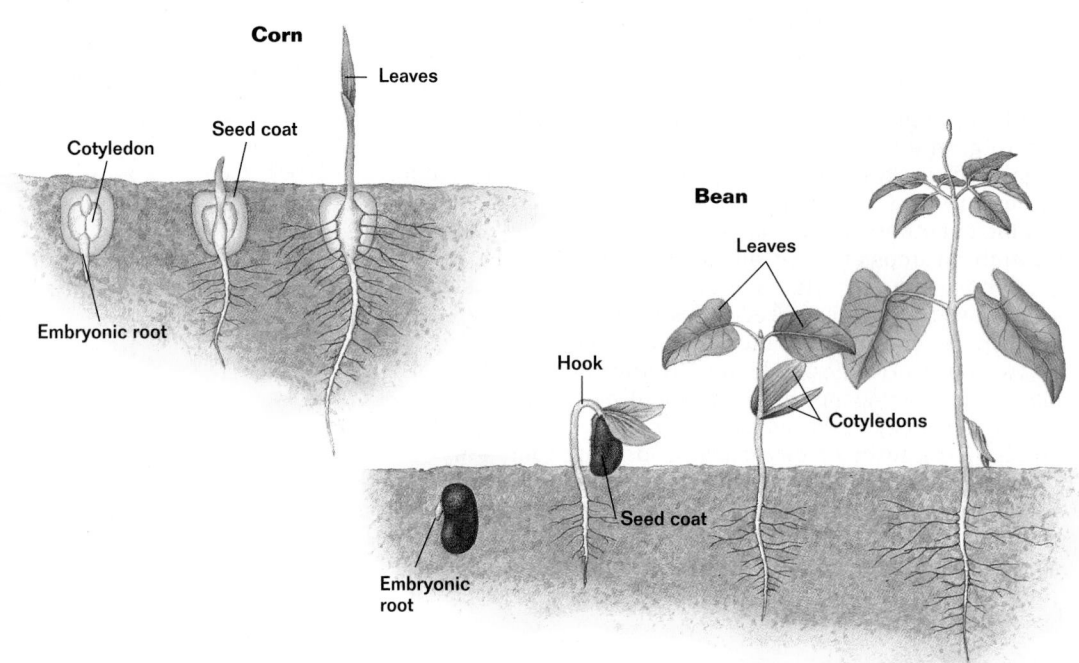

Corn

Cotyledon

Seed coat

Leaves

Embryonic root

Bean

Leaves

Hook

Cotyledons

Seed coat

Embryonic root

☐ CAPSULE SUMMARY

Germination is the resumption of growth by the dormant embryo in a seed. Water and oxygen must penetrate the seed coat before a seed can germinate.

Seed germination cannot take place until water and oxygen penetrate the seed coat. When water penetrates the seed coat, the tissues in the seed swell, and the seed coat breaks. If adequate water and oxygen are available, the young plant, or **seedling,** will continue to grow. Many seeds require exposure to heat or cold to germinate, and some must be exposed to light. The seed coats of other seeds must be damaged to allow water to penetrate the seed. Exposure to fire, passing through the digestive system of an animal, and falling on rocks are several natural ways that seed coats are damaged. ☐

Effective Reading

Defining Terms

Instruct students to write definitions of the boldface terms in the chapter as they encounter them in their reading. Advise students to reread paragraphs that have two or more boldface terms after they have defined the terms to better understand what they have read.

Apical Meristems Produce Primary Growth

All plants grow by cell division that occurs in meristems located at the tips of their shoots and roots. To better understand this pattern of growth, imagine a stack of dishes—the stack can get taller but not wider. Growth that occurs from the formation of new cells at the tips of a plant is called **primary growth.** The primary growth of a seedling occurs in **apical** *(AP ih kuhl)* **meristems.** As Figure 25-3 indicates, apical meristems are located at the tips of stems and at the tips of roots, just behind the root cap. The plant tissues that result from primary growth are called **primary tissues.**

Figure 25-3 A cross section of the shoot tip of a coleus plant, *below left,* shows the shoot apical meristem as well as young leaves and buds. A cross section of the root tip of a radish plant, *below,* shows the root apical meristem and the root cap.

SECTION 25-1

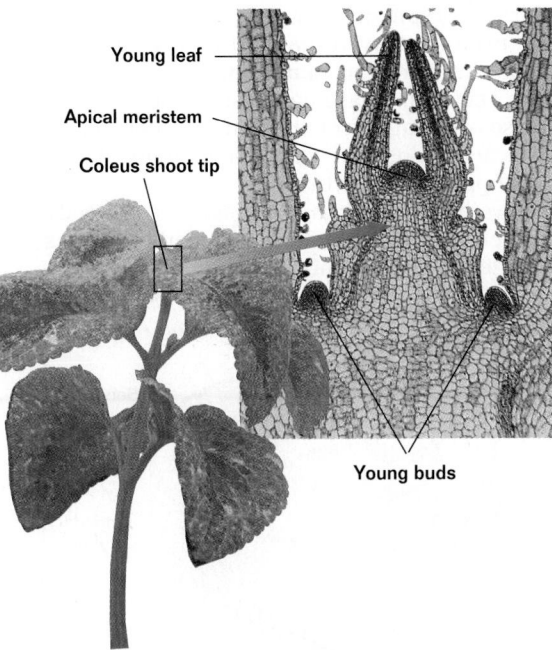

Young leaf

Apical meristem

Coleus shoot tip

Young buds

Radish root tip

Apical meristem

Root cap

During periods of growth, the cells of apical meristems divide and continually add more cells to the tips of a seedling's body. Thus, the seedling's stems and roots lengthen. After new cells are formed, they grow and undergo **differentiation,** the process by which cells become specialized in form and function. The new cells produced by the apical meristem of a stem become part of the primary dermal, ground, and vascular tissues of the lengthening stem and its young leaves. Some of the new cells produced by a root apical meristem elongate and become the primary dermal, ground, and vascular tissues of the lengthening root. Cells produced by the root apical meristem also become part of the root cap, replacing root cap cells that are constantly worn away as the root pushes through the soil.

❏ CAPSULE SUMMARY

Apical meristems located at the tips of shoots and roots produce primary growth. Tissues that result from primary growth are called primary tissues.

CONNECTIONS

Chapter 23

Meristems

Remind students that plants have meristems, which are groups of actively dividing cells that differentiate to form all the specialized cells of the plant body.

Teaching Tips
Prepared Slides

Let students observe prepared slides of the apical meristems of the shoots and roots of several plants. Have them draw what they see and label the parts seen in Figure 25-3. Have students write a brief paragraph describing how the cells in the apical meristems differ from those in the surrounding tissues. *(The cells of the apical meristems are small and unspecialized, while surrounding cells tend to be larger and longer and have thicker walls, and some have specialized structures such as chloroplasts or root hairs.)*

What If the apical meristems of a plant were removed? Removing or damaging the apical meristems of a plant would stop or limit the vertical growth of the stem or root. This frequently occurs in nature, as tips of stems are broken off by wind or animals, killed by disease, or eaten by insects or other animals. Lateral buds that develop at the nodes of a stem often begin to grow after the apical meristem is removed. Thus, the growth of the plant continues, but the plant may not grow as tall.

◈ Did You Know?

You may have noticed that plants have two patterns of primary growth. Some continue to grow taller throughout their lives, and others do not. Plants with stems that stop growing after reaching a certain length are said to have a determinate *(dee TUR mih niht)* growth pattern. Plants with stems that continue to grow longer as long as the plant lives are said to have an indeterminate growth pattern. Many plants with determinate growth patterns stop growing when they produce flowers at the tips of their stems.

Lateral Meristems Produce Secondary Growth

Many vascular plants also increase in width as they grow taller. Growth that causes a plant to increase in width is called **secondary growth.** The tissues that develop as a result of secondary growth are referred to as **secondary tissues.** Although secondary growth occurs in many non-woody plants, its effects are most dramatic in woody plants.

Woody plants have two cylinders of actively dividing cells called lateral meristems. Cell division in **lateral meristems** adds layers of new cells around the outside of a plant's body. Within the bark of a woody stem is the **cork cambium** (*KAM bee uhm*), a lateral meristem that produces the cork cells of the outer bark. Cork cambium originates from primary stem tissues that were produced by the apical meristem. Just beneath the bark is the **vascular cambium,** a lateral meristem that produces secondary vascular tissue. The vascular cambium, which also originates from primary tissues, forms between the xylem and phloem of vascular bundles, which are also the result of primary growth.

Figure 25-4 shows how the secondary tissues of a woody stem develop. Notice that the secondary vascular tissues form on opposite sides of the vascular cambium. Secondary

Figure 25-4 A woody stem first forms by primary growth, *below.* After the vascular cambium forms from primary stem tissues between the primary xylem and phloem, secondary growth begins, *below right.* Layers of secondary xylem and phloem form between the primary xylem and phloem, and the stem grows in diameter, *below center.* The cork cambium forms when the epidermis is stretched and broken as the stem thickens and produces cork cells that become part of the bark.

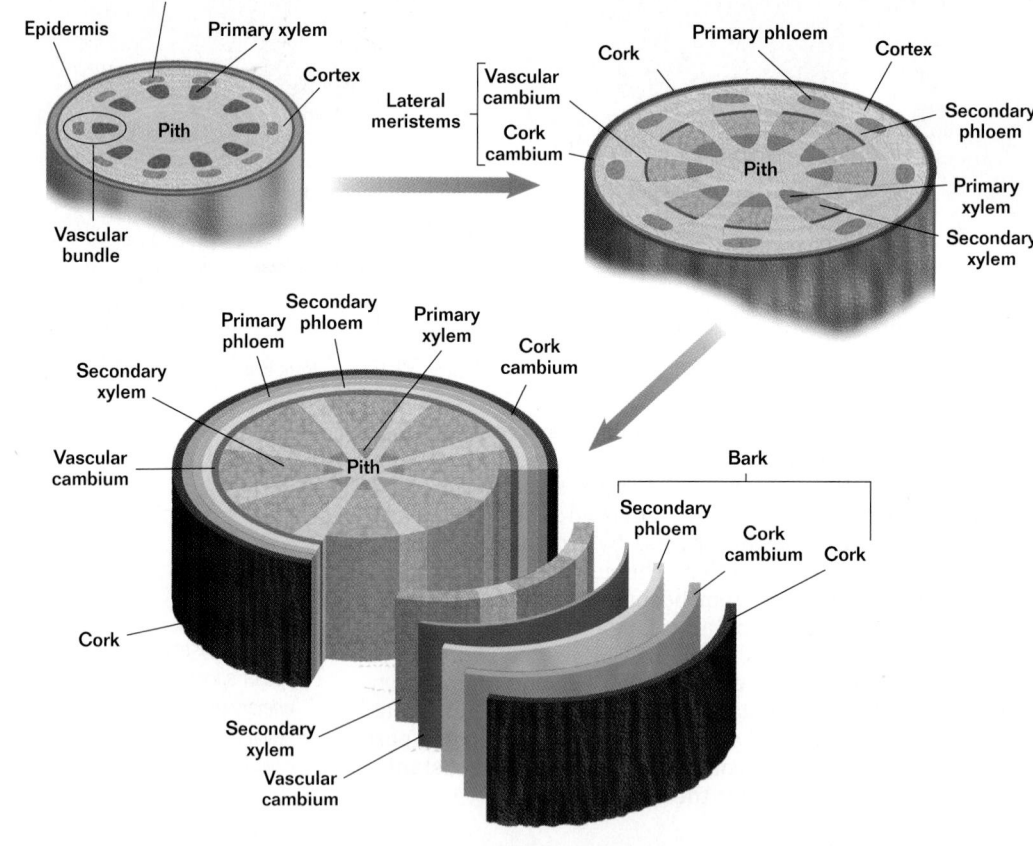

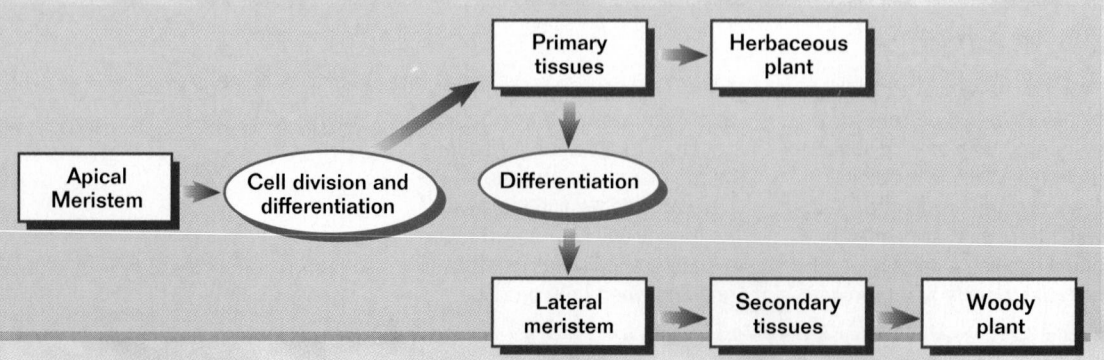

xylem forms in thick layers on the inner side of the vascular cambium and is the main component of wood. Secondary phloem forms on the outer side of the vascular cambium and becomes part of the inner bark. Thus, secondary phloem, which transports sugars from the leaves to the roots, is very close to the outer surface of a woody stem. Removing the bark of a tree damages the phloem and both cambia and may eventually kill the tree. ◻

Plant Development Is Continuous and "Reversible"

Although both plants and animals develop according to a genetic blueprint, their patterns of development are very different. The seedling that emerges from a germinating seed is a small version of an adult plant. As the plant grows, new cells are continuously produced in its apical and lateral meristems. These cells differentiate and replace or add to existing tissues, as seen in Figure 25-5. Thus, a plant continues to develop throughout its life. This developmental pattern is very different from that of animals, which develop until reaching a certain point at which they are considered to be adults.

Plant development and animal development also differ in how they are affected by the environment. The outcome of plant development is greatly influenced by a plant's immediate surroundings. A plant cannot move about, so it is critical that during development the seedling adapt to the particular conditions that the mature plant will have to face. Plants are able to change when the local environment changes because they develop continuously. For example, many trees and shrubs stop growing and drop their leaves during periods of cold weather or drought. When conditions improve, these trees and shrubs produce new leaves and resume growth. Animals, which can move to new environments when conditions change, develop without being as strongly influenced by environment.

Unlike most animal cells, many plant cells retain the functional genetic instructions necessary to produce *all* of the tissues of a mature plant. A different part of the complete set of genetic instructions for an organism is utilized in the cells of each of its tissues. During animal development, critical sets of genes that control development are lost or "turned off" and may not be reactivated. In many plant tissues, however, the unused sets of genes are preserved and can be activated in the future. When separated from the body of a mature plant, the cells of certain plant tissues can begin to undergo cell division and form masses of undifferentiated cells. Thus, in a sense, their development has been reversed. The masses of unspecialized plant cells can later resume the process of differentiation and form all of the tissues of a mature plant.

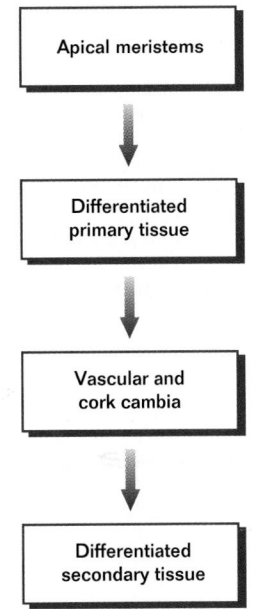

Figure 25-5 This graphic organizer shows the stages of plant cell differentiation. New cells form by cell division in the apical meristems. These cells differentiate to form primary tissues. In woody plants, the vascular and cork cambia differentiate from primary tissues and then give rise to secondary tissues.

Application

Agriculture Hybrid fruit trees are produced by grafting desirable stems onto a hardy root stock. The upper part of the tree that bears the fruit is the hybrid, while the roots are those of a native tree with cold tolerant or disease resistant roots. Grafting requires that the vascular cambium of the hybrid be aligned with the vascular cambium of the native tree. The new tissue is secured in place with a waterproof wrapping material, and it eventually grows and connects with the tissue of the root stock.

Teaching Tips
Finding the Vascular Cambium
Vascular cambium is only one layer of cells. It is found just under the bark of a woody stem and can be located in microscope slides of woody-stem cross sections by finding the phloem cells with their companion cells and the large, thickened xylem cells. Both xylem and phloem develop from the vascular cambium cells, so they develop in rows on either side of this meristematic tissue. Have students locate the layer that separates the two types of cells in a woody stem cross section. This is the row of vascular cambium cells. Have students describe how xylem and phloem originate from vascular cambium. (*through cell division and cell differentiation*)

How Is Plant Development "Reversible"?
When told that plant development is reversible, students may picture mature plants slowly decreasing in size and returning to a seed. Explain that in this case "reversible development" means that fully differentiated cells, such as those of the cortex or phloem, can divide and produce undifferentiated cells like those in a plant's meristems. Those cells can then repeat the steps in plant growth and differentiation, eventually producing a new adult plant.

Overcoming Misconceptions
Some Plants Have Juvenile Forms
Not all plants sprout from a seed with the characteristics of the adult plants. Some plants, such as English ivy, junipers, and philodendrons, have both juvenile and adult forms. The leaves of the juvenile and adult stages are distinctly different.

UNIT 6

CHAPTER 25

Figure 25-6

Walk students through Steward's experiment. Students may be interested to know that Steward used coconut milk as the medium in his tissue culture experiment. Tell students that coconut milk is the endosperm of a coconut seed. Ask: Why do you think Steward chose this medium? *(Endosperm provides nutrients for a plant embryo to grow when a seed germinates, so it should have the nutrients necessary for growing plants in tissue cultures.)*

Application

Agriculture Many commercially grown houseplants are produced by tissue culture. Tissue culture provides a means to inexpensively grow a pest-free plant in a shorter period of time. One plant can be used to produce thousands of plants because only a few cells are needed to make a new plant. A leaf can be cut into many small sections, which are placed on a nutrient medium that contains hormones to stimulate the leaf sections to differentiate and grow quickly. Sterile techniques ensure that the plants grow in disease-free conditions that also lack insect pests.

☐ CAPSULE SUMMARY

Plants develop continuously as new cells form and differentiate. Plant development is considered reversible because cells of certain tissues can produce unspecialized cells that then differentiate into all of the mature plant's tissues.

Figure 25-6 F. C. Steward's experiments with carrots demonstrated that some plant cells are capable of reversing the process of development. All of the instructions necessary for producing a normal carrot plant were present in the phloem cells Steward extracted from a mature carrot.

The ability of specialized plant cells to reverse and resume development was first demonstrated in 1958 by the botanist F. C. Steward, whose famous experiment is illustrated in Figure 25-6. Techniques for growing pieces of living tissue in artificial media, such as those developed by Steward, are known as **tissue culture.** Today, tissue culture is an important tool in plant research and in the commercial propagation (reproduction) of plants such as orchids. Although many types of animal tissues are also grown in tissue cultures, most are unable to reverse and resume the process of development to produce a complete individual. To learn more about how plants can be propagated, look at *Up Close: Kalanchoë* on pages 578–579. ☐

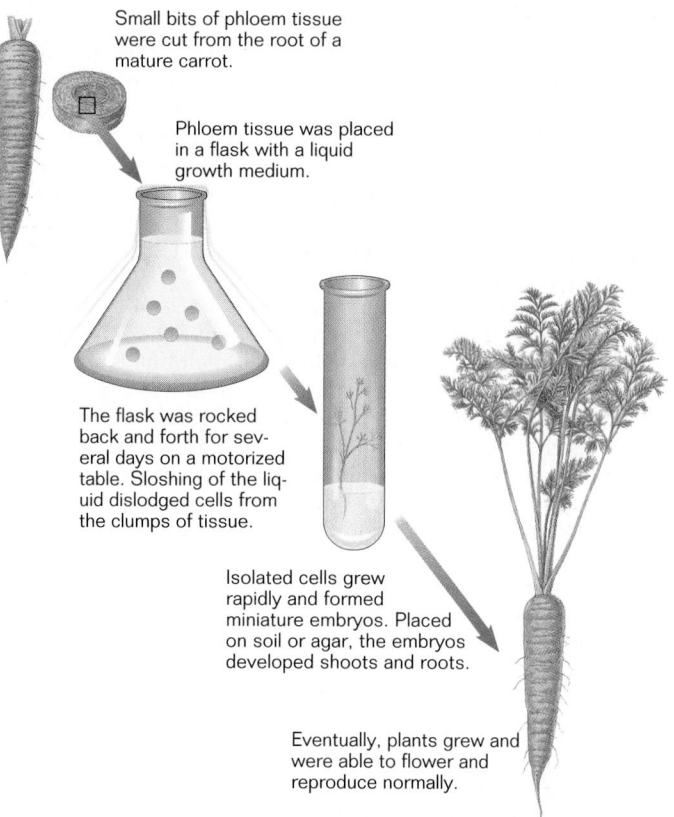

Small bits of phloem tissue were cut from the root of a mature carrot.

Phloem tissue was placed in a flask with a liquid growth medium.

The flask was rocked back and forth for several days on a motorized table. Sloshing of the liquid dislodged cells from the clumps of tissue.

Isolated cells grew rapidly and formed miniature embryos. Placed on soil or agar, the embryos developed shoots and roots.

Eventually, plants grew and were able to flower and reproduce normally.

How Long Does a Plant Live?
....

As you learned in Chapter 23, the oldest known trees are gymnosperms that are estimated to be about 5,000 years old. Some plants, however, live for only a few weeks. Depending on how long they live, all plants can be classified as one of three basic types: annuals, biennials, or perennials. Examples of each of these types of plants can be seen in Table 25-1.

 Did You Know?

Cotton fibers can be produced in a test tube by means of tissue culture, which requires a single cell from the cotton fibers of a cotton plant.

1. **Annuals** An **annual** plant completes its life cycle (grows, flowers, and produces fruits and seeds) within one growing season and then dies. Most annuals are non-woody, or herbaceous, plants that grow rapidly under favorable conditions and increase greatly in size if supplied with adequate water and nutrients. Plants that are annuals include sunflowers, dandelions, lupines, and many weeds.

2. **Biennials** A **biennial** plant takes two years to complete its life cycle. During the first year, biennials produce a short stem and a rosette (circular cluster) of leaves. In the second year, the energy stored in the plant's roots and shoots is used to produce an elongated flowering stalk. After flowering and producing fruits and seeds, the plant dies. Biennial plants include yellow sweet clover, Queen Anne's lace (wild carrot), and parsley.

3. **Perennials** A **perennial** plant lives for more than two years and may produce flowers, fruits, and seeds many times during its life. The majority of vascular plants are perennials, including many herbaceous plants and all woody plants. Herbaceous perennials include plants such as buttercups, morning glories, and evening primroses. The shoots of herbaceous perennials may die each year after a season of growth and food accumulation. Food for the next season's growth is stored in fleshy roots or underground stems. Woody perennials include trees (woody plants with one main stem), shrubs (woody plants with many stems), and many vines. Some woody perennials drop their leaves each year. Trees, shrubs, and woody vines that drop all of their leaves at the end of each growing season, such as elms, roses, and grapevines, are said to be **deciduous** (dee SIHJ oo uhs). Those that drop a few leaves at a time throughout the year, such as pines, junipers, and honeysuckles, are called **evergreens.** □

Table 25-1 Types of Plants

Type	Example
Annual	 Lupine
Biennial	 Yellow sweet clover
Herbaceous perennial	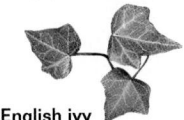 Buttercup
Woody perennial	English ivy

❏ **CAPSULE SUMMARY**

Annuals complete their life cycle in one growing season. Biennials take two years to complete their life cycle. Perennials live for many years and may flower and produce seeds many times.

SECTION 25-1

Teaching Tips

Nature Walk

Take students on a tour around the school grounds or a park, and have them identify plants as annuals, biennials, and perennials. Have them keep a log of the examples they see, using general terms such as flower, tree, grass, or shrub. Have students write a paragraph that answers the following questions based on what they observe. Do you see more of one type of plant than another? If so, might the location or time of year make a difference? (*More annuals tend to be present in spring and summer, while only woody perennials may be present in the winter.*) What could make your observations differ from what you would expect? (*mowing or other gardening procedures, unusual weather*) Why is this information significant? (*It helps people to know when, where, and how to grow these types of plants.*)

Design a Flower Bed

Have students work in groups of four to design a flower garden that will provide color in every season. Let students use nursery catalogs to find examples of annuals, biennials, and perennials as well as to determine where each plant will grow and when it blooms. Assign each group a set of conditions under which their garden must grow (e.g., size, distribution of sun and shade, soil type). Finally, tell them to consider aesthetic factors (e.g., color combinations, taller plants in back, shorter plants in front). Have each student design a graphic that shows the group's garden from the top during one of the four seasons and identifies their plants.

Section Review

1. *How does the germination and initial growth of a corn seedling differ from that of a bean seedling?*

2. *How is secondary growth produced?*

3. *How does plant development differ from animal development?*

4. *What is the difference between a biennial and a perennial?*

5. *Explain how a Kalanchoë can be propagated without seeds.*

Critical Thinking

6. *Why might being an annual be an advantage to a plant?*

Answers to Section Review

1. Corn seedlings grow straight, leaving the cotyledon and seed coat in the soil. Bean seedlings hook as they grow through the soil; as the stem straightens, 2 cotyledons unfold, becoming the seedling's first "leaves."

2. Secondary growth results from cell division in lateral meristems (the cambia).

3. Plant development is continuous and "reversible," while animal development is not.

4. Biennials live for two growing seasons, flower and make seeds once, and then die.

Perennials live for more than two years and may reproduce many times.

5. Kalanchoës can be propagated vegetatively with plantlets, leaf cuttings, and stem cuttings.

6. Answers will vary but might suggest that annuals have an advantage in extreme (cold, hot, or dry) climates because they grow quickly in favorable conditions and produce seeds, which lie dormant until favorable conditions recur.

Instructional Strategies

- Let students asexually propagate *K. daigremontiana* by using plantlets and stem cuttings.

- Show students examples of other *Kalanchoë* species. Be sure they notice that while the size, shape, and color of the leaves vary, the flower structure is very similar.

- For interest, have a horticulturist talk to students about the production of kalanchoës for sale as ornamental plants.

- Tell students that like *Kalanchoë daigremontiana*, other horticulturally important plant species are natives of Madagascar, which is an island off the southeast coast of Africa. One such plant is the rosy periwinkle, *Catharanthus roseus*, which is both a popular bedding plant and the original source of two important cancer treatment drugs. Many such economically important plants have been collected in their native lands, which are often developing countries with very low incomes and standards of living. Have students research the concept of *intellectual property rights* and then debate whether businesses should have the right to use valuable plants for economic gain without compensation to the people of the countries in which the plants are collected.

- Have students make cross sections of *K. daigremontiana* leaves and stems and observe the cross sections with a microscope.

- CAM photosynthesis may be difficult for students to comprehend. Help students understand CAM photosynthesis by having them illustrate the process.

UP CLOSE *KALANCHOË*

- **Scientific Name:** *Kalanchoë daigremontiana*
- **Range:** Native to southwestern Madagascar; cultivated worldwide as an indoor potted plant and as an outdoor perennial in warm climates
- **Habitat:** In nature, semiarid tropical grassland with well-drained, fertile soil and moist summers; in cultivation, warm, brightly lighted area with rich, well-drained soil
- **Size:** Height—30 cm to 1 m (1 ft. to 3 ft.)
- **Importance:** Kalanchoës *(kal an KOH eez)* are members of the Crassulaceae family, a group of succulent plants that are adapted to hot climates. They are valued as ornamentals for their colorful flowers, interesting foliage, and ease of cultivation and propagation.

External Structures

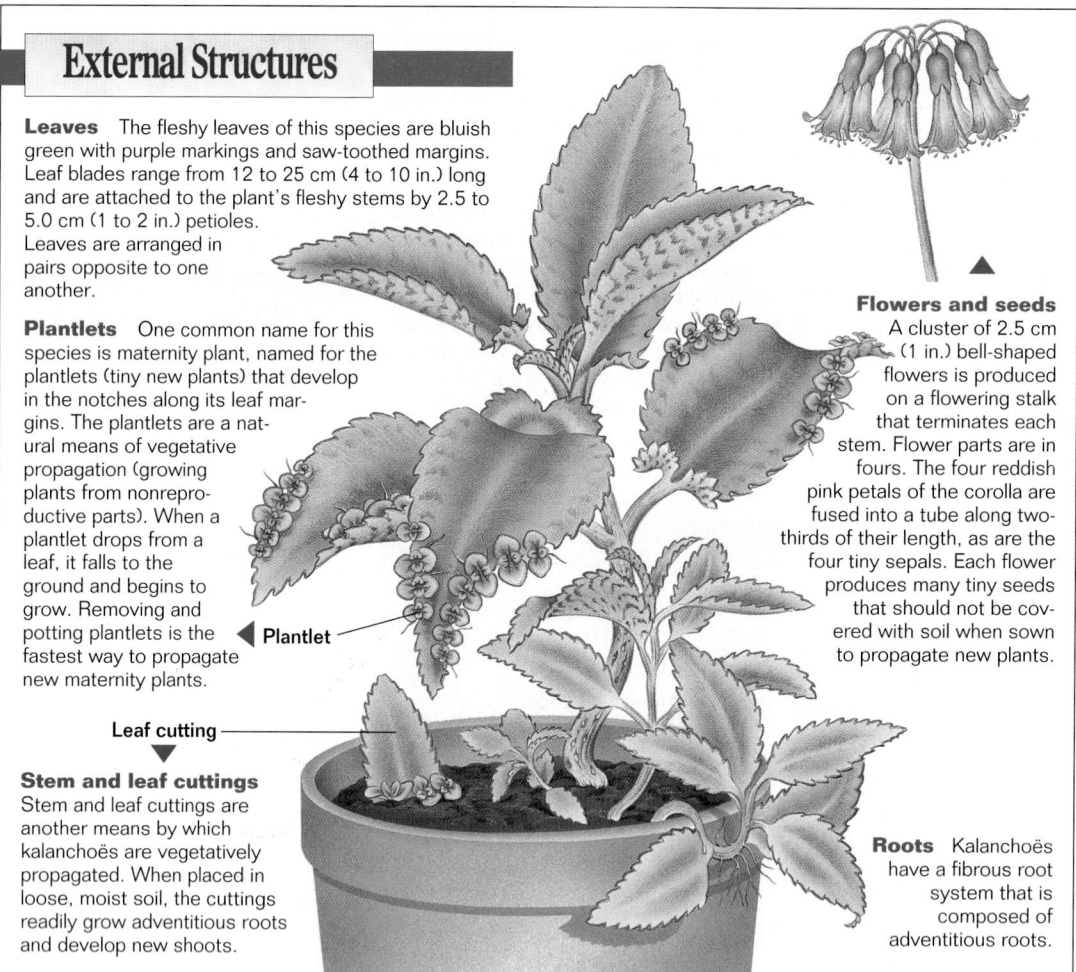

Leaves The fleshy leaves of this species are bluish green with purple markings and saw-toothed margins. Leaf blades range from 12 to 25 cm (4 to 10 in.) long and are attached to the plant's fleshy stems by 2.5 to 5.0 cm (1 to 2 in.) petioles. Leaves are arranged in pairs opposite to one another.

Plantlets One common name for this species is maternity plant, named for the plantlets (tiny new plants) that develop in the notches along its leaf margins. The plantlets are a natural means of vegetative propagation (growing plants from nonreproductive parts). When a plantlet drops from a leaf, it falls to the ground and begins to grow. Removing and potting plantlets is the fastest way to propagate new maternity plants.

◄ **Plantlet**

Leaf cutting ▼

Stem and leaf cuttings
Stem and leaf cuttings are another means by which kalanchoës are vegetatively propagated. When placed in loose, moist soil, the cuttings readily grow adventitious roots and develop new shoots.

▲

Flowers and seeds
A cluster of 2.5 cm (1 in.) bell-shaped flowers is produced on a flowering stalk that terminates each stem. Flower parts are in fours. The four reddish pink petals of the corolla are fused into a tube along two-thirds of their length, as are the four tiny sepals. Each flower produces many tiny seeds that should not be covered with soil when sown to propagate new plants.

Roots Kalanchoës have a fibrous root system that is composed of adventitious roots.

Internal Structures

Leaf structure Kalanchoës are succulents, which means that they have fleshy leaves and stems that store water. A look inside the leaf of a kalanchoë plant discloses how some succulents are adapted for conserving water. Notice that the epidermis consists of several layers of cells covered by a thick waxy cuticle. Relatively few, very small stomata dot the leaf surfaces. The leaf's mesophyll consists of uniformly large cells with little air space between them.

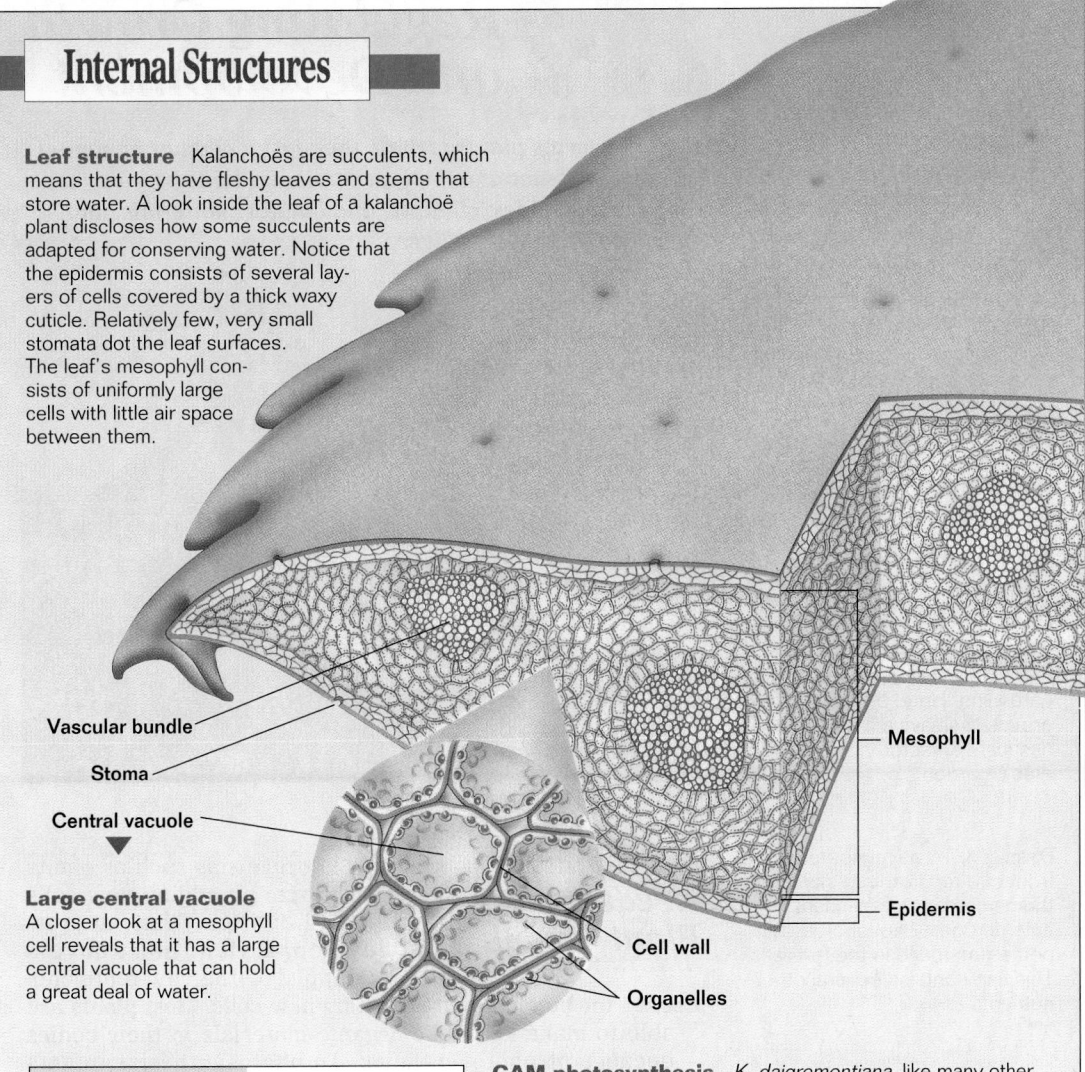

Vascular bundle

Stoma

Central vacuole ▼

Mesophyll

Epidermis

Cell wall

Organelles

Large central vacuole
A closer look at a mesophyll cell reveals that it has a large central vacuole that can hold a great deal of water.

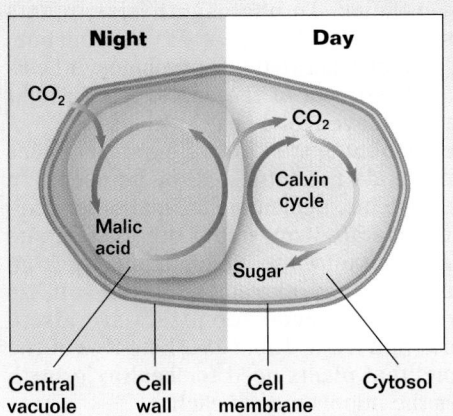

Night	Day
CO₂	

Central vacuole   Cell wall   Cell membrane   Cytosol

CAM photosynthesis *K. daigremontiana*, like many other succulents, fixes carbon by a water-conserving pathway called crassulacean acid metabolism (CAM). Like C$_4$ plants, CAM plants fix carbon by first incorporating carbon dioxide into a compound called malic acid. Later, carbon dioxide is released from malic acid and used by the Calvin cycle to make sugar. Unlike C$_4$ plants, the two processes occur in the same cells but at different times of the day.

Unlike the stomata of other plants, the stomata of CAM plants open only at night. Thus, CAM plants make malic acid only at night. The malic acid is stored in the large central vacuoles of the mesophyll cells. A rising concentration of malic acid causes water to enter the vacuoles by osmosis, thus helping CAM plants to store water. During the day, malic acid diffuses out of the vacuoles, and the carbon dioxide released from it is fixed by the Calvin cycle.

Discussion

Guide the discussion by posing the following questions.

1. How does the appearance of other kalanchoë flowers compare to that of the flowers of *K. daigremontiana*? (*Students should infer that other kalanchoë flowers are similar to those of* K. daigremontiana.)

2. What is one rule (criterion) that you could use to identify an unknown plant as a kalanchoë? (*Flower structure is the best way to identify a plant as a kalanchoë.*)

3. Why can a kalanchoë survive in hot, dry environments? (*Kalanchoës are able to survive in hot, dry environments because they are succulents, which are able to store water in their cells and tissues. They also have very few stomata and thick epidermal and cuticle layers that reduce water loss.*)

4. Why is it an advantage to have stomata that open at night, as CAM plants do? (*If stomata are closed during the day when temperatures are the hottest, the plant will lose less water through transpiration and be able to survive with less water.*)

Vocabulary Preview

- nutrient
- auxin
- hormone
- apical dominance
- cytokinin
- ethylene
- gibberellin
- tropism
- phototropism
- gravitropism
- thigmotropism
- photoperiodism
- short-day plant
- long-day plant
- day-neutral plant
- dormancy

Opening Question

Ask students to make a list of what they consider to be the 10 most important plants and justify each of their choices. Students' lists will include types of trees, herbs, fruits, vegetables, and ornamentals.

Demonstration

Dormant Apple Seeds

Cut an apple in half, and show students the seeds inside. Ask: Why are the seeds not growing inside of the apple? Let students suggest reasons. Help them to derive that chemicals (hormones) produced by the ovary prevent apple seeds from germinating while inside the fruit.

👁 VISUAL STRATEGY

Figure 25-8

Point out the numbers on the bag of fertilizer. Explain that the numbers designate the percentages of N, P, and K in the fertilizer. Tell students that the bag also contains "filler," which accounts for most of its mass.

Section Objectives

- Identify basic requirements for healthy plant growth.
- Explain why nutrients are important to a plant.
- Name four types of plant hormones, and describe how each affects plant growth.
- Describe how plant growth is affected by environment.

Figure 25-7 Many people enjoy gardening. Here, a gardener proudly displays a cabbage she has just harvested from her vegetable garden.

Figure 25-8 Three numbers on the front of a fertilizer bag indicate the percentage of the plant nutrients nitrogen, phosphorus, and potassium found in the fertilizer. This bag contains several other nutrients as well.

Effective Reading

Crossword Puzzle

Have students work in small groups to design crossword puzzles using the boldface vocabulary words in Section 25-2. Groups can exchange puzzles and complete them to review the concepts in this section.

25-2 Regulating Growth and Development

Growing plants is more than just a popular hobby—it is essential to human survival. Plants provide food, medicines, clothing, and building materials, and they also brighten our lives, as Figure 25-7 illustrates. Success in growing plants—whether on an apartment windowsill, in a home garden, or in a wheat field—requires an understanding of the many factors that affect plant growth.

Nutrients Are Needed for Plant Growth

Multicelled organisms such as plants increase in size by adding new cells through cell division. Thus, to grow and develop, a plant must have a steady supply of the raw materials needed for building and operating new cells. Most plants are able to make all of the organic materials in their bodies because of photosynthesis. To photosynthesize, plants require light, water, and carbon dioxide. The carbon, hydrogen, and oxygen incorporated during photosynthesis make up more than 90 percent of the dry weight (weight after water has been removed) of a healthy plant.

However, the carbohydrates that a plant produces during photosynthesis do not satisfy all of its needs. In order to produce proteins, nucleic acids, and other molecules needed for metabolism, many other elements are required. Plants obtain these elements mostly from mineral **nutrients** that they extract from the soil. As Figure 25-8 shows, commercial fertilizers are also a source of mineral nutrients. Table 25-2 lists five of the many nutrients that plants need for healthy growth and describes the importance of each.

Table 25-2 Five Plant Nutrients and Their Importance

Nutrient	Importance
Nitrogen (N)	Part of all proteins, nucleic acids, chlorophylls, and coenzymes
Phosphorus (P)	Part of ATP, ADP, nucleic acids, phospholipids of plasma membranes, and some coenzymes
Potassium (K)	Needed to perform active transport, activate enzymes, regulate osmotic balance, and open stomata
Magnesium (Mg)	Part of chlorophyll; needed for photosynthesis and activation of enzymes
Sulfur (S)	Part of some proteins and coenzyme A; needed for cellular respiration

Plants also require oxygen for cellular respiration. While the green parts of a plant produce oxygen during photosynthesis, most of the oxygen needed by these parts comes from the atmosphere. A plant's roots, which do not usually conduct oxygen-producing photosynthesis, obtain oxygen from air in the spaces between soil particles. If the soil in which a plant is growing becomes compacted or saturated with water, there may not be enough oxygen available for its roots, and the plant could die. Figure 25-9 shows how one plant has adapted to living in places where little oxygen is present in the soil. ❑

❑ **CAPSULE SUMMARY**

Plants are able to produce all of their organic molecules as a result of photosynthesis, but they require several inorganic nutrients to convert carbohydrates into other important organic molecules.

Hormones Control Plant Growth and Development

In 1851, the great biologist Charles Darwin and his son Francis published a book called *The Power of Movement in Plants*. In the book, they reported the results of their experiments on how plants grow toward light. Young grass seedlings bend strongly toward a light source as their shoots elongate. The Darwins found that if they covered the tip of a seedling with material that prevented light from reaching the tip, the shoot would not bend. If they instead covered the tip of a seedling with a cap of gelatin that enabled light to reach the shoot, the shoot would bend. Charles and Francis Darwin concluded that some "influence" arose in the tip of the shoot and was transmitted downward, causing the shoot to bend.

With experiments conducted in the 1920s, the Dutch plant physiologist Frits Went showed that a chemical is the "influence" responsible for making plants bend toward light.

Figure 25-9 Mangroves, which grow in swamps, produce "air roots," which emerge from water and gather oxygen from the air.

☑ RESEARCH Update

The Biotech Tomato: Is Something Fishy?

The taste of home-grown tomatoes tends to be better than that of store-bought tomatoes. Why? A tomato just picked from the vine is fresher than one purchased from a store. Also, tomatoes in the store spent less time on the vine than did tomatoes from a home garden. Time spent ripening on the vine improves a tomato's taste but also increases the likelihood that it will rot.

Calgene, a biotechnology firm in Davis, California, has come up with a new tomato variety called Flavr Savr that has the same qualities as home-grown tomatoes but resists rotting. How? Calgene essentially took a gene from a cold-water fish (halibut) and inserted it into the tomato genome. In cold-water fishes, ethylene (the compound that causes fruit to ripen) is produced as ethylene glycol (antifreeze). By cloning the responsible gene and inserting it into tomato plants, Calgene created tomatoes that produce ethylene glycol instead of ethylene. The result is tomatoes that can stay on the vine longer while resisting rotting.

While FDA officials and many others in the scientific community approve of the tomato, some public-interest groups are leery of genetically engineered food. In the amounts encountered in Flavr Savr fruits, ethylene glycol is not dangerous. Calgene has provided the data on its research and voluntarily submitted to safety trials and specific labeling. The Union of Concerned Scientists' Jane Rissler thinks that the FDA should respond to the public's concerns before approving any genetically engineered food. Calgene looks at biotechnology as helping to create more efficient food—not necessarily making new food. ☑

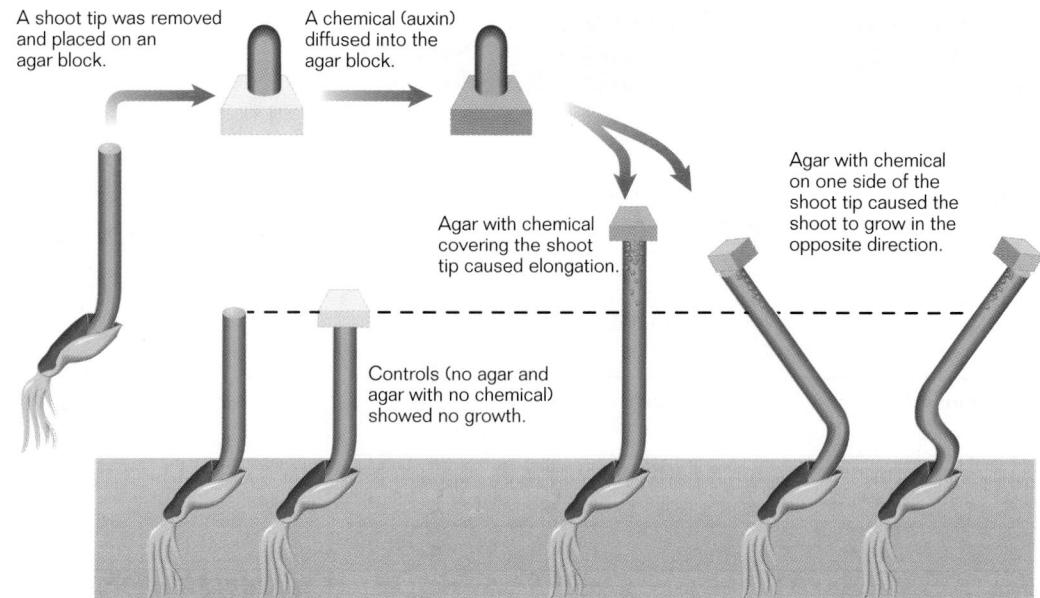

A shoot tip was removed and placed on an agar block.

A chemical (auxin) diffused into the agar block.

Agar with chemical on one side of the shoot tip caused the shoot to grow in the opposite direction.

Agar with chemical covering the shoot tip caused elongation.

Controls (no agar and agar with no chemical) showed no growth.

Figure 25-10 Frits Went's experiments, *above,* showed that a chemical (auxin) produced in the tips of oat (*Avena* sp.) seedlings causes shoots to grow toward light.

Figure 25-11 These coleus plants illustrate the effects of apical dominance and the removal of terminal buds. The terminal buds were not removed from the plant on the left, and thus it is taller and not as bushy as the plant on the right, from which the terminal buds were removed so that lateral buds could grow.

Went's experiments are described in Figure 25-10. Went named the growth-promoting chemical **auxin** (*AWK sihn*). The word *auxin* comes from the Greek word *auxein,* meaning "to increase." Auxin causes the elongation of plant cells by increasing the plasticity of cell walls and enabling them to stretch during active cell growth. Influenced by light, auxin migrates from the lighted side of a stem to the darkened side. As a result of being exposed to a higher concentration of auxin, the stem cells on the darkened side of a seedling elongate faster than the cells on the lighted side. This differential growth causes the stem to grow toward light.

Auxin is one of many growth-regulating chemicals that act as hormones. A **hormone** is a chemical that is produced in one part of an organism and then transported to another part of the organism, where it brings about a response. The word *hormone* comes from the Greek word *horman,* meaning "to set in motion." Your body uses hormones to regulate growth and many other activities. Some of the hormones in your body are produced by specialized glands. In plants, however, all hormones are produced in various tissues throughout the plant. Four major types of plant hormones that have been identified by biologists are discussed below.

1. **Auxin**   Auxin is produced in the tips of stems. In addition to promoting stem elongation, auxin stimulates fruit development, suppresses leaf and fruit drop, and inhibits lateral bud growth. The inhibition of lateral bud growth by auxin is called **apical dominance.** Removal of a terminal bud enables lateral buds to grow. Gardeners use this knowledge to shape the growth of their plants, as Figure 25-11 shows. Synthetic auxins are used in agriculture to control

Historical Note

Agent Orange

Agent Orange, a herbicide made of two auxin-based chemicals, 2,4-D and 2,4,5-T, was used widely as a defoliant in the jungles of Vietnam during the Vietnam War. Dioxin, a contaminant that results from the production of Agent Orange, is a highly toxic chemical that has been linked to many human health problems.

leaf, flower, and fruit drop and are a component of some herbicides (chemicals that kill weeds).

2. **Cytokinins** The **cytokinins** *(seye toh KEYE nihnz)* are a group of chemicals that are produced in root tips. They stimulate cell division, promote lateral bud growth, and inhibit leaf drop. Synthetic cytokinins are used in tissue cultures to stimulate growth and in agriculture to break apical dominance.

3. **Ethylene** The simple gaseous compound **ethylene** is produced in most tissues of the plant body. It stimulates fruit ripening, promotes leaf, flower, and fruit drop, and retards lateral bud growth. Ethylene is used extensively in agriculture to ripen fruits that are harvested before they ripen naturally and to loosen fruits so that they drop more readily for machine harvest.

4. **Gibberellins** The **gibberellins** *(jihb uhr EHL ihnz)* are a group of chemicals that are produced in developing shoots and seeds. They were first discovered by Japanese scientists studying an abnormal elongation of rice seedlings, called "foolish seedling disease," and then were named for the fungus that causes the disease—*Gibberella*. Gibberellins cause stem elongation by stimulating cell division and elongation. They also induce seed germination and fruit development. Synthetic gibberellins are used in agriculture to produce seedless fruits, as seen in Figure 25-12, and to promote uniform seed germination. □

Figure 25-12 Seedless white table grapes are produced by treatment with gibberellin. Compare a cluster of treated grapes, *bottom right,* with a cluster of untreated grapes, *top left,* and you will notice another effect of gibberellins on plants.

□ CAPSULE SUMMARY

Hormones, chemicals produced in one part of an organism and transported to another part of the organism, regulate plant growth and development.

Plant Growth Responds to Environmental Factors

Because plants are anchored in one spot, their growth must be adjusted continually in response to changes in the environment. In most cases, a plant's responses to environmental stimuli are triggered by the hormones that regulate plant growth.

Tropisms

A **tropism** *(TROH pihz uhm)* is a growth response in which the direction of growth is determined by the direction from which the stimulus comes. Auxins are responsible for producing tropisms. If the response is toward a stimulus, it is called a positive tropism. If the response is away from a stimulus, it is called a negative tropism. Three common stimuli to which plants respond are light, gravity, and touch. **Phototropisms** are growth responses to light. The growth of a plant's shoots toward light is a positive phototropism, while the growth of its roots away from light is a negative phototropism. **Gravitropisms** are growth responses to gravity. The upward growth of a germinating seed's shoot is a negative gravitropism, while the downward growth of its root is a

◉ **VISUAL STRATEGY**

Figure 25-12
Be sure students notice that the treated cluster is looser than the untreated cluster. Tell students that this looseness results from elongation of the cluster's stems, which is caused by gibberellin.

Demonstration
Ethylene and Leaf Drop
Materials: Two 1 gal. glass jars with lids, 2 small bedding plants in 2 in. pots, and a small apple.
Procedure: Place one plant in one of the jars, and tightly secure the lid. Place the other plant and the apple in the second jar, and secure the lid. Watch the jars for several days to see if a difference is observed in the plants. (The leaves should drop from the plant in the jar with the apple because of ethylene gas produced by the ripened fruit.)

Application

Agriculture Pineapples in Hawaii are sprayed with ethylene to induce all of the plants in a field to flower at the same time. Ask: Why is this important? *(They can all be harvested at the same time.)*

Demonstration
Plant Responses
Place bean seedlings in various places around the classroom to demonstrate phototropism and gravitropism. Seedlings that are placed near windows with bright light will show some bending toward the light. Those placed farther from a light source will bend strongly toward the light. Try to place some seedlings so that they are oriented sideways or upside down. These seedlings will show a negative gravitropism by bending upward.

Application

Agriculture Poinsettias, the Christmas flower, are subjected to artificial darkness in the greenhouses where they are grown beginning around September. Plants must be exposed to 10-hour days for eight weeks; so during late afternoon, the rooms are darkened to imitate short days. The poinsettias flower in response to the shorter days and longer nights.

Teaching Tips

What If the long night period in a poinsettia greenhouse were interrupted?
After examining Figure 25-14, students should realize that the length of the night period determines whether a short-day plant will bloom. Interrupting the night period in a poinsettia greenhouse, which can occur if lights are turned on or even when bright flashes of light from headlights or some other source come from the outside environment, simulates a short night. As a result, the blooming of the plants will be delayed or reduced. To ensure that their plants will have lots of large, beautiful blooms for Christmas, commercial poinsettia growers cover their greenhouses with heavy black cloth to prevent outside light from entering during a critical time of the year.

Identify the Photoperiod
Use nursery catalogs to provide students with examples of flowering plants. Include information about the blooming of the plant with the picture. Have students determine the photoperiod from the information given about the plant.

Figure 25-13 The coiling of grapevine tendrils, *above,* is a thigmotropism. The bending of an amaryllis toward a window, *above center,* is a positive phototropism. In germinating corn seedlings, *above right,* the upward growth of shoots is a negative gravitropism, and the downward growth of roots is a positive gravitropism.

positive gravitropism. **Thigmotropisms,** from the Greek word *thigma,* meaning "touch," are growth responses to touch. Effects of tropisms can be seen in Figure 25-13.

Photoperiodism

Have you ever wondered why certain plants bloom in the spring and others bloom in the summer or fall? Changes in the length of days (and nights) are responsible for the seasonal patterns of flowering in plants and also for many other seasonal patterns of plant growth and development. The response of a plant to the length of days and nights, or day length, is called **photoperiodism.** Plants can be categorized as one of three types, depending on how they respond to day length. If a response such as flowering occurs when days

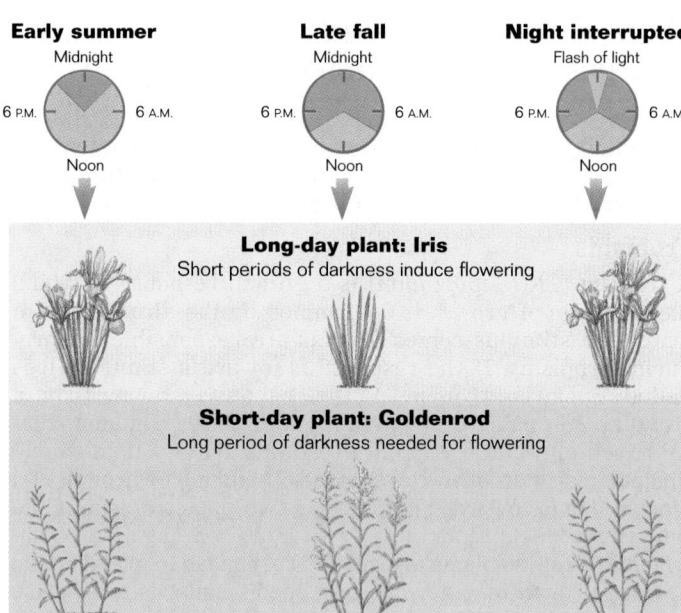

Figure 25-14 Day length (actually night length) controls flowering in many plants. Long-day plants flower when nights are short, and short-day plants flower when nights are long.

 Did You Know?

Tulips are produced from bulbs, which are modified underground stems. Tulip bulbs are refrigerated by commercial nurseries before they are grown because a bud within the tulip bulb requires a cold treatment to overcome dormancy. In the horticulture industry, this procedure is called "forcing."

become shorter than a critical length, the plant is said to be a **short-day plant.** If the response occurs when days become longer than a critical length, the plant is said to be a **long-day plant. Day-neutral plants** are plants that are not affected by day length. However, as Figure 25-14 shows, it is not really the *day length* that initiates a photoperiodic response—it is the length of the nights. Knowledge of photoperiodism is very important to commercial flower growers. By artificially controlling the length of days and nights in greenhouses, plants such as poinsettias, Easter lilies, and chrysanthemums can be forced to flower when they ordinarily would not.

Dormancy

Dormancy is a condition in which a seed or a plant remains inactive for a period of time. It is thought that abscisic acid, another plant hormone, plays a role in initiating dormancy. Seeds often remain dormant for a period of time before germinating. Dormancy may be broken after the seeds have been exposed to a period of low temperatures. This prevents the seeds of plants that live in regions with distinct seasons from germinating before winter has ended. Many perennial plants temporarily stop growing and become dormant when the environment becomes unfavorable for continued growth. The buds of deciduous woody plants usually become dormant in late summer or fall when thick, protective scales form around each bud, as seen in Figure 25-15. Then, a drop in auxin concentration and a rise in ethylene production combine to initiate leaf drop. After a period of exposure to cold temperatures, bud dormancy is broken and plant growth resumes. The shoots of herbaceous perennials, on the other hand, typically die during cold or dry weather. Carbohydrates stored in the dormant underground stems or fleshy roots of these perennials serve as fuel for the growth of new shoots. ▢

Figure 25-15 Thick scales cover these dormant buds on this twig from an apple tree.

▢ **CAPSULE SUMMARY**

Plant hormones are produced in response to stimuli from the environment. Therefore, plant growth and development respond to the environment in many ways.

Section Review

1. *What could you do to promote healthy plant growth?*
2. *How are nutrients important to plant growth?*
3. *List four types of plant hormones, and briefly describe how each affects plant growth and development.*
4. *How do tropisms affect plant growth?*

Critical Thinking
5. *Why is it an advantage for plant growth and development to be regulated by stimuli received from the environment?*
6. *Why is apical dominance an advantage to a young plant?*

SECTION 25-2

👁 **VISUAL STRATEGY**

Figure 25-15
Instruct students to locate and study the buds on the dormant apple twig in the photograph. Ask: How does the structure of the buds enable them to perform their function? *(Thick bud scales insulate the tender tissues inside the bud from the cold and prevent it from drying out or being damaged otherwise.)*

Chapter Closure

Have students write a short story describing the production of a crop on a farm of the future. Tell students to consider how a farm of the future might be different through the use of technology. For example, cotton fibers can now be produced in tissue culture without growing the plant in a field, but this presently takes place only on a small scale. Have students choose a crop and imagine how it might be planted, watered, fertilized, and harvested in the future. Ask them to include other factors that would make their stories more real, such as the atmosphere of the planet, people's attitudes about the new production methods, and other topics you feel would be important.

Answers to Section Review

1. To promote healthy plant growth, provide plants with water, air, sunlight, and fertilizer.

2. Nutrients are important to plant growth because they provide the elements plants need to make proteins, nucleic acids, enzymes, phospholipids, and ATP (needed for building and regulating new cells) from the carbohydrates they produce during photosynthesis.

3. Auxin promotes stem elongation and inhibits lateral bud growth, leaf drop, and fruit drop. Cytokinins stimulate lateral growth and cell division, and inhibit leaf drop. Ethylene ripens fruit and causes leaf, flower, and fruit drop. Gibberellins cause cell elongation, promote seed germination, and stimulate fruit development.

4. Tropisms cause plant growth toward or away from a stimulus.

5. It is an advantage for plant growth to be regulated by environmental stimuli because plants are unable to move to a new environment when conditions become unfavorable.

6. Apical dominance enables a plant to escape shade by extending its leaves toward light.

CHAPTER REVIEW ANSWERS

Each item in the Chapter Review is correlated by text section in the assignment guide that follows.

ASSIGNMENT GUIDE

Section	Review	Themes Review	Critical Thinking
25-1	1–6, 11–14	18, 19	21
25-2	7–10, 15–17	20	22, 23

Review

Multiple Choice

Code in parentheses indicates section number and objective number.

1. a (25-1.1)
2. a (25-1.1)
3. c (25-1.1)
4. d (25-1.2)
5. c (25-1.3)
6. b (25-1.4)
7. b (25-2.1)
8. d (25-2.2)
9. b (25-2.3)
10. c (25-2.4)

Completion

11. cotyledons, endosperm (25-1.1)
12. woody, deciduous (25-1.4)
13. cork, xylem, phloem (25-1.2)
14. vegetative propagation (25-1.5)
15. gibberellins, ethylene (25-2.3)
16. positive phototropism, negative gravitropism (25-2.4)
17. fall, summer (25-2.4)

---

25 CHAPTER REVIEW

Vocabulary

annual (577)	ethylene (583)	phototropism (583)
apical dominance (582)	evergreen (577)	primary growth (573)
apical meristem (573)	germination (572)	primary tissue (573)
auxin (582)	gibberellin (583)	secondary growth (574)
biennial (577)	gravitropism (583)	secondary tissue (574)
cork cambium (574)	hormone (582)	seedling (572)
cytokinin (583)	lateral meristem (574)	short-day plant (585)
day-neutral plant (585)	long-day plant (585)	thigmotropism (584)
deciduous (577)	nutrient (580)	tissue culture (576)
differentiation (573)	perennial (577)	tropism (583)
dormancy (585)	photoperiodism (584)	vascular cambium (574)

Review

Multiple Choice

1. The first sign of germination of a bean seed is the emergence of the embryo's
   a. root.
   b. hooked shoot.
   c. protected shoot.
   d. cotyledons.

2. Providing cells for growth at the tips of the plant is the primary function of the
   a. apical meristems.
   b. root cap.
   c. apical endosperm.
   d. dermal ground tissue.

3. Cell division that results in the increased width of a tree occurs in the
   a. apical meristems.
   b. primary tissues.
   c. lateral meristems.
   d. sieve tube vessels.

4. Plant and animal development differ in that plant development
   a. stops soon after the plant reaches maturity.
   b. is not affected by environmental factors.
   c. is controlled by genes that cannot be reactivated.
   d. is continuous and enables whole plants to develop in tissue cultures.

5. Plants that live three or more years are
   a. annuals.
   b. biennials.
   c. perennials.
   d. evergreens.

6. The fastest way to propagate kalanchoës is to plant
   a. leaf cuttings.
   b. plantlets.
   c. seeds.
   d. stem cuttings.

7. Which of the following is *not* a basic requirement for healthy plant growth?
   a. oxyen
   b. vitamins
   c. carbon dioxide
   d. water

8. Which nutrient is important to plants because of its role in active transport?
   a. nitrogen
   b. sulfur
   c. magnesium
   d. potassium

9. What hormone should be used if lateral bud growth is desired?
   a. ethylene
   b. cytokinin
   c. auxin
   d. gibberellin

10. The response of a plant to the length of days and nights is
    a. negative phototropism.
    b. positive phototropism.
    c. photoperiodism.
    d. dormancy.

Completion

11. In bean seeds, the embryo's energy reserves are stored in the _____ ; in corn seeds, the embryo's energy reserves are stored in the _____ .

12. Oaks and elms are examples of _____ perennials. Because these trees lose their leaves at the end of each growing season, they are said to be _____ .

13. Cork cambium produces _____ cells, and vascular cambium forms _____ and _____ .

14. Growing new plants from a plant's non-reproductive parts, such as stems and leaves, is called _____ .

15. Synthetic _____ are used to inhibit seed formation in grapes, while _____ is used to ripen apples and pears that are picked before they ripen naturally.

16. Plant growth *toward* light is called _____ . Plant growth *away* from gravity is called _____ .

17. Short-day plants tend to flower in the _____ , and long-day plants tend to flower in the _____ .

Themes Review

18. **Levels of Organization** Explain how a plant's tissues form through differentiation, beginning with the cells of the apical meristems and ending with secondary tissues.

19. **Flow of Energy** A carrot plant is a biennial that completes its life cycle in two years. How is the energy required for the plant's second year of growth stored at the end of the first year?

20. **Evolution** The pattern of plant growth seen in the photograph below is caused by apical dominance. What is the most obvious adaptive advantage of apical dominance to a young plant?

Critical Thinking

21. **Making Predictions** A student placed a green banana in each of several plastic bags. The student also placed a ripe pear in half of the bags and then sealed all of the bags. Which group of bananas (with or without pears) do you think will ripen sooner? Justify your answer.

22. **Making Inferences** Suppose that a friend who lives in North Dakota gives you some seeds from a plant that you admired when you saw it growing in your friend's yard. You plant the seeds at your home in Florida, but they fail to germinate. Based on your knowledge of seed germination, what might be preventing the germination of the seeds?

23. **Designing Experiments** Design an experiment that will test the inference you made for question number 22. Be sure to include a control for each variable in your experiment.

Activities and Projects

24. **Cooperative Group Project** Interview or write to people who work at commercial nurseries to find out how they produce the plants that are sold at garden centers and flower shops. Some specific plants that you might want to ask about include chrysanthemums, Easter lilies, fruit and nut trees, lawn grasses, orchids, pansies, and poinsettias. Summarize your findings in a report that can be shared with your class.

25. **Multicultural Perspective** Research the practice of and philosophical basis of bonsai, the Oriental art of growing miniature plants. Find out when and where bonsai originated and how bonsai plants are kept small. Relate your findings in a report that also explains how an understanding of plant growth and development is important to success in bonsai.

Critical Thinking

21. Students should predict that the bananas with ripe pears will ripen sooner. Ripe fruits give off ethlyene, which causes ripening in other fruits. (25-1.3)

22. Answers will vary. Some possibilities are: the seeds did not receive the required cold period; they did not receive the right amount of light; they received too much water; they got too hot. Students should provide a reasonable explanation for the inference they make. (25-2.4)

23. Answers will vary depending on the inference made for Question 22. Some things to look for in a good experimental design include a testable hypothesis based on the inference, use of controls for each variable, use of many test plants (seeds), and repetition of tests. (25-2.4)

Activities and Projects

24. Students can find out how to get information on plant propagation through a local garden center, your local agricultural extension service, or your state's agricultural colleges and universities.

25. In bonsai, plants are shaped by selective pruning and training and are kept small by pruning their roots, potting in small containers, and watering and fertilizing sparingly. The bonsai gardener must have a knowledge of the growth habits and nutritional needs of a plant in order to successfully grow healthy miniature plants.

Themes Review

18. Primary dermal, ground, and vascular tissues of the stem and leaves are produced by the shoot apical meristem, while the root apical meristem produces the primary dermal, ground, and vascular tissues of the root. The lateral meristems, vascular cambium and cork cambium, form from primary tissues. Vascular cambium produces secondary xylem and phloem, and cork cambium produces the cork cells of the outer bark. (25-1.2)

19. The energy for a carrot plant's second year of growth is stored in the familiar and edible root. (25-1.4)

20. Apical dominance enables plants to direct their growth upward, toward sunlight. (25-2.3)

LABORATORY Investigation Chapter 25

Seed Structure and Seedling Development

OBJECTIVE

Observe the structures of dicot and monocot seeds, and compare the development of their embryos.

PROCESS SKILLS

- relating structure to function
- comparing features of monocots and dicots

MATERIALS

- 1 pea seed soaked overnight
- 6 bean seeds soaked overnight
- 6 corn kernels soaked overnight
- stereomicroscope
- scalpel
- Lugol's iodine solution in dropper bottle
- paper towels
- 2 rubber bands
- 150 mL beakers (2)
- glass-marking pen
- metric ruler
- microscope slide
- medicine dropper
- compound light microscope

BACKGROUND

1. What are the parts of a seed?
2. In what ways are seeds like their parent plant?
3. How do monocotyledons and dicotyledons differ?
4. Write your own question to explore in your lab report or notebook.

TECHNIQUE

Part A: Seed Structure

1. Obtain each of the seeds—pea, bean, and corn. Remove the seed coats of the pea and bean seeds. Open the seeds to reveal the two embryonic leaves.

2. Using the stereomicroscope, examine the embryos of the pea seed and the bean seed.

3. In the **Records** section of your report, draw the pea and bean embryos and label all of the parts that you can identify.

4. Examine a corn kernel and find a small, oval, light-colored area that shows through the seed coat. **CAUTION: Use the scalpel carefully to avoid injury.** Use the scalpel to cut the seed in half along the length of this area. Place a drop of iodine on the cut surface.

5. Use the stereomicroscope to examine the corn embryo. In the **Records** section of your report, sketch the embryo and label all the parts that you can identify.

Part B: Seedling Development

6. Set five corn kernels on a folded paper towel. Roll up the paper towel and put a rubber band around the roll. Stand the roll in a beaker with 1 cm of water in the bottom. The paper towel will soak up water and moisten the corn. Keep water at the bottom of the beaker, but do not allow the corn kernels to be covered by water.

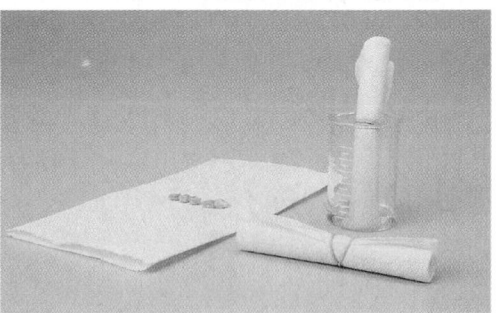

Preparation

Each lab group will need 1 pea seed, 6 bean seeds, 6 corn kernels, 2 rubber bands, two 150 mL beakers, paper towels, and Lugol's iodine solution as well as a scalpel, glass-marking pen, metric ruler, microscope slide and coverslip, medicine dropper, stereomicroscope, and compound light microscope. Soak seeds prior to the investigation. Seeds may be purchased at a nursery, hardware store, or ordered from a biological supply house. Ward's offers the following seed packets: green peas, kidney beans, and corn kernels.

Procedural Notes

- This investigation must be done in three 30-minute periods over 5 days.
- The seed coats of the pea and bean should be easy to remove if the seeds have soaked long enough.
- **Safety:** Caution students to use scalpels carefully to avoid injuring themselves and others.
- **Safety:** Caution students to use the Lugol's iodine solution with care. Remind them that they should not touch the chemical and that if they get the iodine solution on their skin or clothing, they should wash it off at the sink while calling to you.

Background Answers

1. A seed consists of an embryo, a stored food supply, and a seed coat.

2. Like the parent plant, the embryo is a sporophyte.

3. Monocots produce seeds with one cotyledon and usually have flowers with parts in multiples of three, leaves with parallel veins, and stems with scattered vascular bundles. Dicots produce seeds with two cotyledons and usually have flowers with parts in multiples of four or five, leaves with netted veins, and stems with vascular bundles arranged in rings.

4. How does a monocot seed compare with a dicot seed?

Records

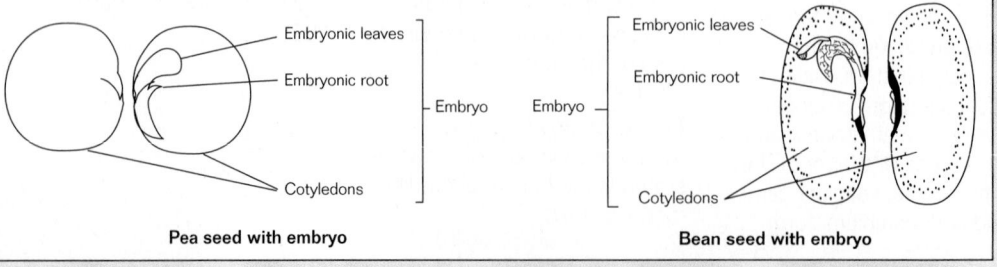

Pea seed with embryo   Bean seed with embryo

7. Repeat step 1 with five bean seeds.

8. After three days, unroll the paper towels and examine the corn and bean seedlings. Use a glass-marking pen to mark the roots and shoots of the developing seedlings. Starting at the seed, mark each 0.5 cm along the root of each seedling. And again starting at the seed, mark each 0.5 cm along the stem of each seedling.

9. Draw a corn seedling and a bean seedling on the **Records** section of your report and record the distance between the marks. Using a fresh paper towel, roll up the seeds, place the rolls in the beakers, and add fresh water to the beakers.

10. After two more days reexamine the seedlings. Measure the distance between the marks.

11. **CAUTION: Use the scalpel carefully to avoid injury.** Using the scalpel, make a cut about 2 cm from the tip of the root of a bean seedling. Place the root tip on a microscope slide and add a drop of water. Using a compound light microscope on low power, observe the root tip. In the **Records** section of your report, draw the root tip.

12. In your report, briefly summarize the procedure you followed.

13. Clean up your materials and wash your hands before leaving the lab.

INQUIRY

1. What types of leaves first appear on the bean seedling?

2. What substance does the black color in the corn kernel indicate? Why might you expect to find this substance in the seed?

3. Has the distance between the marks changed? If it has, where has it changed?

ANALYSIS

1. What parts of the embryo were observed in all seeds on the third day?

2. How does the structure and development of the corn kernel differ from the structure and development of the pea and bean seeds?

3. What was the source of nutrients for each of the seed embryos? What is your evidence?

4. Describe the growth in the seedlings you observed.

5. What is the function of root hairs? How do they improve the function of the root?

6. Corn and beans are often cited as representative examples of monocots and dicots, respectively. Relate the seed structure of each to the terms *monocotyledon* and *dicotyledon*.

7. As they push through the soil, what protects the tips of corn shoots? of bean shoots?

FURTHER INQUIRY

Write a new question that could be explored as a new investigation. The following is an example:

> How do monocots and dicots compare in general plant growth and in the structure of their leaves and flowers?

Inquiry Answers

1. Cotyledons, or seed leaves, are the first leaves to appear on a bean seedling.

2. The black color indicates that starch is contained in the corn kernel. You might expect to find starch in a corn kernel because much of the kernel is a stored food supply for the embryo and starch is the main storage form of carbohydrates produced in photosynthesis.

3. Yes. The distance between the marks has changed at the tips of the stems and the tips of the roots.

Analysis Answers

1. Embryonic leaves and roots are observed in all seeds on the third day.

2. Peas and beans have two cotyledons, while corn has only one. Corn seeds have endosperm, while mature bean and pea seeds do not. The shoots of peas and beans hook as they germinate, while a corn shoot grows straight up.

3. Food is stored in the endosperm of the corn kernel and in the cotyledons of the beans and peas. The iodine test for starch confirms this.

4. The seedlings grow at the tips of their roots and stems.

5. Root hairs absorb water from the soil. They facilitate water absorption by increasing the surface area of a root.

6. Corn seeds have only one cotyledon and thus are monocotyledons; while bean seeds have two cotyledons and thus are dicotyledons.

7. Corn shoots are protected by a sheath as they push through the soil. Bean shoots are protected by a hook in the embryonic stem that pushes through the soil before the cotyledons.

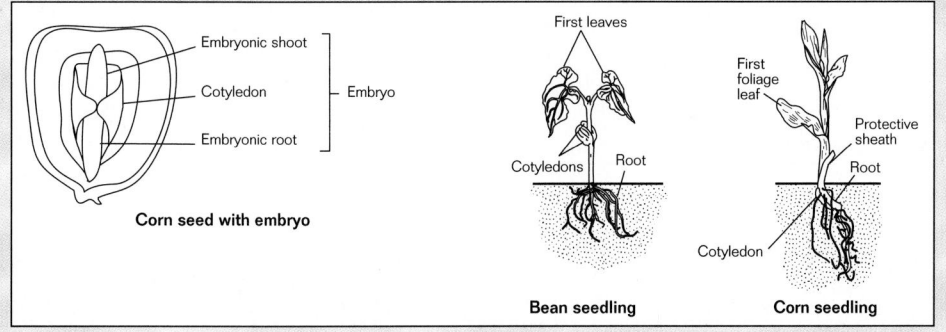

Corn seed with embryo — Embryonic shoot, Cotyledon, Embryonic root (Embryo); First leaves, Cotyledons, Root — Bean seedling; First foliage leaf, Protective sheath, Root, Cotyledon — Corn seedling

UNIT 6

Block Scheduling Guide

Each block represents about 45–50 minutes of instructional time. The resources cited in a block are coded to a specific program component using the Key on the next page.

Lecture Concepts	Lesson Resources

26-1 Plants as Food pp. 591–599

BLOCK 1

a. Cereals such as rice, wheat, and corn are important sources of food. Cereals produce dry fruits called grains, which contain a single seed that is packed with energy-rich endosperm.

b. Legumes such as beans and peas are important foods because they provide essential amino acids that grains lack.

c. Root crops such as potatoes, yams, and cassava are also a major source of calories.

Lecture Resources
- Opening Question, page 591
- Demonstrations: Fruits or Vegetables?, Biodegradable Packing Pellets
- Visual Strategy: Figure 26-3
- Overcoming Misconceptions, page 595
- Research Update, page 599
- Teaching Transparencies: 112, 113, 114
- Holt Biology Videodiscs: Lesson 26-1

Classwork Options
- A Consumer Comparison, page 592
- Brainstorming Crop Engineering, page 594
- Up Close: Bread Wheat, pages 596–597
- Getting the Essential Amino Acids, page 598

Assignment Options
- A Vegetarian's Complete Protein Cookbook, page 598
- Section Review, page 599
- Chapter Review, questions 1–6, 11–14, 18, 19, 21, 22

26-2 Other Uses of Plants pp. 600–603

BLOCKS 2 and 3

a. Wood, an important plant resource, is found in thousands of products. It is cut into lumber for use in building construction, and ground into wood pulp for use in paper and rayon.

b. Plants are the sources of many important medicines used to treat diseases and other ailments.

c. Plants are also sources of cloth (cotton), rubber, and latex.

Lecture Resources
- Opening Question, page 600
- Demonstrations: Uses of Plants, Know Your Lumber, Naturally Colored Cotton
- Products That Come From Trees, page 601
- Holt Biology Videodiscs: Lesson 26-2

Classwork Options
- Effective Reading, page 600
- Observing Paper Types, page 601
- Biotech Plants, page 602

- Closure, page 603
- Laboratory Investigation: Vegetative Propagation, pages 606–607
- Laboratory Techniques and Experimental Design C36: Using Paper Chromatography to Separate Pigments

Assignment Options
- Section Review, page 603
- Chapter Review, questions 7–10, 15–17, 20, 23

Review/Enrichment

BLOCK 4

■ Study Guide: Concept Review and Pretest

Assignment Options
- ■ Activities and Projects, questions 24–26
- ⊙ Teaching Resources CD-ROM: Portfolio Projects, Plant Research Project
- ⊙ Teaching Resources CD-ROM: Portfolio Projects, Plant Focus Worksheet
- ⊙ Teaching Resources CD-ROM: Occupational Applications Worksheets, Forestry Technician

Assessment Options

BLOCK 5

Traditional Assessment
- ■ Chapter 26 Test
- ⊙ Test Generator/Assessment Item Listing: Software item bank for preparing customized chapter tests.

Portfolio Assessment
Select student reports for one or more laboratory experiments from this chapter. *The Direct Observation Checklist* on the *BioSources Teaching Resources CD-ROM* should be completed during a laboratory or other cooperative learning experience.

Holt Biology Videodiscs

Holt Biology Videodiscs gives you a powerful tool for teaching, review, and assessment. *Concepts of Biology* adds interactive lesson plans and extensive tools for customization using CD-ROM technology.

CONCEPTS OF BIOLOGY

Use the following topic from *Concepts of Biology* to help you teach this chapter:

- ■ Topic 16: Plant Structure and Function

For further information regarding the media on the videodiscs, see *Holt Biology Videodiscs Teacher's Correlation Guide to Biology: Principles and Explorations.*

Chapter Themes

Flow of Energy *Plants depend on the sun's energy to make food, and humans depend on plants for the energy necessary for their survival. Humans derive energy from plants eaten as food and use the energy stored in plants for fuel.*

Structure and Function *The structure of plant parts, such as wood and fibers, enables humans to use those parts as building materials and as components of paper and cloth.*

Tapping Prior Knowledge

- Where does the energy stored in plants come from?
- What part of a plant produces a fruit?
- How does a plant produce wood?
- What can happen to a tree if you strip off its bark?

Opening Demonstration

Make some popcorn or grind some dried corn between two stones. Ask students where corn and this practice originated. *(Native Americans developed all of the types of corn known today, such as popcorn and sweet corn, before the time of Columbus. They also developed the processes of popping and grinding corn.)* Introduce this chapter to students by telling them that humans rely on plants for many purposes other than food.

CHAPTER 26
PLANTS IN OUR LIVES

REVIEW

- amino acids and proteins (Section 2-3)
- cellulose (Section 2-3)
- crops (Section 5-1)
- nitrogen-fixation (Section 16-3)
- fruit and endosperm (Section 23-3)
- structure of roots, stems, and leaves (Section 24-1)
- seed structure (Section 25-1)

Plants are an essential source of food.

Authors' Rationale

Students should be aware of all that plants contribute to their lives. Even the most voracious, top-of-the-food-chain carnivore depends on plants or some other autotrophic life-form for food. In addition to food, plants produce wood, medicines, and fibers used for making paper and clothing.

Students have also learned that plants remove carbon dioxide from the air, produce oxygen, prevent soil erosion, and are responsible, ultimately, for any animal product. Challenge your students to come up with any necessity in life that does not involve plants in some way.

26-1 Plants as Food

Humans and other animals depend on the sugar that plants produce during photosynthesis. Certain plants produce far more sugar during photosynthesis than they can immediately use. Such plants store this extra sugar and use it later for activities such as spring growth. Plants with large energy stores are attractive to humans and other animals as food. In this section, you will examine some of the most important food plants.

All Plant Parts Serve as Food

All types of plant parts—roots, stems, leaves, flowers, fruits, and seeds—are eaten as food. For marketing purposes, each of the foods derived from plants is identified by an agricultural commodity term—cereal, fruit, or vegetable—that is registered in Washington, D.C. These terms have different meanings in botany than they do in agriculture, however, and this often causes confusion among the general public. For example, botanically, a fruit is the ripened ovary of a flower, and a **vegetative part** is any nonreproductive part of a plant. The foods that you think of as fruits—such as apples, bananas, and melons—are also fruits in the botanical sense. But the foods known agriculturally as cereals are derived from fruits as well. Vegetables, on the other hand, may be any botanical part of a plant, as you can see in Figure 26-1.

Section Objectives

■ Categorize familiar foods as roots, shoots, stems, leaves, flowers, seeds, or fruits.

■ Identify the three most important cereal grains, and explain how each was derived from its wild ancestors.

■ Describe the role of polyploidy in the origin of bread wheat.

■ Identify nongrain foods derived from plants, and state their dietary importance.

Figure 26-1 Familiar vegetables, *below,* have botanical functions. Broccoli, artichokes, and cauliflower are flowers. Cabbage, celery, and lettuce are leaves. Carrots, radishes, and turnips are roots. Pumpkins, squash, green beans, and tomatoes are fruits. Asparagus and alfalfa sprouts are shoots. Potatoes and onions are modified stems.

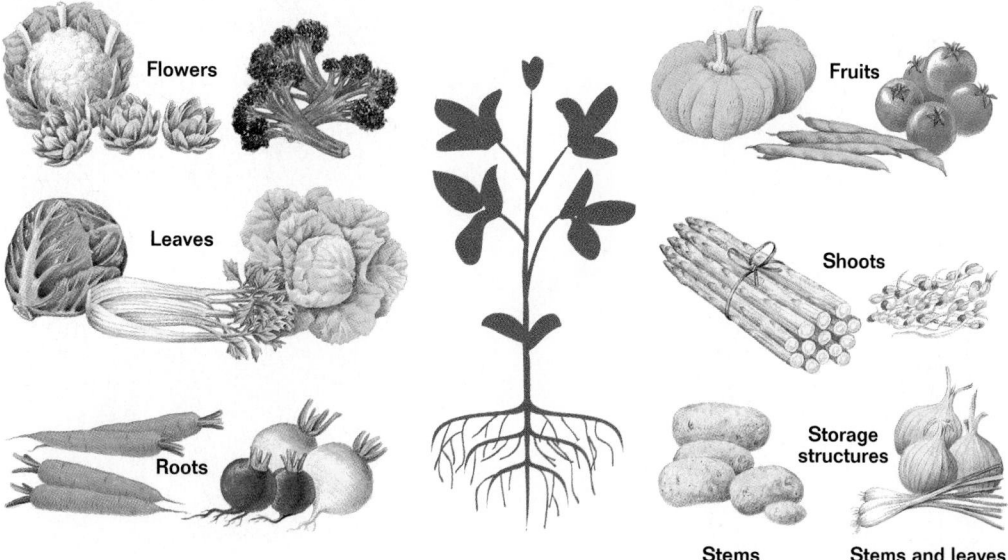

Flowers

Leaves

Roots

Fruits

Shoots

Stems

Storage structures

Stems and leaves

SECTION 26-1

Vocabulary Preview

vegetative part
cereal
grain
bran
wheat germ
whole wheat
essential amino acids
complete protein
incomplete protein
legume
tuber

Opening Question

Ask students to make a list of the different types of plant parts that can be eaten as food. Students should list leaves, roots, stems, fruits, and seeds and may also list flowers, buds, bulbs, tubers, and shoots.

Demonstration
Fruits or Vegetables?

To compare the agricultural and botanical meanings of the word *fruit,* show students a tomato, squash, green pepper, apple, orange, and banana. (Note: Other types of fruits and vegetables could be substituted.) Ask students to classify these foods in any way they wish. (*Students will probably derive the terms* vegetables *and* fruits.) Tell students that they are correct if they are describing an agricultural commodity. Then tell them that these foods can also be placed in one botanical group because they all originate from the ovary of a flower.

 Did You Know?

Approximately 80,000 plant species produce edible parts, but only 50 of these are cultivated as major food sources. Yet out of these, only seven species—rice, wheat, corn, barley, sorghum, potatoes, and cassava—are the the source of 75 percent of the world's food.

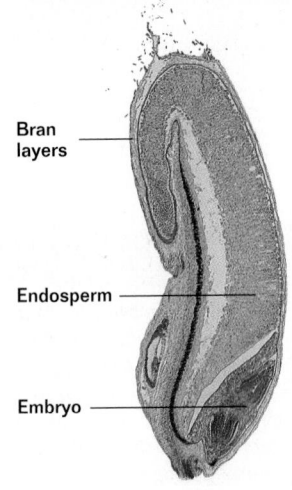

Figure 26-2 A wheat grain is a single-seeded dry fruit.

Figure 26-3 As this map illustrates, the United States is one of the world's major producers of rice, wheat, and corn. What other countries are major producers of these three important cereal crops?

Cereals Are the Most Important Sources of Food

Most of the foods that people eat come directly or indirectly from the fruits of **cereals,** which are grasses that are grown as food for humans and livestock. Cereal grasses produce large numbers of a type of edible, dry fruit called a **grain.** As you can see in Figure 26-2, a grain contains a single seed with a large supply of endosperm. A grain is covered by a dry, papery husk called the **bran,** which includes the wall of the ovary and the seed coat. Grains are rich in carbohydrates and also contain protein, vitamins, and dietary fiber. More than 70 percent of the world's cultivated farmland is used for growing cereal grains. The map in Figure 26-3 shows where the three most important cereal crops—rice, wheat, and corn—are grown.

Rice: An Important Source of Carbohydrates

For more than half of the people in the world, rice, *Oryza sativa*, is the main part of every meal. Although it is low in protein, rice is an excellent source of carbohydrates. While brown rice still has its vitamin-rich bran layers, white rice has been processed to remove the bran layers. This processing helps to prevent spoilage in stored rice. In societies where people eat mainly rice, vitamin-rich sauces such as soy sauce are added to white rice to make meals more nutritious. The white rice you buy at a grocery store is enriched

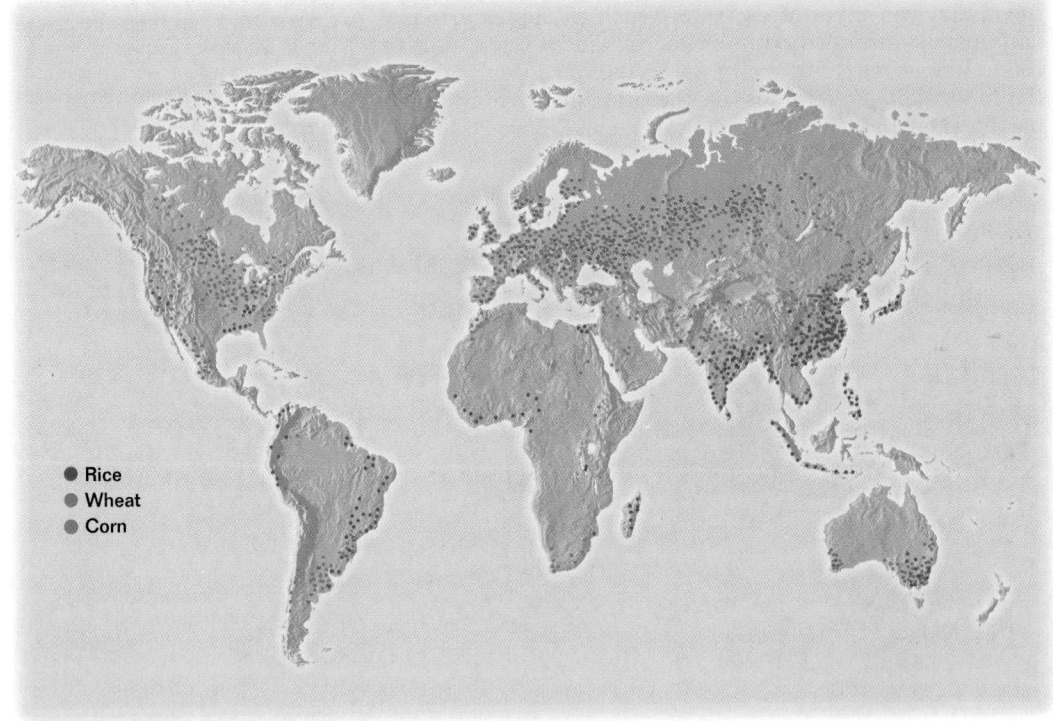

- Rice
- Wheat
- Corn

White rice

Brown rice

with added vitamins. Rice is often added to processed foods such as breakfast cereal, soup, baby food, and flour to increase their energy content.

Rice is native to tropical Asia, where it was gathered and eaten by people more than 10,000 years ago. Archaeologists have found evidence that people cultivated rice in southern China as early as 5000 B.C. From China, rice cultivation spread east to Japan and west to India. Rice thrives in areas with abundant rainfall, such as southeast Asia and the Gulf Coast of the United States, where it is grown in standing water. Rice is pictured in Figure 26-4.

Wheat: The World's Most Widely Grown Crop

For more than one-third of the world's population, wheat is the primary source of food. High in carbohydrates, the endosperm of wheat grains is commonly ground into white flour and used in breads and pasta. Vitamin-rich **wheat germ** consists of the embryos of wheat grains. **Whole wheat** consists of the endosperm plus the germ and bran layers. In the Middle East, wheat grains are often boiled or soaked, dried, and then pounded until they crack. The cracked grains, called bulgur *(BUL guhr)*, are used in dishes such as tabbouleh *(tuh BOO lee)* and pilaf *(pih LAHF)*. Most wheat is grown in temperate regions that have fertile soil and moderate rainfall. One of the world's best wheat-growing areas is the Great Plains region of the United States and Canada—a temperate grassland biome.

Even before the cultivation of wheat began in the Middle East about 11,000 years ago, people gathered and ate the grains of wild wheat plants. However, the grains of the first

Figure 26-4 Rice plants, *left,* are grown in standing water. White rice, *center,* lacks the vitamin-rich bran layers, but is still rich in carbohydrates. Brown rice, *right,* is more nutritious than white rice because it has not been processed to remove the bran layers.

Figure 26-5 The grains of modern bread wheat, *Triticum aestivum*, *top right* and *bottom left*, are much larger than the grains of *Triticum monococcum, top left*, one of three wild ancestors of modern bread wheat.

wild wheats that were cultivated bear little resemblance to the grains of modern wheat, as you can see in Figure 26-5. Early farmers selected grains from their best wheat plants—such as those with bigger and more numerous grains, those with stalks that did not break in the wind, and those with grains that fell off the stalk more easily during threshing—to use as seeds for planting. Over time, selection by farmers resulted in modern bread wheat, or *Triticum aestivum*. To learn more about the history of modern bread wheat, look at *Up Close: Bread Wheat* on pages 596–597.

Corn: The American Grain

Corn, *Zea mays*, is the most widely cultivated crop in the United States. American colonists of the 1600s and 1700s first learned how to grow corn from Native Americans. In the southeastern United States, corn was more widely grown than wheat, which does not grow as well in hot climates. Thus, corn-based foods—corn bread, corn pone, hominy, and grits—are a traditional component of the southeastern American diet. Corn is also one of the world's chief foods for farm animals. About 70 percent of the United States corn crop is consumed by livestock. Other uses for corn include the production of corn syrup, margarine, corn oil, cornstarch, and fuel-grade ethanol. Most of the corn grown in the United States today comes from a region known as the Corn Belt, which includes Iowa, Nebraska, Minnesota, Illinois, and Indiana.

Corn is the only major grain crop that originated in America. Early native farmers in Mexico are thought to have selected seeds from the largest flower spikes of teosinte *(tee oh SIHN tee)*, *Zea mexicana*, a wild annual grass. Teosinte

Corn, *Zea mays*

Tassels (male flower spikes)

Immature ears (female flower spikes)

Teosinte, *Zea diploperennis*

Flower spike

and modern corn are compared in Figure 26-6. Selective breeding by humans eventually produced plants that had flower spikes with many parallel rows of grain, resembling an ear of corn. Corn was later cultivated by the Aztecs of central Mexico, the Mayas of central Mexico and northern Central America, and the Incas of western South America. The cultivation of corn also spread among many other native North American cultures. Columbus and other European explorers introduced corn to many parts of the world. Now one of the principal crops grown in Africa, corn is the third most important source of food for humans. ❑

Figure 26-6 Modern corn, *Zea mays, above left,* was derived from *Zea mexicana,* or teosinte, a native of Mexico. An ear of corn is a thick spike with many rows of grains. Teosintes such as *Zea diploperennis, above right,* a species that is almost identical to *Zea mexicana,* produce narrow spikes containing two rows of grains.

Many Other Plants Supply Important Foods

Although cereals provide most of the calories consumed by humans, they do not contain all of the nutrients needed for a healthy diet. Other significant foods derived from plants are seeds, stems, or roots that provide important vitamins, minerals, and amino acids. For example, your body produces only 11 of the 20 amino acids that are needed to make your proteins. The other 9 amino acids, called **essential amino acids,** must be obtained from the foods you eat. Meat provides all of the amino acids your body needs, and meat protein is therefore said to be a **complete protein.** Plant proteins, which lack one or more of the essential amino acids, are said to be **incomplete proteins.** In societies where there is little meat in the diet, it is particularly important that people eat combinations of grains and other foods that provide a complete protein.

❑ **CAPSULE SUMMARY**

Rice, wheat, and corn are the most important sources of food for humans. These crops, called cereals, produce dry fruits called cereal grains, which contain a single seed that is packed with energy-rich endosperm.

Demonstration
Biodegradable Packing Pellets
Materials: biodegradable packing pellets, a beaker of water, iodine.
Procedure: To show students that biodegradable materials can replace petroleum-based products that cannot be recycled, drop a few of the packing pellets into a beaker of water. The pellets will dissolve. Then add a few drops of iodine to the beaker, and note the color change. Ask students to identify the composition of the pellets based on the color change that occurs. *(Purple indicates the presence of starch.)* Tell students that these pellets, which are replacing the plastic foam pellets derived from petroleum-based products, are made of cornstarch, which is derived from corn.

Overcoming Misconceptions

Vegetarianism Is Not Always Healthy

Many people believe that by not eating animal products, their diet will be more healthy. This is true only to a certain extent. Vegetarians must be sure to eat combinations of foods that provide a complete protein as well as needed vitamins and minerals. One consequence of eating a diet that is deficient in essential amino acids is kwashiorkor, a severe protein deficiency suffered by young children whose diet is composed only of grains. Kwashiorkor is characterized by stunted growth, edema, and swelling of the belly and can result in mental retardation. It can be prevented by adding protein to the diet in the form of meat or legumes.

Instructional Strategies

- Show students specimens of living or preserved wheat plants.

- Let students dissect kernels of wheat and locate the endosperm, bran, and germ of the kernels. Explain that the endosperm serves as a food supply for the embryo when it begins to grow. Tell students that white wheat flour contains only endosperm and that whole wheat flour contains both endosperm and bran.

- For interest, have students make bread so that they notice the elasticity of the dough, which is caused by gluten.

UP CLOSE   BREAD WHEAT

- **Scientific Name:** *Triticum aestivum*
- **Range:** Agricultural regions worldwide
- **Habitat:** Cultivated fields in temperate and subtropical grasslands
- **Size:** 0.3 m (1 ft.) to 0.8 m (2.5 ft.) tall
- **Importance:** Bread wheat is the world's most agriculturally important annual grass. A staple food in temperate regions of the world, the grains of *Triticum aestivum* are high in gluten (a sticky mixture of proteins that makes dough elastic) and are usually ground into flour that is used to make bread.

External Structures

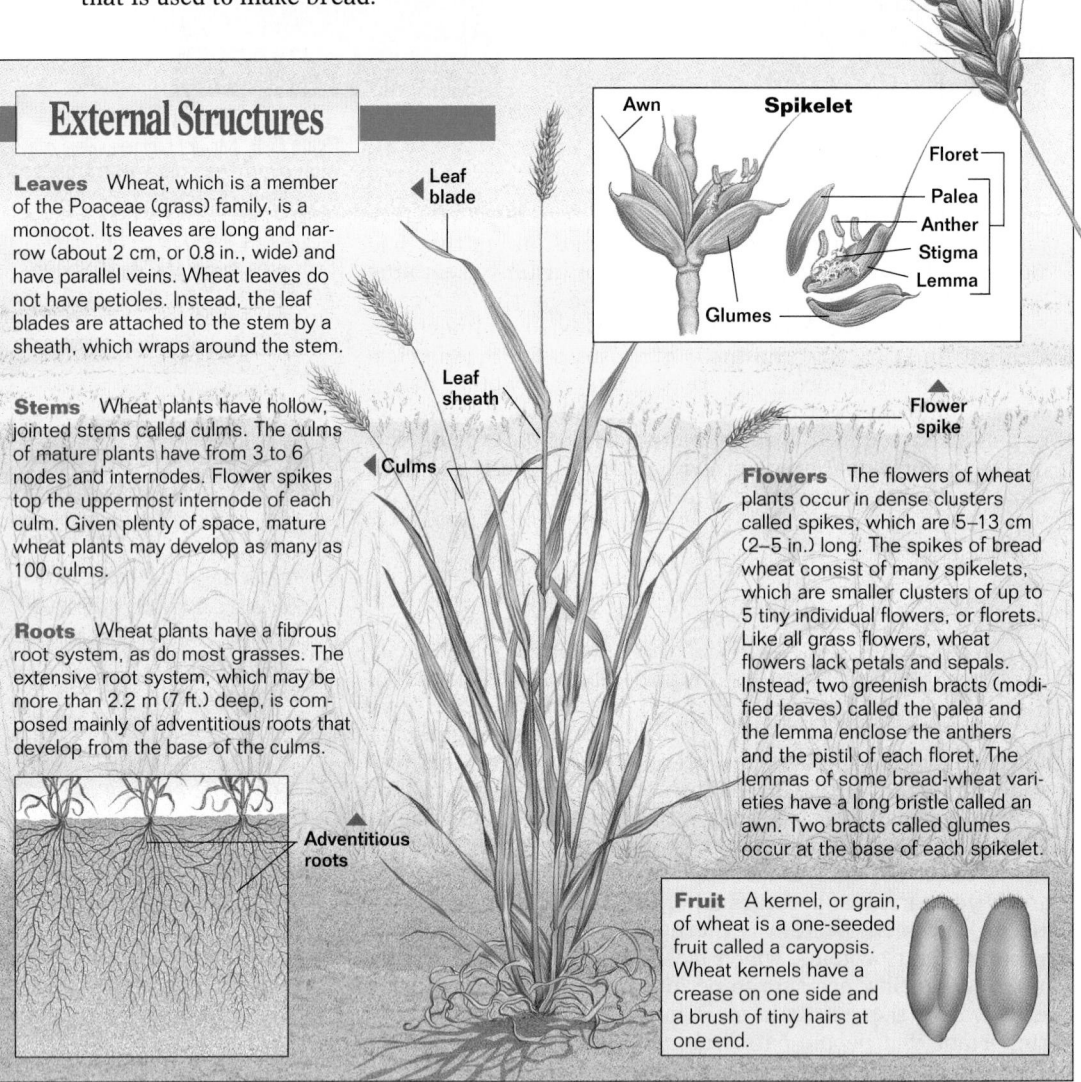

Leaves   Wheat, which is a member of the Poaceae (grass) family, is a monocot. Its leaves are long and narrow (about 2 cm, or 0.8 in., wide) and have parallel veins. Wheat leaves do not have petioles. Instead, the leaf blades are attached to the stem by a sheath, which wraps around the stem.

Stems   Wheat plants have hollow, jointed stems called culms. The culms of mature plants have from 3 to 6 nodes and internodes. Flower spikes top the uppermost internode of each culm. Given plenty of space, mature wheat plants may develop as many as 100 culms.

Roots   Wheat plants have a fibrous root system, as do most grasses. The extensive root system, which may be more than 2.2 m (7 ft.) deep, is composed mainly of adventitious roots that develop from the base of the culms.

Flowers   The flowers of wheat plants occur in dense clusters called spikes, which are 5–13 cm (2–5 in.) long. The spikes of bread wheat consist of many spikelets, which are smaller clusters of up to 5 tiny individual flowers, or florets. Like all grass flowers, wheat flowers lack petals and sepals. Instead, two greenish bracts (modified leaves) called the palea and the lemma enclose the anthers and the pistil of each floret. The lemmas of some bread-wheat varieties have a long bristle called an awn. Two bracts called glumes occur at the base of each spikelet.

Fruit   A kernel, or grain, of wheat is a one-seeded fruit called a caryopsis. Wheat kernels have a crease on one side and a brush of tiny hairs at one end.

Did You Know?

More than 200 varieties of wheat are grown in the United States. Hard red wheats, the varieties grown for making bread flour, belong to the species *Triticum aestivum*. Durum wheats, or macaroni wheats, which are grown for making pasta because they contain less gluten, belong to the species *Triticum dicoccum*. Wheat may be planted in either the fall or the spring. The term *winter wheat* refers to a crop that is planted in the fall, overwinters as seedlings, and is harvested in late spring or summer. The term *spring wheat* refers to a crop that is planted in the spring for harvest in the fall.

Internal Structures

Grain structure A cross section of a wheat kernel shows that it is mainly (about 85 percent) starchy endosperm. The kernel's outer layers, consisting of the ovary wall, seed coat, and aleurone layer (a layer that contains protein and oils), make up the bran (about 12 percent of the kernel). The embryo, or wheat germ, is only a small portion (less than 3 percent) of the kernel.

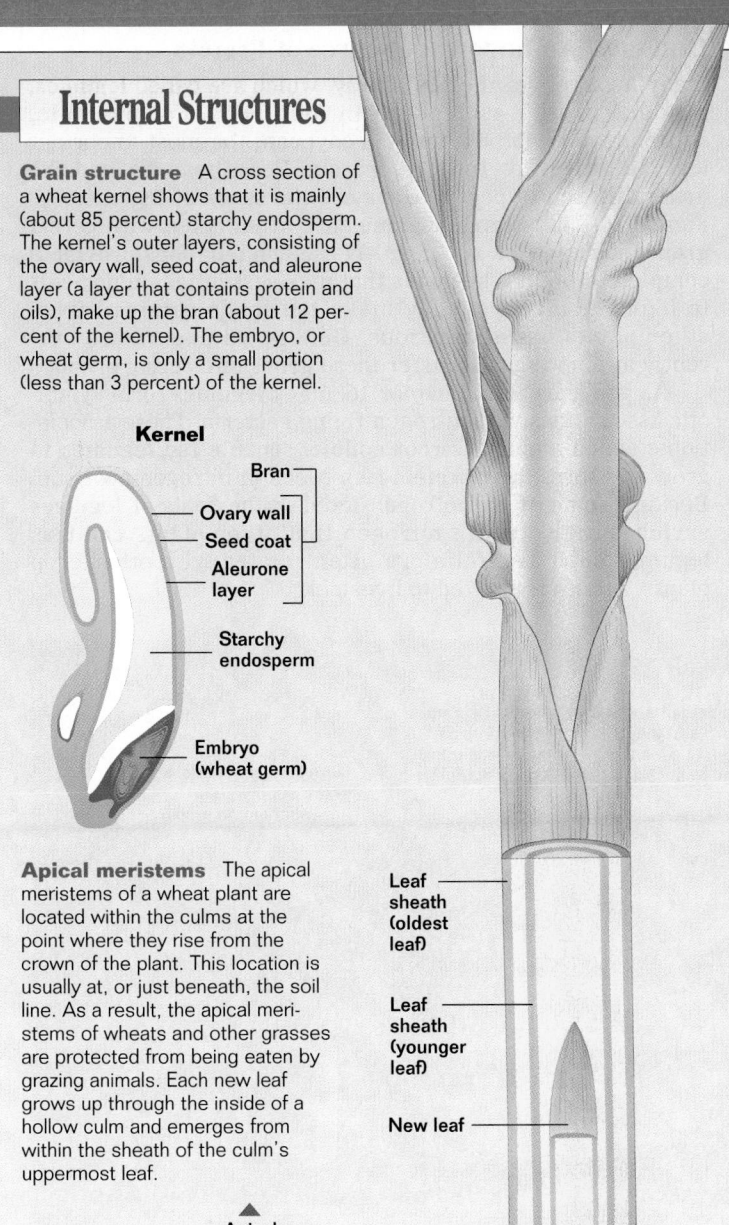

Kernel

- Bran
  - Ovary wall
  - Seed coat
  - Aleurone layer
- Starchy endosperm
- Embryo (wheat germ)

Apical meristems The apical meristems of a wheat plant are located within the culms at the point where they rise from the crown of the plant. This location is usually at, or just beneath, the soil line. As a result, the apical meristems of wheats and other grasses are protected from being eaten by grazing animals. Each new leaf grows up through the inside of a hollow culm and emerges from within the sheath of the culm's uppermost leaf.

Leaf sheath (oldest leaf)

Leaf sheath (younger leaf)

New leaf

Apical meristem

Soil level

Chromosomes The cells of bread wheat have a chromosome number of $2n = 42$, or $n = 21$. Close examination of bread wheat's chromosomes reveals an interesting pattern. Among the 42 chromosomes in a bread-wheat cell, there are actually three distinct sets of 14 chromosomes (7 pairs). Because the chromosomes in each set are slightly different from those in the other two sets, botanists represent these sets of paired chromosomes with the letters AA BB DD. Thus, bread wheat is not a diploid plant but is instead a hexaploid ($6n$) plant—a polyploid plant with 6 of each kind of chromosome.

Wheat cell

■ A   ■ B   ■ D

Karyotype of wheat

Polyploidy Polyploidy, which means having many sets of chromosomes, is common among cultivated food and ornamental plants. Polyploids tend to be larger, or prettier, than their wild ancestors. Among plants, polyploidy is an important evolutionary tool that enables fertile hybrids to result from crosses among different species. Such is the case with bread wheat. Three different closely related species hybridized naturally to produce bread wheat. Polyploids that contain sets of chromosomes from two or more different species are called allopolyploids. The prefix *allo-* comes from the Greek word *allos*, meaning "other."

Discussion

Guide the discussion by posing the following questions.

1. 2, 4-D is a chemical that is used to kill dicots. Will it kill wheat plants? Why or why not? *(No. Wheat is a monocot, which can be determined by observing some typical characteristics of the plant, such as parallel leaf veins, fibrous roots, and seeds with a single cotyledon.)*

2. How did our ancestors help produce the bread wheat varieties that exist today? *(From wild hybrids, they selected plants with desirable characteristics such as higher yields, larger seeds, and more gluten.)*

3. How would a human karyotype differ from the wheat karyotype? *(A human karyotype normally has 23 pairs of chromosomes, while a wheat karyotype has seven sets of six similar chromosomes, in which there are three homologous pairs.)*

4. Is polyploidy beneficial to all living organisms? Why or why not? *(No. Polyploidy may be beneficial in plants because it causes plants to be larger, showier, and more vigorous. However, it can also be detrimental in nature if the plant is sterile, which often occurs in triploids. In most animals, including humans, polyploidy is usually detrimental.)*

Teaching Tips

Getting the Essential Amino Acids

Tell students that foods containing animal protein provide all the essential amino acids. Ask: What is another name for animal protein? *(a complete protein)* What are some examples of foods that provide a complete protein? *(meat, eggs, and milk)* However, many cultures depend on other foods, such as rice, as a main staple. Cereals lack certain amino acids, but those amino acids are found in legumes. Legumes also lack certain amino acids; but when legumes are eaten with cereals, the combination provides all of the essential amino acids.

A Vegetarian's Complete Protein Cookbook

Let students work in teams to make lists of foods from different cultures that contain combinations of two crops that provide a complete protein. Have each group research recipes for the foods they list and compile cookbooks with the recipes. Allow students to work together to gather the recipes, but have each student individually make a copy of the cookbook. Ask each student to write a preface for their cookbook describing how the cultures from which the recipes come might have learned to combine these foods to get complete proteins.

Legumes: An Important Source of Protein

Many members of the pea family, which are called **legumes,** produce protein-rich seeds in long pods. For example, approximately 45 percent of a soybean, the most important legume grown for food, is protein. Peas, peanuts, and the many different types of beans are the seeds of legumes. As Figure 26-7 illustrates, legumes are often eaten with cereal grains such as rice and corn. These combinations provide a complete protein. The genes that make essential amino acids in legumes may someday make the foods obtained from other plants more nutritious. Genetic engineers are currently attempting to transfer these genes into cereal grains.

As you learned in Chapter 16, many legumes form symbiotic associations with nitrogen-fixing bacteria. These associations, which appear as root nodules, enable the legumes to grow and form their protein-rich seeds in nitrogen-poor soil. Because some of the nitrogen fixed in the roots of legumes enriches the soil with nitrogen that other plants can use, legumes such as alfalfa are often rotated with other crop plants. Alfalfa is also fed to livestock.

Figure 26-7 This meal supplies a complete protein by combining foods made from two grains (rice and corn) with a legume (beans).

Tamales (made of corn and meat; supply essential amino acids)

Beans (legume that contains some essential amino acids)

Corn tortillas (made of grain; supply some essential amino acids)

Rice (grain rich in carbohydrates)

Root Crops: Dietary Staples for Many People

Potatoes, *Solanum tuberosum,* are an important food staple in many regions of the world. Although potatoes are thought of as a root crop because they grow underground, they are actually **tubers,** modified underground stems that store starch. Yams, an essential food crop in many tropical parts of the world, are also tubers. Native to the Andean region of

Historical Note

Potatoes and the Irish Immigration

Millions of people immigrated to the United States from Ireland in the late 1840s because of the Irish potato famine. The loss of potatoes to late blight, a fungal disease, caused 400,000 people to die of starvation or of diseases caused by starvation.

Figure 26-8 Cassava, *Manihot esculenta, left,* is a large, fleshy root that is rich in starch. Cassava, also known as Manioc, supplies more than one-third of the calories consumed in Africa. In Central America, cassava flour is used to make a type of bread. This native Panamanian woman, *far left,* is grinding cassava roots to release toxic chemicals found in the roots. Foods made from cassava must be cooked to completely destroy the roots' toxic components.

☑ RESEARCH Update

Should Nicotine Be Controlled?

The tobacco industry has been battling an image problem since tobacco was linked to cancer. In June of 1994, the CEOs of the seven major tobacco firms in the United States testified before a Congressional subcommittee that the nicotine in tobacco products is not an addictive drug. However, scientists tend to agree that nicotine is *very* addictive in a manner similar to cocaine. If this is the case, cigarettes could be regulated as a controlled substance (drug) under the Federal Food, Drug, and Cosmetic Act.

John Dani, a researcher at Baylor College of Medicine, states that nicotine's principal similarity to cocaine is its relationship to dopamine, a "pleasure," or "reward," chemical produced in the brain. Researchers have shown that when enough dopamine is secreted, an organism feels a sense of satisfaction or pleasure. The brain regulates its release of dopamine by taking the chemical back into the tissues (reuptake). Cocaine blocks the reuptake of dopamine, so the sensation of pleasure is prolonged. Dani and other scientists have shown that nicotine stimulates the release of dopamine.

Will nicotine be regulated as a drug? Many people contend that it is a substance that needs to be controlled. The tobacco industry and their substantial number of customers disagree. It should be an interesting fight. ☑

South America, potatoes quickly became an important crop in Europe after they were introduced there by the explorers who followed Columbus. Rich in calories and easy to cultivate, potatoes are an ideal crop for a small farm. A small plot can provide enough calories for an entire family.

Other important root crops include sweet potatoes, carrots, radishes, turnips, beets, and cassava *(kuh SAH vuh).* These vegetables are enlarged roots that store starch. Cassava, seen in Figure 26-8, is the staple food of more than 500 million people, making it the most important noncereal crop. Tapioca pudding is made from the starch extracted from cassava roots. Large quantities of sugar are extracted from sugar beets, which are large, fleshy roots that store sugar rather than starch. Much of the sugar you eat, however, comes from the stems of a grass—sugar cane. ▫

▫ **CAPSULE SUMMARY**

Legumes such as beans and peas are important foods because they provide essential amino acids that grains lack. Root crops such as potatoes, yams, and cassava are a major source of calories in many cultures.

Section Review

1. *Name three common vegetables that are actually fruits and two vegetables that are actually flowers.*
2. *What are the three most important cereal grains, and where did each originate?*
3. *How does bread wheat differ from its wild ancestors?*
4. *How do potatoes and yams differ from other root crops?*

Critical Thinking

5. *How could transferring specific genes from legumes into rice plants help reduce malnutrition?*

Answers to Section Review

1. Answers will vary but might mention tomatoes, peppers, squash, green beans, or any plant part that contains seeds as examples of vegetables that are fruits. Vegetables that are flowers include cauliflower, broccoli, and artichokes.

2. The three most important cereal grains are rice, wheat, and corn. They originated in Asia, the Middle East, and Mexico, respectively.

3. Bread wheat, which produces larger grains than its wild ancestors, also differs genetically from those ancestors because it is a hexaploid plant, while its ancestors are either diploid or tetraploid species.

4. These vegetables differ from other roots crops because they are actually tubers, or underground stems.

5. Transferring genes from legumes into rice plants could result in rice that contains a complete protein, which could help reduce malnutrition by supplying more of the nutrients needed for a healthy diet even when other types of foods are unavailable.

Vocabulary Preview

lumber

wood pulp

aspirin

cortisone

digitalis

latex

Opening Question

Ask students to look around the room and identify everything derived from a plant. Students should be able to identify many materials in the classroom that are derived from plants.

Demonstration

Uses of Plants

Make a list on the board or overhead of student answers to the **Opening Question** on this page. Add to their list things they might have missed, such as starch in their clothes or paint on the wall. Use the list to introduce this section on uses of plants.

26-2 *Other Uses of Plants*

Section Objectives

- Describe several ways that wood is used.
- Explain how plants are used to treat human ailments.
- Describe how plants are used to make rubber and cloth.

Plants are used by people for many purposes other than food. For example, if you were to look at this sheet of paper very closely through a magnifying glass, you would see that it is made of many fibers laid across one another in a thin mat. These fibers are strands of cellulose from the cell walls of plants. Plant fibers are also used to make cloth. Trees can be cut into lumber, burned as fuel for heating and cooking, and processed for use in paper, plastics, and rayon. Plants are also the original sources of many important medicines.

Wood Is the Most Valuable Plant Product Besides Food

Figure 26-9 Wood has many uses. Like this guitar, *top left,* many instruments are made of wood. When burned in an efficient wood-burning stove, *bottom left,* wood provides heat for warming homes and cooking. Most wood is sawed into lumber and used in construction for purposes such as framing a house, *right.*

After food, wood is the single most valuable resource obtained from plants. For more than a quarter of the world's people, wood is still the main source of fuel for heating and cooking. Thousands of products are also made from wood. The wood from trees that have been cut down and sawed into boards and planks is called **lumber.** Nearly 75 percent of the lumber cut in the United States is used for building construction, as Figure 26-9 illustrates. The rest is used to make products such as furniture, or it is ground and moistened to make **wood pulp.** Wood pulp is made into paper, rayon, and many other products. Although most of the cellulose fibers used to make paper are extracted from wood, paper-making fibers come from many different plants. For example, fibers from cotton, flax (from which linen is made), rice, papyrus *(puh PEYE ruhs),* and bamboo are found in many types of stationery.

Effective Reading

Listing Facts

Have students keep a list of the uses of plants as they read this section. When they have finished reading, have them make a table of the items in their list.

❓ Did You Know?

Paper fibers are held together by cornstarch, a natural plant product derived from corn.

Although trees are a renewable resource, they are currently being used faster than they can be replaced. As seen in Figure 26-10, recycling paper is one way that people are trying to limit their consumption of trees. Today, paper is often recycled by dissolving it in water, washing off the inks and dyes, and re-pressing the fibers into new paper. Unfortunately, the expense of ink and dye removal makes the cost of recycling paper exceed the cost of making new paper from wood pulp. As a result, commercial landfills contain more than 60 percent wastepaper. Efforts are now underway to recycle this wastepaper in a different way. By using genetically engineered bacteria that have the enzymes necessary to break down cellulose and ferment the resulting sugar, wastepaper is being used to produce fuel-grade ethanol that is added to gasoline. ☐

Figure 26-10 A worker examines mountains of corrugated paper, *above left*, that will be used to make recycled paper, *above*.

☐ **CAPSULE SUMMARY**

Wood is an important resource that is part of thousands of products. It is cut into lumber for use in building construction and ground into wood pulp for use in paper and rayon.

Many Medicines Are Obtained From Plants

People have always used substances obtained from plants to treat a variety of ailments. Rural and primitive cultures around the world still depend on native plants to ease pain and cure illnesses. By studying the plants traditionally used to treat human ailments, researchers have developed many "modern" medicines. For example, solutions made by soaking the bark of willow trees, *Salix*, were a traditional cure for aches and pains. The ingredient in willows that reduced pain was isolated in 1827 and named salicin (*SAL uh sihn*). Today, **aspirin** (acetylsalicylic acid), a derivative of salicin, is the most widely used drug in the world. Each of the

Teaching Tips
Observing Paper Types
Have students make wet-mount slides of various types of paper and then observe the slides under a microscope. Staining the paper with methylene blue will make the fibers more visible. Have students draw what they see, and then ask them to relate the fibers that they saw to the characteristics of the papers.

Products That Come From Trees
Have students make a concept map illustrating the importance of trees by using the applicable vocabulary words for this section plus any other words that they wish to include. An example of a possible concept map is contained in the *Graphic Organizer* on page 601.

Demonstration
Know Your Lumber
Tell students that different kinds of wood have different degrees of hardness and therefore are used for different purposes. Some of these include: pine (soft, polishes well), oak (hard, impermeable to liquids), maple (hard, resists abrasion), balsa (soft, high buoyancy), ebony (hard, difficult to nail), and cedar (soft, fine grained). Show samples of different types of woods, and let students suggest how they could be used commercially. Ask: What makes some woods hard while others are soft? (*Softwoods are fast-growing trees that form loosely packed tracheids and relatively thin-walled vessel elements. Hardwoods are slow-growing trees that form tightly packed tracheids and relatively thick-walled vessel elements.*)

Graphic Organizer
Use this graphic organizer with *Teaching Tip: Products That Come From Trees* on page 601.

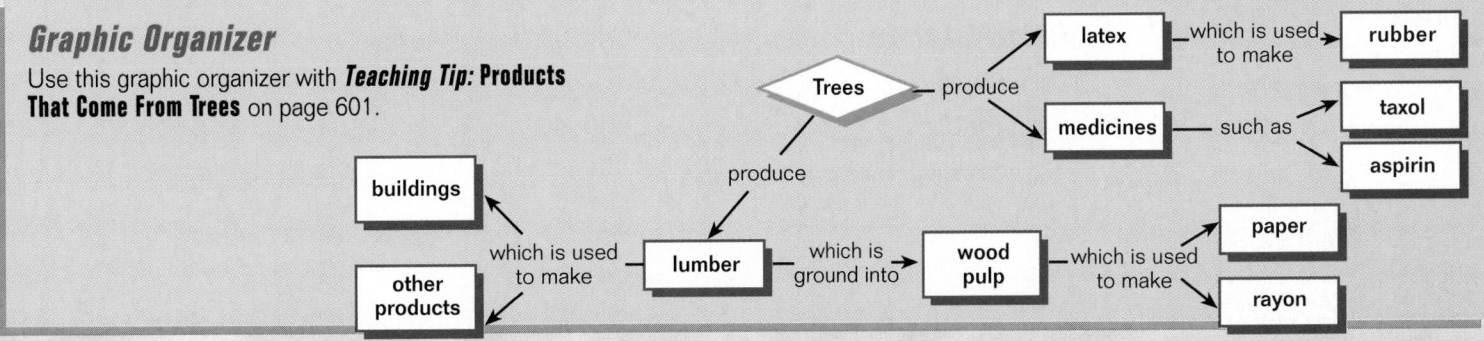

Figure 26-11 The rosy periwinkle, also known as vinca, is a popular bedding plant and the original source of two important drugs that are used to treat cancer.

☐ CAPSULE SUMMARY

Plants are the sources of many important medicines used by humans to treat diseases and other ailments.

Figure 26-12 These workers are stripping the bark from yew trees, a practice that may kill the trees. The bark is processed to obtain taxol, another drug that is used to treat cancer.

20 best-selling drugs in the world is either extracted directly from a plant or linked to a plant in another way.

In the 1960s, two cancer-treatment drugs were discovered in a familiar garden plant—the rosy periwinkle, *Catharanthus roseus*. The rosy periwinkle, seen in Figure 26-11, is the natural source of vinblastine (*vihn BLAS teen*) and vincristine (*vihn KRIHS teen*). Vinblastine is used to treat Hodgkin's disease, a type of cancer that affects the lymph nodes. Vincristine is used to treat childhood leukemia. The use of these drugs, which are now made synthetically, has greatly increased the survival rate of patients with these cancers. For example, the chance that a child who contracts childhood leukemia will live past the age of five has increased from 1 chance in 20 to about 19 chances in 20.

Scientists have recently discovered that yew trees, *Taxus* sp., also produce a chemical with cancer-fighting properties. The chemical, called taxol, is found in several parts of yew trees, including the bark, as Figure 26-12 illustrates. However, only very tiny amounts of taxol are made by a single tree—4 tons of bark are needed to produce 1 lb. of taxol. Leaves may be a more promising source of taxol, a very complicated molecule that was produced synthetically for the first time in 1994.

Today, plants still directly provide many important medicines. For instance, sweet potatoes are the source of **cortisone** (*KAWRT uh sohn*), a drug that is used to treat inflammation and allergies. The leaves of foxglove (an extremely poisonous European plant) yield **digitalis** (*dihj ih TAL ihs*), a drug that is used to stabilize irregular heartbeats and to treat cardiac disorders. Reserpine (*rih SUR pihn*), a drug obtained from the shrub *Rauwolfia serpentina*, is used to control high blood pressure. More than two-thirds of the people in the world obtain most of their medicinal drugs directly from plants. For example, more than 5,000 kinds of natural drugs obtained from plants are traded and sold in China. ☐

Figure 26-13 Cotton cloth is made from the fine fibers that are attached to the seeds in a cotton boll, *left*, which is the fruit of a cotton plant. After it is harvested, most cotton is sent to a gin, *right*, a machine that removes the cotton fibers from the seeds.

Fibers and Rubber Come From Plants

Plant fiber, in addition to its use in making paper, is also used to make cloth. Although synthetic fibers are now used in more than 30 percent of the world's clothing, natural plant fibers are still prized for their durability and comfort. Cloth made of cotton, the world's most important plant fiber, has been worn for centuries. Cotton thread is spun from the strong, fine fibers on cotton seeds, seen in Figure 26-13. The stems of flax, *Linum usitatissimum*, yield softer, more durable fibers that are used to make linen.

Rubber was first obtained from plants. Native Americans of Central and South America made rubber balls and waterproof shoes from **latex,** the milky white sap of tropical trees of the genus *Hevea*. Latex is extracted from "rubber" trees by the method seen in Figure 26-14. Today, natural rubber comes from rubber trees on plantations in Southeast Asia. Guayule *(gwah YOO lee)*, a member of the sunflower family that is a native of the deserts in the southwestern United States, is a promising new source of natural rubber. Most of today's rubber, however, is synthesized from petroleum, a nonrenewable resource.

Figure 26-14 Latex, used to make natural rubber, is harvested from a rubber tree by cutting and removing strips of bark and collecting the milky white sap as it drips out.

Section Review

1. *What is the main way that people use wood?*
2. *What is wood pulp, and how is it used?*
3. *Name five medicines that are derived from plants, and state how they are used.*
4. *What is the most important plant fiber used in clothing?*

Critical Thinking
5. *Why might guayule become a valuable cultivated crop?*

Answers to Section Review

1. The main use of wood is as fuel for cooking and heating.

2. Wood pulp is a product that is derived from moistening ground wood. It is used to make paper, fabric, and other goods.

3. Answers will vary but might include the following: aspirin to reduce pain, vinblastine to treat Hodgkin's disease, vincristine to treat childhood leukemia, taxol to treat cancer, cortisone to reduce inflammation, and digitalis to treat heart disease.

4. Cotton is the most important plant fiber used in clothing.

5. Guayule may become an important cultivated crop because it can be used to make rubber and could replace synthetic rubber produced from petroleum products.

CHAPTER REVIEW ANSWERS

Each item in the Chapter Review is correlated by text section in the assignment guide that follows.

ASSIGNMENT GUIDE

Section	Review	Themes Review	Critical Thinking
26-1	1–6, 11–14	18, 19	21–23
26-2	7–10, 15–17	20	

Review
Multiple Choice
Code in parentheses indicates section number and objective number.

1. d (26-1.1)
2. b (26-1.2)
3. c (26-1.2)
4. d (26-1.3)
5. a (26-1.3)
6. b (26-1.5)
7. b (26-2.1)
8. c (26-2.1)
9. b (26-2.2)
10. a (26-2.3)

Completion

11. fruit, ovary (26-1.1)
12. corn, teosinte (26-1.2)
13. grain, bran (26-1.2)
14. legumes, soybean (26-1.4)
15. lumber, wood pulp (26-2.1)
16. aspirin, digitalis (26-2.2)
17. cotton (26-2.3)

26 CHAPTER REVIEW

Vocabulary

aspirin (601)	essential amino acids (595)	tuber (598)
bran (592)	grain (592)	vegetative part (591)
cereal (592)	incomplete protein (595)	wheat germ (593)
complete protein (595)	latex (603)	whole wheat (593)
cortisone (602)	legume (598)	wood pulp (600)
digitalis (602)	lumber (600)	

Review

Multiple Choice

1. Which item correctly matches a food and the plant part from which it comes?
   a. yam : root
   b. sweet potato : leaves
   c. cassava : stem
   d. banana : fruit

2. Which one of the following cereal grains is eaten by more of the world's people?
   a. wheat   c. corn
   b. rice   d. oats

3. What trait did early farmers select when trying to increase the yield from wheat plants?
   a. fast growth in areas with much rainfall
   b. survival in hot climates
   c. large grains
   d. parallel rows of grain

4. Which of the following statements best describes the origin of bread wheat?
   a. It is a diploid plant essentially identical to the first wheat species cultivated about 11,000 years ago.
   b. It is a haploid plant.
   c. It is a tetraploid plant derived from the hybridization of two closely related species of wheat.
   d. It is a hexaploid plant derived from the hybridization of three closely related species of wheat.

5. A complete protein is provided by which of the following combinations of foods?
   a. beans and rice   c. potatoes and rice
   b. wheat and oats   d. rice and corn

6. An important nongrain food for many African people is
   a. sugarcane.   c. alfalfa.
   b. cassava.   d. corn.

7. What makes paper recycling so costly?
   a. collecting the paper to be recycled
   b. removing the inks and dyes from the paper
   c. growing the bacteria that digest cellulose
   d. paying taxes on recycled paper

8. Besides food, what is the most valuable plant product?
   a. medicine   c. wood
   b. oil   d. latex

9. Drugs derived from the rosy periwinkle are used in the treatment of
   a. heart disease.   c. arthritis.
   b. leukemia.   d. headaches.

10. Trees of the genus *Hevea* are a source of latex, which is used to make
    a. rubber.   c. petroleum.
    b. reserpine.   d. cloth.

Completion

11. As an agricultural commodity, tomatoes are classified as vegetables. But a tomato is actually a _____ because, like an apple, it is the ripened _____ of a flower.

12. The origin of _____ has been traced to the wild annual grass of Mexico, called _____ .

13. The dry fruit of a cereal plant is called a _____ ; it is covered by a dry husk called the _____ .

Themes Review

18. Agricultural crops were derived by selecting certain plants with desirable characteristics to become the parents of the next generation. Crossbreeding individuals with desirable characteristics and selecting the desirable hybrids has further defined the characteristics of modern agricultural crops. In both artificial selection and natural selection, offspring of certain individuals survive to become the parents of the next generation, and thus over time, a population's characteristics become more similar to the characteristics of the individuals selected. (26-1.2)

14. Plants that produce protein-rich seeds in long pods are called _____ . The most important of these plants is the _____ .

15. Wood is sawed into _____ and then used in construction. It is also processed into _____ and used to make paper.

16. The willow tree, *Salix,* was the orginal source of _____. _____ is extracted from the leaves of the foxglove plant.

17. The world's most important plant fiber is _____.

Themes Review

18. **Evolution** How did artificial selection by humans play a role in the origin of agricultural crops? How is artificial selection similar to natural selection?

19. **Levels of Organization** Enzymes are proteins (chains of amino acids) that catalyze chemical reactions in living cells. How might eating a diet that lacks some of the essential amino acids affect the functioning of living cells?

20. **Homeostasis** How do the drugs digitalis and reserpine help the human body maintain homeostasis?

Critical Thinking

21. **Communicating Effectively** Suppose that a friend asks you why corn, which he or she considers to be a vegetable, is listed as a cereal crop in the encyclopedia. To answer this question, write a paragraph that explains why corn is a cereal crop, agriculturally, and why it is also a fruit, botanically. Include other examples of foods that are classified as vegetables or grains but are also fruits.

22. **Making Inferences** Suppose that you discover a new strain of wheat that has much larger spikes and grains than bread wheat. How might this wheat differ from bread wheat? Explain.

23. **Interpreting Data** Recent medical reports suggest that a high-fiber diet may help to prevent colon cancer. The table below shows the approximate percentage of fiber in different parts of a wheat grain. Would you expect to find more cases of colon cancer among people who eat only whole wheat bread (made with endosperm, germ, and bran) or among people who eat only white bread (made with endosperm only)? Why?

Part of wheat grain	Percentage of fiber
Endosperm	4%
Germ	12%
Bran	40%

Activities and Projects

24. **Cooperative Group Project** Interview several people who are vegetarians. Find out what they eat and why they became vegetarians. Then, write an article about being a vegetarian. Include information from the interviews and from other sources such as books and magazines. Publish the article in your school newspaper.

25. **Multicultural Perspective** Native Americans of the Southwest and Great Plains refer to corn, beans, and squash as the "three sisters." Find out why Native Americans have traditionally grown these plants together and why they are important to human nutrition. Relate your findings in a written report.

26. **Research and Writing** Plywood is usually sold in sheets 4 ft. wide by 8 ft. long. Find out how plywood is made and why it is stronger than a regular wooden board of the same thickness. Summarize your findings in a written report.

Critical Thinking

21. Agriculturally, corn is a cereal crop because it is a grass that produces seeds in grains, and botanically, it is a fruit because each grain on an ear of corn is an individual fruit (ripened ovary) with one seed. All other cereals, such as rice, wheat, oats, and barley, as well as vegetables such as tomatoes, bell peppers, hot peppers, eggplants, cucumbers, squash, and green beans, are botanically fruits. (26-1.1)

22. Students might expect to find that the new strain of wheat is a polyploid with eight or more sets of chromosomes, which would be more than the six sets of chromosomes found in bread wheat. Polyploids tend to be larger with more sets of chromosomes. (26-1.3)

23. Students should expect to find more cases of colon cancer among people who eat white bread than among people who eat whole-wheat bread because there is more fiber in whole-wheat bread. (26-1.2)

Activities and Projects

24. Caution students that some of their schoolmates may be sensitive about being labeled vegetarians and therefore may be reluctant to be interviewed. Reasons given by teenagers for becoming vegetarians may include: love of animals, revulsion to the sight of bloody meat, desire to assert nutritional, religious, and economic independence.

25. Native Americans grew corn, squash, and beans together in their gardens because the beans, which are nitrogen-fixing legumes, enriched the soil for the other plants, which provided a place for vines of the beans to grow. Nutritionally, these three foods provide a complete protein when eaten together.

26. Plywood is made of several sheets of veneer glued together. Veneer is a paper-thin sheet of wood shaved off a log by turning the log against a sharp blade. Plywood is strong because each sheet of veneer is placed so that the direction of its wood fibers is at right angles to the direction of those of the sheets below and above it.

19. Without all the essential amino acids required by your body, you would be unable to make all the enzymes necessary for the functioning of cells. If these enzymes were missing, your cells would not be able to function (e.g., convert sugars to ATP, reproduce, make structural materials) and would probably die. (26-1.4)

20. Digitalis is used to stabilize irregular heartbeats, and reserpine is used to control high blood pressure. (26-2.2)

LABORATORY Investigation Chapter 26

Vegetative Propagation

Preparation

Each lab group will need 12 duckweed plants, 5 petri dishes, a stereomicroscope, a glass-marking pen, pond water, Knop's solution, a 0.1% fertilizer solution, and distilled water. Obtain duckweed from a pond or quiet stream. You can also order live duckweed from a biological supply company. Duckweed can be obtained from Ward's by ordering *Lemna*. Obtain 23-19-17 fertilizer from a gardening store or from Ward's (Rapid Gro plant food). To make a 0.1% solution of fertilizer, measure 1 g of fertilizer and add enough distilled water to make 1 L. Do not use less potent fertilizers, such as 5-10-5, which will show no increase in growth compared to distilled water.

Procedural Notes

- Allow students 20 to 30 minutes to begin this investigation, and then allow them 10 minutes each day for the remainder of the two-week period to make their observations.
- Students should see abundant vegetative reproduction in the two-week period.
- Tell students that any population growth observed in the petri dishes is the result of vegetative reproduction.
- **Safety:** Students should wear safety goggles when working with Knop's solution and the 0.1% fertilizer solution.
- **Safety:** Students should wash their hands before leaving the lab.

OBJECTIVES

- Recognize the ability of some plants to reproduce vegetatively.
- Compare vegetative reproduction of duckweed in different nutrient solutions.

PROCESS SKILLS

- identifying the structures of duckweed
- observing vegetative propagation in duckweed
- hypothesizing on conditions that would increase growth of duckweed
- testing a hypothesis by performing experiments to observe optimum growth of duckweed

MATERIALS

- duckweed culture
- 5 petri dishes
- stereomicroscope
- glass-marking pen
- pond water
- Knop's solution
- 0.1% fertilizer solution
- distilled water
- safety goggles

BACKGROUND

1. What are vegetative plant parts?
2. How can a plant reproduce asexually by means of vegetative plant parts?
3. Write your own question to explore in your lab report or notebook.

TECHNIQUE

1. Place a duckweed plant in a petri dish. Add a few drops of water. Observe the duckweed under a stereomicroscope. Record your observations in the **Records** section of your report by sketching what you see.
2. Label four petri dishes as follows:
   A—pond water
   B—Knop's solution
   C—0.1% fertilizer solution
   D—distilled water
3. **CAUTION: Wear safety goggles for this step. Growth solutions are mild eye irritants. Avoid contact with your skin and eyes.** In case of contact, notify your teacher and flush the area with running water. Fill each petri dish three-quarters full with the solution identified on the label of the dish.
4. Place three duckweed plants in each Petri dish. Put the covers on the dishes and place the dishes in a well-lighted area.

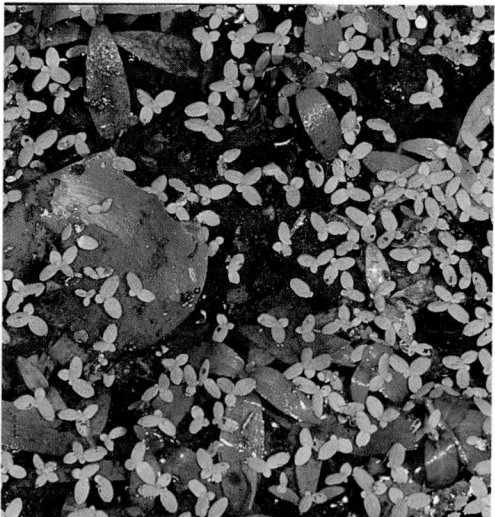

Background Answers

1. Vegetative plant parts are nonreproductive parts—such as leaves, stems (rhizomes, tubers, bulbs, corms), and roots—that can produce new plants through asexual means.

2. Plants reproduce asexually by means of vegetative parts as masses of undifferentiated cells begin to differentiate and develop into new leaves, stems, and roots. A plant may be stimulated to reproduce vegetatively when conditions are ideal and nutrients are abundant.

3. How does the rate of vegetative propagation in duckweed vary with different conditions?

Records

Students should compile their data into a bar graph.

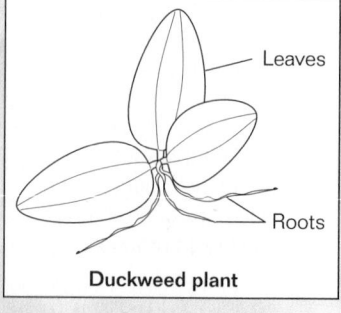

Duckweed plant

5. Hypothesize which solution should produce the largest number of duckweed plants. Record your hypothesis in the **Records** section of your report.

6. Observe the four petri dishes each day for two weeks. In the **Records** section of your report, make a chart to record the number of individual plants in each solution on each day, at the end of one week, and at the end of the two-week period. Record any other observations about the condition of the plants at those times in the **Records** section of your report.

7. In the **Records** section of your report, make a bar graph with the *y*-axis labeled "Number of plants" and the *x*-axis "Days." Make a key for the graph to indicate which lines represent the three solutions and the distilled water. Title your graph. In your report, briefly summarize the procedure you followed.

8. Clean up your materials and wash your hands before leaving the lab.

INQUIRY

1. In which dish did the greatest amount of growth take place?

2. In which dish did the least amount of growth take place?

3. Did the results you observed agree with your hypothesis? If not, how are they different?

4. As the number of new plants increases, what happens to the group of plants?

5. Describe the attachment of the new plants.

ANALYSIS

1. Describe the appearance of plants growing in the Knop's solution. Offer an explanation for your observations.

2. Describe the appearance of plants growing in distilled water. Explain your observations.

3. Based on this investigation, what is the connection between the general health of a plant and vegetative reproduction?

4. What factors regulate the rate of vegetative reproduction in duckweed?

5. Explain what factors accounted for the amount of growth in the dishes containing the fertilizer solution and the pond water.

6. Is there a control in this investigation. If so, what is the control? Give reasons for your answer.

7. Why is a new duckweed plant produced by vegetative reproduction genetically the same as the parent plant?

8. How does an organism benefit from receiving all its genetic information from and being identical to, a single parent? What disadvantages might there be?

FURTHER INQUIRY

Write a new question that could be explored as a new investigation. The following are examples:

How could you check for the effects of other environmental factors, such as light intensity or temperature, on the growth of duckweed?

How could your findings in this lab be applied to raising crop plants by means of vegetative propagation?

Number of Duckweed Plants in Various Nutrient Solutions

Solution	Day											
	1	2	3	4	5	1 Week	8	9	10	11	12	2 Weeks
A												
B												
C												
D												

Inquiry Answers

1. The greatest amount of population growth, or reproduction, occurred in the dish containing Knop's solution.

2. The least amount of population growth, or reproduction, occurred in the dish containing distilled water.

3. Answers will vary.

4. It divides.

5. Answers will vary.

Analysis Answers

1. Answers will vary but the plants were probably darker green, larger, and more numerous. Knop's solution had the most of the necessary plant nutrients and in the proper amounts.

2. Answers will vary but may mention that the plants were fewer in number and showed loss of color, deterioration of parts, or death of plants. These changes occurred because no nutrients were present in the distilled water to stimulate plant growth and reproduction.

3. Plants reproduce vegetatively when they are healthy.

4. Answers will vary but should mention that, according to this experiment, the amount of essential plant nutrients present regulates the rate of vegetative reproduction in duckweed.

5. Answers will vary but may mention that there was some population growth in the fertilizer solution because some nutrients were available and that the plants in the pond water showed more reproduction because it had more necessary nutrients.

6. Yes. Answers will vary. Some students will respond that distilled water is the control because it shows that no reproduction occurs when no nutrients are present. Others will respond that pond water is the control because it shows what happens in a normal duckweed environment.

7. The offspring come from the division of the cells of the parent, not from the union of gametes from two parents.

8. Advantages of vegetative reproduction include the following: if the parent is in an environment where it is flourishing, the offspring should also do well, and the plants can rapidly increase in number when conditions are favorable. A disadvantage of vegetative reproduction is that when the environment changes, there will be less variation in the population, and, as a result, the number of individuals that can survive in the new environment may be few or none.

UNIT 6

SCIENCE, TECHNOLOGY, AND SOCIETY

Point of View: Biotechnology and Agriculture

Objectives

- Develop scientific literacy
- Identify an author's point of view
- Detect bias in a persuasive argument

Background

Each year scientific issues become more important in determining the nature and direction of our society, and yet most citizens have little or no training in evaluating the arguments that surround controversial issues. The purpose of this feature, Point of View, is to provide practice in identifying and evaluating a writer's point of view.

A critical step in this process is detecting bias. Much of the writing on critical issues is persuasive and not objective. A writer will include information that bolsters his or her point of view and suppress evidence that contradicts it.

Instructional Strategies

- As students read this article, remind them that it represents one person's point of view. They must read critically, looking for signs that would indicate biased or slanted writing. In particular, have students look closely for the following:

1. Does the article present both sides of the issue fairly and objectively?

2. Is the writer seeking to shape your opinion on the issues?

3. Does the writer use "loaded" language? Are some words chosen because they have an emotional rather than a logical appeal to the reader?

- On the chalkboard, write the following list of biotechnology research projects currently under-

BIOTECHNOLOGY AND AGRICULTURE

What Will the Harvest Be?

POINT OF VIEW

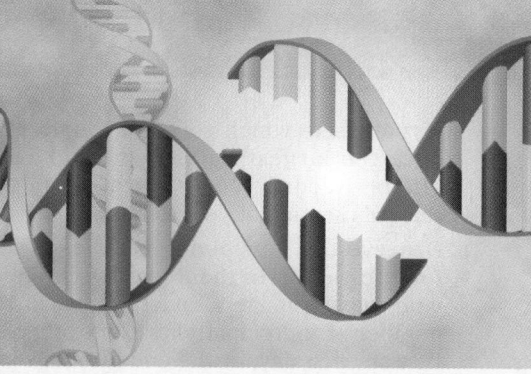

The genetic engineering process for making BST starts with the use of a restriction enzyme to cleave DNA into fragments with sticky ends.

BY TRACEY COHEN

In February 1994 a genetically engineered drug called bovine somatotropin (BST) went on sale to dairy farmers in the United States. The drug, commonly known as recombinant bovine growth hormone (rBGH), increases milk production in cows by as much as 40 percent while increasing their food consumption by only 5 to 15 percent. Monsanto and over 400 other companies are collectively spending close to $2 billion a year on biotechnology research and development. Even federal and state governments are involved. In 1990, for example, federal funding for biotechnology was almost $3.5 billion.

Proponents of BST hailed its introduction as another means of reducing food costs and increasing food production. Not everyone greeted the news enthusiastically, however. Concerned about the possibility of hormone-tainted milk, several consumer groups called for a boycott of BST products and demanded that dairy products from BST cows be labeled as such.

Biotechnological innovations like BST are transforming agriculture. The list of traits that could potentially be genetically engineered into various agricultural organisms is endless. Researchers are working on tomatoes that ripen more slowly and keep their flavor longer, potatoes with less water and more flesh, and low-fat pigs that yield leaner cuts of pork. While these traits may be appealing to consumers, there are broad environmental, social, economic, and ethical issues of agricultural biotechnology that have yet to be adequately addressed.

What Effects Might Agricultural Biotechnology Have on the Environment?

Agricultural biotechnology poses environmental risks because many of the organisms being engineered will be released outdoors. A good example is the effort to develop herbicide-resistant crops (HRC's), which involves more than 30 companies, along with researchers at several universities and the U.S. Department of Agriculture. Scientists are trying to engineer corn, new varieties of wheat, and other major crops that will resist damage from high doses of specific herbicides. Farmers will

then be able to use heavier doses of herbicides to kill all the weeds in a field without harming their crops.

The use of HRCs, however, ignores complex ecological factors. For example, in any population of weeds, some plants will have a natural resistance to whatever herbicide is being applied. (It is the genes for such resistance that are being transferred into crop plants.) Those resistant weeds that are not affected by the herbicide will survive and reproduce, becoming the dominant population.

Eventually, farmers would have to try something new, creating a vicious cycle of applying more and more poison. This would result in the evolution of more resistant weeds. About 100 species of weeds are currently known to be resistant to one or more herbicides.

Will Farm Animals Really Be "Improved" by Genetic Engineering?

Many farm animals today are raised under conditions designed to produce the largest number of animals and animal products rapidly and at the lowest cost. This approach to animal husbandry, often justified as "progress," treats farm animals

way: (a) creating insect predators that are resistant to pesticides; (b) creating nonnative crops that can grow in marshes or deserts; (c) engineering pigs to provide organs for transplant in humans; (d) creating crops that make their own nitrogen compounds and do not need chemical fertilizers. Have students evaluate each project in

terms of its environmental, social, economic, and ethical impacts.

Answers to Analyzing the Issue

1. (a) The writer opposes biotechnology that has not been evaluated in terms of its broad ecological, social, cultural, and

economic consequences. **(b)** The writer advocates becoming involved in the development of standards for biotechnology products. **(c)** Facts include weed resistance to herbicides, risk of udder infection and deformed calves in cows that are treated with BST, and loss of family farms. **(d)** Opinions include the beliefs that biotechnology will as if they were little more than machines for producing milk, eggs, or other foods. Biotechnical research directed at farm animals is likely to make this situation even worse.

For example, the product label for BST shows that it increases the risk of a painful udder infection in cows known as mastitis. Research also indicates that BST may cause cows to have more stillborn and deformed calves than do untreated cows. And the increased stress caused by excessive milk production may bring on early death in BST-treated animals.

Other work on animals involves inserting genes for desired traits directly into animal embryos. For example, the USDA developed transgenic pigs carrying the gene for human growth hormone. The idea was to create a variety of "superpig" that would grow rapidly to huge size. Only one of every 200 embryos survived the procedure. Those that survived had arthritis, vision problems, and an impaired immune system that left them with a tendency to contract pneumonia. There is no way to know for certain what effects genetic engineering will have on an animal and whether the animal will suffer as a result.

Do We Want More Milk?

The use of BST provides a good opportunity to examine the social costs of agricultural biotechnology. Though BST increases total milk production, its use comes at a time when there is already a milk surplus in the United States. A surplus of milk will drive milk prices down. A number of studies, including one by the U.S. Office of Technology Assessment, predict that use of BST will ultimately drive many small dairy farmers out of business. It is expected that the remaining dairy farms will become more efficient, but are society's interests really served by driving small farmers into bankruptcy?

The loss of a small farm has effects that ripple throughout the economy. Family farmers forced out of business often join the ranks of the poor and the unemployed. Rural businesses suffer as the demand for goods and services drops. It is estimated that for every farmer driven out of business, as many as 25 other dairy-related jobs are lost. Data from the USDA show that the unemployment rate is already much higher among agricultural workers than for all other areas of the civilian economy combined: 11.1 percent in agriculture versus 7.0 percent for all other areas in 1993. Poverty in rural America is also higher than in urban areas.

What Should Be Done?

When industry designs a new product, there are important questions to ask. What does society gain? Are real social needs being met? What are the risks? Are potential consequences shared equally or will some areas be threatened more than others? In the case of biotechnology, these questions are mostly unanswered.

It is also important to have a set of social standards for biotechnology products. For a product to be developed, it must be so needed by society that the benefits outweigh any risks. A product with minimal social value is not worth any risk at all.

Each biotechnological innovation must be thoroughly evaluated. It is not enough to look at only the usual business criteria of profit and loss. The broader ecological, social, cultural, and long-term economic implications must also be considered. Only then will this powerful new technology live up to its promises.

Tracey Cohen is a freelance writer specializing in science and environmental issues.

SCIENCE, TECHNOLOGY, AND SOCIETY

mental Protection Agency, and Food Safety and Inspection Service. The U.S. Office of Technology Assessment has found a lack of coordination among the regulatory agencies.

References

1. Beck, C. I. and T. H. Ulrich. "Environmental Release Permits: Valuable Tools for Predicting Food Crop Developments." *Bio/Technology,* Vol. 11, Dec. 1993, pp. 1524–1528.

2. Dekker, J. and G. Comstock. "Ethical Environmental Considerations in the Release of Herbicide Resistant Crops." *Agriculture and Human Values,* Summer 1992, pp. 31–43.

3. Doyle, J. "Farming Genes for Profit." *Coop America Quarterly,* Spring 1992, pp. 15–19.

4. Fox, M. W. *Superpigs and Wondercorn: The Brave New World of Biotechnology and Where It All May Lead.* Lyons & Burford, New York, 1992.

5. Goodno, J. B. "Fields of Misfortune: Colonialism in the Heartland." *Dollars & Sense,* March 1992, pp. 6–9.

6. Kimbrell, A. *The Human Body Shop: The Engineering and Marketing of Life.* Harper, San Francisco, 1993.

7. Kolata, G. "Super Cows and Tasty Tomatoes: When the Geneticists' Fingers Get in the Food." *The New York Times,* February 20, 1994.

8. Krimsky, S., et al. "Controlling Risk in Biotech." *Technology Review,* July 1989, pp. 62–70.

9. LeBaron, H. M. and J. McFarland. *Herbicide Resistance in Weeds and Crops: An Overview and Prognosis.* ACS Symposium Series, American Chemical Society, Washington, D.C., 1990.

10. MacKenzie, D. "Doubts Over Animal Health Delay Milk Hormone." *New Scientist,* Jan. 18, 1992, p. 13.

11. Schneider, K. "Lines Drawn in a War Over a Milk Hormone." *The New York Times,* March 9, 1994.

12. Sharples, F. E. "Regulation of Products From Biotechnology." *Science,* Vol. 235, March 13, 1987, pp. 1329–1332.

Analyzing the Issue

1. **Detecting Bias** This article is an argumentative essay. It attempts to convince a reader to agree with a certain viewpoint, take a specific action, or make a decision about a subject on which many possible opinions and actions are possible.
   a. What is the writer's point of view about biotechnology in agriculture?
   b. What action does the writer advocate?
   c. List the key factors the writer uses to support the argument.
   d. List the statements from the article that are opinions.

2. **Formulating an Opinion** Write an essay expressing your views about biotechnology in agriculture. You may need to do more library research. Before you begin writing, create an outline that includes the following:

   a. Main point or thesis to be proved
   b. Supporting facts, examples, and arguments
   c. Opposing facts and arguments

3. **Get the Facts** Investigate the movement for sustainable agriculture. How does this approach to farming compare with the most widely used current methods? What impact might biotechnology have on farmers' ability to practice alternative agriculture? Should research in biotechnology support the current system or help to change it? Explain your answer.

4. **Take Action** Find out what federal, state, and local laws regulate biotechnology. Contact the U.S. General Accounting Office and the Office of Technology Assessment for their evaluation of the regulatory framework. What changes, if any, need to be made?

worsen the situation for farm animals and that questions must be answered before a product is developed.

2. Students' opinions will vary. Their positions should be clearly presented and supported by statistics and facts. You may wish to give students the list of references used in developing this essay.

3. The sustainable-agriculture movement is concerned with ecologically sound farming. An important component is the use of organic (non-petrochemical) methods to grow crops, maintain soil fertility, and control pests. By contrast, commercial agriculture relies heavily on chemicals and intensive cultivation that eventually ruin the soil

and pollute air and water. Effects of biotechnology may be negative if farmers cannot get seeds for plants that do not need herbicides and other chemicals.

4. At the federal level, four agencies oversee a regulatory patchwork for biotechnology: Animal and Plant Health Inspection Service, Food and Drug Administration, Environ-

> Each block represents about 45–50 minutes of instructional time. The resources cited in a block are coded to a specific program component using the Key on the next page.

Lecture Concepts   Lesson Resources

27-1 Advent of Tissues and Symmetry pp. 611–616

BLOCKS 1 and 2

a. A series of key evolutionary adaptations led to the major animal phyla.
b. Sponges are the simplest kind of animals. They lack tissues and body symmetry and live attached to the sea floor, where they filter food particles from the water.
c. Cnidarians have tissues, tentacles, and radial symmetry. The digest their food extracellularly, in a digestive cavity.

Lecture Resources
- Opening Question, page 611
- Demonstration: Using the Tree to Describe You and Me
- What Good Is a Sponge? page 612
- The Real Thing, page 612
- Visual Strategies: Figures 27-4, 27-7
- Symmetry, page 616
- Teaching Transparencies: 134, 135, 144, 145, 146, 153, 154
- Holt Biology Videodiscs: Lesson 27-1

Classwork Options
- Effective Reading, page 611

- Connecting the Abstract to the Concrete, page 612
- Highlights: Invertebrate Evolution, pages 613, 615
- Unscrambling the Groups, page 613
- Quick Labs A21: Recognizing Patterns of Symmetry

Assignment Options
- Heterotrophic Lifestyles, page 614
- Section Review, page 616
- Chapter Review, questions 1–3, 11, 12, 18, 33
- Highlights Review, questions 22, 23

27-2 Evolution of the Body Cavity pp. 617–623

BLOCK 3

a. Flatworms lack a body cavity and have bilateral symmetry, simple organs, and a digestive tract with a single opening.
b. Roundworms have a body cavity (pseudocoelom), muscle tissue, and a gut that is open at two ends.
c. Mollusks, and all animal phyla that evolved after this group, have a coelom, which enabled the formation of advanced body systems.

Lecture Resources
- Opening Question, page 617
- Demonstrations: Following the Directions, Longitudinal Muscles
- Visual Strategies: Figures 27-10, 27-11, 27-12, 27-13
- Size Limitations, page 619
- Research Update, page 623
- Teaching Transparencies: 157, 158, 159, 160
- Holt Biology Videodiscs: Lesson 27-2

Classwork Options
- Highlights: Invertebrate Evolution, pages 618, 620, 622
- Cell Talk, page 621
- Fluid Movement, page 621

Assignment Options
- Section Review, page 623
- Chapter Review, questions 4, 11, 13–15, 19, 20, 30–32, 34, 35
- Highlights Review, questions 24, 25

27-3 Four Innovations in Body Plans pp. 624–632

BLOCK 4

a. Segmented worms have evolved complex body systems, including a nervous system and a circulatory system.
b. Arthropods, such as spiders, scorpions, crabs, and lobsters, are characterized by jointed appendages and an exoskeleton.

Lecture Resources
- Opening Question, page 624
- Demonstration: Muscle Resistance
- Real Worms, page 625
- Visual Strategies: Figures 27-15, 27-16, and 36-4
- Arthropod Ancestors, page 626

- Teaching Transparencies: 161
- Holt Biology Videodiscs: Lesson 27-3

Classwork Options
- Highlights: Invertebrate Evolution, pages 625, 627
- Worms and Bugs, page 626
- Joint Appreciation, page 627

BLOCKS 5 and 6

a. Echinoderms and chordates have similar patterns of development and very likely share a common ancestor.
b. Echinoderms have an endoskeleton, radial symmetry, a simple nervous system, and a water vascular system.
c. All chordates develop a hollow dorsal nerve cord, notochord, and pharyngeal slits sometime during their lifetime. Vertebrates are chordates that have a cartilaginous or bony endoskeleton that includes a vertebral column.

Lecture Resources
- The Blastopore, page 628
- Predetermination, page 629
- Good Support, page 631
- Teaching Transparencies: 162, 163
- Holt Biology Videodiscs: Lesson 27-3

Classwork Options
- Highlights: Invertebrate Evolution, pages 630, 631

- Laboratory Investigation: Survey of Invertebrates, pages 636–637

Assignment Options
- Section Review, page 632
- Chapter Review, questions 6–10, 16, 17, 21, 36
- Highlights Review, questions 25–29

Review/Enrichment

BLOCK 7

▦ Study Guide: Concept Review and Pretest

Assignment Options
▦ Activities and Projects, questions 37–40
▦ Readings, questions 41, 42

Assessment Options

BLOCK 8

Traditional Assessment
▦ Chapter 27 Test
◉ Test Generator/Assessment Item Listing: Software item bank for preparing customized chapter tests.

Portfolio Assessment
Select student reports for one or more laboratory experiments from this chapter. *The Direct Observation Checklist* on the *BioSources Teaching Resources CD-ROM* should be completed during a laboratory or other cooperative learning experience.

Answer to Concept Map

The following is one possible answer to the Concept Mapping exercise on page 633.

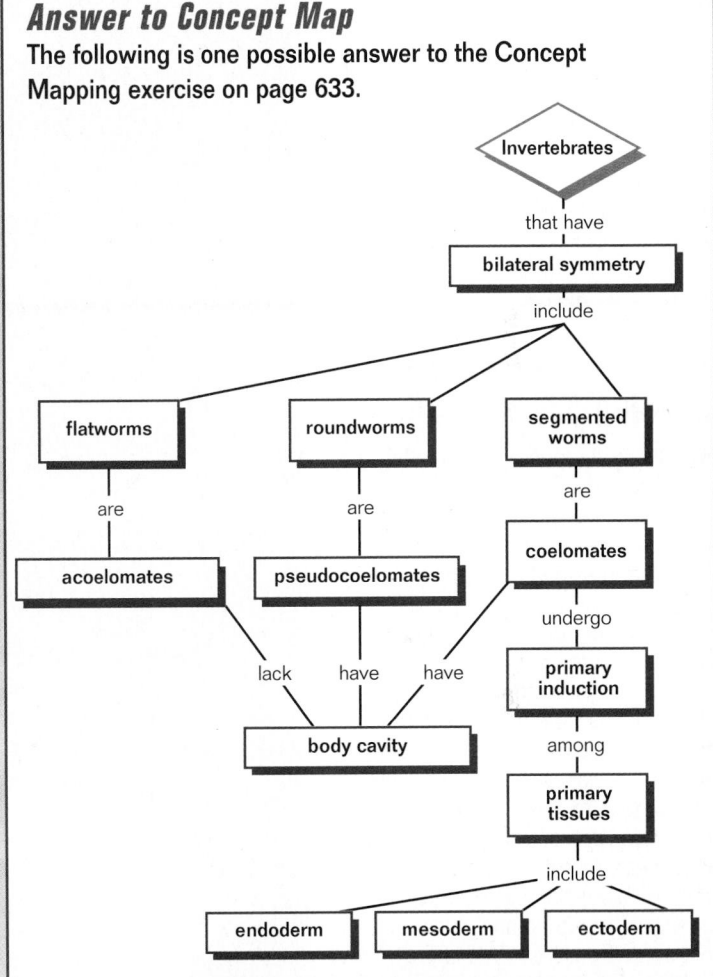

Holt Biology Videodiscs

Holt Biology Videodiscs gives you a powerful tool for teaching, review, and assessment. *Concepts of Biology* adds interactive lesson plans and extensive tools for customization using CD-ROM technology.

CONCEPTS OF BIOLOGY

Use the following topic from *Concepts of Biology* to help you teach this chapter:

■ Topic 18: Invertebrate Diversity

For further information regarding the media on the videodiscs, see *Holt Biology Videodiscs Teacher's Correlation Guide to Biology: Principles and Explorations.*

Chapter Theme

Evolution *The anatomic diversity seen in animals today is the result of major innovations in the body design of animals that have occurred over millions of years of evolution.*

Tapping Prior Knowledge

- What are some general characteristics of animals?
- Compare a cell with a tissue.
- What is an arthropod?
- What significant roles have arthropods played during the history of life on Earth?

Opening Demonstration

Explain to students that one kind of animal they will study in this chapter is the sponge. Then place a potted plant on your desk and ask students what the plant and a sponge have in common. *(Eventually, someone should say both organisms are stationary.)* Next, draw a simple diagram of a sponge that reaches from the bottom to the top of the chalkboard. Explain that some giant sponges can grow this large, and even larger. Then ask them what a sponge this size would eat. *(nothing larger than tiny food particles)* Explain how all sponges (tiny ones and gigantic ones) feed by filtering tiny food particles from sea water.

OVERVIEW

CHAPTER 27

OVERVIEW OF INVERTEBRATES

Moon jellyfish

REVIEW

- diffusion and endocytosis (Section 3-2)
- haploid and diploid cells (Section 6-1)
- natural selection (Section 12-1)
- evolution of arthropods (Section 13-2)
- evolution of vertebrates (Section 13-3)

Authors' Rationale

The detailed exploration of the animal kingdom begins here, with an overview of the extremely diverse group of invertebrates. The animal kingdom is typically divided into vertebrates and invertebrates, so the distinction between both groups is important to make. The invertebrates, obviously, are those animals that do not have a backbone. This group is usually further subdivided, based on body cavity (no body cavity, pseudocoelom, or coelom). From the simplest animals, such as sponges, your students can trace the evolution of arthropods, echinoderms, and, ultimately, chordates.

27-1 Advent of Tissues and Symmetry

*I*n this chapter you will discover how a series of key evolutionary innovations has led to today's animal phyla. These phyla are shown in Figure 27-1 and are described in Highlights: Invertebrate Evolution *features that occur throughout the chapter. You will be provided with a broad overview of the body plan—the shape, symmetry, and internal organization—of different animals and learn how those body plans arose. An animal's body plan results from a pattern of development programmed into the animal's genes by natural selection.*

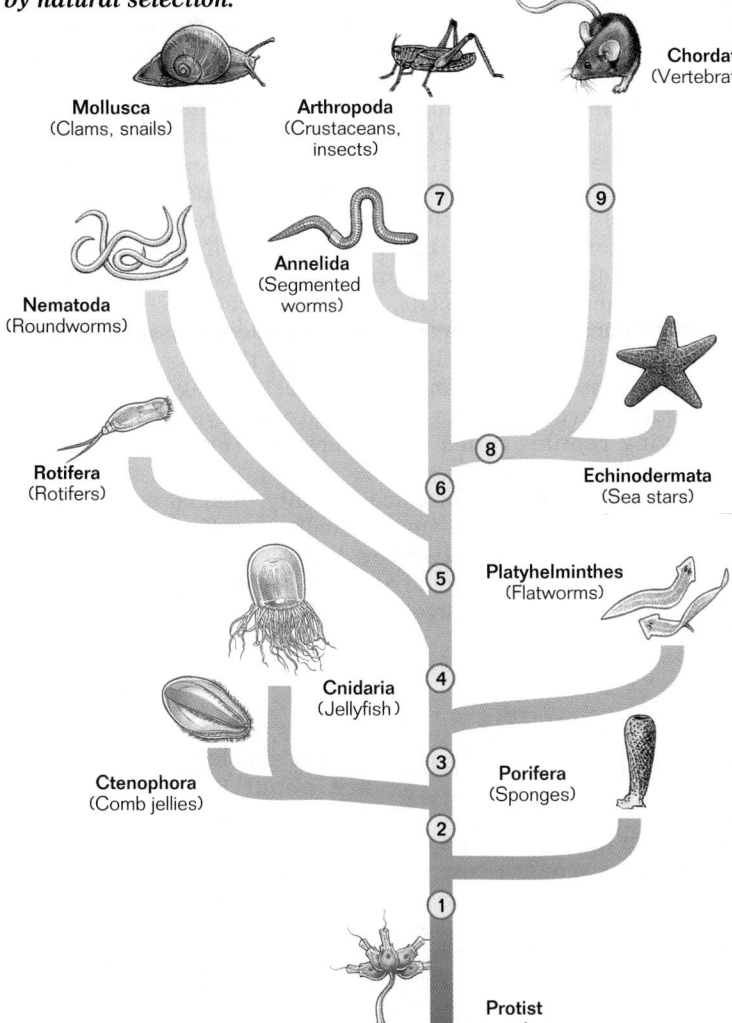

Mollusca (Clams, snails)

Arthropoda (Crustaceans, insects)

Chordata (Vertebrates)

Annelida (Segmented worms)

Nematoda (Roundworms)

Rotifera (Rotifers)

Echinodermata (Sea stars)

Platyhelminthes (Flatworms)

Cnidaria (Jellyfish)

Ctenophora (Comb jellies)

Porifera (Sponges)

Protist ancestors

Figure 27-1 This phylogenetic tree identifies the major animal phyla. Although many animals live in water, some inhabit land. Most terrestrial animals are mollusks, arthropods, or chordates. The circled numbers indicate important evolutionary stages, which are listed in the table below.

Evolutionary stages in the animal body	
Stage	**Milestone**
1	Multicellularity
2	Tissues
3	Bilateral symmetry
4	Body cavity
5	Coelom
6	Segmentation
7	Jointed appendages
8	Deuterostomes
9	Notochord

Teaching Tips

Connecting the Abstract to the Concrete

Terms and phrases such as "heterotroph," "multicellular," "lacking rigid cell walls," and "reproduce sexually," are used to describe features of animals discussed on this page. Help students connect these terms and phrases to their own concrete experiences by asking them to choose any animal discussed in the next eight chapters. Have them write the name of the animal and the page number on which it appears, and then describe evidence that their animal is heterotrophic, multicellular, lacks rigid cell walls, and reproduces sexually. Have a few students explain to the class why they know that their animals have the features common to all animals.

What Good Is a Sponge?

Students need to feel that what they are learning is relevant and useful, so take time to discuss or research the role that sponges play in aquatic ecosystems, as well as their commercial importance to humans.

The Real Thing

Dried real sponges are readily available in stores that carry bathroom accessories. Students gain a greater appreciation for the porous and asymmetrical structure of a sponge when they have the opportunity to touch and examine a real sponge. You might want to provide an artificial sponge for comparison.

CONNECTIONS

Chapter 19

Heterotrophic Organisms

Animals are not the only organisms that are heterotrophs. Many bacteria and protists are heterotrophs, as are all the fungi.

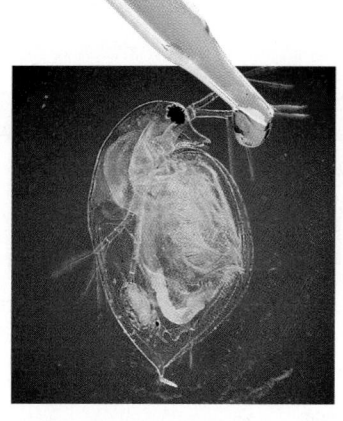

Figure 27-2 Some people may not realize that these tiny organisms, called *Daphnia,* are animals. Also known as water fleas, *Daphnia* make up part of the plankton in freshwater and marine ecosystems.

Figure 27-3 Sponges, like these purple tube sponges, are attached to the sea floor, unable to move around. While they may look inactive, they are actually very busy filtering food particles from sea water.

Some General Features of Animals

Animals are heterotrophs—they ingest their food before digesting it. Most animals move from place to place searching for food, which they then take into their body. In most animals, ingestion of food is followed by digestion in an internal cavity.

All animals are multicellular, and almost all (99 percent) are invertebrates (animals without backbones). Scientists estimate that there are between 5 million and 10 million living animal species. Only 42,500 of these species have a backbone. Animals range in size from organisms too small to see clearly with the naked eye, like the *Daphnia* in Figure 27-2, to enormous whales and giant squids. The animal kingdom includes about 35 phyla; most occur in the sea, with far fewer in fresh water and fewer still on land. Members of three phyla, Arthropoda (spiders, insects, and crustaceans), Mollusca (snails), and Chordata (vertebrates), dominate animal life on land.

Animal cells are distinct among multicellular organisms because animal cells lack rigid cell walls and are usually quite flexible. The cells of all animals except sponges are organized into structural and functional units called tissues.

The ability of animals to move, more rapidly and in more complex ways than the members of other kingdoms, is perhaps their most striking characteristic. Animals move by means of muscle cells, specialized cells able to contract with considerable force. A remarkable form of movement unique to animals is flying, an ability that is well developed among both insects and vertebrates. Among today's vertebrates, birds and bats are both strong flyers. At one time, flying reptiles called pterosaurs, now extinct, dominated the skies.

Most animals reproduce sexually. Animal egg cells are much larger than the small, usually flagellated sperm cells, and, unlike sperm cells, they do not swim. With few exceptions, animals are diploid and the gametes are the only haploid cells in their life cycles.

Sponges: The Simplest Animals

Sponges, shown in Figure 27-3, are the simplest animals. There are about 9,000 species of sponges, almost all of which live in the sea (a few species live in fresh water). The bodies of most sponges completely lack **symmetry**—they do not have body parts that grow around a central point or a central axis as do all other animals. The cells of sponges are not organized into tissues or organs. The bodies of sponges consist of little more than masses of specialized cells embedded in a gel-like substance, called matrix, like chopped fruit in gelatin. However, sponge cells do have a key property of all

Highlights
Invertebrate Evolution

Animals Without Tissues

Highlights
Invertebrate Evolution

Stage 1 Multicellularity

The bodies of *all* animals, including sponges (phylum Porifera), are **multicellular**—made of many cells. Sponges are composed of several different cell types whose activities are coordinated with each other.

The Body of a Sponge Is Built for Filter Feeding

The outside of a sponge is covered with epithelial cells that protect the sponge. Water, *blue arrows,* enters the sponge's central cavity by passing through many tiny pores that penetrate the skin. The beating action of many choanocyte flagella creates a water current inside the sponge. This current pulls a steady stream of water into the sponge. The choanocytes trap and ingest food particles that are suspended in the water, and the water eventually exits through the osculum.

Protist Ancestors

The choanocytes of sponges very closely resemble a kind of protist called a choanoflagellate, and ancient choanoflagellates are thought to have been the ancestors of sponges. Other, free-swimming colonial flagellates more closely resemble sponge larvae, however, and some believe them to be the true ancestors of sponges.

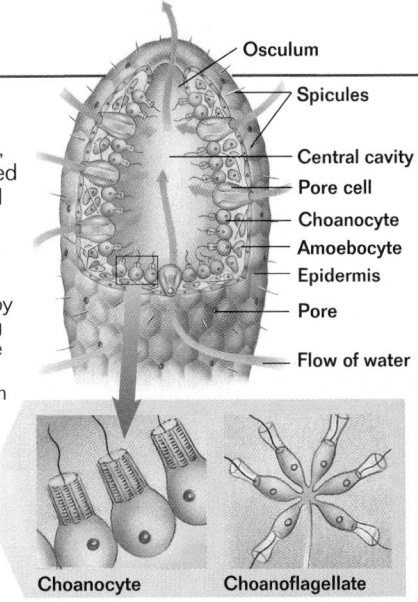

Osculum
Spicules
Central cavity
Pore cell
Choanocyte
Amoebocyte
Epidermis
Pore
Flow of water

Choanocyte   Choanoflagellate

---

animal cells—cell recognition. The ability of a sponge cell to recognize another sponge cell can be demonstrated during lab experiments in which a sponge is passed through a fine silk mesh. Individual cells separate and then recombine on the other side of the mesh to re-form the sponge.

The body of an adult sponge is anchored in place on the sea floor and functions as a water-filtering machine. As sea water filters through the sponge, the sponge is able to trap protists and tiny animals that live in the water. The body of the sponge is perforated by tiny holes or pores. The name of the phylum, Porifera, refers to this system of pores. Inside the sponge is an internal cavity. Facing into the internal cavity are unique flagellated cells called **choanocytes** (*koh AN oh seyets*), also known as collar cells. The beating of the flagella of the many choanocytes that line the body cavity of the sponge draws water in through the pores and drives the water through the cavity. The water exits the sponge through one or more large openings in the sponge's body wall.

One cubic centimeter of a sponge can propel more than 20 L (5.3 gal.) of water a day into and out of the sponge's body. Why all this moving of water? The sponge is a filter feeder. The beating of each choanocyte's flagellum draws water down through its collar. The collar is made of small hairlike projections resembling a picket fence. Food particles in the water are trapped in the collar and later ingested by the choanocyte or by the sponge's other cells. ◻

CONTENT LINK

Information on sponge classification and reproduction can be found in **Chapter 28.**

◻ CAPSULE SUMMARY

Sponges are the simplest animals. Lacking tissues and body symmetry, sponges live attached to the sea floor, where they filter food particles from the water. The food is digested within the sponge's cells.

Instructional Strategy

Have students study the pictures of the choanocytes and the choanoflagellates to find three things they have in common and three ways that they are different.

Discussion

Guide the discussion by posing the following questions.

1. Which protists were probably the ancestors of sponges? *(choanoflagellates)*

2. Are sponges the ancestors of all multicellular animals? *(It appears they are.)*

3. Which types of cells are found in a sponge, and why is each needed for the whole sponge to live? *(Choanocytes enable the sponge to feed, amoebocytes transport nutrients and wastes, and epidermal cells protect the sponge.)*

Teaching Tip
Unscrambling the Groups

Write a *Graphic Organizer* similar to the one located at the bottom of page 613 on the chalkboard or overhead projector but without the names of the animal groups. Scramble the order of the animal groups and list them separately next to the spaces. Beginning with "Multicellularity," have a student come up to the board and write the correct animal group next to this description. Then ask the student to explain what the description means. Have other students come up and try the other descriptions.

Graphic Organizer

Use this graphic organizer in **Teaching Tip: Unscrambling the Groups** on page 613

Animal Group	Evolutionary advances
Chordates	Improved skeleton
Echinoderms	Deuterostome pattern of embryo development
Arthropods	Jointed appendages
Annelids	Body plan based on segmentation
Mollusks	Formation of a better body cavity
Roundworms	Formation of a body cavity
Flatworms	Formation of organs
Cnidarians	Tissues and symmetry
Sponges	Multicellularity

VISUAL STRATEGY

Figure 27-4

Have students study Figure 27-4; then name an organ, and ask them which embryonic tissue layer gave rise to the organ. You might name organs of ectodermal origin such as the skin, hair, fingernails, brain, spinal cord, and nerves. Organs of mesodermal origin could include the heart, arteries, veins, muscles, kidneys, bladder, uterus, testes, lungs, and thyroid gland. Organs of endodermal origin could include the stomach, small intestine, gall bladder, pancreas, esophagus, and colon.

Teaching Tips
Heterotrophic Lifestyles

Since animals must obtain their energy from food molecules from outside their bodies, they have evolved an amazing array of adaptations for detecting, capturing, and digesting food. Ask students to compare some of the ways invertebrates obtain their food, paying special attention to their sense organs and ways that they trap and digest food. Have students compare filter feeders, carnivores, herbivores, and parasites.

What If you were to prick one of the tentacles of a hydra with a pin? The hydra would respond by simultaneously contracting all of its tentacles, not just the one that was pricked. This is because cnidarians have a diffuse nervous system called a nerve net. This kind of very simple nervous system doesn't allow cnidarians to have any kind of fine control over their body movements.

CONTENT LINK

More information on cnidarian classification and life cycles can be found in **Chapter 28**.

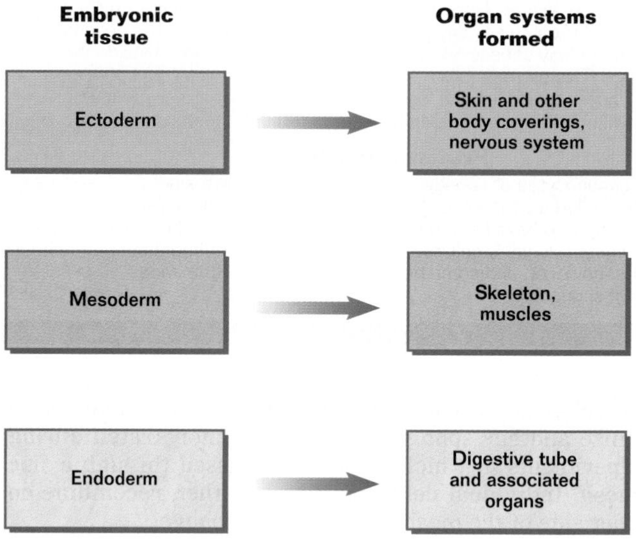

Figure 27-4 All of the organ systems of an animal develop from three embryonic tissue layers: ectoderm, mesoderm, and endoderm. The eyes you are using to read these words developed from embryonic ectoderm.

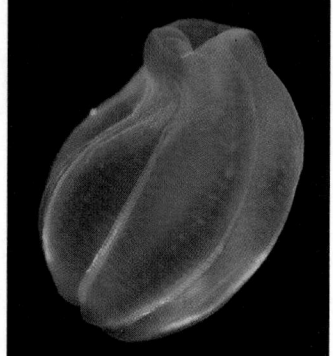

Figure 27-5 Ctenophores are marine animals that are found mostly in deep, open oceans. This particular species of ctenophore likes to feed on other ctenophores, so it is possible that the red object in its center is a relative.

Cnidarians: Radially Symmetric Animals

All animals other than sponges have both tissues and symmetry and are called "true animals," or **eumetazoans** *(YOO meht uh zoh uhns)*. Early in their development, the embryos of most eumetazoans develop three distinct cell layers. These three embryonic tissues give rise to the many tissues of the adult body, as shown in Figure 27-4. An outer **ectoderm** layer gives rise to the outer coverings of the body and the nervous system; a middle **mesoderm** layer gives rise to the skeleton and muscles; and an inner **endoderm** layer gives rise to the digestive organs.

Embryonic tissue → **Organ systems formed**

Embryonic tissue	Organ systems formed
Ectoderm	Skin and other body coverings, nervous system
Mesoderm	Skeleton, muscles
Endoderm	Digestive tube and associated organs

The two most primitive eumetazoan phyla are Cnidaria *(nye DAIR ee uh)* and Ctenophora *(TEHN uh fawr ah)*. The phylum Cnidaria includes jellyfish, hydra, sea anemones, and corals. The phylum Ctenophora is a minor phylum that includes the comb jellies, like the one shown in Figure 27-5. Comb jellies are delicate marine animals that look like tiny jellyfish and swim by moving clusters of cilia. Both cnidarians and ctenophores are radially symmetric. **Radially symmetric** animals have body parts arranged around a central point, as shown in Figure 27-6.

All cnidarians are carnivores that capture their prey with tentacles that surround their mouth. These tentacles bear unique stinging cells called **cnidocytes** *(nih DOH seyets)*, which occur in no other organism and which give the phylum its name. Within each cnidocyte is a small barbed harpoon called a **nematocyst** *(NEHM uh toh sist)*, which cnidarians use to spear their prey. The captured prey is then drawn back to the tentacle containing the cnidocyte.

 Did You Know?

Some nudibranchs, or sea slugs, feed on cnidarians and save the untriggered nematocysts within their own bodies. When other animals try to eat the nudibranchs, they trigger the nematocysts. As a result, they learn to avoid the painful mollusks in subsequent encounters.

Highlights
Invertebrate Evolution

Tissues Lead to Greater Specialization

Stage 2 Tissues

The cnidarian body is more complex than that of sponges. Cnidarians (phylum Cnidaria) have specialized **tissues** that carry out particular functions. Cnidarians also exhibit **radial symmetry,** with their parts arranged around a central axis like the petals of a daisy. An interior digestive cavity is the site for **extracellular digestion**—digestion outside of cells.

Hydras Are Common Freshwater Cnidarians

The body wall of hydras, like those of other cnidarians, is composed of an outside ectodermal layer, an inside endodermal layer, and a middle, gel-like, mesogleal layer. The tentacles of cnidarians contain many stinging cells called cnidocytes, each of which houses a harpoon-like nematocyst. Cnidarians use their nematocysts to spear prey.

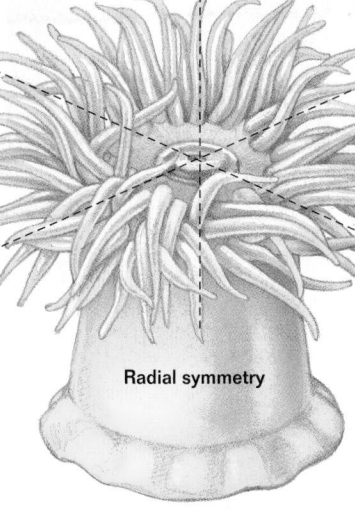

Nerve cell · Nematocyst (discharged) · Tentacles · Cnidocyte · Nematocyst (coiled) · Digestive cavity · Mesoglea · Ectoderm · Endoderm

A Lethal Weapon

This scanning-electron micrograph shows a barbed nematocyst bursting out of its cnidocyte. The cnidocyte builds up a very high internal osmostic pressure and pushes the nematocyst outward so explosively that it can penetrate even the hard shell of a crab!

A major evolutionary innovation that occurred among the cnidarians is the extracellular digestion of food. Recall that food trapped by a sponge choanocyte is taken directly into that cell or another cell by endocytosis and digested intracellularly (within the cell). In a cnidarian, food is digested extracellularly (outside the cell), in a digestive cavity. Extracellular digestion is the same heterotrophic strategy used by fungi, except that fungi digest food outside their bodies. The extracellular

Radial symmetry

Figure 27-6 The body parts of this sea anemone, and of all other cnidarians, are arranged around a central point. This kind of body organization is called radial symmetry. Can you name another commonly known animal that has radial symmetry?

Instructional Strategy

Have students compare the anatomy and lifestyle of a hydra with that of a sponge. Ask students why both organisms are considered to be animals, and why a hydra's body is more complex than a sponge's body. Have students pay special attention to the feeding styles of both animals. You might begin a wall chart to record the evolutionary advances of each group of animals as they are encountered in this chapter.

Discussion

Guide the discussion by posing the following questions.

1. How does the body of a hydra show better organization than the body of a sponge? *(The cells of the hydra are organized into tissues, while those of the sponge are not.)*

2. How is the interior digestive cavity of the hydra different from the interior cavity of the sponge? *(Sponges do not digest food in their interior cavity.)*

3. What provides the power for firing a nematocyst out of a cnidocyte? *(The cnidocyte builds up a very high internal osmotic pressure.)*

4. If one tentacle catches a small animal, the food is shared with the entire hydra. Why? *(Because the animal is digested extracellularly, and the resulting nutrients diffuse into the hydra's cells.)*

Teaching Tip
Symmetry

Use some common classroom objects to illustrate different types of symmetry. For instance, a crumpled piece of paper is asymmetrical, a roll of tape is radially symmetric, and a stapler is bilaterally symmetric. Point out that the symmetry doesn't have to be perfect in every detail to be considered a type of symmetry. A hydra is still radially symmetric even though all its tentacles aren't exactly the same size. Then give students a series of food items to classify as asymmetrical (*scrambled eggs, mashed potatoes, popcorn*), radially symmetric, (*pizza, doughnut, bagel*) or bilaterally symmetric (*taco, hot dog, loaf of bread*).

 VISUAL STRATEGY

Figure 27-7

Have students study Figure 27-7 to compare and contrast the medusa and polyp stages of a jellyfish. Ask students to name any other animals that have different bodies for different stages of their life. (*Answers may include examples from these groups of animals: amphibians, insects, mollusks, echinoderms.*) Then have your students find an example of a medusa and of a polyp elsewhere in this textbook. (*Many examples are in Section 28-2.*)

☐ CAPSULE SUMMARY

Cnidarians are eumetazoans (true animals) that have tissues, tentacles, and radial symmetry. They digest their food extracellularly, in a digestive cavity.

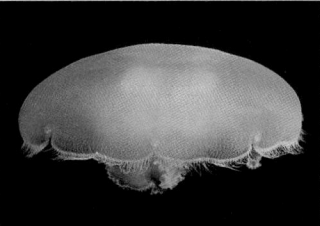

Figure 27-7 Many cnidarians, like *Aurelia*, have life cycles that include both a polyp and medusa stage. Medusae, *top*, are free-swimming. The black-and-white photograph of the polyp, *bottom*, shows that polyps are stationary.

digestion of food has been retained by all of the more advanced groups of animals.

Cnidarians have two basic body forms. **Medusae** are free-floating, gelatinous, and often umbrella shaped. **Polyps** are cylindrical, pipe-shaped animals that are usually attached to a rock. Many cnidarians exist only as medusae, others only as polyps, and still others alternate between these two phases during the course of their life cycle, as shown in Figure 27-7. ☐

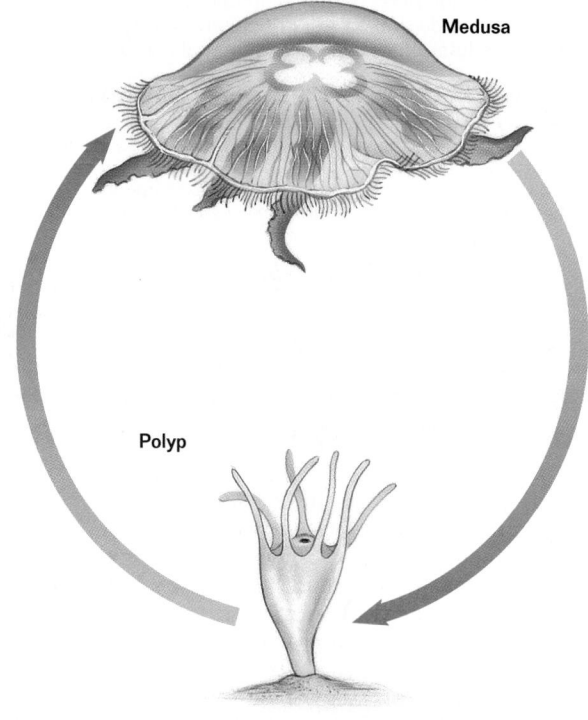

Medusa

Polyp

Section Review

1. *What characteristic of sponges could lead to their being mistakenly classified as plants instead of animals?*
2. *Describe the symmetry of a jellyfish.*
3. *Compare the way sponges capture and digest food with that of jellyfish.*

Critical Thinking
4. *Imagine a sponge that is living on the ocean floor and is covered by a watertight plastic bag. After a period of time, the ATP level in the sponge's cells begins to decrease. Explain why.*

Answers to Section Review

1. Sponges are anchored in place, as are most plants. Animals usually aren't anchored.

2. A jellyfish is radially symmetric. Its body is round and bell-shaped, with tentacles that dangle downward around the edges, so its parts are arranged in a circle around a central axis.

3. Sponges filter food particles from water by trapping them in their choanocytes. The food is digested inside the cells of sponges. Jellyfish capture food with their nematocysts, and digest the food in their digestive cavity.

4. The sponge continues to filter the same water over and over inside the bag. After a period of time, it has filtered all the food out of the water, so its cells are no longer recieving food. Without food, the sponge's cells cannot produce ATP molecules.

27-2 Evolution of the Body Cavity

No process more clearly illustrates the way the animal body plan has evolved than the development of the body cavity. As you will see in this section, the fundamental architecture of the animal body is largely determined by the nature of the body cavity.

Flatworms: Worms With Solid Bodies

Although cnidarians and ctenophores have radial symmetry, all other eumetazoans have bilateral symmetry. **Bilaterally symmetric** animals have left and right halves that mirror each other when they are divided by an imaginary plane passing through their longitudinal center, as shown in Figure 27-8. In a bilaterally symmetric animal, the top surface of the animal is referred to as **dorsal** and the bottom surface as **ventral**. The front end of the animal is **anterior** and the back end is **posterior**. Bilateral symmetry was a major evolutionary advancement among animals because it enabled different parts of the body to become specialized in different ways. For example, most bilaterally symmetric animals have evolved a definite head end, a process called **cephalization.** Animals with heads are often active and mobile, moving through their environment headfirst, with sensory organs concentrated in front so the animal can sense food, danger, and potential mates as it enters new surroundings.

Section Objectives

■ *Define bilateral symmetry.*
■ *Describe the evolution of the body cavity in animals.*
■ *Compare and contrast flatworms with roundworms.*
■ *Describe the basic body plan of a mollusk.*

Figure 27-8 Most animals have bilateral symmetry, in which the right and left halves of the animal are mirror images of each other, *left.* Bilaterally symmetric animals have surfaces that are identified as anterior (head), posterior (rear), dorsal (back), and ventral (stomach), *right.*

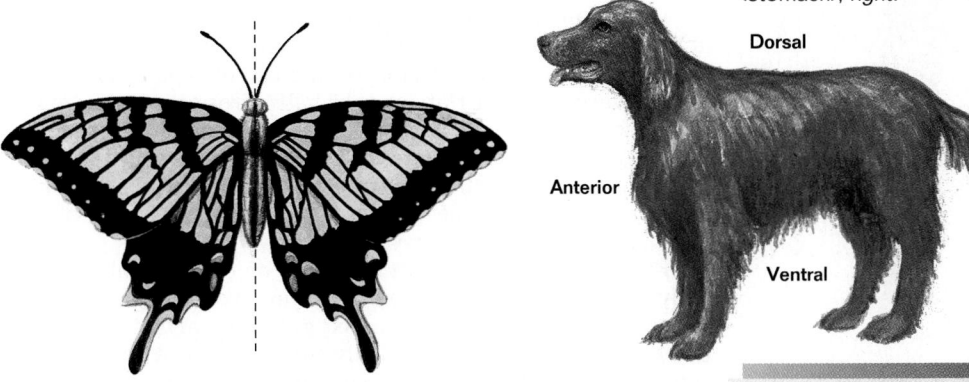

Bilateral symmetry

Dorsal

Posterior

Anterior

Ventral

The simplest of all bilaterally symmetric animals are the solid worms. The largest phylum is Platyhelminthes, with about 20,000 species. The animals in this phylum are commonly called flatworms. Flatworms are the simplest animals

CONTENT LINK

Learn more about flatworm classification and the life cycles of several parasitic flatworms in **Chapter 28.**

Vocabulary Preview

bilaterally symmetric
dorsal
ventral
anterior
posterior
cephalization
organ
acoelomate
pseudocoelom
coelom
primary induction
circulatory system
mollusk
visceral mass
mantle
gill
radula

Opening Question

Have students call on their own experiences with animals to identify what cues they use to identify the body orientation and normal direction of movement of different kinds of animals. How can you tell what is the front and what is the back of an animal? Tell them that sensory organs are usually concentrated at the front end of an animal. If a tail is present, it will be located at the back end. How can you tell what are the upper and lower surfaces of an animal? Tell them that if an animal is legless, like a worm, then it will crawl on its lower surface. If the animal has legs, the surface closest to the ground when it walks is the lower surface.

Highlights
Invertebrate Evolution

Organs and Bilateral Symmetry

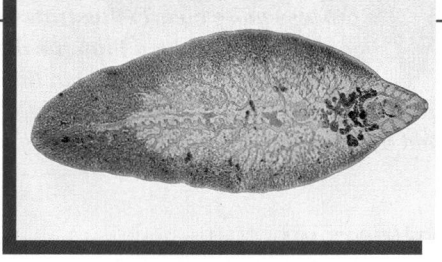

Stage 3 Bilateral Symmetry

The evolution of a middle tissue layer called mesoderm in the solid-bodied flatworms (phylum Platyhelminthes) made possible the formation of organs such as testes and ovaries. Flatworms were the first animals to be **bilaterally symmetric**, with left and right halves that mirror each other. They were also the first animals to develop a distinct head region.

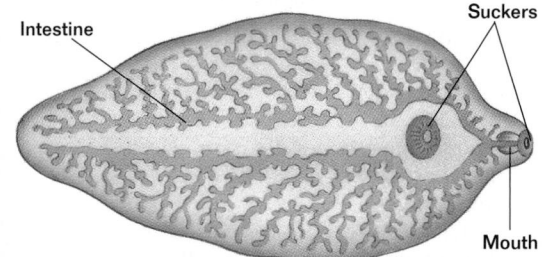

Intestine

Suckers

Mouth

A Parasitic Flatworm

Many flatworms, like these parasitic sheep **liver flukes,** have flat bodies with a mouth that is located on their undersurface. Adult flukes use two sucker-like mouths to attach themselves to the inside of a sheep's liver or gallbladder. The mouth of a flatworm is the only opening to their gut, and all food and wastes pass through it. Their intestine has many branches, ensuring that nutrients are brought close to all the worm's tissues.

Instructional Strategy

Have students study the information about the sheep liver fluke in this *Highlights* feature and compare it to information about the hydra in the *Highlights* feature on page 615. Direct students' understanding by asking questions such as: How is the symmetry of a fluke different from that of a hydra? How is their digestion different? Why did cephalization evolve in the fluke?

Discussion

Guide the discussion by posing the following questions.

1. What advances does the flatworm have over the hydra and the sponge? *(The flatworm has bilateral symmetry, organs, cephalization, and a solid body.)*

2. Why is it important to have a head, or ovaries and testes? *(head—for a region where sense organs are located; ovaries and testes—to produce gametes for reproduction)*

Demonstration
Following the Directions

After introducing the terms anterior, posterior, ventral, and dorsal to your students, you may wish to reinforce this anatomical terminology by having students sketch the outline of a worm, then direct them to draw different structures on their worm. Examples of directions could be the following: Draw bristles on the anterior end of your worm; shade the dorsal surface black; draw rings around the posterior end of the worm; draw spots on the ventral surface of the worm; draw eyes on the anterior end of your worm. Finish by drawing on the chalkboard what the completed worm should look like.

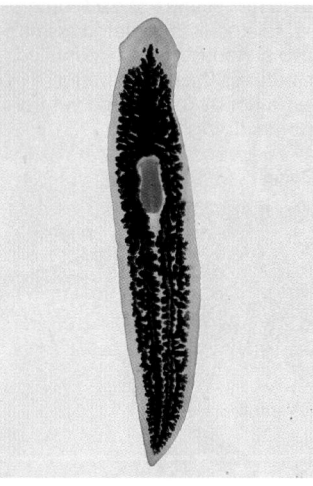

Figure 27-9 This common freshwater flatworm is called a planarian. The two dark spots located on the animal's shovel-shaped anterior end are light-sensitive structures called eyespots. The darkly stained interior portion is the planarian's branched intestine.

that have organs. An **organ** is a collection of different tissues that work together as a unit to perform a particular function. The dark spots on the head of the flatworm shown in Figure 27-9 are sense organs called eyespots. Eyespots can detect light, but they cannot focus an image like your eyes can.

Flatworms lack any internal body cavity other than the gut (digestive tube). They are soft-bodied animals that are flattened from top to bottom, like a piece of tape or ribbon. If you were to cut a flatworm in half across its body, you would see that the gut is completely surrounded by tissues and organs. This solid body construction, shown in Figure 27-10, is called **acoelomate** *(ay SEEL oh mayt)*. The term *acoelomate* is from the Greek *a*, meaning "without," and *koilia*, meaning "body cavity."

Flatworms must be thin because of their acoelomate body design. Since flatworms have no circulatory system, dissolved substances such as oxygen and carbon dioxide must pass through the solid body by diffusion. A thin body shortens the distance that these substances must travel to reach each cell. The gut of a flatworm is highly branched. Portions of it run close to practically all of the flatworm's tissues, giving each cell access to food molecules. The gut has only one opening—the mouth. This means that material must move through the mouth in two directions: foods enter and wastes exit.

 Did You Know?

Planarians have extraordinary powers of regeneration. Any part of their body that is destroyed by injury may be replaced, and if a planarian is cut into two or more pieces, each piece will eventually grow into a complete new individual.

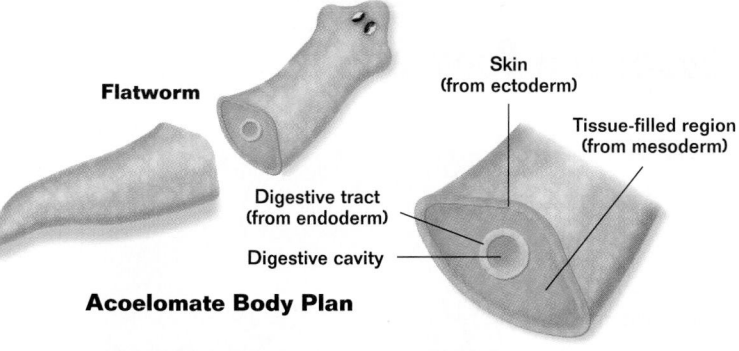

Flatworm

Skin
(from ectoderm)

Tissue-filled region
(from mesoderm)

Digestive tract
(from endoderm)

Digestive cavity

Acoelomate Body Plan

Some flatworms are free-living, but many species are parasitic. Flatworms range in size from free-living forms less than 1 mm in length to parasitic intestinal tapeworms that grow to several meters long.

Roundworms: Nematodes

All bilaterally symmetric animals other than the solid worms have an internal body cavity. A **pseudocoelom** (*SOO duh see luhm*) is a body cavity located between the endoderm and the mesoderm, as shown in Figure 27-11. The evolution of a body cavity was an important improvement in animal body design for several reasons.

1. **Circulation**   Fluids that move within the body cavity can serve the function of a circulatory system, permitting the rapid passage of materials from one part of the body to another.

2. **Movement**   Fluid in the body cavity makes the animal's body rigid—like a balloon filled with water. A rigid body offers resistance to contracting muscle cells, enabling muscle-driven body movement.

3. **Organ function**   Body organs that are surrounded by a body cavity can function without being distorted by surrounding muscles. For example, food can pass freely through a gut suspended within a cavity. The rate at which the food moves through the digestive tube is not affected when the animal contracts its body muscles to move.

Roundworm

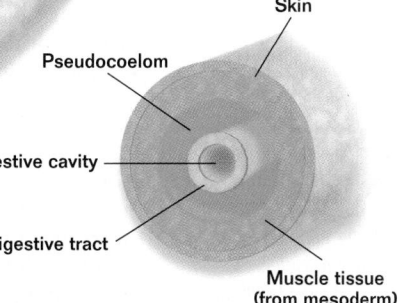

Skin

Pseudocoelom

Digestive cavity

Digestive tract

Muscle tissue
(from mesoderm)

Pseudocoelomate Body Plan

Figure 27-10 Flatworms are called solid-bodied worms because they have no body cavity between their digestive tract and skin. An animal without a body cavity has a limited amount of organ development.

CAPSULE SUMMARY

Flatworms have bilateral symmetry and no body cavity. Simple organs and a digestive tract with a single opening characterize these invertebrates.

CONTENT LINK

More information on roundworms and on rotifers, other pseudocoelomates, can be found in **Chapter 28**.

Figure 27-11 Roundworms are called pseudocoelomates because they have a pseudocoelom, a body cavity that develops between their digestive tract and their mesoderm layer. The fluid-filled pseudocoelom serves as a simple circulatory system in roundworms.

❓ Did You Know?

Caenorhabditis elegans may be small but it is probably the most famous nematode in modern biology. What *E. coli* was to microbiologists and *Drosophila* was to geneticists, *C. elegans* is to modern developmental biologists. Tracing the development of every cell in this famous nematode has shed light on many processes and problems of embryological development in higher animals, including *H. sapiens*.

Highlights   *Advent of a Body Cavity*

Invertebrate Evolution

Instructional Strategy

Ask students to compare the digestive system of a roundworm with the digestive systems of less complex animals. If students don't see any problem with animals such as the hydra or sheep liver fluke having only one opening to their digestive cavity, have students imagine how inefficient it would be if they were to go through a cafeteria line picking up a salad, a sandwich, and a drink, then back through the line in the opposite direction to leave. They would have to make their way past people choosing drinks, choosing sandwiches, and choosing salads. If there aren't many people, it's not a big problem, but for efficient handling of crowds, it would be gridlock. Simpler animals have lower energy demands, so they can live with inefficient body systems. More complex animals cannot.

Discussion

Guide the discussion by posing the following questions.

1. Name two anatomical features present in roundworms but not flatworms. *(a pseudocoelom and a one-way gut)*

2. What internal organs of a roundworm are bilaterally symmetric? *(muscles, oviducts)*

Demonstration
Longitudinal Muscles
Obtain about six inches of thick, large diameter flexible tubing and attach rubber bands to the outside of the tubing with paper clips. Pull the rubber bands on one side, then on the other side, to demonstrate the whipping action that results when muscles contract on opposite sides of the roundworm.

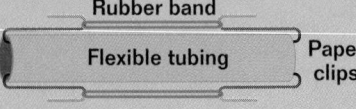
Rubber band
Flexible tubing
Paper clips

Stage 4 Body Cavity

Roundworms (phylum Nematoda) are bilaterally symmetric, cylindrical worms. The major innovation in their body design is a body cavity called a **pseudocoelom**, which forms between the gut and the body wall.

Most Roundworms Are Tiny
Some free-living roundworms are very small—less than 1 mm (0.04 in.) long. *Caenorhabditis elegans* is composed of only about 1,000 cells and is the only animal whose complete cellular anatomy is known.

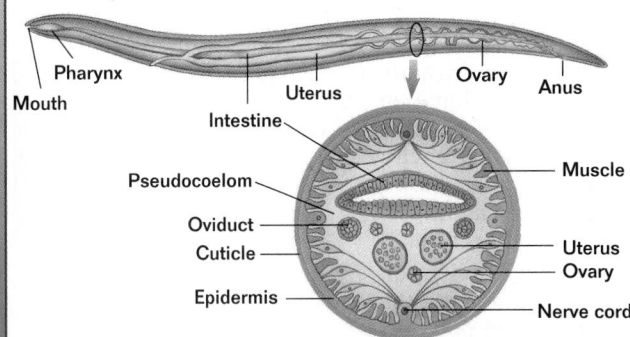

Pharynx
Mouth
Uterus
Ovary
Anus
Intestine
Pseudocoelom
Muscle
Oviduct
Cuticle
Uterus
Ovary
Epidermis
Nerve cord

Why Roundworms Wriggle
Roundworms, like these vinegar eels, have muscles that extend along the length of their bodies, rather than encircling them, enabling the worms to wriggle. The nematode body is covered with a flexible, thick cuticle that is shed as the worm grows. The nematode has a one-way digestive tube. Food enters the mouth at one end of the worm, and waste exits through the anus at the other end.

☐ CAPSULE SUMMARY

Roundworms were the first animals to have a body cavity. Called a pseudocoelom, this cavity is positioned between mesoderm and endoderm. Roundworms have muscle tissue and a gut that is open at both ends.

All pseudocoelomates (animals with a pseudocoelom) have a one-way gut through which food passes into the mouth and out the anus. This is a major improvement in body design, as it permits far greater specialization of the gut. A one-way digestive tube can function like an assembly line, with food being acted on in different ways in each section as it passes through.

Seven animal phyla have a pseudocoelom. Only one of the seven pseudocoelomate phyla contains a large number of species, the phylum Nematoda. Nematodes are commonly called roundworms. There are estimated to be between 500,000 and 600,000 nematode species with only about 13,000 species actually named, most of which are tiny. Most nematodes live in soil. It has been estimated that one spadeful of soil contains an average of 1 million nematodes! There are also some parasitic species, several of which infect humans. In nematodes, a layer of muscle extends along the length of the worm beneath a flexible, thick cover of epidermis and cuticle (protective layer). These long muscles push against the cuticle and the pseudocoelom, whipping the worm's body from side to side as the animal moves. ☐

Mollusks: Animals With Three-Part Bodies

Even though acoelomates and pseudocoelomates have flourished, a third way of organizing the animal body evolved among the mollusks. Mollusks and all animals that evolved after them, such as the earthworm shown in Figure 27-12, have a body cavity called a coelom and hence are called coelomates. A **coelom** (*SEE luhm*), is a fluid-filled body cavity that develops not between endoderm and mesoderm, but entirely within the mesoderm.

What is the functional difference between a pseudocoelom and a coelom, and why has the latter type of body cavity been so successful? The answer has to do with the nature of animal embryonic development. In animals, the development of specialized tissues involves a process called primary induction. In **primary induction,** one of the three primary tissues (endoderm, mesoderm, or ectoderm) interacts with another. The interaction requires physical contact. A major advantage of the coelomate body plan is that it allows contact between mesoderm and endoderm, which enables primary induction to occur during development. For example, contact between mesoderm and endoderm permits localized portions of the digestive tract to develop into complex, highly specialized regions, like the stomach. In pseudocoelomates, mesoderm and endoderm are separated by the body cavity, limiting developmental interactions between these tissues.

CONTENT LINK

Learn more about mollusk organ systems and mollusk classification in **Chapter 29**.

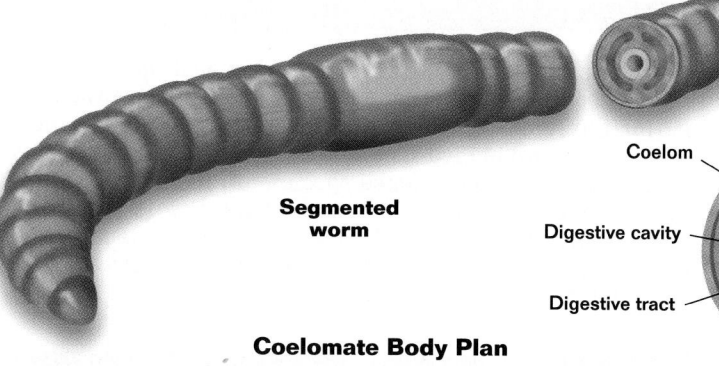

Segmented worm

Coelomate Body Plan

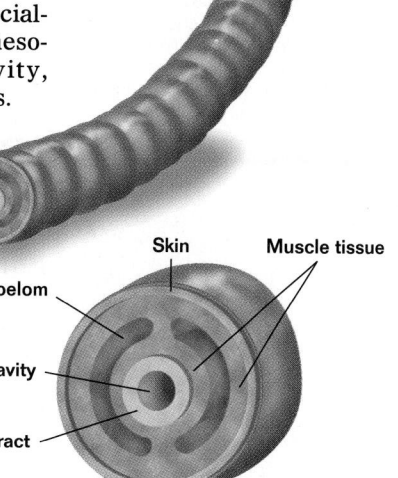

Skin

Muscle tissue

Coelom

Digestive cavity

Digestive tract

Figure 27-12 This earthworm, like most other animals, is called a coelomate because it has a coelom, a body cavity that develops within the mesoderm. The tissue interactions that occur in coelomates have enabled them to develop complex organ systems.

The formation of the coelom does re-create one old problem that was solved among pseudocoelomates—the circulation of nutrients, oxygen, and wastes. In coelomates, the digestive tube is again surrounded by tissue that presents a barrier to diffusion, just as in flatworms. In coelomates, however, the efficient circulation of nutrients, oxygen, and wastes is accomplished by a circulatory system. A **circulatory system** is a network of vessels that carries fluids to all parts of the body. The circulating fluid, called blood, carries nutrients and oxygen to the tissues and removes waste and carbon dioxide. Blood is usually pushed through the circulatory system by the contraction of one or more muscular hearts.

SECTION 27-2

Teaching Tips
Cell Talk

The interaction of the three primary tissues in primary induction results from the secretion of chemical messengers and from the interaction of the receptors in the plasma membranes of certain cells with those of other cells. These receptor interactions activate or suppress different genes. These interactions can be very complex and are an important area of biological study. To help students appreciate how important these interactions are, refer students back to Figure 27-4 to see some of the organs that develop from different cell layers. Ask students which two tissue layers interacted to form the stomach (*mesoderm and endoderm*) and which two tissue layers interacted to form the backbone (*ectoderm and mesoderm*).

Fluid Movement

The coelomates have a circulatory system that transports fluids throughout the body. As animals become larger and more complex, vessels and pumps become more important for fluid circulation. To help students appreciate how important circulation is to survival, ask them how long a person can live after their heart stops. (*minutes*) Then ask why a person dies when their heart stops. (*The most critical need is moving oxygen to tissues.*)

◉ VISUAL STRATEGY

Figures 27-11 and 27-12

Have students compare the coelom in Figure 27-12 to the pseudocoelom in Figure 27-11 on page 619. Draw students' attention to the placement of the three primary tissues in each figure, then relate the reduced mesodermal layer in pseudocoelomate animals to their lack of complex organs. Without much mesoderm tissue, the muscular system of a pseudocoelomate animal is reduced, and its gut lacks an outer layer of muscle tissue.

Instructional Strategy

Have students compare the diagram of the anatomy of the snail with the three primary tissue layers in Figure 27-12 on page 621. Then have students classify each structure in the snail with its primary tissue layer(s).

Discussion

Guide the discussion by posing the following questions.

1. How are a coelom and the development of specialized organs related to each other? *(The position of the coelom enables mesoderm and endoderm to interact, leading to the formation of complex organs, such as a highly modified gut.)*

2. Where is the coelom in an adult snail? *(around the heart)*

3. What are some of the specialized organs of a snail, and why are these organs necessary for a snail to live? *(gills—for breathing; kidney—for removing nitrogen waste; heart—for pumping blood)*

👁 VISUAL STRATEGY

Figure 27-13
Have students associate the photographs of the mollusks with the groups of mollusks described on page 623. Stress the common body design of each of these groups, then check for student understanding by asking questions such as: Does a scallop have endoderm/mesoderm/ectoderm? *(yes)* What kind of symmetry does a squid have? *(bilateral)*

Highlights
Invertebrate Evolution

Building a Better Body Cavity

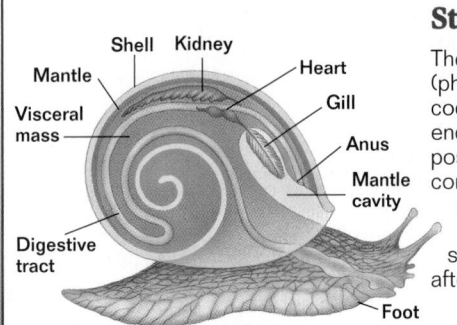

Stage 5 Coelom

The body cavity of a mollusk (phylum Mollusca) is called a coelom and is completely enclosed within mesoderm. The position of the coelom allows physical contact between mesoderm and endoderm, permitting tissue interactions that lead to the development of highly specialized organs in mollusks, such as a heart and stomach. All animal phyla that evolved after mollusks have a coelom.

Inside a Snail

The coelom of a snail is confined to a small zone around the heart. The **mantle** is a heavy fold of tissue that wraps around the snail's organs, or **visceral mass.** In aquatic snails, the cavity between the mantle and visceral mass contains the **gills.** The gills capture oxygen from water as it passes through the mantle cavity. In snails, the mantle secretes a hard outer shell. A simple kidney gathers wastes from the coelom and discharges them into the mantle cavity. Within the mouth of a snail is a unique rasping tongue called a **radula,** shown in the photo to the right. Snails creep along the ground on a muscular **foot.**

A radula's surface magnified many times

The least advanced of the coelomates, and thus probably the first to evolve, are the **mollusks.** Mollusks are the only major phylum of coelomates without segmented bodies. Mollusks, shown in Figure 27-13, are divided into three classes and, with over 100,000 named species, are the largest animal phylum except for the arthropods. Mollusks occur almost everywhere. Next to insects, they are the most successful land animals. There are more terrestrial mollusk species (35,000) than there are terrestrial vertebrate species (20,000).

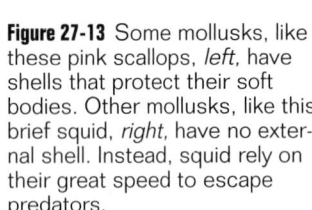

Figure 27-13 Some mollusks, like these pink scallops, *left,* have shells that protect their soft bodies. Other mollusks, like this brief squid, *right,* have no external shell. Instead, squid rely on their great speed to escape predators.

 Did You Know?

Scientists have traditionally thought of squids as creatures of the oceans' middle depths, constantly swimming about or floating in a state of neutral buoyancy. But recently, researchers have learned that some squid species actually rest on the bottom, using their tentacles to prop up their bodies, so that their breathing tubes stay clear.

The bodies of all mollusks are composed of three distinct regions. There is a **visceral mass,** a central section that contains the body's organs. Wrapped around the visceral mass like a cape is a heavy fold of tissue called the **mantle.** In aquatic snails, gills are positioned on the inner surface of the mantle. **Gills** are filamentous projections of tissue, rich in blood vessels, that capture oxygen from the water. The water circulates between the mantle and visceral mass, and the oxygen diffuses into the gills. Carbon dioxide is released from the gills into the water. Terrestrial snails, which have no gills, breathe air with simple lungs. Finally, every mollusk has a muscular region called a foot.

The three classes of mollusks are different variations upon this same basic body design.

1. **Gastropods**   Gastropods (snails and slugs) use their muscular foot to crawl, and their mantle often secretes a hard protective shell. All terrestrial mollusks are gastropods.

2. **Bivalves**   Bivalves (clams, oysters, and scallops) secrete a two-part shell (also called a valve) with a hinge. They feed by drawing water into their shells and filtering out food particles.

3. **Cephalopods**   Under their mantle, cephalopods (octopuses and squids) have a modified cavity that creates a jet propulsion system to propel them rapidly through the water.

A characteristic feature of mollusks is a structure called a **radula** (RAJ u luh), a rasping tongue-like organ. All mollusks except bivalves have a radula. With rows of pointed, backward-curving teeth, the radula is used by some snails to scrape algae off rocks. Cephalopods and many gastropods are active predators that use their radula as a weapon to puncture their prey. The small holes often seen in oyster shells are produced by the radula of certain gastropods. Once the hole is made, the snail extracts the oyster's body for food. ▢

❏ CAPSULE SUMMARY

The development of a coelom occurred first in the mollusks and is found in all animal phyla that evolved after this group. Embryos with a coelom have greater tissue interaction during development, enabling the formation of advanced body systems.

☑ **RESEARCH Update**

What Happens If a Link in the Food Chain Is Removed?

UPDATE

The Everglades National Park in Florida is in trouble. The overall size of the park has been reduced by some 2,000 sq. mi. over the last 100 years, and the parts of the park that remain have been further sectioned, drained, flooded, or otherwise tampered with to suit the needs of a growing human population. Dr. Mark Maffei is a refuge biologist at the Everglades and points to one example of a "biological domino effect" that has occurred in the park. The snail kite is a bird that preys solely on the apple snail, a once abundant inhabitant of the Everglades. With the increased manipulation of water levels, however, the snail started to disappear. When the snail disappeared, the snail kites disappeared.

Dr. Maffei arranged for sufficient irrigation of the area to support a population of apple snails, which require 15 to 18 months of flooded conditions to survive. Once the snail population rebounded, Maffei started seeing snail kites—one nest in 1992 and four nests in 1993. At least 17 snail kites were counted in 1994, the most snail kites seen since 1974. Maffei is optimistic about the snail kite data. He feels that it represents a universal model that will demonstrate what happens when humans tinker too much with a marsh ecosystem. ☑

Section Review

1. *A dog is rolling on its back, wagging its tail. What kind of symmetry does the dog have? Name the body surface of the dog that is rolling on the ground, and identify the end with the wagging tail.*

2. *Compare and contrast the body plan of flatworms with that of roundworms.*

3. *Describe a* coelom *and identify the group of animals in which it first developed.*

4. *What are three major mollusk characteristics?*

Critical Thinking

5. *Explain why the wall of your digestive tube (intestine) contains muscle tissue but the wall of a roundworm's digestive tube does not.*

Answers to Section Review

1. The dog has bilateral symmetry. Its body surface in contact with the ground is its dorsal side, and the dog is wagging its posterior end.

2. Flatworms and roundworms both have three tissue layers, are bilaterally symmetric, and have a head region. The roundworm has a body cavity, fluids that circulate through its body, and a one-way digestive system with both a mouth and an anus. A flatworm has none of these characteristics.

3. A true coelom is a fluid-filled body cavity that develops entirely within the mesoderm. The mollusks were apparently the first group of animals to develop a coelom.

4. Three major mollusk characteristics are a visceral mass, a mantle, and a muscular foot.

5. Muscle is mesodermal in origin, so a layer of muscle surrounds the endoderm of the gut in animals with a coelom. A roundworm's endoderm is not surrounded by mesoderm, so its gut doesn't contain any muscle.

Vocabulary Preview

cerebral ganglion

larva

exoskeleton

protostome

deuterostome

endoskeleton

water vascular system

vertebrate

Opening Question

Ask students why the segmented worms are considered to be more advanced than flatworms or roundworms. The most obvious reason is that segmented worms have segmented bodies and the other kinds of worms don't. Segmented worms also have an advanced body cavity called a coelom and organs that are more complex than the organs of flatworms or roundworms.

Demonstration

Muscle Resistance

You can use balloons to simulate contracting muscles with and without internal resistance. Partially fill one long balloon with water, and tie three other long balloons loosely around the water-filled balloon. Tightly constrict the surrounding balloons, starting at one end, to simulate the fluid movement and body elongation that accompanies circular muscle contraction. Contrast this with what would happen if there was little water in the balloon.

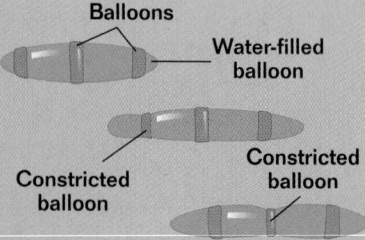

Balloons

Water-filled balloon

Constricted balloon

Constricted balloon

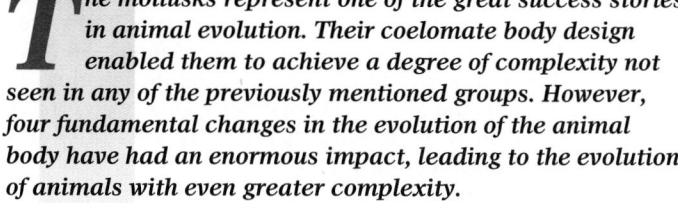

27-3 Four Innovations in Body Plan

*T**he mollusks represent one of the great success stories in animal evolution. Their coelomate body design enabled them to achieve a degree of complexity not seen in any of the previously mentioned groups. However, four fundamental changes in the evolution of the animal body have had an enormous impact, leading to the evolution of animals with even greater complexity.***

Section Objectives

- List the characteristics of an annelid.
- List the characteristics of an arthropod.
- Contrast protostome development with deuterostome development.
- Compare and contrast echinoderms with chordates.

CONTENT LINK

*More information on annelid organ systems and annelid classification can be found in **Chapter 29**.*

Figure 27-14 Both of these animals are segmented worms. The marine polychaete, *left,* is called a fireworm. If you touch it, tiny hair-like structures break off in your skin and cause excruciating pain. An earthworm, *right,* is a terrestrial segmented worm. Earthworms are important because they aerate and fertilize the soil.

Annelids: Segmented Worms

One of the early key innovations in body plans to arise among the coelomates was segmentation, the building of a body from a series of similar segments. The first segmented animals to evolve were the annelid worms. Annelids, like the polychaete (*PAHL ih keet*) shown in Figure 27-14, are composed of a chain of nearly identical segments, like the boxcars that make up a train. The great advantage of such segmentation is the evolutionary flexibility it offers. A small change in an existing segment can produce a new kind of segment with a different function. Thus, in an annelid worm, some segments are modified for reproduction, some for feeding, and others for eliminating wastes.

Two-thirds of all annelids (about 10,000 species) live in the sea, and most others (about 5,000 species) are earthworms. The basic body plan of an annelid is often described as "a tube within a tube." The gut is a tube suspended within the coelom, which is itself a tube running from mouth to anus.

All annelids share three characteristics.

1. **Repeated segments** The body segments of an annelid are visible as a series of ringlike structures running the length of the body. The segments are divided from one another internally by partitions. In each of the cylindrical

Stage 6 Segmentation

The marine polychaetes, freshwater leeches, and terrestrial earthworms are all annelids (phylum Annelida). They were the first organisms to evolve a body plan based on repeated body segments. Most segments in an annelid look alike, and each segment is separated from the next by an internal partition.

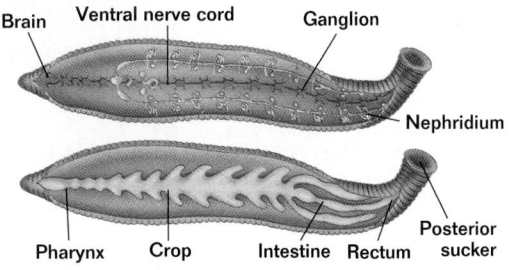

Leeches Have a Blood Diet

Many leeches are bloodsucking parasites. Like most annelids, they have a well-developed gut, *above*, that includes a muscular pharynx, a food storage area called a **crop,** and an intestine. Each segment of an annelid has a pair of excretory organs called **nephridia**, *top*, and a nerve center called a ganglion. A well-developed brain is located in an anterior segment.

segments, the digestive, excretory, and locomotor (movement) organs are repeated. The fluid within the coelom of each segment creates a hydrostatic (liquid-supported) skeleton that gives the segment rigidity. Muscles within each segment contract against the resistance offered by the fluid in the coelom. Because each segment is separate, each is able to expand or contract independently. This enables an annelid to move in ways that are quite complex. When an earthworm crawls on a flat surface, for example, it lengthens some parts of its body while shortening others.

2. **Specialized segments** The anterior (front) segments of annelids are modified and contain the sensory organs of the worm. Some of these organs are sensitive to light; elaborate eyes with lenses and retinas have evolved in some annelids. A well-developed **cerebral ganglion,** or brain, is contained in one anterior segment.

3. **Connections between segments** Although partitions separate the segments, materials and information must still pass between segments. A circulatory system carries blood from one segment to another, while a nerve cord connects the nerve centers, or ganglia, located in each segment with each other and with the brain. These nerve connections enable the brain to coordinate the worm's activities.

   Segmentation underlies the body organization of all advanced coelomate animals. Not only annelids but also arthropods (crustaceans, spiders, and insects) and chordates

Highlights
Invertebrate Evolution

Instructional Strategy

A segmented worm is separated in some ways and joined in other ways. Ask students to read this *Highlights* section, and then describe how the segments of a leech are independent from each other and how they are joined.

Discussion

Guide the discussion by posing the following questions.

1. How can you tell that a leech is segmented? *(The outside of its body looks segmented, and internally it has pairs of organs that occur in each segment.)*

2. How can you tell where the head of a leech is? *(Its brain is located in the head.)*

3. How is the body of a leech adapted to its parasitic lifestyle, compared to the body of other annelids? *(It has suckers that it uses to attach itself to its host.)*

4. Why does a leech need a well-developed brain and nerve cord? *(so that it can coordinate its body movements to crawl)*

Teaching Tip
Real Worms

Living, functioning animals are highly motivating for students studying invertebrate anatomy and physiology. Earthworms are cheap and easy to obtain in good soil or at bait shops. The transparent nature of their skin allows observation of a working circulatory and digestive system. Earthworms are easily maintained in a container of soil in a refrigerator for months, or you may wish to release them to a garden in your neighborhood. For increased transparency, you may wish to culture the worms for a few days on ordinary agar gel punched with holes. As the worms consume the transparent agar and expel opaque dirt, their internal organs become visible.

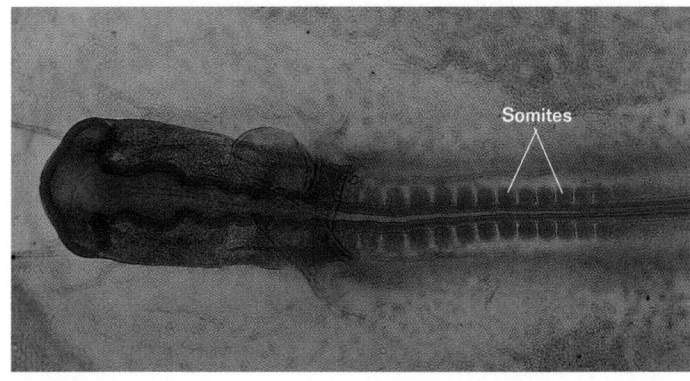

Somites

 VISUAL STRATEGY

VISUAL STRATEGY

Figures 27-15, 27-16, and 36-4

Have students look at Figure 27-15, Figure 27-16, and Figure 36-4 on page 858 (a human skeleton) to identify repeated segments in the photographs and drawing.

Teaching Tips
Worms and Bugs

Have students compare the anatomy and movement of an earthworm with that of a familiar higher arthropod, like a caterpillar or millipede. Point out that even though both animals are wormlike, their bodies are very different. Be sure students notice that both are segmented, but only the arthropod has jointed legs and a high degree of cephalization. Have the students compare how both kinds of animals eat, move, and respond to stimuli.

Arthropod Ancestors

The annelids are considered to be the ancestors of arthropods, yet adult arthropods may not resemble annelids at all. Point out that sometimes biologists have to look at the embryos or larval stages of animals to find clues about their evolutionary history. For example, caterpillars, the larvae of moths and butterflies, provide evolutionary clues not evident in the adults.

Figure 27-15 Your body's muscles formed from blocks of mesoderm tissue called somites when you were an embryo. The somites of this 33-hour-old chicken embryo are cleary visible.

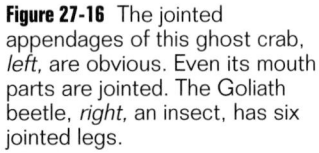

□ **CAPSULE SUMMARY**

Segmented worms have evolved complex body systems, including a nervous system and a circulatory system.

Figure 27-16 The jointed appendages of this ghost crab, *left,* are obvious. Even its mouth parts are jointed. The Goliath beetle, *right,* an insect, has six jointed legs.

(mostly vertebrates) exhibit some degree of body segmentation. Sometimes the segmentation is not obvious when viewing an adult arthropod or vertebrate. In many arthropods the segments are fused, and the underlying pattern is difficult to perceive. In the adult human, segments are not apparent, although they can be seen clearly during embryonic development. Vertebrate muscles develop from repeated blocks of tissue called somites, shown in Figure 27-15, that occur in the embryo. Another example of vertebrate segmentation is the vertebral column, which is a stack of very similar vertebrae. □

Arthropods: Walking Animals

The first animals with jointed appendages were the arthropods, representatives of which are shown in Figure 27-16. Arthropods almost certainly evolved from the annelids. Arthropod bodies are segmented like those of annelids. Individual segments of arthropods often exist only during their larval stages, however. A **larva** is an immature organism that has a body form different from that of the adult. In the adult arthropod, the segments are fused into functional units. For example, the

 Did You Know?

Using ancient Chinese, German, and Russian texts, Conrad C. Labandeira, paleoentomologist at the National Museum of Natural History of the Smithsonian Institution, and his colleague J. John Sepkoski Jr. of the University of Chicago, have studied the fossils of over 1,263 insect families since the 1980s. What did they conclude after all this research? Insects rarely become extinct even when large masses of other groups of animals perish.

Highlights
Invertebrate Evolution

Invention of Jointed Appendages

Stage 7 Jointed Appendages

Arthropods (phylum Arthropoda) have a coelom, segmented bodies, and **jointed appendages.** Each body region of an arthropod (head, thorax, and abdomen) is composed of a number of individual segments that fused during development. All arthropods have a strong **exoskeleton** made of **chitin.** Only one class of arthropods, the insects, has evolved wings, which permit them to fly rapidly through the air.

What an Aphid Left Behind

As arthropods grow, they shed their old exoskeletons in a process called **molting.** Here, insects called aphids voraciously feed on a plant. The white, paperlike objects are exoskeletons that the aphids have recently shed.

A Hovering Dragonfly

The jointed legs of insects attach to the central body region called the thorax. Insects have three pairs of legs and usually two pairs of wings (some insects, like flies, have retained only one wing pair).

larva of a butterfly, called a caterpillar, has many segments, while the adult butterfly has only three functional body units—a head, thorax, and abdomen. Each unit is composed of several fused segments. Having a head, thorax, and abdomen is characteristic of all adult insects.

All arthropods have jointed appendages. The name *arthropod* is from the Greek *arthron*, meaning "joint," and *podos*, meaning "foot." To gain some idea of the importance of jointed appendages, imagine yourself without them. Without jointed appendages—hips, knees, ankles, shoulders, elbows, wrists, knuckles—you could not walk or grasp any object. Arthropods use jointed appendages as legs for walking, as wings for flying, as antennae to sense their environment, and as mouthparts for sucking, ripping, and chewing their food. The scorpion, shown in Figure 27-17, seizes and tears apart its prey with jointed mouthparts that have been modified into large pincers. Jointed appendages have proven to be very successful—more than half of all named species on Earth are arthropods. Scientists estimate that 10^{18} insects are alive at any one moment!

CONTENT LINK

Learn more about arthropod body plans and classifications in **Chapter 30**.

Figure 27-17
The stinger of this desert hairy scorpion from Arizona is found at the tip of its tail. Scorpions, along with spiders and mites, belong to a group of arthropods called arachnids.

Multicultural Perspective

Native American Mythology

Beetles of the genus *Eleodes* have an odd behavior. When the beetle is threatened, it will lower its head to the ground and release a foul-smelling fluid. The Cochiti of the Southwestern United States have incorporated the beetle's behavior into one of their creation stories. The story states that long ago the beetle was told to place the stars in the sky. But the beetle became careless and dropped the stars, causing them to scatter. As a result, the Milky Way was formed. To this day the beetle is so ashamed of the mishap, it lowers its head in disgrace whenever someone approaches.

Highlights
Invertebrate Evolution

Instructional Strategy

Have students study the photos of the aphids and the dragonfly. Ask students to identify the head, thorax, and abdomen of these insects, as well as the number of legs they have. Then have them look at the photos of the other arthropods on page 626 to find another insect with the same body parts and number of legs. *(goliath beetle)*

Discussion

Guide the discussion by posing the following questions.

1. Why does an insect have to shed its exoskeleton as it grows? *(because the exoskeleton is made of nonliving chitin, which does not grow)*

2. Why don't dragonflies grow to the size of chickens? *(To support that much body weight, their exoskeleton would have to be unrealistically thick.)*

3. How do aphids and dragonflies use their jointed legs? *(They use them to walk.)*

Teaching Tip
Joint Appreciation

Have students try to walk, pick up objects, and sit down without bending their elbows, knees, fingers, wrists, or ankles. Then have them perform the same activities using their joints. Have them closely observe how they use their muscles and joints to coordinate their actions.

Another feature of the arthropod body plan is an **exoskeleton,** a rigid external shell made of a substance called chitin. In any animal, a skeleton provides places for muscle attachment. In arthropods, the muscles attach to the interior surfaces of their exoskeletons. An exoskeleton also protects an arthropod from predators and impedes water loss.

The arthropod exoskeleton does have one great limitation. Though chitin is tough, it is also brittle and cannot support great weight. To bear the pull of their more powerful muscles, the exoskeletons of large arthropods must be much thicker than the exoskeletons of small arthropods. The reason why you don't see beetles as big as birds or crabs the size of cows is that their exoskeletons would have to be so thick that the arthropod couldn't move its great weight. Because this size limitation is inherent in the body design of arthropods, no arthropods have ever grown to a great size. ▢

Echinoderms: Animals With a Five-Part Body Plan

Figure 27-18 Animal embryos develop in one of two ways. In protostome embryo development, cell divisions occur in a spiral pattern. The mouth of the animal forms from the blastopore, and the anus forms from a secondary opening, *left.* In deuterostome embryo development, cell divisions occur in a radial pattern. The anus forms from the blastopore, and the mouth forms from the secondary opening, *right.*

There are two major patterns of embryological development that occur among coelomate animals. The embryos of these animals begin as hollow balls of cells, as shown in Figure 27-18. The balls of cells then indent to form a ball two cell layers thick with an opening called a blastopore to the outside. In annelids, mollusks, and arthropods, the mouth develops from or near the blastopore. This same pattern of development is seen in all acoelomate animals. An animal whose mouth develops from the blastopore is called a **protostome.** The term *protostome* is from the Greek *protos,* meaning "first," and *stoma,* meaning "mouth."

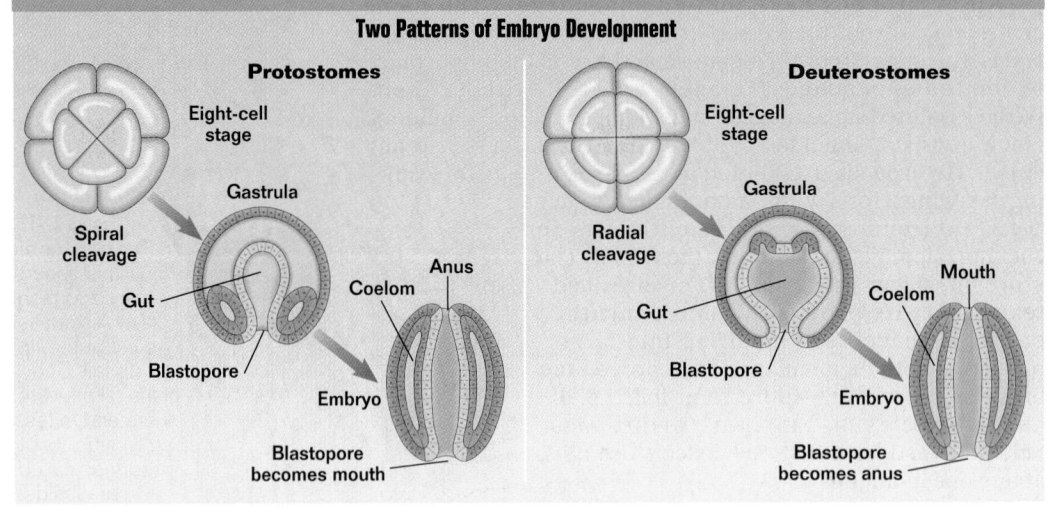

Two Patterns of Embryo Development

Protostomes

Eight-cell stage
Gastrula
Spiral cleavage
Gut
Blastopore
Coelom
Anus
Embryo
Blastopore becomes mouth

Deuterostomes

Eight-cell stage
Gastrula
Radial cleavage
Gut
Blastopore
Coelom
Mouth
Embryo
Blastopore becomes anus

A second kind of coelomate animal developed after the arthropods evolved. In these animals the basic pattern of embryological development differs from the pattern seen in protostome development. In the echinoderms, chordates, and a few other small phyla, the anus—not the mouth—develops from or near the blastopore. The mouth forms later, on another part of the embryo. These animals are called **deuterostomes** because the mouth forms second, from an opening other than the blastopore. The term *deuterostome* is from the Greek *deuteros,* meaning "second," and *stoma,* meaning "mouth." Echinoderms and chordates are clearly related to each other, as demonstrated by their shared pattern of embryonic development.

Deuterostomes represent a revolution in embryonic development. While the pattern of cell division in protostomes is spiral, it is radial in deuterostomes, as shown in Figure 27-18. Furthermore, in protostomes, the developmental fate of each embryonic cell is fixed, meaning that each cell is destined to develop into a particular part of the animal and no other. Because the chemicals that act as developmental signals to the cells are localized within the egg, there is no such developmental control programmed into each individual cell of the protostome embryo. However, if the cells of a deuterostome embryo are separated from each other at an early stage, each cell can develop into a complete organism. This is because the chemical developmental signals in deuterostome embryos are generated by the chromosomes contained in each individual cell. In protostomes, cell position is the key to the developmental fate of embryo cells, while in deuterostomes, whole groups of cells move around during the course of development to form new tissue associations. ▢

The first deuterostomes, marine animals called echinoderms, evolved more than 650 million years ago. Echinoderms, shown in Figure 27-19, were the first animals to develop an **endoskeleton,** an internal skeleton. The term

▢ *CAPSULE SUMMARY*

Because echinoderms and chordates have similar patterns of embryological development, it is likely that they share a common ancestor.

Figure 27-19 The vermillion biscuit sea star, *left,* and purple sea urchin, *right,* are echinoderms. Most sea stars have five arms, but there are some species that have many more. The long, movable spines of the sea urchin are used to push the animal along the ocean bottom.

SECTION 27-3

Teaching Tip
Predetermination
Emphasize the fixed developmental fate of every cell in a protostome embryo, compared to the flexible developmental fate of the cells in a deuterostome embryo. Tell students that developmental biologists sometimes separate the cells of early deuterostome embryos so that they can obtain new individuals that are genetic duplicates.

Application
℞ **Medicine** The importance of chemical signals in the development of deuterostomes has important medical implications for the regeneration of lost body parts. Scientists are working on discovering factors that cause cells to change and grow so that someday it might be possible to grow back a severed spinal cord or an amputated limb.

Historical Note

Embryo Development
A German surgeon named Kaspar Friedrich Wolff (1733–1794) is regarded as the "father" of embryology. Wolff introduced the idea that the cells of a plant or animal embryo initially are undefined but later they differentiate to form tissues, organs, and organ systems. Prior to Wolff's theory, it was generally thought that a living organism developed from an exact miniature of the adult that resided in the sperm.

Instructional Strategy

Ask students to explain the evolutionary link between sea stars and chordates. Point out that although the relationship between adult echinoderms and chordates may not be apparent, the larvae/embryos of both groups have certain developmental stages in common. Echinoderms not only share our deuterostome pattern of embryo development, but also have larvae with bilateral symmetry and adults with endoskeletons.

Discussion

Guide the discussion by posing the following questions.

1. What makes a tube foot extend? *(The water-filled ampulla contracts.)*

2. What makes it withdraw? *(The ampulla relaxes and refills with water.)*

3. Why can't sea stars live in terrestrial environments? *(They wouldn't be able to move about, since they need water to operate their water vascular system.)*

4. Why doesn't a sea star have a head? *(Since there is no brain, there has been no cephalization in these animals.)*

Highlights
Invertebrate Evolution

A New Pattern of Embryo Development

Stage 8 Deuterostomes

Echinoderms (phylum Echinodermata), like sea stars, are coelomates that have a **deuterostome pattern of embryo development**. This same pattern of development occurs in the chordates. A delicate skin stretches over the internal endoskeleton of echinoderms. Their **endoskeleton** is composed of calcium-rich plates that often are fused together.

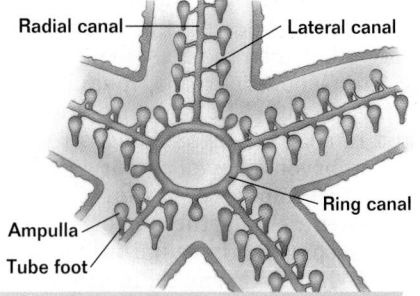

How a Sea Star Crawls

Sea stars crawl by using thousands of tiny tube feet and a hydraulic **water vascular system** made up of water-filled canals. Each tube foot has a water-filled sac at its base called an ampulla. When the sac contracts, the tube foot is extended in the same way that a water-filled balloon extends when it is squeezed. Hundreds of tube feet extend from the bottom of each arm and attach to the sea floor. The sea star's body muscles can then pull against them, enabling the sea star to haul itself along.

CONTENT LINK

*More information about echinoderm structure and classification can be found in **Chapter 30**.*

☐ CAPSULE SUMMARY

The endoskeleton first appeared in the echinoderms. Radial symmetry, a simplified nervous system, and a water vascular system characterize these marine invertebrates.

echinoderm is from the Greek *echinos*, meaning "spiny," and *derma*, meaning "skin." This "spiny skin" refers to the echinoderm's endoskeleton, which is composed of hard calcium-rich plates called ossicles that are just beneath the delicate skin. When ossicles first form, they are enclosed in living tissue and so are truly an endoskeleton. In adults, the ossicles fuse, forming a hard shell. Today there are about 6,000 echinoderm species, almost all of which live on the ocean bottom. Many of the most familiar animals seen along the seashore are echinoderms, including sea stars (starfish), sea urchins, sand dollars, and sea cucumbers.

The body plan of echinoderms undergoes a fundamental shift during development. All echinoderms are bilaterally symmetric as larvae, but they become radially symmetric as adults. Adult echinoderms have a five-part body plan, easily seen in the five arms of a sea star. The nervous system consists of a central ring of nerves from which five branches arise. While the animal is capable of complex response patterns, there is no centralization of nerve function—no "brain." Apparently, the development of a central nervous system is not feasible in animals with radial symmetry.

A key evolutionary innovation of echinoderms is the development of a hydraulic system to aid movement. Called a **water vascular system,** this fluid-filled system is made up of interconnected internal canals and thousands of tiny hollow tube feet. ☐

? Did You Know?

The tube feet of sea stars appear delicate, and they actually can be individually pulled off an object they are attached to, but when they are all pulling together, a sea star can exert tremendous force—enough to pull apart the valves of a clam.

People who shell clams have to use a knife to cut the muscles of the clam to open up its valves.

Highlights
Invertebrate Evolution

Improving the Skeleton

Stage 9 Notochord

Tunicates, lancelets, and all the vertebrates are chordates. Chordates (phylum Chordata) are coelomate animals that have a flexible dorsal rod called a **notochord**. Chordates also have **pharyngeal slits** and a **dorsal hollow nerve cord**. In the vertebrates, the notochord is replaced during embryonic development by a vertebral column.

Seeing Through a Lancelet
Unlike the skin of other vertebrates, the skin of a lancelet is transparent.

How Lancelets Move

In a lancelet, a simple chordate, the flexible notochord persists throughout life. It aids swimming by giving the lancelet's muscles an anchor point to pull against. These muscles form a series of discrete blocks that can easily be seen.

How Lancelets Feed

Lancelets feed on microscopic protists that are caught as they filter through cilia and gills on the pharyngeal slits. The beating of cilia that line the front end of the digestive tract draws water, *blue arrow*, through the mouth, through the pharynx, and out the slits.

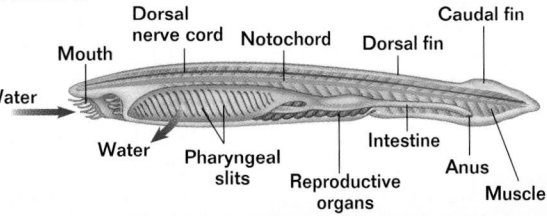

Chordates: Animals With a Notochord

Functionally, the endoskeleton of echinoderms is similar to the exoskeleton of arthropods. It is a hard shell that encases the body and serves as an attachment site for muscles. Members of the second major group of deuterostomes, the chordates, have a very different kind of endoskeleton. The chordate endoskeleton is completely internal and is characterized by a stiff rod that develops along the back of the organism during embryological development. Muscles attached to this rod allowed early chordates to swing their backs from side to side, enabling them to swim through the water. The attachment of muscles to an internal skeleton was an important evolutionary advancement. Endoskeletons were the first step along an evolutionary path that led to the vertebrates, and they made it possible for animals to grow large and move quickly.

The approximately 42,500 species of chordates share three characteristics.

1. **Nerve cord** Chordates have a single, hollow, dorsal nerve cord. Nerves attached to the nerve cord travel to different parts of the body.
2. **Notochord** Chordates have a long, stiff rod called a notochord that forms beneath the nerve cord. The notochord is positioned between the nerve cord and the developing gut in the early embryo.

> CONTENT LINK
>
> *More detail about different kinds of invertebrate chordates is found in* **Chapter 30**.

Teaching Tips

Demonstrate Your Own Vertebral Column

Emphasize the combination of strength and flexibility that a vertebral column gives by bending in different directions while explaining to your students that a backbone was an innovation that made it possible for large animals to move successfully on land.

Change the Rules

Have your students pretend that they wake up one morning with no bones. What would they look like? Would they be able to get out of bed? Eat? Watch TV? Encourage students to be creative yet logical as they speculate on the consequences of a lack of bones in a vertebrate.

Back to Basics

This is a good time to review the basic characteristics of animals. Have students relate those characteristics to the animals they have studied in this chapter.

Chapter Closure

Ask students to list the nine phyla of animals that were discussed in the chapter. Next to each phylum name they should write two distinguishing characteristics common to all the animals contained in that phylum.

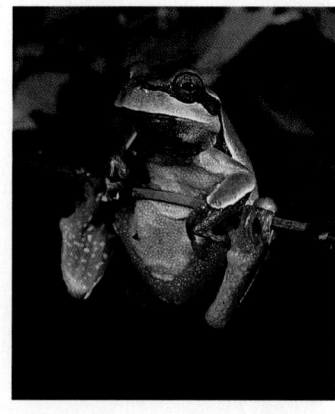

Figure 27-20 This pine-barrens tree frog is found in New Jersey. Frogs belong to a group of vertebrates called amphibians. Amphibians were the first vertebrates to live successfully on land.

❏ CAPSULE SUMMARY

All chordates develop a hollow dorsal nerve cord, notochord, and pharyngeal slits sometime during their lifetime. Vertebrates are chordates that have a cartilaginous or bony endoskeleton that includes a vertebral column.

3. Pharyngeal slits Chordates have a series of slits that develop in the wall of their pharynx. The pharynx, located behind the mouth, is a muscular tube that connects the mouth to the digestive tract and windpipe.

All chordates have all three of these characteristics at some time in their life. For example, humans have a nerve cord, a notochord, and pharyngeal slits as embryos. As adults, humans retain only the nerve cord and the remnants of one pair of pharyngeal slits, which exist as the eustachian tubes that connect the throat to the middle ear.

With the exception of a group of animals called tunicates, which live attached to the ocean bottom, and a small group of fishlike marine animals called lancelets, all chordates are vertebrates. **Vertebrates,** like the frog shown in Figure 27-20, are chordates that have two important evolutionary advancements, a vertebral column and a head.

1. Vertebral column During embryological development, the notochord becomes replaced by a hollow, bony vertebral column, also called a backbone. The vertebral column is composed of many hollow bones called vertebrae that are stacked on top of one another. The vertebral column surrounds and protects the dorsal nerve cord.

2. Head In all vertebrates except the earliest fishes, there is a distinct and well-differentiated head containing a skull and a brain. For this reason, the vertebrates are sometimes called the craniate chordates. The term *craniate* is from the Greek *kranion,* meaning "skull."

All vertebrates have an internal skeleton made of bone or cartilage. The vertebrate skeleton is a solid framework against which attached muscles can contract. A strong supporting endoskeleton makes possible the great size and extraordinary powers of movement that characterize the vertebrates. ❏

Section Review

1. *Identify and explain the significance of a major evolutionary development that first evolved in annelids.*
2. *What limits the maximum size of an arthropod?*
3. *State the evidence that enables scientists to conclude that you are more closely related to echinoderms than to arthropods.*

Critical Thinking
4. *The* law of recapitulation *was proposed by the German scientist Ernst Haeckel in 1866. It states that a species' evolutionary history can be deciphered by studying the stages of its embryological development. Describe evidence that supports this law.*

Answers to Section Review

1. The annelids were the first animals to show segmentation, which gives anatomic flexibility. A small change in an existing segment can produce a new specialized segment for a different function.

2. The arthropod's exoskeleton limits its size. For example, the thickness and weight needed to support an arthropod the size of human would make it too heavy to move in terrestrial environments.

3. Humans and echinoderms share the same deuterostome pattern of embryological development. Arthropods exhibit a protostome pattern of embryological development.

4. Humans have a notochord and pharyngeal slits during development, similar to structures of ancient chordates.

Vocabulary

acoelomate (618)
anterior (617)
bilaterally symmetric (617)
cephalization (617)
cerebral ganglion (625)
choanocyte (613)
circulatory system (621)
cnidocyte (614)
coelom (621)
deuterostome (629)
dorsal (617)
ectoderm (614)

endoderm (614)
endoskeleton (629)
eumetazoan (614)
exoskeleton (628)
gill (623)
larva (626)
mantle (623)
medusa (616)
mesoderm (614)
mollusk (622)
nematocyst (614)
organ (618)

polyp (616)
posterior (617)
primary induction (621)
protostome (628)
pseudocoelom (619)
radially symmetric (614)
radula (623)
symmetry (612)
ventral (617)
vertebrate (632)
visceral mass (623)
water vascular system (630)

CHAPTER REVIEW ANSWERS

Each item in the Chapter Review is correlated by text section in the assignment guide that follows.

ASSIGNMENT GUIDE

Section	Review	Themes Review	Critical Thinking
27-1	1–3, 11, 12, 18		33
27-2	4, 11, 13–15, 19, 20	30–32	34, 35
27-3	6–10, 16, 17, 21		36

Concept Mapping

Construct a concept map that shows the similarities and differences among flatworms, roundworms, and segmented worms. Use as many terms as needed from the vocabulary list. Try to include the following items in your map: bilateral symmetry, ectoderm, mesoderm, endoderm, acoelomates, pseudocoelomates, and coelomates. Include additional terms in your map as needed.

Review

Multiple Choice

1. Animals differ from other eukaryotes because they typically
   a. reproduce sexually.
   b. are organized into tissues.
   c. use muscle cells to move.
   d. are heterotrophs.

2. Which one of the following animals exhibits radial symmetry?
   a. dog   c. giant squid
   b. sponge   d. sea anemone

3. From what kind of embryonic tissue do adult digestive organs develop?
   a. ectoderm   c. protostome
   b. endoderm   d. deuterostome

4. Which of the following is a coelomate?
   a. octopus   c. sponge
   b. nematode   d. jellyfish

5. Flatworms and roundworms are different because flatworms
   a. are bilaterally symmetric.
   b. have a gut with one opening.
   c. may be parasitic.
   d. exhibit cephalization.

6. What specialized body parts evolved after animals developed segmentation?
   a. tentacles   c. wings
   b. flagella   d. blood vessels

7. Which is *not* a characteristic of annelids?
   a. jointed appendages
   b. repeated segmentation
   c. a coelomate body plan
   d. specialized segments

8. Which is a feature of the arthropod body plan?
   a. a hydrostatic support system
   b. pharyngeal slits
   c. an exoskeleton
   d. a nonsegmented body

9. Which is *not* a deuterostome and does not have an exoskeleton?
   a. a butterfly
   b. an earthworm
   c. a crab
   d. a starfish

Concept Mapping

Map answer is shown on page 609B.

Review

Multiple Choice

Code in parentheses indicates section number and objective number.

1. c (27-1.1)	**6.** c (27-3.1)
2. d (27-1.2)	**7.** a (27-3.2)
3. b (27-1)	**8.** c (27-3.2)
4. a (27-2.2)	**9.** b (27-3.1)
5. b (27-2.3)	**10.** b (27-3.4)

Completion

11. bilateral, radial (27-1.2, 27-2.1)
12. eumetazoans, tissues (27-1.3)
13. Platyhelminthes, one opening (27-2.3)
14. Nematoda, circulatory (27-2.3)
15. foot, visceral mass, mantle (27-2.4)
16. annelid, anterior (27-3.1)
17. internal, bilateral, radial (27-3.4)

Short Answer

18. Medusae are free-floating, gelatinous, and often umbrella shaped. Polyps are cylindrical, pipe-shaped animals that are usually attached to rocks. (27-1.3)

CHAPTER REVIEW ANSWERS *continued*

Short Answer *continued*

19. Cephalization is the process by which animals evolve a definite head end. Having a head, in which sense organs are centralized, enables an animal to sense what is in a new environment before entering it. (27-2.1)

20. Primary induction is the interaction of one primary tissue with another by physical contact. Contact between the mesoderm and endoderm results in the development of the stomach. (27-2.2, 27-2.4)

21. No. Like humans, the embryos of cats have a notochord. (27-3.4)

Highlights Review

22. The choanocytes of sponges closely resemble protists called choanoflagellates. (27-1)

23. Brushing up against the tentacles would cause a barrage of nematocyst barbs to penetrate my body. The nerves associated with the cnidocytes would signal the tentacles to move me toward the mouth. Once inside the gut, digestive enzymes would destroy my body. (27-1)

24. The digestive tract of a flatworm has many branches, thus bringing nutrients close to all the flatworm's body cells. (27-2)

25. Earthworms are segmented, coelomate worms that live in soil. Roundworms are unsegmented, pseudocoelomate worms, many of which also live in soil. (27-2, 27-3)

26. The mantle cavity contains the gills, which are used to remove oxygen from the water. The mantle cavity also is the dumping site for wastes before they exit the mollusk's body. (27-2)

27. Similar to annelids, arthropods have a coelom and a body divided into segments. (27-3)

28. The sea star would attach its many tube feet to both shells of the clam. Then the sea star would use the force generated by its body mus-

---

10. Which of the following do adult chordates and adult echinoderms have in common?
   a. a nonsegmented body
   b. an internal skeleton
   c. a water vascular system
   d. bilateral symmetry

Completion

11. Crabs exhibit _____ symmetry, and hydras exhibit _____ symmetry.

12. Unlike sponges, cnidarians are _____, or true animals. This means that cnidarians have body symmetry and _____.

13. Flukes and tapeworms are members of the phylum _____. These animals have a digestive cavity with _____ for taking in food and eliminating waste.

14. The first animals to have body cavities were members of phylum _____. Fluids moving in the body cavity function as a _____ system to transport nutrients, oxygen, and waste.

15. All mollusk bodies contain a muscular _____ that may be used for movement, a _____ that contains internal organs, and a fold of tissue called the _____ that may secrete a shell.

16. An earthworm is a terrestrial segmented worm, or _____. Its _____ segments are modified and contain cerebral ganglia and other sensory organs.

17. A starfish has an _____ skeleton. It exhibits _____ symmetry as a larva and _____ symmetry as an adult.

Short Answer

18. Name and describe the two body forms of cnidarians.

19. What is cephalization? What is the main advantage of cephalization?

20. What is primary induction? Give an example of the outcome of primary induction.

21. Adult cats do not have a notochord. Does this mean that cats are not chordates?

Highlights Review

22. What is the evidence that suggests sponges evolved from protists?

23. Suppose you were a small floating animal that happened to brush against the tentacles of a hydra. Describe the action of the hydra that would result in your being pulled through its mouth and into its gut.

24. In animals, nutrients are transferred from the digestive tract to body cells. How does the digestive tract of a flatworm facilitate this process?

25. Compare the body plans and habitats of earthworms with those of roundworms.

26. Why might the mantle cavity of a snail be considered a "multifunctional" structure?

27. What is the evidence that suggests annelids are the ancestors of arthropods?

28. How would a sea star use its water vascular system to pry open the valves of a clam?

29. How is the food gathering mechanism used by lancelets similar to that used by sponges? How is it different?

Themes Review

30. Evolution What evolutionary trend do the body cavities of flatworms, roundworms, and segmented worms show?

31. Levels of Organization Describe the relationship between cells, tissues, and organs using an example from the basic flatworm body plan.

32. Structure and Function What is a radula? How do gastropods use the radula to obtain food?

Critical Thinking

33. Making Inferences How is bilateral symmetry better suited to terrestrial animals than radial symmetry?

---

cles to pull the shells apart. (27-3)

29. Both lancelets and sponges are filter feeders that remove tiny food particles from water. The filter-feeding mechanism used by the two animals is different, in that the lancelets filter the protists through cilia and gills located on their pharyngeal slits. Sponges filter food parti-

cles through the collars of choanocytes that line their body cavity. (27-3)

Themes Review

30. Increasing complexity, from a solid body construction, to a body cavity located between the endoderm and mesoderm, and finally to a fluid-filled body cavity

that develops entirely within the mesoderm. (27-2.2)

31. The eyespot on the head of a planarian is an organ that is composed of a collection of different tissues. Each of the tissues is made of different kinds of cells. (27-2.3)

32. The radula is a rasping, tongue-like organ that appears

34. Communicating Effectively The diagram below shows a cross section of a bilaterally symmetric animal. Name the primary germ layers and structures indicated by the lines labeled *a* through *d*. Is the animal an acoelomate, a pseudocoelomate, or a coelomate? Name the phylum whose representatives have this body plan.

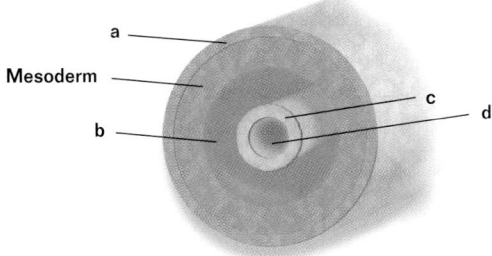

a

Mesoderm

b

c

d

35. Interpreting Data The shell pictured to the right was found on a beach. To which class of mollusks did the organism that occupied the shell belong? What is the probable cause of the organism's death?

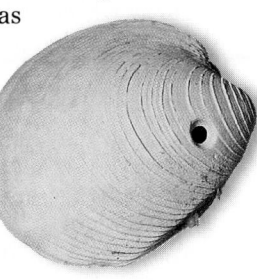

36. Responding Critically A boy who was hospitalized after being stung by bees was interviewed on the nightly news. During the interview, the boy said that the bees were as large as sparrows and seemed to fly at speeds nearing 30 mph. Do you think the boy's story is accurate or exaggerated? Explain.

Activities and Projects

37. Research and Writing Find out about vinegar eels and the conditions under which they thrive. Write a report detailing how to maintain a culture of vinegar eels and describing what we can learn from studying them.

38. Multicultural Perspective Working in groups of three to four students, give a class presentation on the ceremonial use of invertebrates by the indigenous people of another culture. You might consider types of musical instruments, art forms, diet, oral stories, and seasonal celebrations passed down over many generations.

39. Cooperative Group Project Collect clam chowder recipes from different cookbooks. After studying several of the recipes, write your own recipe for clam chowder. Then prepare a pot of clam chowder using your recipe. Allow members of the class to taste and critique your chowder. Different group members may take responsibility for collecting recipes, writing the group's recipe, preparing the chowder, and arranging the chowder tasting.

40. Research and Writing Use library resources and interview medical professionals to identify the species of the "medicinal leech." Find out how it was used in medicine during the eighteenth and nineteenth centuries and how it is used today. In your report, describe why medical professionals stopped using the leech at the turn of the century and why they have resumed medical use of the leech.

Readings

41. Read "Luck Be a Ladybug," in *National Wildlife*, June–July 1994, pages 30–32. Why are "bug-pickers" traveling to the foothills of California's Sierra range each fall season?

42. Read "Dusk and Dawn Are Rush Hours on the Coral Reef," in *Smithsonian*, October 1993, pages 104–115. Describe the daily activity of three different invertebrates that live in the reef along the Gulf of Aqaba.

Activities and Projects

37. Vinegar eels are nonparasitic roundworms that feed on fungi growing in nonpasteurized vinegar. They may be cultured in pure cider vinegar in which small cubes of raw apple have been placed. Vinegar eels can be used to study roundworm anatomy and general characteristics.

38. Answers will vary. Suggest that students use the *Cultural Atlas* series published by Facts on File in addition to the standard encyclopedias and reference books in the school library.

39. Frozen clams can be found in most grocery stores. If fresh clams are used, insist that students read instructions about how to prepare clams for cooking that can be found in most seafood cookbooks.

40. The medicinal leech, *Hirudo medicinalis*, was used for bloodletting because it was believed that the cause of fevers and other medical problems was the accumulation of excess blood in the body. Today, the leech is used to aid the reattachment of severed fingers, toes, and other body parts. Unlike veins, arteries can be reconnected surgically. Blood that would accumulate in the digit is removed by the leech until the veins grow back into the digit.

Readings

41. They are collecting *Hippodamia convergens,* a species of ladybird beetle that is sold to gardeners. The beetles can bring the collectors several hundred dollars for a day's work.

42. Almost all of the invertebrates that live on the reef hide from predators during the day. For example, corals retract their tentacles and bodies deep into their limestone skeletons, and sea lilies and sea snails hide in the cracks and crevices of the reef. At night, the corals inflate their bodies with water and extend their tentacles, the sea snails come out and graze, and the sea lilies crawl atop coral branches to feed.

---

to have rows of pointed, backward-curving teeth. It is used by gastropods to scrape algae or to puncture prey. (27-2.4)

Critical Thinking

33. Radial symmetry is beneficial for sessile animals or animals that float freely in water because food may come from any direction. Bilateral symmetry is better

suited to terrestrial animals that must forage or hunt for food. (27-1.2, 29-2.1)

34. (a) ectoderm, (b) pseudocoelom, (c) endoderm, (d) cavity of digestive tract; pseudocoelomate; Nematoda (27-2.2)

35. The shell is that of a bivalve. The hole in the shell suggests that the organism was eaten by a gastropod. (27-2.4)

36. Bees could not be as large or fly as fast as the boy reported. The thick exoskeleton of a bee the size of sparrow would make flight at any speed impossible. (27-3.2)

Preparation

If you use specimens that have been preserved in formalin, be sure to conduct the lab in a well-ventilated room. If live specimens are used, remind students of the proper care and treatment of live animals. Whether live or preserved animals are used, remind students to wash their hands after working with specimens. Release only native animals into your local environment.

Procedural Notes

- Students should work in groups of 2–4 and rotate in clockwise order as they move from station to station.
- Have students set up the **Records** section of their lab report by numbering before starting the lab. This will make it easier for students to answer the questions in the stations since not all students will be starting with Station 1.
- **Safety:** Students need to use extreme care when observing the preserved specimens. The preservative can leak if the jars are tilted. Since some animals may be preserved in glass jars, students should be careful not to break the jars. Be sure to tell students to wash their hands if they should come into contact with the preservative. Also remind students to wash hands before leaving the laboratory.

Background Answers

1. polyp, medusa

2. cephalization

3. visceral mass, mantle, and foot

4. head, thorax, and abdomen

5. water vascular system

6. What are some similarities and differences among invertebrates?

Survey of Invertebrates

OBJECTIVES

- Compare similarities and differences among invertebrates.
- Relate structural adaptations to the evolution of invertebrates.

PROCESS SKILLS

- classifying invertebrates
- relating structural features to functions
- analyzing data for trends from simple to complex development

MATERIALS

- preserved or living specimens of invertebrate animals
- blunt probes
- hand lens or stereomicroscope

BACKGROUND

1. What type of body form does a sea anemone have? a jellyfish?
2. What term describes concentration of nervous tissue at the anterior end of an animal's body?
3. What are the three body regions of a mollusk?
4. Name the three body regions in adult insects.
5. What is the name of the system of movement in echinoderms?
6. Write your own question to explore in your lab report or notebook.

TECHNIQUE

 You will travel to seven stations to observe organisms that are representative of the invertebrate groups, or phyla. In the **Records**

section of your report, construct and complete a table. The headings are: Group Name, Phylum Name, Type of Symmetry, Type of Body Plan, Distinguishing Characteristics, and Animal Examples. Under "Group Name" list the name from each "Station" number. Record the answers to the following questions in the **Records** section of your report.

Station 1: Sponges

1. Describe the shape of the sponge. Why is the skeleton filled with holes or pores?
2. Since a sponge is sessile or permanently attached, how does it obtain food?

Station 2: Cnidarians

3. What body form do hydra and coral exhibit?
4. How do they paralyze their prey?
5. These organisms have a medusa body form which allows them to be free-floaters. What part of their body enables them to float?

Station 3: Worms

6. What is the shape of the planarian and sheep liver fluke? Relate this shape to their method of obtaining and distributing oxygen.
7. To what group of invertebrates do these animals belong?
8. How many body openings do these animals have?
9. Describe the outside and the shape of the body of *Ascaris*.

Station 4: Mollusks

10. Of the snail, clam, and squid, which have a visible shell?
11. How do each of these animals obtain food?

Station 5: Segmented Worms

12. Observe and compare the earthworm and the leech. What features do they have in common? To what group of worms do these two specimens belong?

Records

Students should construct and complete a table as described under Technique.

Station 1: Sponges

1. No shape; to allow water carrying food to enter

2. The water current created by choanocytes brings in food. The food lands on the collars of the choanocytes and is then digested.

Station 2: Cnidarians

3. polyp

4. They have cells called cnidocytes. They shoot out a harpoon called a nematocyst that paralyzes the prey.

5. the top part called the bell

Station 3: Worms

6. flat and thin; oxygen diffuses from the outside environment,

through their skin, and into all their cells

7. flatworms or Platyhelminthes

8. one

9. They have a smooth, tough surface and are round.

Station 4: Mollusks

10. the snail and clam

11. The snail and squid use a tongue-like structure called a radula. The clam gets food by

Station 6: Arthropods

13. Observe and compare arthropod specimens such as crayfish, pill bug, grasshopper, scorpion, millipede, and centipede. Describe the outer covering of these animals. Are the appendages straight and stiff or are they jointed?

14. Look closely at the antennae of the grasshopper and crayfish. Speculate about their function.

15. Compare and contrast the walking legs of the millipede and the centipede.

16. Describe how each of the arthropod species shown at this station obtains oxygen.

Station 7: Echinoderms

17. Observe the sea star (starfish), sand dollar, and sea urchin specimens and compare the spiny skin of these specimens.

18. Describe the symmetry of echinoderms.

19. These animals move by a water vascular system. Look at the ventral surface of the sea star, and sketch the tube feet that make up this system.

20. Clean up your materials and wash your hands before leaving the laboratory.

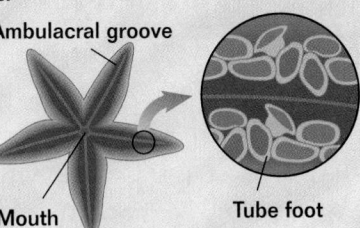

19.

Ambulacral groove

Mouth

Tube foot

Inquiry Answers

1. No. Since cnidarians lack a centralized collection of nerve cells called a brain, there has not been the evolution of a head region in these animals.

2. Advantages are they are attachment sites for muscles, impede water loss, and protect. A disadvantage is that they are brittle and can't support much weight.

3. An endoskeleton gives more support to an animal's body than an exoskeleton and acts as a better anchor site for the attachment of muscles.

Analysis Answers

1. This covering enables *Ascaris* to live inside its host and withstand the chemical climate in which it lives.

2. They are more developed, with a higher degree of specialization. Segmented body regions are apparent in the more advanced animals such as the chordates.

3. They are protected by an exoskeleton and have well-developed sense organs. All are equipped with jointed appendages used for walking, and many can fly.

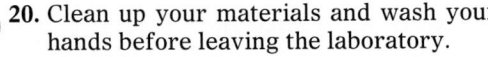

using a siphon to suck it into its mantle cavity.

Station 5: Segmented worms

12. They have body segments. segmented worms or Annelida

Station 6: Arthropods

13. They have a tough exoskeleton of chiton, which appears brittle. They are jointed.

14. Antennae sense objects in the environment.

15. Millipedes and centipedes have jointed legs, but millipede legs appear to be smaller than centipede legs. Millipedes have two pairs of legs per body segment, and centipedes have one pair of legs per body segment.

16. The crayfish and pill bug breathe with gills. The scorpion breathes with book lungs. All the rest breathe with spiracles and tracheae.

Station 7: Echinoderms

17. The starfish looks rough with raised dots on its body; the sand dollar appears velvety; and the sea urchin has sharp spines that radiate out from its body.

18. Echinoderm larvae are bilaterally symmetric, and the adults have radial symmetry and a five part body plan.

Block Scheduling Guide

Each block represents about 45–50 minutes of instructional time. The resources cited in a block are coded to a specific program component using the Key on the next page.

Lecture Concepts	Lesson Resources

28-1 Sponges pp. 639–641

BLOCK 1

a. Sponges are simple filter-feeding animals that have a supportive skeleton composed of soft spongin fibers or hard spicules, or a combination of both.

b. Sponges are capable of both asexual and sexual reproduction.

Lecture Resources
- Opening Questions, page 639
- Demonstration: Choanocytes in Your Kitchen
- Motile and Sessile, page 641
- It Takes Only One, page 641
- The Logic of Sexual Reproduction, page 641
- Teaching Transparencies: 135, 144, 145

- Holt Biology Videodiscs: Lesson 28-1

Classwork Options
- Effective Reading, page 639
- Gemmule Survival Pods, page 640
- Know Those Animals, 640

Assignment Options
- Section Review, page 641
- Chapter Review, questions 1, 2, 11, 12, 19, 20, 22

28-2 Cnidarians pp. 642–647

BLOCK 2

a. Cnidarians, such as hydras, jellyfish, corals, and sea anemones, are soft-bodied animals with tentacles armed with stinging cells called cnidocytes.

b. Cnidarians are grouped into three classes: Hydrozoa, Scyphozoa, and Anthozoa.

c. Sexual reproduction among many cnidarians leads to the development of an intermediate larval stage called a planula.

Lecture Resources
- Opening Questions, page 642
- Demonstration: Looking and Learning
- Visual Strategies: Figures 28-5, 28-6, 28-7, 28-9, 28-12
- Cnidarian Locomotion, page 643
- The Problems and the Pluses of Planualae, page 644
- Finding Coral Reefs, page 646
- Research Update, page 647
- Teaching Transparencies: 134, 147, 148, 153, 154, 155

- Holt Biology Videodiscs: Lesson 28-2

Classwork Options
- Hermaphrodite Advantages, page 644
- The Big and the Small, page 645
- A Carnivore's Venom, page 645
- Plant or Animal? page 646

Assignment Options
- Section Review, page 647
- Chapter Review, questions 3–6, 13, 14, 21, 23

28-3 Kinds of Simple Worms pp. 648–653

BLOCKS 3 and 4

a. Flatworms have flattened bodies that lack a body cavity. Many are parasitic, but some are free-living.

b. Ribbon worms (acoelomates) and round-worms (pseudocoelomates) have digestive tubes open at both ends.

Lecture Resources
- Opening Question, page 648
- Demonstration: Observing Planaria
- Visual Strategies: Figures 28-14, 28-15, 28-16
- Adapting to a New Environment, page 649
- The Predator's Poison, page 652
- Teaching Transparencies: 134, 149, 150, 156, 157, 158
- Holt Biology Videodiscs: Lesson 28-3

Classwork Options
- Up Close: Planarian, page 651
- Create a Life Cycle, pages 652
- The Small End of the Food Chain, page 653
- Closure, page 653
- Laboratory Investigation: Hydra Behavior, pp. 656–657
- Inquiry Skills Development B21: Flatworm Behavior

Assignment Options
- Section Review, page 653
- Chapter Review, questions 7–10, 15–18, 24

Review/Enrichment

BLOCK ⑤

■ Study Guide: Concept Review and Pretest

Assignment Options
■ Activities and Projects, questions 25, 26

Assessment Options

BLOCK ⑥

Traditional Assessment
■ Chapter 28 Test
⊙ Test Generator/Assessment Item Listing: Software item bank for preparing customized chapter tests.

Portfolio Assessment
Select student reports for one or more laboratory experiments from this chapter. *The Direct Observation Checklist* on the *BioSources Teaching Resources CD-ROM* should be completed during a laboratory or other cooperative learning experience.

Holt Biology Videodiscs

Holt Biology Videodiscs gives you a powerful tool for teaching, review, and assessment. *Concepts of Biology* adds interactive lesson plans and extensive tools for customization using CD-ROM technology.

CONCEPTS OF BIOLOGY

Use the following topic from *Concepts of Biology* to help you teach this chapter:

■ Topic 18: Invertebrate Diversity

For further information regarding the media on the videodiscs, see *Holt Biology Videodiscs Teacher's Correlation Guide to Biology: Principles and Explorations.*

UNIT 7
CHAPTER 28

CHAPTER 28 — SIMPLE INVERTEBRATES

Chapter Theme

Structure and Function

The varied structural adaptations of simple invertebrates make it possible for them to successfully inhabit and function in a wide variety of water-rich habitats, from the ocean floor to the human intestine.

Tapping Prior Knowledge

- Why are all the organisms described in this chapter considered to be heterotrophs?
- How is the body of a sponge built for filter feeding?
- Why are cnidarians sometimes described as hollow-bodied animals?
- Why is the body design of a segmented worm considered to be more complex than that of a flatworm?

Opening Demonstration

Show students the skeleton of a natural sponge. Ask them what kinds of information they can gain about the structure of living sponges by observing the skeleton of the dead sponge. *(They are asymmetrical and contain many pores.)* Next, soak up and release water from the sponge to show the large amount of water that the sponge can hold. Ask students how this feature is helpful to a living sponge. *(It allows more oxygen and food to surround the sponge's cells.)*

REVIEW

- osmosis and endocytosis (Section 3-2)
- choanocytes and filter feeding (Section 27-1)
- radial symmetry and cnidocytes (Section 27-1)
- polyps and medusae (Section 27-1)
- cephalization (Section 27-2)
- solid-bodied and pseudocoelomate worms (Section 27-2)

Azure vase sponge

Authors' Rationale

This chapter concentrates on the simplest invertebrates—the sponges, cnidarians, flatworms, and roundworms. These invertebrates, which made their appearance early during the history of life on Earth, are either acoelomate or possess a pseudocoelom. Sponges are a pivotal group because they practically scream "missing link." Sponges are animals, but don't they also appear to be colonial protists? The simple acoelomate cnidarians can be some of the largest animals encountered, for example, coral reefs, and jellyfish that have 80-ft.-long tentacles!

28-1 Sponges

***W**hatever their shape or size, all sponges have two characteristics in common: they have a body wall penetrated by many pores, and they are sessile.* **Sessile** *animals are not mobile. Early in their lives, sponges attach themselves firmly to the sea bottom or some other submerged surface, like a rock or coral reef. There they remain for their entire life.*

Section Objectives

- *Describe how sponge cells receive nutrients.*
- *Describe how a sponge's body is structurally supported.*
- *Compare and contrast asexual and sexual reproduction in sponges.*

Sponges Are Filter Feeders

A sponge has the same basic structure as a bag—a large internal cavity with an opening at the top. As you learned in Chapter 27, facing into the internal cavity of a sponge is a layer of choanocytes, or collar cells. By beating their flagella, the collar cells draw water through the sponge's many pores and into the internal cavity of the sponge. As water circulates through a sponge, the collar cells trap food particles by functioning as sieves, as shown in Figure 28-1.

How do the other cells of the sponge, like the epithelial cells, survive if the collar cells take in all of the food? The collar cells release nutrients into the mesenchyme, the jelly-like middle layer of the sponge body. Here, the nutrients are picked up by other specialized cells, called amoebocytes *(uh MEE boh seyets)*. **Amoebocytes,** cells that have irregular amoeba-like shapes, wander about the mesenchyme, supplying the rest of the sponge's cells with nutrients, and carrying away their wastes.

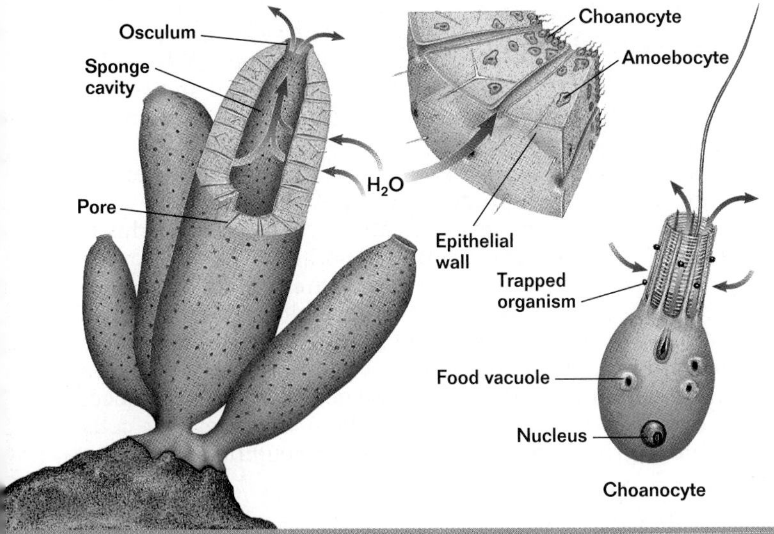

Figure 28-1 Water enters a sponge by passing through pores in the sponge's body wall, *left.* The inside of a sponge is lined by cells called choanocytes, *center,* which trap tiny organisms in the water. The collar of a choanocyte is made of small, hairlike projections resembling a picket fence, *right.* The beating of each choanocyte's flagella draws water, *blue arrows,* down through its collar. Organisms like bacteria and algae are trapped in the collar. These organisms then move toward the cell's cytoplasm, *red arrows,* where they are ingested through endocytosis.

SECTION 28-1

Vocabulary Preview

amoebocyte
spongin
spicule
gemmule
hermaphrodite

Opening Questions

Have students begin to think about different structures for different lifestyles by asking them how an animal that can't move is able to feed and protect itself. Have them imagine that they are stuck to the floor and can't ever move again. How will they get food? How will they protect themselves from large motile predators? Answers may vary, but some might say that they could just wait until food came close enough, then they could reach out and grab it. For protection, they might curl up into a ball if a predator comes too close. This would serve to protect their head, which contains their brain and important sense organs.

Demonstration

Choanocytes in Your Kitchen

Pour a mixture of marbles and water through a spaghetti strainer, trapping the marbles inside the strainer (do this over a sink, of course). Show the trapped marbles to your students and tell them to imagine that the marbles are food particles. Explain to them that the collars of the choanocytes that line the inside of a sponge trap food particles in water in a similar manner.

Effective Reading

Keeping Track

Have students use removable tags or stickers to flag the parts of this chapter that they understand and the parts with which they need help. It helps to color-code these "flags." Green could indicate a section that is understood, and red could signify a part that needs further explanation. After students flag their chapters, they can work together in small groups and help each other understand the parts they are having difficulty with.

UNIT 7
CHAPTER 28

Teaching Tips
Gemmule Survival Pods

Survival pods have been used for submarines and spacecraft, and they should be familiar to your students. Ask them what a good survival pod on a submarine should contain. Students should come up with ideas such as food, water, oxygen, the means to get rid of wastes, and protection from the outside environment. List their ideas as they volunteer them, then take each item on their list and show how a gemmule provides all these same necessities of life for survival in a hostile environment.

Know Those Animals

On the chalkboard, draw a chart similar to the abbreviated version located in the *Graphic Organizer* at the bottom of page 640, but leave the *Structures/Functions* column empty. Extend your chart so that five new rows are created. Begin each new row with a different animal group name. The names should be Cnidarians, Flatworms, Ribbon worms, Roundworms, and Rotifers. Also fill in the *Lifestyles* column for each animal group. Then have a student come up and fill in the *Structure/Function* column for sponges. If he or she is having trouble, ask for help from the class. At least three major structures and their functions should be filled in for sponges, but encourage the student and class to list as many as possible. Have other students come up to the board for the other animal groups.

Figure 28-2 Calcareous sponges, *above left,* have spicules composed of calcium carbonate. In glass sponges, *above center,* the spicules are made of silica. Demosponges, *above right,* may have skeletons made of silica, spongin fibers, or both.
Eighty percent of all sponge species are demosponges.

Sponges Have Simple Skeletons of Varied Composition

For the sponge's body to function effectively, the body wall must be rigid enough to prevent the sponge from collapsing in on itself. Imagine a hollow ball made of gelatin, and you will see the point. The skeleton of a sponge is not a fixed framework like your skeleton, but rather a diffuse network of fibers or minerals. The skeletons of most sponges are made of a resilient flexible protein fiber called spongin, which gives sponges their name. A few sponges have more brittle skeletons composed of tiny hard needles of silica or calcium carbonate called spicules, often embedded within spongin. Taxonomists group sponges into three classes based on the composition of their skeletons. A representative from each class is shown in Figure 28-2.

Asexual and Sexual Reproduction Occur in Sponges

One of the most remarkable properties of sponges is that they will regenerate when cut into pieces. When the pieces are returned to the ocean, each bit of sponge, however small, will grow into a complete new sponge.

As you might suspect from this ability to regenerate, sponges frequently reproduce by simply breaking into fragments. Each fragment develops into a new individual. This is a form of asexual reproduction. Another form of asexual reproduction occurs in some freshwater sponges (a relatively small group). When living conditions become harsh (cold or very dry), some freshwater sponges ensure their survival by forming **gemmules,** clusters of amoebocytes encased within protective coats. Sealed in with ample food, the cells survive even if the rest of the sponge dies. The cells are able to grow into a new sponge when conditions improve.

Graphic Organizer

Use this graphic organizer in *Teaching Tip:* **Know Those Animals** on page 640.

Animal Groups	Lifestyles	Structures/Functions
Sponges	Sessile filter feeders	Collar cells/Filter food & intracellular wastes Spicules/Support & protection Spongin/Support

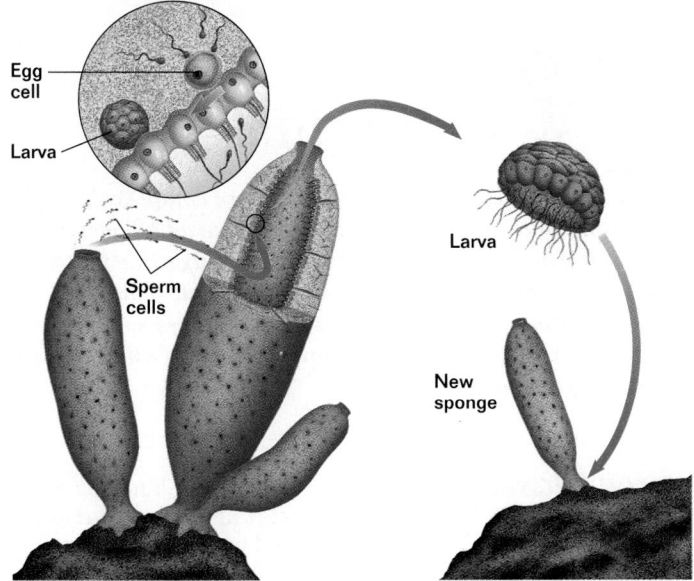

Egg cell
Larva
Sperm cells
Larva
New sponge

Figure 28-3 Sperm from one sponge enter another sponge through its pores, *left.* The sperm are taken in by the receiving sponge's choanocytes and pass into the mesenchyme layer, where egg cells reside. After fertilization, larvae develop. The larvae swim out of the sponge and eventually grow into a new sponge, *right.*

Sexual reproduction among sponges, shown in Figure 28-3, is also common. Although sponges may reproduce sexually, most sponges are not exclusively male or female. Instead, each individual produces both eggs *and* sperm. An organism that produces both eggs and sperm is called a **hermaphrodite** *(huhr MAF roh deyet).* Can eggs from a certain individual be fertilized by sperm from the same individual? No. Eggs and sperm are produced at different times, so self-fertilization is avoided. The advantage of hermaphroditism lies in the fact that the sperm of any individual can fertilize the egg of any other individual. You can see that the probability of one sponge successfully fertilizing another sponge is much greater than if only half of the population produced eggs. Many types of animals that rarely encounter members of their own species—such as sessile invertebrates and even some vertebrates like deep-sea fishes—are hermaphroditic. ◻

❏ CAPSULE SUMMARY

Sponges are simple filter-feeding animals that have a supportive skeleton composed of soft spongin fibers or hard spicules, or a combination of both. Sponges are capable of both asexual and sexual reproduction.

Section Review

1. *What are amoebocytes, and what are their functions?*
2. *Explain why not all sponges would make good bath sponges.*
3. *In what way is sponge reproduction and human reproduction similar?*

Critical Thinking
4. *What evolutionary advantage is there to a free-swimming larval stage in sponges?*

SECTION 28-1

Teaching Tips
Motile and Sessile
There are many examples of motile and sessile organisms in this chapter. Have students look at the pictures of the following organisms and classify them as motile, sessile, or both: calcareous sponge, glass sponge, demosponge, hydra, Portuguese man-of-war, *Obelia,* sea wasp, *Aurelia,* sea fan, sea anemone, coral, planaria, blood fluke, beef tapeworm, ribbon worm, *Ascaris,* and rotifer.

It Takes Only One
Emphasize that asexual reproduction requires only one individual and will create exact copies of that individual. Ask students to explain why it is sometimes good to make numerous copies of a single organism. *(It is an advantage when the individual is highly successful.)* Then ask why it is sometimes good for a sessile animal to be able to reproduce by itself. *(It is sometimes difficult for two animals of opposite sexes to get together.)*

The Logic of Sexual Reproduction
Tell students which form of reproduction produces new organisms identical to the original animal *(asexual)* and which produces variable offspring sharing characteristics of both parents *(sexual).* Then help students relate each form of reproduction to environmental pressures. Ask about situations that require new varieties of an animal (produced by sexual reproduction) in order for the species to survive. *(Species-threatening situations may include change in climate, new predators, loss of food source, competition, and new diseases.)*

Answers to Section Review

1. Amoebocytes are specialized cells in a sponge that pick up nutrients in the mesenchyme and deliver the nutrients to other cells in the sponge. Amoebocytes also pick up and get rid of cellular waste.

2. Many sponges have skeletons made up of tiny, hard needles of silica or calcium carbonate called spicules.

3. Both humans and sponges sexually reproduce with eggs and sperm.

4. The free-swimming larval stage allows sponges to spread out into new areas before they settle down to their sessile life as an adult. If there were no free-swimming stage, all the adult sponges in an area could be wiped out by a local disaster, and the sponges could become extinct.

Opening Questions

Ask students about their own experiences with cnidarians. Have they ever seen a jellyfish, a Portuguese man-of-war, a coral, or a sea anemone? Have they ever touched a cnidarian? What did it feel like? Where have they seen cnidarians, in fresh water or salt water? Take this opportunity to explain that while most cnidarians are marine, some, like hydras, live in fresh water.

Demonstration
Looking and Learning

Show students several colorful pictures of sea anemones, as well as some preserved specimens. Ask them to use what they have learned about sponges to answer the following questions about sea anemones. Are sea anemones plants or animals? (Although they may appear plantlike because they are attached to the ocean bottom, they are really animals.) How do the sea anemones differ from sponges? (They don't have pores in their bodies, and sea anemones have tentacles.) Are the sea anemones filter feeders, like sponges? (no) If not, then how do they capture food? (with their tentacles)

Section Objectives

- Compare and contrast a hydra and a Portuguese man-of-war.
- Describe the life cycle of Obelia *and* Aurelia.
- Name several kinds of anthozoans.

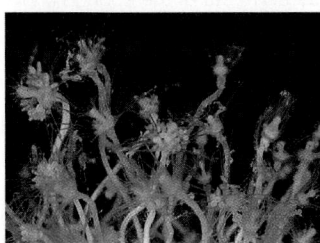

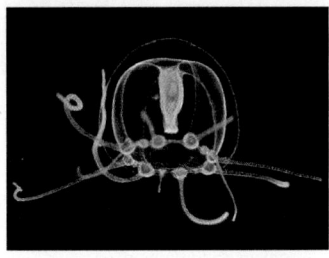

Figure 28-4 Polyps, *top,* are specialized for a sessile existence. They attach to a surface with their mouth facing upward. A medusa, *bottom,* is specialized for swimming. Its mouth faces downward, with its tentacles dangling down around it.

 Did You Know?

There are polyps other than the invertebrate animal kind. Sometimes people have a polyp removed from their colon. The polyp obviously isn't a cnidarian living in the colon—it is a small tumor that grows in the shape of a cnidarian polyp.

28-2 *Cnidarians*

The body plan of a cnidarian is more complex than that of a sponge. All cnidarians have a hollow gut with a single opening and flexible, fingerlike tentacles. Located on the tentacles are stinging cells called cnidocytes that contain harpoon-like nematocysts. Tiny freshwater hydra, jellyfish, and flowerlike coral all belong to the phylum Cnidaria.

Three Classes of Cnidarians

Two very different kinds of body forms exist among cnidarians: vase-shaped polyps and umbrella-shaped medusae, each shown in Figure 28-4. Taxonomists divide cnidarians into three classes based on their life cycles. The most primitive cnidarians are members of the class Hydrozoa. Most hydrozoans spend part of their lives as medusae and part as polyps. The other two classes of cnidarians, which evolved from hydrozoans, emphasize one or the other body form. The jellyfish (class Scyphozoa) pass briefly through a polyp stage but spend most of their lives as medusae. The other class, corals and sea anemones (class Anthozoa), spend all of their lives as polyps and have no free-swimming medusa stage.

Hydrozoans Are Found in Fresh and Salt Water

The class Hydrozoa includes about 2,700 named species. Most species are colonial marine organisms with both polyp and medusa stages in their life cycles. But the freshwater hydrozoans, so often studied in school laboratories, are probably most familiar to us. The abundant freshwater genus *Hydra* is unique among hydrozoans because it has no medusa stage and exists only as a solitary polyp. Hydras live in quiet ponds, lakes, and streams. Hydras attach themselves to rocks or water plants by means of a sticky secretion produced by an area called the **basal disk**. Hydras can glide around by decreasing the stickiness of the material secreted by their basal disk. They can also move by tumbling, as shown in Figure 28-5. Most hydras are white or brown, although some appear green because of the algae living beneath their outer cells.

Marine hydrozoans are typically far more complex than freshwater hydrozoans. They often form colonies, with many individuals living together. While the cells of the

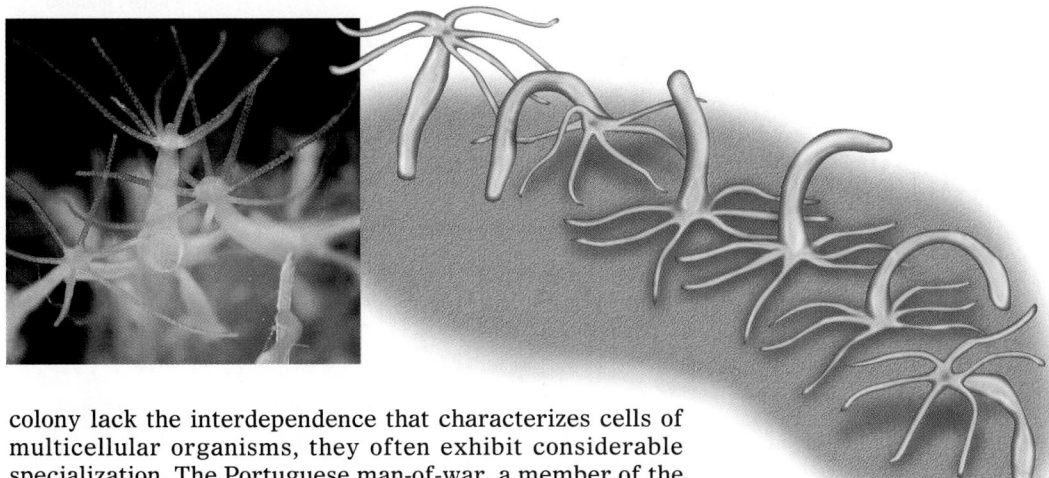

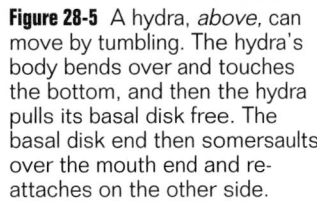

colony lack the interdependence that characterizes cells of multicellular organisms, they often exhibit considerable specialization. The Portuguese man-of-war, a member of the marine genus *Physalia,* incorporates both medusae and polyps. As shown in Figure 28-6, a gas-filled medusa of *Physalia* floats like a balloon on the surface of the water. Dangling below, nematocyst-studded tentacles that can be 15 m (50 ft.) long in large specimens stun and entangle prey. Several different kinds of polyps are attached to the tentacles, each carrying out a different function, such as feeding, sexual reproduction, and defense.

Figure 28-5 A hydra, *above,* can move by tumbling. The hydra's body bends over and touches the bottom, and then the hydra pulls its basal disk free. The basal disk end then somersaults over the mouth end and re-attaches on the other side.

Figure 28-6 As many as 1,000 individual medusae and polyps may compose a single Portuguese man-of-war colony, *far left.* Ocean currents and winds often direct Portuguese man-of-war colonies onto beaches in the southern United States, where they become a hazard to swimmers, *left.*

Reproduction in Hydrozoans

Most hydrozoans are colonial organisms whose polyps reproduce asexually by forming small buds on the outside of their body. These buds develop into miniature polyps. Eventually the offspring separate from the colony and begin living independently. Many hydrozoans are also capable of sexual reproduction. The genus *Obelia* is a typical colonial hydrozoan. The body of *Obelia* is branched like deer antlers, with various polyps attached to the branched stalks. As in the Portuguese

Multicultural Perspective

A Mythological Hydra
The name *hydra* is a reference to the nine-headed serpent that was slain by Hercules, a figure in Greek and Roman mythology. According to the myth, the slaying of the Hydra was one of the twelve labors that Hercules was assigned to perform. Killing the Hydra was not an easy task for Hercules, because when one head of the Hydra was cut off, two others would grow in its place.

Figure 28-7
Check for student understanding of the life cyle of *Obelia* in the diagram on this page. Ask which stages are motile and which are sessile. In which stage do males and females exist?

Teaching Tips
The Problems and the Pluses of Planulae

There are both negative and positive aspects to the existence of the free-swimming planulae of *Obelia*. Ask students about the dangers to free-swimming larvae in the ocean, and why it is an advantage to the species that the larvae can swim far from their parents. *(Larvae may starve, be eaten, wash up on shore, or not find a suitable place to settle down. Motile larvae allow* Obelia *to spread to new places.)*

Hermaphrodite Advantages

Ask students why it is an advantage for a sessile organism to be a hermaphrodite. Stress that hermaphrodites don't usually fertilize their own eggs, because an advantage of sexual reproduction is the increased variety of offspring that comes with combining the genetics of two different individuals. Guide students to the idea that slow or sessile animals may not come in contact with many others of their species, so it is an advantage for them to be able to reproduce with any other species member they encounter.

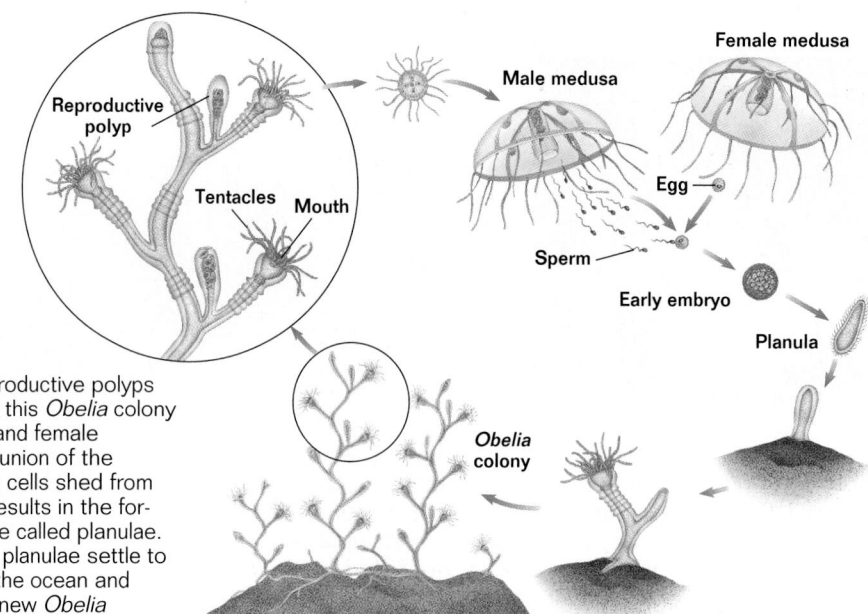

Figure 28-7 Reproductive polyps that grow from this *Obelia* colony produce male and female medusae. The union of the sperm and egg cells shed from the medusae results in the formation of larvae called planulae. Eventually, the planulae settle to the bottom of the ocean and develop into a new *Obelia* colony.

man-of-war, some of the polyps are specialized for feeding, and others are for reproduction. Reproductive polyps give rise to male and female medusae. The medusae leave the polyps and release eggs and sperm into the water. The gametes fuse and produce zygotes that develop into free-swimming larvae called **planulae** *(PLAN yoo lee)*. The planulae eventually settle on the ocean bottom and develop into new polyps, shown in Figure 28-7. Sexual reproduction in *Obelia* usually occurs in the fall, when low water temperatures trigger the development of eggs and sperm in the medusae.

Like sponges, some species of *Hydra* are hermaphrodites; a single individual is capable of producing both eggs and sperm. The gametes are produced in swellings along the hydra's body wall. In other species, there are separate sexes. During sexual reproduction, sperm released into the water swim to eggs contained in a nearby hydra. Fertilization typically occurs in the fall. The resulting zygote forms a cell mass with a hard covering that protects it through the winter. In the spring the embryo bursts through the cover and develops into a new hydra.

Scyphozoans Are Marine Jellyfish

The best known cnidarians are the jellyfish, members of the class Scyphozoa *(seye fuh ZOH uh)*. The name Scyphozoa is from the Greek *skyphos,* meaning "cup," and *zoia,* meaning "animal." Their name refers to the fact that members of this class spend most of their lives as medusae, which have

Figure 28-8 The diver is photographing a sea wasp jellyfish in the waters along the coast of Australia. The stings from a large sea wasp jellyfish can cause a human to die in a matter of minutes.

the shape of an inverted cup. There are about 200 species of jellyfish. Some are as small as a thimble, and others are as large as a queen-sized mattress. Jellyfish are active predators that ensnare and sting prey with their tentacles. The toxins contained within the nematocysts of some species are extremely potent. The sea wasp jellyfish, shown in Figure 28-8, lives in the sea along the tropical northern coast of Australia and can inflict severe pain and even death on humans.

Although the jellyfish that you see in the ocean are medusae, most species also go through a small, inconspicuous polyp stage at some point in their life cycle. Most jellyfish reproduce sexually, and their life cycle often includes the development of larval planulae. One of the most familiar jellyfish is the stinging nettle, *Aurelia*. The life cycle of *Aurelia*, shown in Figure 28-9, includes both medusa and polyp stages.

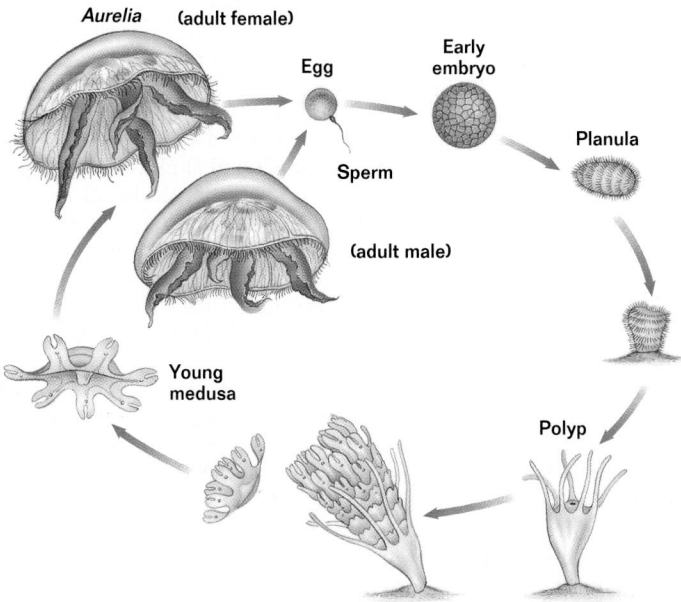

 Figure 28-9 The adult male and female medusae of *Aurelia* produce gametes. The early embryos that result from the fusion of the gametes develop into planulae that settle and attach to the ocean bottom. The unattached end of the planula develops a mouth and tentacles, becoming a polyp. As the polyp grows, it forms medusae, which eventually bud off and develop into jellyfish.

Teaching Tips
The Big and the Small
Find how small the smallest cnidarians are and how large the largest are. (*Hydras are barely visible to the naked eye, while box jellyfish may be the size of a large mattress.*)

A Carnivore's Venom
Jellyfish have toxins in their nematocysts that paralyze and kill fish. Ask students why it is so important for a simple floating animal like a jellyfish to immobilize its prey. (*A jellyfish can't stalk or chase prey, and it has no jaws or claws to capture prey. It must catch and immobilize any fish that accidentally swims into its tentacles.*)

Application
Health When you are swimming in or walking near the ocean, be aware of stinging jellyfish so that you can avoid them. But if you do happen to get stung, some people recommend treating the wound by washing it with sea water mixed with a meat tenderizer. Meat tenderizer contains an enzyme called papain that is thought to break down the protein toxins released by the jellyfish.

VISUAL STRATEGY

Figure 28-9
Have students compare and contrast the anatomy and life history of *Aurelia* with that of *Obelia*. On the basis of their answers, have them criticize or defend the placement of *Aurelia* in the same class of cnidarians as *Obelia*.

Teaching Tips
Plant or Animal?
Some students find it difficult to believe that cnidarians such as sea anemones and sea pansies are animals. Furthermore, the anthozoans are sessile and may contain photosynthetic dinoflagellates. Ask students to defend the classification of anthozoans in the animal kingdom.

Finding Coral Reefs
Obtain several good world maps showing details of the oceans, and have students find some of the great reefs of the world. Students should find the Florida Keys Reef, Australia's Great Barrier Reef, and the Eniwetok Atoll in the Marshall Islands, as well as others.

 VISUAL STRATEGY

Figure 28-12
Direct students' attention to Figure 28-12 on page 647, and have them identify as many animals as they can. Ask them how many are cnidarians, and have them classify the animals that are anthozoans, scyphozoans, or hydrozoans.

Figure 28-10 The flat, lacelike branches of this sea fan from Micronesia are penetrated by many tiny holes. A polyp lives in each hole.

Figure 28-11 Once the sea star is paralyzed by the sea anemone's stinging tentacles, it will be directed through a centrally located, slitlike mouth and dissolved in the anemone's digestive cavity.

Anthozoans Are Marine Polyps

The largest class of cnidarians is the anthozoans. The class name Anthozoa is from the Greek *anthos*, meaning "flower," and *zoia*, meaning "animal." There are approximately 6,200 species in this class. The most familiar anthozoans are the brightly colored sea anemones and corals. Other members are known by such fanciful names as sea pansies, sea fans, and sea whips. A sea fan is shown in Figure 28-10. Anthozoans typically have a thick, stalklike body topped by a crown of tentacles that occur in groups of six. Nearly all of the shallow-water species contain symbiotic dinoflagellates. The anthozoans provide a place for the dinoflagellates to live in exchange for some of the food that the dinoflagellates produce. The fertilized eggs of anthozoans develop into planulae that develop into polyps. No medusae are formed by anthozoans.

Sea Anemones
Sea anemones are a large group of soft-bodied polyps found in coastal areas all over the world. Most do not grow very large, only from 5 mm (0.2 in.) to 100 mm (4.0 in.) in diameter. But one species on the northern Pacific coast of the United States may reach up to 1 m (3.3 ft.) in diameter. Sea anemones feed on fish and other marine life that happen to swim within reach of their tentacles. When they are touched, most sea anemones retract their tentacles into their body cavity and contract into a tight ball. Sea anemones are highly muscular and relatively complex animals with many divided internal cavities. Many species, such as the one shown in Figure 28-11, are quite colorful.

Corals
Most corals live in colonies. Each small polyp secretes a tough, stonelike outer skeleton. Coral reefs form when these calcium-rich skeletons become cemented to those of their neighbors. When a polyp dies, its hardened skeleton remains. Like the cement foundation of a house, the old coral skeletons provide a foundation for new coral polyps growing above them. Eventually, the reef may grow all the way to the surface. Hundreds of thousands of polyps live together on the upper part of the reef at any one time. A coral reef, shown in Figure 28-12, provides food and shelter for an enormous variety of fishes and invertebrates.

There are three kinds of coral reefs. **Fringing reefs** form close to a beach. The reefs off the coastal beaches of Florida are fringing reefs. **Barrier reefs** form in much deeper water farther out from the shore. The Great Barrier Reef, off the coast of eastern Australia, is the largest reef in the world, easily visible from outer space. Even farther out from land are **atolls,** coral islands that form far out at sea and grow into a ring shape, with a lagoon in the center.

 Did You Know?

One interesting example of symbiosis is a clown fish living in the tentacles of a sea anemone. Usually a sea anemone catches and eats fish, yet the clown fish seems to be immune to the barbs of the sea anemone's nematocysts. The clown fish gets protection by hiding in the tentacles, and the sea anemone gets cleaned as the clown fish eats undigested debris. Clown fish do not swim straight into the tentacles of a strange sea anemone; rather, they brush lightly against the tentacles several times, apparently building immunity to the sting of the nematocysts before taking up residence in a new sea anemone.

Figure 28-12 A growing coral reef in Fiji teems with life, *left*. The many creatures that live around a coral reef interact to form complex food webs. These feeding corals, *above*, may appear to be individual organisms. Actually, the body walls of all of them are connected.

Coral reefs are usually found in tropical seas in shallow, clear water. The symbiotic algae that live inside the cells of the coral require sufficent light and warmth for photosynthesis. Given this fact about coral reef growth, naturalists were long puzzled by how an atoll could form. The deepest corals of atolls are far below the surface light. Darwin solved the puzzle over a century ago. He suggested that atolls form from coral colonies that are attached to the tops of undersea mountains or volcanoes that were formerly at sea level but have slowly sunk beneath the sea. The coral colonies continued to grow upward at the same speed that the mountains sank, keeping the coral colony's "living zone" near the surface. ☐

☐ **CAPSULE SUMMARY**

Hydras, jellyfish, corals, and sea anemones are members of the phylum Cnidaria. These soft-bodied animals have tentacles armed with stinging cells (cnidocytes). Cnidarians are grouped into the three classes Hydrozoa, Scyphozoa, and Anthozoa. Sexual reproduction among many cnidarians leads to the development of an intermediate larval stage called a planula.

Section Review

1. *Explain why a Portuguese man-of-war is not considered a jellyfish.*
2. *What is a planula, and what does it become?*
3. *Which body form of* Aurelia *dominates its life cycle?*
4. *Compare and contrast sea anemones with corals.*
5. *Describe the formation of an atoll.*

Critical Thinking

6. *Damage to coral reefs by human activity is increasing worldwide. Identify some activities that may be causing this damage, and explain how the damage could affect humans and other animals.*

☑ **RESEARCH Update**

Killer Comb Jellies

UPDATE

Comb jellies, or ctenophores, don't sting as their jellyfish cousins do, but they can still be enormous pests. *Mnemiopsis leidyi*, in particular, is a walnut-sized, compact lump of jelly with a voracious appetite for plankton and the eggs of fish. It has invaded the Black and Azov Seas, sneaking in as a stowaway in the ballast water of commercial ships from the coast of the Americas. *Mnemiopsis* has caused an estimated $250 million in damage to the fishing industry of the Black Sea. In the Azov, fisheries have closed down.

European and Mediterranean fisheries, along with experts from the United States and the United Nations, are trying desperately to come up with a remedy that will save the flagging fishing industries. Other countries are interested in the fight because, as James Carlton, an "invasion expert" from Williams College in Connecticut points out, "This is not the last ctenophore invasion in the world." Despite the obvious risks of introducing another foreign species into the seas to control the pest, most of the researchers working on this problem propose the introduction of a predatory fish. Right now there are no natural predators of the ctenophore in Europe. If they can find a specific predatory fish, maybe it can keep the ctenophore population down *and* provide another species for the fishing industry. ☑

Answers to Section Review

1. The Portuguese man-of-war is a hydrozoan that is a complex colony containing both medusae and polyps, while jellyfish, which are scyphozoans, spend most of their lives as medusae.

2. Planulae are the free-swimming larvae that result from sexual reproduction in hydrozoans and scyphozoans. The planulae develop into polyps.

3. *Aurelia* spends most of its life as a medusa.

4. Sea anemones and corals are both Anthozoans and polyps. The sea anemone is much larger than a single coral animal and doesn't produce a hard skeleton.

5. Atolls form from coral colonies that grew around a mountain that sank into the ocean while the corals continued to grow upward.

6. Some of the human activities that cause damage to coral colonies are water pollution and collecting coral for selling. This damage affects a huge array of fishes and invertebrates, some of which are used as food by humans.

Vocabulary Preview

fluke

endoparasite

ectoparasite

proglottid

tegument

Opening Question

The word *worm* can refer to a wide variety of organisms. Ask students to think about what a worm is, and ask them to name or describe all the kinds of worms they can remember. Then ask them to describe where worms live and what they eat. Have volunteers share their descriptions with the rest of the class.

Demonstration

Observing Planaria

Have students observe living planaria through a stereoscope or hand lens. Encourage the students to observe how the planaria move and respond to touch and light. Have students draw a detailed sketch of a planarian and list four major flatworm characteristics next to their drawing.

👁 VISUAL STRATEGY

Figure 28-14

Have students look at Figure 28-14 and identify the head of the worm. Ask them how they can tell it is the head. Point out that sponges and cnidarians have no identifiable head, so the cephalization of worms is an anatomical advancement.

28-3 Kinds of Simple Worms

Section Objectives

- Compare and contrast turbellarians with trematodes.
- Describe the life cycle of a beef tapeworm.
- List the major characteristics of the ribbon worms.
- Contrast the life cycles of several different parasitic roundworms.
- Describe the characteristics of a rotifer.

*T*he remaining major phyla of simple invertebrates all have basically tubular bodies—many are what you call worms. Here, you will consider four phyla: flatworms (phylum Platyhelminthes), ribbon worms (phylum Rhynchocoela), roundworms (phylum Nematoda), and rotifers (phylum Rotifera). These organisms, although all simple in body plan, are a great deal more complex than sponges or cnidarians.

Flatworms Have a Solid Body

Flatworms, like the one shown in Figure 28-13, are bilaterally symmetrical with simple bodies that lack both respiratory and circulatory systems. Flatworms have no need for these systems because each cell of a flatworm's body is close to the animal's exterior. A flatworm's body is solid because it has no body cavity. There are approximately 20,000 named species of flatworms, which compose three major classes: Turbellaria, Trematoda, and Cestoda.

Figure 28-13 Many of the free-living flatworms are beautiful marine species that swim with wavelike movements of their flattened body.

Figure 28-14 This small planarian is common in clear lakes and streams. Since planarians avoid light, they can most often be found clinging to the undersides of rocks or logs in the water.

Turbellaria

There are over 3,000 species in the class Turbellaria, almost all of them free-living marine flatworms. Marine flatworms are rarely studied by students, however, as they are difficult to raise. Instead, students usually study a freshwater flatworm like the planarian *Dugesia*, shown in Figure 28-14 and in *Up Close: Planarian* on page 651. Like all platyhelminthes, its body is flat like a piece of tape. Dugesia's rear is tapered, while its front end is rounded like a shovel. It swims by flexing its body in a wavelike motion that somewhat resembles the butterfly stroke.

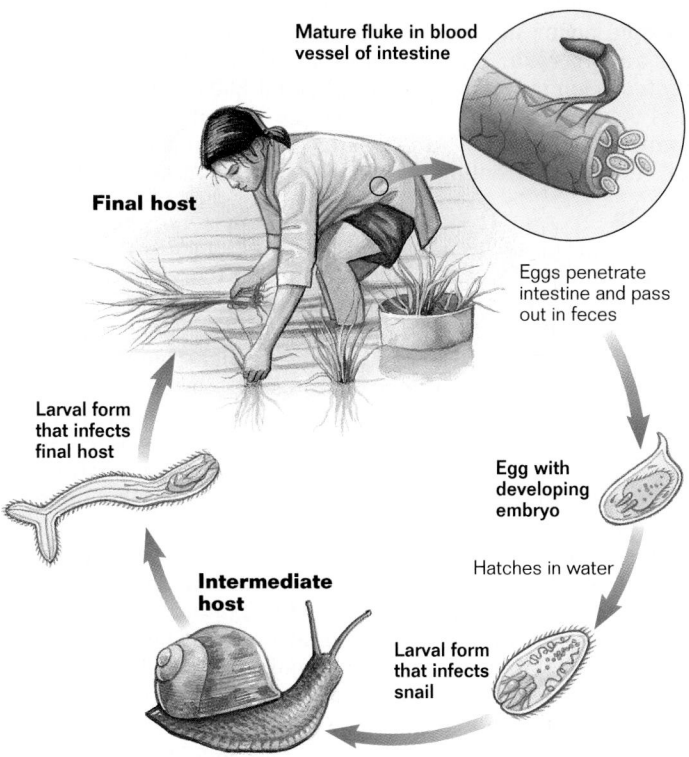

Mature fluke in blood vessel of intestine

Final host

Eggs penetrate intestine and pass out in feces

Larval form that infects final host

Egg with developing embryo

Hatches in water

Intermediate host

Larval form that infects snail

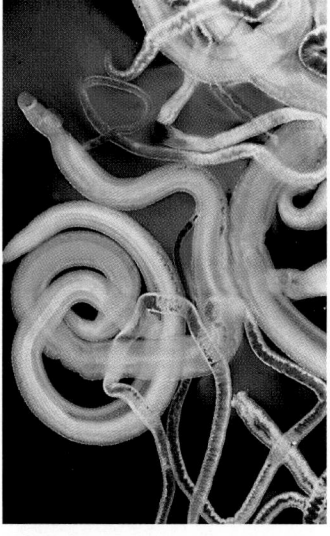

VISUAL STRATEGY

Figure 28-15

Have students analyze Figure 28-15 from the perspective of a public health investigator. Divide the class into small groups of two to three students, and have them devise a plan to stop the infection of people by the blood fluke.

Teaching Tip
Adapting to a New Environment

Parasitic flatworms have a thick protective covering, and a digestive tract may be completely absent. Ask students why parasites have these anatomical characteristics. Then ask what other adaptations the worms must have to burrow through skin, migrate around the body to specific areas, stay attached to blood vessels and intestine, and reproduce inside a human. *(Students' answers might include specialized mouthparts, suckers and hooks to hang on to a host, good sense of smell or taste to detect a host or mate, and the ability to swim or crawl around a host's body.)*

Trematoda

Most flatworm species are parasitic. The largest flatworm class, Trematoda, consists of about 6,000 species of parasitic worms called **flukes**. Some flukes are endoparasites. **Endoparasites** live *inside* their hosts. Others are ectoparasites. **Ectoparasites** live on the outside of their hosts. To avoid being digested by their host, endoparasites often have a thick protective covering of cells called the **tegument**.

Flukes and the other parasitic flatworms have extremely simple bodies with few organs. Biologists think that parasitic flatworms evolved from free-living forms. After adopting their parasitic lifestyle, they no longer needed many of the organs necessary for independent living, like well-developed mouths or digestive systems. Other organs became modified. For example, many parasitic flatworms have one or more suckers that they use to attach themselves to their host.

Most flukes have complex life cycles involving more than one host. Blood flukes of the genus *Schistosoma* are responsible for schistosomiasis *(shihs tuh soh MEYE uh sihs)*, a major public health problem in many parts of the tropics. These parasites, shown in Figure 28-15, live in the blood vessels of infected individuals and cause bleeding of the intestinal wall and decay of the liver due to blocked blood passages. The life cycle of blood flukes includes a particular species of snail as an intermediate host. Most blood fluke infestations

Figure 28-15 Adult blood flukes, *above,* normally live as male-female pairs in the blood vessels of the small intestine. The males are the thick-bodied worms, and females are the thinner, thread-like worms. Workers may be exposed to blood fluke larvae when they wade in fields contaminated with human feces, *left.* The larvae are able to bore through the workers' skin and enter their bloodstream.

CONNECTIONS
....

Chapter 17
Parasitism

Parasitism is a special type of predation in which one organism (the parasite) feeds on and typically lives on or in another organism (the host). A successful parasite doesn't significantly harm its host. It is not to a parasite's advantage to kill the host on which it depends. The longer it can live in the host, the more eggs it produces, and the more successful its species is.

Application

Health Undercooked pork and meat from game animals may contain infective roundworms, and undercooked beef may contain infective tapeworms, so it's a good health habit to cook *all* meat until it is well done.

◆ VISUAL STRATEGY

Figure 28-16

Ask students to look at Figure 28-16 to discover how cows become infested with tapeworms. Point out that *they* aren't eating undercooked beef, so how do they become infested? *(Infested cow feces are on the grass they eat.)*

❑ CAPSULE SUMMARY

Flatworms (phylum Platyhelminthes) have flattened bodies that lack a body cavity (they are acoelomate). Many flatworms are parasites (flukes and tapeworms), but some are free-living (planarians and marine flatworms).

of people occur in tropical countries, particularly those in Asia and Africa. Hundreds of millions of people are affected. Schistosomiasis will kill about 800,000 people this year! A second kind of fluke is the human liver fluke, *Clonorchis sinensis*. This parasite passes from humans to snails to fish, reinfecting humans who eat the infected fish.

Cestoda

A second group of parasitic flatworms belongs to the class Cestoda, commonly called tapeworms. There are about 1,500 tapeworm species. In contrast to flukes, tapeworms permanently attach themselves to the inner intestinal wall of their host. Tapeworms do not have mouths or digestive systems. They absorb food from the host's intestine directly through their skin. They grow by producing a string of rectangular body sections called **proglottids** *(proh GLAHT ihds)* immediately behind their head. These sections are added continuously during the life of the tapeworm. The long, ribbonlike body of a tapeworm may grow up to 12 m (40 ft.) long!

Most tapeworms occur in vertebrates, and about a dozen different kinds infect humans. One tapeworm that infects humans is the beef tapeworm, *Taenia saginata*, whose life cycle is shown in Figure 28-16. Humans can become infected if they eat beef that has not been cooked to a temperature high enough to kill the larval tapeworms that occur as cysts inside the meat. About 1 percent of the cattle in the United States are infected by tapeworms. Since approximately 20 percent of the beef consumed in the United States is not federally inspected, meat from infected cattle may reach the marketplace and be eaten. As a result, the beef tapeworm is a frequent human parasite in the United States. ❑

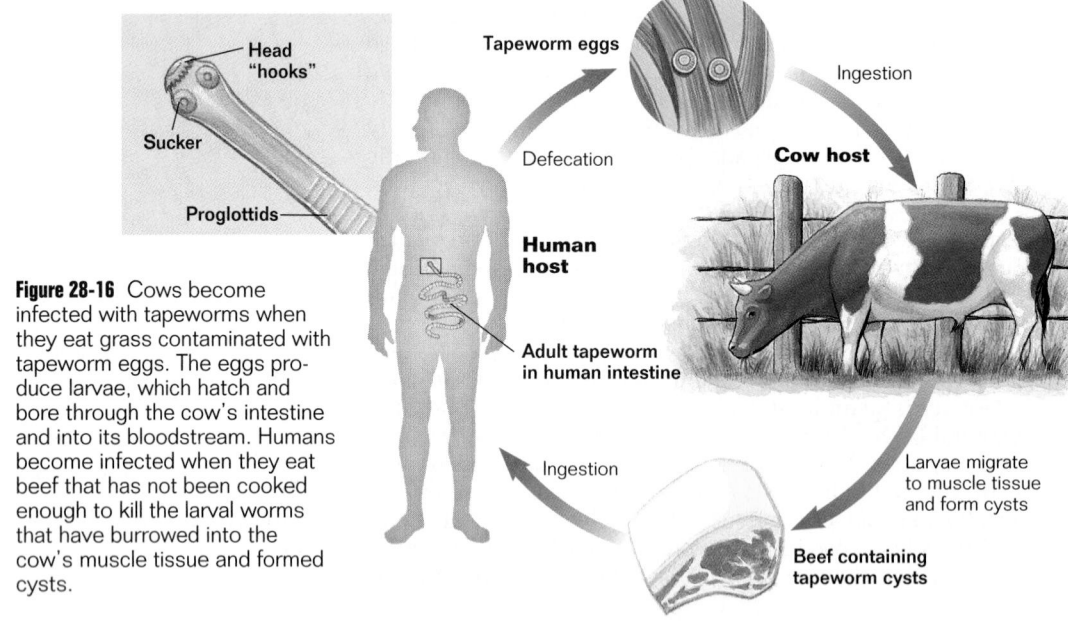

Figure 28-16 Cows become infected with tapeworms when they eat grass contaminated with tapeworm eggs. The eggs produce larvae, which hatch and bore through the cow's intestine and into its bloodstream. Humans become infected when they eat beef that has not been cooked enough to kill the larval worms that have burrowed into the cow's muscle tissue and formed cysts.

Historical Note

The History of Meat Inspection

The meatpacking industry in the United States began when the American colonies still existed. But it wasn't until 1890 that Congress passed a meat inspection law for meats that were to be exported. Next, in 1906, a law was passed that required the inspection of meats being transported across state lines. Today, the residents of every state are protected. A law passed in 1967 requires meatpackers in each state to meet inspection standards equal to federal government standards.

Up Close PLANARIAN

- **Scientific name:** *Dugesia* sp.
- **Range:** Worldwide
- **Habitat:** Cool, clear, permanent lakes and streams
- **Size:** Average length of 3–15 mm (0.1–0.6 in.)
- **Diet:** Protozoa and dead and dying animals

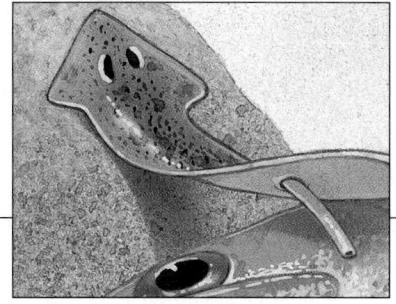

Feeding *Dugesia*, a free-living flatworm, must extend its muscular **pharynx** out of its centrally located mouth in order to feed.

Digestion and Excretion Food drawn into the pharynx passes into a closed, branched intestine. The branching of a flatworm's intestine enables nutrients to pass close to all of the worm's tissues. The nutrients are absorbed through the intestinal wall, and undigested food is expelled through the mouth.

External Structures

Nervous System Sensory information gathered by the brain is sent to the muscles by two main nerve cords that are connected by cross branches. Planarians show an ability to learn complex tasks, such as how to find their way through an experimental maze. Interestingly, their memory of learned tasks appears to be stored chemically.

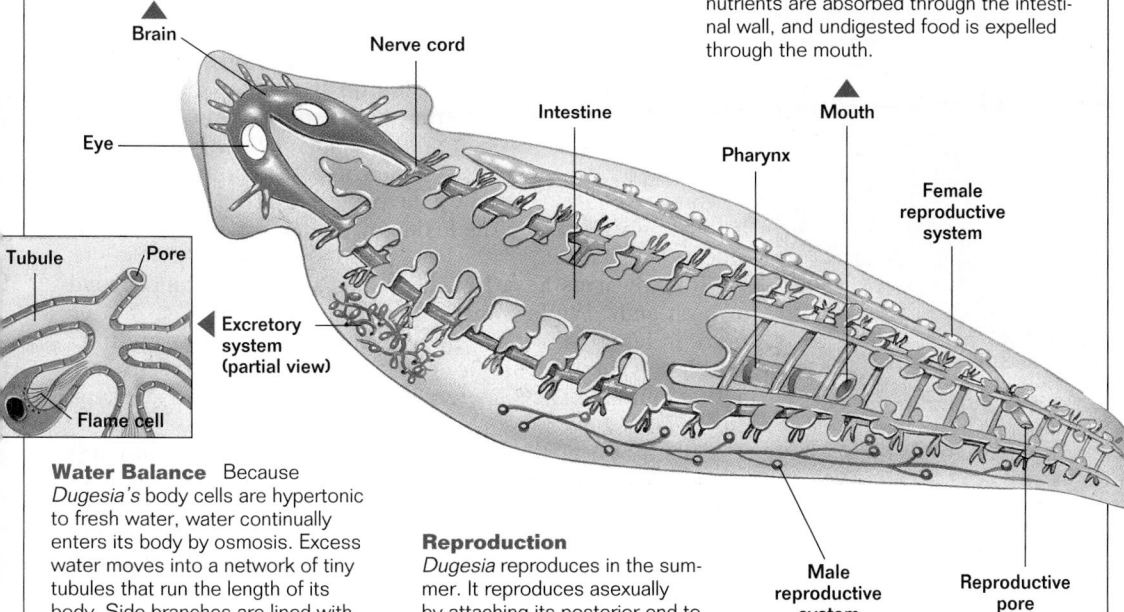

Brain · Nerve cord · Eye · Intestine · Mouth · Pharynx · Female reproductive system · Tubule · Pore · Excretory system (partial view) · Flame cell · Male reproductive system · Reproductive pore

Water Balance Because *Dugesia's* body cells are hypertonic to fresh water, water continually enters its body by osmosis. Excess water moves into a network of tiny tubules that run the length of its body. Side branches are lined with many **flame cells,** specialized cells with beating tufts of cilia that resemble a candle flame. The beating cilia draw water through pores to the outside of the worm's body. Most flatworms, including flukes and tapeworms, have a similar water-removal system.

Reproduction
Dugesia reproduces in the summer. It reproduces asexually by attaching its posterior end to a stationary object and stretching until it tears in two. Each half then regenerates another complete animal. *Dugesia* is also capable of sexual reproduction. As **hermaphrodites**—individuals that have both male and female sex organs—two individuals simultaneously fertilize each other, each transferring sperm to the other. Protective capsules enclose groups of several fertilized eggs. The capsules are laid in bunches, and the eggs inside hatch in two to three weeks.

Instructional Strategy

Have your students imagine a day in the life of this planarian, and have them write a story about the planarian's adventures. Included in their story should be the text's descriptions of the nervous system, water balance, reproduction, feeding, digestion, and excretion. Caution them to avoid anthropomorphic descriptions of their worm, though a little "creative license" should be allowed.

Discussion

Guide the discussion by posing the following questions.

1. Instead of a circulatory system that delivers nutrients to tissues, what does a planarian have? *(Branches of the digestive tract reach the tissues directly.)*

2. Water continually flows into a planarian because it lives in fresh water. What process causes the water flow, and how does a planarian counteract the flow? *(Osmosis causes water to flow into the planarian, and it has beating cilia to push excess water out through pores in its body.)*

3. How do planaria reproduce asexually? *(They tear themselves in two, and each half regenerates to form a complete worm.)*

4. How do planaria reproduce sexually? *(They are hermaphrodites that fertilize each other's eggs. Protective capsules surround groups of fertilized eggs, which hatch in two to three weeks.)*

5. Planaria are often found under rocks. How do they detect dark places and how do they move to those places? *(They have eyes that detect light and dark, and they have a brain and nerve cord that contract their muscles so that they can crawl.)*

Teaching Tips

The Predator's Poison

Ask students why many ribbon worms poison their prey before eating them, while planarians have no poison. You might encourage students to think about a ribbon worm's need to subdue active prey, while planaria eat rotting organic matter that doesn't need to be subdued.

Create a Life Cycle

Have students pick either *Ascaris* or *Necator*. Ask them to research their worm in a zoology college textbook and then diagram a life cycle for their worm similar to the life cycles diagramed earlier in the chapter.

Application

Agriculture When you buy plants for your vegetable garden, they are sometimes marked with the letters *V, F,* or *N.* The *N* stands for "nematode resistant." Since roundworms are so successful eating plants in some areas of the country, it is often necessary to breed plants that are resistant to roundworms. By the way, the *V* stands for "virus resistant," and the *F* stands for "fungus resistant."

Figure 28-17 This beautiful ribbon worm is from Panama. Some ribbon worms burrow beneath sand. Others live under rocks or in algae.

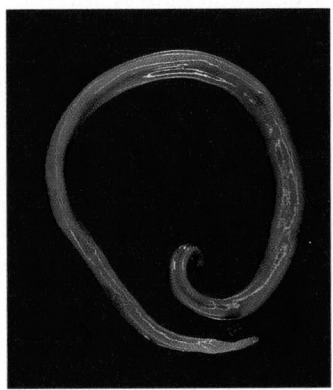

Figure 28-18 This adult *Ascaris* roundworm is a human parasite that may wander into the ducts of the pancreas or gallbladder, causing a blockage. Other, related species infect dogs, cats, pigs, and other vertebrates.

Ribbon Worms Have a Digestive Tube With Two Openings

The longest worms by far are the free-living, flat-bodied, marine ribbon worms, members of the phylum Rhynchocoela *(RIHNG koh seel ah)*. There are about 650 species of marine ribbon worms. Individual worms are typically 0.31 m (1 ft.) long, but some may reach more than 30 m (100 ft). They are characterized by a long proboscis *(proh BAHS kihs)*, a muscular tube that they quickly thrust out to capture prey. Ribbon worms, like the one shown in Figure 28-17, share many characteristics with the free-living flatworms; their bodies are flattened, and most biologists consider them to be solid-bodied (acoelomate). At the same time, the body structure of a ribbon worm is far more complex than that of a flatworm. For example, they are the simplest animals that have a digestive tube open at both ends. They also have a circulatory system. The structure of ribbon worms provides early indications of important evolutionary trends that became fully developed in the more advanced animals.

Roundworms Are the Simplest Animals With a Body Cavity

Roundworms are members of the phylum Nematoda. There are about 12,000 known species, although it has been estimated that between 500,000 and 750,000 species may actually exist. As you learned in Chapter 27, roundworms are characterized by the presence of a pseudocoelom, a body cavity lined on the inside by endoderm and on the outside by mesoderm.

Roundworms have bodies shaped like pencils sharpened at both ends. While some roundworms grow to be a foot or more in length, most are microscopic or only a few millimeters long. The vast majority of roundworms are free-living, active hunters. They are found on land, in lakes and streams, and in all oceans.

About 50 roundworm species are plant and animal parasites that cause considerable economic damage and human suffering. One roundworm that infects humans is the intestinal roundworm *Ascaris*, shown in Figure 28-18. Because the eggs of *Ascaris* are carried to soil by way of human waste, infestation is greatest in those areas without modern plumbing. The eggs of *Ascaris* can live in the soil for years and may enter the body when aspects of personal hygiene, such as hand washing, are not properly practiced. In the intestine, the eggs develop into larvae that bore through blood vessels and enter the bloodstream. The blood carries the larvae to the lungs, where they cause respiratory distress. Eventually, they return to the intestine, where they mature and mate.

Did You Know?

Parasites are a leading health problem in the world today. We just happen to live on a continent that has few parasitic worms. Public-health laws have stopped widespread parasitic worm infestations in the United States. You see evidence of one of those laws when you enter a store or restaurant that has a sign reading "No shoes, no service!" Ask students which parasites can burrow straight through human flesh when people walk on infested ground with bare feet. *(hookworms)*

Adult *Ascaris* may grow up to 0.3 m (1 ft.) in length while in the intestine.

Another roundworm that infects people is the hookworm *Necator*. Members of this genus live mostly in the warm, moist soils of the tropics. When people step barefooted on infected soil, the hookworms penetrate the soles of their feet and attach to the sides of blood vessels. They consume blood cells, causing anemia in the infected person. Their eggs are shed in human feces and hatch in the soil, where they await another barefooted human to begin the cycle again. Hookworm infections can be prevented by wearing shoes and properly disposing of waste matter.

Trichinella, a parasitic roundworm that infects pigs, causes a disease called trichinosis (*trihk ih NOH sihs*) in humans. The disease occurs when infected, undercooked pork is consumed. Research has shown that freezing as well as proper cooking kills the parasite. Trichinosis is now rare in the United States. ◻

Most Rotifers Are Microscopic

Rotifers are complex, free-swimming hunters too tiny to see with the naked eye. There are over 1,750 named species, some marine and others living in fresh water. Rotifers, shown in Figure 28-19, are characterized by a row of cilia surrounding their mouth. The action of the cilia sweeps food into the mouth. Under a microscope, the beating of the cilia looks much like a rotating wheel, which explains why these animals are called rotifers. Rotifers feed on bacteria and protists. Like roundworms, they have a well-developed, one-way gut and a pseudocoelom.

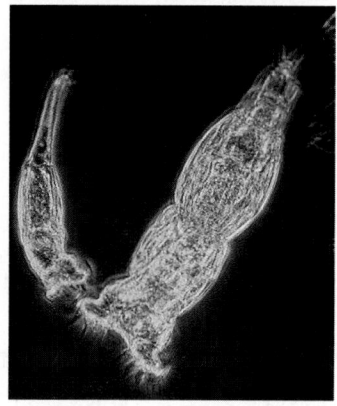

Figure 28-19 Rotifers are often found living in debris at the bottom of lakes and ponds and feed mainly on protozoans.

Section Review

1. *How does the internal anatomy of a fluke differ from that of a planarian?*

2. *Why is it important to properly cook beef?*

3. *Why might it be accurate to describe ribbon worms as a "transitional" group between flatworms and roundworms?*

4. *Compare and contrast* Ascaris *and* Necator.

Critical Thinking

5. *A student observing pond water through a microscope identifies a paramecium and a rotifer, both ciliated organisms. The student classifies both as protists. Is he correct? Explain your answer.*

CHAPTER REVIEW ANSWERS

Each item in the Chapter Review is correlated by text section in the assignment guide that follows.

ASSIGNMENT GUIDE

Section	Review	Themes Review	Critical Thinking
28-1	1, 2, 11, 12	19, 20	22
28-2	3–6, 13, 14	21	23
28-3	7–10 15–18		24

Review

Multiple Choice

Code in parentheses indicates section number and objective number.

1. b (28-1.2)

2. a (28-1.3)

3. c (28-2.1)

4. a (28-2.2)

5. d (28-2.3)

6. c (28-2.3)

7. b (28-3.1)

8. c (28-3.2)

9. a (28-3.3)

10. a (28-3.4)

Completion

11. nutrients (28-1.1)

12. hermaphrodites, self-fertilizing (28-1.3)

13. medusae, polyps (28-2.1, 28-2.2, 28-2.3)

14. polyps, planulae (28-2.2)

15. ectoparasites (28-3.1)

16. freezing, cooking (28-3.4)

17. circulatory, digestive tract (28-3.3)

18. Animalia (28-3.5)

---

28 CHAPTER REVIEW

Vocabulary

amoebocyte (639)	endoparasite (649)	planula (644)
atoll (646)	fluke (649)	proglottid (650)
barrier reef (646)	fringe reef (646)	spicule (640)
basal disk (642)	gemmule (640)	spongin (640)
ectoparasite (649)	hermaphrodite (641)	tegument (649)

Review

Multiple Choice

1. A protein sponge skeleton is composed of
a. spicules.
b. spongin.
c. mesenchyme.
d. amoebocytes.

2. What prevents self-fertilization among sponges?
a. Eggs and sperm are released at different times.
b. Few male sponges exist.
c. Sponges are hermaphrodites.
d. Encounters between members of the same species are rare.

3. A Portuguese man-of-war and a hydra are alike because both
a. are colonial.
b. contain medusae and polyps.
c. are hydrozoans.
d. produce planulae.

4. Which sequence reflects the life cycle of *Aurelia*?
a. polyp → medusa → planula
b. medusa → polyp → planula
c. planula → medusa → polyp
d. polyp → planula → medusa

5. Which is an anthozoan?
a. hydra
b. jellyfish
c. Portuguese man-of-war
d. sea anemone

6. Sinking volcanoes explain the existence of
a. sponges.
b. barrier reefs.
c. atolls.
d. ocean ridges.

7. The covering that protects endoparasites from the actions of digestive enzymes is called the
a. osculum.
b. tegument.
c. proglottid.
d. basal disk.

8. Which is *not* true about the life cycle of a beef tapeworm?
a. Humans can become infected by eating raw steak.
b. Cattle are not the only hosts of the parasite.
c. The adult tapeworm attaches to the human liver.
d. Eggs are deposited in human feces.

9. The most characteristic physical feature of members of the phylum Rhynchocoela is their
a. proboscis.
b. parasitic lifestyle.
c. coelomate body.
d. ciliated mouth.

10. Humans living in an area that lacks modern plumbing are likely to be parasitized by
a. *Ascaris*.
b. rotifers.
c. *Dugesia*.
d. *Trichinella*.

Completion

11. In sponges, amoebocytes are responsible for the movement of _____ from choanocytes to other sponge cells.

12. Sponges are usually _____ , producing both egg and sperm. But they are not _____ , because egg and sperm are produced at different times.

13. Some cnidarians live as free-floating _____ , some only as sessile _____ . Others pass through both in their life cycles.

14. In *Obelia*, reproductive _____ give rise to both male and female medusae. The zygotes produced from the union of the sperm and eggs released by the medusae develop into free-swimming larvae called _____ .

15. Flukes that live on the external surface of their hosts are called _____ .

16. Trichinosis in humans can be prevented by _____ or properly _____ pork.

17. Ribbon worms are more highly organized than flatworms, possessing a simple _____ system and a separate exit for the _____ tract.

18. Even though rotifers are microscopic, they are classified in kingdom _____ rather than in kingdom Protista.

Themes Review

19. **Structure and Function** Describe the structure of gemmules. How do gemmules ensure the survival of sponges?

20. **Evolution** Porifera has been called a dead-end phylum. List several reasons why no animal group evolved from the sponges.

21. **Flow of Energy** How does seasonal variation in light energy (and therefore heat) play a role in the life cycle of hydrozoans?

Critical Thinking

22. **Making Inferences** A single species of sponge may assume different appearances due to differences in substrate, availability of space, and the velocity and temperature of water currents. How might these factors make the classification of sponges confusing?

23. **Interpreting Data** Two living coral samples of equal size were placed in identically lighted, 100 gal. aquariums. The symbiotic algae were removed from one coral sample. The graph below shows the rate at which coral skeletons were deposited each year for four years. What do the data show about how the presence of the symbiotic algae affected the deposit of coral skeletons?

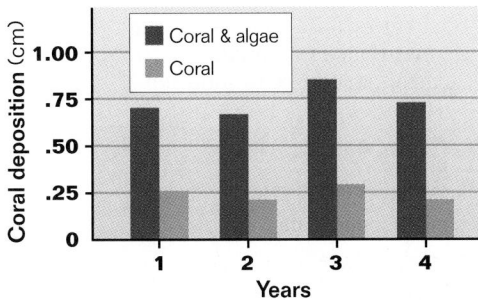

24. **Designing Experiments** Freshwater planarians reproduce sexually only during the fall. Design an experiment to determine whether day length or lower water temperature is the stimulus for sexual reproduction.

Activities and Projects

25. **Research and Writing** Look in the library for information about three of the parasitic roundworms that are not discussed in the text. Write a brief report about them, including information about their life cycles (include a diagram of each) and their economic impact.

26. **Multicultural Perspective** Compare and contrast the hydra of Greek mythology with the invertebrate hydra that you learned about in this chapter.

Themes Review

19. Gemmules are protective capsules that contain amoebocytes and a nutrient supply. Gemmules protect the amoebocytes from harsh environmental conditions that would kill adult sponges. (28-1.3)

20. Unlike other animals, sponges have no mouth or digestive tract. The structure of the sponge body is built around a water canal system and the outer body layer is poorly developed. (28-1)

21. The low water temperatures of autumn trigger the development of eggs and sperm. (28-2.1)

Critical Thinking

22. Differences in appearance may cause members of the same species to be classified as different species. (28-1.2)

23. The symbiosis facilitates the deposition of coral skeletons. (28-2.3)

24. Planarians can be placed in culture dishes where day length is varied and water temperature is kept constant. Other planarians can be placed in culture dishes where water temperature is varied and day length is kept constant. From these two experiments, it can be determined if one of the factors is responsible for the onset of sexual reproduction. (28-3.1)

Activities and Projects

25. Any college-level parasitology textbook will provide several examples of economically important parasitic roundworms.

26. Both hydras are animals. But the hydra that you studied is a tiny invertebrate, and the Greek Hydra was an enormous nine-headed serpent (a vertebrate) that lived in a lake on the Peloponnesus peninsula.

LABORATORY Investigation Chapter 28

Hydra Behavior

Preparation
Hydras are very sensitive to metals, such as copper, so do not use tap water unless it has been treated with a chelating agent, such as Ward's water conditioner. You may also use filtered, sterile pond water or bottled spring water. Hydras can be fed small live crustaceans like *Daphnia* and brine shrimp daily or every other day. Do not feed hydras for one or two days prior to the lab. Hydras that have recently been fed will not discharge nematocysts. Hydras will not survive if their water becomes too dirty, so the aquarium should be cleaned and the water changed at least every other day.

Procedural Notes
- Part A will take about 30 minutes to complete. Part B will take 10 to 15 minutes.
- Prior to the investigation, you might have students cut the filter paper in the shape of small pennants.
- Demonstrate to students how to pick up hydras and daphnias with a medicine dropper. If the tip of the dropper is too small, the dropper can be inverted.
- Instruct students to handle the hydras gently and watch patiently—it may take time for certain behaviors to occur. Encourage students to observe the specimens of other groups and share information.
- You may wish to have students observe prepared slides of hydras to study their internal structure. Prepared slides can be purchased from a biological supply house, such as Ward's.
- Discuss some of the basic principles of sensory cells, neurons, and effector cells, relating the information to the way hydras capture prey.

Background Answers
1. Animals respond to stimuli by moving toward or away from the

OBJECTIVE
Observe hydras to determine how they respond to different stimuli and how they capture and feed on prey.

PROCESS SKILLS
- observing an animal's feeding behavior and response to stimuli
- relating structure to function

MATERIALS
- silicone culture gum
- microscope slide
- hydra culture
- 2 medicine droppers
- stereomicroscope
- filter paper cut into pennant shapes
- forceps
- concentrated beef broth
- *Daphnia* culture

BACKGROUND
1. How do animals respond to stimuli in their environment?
2. How does a sessile animal, such as a hydra, obtain food?
3. Write your own question to explore in your lab report or notebook.

TECHNIQUE
Part A: Response to Stimuli
1. Using a long piece of silicone culture gum, make a circular "well" on a microscope slide, as shown in the illustration.

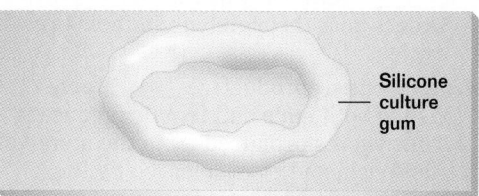
Silicone culture gum

2. With a medicine dropper, gently transfer a hydra from the culture dish to the well on the slide, making sure the hydra is in water. Allow the hydra to settle. Then examine it under the high power of a stereomicroscope. Identify and draw the hydra's body stalk, mouth, and tentacles in the **Records** section of your report.

3. In the **Records** section of your report, make a table like the one shown on the next page. Observe whether a hydra responds to a chemical stimulus, in this case a nutrient. First, hold a pennant-shaped piece of filter paper with forceps and move the long tip of the pennant near, but not touching, the hydra's tentacles. Observe and record the hydra's response to the filter paper in the **Records** section of your report. Next, dip the same piece of filter paper in beef broth and repeat the procedure. Record the hydra's response to the beef broth in the **Records** section of your report.

stimulus, by releasing chemicals, or by exhibiting other behaviors.
2. Sessile animals obtain food by taking in food that passes by them or by actively capturing the food.
3. How do hydras respond to different stimuli?

Records

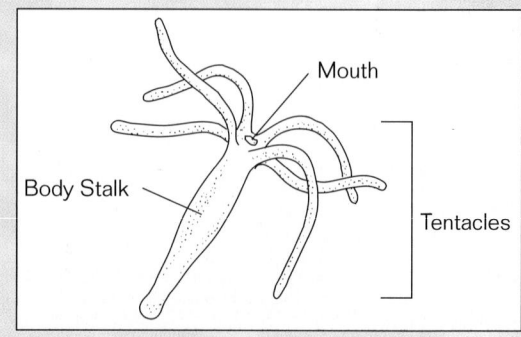
Mouth
Body Stalk
Tentacles

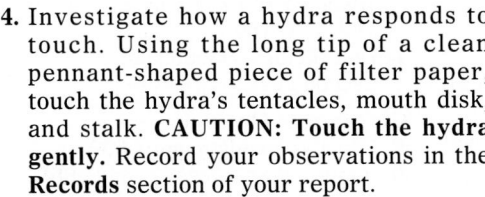

Observations of Hydra	
Response to filter paper	
Response to beef broth	
Response to touch	
Feeding behavior	

4. Investigate how a hydra responds to touch. Using the long tip of a clean pennant-shaped piece of filter paper, touch the hydra's tentacles, mouth disk, and stalk. **CAUTION: Touch the hydra gently.** Record your observations in the **Records** section of your report.

Part B: Feeding Behavior

5. Hydras eat small crustaceans, such as *Daphnia*. Use a medicine dropper to transfer live *Daphnia* to the well with the hydra on the microscope slide. Observe the hydra carefully under the stereomicroscope. Watch for threadlike nematocysts shooting out from the hydra. Some nematocysts release a poison that paralyzes the prey. If the hydra does not respond after a few minutes, it may not be hungry. In that case, obtain another hydra from the culture dish.

6. Observe the way the hydra captures and ingests the *Daphnia*, and record your observations in the **Records** section of your report. Also, briefly summarize the procedure you followed.

7. Clean up your materials and wash your hands before leaving the lab.

INQUIRY

1. Based on your observations, how do you think a hydra behaves when threatened in its natural habitat?
2. Describe a hydra's feeding behavior.
3. What happens to food that has not been digested by a hydra?
4. What was the purpose of using the untreated filter paper in step 3?
5. Did the hydra show a feeding response or a defensive response to the beef broth? Explain.

ANALYSIS

1. How is a hydra adapted to a sedentary lifestyle?
2. How is the feeding method of a hydra different from that of a sponge?

FURTHER INQUIRY

Write a new question that could be explored as a new investigation. The following are examples:

How does a hydra respond to light?

What kinds of food does a hydra eat?

Inquiry Answers

1. Based on the hydra's response to touch, students should conclude that a hydra contracts its body when it is threatened.

2. A hydra feeds by capturing prey with its tentacles and nematocysts, which sting the prey. The tentacles then push the food into the hydra's expanded mouth.

3. Undigested food is released from the hydra's mouth.

4. The untreated filter paper served as a control to show that the hydra was responding to the beef broth rather than to the filter paper.

5. The hydra should show a feeding response to the beef broth, which includes expansion of its mouth, movement of its tentacles, elongation of its body, and the release of nematocysts.

Analysis Answers

1. Tentacles and nematocysts enable a hydra to capture prey that drift past it in slow-moving water. Its ability to contract when touched protects the hydra from predators.

2. Both sponges and hydras are sessile. However, sponges obtain food passively by filter feeding. They have no specialized structures for capturing food. Hydras are active feeders that have tentacles and nematocysts for capturing food.

Observations of Hydra	
Response to filter paper	no response
Response to beef broth	moves toward filter paper
Response to touch	coils up
Feeding behavior	Nematocysts shoot out from the hydra, grab the *Daphnia*, and pull it to the hydra's mouth.

CHAPTER 29 · MOLLUSKS AND ANNELIDS

Block Scheduling Guide

> Each block represents about 45–50 minutes of instructional time. The resources cited in a block are coded to a specific program component using the Key on the next page.

Lecture Concepts	Lesson Resources

29-1 Mollusks pp. 659–667

BLOCK 1

a. Most mollusks breathe with gills and have an open circulatory system and nephridia, which enable mollusks to recover substances from wastes prior to excretion.
b. Bivalves, such as oysters, mussels, and scallops, are aquatic and have hard valves (shells) made of calcium carbonate that surround and protect their soft bodies.

Lecture Resources
- Opening Question, page 659
- Demonstrations: Mollusks on the Menu, Open and Closed Circulatory Systems
- Oxygen Delivery Systems, page 660
- Excretory Strategies, page 661
- The Root *Nephr-*, page 661
- Visual Strategy: Figure 29-5

- Teaching Transparencies: 135, 141, 142, 151, 159
- Holt Biology Videodiscs: Lesson 29-1

Classwork Options
- Effective Reading, page 659
- Picture of an Ancestor, page 661

BLOCK 2

a. Gastropods, such as snails, slugs, and nudibranchs, live in oceans, in fresh water, and on land. Gastropods are unique among mollusks because their bodies undergo torsion.
b. Cephalopods, such as octopuses, squids, and nautiluses, have a well-developed head region and many tentacles equipped with suction cups.

Lecture Resources
- Research Update, page 663
- A Good Defense, page 664
- Demonstrations: Snails, Suction Cups, Fossil Mollusks
- The Versatile Mantle, page 666
- Teaching Transparencies: 159
- Holt Biology Videodiscs: Lesson 29-2

Classwork Options
- Mollusks and Humans, page 665

- Major Characteristics, page 665
- Design a Mollusk, page 667
- Inquiry Skills Development B24: Snails

Assignment Options
- What's My Name? page 667
- Section Review, page 667
- Chapter Review, questions 1–5, 9, 11–14, 16, 18–20

29-2 Annelids pp. 668–673

BLOCK 3

a. Annelids are coelomate worms that have segmented bodies, a closed circulatory system, and carry out respiration through their skin.
b. Annelids include polychaetes, earthworms, and leeches. Marine polychaetes and terrestrial earthworms have external appendages (parapodia or setae). Leeches are both aquatic and terrestrial and lack such appendages.

Lecture Resources
- Opening Question, page 668
- Demonstration: Which End Is the Head?
- Overcoming Misconceptions, page 668
- Closed Circulatory System, pages 669
- Visual Strategy: Figure 29-15
- Nephridia, page 670
- Teaching Transparencies: 134, 152, 160, 136A
- Holt Biology Videodiscs: Lesson 29-2

Classwork Options
- Annelid Characteristics, page 669
- Up Close: Earthworm, page 672
- Closure, page 673
- Laboratory Investigation: Mollusks, pages 656–657
- Inquiry Skills Development B22: Earthworm Dissection
- Inquiry Skills Development B23: Live Earthworms

Assignment Options
- Section Review, page 673
- Chapter Review, questions 6–10, 15, 17, 21

Review/Enrichment

BLOCK 4

■ Study Guide: Concept Review and Pretest

Assignment Options
■ Activities and Projects, questions 22, 23

Assessment Options

BLOCK 5

Traditional Assessment
■ Chapter 29 Test
◉ Test Generator/Assessment Item Listing: Software item bank for preparing customized chapter tests.

Portfolio Assessment
Select student reports for one or more laboratory experiments from this chapter. *The Direct Observation Checklist* on the *BioSources Teaching Resources CD-ROM* should be completed during a laboratory or other cooperative learning experience.

Holt Biology Videodiscs

Holt Biology Videodiscs gives you a powerful tool for teaching, review, and assessment. *Concepts of Biology* adds interactive lesson plans and extensive tools for customization using CD-ROM technology.

CONCEPTS OF BIOLOGY

Use the following topic from *Concepts of Biology* to help you teach this chapter:

■ Topic 18: Invertebrate Diversity

For further information regarding the media on the videodiscs, see *Holt Biology Videodiscs Teacher's Correlation Guide to Biology: Principles and Explorations.*

Chapter Themes

Structure and Function

The development of a coelom in animals is a major evolutionary adaptation. The structure of the coelom led to the development of a closed circulatory system, provided a fluid environment within which digestive, sexual, and other organs could be suspended, and facilitated muscle-driven body movement.

Evolution
Mollusks and annelids are diverse groups of invertebrates that share common anatomical features. The shared pattern of development of trochophore larvae in mollusks and annelids is evidence of their common ancestry.

Tapping Prior Knowledge

- What are three major mollusk characteristics?
- Name a major evolutionary development that first evolved in annelids.

Opening Demonstration

Invite students to bring any shell collections they might have to class. Display the specimens with mollusks or mollusk shells you are able to acquire. Examples include clams, mussels, scallops, oysters, octopuses, squids (available from fish markets), and snails and slugs (found in gardens). Ask students to look for similarities among the mollusks and shells on display. Point out the variety of shapes and sizes among the shells.

CHAPTER 29

MOLLUSKS AND ANNELIDS

REVIEW

- development of the coelom (Section 27-2)
- cephalization (Section 27-2)
- mantle, visceral mass, foot, radula (Section 27-2)
- mollusk classification (Section 27-2)
- segmentation (Section 27-3)
- annelid classification (Section 27-3)

Blue-ringed octopus

Authors' Rationale

Mollusks and annelids are two invertebrate groups of exquisite design and diversity. Section 29-1 discusses how clams, squids, slugs, and snails are related to each other and how their structures and functions are suited to each mollusk's environment. Section 29-2 presents the characteristics, organ systems, and diversity of annelids. Students will learn that the first example of segmentation can be found in annelids.

29-1 Mollusks

A snail may not seem to have much in common with an earthworm, but in fact, these two very different-looking animals are related. Mollusks and annelids were probably the first major groups of organisms to develop a true coelom. Also, the fertilized eggs of both groups develop into a distinct larval form called a *trochophore* (TRAHK oh fawr), *shown in Figure 29-1.*

Characteristics of Mollusks

Mollusks are one of the most successful of all animal phyla. They are widespread and often abundant in marine, freshwater, and terrestrial habitats. They are the largest animal phylum, except for the arthropods, with over 100,000 named species. Members include a wide variety of animals, such as snails, oysters, octopuses, and cuttlefishes like the one shown in Figure 29-2. Despite their varied appearance, all mollusks share the following characteristics.

1. **Body cavity** All mollusks have a true coelom, although in most it is reduced to a small area immediately surrounding the heart.
2. **Symmetry** Most mollusks have bilateral symmetry, and many have one or more shells called valves.
3. **Organ systems** Mollusks have organ systems for circulation, respiration, digestion, and excretion.
4. **Three-part body plan** The body of every mollusk has three distinct parts: the muscular foot, the head, and the visceral mass.

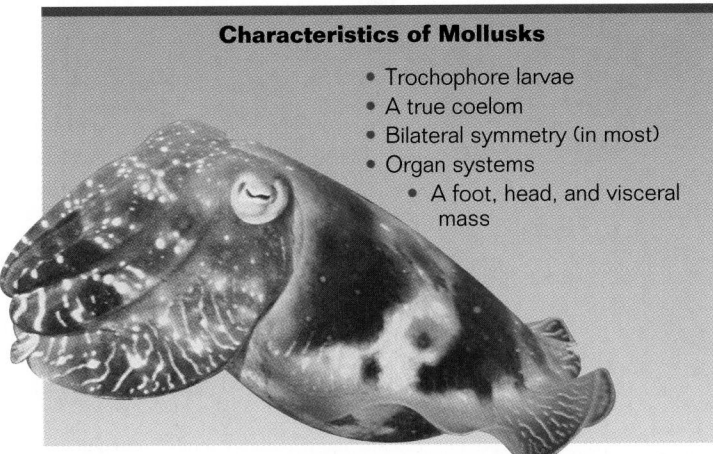

Characteristics of Mollusks

- Trochophore larvae
- A true coelom
- Bilateral symmetry (in most)
- Organ systems
  - A foot, head, and visceral mass

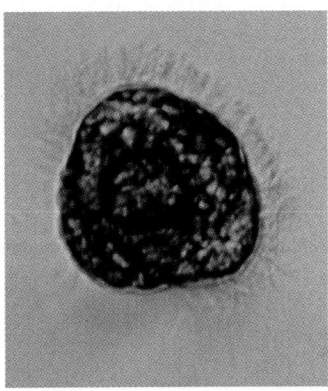

Figure 29-1 A trochophore larva has a belt of cilia that circles its body. The beating of the cilia propels the larva through the water. The cilia may also function to trap the tiny plankton that the trochophore feeds on.

Figure 29-2 This cuttlefish has all of the major mollusk characteristics, although the foot has been modified into tentacles. Extremely agile swimmers, most species of cuttlefish hunt small fishes and crustaceans at night.

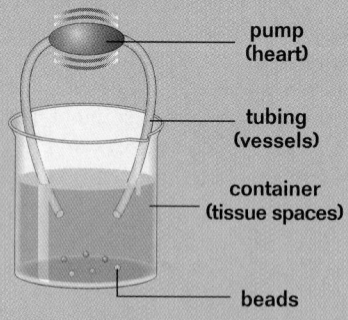
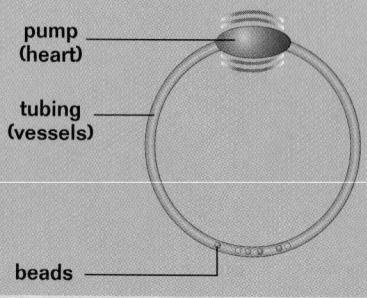

Organ Systems of Mollusks

As you learned in Chapter 27, the evolution of a coelom in mollusks enabled the development of complex organ systems in these animals. Mollusks are one of the earliest evolutionary lines to have developed an efficient excretory system.

Respiration

Most mollusks breathe with ciliated gills located in their mantle cavity. The **mantle cavity** is a space between the mantle and the visceral mass. The constant beating of the cilia causes a continuous stream of water to pass over the gills. Mollusk gills are very efficient and may extract 50 percent or more of the dissolved oxygen from the water that passes over them. Some terrestrial snails lack gills. Instead, their mantle cavity functions as a simple lung.

Circulation

Most mollusks have a three-chambered heart and an open circulatory system. In an **open circulatory system,** the blood leaks out of blood vessels and bathes the body's tissues directly, as shown in Figure 29-3. Two of the heart's chambers collect blood from the gills. The third chamber pumps the oxygenated blood out of the vessels and into spaces in the mollusk's tissues, where nutrients, oxygen, and carbon dioxide are exchanged between the blood and tissues. The blood then returns to the heart via the gills. Among mollusks, only octopuses and squids have a **closed circulatory system,** in which the blood never leaves the blood vessels. In a closed circulatory system, materials pass into and out of the blood by diffusing across the walls of the vessels.

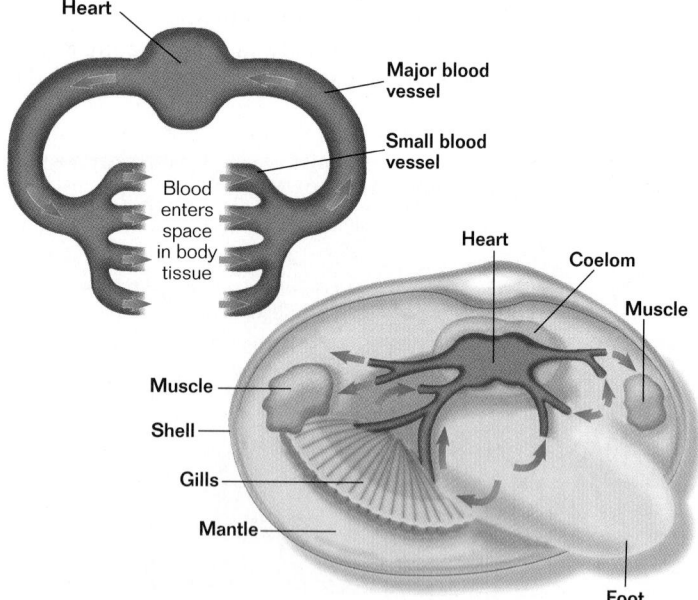

Figure 29-3 Most mollusks have an open circulatory system. The diagram, *top*, shows that in an open circulatory system blood leaks out of vessels and bathes tissue directly. In a bivalve's open circulatory system, *bottom*, blood flows out of some blood vessels and into others.

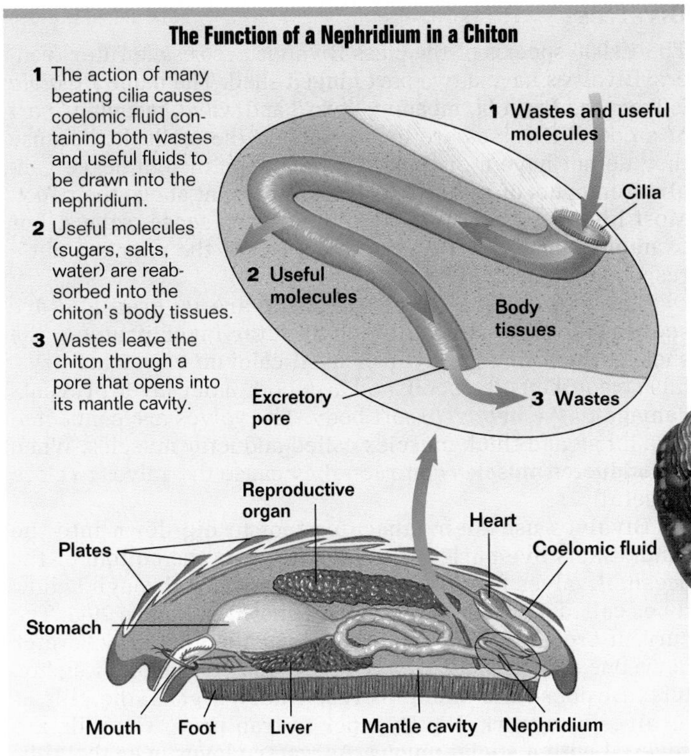

The Function of a Nephridium in a Chiton

1 The action of many beating cilia causes coelomic fluid containing both wastes and useful fluids to be drawn into the nephridium.

2 Useful molecules (sugars, salts, water) are reabsorbed into the chiton's body tissues.

3 Wastes leave the chiton through a pore that opens into its mantle cavity.

1 Wastes and useful molecules

Cilia

2 Useful molecules

Body tissues

Excretory pore

3 Wastes

Reproductive organ

Heart

Coelomic fluid

Plates

Stomach

Mouth   Foot   Liver   Mantle cavity   Nephridium

Figure 29-4 Chitons, *below* and *bottom left,* are mollusks that are commonly found attached to rocks in shallow water along shorelines. As in all mollusks, chitons use structures called nephridia, *top left,* to recycle useful molecules from their wastes before the wastes are eliminated from the chiton's body.

Excretion

Mollusks use their coelom as a refuse dump in which waste-laden body fluids are collected. Nitrogen-rich wastes are filtered from the coelomic fluid by tiny tubular structures called **nephridia** (*nee FRIHD ee ah*). Nephridia, diagramed in Figure 29-4, are found in all coelomates except arthropods and chordates. They are a highly successful evolutionary innovation that allow useful molecules to be efficiently recovered from body fluids before wastes are discharged from the animal. ▢

▢ CAPSULE SUMMARY

Most mollusks breathe with gills and have an open circulatory system and nephridia. Nephridia enable mollusks to recover the useful substances from wastes before they are excreted from the animal's body.

Mollusk Diversity

There are seven classes of mollusks. By studying four of the minor classes, taxonomists have learned a great deal about the probable ancestor of this phylum. Research has indicated that the ancestor was a flattened, unsegmented, wormlike animal that glided along its ventral surface. Chitons, shown in Figure 29-4, are members of the class Polyplacophora. This class is one of the smaller groups of mollusks and still has many of the characteristics of its mollusk ancestors. As you read in Chapter 27, the three major mollusk classes are the bivalves, gastropods, and cephalopods.

SECTION 29-1

Teaching Tips
Excretory Strategies
Relate the excretion of wastes in nephridia to cleaning up a table after a meal. Nephridia get rid of wastes as well as substances that the mollusk needs; thus, some substances must be reabsorbed before the wastes are discharged. The process is similar to throwing away everything on a table after a meal, then going through the trash to recover plates and silverware that you want to keep.

The Root *Nephr-*
In medicine, the root *nephr-* is often found, in reference to the human excretory organ, the kidney. Nephritis is inflammation of the kidneys, a nephrectomy is removal of a kidney, a nephron is a structure in the kidney that filters blood, and nephrosis is degeneration of kidneys. Ask students to think of a likely word for a kidney doctor and the field of medicine a kidney doctor is likely to practice. (*A nephrologist practices nephrology.*)

Picture of an Ancestor
The text describes the probable ancestor of mollusks as "a flattened, unsegmented, wormlike animal that glided on its ventral surface." Ask students to look through this section and pick out a modern mollusk that most closely matches this description. (*chiton*)

Application

 Health Eating raw mollusk filter feeders such as oysters or clams can be a health risk because they filter pathogens from the water. If poorly processed sewage is dumped near shellfish beds, mollusks can collect pathogens that cause cholera, hepatitis, and other serious diseases. Cooking mollusks kills the pathogens.

👁 VISUAL STRATEGY

Figure 29-5

Use this figure to point out that mollusks have advanced organ systems. Ask students to categorize each of the quahog clam's organs as one of the following systems: digestive, circulatory, respiratory, or excretory. Some organs, such as the siphon and gills, may be listed under more than one system. *(digestive system—incurrent siphon, gills, mouth, esophagus, stomach, intestine, excurrent siphon; circulatory system—heart; respiratory system—gills, incurrent siphon, excurrent siphon; excretory system—kidney, excurrent siphon)* Have students explain why each system is necessary for the mollusk to live.

Bivalves

The 10,000 species of the class Bivalvia are sessile filter feeders. Bivalves have a two-part hinged shell. The name *Bivalvia* is from the Latin *bi*, meaning "two," and *valva*, meaning "part of a door." Bivalves are unique among the mollusks because they do not have a distinct head region, although nerve ganglia (remnants of a simple brain) are present above their foot. Most bivalves have at least rudimentary sense organs. For example, sensory cells along the edge of the clam's mantle respond to light and touch.

The valves (shells) of a bivalve are layered and are secreted by the mantle. A tough outer layer protects the shell, a thick middle layer of hard calcium carbonate crystals strengthens the shell, and a smooth inner layer prevents damage to the bivalve's soft body. The valves are connected by a hinge and thick muscles called adductor muscles. When the **adductor muscles** contract, they cause the valves to close forcefully.

Bivalves use their muscular foot to dig down into the sand. Clams live buried in mud or sand on the bottom of the ocean. Bivalves feed by sucking in sea water through hollow tubes called **siphons** (*SEYE fuhns*), shown in Figure 29-5. The current created by the cilia that cover the gills draws water down one siphon tube, over the gills, and out the other siphon tube. Besides functioning as respiratory organs, the gills of bivalves also work like flypaper to trap prey. The gills are covered with a sticky mucus. As water moves over the gills, small marine animals, protists, and organic material become trapped in the mucus. The cilia then direct the food-laden mucus to the bivalve's mouth.

Figure 29-5 Many bivalves, like this quahog clam, *above,* burrow into sand or mud on the ocean bottom. They feed by drawing sea water in one siphon and expelling it out the other. Bivalves, like all mollusks, have all of the major organ systems, *right.*

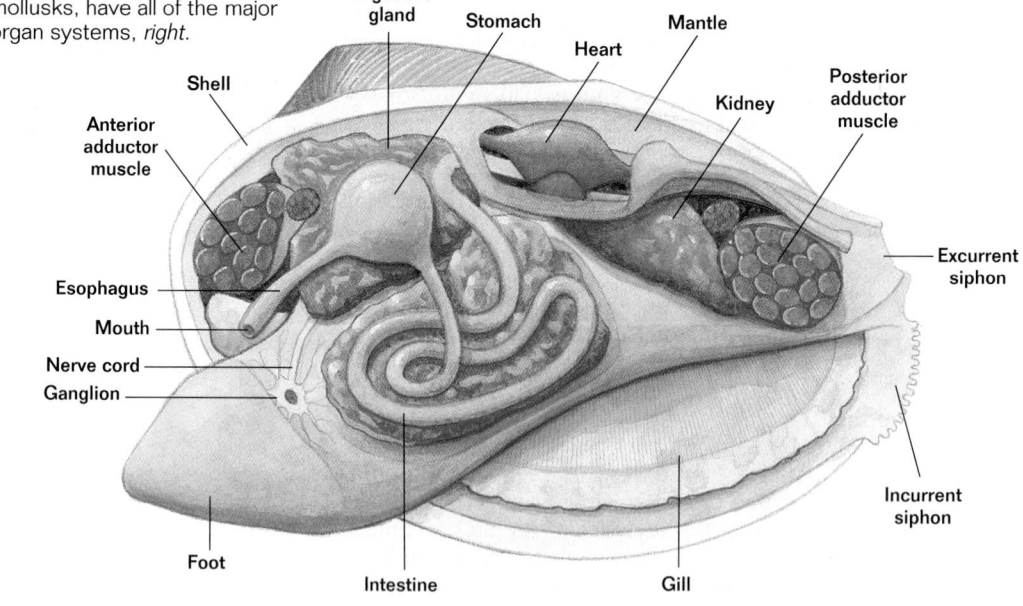

Different kinds of bivalves feed in various ways. Oysters feed in the open water with their shells permanently attached to rocks, while scallops don't attach themselves to anything. Instead, water passes over their gills as they swim. A swimming scallop looks like it might be eating, its valves rapidly opening and shutting like jaws. Actually, it is pushing itself through the water with the jets of water it expels when its valves snap shut. One bivalve does not filter-feed at all. The teredo, or shipworm, shown in Figure 29-6, digests the cellulose in wood by using symbiotic protists that live inside its intestine, much as a termite does.

Bivalves reproduce sexually. Most are either male or female, but a few species are hermaphroditic. They reproduce by shedding sperm and eggs into the water, where fertilization occurs. The fertilized eggs develop into free-swimming trochophore larvae. The trochophore larvae of marine bivalves, as well as those of marine gastropods, develop into a second free-swimming stage called a **veliger** (*VEE lah jur*). Veligers drift in ocean currents and are dispersed far and wide. In the veliger stage, the beginnings of the foot, shell, and mantle become evident. Eventually, veligers settle to the ocean bottom and grow into adults.

Gastropods

The 80,000 species of the class Gastropoda are snails and slugs. Gastropods are primarily a marine group that has very successfully invaded freshwater and terrestrial habitats. As you learned in Chapter 27, their bodies are generally divisible into a conspicuous head, foot, and visceral mass. The foot of gastropods is adapted for locomotion. Terrestrial species secrete mucus from the base of their foot, forming a slimy path that they can glide along. Most gastropods have a pair of tentacles, on which the eyes are often located, on their head. Many gastropods have a single shell. During the evolution of slugs and nudibranchs (*NOO dih brangks*), or sea slugs, shown in Figure 29-7, the shell was lost. The shells of most marine snails can be closed by a plate that the animal pulls into place like a door; most land snails lack this plate.

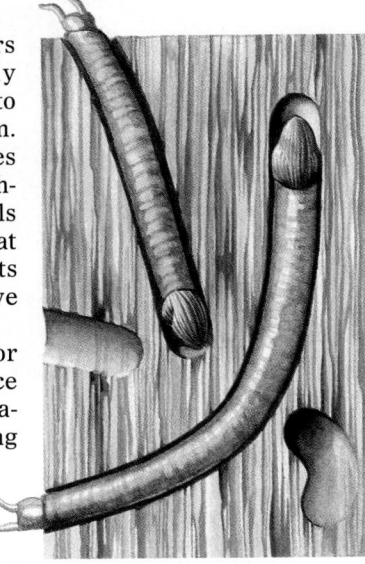

Figure 29-6 A shipworm's valves are highly reduced and function as a drill, which enables the animal to bore into wood. Shipworms can cause extensive damage to wooden ship bottoms.

Figure 29-7 Many sea slugs, *below left,* secrete toxic substances from their skin. Their bright coloration acts as a warning sign to other animals. Many terrestrial slugs, *below,* breathe air. Their mantle cavity has been modified into a simple lung.

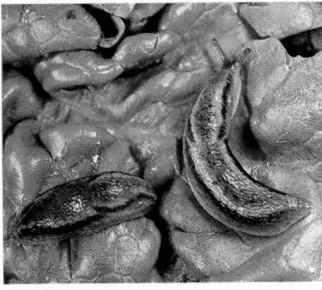

☑ RESEARCH Update

UPDATE

Mini Mussel Makes Major Mess

Dreissena polymorpha, the zebra mussel, is a small freshwater bivalve mollusk. In Europe, natural predators control the zebra mussel population. However, when zebra mussels hitchhiked across the Atlantic Ocean in the ballast water of ships, they started colonizing freshwater environments in the United States that contained few natural predators of the mussels. Not found in this country prior to 1985, zebra mussels have already exploded in population, occluding industrial water supply intakes and threatening established aquatic populations.

Zebra mussels attach with tough byssal threads to any firm structure and proceed to attach to each other, forming thick blankets of mussels. The Detroit Edison Monroe Power Plant noticed that the number of mussels per square meter of trash bars, which screen incoming water, increased from between 500 and 1,000 to more than 7,000 during a six-month period.

Each zebra mussel siphons about 1 qt. of water a day through its filter-feeding apparatus, feeding on plankton that would have been eaten by indigenous zooplankton, hence damaging native food chains. Considering the number of freshwater systems interconnected by open canals and the shipping industry, it isn't likely that "accidental immigration" such as that of the zebra mussel can be prevented. Marine biologists at the University of North Carolina found 367 kinds of marine organisms in ballast water samples from 159 Japanese cargo ships entering Coos Bay, Oregon. Because 39,000 merchant ships sail the world's oceans, the problem is likely to grow. ☑

❓ Did You Know?

One of the ways that biological oceanographers study the physical dynamics of organisms is to examine how the special formation of stacks of mussels improves their ability to feed. As a biological oceanographer at the Woods Hole Oceanographic Institution, where Rachel Carson once studied, Dr. Cheryl Ann Butman's research activities include placing special equipment on the ocean floor to monitor the growth and development of marine invertebrates.

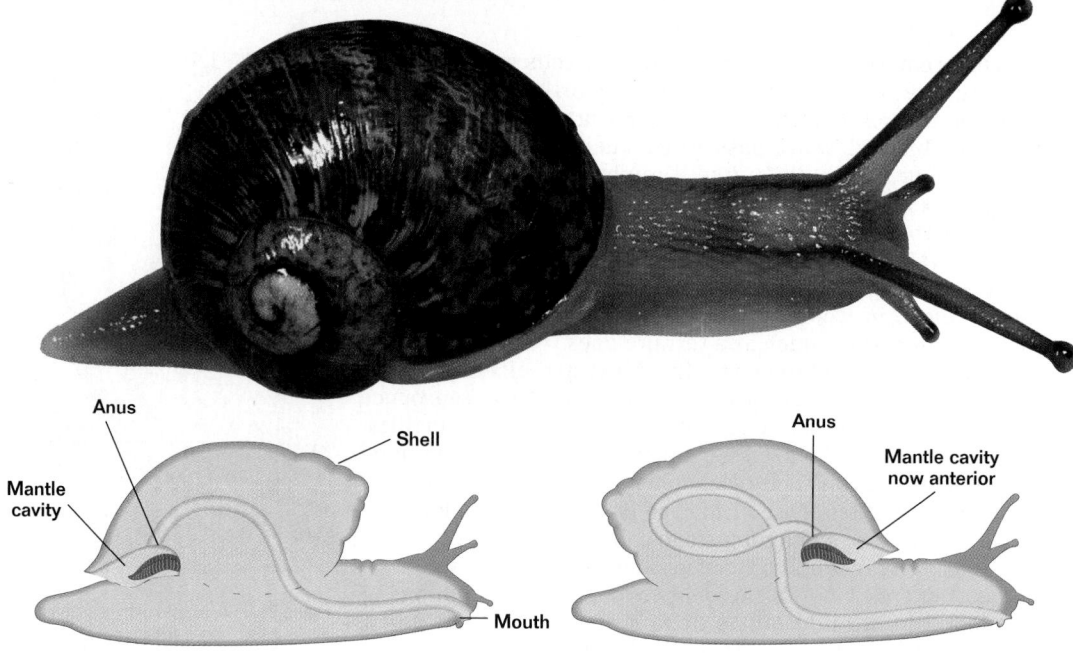

Teaching Tip
A Good Defense
Ask students to list some ways that mollusks can escape from predators. *(Snails can close their shells with a plate, squids can speed away on a jet of water, sea slugs can excrete toxic substances, etc.)*

Demonstration
Snails
Allow students to explore snail behavior. Obtain several land snails from gardens, wooded areas, or a biological supply house. Have students devise careful ways to test the snails' responses to touch, light, moisture, and gravity. Store snails in a cool, moist terrarium with pieces of lettuce.

Anus

Shell

Mantle cavity

Mouth

Anus

Mantle cavity now anterior

Figure 29-8 Torsion, a 180° twisting of a gastropod's visceral mass, occurs during the gastropod's embryonic development. Before torsion, *above left,* the snail's mantle cavity is posterior. After torsion, *above right,* the mantle cavity is anterior, giving the snail a space to pull its head into when it is threatened.

❏ CAPSULE SUMMARY

Bivalves (oysters, mussels, and scallops) are aquatic and have hard valves (shells) made of calcium carbonate that surround and protect their soft bodies. Gastropods (snails, slugs, and nudibranchs) live in oceans, in fresh water, and on land. Gastropods are unique among mollusks because their bodies undergo torsion.

The body plan of gastropods has undergone a significant change from that of their mollusk ancestors. In gastropods, the visceral mass rotates 180° during development. This twisting, called **torsion** *(TAWR shuhn)*, causes a rearrangement of organs and moves the mantle cavity from the back to the front of the animal, as shown in Figure 29-8. The spiraling of a gastropod's shell is not caused by torsion. It occurs before torsion begins.

Respiration among gastropods is carried out in a variety of ways. Aquatic snails breathe with gills that are located in the mantle cavity. Nudibranchs (marine snails) lost their gills as well as their shells through evolution—gas exchange takes place directly through their skin. Gills, which can function only when the delicate filaments of gill tissue are supported by water, have been lost in terrestrial snails as well. The empty mantle cavity of terrestrial snails acts as a primitive lung. Oxygen in the air diffuses across the thin membrane that lines the cavity. Because this membrane must be kept moist for respiration to occur, terrestrial snails are most active when air has a high moisture content, such as at night or after it rains. During dry weather, a terrestrial snail avoids water loss by creeping back into its shell and plugging the opening with a wad of mucus to keep water in. ❏

Gastropods display extremely varied feeding habits. Many are herbivores that scrape algae off rocks with their radula. Remember from Chapter 27 that a radula is a tongue-like scraping organ. Some terrestrial snails can be serious garden and agricultural pests by using their radula to saw off leaves. Nudibranchs and many other gastropods are active

 Did You Know?

Not all snail shells twist the same way. Some shells twist to the left and some twist to the right, depending on the species.

predators. Whelks and oyster drills, for example, use their radula to bore holes in the shells of other mollusks. Then the tissue of their prey is sucked out. In gastropods called cone shells, shown in Figure 29-9, the radula has been modified into a kind of poison-tipped harpoon that is shot into prey.

Cephalopods

The more than 600 species of the class Cephalopoda include squids, octopuses, cuttlefishes, and nautiluses. The name *cephalopod* is from the Greek *kephalicos*, meaning "head," and *pous*, meaning "foot." Appropriately named, most of their body is a large head attached to tentacles (the foot divided into numerous parts), as shown in Figure 29-10. The

Figure 29-9 The eastern Pacific cone shell sweeps its long proboscis back and forth over the ocean bottom in search of prey, *left*. When a fish gets close, the cone shell quickly paralyzes the fish by stabbing it with the poison-tipped radula located at the tip of the proboscis. The fish is then swallowed whole, *right*.

Teaching Tips
Mollusks and Humans
Have students choose a mollusk that affects humans. Then have them research and report on how the mollusk has been affected by humans and how humans have been affected by the mollusk. Topics could include edible mollusks, mollusk pests, or mollusks used for commercial products such as pearls.

Major Characteristics
Ask students to develop a table like the *Graphic Organizer* at the bottom of this page that compares the major characteristics of bivalves, gastropods, and cephalopods.

Figure 29-10 Most octopuses are bottom dwellers that live in crevices among rocks and corals. They move around mainly by crawling, and they swim only to escape from enemies.

Graphic Organizer
Use this graphic organizer with *Teaching Tip:* **Major Characteristics** on this page.

	Bivalves	Gastropods	Cephalopods
Have head, foot, and visceral mass	yes; head undeveloped	yes; torsion of visceral mass	yes; foot divided into tentacles
Have true coelom	yes	yes	yes
Have shells	two shells	most have one shell	no, except chambered nautilus
Use locomotion	most do not	yes	yes

Figure 29-11 A nautilus, *above left*, swims backward, with its coiled shell positioned over its head. The animal lives only in the outermost compartment of its partitioned shell, *above right*. The inner chambers of the shell are filled with gas. By regulating the amount of gas in the chambers, the nautilus is able to adjust its buoyancy and thus control its depth in the water.

tentacles of cephalopods are equipped with suction cups or hooks for seizing prey. Squids have 10 tentacles, octopuses have 8, and the nautiluses have 80 to 90. All cephalopods are active marine predators. They feed on fish, mollusks, crustaceans, and worms. Once the prey has been snared by the tentacles, it is pulled to the mouth, where it is bitten by strong, beaklike jaws. The cephalopod's radula then pulls the prey into the mouth.

Like most aquatic mollusks, cephalopods draw water into their mantle cavity and expel it through a siphon. The squids and octopuses, however, have modified this system into a means of jet propulsion. When threatened, they quickly close their mantle cavity, causing water to shoot forcefully out of the siphon. Both squids and octopuses can also release a dark fluid that clouds the water and thus disguises the direction of their escape.

Although they evolved from shelled ancestors, most modern cephalopods lack an external shell. The nautilus, shown in Figure 29-11, is the only living cephalopod that has retained its outer shell.

Cephalopods are the most intelligent of all invertebrates. In contrast to the other mollusks, cephalopods have a complex nervous system that includes a well-developed brain, and they are capable of exhibiting complex behaviors. Octopuses can easily be trained to distinguish among classes of objects.

The structure of a cephalopod eye is similar to that of a vertebrate eye, and some cephalopod eyes grow quite large. A giant squid that washed up on a beach in New Zealand in 1933 had eyes that were 40 cm (about 16 in.) across, the largest eyes ever measured in an animal.

Figure 29-12 During reproduction, the male octopus, *left*, removes a sperm packet from inside his mantle cavity and transfers it into the mantle cavity of a female, *far left*. The eggs are fertilized as the female lays her eggs on the ocean bottom, and they are guarded until they hatch.

Cephalopods exhibit both sexes. In males, sperm are stored in sacs that open into the mantle cavity. During breeding, the male uses a modified tentacle to transmit a packet of sperm from its own mantle cavity into the female's mantle cavity, as shown in Figure 29-12. The eggs are fertilized as they leave her body and are attached to stones or other objects. ❑

❑ CAPSULE SUMMARY

Cephalopods (octopuses, squids, and nautiluses) have a well-developed head region and many tentacles equipped with suction cups. An effective propulsion system and a complex nervous system have enabled these mollusks to become effective predators.

Section Review

1. *What evidence suggests that mollusks and annelids share a common ancestor?*
2. *Sketch a nephridium, and describe how it functions.*
3. *Why would you expect the blood pressure inside a mollusk's blood vessels to be quite low?*
4. *Define* torsion, *and identify the mollusk group in which it occurs.*
5. *Contrast the feeding habits of cephalopods with those of gastropods.*

Critical Thinking

6. *A chemical pollutant accidentally spills into a bay. One of the effects of the chemical is that it paralyzes cilia. The next day almost all of the oysters in the bay are dead. Explain why.*
7. *Squids are the fastest swimmers of all aquatic invertebrates. Name two structural adaptations in squids that may have enhanced their swimming ability.*

Opening Question

Ask students to list the anatomical features that make earthworms more advanced than flatworms or roundworms. Help students recall that earthworms have a segmented body, a true ceolom, a highly specialized gut, a closed circulatory system, and external bristles.

Demonstration

Which End Is the Head?

Obtain earthworms from garden soil or a bait shop, and have students observe their behavior. Ask students to describe how they can tell which end of the earthworm is the head. If students say that the head is the leading end, remind them that a squid sometimes goes tail first. Direct their attention to the earthworm's mouth. Remind students to keep their worm moist. Ask them why dry worms die. *(They suffocate.)*

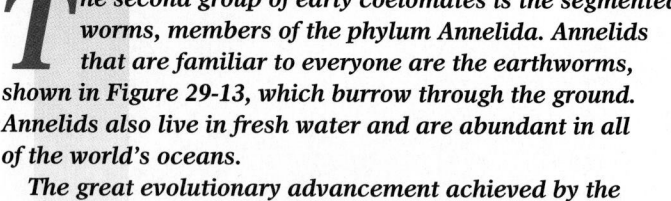

29-2 Annelids

The second group of early coelomates is the segmented worms, members of the phylum Annelida. Annelids that are familiar to everyone are the earthworms, shown in Figure 29-13, which burrow through the ground. Annelids also live in fresh water and are abundant in all of the world's oceans.

The great evolutionary advancement achieved by the annelids is segmentation. Many scientists believe that segmentation first evolved as an adaptation for burrowing. Segmentation enables the worm to produce strong waves of muscular contraction along the length of its body, making burrowing easier and faster.

Section Objectives

- State the major annelid characteristic that distinguishes annelids from mollusks.
- Describe the circulatory system of an annelid.
- Describe three classes of annelids.
- Identify the internal structures of an earthworm.

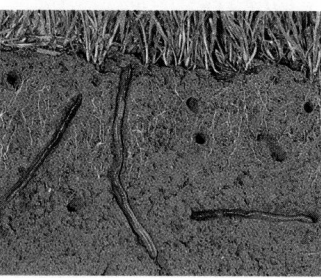

Figure 29-13 Earthworms come to the surface only at night or during heavy rains. During dry or cold weather, they burrow deep into the soil and become inactive.

Characteristics of Annelids

Annelids are easily recognized by their segments, which are visible externally as a series of ringlike structures along the length of their body. The name *annelid* is from the Latin word *annellus*, meaning "ring." At first glance the annelids that burrow beneath the ground seem to have little in common with the fierce predatory annelids of the open ocean. But in fact, all annelids, like the one shown in Figure 29-14, exhibit certain basic characteristics.

1. **Body cavity** The body cavity in annelids is a true coelom.
2. **Segmentation**   In all annelids, the body is segmented, divided into many nearly identical units. Some segments fuse during development, but many segments remain separate.

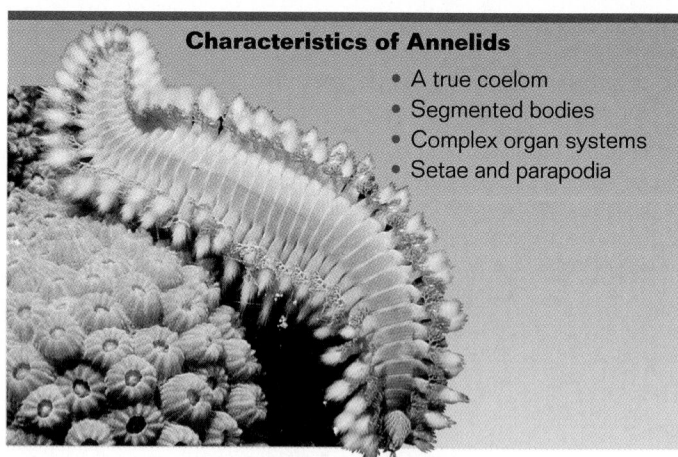

Characteristics of Annelids

- A true coelom
- Segmented bodies
- Complex organ systems
- Setae and parapodia

Figure 29-14 Notice the paired parapodia and numerous setae of this polychaete. The term *polychaete* is from the Greek *poly*, meaning "many," and *chaite*, meaning "hair."

Overcoming Misconceptions

Grubs and Caterpillars

Many people mistake grubs and caterpillars for worms. Have students list the annelid characteristics listed in Figure 29-14 that a grub or a caterpillar has. *(Answers will vary, but some students may find indications* *for all four characteristics.)* Make sure students understand that grubs and caterpillars are not annelids but are arthropods. Remind students that grubs and caterpillars are insects in their larval stages.

3. **Organ systems** The major organ systems of annelids include a highly specialized gut, a closed circulatory system, and many nephridia.

4. **Bristles** Most annelids have external bristles called **setae** *(SEET ee)*. Marine annelids also have many fleshy appendages called **parapodia** *(par uh POH dee uh)*.

Organ Systems of Annelids

The organ systems of annelids display a high degree of specialization. Cephalization has led to the development of anterior sense organs and a brain. A ventral nerve cord and pairs of segmental ganglia make the coordinated movement of each body segment possible. The digestive tube is modified into specialized regions: the crop, stomach, and intestine.

Circulation and Respiration

Annelids and many other coelomates have a closed circulatory system, like the one shown in Figure 29-15. Only the two largest animal phyla, arthropods and mollusks, have an open circulatory system. Blood moves through a closed system of blood vessels faster than through an open system because the blood is under greater pressure. In several

Figure 29-15 All annelids have a closed circulatory system. The diagram, *top*, shows that in a closed circulatory system, the blood never leaves the hearts or blood vessels. The locations of the hearts and major blood vessels in an earthworm, *bottom*, are also shown.

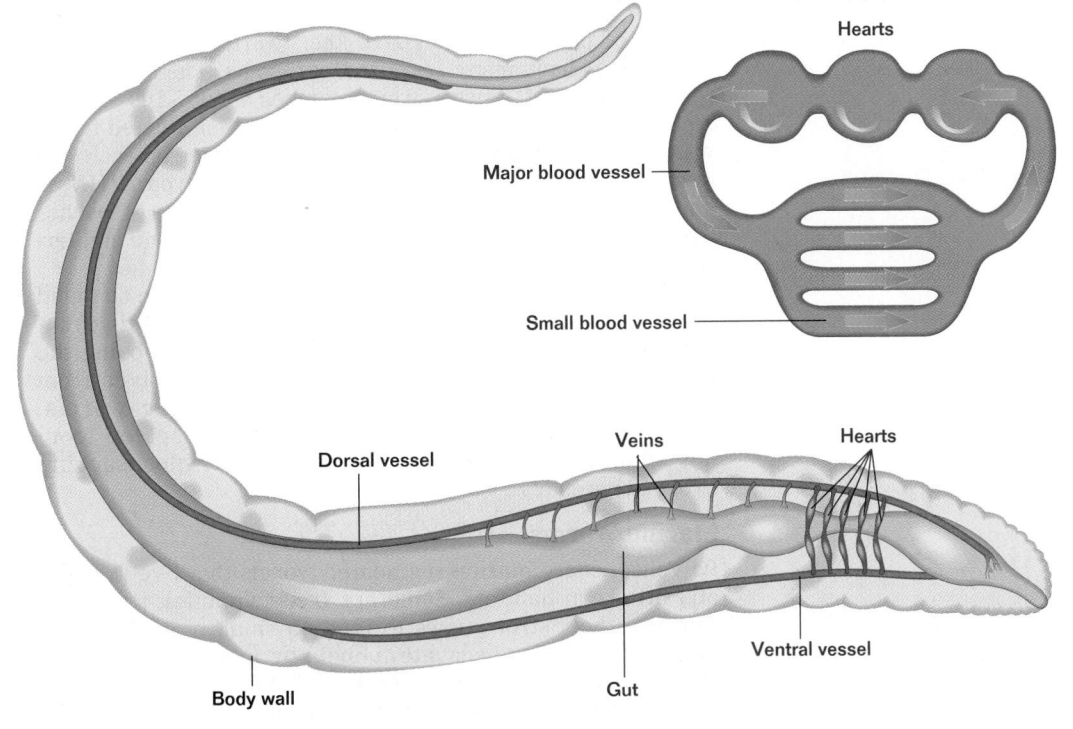

Hearts

Major blood vessel

Small blood vessel

Dorsal vessel

Veins

Hearts

Ventral vessel

Body wall

Gut

Teaching Tips
Annelid Characteristics
Have students defend the classification of an earthworm as an annelid, using the characteristics outlined in Figure 29-14 on page 668. Ask students to describe any evidence they see that an earthworm has a body cavity, segmentation, organ systems, and bristles.

Closed Circulatory System
Explain that a closed circulatory system is essential for efficiently transporting materials in complex animals. A closed system allows greater regulation of the composition of circulating fluids.

VISUAL STRATEGY

Figure 29-15
Use this figure to discuss the closed circulatory system of annelids. Ask the following questions about how the anatomical structures work. What parts produce pressure in the system? *(hearts)* What parts are in close contact with body tissues? *(small blood vessels)* How do oxygen and nutrients get to the tissues if the blood doesn't touch body tissues directly? *(The oxygen and nutrients pass through the vessel walls.)*

Teaching Tips
Earthworm Circulation

Ask students whether an earthworm could die of a heart attack. *(Students should guess that one defective heart won't stop circulation of blood.)* Point out that defective parts in a human heart do not necessarily stop circulation of blood either.

Nephridia

Point out that liquid waste that is excreted by nephridia serves another important function. The liquid helps keep the worm's surface moist.

CONTENT LINK

You can learn about how hemoglobin functions in human blood in Chapter 37.

❏ CAPSULE SUMMARY

Annelids are coelomate worms that have segmented bodies. Respiration in annelids occurs through their skin, and their blood is pumped through a closed circulatory system.

segments of an annelid, the blood vessels are enlarged and heavily muscled. These enlarged vessels serve as simple hearts that pump the blood. Earthworms, for example, have five pairs of hearts. Annelids do not have gills, lungs, or other specialized respiratory organs. They exchange oxygen and carbon dioxide with the environment directly through their body surfaces. Oxygen then diffuses into blood vessels that lie under the skin. An annelid's blood typically contains hemoglobin, a protein that binds to oxygen molecules. In some annelids, other respiratory molecules with a similar function are found.

Excretion

The excretory system of annelids is very similar to that of mollusks, consisting of ciliated, funnel-shaped nephridia. Each segment of an annelid has a pair of nephridia that collect waste products. The wastes are then transported out of the annelid's body through pores that open on the sides of each segment. ❏

Annelid Diversity

There are roughly 12,000 known species of annelids. They range in size from less than 1 mm (0.04 in.) long to more than 3 m (10 ft.) long. Annelids are classified into three classes based on the number of setae they have and the presence or absence of parapodia. The approximately 8,000 species of polychaete *(PAHL ih keet)* worms are all marine. Polychaetes have many setae and parapodia. The oligochaete *(AHL ih goh keet)* worms, which number about 3,100 species, include terrestrial earthworms as well as some related freshwater worms. Oligochaetes have only a few setae on each segment and no parapodia. The hirudinean *(hihr yoo DIHN ee ahn)* worms are the leeches. Most of the 600 species of leeches live in fresh water, although there are a few marine and terrestrial species. Leeches lack both setae and parapodia.

Annelids are an ancient phylum. Their fossils can be found in rock that is 530 million years old. Scientists think that annelids evolved in the sea, with the polychaetes being the ancestral group. Oligochaetes appear to have evolved from polychaetes, perhaps by way of freshwater worms. Taxonomists generally agree that leeches evolved from oligochaetes, some becoming specialized as bloodsucking external parasites.

Marine Worms

Polychaetes are marine segmented worms that live in virtually all ocean habitats. Some are free-swimming predators that use their strong jaws to feed on small animals. Some burrowing species excavate tunnels by ingesting sediment. Others feed by pumping water through their bodies or scouring the ocean bottom with their tentacles.

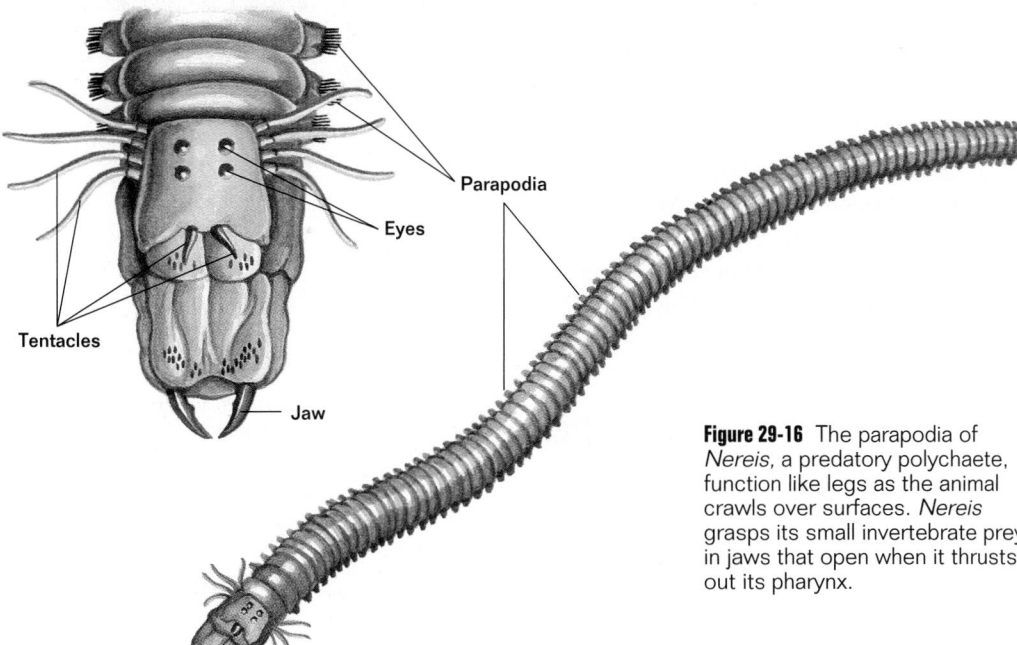

Parapodia

Eyes

Tentacles

Jaw

Figure 29-16 The parapodia of *Nereis,* a predatory polychaete, function like legs as the animal crawls over surfaces. *Nereis* grasps its small invertebrate prey in jaws that open when it thrusts out its pharynx.

Teaching Tip
Pet Show
Have students pick an annelid or mollusk in this chapter and pretend that the animal is a pet. Ask them to provide a picture of the pet and prepare a report about it. In their report, students should give each mollusk a name, describe the kind of habitat necessary to keep the pet, describe what and how they would feed it, and provide warnings for anyone handling it.

Polychaetes, unlike earthworms, have a well-developed head, as shown in Figure 29-16. Many have antennae, specialized mouthparts, and sense organs, often including stalked eyes. Polychaetes are often beautiful; some have unusual forms and iridescent colors. Spectacular polychaetes called feather dusters are shown in Figure 29-17. A distinctive characteristic of polychaetes is the pair of fleshy, paddle-like flaps called parapodia that occur on most of their segments. The parapodia, which usually have setae, are used to swim, burrow, or crawl. They also greatly increase the surface area of the body, making gas exchange between the animal and the water more efficient.

Individual polychaetes are either male or female, and fertilization is usually external. There are no distinct male sex organs that produce sperm. Instead, sperm are produced from cells that line the male's coelom. Eggs are typically attached to a rock by the female and are fertilized by sperm released into the water by the male. Fertilization results in the growth of ciliated trochophore larvae. After a long period of development, the larvae begin to add segments and thus change to juvenile polychaetes that more closely resemble the adult worms.

Earthworms

Earthworms are terrestrial worms that literally eat their way through the soil. As highly specialized scavengers, earthworms take organic matter and other soil material into their mouths as they tunnel through the soil. The ingested soil moves through a long, straight digestive tube. In one

Figure 29-17 Feather dusters live in hard, narrow tubes that they make with their body secretions. They filter-feed by trapping food particles in the spiral of feather-like head structures that extend from the end of their burrow. The beating of the cilia located inside their feathery cone directs the food to their mouth.

Historical Note

The Service of Earthworms
Charles Darwin (1809–1882), a British naturalist famous for his theories of evolution and natural selection, was fascinated with the ways that earthworms aerate and enrich soil. In his last botanical book, *The*

Formation of Vegetable Mould Through the Action of Worms, Darwin noted the service of earthworms in recycling organic matter in soil. He also observed that an earthworm could ingest its weight in soil each day, and that in one year the earthworms

inhabiting one hectare could digest 22 to 40 metric tons of soil. The work was published only six months prior to Darwin's death and was called a pioneering study in quantifying ecology.

Instructional Strategy

Earthworms, which have no teeth, eat their way through soil and grind it up for nutrients. Ask students what structure earthworms use to grind up soil. (*gizzard*) Ask students which other toothless animals use gizzards to grind food. (*birds*) Finally, ask students why food needs to be ground up. (*Smaller food particles are digested more easily.*)

Discussion

Guide the discussion by posing the following questions.
1. How might an earthworm try to escape if its anterior end were being pulled out of the ground by a bird? (*The earthworm would dig its posterior setae into the dirt and pull its anterior section back into the ground.*)
2. How do earthworms loosen soil and help recycle soil nutrients? (*Earthworms make holes in soil by moving through it, and their digestive system breaks down soil matter to simpler components.*)
3. Why is the clitellum necessary for earthworm reproduction? (*It secretes a protective layer of chitin for the fertilization and incubation of eggs.*)

UP CLOSE EARTHWORM

■ **Scientific Name:** *Lumbricus terrestris*
■ **Range:** Europe and eastern North America
■ **Habitat:** Damp soil
■ **Size:** Grows up to 30 cm (12 in.) long
■ **Diet:** Organic matter contained in soil

Characteristics

Respiration Oxygen and carbon dioxide diffuse through the earthworm's skin, as in all annelids. This exchange can take place only if the worm's skin is kept moist.

Digestion Soil is taken into the digestive passage by the muscular throat, or **pharynx.** The soil then passes through the **esophagus** to a temporary storage area called a **crop,** and from there it passes to the **gizzard.** The thick, muscular gizzard walls contract and grind the soil, breaking up the organic matter contained in it. Food molecules pass across the walls of the **intestine** and are absorbed into the bloodstream. The blood is pumped throughout the earthworm's closed circulatory system by a series of muscular hearts.

Movement As in most annelids, circular and longitudinal muscles line the interior body wall of an earthworm. In order to crawl, an earthworm first anchors several of its segments by sinking stiff bristles, called **setae,** into the ground (1). The worm then contracts the circular muscles in front of the anchored segments. This causes the anterior segments to elongate (2). The worm then retracts the rear set of setae and grips the ground again with setae that are in front of the stretched region (3). The rear segments are then pulled forward (4).

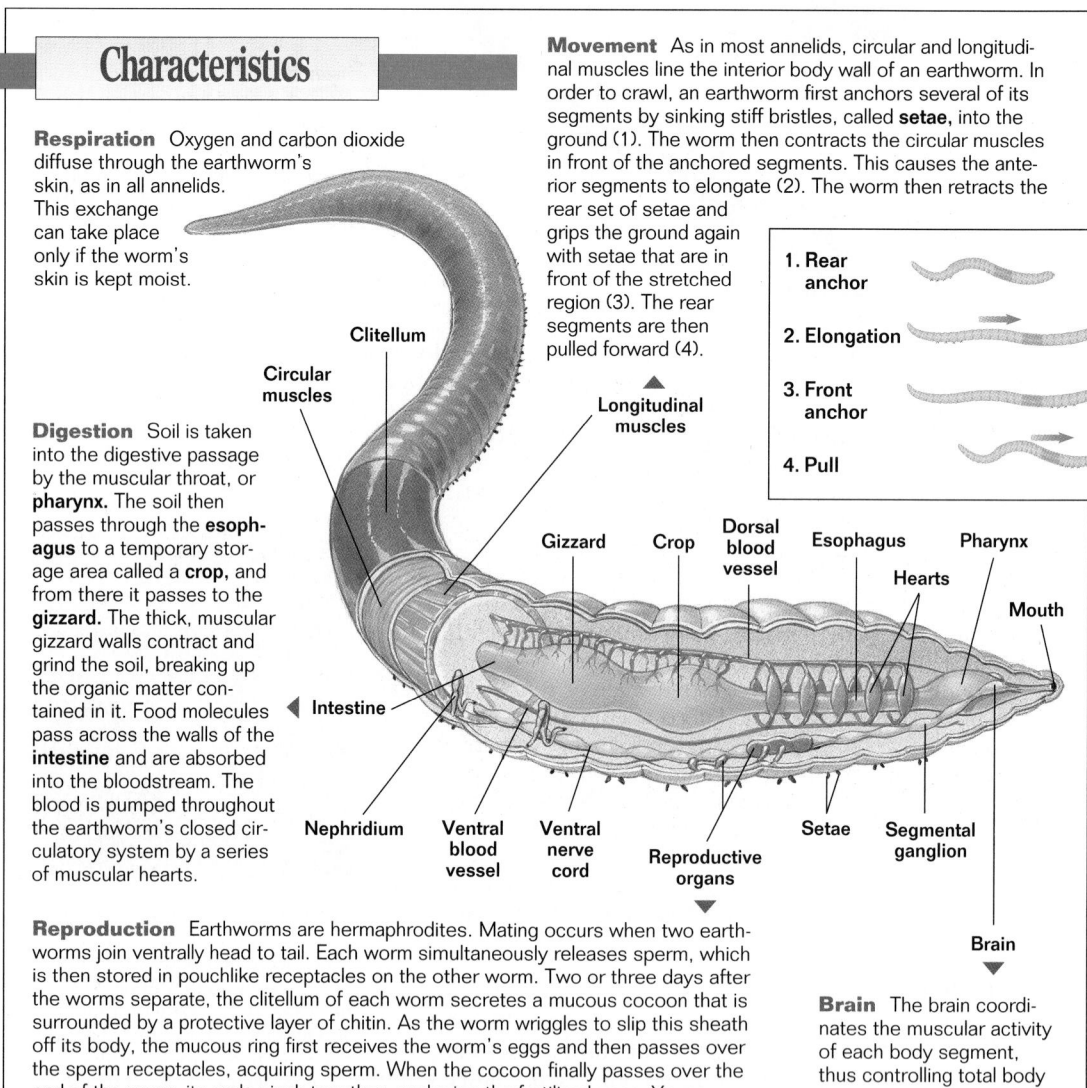

1. Rear anchor
2. Elongation
3. Front anchor
4. Pull

Clitellum
Circular muscles
Longitudinal muscles
Gizzard
Crop
Dorsal blood vessel
Esophagus
Pharynx
Hearts
Mouth
Intestine
Nephridium
Ventral blood vessel
Ventral nerve cord
Reproductive organs
Setae
Segmental ganglion
Brain

Reproduction Earthworms are hermaphrodites. Mating occurs when two earthworms join ventrally head to tail. Each worm simultaneously releases sperm, which is then stored in pouchlike receptacles on the other worm. Two or three days after the worms separate, the clitellum of each worm secretes a mucous cocoon that is surrounded by a protective layer of chitin. As the worm wriggles to slip this sheath off its body, the mucous ring first receives the worm's eggs and then passes over the sperm receptacles, acquiring sperm. When the cocoon finally passes over the end of the worm, its ends pinch together, enclosing the fertilized eggs. Young worms emerge from the cocoon after several weeks.

Brain The brain coordinates the muscular activity of each body segment, thus controlling total body movement.

Graphic Organizer

Use this graphic organizer with the **Chapter Closure** on page 673.

	Body cavity	Body plan	Organ systems	Larvae
Mollusks	true coelom	muscular foot, head, visceral mass	circulatory (usually open system), respiratory, digestive, excretory	trochophore
Annelids	true coelom	segmented with bristles	circulatory (closed), respiratory, digestive, excretory	trochophore

modified portion of the tube, called the **gizzard,** strong muscles grind up the organic material in the soil. The material that passes all of the way through the digestive system without being absorbed exits the worm through the anus and is deposited outside its burrow in the form of characteristic castings. The tunneling activity of earthworms aerates the soil, and their castings fertilize it. An earthworm eats its own weight in soil every day. Rich, organic soil may contain thousands of earthworms per acre.

Earthworms lack the distinctive head region of polychaetes, and they have no eyes. This is not surprising when you consider an earthworm's underground lifestyle. However, earthworms do have light-sensitive and touch-sensitive organs located at each end of their body. Earthworms also have sensory cells that detect moisture. Learn more about the structure and lifestyle of earthworms in *Up Close: Earthworm* on page 672. ❏

Leeches

The body of a leech is flattened. Most species are 2.5–5.0 cm (1–2 in.) long, although one tropical species grows up to 30.5 cm (1 ft.) long. The coelom of leeches is not segmented as in other annelids. A leech, shown in Figure 29-18, has suckers at both ends of its body. Leeches move by attaching first one sucker and then the other and pulling themselves forward. Most species of leeches are predators or scavengers, but some have evolved into parasites. Parasitic leeches suck the blood from mammals and other vertebrates. A few species even suck blood from crustaceans. Many parasitic freshwater leeches remain on their hosts for extended periods. Most of the terrestrial parasitic leeches position themselves on low-growing plants, waiting to attach themselves to a suitable endothermic host. Some leeches, however, climb trees to seek out mammals and birds.

Figure 29-18 Well-developed, powerful muscles in the body wall of a leech enable it to carry out the complex body movements that are necessary for it to crawl along.

Section Review

1. *If you have two worms with different external characteristics, how can you tell if either is an annelid?*

2. *How are the circulatory systems of a human and an earthworm similar?*

3. *Contrast polychaetes with leeches.*

Critical Thinking

4. *A mutation results in the birth of an earthworm that lacks moisture-sensing cells in its skin. Explain why this earthworm is less likely to survive than one with the sensory cells.*

Answers to Section Review

1. The bodies of all annelids are segmented, and most have external bristles.

2. Both annelids and humans have a closed circulatory system consisting of a heart (or hearts) that pumps blood to the body through a system of vessels.

3. Polychaetes are either free-swimming predators with strong jaws or burrowing filter feeders, while leeches are either ectoparasites that feed on blood, or scavengers. Polychaetes have parapodia, and leeches have suckers at both ends of their bodies.

4. Earthworms must stay moist in order to breathe since oxygen and carbon dioxide diffuse through moist skin surfaces. A worm without the ability to sense moisture is more likely to dry out and suffocate.

29 CHAPTER REVIEW ....

CHAPTER REVIEW ANSWERS

Each item in the Chapter Review is correlated by text section in the assignment guide that follows.

ASSIGNMENT GUIDE

Section	Review	Themes Review	Critical Thinking
29-1	1–5, 9, 11–14	16, 18	19, 20
29-2	6–10, 15	17	21

Review

Multiple Choice

Code in parentheses indicates section number and objective number.

1. a (29-1.1)

2. c (29-1.2)

3. a (29-1.2)

4. a (29-1.3)

5. c (29-1.3)

6. d (29-2.3)

7. b (29-2.1)

8. d (29-2.2)

9. d (29-1.2, 29-2.3)

10. b (29-2.4)

Completion

11. trochophore, veliger (29-1.1, 29-1.3)

12. open, closed (29-1.2, 29-1.3)

13. three-part, Cephalopoda, Gastropoda (29-1.3)

14. chitons (29-1.3)

15. pharynx, crop, gizzard (29-2.4)

Vocabulary

adductor muscle (662)	mantle cavity (660)	setae (669)
closed circulatory system (660)	nephridia (661)	siphon (662)
	open circulatory system (660)	torsion (664)
gizzard (673)	parapodia (669)	veliger (663)

Review

Multiple Choice

1. Mollusks and annelids are alike in that they both
 a. are coelomates.
 b. undergo torsion.
 c. have veliger larvae.
 d. have a visceral mass.

2. Cephalopods have all of the following characteristics except
 a. bilateral symmetry.
 b. a three-part body plan.
 c. an open circulatory system.
 d. a true coelom.

3. Which reflects the circulatory path exhibited by the bivalves?
 a. heart→tissue sinus→gills→heart
 b. heart→gills→heart→tissue sinus
 c. heart→lungs→arteries→veins
 d. heart→tissue sinus→lungs→heart

4. Which mollusk has adductor muscles?
 a. clam   c. squid
 b. snail   d. nudibranch

5. The visceral mass of gastropods twists during the larval stage, bringing the mantle cavity toward the front of the animal. This is called
 a. spiraling.
 b. segmentation.
 c. torsion.
 d. radial symmetry.

6. Annelids are divided into three classes. This classification is based on the number of setae and the presence or absence of
 a. segments.   c. sauropodia.
 b. hearts.   d. parapodia.

7. Annelids are most easily recognized by their
 a. cephalization.   c. nephridia.
 b. segmentation.   d. body cavity.

8. Blood in the circulatory system of an annelid
 a. flows in open sinuses.
 b. moves very slowly.
 c. passes through gills.
 d. transports oxygen.

9. The nephridia of annelids and mollusks function in
 a. respiration.   c. digestion.
 b. circulation.   d. excretion.

10. Earthworm movement requires all of the following except
 a. circular muscles.
 b. secretion of a mucus layer.
 c. muscle contractions.
 d. traction provided by setae.

Completion

11. Mollusks and annelids commonly develop a(n) _____ larva, but only certain mollusks develop a second larva called a(n) _____ .

12. Oysters have a(n) _____ circulatory system in which blood leaves the vessels, but earthworms have a(n) _____ circulatory system in which blood circulates entirely within vessels.

13. All mollusks exhibit a(n) _____ body plan that includes the foot, head, and visceral mass. The head is most specialized among members of the class _____ . Torsion of the visceral mass is seen among members of the class _____ .

14. Ancestral characteristics of the phylum Mollusca are shown by the flattened, wormlike animals called _____ .

15. Soil taken in by an earthworm passes through a muscular throat called a _____ before it enters the esophagus. Upon leaving the esophagus, the soil is stored for a short time in the _____ . Then it is ground up by the contracting walls of the _____ . Food extracted from the soil is absorbed by the blood as the soil passes through the intestine.

Themes Review

16. **Homeostasis** Mollusks and annelids possess nephridia. How do nephridia function to regulate the balance of sugar, water, salt, and other useful molecules in the bodies of these animals? What other function is performed by nephridia?

17. **Evolution** The earthworm gizzard, which is used for grinding food, is lined with a cuticle and is very muscular. In comparison, the gizzards of annelids that have returned to an aquatic environment are much smaller and less muscular. How do the differences between the gizzards represent adaptations to terrestrial and aquatic feeding?

18. **Structure and Function** Most gastropods are marine and breathe with gills, but terrestrial snails are an exception. What modification to the mantle cavity enables these snails to live on land?

Critical Thinking

19. **Controlling Variables** As a science-fair project, a student decided to determine if a new, experimental paint reduces the boring by shipworms on ocean pier pilings more than other paints do. What variables must the student control for if the test of the new paint is to be a fair one?

20. **Making Inferences** Classification of organisms has traditionally been based on physical similarities. Given the physical differences of adult chitons, gastropods, bivalves, and cephalopods, how did they come to be grouped into the phylum Mollusca?

21. **Making Inferences** Examine the photograph below. Is the annelid shown in the photograph a polychaete, oligochaete, or hirudinean? What observable characteristics support your choice?

Activities and Projects

22. **Science-Technology-Society** *Nautilus* was the name given to the U.S. Navy's first nuclear submarine. Compare the shell structure of the cephalopod nautilus and the functioning of its siphuncle with the structures that enable a submarine to submerge and surface. Why do you think the Navy chose to name the submarine *Nautilus?* Support your explanation with cut-away sketches of the cephalopod and a submarine.

23. **Multicultural Perspectives** Many marine seashells are named after Greek or Roman goddesses. Using the *Audubon Society Field Guide to North American Seashells* (published by Alfred A. Knopf, Inc., 1992) and sources of Greek and Roman mythology, compile a list of 5–10 shells named after mythological figures.

Themes Review

16. Nephridia allow the reabsorption of useful molecules (such as sugar, water, and salt) as well as the storage and elimination of nitrogen-rich wastes. (29-1.2, 29-2)

17. The cuticle and strong muscles of the terrestrial earthworm are well suited for grinding up the organic material contained in soil. Abundant water in the aquatic environment makes the reduction of organic material a much easier task, thus the stronger muscles found in the terrestrial earthworm are not required among aquatic species. (29-2.4)

18. The mantle cavity is a modified lung in which oxygen from air diffuses across the thin, moist membrane that lines the cavity. (29-1.2)

Critical Thinking

19. Answers will vary but should address such variables as the age of the wood, wood type and quality, previous shipworm damage to wood, location of pier, amount of paint used, how paint is applied, etc. (29-1.3)

20. While they are different, they all have a true coelom, produce a trochophore larva (with bivalve and gastropod trochophores developing into veligers), and have organ systems. Classification into one phylum was based on these characteristics. (29-1.3)

21. The worm is a polychaete. The well-developed head and parapodia are distinguishing characteristics of polychaetes. (29-2.3)

Activities and Projects

22. The shell of the *Nautilus* is divided into chambers, and only the last chamber is occupied by the animal. The wall separating each chamber is perforated by a thread of body tissue called the siphuncle, which secretes gas into the chambers and makes the shell buoyant. A submarine also has chambers along its length that are filled with gas to make it buoyant.

23. The list should include the names Venus, Junonia, Margarite, Arene, Pandora, Semel, Thracia, and Astarte.

LABORATORY Investigation Chapter 29

Mollusks

OBJECTIVES

- Observe the behavior of a live clam.
- Examine the structure and composition of a clam shell.

PROCESS SKILLS

- identifying structures
- relating structure and function

MATERIALS

- safety goggles
- lab apron
- disposable gloves
- live clam
- small beaker or dish
- food coloring in a dropper bottle
- glass stirring rod
- clam shell
- petri dish
- scalpel
- stereomicroscope
- 10% HCl in a dropper bottle

BACKGROUND

1. What kinds of mollusks do people eat?
2. What makes a clam a bivalve?
3. What kinds of food do bivalves eat?
4. What keeps the two shells of a clam together?
5. Write your own question to explore in your lab report or notebook.

TECHNIQUE

Part A: Live Clam

1. CAUTION: **Put on safety goggles, a lab apron, and disposable gloves.** Place a live clam upright in a small beaker or dish. Using a medicine dropper, apply two drops of food coloring between the two shells, as shown below. Observe what happens to the food coloring. Record your observations in the **Records** section of your report.

2. Using the stirring rod, gently touch the clam's mantle. The mantle is the internal membrane that lines the shells. Describe the clam's response in the **Records** section of your report.

Part B: Examining a Clam Shell

3. Observe a clam shell. Look at the concentric growth rings on the shell. As a clam grows, its mantle secretes a layer of shell. Locate the knob-shaped umbo on the shell. The umbo is the oldest part of the shell. Each successive ring of shell formed beyond the umbo is younger than the one before it. In the **Records** section of your report, record the number of growth rings on the clam shell.

Procedural Notes

- This investigation can be completed during one 40-minute period.
- Keep the clams in a freshwater aquarium with green algae and a layer of sand until lab time.
- **Safety:** Students must wear safety goggles, lab aprons, and disposable gloves throughout this investigation.
- In step 4, remind students to use extreme care when chipping clam shells with their scalpel. Have students use a scalpel that has a permanent blade.
- **Safety:** In step 5, remind students to keep hydrochloric acid (HCl) away from skin and clothing. If acid is spilled, thoroughly flush all affected areas with running water and mop the spill up with cloths designated for spill cleanup. Avoid having students handle acid spills.

Preparation

Obtain enough live clams and empty clam shells for each lab group to have one of each. Live clams can be purchased from a local seafood market or from a biological supply house such as Ward's. Clam shells can be obtained from a seafood market or restaurant. Keep the live clams in a freshwater aquarium with green algae. Provide a layer of sand as a substrate. In addition to a clam and a clam shell, each lab group will need a small beaker or dish, a dropper bottle with food coloring, a glass stirring rod, a petri dish, a scalpel, and a bottle of 10% HCl. Have students use a scalpel that has a permanent blade as opposed to one that has a thin, disposable blade. If 10% HCl is unavailable, have students use a 1.0 M HCl solution (made from 42 mL HCl diluted to 1 L with distilled water).

To clean up an acid spill, dilute the spilled acid with water, and mop up the spill with wet cloths designated for spill cleanup. Avoid having students handle acid spills themselves.

Students will also require access to a stereomicroscope.

Background Answers

1. People eat a variety of mollusks, including clams, oysters, mussels, snails, octopuses, and squids.

2. Clams are bivalves because they have two shells.

3. Bivalves feed on microscopic organisms in water, such as bacteria, protozoans, algae, and tiny animals.

4. The shells are connected by a hinge as well as by thick adductor muscles.

5. How do materials move into and out of a clam? What is the structure and composition of a clam shell?

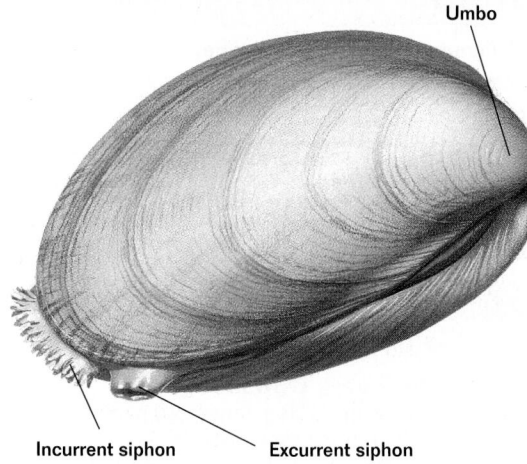

Umbo

Incurrent siphon   Excurrent siphon

4. CAUTION: Use extreme care when working with sharp objects. Place the clam shell in a petri dish and use a scalpel to chip away part of the shell to expose its three layers. Observe the layers under a stereomicroscope. The outermost layer of shell protects the clam from acids in the water. The innermost layer is a pearly material called mother-of-pearl. It is this material that forms a pearl inside a shell.

5. The middle layer of the shell contains crystals of calcium carbonate. **CAUTION: Hydrochloric acid can burn skin and damage clothing.** Test for the presence of this compound by placing one drop of 10% HCl on the middle layer of the shell. If calcium carbonate is present, it will form bubbles in the shell. Record your observations in the **Records** section of your report. Also, briefly summarize the procedure you followed.

6. Clean up your materials and wash your hands before leaving the lab.

INQUIRY

1. Find the incurrent and excurrent siphons of the clam in the diagram on this page. The incurrent siphon draws water into the clam, and the excurrent siphon expels the water from the clam. Using this information, explain your observations in step 1.

2. What is the purpose of a clam's shell?

ANALYSIS

1. Based on your observations, how do you think clams respond when they are touched or threatened in their natural habitat?

2. What does a clam take in from water that passes through it?

3. Water that enters a clam's incurrent siphon passes over the clam's gills. How does this help the clam to breathe?

4. How does a clam get rid of its wastes? What kinds of wastes are removed?

5. Look at the picture of the clam. Where would you expect the clam's foot to appear if it was extended?

6. Bivalve fossils are one of the most common kinds of fossils. Explain why.

FURTHER INQUIRY

Write a new question that could be explored as a new investigation. The following is an example:

How similar are the layers of the shells of univalves and bivalves?

Inquiry Answers

1. The food coloring was drawn into the clam through the incurrent siphon and released through the excurrent siphon.

2. The shell protects the clam from injury and from predators.

Analysis Answers

1. Clams draw in their siphons and foot and close their shells tightly when threatened or touched.

2. Food and oxygen are taken in from the water.

3. Mollusk gills can extract 50 percent or more of the dissolved oxygen from water that passes over them.

4. Wastes, such as carbon dioxide, undigested food, and sediments, are removed with water through the excurrent siphon.

5. The clam's foot would be extended on the side of the shell across from the incurrent and excurrent siphons.

6. As in bones, calcium carbonate enables bivalve shells to avoid deterioration over long periods and to fossilize well.

Records

1. The clam should draw in the food coloring through its incurrent siphon and expel the food coloring through its excurrent siphon.

2. The clam should draw in its siphons and foot and close its shells tightly when it is disturbed by the stirring rod.

3. The number of growth rings will depend on the age of the clam.

4. Students should observe bubble formation, indicating the presence of calcium carbonate in the middle layer of the shell.

Each block represents about 45–50 minutes of instructional time. The resources cited in a block are coded to a specific program component using the Key on the next page.

Lecture Concepts

Lesson Resources

BLOCK 1

a. Arthropods, the most abundant animals living on Earth, evolved from annelids and are characterized by jointed appendages, a segmented body, and an exoskeleton.

b. An arthropod must shed its exoskeleton to grow.

Lecture Resources
- Opening Question, page 679
- Demonstration: Comparing Arthropods
- Visual Strategies: Figures 30-3, 30-5
- Wings, page 680
- The Dangers of Molting, page 682
- Teaching Transparencies: 138, 139, 161, 134A

- Holt Biology Videodiscs: Lesson 30-1

Classwork Options
- Arthropod Segmentation, page 680
- Excretion, page 681
- Inquiry Skills Development B25: Crayfish Dissection

Assignment Options
- Section Review, page 682
- Chapter Review, questions 1, 2, 4, 9, 13–15

30-2 Arthropod Diversity pp. 683–693

BLOCK 2

a. Arachnids (spiders, scorpions, and mites) are characterized by four pairs of legs, fused body regions, and mouthparts called chelicerae.

b. Crustaceans have hardened exoskeletons and mouthparts called mandibles. Most crustaceans (crabs, lobsters, and crayfish) are aquatic and have greatly modified appendages. Other crustaceans (pill bugs and sand fleas) are terrestrial.

Lecture Resources
- Opening Question, page 683
- Demonstration: Arthropod Collection
- Venom, page 684
- Hunting Techniques, page 684
- Research Update, page 685
- Demonstration: Crustaceans, page 686
- Demonstration: Sea Monkeys, page 686

- Visual Strategy: Figure 30-14
- Crustaceans and Humans, page 687
- Teaching Transparencies: 134A
- Holt Biology Videodiscs: Lesson 30-2

Classwork Options
- Effective Reading, page 683
- Inquiry Skills Development B20: Life in a Pine Cone

BLOCK 3

a. Insects, the most numerous of all animals, have jaws and three body regions: head, thorax, and abdomen. They are the only invertebrates capable of flight.

b. Metamorphosis in insects may be either complete or incomplete.

c. Some insects, such as bees and termites, have evolved complex social systems.

Lecture Resources
- Essay on Arthropods, page 688
- Visual Strategies: Figure 30-17; Table 30-1
- Demonstration: Twin Silkworms
- Social Insects, page 693
- Teaching Transparencies: 136, 137, 140, 134A, 135A,
- Holt Biology Videodiscs: Lesson 30-2

Classwork Options
- Orders of Insects, page 689

- Up Close: Eastern Lubber Grasshopper, pages 690–691
- Silk, page 692
- Life Cycles, page 692
- Quick Labs A24: Observing Insect Behavior
- Laboratory Techniques and Experimental Design C39: Response in the Fruit Fly

Assignment Options
- Section Review, page 693
- Chapter Review, questions 3, 5, 10, 11, 13–15

30-3 Echinoderms and Invertebrate Chordates pp. 694–697

BLOCK 4

a. Echinoderms (sea stars, urchins, and sea cucumbers) are deuterostome marine invertebrates that are radially symmetric as adults and have a water-vascular system.

b. Tunicates and lancelets are invertebrate chordates that have a body cavity, nerve cord, and notochord during their larval stages.

Lecture Resources
- Opening Questions, page 694
- Demonstration: Observing Echinoderms
- Visual Strategy: Figure 30-22
- Overcoming Misconceptions, page 695
- Teaching Transparencies: 162, 163
- Holt Biology Videodiscs: Lesson 30-3

Classwork Options
- Chordates, page 697
- Closure, page 697
- Laboratory Investigation: Arthropod Responses, pages 700–701

Assignment Options
- Section Review, page 697
- Chapter Review, questions 6–8, 12, 14, 16

Review/Enrichment

BLOCK ⑤

- ■ Study Guide: Concept Review and Pretest

Assignment Options
- ■ Activities and Projects, questions 17, 18
- ◉ Teaching Resources CD-ROM: Supplemental Reading Worksheets and Test, *Journey to the Ants: A Story of Scientific Exploration*

Assessment Options

BLOCK ⑥

Traditional Assessment
- ■ Chapter 30 Test
- ◉ Test Generator/Assessment Item Listing: Software item bank for preparing customized chapter tests.

Portfolio Assessment
Select student reports for one or more laboratory experiments from this chapter. *The Direct Observation Checklist* on the *BioSources Teaching Resources CD-ROM* should be completed during a laboratory or other cooperative learning experience.

Holt Biology Videodiscs

Holt Biology Videodiscs gives you a powerful tool for teaching, review, and assessment. *Concepts of Biology* adds interactive lesson plans and extensive tools for customization using CD-ROM technology.

CONCEPTS OF BIOLOGY

Use the following topic from *Concepts of Biology* to help you teach this chapter:

- ■ Topic 18: Invertebrate Diversity

For further information regarding the media on the videodiscs, see *Holt Biology Videodiscs Teacher's Correlation Guide to Biology: Principles and Explorations.*

Chapter Theme

Evolution *Arthropods and echinoderms are diverse groups of animals that have survived millions of years of change on Earth. Adaptations of arthropods have enabled crustaceans, arachnids, and insects to live in many different environments and in greater numbers than any other animal phylum. Echinoderms are among the oldest phyla of animals. Fossil and embryological evidence shows that echinoderms are relatively close to chordates in evolutionary development.*

Tapping Prior Knowledge

- Contrast protostome and deuterostome development.
- What are some characteristics of arthropods?
- Name some similarities and differences between echinoderms and chordates.

Opening Demonstration

Display a phylogenetic tree of the major animal phyla like the one shown on page 611. Review the invertebrate animal phyla that have been discussed so far in Unit 7, including Cnidaria, Platyhelminthes, Nematoda, Mollusca, and Annelida. Point out Arthropoda and Echinodermata on the phylogenetic tree. Explain that arthropods are protostomes and echinoderms are deuterostomes. Ask students to name some animals from each phylum. *(Arthropods include insects, spiders, shrimps, mites, etc. Echinoderms include sea stars, sea urchins, sea cucumbers, etc.)*

CHAPTER 30

ARTHROPODS AND ECHINODERMS

European swallowtail

REVIEW

- arthropod evolution (Section 27-3)
- jointed appendages, exoskeleton (Section 27-3)
- protostome and deuterostome embryo development (Section 27-3)
- endoskeleton (Section 27-3)
- nerve cord, notochord, pharyngeal slits (Section 27-3)

Authors' Rationale

Arthropods are the largest and most diverse group of invertebrate animals. Their impact on human society ranges from beneficial (pollination, pest control, food sources) to devastating (crop destruction, pests, venomous bites). Echinoderms, which are exclusively marine, have symmetry (usually five-part radial symmetry) and a unique physiology. This chapter presents the distinguishing characteristics and diversity of Arthropoda and Echinodermata and briefly discusses invertebrate chordates.

30-1 Features of Arthropods

*T*here are more arthropods living on Earth than any other kind of animal, and many are of enormous importance to human life. The largest arthropod group is the class Insecta. Insects eat virtually every kind of plant, including most agricultural crops. On the other hand, insects are beneficial to agriculture. They eat the weeds that compete with the crops, and they are primary crop pollinators.

Arthropods Evolved From Annelids

The ancestor of arthropods is undoubtedly some kind of annelid. Like annelids, arthropods have a coelom and a segmented body. Scientists are not quite sure which annelid gave rise to arthropods. Until recently, experts thought velvet worms, shown in Figure 30-1, might be the "missing link" between annelids and arthropods. Velvet worms appear to have a combination of both annelid and arthropod characteristics—segmented, wormlike bodies with paired legs. However, Australian researchers in 1993 analyzed the DNA of velvet worms and discovered that they are in fact true arthropods and are closely related to scorpions.

Arthropod fossils are among the oldest, best-preserved fossils of multicellular animals; some are 630 million years old. Among the most successful of the early arthropods were the trilobites—highly segmented, plate-shaped animals. Trilobites were the first animals with eyes capable of forming images. Trilobites, shown in Figure 30-2, were very abundant in the seas until they became extinct 250 million years ago. The first terrestrial arthropods were probably scorpions, whose fossils date back 425 million years.

Figure 30-1 Velvet worms live in the leaf litter of tropical and subtropical forests. This species is from the rain forests of Costa Rica. Active mainly at night, velvet worms capture insects by trapping them in a sticky substance secreted near their mouth.

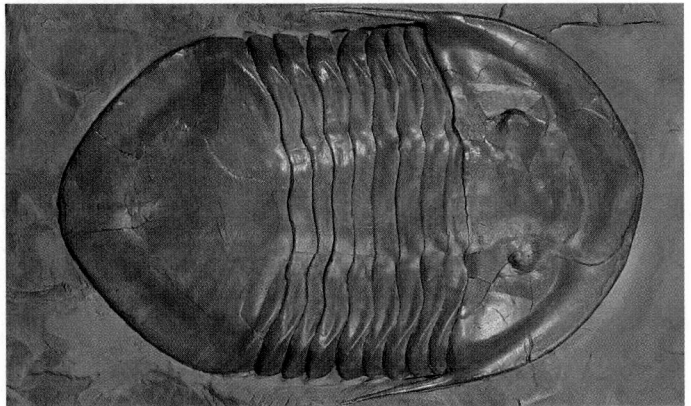

Figure 30-2 Although trilobites no longer exist, their hard exoskeletons made fossilization easy. As a result, trilobites are extremely common in the fossil record. By studying the fossils, scientists have been able to identify nearly 4,000 different trilobite species.

Figure 30-3

Use this figure to point out mandibles and chelicerae. Beetles use mandibles to eat plants, other insects, and dead plants and animals. Spiders use chelicerae to inject enzymes into their prey to liquefy and eat them. Help students understand the classification of arthropods as mandibulates and chelicerates by relating the diets and feeding styles of different animals to their mouthparts.

Teaching Tips
Arthropod Segmentation

Have students look through this chapter and find examples of arthropod segmentation in the figures. Ask students to summarize any trends that they notice. Which are more segmented, larvae or adults? *(larvae)* Which types of arthropods show more segmentation? *(Answers will vary. Millipedes and centipedes are highly segmented arthropods.)*

Wings

Have students compare insect flight to that of birds, kites, and airplanes. Which have movable wings? *(Insects and birds move their wings; kites and airplanes have fixed wings.)* Which can take off from a standing start without gathering speed? *(Insects and birds can fly from a standing start; kites must have wind; and airplanes must gather speed for take-off.)* Is there a connection between the type of wing and the ability to take off from a standing start? *(Movable wings make a takeoff from a standing start possible.)* Compare insect wings with bird wings. *(Answers will vary but should mention that bird wings have feathers and bones and are heavier than insect wings.)*

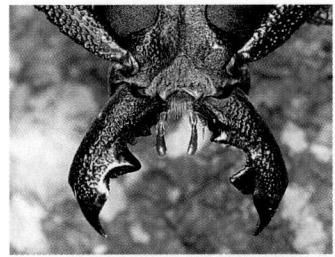

Figure 30-3 Notice the mandibles (jaws) of the wood borer beetle, *top,* a mandibulate arthropod. The baboon spider, *bottom,* is a chelicerate arthropod because it has chelicerae (fangs) as its mouthparts.

Figure 30-4 The compound eyes of this fruitfly, an insect, are not capable of forming images as clear as those formed by vertebrate eyes, but insect eyes are much better at detecting motion. That is why it is difficult to sneak up on a fly.

The Body Plan and Characteristics of Arthropods

As you read in Chapter 27, arthropods have many kinds of jointed appendages, including legs, antennae, and mouthparts. Living arthropods are traditionally separated into two large groups. This separation is based on the kind of mouthparts they have. Arthropods with jaws (crustaceans, insects, centipedes, and millipedes) are called **mandibulates**. The scientific word for jaw is mandible. Arthropods with fangs or pincers (spiders, mites, and scorpions) are called **chelicerates** *(kuh LIHS uhr ayts)*. The scientific name for their mouthparts is chelicerae *(kuh LIHS uhr ee)*. A mandibulate and a chelicerate arthropod are shown in Figure 30-3.

There are nine arthropod characteristics of particular importance.

1. **Jointed appendages**   All arthropods have jointed legs and other jointed appendages. Joints enable an appendage to be far more flexible.

2. **Segmentation**   All arthropods are segmented, although the adults of the more advanced species have many of the segments fused together to form regions called the head, thorax, and abdomen.

3. **Head**   Most arthropods have a distinct head, although in some (lobsters and spiders) the head fuses with the thorax (the mid-body region) to form a body region called the **cephalothorax**. This sort of fusion is unique and does not occur in any other animal phylum.

4. **Exoskeleton**   As you read in Chapter 27, the bodies of arthropods are encased in a shell-like **exoskeleton.**

5. **Compound eyes**   Many arthropods have compound eyes, like those shown in Figure 30-4. A compound eye is made of thousands of individual visual units called **ommatidia** *(ahm uh TIHD ee ah)*, each with its own lens and retina. An insect's brain receives the many inputs of the individual ommatidia and composes a detailed image of the object. Some arthropods also have simple, single-lens eyes called **ocelli** *(oh SEHL eye)* that do not form images. Ocelli can only distinguish light from dark.

6. **Spiracles**   Crustaceans breathe with gills, and spiders have tiny lungs, but the majority of terrestrial arthropods breathe with a network of fine tubes called **tracheae.** Air from outside the animal passes into the tracheae by way of special openings called **spiracles** that are opened and closed by valves. Tracheae, shown in Figure 30-5, transmit oxygen throughout the body of the arthropod. The ability to close their spiracles and prevent water loss was a key adaptation for insects when they invaded land.

7. **Circulation**   The arthropod circulatory system is open, like that of mollusks, with a heart along the top of the body that squeezes blood out into internal body spaces.

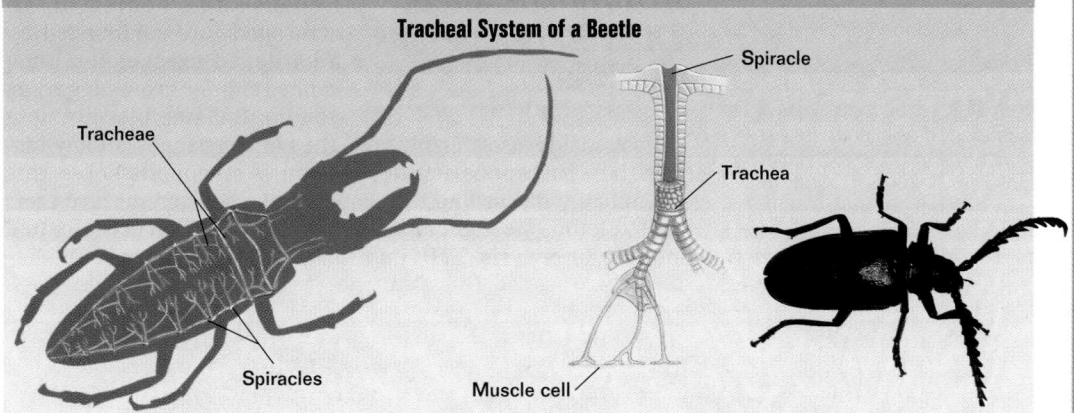

Tracheal System of a Beetle

Tracheae

Spiracle

Spiracles

Trachea

Muscle cell

👁 *VISUAL STRATEGY*

Figure 30-5
Use this figure to compare the respiratory systems of terrestrial vertebrates with those of insects. Ask students the following questions: Is it easier for a vertebrate to choke than it is for an insect to choke? *(It is easier for a vertebrate to choke because vertebrates have only one opening to the respiratory system.)* Is there a disadvantage to having many openings to the respiratory system? *(Answers will vary but may include increased chance of infection and water loss.)*

8. **Malpighian tubules** Terrestrial arthropods have a unique excretory system that efficiently conserves water and eliminates metabolic wastes. This system is composed of excretory units called **Malpighian tubules.** Malpighian *(mal PIHG ee an)* tubules, shown in Figure 30-6, are slender, fingerlike structures that extend from the arthropod's gut and are bathed by the surrounding blood. The fluid component of the blood (consisting of water and small dissolved particles) moves through the tubule and into the arthropod's midgut. As this fluid flows down to the hindgut, most of the water, valuable ions, and metabolites from the fluid are reabsorbed into the arthropod's body tissues by osmosis. Metabolic wastes remain in the gut and eventually leave the body through the anus.

9. **Wings** Insects, the most abundant of the arthropods, were the first animals to evolve wings. For more than 100 million years, until flying reptiles appeared, insects were the only flying organisms in existence.

Figure 30-5 A complex series of hollow tubes called tracheae run throughout the bodies of terrestrial arthropods like this long-horned beetle, *above right.* Air enters the tracheal system through spiracles that open to the outside, and it is delivered to all of the beetle's cells, *above left.*

Teaching Tip
Excretion

Ask students to draw a *Graphic Organizer* like the one at the bottom of this page to explain the excretory process of terrestrial arthropods. Ask students to summarize the excretory process of terrestrial arthropods in a short paragraph. Summaries should include that the excretory system of arthropods removes not only waste products to be excreted, but also removes ions, water, and nutrients to be reabsorbed into body tissues.

✉

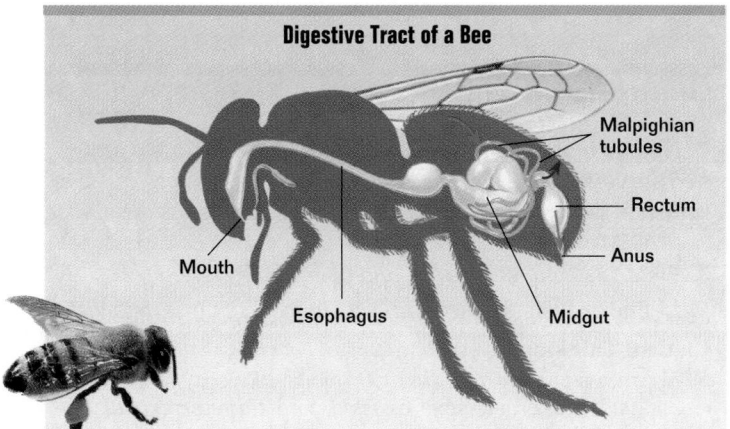

Digestive Tract of a Bee

Malpighian
tubules

Rectum

Anus

Mouth

Esophagus

Midgut

Figure 30-6 Most terrestrial arthropods, like bees, have excretory organs called Malpighian tubules that extend from their gut. A mixture of both wastes and useful molecules passes into the tubules from the surrounding blood that bathes them (red arrow). As this mixture travels down the gut, the useful molecules are reabsorbed back into the blood (blue arrow). The wastes remain in the gut and leave the bee's body through its anus (green arrow).

Graphic Organizer

Use this graphic organizer with
Teaching Tip: Excretion on this page.

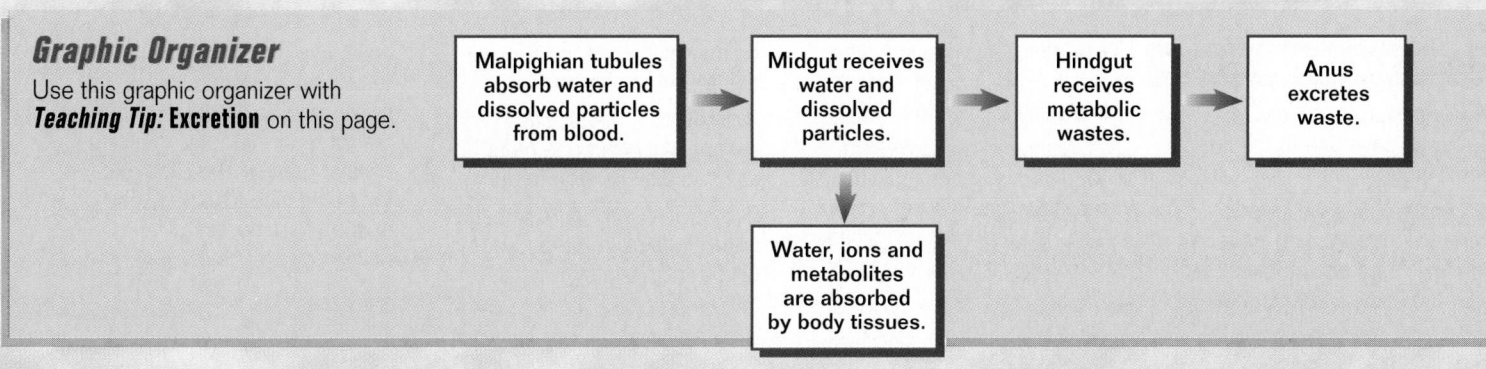

| Malpighian tubules absorb water and dissolved particles from blood. | → | Midgut receives water and dissolved particles. | → | Hindgut receives metabolic wastes. | → | Anus excretes waste. |

Water, ions and metabolites are absorbed by body tissues.

Teaching Tips
The Dangers of Molting

Encourage students to think about some of the problems inherent in molting. Emerging from an exoskeleton can be hazardous to an arthropod. For example, a spider may lose a leg while molting. For one or two hours after emerging from an outgrown exoskeleton, an arthropod without a hard exoskeleton for protection is vulnerable to predators and the environment. Ask students how animals can protect themselves during molting. (*Animals usually hide during molting until their new exoskeletons have hardened.*)

What If an arthropod's new exoskeleton failed to harden? The animal would probably soon become prey to one of its predators or die from desiccation. In the case of crayfish, which are cannibals, the animal is likely to be eaten by another crayfish.

Figure 30-7 As a lobster grows, it must periodically shed its exoskeleton. First, the exoskeleton ruptures, *below left*. Next, the carapace opens, *below center*. Finally, the lobster emerges from its old exoskeleton, *below right*.

Arthropods Shed Their Exoskeletons to Grow

Because the bodies of all arthropods are surrounded by a hardened exoskeleton, they cannot simply grow bigger, as you do. Imagine blowing up a balloon inside a soft drink can—the balloon cannot get any bigger than the space inside the can. Arthropods solve this dilemma by discarding the "can." They shed and discard their exoskeletons periodically as they grow in a process called molting, or **ecdysis** (*EHK duh sihs*), shown in Figure 30-7.

CAPSULE SUMMARY

Arthropods evolved from annelids and are characterized by jointed appendages, a segmented body, and an exoskeleton. An arthropod must shed its exoskeleton to grow.

Molting is signaled by hormones. Just prior to molting, a new exoskeleton forms underneath the old one. When the new exoskeleton is fully developed, it becomes separated from the old one by fluid. This fluid dissolves the chitin and, in a crustacean, the calcium carbonate of the old exoskeleton. The weakened exoskeleton then cracks open and is shed, and the arthropod emerges in its new, still-soft exoskeleton. Almost immediately, the new exoskeleton begins absorbing air or water and expands to fit the larger size of the animal. Within an hour or two after exposure to air or water, the new exoskeleton is hardened. Most insects molt four to eight times during their development, but some may molt as many as thirty times. ◻

Section Review

1. *Name the invertebrate phylum that gave rise to the first arthropods.*
2. *Name three important structural characteristics of arthropods.*
3. *What are compound eyes, and how do they function?*
4. *Define* ecdysis, *and explain why it occurs in arthropods.*

Critical Thinking

5. *Hypothesize why aquatic arthropods, like crayfish, evolved thicker, stronger exoskeletons than terrestrial arthropods, like insects.*

Answers to Section Review

1. The invertebrate phylum Annelida gave rise to the first arthropods.

2. Answers should include three of the following characteristics: jointed appendages, segmented bodies, distinct head, exoskeleton, compound eyes, spiracles, open circulation, Malpighian tubules, and wings.

3. Compound eyes are composed of thousands of visual units that contain a lens and a retina. The individual units send information to the arthropod's brain, which uses the information to form an image.

4. Ecdysis is the process of molting. An arthropod must shed its exoskeleton to grow.

5. Answers may vary. Aquatic arthropods have the buoyancy of water to help support a heavier exoskeleton, while terrestrial arthropods must move their bodies around with the force of their own muscles against gravity. The exoskeletons of insects must be light enough for insects to fly.

30-2 Arthropod Diversity

*T**he arthropods are the most diverse of all animal phyla. The great majority of arthropods are small, about 1 mm (0.04 in.) in length—the size of a grain of rice. The very smallest are parasitic mites only 80 μm (0.003 in.) long. The largest arthropods are gigantic crabs 3.6 m (12 ft.) across, found in the sea off the coast of Japan.*

Section Objectives

- List the three subphyla of arthropods.
- Describe the characteristics of arachnids and crustaceans.
- Compare and contrast millipedes, centipedes, and insects.
- Identify the external and internal structures of the Eastern lubber grasshopper.
- Compare and contrast complete and incomplete metamorphosis.

Arachnids Have Eight Legs

Arthropods are traditionally grouped into three separate subphyla, each with a distinct evolutionary line, as shown in Figure 30-8. They are Chelicerata (spiders, scorpions, and their relatives), Crustacea (crabs, lobsters, and their relatives), and Uniramia (insects, millipedes, and centipedes).

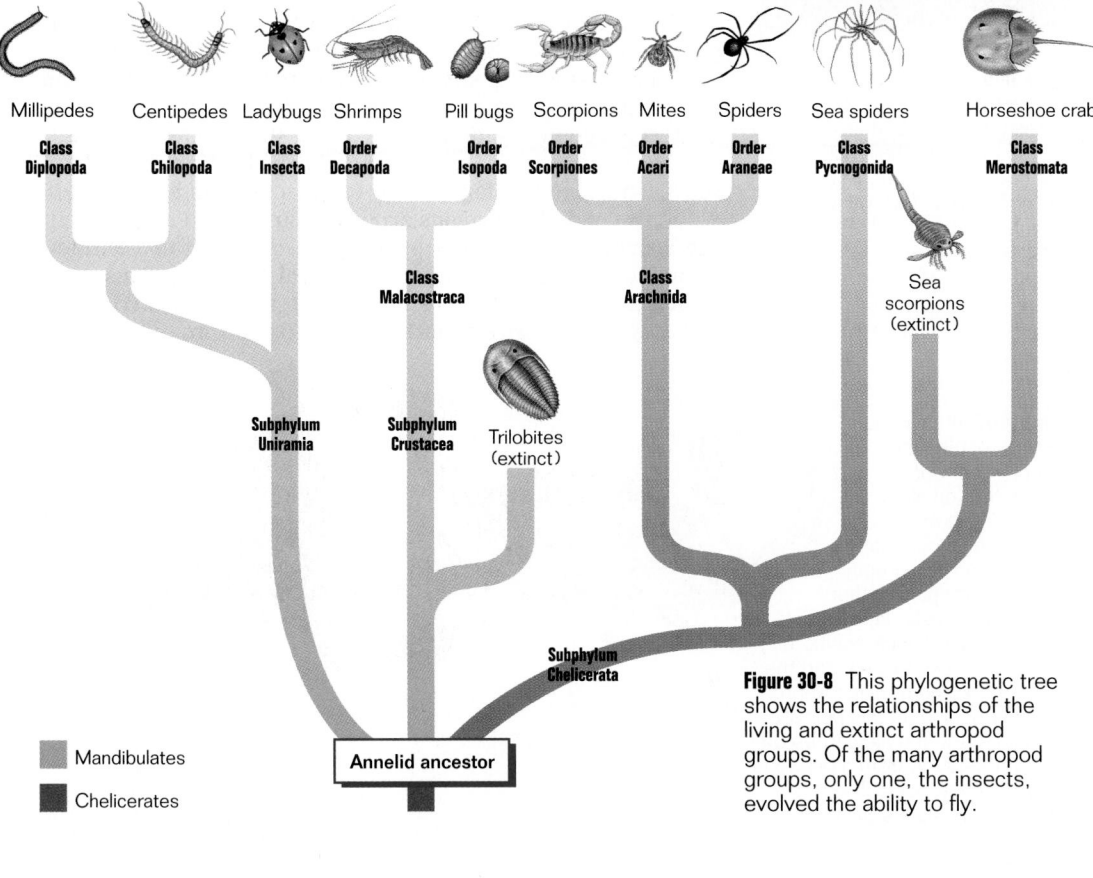

Figure 30-8 This phylogenetic tree shows the relationships of the living and extinct arthropod groups. Of the many arthropod groups, only one, the insects, evolved the ability to fly.

Vocabulary Preview

- arachnid
- palp
- carapace
- swimmeret
- uropod
- telson
- isopod
- metamorphosis
- chrysalis
- nymph
- caste

Opening Question

Before students read this section, write the names of the organisms in Figure 30-8 on the chalkboard, and ask students to divide them into three groups on the basis of common characteristics. After they have classified the organisms, have students look at Figure 30-8. Explain that taxonomists classify arthropods into three subphyla: Chelicerata (spiders, scorpions, mites, sea spiders, horseshoe crabs, etc.), Crustacea (shrimps, pill bugs, lobsters, etc.), and Uniramia (insects, millipedes, centipedes, etc.). Stress that DNA studies surprised some biologists who at one time considered horseshoe crabs a type of crab.

Demonstration

Arthropod Collection

Bring examples of live or preserved arthropods to class for display, or bring detailed photographs of arthropods from a nature magazine. Ask students to collect various harmless arthropods that they can find in the community and bring them to class. (Students may prefer to collect live animals that they can later release.) Allow students to compare the animals. Ask them to point out examples of the arthropod characteristics discussed in Section 30-1. Students should be able to find examples of jointed appendages, segmentation, exoskeletons, cephalothoraxes, wings, and spiracles among the displayed specimens.

Effective Reading

Arthropod ID Cards

This chapter describes many different animals and their characteristics. Ask students to organize the information about each animal as they read. Encourage students to use their notes to make identifica-tion cards for animals in each of the subphyla of Arthropoda. On one side of each card, students can include a photograph or sketch of the animal and a list of the animal's characteristics. On the other side of the card, students should identify the organism and its subphylum. Students can then use their cards as flash cards. Allow students time to quiz each other using each other's cards.

Application

◆ **Anthropology** Arachnids are named for Arachne, a weaver in Greek mythology. Arachne boastfully challenged the goddess Athena to a weaving contest. Arachne produced a beautiful tapestry. Athena was so enraged by Arachne's skill that she changed Arachne into a spider. Arachne's skill can still be found in the beautiful webs of spiders.

Teaching Tips
Venom

Ask students what the selective advantage of venom production might be in bees, ants, scorpions, spiders, and centipedes. (*Bees and ants use venom for defense. Scorpions, spiders, and centipedes use venom to capture prey.*) Students might extend their study by researching different kinds of venom, different ways venom is injected, or different effects of venom. (*Venom is injected via piercing devices, such as stingers or fangs. Venoms can slow heart rate, interrupt transmission of nerve impulses, degrade capillaries, cause blood clotting, or prevent blood clotting.*)

Hunting Techniques

Ask students to compare the hunting techniques of spiders to those of humans. Encourage students to give examples of human hunters who track their prey, hide and wait for their prey, or capture their prey with a web (net). Point out that not all spiders hunt alone. Some species, such as *Aglena consociata* of Gabon and *Stegodypgus* of Pakistan, build a communal web and hunt in packs.

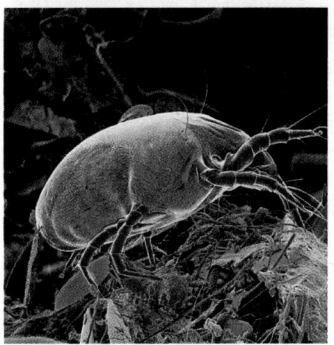

Figure 30-9 A tarantula, *top*, does not capture its prey in a web. Instead, tarantulas stalk, lunge at, and then bite their victims. The house dust mite, *bottom*, is a major cause of allergies in humans when it is inhaled in household dust.

Arachnids, such as those in Figure 30-9, compose the largest class in the subphylum Chelicerata. Two additional minor classes of chelicerates, horseshoe crabs and sea spiders, are marine. There are over 57,000 named species of arachnids, including spiders, ticks, mites, scorpions, and daddy longlegs. Arachnids have a pair of chelicerae as their foremost appendage. A second pair of appendages, called **palps,** are used to catch and handle prey, although they are sometimes specialized for sensory or even reproductive functions. The palps are followed by four pairs of walking legs.

All arachnids, except mites, are carnivores. Arachnids have no jaws. Because they are able to ingest only liquid food, spiders and other carnivorous arachnids must digest their prey externally. They accomplish this by injecting their prey with powerful enzymes, which cause the prey's tissues to liquefy. The liquid contents are then sucked up through the arachnid's muscular pharynx.

Arachnids occur primarily in terrestrial habitats. Three major orders of arachnids are discussed below.

Scorpions

Scorpions are arachnids whose palps have evolved into large, grasping pincers. Scorpions use these pincers for handling and tearing apart their food and for reproducing, as shown in Figure 30-10. They are not used as weapons. Scorpions have long, slender, segmented abdomens that end in a venomous stinger used to stun their prey. A scorpion holds its abdomen folded forward over its body as it walks.

Scorpions are thought to have been the first animals to successfully invade land; some fossil scorpions are over 425 million years old. Scorpions evolved from eurypterids, ancient sea scorpions. The sea scorpions, all of which are extinct, breathed with gills. Some grew several feet long. There are some 1,200 species of scorpions living today, all terrestrial, and they range in size from 1 to 18 cm (0.4 to 7 in.).

Figure 30-10 In this courtship ritual, the male scorpion, *far right,* grasps the female's pincers in his and dances her around until he positions her directly over a sperm packet that he has previously deposited on the ground. Once positioned, the packet bursts open, and the sperm enter the female's reproductive opening. In this pair of scorpions, the female carries offspring, the result from a previous mating, on her back. The red spots on the male's back are mites.

Figure 30-11 This garden spider has spun what is called an orb web. The spider begins building its web by laying down a foundation of spokes that radiate from a center point. Next, it spins a continuous spiral outward from the center.

Spiders

Spiders have poison glands that secrete toxin through chelicerae that have been modified into fangs. Spiders use their fangs to bite and paralyze prey. Only two species of spiders living in the United States, the black widow and brown recluse, are dangerous to humans. There are about 35,000 named species of spiders, but many more await discovery.

Spiders are important insect predators in virtually every terrestrial ecosystem. Many, like the familiar wolf spiders, actively hunt their prey by running them down. Others, like the trapdoor spiders, construct silk-lined burrows with lids that open easily. The spider hides underneath the lid and then leaps out to seize its prey as it passes by the burrow. Many spiders trap their prey in sticky webs of remarkable diversity, as shown in Figure 30-11. The strands of the webs are made of pure silk. The silk is formed from a liquid protein forced out of specially modified appendages called spinnerets that are located at the end of the spider's abdomen.

Mites

The mites are by far the largest and most diverse group of arachnids. Few scientists study mites, so many of the members of this group are not well known. There are about 35,000 identified species, but taxonomists estimate that there may be up to a million species that exist. Some mites, including chiggers and ticks, are well known to humans because of their irritating bites.

In mites, the head, thorax, and abdomen are fused into a single, unsegmented body. Most mites pass through a series of larval phases, changing from a six-legged to an eight-legged animal. As adults, most mites are quite small, typically less than 1 mm (0.04 in.) long. A tick, shown in Figure 30-12, grows larger. Some can be up to 3 cm (1.2 in.) long. Mites are found in virtually every terrestrial, freshwater, and marine habitat. Many marine and freshwater mites are herbivores, while terrestrial mites are usually predators. Many plants have special pits on their leaves inhabited by predaceous mites that protect the plants from herbivores. ❏

Figure 30-12 Some ticks carry diseases that can be transmitted to humans when the tick bites them. This particular species, commonly called a deer tick, harbors the bacteria that cause Lyme disease.

❏ CAPSULE SUMMARY

Arachnids include spiders, scorpions, and mites. Arachnids are characterized by four pairs of legs, fused body regions, and mouthparts called chelicerae.

SECTION 30-2

UPDATE

☑ RESEARCH Update

Campers Stay Home

James Olson, Jacquiline Dawson, and Daniel Fishbein, researchers at the Centers for Disease Control and Prevention in Atlanta, Georgia, have been studying ehrlichosis. Ehrlichosis is a tick-borne disease similar to Rocky Mountain spotted fever that is caused by a ricketsial bacterium. Although the disease is probably not new, it has only recently been distinguished from others like it. Symptoms of infection include fever, headache, muscle pain, nausea, depressed white blood cell count, and impaired liver function. Successful treatment must be immediate—often before the bacterium can be clinically detected in the blood.

Cases of ehrlichosis are becoming more common, Olson contends, because more people are hiking in the wilderness where ticks abound. He and his colleagues recommend a good personal tick check after spending time outdoors. The researchers think that the disease cannot be transmitted unless the tick has been attached for 24 hours, so the sooner you inspect your body for ticks, the better the chances of preventing ehrlichosis. But remember, as unpleasant as having an arachnid imbedded in your skin can be, it's the *bacterium* that causes the disease. Olson says, "The tick is an innocent reservoir." ☑

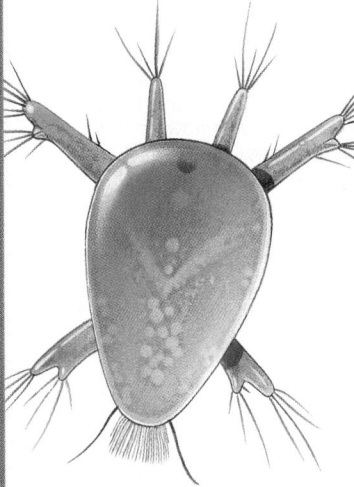

Figure 30-13 All crustaceans pass through a free-swimming larval stage called a nauplius. This is a nauplius of a tiny crustacean called a copepod.

Most Crustaceans Are Aquatic

The crustaceans are a large group of primarily aquatic arthropods that includes crabs, lobsters, crayfish, shrimps, barnacles, water fleas, and pill bugs. There are eight separate crustacean orders, which include some 35,000 species. Crustaceans are often very abundant in marine and freshwater habitats and are a major food source for humans.

Like insects, crustaceans have jaws and are considered mandibulates. However, crustaceans differ from insects in a number of important respects.

1. Most appendages of crustaceans are branched at the ends, although some have become unbranched in the course of evolution. The appendages of insects are unbranched.
2. All crustaceans have a distinctive larval form, called a nauplius, that has three pairs of branched appendages, as shown in Figure 30-13.
3. Aquatic crustacean exoskeletons are hardened with calcium carbonate.
4. Crustaceans are the only arthropods with two pairs of antennae. Insects have only one pair.
5. Most crustaceans have three pairs of chewing appendages, while insects have only one pair.
6. Crustaceans, unlike insects, have legs attached to their abdomen as well as to their thorax.
7. All crustaceans breathe with gills.

Decapods

Large marine crustaceans such as shrimps, lobsters, and crabs, along with the freshwater crayfish shown in Figure 30-14, have five pairs of walking legs. These crustaceans are often referred to as decapods. They are all members of the order Decapoda. The term *decapod* is from the Greek *deka*, meaning "ten," and *pous*, meaning "foot." The head and thoracic body segments of decapods are fused into a single cephalothorax. The cephalothorax is covered on top by a shield called a **carapace**. In lobsters and crayfish, appendages called **swimmerets** are attached to the underside of the abdomen. The swimmerets are used in swimming and in reproduction. Flattened, paddle-like appendages, called **uropods,** are at the end of the abdomen, and many decapods have a **telson,** or tail spine. Decapods such as crayfish, lobsters, and shrimps can propel themselves through the water by forcefully flexing their abdomen.

Unlike decapods, the members of the other seven orders of crustaceans are quite small. Minute crustaceans smaller than the tip of a pencil are abundant in marine and freshwater habitats all over the world. Tiny copepods *(KOH puh pahds)* are among the most abundant multicellular organisms on Earth and are a key food source in the marine food chain. Also common are the minute ostracods, fairy shrimps, and water fleas.

 Did You Know?

Whales often have barnacles attached to their skin. In fact, some whale researchers distinguish the whales they are studying by the pattern of barnacle "passengers" on the whales' skin.

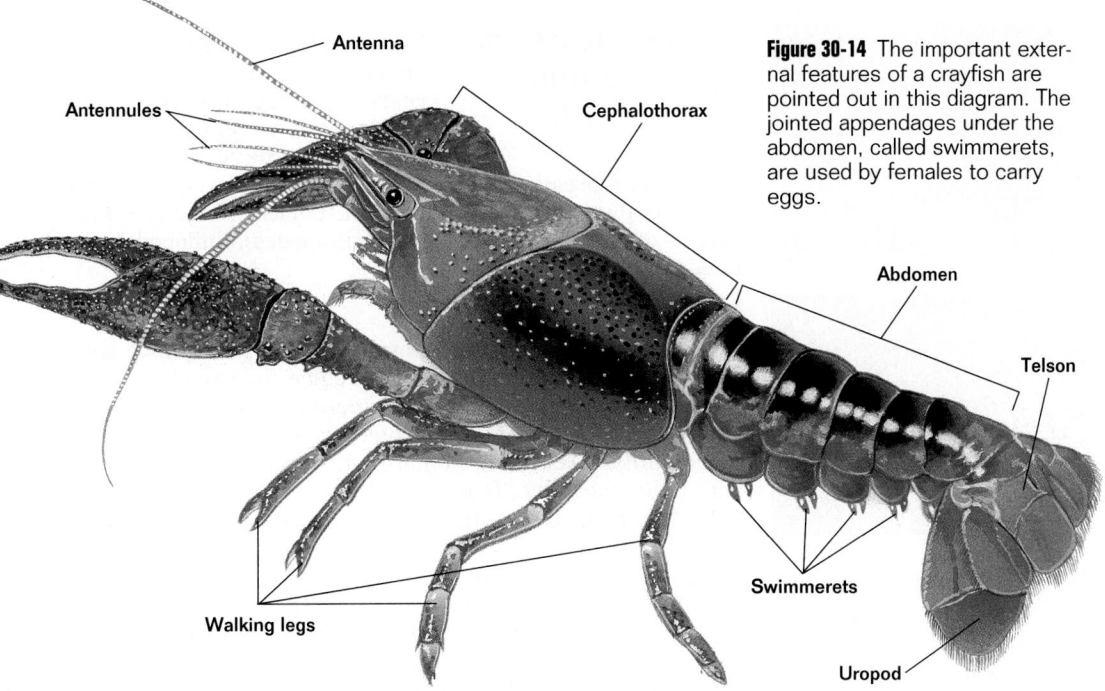

Antenna

Antennules

Cephalothorax

Abdomen

Telson

Walking legs

Swimmerets

Uropod

Figure 30-14 The important external features of a crayfish are pointed out in this diagram. The jointed appendages under the abdomen, called swimmerets, are used by females to carry eggs.

One marine crustacean, the barnacle, is sessile (permanently attached to a rock or other object) as an adult, although the larvae are free swimming. Barnacles attach themselves upside down to a rock or other object. With their bodies protected by hard, sharp plates, barnacles stir food into their mouths with their feathery legs.

Terrestrial Crustaceans

Only a few crustaceans are terrestrial. Other than a few crab species, there are two principal groups of terrestrial crustaceans. One group is composed of the pill bugs and sow bugs, which belong to a large order known as the **isopods.** A pill bug is shown in Figure 30-15. About half of the 4,500 species of isopods are terrestrial, and the rest are marine. The other group of terrestrial crustaceans is composed of the sand fleas, with several thousand terrestrial species typically found along beaches. ◻

◻ CAPSULE SUMMARY

Crustaceans have hardened exoskeletons and mouthparts called mandibles (jaws). Most crustaceans (crabs, lobsters, and crayfish) have greatly modified appendages and are aquatic. Other crustaceans (pill bugs and sand fleas) are terrestrial.

Figure 30-15 Although this pill bug is a terrestrial arthropod, it still breathes with gills that must be kept moist to function. Consequently, pill bugs are usually found in damp environments and are most active at night, when the humidity is high.

Figure 30-16 Centipedes, *top,* are carnivorous predators that inject poison into their prey through a pair of hollow fangs. Millipedes, *bottom,* are harmless herbivores that feed on decayed vegetation. Millipedes avoid predators by releasing a foul-smelling fluid from the side of each body segment.

Most Uniramians Live on Land and Can Fly

The subphylum Uniramia is an enormous group of mostly terrestrial arthropods that differ structurally from crustaceans. Uniramians have unbranched appendages, breathe with tracheae and spiracles, and excrete waste products through Malpighian tubules. There are three classes of uniramians: Diplopoda (millipedes), Chilopoda (centipedes), and Insecta (insects).

Millipedes and Centipedes

Millipedes and centipedes, shown in Figure 30-16, have similar body designs. Each has a head region followed by numerous segments (usually 15, but sometimes more) that are all similar. Each segment bears one or two pairs of legs. The name *millipede* is from the Latin *mille,* meaning "thousand," and *pedis,* meaning "foot." The name *centipede* is from the Latin *centum,* meaning "hundred," and *pedis,* meaning "foot." Actually, millipedes have two pairs of legs on each body segment, so they typically have 60 legs in all. Centipedes have one pair of legs on each segment, so they have a total of 30 legs. Each segment of modern millipedes evolved from two segments of their ancestors. Approximately 2,500 species of centipedes and over 10,000 species of millipedes have been named.

Insect Diversity

The insects are by far the largest group of organisms on Earth, as shown in Figure 30-17. There are over 700,000 named insect species, comprising more than 50 percent of all named animal species. Most scientists agree that there may be several million insect species in existence. Most of the undiscovered species live in the tropics. There are more kinds of beetles (350,000 species) than all the kinds of non-insect animals put together! Insects are primarily a terrestrial group, and it is thought that those that are aquatic probably had terrestrial ancestors. Seven of the most common orders of insects are described in Table 30-1. You can learn more about the structure of one particular insect, the grasshopper, in *Up Close: Eastern Lubber Grasshopper* on pages 690–691.

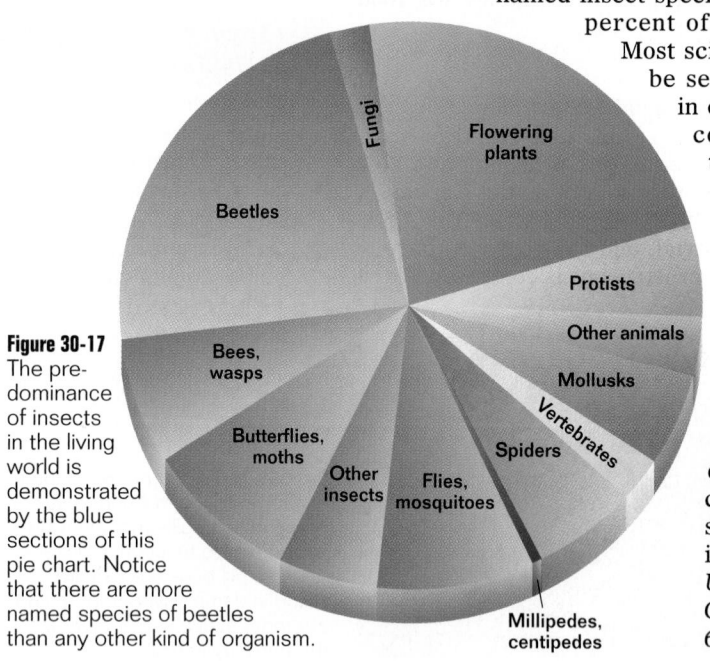

Figure 30-17
The predominance of insects in the living world is demonstrated by the blue sections of this pie chart. Notice that there are more named species of beetles than any other kind of organism.

Table 30-1 Major Orders of Insects

Order	Number of Species	Main Characteristics	Examples
Coleoptera "Shield winged"	350,000	Two pairs of wings (front pair covers transparent hind pair); heavy, armored exoskeleton; biting and chewing mouthparts; complete metamorphosis	Beetles, weevils
Diptera "Two winged"	120,000	Transparent front wings; hind wings reduced to knobby balancing organs; sucking, piercing, and lapping mouthparts; complete metamorphosis	Flies, mosquitoes
Lepidoptera "Scale winged"	120,000	Two pairs of broad, scaly wings; hairy bodies; tubelike, sucking mouthparts; complete metamorphosis	Butterflies, moths
Hymenoptera "Membrane winged"	100,000	Two pairs of transparent wings; mobile head; well-developed eyes; chewing and sucking mouthparts; stinging; many social species; complete metamorphosis	Ants, bees, wasps
Hemiptera "Half winged"	60,000	Two pairs of wings or wingless; piercing, sucking mouthparts; incomplete metamorphosis	Giant water bug, bedbug, chinch bug
Orthoptera "Straight winged"	20,000	Two pairs of wings or wingless; biting and chewing mouthparts in adults; incomplete metamorphosis	Grasshoppers, crickets, cockroaches, mantids
Odonata "Toothed"	5,000	Two pairs of transparent wings; chewing mouthparts; incomplete metamorphosis	Dragonflies, damselflies

SECTION 30-2

👁 VISUAL STRATEGY

Table 30-1
Correlate this table with the pie chart in Figure 30-17. Point out that beetles are the predominant order of insects. Ask students to explain how the insect orders are arranged in Table 30-1. (*They are arranged in order of descending number of species.*)

Teaching Tip
Orders of Insects
Play an insect identification game with your students to help them distinguish characteristics of the different orders of insects. Divide your class into about five groups, and have each group choose an insect order from Table 30-1. Then have the other groups take turns asking one question at a time about the unknown insect order, such as "Does it have transparent wings?" The questions should be phrased so that they can be answered yes or no. The group that guesses the insect order should be allowed to take the next turn.

Multicultural Perspective

Crickets

To the British, a cricket on the hearth is a sign of good luck. To the Japanese, crickets are considered pets and a delightful source of musical and artistic pleasure. Favorite pet crickets are often given elaborate cages and beautiful porcelain water dishes. Japanese cricket owners even "tickle" their crickets with delicate hand-carved brushes to encourage them to sing.

Instructional Strategies

- Have students look up grass-hoppers in field guides that are used for insect identification. Ask students to compare the external structures of the Eastern lubber grasshopper with those of other grasshoppers. *(The Eastern lubber grasshopper is a short-horned grasshopper. Long-horned grasshoppers, such as katydids and Mormon crickets, have long, threadlike antennae.)*

- Ask students to choose one of the organ systems discussed in *Up Close: Eastern Lubber Grass-hopper* to illustrate in a cartoon or flowchart. Students can choose from the reproductive system, circulatory system, digestive system, or nervous system of the grass-hopper. Cartoons and flowcharts should show all of the organs involved in the system.

- Ask students to list the parts of the digestive tract that help to physically break down the food and to list the parts that chemically break down food. *(Physical digestion includes labrum, labium, mandible, maxilla, and gizzard. Chemical digestion includes midgut and coelom.)*

UP CLOSE EASTERN LUBBER GRASSHOPPER

- **Scientific name:** *Romalea microptera*
- **Range:** Eastern United States
- **Habitat:** Fields and meadows
- **Size:** 5 cm (2 in.) to 6.5 cm (2.6 in.) in length
- **Diet:** Grasses and other leafy vegetation

External Structures

Head   Two antennae contain sense organs for both touch and smell. On the sides of the head are a pair of very large compound eyes, each containing hundreds of six-sided lenses. Located high on the forehead are three light-detecting ocelli. The mouthparts are composed of four kinds of appendages. The stiff upper labrum and lower labium are "lips" that hold a leaf or blade of grass. The mandibles (jaws), assisted by maxillas (graspers), are then used to tear off pieces of the plant.

Wings   Like most insects, grasshoppers have two pairs of wings. A pair of leathery forewings protects the more delicate flying wings.

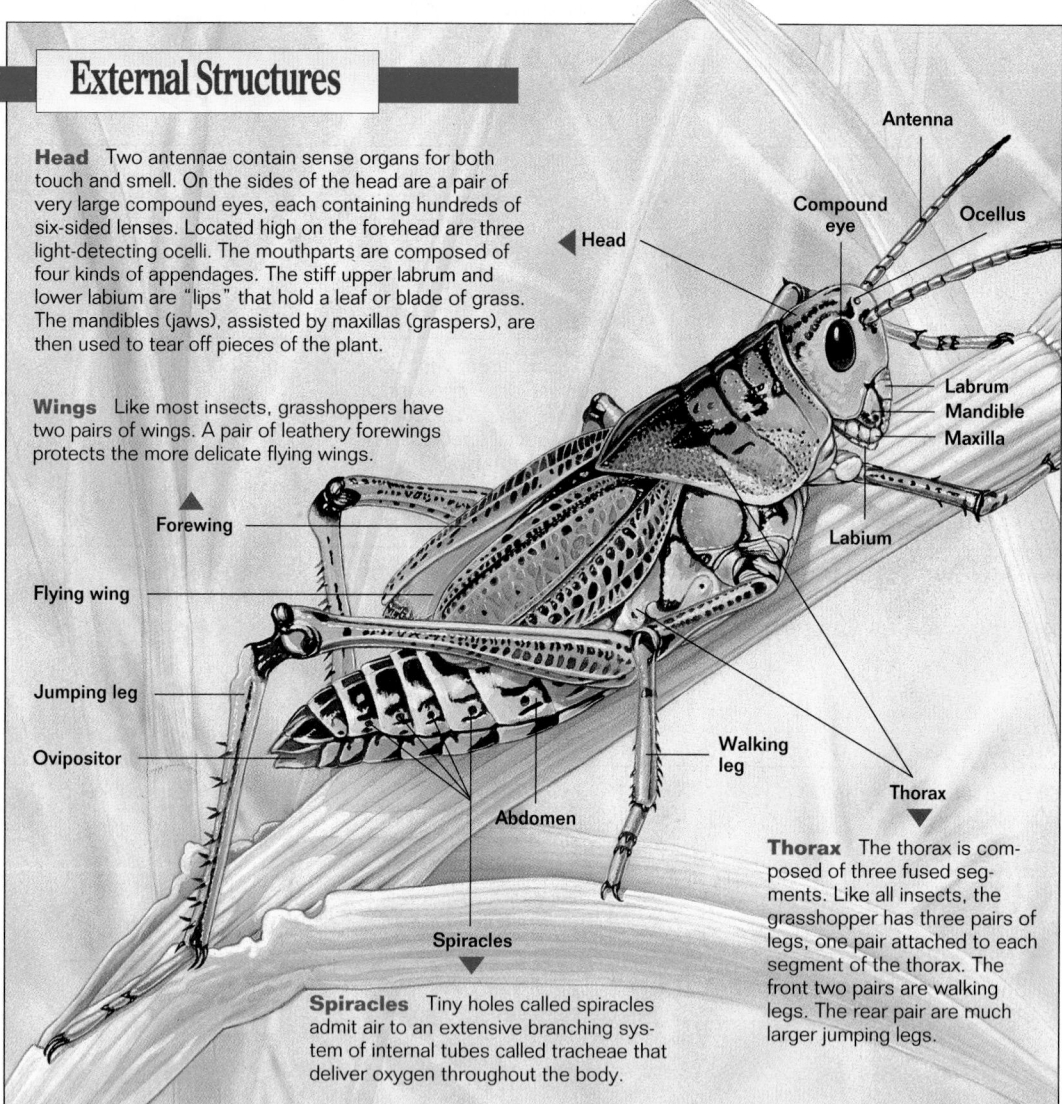

Spiracles   Tiny holes called spiracles admit air to an extensive branching system of internal tubes called tracheae that deliver oxygen throughout the body.

Thorax   The thorax is composed of three fused segments. Like all insects, the grasshopper has three pairs of legs, one pair attached to each segment of the thorax. The front two pairs are walking legs. The rear pair are much larger jumping legs.

Multicultural Perspective

Insect Behavior
Dr. Charles Turner (1867–1923), an entomologist, was among the first scientists to prove that insects can hear and distinguish pitch. He also showed that cockroaches can learn by trial and error. This African American pioneer in the field of insect behavior (and author of 50 research articles) was devoted to teaching his students the art of observing nature. Dr. Turner obtained his Ph.D. from the University of Chicago in 1907.

Internal Structures

Reproductive system The reproductive organs (testes and ovaries) of grasshoppers are located in the abdomen. During the summer mating season each male "sings" to potential mates. A "song" is produced when he rubs the row of pegs on the inside of a jumping leg against ridges on a forewing. During mating, the female collects the male's sperm in a storage pouch called a seminal receptacle. Later that summer, the female digs a hole using two pairs of pointed organs called ovipositors. The eggs are fertilized by the stored sperm as she releases the eggs into the hole. They stay dormant over the winter and hatch the following spring.

Circulatory system The grasshopper's circulatory system is composed of a long blood vessel that runs along its back, with a series of muscular "hearts" located over the abdomen. The circulatory system is open; that is, the blood does not stay in the vessel, but is pumped out and bathes the tissues directly.

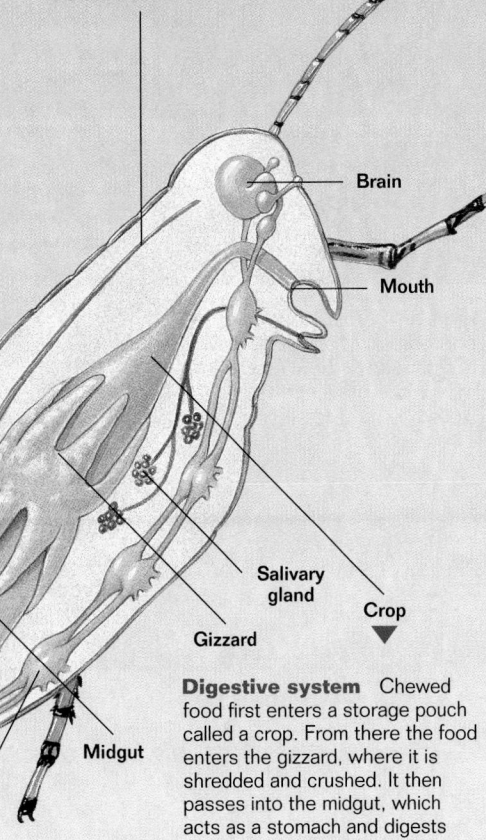

Ovary

Hearts

Seminal receptacle

Dorsal blood vessel

Brain

Mouth

Salivary gland

Crop

Gizzard

Midgut

Ganglia

Nerve cord

Anus

Digestive system Chewed food first enters a storage pouch called a crop. From there the food enters the gizzard, where it is shredded and crushed. It then passes into the midgut, which acts as a stomach and digests the food with the aid of enzymes. Food molecules then pass across the wall of the midgut into the fluid of the coelom. This fluid eventually enters the circulatory system, and the nutrients are delivered to body tissues.

Nervous system The nervous system is composed of a major ventral nerve cord with ganglia located in each body segment. Three fused ganglia in the head serve as the brain. The brain coordinates responses such as the very rapid leap that the grasshopper makes when threatened.

SECTION 30-2
UP CLOSE
EASTERN LUBBER GRASSHOPPER

Discussion

Guide the discussion by posing the following questions.

1. How do grasshopper eyes make it almost impossible to sneak up on this insect? *(Grasshoppers have compound eyes on each side of their head and three light-detecting ocelli on their forehead.)*

2. How can you tell that this grasshopper is a female from its external structures? *(This grasshopper has ovipositors, which are used to deposit fertilized eggs.)*

3. What structures do grasshoppers use to move slowly? How do they make quick escapes? *(Grasshoppers use their walking legs to move slowly. When they need to escape quickly, they use their jumping legs or wings.)*

4. What are the three sections of the grasshopper's body, and what is a major body function that occurs in each section? *(The grasshopper has a head, thorax, and abdomen. Various body functions occur in each of the body sections. For example, touching, smelling, seeing, and chewing occur in the head; walking and jumping occur in the thorax; and respiration, reproduction, and excretion occur in the abdomen.)*

5. What would happen if every grasshopper were to freeze during a cold winter? *(The species would continue to survive because grasshopper eggs stay dormant in winter and hatch in the spring.)*

Historical Note

Fear of Insects

Dr. May Roberta Berenbaum, head of the entomology department at the University of Illinois, was terrified of insects as a child. While at Yale University, she enrolled in a course on terrestrial arthropods and was determined to learn about the insects that she had grown up fearing. Dr. Berenbaum has founded several programs to promote learning about insects, including an annual Insect Fear Film Festival, a radio commentary called *Those Amazing Insects*, and two books, *Ninety-Nine Gnats, Nits and Nibblers* and *Ninety-Nine More Maggots, Mites and Munchers*. Dr. Berenbaum has won several awards for her research, teaches some of the most popular courses at the University of Illinois, and was recently elected to the National Academy of Sciences.

Application

Agriculture
Silk is made today with cultivated silkworms, usually caterpillars of the genus and species *Bombyx mori*. The female lays 300 to 500 eggs on special paper provided by silk farmers. When fully grown, the silkworms spin a cocoon made of one continuous silk thread, in which the silkworm will develop into a pupa and then emerge as a moth. Silk farmers do not allow most pupas to become moths because a moth will shatter the silk cocoon into many pieces as it leaves the cocoon. Instead, farmers kill most pupas and process the silk cocoons. First, the cocoon is unwound, and the threads are strengthened by several processes. Then the thread is dyed and woven into fabric.

Teaching Tips
Silk

Encourage students to research the history of the discovery of silk. China was the first to produce silk fabrics and for about 3,000 years was the only country that knew how to cultivate silkworms. Chinese legends colorfully tell of the discovery of silk by Emperor Huangdi's wife in about 2700 B.C.

Life Cycles

Ask students to compare complete and incomplete metamorphosis in a *Graphic Organizer* like the one at the bottom of this page. Students can use the life cycles of specific insects in their comparisons or use generalizations. Ask students to write a brief paragraph to accompany and explain their chart.

Figure 30-18 This stop-action photograph of a green lacewing in flight shows how its wings move. The movement is controlled by muscles inside the insect's thorax.

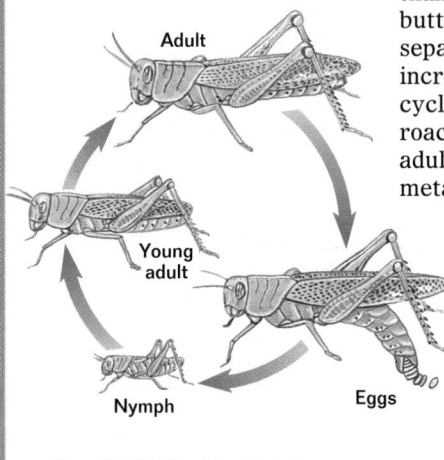

Figure 30-19 The Monarch butterfly, *right*, undergoes complete metamorphosis. During complete metamorphosis, each developmental stage appears markedly different. The Alutacea bird grasshopper, *above*, undergoes incomplete metamorphosis, during which the developmental differences between stages are not nearly as great.

Insect Structure and Life Cycles

Most insects are small, only a few centimeters in length. All insects have the same general body plan: three body sections (a head, a thorax, and an abdomen); three pairs of legs, all attached to the thorax; and one pair of antennae.

As shown in Figure 30-18, flight is one of the great evolutionary achievements of insects. While some insects, like fleas, lice, and silverfish, lack wings, all other insects have one or two pairs. The insect thorax consists of three fused segments. The wings are attached to the middle and rear segments. Each wing is a solid sheet of chitin that grows from a saclike outgrowth of the body wall. The wings are strengthened by veins made of tubes of chitin.

The life cycles of most insects are complex, often requiring several molts before the adult stage is reached. In the final molt to adult, the juvenile undergoes a process of physical change called **metamorphosis**. Flies, beetles, ants, bees, wasps, butterflies, and moths—90 percent of all insect species—undergo "complete" metamorphosis, shown in Figure 30-19. In complete metamorphosis, the juvenile changes from a wingless, wormlike larva (called a caterpillar in butterflies and moths) to a winged adult by becoming a pupa enclosed within a protective capsule called a **chrysalis** (*KRIHS uh lihs*). What is the evolutionary advantage of this complicated life cycle? During development, the larvae exploit different habitats and different food sources than the adult. For example, the larvae of nectar-drinking butterflies are caterpillars that eat leaves! This ecological separation of young from adults eliminates competition, thus increasing the chance of survival for each phase of the life cycle. In a smaller number of species (grasshoppers, cockroaches, and mosquitoes), metamorphosis from juvenile to adult is much less dramatic and is described as "incomplete" metamorphosis. In incomplete metamorphosis, also shown in

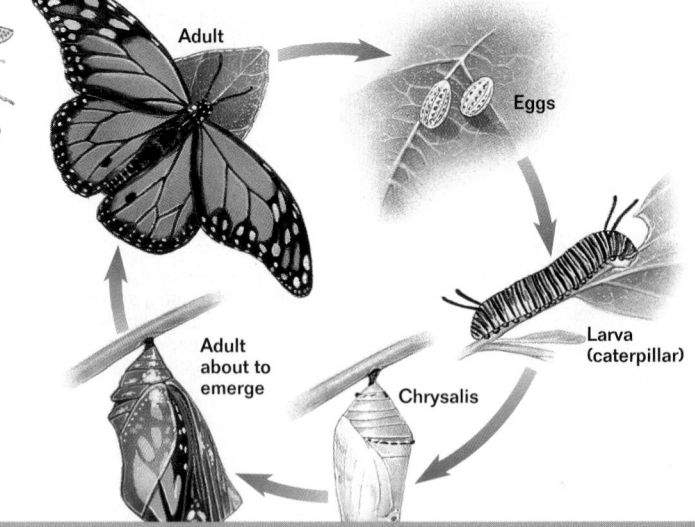

Graphic Organizer

Use this graphic organizer with *Teaching Tip:* **Life Cycles** on this page.

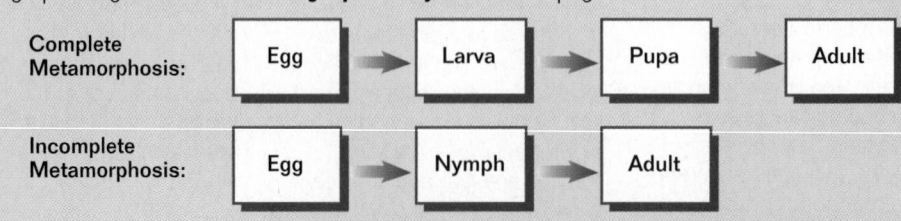

| Complete Metamorphosis: | Egg | → | Larva | → | Pupa | → | Adult |

| Incomplete Metamorphosis: | Egg | → | Nymph | → | Adult |

Figure 30-19, the juveniles have wings, and there is usually no pupa before the last molt. The juvenile, called a **nymph** (NIHMF), is essentially a smaller version of the adult.

Two orders of insects, Hymenoptera (ants, bees, and wasps) and Isoptera (termites), have evolved elaborate social systems. These insects often live in highly organized societies of genetically related individuals. The societies are characterized by marked division of labor, meaning that different kinds of individuals serve the group by performing specific functions. In the termite colony shown in Figure 30-20, for example, small, active members called workers gather the food, raise the young, and excavate tunnels. Other, larger termites, called soldiers, defend the colony with their immense jaws. Both workers and soldiers are sterile. Only the queen and king reproduce. The role played by an individual in a colony is called its **caste**. Social insects have no choice about their caste. Their genes determine their caste, just as your genes determine your sex. □

□ **CAPSULE SUMMARY**

Insects, the most numerous of all animals, have jaws and three body regions: head, thorax, and abdomen. They are the only invertebrates capable of flight. Metamorphosis in insects may be either complete or incomplete. Some insects, like bees and termites, have evolved complex social systems.

Figure 30-20 Most of the members of this termite colony are sterile (unable to reproduce). The queen, with her distended abdomen, is the egg-laying machine of the colony.

Section Review

1. *What is the major difference between chelicerate and mandibulate arthropods?*
2. *Compare the body segmentation of a crayfish with that of an insect.*
3. *Explain why a centipede is more closely related to a housefly than to a scorpion.*
4. *Describe the digestive system of the Eastern lubber grasshopper.*
5. *Describe the life cycle of a butterfly.*

Critical Thinking

6. *Arthropods first invaded land about 400 millions years ago. They have endured several mass extinctions in which many other kinds of organisms became extinct. What characteristics have enabled arthropods to thrive?*

Answers to Section Review

1. Mandibulates have jaws, and chelicerates have fangs or pincers.
2. The head and thorax of a crayfish is fused into a part called the cephalothorax, while an insect has a separate head and thorax. Both an insect and a crayfish have an abdomen, which shows further segmentation in a crayfish and may be segmented in an insect.
3. Centipedes and houseflies are uniramians, while the scorpion is a chelicerate. Centipedes and insects have unbranched appendages, tracheae, spiracles, and Malpighian tubules.
4. See *Up Close: Eastern Lubber Grasshopper* on page 691.
5. A butterfly begins as an egg, which hatches into a larva. The larva forms a chrysalis, from which the adult butterfly emerges to mate and lay eggs.
6. Answers might include the efficient use of food sources during different stages of arthropod development, rapid reproduction, and widespread populations in diverse aquatic and terrestrial habitats.

Vocabulary Preview

skin gill

Opening Questions

Ask students whether there are any terrestrial echinoderms. They should recall from Chapter 27 that echinoderms are marine animals, most of which live on the ocean bottom. Challenge students to explain why echinoderms are strictly aquatic organisms by asking them about loco- motion, circulation, and respiration in echinoderms. Lead students to con- clude that echinoderms are better adapted to marine environments.

Demonstration

Observing Echinoderms

Display any echinoderms you or your students may have collected, such as sea stars or sand dollars. Emphasize the five-part radial sym- metry of the specimens. Point out that during drying, the skin of sand dollars goes away, and only the hard internal skeleton remains. Ask students how echinoderms differ from arthropods. *(Echino- derms have endoskeletons, no body segments, and five-part sym- metry, while arthropods have exoskeletons, body segments, and bilateral symmetry.)*

CONNECTIONS

Chapter 27

Deuterostome Development

In deuterostome embryo devel- opment, cell division occurs in a radial pattern. The anus forms from the blastopore, and the mouth forms from the second- ary opening.

30-3 Echinoderms and Invertebrate Chordates

Section Objectives

- Describe the major charac- teristics of echinoderms.
- Compare and contrast the lifestyles of the organisms in each of the five echinoderm classes.
- Describe how sea stars feed.
- Compare and contrast tunicates and lancelets.

Mollusks, annelids, and arthropods follow a proto- stome pattern of development. In echinoderms (sea urchins and sea stars) and chordates (tuni- cates, lancelets, and vertebrates), however, development is organized very differently. Their deuterostome pattern of development represents one of the most fundamental changes in body plan in the history of animal evolution. The developmental similarity between echinoderms and chordates unites these two seemingly dissimilar animal phyla and leads scientists to believe that both groups were derived from a common ancestor, as shown in Figure 30-21.

Echinoderm Larvae and Adults Have Different Body Plans

Echinoderms are familiar to all of us as "starfish." They are not really fish and are more properly called sea stars. Most people rec- ognize the obvious radial symmetry of adult sea stars. It may be less obvious, however, that like all the advanced animal phyla, echinoderms are also *bilaterally* symmetric. In fact, it is the larvae of all echinoderms that are bilaterally symmet- ric. Biologists believe that the echinoderms were once far more mobile than they are today. The radial symmetry of

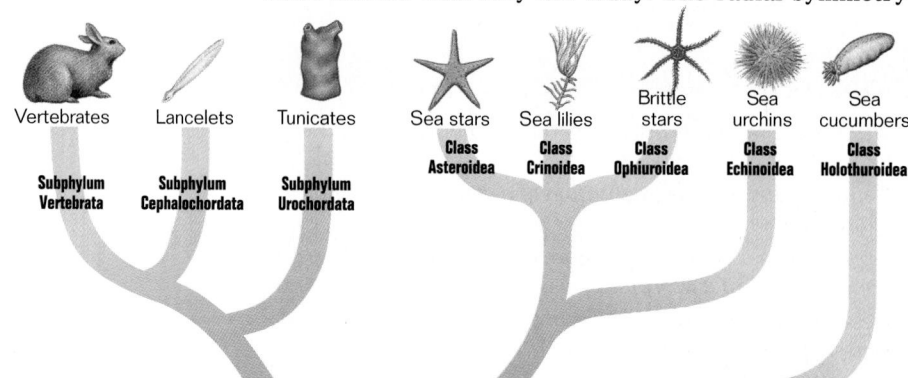

Figure 30-21 This phylogenetic tree shows the evolution of chordates and echinoderms from a common ancestor, as well as the major chordate and echinoderm groups.

adults evolved later, when echinoderms adopted a more sessile style of living.

The adult echinoderm body has no head or brain. Instead, the nervous system consists of a central ring of nerves with branches extending into the arms. Although the animal is capable of complex response patterns, each arm acts more or less independently. Apparently, centralization of the nervous system is not feasible in radially symmetric bodies.

All Echinoderms Share Four Major Characteristics

The different kinds of echinoderms vary considerably in the details of their body design and in how they go about acquiring food. Despite the apparent differences among echinoderms, all share four fundamental characteristics.

1. **Endoskeleton** Echinoderms have an endoskeleton composed of individual plates called ossicles. In sea stars and many other echinoderms, a large number of these plates are fused together. The fused plates function much as an arthropod exoskeleton by providing muscle attachment sites and shell-like protection. In most echinoderms, the endoskeletal plates bear spines that project upward through their skin. The term *echinoderm* is from the Greek *echinos*, meaning "spiny," and *derma*, meaning "skin."

2. **Five-part radial symmetry** Most echinoderms have five arms extending radially from a central point.

3. **Water-vascular system** Echinoderms have a water-filled system of interconnected canals and tube feet called a water-vascular system. Echinoderms use their many tube feet to crawl across the sea floor.

4. **Coelomic circulation and respiration** The echinoderm body cavity functions as a simple circulatory system. Particles move freely throughout the large, fluid-filled coelom. In many echinoderms, respiration and waste removal are performed by **skin gills,** shown in Figure 30-22. Skin gills are small, fingerlike projections that grow between the spines.

Figure 30-22 The extensions of skin on the surface of this sea star are called skin gills. Skin gills create an increased surface area through which respiratory gases (oxygen and carbon dioxide) can be exchanged. Skin gills also function as excretory organs. Wastes that accumulate in the skin gills are released into the surrounding water.

❏ CAPSULE SUMMARY

Echinoderms (sea stars, sea urchins, and sea cucumbers) are deuterostome marine invertebrates that are radially symmetric as adults and have a water-vascular system.

Echinoderms Are a Diverse Group

Echinoderms are one of the most successful of all marine phyla, although they were far more so in the past than they are now. There are more than 20 extinct classes of echinoderms and an additional 5 classes of living members. The 5 living classes are sea stars, feather stars and sea lilies, brittle stars, sea urchins and sand dollars, and sea cucumbers.

Teaching Tip

What If a sea star is cut in half? Each half of the sea star will become a new sea star. This process of regeneration surprised some fishermen who decided to cut up sea stars that were feasting on their oyster beds. The fishermen's problem multiplied because each piece of a sea star that is connected to the central ring of the original echinoderm has the ability to develop into a new sea star.

 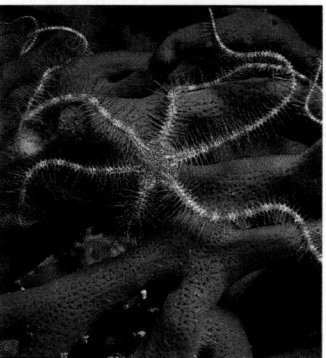 

Figure 30-23 These three representative echinoderms each belong to a separate class. Feather stars, *above left,* along with sea lilies, are primitive echinoderms. Brittle stars, *above center,* are the largest of the major groups of echinoderms. Sea urchins, *above right,* often prefer to live on rocky ocean bottoms.

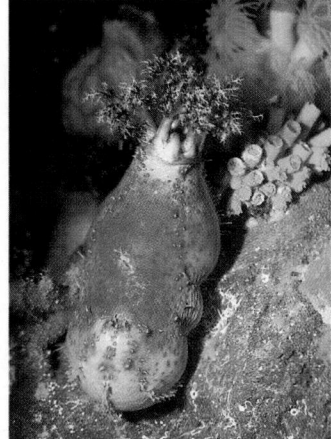

Figure 30-24 The tentacles of a sea cucumber are covered with sticky mucus, to which plankton adhere. Periodically, the tentacles are inserted into the mouth, cleaned of the organism-bearing mucus, and recoated with a fresh supply of mucus.

Sea stars are the echinoderms most familiar to people. All of the roughly 1,500 species of sea stars are carnivores. They are among the most important predators in many marine ecosystems. For example, the crown-of-thorns sea star consumes the cnidarians that make coral reefs. In one hour, these sea stars are able to graze along 20 m of a reef and can quickly destroy entire coral reef ecosystems. Other sea stars specialize in hunting bivalve mollusks—clams and oysters.

The feather stars and sea lilies differ from all other living echinoderms because their mouth is located on their upper surface. Sea lilies, the most primitive living echinoderms, are sessile. They are attached to the ocean floor by a stalk that can be 1 m (3.3 ft.) long. More than 600 fossil species of sea lilies are known, but there are only 80 living species. Also in this class are 520 living species of feather stars. Feather stars, shown in Figure 30-23, have stalks initially but lose them early in their development. The adult has a centrally located disk with hooklike projections that it uses to attach itself directly to the ocean bottom or a coral reef, but it is not fully sessile. It can crawl along the surface or even swim short distances.

Brittle stars have slender branched arms that they move in pairs from side to side, rowing along the ocean floor. Close relatives of the sea stars, the 2,000 species of brittle stars are sometimes grouped with them by taxonomists.

The sea urchins and sand dollars lack distinct arms but have the same basic five-part body plan as all echinoderms. Both sea urchins and sand dollars have hard endoskeletons of fused plates. There are about 900 living species of sea urchins and sand dollars.

The 1,500 species of sea cucumbers differ from other echinoderms in that their ossicles are small and not connected. Because they do not have a fused skeleton, the bodies of sea cucumbers are soft. Sea cucumbers, shown in Figure 30-24, are sluggish animals that often have a tough, leathery epidermis. They lie on their side on the ocean floor with their mouth at one end. The mouth is surrounded by several dozen tube feet that have been modified into tentacles. The tentacles capture the small organisms that the animal eats.

Graphic Organizer

Use this graphic organizer with the **Chapter Closure** on page 697.

Arthropod characterisitics

- Exoskeleton
- Segmentation
- Cephalothorax
- Open circulation
- Wings
- Jointed appendages
- Compound eyes
- Spiracles
- Malpighian tubules

Invertebrate animals

Echinoderm characterisitics

- Endoskeleton
- Five-part radial symmetry
- Water-vascular system
- Coelomic circulation and respiration

Protostome development   Deuterostome development

Invertebrate Chordates

The phylum Chordata is divided into three subphyla containing some 42,500 species. One subphylum, the vertebrates (subphylum Vertebrata), contains almost all chordate species. The other two, much smaller invertebrate subphyla are the tunicates (subphylum Urochordata) and the lancelets (subphylum Cephalochordata).

The tunicates are a group of about 1,250 species of sessile, filter-feeding marine animals. An adult tunicate, shown in Figure 30-25, exhibits few chordate characteristics. Only tunicate larvae have a body cavity, nerve cord, and notochord. All of these features are lost during development of the adult body form. The adult develops a tough sac, called a tunic, around its body. The phylum is named for this structure. Interestingly, the tunic is made of cellulose, a substance commonly found in plants and algae but rarely in animals.

Lancelets are scale-less, fishlike chordates a few centimeters long that live in shallow water throughout the oceans of the world. There are only 23 species. Lancelets spend most of their time buried in mud or sand with only their mouths protruding. They create a current with cilia that line their tentacled mouths and filter microscopic plankton from the water. They have no head or eyes. Scientists have determined that lancelets are the direct ancestors of the fishes, the first vertebrates.

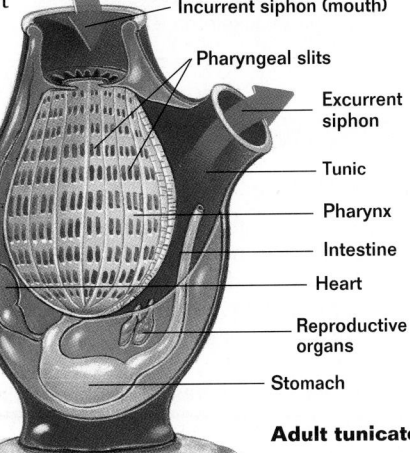

Incurrent siphon (mouth)

Pharyngeal slits

Excurrent siphon

Tunic

Pharynx

Intestine

Heart

Reproductive organs

Stomach

Adult tunicate

Figure 30-25 Tunicates are found in shallow- and deep-water marine environments. They are often called "sea squirts," because many species shoot water out from their siphons when they are touched. The species shown is a blue sea squirt from the Philippines.

Section Review

1. List the major echinoderm characteristics.
2. Explain why some echinoderms have bodies that are softer than others.
3. Compare and contrast the feeding habits of a sea star and a sea cucumber.
4. Name the echinoderms that are completely sessile, and describe their basic structure.
5. Name two groups of invertebrate chordates, and describe how they feed.

Critical Thinking

6. A scientist collects several specimens of an animal that he has never seen before. After conducting an in-depth study of the organisms, the scientist observes that they have tube feet, an endoskeleton, and a protostome pattern of embryonic development. Why is the classification of this organism difficult?

CHAPTER REVIEW ANSWERS

Each item in the Chapter Review is correlated by text section in the assignment guide that follows.

ASSIGNMENT GUIDE

Section	Review	Themes Review	Critical Thinking
30-1	1, 2, 4, 9	13, 14	15
30-2	3, 5, 10, 11	13, 14	15
30-3	6–8, 12	14	16

Review

Multiple Choice

Code in parentheses indicates section number and objective number.

1. c (30-1.1)
2. a (30-1.2)
3. d (30-2.3)
4. a (30-1.2)
5. c (30-2.5)
6. c (30-3.1)
7. a (30-3.2)
8. c (30-3.3)

Completion

9. ommatidia, ocelli (30-1.2)
10. Chelicerata, Crustacea, and Uniramia (30-2.1)
11. ventral, ganglia (30-2.4)
12. larval (30-3.4)

30 CHAPTER REVIEW

Vocabulary

arachnid (684)
carapace (686)
caste (693)
cephalothorax (680)
chelicerate (680)
chrysalis (692)
ecdysis (682)
exoskeleton (680)

isopod (687)
Malpighian tubule (681)
mandibulate (680)
metamorphosis (692)
nymph (693)
ocelli (680)
ommatidia (680)
palp (684)

skin gill (695)
spiracle (680)
swimmeret (686)
telson (686)
trachea (680)
uropod (686)

Review

Multiple Choice

1. What evidence suggests that arthropods evolved from annelids?
   a. Arthropods and annelids have gills.
   b. Both groups have marine species.
   c. Segmentation is present in both groups.
   d. Arthropods have vestigial parapodia.

2. Lobsters, barnacles, and isopods
   a. breathe with gills.
   b. have only calcium carbonate in their exoskeleton.
   c. are scavengers.
   d. have unbranched appendages.

3. Millipedes and centipedes differ in that millipedes
   a. are terrestrial and segmented.
   b. have one pair of legs on each segment.
   c. have poisonous fangs.
   d. are herbivores.

4. The chief organ of excretion in insects is the
   a. Malpighian tubule.
   b. palp.
   c. nephridium.
   d. spiracle.

5. Which sequence shows the life cycle of an insect that undergoes incomplete metamorphosis?
   a. egg → larva → adult
   b. larva → pupa → adult
   c. egg → nymph → adult
   d. nymph → pupa → adult

6. Which of the following characteristics is typical of echinoderms?
   a. nauplius larva
   b. notochord
   c. water-vascular system
   d. exoskeleton present, endoskeleton absent

7. Which pair shows a correct match between an adult echinoderm and one of its unique characteristics?
   a. sea cucumber: leathery epidermis
   b. sand dollar: has five arms
   c. sea lily: free-swimming
   d. sea star: sessile

8. Which echinoderm group contains species that specialize in hunting bivalves?
   a. sea cucumbers   c. sea stars
   b. sand dollars   d. sea lilies

Completion

9. The compound eyes of insects form images and are made up of visual units called _____ , while some arthropods have simple eyes, called _____ , that do not form images.

10. Arthropods are classified into three subphyla: _____ , which contains spiders and scorpions; _____ , which includes crabs and lobsters; and _____ , which includes the millipedes and centipedes.

11. The nervous system of the Eastern lubber grasshapper is composed of a major _____ nerve cord with _____ located in each body segment.

12. Only the _____ stage of urochordates possesses the distinctive chordate characteristics: a nerve cord, notochord, and pharyngeal slits.

Themes Review

13. **Evolution** Included among the mandibulates, or jawed arthropods, are the crustaceans, insects, millipedes, and centipedes. There is growing acceptance among zoologists of the view that the crustacean mandibles and the mandibles of insects, millipedes, and centipedes reflect convergent evolution. What anatomical evidence suggests that the modern crustaceans are rather distantly related to other mandibulates?

14. **Structure and Function** Gas exchange is essential to crustaceans, insects, and echinoderms. Compare the manner in which gas exchange occurs in most crustaceans, insects, and echinoderms.

Critical Thinking

15. **Interpreting Data** As part of a study of North American spiders, data were collected on spider size and the number of molts undergone by spiders before reaching sexual maturity. Spider size was determined at time of sexual maturity by measuring from the tip of the spider's head to the end of its abdomen. The graph shows data collected from

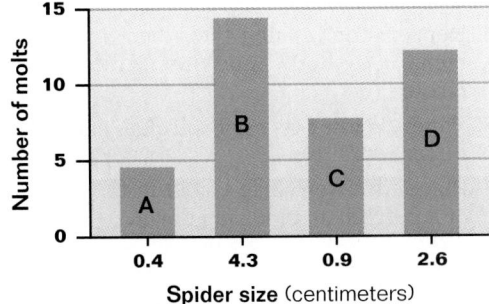

Spider size (centimeters)

100 individuals of four species. What is the relationship between spider size and number of molts that a spider undergoes before reaching sexual maturity?

16. **Designing Experiments** Two students agree that sea stars can regenerate but disagree on what body parts are required for regeneration to occur. One student claims that a complete sea star will develop from one arm. The other student insists that complete development will occur only when a part of the central disk is attached to the arm. Design an experiment to test the students' hypotheses and find out if one is correct.

Activities and Projects

17. **Cooperative Group Project** Work in teams to collect a variety of insects. Pool your team's collections and use field guides to group the insects into orders. Construct a bar graph that shows the number of individuals from each insect order collected by your team. Attach a photograph of a representative insect next to the portion of the graph that shows the number of individuals of that order. Speculate about how your team's graph might be different if insects were collected from a different location or during a different season.

18. **Multicultural Perspective** People from the Russian steppes and the plains of central Europe invaded the Indus Valley (of present day India) about 1300 B.C. Afterwards, the native people, the Dasa, were separated from the invaders, the Aryans. This separation evolved into a caste system that affects Indian culture today. What are the castes that evolved from the separation of the Dasa and Aryans? How do people who are members of one caste interact with people from a different caste? How is this caste system similar to the caste systems that exist among social insects?

Themes Review

13. Crustaceans have a second pair of antennae, appendages that are typically branched at the end, legs attached to the abdomen, and gills (among other differences). (30-1.1, 30-2.3)

14. Gas exchange among crustaceans takes place in gills. Gas exchange in insects occurs directly between cells of the animal and the branching tracheae. The movement of gas into and out of the tracheae takes place through the spiracles. In most echinoderms, gas exchange takes place in fingerlike projections called skin gills. (30-1, 30-2.3, 30-3.1)

Critical Thinking

15. The larger the species, the more molts an individual undergoes before reaching sexual maturity. (30-1.3, 30-2.2)

16. Sea-star arms and arms attached to pieces of the central disk should be placed in identically prepared aquariums and observed over time for changes. All sea-star parts should come from healthy individuals of the same species. (30-3.1)

Activities and Projects

17. Make available insect field guides that include pictures.

18. Castes include priests called Brahmins, warriors called Ksatriyas, traders and landowners called Vaisyas, and beggars called Sudras. A person cannot change caste or interact socially with members of another caste. India's caste system is similar to the caste systems of social insects in that membership within a caste is determined by birth, each caste has a specific function in society, individuals rarely change caste, and a hierarchy exists among the castes.

LABORATORY Investigation Chapter 30

Arthropod Responses

Preparation

Each group of students will require four pill bugs, which can be found outdoors under logs and rocks or ordered from Ward's. Keep the pill bugs in a plastic container with crumpled, moist paper towels and a slice of potato until lab time. Lab groups will also need two petri dishes, a hand lens (or stereomicroscope), a blunt probe, a pair of scissors, transparent tape, a clock or watch with a second hand, and four types of fabric that differ in texture, such as wool, polyester, silk, and flannel.

Procedural Notes

- This investigation can be completed during two class periods.
- Distribute the pill bugs to students in small paper cups.
- **Safety:** In step 3, remind students to be careful when using scissors.
- To extend the investigation, students can tape together the four fabrics in all possible combinations, which would result in six different kinds of two-fabric circles. Then have students calculate the total time pill bugs spend on each type of fabric.

Background Answers

1. Arthropods have jointed appendages, segmentation, a cephalothorax, an exoskeleton, compound eyes, spiracles, open circulation, and Malpighian tubules.

2. Crustaceans include crabs, lobsters, crayfish, shrimps, barnacles, water fleas, and pill bugs. They are mandibulates that breathe with gills.

3. Pill bugs live in moist terrestrial enviroments such as in soil and under rocks and logs.

4. Pill bugs respond to stimuli, such as light, moisture, and touch, by moving toward or away from the stimulus or by curling into a ball.

5. How do pill bugs respond to surfaces that have different textures?

OBJECTIVES

- Recognize arthropod characteristics in the pill bug, a representative arthropod.
- Observe the behavior of pill bugs on surfaces with different textures.
- Infer the type of texture pill bugs prefer.

PROCESS SKILLS

- relating structure to function
- inferring adaptive advantages of certain behaviors

MATERIALS

- 4 adult pill bugs
- 2 petri dishes
- stereomicroscope or hand lens
- blunt probe
- 4 fabrics of different texture (wool, polyester, silk, flannel, etc.)
- scissors
- transparent tape
- clock or watch with second hand

BACKGROUND

1. What are some characteristics of arthropods?
2. What is a crustacean? How do crustaceans obtain oxygen?
3. In what kind of environment do pill bugs live?
4. What kinds of stimuli do pill bugs respond to in their environment? How do they respond to the stimuli?
5. Write your own question to explore in your lab report or notebook.

Records

1. The pill bug can be recognized as an arthropod by its jointed appendages, segmentation, and exoskeleton.

2. The pill bug should curl into a ball when touched with a probe.

TECHNIQUE

Part A: Pill Bug Characteristics

1. Place a pill bug in a petri dish. Use the stereomicroscope or hand lens to observe the pill bug. In the **Records** section of your report, list the characteristics that enable you to recognize that the pill bug is an arthropod.

2. Using the blunt probe, gently touch the pill bug. In the **Records** section of your report, describe your observations.

Part B: Pill Bug Behavior

3. On one of the fabrics, trace the outline of the bottom of the petri dish. **CAUTION: Be careful when using the scissors.** Cut out the circle from the fabric and fold it in half. Then cut along the fold to produce two half-circles.
4. Repeat step 3 using the other three fabrics. You should now have eight half-circles.
5. Tape together two half-circles, each of a different fabric.
6. Place the two-fabric circle in the bottom of a petri dish, tape side down.

7–10. Students should draw a total of six circles. In each circle, the path that the pill bug travels should be drawn. Below each half-circle, the type of fabric and the amount of time the pill bug spends on each type of fabric should be recorded.

7. In the **Records** section of your report, draw the circle you made in step 5. Label the two types of fabric that make up the circle.

8. Place a pill bug in the center of the circle. Observe the movement of the pill bug for five minutes. Have one member of the group keep track of the amount of time the pill bug spends on each fabric. Have another member draw in the **Records** section the path the pill bug travels in the circle. Below the circle, record the amount of time spent on each type of fabric.

9. Repeat steps 5–8 using the other two fabrics.

10. Repeat the investigation twice, each time using a different pill bug. Reuse the circles you made. Record your observations in the **Records** section of your report. Also, briefly summarize the procedure you followed.

11. Clean up your materials and wash your hands before leaving the lab.

1. Why were three pill bugs tested in this investigation instead of just one?
2. Did the three pill bugs show a similar pattern of movement? Explain.
3. Rank the fabrics according to the total amount of time spent on them by the three pill bugs.

ANALYSIS

1. In what way is the pill bug's response to disturbance advantageous?
2. In Part B, which fabric did the pill bugs prefer? Describe the texture of that fabric.
3. What stimulus did the pill bugs respond to? What was their response?
4. How is being able to detect surface texture a good adaptation for pill bugs in their natural habitat?

FURTHER INQUIRY

Write a new question that could be explored as a new investigation. The following are examples:

Do pill bugs prefer moist or dry environments?

Do pill bugs prefer light or dark environments?

Inquiry Answers

1. Three pill bugs were tested to improve the reliability of the data.

2. Answers may vary. The pill bugs should show a similar pattern of movement.

3. Answers will depend on the types of fabrics used. Pill bugs should spend more time on rough-textured fabrics, such as wool, than on smooth-textured fabrics, such as silk.

Analysis Answers

1. The pill bug's response to disturbance allows it to protect its gills and underside by rolling into a ball and to distract any predators by pretending to be dead.

2. Answers will vary. The pill bugs should prefer rough-textured fabrics.

3. The pill bugs responded to the textured surface of the cloth by either moving on the surface or by moving away from it.

4. Being able to distinguish between rough textures and smooth textures may help pill bugs find diggable substrates and avoid crawling over stones that might expose them to predators.

Block Scheduling Guide

Each block represents about 45–50 minutes of instructional time. The resources cited in a block are coded to a specific program component using the Key on the next page.

Lecture Concepts | Lesson Resources

31-1 Story of Vertebrate Evolution pp. 702–710

BLOCK 1

a. Jawless fishes were the first vertebrates. Reptiles were the first terrestrial vertebrates. Birds and mammals have reptile ancestors.
b. Lampreys and hagfishes are the only surviving jawless fishes. Sharks have a lightweight skeleton. Bony fishes have a heavier skeleton and use a swim bladder to provide buoyancy.
c. Amphibians were the first vertebrates to live on land. Because they lack watertight eggs, they must return to the water to reproduce and to keep their eggs from drying out.
d. The dry skin of a reptile prevents water loss. Reptile eggs protect the embryo from drying out on land.
e. Mammals are classified into three groups based on reproductive differences. Monotremes lay eggs. Marsupials bear underdeveloped young that mature in a pouch. Placental mammals nourish developing embryos by means of a placenta.

Lecture Resources
- Opening Question, page 703
- Demonstration: Exploring the Backbone
- Visual Strategy: Figure 31-3
- Teaching Transparencies: 134, 164, 137A, 138A, 139A, 140A
- Holt Biology Videodiscs: Lesson 31-1

Classwork Options
- Overcoming Misconceptions, page 708

- Mammals and Milk, and Graphic Organizer, page 709
- Quick Labs: Lab A23 Observing Some Major Animal Groups
- Laboratory Techniques and Experimental Design C20: Observing Animal Behavior

Assignment Options
- Section Review, page 710
- Chapter Review, questions 1, 2, 8, 9, 17, 28, 30

31-2 Challenge of Obtaining Oxygen pp. 711–716

BLOCK 2

a. In the gill of a bony fish, water and blood flow in opposite directions, maximizing the amount of oxygen that can be extracted from water.
b. The efficiency of gas exchange in the lung increases with increased surface area. The amphibian lung consists of folded sacs. Alveoli in reptiles provide more surface area than is provided by those of the amphibian lung. Mammals have more alveoli than reptiles.
c. Bird lungs consist of a series of air sacs connected to the lungs. This system, which provides air flow and blood flow similar to those of fish, provides maximum lung efficiency.

Lecture Resources
- Opening Question, page 711
- Demonstrations: Breathing Goldfish, Modeling the Increase in Lung Surface Area
- Teaching Transparencies: 166, 180, 181, 182

- Visual Strategies: Figures 31-11, 31-13
- Holt Biology Videodiscs: Lesson 31-2

Assignment Options
- Section Review, page 716
- Chapter Review, questions 3, 11, 12, 19, 32

31-3 Evolution of a Better Heart pp. 717–722

BLOCKS 3 and 4

a. The fish heart is a tube of four connected chambers that collects deoxygenated blood from the body and pumps it to the gills.
b. The amphibian atrium is divided by a septum that sends deoxygenated blood to the lungs and oxygenated blood to the body, with some mixing of the two bloodstreams.
c. The reptile heart has a partial division of the ventricle, which reduces mixing of oxygenated and deoxygenated blood.
d. Crocodiles, birds, and mammals have completely separate ventricles, which prevents any mixing of blood.

Lecture Resources
- Opening Question, page 717
- Demonstrations: The Blood Pump, One Way Flow and Blood Flow in a Goldfish Tail
- Visual Strategies: Figures 31-15, 31-16
- Effective Reading, page 718
- Teaching Transparencies: 171, 172

- Research Update, page 721
- Holt Biology Videodiscs: Lesson 31-3

Classwork Options
- Concept Mapping, page 722
- Laboratory Techniques and Experimental Design C21: Observing Animal Behavior — Grant Application

Assignment Options
- Section Review, page 722
- Chapter Review, questions 4, 10, 16, 18

31-4 Challenge of Retaining Water pp. 723–726

BLOCK 5

a. The excretory system of vertebrates increases in complexity from fish to mammals, enabling reptiles, birds, and mammals to handle the demands of living in a dry environment.
b. The evolution of protective coverings for eggs has enabled reptiles, birds, and mammals to reproduce in dry environments.

Lecture Resources
- Opening Question, page 723
- Demonstration: Fresh Kidneys

- Visual Strategy: Figure 31-22
- Teaching Transparencies: 183
- Holt Biology Videodiscs: Lesson 31-4

31-5 Reproduction and Development pp. 727–731

BLOCK 6

a. Fertilization of eggs is external in most fishes and amphibians. After hatching, the larval amphibian feeds and grows until it reaches a certain size and transforms into an adult.

b. Fertilization is internal in birds and reptiles. Most reptiles and all birds are oviparous.

c. The placenta provides nourishment for a developing embryo while keeping the bloodstreams of the embryo andof the mother separate.

Lecture Resources
- ■ Opening Question, page 727
- ■ Demonstration: Stages of Development
- ■ Overcoming Misconceptions, page 728
- ■ Visual Strategy: Figure 31-25
- ◉ Holt Biology Videodiscs: Lesson 31-5
- 🔖 Teaching Transparencies: 228

Classwork Options
- ■ Laboratory Investigation: Comparing Vertebrate Characteristics, pages 742–743
- ■ Closure, page 730

Assignment Options
- ■ Section Review, page 730
- ■ Chapter Review, questions 7, 9, 15, 33

Review/Enrichment

BLOCK 7

- ■ Highlights: Vertebrate Evolution, pages 731–738
- ■ Highlights Review, questions 21–27
- ■ Study Guide: Concept Review and Pretest

Assignment Options
- ◉ Teaching Resources CD-ROM: Supplementary Reading Worksheets and Test, *The Dinosaur Heresies*
- ■ Activities and Projects, questions 34–36
- ■ Readings, questions 37–38

Assessment Options

BLOCK 8

Traditional Assessment
- ■ Chapter 31 Test
- ◉ Test Generator/Assessment Item Listing: Software item bank for preparing customized chapter tests.

Performance Assessment
- ◉ Test Generator : The Challenge of Retaining Water

Portfolio Assessment
Select student reports for one or more laboratory experiments from this chapter. *The Direct Observation Checklist*, on the *BioSources Teaching Resources CD-ROM*, should be completed during a laboratory or other cooperative-learning experience.

Answer to Concept Map
The following is one possible answer to the Concept Mapping exercise on page 739.

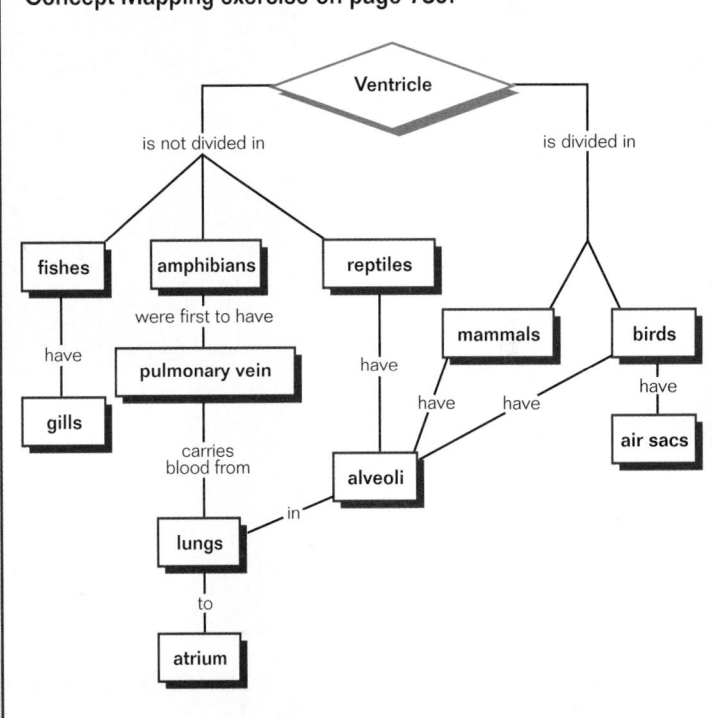

Holt Biology Videodiscs   CONCEPTS OF BIOLOGY

Holt Biology Videodiscs gives you a powerful tool for teaching, review, and assessment. Concepts of Biology adds interactive lesson plans and extensive tools for customization using CD-ROM technology.

Use the following topics from *Concepts of Biology* to help you teach this chapter:

- ■ *Topic 19: Evolutionary Trends in Vertebrates*
- ■ *Topic 20: Vertebrate Diversity*

For further information, see *Holt Biology Videodiscs Teacher's Correlation Guide to Biology: Principles and Explorations.*

CHAPTER 31 OVERVIEW OF VERTEBRATES

Chapter Theme

Evolution *Vertebrate evolution is the story of a series of changes in vertebrates that have enabled them to survive in new habitats and adopt new ways of life. Vertebrates have become more complex in structure in response to pressures and opportunities in the environment. This chapter summarizes many of these changes and their origins.*

Tapping Prior Knowledge

- What are some evolutionary trends in invertebrates?
- Based on our knowledge of invertebrate evolution, what are some questions we can ask about evolutionary trends in vertebrates?

Opening Question

Ask students what good a backbone is. Students' answers might include that a backbone helps us walk, stand, and sit or that it protects the spinal cord. List their answers and prompt students with more questions: What does a backbone support? What does a backbone protect? What kind of movements require a backbone?

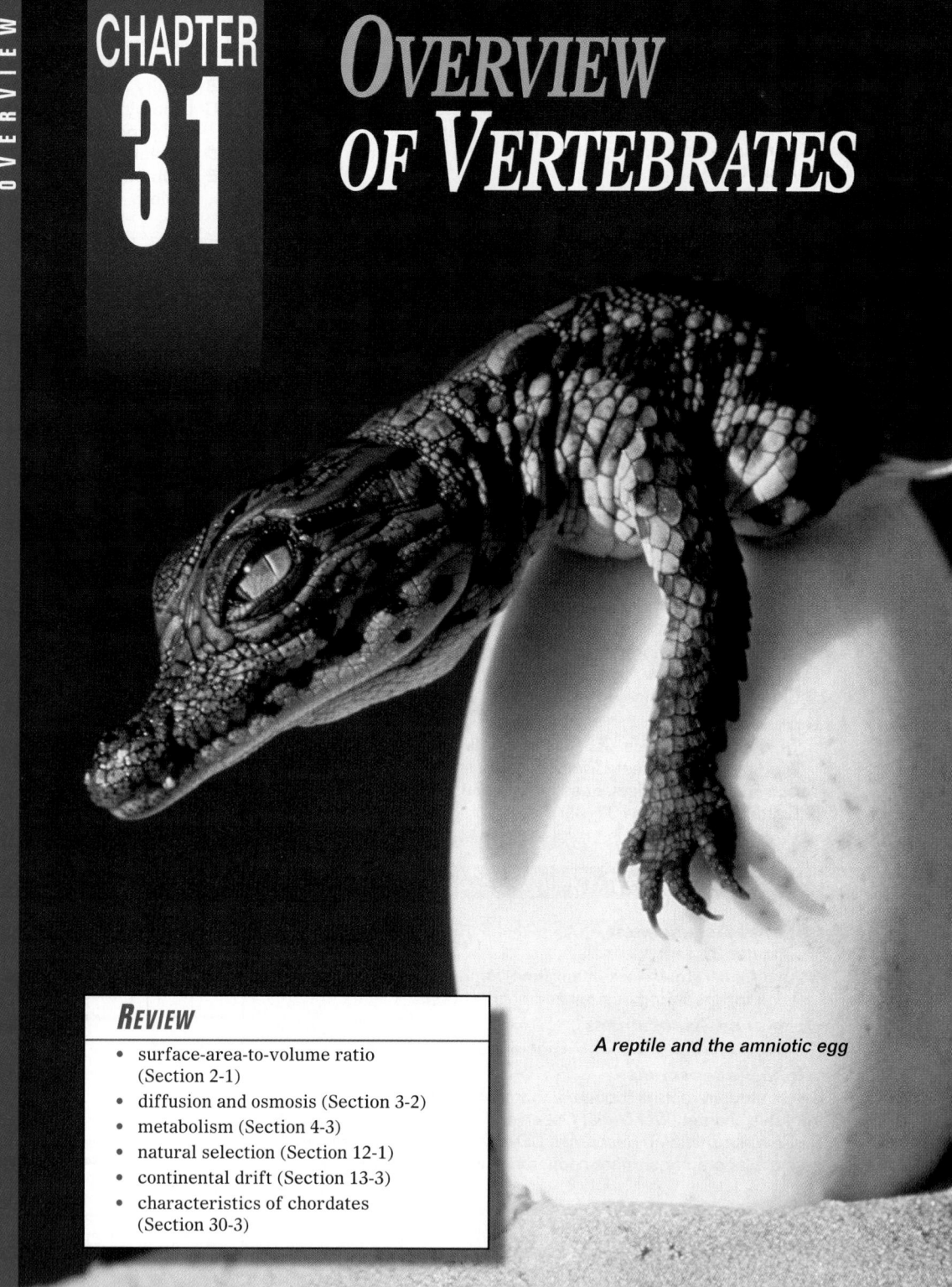

A reptile and the amniotic egg

REVIEW

- surface-area-to-volume ratio (Section 2-1)
- diffusion and osmosis (Section 3-2)
- metabolism (Section 4-3)
- natural selection (Section 12-1)
- continental drift (Section 13-3)
- characteristics of chordates (Section 30-3)

Authors' Rationale

Your students should be familiar with the vertebrates, comprising the fish groups, amphibians, reptiles, birds, and mammals. This chapter is an overall discussion of the entire group, including how they evolved and problems they had to overcome to adapt to life on land: support, dehydration, gas exchange, and reproduction. Before learning about the individual vertebrate classes in more detail, students should understand how a vertebrate is organized, how the classes are related to each other, when they first appeared in the evolutionary history of the Earth, and the evolutionary forces behind their appearance.

31-1 Story of Vertebrate Evolution

Vertebrates are chordates with a backbone. These animals are named for the individual segments, called vertebrae, that make up the backbone. Vertebrae form a central axis for muscle attachment. They also protect the dorsal nerve cord. These functions paved the way for the development of the brain and an internal skeleton that allowed vertebrates to grow larger than their invertebrate ancestors.

Figure 31-1 Like all of the earliest vertebrates, this ancient fish had no jaws.

History of Vertebrates

The first chordates probably evolved about 600 million years ago. At that time many different groups of organisms appeared in the shallow seas covering much of Earth's continents. The oldest vertebrate fossils, dating back to about 500 million years ago, are of fishes similar to the one shown in Figure 31-1. Unlike most of the fishes with which you are familiar, the earliest fishes had neither jaws nor paired fins. Many of them looked something like a flattened hot dog with a hole at one end and a fin at the other.

For more than 100 million years, fishes were the only vertebrates. They became the dominant animals in the sea, diversifying into a great variety of species. Some were just a few centimeters long; others were bigger than a car. Amphibians, which evolved from fishes, were the first vertebrates to invade the land. Frogs and salamanders are the descendants of those early amphibians. Like their ancestors, frogs and salamanders live successfully on land but must reproduce in water or very moist environments.

Teaching Tips

Make a Tree of Vertebrates

Have students create a phylogenic tree out of craft paper or rolled-up newspaper. Divide the class into six groups. One group should create the tree, while the other five groups make diagrams and labels of major groups of vertebrates—fishes, amphibians, reptiles, birds, and mammals. Diagrams may be drawn freehand or traced, and the artists should sign their work. If you have several classes working on this assignment, have groups in each class add their own unique vertebrate species to each branch of the phylogenic tree. Students may wish to consult library references to learn more about early vertebrates.

Diversification

Ask students the following questions about reptile and mammal diversification. Why didn't reptiles re-emerge as the dominant life-form after the extinction of dinosaurs? *(A hot and dry climate favored reptiles. After the extinction of the dinosaurs, the cold climate favored mammals.)* Why do animal species diversify? That is, why do they become adapted to new ecological niches, as the marsupials did in Australia? *(They evolve to take advantage of places vacated by other species, or of new environments that have appeared.)* Why do animals evolve into forms that can take advantage of new food sources? *(The competition for food favors those animals that can use alternative sources when one kind of food disappears.)* Have students look at page 731. Ask them to list as many reasons as they can to explain why there are so many species of bony fishes today. Compare and consolidate the lists. *(Their reasons should include a variety of adaptations to the physical and biotic environments of the fish.)*

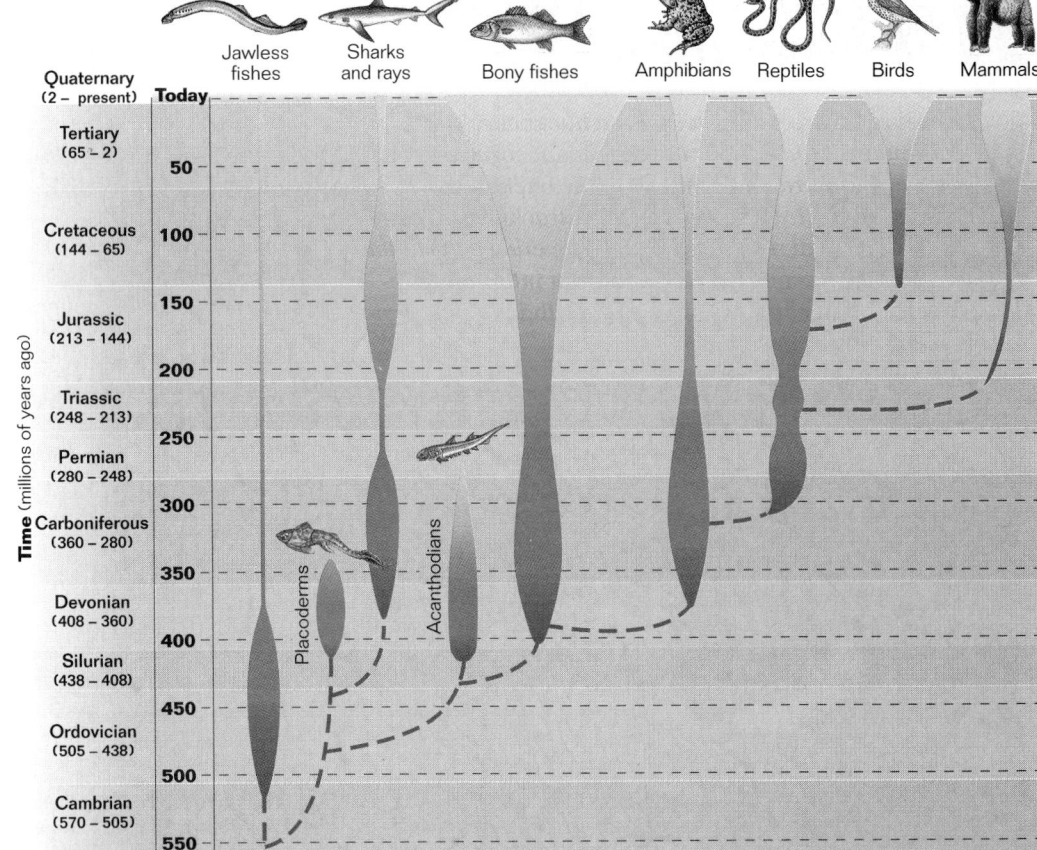

Figure 31-2 This diagram shows a simplified history of the vertebrates and the relationships among the major groups. Dotted lines indicate the origin of one group from another. Each group shown here is a class. The width of the colored bar representing the class indicates its approximate diversity (number of species).

CONTENT LINK
Curious about what caused the extinction of the dinosaurs? Part of **Chapter 33** is devoted to this puzzle.

Amphibians, in turn, gave rise to reptiles, which were better suited to living out of water. Within 50 million years of their origin, reptiles replaced amphibians as the dominant land vertebrates. Because reptiles are **terrestrial**, able to live their whole lives on land, they were able to spread out and diversify. Many different species of reptiles evolved. Some were smaller than a chicken, while others were as big as a truck.

From these early reptiles, the two great lines of terrestrial vertebrates evolved—dinosaurs and mammals. Dinosaurs and mammals appear at about the same time in the fossil record, 220 million years ago. But the dinosaurs quickly won out in the evolutionary competition and dominated the Earth for 150 million years. Over this long period, the largest mammal was no bigger than a cat. When the dinosaurs abruptly disappeared 65 million years ago, mammals quickly took their place. The reason for the disappearance of the dinosaurs is still hotly debated, but there is no doubt that their extinction allowed mammals to become abundant and diverse. Figure 31-2 summarizes the evolutionary history of the vertebrates.

Historical Note

Pierre Teilhard de Chardin (1881–1955)

The famous French philosopher and Jesuit priest Pierre Teilhard de Chardin first began studying vertebrate paleontology when he went to live at a Jesuit college at age 10. Chardin's career included teaching science in Egypt, studying human paleontology in Paris, and serving as a stretcher-bearer during World War I (for which he won the Legion of Honor). He also worked in India, Burma, Java, and China, where he helped to unearth the skull of Peking Man in 1929.

The history of vertebrates has been a series of evolutionary advances that allowed these animals to diversify first in the sea and then on the land. In this section, you will look briefly at some evolutionary changes in the vertebrate body plan that contributed to their success. You can read more about key steps in vertebrate evolution in *Highlights: Vertebrate Evolution* on pages 731–738. In the remaining sections of this chapter, you will examine the key adaptations that permitted vertebrates to invade the land. This invasion was a staggering evolutionary achievement that involved fundamental changes in many body systems.

Fishes Dominate the Sea

Jawless fishes were the first vertebrates. As they swam along the ocean bottom, these animals fed by filtering tiny invertebrates from water passing through their mouths. Their bodies were covered with thick, bony plates and smaller scales.

Jawless fishes dominated the world's oceans for about 100 million years, until they were eventually replaced by new kinds of fishes that were hunters. Only 63 species of jawless fishes exist today, the hagfishes and lampreys. Figure 31-3 shows a lamprey.

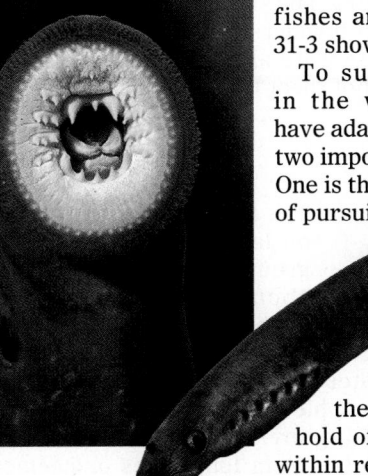

To survive as predators in the water, fishes must have adaptations that meet two important challenges. One is the prolem of pursuing prey through the water. The second is the problem of grabbing hold of prey once they are within reach. A new group of fishes that had met these challenges appears in the fossil record about 430 million years ago. These fishes were the acanthodians (*uh KAN thoh dee uhns*). They had strong biting jaws with jagged bony edges that served as teeth. Such adaptations made it possible to seize prey. Chasing prey through the water was possible because of several other changes. For instance, the bodies of these early fishes had become streamlined and flattened sideways. This made it easier to move through the water. Paired fins provided better control of fast swimming. Acanthodians had small bodies covered with protective spines.

Figure 31-3 This lamprey is a parasite. Its mouth, *left,* has a suction-cup-like structure. After attaching to a host fish, the lamprey gouges out a wound with its rough tongue and then feeds on blood and bits of flesh from the wound.

Teaching Tips
Sharks Aren't Boneheads

Remind students about the characteristics of cartilage and bone by finding examples of each kind of material on their own head. Why is it good that the end of the nose is cartilage instead of bone? *(It would bump into things and break easily.)* Why is the skull made of bone? *(The brain is delicate and needs a hard, protective covering.)* Why does a cartilaginous skeleton give sharks a maneuverability advantage? *(Cartilaginous skeletons are more flexible and lightweight than those of bone.)*

Ocean Emigration

Why did vertebrates invade the land? What was so unfavorable about a water environment or so favorable about a terrestrial one that the first amphibious vertebrates found land a better environment than water? Split your class into small groups of two to three students, and have them discuss their answers to these questions. Follow their discussions with a collaboration of everyone in the class to assemble a list of advantages offered by a terrestrial environment. *(Reasons should include new plant and animal food sources on land, less competition, fewer predators, and more room.)*

Figure 31-4 Sharks were among the first vertebrates to have jaws, which allow them to bite. Sharks also have streamlined bodies and paired fins, which allow for fast swimming with better control.

❑ CAPSULE SUMMARY

Jawless fishes were the first vertebrates. Lampreys and hagfishes are the only surviving jawless fishes. Sharks have a lightweight skeleton of cartilage. Bony fishes have a heavier skeleton of bone and a swim bladder, which provides buoyancy.

CONTENT LINK

*See **Chapter 32** for more information on lungfishes and coelacanths.*

About 410 million years ago, larger jawed fishes called placoderms (PLAK uh durms) appeared. These fishes had massive heads armored with bony plates. Placoderms and acanthodians are now extinct. Soon after the appearance of the placoderms, two other groups of jawed fishes evolved—sharks and bony fishes. Each group had somewhat different features for fast and maneuverable swimming. In sharks, such as *Hybodus* shown in Figure 31-4, the skeleton is composed of **cartilage,** a lightweight, strong, and flexible material. In bony fishes, represented by the coelacanth (SEE luh kanth) shown on page 731, the skeleton is made of bone, which is heavier and less flexible than cartilage. However, bony fishes compensate for their heavier skeletons with a **swim bladder,** a gas- or fat-filled sac that provides buoyancy. There are more than 20,000 species of fishes in the world today, and the vast majority of them are bony fishes with swim bladders. In fact, if you could watch every species of vertebrate alive today pass by you, one after another, one-half of them would be bony fishes. ❑

Amphibians Invade the Land

The first vertebrates able to live on land were amphibians. This group of animals first appeared about 370 million years ago. Amphibians evolved from lobe-finned fishes, a group of bony fishes that includes coelacanths and lungfishes. Lungfishes live in water that often has a low oxygen content, and their lungs enable them to supplement their oxygen intake by breathing air. Most scientists think that amphibians didn't evolve from lungfishes or coelacanths, but from an extinct group of lobe-finned fishes that had fins more like amphibian limbs.

Life on land is quite different from life in the water. Thus, the successful invasion of land by vertebrates involved a number of major innovations. Look at the characteristics of *Ichthyostega*, an early amphibian shown on page 732, as an example. Air is less buoyant than water, so legs were necessary to support body weight as well as to allow movement from place to place. To live on land, amphibians also needed lungs. The delicate structure of a fish's gills depends on the buoyancy of water for support. Out of water, gills stick together, reducing the surface area available for gas exchange. That is why a fish will suffocate on land, even though there is more oxygen in air than in water. Walking around on land requires a higher metabolism, which in turn uses greater amounts of oxygen. Thus, the heart also underwent change.

Some problems remained, however. For example, the eggs of amphibians are not watertight. Thus, amphibians generally seek out water or damp areas in which to reproduce and in which their young can live as they grow and mature. But evolution does not result in perfect solutions, only workable ones. The adaptations that allowed the first amphibians to climb out onto land have enabled their descendants to survive for over 350 million years. Today there are approximately 4,200 species of amphibians, including the familiar frogs, toads, and salamanders. ❏

Reptiles Conquer the Land

If you think of amphibians as the "first draft" of a manuscript about survival on land, then reptiles are the finished book. Each of the adaptations that allowed amphibians to lead a terrestrial existence were refined in reptiles. Their legs, for example, were positioned to support the body more effectively. So reptiles not only were bigger, but also could run. Changes in the lungs and heart made these organs more efficient. The most significant adaptations, though, were internal fertilization and watertight eggs. Watertight eggs made reptiles the first completely terrestrial vertebrates. Today there are about 7,000 species of reptiles, mostly snakes and lizards, found in practically every habitat on Earth. The chameleon shown in Figure 31-5 is one of these species.

When reptiles first evolved, about 320 million years ago, Earth was entering a long, dry period. Early reptiles, as represented by *Dimetrodon* shown on page 733, were well suited to these conditions and quickly diversified. In particular, their ability to conserve water allowed reptiles to have large bodies in dry conditions. This was something amphibians could not do. Within 50 million years, reptiles had replaced amphibians as the large, terrestrial vertebrates. All land vertebrates larger than a chicken were reptiles.

Living reptiles are **ectothermic**. Their metabolism is too slow to produce enough heat to warm their bodies. They must absorb heat from the environment. **Endothermic** animals, such as mammals and birds, maintain a high, constant body temperature because they produce heat internally through a faster metabolism. Therapsids, an order of extinct reptiles believed to be the ancestors of mammals, were probably endotherms. They were replaced by ectothermic thecodonts and then by dinosaurs.

CONTENT LINK

More information on characteristics, classification, and anatomy of reptiles can be found in **Chapter 33**.

Figure 31-5 Unlike the moist skin of most present-day amphibians, reptilian skin is covered with scales that keep the body from drying out.

Demonstration
Breathing Lessons
Pour a liter of water into a 1 L bottle, and set it beside another 1 L bottle full of air. Ask students the following: Which bottle is heavier? Which bottle contains more oxygen? Make the point that air is about 21 percent oxygen, while water contains less than 1 percent. Stagnant, warm water has the least amount of oxygen. If fish can only get oxygen by water passing through their gills, they can suffocate in stagnant water. A goldfish or Siamese fighting fish in a fish bowl is interesting to show to your class. These fish will come to the surface to gulp air because the warm, still water in their fish bowl is so oxygen-poor. (Don't place a Siamese fighting fish with any other fish because it may attack and kill the other fish.)

Teaching Tip
Letter of Recommendation
Have your class pretend that your school is preparing a desert exhibit and is deciding whether to place reptiles or amphibians in the exhibit. Have each student pick a different animal to recommend for the exhibit, describing in the detail how well-adapted the reptile is for a warm, terrestrial environment compared to an amphibian. Have your class share some of the reptile characteristics identified in their descriptions that make the reptiles ideal candidates for the exhibit. Even though students will describe different reptiles, they should recommend similar characteristics, such as watertight eggs, watertight skin, feeding behaviors, and better locomotion.

Teaching Tips

Reptiles and Therapsids

Have your students compare the early reptiles and therapsids on page 734. What advantage did the therapsid jaw and stance give this group of animals? *(The jaws were powerful and had efficient chewing surfaces. Their improved stance resulted in less awkward, more rapid locomotion.)* Why was their endothermy an advantage in cold weather? *(They could stay active.)* How was their endothermy a disadvantage? *(They had to eat 10 times more than their ancestors to maintain their body temperature.)*

Warm Bodies for Cold Places

Ask your class to individually list five vertebrates that may be found in polar regions *(examples: polar bears, musk oxen, Arctic foxes, seals, walruses, penguins)* and five vertebrates that are found in tropical regions, deserts, or forests *(examples: snakes, lizards, monkeys, frogs, parrots, sloths, toucans, jaguars, crocodiles).* Classify the listed animals by vertebrate group. Students should notice that there is a preponderance of birds and mammals, both endotherms, at the polar regions, but other kinds of vertebrates predominate in the tropics. Lead students to conclude that the Arctic is too cold for animals that cannot maintain a constant body temperature.

Not Just for the Birds and Bears

People use the same insulation that birds and other mammals use to maintain body temperature. Have your class name some of the ways that humans use leather, fur, and feathers to keep warm. If students in your class are opposed to wearing animal furs, suggest that they think of animal products we use without killing the animals *(different kinds of wool).*

❑ CAPSULE SUMMARY

One reason that reptiles were successful on land is that their dry skin prevented water loss. The reptilian egg also protected the embryo against drying.

Figure 31-6 *Eozostrodon* was an early mammal that was only about 10 cm (4 in.) long. It lived during the late Triassic period. All of the earliest known mammals were very small.

Figure 31-7 Feathers on birds and fur on mammals are adaptations that enabled these groups to survive in colder climates.

Evidence indicates that at least some, perhaps many, dinosaurs were endothermic. Studies of radioactive isotopes show that some bones at the tip of the tail of large dinosaurs appear to have formed at the same temperature as the rib bones in the body's center. In an ectothermic reptile, the extremities are much cooler than the body's center. Generally speaking, endothermic animals are more active and are able to live in a wider range of habitats than ectothermic animals. The distribution of dinosaur fossils provides additional evidence that some of these animals were endothermic. ❑

Mammals and Birds Adapt to Colder Climates

Sixty-five million years ago, global temperatures cooled dramatically. Many scientists think that this cooling was caused, at least in part, by huge clouds of dust thrown into the atmosphere by the impact of a huge meteorite or comet. If dinosaurs were endothermic, they may have been particularly vulnerable to the cooler temperatures because, unlike mammals and birds, dinosaurs had no insulation. No land vertebrate bigger than a cat survived the cold.

When the dinosaurs disappeared, their ecological roles as the large land vertebrates were taken over by mammals. With bodies insulated by fur, mammals were better suited to the colder climates that were typical then. Look at *Megazostrodon*, shown on page 735, and *Eozostrodon*, shown in Figure 31-6, as representative examples of early mammals.

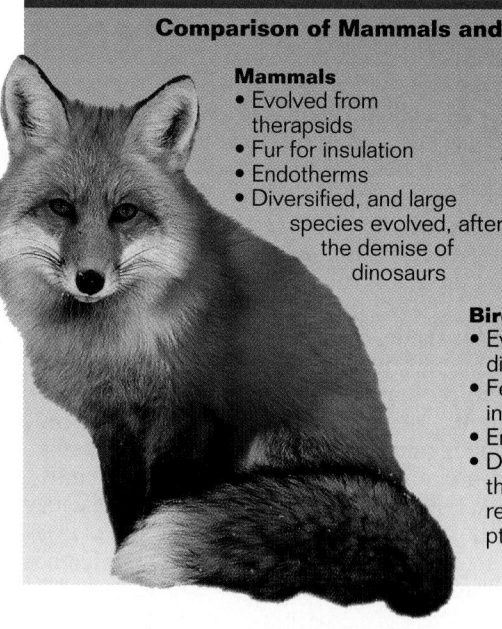

Comparison of Mammals and Birds

Mammals
- Evolved from therapsids
- Fur for insulation
- Endotherms
- Diversified, and large species evolved, after the demise of dinosaurs

Birds
- Evolved from dinosaurs
- Feathers for insulation
- Endotherms
- Diversified after the demise of flying reptiles known as pterosaurs

Overcoming Misconceptions

The First Bird

Many students think that modern birds evolved from pterodactyls, extinct flying reptiles. However, fossil pterodactyls have a wing structure unlike any living birds. Be sure that your students understand that birds probably evolved from other reptiles, not from pterodactyls. *Archaeopteryx,* the oldest known bird, had characteristics of both birds and reptiles.

Birds, which evolved from dinosaurs early in their history, had feathers for insulation. Once the pterosaurs were gone, birds were able to diversify. Look at *Archaeopteryx*, shown on page 738, which is the oldest known bird, to see some characteristics of birds. Today over 8,000 species of birds fly the world's skies. Figure 31-7 contrasts birds and mammals.

The mammals that evolved 220 million years ago alongside the dinosaurs would look strange to you. These early mammals were quite small and probably in some ways resembled the most primitive mammals living today—the monotremes. Monotremes, such as the duckbill platypus, lay shelled eggs and have a shoulder structure similar to that of early reptiles. Like all mammals, monotremes nurture their young with milk produced by mammary glands (hence the name *mammal*). Only three species of monotremes exist today: the duckbill platypus and two species of spiny anteaters.

Monotremes were replaced 100 million years ago by marsupials. Unlike monotremes, marsupials do not lay shelled eggs. The embryos develop within the mother's body and are born live. Upon birth, the young marsupials, which are tiny and immature, crawl into an external pouch to nurse and complete their development. About 280 species of marsupials survive today, mostly in Australia and New Guinea.

Placental mammals evolved at about the same time as marsupials and eventually replaced them throughout most of the world. Marsupials survived in Australia and New Guinea because continental drift separated these land masses from the other continents before placental mammals arrived. Placental mammals, such as the blue whale shown in Figure 31-8, invest even more time in nurturing their young than marsupials do. Embryos spend a much longer period of

Figure 31-8 The blue whale is the largest living mammal. An adult may be up to 30 m (99 ft.) long and weigh 136,000 kg (150 tons).

Teaching Tips
Mammals and Milk
Have students make a chart comparing monotremes, marsupials, and placental mammals. The chart headings should be Type of Mammals, Examples, Type of Reproduction, Food Source for Growing Embryo, and Food Source for Young. The answers will vary according to type of mammal, except for the food source for the young, which is milk in every case.

How Does a Whale Stay Warm?
Students already know that whales aren't covered with fur as other mammals are, so ask them how whales do stay warm. If necessary, prompt students to arrive at the idea that fat or blubber is an excellent insulator. Have them name other animals that use fat for insulation (*seals, walruses, bears*).

Major Characteristics
Help students recognize the major characteristics of early fishes, amphibians, reptiles, mammals, and birds by making a flowchart similar to the *Graphic Organizer* shown at the bottom of the page. Ask students to include a brief summary of the changes outlined in their charts.

Graphic Organizer
Use this graphic organizer with *Teaching Tip:* **Major Characteristics** on page 709.

Major characteristics of fishes: jaws, paired fins	→	Major characteristics of amphibians: legs, lungs	→	Major characteristics of reptiles: watertight eggs, scales

Major characteristics of mammals: constant body temperature, fur, placenta, mammary glands

Major characteristics of birds: constant body temperature, feathers, wings

Teaching Tip
How Do They Know That?
Help students understand how the history of mammals has been assembled using the evidence of fossil skeletons of mammals, often their skulls. Ask students if they would be able to distinguish the skull of a dog from that of a cow. What are some of the differences between these skulls? (*Answers should include such characteristics as type of teeth, skull size, and presence or absence of horns.*) How would you identify a rodent skull? (*presence of large chisel-shaped incisors*) It is particularly helpful to display diagrams of different vertebrate skulls while asking these questions. Make the point that if beginning biology students can figure out the type of animal and its diet from skulls, then paleontologists and biologists can make even more extensive conclusions based on their knowledge of mammal anatomy.

Figure 31-9 *Smilodon* was a little over 1m (about 4 ft.) in length, roughly the size of a mountain lion.

□ CAPSULE SUMMARY

There are three groups of mammals. Monotremes lay eggs. Marsupials suckle their young, which are born after a short period of development, in an external pouch. Placental mammals nourish their developing embryos by means of a placenta.

development within the mother's body. This long period of growth and development is made possible by the placenta. The **placenta** is a structure that transfers nutrients from the mother's blood supply directly to the growing embryo. You will read more about its function in Section 31-5.

Today there are over 4,000 species of placental mammals, ranging in size from 1.5 g (less than 0.1 oz.) shrews to 136,000 kg (150 tons) whales. Almost half of these species are rodents (mice and their relatives), and another quarter are bats, the only flying mammals. Mammals have also invaded the seas, just as the reptiles known as plesiosaurs and ichthyosaurs did so successfully millions of years earlier. There are 79 living species of whales and dolphins. Primates, the order to which humans belong, are not a major group in terms of species diversity. There are only 233 known species. Humans evolved very recently—less than 2 million years ago. There have been at least three species of humans, but *Homo sapiens* is the only one that survives. Humans are notable among primates for lacking a complete body covering of fur. The climate in which humans evolved was so much warmer, that fur—so central to the success of mammals—was no longer essential.

Today, all very large land animals are mammals. However, the largest land mammals reached their peak during past ice ages, when the saber-toothed cat shown in Figure 31-9 lived in North America. The warming of Earth's climate in recent times has favored smaller bodies that are easier to cool. ▢

Section Review

1. *From which group of vertebrates did the reptiles evolve? What groups of vertebrates evolved from reptiles?*
2. *Explain the advantages of jaws and paired fins.*
3. *Describe the characteristics that enable amphibians to survive on land.*
4. *Name two adaptations that allowed reptiles to live and reproduce completely on land.*

Critical Thinking

5. *What role have mass extinctions played in the evolution of mammals?*
6. *It is conventional to view mammals as superior to or "more advanced" than reptiles. What evidence from the history of vertebrates contradicts this view?*

Answers to Section Review

1. Reptiles evolved from amphibians, and mammals and birds evolved from reptiles.

2. Jaws make it possible to capture and hold onto prey, while paired fins make it

possible to maneuver through water in pursuit of prey.

3. Amphibians have lungs to breathe air and legs to move on land, and their heart circulates blood through the lungs.

4. Two adaptations were internal fertilization and watertight eggs. Others include improved

heart and lungs, better leg structure, and watertight skin.

5. Mammals and reptiles appeared about the same time, but reptiles were more successful. It wasn't until a mass extinction occurred, killing all of the large reptiles, that mammals diversified into Earth's

dominant life-forms.

6. Mammals coexisted with reptiles for more than 150 million years without growing large or diversifying. Only after the most successful reptiles, the dinosaurs, became extinct did the mammals diversify and give rise to large species.

31-2 Challenge of Obtaining Oxygen

*O*ne of the defining challenges faced by animals has been the need to get enough oxygen to fuel their metabolism. Sponges, cnidarians, many flatworms and roundworms, and some annelids obtain oxygen by diffusion through the body surface. The more advanced marine invertebrates—mollusks, arthropods, and echinoderms—have gills, which are specialized respiratory organs. Fishes also obtain oxygen with gills. But vertebrate gills are useless on land. In this section you will learn how gills function and how terrestrial vertebrates obtain oxygen without gills.

Section Objectives

■ Describe how countercurrent flow increases oxygen absorption.

■ Explain how vertebrates obtain oxygen from the air.

■ Identify the key changes in the structure of the vertebrate lung.

Fishes Use Gills to Get Oxygen From Water

Have you ever looked closely at the face of a swimming fish? If so, you've probably noticed that as it swims, a fish continuously opens and closes its mouth, as if it were trying to eat the water. What looks like eating is actually breathing. The major respiratory organ of a fish is the gill. A gill is a thin sheet of tissue that increases the surface area available for the diffusion of oxygen. The gills of vertebrates, specifically those of bony fishes, are constructed in a way that makes them the most efficient oxygen-gathering organs ever to have evolved. In a fish, gills hang like a curtain between the mouth and the cheeks. At the rear of the cheek cavity is an opening called a gill slit. The whole point of a fish's "swallowing" water is to force the water from the mouth, over the gills, and out through the gill slits.

This swallowing procedure is the core of a great advance in gill design shown by fishes. To understand the importance of this process, first think about the way a mollusk gets oxygen with its gills. In a mollusk, the gills are in an internal cavity, but the water simply sloshes in and out, washing back and forth over the gills. In a fish, the swallowing process forces water to flow only in one direction. One-way flow of water, combined with a specific arrangement of gill tissue, permits countercurrent flow, which is a very efficient way of extracting oxygen.

Here is how countercurrent flow works. Each gill is made of two rows of **gill filaments**, which are fingerlike projections composed of thin membranous plates stacked on top of one another, as shown in Figure 31-10. These filaments stick out into the flow of the water. The water entering the gills is higher in oxygen content than the water leaving. Within each filament, blood flows from the back to the front,

Vocabulary Preview

gill filament
lung
alveoli

Opening Questions

Ask students why animals die if they stop breathing. Guide students' answers to the need for a steady supply of oxygen. If fishes need oxygen, how can you tell if a fish is breathing? Have students watch real fishes on video or in an aquarium to observe the breathing (gill) movements of fishes.

Demonstration
Breathing Goldfish

Carefully place a goldfish in a jar or beaker containing about 500 mL of dechlorinated water. Allow the fish time to settle down. Measure the temperature of the water, then count how many times the gill covering opens and closes per minute (operculum beats). Pour about 100 mL of water out of the jar, and add 100 mL of cold dechlorinated water. Record the temperature of the water again, and count the operculum beats per minute for the fish at the colder temperature. Continue the same procedure until you have five separate readings for increasingly colder temperatures. (All temperature changes should be gradual. Never add water warmer than room temperature to the container.) Graph the beats per minute as a function of temperature. How are water temperature and breathing rate related? (*As the water temperature decreases, so does the rate of operculum beating.*) Why does the breathing rate vary with temperature? (*Answers should include a reduced need for oxygen as the metabolic rate falls with temperature.*)

Teaching Tips
More Than One Way to Breathe

Remind students that some terrestrial animals use structures other than lungs to get the oxygen they need. Students should remember from earlier chapters that some animals use gills (sow bugs), skin (earthworms), spiracles and tracheae (grasshoppers), and book lungs (spiders).

What If a frog's skin dries out? Frogs that depend on respiration through their skin will suffocate and die if their skin is not kept moist. Ask students some of the things a frog could do to keep its skin moist. List their suggestions, which may include sitting in water, hiding in the shade, hopping in and out of water, having a waterproof covering on its skin, and burying itself in mud. Different kinds of frogs actually do all of these things.

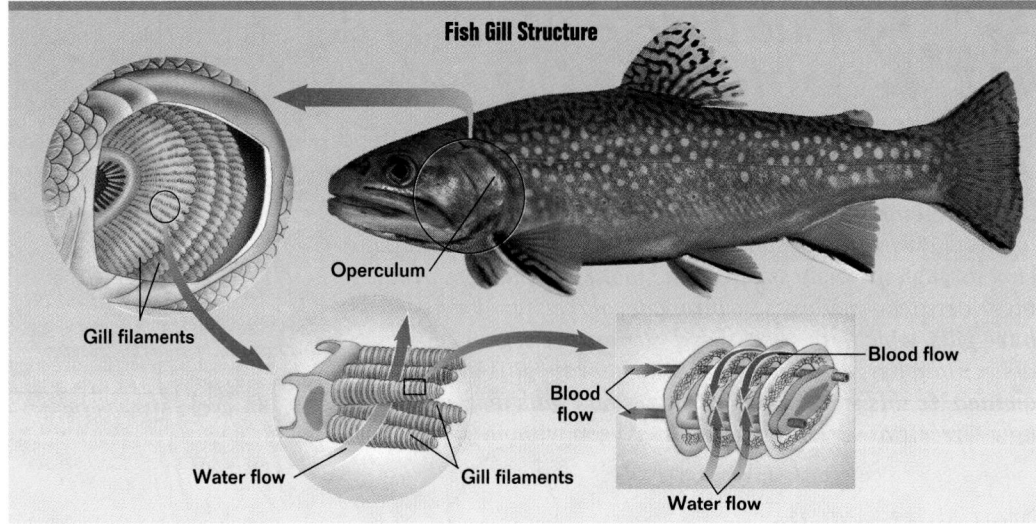

Fish Gill Structure

Figure 31-10 The gills of a bony fish are located behind the head, underneath a flap called the operculum. In the filaments of the gills, oxygen from water diffuses into the blood vessels carrying oxygen-poor blood to be transported throughout the body.

CAPSULE SUMMARY

The structure of a bony fish's gills ensures that water and blood flow in opposite directions. This arrangement maximizes the amount of oxygen that can be extracted from water.

as shown in Figure 31-10. This blood is low in oxygen (indicated by blue) when it enters the gills and high in oxygen (indicated by red) when it leaves. Water, however, flows past the filaments in the opposite direction—from front to back.

Why is the countercurrent arrangement important? Countercurrent flow ensures that oxygen diffuses into the blood over the whole length of the blood vessels in the gills. If blood and water flowed in the same direction, the amount of oxygen in the blood would increase while that in the water decreased until an equilibrium was reached and diffusion stopped. The gills of bony fishes are the most efficient respiratory organs possible. They are able to extract up to 85 percent of the available oxygen from water. ◻

Amphibians Use Lungs to Get Oxygen From Air

Air contains about 20 times as much oxygen as sea water does. Yet all of this oxygen won't help a land-dwelling animal much if the structures for gathering oxygen don't work. And gills, which work very well in water, do not work in the air. Out of water the delicate gill membranes have no support. Therefore, they collapse into a soggy mass. The surface area for gas exchange is so dramatically reduced that a fish out of water suffocates from lack of oxygen. Thus, one of the major challenges that faced the first land vertebrates was obtaining oxygen from air.

The evolutionary solution to the problem of getting oxygen from air is the lung. A **lung** is basically a baglike respiratory organ that allows gas exchange between the air and the blood. The amount of oxygen a lung can absorb depends on

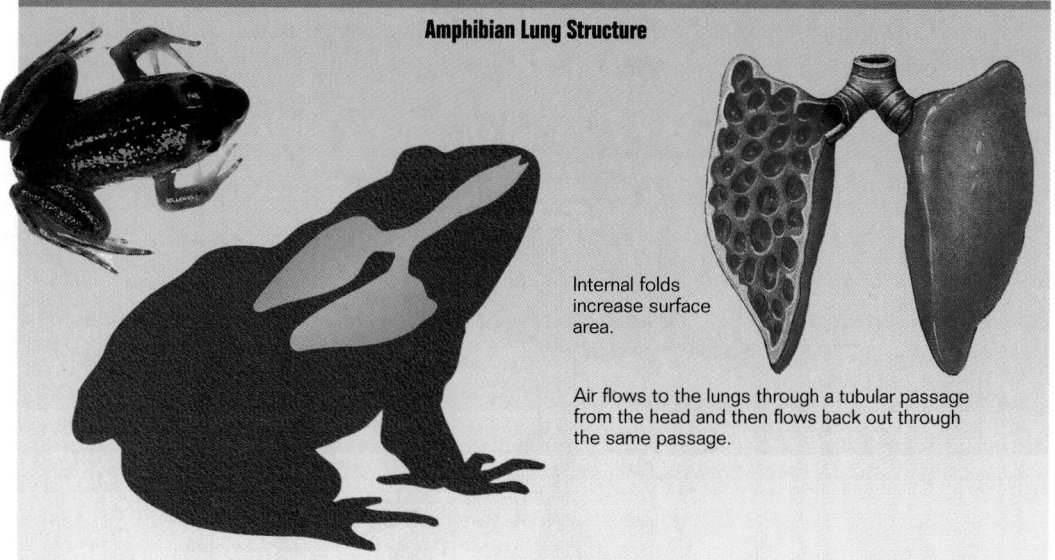

Amphibian Lung Structure

Internal folds increase surface area.

Air flows to the lungs through a tubular passage from the head and then flows back out through the same passage.

its internal surface area. The greater the surface area, the more oxygen that can be absorbed. To get as much oxygen as possible, the interior surface of the lung came to be highly folded, which greatly increased the surface area. Although amphibians could absorb oxygen through their skins, lungs eventually proved to be far more efficient. In amphibians, the lungs are hardly more than sacs with a wrinkled inner membrane, as shown in Figure 31-11. With each breath, fresh air rich in oxygen is drawn into the lungs, where it mixes with a small volume of air that has already given up its oxygen. Because the diffusion surface of the lung is exposed to a mixture of fresh and partly depleted air, the respiratory efficiency of lungs is much less than that of gills. But because there is so much more oxygen in air, amphibian lungs don't have to be as efficient as gills. This arrangement works perfectly well for amphibians, many of which also obtain some oxygen through their thin, moist skin.

Figure 31-11 The lungs of amphibians are sacs with a folded internal membrane that provides a large surface area for gas exchange.

Lung Surface Area Increases in Reptiles and Mammals

Reptiles are far more active than amphibians, so they have much greater metabolic demands for oxygen. An amphibian's lungs cannot provide that much oxygen. However, reptiles cannot rely on their skin for additional respiration in the way that many amphibians can. Remember that reptiles have dry, scaly skin to prevent water loss; they don't have a moist outer membrane for gas exchange.

Teaching Tip
Watching Frogs Breathe
Use videos or real frogs to observe the breathing motions of an amphibian. Have students compare their observations of the way a frog breathes to the way a human breathes. (*Answers should focus on the entrance and exit of the air or water, the breathing rate, and anatomical movements associated with breathing.*) Have students organize their observations on a chart or diagram.

 VISUAL STRATEGY

Figure 31-11
Have your class compare the lungs of the frog in this figure to the lungs of the turtle (Figure 31-12), human (Figure 31-13), and bird (Figure 31-14) diagramed on the following pages. What trends are there in the efficiency and structure of the vertebrate lung? (*more efficient, more internal folds*) How does the number and size of alveoli change? (*more alveoli, smaller size*)

Demonstration
Modeling the Increase in Lung Surface Area
Use a paper towel to line the inside of a 500 mL beaker labeled "swim bladder." Then crumple up towels, and stuff them into a beaker of equal size labeled "lung." Pull the paper towel out of the lined beaker, and have a student calculate the area of the towel in sq. cm. Then, pull the towels out of the other beaker and have a couple of students calculate the area of the towels in sq. cm. This demonstration should yield dramatic evidence of the increase in surface area of an organ that contains internal folds, such as the vertebrate lung.

Teaching Tips
Lung Adaptation to Habitat
Vertebrates with lungs are found from high altitudes to the depths of the ocean. Have your students pick a vertebrate that can survive in an oxygen-poor environment, research its respiratory system, and report on the special adaptations that the vertebrate needs to survive. Choices may include seals, whales, ptarmigans, mountain sheep, mountain goats, marmots, and pikas.

Lung Surface Area and Metabolism
Check for your students' understanding of the relationship between the number of alveoli and lung volume and the metabolic rate of vertebrates. Name pairs of vertebrates and have your class name the member of the pair that probably has more alveoli and greater lung volume. Pairs could be a mouse and a lizard (*mouse has more because of a higher metabolic rate*), human and an alligator (*human has more because of a higher metabolic rate*), sloth and a monkey (*monkey has more because it is more active*). Give students a "brain check" by including a pair such as a goldfish and a turtle. Your class should recognize quickly that goldfish have no alveoli because they have gills, not lungs.

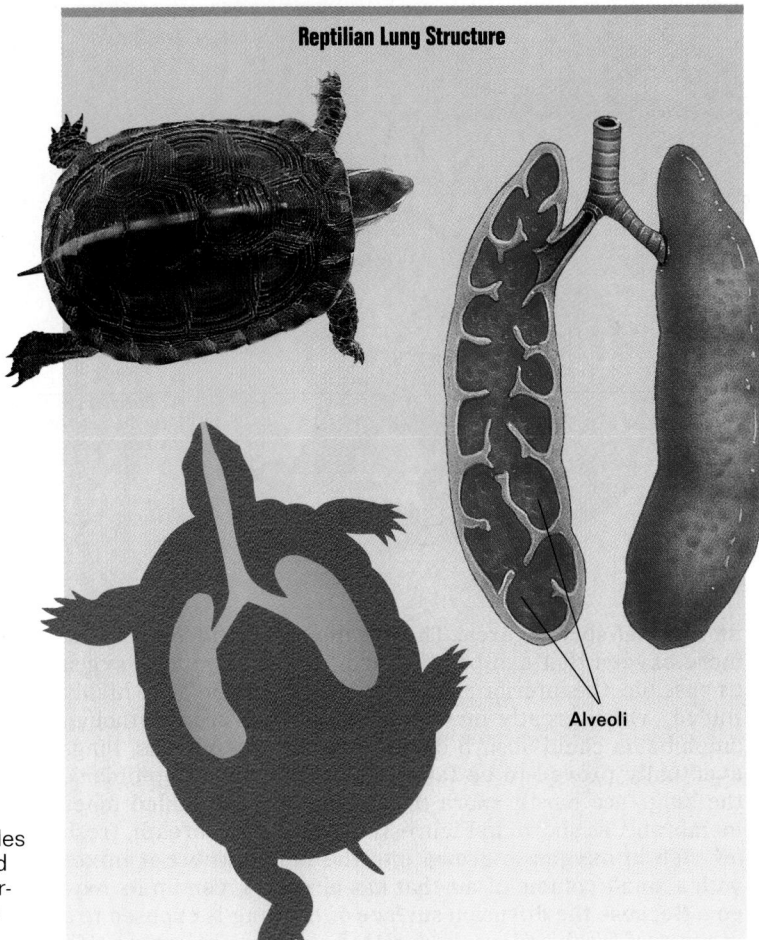

Reptilian Lung Structure

Alveoli

Figure 31-12 The lungs of reptiles contain small chambers called alveoli that create a larger surface area than do the folds in amphibian lungs.

☐ CAPSULE SUMMARY

Efficiency of gas exchange in the lung increases with increasing surface area. Reptilian lungs have many alveoli that greatly increase internal surface area. Mammalian lungs have even more alveoli and so have an even larger internal surface area.

In reptiles, increased oxygen demand is met by greatly enlarged surface area of the lung available for diffusion. The inner surface of reptile lungs consists of many small chambers called **alveoli** (singular, alveolus) that are clustered together like grapes, as shown in Figure 31-12.

Mammals have an even greater demand for oxygen than reptiles do because mammals maintain a constant body temperature. The problem of harvesting more oxygen is solved by increasing the diffusion surface area within the lung even further, as shown in Figure 31-13.

Some mammals are far more active than others, but the more active ones do not have proportionally larger lungs. Instead, in active mammals the individual alveoli are smaller and more numerous, which further increases the surface area for diffusion. Throughout the long history of evolutionary refinement of the vertebrate lung, increases in efficiency have been achieved by increases in the lung's internal surface area. ☐

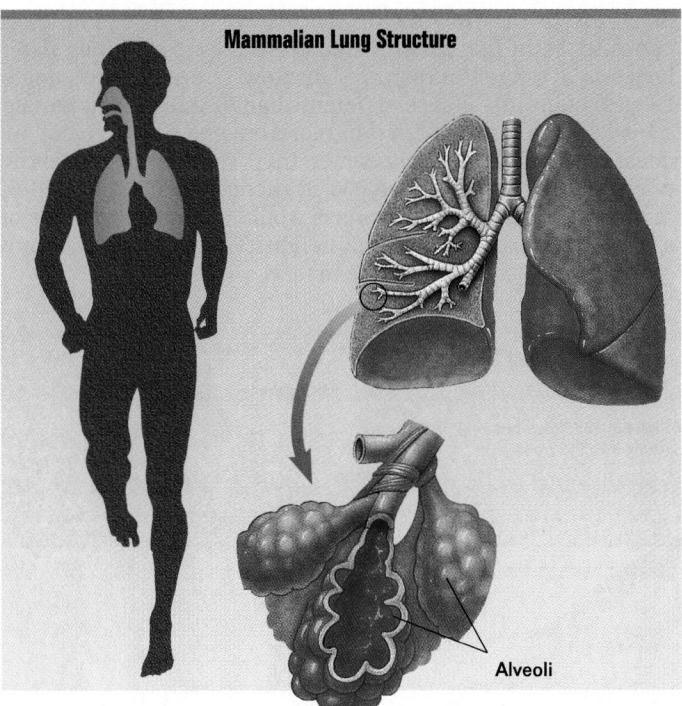

Mammalian Lung Structure

Figure 31-13 The lungs of mammals contain many clusters of alveoli that increase surface area even more than the alveoli in reptilian lungs. Humans have about 300 million alveoli in their lungs. In each lung, the total surface area devoted to diffusion is about 80 sq. m (860 sq. ft.)—slightly less than one-fourth the area of a full-sized basketball court.

Alveoli

The Lungs Are Most Efficient in Birds

There is a limit to how much efficiency can be improved by increasing the surface area of the lung. The more active mammals have, in fact, already reached this limit. Birds, however, have a respiratory demand for oxygen that exceeds the capacity of the lungs of even the most active mammal. Bird flight uses a lot of energy very quickly. Unlike bats, whose flight involves considerable gliding, most birds beat their wings rapidly as they fly. And they often fly this way for a long time. Thus, flying birds must carry out very rapid oxidative respiration within their cells to replace the ATP expended by their contracting flight muscles. The amount of oxygen needed is much greater than any mammalian lung can deliver.

If the oxygen demands of flight cannot be met by an increase in lung surface area, how do birds get the energy they need? In a sense, they do what fishes do. Just as water flows over a fish's gills in one direction, air flow in birds occurs in only one direction. One-way air flow was made possible by the evolution of a series of air sacs connected to the bird's lungs, shown in Figure 31-14.

There are two important advantages to this complicated system. First, there is no oxygen-poor air left in the lungs, as there is in mammalian lungs, so the diffusion surfaces of a bird's lungs are exposed only to fully oxygenated air. Second, the flow of blood in the lungs runs in a different

 VISUAL STRATEGY

Figure 31-13
Use this figure to emphasize the surface area of human lungs. Have students measure an area with chalk or tape that is 8 m by 10 m, the surface area of human lungs, on the classroom floor. How does this space fit into the lungs? *(The shape and number of alveoli increase the surface area.)*

Teaching Tip
Efficient Respiration

Ask students to discuss efficiency in the context of fuel efficient cars. Relate fuel efficiency to the efficient use of air in birds. Ask students how the efficiency of a car differs from the efficiency of a lung. *(Cars are considered more efficient when they use less gas. Lungs, in contrast, are more efficient when they capture more oxygen.)* Birds can get more oxygen into their bodies than any other vertebrate, without increasing lung surface area.

direction to the flow of air. The flows are not completely opposite, as in fish gills. Instead, the network of lung capillaries is arranged across the air flow at a 90-degree angle. Even though this is less efficient than in gills, blood leaving a bird's lung can still contain more oxygen than exhaled air does. No mammalian lung can do this. That is why a sparrow can fly at an altitude of 6,000 m (about 20,000 ft.), while a mouse with the same body mass would be unable to move on a mountain peak at the same height. The sparrow is simply getting more oxygen than the mouse.

Figure 31-14 Birds can fly at high altitudes where there is limited oxygen because there is always fully oxygenated air in their lungs and because their lungs are very efficient at absorbing oxygen. There is no gas exchange in the air sacs; they act simply as holding tanks.

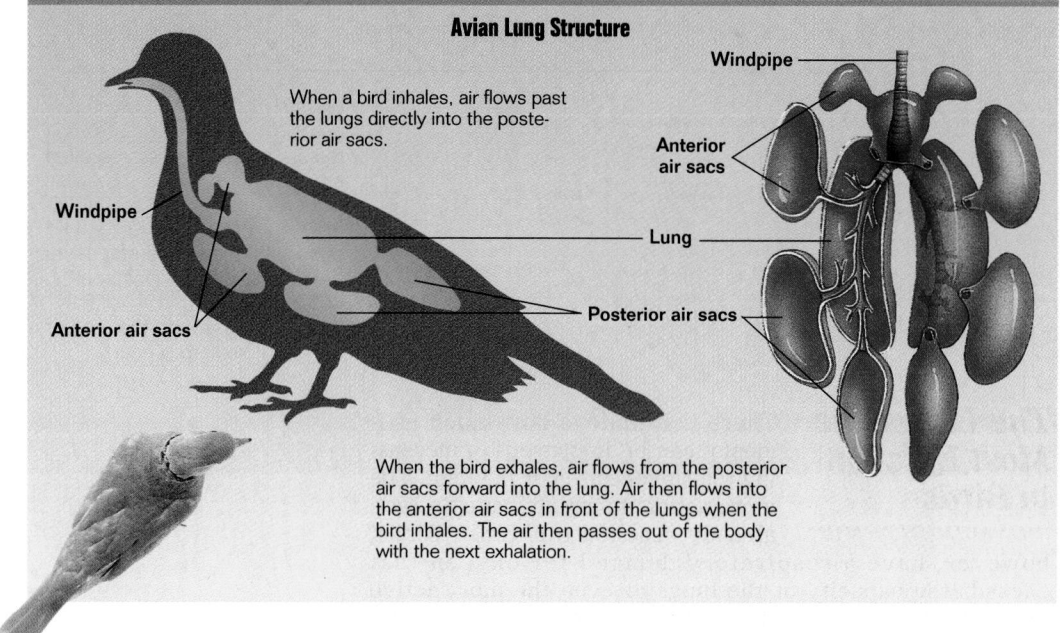

Avian Lung Structure

When a bird inhales, air flows past the lungs directly into the posterior air sacs.

Windpipe

Anterior air sacs

Windpipe

Anterior air sacs

Lung

Posterior air sacs

When the bird exhales, air flows from the posterior air sacs forward into the lung. Air then flows into the anterior air sacs in front of the lungs when the bird inhales. The air then passes out of the body with the next exhalation.

Section Review

1. *How does a fish obtain the maximum amount of oxygen from the water flowing over its gills?*
2. *What function do lungs serve in terrestrial vertebrates?*
3. *How does the structure of the lung in reptiles and mammals allow these animals to obtain sufficient oxygen?*
4. *What adaptations improved the efficiency of bird respiration?*

Critical Thinking

5. *In addition to capturing oxygen, a bird's unusual respiratory system plays a role in flight. Explain how.*

31-3 Evolution of a Better Heart

A long with the need to obtain oxygen, vertebrates need to deliver it efficiently to the body's tissues. Delivering oxygen to and removing carbon dioxide from the body's tissues are the functions of the vertebrate circulatory system. The circulatory systems of all vertebrates share two basic features: a network of vessels through which blood can circulate, and a heart to pump that blood. Much of the evolution of the vertebrate circulatory system involved changes in the heart.

The Chordate Heart Is a Simple Pump

Chordates that were ancestral to the vertebrates had simple tubular hearts similar to those seen in lancelets. This heart was little more than a specialized zone of the ventral (lower) artery that was more heavily muscled than the rest of the arteries. This artery-pump beat in simple peristaltic waves. A peristaltic wave starts at one end of a tube and moves progressively along toward the other end. Suppose you take a soft-drink straw and squeeze one end, and then move your squeezing fingers down the length of the straw. That's a peristaltic wave. If you squeezed a straw in the middle, any liquid in it would squirt in both directions. The same thing happens with peristaltic pumps such as a chordate heart. Blood is pushed in both directions as the heart contracts, so these organs are not very efficient.

The evolution of the vertebrate heart reflects two important transitions in vertebrate behavior. The first was the shift from filter-feeding to active prey capture that accompanied the evolution of jaws. The second was the increased activity that accompanied the invasion of land and the control of body temperature.

A Heart With Chambers Evolves in Fishes

The evolution of gills in fishes created a serious problem. As the diameter of a tube gets smaller, resistance to the flow of liquid increases dramatically. So the tiny diameters of the many small blood vessels in a fish's gills create enormous resistance to the flow of blood. This resistance is too great for any peristaltic pump to overcome. In fishes, the peristaltic pump of early chordates has been replaced with a pump of a very

 Did You Know?
The differences among the hearts of different species of vertebrates may seem to have appeared mysteriously from nowhere, but the genetic variability that provides the raw material for natural selection is with us even today. There are many variations even in the human heart, from location and size of blood vessels to orientation and function of structures. Genetic changes are usually not beneficial, however. Many of these variations are identified as congenital defects of the human heart.

UNIT 8
CHAPTER 31

Figure 31-15

Use this figure to help students figure out the pumping sequence in the heart of a fish. Start with the oxygenated blood in the gills. Where should the blood go next (what needs oxygen)? *(the body cells)* After the oxygen has been delivered, the blood pressure is low, and the blood moves sluggishly. Where should the blood go next? *(the heart)* What is the collection chamber for blood entering the heart? *(sinus venosus)* What holds the blood about to enter the pumping chamber? *(atrium)* What is the main pumping chamber of the heart? *(ventricle)* Why is the blood pressure here so high? *(It takes high blood pressure to push the blood through the many narrow vessels of the gills.)* Where does the blood go next? *(It goes through the cycle again.)*

Demonstration
One Way Flow

Demonstrate the way a valve works by using your hands to model the valve that prevents backward blood flow from the heart into the sinus venosus. Student assistance and props such as red ribbons to represent blood can add to the effect of the demonstration.

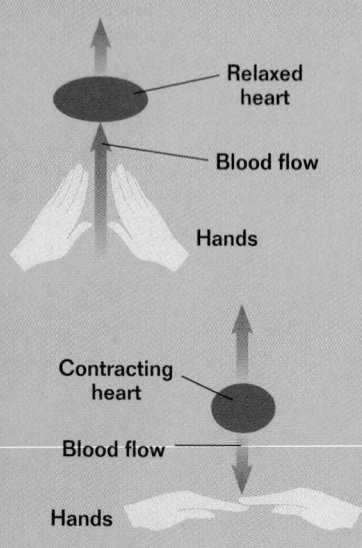

Fish Heart Structure

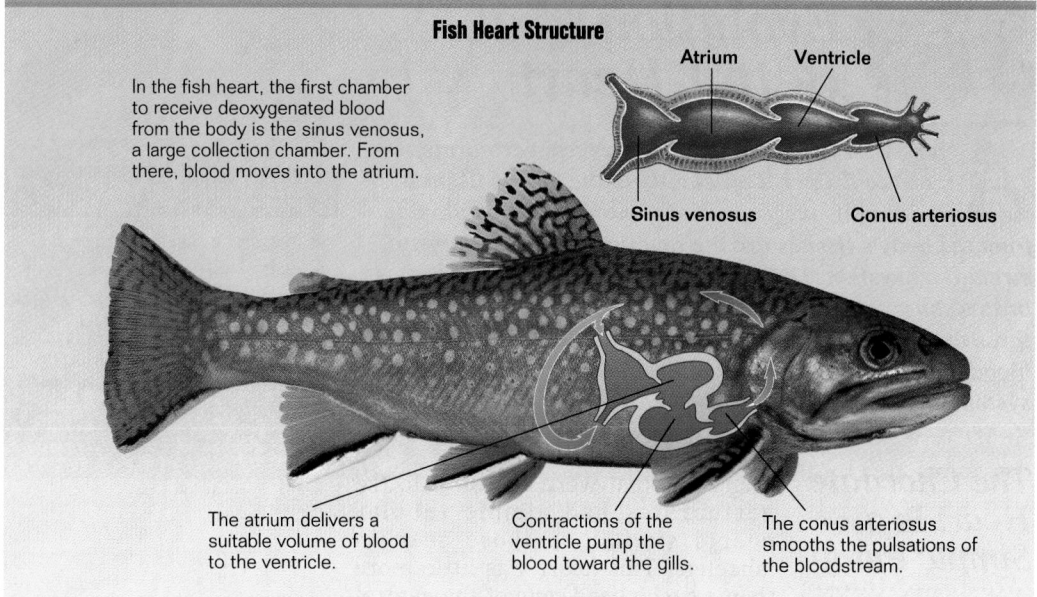

In the fish heart, the first chamber to receive deoxygenated blood from the body is the sinus venosus, a large collection chamber. From there, blood moves into the atrium.

Atrium   Ventricle
Sinus venosus   Conus arteriosus

The atrium delivers a suitable volume of blood to the ventricle.

Contractions of the ventricle pump the blood toward the gills.

The conus arteriosus smooths the pulsations of the bloodstream.

Figure 31-15 The heart of a fish is a tube of four connected chambers.

different nature—a chamber-pump heart, which is shown in Figure 31-15.

The heart of a gill-breathing fish can be thought of as a tube with four chambers in a row. The first two chambers are collection chambers. The second two are pumping chambers.

1. **Sinus venosus** *(SEYE nuhs vuh NOH suhs)* This chamber is a large collection chamber that acts to reduce the resistance of blood flow into the heart. A one-way valve at its entrance prevents blood from flowing backward out of the chamber when its walls contract.
2. **Atrium** Blood from the sinus venosus fills this chamber, which is large and has thin, muscular walls.
3. **Ventricle** To provide enough force to push the blood through the gills, the third chamber is a thick-walled pump with enough muscle to contract strongly.
4. **Conus arteriosus** *(KOH nuhs ahr TIHR ee oh suhs)* A second, more elongated pump, the fourth chamber smooths the pulsations and adds still more force.

You can see the ancestry of fish in how their hearts operate. As in their chordate ancestors, the sequence of contraction is a peristaltic wave starting at the rear and moving to the front. The first of the four chambers to contract is the sinus venosus, which loads the atrium. Then the atrium contracts and fills the ventricle with blood. Next the ventricle contracts, forcing blood into the conus arteriosus. This chamber contracts last, smoothing the pulsations of the bloodstream. Although the relative positions of the heart chambers have changed throughout vertebrate evolution, this heartbeat sequence has remained the same. ▫

Relaxed heart

Blood flow

Hands

Contracting heart

Blood flow

Hands

🞑 **CAPSULE SUMMARY**

A fish's heart collects deoxygenated blood from the body and pumps it to the gills.

Effective Reading

Making Sense of New Words

Talk about the origins of the terms for heart anatomy. Here are some of the meanings given in a dictionary.

sinus—a hollow space formed by a curved surface

venosus—a vein; a vessel that brings blood back to the heart

atrium—a hall or entrance court (originally named for central court of Roman homes)

ventricle—derived from a word for stomach; means a hollow organ

conus—cone shape

arteriosus—an artery; a vessel that carries blood from the heart to the body

Have students make a diagram or icon to represent each word's original meaning.

The fish heart represents one of the great evolutionary innovations in the vertebrates. Because it pumps blood first through the gills and then to the rest of the body, the blood delivered to the tissues is fully oxygenated. This arrangement has one important limitation, however. After passing through the fine network of tiny blood vessels in the gills, the flow of blood has lost much of the force applied by the contraction of the heart. Hence, the circulation of blood from the gills to the rest of the body is sluggish. This means that oxygen cannot be quickly delivered to body muscles.

The Pulmonary Vein Evolves in Amphibians

With the evolution of amphibians, the limitations of pumping blood through the lungs to the rest of the body presented an even greater problem. The first lungs were probably quite inefficient, but to be active on land, amphibians needed even more oxygen to be delivered to their muscles. So in this group of organisms, the circulatory system underwent another change—the addition of another pair of major blood vessels in the heart. These vessels, the **pulmonary veins**, carry blood from the lungs to the heart. Therefore, after picking up oxygen in the lungs, blood returns to the heart to be pumped to the rest of the body, as shown on the right in Figure 31-16.

Not only did the path of circulation in amphibians change, but so did the heart itself. A dividing wall known as

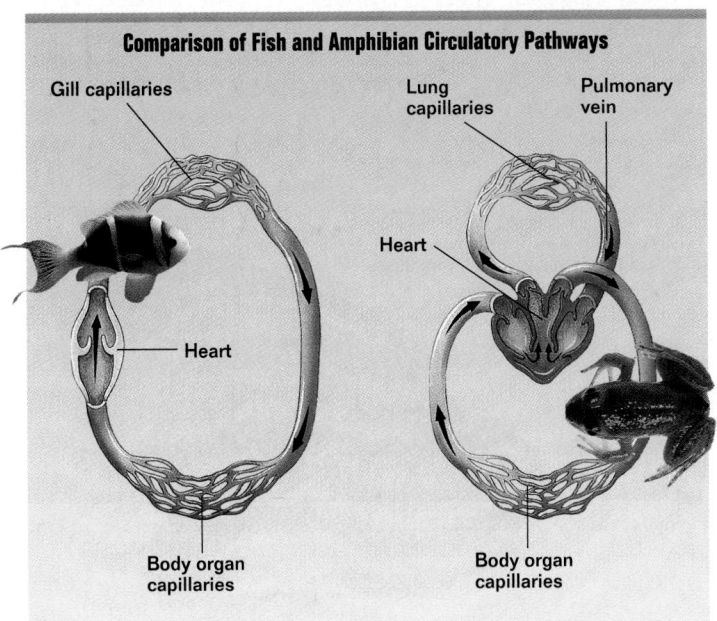

Comparison of Fish and Amphibian Circulatory Pathways

Gill capillaries

Lung capillaries

Pulmonary vein

Heart

Heart

Body organ capillaries

Body organ capillaries

Figure 31-16 Circulation in fishes differs from amphibian circulation in that amphibians have a pulmonary circulation loop. This loop moves blood from the heart to the lungs and back to the heart.

SECTION 31-3

Teaching Tip
Oxygen Requirements
Point out that fishes live in an environment that is low in oxygen (compared to air) and they have sluggish circulation compared to that of mammals and birds. How, then, do fish survive? (*They have a low demand for oxygen compared to other endothermic vertebrates.*) Students should realize that the fish's circulatory system, though not delivering oxygen as rapidly as that of a mammal or bird, still provides enough oxygen for the fish's needs.

Demonstration
Blood Flow in a Goldfish Tail
Have students observe real blood flow in the capillaries of a goldfish's tail. Wrap wet cotton balls around the gills of a goldfish, very gently spread its tail out on a slide, and observe at low power under a microscope. Once focused, you can go to high power and actually see individual blood cells. Don't leave the fish out of water more than a few minutes (less time if it flops around in distress). Be sure that it doesn't dry out.

VISUAL STRATEGY

Figure 31-16
Use this figure to compare the circulation of fish and amphibians. Which system is single and which is double? (*The fish system is single, and the amphibian system is double.*) Why does the amphibian need an extra loop in its circulation? (*It has to send blood to lungs.*) Name a variety of vertebrates, and have individual students say whether the animal yes, has a double circulatory system or no, doesn't have a double system. (*All of the fish you name should be no, and all of the amphibians, reptiles, birds, and mammals you name should be yes.*) If students have trouble, ask them whether the animal has gills or lungs.

Teaching Tips
Amphibian Circulation

Amphibians pump blood through their lungs, then they combine the newly oxygenated blood with oxygen-poor blood. This apparently inefficient system works for several reasons. Have students review what they know about amphibians in general, and frogs in particular, to suggest reasons why this circulatory system works for amphibians. (*Reasons may include that their metabolic rate is slow, their skin supplies an extra respiratory surface, and their oxygenated blood tends not to mix with oxygen-poor blood.*)

Comparative Vertebrate Anatomy

Have students compare the hearts of a fish and an amphibian. Ask them to list the similarities and differences between the two types of hearts. Have students compare their lists.

Demonstration
Separating Blood

To show that oxygenated and deoxygenated blood tend to stay separate in the ventricle, even though there is no wall in the ventricle of the amphibian heart, demonstrate how liquids with different densities tend to separate. Fill about half of one graduated cylinder with corn syrup. Ask what will happen if you add colored water. Will the two liquids mix? Fill the rest of the cylinder with colored water. The clear syrup should remain at the bottom with the water on the top. Reverse the order of adding the liquids. Fill about half the cylinder with colored water, and ask your students to predict what will happen when you add the syrup. When you add the syrup, it should sink to the bottom. Explain that there is also a slight density difference between oxygenated and deoxygenated blood; thus, they stay separate in the ventricle.

a **septum** separates the atrium into right and left halves. The pulmonary veins, carrying oxygen-rich blood from the lungs, empty into the left atrium. The sinus venosus, carrying oxygen-poor blood from the rest of the body, empties into the right atrium. Since no wall divides the ventricle, some mixing of oxygen-rich and oxygen-poor blood does occur in that chamber. A number of amphibians also have a spiral valve that divides the conus arteriosus.

In effect, the changes in the amphibian heart match the two circulation paths created by the pulmonary veins. Blood entering from the lungs tends to stay on the side of the heart that will exit toward the rest of the body. Blood entering from the rest of the body tends to stay on the other side of the heart, and it exits toward the lungs.

The advantage of this new arrangement is that oxygenated blood can be pumped to the muscles at much higher pressures. The disadvantage is that oxygenated blood from the lungs is mixed in the heart with blood that has already given up its oxygen as it circulated through the rest of the body. So the amphibian heart pumps out a mixture of oxygen-rich and oxygen-poor blood, as shown in Figure 31-17.

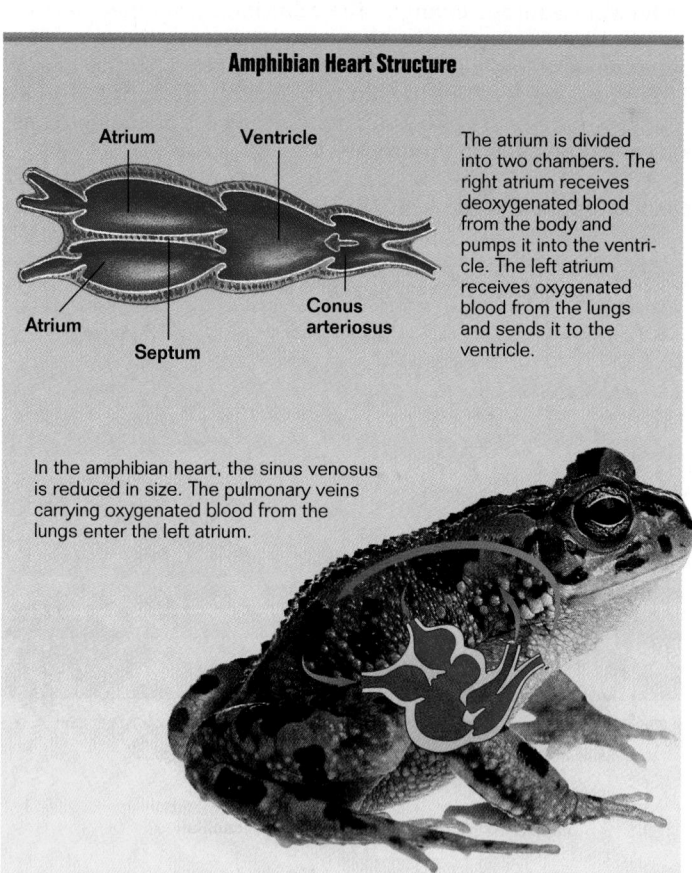

Figure 31-17 The amphibian atrium is divided by a septum that helps limit the mixing of oxygenated and deoxygenated blood.

The Heart Is Divided in Terrestrial Vertebrates

In the amphibian heart, oxygen-poor blood dilutes the oxygen-rich blood that has returned from the lungs. It is no accident, then, that amphibians are usually sluggish. Among reptiles, however, the septum that completely divides the atrium into right and left halves extends into the ventricle. There is therefore much better separation of oxygen-rich and oxygen-poor blood. The conus arteriosus is also divided completely in two. It is no longer a heart chamber. Instead, it forms the trunks of the two large arteries leaving the heart.

Because the reptilian heart achieves a better separation of blood, it works more efficiently than the amphibian heart at driving the unique "double" circulation that first evolved in amphibians. However, the efficiency of the system is still somewhat limited since the separation of the ventricle is not complete. Mixing of blood still occurs, as shown in Figure 31-18.

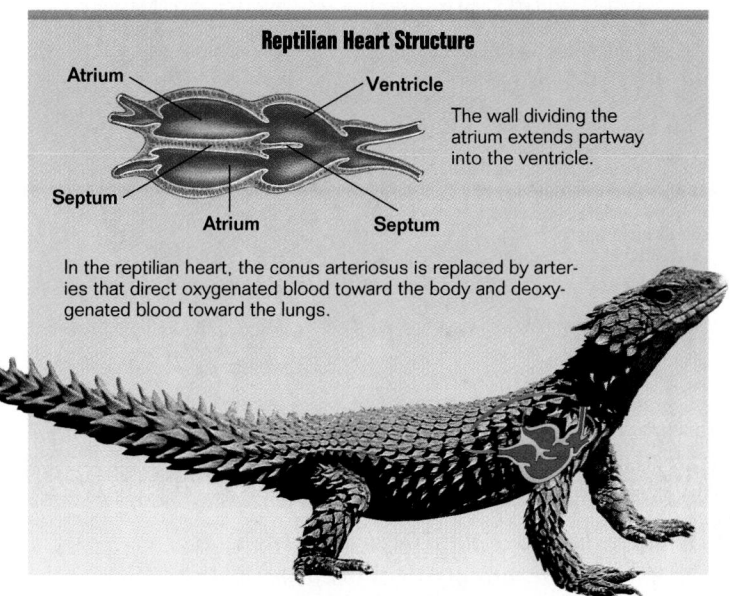

Reptilian Heart Structure

Atrium

Ventricle

The wall dividing the atrium extends partway into the ventricle.

Septum

Atrium

Septum

In the reptilian heart, the conus arteriosus is replaced by arteries that direct oxygenated blood toward the body and deoxygenated blood toward the lungs.

Figure 31-18 In all reptiles but crocodiles, the ventricle is partly divided by a septum.

Only a slight change was needed to complete the evolutionary transition of the vertebrate heart begun in amphibians. Mammals, birds, and crocodiles all have completely separated ventricles. In these animals, the septum completely divides the ventricle into two pumping chambers, as illustrated in Figure 31-19. The closing of the ventricular septum solves the problem originally created when pulmonary veins evolved in amphibians. With total separation of the lung and body circulations, blood pumped to the tissues is now fully oxygenated.

UPDATE

☑ **RESEARCH Update**

Who's Afraid of the Big Bad Wolf?

Ranchers were and still are. By the 1920s in Yellowstone National Park, the entire wolf population had been killed off by ranchers defending their livestock. Wildlife conservation groups have been pushing since the early 1980s for the reintroduction of the gray wolf to Yellowstone. In June of 1994, Secretary of the Interior Bruce Babbit finally approved a measure that would import 30 wolves from Canada to Yellowstone. The ultimate goal of the government plan is to have 100 wolves in Yellowstone by 2002. This mandate has drawn mixed reviews. The ranchers are upset. They fear the loss of their livestock, particularly young animals. They already have to fight off mountain lions and coyotes, at considerable expense, and don't want to have to contend with another major predator.

Conservation groups have stated that while the reintroduction of wolves to the park is certainly a step in the right direction, it isn't enough. The government plans to reintroduce the wolves but not give them endangered species protection, as a concession to the ranchers. This means that the ranchers are entitled to kill any wolf on their property that is found killing their livestock.

Wildlife biologists applaud the measure because, as soon as the population of wolves disappeared, their prey flourished—too much. Elk and deer populations exploded without any predatory control, and went on to overbreed and tax the forest ecosystem, destroying grasses and herbs. ☑

UNIT 8

CHAPTER 31

Teaching Tips

Concept Mapping

Have students organize the information they are learning in a chart, concept map, or picture using the following headings: Bony fishes, Amphibians, Reptiles, Birds, and Mammals. Students should find connections between the type of circulatory system and the type of respiratory system for each class of vertebrate. They should also see a relationship between the type of circulatory system and the metabolic demands of the vertebrate class. In addition, students should be able to trace changes in the structures of the heart in different vertebrate classes.

Independent Evolution

Direct student attention to the observation that a septum between the ventricles evolved independently several times. When the same structure evolves in different groups of animals, there is usually a great advantage associated with the structure. What advantage does a septum between the ventricles give to a crocodile, bird, or mammal? *(The separation of oxygenated and deoxygenated blood ensures that only fully oxygenated blood is pumped to the tissues, improving the efficiency of the system.)*

Application

℞ Medicine If a child is born before the septum between the atria has grown together, the hole in the septum allows the oxygenated and deoxygenated blood to mix. Deoxygenated blood is pumped out to the body causing the babies to have a cyanotic, or blue, appearance. This mixing of blood may work in amphibians, but humans born this way must go through surgery to repair the defect if the septum does not grow together naturally.

❏ CAPSULE SUMMARY

A fish's heart only pumps deoxygenated blood to the gills. The amphibian heart sends deoxygenated blood to the lungs and oxygenated blood to the body, with some mixing of the two bloodstreams. A partial division of the ventricle in most reptiles reduces this mixing. No mixing occurs in the hearts of crocodiles, birds, and mammals because they have two completely separate ventricles.

Figure 31-19 The avian heart shows a complete division of the ventricle by the septum. Oxygenated and deoxygenated blood are kept completely separate, meaning that oxygen is moved through the body more efficiently. This type of heart is also found in mammals and crocodiles.

What has happened to the sinus venosus over the course of vertebrate evolution? Remember that this is a major chamber in the fish heart. In amphibians the sinus venosus is reduced in size, and it is even smaller in reptiles. Mammals and birds no longer use the sinus venosus as a separate chamber. However, some tissue from it remains and plays an important role.

Throughout the evolutionary history of the heart, the sinus venosus has had two functions. One function is to act as a collection chamber for blood. The other function is as the site from which the heartbeat begins. The second function is indispensable. Thus, mammals and birds have retained some of the excitatory tissue of the sinus venosus in the wall of the right atrium, at the point where it was located in the fish heart. The sinus venosus has become the pacemaker in the hearts of mammals and birds. It is the point of origin of each heartbeat. ❏

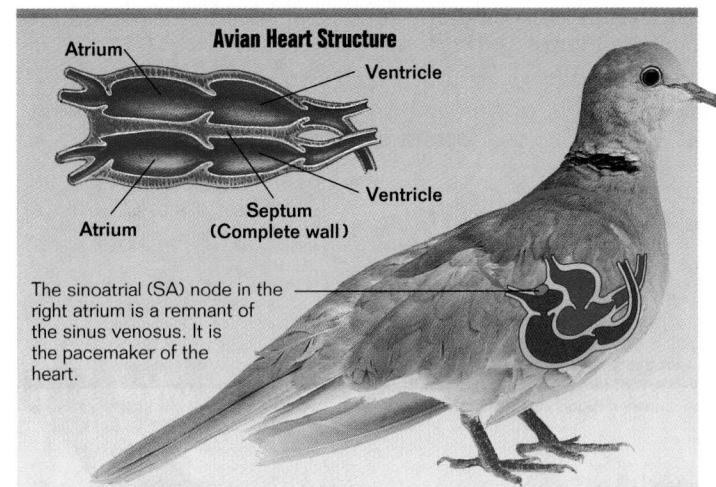

Avian Heart Structure

Atrium — Ventricle — Ventricle — Atrium — Septum (Complete wall)

The sinoatrial (SA) node in the right atrium is a remnant of the sinus venosus. It is the pacemaker of the heart.

1. *Trace the flow of blood through a fish's heart.*
2. *How do the hearts of fish, amphibians, and reptiles differ? How are they alike?*
3. *Describe how the circulatory system of a mammal is an improvement on the double-loop circulation of an amphibian?*
4. *What is the advantage of separating oxygen-rich and oxygen-poor blood during circulation?*

Critical Thinking
5. *In your heart, the left ventricle, which pumps blood to the body, is much stronger than the right ventricle, which pumps blood to the lungs. Explain this difference.*

Answers to Section Review

1. Sinus venosus→atrium→ventricle→conus arteriosus

2. Their hearts differ in the number of atria and in the presence of a septum, sinus venosus, and conus arteriosus. For further information, see pages 718, 720, and 721.

3. The heart of a mammal has complete separation of the ventricle into right and left parts.

4. Separating the blood in the ventricles results in oxygenated blood being sent to the tissues that need oxygen and oxygen-poor blood being sent to the lungs to absorb oxygen.

5. The left ventricle has to provide enough pumping pressure to move blood through the entire body, except for the lungs.

31-4 Challenge of Retaining Water

*V*ertebrates evolved in water. Though vertebrates now also live on land, no vertebrate can do without water for long. About two-thirds of a vertebrate's body is water, and if the amount of water falls much lower than this, the animal will die. Losing water is called dehydration. Preventing dehydration has been a key evolutionary challenge facing vertebrates in all environments. Even marine fishes, which live in sea water, must cope with the problem of water loss. If this seems strange to you, remember the process of osmosis. Osmosis causes a net movement of water through membranes toward regions of higher ion concentration. Sea water is three times as salty as the tissues of a marine bony fish. As a result, water tends to leave its body by osmosis. How, then, do marine fishes avoid dehydrating in sea water?

Fishes Use Kidneys to Balance Water and Salt

Marine bony fishes lose water continuously by osmosis to the saltier water in which they swim. To make up for this water loss, marine fishes drink a lot of water. But because this water contains high levels of salt, marine fishes have an osmotic problem. How can they avoid losing water and taking in too much salt?

Freshwater fishes have the opposite osmotic problem. Their bodies have more salt than the surrounding water. So freshwater fishes need to avoid taking in too much water and losing too much salt. The evolutionary solution for both marine and freshwater fishes is a pair of kidneys.

The **kidney**, shown in Figure 31-20, regulates salt and water balance and removes metabolic wastes from the blood. It is a complex organ made up of thousands of disposal units called nephrons. Each **nephron** has the structure of a bent tube, as shown in Figure 31-21 on the next page. Blood pressure forces the fluid in blood past a filter at the top of each nephron. This filter keeps blood cells, proteins, and other useful large molecules in the blood. However, water and the small molecules and wastes dissolved in it pass through the filter and into the bent part of the tube. As the filtered fluid passes through the nephron tube, useful sugars and ions are recovered by active transport. Some of the water is also reabsorbed. The remaining water and dissolved metabolic wastes (such as urea) that are left behind form a fluid called **urine**.

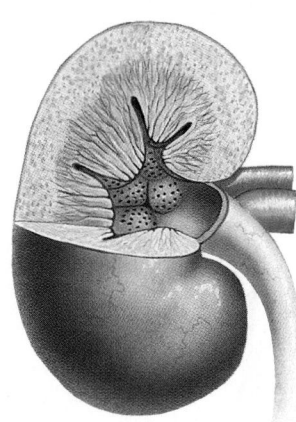

Figure 31-20 The kidney is the organ that filters waste from the vertebrate body. Kidneys differ among fish, amphibians, reptiles, birds, and mammals to address different needs in maintaining water balance. As vertebrates moved onto land, kidneys became more specialized in concentrating urine to prevent water loss.

Teaching Tips

Teaching Tips
Animal Urinalysis

To discuss what urine reveals about the animal that produces it, separate your class into groups of two, and ask what environment the following animals might inhabit. Where would you find a fish that produces watery (hypotonic) urine? *(freshwater)* a fish that produces salty (hypertonic) urine? *(saltwater)* a frog that produces watery (hypotonic) urine? *(freshwater)* a rodent that produces salty (hypertonic) urine? *(desert)*

Visual Thinking

Have students visualize osmotic water flow by thinking of an analogy. For example, they might compare the percent of water with a mountain—the higher the percent of water, the higher the elevation. Students should be able to describe water flow into and out of a fish with their analogy. Have students draw a diagram of their analogy. Students could include names of representative species in their analogies. Another analogy could relate the percent of water found in urine to a temperature scale.

How concentrated the urine is depends on the environment in which the animal lives. Freshwater fishes produce large amounts of dilute urine. In marine fishes, ions and nitrogenous wastes are excreted by the gills, which are their major organs for regulating ion balance. Also, the ion channels in the nephron membranes of marine fishes are reversed in orientation. This allows the nephron to actively excrete excess salts into the urine.

Sharks are an exception. These marine fishes do not actively pump ions out of their bodies through their kidneys. Instead, sharks maintain the same ion concentration as the surrounding water. So a shark does not gain or lose water by osmosis, since there is no osmotic difference between its fluids and the sea water in which it is swimming.

The vertebrate kidney evolved in fresh water. But over millions of years of evolution, it has become adapted for life in a wide range of environments.

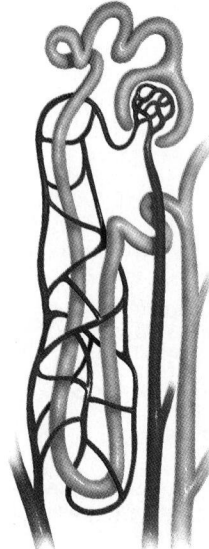

Figure 31-21 The nephron, *above*, is the collecting device in the kidney. The bend in the tube, which evolved in mammals, helps concentrate urine and further reduce water loss. The excretory systems of bony fishes, birds, and mammals, *right*, show increasing complexity in the structure of the kidney.

Urine Is Concentrated in Terrestrial Vertebrates
....

Because they spend so much time in or near fresh water, amphibians, like freshwater fishes, produce dilute urine. Reptiles, however, face not an osmotic problem but an evaporation problem. Thus, reptiles reabsorb much more of the water in their kidneys and produce a far more concentrated urine than amphibians do. This urine, though, cannot become any more concentrated than a reptile's blood plasma. Otherwise, water from the animal's body would simply flow into its urine by osmosis while the urine was still in the kidneys.

Mammals do an even better job of retaining water than do reptiles because mammalian kidneys have evolved to remove far more water from urine. Human urine may be as

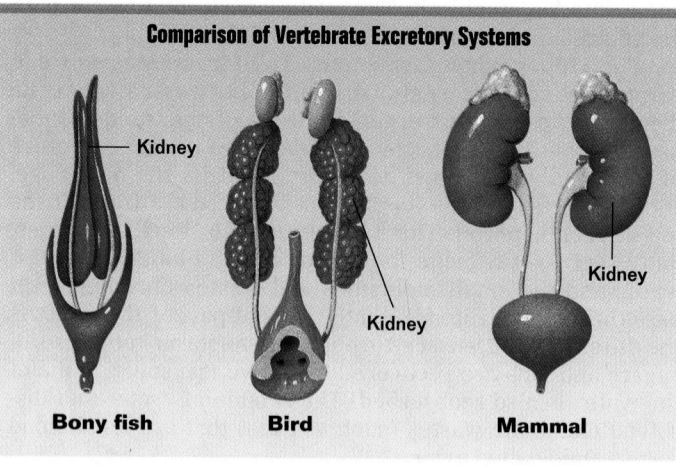

Comparison of Vertebrate Excretory Systems

Kidney

Bony fish

Kidney

Kidney

Bird

Kidney

Mammal

much as four times as concentrated as blood plasma. In desert mice, the urine is over 20 times as concentrated. How do mammalian kidneys avoid the osmotic trap of having body water flow back into the urine within the kidney? In a remarkably simple and powerful innovation, the nephron tube is bent back on itself, as you can see in Figure 31-21. This structure places the end segment of the nephron in a region of the kidney with a very high local concentration of ions and metabolic wastes. The high salt concentration draws water out of the tube by osmosis. Thus, a very highly concentrated urine solution is left behind.

CONTENT LINK

Suppose you drink a large glass of water. How will your kidneys respond? You can find out in **Chapter 38.**

Watertight Skin Evolves in Reptiles

On land, water tends to evaporate from surfaces into the air. So a terrestrial animal faces the problem of water loss through its skin. The first amphibians had bodies covered with the same kind of small bony scales as their fish ancestors. Most amphibians that survive today have moist skins, however. To slow water loss, many present-day amphibians secrete a slippery mucus that tends to retain water. This mucus is what makes these animals feel slimy. Even with this coating, though, the skin of amphibians is not very watertight.

Reptiles had a different evolutionary solution to the problem of water loss through the skin. Instead of bony scales, the earliest reptiles were covered with lighter, flexible scales made of protein. These scales overlap and form a watertight, dry covering that minimizes evaporation of body water through the skin. This dry skin helped to free reptiles from the necessity of living in a wet environment. Mammals and birds, which evolved from reptiles, also have skin that is dry and virtually watertight.

Watertight Eggs Evolve in Reptiles

Water loss is not a significant problem for the eggs of fishes and amphibians. The eggs of most species of amphibians are released by the female into water, where they are fertilized by sperm released by a male. Once fertilized, the eggs go on to develop in the water.

For a reptile living on dry land, however, reproduction presents a serious water-loss problem. Without a watery environment, both eggs and sperm will dry out. Also, sperm need water in which to swim. And finally, the fertilized egg needs a moist environment in which to develop. In the eggs of reptiles, the embryo develops surrounded by a pair of protective membranes, as shown in Figure 31-22. The **amnion**

SECTION 31-4

Application

Health Reptiles, birds, and mammals have watertight skin. However, humans and many other mammals lose water as sweat, which is released by sweat glands in the skin to help cool the body. Humans have to beware of heatstroke when they are active during the summer because they can lose large amounts of water through sweating. It is possible to lose so much water that the body cannot cool itself, raising the body temperature to life-threatening levels.

Teaching Tips
Life Without Free Water

Help students relate water retention in the mammalian kidney to the lifestyle of different mammals. There are desert rodents that never drink water, but derive all of their water from the food they eat. Ask students what they would choose to eat if there were no water for drinking. (Answers might include fruits, vegetables, or succulents that have a high water content.)

Incubating Eggs

If you have an incubator, fertile eggs, and a home for the hatchlings, incubating eggs can be a highly motivating experience for students. If you don't have access to these materials, you may be able to obtain a video of egg development and hatching from a local zoo or library. Have your students pay special attention to the water requirements of eggs. Eggs of reptiles and birds don't need to be in water like the eggs of amphibians, but they do need humid air.

Figure 31-22

Use this figure to have students identify the parts and functions of egg anatomy. Every part except the shell is important in the development of placental mammal embryos. Have students compare the hatching eggs in this figure to regular chicken eggs. How are the shells different? (*These eggs are flexible, not hard like chicken eggs usually are.*) Why aren't reptile eggs hard? (*They are easier for the mother to lay, and the parents don't sit on the eggs.*)

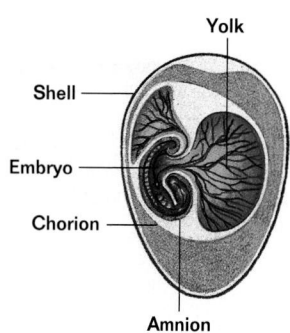

Figure 31-22 This cross section, *above*, shows the internal structure of an amniotic egg. The evolution of the amniotic egg allowed reptiles, such as these hognose snakes, *right*, to make the complete transition to land.

❑ *CAPSULE SUMMARY*

The transition from life in water to life on land required a more complex excretory system to maintain water and salt balance. It also required the evolution of protective coverings to prevent the body and eggs from losing water.

(*AM nee awn*) encloses the embryo within a watery environment. In a sense, this membrane creates a little pond that substitutes for the water in which amphibians lay their eggs. Surrounding the amnion is another membrane, the **chorion** (*KAWR ee awn*). This membrane allows oxygen to enter the egg and carbon dioxide to leave, but it is impermeable to water. This kind of watertight, fluid-filled egg is called an **amniotic egg**. Reptiles, birds, and mammals produce amniotic eggs. Most reptiles, all birds, and three species of mammals (the monotremes) lay amniotic eggs surrounded by a tough shell. ❑

Section Review

1. *Describe two ways marine bony fishes maintain their salt and water balance.*
2. *How are mammals able to produce a urine that is more concentrated than that of reptiles?*
3. *How does a reptile's dry skin help conserve body water?*
4. *What are the functions of the amnion and chorion in a bird's egg?*

Critical Thinking

5. *A few species of sharks live in fresh water. Would you expect these species to have the same internal ion concentration as marine sharks? Explain your answer.*

Answers to Section Review

1. Marine bony fishes drink large quantities of water and excrete excess salt from their gills.

2. Mammals have a loop in their nephrons that makes it possible for them to have very high local concentrations of ions and metabolic wastes, which create concentrated urine.

3. Reptiles are covered with flexible, overlapping scales that are watertight.

4. The amnion encloses a watery environment in which the embryo develops. The chorion allows the embryo to exchange gases without losing water.

5. Marine sharks have the same internal ion concentration as sea water, so freshwater sharks probably have the same ion concentration as their freshwater environment. Therefore, marine sharks and freshwater sharks probably do not have the same internal ion concentration.

31-5 Reproduction and Development

*I**n Section 31-4, you saw that one of the most severe problems faced by vertebrates as they began living on land was the danger of drying out. This problem was particularly severe for their small and vulnerable gametes. If released on land, the gametes of fishes would dry out and perish. As vertebrates adapted to life on land, different ways of protecting gametes and embryos from drying out evolved in each vertebrate class.*

Section Objectives

■ List the advantages of internal fertilization over external fertilization.

■ Contrast oviparity, ovoviviparity, and viviparity.

■ Contrast the parental care provided by reptiles, birds, and mammals.

Fishes Fertilize Their Eggs Externally

Most fishes fertilize their eggs like other aquatic animals, by simply releasing male and female gametes near one another in the water. This approach is called **external fertilization**. A fish egg contains only enough yolk to nourish the developing zygote for a short time. A hatchling fish has a very short larval stage during which the yolk sac of the egg is still attached. When the yolk is depleted, the growing fish must get along on its own, seeking food in the waters it lives in. Young, growing fishes are unlikely to survive because food is often hard to come by, infection by microbes is common, and predators abound. Thus, while many thousands of eggs may be fertilized in a single mating, few of them survive this harsh introduction to life and grow to maturity. As you might expect, natural selection has favored changes that make this dangerous period as short as possible. The fertilized egg develops very quickly, and the young that survive the dangers of early growth mature quickly.

Amphibians Are Still Tied to Water

As you have seen repeatedly in this chapter, amphibians, the first vertebrates to successfully invade land, did not make a full transition to terrestrial life. The reproductive cycle of most living amphibians is still dependent on the presence of free water. Among most amphibians today, fertilization is still external, just as it is among most fishes. Where do amphibians find water in which to carry out external fertilization? Many female frogs and toads lay their eggs in puddles or ponds. The male grasps the female and discharges fluid containing sperm onto the eggs as she releases them into the water, as shown in Figure 31-23.

Vocabulary Preview

external fertilization
tadpole
metamorphosis
internal fertilization
semen
penis
oviparous
ovoviviparous
milk
viviparous

Opening Question

Ask students to discuss the roles that parents play in raising the following types of offspring: birds, turtles, kangaroos, and humans. Lead students to realize that offspring are born at different stages of development. The young of some animals rely on parental protection and nurturing until their development is complete.

Demonstration

Stages of Development

Use videos, still pictures, or preserved specimens to demonstrate larvae and embryos. Have students guess whether a specimen is a larva or an embryo and which kind of vertebrate it is. Also, have students match pictures of adult animals to their embryos.

Figure 31-23 The male frog, *above left,* has grasped a female and is about to fertilize her eggs. Each egg is enclosed in a jelly-like coating, *above center.* The tadpole, *above right,* is a larval stage of the developing frog.

☐ **CAPSULE SUMMARY**

Fertilization of eggs is external in most fishes and amphibians. After hatching, the larval amphibian feeds and grows until it reaches a certain size, then transforms into an adult.

The body of an amphibian is far more complex than that of a fish, and, as you might expect, it takes an amphibian far longer to develop. However, amphibian eggs have about the same amount of yolk as fish eggs, so the amphibian embryo isn't protected in the egg any longer than a fish embryo. Instead, amphibian development takes place in two phases. First, the egg hatches into a larval stage, like some of the larvae found among insects. The development of the embryo into a larva is rapid because it uses yolk from the egg. The larva then functions, often for a considerable period of time, as an independent food gatherer, living in water and getting oxygen through gills. Larval amphibians typically grow quite rapidly. **Tadpoles**, which are frog larvae, can grow in a few days from the size of a grain of rice to the size of a goldfish. Only when an individual has grown to a sufficient size does it undergo the second phase of its development—**metamorphosis**, a radical developmental transformation into the terrestrial adult form. ☐

Reptiles, Birds, and Mammals Have Internal Fertilization

Reptiles were the first class of vertebrates to completely abandon aquatic habitats. Unlike the eggs of most amphibians, reptilian eggs are fertilized within the female before they are laid, a process called **internal fertilization**. The male introduces his **semen**, a fluid containing sperm and fluid secretions, directly into the female's body. In this way, fertilization takes place in a wet environment and the gametes are protected from drying out, even though the adult animals are fully terrestrial. Most vertebrates that fertilize internally use the **penis**, a tubelike organ, to inject semen into the female. Composed mostly of tissue that can become rigid and erect, the penis penetrates far into the female reproductive tract.

Most birds do not have a penis (swans are an exception). Birds achieve internal fertilization simply by the male pressing his reproductive opening against the female's reproductive opening and releasing sperm.

Many reptiles are **oviparous**, meaning their young hatch from eggs laid outside the mother's body. Other reptiles are **ovoviviparous**, meaning their young are born live from eggs that hatch within the mother's body. Reptile eggs contain a considerable amount of yolk, which provides a rich food supply for the developing embryo. The shells of most reptile eggs are leathery, but in birds the shells are hard. Because adult birds sit on the eggs to maintain the temperature needed for the embryo to develop, the shells must be strong enough to resist cracking. Both reptiles and birds produce amniotic eggs.

Young that hatch from reptile eggs are typically fully formed at birth and able to fend for themselves. Bird hatchlings, on the other hand, are not able to survive unaided, since their development is still incomplete. While most young reptiles are ignored by their parents (crocodiles and alligators are exceptions), young birds are fed and nurtured by their parents, growing to maturity only gradually. ◻

Mammals Nourish Their Young

Mammals are unique among the vertebrates in nourishing their young after birth with a nutrient-rich fluid called **milk**, which is also a rich energy source. There are three types of reproduction among mammals, as described in Figure 31-24. The first mammals to evolve, monotremes, had a reproductive strategy similar to the reptiles from which they evolved—they were oviparous. No other mammals lay eggs. All other members of this class are **viviparous**, which describes the situation in which the young are born live from egg cells that develop within the mother's body and are nourished from nutrients that pass from the mother to the embryo.

Just as among birds, the young of viviparous mammals are nourished and protected by their parents. Mammals

> ◻ **CAPSULE SUMMARY**
>
> *Fertilization is internal in birds and reptiles. Most reptiles and all birds are oviparous, meaning they lay shelled eggs*

Figure 31-24 The duckbill platypus, *below left,* lays shelled eggs. When the young hatch, they are nursed by the mother. Marsupials, such as the kangaroo, *below center,* give birth to immature live young that continue development within the mother's pouch. Placental mammals, such as the house cat, *below right,* nourish their developing offspring through the placenta, then give birth to live young that receive milk from the mother.

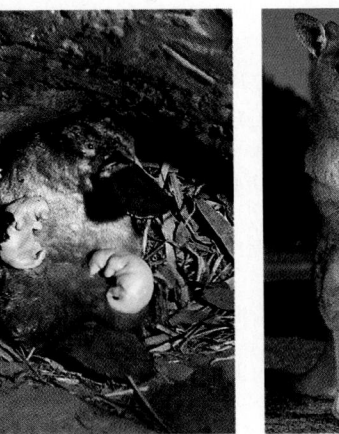  

Teaching Tips
Ovovivi-What?
Point out that *ovo* or *ovi* means "egg," *vivi* means "alive," and *parous* means "to produce." Check for student understanding by describing the reproduction of different vertebrates and asking students to apply the correct term. Examples could include the animals pictured at the bottom of this page, as well as the following: a boa constrictor that retains its eggs to give birth to live young (*ovoviviparous*); a dog that gives birth to live puppies (*viviparous*); an alligator that lays eggs in a nest (*oviparous*).

Milk Is More Than Food
Mammalian milk is a rich source of food, but it contains other substances that affect the life of young mammals. Milk can contain substances important to the immunity of young mammals. It can also contain infectious organisms and pollutants to which the mother has been exposed. Bring empty cartons of milk to class, and have students read the carton labels that list some nutrients in the milk. Finish by asking what kinds of substances sometimes get into milk, and how these substances might affect young mammals. *(Answers might include pollutants from the air, water, or food that the mother ingests or substances such as prescription drugs, alcohol, hormones, or illegal drugs.)*

Multicultural Perspective
Cultural Uses of Mammals
Apart from their role as pets and food, some mammals have played special roles in cultural celebrations for thousands of years. For example, Jewish people blow a ram's horn known as a *shofar* during Rosh Hashanah and Yom Kippur. The shofar is a reminder of the ancient means of communication used at Mount Sinai. On Rosh Hashanah, the shofar is blown 100 times in the synagogue.

Teaching Tip
The Ancient Times

Teaching Tip
The Ancient Times

Have students pick any section title or subheading in Chapter 31 and use it as the title for an article in the *Ancient Times* newspaper. Students should write the article as if the information were appearing in a modern commercial newspaper. Assemble the articles together so that they appear to be a full-spread newspaper, including advertisements for extinct vertebrates and illustrations similar to those in the chapter.

Figure 31-25

Use this figure to show students what an amazing organ the placenta is. Have students pay attention to the intimate association of maternal and fetal blood vessels. What substances must diffuse from the mother's blood to the fetal blood? *(water, oxygen, simple sugars, amino acids)* What substances must diffuse from the fetal blood to the maternal blood? *(wastes such as carbon dioxide)*

Chapter Closure

Have students construct a concept map to summarize the major evolutionary adaptations and systems of fishes, amphibians, reptiles, birds, and mammals discussed in this chapter. Students should try to incorporate many of the boldface vocabulary words in the chapter as well as any additional concepts as needed to complete the map.

other than monotremes show two patterns of development and parental care.

1. Marsupials give birth to live embryos at a very early stage of development. Born just days or weeks after fertilization, newborn marsupials are no bigger than the tip of your finger. The tiny animals crawl to and enter a pouch on the mother's body, where they attach to a nipple and continue their development for many months. They eventually emerge when they are able to function on their own.

2. Placental mammals retain their young for a much longer time within the body of the mother. To nourish the developing fetus, placental mammals have a placenta, through which nutrients are channeled to the embryo from the blood of the mother, as described in Figure 31-25.

CONTENT LINK

*More information about the structure and function of the placenta is found in **Chapter 42**.*

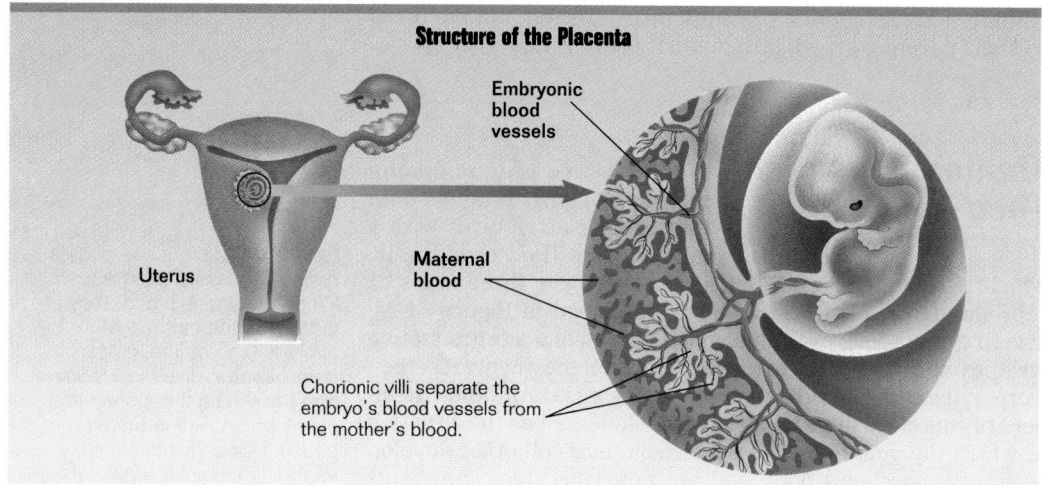

Structure of the Placenta

Embryonic blood vessels

Uterus

Maternal blood

Chorionic villi separate the embryo's blood vessels from the mother's blood.

Figure 31-25 The placenta is a composite structure containing tissue from both mother and embryo. Chorionic villi, outgrowths from the chorion, penetrate into the wall of the uterus and absorb nutrients from pools of maternal blood that form around them. As a result of this structure, the blood of the developing embryo does not mix with that of the mother.

Section Review

1. List the advantages of internal fertilization over external fertilization.
2. How does viviparity increase the likelihood that offspring will survive?
3. How might the amount of parental care provided be related to the number of eggs laid or offspring produced? Explain your answer.

Critical Thinking

4. Describe two disadvantages to the parents of internal fertilization.

Answers to Section Review

1. Internal fertilization does not require a body of water, and it increases the odds of survival for the sperm and developing young.

2. When a female retains her young in her body, she provides an ideal physical environment for their development. Also, the female body is safer than the external environment.

3. The greater the parental care, the fewer the eggs laid or offspring produced. Fish and amphibians, which produce thousands of eggs, deposit them and leave the young to fend for themselves. Birds and mammals, which have few offspring, invest a great deal of time incubating young and raising them.

4. Answers might include being infected with disease during mating, the possibility of increased risk from predators while the mother is carrying the developing young or eggs, and danger to the mother while laying eggs or giving birth.

Bony Fishes

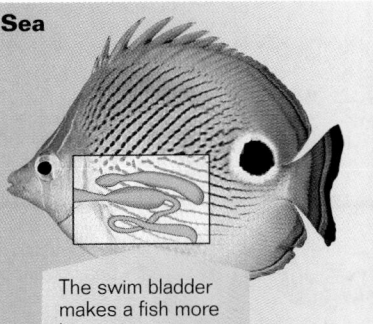

Control of Buoyancy

Bony fishes, such as this lobe-finned coelacanth, evolved a **swim bladder** that gave them active control of buoyancy in the water. This control allowed them to float at any level in the water without swimming, which expends energy. Bony fishes use less energy to maintain their positions than sharks, which sink when they stop swimming. Lobe-finned fishes evolved **lungs**, which let them extract oxygen from air and enabled them to remain out of the water for long periods. Lobe-finned fishes are the direct ancestors of amphibians.

Latimeria
Length: 1.8 m (6 ft.)
Present day

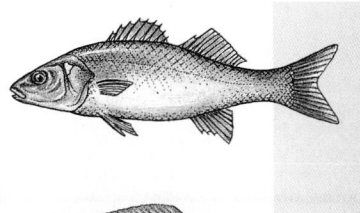

Conquering the Sea

The "slimy" surface of a fish reduces water friction by more than 66 percent.

In almost all bony fishes, fertilization of eggs by sperm occurs outside the female's body.

The swim bladder makes a fish more buoyant.

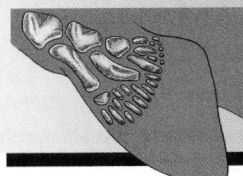

The Road to Land

Bony fishes have paired pectoral (shoulder) and pelvic (hip) fins. In most species, fins are fan shaped and supported by thin bony rays.

Pectoral fin — Pelvic fin

Lobe-finned fishes have fleshy fins supported by central bones, from which the limbs of amphibians are thought to have evolved.

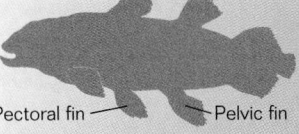

Bony fishes are the most numerous vertebrates. There are more than 18,000 species.

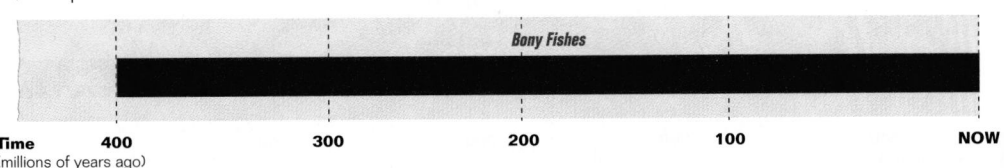

Bony Fishes

Time	**400**	**300**	**200**	**100**	**NOW**
(millions of years ago)					

Instructional Strategy

Provide several reference materials for students to use in researching the major groups of bony fishes. Point out that bony fishes fall into two main groups: ray-finned fishes and lobe-finned fishes. The fins of ray-finned bony fishes have thin rays and are fan shaped, while the fins of lobe-finned fishes have central bones and are fleshy. Encourage students to find examples of ray-finned and lobe-finned fishes. Ask students which group of bony fishes amphibians most likely evolved from. (*Amphibians are thought to have evolved from lobe-fined fishes because these fishes have sturdy fins and mitochondrial DNA that closely matches that of amphibians.*)

Discussion

Guide the discussion by posing the following questions.

1. What are some features that make a bony fish successful in the water? (*Most bony fishes have a swim bladder that allows them to conserve energy and float at any level without swimming. They also have a slimy outer surface that reduces water friction.*)

2. What is one feature that makes it possible for some lobe-finned fishes to survive out of water for long periods of time? (*Some lobe-finned fishes have functional lungs that allow them to breath air.*)

3. How are competition swimmers who shave their body hair and smear their body with oil similar to bony fishes? (*These swimmers have a smooth outer surface that reduces water friction just as the slimy outer surface of fishes does.*)

4. Submarines can dive by allowing in water, and they can resurface by expelling water. How are submarines that pump water in and out to regulate their depth similar to bony fishes? (*Submarines can regulate their depth, as bony fishes do, by controlling their buoyancy. The pumping in and out of water of a submarine mimics a bony fish's swim bladder.*)

Instructional Strategy

Guide students to appreciate how important environmental moisture is to the life of amphibians by comparing the amphibians on this page with the reptiles, mammals, and birds on the following pages. Encourage students to consider the importance of moisture in reproduction, breathing, and habitat for each type of animal.

Discussion

Guide the discussion by posing the following questions.

1. How does the skin of modern amphibians differ from that of early amphibians? *(Early amphibians had a dry skin that prevented water loss. Modern amphibians have a moist, scaleless skin that aids gas exchange.)*

2. What kind of eggs do amphibians lay? How do the eggs stay moist? *(Amphibians lay eggs that are not watertight and that do not have hard shells. The eggs stay moist because they are laid in moist habitats.)*

3. How might amphibians living in a temperate climate survive the winter? *(During winter, most amphibians hiberate in soft mud at the bottom of ponds.)*

Highlights
Vertebrate Evolution

Early Amphibians

Moving on Land

Ichthyostega and other early amphibians were the first vertebrates to live largely on land, beginning about 370 million years ago. **Pectoral and pelvic legs** supported their body weight on land. A **pulmonary vein** sent oxygenated blood from the lungs back to the heart for repumping (see Figure 31-16). Early amphibians had a dry skin to avoid water loss (the moist skins of today's amphibians evolved much later). Many early amphibians had body armor and grew to be quite large. Large amphibians were the dominant land animals for 70 million years, until they were slowly replaced by reptiles. *Ichthyostega* became extinct 360 million years ago. Most other large amphibians had died out by 160 million years ago.

Ichthyostega
Length: 1 m (39 in.)
Late Devonian period

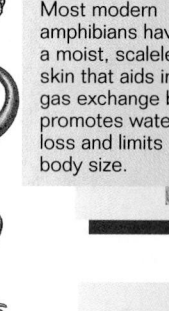

Tied to Water

Most modern amphibians have a moist, scaleless skin that aids in gas exchange but promotes water loss and limits body size.

The majority of amphibians lay their eggs in the environment. The eggs are not watertight, so amphibians must reproduce in water or moist habitats. Most amphibians have an aquatic larval phase.

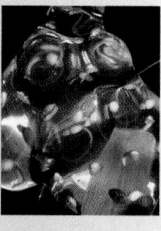

And Adjusting to Life on Land

The amphibian heart is partly divided, and mixing of oxygenated and deoxygenated blood occurs within it.

Amphibians are ectotherms (their body temperature is determined by their surroundings). During winter, most hibernate in soft mud at

Over 70 extinct families of ancient amphibians are known, compared with 37 families of today's amphibians, which fall into three groups: frogs and toads, salamanders, and caecilians.

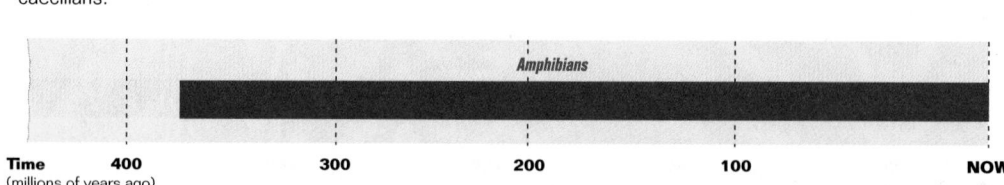

Amphibians

Time	400		300		200		100		NOW

(millions of years ago)

Early Reptiles

Becoming a Better Predator

Early reptiles, such as this pelycosaur called *Dimetrodon,* had powerful jaws. With large teeth and powerful jaws, pelycosaurs could devour much larger prey than their ancestors. Dominant for 50 million years, they were better adapted to life on dry land than amphibians, partly because they had **watertight eggs**. Pelycosaurs once composed 70 percent of all land vertebrates. They died out about 250 million years ago, replaced by their direct descendants, the therapsids.

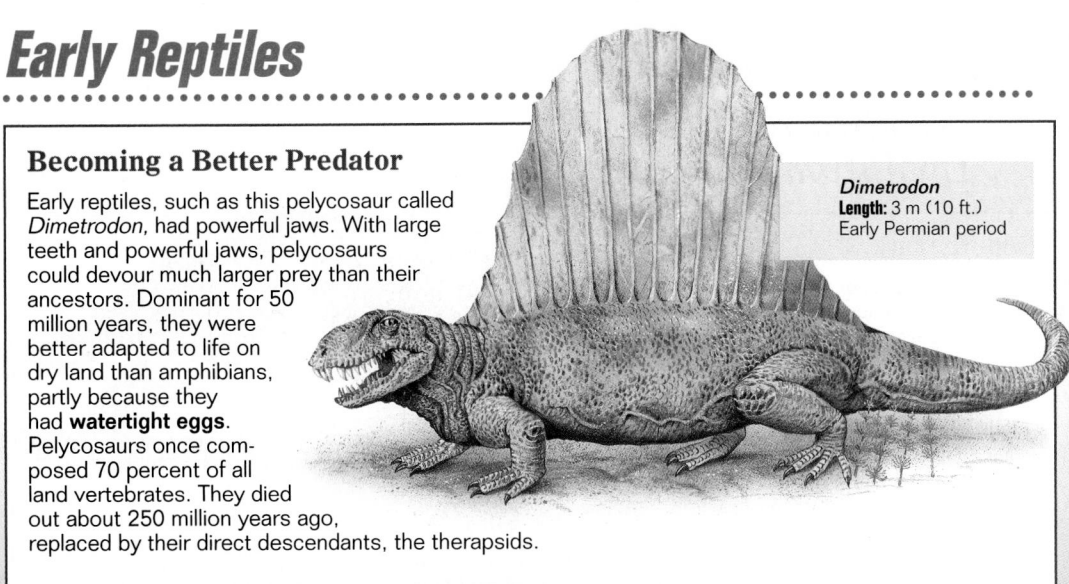

Dimetrodon
Length: 3 m (10 ft.)
Early Permian period

Pelycosaur skulls have a single, large opening behind each eye socket, allowing long jaw muscles to attach to the rear of the skull. This kind of skull is called a synapsid skull. The lower pelycosaur jaw is composed of several bones.

Eating

With long, sharp, "steak knife" teeth, *Dimetrodon* was the first land vertebrate able to kill animals its own size.

Pelycosaurs appear to have been the first land vertebrates with teeth of different sizes and shapes.

Standing

Pelycosaurs stood in a sprawling stance: legs bent, knees and elbows stuck out, and feet flat on the ground, pointed outward.

The spectacular "sail" on the back of many pelycosaurs earned them the nickname "fin backs." The framework of the sail is made of spines up to 1 m (39 in.) long that are extensions of the back vertebrae. The sail may have helped control body temperature.

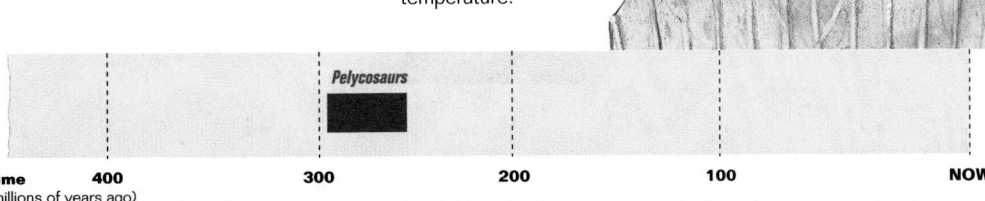

Pelycosaurs

Time   400   300   200   100   NOW
(millions of years ago)

Instructional Strategy

Therapsids are considered to be the reptiles from which mammals evolved. Ask students to make a list of reptilian and mammalian characteristics of therapsids. (*Like pelycosaurs, therapsids had a synapsid skull and powerful jaws. Like mammals, therapsids were endothermic, ran quickly on four legs because of their more upright stance, and had specialized teeth that were replaced only once or twice in a lifetime.*)

Discussion

Guide the discussion by posing the following questions.

1. What were some advantages and disadvantages of endothermy for therapsids? (*Therapsids could be active in cold weather, but they also required food 10 times more frequently than ectotherms.*)

2. What adaptations did therapsids have that improved the way they ate? (*The long tooth life of their teeth created an effective chewing surface. A secondary palate allowed therapsids to chew and breathe simultaneously.*)

3. Why might therapsids have moved more quickly than the pelycosaurs did? (*Therapsids walked on all fours with a more upright stance than that of pelycosaurs.*)

Highlights
Vertebrate Evolution

Therapsids

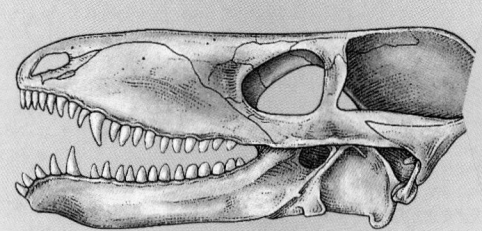

Endothermy

Like their pelycosaur ancestors, therapsids such as *Cynognathus* were synapsids with powerful jaws. But therapsids ate 10 times more frequently. This frequent eating provided the fuel to produce body heat. Therapsids probably were **endotherms,** meaning they maintained a constant, high body temperature. Endotherms could be far more active during prolonged cold periods than other vertebrates of that time. For 20 million years therapsids were the dominant land vertebrates, until largely replaced 230 million years ago by ectothermic thecodonts, ancestors of dinosaurs. The last therapsid became extinct 170 million years ago.

Cynognathus
Length: 1 m (39 in.)
Early Triassic period

Eating

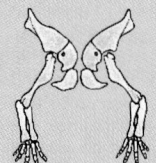

The lower jaw of a therapsid is made of fewer bones than a pelycosaur's jaw. One dominant bone bears the teeth, and another occupies the rear of the jaw, forming the joint with the skull.

Therapsid teeth were replaced only once or twice in a lifetime, unlike reptile teeth, which are lost and replaced periodically. This long tooth life allows opposing teeth to mesh, creating a very effective chewing surface. Therapsids also had a secondary palate, a sheet of bone that separates the nasal passages from the mouth, making simultaneous breathing and chewing possible.

Improved Stance

Therapsids walked on all fours, but adopted a more upright stance than pelycosaurs, with their legs forming a "vee" ($\wedge$) rather than a horizontal "chair" ($\sqcap$).

The rib cage is greatly reduced in the lower body, which is thought to reflect the closing off of the front of the body cavity (which houses the lungs and heart) by a muscular sheet of tissue, the diaphragm. The diaphragm helps ventilate the lungs.

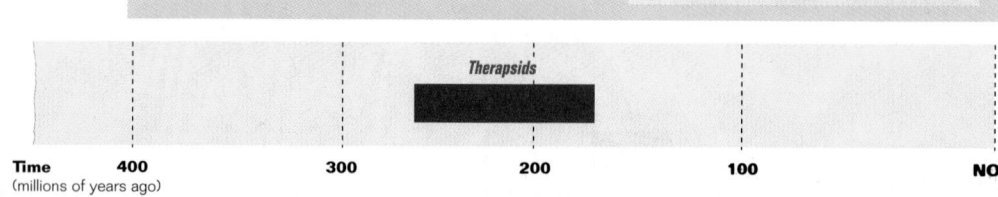

Therapsids

Time (millions of years ago)	400	300	200	100	NOW

Mammals

Fur for Insulation

Mammals, such as *Megazostrodon*, evolved from therapsids at the same time that the dinosaurs first appeared. Throughout the 150-million-year reign of the dinosaurs, all mammals were small, the largest no bigger than a cat. Like their therapsid ancestors, mammals are endotherms and devote 90 percent of their food energy to making heat. Mammals are covered with insulating **fur** to retain body heat. Females produce **milk** to feed their young. When the dinosaurs disappeared and world climates turned progressively colder, mammals became the world's dominant large land animals, as they still are.

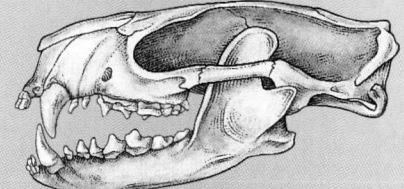

Megazostrodon
Length: 12 cm (5 in.)
Late Triassic period to early Jurassic period

Eating

Mammals have a characteristic arrangement of four kinds of teeth: chiseling incisors in front, stabbing canines at the corners, and cutting or grinding premolars and molars on the sides.

The mammalian jaw is able to move sideways as it closes, permitting very effective grinding of food by molar teeth.

The mammalian jaw is a single bone; the other bones at the rear of the therapsid jaw evolved to become the bones of the mammalian middle ear.

Past Their Prime?

Mammals reached their maximal diversity 15 million years ago. Many large mammals flourished 2 million year ago, but most disappeared as humans became common over the last 12,000 years.

Most mammals living today are small, no larger than during the long reign of the dinosaurs. Seventy-eight percent of mammal species are rodents, bats, or insectivores, most of which are small enough to hold in your hand.

Mammals

| Time | 400 | | 300 | | 200 | | 100 | | NOW |
(millions of years ago)

Instructional Strategy

As endotherms, mammals use about 90 percent of their food energy to make heat. Encourage students to consider the relationships between a high metabolic rate and the need for a constant food supply, efficient respiratory and circulatory systems, and a complex brain. Students may also consider types of insulation and behavioral traits that conserve energy in mammals.

Discussion

Guide the discussion by posing the following questions.

1. Which are the most successful mammals according to number of species? (*Small animals, such as rodents, bats, or insectivores, make up 78 percent of mammal species.*)

2. How could the success of humans have driven large mammals to extinction? (*Answers will vary. Students may suggest that as humans have grown in population they have required more environmental resources.*)

Instructional Strategy

Dinosaurs have been portrayed as large, stupid, sluggish and doomed to extinction. Have students evaluate popular images of dinosaurs. Lead students to conclude that dinosaurs were the most successful of all land vertebrates. Remind students that while dinosaurs and mammals coexisted, dinosaurs were able to diversify and dominate the Earth, while mammals remained small and mainly nocturnal. It was not until the extinction of dinosaurs that mammals began to flourish.

Discussion

Guide the discussion by posing the following questions.

1. What effect did having legs positioned directly under the body have on dinosaurs? (*Dinosaurs could grow to larger sizes and could run faster with greater agility.*)

2. What is the difference between a lizard-hipped and a bird-hipped dinosaur? (*In lizard-hipped dinosaurs, one of the bones of the pelvis points forward. In bird-hipped dinosaurs, the pubis is directed backward.*)

3. How do scientists think that dinosaurs became extinct? (*Most scientists agree that dinosaurs could not survive the cold that resulted from debris in the atmosphere caused by the impact of a meteorite or comet.*)

Highlights
Vertebrate Evolution

Dinosaurs
....

An Upright Stance

Dinosaurs are reptiles. They evolved from the crocodile-like reptiles known as **thecodonts** and had a significant improvement in body design—**legs** positioned **directly underneath** the body. This placed the weight of the body directly over the legs, which could bear far more weight, and allowed some dinosaurs to run with great speed and agility. Scientists now think many dinosaurs were endotherms, maintaining a high, constant body temperature in all environments. Dinosaurs were the most successful of all land vertebrates—the dominant large animals on Earth for 150 million years. They disappeared abruptly 65 million years ago. Most scientists now agree that dinosaurs became extinct as the result of the impact of a meteorite or comet. Perhaps the endothermic but uninsulated dinosaurs could not survive the intense cold produced by debris in the atmosphere.

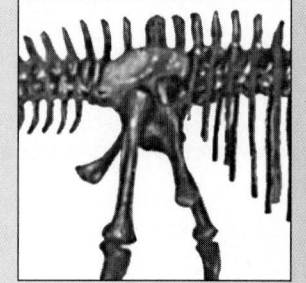

Apatosaurus
Length: 20 m (66 ft.)
Mass: 30,000 kg (33 tons)
Late Jurassic Period

Standing Upright

The key trait that distinguishes a dinosaur from a thecodont is the presence of a hole in the side of the hip socket. Because the dinosaur leg is positioned underneath the socket, force is directed upward, not inward, so there was no need for bone on the side of the socket.

Dinosaurs can be divided into two groups based on pelvic structure. In the lizard-hipped dinosaurs, one of the bones of the pelvis, the pubis, points forward. Lizard-hipped dinosaurs include *Apatosaurus*, *Deinonychus*, and the famous predator *Tyrannosaurus rex.*

Deinonychus
Length: 3-4 m (10-13 ft.)
Height: 1.8 m (6 ft.)
Early Cretaceous period

Pubis

A lizard-hipped dinosaur

				Dinosaurs			

Time   400   300   200   100   NOW
(millions of years ago)

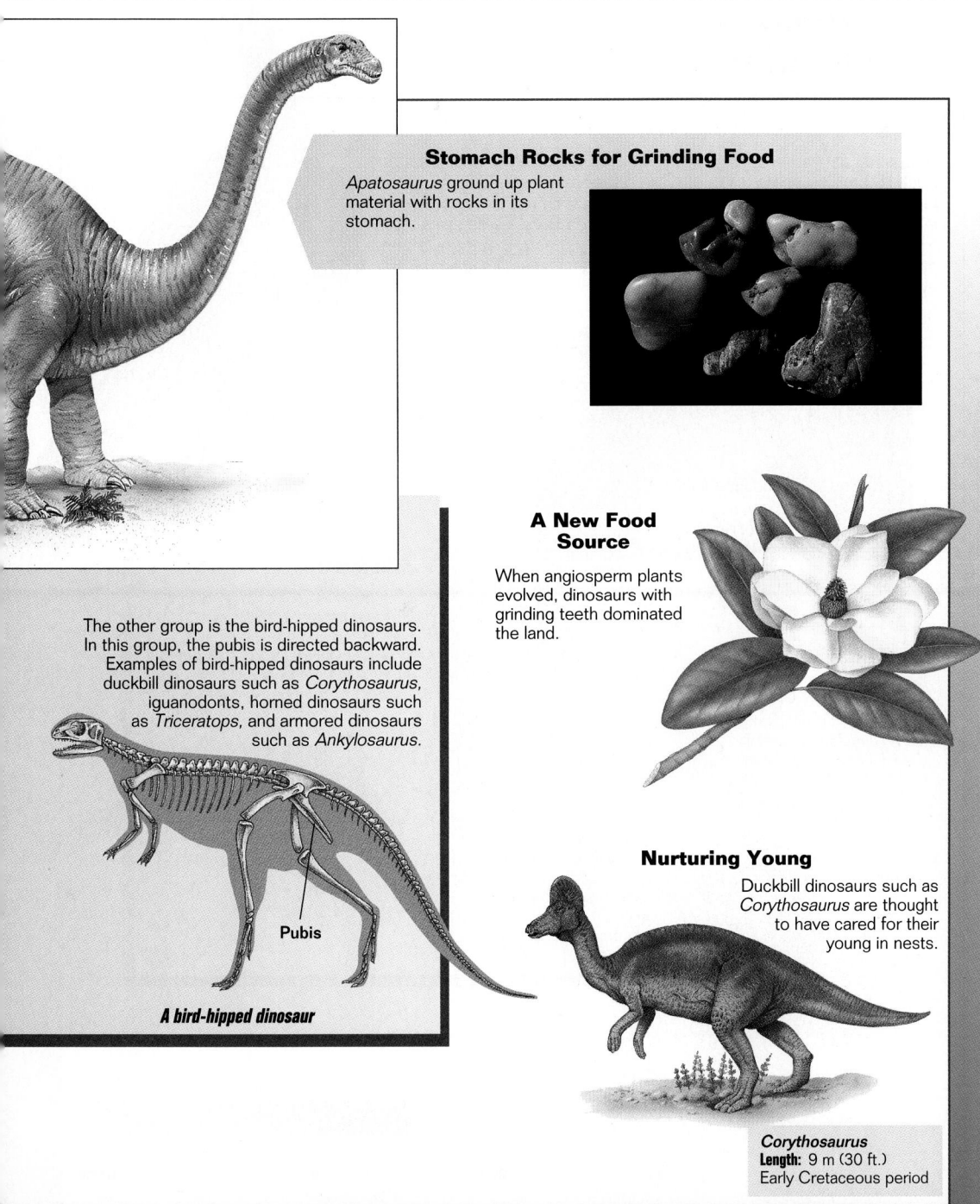

Stomach Rocks for Grinding Food

Apatosaurus ground up plant material with rocks in its stomach.

A New Food Source

When angiosperm plants evolved, dinosaurs with grinding teeth dominated the land.

The other group is the bird-hipped dinosaurs. In this group, the pubis is directed backward. Examples of bird-hipped dinosaurs include duckbill dinosaurs such as *Corythosaurus*, iguanodonts, horned dinosaurs such as *Triceratops*, and armored dinosaurs such as *Ankylosaurus*.

Pubis

A bird-hipped dinosaur

Nurturing Young

Duckbill dinosaurs such as *Corythosaurus* are thought to have cared for their young in nests.

Corythosaurus
Length: 9 m (30 ft.)
Early Cretaceous period

Instructional Strategy

Write the word *flying* on the chalkboard, and ask students to consider the impact of flight on the life and body of birds. Encourage students to consider anatomical adaptations, behavioral adaptations, and physiological adaptations to life in the air.

Discussion

Guide the discussion by posing the following questions.

1. Special adaptations found in flying birds give them more power and less weight. List examples of special features that contribute to these two ends. (*Special features found in flying birds include feathers, hollow and light bones, a well-developed collarbone, very efficient lungs, and one-way air flow through the lungs.*)

2. Why do some scientists consider birds to be feathered dinosaurs? (*Birds are structurally similar to dinosaurs in most respects except in the presence of insulating feathers.*)

3. What are two functions of feathers? (*Feathers are used for flight and insulation.*)

4. What is the purpose of a wishbone in birds? (*It acts as a torque-absorbing strut, which is essential to flight.*)

Highlights   Birds
Vertebrate Evolution

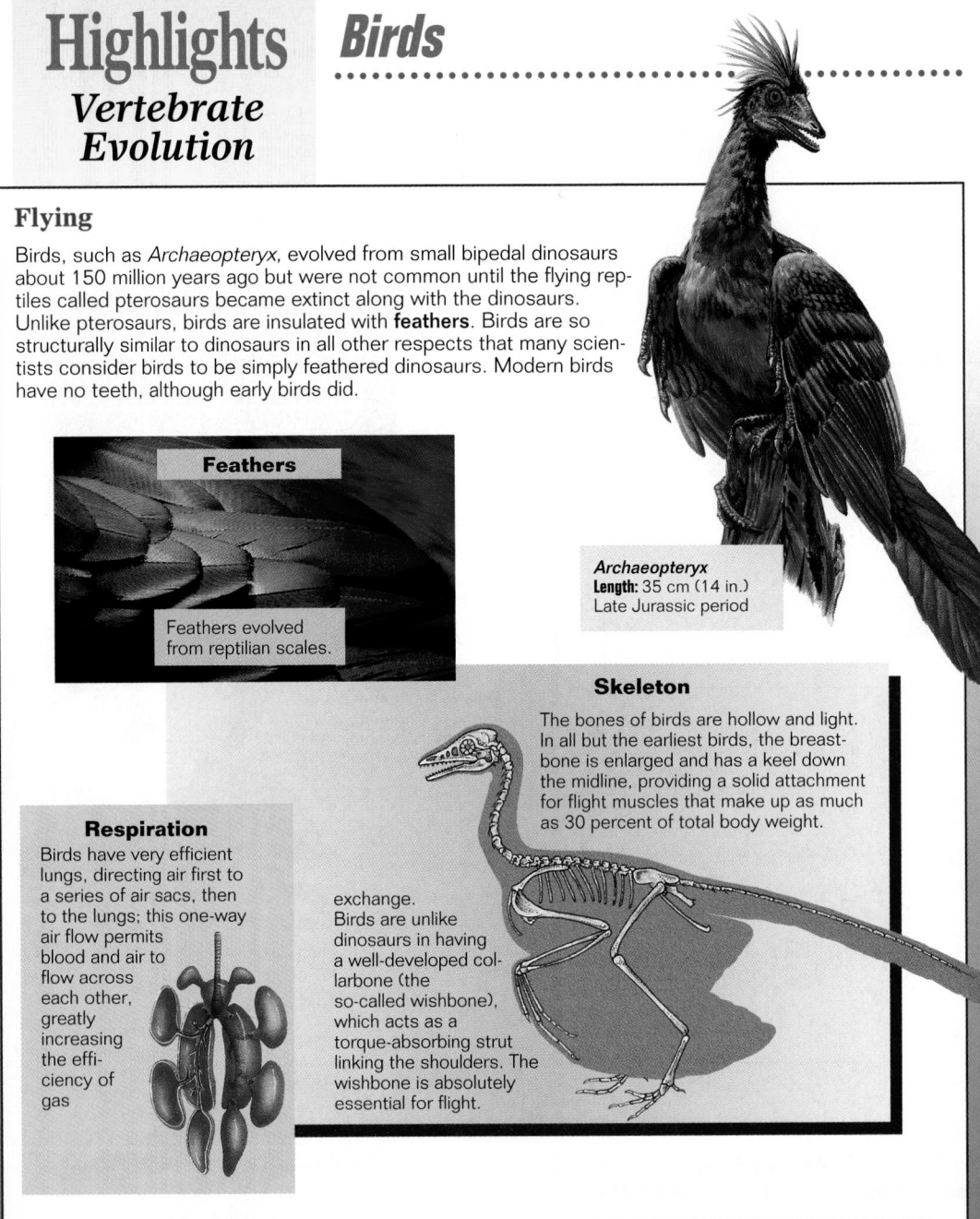

Flying

Birds, such as *Archaeopteryx*, evolved from small bipedal dinosaurs about 150 million years ago but were not common until the flying reptiles called pterosaurs became extinct along with the dinosaurs. Unlike pterosaurs, birds are insulated with **feathers**. Birds are so structurally similar to dinosaurs in all other respects that many scientists consider birds to be simply feathered dinosaurs. Modern birds have no teeth, although early birds did.

Feathers

Feathers evolved from reptilian scales.

Archaeopteryx
Length: 35 cm (14 in.)
Late Jurassic period

Skeleton

The bones of birds are hollow and light. In all but the earliest birds, the breastbone is enlarged and has a keel down the midline, providing a solid attachment for flight muscles that make up as much as 30 percent of total body weight.

Respiration

Birds have very efficient lungs, directing air first to a series of air sacs, then to the lungs; this one-way air flow permits blood and air to flow across each other, greatly increasing the efficiency of gas exchange.

Birds are unlike dinosaurs in having a well-developed collarbone (the so-called wishbone), which acts as a torque-absorbing strut linking the shoulders. The wishbone is absolutely essential for flight.

Birds

| Time | 400 | | 300 | | 200 | | 100 | | NOW |

(millions of years ago)

31
CHAPTER REVIEW

Vocabulary

alveolus (714)
amnion (725)
amniotic egg (726)
atrium (718)
cartilage (706)
chorion (726)
conus arteriosus (718)
ectothermic (707)
endothermic (707)
external fertilization (727)
gill filament (711)

internal fertilization (728)
kidney (723)
lung (712)
metamorphosis (728)
milk (729)
nephron (723)
oviparous (729)
ovoviviparous (729)
penis (728)
placenta (710)
pulmonary vein (719)

semen (728)
septum (720)
sinus venosus (718)
swim bladder (706)
tadpole (728)
terrestrial (704)
urine (723)
ventricle (718)
viviparous (729)

CHAPTER REVIEW ANSWERS

Each item in the Chapter Review is correlated by text section in the assignment guide that follows.

ASSIGNMENT GUIDE

Section	Review	Themes Review	Critical Thinking
31-1	1, 2, 8, 9, 17	28, 30	
31-2	3, 11, 12, 19		32
31-3	4, 10, 16, 18		
31-4	5, 6, 13, 14, 20	29	31
31-5	7, 9, 15		

Concept Mapping

Construct a concept map that shows comparisons of the circulatory and respiratory systems of the various groups of vertebrates. Use as many terms as needed from the vocabulary list. Try to include the following items in your map: pulmonary vein, lungs, gills, ventricle, atrium, air sacs, and alveoli. Use additional concepts in your map as needed.

Review

Multiple Choice

1. Which sequence reflects the order in which the major groups of vertebrates are thought to have evolved?
   a. bony fishes → reptiles → amphibians
   b. bony fishes → amphibians → reptiles
   c. amphibians → mammals → reptiles
   d. birds → mammals → reptiles

2. Mammals are thought to have survived the climatic changes that caused the extinction of dinosaurs because the bodies of mammals
   a. had little insulation.
   b. are ectothermic.
   c. maintain a constant temperature.
   d. had insulation.

3. Fish are much more efficient in getting oxygen from the water than oysters because
   a. fish gills are in an internal cavity.
   b. oxygen diffuses directly into the blood of fish.
   c. water flows in one direction over fish gills.
   d. gill slits permit fish to swallow a lot of water.

4. The heart of an amphibian
   a. has a fully divided atrium.
   b. pumps blood to the lungs through the left atrium.
   c. completely separates oxygenated and deoxygenated blood.
   d. receives oxygen-poor blood from the lungs.

5. Compared with reptilian kidneys, mammalian kidneys are more effective water conservation organs because they
   a. are paired.
   b. remove more water from urine.
   c. produce a very dilute urine solution.
   d. allow urine to diffuse into the kidney.

6. Which is *not* an adaptation of reptiles for life on land?
   a. watertight skin
   b. external fertilization
   c. amniotic egg
   d. kidneys

Concept Mapping

Map answer is shown on page 701B.

Review

Multiple Choice

Code in parentheses indicates section number and objective number.

1. b (31-1.1)
2. d (31-1.4)
3. c (31-2.1)
4. a (31-3.2)
5. b (31-4.2)
6. b (31-4.2)
7. c (31-5.3)

Completion

8. jaws, streamlined (31-1.2)
9. moist, external (31-1.3, 31-5.1)
10. pulmonary, oxygenated (31-3.2, 31-3.3)
11. lungs, increasing (31-2.2, 31-2.3)
12. alveoli, oxygen (31-2.3)
13. loss, water (31-4.2)
14. amniotic, losing water (31-4.3)
15. viviparity (31-5.2)

CHAPTER REVIEW ANSWERS *continued*

Short Answers

16. In fish, the sinus venosus is a large collection chamber. In mammals, the sinus venosus is no longer a separate chamber, although some traces of it remain and control heartbeat. (31-3.2)

17. Snakes are better suited for life on land than frogs because they have more efficient lungs and hearts. They also have internal fertilization and watertight eggs. (31-1.4)

18. Amphibians are more sluggish than reptiles because their ventricles allow oxygen-rich and oxygen-poor blood to mix. The reptilian heart has a septum that partially divides the ventricle. Moreover, in reptiles the conus arteriosus is replaced by arteries that direct oxygenated blood toward the body and deoxygenated blood toward the lungs. (31-3.2)

19. A bird would be able to achieve a higher maximal altitude in flight than a bat because airflow in birds is one-way. Thus, the diffusion surfaces of a bird's lungs are exposed only to fully oxygenated air. (31-2.3)

20. Freshwater fishes produce large amounts of dilute urine. Marine fishes excrete nitrogenous wastes by the gills and produce small amounts of concentrated urine. Also, the ion channels in the nephron membranes of marine fishes are reversed in orientation, allowing the nephron to excrete excess salts in the urine. (31-4.1)

Highlights *Review*

21. Lobe-finned fishes have fleshy fins with sturdy supporting bones from which the limbs of amphibians are thought to have evolved.

22. Early amphibians had dry skin, but most modern amphibians have a moist, scaleless skin.

23. Pelycosaurs could devour large prey because they had powerful jaws. Large openings in the skull behind the eye sockets allowed long jaw muscles to attach to the rear of the skull.

7. What makes marsupial reproduction unique among mammals?
   a. marsupials are viviparous
   b. the mother nourishes her young with milk
   c. the young develop in their mother's pouch
   d. the young hatch from eggs

Completion

8. The early fishes that replaced jawless fishes were successful because they had _____ to capture prey and _____ bodies to aid their movement.

9. Characteristics that make amphibians transitional land-dwellers include _____ skin, _____ fertilization, and eggs that develop in water.

10. In amphibians, the _____ veins carry _____ blood from the lungs to the heart.

11. In terrestrial vertebrates, gas exchange between the air and blood occurs in the _____ . The efficiency of this baglike structure can be improved by _____ its internal surface area.

12. Small chambers called _____ are present in the lungs of reptiles and mammals. They are where _____ diffuses into the blood.

13. The kidney is the organ in vertebrates that works to prevent water _____ . The fluid discharged from the kidney, called urine, contains dissolved metabolic wastes and _____ .

14. Reptiles and birds produce _____ eggs. Eggs of this type keep the developing embryo from _____ .

15. An advantage of _____ is that the embryos are protected and nourished within the mother's body while they develop.

Short Answer

16. How does the structure and function of the sinus venosus in fish compare with its structure and function in mammals?

17. In what ways is a snake better suited for life on land than a frog is?

18. In foot races staged between reptiles and amphibians, the reptiles won 9 out of 10 times. Explain the race results in terms of the structure of reptilian and amphibian hearts.

19. By comparing the structure of mammalian and avian lungs, predict whether a bat or a bird would be able to achieve a higher maximal altitude in flight. Explain your prediction.

20. Contrast the ways in which saltwater and freshwater fishes maintain salt and water balance.

Highlights *Review*

21. Explain why lobe-finned fishes, rather than other kinds of bony fishes, are thought to be the ancestors of amphibians.

22. How does the skin of modern amphibians differ from the skin of the first amphibians?

23. How is the synapsid skull related to the feeding habits of pelycosaurs?

24. The diaphragm increases the efficiency of breathing. How would this structure benefit an endothermic therapsid?

25. Describe two similarities between mammals and therapsids. What conclusion about the relationship between these two groups have scientists drawn from these similarities?

26. Scientists have discovered piles of smooth stones within the ribcages of some dinosaurs. What do these stones suggest about the feeding habits of these dinosaurs?

27. In what ways does a bird's skeleton differ from that of other vertebrates? How are these differences related to flying ability?

24. An endotherm requires more oxygen which the diaphragm helps supply.

25. Both have synapsid skulls and diverse teeth. Scientists have concluded that mammals probably evolved from therapsids.

26. The stones suggest that these dinosaurs used the rocks

in their stomachs to grind up plant material.

27. A bird's skeleton is made of hollow, light bones. The breastbone is enlarged and has a keel down the middle to provide attachment for flight muscles. The collarbone is well developed and acts as a torque-absorbing strut.

Themes Review

28. Bottom-dwelling fishes do not need a swim bladder which enables a fish to adjust its position in the water. (31-1.2)

29. Salt water has a higher ion concentration than a fish's tissues, and so they tend to lose water. (31-4.2)

Themes Review

28. Structure and Function Bottom-dwelling fishes often lack a swim bladder. Explain the adaptive advantage of this condition.

29. Homeostasis Saltwater fishes drink more water and produce less urine than freshwater fishes. How do you account for this difference?

30. Evolution In what ways are the adaptations of reptiles to land similar to the adaptations of vascular plants to land?

Critical Thinking

31. Making Inferences When a female leatherback turtle comes onto a beach to lay eggs, she digs a deep hole, lays her eggs, and covers them with sand. Then she crawls about 100 m and digs another hole. This time she lays no eggs, and then covers the hole with sand. Suggest a possible explanation for this behavior.

32. Making Predictions How would having the same lung structure as birds affect the performance of a marathon runner?

33. Making Inferences Fossil evidence collected in the 1980s in Arctic Alaska suggests that some dinosaurs were year-round residents of areas that reached freezing temperatures and were in total darkness during the winter months. Does this evidence of "Arctic dinosaurs" support or contradict the hypothesis that the extinction of dinosaurs was due to a period of intense cold produced by debris in the Earth's atmosphere? Explain your answer.

Activities and Projects

34. Research and Writing Research the Alvarez hypothesis, proposed in 1980 by Walter Alvarez and his father, Luis, to explain what triggered the mass extinction of dinosaurs near the end of the Mesozoic era. What evidence supports the hypothesis?

35. Research and Writing Find out how marine mammals differ structurally from fish. In your report, include structural diagrams showing similarities and differences between these animals.

36. Multicultural Perspective Research the importance of fish in the Japanese culture. Find out about the use of carp in Japan and how it differs from that in the United States.

Readings

37. Read the book *Dinosaurs Rediscovered* by Don Lessem. Describe two recent discoveries that have changed how scientists view dinosaurs.

38. Read the article "Frogs and Toads in Deserts," in *Scientific American*, March 1994, pages 82–88. Why are frogs and toads such unlikely desert inhabitants? Describe two adaptations that enable these animals to survive or reproduce in deserts.

Activities and Projects

34. The Alvarez hypothesis proposes that a large meteorite struck the Earth in northern Yucatan, Mexico, about the time that dinosaurs disappeared. The collision would have produced thick dust clouds that would have blocked out sunlight for several months. The lack of sunlight would have adversely affected the climate and plant life that dinosaurs depended on for survival. There is a layer of iridium, an element rare on Earth but common in meteorites, in rocks laid down at the end of the Cretaceous period.

35. Marine mammals have a vestigial pelvis as evidence of their evolution from four-legged mammals. They also have lungs and hair and are endothermic.

36. Although rice is the principal food in the Japanese diet, fish is the main source of protein. Favorite ways of eating fish are raw, as in sashimi and sushi, or salted and broiled over an open flame. Japan has the world's largest fishing industry and the most fishing vessels of any country. Carp is a very popular food in Japan; however, it does not appeal to people in the United States, who often regard the fish as a nuisance and an unappetizing competitor of more desirable fish.

Readings

37. Answers will vary.

38. Frogs and toads require water to keep their skins moist for breathing. Some amphibians hibernate during dry periods. Others have a waxy coating that limits water loss.

30. To prevent water loss, reptiles developed watertight eggs, and vascular plants developed outer jackets of cells surrounding gametes. Plants and reptiles also have a watertight outer coating. (31-1.4)

Critical Thinking

31. The leatherback turtle protects her eggs by creating a decoy nest. (31-4.3)

32. A marathon runner having the same lung structure as birds would be able to absorb more oxygen from the air and would have an advantage over competitors at higher altitudes. (31-2.3)

33. This evidence contradicts the hypothesis that dinosaurs became extinct 65 million years ago as a result of intense cold. If dinosaurs normally lived in cold climates, then a period of cold could not have caused their extinction. (Highlights)

Preparation

If you use specimens that have been preserved or fixed in formalin, be sure to conduct the lab in a well-ventilated room. If live specimens are used, remind students of the proper care and treatment of live animals. Remind students to thoroughly wash their hands after working with the animals. Release into your local environment only native animals.

Procedural Notes

- This activity may take more than one class period.
- Students can work in groups of two to four students.
- Students can begin this investigation at any available station. They should rotate from one station to another until they have completed the activities of each station. Encourage students to keep specimens at their appropriate stations.
- Have students wash their hands and clean up laboratory materials at the end of each class period.

Background Answers

1. Vertebrates have a backbone that forms a central axis for muscle attachment and protects the dorsal nerve cord.

2. Amphibians reproduce in water or very moist environments.

3. Answers may vary but should include internal fertilization and watertight eggs and skin.

4. Mammals and birds have higher metabolic rates because they maintain a high, constant body temperature.

5. How are vertebrate classes similar and different?

Records

1. (a) Gills move rhythmically, expanding and contracting. (b) Fish move using fins and whole-body movements. They eat by opening their mouths, sucking in food, and moving away quickly. (c) The bodies

Comparing Vertebrate Characteristics

OBJECTIVE

Categorize similarities and differences among vertebrate classes by observing representative organisms.

PROCESS SKILLS

- classifying organisms
- comparing and contrasting structural features
- relating structural features to functions
- organizing observations

MATERIALS

- field guides for fishes, amphibians, reptiles, birds, and mammals
- live or preserved specimens of fishes, amphibians, reptiles, birds, and mammals
- aquarium with fish
- frog or toad
- cricket or other food for the amphibian
- bird feather
- compound light microscope
- chicken or turkey bones
- beef bones or pork bones

BACKGROUND

1. What characteristic distinguishes vertebrates from other phyla?
2. Why do all amphibians have to be near water at some times in their lives?
3. What factors allowed reptiles to live exclusively on land?
4. Why do mammals and birds have higher metabolic rates than other vertebrates?

5. Write a question to explore in your lab report or notebook.

TECHNIQUE

You will travel to five stations to observe organisms that are representative of the most common vertebrate classes. Record your observations and answer the following questions in the **Records** section of your report.

Station 1: Fishes

1. Observe a freshwater tank containing representative bony fishes.
   a. What do you notice about the motion of their gills?
   b. Describe how they behave and eat.
   c. Describe their body coverings.
2. Using the field guide to fishes, look up two very different bony fishes and describe their similarities and differences.

Station 2: Amphibians

3. Observe a frog or toad in an aquarium. Pay particular attention to the skin of the animal.
   a. Describe the animal's behavior.
   b. How does it breathe?
4. Feed it a cricket. Describe how it eats.
5. Using the field guide to amphibians, find frogs, toads, and salamanders and compare their similarities and differences.
   a. Relate the way each moves to the structure of its legs.

Station 3: Reptiles

6. Observe a live or preserved turtle. Describe the design of the shell. What are the advantages of having a shell? What are the limitations?
7. Using the field guide to reptiles, look up snakes. What adaptations enhance the survival of snakes?

of fish are covered with shiny, slimy scales.

2. Students may choose to compare fin structure and shape, habitat, mating habits, or size and shape of the fishes.

3. (a) The animal's behavior will vary. (b) The amphibian breathes through its mouth and skin.

4. The animal eats by quickly catching the cricket with its tongue and putting it into its mouth.

5. Frogs and toads jump; salamanders "waddle" and are longer and thinner than frogs and toads. Frogs and salamanders have moist skin; toads have dry skin. a. The large,

strong back legs of frogs are best suited for jumping. The smaller legs of salamanders are more appropriate for walking or waddling.

6. The turtle's shell has a symmetrical, geometric pattern. The advantage of a shell is its pro-

Station 4: Birds

8. Observe a live or preserved bird specimen. List two qualities birds have in common with reptiles.

9. Run your finger and thumb along a feather in one direction. Now examine the feather under the microscope. Run your finger and thumb along the feather in the opposite direction. Look at the feather under the microscope. How has it changed? Predict what could happen if a bird never groomed itself.

10. Examine the *cleaned, broken* wing and leg bones from a chicken or turkey. Describe the interior of the bones. Look at cleaned beef bones or pork bones. How do they differ from those of the birds?

11. Using the field guide to birds, find five examples of how a bird's feet relate to its feeding habits and habitat.

Station 5: Mammals

12. Observe live or preserved mammal specimens. What two characteristics distinguish mammals from other vertebrates?

13. Using the field guide to mammals, find the only true flying mammal.

14. Wash your hands thoroughly before leaving the laboratory.

INQUIRY

1. Describe the differences between snakes and amphibians.

2. How might the characteristics of the bird bones you examined in step 10 be beneficial to birds?

3. What quality do mammals share with birds?

ANALYSIS

1. Construct a table of vertebrate characteristics using your observations from Stations 1–5. Put the characteristics down the left column and the five common classes of vertebrates along the top of the table. Check off the characteristic in each column if it is common to that vertebrate class.

2. The echidna and crocodile are alike in that they both lay eggs, have similarly shaped pelvises, and have a single opening through which feces, urine, and reproductive products pass. Why is the echidna in a class different from the crocodile?

FURTHER INQUIRY

Write a new question that could be the focus of a new investigation. The following is an example:

A bird's wing is made of feathers. A bat's wing is a thin membrane of skin stretched across elongated fingers. Could a functional wing be made of hair? Why or why not?

Inquiry Answers

1. Amphibians have pectoral and pelvic legs; snakes do not have legs. Snakes have dry skin; amphibians have moist skin.

2. Bird bones are hollow and light. These characteristics provide strength without adding a lot of weight, enabling birds to fly.

3. Both birds and mammals are endotherms. Mammals have insulating fur or hair, and birds have insulating feathers.

Analysis Answers

1. Tables will vary but should include categories for behavior, bone type, body covering, respiration, and endothermy as well as others.

2. Echidnas have milk-secreting mammary glands and hair.

tection; the disadvantage is its inflexibility.

7. Adaptations include the ability to move (rectilinear movement), locate and kill prey (venom, constriction), swallow and digest (unhinged jaw), and defend itself (camouflage).

8. Birds and reptiles have scales and lay eggs.

9. The feather loses its smooth contour when it is rubbed in the opposite direction. Birds have to preen their feathers to keep them contoured and smooth for flight and insulation.

10. Poultry bones are hollow and light; pork and beef bones contain marrow.

11. Common types of bird feet include those best suited for grasping, scratching, swimming, perching, running, or climbing.

12. Mammals have hair and milk-secreting mammary glands.

13. Bats are the only flying mammals.

UNIT 8

FISHES AND AMPHIBIANS

Block Scheduling Guide

Each block represents about 45–50 minutes of instructional time. The resources cited in a block are coded to a specific program component using the Key on the next page.

Lecture Concepts
Lesson Resources

32-1 History and Diversity of Fishes pp. 745–750

BLOCK 1

a. Jawless fishes, the first vertebrates, evolved over 500 million years ago. The only species that survive today are lampreys and hagfishes.

b. Jaws, derived from gill arches, first evolved in the spiny fishes.

c. Bony fishes and sharks evolved 400 million years ago. Bony fishes, the most diverse and abundant group of fishes today, have powerful muscles, fins, and a swim bladder.

Lecture Resources
- Opening Question, page 745
- Demonstrations: Fish Diversity, Neutral Buoyancy
- The Price of Protection, page 746
- Get a Grip, page 746
- Natural Selection of Body Shape, page 748
- Visual Strategy: Figure 32-8
- The Living Fossil, page 750
- Teaching Transparencies: 164

- Holt Biology Videodiscs: Lesson 32-1

Classwork Options
- Effective Reading, page 745
- Time Travel, page 747
- Organizing Information, page 747
- Design a Fish, page 750

Assignment Options
- Section Review, page 750
- Chapter Review, questions 1–4, 8–10, 16, 19

32-2 Key Adaptations of Fishes pp. 751–755

BLOCKS 2 and 3

a. All fish share four characteristics: gills, vertebral column, single-loop blood circulation, and an inability to make specific amino acids.

b. Bony fishes have additional adaptations such as a swim bladder, a lateral line system, and a gill cover called an operculum.

Lecture Resources
- Opening Question, page 751
- Demonstrations: Watch a Real Fish, Swim Bladders and Buoyancy Compensators
- Fish Kills, page 751
- Train Your Fish, page 754
- Different Sense for Different Places, page 754
- Teaching Transparencies: 165, 166, 168, 138A, 139A, 140A
- Holt Biology Videodiscs: Lesson 32-2

Classwork Options
- Up Close: Yellow Perch, pages 752–753
- Advantages of an Operculum, page 755
- Inquiry Skills Development B26: Perch Dissection
- Inquiry Skills Development B27: Schooling Behavior in Fishes

Assignment Options
- Section Review, page 755
- Chapter Review, questions 5, 11, 17, 18

32-3 History and Diversity of Amphibians pp. 756–759

BLOCK 4

a. Amphibians were the dominant land vertebrates for 100 million years. Their numbers began declining when the numbers of reptiles began rising.

b. Reptiles halted the expansion of amphibians into terrestrial habitats, greatly diminishing amphibian diversity and numbers.

Lecture Resources
- Opening Question, page 756
- Demonstration: Learning From Larvae
- Tracing a Family Tree, page 756
- Visual Strategy: Figure 32-15
- First on Land, page 757
- Research Update, page 758
- Adios Amphibians, page 759
- Teaching Transparencies: 167

- Holt Biology Videodiscs: Lesson 32-3

Classwork Options
- The First Amphibian, page 757
- The Rise, Fall, and Rise of Amphibians, page 759
- Less in More, page 759

Assignment Options
- Section Review, page 759
- Chapter Review, questions 6, 12, 13

32-4 Key Adaptations of Amphibians pp. 760–765

BLOCK 5

a. Modern amphibians (frogs, toads, salamanders) have legs, lungs, cutaneous respiration, pulmonary veins, and a partially divided heart. Caecilians are legless, burrowing amphibians.

Lecture Resources
- Opening Question, page 760
- Surface-Area-to-Volume Ratio, page 760

- Overcoming Misconceptions, page 760
- Lifestyles of the Cold and Slimy, page 761
- An Extra Eyelid, page 761

32-4 Key Adaptations of Amphibians pp. 760–765 (continued)

b. Most amphibian species live in moist or aquatic habitats, eat small invertebrates, and lay eggs that are fertilized externally.

- ■ Visual Strategies: Figures 32-20, 32-21
- Teaching Transparencies: 169, 170, 171, 172, 180
- ◉ Holt Biology Videodiscs: Lesson 32-4

Classwork Options
- ■ Up Close: Leopard Frog, pages 762–763
- ■ What's the Difference? page 765
- ■ Closure, page 765
- ▲ Quick Labs A25: Observing a Frog

▲ Quick Labs A34: Culturing Frog Embryos
▲ Inquiry Skills Development B28: Frog Dissection
- ■ Laboratory Investigation: Live Frogs, pages 768–769

Assignment Options
- ■ Section Review, page 765
- ■ Chapter Review, questions 7, 14, 15, 20

Review/Enrichment

BLOCK 6

- ■ Study Guide: Concept Review and Pretest

Assignment Options
- ■ Activities and Projects, questions 21–23

Assessment Options

BLOCK 7

Traditional Assessment
- ■ Chapter 32 Test
- ◉ Test Generator/Assessment Item Listing: Software item bank for preparing customized chapter tests.

Portfolio Assessment
Select student reports for one or more laboratory experiments from this chapter. *The Direct Observation Checklist* on the *BioSources Teaching Resources CD-ROM* should be completed during a laboratory or other cooperative learning experience.

Holt Biology Videodiscs

CONCEPTS OF BIOLOGY

Holt Biology Videodiscs gives you a powerful tool for teaching, review, and assessment. *Concepts of Biology* adds interactive lesson plans and extensive tools for customization using CD-ROM technology.

Use the following topics from *Concepts of Biology* to help you teach this chapter:

- ■ Topic 19: Evolutionary Trends in Vertebrates
- ■ Topic 20: Vertebrate Diversity

For further information regarding the media on the videodiscs, see *Holt Biology Videodiscs Teacher's Correlation Guide to Biology: Principles and Explorations.*

Chapter Theme

Evolution *Changes in their skeletons and in their body systems gave fish an advantage in the water and amphibians an advantage on land.*

Tapping Prior Knowledge

- Name the structure that identifies fishes and amphibians as being vertebrates.
- Describe the function of a swim bladder, and identify the group of fishes in which it first evolved.
- Name two amphibian structural adapations that enabled them to live on land.

Opening Demonstration

Bring several sizes and types of pliers or wrenches to class. Have students pick up different objects and analyze which tools exert the most force and which are best for picking up small objects. Relate the function of the tools to jaw function. Ask students to name the vertebrate group in which jaws first evolved. *(fishes)*

CHAPTER 32 FISHES AND AMPHIBIANS

A green frog and koi (fancy carp)

REVIEW

- vertebrate evolution (Sections 13-3 and 33-1)
- Highlights: Bony Fishes on page 731 and Early Amphibians on page 732
- vertebrate respiration (Section 31-2)
- vertebrate circulation (Section 31-3)
- vertebrate excretion (Section 31-4)
- vertebrate reproduction (Section 31-5)

Authors' Rationale

The fishes and amphibians were the earliest vertebrates, and for that reason they are very significant. Fishes were the first animals with a backbone. Why? The transition of fish groups from cartilaginous fishes to lobe-finned fishes is particularly significant. The significance of fin-to-limb evolution is critical to the understanding of vertebrate evolution. Amphibians are strictly freshwater animals. Why is that evolutionarily significant? The amphibians present us with an extant "missing link" group—crawling about on land, yet still tied to the water.

32-1 History and Diversity of Fishes

What is a fish? All of us have a pretty good idea. A fish is a vertebrate (animal with a backbone) that lives in water, breathes with gills, and swims with fins. More than one-half of all living vertebrates are fishes. The largest and most diverse vertebrate group, fishes are the evolutionary base from which amphibians, the first land vertebrates, arose. In many ways an amphibian can be thought of as a "fish out of water"—a vertebrate able to live on land but still tied to water. For this reason, fishes and amphibians will be considered together in this chapter so that the similarities between them are not lost among the host of obvious differences.

Jawless Fishes Were the First Vertebrates

The story of vertebrate evolution started over 500 million years ago during the Cambrian period, as indicated in Figure 32-1. It was then that the first animals with backbones, the fishes, appeared in the ancient seas. Wriggling through the water, jawless and toothless, these

Section Objectives

- Describe the first vertebrates.
- Describe the evolution of jaws.
- Compare and contrast ostracoderms with placoderms.
- Give examples of two modern agnathans.
- Identify several important characteristics of sharks.
- Describe the evolution of bony fishes.

Figure 32-1 This phylogenetic tree shows the evolutionary relationships among different groups of fishes as well as between fishes and amphibians. Living species are at the top of the tree, and extinct forms are at the bottom.

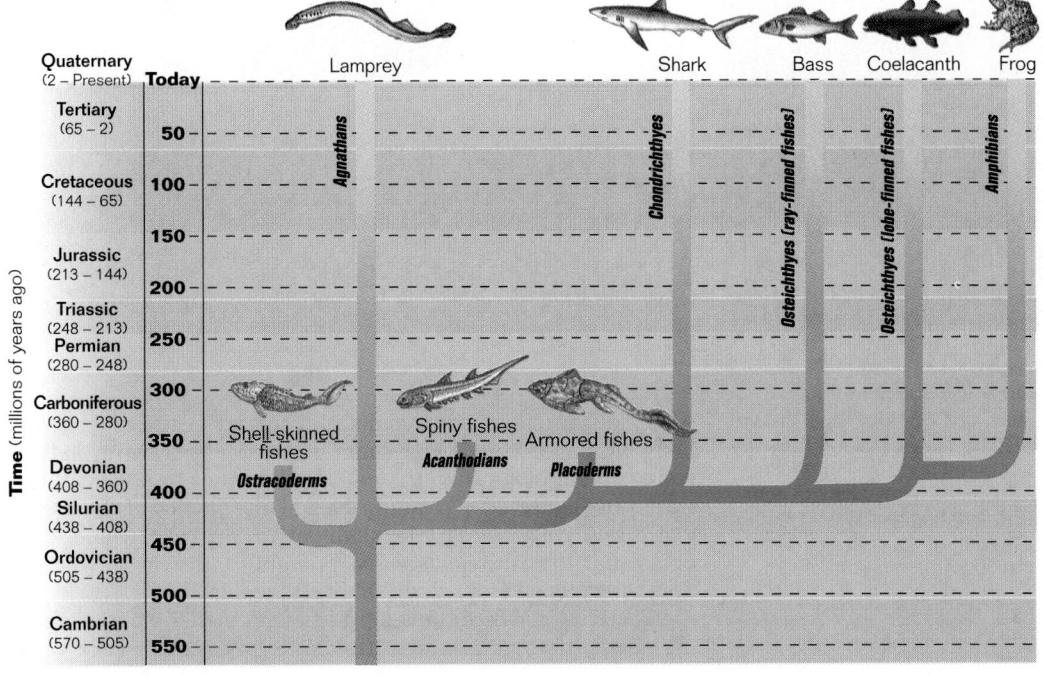

Opening Question

Pose the first question in this section, "What is a fish?" Invite answers from your students, then ask them some questions to help them refine their definition. Ask if a starfish is a fish, then discuss why it is not. Ask if a dolphin is a fish, then discuss why it is not. Ask about some of the more unusual fishes such as sea horses, basking sharks, lampreys, and mudskippers. Finish by discussing characteristics that are specific to particular groups of fish described in this chapter, and then discuss characteristics that are shared by all fish.

Demonstration
Fish Diversity

Provide students with approximately 12 specimens or pictures representing the three classes of fishes. Have them note how the fishes are alike and how they differ. (*Eels and hagfish are both long and slender, but hagfish have no paired fins; skates and flounders are flat, but flounders lie on their sides; sharks and tuna are streamlined, but sharks have multiple gill slits.*) Ask students to group the fishes by identifying common attributes. Prompt students by asking the following questions: Which fishes should be grouped together? What criteria can be used? How are the groups alike and how do they differ? Have students compare their groupings to the three fish classes, Agnatha, Chondrichthyes, and Osteichthyes. Notice how each group feeds and is adapted to living in the water.

Effective Reading

For each paragraph in this chapter, have your students write a question and an answer to the question. Ask students to pose questions on the major points made in each paragraph rather than on small details.

Figure 32-2 A predatory sea scorpion swims above three bottom-dwelling ostracoderms. Only the head shields of ostracoderms were made of bone. Their internal skeleton was constructed of cartilage. Cartilage weighs less than bone, but it is not as strong.

Figure 32-3 Most lamprey species are parasitic on other living fishes, *above.* Hagfishes, *above right,* are scavengers of dead and dying fishes on the ocean bottom. The slimy skins of lampreys and hagfishes have neither the plates nor scales of their ostracoderm ancestors.

were the first members of the vertebrate class Agnatha. The term *agnatha* is from the Greek *a,* meaning "not," and *gnathos,* meaning "jaw." These first fishes breathed with gills but had no fins. They sucked up small food particles from the ocean floor like miniature vacuum cleaners.

For 50 million years, throughout the Ordovician period, simple jawless fishes were the only vertebrates that existed. Few reached more than 35 cm (1 ft.) long. By the end of this period, they had developed primitive fins that helped them swim. They had also developed massive shields of bone that may have protected them from attacks by the giant sea scorpions that were fearsome predators. These armored fishes, called ostracoderms *(AHS truh kah durms),* are shown in Figure 32-2. The term *ostracoderm* is from the Greek *ostrakon,* meaning "shell," and *dermis,* meaning "skin."

At one time there were five ostracoderm orders containing nearly 40 families. Most had become extinct by the close of the Devonian period. The descendants of one of the ancient ostracoderm orders survive today as lampreys and hagfishes. Lampreys and hagfishes, shown in Figure 32-3, are the only remaining agnathans.

The Evolution of Jaws

A key evolutionary advancement occurred in fishes 440 million years ago—the development of jaws. The transformation from cartilaginous supports called gill arches to jaws is illustrated in Figure 32-4. The top half of the jaw in early fishes was directly attached to the skull only at the rear, which indicates that the first jaws were probably not very powerful. Teeth developed from skin that lined the mouth.

The first fishes to develop jaws were called spiny fishes, members of the class Acanthodia. Spiny fishes were very

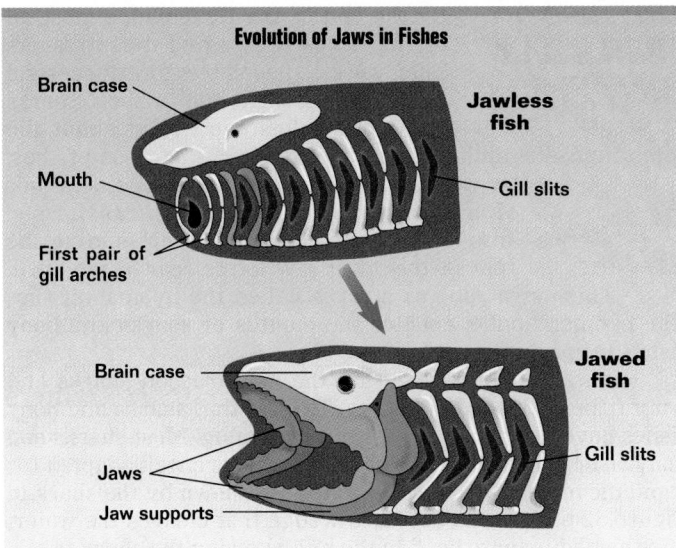

Evolution of Jaws in Fishes

Brain case

Mouth

First pair of gill arches

Jawless fish

Gill slits

Brain case

Jawed fish

Gill slits

Jaws

Jaw supports

Figure 32-4 Jaws evolved from a front pair of a series of structures called gill arches. A gill arch is the cartilage support that reinforces the tissue between two gill slits (openings between each gill) and holds the slits open.

common during the early Devonian period and replaced most of the ostracoderms. By the end of the Devonian, spiny fishes were extinct. The spiny fishes had internal skeletons made of cartilage, although some fossils indicate that their skeletons also contained some bone. Their scales also contained small plates of bone. The presence of bone in the spiny fishes foreshadows the much larger role that bone would play in their descendants.

By the mid-Devonian, spiny fishes were being replaced by fishes with stronger, more efficient jaws—the heavily armored placoderms. A placoderm is shown in Figure 32-5. ❑

❑ **CAPSULE SUMMARY**

The first vertebrates, the jawless fishes, evolved in the sea over 500 million years ago. The only jawless fishes that survive today are lampreys and hagfishes. Jaws, derived from gill arches, first evolved in the spiny fishes.

Figure 32-5 *Dunkleosteus,* a placoderm, grew to more than 3 m (10 ft.) long. The placoderm jaw was fused to the skull, and the skull was hinged on the back, in the "shoulder" area. Only the front of the placoderm body was armored. Most lived near the sea floor because of their great weight.

SECTION 32-1

CONNECTIONS

Chapter 27
Human Slits
You once had slits in your throat that were similar to a fish's gill slits. The formation of pharyngeal slits occurs during human embryonic development. In the early stages of human development, pharyngeal slits develop into parts of the ear and throat rather than into functioning gills.

Teaching Tips
Time Travel

Have your students create a mural of the environment and organisms in a Devonian sea based on the pictures in this chapter or other reference books. Have different groups work on different parts of the mural—the background, the plant life, invertebrates, and different vertebrates that lived in the ancient ocean. Sections of the mural can be drawn, cut out of paper, or traced. Be sure that students label or classify the vertebrates in the mural.

Organizing Information

Have students create a Venn diagram similar to the one shown in the *Graphic Organizer* at the bottom of page 747 (but leave the subheads out). As students complete each section in the chapter, have them list in the appropriate regions of the diagram the characteristics that are exclusive to fish, exclusive to amphibians, and common to both groups. After their diagram is complete, have each student use the information in their diagram to write a paragraph that either supports or rejects the placement of fish and amphibians into two separate vertebrate groups.

Graphic Organizer
Use this graphic organizer in *Teaching Tip:* **Organizing Information** on page 747.

Fish characteristics

exclusive characteristics

characteristics *in common*

Amphibian characteristics

exclusive characteristics

Teaching Tip
Natural Selection of Body Shape
To look for similarities in body shape, have students diagram the profile of several living and nonliving things that move quickly through water. The examples you might give to students are a tiger shark, a tuna, a bottle-nose dolphin, a submarine, a torpedo, and a seal. All these things have the basic shape shown below. Ask students why this same shape occurs in things that need to move rapidly through water? Direct students to think about water resistance and the need to navigate through water using fins.

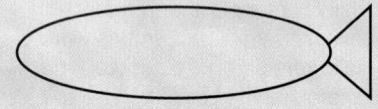

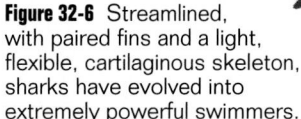

Figure 32-6 Streamlined, with paired fins and a light, flexible, cartilaginous skeleton, sharks have evolved into extremely powerful swimmers.

☐ CAPSULE SUMMARY

A reinforced jaw, a streamlined body, and functional fins have made sharks and bony fishes the dominant fishes of today.

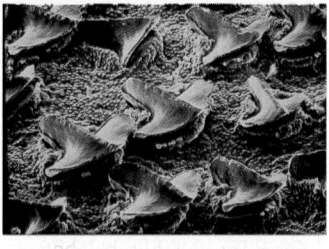

Figure 32-7 The teeth of sharks, *top*, are not set into the jaw as yours are, but rather sit atop it. Therefore, their teeth are easily lost. These shark scales, *bottom*, have been magnified 500 times. The scales give the shark's skin a rough "sandpaper" texture.

The Rise of Sharks and Bony Fishes

By the end of the Devonian, almost all of the early fishes had disappeared. They were replaced by more successful forms—sharks and bony fishes. Sharks and bony fishes evolved at about the same time, 400 million years ago. In sharks and bony fishes, the jaw was improved even further. The next pair of arches, behind the jaws, was transformed into a supporting strut, or prop, joining the rear of the lower jaw to the rear of the skull. This extra support strut is called the hyomandibular. The hyomandibular enables the mouths of sharks and bony fishes to open very wide.

Scientists hypothesize that the main reason sharks and bony fishes replaced primitive fishes is that sharks and bony fishes have a superior design for swimming. Most sharks and bony fishes have streamlined bodies that are well adapted for rapid movement through the water. As shown by the shark in Figure 32-6, the head acts as a wedge that cleaves the water, and the body tapers back to the tail, allowing the shark to slip through the water with minimal resistance. In addition, sharks and bony fishes have an assortment of movable fins that greatly aid their swimming. ☐

Sharks Became Top Predators

In the period following the Devonian, the Carboniferous period, sharks became the dominant vertebrates in the sea. Sharks are members of the class Chondrichthyes, the cartilaginous (*KAHRT'l aj uh nuhs*) fishes. These fishes have skeletons made completely of cartilage. Many of the early sharks died out during a worldwide mass extinction that occurred at the end of the Permian period, 250 million years ago. Those sharks that survived this extinction underwent a great burst of evolution during the age of dinosaurs. Most of the modern groups of sharks evolved during this time. Competing successfully with the marine reptiles of that time, sharks became—and remain today—major predators in the seas.

Like many earlier fishes, sharks have a skeleton made of cartilage, but the shark's skeleton is different because it is calcified. It is strengthened by the mineral calcium carbonate (the material oyster shells are made of). The calcium carbonate is deposited in the outer layers of cartilage, and a thin layer of bone covers this reinforced cartilage. The result is a very light but strong skeleton.

Sharks were among the first vertebrates to develop bony teeth, shown in Figure 32-7. In a shark's mouth, the teeth are arranged in 6 to 10 rows. The teeth in front are pointed and sharp to do the work of biting and cutting, while behind them rows of immature teeth are growing. When a functional tooth breaks or is worn down, a replacement tooth moves forward. One shark may use more than 20,000 teeth during its lifetime! This system of tooth replacement guarantees that the teeth being used are always new and sharp.

 Did You Know?

The largest fish that exists today is the whale shark. These giants can grow to more than 40 ft. (12.5 m) long and can weigh up to 16.5 tons (15 tonnes). A whale shark is not considered a dangerous predator like some of the other sharks. Whale sharks are filter feeders. They hold open their wide, gaping mouth as they swim and filter tiny plankton from the water.

The method of reproduction among sharks is quite advanced. Shark eggs are fertilized internally. During mating, the male grasps the female with modified pelvic fins called claspers. Sperm runs from the male into the female through grooves in the claspers. A few shark species lay eggs that are enclosed in leathery cases. But the eggs of most species develop within the female's body, and the offspring (commonly called pups) are born alive.

Skates and rays, shown in Figure 32-8, are cartilaginous relatives of sharks. They evolved about 200 million years after the first sharks appeared. Today there are 275 species of sharks, apparently more kinds than existed during the Carboniferous period.

Bony Fishes Dominate the Waters

Bony fishes are members of the class Osteichthyes. They evolved at the same time as sharks, about 400 million years ago. The early bony fishes took quite a different evolutionary road from sharks. Instead of gaining speed through lightness, as sharks did so successfully, bony fishes evolved a heavy internal skeleton made completely of bone. Such a skeleton is very strong and provides a solid base against which powerful swimming muscles can pull. The process of ossification (the replacement of cartilage by bone) is complete in the bony fishes.

Unlike sharks, bony fishes evolved in fresh water. The first bony fishes were small and had paired air sacs connected to the back of the throat. These air sacs could be inflated with gas to make the fish more buoyant in the water.

Ray-finned bony fishes, shown in Figure 32-9, comprise the vast majority of living fishes. In **ray-finned fishes** the primitive air sacs were transformed into a structure called a swim bladder. By adjusting the amount of air in its swim bladder, a ray-finned fish can easily adjust its depth in the water. A shark, however, must swim through the water or sink, because it has no swim bladder and its body is denser than water.

Teleosts *(TEL ee ahsts)* are the most advanced of the ray-finned bony fishes. Teleosts have highly mobile fins, very thin scales, and completely symmetrical tails. About 95 percent of all living fish species are teleosts.

The first teleosts evolved during the Jurassic period and included forms that were similar to modern sardines, tarpons, and eels. In the mid-Cretaceous period, a second burst of teleost evolution occurred, producing fishes with heavily muscled bodies and a second dorsal fin. Examples include salmon, trout, carp, cod, and haddock. A third, and particularly intense, burst of evolution occurred in the early Tertiary period and produced highly specialized fishes with

Figure 32-8 Rays, *top*, and skates, *bottom*, have flattened bodies and live on the sea floor. Most species have flattened teeth that are used to crush their prey, mainly mollusks and crustaceans.

Figure 32-9 This large sturgeon, *top*, is a modern survivor from an early evolutionary line of ray-finned fishes. The characteristic feature of all ray-finned fishes is an internal skeleton consisting of parallel bony rays that support and stiffen each fin. There are no muscles within the fins; the fins are moved by muscles within the body. The bluegill, *bottom*, evolved much more recently than the sturgeon.

 Did You Know?

The buoyancy of human swimmers varies. A swimmer's body has a tendency to sink or rise in the water, depending on the salinity of the water and the fat content of the swimmer's body. The greater the water's salinity and the swimmer's fat content, the more readily he or she floats. It is interesting that some humans can actually swim faster than a fish. Humans have been clocked at swimming speeds of 5.2 mph, while a fish called blenny swims only 5 mph.

04:05# UNIT 8

CHAPTER 32

Teaching Tips
Design a Fish

Help students appreciate the diversity of the teleosts by first showing them pictures of different bony fishes. Then have them design a fish for a specific habitat, such as one that is able to leave the water and fly through the air or a fish that eats coral. Be sure students explain how the shape of the mouth, body shape, fins, and body coloration help the fish survive in its habitat. Have students name their fish on the basis of its characteristics.

The Living Fossil

Tell students that the coelacanth has been called a living fossil because it was identified from fossil specimens before a living specimen was ever caught. It is not known how similar ancient coelacanths were to those that live today, just that their skeletons are similar.

❏ **CAPSULE SUMMARY**

The bony fishes evolved along with sharks 400 million years ago. Powerful muscles for swimming, fins capable of subtle movements, and a fully functional swim bladder have enabled the bony fishes to become the most diverse and abundant group of fishes today.

Figure 32-10 Lungfishes are found in streams and rivers in Australia, South America, and Africa. The lungfish shown is an Australian species that grows up to 1.5 m (5 ft.) long.

more flattened, perchlike bodies. This very diverse group includes perches, sunfishes, bass, snappers, flounders, barracudas, swordfish, and many others. To learn about a common freshwater teleost, read *Up Close: Yellow Perch* on pages 752–753. ❏

Lobe-Finned Fishes Paved the Way to Land

The other major group of bony fishes, the **lobe-finned fishes**, evolved 390 million years ago, shortly after the first bony fishes appeared. Only seven species of lobe-finned fishes survive today. One species is the coelacanth (*SEE luh kanth*). The other six species are called lungfishes. One species of lungfish is shown in Figure 32-10. The lobe-finned fishes have paired fins that are structurally very different from the fins of ray-finned fishes. In many lobe-finned fishes, each fin consists of a long, fleshy, muscular lobe (hence the name of the group) that is supported by a central core of bones. The bones form fully articulated joints with one another, like the joints between the bones in your hand. Bony rays are found only at the tips of each lobed fin. Muscles within each lobe can move the fin rays independently of each other, a feat no ray-finned fish can accomplish.

Although rare today, lobe-finned fishes have played a particularly important role in the evolutionary history of vertebrates. It is from lobe-finned fishes that amphibians, the first land vertebrates, almost certainly evolved.

Section Review

1. *When and where did the first vertebrates evolve?*
2. *Explain the meaning of the following sentence: The evolution of gill arches is directly related to the evolution of predatory fishes.*
3. *The appearance of sharks probably caused the extinction of placoderms. Why?*
4. *Compare and contrast a sunfish and a coelacanth in terms of time of evolutionary appearance, classification, and structural characteristics.*

Critical Thinking

5. *How is an ostracoderm more similar to you than to a lamprey?*

Answers to Section Review

1. The first vertebrates evolved in ancient seas about 500 million years ago.

2. The jaws of predatory fishes evolved from gill arches.

3. The sharks were better swimmers and more successful predators than the placoderms, which had to live on the bottom of the ocean due to their great body weight. Sharks probably out-competed the placoderms in the search for food.

4. The sunfish evolved recently, while the coelacanth belongs to an ancient line of fish. The sunfish is a ray-finned bony fish, and the coelacanth is a lobe-finned bony fish. The sunfish has mobile fins with parallel bony rays that support and stiffen each fin. The fin of a coelacanth is fleshy, with fully articulated bones that allow it to move the fin rays independently of each other, which the sunfish cannot do.

5. An ostracoderm had bony shields that a lamprey doesn't have. The lampreys' lack of bony tissue makes them unlike humans.

32-2 Key Adaptations of Fishes

The great diversity of fishes reflects the many ways that fishes can live in the oceans and fresh waters around the world. Fishes vary in size from whale sharks, 18 m (59 ft.) giants that feed on plankton, to tiny cichlids no larger than your fingernail. Some fishes live in freezing Arctic seas, while others live in warm freshwater lakes. Mudskippers, such as the one shown in Figure 32-11, are found in Asia and Africa and spend a lot of time out of water. As you have learned, the majority of living fishes are bony fishes. In this section you will first look at the characteristics common to all fishes, and then you will learn about important adaptations of bony fishes in particular.

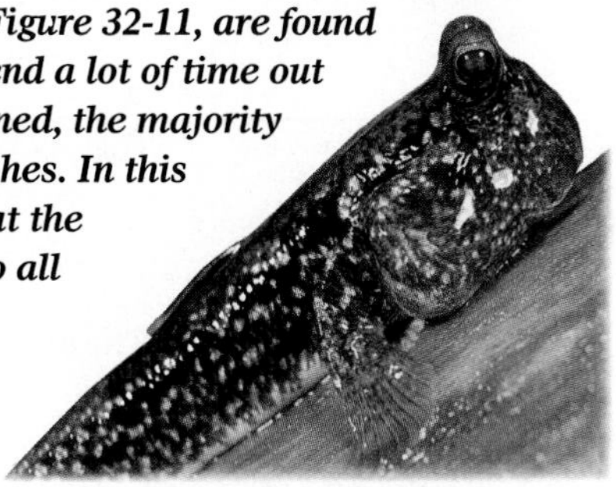

Figure 32-11 The mudskipper is found in shallow bodies of fresh water in Africa and Asia. The modified front fins of this teleost fish enable it to creep out of the water on a regular basis.

Section Objectives

- *Describe the characteristics common to all fishes.*
- *Summarize the unique features of bony fishes.*
- *Identify and describe the major external and internal characteristics of the yellow perch.*

Characteristics of Fishes

However varied their appearance, all fishes share four characteristics. For the most part, these characteristics enable fishes to live more successfully in their aquatic environments.

1. **Gills** Fishes are water-dwelling organisms and must obtain the oxygen required for metabolism from the oxygen gas dissolved in the water around them. They do this by pumping a great deal of water through their mouths. The water passes over fine filaments of tissue called gills in the back of the mouth, and then exits the body through slits in the side of the throat.
2. **Vertebral column (backbone)** All fishes have an internal skeleton made of either cartilage or bone, with a vertebral column surrounding the spinal cord. The brain is fully encased within a protective covering called the skull or cranium.
3. **Single-loop blood circulation** Blood is pumped from the heart to the gills, where the blood is oxygenated. From the gills, the oxygenated blood passes to the rest of the body and then returns to the heart.
4. **Nutritional requirements** Fishes are unable to synthesize the aromatic (ring-structured) amino acids and must consume these important protein-building amino acids in their diet. This inability to synthesize one or more aromatic amino acids has been inherited by all the vertebrate descendants of fishes, including humans.

SECTION 32-2

Vocabulary Preview

operculum

Opening Question

Divide the class into groups of two to three students and have them make a list of every fish they know. After they make their lists, make sure they haven't included invertebrates such as starfish and mammals such as whales. Have them go back and circle every fish that is a bony fish. It should be apparent that most fish in the world are bony fish.

Demonstration

Watch a Real Fish

Give students time to observe fish in an aquarium, a goldfish in a bowl, or videos of fish. Give your students specific behaviors to look for and describe. How is the fish breathing? What does it respond to? What does it use to move up and down? How does it use its fins to navigate through the water? How does the fish feed? What startles the fish? Have the students write an accurate and creative account of their observations.

Teaching Tip

Fish Kills

Tell students that when weather, pollution, or a cyanobacteria bloom depletes the oxygen content in water, fish gills are no longer able to absorb enough oxygen, and the fish suffocate. When this happens over a widespread area, thousands of fish bodies float to the surface of the water or wash ashore.

Multicultural Perspective

Fish in Asia—Art and Food

Asian people have celebrated the elegant beauty of carp and goldfish for many centuries. Elaborately decorated fishbowls and kites with carp designs can be found throughout China, Japan, and Korea, as can serene water gardens with fish swimming among the willows and water lilies. Even a fish dinner becomes a celebration of beauty as Asian chefs strive to create works of edible art.

Instructional Strategies

- Have students relate the perch's external structures to its lifestyle and habitat. Have students pay special attention to the specialized senses and swimming adaptations of this successful fish.

- Have students classify every labeled part of the internal structures of the yellow perch according to these body systems: nervous system, circulatory system, reproductive system, digestive system, respiratory system, excretory (urinary) system, muscular system, and skeletal system.

Discussion

Guide the discussion by posing the following questions:

1. How does the yellow perch control roll, pitch, and yaw when it swims? *(The dorsal fins prevent roll, the caudal fin controls pitch, and the anal fin controls yaw.)*

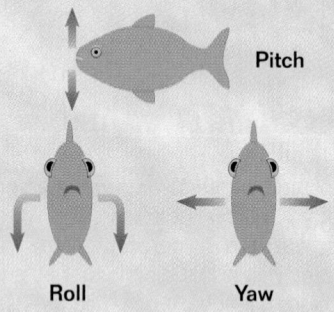

Pitch

Roll   Yaw

2. How would a perch find food if it were kept in a dark room? *(It could smell the food and sense the food's movement with its lateral line.)*

3. Why are the mouth and operculum needed for a perch to breathe? *(Movements of the opercula draw water into the mouth and force the water over the perch's gills, where oxygen is removed from the water.)*

UP CLOSE YELLOW PERCH

- **Scientific name:** *Perca flavescens*
- **Range:** Found in lakes and rivers from the Great Lakes to the Atlantic coast and as far south as South Carolina
- **Habitat:** Lives concealed among vegetation or submerged tree roots
- **Size:** Grows to about 0.30 m (1 ft.) long and up to 2.3 kg (5 lb.)
- **Diet:** Feeds on insect larvae, crustaceans, and other fishes

External Structures

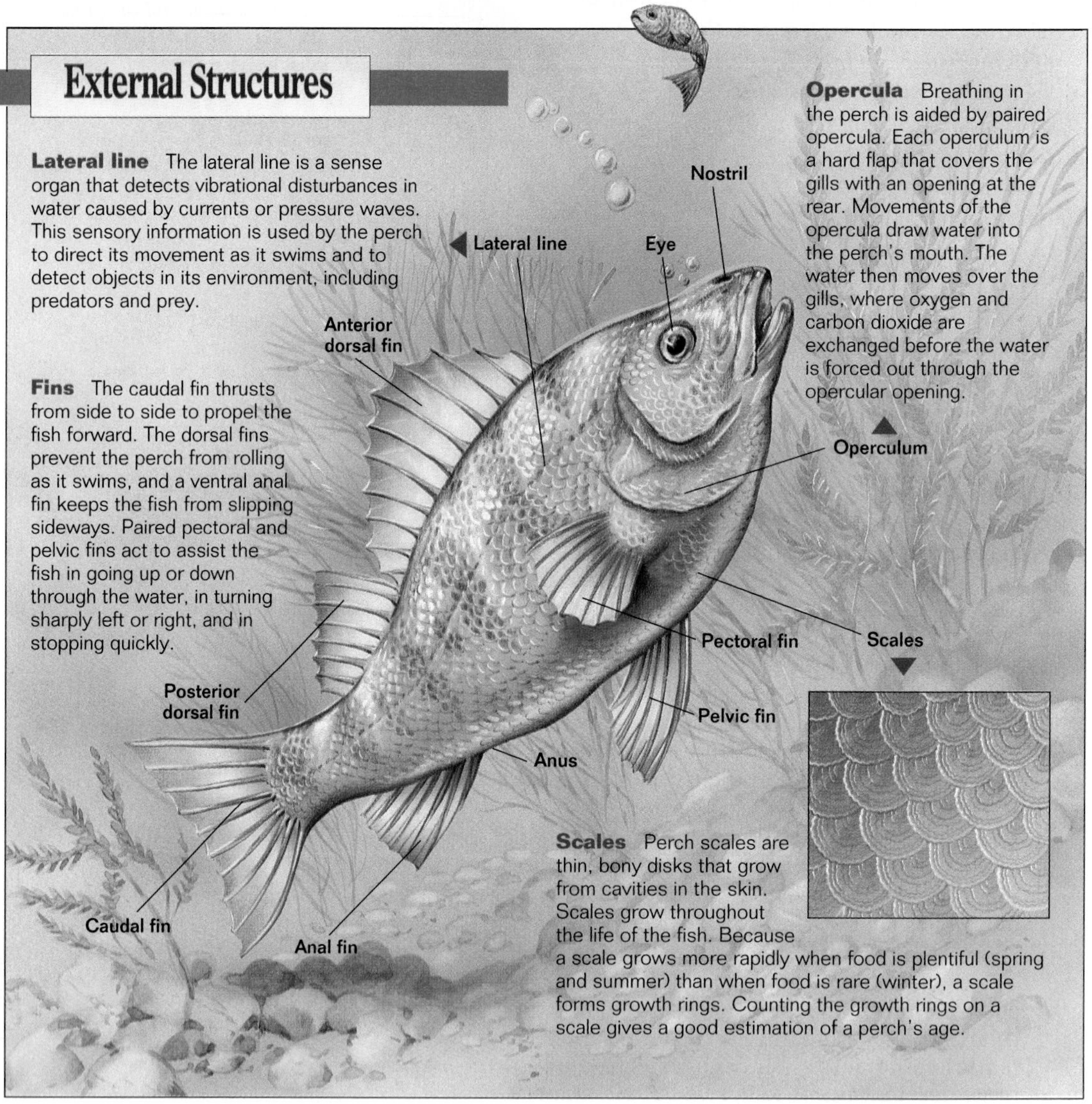

Lateral line The lateral line is a sense organ that detects vibrational disturbances in water caused by currents or pressure waves. This sensory information is used by the perch to direct its movement as it swims and to detect objects in its environment, including predators and prey.

Fins The caudal fin thrusts from side to side to propel the fish forward. The dorsal fins prevent the perch from rolling as it swims, and a ventral anal fin keeps the fish from slipping sideways. Paired pectoral and pelvic fins act to assist the fish in going up or down through the water, in turning sharply left or right, and in stopping quickly.

Opercula Breathing in the perch is aided by paired opercula. Each operculum is a hard flap that covers the gills with an opening at the rear. Movements of the opercula draw water into the perch's mouth. The water then moves over the gills, where oxygen and carbon dioxide are exchanged before the water is forced out through the opercular opening.

Nostril

Lateral line

Eye

Anterior dorsal fin

Operculum

Posterior dorsal fin

Pectoral fin

Scales

Pelvic fin

Anus

Caudal fin

Anal fin

Scales Perch scales are thin, bony disks that grow from cavities in the skin. Scales grow throughout the life of the fish. Because a scale grows more rapidly when food is plentiful (spring and summer) than when food is rare (winter), a scale forms growth rings. Counting the growth rings on a scale gives a good estimation of a perch's age.

Internal Structures

Reproductive organs Yellow perch produce gametes (sperm cells and egg cells) during their breeding season in the spring. During this period the testes of males produce enormous numbers of tiny sperm cells, and the ovaries of females become swollen with egg cells. Eggs are fertilized externally. The male deposits milt, a fluid containing the sperm, on strings of eggs laid by the female that are wound in and out of weeds and twigs in the water. The young hatch from fertilized eggs within days in warm water (in cold water it may take much longer) and grow quickly.

Brain The brain is divided into anterior (front), middle, and posterior (back) regions. Optic lobes receive sensory information from the eyes. The large size of these lobes indicates the importance of vision. The olfactory bulbs are devoted to receiving information concerning smell (another important sense in perch) from chemical-sensing cells. The cerebrum processes mainly sensory information. The cerebellum coordinates muscle activity, and the medulla oblongata controls the function of many internal organs.

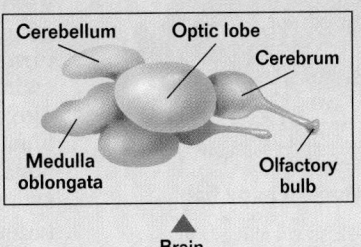

Cerebellum · Optic lobe · Cerebrum · Medulla oblongata · Olfactory bulb

Brain

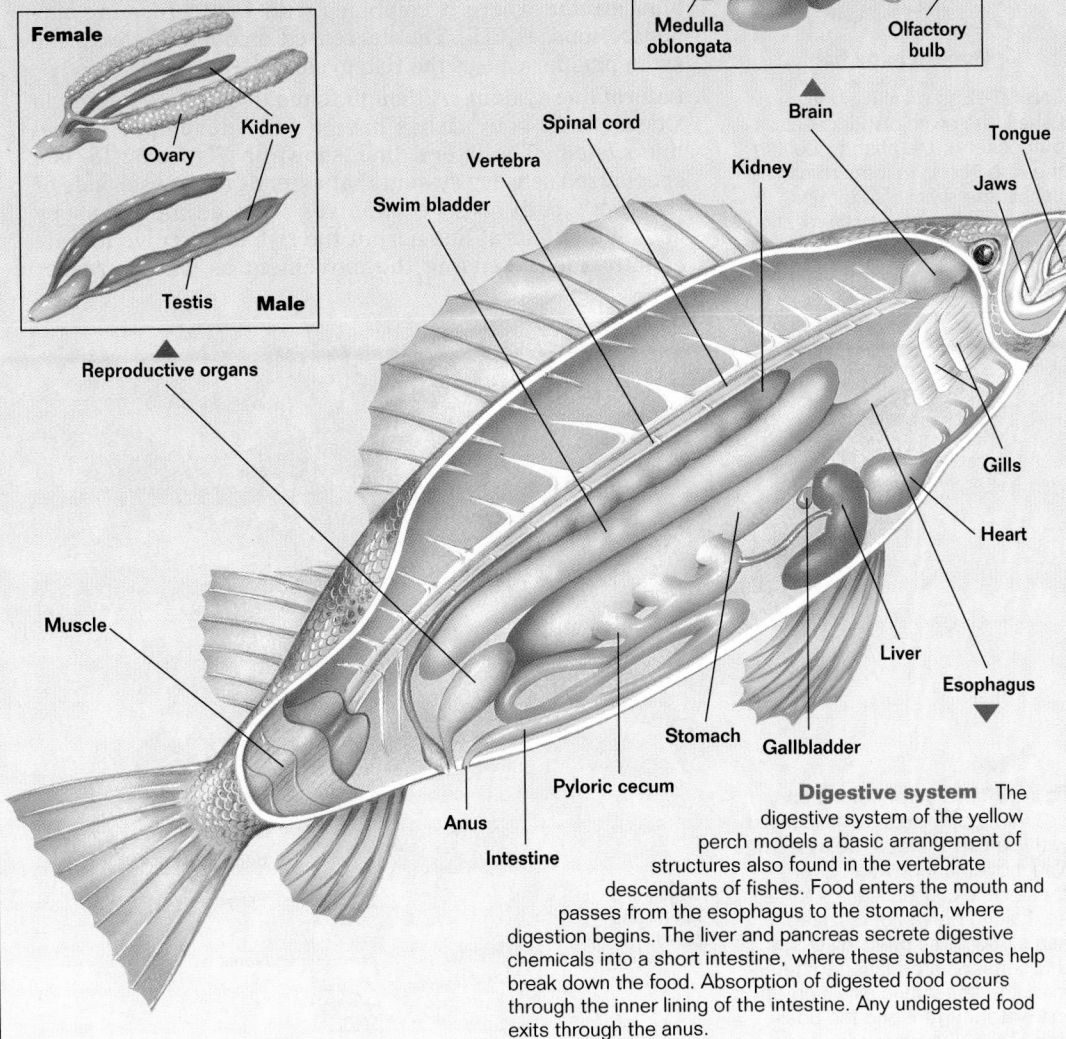

Female

Ovary · Kidney

Testis · **Male**

Reproductive organs

Spinal cord · Vertebra · Swim bladder · Kidney · Brain · Tongue · Jaws · Gills · Heart · Liver · Esophagus · Gallbladder · Stomach · Pyloric cecum · Intestine · Anus · Muscle

Digestive system The digestive system of the yellow perch models a basic arrangement of structures also found in the vertebrate descendants of fishes. Food enters the mouth and passes from the esophagus to the stomach, where digestion begins. The liver and pancreas secrete digestive chemicals into a short intestine, where these substances help break down the food. Absorption of digested food occurs through the inner lining of the intestine. Any undigested food exits through the anus.

4. Pick a body system and ask students to describe what would happen to a yellow perch that was missing that system. *(nervous system—it could not move or sense its environment; circulatory system—its cells could not get oxygen and nutrients or get rid of wastes; reproductive system—it could not produce offspring; respiratory system—it could not breathe; digestive system—it could not break down food; excretory (urinary) system—it could not get rid of nitrogen wastes; muscular system—it could not move; skeletal system—its body would collapse)*

5. Why would a perch without a swim bladder probably have to eat more than a perch that had a swim bladder? *(because it would use more energy swimming from place to place)*

6. Name the dorsally located structure that identifies the perch as a chordate. *(spinal cord or nerve cord)*

7. If you caught, cooked, and ate a yellow perch, which labeled structure did you eat? *(muscle)*

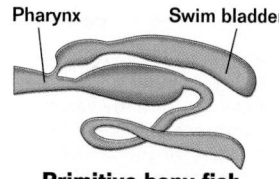

Primitive bony fish

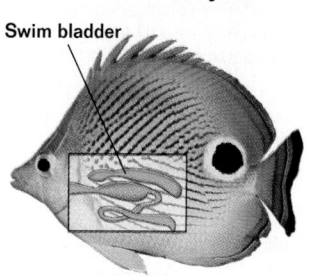

Modern bony fish

Figure 32-12 In primitive bony fishes, the swim bladder is an outpocket of the pharynx behind the throat, *top*. In modern bony fishes, the swim bladder has separated from the pharynx and is an independent organ, *bottom*.

Figure 32-13 Clusters of sensory cells are contained in pits located within a fish's lateral line system. The movement of water over these cells causes signals to be sent along nerve fibers to the fish's brain. The brain assesses this sensory information to determine the speed and direction of the water current and the position of the objects in the water.

Bony Fishes Have Additional Adaptations

Although it is correct to consider jawless and cartilaginous fishes successful in their own right, they in no way approach the numbers or diversity seen in the bony fishes. The remarkable success of bony fishes is due to a series of unique structural adaptations.

1. **Swim bladder** Early bony fishes had to gulp air to fill their swim bladder. In modern bony fishes, the swim bladder, shown in Figure 32-12, is able to secrete and absorb its own gases. The cells that make up the swim bladder generate carbon dioxide, CO_2, by carrying out cellular respiration, which fills the bladder with CO_2 gas. The filled swim bladder makes the fish rise in the water. Carbon dioxide is absorbed from the swim bladder into the fish's bloodstream where it combines with water, forming carbonate ions, H_2CO_3. The decreased amount of gas in the swim bladder causes the fish to sink.

2. **Lateral line system** Although found to a limited degree in sharks, only bony fishes have a fully developed lateral line system. The lateral line, shown in Figure 32-13, is a specialized sensory system that extends along each side of the fish's body. Nerve impulses from ciliated sensory cells in the lateral line permit the fish to perceive its rate of movement, sensing the movement as water presses

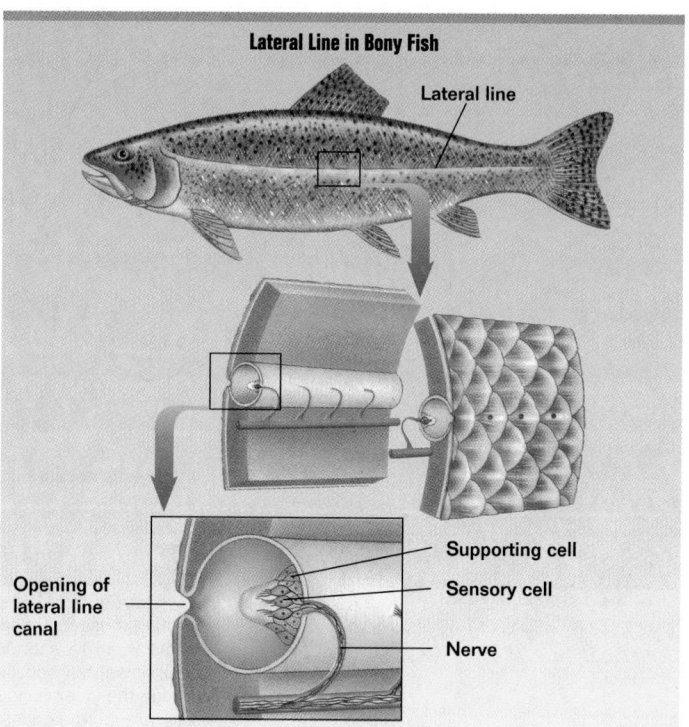

Lateral Line in Bony Fish

Lateral line

Opening of lateral line canal

Supporting cell

Sensory cell

Nerve

against its lateral line. A trout uses its lateral line system to obtain the sensory information it needs to orient itself with its head upstream.

The lateral line system also enables a fish to detect a motionless object by the movement of water reflected off that object. In a very real sense, this is equivalent to hearing. The way that a fish detects an object with its lateral line and the way that you hear a symphony with your inner ear share the same basic mechanism—cilia deflected by waves of pressure. Like fishes, you detect patterns in the waves of pressure in the medium (air, in your case) around you. The sound receptors within the ears of terrestrial vertebrates are thought to have evolved from lateral line receptors.

3. **Gill cover**   What is a goldfish doing when it hovers at some depth in its fish tank, periodically taking in a mouthful of water? If it is not eating, then it is probably breathing. Most bony fishes have a hard plate called an **operculum** that covers the gills on each side of the head. Movements of the opercula, shown in Figure 32-14, permit a bony fish to pump water over the gills, enabling the fish to breathe. By using this very efficient pump, bony fishes can move water over their gills while remaining stationary in the water. A bony fish doesn't have to swim forward with its mouth open to move water over its gills as fishes without opercula, like sharks, must do. This ability to breathe without swimming enables a bony fish to conserve energy—energy that can be spent chasing after prey and escaping from predators.

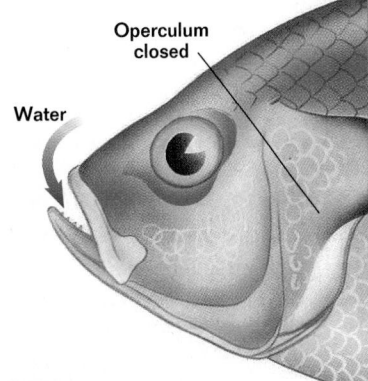
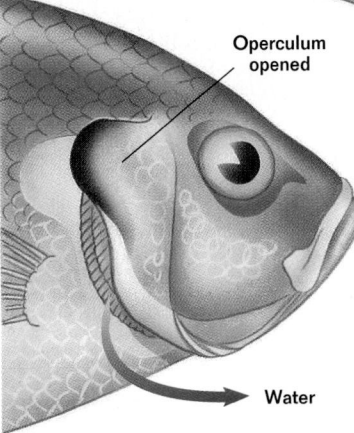

Figure 32-14 A fish breathes by moving its operculum. When the fish's mouth opens, the operculum closes over the gills, sealing off the body wall opening. This increases the volume of the mouth cavity, causing water to be drawn into the mouth, *top*. When the mouth is closed, the operculum moves away from the body wall, revealing the opening. As a result, the volume of the mouth cavity is decreased, and water is forced over the gills and out of the fish, *bottom*.

Section Review

1. *Of the characteristics that are common to all fishes, which one could affect a fish's ability to produce enzymes?*

2. *An unidentified species of fish is caught. It has rough skin, several rows of teeth, and no opercula. Based on this external description of the fish, would you expect it to have a swim bladder? Explain why or why not.*

3. *Describe the digestive system of the yellow perch.*

4. *What are the similarities between a yellow perch's lateral line and your ear?*

Critical Thinking

5. *Two identical goldfish bowls each contain one goldfish of the same type and size. A student observes that the opercula of one fish are moving at a much slower rate than the opercula of the other fish. Hypothesize about what might cause the different rates.*

SECTION 32-2

Teaching Tip
Advantages of an Operculum
Ask students to think of reasons why it is an advantage for bony fishes to have gill coverings, and have them list their ideas. Have students first read the section in the text on the gill cover, which mentions the ability to breathe without swimming and wasting energy. Then encourage students to think in terms of protection (from injury and parasites), communicating with other fish (Siamese fighting fish flare their gill covers when they are aggressive), and keeping gills moist (mudskippers hold water under their gill covers while out of water).

Application
Health Some fish oils, like Omega-3, seem to have a beneficial effect on human health by lowering the levels of certain harmful cholesterols in the blood and by slowing the spread of some cancers. But more research is needed before these effects can be documented with any great degree of certainty.

Answers to Section Review

1. All fish are unable to synthesize aromatic amino acids that might be needed to make enzymes, since enzymes are proteins and therefore made of amino acids.

2. No, it has the characteristics of a shark, which lacks a swim bladder.

3. The digestive system of a yellow perch is the basic arrangement of the digestive structure found in all vertebrate descendants of fish: food enters the mouth, passes through the esophagus to the stomach and small intestine, and exits through the anus.

4. Both the lateral line of the yellow perch and the human ear can detect pressure changes as sensory information. Both systems share the same basic mechanism of movement of cilia in response to pressure changes.

5. Answers may vary but may include that one fish is swimming in cooler water, one fish is swimming rapidly while the other is resting, or the water in one bowl contains less oxygen than the water in the other bowl.

Opening Question

Ask students to read Linnaeus's description of amphibians in the introduction to this section and list the things he wrote that were wrong as well as the things he wrote that were right.

Vocabulary

(None)

Demonstration

Learning From Larvae

Show students a living leopard frog or some other common frog species. Also have on hand an aquarium that contains some living, young frog tadpoles. Ask the students to identify which is in the adult stage and which is in the larval stage. *(frog—adult, tadpoles—larval)* Next, have the students describe the habitat and physical appearance of a tadpole. *(It is aquatic and has gills, a tail, and no limbs.)* Ask them to name another vertebrate that also seems to fit their description. *(fishes)* Explain to them that the ancestors of amphibians were undoubtedly lobe-finned fishes. End the demonstration by stating that although a tadpole is similar to a fish, it is also different. Ask the students what makes a fish different from a tadpole? *(Answers may vary but may include that fish have scales and paired fins, and swim better.)*

Teaching Tip

Tracing a Family Tree

This section is a good place to remind students that since Linnaeus's time, anatomical features have been the primary basis of establishing degree of relatedness, but that now we can also use biochemical evidence of similarity. All vertebrates have DNA, so it is a good chemical to compare when looking for relatedness among modern-day vertebrates.

Section Objectives

- Define the term amphibian.
- Identify and describe the direct ancestors of the first amphibians.
- Discuss the major changes that have occurred in amphibians throughout their evolutionary history.

Frogs, toads, salamanders, and newts—these air-breathing, damp-skinned vertebrates are the direct descendants of fishes. They are the sole survivors of a very successful group, the amphibians, the first vertebrates to walk on land. Most modern amphibians are small and live unnoticed by humans. When describing amphibians in 1758, the biologist Carolus Linnaeus said, "These foul and loathsome animals are abhorrent because of their cold body, pale color, filthy skin, fierce aspect, calculating eye, offensive smell, harsh voice, squalid habitation, and terrible venom; and so their Creator has not exerted his powers to make many of them."

Many of Linnaeus's observations were inaccurate, particularly his claim that amphibians were few in number. Modern biologists have discovered that amphibians are among the most numerous of terrestrial vertebrates. There are as many amphibian species as mammalian species, if not more, and throughout the world amphibians play key roles in terrestrial as well as aquatic food chains.

Amphibians Evolved From a Type of Lobe-Finned Fish

The term *amphibian* is from the Greek *amphi*, meaning "double," and *bios*, meaning "life." The term nicely describes the essential characteristic of today's amphibians, reflecting their ability to live in two worlds—the aquatic world of their fish ancestors and the terrestrial world that their ancestors first invaded.

Fossil evidence indicates that amphibians evolved from lobe-finned fishes about 370 million years ago. For many years there was considerable disagreement about which kind of lobe-finned fish was the direct ancestor of amphibians. Some scientists argued that the coelacanths were the direct ancestors, while others insisted it was the lungfishes. A third group of scientists argued that a different group of lobe-finned fishes called rhipidistians (*RIHP uh DIST tee uhns*), all of which are now extinct, were the true ancestors of amphibians. Good arguments can be made to support each of these possibilities. Recent analysis of the DNA of coelacanths and lungfishes indicates lungfishes are in fact far more closely related to amphibians than are coelacanths. But due to similarities in the skeletal features of amphibians and rhipidistian fishes, most paleontologists think that

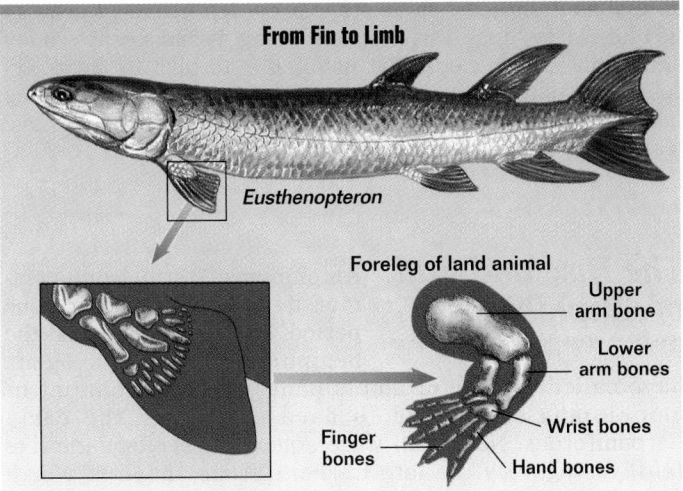

From Fin to Limb

Eusthenopteron

Foreleg of land animal

Upper arm bone

Lower arm bones

Wrist bones

Finger bones

Hand bones

Figure 32-15 The pattern of bones in an amphibian's limbs bears a remarkable resemblance to that of the fin bones of *Eusthenopteron*, a rhipidistian fish. This evidence leads most biologists to think that rhipidistian fishes are the direct ancestors of amphibians.

amphibians evolved directly from the rhipidistian fishes rather than from lungfishes. Figure 32-15 compares the fin bones of *Eusthenopteron*, an extinct rhipidistian fish, to the front leg bones of a typical land animal.

The earliest known amphibian fossil is that of an extinct amphibian called *Ichthyostega*. This fossil was found in 370-million-year-old rock in Greenland. At the time that this animal lived, Greenland was part of the North American continent and lay near the equator. Fossils from amphibians that lived during the next 100 million years have been found only in North America. Only when Asia and the southern continents merged with North America 250 million years ago to form one giant supercontinent did amphibians spread throughout the world.

Ichthyostega was a strongly built animal, with four sturdy legs well supported by hip and shoulder bones, as shown by the skeleton in Figure 32-16. Like its fish ancestors, *Ichthyostega* had bony scales, although the scales were set only in the skin of its belly and flattened tail. It also had a long fin that extended along the length of the tail. In the water the tail could be thrust from side to side to propel the

Figure 32-16 *Ichthyostega* had long, broad ribs that overlapped each other and formed a solid cage for the lungs and heart. Because its rib cage was not expandable, *Ichthyostega* probably breathed with mouth movements similar to those of a fish—lowering the floor of the mouth to draw in air and raising it to push air down the windpipe into the lungs.

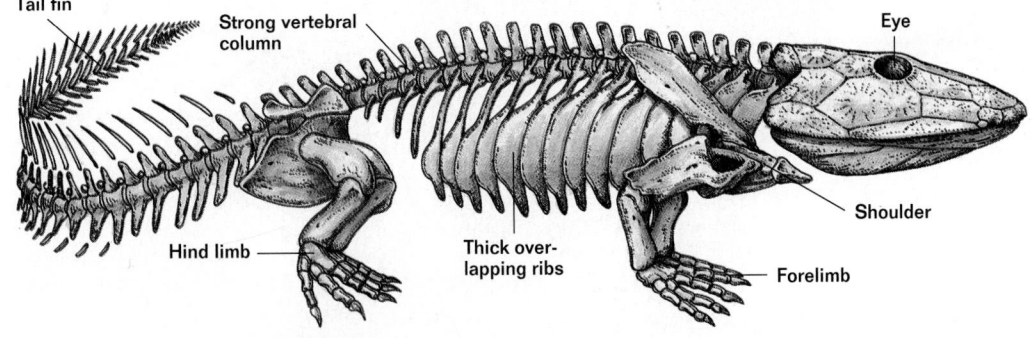

Tail fin

Strong vertebral column

Eye

Hind limb

Thick overlapping ribs

Shoulder

Forelimb

☑ RESEARCH Update

Frogs, Toads, and Deserts

Amphibians are typically associated with aquatic, or at least damp, environments. However, three researchers, two from the University of California at Riverside and one from California State University at Fullerton, have focused their studies on the hardy amphibians that live in the desert. The desert is not known for its aquatic habitats, and Dr. Lon L. McClanahan, Dr. Rodolfo Ruibal, and Dr. Vaughan H. Shoemaker have made some recent conclusions that challenge the conventional understanding of amphibian physiology.

How can frogs and toads live in a desert? The researchers have studied species from three continents and compiled an impressive inventory of heat-coping strategies. As with many other desert animals, the scientists found that the amphibians coped with desert heat either behaviorally or physiologically. Behaviorally, frogs and toads would do such things as sit in the shade, burrow in the soil, or construct cocoons around themselves to minimize water loss. Physiologically, they would cool themselves through evaporation, secrete waterproof waxes from their skin, or even resorb water from urine in their bladders.

The researchers have issued a warning about the global trend toward decreased numbers of amphibian populations. While man-made endeavors certainly hasten the demise of some populations, other populations untouched by humans are also disappearing. The scientists suggest that perhaps there is a subtle environmental change occurring on a global scale. Regardless of the cause of decreased amphibian numbers, protecting and maintaining amphibian populations will require an understanding of how frogs and toads adapt to their diverse environments. ☑

❏ *CAPSULE SUMMARY*

The earliest amphibians, like the semiaquatic Ichthyostega, *probably evolved from rhipidistian lobe-finned fishes about 390 million years ago.*

Figure 32-17 *Eryops* grew to 1.8 m (6 ft.) long and had a heavy, flat skull bearing strong teeth. A thick, strong backbone indicates that *Eryops* was well adapted for life on land. Bony lumps in the skin formed a heavy armor. *Eryops* probably lived like a modern alligator, in and out of the water of upland streams, rivers, and lakes.

animal as it chased fish or other prey. *Ichthyostega* grew to be quite large, up to 1 m (3 ft. 3 in.) long. It had a heavy, bony skull and a short neck that enabled it to turn its head. Its four legs splayed out from the sides of its body, and each leg had a foot with five toes. On land *Ichthyostega* would have walked with an awkward sprawling gait. ❏

The Rise and Fall of Amphibians

Amphibians first became common during the Carboniferous period. This period marked the beginning of a time biologists have called the age of amphibians. Fourteen families of amphibians are known to have existed in the early Carboniferous. Nearly all were aquatic or semiaquatic like *Ichthyostega*. By the late Carboniferous, much of North America was covered by low-lying tropical swamplands. Thirty-four families of amphibians thrived in this wet terrestrial environment, sharing it with pelycosaurs and other early reptiles. In the Permian period that followed, amphibians reached their greatest diversity, increasing to 40 families. In the early Permian a remarkable change occurred among amphibians—many of them began to leave the marshes for dry uplands. Many of these terrestrial amphibians had bony plates and armor covering their bodies and grew to be very large, some as big as a pony. Both their large size and shielded bodies suggest that the skin of these amphibians did not play a respiratory role, as does the skin of present-day amphibians. Rather, these early terrestrial amphibians had an impermeable leathery skin that prevented water loss. By the mid-Permian, 60 percent of all amphibian species were completely terrestrial. *Eryops*, shown in Figure 32-17, was typical of the terrestrial amphibians.

The middle Permian period marked the peak of amphibian success. By the end of this period a new kind of terrestrial reptile, the therapsid, had become common. Gradually the therapsids would be successful in replacing all terrestrial amphibians. Over half of all remaining amphibians were aquatic by the end of the Permian, and this trend continued into the Triassic period. By the end of the Triassic there were only 15 families of amphibians, including the first frogs. Almost without exception, these families were aquatic. Some contained species that grew to enormous sizes, up to 4 m (13 ft.) long. By the Jurassic period only two families of amphibians survived. One was the anurans *(uh NUR uhns)*, frogs and toads, and the other was the urodeles *(YOOR oh dehls)*, salamanders and newts. The age of amphibians was over. ❑

The Success of Amphibians Today

All of today's amphibians descended from the anurans and urodeles that survived the Age of Reptiles, which lasted throughout the Mesozoic era. During the Tertiary period these remaining moist-skinned amphibians underwent a highly successful invasion of wet habitats all over the world. Today there are over 4,200 species of amphibians in 37 different families. They consist of frogs, toads, salamanders, and caecilians. While amphibians are not nearly as diverse in form or habitat as they were during the Permian period, modern amphibians have been very successful in occupying their particular niches.

Section Review

1. *Why is the term* amphibian *very appropriate for the members of this class?*
2. *Why would the capture of a rhipidistian fish today be considered a monumental scientific event?*
3. *Describe the physical appearance of* Ichthyostega, *and state when it lived.*
4. *State the scientific evidence suggesting that amphibians did not always live in aquatic or moist environments.*
5. *What kind of animal is a therapsid, and what influence did they have on early amphibians?*

Critical Thinking

6. *Offer a possible reason why amphibians evolved when they did, rather than 50 million years earlier. To answer this question, it may be helpful to refer to the timeline on pages 274–287 in Chapter 13.*

❑ **CAPSULE SUMMARY**

Amphibians were the dominant land vertebrates for 100 million years. The rise of reptiles marked the beginning of the decline of amphibians. Reptiles halted the expansion of amphibians into terrestrial habitats, greatly diminishing amphibian diversity and numbers.

SECTION 32-2

Teaching Tips

The Rise, Fall, and Rise of Amphibians

Have students read the last two pages of this section to find how many families of amphibians there were during the following periods: Permian, Triassic, Jurassic, present-day *(40, 15, 2, 37)*. Then ask students during which period amphibians were possibly headed for extinction? *(Jurassic)* Finish by comparing amphibian diversity today with amphibian diversity during the Permian period. *(They are less diverse today than they were in the Permian.)*

Adios Amphibians

During the Permian period, terrestrial amphibians were apparently outcompeted by an early reptile. Have students look at the picture of *Eryops* in Figure 32-17 and explain why this kind of amphibian couldn't successfully compete with the new reptiles. *(Answers might include that it moved more awkwardly on land, it didn't have a respiratory system that could support much activity, and its reproduction was tied to water.)*

Less Is More

Ask students to think about how different today's amphibians are from the sprawling giants in the Carboniferous and Permian periods. Why are today's small amphibians successful and the 6-ft.-long Eryops extinct? Encourage students to think about escaping enemies, capturing prey, finding places to live, and the amount of food needed.

Answers to Section Review

1. The word *amphibia* comes from Greek root words meaning "double life," which is what amphibians lead. They begin their life in water with fins and gills, then metamorphose into a terrestrial body type with legs and lungs.

2. Studying the DNA of a captured rhipidistian fish could settle the uncertain evolutionary history of amphibians.

3. *Ichthyostega,* which lived during the late Devonian and early Carboniferous periods, was strongly built and over 3 ft. long. It had four legs and a long, flattened tail that had a fin along its upper edge.

4. During the Permian period, amphibians had bony plates and armor, indicating that their skin was tough and leathery to prevent water loss.

5. Therapsids were terrestrial reptiles that out-competed the terrestrial amphibians that lived during the late Permian period.

6. Answers may vary but might include that 50 million years earlier, before the invasion of land by arthropods, there would not have been a terrestrial food supply for the carnivorous amphibians.

Vocabulary

(None)

Opening Question

Two characteristics of amphibians are the presence of legs and lungs. What are some exceptions? Tell students to look at the characteristics of modern amphibians in their textbook to find the answers. *(Caecilians are legless, and some salamanders do not have lungs.)*

Demonstration

Surface-Area-to-Volume Ratio
Caution: iodine can be toxic if consumed, and some people are allergic to it, so use forceps and a knife to handle the cubes.
Materials: Whole raw potato, knife, beaker, metric ruler, iodine solution (available from biological supply companies) **Procedure:** Cut three cubes out of a raw potato; make one 4 cm on a side, one 2 cm on a side, and one 1 cm on a side. Place all three cubes in an iodine solution for about 5 minutes, then cut the cubes open and examine them. Ask students in which cube the iodine reached the middle. Have students imagine that the potato is an amphibian and the iodine is oxygen. Which "amphibian" would have "oxygen" reach every part of its body? *(the smaller ones)*

Teaching Tip

Why Are There No Flying Frogs?
Tell students that flying requires a high metabolic rate. Then ask them why there aren't any flying frogs? Direct student attention to the inefficient respiratory and circulatory systems of frogs for an explanation.

32-4 Key Adaptations of Amphibians

Biologists classify today's species of amphibians into three orders. The order **Anura** is composed of frogs and toads. The term **Anura** is from the Greek **a**, meaning "without," and **oura**, meaning "tail." The order **Urodela** is composed of salamanders and newts. The term **Urodela** is from the Greek **oura**, meaning "tail," and **delos**, meaning "visible." The order **Apoda** is made up of wormlike, nearly blind organisms called caecilians. The term **Apoda** is from the Greek **a**, meaning "without," and **pous**, meaning "foot."

Section Objectives

■ Describe the characteristics common to all modern amphibians.
■ Contrast the three orders of modern amphibians.
■ Identify and describe the major external and internal characteristics of the leopard frog.

Characteristics of Modern Amphibians

Even though each amphibian order evolved at a different time during the Tertiary period, all modern amphibians have certain key characteristics in common.

1. **Legs** Frogs and salamanders have four legs and can move about on land quite well. The evolution of legs was one of the key adaptations for vertebrates that live on land. Caecilians gradually lost their legs during the evolutionary course of adapting to a burrowing existence.

2. **Lungs** Most larval amphibians have gills, but by adulthood the gills have usually disappeared and breathing is accomplished with a pair of lungs (lungless salamanders are an exception). The internal surfaces of amphibian lungs are poorly developed, with much less surface area than reptilian, bird, or mammalian lungs. Modern amphibians still breathe as *Ichthyostega* did, by lowering the floor of the mouth to suck in air and then raising it back up to force the air down into the lungs. Figure 32-18 demonstrates the breathing movements of a frog.

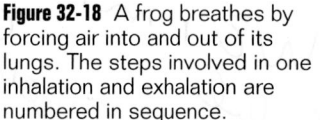

Figure 32-18 A frog breathes by forcing air into and out of its lungs. The steps involved in one inhalation and exhalation are numbered in sequence.

Lung   Mouth

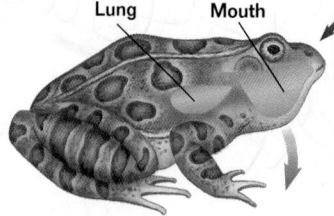 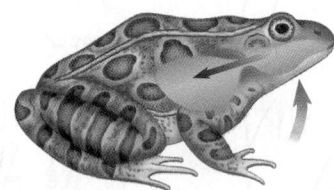 

1 The floor of the mouth is lowered, causing air to be sucked in through the nostrils.

2 The nostrils close and the floor of the mouth is elevated, forcing air down into the lungs.

3 Contraction of muscles in the body wall forces air out through the mouth.

Overcoming Misconceptions

Toads Don't Give You Warts
Some students may believe that they can get warts from toads. Be sure to explain that warts are caused by viruses that are not carried on toads. Have students look at the photograph of a toad on page 761 and suggest a reason why people might associate warts with toads. *(The skin of a toad looks pebbly and "warty.")* The rough skin of a toad is an adaptation for living successfully in dry habitats and is not due to skin problems.

3. **Cutaneous respiration**   Frogs, salamanders, and caecilians supplement the use of their lungs by respiring directly through their skin. The skin of amphibians is moist and provides an extensive surface area. Cutaneous respiration ("skin breathing") limits the maximum body size of amphibians because it is efficient only when there is a high ratio of skin surface area to body volume.

4. **Pulmonary veins**   After blood is pumped through the lungs, two large veins called pulmonary veins return the oxygenated blood to the heart for repumping. This allows the oxygenated blood to be pumped to the tissues at a much higher pressure than it has when it leaves the lungs. Under high pressure, the blood travels quickly through the circulatory system, ensuring that the body tissues receive the oxygen they need.

5. **Partially divided heart**   As you learned in Chapter 31, the amphibian heart is partitioned internally to form chambers. Because the division is incomplete (the atrium is divided into left and right sides, but the ventricle is not), a mixture of oxygenated and deoxygenated blood is delivered to the amphibian's body tissues.

Frogs and Toads Are Tailless
....

Frogs and toads, shown in Figure 32-19, make up the order Anura. There are 3,680 species of frogs and toads in 22 families. They live in environments ranging from deserts to mountains and ponds to puddles. All adult anurans are carnivores, eating a wide variety of insects. To learn about an anuran called the leopard frog, see *Up Close: Leopard Frog* on pages 762–763.

Figure 32-19 Frogs, like the green frog, *below left,* and the tree frog, *below center,* have smooth, moist skin, a broad body, and long hind legs that make them excellent jumpers. Most frogs live in or near water, although some tropical species live in holes in trees. Unlike frogs, toads, like the common Asiatic toad, *below right,* have dry, bumpy skin, short legs, and are well adapted to dry environments.

SECTION 32-3

Teaching Tips

Lifestyles of the Cold and Slimy

All adult anurans are carnivores that eat mostly insects. Ask students why a frog in a pond might be an important part of a food chain. Ask students what the frog might eat and what might eat the frog. Finish by asking what would happen to the pond if there were no more frogs?

Classroom Visitors

Amphibians are sometimes difficult to keep alive for long periods of time, but they make good visitors to your room. Tadpoles may be kept in an ordinary aquarium and fed fish food. Frogs do well in a terrarium with a bowl of water, and they eat crickets or other small insects. Animals are wonderful teaching tools, and they give students many opportunities to watch behaviors.

An Extra Eyelid

Frogs have a transparent membrane to cover their eyes underwater. Humans have the remnants of that membrane in the inner corner of the eye in the form of a little bulge of tissue, as shown below. Ask students to give examples of animals they know that have another eyelid. (*They may mention cats, reptiles, or birds.*)

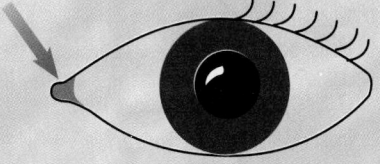

Instructional Strategies

- Have students look at the diagram on page 762 and explain how the frog's head sticking out of the water is well adapted to air, while the body sitting in the water is well adapted to living in water.
- Have students compare the organ systems of the frog with the organ systems of the yellow perch, and have them describe the similarities and differences.

Discussion

Guide the discussion by posing the following questions:

1. How do you think the leopard frog got its name? *(because of its spotted appearance)*

2. How is the ear of the frog similar to the lateral line of the yellow perch? *(Both structures contain ciliated sensory cells.)*

3. Why do leopard frogs produce so many eggs and sperm? *(because their eggs and larvae are eaten by many predators)*

4. How do leopard frogs breathe? *(with lungs and through their skin)*

5. Why is a fish more agile in the water than a frog? *(A more developed cerebellum in the fish enables it to have better muscle coordination.)*

6. Why will a fish suffocate if it is taken out of water. *(Its gills will dry out.)*

7. How does a frog change its depth in water? *(It swims using its webbed feet.)*

8. Why is it vital for male frogs to call to females? *(Frogs are camouflaged and may not find each other. And each species of frog has its own mating call so females won't come to the wrong kind of male.)*

UP CLOSE   LEOPARD FROG

- **Scientific name:** *Rana pipiens*
- **Range:** From northern Canada to southern New Mexico and from eastern California to the Atlantic coast
- **Habitat:** Lives in the short grass of meadows and around ponds
- **Size:** Body length (legs excluded) of 5–9 cm (2–3.5 in.)
- **Diet:** Feeds on crickets, mosquitoes, and other insects

External Structures

Skin Numerous mucous glands embedded within the skin supply a lubricant that keeps the leopard frog's skin moist. A moist surface is necessary for respiration. Unlike many other frogs and toads, the leopard frog does not have skin glands that secrete poisonous or foul-tasting substances. Instead, the leopard frog must rely on its protective coloration and speed to evade predators, which include fishes, birds, turtles, and small mammals.

Eye Because its eyes bulge out from the head, the leopard frog can stay almost fully submerged while literally "keeping an eye out" for predators. Its eyes work equally well in or out of water. Eyelids that blink protect the eyes from dust. In addition, a transparent membrane covers each eyeball, keeping it moist and protecting it when the frog is underwater.

Tympanic membrane When sound causes the tympanic membrane (eardrum) to vibrate, a tiny bone transmits the vibrations to the middle ear. Within the middle ear are ciliated sensory cells (similar to those found in the lateral line of a fish) that are able to detect sound and help the frog maintain balance. Leopard frogs hear well in both water and air.

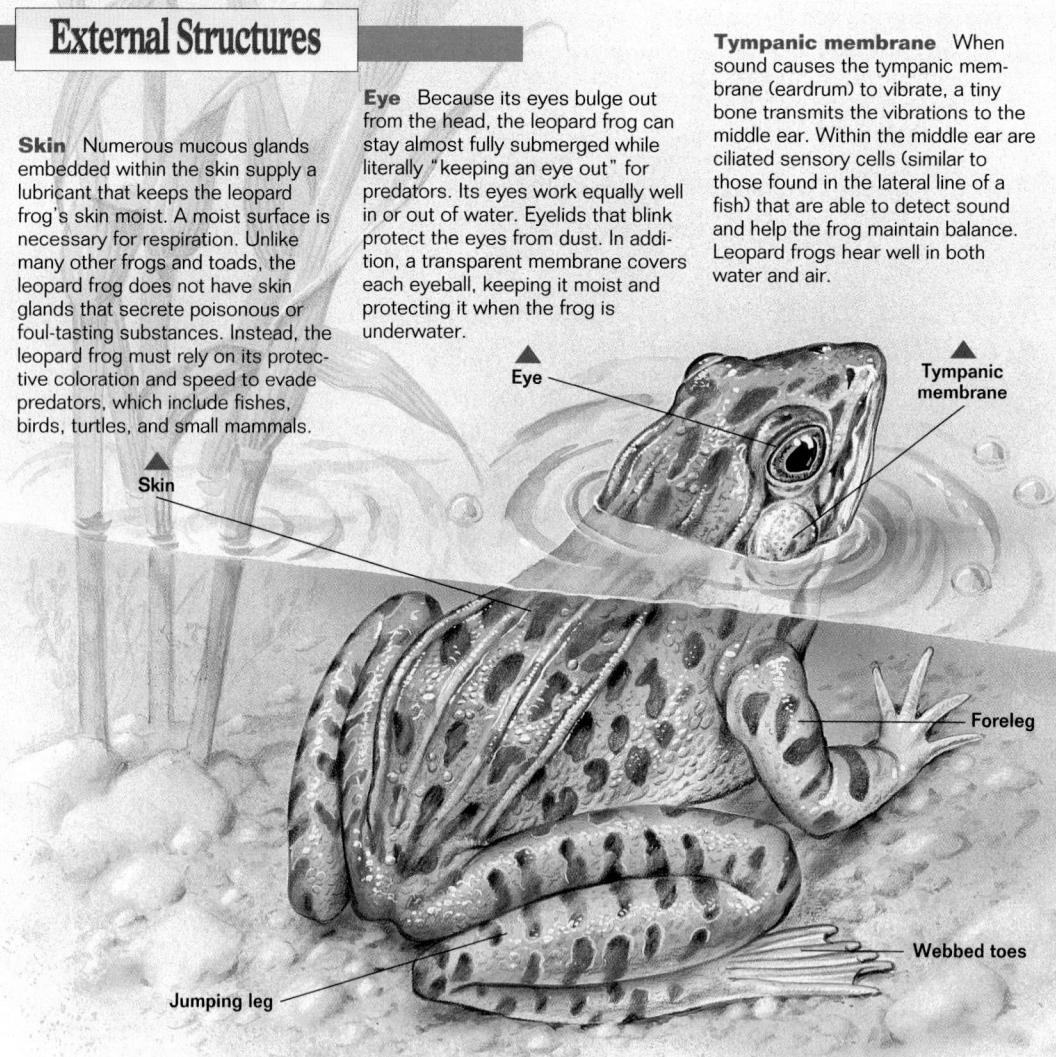

Internal Structures

Reproductive organs Prior to breeding, the reproductive organs of male and female leopard frogs produce enormous numbers of sperm and egg cells, respectively. Next, males establish breeding territories and call to females using prominent vocal sacs that amplify their croak. When a female approaches, the male frog climbs on her back and grasps her firmly just behind the forelegs. When the female leopard frog releases a cluster of eggs into the water, the male discharges his sperm over them, fertilizing them externally.

Brain What makes the frog's brain different from the fish's brain is the degree of development of each brain component. For example, the larger, more complex cerebrum of a frog is able to process a wider assortment of sensory information than that of a fish.

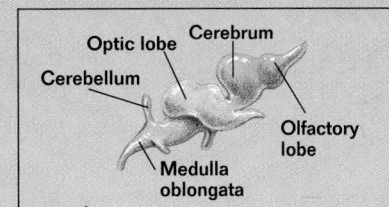

Optic lobe
Cerebrum
Cerebellum
Olfactory lobe
Medulla oblongata

▲ Brain

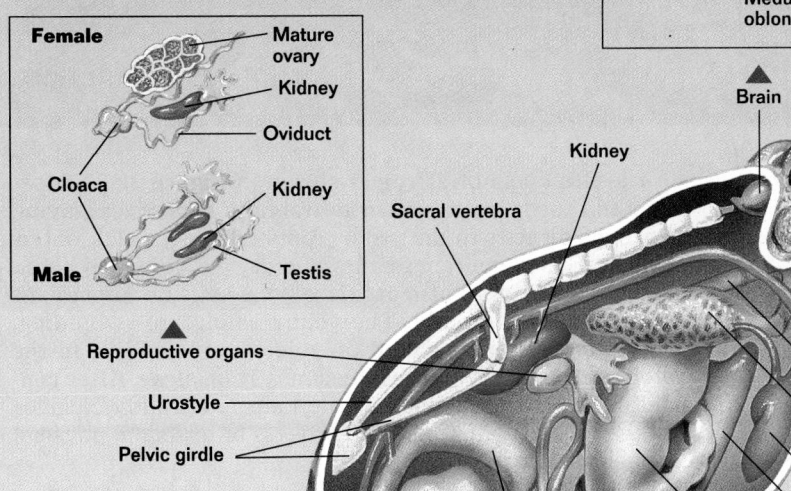

Female
Mature ovary
Kidney
Oviduct
Cloaca
Kidney
Male
Testis

▲ Reproductive organs

Urostyle
Pelvic girdle
Sacral vertebra
Kidney
Teeth
Esophagus
Lung
Heart
Liver
Tongue ▼
Stomach
Intestine
Urinary bladder
▲ Cloaca

Tongue The tongue flicks out at great speed, curls around the prey, then flicks back into the mouth with the insect. To prevent escape of its prey, the frog's upper jaw is lined with small, sharp teeth. In addition, two larger teeth project inward from the roof of the mouth to impale struggling prey. Food is swallowed whole.

Cloaca The excretory system of the leopard frog is very straightforward—there is only one way out. From the intestine, undigested foods are pushed into a cavity called the cloaca. Urine from the kidneys and bladder also passes into the cloaca, as do either egg cells or sperm cells from the reproductive organs. All of these materials exit the body through the cloacal opening.

Skeleton The skeletal system of the leopard frog (and all modern frogs) is highly reduced. It has only nine vertebrae and no ribs. From the sacral vertebrae, a long slender bone called the urostyle extends back to the center of the pelvic (hip) girdle. The bones of the frog's hind legs insert directly into sockets in the pelvic girdle. The two bones that make up the pelvic girdle extend forward to meet the sacral vertebrae. This three-strut support structure acts as a shock absorber for the long leg bones when the frog lands.

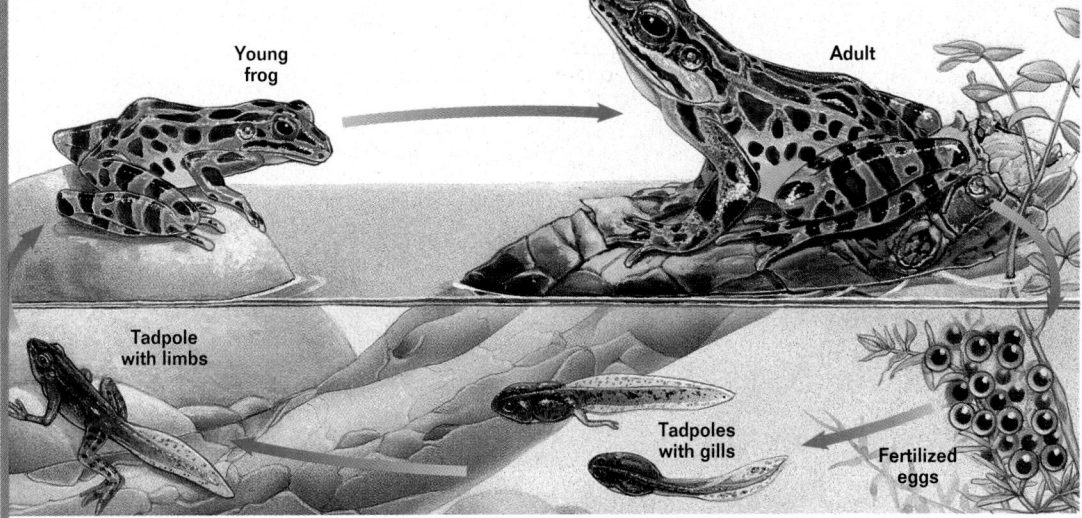

CONNECTIONS
....

Chapter 2
Cellular Digestion

Think about what must happen when a tadpole changes into a frog. Some structures must disappear. These changes occur as a result of cellular processes. To absorb a tail, the tadpole's lysosomes must first digest the tail cells.

 **VISUAL STRATEGY**

Figure 32-20, Figure 32-21

Ask students to compare the photographs of the salamanders with the picture of the tadpoles with gills. Then have them tell you their observations about the similarities and differences between the salamanders and the tadpoles.

Figure 32-20 The transition from a larval frog (tadpole) to an adult involves a complex series of external and internal body changes. The tadpole's tail, gills, and lateral line system all disappear, and legs grow from the body. A saclike bladder in the throat (the position of the air sacs in its fish ancestors) divides into two sacs that become lungs. The pulmonary vein develops, and the heart develops its internal walls.

The life cycle of a frog is shown in Figure 32-20. Most frogs and toads must return to water to reproduce, laying their eggs directly in the water. Amphibian eggs lack watertight external membranes and would dry out quickly if not in water. Like the eggs of most fishes, the eggs of amphibians are fertilized externally. The young hatch into swimming, fishlike larval forms called tadpoles. Tadpoles live in the water and, being herbivores, feed mostly on algae. After considerable growth, the body of the tadpole abruptly changes into that of an adult frog. This process of dramatic physical change is called metamorphosis.

Figure 32-21 Most salamanders, like this marbled salamander, *left,* live in damp places, such as under stones or logs or among the leaves of plants. Some aquatic salamanders, like the Texas spring salamander, *right,* spend their entire life in water.

Salamanders Have Tails
....

Salamanders, shown in Figure 32-21, are members of the order Urodela. Salamanders have elongated bodies, long tails, and smooth, moist skin. There are about 369 species of salamanders in 9 families. They typically range from 10 cm to 0.3 m (4 in. to 1 ft.) in

length, although giant Asiatic salamanders of the genus *Andrias* grow to as long as 1.5 m (5 ft.) and weigh up to 41 kg (90 lb.). In general, salamanders are unable to remain away from water for long periods as toads do, although some salamander species manage to live in dry areas by remaining inactive during the day.

Salamanders lay their eggs in water or in moist places. Fertilization is usually external, although a few species practice a type of internal fertilization in which the female picks up sperm packets that have been deposited by the male. Unlike anuran larvae, salamander larvae do not undergo a dramatic metamorphosis. The young that hatch from salamander eggs are carnivorous and resemble small versions of the adults, except that the young have gills.

Caecilians Are Legless

Caecilians, members of the order Apoda, are a highly specialized group of tropical, burrowing amphibians. This order is made up of about 168 species in 6 families. These legless, wormlike animals, shown in Figure 32-22, grow to about 0.3 m (1 ft.) long, but some species can be up to 1.2 m (4 ft.) long. During breeding, the male caecilian deposits sperm directly into the female. Depending on the species, the female may bear live young or lay eggs that develop externally. ☐

Figure 32-22 This caecilian, *left*, is from Colombia, in South America. A caecilian has very small eyes and is often blind. As it burrows through the soil, it searches for small invertebrate prey. A unique group of amphibians, caecilians have small bony scales embedded in their skin.

☐ **CAPSULE SUMMARY**

Modern amphibians include frogs, toads, salamanders, and the legless caecilians. Most amphibian species live in moist or aquatic habitats, eat small invertebrates, and lay eggs that are fertilized externally.

Section Review

1. *Offer an explanation for the fact that there are no frogs as large as elephants.*
2. *Although most amphibians have lungs, a few salamander species are lungless. In what kinds of environments would you expect to find these lungless salamanders?*
3. *Compare and contrast the anurans and apodans.*
4. *Explain why it is difficult to "sneak up" on a leopard frog.*

Critical Thinking

5. *Would you expect the digestive system of a tadpole to look and function the same way as the digestive system of an adult frog? Explain your answer.*

SECTION 32-3

Teaching Tips
Bait-Shop Salamanders
It is easy to get real salamanders to keep in your classroom if you have access to a bait shop. Ask if they have any "mud puppies," a type of salamander. The bait shop can tell you how to keep them and what to feed them.

What's the Difference?
Challenge students to tell you why a caecilian isn't a worm. Then challenge them to tell you why a caecilian isn't a snake.

Chapter Closure
Ask students to draw a concept map that shows the different groups of fishes and amphibians and their major characteristics.

CHAPTER REVIEW ANSWERS

Each item in the Chapter Review is correlated by text section in the assignment guide as follows.

ASSIGNMENT GUIDE

Section	Review	Themes Review	Critical Thinking
32-1	1–4, 8–10	16	19
32-2	5, 11,	17, 18	
32-3	6, 12, 13		
32-4	7, 14, 15	17	20

Review
Multiple Choice
Code in parentheses indicates section number and objective number.

1. c (32-1.1)
2. b (32-1.2)
3. a (32-1.3)
4. a (32-1.5)
5. d (32-2.1, 32-2.2, 32-2.3)
6. d (32-3.2)
7. a (32-4.2)

Completion
8. fish, Cambrian (32-1.1)
9. lampreys, hagfishes (32-1.4)
10. skeleton, teeth (32-1.5)
11. rise, sink (32-2.2)
12. water, land (32-3.1)
13. Permian, respiration (32-3.3)
14. gills, lungs (32-4.1, 32-4.3)
15. Anura, Urodela (32-4.2)

---

32 CHAPTER REVIEW

Vocabulary

lobe-finned fish (750)
operculum (755)
ray-finned fish (749)
teleost (749)

Review

Multiple Choice
1. Modern-day lampreys and hagfish evolved from an ancient order of
   a. sharks.
   c. ostracoderms.
   b. bony fishes.
   d. coelacanths.
2. The jaws of spiny fishes evolved from
   a. plates of bone.
   c. muscles.
   b. gill arches.
   d. ray fins.
3. Placoderms differed from ostracoderms in that placoderms had
   a. a jaw fused to the skull.
   b. full body armor.
   c. a swim bladder.
   d. two dorsal fins.
4. What characteristic did not help sharks become successful ocean predators?
   a. internal fertilization of eggs
   b. streamlined body
   c. powerful jaws
   d. sharp, replaceable teeth
5. Yellow perch and sharks share all of the following characteristics except
   a. gills.
   b. an internal skeleton.
   c. a single-loop circulatory system.
   d. a swim bladder.
6. The first amphibian evolved from
   a. chondrichthyes.
   c. agnathans.
   b. reptiles.
   d. rhipidistians.
7. Caecilians are different from other amphibians in that they
   a. are legless.
   b. can breathe through their skin.
   c. have a partially divided heart.
   d. mix oxygenated and deoxygenated blood.

Completion
8. The first vertebrates were jawless _____ that evolved during the _____ period.
9. Living agnathans include the parasitic _____ and scavenger _____ .
10. Sharks have a cartilaginous _____ that is covered by a thin layer of bone. Sharks also have bony _____ that are used for biting and tearing their food.
11. A fish will _____ when its swim bladder is filled with gas, but it will _____ when its swim bladder is emptied.
12. A typical amphibian lives a double life. It spends part of the time in the _____ and part of the time on _____ .
13. Amphibians reached their greatest diversity during the _____ period. The skin of the terrestrial amphibians that lived during this time was impermeable to water and was not used for cutaneous _____ .
14. During the tadpole stage, most amphibians respire by means of _____ . However, as adults they respire using simple, saclike _____ .
15. The order _____ includes frogs and toads. Amphibians that have tails are members of the order _____ .

Themes Review
16. **Evolution** Most sharks bear live young rather than laying eggs. How is bearing live young an adaptive advantage for these fishes?
17. **Structure and Function** Contrast the location and function of the swim bladder of a yellow perch with the urinary bladder of a leopard frog.
18. **Levels of Organization** What organs comprise the digestive system of the yellow perch?

Critical Thinking

19. Responding Critically A guide who leads tours of a large aquarium bases his discussion of the evolution of fishes on the idea that all bony fishes evolved from cartilaginous sharks. At the end of the tour, audience members are asked to rate and comment on various aspects of the guide's presentation. Suppose that you are a member of the audience. You are asked to rate and comment on the accuracy of the presentation. What rating on a scale of 1 to 10 (10 being the best) would you give the presentation's accuracy? What comments would you make to justify your rating?

20. Interpreting Data An ecologist collected data on the average level of the pesticide DDT contained in the tissues of insects eaten by leopard frogs in a pond. Additional data showed the number of leopard frog tadpoles captured from and released to the pond for six years. Based on the following data, what conclusions can you draw about the relationship between DDT levels and the number of tadpoles in the pond? Imagine that the average DDT level in insects is 10 parts per million in year 7. Will the ecologist likely see more or fewer tadpoles than in year 6?

Year	DDT in parts per million	Number of tadpoles
1	5	167
2	11	75
3	9	123
4	14	51
5	8	146
6	6	159

Activities and Projects

21. Multicultural Perspectives Research how dart-poison frogs are used by the native hunters of Central and South America. Find out why North American scientists are interested in studying the poison that is secreted by the skin of these frogs.

22. Research and Writing Do research to learn what caviar is and where it comes from. Also, contact a grocery store or market to find out the current price of caviar. Make a poster that presents the findings of your research in both words and pictures.

23. Cooperative Group Project Hundreds of thousands of frogs are killed and preserved each year by biological supply companies for use as specimens in laboratories. Organize a debate regarding the pros and cons of this practice. After forming debate teams, research various perspectives of the issue, including classroom demand for preserved specimens and the practices of the supply companies. Debate the issue before a class of students, following agreed-on rules of conduct. Allow the class to select a winner using predetermined criteria that are approved by your teacher.

Themes Review

16. Live young are mobile and are able to escape from some would-be predators. Eggs are easy prey to all would-be predators. (32-1.5)

17. The swim bladder in a yellow perch is located on its dorsal side. By increasing and decreasing the amount of gas in the bladder, the perch can adjust its depth in the water. In contrast, the leopard frog's urinary bladder is located ventrally and posteriorly, and functions in the storage of urine. (32-2.2, 32-2.3, 32-4.3)

18. Digestive organs include the esophagus, stomach, intestine, liver, and pancreas. (32-2.3)

Critical Thinking

19. Student ratings will vary but should be on the low end of the scale. Low ratings should be based on knowledge of the fossil evidence that indicates that fishes had bone before there were sharks, and that sharks and the "bony fishes" arose at about the same time. (32-1.6)

20. The higher the level of DDT, the fewer tadpoles in the pond. Fewer tadpoles will be seen in year 7. (32-4.1)

Activities and Projects

21. Native hunters dip the points of their arrows in the toxic sweat of dart-poison frogs to paralyze their prey. North American scientists are studying the frog sweat as a possible drug to stimulate the heart and brain, control pain, and relieve muscle spasms.

22. Caviar is the processed, salted eggs of a large fish, usually a sturgeon. Caviar is sold by the ounce and is quite expensive.

23. Encourage debaters to base their arguments on data and not on feelings or hearsay. Review rules of conduct and criteria used to select a winner.

UNIT 8
CHAPTER 32

LABORATORY Investigation Chapter 32

Preparation

To dechlorinate tap water, pour the water into a tank, and allow it to stand for two days. Pond water may be used in place of dechlorinated tapwater. You can purchase frogs from Ward's. Small, clear-plastic terrariums in which to examine frogs can also be purchased. Many pet stores stock live crickets or mealworms, which can be fed to the frogs. Release only native animals into your local environment. Consult the NABT *Guidelines for the Use of Live Animals.* Students should wash their hands after handling frogs.

Procedural Notes

- This investigation can be completed during one 40-minute period.
- Have students follow guidelines from the National Association of Biology Teachers (NABT) for handling vertebrate animals.
- Have groups take turns releasing their frog into the freshwater aquarium. After students have completed their observations, remove each frog from the aquarium yourself to avoid injury to the frogs.
- If you use frogs that are not native to your area, do not release them into the local environment when the investigation is completed. Students may wish to take the frogs home to keep as pets. Indigenous frogs may be released into an appropriate habitat.

Background Answers

1. *Amphibious* means having two natures or qualities.

2. Adult amphibians vary in the amount of time they spend on land. Most amphibians lay their eggs in water, where they hatch and develop into adults.

3. Amphibians are cold-blooded vertebrates that have a three-chambered heart, breathe both through their skin and with lungs, and have scaleless skin through which they take in and lose water. Most amphibians have appendages and lay eggs in water.

4. How are a frog's external features and behavior adapted to life on land and in water?

Live Frogs

OBJECTIVES

- Examine the external features of a frog.
- Observe the behavior of a frog.
- Explain how a frog is adapted to life on land and in water.

PROCESS SKILLS

- relating structure to function
- recognizing the relationship between structure and evolutionary success

MATERIALS

- live frog in a terrarium
- aquarium half-filled with dechlorinated water
- live insects (crickets or mealworms)
- 600 mL beaker

BACKGROUND

1. What does *amphibious* mean?
2. How do amphibians live part of their life on land and part in water?
3. What are some major characteristics of amphibians?
4. Write your own question to explore in your lab report or notebook.

TECHNIQUE

1. Observe a live frog in a terrarium. Closely examine the external features of the frog. Make a drawing of the frog in the **Records** section of your report. Label the eyes, nostrils, tympanic membranes, front legs, and hind legs. The tympanic membrane, or eardrum, is a disklike membrane behind each eye.

2. Make a table like the one below in the **Records** section of your report to note all your observations of the frog in this investigation.

Observations of Frog	
Breathing	
Eyes	
Legs	
Response to food	
Response to noise	
Skin	
Swimming behavior	

3. Watch the frog's movements as it breathes air with its lungs. Record your observations in the **Records** section of your report.

4. Look closely at the frog's eyes and note their location. Examine the upper and lower eyelids, as well as a third transparent eyelid called a nictitating membrane. The upper and lower eyelids do not move. The nictitating membrane moves upward over the eye. This eyelid protects the eye when the frog is underwater and keeps it moist when the frog is on land.

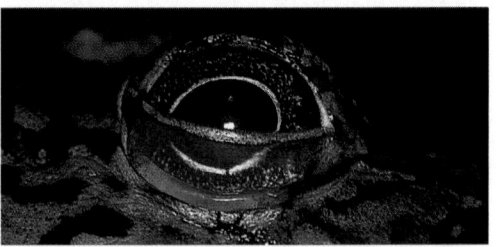

5. Study the frog's legs, noting the difference between the front and hind legs.

Records

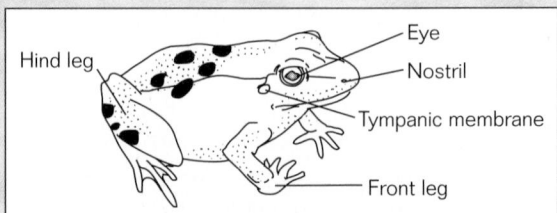

6. Place a live insect, such as a cricket or a mealworm, into the terrarium. Observe how the frog reacts.

7. Tap the side of the terrarium farthest from the frog and observe the frog's response.

8. **CAUTION: You will be working with a live animal. Handle it gently and follow instructions carefully.** Frogs are slippery! Do not allow the frog to injure itself by jumping from the lab bench to the floor. Place a 600 mL beaker in the terrarium. Carefully pick up the frog and examine the skin. How does it feel? The skin of a frog acts as a respiratory organ, exchanging oxygen and carbon dioxide with the air or water. A frog also takes in and loses water through its skin. Place the frog in the beaker. Cover the beaker with your hand and carry it to a freshwater aquarium. Tilt the beaker and gently submerge it beneath the surface of the water until the frog swims out of the beaker.

9. Closely watch the frog float and swim in the aquarium. How does the frog use its

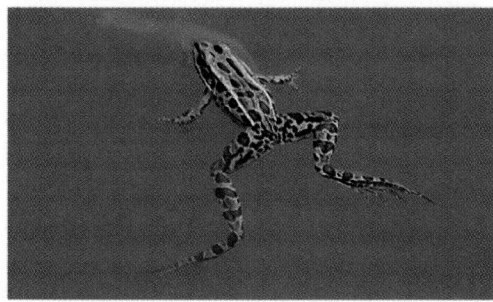

legs to swim? Notice the position of the frog's head. As the frog swims, bend down and look up into the aquarium so that you can see the underside of the frog. Then look down on the frog from above. Compare the color on the dorsal and ventral sides of the frog. When you are finished observing the frog, your

teacher will remove the frog from the aquarium. In your report, briefly summarize the procedure you followed.

10. Clean up your materials and wash your hands before leaving the lab.

INQUIRY

1. From the position of the frog's eyes, what can you infer about the frog's field of vision?

2. How does the position of the frog's eyes benefit the frog while it is swimming?

3. How does a frog hear?

4. How can a frog breathe air while it is swimming in water?

5. Why must a frog keep its skin moist while it is on land?

ANALYSIS

1. How are the hind legs of a frog adapted for life on land and in water?

2. What adaptive advantage do frogs have in showing different coloration on their dorsal and ventral sides?

3. What features provide evidence that an adult frog has an aquatic life and a terrestrial life?

FURTHER INQUIRY

Write a new question that could be explored as a new investigation. The following is an example:

How are other amphibians adapted to life on land and in water?

CHAPTER 32

Inquiry Answers

1. A frog can see in almost all directions.

2. Having eyes on the top of its head enables a frog to see above the waterline while submerged in water, thus allowing the frog to remain camouflaged.

3. Sound waves hit the tympanic membranes, causing them to vibrate. The vibrations send signals to the brain, allowing the frog to hear the sounds.

4. Keeping the nostrils above the waterline allows the frog to breathe with its lungs while it is swimming.

5. Moisture is necessary for the diffusion of oxygen and carbon dioxide through the skin. It also keeps the frog from drying out.

Analysis Answers

1. The webs of the hind legs push back water as the frog swims. The large muscles of the hind legs allow the frog to jump on land.

2. The dark, dorsal skin blends in with the color of the water as seen from above, thus making the frog less visible to predators above water. The light, ventral skin blends in with the color of the sky above the water, making the frog less visible to predators under water.

3. aquatic life: webbed feet, eyes on top of head, breathes through skin; terrestrial life: jumping legs, breathes with lungs

Observations of Frog	
Breathing	The floor of the frog's mouth moves up and down as it breathes.
Eyes	The frog's eyes bulge out from the top sides of its head. It has three eyelids. The upper and lower eyelids do not move. The transparent nictitating membrane moves upward over the eye to protect it when the frog is submerged under water.
Legs	The frog has four legs. Each front leg typically has four toes, and each back leg typically has five toes. Toes on the back legs are webbed.
Response to food	The frog eats by quickly capturing a cricket or mealworm with its tongue and pulling it into its mouth.

Observations of Frog	
Response to noise	Answers may vary. The frog should appear to be startled. It may jump or move around.
Skin	Most frogs have smooth, moist skin. Coloration varies among frog species.
Swimming behavior	The frog uses its hind legs and webbed back feet to move through the water. The frog's head stays partially above water. The ventral side of the frog is light colored, and the dorsal side of the frog is dark colored. Thus, the frog's coloring blends into the environment from below and from above.

UNIT 8

Block Scheduling Guide

Each block represents about 45–50 minutes of instructional time. The resources cited in a block are coded to a specific program component using the Key on the next page.

Lecture Concepts	Lesson Resources

33-1 History of Today's Reptiles pp. 771–773

BLOCK ❶

a. Present-day reptiles are classified into four orders: Chelonia (turtles), Rhynchocephalia (tuataras), Squamata (lizards and snakes), and Crocodilia (crocodilians).
b. Reptilian classification is based on the following three types of skull structures: anapsid, diapsid, and synapsid.

Lecture Resources
- Opening Questions, page 771
- Demonstration: Tied To Land
- Origin of Snakes, page 772
- Visual Strategy: Figure 33-3
- Anapsid, Diapsid, and Synapsid Skulls, page 773
- Teaching Transparencies: 173, 141A
- Holt Biology Videodiscs: Lesson 33-1

Classwork Options
- Effective Reading, page 771
- Reptile Keeper, page 772

Assignment Options
- Section Review, page 773
- Chapter Review, questions 1, 2, 12, 13, 23

33-2 Characteristics and Diversity of Living Reptiles pp. 774–781

BLOCK ❷

a. Reptiles are characterized by their dry, scaly skin (which minimizes water loss), a partially divided ventricle (completely divided in crocodiles), and lungs with a larger internal surface area than those of amphibians.
b. Reptiles lay watertight eggs that contain their own water supply and food source.
c. Present-day reptiles are ectothermic. Many species regulate their body temperature by controlling their exposure to sunlight.

Lecture Resources
- Opening Questions, page 774
- Demonstrations: Behavioral Temperature Regulation, Snakelike Lizards
- Visual Strategies: Figures 33-5, 33-7, 33-11, 33-12
- Overcoming Misconceptions, page 775
- Turtles vs. Tortoises, page 776

- Teaching Transparencies: 174, 179, 141A
- Holt Biology Videodiscs: Lesson 33-2

Classwork Options
- Up Close: Timber Rattlesnake, pages 778–779

Assignment Options
- Section Review, page 781
- Chapter Review, questions 3, 4, 12, 14, 20, 21

33-3 Dinosaurs pp. 782–787

BLOCK ❸

a. Dinosaurs are considered the most successful of terrestrial vertebrates. They first appeared in the Triassic period (248–213 million years ago) as small, bipedal carnivores.
b. During the Jurassic period (213–144 million years ago), dinosaurs such as sauropods, stegosaurs, and therapods were diverse and abundant.
c. The Cretaceous period (144–65 million years ago) was dominanted by iguanodonts, which were later replaced by duckbill dinosaurs, horned dinosaurs, and armored dinosaurs.
d. Dinosaurs disappeared at the end of the Cretaceous period (65 million years ago). The most likely cause was the impact of a meteorite off the coast of the Yucatan peninsula.

Lecture Resources
- Opening Question, page 782
- Demonstration: Dinosaur Skeleton
- How Do They Know That? page 783
- Overcoming Misconceptions, page 783
- Visual Strategy: Figure 33-15
- Research Update, page 785
- Demonstration: New Diet, New Dinosaurs
- A Worldwide Disaster, page 787
- Teaching Transparencies: 142A

- Holt Biology Videodiscs: Lesson 33-3

Classwork Options
- The Age of Dinosaurs, page 783
- Actual Size, page 784
- Were Dinosaurs Endothermic? page 787

Assignment Options
- Section Review, page 787
- Chapter Review, questions 5–7, 15, 18, 19

33-4 History of Birds pp. 788–789

BLOCK ❹

a. The oldest known bird is *Archaeopteryx*, which dates from 150 million years ago.
b. Most of the modern orders of birds are thought to have arisen before the extinction of the dinosaurs.

Lecture Resources
- Opening Questions, page 788
- Demonstration: Bird Bones
- Are Birds Reptiles? page 789
- Holt Biology Videodiscs: Lesson 33-4

Assignment Options
- Section Review, page 789
- Chapter Review, questions 8, 9

33-5 Characteristics and Diversity of Birds pp. 790–797

BLOCK ⑤

a. A bird's feathers, lightweight skeleton, and rapid metabolism help make flight possible.
b. Birds can be classfied according to the shapes of their beaks and feet.

Lecture Resources
- ■ Opening Questions, page 790
- ■ Demonstrations: Types of Feathers, Wings of Birds and Bats
- ■ White or Dark Meat? page 790
- ■ Preening, page 791
- ■ Heat Source, page 794
- ■ Numbers of Species, page 794
- 📖 Teaching Transparencies: 175, 176, 182, 143A
- ◢ Holt Biology Videodiscs: Lesson 33-5

Classwork Options
- ■ Up Close: Bald Eagle, pages 792–793

- ■ Summarizing the Differences Between Reptiles and Birds, page 796
- ■ Closure, page 797
- ◢ Laboratory Techniques and Experimental Design C40: Conducting a Bird Survey
- ■ Laboratory Investigation: Chicken Anatomy, pages 800–801

Assignment Options
- ■ Endangered Parrots, page 795
- ■ Section Review, page 797
- ■ Chapter Review, questions 10, 11, 16, 17, 22

BLOCK ⑥

Review/Enrichment

- ■ Study Guide: Concept Review and Pretest

Assignment Options
- ■ Activities and Projects, questions 24, 25
- ◉ Teaching Resources CD-ROM: Supplemental Reading Worksheets and Test, *The Dinosaur Heresies*

BLOCK ⑦

Assessment Options

Traditional Assessment
- ■ Chapter 33 Test
- ◉ Test Generator/Assessment Item Listing: Software item bank for preparing customized chapter tests.

Portfolio Assessment
Select student reports for one or more laboratory experiments from this chapter. *The Direct Observation Checklist* on the *BioSources Teaching Resources CD-ROM* should be completed during a laboratory or other cooperative learning experience.

Holt Biology Videodiscs · HRW · CONCEPTS OF BIOLOGY

Holt Biology Videodiscs gives you a powerful tool for teaching, review, and assessment. *Concepts of Biology* adds interactive lesson plans and extensive tools for customization using CD-ROM technology.

Use the following topic from *Concepts of Biology* to help you teach this chapter:

- ■ Topic 20: Vertebrate Diversity

For further information regarding the media on the videodiscs, see *Holt Biology Videodiscs Teacher's Correlation Guide to Biology: Principles and Explorations.*

Chapter Theme

Evolution *Students may have the misconception that reptiles are a "has-been" group that has been in decline since the dinosaurs became extinct. In fact, the class Reptilia is still a successful and diverse group, inhabiting all of the continents except Antarctica and containing about 7,000 species. In diversity among terrestrial vertebrates, reptiles are second only to birds.*

Tapping Prior Knowledge

- When did reptiles first appear?
- Explain how the skin and eggs of reptiles are well suited for life on land.
- From which reptiles did birds evolve?
- What are the unique characteristics of birds?

Opening Demonstration

Have students make a list of some of their preconceptions about snakes. For instance, many students believe that a snake's skin is slimy, like that of a frog. Another misconception is that the tongue of a snake is poisonous. Show the class a living snake, such as a garter snake, to dispel these misconceptions. Allow the students to examine and touch its skin, observe the flicking of its tongue, and watch its movements. Discuss with students how misconceptions contribute to the fear of snakes.

CHAPTER 33 REPTILES AND BIRDS

REVIEW

- continental drift (Section 13-3)
- angiosperms (Section 23-2)
- main characteristics of reptiles and birds (Section 31-1)
- ectothermy and endothermy (Section 31-1)
- differences between reptilian and avian circulatory and respiratory systems (Sections 31-2 and 31-3)
- structure of the amniotic egg (Section 31-4)
- oviparous, ovoviviparous, and viviparous reproduction (Section 31-5)
- *Highlights: Therapsids* on page 734

New Zealand gecko

Authors' Rationale

The reptiles truly conquered land, becoming independent of water with the evolution of the amniotic egg. During the Mesozoic era, dinosaurs predominated. But then they became extinct. Or did they? Reptiles are a diverse group—so diverse, in fact, that some scientists place not only the "typical" reptiles such as snakes, lizards, and dinosaurs in this group, but also birds and even mammals. It is not uncommon to hear heretical statements like "the dinosaurs aren't extinct—birds are dinosaurs." If any group is ripe for a debate on classification, this is the one.

33-1 History of Today's Reptiles

More than 7,000 species of reptiles (Class Reptilia) currently live on Earth. They are a highly successful group, more diverse than even the mammals (there are three species of reptiles for every two species of mammals). Although it is traditional to think of reptiles as more primitive than mammals, the great majority of reptiles that live today belong to groups that appeared *after* the therapsids, the group from which mammals evolved. This section examines the history of the living reptiles.

Section Objectives

■ *Identify the four orders of living reptiles.*
■ *Recognize the close relationship between crocodiles and birds.*
■ *Describe the evolution of the reptilian skull.*

Origin of Present-Day Reptiles

Figure 33-1 shows the relationship of the living reptiles to some of the extinct groups of reptiles, such as the dinosaurs, and to the groups that evolved from reptiles, the mammals and birds. Of the 16 orders of reptiles known to have existed, only 4 survive. The most ancient surviving group is the turtles (order

Figure 33-1 There are four orders of living reptiles: turtles, crocodilians, tuataras, and lizards and snakes. This phylogenetic tree shows how these four orders are related to each other and to dinosaurs, birds, and mammals.

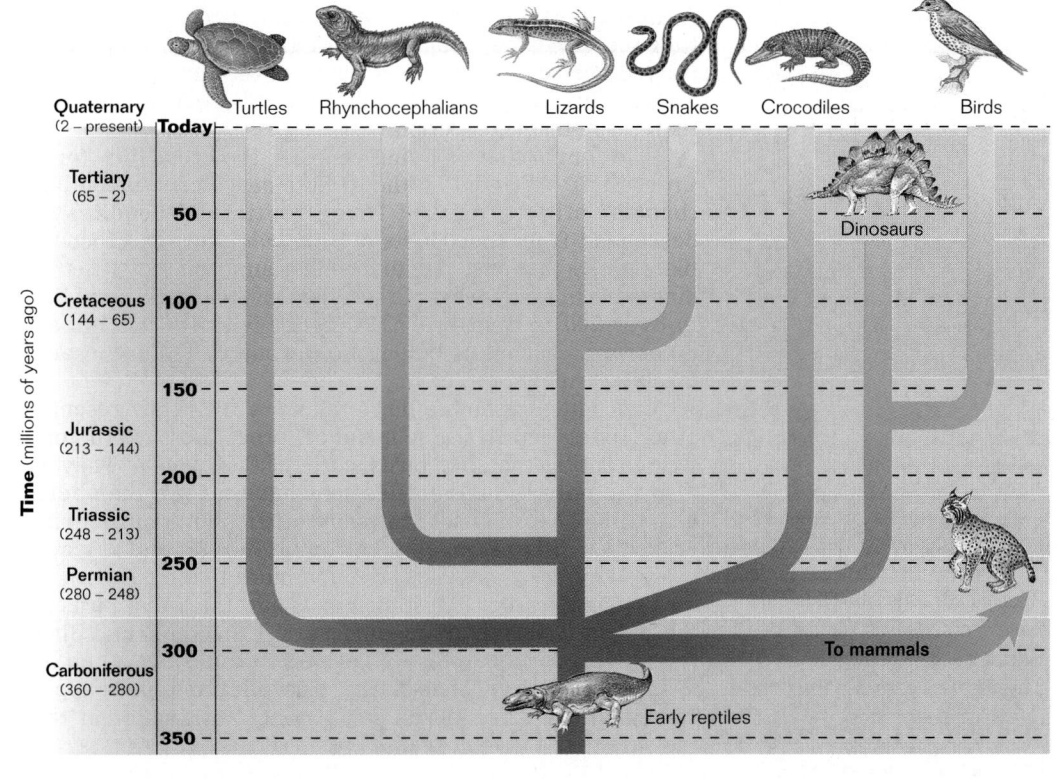

SECTION 33-1

Vocabulary Preview

anapsid
diapsid
synapsid

Opening Questions

Help students recall what they learned about the origins of reptiles in Chapter 31 by asking the following questions. From which animals did reptiles evolve? *(amphibians)* List three differences between reptiles and present-day amphibians. *(Reptiles have dry, scaly skin, produce amniotic eggs, have larger lung surface area than amphibians, and have hearts with a partially divided ventricle or, in the case of crocodiles, completely separated ventricles.)*

Demonstration
Tied To Land
Show the class a picture of an aquatic reptile such as a turtle or crocodile that is laying its eggs on land. Point out that just as most amphibians are tied to water for reproduction, egg-laying reptiles are tied to land. The amniotic, shelled egg enables reptiles to reproduce on land, but the embryo inside the egg will drown if the egg is laid in water. Tell students that even permanently aquatic reptiles such as sea turtles must return to land to lay eggs.

Effective Reading
Who, What, When, Where, and Why
Have students phrase the heads within this chapter as who, what, when, where, or why questions. Then they should read the chapter and answer the questions they created. For example, "Origin of Present-Day Reptiles" could be used to create the question, "When did present-day reptiles originate?" The answer would be that turtles and lizards appeared in the Permian period, tuataras and crocodiles in the Triassic period, and snakes in the Cretaceous period.

Application

 Language The term *reptile* is derived from the Latin word *repere*, meaning "to crawl."

Teaching Tips
Origin of Snakes

Point out that although the line leading to lizards and snakes diverged in the Permian period, there is no fossil evidence of snakes until the Cretaceous period. It is still a matter of debate as to which group of lizards contains the ancestors of snakes.

Reptile Keeper

Have students pick one of the present-day reptiles mentioned in this chapter and write a report describing what would be needed to maintain the reptile in good health. Students may use library books, pet store pamphlets, or observations of zoo exhibits. In their reports, students should explain why most reptiles don't make good pets and why the pet trade is endangering many reptile species.

Figure 33-2 The indigo snake is one of the largest snakes in the United States, reaching a length of 262 cm (100 in.). It lives in the southeastern states, Texas, and Mexico and feeds on mammals, birds, and other snakes.

☐ CAPSULE SUMMARY

Present-day reptiles are classified into four orders: Chelonia (turtles), Rhynchocephalia (tuataras), Squamata (lizards and snakes), and Crocodilia (crocodilians).

Chelonia). Turtles have skulls much like those of the first reptiles and have changed little in basic structure since before the time of the dinosaurs.

The vast majority of living reptilian species, including the indigo snake shown in Figure 33-2, belong to the second group to evolve, the lizards and snakes (order Squamata). About 6,800 species of this order are alive today and inhabit every continent except Antarctica. Lizards and snakes are descended from an ancient line of lizardlike reptiles that branched off the main line of reptilian evolution in the late Permian period, about 250 million years ago. Throughout the time of the dinosaurs, these reptiles survived as minor elements of the landscape, much as mammals did. Also like mammals, lizards and snakes became diverse and common only after the disappearance of the dinosaurs.

The third group of surviving reptiles to evolve were members of the order Rhynchocephalia. Rhynchocephalians (*RIHN koh suh FAYL ee uhnz*) are small reptiles that appeared shortly before the dinosaurs. They lived throughout the time of the dinosaurs and were common in the Jurassic period. They began to decline in abundance during the Cretaceous period, however, possibly due to competition from lizards. Rhynchocephalians were already rare by the time the dinosaurs disappeared. Today only two very closely related species of this order survive, the tuataras (*TOO uh TAHR uhz*) of New Zealand. You will read more about tuataras in the next section.

The fourth line of living reptiles, the crocodiles (order Crocodilia), appeared on the evolutionary scene much later than the other groups of living reptiles. Crocodiles are descended from thecodonts (*THEE kuh dawntz*), crocodile-like reptiles that gave rise to the dinosaurs, and they resemble dinosaurs in many ways. Crocodiles have changed very little in over 200 million years. Together with thecodonts and dinosaurs, crocodiles belong to a group called archosaurs, meaning "ruling reptiles."

Crocodiles resemble birds far more than they resemble other living reptiles. For instance, crocodiles are the only living reptiles that, like birds, care for their young. They are also the only living reptiles that have a four-chambered heart like that of birds. In many other points of anatomy, crocodiles differ from all other living reptiles and resemble birds. Why are crocodiles so much more like birds than are other living reptiles? Most biologists now think that birds are the direct descendants of dinosaurs. Crocodiles and birds, then, are far more closely related to dinosaurs than are lizards and snakes. That is why crocodiles and birds appear so similar; they are more closely related to each other than they are to lizards and snakes. ☐

Did You Know?

Until recently, most textbooks and experts recognized only one species of tuatara, *Sphenodon punctatus*. Though first described in the late 1800s, the second species of tuatara, *S. guntheri*, was overlooked or classified as a sub- species of *S. punctatus* until the 1990s. Molecular studies published in 1990 confirmed that *S. guntheri* differed enough from *S. punctatus* to be considered a separate species. Recognition of this second kind of tuatara may have come just in time to help save the species. Presently, its population numbers less than 300 individuals, all living on one 4-hectare island.

Evolution of the Reptilian Skull

The classification of reptiles into major groups is based largely on the structure of the skull. Early reptiles, like their amphibian ancestors, had **anapsid** skulls. This type of skull has openings only for the eyes and nostrils, as illustrated in Figure 33-3. The muscles that move the jaws lie underneath this solid covering. Of today's reptiles, only turtles have anapsid skulls. Because of this characteristic, turtles are thought to be closely related to the early reptiles.

As you learned in Chapter 31, reptiles evolved openings in the skull behind the eyes. **Diapsid** skulls have two openings on each side and characterize tuataras, crocodiles, and dinosaurs. In a living animal, these openings are covered by connective tissue. In a fossil, of course, the connective tissue has long since decayed, leaving "windows" in the skull. One advantage of these openings may have been a reduction in skull weight. They may also have permitted larger, stronger jaw muscles to evolve, since these muscles would no longer be confined by the solid skull wall and could expand into the opening when contracting.

The skulls of lizards and snakes are modified versions of the diapsid skull. Lizards have lost the arch of bone below the lower of the two openings. This arrangement increases the flexibility of the skull and allows the mouth to open wider. Snakes have even more flexible jaws because they have lost the arches below both openings, resulting in one large opening.

Very different skulls evolved among the reptiles that led to mammals. Therapsids had **synapsid** skulls with a single opening on each side. Mammals have the same type of skull, except that the opening is much larger and has merged with the eye socket.

Anapsid skull

Diapsid skull

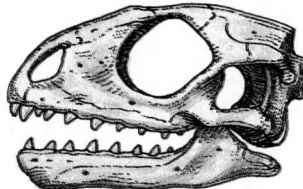
Modified diapsid skull of a lizard

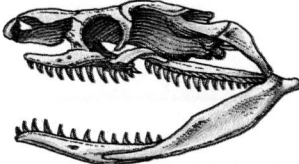
Modified diapsid skull of a snake

Synapsid skull

Figure 33-3 Skull structure is an important characteristic for classifying reptiles. One reason this characteristic is informative is that it is evolutionarily conservative, changing little within major groups. Characteristics such as tooth and limb structure tend to be more flexible and therefore provide less information for identifying major groups.

Section Review

1. *Which reptilian order is most closely related to the early reptiles?*

2. *Describe two similarities between crocodiles and birds.*

3. *If you found a reptilian skull, what characteristics would you look for to identify what type of reptile the skull belonged to?*

Critical Thinking

4. *Explain why a lighter skull would be particularly advantageous for a terrestrial vertebrate. In what ways would a lighter skull be disadvantageous?*

Vocabulary Preview

carapace

plastron

Opening Questions

Have students imagine that they are making a list of reptilian species and want to see as many as possible. Where in the United States could they see the most species of reptiles? *(the southern states)* Where could they see the fewest? *(the northernmost states)* Why are reptiles not found in very cold climates? *(As ectotherms, they need heat from the environment to warm their bodies. They cannot obtain enough heat in cold climates.)*

Demonstration

Behavioral Temperature Regulation

Show students a photograph of a lizard that is basking in the sun. Point out the behaviors that enable the lizard to absorb more solar energy. The lizard usually flattens its body against the surface, thereby exposing more of its surface to the sun. When cold, the lizard is also typically dark so that it absorbs more solar radiation.

Section Objectives

- Describe the structure of a turtle's shell.
- Identify two differences between lizards and snakes.
- Describe the timber rattlesnake's adaptations for locating and capturing prey.
- Contrast parental care in crocodilians with that in other reptiles.

Figure 33-4 The thorny devil, a fierce-looking but harmless lizard of arid Australia, is a representative reptile. Like North American horned lizards, the thorny devil feeds on ants.

*R*eptiles occur worldwide except in the coldest regions, where it is impossible for ectotherms to survive. In recent times, humans have had an adverse impact on the number and distribution of reptiles. Many species of turtles, for example, are prized for food and have been hunted almost to extinction. Despite overhunting and the destruction of many of their natural habitats, reptiles remain among the most numerous and diverse of terrestrial vertebrates. In this section you will study the major characteristics of reptiles and the four surviving reptilian orders, which contain about 7,000 species.

Key Characteristics of Living Reptiles

All living reptiles share certain fundamental characteristics, features they retain from the time when they replaced amphibians as the dominant terrestrial vertebrates. Figure 33-4 summarizes these key features, several of which are described in more detail in this section.

Amniotic Eggs Make Terrestrial Reproduction Possible

Amphibians never succeeded in becoming fully terrestrial because their eggs dry out and die unless laid in water or in moist areas. In contrast, reptilian eggs can be laid on

Characteristics of Living Reptiles

- Amniotic eggs, with embryo surrounded by two membranes, the chorion and the amnion
- Internal fertilization of eggs
- Dry, watertight, scaly skin that minimizes water loss
- Partially divided ventricle of heart, reducing dilution of oxygen-rich blood by oxygen-poor blood; in crocodiles, completely divided ventricle
- Lungs have larger internal surface area than lungs of amphibians
- Ectothermic

dry land because they are watertight and contain their own supply of water. A reptilian egg contains a food source (the yolk) and a series of four membranes: the yolk sac, the amnion, the allantois *(uh LAHN toh ihs)*, and the chorion. Each membrane plays a role in making the egg an independent life-support system. The outermost membrane of the egg is the chorion. This membrane allows oxygen to enter, but it retains water within the egg. The amnion encloses the developing embryo within a fluid-filled cavity. The yolk sac contains the yolk, which the embryo absorbs through blood vessels connected to its gut. The allantois surrounds a cavity into which waste products from the embryo are excreted. All modern reptiles show this pattern of membranes within the egg.

The majority of reptiles are oviparous. Their eggs are surrounded by a protective shell and are deposited in a suitable place in the environment. Heat from the environment incubates the eggs, which are usually left unprotected by the parents. However, several species of lizards and snakes, including rattlesnakes and some horned lizards, are ovoviviparous or viviparous and give birth to live, fully formed young.

Reptiles Are Ectothermic

The metabolism of present-day reptiles is too slow to generate enough heat to warm their bodies, so they must absorb heat from their surroundings. A reptile's body temperature is largely determined by the temperature of its environment. Many reptiles regulate their temperature by their behavior. They bask in the sun to warm up and seek shade to prevent overheating. Figure 33-5 shows that a lizard can maintain a fairly constant body temperature throughout the day by moving between sunlight and shade. You can also see why it is inaccurate to call ectotherms "cold blooded." The lizard's body temperature is higher than yours for part of the day. ☐

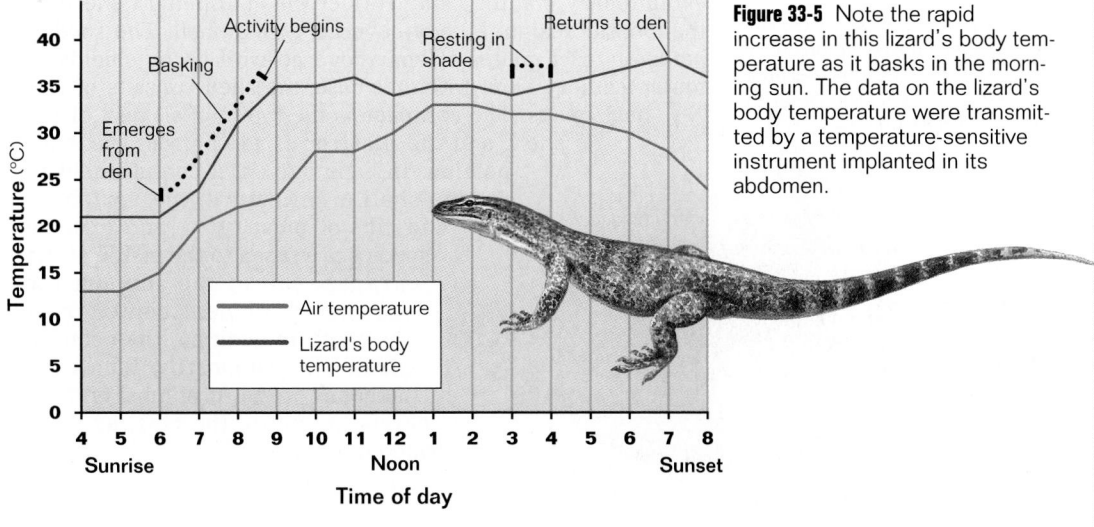

Figure 33-5 Note the rapid increase in this lizard's body temperature as it basks in the morning sun. The data on the lizard's body temperature were transmitted by a temperature-sensitive instrument implanted in its abdomen.

At very cold temperatures, reptiles become sluggish and unable to function. Intolerance of cold limits reptiles' geographical range and forces them to hibernate through the winter in temperate climates. Furthermore, the "low-power" metabolism of an ectotherm cannot support sustained, energy-demanding activity. Lizards are often swift sprinters, but they tire quickly. However, an ectotherm has one significant advantage over an endotherm, especially in environments where food is scarce. Because an endotherm uses 80 percent of its food energy to maintain a high body temperature, it must eat about 10 times more food than an ectotherm of the same size.

The thecodont ancestors of crocodiles were ectothermic, as crocodiles are today. The later dinosaurs, from which birds evolved, were probably endothermic, which is why crocodiles and birds differ in this one important aspect. This difference is a principal reason why crocodiles are classified as reptiles, while birds are not.

Figure 33-6 Like other sea turtles, this hawksbill turtle, *top*, spends virtually its entire life in the sea. Females leave the water only to lay their eggs on sandy beaches. Darwin was amazed by the diversity of shell shapes among the Galápagos tortoises, *bottom*. Each island has a different variety of tortoise with a uniquely shaped shell.

Figure 33-7 This cutaway view of a turtle's shell shows the unique arrangement of the ribs, pelvis, and pectoral girdle. The plastron has been removed.

Turtles and Tortoises

There are about 250 species of turtles (which generally live in water) and tortoises (which live on land), all classified in the order Chelonia. They differ from other reptiles in that their bodies are encased within a protective shell. Many of them can pull their head and legs into the shell for effective protection from predators. While most tortoises have a dome-shaped shell, water-dwelling turtles have a streamlined, disk-shaped shell that permits rapid turning in water. Turtles and tortoises lack teeth but have a sharp beak. Figure 33-6 shows a turtle and a tortoise.

Today's turtles and tortoises differ little from the first turtles that appeared more than 200 million years ago. This evolutionary stability may reflect the continuing benefit of their basic design, a body covered with a shell. The shell is made of fused plates of bone covered with horny shields or tough leathery skin. In either case, the shell consists of two basic parts. The **carapace** is the top (dorsal) part of the shell, and the **plastron** is the bottom (ventral) portion. In a fundamental commitment to this shell architecture, the vertebrae and ribs of most turtle and tortoise species are fused to the inside of the carapace; all of the support for muscle attachment comes from the shell. In addition, the pectoral girdle (the supporting bones for the bones of the forelimbs) and the pelvis lie within the ribs, as illustrated in Figure 33-7.

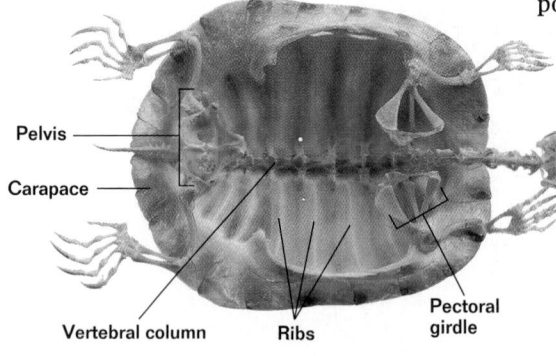

Pelvis

Carapace

Vertebral column   Ribs

Pectoral girdle

Tuataras Live Only in New Zealand

Today the order Rhynchocephalia contains only two very closely related species, the tuataras, members of the genus *Sphenodon*. *Sphenodon punctatus*, the more common and widespread species, is shown in Figure 33-8. Tuataras are lizardlike reptiles up to 70 cm (2 ft.) long. Unlike most reptiles, tuataras are most active at low temperatures. They burrow or bask in the sun during the day and feed on insects, worms, and small animals at night.

Figure 33-8 Tuataras formerly lived on several of the islands of New Zealand. When humans colonized New Zealand, they destroyed most of the forests and introduced rats, cats, dogs, weasels, and other animals that preyed on or competed with tuataras. Tuataras survived this onslaught only on a few small islands that were largely spared from human interference.

Lizards and Snakes

The order Squamata consists of about 3,800 species of lizards and about 3,000 species of snakes. The distinguishing characteristics of this order are the paired reproductive organs of males and a lower jaw that is loosely connected to the skull. This loose connection allows the mouth to open wide enough to accommodate large prey. In addition, the loss of the lower arch of bone below the lower opening in the skull of lizards makes room for large muscles to move the jaws. Most lizards and all snakes are carnivores. These improvements in jaw design have made a major contribution to their success as predators.

Lizards appeared in the late Permian period, about 250 million years ago. Common present-day lizards include iguanas, chameleons, geckos, horned lizards (often mistakenly called "horny toads"), and anoles. Several lizard species are shown in Figure 33-9. Most lizards are small, measuring less than 30 cm (1 ft.) in length. The largest lizards belong to the monitor family. The largest of these is the Komodo dragon of Indonesia, which reaches 3 m (10 ft.) in length and weighs up to 250 kg (550 lb.).

Figure 33-9 This gecko, *bottom*, like many species of lizards, has the ability to lose its tail when seized by a predator and to then grow a replacement. This individual's tail apparently did not completely detach, and a second tail has begun to grow. The Gila monster, *top right*, of the southwestern United States and northern Mexico is one of only two species of venomous lizards. Because it lacks the venom-injecting fangs found in poisonous snakes, the Gila monster is not dangerous to humans unless harassed. A common lizard throughout the western United States is the western fence lizard, *top left*, named for its habit of perching on fences.

Application

Geography Have students use a globe or world map to locate New Zealand. Point out that before humans colonized New Zealand, tuataras lived on both of the main islands of New Zealand (North Island and South Island). Humans and their introduced animals have eliminated tuataras from all but a few small islands.

CONNECTIONS

Chapter 19
Scientific Names
One advantage of scientific names over common names is that common names for organisms can be misleading. The common names for horned lizards—horned toads, or horny toads—are good examples.

Historical Note

New Zealand's Disappearing Flora and Fauna

New Zealand separated from Australia about 80 million years ago. During this long isolation, a variety of unique plants and animals evolved in New Zealand, including tuataras, huge flightless birds called moas, flightless parrots, kauri trees, and kiwis. The first Polynesian settlers, who probably arrived around the eighth century A.D., began the destruction of this unique flora and fauna. By the time Europeans sighted New Zealand in the 1640s, all 12 species of moas had been hunted to extinction. European settlers accelerated and expanded the assault on the environment. Today, over 50 percent of the native species of birds have disappeared from New Zealand, and many more are endangered.

Instructional Strategies

- Point out to students that only a small minority of snake species are venomous. In the United States, only 17 of 115 species are poisonous. These snakes inflict about 1,500 bites a year, about 45 of which are fatal.

- Help students understand the rattlesnake's ability to sense by using a warm object such as a heating pad. Place the heating pad on a table and allow it to warm the surface. Remove the heating pad and have students move their hands above the table top without touching the table itself. They should be able to feel the heat radiating from the warmed spot without actually touching the table. Have students determine how close their hand must be to detect the heat. A snake can locate warm prey from a distance of 1 m (39 in.).

- There are many misconceptions about why a snake flicks its tongue out. Have students pay special attention to the section "Jacobson's organs." Students should note that snakes flick their tongue in order to sample the air. Particles picked up from the air are transferred to the Jacobson's organs, which "taste" (smell) the air.

- The survival advantage of having a rattle is the source of much conjecture. Why would a predator warn other animals of its presence? Some biologists think that the rattle is an advantage to snakes that might be injured when stepped on by large herbivores. A warning rattle benefits both the snake and the bison, deer, or elk that doesn't notice a camouflaged snake on the ground.

UP CLOSE TIMBER RATTLESNAKE

- **Scientific name**: *Crotalus horridus*
- **Range:** Eastern and central United States, from northern New York to northern Florida and central Texas
- **Habitat:** Prefers areas of thick brush, dense woodland, or swamp
- **Size:** Typically 90–150 cm (36–60 in.); maximum 189 cm (74 in.)
- **Diet:** Mainly mammals

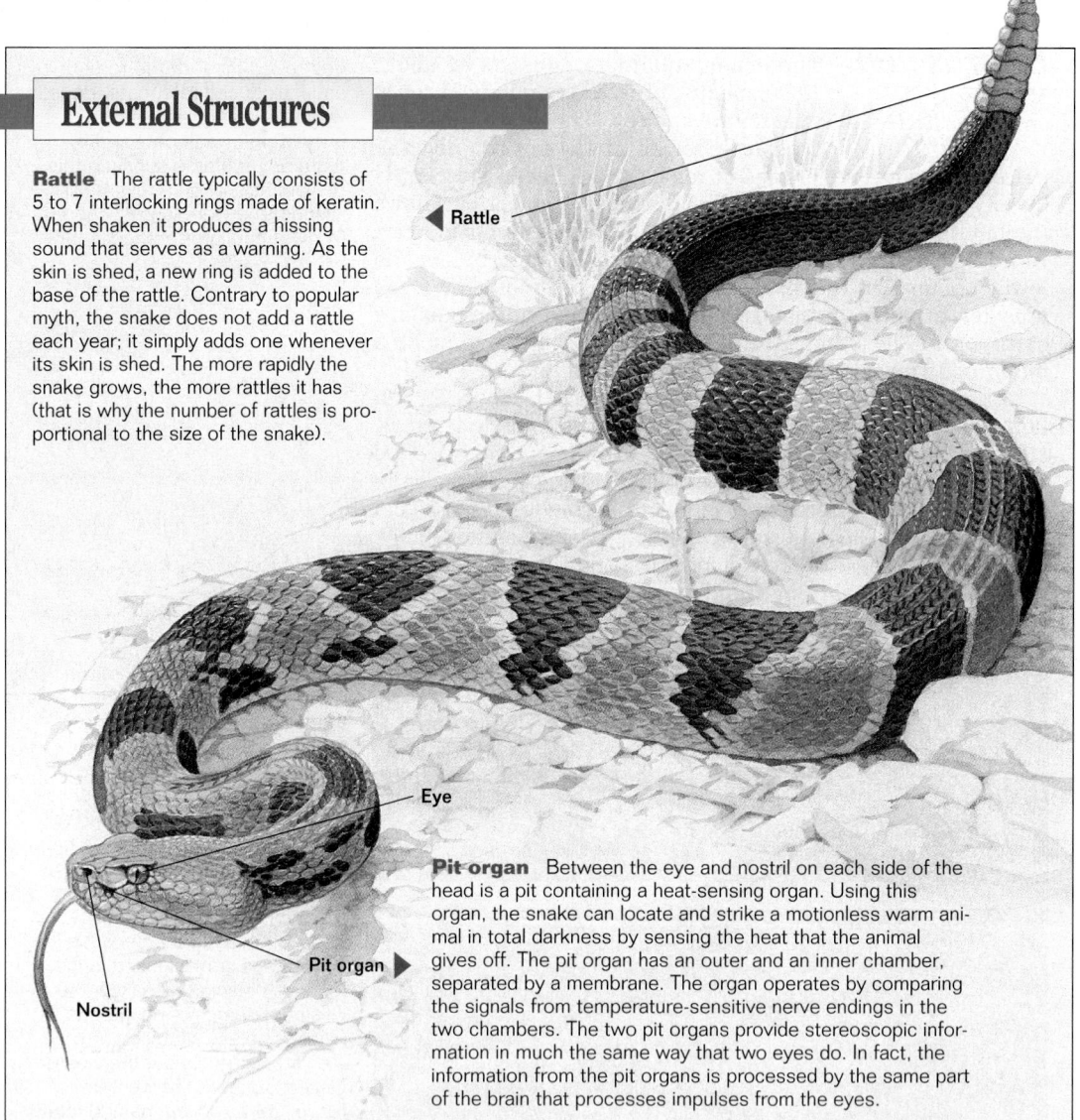

External Structures

Rattle The rattle typically consists of 5 to 7 interlocking rings made of keratin. When shaken it produces a hissing sound that serves as a warning. As the skin is shed, a new ring is added to the base of the rattle. Contrary to popular myth, the snake does not add a rattle each year; it simply adds one whenever its skin is shed. The more rapidly the snake grows, the more rattles it has (that is why the number of rattles is proportional to the size of the snake).

Rattle

Eye

Pit organ Between the eye and nostril on each side of the head is a pit containing a heat-sensing organ. Using this organ, the snake can locate and strike a motionless warm animal in total darkness by sensing the heat that the animal gives off. The pit organ has an outer and an inner chamber, separated by a membrane. The organ operates by comparing the signals from temperature-sensitive nerve endings in the two chambers. The two pit organs provide stereoscopic information in much the same way that two eyes do. In fact, the information from the pit organs is processed by the same part of the brain that processes impulses from the eyes.

Pit organ

Nostril

Internal Structures

Venom The timber rattlesnake injects its prey with toxic venom. Its upper front teeth are large hollow fangs. When the rattlesnake strikes, these hinged fangs swing forward from the roof of the mouth and inject venom deep into the prey. The venom contains hemotoxins, proteins that attack the circulatory system, destroying red blood cells and causing internal hemorrhaging. Modified salivary glands in the upper jaw produce the venom.

Venom gland

Fang

Jacobson's organs

Tongue

Reproductive organs This male rattlesnake produces sperm in his testes. Female timber rattlesnakes are ovoviviparous. A female carries her fertilized eggs in her body throughout development. Each egg has a thin membrane through which water and oxygen pass from the mother to the embryo, although all nourishment is provided by the egg's yolk. After the eggs hatch in the mother's body, the hatchlings are ejected to fend for themselves.

Small intestine

Testes

Large intestine

Kidneys

Pancreas

Gallbladder

Stomach

Unique internal anatomy The internal organs are elongated, matching the snake's body shape. The left lung is nonfunctional.

Liver

Cloaca

Right lung

Left lung

Esophagus

Trachea

Heart

Jacobson's organs
Flicking its forked tongue into the air, the rattlesnake gathers chemicals from the environment. These chemicals are transferred to two depressions in the roof of the mouth called Jacobson's organs, which act as very sensitive taste buds.

Movement without legs The rattlesnake moves by slithering—undulating gracefully and with surprising speed. This motion is made possible by the unique anatomy of its backbone and muscles. The backbone is made up of several hundred vertebrae, each with its own pair of attached ribs. These bones provide the framework for thousands of muscles. The muscles manipulate not only the skeleton but also the snake's skin, causing this fabric of overlapping scales to extend and contract. To move, the timber rattlesnake slides its head to one side, which initiates a wave of muscular contractions along the vertebrae. As this lateral motion moves down its trunk, the rattlesnake uses the sides of its body to push off against pebbles, twigs, and small irregularities of the ground, moving forward in an S-shaped path.

Discussion

Guide the discussion by posing the following questions.

1. Why is it advantageous for the rattlesnake to inject its prey with venom? *(It subdues and starts digesting the prey.)*

2. In what ways is the rattlesnake a typical reptile? *(It has dry, scaly skin, produces amniotic eggs, breathes with lungs, and has a partially divided ventricle in the heart.)*

3. What are some of the atypical features of the timber rattlesnake? *(It has no legs, has only one functional lung, produces venom, and is ovoviviparous.)*

Figure 33-10 A South American emerald boa rests on a branch, *left.* The copperhead, *right,* is a common venomous snake in eastern and southern North America.

Figure 33-11 A snake's very flexible jaws allow it to swallow prey much larger than its head. This copperhead is swallowing a mouse.

Figure 33-10 shows a small sample of the diversity among snakes. Snakes probably evolved from lizards during the Cretaceous period. The close relationship between lizards and snakes is reflected in their many similarities. In fact, it is often difficult to distinguish between lizards and snakes. All snakes are limbless, but so are a number of lizards, such as the glass lizard of the United States. Snakes lack movable eyelids and external ears, as do several species of lizards. However, the internal anatomy of snakes is unique. No snake has any trace of a pectoral girdle, which is found even in legless lizards. Recall also that snakes have lost both arches of bone behind the eye, while lizards have lost only the lower arch. The snake jaw is also distinctive. It is very flexible because it has five points of movement (your jaw has only one movement point). One of these points is the chin, where the halves of the lower jaw are connected by a ligament and can spread apart when a large meal is being swallowed, as shown in Figure 33-11.

Very few species of snakes are poisonous. The poison is produced by modified salivary glands and is injected into the victim through grooved or hollow teeth. Snakes dangerous to humans are concentrated in four families: cobras, kraits, and coral snakes; sea snakes; adders and vipers; and rattlesnakes, water moccasins, and copperheads.

You can read more about the biology of snakes in *Up Close: Timber Rattlesnake* on pages 778–779.

Crocodiles and Alligators

The order Crocodilia is composed of 25 species of large, aquatic reptiles. In addition to crocodiles and alligators, this order also includes the alligator-like caimans and the fish-eating gavial. You can see two species of crocodilians in Figure 33-12.

All crocodilians are aggressive carnivores. They generally capture prey by stealth, often floating just beneath the water's surface near the shore. If an animal comes to the water to drink, the crocodilian explodes out of the water, seizes the prey, and hauls it back into the water to be

Teaching Tip
Endangered Crocodiles
Although crocodilians are fierce predators and some species occasionally eat humans, many of the 25 species of crocodilians are endangered or threatened. Overhunting—for their hides, which are used to make leather goods—is the chief cause of the decline in numbers of crocodilians.

◉ VISUAL STRATEGY

Figure 33-12
Have students note the crocodiles basking in Figure 33-12. Ask students why crocodilians are found only in tropical or subtropical climates. (*Because they are large, crocodilians require large amounts of heat to warm their bodies. They would warm up too slowly in temperate climates.*)

drowned and eaten. The bodies of crocodilians are well adapted for this form of hunting. Their eyes are high on the sides of the head, and their nostrils are on top of the snout; this enables them to see and breathe while lying nearly submerged in the water. They have an enormous mouth studded with sharp teeth and have a very strong neck. A valve in the back of the mouth prevents water from entering the air passages when crocodilians feed underwater.

Unlike other living reptiles, crocodilians care for their young after hatching. For instance, a female American alligator builds a nest for her eggs from rotting vegetation. Heat from the decaying vegetation helps to incubate the eggs. After the eggs hatch, the mother may tear open the nest to free the hatchlings. The young alligators remain under her protection for up to a year.

Figure 33-12 The American alligator, *above left,* lives throughout the southern United States from Texas to Florida. Twenty years ago, the alligator was an endangered species, but careful management has restored its numbers. These crocodiles, *above right,* are basking alongside a river in India.

Section Review

1. Contrast the position of a turtle's shoulder with that of yours.

2. Why is the lack of legs not a good characteristic for distinguishing a lizard from a snake?

3. A blinded rattlesnake can still strike its prey. Explain how.

4. Explain how the parental care shown by alligators differs from that shown by most other reptiles.

Critical Thinking

5. Many viviparous snakes and lizards live in cold climates. Why might viviparity be advantageous in such environments?

Opening Question

Ask students to explain how the word *dinosaur* is used in everyday conversation. *(It is usually used to describe something that is large and lumbering or something that is old and nearing [or deserving] extinction.)* Have students make a list of the attributes they would expect dinosaurs to have, based on the everyday usage of the term. Point out that dinosaurs were not necessarily lumbering and slow, and that they certainly were not destined for extinction. Dinosaurs were a very successful group that probably prevented the mammals from diversifying.

Demonstration

Dinosaur Skeleton

Show students a photograph of a nearly complete dinosaur skeleton. Ask them what inferences they can draw regarding the diet and habits of the dinosaur just by looking at the skeleton. For instance, the teeth will probably give away the diet. Sharp, pointed teeth indicate a carnivore, while flattened grinding teeth indicate a herbivore. Long limbs may indicate a fast runner. Point out that everything scientists know about dinosaurs has been inferred from fossils such as skeletons, footprints, and impressions of skin.

33-3 Dinosaurs

Section Objectives

- Recognize three factors that contributed to the success of the dinosaurs in the Triassic period.
- Explain the role of continental drift in the evolution of dinosaurs.
- Describe how plant evolution affected the evolution of dinosaurs.
- Explain the meteorite-impact hypothesis for the extinction of the dinosaurs.

Of the major groups of terrestrial vertebrates, dinosaurs are considered the most successful. They dominated life on land for roughly 150 million years. During their long history, dinosaurs changed a great deal because the world they inhabited changed. One reason they changed was that the continents moved and radically altered the Earth's climates. Thus, you cannot study dinosaurs as if they were a particular kind of animal, with one type representing the group. Rather, you have to look at dinosaurs more as a story, a long parade of change and adaptation. This section will examine a variety of very different dinosaurs—animals that lived at different times and were adapted to very different worlds.

Figure 33-13 How do paleontologists know what dinosaurs were like, what they ate, and how they walked? Paleontologists must draw their conclusions from traces the dinosaurs left behind, such as this skeleton of a small, carnivorous dinosaur.

The Triassic Period: Origin of Dinosaurs

Dinosaurs are reptiles that first appeared in the Triassic period (248–213 million years ago). The first dinosaurs evolved from thecodonts, a group of crocodile-like, carnivorous reptiles that are now extinct. Fossils of the oldest dinosaurs for which there is clear evidence were found in Argentina in early Triassic rock some 235 million years old. Only about 30 cm (1 ft.) long, the oldest known dinosaur is a bipedal carnivore named *Eoraptor*, illustrated in Figure 33-14. Almost as old is *Herrerasaurus*, a bulky carnivore 4 m (13 ft.) long. *Herrerasaurus* had sharp, pointed teeth and a sliding jaw joint that enabled it to slice its victims as it bit into them.

These early dinosaurs were the first vertebrates to have a key improvement in body structure: their legs were positioned directly under the body, allowing them to run swiftly after prey. By the end of the Triassic period, small, carnivorous dinosaurs were very common; most of them were lightly built and bipedal.

In the late Triassic period, all of the continents were joined in a single supercontinent called **Pangaea** *(pan GEE uh)*. There were few mountain ranges over this enormous stretch of land, and the interior was arid. Coastal climates were much the same all over the world—quite warm, with a dry season followed by a very wet monsoon season. By the end of the Triassic period, some 22 million years after the oldest known dinosaur fossils appear in the fossil record, dinosaurs had become common and had largely replaced the thecodonts.

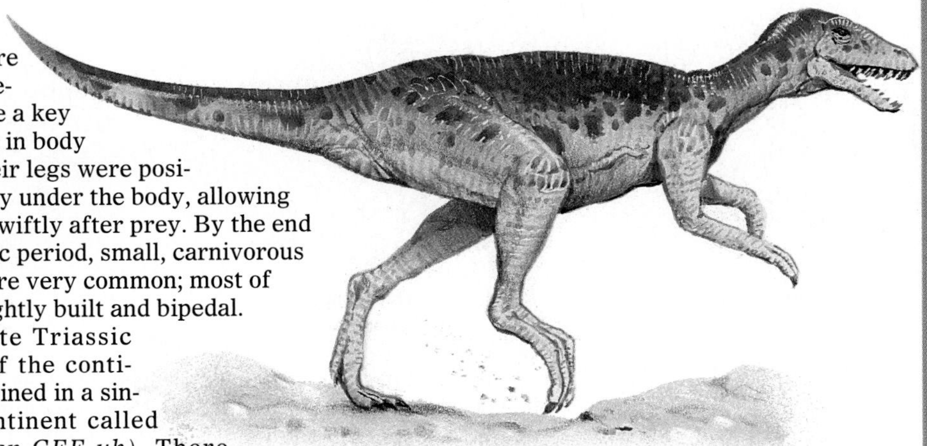

Figure 33-14 *Eoraptor* was about the size of a chicken. Fossils of this dinosaur, the oldest species yet unearthed, were discovered in Argentina in 1989. Since no one has seen a living dinosaur, the coloration of the dinosaurs shown in this section is the artist's speculation.

There are at least three reasons why dinosaurs were so successful.

1. **Leg structure** Legs positioned directly under the body enabled the dinosaurs to be faster and more agile runners than the thecodonts.
2. **Drought resistance** In the late Triassic period, Pangaea's interior was dry, and dinosaurs were superbly adapted to arid conditions. The mammals of the late Triassic period, by contrast, were not as effective at water conservation. They lost water by sweating, their chief means of releasing excess body heat.
3. **Luck** At the end of the Triassic period, a large meteorite landed in northeastern Canada (the site, the Manicuoagan Crater, is still visible today), and it might have been responsible for the great loss of diversity that occurred at the end of the Triassic period. Thecodonts and many other species became extinct, but the dinosaurs survived.

The Jurassic Period: The Golden Age of Dinosaurs

At the beginning of the Jurassic period (213–144 million years ago), vast deserts still covered much of Pangaea. What was to become western North America was covered by a sea of sand during much of this time. The dominant trees were cycads, which resemble palm trees and are well adapted to arid climates.

SECTION 33-3

Teaching Tips

Fred Flintstone Paleontology
Cartoons often portray cave men with dinosaurs, so it is important to stress that there is no credible fossil evidence of humans coexisting with dinosaurs.

How Do They Know That?
Students may wonder how some of our knowledge of dinosaurs was obtained. For example, how do scientists know that dinosaur limbs were positioned underneath the body, as in mammals? Dinosaur footprints, preserved when mud or wet volcanic ash hardened into rock, indicate that dinosaurs walked with their legs underneath their bodies—the prints of the left and right feet are close together.

The Age of Dinosaurs
Have students create a time line illustrating when different dinosaurs lived on Earth. Divide the class into groups of two to three students, and have one group make a time line across the wall for the Triassic, Jurassic, and Cretaceous periods. Have the other groups pick a dinosaur to diagram, label the dinosaur, and indicate its correct place on the time line. If you give students a scale to follow, it is possible to compare the sizes of the dinosaurs. A scale of 1 cm of diagram to 1 m of dinosaur works well if the time line has a scale of 1 cm to 1 million years, and if students place the dinosaur diagrams above or below the time line. A time line of this scale will be about 2 m long.

Overcoming Misconceptions

Dinosaur Contemporaries
In common parlance, all large reptiles of the Mesozoic era were dinosaurs. This is not accurate. Many other kinds of large reptiles coexisted with the dinosaurs, and these reptiles are not dinosaurs. Examples of these dinosaur contemporaries include the flying reptiles known as pterosaurs, the dolphinlike ichthyosaurs, and the long-necked plesiosaurs. Like the dinosaurs, these large reptiles did not survive past the end of the Cretaceous period.

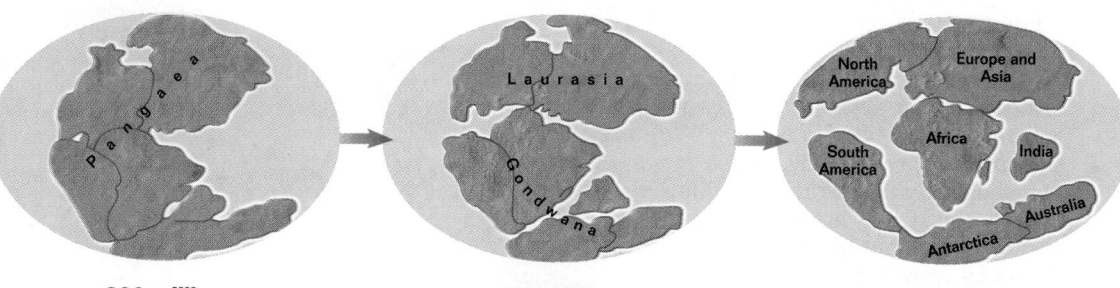

200 million years ago

180 million years ago

65 million years ago

Figure 33-15 The continents were not in their present positions when the dinosaurs lived. In the Triassic period, *above left,* all of the continents were connected into a huge supercontinent, Pangaea. This landmass began to break up during the Jurassic period, *above center.* By the Cretaceous period, *above right,* today's continents began to be recognizable, although Australia and Antarctica did not separate until 50 million years ago, 15 mil-lion years after the dinosaurs became extinct.

The great supercontinent of Pangaea was beginning to break up, as shown in Figure 33-15. Long fingers of ocean began to separate the northern part, called Laurasia (the future continents of North America, Europe, and Asia), from the southern part, called Gondwana (India and the future con-tinents of South America, Africa, Australia, and Antarctica). These two landmasses were fully separated by the end of the Jurassic period. World sea levels began to rise, and much of Laurasia and Gondwana were flooded by sea water, forming shallow inland seas. Because so much of the land was nearer to the oceans, conditions became progressively less arid. Also, the world's climate became even warmer.

The Jurassic period is called the golden age of dinosaurs because of the variety and abundance of dinosaurs that lived during this time, including the largest land animals of all time, the sauropods (*SAWR oh pawdz*). Sauropods, such as *Brachiosaurus* and *Diplodocus* shown in Figure 33-16, were the dominant herbivores of the Jurassic period. All sauropods had enormous barrel-shaped bodies, heavy columnlike legs, and very long necks and tails.

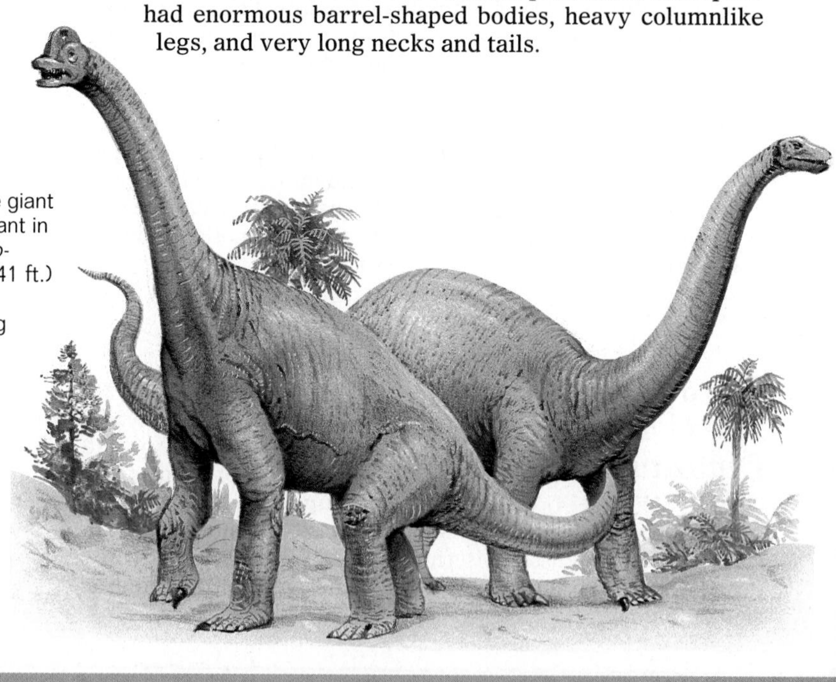

Figure 33-16 Sauropods were giant herbivores that were abundant in the Jurassic period. *Brachio-saurus, left,* stood 12.5 m (41 ft.) tall and was 23 m (75 ft.) in length. It weighed 81,000 kg (89 tons), more than 14 African elephants or 1,500 students. *Diplodocus, right,* was longer and slimmer, reaching 26 m (86 ft.) in length and weighing 10,000 kg (11 tons). *Apatosaurus,* which is commonly but incorrectly known as *Brontosaurus,* is closely related to *Diplodocus.*

Historical Note

Brontosaurus's Name Change
Most students probably recog-nize the name *Brontosaurus,* but few realize that *Apato-saurus* is the accepted name for this genus. In the late 1800s, the paleontologist O. C. Marsh named both genera based on fragmentary fossils. Later research showed the two genera to be one. By the rules of priority, the first name Marsh coined, *Apatosaurus,* was applied to the genus, although this name (meaning "deceptive lizard") is certainly less evoca-tive than *Brontosaurus* (mean-ing "thunder lizard").

The second largest Jurassic herbivores, about the size of a big pickup truck, were the stegosaurs (STEHG uh sawrz). For example, *Stegosaurus*, illustrated in Figure 33-17, weighed about 900 kg (1 ton) and was 4.5 m (15 ft.) long. Stegosaurs fed on plants that grew close to the ground. They had a row of narrow plates along their back and sharp spikes at the tip of their tail.

By the late Jurassic period, very sophisticated carnivorous dinosaurs had evolved. These dinosaurs, known as theropods (THEHR uh pawdz), were descendants of *Herrerasaurus* and preyed on the large herbivorous dinosaurs. Theropods were the dominant terrestrial predators until the dinosaurs disappeared at the end of the Cretaceous period. Figure 33-18 shows three representative theropods, all of which had the typical body structure of this group: bipedal stance, powerful legs, short arms, and a large head. This anatomy was well suited for rapid running and quick, slashing attacks.

Figure 33-17 What was the function of the plates along *Stegosaurus*'s back? Like the spikes on the tail, they may have been defensive weapons. However, the plates were interlaced with numerous blood vessels, which suggests they played a part in heating and cooling the animal.

Figure 33-18 There were three types of theropods. Coelurosaurs (suh LUHR uh sawrz) were small and swift with hollow bones. Birds are thought to have descended from this group, which is typified by *Composgnathus, bottom.* Raptors (RAHP turz), typified by *Velociraptor, top,* had lethal, sickle-shaped claws on each foot for ripping open their prey. Most raptors were about the size of a human. The largest theropods were the carnosaurs (KAHRN uh sawrz), typified by *Tyrannosaurus rex, above left. T. rex* was 6 m (20 ft.) tall and weighed 7,300 kg (8 tons). Each of its sharp teeth was 15 cm (6 in.) long.

The Cretaceous Period: Triumph of the Chewing Dinosaurs

The Jurassic period ended about 144 million years ago and was followed by the Cretaceous period (144–65 million years ago), a time of profound change for dinosaurs. During the Cretaceous period, Laurasia and Gondwana for the most part split into the continents we know today. Sea levels continued to rise. By the mid-Cretaceous period, sea levels had reached an all-time high, and the interior of North America was a vast inland sea. The climate of much of the world was tropical—hot and wet, like in a greenhouse. Most important, the Cretaceous period saw the rise to dominance of

☑ **RESEARCH Update**

Tyrannosaurus rex: Coldblooded Killer or Warm and Cuddly Killer?

Many researchers believe that at least some dinosaurs might have been endothermic. Reese Barrick and William Showers, a paleontologist and a geochemist, respectively, from North Carolina State University are in the endothermic camp. They are attempting to convince the rest of the highly skeptical paleontological community that *Tyrannosaurus rex* was endothermic. These collaborators have been studying oxygen isotopes in the bones of *T. rex* fossils, including a *T. rex* fossil whose bones are not completely mineralized. Oxygen isotopes are temperature-sensitive, and if they become locked in a fossil, contend the NCSU scientists, they may be able to give information about the body temperature of the original animal.

Barrick and Showers made repeated and improved tests on *T. rex* bones and concluded that body temperature did not vary significantly (less than 4°C) in any region of the body, so the dinosaur was probably endothermic. Critics of this hypothesis point to two potential flaws: inaccuracy of the oxygen isotope test, and that the ratio of oxygen isotopes may not be an indication of endothermy (high metabolism). Other scientists contend that even if the oxygen isotope readings do indicate constant temperature, temperature would have been constant across the animal due to its sheer size. ☑

CONNECTIONS

Chapter 23
Angiosperms
Remind students that angiosperms are flowering plants such as daisies, apple trees, and grasses. Also point out that the majority of today's plant species are angiosperms.

Chapter 16
Rare Predators
Point out to students that large predators are usually rare even today. This rarity is a consequence of the loss of energy as it passes from trophic level to trophic level. Therefore, *Tyrannosaurus rex*'s rarity is not unexpected.

Demonstration
New Diet, New Dinosaurs
Simulate the physical digestion of sauropods and stegosaurs by filling a strong, clear, plastic bag with egg-sized rocks and several tufts of grass. Shake the bag for a couple of minutes, then open it and examine the grass. Simulate the grinding action of iguanodont teeth by placing several tufts of grass in a mortar and grinding it with a pestle for the same amount of time. Compare this grass with the first sample, and ask students which method breaks grass down more effectively *(grinding)*. Ask them why grinding food is important for efficient digestion. *(It breaks down the structure of the plant material, allowing digestive enzymes access to more of the macromolecules it contains.)* Then have students apply their observations to explain the success of the iguanodonts at a time when angiosperms were becoming the most common terrestrial plants.

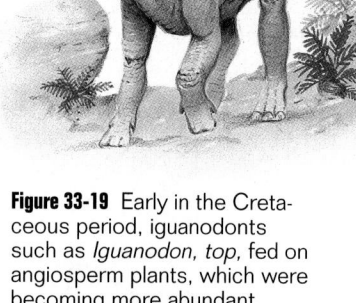

Figure 33-19 Early in the Cretaceous period, iguanodonts such as *Iguanodon, top,* fed on angiosperm plants, which were becoming more abundant. Fossils of *Iguanodon* were among the first dinosaurs discovered. Later in the Cretaceous period, duckbill dinosaurs such as *Maiasaura, bottom,* replaced the iguanodonts.

flowering plants, angiosperms. This event had a profound effect on herbivorous dinosaurs. Stegosaurs and most sauropods became extinct, replaced by a totally different kind of plant eater that was better adapted for consuming these tough, versatile plants.

Sauropods and stegosaurs did not chew the plants they ate. Their teeth had no grinding surfaces. Sauropods and stegosaurs shredded leaves and stems and swallowed the shreds whole. Rocks within their stomachs then battered the plant material to a pulp. Though a good strategy with cycads, which have soft, pulpy interiors, this approach didn't work as well with angiosperms, which are much tougher. With the rise of angiosperms during the Cretaceous period, sauropods and stegosaurs were replaced first by iguanodonts *(ih GWAHN uh dawntz),* which had chewing teeth. The jaws of iguanodonts contained enormous grinding teeth that could shred, pound, and grind even the toughest angiosperms. Even bigger than the stegosaurs, *Iguanodon,* the iguanodont illustrated in Figure 33-19, was as heavy as an elephant.

Later in the Cretaceous period, iguanodonts were replaced by three very successful groups of large chewing herbivores. The first group was the duckbill dinosaurs, typified by *Maiasaura,* illustrated in Figure 33-19. Many of these dinosaurs had bony crests on their heads. The second group was the horned dinosaurs, which had horns and bony head frills. *Triceratops,* shown in Figure 33-20, is a well-known member of this group. The third group, the armored dinosaurs, is typified by *Ankylosaurus,* also shown in Figure 33-20. Armored dinosaurs were the most diverse of all dinosaurs.

The flesh-eating theropods of the Cretaceous period were more diverse and formidable than those of the Jurassic. Among the largest was *Tyrannosaurus rex.* Only a few dozen *Tyrannosaurus rex* skeletons have been discovered, indicating that this large predator was probably never common.

Figure 33-20 Two Cretaceous herbivores were *Triceratops, left,* and *Ankylosaurus, right.*

Extinction of the Dinosaurs

Toward the end of the Cretaceous period, sea levels began to fall and the climate began to cool. Many kinds of dinosaurs became less common, and then suddenly, at the end of the Cretaceous period 65 million years ago, all dinosaurs disappeared. What caused the sudden extinction of the dinosaurs after 170 million years of existence? Most scientists now agree that the most likely cause was the impact of a gigantic meteorite 8–16 km (5–10 mi.) in diameter off the coast of the Yucatan peninsula in Mexico, as illustrated in Figure 33-21. The thin line of sediment that marks the end of the Cretaceous period in rocks is rich in iridium (a mineral rare in the Earth's crust but common in meteorites), tiny spheres of cooled molten rock, and bits of quartz shocked by a high-velocity impact. The impact created a huge crater 300 km (185 mi.) in diameter and threw massive amounts of material into the atmosphere that would have blocked out all sunlight for a considerable period of time, creating a worldwide period of low temperature. The endothermic birds and mammals, insulated with feathers or fur, survived. The ectothermic reptiles and amphibians also survived, because they could simply lower their activity levels.

No one can be sure why the dinosaurs did not live through the deep cold. Disease might have killed them, or massive volcanic eruptions might have led to their extinction. However, the most reasonable and widely accepted explanation is that the cold itself killed them. Most, if not all, Cretaceous dinosaurs appear to have been endothermic. Unlike birds and mammals, however, dinosaurs had no insulation, no way to retain body heat. Endothermy was a great contribution to the success of the dinosaurs, but it created an evolutionary dead end from which they could not emerge.

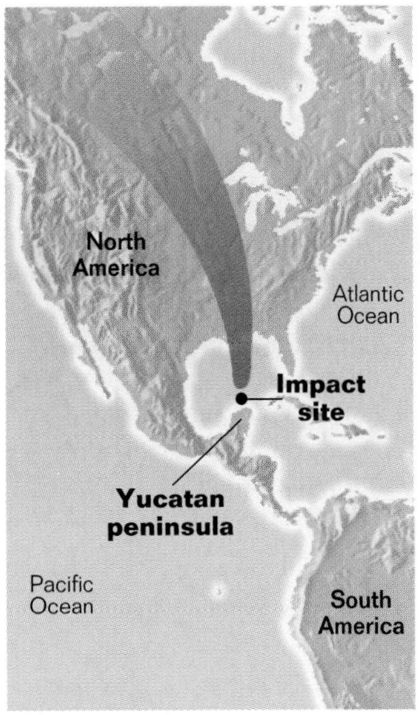

Figure 33-21 A very large impact crater lies off the coast of the Yucatan peninsula in Mexico. The age of this crater coincides with the extinction of the dinosaurs at the end of the Cretaceous period. Scientists have calculated that an impact large enough to create this crater would have thrown enough dust into the atmosphere to drastically reduce the amount of sunlight reaching the Earth's surface.

SECTION 33-3

Teaching Tips
A Worldwide Disaster

Be sure students understand that the dinosaurs weren't the only organisms to disappear at the end of the Cretaceous period. There were massive extinctions in the oceans as well as on land, so any explanation for the extinction of the dinosaurs has to also explain the disappearance of other organisms. Also emphasize that extinction had been happening throughout the Age of Dinosaurs. *Brachiosaurus,* for instance, died out in the Jurassic period. The end of the Cretaceous period saw the extinction of the dinosaur species alive at the time, not the extinction of all species that had ever existed.

Were Dinosaurs Endothermic?

This question is still the subject of vigorous debate. Have students research the controversy about endothermic dinosaurs. Have students write a report that summarizes the evidence for and against endothermy in dinosaurs. In their reports, students should state which side they believe has made the strongest case.

Section Review

1. *Describe two of the characteristics of dinosaurs that gave them an advantage in the late Triassic period.*

2. *How did the breakup of Pangaea alter the global climate?*

3. *What effect did the rise to dominance of the angiosperms have on the evolution of dinosaurs?*

4. *What combination of features made the dinosaurs particularly vulnerable to cold periods?*

Critical Thinking

5. *Describe two pieces of evidence that would disprove the meteorite-impact hypothesis for the extinction of the dinosaurs.*

Answers to Section Review

1. Dinosaurs were faster and more agile due to a new leg structure that positioned the legs under the body. Dinosaurs were able to live successfully in drought conditions. A meteorite caused widespread weather changes that killed off many animals and created new opportunities for dinosaurs.

2. Shallow inland seas formed, the land became less arid, and the climate became warmer.

3. Angiosperms often have tougher leaves and are harder to chew, so dinosaurs with the ability to chew and grind food had an advantage in using this new food source.

4. The dinosaurs were endothermic and lacked insulation, so they couldn't stay warm in the cold weather and couldn't reduce their body temperature to conserve energy.

5. Evidence that dinosaurs died out before the impact would disprove the hypothesis, as would evidence that they survived for a substantial period after the impact.

Vocabulary Preview

(none)

Opening Questions

Ask students about the origin of birds. What living reptiles are most closely related to birds? *(crocodiles)* Name two characteristics crocodiles and birds share. *(heart with completely divided ventricles, parental care of young, shelled eggs)* Why are birds and crocodiles more similar than birds and lizards? *(Birds evolved from dinosaurs, which evolved from thecodonts. Crocodiles also evolved from thecodonts.)*

Demonstration

Bird Bones

Bring clean chicken bones to class for students to examine. Cut a few bones in half so the hollow structure can be seen. If you have a wishbone, allow students to pull it gently to see how flexible it is. The cartilage keel on the breast of a chicken is also interesting to examine because it provides an attachment surface for flight muscles.

Section Objectives

- Recognize the position of *Archaeopteryx* in avian evolution.
- Identify two characteristics that *Archaeopteryx* shares with modern birds.

Figure 33-22 Macaws, *far right,* are fruit-eating parrots of Central America and South America.

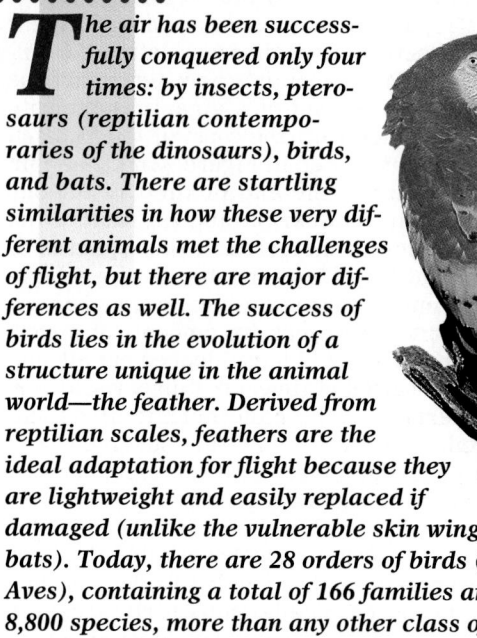

33-4 History of Birds

The air has been successfully conquered only four times: by insects, pterosaurs (reptilian contemporaries of the dinosaurs), birds, and bats. There are startling similarities in how these very different animals met the challenges of flight, but there are major differences as well. The success of birds lies in the evolution of a structure unique in the animal world—the feather. Derived from reptilian scales, feathers are the ideal adaptation for flight because they are lightweight and easily replaced if damaged (unlike the vulnerable skin wings of bats). Today, there are 28 orders of birds (Class Aves), containing a total of 166 families and about 8,800 species, more than any other class of terrestrial vertebrates.

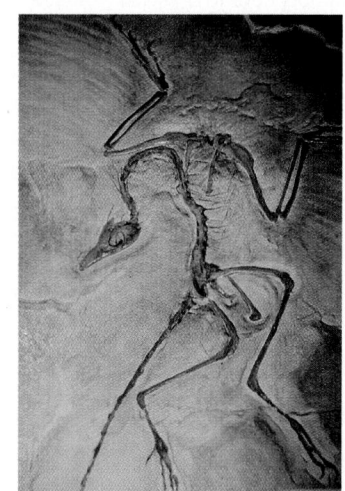

Figure 33-23 Notice the clear impressions of feathers on this specimen of *Archaeopteryx.*

Origin of Birds

The earliest known bird is *Archaeopteryx* (meaning "ancient wing"), which is shown in Figure 33-23. The first specimen of *Archaeopteryx* was found in a limestone quarry in Bavaria (in southern Germany) and is about 150 million years old. *Archaeopteryx* was about the size of a crow and shared many features with small theropod dinosaurs. For example, it had teeth and a long reptilian tail, and very few of its bones were fused to each other. These are features of dinosaurs, not of birds. Also, it had no breastbone such as modern birds have to anchor flight muscles. And unlike the hollow bones of present-day birds, its bones were solid. Finally, it had the forelimbs of a dinosaur. Because of these dinosaurian features, several *Archaeopteryx* fossils were originally classified as *Compsognathus,* a dinosaur of similar size, until impressions of feathers were discovered on the fossils. What makes *Archaeopteryx* distinctly avian is the presence of feathers on its wings and tail. It also had other avian features, notably a wishbone (dinosaurs had no wishbone).

Today almost all biologists agree that *Archaeopteryx* is very closely related to *Compsognathus.* Indeed, some biologists go so far as to classify *Archaeopteryx* and other birds as "feathered dinosaurs" and speak jokingly of "carving the dinosaur" at Thanksgiving dinner. However, most biologists continue to classify birds in a separate class, Aves, because of their key evolutionary novelties: feathers, hollow bones, and physiological mechanisms—such as super-efficient lungs—that permit sustained, powered flight. This judgment should not conceal the agreement among almost all biologists that birds are the direct descendants of theropod dinosaurs.

By the early Cretaceous period, only 15 million years after *Archaeopteryx*, a variety of birds with many of the features of modern birds had evolved. Fossils discovered within the last few years in Mongolia, Spain, and China reveal a diverse collection of toothed birds with the hollow bones and breastbones necessary for sustained flight. Other birds were highly specialized for a flightless, diving existence. The diverse birds of the Cretaceous period shared the skies with pterosaurs for 70 million years.

Since the impressions of feathers are rarely fossilized, and since modern birds have hollow, delicate bones, the fossil record of birds is incomplete. Relationships among the families of modern birds are mostly inferred from studies of the degree of DNA similarity among living birds. These studies suggest that the ostrich and its relatives belong to the oldest group of living birds. Ducks, geese, and other waterfowl arose next, in the early Cretaceous period, followed by a diverse group of woodpeckers, parrots, swifts, and owls. The largest of the bird orders, the Passeriformes, or songbirds (containing 60 percent of present-day bird species), appeared in the mid-Cretaceous period. The more specialized orders of birds, such as shorebirds, birds of prey, flamingos, and penguins, did not evolve until the late Cretaceous period. All but a few of the modern orders of birds are thought to have arisen before the disappearance of the pterosaurs and dinosaurs 65 million years ago. ◻

□ **CAPSULE SUMMARY**

The oldest known bird is Archaeopteryx, which dates from 150 million years ago. Most of the modern orders of birds are thought to have arisen before the extinction of the dinosaurs.

SECTION 33-4

Teaching Tip
Are Birds Reptiles?
The answer to this question depends on who you ask. For cladists, birds are reptiles, albeit feathered ones. Cladistic classifications only accept groups that contain all the descendants of a common ancestor. Birds evolved from dinosaurs, so they must be included in the same group as the dinosaurs, the reptiles. For more traditional biologists, the unique features of birds—feathers, lightweight skeleton, efficient respiratory system—justify placing them in a group of their own.

CONNECTIONS
....

Chapter 14
DNA and Phylogeny
Remind students that the degree of DNA similarity has also been used to infer the relationships among apes and humans.

Section Review

1. *In later editions of* The Origin of Species, *Darwin cited* Archaeopteryx *as an example of a transitional form. Is* Archaeopteryx *a good candidate for a transitional form between reptiles and birds? Explain your answer.*

2. *Explain why most biologists classify birds in a separate class rather than with the reptiles.*

Critical Thinking

3. *What is an adaptive advantage of toothlessness for birds?*

Answers to Section Review

1. *Archaeopteryx* is a good candidate for a transitional form because it had both reptilian and avian characteristics. Its reptilian characteristics are teeth, a long tail, solid bones, no breastbone like that of modern birds, and few fused bones. Its avian features are feathers and a wishbone.

2. Birds are classified separately because they have features that no reptile has: feathers, hollow bones, and very efficient lungs.

3. Because teeth are heavy, they are disadvantageous for a flying animal.

Vocabulary Preview

(none)

Opening Questions

Help students think about the diversity of birds by asking which is the largest bird in the world. *(ostrich)* Which are the smallest? *(hummingbirds)* Where do birds live? *(on all continents)* Do all birds fly? *(No. There are several flightless species, including ostriches and kiwis.)*

Demonstration
Types of Feathers

Show students a contour feather and a down feather. Allow students to examine each kind of feather under a microscope and compare their texture. On the contour feather, students should see the interlocking hooks and barbs pictured in Figure 33-25. Students should notice that, because of the interlocking hooks and barbs, the contour feather keeps its shape even when touched. Contour feathers provide the surface area for the wings and tail, so they must be smooth and aerodynamic. Point out that hooks and barbs are not found on down feathers, which serve only as insulation.

Teaching Tip
White or Dark Meat?

The dark meat on a chicken is partly due to the presence of myoglobin, a hemoglobin-related molecule that helps provide oxygen to muscle that must contract vigorously and repeatedly. Ask students why a duck has dark breast meat, but a chicken doesn't. *(The duck flies long distances, the chicken just flies short distances.)* Ask what chickens usually use for locomotion and what color that meat is. *(They usually use their legs to walk, and the legs are composed of dark meat muscle.)*

33-5 Characteristics and Diversity of Birds

Modern birds lack teeth and have only a vestigial tail, but they retain many other reptilian characteristics. For instance, birds lay amniotic eggs. Also, reptilian scales are present on the feet and lower legs of birds. What makes birds unique? What traits distinguish them from living reptiles? This section will answer these questions.

Section Objectives

■ Describe two functions of feathers.
■ Identify two avian adaptations for weight reduction.
■ Describe the digestive system of the bald eagle.

Main Characteristics of Birds

Figure 33-24 lists several distinguishing features of birds. Some of these features are explained in more detail following the figure.

In addition, more information on the anatomy and habits of birds is presented in *Up Close: Bald Eagle* on pages 792–793.

Figure 33-24 This tern is a representative bird.

Characteristics of Birds

- Feathers
- Hollow bones
- No teeth
- Super-efficient respiratory system
- Completely divided ventricle of heart; no mixing of deoxygenated and oxygenated blood
- Endothermic, maintaining a higher body temperature than mammals
- Oviparous, laying hard-shelled eggs
- Parental care of hatchlings

Feathers Insulate and Assist in Flight

Feathers are modified reptilian scales that serve two functions: providing lift for flight and conserving heat. The structure of a feather, illustrated in Figure 33-25, combines maximal flexibility and strength with minimal weight. Feathers develop from tiny pits, called follicles, in the skin. A shaft emerges from the follicle. Pairs of vanes develop from opposite sides of the shaft. At maturity each vane has many branches called barbs. The barbs in turn have many projections called barbules that are equipped with microscopic

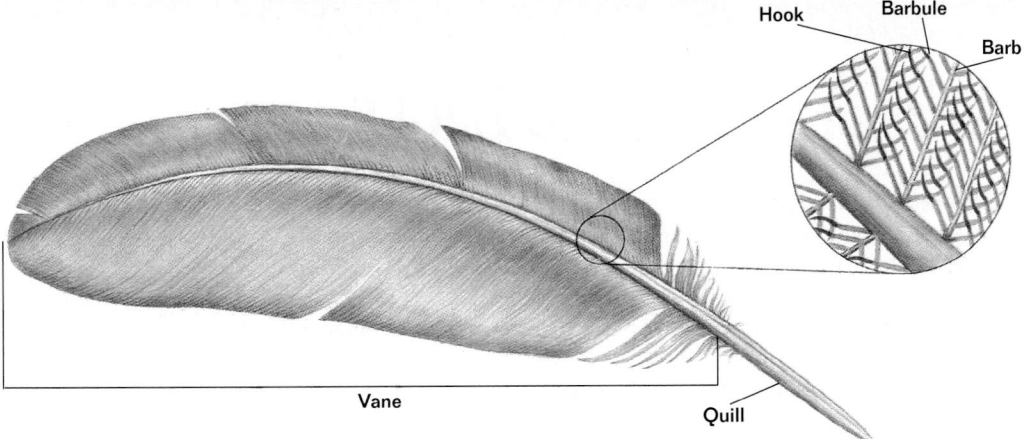

Figure 33-25 The microscopic structure of a feather helps create a smooth, aerodynamic surface. Interlocking hooks and barbs form a continuous surface.

hooks. These hooks link the barbs to one another, giving the feather a continuous surface and a sturdy but flexible shape. Like scales, feathers can be regrown.

Birds Have a Strong, Lightweight Skeleton

The bones of birds are thin and hollow. Many of the bones are fused, making a bird's skeleton more rigid than a reptile's. The fused sections of the backbone and of the pectoral girdle and pelvis form a sturdy frame that anchors muscles during flight. The power for active flight comes from large breast muscles that can make up 30 percent of a bird's total body weight. These muscles stretch from the wing to the breastbone, which is greatly enlarged and bears a prominent keel for muscle attachment, as illustrated in Figure 33-26. Muscles also attach to the fused collarbones that form the so-called wishbone. No other living vertebrates have a fused collarbone or a keeled breastbone.

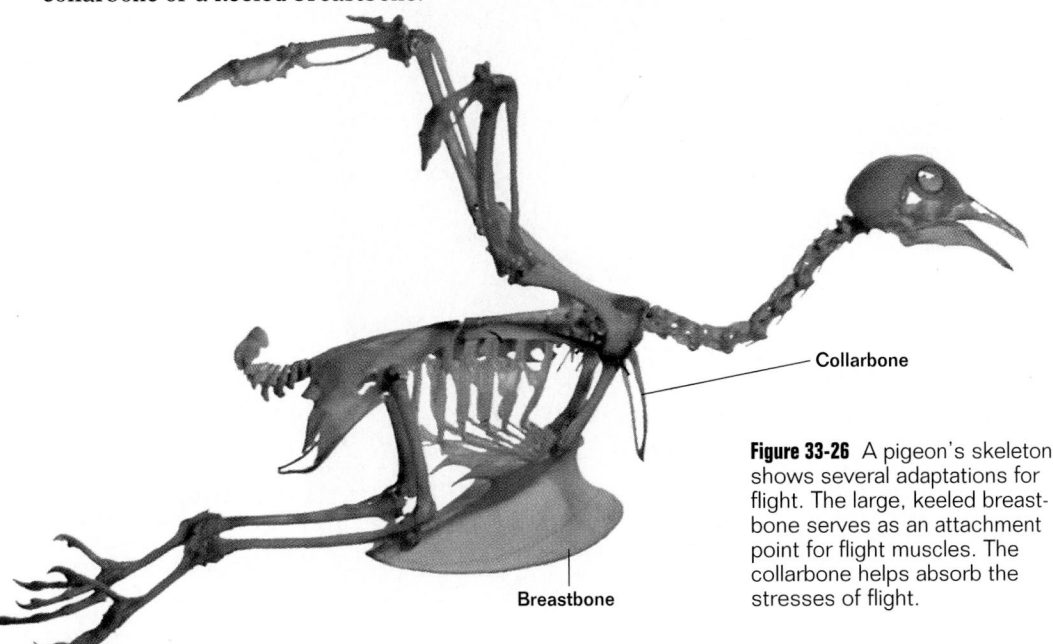

Figure 33-26 A pigeon's skeleton shows several adaptations for flight. The large, keeled breastbone serves as an attachment point for flight muscles. The collarbone helps absorb the stresses of flight.

SECTION 33-5

Teaching Tips
Preening

Students may have seen birds preening their feathers. Have students hypothesize why birds spend so much time in this activity. Point out that preening smooths the feathers, making them more aerodynamic. Preening also spreads oil to make the feathers waterproof.

What If a bird lost all its feathers? Could it still fly? No. Feathers make up most of the surface area of the wing. Without feathers, the bird could not create enough lift to get off the ground. Point out that birds do shed and replace their feathers. However, in most species only a few feathers are lost at a time, so the ability to fly is not affected.

Demonstration
Wings of Birds and Bats

Show students a photo of a bat with its wing extended. Point out that the bat's wing is largely composed of skin that is stretched across a framework composed of the bones of the arm and the elongated bones of four fingers. Contrast the structure of the bat's wing with that of a bird. Point out that feathers make up most of the surface area of a bird's wing. Instead of being elongated, the bones of the bird's hand are greatly reduced in size.

Instructional Strategies

- Stretch a string 2 m (6.5 ft.) long to show the wingspan of a bald eagle, then have students hold a 7 kg (15 lb.) bag by straightening out their arm at the shoulder to feel what it would be like to hold an eagle.

- Eagles have better eyes than humans do. Ask students why vision is so important to an eagle, and ask them if it would be possible to locate prey half a mile away by sound or smell? To simulate how much better an eagle's eyes are, tape a worksheet on the wall and have students stand about 3 m (10 ft.) away and try to read it. Then have them move 1 m (39 in.) away and read the worksheet. An eagle could see the paper as well at 3 m as humans can at 1 m.

- After World War II, the numbers of bald eagles fell drastically due to the widespread use of DDT as a pesticide. DDT accumulated in the food chain and caused the egg shells of bald eagles and many other birds to become thin and fragile. By 1978 almost all populations of bald eagles in the United States (except Alaska) were listed as endangered, and fewer than 500 breeding pairs remained. More than two decades after DDT was banned, eagles are on the increase. Over 22,000 eagles now live in the lower 48 states, and about twice as many occur in Alaska.

- Use the diagram on page 793 to compare the advantage of a gizzard over teeth as a means for an eagle to physically break down its food. Emphasize that teeth are very heavy and would weigh down the bird. Remind students that the crop may moisten and soften food before it goes to the gizzard, and that an eagle uses its beak and talons to rip its meat before swallowing.

UP CLOSE *BALD EAGLE*

- **Scientific name:** *Haliaeetus leucocephalus*
- **Range:** Nearly all of North America, from Florida to northern Alaska
- **Habitat:** Forested areas near water that have tall trees for perching and nesting
- **Size:** Wingspan is typically over 2 m (6.5 ft.), and body mass often exceeds 7 kg (15 lb.)
- **Diet:** Fish, small mammals, birds, carrion

External Structures

Eye Vision is the most important sense of the bald eagle. Keen eyesight allows it to see prey at great distances. Its eyes are so large that they occupy most of the space in the head. The bald eagle's visual acuity is 3–4 times higher than yours.

Eye

Feathers

Feathers The body of the bald eagle is covered with feathers everywhere except the feet, which are bare. Both sexes develop the characteristic white head and neck at maturity.

Nostril

Beak

Grasping feet The bald eagle has large feet and talons—the hind claw may be 5 cm (2 in.) long. The talons are used to snatch fish from the water while the eagle is on the wing. When the muscles of the legs contract, the tendons in the lower legs contract, and the talons lock together around the fish.

Grasping feet

Beak The beak is massive, with an elongated, sharp, downward-curving tip. Having no teeth, bald eagles do not chew their food. Instead, they use their beak to tear their prey into manageable portions that are swallowed whole.

Internal Structures

Brain In ratio of brain size to body size, birds rank second among vertebrates, behind only mammals. The large cerebellum receives and integrates information from the muscles, eyes, and inner ears, making possible the precise control of movement and balance necessary for flight. Since the optic lobe processes input from the eagle's most important sense organs—the eyes —it is large. The cerebrum performs many functions, including evaluation of sensory information, control of behavior, and learning.

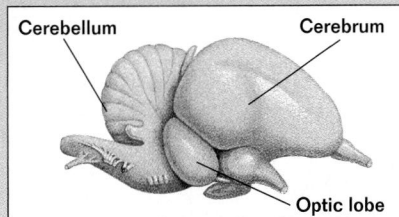

Brain

Excretory system The bald eagle's excretory system is efficient and lightweight. It does not store waste liquids in a bladder as you do. Your excretory system cannot concentrate the urea that is the byproduct of metabolism because urea is very toxic, so you must expel it in dilute form, urine. The bald eagle (and other birds and reptiles) instead converts its nitrogenous wastes to a nontoxic form called uric acid, which is concentrated into a harmless white paste. Undigested food and uric acid travel to the terminal portion of the gut, the cloaca, and are eliminated.

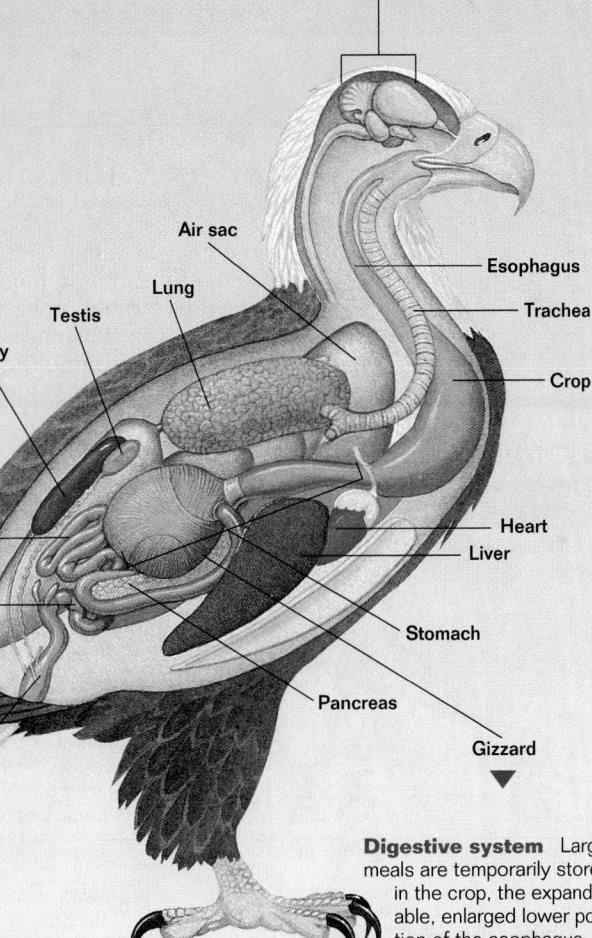

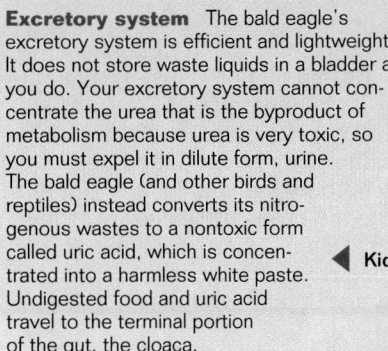

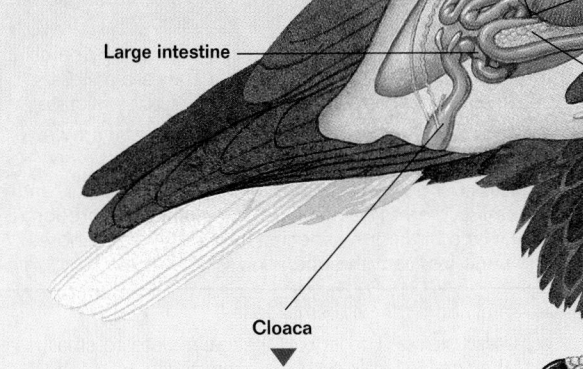

Cloaca The cloaca also serves a reproductive function. As in most other species of birds, male bald eagles do not have a penis. Sperm produced in the testes pass into the male's cloaca. During mating, the male presses his cloaca against the female's cloaca and releases sperm.

Digestive system Large meals are temporarily stored in the crop, the expandable, enlarged lower portion of the esophagus. The food then passes into a two-part stomach. In the first chamber, stomach acids begin breaking down the food. The partially digested food is then passed to a second chamber, the gizzard, where it is ground and crushed. The gizzard often contains small stones that the bird has swallowed.

Discussion

Guide the discussion by posing the following questions.

1. Identify three differences between the internal structures of the timber rattlesnake and the bald eagle. *(The rattlesnake has one functional lung, lacks air sacs, has neither a crop nor a gizzard, has venom-producing glands, and does not have hollow bones.)*

2 Would the heat-sensitive pits of the timber rattlesnake be an effective way of sensing prey for a bald eagle? *(No. The rattlesnake's pits are only sensitive to nearby heat sources, and the bald eagle spots its prey from far off.)*

3. Why would it be a disadvantage for the eagle to produce urine, as we do? *(Urine contains much liquid and is therefore heavy.)*

Teaching Tips

Heat Source

The high temperature of birds is a byproduct of their rapid metabolism. You may wish to use the metaphor of a car engine for metabolism. If the car is running, heat is produced. Ectothermic reptiles are like cars whose engines run very slowly all the time. In contrast, birds (and mammals) keep their engines running at high speed at all times.

Numbers of Species

Caution students that the numbers of species for each order in this table are not set in stone. The numbers of species can change as more species are discovered and if known species are reclassified.

Introduced Pests

House sparrows and starlings are two passerines that students probably are familiar with. Inform students that these two species were introduced into the United States from Europe. Both species have become serious pests. Starlings outcompete native songbirds for food and nesting sites. For example, starlings are thought to be at least partly responsible for the decline in numbers of bluebirds.

☐ CAPSULE SUMMARY

A bird's feathers, lightweight skeleton, and rapid metabolism help to make flight possible.

Birds Are Endothermic

Birds, like mammals, are endothermic. They generate enough heat through metabolism to maintain a high body temperature. Many paleontologists think the later dinosaurs, from which birds evolved, were endothermic. Birds maintain body temperatures significantly higher than those of most mammals, ranging from 40°C to 42°C (104°F to 108°F). For comparison, your body temperature is 37°C (98°F). The high temperatures maintained by endothermy permit metabolism in the bird's flight muscles to proceed rapidly. A rapid metabolism is necessary to satisfy the large energy requirements of flight. Feathers provide excellent insulation, helping to conserve body heat. ☐

The Major Orders of Birds

You can tell a great deal about the habits and diet of a bird by examining its beak and feet. For instance, carnivorous birds such as hawks have curved talons for seizing prey and a sharp beak for tearing apart their meal. The beaks of ducks are flat for shoveling through mud, while the beaks of finches are short and thick for crushing seeds. Of the 28 orders of birds, the dozen orders containing more than 100 species are briefly described in Table 33-1. Pay particular attention to the feet and beaks of these birds.

Table 33-1 Major Orders of Birds

Order	Number of Species	Description
Passeriformes: Songbirds	5,276	By far the largest order of terrestrial vertebrates, comprising 60 percent of the species of birds **Key Characteristics** Passerines are characterized by perching feet and well-developed vocal cords. They are generally small; the largest are the crows. Many species migrate. **Important Examples** Sparrows, robins, warblers, crows, starlings, and mockingbirds **Distribution** Found worldwide, except polar regions, and in nearly all terrestrial habitats, passerines are predominantly land birds. Few species frequent the seashore, and none frequent the open sea.
Apodiformes: Hummingbirds and swifts	428	Streamlined birds with small legs **Key Characteristics** The legs of these birds are so small that they cannot walk on the ground. Swifts are high-speed fliers that catch insects in the air. Hummingbirds, the smallest birds, beat their wings very rapidly, hovering over flowers to suck nectar. **Important Examples** Ruby-throated hummingbird and chimney swift **Distribution** Swifts are found worldwide, except in polar regions. Hummingbirds live only in the Western Hemisphere.

Order	Number of Species	Description
Piciformes: Woodpeckers	383	Birds with highly specialized beaks **Key Characteristics** These are forest birds that nest in holes in trees and have grasping feet. Many have strong, chisel-like beaks and feed on insects they dig from under bark by pounding into it with their beaks. Most species are solitary and do not migrate. **Important Examples** Woodpeckers, toucans, and honeyguides **Distribution** Woodpeckers occur on all continents except Australia; toucans and honeyguides are tropical.

Psittaciformes: Parrots	340	Brightly colored vegetarians of the tropics **Key Characteristics** All members of this distinctive order are similar, with short necks, compact bodies, and short, stout, hooked bills. All are vegetarians and are very fond of fruits and berries. They nest in holes in trees and are very gregarious. In captivity, some can be taught to imitate human speech. **Important Examples** Parrots, parakeets, cockatoos, and cockatiels **Distribution** Mainly tropical, but also New Zealand and temperate areas of Australia

Charadriiformes: Shorebirds	331	A varied group of shorebirds **Key Characteristics** Plovers and sandpipers have long legs and long, slender bills used to probe for marine animals in the sand. Terns and gulls have streamlined, pointed wings and fly over water in search of fish. Most shorebirds are very gregarious, and many migrate long distances. **Important Examples** Terns, gulls, plovers, and sandpipers **Distribution** Marshes and seashores throughout the world

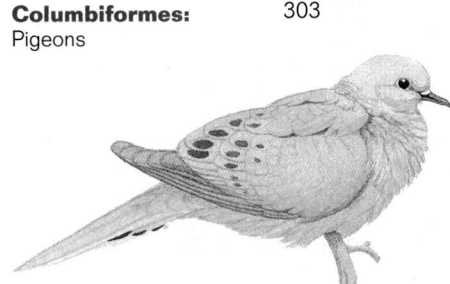

Columbiformes: Pigeons	303	Pigeons and doves **Key Characteristics** These are compact, plump birds with small heads, short beaks, perching feet, and soft, dense plumage. Most are strong fliers and gather in flocks. They feed on seeds, fruit, and berries. They vary greatly in size, from smaller than a sparrow to the size of a small turkey. **Important Examples** Mourning dove and common pigeon **Distribution** Worldwide, except polar regions

SECTION 33-5

Teaching Tip
Endangered Parrots

Parrots are sought after as pets because of their intelligence and ability to speak. However, parrots often reproduce slowly, and their populations cannot quickly rebound when large numbers of individuals are collected. Have students write a report on the effects the pet trade is having on populations of parrots throughout the world. Students should also describe some of the efforts to conserve parrots.

Historical Note

The Carolina Parakeet

There are no native parrots or parakeets in the United States today. Three hundred years ago, the Carolina parakeet lived in the central and southern states. Carolina parakeets were slaughtered for their feathers, for sport, and because they were considered agricultural pests. The last individual of this species died in captivity in 1914.

Order	Number of Species	Description
Falconiformes: Birds of prey	288	Day-active carnivores, the lions of the air **Key Characteristics** Strong fliers with keen vision, they have curved, pointed beaks for tearing flesh and powerful, curved talons for seizing prey. Birds of prey range in size from small, insect-eating sparrow hawks to the South American harpy eagle, which has a wingspan of 3 m (10 ft.). **Important Examples** Eagles, hawks, falcons, ospreys, and vultures **Distribution** Worldwide

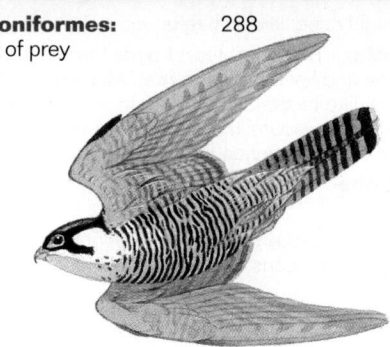

Galliformes: Fowl	268	Small, rapid-running ground feeders **Key Characteristics** Sometimes called "gamebirds" because many of them are hunted for sport, these birds typically have rounded bodies and often only limited flying ability. Most fowl are browsers or grazers and live in forests and fields. **Important Examples** Quail, grouse, turkeys, partridges, pheasants, and domestic chickens **Distribution** Worldwide

Gruiformes: Marsh birds	209	Ground-living marsh dwellers **Key Characteristics** Marsh birds have small bodies and long bills. Rails are small, have long legs, and bob their heads as they walk. Many rails are nocturnal, solitary, and rarely fly. Cranes are the exact opposite—powerful fliers that mate for life and often migrate long distances. **Important Examples** Rails, coots, and cranes **Distribution** Worldwide

Anseriformes: Waterfowl	150	Aquatic diving birds **Key Characteristics** Waterfowl have long necks and blunt, flat tails. Three of the toes are linked by webs for improved swimming. Flight feathers are molted after the breeding season, and individuals pass through a flightless period of up to a month. Three-quarters of the species are the so-called dabbling ducks—including mallards, teals, and pintails—which tip down to feed on vegetation a few feet below the water's surface. Waterfowl migrate seasonally along established flyways. **Important Examples** Ducks, geese, and swans; all domestic ducks are descended from the mallard **Distribution** Worldwide

Multicultural Perspective
Eagle Feathers
The only people in North America who can legally own an eagle feather are Native Americans. The eagle is so highly valued by tribes throughout the United States that its feathers must be earned through personal sacrifice and then used only in special ceremonies. For instance, if a Winnebago pow-wow dancer accidentally drops an eagle feather during a performance, the dance is stopped until the feather is purified by an elder and then reclaimed by the dancer, who is not allowed to dance again for a year.

Order	Number of Species	Description
Strigiformes: Owls 	146	Nocturnal carnivores; the nighttime equivalents of birds of prey

Strigiformes: Owls — 146

Nocturnal carnivores; the nighttime equivalents of birds of prey

Key Characteristics   Owls have soft plumes, short tails, and large heads with hooked bills and large eyes that are directed forward. Owls have extraordinarily good hearing and excellent night vision. They cannot see in absolute darkness, but they have a very large number of light-sensitive rods in the retinas of their eyes that provide exceptional vision in poor light, such as moonlight.

Important Examples   Barn owl, great horned owl, and snowy owl

Distribution   Worldwide

Ciconiiformes: Herons — 114

Long-legged waders

Key Characteristics   Often confused with the marsh birds, the members of this order have large bodies and long necks. Herons have long, sharp bills that they use to spear or grasp fish or other prey.

Important Examples   Herons, bitterns, egrets, and flamingos

Distribution   Worldwide, except polar regions

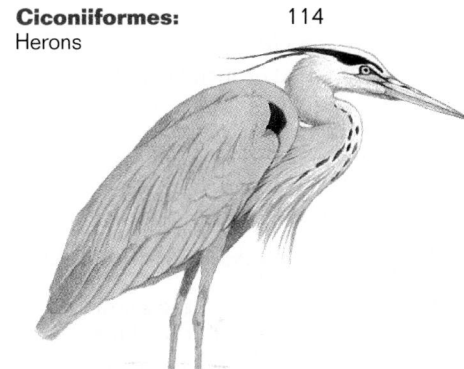

Section Review

1. *Could a bird fly without feathers? Explain your answer.*
2. *Describe two skeletal features of birds that help reduce weight.*
3. *What is the function of the gizzard? Explain why this function is more critical for a seed-eating bird than for a carnivorous bird.*

Critical Thinking

4. *In the majority of species of birds, both parents care for the young. In contrast, only the mother provides parental care in most mammalian species. Propose a hypothesis to explain this difference. How would you test your hypothesis?*

Answers to Section Review

1. No. Feathers provide most of the surface area of the wings and are necessary for maneuvering.

2. Bird bones are thin and hollow.

3. The gizzard grinds up food. The gizzard is particularly important to seed-eating birds because they swallow seeds whole, while meat-eating birds tear their food up before swallowing it.

4. Accept any logical and testable hypothesis. One hypothesis might be that bird offspring demand more parental care than can be supplied by just one parent. Answers will vary for the test of the hypothesis but may include observing the success rate of "single-parent" birds raising their young versus the success rate of two-parent nests.

Teaching Tip
Owl Eyes

Owls have very large eyes for their body size. Ask students why large eyes would be an adaptation for nocturnal hunting. (*Larger eyes can receive more light in low-light conditions.*) Tell students that although bats hunt at night, they do not have especially large eyes. Bats navigate and locate prey with a sonar system that uses high-pitched sounds, as described in the next chapter.

Chapter Closure

Birds are adapted to a variety of habitats and are very diverse in size and anatomy. Some are aquatic, some are terrestrial and flightless, and some cannot land on the ground because their legs are too small. Have students pick one species of bird and write an account of its adaptations to its environment. Students should explain how these adaptations enable the bird to cope with the challenges of its environment.

CHAPTER REVIEW ANSWERS

Each item in the Chapter Review is correlated by text section in the assignment guide that follows.

ASSIGNMENT GUIDE

Section	Review	Themes Review	Critical Thinking
33-1	1, 2, 12, 13		
33-2	3, 4, 12, 14	20	21, 23
33-3	5–7, 15	18, 19	
33-4	8, 9		
33-5	10, 11, 16, 17		22

Review

Multiple Choice

Code in parentheses indicates section number and objective number.

1. b (33-1.3)
2. d (33-1.2)
3. c (33-2.2)
4. b (33-2.4)
5. c (33-3.1)
6. d (33-3.1)
7. b (33-3.4)
8. a (33-4.2)
9. b (33-4.2)
10. c (33-5.3)
11. b (33-5.1)

Completion

12. crocodiles, young (33-1.2)
13. crocodiles, lizards, or snakes; anapsid (33-1.3)
14. Chelonia, plastron (33-2.1)
15. *Triceratops, Tyrannosaurus rex* (33-3.3)
16. high, larger (33-5.1)
17. bones, light (33-5.2)

Vocabulary

anapsid (773)
carapace (776)
diapsid (773)
Pangaea (783)
plastron (776)
synapsid (773)

Review

Multiple Choice

1. The skulls of the earliest reptiles are most similar to those of
a. crocodiles.
b. turtles.
c. snakes.
d. lizards.

2. Which group of living reptiles is most closely related to birds?
a. snakes
b. turtles
c. rhynchocephalians
d. crocodiles

3. Snakes are different from lizards because snakes do not have
a. a tail.
b. reproductive organs.
c. a pectoral girdle.
d. a loosely connected lower jaw.

4. Compared with other reptiles, crocodiles
a. are harmless and tame.
b. provide better care for their young.
c. have weak jaw muscles.
d. can tolerate drier conditions.

5. The oldest known dinosaur is
a. *Brachiosaurus.*
b. *Dimetrodon.*
c. *Eoraptor.*
d. *Herrerasaurus.*

6. Which factor did *not* contribute to the success of dinosaurs?
a. the ability to live in arid conditions
b. legs positioned directly under the dinosaur body
c. a meteorite striking the Earth at the end of the Triassic period
d. the break up of Pangaea

7. What evidence best supports the hypothesis that the impact of a meteorite caused the extinction of the dinosaurs?
a. fossils
b. iridium sediments
c. continental drift
d. plant evolution

8. What evidence convinced scientists that fossils once classified as *Compsognathus* should be reclassified as *Archaeopteryx*?
a. feathers
b. fused bones
c. teeth
d. hard-shelled eggs

9. The fossil record of birds is incomplete because
a. reptiles gave rise to birds.
b. birds have thin, hollow bones.
c. mammals prey on birds.
d. bird cells do not contain carbon.

10. What structure has taken over the function of teeth in birds?
a. the beak
b. the crop
c. the gizzard
d. the claws

11. The feathers of most birds are well adapted for
a. swimming and repelling water.
b. flying and insulating.
c. flying and conducting.
d. expelling heat and feeding.

Completion

12. Like dinosaurs, _____ evolved from thecodonts. They are aggressive carnivores and, unlike other reptiles, care for their _____ .

13. Living reptiles that have diapsid skulls include tuataras and _____ . These reptiles are thought to have appeared more recently than reptiles with _____ skulls, such as turtles.

14. Turtles and tortoises are members of the order _____ . Their bodies are covered with a shell that consists of two parts: the carapace and the _____ .

15. Herbivores of the Cretaceous period include the horned _____ and the duck-billed *Maiasaura.* A very large carnivore of the same period is _____ .

16. Due to their _____ rate of metabolism, birds probably eat a _____ amount of food for their size.

17. The feathers and _____ of birds are strong yet _____ . These adaptations are important for flight.

Themes Review

18. **Evolution** Coevolution occurs when interacting groups of organisms evolve together. How is the evolution of modern angiosperms and herbivorous dinosaurs an example of coevolution?

19. **Homeostasis** The plates along *Stegosaurus's* back contained many blood vessels. How did these blood vessels help *Stegosaurus* maintain a constant body temperature?

20. **Structure and Function** The young of birds and reptiles develop in an amniotic egg. Name the four membranes contained in an amniotic egg. Describe how they help make the egg an independent life-support system.

Critical Thinking

21. **Designing an Experiment** Rattlesnakes and other pit vipers have special sensory pits on each side of the head near the nostril. Scientists think that these sensory pits help snakes locate and bite objects that have a higher temperature than the surrounding environment. Design an experiment to test this hypothesis.

22. **Making Predictions** A hummingbird has a four-chambered heart. How would having a three-chambered heart, like that of most reptiles, affect a hummingbird in flight?

23. **Interpreting Data** The graph below shows the body temperature range of the members of the major groups of reptiles.
    a. Which two groups show a much narrower range of temperatures than the other groups? What might account for this? (Hint: Consider the number of different species in each group.)
    b. Which groups do you think could best tolerate the temperature extremes found in deserts?

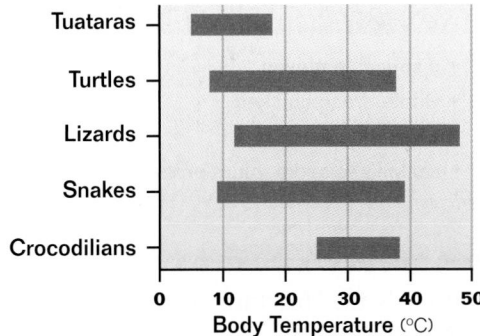

Body Temperature (°C)

Activities and Projects

24. **Cooperative Group Project** Organize a bird-watching trip. Have different members of your group visit different ecosystems in your area, such as woods, lakes, apartment houses, and fields. Using a field guide to birds, identify as many birds in your ecosystem as you can. Make a list of the different species of birds you see, and record the number of times you see each species. Share your findings with your class.

25. **Multicultural Perspective** Different cultures have different attitudes toward snakes. Research the important cultural roles of snakes in India and Africa. Write a report that summarizes what you have learned.

CHAPTER 33

Themes Review

18. Early in the Cretaceous period, sauropods and stegosaurs became extinct and were succeeded by the iguanodonts as the most successful herbivores. This was due to the rise of angiosperms and the ability of the iguanodonts to chew the tougher angiosperm leaves. (33-3.3)

19. The blood circulating through the vessels in the plates could be heated or cooled by coming in close contact with air. By moving in and out of direct sunlight, a *Stegosaurus* could regulate its body temperature by heating or cooling the blood flowing through the vessels. (33-3.1)

20. The chorion is the outermost membrane; it keeps water in the egg and allows oxygen to enter. The amnion encloses the developing embryo within a fluid-filled cavity. The yolk sac contains the yolk that serves as a food source for the developing embryo. The allantois surrounds a cavity into which waste products from the embryo are excreted. (33-2.4)

Critical Thinking

21. An adequate test of this hypothesis would involve destroying or blocking all other major sensory organs of rattlesnakes and then seeing if the snakes could strike a target with accuracy. (33-2.3)

22. A three-chambered heart would allow dilution of oxygen-rich blood by oxygen-poor blood. The diluted blood would not provide enough of the oxygen needed for the bird's very rapid metabolism, and the bird would not have the energy to fly. (33-5.1)

23. Tuataras and crocodilians show a much narrower temperature range. This difference may be accounted for by the fact that these groups contain few species, and these are adapted to similar environments. Snakes and lizards are best able to tolerate desert temperatures. (33-1.1)

Activities and Projects

24. Answers will vary.

25. Answers will vary but may include the worship of snakes in some Indian cultures.

Chicken Anatomy

OBJECTIVE

Analyze how the muscular and skeletal systems of a bird are adapted for flight.

PROCESS SKILLS

- observing the skeletal and muscular structure of an organism
- identifying anatomical features

MATERIALS

- disposable gloves
- whole, fresh chicken
- dissecting pan
- scalpel, scissors, blunt probe

BACKGROUND

1. How is the structure of a bird typical of all vertebrates?
2. How are hollow bones and fused bones adaptations for flight?
3. What muscles of a bird are especially large, allowing flight?
4. Write your own question to explore in your lab report or notebook.

TECHNIQUE

1. Put on disposable gloves. **CAUTION: Chicken is sometimes contaminated with *Salmonella* bacteria. Wear gloves throughout this investigation and keep your hands away from your face and mouth.**
2. **CAUTION: Be careful not to cut yourself with the scalpel or scissors.** Use the scalpel and scissors to remove the skin from the chicken, exposing as many muscles as possible. Record your observations in the **Records** section of your report.

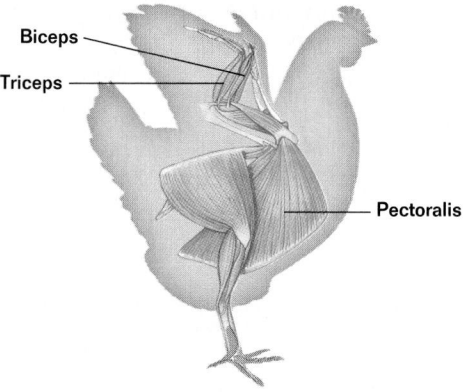

Biceps
Triceps
Pectoralis

3. Compare the size and color of the muscles of the wing with those of the leg. How do they differ?
4. Compare the muscles on the back and on the chest. How do they differ?
5. The main muscle of the chest is the pectoralis. This muscle originates on the breastbone and the fused clavicle bones, or "wishbone," and attaches to the underside of the humerus at some distance from the shoulder joint. When this muscle contracts, it pulls the wing down. Use the blunt probe to pull on the pectoralis.
6. Cut the pectoralis from the humerus and peel back the muscle. This will expose the supracoracoideus muscle. It originates on the coracoid and the breastbone and attaches on the upper side of the humerus. Pull on this muscle to find out how it functions.
7. Identify the biceps and triceps of the bird. How do these muscles function?
8. Use the scissors and scalpel to remove muscles from the breast, back, and thigh of the chicken.

Preparation

Purchase fresh whole chickens at a supermarket or butcher shop. Keep the chickens refrigerated and covered with plastic wrap until lab time. Each group of 2–4 students should have their own chicken. Students will also need a dissecting pan, scissors, a blunt probe, and a scalpel with a permanent blade. Caution students to use their dissecting tool carefully to avoid cutting themselves. Because chicken is sometimes contaminated with *Salmonella* bacteria, students should wear disposable gloves, safety goggles, and a lab apron throughout this investigation. Remind students to keep their hands, dissecting tools, and chicken parts away from their face and mouth. After the investigation, have students use soap and hot water to thoroughly wash their hands, lab area, and any surfaces that came into contact with the chicken or its juices. Dispose of chickens promptly.

Procedural Notes

- Students may work on the dissection in groups of two to four. This lab should be completed in one class period for sanitation purposes.
- **Safety:** Because chicken is so ubiquitous, students may not appreciate the danger of *Salmonella* poisoning. Encourage them to be very careful handling the chicken and to wash all surfaces with soap and hot water at the end of the lab. Students should wash their hands before leaving the lab.
- **Safety:** Remind students to be careful with the sharp dissecting tools, and caution them to keep the tools, their hands, and the chicken away from their faces.

Background Answers

1. Like other vertebrates, birds have a backbone, paired appendages, well-developed organ systems, and are bilaterally symmetric.

2. Hollow bones and fused bones are light yet strong, cutting down on mass while tolerating the stresses of flight.

3. the breast muscles

4. How does the structure of the chicken's skeleton and bones relate to its ability to fly?

Records

Observations will vary. However, students should note the differences in color and size between the muscles of the leg and those of the breast. Drawings of bones and muscles should be clear and labeled.

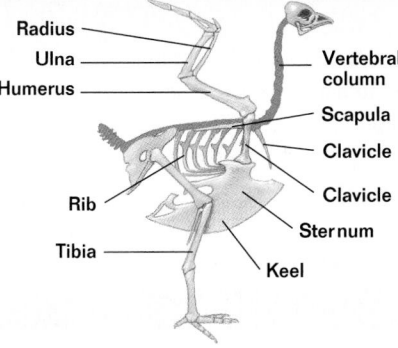

Radius
Ulna
Humerus
Vertebral column
Scapula
Clavicle
Clavicle
Rib
Sternum
Tibia
Keel

9. Examine the vertebral column of the chicken, which is shown above in pink. Unlike most other vertebrates, birds have a rigid backbone. This provides support for flight. In contrast, the relatively long neck is flexible, allowing a bird to use its beak as a tool.

10. The ribs come together in a large shield-shaped sternum, or breastbone. The backbone, ribs, and sternum together form a flexible but strong box that houses the heart, lungs, and visceral organs. Find the sternum and the keel that extends from it. The pectoralis and supracoracoideus attach to the keel.

11. Follow the sternum up to the coracoid. The coracoid, scapula (shoulder blade), and clavicle (collarbone) form the shoulder. The clavicles, one from each shoulder, join together in the front to form the "wishbone." Find these bones.

12. Next find the femur, or thighbone. Break the thighbone and examine the broken ends. Notice that it is not solid, but is instead filled with air pockets. In the lower leg, the tibia is the main bone and the fibula is reduced to a splint. Because they have only one bone in this part of the leg, birds cannot twist their legs or step sideways; they can only walk forward or backward.

13. In your report, briefly summarize the procedure you followed. In the **Records** section of your report, sketch the bones and muscles you have located in this lab.

14. Follow your teacher's instructions to dispose of the chicken and clean up your materials. Before leaving the lab, use soap and water to thoroughly wash your hands and any surfaces touched by the chicken.

INQUIRY

1. How does the color of the wing muscles differ from that of the leg muscles?

2. What is the chief function of the chest muscles?

ANALYSIS

1. What adaptations for flight did you find in the chicken skeleton and musculature?

2. Muscles that have a large blood supply are darker than those that have a smaller amount of blood flow. Generally, those muscles that are the most active have the greatest supply of blood. Relate these facts to your observations during this investigation.

3. The keel increases the area available for the attachment of flight muscles. Compared with that of other birds, the keel of the chicken is small. Why?

FURTHER INQUIRY

Write a new question that could be explored as a new investigation. The following is an example:

How do the structures of flightless birds and songbirds differ?

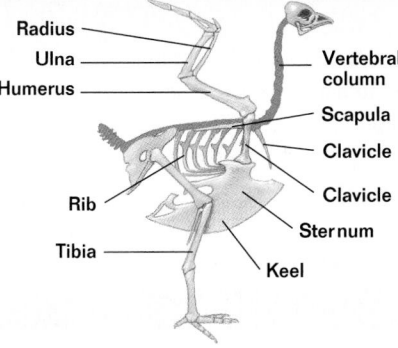

CHAPTER 33

Inquiry Answers

1. Muscles of the wings are light; those of the leg are dark.
2. They move the wings during flight.

Analysis Answers

1. Answers may include hollow bones, rigid backbone, fused bones, and front limbs adapted as wings.

2. Chickens can only fly short distances and have a relatively small blood supply to the chest muscles. Therefore, these muscles are light colored. Chickens rely on their leg muscles for running and walking. These muscles receive more blood and are dark colored.

3. The chicken doesn't need such large flight muscles because it flies for short distances. Therefore, the keel on its breastbone is small.

Each block represents about 45–50 minutes of instructional time. The resources cited in a block are coded to a specific program component using the Key on the next page.

Lecture Concepts

Lesson Resources

34-1 History of Mammals pp. 803–805

BLOCK ①

a. Mammals are the descendants of therapsids, animals that had diverse teeth and a secondary palate. Some therapsids may have been endothermic.

b. The first mammals appeared about 220 million years ago, just as the first dinosaurs were evolving from thecodonts.

c. Mammals survived the mass extinctions that took place at the end of the Cretaceous period and rapidly diversified during the Tertiary period (65–2 million years ago).

Lecture Resources
- Opening Questions, page 803
- Demonstrations: Times of Origin, The Secondary Palate
- Visual Strategy: Figure 34-3
- Holt Biology Videodiscs: Lesson 34-1

Classwork Options
- Effective Reading, page 803
- Rare Fossils, page 805
- Viviparity and Oviparity, page 805

Assignment Options
- Section Review, page 805
- Chapter Review, questions 1–5, 10, 16, 21

34-2 Key Mammalian Adaptations pp. 806–812

BLOCKS ② and ③

a. Mammals can be characterized by the presence of hair; milk-producing mammary glands; diverse, specialized teeth; and a high metabolic rate (necessary for endothermy).

b. Present-day mammals can be divided into three groups based on styles of reproduction: monotremes, marsupials, and placental mammals.

Lecture Resources
- Opening Questions, page 806
- Demonstration: A Mammal's Coat
- Parental Care in Mammals, page 807
- Learning, page 807
- Overcoming Misconceptions, page 810
- Visual Strategy: Figure 34-6
- Natural Selection at Birth, page 811
- Placental Marsupials? page 812
- Teaching Transparencies: 144A
- Holt Biology Videodiscs: Lesson 34-2

Classwork Options
- Up Close: Siberian Tiger, pages 808–809
- Natural Selection at Birth, page 811
- Drawing a Cladogram, page 811
- Inquiry Skills Development B29: Fetal Pig Dissection

Assignment Options
- Section Review, page 812
- Chapter Review, questions 6–8, 11, 13, 17–20, 22, 23

34-3 Mammalian Diversity pp. 813–819

BLOCKS ④ and ⑤

a. Monotremes are egg-laying mammals and consist of the duckbill platypus and two species of echidnas.

b. Marsupials are pouched mammals such as kangaroos, opossums, koalas, and wombats.

c. More than 90 percent of mammal species are placental mammals.

Lecture Resources
- Opening Question, page 813
- Demonstration: Mammalian Diversity
- More Oddities of Monotreme Reproduction, page 814
- Continental Drift and Mammalian Distribution, page 814
- Research Update: Dealing With Crowding: Fighting or Coping? page 816
- Teaching Transparencies: 177, 178, 181

- Holt Biology Videodiscs: Lesson 34-3

Classwork Options
- Misconception About Bats, page 815
- Closure, page 819
- Laboratory Investigation: Mammalian Behavior: Human Postural Signs, pages 822–823

Assignment Options
- Section Review, page 819
- Chapter Review, questions 9, 12, 14, 15

Review/Enrichment

BLOCK ⑥

■ Study Guide: Concept Review and Pretest
◉ Teaching Resources CD-ROM: Occupational Applications Worksheets, Veterinary Technician

Assignment Options
■ Activities and Projects, questions 24–26

Assessment Options

BLOCK ⑦

Traditional Assessment
■ Chapter 34 Test
◉ Test Generator/Assessment Item Listing: Software item bank for preparing customized chapter tests.

Portfolio Assessment
Select student reports for one or more laboratory experiments from this chapter. *The Direct Observation Checklist* on the *BioSources Teaching Resources CD-ROM* should be completed during a laboratory or other cooperative learning experience.

Holt Biology Videodiscs

Holt Biology Videodiscs gives you a powerful tool for teaching, review, and assessment. *Concepts of Biology* adds interactive lesson plans and extensive tools for customization using CD-ROM technology.

CONCEPTS OF BIOLOGY

Use the following topic from *Concepts of Biology* to help you teach this chapter:

■ Topic 20: Vertebrate Diversity

For further information regarding the media on the videodiscs, see *Holt Biology Videodiscs Teacher's Correlation Guide to Biology: Principles and Explorations.*

Chapter Theme

Evolution *Mammals are very diverse anatomically and occupy a variety of habitats. But for most of their history, mammals were small and not very diverse. The diversification of mammals occurred only after the extinction of the dinosaurs.*

Tapping Prior Knowledge

- When did the first mammals appear?
- What group of reptiles is ancestral to the mammals?
- What effect did the extinction of the dinosaurs have on mammalian evolution?

Opening Question

Ask students to make two lists. On one list they should write down as many large terrestrial herbivores as they can think of. On the other list they should write down as many large terrestrial carnivores as possible. Have students determine what percentage of the animals on their lists are mammals. Point out that today these two ecological roles are dominated by mammals. During the age of dinosaurs, however, reptiles dominated these roles.

CHAPTER
34 MAMMALS

REVIEW

- continental drift (Sections 13-3 and 33-3)
- cladogram (Section 19-4)
- mammalian characteristics (Section 31-1)
- endothermy (Section 31-1)
- differences among marsupials, monotremes, and placental mammals (Section 31-1)
- structure of mammalian lungs (Section 31-2)
- structure of mammalian heart (Section 31-3)
- viviparous and oviparous reproduction (Section 31-5)
- *Highlights: Therapsids on page 734*

Female jaguar

Authors' Rationale

Mammals are an evolutionary success. The age of dinosaurs gave way to the Age of Mammals about 65 million years ago. This chapter discusses the attributes that made mammals so successful as a class, the uniquely mammalian characteristics and variations, and mammalian diversity. The reptilian ancestry of mammals is suggested by the egg-laying monotremes. Yet the monotremes are definitely mammals—they are covered by hair, nurse their young, and are endothermic. In order to understand human biology, a thorough comprehension of general mammalian history and biology is necessary.

34-1 History of Mammals

T oday almost all large, land-dwelling vertebrates are mammals, and they therefore tend to dominate terrestrial communities. When you look out over an African plain, for example, you see the big mammals—the lions, elephants, antelopes, and zebras. Your eye does not as readily pick out the many birds, lizards, snakes, and frogs that also live there. But only within the last 65 million years have mammals become dominant. For most of their history, mammals were small animals living in the shadow of the dinosaurs. In this section you will learn about the history of mammals (class Mammalia).

Ancestors of Mammals

The first step in the evolutionary journey leading to mammals occurred in the warm, moist tropical forests that covered most of the globe 300 million years ago. This step was the evolution of reptiles known as pelycosaurs (*PEHL uh kuh SAWRZ*), including *Archaeothyris*, shown in Figure 34-1. Early pelycosaurs, which were small and lizardlike, seem very different from mammals. However, both groups have the same type of skull, the synapsid skull. Recall from Chapter 33 that a synapsid skull has a single opening behind each eye socket. This arrangement enables larger, stronger muscles to attach to the jaw, increasing biting power. In their powerful jaws, pelycosaurs had teeth differing in size and shape, another characteristic shared with mammals. The degree of difference between teeth in pelycosaurs is much less than in mammals, however. For 40 million years, over 70 percent of the land vertebrates were pelycosaurs. Some were large herbivores; others, such as *Dimetrodon*, which you saw in Chapter 31, were ferocious predators.

About 260 million years ago, the therapsids (*thuh RAP sihdz*) evolved from and replaced the pelycosaurs. As you learned in Chapter 31, therapsids, such as *Cynognathus* illustrated in Figure 34-2, are the direct ancestors of mammals. Therapsids had

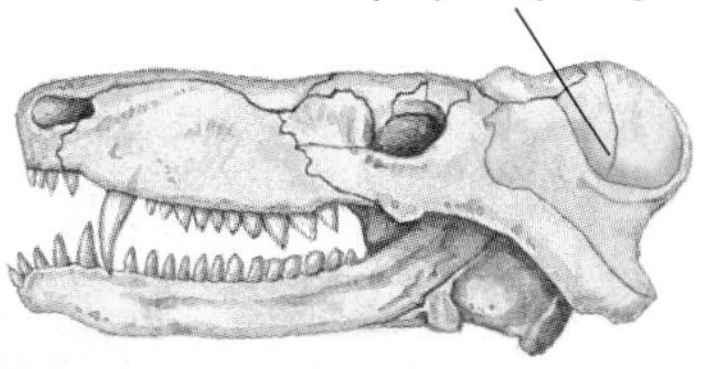

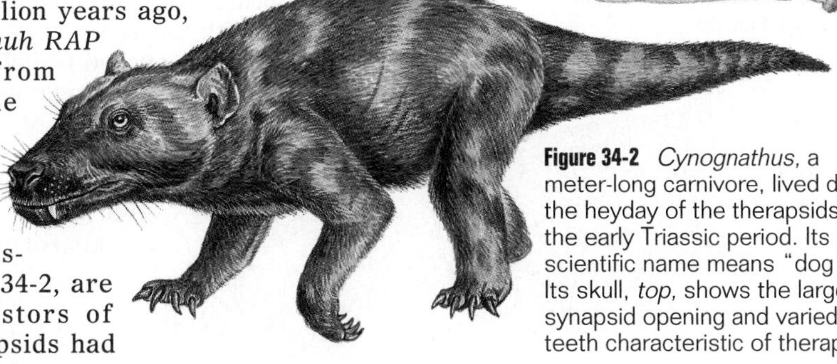

Figure 34-1 *Archaeothyris* is the oldest known pelycosaur. About 50 cm (20 in.) from nose to tail, it preyed on other early reptiles.

synapsid opening

Figure 34-2 *Cynognathus*, a meter-long carnivore, lived during the heyday of the therapsids in the early Triassic period. Its scientific name means "dog jaw." Its skull, *top*, shows the large synapsid opening and varied teeth characteristic of therapsids.

Section Objectives

- Recognize the role of pelycosaurs in mammalian evolution.
- Describe three similarities between therapsids and mammals.
- Identify three characteristics of the earliest mammals.

Vocabulary Preview

(none)

Opening Questions

"Dinosaurs suppressed the evolutionary potential of mammals, not the other way around." This is the judgment of paleontologist Robert T. Bakker. Ask students to cite evidence that supports Bakker's statement. (*The mammals appeared at about the same time as the dinosaurs, but they remained small and did not diversify much until the disappearance of the dinosaurs.*) Ask students what view of mammals Bakker wanted to counter with this statement. (*the view that mammals were superior to dinosaurs and diversified at their expense*)

Demonstration

Times of Origin

To help students understand the times of origin and extinction mentioned in this section, on the board draw a geological time line that begins with the opening of the Carboniferous period about 360 million years ago. On the time line, draw bars to represent the "lifetimes" of the major groups mentioned in this section: pelycosaurs—300 million years ago to 255 million years ago; therapsids—260 million years ago to 150 million years ago; dinosaurs—235 million years ago to 65 million years ago; mammals—220 million years ago to present. Point out that the Age of Mammals began with the extinction of the dinosaurs 65 million years ago.

Effective Reading

Reading With a Plan

Tell students that they can better understand the chapter if they make a chart to organize the information they read. For example, column one could contain the heads in each section of the text, and column two could contain explanations, descriptions, or examples. The first row of the chart could be like the following.

Section Subheads	Explanations and Examples
Ancestors of Mammals	Probable reptile ancestors of mammals were called pelycosaurs, followed by therapsids. Both groups share the synapsid skull with mammals.

Application

℞ **Medicine** Have students run their tongue along the roof of their mouth to feel the seam where their secondary palate grew together before birth. Some babies are born with a hole in the roof of their mouth where the secondary palate didn't close as it should, a condition called a cleft palate. Babies with this condition have trouble feeding and may choke on their milk. Cleft palates can be repaired surgically.

Demonstration
The Secondary Palate

Show students the secondary palate on a photograph or drawing of a cutaway human skull or on a sectioned real skull. The secondary palate is the "ceiling" of the mouth cavity. Show students how the secondary palate diverts air inhaled through the nostrils toward the back of the mouth. They should see how a secondary palate makes simultaneous chewing and breathing possible. Point out that most reptiles lack a secondary palate.

👁 VISUAL STRATEGY

Figure 34-3
Students should note *Eozostrodon's* relatively large eyes. Scientists infer that *Eozostrodon* was nocturnal because so many of today's vertebrates with very large eyes—owls, for instance—are nocturnal. Large eyes can receive more light than small eyes, an advantage in low-light conditions such as twilight and night.

❏ **CAPSULE SUMMARY**

Therapsids were the direct ancestors of mammals. Like mammals, they had diverse teeth and a secondary palate. At least some therapsids may have been endothermic.

Figure 34-3 Only 12 cm (5 in.) in total length, *Eozostrodon* is typical of the small, nocturnal early mammals. Like modern shrews, *Eozostrodon* fed on insects and other small animals.

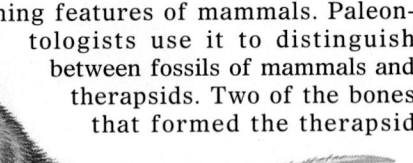

several adaptations for more efficient feeding, as evidenced by the skull of *Cynognathus*. Therapsid teeth, for instance, are much more diverse than the teeth of pelycosaurs. In addition to sharp, stabbing teeth, some therapsids had teeth with complex surfaces for crushing, grinding, or tearing food. The secondary palate, a sheet of bone that separates the mouth cavity from the nasal passages, is another adaptation that appears in therapsids. An animal with a secondary palate can breathe and chew simultaneously. (Your secondary palate is the roof of your mouth.) In therapsid skulls, the synapsid opening is much larger, making room for even bigger jaw muscles. These three adaptations—complex teeth, secondary palate, and large synapsid openings—are also characteristic of mammals.

Therapsids were active, and some were probably endothermic. How is it possible to know the metabolic state of these animals without examining a living specimen? For one thing, the therapsid adaptations for efficient feeding suggest animals with a high demand for energy. In addition, therapsid fossils have been found in areas that had cold winters in the Permian period. Evidence also suggests that at least some therapsids had fur to insulate their endothermic bodies.

Therapsids were the dominant terrestrial animals from the middle of the Permian period until the middle of the Triassic period. As climates warmed in the mid-Triassic period, however, therapsids entered a period of decline, perhaps because of competition from thecodonts. It was during this decline that the mammals evolved. Therapsids became extinct about 150 million years ago. ❏

Origin of Mammals

The first mammals appeared about 220 million years ago, just as the first dinosaurs were evolving from thecodonts. These early mammals, such as *Eozostrodon* illustrated in Figure 34-3, were small (about the size of mice), shrewlike, insectivorous tree dwellers. Their relatively large eye sockets indicate they were active at night. The early mammals had a number of improvements on the therapsid feeding adaptations. Larger jaw muscles were possible because the synapsid opening had merged with the eye socket. The jaw itself was stronger because it was composed of only one bone, rather than several as in therapsids. The single jawbone is one of the defining features of mammals. Paleontologists use it to distinguish between fossils of mammals and therapsids. Two of the bones that formed the therapsid jaw joint became part of the chain of tiny bones that transmits sound in the middle ear of mammals.

For 155 million years, while the dinosaurs flourished, mammals were a minor group that changed little. Only five orders of mammals arose in that time, and their fossils are scarce, indicating that mammals were not abundant. However, the two groups to which present-day mammals belong did appear. The most primitive mammals, direct descendants of therapsids, were members of the subclass Prototheria *(PROH toh THIHR ee uh)*. Most prototherians were small and resembled modern shrews. All prototherians laid eggs, as did their reptilian ancestors. The only prototherians surviving today are the monotremes—the duckbill platypus and the echidnas, or spiny anteaters. The other major mammalian group is the subclass Theria *(THIHR ee uh)*. All of the mammals you are familiar with, including humans, are therians. Therians are viviparous. The two major therian groups are marsupials, or pouched mammals, and placental mammals. Kangaroos, opossums, and koalas are marsupials. Dogs, cats, humans, horses, and most other mammals are placentals. You will learn more about these groups later in the chapter.

The Age of Mammals

At the end of the Cretaceous period, 65 million years ago, the dinosaurs and numerous other land and marine animals became extinct, but mammals survived. In the Tertiary period (lasting from 65 million years ago to 2 million years ago), mammals rapidly diversified, taking over many of the ecological roles once dominated by dinosaurs. Mammals reached their maximal diversity late in the Tertiary period, about 15 million years ago. At that time, tropical conditions existed over much of the world. During the last 15 million years, world climates have deteriorated, and the area covered by tropical habitats has decreased, causing a decline in the total number of mammalian species.

Section Review

1. *Identify two mammalian features shown by pelycosaurs.*
2. *Name two features you share with therapsids.*
3. *How could you distinguish between a therapsid fossil and a fossil of an early mammal?*

Critical Thinking
4. *How might mammalian evolution have been different if the dinosaurs had not become extinct?*

SECTION 34-1

Teaching Tips
Rare Fossils
The text states that mammal fossils were rare from the time of the dinosaurs because mammals were not abundant. Have your students suggest other reasons why small, delicately built, terrestrial animals might not leave many fossils. *(Mammals were either caught and consumed by predators or their skeletons were destroyed by scavengers. Delicate bones and teeth do not fossilize well, and terrestrial environments do not favor the quick burial needed for fossilization.)*

Viviparity and Oviparity
Remind students that oviparous animals lay eggs. Birds, most reptiles, and monotremes are oviparous. The young of viviparous animals receive nourishment from their mother before birth and are born live. Marsupials, placental mammals, and a few reptiles are viviparous.

CONNECTIONS
Chapter 17
Disappearing Mammals
There are more species of organisms in the tropics than in the temperate regions. This trend holds true for mammals. As tropical rain forests are destroyed, therefore, mammalian species disappear along with the plants and insects. Humans are contributing to the decline in mammalian species that has been occurring over the last 15 million years.

Answers to Section Review

1. Pelycosaurs had the mammalian features of a synapsid skull and teeth of different sizes and shapes.

2. Answers should include two of the following features that humans share with therapsids: complex teeth, secondary palate, synapsid skull, and large synapsid openings.

3. Early mammal fossils have a strong jaw composed of only one bone, while the therapsid jaw is composed of many bones.

4. Answers will vary but should include a description of the lack of diversity in mammals for millions of years while the dinosaurs lived on Earth. If there hadn't been a mass extinction at the end of the Cretaceous period, dinosaurs might still dominate the Earth, and mammals might still be small and rare.

Opening Questions

Mammals provide extensive care to their offspring. List two benefits the offspring receive. (*The young are protected from predators and environmental extremes. They receive food, and they can learn from their parent or parents.*) List two disadvantages for the parents. (*By putting more effort and resources into each offspring, they can have fewer offspring in total. Dependent offspring make the parents more vulnerable and more obvious to predators. Resources given to offspring are lost to the parents.*)

Demonstration

A Mammal's Coat

Show students a pelt of a mammal. Allow them to examine the texture and color of the fur. Point out that two major types of hair make up the coat: guard hairs and underhair. Guard hairs are the thick, coarse, long outer hairs on the coat. Underhair is the denser, shorter hair that grows underneath the guard hairs. Students can see the underhair by brushing back the guard hairs. The coloration of the coat comes primarily from the guard hairs. Underhair serves primarily as insulation.

One hundred million years ago dinosaurs were the large land animals, and mammals were small, insectivorous, and nocturnal. No mammal was larger than a cat until after the dinosaurs became extinct. Today, mammals have succeeded dinosaurs as the dominant land animals. Dinosaurs were successful in diverse habitats because they had adaptations such as horns, claws, expandable jaws, and specialized kinds of teeth. In this section you will explore a few of the adaptations that contributed to the success of mammals.

Section Objectives

■ Identify three functions of hair.

■ Contrast parental care in reptiles and mammals.

■ Describe two hunting adaptations of the Siberian tiger.

■ Compare patterns of reproduction in monotremes, marsupials, and placental mammals.

Hair Is a Unique Mammalian Feature

Mammals are hairy. Even whales and dolphins, which appear to be hairless, have a few sensitive bristles on the snout. No other living animals have hair. A hair is a filament composed mainly of dead cells filled with the protein keratin. Each hair is anchored in and produced by a bulb-shaped structure, the hair follicle, which lies beneath the surface of the skin. The evolutionary origin of hair is unknown, but it is probably not derived from reptilian scales.

The primary function of hair is insulation. Mammals, such as the polar bear in Figure 34-4, tend to lose body heat because they typically maintain body temperatures higher than the temperature of their surroundings. Most mammals are covered with a dense coat of hair that reduces the amount of body heat escaping into the environment. We humans need clothes in most climates because our hair is too sparse to be adequate insulation.

Hair has many functions besides insulation. The coloration and pattern of a mammal's coat often make very effective camouflage. A little brown mouse is practically invisible against the brown leaf litter of the forest floor, and the orange and black stripes of a Bengal tiger cause it to blend in with the tall, orange-brown grass in which it hunts. A mammal's coat may also be a conspicuous signal. The black and white fur of a skunk, for instance, warns would-be predators to stay away.

In some mammals, hairs serve a sensory function. The whiskers of cats and dogs are stiff hairs that are very sensitive to touch. Mammals that are active at night or that live underground often rely on their whiskers for information about the environment. Hair can also be a defensive weapon. Porcupines and hedgehogs are protected by long, sharp, stiff hairs called quills.

Figure 34-4 A thick coat of fur, coupled with a layer of blubber beneath the skin, insulates this polar bear in its frigid arctic habitat. The coloration of the coat helps conceal the bear as it stalks seals, its main prey.

Nursing and Caring for the Young

Female mammals have mammary glands, the unique feature for which the class Mammalia was named. The word *mammary* is derived from the Latin word *mamma*, meaning "breast." These glands, located on the chest or abdomen, produce milk. Newborn mammals grow rapidly, and milk provides the nutrition to support this growth. Milk is rich in protein, carbohydrates (chiefly the sugar lactose), and fat, which accounts for 50 percent of the energy in milk. It also contains water to prevent dehydration and contains minerals, such as calcium, that are critical to early growth. Young mammals are nourished on milk from birth (or hatching, in the case of monotremes) until weaning, when the mother stops nursing them.

Figure 34-5 By watching their mother stalk and capture prey, these lion cubs will learn to hunt. Though weaned at the age of one year, the cubs will not be skilled enough to survive on their own until they are about three years old.

Reptiles seldom provide parental care to their offspring. Typically, a female reptile buries and leaves her eggs. The hatchlings must fend for themselves. In contrast, mammalian young are dependent on parental care for a relatively long period, receiving milk and other food, protection, and shelter. Like the lion cubs shown in Figure 34-5, young mammals often learn necessary skills, such as hunting, during their period of dependence. Learning is especially important for primates, which have particularly long periods of dependence.

You can read more about the habits and anatomy of mammals in *Up Close: Siberian Tiger* on pages 808–809.

Teaching Tips

Parental Care in Mammals

Inform students that in birds, both parents typically help raise the offspring. In most mammalian species, however, only the female provides parental care. The male's sole contribution to his offspring is genetic. The reason for this difference between birds and mammals is not known. Students were asked to form a hypothesis on this subject in Chapter 33.

Learning

Students should recognize that mammals depend on learning and that humans probably rely more on learning than do any other mammals. Ask students to list the advantage of learned behaviors over instinctive behaviors. (*Learned behaviors are more flexible, so an animal can adjust its response to match the situation.*) Ask students to list some disadvantages of learned behavior. (*The wrong response may be learned. The correct response may not be learned at all or may be forgotten.*)

Instructional Strategies

- Inform students that all tigers belong to one species, *Panthera tigris*. Only about 5,000 to 7,500 tigers remain in the wild, mostly in India, and their populations are rapidly shrinking. The Siberian tiger is the most northerly subspecies, and it is one of the most endangered. Probably fewer than 200 Siberian tigers exist outside of zoos. Poaching is rapidly reducing this number. According to Russian officials, as many as 96 tigers were killed in the winter of 1993–94.

  The collapse of the Soviet Union has been disastrous for Siberian tigers. Under the Soviet government, the Siberian tiger was strictly protected, and its population increased from around 30 in the 1930s to about 400 in the 1980s. In the economic and social chaos of post-Soviet Russia, wildlife conservation has been a low priority. Moreover, economic hardship has made the tiger a more attractive target for poachers.

- The tiger has a dark upper surface and a creamy white underside. This contrasting pattern, known as countershading, is widespread in animals. This double camouflage helps make an animal inconspicuous from below and above. If possible, take your students outside and have them look up at the sky and down at the ground to appreciate how different the light intensity is depending on the view. They should then see that a light surface is less conspicuous against the bright sky than is a dark surface. Similarly, when seen from above, a dark surface is less conspicuous against the dark ground than is a light surface.

UP CLOSE SIBERIAN TIGER

- **Scientific name:** *Panthera tigris altaica*
- **Range:** Far northeastern China and southeastern Russia
- **Habitat:** Undisturbed forests with ample cover; fewer than 200 Siberian tigers remain in the wild
- **Size:** Up to 300 kg (660 lb.) and 3 m (10 ft.) from nose to tip of tail
- **Diet:** Large mammals such as wild boars and red deer

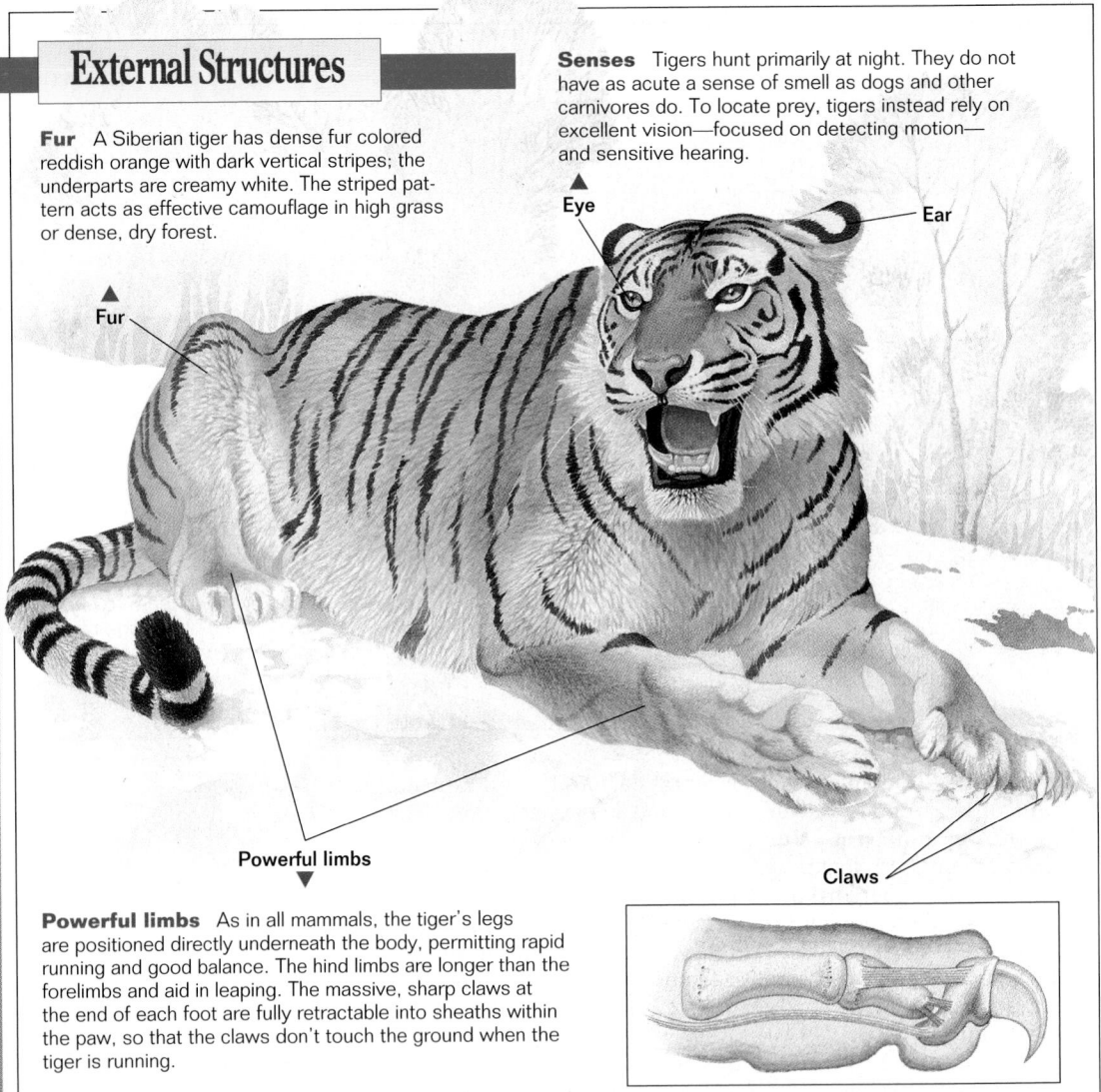

External Structures

Fur A Siberian tiger has dense fur colored reddish orange with dark vertical stripes; the underparts are creamy white. The striped pattern acts as effective camouflage in high grass or dense, dry forest.

Senses Tigers hunt primarily at night. They do not have as acute a sense of smell as dogs and other carnivores do. To locate prey, tigers instead rely on excellent vision—focused on detecting motion—and sensitive hearing.

Eye

Ear

Fur

Powerful limbs

Claws

Powerful limbs As in all mammals, the tiger's legs are positioned directly underneath the body, permitting rapid running and good balance. The hind limbs are longer than the forelimbs and aid in leaping. The massive, sharp claws at the end of each foot are fully retractable into sheaths within the paw, so that the claws don't touch the ground when the tiger is running.

Internal Structures

Reproductive system Like all placental mammals, Siberian tigers have internal fertilization and nourish their embryos through the placenta. Among wild Siberian tigers, reproduction occurs only every 4–6 years. Litters are small, only 2 or 3 cubs, each weighing only about 1 kg (2 lb.) at birth. Born blind and helpless, the cubs open their eyes after 6–12 days and nurse for about 6 months. The mother guards the cubs as they grow and teaches them to hunt. The cubs will hunt with their mother for over 2 years before striking off on their own.

Skull The skull is particularly heavy and anchors massive jaw muscles. The tiger's head is large and foreshortened, providing leverage for its powerful jaws. In the lower jaw there is a gap between the canines and the premolars, allowing the long, sharp upper canines to pierce deeply into prey during a bite. Tigers usually kill their prey by biting its upper neck and then using their powerful muscles to twist and snap its neck.

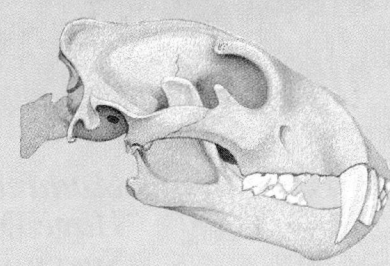

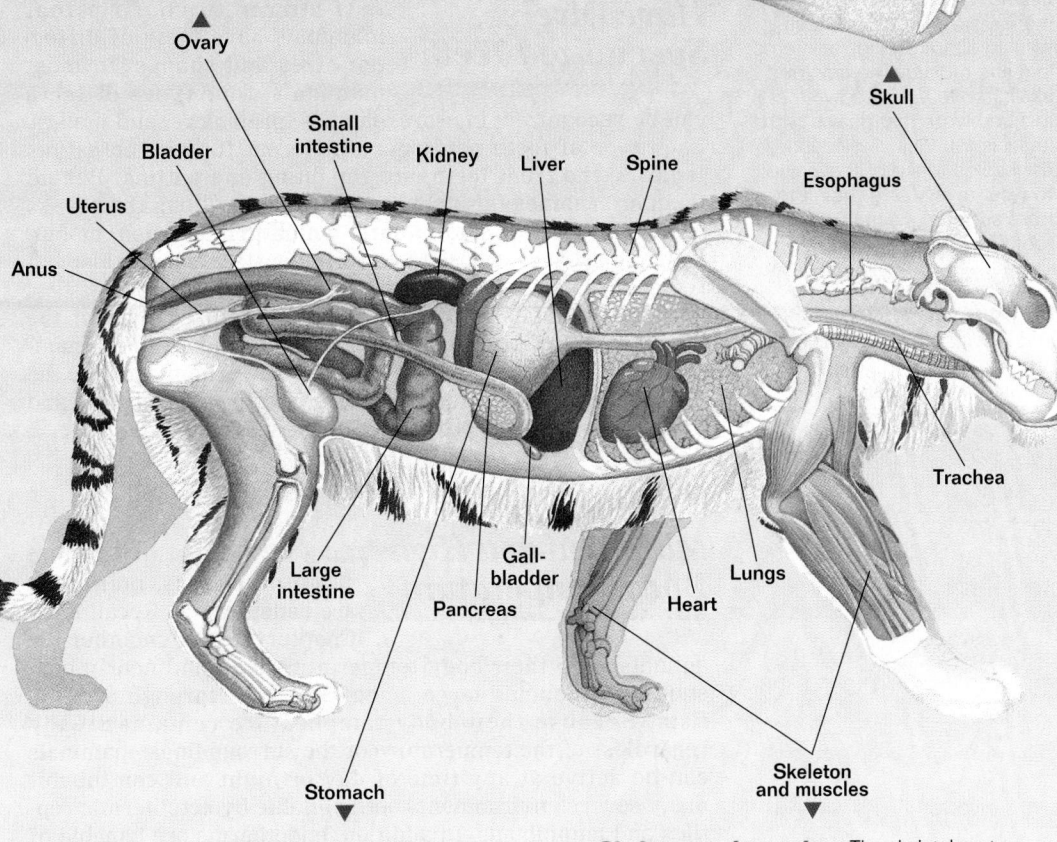

Digestive system A tiger is pure carnivore. Because a tiger's largely protein diet is more easily digested than grass, its digestive system is much shorter than that of a herbivore. A tiger in a zoo will eat 6 kg (13 lb.) of meat per day. In the wild, tigers eat far less frequently but eat more—up to 40 kg (90 lb.)—in each meal.

Skeleton and muscles The skeletal system of a tiger, like that of all mammals, is made of bone. A tiger's body is compact and has the most powerful muscular system of any cat.

Discussion

Guide the discussion by posing the following questions.

1. Identify several differences between the internal organs of the Siberian tiger and the bald eagle, shown in Chapter 33. *(The bald eagle has a crop for storing food, a two-part stomach, air sacs connected to the lungs, and a cloaca.)*

2. There is a gap between the canines and the premolars in the lower jaw. How does this gap affect the tiger's bite? *(The long upper canines fit into this gap, so the tiger can bite deeply into its prey.)*

3. Why might a wild tiger eat more in each meal than a tiger dwelling in a zoo? *(The wild tiger is likely to find food less often, so it eats as much as possible when food is available.)*

👁 VISUAL STRATEGY

Figure 34-6

Point out that the coyote's teeth function like a pair of scissors; two sharp surfaces slide across one another as the coyote closes its mouth. Students should see that the coyote's teeth are unsuited for grinding plant material. The beaver's molars are suited for this task because they function like a grinding mill; two rough surfaces come together and slide across one another when the beaver closes its mouth.

Application

Health Adult humans usually have 12 molars—six in each jaw. The rearmost molars, the so-called wisdom teeth, are the last to erupt, typically emerging between the ages of 17 and 25. Not everyone has a full complement of wisdom teeth. Some people have two or none. If the wisdom teeth grow in at an angle, they can force the other teeth out of alignment and may have to be removed.

Teaching Tip

Metabolic Rate and Body Size

All mammals use energy from the breakdown of food to maintain their body temperature. The rate at which mammals use food depends on their body size, which determines their surface-area-to-volume ratio. Small mammals such as shrews have a high surface-area-to-volume ratio and lose heat so rapidly that they use 80 to 90 percent of their energy just trying to stay warm. Large mammals such as elephants have the opposite problem. They have trouble dissipating excess body heat because their surface-area-to-volume ratio is so low. Ask students which animal they would expect to have the higher metabolic rate. *(shrew)* Because the shrew loses so much heat to the environment, it must replace that heat by consuming more fuel. In fact, a shrew needs so much energy that it must eat its mass in food each day.

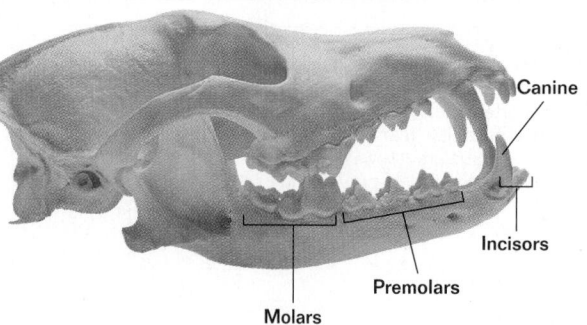
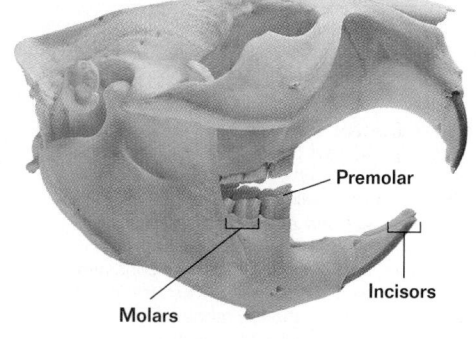

Figure 34-6 The coyote's skull, *above left,* reveals it to be a predator. Long canine teeth enable it to bite and hold prey; it shears off chunks of flesh with its sharp, triangular premolars and molars. The beaver's skull, *above right,* shows that it is a herbivore that gnaws with its incisors. Gnawing causes the back surface of the incisor to wear away faster than the front surface, maintaining the tooth's sharp, chisel-like edge. The incisors continue to grow throughout the beaver's life. The surfaces of its premolars and molars show a complex pattern of ridges and grooves that forms a grinding surface.

Mammals Have Diverse, Specialized Teeth

You need only look at your teeth in a mirror to confirm that mammals have teeth of different sizes and shapes. In most mammals, four types of teeth can be recognized: incisors, canines, premolars, and molars. Each type of tooth performs a different function in eating. Incisors, the front teeth, are for biting and cutting. Behind them are canines for stabbing and holding. Along the cheeks are the premolars and molars, which grind, crush, or cut. Unlike dinosaurs, whose teeth were constantly being lost and replaced, an adult mammal keeps its teeth all its life.

The teeth of a mammal are highly specialized to match the food it eats. In fact, it is usually possible to determine a mammal's diet just by examining its teeth. For example, notice the differences between the teeth of a coyote (a carnivore) and those of a beaver (a herbivore), shown in Figure 34-6.

Maintaining a High Body Temperature

Like birds, therapsids, and some dinosaurs, mammals are endothermic. Recall from Chapter 31 that endothermic animals keep their body temperature high and nearly constant by producing large amounts of heat through metabolism. Because their body temperature remains steady regardless of the temperature of the surroundings, mammals can be active at any time of day or night and can inhabit many severe environments not habitable by ectothermic reptiles and amphibians. In addition, endotherms are capable of sustained activity, such as running or flying long distances.

To maintain the high metabolic rate necessary for endothermy, a mammal must eat about 10 times as much food as an ectotherm of similar size. Endothermy also requires rapid delivery of oxygen to metabolizing tissues. The mammalian respiratory and circulatory systems are very efficient at acquiring and distributing oxygen.

Overcoming Misconceptions

Long Canines in Herbivores

Students looking at the skull of a gorilla or chimpanzee might get the impression that these animals are carnivores. After all, they have long, large canines. Scientists have carefully observed the behavior of both primates in the wild; chimpanzees rarely eat meat, and gorillas never do. How can this apparent contradiction be explained? Male gorillas and chimpanzees display their long canines to frighten away rival males. Long canines have evolved because they produce a more threatening and impressive display.

In Chapter 31 you learned that mammals have a four-chambered heart in which oxygenated and deoxygenated blood never mix. Only oxygen-rich blood is delivered to the tissues. Recall also that mammalian lungs, because of their large internal surface area, are much more effective at absorbing oxygen than are reptilian and amphibian lungs. Efficient respiration in mammals is aided by the **diaphragm,** a sheet of muscle at the bottom of the rib cage. Contraction of the diaphragm helps draw air into the lungs.

Three Styles of Reproduction

In the previous section you learned that present-day mammals are divided into three groups: monotremes, marsupials, and placental mammals. These three groups differ in their manner of reproduction. Monotremes are oviparous, laying shelled eggs in which yolk nourishes the embryo. After hatching, a young monotreme feeds on its mother's milk for several months.

Placental mammals are viviparous. The key feature of embryonic development in placental mammals is the placenta, which you studied in Chapter 31. Recall that the placenta brings the bloodstreams of the mother and offspring into close proximity. Nutrients and oxygen pass from mother to offspring; wastes pass in the opposite direction and are eliminated by the mother's excretory system. Nourished by nutrients from the placenta, the offspring remains in the uterus until development is essentially complete. The length of time between fertilization and birth is the **gestation period** (jeh STAY shuhn). Compared with marsupials, placental mammals have long gestation periods.

Marsupials, such as the Virginia opossum shown in Figure 34-7, are also viviparous, but only a few species have a placenta. Instead, much of the embryonic marsupial's nutrition is provided by a rich fluid secreted by the uterus. The gestation period in marsupials is very short—as little as eight days in some species. A newborn marsupial is tiny, hairless, and

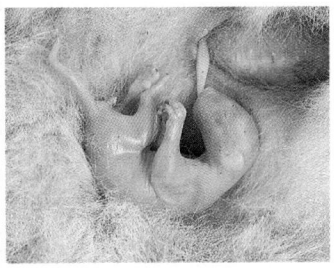

Figure 34-7 Young Virginia opossums, *left,* ride on their mother's back. A newborn opossum, *above,* is smaller than your thumb. It must crawl through the mother's fur to reach her pouch, where the nipples are located.

Graphic Organizer

Use this Graphic Organizer in **Teaching Tip: Drawing a Cladogram** on page 811.

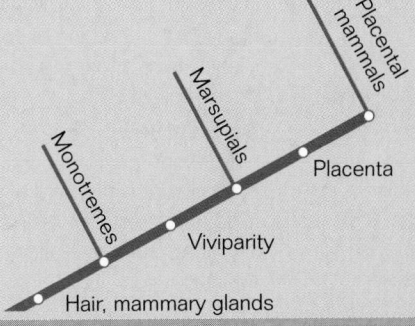

Teaching Tip
Placental Marsupials?

Students may wonder why the few marsupials that have a placenta aren't considered placental mammals. The "placental" marsupials are the bandicoots (family Paramelidae), insectivorous inhabitants of New Guinea and Australia that range from the size of a rat to the size of a small dog. Though bandicoots have a placenta, it is neither as efficient nor as intimately associated with the mother's tissues as the placenta of a placental mammal. In bandicoots, the fertilized egg adheres to the lining of the uterus. Blood vessels in the lining of the uterus exchange materials with blood vessels from the allantois of the egg. In placental mammals, by contrast, the fertilized egg buries itself in the lining of the uterus. Fingerlike outgrowths from the chorion, called chorionic villi, penetrate into the uterine lining and greatly increase the surface area through which nutrients can be absorbed. No chorionic villi form in bandicoots. Their placental arrangement cannot provide all the embryo's nutritional requirements, so supplementary nutrition is furnished by the secretions of the uterus, as in other marsupials. Moreover, the gestation period in bandicoots is short, as in other marsupials.

CONTENT LINK

Chapter 42 *further describes the structure and function of the placenta.*

incompletely developed, as you can see in Figure 34-7. Without any help from its mother, the newborn struggles to the nipples, which are usually located in a pouch on the mother's abdomen. It attaches to a nipple and continues its growth and development, which may take several months. Table 34-1 summarizes some of the similarities and differences among monotremes, marsupials, and placental mammals.

Table 34-1 Comparison of Monotremes, Marsupials, and Placental Mammals

Group	Mode of Reproduction	Placenta Present?	Distribution	Number of Species
Monotremes	Oviparity	No	Confined to Australia and New Guinea	3
Marsupials	Viviparity	Only in a few species	Concentrated in Australia and New Guinea, but also in nearby Asian islands and in the Americas	280
Placental Mammals	Viviparity	Yes, in all species	Worldwide	4,100+

Section Review

1. *Explain two ways a cat would be hampered by the removal of all its hair.*
2. *How does the parental care shown by mammals and crocodilians differ?*
3. *What adaptations enable the Siberian tiger to hunt at night?*
4. *Name two differences between marsupials and placental mammals.*

Critical Thinking

5. *Reptiles swallow their food without chewing. Mammals usually chew their food. Why is chewing an advantage for an endotherm?*
6. *How is the ability to chew related to the replacement pattern of mammalian teeth?*

Answers to Section Review

1. The cat would lack insulation, camouflage, and sensory input from sensitive hairs.

2. Crocodiles protect their eggs and young. Mammals protect and nurse their young and help them learn survival skills.

3. The Siberian tiger has whiskers for sensing its position in the dark, sensitive hearing, and excellent night vision.

4. After a short gestation period, marsupials give live birth to an immature newborn that must crawl to the mother's pouch to reach the mother's nipples and continue development. Placental young are more developed at birth.

5. Endotherms need to rapidly release the energy contained in food. Chewing breaks down food into smaller particles for more rapid and efficient digestion.

6. Most mammals retain their adult teeth for life. Opposing teeth keep the same position and can work against each other to form a grinding surface.

34-3 Mammalian Diversity

*T*here are about 4,400 living species of mammals. In number of species, mammals are not the most successful vertebrate class; there are more species of birds, reptiles, and bony fishes. In diversity of anatomy and habitat, however, mammals far surpass the other vertebrate groups. For instance, consider the differences between a bat and a whale, some of which are listed in Figure 34-8. In this section, you will get an idea of the diversity within the class Mammalia by studying the 21 mammalian orders recognized by most biologists.

Section Objectives

- Recognize the limited distribution of monotremes and marsupials.
- Recognize the diversity of placental mammals.

Vocabulary Preview

(none)

Opening Question

Students are probably aware that mammals live in a great variety of habitats. Ask students to list some habitats in which they think mammals *cannot* be found. *(Acceptable answers are environments of extreme cold, such as Antarctica and the peaks of tall mountains, the deepest parts of the ocean, and very hot environments, such as boiling springs.)*

Comparison of Bats and Whales

Bats
- Forelimbs are wings of leathery skin
- Elongated bones of forelimbs and four fingers form framework of wing
- Most species fly at night and navigate by using echoes of high-pitched sounds emitted from the nose or mouth

Whales
- Nearly hairless, streamlined body
- Forelimbs are flattened and paddle-shaped; no hind limbs.
- Propelled by flattened tail
- Many species use sound to communicate and perceive the environment

Figure 34-8 Bats and whales live in very different environments for which they have different adaptations. All bats fly, and most are active at night. Whales are permanently aquatic. However, both groups face one similar challenge—how to navigate in an environment with poor visibility. Both groups evolved a similar solution to this problem: they use sound.

Demonstration
Mammalian Diversity
To demonstrate some of the diversity found among the mammals, show the class photographs of several kinds of mammals that live in varied habitats. Point out some of the adaptations that enable each mammal to live where it does. For instance, whales have flippers and powerful, flattened tails for swimming. They use sonar to communicate and navigate, and are insulated against the cold of the ocean by a thick layer of blubber under the skin.

Monotremes Are Egg-Laying Mammals

The duckbill platypus and two species of echidnas *(ee KIHD nuhz),* or spiny anteaters, are the only living monotremes (order Monotremata). The Australian echidna is shown in Figure 34-9. Recall from Chapter 31 that mammals evolved from reptiles. Because monotremes

Figure 34-9 Termites and ants are the main foods of echidnas.

Historical Note

Mysteries of the Platypus
When the first specimens of the platypus arrived in Europe in the 1790s, the first scientists to examine them suspected a hoax. The platypus seemed to be composed of parts of two very different animals. Like known mammals, it had hair, a single jawbone, and three middle-ear bones. But it also had a bill like a duck's. As they could find no evidence of stitching or tampering, the scientists concluded that the platypus, however odd, was a mammal. Its reproductive habits were even more unusual. The reproductive tract was similar to that of a bird, suggesting that platypuses were oviparous like birds and most reptiles. That platypuses laid eggs was not conclusively established until 1883, when platypus eggs were discovered.

share several characteristics with reptiles, biologists think monotremes are more closely related to the early mammals than are any other living mammals. For instance, monotremes, like reptiles, lay shelled eggs. The monotreme shoulders and forelimbs are also quite reptilian. Monotremes, like reptiles and birds, have a cloaca, a common passageway for the digestive, reproductive, and urinary systems. Despite their reptilian features, monotremes are definitely mammals; they have hair, and the females produce milk. All monotremes live in Australia and New Guinea. This limited distribution resulted from continental drift, which you studied in Chapter 33. Australia and New Guinea moved away from the other landmasses about 50 million years ago, before placental mammals had a chance to reach them.

The platypus inhabits lakes and streams in eastern and southern Australia. With its soft, sensitive bill, it searches through mud and gravel for worms and other soft-bodied animals. Echidnas are terrestrial, and they have very strong, sharp claws, which they use for burrowing and for digging out insects.

Marsupials Are Pouched Mammals

The order Marsupialia contains 280 species of kangaroos, opossums, wombats, and a variety of other animals, including the koala shown in Figure 34-10. As you learned in Section 34-2, marsupials differ from placental mammals in their pattern of development. The gestation period is very short, and the young are born incompletely formed. They crawl to the pouch, attach to nipples, and complete their development. Only one species of marsupial, the Virginia opossum, lives in North America today. Australia, New Guinea, and a few nearby islands contain the majority of marsupial species. Most mammalian species in Australia and New Guinea are marsupials.

Figure 34-10 The koala is one of nearly 100 species of marsupials found in Australia. Its cuddly, "teddy bear" appearance belies a testy, pugnacious disposition. Koalas have a very specialized diet, feeding only on the leaves of 17 species of eucalyptus trees.

Most Mammals Are Placental Mammals

Nineteen orders of mammals, comprising more than 90 percent of the species, contain only placental mammals. Recall that the offspring of placental mammals remain in the uterus through a relatively long gestation period, during which they receive nourishment through the placenta. At birth, placental young are further along in development than are marsupial young. Terrestrial placental mammals inhabit all continents except Antarctica, and aquatic placental mammals inhabit all oceans. Table 34-2 summarizes the orders of placental mammals.

Table 34-2 Orders of Placental Mammals

Order	Number of Species	Main Characteristics
Rodentia: Gnawing mammals 	1,814	Over 40 percent of mammalian species are rodents. **Key Characteristics** While most mammals have four incisors (front biting teeth) in each jaw, a rodent has only two, which continue to grow throughout its lifetime. Famous for their reproductive powers, rodents can give birth to a new litter every three weeks. **Important Examples** The families Muridae (mice and rats), Sciuridae (squirrels and chipmunks), and Cricetidae (voles, lemmings, deer mice, and hamsters) **Distribution** Worldwide, in almost all habitats
Chiroptera: Flying mammals 	986	Bats are the only mammals capable of powered flight. **Key Characteristics** The wing of a bat is a leathery membrane stretching across the forelimb and four elongated fingers. Most bats feed on night-flying insects and navigate in the dark with a sonar system that employs very high-pitched sounds. **Important Examples** Little brown bat *(Myotis lucifugus),* freetail bats *(Tadarida),* and flying foxes *(Pteropus)* **Distribution** Worldwide, except in polar environments
Insectivora: Insect eaters 	390	Insectivores are small and have a high metabolic rate. They are the mammals most similar to the ancestors of the placental mammals. **Key Characteristics** Most have long pointed snouts that they use to root for insects and worms. Their sharp teeth are adapted for grabbing and piercing prey. **Important Examples** Shrews, among the smallest of mammals, feed above ground by sweeping invertebrates into their mouths with clawed paws. Moles, which burrow, have small eyes and no external ears. **Distribution** All continents except Australia
Carnivora: Flesh eaters	240	The large predators of the animal kingdom, the carnivores include some of the strongest and most intelligent of animals. **Key Characteristics** Carnivores have long canine teeth, strong jaws, and clawed toes that aid in seizing and holding prey. Most have keen senses of sight and smell and can run quickly. **Important Examples** The families Canidae (domestic dogs, wolves, foxes, and coyotes), Felidae (the cats, including tigers, lions, leopards, and domestic cats), Mustelidae (weasels, minks, otters, and skunks), Procyonidae (raccoons), and Ursidae (bears, many of which are omnivores) **Distribution** Native to all continents except Australia

SECTION 34-3

CONNECTIONS

Chapter 15
r-strategists
Remind students of the characteristics of r-strategists: small, fast-growing, reproduce early in life, have many offspring, provide little care to each offspring. Point out that these characteristics describe most rodents.

Teaching Tips

What Order Are the Orders In?
Have students look at the orders of placental mammals and figure out by what criterion they are ordered. *(descending number of species).*

What If a bat is blinded? Could it still fly? The Italian scientist Lazzarro Spallanzani answered this question in the late 1700s. He found that blinded bats could still fly and capture prey. However, deafened bats could not avoid objects in their path, and they could not capture prey. He concluded that bats depend on sound to navigate.

Misconceptions About Bats
Bats have been burdened with more misconceptions, superstitions, and old wives' tales than any group of animals apart from snakes. Have students research these misconceptions and write a report summarizing their findings. Students should address the misconceptions, such as the belief that bats suck blood, and explain how they may have originated.

Multicultural Perspective

Folk Medicine Threatens the Tiger's Survival

Tigers are disappearing because humans are encroaching on their habitat and because of overhunting. Poachers kill tigers for the pelt, which can be worth as much as $15,000. But tigers are also slaughtered to feed the large market for tiger-based folk medicines, which are used mainly in Korea, China, and Taiwan. Potions made from the bones, eyes, and other parts of the tiger purportedly lengthen life, instill vigor, and cure diseases and impotence.

CONNECTIONS
Chapter 14
Opposable Thumb
Remind students that the opposable thumb can be bent towards the other fingers so that objects can be gripped. Most primates have opposable big toes, but humans do not.

☑ RESEARCH Update

Dealing With Crowding: Fighting or Coping?

For the past three decades, behavioral scientists and psychologists have held firm to the theory that crowded conditions lead to aggressive behavior. These conclusions were based on studies of rats and mice that were packed into cages. The rodents turned vicious, attacking, biting, and even killing others at the peak of their stress-induced frenzy. The overcrowded-rats model has frequently been applied to human situations as well. Researchers Peter Judge and Frans de Waal of the Yerkes Regional Primate Research Center in Atlanta decided to apply this theory to non-human primates. For their study they chose rhesus monkeys, which have a reputation for feistiness and aggression.

These researchers found that in crowded conditions rhesus monkeys didn't behave as aggressively as the rats had. Instead, the monkeys tried coping with the added stress by doing such things as grooming, huddling together in groups, avoiding confrontational situations, and establishing pecking orders. While the frequency of aggressive acts did go up slightly in more cramped quarters, the increase was not particularly significant. ☑

Order	Number of Species	Main Characteristics
Primates: Large-brained mammals	233	Primates have the largest brains for their body size of any animals. **Key Characteristics** The only mammals with binocular vision and opposable thumbs, primates have five digits on all four limbs. None have horns or hoofs, and most have flat fingernails rather than claws. Hair covers the bodies of all primates except humans. **Important Examples** Prosimians (lemurs, lorises, and tarsiers), monkeys (baboons, squirrel monkeys, and howler monkeys), and hominoids (apes and humans). The hominoids include the families Hylobatidae (gibbons), Pongidae (orangutans, chimpanzees, and gorillas), and Hominidae (australopithecines and humans). **Distribution** Humans live on all continents; most other species live in tropical regions of Africa, Asia, and the Western Hemisphere.
Artiodactyla: Even-toed, hoofed mammals	211	Fast-running antelopes and other grazers and browsers **Key Characteristics** On each foot, artiodactyls have an even number of toes, each encased in a horny hoof. All are herbivores, with large flat molars for grinding plant material. Many have stomachs with a storage chamber called the rumen, where bacteria break down cellulose. **Important Examples** There are nine families, grouped into three suborders: Suina (pigs and hippopotamuses), Tylopoda (camels), and Ruminantia (deer, antelope, cattle, sheep, and giraffes). **Distribution** All continents except Australia
Cetacea: Fully marine mammals	79	Cetaceans are permanently aquatic and have fishlike bodies. **Key Characteristics** Cetaceans have forelimbs modified as flippers, no hind limbs, and broad, flat tails for swimming. They breathe through a nostril, or blowhole, on top of the head. A thick layer of blubber beneath the skin serves as insulation. Except for a few bristles on the muzzle, cetaceans are hairless. **Important Examples** Cetaceans are divided into two groups: the predatory toothed whales (sperm whales, killer whales, dolphins, and porpoises) and the filter-feeding baleen whales (blue whales, gray whales, and humpback whales). The blue whale is the largest animal that ever lived. **Distribution** All oceans; some species of dolphins live in freshwater rivers and lakes.
Lagomorpha: Rabbits	69	Small, swiftly running herbivores with powerful hind legs **Key Characteristics** Like rodents, they have one pair of long, continually growing incisors; unlike rodents, they have an additional pair of peglike incisors growing behind the front pair. **Important Examples** Rabbits and hares **Distribution** Native to all continents except Australia

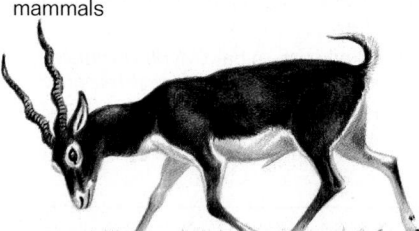

Order	Number of Species	Main Characteristics
Pinnipedia: Marine hunters	34	Partially marine carnivores that feed in the sea but mate and sleep on land **Key Characteristics** All four limbs have been modified as flippers for swimming. Their bodies are streamlined for rapid movement through the water. **Important Examples** Seals, sea lions, and walruses **Distribution** Temperate and polar oceans

Edentata: Toothless mammals	30	Mammals with poorly developed or no teeth **Key Characteristics** The name of this order means "without teeth," and its members are toothless or have simple, peglike, rootless molars without enamel. **Important Examples** Anteaters use a long sticky tongue to lap up insects; sloths are sluggish, tree-dwelling herbivores; armadillos feed on any kind of meat, including carrion. **Distribution** Western Hemisphere only, primarily South America and Central America

Macroscelidea:	19	Hopping shrews **Key Characteristics** Elephant shrews are insect eaters that have a long, flexible snout. They resemble true shrews but hop about somewhat like small kangaroos. **Distribution** Africa

Perissodactyla: Odd-toed, hoofed mammals	17	Horses and rhinos **Key Characteristics** Hoofed herbivores with an odd number of toes per foot. They have a cecum, a pouch branching from the intestine in which bacteria digest cellulose. **Important Examples** Members of the family Equidae (horses, asses, and zebras) have one toe per foot. Rhinoceroses have three toes per foot. **Distribution** Most species are native to Asia and Africa.

Teaching Tips
Comparing Whales With Pinnipeds

Whales and pinnipeds (seals, sea lions, walruses) are adapted for life in the water. Have students create a table that lists similarities and differences between these aquatic mammals. *(Similarities: forelimbs modified as flippers, feed in the water, breathe air, streamlined bodies. Differences: whales lack hind limbs; pinnipeds have paddlelike hind limbs; whales cannot leave the water; pinnipeds mate and sleep on land.)*

Advancing Armadillos

The nine-banded armadillo *(Dasypus novemcinctus)* is one of the few mammalian species whose range is expanding. In the late 1800s, armadillos lived only as far north as central Texas. Now they inhabit most of Texas and all of Louisiana, and have spread to parts of Oklahoma, Arkansas, Mississippi, New Mexico, Georgia, Florida, and Alabama. Isolated sightings have been reported in Tennessee and South Carolina.

Historical Note

Wild Horses

Herds of wild horses are part of the standard picture of the West. However, these horses are recent arrivals. All are descendants of horses that were released by or escaped from European colonists, beginning after Columbus's arrival in 1492. Ironically, horses probably evolved in North America about 4 million years ago. They migrated to Asia and Europe but became extinct in North America about 12,000 years ago.

Teaching Tip
Adaptations of Pangolins
Point out some of the food-gathering adaptations of pangolins. The fore-limbs are powerful and tipped with sharp, sturdy claws for tearing open ant nests and termite mounds. The tongue is sticky and very long—together, the tongue and its muscles are longer than the head and body. The tongue can be inserted into a nest or mound to ensnare prey.

Application

Foreign Language The order Sirenia is named for the sirens of Greek and Roman mythology. Sirens had the head and upper body of a woman and the lower body and limbs of a bird. Their sweet songs lured sailors to their death. Manatees and dugongs were thought to resemble sirens in their manner of nursing their young.

Order	Number of Species	Main Characteristics
Scandentia: Tree shrews	16	Small, squirrel-like mammals of the rain forest **Key Characteristics** They have a shrewlike body, a long snout, and sharp teeth. Despite their name, tree shrews live mainly on the ground. They feed on animals and fruit. **Distribution** Tropical rain forests of southern Asia
Pholidota: Pangolins	7	Armored anteaters **Key Characteristics** The word *pholia,* the root word of the name for this order, means "horny scale." The bodies of pangolins are covered with overlapping scales formed from fused bundles of hair. Pangolins have no teeth and feed on termites and ants they capture with a very long tongue. **Distribution** Tropical Asia and Africa
Hyracoidea: Hyraxes	7	Rabbitlike desert dwellers **Key Characteristics** Hyraxes have rabbitlike bodies, short ears, and hoofs. They have four hoofed toes on the front feet and three on the back. **Distribution** Africa and the Middle East
Sirenia: Manatees	4	Tubby, clumsy-looking aquatic herbivores **Key Characteristics** Like whales, sirenians have front limbs modified as flippers, no hind limbs, and a flattened tail used for propulsion. **Important Examples** Manatees, dugongs, and sea cows **Distribution** Tropical coastlines and estuaries of rivers

Order	Number of Species	Main Characteristics
Proboscidea: Elephants	2	The largest land animals alive today, elephants can weigh up to 5,400 kg (6 tons). **Key Characteristics** Elephants have a long, boneless, trunked nose. Their two upper incisors are long, curved tusks. **Important Examples** The African (large-eared) elephant and the much smaller Asian (small-eared) elephant **Distribution** Africa and Asia
Dermoptera: Flying lemurs	2	Large, squirrel-like gliders **Key Characteristics** Flying lemurs are misnamed on both counts: they cannot fly, and they are not lemurs (which are primates). Like flying squirrels, they glide on a sheet of skin stretching between their forelegs and hind legs. **Distribution** Southeastern Asia
Tubulidentata: Aardvark	1	The most unusual mammalian order **Key Characteristics** The name *aardvark* means "earth pig" in Afrikaans. The nocturnal aardvark has a pig-like body, big ears, and a long snout. Like pangolins, they feed on ants and termites. **Distribution** Southern Africa

SECTION 34-3

Chapter Closure

Have students make a dichotomous key that could be used to classify a specimen into one of the 21 mammalian orders. Tell students that they can use the information in the chapter and consult other references.

Section Review

1. *To which country would you have to travel to see a monotreme? to see a marsupial?*

2. *Like reptiles in the Mesozoic era, mammals have taken to the air and live in the ocean. Describe one adaptation that enables mammals to live in each habitat.*

Critical Thinking

3. *Few species of placental mammals have been able to reach Australia without human help. Most that did are rodents and bats. Why do you think these mammals were able to make this journey, while members of other orders could not?*

Answers to Section Review

1. One would have to travel to Australia or New Guinea to see a monotreme; one could see a marsupial in the United States.

2. Mammalian adaptations for flight include a sheet of skin between the forelimbs and fingers and a sonar system for navigation. Adaptations for swimming include flippers, blowhole on top of head, streamlined bodies, and flattened tails for propulsion.

3. Rodents tend to be small and could hitch a ride on floating debris more easily than could a large mammal. Bats can fly or be blown across the sea.

CHAPTER REVIEW ANSWERS

Each item in the Chapter Review is correlated by text section in the assignment guide that follows.

ASSIGNMENT GUIDE

Section	Review	Themes Review	Critical Thinking
34-1	1–5, 10	16	21
34-2	6–8, 11, 13	17–20	22, 23
34-3	9, 12, 14, 15		

Review

Multiple Choice

Code in parentheses indicates section number and objective number.

1. b (34-1.2)
2. c (34-1.2)
3. d (34-1.3)
4. a (34-1.3)
5. d (34-1.3)
6. b (34-2.1)
7. c (34-2.2)
8. a (34-2.4)
9. b (34-3.2)

Completion

10. pelycosaurs (34-1.1)
11. mammary, milk (34-1.3)
12. Monotremata, reptiles (34-2.4)
13. oviparous, viviparous (34-2.4)
14. Artiodactyla (34-3.2)
15. the air, the ocean (34-3.2)

Vocabulary

diaphragm (811)
gestation period (811)

Review

Multiple Choice

1. What shared characteristics suggest that therapsids are the ancestors of mammals?
   a. secondary palate and milk production
   b. complex teeth and a large synapsid opening
   c. endothermic body and primary palate
   d. fur and large eye sockets

2. Evidence that therapsids were efficient feeders and lived in cold climates indicates that they may have been
   a. covered with scales.
   b. ectothermic.
   c. endothermic.
   d. carnivores.

3. A paleontologist finds a fossilized jaw believed to be that of a mammal. If it is that of a mammal, how many bones will compose the jaw?
   a. 12   c. 2
   b. 4   d. 1

4. After the dinosaurs became extinct, mammalian species flourished because
   a. many new habitats became available to them.
   b. the climate became warmer and dryer.
   c. they have four-chambered hearts.
   d. young mammals can learn necessary skills.

5. A dolphin is classified as a mammal because it
   a. has a bony jaw with teeth.
   b. breathes air.
   c. has a four-chambered heart.
   d. feeds milk to its offspring.

6. Sheep are sheared annually for their hair, called wool. The function served by this hair is
   a. camouflage.   c. defense.
   b. insulation.   d. sensory.

7. The care given by mammals to their young
   a. begins after the young are weaned.
   b. is similar to that given by reptiles to their young.
   c. involves nursing and teaching survival skills.
   d. ends soon after the young are born or hatched.

8. Monotremes differ from marsupials in that monotremes
   a. lay eggs.
   b. are viviparous.
   c. nourish unborn young via the placenta.
   d. do not have mammary glands.

9. Members of the order Rodentia include
   a. polar bears and bats.
   b. chipmunks and gerbils.
   c. deer mice and wolves.
   d. deer and cats.

Completion

10. Scientific evidence suggests that mammals evolved from reptiles. Reptiles called _____ are considered to be the ancestors of therapsids, which are the direct ancestors of mammals.

11. The class Mammalia is named for the _____ glands found on female mammals. These glands produce _____ .

12. The spiny anteater and duckbill platypus belong to the order _____ . They share several characteristics with _____ , including forelimb structure and a cloaca.

13. Monotremes are _____ , meaning they lay eggs. Marsupials and placental mammals are _____ .

14. The hippopotamus and camel belong to the order _____ .

15. Not all placental mammals are active solely on land. For example, members of the order Chiroptera are active primarily in _____ , while members of the order Cetacea are active primarily in _____ .

Themes Review

16. **Evolution** Fewer species of mammals live today than did during the Tertiary period. What contributed to the decline in mammalian diversity?

17. **Structure and Function** The photograph shows a polar bear in its habitat. What visible characteristics enable the polar bear to survive in this habitat?

18. **Homeostasis** How do hair and a high rate of metabolism help mammals maintain homeostasis?

19. **Structure and Function** How does the shape of a beaver's teeth relate to its diet?

20. **Structure and Function** Explain why a tiger has a shorter digestive system than a herbivore of the same size.

Critical Thinking

21. **Making Inferences** The fossil record indicates that early mammals had large eye sockets. What does this evidence suggest about how early mammals survived competition with dinosaurs?

22. **Making Predictions** The gestation period of a mouse is about 21 days, whereas the gestation period of a moose is about 8 months. If you were a scientist looking for a laboratory animal for experiments dealing with mammalian development and heredity, which animal would you select for your experiments? Explain your choice.

23. **Making Comparisons** Some mammal species must care for their young for many years until the young reach maturity and can survive on their own. This pattern of rearing young has advantages over the way reptiles rear their young. Longer periods of parental care offer the young food, protection, and time to learn necessary survival skills. What are some disadvantages of this pattern of rearing young, compared with the reptilian pattern?

Activities and Projects

24. **Research and Writing** Find out about the theory of continental drift and how the theory can explain the distribution of mammals on Earth. In your report, describe how the breakup of Pangaea led to the predominance of marsupials in Australia and the predominance of placental mammals on other continents.

25. **Cooperative Group Project** Obtain a live female mammal, such as a mouse, hamster, or gerbil, that has just given birth. Get instructions for the care and feeding of the mammal. Then observe the mother and her babies for several weeks, or more if possible. Take notes on the feeding, sleeping, and interaction patterns of the mother and her babies. Share what you learn with your class.

26. **Multicultural Perspective** Research the importance of bison to the Native Americans that lived on the Great Plains. Write a report that summarizes what you have learned.

Critical Thinking

21. Large eye sockets indicate that the animals had large eyes and could see well at night. By hunting for food at night and hiding during the day, they could avoid predation by dinosaurs. (34-1.3)

22. Mice. Because of their short gestation period, they reproduce rapidly. Many generations of mice can be studied in a short time. (34-2.4)

23. Answers will vary, but students should mention one or more of the following disadvantages: adults must protect young, they must gather food for young, and the reproductive rate is lower. (34-2.2)

Activities and Projects

24. Student reports should mention that the movement of the continents away from each other prevented placental mammals from inhabiting all continents. Placental mammals did not reach Australia before it separated from the other continents.

25. Answers will vary. Emphasize that the students' notes should include both observations and inferences drawn from the observations. "Mother ate the carrot" is an example of an observation, while "Mother seemed to like carrots" is an inference based on the observation.

26. Native Americans worshiped, celebrated, and hunted bison. They were used to satisfy an amazing variety of needs. The meat was eaten, the skins provided clothing and shelter, the organs served as containers, the blood became paint, and the bones were used for utensils and ceremonies. Even the dried dung was used for fuel.

Themes Review

16. Climatic changes caused the area covered by tropical habitats to decrease. (34-1.3)

17. The white hair camouflages the bear and insulates it from the cold. (34-2.1)

18. A high rate of metabolism ensures that mammals produce enough heat to maintain a high body temperature. Hair helps prevent some of this heat from escaping. (34-2.1)

19. Beavers feed on plants. With their long, continually growing incisors, they strip plant material such as bark. They grind the food with the rough surfaces of their large, flattened molars and premolars. (34-2.2)

20. Meat is easier to digest than plants, which contain large amounts of indigestible cellulose. Therefore, herbivores must subject their food to more digestion and have longer digestive tracts. (34-2.3)

Preparation

Students will observe at least three conversations of 45 seconds to 5 minutes long. Divide the class into groups of two to three students. It is important that members of each group have time together outside of class to conduct their observations. Each observation will require the presence of at least two members of the group. Students will need a stopwatch or watch with a second hand, paper, and pencil.

Procedural Note

- Allow students to organize themselves into teams of two to three to conduct observations outside the lab period.

Background Answers

1. Mammals communicate with sound, visual signals such as colors and postures, direct contact, and chemicals.

2. language

3. Answers will vary, but may include facial expressions such as smiling and frowning, gestures such as waving and shaking a fist, and body posture.

4. How do humans communicate the intention to depart from a dyadic exchange?

Records

Observations will vary. Students should compile their observations in tables and summarize them in bar graphs.

LABORATORY Investigation Chapter 34

Mammalian Behavior: Human Postural Signals

OBJECTIVE

Observe human postural cues, a type of nonverbal communication.

PROCESS SKILL

- observing and identifying behaviors

MATERIALS

- stopwatch or watch with second hand
- paper and pencil

BACKGROUND

1. What are some types of communication among mammals?

2. What is the most common way in which humans communicate?

3. What are some types of nonverbal communication in humans?

4. Write your own question to explore in your lab report or notebook.

TECHNIQUE

1. Work in a group of two or three to observe pairs of people in conversation. It is important that your subjects are unaware of being observed.

2. You will observe at least three conversations. Each conversation must last between 45 seconds and 5 minutes, and it must involve only two people. Data for conversations that last less than 45 seconds will not give significant results for this investigation.

3. People who study behavior give specific definitions to certain actions. An interaction between two people is called a *dyadic exchange*. The position of the body while standing is called a *stance*. In an *equal stance*, the body weight is supported equally by both legs. And in an *unequal stance* more weight is supported by one leg than by the other.

4. For each dyadic exchange you will record the actions of only one person. Another member of your team will be the timekeeper and will alert you to 15-second intervals. Within each 15-second interval, you will make note of all the changes of stance by the person you are observing.

5. How often do you think people change their stance while they are in conversation? How might the gender of the people in conversation influence the number of stance shifts? How might the timing of the conversation—beginning, middle, or end—affect the number of stance shifts? Record your expectations in the **Records** section of your report.

6. In the **Records** section, set up a table for your observations. You may choose to use a table like the one shown below. Record the sexes of both people in the dyad: male/female, male/male, or female/female. Then record the sex of the person you are observing (*s.o.*). Each of the numbered boxes represents a 15-second interval. Notice that the table is reduced here and that you will need to allow extra space for additional 15-second intervals on your own table.

Observations of Dyadic Exchanges									
	sexes	s.o.	1	2	3	...	18	19	20
Dyad 1									
Dyad 2									
Dyad 3									

7. Record every time your subject shifts from an equal stance to an unequal one, or vice versa. When the subject assumes an unequal stance, record the number of discernible weight shifts from one foot to the other. To record this simply, you may write *E* or *U* to identify an equal or unequal stance, and *W* to identify a weight shift. Record all of the changes within a single 15-second block until the timekeeper signals you to go on to the next block.

8. At the end of the conversation, write down whether the pair departed together or separately. Record your observations in the **Records** section of your report.

9. After you have completed each observation, tally the total number of shifts within each 15-second block.

   IMPORTANT! Only retain data for conversations that last at least 45 seconds. If you have data for 30 seconds and the pair depart, you must observe a new dyadic exchange.

10. After you have completed your observations, share your data with other teams in your class. Combine and analyze the data from all the teams. Find the most common stance during the first 15 seconds of an exchange, the middle intervals, and the last 15 seconds.

11. Average the number of weight shifts for the beginning, middle, and end intervals. Record these numbers in the **Records** section of your report.

12. Analyze the data by the gender of the subject you observed and by the gender of the partner. Compare these with the averages not related to gender, and record any differences in the **Records** section.

13. Construct bar graphs to express data in the **Records** section of your report. Also, briefly summarize the procedure you followed.

INQUIRY

1. Which is the most common stance in a dyadic exchange?

2. Which is seen more often as the members of the dyad prepare to depart: stance change or weight shift?

3. What happens when a member of the dyad prepares to leave?

ANALYSIS

1. Were your observations what you had expected? If not, how were they different?

2. Are there differences in the departure signals of males and females? If so, what are they?

3. What do you think might be an adaptive significance of a departure signal?

4. What other types of behaviors might serve as departure signals for humans?

5. What other nonverbal communication did you observe during this investigation?

FURTHER INQUIRY

Write a new question that could be explored as a new investigation. The following is an example:

> Do people use body language even when their actions cannot be seen? For example, does a person speaking at a public pay phone shift stances as the conversation comes to a close?

Inquiry Answers

1. During a human dyadic exchange, the most common stance is equal weight on each leg.

2. Answers will vary.

3. When one of the dyad intends to depart, he or she signals by changing stance or shifting weight.

Analysis Answers

1. Answers will vary.

2. While researchers have not found significant differences between the sexes in these signals, your students may see some trends due to the small sample size. This could lead to an interesting discussion about statistics.

3. Answers will vary but may include the idea that nonverbal cues may make the departure expected and less abrupt. This may ensure that the leaving person does not insult the other member of the dyad.

4. Answers will vary but may include breaking eye contact or more frequent movements.

5. Answers will vary.

Block Scheduling Guide

> Each block represents about 45–50 minutes of instructional time. The resources cited in a block are coded to a specific program component using the Key on the next page.

Lecture Concepts	Lesson Resources

35-1 The Human Body Plan pp. 825–828

BLOCK 1

a. Humans are vertebrates with a jointed skeleton that is cushioned by cartilage and held together by ligaments.

b. The human nervous system is composed of two functional units: a central nervous system and a peripheral nervous system.

c. The human body has a closed circulatory system and a large body cavity containing visceral organs. It carries out extracellular digestion within a gut and maintains a fairly constant internal temperature of 37°C.

Lecture Resources
- Opening Questions, page 825
- Demonstration: The Skeleton
- Visual Strategy: Figure 35-2
- Building Bones and Bone Reabsorption, page 826
- Research Update, page 827
- Temperature Variation, page 828
- Teaching Transparencies: 184, 187

- Holt Biology Videodiscs: Lesson 35-1

Classwork Options
- Effective Reading, page 825

Assignment Options
- Section Review, page 823
- Chapter Review, questions 1, 2, 9, 10, 16, 28

35-2 How the Body is Organized pp. 829–835

BLOCK 2

a. Cells in the body are organized into four kinds of tissue: epithelial, connective, nervous, and muscular.

b. Tissues are organized into organs, which, in turn, belong to organ systems.

Lecture Resources
- Opening Questions, page 829
- Demonstration: Specialized Cells
- Other Functions of Epithelium, page 830
- Visual Strategies: Figures 35-6, 35-9
- Connective Tissue Has Many Functions, page 831
- Overcoming Misconceptions, page 831
- Overcoming Misconceptions page 833

- Teaching Transparencies: 189, 191, 193, 194
- Holt Biology Videodiscs: Lesson 35-2

Classwork Options
- Homeostasis and the Organization of the Body, page 834

Assignment Options
- Section Review, page 835
- Chapter Review, questions 3–6, 11–13, 17, 18, 24, 25, 27, 29

35-3 Homeostasis pp. 836–840

BLOCKS 3 and 4

a. In body systems controlled by negative feedback, deviations from the set point are opposed, and therefore changes tend to be minimized.

b. In body systems controlled by positive feedback, deviations from the set point are magnified, resulting in rapid change from the set point.

c. The hypothalamus, a region of the brain, monitors and regulates the body's condition. Information is transmitted from one part of the body to another by the nervous system and by hormones.

Lecture Resources
- Opening Question, page 836
- Demonstration: Positive and Negative Feedback
- Overcoming Misconceptions, page 837
- Countering Cold Temperatures, page 839
- Electrical Signals, page 839
- Teaching Transparencies: 177, 178, 181
- Holt Biology Videodiscs: Lesson 34-3

Classwork Options
- Negative Feedback and Room Temperature, page 837
- Contrasting Negative Feedback With Positive Feedback, page 838
- Laboratory Investigation: Thermoregulation, pages 852–853
- Closure

Assignment Options
- Section Review, page 840
- Chapter Review, questions 7, 8, 14, 15, 19, 26, 30

Review/Enrichment

BLOCK ⑤

■ Highlights: Organ Systems, pages 841–848
■ Highlights Review, questions 20–23
■ Study Guide: Concept Review and Pretest

Assignment Options
⊙ Teaching Resources CD-ROM: Occupational Applications Worksheets, Emergency Medical Technician and Nurse Practitioner
■ Activities and Projects, questions 31, 32
■ Readings, questions 33–35

Assessment Options

BLOCK ⑥

■ Chapter 35 Test
⊙ Test Generator/Assessment Item Listing: Software item bank for preparing customized chapter tests.

Portfolio Assessment
Select student reports for one or more laboratory experiments from this chapter. *The Direct Observation Checklist* on the *BioSources Teaching Resources CD-ROM* should be completed during a laboratory or other cooperative learning experience.

Answer to Concept Map

The following is one possible answer to the Concept Mapping exercise on page 849.

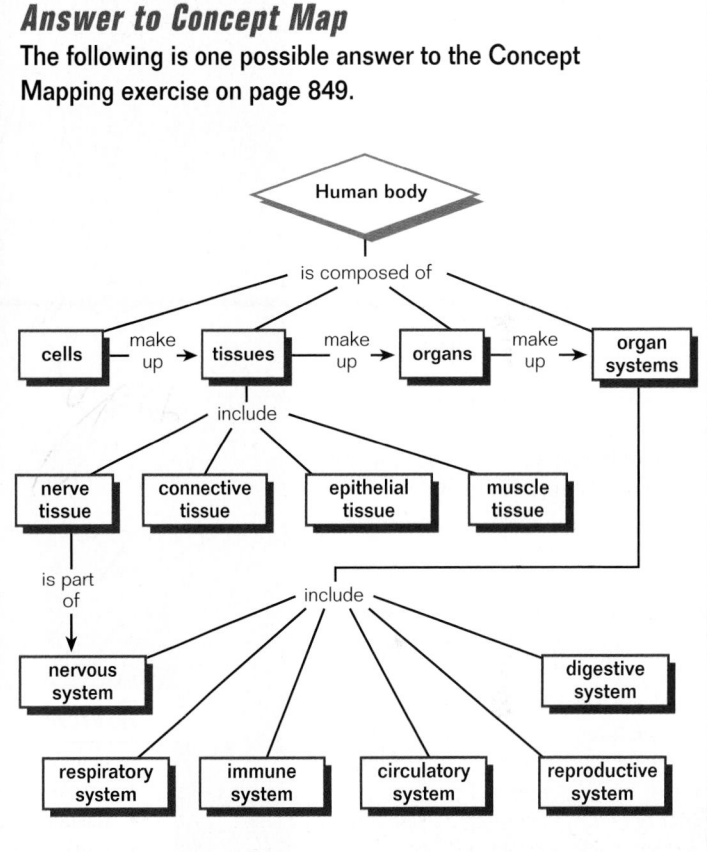

Holt Biology Videodiscs   CONCEPTS OF BIOLOGY

Holt Biology Videodiscs gives you a powerful tool for teaching, review, and assessment. *Concepts of Biology* adds interactive lesson plans and extensive tools for customization using CD-ROM technology.

Use the following topic from *Concepts of Biology* to help you teach this chapter:

■ Topic 21: Body Systems

For further information regarding the media on the videodiscs, see *Holt Biology Videodiscs Teacher's Correlation Guide to Biology: Principles and Explorations.*

UNIT 9

CHAPTER 35

Chapter Theme

Levels of Organization

The human body is composed of some 100 trillion cells. None of these cells can function alone; each depends on interactions with other cells of the body. The organization of the human body enables cells to exchange materials and information and thereby to function in concert to meet the requirements of living. Thus, the human body is a highly integrated whole, dependent on the organization of its component systems to maintain its function.

Tapping Prior Knowledge

- Name the major properties shared by all living things.
- What is homeostasis, and why is it crucial for survival?
- What are the main characteristics of vertebrates?

Opening Demonstration

Illustrate the interdependence of the systems of the human body by showing students flip charts of the major organ systems. Point out that each organ system carries out a major task—breaking down food, sensing the environment, and so on—and that each system depends on the others. For instance, cells of the nervous system cannot transmit information unless they have oxygen and nutrients, which are obtained by the respiratory and digestive systems and transported by the circulatory system. Because of the interdependence of body systems, total failure of one system is usually fatal.

CHAPTER 35

OVERVIEW OF THE HUMAN BODY

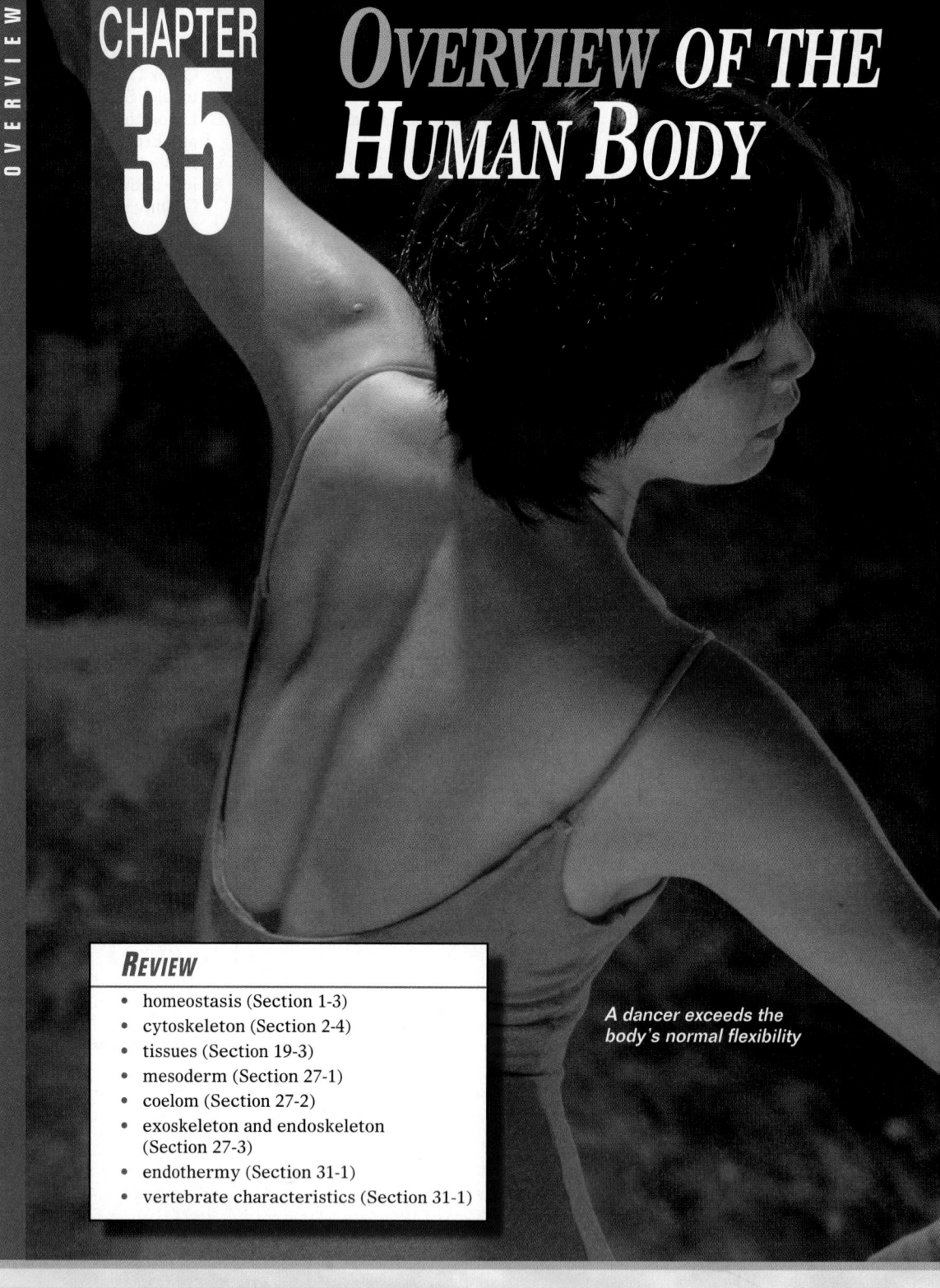

A dancer exceeds the body's normal flexibility

REVIEW

- homeostasis (Section 1-3)
- cytoskeleton (Section 2-4)
- tissues (Section 19-3)
- mesoderm (Section 27-1)
- coelom (Section 27-2)
- exoskeleton and endoskeleton (Section 27-3)
- endothermy (Section 31-1)
- vertebrate characteristics (Section 31-1)

Authors' Rationale

Our investigation of life culminates with an in-depth look at the human body. This chapter serves as an introduction to the human organism, approaching the human body in terms of overall layout—body plan, organization, the systems that maintain life, and regulation, or homeostasis. The need for this kind of information is obvious—human bodies do not come with owner's manuals. Students need to understand that when homeostasis is disrupted, disease results. Furthermore, the best route to health is through prevention, which is based on detailed knowledge about the intricate workings of the body.

35-1 The Human Body Plan

*I*n this unit you will study the human body and how it functions. You will begin your study by exploring the general nature of the human body plan. To understand how the human body functions, you must first look at its overall architecture. Only then can you properly appreciate how the various parts function. Imagine trying to explain an automatic transmission to someone who doesn't know what a car is. In this section you will examine six general functions of the human body plan. Just as in the discussion of the animal body plan in Chapter 27, the use of the term **plan** in no way implies a conscious design. Rather, it refers to the organization of the human body that has been programmed into its genes by thousands of centuries of evolution.

Section Objectives

- Describe the role of joints in the skeleton.
- Contrast the functions of the central and peripheral nervous systems.
- Identify two advantages of maintaining a constant body temperature.

Supporting the Body

Perhaps the most important feature of vertebrates is their endoskeleton, or internal skeleton. Instead of a rigid exterior skeleton like that of arthropods, vertebrates have soft, flexible skin that stretches to accommodate the body's movement. When you bend your arm, the skin covering the joint of your elbow stretches; if it didn't, it would tear. The human endoskeleton, shown in Figure 35-1, is composed of bone, a strong and nonbrittle material. There are 206 individual bones in the skeleton.

Like insects, vertebrates have jointed appendages. In the human body, the jointed appendages are the arms, hands, legs, and feet. The movement of an appendage such as the arm occurs when its bones move relative to the rest of the skeleton. The movement of the skeleton occurs at **joints,** where one bone meets another. The freely movable joints of the skeleton are cushioned by pads of cartilage and held together by bands of elastic tissue called **ligaments.** The pads of cartilage at the tips of bones do not touch

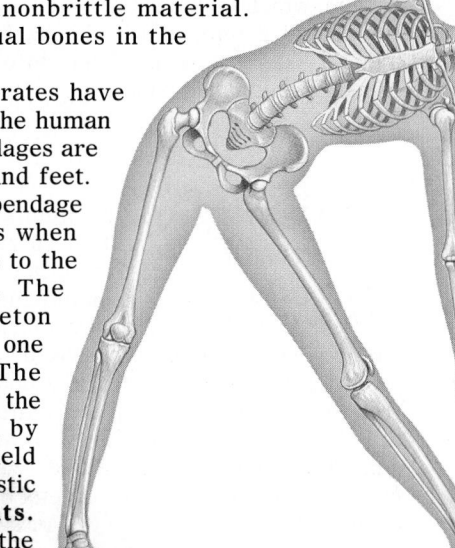

Figure 35-1 Your skeleton supports and protects your body.

SECTION 35-1

Vocabulary Preview

joint
ligament
central nervous system
peripheral nervous system

Opening Questions

Ask students what is meant by the term *body plan.* Lead them to conclude that an organism's body plan is its architecture, or the way it is organized. Ask students to list several ways that the human body plan is similar to that of other vertebrates. For example, humans are bilaterally symmetric, have a definite head, and have a vertebral column.

Demonstration

The Skeleton

Show students an articulated skeleton, full-sized if possible. Point out the range of movements possible at some of the major joints, such as the knee, elbow, shoulder, and hip. Also point out some of the immovable joints, such as those in the skull. Ask students to list some of the advantages of an internal skeleton over an external skeleton such as those of insects and crustaceans. *(An internal skeleton can grow with the body. It is also lighter, since it does not cover all of the body.)* Ask students to list some disadvantages of an internal skeleton. *(It does not protect the entire body.)*

Effective Reading

Expanding Knowledge

Have students prepare a list of questions that they would like to see answered in this chapter. Encourage them to keep a log of their questions, the answers, and the page numbers where they find each answer.

VISUAL STRATEGY

Figure 35-2
Point out that the sac surrounding the joint holds the lubricating fluid in place. Prolonged kneeling can keep the fluid from circulating through the joint, resulting in the "squeaky" knees often experienced when one rises from a kneeling position.

Teaching Tips
Bone Building and Bone Reabsorption
Tell students that bone is the body's storehouse for calcium. If calcium is needed elsewhere in the body, it may be released from the bones (a process called bone reabsorption). Conversely, if calcium is abundant in the body, it may be added to the bones (a process called bone building). The amount of calcium in the body is monitored by the hypothalamus and is controlled by hormones released by the parathyroid glands and the thyroid gland.

What If bone is reabsorbed faster than it is formed? Imbalance between bone formation and reabsorption can result in several diseases. The most common bone disease that can result from this imbalance is osteoporosis. Osteoporotic bones become more porous and lighter due to a more rapid rate of bone reabsorption than deposition. This loss of bone mass leads to fractures and deformities, such as compression of the vertebrae and fractures of the hip.

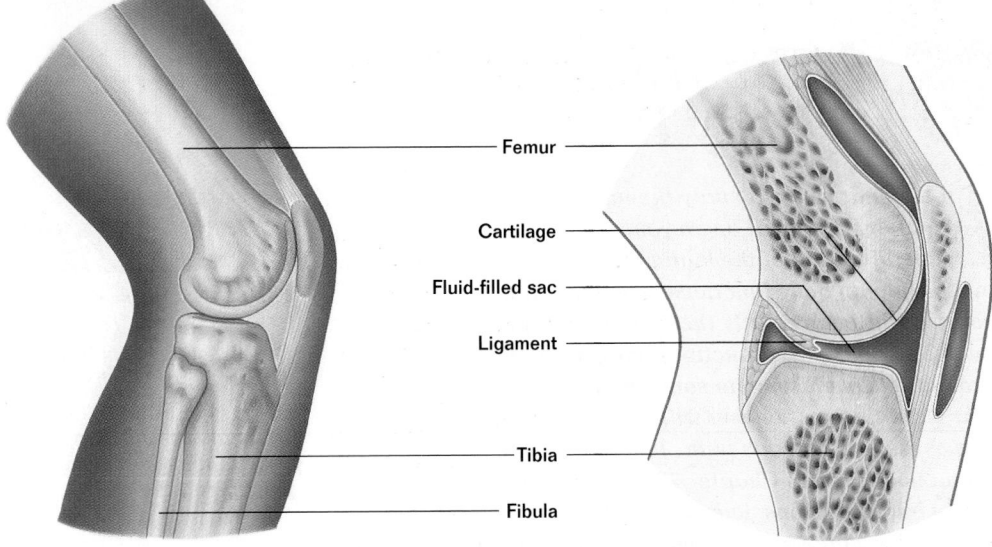

Figure 35-2 Movement in the skeleton is possible at joints, where bones meet. This diagram shows the structure of the knee.

CONTENT LINK
Rheumatoid arthritis is a very painful disease in which, among other things, the cartilage of the joints is destroyed. To find out more about this disease, turn to **Chapter 36.**

at the joints. As shown in Figure 35-2, membranes filled with fluid occur where the bones meet, and this fluid lubricates the movement of these pads across one another.

Collecting and Evaluating Information

Humans, like all vertebrates, have a brain that coordinates and directs the activities of the body. The human nervous system is composed of two contrasting functional groups. The first is the central processing region, the **central nervous system,** which includes the brain and spinal cord. The second functional group, called the **peripheral nervous system,** includes the nerves that bring information to the brain and transmit commands from it.

Transporting Oxygen and Nutrients

Most animals have a circulatory system that carries oxygen and nutrients to the cells of the body and removes carbon dioxide and water. Humans and other vertebrates have a closed circulatory system. Blood is enclosed within blood vessels that separate it from the rest of the body's fluids and prevent it from mixing freely with them. Materials pass into and out of the blood by diffusion. The great advantage of a closed circulatory system is that the body can maintain different circulation rates in different organs by changing the diameter of blood vessels. A second advantage of a closed circulatory system is that it allows the blood to be pumped under

 Did You Know?

Physicians classify many types of disorders under the umbrella of "arthritis." Two of the most common forms of arthritis are osteoarthritis and rheumatoid arthritis. Osteoarthritis, often known as "wear-and-tear" arthritis, occurs when the cartilage in joints degenerates, leaving bone rubbing against bone. Rheumatoid arthritis, on the other hand, is an autoimmune disorder in which the immune system attacks the body's own tissues, including those at the joints. Of these two types of arthritis, osteoarthritis is the more common.

pressure. The heart pushes the blood out through the body's arteries, sending the blood through the body much faster than would be possible in an open system.

Protecting the Heart, Lungs, and Digestive Organs

Like all vertebrates, humans are coelomates. Every human body contains a large body cavity, the coelom, that develops entirely within mesoderm tissue in the embryo. As you can see in Figure 35-3, within this cavity are found the large organs of the body, which are suspended in fluid that supports their weight and prevents them from being deformed by body movements. If your heart were compressed every time you turned at the waist or bent over, it could not function correctly. Why doesn't it get deformed by body movements? Imagine a balloon full of water that is floating within another, larger balloon, also full of water. Pushing your thumb into the outer balloon doesn't deform the inner one. The human coelom protects the body's organs in much the same way.

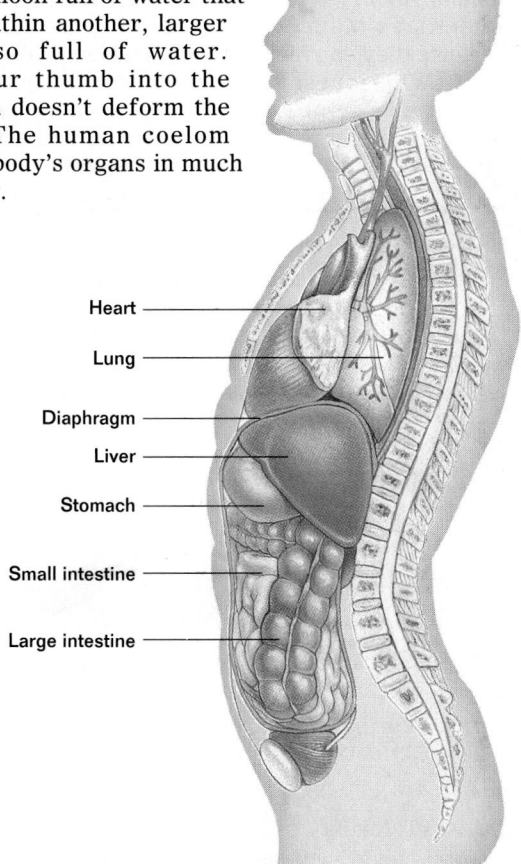

Heart
Lung
Diaphragm
Liver
Stomach
Small intestine
Large intestine

Figure 35-3 The heart, lungs, and digestive organs lie within the coelom. This fluid-filled cavity protects and supports the organs. The diaphragm separates the body cavity into the thoracic cavity, which houses the lungs and heart, and the abdominal cavity, which houses the digestive organs.

Historical Note

Observing Digestion

William Beaumont, a U.S. Army surgeon, discovered many of the processes of digestion in humans through the study of a young French Canadian, Alexis St. Martin, who had accidentally shot himself in the abdomen with a shotgun. St. Martin's wound didn't completely heal, leaving an opening through which Beaumont could observe the interior of the stomach and experiment with gastric secretions. Beaumont published his studies in *Experiments and Observations on the Gastric Juice and the Physiology of Digestion* in 1838.

☑ **RESEARCH Update**

How Do We Get an Overview of the Human Body?

Medical science frequently uses noninvasive diagnostic techniques such as positron emission tomography (PET), magnetic resonance imaging (MRI), computerized axial tomography (CAT), and X-ray analysis. Researchers are working on an exciting new noninvasive diagnostic technique called *optical tomography*. Robert Alfano and his collaborators from City University of New York were the first to extract information using this technique, which relies on the differential scattering, absorption, and transmission of photons by living tissue.

While the technology is not yet as good as existing techniques, David Benaron of Stamford University sees optical tomography as a potential *bedside* diagnostic tool. Researchers feel that once the technique is refined and amplified, it may offer patients a test that is inexpensive, no-risk, high-resolution, and fast. The promise this new technique offers has not been missed by the National Institutes of Health (NIH), which has funded a number of unique research projects in the field. One project is to use the technique to detect breast cancer, particularly in younger women. Present-day mammography techniques are often inadequate to detect this type of cancer in young women. Benaron and his colleagues at Stamford are working on an NIH-supported device that will measure oxygen levels in the brain tissues of infants. How? The babies will wear a comfortable headband loaded with the latest in fiber optic technology, which will send streams of light across the skull and interpret absorptive differences to calculate areas where oxygen levels are low in the brain. The day may come when "going to the hospital for some tests" will not instill fear in even the most faint-hearted of individuals. ☑

CONNECTIONS

Chapter 4
Enzymes

Although a fever can be helpful by inactivating the enzymes of an invading pathogen, it can also rise to a point at which it begins to inactivate the enzymes of the body. Moreover, the body's homeostatic mechanisms may not be able to reestablish control over body temperature once it rises above a certain level. Uncontrolled high fever may result in serious tissue damage or death.

Teaching Tip
Temperature Variation

Tell students that body temperature is not always kept precisely at 37°C. For example, at night the body lowers its temperature by 1°C or more, and activity or intense emotion may cause a rise in body temperature of about 1°C. Be sure students recognize that these variations are normal.

CONTENT LINK

To find out what role each organ of the digestive system plays in extracting nutrients from the food you eat, turn to **Chapter 38**.

Obtaining Nutrients From Food

The human body, like that of nearly all animals, is organized to carry out extracellular digestion within the gut. As in most animals, the flow of materials through the gut proceeds in one direction. Food enters the body through the mouth and passes through the digestive tract and out the anus. In humans, the digestive tract is a long tube that extends from mouth to anus and is suspended within the coelom.

Maintaining a Relatively Constant Temperature

Like all mammals, human beings are endothermic, maintaining a fairly constant internal temperature of 37°C (98°F). A large percentage of the calories you consume are devoted to maintaining your body temperature, and you cannot live if it falls far below normal for long. Very high temperatures, such as fevers sometimes induced by infections, are also dangerous because they can inactivate critical enzymes.

Why invest so much energy keeping your body warm? One great advantage of endothermy is that it permits the body to maintain its activity at all times and in many different places, regardless of the temperature of the surroundings. Mountain butterflies, which use the sun's rays to heat their bodies, cannot fly if the sun doesn't shine, just as you could not walk on a cold day if you were not endothermic. In addition, endothermy permits you to sustain strenuous activity, while ectotherms such as lizards and frogs are limited to short periods of exertion.

Figure 35-4 Running a marathon requires more than two hours of exertion. Strenuous exercise of this duration, though beyond the capabilities of most people, depends on the large amount of energy supplied by an endothermic metabolism. No ectotherm could run for even a tiny fraction of this time.

Section Review

1. *Explain how the joints of the skeleton both permit and limit movement.*
2. *Describe the relationship between the central and peripheral nervous systems.*
3. *Why are very high body temperatures often fatal?*

Critical Thinking
4. *Explain why your body temperature usually rises slightly after you eat.*

Answers to Section Review

1. Because bone is rigid, the skeleton can move only at the places where bones meet—the joints. However, joints only permit bones to move in certain ways, thus limiting movement.

2. The central nervous system evaluates information and initiates and controls responses. The peripheral nervous system gathers information, relays it to the central nervous system, and transmits the "commands" of the central nervous system to the muscles and other effectors.

3. High temperatures can change the shape of enzymes, thus inactivating them.

4. Digestion breaks down large molecules into smaller ones, releasing energy. Also, after a meal the cells of the body have more energy and can metabolize at a higher rate.

35-2 How the Body Is Organized

*T*he human body is composed of over a hundred differ- ent kinds of cells. As illustrated in Figure 35-5, cells within the body are organized into larger functional units. Recall from Chapter 19 that a tissue is a group of simi- lar cells that work together to perform a common function. The diverse cell types of the human body are traditionally grouped by function into four basic types of tissues: epithelial (ehp uh THEE lee uhl), connective, nerve, and muscle tis- sues. These four kinds of tissues are the building blocks of the human body. Each **organ** of the human body is composed of a mixture of these tissues organized in various ways. An **organ system** is a group of organs that function together to carry out a major activity of the body.

Section Objectives

- Contrast the characteristics and functions of the four basic types of tissue.
- Contrast smooth, skeletal, and cardiac muscle.
- Describe the function of each organ system.

Vocabulary Preview

organ

organ system

epithelium

macrophage

lymphocyte

erythrocyte

neuron

supporting cell

glial cell

dendrite

axon

actin

myosin

Opening Question

Ask students to explain how "divi- sion of labor" applies to the organi- zation of the human body. Lead students to conclude that tasks are allocated among different systems of the body, so each system carries out only certain tasks.

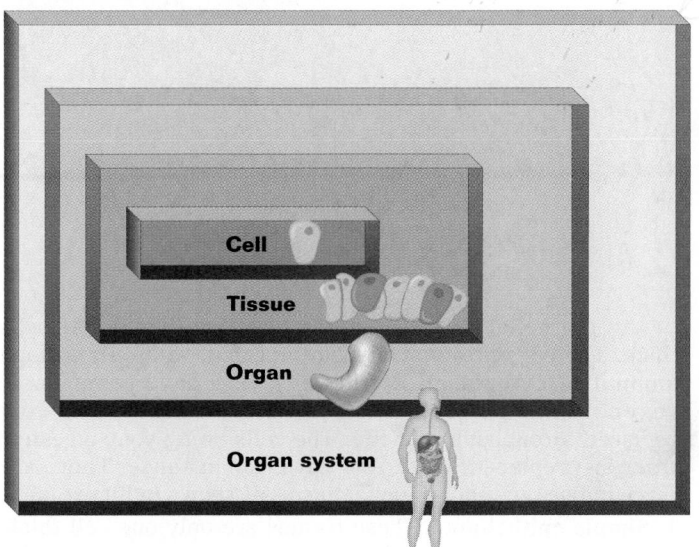

Figure 35-5 Within your body, cells belong to functional units called tissues. Tissues are organized into organs, and organs belong to groups of organs called organ systems.

Demonstration

Specialized Cells

Point out that specialization is a key feature of the cells of the human body by showing students a photograph of a highly special- ized cell—a red blood cell (erythro- cyte). Point out some of the specialized features of the cell. For instance, mature erythrocytes have no nucleus. Therefore, they cannot produce protein during their functional lifetime, which is only about 120 days. During its devel- opment, however, each erythro- cyte is stuffed with 300 million molecules of oxygen-carrying hemoglobin. The flattened shape of the erythrocyte has a large sur- face area through which oxygen can be absorbed. Ask students to think of other highly specialized cells of the body, such as neurons.

Epithelial Tissue Is Protective Tissue

Epithelial cells are the guards of the body. Epithelial tissue, or **epithelium,** covers the body's surfaces, protect- ing the tissues beneath from dehydration and physical dam- age. Because epithelium covers the body's outer surface and lines the gut and lungs, every substance that enters or leaves the body must cross an epithelial layer. Therefore, epithelial cells control which substances enter and leave the cells of the body. Many secretory glands are derived from epithelium, as are many of the body's sensory organs.

Multicultural Perspective

The Asian Theory of *Chi*

Asian people have a complex theory of *chi*, an energy field that flows along the body's meridians, which are somewhat parallel to the circulatory and nervous systems. This system, which was first illustrated with elaborate charts thousands of years ago, is still used exten- sively in Asia. The concept of *chi* is becoming more popular in the West thanks to the use of acupuncture and acupressure in treating a variety of illnesses. Western scientists have demonstrated that endorphins, the body's own painkillers, are released during acupuncture and acupressure.

Teaching Tip
Other Functions of Epithelium

Inform students that epithelial tissue has functions other than protection. For example, the function of the cuboidal epithelial tissue that lines the tubules of the kidneys is secretion. Any tissue having the characteristics of epithelium is classified as epithelial tissue, no matter what function it serves.

 VISUAL STRATEGY

Figure 35-6

Ask students why it is advantageous for the epithelial tissues of the lungs to be only one cell layer thick. *(Oxygen must cross only a thin layer to reach the blood.)* Why isn't the skin composed of epithelium that is a single layer thick? *(The skin must be thick so that it is able to resist abrasion and wear. Its main function is to keep pathogens out and the body's fluids in.)*

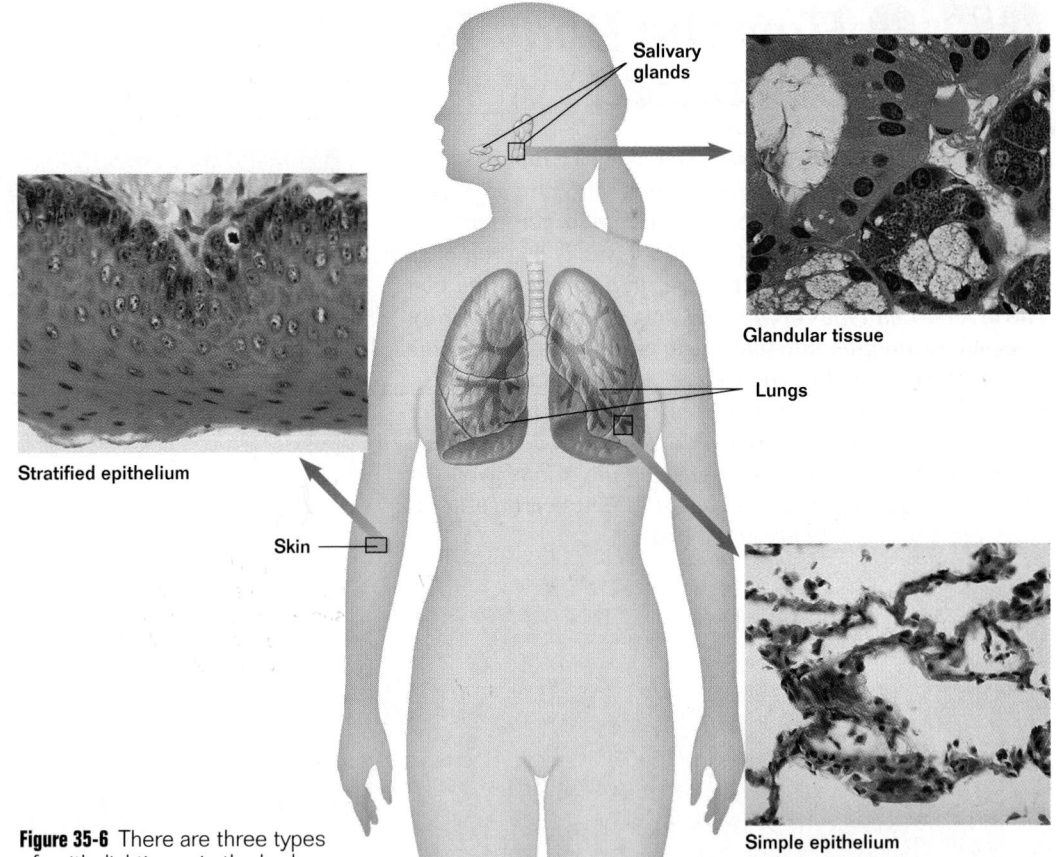

Figure 35-6 There are three types of epithelial tissue in the body.

An epithelial layer is usually no more than a few cells thick. The cells are typically flat and thin, with only a small amount of cytoplasm. Epithelial layers have remarkable regenerative powers—their cells are constantly being replaced throughout your life. The cells lining your digestive tract are replaced every few days, for instance. There are three major types of epithelial tissue, as shown in Figure 35-6.

1. **Simple epithelium**   These tissues are only one cell thick. They line the respiratory tract, lungs, and major cavities of the body.

2. **Stratified epithelium**   These tissues are several cell layers thick. Stratified epithelium is most abundant in the skin, where it forms the outer layer. A characteristic property of stratified epithelial cells is their ability to produce keratin, a very strong, fibrous protein. The calluses that occur on your hands are composed of keratin, and so is hair.

3. **Glandular tissue**   The human body contains many glands, all composed of stratified epithelia. These glands produce sweat, milk, saliva, digestive enzymes, hormones, and many other substances.

 Did You Know?

Many of the most familiar glands are epithelial. Sweat glands and sebaceous glands, which release oils to protect the skin, are good examples. Mammary glands are also epithelial in origin.

Connective Tissue Supports and Defends the Body

The cells of connective tissue provide the body with its structural building blocks and its most potent defenses. Connective tissue is the most diverse of the four principal types of tissue. Some connective tissue cells, such as those in bone, are densely packed together. Others, such as blood cells, are spaced well apart from each other. Connective tissue cells fall into three functional categories: cells of the immune system, cells of the skeleton, and cells that accumulate and store molecules.

Cells of the Immune System

The immune system is composed of cells that defend the body from infection and cancer. Immune system cells are typically small, and they roam the bloodstream searching for invaders. Some immune system cells, called **macrophages** *(MAK roh FAY juhz)*, are mobile cells that engulf and digest bacteria and other microbes. Other cells, called **lymphocytes** *(LIM foh seyets)*, make antibodies or attack virus-infected cells and cancer cells. Both types of cells are shown in Figure 35-7.

Cells of the Skeletal System

The human skeleton is composed of three kinds of tissue, shown in Figure 35-8, that are distinguished by the nature of the material that is laid down between the cells.

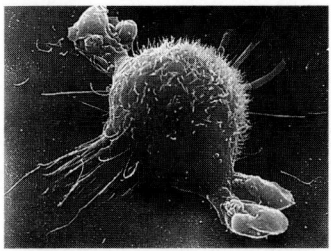
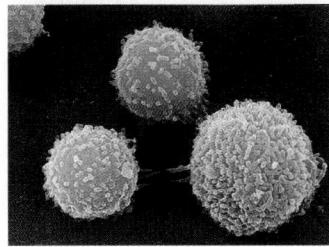

Figure 35-7 Macrophages, *top*, are immune system cells that consume invaders. B cells, *bottom*, are lymphocytes that produce defensive proteins.

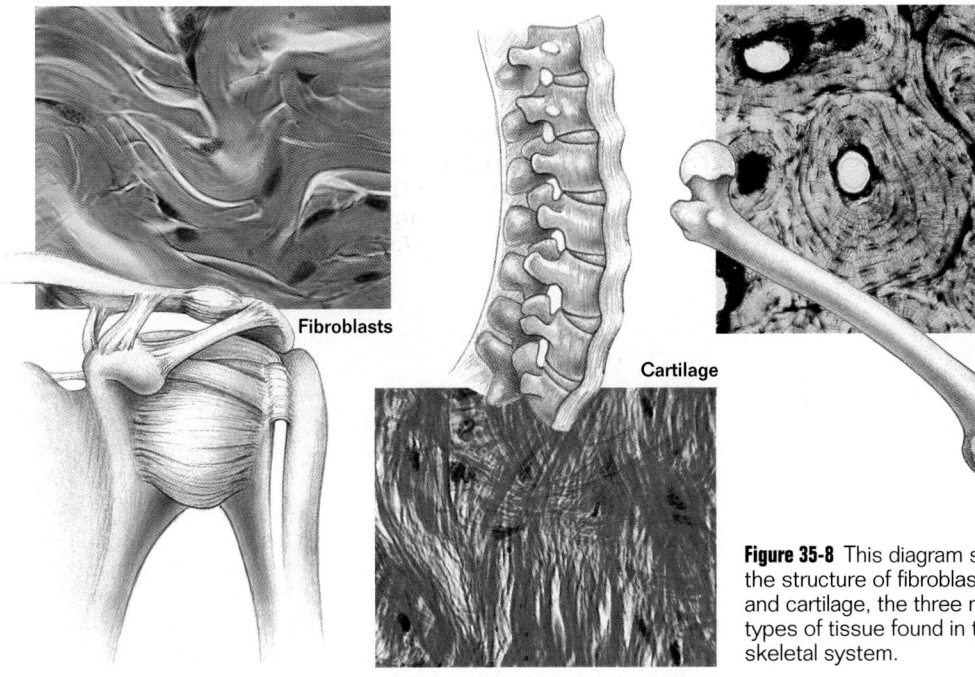

Fibroblasts

Cartilage

Bone

Figure 35-8 This diagram shows the structure of fibroblasts, bone, and cartilage, the three major types of tissue found in the skeletal system.

1. **Fibroblasts** (*FEYE broh blasts*) These cells are the most common connective tissue cells. They are flat, irregularly shaped cells that secrete the strong structural protein **collagen** into the spaces between the cells. Collagen makes up one-fourth of the protein in the body. Fibroblasts also produce scar tissue.
2. **Cartilage** This material is firm, flexible, and very strong. It is formed at positions of mechanical stress by fibroblasts, which lay down long parallel strands of collagen fibers along the lines of stress. Cartilage also covers the ends of bones at joints.
3. **Bone** In bone, the collagen fibers between cells are coated with a calcium phosphate salt, making bone more rigid than collagen, but not brittle. The human endoskeleton is composed mainly of bone.

Cells That Accumulate and Transport Molecules

Many of the body's connective tissues are specialized for storing and transporting substances; examples include pigment-containing cells and the fat-storing cells of adipose (fat) tissue. The most important storage and transportation cells are red blood cells, called **erythrocytes** (*eh REE throh seyets*), which are shown in Figure 35-9. Erythrocytes circulate in the blood, carrying oxygen from the lungs to the body's cells and transporting carbon dioxide from the cells back to the lungs. There are about 5 billion erythrocytes in every milliliter (0.03 oz.) of human blood. Erythrocytes lose their cell nucleus during development and are packed with the oxygen-binding protein hemoglobin. Each erythrocyte contains about 300 million molecules of hemoglobin.

> CONTENT LINK
>
> An erythrocyte lives for only about 120 days. In **Chapter 37** you will learn how and where new erythrocytes are produced.

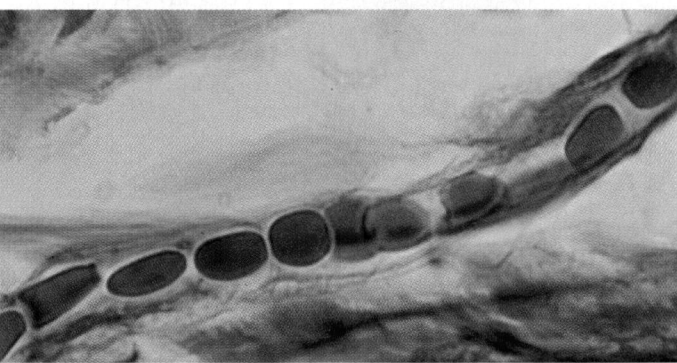

Figure 35-9 These red blood cells are moving single file through a capillary, a very small blood vessel. Red blood cells release their load of oxygen and nutrients while in capillaries.

Nerve Tissue Conducts Signals Rapidly

Nerve tissue, the third major class of human tissue, is composed of two kinds of cells: **neurons,** which are specialized for the transmission of nerve impulses, and **supporting cells,** also called **glial** (*GLEE uhl*) **cells,** which insulate the neurons and provide them with nutrients.

Neurons have a central cell body, which contains the nucleus, and two kinds of cytoplasmic extensions, as you can see in Figure 35-10. The first kind of extension consists of threadlike protrusions called **dendrites,** which act as antennae for the reception of nerve impulses from other cells or sensory systems. A single neuron may have many dendrites. The second kind of cytoplasmic extension that projects from the cell body of a neuron is the **axon.** An axon is a long, tubular extension of the neuron that carries the nerve impulse away from the cell body, often for considerable distances. A single axon connects your heel with the base of your spine, a distance of more than 1 m (39 in.). Another neuron runs from your spine all the way across your shoulder to your thumb. Single neurons more than 1 m long are common. Nerves in the human body appear as fine white threads, but they actually are composed of hundred of axons bunched together like the wires of a telephone cable.

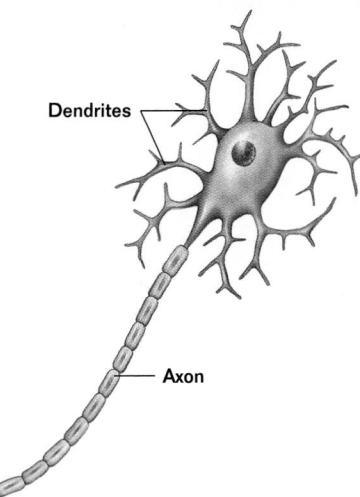

Dendrites

Axon

Figure 35-10 The small, hairlike projections on this neuron are dendrites. The larger projection is an axon.

Muscle Tissue Permits Movement

Muscle cells, formed from mesoderm early in development, are the workhorses of the human body. The key property of muscle cells is the relative abundance of the protein filaments **actin** and **myosin,** which enable the muscle cell to contract. These protein filaments are present as part of the cytoskeleton in all eukaryotic cells, but they are far more abundant in muscle cells. The three kinds of muscle cells in the human body are illustrated in Figure 35-11.

1. **Smooth muscle**   Smooth muscle cells are long, spindle shaped, and packed with myosin and actin filaments. The individual filaments are not aligned with each other. Smooth muscle cells are typically organized into sheets of cells. Some smooth muscles, such as those lining the walls of blood vessels, contract only when stimulated by nerves or hormones. Other smooth muscles, such as those lining the wall of the gut, contract spontaneously, leading to a slow, steady contraction of the tissue.

2. **Skeletal muscle**   Skeletal muscles (also called striated muscles) move the endoskeleton. Each muscle is a tissue made of many individual muscle cells acting together. The cells fuse end to end during development to form a long muscle fiber with a central cable of filaments and many nuclei pushed out to the edges. Because the filaments are aligned, skeletal muscles are much stronger than smooth muscles.

3. **Cardiac muscle**   The muscles of the human heart, called cardiac muscles, are organized very differently from skeletal muscles. Instead of long cells with many nuclei, cardiac muscle is composed of chains of single cells, each cell having its own nucleus. These chains are organized

UNIT 9
CHAPTER 35

Teaching Tips

Contraction of Smooth Muscles

Point out to students that the contraction of the smooth muscles lining the gut is not under conscious control. These muscles produce peristalsis, the wavelike contraction that forces food through the gut. Ask students whether they can feel peristalsis occurring when they place their hand on their stomach. They cannot, because the gut is insulated from the surface of the body by the fluid-filled coelom and the body wall.

Homeostasis and the Organization of the Body

Emphasize to students that maintaining homeostasis in a multicellular organism is much more difficult than doing so within a single cell. Each cell involved must contribute its share of the work. To ensure that the balance among body cells is not upset, each cell must coordinate its activities not only with those of the cells belonging to the same tissue, organ, and organ system, but also with the activities of cells of other organ systems. Disease is often the result of disruption of homeostasis in the body. Have students select one disease that is caused by a breakdown in homeostasis and write a report about it. Possible diseases include diabetes (Type I or Type II), Graves' disease, and porphyria. In their report, students should address the cause of the disease, its effects on the body, and possible treatments.

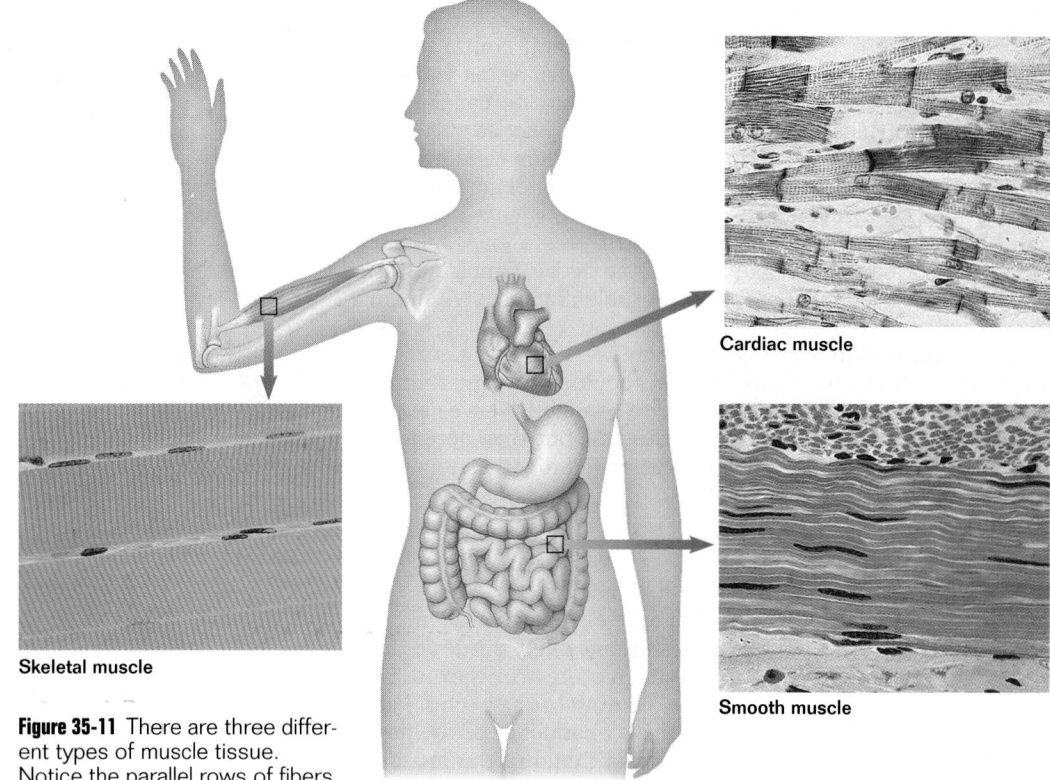

Cardiac muscle

Skeletal muscle

Smooth muscle

Figure 35-11 There are three different types of muscle tissue. Notice the parallel rows of fibers in skeletal muscle.

into fibers that branch and interconnect, forming a latticework. When two fibers touch in the lattice, they make an electrical junction. When the heart starts to contract, the electrical wave that initiates contraction rapidly spreads over the whole lattice. Thus, a group of neighboring cells tends to contract all at once, rather than individually.

❑ **CAPSULE SUMMARY**

Cells in the body are organized into tissues. There are four kinds of tissue: epithelial, connective, nervous, and muscular. Tissues are organized into organs, which, in turn, belong to organ systems.

Organs and Organ Systems
....

All of the organs in your body are composed of combinations of the four types of tissue just described. The heart, for example, is composed of cardiac muscle tissue and connective tissue, and is connected with many nerves. All of these tissues work together to pump blood through your body. Each organ, in turn, belongs to an organ system that carries out a critical function. For example, the digestive system is composed of individual organs that break up and digest food, absorb nutrients, and compact and expel the unabsorbed residue. The human body contains 11 principal organ systems that are briefly described in Table 35-1. *Highlights: Organ Systems* on pages 841-848 further examines the structure and function of these organ systems. ❑

Table 35-1 Major Organ Systems of the Human Body

System	Major Components	Functions
Systems That Cover, Support, or Produce Movement		
Integumentary	Skin and associated structures	Protects against injury, infection, and fluid loss
Muscular	Skeletal and smooth muscles	Movement
Skeletal	Bones of the skeleton	Protects and supports the body
Systems That Regulate Body Functions		
Endocrine	Various glands, including pituitary gland, thyroid, adrenal glands, and pancreas	Regulates and coordinates the body's functions
Nervous	Brain, spinal cord, nerves, sense organs	Collects and evaluates information; issues commands
Reproductive	Testes, penis, and associated ducts (males); ovaries, uterus (females)	Produces gametes; delivers gametes (males); nourishes and shelters fertilized egg (females)
Systems That Transport or Protect		
Circulatory	Heart, blood vessels, lymphatic vessels	Transports oxygen, carbon dioxide, nutrients, and cells
Immune	White blood cells	Defends against pathogens and cancer
Systems Involved in Metabolism or Excretion		
Respiratory	Lungs, air passages	Obtains oxygen; releases carbon dioxide
Digestive	Mouth, esophagus, stomach, small intestine, large intestine, liver, pancreas	Extracts and absorbs nutrients from food
Urinary	Kidneys, urinary bladder, ureters, urethra	Removes wastes from blood; regulates concentration of the body's fluids

Section Review

1. Why is it advantageous to be able to rapidly replace epithelial cells?

2. Describe two differences between cardiac and skeletal muscle.

3. Complex machines often have redundant systems so that if one system fails, another can take over its function. Are there any redundant organ systems in the human body? Explain your answer.

Critical Thinking

4. Some people living at high elevation have more erythrocytes in their blood than do people living at lower elevations. Propose a hypothesis to account for this observation.

SECTION 35-2

Teaching Tip

Interdependence of Major Organ Systems

As students read Table 35-1, encourage them to think about the ways the systems in the body interact and depend on each other. For instance, the cells of the immune system move through the body in the circulatory system. Similarly, the respiratory system captures oxygen that is used by all the cells of the body.

Answers to Section Review

1. Epithelial cells are constantly lost. Rapid replacement maintains the integrity of the epithelial tissues.

2. Cardiac muscle is composed of cells that do not fuse and that are organized into a branching network. Skeletal muscle is composed of fibers formed by the fusion of many cells end to end. There are many nuclei per fiber.

3. No. Each system is so specialized that it cannot take over the duties of a failed system. For instance, if the heart stops pumping blood, no other organ can supply the cells of the body with oxygen and nutrients, and the result is death.

4. Erythrocytes carry oxygen to the tissues. A liter of air at high altitude contains less oxygen than a liter of air at sea level. Therefore, at high altitude the body needs more red blood cells to carry the same amount of oxygen as at low altitude.

Vocabulary Preview

feedback loop

negative feedback

set point

positive feedback

hypothalamus

electrical signal

chemical signal

hormone

Opening Question

Ask students to list as many examples of homeostasis from everyday life as possible. Examples include climate control in buildings, temperature in a refrigerator, and water levels in a water heater. Students should recognize that feedback control mechanisms operate in everyday life as well as in living things, and that humans and living systems rely on some of the same control strategies.

Demonstration

Positive and Negative Feedback

To help students better understand positive and negative feedback, draw a graph that depicts the growth of a population of bacteria that begins with one individual and doubles in size every hour. Students should see not only that the population increases rapidly, but that the amount by which the population grows increases as the population grows. As a result, the population moves farther from its starting point with each new generation. Rapid deviation from starting conditions is characteristic of positive feedback. Ask students how negative feedback differs from positive feedback. (*Negative feedback prevents a variable from deviating from its set point.*)

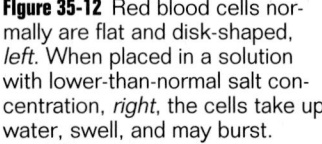

Section Objectives

- Recognize why it is essential for the body to maintain homeostasis.
- Contrast positive feedback with negative feedback.
- Contrast chemical signals with electrical signals.

Every cell in your body is continuously bathed in fluid that supplies it with nutrients and carries away its wastes. No human cell can live for long without this supply line; a brain cell will die in minutes if its supply of oxygen and nutrients is cut off. There is nothing unusual about this. All living cells, whether single-celled paramecia or your own cells, face this same stringent requirement. In your body, however, there are over 100 trillion cells. Each cell must get nutrients from and release wastes into the same 15 L (4 gal.) of body fluid.

Why Homeostasis Is Critical

The body fluid not inside your cells is called extracellular fluid. It occupies the spaces between cells and forms the plasma, the fluid portion of the blood. Because so much exchange occurs across the membranes of cells, any extreme change in the composition or volume of the extracellular fluid can have very serious effects on the activities inside cells, as shown in Figure 35-12. The pH level, the concentration of key ions such as potassium, sodium, and calcium, and the level of sugar in the extracellular fluid are all critical to the body's functions. If any of these factors vary outside the range that individual cells require, death may result. A stable fluid environment is one of the key achievements of multicellular organisms, and it is maintained by a complex set of physiological mechanisms. As you learned in Chapter 1, maintaining a relatively constant internal environment is called homeostasis.

Figure 35-12 Red blood cells normally are flat and disk-shaped, *left.* When placed in a solution with lower-than-normal salt concentration, *right,* the cells take up water, swell, and may burst.

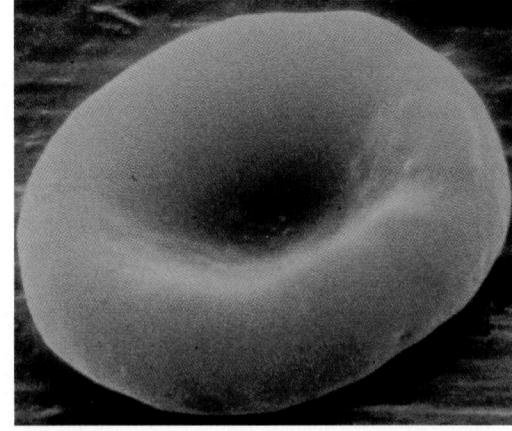

The Feedback of Information

To maintain homeostasis, your body must keep track of what is going on in all of its organs and tissues. In particular, the concentrations of ions and other chemicals in the extracellular fluid must be monitored. To do this, the central nervous system gathers information about the body's condition, evaluates it, and issues commands to counteract any deviation from normal conditions. Such a process of surveillance and response is called a **feedback loop.** Homeostasis is maintained by feedback loops.

Many of the most important homeostatic feedback loops involve **negative feedback.** Negative feedback prevents a variable such as pH or temperature from deviating from its normal value, called its **set point.** A good everyday example of negative feedback is driving a car. The variable the driver wants to keep at a constant value is the position of the car in its lane; the set point is the center of the lane. The eyes of the driver serve as sensors that feed information to the driver's brain about the car's position and constantly compare the car's actual position with the center of the lane. Deviations from the set point, caused by bumps or curves in the road, are recognized by the brain, which issues signals to the driver's muscles to use the car's steering system to correct the deviations. Figure 35-13 illustrates a generalized negative feedback loop.

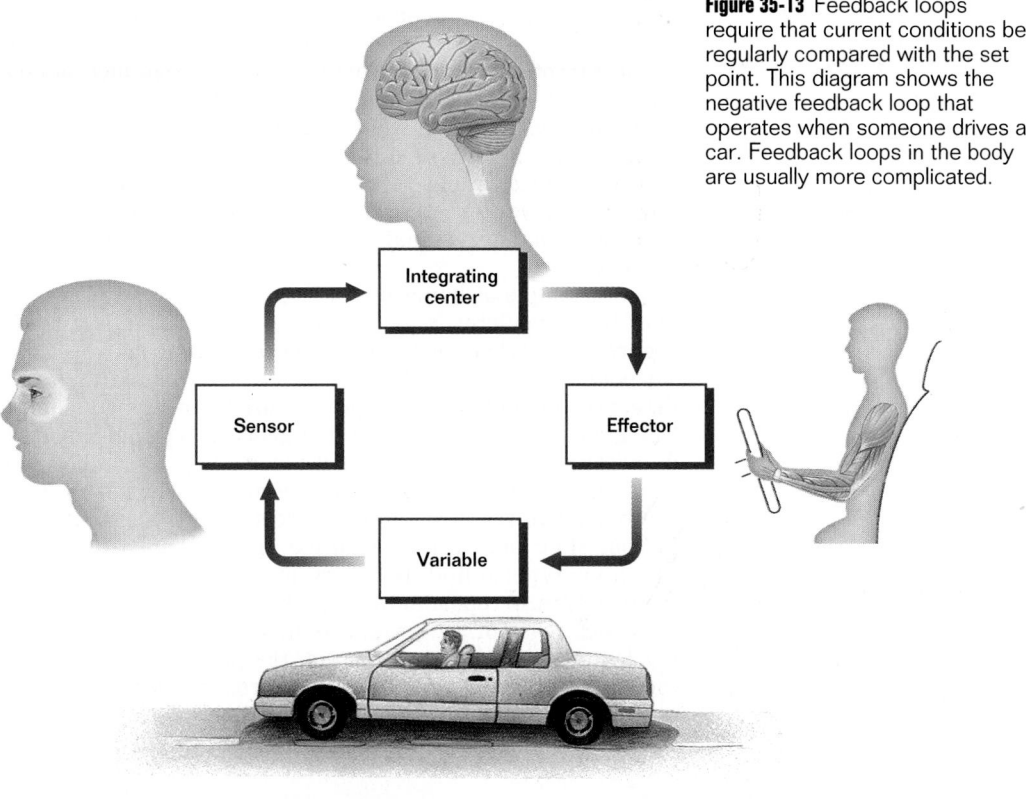

Figure 35-13 Feedback loops require that current conditions be regularly compared with the set point. This diagram shows the negative feedback loop that operates when someone drives a car. Feedback loops in the body are usually more complicated.

Teaching Tip
Contrasting Negative Feedback With Positive Feedback

The level of glucose in the blood is under precise control. Have students predict what would happen under the following conditions. The level of glucose in the blood of an individual is above the set point. What would happen to the glucose level if it were controlled by a negative feedback loop? *(It would return to the set point.)* by a positive feedback loop? *(It would increase away from the set point.)* Have students create a *Graphic Organizer* in the form of two graphs that show the change in blood glucose level over time in each of these situations. A completed *Graphic Organizer* is shown at the bottom of the page.

CONNECTIONS

Chapter 33
Control of Body Temperature

The hypothalamus regulates body temperature so precisely that it normally varies only about 4°C, even between the hottest and coldest environments. Have students contrast this small variation with the large daily variation shown by the lizard depicted in Figure 33-5 on page 775. The lizard depends on external heat, so its body temperature follows that of the environment.

Application

Health Sometimes a substance in the environment can trigger a positive feedback loop in the body. The body overreacts to the substance, often a foreign protein, and causes damage to itself, which triggers a greater reaction. Allergies are an example of such an overreaction. Corticosteroids, such as cortisone, are hormones that reduce inflammation, and thus break the feedback loop.

☐ CAPSULE SUMMARY

In body systems controlled by negative feedback, deviations from the set point are opposed, and therefore changes tend to be minimized. In systems controlled by positive feedback, deviations from the set point are magnified, resulting in rapid change from the set point.

CONTENT LINK

*In **Chapter 41** you will learn more about the hypothalamus's regulatory functions.*

In the negative feedback controls of the human body, the values of variables such as body temperature, blood pressure, and pH are continuously compared with their set point values. Any changes that increase the difference between the variable and its set point initiate responses that tend to oppose the changes and restore the variable to its set point. That is why it is called negative feedback—it *reduces* the difference between variable and set point. However, the human body may change its physiological set points from time to time. For example, body temperature decreases during sleep.

Most physiological feedback is negative, but there are exceptions. **Positive feedback** refers to the condition in which a change in a variable causes the body to drive the variable even farther from the initial value. Positive feedback systems tend to be highly unstable—an explosion is an example of positive feedback. In humans, positive feedback plays an important role in childbirth; the pressure of the baby's head on the lower part of the uterus increases the frequency and intensity of uterine contractions. ☐

Homeostasis in Action

Using sensors scattered throughout the body, the central nervous system constantly monitors temperature, pH, blood pressure, ion concentrations, and many other factors. Much of this surveillance is centered in a small, marble-sized region of the brain called the **hypothalamus.** When the hypothalamus detects a disturbance in the body's condition, it issues orders to correct the disturbance. Sometimes these orders alter body functions, such as how rapidly you breathe. At other times, they call for production of chemical signals, which you will study later in this section.

An example of homeostasis in action is the way in which your body regulates its temperature, a process illustrated in Figure 35-14. When the temperature of your blood exceeds its set point, which is usually 37°C (98°F), neurons in the brain detect the temperature change and inform the hypothalamus. The hypothalamus responds by triggering mechanisms, such as sweating and the expansion of blood vessels, that dissipate heat. In addition to neurons in the central nervous system, there are two types of temperature-sensitive nerve endings in your skin. One type is sensitive to low temperatures, and the other type to high temperatures. These "surface thermometers" also report to the hypothalamus.

Other parts of the brain are also important in maintaining homeostasis. The brain stem, the part of the brain that joins to the spinal cord, keeps blood pressure constant. Blood pressure is measured at sites in the major arteries where the wall of the artery is thin and contains a highly branched

---

Graphic Organizer

Use this graphic organizer in *Teaching Tip:* **Comparing Negative Feedback With Positive Feedback** on page 838.

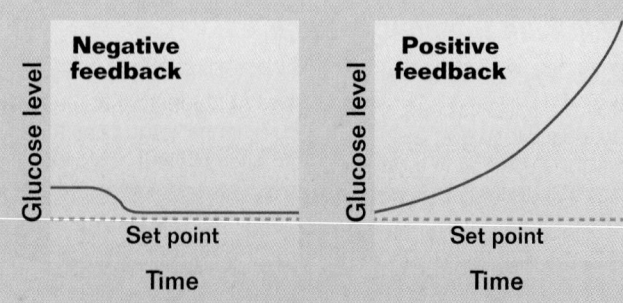

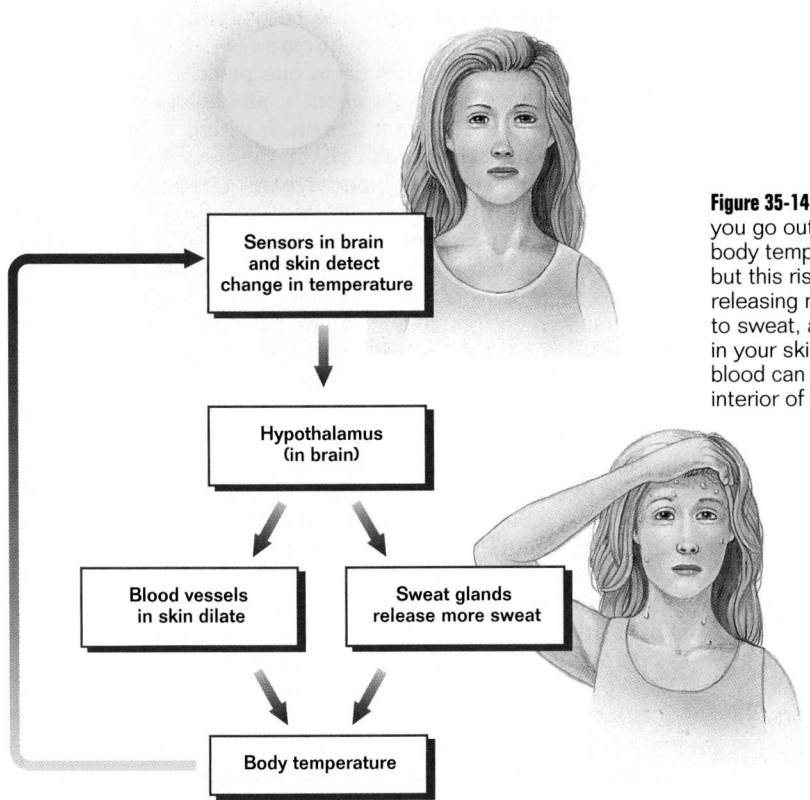

Figure 35-14 What happens when you go out on a hot day? Your body temperature initially rises, but this rise is countered by heat-releasing mechanisms. You begin to sweat, and the blood vessels in your skin expand so that the blood can transfer heat from the interior of the body to its surface.

network of nerve endings. When blood pressure increases, the thin wall of the artery bulges out, stretching the nerve endings, which relay signals to the brain stem. The brain stem responds by adjusting the rate at which the heart beats, the force of its contractions, and the diameter of some blood vessels. This negative feedback loop moves blood pressure back toward its set point.

The Body's Signaling System

Maintaining homeostasis requires a great deal of signaling back and forth between the central nervous system and the rest of the body. There are two types of signals. Some are rapidly transmitted **electrical signals,** which are carried by neurons from one place in the body to another much as wires carry telephone messages. An electrical signal's effects usually last only for brief periods. **Chemical signals** are more slowly transmitted and are carried by the bloodstream from one place in the body to another. Unlike the effects of electrical signals, the effects of chemical signals tend to last for a long time.

Chapter Closure

On large pieces of butcher paper, have students make models of the various organ systems discussed in this chapter. Have them draw the systems within the outline of a human figure. They should put one system on each sheet so that the sheets form a flip chart of the human body. Also have students label the components of each system. Each sheet should also include a description of the function of each system. Have students save their sheets and add more information as they read the remaining chapters in this unit.

CONTENT LINK

How does a hormone change the functions of its target tissue? You will study this topic in Chapter 41.

☐ CAPSULE SUMMARY

The hypothalamus, a region of the brain, monitors and regulates the body's condition. Information is transmitted from one part of the body to another by the nervous system and by hormones.

To maintain homeostasis, the body employs a battery of specific molecules called **hormones.** A hormone is a chemical messenger produced in one place and transported to another, where it brings about a physiological response. One example is the hormone insulin, which is produced by the pancreas and stimulates cells in the liver and muscles to absorb glucose from the bloodstream. Chemical messages obviously can be effective only if they are recognized. How does a tissue recognize a hormone with a particular shape? It does so by having a specific receptor protein that matches the shape of the hormone. Using such a receptor, the tissue can recognize a hormone with precision and select one molecule from billions of others. Specificity is the essence of a receptor protein's function.

The great advantage of a hormone as a body messenger is that it can be directed at a particular protein receptor on the target cells, and the receptor recognizes only this molecule. In every case the operating principle is the same: only cells whose membranes contain an appropriate receptor protein will respond to the chemical message.

Why aren't all communications between cells in the body handled by chemical signals? Their transmission is very slow. If the message to be delivered to your leg muscles is "Contract quickly; we are being pursued by a leopard," a quicker means of communication than hormones is required. To answer this need, humans and all complex animals have specialized cells called neurons, which transmit electrical signals over considerable distances with great speed. ☐

Section Review

1. *Why is homeostasis crucial to the body?*
2. *Explain why most body functions are regulated by negative, rather than positive, feedback loops.*
3. *Identify one advantage and one disadvantage of chemical signals, compared with electrical signals.*

Critical Thinking

4. *A thermostat plays the same role in regulating the temperature of a room as your brain does in regulating the temperature of your body. Which situation would produce a more pleasant room temperature: a thermostat controlling a positive feedback loop or one controlling a negative feedback loop? Explain your answer.*
5. *Why is it advantageous for uterine contractions during birth to be regulated by a positive feedback loop?*

Answers to Section Review

1. The chemical reactions in the body require fairly constant conditions.

2. Negative feedback keeps the variable near its set point, while positive feedback drives the variable to extreme values.

3. Chemical signals are long lasting and specific, but they are slow.

4. Negative feedback would keep the temperature moderate. Positive feedback would produce either a very hot or a very cold temperature.

5. If uterine contractions were controlled by a negative feedback loop, the pressure of the baby's head on the lower part of the uterus would decrease the force of contractions, and the baby would not be forced out of the uterus.

Highlights
Organ Systems

Skeletal System

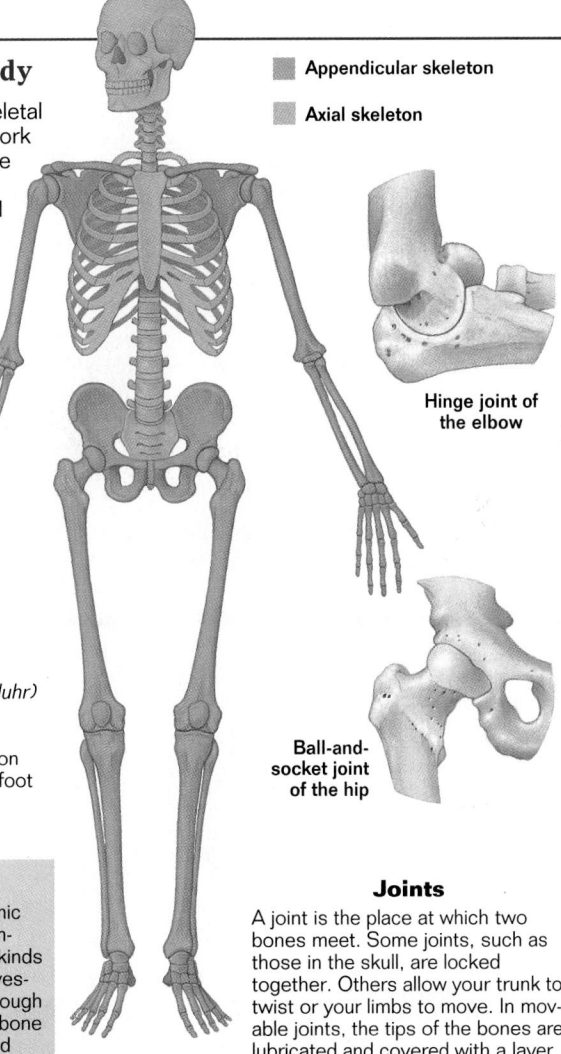

- ■ Appendicular skeleton
- ▨ Axial skeleton

Protecting and Supporting the Body

The bones, cartilage, and ligaments of the skeletal system provide a strong, rigid internal framework against which the body's muscles can pull. The skeletal system also protects the body by encasing the central nervous system in a shell of bone and shielding the thoracic cavity with a framework of ribs. A flexible endoskeleton of bone allows the body a wide range of motion while supporting a body of considerable size. There are 206 individual bones in the human body.

The Axial Skeleton

The bones of the **axial** *(AX ee uhl)* **skeleton** support and protect the trunk. Twenty-eight of these bones form the skull, which protects the brain; 26 bones compose the backbone, which protects the spinal cord; and 24 bones form the rib cage (12 pairs of ribs attached to the breastbone and the backbone), which protects the heart and lungs.

The Appendicular Skeleton

The bones of the **appendicular** *(ap ehn DIHK yoo luhr)* **skeleton** form the limbs and the points where the limbs attach to the axial skeleton. The 30 bones of each arm and hand are attached to the axial skeleton at the shoulder, and the 30 bones of each leg and foot are attached to the axial skeleton at the hip.

Hinge joint of the elbow

Ball-and-socket joint of the hip

Bone

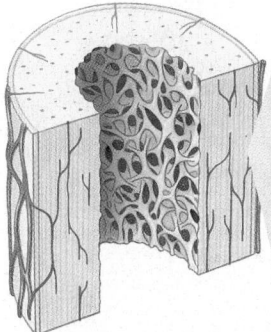

Bone is a dynamic living tissue composed of many kinds of cells. Blood vessels passing through channels in the bone bring oxygen and nutrients to these cells. Bone is hard because some of its cells form crystals of calcium phosphate.

Joints

A joint is the place at which two bones meet. Some joints, such as those in the skull, are locked together. Others allow your trunk to twist or your limbs to move. In movable joints, the tips of the bones are lubricated and covered with a layer of smooth cartilage, allowing them to slide easily against each other. The elbow and knee are hinge joints. They can swing up and down, but not side to side. The hip and shoulder are ball-and-socket joints. In this type of joint, a long bone that ends in a ball fits into a hollow socket. The long bone can move in almost any direction by swiveling within its socket.

Ligaments and Tendons

Strong fibers called **ligaments** hold bones together. They act as cables to hold bones flexibly within their sockets. Muscles are attached to bones by straps of tough cartilage called **tendons.**

Instructional Strategy

Students learn about the skeletal system more effectively if they can examine a life-size skeleton or model of a skeleton. Encourage them to observe the motions of their own joints and then correlate these motions to structure of the joints. For instance, the knee allows back-and-forth movement but not pivoting. Students can understand knee movement by examining the articulation of the bones at the knee.

Discussion

Guide the discussion by posing the following questions.

1. What bones compose the axial skeleton? *(spine, ribs, and skull)* What are the functions of the axial skeleton? *(It protects the brain, thorax, and spinal cord. It also serves as an attachment point for the appendicular skeleton.)*

2. What bones compose the appendicular skeleton? *(bones of the legs, arms, feet, and hands; the pelvis; and the pectoral girdle)*

3. How do the cells in bone get nutrients and oxygen? *(from blood vessels passing through canals in the bone)*

4. Distinguish between tendons and ligaments. *(Ligaments attach bones to other bones; tendons connect muscles to bones.)*

Instructional Strategy

Review the structures of skeletal and smooth muscle. Skeletal muscle is composed of fibers formed when several cells fuse. The protein filaments are aligned, producing strong contraction. Smooth muscle cells do not fuse, and their filaments are not aligned. Smooth and skeletal muscles also differ in how they are controlled. Skeletal muscles are voluntary, while smooth muscles are involuntary. Ask students to explain this difference based on the function of each kind of muscle.

Discussion

Guide the discussion by posing the following questions.

1. Where are smooth muscles found? *(in the walls of blood vessels and hollow internal organs)*

2. Name two ways that smooth muscles can be stimulated. *(by nerves and hormones)*

3. What is a motor unit? *(All the muscle fibers controlled by one neuron make up a motor unit.)*

4. Compare summation with recruitment. *(Summation is the repetitive firing of a motor neuron, producing stronger contraction of muscle fibers. Recruitment is the activation of additional motor units, also producing stronger contraction.)*

5. Why is the lifetime of a skin cell so short? *(Skin cells wear out because they are exposed to the environment.)*

Highlights
Organ Systems

Muscular and Integumentary Systems

Producing Movement

The **muscular system** is composed of
1. smooth muscles that line the blood vessels and hollow internal organs, such as those of the digestive tract, and
2. skeletal muscles that move the head, chest, and appendicular

Kinds of Smooth Muscles

Smooth muscles are organized into sheets of tissue. In some smooth muscles, such as those that line the walls of many blood vessels, the cells contract simultaneously when the sheet is stimulated by a nerve or hormone. In other smooth muscles, such as those that line the wall of the gut, individual cells contract spontaneously, producing a slow, steady contraction.

Kinds of Skeletal Muscles

Skeletal muscles have different functions. Muscles of the fingers and eyes must contract rapidly, and they fatigue with continued use. The muscles that control posture, such as the long muscles of the back, contract slowly and do not fatigue. Skeletal muscles can function in these different ways because they have different kinds of fibers. Muscles that contract rapidly have a fast but inefficient metabolism, while muscles that contract slowly have a slower but more efficient metabolism.

Covering the Body

The **integumentary system** covers and protects the body. Human skin is from 10 to 30 cells thick, about as thick as this page; the outer layer of the skin is epithelial tissue.

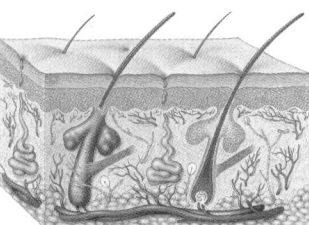

Maintaining the Skin

Cells from the outer layer are continually being injured and worn; they also lose moisture and dry out. The integumentary system deals with this damage not by repairing cells but by replacing them. Cells from the outer layer are shed continuously and replaced by new cells that have been produced deeper within the skin. A cell normally lives on the surface of the skin for about a month before it is shed.

Motor Units

Each cell of a skeletal muscle is called a **muscle fiber.** Nerve cells called motor neurons control the contraction of muscle fibers. Motor neurons usually branch at their tips and contact several muscle fibers. All of the fibers controlled by one motor neuron, called a **motor unit,** contract together when the motor neuron fires. Large muscles, such as those of the leg, have more than 100 muscle fibers per motor unit, while small muscles, such as those of the eye, have less than 10.

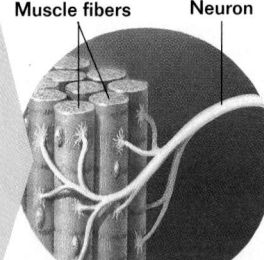

Motor Unit

Muscle fibers   Neuron

How Much Force?

The total amount of force developed by a muscle depends on two things:
1. how often the motor neuron leading to the muscle fires—repetitive firing produces stronger contraction of individual muscle fibers, an effect called **summation;** and
2. the number of muscle fibers involved in the contraction—stronger contractions occur when additional motor neurons fire, activating additional groups of muscle fibers, an effect called **recruitment.**

Circulatory and Immune Systems

The Body's Transportation System

The **circulatory system** contains the network of blood vessels that connects the organs and muscles of the body, the muscular heart that pumps blood through the vessels, and the lymphatic vessels that collect excess fluid. The circulatory system transports food and nutrients, oxygen-bearing red blood cells, and defensive immune system cells. It also helps maintain a uniform body temperature by transferring heat from one part of the body to another.

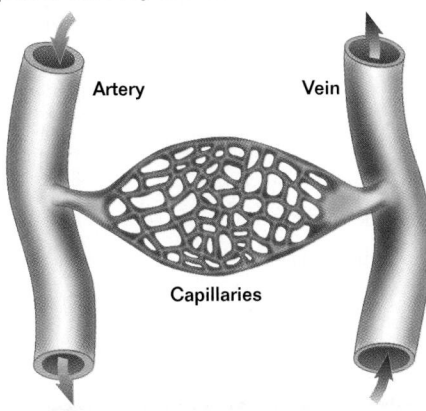

Arteries and Veins

Arteries carry blood away from the heart. Because they have elastic walls, arteries can expand their volume when the heart contracts. Their walls also contain a layer of smooth muscle that allows the body to control the amount of blood flowing to particular tissues. Veins return blood to the heart. The walls of veins are much thinner than those of arteries—when a vein is empty, its walls collapse.

Capillaries

Capillaries connect arteries to veins. They are very short and narrow. The average capillary is only 1 mm (0.04 in.) long with an internal diameter of 8 μm (slightly larger than the diameter of a red blood cell). However, there are many capillaries in the body. Laid end to end, the capillaries in your body would extend across the United States. Oxygen and food are transferred from the blood to the body's cells, and wastes are collected in capillaries. No cell of the human body is more than 100 μm from a capillary.

The Lymphatic System

Because some liquid is always being lost from the blood supply by diffusion through the walls of the smallest blood vessels, it is necessary to constantly re-collect this fluid. This recycling task is carried out by the lymphatic system, an open series of vessels that collects fluid and returns it to the blood vessels.

Defending the Body Against Infection and Cancer

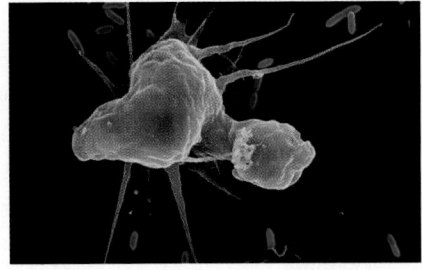

The **immune system** is composed of defensive white blood cells that circulate in the bloodstream and line the lymphatic system. The cells of the immune system constantly monitor the bloodstream and tissues to detect invading bacteria, fungi, protists, or viruses; to identify body cells infected with viruses; and to screen for cancer cells. When an infection occurs or an infected or cancerous cell is identified, the cells of the immune system marshal an attack and destroy the invading microbes or abnormal cell. No one can live for long

Blood

Blood is a protein-rich fluid in which many kinds of cells are suspended—cells are 40 percent of the total volume. Each milliliter of blood contains about 5 million oxygen-carrying **erythrocytes** (red blood cells) and 10,000 body-defending **leukocytes** (white blood cells), as well as cells and cell fragments involved in blood clotting and waste removal.

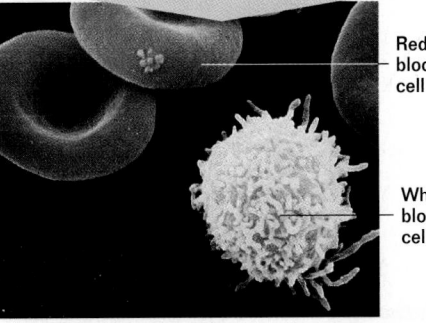

Red blood cell

White blood cell

Instructional Strategy

Remind students of the exchange of gases that occurs in the lungs—carbon dioxide diffuses from the blood into the air in the lungs, while oxygen diffuses from this air into the bloodstream. Remind students that more oxygen can diffuse across a large surface area than across a small one. This fact accounts for the large internal surface area of the lungs.

Discussion

Guide the discussion by posing the following questions.

1. Trace the path an inhaled molecule of oxygen would follow to the lungs. *(nose or mouth, trachea, bronchus, bronchiole, alveolus)*

2. What is the function of mucus in the respiratory system? *(It traps dust and other foreign particles before they reach the lungs.)*

3. In which structures does gas exchange take place? *(in the alveoli)*

4. Which muscles are involved in breathing? *(the diaphragm and the muscles between the ribs)*

Exchanging Gases

The respiratory system captures oxygen and releases carbon dioxide. This system is composed of a pair of **lungs,** which hang free within the fluid-filled chest cavity, and the passages that link the lungs to the nose and mouth.

Trachea

Air travels from the nose or mouth to the lungs through a long tube called the **trachea,** which branches near the lungs into the left and right **bronchi** (singular, bronchus). At the lung, each bronchus branches into many bronchioles, each connected to a large number of alveoli. Mucus produced by cells lining the trachea traps dust particles before they reach the lungs. Cilia on cells lining the trachea sweep the mucus upward so that it can be swallowed.

Alveoli

Each lung consists of some 300 million tiny air-filled sacs called **alveoli.** Each alveolus is surrounded by a network of about 100 tiny blood vessels. Gases diffuse across the short distance from alveolus to blood and back. The key to the structure of the lung is the enormous surface area that this arrangement provides for gas exchange—60 to 80 sq. m (650 to 860 sq. ft.) per lung. The total inner surface area of your lungs is about one-half the area of a tennis court.

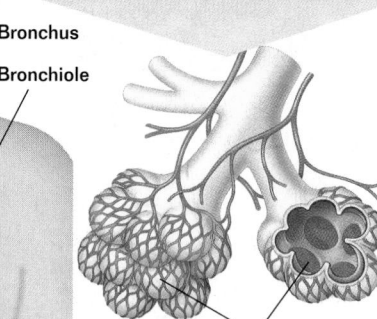

Trachea
Lungs
Bronchus
Bronchiole
Alveoli
Diaphragm

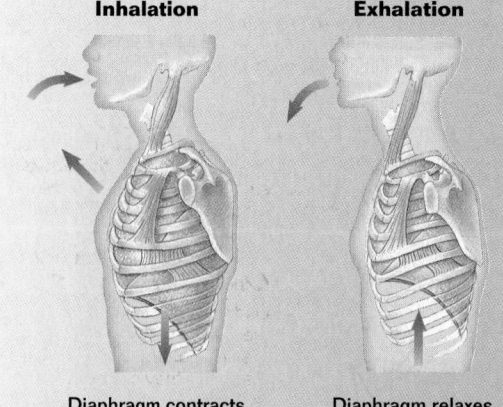

Inhalation

Exhalation

Diaphragm contracts

Diaphragm relaxes

Moving Air Into and Out of the Lungs

A thick layer of muscle called the **diaphragm** separates the thoracic cavity, containing the lungs, from the abdominal cavity, containing the digestive organs. The diaphragm is dome-shaped when relaxed. During inhalation, the diaphragm contracts, causing it to flatten. Muscles on the ribs raise and expand the rib cage. The actions of these muscles enlarge the thoracic cavity and reduce the pressure in the lungs, allowing air to rush inward. During exhalation, the diaphragm relaxes and the muscles on the ribs lower the rib cage, compressing the thoracic cavity and driving air outward.

Digestive and Urinary Systems

Obtaining Nutrients From Food

The **digestive system** is a series of organs involved in breaking down food and absorbing nutrients. The digestive system includes the organs through which food passes and also includes the liver and pancreas, which manufacture many digestive enzymes and hormones.

The function of the digestive system is to break up the large molecules in food (proteins, carbohydrates, and fats) into small subunits (amino acids, sugars, glycerol, and fatty acids). These molecules are then absorbed into the bloodstream and carried to the cells of the body.

Regulating the Body's Salt and Water Levels

The **urinary system** filters metabolic wastes from the bloodstream and controls the ionic composition of the blood. This system is composed of the kidneys, urinary bladder, and associated ducts.

Kidneys

The kidneys are about the size of a small fist and are located in the lower back region. The kidneys play a crucial role in waste removal and water retention. Your blood passes through the kidneys for cleansing about once every four minutes; in the course of a full day, about 2,000 L (530 gal.) of fluid pass out of the bloodstream and into the kidneys, and all but a very small portion of that fluid is reabsorbed.

Urinary Bladder

Urine, the residual unabsorbed fluid that contains the body's wastes, is stored in a hollow, muscular sac called the bladder. The bladder can hold over 0.5 L (about 0.5 qt.) when full.

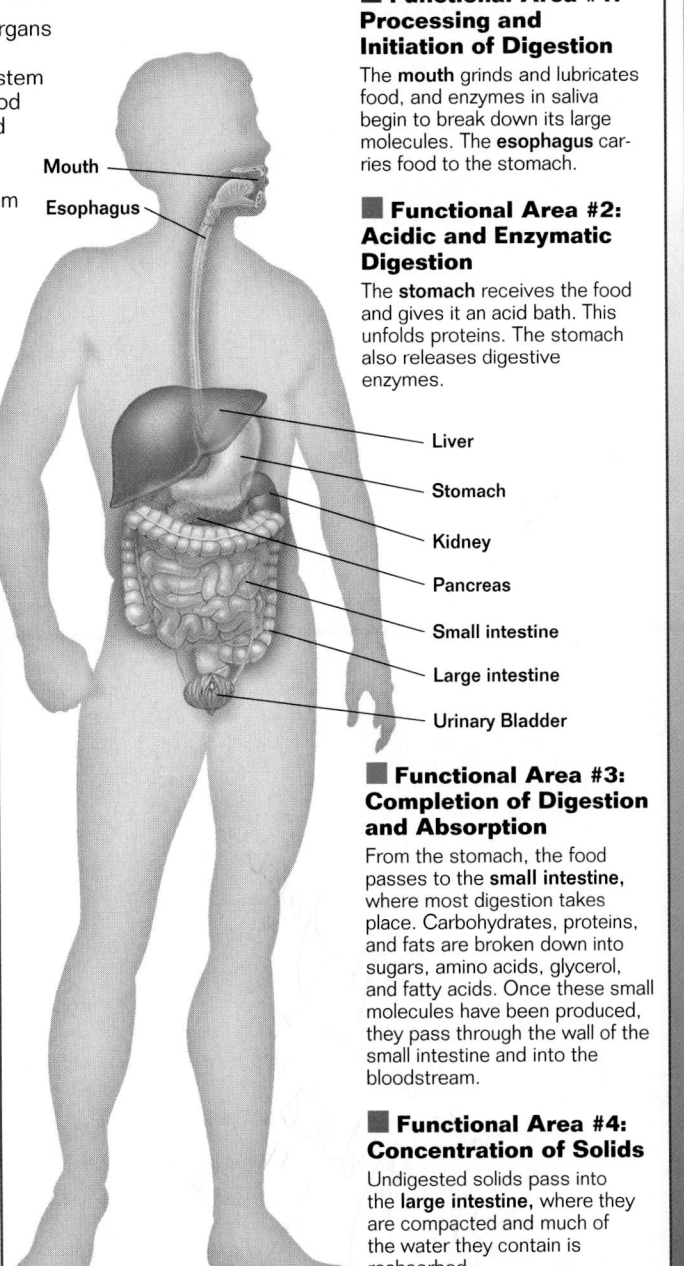

Mouth
Esophagus
Liver
Stomach
Kidney
Pancreas
Small intestine
Large intestine
Urinary Bladder

■ Functional Area #1: Processing and Initiation of Digestion

The **mouth** grinds and lubricates food, and enzymes in saliva begin to break down its large molecules. The **esophagus** carries food to the stomach.

■ Functional Area #2: Acidic and Enzymatic Digestion

The **stomach** receives the food and gives it an acid bath. This unfolds proteins. The stomach also releases digestive enzymes.

■ Functional Area #3: Completion of Digestion and Absorption

From the stomach, the food passes to the **small intestine,** where most digestion takes place. Carbohydrates, proteins, and fats are broken down into sugars, amino acids, glycerol, and fatty acids. Once these small molecules have been produced, they pass through the wall of the small intestine and into the bloodstream.

■ Functional Area #4: Concentration of Solids

Undigested solids pass into the **large intestine,** where they are compacted and much of the water they contain is reabsorbed.

Instructional Strategy

The digestive system chemically and physically transforms food and extracts its nutrients. Remind students that digestion begins in the mouth. Teeth crush and grind food, and enzymes in saliva begin the process of chemical breakdown. The stomach continues digestion, but the site of most digestion and of the absorption of nutrients is the small intestine. In concert with the endocrine system, the urinary system regulates the composition of the body's fluids. It removes wastes from the blood and eliminates them from the body.

Discussion

Guide the discussion by posing the following questions.

1. Food does not pass through two of the organs of the digestive system shown here. Name them. *(liver, pancreas)*

2. List in order the organs through which food passes on its journey through the digestive system. *(mouth, esophagus, stomach, small intestine, large intestine)*

3. What are the two main functions of the urinary system? *(It filters metabolic wastes from the blood and regulates the ionic composition of the body's fluids.)*

4. What is the function of the urinary bladder? *(It stores urine.)*

Instructional Strategy

The body has two different control systems, each fulfilling different needs. The endocrine system, though relatively slow acting, exerts long-term control over the body. Students should recognize that specificity is the essence of endocrine control. Endocrine glands communicate with each other and with their target tissues by means of hormones, chemical messengers that circulate in the blood. Hormones act on very specific targets; only cells with the appropriately shaped receptors will respond to the hormone.

Discussion

Guide the discussion by posing the following questions.

1. What is the relationship between the hypothalamus and the pituitary gland? *(The hypothalamus releases hormones that stimulate the pituitary to release hormones into the bloodstream.)*

2. Which endocrine gland also functions in the digestive system? *(pancreas)*

3. How are the effects of hormones different from those of the chemicals that function as messengers in the nervous system? *(The effects of hormones tend to persist.)*

4. What is the function of adrenaline, and which gland produces it? *(It prepares the body for activity and is released by the adrenal medulla.)*

Highlights *Endocrine System*

Organ Systems

Regulating the Body's Functions

The endocrine system uses chemical signals called **hormones** to coordinate and integrate the activities of the body. This system is composed of ductless glands called endocrine glands, such as the pituitary, thyroid, and adrenal glands. Hormones regulate growth and sexual development and maintain physiological conditions within narrow bounds. The proper balance of sodium, potassium, and calcium ions is under careful hormonal control, as is the level of glucose in the blood.

Hypothalamus
Pituitary
Thyroid
Parathyroids

Adrenal glands
Pancreas

Adrenal Glands

The interior of the adrenal gland, the **adrenal medulla,** releases adrenaline, a hormone that enables you to act in an emergency. Among adrenaline's effects are increased heart rate, heightened awareness, expansion of the airways leading to the lungs, and an increase in metabolic rate. Secretions of the outer layer of the adrenal gland, the **adrenal cortex,** maintain the levels of sodium and potassium in the body.

Pancreas

Controls levels of glucose in the blood

Hypothalamus

A small area in the lower portion of the brain called the hypothalamus directs much of the endocrine system. Special brain hormones diffuse from the hypothalamus to a nearby endocrine gland called the **pituitary gland,** which responds by producing one of eight pituitary hormones.

Parathyroids

Regulate calcium metabolism and bone building

Thyroid

Regulates metabolism

Hormones

Hormones are chemical messenger molecules that are manufactured in a small quantity in one part of the body and then transported through the bloodstream to another location, where they bring about a physiological response. Unlike the effects of chemicals used as messengers in the nervous system, the effects of hormones tend to persist for a long time.

Pituitary Hormones

Pituitary hormones are action orders. They pass to endocrine glands located in other parts of the body and cause these glands to begin production of particular hormones. The hypothalamus controls the commands that the pituitary issues to the endocrine system like an army general controls the orders that a captain issues to the troops.

Nervous System

Receiving and Interpreting Information

The nervous system is composed of

1. the body's **sense organs**, which provide information about the body and its environment,
2. **sensory nerves**, which relay the sensory information to the central nervous system,
3. the **brain** and **spinal cord**, which process and integrate sensory information and issue commands in the form of nerve impulses, and
4. **motor nerves**, which transmit commands to the body's muscles, organs, and glands.

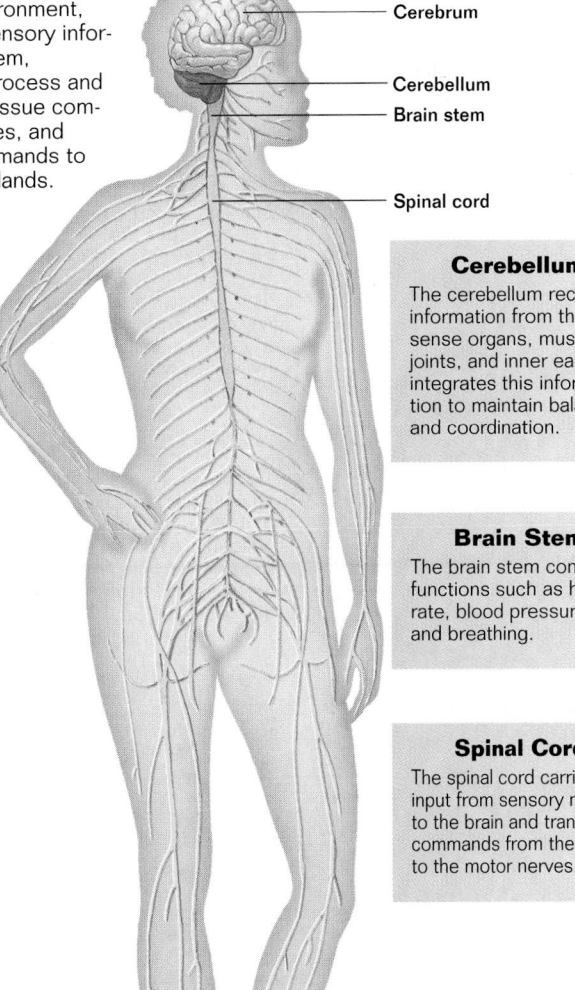

Cerebrum

Cerebellum

Brain stem

Spinal cord

Cerebrum

The cerebrum is responsible for learning, memory, and thought. It evaluates sensory information and issues commands. The cerebrum is the largest part of the brain. The cerebrum is so large that it envelops much of the rest of the brain. A human brain weighs about 1.5 kg (3 lb.), and over 1 kg (2 lb.) of it is the cerebrum.

Cerebral Cortex

Much of the learning activity that occurs in the cerebrum takes place within a thin layer on its outer surface called the cerebral cortex. The human cerebral cortex contains over 10 billion nerve cells, amounting to roughly 10 percent of all the neurons in the brain.

Nerve Impulses

A nerve impulse is a short-lived electrical disturbance that passes rapidly along a neuron. Every nerve impulse in the nervous system is the same, involving the same magnitude of electrical change in the neuron cell membrane. The frequency (pattern) and the point of origin of nerve impulses are what determine what information is carried from one part of the body to another.

Cerebellum

The cerebellum receives information from the sense organs, muscles, joints, and inner ears. It integrates this information to maintain balance and coordination.

Brain Stem

The brain stem controls functions such as heart rate, blood pressure, and breathing.

Spinal Cord

The spinal cord carries input from sensory nerves to the brain and transmits commands from the brain to the motor nerves.

Highlights
Organ Systems

Instructional Strategy

While the endocrine system is slow-acting but has long-term effects, the body's other major regulatory system, the nervous system, is fast-acting and has short-term effects. Messages in the nervous system are transmitted by short-lived electrical disturbances across the membranes of nerve cells. Point out that the nervous system can be divided into two units: the central nervous system, which is the command and control center, and the peripheral nervous system, which transmits information to the central nervous system and carries its commands to the muscles, organs, and glands.

Discussion

Guide the discussion by posing the following questions.

1. What are the three main parts of the brain? (cerebrum, cerebellum, brain stem)

2. Damage to the cerebellum would impair which functions? (balance and coordination)

3. Describe the relationship between the brain and the spinal cord. (The spinal cord carries messages from sensory nerves to the brain and relays the brain's messages to motor nerves.)

4. What is the largest part of the human brain? (cerebrum)

Instructional Strategy

Emphasize the strong connection between the reproductive and endocrine systems. The testes and ovaries produce hormones that control sexual development and maturation. The ovaries also produce hormones that influence the uterus during pregnancy. The endocrine activities of the ovaries and testes are controlled by the pituitary gland. The pituitary gland also releases hormones that control the production of gametes.

Discussion

Guide the discussion by posing the following questions.

1. How do the sexes differ in their production of gametes? *(In males, gamete production begins at puberty and continues throughout adult life. A female is born with all the eggs she will ever have. Typically, one egg matures and is released each month.)*

2. What are oocytes? *(They are the cells that may eventually become eggs.)*

3. What is the function of the fallopian tubes? *(They conduct the egg to the uterus and serve as the site of fertilization.)*

Highlights   *Reproductive System*
Organ Systems

Producing Gametes

The female and male reproductive systems act to join **egg cells** from a female with **sperm cells** from a male. As in all mammals, the egg is fertilized within the female. The fertilized egg develops into a mature fetus in the uterus.

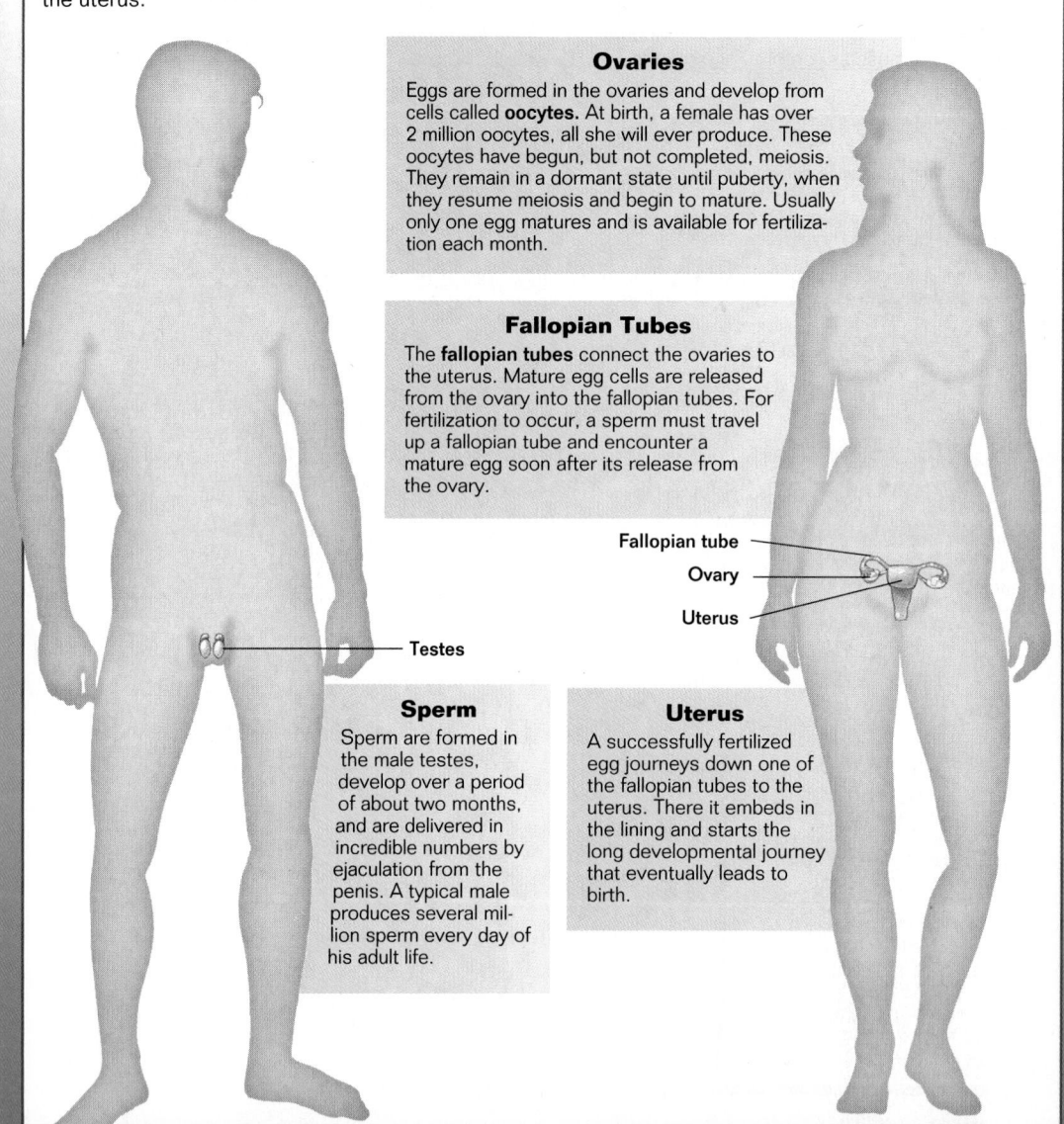

Ovaries

Eggs are formed in the ovaries and develop from cells called **oocytes.** At birth, a female has over 2 million oocytes, all she will ever produce. These oocytes have begun, but not completed, meiosis. They remain in a dormant state until puberty, when they resume meiosis and begin to mature. Usually only one egg matures and is available for fertilization each month.

Fallopian Tubes

The **fallopian tubes** connect the ovaries to the uterus. Mature egg cells are released from the ovary into the fallopian tubes. For fertilization to occur, a sperm must travel up a fallopian tube and encounter a mature egg soon after its release from the ovary.

Fallopian tube
Ovary
Uterus

Testes

Sperm

Sperm are formed in the male testes, develop over a period of about two months, and are delivered in incredible numbers by ejaculation from the penis. A typical male produces several million sperm every day of his adult life.

Uterus

A successfully fertilized egg journeys down one of the fallopian tubes to the uterus. There it embeds in the lining and starts the long developmental journey that eventually leads to birth.

Vocabulary

actin (833)	feedback loop (837)	negative feedback (837)
axon (833)	glial cell (832)	neuron (832)
central nervous system (826)	hormone (840)	organ (829)
chemical signal (839)	hypothalamus (838)	organ system (829)
collagen (832)	joint (825)	peripheral nervous system (826)
dendrite (833)	ligament (825)	positive feedback (838)
electrical signal (839)	lymphocyte (831)	set point (837)
epithelium (829)	macrophage (831)	supporting cell (832)
erythrocyte (832)	myosin (833)	

Concept Mapping

Construct a concept map that shows how the human body is organized. Use as many terms as needed from the vocabulary list. Try to include the following items in your map: cells, tissues, organs, organ systems, epithelial tissue, connective tissue, nerve tissue, muscle tissue, endocrine system, nervous system, reproductive system, circulatory system, immune system, respiratory system, and digestive system. Include additional terms in your map as needed.

Review

Multiple Choice

1. Movement of the skeleton is made possible by
   a. bones.
   c. cartilage.
   b. joints.
   d. skin.

2. What enables chemical reactions occurring in your body to proceed at a constant rate?
   a. endothermy
   b. fat content of muscles
   c. body movement
   d. the coelom

3. What tissue type functions to cover and protect the body?
   a. connective
   c. muscle
   b. nerve
   d. epithelial

4. Cells of the nervous system that surround and insulate neurons are called
   a. axons.
   c. glial cells.
   b. dendrites.
   d. myosin.

5. What muscle type lines the walls of the blood vessels and gut?
   a. cardiac muscle
   c. smooth muscle
   b. skeletal muscle
   d. rough muscle

6. Which organ system concentrates and removes waste from the body?
   a. the urinary system
   b. the immune system
   c. the skeletal system
   d. the endocrine system

7. Which medical test is least likely to reveal information about homeostasis in the human body?
   a. chest X-ray
   b. urinalysis
   c. blood test
   d. measurement of body temperature

8. Which is an example of positive feedback?
   a. the operation of a home thermostat
   b. an explosion
   c. upward and downward trends in the stock market
   d. fluctuations in pH of body fluids

Completion

9. The site at which two bones meet is called a(n) _____ . Tough bands of connective tissue called _____ hold the bones in place.

CHAPTER REVIEW ANSWERS

Each item in the Chapter Review is correlated by text section in the assignment guide that follows.

ASSIGNMENT GUIDE

Section	Review	Themes Review	Critical Thinking
35-1	1, 2, 9, 10, 16		28
35-2	3–6, 11–13, 17, 18	24, 25, 27	29
35-3	7, 8, 14, 15, 19	26	30

Concept Mapping

Map answer is shown on page 823B.

Review

Multiple Choice
Code in parentheses indicates section number and objective number.
1. b (35-1.1)
2. a (35-1.3)
3. d (35-2.1)
4. c (35-2.1)
5. c (35-2.2)
6. a (35-2.3)
7. a (35-3.1)
8. b (35-3.2)

Completion
9. joint, ligaments (35-1.1)
10. central, peripheral (35-1.2)
11. simple, stratified (35-2.1)
12. organs (35-2.3)
13. immune, white blood (35-2.3)
14. hypothalamus (35-3.1)
15. electrical, chemical (35-3.3)

Short Answer
16. Endothermic animals can be active at cold temperatures and in cold habitats and can perform strenuous activities for long periods. Endothermy requires large amounts of food. (35-1.3)

Short Answer *continued*

17. Cardiac muscle. The cells are connected in a latticework structure so that electrical impulses can be transmitted from cell to cell. Neighboring cells can contract almost simultaneously. (35-2.2)

18. Bone is a supportive and protective tissue that is composed of cells surrounded by collagen fibers that are hardened by calcium salts. Cartilage is produced along lines of stress in the skeleton and covers the ends of bones. It is composed of parallel collagen fibers. Fibroblasts are generalized connective tissue cells that make up much of the connective tissue in the body. They produce collagen. (35-2.1)

19. The body initiates heat-releasing mechanisms such as increased sweating and dilation of blood vessels near the skin. The hypothalamus senses increases in body temperature and stimulates heat-releasing mechanisms. (35-3.1)

Highlights Review

20. Bones of the skull, ribs, and backbone form the axial skeleton; they protect the brain, spine, and organs of the chest, including the heart and lungs. The bones of the limbs and their points of attachment to the axial skeleton make up the appendicular skeleton; these bones provide points of muscle attachment, enabling movement of the limbs.

21. The lymphatic system collects the body fluids that constantly leak from the capillaries and delivers them back to the circulatory system through a system of vessels.

22. Air is moved by the action of the diaphragm and the muscles of the ribs. During inhalation, the diaphragm contracts and the muscles of the ribs raise and expand the chest cavity. During exhalation, the diaphragm relaxes and the muscles of the ribs contract the chest cavity.

10. The brain and spinal cord make up the _____ nervous system, while the _____ nervous system consists of nerves that connect the brain and the spinal cord to the rest of the body.

11. The lungs are lined by _____ epithelium that is one cell thick. The skin is partly composed of _____ epithelium that is several cells thick.

12. Organ systems are made up of _____ that, in turn, are composed of different types of tissues.

13. The _____ system defends the body against infection. The functional units of this system are _____ cells that make antibodies or attack infected cells or pathogens.

14. Homeostasis of the body is regulated by a region of the brain called the _____ , which detects and initiates changes in the body's condition.

15. Maintaining homeostasis in the human body involves the operation of _____ signals carried by neurons and _____ signals transmitted by hormones.

Short Answer

16. List two benefits of endothermy. What is the major disadvantage of endothermy?

17. What type of muscle tissue is shown in the photograph below? How does the structure of this kind of tissue reflect its function?

18. Contrast the functions and characteristics of each kind of connective tissue.

19. Explain how your body responds to an increase in its temperature. What role does the hypothalamus play in the response?

Highlights Review

20. The human skeleton is composed of two major parts: the axial and the appendicular skeletons. What bones make up the axial skeleton? What bones make up the appendicular skeleton? What are the functions of these two parts of the human skeleton?

21. The lymphatic system is sometimes called a backup circulatory system. What does the lymphatic system do that has earned it this title?

22. The main organs of the respiratory system are the lungs. Air is drawn into and forced out of the lungs, but they do not have muscles. How is air moved into and out of the lungs?

23. Serious diarrhea is associated with failure of the large intestine to reabsorb water. How can serious diarrhea affect water balance in the body?

Themes Review

24. Levels of Organization What is the relationship between an organ and its organ system?

25. Structure and Function Blood vessels have one layer of smooth muscle that encircles the vessel wall. The stomach and intestine have one layer of circular smooth muscle and one layer of longitudinal (running lengthwise) smooth muscle. How does this muscle arrangement reflect the function of each structure?

26. Homeostasis How does a feedback loop that involves negative feedback help the body to maintain homeostasis?

23. A person with diarrhea could become dehydrated and possibly die from dehydration.

Themes Review

24. An organ system contains various organs that work together. (35-2.3)

25. Vessels need to constrict in only one direction to push blood through the system. The stomach and small intestine constrict in two directions to break up and move food. (35-2.2)

26. A feedback loop that involves negative feedback prevents important variables from deviating from their set points. (35-3.2)

27. Smooth muscles are organized into sheets, and the filaments of smooth muscles are not aligned. The cells of skeletal muscles form a long fiber with a central cable of aligned filaments. Cardiac muscles are organized into branched and interconnected fibers. The lack of alignment in smooth muscles allows for slow and

27. **Structure and Function** Compare and contrast the structures of smooth muscle, skeletal muscle, and cardiac muscle. How is the structure of each muscle type related to its function?

Critical Thinking

28. **Interpreting Data** The graph below shows the relationship between body temperature and environmental temperature for two animals. Which line, A or B, best represents the human condition? Explain your answer.

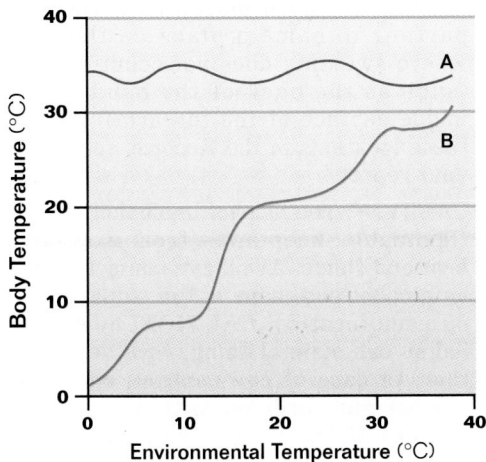

29. **Making Inferences** Polio is a disease that paralyzes muscles by affecting the nerves that make them move. Before polio vaccines were developed, many people were stricken by polio, and they sometimes died because they could not breathe. From what you know about the respiratory system, why might this happen?

30. **Making Inferences** The hormone insulin travels throughout the body in the bloodstream, but only certain types of cells respond to its presence by taking in glucose. How can this selective influence be explained?

Activities and Projects

31. **Research and Writing** Do library research and talk to medical doctors to find out about arthroscopic surgery. If possible, view a videotape of a surgical procedure to gain additional information. Write a report including a description of the nature of the surgical procedure, the types of injuries best suited for arthroscopic surgery, and the benefits associated with this type of surgery.

32. **Science-Technology-Society** Medical care for AIDS patients is very expensive. For this reason, insurance companies routinely require blood tests for persons applying for medical insurance and deny insurance to applicants carrying HIV, the virus that causes AIDS. Is this practice ethical? Construct an answer to this question by surveying students in your school, insurance agents, and medical workers who care for AIDS patients. Share your findings with your class.

Readings

33. Read the article "A vaccine for AIDS," in *The American Biology Teacher*, April 1993, pages 198–202. Describe one AIDS vaccine in development, and explain its disadvantages. What problems hamper the development of all AIDS vaccines?

34. Read the article "The Inside Dope on Runner's High," in *Runner's World*, August 1994, pages 60–64. What triggers runner's high and what physical sensations are associated with it? How far must a person run to experience runner's high?

35. Read the article "On the trail of lethal DNA," in *Newsweek*, May 17, 1993, page 63. About how many people die of colon cancer each year in the United States? What have new discoveries revealed about the cause of colon cancer?

Activities and Projects

31. Arthroscopic surgery involves the use of fiber optic technology and miniaturized instruments. Arthroscopic techniques are best suited for injuries to joints such as the knee and elbow. Arthroscopic surgery can usually be performed on an outpatient basis, usually nothing more than a local anesthetic is required, and rehabilitation and healing times are shorter than when traditional surgical procedures are used.

32. A clear-cut answer to the question is unlikely. Students will find that people's answers are based on their past experiences, including experiences with insurance companies and AIDS patients, and how the practice of medical screening has affected them personally.

Readings

33. Variable-vector vaccine is one in development. It has a quick and substantial immune response but involves the active cowpox virus. Problems hampering vaccine development include the following: (1) when HIV is incorporated into a host genome, it escapes antibodies; (2) the envelope proteins of HIV mutate; and (3) trial volunteers will test positive for HIV antibodies.

34. Runner's high is triggered by the physical stress caused by running and leaves a person feeling comfortable and relaxed. How far a person must run to experience runner's high varies. In fact, some people may never experience runner's high, no matter how far they run.

35. About 57,000 people in the United States die from colon cancer each year. A gene that is believed to cause colon cancer prevents DNA from making faithful copies of itself during cell division.

steady contraction, while the aligned filaments of skeletal muscles permit faster and more powerful contraction. The synchronized contraction of cardiac muscles is made possible by fibers that branch and interconnect. (35-2.2)

Critical Thinking

28. Line A best represents the human condition, one in which a fairly constant internal temperature is maintained. (35-1.3)

29. Polio can affect the muscles of the chest and diaphragm that control breathing. If these muscles cannot move, the polio victim cannot breathe. (35-2.2)

30. The selective influence is due to the presence of receptor molecules on only certain types of cells. (35-3.3)

LABORATORY Investigation Chapter 35

Thermoregulation

OBJECTIVES

- Demonstrate the distribution of sweat glands in skin.
- Determine the sweat gland activity with a change of environmental temperature and body temperature.

PROCESS SKILLS

- organizing data in tables
- comparing effects of different experimental conditions

MATERIALS

- safety gloves
- pre-moistened alcohol pads
- antibacterial soap (with iodine)
- cotton swabs
- 2 in. squares of erasable bond paper (no substitution)
- heat source lamp (no sun lamps)
- towel
- metric ruler

BACKGROUND

1. What is homeostasis?
2. How is internal body temperature regulated in humans?
3. How does the internal body temperature of mammals and birds compare to that of reptiles and amphibians?
4. What are the sensory organs that communicate with the brain's temperature control center in the hypothalamus?
5. Write your own question to explore in your lab report or notebook.

TECHNIQUE

CAUTION: Wear safety goggles during this investigation. Students should be careful when handling any solutions they use to swab the skin, especially on the face.

Part A: Heat and Sweat Gland Activity

1. Either you or your lab partner will be the experimental subject and the other will be the experimenter. Ask your partner to name an area on the skin where sweating occurs rapidly when he or she gets nervous or too warm (the forehead near the hairline is a good place). Also ask your partner to name a place on the skin where sweating does not occur rapidly (such as the back of the hand or the inside surface of the forearm). Record these locations in the **Records** section of your report.

2. **CAUTION: The alcohol in alcohol pads is flammable; keep away from excessive heat and flame. Avoid eye contact.** Using an alcohol pad, wipe a 3 in. square area on each location. **CAUTION: Soap with iodine can stain clothing. Avoid eye contact. In case of eye contact, call your teacher and flush eye with water for 15 minutes.** Using the soap with iodine and a cotton swab, carefully cover a 2 in. square in each of those areas. Blot them dry. After they have dried, gently hold a 2 in. square of the bond paper on each area for 30 seconds.

3. Take the paper from the skin and observe it closely for any blue dots. The bond paper contains starch, which reacts with the iodine to form a blue-black color.

4. Find the area of the bond paper with the greatest concentration of dots and mark off a 1 cm square. In the **Records** section of your report, make a table similar to

Preparation

Each pair of students will need 6 pre-moistened alcohol pads, 6 cotton swabs, a liquid antibacterial soap (with iodine), six 2 in. squares of erasable bond paper, a towel, a metric ruler, and a heat source lamp. Students should be careful when using the alcohol and iodine solutions. Goggles must be worn throughout the investigation. In case of eye contact with the solutions, direct students to flush the affected eye with water for 15 minutes and to notify you. Iodine can stain clothing. Do not allow students to use sun lamps. Caution students to avoid burns by keeping the lamps at least 6 in. from skin surfaces and avoiding extended exposure of skin to the lamp's heat. Students should shield exposed skin areas with a towel.

Procedural Notes

- This investigation should take 40 minutes to complete.
- The purpose of the alcohol is to remove skin oils so that the sweat glands can be mapped.
- **Safety:** Caution students not to get the alcohol or iodine soap in their eyes. If they do, have them flush their eyes with water, and then consult a physician. Safety goggles should be worn when applying both types of chemicals to the skin. Remind students that iodine stains.
- Since some students may be embarrassed to exercise in front of their peers, you may wish to ask for volunteers to serve as participants for Part B. Volunteers should not have health conditions that preclude vigorous exercise.

Background Answers

1. Homeostasis is the maintenance of fairly constant internal conditions by organisms.

2. If blood temperature changes, messages are sent to the hypothalamus, which then triggers a reaction, such as shivering or sweating, to bring the body temperature back to normal.

3. Mammals and birds are endothermic; they maintain a high, fairly constant internal body temperature. Reptiles and amphibians are ecto-therms; their body temperature varies with that of their surroundings.

4. temperature-sensitive nerve endings in the skin

5. What are the effects of heat and exercise on sweat gland activity?

the one below. Record the number of dots in your table under the column labeled "No stimulus."

Sweat Gland Activity

Number of dots

	No stimulus	After heat	After exercise
Forehead			
Back of hand			

5. Wipe the areas of the skin with a fresh alcohol pad and reswab them with iodine using a fresh cotton swab. Cover those areas with a towel or other cloth to prevent heat from being directed onto bare skin.

6. After you turn on the lamp, your lab partner should tell you the instant the heat is actually felt. DO NOT bring the lamp closer than 6 in. from the skin. DO NOT touch the lamp. Gently hold a 2 in. square of bond paper on each area, asking the subject to help hold it in place. From that point, time the heat application for 10 minutes. Record the resulting number of dots in the column labeled "After heat" in your table in the **Records** section of your report.

Part B: Exercise and Sweat Gland Activity

7. Repeat steps 5 and 6 with exercise rather than heat as the variable. Have the subject jog in place or do jumping jacks, push-ups, sit-ups, or another form of moderate exercise for 10 minutes, allowing brief rest breaks as needed. Record the number of dots under the column labeled "After exercise" in your table in the **Records** section. In your

report, also briefly summarize the procedure you followed.

8. Clean up your materials and wash treated areas with soap and water before leaving the lab.

INQUIRY

1. Where do you hypothesize that the greatest number of sweat glands are located: on the back of the hand or forearm or on the forehead?

2. What do the dots on the bond paper represent?

3. Were there more dots on the paper before or after the heat exposure?

ANALYSIS

1. What does sweating accomplish for the body?

2. How is sweating an adaptation that maintains homeostasis?

3. As your subject reacted to heat and exercise, what changes were observed in the sweat glands?

4. How do the sweat glands fit into the analogy of the body as a human thermostat?

FURTHER INQUIRY

Write a new question that could be explored as a new investigation. The following is an example:

Will the placement of cold compresses near a skin area reduce the activity of sweat glands there?

Inquiry Answers

1. Answers will vary.
2. The dots show the location of active sweat glands.
3. Answers should reflect an increase in sweat gland activity after heat exposure and exercise.

Analysis Answers

1. It cools the body.
2. It helps return body temperature to normal.
3. Their activity increased.
4. When body temperature exceeds the temperature at which the "thermostat" is set, sweat gland activity increases, perspiration evaporates from the body, and body temperature decreases toward the normal level.

Records

Answers will vary. Students should see increased sweat gland activity after exposure to heat and after exercise.

Sweat Gland Activity

Number of dots

	No stimulus	After heat	After exercise
Forehead			
Back of hand			

UNIT 9

SKELETON, MUSCLES, AND SKIN

Block Scheduling Guide

Each block represents about 45–50 minutes of instructional time. The resources cited in a block are coded to a specific program component using the Key on the next page.

Lecture Concepts	Lesson Resources

36-1 The Skeletal System pp. 855–860

BLOCK 1

a. Compact bone consists of concentric rings deposited around a Haversian canal. Spongy bone is porous and is partially filled with red marrow.
b. Blood vessels run through the peristeum and Haversian canals and supply bone cells with blood and oxygen.
c. Osteoporosis is a loss of bone density that causes bones to become brittle and break easily.
d. The axial skeleton is made up of the skull, spine, and rib cage. The appendicular skeleton consists of the bones of the arms and legs and of the pectoral and pelvic girdles.

Lecture Resources
- Opening Questions, page 855
- Overcoming Misconceptions, page 855
- Demonstrations: Bone Structure, Freely Movable Joints
- Visual Strategies: Figures 36-2, 36-3
- Teaching Transparencies: 185, 186, 198, 184A
- Holt Biology Videodiscs: Lesson 36-1

Classwork Options
- A Skeleton Game, page 858
- A Skeleton Puzzle, page 858
- Bone Diseases, page 859
- Quick Labs A26: Vertebrate Skeletons
- Quick Labs A27: Comparing Skeletal Joints

Assignment Options
- Section Review, page 860
- Chapter Review, questions 1–3, 11, 12, 18, 21, 22

36-2 The Muscular System pp. 861–865

BLOCK 2

a. Skeletal muscles consist of bundles of fibers that contain bundles of smaller fibers called myofibrils. Muscles contract when myosin and actin fibers arranged in repeating units called sarcomeres slide together and shorten the sarcomeres along a myofibril.
b. Exercise can increase the efficiency of muscles by increasing the body's ability to supply oxygen to its muscle cells. It can also increase the size and strength of muscles.

Lecture Resources
- Opening Questions, page 861
- Muscle Contraction, page 861
- Demonstration: A Chicken's Wing
- Visual Strategies: Figures 36-7, 36-8
- Overcoming Misconceptions, page 862
- The Role of ATP, page 863
- Research Update: Muscle Fatigue, page 864
- Teaching Transparencies: 190, 192, 193

- Holt Biology Videodiscs: Lesson 36-2

Classwork Options
- Muscle Diseases, page 865
- Inquiry Skills Development C41: Evaluating Muscle Exhaustion

Assignment Options
- Section Review, page 865
- Chapter Review, questions 4–7, 13–15, 19, 23

36-3 The Skin pp. 866–869

BLOCK 3

a. Human skin, the largest organ of the body, consists of three layers: the epidermis, dermis, and subcutaneous layer.
b. The skin and its specialized structures help protect the body from disease and help maintain the body's homeostasis.
c. Carcinomas and melanomas are skin cancers, which result from mutations in skin cells.

Lecture Resources
- Opening Question, page 866
- Demonstration: Examining Skin
- Straight, Wavy, or Curly, page 867
- Overcoming Misconceptions, page 867
- Visual Strategies: Figures 36-12, 36-13
- Teaching Transparencies: 188
- Holt Biology Videodiscs: Lesson 36-3

Classwork Options
- Effective Reading, page 866

- Regulating Body Temperature, page 868
- Acne Advice, page 869
- Closure, page 869
- Laboratory Investigation: Muscle Contraction, pages 872–873
- Inquiry Skills Development B30: Touch Receptors in the Skin

Assignment Options
- Section Review, page 869
- Chapter Review, questions 8–10, 16, 17, 20

Review/Enrichment

BLOCK 4

- ■ Study Guide: Concept Review and Pretest

Assignment Options
- Teaching Resources CD-ROM: Occupational Applications Worksheets, Medical Sonographer and Physical Therapist
- ■ Activities and Projects, questions 24–26

Assessment Options

BLOCK 5

Traditional Assessment
- ■ Chapter 36 Test
- Test Generator/Assessment Item Listing: Software item bank for preparing customized chapter tests.

Portfolio Assessment
Select student reports for one or more laboratory experiments from this chapter. *The Direct Observation Checklist* on the *BioSources Teaching Resources CD-ROM* should be completed during a laboratory or other cooperative learning experience.

Holt Biology Videodiscs

Holt Biology Videodiscs gives you a powerful tool for teaching, review, and assessment. *Concepts of Biology* adds interactive lesson plans and extensive tools for customization using CD-ROM technology.

CONCEPTS OF BIOLOGY

Use the following topic from *Concepts of Biology* to help you teach this chapter:

- ■ Topic 21: Body Systems

For further information regarding the media on the videodiscs, see *Holt Biology Videodiscs Teacher's Correlation Guide to Biology: Principles and Explorations.*

CHAPTER 36 — SKELETON, MUSCLES, AND SKIN

Chapter Themes

Homeostasis *Any classification system is limited by artificial divisions made for convenience. By the nature of homeostasis, all of the systems work as an integrated whole. The skeletal system, muscular system, and integumentary system share the common functions of protection, support, and movement; they also interact with other systems of the body to maintain homeostasis.*

Levels of Organization *The organization of skeletal muscle tissue enables the contraction of muscle cells and the overall contraction of a muscle. Movement depends on the proper functioning of the entire system of muscles, including the heart (cardiac) muscle, which pumps needed oxygen to cells.*

Tapping Prior Knowledge

- What is the advantage of an endo-skeleton over an exoskeleton?
- What is the major function of the integumentary system (the skin)?

Opening Demonstration

Bring skeletons of a human and another vertebrate to class for students to study. These can be purchased from a biological supply company or borrowed from a local university. Ask students to compare the skeletons. *(Students should see similarities among bones of the vertebral column and appendages. They should find differences in the size and shape of many bones, including the skull and mandible.)* Discuss how the parts of the skeletons provide movement and protection by pointing out the range of motion of the appendages and the placement of organs.

A hurdler expending energy

> ## REVIEW
> - ATP as an energy source (Section 4-3)
> - cellular respiration (Section 5-3)
> - endoskeletons and exoskeletons (Section 27-3)
> - bone, cartilage, and collagen (Section 35-2)
> - actin, myosin, and muscle tissue (Section 35-2)
> - epithelial, glandular, and nerve tissue (Section 35-2)
> - *Highlights: Skeletal System (p. 841)*
> - *Highlights: Muscular and Integumentary Systems (p. 842)*

Authors' Rationale

The skeletal, muscular, and integumentary systems provide the body with protection, support, and a means of movement. Although students know what bones, muscles, and skin are, they probably do not know how they work or how they interact with other systems. This chapter presents the structure of bones, the skeleton, skeletal muscles, and skin. Students will learn how bones and muscles create body movement as well as about disorders of the skeletal, muscular, and integumentary systems.

36-1 The Skeletal System

An internal framework of calcium-hardened bones shapes and supports the human body. This endo-skeleton is one of the hallmarks of the vertebrate body. All of the skeletons discussed in previous chapters have a serious limitation. Chitin, a carbohydrate found in arthropod exoskeletons, is brittle and is not very strong. Calcium alone is hard and strong but is also brittle and likely to fracture. The calcium-hardened bone endoskeleton of vertebrates, however, has enabled them to become the most mobile animals. The human skeleton provides protection for internal organs and acts as a versatile system of levers and joints. The body's muscles pull against these levers and joints, enabling the arms and legs to move.

Section Objectives

- Describe the structure of a bone.
- Differentiate between the axial skeleton and the appendicular skeleton.
- Name the three main types of joints, and explain how an example of each works.
- Discuss causes and effects of osteoporosis, and explain how it can be prevented.

SECTION 36-1

Vocabulary Preview

compact bone
spongy bone
red marrow
yellow marrow
periosteum
osteoblast
Haversian canal
osteocyte
osteoporosis
axial skeleton
appendicular skeleton
cranium
spine
sternum
rib cage
pectoral girdle
pelvic girdle
joint
immovable joint
slightly movable joint
freely movable joint
rheumatoid arthritis

Opening Question

Ask students what their bones are made of. Students may suggest calcium, bone marrow, blood vessels, etc. Emphasize that bone is living tissue made of cells and minerals. Point out the parts of the bone (compact bone, spongy bone, yellow marrow, red marrow, and periosteum) shown in Figure 36-1.

The Human Skeleton Is Mainly Bone

Your skeleton is made mostly of bone, a dynamic living tissue that is constantly formed and replaced as long as you live. Bone has a great advantage as a structural material. Its calcium-hardened collagen fibers make it strong but flexible, unlike the brittle chitin of the arthropod exoskeleton. As Figure 36-1 shows, bone tissue occurs in two forms. **Compact bone** is a dense, almost solid tissue that provides a great deal of support. **Spongy bone,**

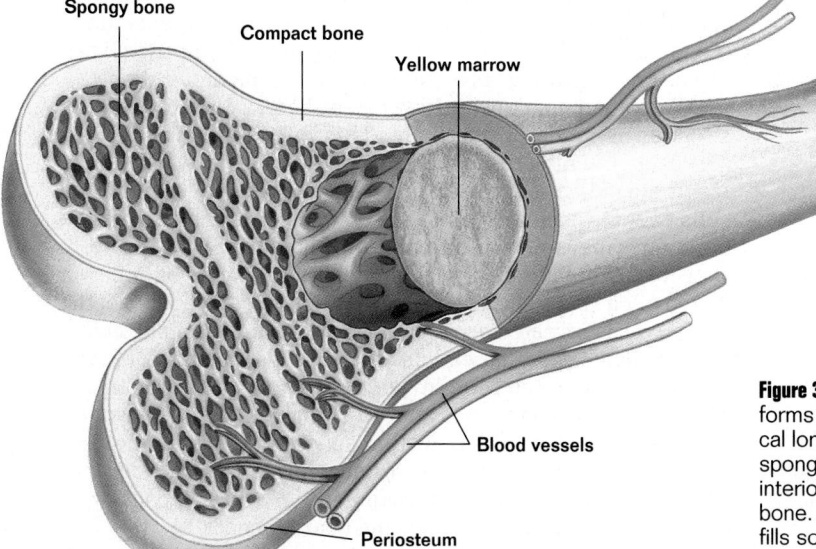

Spongy bone
Compact bone
Yellow marrow
Red marrow
Blood vessels
Periosteum

Figure 36-1 Dense compact bone forms most of the shaft of a typical long bone, while porous spongy bone forms most of the interior portion at the ends of the bone. A soft tissue called marrow fills some spaces in bone tissue.

Overcoming Misconceptions

Dry as a Bone

Many students think of bone as a nonliving, rocklike substance because their experience with bones is with skeletal remains. Explain that bones are living tissue. Remind students that a bone can repair itself after a fracture because cells and blood vessels regenerate in bone. Inorganic mineral salts, such as calcium phosphate, calcium hydroxide, and calcium carbonate, enable bones to maintain their shape and size for centuries after an animal's death.

VISUAL STRATEGY

Figure 36-2
Use this figure to correlate the scanning electron micrograph of compact bone with the cross sections of compact bone. Point out the Haversian canal(s) in each picture. Ask students to visualize the layers of osteoblasts, bone, and osteocytes surrounding the Haversian canal shown in the micrograph.

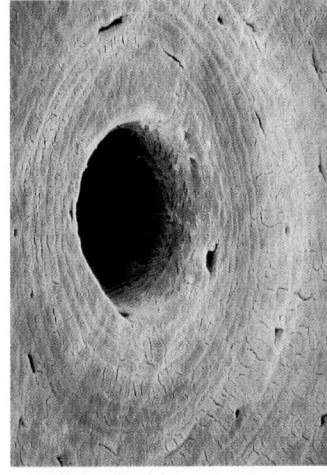

Figure 36-2 A scanning electron micrograph (SEM) of a section of compact bone from which all organic tissue was removed, *above,* shows the concentric rings of bone that surround a Haversian canal. The diagrams, *right,* show the structure of compact and spongy bone, including the many living cells and blood vessels that are an important part of bone tissue.

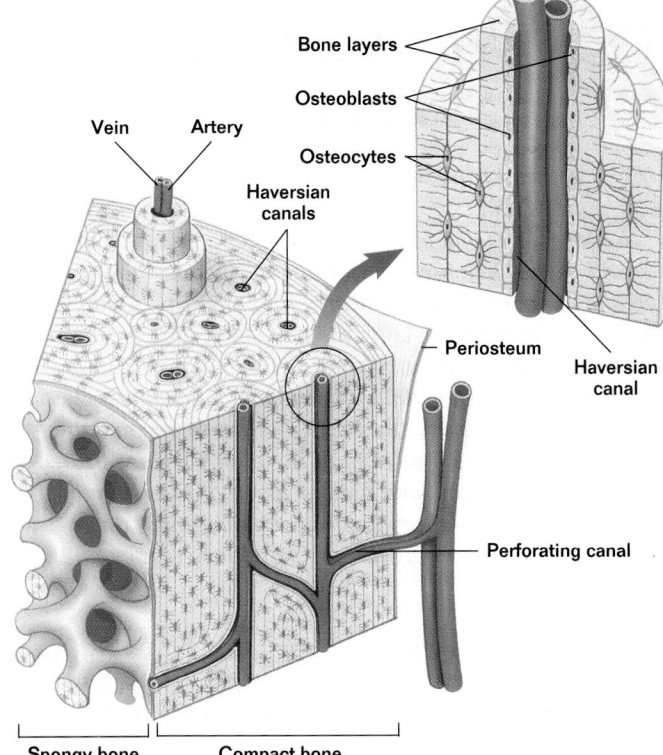

CAPSULE SUMMARY

Compact bone consists of concentric rings deposited around a Haversian canal. Red marrow fills some spaces in the porous spongy bone. Blood vessels that run through the periosteum and Haversian canals supply bone cells with oxygen and nourishment.

which provides lightweight support, consists of hardened fibers interspersed with many spaces. Some of these spaces are filled with a soft tissue called **red marrow,** which produces red blood cells in some bones. The hollow interiors of the arm and leg bones are filled with **yellow marrow,** which stores fat. Individual bones are surrounded by a membrane, the **periosteum** *(pehr ee AHS tee uhm),* which contains many blood vessels. The word *periosteum* comes from the Greek words *peri,* meaning "around," and *osteon,* meaning "bone."

When the human body first takes shape as an embryo, the skeleton is made of cartilage. This cartilaginous skeleton serves as a template for bone formation. As you grow, bone forms along lines of stress. New bone is formed by young bone cells called **osteoblasts** *(AHS tee oh blasts),* which secrete the collagen fibers on which calcium phosphate is deposited. Osteoblasts build bone by laying down thin layers of collagen on top of one another, like layers of paint. In compact bone, the first layer is secreted around a narrow, hollow channel, called a **Haversian canal,** that extends the length of the bone. As you can see in Figure 36-2, additional layers are laid down, creating a series of concentric tubes of bone. Eventually, young bone cells become trapped within spaces in the bone they lay down, and they are then called **osteocytes** *(AHS tee oh seyets).* The blood vessels that run through each Haversian canal supply the osteocytes with oxygen and nourishment. ❑

Did You Know?
Bones have the tensile strength of cast iron but only one-third of its weight. They can resist 25,000 lb./sq. in. of compression and 15,000 lb./sq. in. of tension.

Osteoporosis Results From Bone Loss

Bone tissue, which also serves as a storehouse for minerals needed by the body, is continuously broken down and replaced. In babies, new bone is added more rapidly than it is broken down, and bones grow larger and denser. In young adults, bone tissue is broken down and replaced at the same rate, so the density of bone usually remains relatively constant. However, during middle age, bone replacement gradually becomes less efficient. As a result, bones become less dense and store fewer minerals. When bone loss is severe, as seen in Figure 36-3, the condition is called **osteoporosis** (*ahst ee oh puh ROH sihs*), which means "porous bone." Bones that are affected by osteoporosis become brittle and are easily broken. In the United States, more than 600,000 fractures result from osteoporosis each year.

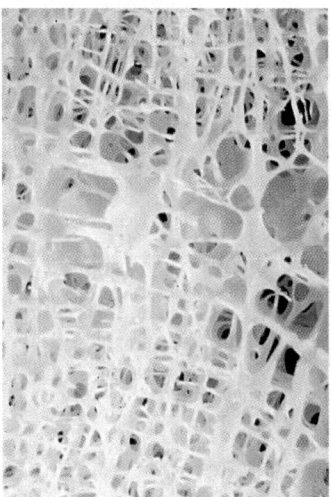 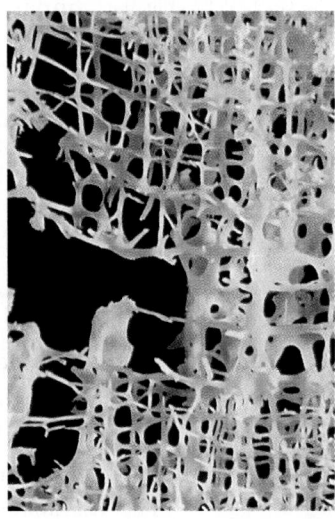 

Figure 36-3 In the photos of bone tissue, *below left*, compare the density of bone tissue in a normal bone, *left*, with that in a bone weakened by osteoporosis, *right*. An active lifestyle, *below*, and regular exercise help prevent or delay osteoporosis.

Although both women and men lose bone as they age, many more women are affected by osteoporosis. Because their bones are smaller, women cannot afford to lose as much bone tissue as men. Women also lose a lot of calcium each month during the menstrual cycle. Moreover, the production of sex hormones, which help maintain bone density, declines rapidly during menopause, resulting in an increased rate of bone loss in women.

To avoid the consequences of osteoporosis, you must act at an early age. Bone density can be increased only during the teens and twenties with regular exercise and a calcium-rich diet that includes dairy products such as milk and yogurt. Although regular exercise throughout your life will help maintain bone density, strengthening your bones now will make you less likely to be affected by osteoporosis later in life.

☐ CAPSULE SUMMARY

Osteoporosis, which affects many more women than men, is a loss of bone density that causes bones to become brittle and to break easily. Osteoporosis can be delayed or prevented by a calcium-rich diet and regular exercise starting at an early age.

? Did You Know?

As bone grows, its shape changes. A long bone does not grow from its center in the direction of its ends. Instead, a long bone lengthens through the growth of plates near its ends. This process requires extensive remodeling of the bone's shape as it grows.

Teaching Tips
A Skeleton Game

Display a model, or a large picture, of a human skeleton. Make a game of learning the scientific names of the major bones. Starting with the cranium, pronounce the name of the bone, then have the first student in a row repeat it as you point to the cranium. Naming the mandible, have the next student in the row repeat "mandible" and "cranium" as you point to the bones. Go down the rows, naming a new bone with each student. As an extension of the activity, reward students who can name each bone on their own.

A Skeleton Puzzle

Make copies of a full-page picture of the human skeleton, and give one copy to each student. Tell students to carefully cut out the bones of the skeleton with a pair of scissors. (You may have students group some of the bones together, such as the phalanges, metacarpals, and carpals.) Ask students to take a blank sheet of paper and piece the skeleton back together, gluing each bone in place. Finally, ask students to label the major bones on their completed puzzles.

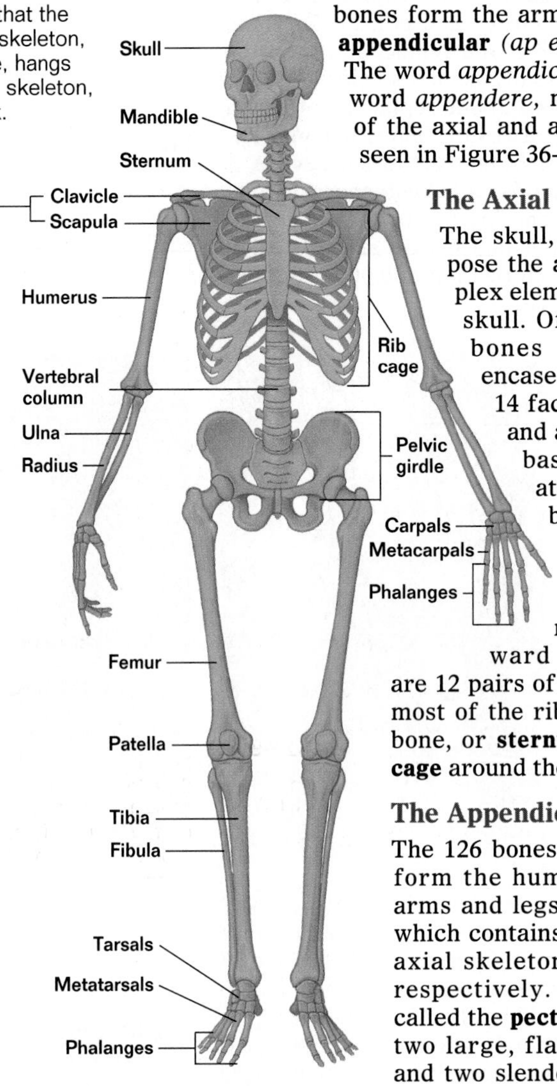

Figure 36-4 Some major bones of the human skeleton are identified here. Notice that the appendicular skeleton, shown in blue, hangs from the axial skeleton, shown in pink.

Skull — Mandible — Sternum — Pectoral girdle — Clavicle — Scapula — Humerus — Vertebral column — Ulna — Radius — Rib cage — Pelvic girdle — Carpals — Metacarpals — Phalanges — Femur — Patella — Tibia — Fibula — Tarsals — Metatarsals — Phalanges

The Human Skeleton Has Two Main Parts

The human skeleton is made up of 206 individual, calcium-hardened bones. Of these, 80 bones form the main body axis and are called the **axial skeleton.** The other 126 bones form the arms and legs and are called the **appendicular** (ap ehn DIHK yoo luhr) **skeleton.** The word *appendicular* is derived from the Latin word *appendere,* meaning "to hang." The bones of the axial and appendicular skeletons can be seen in Figure 36-4.

The Axial Skeleton

The skull, backbone, and rib cage compose the axial skeleton. The most complex element of the axial skeleton is the skull. Of the 29 bones in the skull, 8 bones form the **cranium,** which encases the brain. The skull also has 14 facial bones, 6 middle-ear bones, and a single bone that supports the base of the tongue. The skull is attached to the top of the backbone, or **spine,** which is a flexible stack of 26 vertebrae. The spinal cord passes through all but the lowermost vertebrae. Curving forward from the middle vertebrae are 12 pairs of ribs. At the front of the body, most of the ribs are attached to the breastbone, or **sternum,** forming a protective **rib cage** around the heart and lungs.

The Appendicular Skeleton

The 126 bones of the appendicular skeleton form the human body's appendages—the arms and legs. The arms and legs, each of which contains 30 bones, are attached to the axial skeleton at the shoulders and hips, respectively. The shoulder attachment is called the **pectoral girdle** and is composed of two large, flat shoulder blades (scapulas) and two slender, curved collarbones (clavicles). The collarbones, which connect the shoulder blades to the sternum, hold the shoulders apart. This arrangement permits the full rotation of the arms about the shoulder joints. When you fall on an outstretched arm, much of the force is transmitted to the collarbones—the most frequently broken bones in the human body. The hip attachment is called the **pelvic girdle** and is composed of two large pelvic bones that form a bowl. The pelvic bones transmit the weight of the body squarely down the legs. ❑

❑ CAPSULE SUMMARY

The skull, spine, and rib cage make up the axial skeleton. The appendicular skeleton consists of the bones of the arms and legs and of the pectoral and pelvic girdles.

Did You Know?

The fibula, a sticklike bone found in the lower leg, has been used as a source of bone tissue for bone grafting because this bone does not bear weight in humans and is a nonessential bone. Its main function is to stabilize the ankle. Portions of the fibula and the blood vessels feeding it can be removed and used to replace large bone segments elsewhere in the body.

Joints Fasten Bones Together

The junction of two bones is called a **joint.** Most joints permit movement of the bones they join. These movements are aided by the structure of the joint. For example, pads of cartilage cushion the ends of both bones in a joint. The axial and appendicular skeletons contain three different kinds of joints that enable varying degrees of movement. Examples of the three basic types of joints can be seen in Figure 36-5.

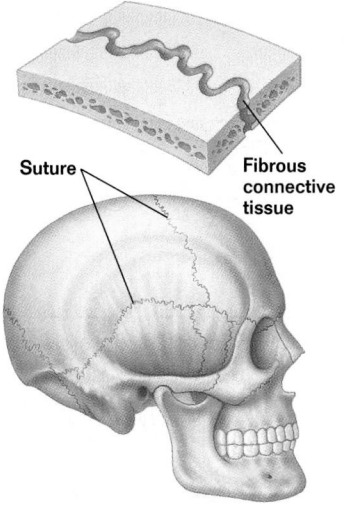

Suture
Fibrous connective tissue

Immovable joint

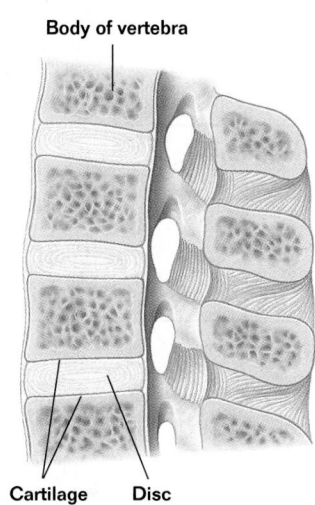

Body of vertebra

Cartilage   Disc

Slightly movable joint

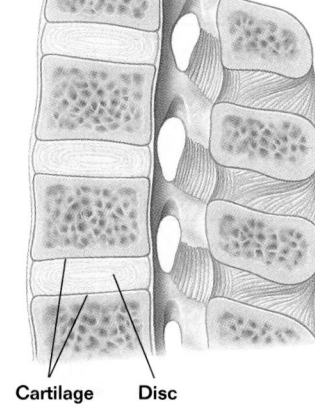

Cartilage

Membrane

Lubricating fluid

Freely movable joint

Figure 36-5 Individual bones in the skull are joined by sutures, *left,* a type of immovable joint. The vertebrae in the spine are joined by cartilaginous joints, *center,* a type of slightly movable joint. The knee, *right,* is an example of a hinge joint, a type of freely movable joint.

1. **Immovable joints** Very tight joints that permit little or no movement of the bones they join are called **immovable joints.** The cranial bones of the skull are joined by sutures, a type of immovable joint in which the bones have only a thin layer of connective tissue between them.

2. **Slightly movable joints** Joints that permit limited movement of the bones they join are called **slightly movable joints.** The vertebrae in the spine are joined by cartilaginous joints, a type of slightly movable joint in which a bridge of cartilage joins the two bones. The pads of cartilage, or disks, that separate the vertebrae also act as shock absorbers.

3. **Freely movable joints** The joints that permit the greatest degree of movement are called **freely movable joints.** Recall that ligaments hold the bones of a freely movable joint together and that a membrane surrounding the joint contains a lubricating fluid that separates the ends of the bones. **Rheumatoid arthritis,** a very painful degeneration of movable joints, occurs when cells of the immune system attack these membranes, weaken the cartilage, and deposit bone in its place. The hinge joints in elbows and

Multicultural Perspective
Bone China

Bone china is a form of porcelain made from burned animal bones. Bone ash is mixed with kaolin, a white clay found in Jiangxi Province, China, and petuntse, a type of feldspar found only in China. The bone ash increases the porcelain's translucence. Bone china is not as hard as other forms of porcelain, but it is more durable. Porcelain was first made in China during the Tang dynasty (618–907). For centuries Chinese porcelain was considered the finest in the world. It was not until the 1700s that Europe began producing its own porcelain. Around 1750, the English found a new way of making porcelain—with bone ash. Today England produces most of the world's bone china.

knees are perhaps the most familiar freely movable joints. Table 36-1 lists some other freely movable joints found in the appendicular skeleton.

Table 36-1 Types of Freely Movable Joints

Name	Type of Movement	Examples
Ball-and-socket joint		Shoulders and hips
Pivot joint		Elbows
Plane joint		Carpals of hands and tarsals of feet
Saddle joint		Thumbs

Section Review

1. *How does compact bone differ from spongy bone?*
2. *Where are the axial and appendicular skeletons attached to each other?*
3. *Which type of freely movable joint allows the widest range of motion? Give an example of this kind of joint.*
4. *What are two reasons that women are more likely than men to develop osteoporosis?*

Critical Thinking
5. *The bones of a newborn baby are made mostly of cartilage. Why is this is an advantage?*

36-2 The Muscular System

W hile smooth muscle lines your organs and blood vessels, your body is composed largely of skeletal muscle that is devoted to moving the parts of your body. If you have ever tried to lift a heavy object, you are familiar with a basic problem of movement—force must be applied to overcome the forces that tend to hold the object in place. By splitting ATP molecules into ADP and inorganic phosphate, muscle cells utilize the body's store of chemical energy to contract (shorten their length). Thus, muscle cells exert force by pulling on surrounding tissue. When a lot of muscle cells contract at one time, they are able to exert a great deal of force.

Section Objectives

■ Describe the action of flexors and extensors in moving the parts of the body.

■ Explain how muscle contractions are produced.

■ Discuss the two factors that determine the strength of a muscle contraction.

■ Discuss the importance of exercise in increasing endurance and maintaining muscle strength.

Muscles Make the Skeleton Move

For muscles to move the parts of the body, they must be attached to something they can pull against. The bones of the skeleton provide these points of attachment. Most skeletal muscles are attached to bones by strips of dense connective tissue called **tendons.** One end of the muscle, the **origin,** is attached to a bone that remains stationary during a muscle contraction. This gives the muscle something to pull against. The other end of the muscle, the **insertion,** is attached to a bone that moves when the muscle contracts. A muscle's insertion always moves towards its origin.

Skeletal muscles are attached to the bones of the appendicular skeleton in opposing pairs called flexors and extensors. As Figure 36-6 illustrates, **flexors** cause the limbs to

Figure 36-6 A flexor and an extensor work together to enable you to bend your arm. Contraction of the biceps muscle, a flexor, bends the arm at the elbow. Contraction of the triceps muscle, an extensor, straightens the arm.

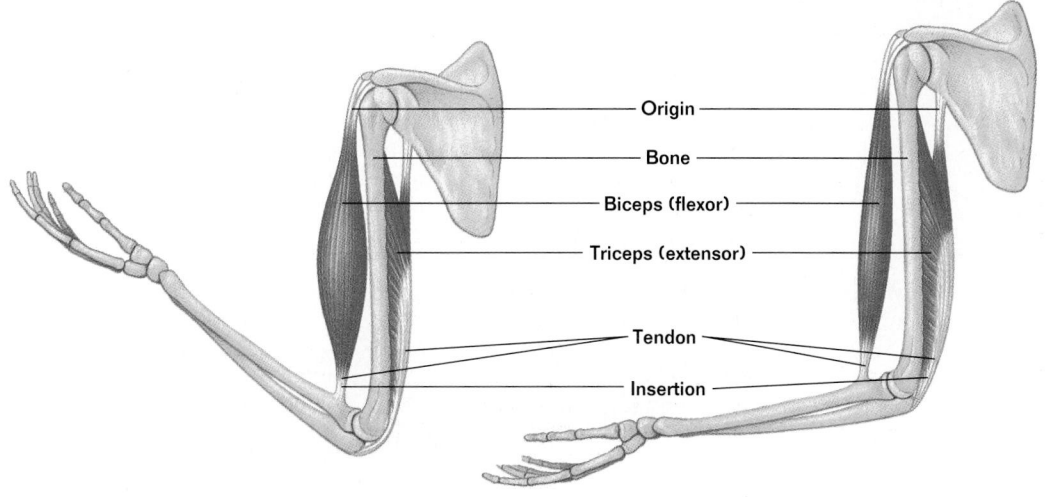

- Origin
- Bone
- Biceps (flexor)
- Triceps (extensor)
- Tendon
- Insertion

❓ Did You Know?

Because ATP plays such an important role in muscle contraction, active muscles use larger amounts of ATP than resting muscles or other tissues. In fact, an exercising muscle can use as much as 200 times more ATP than the same muscle at rest.

SECTION 36-2

Vocabulary Preview

tendon
origin
insertion
flexor
extensor
myofibril
Z line
sarcomere
sliding filament theory
anabolic steroid
muscle strain
sprain
tendinitis

Opening Questions

Ask students to describe the basic mechanism of ATP production and to list the products, reactants, and location of the process. Students should recall that ATP is a product of cellular respiration, which takes place in mitochondria. Glucose and oxygen are reactants; carbon dioxide and ATP are the major products. Ask students which organelle they would expect to find in great numbers in muscle cells and why. Lead students to realize that muscle cells contain many mitochondria because they produce ATP needed for muscle contraction.

Teaching Tip
Muscle Contraction

Have students imagine a racing shell (boat) with a full crew on a broad, straight river suitable for rowing races. To maintain the shell's position in the river, the crew must continually row; otherwise, the shell will be carried downstream. Simply having the oars in the water will not maintain the shell's position. Similarly, a muscle that has contracted to its full extent must use ATP to maintain its contraction.

Demonstration

A Chicken's Wing

Caution: Raw chicken may carry *Salmonella*. Soak in 70 percent ethanol and refrigerate overnight. Wear disposable gloves while handling raw chicken and keep hands and chicken away from your face during the demonstration. Wash your hands and any materials that come into contact with the chicken or its juices with soap and hot water after handling it. Show students a fresh, raw chicken wing from a local grocery store. Show the muscles in the upper and lower parts of the wing. Ask students to pay attention to where and how the muscles are attached to the wing. Bend the wing, simulating muscle movement.

Teaching Tip

What If both the extensor muscle and the flexor muscle in an antagonistic pair were to contract with the same force at the same time? In such a case, the limb would not move, and the joint would be frozen.

👁 VISUAL STRATEGY

Figure 36-7

Use this figure to identify the structure of a skeletal muscle. Point out the muscle fibers, motor neuron, myofibrils, sarcomere, Z line, actin, and myosin. Make sure that students understand the separate levels of organization shown in the figure. Help students understand that muscles comprise many bundles of muscle fibers, that a muscle fiber contains many myofibrils, and that a myofibril consists of actin and myosin. Remind students that muscle contractions begin at the lowest level—with actin and myosin.

Figure 36-7 Bundles of muscle fibers (muscle cells) make up a skeletal muscle. Each muscle fiber is a bundle of many smaller fibers (myofibrils) that are bundles of protein (myosin and actin) fibers. Muscle contractions begin with the movements of myosin and actin fibers within sarcomeres.

bend at a joint, while **extensors** cause them to straighten. Thus, one muscle in a pair pulls a bone in one direction, and the other pulls it back. When you move, the command to contract goes from the nervous system to one muscle or the other, but not to both muscles at the same time. If both muscles were commanded to contract, they would simply pull against each other, and the limb would not move. Thus, to produce movement, muscle contraction and relaxation are carefully coordinated and controlled by the nervous system. ❏

Myosin and Actin Cause Muscle Fibers to Shorten

The structure of a muscle, such as that of the skeletal muscle seen in Figure 36-7, enables it to contract. Recall that skeletal muscles are made of many muscle fibers, which are strands of muscle cells joined end to end. Within each muscle fiber are many bundles of smaller

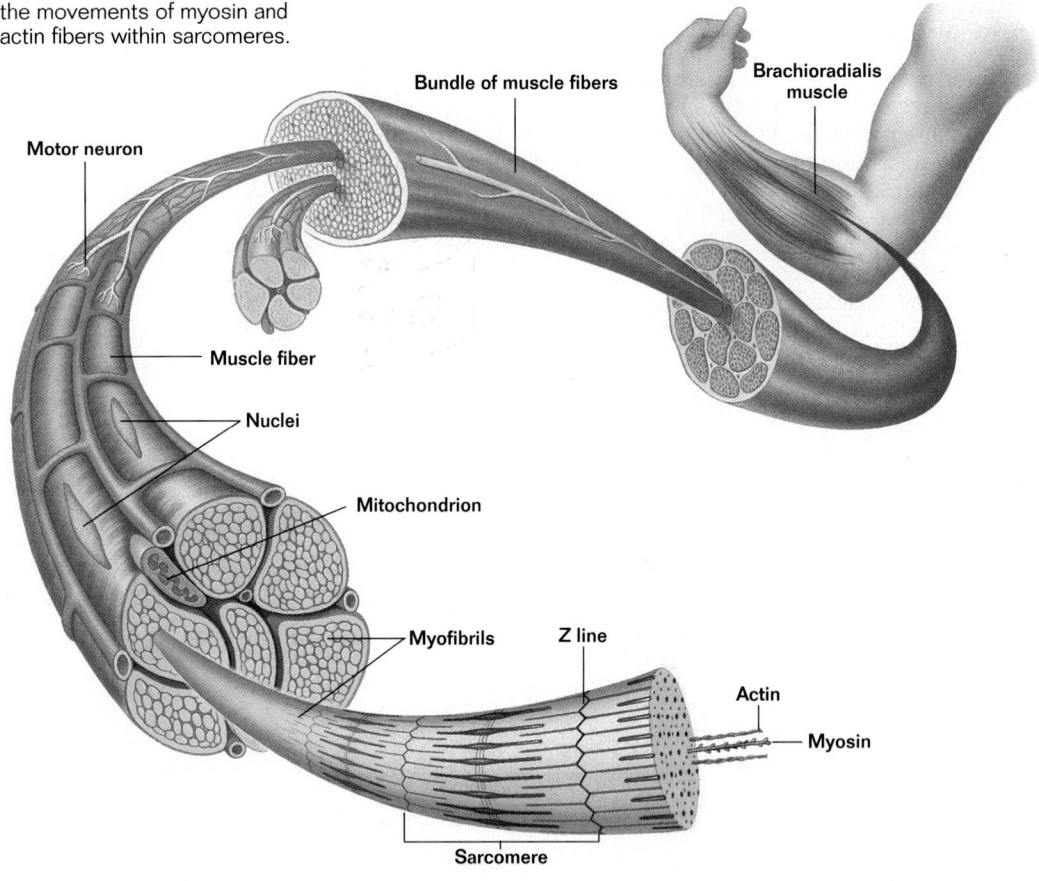

Bundle of muscle fibers

Brachioradialis muscle

Motor neuron

Muscle fiber

Nuclei

Mitochondrion

Myofibrils

Z line

Actin

Myosin

Sarcomere

Overcoming Misconceptions

Fine Motor Control

Students may believe that only one muscle contracts to produce a specific movement. However, both muscles in a flexor-extensor pair commonly contract with different degrees of force during the movement of a joint. The slight contraction of the antagonist muscle during movement helps keep the muscle exerting the greater force from overshooting the desired position of the joint. Other muscles not directly involved with the desired movement also contract in order to stabilize the joint, thus contributing to what is commonly called fine motor control.

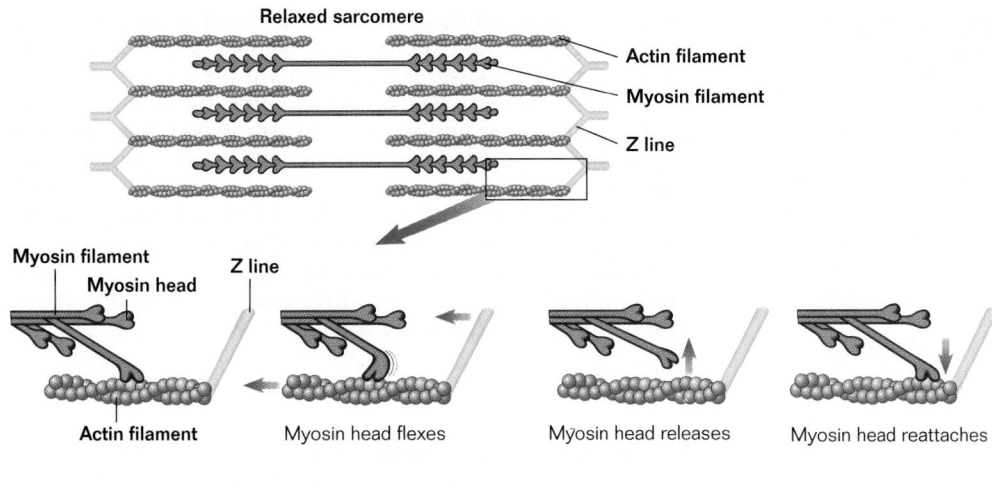

Relaxed sarcomere

Actin filament
Myosin filament
Z line

Myosin filament
Myosin head
Z line
Actin filament

Myosin head flexes

Myosin head releases

Myosin head reattaches

Contracted sarcomere

fibers called **myofibrils,** which in turn contain many filaments of the proteins myosin and actin. Notice that each myosin filament consists of a cluster of myosin molecules, each of which looks like a long rod with a large "head" at one end. The thinner actin filaments resemble two strings of beads twisted together. One end of each actin filament is anchored at a **Z line,** which is the location of a protein that crosses a myofibril. Myosin and actin filaments lie parallel to one another and are arranged in **sarcomeres** (SAHR koh mihrz), repeating units that are bounded by two Z lines. The characteristic light and dark bands of striated muscle result from the orderly and slightly overlapping arrangement of the myosin and actin filaments in the sarcomeres.

Sarcomeres are the functional units of contraction in a myofibril. In a relaxed muscle fiber, the ends of the myosin and actin filaments in each sarcomere barely overlap. When a muscle fiber contracts, adjacent myosin and actin filaments slide along one another so that they overlap a great deal and shorten each sarcomere. This explanation of muscle contraction is called the **sliding filament theory.**

How do actin filaments slide along myosin filaments, and how does this lead to the contraction of a muscle? As illustrated in Figure 36-8, myosin and actin filaments are so close to each other that the myosin heads can touch the adjacent actin filaments. Muscles contract when the myosin heads attach to the adjacent actin filaments and bend inward, pulling the actin filaments along with them. Each myosin head then releases the actin filament, "recocks" (flexes back

Figure 36-8 According to the sliding filament theory, muscle contraction occurs when myosin fibers within a sarcomere "walk" along adjacent actin fibers. This movement pulls the Z lines (to which actin is attached) of a sarcomere together, shortening the sarcomere. The shortening of the sarcomeres along a myofibril causes the myofibril to contract.

Historical Note

The Sliding Filament Theory

The English molecular biologist Hugh Huxley (1924–) studied the changing widths of the bands in sarcomeres using X-ray diffraction and electron microscopy. On the basis of his observations, Huxley proposed a mechanism for muscle contraction that explained the conversion of chemical energy (ATP) to mechanical energy, known as the sliding filament theory. Huxley's theory was the first formulation of the molecular mechanism for muscle contraction and is accepted by most scientists today.

SECTION 36-2

👁 VISUAL STRATEGY

Figure 36-8
Use this figure to trace the events underlying muscle contraction: attachment of myosin heads to actin, bending of myosin heads, release of bonds between myosin and actin, and recocking of myosin heads for the next cycle. Emphasize that when a muscle cell contracts it becomes shorter as its sarcomeres and the distance between Z lines shorten.

CONNECTIONS

Chapter 4
ATP Supplies Energy
ATP, which is produced by cellular respiration, is used during the cocking of myosin heads. As a myosin head cocks and prepares to attach to actin, one molecule of ATP is split. The hydrolysis (splitting) of ATP releases the energy needed to detach a myosin head from an actin filament at the end of the next cycle.

Teaching Tip
The Role of ATP
Ask students to draw a diagram to illustrate the role of ATP in the sliding filament theory. Diagrams should show each step in the cycle, including (1) attachment of myosin head to actin, (2) pulling of actin by myosin head, (3) detachment of myosin head from actin, and (4) recocking of myosin head for the next cycle. Ask students to write brief explanations to accompany their diagrams. Myosin uses the energy to attach to and pull actin in steps 1 and 2 and to detach from actin in step 3. ATP is split into ADP and inorganic phosphate (P_i) as the myosin head recocks and prepares for the next cycle in step 4.

☑ RESEARCH Update

Muscle Fatigue

Everyone has experienced muscle fatigue at some point. Muscle physiologists consider fatigue to be the inability to maintain force in a sustained contraction (you can't get the bag of groceries all the way to the house) or the inability to sustain repeated muscle contractions of the same force (your arms give out when you're doing push-ups). Although scientists know that muscles don't work without calcium or ATP, they don't know what causes muscle fatigue. In light of modern ailments like chronic fatigue syndrome, understanding the causes of muscle fatigue has become increasingly important.

John Lee of Sheffield University in England recently reviewed several muscle fatigue hypotheses. One group of scientists claims that something happens in the muscle during fatigue; another group says that something happens in the brain, which tells the muscle to slow or stop its contraction. Physiologically, an unequal distribution of calcium, an insufficient supply of ATP, or chemical byproducts of muscular contraction (such as lactic acid) could interfere with the attachment of myosin heads to actin filaments. Behavioral scientists have begun to focus on proteins that seem to cause lethargy (acute phase proteins) as a cause of muscle fatigue, because running on a treadmill may fatigue more than a person's muscles. Moreover, the relationship between fatigue and illness has prompted many researchers to think that a problem with the hypothalamus may cause chronic fatigue syndrome. ☑

❑ CAPSULE SUMMARY

Skeletal muscles consist of bundles of fibers that contain bundles of smaller fibers called myofibrils. Muscles contract when myosin and actin fibers arranged in repeating units called sarcomeres slide together and shorten the sarcomeres along a myofibril.

to where it was), and reattaches to it, ready to bend and pull again. Each flex and recock uses a molecule of ATP. With each flex of the myosin heads, the actin filaments slide along the myosin filaments. The actin filaments drag their attached Z lines with them, pulling the Z lines closer together and shortening the sarcomeres. When stimulated to contract, this shortening of the sarcomeres occurs along the entire length of a muscle fiber. The contraction of an individual muscle fiber is an all-or-nothing response. Thus, the total amount of force that a muscle can exert depends on two factors: how often individual muscle fibers are stimulated to contract and how many muscle fibers contract in a given muscle. ❑

Exercise Increases Muscle Size and Efficiency

The ATP used to fuel muscle contraction is usually supplied by the aerobic pathway of cellular respiration. However, in rapidly contracting muscles, the oxygen supply soon becomes inadequate, and anaerobic processes take over. In the absence of oxygen, fermentation follows glycolysis, and the muscle has only glycolysis as a source of ATP. The lactic acid produced during fermentation causes muscle fatigue (the more acidic conditions lower the activity of glycolytic enzymes and interfere with the action of the myosin heads). Because the production of ATP by glycolysis is so inefficient, the body begins using the glycogen stored in muscles as a source of glucose. When the glycogen is depleted, the body must begin using fat as its only source of energy. However, energy production from fat occurs at only about half of the rate of energy production

Figure 36-9 Aerobic exercises, such as jogging, *below,* improve your body's ability to supply oxygen to its muscles and increase the efficiency of ATP production. Resistance exercises, such as chin-ups, *below right,* increase muscle size and strength.

Historical Note

When Is ATP Expended?

ATP is the energy currency of all cells, including muscle cells. Scientists might assume that ATP is consumed at the moment muscles contract since movement requires energy. A key discovery made by the British physiologist and biophysicist Archibald V. Hill helped biologists determine the sequence of events. Hill discovered that muscle cells use oxygen after, not during, contraction. Since cellular respiration uses oxygen to produce ATP, ATP cannot cause contraction directly. Instead, ATP prepares the muscle for the next contraction. In 1922 Hill received a Nobel Prize in medicine and physiology for his work on muscles. It was left to Hugh Huxley to formulate a theory that explained the role of ATP in cocking myosin heads prior to the contraction of the sarcomere.

from glucose, so the depletion of glycogen stores is marked by a substantial decrease in muscle performance. Long-distance runners refer to this as "hitting the wall."

With aerobic training and exercise, chest muscles can be strengthened so that more air (and oxygen) enters the body with each breath. Such exercises also make the heart pump more efficiently and increase the number of blood vessels in the muscles. These changes expand your body's ability to supply oxygen to its muscles and thus make ATP production more efficient. The increase in muscle efficiency results in greater endurance (ability to continue exercising).

Although exercise cannot increase the number of muscle cells, resistance exercises such as the chin-ups shown in Figure 36-9 can increase muscle size and strength. However, because these exercises do not significantly improve your body's ability to deliver oxygen to your muscles, they do not increase your endurance. Some athletes are tempted to use drugs called **anabolic steroids** to increase the size (and thus the strength) of their muscles. Anabolic steroids are powerful synthetic chemicals that resemble the male sex hormone testosterone and that trick the muscles into growing larger. Unfortunately, their use has many serious, often irreversible side effects, including cancer, heart disease, and altered sexual development. Anabolic steroids are dangerous drugs, and their use is illegal in both amateur and professional sports.

Overusing your muscles by exercising more than you usually do can lead to muscle injury. A **muscle strain,** commonly called a "pulled muscle," is the overstretching or tearing of a muscle. Muscle strain occurs when a muscle is overused or when strenuous exercise is done without warm-up exercises. A **sprain** is a torn or overly stretched muscle, ligament, or tendon. If excessive stress causes the tendons that attach the muscles to bone to become inflamed, a painful condition called **tendinitis** results. ❑

CONTENT LINK
Steroids are described in greater detail in Chapter 41.

❑ CAPSULE SUMMARY
Exercise can increase the efficiency of your muscles by increasing your body's ability to supply oxygen to its muscle cells. Exercise can also increase the size and strength of your muscles.

Application
R͟x͟ **Medicine** Muscles require ATP to break connections between myosin and actin. When the supply of ATP is interrupted at death, muscles stiffen into rigor mortis within three to four hours. Peak rigidity occurs at 12 hours, gradually slackening over the next 48 to 60 hours due to decomposition of muscle tissue. Coroners and police examiners use these facts to estimate a murder victim's time of death.

Teaching Tip
Muscle Diseases
Have students prepare newspaper-style articles about diseases or injuries that affect muscles. Diseases include tetanus, Cushing's syndrome, muscular dystrophy, myasthenia gravis, poliomyelitis, McArdle's disease, etc. Articles should cover causes, symptoms, and treatments of the condition. Post the articles around the room, or have students present their work to the class.

Section Review

1. *How do flexors and extensors work together to move the limbs of the body?*
2. *How are the Z lines of a myofibril pulled closer together?*
3. *How can the force that a muscle exerts be increased?*
4. *What type of exercise helps to increase muscle size and strength?*

Critical Thinking
5. *After looking back at Figure 5-16 on page 111, explain why an increase in muscle efficiency increases your endurance.*

Answers to Section Review

1. They oppose each other in order to move a joint in opposite directions. Flexors cause bending of a joint and extensors cause straightening of a joint.

2. Z lines of a myofibril are pulled closer as adjacent myosin and actin filaments slide along one another and shorten each sarcomere.

3. The force that a muscle exerts can be increased by increasing either how often or how many muscle fibers are stimulated in a given muscle.

4. Resistance exercises, such as chin-ups, increase muscle size and strength.

5. Increased muscle efficiency, which is the result of aerobic exercise that makes more oxygen available to muscles, enables muscles to produce more ATP for each glucose molecule consumed and thus supplies the energy the muscle needs to work for a longer period of time without rest.

Opening Question

Ask students what the largest single organ of the body is. Tell them it's the skin. Then have students list the functions of skin. Lists should include that skin protects and cushions the body, acts as a sense organ, eliminates body wastes through sweat, controls body temperature, and maintains body fluids inside the body.

Demonstration

Examining Skin

Give students hand lenses to examine their skin. As students make their observations, have them answer the following questions: How does skin look? (*rough and leathery*) What structures are in skin? (*pores and hairs*) What does their skin feel like? (*Answers will vary. Skin may feel smooth and soft, dry and leathery, hard and calloused, etc.*)

Section Objectives

- Name the three layers of the skin, and describe the main functions of each.
- Explain how the epidermis, hair, and nails are formed.
- Explain how the dermis helps regulate body temperature.
- Identify the causes of skin cancer and acne, and describe how you can minimize your chances of getting skin cancer and how you can minimize the symptoms of acne.

36-3 The Skin

The largest organ of the human body—accounting for about 15 percent of your total body weight—is the skin. Many kinds of specialized structures are found in the skin, which forms the integumentary system. Among the stratified epithelial cells in 1 sq. cm of your skin (an area about the size of a dime) are about 200 nerve endings, 10 hairs and muscles, 100 sweat glands, 15 oil glands, 3 blood vessels, 12 heat sensors, 2 cold sensors, and 25 pressure sensors.

The Skin Is Composed of Three Layers

Skin is the outermost layer of the human body. It protects the body from injury, provides the first line of defense against disease-causing microbes, helps regulate body temperature, and prevents the body from drying out through evaporation. Human skin is a very complex organ that is composed of three layers: an outer epidermis, a lower dermis, and an underlying layer of subcutaneous tissue. The structure of the skin is seen in Figure 36-10.

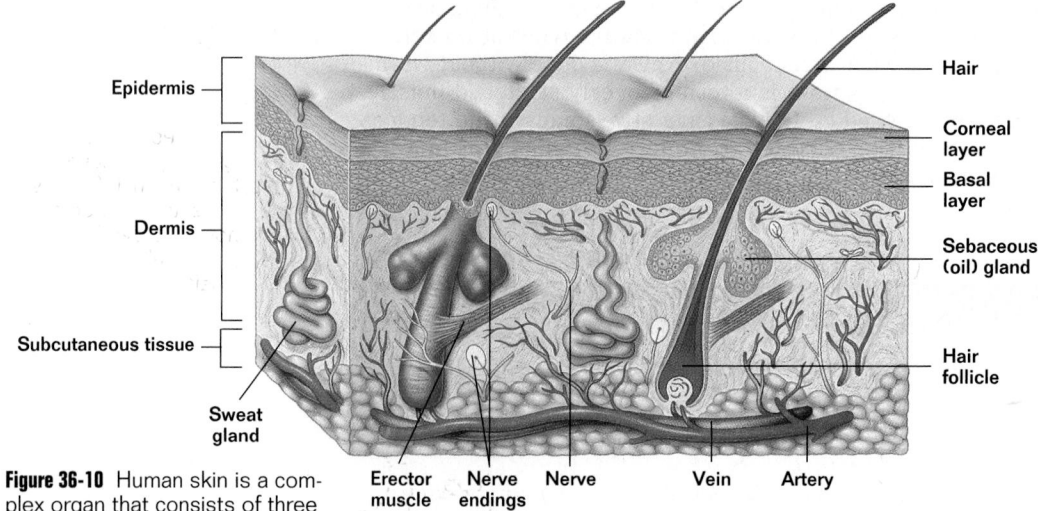

Figure 36-10 Human skin is a complex organ that consists of three layers as well as many blood vessels, nerve endings, muscles, hairs, and glands.

Epidermis

The **epidermis,** which is from 10 to 30 cells thick, is the outermost layer of the skin. About as thick as this page, the epidermis consists of several layers of stratified epithelial cells. The part of the epidermis that you see when you look in a mirror is the **corneal layer,** a thin layer of flattened,

Effective Reading

Personal Hygiene

High school students are particularly conscious of their personal appearance. Ask students to formulate an answer to the following question as they read this section: How will information in this section enable you to take better care of your skin? Then have students share their answers in small group discussions.

mostly dead cells that are filled with the protein keratin, an excellent waterproofing agent. The cells of the corneal layer are continuously damaged by encounters with the outside world. They are scraped, ripped, worn away by friction, and dried out by moisture loss. Your body deals with this damage not by repairing the cells but by replacing them.

The outermost cells of the epidermis are continuously shed and replaced by cells from the **basal layer,** a layer of actively dividing cells that lies at the base of the epidermis. As new skin cells form, they migrate upward and produce large amounts of keratin, the protein that makes the skin tough. After reaching the surface, a skin cell normally lives in the corneal layer for about a month before it is shed. The basal layer also contains **melanocytes** (muh LAN uh seyets), which are cells that produce the brown pigment melanin (MEHL uh nihn). Human skin color results from melanin, which provides the skin with protection from the sun's ultraviolet rays. The more these cells are exposed to the sun, the more melanin they produce. That is why skin "tans."

Hair and nails are also produced by the cells of the epidermis. Specialized epidermal structures called **hair follicles** produce individual hairs. As Figure 36-11 shows, hair is composed of dead, keratin-filled cells stacked on top of one another like a pile of roof shingles. Each hair on your head grows for several years; then the follicle enters a resting phase for several months, and the hair is shed. Nails are produced by specialized epidermal cells located in the light "half-moon" area at their base. These cells fill with tough keratin as they are pushed outward by the production of new cells.

Dermis

The **dermis** is the chief framework of the skin. It is 15 to 40 times thicker than the epidermis. The leather used to make belts and shoes is made from very thick animal dermis. The dermis serves as a structural support and as a matrix for the many nerve endings, blood vessels, and specialized cells of the skin. For example, the human sense of touch originates with the nerve endings in the dermis. These nerve endings, which are sometimes coupled to simple sensory receptors, enable you to sense pressure, pain, and temperature.

The dermis has tiny muscles that are attached to the hair follicles in your skin. When you are cold or afraid, these muscles contract and pull the hairs upright. Similar muscles cause a cat's fur to stand up when it is frightened (making it look bigger and more dangerous) or cold (trapping air around its body to insulate it from heat loss). In humans, the muscles that pull hairs upright also cause goose bumps.

The dermis is crisscrossed by a network of blood vessels that provide nourishment to the living cells of the skin. These blood vessels also help regulate body temperature either by radiating heat into the air or by helping to insulate the body. If your body gets too hot, muscles around the skin's blood vessels relax. When the muscles relax, the blood vessels enlarge

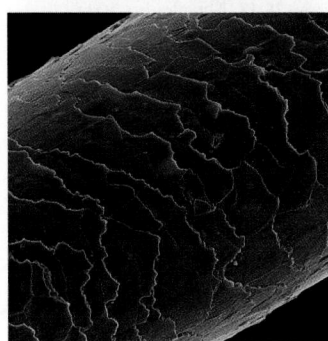

Figure 36-11 Human hair, *top,* grows from hair follicles in the skin. Hair color results from melanin pigments produced by melanocytes in the hair follicles. A scanning electron micrograph, *bottom,* shows the overlapping, flattened epidermal cells that cover the shaft of a hair.

UNIT 9

CHAPTER 36

 VISUAL STRATEGY

Figure 36-12
Compare the capillaries of the skin's surface to the radiator of an automobile. The blood carries heat from the body's core to the surface where it radiates to the environment. This function is enhanced by the evaporation of sweat from the skin's surface.

Teaching Tip
Regulating Body Temperature

Ask students to draw a sequential diagram that shows how skin helps regulate body temperature. The *Graphic Organizer* at the bottom of this page is one example. Have students include a brief paragraph that describes a situation in which events outlined in their diagrams might occur.

Application

Anthropology In ancient Greece, a pale complexion was an indication that one did not have to work in the fields. Until the recent discovery that overexposure to sun caused skin cancer, many people spent hours "baking" in the sun trying to achieve a "perfect tan" because a tan gave the impression that one had plenty of leisure time to relax in the sun. Today the American Cancer Society encourages people to wear protective sunscreens outdoors. Thus, suntans are a reflection of cultural and medical values.

Figure 36-12 The flushed (reddened) skin of this cyclist is a sign that her body is trying to cool itself.

CAPSULE SUMMARY

Human skin, the largest organ of the body, consists of three layers: the epidermis, dermis, and subcutaneous layer. The skin and its specialized structures help protect the body from disease and help maintain the body's homeostasis.

---

so that more blood flows near the skin's surface and more heat is dissipated. This is why the skin of people with light complexions may become reddish during strenuous exercise, as seen in Figure 36-12. If your body gets too cold, the muscles around the the skin's blood vessels contract, making the blood vessels constrict. This constriction keeps the blood deeper in the skin and helps insulate the body to reduce heat loss.

The dermis has another way to remove excess body heat. It contains about 100 sweat glands per square centimeter. The evaporation of sweat from the surface of your skin removes heat much more efficiently than by dissipation through the skin from the blood. Most sweat is about 99 percent water and 1 percent dissolved salts and acids. Certain sweat glands located in body areas with dense hair, such as the armpits, also secrete proteins and fatty acids. Because these substances provide a rich food source for bacteria, stale sweat often has the rank odor of bacterial waste products.

Subcutaneous Tissue

The **subcutaneous tissue** is a layer of fat-rich cells lying just beneath the dermis. These cells act as shock absorbers, provide additional insulation to conserve body heat, and store energy and fat-soluble vitamins. Different parts of the body have very different thicknesses of subcutaneous tissue—from the eyelids, which have none, to the buttocks and thighs, which may have a lot. The pads of subcutaneous tissue in the soles of your feet may be one-fourth of an inch thick or more. ❑

Skin Disorders Are Common

The skin is the most exposed part of the body and is therefore continually exposed to damaging agents such as insects, microorganisms, and sunlight. Injuries such as insect bites, scrapes, and blisters are often minor and usually heal rapidly without permanent scarring. Burns, however, may be very serious and may result in permanent scarring or even death. Some skin disorders are the result of changes that occur within your body.

Skin cancer may result from mutations caused by years of exposure to the sun's ultraviolet rays. The most common types of skin cancer are **carcinomas,** which originate in the non-pigment-producing cells of the epidermis. If detected early, carcinomas have a very high cure rate. About 1 percent of skin cancers, however, result from mutations that occur in the pigment-producing melanocytes. These cancers, called **malignant melanomas,** grow very fast and spread easily to other parts of the body. People with malignant melanomas have a very low survival rate. A carcinoma and a melanoma can be seen in Figure 36-13. The best way to minimize your risk of skin cancer is to avoid overexposure to sunlight.

Graphic Organizer

Use this graphic organizer with *Teaching Tip:* Regulating Body Temperature on this page.

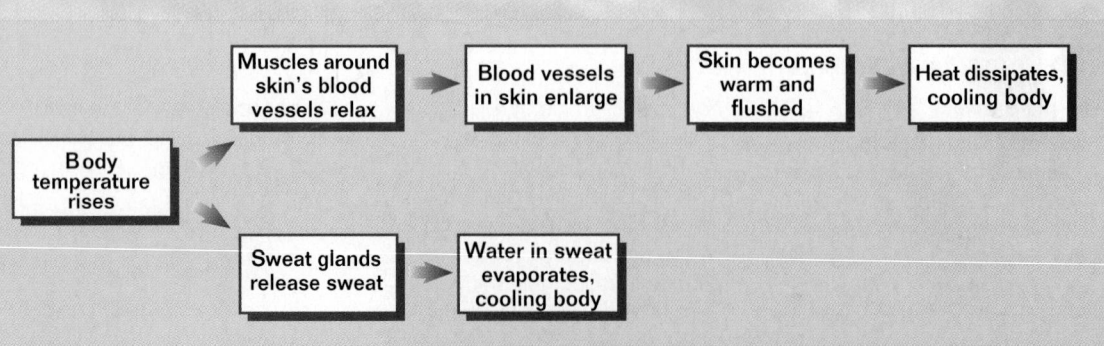

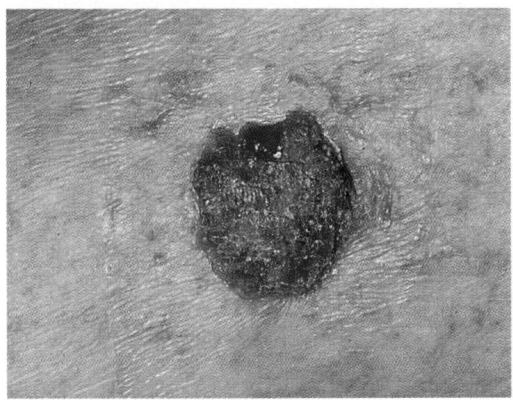
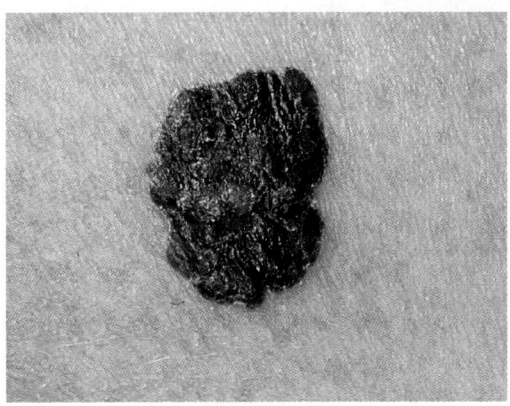

In many teens, high levels of sex hormones produced during adolescence increase the activity of the skin's oil glands. Like sweat glands, oil glands are **exocrine glands,** which release their products through ducts. If the oil ducts become clogged, **acne** results. In the first stage of acne, the buildup of oil in a clogged oil duct causes a swelling called a whitehead. When exposed to air, a whitehead becomes a blackhead. If the oil buildup continues until the gland bursts, the surrounding skin becomes red and inflamed. The swollen pimple that forms is the result of your body's attempts to fight both the tissue damage and the bacteria that infect the damaged area. Although acne cannot be prevented, it can usually be controlled with proper skin care.

Psoriasis (*suh REYE uh sihs*), which affects about 4 million Americans, is an inherited skin disorder in which new epidermal cells are produced rapidly in the basal layer and reach the skin's surface every three or four days—about eight times faster than normal. The cause of this rapid cell production is still unknown. Though the raised red patches and scaly white skin caused by psoriasis may look unpleasant, the health of most people with this condition is not affected. ☐

Figure 36-13 In its early stages, a carcinoma such as a squamous cell carcinoma, *above left*, may look like a wart. If what you think is merely a wart grows in size and begins to bleed, you should consult a doctor because it could be a carcinoma. A malignant melanoma, *above*, often resembles a mole. If you have a mole that changes in size, shape, or color, you should immediately consult a doctor because it could be a malignant melanoma.

☐ CAPSULE SUMMARY

Carcinomas and melanomas are skin cancers, which result from mutations in skin cells. Acne is a common skin condition that results from clogged oil ducts.

Use this figure to emphasize the necessity of detecting skin cancer through self-examination. Although skin carcinomas are slow growing, inattention to them can lead to serious problems and even death. Even though people with malignant melanomas have a low survival rate, early detection improves the chances for successful treatment.

Teaching Tip
Acne Advice

Tell students that a recent article in a fashion magazine recommends a way to avoid acne. Their advice: request an antibiotic such as tetracycline or erythromycin from your physician and apply it every day regardless of whether blemishes appear. Is this good advice? Ask students to write a brief essay explaining their viewpoints. *(Essays should weigh some of the following factors: antibiotic resistance, alternative acne treatments, severity of acne, etc.)*

Chapter Closure

Ask students to make an instruction manual for the three systems covered in this chapter. Manuals should have one chapter for each of the systems: skeletal, muscular, and integumentary. Each chapter can be divided into two sections: structure and maintenance. The sections on structure should include labeled diagrams of the structures of skeleton and bones, muscles, and skin. The sections on maintenance should present ways of taking care of bones, muscles, and skin.

Section Review

1. *What is the function of the skin's subcutaneous layer?*
2. *Where does hair form, and how does it grow in length?*
3. *Describe the two main ways that the dermis helps regulate body temperature.*
4. *What is the most common cause of skin cancer, and how can skin cancer best be prevented?*

Critical Thinking

5. *Why is a third-degree burn, which destroys the epidermis and dermis of the skin, such a serious injury?*

Answers to Section Review

1. The skin's subcutaneous layer acts as a shock absorber, provides insulation to conserve body heat, and stores energy and fat-soluble vitamins.

2. Hair forms in follicles in the dermis. It grows by stacking dead, keratinized cells on top of one another.

3. Muscles around the blood vessels in the dermis regulate body temperature by relaxing to enhance blood flow and enable heat to radiate from the blood into the air or by contracting to constrict blood flow and thus help to insulate the body. Dermis also contains sweat glands that dissipate heat through the secretion and evaporation of sweat.

4. Mutations that result from exposure to the sun's ultraviolet rays are the greatest cause of skin cancer. Avoiding the sun is the best prevention method.

5. Serious burns to these layers can result in dehydration, inability to help cool or heat the body, loss of protection from disease-causing agents, and even death.

36 CHAPTER REVIEW

CHAPTER REVIEW ANSWERS

Each item in the Chapter Review is correlated by text section in the assignment guide that follows.

ASSIGNMENT GUIDE

Section	Review	Themes Review	Critical Thinking
36-1	1–3, 11, 12	18	21, 22
36-2	4–7, 13–15	19	23
36-3	8–10, 16, 17	20	

Review
Multiple Choice
Code in parentheses indicates section number and objective number.

1. d (36-1)
2. c (36-1.2)
3. b (36-1.2)
4. c (36-2.1)
5. a (36-2.2)
6. b (36-2.3)
7. c (36-2.4)
8. b (36-3.1)
9. a (36-3.3)
10. d (36-3.4)

Completion

11. osteocytes, Haversian canal (36-1.1)
12. immovable, rheumatoid arthritis (36-1.3)
13. Flexors, extensors (36-2.1)
14. muscle fibers, contract (36-2.2)
15. Resistance, aerobic (36-2.4)
16. corneal, basal, hair, nails (36-3.1)
17. malignant melanoma, Carcinomas (36-3.4)

Vocabulary

acne (869)	Haversian canal (856)	red marrow (856)
anabolic steroid (865)	immovable joint (859)	rheumatoid arthritis (859)
appendicular skeleton (858)	insertion (861)	rib cage (858)
axial skeleton (858)	joint (859)	sarcomere (863)
basal layer (867)	malignant melanoma (868)	sliding filament theory (863)
carcinoma (868)	melanocyte (867)	slightly movable joint (859)
compact bone (856)	muscle strain (865)	spine (858)
corneal layer (866)	myofibril (863)	spongy bone (855)
cranium (858)	origin (861)	sprain (865)
dermis (867)	osteoblast (856)	sternum (858)
epidermis (866)	osteocyte (856)	subcutaneous tissue (868)
exocrine gland (869)	osteoporosis (857)	tendinitis (865)
extensor (862)	pectoral girdle (858)	tendon (861)
flexor (861)	pelvic girdle (858)	yellow marrow (856)
freely movable joint (859)	periosteum (856)	Z line (863)
hair follicle (867)	psoriasis (869)	

Review

Multiple Choice

1. Which of the following is *not* a role of the skeletal system?
   a. support
   b. protection
   c. movement
   d. homeostasis

2. The outer membrane of the bone that contains blood vessels is called the
   a. marrow.
   b. spongy bone.
   c. periosteum.
   d. Haversian canal.

3. The bones of the arms and legs form the
   a. axial skeleton.
   b. appendicular skeleton.
   c. compact bone system.
   d. spongy bone system.

4. The end of a muscle that does not move during muscle contraction is called the
   a. tendon.
   b. insertion.
   c. origin.
   d. flexor.

5. Synchronized shortening of sarcomeres along the length of a muscle fiber causes
   a. muscles to contract.
   b. Z lines to move apart.
   c. muscles to relax.
   d. myofibrils to slide.

6. The factor that affects the amount of force exerted when a muscle contracts is the
   a. mass of the object being moved.
   b. rate of the muscle fiber contractions.
   c. distance between the Z lines.
   d. length of the sarcomeres.

7. Resistance exercises
   a. decrease endurance.
   b. decrease steroid uptake.
   c. increase muscle size.
   d. increase muscle cell number.

8. Which one of the following layers of skin provides structural support?
   a. epidermis
   b. dermis
   c. subcutaneous
   d. corneal

9. The dermis helps to regulate body temperature by producing
   a. sweat.
   b. goose bumps.
   c. oil.
   d. melanocytes.

10. The risk of developing skin cancer is greatly increased by
    a. eating oily foods.
    b. using suntan lotion.
    c. exercising.
    d. sunbathing.

Completion

11. Living bone cells called _____ get trapped between concentric layers of bone that surround a tube called a(n) _____ .

12. The sutures of your skull are _____ joints. A painful condition that affects movable joints is called _____ .

13. _____ are muscles that cause limbs to bend at joints, and _____ are muscles that cause limbs to straighten.

14. The strength of a muscle contraction depends on the number of _____ that contract and on how often they _____ .

15. _____ exercises increase strength, and _____ exercises increase endurance.

16. Cells of the _____ layer of the epidermis are filled with keratin. These cells are constantly shed and replaced by cells from the _____ layer, which also makes cells for the _____ and _____ .

17. A _____ is a very serious type of skin cancer that results from mutations in pigment-producing cells. _____ are the most common type of skin cancer.

Themes Review

18. **Structure and Function** Give an example of a ball-and-socket joint and a hinge joint. What type of movement is permitted by each of these joints?

19. **Levels of Organization** How are the actions of myosin and actin fibers translated into the contractions of muscles attached to the bones of the arm?

20. **Evolution** How do the tiny muscles in the dermis benefit dogs and cats during cold weather? What vestige (sign) of this benefit is observed in humans?

Critical Thinking

21. **Communicating Effectively** Write a letter about the consequences of osteoporosis to a mature female friend or relative, and persuade her to start a diet and exercise program to delay osteoporosis.

22. **Making Inferences** Look at the X rays below of two different hands. Which is an adult's hand? Which is a child's hand? Explain your decision.

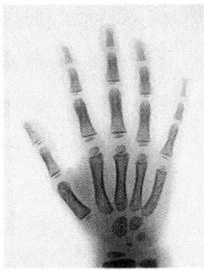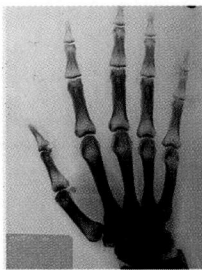

23. **Making Predictions** Leg muscles have a rich supply of blood and energy stored in the form of fat and glycogen. In contrast, arm muscles have a smaller supply of blood and energy, which comes from stored glycogen and blood sugar. Which muscles will tire more quickly? Explain.

Activities and Projects

24. **Cooperative Group Project** Visit a gym that has exercise machines. Test several of the machines, and talk to exercise consultants to determine which muscles are worked by each machine. Document the visit with photographs or a videotape. Identify other activities that work the same muscles as the machines tested. Present your findings to your class.

25. **Multicultural Perspective** Research the incidence of skin cancer in several cultures from different parts of the world. Look for differences and similarities in skin cancer type, prevalence, and occurrence in males and females among several ethnic groups. Relate your findings in a written or oral report.

26. **Research and Writing** Research the link between Accutane® and birth defects. Find out how its effectiveness compares with that of other prescription and over-the-counter acne treatments. Relate your findings in a written report.

Critical Thinking

21. Answers will vary. Students should mention that osteoporosis is a loss of bone that occurs with age, particularly in women. It causes bones to become brittle and to break easily. A calcium-rich diet and low-impact exercises help to retard bone loss. (36-1.4)

22. The X ray on the right with the most dark areas (which indicate bone) is of an adult's hand. The X ray on the left with the most light areas (which indicate cartilage) is of a child's hand. The human skeleton is mostly cartilage at birth. Much of this cartilage is later replaced by bone. (36-1.1)

23. Muscles tire when energy stored in glycogen, fat, and sugar is used up. Arm muscles will tire more quickly than leg muscles because leg muscles receive more blood, which contains blood sugar, and they store energy in glycogen and fat. (36-2.3)

Activities and Projects

24. Encourage students to contact a gym or health club a week or two before visiting and ask permission to videotape or take photographs in the club. Other activities may include walking, swimming, playing tennis, dancing, etc.

25. The incidence and type of skin cancer varies throughout the world. The most current global statistics and information about skin cancer are available from the American Cancer Society.

26. Accutane® dermatological preparation is considered by many medical specialists to be the most effective treatment for acne available; however, its use has been linked to lethal birth defects. For this reason, women who want to use Accutane® are required to take pregnancy tests before, during, and after treatment.

Themes Review

18. The hip is a ball-and-socket joint, and the knee is a hinge joint. The ball-and-socket joint permits movement in all directions, while the hinge joint permits only back-and-forth movement. (36-1.3)

19. Muscles contract when myosin heads attach to adjacent actin filaments and bend inward, pulling the actin filaments with them and shortening the sarcomeres, which in turn causes muscle fibers to shorten and muscles to contract. (36-2.2)

20. The muscles of the dermis cause hairs in the fur of dogs and cats to rise, thus insulating the animals from cold. When the hairs of humans stand up, humans get goose bumps instead of getting fluffier because they have fewer hairs than dogs and cats. (36-3.3)

LABORATORY Investigation Chapter 36

Muscle Contraction

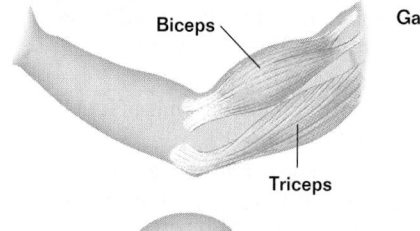

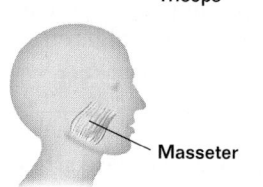

Biceps
Gastrocnemius
Triceps
Masseter

OBJECTIVES

- Demonstrate muscle tone.
- Determine muscle fatigue.

PROCESS SKILLS

- measuring changes in muscle size
- organizing data in tables

MATERIALS

- tape measure
- 5–7 lb. weight
- used tennis ball or hollow rubber bouncing ball
- stopwatch or second hand

BACKGROUND

1. What are the two protein filaments that are abundant in muscle cells and allow them to contract?
2. What is the structure of skeletal muscle, and how does it function?
3. How do smooth and cardiac muscle differ from skeletal muscle?
4. Write your own question to explore in your lab report or notebook.

TECHNIQUE

Part A: Muscle Tone

1. Each pair of students in your class will take turns performing the following tasks. In each case, note the degree of biceps and triceps contraction by feeling each of these muscles to check for hardening and/or tightening.

a. Sit comfortably with your forearms on the table, palms down.
b. Sit comfortably in the same position, but this time with palms up. Note differences, if any.
c. Sit comfortably, grasp the weight and lift it with your palms facing up. This is called a weight curl.
d. Stand up straight and let both arms hang down at your sides.
e. Stand up straight and hold the weight with one hand at the side of the body.

Record all observations in the **Records** section of your report.

2. Take turns doing the following tasks and note the degree of contraction of the gastrocnemius muscle, recording your observations in the **Records** section of your report.

a. Sit on the edge of a chair with both feet on the floor.
b. Sit on the edge of a table with feet dangling.
c. Stand up straight and lock knees for 5 seconds.
d. Stand up straight with legs relaxed.
e. Stand on toes, extend arms overhead, and reach for the ceiling.

Records

1. In general, the biceps muscle should feel tight when the elbow is bent, and the triceps muscle should feel tight when the arm is straightened. Both muscles may feel tight when the weight is added.

2. The gastrocnemius muscle should feel least tight in a and tightest in e.

3. The masseter muscle bulges out of the jaw slightly when it is tightened.

4–7. Answers will vary. The dominant hand should resist muscle fatigue longer than the weaker hand.

3. Place your hands on the masseter muscles of your lab partner's face and notice what happens when your partner relaxes and then tightens these muscles. Record your observations in the **Records** section of your report.

Part B: Muscle Fatigue

Each member of each team may do Part B in turn.

4. Hold the ball in your right hand and squeeze the ball repeatedly as quickly as you can. Count the number of complete contractions during each 10 second interval for a total period of 3 minutes or until fatigue is complete. Record the time of fatigue as well as the number of contractions during each 10 second interval the in a table similar to the one below in the **Records** section in your report.

Muscle Contraction During Ten-Second Intervals			
Seconds	**Contractions**	**Seconds**	**Contractions**
10		100	
20		110	
30		120	
40		130	
50		140	
60		150	
70		160	
80		170	
90		180	

5. Relax for 1 minute and repeat step 4. Record the number of contractions and the time of complete fatigue in the **Records** section of your report.

6. Wait 5 minutes, repeat step 4 again, and record your data.

7. Now repeat steps 4–6 with your left hand. Record all data in the **Records** section of your report. Graph your data as well.

8. Each lab partner should keep records of his or her personal data. In your report, briefly summarize the procedure you followed.

9. Clean up your materials and wash your hands before leaving the laboratory.

INQUIRY

1. Compare the muscle tone and contractions of the muscles studied in the lab.
2. Which tasks were harder to complete than the others? Why?
3. Were differences in muscle fatigue found between the right and left hands?

ANALYSIS

1. In which tasks was the muscle tension evident?
2. What variables might have influenced the differences between the left and right arm?
3. What is muscle fatigue?

FURTHER INQUIRY

Write a new question that could be explored as a new investigation. The following is an example:

> What is the relationship of muscle tone to fatigue among people who exercise regularly compared with those who do not?

Inquiry Answers

1. Answers will vary. Generally, the better muscle tone a muscle has, the more prepared it is to contract.

2. Answers will vary. Tasks requiring strength or endurance should be harder to complete than others.

3. The dominant hand should resist muscle fatigue longer than the weaker hand.

Analysis Answers

1. Muscle tension was evident in tasks in which the load was heavy.

2. One arm is likely to be dominant, and this arm will have more capable muscles. Other variables include differences in how quickly and how tightly the ball was squeezed by each hand.

3. Muscle fatigue occurs when muscles can no longer repeat muscle contractions of the same force.

Block Scheduling Guide

Each block represents about 45–50 minutes of instructional time. The resources cited in a block are coded to a specific program component using the Key on the next page.

Lecture Concepts	Lesson Resources

37-1 The Circulatory System pp. 875–880

BLOCKS ① and ②

a. Arteries have thick, muscular walls and carry blood away from the heart.
b. Capillaries have walls only one cell thick and are the sites where materials are exchanged between blood and tissue.
c. Veins, with walls less muscular than an artery's, carry blood back to the heart.
d. Plasma, the liquid component of blood, is composed of water and dissolved solutes. The cellular components of blood consist of red blood cells, white blood cells, and platelets.
e. Blood type (A, B, AB, O) is determined by the presence of A and B antigens located on the surface of red blood cells.

Lecture Resources
- Opening Question, page 875
- Demonstrations: The Sea Inside Your Body, Valves in Veins
- A Closed System, page 877
- Visual Strategies: Figures 37-2, 37-4, 37-4, 37-5, 37-6; Table 37-2
- Overcoming Misconceptions, page 880
- Teaching Transparencies: 211, 212, 213, 214, 186A, 187A, 189A
- Keeping Track, page 880
- Looking at Cells, page 881
- The Role of Platelets, page 881
- Holt Biology Videodiscs: Lesson 37-1

Classwork Options
- Effective Reading, page 876
- Interactive Explorations in Biology Laboratory Manual E6: Hemoglobin
- Laboratory Techniques and Experimental Design C42: Blood-Typing
- Laboratory Techniques and Experimental Design C43: Blood Typing—Whodunit?
- Laboratory Techniques and Experimental Design C44: Blood Typing—Pregnancy and Hemolytic Disease

Assignment Options
- Section Review, page 882
- Chapter Review, questions 1–4, 9, 12, 15

37-2 The Heart pp. 883–887

BLOCK ③

a. Oxygen-rich blood from the lungs enters the left atrium and is pumped from the left ventricle to the body's tissues.
b. After returning to the right atrium, the oxygen-poor blood is pumped from the right ventricles to the lungs.
c. Blood traveling through the heart does not flow backward because the heart contains one-way valves.
d. Cardiovascular diseases, like artherosclerosis and arteriosclerosis, are the leading causes of death in America.

Lecture Resources
- Opening Questions, page 883
- Demonstration: Blood Pressure
- Visual Strategies: Figures 37-8, 37-9, 37-11, 37-12
- Teaching Transparencies: 210, 185A
- Holt Biology Videodiscs: Lesson 37-2

Classwork Options
- Learn About Pacemakers, page 885
- The Silent Killer, page 886

Assignment Options
- Section Review, page 887
- Chapter Review, questions 5, 10, 14

37-3 The Respiratory System pp. 888–893

BLOCKS ④ and ⑤

a. Air is carried to the lungs by way of a branched system of tubes and air sacs.
b. Oxygen is transported to tissues by combining with hemoglobin molecules inside red blood cells.
c. Most carbon dioxide is transported to the lungs as bicarbonate ions inside the cytoplasm of red blood cells.

Lecture Resources
- Opening Question, page 888
- Visual Strategies: Figures 37-13, 37-16, 37-17, 37-18
- Demonstrations: The Breathing Tube, Building Lungs
- Research Update, page 890
- Diffusion in the Body, page 891
- Teaching Transparencies: 215, 217
- Holt Biology Videodiscs: Lesson 37-3

Classwork Options
- Interconnected Systems, page 892
- Closure, page 893
- Quick Labs A31: Determining Lung Capacity
- Laboratory Investigation: Lung Capacity and Carbon Dioxide Production, pages 896–897

Assignment Options
- Section Review, page 893
- Chapter Review, questions 6–8, 11, 13, 16

Review/Enrichment

BLOCK ⑥

- ■ Study Guide: Concept Review and Pretest

Assignment Options
- ◉ Teaching Resources CD-ROM: Occupational Applications Worksheets, Respiratory Therapist and Blood-Bank Technologist
- ◉ Teaching Resources CD-ROM: Problem-Solving Worksheets, Pressure Gradients in Breathing and Calculating Respiratory Volume, Computing Rates and Heart Efficiency
- ■ Activities and Projects, questions 17, 18

Assessment Options

BLOCK ⑦

Traditional Assessment
- ■ Chapter 37 Test
- ◉ Test Generator/Assessment Item Listing: Software item bank for preparing customized chapter tests.

Portfolio Assessment
Select student reports for one or more laboratory experiments from this chapter. *The Direct Observation Checklist* on the *BioSources Teaching Resources CD-ROM* should be completed during a laboratory or other cooperative learning experience.

Holt Biology Videodiscs

Holt Biology Videodiscs gives you a powerful tool for teaching, review, and assessment. *Concepts of Biology* adds interactive lesson plans and extensive tools for customization using CD-ROM technology.

CONCEPTS OF BIOLOGY

Use the following topic from *Concepts of Biology* to help you teach this chapter:

- ■ Topic 22: Circulation and Respiration

For further information regarding the media on the videodiscs, see *Holt Biology Videodiscs Teacher's Correlation Guide to Biology: Principles and Explorations.*

Chapter Theme

Structure and Function

There is a close relationship between the circulatory and respiratory systems. To conveniently distinguish the two systems, this chapter concentrates on the gas exchange function of the respiratory system and the transport function of the circulatory system. The respiratory system captures and releases gases and provides a moist surface across which gases are exchanged between the lungs and the blood. The circulatory system transports gases and other important solutes throughout the body.

Tapping Prior Knowledge

- Why do your cells need oxygen?
- What are the advantages of a closed circulatory system?

Opening Demonstration

Have one student take the resting pulse (in beats per minute) of a student volunteer and another student count the volunteer's breathing rate (in breaths per minute). Record both rates on the chalkboard. Have the volunteer do some jumping jacks or other aerobic exercise, then remeasure and record his or her pulse and breathing rate. Ask students the following questions. How did the volunteer's pulse and breathing rate vary before and after exercise? *(Both should increase after exercise.)* Why would pulse rate increase during exercise? *(The heart must pump faster to ensure that active muscles get enough food and oxygen.)* Why does the breathing rate increase during exercise? *(The body needs more oxygen and needs to get rid of more carbon dioxide.)*

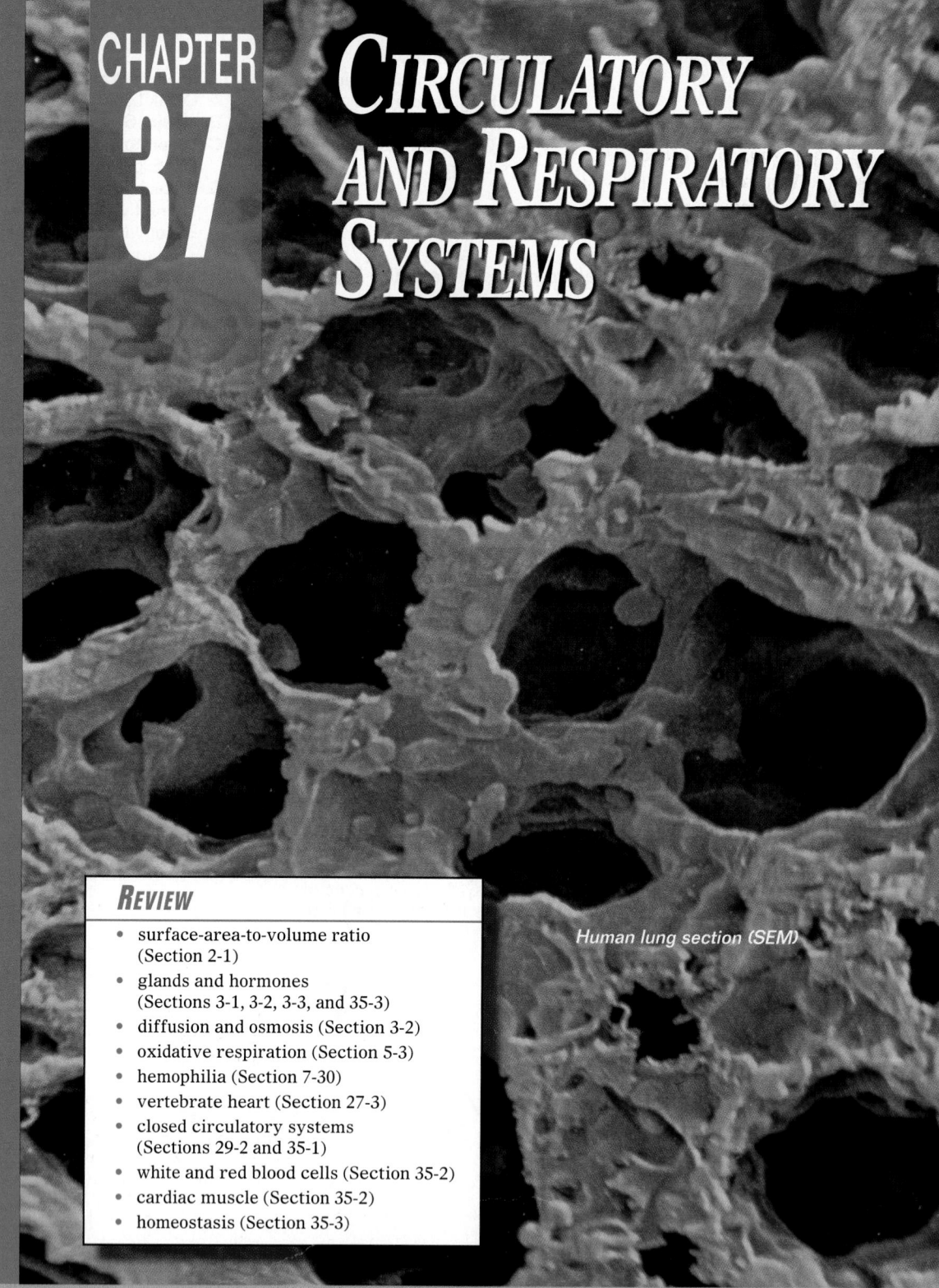

CHAPTER 37 CIRCULATORY AND RESPIRATORY SYSTEMS

Human lung section (SEM)

REVIEW

- surface-area-to-volume ratio (Section 2-1)
- glands and hormones (Sections 3-1, 3-2, 3-3, and 35-3)
- diffusion and osmosis (Section 3-2)
- oxidative respiration (Section 5-3)
- hemophilia (Section 7-30)
- vertebrate heart (Section 27-3)
- closed circulatory systems (Sections 29-2 and 35-1)
- white and red blood cells (Section 35-2)
- cardiac muscle (Section 35-2)
- homeostasis (Section 35-3)

Authors' Rationale

It is easy to understand that we breathe and that we have a heart that pumps blood. However, understanding the mechanics of how these two systems interact to make a continuous supply of oxygen available to the tissues is more abstract. How, exactly, do the lungs extract oxygen from the air and get it to the blood? Where does carbon dioxide come from? And how is it eliminated from the body? These and other questions you may have about the circulatory and respiratory systems will be answered in this chapter.

37-1 The Circulatory System

Humans, like all vertebrates, are mobile animals. We are able to walk, run, and swim because of an extensive muscular system and an elaborate internal skeleton. All of this muscle and bone would be of little use, however, if there were no way to provide it with food and no way to take away its wastes. This need is met in humans by a highly efficient closed circulatory system, a network of vessels that extends to every tissue of the body.

The Circulatory System Transports Materials and Distributes Heat

The human circulatory system functions like a network of highways. It connects the various muscles and organs of the body with one another. Four kinds of traffic move along this highway, as shown in Figure 37-1.

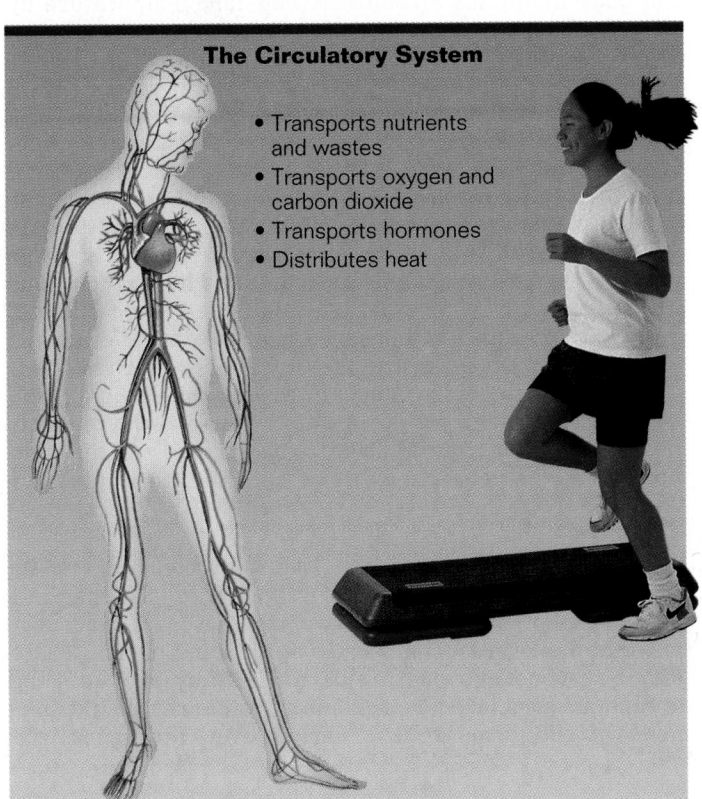

The Circulatory System

- Transports nutrients and wastes
- Transports oxygen and carbon dioxide
- Transports hormones
- Distributes heat

Figure 37-1 The circulatory system transports a variety of materials throughout the body. The circulatory system also helps maintain a constant body temperature by releasing heat into the environment when you are hot and retaining heat when you are cold. The girl doing step aerobics releases her excess body heat through blood vessels under her skin.

Historical Note

William Harvey

Prior to the mid-1500s, no one had any conception of blood circulating through the organs of the body. William Harvey, an English physician and anatomist, postulated the muscular nature of the heart and its function in pumping blood into the arteries. Fabricius ab Aquapendente, an accomplished Italian surgeon and anatomist, and Harvey's mentor, had discovered valves in veins. Building upon this knowledge, Harvey used tourniquets to reveal the direction of blood flow in the arteries and veins, and demonstrated that venous blood can flow only toward the heart. In 1628, Harvey published his theory describing the circulation of blood.

CONNECTIONS

Chapter 4

Energy and Metabolism

The circulatory system must deliver the raw materials of cellular respiration, including glucose and oxygen, to the cells while eliminating the waste product, carbon dioxide. Buildup of wastes or deficits in the delivery of raw materials can be dangerous, even fatal. For example, the human heart needs a constant supply of oxygen. Cardiac muscle cells completely deprived of oxygen for as little as 30 seconds will cease to contract, with heart failure as a result.

Teaching Tip

What If a person lost too much blood or had an obstructed airway? The result can be shock. This condition is a direct result of an oxygen deficit to the tissues. Shock can become irreversible when the oxygen deficit becomes very great. Even hyperoxygenation or blood transfusions will not restore the homeostasis of the body's organs, especially of the heart and brain. This kind of irreversible shock often leads to death.

Nutrients and Wastes

Food molecules are the nutrients that fuel muscle contraction and other cell activities. After nutrients pass from the small intestine into the bloodstream, they are transported to the cells of the body by the circulatory system.

Cells that metabolize (break down) proteins dump the resulting nitrogen-containing wastes into the circulatory system. The circulatory system carries wastes to excretory organs called kidneys, where the wastes are filtered from the blood.

Oxygen and Carbon Dioxide

Our cells must have a way of getting oxygen and disposing of carbon dioxide. Both of these gases are transported to and from cells by the circulatory system. In a process known as gas exchange, oxygen is carried to cells, and carbon dioxide is carried away from cells. The transportation provided by the circulatory system is vital. It has been estimated that it would take a molecule of oxygen three years to travel from your lung to your toe if the transport depended only on diffusion rather than the circulatory system.

Heat

The body maintains a relatively constant temperature by using metabolic energy to make heat. The circulatory system distributes this heat more or less uniformly to all parts of the body.

But environmental (outside) temperature also affects body temperature. Passing just beneath the skin, the circulatory system absorbs heat from a hot environment and gives off heat to a cold one. Distribution of heat by the circulatory system thus tends to adjust the entire body's temperature toward that of the environment. However, your body has several mechanisms that limit how much your body temperature can fluctuate. When it is cold outside, your body attempts to conserve heat by constricting the blood vessels that lie near the surface so that less heat escapes into the environment. In the opposite manner, the blood vessels under your skin dilate when it is hot outside, thereby delivering more blood to the surface and allowing more heat to escape.

CONTENT LINK

*You can learn more about glands and the hormones they release into the bloodstream in **Chapter 41**.*

Hormones

As you learned in Chapter 35, the body coordinates the activities of its many organs with hormones, the chemical messengers of the endocrine system. The brain sends signals that activate special hormone-producing glands. These glands respond by releasing particular hormones into the circulatory system. The circulatory system carries the hormones to the target cells at various sites throughout the body.

Effective Reading

Studying Vocabulary

Encourage students to keep a vocabulary notebook, a stack of vocabulary cards, or some other way of recording and reviewing unfamiliar vocabulary. Each entry should include not only the correct spelling of the word, but also a definition, preferably one in their own words. This strategy will be especially beneficial in a chapter like this one, in which students encounter many new words.

Components of the Circulatory System

There are four major components of the human circulatory system. Together, they form the transportation network that delivers essential materials to every cell in your body.

1. **Blood vessels** Blood vessels are a network of tubes through which blood moves.
2. **Lymphatic vessels** Lymphatic vessels intertwine with blood vessels. They recover fluid that leaks out of the blood vessels.
3. **Blood** Blood is a complex mixture of specialized cells and fluid.
4. **Heart** The heart is a muscular pump that propels blood through the blood vessels of the circulatory system.

Circulation Pathway

The general path of blood through the circulatory system is shown in Figure 37-2. Blood leaves the heart through vessels known as **arteries.** From the arteries, the blood passes into a large network of **arterioles** (*ahr TIHR ee uhls*), or small arteries. Eventually, the blood is pushed through the **capillaries,** an extensive network of very narrow tubes. The term *capillary* is from the Latin *capillaris,* meaning "hair." It is while the blood is passing through the capillaries that gases and metabolites are exchanged with the cells of the body. After leaving the capillaries, the blood flows into **venules** (*VEHN yools*), or small veins. The network of venules empties into larger **veins,** which carry the blood back to the heart.

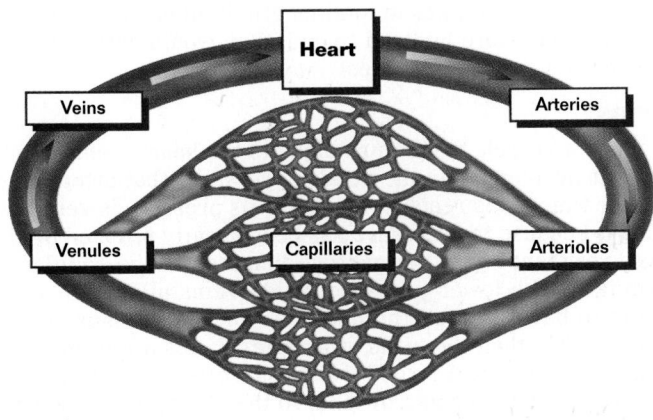

Arteries

Arteries are the tubes that carry blood away from the heart, but they must be more than just tubes to do their job. Blood leaves the heart not in a smooth flow but rather in pulses as the heart forcefully ejects it with contractions. To accommodate each forceful pulse of blood, an artery's wall must be able to expand.

Figure 37-2 Blood leaves the heart through blood vessels called arteries, which branch into many smaller arterioles. The smallest and most numerous blood vessels are capillaries. Blood returns to the heart through venules and veins.

SECTION 37-1

Teaching Tip
A Closed System
Remind students that the human circulatory system is a closed system, one in which the blood never leaves the heart and blood vessels. Materials like nutrients, oxygen, and wastes are transferred between the blood and tissues by passing through the walls of the smallest blood vessels, called capillaries.

◉ VISUAL STRATEGY

Figure 37-2
Name the different vessels that blood travels through as it flows from one side of the heart to the other side of the heart. Explain that numerous capillaries are found within every tissue of the body. Ask students why. (*because each tissue requires nutrients and oxygen from the blood inside the capillaries*)

 Did You Know?

Arteries can be classified as either *conducting arteries* or *distributing arteries*. While both kinds have substantial amounts of smooth muscle in their walls, the conducting arteries are large, thick-walled arteries, like the ones that lead out of the heart. Conducting arteries are relatively inactive during vasoconstriction (narrowing of the blood vessels). Instead, they serve as low-resistance conduits that carry blood to the more numerous, and smaller, distributing arteries. The distributing arteries play a role in vasoconstriction. For example, they attempt to bolster the blood pressure by contracting after large-scale blood loss occurs.

Demonstration
Valves in Veins

Tell students that they can demonstrate the existence of valves in their own veins. Have students form into groups of two and designate one student of each pair as an observer. Instruct the other student in the pair to stand with arms hanging by his or her sides until the veins of the back of the hands become distended. Then have the observer place both of his or her index fingers (side by side) on one of the veins of one of the partner's hands. Keeping the finger closest to the ground stationary, the observer should move the finger closest to the heart slowly along the vein for 2–4 cm while pressing firmly. The observer should then release the finger nearest the heart while keeping the finger closest to the floor in place. Finally, the observer should release the finger closest to the floor. The pair should repeat the demonstration after switching roles. Then ask students the following questions: How did the vein appear when your partner moved a finger toward the heart? Why? *(It seemed to disappear because the blood was pushed out of it.)* How did the vein appear after your partner removed the finger closest to your heart? *(Student answers will vary, depending on how far they stripped the vein of blood. Some will see no change, others will see the blood return to part of the vein until a valve is reached.)* Why did blood not return to the vein at this time? *(A valve in the vein restricted its backward flow.)* How did the vein appear when your partner removed the finger closest to the floor? *(It became distended because it filled with blood.)*

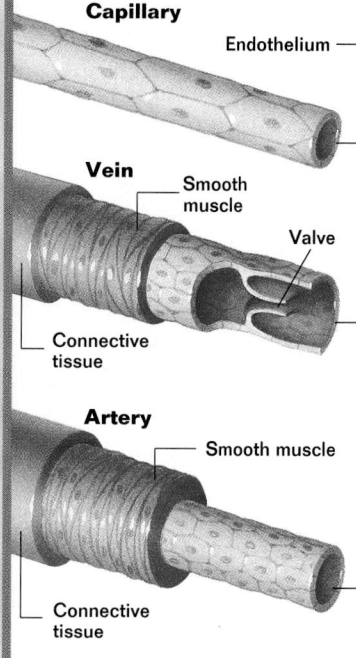

Figure 37-3 The thick, muscular wall of an artery, *bottom*, enables the artery to withstand the high blood pressure that develops inside it every time the heart contracts. Veins, *center*, have thinner walls than arteries since the blood pressure inside veins is much lower. Capillaries, *top*, have the thinnest walls, allowing materials like oxygen and carbon dioxide to easily pass through.

❏ CAPSULE SUMMARY

Arteries, capillaries, and veins are three kinds of blood vessels. Arteries have thick, muscular walls and carry blood away from the heart. Capillaries have walls only one cell thick and are the sites where materials are exchanged between blood and tissue. Veins, with walls less muscular than an artery's, carry blood back to the heart.

How is an artery able to expand? The wall of an artery is made up of three layers of tissue, as shown in Figure 37-3. The innermost layer is a thin skin of endothelial cells. Surrounding this is a layer of elastic, smooth muscle tissue. Finally, a protective layer of connective tissue wraps around the smooth muscle tissue. Just as a balloon expands when you blow more air into it, so the elastic artery expands when the blood is pumped into it each time the heart contracts.

Capillaries

In capillaries, food and oxygen molecules are transferred from the blood to the body's cells, and carbon dioxide and wastes are picked up. This extensive back-and-forth traffic is possible only because of two key properties of capillaries.

1. **Thin walls**   Built like a soft-drink straw, capillaries are simple tubes with walls that are only one cell thick. The walls of capillaries are so thin that food and gas molecules easily pass through them.

2. **Narrow diameter**   All capillaries are very narrow, with an internal diameter of about 8 µm (0.0003 in.). Because this is only slightly larger than the diameter of a red blood cell, blood cells passing through a capillary slide along the capillary's inner wall. This tight fit makes it easy for oxygen to move through the wall as it diffuses from the blood cells to body cells.

No cell in your body is more than about 100 µm (0.004 in.) away from a capillary. At any one moment, about 5 percent of your blood is in capillaries. The network of capillaries is several thousand miles in total length; if all of the capillaries of your body were laid end to end, they would extend all the way across the United States!

Veins

Veins are vessels that return blood to the heart. Veins do not have to accommodate the pulsing pressures that arteries do. By the time blood reaches the veins, its pressure is very low. This is because the pressure has been greatly reduced by the high resistance that the narrow diameter of the capillaries provides. The lower pressure inside veins allows for their walls to have a much thinner layer of smooth muscle and elastic fiber than the walls of arteries. Veins are often quite large in diameter. A large blood vessel offers less resistance to blood flow than a narrow one, so the blood can move more quickly through large veins. The largest vein in the human body is the vena cava *(VEE nuh KAYE vuh)*, the vein that leads into the heart. The vena cava is about 3 cm (1.2 in.) in diameter—you could easily slide your thumb into it. Most veins have internal, one-way valves spaced at regular intervals. These valves prevent the blood from flowing backward during its trip to the heart. ❏

Lymphatic Vessels

The heart and the blood vessels, collectively, are called the cardiovascular system. The cardiovascular system is very leaky. Fluids are forced out of the thin walls of the capillaries by the pressure generated every time the heart pumps. This loss of fluid is unavoidable because the capillaries could not function without very thin walls. But the loss must be made up or the cardiovascular system would soon dry up. About 3 L (3.2 qt.) of fluid leak out of your cardiovascular system each day—more than half of your total supply of about 5.6 L (5.9 qt.) of blood. To collect and recycle this fluid, the body utilizes another circulatory system called the **lymphatic** *(lihm FAT ihk)* **system,** shown in Figure 37-4. The fluid that leaks out of the capillaries accumulates in the spaces around the body's cells and diffuses into vessels called lymphatic capillaries. Once the fluid enters the lymphatic capillaries, it is called **lymph.** The recovered fluid passes through a series of progressively larger lymphatic vessels into two large lymphatic ducts that drain into veins in the lower part of the neck.

Blood Has Liquid and Solid Components

The blood that circulates through the cardiovascular system is composed of water, a variety of molecules dissolved in the water, and three kinds of cells, as shown in Table 37-1.

Blood Plasma

The noncellular portion of blood is called **plasma.** Plasma is a complex solution of 90 percent water and 10 percent solute. Three very different kinds of substances compose the solute component of plasma.

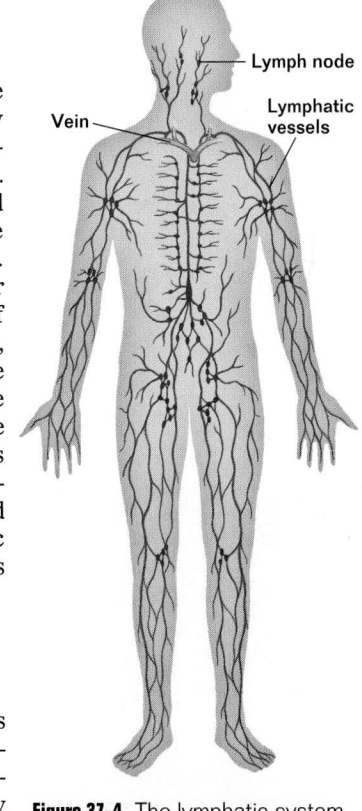

Figure 37-4 The lymphatic system, in green, returns the fluid that collects around tissues to veins in the neck. The fluid, called lymph, is driven through lymphatic vessels as the vessels are squeezed by the movements of the body's muscles. Structures called lymph nodes filter foreign substances from the lymph.

SECTION 37-1

👁 VISUAL STRATEGY

Figure 37-4
Point out the major features of the lymphatic system, including the lymphatic vessels and the lymph nodes. The lymph returns to the bloodstream via the left and right subclavian veins, just beneath the collar bones. Lymph vessels contain valves that are similar to the ones in veins.

CONNECTIONS

Chapter 3
Solutions
Tell students that plasma, the liquid portion of blood, is a solution. Like all solutions, plasma contains solutes that are dissolved in a solvent. The solutes in plasma are metabolites, salts, ions, proteins, and wastes, and the solvent is water.

Table 37-1 Composition of Blood

Components of Blood	Function
Plasma portion (60% of total blood volume)	
Water	Acts as solvent
Metabolites and wastes, salts and ions, proteins	Play diverse roles (nourish cells, catalyze chemical reactions, act as chemical messengers, maintain blood volume, fight infection, etc.)
Cellular portion (40% of total blood volume)	
Red blood cells	$O_2 + CO_2$ transport
White blood cells	Produce antibodies, ingest foreign materials
Platelets	Aid in clotting blood

 Did You Know?

Our bodies have many types of lymphoid organs that contain lymphatic tissue. They include the lymph nodes, the spleen, the thymus gland, and the tonsils. Specialized lymph vessels called lacteals in the intestines function to absorb fats from digested foods. They deliver their contents to other lymph vessels, which then conduct the fats to the bloodstream.

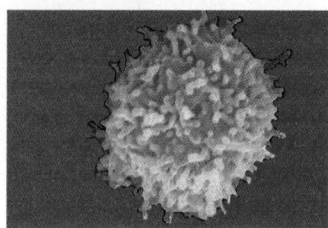

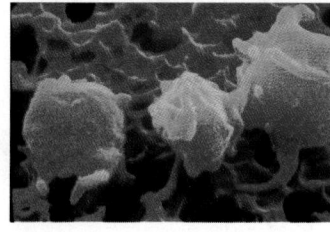

Figure 37-5 Red blood cells, *top,* get their red color from the hemoglobin molecules inside them. Lymphocytes, *center,* are specific kinds of white blood cells. Lymphocytes, slightly larger than red blood cells, have large nuclei with very little cytoplasm. Platelets, *bottom,* are essentially membrane-bound "enzyme packets." The enzymes are used for blood clotting.

1. **Metabolites and wastes** Dissolved within the plasma are glucose and other food molecules, as well as vitamins, hormones, and nitrogen-containing wastes.

2. **Salts and ions** The chief plasma ions are sodium, chloride, and bicarbonate. Plasma also contains trace amounts of calcium, magnesium, and metallic ions such as copper, potassium, and zinc.

3. **Proteins** If not for the high concentration of protein in the plasma, the cells of your body would soak up much of the water in plasma. Remember from Chapter 3 that water tends to move from a region of lower solute concentration toward a region of higher solute concentration in a process called osmosis. Plasma avoids losing its water to cells because it is rich in dissolved proteins. In fact, the total amount of protein inside cells and in plasma is the same, making cytoplasm and plasma essentially isotonic to each other.

Blood Cells

About 40 percent of the total volume of blood is not plasma; it consists of cells and cell fragments that are suspended in the plasma. There are three principal types of cells in human blood: red blood cells, white blood cells, and pieces of cells called platelets.

1. **Red blood cells** Each milliliter of your blood contains about 5 million oxygen-carrying red blood cells, also called **erythrocytes** *(eh RIHTH roh seyets).* Each red blood cell has the shape of a flat disk with a collapsed center, as shown in Figure 37-5. Almost all of the interior of a red blood cell is packed with hemoglobin, a protein that binds to oxygen in the lungs and transports it to the tissues of the body. A red blood cell is simply a hemoglobin container; it has no nucleus, and it cannot make proteins or repair itself. As a result, red blood cells have a short life span—about four months. New red blood cells are produced constantly by stem cells, specialized cells in bone marrow. If the production of new red blood cells slows, the blood's population of them soon decreases, and a condition called **anemia** develops.

2. **White blood cells** A small proportion of the cells in your blood are white blood cells, or leukocytes *(LOO koh seyets).* There are only one or two white blood cells for every 1,000 red blood cells. White blood cells are larger than red blood cells and contain no hemoglobin. **Leukocytes** are the primary cells of the immune system, your body's defense against disease.

   There are different kinds of white blood cells. Each kind has a different function in the immune system. Some are called lymphocytes *(LIM foh seyets),* many of which produce antibodies. Others, called macrophages *(MAK roh fayj ehs),* are unique among blood cells in that they do not always stay within the vessels of the cardiovascular system. They are mobile soldiers that migrate into the fluid surrounding the body's cells. Sometimes cancer occurs among the cells that give rise to white blood cells. As a result, too many white blood cells are produced, and a disease called **leukemia** *(loo KEE mee uh)* develops. Leukemia is often fatal.

3. **Platelets**   Certain large cells in bone marrow called megakaryocytes *(mehg uh KAR ee oh seyets)* regularly pinch off bits of their cytoplasm. These unnucleated cell fragments, called **platelets** *(PLAYT lihts)*, play a key role in blood clotting. Remember that humans, like other vertebrates, have a closed circulatory system. If a blood vessel gets a hole in its wall, the hole must be plugged quickly. If the hole is not plugged, all of the blood will leak out of the system, and death will occur.

Circulating platelets start the clotting process when they encounter chemicals released by damaged blood vessel cells. The platelets then release a clotting protein into the blood that initiates a series of chemical reactions resulting in the formation of a protein called **fibrin**. The fibrin begins to form a netlike covering over the damaged site, as shown in Figure 37-6. Very quickly, the gluey mesh of fibrin and platelets develops into a mass, or clot, that plugs the hole in the blood vessel. Because of its initially gluey nature, the clot fits itself to the shape of the rupture in the blood vessel and provides a tight, strong seal. As you learned in Chapter 7, the lack of one of the clotting proteins causes hemophilia. ❑

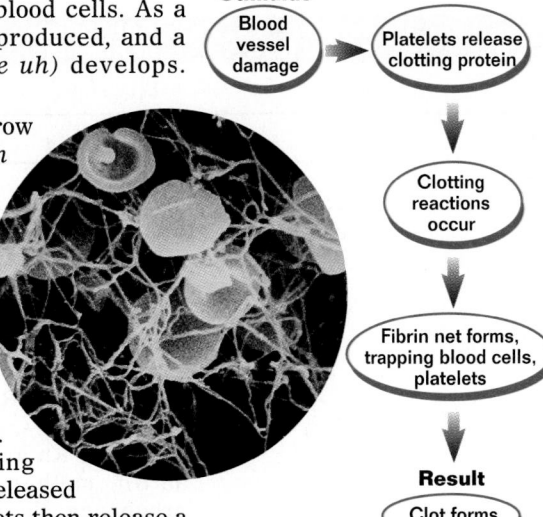

Stimulus

Blood vessel damage → Platelets release clotting protein

Clotting reactions occur

Fibrin net forms, trapping blood cells, platelets

Result

Clot forms

Figure 37-6 The release of enzymes from platelets at the site of a damaged blood vessel initiates a "clotting cascade." During this cascade, a series of chemical reactions occurs, resulting in the formation of a fibrin net in which blood cells are trapped, *above left*. The net eventually forms into a clot in the hole of the damaged vessel.

Surface Proteins on Red Blood Cells Determine Blood Type

Before a blood transfusion can be performed, it is important to know the blood type of the donor and recipient. If the blood of each is not compatible, a life-threatening situation can develop. Blood compatibility is determined by proteins called antigens that coat the outside of red blood cells.

The system used to classify human blood is called the **A-B-O system.** Your red blood cells have either A antigens, B antigens, both A and B antigens, or no antigens on their surface. The kinds of antigens present on red blood cells determine your blood type. The danger of mixing bloods of different blood types comes from the presence of other proteins, called antibodies, that are present in the blood plasma. Antibodies seek out specific foreign antigens and attack them. For example, type B blood contains antibodies that will attack A antigens. If a person with type B blood receives type A or AB blood, the recipient's antibodies will react with

SECTION 37-1

Teaching Tips
Looking at Cells
Show students prepared slides of human red blood cells. Ask students to describe the shape of the cells and any other unique physical characteristics the cells exhibit. Ask students to infer relationships between the red cell's unusual disklike shape and its function.

The Role of Platelets
Discuss with students the function of platelets and their structure. On the chalkboard, review the step-by-step process of blood clotting. Ask students to make predictions about what kind of problems a person might encounter if his or her blood suddenly lost the ability to clot.

◉ **VISUAL STRATEGY**

Figure 37-6
Trace the blood-clotting process depicted in this figure. This is a simplified version of a complex, multi-step pathway involving as many as 12 chemical factors either found in plasma or released from megakaryocytes or damaged tissues. The factors range from simple calcium ions to lipoprotein complexes to large protein components. Emphasize that the fibrin net is the end result of the blood-clotting process.

Historical Note

Blood Groups

In 1900, the Austrian physician Karl Landsteiner demonstrated that there are three blood groups. He distinguished the three groups as A, B, and O (for zero), based upon incompatibilities between the plasmas and erythrocytes of the different groups. Landsteiner's discovery of a fourth blood group, AB, in 1902 earned him a Nobel Prize for physiology or medicine in 1930.

VISUAL STRATEGY

Table 37-2

Go over this table with students to reinforce the material presented in the text. Stress that antigens are found on the surface of the erythrocytes (RBCs), while antibodies are dissolved in the plasma.

Table 37-2 Blood Types

Type	Antigen on the RBC	Antibodies in Plasma	Can Receive Blood From	Can Donate Blood to
A	A	B	O, A	A, AB
B	B	A	O, B	B, AB
AB	A, B	None	O, A, B, AB	AB
O	None	A, B	O	O, A, B, AB

CAPSULE SUMMARY

Blood type (A, B, AB, O) is determined by the presence or absence of A and B antigens located on the surface of red blood cells. Another important antigen found on red blood cells is called Rh factor. Rh+ blood has Rh antigens, and Rh- blood does not.

the donor's blood and cause the blood to clump. Table 37-2 lists the four different blood types and the antigens and antibodies found in each.

Another important antigen on the surface of red blood cells is called **Rh factor.** People who have this protein are said to be Rh+, and those who lack it are Rh–. The greatest danger of Rh incompatibility arises in pregnancies in which an Rh– mother is carrying an Rh+ child. Throughout the pregnancy, the mother's blood remains separate from the child's, but during delivery some mixing usually occurs. As a result of this mixing, the mother's immune system responds to the Rh antigens in the child's blood by producing Rh antibodies. Because it takes some time for the mother's body to produce the antibodies, the first child is usually born unharmed. But if the next child that the mother carries is also Rh+, the antibodies that remain in the mother's blood will attack this child's blood cells, causing severe complications for the child. Fortunately, a drug can be given to the mother shortly after delivery of her first Rh+ child that will destroy any Rh antigens in her blood before her body has a chance to produce antibodies to them, thus making it safe for the next Rh+ child. ◼

Section Review

1. *Why might a person's face become flushed and red when the weather is hot?*

2. *Why is a cut artery usually considered more life threatening than a cut vein?*

3. *How does the total volume of blood in the cardiovascular system remain constant?*

4. *How is oxygen transported in the blood?*

5. *A person with type O blood is considered a universal blood donor. Explain why.*

Critical Thinking

6. *Standing in one place for a long period of time often results in swollen ankles and feet. Explain why this happens. How could the swelling be relieved?*

Answers to Section Review

1. An increased blood flow to the surface of the body radiates heat into the environment, which cools the body's central core.

2. Arteries contain blood under higher pressure than the blood in veins. As a result, more blood flows per second through an artery of a given diameter than through a vein of comparable size. Severed arteries result in quicker blood loss.

3. Blood volume is kept constant because blood plasma lost to the tissues is returned to the blood vessels by the lymphatic system.

4. The majority of the oxygen carried in blood is bound to hemoglobin in the erythrocytes.

5. Since type O blood has no antigens on its erythrocytes, it can usually be transfused into anyone without harm.

6. Standing in one place causes blood to pool in the lower extremities because inactivity does not allow the skeletal muscles to help push the blood back up toward the heart. Two remedies for this pooling are exercise and propping the feet up.

37-2 The Heart

*T*he human heart is a double *pump. One side of the heart powers an oxygen-acquiring circulatory phase, while the other side powers an oxygen-delivering circulatory phase. But oxygen is not the only gas carried by the circulatory system. When oxygen is acquired or delivered, it is exchanged with carbon dioxide.*

Section Objectives

- Describe pulmonary and systemic circulation.
- Identify the chambers, vessels, and valves of the heart.
- Trace the path of blood through the heart.
- Describe two ways that medical technicians can monitor cardiovascular performance.
- Describe two different blood vessel diseases.

The Heart Pumps Blood in Two Separate Loops

The pattern of circulation in the human body includes two separate circulatory loops, as shown in Figure 37-7. The right side of the heart is responsible for driving the pulmonary circulation loop, which pumps oxygen-poor blood through pulmonary arteries to the lungs. In the lungs, gas exchange occurs: carbon dioxide is released, and the blood receives oxygen. This oxygenated blood is then returned to the left side of the heart through pulmonary veins. The left side of the heart is responsible for driving the systemic circulation loop, which pumps oxygen-rich blood through a network of arteries to the tissues of the body. The deoxygenated blood is then returned to the right side of the heart by veins.

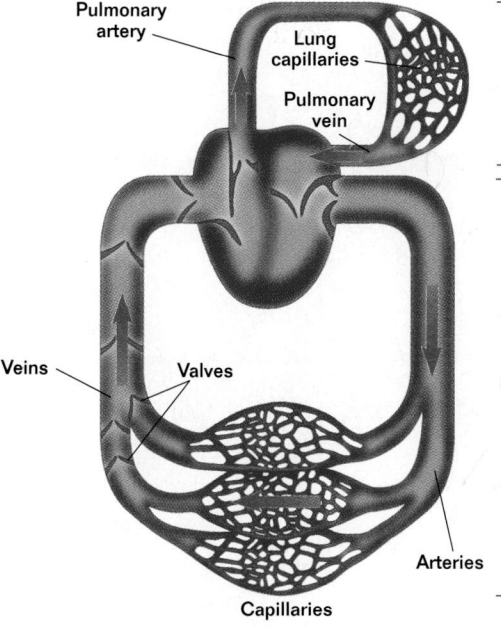

Figure 37-7 The pulmonary circulatory loop transports blood from the right side of the heart to the lungs and then to the left side of the heart. The systemic circulatory loop transports blood from the left side of the heart to all the body's tissues and then to the right side of the heart. The designation of the left and right sides of the heart in the diagram is based on the left and right sides of an intact heart that is still in a body.

SECTION 37-2

Vocabulary Preview

pulmonary vein
mitral valve
aorta
aortic valve
coronary artery
superior vena cava
inferior vena cava
tricuspid valve
pulmonary valve
sinoatrial node
diastolic pressure
systolic pressure
atherosclerosis
arteriosclerosis

Opening Questions

Ask students to name the type of muscle found in the heart, and then ask them what makes heart muscle different from the other kinds of muscle. They should remember that the muscle in the heart, called cardiac muscle, is composed of chains of single cells, each with its own nucleus. The chains form fibers that branch and interconnect to form a latticework.

 Did You Know?

The heart cannot get food and oxygen from the blood in its own chambers because the heart muscle is too thick for diffusion to be a practical means of distribution. Instead, the heart must rely on the coronary arteries to sustain it. These two arteries lie in grooves that spiral around the heart. An obstruction in one of their branches often necessitates bypass surgery to restore proper blood flow to the heart. Restricted blood flow can result in angina pectoris, which is severe chest pain, or a myocardial infarction, which is a heart attack. Heart cells die during an infarction, which limits the heart's pumping efficiency.

Demonstration
Blood Pressure

Take a student volunteer's resting blood pressure, and record the results on the chalkboard. (Average is 120 mm Hg systolic and 70–80 mm Hg diastolic.) Have the volunteer do some jumping jacks or other aerobic exercise. As soon as the student stops exercising, measure and record the blood pressure again. Then ask the class the following questions. What differences did you see between the volunteer's at-rest blood pressure and after-exercise blood pressure? *(Blood pressure after exercise is higher.)* Why does the blood pressure increase with exercise? *(The heart must contract more strongly to deliver oxygen to the tissues at a greater rate.)*

 VISUAL STRATEGY

Figure 37-8

Ask students to describe the functions of the following structures: right atrium, left atrium, right ventricle, left ventricle, tricuspid valve, mitral valve, pulmonary valve, aortic valve, aorta, pulmonary artery, superior vena cava, pulmonary veins, and inferior vena cava. Have students apply their knowledge by tracing the pathway of blood through the heart.

Tracing the Path of Blood Through the Heart
....

Let's follow the journey of blood through the human heart, shown in Figure 37-8, starting with the entry of oxygen-rich blood from the lungs into the heart. This oxygenated blood enters the left side of the heart through the **pulmonary veins,** emptying directly into the heart's collection chamber, the left atrium. From the left atrium, the blood flows down into the left ventricle. Most of this flow occurs before a contraction starts. When the heart starts to contract, the atrium contracts first, pushing its remaining blood into the ventricle.

After a slight delay that permits the atrium to empty fully, the ventricle contracts. The walls of the ventricle are far more muscular than those of the atrium, so the ventricle's contraction is much more forceful than that of the atrium. As the left ventricle contracts, the blood is prevented from going back into the left atrium by a one-way valve called the **mitral** *(MEYE truhl)* **valve**. All four valves in the heart are one-way valves. Each has flaps that act as doors that open in only one direction, ensuring that the blood that passes through them will not flow back.

Prevented from reentering the atrium, the blood within the contracting left ventricle enters the largest artery of the body, the **aorta** *(ay AWR tuh)*. Once inside the aorta, the blood is prevented from reentering the left ventricle by another large

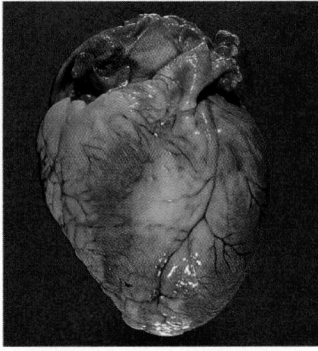

Figure 37-8 The ventricles (lower chambers) of the human heart have thicker walls than the atria (upper chambers) because the ventricles must pump the blood inside them a long distance. The arrows indicate the pathway that the blood takes as it travels into, through, and out of the heart. Oxygen-rich blood from the lungs, red arrows, enters the left side of the heart through pulmonary veins. Oxygen-poor blood from the body, blue arrows, enters the right side of the heart through two large veins, the superior vena cava and the inferior vena cava.

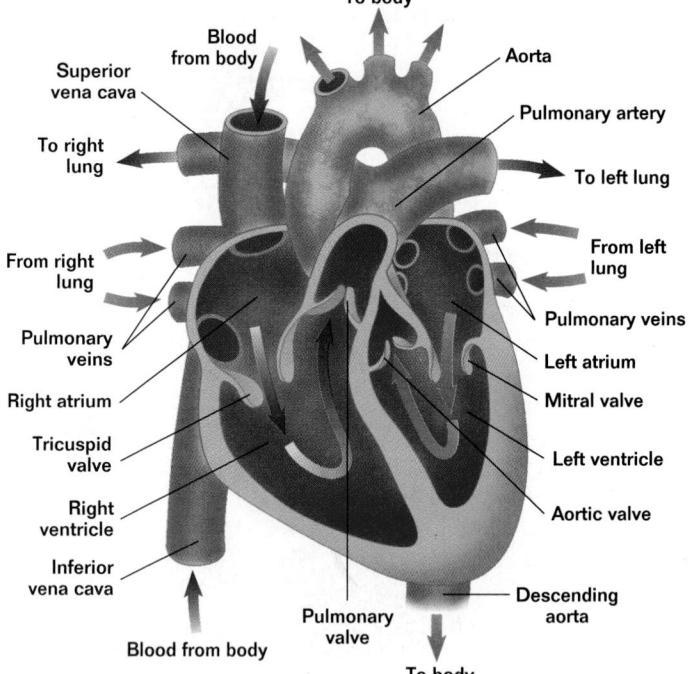

valve, the **aortic valve.** Many major arteries branch from the aorta and carry oxygen-rich blood to all parts of the body.

The first arteries to branch from the aorta are the **coronary** *(KAWR uh neh ree)* **arteries,** which carry freshly oxygenated blood to the heart muscle itself. Many other arteries also branch from the aorta. For example, two renal *(REEN uhl)* arteries leave the aorta and carry blood to the kidneys, where nitrogen wastes are filtered from the blood.

After delivering its cargo of oxygen to the cells of the body, the blood makes its way back to the heart through the body's many veins. Two large veins collect all the oxygen-poor blood from the systemic circulatory system. The **superior vena cava** *(VEE nuh KAY vuh)* drains blood from the upper body, while the **inferior vena cava** drains blood from the lower body. These two veins empty directly into the right atrium of the heart. Blood passes from the right atrium into the right ventricle through the **tricuspid** *(treye KUHS pihd)* **valve.** As the right ventricle contracts, it sends the blood through the **pulmonary valve** and into the pulmonary arteries, which carry the blood to the lungs. The blood then returns from the lungs to the left atrium with a new cargo of oxygen, and the cycle begins again. ◾

◻ **CAPSULE SUMMARY**

Oxygen-rich blood from the lungs enters the left atrium and is pumped from the left ventricle to the body's tissues. After returning to the right atrium, the oxygen-poor blood is pumped from the right ventricle to the lungs. Blood traveling through the heart does not flow backward because the heart contains one-way valves.

A Wave of Contraction Spreads Over the Heart

Contraction of the heart is initiated by a small cluster of cardiac muscle cells embedded in the upper wall of the right atrium, as shown in Figure 37-9. Called the **sinoatrial** *(SEYE noh ay tree ahl)* **node,** or SA node for short, these cells act as the pacemaker of the heart, spontaneously starting contractions with a regular rhythm. Each contraction initiated at the SA node travels quickly in a wave that causes both the right and the left atria to contract almost simultaneously.

The wave of contraction does not immediately spread to the ventricles, however. Almost one-tenth of a second passes before the lower half of the heart starts to contract. The delay is critical to proper functioning of the heart, as it permits the atria to finish emptying blood into the ventricles before the ventricles contract. The reason for the delay is that the two upper chambers of the heart are separated from the two lower chambers by connective tissue. Connective tissue cannot transmit the contraction. The wave of contraction would not pass to the ventricles at all except for a slender bridge of cardiac muscle cells that connects the upper chambers of the heart to the lower chambers. This bridge is called the atrioventricular node, or AV node. From the AV node, the wave of contraction is conducted rapidly over both ventricles by a network of fibers called the Bundle of His. Both ventricles contract almost simultaneously.

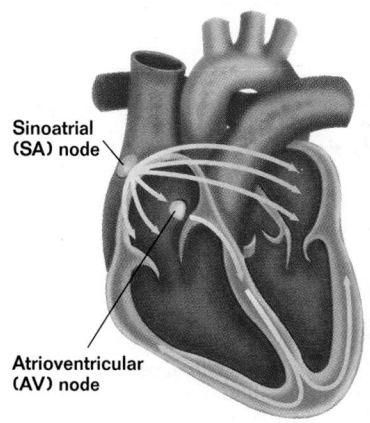

Sinoatrial (SA) node

Atrioventricular (AV) node

Figure 37-9 The sinoatrial node, or pacemaker, initiates each heart contraction. After initiation, the wave of contraction spreads across both atria, delays for an instant at the atrioventricular node, travels to the bottom of the heart, and then moves back up and across both ventricles. On average, heart contractions are initiated at a rate of about 72 times per minute. During sleep the rate decreases, and during exercise it increases.

SECTION 37-2

Application

℞ **Medicine** Some people have a heart that beats abnormally. Without proper control of the SA node as mediated by the AV node, the ventricles begin to beat at their own rate, a rate too slow to maintain adequate circulation. A patient suffering from this malady receives a device called a fixed-rate artificial pacemaker that electrically stimulates the ventricles to pump faster. Sometimes a patient's AV node transmits the contractile wave intermittently. These patients are given demand-type pacemakers that deliver stimulation only when it is needed.

Teaching Tip
Learn About Pacemakers
Have students write a research paper on artificial pacemakers. Suggest the following areas of focus: the kinds of medical conditions that call for pacemakers, how pacemakers are implanted and maintained, how pacemakers have changed as technology has improved, and the number of people who have pacemakers.

✉

◉ VISUAL STRATEGY

Figure 37-9
Point out the locations of the SA node (also called the pacemaker) and AV node. Then ask students about the consequences of damaging either the SA node or the AV node. *(Student answers may vary. Irregular heart beat would result, possibly leading to death.)* How might such damage occur? *(Heart attacks can damage the nodes.)*

 Did You Know?

Cardiac muscle cells have an intrinsic (self-initiated) beat. Contractions of the heart muscle are not produced by stimulation from nerves. Nerves control only the rate of the heart's contractions. Additionally, cardiac muscle cells influence each other. They will synchronize their rhythms in response to other cardiac muscle cells with which they have contact.

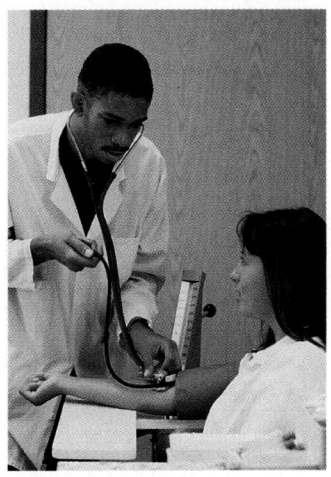

Figure 37-10 Blood pressure is quickly and easily measured with a sphygmomanometer, or blood pressure cuff, and a stethoscope. The blood pressure readings obtained from females are generally slightly lower than the readings from males.

Figure 37-11 The electrical changes that occur in the heart each time it contracts are measured with a device called an electrocardiograph, *below right*. Each time the heart beats, a recording pen graphs characteristic up-and-down waves, *below*. These waves are analyzed by medical professionals to assess the health of the heart.

Cardiovascular Functions Can Be Monitored

As you have seen, the contraction of the heart is not simply a continuous squeeze-release cycle, but rather a series of events that occurs in a predictable order. Medical technicians can monitor these events with special instruments.

Blood pressure is measured with a device called a sphygmomanometer *(sfihg moh muh NAHM uht uhr)*, shown in Figure 37-10. During the first part of the heartbeat, the atria are filling, so the pressure in the arteries leading out to the tissues of the body falls slightly. This low pressure that occurs during relaxation of the heart is called the **diastolic** *(DEYE uh stahl ihk)* **pressure.** Then, with the contraction of the ventricles, a pulse of blood is forced into the systemic arterial system, immediately raising the blood pressure. This higher blood pressure is called the **systolic** *(sihs TAHL ihk)* **pressure.** A blood pressure reading is usually reported as the systolic pressure written over the diastolic pressure. Normal blood pressure values are from 100 to 130 for systolic and from 70 to 90 for diastolic, so an example of a normal reading would be written 120/80. Many Americans suffer from a condition called high blood pressure, or hypertension. People with this condition have elevated systolic and diastolic blood pressures. Left untreated, hypertension can lead to heart damage, a stroke (rupture of a blood vessel in the brain), or kidney failure.

Another way to monitor the heartbeat is to measure the tiny electrical impulses produced by the heart muscle when it contracts. Because the human body is composed mostly of water, it conducts electrical currents rather well. A wave of muscle contraction passing over the surface of the heart generates an electrical current that passes in a wave through the entire body. Although the magnitude of this electrical pulse is tiny, it can be detected with special sensors that are placed on the skin. A recording of these impulses is called an electrocardiogram *(ee LEHK troh kahr dee ah graym)*. In one normal heartbeat, three successive electrical impulses are recorded, as shown in Figure 37-11.

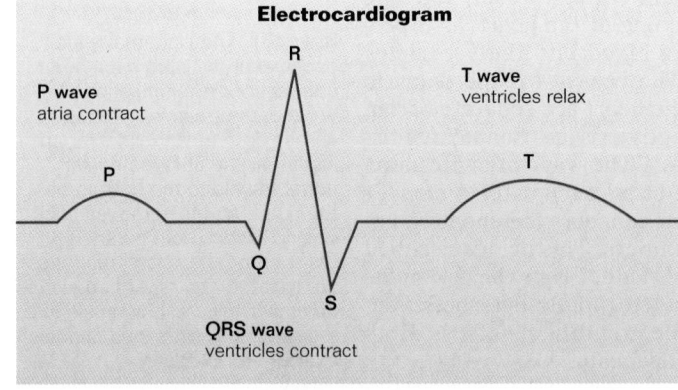

Electrocardiogram

P wave
atria contract

T wave
ventricles relax

QRS wave
ventricles contract

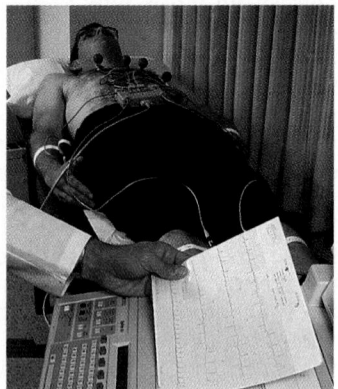

Heart and Blood Vessel Diseases Kill Many Americans

Diseases of the heart and blood vessels are the leading cause of death in the United States. Over 16 percent of the population, or about 42 million Americans, are affected by some form of cardiovascular disease.

Heart attacks are the main cause of death due to cardiovascular disease. They result when an area of heart muscle does not receive enough blood (and consequently, oxygen), leading to the death of cardiac muscle cells. Heart attacks are sometimes caused by a blood clot that has formed somewhere else in the body. The clot can be carried through the heart and into the coronary arteries, where it blocks the passage of blood to part of the heart. Heart attacks may also occur if the coronary arteries become blocked by deposits of fatty materials on their inner walls.

Atherosclerosis (ath uhr oh skluh ROH sihs) is a buildup of fatty deposits on the inner walls of arteries, as shown in Figure 37-12. The deposits are composed of cholesterol and cellular debris of various kinds. When this condition is severe, blood flow through the artery becomes greatly restricted. Exercise, along with diets low in cholesterol and saturated fats are prescribed to help prevent atherosclerosis. **Arteriosclerosis** (ahr tihr ee oh skluh ROH sihs), or hardening of the arteries, occurs when calcium is deposited in the fatty buildup caused by atherosclerosis. Hardened arteries cannot expand to accommodate the volume of blood that enters them every time the heart contracts. As a result, pressure builds up in the artery and feeds back to the heart, causing it to work harder. ◻

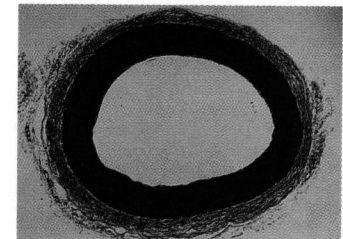

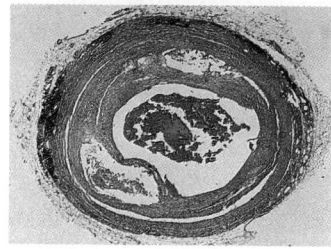

Figure 37-12 A normal artery, *top,* has smooth interior walls and a large diameter through which blood can flow. An atherosclerotic artery, *bottom,* has accumulated deposits on its interior walls. As a result, blood flow through the artery is greatly diminished.

◻ CAPSULE SUMMARY

Monitoring blood pressure and the electrical activity of the heart enables medical professionals to assess the condition of the heart. Cardiovascular diseases, like atherosclerosis and arteriosclerosis, are the leading cause of death in America.

SECTION 37-2

◉ VISUAL STRATEGY

Figure 37-12
Compare the two photographs. Point out the plaque (accumulated deposits of fat and calcium) on the walls of the atherosclerotic vessel. Explain the increase in pressure due to the reduced inner diameter of the atherosclerotic vessel with the following model. Tell students to imagine water flowing through a large-diameter hose that is connected to one with a smaller diameter. Water in the smaller hose will have to flow with a greater velocity in order to accommodate the volume delivered to it by the larger hose. Friction between the water and the wall of the hose will also be greater in the smaller hose than in the larger one, so pressure in the smaller hose will be higher. Ask students why restricted blood flow is a disadvantage. (*Tissues need a constant supply of materials, especially oxygen, in order to survive.*)

Section Review

1. *Is the blood that your heart pumps to your stomach part of the systemic or pulmonary circulatory loop?*

2. *A red blood cell loaded with oxygen must be located in which side of the heart?*

3. *Beginning with the excitation of the SA node, list the sequence of events that results in atrial and ventricular contraction.*

4. *If a person has severe atherosclerosis, would you expect their blood pressure to be low, normal, or high? Explain your answer.*

Critical Thinking

5. *The pulmonary artery is the only artery in the body that carries oxygen-poor blood. Explain why this blood vessel cannot be a vein.*

Answers to Section Review

1. It is part of systemic circulation.

2. It must be located in the left side of the heart.

3. First, the SA node initiates a wave of contraction in the top part of the heart (the atria).

Second, after a delay of about 0.1 second at the AV node, another wave of contraction spreads through fibers in the ventricles, causing them to contract.

4. Their blood pressure would be high due to the smaller inner diameter of the vessels.

5. The pulmonary artery takes blood away from the heart. All vessels conducting blood away from the heart have thicker walls that are adapted to withstand the higher blood pressures generated close to the heart.

Vocabulary Preview

respiration
pharynx
larynx
trachea
epiglottis
bronchi
thoracic cavity
diaphragm
alveoli
bronchiole
Bohr effect
carbonic anhydrase
asthma
emphysema
lung cancer

Opening Question

Ask students to describe the function of the respiratory system. It is important for students to understand that oxygen and carbon dioxide move in opposite directions in the lungs. Explain to them that the lungs obtain oxygen from the atmosphere and release it into the blood. The lungs also obtain carbon dioxide from the blood and release it into the atmosphere.

👁 VISUAL STRATEGY

Figure 37-13

Trace the path of air through the passages of the respiratory system. Point out to students how air can move into and out of the lungs by passing through either the nose or the mouth. Tell them that it is better to breathe in through the nose and out through the mouth. Air coming in through the nose is cleaned, warmed, and humidified before it enters their lungs.

37-3 The Respiratory System

One of the major tasks of the circulatory system is to transport oxygen to the cells of the body. Air is about 21 percent oxygen gas. When you breathe, air moves into the large, saclike organs in your chest called lungs. The oxygen in the air then moves into lung capillaries and is carried to the heart by the pulmonary veins. The simultaneous uptake of oxygen and release of carbon dioxide by your lungs is called **respiration**.

The Respiratory System Is a Network of Tubes and Air Sacs

The human respiratory system, shown in Figure 37-13, is composed of the respiratory passages, the lungs, and the thoracic cavity. Together, these structures function to provide your body with a constant supply of oxygen.

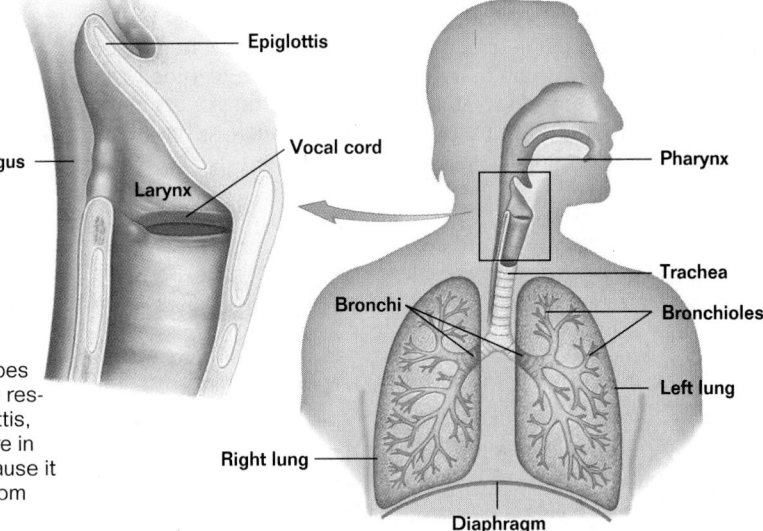

Figure 37-13 A network of tubes and two lungs compose the respiratory system. The epiglottis, *left,* is an important structure in the respiratory system because it prevents food and liquids from entering the trachea.

Respiratory Passages

Air normally enters your respiratory system through the nostrils. Hairs in your nose filter dust and other particles out of the air, and the epithelial tissue that lines the nasal cavity moistens and warms it.

The air then passes through the **pharynx** (FAR ingks), or upper throat, and the **larynx** (LAR ingks), or voice box. Next, the air enters the long, straight **trachea** (TRAY kee uh), the tube that carries air down to the lungs. A flap of tissue called the **epiglottis** (ehp uh GLAHT ihs) covers the opening to the trachea when you swallow. Without an epiglottis, food and liquids could easily pass into your lungs. Air moves down the trachea and then through two branches called **bronchi** (BRAHNG keye). Each bronchus enters a lung. Ciliated cells in the bronchi and trachea secrete mucus that traps any foreign particles that may remain in the air. This mucus is directed upward by the beating of the cilia. When it reaches the epiglottis, it is swallowed and digested. Bacteria and other microbes are destroyed by the strong acids and enzymes in the stomach.

Lungs

The lungs are suspended in the **thoracic cavity,** bounded on the sides by the ribs and on the bottom by the **diaphragm** (DEYE uh fram). A protective double membrane, called the pleural membrane, surrounds both lungs. The outermost membrane is attached to the the wall of the thoracic cavity, and the inner membrane is attached to the surface of the lungs. Between both membranes is a small space called the pleural cavity. Your lungs are among the largest organs in your body. The interior of each lung is not an open cavity like a bag. Instead, each lung is subdivided into about 300 million small chambers called **alveoli** (al VEE uh leye).

The alveoli, shown in Figure 37-14, are clustered together in groups, like bunches of grapes. Alveoli increase the surface area of your lungs to as much as 80 sq. m, or 42 times the surface area of your body. The alveoli are surrounded by many capillaries, enabling blood to flow over them in a nearly continuous sheet.

The alveoli are connected to the bronchi by a network of tiny tubes called **bronchioles** (BRAHNG kee ohls) within each lung. Most of the bronchioles have smooth muscle in their walls. ❑

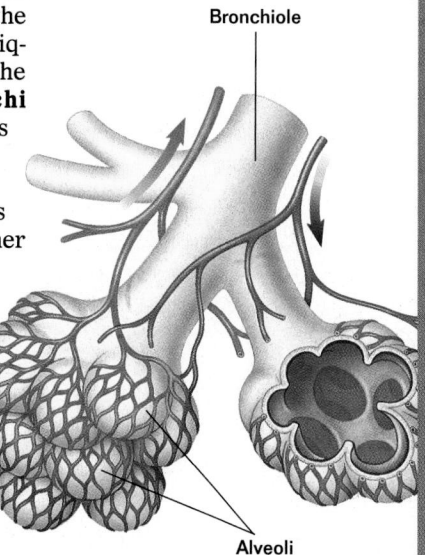

Bronchiole

Alveoli

Figure 37-14 The walls of the alveoli inside the lungs are extremely thin, enabling oxygen and carbon dioxide to easily pass through. Oxygen diffuses from the alveoli into the surrounding capillaries, and carbon dioxide diffuses from the capillaries into the alveoli.

❑ CAPSULE SUMMARY

Air is carried to the lungs by way of a branched system of tubes that looks much like an upside-down tree. The ends of the smallest branches inside the lungs are connected to microscopic air sacs called alveoli.

Inhalation and Exhalation Are Caused by Pressure Changes

Breathing is the result of pressure changes that occur inside the pleural cavity and lungs. Air moves into the lungs when the air pressure inside them is reduced by the expanding walls of the chest cavity, which contains a great deal of skeletal muscle. During inhalation, the diaphragm contracts and moves downward, and the rib cage moves upward and outward. In effect, this enlarges the volume of the chest cavity as it is pulled in all directions. The air pressure inside the pleural cavity and lungs decreases, causing air to rush into the lungs from outside the body.

Demonstration
The Breathing Tube
Have students gently rub the front of their neck to feel their trachea. Ask them to describe how it feels. *(They should say it feels rigid.)* Ask them why their trachea is not soft and muscular like their esophagus, the tube that leads to their stomach. *(The trachea must be rigid so that it does not collapse during the air pressure changes that occur in their respiratory system as they breathe.)*

Demonstration
Building Lungs
Demonstrate the role of the diaphragm in altering air pressure within the body. Place a one-hole stopper in the neck of a bell jar. Use petroleum jelly as a seal. Insert a Y-shaped tube into the stopper so that the two branches of the Y are inside the jar. Attach a small balloon securely to the end of each Y-tube branch. Firmly attach a thin rubber membrane over the mouth of the bell jar. Gently pull the membrane down. This action decreases air pressure within the bell jar. Air flows into the balloons in response to the reduced pressure inside the bell jar. Lead students in a discussion of how the action of this model approximates what happens to the lungs when the diaphragm contracts. Let the membrane return to its normal position and the balloons will deflate.

During exhalation, the diaphragm and ribs return to their
original resting position. The compression exerted by the ribs
and diaphragm forces air out of the lungs. The air pressure
changes that occur in the pleural cavity and lungs during
inhalation and exhalation are described in Figure 37-15.

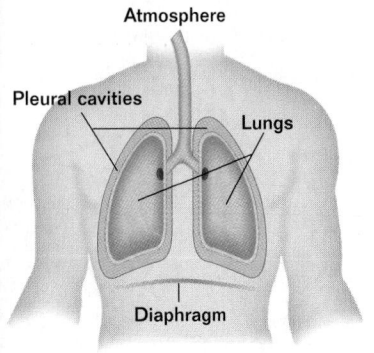
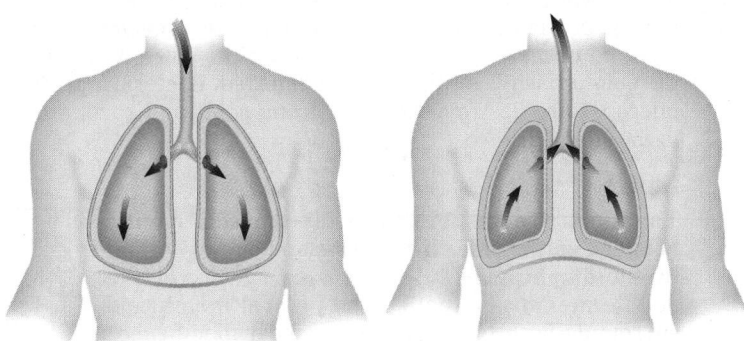

Figure 37-15 Immediately before
inhalation, *left,* the pressure
inside the lungs is equal to
atmospheric pressure. Once the
chest expands, *center,* the pressure inside the pleural cavity
decreases, causing the lungs to
push outward. As a result, the
pressure inside the lungs drops
below atmospheric pressure, and
air enters. During exhalation,
right, the pressure inside the
pleural cavity increases, compressing the lungs. Air is forced
from the lungs until the pressure
inside the lungs equals atmospheric pressure.

Breathing Exchanges O_2 and CO_2

When oxygen molecules diffuse from
the air into the cells of your alveoli,
their journey has just begun, as shown
in Figure 37-16. Passing into the plasma
of the bloodstream, the oxygen is picked
up by red blood cells that contain an oxygen-carrying protein, hemoglobin.

Oxygen Transport

Each hemoglobin molecule contains an atom of iron that
binds reversibly with oxygen. Therefore, at the appropriate
time, the oxygen can be released elsewhere in the body and
be taken up by the cells that need it. Hemoglobin is manufactured within red blood cells, giving them their red color.
Oxygen is simply loaded and unloaded from the hemoglobin
inside red blood cells.

Hemoglobin molecules act like little sponges, soaking up
the oxygen that diffuses into the red blood cells and causing
more oxygen to diffuse in from the blood plasma. At the high
oxygen levels that occur in the blood inside the lungs, most
hemoglobin molecules carry a full load of oxygen. Later, in
the tissues, oxygen levels are much lower, causing the hemoglobin to release its bound oxygen.

In tissues, the presence of carbon dioxide produced by
cellular respiration makes the blood more acidic and causes
the hemoglobin molecules to assume a different shape, one
that gives up oxygen more easily. This speeds the unloading
of oxygen from hemoglobin even more. The effect of carbon
dioxide on oxygen unloading is called the **Bohr effect.** The

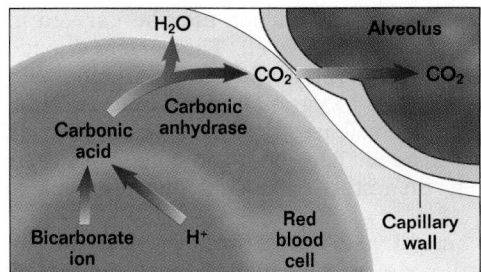

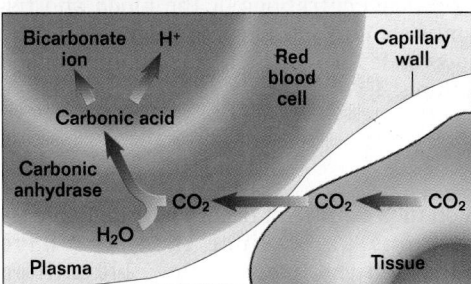

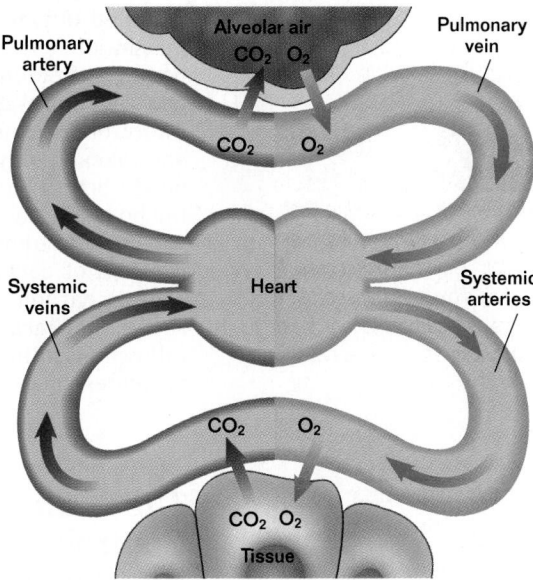

Figure 37-16
Students often find it difficult to remember the directions that oxygen and carbon dioxide molecules move within the lungs and within the tissues. Use this overview to help explain these events to them. It may be a good idea to begin by reviewing the pulmonary and systemic circulatory loops. Be sure to point out that, in the lungs, oxygen moves from the alveoli into the blood and carbon dioxide moves from the blood into the alveoli. In the tissues, oxygen moves from the blood into the tissues and carbon dioxide moves from the tissues into the blood. Tell students that the red blood cell shown in the bottom left part of the figure has just picked up carbon dioxide from the tissues and that the top left part of the figure shows that the red blood cell has traveled to the lungs, where it releases its carbon dioxide. Point out that most of the carbon dioxide transported by the blood travels in the form of bicarbonate ion.

Teaching Tip
Diffusion in the Body
Briefly review with students the process of diffusion. Have the students relate diffusion to the processes that occur with oxygen and carbon dioxide within the alveoli and in the tissues of the body.

Bohr effect is of great importance because it means that blood unloads oxygen more readily within those tissues that undergo a high rate of cellular respiration. These are the very tissues that need oxygen the most.

Carbon Dioxide Transport

At the same time that the red blood cells are unloading oxygen to tissues, they are also absorbing carbon dioxide. Only a tiny fraction of the carbon dioxide that the blood carries is dissolved in plasma (about 7 percent). About 23 percent is carried by the hemoglobin molecules inside red blood cells. The remaining 70 percent is carried within the cytoplasm of red blood cells.

How do the red blood cells manage to keep all of this carbon dioxide inside their cytoplasm? Why doesn't it simply diffuse back into the plasma, where carbon dioxide levels are lower? An enzyme within the red blood cells, called **carbonic anhydrase,** combines carbon dioxide molecules with water to form carbonic acid, H_2CO_3. The carbonic acid then dissociates (breaks apart) to form bicarbonate ions, HCO_3^-.

$$H_2O + CO_2 \longrightarrow H_2CO_3 \longrightarrow HCO_3^- + H^+$$

Bicarbonate ions are unable to pass through the membranes of red blood cells. The red blood cells carry their cargo of bicarbonate ions back to the lungs, where the low carbon dioxide concentration of the air causes the carbonic anhydrase reaction to proceed in the reverse direction.

$$HCO_3^- + H^+ \longrightarrow H_2CO_3 \longrightarrow H_2O + CO_2$$

As a result, gaseous carbon dioxide is released and diffuses from the blood into the alveoli. The carbon dioxide is then

Figure 37-16 This overview of respiration, *above,* shows that in the lungs, oxygen diffuses into the blood and carbon dioxide diffuses into alveoli. In the tissues, oxygen diffuses into the tissue's cells and carbon dioxide diffuses into the blood. Most carbon dioxide is transported in the form of bicarbonate ions inside red blood cells. Bicarbonate ions are chemically produced from carbon dioxide that has diffused from the tissue, through a capillary wall, and into a red blood cell, *bottom left.* In the lung, the chemical reaction is reversed. The carbon dioxide that is produced diffuses out of the red blood cell, through a capillary wall, and into an alveolus, *top left.*

 Did You Know?

Carbon dioxide does not bond to the iron in hemoglobin as does oxygen. Instead, it bonds to the polypeptide portion of hemoglobin. The uptake of carbon dioxide by the polypeptides reduces the effect that carbon dioxide has on the pH (acidity) of the blood. In other words, hemoglobin acts as a buffer. This effect is called the Haldane effect.

Teaching Tip
Interconnected Systems
Remind students of how the organ systems are interrelated. Ask them to create a *Graphic Organizer* like the one shown at the bottom of page 892 that demonstrates the relationship between the circulatory, cardiovascular, immune, lymphatic, and respiratory systems. If students are having difficulty, you can put an empty skeleton of the graphic organizer on the chalkboard or overhead projector, and have them fill in the boxes with the correct organ systems.

◀ VISUAL STRATEGY

Figure 37-17
Tell students that the respiratory control center is located in the medulla oblongata region of the brain. Normally, the center is stimulated by changes in the pH of the cerebrospinal fluid that surrounds the brain. When carbon dioxide levels in the blood are high, carbon dioxide can diffuse into the cerebrospinal fluid. Since cerebrospinal fluid has no buffer in it, small changes in carbon dioxide concentration quickly affect its pH. The medulla responds to the low pH by increasing the rate of respiration.

❑ CAPSULE SUMMARY

Oxygen is transported to tissues by combining with hemoglobin molecules inside red blood cells. Most carbon dioxide is transported to the lungs as bicarbonate ions inside the cytoplasm of red blood cells.

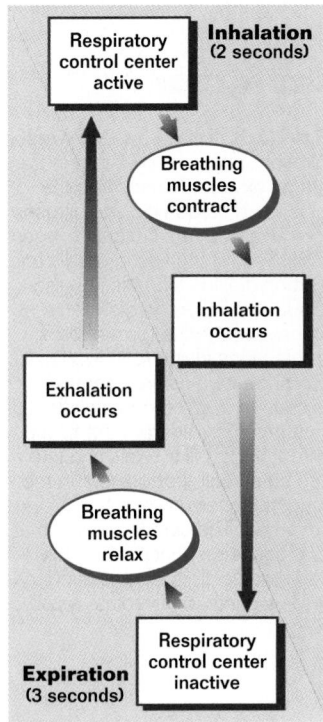

Figure 37-17 The respiratory control center in the brain initiates each breath. When the center sends out a nerve impulse, inhalation occurs. When the center is inhibited, exhalation occurs.

expelled during exhalation. The carbon dioxide carried in plasma diffuses into the alveoli as well. The carbon dioxide bound to hemoglobin is also released because hemoglobin has a greater affinity for oxygen than carbon dioxide. Therefore, the hemoglobin releases its bound carbon dioxide and takes up oxygen instead. The red blood cells, with their newly bound oxygen, then start their next journey back to the body's tissues.

The carbonic anhydrase reaction is critical to the removal of carbon dioxide from tissues because the difference in carbon dioxide concentrations of the blood and tissues is not large (only about 5 percent). We will see in the following section how the level of carbon dioxide in the blood regulates your breathing. ❑

The Brain Controls Breathing

You took your first breath within moments of being born. Since then, you have repeated the process over 200 million times. Every one of these breaths was initiated by the respiratory control center of the brain, as is shown in Figure 37-17.

When the body is at rest, the respiratory control center sends nerve signals to the diaphragm to initiate inhalations. During vigorous breathing, such as during exercise, the level of carbon dioxide in the blood increases, making the blood more acidic. The respiratory control center responds to this increased acidity by sending signals to the muscles between your ribs as well as to the diaphragm. The contraction of the rib muscles causes the chest cavity to expand even further, enabling you to take in more air. Right after inhalation occurs, other neurons from the respiratory center inhibit the stimulation of the diaphragm and the muscles between the ribs so that they relax, and the body exhales.

The respiratory control center also regulates breathing rate (how many breaths you take per minute). Again, blood pH is the controlling factor. When you are sleeping, your muscles are not being used very much, so they don't produce much carbon dioxide. As a result, blood pH remains relatively constant, and you breathe at a slow rate. When you run, intense muscle activity increases the level of carbon dioxide in the blood, and blood pH drops. In response, you breathe more rapidly.

The signals that travel from the breathing center of the brain are not subject to voluntary control. You cannot simply decide to stop breathing. You can hold your breath for a while, but ultimately your respiratory control center will take over and force your body to breathe.

Graphic Organizer
Use this graphic organizer in *Teaching Tip:* **Interconnected Systems** on page 892.

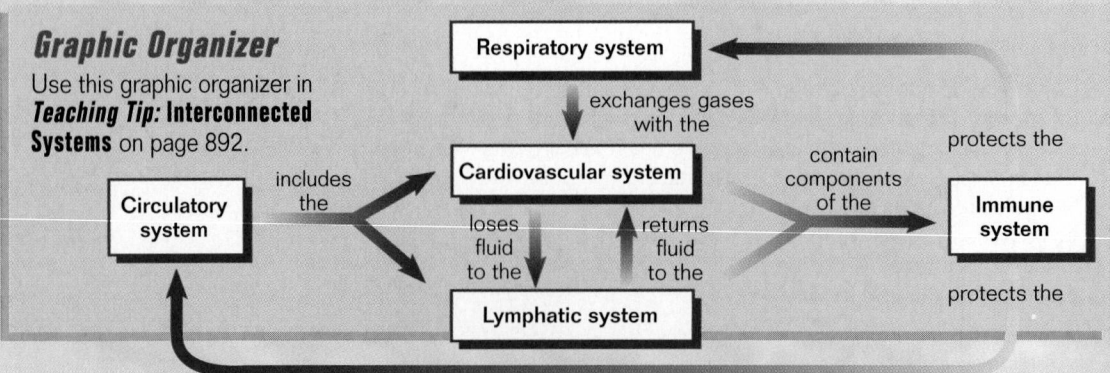

Respiratory Diseases Limit Lung Function

Respiratory diseases affect millions of Americans. **Asthma** is a respiratory disease in which the bronchioles of the lungs become constricted (narrowed) because of their sensitivity to certain stimuli in the air. The narrowing of the airways restricts air flow, making it difficult to exhale air from the alveoli. Left untreated, asthma can be dangerous. In severe asthma attacks, the alveoli may swell enough to rupture. Fortunately, inhalant medicines can counteract an asthma attack by dilating (expanding) the bronchioles.

Cigarette smoking has been linked to emphysema (*ehm fuh SEE muh*) and lung cancer, two respiratory diseases that claim millions of lives annually. Over 2 million American teenagers smoke, a pattern of behavior that may someday kill them. Almost no Americans begin smoking as adults. In people who develop **emphysema,** the lungs' alveoli lose their elasticity, making it difficult for the alveoli to release their air during exhalation. There is also a great reduction in the efficiency of gas exchange. Severely affected individuals must breathe from tanks of pure oxygen in order to live. Carcinogens present in cigarette smoke can also cause **lung cancer.** As you read in Chapter 9, cancer is a disease characterized by abnormal cell growth. Once the lung cancer, shown in Figure 37-18, is detected, the affected lung is usually removed surgically. Even with such drastic measures, fewer than 10 percent of lung cancer victims live more than five years after diagnosis.

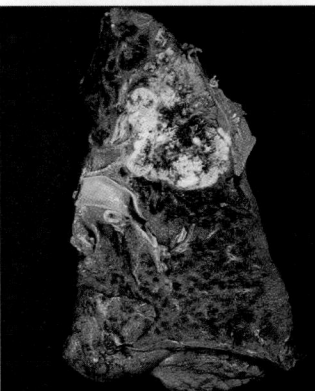

Figure 37-18 A healthy lung, *top,* contains normal cells and tissues. A cancerous lung, *bottom,* contains cells that divide uncontrollably. Scientific studies indicate that between 85 and 90 percent of all lung cancers are the direct result of cigarette smoking.

SECTION 37-3

👁 *VISUAL STRATEGY*

Figure 37-18
Have students compare and contrast the characteristics evident in the two photographs. Have them record their observations on paper and then answer the following question. What evidence in this figure suggests that a victim of lung cancer might be required to breathe pure oxygen? *(Student answers will vary but may include obstruction of airways within the lung by cigarette residues, breakdown of alveoli, and reduced oxygen transfer across damaged tissues.)*

Chapter Closure

Have students contact agencies such as the American Cancer Society and the American Heart Association for pamphlets concerning respiratory and circulatory diseases. Have them prepare short oral presentations about the information they discovered.

✉

Section Review

1. *Draw a diagram of the respiratory system.*

2. *After an exhalation is completed, in what direction will the diaphragm move next?*

3. *Explain the role that carbonic anhydrase plays in transporting carbon dioxide in the blood.*

4. *If the nerves going to the diaphragm were severed, would breathing be affected? Explain your answer.*

5. *Explain why a person with emphysema would have trouble climbing stairs.*

Critical Thinking

6. *Researchers have determined that cigarette smoke paralyzes the cilia in the trachea. Knowing this, explain why smokers often suffer from severe bouts of coughing.*

Answers to Section Review

1. See Figure 37-13.

2. downward

3. Carbonic anhydrase is an enzyme that is involved in the formation of bicarbonate ion from carbon dioxide and water. Most of the carbon dioxide in the blood is transported in the form of bicarbonate ions.

4. Yes, the medulla oblongata controls breathing through its nervous system connections with the diaphragm.

5. Emphysema causes a breakdown in the alveoli, resulting in reduced surface area for gas transfer. Exertion makes patients with emphysema unable to meet their oxygen demands.

6. Cilia in the trachea constantly sweep mucus upward. When the cilia are paralyzed, the mucus accumulates until the body tries to eject it by coughing.

CHAPTER REVIEW ANSWERS

Each item in the Chapter Review is correlated by text section in the assignment guide as follows.

ASSIGNMENT GUIDE

Section	Review	Themes Review	Critical Thinking
37-1	1–4, 9	12	15
37-2	5, 10		14
37-3	6–8, 11	13	16

Review

Multiple Choice

Code in parentheses indicates section number and objective number.

1. d (37-1.1)
2. b (37-1.3)
3. c (37-1.5)
4. a (37-1.5)
5. c (37-2.4)
6. a (37-3.2)
7. c (37-3.4)
8. c (37-3.5)

Completion

9. arterioles, veins (37-1.2)
10. atrium, aorta, pulmonary artery (37-2.2, 37-2.3)
11. bronchi, alveoli (37-3.1)

37 CHAPTER REVIEW

Vocabulary

A-B-O system (881)	diaphragm (889)	pharynx (888)
alveoli (889)	diastolic pressure (886)	plasma (879)
anemia (880)	emphysema (893)	platelet (881)
aorta (884)	epiglottis (889)	pulmonary valve (885)
aortic valve (884)	erythrocyte (880)	pulmonary vein (884)
arteriole (877)	fibrin (881)	respiration (888)
arteriosclerosis (887)	inferior vena cava (885)	Rh factor (882)
artery (877)	larynx (888)	right atrium (885)
asthma (893)	left atrium (884)	sinoatrial node (885)
atherosclerosis (887)	left ventricle (884)	superior vena cava (885)
Bohr effect (890)	leukemia (881)	systolic pressure (886)
bronchi (889)	leukocyte (880)	thoracic cavity (889)
bronchiole (889)	lung cancer (893)	trachea (888)
capillary (877)	lymph (879)	tricuspid valve (885)
carbonic anhydrase (891)	lymphatic system (879)	vein (877)
coronary artery (885)	mitral valve (884)	venule (877)

Review

Multiple Choice

1. Which is not a function of the circulatory system?
   a. gas exchange
   b. temperature regulation
   c. hormone transport
   d. action potential transport

2. Lymphatic vessels
   a. transport blood.
   b. return fluid to the blood.
   c. produce antibodies.
   d. control blood clotting.

3. Blood that contains A antibodies (but not B antibodies) in the plasma and lacks Rh antigens is labeled
   a. AB negative.   c. B negative.
   b. A positive.   d. O positive.

4. Blood traveling through the pulmonary veins is
   a. oxygenated.   c. oxygen-poor.
   b. iron-poor.   d. calcium-rich.

5. A sphygmomanometer is used to measure
   a. heart rate.   c. blood pressure.
   b. lung volume.   d. breathing rate.

6. The diaphragm contracts and the pressure in the pleural cavity decreases during
   a. inhalation.   c. respiration.
   b. exhalation.   d. asthma attacks.

7. Breathing rate will automatically increase when
   a. blood pH is high.
   b. the amount of carbon dioxide in the blood decreases.
   c. blood acidity increases.
   d. hemoglobin is unloaded.

8. A disease in which the alveoli lose their elasticity is called
   a. atherosclerosis.   c. emphysema.
   b. arteriosclerosis. d. asthma.

Completion

9. Blood flows from arteries to _____ and then capillaries. From capillaries, blood flows through venules and into _____ before returning to the heart.

10. In each side of the heart, blood flows from the _____ to the ventricle. Blood flows into the _____ from the left ventricle and into the _____ from the right ventricle.

11. During inhalation, air flows through the trachea, into the right and left _____ , and then into the bronchioles before reaching the _____ of the lungs.

Themes Review

12. **Homeostasis** How is body temperature regulated by blood vessel constriction and dilation?

13. **Levels of Organization** The site of gas exchange in the human body is the lungs. How do alveoli function to increase the overall efficiency of the lungs?

Critical Thinking

14. **Interpreting Data** Salt intake and blood pressure data were collected from people representing more than 20 cultures, ranging from factory workers in France to natives of Colombia. The data are displayed as a best fit curve on the graph below. What is the relationship between salt intake and blood pressure? If an American businessman averages a daily salt intake of 26 g, what would you predict his systolic pressure to be? Knowing that high blood pressure can lead to a stroke or kidney failure, what advice would you give this businessman?

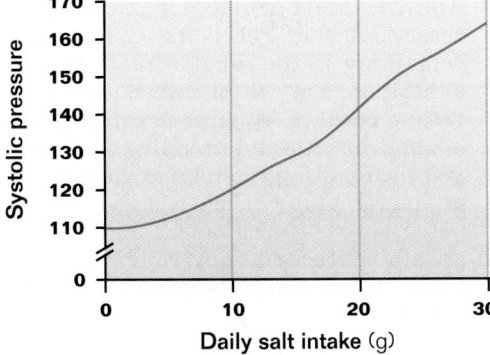

15. **Making Inferences** The frequency of blood clots and heart attacks is much lower among the Inuit, the nomadic hunters of the North American Arctic, than it is among other North Americans and Europeans. This difference is due to fish oils in the Inuit diet that cause their platelets to be more slippery. How do you think the clotting ability of the Inuit's blood is affected by the slippery platelets?

16. **Making Predictions** As altitude increases, the gases that make up the atmosphere, including oxygen, become more scarce. When a runner who trained at sea level competes at a location 500 m above sea level, how will his performance at this altitude compare with his performance during training?

Activities and Projects

17. **Cooperative Group Project** With a small group of classmates, obtain training in first aid procedures from the American Red Cross. Be sure your training includes the Heimlich maneuver and cardiopulmonary resuscitation (CPR). After receiving first aid training, organize training sessions with Red Cross officials for students and teachers at your school. Build interest in the first aid training by constructing and displaying posters that describe or picture medical emergencies and ask if the reader is prepared to help save a life.

18. **Multicultural Perspective** Dr. Charles Drew was an African American physician who lived during the first half of the twentieth century. Research the life of Dr. Drew and his many accomplishments, and how his work with blood helped save many lives.

Themes Review

12. Blood vessel constriction reduces heat loss to the environment. Blood vessel dilation allows heat to escape to the environment. (37-1.1)

13. Alveoli, the tiny air sacs, increase the surface area of the lung. (37-3.1)

Critical Thinking

14. As salt intake increases, blood pressure also increases. The businessman's systolic pressure would be about 150. He should reduce his daily salt intake. (37-2.4)

15. The slippery platelets would be less likely to get caught on the damaged site of a blood vessel. Also, if a clot does begin to form, slippery platelets could pass through the fibrin net, thus slowing the clotting process. (37-1.4)

16. Due to the altitude, he will likely experience shortness of breath and fatigue and his performance will be poorer. (37-3)

Activities and Projects

17. A telephone number for the American Red Cross should be listed in your local telephone directory. So as not to frighten people away from first-aid training, encourage students not to make the posters extremely graphic.

18. In 1931, Dr. Charles Drew graduated second in his class of 137 at McGill University Medical School in Canada. He went on to become the first African American physician to receive a Doctor of Science (Sc.D) from Columbia University in 1940. Dr. Drew directed blood bank projects for England and the American Red Cross during World War II, and his work in blood research led to the use of plasma to treat surgical shock.

*L*ABORATORY *Investigation* Chapter 37

Procedural Notes

- This investigation should take 50 minutes to complete. Have half of the students perform Part A while the other half perform Part B, then switch.
- To prevent the embarrassment of students not comfortable with exercising in front of their peers, you may wish to ask for students to serve as volunteer participants for Part B. These students should not have health conditions that would preclude sustained vigorous exercise.
- Review the use of a spirometer and the terms *tidal volume, vital capacity,* and *expiratory reserve volume.*
- **Safety:** Remind students to wear safety goggles for Part B and to avoid skin and eye contact with methyl red and sodium hydroxide.
- You may wish to have students record their data in a class data table on the chalkboard. They need not identify themselves by name, only by gender, athlete or nonathlete, smoker or non-smoker. Calculate means for males, females, smokers, non-smokers, etc.

Preparation

Inexpensive spirometers are available from Ward's. Provide plenty of clean spirometer mouth pieces for students to use. Direct students not to share spirometer mouth pieces. To make 0.5 L of 0.1% methyl red acid-base indicator, measure 0.5 g of methyl red and add enough distilled water to make 500 mL in a beaker. Prepare several dropper bottles of 0.1% methyl red solution for each class. Caution students not to inhale methyl red as they blow through their straws into the solution. Work in a ventilated hood and wear goggles, a face shield, impermeable gloves, and a lab apron when you prepare the 0.04% sodium hydroxide solution. In case of a spill, dilute first with water. Then mop up

Lung Capacity and Carbon Dioxide Production

OBJECTIVES

- Indirectly measure lung capacity.
- Measure the amount of carbon dioxide produced before and after exercise.

PROCESS SKILLS

- observing and comparing differences
- charting effects of experimental results

MATERIALS

- spirometer
- safety goggles
- marker
- 250 mL flasks
- 0.1% methyl red in dropper bottle
- plastic wrap
- drinking straws
- 10 mL and 100 mL graduated cylinders
- graduated pipettes with rubber bulbs
- 0.04% sodium hydroxide in dropper bottle

BACKGROUND

1. Where does the carbon dioxide we exhale come from?
2. What is lung capacity?
3. How does exercise influence the amount of carbon dioxide you produce?
4. Write your own question to explore in your lab report or notebook.

TECHNIQUE

Part A: Determining Lung Capacity

CAUTION: Do not share spirometer mouth-pieces; change them as your teacher directs.

1. Tidal volume is one of several measurements that can be made by a spirometer. Tidal volume is the amount of air inhaled or exhaled in a normal breath. Remember that your lung capacity is influenced by many factors (age, diaphragm and chest muscle strength, disease, gender, and body position while testing).

2. With the tube of the spirometer near your mouth, inhale a normal breath. Then exhale a normal breath into the spirometer tube and take your reading. In the **Records** section of your report, record your results in a table similar to the one below.

Measurements of Lung Capacity

	Avg. for young adult male	Your readings
Tidal volume	500 mL	
Expiratory reserve volume	100 mL	
Vital capacity	4600 mL	

3. Another measurement is expiratory reserve volume. This is the amount of air remaining in the lungs after a normal exhalation. You will measure your expiratory reserve by first breathing in a normal breath and exhaling normally and then putting the tube to your mouth as you forcefully exhale whatever is left, being sure to force out as much air as possible. Record your reading in the table in the **Records** section of your report.

the spill with wet cloths designated for spill cleanup. Avoid having students handle base spills themselves. To make 1 L of 0.04% sodium hydroxide solution, add 100 mL of 0.4% sodium hydroxide to about 800 mL of distilled water in a beaker, and dilute to 1 L. Have students wear goggles, impermeable gloves, and a lab apron when handling sodium

hydroxide. Students should avoid skin and eye contact with methyl red and sodium hydroxide. Students will also need a marker, plastic wrap, two drinking straws, three 250 mL flasks, a 100 mL graduated cylinder, and a graduated pipette with a rubber bulb. After the investigation, collect the sodium hydroxide solutions in a single container, and neutralize to

a pH of 7 with a dilute acid before discarding in the sink.

Background Answers

1. Carbon dioxide is a product of cellular respiration that enters the bloodstream and diffuses through the membrane of the alveoli of the lungs; the lungs then exhale the carbon dioxide.

4. Another measurement determined with a spirometer is vital capacity. This is the maximum amount of air that you can possibly inhale or exhale. Inhale the biggest breath you can, and then exhale all of the air possible into the spirometer tube. Record your reading in the table in the **Records** section.

5. The table on the previous page includes the values for young adult males. The average volumes for young adult females are 20–25% lower. Athletes can have volumes 30–40% greater than the average for their gender.

Part B: Carbon Dioxide Production

6. **CAUTION: Wear safety goggles at all times during this procedure.** Label and mark three flasks as 1, 2, and "Control."

7. **CAUTION: Avoid skin and eye contact with methyl red and sodium hydroxide.** Add 100 mL of tap water into each flask. Add 10 drops of methyl red to each and swirl to mix. Cover the mouth of each flask with plastic wrap.

8. Remove the plastic wrap from flask 1. Blow gently through one straw into flask 1 for exactly 2 minutes, exhaling slowly so that the solution does not bubble up. Be careful not to inhale the solution.

9. Using your pipette with a bulb, add NaOH 1 mL at a time, swirling with each addition and counting the milliliters you add to get the same pink color as in the control flask. In the **Records** section of your report, record the number of milliliters of NaOH added.

10. Exercise by jogging in place or doing jumping jacks for 2 minutes. Immediately blow gently through a new straw into flask 2 for exactly 2 minutes.

11. Repeat step 9 on flask 2 until you get the same pink color as in the control flask.

Record the number of milliliters of NaOH added to flask 2 in the **Records** section. In your report, also briefly summarize the procedure you followed.

12. Clean up your materials and wash your hands before leaving the lab.

INQUIRY

1. How did your tidal volume compare with that of your classmates?
2. What color change was observed in Part B when you blew into the flask?
3. What part of respiration is measured indirectly in Part B?

ANALYSIS

1. Why would males and athletes have greater vital capacities than females?
2. If a person's vital capacity was extremely low, how could it be increased?
3. How do you know if you produced more carbon dioxide before or after you exercised? What is your evidence from this lab?
4. Why were the flasks covered with plastic wrap?

FURTHER INQUIRY

Write a new question that could be explored as a new investigation. The following is an example:

How do asthma and emphysema reduce the efficiency of the respiratory system as measured by a spirometer?

Inquiry Answers

1. Answers will vary.
2. The solution becomes colorless.
3. The cell respiration rate is indirectly measured. It may be slowly or rapidly producing energy for the body.

Analysis Answers

1. Vital capacity depends mostly on body size. Respiratory muscle strength and the expansion capacity of the lungs are also important. Males are generally larger than females. Athletes have stronger respiratory muscles and greater expansion capacity.
2. Vital capacity can be increased by consistent rigorous exercise to strengthen the diaphragm and internal and external intercostal muscles.
3. Greater CO_2 is put out after exercise, as indicated by more NaOH being needed to return flask #2 to its original color.
4. This prevented any CO_2 in the room air from reacting with the water in the flask.

2. Lung capacity is the total amount of air contained in the lungs at the end of a forced inhalation.

3. Exercise increases the amount of CO_2 produced, since it increases cellular respiration.

4. What differences in lung capacity exist among individuals?

Records

Table 1 Measurements of Lung Capacity

	Avg. for young adult male	Your readings
Tidal volume	500 mL	Answers will vary.
Expiratory reserve volume	100 mL	Answers will vary.
Vital capacity	4600 mL	Answers will vary.

Table 2 Carbon Dioxide Production

	Volume of 0.04% NaOH added to balance acidity (index of CO_2 production)
Flask 1 (rested)	
Flask 2 (after exercise)	

Block Scheduling Guide

Each block represents about 45–50 minutes of instructional time. The resources cited in a block are coded to a specific program component using the Key on the next page.

Lecture Concepts

Lesson Resources

38-1 Diet: What We Need to Eat and Why pp. 899–903

BLOCKS 1 and 2

a. Carbohydrates, proteins, and fats are energy-rich compounds present in food. Cells usually use carbohydrates as an energy source, although they are capable of using proteins and fats.

b. Essential amino acids, vitamins, and trace elements are necessary for proper health. Essential amino acids are used as building blocks for proteins. Vitamins and trace elements participate in many cellular chemical reactions.

Lecture Resources
- Opening Questions, page 899
- A Three-Dimensional Model of a Fat, page 899
- Visual Strategies: Figure 38-2, Table 38-1
- Food Preparation, page 900
- Diagosing Deficiencies, page 902
- Overcoming Misconceptions, page 902
- Research Update, page 903
- Teaching Transparencies: 222, 190A, 191A, 192A

- Holt Biology Videodiscs: Lesson 38-1

Classwork Options
- Effective Reading, page 901
- World Diets, page 901
- Quick Labs A33: Reading Labels: Nutritional Information
- Laboratory Techniques and Experimental Design C46: Identifying Food Nutrients

Assignment Options
- Section Review, page 903
- Chapter Review, questions 1, 2, 8, 9, 14

38-2 Digestion pp. 904–909

BLOCKS 3 and 4

a. Amylases, lipases, and proteases are three classes of digestive enzymes.

b. Carbohydrate digestion begins in the mouth, and protein digestion begins in the stomach. Digestion is completed in the small intestine.

c. The end products of digestion are absorbed through intestinal villi and pass into the bloodstream. The waste products of digestion are stored in the large intestine before passing out of the body.

Lecture Resources
- Opening Questions, page 904
- Demonstration: Peristaltic Action
- Bile and Fats
- Visual Strategies: Figures 38-7, 38-8
- Organizing Information, page 907
- Teaching Transparencies: 216, 219, 193A
- Holt Biology Videodiscs: Lesson 38-2

Classwork Options
- Observing and Drawing, page 906
- Laboratory Techniques and Experimental Design C47: Identifying Food Nutrients—Food Labeling

Assignment Options
- Section Review, page 909
- Chapter Review, questions 3, 4, 10, 15

38-3 Excretion pp. 910–915

BLOCKS 5 and 6

a. The major waste products excreted by humans are carbon dioxide, water, and urea.

b. Carbon dioxide and water are excreted by the lungs when you exhale.

c. The kidneys remove water and urea (and some salts) from the blood to form urine. The microscopic blood-filtering units of the kidneys are nephrons.

Lecture Resources
- Opening Questions, page 910
- Demonstration: Inside a Kidney
- A Magnified View, page 911
- Visual Strategies: Figures 38-14, 38-15, 38-16
- Summarizing a Function, page 914
- Teaching Transparencies: 218, 220, 221
- Holt Biology Videodiscs: Lesson 38-3

Classwork Options
- Note Taking, page 912

- Learning Tool, page 913
- Using the Library, page 915
- Closure, page 915
- Laboratory Investigation: Lactose Digestion, pages 918–919
- Laboratory Techniques and Experimental Design C48: Urinalysis Testing

Assignment Options
- Section Review, page 915
- Chapter Review, questions 5–7, 11–13

Review/Enrichment

BLOCK 7

■ Study Guide: Concept Review and Pretest

Assignment Options

◉ Teaching Resources CD-ROM: Occupational Applications Worksheets, Dietitian

◉ Teaching Resources CD-ROM: Problem-Solving Worksheets, Using Food Labels to Calculate Percentage of Nutrients and Calories

■ Activities and Projects, questions 16–18

Assessment Options

BLOCK 8

Traditional Assessment

■ Chapter 38 Test

◉ Test Generator/Assessment Item Listing: Software item bank for preparing customized chapter tests.

Portfolio Assessment

Select student reports for one or more laboratory experiments from this chapter. *The Direct Observation Checklist* on the *BioSources Teaching Resources CD-ROM* should be completed during a laboratory or other cooperative learning experience.

Holt Biology Videodiscs

Holt Biology Videodiscs gives you a powerful tool for teaching, review, and assessment. *Concepts of Biology* adds interactive lesson plans and extensive tools for customization using CD-ROM technology.

CONCEPTS OF BIOLOGY

Use the following topic from *Concepts of Biology* to help you teach this chapter:

■ Topic 23: Digestion and Excretion

For further information regarding the media on the videodiscs, see *Holt Biology Videodiscs Teacher's Correlation Guide to Biology: Principles and Explorations.*

Chapter Theme

Homeostasis *The human body requires a very special type of balance. It is a balance that depends on all of the systems of the body to maintain their own dynamic equilibrium. The metabolic processes that each system performs require materials and energy ultimately obtained from food that has been digested. Likewise, the same metabolic processes often generate toxic waste products. Ultimately, these materials are removed from the body by the process of excretion.*

Tapping Prior Knowledge

- What type of digestion does the human body use, intracellular or extracellular?
- Movement of materials in the human digestive system can be described as one-way traffic. Why is this description appropriate?
- What is exocytosis? Why is this process especially important in glandular tissues?

Opening Demonstration

Show students a model or a life-sized diagram of the human digestive system. As a visual comparison of the relative length of the parts of the digestive system, cut pieces of string in the following lengths to represent the following organs: esophagus—25 cm, stomach—30 cm, small intestine—7 m, and large intestine—1.5 m. Hang the pieces of string near the chalkboard and refer to their relative lengths as you discuss the parts of the digestive system.

CHAPTER 38

DIGESTIVE AND EXCRETORY SYSTEMS

REVIEW

- pH scale (Section 2-2)
- carbohydrates, fats, and proteins (Section 2-3)
- diffusion and osmosis (Section 3-2)
- homeostasis (Sections 3-2 and 35-3)
- cellular respiration (Sections 5-1 and 5-3)
- water and salt balance in vertebrates (Section 31-4)
- negative feedback (Section 35-3)
- antigens (Section 37-1)
- carbon dioxide transport (Section 37-3)
- *Highlights: Digestive and Urinary Systems (p. 845)*

Villi in human intestine

Authors' Rationale

Students are not likely to think about ATP when they eat a pizza, but now is the time to remind them about the energy requirements of living things. Humans are heterotrophs—they have to eat food to get energy. What elements should compose the human diet? Why? How can eating the proper foods play a significant role in maintaining health? The excretory system is responsible for removing nitrogen wastes from the body, not solid wastes, as is often assumed. Plasma levels of salts, acids, bases, and water are also regulated by the excretory system, and so this system plays a crucial role in maintaining homeostasis.

38-1 Diet: What We Need to Eat and Why

You obtain energy from the foods you eat to fuel your every activity. Your body uses energy to move, to grow, and to do countless other things—like read these words. The energy in food is stored in its chemical bonds. As you learned in Chapter 5, your cells break the chemical bonds of food molecules and harvest their energy to make ATP.

In addition to providing energy, the food you eat provides raw materials that your body uses to manufacture molecules for its own use. For example, calcium is used to make bone, and amino acids are used to make proteins. Every molecule in your body, every hair and bone and cell, is built from raw materials that came from food you ate.

Section Objectives

- *Describe how the body uses carbohydrates, fats, and proteins.*
- *Explain what essential amino acids are.*
- *Compare and contrast vitamins and trace elements.*

Energy-Rich Compounds Are Used to Make ATP

Much of your need for food is the result of your body's constant use of and need for energy. Everything that you eat—meals and snacks and soda at the movies—is your diet, a record of what you consume. An optimal diet contains more carbohydrates than fats and also contains a significant amount of protein. To help educate consumers about the food choices they make, the United States government has set new standards for the nutrition labels that appear on food products. Figure 38-1 shows you one of these labels and explains how to interpret it.

Figure 38-1 The Food and Drug Administration's new system for labeling foods closely reflects the dietary guidelines developed by the American Heart Association. The label is not as complex as it looks. Study this sample label from a frozen dinner to learn how to interpret the new system.

1. **Serving Size** Serving sizes formerly were unrealistically small. New serving sizes set by the FDA are closer to the amounts that people actually eat.
2. **Calories From Fat** Beware of any food that gets more than one-third of its Calories from fat.
3. **% Daily Value (DV)** This tells you what percentage of the daily requirement for a nutrient you are getting.
4. **Total Fat** Keep a close eye on saturated fats; they cause clogged arteries.
5. **Cholesterol** High levels of fatlike cholesterol molecules in the blood can also lead to clogged arteries. Junk foods made from vegetables, like corn chips, may be advertised as cholesterol-free, but they are often high in other fats.

Nutrition Facts

1	Serving Size 1 Package (258 g)	
	Servings Per Container 1	
	Amount per Serving	
2	**Calories** 270	From Fat 70
3		**% Daily Value**
4	**Total Fat** 8 g	**12%**
	Saturated Fat 3.5 g	**15%**
	Polyunsaturated Fat .5 g	
	Monosaturated Fat 1.5 g	
5	**Cholesterol** 30 mg	**9%**
6	**Sodium** 500 mg	**20%**
7	**Total Carbohydrate** 28 mg	**9%**
	Dietary Fiber 3 g	**13%**
	Sugars 5 g	
8	**Protein** 21 g	
9	Vitamin A 8% • Vitamin C 20%	
	Calcium 35% • Iron 6%	

6. **Sodium** Sodium is abundant in table salt and many food products. Too much sodium can cause high blood pressure.
7. **Total Carbohydrate** Getting enough carbohydrates, especially complex ones, like starches, is important. Carbohydrates are used by your cells as a source of energy.
8. **Protein** Although the government has not set a percent DV for proteins, it is generally agreed they should compose no more than about 15 percent of your total daily Calories.
9. **Vitamins and Minerals** A DV of about 10 percent indicates that a food is a good source of these essential nutrients.

Historical Note

The Discoverer of ATP

ATP is the energy currency of all living cells, including yours. Fritz Lipmann isolated ATP from muscle tissue in 1929. In research beginning in 1941, he discovered that ATP and similarly structured compounds shuttle energy around the cell once the energy is captured from carbohydrates, fats, and proteins during cellular respiration. In 1953 Lipmann shared the Nobel Prize in physiology or medicine for his discoveries.

SECTION 38-1

Vocabulary Preview

carbohydrate
Calorie
protein
fat

Opening Questions

Ask students to name the one compound that their cells prefer to use as an energy source. Tell them that glucose, the molecule that supplies energy to cells, is a type of carbohydrate. Then ask them to name some of the high-carbohydrate foods that they have eaten during the past day or two. Stress to them that foods that contain complex carbohydrates, like fruits, vegetables, and grains, are very healthy for their bodies. These kinds of foods supply them with energy, vitamins, and minerals and may even help prevent certain diseases, like colon cancer. Finish by telling them that even though carbohydrates in the diet are important, it is also important that they consume some fat and protein. Then ask them why. If they have difficulty coming up with an answer, ask them to identify the two major components of the cell membrane.

Demonstration

A Three-Dimensional Model of a Fat

Using gumdrops as atoms and toothpicks as chemical bonds, construct a glucose molecule and a fat molecule. Review the concept of covalent bonds with students. Reiterate that most of the energy we obtain from food is stored in the food's carbon-to-hydrogen bonds. It should be obvious to students why fats contain so much more energy per gram than carbohydrates when they see the ratio of hydrogen atoms to carbon atoms in both types of compounds.

VISUAL STRATEGY

Figure 38-2

Have students study Figure 38-2, and then ask them the following questions. Why are carbohydrate-rich foods an especially good choice for weight-conscious people? *(Carbohydrate-rich foods take the body longer to digest, while releasing only 4 Calories per gram of carbohydrate. Longer digestion time means your body can use more of the energy stored in carbohydrates for cellular processes and store less in the form of fat.)* Comparing the three lists of foods, why are carbohydrate-rich foods a better choice to eat than protein-rich foods? *(Protein-rich foods tend to contain hidden fat. See the foods which appear on both the fat and protein lists.)*

Teaching Tip
Food Preparation

Bring some French fries and a baked potato to class, or show students a picture of these foods. Have students examine the baked potato and the French fries carefully, and ask them to brainstorm differences between the two foods—in their preparation, cooking, taste, and nutritional value. *(Unlike the baked potato, French fries are peeled, sliced, fried in oil or fat, and salted.)* Ask students how the body is affected by a diet high in oils, fats, and salt, that is, foods processed like French fries. *(Oils and fats increase caloric intake. Saturated fats may increase cholesterol levels. A diet high in salt can cause fluid to be retained in the body. Over a long period of time, high salt intake can contribute to hypertension.)*

Figure 38-2 Carbohydrate-rich foods, *left,* should compose a major portion of your diet (50 to 55 percent). The majority of these foods should be foods such as breads, pastas, fruits, and vegetables, which contain complex carbohydrates and natural sugars. Foods containing refined sugars like candies and soft drinks, should be consumed sparingly. Protein-rich foods, *center,* should compose only about 15 percent of your diet, and fatty foods, *right,* no more than 30 percent of your diet. Fats contain more than twice as many Calories per gram as carbohydrates or proteins, so fat-rich foods should be avoided if you are trying to lose weight.

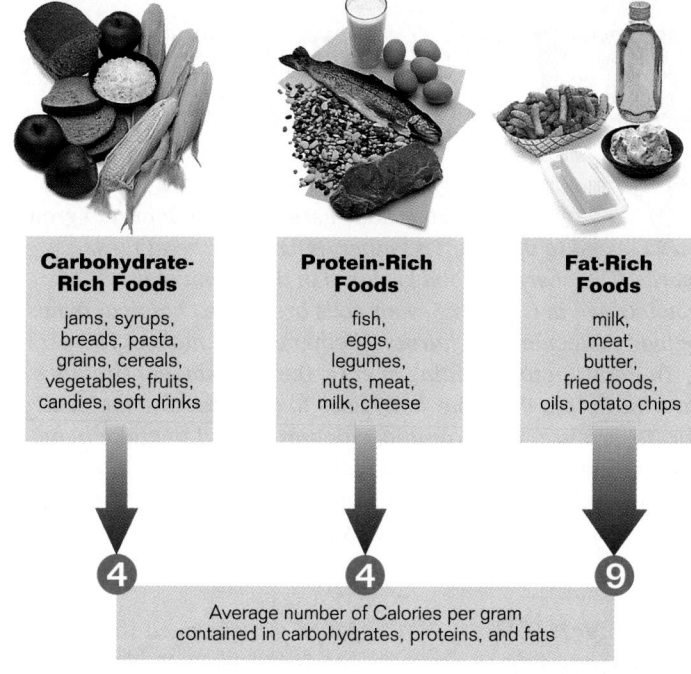

Carbohydrate-Rich Foods	Protein-Rich Foods	Fat-Rich Foods
jams, syrups, breads, pasta, grains, cereals, vegetables, fruits, candies, soft drinks	fish, eggs, legumes, nuts, meat, milk, cheese	milk, meat, butter, fried foods, oils, potato chips

4   4   9

Average number of Calories per gram contained in carbohydrates, proteins, and fats

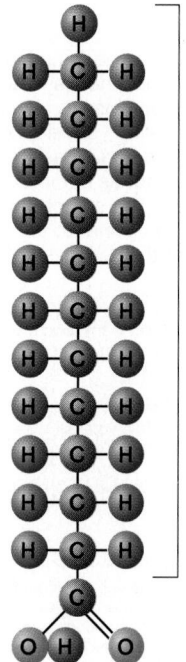

Long carbon-hydrogen chain

Figure 38-3 Fats are composed of molecules called fatty acids. The many carbon-hydrogen bonds of a fatty acid molecule store a great deal of energy.

Carbohydrates are a major energy source for your cells. The carbohydrates in a typical human diet are obtained primarily from cereals, grains, breads, fruits, and vegetables like the ones shown in Figure 38-2. On average, carbohydrates contain 4 Calories per gram. A **Calorie** is a unit of energy that indicates the amount of energy contained in food. Specifically, a Calorie is the amount of heat energy required to raise the temperature of 1 kg (2.2 lb.) of water 1°C (1.8°F). The greater the number of Calories in a quantity of food, the more energy it contains.

Proteins, like carbohydrates, contain 4 Calories per gram. Proteins are obtained from dairy products, poultry, fish, meat, and grains. Although cells can use proteins as an energy source, your body uses them mainly as building materials for cell structures, enzymes, hormones, muscles, and bones.

Fats are obtained from oils, margarine, and butter. They are abundant in fried foods, meats, and processed snack foods such as potato chips and crackers. Your body uses fats to construct cell membranes and other cell structures, to insulate nervous tissue, and as an energy source. Fats also contain certain fat-soluble vitamins that are essential for proper health.

A fat molecule with its long fatty acids, like the one shown in Figure 38-3, contains far more energy-rich carbon-hydrogen chemical bonds than a carbohydrate or protein molecule does. Therefore, fats contain much more energy per gram than carbohydrates or proteins do, about 9 Calories per gram. Although fats may be used as an energy source by cells, fats are also a very efficient way to store energy in the body.

 Did You Know?

Until recently, scientific research literature used the calorie (1/1000th of a Calorie) to describe energy changes. Since the advent of the International System of Units (abbreviated SI, for the French *le Système International d'Unités*), the preferred unit to use is the joule. However, since we cannot go back and change any of the older research literature, scientists must sometimes convert back and forth between units. One calorie equals 4.184 joules, so one Calorie equals 4,184 joules.

Calories that are not used by your body are converted into fat tissue. When you eat a meal, your body can do only two things with the energy it gets from the food. The energy is either used by the muscles and other cells of the body, or it is converted into fat and stored in fat cells. This relationship can be expressed by the following simple equation.

energy obtained from food
− energy used by body's cells

= energy stored in fat

In wealthy countries like ours, obesity, being more than 20 percent heavier than the average person of the same sex and height, is common. Obesity usually results from habitually overeating an unbalanced diet that is too high in fat. In the United States, about 30 percent of middle-aged women and 15 percent of middle-aged men are classified as obese. Obesity significantly increases an individual's risk of diabetes, coronary heart disease, and many other disorders. ◻

☐ **CAPSULE SUMMARY**

Carbohydrates, proteins, and fats are energy-rich compounds present in food. Your cells usually use carbohydrates as an energy source, although they are also capable of using proteins and fats.

Other Compounds Serve as Raw Materials

Over the course of evolution, many animals have lost the ability to manufacture some of the substances they need. These substances often play critical roles in metabolism. Typically, this ability is lost because a substance has become plentiful in the animal's diet. For example, mosquitoes and many other bloodsucking insects cannot manufacture cholesterol (a critical component of cell membranes). But that does not matter because the blood they consume is rich in cholesterol. There are essential substances that humans can get only from the foods they eat. These substances include certain amino acids, vitamins, and trace elements.

Amino Acids

Humans are unable to manufacture 8 of the 20 amino acids used to make proteins: lysine, tryptophan, threonine, methionine, phenylalanine, leucine, isoleucine, and valine. These eight amino acids are called essential amino acids because without them in your diet, your cells would be unable to manufacture the many proteins they need. For this reason, it is important to eat complete proteins, those containing all of the essential amino acids.

Vitamins

Vitamins are essential organic substances that the body requires in tiny amounts for normal growth and activity. Just as a computer uses only a tiny amount of grease (to lubricate the spinning hard disk) but cannot function without it, so our bodies use minute traces of vitamins to maintain good health.

Teaching Tip
World Diets
Divide the class into groups of three to five members. Explain that members of each group will work cooperatively to develop a report on dietary conditions around the world. Have members of each group work together to decide which countries or regions of the world they would like to investigate. Each member should then research his or her assigned country or region, ascertaining the foods that are eaten, the nutritional strengths and weaknesses of the diet, and nutritional disorders associated with the diet.

After some information has been gathered, the group should meet and discuss the focus of the report. Each member should then write and illustrate his or her segment, which will then be combined into a group report to be shared with the class.

CONNECTIONS

Chapter 2
Protein Structure and Function
There are an enormous number of different proteins in the human body, and each kind has a unique function. Many proteins are enzymes that speed up chemical reactions. Others help build tissues or protect you from disease. Ultimately, the function of any protein is determined by its primary structure, the sequence of its amino acids.

Effective Reading

Making Tables to Summarize Information
Teachers need to share their knowledge of *how* to learn as well as what to learn. Students need to be encouraged to study the tables in the text.

They should also be encouraged to make their own tables to summarize information they have read. Making tables will accomplish two things for students: first, the act of writing the information will help them remember it; second, tables organize information, making the information more accessible to the students when they start preparing for an assessment.

VISUAL STRATEGY

Table 38-1
Pass a bottle of vitamin tablets around the class. Have students examine the label and identify vitamins that are listed on the label but not on this table. List these additional vitamins on the chalkboard. Have students make Table 38-1 more complete by checking references to determine the sources, functions, and deficiency information for each of the vitamins listed on the chalkboard.

CONNECTIONS

Chapter 4
Enzymes and Coenzymes
Enzymes are biological catalysts that increase the rate of chemical reactions. Some enzymes cannot function unless they are assisted by coenzymes, which often are vitamins. Enzyme molecules can be reused many times by cells. Ask students why vitamins that function as coenzymes are needed in only small amounts. (*because like enzymes, coenzyme molecules are used over and over again*)

Teaching Tip
Diagnosing Deficiencies
Tell students to imagine that they are physicians. Based on their knowledge of vitamins, how would they answer the following questions. If a child suffers from frequent internal infections and has difficulty seeing in dim light, what would you suspect to be the problem? (*The child is suffering from a vitamin A deficiency.*) What foods would you prescribe to someone who has sore gums and tends to bruise easily? (*fruit, especially citrus fruits*)

Many vitamins function as integral parts of cellular enzymes. Other vitamins are essential components of coenzymes, organic molecules that take part in metabolic reactions. In Chapter 5, you learned that FAD accepts hydrogen atoms during the Krebs cycle. FAD is a coenzyme that has riboflavin, vitamin B_2, as an essential component. Vitamins must be obtained from the foods we eat because we are unable to manufacture them.

Some vitamins are soluble in water, while others are soluble in fat. Ingesting too much of a fat-soluble vitamin can be dangerous. Excessive amounts accumulate in the body's fatty tissues and may reach toxic levels. Table 38-1 lists several important water-soluble and fat-soluble vitamins and describes their food sources and biological roles.

Table 38-1 Vitamins

Vitamin	Food Sources	Role	Effects of Deficiency
Water-Soluble			
Vitamin B_1 (thiamin)	Most vegetables, nuts, organ meats	Carbohydrate metabolism, helps nerves and heart to function properly	Digestive disturbances, impaired senses
Vitamin B_2 (riboflavin)	Fish, poultry, cheese, yeast, green vegetables	Needed for healthy skin and tissue repair, carbohydrate metabolism	Blurred vision, cataracts, cracking of skin, lesions of intestinal lining
Vitamin B_3 (niacin)	Whole grains, fish, poultry, liver, tomatoes, legumes, potatoes	Keeps skin healthy, carbohydrate metabolism	Mental disorders, diarrhea, inflamed skin
Vitamin B_{12} (cobalamin)	Meat, poultry, green vegetables, milk, dairy products	Needed for formation of red blood cells	Reduced number of red blood cells
Vitamin C (ascorbic acid)	Citrus fruits, strawberries, potatoes	Needed for wound healing, healthy gums and teeth	Swollen bleeding gums, loose teeth, and slow-healing wounds
Fat-Soluble			
Vitamin A (retinol)	Carrots, green leafy vegetables, butter, eggs, liver, sweet potatoes	Keeps eyes and skin healthy, needed for strong bones and teeth	Infections of urinary and digestive systems, night blindness
Vitamin D (cholecalciferol)	Salmon, tuna, fish liver oils, fortified milk	Calcium and phosphorus metabolism, needed for strong bones and teeth	Bone deformities in children, loss of muscle tone
Vitamin E (tocopherol)	Many foods, especially wheat germ oil and olives	Protects cell membranes from damage by reactive oxygen compounds	Reduced number of red blood cells; nerve tissue damage in infants
Vitamin K	Leafy green vegetables, liver, cauliflower	Necessary for normal blood clotting	Bleeding caused by a prolonged clotting time

Overcoming Misconceptions

Vitamin Overdoses
Some people have the idea that "If a little is good, then a whole lot ought to be better." Unfortunately, this is not the case with vitamins. For example, severe overdoses of vitamin A can cause enlargement of the liver and spleen. Overdoses of vitamin D can cause kidney damage and calcification of soft tissues.

While vitamins A and D are both fat-soluble, making them more difficult for the body to eliminate, even the water-soluble vitamins have some negative effects if overused. Massive doses of vitamin C can cause kidney stones and enhance blood coagulation, and megadoses of niacin, one of the B vitamins, can cause liver damage and gout.

Trace Elements

Trace elements are minerals that are required by your body. Unlike calcium, sodium, and other inorganic elements that are present in your body in large quantities, trace elements are present in only minute amounts. Most trace elements are also essential for plant growth. Therefore, humans usually obtain adequate amounts of the required trace elements directly from the plants we eat or indirectly from animals that have eaten plants. Some important trace elements, their food sources, and their biological roles are presented in Table 38-2. ■

Table 38-2 Trace Elements

Trace Element	Best Sources	Role
Iodine	Iodized salt, seafood, plants grown in high-iodine soil	Synthesis of thyroid hormone
Cobalt	Leafy vegetables, liver, kidney	Synthesis of vitamin B_{12}
Zinc	Meat, shellfish, dairy products	Synthesis of digestive enzymes
Molybdenum	Legumes, cereals, milk	Protein synthesis
Manganese	Whole grains, nuts, legumes	Hemoglobin synthesis, urea formation
Selenium	Meat, seafood, cereal grains	Preventing chromosome breakage

Section Review

1. *State several reasons why you need more carbohydrates than fats in your diet.*

2. *Why would a diet without any protein be dangerous to your health?*

3. *Explain why zinc and manganese are not considered vitamins.*

Critical Thinking

4. *Each week for a month, an overweight person used more Calories exercising than were obtained from the food he or she had eaten that week. Contrast the amount of fat tissue in the individual's body at the beginning of the month with the amount present at the end of the month, and explain any difference.*

5. *Explain why a physician would prescribe an injection of several B vitamins to a person whose major symptoms are persistent fatigue and lack of energy.*

Answers to Section Review

1. Student answers will vary but may include: Fats can deliver *too* much energy, in which case the excess is simply stored as body fat. Fats may contribute to the development of atherosclerosis.

2. The body would have no way to grow, repair itself, or produce new enzyme molecules.

3. Vitamins are organic compounds. Both zinc and manganese are inorganic.

4. Exercise created an energy deficit in the person. The energy released from stored fat made up for this deficit. Therefore, the person would have less fat in his or her body at the end of the month than at the beginning of the month.

5. These might be prescribed to increase the person's carbohydrate metabolism efficiency and to fight anemia.

□ **CAPSULE SUMMARY**

Essential amino acids, vitamins, and trace elements are necessary for proper health. Essential amino acids are used as building blocks for proteins, and vitamins and trace elements participate in many cellular chemical reactions.

☑ **RESEARCH Update**

What's for Dinner?

Fat is bad, fiber is good, eat more greens, eat less meat—we are constantly bombarded by nutritional advice. Where does all this advice come from? It comes from scientific literature. Scientists who make these and other nutritional claims are basing their comments on many nutritional experiments and surveys. Walter C. Willet, of the Department of Nutrition and Epidemiology at Harvard University, has recently analyzed all of the recent nutritional data and produced a definitive summary of the current status of nutritional recommendations. Willet points to evidence collected by numerous researchers that indicates diet is the largest controllable aspect one has over one's life in the prevention of serious diseases like cancer, coronary heart disease, and birth defects. Willet's review of the current data has led him to support the following summary of recommendations proposed by the National Research Council (NRC) in 1989. (1) Decrease the total amount of fat ingested to 30 percent or less of calories, and decrease saturated fatty acid ingestion to below 10 percent of calories. (2) Eat five or more servings of a combination of vegetables and fruits daily, especially green and yellow vegetables and citrus fruits, and eat six or more daily servings of complex carbohydrates. (3) Consume moderate amounts of protein. (4) Balance food intake and exercise to maintain healthy body weight. (5) Limit alcohol to an average of 1 oz. per day, but it is better to drink no alcohol. (6) Limit daily salt intake to 6 g or less. (7) Consume adequate amounts of calcium. (8) Avoid taking dietary supplements in amounts that exceed the RDA. (9) Maintain optimal intake of fluoride (pertains especially to children). ☑

UPDATE

Vocabulary Preview

digestion

amylase

emulsification

bile salt

lipase

protease

pharynx

esophagus

peristaltic contraction

chyme

gastrin

ulcer

pancreas

duodenum

villi

microvilli

colon

rectum

liver

Opening Questions

Ask students which animals have a digestive system and which do not. Since they have probably already studied the invertebrate unit, they should know that any animal that is more complex than a cnidarian will have a digestive system. Then ask them to describe the relationship between a digestive system and the cell membranes of an animal's cells. They should tell you that a digestive system breaks down foods into individual molecules that are small enough to pass through cell membranes.

38-2 Digestion

Your cells obtain the energy they need by extracting it from sugars, fatty acids, and amino acids during cellular respiration. Eating a plant or an animal provides you with a rich source of complex starches, fats, and proteins. These molecules occur as long chains composed of individual sugars, fatty acids, or amino acids, like strings of beads. Your cells cannot extract energy from these large chains. First, these large molecules must be broken down into their individual components during a process called digestion.

Section Objectives

■ Trace the path of food through the digestive system.

■ Name the enzymes that digest carbohydrates, fats, and proteins.

■ Describe the major digestive processes that occur in the stomach, small intestine, and large intestine.

■ Describe the role of the pancreas and liver in digestion.

■ Explain how nutrients are absorbed into the bloodstream from the digestive system.

Food Molecules Are Broken Down by Enzymes

Each of the three types of large food molecules—starch, fat, and protein—presents a different digestive challenge and requires a unique strategy, as shown in Figure 38-4. For example, starches, which are chains of sugar molecules, are readily broken down into sugars by enzymes called **amylases** (*AM uh lays uhs*).

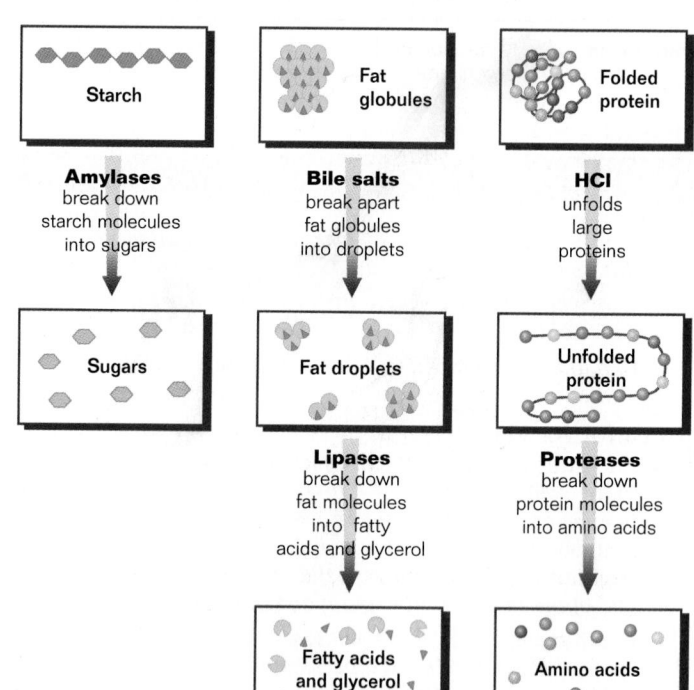

Figure 38-4 Starches, *top left,* are chemically digested into sugars by enzymes called amylases. Globules of fat, *top center,* must be broken apart by bile salts before the individual fat molecules can be digested into fatty acids (yellow) and glycerol (orange) by enzymes called lipases. Large, folded proteins, *top right,* must be unfolded by hydrochloric acid (HCl) before the proteins can be broken into amino acids by enzymes called proteases.

Fats are insoluble in water and tend to aggregate into large globules consisting of many fat molecules. As a result, many of the individual fat molecules are not easily attacked by enzymes. Before fats can be digested, they must first be treated with a detergent to make them water soluble, a process that is called emulsification. **Emulsification** *(ee MUHL suh fih kay shuhn)* breaks up fat globules into many tiny fat droplets, thus exposing many more fat molecules to enzymes. Your liver produces detergent molecules called **bile salts,** which emulsify the fats contained in the foods you eat. After these fats have been emulsified, enzymes called **lipases** *(LEYE pays uhs)* break the fat molecules into fatty acids and glycerol.

Proteins, which are chains of amino acids, present a particularly tough challenge to the digestive system. Almost all proteins are either folded into tight balls or wound together into tough fibers. Enzymes cannot break down these balls and fibers because they cannot get at the individual protein chains. The human body solves this problem by carrying out protein digestion in two steps. First, hydrochloric *(HEYE droh klawr ihk)* acid, HCl, in the stomach is used to unfold large proteins into single polypeptide strands; then enzymes called **proteases** *(PROHT ee ays uhs)* attack the strands, cutting them into smaller fragments.

Most digestive enzymes cannot tolerate extremely acidic conditions, so human digestion is carried out in phases, each of which occurs in a different part of the digestive system. The digestion of proteins by acid, as well as some enzymatic digestion of proteins, takes place in the stomach. Next, the food moves to the small intestine, where the acid from the stomach is neutralized, thus allowing the intestinal enzymes to function. In the small intestine fats are emulsified, and carbohydrates, fats, and proteins are completely dismantled by amylases, lipases, and proteases. The products of this digestion—amino acids, fatty acids, glycerol, and sugars—then pass through the wall of the small intestine and enter the bloodstream. ◼

The Digestive Journey Begins in the Mouth

Humans regularly eat a wide variety of plants and animals, and the structure of our teeth, shown in Figure 38-5, reflects our omnivorous *(ahm NIHV uh ruhs)* diet. The front teeth of humans are structurally similar to the pointed, cutting, and ripping teeth characteristic of meat eaters (carnivores). Our back teeth resemble the flat, grinding teeth characteristic of plant eaters (herbivores).

After food has been ripped or chewed into shreds by the teeth, the tongue mixes it with a watery solution called saliva. Saliva is secreted into your mouth by three pairs of salivary

◻ **CAPSULE SUMMARY**

Amylases, lipases, and proteases are three classes of digestive enzymes. Amylases break down starches into sugars. After fat globules have been emulsified by bile salts, lipases break down individual fat molecules into fatty acids and glycerol. After compacted proteins have been unfolded by HCl, proteases break down the proteins into polypeptides and amino acids.

Figure 38-5 Like most mammals, humans have teeth that are differentiated. Canine and incisor teeth, used for cutting and tearing food, are located toward the front of both the upper, *top,* and lower, *bottom,* jaws. The molars, located toward the back of the jaws, are used to grind food.

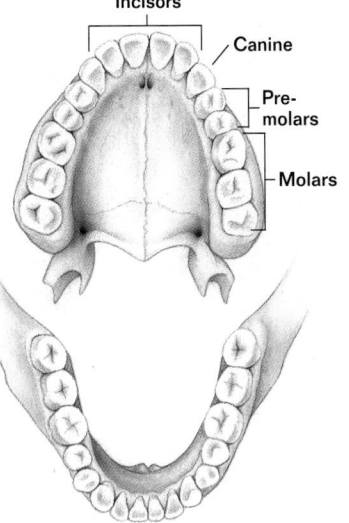

Incisors

Canine

Pre-molars

Molars

Demonstration
Peristaltic Action

Fill a 2 ft. section of rubber tubing with water. Then have an assistant tightly pinch the top and bottom of the tubing shut between thumb and forefinger. Starting at the bottom of the tube, pinch the tubing just above your assistant's fingers with one hand. While still pinching with one hand, pinch just above your own fingers with your other hand. Then release your bottom hand and use it to pinch just above what was your top hand. Continue moving your hands alternately up the length of the tubing, all the while encouraging your assistant to hold tightly to both ends. When you get near the top, have your assistant aim the tube out toward a sink or a bucket, releasing just his or her top hand. Depending on how well your assistant did the job, a stream of water should spurt out of the tube some distance. Explain to students that their digestive tract operates in a similar manner. Tell them that the muscular wall of their esophagus contracts and pushes food down into their stomach.

Teaching Tip
Bile and Fats

Pour some oil into water. Have students note how the two are immiscible. Then slowly add a liquid detergent while stirring. Tell students that bile acts like the detergent—both emulsify the fat into smaller droplets. Once again, emphasize how increased surface area is important to the digestive process.

Teaching Tip
Observing and Drawing

Set up microscope stations around the classroom and have students observe prepared slides of digestive organs, including the stomach, small intestine, liver, and pancreas. Have students make drawings of what they see. Discuss as a class how the structures in the slides relate to function, such as how the distribution of villi of the small intestine relates to nutrient absorption.

 VISUAL STRATEGY

Figure 38-7

Be sure students know where the esophagus is located in the body. Ask them where the mouth is located in relation to the esophagus. (*at its top*) Then ask them where the stomach is located. (*at its bottom*) Finally, ask them to name the process depicted in the figure. (*peristalsis*)

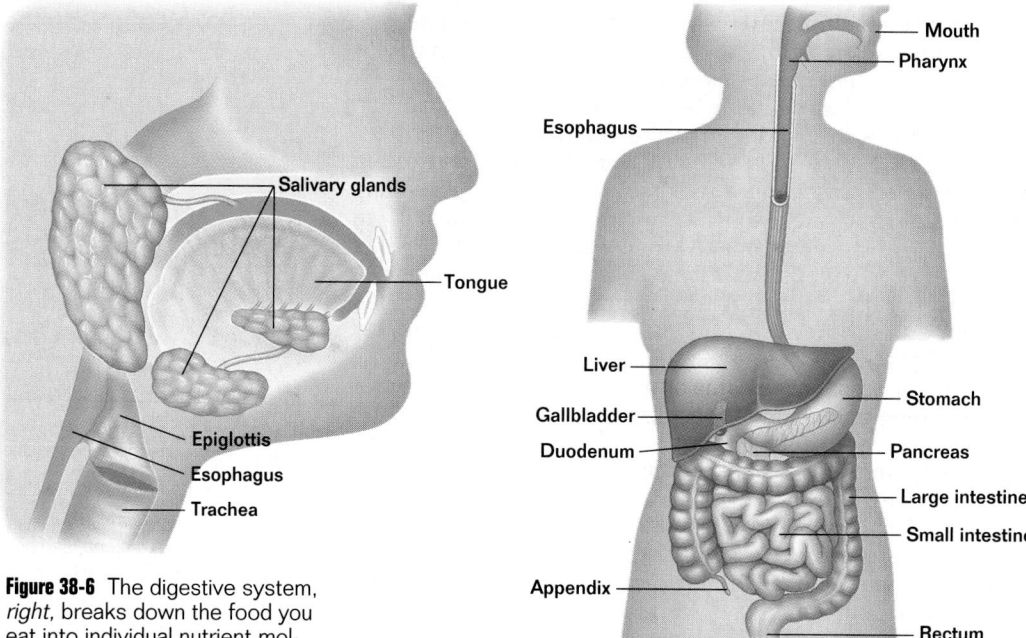

Figure 38-6 The digestive system, *right*, breaks down the food you eat into individual nutrient molecules that can be absorbed into the bloodstream. The salivary glands, *left*, secrete saliva through ducts that open into the mouth.

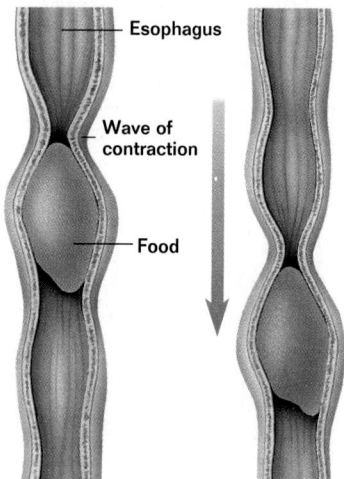

Figure 38-7 Food enters the esophagus from the pharynx. The food is pushed down the esophagus and toward the stomach by a wave of muscular contraction that occurs in the wall of the esophagus.

glands, shown in Figure 38-6, located above the upper jaw and below the lower jaw. Saliva moistens and lubricates the food so that it can be swallowed more easily. Saliva also contains amylase, which begins the breakdown of starch.

After passing through the region in the back of the throat called the **pharynx** (*FAIR ihnks*), the food enters the esophagus. The **esophagus** (*ih SAHF uh guhs*) is a long tube that connects the mouth to the stomach. No digestion takes place in the esophagus. Its role is to act as a kind of descending escalator, moving food down to the stomach. Your esophagus is about 25 cm (10 in.) long. The lower two-thirds of the esophagus is wrapped in sheets of smooth muscle. Food doesn't just fall down into the stomach—it is "pushed" down, as shown in Figure 38-7. Successive rhythmic waves of contraction of the smooth muscle in the wall of the esophagus, called **peristaltic** (*pehr uh STAHL tihk*) **contractions,** move the food along toward the stomach.

Proteins Are Dismantled in the Stomach

Food exits the esophagus and passes into the stomach through a muscular valve called a sphincter (*SFIHNGK tuhr*). The sphincter prevents acid-soaked food in the stomach from making its way back into the esophagus. The stomach is a saclike

---

 Did You Know?

The condition known as heartburn is caused by the reflux (backward movement) of stomach fluid into the esophagus. The esophagus, lacking the mucous coating of the stomach, becomes irritated by the acidic fluid, and a burning sensation results. Occasional indigestion may be caused by overeating, upset stomach, or stress, but it is usually not serious. A physician should evaluate cases of persistent heartburn.

organ located just beneath the diaphragm. Besides temporarily storing food, the stomach also mechanically breaks down food and chemically unravels and breaks down proteins.

When food enters the stomach, "gastric juice," a combination of HCl and acid-stable proteases, is secreted by the stomach's epithelial lining. In the upper portion of the stomach's inner wall are deep depressions called gastric pits. These pits contain cells that secrete HCl. The proteases contained in gastric juice are secreted by other cells in the gastric pits. The stomach mixes its contents with a churning action caused by the contraction and relaxation of its muscular wall. The mixture of food, HCl, and enzymes eventually forms a soupy, semisolid material called **chyme** (KEYEM).

Your stomach secretes about 2 L (2.11 qt.) of HCl every day, creating a very concentrated acid solution—about 3 million times more acidic than your bloodstream. HCl unfolds proteins because its low pH (between 1.5 and 2.5) breaks the molecular attractions that hold the polypeptide chains of proteins together. Once the chains have been unfolded, the proteases cut them into smaller polypeptide fragments.

It is important that your stomach not produce *too* much acid, for if it did, it would be impossible for your body to neutralize the acid later in the small intestine. Acid production is controlled by hormones produced in endocrine cells that are scattered within the walls of the stomach. The hormone **gastrin** regulates the synthesis of HCl, permitting it to be made only when the pH in the stomach is higher than about 1.5. This negative feedback loop is shown in Figure 38-8. Some people oversecrete gastrin, causing their stomachs to produce too much HCl. The excessive acid may eventually eat holes in the walls of the stomach and small intestine. These holes are called ulcers (UHL suhrs).

Digestion and Absorption Occur in the Small Intestine

From the stomach, food passes into the small intestine. Only small portions of acidic chyme are introduced into the small intestine, thereby ensuring that the acid will be neutralized and that the intestinal enzymes will be able to function.

Some of the digestive enzymes in the small intestine are secreted by the cells of the intestinal wall. Most of the digestive enzymes, however, are produced in a gland called the pancreas, shown in Figure 38-9. The **pancreas** (PAN kree uhs) sends enzymes through a duct into the first part of the small intestine, called the **duodenum** (doo oh DEE nuhm). Your small intestine is approximately 6 m (19.8 ft.) long. Unwound and

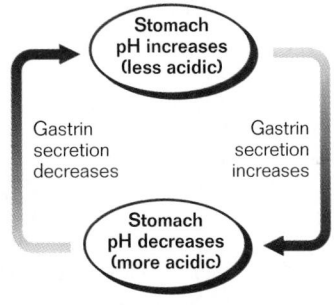

Figure 38-8 The pH in the stomach remains relatively constant (about 2) due to a hormonal negative feedback loop. If the pH in the stomach begins to rise too high, the stomach secretes the hormone gastrin, which stimulates the stomach to secrete more HCl, thus decreasing the pH inside the stomach. If the pH begins to drop too low, gastrin secretion decreases, less HCl is secreted, and the pH increases.

Figure 38-9 The pancreas and liver are often called accessory digestive organs because the food within the digestive tract never actually enters them. Instead, bile from the liver and digestive enzymes from the pancreas pass through ducts into the first part of the small intestine, the duodenum.

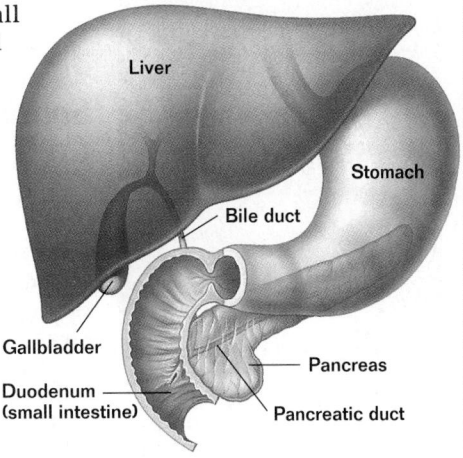

Did You Know?

The pancreas is an endocrine gland as well as a digestive organ. The pancreas secretes two different hormones, insulin and glucagon, each with an opposing effect on the amount of glucose in the bloodstream. Insulin causes the glucose level in the blood to decrease, and glucagon causes the level to increase.

SECTION 38-2

 VISUAL STRATEGY

Figure 38-8
Ask students why this figure demonstrates a *negative* feedback loop and not a *positive* feedback loop. *(because the release of gastrin causes a negative, or reverse, effect—in this case a decrease in the pH of the stomach)* Ask students where gastrin must first go before it stimulates the stomach to secrete HCl. *(Since gastrin is a hormone, it must first enter the bloodstream.)*

Teaching Tip
Organizing Information
To help students comprehend the sources and functions of the various digestive enzymes, have them make a table with the following headings down the left side: Enzyme, Source, Site of Enzymatic Action, and Nutrient Acted Upon. Across the top of the table, have students write the following headings: Mouth, Stomach, Small Intestine, Pancreas, and Large Intestine. Then have students supply the information to complete the table.

Application
Medicine For years, physicians have associated the cause of most stomach ulcers with poor lifestyle habits—too much stress, too much alcohol, and too many spicy foods. But recent research findings are forcing physicians to reconsider the cause and treatment of stomach ulcers. Although stress and diet may somehow contribute to the formation of an ulcer, it seems the major cause is a bacterium called *Helicobacter pylori*. So instead of prescribing a job change, a different diet, and antacids to patients with ulcers, many physicians are now giving their patients antibiotics to kill off the bacteria that live in their stomachs. The new therapy seems to be working. One study showed up to a 90 percent cure rate when antibiotics were used.

VISUAL STRATEGY

Figure 38-10
Correlate the diagram with the photo. Explain to students that the photograph shows many whole villi are attached to the small intestine's inner wall. The drawing depicts several villi that have been sectioned longitudinally, thus displaying their internal anatomy. Ask students the function of the capillaries inside the villi. *(They absorb food molecules.)*

Application

℞ **Medicine** The colon pushes wastes to the end of the digestive tract and absorbs ions, vitamins, and excess water. Although these are important functions, none are essential for life. In fact, many people who have colectomies (surgical removal of all or part of their colons) live full, rewarding lives. If the colon is removed, surgeons reattach the end of the intestine so that it empties wastes through the abdominal wall into a specially designed bag.

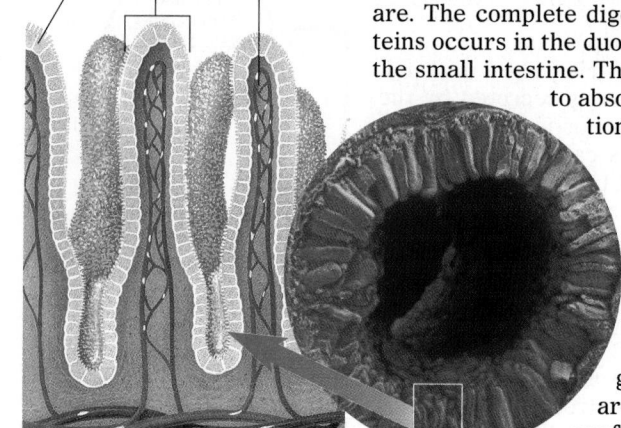

Microvilli Villus Capillaries

Figure 38-10 This cross section of the small intestine, *right,* has been magnified about 30 times. The fingerlike projections called villi, *left,* are the sites where food molecules are absorbed from the intestine into the bloodstream. Hairlike extensions called microvilli on the surface of the villi greatly increase the internal surface area of the small intestine.

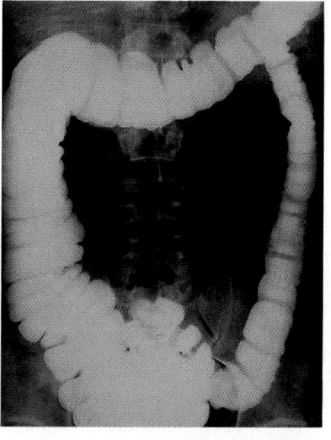

Figure 38-11 This X ray of the large intestine shows its location in the abdomen. The diameter of the large intestine is about three times that of the small intestine.

held by one end, it would be over three times taller than you are. The complete digestion of carbohydrates, fats, and proteins occurs in the duodenum, about the first 25 cm (10 in.) of the small intestine. The rest of the small intestine is devoted to absorbing water and the products of digestion into the bloodstream.

The lining of the small intestine is covered with fine fingerlike projections called **villi,** shown in Figure 38-10, each too small to see with the naked eye. In turn, the cells covering each villus have cytoplasmic projections on their outer surface called **microvilli.** The villi and microvilli greatly increase the absorptive surface area of the small intestine. The average surface area of an adult human's small intestine is about 300 sq. m (3,229 sq. ft.)— greater than the surface area of many large swimming pools.

The volume of material passing through the small intestine is surprisingly large. An average human consumes about 0.85 L (0.90 qt.) of solid food and 1.2 L (1.27 qt.) of water a day, for a total volume of about 2 L (2.11 qt.). To this amount are added secretions from the salivary glands, stomach, liver, pancreas, and small intestine, bringing the total up to a remarkable 9 L (9.51 qt.)—more than 10 percent of the total volume of your body!

Although the volume of material that flows through the digestive system is great, the amount that leaves the body as waste is small. This is because almost all of these fluids and solids are absorbed during their passage through the small intestine. Only a small quantity is absorbed through the wall of the large intestine. Of the 9 L (9.51 qt.) of material that enter the digestive tract each day, only about 0.05 L (0.06 qt.) of solid and 0.1 L (0.11 qt.) of liquid leave the body as wastes. The fluid absorption efficiency of the intestine approaches 99 percent.

Solids Are Compacted in the Large Intestine

Food passes from the small intestine to the large intestine, shown in Figure 38-11. The large intestine, or **colon,** is much shorter than the small intestine, about 1 m (3.3 ft.) long. The large intestine is not coiled up like the small intestine. Instead, it is in three relatively straight segments. No digestion takes place within the large intestine.

The inner surface of the large intestine does not have villi. Consequently, the large intestine has only one-thirtieth of the absorptive surface area of the small intestine. Although a small amount of fluid, along with sodium and

Did You Know?

When the intestine rushes its contents through without sufficient time to absorb the water from the wastes, diarrhea results. Irritations from drugs, parasites or bacteria, and stress can all cause diarrhea. Prolonged diarrhea, such as occurs from amoebic dysentery or cholera, can cause lethal dehydration and salt imbalance. Children and infants are especially vulnerable since they cannot store as much water in their tissues as adults.

vitamin K, are absorbed through its walls, the primary function of the large intestine is to act as a trash compactor. Within it, undigested material, including large amounts of plant fiber and cellulose, is compacted and stored. Many bacteria live and actively divide within the large intestine, where they play a role in processing undigested material into the final excretory product, feces.

The final segment of the digestive tract is a short extension of the large intestine called the **rectum.** Compacted solids within the colon pass into the rectum as a result of the peristaltic contractions of the muscles in the wall of the large intestine. From the rectum, the solid feces pass out of the body through the anus. ☐

How Nutrients Are Delivered to Cells

Blood that leaves the small intestine, rich with the products of digestion, is collected in the portal vein and carried to the liver.

Your **liver,** a very large organ, is the size of a football and weighs over 1.4 kg (3.1 lb.). The liver performs a wide variety of metabolic functions. For example, the liver supplies quick energy to the body by releasing sugar into the blood, and it builds complex carbohydrates from sugar. It also makes proteins from amino acids, stores vitamins and minerals, produces bile salts, regulates blood clotting, monitors the production of cholesterol, and detoxifies alcohol and poisons.

After flowing through the many fine passages of the liver, the blood is collected into the hepatic vein, which carries the blood back toward the heart to be circulated to the rest of the body.

❏ CAPSULE SUMMARY

Carbohydrate digestion begins in the mouth, and protein digestion begins in the stomach. The pancreas secretes amylases, lipases, and proteases into the small intestine, where carbohydrates, fats, and proteins are completely digested. The end products of digestion are absorbed through intestinal villi and pass into the bloodstream. The waste products of digestion are stored in the large intestine before passing out of the body.

CONTENT LINK

You can learn about the hormones that cause the liver to store and release sugar in **Chapter 41.**

Section Review

1. *Why does protein digestion begin in the stomach?*
2. *Define the term* emulsification, *and explain why it occurs during digestion.*
3. *When during the digestive process does food enter the liver and pancreas?*
4. *Describe digestion in the small intestine, and explain how food molecules enter the bloodstream.*
5. *What is the function of the large intestine?*

Critical Thinking

6. *A person has a small intestine that has villi but lacks microvilli. Would you expect this person to be underweight or overweight? Explain your answer.*

SECTION 38-2

Application

℞ Health The modern image of beauty has produced two scourges: anorexia nervosa and bulimia. Though eating disorders often affect females of high school or college age, sufferers may be male and of any age group.

People who suffer from anorexia nervosa deliberately starve themselves, often while undertaking a strenuous exercise regimen, in an attempt to gain an "ideal" body image. However, the anorexic never can attain that ideal, since a poor self-image is often the true problem.

Those who suffer from bulimia also often have a poor self-image. Instead of starving themselves, bulimics binge on huge quantities of food and then purge themselves through vomiting, the use of huge doses of laxatives, or both. While a bulimic may look healthy, the acid from constant vomiting damages their esophagus and teeth, and many become laxative dependent. They may also suffer from life-threatening electrolyte (salt) imbalances and internal hemorrhages.

Treatment for both conditions is prolonged, often involving hospitalization during which the individual's health is monitored and new behavior patterns are taught. Treatment may also include drug and psychological therapy.

Answers to Section Review

1. Chemicals that are able to digest proteins (pepsin and HCl) are first encountered in the stomach.

2. Emulsification is the breaking up of large fat globules into small fat droplets.

Emulsification serves to expose fats to lipases, thus speeding their enzymatic digestion and absorption.

3. Food never enters the liver and pancreas during the digestive process.

4. The small intestine secretes some of its own digestive enzymes and receives others from the pancreas. In addition, the liver and gallbladder deliver bile to the duodenum for the emulsification of fat. Food molecules enter the bloodstream by passing through the walls of capillaries that are inside villi.

5. The large intestine functions primarily in compacting wastes.

6. The person would be underweight due to the inefficient absorption of food molecules caused by the reduced surface area of the intestinal lining.

Vocabulary Preview

excretion

urea

nephron

Bowman's capsule

glomerulus

filtrate

renal tubule

loop of Henle

collecting duct

ureter

urinary bladder

urethra

hemodialysis

histocompatibility
  antigen

Opening Questions

Remind students that all cells produce nitrogen-containing waste products as a result of protein metabolism. Then ask them where those waste products go after the cells release them. They should answer that they enter the bloodstream. Finally, ask the students which system of the body removes nitrogen waste products from the blood. They should know that the urinary system carries out this function.

Section Objectives

■ Identify the major wastes produced by humans, and describe how they are eliminated from the body.

■ Draw a simple sketch of a nephron.

■ Explain how a nephron forms urine.

■ Trace the flow of urine through the human urinary system.

■ Compare and contrast kidney dialysis with kidney transplants.

38-3 Excretion

*I*f you live to be 70 years old, you will, in the course of your life, eat some 45,000 lb. of food and drink over 7,250 gal. of fluid—enough to fill a tanker truck. Did you ever wonder what happens to all of this food and fluid? Every atom of it still exists, if not in your body, then in the wastes that have been eliminated from it. Humans eliminate wastes in a process called excretion.*

Carbon Dioxide, Nitrogen Compounds, and Water Are Excreted

Excretion rids the body of toxic chemicals, excess water, salts, and carbon dioxide. Among the many kinds of substances you excrete, the most important are carbon dioxide, nitrogen wastes, and water, as shown in Figure 38-12.

Almost all of the energy in the food you eat is contained in the food's carbon-hydrogen bonds. During cellular respiration this energy is extracted from the food molecules. The carbon and hydrogen atoms that are left behind combine with oxygen to form carbon dioxide and water. The carbon dioxide is transported to your lungs by the circulatory system and excreted from your body every time you exhale.

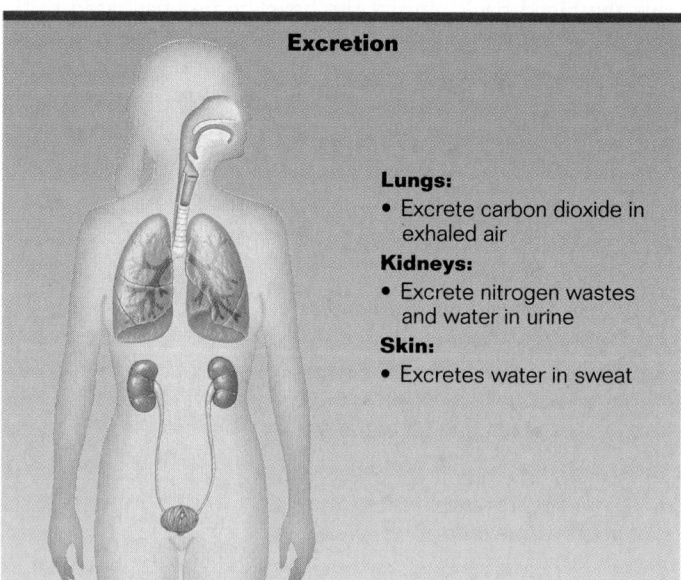

Excretion

Lungs:
• Excrete carbon dioxide in exhaled air

Kidneys:
• Excrete nitrogen wastes and water in urine

Skin:
• Excretes water in sweat

Figure 38-12 The lungs, the kidneys (along with their associated system of tubes), and the skin all function as excretory organs. For example, every time you exhale, you release carbon dioxide, a waste product of cellular respiration, into the atmosphere.

The removal of the amino groups from the amino acids of proteins produces ammonia, NH_3. Ammonia is extremely toxic to cells. Getting rid of this metabolic nitrogen waste product is a problem that every organism must solve. In aquatic animals like fishes, the ammonia is excreted through gills, directly into the surrounding water. Land animals, like humans, must solve the problem of ammonia accumulation a different way. In the liver, a complex series of chemical reactions combines pairs of amino groups with carbon dioxide to form a much less toxic nitrogen waste called **urea** (*yoo REE uh*). Urea is a principal component of urine, the fluid formed by the urinary system.

The average person drinks over 1 L (1.1 qt.) of fluid every day. You would swell up like a balloon if your body did not lose an equal amount of water through sweat and urine. As you learned in Chapter 35, sweating plays an important role in controlling body temperature, while urine is a vehicle for discharging urea and excess salts from the body.

Your body eliminates urea and excess water by means of a pair of bean-shaped, reddish brown kidneys located in the lower back. Kidneys, shown in Figure 38-13, are the size of a small fist. Urea is carried by the bloodstream to the kidneys, where it is removed from the blood. The kidneys receive a flow of about 2,000 L (2,114 qt.) of blood each day—more than the volume of a car! Since your body holds only 5.6 L (5.9 qt.) of blood, you can see that your blood must circulate through the kidneys many times during the day (about 350 times per day, or once every four minutes).

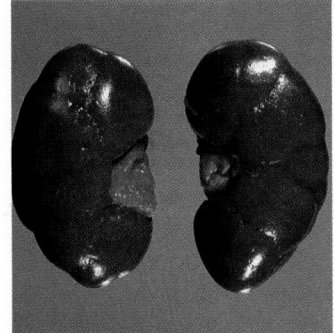
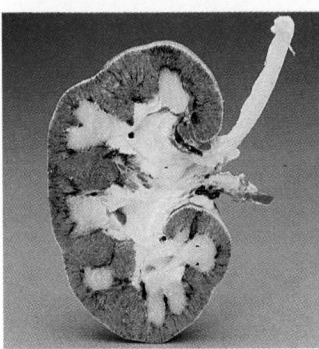

Figure 38-13 Human kidneys, *top*, filter wastes (mostly urea and excess water) from the blood that circulates through them. The blood supply to a kidney is enormous, as can be seen in this dissected kidney, *bottom*. This kidney has been injected with a dye that stains arteries red and veins blue.

The Kidneys Regulate Blood Plasma Composition

The kidneys are extremely important organs because of the role they play in homeostasis—maintaining a balanced state within the body. As you learned in Chapter 35, the chemical composition of extracellular fluid (which includes blood plasma) cannot vary to any great degree without causing serious harm to the body's cells and organ systems. The kidneys play a key role in regulating the amount of water and salt contained in blood plasma.

If you were to drink a large amount of water, your blood would become diluted and your blood pressure would sky-rocket unless you somehow got rid of the extra water. Your body uses the kidneys to remove the excess water by increasing urine production. Besides removing excess water from the blood, your kidneys also conserve water. Being a terrestrial organism, your body has a much higher concentration of water than the surrounding air does, so you constantly lose water to the air by evaporation. Water is also lost as water vapor in the air that you exhale. In order to

Did You Know?

The kidneys develop from the embryonic tissue called mesoderm. The average human kidney weighs about 150 g (5 oz.) and is about 12.5 cm (5 in.) long, 7.5 cm (3 in.) wide, and 2.5 cm (1 in.) thick. Both kidneys lie in a vertical position that spans from about the twelfth thoracic vertebra to the third lumbar vertebra, high up against the back wall of the abdomen.

Demonstration
Inside a Kidney

Purchase a sheep or beef kidney from your supermarket. Slice the kidney from top to bottom, separating it into two mirror-image halves. (Imagine the kidney as a large lima bean: separate the two halves.) Allow students to observe both the internal and external features of the kidney. Have a hand lens on hand and encourage them to use it. Have students make drawings and help them to identify the following structures so that they can label them: renal pelvis, renal medulla, renal cortex, and renal blood vessels. Ask the students how similar the kidney they have drawn is to their own. (*very similar*) Ask them why. (*because the kidney they have drawn is from a mammal, and they are also mammals*)

Application

Health It is ironic that as food conscious as Americans are, they often pay little attention to the other side of the coin—their fluid intake. In fact, people can survive without food much longer than they can survive without water. On average, an adult needs 1.5–3 L of water per day, some of it being supplied by moist food. Any combination of foods and beverages that provides the needed amount of water is acceptable, but most nutritionists recommend 6–8 glasses of liquid a day over and above the amount contained in food.

Teaching Tip
A Magnified View

Provide students with microscopes and prepared slides of tissue from the kidney. Have students observe a Bowman's capsule surrounding a glomerulus. Ask students to make a drawing of what they see. When students have completed their observations, have them label the parts.

Teaching Tip
Note Taking

Have students practice their note-taking skills while reading this section. Remind them that they should always use their own words and write their notes in words and phrases (not in complete sentences) in order to save time.

👁 VISUAL STRATEGY

Figure 38-14

To help students better understand how the kidneys are able to carry out their function, explain to them the arrangement of blood vessels and nephrons within a kidney. Each kidney receives a branch from the descending aorta called the renal artery. Within the kidney, branches from the artery radiate outward through the renal medulla to the renal cortex. The smallest of these branches supply blood to the kidney's million or so nephrons. Blood leaves the nephrons through tiny veins, which gradually converge until they form a large renal vein that exits the kidney near where the renal artery enters. The renal vein from each kidney drains into the inferior vena cava. The glomeruli and their surrounding Bowman's capsules are found in the renal cortex. In general, the renal tubules extend from the cortex down into the medulla and back up into the cortex again where they connect with the collecting ducts. The collecting ducts then plunge back into the medulla where they fuse and form the renal pelvis, which connects to the ureter. Review the drawing on the far right with students to help explain how the processes of filtration and reabsorption lead to the formation of urine within a nephron.

Figure 38-14 A kidney, *left,* contains microscopic filtering units called nephrons. Each nephron, *center,* is composed of a hollow tubule system with blood vessels wrapped around it. The processes of filtration and reabsorption that occur in the nephron lead to the formation of urine, *right.* During filtration, urea, water, sodium, and glucose from the blood pass into a Bowman's capsule. During reabsorption, all of the glucose, and most of the sodium and water, pass into the capillaries that surround the loop of Henle. Urine is composed of the urea and excess water and sodium that remain in the collecting duct.

minimize water loss, your kidneys tend to concentrate the urine they produce.

Your body also monitors the levels of sodium (salt) in your blood. When the number of sodium ions falls below normal, your kidneys direct the adrenal glands to send out hormones that cause the kidneys to remove more sodium ions from the urine forming within them. This sodium then returns to the blood. On the other hand, if levels of sodium in the bloodstream rise too high, hormone levels are decreased so that more sodium is excreted in the urine. ◻

Kidneys Are Composed of Tiny Blood Filters

The kidney is a complex organ composed of roughly 1 million microscopic blood-filtering units called **nephrons** (*NEHF rahns*), shown in Figure 38-14. Each nephron produces urine. A nephron is composed of three elements. A different phase of urine production occurs in each element.

1. **Filtration** The filtration device at the top of each nephron is called a **Bowman's capsule.** Within each Bowman's capsule, an arteriole enters and splits into a fine network of capillaries called a **glomerulus** (*gloo MEHR yoo luhs*). The glomerulus acts as a filtration device. The blood pressure inside the glomerulus forces fluid through the capillary walls. The walls prevent blood cells, proteins, and other large molecules from leaving the blood but allow water

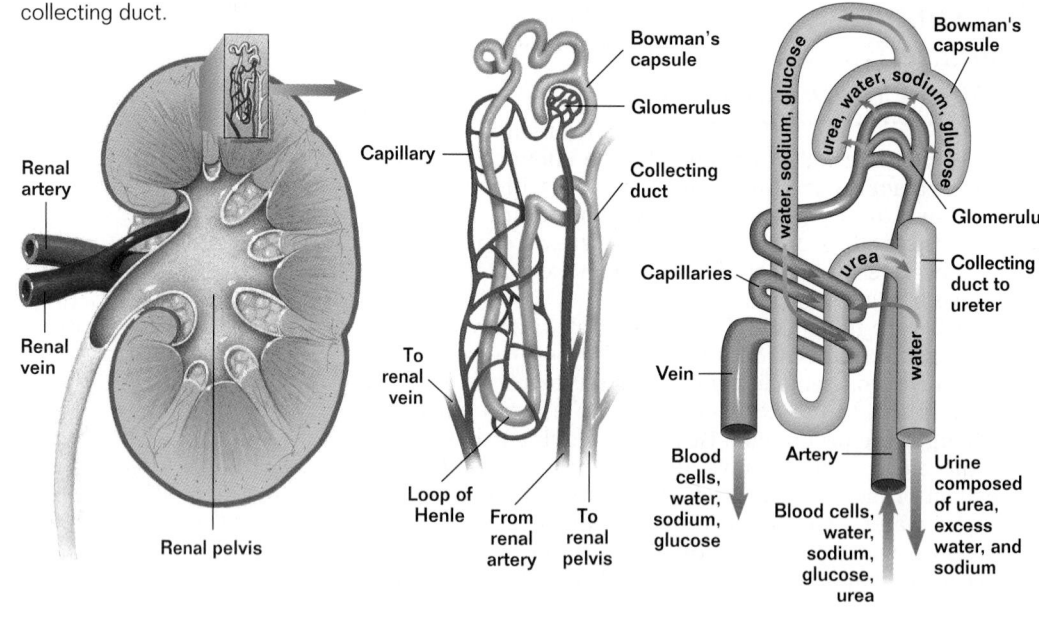

and small molecules such as urea, sodium ions, and glucose to pass through and enter the hollow interior of a Bowman's capsule. The fluid inside a Bowman's capsule is called the **filtrate.**

2. **Reabsorption** A Bowman's capsule is connected to a long narrow tube called the **renal tubule.** The renal tubule bends back on itself in its center, forming what is called the **loop of Henle.** The renal tubule is a reabsorption device. It extracts a variety of useful molecules, including glucose, ions, and some water, from the filtrate that passes through it. These substances reenter the bloodstream by passing into capillaries that wrap around the tubule. If it were not for the reabsorption that occurs in this portion of the nephron, these molecules would be eliminated from the body in the urine.

3. **Urine formation** The renal tubule empties into a larger tube called a **collecting duct.** The collecting duct removes much of the water from the filtrate that passes through it. As a result, human urine is four times more concentrated than blood plasma. Your kidneys achieve this remarkable degree of water reabsorption by a simple but superbly efficient mechanism. Because the collecting duct is slightly permeable to urea, some urea diffuses out and accumulates in the tissue around the duct. In a similar manner, some sodium also diffuses out from the portion of the tubule that is directly after the loop of Henle. Because of the high concentration of solute that exists around the collecting duct, an osmotic gradient is created, and water is drawn out of the collecting duct into the surrounding tissue, as is shown in Figure 38-15. The tissue absorbs water from the filtrate and passes it on to blood vessels that carry it out of the kidneys and back into the bloodstream. The filtrate that remains after salts, nutrients, and water have been removed is called urine. ■

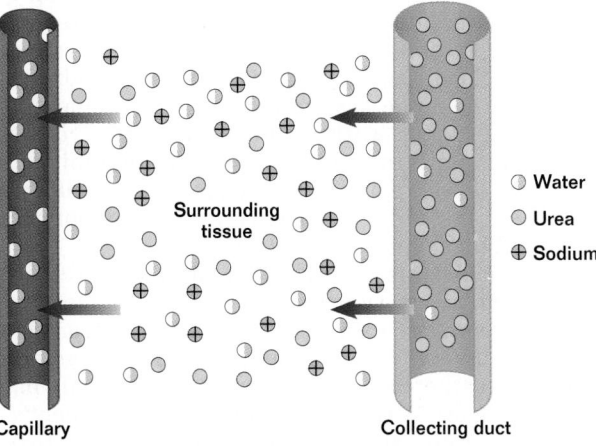

Water
Urea
Sodium

Surrounding tissue

Capillary

Collecting duct

□ CAPSULE SUMMARY

The microscopic blood-filtering units of the kidneys are called nephrons. Water, urea, sugars, amino acids, and salts enter one end of a nephron. As these materials move through the nephron's tubule and collecting duct, most of the water, some of the salts, and all of the sugar and amino acids are reabsorbed into the bloodstream. The mixture of water, urea, and salts that remains in the nephron is called urine.

Figure 38-15 The high concentration of urea and sodium that surrounds the collecting duct of a nephron creates a hypertonic environment, which draws water out of the collecting duct. The water then diffuses into capillaries and is returned to the body. The capillaries are impermeable to urea. Although some sodium enters the capillaries, equal quantities of sodium exit. Thus, the hypertonic nature of the surrounding tissue is maintained.

SECTION 38-3

Application

Rx **Medicine** Kidney stones are caused by the formation of crystals of calcium, magnesium, or uric acid salts in the kidney. This relatively rare occurrence causes excruciating pain as the body attempts to pass the sharp crystals through the renal pelvis and ureter. In the past, surgery was the only remedy, but a new treatment uses sound waves to shatter the crystals into tiny, nonobstructive pieces.

Teaching Tip
Learning Tool

It is important for students to understand the pathway that blood takes in the kidneys, when filtrate forms, and where the filtrate goes. Guide students through the making of a *Graphic Organizer* like the one shown at the bottom of the page. If you wish, you can put the diagram on the chalkboard (minus all but two or three of the labels). Then have a student volunteer come up to the chalkboard and try to fill in the empty boxes. If he or she is having difficulty with labeling any of the boxes, ask the rest of the class for help.

◉ VISUAL STRATEGY

Figure 38-15

Review the concept of osmosis with students and ask them to define the terms *hypertonic, hypotonic,* and *isotonic.* Also correlate this figure with the far-right diagram in Figure 38-14. Explain to students that Figure 38-15 demonstrates a process that is occurring in one portion of a nephron, which is the collecting duct and the capillaries that wrap around the renal tubule.

Graphic Organizer

Use this graphic organizer in *Teaching Tip:* **Learning Tool** on page 913.

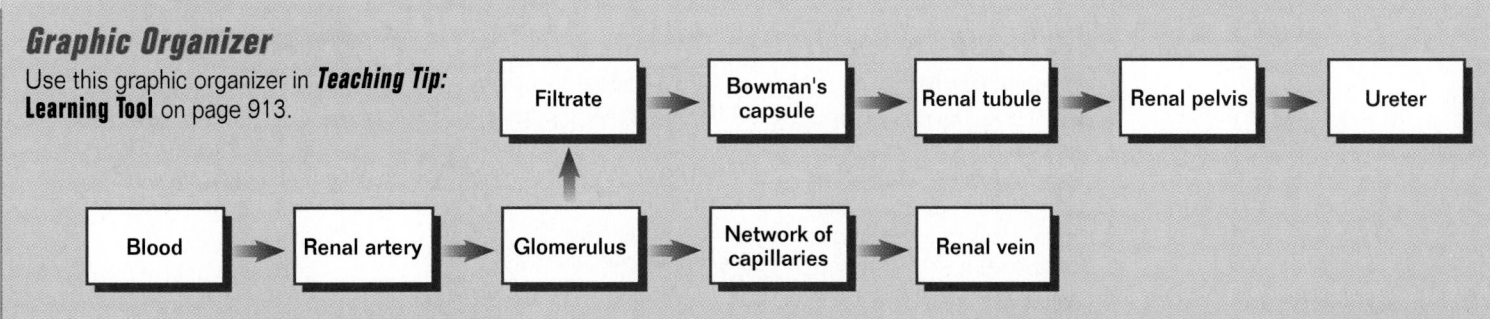

Filtrate → Bowman's capsule → Renal tubule → Renal pelvis → Ureter

Blood → Renal artery → Glomerulus → Network of capillaries → Renal vein

Glomerulus → Filtrate

👁 VISUAL STRATEGY

Figure 38-16

Help students trace the path of urine from the kidneys through the ureter to the urinary bladder and through the urethra. Tell students that the right kidney is slightly lower than the left because of the large amount of space taken up by the liver.

Teaching Tips
Summarizing a Function

Ask students to summarize the function of the urinary system in a paragraph or two, using their own words. Tell them not to look in their textbook or notes for information, so that they can see how much they *really* know. Tell them to use as many science terms as possible in their summary. They may want to begin by making a list of the terms that relate to the urinary system and the formation of urine.

🔲 What If the kidneys were to malfunction? Nitrogen

wastes would build up in the blood, eventually reaching poisonous levels. The condition in which poisonous nitrogen wastes build up in the blood is called uremia.

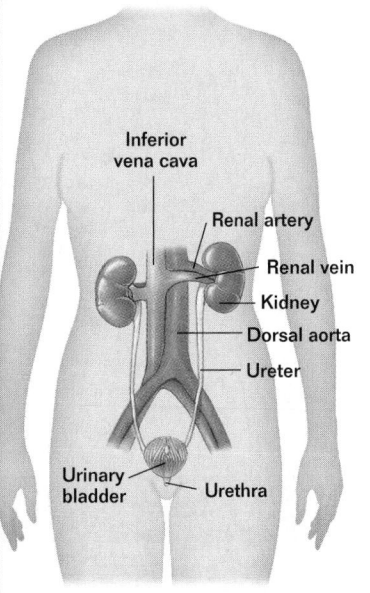

Figure 38-16 Urine exits the kidneys by way of two ureters that empty into a storage organ called the urinary bladder. Urine exits the body through the urethra.

CONTENT LINK

You can learn more about the structural relationship between the urinary and reproductive systems in **Chapter 42**.

Urine Leaves the Body Through a System of Tubes

Urine produced in the kidneys passes into the urinary bladder through tubes called **ureters** *(yoo REET uhrs)*, shown in Figure 38-16. The ureters have smooth muscle in their walls. The slow, rhythmic contractions of this muscle move the urine through the ureters.

The ureters direct the urine into the **urinary bladder,** a hollow, muscular sac that gradually expands as it fills. The bladder can hold up to 0.6 L (0.63 qt.) when full. Urine leaves the bladder and exits the body through a tube called the **urethra** *(yoo REE thruh)*. In males the urethra passes through the penis. In females the urethra lies in front of the vagina and is only about 2.5 cm (1 in.) long. Such a short length makes it easy for bacteria and other pathogens to invade the urinary system, which explains why females are more prone to urinary infections than males are. There is no connection between the urethra and the genital (reproductive) system in females. In males, both sperm and urine exit the body through the urethra.

The elimination of urine from the body through the urethra is called urination. When the bladder fills with urine, stretch receptors in its wall send nerve impulses to the spinal cord. In response, the spinal cord returns impulses to the bladder, causing its muscular walls to contract and the tight rings of muscle closing off the urethra to relax. The bladder then empties its contents through the urethra. In older children and adults, the brain controls this urination reflex, delaying the release of urine until a convenient time.

Kidney Failure Is Life Threatening

Because of the important role that the kidneys play in maintaining homeostasis of body fluids, kidney failure can quickly become life threatening. The most common causes of kidney failure are infection, long-term diabetes, untreated long-term high blood pressure, and damage to the kidneys by the body's own immune system (autoimmune kidney disease). When kidneys stop working, toxic waste materials such as urea accumulate in the blood plasma. In addition, blood plasma ion levels rapidly depart from their normal values. If the kidneys fail, there are only two treatment options, dialysis *(deye AL uh sihs)* or transplant.

In **hemodialysis**, tubes called catheters are surgically inserted into an artery and a vein, usually on the lower arm. These catheters are equipped with valves that can be opened and closed. Every few days the individual must go to a clinic where the catheters are connected to a dialysis machine, shown in Figure 38-17. Blood passes from the patient's artery into the machine and then back into the vein. Inside the

Multicultural Perspective

A World-Class Doctor

Dr. Andrew Thomas, an African American urologist who practices medicine in Chicago, has dedicated his career to world health. He has studied African diseases as well as how medicine is practiced in Africa, and in 1975 he helped coordinate America's first delegation to the People's Republic of China. In this country, Dr. Thomas founded "Project 75," an organization with the goal of increasing the number of minority physicians in the United States.

❓ Did You Know?

The first successful organ transplant, performed in 1954, was that of a kidney. Since the donor and recipient were identical twins, rejection was not a problem.

dialysis machine, the blood travels through a unit of many hollow tubes. The tubes are surrounded by a thin permeable membrane. Waste materials and ions that have accumulated in the person's blood plasma diffuse through the membrane into a fluid that bathes the outside of the membrane. The fluid has the same composition as normal blood plasma and is free of wastes. Dialysis patients must carefully manage their salt and water intake because the dialysis machine, unlike the kidney, cannot regulate blood volume and sodium levels.

Dialysis is not a permanent solution to kidney failure. A single healthy kidney can meet all of the homeostatic needs of the body, but no dialysis machine can. A more permanent solution to kidney failure is transplantation of a kidney from a healthy donor.

One great problem with kidney transplants is common to all organ transplants—rejection of the transplanted organ by the recipient's immune system. All of the cells of your body have "self-markers" on their surfaces to identify them so that they are not attacked by your body's immune system. These markers are called **histocompatibility** *(hihs toh kehm pat uh BIHL uh tee)* **antigens.** The combination of these antigens displayed on your body's cells is as unique as your fingerprints. Only identical twins have the same histocompatibility antigens. The more closely related two individuals are to one another, the more likely they are to have some common histocompatibility antigens. This is why tissue transplants are more likely to succeed if the donor and recipient are closely related. But even in close matches, there is some chance of transplant rejection. To reduce chances of rejection, the recipient is treated with drugs like cyclosporin, which suppresses the activity of the immune system. However, when such drugs are given, there is an increased risk of infection.

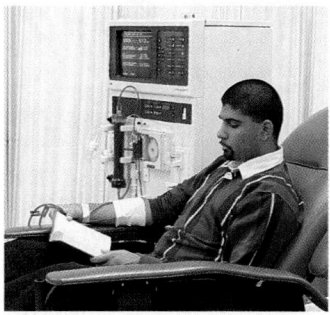

Figure 38-17 Hemodialysis has prolonged the lives of many people with damaged or diseased kidneys. The dialysis machine functions like a kidney in that it filters urea and excess ions from the blood.

CONTENT LINK

You will learn more about why the body rejects foreign tissues in Chapter 39.

SECTION 38-3

Teaching Tip
Using the Library
Have students use library references to research kidney diseases and disorders. Have students write and illustrate a report focusing on the causes of these illnesses, their symptoms, and their treatment. You may wish to have students share their reports with the class.

Chapter Closure
Have students write down all the major heads that are in the chapter. Next to each head, ask students to write a sentence that summarizes an important point made in the text that follows the head.

Section Review

1. *How is the carbon dioxide in your body produced and excreted?*
2. *Name the liquid that would be found inside the collecting duct of a nephron.*
3. *How is urine stored and eliminated from the body?*
4. *What are the benefits and risks involved in kidney transplants?*

Critical Thinking
5. *Suppose that a urine sample contains a high concentration of protein. Explain why this may indicate damaged kidneys.*

Answers to Section Review

1. Carbon dioxide production is a result of cellular respiration. It is excreted by the lungs and by the kidneys as a component of urea.

2. urine

3. Urine is stored in the urinary bladder and eliminated from the body through the urethra.

4. The benefits of transplantation include the patient's mobility and the constant cleansing of the blood afforded by the transplant. The disadvantages are the cost, the surgical risks, and the need for constant immunosuppression therapy.

5. In a healthy kidney, blood proteins never leave the blood and enter the kidney's tubule systems, thus the urine should contain no proteins.

38 CHAPTER REVIEW ....

CHAPTER REVIEW ANSWERS

Each item in the Chapter Review is correlated by text section in the assignment guide that follows.

ASSIGNMENT GUIDE

Section	Review	Themes Review	Critical Thinking
38-1	1, 2, 8, 9		14
38-2	3, 4, 10		15
38-3	5–7, 11	12, 13	

Review
Multiple Choice
Code in parentheses indicates section number and objective number.

1. c (38-1.1)
2. b (38-2.1)
3. c (38-2.2)
4. a (38-2.5)
5. d (38-3.1)
6. b (38-3.2)
7. d (38-3.5)

Completion
8. 4, 4, 9 (38-1.1)
9. 8 (38-1.2)
10. colon, K (38-2.3)
11. kidney, ureter. (38-3.4)

Themes Review
12. The concentration of ammonia would increase. Ammonia is converted to urea in the liver. (38-3.1)
13. The loop of Henle is shorter in beavers. Because beavers live in a watery environment, they do not need to conserve water. (38-3.3)

Vocabulary

amylase (904)	filtrate (913)	pharynx (906)
bile salt (904)	gastrin (907)	portal vein (909)
Bowman's capsule (912)	glomerulus (912)	protease (905)
Calorie (900)	hemodialysis (914)	protein (900)
carbohydrate (900)	hepatic vein (909)	rectum (909)
chyme (907)	histocompatibility antigen (915)	renal tubule (913)
collecting duct (913)	lipase (905)	ulcer (907)
colon (908)	liver (909)	urea (911)
duodenum (907)	loop of Henle (913)	ureter (914)
emulsification (904)	microvilli (908)	urethra (914)
esophagus (906)	nephron (912)	urinary bladder (914)
excretion (910)	pancreas (907)	villi (908)
fat (900)	peristaltic contraction (906)	

Review

Multiple Choice

1. Which compound is used by the body mainly to build muscle and bone?
a. fat
c. protein
b. carbohydrate
d. iodine

2. Select the sequence that most accurately reflects the order of movement and digestive activity as food passes through the digestive tract.
a. duodenum, pancreas, stomach, small intestine, large intestine
b. stomach, small intestine + pancreas, large intestine
c. duodenum, esophagus, small intestine + pancreas, large intestine
d. esophagus, small intestine, large intestine, pancreas, colon

3. Which pair matches the enzymes with the food molecules they digest?
a. lipases: starches
b. amylases: fats
c. lipases: fats
d. proteases: starches

4. Nutrients in the small intestine enter the bloodstream by passing through structures called
a. villi.
c. gastric pits.
b. macrovilli.
d. both a and b.

5. Which substance is *not* a waste eliminated from the body through the kidneys?
a. urea
c. salts
b. water
d. oxygen

6. The fine capillary network of the nephron system is called the
a. collecting duct.
b. glomerulus.
c. loop of Henle.
d. Bowman's capsule.

7. A kidney transplant is likely to be most successful if the kidney donor is the recipient's
a. friend.
c. wife.
b. uncle.
d. sister.

Completion

8. Studies have shown that 1 g of carbohydrate yields about _____ Calories; 1 g of protein yields about _____ Calories; and 1 g of fat yields approximately _____ Calories.

9. Of the 20 amino acids used to make proteins in the human body, _____ must be obtained through food.

10. The large intestine, or _____ , is much shorter than the small intestine. It functions to compact undigested materials and reabsorb vitamin _____ .

11. Urine is produced in the _____ , each of which drains into the urinary bladder through a tube called the _____ .

Themes Review

12. **Homeostasis** How would liver failure affect the concentration of ammonia within the body? What is the relationship of ammonia to urea?

13. **Evolution/Structure and Function** The loop of Henle functions to conserve water by reabsorbing it. Its length varies among mammal species. Would you expect the loop of Henle of an animal like the beaver, which lives in a watery environment, to be longer or shorter than that found in humans? Explain.

Critical Thinking

14. **Interpreting Data** The table below shows nutrition data for a single serving of two breakfast cereals. If you were most concerned with preventing colon cancer, keeping your blood pressure low, and supplying your body's cells with enough energy, which cereal, A or B, would you choose for breakfast? Explain.

Cereal A		Cereal B	
Serving Size 1 box (19 g)		Serving Size 1 box (32 g)	
Amount/serving		Amount/serving	
Calories	70	Calories	100
Fat Calories	0	Fat Calories	5
	% DV		% DV
Total Fat 0 g	0%	Total Fat 0.5 g	1%
Saturated Fat 0 g	0%	Saturated Fat 0 g	0%
Cholesterol 0 mg	0%	Cholesterol 0 mg	0%
Sodium 210 mg	9%	Sodium 170 mg	7%
Total Carbohydrates 17 g	6%	Total Carbohydrates 25 g	8%
Fiber 1 g	4%	Fiber 4 g	16%
Sugars 1 g		Sugars 10 g	
Protein 1 g		Protein 2 g	

15. **Making Inferences** Pancreatic fluid contains sodium bicarbonate, which changes the pH of the chyme from acid to base as it moves into the duodenum. How would the absence of sodium bicarbonate in the pancreatic fluid affect the functioning of the many enzymes also carried in this fluid?

Activities and Projects

16. **Research and Writing** Write an article for the health section of your school or local newspaper that discusses diuretics. You may wish to address these questions in your article: What is a diuretic and how does it affect kidney function? Why is thirst often intensified by a diuretic? How do diuretic drugs function to lower blood pressure? To capture your reader's interest, why not begin your article by identifying diuretics that most people have heard of, such as the caffeine in coffee and soft drinks.

17. **Cooperative Group Project** Survey students in your school to find out what they know about vitamins. Ask about vitamin use in general and about specific vitamins, such as vitamin A and vitamin C. Organize your survey questions into four sections: (1) general vitamin facts, (2) sources of specific vitamins, (3) benefits of specific vitamins, and (4) deficiency diseases and symptoms associated with specific vitamins. Compile the survey results and present them on a poster. Display the poster where it can be seen by students and teachers.

18. **Multicultural Perspectives** Look at your library's reference books that describe the cultures and customs of other nations. Find information relating to ideas about ideal body weight and eating habits for at least five different countries. How do these ideas differ from those in America?

Critical Thinking

14. Cereal B. Cereal B contains more fiber, less sodium, and more carbohydrate than cereal A. Fiber may prevent colon cancer, too much sodium elevates blood pressure, and cells prefer to use carbohydrates as a source of energy. (38-1.1)

15. The enzymes would not function since the environment would remain acidic. (38-2.4)

Activities and Projects

16. A diuretic is a substance that increases the volume of urine. Thirst is often intensified by a diuretic because the diuretic causes an excessive amount of water to be lost by the body. Diuretic drugs function by decreasing the reabsorption of sodium ions, which leads to an increase in urine volume. Blood pressure is lowered because there is less fluid in the blood.

17. General questions may deal with issues such as the need for vitamin supplements, appropriate vitamin dosage, and the use of vitamins by the body. Questions about specific vitamins can be developed from information presented in Table 38-1.

18. Answers will vary; however, point out to students that not every nation believes "thin is beautiful." In some countries, people who are relatively thin are considered unhealthy. You might also compare America's "fast-food" and "junk food" compulsions with other eating habits and dietary customs around the world: for example, in many European, Middle-Eastern, and Asian countries, it is still considered impolite to rush through a meal.

LABORATORY *Investigation* *Chapter 38*

Lactose Digestion

OBJECTIVES

- Identify the types of large molecules present in milk products and other foods.
- Observe the action of enzymes.
- Recognize the relationship between enzymes and the digestion of food molecules.

PROCESS SKILLS

- designing and performing an experiment
- collecting data

MATERIALS

- milk-treatment product (liquid)
- toothpicks
- depression slides
- droppers
- whole milk
- glucose
- glucose test strips

BACKGROUND

1. What are the three types of large molecules found in foods?
2. How does the human body break down food molecules?
3. Describe the chemical composition of carbohydrates.
4. What is milk sugar?
5. What is milk intolerance? What happens to people who drink milk but cannot digest it?
6. Write your own question to explore in your lab report or notebook.

TECHNIQUE

1. Working with a lab partner, prepare a list of at least 15 foods that contain milk. List the number of times each week that you and your lab partner are likely to eat these foods.

2. Read the information sheet about the milk-treatment product. Discuss with your lab partner what the product is and what it does. Summarize your discussion in the **Records** section of your report.
3. Using the materials listed, design a control experiment that will test the effectiveness of the milk-treatment product. In the **Records** section of your report, describe the experiment that you have designed. Be sure to include one or more data tables.
4. Show your experimental design to your teacher. You must receive approval from your teacher before you can perform the actual experiment.

Preparation

Commercial milk-treatment products that contain yeast derived lactase are available from most pharmacies. Avoid buying this product in solid form. Each lab group will need copies of the information sheet that accompanies the product. You will also need several boxes of glucose test strips, which can be obtained from a pharmacy. Buy several boxes so the lab groups each have a color guide. Glucose solution can be purchased from Ward's. Groups will also need toothpicks, 2–5 depression slides, and droppers. Provide reference materials on lactose intolerance as a help to students. Students' experiments should include testing the milk for the presence of glucose before and after treatment with lactase.

Procedural Notes

- This investigation will take one or two periods, depending on how much time designing the experiment will take.
- When you read the student's protocols for approval, direct students to use only drops of milk and milk-treatment product instead of full quarts of milk. Some students may want to perform a Benedict's test for the presence of glucose. Convince them that glucose test strips are much more convenient, in that they do not require a hot water bath. They also give faster results.
- Some students will be unclear about using a control in their experiment. They will know to use milk with, and without, milk-treatment product, but they may not think about testing the milk-treatment product itself for sugar. Milk-treatment product tablets will give positive glucose results, so they should not be used for this investigation. Instruct students to test the glucose test strip for a positive glucose test in a known glucose solution.

Background Answers

1. carbohydrates, fats, and proteins
2. Most food molecules are broken down with enzymes.
3. Carbohydrates are composed of carbon, hydrogen, and oxygen atoms in the proportion of 1:2:1.
4. Milk sugar is lactose, a disaccharide that contains glucose and galactose.
5. Milk intolerance occurs when a person's digestive system does not produce lactase, the enzyme that breaks down lactose. They may experience gastrointestinal pain, vomiting, and diarrhea.
6. How effective is a milk-treatment product in breaking down lactose?

Records

Answers will vary. One possible data table is shown on the next page.

100% Lactose Reduced
Nonfat Milk

5. After your teacher has approved your procedure, change the measures from the ones given by the product to the following: quart of milk = drop of milk in a depression slide; teaspoon of milk-treatment product = drop of milk-treatment product.

6. To use the glucose test strips, follow the instructions printed on their container.

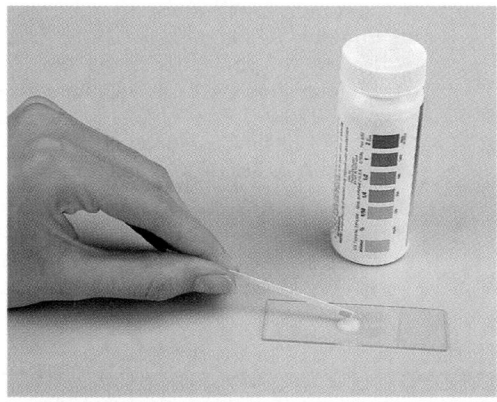

7. Fill in your self-designed data table in the **Records** section of your report.

8. Clean up your materials and wash your hands before leaving the lab.

INQUIRY

1. What are the ingredients of the milk-treatment product?
2. What does the product do when added to milk?
3. Why was it important for your experiment to include glucose?

ANALYSIS

1. What is the definition of milk intolerance?
2. What are lactose and lactase, and how do they interact with each other?
3. What do the results of this lab imply about treatments for other medical problems resulting from enzyme deficiencies?
4. As a person grows older, do you think the likelihood of that individual exhibiting milk intolerance increases or decreases? Explain your answer.
5. Do you think milk intolerance might be inherited? Why or why not?

FURTHER INQUIRY

Write a new question that could be explored as a new investigation. The following is an example:

> In a hospital study of people with milk intolerance, how would blood glucose levels in human subjects receiving treated milk compare with those of subjects receiving untreated milk?

Inquiry Answers

1. It is a commercial product made of yeast-derived lactase.
2. The lactase digests the lactose in the milk.
3. The test shows that the glucose strip actually tests for glucose.

Analysis Answers

1. Milk intolerance is the gastrointestinal pain, vomiting, and diarrhea that results from the body's inability to digest lactose, also called milk sugar. This inability to digest lactose is caused by a lower than normal level of lactase activity.
2. Lactose is the disaccharide found in milk; lactase is the enzyme that catalyzes the breakdown of lactose into glucose and galactose.
3. Other products that contain the missing enzyme might be used to treat other enzyme deficiencies.
4. It increases, because lactase is required most during infancy and early childhood.
5. yes, because certain populations have a higher incidence of lactose intolerance than others

Lactose Breakdown As Evidenced by the Presence of Glucose		
Solution tested for glucose	Results	Glucose present?
Milk	–	no
Milk-treatment product	–	no
Milk treated with milk-treatment product	+	yes
Glucose solution	+	yes

UNIT 9

THE BODY'S DEFENSES

Block Scheduling Guide

> Each block represents about 45–50 minutes of instructional time. The resources cited in a block are coded to a specific program component using the Key on the next page.

Lecture Concepts

Lesson Resources

39-1 Defending Against Infection pp. 921–924

BLOCK 1

a. Skin and mucous membranes help exclude pathogens from the body. Any pathogens that do enter may be attacked by white blood cells, such as macrophages and neutrophils.

b. Pathogens may also stimulate an inflammatory response or an increase in body temperature.

Lecture Resources
- Opening Questions, page 921
- Demonstration: Microbes Around Us
- Neutrophils Are Also Active Phagocytes, page 922
- Natural Killer Cells, page 923
- Inflammation: Isolating an Infection, page 923

- Visual Strategies: Figures 39-5, 39-6
- Teaching Transparencies: 235
- Holt Biology Videodiscs: Lesson 39-1

Assignment Options
- Section Review, page 924
- Chapter Review, questions 1, 2, 8, 9, 16, 19

39-2 The Immune System pp. 925–928

BLOCKS 2 and 3

a. The immune system consists of an army of various kinds of white blood cells: macrophages, killer T cells, helper T cells, and B cells.

b. Helper T cells stimulate killer T cells, which attack pathogens and infected cells. Helper T cells also stimulate B cells, which release antibodies that mark pathogens for destruction.

Lecture Resources
- Opening Questions, page 925
- Demonstration: Viruses
- B Cells and T Cells, page 925
- Nonspecific Diseases, page 926
- Visual Strategies: Figures 39-7, 39-8
- B Cell Production, page 927
- Vaccination, page 928
- Teaching Transparencies: 236, 237, 202A
- Holt Biology Videodiscs: Lesson 39-2

Classwork Options
- Effective Reading, page 926
- History of Vaccination, page 928
- Summarizing the Body's Defenses, page 928
- Inquiry Skills Development B32: Antigen-Antibody Interaction

Assignment Options
- Section Review, page 928
- Chapter Review, questions 3, 4, 10–12, 17, 18

39-3 Malfunctions and Failures of the Immune System pp. 929–933

BLOCK 4

a. In autoimmune diseases, the body fails to distinguish its own cells from those of a pathogen and thus attacks its own cells.

b. In allergies, the body responds to a harmless substance such as pollen or dust as if it were a pathogen.

c. AIDS is caused by the human immunodeficiency virus (HIV), which destroys helper T cells in large numbers, thereby disabling the immune system.

Lecture Resources
- Opening Question, page 929
- Demonstration: Structure of HIV
- The Rapidly Evolving Influenza Virus, page 930
- The Origin of HIV, page 931
- Mutation and HIV, page 931
- Overcoming Misconceptions, page 932
- Teaching Transparencies: 238, 203A
- Holt Biology Videodiscs: Lesson 39-3

Classwork Options
- Keeping Ahead of Flu Epidemics, page 930
- Closure, page 933
- Laboratory Disease Transmission, pages 936–937
- Inquiry Skills Development B33: Transmission of a Communicable Disease

Assignment Options
- Section Review, page 933
- Chapter Review, questions 5–7, 13–15, 20

Review/Enrichment

BLOCK 5

- Study Guide: Concept Review and Pretest

Assignment Options
- Activities and Projects, questions 21–23

Assessment Options

BLOCK 6

Traditional Assessment
- Chapter 39 Test
- Test Generator/Assessment Item Listing: Software item bank for preparing customized chapter tests.

Portfolio Assessment

Select student reports for one or more laboratory experiments from this chapter. *The Direct Observation Checklist* on the *BioSources Teaching Resources CD-ROM* should be completed during a laboratory or other cooperative learning experience.

Holt Biology Videodiscs

Holt Biology Videodiscs gives you a powerful tool for teaching, review, and assessment. *Concepts of Biology* adds interactive lesson plans and extensive tools for customization using CD-ROM technology.

CONCEPTS OF BIOLOGY

Use the following topics from *Concepts of Biology* to help you teach this chapter:
- Topic 13: Bacteria and Viruses
- Topic 25: Drugs and the Human Body

For further information regarding the media on the videodiscs, see *Holt Biology Videodiscs Teacher's Correlation Guide to Biology: Principles and Explorations*.

Chapter Theme

Structure and Function

Cells of the immune system continuously monitor the body for infection by pathogens and for cancerous cells. This monitoring is known as immune surveillance. How can immune-system cells distinguish invaders from the cells of the body? The body's cells have protein markers on their surface that signal their identity. Immune-system cells have protein receptors that match these marker proteins. The surface proteins of pathogens will differ from the body's, allowing the immune system to recognize and attack these invaders.

Tapping Prior Knowledge

- Where is the immune system located?
- What type of cells composes the immune system?
- Describe the structure of HIV, the human immunodeficiency virus.

Opening Demonstration

If the immune system fails, pathogens or cancers will quickly overwhelm the body. To illustrate our dependence on a functional immune system, show students a picture of David, the "boy in the bubble." David was born with Severe Combined Immune Deficiency (SCID), a disease characterized by the inability to produce B cells and T cells, which play key roles in the immune system. David had to live in a sterile, sealed plastic enclosure because he could not fight off pathogens. He chose to leave this sterile environment at age 12, and he died shortly thereafter—not from bacterial or viral infections as might be expected, but from cancers that had been developing during his stay in the "bubble."

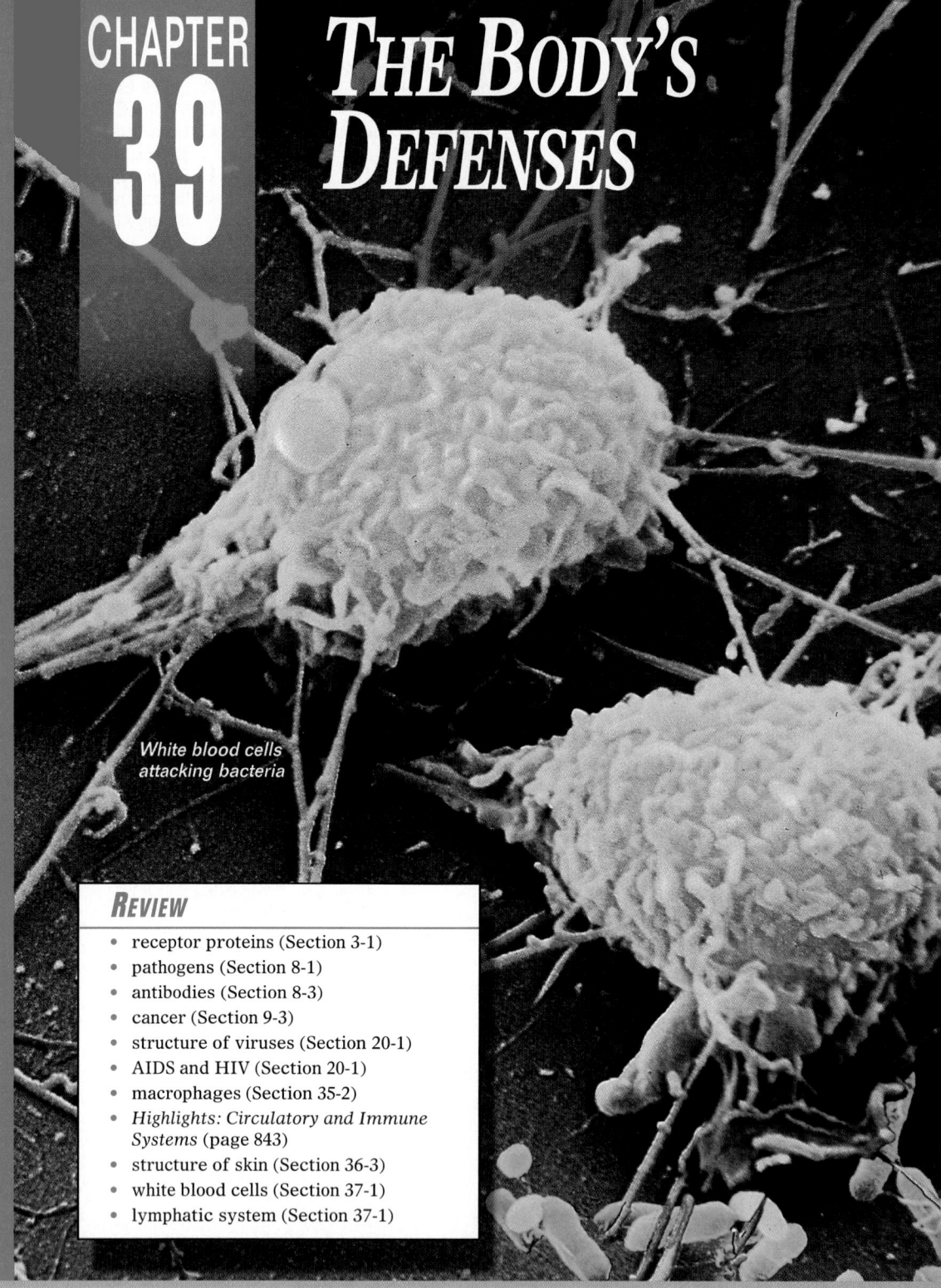

CHAPTER 39
THE BODY'S DEFENSES

White blood cells attacking bacteria

REVIEW

- receptor proteins (Section 3-1)
- pathogens (Section 8-1)
- antibodies (Section 8-3)
- cancer (Section 9-3)
- structure of viruses (Section 20-1)
- AIDS and HIV (Section 20-1)
- macrophages (Section 35-2)
- *Highlights: Circulatory and Immune Systems* (page 843)
- structure of skin (Section 36-3)
- white blood cells (Section 37-1)
- lymphatic system (Section 37-1)

Authors' Rationale

Awareness of and education about AIDS and other sexually transmitted diseases have skyrocketed recently. Why? Antibiotics, the wonder drugs of the mid-twentieth century, are becoming less and less effective. Understanding how the immune system works is essential to good health. Students need to understand that there are many diseases over which they have no control, but there are many, such as AIDS, that can be avoided through education, intelligent behavior, and prevention.

39-1 Defending Against Infection

When you think of how animals defend themselves, it is natural to think of armor, such as the hard shells of turtles, clams, and armadillos. However, even armor would offer no protection against the most dangerous enemies that the human body faces: pathogenic bacteria, viruses, fungi, and protists. The world is filled with pathogens, and you could not live for long unprotected. You survive because your body has a variety of very effective defenses against this constant attack, as summarized in Figure 39-1.

Section Objectives

- Describe two ways that skin and mucous membranes defend the body.
- Explain the roles of macrophages, neutrophils, and natural killer cells in combating pathogens.
- Contrast the inflammatory response with the temperature response.

Skin and Mucous Membranes: The First Line of Defense

The outermost layer of the human body, the skin, provides the first defense against invasion by pathogens. Skin acts as a wall that keeps pathogens out of the body. Cells of the outer layer of skin, the epidermis, are continually worn away. These cells are quickly replaced by new cells that move up from the lower layers, where cell division occurs. In only 40 minutes, your body loses and replaces approximately 1 million skin cells.

Your skin defends your body not only by providing a nearly impermeable barrier, but also by reinforcing this defense with chemical weapons. Secretions of the oil and sweat glands within the dermis make the skin's surface very acidic, inhibiting the growth of many pathogens. Sweat also contains the enzyme lysozyme, which attacks and digests the cell walls of many bacteria.

Internal surfaces of the body through which pathogens could pass are covered by **mucous membranes**, epithelial layers that produce the sticky fluid called mucus. Mucous membranes line the digestive system, nasal passages, lungs, respiratory passages, and reproductive tract. Like the skin, mucous membranes not only serve as a barrier to pathogens, but also produce chemical defenses. Cells lining the bronchi and bronchioles in the respiratory tract secrete a layer of sticky mucus that traps microorganisms before they can

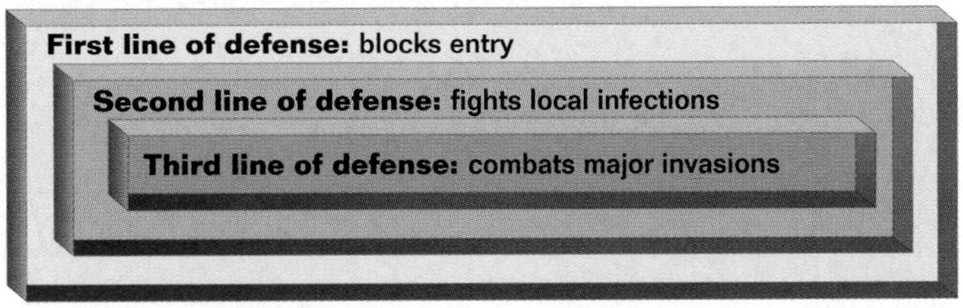

First line of defense: blocks entry

Second line of defense: fights local infections

Third line of defense: combats major invasions

Figure 39-1 Pathogens that attack your body are opposed by three lines of defense. The first defense consists of the skin and the epithelial layers that line the body's interior surfaces. This defensive line prevents pathogens from entering the body. The cells, proteins, and physiological responses of the second line of defense generally deal with minor infections, such as those resulting from small breaks in the skin. The third line of defense, the immune system, is activated only when your body is facing a major invasion by pathogens that have broken through the first two lines.

Vocabulary Preview

mucous membrane

neutrophil

natural killer cell

complement

inflammatory response

Opening Questions

Have each student list five diseases and the means by which each is transmitted. Tables of viral and bacterial diseases are shown in Chapter 20, and a table of protist-caused diseases is shown in Chapter 21. Ask students to explain how each pathogen's means of transmission enables it to penetrate the body's defenses.

Demonstration

Microbes Around Us

Materials: sterile agar solution, five petri dishes, sterile swabs, sterile distilled water

Procedure: To demonstrate that microbes are ubiquitous, prepare a sterile agar solution and add it to five sterile petri dishes. Add a sample of human saliva to one petri dish. To the other four dishes add samples taken from various places in the classroom. You can collect microbes from surfaces such as desktops by sweeping them with sterile swabs moistened with sterile distilled water. Seal each petri dish with masking tape, label each petri dish, and place all five in a warm, dark place or an incubator for several days. After this time period, show students the colonies of microbes that have grown on the agar. Point out that most of the microorganisms growing on the dishes are harmless but that some may be pathogens. **Caution: Students should not open the sealed petri dishes, in case pathogens are growing there. Dispose of the swabs and petri dishes safely.**

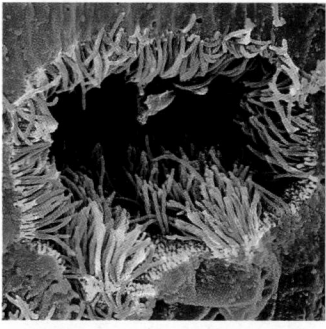

Figure 39-2 Cells lining the respiratory tract are covered by beating cilia. Movement of the cilia carries pathogens and foreign objects up to the esophagus to be swallowed. This cross section of a respiratory passage in the lung shows the hairlike cilia.

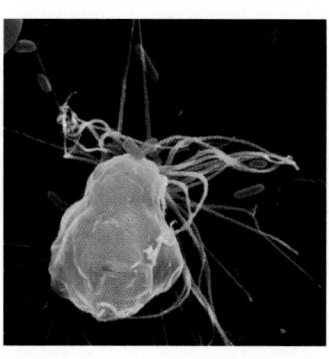

Figure 39-3 The extensions of this macrophage (shown here in yellow) are capturing bacteria.

Figure 39-4 The large cell with two dark blotches is a neutrophil. Neutrophils are white blood cells that release chemicals that are lethal to pathogens—and themselves.

reach the warm, moist lungs, which are ideal breeding grounds for microbes. Cilia on the cells of the respiratory tract, shown in Figure 39-2, continually sweep the mucus upward so that it can be swallowed, sending potential invaders to the stomach, where they are destroyed by acid.

The surface defenses of your body are very effective, but they are occasionally penetrated. You can take pathogens into your body when you eat and breathe. They can also enter through wounds or open sores. When invaders reach deeper tissue, a second line of defense comes into play.

Counterattacks: The Second Line of Defense

When the body's interior is invaded, a host of cellular and chemical defenses swing into action. Four are of particular importance: (1) cells that kill invading microbes; (2) proteins that kill invading microbes; (3) the inflammatory response, which sends defensive cells to the point of infection; and (4) the temperature response, which elevates body temperature to slow the growth of invading bacteria.

Cells That Kill Invading Microbes

The most important of the counterattacks against infection are those carried out by cells that attack invading microbes. These cells patrol the bloodstream and wait within the tissues for invaders. There are three kinds of these cells. Each kind kills invading microbes differently.

1. **Macrophages** White blood cells called macrophages, illustrated in Figure 39-3, kill bacteria one at a time by ingesting them. Although some macrophages stay within particular organs, particularly the spleen, most of the body's macrophages travel throughout the body in the blood, lymph, and fluid between cells.

2. **Neutrophils** White blood cells called **neutrophils** (*NOO truh fihlz*), shown in Figure 39-4, sacrifice themselves to defend the body. Neutrophils release chemicals that are

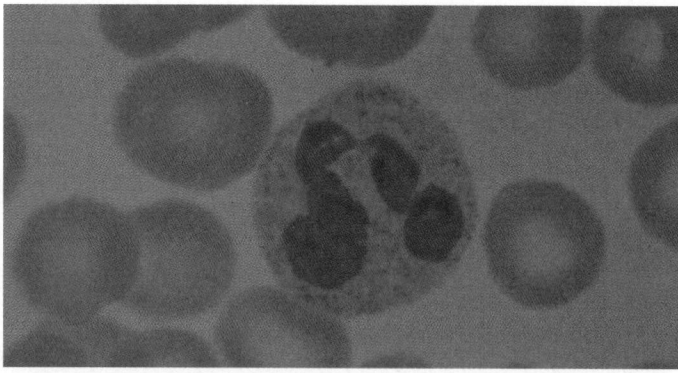

 Did You Know?

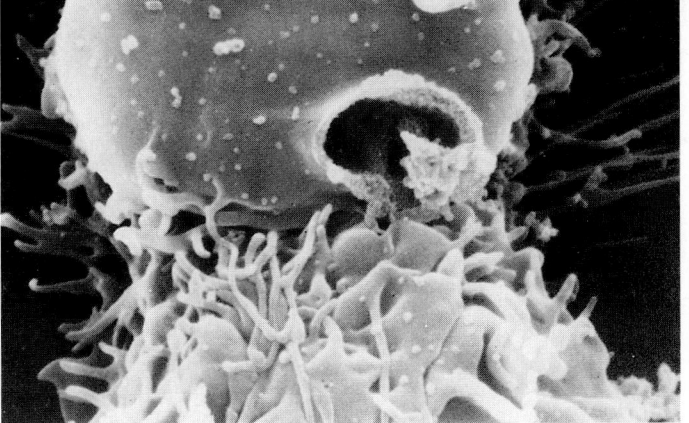

Figure 39-5 Natural killer cells monitor the body for the presence of cancer cells. The cell membrane of this cancer cell, *top,* has been punctured by a natural killer cell, *bottom;* the wound is fatal to the cell.

identical to household bleach, killing any nearby bacteria—and themselves in the process.

3. **Natural killer cells** Unlike macrophages, **natural killer cells** attack only cells that have been infected by microbes, not the microbes themselves. Natural killer cells are particularly effective at detecting and attacking body cells infected with viruses. Natural killer cells destroy a cell by puncturing its membrane, as shown in Figure 39-5. This allows water to rush into the cell, causing it to burst. Natural killer cells are also able to detect cancer cells, which they often kill before the cancer cells can develop into a tumor. Immune surveillance by natural killer cells is one of the body's most potent defenses against cancer.

Proteins That Kill Invading Microbes

Your body also uses a very effective chemical defense to assist its cellular defenses. This defense, called **complement,** consists of proteins that circulate in the blood plasma. Their defensive activity is triggered when they encounter the cell walls of bacteria or fungi. Then the complement proteins interact to form a membrane attack complex (MAC). The MAC inserts itself into the pathogen's cell membrane, creating a hole.

The Inflammatory Response

An injury or local infection causes an **inflammatory response,** a series of events that suppresses infection and speeds healing. For example, imagine that you have cut your finger, creating an entrance for pathogens. Infected or injured cells in your finger release chemical alarm signals that cause expansion of local blood vessels, increasing the flow of blood to the site of infection. Increased blood flow produces the redness and swelling so often associated with infections. It also promotes the migration of macrophages and neutrophils to the infection site, where they can attack invading microbes. Neutrophils arrive first, spilling out chemicals that kill the invading microbes and the tissue cells

Teaching Tips
Natural Killer Cells

Lymphocytes, such as killer T cells, and natural killer cells (NK cells) are very similar in shape. NK cells, once thought to be special lymphocytes, can spontaneously lyse (burst) any virus-infected cell or tumor cell. Lymphocytes are more restricted in their actions; they must recognize a specific antigen (a foreign molecule) before they act.

Inflammation: Isolating an Infection

What causes the area around a cut to swell? First, local blood vessels expand, which brings more blood to the site of the cut. Second, the vessels also become leakier, so more fluid leaves the bloodstream and enters the tissues. Third, the lymphatic vessels and the spaces between cells become filled with clots of fibrin, causing a reduction in the drainage of fluid from the area of the cut. This isolation delays the spread of bacteria or other toxic products. In effect, inflammation results in a blockade of the site of an infection.

 VISUAL STRATEGY

Figure 39-5

Ask students what would happen to the cancer cell in this photograph after its membrane is punctured? *(Without an intact membrane, the cell could not regulate its internal conditions. Water would move into the cell, causing it to burst.)*

 Did You Know?

Viruses can't reproduce or metabolize on their own. They must invade a cell, commandeer its reproductive and energy-releasing machinery, and force the infected cell to replicate new viral nucleic acids and protective envelopes.

Although the unfortunate cells, once overrun, cannot save themselves, they can defend other cells by sending out chemicals called interferons. Interferons fasten to the cell membranes of uninfected cells and interfere with the virus's

ability to bind, rupture, and enter these membranes. Interferons probably stimulate protein synthesis in the target cells to inhibit production of the virus's genetic material or its coat proteins.

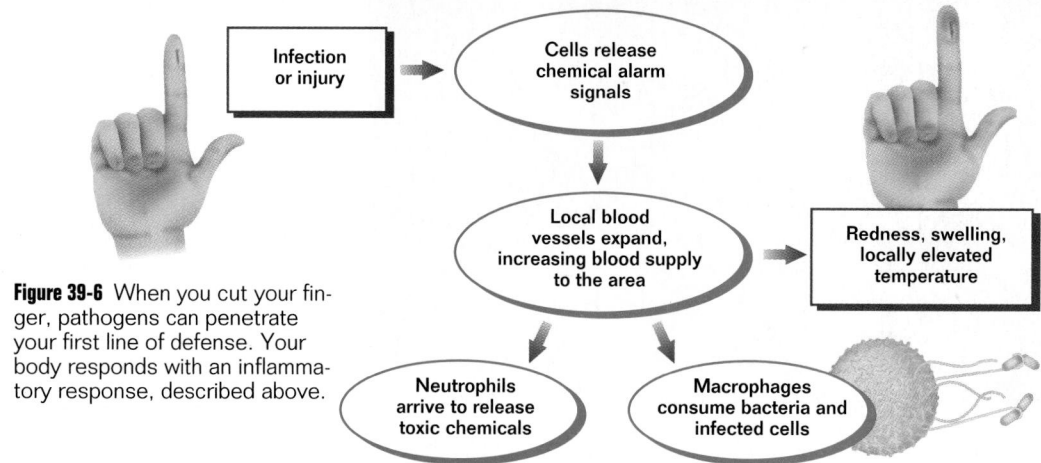

Figure 39-6 When you cut your finger, pathogens can penetrate your first line of defense. Your body responds with an inflammatory response, described above.

CAPSULE SUMMARY

Skin and mucous membranes help exclude pathogens from the body. Any pathogens that do enter may be attacked by white blood cells, such as macrophages and neutrophils. Pathogens may also stimulate an inflammatory response or an increase in body temperature.

in the immediate area. After the neutrophils come macrophages, which clean up the remains of the dead cells. The pus associated with some infections is a mixture of dead or dying neutrophils, broken-down tissue cells, and dead pathogens. Figure 39-6 summarizes the inflammatory response.

The Temperature Response

When macrophages initiate their counterattack, they increase the odds in their favor by sending a message to the brain to raise the body's temperature several degrees above the normal value of 37°C (98°F). This higher temperature is called a fever. Fever is helpful because bacteria that cause human diseases do not grow well at high temperatures. Although fever inhibits microbial growth, very high fevers are dangerous because excessive heat can destroy important cellular enzymes. In general, temperatures greater than 39°C (103°F) are considered dangerous, and those greater than 41°C (105°F) are often fatal.

Section Review

1. *Describe the skin's chemical defenses.*
2. *Why are neutrophils used to combat only local infections and not body-wide diseases?*
3. *Aspirin reduces fever. Explain why taking aspirin for a mild fever might slow rather than hasten your recovery from a bacterial infection.*

Critical Thinking

4. *Would you expect the temperature response to be as effective against a viral infection as against a bacterial infection? Explain your answer.*

39-2 The Immune System

Only occasionally do bacteria or viruses overwhelm your body's second line of defense. When this happens, the invaders face a third line of defense, the immune system, which is more difficult to evade than the first two lines of defense. Your immune system is not localized in any one place in the body, nor is it controlled by any one organ, such as the brain. Rather, it is an army of individual cells that rush to the site of an infection to combat invading microorganisms.

The Key Players in the Immune System

White blood cells, which are produced in the bone marrow and circulate in blood and lymph, constitute the immune system. These cells are very numerous; of the 100 trillion cells in your body, 2 trillion are white blood cells. Four kinds of white blood cells participate in the immune system's attack on pathogens: macrophages, killer T cells, helper T cells, and B cells. Each kind of cell performs a different function. **Killer T cells** attack and kill infected cells. **B cells** label invaders for later destruction by macrophages. **Helper T cells** activate killer T cells and B cells. And, as you have already seen, macrophages consume pathogens and infected cells. These four kinds of white blood cells exchange information and act in concert as a functional, integrated system.

To understand how this third line of defense works, imagine that you have just come down with the flu. Influenza viruses have entered your body in small water droplets inhaled into your respiratory system. They were not ensnared in the mucus covering the respiratory membranes (first line of defense), they slipped past patrolling macrophages (second line of defense), and they have begun to infect and kill mucous membrane cells. You feel sick because large numbers of the cells lining your respiratory tract are dying.

At this point, macrophages initiate the immune defense by releasing an alarm signal, the protein interleukin-1 *(IHN tuhr LOO kihn)*. This protein activates helper T cells. However, helper T cells do not attack pathogens. Instead, they serve as the "generals" of the immune system. Helper T cells respond to the alarm broadcast from macrophages by simultaneously activating two different types of immune system cells: killer T cells, which attack infected cells, and B cells, which produce defensive proteins.

Vocabulary Preview

killer T cell

B cell

helper T cell

antibody

Opening Questions

How do viruses reproduce? *(They must invade cells and convert them to the production of viruses.)* What are antibodies? *(defensive proteins produced by the body)* Where are the cells of the immune system located? *(in the blood and lymph and within some organs)*

Demonstration
Viruses

To emphasize the damage that pathogens can cause to the body, show the class a photograph of viruses bursting out of an infected cell. Point out that viruses often kill their host cell. Students should recognize that many cells in the body are infected and destroyed, making the destructive effects of the virus widespread. Students should also see that each cell produces many virus particles, each of which can go on to infect another cell, multiplying the damaging effects of the virus.

Teaching Tip
B Cells and T cells

B cells and T cells are formed in the bone marrow, but the two types of cells mature in different parts of the body. T cells are so named because they mature in the thymus, a small gland located in the neck. B cells were so named because in chickens the cells mature in the bursa of Fabricus, an organ of the lymphatic system. Humans do not have a bursa of Fabricus, and the B cells may mature in the liver or bone marrow.

The first and second lines of defense are called the nonspecific defenses because they are not directed at a specific pathogen. The cells of the immune system, by contrast, are referred to as the specific defenses. Inform students that each time the immune system is activated, it responds only to a particular pathogen. Each invasion by a different pathogen stimulates a new response from the immune system.

 VISUAL STRATEGY

Figure 39-7

Remind students that all cells in the body have receptor proteins and marker proteins on their surface. Killer T cells have receptors that recognize the proteins of pathogens. The body makes millions of types of killer T cells, and each type has uniquely shaped receptor proteins. Some killer T cells have receptors that match the shape of the proteins of the influenza virus. These proteins appear not only on the viruses themselves, but also on the membranes of infected cells, allowing killer T cells to recognize and destroy cells that have been taken over by the virus.

 VISUAL STRATEGY

Figure 39-8

Be sure students recognize that the lower-right portion of Figure 39-8 shows the same process as Figure 39-7—recognition and destruction of an infected cell by a killer T cell. Also point out the central role of helper T cells. They stimulate the activities of killer T cells and B cells. The activities of the killer T cell arm of the immune system are often called cell-mediated immunity, while the activities of the B cells are called antibody-mediated immunity or cellular immunity.

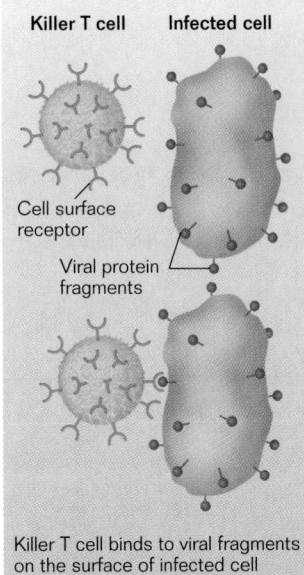

Killer T cell   Infected cell

Cell surface receptor

Viral protein fragments

Killer T cell binds to viral fragments on the surface of infected cell

Figure 39-7 How does a killer T cell recognize an infected cell? The infected cell will have some of the pathogen's proteins on its surface, and the killer T cell has receptors that detect these foreign proteins.

Killer T Cells Attack Infected Body Cells

Stimulation by the interleukin-1 signal causes helper T cells to immediately unleash a very potent attack against the virus, which now is located mainly inside the cells of your respiratory tract. Using a second chemical signal called interleukin-2, the helper T cells call into action killer T cells, which recognize and destroy body cells that have been infected with the virus. This attack is illustrated in Figure 39-7. Infected body cells display little bits of viral protein on their surface, and it is these telltale traces that the killer T cells recognize. Any of your body's cells that bear traces of viral infection are destroyed. The method used by killer T cells to kill infected cells is similar to that used by natural killer cells and complement—they puncture the cell membrane of the infected cell.

How do killer T cells recognize viral proteins? As explained in Figure 39-7, T cells have receptor proteins scattered over their cell membranes. Your body makes millions of different types of T cells. Each type of T cell bears a unique kind of receptor protein on its cell membrane, and this receptor is able to bind to a particular viral or bacterial protein. When you were still a fetus, any cells with receptors that recognized your own proteins were eliminated. ▫

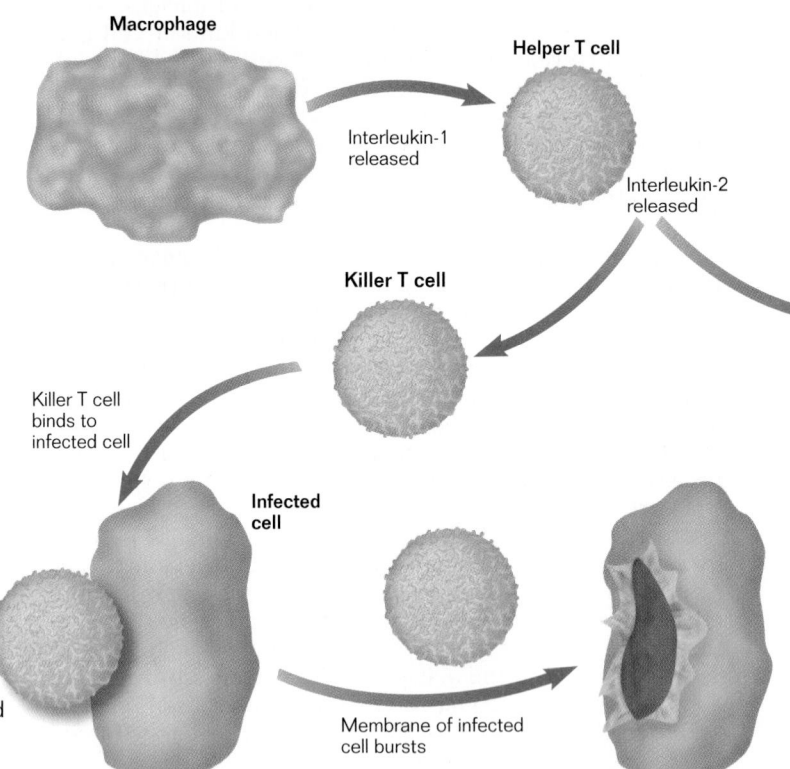

Macrophage

Helper T cell

Interleukin-1 released

Interleukin-2 released

Killer T cell

Killer T cell binds to infected cell

Infected cell

Membrane of infected cell bursts

Figure 39-8 When stimulated by macrophages, helper T cells activate the two arms of the immune system: killer T cells and B cells. For simplicity, the receptors on the killer T cell and the viral fragments on the infected cell are not shown.

Effective Reading

A Summary of the Immune Response

Have students carefully read the information on pages 926–928, paying particular attention to Figures 39-7 and 39-8. Then have students translate the information in this sub- section into a concise, step-by-step description of the events that follow infection by the influenza virus.

B Cells Attack Invading Microbes

The interleukin-2 released by helper T cells simultaneously activates the second kind of defensive white blood cells, B cells. Like killer T cells, B cells have unique receptor proteins on their surface. These receptor proteins are called **antibodies.** B cells can release copies of their antibodies into the bloodstream or attach them directly to pathogens.

B cells do not directly attack pathogens or infected cells. Instead, they mark the pathogen for destruction by macrophages and natural killer cells. When a B cell encounters a foreign microbe with a surface protein that matches the shape of its antibodies, it simply sticks an antibody onto the microbe. The antibody acts as a flag to attract macrophages and natural killer cells. In addition, the B cell is stimulated by its encounter to divide repeatedly, forming a large population (known as a "clone") of identical B cells. Then all these B cells begin to make large amounts of antibodies, which they secrete into the bloodstream. Antibodies attach to any invading pathogens that might be present, marking them for destruction. Figure 39-8 summarizes the roles B cells and killer T cells play in fighting pathogens.

The B-cell defense is very powerful because it amplifies the reaction to an initial pathogen encounter a millionfold. It

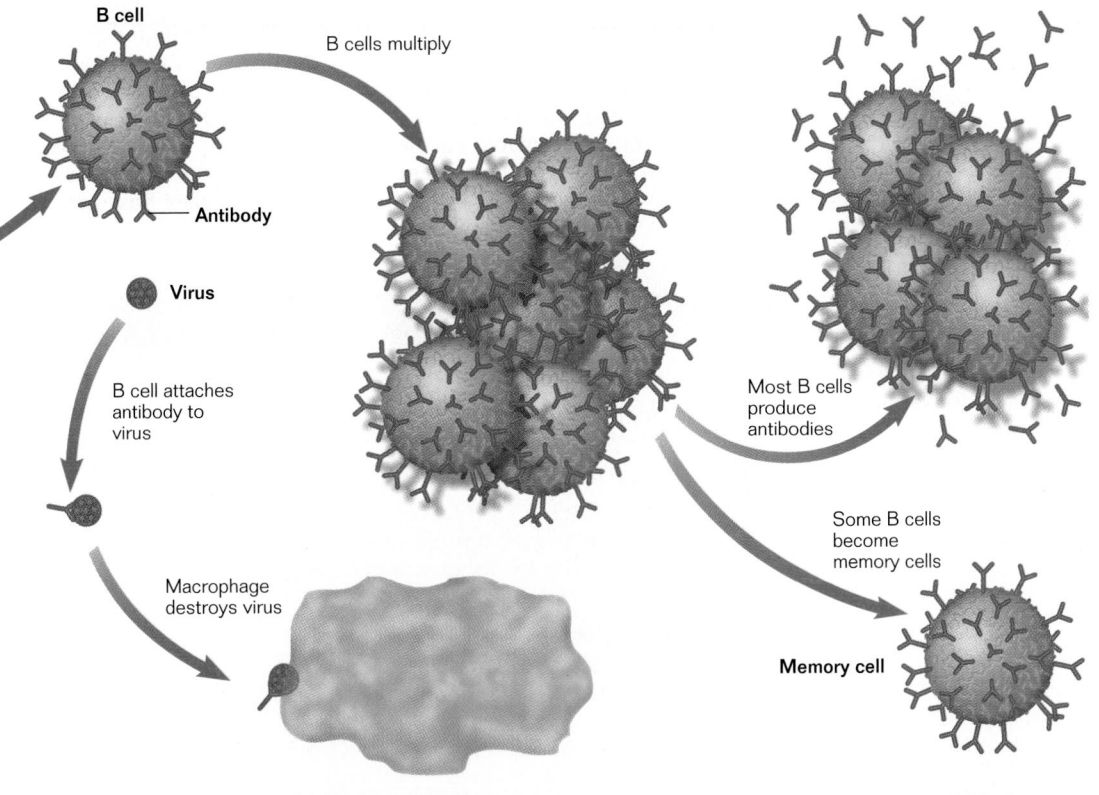

B cell

B cells multiply

Antibody

Virus

B cell attaches antibody to virus

Macrophage destroys virus

Most B cells produce antibodies

Some B cells become memory cells

Memory cell

Vaccination

Emphasize that vaccination is effective because it exposes the body to the surface proteins of a pathogen that have been stripped of disease-causing ability. AIDS, malaria, and influenza are among the diseases that have defeated attempts to produce vaccines. The pathogens that cause these diseases (AIDS and influenza are caused by viruses, malaria by a protist) have rapidly evolving surface proteins. Ask students why such proteins make vaccination development difficult. *(The immune system only produces defenses against those surface proteins to which it was exposed by the vaccine. The immune system probably will not recognize the altered proteins of a mutant pathogen.)*

History of Vaccination

Have students research the contributions of the English physician Edward Jenner to the development of vaccination. Students should write a report summarizing what they have learned.

Summarizing the Body's Defenses

Have students create a *Graphic Organizer* in the form of a table. Along the top of the table they should list the following categories: Unique Characteristics, Function, Location in Body. Along the margin they should list all the types of cells that play a role in the body's defenses: macrophage, neutrophil, natural killer cell, helper T cell, killer T cell, B cell.

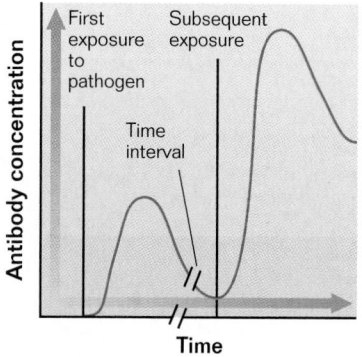

Figure 39-9 The first time you are exposed to a pathogen, your immune system responds slowly. At the second exposure, the response is faster and results in more antibodies.

❏ CAPSULE SUMMARY

Helper T cells stimulate killer T cells, which attack pathogens and infected cells, and B cells, which release antibodies that mark pathogens for destruction.

is also a very long-lasting defense. A few of the B cells resulting from a pathogen encounter do not become antibody producers. Instead, they become specialized memory cells that continue to patrol your body's tissues, circulating through your blood and lymph for a long time—sometimes for the rest of your life. If the pathogen that they are specialized to recognize ever appears again, those memory cells start the production of a new generation of antibody-producing cells directed against that particular pathogen. Your body does not wait until an infection is well underway to begin to fight back. The new generation of B cells produces large amounts of antibodies, as shown in Figure 39-9, and macrophages destroy the pathogen *before* you become ill. You are not even aware of the battle going on in your body and are said to be "immune" to the pathogen. ❏

Vaccination Prepares the Immune System

Vaccination is the introduction into your body of a dead or disabled pathogen or of a harmless microbe with the protein of a pathogen on its surface. Vaccination triggers an immune system response against the pathogen without an infection ever occurring. Afterward, the bloodstream of the vaccinated person contains memory cells that are directed against that specific pathogen. The vaccinated person is said to be "immunized" against the disease. Vaccination has dramatically reduced the incidence of many bacterial and viral diseases, including polio, tetanus, and diphtheria. More significant, an intensive vaccination program led to the elimination of the deadly disease smallpox in the 1970s. However, as you will see in the next section, some diseases have not responded so favorably.

Section Review

1. *Contrast B cells with killer T cells.*
2. *Could the immune system respond to an infection without helper T cells? Explain your answer.*
3. *How do macrophages assist B cells in fighting pathogens?*
4. *In what way is the immunity you acquire from a vaccination similar to that acquired from actually having the disease? In what way is it different?*

Critical Thinking
5. *What would have happened if the killer T cells that recognized your tissues hadn't been weeded out before birth?*

Answers to Section Review

1. Killer T cells are produced in the bone marrow, mature in the thymus, and kill infected cells by puncturing their membranes. B cells are produced in the bone marrow and produce antibodies.

2. The immune system would still react, but to a very limited degree; macrophages would still attack and consume pathogens, but killer T cells and B cells would not be activated, and no antibodies would be produced.

3. Macrophages envelop pathogens, especially those tagged with antibodies by B cells.

4. They are similar in that immunity is stimulated by the reaction to the pathogen's surface proteins. They differ in that a full-blown case of the disease is not necessary with vaccination.

5. The body might attack its own tissues, resulting in autoimmune diseases.

39-3 Malfunctions and Failures of the Immune System

Though the immune system is one of the most sophisticated systems of the human body, it is not perfect. Many of today's major diseases, and other health concerns, are the result of failures or malfunctions of the immune system.

Section Objectives

- Compare autoimmune diseases with allergies.
- Explain how HIV cripples the immune system.
- Recognize the ways HIV can be transmitted.

When the Body Attacks Its Own Tissues

The ability of killer T cells and B cells to distinguish cells of your own body from foreign cells is crucial to the fight against pathogens. In **autoimmune diseases,** this ability breaks down, causing the body to attack its own cells. Multiple sclerosis is an autoimmune disease that usually strikes people between the ages of 20 and 40. In multiple sclerosis, the immune system attacks and destroys the sheath of myelin that insulates motor nerves, which carry commands from the brain to the muscles and organs. Degeneration of the myelin sheath interferes with the transmission of nerve impulses, ultimately stopping their transmission altogether. Voluntary functions, such as the movement of limbs, and involuntary functions, such as bladder control, are lost, leading to paralysis and, eventually, death. Scientists do not know what stimulates the immune system to attack myelin. Table 39-1 describes several other autoimmune diseases and the tissues they affect.

Figure 39-10 Multiple sclerosis is a slow, degenerative disease in which transmission of nerve impulses is slowed and blocked. This teacher and multiple sclerosis patient requires a cane for support.

Table 39-1 Some Autoimmune Diseases

Disease	Areas affected	Symptoms
Systemic lupus erythematosus	Connective tissue, joints, kidney	Facial skin rash, painful joints, fever, fatigue, kidney problems, weight loss
Type 1 diabetes	Insulin-producing cells in pancreas	Excessive urine production, blurred vision, weight loss, fatigue, irritability
Graves' disease	Thyroid	Weakness, irritability, heat intolerance, increased sweating, weight loss, insomnia
Rheumatoid arthritis	Joints	Crippling inflammation of the joints

Vocabulary Preview

autoimmune disease

allergy

histamine

HIV positive

Opening Question

Ask students to consider Figure 39-8 and imagine they are designing a drug to suppress the immune system. Which step in the immune response, if inhibited, would have the most damaging effects? Have students justify their answers. Point out that HIV, the virus that causes AIDS, destroys helper T cells. Without these cells to stimulate B cells and killer T cells, the body is incapable of fighting off pathogens and cancers.

Demonstration
Structure of HIV

Show students a drawing of HIV. Point out its main components: RNA, reverse transcriptase, surface proteins, membrane, protein coat. Describe the function of each component. Tell students that this virus contains only about 10,000 nucleotides.

Application

Health Allergies can be life threatening, particularly those allergies to food, drugs, or other ingested chemicals that lead to ana- phylactic shock, a severe, system- wide reaction to an allergen. Anaphylactic shock results in con- striction of the bronchioles, possible circulatory collapse, and loss of con- sciousness within minutes.

Teaching Tips
The Rapidly Evolving Influenza Virus

Why does the influenza virus mutate so rapidly? One reason is that, unlike eukaryotic cells, it has no enzymes that correct errors in its genetic material (which is RNA). Also, influenza viruses infecting different species swap genetic material in a complex reshuffling that occurs among pigs, ducks, and other domestic animals living in Asia, the source of most new strains of influenza. Occasionally, as in the worldwide flu pandemic of 1918–19, new and very deadly strains of human influenza emerge from Asia and spread throughout the world.

Keeping Ahead of Flu Epidemics

Have students research the United States government's program to monitor the appearance of new influenza strains and anticipate severe outbreaks. Students should write a report about what they have learned. In their reports, students should include an account of the worst failure of this early-warning system—the swine flu episode of 1976.

Figure 39-11 If your eyes water and your nose runs when you encounter dust, blame this microscopic mite that lives in the dust. You are allergic to proteins in its feces.

☐ CAPSULE SUMMARY

Autoimmune diseases and allergies are both malfunctions of the immune system. In autoimmune diseases, the body fails to distin- guish its own cells from those of a pathogen and thus attacks its own cells. In allergies, the body responds to a harmless substance such as pollen or dust as if it were a pathogen.

Allergy: Attacking a Harmless Substance

Although your immune system provides very effective protec- tion against bacteria, viruses, fungi, and protists, it sometimes does its job too well and mounts a major defense against a harmless substance. Such an immune system response is called an **allergy**. Hay fever, the sensitivity that many people experience to even tiny amounts of plant pollen, is a familiar example of an allergy. Many peo- ple are allergic to proteins in the feces of the tiny mite shown in Figure 39-11, which lives on grains of house dust.

What makes an allergic reaction uncomfortable, and sometimes dangerous, is the involvement of antibodies that are attached to a kind of white blood cell called a mast cell. When mast cells encounter something that matches their antibodies, they initiate an inflammatory response. The inflammatory response begins when mast cells release **histamines** (*HIHST uh meens*) and other chemicals that cause capillaries to swell. Histamines also increase mucus produc- tion by the mucous membranes, resulting in the runny nose and nasal congestion that are symptoms of hay fever. Most allergy medicines relieve these symptoms with antihista- mines, chemicals that block the action of histamines.

Asthma is a form of allergic response that takes place in the lungs. In addition to the reactions already described, histamines cause the narrowing of air passages in the lungs of people who have asthma. These people have trouble breathing when exposed to substances to which they are allergic. ☐

The Surface Proteins of Influenza Viruses Evolve Rapidly

If the activities of mem- ory cells provide such an effective defense against future infection, why can you catch some diseases, such as the flu, more than once? Even if you are immunized against influenza, vaccination does not provide long-term

Historical Note

Retrospectively Diagnosed AIDS Cases

AIDS was first recognized as a disease in 1981. Several mys- terious deaths in the late 1970s and in 1980 were easily attrib- uted to AIDS. However, using recently developed tests for HIV, scientists have been able to solve some unexplained deaths from the previous decades. For instance, a British sailor died in 1959, his immune system apparently nonfunc- tional. His doctors saved tissue samples that, 30 years later, were found to contain HIV. Similar studies also implicated AIDS in the death of a Norwegian sailor in 1966.

protection against this disease. The reason you don't remain immune to flu is that the influenza virus has evolved a way to evade the human immune system. The virus changes its surface, the part that white blood cells recognize. How does it do this? The genes encoding the surface proteins of the influenza virus mutate very rapidly. Thus, the influenza virus continually makes new varieties of its surface proteins, and the immune system does not recognize these new varieties as belonging to a pathogen it has previously encountered. For each new variety of influenza virus, therefore, your body needs to mount an entirely new defense.

AIDS: Immune System Collapse

AIDS was first recognized as a disease in 1981. In Figure 39-12, you can see how rapidly the number of AIDS cases has increased since then. The World Health Organization estimates that 40 million people throughout the world will be infected with HIV (human immunodeficiency virus), the virus that causes AIDS, by the year 2000. By 1995, more than 250,000 Americans had died of AIDS, and more than 1.5 million Americans were thought to be infected with HIV. This destructive virus apparently evolved from a very similar virus that infects chimpanzees in Africa. A mutation arose in the chimpanzee virus that allowed it to recognize a human cell surface receptor called CD4. This receptor is present in the human body on certain immune system cells, notably macrophages and helper T cells.

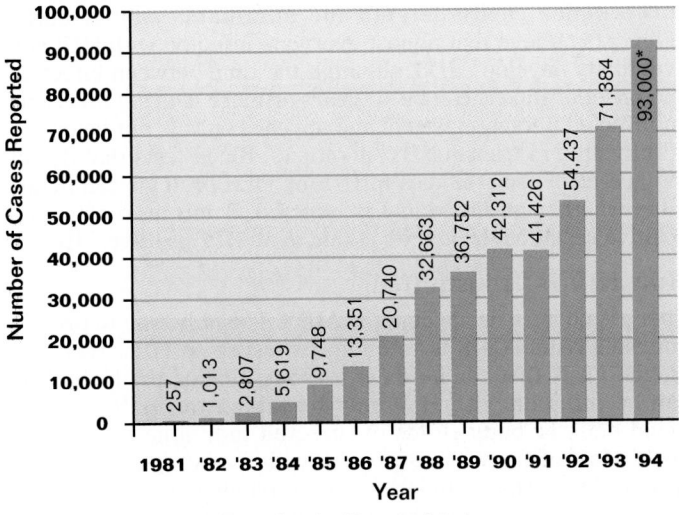

*Centers for Disease Control and Prevention Estimate

Figure 39-12 The number of cases of AIDS in the United States has grown rapidly since the disease was first described in 1981.

Teaching Tips
The Origin of HIV

The apparently sudden appearance of a virus as deadly as HIV inspires many questions. Where and when did HIV originate? How did it spread from its point of origin? Do any similar viruses infect animals? Scientists do not have answers for all these questions. For one thing, HIV did not appear suddenly. Using retrospective studies of stored tissues, scientists have shown that the virus was present in the United States at least a decade before AIDS was described. For example, samples of blood drawn from intravenous drug users in the early 1970s showed an infection rate of about 1 percent in this group. And a blood sample taken in the Belgian Congo (now Zaire) in 1959 had antibodies to HIV. Most scientists think HIV arose in Africa because a similar virus, the simian immunodeficiency virus (SIV), infects monkeys and apes in Africa. No HIV-like viruses are found in wild primates in Asia or the New World. SIV, therefore, is probably the ancestor of HIV. HIV seems to be most closely related to a strain of SIV found in chimpanzees. No one is sure how humans were first infected by this ape virus, where in Africa this occurred, or what allowed the virus to spread throughout the rest of the world.

Mutation and HIV

The ability of viruses to undergo mutations in the genes controlling their surface proteins is not just an interesting sidelight to viral study. Efforts to produce a vaccine against HIV have been thwarted because it quickly mutates, making it impossible (so far) for medical science to produce an enduring vaccine or drug. The rapid mutation rate also quickly renders the virus's surface proteins unrecognizable to the antibodies initially produced by infected patients, leaving them without this vital defense. As the next subsection points out, this is just one step in a cascade of immune-system failures they experience.

Teaching Tip

AIDS and Opportunistic Infections

AIDS results in one or more opportunistic infections or cancers that are usually not harmful to individuals with healthy immune systems. These infections and cancers include *Pneumocystis carinii* pneumonia, Kaposi's sarcoma, candidiasis, and others. *Pneumocystis carinii* pneumonia is the most common of the opportunistic diseases, infecting more than 50 percent of all AIDS patients. The patients have a persistent cough, fever, and tightness in the chest with shortness of breath. The diagnosis depends on the difficult identification of the *Pneumocystis* cysts in the lungs. Treatment is with strong antibiotics. Kaposi's sarcoma is a cancer of the skin and mucous membranes that often spreads through other organs. Initially, the patient experiences red or purplish sores, most commonly on the face or in the mouth. Kaposi's sarcoma can affect the lungs, causing rapid deterioration of the patient's condition and death. Treatment is by cancer chemotherapy drugs. Candidiasis, commonly known as thrush, is a fungal infection producing a thick, white coating of the mouth, tongue, and other components of the digestive system. It is treated with antifungal drugs. Other opportunistic infections include cytomegalovirus, which produces inflammation of the retina, colon, and adrenal glands; and tuberculosis and *Salmonella* infection, both caused by bacteria. Tuberculosis in AIDS patients usually responds to standard treatments (although the incidence of drug-resistant tuberculosis is rising), but *Salmonella* infection, a cause of gastrointestinal problems, does not. No drugs are effective against cytomegalovirus, although several are in experimental testing.

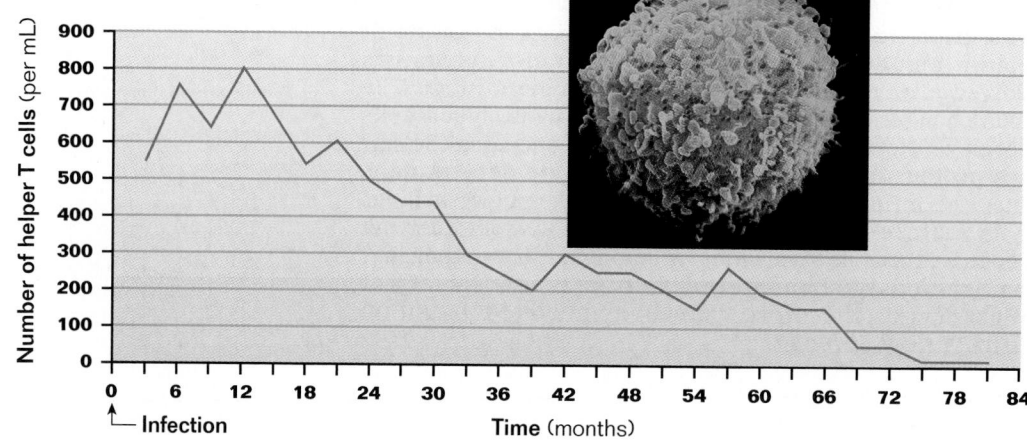

Figure 39-13 HIV destroys helper T cells. The small, green spheres, *above right*, are particles of HIV emerging from a helper T cell. These particles will infect other helper T cells, resulting in a rapid decrease in the number of these cells in the infected person. The graph shows the decline in helper T cells in the blood of an AIDs patient.

❑ CAPSULE SUMMARY

AIDS is caused by the human immunodeficiency virus (HIV). HIV destroys helper T cells in large numbers, thereby disabling the immune system. Without a functional immune system, AIDS patients cannot ward off infections and cancers.

How HIV Disables the Immune System

HIV attacks and cripples the immune system. The virus invades macrophages, which carry it throughout the body. It also attacks helper T cells and causes a drastic reduction in their numbers as the infection progresses, as shown in Figure 39-13. When the number of helper T cells in the blood of an infected person has fallen below 200/mL (a normal count is 800–1,000/mL) the person is said to have AIDS. Without enough helper T cells to activate and direct B cells and killer T cells, the immune system cannot respond to infections and cancer. As a result, the body is soon overwhelmed by pathogens and cancers that it would normally detect and destroy. Pneumocystic pneumonia, a common cause of death in AIDS patients, is an infection caused by a fungus that healthy immune systems easily defeat. Kaposi's sarcoma, another major cause of death in AIDS patients, is a form of cancer that is rare (except in very old people) because normal immune surveillance easily detects and eliminates cancer cells. Scientists believe that almost everyone infected with HIV will eventually develop AIDS, although the time between infection and onset of illness can be 10 years or more. During this time, an infected person may feel and appear healthy, but he or she is fully able to transmit HIV to others. Blood tests that detect the presence of antibodies to HIV or viral proteins can determine whether an individual is infected. Someone receiving a positive result on these tests is said to be **HIV positive**. ❑

How HIV Is Transmitted

There is no cure or vaccine for AIDS. The only way to protect yourself from AIDS is to avoid exposure to HIV. HIV is a fragile virus that cannot exist for long outside the body. You can be infected with HIV only by taking the HIV-infected blood cells or body fluids of infected individuals into your body. Most of the virus is present within macrophages rather than as free virus particles. Because semen and vaginal secretions are rich in macrophages, a person can become infected with HIV through sexual intercourse with an infected person. Worldwide, most HIV infections are spread this way. The

Overcoming Misconceptions

The Difference Between HIV Infection and AIDS

A common misconception is that a person who is HIV positive has AIDS. In fact, AIDS is the last phase of a long HIV infection. A person is declared HIV positive when antibodies to HIV or HIV itself are found in his or her blood or tissues. However, an HIV-positive person may be healthy and disease-free for a decade or more after infection. During this asymptomatic stage, the HIV-positive person is fully infectious and can transmit the virus to others. AIDS is diagnosed when HIV has destroyed enough of the infected person's helper T cells that severe immune suppression is occurring.

virus can be transmitted by men and women during sexual intercourse (the worldwide frequency of AIDS is about the same in men and women). Vaginal, anal, and oral sex can transmit the virus, because macrophages can easily cross any mucous membrane. Use of a latex condom during sex greatly reduces, but does not eliminate, the risk of getting HIV.

Because blood contains a great many macrophages, it provides a vehicle for transmitting AIDS. People who inject intravenous drugs with needles or hypodermic syringes that have become contaminated with HIV-containing blood are at high risk of becoming infected with HIV. The majority of infections in the United States in the late 1980s were transmitted in this manner. In the late 1970s and early 1980s, many people were infected after receiving blood transfusions or injections of blood products, such as the Factor VIII used by hemophiliacs to help their blood to clot. The introduction of tests for HIV has almost completely eliminated this route of infection in the industrialized countries—but not in the developing countries, where such tests are often unavailable or too expensive.

HIV is not transmitted through the air, on toilet seats, or by any other medium where a macrophage would not survive. HIV cannot be contracted through shaking hands, sharing food, or drinking from a water fountain used by an infected person, for the simple reason that macrophages cannot be transmitted by casual contact. Although HIV is found in saliva, tears, and urine, these fluids contain too few virus particles to easily initiate an infection. This is why there are no known instances of anyone being infected with HIV from the small amount of saliva exchanged while kissing. Biting insects such as mosquitoes and ticks do not transmit HIV because they do not transmit macrophages. AIDS is not an easy disease to contract. There are only two common routes of infection: exposure to HIV-contaminated syringes or hypodermic needles and unprotected sexual contact with infected individuals.

Figure 39-14 The impact of AIDS is global. This clinic in Kenya counsels people on how they can avoid being infected with HIV. Africa has been hardest hit by AIDS of any continent. The World Health Organization estimates that over 6 million Africans are already infected by HIV.

Teaching Tip
Avoiding HIV Infection

How to prevent HIV infection may be the most important information students receive from this book. Remind students that there is currently no cure for AIDS and no vaccine for HIV. Furthermore, since HIV mutates so rapidly, the prospects for a cure or vaccine in the near future are bleak. Emphasize that HIV is not a very infectious virus. It is not spread in respiratory droplets like influenza virus, nor is it spread by contaminated food or water like the hepatitis A virus. However, sexual intercourse with an infected individual and use of HIV-contaminated syringes or hypodermic needles are common routes of transmission.

Chapter Closure

Assign short (2–3 pages) research reports on various topics dealing with immunity. AIDS can provide the bulk of the topics, but other topics such as autoimmune diseases and immunosuppression during organ transplantation are also relevant. Encourage students to relate information from this chapter to information they discover in their library research.

Section Review

1. *Describe the differences and similarities between allergies and autoimmune diseases.*

2. *Would HIV's effects on the immune system be as damaging if it invaded only killer T cells? Explain your answer.*

3. *Describe two ways that HIV can be transmitted and two ways that it cannot.*

Critical Thinking

4. *When a pathogen initially invades a new host—such as when HIV first infected humans—it is usually deadly. After some time, however, the pathogen usually evolves to become less deadly. Explain why natural selection would favor a less-harmful pathogen.*

Answers to Section Review

1. An allergy is a reaction to a harmless substance entering the body. An autoimmune disease is an attack on the body's own cells. Both are inappropriate overreactions by the immune system.

2. The effects would not be as great since the body would still have macrophages to attack pathogens, and helper T cells would be free to activate B cells, resulting in antibody production.

3. HIV can be transmitted by transfer of blood through use of contaminated needles or syringes and by unprotected sex with an infected partner. HIV cannot be transmitted by casual contact such as kissing, by toilet seats, or by insect bites.

4. It is advantageous for a pathogen to become more benign, since the death of the host is usually a dead end for the pathogen—its spread is usually hindered or stopped with the death of its host.

CHAPTER REVIEW ANSWERS

Each item in the Chapter Review is correlated by text section in the assignment guide that follows.

ASSIGNMENT GUIDE

Section	Review	Themes Review	Critical Thinking
39-1	1, 2, 8, 9	16	19
39-2	3, 4, 10–12	17, 18	
39-3	5–7, 13–15		20

Review

Multiple Choice

Code in parentheses indicates section number and objective number.

1. b (39-1.1)
2. a (39-1.3)
3. c (39-2.2)
4. a (39-2.4)
5. b (39-3.1)
6. b (39-3.2)
7. a (39-3.3)

Completion

8. macrophages, neutrophils (39-1.2)
9. inflammatory, temperature (39-1.3)
10. macrophages, helper (39-2.1)
11. killer, B (39-2.3)
12. dead, memory (39-2.4)
13. autoimmune, multiple sclerosis or any of the diseases listed in Table 39-1 (39-3.1)
14. histamines, antihistamines (39-3.1)
15. cancers (39-3.2)

Vocabulary

allergy (930)
antibody (927)
autoimmune disease (929)
B cell (925)
complement (923)

helper T cell (925)
histamine (930)
HIV positive (932)
inflammatory response (923)

killer T cell (925)
mucous membrane (921)
natural killer cell (923)
neutrophil (922)

Review

Multiple Choice

1. A mucous membrane defends against infection by
   a. ejecting lysozyme, which digests bacterial cell walls.
   b. secreting a sticky fluid that traps pathogens.
   c. preventing blood clots.
   d. engulfing and destroying bacteria.

2. An increased body temperature helps the body fight bacterial infection because it
   a. inhibits bacterial growth.
   b. stimulates antibody production.
   c. causes a membrane attack complex to form.
   d. produces redness and swelling.

3. B cells and killer T cells are stimulated to act by a substance called interleukin-2, which is released by
   a. macrophages.   c. helper T cells.
   b. neutrophils.   d. natural killer cells.

4. A flu shot must be taken each year to be even marginally effective because
   a. the influenza virus mutates rapidly.
   b. flu is caused by a retrovirus.
   c. very few memory cells are produced.
   d. macrophages cannot engulf the flu virus.

5. Rheumatoid arthritis, a disease that involves an immune system attack on the membranes around the joints, is an example of a(n)
   a. allergy.
   b. autoimmune disease.
   c. AIDS-related infection.
   d. passive immunity.

6. HIV disables the immune system by
   a. blocking the action of macrophages.
   b. destroying helper T cells.
   c. activating the production of B cells.
   d. triggering genetic mutations in CD5 surface receptors.

7. HIV can be transmitted by
   a. sexual intercourse.
   b. mosquito bites.
   c. shaking hands.
   d. vaccination only.

Completion

8. Your body's second line of defense against invading pathogens involves both _____ , which consume pathogens, and _____ , which destroy pathogens and themselves with chemicals.

9. The _____ response is recognized by the redness and swelling of affected tissue, while evidence of the _____ response is fever.

10. The immune system is activated by _____ that secrete the protein interleukin-1. This protein activates _____ T cells.

11. Infected cells are attacked directly by _____ T cells, while microbes are labeled by _____ cells for later attack by macrophages and natural killer cells.

12. Vaccines may be prepared from a weakened or _____ pathogen. Once vaccinated, a person's blood contains _____ cells that provide protection from the disease caused by that pathogen.

13. Destructive responses of the immune system to cells of the person's own body are called _____ diseases. One such disease is _____ .

14. The runny nose and nasal congestion associated with hay fever are due to the release of _____ . These hay fever symptoms can be relieved by using medicines that contain _____ .

15. AIDS patients die from infections and _____ because their immune systems have been crippled by HIV.

Themes Review

16. **Structure and Function** What are the outward signs of the inflammatory response? How do the changes that accompany the outward signs of the inflammatory response help to destroy pathogens?

17. **Structure and Function** When a B cell encounters a recognizable pathogen protein, it divides rapidly and produces clones that make large amounts of antibodies. These antibody-producing cells contain many Golgi apparatuses and large amounts of endoplasmic reticulum. How is the presence of these components related to the function of these cells?

18. **Evolution** Scientists have been successful in developing vaccines that provide long-term protection against a number of diseases, including polio and smallpox. Yet, they have been unsuccessful in developing a vaccine that protects against the influenza virus responsible for flu. Why has developing a flu vaccine been so difficult?

Critical Thinking

19. **Making Inferences** People who receive severe burns often die from infection. Given what you know about the body's defenses, explain how this is possible.

20. **Making Inferences** A government agency is reviewing two proposals for HIV research, but can fund only one. Suppose you are asked to provide input into the decision. Which proposal would you recommend that the agency fund? You should consider not only the likely effectiveness of the treatment but also possible side effects. Explain how you made your choice.

Proposal 1: Develop a drug that interferes with protein synthesis.

Proposal 2: Develop a substance that binds to the CD4 receptors on macrophages and helper T cells.

Activities and Projects

21. **Research and Writing** Do library research and talk to doctors and nurses to find out how the drug cyclosporin has affected the success of organ transplants. In your report include information about the source of the drug and how it works.

22. **History** The German physician Robert Koch was the first person to develop a step-by-step procedure for identifying the pathogen responsible for a given disease. What are the steps of Koch's procedure and what was the first pathogen identified using the procedure? What is the advantage of using Koch's procedure to identify the pathogen responsible for a given disease?

23. **Science-Technology-Society** Several AIDS vaccines are under development. The final test of these vaccines will be to inject them into healthy people and see if they develop AIDS or begin to build up antibodies to fight HIV, the virus that causes AIDS. Organize your group into two teams to debate the pros and cons of this practice.

Critical Thinking

19. The first line of defense is weakened because skin damaged by burns does not provide the same protection against pathogens as healthy skin. (39-1.1)

20. Fund Proposal 2 because HIV is known to recognize and attach to CD4 surface receptors on macrophages and helper T cells. By preventing attachment to CD4 receptors, the virus could not send its RNA into the macrophages and helper T cells. It is highly likely that a drug that interferes with synthesis of proteins will interfere with synthesis of proteins in the host cells and cause severe side effects. (39-3.2)

Activities and Projects

21. Cyclosporin is derived from a soil fungus. It increases the success of organ transplants by suppressing the killer T cells that reject foreign tissue. Cyclosporin does not suppress the action of the helper T cells that stimulate antibody production in B cells or the killer T cells that attack cancer cells.

22. The steps include: (1) Isolate the organism suspected of causing the disease. (2) Grow the organism in laboratory culture. (3) Inoculate a healthy animal with the cultured organisms. Observe whether the animal contracts the disease. (4) If the animal contracts the disease, isolate the organism that caused it. Compare this organism with the one isolated in step 1. The procedure was first used to identify the bacterium that causes anthrax. The step-by-step procedure is a systematic way that can lead to the identification of disease-causing pathogens.

23. The practice of injecting healthy people with an experimental vaccine was used by Jenner in developing a smallpox vaccine and by Salk in developing one for polio. Provide students with time to survey the AIDS literature in preparation for the debate.

Themes Review

16. The outward signs of the inflammatory response are redness and swelling of tissue. The redness and swelling are associated with increased blood flow that promotes the migration of macrophages and neutrophils to the infection site. (39-1.3)

17. Golgi apparatuses and the endoplasmic reticulum are involved in producing proteins, such as the antibodies that are secreted by these cells into the bloodstream. (39-2.3)

18. The genes of the influenza virus evolve rapidly. Therefore, its surface proteins change rapidly. Changing its surface proteins allows the virus to evade the immune system. (39-2.4)

LABORATORY Investigation Chapter 39

Preparation

To prepare unknown solutions, prepare a stock dropper bottle of distilled water for each student, minus one, and a dropper bottle containing 10% ascorbic acid for the remaining student. Both of these solutions are clear. The ascorbic acid solution should be pink or clear after indophenol has been added. Distilled water should become blue when indophenol is added. Each student will need a dropper bottle of unknown solution, a large test tube, and indophenol indicator. Indophenol indicator and ascorbic acid are available from a biological supply house such as Ward's. Students whould not allow solutions to touch skin or clothing. They should wear a lab apron, goggles, and disposable gloves throughout this investigation. After completing the investigation, students should rinse the test tubes well. The solutions retain their potency for at least a month.

Procedural Notes

- This investigation should take 15 minutes to complete.
- If the procedure is precisely followed, the route of transmission can be easily traced.
- Before students begin, remind them of safety practices and to be careful not to allow the solutions to touch their clothes or skin. If this happens, flush the area with water.

Background Answers

1. The viruses that cause colds and flu are spread by droplets expelled into the air by sneezing and coughing.
2. The skin and mucous membranes help keep viruses out. Viruses that get into the body will stimulate the immune system. Killer T cells destroy cells infected by viruses. B cells release antibodies that mark the virus for attack by macrophages.
3. AIDS is an incurable disease that has spread rapidly in the past several decades. Knowledge of its means of

Simulating Disease Transmission

OBJECTIVE

Simulate the transmission of a disease.

PROCESS SKILLS

- testing unknown solutions with chemical indicator
- recording data in tables

MATERIALS

- lab apron
- safety goggles
- disposable gloves
- dropper bottle of unknown solution
- large test tube
- indophenol indicator

BACKGROUND

1. How does a cold or flu spread from person to person?

2. How does the body fight invading viruses?

3. Why has the transmission of AIDS become a great concern worldwide?

4. Why is a person with HIV less able to combat infections?

5. Write your own question to explore in your lab report or notebook.

TECHNIQUE

1. CAUTION: Do not allow any solutions to touch your skin or clothing. Put on a lab apron, goggles, and disposable gloves. This investigation will involve the class in a simulation of disease transmission. After the simulation, you will try to identify the original infected person in the closed class population.

2. You have been given a dropper bottle of unknown solution and a clean test tube. Handle the unknown solutions with care because they are not simply water. When your teacher says to begin, transfer 3 dropperfuls of your solution to your clean test tube.

3. Randomly select one person to be your partner. Let one partner pour the contents of his or her test tube into the other partner's test tube. Then pour half of the solution back into the first test tube. You and your partner now share pathogens of any possible transmittable disease that either of you might have had. Record the name of your first partner (Round 1) in the **Records** section of your report.

4. For Round 2, wait for your teacher's signal and then find a different student partner and exchange solutions in the same manner as in step 3. Record the name of your second partner (Round 2) in the **Records** section of your report. Do not exchange solutions with the same person more than once. Repeat again for Round 3.

5. After all rounds are finished, your instructor will ask you to add one dropperful of indophenol indicator to your test tube to see if the fluids in your test tube have become infected. Infected solutions will be colorless or light pink in color. All uninfected solutions will appear blue. Record the outcome of your test in the **Records** section of your report.

Path of Disease Transmission

Infected Person	Names of Partners		
	Round 1	Round 2	Round 3

transmission can be used to prevent further spread.

4. HIV kills helper T cells, without which killer T cells and B cells cannot be activated to fight pathogens.

5. How can disease transmission be traced to the original carrier in a closed population?

Records

Answers will vary. One sample answer follows.

6. If you are an "infected" person, give your name to your teacher. As names of "infected" people are written on the chalkboard or on the overhead projector, record them in the **Records** section of your report in a table similar to the one above.

7. Try to trace the original source of the infection, then determine the transmission route of the disease. In your table, cross out the names of all the uninfected partners in Rounds 1, 2, or 3. There should be only two people in Round 1 who were infected. One of these was the original carrier. Devise a method to determine the route of transmission by constructing a diagram similar to the one below. Draw your diagram on the back of your report. Insert the names of the two people in Round 1 who were infected and the names of their partners in Round 2 and Round 3.

Transmission Route

Round 1	Round 2	Round 3
____ →	____ →	____
	____ →	____
↓ or ↑		
____ →	____ →	____
	____ →	____

8. To test which person was the original disease carrier, pour a sample from his/her dropper bottle into a clean test tube and test it with indophenol indicator.

9. Clean up your materials and wash your hands before leaving the lab. In your report, briefly summarize the procedure you followed.

Inquiry Answers

1. aerosol droplets from a cough or sneeze, blood, or other bodily fluids
2. No. The fluid is clear.
3. indophenol
4. clear to pink color
5. blue
6. Answers will vary.

Analysis Answers

1. Answers will vary, depending on class size.
2. They might post signs, alert the media, notify schools and other places where many people congregate each day, and encourage people with symptoms to stay home. They might also begin a vaccination or treatment program.

Path of Disease Transmission

Infected Person	Names of partners		
	Round 1	Round 2	Round 3
Kendra*	Jason	Mike*	Missy*
Missy*	Than*	Andra*	Kendra*
Than*	Missy*	Otis*	Andrew*
Greg*	Shayne	Jennifer	Doug*
Otis*	Dina	Than*	Greg*
Andra*	Elena	Missy*	Mike*
Mike*	Anne	Kendra*	Andra*
Andrew*	Earl	Dierdre	Than*

Transmission Route

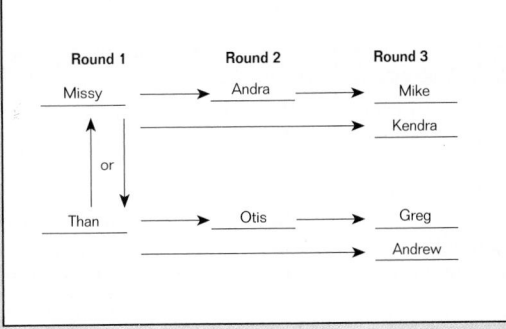

CHAPTER 40 NERVOUS SYSTEM

Block Scheduling Guide

> Each block represents about 45–50 minutes of instructional time. The resources cited in a block are coded to a specific program component using the Key on the next page.

Lecture Concepts	Lesson Resources

40-1 Neurons and Nerve Impulses pp. 939–943

BLOCK 1

a. A neuron consists of a cell body, dendrites that carry impulses to the cell body, and an axon that transmits impulses to other cells.

b. When a nerve impulse reaches the end of an axon, neurotransmitters are released into the synapse. They cross the synapse and bind to receptors in the neighboring cell membrane.

Lecture Resources
- Opening Questions, 939
- Demonstration: Measuring Electrical Activity
- Overcoming Misconceptions, page 939
- Visual Strategies: Figures 40-1, 40-2, 40-3, 40-4
- Research Update, page 943
- Teaching Transparencies: 194, 195,
- Holt Biology Videodiscs: Lesson 40-1

Classwork Options
- Observing Neural Structure, page 940
- Producing Chemicals Economically, page 942

Assignment Options
- Section Review, page 943
- Chapter Review, questions 1–3, 9, 13, 14, 16

40-2 Architecture of the Nervous System pp. 944–949

BLOCK 2

a. The central nervous system is made up of the brain and spinal cord, which work together to collect and process information.

b. The peripheral nervous system is made up of branches of nerves that carry information to and from the brain and spinal cord.

c. The autonomic nervous system controls activities in maintaining homeostasis.

Lecture Resources
- Opening Questions, page 944
- Demonstration: Reflexes
- Visual Strategies: Figures 40-6, 40-7, 40-8
- Pick Your Brain, page 945
- Brain Evolution, page 945
- Blink Reflex, page 948
- Role of Sympathetic Division, page 949

- Teaching Transparencies: 196, 197, 199, 197A
- Holt Biology Videodiscs: Lesson 40-2

Assignment Options
- Section Review, page 949
- Chapter Review, questions 4, 5, 10, 15, 17

40-3 The Sense Organs pp. 950–954

BLOCK 3

a. Receptors for equilibrium and hearing are located in the inner ear.

b. Receptors for sensing light are rods and cones, which are located in the retina in the back of the eye.

c. Receptors located throughout the body sense its internal condition.

Lecture Resources
- Opening Question, page 950
- Demonstration: Seeing Spots Before Your Eyes
- Visual Strategies: Figures 40-9, 40-10; Table 40-1
- Motion Sickness, page 951
- Demonstration: The Nose Knows, page 954
- Teaching Transparencies: 200, 201, 202, 203

- Holt Biology Videodiscs: Lesson 40-3

Classwork Options
- Effective Reading, page 950
- Eye Structure, page 952
- Inquiry Skills Development B31: Exploring Vision

Assignment Options
- Section Review, page 954
- Chapter Review, questions 6, 11

40-4 Drugs and the Nervous System pp. 955–961

BLOCK 4

a. Psychoactive drugs can change neurotransmitter levels in a synapse. Addiction occurs when the number of receptor proteins in a neuron's membrane adjusts in response to these changes.

b. Narcotics are highly addictive painkillers. They have a molecular structure similar to that of enkephalins.

Lecture Resources
- Opening Question, page 955
- Demonstration: Effect of Sidestream Smoke
- Visual Strategies: Figures 40-13, 40-14, 40-16; Table 961
- Cocaine and Depression, page 957

- Demonstration: Healthful Cigarettes?
- Alcoholic Beverages, page 960
- Overcoming Misconceptions, page 960
- Teaching Transparencies: 233, 234, 201A

40-4 Drugs and the Nervous System pp. 955–961 (continued)

c. Cocaine blocks the reabsorption of dopamine, a neurotransmitter that stimulates the pleasure center in the brain.

Classwork Options
- ■ Social Drugs and the Media, page 959
- ■ Closure, page 961
- ■ Laboratory Investigation: Neuron Model, pages 964–965

Assignment Options
- ■ Drug Action, page 956
- ■ Opium, page 957
- ■ The Perfect Drug, page 958
- ■ Section Review, page 961
- ■ Chapter Review, questions 7, 8, 12

Review/Enrichment

BLOCK ⑤

- ◉ Teaching Resources CD-ROM: Occupational Applications Worksheets, Pharmacist and Substance-Abuse Counselor

Portfolio Assessment
- ■ Study Guide: Concept Review and Pretest

Assignment Options
- ■ Activities and Projects, question 18

Assessment Options

BLOCK ⑥

Traditional Assessment
- ■ Chapter 40 Test
- ◉ Test Generator/Assessment Item Listing: Software item bank for preparing customized chapter tests.

Portfolio Assessment
Select student reports for one or more laboratory experiments from this chapter. *The Direct Observation Checklist* on the *BioSources Teaching Resources CD-ROM* should be completed during a laboratory or other cooperative learning experience.

Holt Biology Videodiscs · HRW · CONCEPTS OF BIOLOGY

Holt Biology Videodiscs gives you a powerful tool for teaching, review, and assessment. *Concepts of Biology* adds interactive lesson plans and extensive tools for customization using CD-ROM technology.

Use the following topics from *Concepts of Biology* to help you teach this chapter:
- ■ Topic 24: Nervous and Endocrine Systems
- ■ Topic 25: Drugs and the Human Body

For further information regarding the media on the videodiscs, see *Holt Biology Videodiscs Teacher's Correlation Guide to Biology: Principles and Explorations.*

Chapter Theme

Homeostasis *The nervous system helps control the human body by transmitting information from muscles, organs, and sensory receptors to the brain. Here the information is integrated and used to send signals to muscles and glands. This function is the crux of homeostasis. The nervous system's workings are fast-acting but of short duration. Through the transmission of nerve impulses, an organism can respond quickly to a stimulus, either internal or external, without experiencing any long-term disturbance in its systems.*

Tapping Prior Knowledge

- What kind of cell is capable of sending electrical messages throughout the body?
- Name the three main parts of a neuron.

Opening Demonstration

Ask for a student volunteer to demonstrate an activity in which he or she engages, such as playing basketball, swimming, reading, listening to music, or eating. Lead a brainstorming discussion on all the things the nervous system needs to control and coordinate in order to perform the activity. Help students understand the profound complexity of the nervous system's functions. Encourage students to name not only the obvious movements of the activity (for example, dribbling a basketball), but also everything else that the system keeps under control, including internal functions (such as heart rate and breathing).

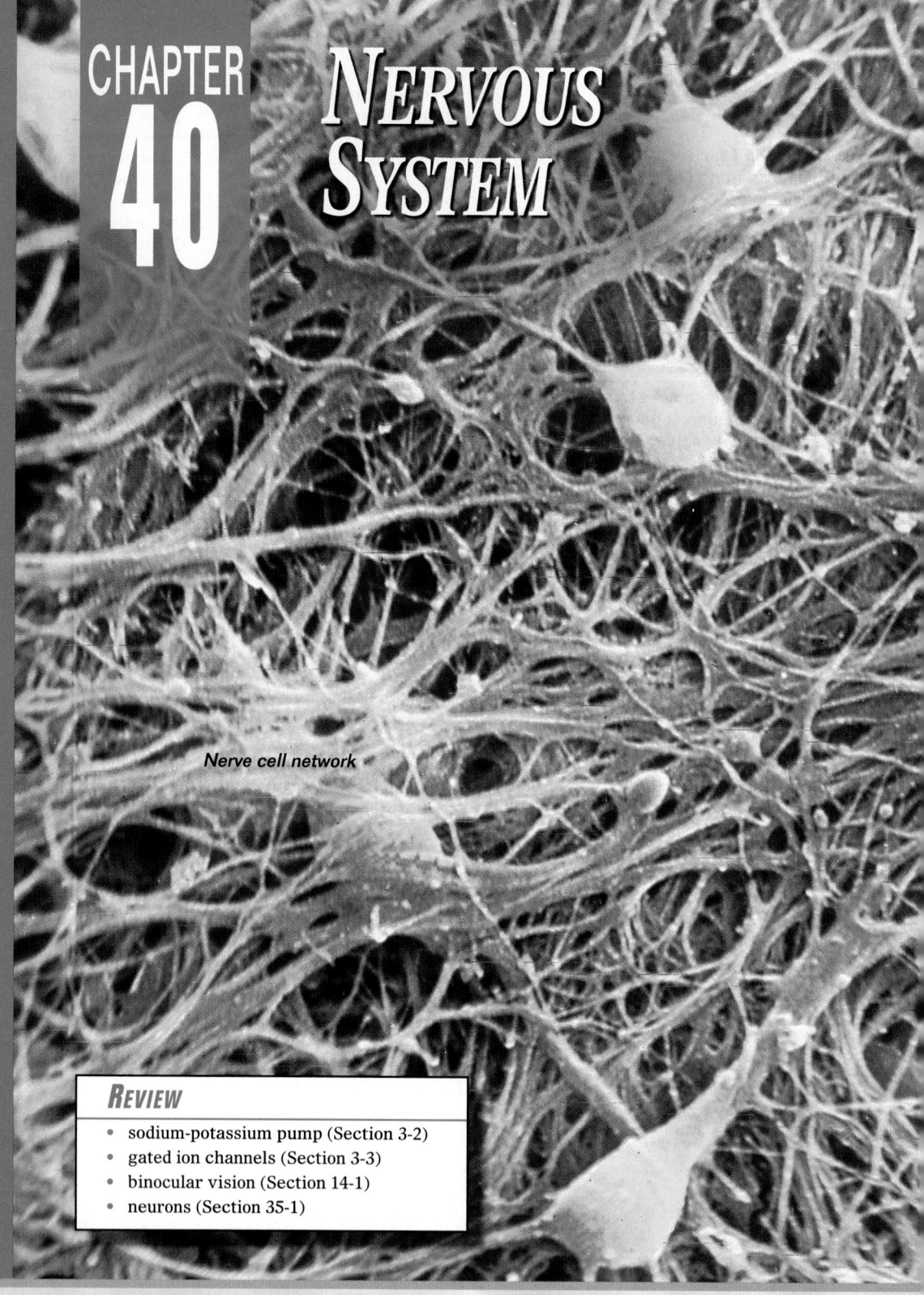

CHAPTER 40
NERVOUS SYSTEM

Nerve cell network

REVIEW

- sodium-potassium pump (Section 3-2)
- gated ion channels (Section 3-3)
- binocular vision (Section 14-1)
- neurons (Section 35-1)

Authors' Rationale

Life consists of perceiving the environment, processing information, and reacting accordingly. In a nutshell, this is the function of the nervous system. Understanding the details of this system enables students to answer questions about homeostasis such as How is information perceived? and How is information sent electrically through the body? Students will also learn how homeostasis is affected by the presence of drugs in the body.

40-1 Neurons and Nerve Impulses

Cells in the human body could communicate using only chemical signals, except for one problem: circulating chemicals is a very slow form of communication. A quicker means of communication is needed, especially if your brain has an urgent message for the muscles in your leg, such as "Contract quickly, a speeding car is headed this way!" To solve this problem, humans and most other animals have neurons, specialized cells that can quickly transmit messages throughout the body.

Section Objectives

- Describe the structure and function of a neuron.
- Explain the importance of myelin sheaths.
- Define resting potential, and describe its physiological basis.
- Explain how action potentials are generated and propagated along a neuron.
- Define synapse, and describe the events that occur in the transmission of an impulse.

Neurons Transmit Electrical Signals

Recall from Chapter 35 that a neuron is a cell specialized for transmitting electrical signals called nerve impulses. A neuron has an electrical charge because sodium-potassium pumps transport ions across the plasma membrane. This activity creates a difference in electrical charge on both sides of the membrane, which is said to be polarized. Loss of this charge difference is called **depolarization**. A neuron transmits a nerve impulse as a wave of depolarization that travels along its membrane like a rapidly burning fuse.

The human body contains many different types of neurons. Some neurons are tiny with few projections. Others are bushy with many projections, like the neuron shown in Figure 40-1, and still others have a single projection several meters long. Despite these differences, all neurons have the same functional architecture. Cytoplasmic extensions called **dendrites** (DEHN dryts) extend from the body of the neuron. Dendrites are the antennae of the neuron, enabling it to receive information simultaneously from many different sources. The surface of the **cell body** collects the information arriving from the many different dendrites. The information then travels as a nerve impulse from the cell body

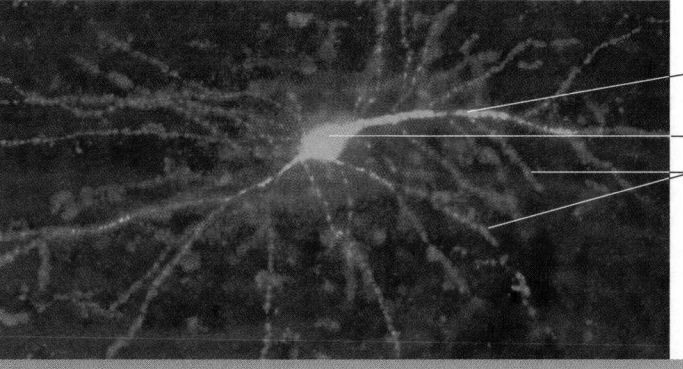

Axon

Cell body

Dendrites

Figure 40-1 A neuron consists of a cell body, dendrites and an axon.

Vocabulary Preview

depolarization
dendrite
cell body
axon
nerve
Schwann cell
myelin sheath
multiple sclerosis
resting potential
action potential
repolarization
synapse
neurotransmitter
integration

Opening Question

Ask students to describe the two kinds of signals the body uses to send messages. Students should describe nerve impulses and chemical messengers.

Demonstration
Measuring Electrical Activity
Talk with a physics teacher to arrange a simple demonstration that shows how electrodes conduct electrical activity detectable by a voltmeter or an oscilloscope.

👁 VISUAL STRATEGY

Figure 40-1
Review the names of the labeled structures of the neuron. Have students draw a neuron with arrows indicating the route that an electrical impulse travels along a neuron. Also have students add the information from this figure to the life-sized diagrams they made for the **Chapter Closure** for Chapter 35 on page 840.

Overcoming Misconceptions

Brain Tumors
Because neurons are incapable of mitosis, students may think that a brain cannot develop a tumor. However, brain tumors are not formed from neuronal tissue. They arise from the supportive glial cells in the nervous system which do undergo mitosis.

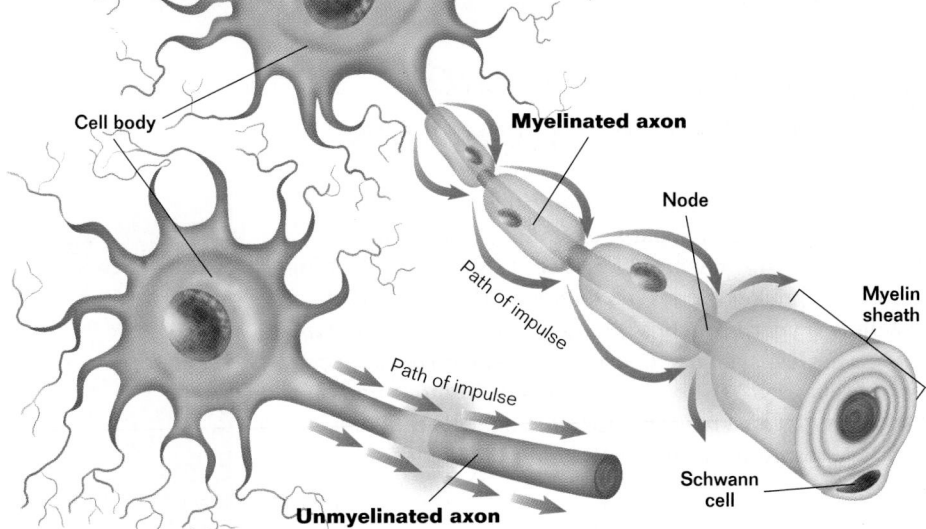

Figure 40-2 Myelin sheaths cover the axons of nerves that carry information to and away from the brain. A Schwann cell wraps its cell membrane around the axon to form the insulating myelin sheath. Impulses hop from node to node, which is faster than traveling the full length of an unmyelinated axon.

❑ CAPSULE SUMMARY

A neuron consists of a cell body, dendrites that carry impulses to the cell body, and an axon that transmits impulses to other cells. Many nerves have myelin sheaths, which expedite impulses.

◀ VISUAL STRATEGY

Figure 40-2

Use this figure to have students answer the following questions. Which kind of neurons carry impulses faster—myelinated or unmyelinated? Why? (*myelinated neurons, because the impulse can skip from node to node without having to travel along the entire length of the axon*)

Teaching Tip
Observing Neural Structure

Have students observe prepared slides of myelinated and unmyelinated neurons. Ask that they look for cell bodies, axons, and dendrites. Have students draw and label what they see and compare their drawings with the neurons shown in Figure 40-2.

along an **axon.** Most neurons have only a single axon, which may be quite long.

Bundles of neurons are called **nerves,** which appear as fine, white threads when viewed with the naked eye. Like a telephone line, nerves contain a large number of independent communications channels. In addition, they are composed of many supporting cells that form the structure of nervous tissue and assist neurons. In some neurons, a form of supporting cell called a **Schwann cell** wraps around the axon, forming a fatty, insulating covering called a **myelin** (*MY uh lihn*) **sheath.** As you can see in Figure 40-2, the myelin sheath is interrupted at intervals, leaving exposed gaps called nodes. A nerve impulse traveling down a myelinated axon jumps from node to node, which is much faster than traveling along the full length of a bare, unmyelinated axon. Destruction of large patches of myelin characterizes a disease called **multiple sclerosis.** In multiple sclerosis, small, hardened scars appear throughout the myelin sheath and interfere with the transmission of impulses. Normal nerve function is impaired, and symptoms such as double vision, muscular weakness, loss of memory, and paralysis result. ❑

A Reversal in Voltage Triggers an Impulse

When a neuron is not transmitting an impulse, the sodium-potassium pumps in its cell membrane are transporting sodium ions (Na^+) out of the cell and potassium ions (K^+) into it. Once they are pumped out, sodium ions cannot easily move back into the cell; therefore, the concentration of sodium ions is higher outside the cell. At the same time, potassium ions build up inside the cell. However, potassium ions are able to diffuse out through open channels. As a result, the outside of the neuron has a higher positive charge

◆ Did You Know?

Nerve fibers are classified as group A, B, or C fibers according to their diameter, degree of myelination, and speed of conduction. Group A fibers have the largest diameters. They are mostly sensory and motor fibers in the skin, skeletal muscles, and joints with an impulse speed of 15–130 m/s (50–430 ft./s). Nerve fibers serving the autonomic nervous system belong in groups B and C. Group C fibers have the smallest diameters. They are unmyelinated and therefore conduct impulses very slowly, at 1 m/s (3 ft./s) or less. Group B fibers, including autonomic nervous system motor fibers, visceral sensory fibers, and smaller pain and touch fibers, carry impulses at 3–15 m/s (10–49 ft./s). Their myelinated fibers have a diameter between those of the A and C fibers.

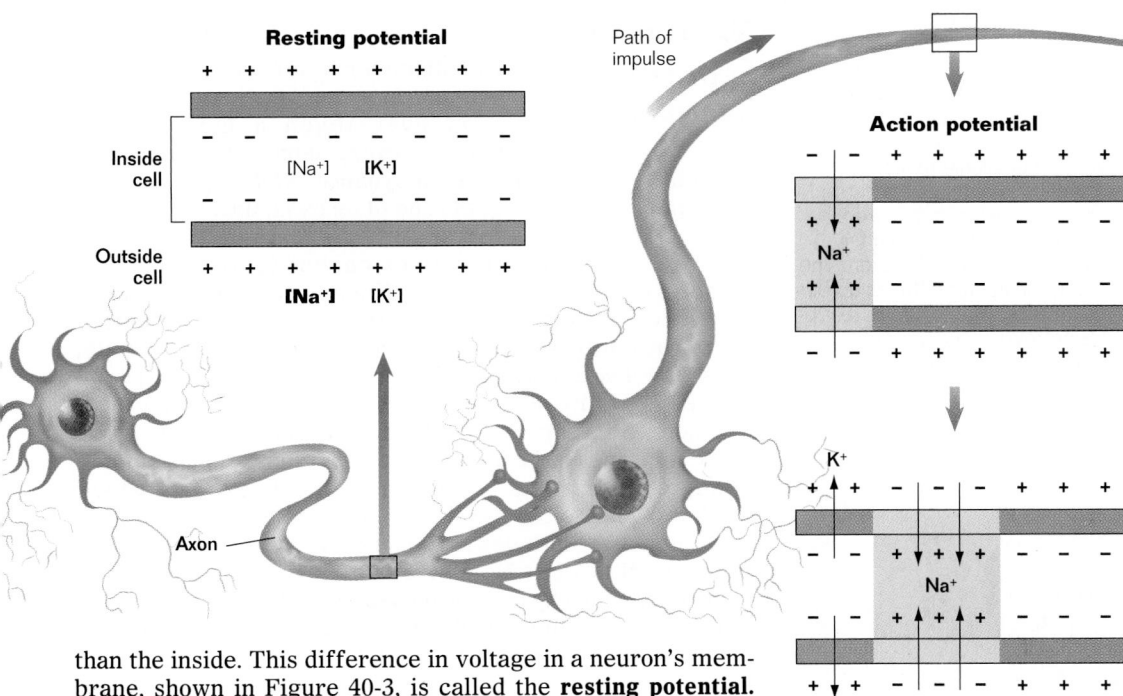

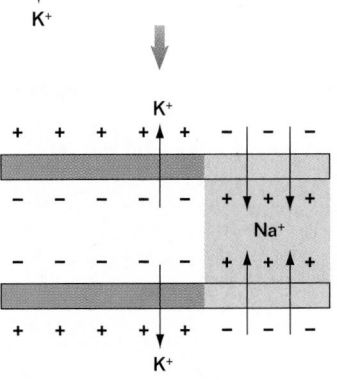

Figure 40-3 At rest, sodium-potassium pumps in a neuron's membrane keep a higher concentration of sodium ions outside the cell and a higher concentration of potassium ions inside, creating a voltage difference called the resting potential, *above left*. When an impulse moves down an axon, sodium ions rush into the cell, creating a reversal in voltage called an action potential, *above right*.

than the inside. This difference in voltage in a neuron's membrane, shown in Figure 40-3, is called the **resting potential.** The resting potential is the starting point for the transmission of a nerve impulse.

A nerve impulse starts when pressure, or other stimuli, disturbs a neuron, opening sodium channels in a small patch of membrane. As a result, sodium ions outside the cell flood into the neuron, depolarizing that patch of the membrane. For a brief moment the inside of that section of membrane becomes more positively charged than the outside. This sudden local reversal of voltage across the neuron membrane, shown in Figure 40-3, is called an **action potential.**

The sodium channels in the small patch of the membrane with the action potential remain open for only about half of a millisecond. The change in voltage causes other nearby voltage-gated sodium channels to open, reversing the voltage in that patch of membrane. The action potential that results causes more channels to open. The action potential moves down the neuron like a chain of falling dominoes as the reversal of voltage causes adjacent gated channels to open.

When the action potential has passed, the voltage-gated sodium channels snap shut, and the resting potential is restored. Why does the potential return to normal? The depolarization briefly opens voltage-gated potassium channels and allows potassium ions to leave the cell, making the inside of the cell less positively charged again. The full potential difference is then re-created by the sodium-potassium pump, and the resting potential is restored. Restoration of the resting potential is called **repolarization.** The neuron cannot transmit another signal for a short period of time until this recovery is complete.

SECTION 40-1

👁 **VISUAL STRATEGY**

Figure 40-3
Use this figure to show the differences between resting potential and action potential as well as the events accounting for the differences. Emphasize the changes in the ion channels and sodium-potassium pumps. Have students answer the following question. How would a neuron's function be affected by a chemical that opened more potassium channels? *(It would be less sensitive to depolarization.)*

Teaching Tip

What If blood flow is cut off from a region of the body? Restricting the blood flow to an area reduces the oxygen and nutrient delivery to the neurons, inhibiting nerve impulses. The numbness produced by exposure to the cold and found in the hand or foot that has "gone to sleep" is an example of inhibited neuronal activity resulting from reduced blood flow. The feeling of "pins and needles" once the blood flow is restored is evidence of the reestablishment of impulse transmission.

Did You Know?

A voltmeter measuring the voltage across a resting neuron's membrane records approximately –70 mV. The minus sign indicates that the inside of a neuron is negatively charged with respect to the outside. An action potential changes the voltage across the neuron's membrane about 100 mV, from –70 mV to +30 mV.

VISUAL STRATEGY

Figure 40-4
Review the activity that occurs at the synapse when an impulse passes from one neuron to another. Have students trace the path of this activity in the figure. Ask students the following question. What is the name of the process in which particles from within a cell are released to the environment? (*exocytosis*)

Teaching Tip
Producing Chemicals Economically
Discuss with the class ways in which the body has evolved conservative means of producing the myriad of chemicals it needs. One example is many neurotransmitters are simply intermediates in a metabolic pathway, as the *Graphic Organizer* below illustrates. Here the amino acid tyrosine is the common starting point for each of the neurotransmitters shown. A single enzyme converts tyrosine to L-dopa, which can in turn be transformed into dopamine by a different enzyme, and so on, finally ending with epinephrine (adrenaline). Each of the intermediates may either function as an independent neurotransmitter or enter the conduit to produce another neurotransmitter later in the pathway.

Figure 40-4 Most neurons are separated from each other by a gap called a synapse. A nerve impulse is unable to jump across a synapse. Instead, it is carried across by neurotransmitters, chemicals released from tiny sacs in the end of the axon. The neurotransmitter diffuses across the synapse and binds to receptor proteins in the adjacent neuron, triggering an impulse.

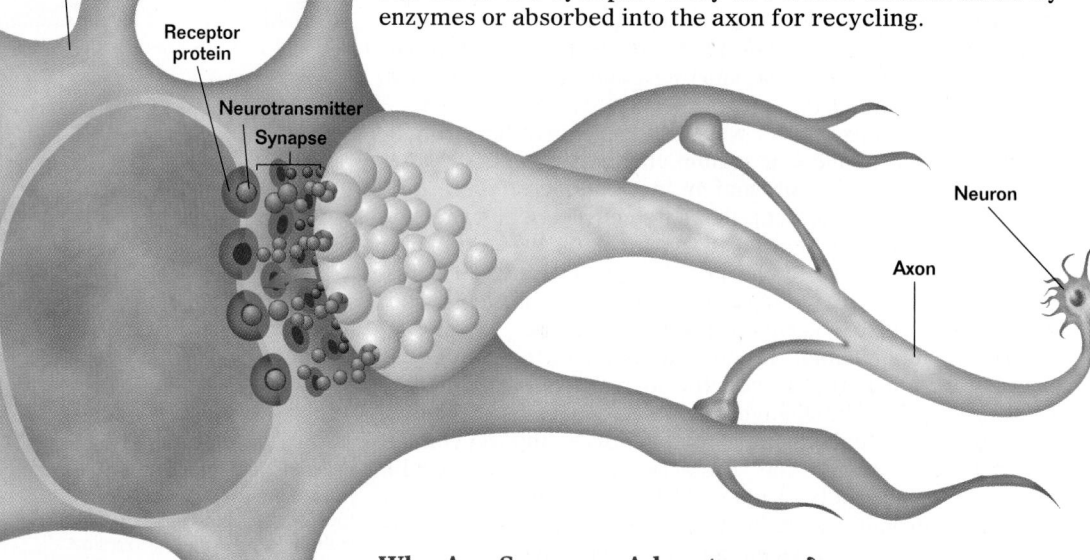

Neurons Are Separated by Synapses

If you examine a nerve closely, you will discover a surprising fact—no two neurons ever touch. Nor does an axon ever touch a muscle cell. Between every axon and its neighboring cell is a tiny gap called a **synapse** (SIHN aps). When a nerve impulse arrives at the end of an axon, the action potential must cross the synapse for the message to continue. However, electrical impulses cannot jump across the synapse; they must be carried. The end of an axon contains tiny sacs filled with chemical messengers called **neurotransmitters.** When a nerve impulse reaches the tip, the sacs release neurotransmitters into the synapse. As you can see in Figure 40-4, the neurotransmitter molecules diffuse across the synapse and bind to receptor proteins in the cell membrane on the other side of the synapse. This binding of neurotransmitters causes ion channels to open, which results in voltage changes in the membrane. As a result, the nerve impulse continues in the adjacent neuron, passing the signal to that cell. Neurotransmitter molecules in the synapse are then quickly removed so that another message can cross the synapse. They are either broken down by enzymes or absorbed into the axon for recycling.

Why Are Synapses Advantageous?
Electrical signals travel along an axon much faster than neurotransmitters cross a synapse. If impulses travel faster along axons, what is the evolutionary advantage of synapses? Why don't neurons have continuous electrical contact? The great advantage of a chemical junction like the synapse is that the nature of the chemical signal can be different in different junctions, permitting different kinds of responses. Over 60 different chemicals have been identified that either act as specific neurotransmitters or modify the activity of neurotransmitters. ☐

☐ CAPSULE SUMMARY
When a nerve impulse reaches the end of an axon, neurotransmitters are released into the synapse. They cross the synapse and bind to receptors in the neighboring cell membrane.

Graphic Organizer
Use this graphic organizer in *Teaching Tip:* **Producing Chemicals Economically** on page 942.

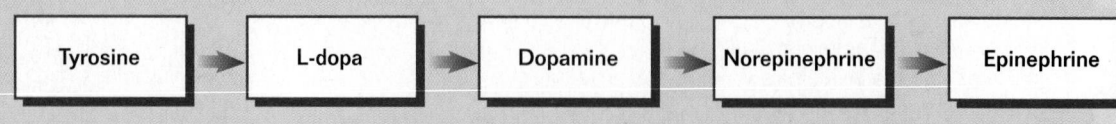

Neurons Form an Integrated Circuit

Synapses fall into two general classes depending on what happens when their cell membrane receptors bind neurotransmitters. In an excitatory synapse, the receptor is a gated sodium channel that is closed while at rest. When a neurotransmitter binds to it, the sodium channel opens, allowing sodium ions to flood inward and depolarize the membrane. If enough sodium channels are opened by neurotransmitters, an action potential that initiates a new nerve impulse in the neuron is created. In an inhibitory synapse, the receptor is a gated potassium channel. When a neurotransmitter binds to it, the potassium channel opens, allowing positively charged potassium ions to leave the cell, giving it a more negatively charged interior. This change in voltage inhibits nerve impulses because even more sodium channels must be opened to initiate an action potential.

An individual cell can have both excitatory synapses and inhibitory synapses. When signals from both excitatory and inhibitory synapses reach the neuron cell body, the excitatory and inhibitory effects interact with one another. The result is a process of **integration** in which the various excitatory and inhibitory electrical effects tend to cancel or reinforce one another. Individual neurons often receive many inputs. A single motor neuron in the spinal cord may have as many as 50,000 synapses. The combined influences of all of the inputs determine whether the cell membrane will be sufficiently depolarized to initiate a nerve impulse along the axon. Your every thought and feeling result from such integration.

Section Review

1. *Diagram and label a neuron. Next to each part, list its function.*

2. *What is a myelin sheath? What is its function?*

3. *Define the following terms: resting potential, action potential, depolarization, and repolarization.*

4. *Outline the major steps in the transmission of a nerve impulse.*

5. *What events are involved in the transmission of a nerve impulse across a synapse?*

Critical Thinking

6. *How are nerve impulses inhibited? Of what advantage is this to the body?*

7. *How are synapses involved in integration?*

☑ RESEARCH Update

Alzheimer's Disease

A substance called beta amyloid protein (βAP) is the main constituent of dense plaques in the brain tissue of individuals suffering from Alzheimer's disease (AD). The presence of these plaques is the key factor used to diagnose Alzheimer's disease, which can only be made after an autopsy. Alzheimer's disease researchers are now looking for the mechanism by which βAP might produce early symptoms of AD such as memory loss, which can occur prior to plaque formation.

René Etcheberrigaray, a researcher at the National Institutes of Health, and his colleagues have performed experiments that show a serious effect of βAP. Using normal cultured fibroblast cells, they were able to re-create an existing physiological condition prevalent in cells from patients with Alzheimer's disease (AD cells). One of the important potassium ion channels in AD cells is inoperative. The NIH team re-created the same pathology in normal cells by exposing them to high amounts of βAP. The βAP impaired the potassium channel function in the normal cells, creating conditions similar to those encountered in AD cells. This has led the scientists from the NIH to suggest that "βAP might alter potassium channels and thus impair neuronal function to produce symptoms such as memory loss by a means other than plaque formation." ☑

UPDATE

Answers to Section Review

1. See page 939.

2. A myelin sheath is a segmented layer of insulation covering an axon. It speeds up impulses along an axon.

3. Resting potential: the difference in the charge across a neuron's membrane. Action potential: a reversal of voltage across the membrane. Depolarization: the loss of the charge difference across the membrane. Repolarization: the restoration of the membrane's resting potential.

4. Stimulation opens sodium channels, allowing sodium to enter the neuron and reverse its charge, which opens other sodium channels in a chain reaction down the axon.

5. A nerve impulse arrives at the tip of an axon where neurotransmitters are then released, into the synapse. Neurotransmitters diffuse across the gap, and trigger an impulse in the adjacent neuron.

6. They are inhibited when neurotransmitters bind to gated potassium channels. This process can inhibit messages for pain.

7. Signals from inhibitory and excitatory synapses cancel or reinforce one another.

Vocabulary Preview

sensory nerve

motor nerve

somatic nervous system

autonomic nervous system

cerebrum

cerebral cortex

thalamus

hypothalamus

limbic system

cerebellum

brain stem

reticular formation

spinal cord

reflex

sympathetic division

parasympathetic division

Opening Questions

Ask students to state the function of the nervous system and to name its main components. They should recall that the nervous system receives and interprets information and is composed of the brain, spinal cord, sensory nerves, and motor nerves.

Demonstration

Reflexes

Demonstrate the familiar knee-jerk reflex on a student volunteer. Gently strike his or her patellar tendon (connective tissue connecting knee cap to the tibia) with a rubber reflex hammer or the edge of a wooden ruler. Ask students to recall what they know about the central nervous system and to devise a possible electrical circuit that would elicit this kind of reflex.

Section Objectives

■ Explain the structural and functional divisions of the nervous system.

■ Name the major regions of the brain, and explain their functions.

■ Describe the peripheral nervous system, and distinguish between sensory and motor nerves.

■ Explain the action of a reflex.

■ Compare the functions of the two divisions of the autonomic nervous system.

Figure 40-5 The nervous system is divided into the central nervous system and the peripheral nervous system, which consists of two other divisions of nervous systems.

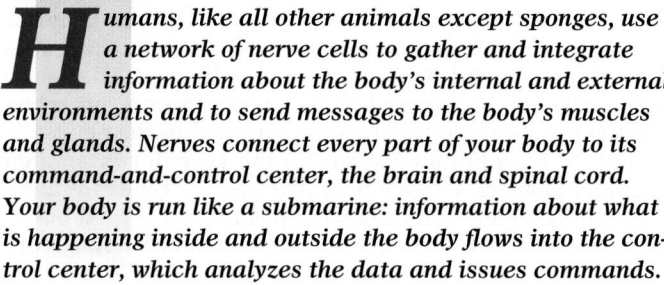

40-2 Architecture of the Nervous System

*H*umans, like all other animals except sponges, use a network of nerve cells to gather and integrate information about the body's internal and external environments and to send messages to the body's muscles and glands. Nerves connect every part of your body to its command-and-control center, the brain and spinal cord. Your body is run like a submarine: information about what is happening inside and outside the body flows into the control center, which analyzes the data and issues commands.

The Nervous System Has Two Divisions

As you learned in Chapter 35, the human nervous system is divided into two principal parts—the central nervous system and the peripheral nervous system. The central nervous system, composed of the brain and the spinal cord, is the body's main processing center. It receives and sends information through the branches of nerves that make up the peripheral nervous system.

The peripheral nervous system consists of two types of nerves. **Sensory nerves** gather information about your environment and your body's condition and deliver it to the central nervous system. **Motor nerves** transmit commands from the central nervous system to muscles and glands all over your body. Motor nerves that conduct impulses to skeletal muscles under our conscious control make up the **somatic nervous system;** other motor nerves that regulate the activity of cardiac muscles, smooth muscles, and glands make up the **autonomic nervous system.** Figure 40-5 shows how the different parts of the nervous system are integrated.

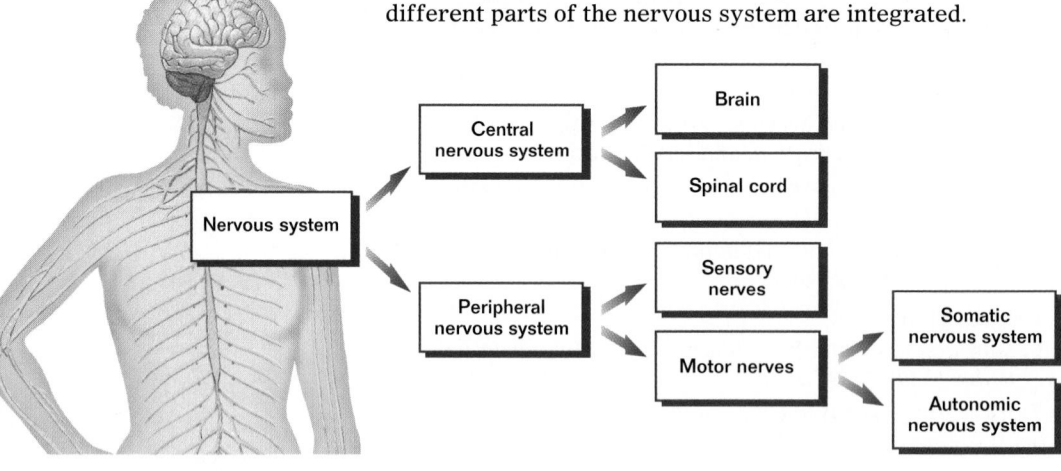

 Did You Know?

Association neurons, or *interneurons*, lie between motor and sensory neurons, often directing signals through complex pathways resulting in diverse integration of information. Interneurons account for over 99 percent of all of the

body's nerve cells and are especially plentiful in the central nervous system.

The Central Nervous System Directs and Coordinates Information

In an average adult, the brain is one of the largest organs of the body, weighing about 1,500 g (approximately 3 lb.). The most conspicuous part of the brain is its rounded, wrinkled outer layer, the **cerebrum,** as shown in Figure 40-6. A long, narrow cleft, or fissure, divides the cerebrum into right and left halves, or hemispheres, which communicate with each other through a connecting tract of fibers. Much of the activity in the cerebrum occurs in the **cerebral cortex,** an outer layer only 2–4 mm thick. The cerebral cortex contains thick layers of unmyelinated neurons that appear gray and are therefore referred to as gray matter. Beneath the cerebral cortex lies a region of myelinated neurons (white matter) that shuttles information between the cortex and the rest of the brain.

Like a hand covering a fist, the cerebrum surrounds the rest of the brain. As shown in Figure 40-6, directly beneath the cerebrum are the thalamus and hypothalamus, important centers of information processing. The egg-shaped **thalamus** is the main site of sensory processing in the brain. Most of

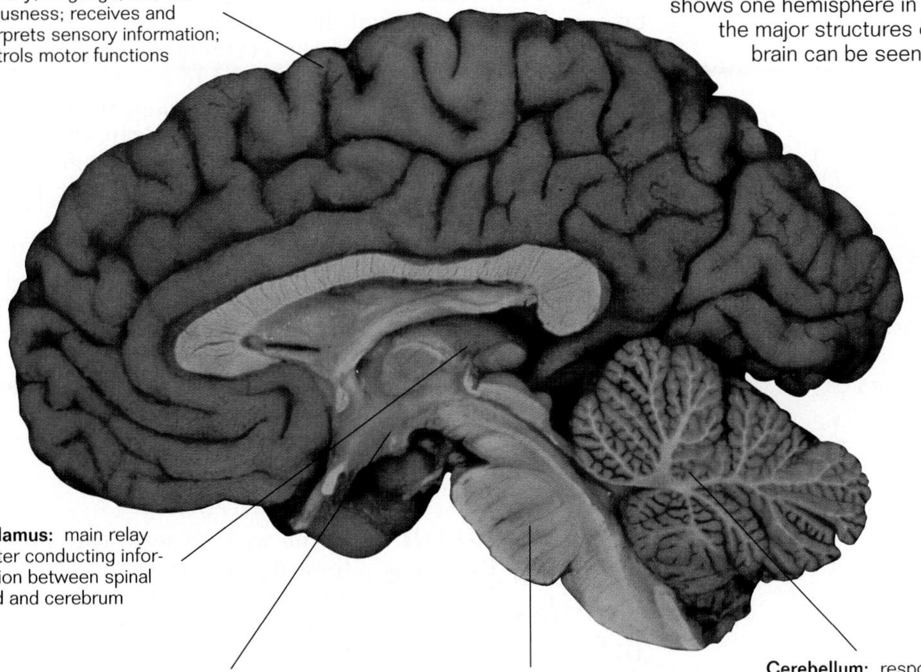

Figure 40-6 A midsagittal cut through a brain divides it into two mirror image halves. This midsagittal view of a human brain shows one hemisphere in which the major structures of the brain can be seen.

Cerebrum: center of intellect, memory, language, and consciousness; receives and interprets sensory information; controls motor functions

Thalamus: main relay center conducting information between spinal cord and cerebrum

Hypothalamus: center for control of homeostasis (body temperature, appetite, water balance); links nervous and endocrine systems

Brain stem: regulates heartbeat, respiration, and blood pressure; helps control swallowing, coughing, sneezing and vomiting

Cerebellum: responsible for smooth, coordinated movement; helps maintain posture, muscle tone, and equilibrium

 VISUAL STRATEGY

Figure 40-6
Use this figure to teach the five major regions of the brain described in the text and their functions. Have students add this information to the life-sized drawings they made for the **Chapter Closure** for Chapter 35.

Teaching Tips
Pick Your Brain
Display an anatomical model of a human brain, and have students compare it with the photograph in Figure 40-6. Help them locate the cerebrum, cerebellum, and brain stem. Then have students relate the different parts of the brain to their functions.

Brain Evolution
Explain the evolution of the vertebrate brain to students. Three structures form the evolutionary basis for the larger brain found in humans and other mammals: the forebrain, the midbrain, and the hindbrain. The forebrain consists of the olfactory bulbs, the cerebrum, and the thalamus/hypothalamus. The midbrain in lower vertebrates functions chiefly in vision, being dominated by its optic lobes. The hindbrain is composed of the medulla oblongata, the cerebellum, and the part of the brain stem known as the pons.

In fish and amphibians, the cerebral cortex is just a smooth covering for the tiny cerebrum. Reptiles have a larger cerebral cortex, which begins to take over some of the midbrain duties such as vision. In birds, the transfer of control to the forebrain is complete. In primitive mammals, the cerebral cortex covers virtually the entire cerebral surface, gradually becoming more and more dominant in its control of brain functions. Eventually, the cerebral cortex grows so large that it can no longer cover the surface without forming folds, or convolutions, in the cerebrum itself in order to increase its surface area. The convolutions are a mark of the complexity and capacity of the brain of a higher mammal.

UNIT 9
CHAPTER 40

Application

Health Medical scientists think psychosomatic illness may be mediated by the hypothalamus because of its connections to the autonomic nervous system and its control of emotions. Many stress-related psychosomatic illnesses such as ulcers, asthma, and high blood pressure are clearly connected to hypothalamic functions.

Teaching Tip

What If electrical activity cannot be registered in the brain? When no electrical activity can be registered in the brain, especially in the region of the brain stem, a person is considered to be "brain dead." In such cases, the vital body processes controlled unconsciously by the brain stem must be sustained by a life-support system.

CONTENT LINK

In *Chapter 41*, you will learn why the pituitary gland is sometimes called "the master gland."

□ *CAPSULE SUMMARY*

The central nervous system is made up of the brain and spinal cord, which work together to collect and process information.

the sensory nerves from all parts of the body converge on the thalamus, which sorts and relays information to appropriate areas in the cerebral cortex. Below the thalamus is the **hypothalamus,** a slender thread of tissue. Despite its small size, the hypothalamus controls many body activities related to homeostasis, such as body temperature, blood pressure, respiration, and heart rate. It also directs the secretions of the brain's major hormone-producing gland, the pituitary gland. The hypothalamus is linked to some areas of the cerebral cortex by an extensive network of neurons. This network, along with the hypothalamus, is called the **limbic system.** The limbic system, sometimes referred to as the emotional brain, is responsible for many basic drives and emotions, such as pain, pleasure, anger, sex, hunger, and thirst.

Extending back from the base of the brain, as shown in Figure 40-6, is a region known as the **cerebellum.** The cerebellum controls balance, posture, and voluntary muscle contractions. This small cauliflower-shaped structure, while well developed in humans and other mammals, is even more developed in birds. Birds perform more complicated feats of balance than do humans. Imagine the kind of balance and coordination needed for a bird to land on a branch.

The base of the brain, called the **brain stem,** connects the rest of the brain to the spinal cord. This stalklike structure contains nerves that control breathing, swallowing, digestive processes, heartbeat, and the diameter of blood vessels. A network of nerves called the **reticular formation** runs through the brain stem and connects to other parts of the brain. Their widespread connections make these nerves essential to consciousness, awareness, and sleep. One part of the reticular formation filters sensory input, enabling you to sleep through repetitive noises such as traffic, yet awaken instantly when a telephone rings.

The **spinal cord** is a cable of nerve tissue that extends from the base of the brain through the backbone to the level just below the ribs. Messages from the brain and the rest of the body run up and down the spinal cord, making it the "information highway" of the central nervous system. The center of the spinal cord consists of a column of gray matter covered by a sheath of white matter. □

The Peripheral Nervous System Shuttles Information

If you were to see a car speeding toward you, your brain would send messages through motor neurons to glands that secrete the hormone adrenaline. Adrenaline would increase your heart rate and breathing rate. Your brain would also send messages through motor neurons to muscles in your legs, which would quickly contract and get you out of

 Did You Know?

Scientists have identified regions of the hypothalamus associated with the perception of pain, pleasure, fear, and rage. Even elements of the sex drive have been localized in the hypothalamus. Most physical expressions of emotion are initi- ated by the hypothalamus, as are the pounding heart and cold sweats that can result from extreme fear.

the way fast. Without these nerve pathways, your brain would be useless because it would never be able to receive or send messages.

As you can see in Figure 40-7, peripheral nerves arise from either the brain or the spine. Twelve pairs of nerves from the brain are associated with motor and sensory functions. Thirty-one pairs of spinal nerves supply the communication links between the central nervous system and the neck, trunk, arms, and legs. This is why injuries to the spinal cord often paralyze the lower part of the body.

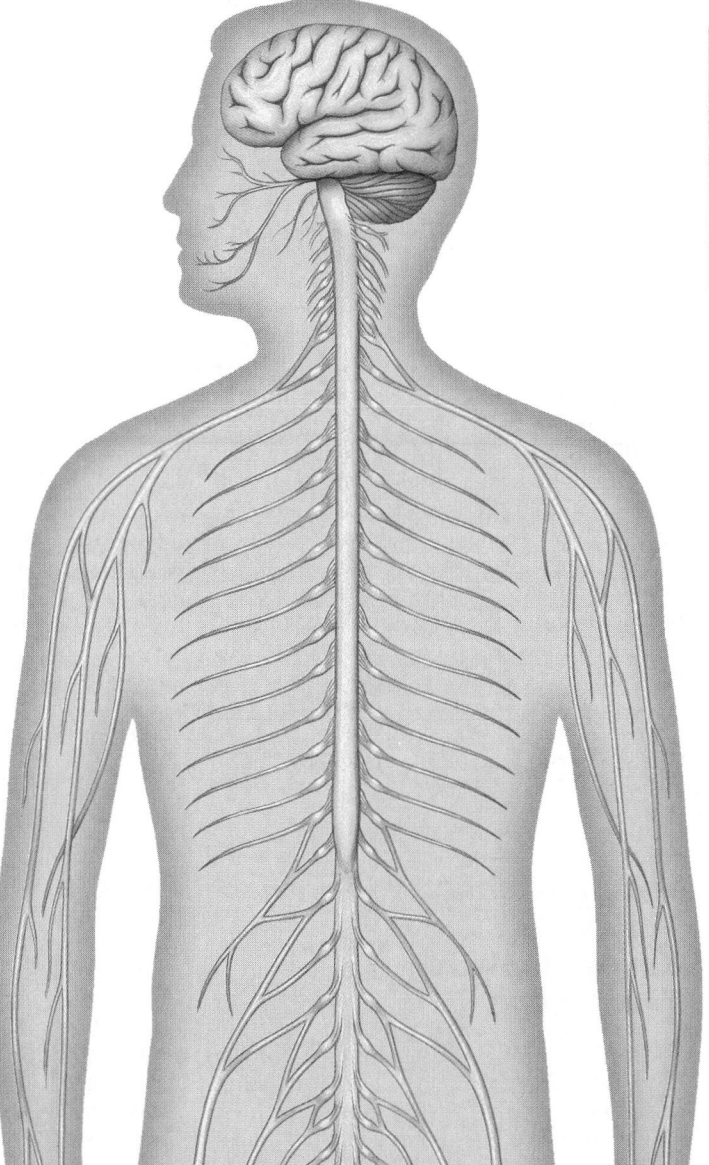

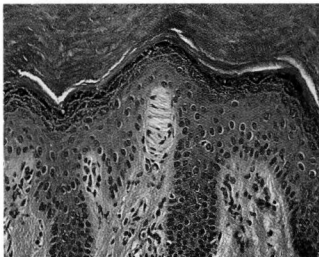

Sensory neuron

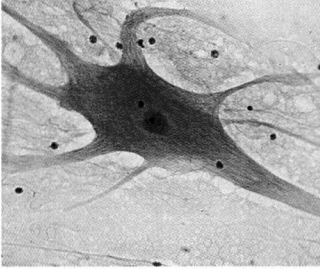

Motor neuron

Figure 40-7 The peripheral nervous system, *left,* has two main types of neurons: sensory neurons and motor neurons. Sensory neurons send messages to the brain; motor neurons deliver messages to muscles and glands.

 VISUAL STRATEGY

Figure 48-7
Use this figure to have students trace the path that an impulse takes along a sensory neuron to the spinal cord and from the spinal cord to a motor neuron. Emphasize that most nerves contain both sensory and motor neurons running side by side.

 Did You Know?

The 12 cranial nerves listed in order from the front of the brain to the back are the *olfactory, optic, oculomotor, trochlear, trigeminal, abducens, facial, vestibulocochlear, glossopharyngeal, vagus, accessory,* and *hypoglossal* nerves. In most cases their names reveal their functions.

The 31 pairs of spinal nerves are named according to their point of origin from the spinal cord: cervical nerves (C_1–C_8), thoracic nerves (T_1–T_{12}), lumbar nerves (L_1–L_5), sacral nerves (S_1–S_5), and the coccygeal nerve (C_0).

Figure 40-8

Ask students why it is advantageous to bypass the brain in a reflex arc. *(It results in a shorter pathway and therefore produces faster reaction times.)*

Demonstration
Blink Reflex

Discuss with students how the involuntary blink reflex protects the eyes. Demonstrate this reflex by tossing a wad of paper at the eyes of a student standing behind a window.

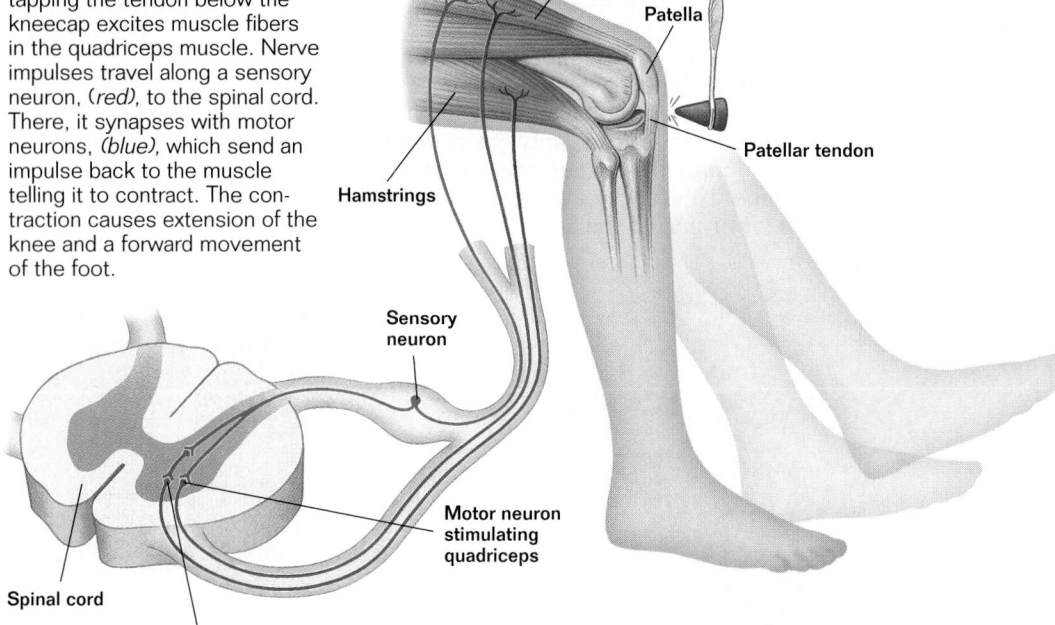

Figure 40-8 In the patellar reflex, tapping the tendon below the kneecap excites muscle fibers in the quadriceps muscle. Nerve impulses travel along a sensory neuron, *(red)*, to the spinal cord. There, it synapses with motor neurons, *(blue)*, which send an impulse back to the muscle telling it to contract. The contraction causes extension of the knee and a forward movement of the foot.

□**CAPSULE SUMMARY**

The peripheral nervous system is made up of branches of nerves that carry information to and from the brain and spinal cord.

Many motor nerves in your peripheral nervous system are wired to give your body the ability to respond quickly and involuntarily to a stimulus. This sudden, rapid response is called a **reflex.** A reflex produces a rapid motor response to a stimulus because the sensory neuron carrying the information connects directly to a motor neuron. The nerve impulse never reaches the brain; it travels only as far as the spinal cord and then comes right back as a motor response. For example, when you strike a muscle below your knee with a reflex hammer, as shown in Figure 40-8, your leg kicks out for a moment. The sudden jolt to the muscle stimulates a nerve impulse that travels to the spinal cord where it synapses with motor neurons. The motor neurons send impulses back to the muscle, causing it to contract. As a result, your leg kicks forward. □

The Autonomic Nervous System Keeps the Body Functioning

The autonomic nervous system is the command network that the central nervous system uses to maintain the body's homeostasis. By using it, the central nervous system regulates heart rate and controls muscle contractions in the walls of your blood vessels and digestive, urinary, and reproductive tracts. It also carries messages that stimulate glands to secrete tears, mucus, and digestive enzymes.

 Did You Know?

While the simplest reflexes may involve single synapses between a sensory neuron and a motor neuron, association neurons are active in more complex reflexes. These association neurons allow you to be aware of the pain of a hot pot handle. Awareness of the temperature follows the release by a measurable time delay, attributable to the added distance between the site of the reflex synapse and the brain centers responsible for the sensory awareness of pain and heat.

Some motor neurons in the autonomic nervous system are active all the time, even when you are asleep. These neurons carry messages from the central nervous system to keep the body functioning even when it is not active. For example, they direct the muscles that control blood pressure, breathing, and the movement of food through the digestive system.

The division of the autonomic nervous system that is referred to as the **sympathetic division** dominates in times of stress. It controls the "fight or flight" reaction—the feeling you experience when you find yourself in an emergency or a dangerous situation, such as being in the path of a speeding car. Its activity increases blood pressure, heart rate, breathing rate, and blood flow to the muscles—things your body needs in order to deal with a crisis. Another division of the autonomic nervous system, called the **parasympathetic division,** has the opposite effect, one that you feel when you relax after eating a heavy meal. The parasympathetic division conserves energy by slowing the heartbeat and breathing rate and by promoting digestion and elimination. Most glands, smooth muscles, and cardiac muscles receive information from *both* the sympathetic and parasympathetic divisions of the body.

Although the autonomic nervous system can carry out its tasks automatically, it is not completely independent of voluntary control. For instance, it enables you to breathe involuntarily, but you can decide to stop breathing for a short time. However, any voluntary control of the autonomic nervous system that endangers life will disturb the homeostasis of the brain tissue and cause unconsciousness. Then the autonomic nervous system takes over and restores normal functioning, which explains why you cannot hold your breath indefinitely. ▢

Teaching Tip
Role of Sympathetic Division
Inform students that the sympathetic division promotes a number of adjustments in the body. It constricts blood vessels in the skin and digestive tract and dilates blood vessels in the heart and skeletal muscles during exercise. This enables these hard-working muscles to get an increased amount of oxygen.

▢ **CAPSULE SUMMARY**

The autonomic nervous system is the division of the peripheral nervous system that controls activities involved with maintaining homeostasis.

Section Review

1. *What are the principal components of the nervous system? What are their general functions?*
2. *What are six major parts of the brain? Describe the function of each part.*
3. *How do sensory nerves and motor nerves differ?*
4. *What is a reflex? Why is a reflex quicker than other responses?*
5. *How does the autonomic nervous system affect your body when you become frightened?*

Critical Thinking
6. *Basic reflexes are unlearned and said to be built into neural anatomy. What is the evolutionary advantage of this?*

Answers to Section Review

1. See page 944.
2. See Figure 40-6 on page 945.
3. Sensory nerves deliver information to the central nervous system, while motor nerves deliver control signals to the muscles and glands.
4. A reflex is an involuntary response. It is quicker than other responses because its sensory neuron connects directly to a motor neuron.
5. The sympathetic nervous system increases the activity of stress-adaptive reactions, while the parasympathetic depresses them.
6. Unlearned basic reflexes enable us to respond to painful or harmful stimuli quickly before physical damage can be done.

Vocabulary Preview

sensory receptor

sensory organ

cochlea

cornea

iris

pupil

lens

retina

rod

cone

taste bud

Opening Question

Ask students to describe the role of sensory nerves in the body. Remind students that sensory nerves relay information to the central nervous system.

Demonstration

Seeing Spots Before Your Eyes

Have students close their eyes and press lightly on their eyelids until they see an effect. Ask what they see. (patterns of colors) These patterns are called phosphenes. The nerves in the eyes send messages to the brain, which interprets these messages as colors. Point out to students that they are not really "seeing" anything. Then ask what this phenomenon tells them about the sense of sight. (The eyes receive information that is sent via nerve impulses to the brain, which analyzes and interprets these

👁 VISUAL STRATEGY

Figure 40-9

Use this figure to answer the following question. The ear can be called a "double organ." Why is this an appropriate explanation? (It is an organ of hearing as well as of balance.)

Section Objectives

- Describe the structure of the inner ear, and explain how it functions to maintain balance and to sense sound waves.
- Trace the pathway of light through the eye to the retina, and describe the events involved in sensing light waves.
- List four kinds of sensory receptors that help maintain homeostasis.
- Describe how taste and smell are sensed.

Figure 40-9 A view of the anatomy of the ear shows the structures that are responsible for sensing balance, *top arrows,* and the structures that enable you to hear, *bottom arrows.*

40-3 The Sense Organs

Over a dozen kinds of sensory cells carry impulses to the central nervous system. These specialized cells, called *sensory receptors,* are capable of detecting various forms of stimuli such as changes in blood pressure, strain on ligaments, and odors in the air. Organs that contain sensory receptors, such as your eyes and ears, are composed of many cell and tissue types and are called *sensory organs.*

Receptors in the Ear Sense Equilibrium and Sound

Although you may think of ears as flattened cups of flesh on the sides of your head, the ear is actually an intricate sense organ, responsible for sensing equilibrium and hearing. The sensory cells for equilibrium are located in chambers within the inner ear. As you can see in Figure 40-9, clusters of hairs in these chambers respond to changes in head position with respect to gravity. The hairs project into a jellylike fluid that contains tiny pieces of calcium carbonate called otoliths, meaning ear stones. Gravity always pulls downward on the otoliths and hairs, sending a steady signal to the brain. When the head moves, the fluid slides over the hairs and bends them in the opposite direction, changing the signal to the brain. The brain interprets the changes in impulses to determine the position of the head. In a similar manner, the three semicircular canals in the inner ear detect rotation of the head.

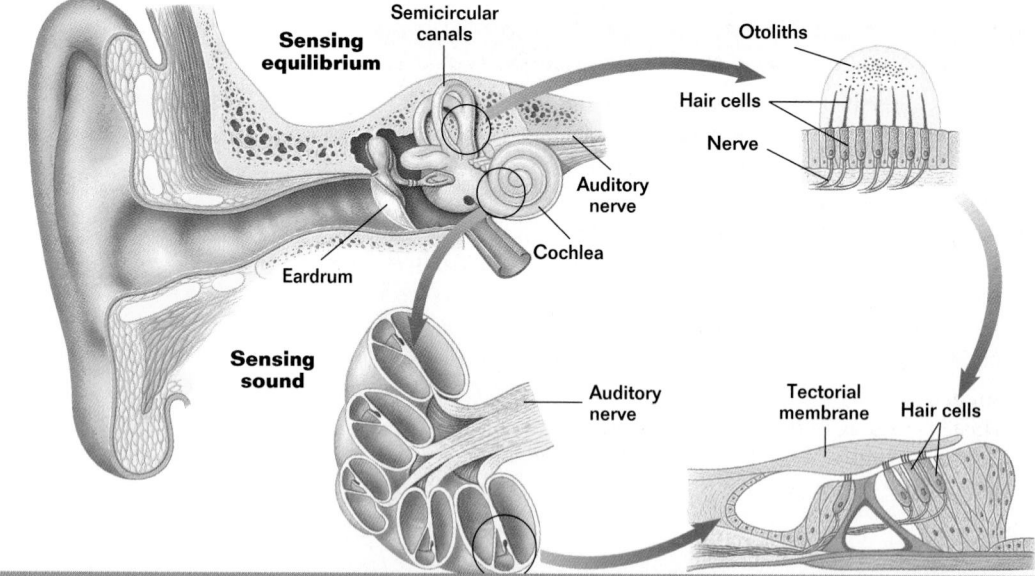

Effective Reading

Visual Learning

Encourage students to make graphic representations of textual information similar to the graphic organizers encountered throughout this text. These learning aids not only summarize key concepts, but also convert the concepts into a visual cue that is often more easily remembered. Additionally, the creation of a graphic organizer is often an interesting way to practice new knowledge.

Responding to Sound Waves

Look again at the structure of the human ear shown in Figure 40-9. When a sound wave enters the ear canal, it strikes the eardrum and causes it to vibrate. Behind the eardrum, three small bones called ossicles transfer the vibration to a fluid-filled chamber within the inner ear. This chamber, called the **cochlea** (KAHK lee uh), is shaped like a tightly coiled snail shell. The sound receptors within the cochlea are hair cells that rest on a membrane separating the chamber into two halves. The hair cells do not project into the fluid filling the cochlea; instead, they are covered by another membrane, called the tectorial membrane. When a sound enters the cochlea, the sound waves cause this membrane "sandwich" to vibrate, bending the hairs pressed against the outer membrane and causing them to send nerve impulses to sensory neurons that travel along the auditory nerve to the brain.

Sounds of different frequencies cause different parts of the membrane to vibrate and thus fire different sensory neurons. Our ability to hear depends upon the flexibility of the membranes within the cochlea. ◻

Photoreceptors in the Eye Sense Light

No other stimulus provides as much detailed information about the environment as light. Vision depends on a special sensory apparatus called an eye. All of the sensory receptors described so far respond to chemical or mechanical stimuli. Eyes contain sensory receptors that respond to photons of light. The light energy is absorbed by pigments that trigger nerve impulses in sensory neurons. Humans have extremely good eyesight. We see in color and can distinguish fine details and movement. Birds are the only animals with better eyesight than humans.

The human eye can be compared to a camera. Light first passes through a transparent, protective covering called the **cornea,** which begins to focus the light onto the back of the eye. The amount of light entering the eye is controlled by a shutter called the **iris,** which consists of tiny muscles arranged in a ring. The black hole in the center of the iris is the **pupil,** the area through which light enters the eye. The pupil gets larger in dim light and smaller in bright light. The beam of light then passes through the **lens,** which completes the focusing. The lens is a fat disk that resembles a flattened balloon. It is attached to ciliary muscles by suspended ligaments. When these muscles contract, they change the shape of the lens and thus the point of focus on the rear of the eye.

◻ **CAPSULE SUMMARY**

The receptors for equilibrium and hearing are located in the inner ear.

Teaching Tip
Motion Sickness
Ask students to recall a time when they experienced motion sickness. Select volunteers to describe their experience and to explain why they think they became sick. After discussing possible causes, explain to them that motion sickness may be due to sensory input mismatch. For example, if you are reading a book in a moving car, your eyes tell your brain that your body is fixed in reference to a stationary environment (the car). But as the car moves forward, impulses from the inner ear disagree with the visual information. The discrepancy between signals may cause excessive salivation, sweating, pallor, hyperventilation, nausea, and vomiting. Some motion sickness medications work by depressing the signals from the otoliths in the inner ear.

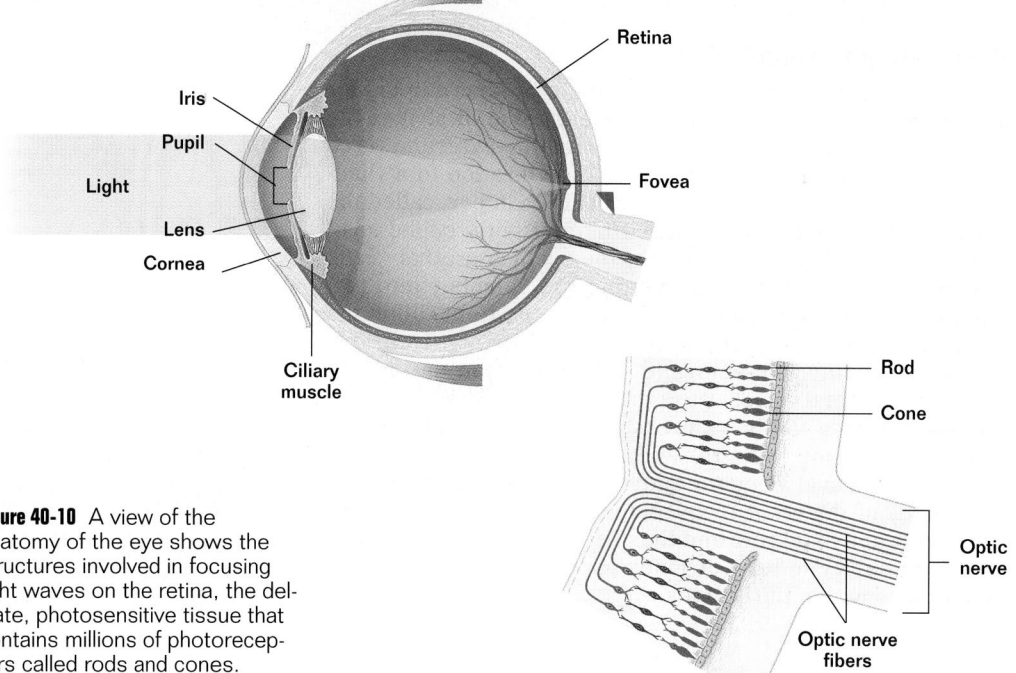

Figure 40-10 A view of the anatomy of the eye shows the structures involved in focusing light waves on the retina, the delicate, photosensitive tissue that contains millions of photoreceptors called rods and cones.

VISUAL STRATEGY

Figure 40-10

Use this figure to explain how light passes through the eye. Tell students that the cornea is a transparent structure that serves as a covering for the eye. Ask what happens to light after it passes through the cornea. (*It passes through the hole in the iris called the pupil and is focused by the lens onto the retina.*) What are the two kinds of photoreceptors in the retina? (*rods and cones*) Ask students what structure sends impulses to the brain from the retina. (*the optic nerve*)

Teaching Tip

Eye Structure

Give each pair of students a hand lens. Then ask the partners to observe each other's eyes. Ask that they first make observations in a well-lit area and then repeat these observations in a dimly lit area. Ask that the students make and label drawings of what they observe. Once all pairs have concluded their observations, discuss how levels of light affect the eye. (*The pupil narrows in bright light and widens in dim light.*)

As light enters the eye, as shown in Figure 40-10, it passes through several transparent materials that focus it onto a delicate layer of tissue called the **retina.** The retina is the light-sensing portion of the eye. It contains about 1 billion light-sensitive receptor cells that, when stimulated by light, generate nerve impulses. **Rods** are receptor cells that are extremely sensitive to light and can detect various shades of gray even in dim light. However, they cannot distinguish colors, and because they do not detect edges well, they produce poorly defined images. **Cones** are receptor cells that detect color and are sensitive to edges, so they produce sharp images. The center of the retina contains a tiny pit densely packed with some 3 million cones. This area, called the fovea, produces the sharpest image, which is why we tend to move our eyes so that the image of an object we want to see clearly falls on this area.

A photoreceptor in your eye is able to detect a single photon of light. The primary sensing event of vision is the absorption of a photon of light by pigment complexes called rhodopsin (*roh DAHP sihn*) in rods and photopsins in cones. When photoreceptors absorb a photon of light, the pigment changes shape and initiates a chain of events that leads to the generation of an action potential. The nerve impulse travels along a short, thick nerve pathway called the optic nerve to the brain, which translates the information into a meaningful image.

In humans, the field of vision of the two eyes overlaps; each eye sees about one-third of what the other sees. The image that each eye sees of the same object, however, is

Did You Know?

Glaucoma is an eye disease in which increased pressure inside the eye causes damage to the retina and the optic nerve. The cause of the increased pressure is a buildup of the aqueous humor, which presses the lens inward.

Glaucoma is the second-most common cause of blindness.

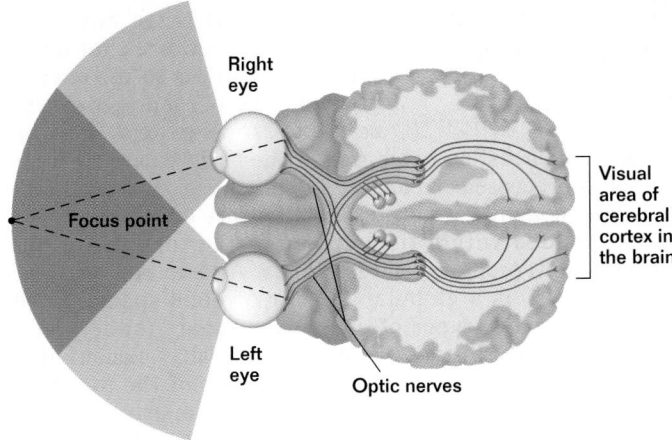

Right eye

Focus point

Left eye

Optic nerves

Visual area of cerebral cortex in the brain

Figure 40-11 Binocular vision results from the overlap of the visual fields from both eyes. In addition, approximately half the optic nerves from each eye cross over and provide each visual area of the cerebral cortex with information from each eye. Binocular vision provides a small visual field, but it enables us to perceive depth.

slightly different because the two eyes view the object from different angles, as shown in Figure 40-11. This slight displacement of images permits the brain to figure out how far away an object is. Interestingly, we are not born with this ability to perceive depth and distance. We learn it as babies by trial and error as the brain explores and remembers how the physical distance to an object compares with the difference between the images. ▢

▢ **CAPSULE SUMMARY**

The receptors for sensing light are rods and cones, which are located in the retina in the back of the eye.

Receptors Throughout the Body Sense Its Internal Condition

Sensory receptors inform the central nervous system about the internal conditions of your body. Much of this information passes to the hypothalamus to help maintain homeostasis—that is, to keep the body's internal environment constant. Table 40-1 lists several stimuli and describes the body's receptors for each and how they work.

Table 40-1 Sensory Receptors

Stimulus	Receptor	Location	Process
Temperature	Heat receptors and cold receptors	Skin, hypothalamus	Changes in temperature alter activity of ion channels
Pain	Nociceptors	Throughout all tissues and organs except the brain	Changes in temperature and pressure open membrane channels
Touch	Mechanoreceptors	Skin surface	Changes in pressure deform nerve
Muscle contraction	Stretch receptors	Wrapped around muscle fibers	Motion of muscle fiber changes nerve

❓ **Did You Know?**

We've all heard of the five traditional senses: hearing, smell, taste, sight, and touch. But in fact, these are misleading categories. Touch actually consists of hot, cold, pressure, and pain, all of which are separate senses with distinctly different receptors. Another sense, equilibrium or balance, is equally important.

The senses may be classified as general or special depending on the distribution of their sensory receptors. *Special senses* are those composed of either large, complex sensory organs such as the eyes and ears or those with small clusters of receptors such as the taste buds and olfactory (smell) organs.

Demonstration

The Nose Knows

Give a pair of students a blindfold and some slices of raw potato and apple. Have one student put on the blindfold while the other directs him or her to the two foods. The blindfolded student should try to identify the foods by tasting, first with nostrils pinched closed, then with nostrils open. Then have the pair switch roles. Repeat with as many pairs of students as want to participate. When all students have tasted the foods, discuss the results and the importance of smell to taste.

CONNECTIONS

Chapter 3

Shapes of Chemicals

Our perceptions of taste and smell depend on the three-dimensional shape of chemicals and their functional groups. Many of the tastes and smells we commonly associate with foods such as mint, bananas, pears, and oranges belong to a class of compounds called esters, each having a similar shape. The interaction of the chemical shape of alkaloids with certain taste buds creates a bit-ter taste, but simple sugars "fit" into other receptors, triggering a sweet sensation.

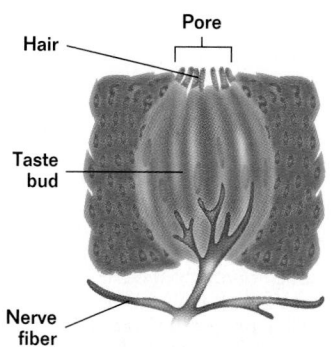

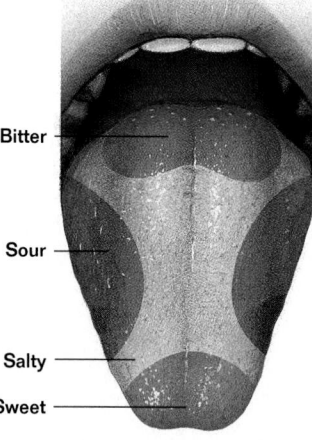

Figure 40-12 A taste bud, *above,* is a cluster of cells surrounding a pore. Several hairs protrude from the tips of the cells through the pore to the surface of the taste bud, where they are bathed in saliva. Taste buds are located all over the tongue, but they tend to be grouped into regions according to which "flavor" they sense, *below.*

Taste and Smell Are Chemical Senses

The senses of taste and smell depend on receptors that detect specific chemicals in the environment. In the presence of chemicals, these sensory neurons transmit information to an area of the brain where the information is processed and analyzed. These neurons are far more sensitive in many vertebrates than they are in humans.

Embedded within the surface of the tongue are over 10,000 taste receptors, or **taste buds.** A taste bud, shown in Figure 40-12, is a globular cluster of cells specialized to detect four basic types of chemicals: sugars (sweet), acids (sour), alkaloids (bitter), and metal ions (salty). Sensitivity to these chemicals varies on different parts of the tongue. In general, the tip of the tongue is most sensitive to sweet tastes, the sides to sour tastes, and the back to bitter tastes. Areas sensitive to salty tastes seem to be distributed evenly over the tongue. A taste bud is stimulated when a chemical dissolved in saliva binds to small hairs that protrude from the tip of the taste bud. The binding generates an action potential that travels along associated sensory neurons to the brain, where taste is perceived. The "hot" sensation of foods such as chili peppers is detected by pain receptors, not chemical receptors.

The receptors that sense smell are in a patch of membrane located in the roof of the nasal passage. There are approximately 5 million receptor cells in this area that respond to chemicals in a gaseous state. The sense of smell, in addition to the sense of taste, is very important in telling us about our food. That is why when you have a bad cold and your nose is stuffed up, your food seems to have little taste.

Section Review

1. *What are two functions of the inner ear?*
2. *Describe the mechanism in the inner ear responsible for equilibrium.*
3. *How are sound vibrations transmitted through the ear?*
4. *Trace the path of light as it passes through the eye. What roles do rods and cones play in the perception of light?*
5. *How are taste buds stimulated?*

Critical Thinking

6. *Describe a possible cause for colorblindness, a condition in which an individual cannot distinguish some colors from others.*
7. *Why is the brain an integral part of any sensation?*

Answers to Section Review

1. balance (equilibrium) and hearing

2. See page 950.

3. Sound waves are picked up by sound receptors in the cochlea. Membranes in the cochlea vibrate in response to the sound waves, making other tiny hairs move. This movement sends action potentials to the brain, which interprets them as sounds of varying pitch and magnitude.

4. Rhodopsin in a cone or photopsin in a rod absorbs a photon of light, changing the pigment's three-dimensional shape. This change initiates an action potential in the optic nerve, sending information to the brain, which translates the information into a meaningful image.

5. See page 954.

6. An individual who is color-blind may have a disorder that involves the normal functioning of the cones in their retina.

7. The brain is needed to inter-pret the information it receives from the sense organ.

40-4 Drugs and the Nervous System

We live in a drug-oriented society. Television commercials and other forms of advertising tell you about pain relievers, antacids, cough syrups, and other medications you can buy to help you feel better. Many drugs can prevent, treat, or cure illnesses that used to be deadly. However, when drugs are taken for nonmedical reasons drug abuse can occur. Few social problems in this country have had a greater impact on people's lives than the spreading abuse of addictive drugs.

Psychoactive Drugs Affect the Nervous System

In the broadest sense, a drug is a chemical that can alter biological functions and structures. Drugs most commonly abused are **psychoactive drugs,** substances that produce their effects on nervous system tissue and are often addictive. Scientists have recognized an important fact: addiction to psychoactive drugs is a physiological response that involves drug molecules and receptors in neuron membranes. Addiction is the body's attempt to cope with the chemical disruption a drug inflicts on a neuron's signaling systems.

To understand the nature of drug addiction, keep in mind how nerves communicate with one another because psychoactive drugs interfere with this process. Recall that neurotransmitters are released from the end of an axon, cross a synapse, and bind to receptor proteins on the adjacent neuron, triggering an impulse. Most psychoactive drugs produce their effects in the synapse. For example, some drugs have chemical structures similar to those of neurotransmitters. When a molecule of such a drug reaches the cell membrane of a neuron, it might be able to fit into a receptor protein for a particular neurotransmitter, thus causing the neuron to react as if a neurotransmitter were present. This process occurs with a number of drugs, such as morphine and heroin. Other drugs work by preventing neurotransmitters from being destroyed or recycled in the synapse, as shown in Figure 40-13.

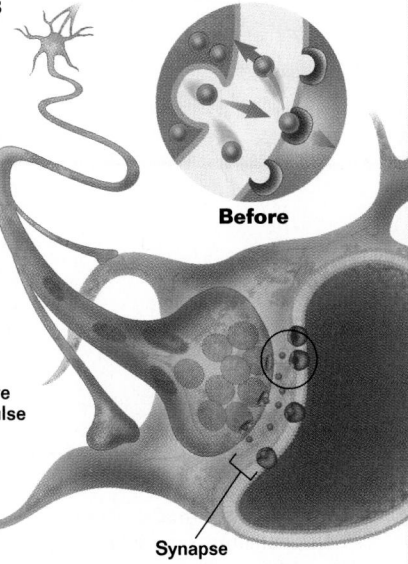

Serotonin

Nerve impulse

After

Before

Synapse

Figure 40-13 After the neurotransmitter serotonin carries a signal across a synapse, it is reabsorbed or broken down, *before*. When a drug that blocks the removal of a neurotransmitter such as Prozac® is taken, the neurotransmitter serotonin remains in the synapse, asserting its effects longer, *after*. This drug is helpful in treating depression, which can result from a shortage of serotonin.

Multicultural Perspective

Tobacco

The word *tobacco* is a Spanish adaptation of the term that the people of the Caribbean used for *cigar.* The majority of the people in South America used the term before Columbus arrived in the Americas. Tobacco was chewed or smoked for centuries to mark events such as war, peace, puberty, the harvest, or death. Tobacco was also burned as incense, sprinkled in leaf form, or buried with the dead. Most uses suggested the notion of sacrificial offerings.

Secular use of tobacco was also popular. It was used to relieve stress from work or war. It was also used to cure disease and as an anesthetic to treat wounds. Tobacco contains the physiologically active alkaloid nicotine.

Addiction Is Due to a Change in Receptor Protein Numbers

When a cell is exposed to a chemical signal for a prolonged period of time, it tends to lose its ability to respond to the stimulus with its original intensity. You are familiar with this loss of sensitivity—when you put on a wristwatch, how long are you aware you are wearing it? Neurons are particularly affected by this loss of sensitivity. If receptor proteins within synapses are exposed to high levels of neurotransmitter molecules for prolonged periods of time, a neuron will often respond by building and inserting fewer receptor proteins into the membrane. This feedback process is a normal part of the functions of all neurons. It is a mechanism that has evolved to adjust the number of receptor proteins in a cell to a certain level of neurotransmitter, enabling a cell to use its energy more efficiently.

Figure 40-14 shows an example of this mechanism, in which a kind of drug that prevents neurotransmitter uptake is present in a synapse. If this kind of drug is taken, large amounts of a neurotransmitter will remain in the synapses for a long time. In response to this surplus, a neuron will produce fewer receptor proteins for that neurotransmitter. In effect, the decreased number of receptor proteins creates a less sensitive neuron. If it didn't, the overabundance of neurotransmitter would cause a neuron to receive information over and over.

The decrease in production of receptor proteins results in addiction because a neuron cannot function normally unless the drug is present. When the drug is removed from the synapses, the neurotransmitter can once again be removed from the synapses. Suddenly there is no longer enough neurotransmitter to enable a neuron to receive information. The neuron cannot function normally until its number of receptor

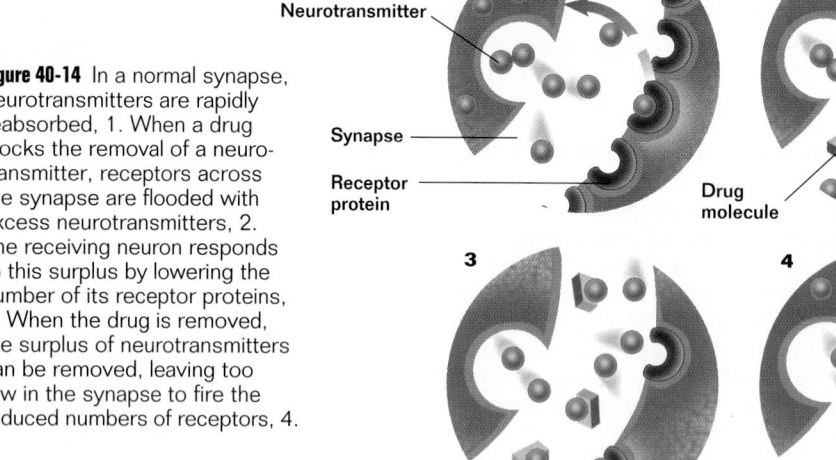

Figure 40-14 In a normal synapse, neurotransmitters are rapidly reabsorbed, 1. When a drug blocks the removal of a neurotransmitter, receptors across the synapse are flooded with excess neurotransmitters, 2. The receiving neuron responds to this surplus by lowering the number of its receptor proteins, 3. When the drug is removed, the surplus of neurotransmitters can be removed, leaving too few in the synapse to fire the reduced numbers of receptors, 4.

proteins have adjusted to the decrease in neurotransmitter. Sometimes in drug treatment programs, a drug is withdrawn slowly to allow the number of receptor proteins to increase gradually, minimizing withdrawal symptoms such as anxiety, depression, and cravings for the drug. As you can see, addiction is not simply a psychological state to be overcome by willpower; addiction is a physiological dependence caused by a change in a neuron's receptor proteins. ☐

❏ CAPSULE SUMMARY

Drugs can change the levels of neurotransmitter in a synapse. Addiction occurs when the number of receptor proteins in a neuron's membrane adjusts in response to these changes.

Narcotics and Cocaine Quickly Produce Addiction

Narcotics are powerful drugs that are used to relieve pain and induce sleep. They are **depressants,** substances that decrease the activity of the central nervous system. Many of the most potent narcotics are derived from chemicals extracted from *Papaver somniferum*, a species of the poppy plant, shown in Figure 40-15. The sap that oozes from the cut seed pod forms a thick, gummy substance called opium. Derivatives of opium are called opiates. Throughout history, the remarkable pain-relieving property of opium made it seem helpful for many illnesses. However, the price for this property is high—opiates produce addiction.

The major active ingredient in opium is morphine, a derivative that is about 10 times more potent than opium and highly addictive. Morphine is one of the most effective pain-relieving drugs known. It was widely used during the American Civil War to treat severely wounded soldiers, a practice that produced an addiction so common that it was often called "soldier's disease." A relatively simple chemical modification of morphine produces heroin, a powerful drug that is even more likely than morphine to produce addiction. Heroin abuse is among the most serious drug problems in society today. To understand how narcotics affect the nervous system and cause addiction, you first need to understand how the body perceives pain.

Narcotics Mimic Natural Painkillers

As uncomfortable as it may feel, pain plays a very important role in the body. It notifies you of damaged tissue and other injury. Imagine how your body would look and function today if you did not have the ability to sense pain. Pain begins as a signal at damaged nerve endings and travels up the spinal cord to the brain. After reaching the brain, a pain signal is shut off when the central nervous system secretes a class of neurotransmitter called **enkephalins** (*ihn KEHF uh lihnz*). Enkephalins are one of several kinds of natural pain relievers your body releases in response to pain and stress. When enkephalins bind to receptor proteins in spinal neurons, potassium channels open, inhibiting pain messages from traveling to the brain.

Figure 40-15 Opium is derived from the poppy *Papaver somniferum.* When scored with a knife, the seed pod oozes a milky white sap. After drying, the sap forms thick, gummy, brown opium.

Teaching Tips
Cocaine and Depression

Discuss with students the reasons why cocaine addiction is so severe and difficult to treat. Explain to them that cocaine addiction causes a severe craving that is notoriously difficult to supervise or treat. Addicts become unable to experience pleasure without more and more cocaine and often experience deep depression. Anxieties increase as sleeplessness and physical deterioration set in.

Treatment of cocaine addicts with drugs has had only moderate success. Some of the drugs stimulate the dopamine receptors to give addicts some relief from their depression, but these drugs do not replace the "high" produced by the cocaine. One promising treatment combines a craving-suppressing drug with amino acid therapy. The amino acids promote dopamine synthesis to replace the addict's low supply.

Opium

Have students conduct library research to identify regions of the world in which opium is grown. Also have them determine what role opium plays in each nation's economy. Ask students to summarize the information they find in a brief report.

☐ CAPSULE SUMMARY

Narcotics are highly addictive painkillers. They have a molecular structure similar to that of enkephalins, a class of neurotransmitters that work to block pain messages from reaching the brain.

Narcotics function by imitating enkephalins. Their similar molecular structure makes it easy for them to bind to the receptor proteins for enkephalins. These enkephalin receptor proteins are called opiate receptors because scientists observed opiates binding to them before enkephalins were ever discovered. Morphine and heroin are potent pain-blocking drugs because they act to block pain signals traveling up the spine to the brain. Narcotics also interact with the brain's limbic system, the "pleasure center," producing a perception of intense well-being. With prolonged use, neurons adjust their internal chemistry to make themselves less sensitive to stimulation. The neurons become so desensitized that the number of opiate receptors may actually increase.

What happens when an addict stops taking a narcotic? Removing the inhibitory action of the narcotic causes the neuron to become extremely sensitive. Unpleasant symptoms result, including extreme anxiety, tremors, and heightened sensitivity to pain. Sensitivity to pain results because the changes in potassium ion levels caused by enkephalins are no longer enough to inhibit the spinal nerves. As a result, pain signals travel freely up the spine, making the body very receptive to pain. The narcotic is required in significant amounts just to make the person's body feel normal. ☐

Cocaine Is a Stimulant

Unlike narcotics, cocaine is a **stimulant,** a substance that excites the central nervous system and speeds up body processes. Cocaine is found in the leaves of coca plants that grow at high altitudes in the mountains of South America. Even though many South Americans had chewed the coca leaf for centuries, the process for extracting cocaine was not developed until the mid-1800s. Many physicians at first con-

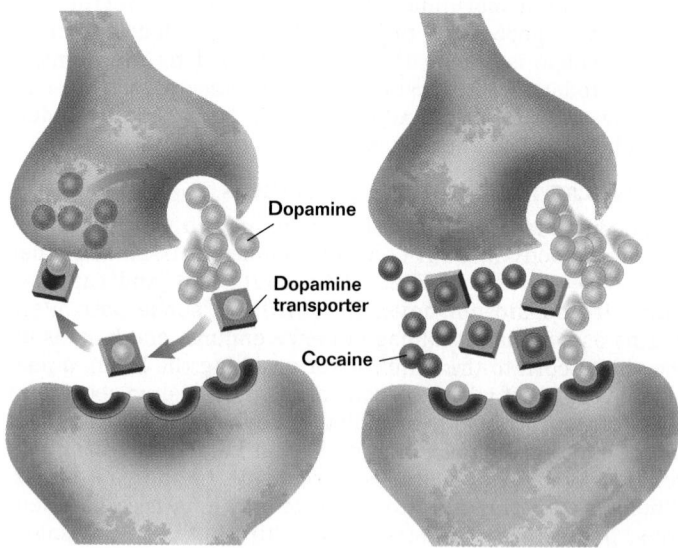

Figure 40-16 Dopamine is a neurotransmitter that helps send pleasure messages to the brain. Cocaine binds with dopamine transporter to block the reabsorption of dopamine so that it remains in the synapse longer.

sidered it a miracle drug and prescribed it for all sorts of physical and mental ailments; it was even added to soft drinks. However, cocaine is highly addictive. Today, laws in the United States forbid the importation, manufacture, and use of cocaine for nonmedical purposes, and medical use is extremely limited.

Despite being illegal and highly addictive, cocaine is still abused by many people. They risk becoming addicted because of the changes cocaine makes in the brain. As shown in Figure 40-16, cocaine works by preventing the reabsorption of dopamine, a neurotransmitter that stimulates the limbic system. The trapped dopamine repeatedly stimulates neurons, producing an intense feeling of well-being. Over a short period of time, neurons adjust to the presence of cocaine by decreasing their number of dopamine receptors. As a result, the neurons in the limbic system become less sensitive. As more time passes, more and more cocaine is needed to stimulate the limbic system. As you can see, cocaine addiction is a result of changes in the number of receptor proteins as neurons desensitize in response to the drug. Even more potent and addictive than cocaine is crack, a smokable form of the drug. When crack is smoked, it is absorbed very rapidly in the lungs and produces addiction very quickly. ❑

Dangerous Social Drugs

Not all dangerous drugs are illegal. More deaths and suffering are caused by familiar, legal drugs such as nicotine and alcohol. **Nicotine** is a highly addictive stimulant found in the leaves of the tobacco plant. Tobacco has many other constituents, but nicotine has the broadest and most immediate effects in the body. The other substances in tobacco, mostly tars, are highly mutagenic, causing changes in DNA that can lead to cancer.

Nicotine is extremely toxic; only 60 mg is lethal in humans. When a smoker inhales a cigarette, the nicotine in the tobacco reaches the brain within 10 seconds. Nicotine binds to specific receptors in the brain that are similar to those for acetylcholine (as ee tihl KOH leen), a neurotransmitter that stimulates skeletal muscles. The binding of these receptors produces a mild, short-lived pleasurable sensation. Nicotine addiction is similar to cocaine addiction. Neurons in the brain adjust to prolonged exposure to nicotine by making fewer receptor proteins, and eventually the body requires nicotine to maintain a "normal" feeling.

You do not have to smoke tobacco to feel the effects of nicotine. Nicotine is also absorbed by the body from smokeless tobacco products, such as chewing tobacco and snuff. These tobacco products are just as addictive as cigarettes. In addition, the tars in chewing tobacco and snuff lead to

❑ CAPSULE SUMMARY

Cocaine blocks the reabsorption of dopamine, a neurotransmitter that stimulates the pleasure center in the brain.

🔍 Did You Know?

Nicotine Is a Powerful Poison
As the text says, nicotine is extremely toxic. In fact, a sulfate derivative of nicotine has been used as a commercial insecticide for years. Nicotine's effects in the lower dosages experienced by smokers are not immediately noticeable as toxic, however.

Initially, nicotine elevates blood pressure by constricting blood vessels and increasing the heart rate. It stimulates the release of free fatty acids into the bloodstream. The added carbon monoxide in the blood causes these fatty acids to be deposited on the inside of the arteries and thus is a major contributor to cardiovascular disease. The combined effect of the strain of cardiovascular disease and the lack of oxygen contributes heavily to the damage seen in smokers' hearts.

Teaching Tip
Alcoholic Beverages

Point out that beer, wine, and liquor differ in their ethanol concentrations. However, a 12 oz. can of beer, a 4 oz. glass of wine, and a 1 oz. shot of liquor all contain about the same amount of ethanol. Tell students that "proof" reflects the alcohol concentration in a beverage. Alcohol content is determined by dividing the proof number in half—80 proof liquor has a 40 percent alcohol content.

CONNECTIONS

Chapter 5
Fermentation
Review how yeast and some other microorganisms, in the absence of oxygen, convert organic compounds such as glucose into ethyl alcohol. Ask students what glucose is converted into by human muscle cells in the absence of oxygen. (*lactic acid*)

CAPSULE SUMMARY

Nicotine acts like a stimulant because it binds to receptor proteins similar to those for acetylcholine, a neurotransmitter that stimulates skeletal muscles.

CAPSULE SUMMARY

Alcohol alters neuron membrane structure, which changes the shape of receptor proteins.

increased risk of mouth and throat cancer. Chewing tobacco instead of smoking it simply trades one form of deadly cancer for another.

It was once thought that cigarette smokers were only harming themselves. However, scientists now know that just being in the same vicinity of people smoking will introduce nicotine, carbon monoxide, and other elements of tobacco smoke into your body. A study of smokers and nonsmokers confined in airplanes revealed that blood samples from nonsmokers contained high levels of nicotine. And a 1992 study of women who died in accidents revealed that almost all of those married to smokers had precancerous lesions in their lungs. Those women married to nonsmokers did not. For these reasons and others, smoking is banned from many public places. ◻

Alcohol

Of all the psychoactive drugs, alcohol (ethanol) is one of the most widely used and abused. Consumed for centuries as wine from fermented grapes or as beer from fermented grain, alcohol reduces inhibitions and produces a sense of well-being. Unfortunately, consuming alcohol can alter both judgment and reaction time, a condition known as being "drunk." Every year, many high school students die in automobile accidents resulting from drunken driving, making alcohol a truly dangerous drug.

Alcohol is an unusual psychoactive drug. Unlike narcotics and cocaine, it has no receptor. Instead, alcohol is able to alter the structure of the membrane bilayer, producing changes in the shape of receptor proteins. An altered receptor protein may become more sensitive to a stimulus or may be, in effect, switched off. As you can imagine, each change in a receptor protein may have complex effects on normal brain functions. Alcohol also inhibits the nerves that repress the limbic system, causing feelings of pleasure and inhibition.

Addiction to alcohol, or alcoholism, is the major drug abuse problem in the United States. People who drink excessive amounts of alcohol over long periods of time develop serious health problems. For example, many alcoholics suffer from a lack of vitamins because they do not eat properly when drinking heavily. This can lead to malnutrition, abnormalities in the circulatory system, and inflammation in the stomach lining. In addition, alcohol is readily converted to energy, preventing the body from breaking down other nutrients, such as sugars, amino acids, and fatty acids. These nutrients are stored as fat in the liver. After several years of drinking, liver cells are filled with fat and begin to die. If heavy drinking continues, a liver condition called **cirrhosis** may develop. In cirrhosis, liver cells are replaced with useless scar tissue, and the liver gradually shrinks into a small, hard mass. In this form, the liver can no longer eliminate body wastes, produce blood clotting, or carry out its other functions. ◻

Overcoming Misconceptions

Marijuana and Safety
A common misconception about marijuana is that it is less harmful than cigarettes. The fact is, that all smoke is harmful due to the tars it contains. Likewise, all smoke contains carbon monoxide. Carbon monoxide lingers in the blood, reducing its oxygen-carrying capacity for hours. A marijuana smoker's explosive cough after a "hit" would seem to be the best indication of its effect on the respiratory system: marijuana smoke is extremely irritating.

As for other, more insidious effects, marijuana use can cause males to produce highly abnormal sperm. Additionally, toxic chemicals carried in a man's semen can be transferred to a woman, increasing the risk of birth defects, miscarriages, and stillbirths.

Psychoactive drugs include a variety of other substances. The caffeine found in coffee and soft drinks is a psychoactive drug, as is the drug in marijuana. Table 40-2 lists and describes additional psychoactive drugs that are commonly abused and that can lead to many serious health problems.

Table 40-2 Additional Psychoactive Drugs

Drug	Examples	Effects	Risks Associated With Abuse
Depressants	Barbiturates, tranquilizers, methaqualone, phencyclidine hydrochloride (PCP)	Slow down the action of the central nervous system	Drowsiness, depression, emotional instability
Stimulants	Amphetamines, caffeine	Speed up the central nervous system, metabolism, blood pressure, heartbeat, and respiratory rate	Irregular heartbeat, high blood pressure, headaches, stomach disorders, exhaustion, violent behavior
Inhalants	Nitrous oxide, ether, paint thinners, glue, cleaning fluids, correction fluids	Disorientation, confusion, memory loss	Hallucinations; permanent damage to brain, kidneys, liver; death
Hallucinogens	Lysergic acid diethylamide (LSD), mescaline, peyote	Distortion in the way the brain translates impulses	Dangerous hallucinations, unpredictable behavior
Marijuana	Derived from dried leaves, flowers, and stems of *Cannabis sativum*	Wide range of effects; short-term memory loss, disorientation, impaired judgment	Lung damage, loss of motivation

Section Review

1. *What is a psychoactive drug?*
2. *What role do enkephalins play in the body?*
3. *Why are narcotics so addictive?*
4. *How does cocaine addiction differ from heroin addiction?*
5. *How does nicotine affect the nervous system?*
6. *How does alcohol differ from other drugs in its effect on the nervous system?*

Critical Thinking

7. *Why is addiction defined as a physiological response?*

SECTION 40-4

 VISUAL STRATEGY

Table 40-2
Review the table with students. Encourage questions and discussion, especially about responsible behavior.

Chapter Closure

Assign three-page research papers that explore a disorder of the nervous system. Students should use references other than their textbook. Topics can include drug effects, metabolic disorders, genetic disorders, injuries, and infectious diseases. Each paper should describe the pertinent aspects of normal functioning of the nervous system and then describe the disorder, its causes and symptoms, and its treatments. Students can also make oral presentations of their papers.

Answers to Section Review

1. A psychoactive drug is a substance that produces its effect on the nervous system.

2. Enkephalins are natural pain relievers.

3. Narcotics are so addictive because they mimic the body's natural painkillers. Their effects on the limbic system make addiction to them especially difficult to treat.

4. Cocaine addiction is a result of the inhibition of the normal reabsorption of dopamine, while heroin (and all opiate) addiction is a result of the binding of heroin to opiate receptors on neurons.

5. Nicotine acts as a nervous system stimulant.

6. Alcohol has a generalized effect on all neurons, not just ones with specific receptors to the drug.

7. Addiction is defined as a physiological response because it interferes with the biological functioning of cells.

CHAPTER REVIEW ANSWERS

Each item in the Chapter Review is correlated by text section in the assignment guide that follows.

ASSIGNMENT GUIDE

Section	Review	Themes Review	Critical Thinking
4-1	1–3, 9	13, 14	16
4-2	4, 5, 10	15	17
4-3	6, 11		
4-4	7, 8, 12		

Review

Multiple Choice

Code in parentheses indicates section number and objective number.

1. c (40-1.2)
2. a (40-1.3)
3. d (40-1.4)
4. b (40-2.2)
5. b (40-2.4)
6. a (40-3.2)
7. c (40-4.1)
8. a (40-4.2)

Completion

9. synapse, neurotransmitter (40-1.1)
10. autonomic, stress (40-2.5)
11. cones, rods (40-3.2)
12. nicotine, alcohol, acetylcholine, limbic (40-4.5)

40 CHAPTER REVIEW

Vocabulary

action potential (941)
autonomic nervous system (944)
axon (939)
brain stem (946)
cell body (939)
cerebellum (946)
cerebral cortex (945)
cerebrum (945)
cirrhosis (960)
cochlea (951)
cone (952)
cornea (951)
dendrite (939)
depolarization (939)
depressant (957)
enkephalin (957)

hypothalamus (946)
integration (943)
iris (951)
lens (951)
limbic system (946)
motor nerve (944)
multiple sclerosis (940)
myelin sheath (940)
nerve (940)
neurotransmitter (942)
nicotine (959)
parasympathetic division (949)
psychoactive drug (955)
pupil (951)
reflex (948)
repolarization (941)

resting potential (941)
retina (952)
rods (952)
Schwann cell (940)
sensory nerve (944)
sensory organ (950)
sensory receptor (950)
somatic nervous system (944)
spinal cord (946)
stimulant (958)
sympathetic division (949)
synapse (942)
taste bud (954)
thalamus (945)

Review

Multiple Choice

1. Compared with the rate at which nerve impulses travel down a myelinated axon, the rate of travel down an unmyelinated axon is
   a. faster.   c. slower.
   b. much faster.   d. no different.

2. Which of the following describes the condition in which there is both a high concentration of sodium ions outside the membrane of the neuron and a high concentration of potassium ions inside the membrane?
   a. resting potential
   b. action state
   c. synapse
   d. neurotransmission

3. Which of the following describes an action potential?
   a. sodium ions flood into a neuron
   b. sodium ions diffuse out of a neuron
   c. potassium ions flood into a neuron
   d. potassium ions reach their resting potential

4. Which of the following might result from damage to the cerebellum?
   a. insatiable thirst
   b. loss of balance
   c. indigestion
   d. dilation of blood vessels

5. A reflex is a very rapid response because the nerve impulse
   a. is passed along an unmyelinated axon.
   b. never reaches the brain.
   c. increases heart rate.
   d. involves only motor neurons.

6. Identify the pathway of light from the environment to the retina.
   a. → cornea → lens → receptor cells
   b. → receptor cells → cornea → lens
   c. → lens → receptor cells → cornea
   d. → lens → cornea → receptor cells

7. Addiction to most psychoactive drugs is considered to be a physiological response because the drugs
   a. can be purchased illegally.
   b. cause long-term suffering among addicts, family, and friends.
   c. affect the operation of neurotransmitters in the synapse.
   d. cause hallucinations, depression, and brain damage.

8. What is the function of enkephalins?
   a. pain relief
   b. increasing the flow of blood to infected areas
   c. restoration of equilibrium
   d. alleviate depression

Completion

9. The small space between an axon of one neuron and a dendrite of another neuron is the _____ . Impulses are carried across this space by chemical messengers called _____ .

10. The _____ nervous system is subdivided into the sympathetic and the parasympathetic nervous system. The sympathetic system dominates in times of _____ , and the parasympathic system dominates in times of rest.

11. Photoreceptor cells of the eye that detect color and produce sharp images are called _____ . Photoreceptor cells that are especially light-sensitive while producing poorly defined images are called _____ .

12. Two legal drugs considered potentially dangerous are _____ and _____ . One of these is found in tobacco and interacts with the _____ receptors on neurons. The other drug inhibits the nerves that repress the _____ system.

Themes Review

13. **Evolution** Many of the axons of humans are wrapped with myelin sheaths. Why are myelinated axons considered an evolutionary advancement?

14. **Structure and Function** Useful information travels in only one direction along a neuron: toward the central nervous system in sensory neurons and away from the central nervous system in motor neurons. What structures of the neuron ensure that this pattern is always followed?

15. **Levels of Organization** The nervous system is divided into various subunits. The two motor subunits that make up the peripheral nervous system are the somatic system and the autonomic system. How do the functions performed by the motor nerves of these two systems compare?

Critical Thinking

16. **Making Inferences** For some people too little potassium in the bloodstream is a serious medical problem. Symptoms of this condition include an increased heart rate and headaches. How might you explain these symptoms?

17. **Interpreting Data** Look at the diagram of the brain of a goose. Using what you know about the human brain, what would you assume the size of the portions of this brain indicate about their relative importance?

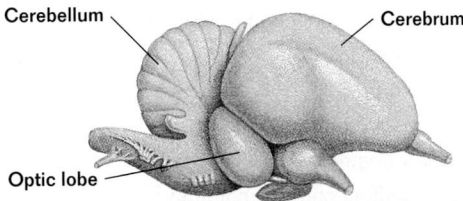

Cerebellum
Cerebrum
Optic lobe

Activities and Projects

18. **Research and Writing** A number of potentially addictive substances, including opiates and cocaine, have now been linked to specific receptors in the brain. Find out how the work of neurochemist Solomon H. Snyder led to a greater understanding of how potentially addictive substances affect the operation of receptors in the brain. In your report, be sure to describe the evidence that led Snyder to infer the existence of endorphins, the naturally occurring opiate-like substances in brain tissue.

CHAPTER 40

Themes Review

13. The myelin sheath increases the conduction rate of nerve impulses, enabling some animals to respond quicker to their environment. (40-1.2)

14. Only an axon terminal is able to release a neurotransmitter. Neurotransmitters move in only one direction across a synapse, from axon terminal to receptor protein. (40-1.5)

15. The motor nerves of the somatic nervous system conduct impulses to skeletal muscles that are under conscious control. In contrast, the motor nerves of the autonomic nervous system regulate the activity involuntary muscles (cardiac muscles, smooth muscles, and glands). (40-2.1, 40-2.3)

Critical Thinking

16. Potassium is necessary for the nerves to conduct impulses. If the potassium level is low, impulses may not be easily inhibited and, in turn, may speed up the functioning of nerves in the brain, heart, and other organs. (40-1.3, 40-1.4)

17. Answers will vary. The size of the cerebellum indicates that the movements of the goose are intricate. The relatively small size of the cerebrum suggests that higher thought processes are not greatly developed. The optic lobe of the goose is very large, indicating that eyesight is important. (40-2.2)

Activities and Projects

18. Snyder found receptors in the brain for opiates and inferred that the brain produces opiate-like substances that bind to these receptors. Later tests by Snyder and colleagues found endorphins in brain tissue.

*L*ABORATORY Investigation Chapter 40

Neuron Model

Procedural Notes

- This investigation should take 40 minutes to complete.
- When constructing the model, do *not* allow the alligator clips to touch the NaCl solution.
- **Safety:** Avoid eye contact with sodium chloride and potassium chloride. If contact does occur, flush eyes with running water for 15 minutes, including under the eyelids.
- Allow the dialysis tubing to remain in the NaCl solution for 1 minute or so to build up a potential. When the KCl begins to diffuse out of the tubing, the potential should rise to about 8 mV.

Preparation

Obtain enough dialysis tubing (available from a pharmacy or medical supply store), nickel-chromium alloy wire, and copper wire (both available from a hardware store) for each lab group to construct a neuron model. Each group of students constructing a neuron model will need a 20 cm strip of dialysis tubing, two 15 cm strips of nickel-chromium alloy wire, and two 40 cm pieces of copper wire. You can purchase 3 M sodium chloride and 3 M potassium chloride solutions from Ward's. Sodium chloride and potassium chloride are both strong eye irritants. Students should wear goggles throughout this investigation and keep the sodium chloride and potassium chloride solutions away from their faces. Students will need to use a DC millivoltmeter that measures 8–10 mV and deflects both negative and positive charges to measure the resting potential of their models. Lab groups will also need scissors, two 150 mL beakers, two glass stirring rods, two alligator clips, a screwdriver, string, and an elastic band for the model.

Unlike an actual neuron, the neuron model in this investigation does not have a sodium-potassium pump to maintain the ion distribution. If left standing, the concentrations of ions in the neuron model will become

OBJECTIVES

- Construct a model to simulate the electrical communication between nerve cells.
- Observe the distribution of ions in a model neuron.

PROCESS SKILLS

- assembling a model
- measuring voltage

MATERIALS

- dialysis tubing
- scissors
- 150 mL beakers (2)
- 15 cm pieces of nickel-chromium alloy wire (2)
- glass stirring rods (2)
- 40 cm pieces of copper wire (2)
- alligator clips (2)
- screwdriver
- DC millivoltmeter
- safety goggles
- 3 M sodium chloride solution
- 3 M potassium chloride solution
- string
- elastic band

BACKGROUND

1. How do nerve cells communicate with each other?
2. Where does the electricity in nerve cells come from?
3. How is the resting potential of nerve cells like a live battery?
4. What is an action potential?
5. Write your own question to explore in your lab report or notebook.

TECHNIQUE

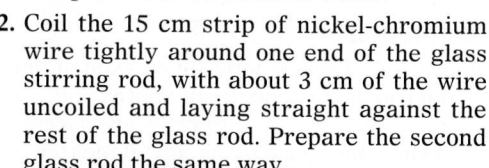

1. **CAUTION: Be careful when using scissors.** Cut a 20 cm strip of dialysis tubing and place it in a beaker of water.
2. Coil the 15 cm strip of nickel-chromium wire tightly around one end of the glass stirring rod, with about 3 cm of the wire uncoiled and laying straight against the rest of the glass rod. Prepare the second glass rod the same way.
3. Pull the copper wire through the hollow end of the alligator clip, loop it around the screw, and tighten the screw with a screwdriver. Repeat with the second clip and copper wire.
4. Clip each alligator clip to each stirring rod with the clip touching the uncoiled part of the nickel-chromium wire, as shown below.

Nickel-chromium wire   Cu++ wire   Stirring rod   Alligator clip

5. Attach the copper wire of one clip to the positive terminal of the meter and the copper wire of the other clip to the negative terminal of the meter by twisting the end of the copper wire around the terminal post and tightening the nut.
6. **CAUTION: Sodium chloride and potassium chloride solutions are strong eye irritants; avoid eye contact and wear safety goggles for the remainder of the procedure.** Fill a 150 mL beaker with 75 mL of 3 M sodium chloride (NaCl) solution.

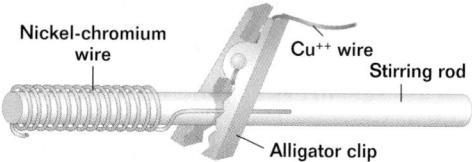

7. Remove the dialysis tubing from the beaker and tie off one end of it with string to make a sack. Fill this sack halfway with 3 M potassium chloride (KCl) solution. This sack is the model of a neuron.

---

equal on both sides of the tubing, causing the resting potential to disappear.

Background Answers

1. Nerve cells communicate by releasing chemicals called neurotransmitters into their synapses (gaps between neurons).

2. The electricity comes from the ions of sodium and potassium.

3. Sodium-potassium pumps transport ions across the plasma membrane of the neuron, causing a difference in charge on either side of the membrane (resting potential).

4. When the neuron is stimulated to open sodium channels,

sodium ions flood into the cell, causing for a short time a reversal in the difference in charge across the membrane. This temporary reversal in charge is called an action potential.

5. How does a nerve cell, represented by a model, develop a resting potential?

8. Place the stirring rod of the apparatus attached to the negative terminal into the potassium chloride solution in the sack and use the elastic band to close the open end of the tubing around the stirring rod as you see in the drawing below. **DO NOT** let any wires touch each other.

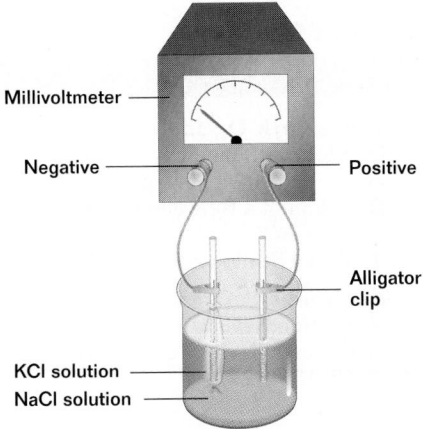

Millivoltmeter
Negative
Positive
Alligator clip
KCl solution
NaCl solution

9. The sodium chloride solution represents the fluid surrounding a neuron when it is not transmitting an impulse. Certain ions can pass through the dialysis tubing (neuron cell membrane) much more easily than others, so a difference in net charge develops across the membrane. This difference in charge, or voltage, is measured in volts or millivolts. Observe the millivoltmeter reading and record it in the **Records** section of your report. This is the resting potential of your model neuron. In your report, briefly summarize the procedure you followed.

10. Clean up your materials and wash your hands before leaving the lab.

INQUIRY

1. What reading was recorded from the voltmeter?

2. Compare your reading with the other lab groups. Were they the same?

ANALYSIS

1. Because the millivoltmeter indicator moves in the direction of the flowing electrons, which way with respect to the potassium chloride solution are electrons flowing in the model?

2. How does the charge on the inside of the nerve cell model compare to the charge on the outside?

3. The potassium (K^+), chloride (Cl^-), and sodium (Na^+) ions are all moving, but perhaps with different ease through the dialysis tubing. Which must be moving more easily to create a net negative charge inside the tubing (neuron)?

4. How is this set up like an unused battery?

5. Why are the sodium and potassium ions rather than the chloride ions responsible for the difference in charge?

FURTHER INQUIRY

Write a new question that could be explored as a new investigation. The following is an example:

How do the concentrations of the sodium chloride and potassium chloride solutions affect the amount of voltage made by the model? Is this important to the study of a functioning neuron?

Inquiry Answers

1. Answers will vary. The potential should rise to about 8 mV.

2. Answers will vary.

Analysis Answers

1. Electrons tend to flow away from the potassium chloride solution.

2. The nerve cell model has a negative charge on the inside and a positive charge on the outside.

3. Sodium ions do not move through the tubing as easily as the potassium ions do. Therefore, more positive ions leave than enter, resulting in a negative net charge on the inside.

4. The neuron not in use still has a resting potential; otherwise it could not transmit impulses. It has to be ready to function.

5. There is no net movement of chloride ions because the chloride concentration is at equilibrium on both sides of the membrane.

Records

Answers will vary but should approximate 8 mV.

Block Scheduling Guide

Each block represents about 45–50 minutes of instructional time. The resources cited in a block are coded to a specific program component using the Key on the next page.

Lecture Concepts	Lesson Resources

41-1 Hormones pp. 967–972

BLOCK 1

a. Hormones regulate the body's ongoing activities. They are slower-acting but longer-lasting than nerve signals.
b. The ductless glands called endocrine glands produce most of the body's hormones.
c. The hypothalamus regulates many activities of the endocrine system by producing hormones that stimulate the pituitary gland to release its hormones.

Lecture Resources
■ Opening Question, page 967
■ Demonstration: The Flight-or-Flight Response
■ Visual Strategies: Figures 41-1, 41-2, 41-3, 41-5
■ Overcoming Misconceptions, page 969
▢ Teaching Transparencies: 204, 205, 209, 229, 230
◉ Holt Biology Videodiscs: Lesson 41-1

Classwork Options
■ Effective Reading, page 967
■ Molecular Models, page 969
■ Action of a Releasing Factor, page 971
■ Visual Strategy: Figure 41-6

Assignment Options
▪ Section Review, page 972
▪ Chapter Review, questions 1–3, 11, 12, 18, 21, 22

41-2 How Hormones Work pp. 973–976

BLOCK 2

a. Peptide hormones work from outside a cell by binding to a receptor protein on the membrane's surface. This typically causes the production of a second messenger that alters the enzymes produced by the cell.
b. Steroid hormones work by passing through a cell's plasma membrane and binding to protein receptors located in the cytoplasm. The hormone-receptor complex then enters the nucleus and attaches to DNA, altering the cell's production of enzymes.

Lecture Resources
■ Opening Question, page 973
■ Demonstration: Solubility of Proteins and Lipids
■ Visual Strategies: Figures 41-8, 41-9, 41-10
▢ Teaching Transparencies: 206, 207
◉ Holt Biology Videodiscs: Lesson 41-2

Classwork Options
■ Modes of Hormonal Action, page 974

Assignment Options
▪ Section Review, page 976
▪ Chapter Review, questions 4–6, 13

41-3 The Body's Endocrine Glands pp. 977–985

BLOCK 3

a. The pituitary gland controls many of the body's other glands.
b. The thyroid gland releases thyroxine (regulates growth and metabolism) and calcitonin (regulates calcium in the body).
c. The parathyroid gland produces parathyroid hormone, which plays a vital role in regulating the body's blood-calcium supply.
d. The adrenal glands produce steroid hormones and the "fight-or-flight" hormones.
e. The pancreas has cells that secrete insulin, a hormone that enables cells in the body to absorb glucose from the blood.

Lecture Resources
■ Opening Question, page 977
■ Demonstration: Locations of the Endocrine Glands
■ Visual Strategies: Figures 41-16, 41-18, 41-19, 41-20
■ Research Update, page 984
■ Problem Solving, page 985
▢ Teaching Transparencies: 208, 209, 194A, 195A
◉ Holt Biology Videodiscs: Lesson 40-3

Classwork Options
■ Problem Solving, page 979
■ Antagonistic Hormones, page 983
■ Closure, page 985
▪ Laboratory Investigation: Effect of Thyroxine on Frog Metamorphosis, pages 988–989

Assignment Options
▪ Section Review, page 985
▪ Chapter Review, questions 7–10, 14–22

Review/Enrichment

BLOCK ④

■ Study Guide: Concept Review and Pretest

Assignment Options

■ Activities and Projects, question 23, 24

Assessment Options

BLOCK ⑤

Traditional Assessment

■ Chapter 41 Test

⊙ Test Generator/Assessment Item Listing: Software item bank for preparing customized chapter tests.

Portfolio Assessment

Select student reports for one or more laboratory experiments from this chapter. *The Direct Observation Checklist* on the *BioSources Teaching Resources CD-ROM* should be completed during a laboratory or other cooperative learning experience.

Holt Biology Videodiscs

Holt Biology Videodiscs gives you a powerful tool for teaching, review, and assessment. *Concepts of Biology* adds interactive lesson plans and extensive tools for customization using CD-ROM technology.

CONCEPTS OF BIOLOGY

Use the following topic from *Concepts of Biology* to help you teach this chapter:

■ Topic 24: Nervous and Endocrine Systems

For further information regarding the media on the videodiscs, see *Holt Biology Videodiscs Teacher's Correlation Guide to Biology: Principles and Explorations.*

Hormones and the Endocrine System **965B**

Chapter Theme

Homeostasis *The dynamic balance between the body's trillions of cells depends on a complex control system. The close association between the nervous system and the endocrine system enables an organism to capitalize on the advantages of each to maintain homeostasis. Figures 41-1, 41-6, 41-14, 41-16, and 41-18 illustrate this theme.*

Tapping Prior Knowledge

- What is a hormone?
- What mechanism of surveillance and response maintains homeostasis in the body?
- What type of feedback does the body use most?

Opening Demonstration

To demonstrate how hormones affect the body, have students record their pulse and breathing rate a few minutes into the class period. Then announce and administer a pop quiz or some other unexpected and stressful assignment. Immediately after the assignment is completed, have students again record their pulse and breathing rate. Ask: How did your pulse and breathing rate differ before and after the assignment? *(Both should be elevated following the assignment.)* Tell students that hormones produced by the endocrine system are responsible for the changes that occurred in their pulse and breathing rate. Ask: What do an elevated pulse and breathing rate do for your body? *(help it take in and deliver more oxygen to the cells so that you can produce more ATP)*

CHAPTER 41 HORMONES AND THE ENDOCRINE SYSTEM

Speed skater ready to race

REVIEW

- action of enzymes (Section 2-3)
- role of DNA and mRNA in protein synthesis (Section 9-1)
- negative feedback (Section 35-3)
- *Highlights: Endocrine System* (p. 846)
- nature of nerve messages (Section 40-1)
- hypothalamus (Section 40-2)

Authors' Rationale

Hormones are involved in the regulation of almost all human systems. The endocrine system assists the nervous system in responding to the environment—in a slower, more sustained way. This chapter emphasizes how hormones work to regulate body functions and maintain homeostasis.

41-1 Hormones

*T*he tissues and organs of your body carry out a multitude of activities. Activities such as those in Figure 41-1 must be coordinated to prevent them from conflicting with one another. This coordination, plus a smoothly functioning set of organs, is what truly makes an organism alive. As you learned in Chapter 35, the central nervous system coordinates the body's activities. Fast-acting but short-lived nerve messages are sent by electrical signals that travel along neurons, much as telephone messages travel along wires. For a longer-lasting effect, the central nervous system signals the release of hormones—chemical messengers that are produced in one place and transported to another by the bloodstream.

Figure 41-1 Combining two activities, such as walking and eating an apple, requires the coordination of many body processes. Such coordination is maintained by hormones, the messengers of the endocrine system.

Hormones Have Advantages Over Nerve Signals

The hormones produced by the endocrine system are the body's chief means of regulating its ongoing activities. Although hormones are slow-acting, their effects tend to persist for a long time. In addition to their longer-lasting effects, hormones have several other distinct advantages over the electrical signals transmitted by nerves.

1. **Scope** For an electrical signal to be effective, it must be transmitted directly to an individual cell through a neuron. To maintain the homeostasis of a group of tissues, organs, or organ systems, electrical signals would have to reach all the cells involved. This would require an enormous number of nerves—far too many to be practical. By contrast, hormones readily spread to all cells and tissues via the blood.

VISUAL STRATEGY

Figure 41-2

Walk students through the diagram, discussing how the two systems compare in their scope, specificity, and flexibility. Ask: What other system is essential for the endocrine system to carry out its function? *(the circulatory system)* Why?*(It delivers the hormones produced by the endocrine system to their widely ranging targets.)*

Teaching Tip

What If a nerve message were to take the wrong path through a nerve?

Sometimes nerve messages intended for one part of the body are routed to another part of the body. Something similar happens in people with multiple sclerosis. Because all nerve messages use the same chemical transmitter to cross a synapse, a response to a nerve message can occur in a tissue that should not have been affected by a particular stimulus. For example, if a nerve message that should have stimulated a gland cell to secrete a chemical is delivered to a muscle cell, the muscle cell will contract. The result is a loss of coordination or an inappropriate response. Point out that such an event cannot occur when a hormone message is sent because only certain types of cells have receptors for particular hormones. If a cell does not have the receptor for a particular hormone, it cannot respond to it.

CONNECTIONS

Chapter 38

The Pancreas and Liver

Remind students that both the pancreas and liver are exocrine glands that function in the process of digestion. Exocrine glands deliver their products to the organs that use them through ducts.

CAPSULE SUMMARY

Hormones, which regulate the body's ongoing activities, are slower-acting but longer-lasting than nerve signals. Hormones also have a greater scope in their effect, yet they are highly specific and are more flexible than nerve signals.

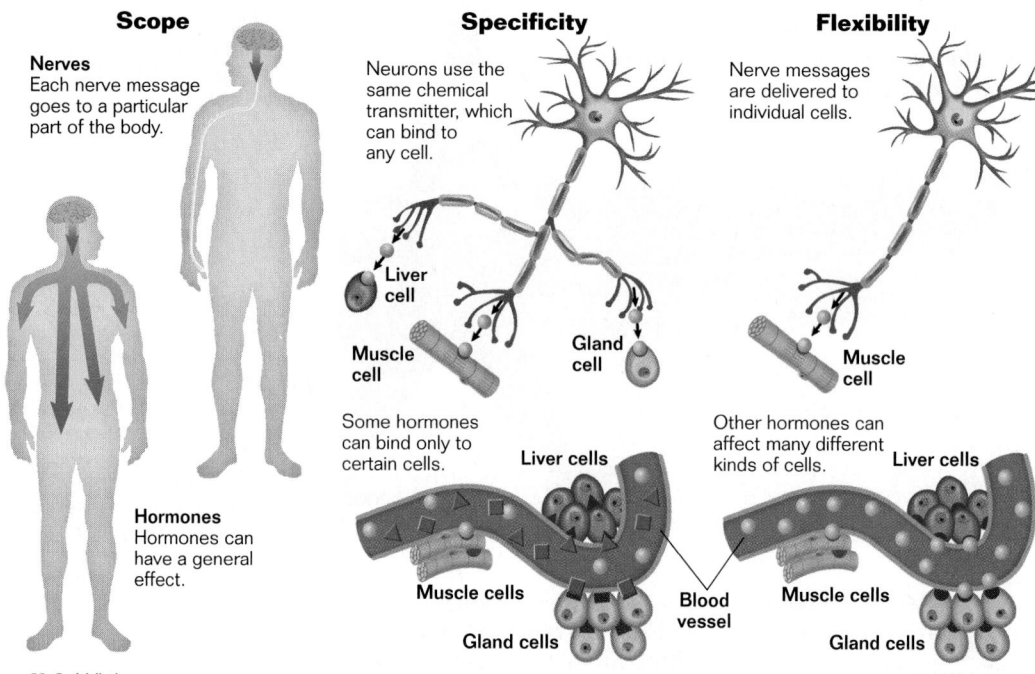

Figure 41-2 While nerve messages are faster-acting, messages carried in the bloodstream by hormones have a greater scope and more flexibility, and they affect only those cells with receptors that match specific hormones.

2. **Specificity** Like letters and postcards, hormones are sent to specific addresses. Each kind of hormone molecule has a shape unlike any other, much as every human face is unique. A hormone's shape matches a particular receptor protein on its target cells. Thus, a hormone will bind only to cells that have a particular receptor protein, ignoring all other cells. In this way, the body ensures that a hormone signal will affect only certain cells.

3. **Flexibility** A particular receptor protein can be present on different kinds of cells in different organs. Therefore, the same hormone can be used by the body to achieve different effects in different tissues. Like the blowing of an official's whistle in a sporting event, what happens after the signal is released depends on the context. Figure 41-2 illustrates the advantages of using hormones to regulate ongoing body processes. ◻

Endocrine Glands Produce Most Hormones

Throughout the body, a variety of tissues produce hormones and other similar chemical messengers. Most hormones are produced by **endocrine glands,** ductless glands that release their products directly into the bloodstream. The thyroid gland and the adrenal glands are examples of endocrine glands. In contrast to endocrine glands, exocrine glands, such as the salivary glands, deliver their products directly to where they are needed through ducts. The pancreas is both an exocrine gland and an endocrine gland. Acting as an exocrine gland, the pancreas produces digestive enzymes and delivers them to the small intestine through ducts. Clusters of cells within the pancreas secrete hormones into the bloodstream, making the pancreas an endocrine gland. In addition to the body's endocrine glands, several other body organs also contain cells that secrete hormones. These organs include the brain, stomach, small intestine, and heart.

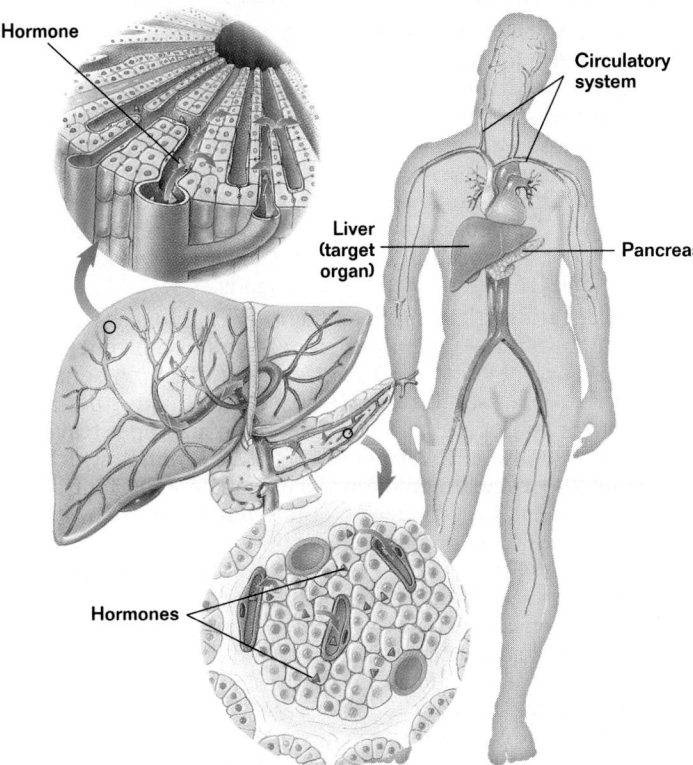

Figure 41-3 The liver is a target organ of hormones produced in the pancreas. Even though the two organs are very near one another, the hormones secreted by endocrine cells within the pancreas travel to the liver through the bloodstream.

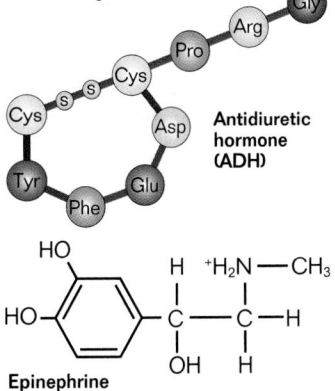

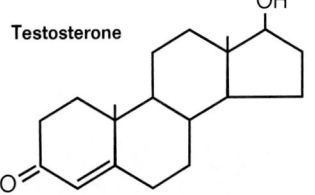

Figure 41-4 These are the structures of some representative hormones. ADH, *top*, is a peptide hormone. Epinephrine (also called adrenaline), *center*, contains a single amino acid. Testosterone, *bottom*, is a steroid hormone.

CONNECTIONS

Chapter 2
Peptides and Steroids
Remind students that chains of many amino acids are called polypeptides, a name that stems from the peptide bond, which holds two amino acids together. Also remind students that steroids are a type of lipid molecule.

Chapter 3
Transport Across the Plasma Membrane
Review the process of diffusion and the structure of the plasma membrane. Remind students that lipid-based molecules, such as steroid hormones, can dissolve in the phospholipids that form the matrix of the double-layered membrane, and thus they can pass through it by diffusion.

Once they are in the bloodstream, hormones travel to a **target cell** where they produce their effect, as illustrated in Figure 41-3. Because they must often be transported in their active form over long distances in the body, hormones are very stable molecules. Most hormone molecules are either peptides or steroids. **Peptide hormones,** which consist of chains of amino acids, are water-soluble. Insulin is an example of a peptide hormone. A few important hormones, such as adrenaline (epinephrine) and norepinephrine, are derived from a single amino acid. **Steroid hormones,** which are lipid molecules similar to cholesterol, are fat-soluble. The sex hormones, estrogen and testosterone, are examples of steroid hormones. Examples of peptide and steroid hormones are illustrated in Figure 41-4.

Teaching Tip
Molecular Models
Have students construct models of the hormones shown in Figure 41-4. Suitable materials might include miniature gumdrops and toothpicks. Tell students that testosterone and epinephrine molecules are relatively close in size. ADH is considerably larger because it is composed of nine amino acids, each of which is approximately the same size as an epinephrine molecule.

Nonendocrine Chemical Signals Also Regulate Cell Activities

The human body also employs many nonendocrine chemical signals to regulate the activities of cells. These substances affect cells in a manner similar to hormones but are not usually referred to as hormones because they act locally instead of at a distance. Nonendocrine chemical signals include many chemicals secreted by the brain and nerves, as well as chemicals that are secreted by cells in a variety of the body's other tissues.

1. **Neurotransmitters** One group of nonendocrine chemical signals produced by the brain and nervous tissue is the neurotransmitters. As you learned in Chapter 40, neurotransmitters take nerve signals across the gaps between individual nerve cells. Unlike hormones, neurotransmitters act only from one neuron to the next. Most neurotransmitters are derived from amino acids. Some neurotransmitters, such as norepinephrine, are chemically identical to hormones secreted by endocrine glands.

2. **Neuropeptides** The brain and nervous tissue also secrete a large number of chemical signals called **neuropeptides.** There are several different groups of neuropeptides. Enkephalins, which were discussed in Chapter 40, are a group of neuropeptides that inhibit pain messages traveling toward the brain. **Endorphins** *(ehn DAWR fihnz)*, which are thought to regulate emotions, are another important group of neuropeptides produced by the brain. Unlike neurotransmitters, enkephalins and endorphins tend to affect many cells near the nerve cells that produce them. Because enkephalins and endorphins often alter a cell's response to a neurotransmitter, they are called **neuromodulators.** Many of the brain's neuropeptides are true hormones. The peptide hormones produced in the brain are secreted by the hypothalamus and are then delivered by the bloodstream to other parts of the body such as the pituitary gland. You will learn more about the relationship between the hypothalamus and the pituitary gland later in this chapter.

3. **Prostaglandins** Among the most important nonendocrine chemical signals are the **prostaglandins** *(prahs tuh GLAN dihnz)*, modified lipids that are made from phospholipids by virtually all cells. Instead of circulating in the blood as hormones do, prostaglandins tend to accumulate in areas where tissue is disturbed or injured. Dozens of different prostaglandins produce a variety of effects. Some prostaglandins stimulate smooth muscle contractions that cause the constriction of blood vessels. The constricted blood vessels in turn affect blood pressure and body temperature. Other prostaglandins cause blood vessels to dilate, which causes inflammation. A headache may result when blood vessels swell and their walls press against nerves in the brain. Aspirin relieves headaches and reduces fever and inflammation by inhibiting prostaglandin production. ☐

☐ CAPSULE SUMMARY

The ductless glands called endocrine glands produce most of the body's hormones. Many nonendocrine chemical signals are produced in the brain and in many cells and tissues throughout the body.

 Did You Know?

The lysosomes of neutrophils (a type of white blood cell) and some other cell types generate prostaglandins through the action of enzymes. Prostaglandins, which are fatty acid molecules, sensitize blood vessels to the effects of other inflammatory chemicals, stimulate the smooth muscles of arterioles, enhance blood clotting, induce fever, and increase pain. They also stimulate uterine muscle contractions during labor.

The Hypothalamus Initiates Most Chemical Signals

A part of the brain controls most of the glands of the endocrine system. In fact, the endocrine system and the nervous system work together so closely that they can be thought of as one system—the **neuroendocrine system.** Until recently, however, one of the great mysteries of medicine was how the brain regulates the endocrine system. Scientists understood that the hypothalamus of the brain issues commands to the **pituitary gland**, an endocrine gland that is located at the base of the brain. They also knew that the pituitary gland in turn sends chemical signals to the body's other endocrine glands. But the way that the hypothalamus issues its commands was not clear. The pituitary gland is very close to the hypothalamus—actually suspended from it by a short stalk, as Figure 41-5 shows. However, no nerves connect the pituitary gland to the hypothalamus or to any other part of the brain.

In the 1930s, researchers discovered a network of tiny blood vessels that spans the short distance between the hypothalamus and the pituitary. This discovery suggested to researchers that perhaps chemical messages carried by hormones passed from the hypothalamus to the pituitary. In 1969, thyrotropin-releasing hormone (TRH), the first of several hormones produced by the hypothalamus to be isolated, was isolated from the brains of pigs. The release of TRH from the hypothalamus triggers the pituitary to release thyrotropin. This hormone then travels to the thyroid gland and causes the release of the thyroid hormones. Six other hypothalamic regulatory hormones that govern the pituitary have since been isolated.

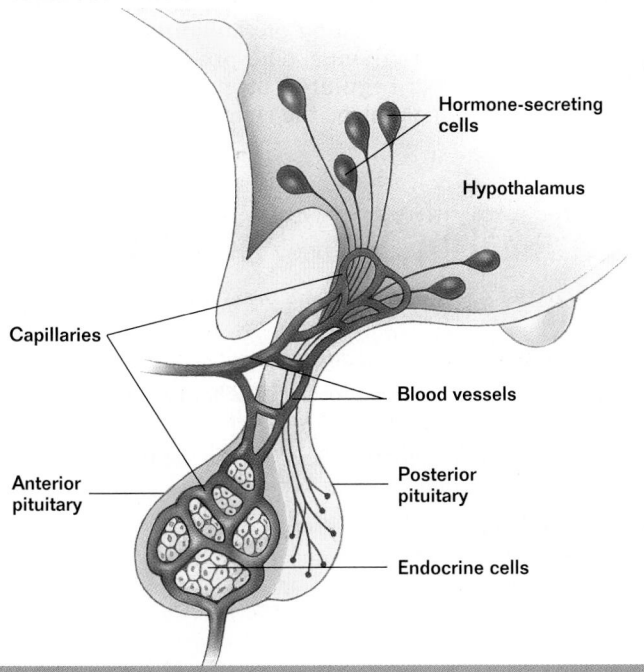

Figure 41-5 A short stalk connects the hypothalamus of the brain to the pituitary gland. Blood vessels run from the hypothalamus to the anterior part of the pituitary gland, which contains hormone-secreting endocrine cells. These blood vessels carry releasing hormones produced by the hypothalamus to the endocrine cells. Hormone-secreting cells in the hypothalamus extend into the posterior part of the pituitary gland, which stores the hormones these cells secrete until it receives a signal to release them.

SECTION 41-1

Teaching Tip
Action of a Releasing Factor

Have students use their textbook to develop a sequential diagram showing how the release of thyrotropin-releasing hormone (TRH) leads to the production of thyroxine. A completed sample of the sequential diagram is found in the *Graphic Organizer* on page 971.

👁 VISUAL STRATEGY

Figure 41-5

Have students compare the internal structure of the two lobes of the pituitary gland. Ask: Which part of the pituitary gland is directly attached to the hypothalamus? (*the posterior lobe*) How does this structure relate to the function of the posterior lobe? (*It enables hormones produced in the hypothalamus to pass directly to the posterior lobe, where they are stored.*) What does the anterior lobe of the pituitary gland contain? (*many endocrine cells and blood vessels*) How does this structure relate to the function of the anterior lobe? (*Hormones are produced in this lobe and enter the bloodstream, where they can be distributed to the rest of the body.*) Tell students that other blood vessels (not shown) enter the posterior lobe of the pituitary gland and distribute the hormones this lobe stores to other parts of the body.

Graphic Organizer

Use this graphic organizer with *Teaching Tip:* **Action of a Releasing Factor** on page 971.

hyphothalamus → TRH → pituitary gland → TSH → thyroid gland → thyroxine

VISUAL STRATEGY

Figure 41-6

Have students write two paragraphs, one describing each of the pathways diagramed in this figure.

CONNECTIONS
....

Chapter 35

Feedback Control of Hormone Release

Review the process of feedback control. Remind students that in negative feedback, the production of a substance such as a hormone is inhibited by an increase in that chemical. The increase in the substance causes the pathway that produces the substance to be shut down.

Chapter 40

Hypothalamus and the Nervous System

Remind students that other human-body responses are mediated by the hypothalamus through the nervous system. These responses include emotions, biorhythms, water balance, and body temperature.

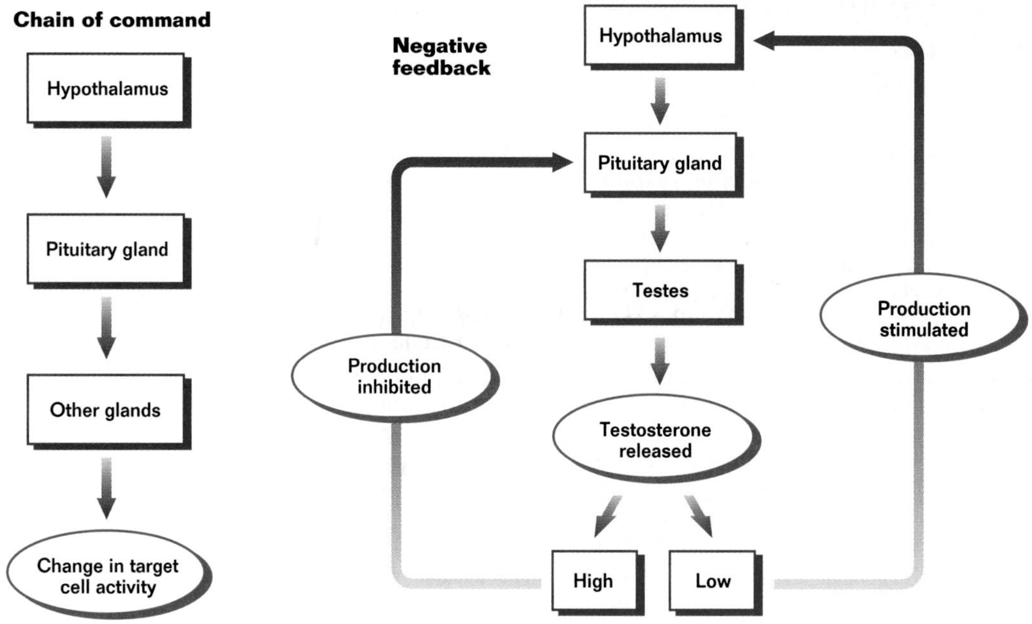

Chain of command

Hypothalamus → Pituitary gland → Other glands → Change in target cell activity

Negative feedback

Hypothalamus → Pituitary gland → Testes → Testosterone released → High / Low

Production inhibited / Production stimulated

Figure 41-6 The hypothalamus regulates many of the body's activities through a chain of command, *above*, in which the pituitary is ordered to release hormones that in turn stimulate other endocrine glands to release their hormones. A negative-feedback system that involves both the pituitary gland and hypothalamus, *above right*, inhibits the secretion of a hormone when a rise in its concentration is detected, or stimulates the hormone's secretion when a drop in its concentration is detected.

Scientists have learned through intensive research that the central nervous system regulates the body's hormones through a chain of command, as illustrated in Figure 41-6. Each of the "releasing" hormones made by the hypothalamus causes the pituitary to synthesize a corresponding pituitary hormone. This hormone then travels to a distant endocrine gland and causes that gland to begin producing its particular hormone or hormones. Some of the hormones released by the pituitary are regulated by an "inhibiting" hormone produced by the hypothalamus. The production of most hormones, however, is regulated by negative-feedback mechanisms like the one seen on the right in Figure 41-6. ▢

□ **CAPSULE SUMMARY**

The hypothalamus regulates many activities of the endocrine system by producing releasing hormones that stimulate the pituitary gland to release its hormones. Many of the pituitary gland's hormones in turn stimulate other endocrine glands to release their hormones.

Section Review

1. *Compare the scope of a chemical messenger with that of a nerve impulse.*
2. *How do peptide hormones differ from steroid hormones?*
3. *Where are neuropeptides produced? What are some functions of neuropeptides?*
4. *What role does the hypothalamus play in the transmission of chemical messages by hormones?*

Critical Thinking

5. *Why are nonendocrine chemical signals not considered to be hormones?*

Answers to Section Review

1. A chemical messenger's scope is broader than that of a nerve impulse, potentially affecting cells throughout the body for longer periods of time.

2. Peptide hormones are made of chains of amino acids and are water-soluble. Steroid hormones are based on the lipid cholesterol and are fat-soluble.

3. Neuropeptides are produced in the brain and other nervous tissue. The hormones produced by the hypothalamus of the brain are neuropeptides, which have a variety of effects that include controlling the release of hormones by other endocrine glands. Enkephalins inhibit pain, while endorphins regulate emotions.

4. The hypothalamus produces releasing hormones that stimulate the pituitary gland to release its hormones, which in turn stimulate other endocrine glands to release their hormones.

5. Nonendocrine chemical signals are not classified as hormones because their effects are more localized.

41-2 How Hormones Work

A fter hormones are produced, they travel through the bloodstream to their target cells, where they cause a change in the activity of the cells. Hormones cause an effect in a cell by attaching to a receptor protein in one of three ways. Some hormones enter the cells they affect, while others do not. One hormone goes directly to the nucleus of a cell. The location of the receptor proteins for a hormone determines its mode of action. In any case, the hormone's message must cross a plasma membrane in order to cause a response, such as the one shown in Figure 41-7. How does the binding of a hormone to a receptor cause a change in a cell's activity? The main way that hormones affect cells is by altering the activity or amounts of certain enzymes, which in turn alters chemical reactions that occur in the cells.

Figure 41-7 A cat's response to danger is triggered by a hormone that causes changes to occur inside of cells. In order for these changes to occur, the hormone must send a message across the plasma membrane.

Peptide Hormones Remain Outside of Cells

A peptide hormone produces its effect from outside a cell. Because peptide hormones are not fat-soluble, they cannot pass through the plasma membrane of a cell. The receptor proteins for peptide hormones are embedded in the plasma membranes of target cells. When a peptide hormone binds to a receptor on the surface of a target cell, a change in the receptor protein's shape is triggered. This change unleashes a series of events in the cell's cytoplasm. One of these events is the production of a **second messenger,** which is a chemical that is produced in response to the binding of a chemical signal on the outside of a cell. **Cyclic AMP,** which is produced by the removal of two phosphate groups from ATP, is a common second messenger.

Vocabulary Preview

- second messenger
- cyclic AMP
- glucagon
- hormone-receptor complex
- cortisol
- thyroxine

Opening Question

Ask students why the receptors on a cell membrane are important. Students should recall that receptors enable communication with the surroundings by binding with chemical messengers and causing a response within a cell.

Demonstration
Solubility of Proteins and Lipids

Materials: 4 small glass beakers or flasks, water, plain gelatin (a protein), cooking oil (very light colored), and vitamin A or vitamin E capsules (fat-soluble vitamins).
Procedure: To demonstrate that lipids are fat-soluble and proteins are not, place a small amount of gelatin in each of two flasks, one half filled with water and one half filled with cooking oil. Repeat the procedure with the vitamin capsules. Swirl the liquids in the flasks until each substance has dissolved in one of the liquids. Ask: Which substance is fat-soluble? *(the vitamin)* Which is not? *(the protein, which is water soluble)* Which type of substance can dissolve in a plasma membrane and pass through it by diffusion? *(a fat-soluble substance)* Why? *(The plasma membrane has a double layer of phospholipids.)* Could the gelatin enter a cell? *(no)* Tell students that like gelatin, protein-based peptide hormones must remain outside of cells, while fat-soluble steroid hormones can enter cells.

? Did You Know?

Cyclic AMP, once triggered, can activate many chemical reactions. The effect is multiplicative: one activated enzyme in a pathway can catalyze hundreds of reactions, resulting in a geometric increase of product molecules at each step of the pathway. Theoretically, a single hormone molecule could generate millions of molecules of the desired final product. The actual action of the cyclic AMP second messenger depends on the nature of the target cell, its enzymes, and the nature of the hormone acting as the first messenger. Thyroid-stimulating hormone (TSH), for example, causes cyclic AMP in the cells of the thyroid gland to promote thyroxine production. Growth hormone triggers cyclic AMP to activate protein synthesis in bone cells.

Chapter 2
Structure of ATP

Remind students that ATP, the cell's main energy currency, is adenosine triphosphate. The ATP molecule is actually a nucleotide (a sugar, a nitrogenous base, and a phosphate group) to which two extra phosphate groups have been attached. When one of the groups is removed, the molecule becomes adenosine diphoshphate (ADP). When two of the groups are removed, the molecule becomes adenosine monophosphate (AMP).

Chapter 4
Action of Enzymes

Review the role of enzymes as catalysts in biochemical reactions. Remind students that the shape of an enzyme enables it to bind with its substrate.

 VISUAL STRATEGY

Figure 41-8

Walk students through the steps called out in this diagram. Ask: Why is glucagon important to the maintenance of homeostasis in a cell? (*Glucagon causes the release of glucose through the breakdown of glycogen and therefore plays a key role in the liberation of energy needed for cell activities.*)

Teaching Tip
Modes of Hormonal Action

Put the diagram seen in the *Graphic Organizer* on page 974 on the board or overhead projector. Have students copy the diagram as you briefly discuss its significance. Have students fill in the blanks in the diagram with the names of each type of hormone as they encounter the answers in the text.

Figure 41-8 Peptide hormones, such as glucagon, bind with receptor proteins located on the surfaces of their target cells. Even though glucagon cannot cross the plasma membrane, this binding initiates a series of events that results in the activation of enzymes and the release of glucose into the bloodstream.

☐ **CAPSULE SUMMARY**

Peptide hormones work from outside a cell by binding to a receptor protein on the membrane's surface, typically causing the production of a second messenger that in turn alters the enzymes produced by the cell.

A second messenger activates or deactivates certain enzymes, altering the chemical activity of a peptide hormone's target cells. Because many molecules of the second messenger are produced by the binding of a single hormone molecule, second messengers amplify the effect of a peptide hormone. The effect of a peptide hormone may be further amplified because a single second-messenger molecule may stimulate the activation or production of many molecules of a particular enzyme. Each of these molecules may in turn activate many molecules of another enzyme, in a cascade of reactions that greatly amplifies the original signal. For discovering how this pathway works, Alfred Gilman and Martin Rodbell won the Nobel Prize for medicine in 1994.

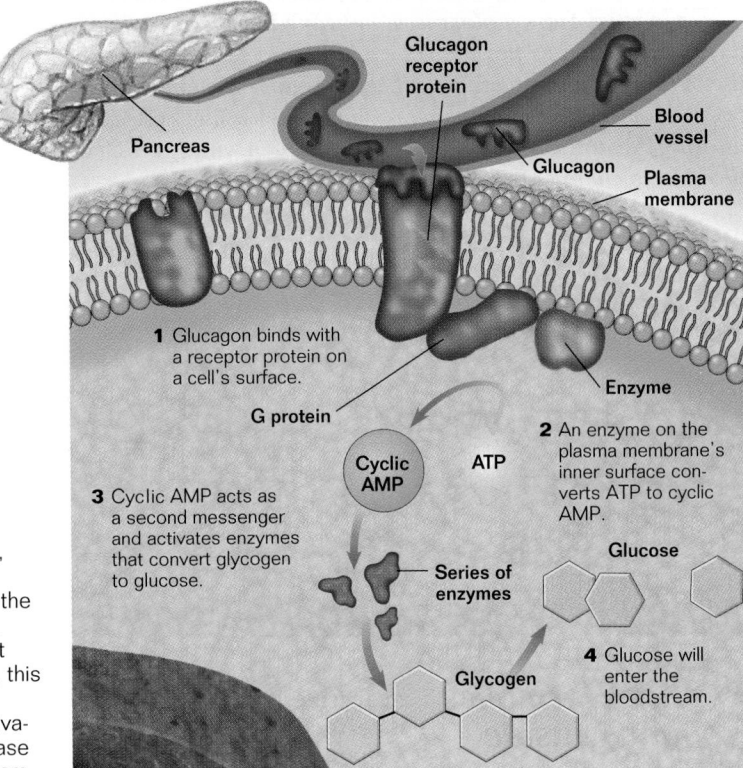

The mechanism by which peptide hormones produce a response is summarized in Figure 41-8. **Glucagon** (*GLOO kuh gahn*) is a peptide hormone that is produced in the pancreas. It travels through the bloodstream and binds to receptor proteins located on the surfaces of liver cells. When glucagon molecules bind to their receptor proteins, the receptor proteins change shape and cause the production of cyclic AMP. Acting as a second messenger, the cyclic AMP molecules alter the activity of liver cells, causing them to activate a series of enzymes that converts glycogen into glucose. ☐

Graphic Organizer

Use this graphic organizer with *Teaching Tip*: **Modes of Hormonal Action** on page 974.

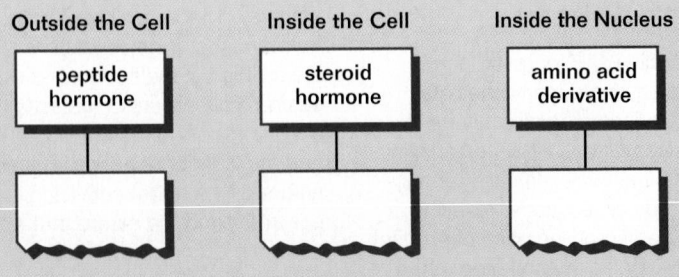

Steroid Hormones Work Inside of Cells

Because steroid hormones are fat-soluble, they can pass readily through the plasma membranes of their target cells and produce a response from inside the cells. A steroid hormone binds to a receptor protein located in a target cell's cytoplasm, which produces a **hormone-receptor complex.** This complex then enters the nucleus of the cell and binds to DNA, causing a change in a gene's activity and thus a change in the cell's activity. Some steroid hormones stimulate genes to synthesize certain proteins, and some repress the synthesis of other proteins.

The mechanism by which steroid hormones produce a response is summarized in Figure 41-9. **Cortisol** *(KAWRT uh sahl)* is a steroid hormone that is produced by the adrenal glands in response to stress. When cortisol molecules reach a target cell, the molecules diffuse through the plasma membrane and bind to receptor proteins located in the cell's cytoplasm. The hormone-receptor complex then enters the cell's nucleus, binds to DNA, and activates the genes that produce certain enzymes. These enzymes catalyze reactions that break down fats and proteins into fatty acids and amino acids, respectively. Other enzymes produced as a result of cortisol's action catalyze the conversion of fatty acids and amino acids to glucose. These changes help you deal with stress by providing extra sources of energy and building blocks for repairing damaged tissues and for making enzymes. ◻

◻ CAPSULE SUMMARY

Steroid hormones work by passing through a cell's plasma membrane and binding to protein receptors located in the cytoplasm. The hormone-receptor complex then enters the nucleus and attaches to DNA, altering the cell's production of enzymes and thus changing the cell's activity.

SECTION 41-2

CONNECTIONS

Chapter 9
Control of Gene Expression
Review how gene expression is controlled with the help of inducers and repressors. Remind students that inducers switch genes on and that repressors switch them off.

👁 VISUAL STRATEGY

Figure 41-9
Walk students through the steps called out in this diagram. You may also want to discuss the action of cortisol using the figure as a guide. Tell students that cortisol is sold in drug stores and supermarkets as hydrocortisone. Used topically, it reduces inflammation. Ask: Why do you suppose instructions on a tube of hydrocortisone limit the drug's use to no more than three applications per day and warn against accidental ingestion? *(Because the drug masks the inflammatory response, it could mask symptoms of a serious infection. Ingesting the drug exposes the entire body to its effects rather than limiting effects to the area of application.)*

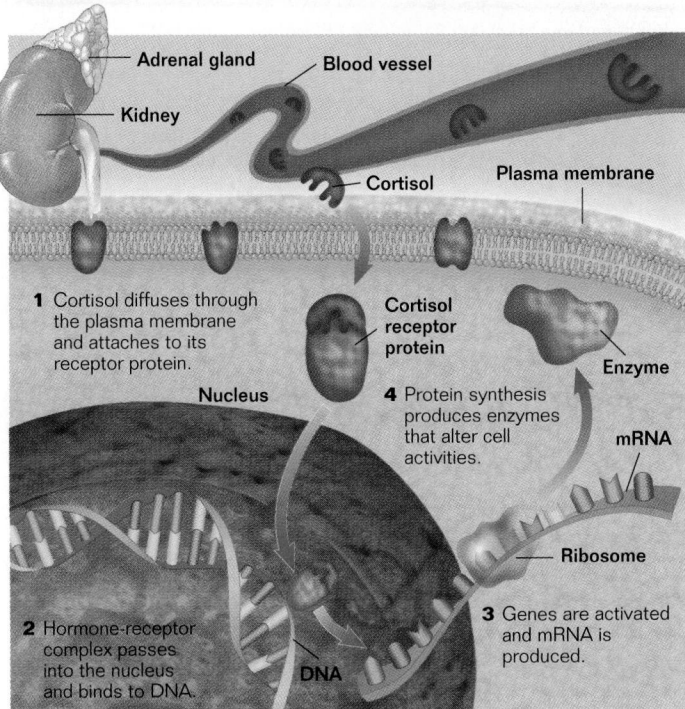

1 Cortisol diffuses through the plasma membrane and attaches to its receptor protein.

2 Hormone-receptor complex passes into the nucleus and binds to DNA.

3 Genes are activated and mRNA is produced.

4 Protein synthesis produces enzymes that alter cell activities.

Figure 41-9 Steroid hormones, such as cortisol, enter their target cells and bind to receptor proteins in the cytoplasm. This binding activates certain genes, which produce enzymes that alter cell activity.

Historical Note

Corticosteroid Drugs

Edward Kendall synthesized the first artificial corticosteroid in 1944. For 20 years prior to this time, a physician named Philip Showalter Hench had been studying rheumatoid arthritis at the Mayo Clinic. He observed that symptoms of rheumatoid arthritis often abate when the body is subjected to stresses such as pregnancy or jaundice. After further research, he suspected that corticosteroids were responsible for the relief some arthritis patients experience during times of stress. In 1949, Hench began using cortisone as a treatment for rheumatoid arthritis. In 1950, Kendall, Hench, and a third contributor to corticosteroid research, Tadeus Reichstein, received a Nobel Prize for medicine and physiology.

VISUAL STRATEGY

Figure 41-10

Walk students through the steps called out in this diagram. Point out that although thyroxine is a peptide hormone made from only two amino acids, it is small enough to pass through openings in the plasma membrane.

CAPSULE SUMMARY

Thyroxine passes through the plasma membrane and cytoplasm of a cell and enters the nucleus, where it attaches to a receptor protein on a DNA molecule and initiates the production of enzymes that change the cell's activity.

One Hormone Works Inside the Nucleus

The receptor proteins for one important hormone are located not on the plasma membrane or in the cytoplasm, but inside the nucleus, as seen in Figure 41-10. **Thyroxine** (*theye RAHKS ihn*), which is produced by the thyroid gland, consists of an amino acid and four attached iodine atoms. Thyroxine molecules are small enough to pass directly through the plasma membranes of their target cells. After diffusing through the cytoplasm, they enter the nucleus and bind to a receptor protein that is attached to a DNA molecule. The binding of a thyroxine molecule to its receptor protein initiates the production of mRNA molecules. These mRNA molecules cause the production of specific enzymes that stimulate cell metabolism and promote growth. ◻

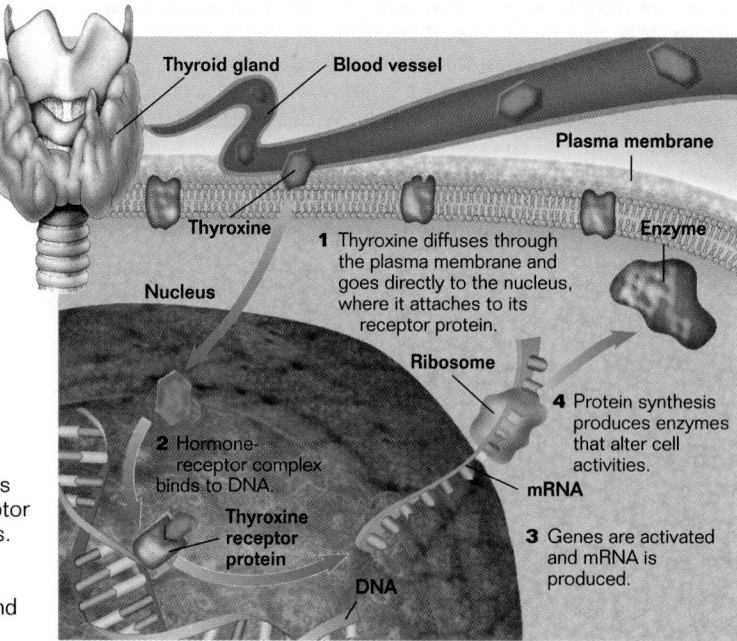

Figure 41-10 Thyroxine enters its target cells and binds to receptor proteins located in the nucleus. This binding activates certain genes that produce enzymes which stimulate metabolism and growth.

(labels in figure): Thyroid gland · Blood vessel · Plasma membrane · Thyroxine · Nucleus · Enzyme · Ribosome · mRNA · Thyroxine receptor protein · DNA

1 Thyroxine diffuses through the plasma membrane and goes directly to the nucleus, where it attaches to its receptor protein.

2 Hormone-receptor complex binds to DNA.

3 Genes are activated and mRNA is produced.

4 Protein synthesis produces enzymes that alter cell activities.

Section Review

1. *How does a peptide hormone produce its response?*
2. *How does a steroid hormone produce its response?*
3. *How does the action of thyroxine differ from that of either a peptide hormone or a steroid hormone?*

Critical Thinking

4. *The excessive use of steroid hormones has been linked to cancer. Why are these hormones likely to cause cancer?*

Answers to Section Review

1. A peptide hormone works through the action of a second messenger, which is released inside a target cell in response to the binding of the hormone to a receptor on the cell membrane's surface. The second messenger activates enzymes that alter the cell's activity.

2. A steroid hormone enters a cell and binds with a receptor to form a hormone-receptor complex. This complex enters the nucleus and activates a gene that produces an enzyme which alters cell activity.

3. Thyroxine works by going directly to a target cell's nucleus and attaching to receptor proteins on DNA.

4. Answers will vary but could suggest that hormones may induce cancer by activating genes that cause the uncontrolled proliferation of cells.

41-3 The Body's Endocrine Glands

About a dozen major endocrine glands collectively make up your endocrine system, which is illustrated in Figure 41-11. Table 41-2 on pages 984–985 lists the major endocrine glands, the hormones they release, the target tissues for each hormone, and the effects that these hormones produce.

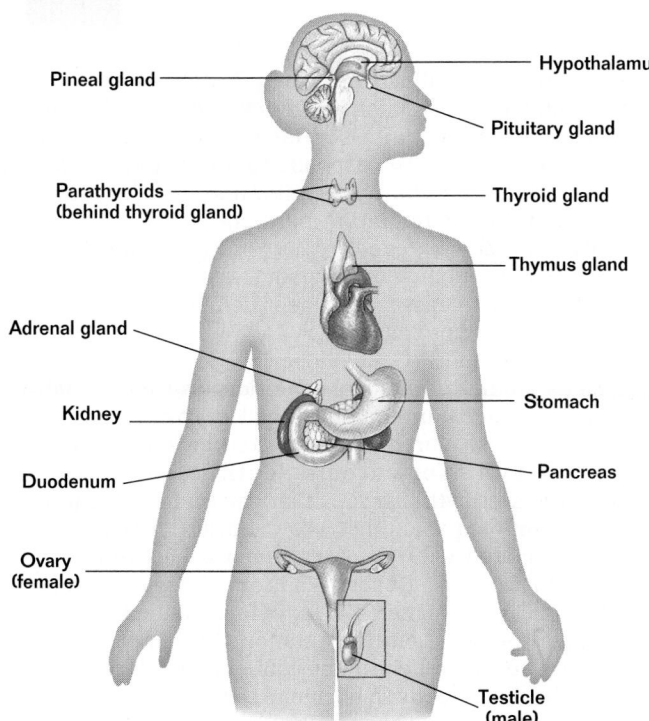

- Pineal gland
- Hypothalamus
- Pituitary gland
- Parathyroids (behind thyroid gland)
- Thyroid gland
- Thymus gland
- Adrenal gland
- Kidney
- Stomach
- Duodenum
- Pancreas
- Ovary (female)
- Testicle (male)

Figure 41-11 Major endocrine glands are located throughout the human body. As research continues, more and more of the body's other organs are being added to the list of endocrine glands because they contain cells that secrete hormones.

Section Objectives
- Identify the major endocrine glands of the body, and name the major hormones produced by each.
- Discuss the effects of the overproduction and underproduction of the hormones thyroxine and insulin.
- Explain why you could not survive without parathyroid hormone and aldosterone.

SECTION 41-3

Vocabulary Preview

- anterior lobe
- posterior lobe
- acromegaly
- gigantism
- goiter
- hypothyroidism
- hyperthyroidism
- calcitonin
- parathyroid gland
- parathyroid hormone
- adrenal medulla
- adrenal cortex
- adrenaline
- norepinephrine
- prednisone
- aldosterone
- islets of Langerhans
- insulin
- diabetes mellitus
- Type I diabetes
- Type II diabetes
- estrogen
- progesterone
- testosterone
- pineal gland
- melatonin

Opening Question

Ask students how the different hormones circulating in the blood produce their effects only on certain target cells. Students should recall that different types of target cells have receptors that are specific to certain hormones.

The Pituitary Gland Regulates Other Endocrine Glands

The pituitary gland is a small endocrine gland that is located just beneath the hypothalamus within a bony recess at the base of the skull. One of the most important endocrine glands in the body, the pituitary releases nine major hormones. Because many of these hormones act principally to influence other endocrine glands, it was fashionable until recently to regard the pituitary as the "master gland" of the endocrine system. However, that role actually belongs to the hypothalamus, because it controls the pituitary.

Multicultural Perspective

Choh Hao Li

The Chinese American endocrinologist Choh Hao Li isolated and identified five pituitary hormones. He also discovered that Somatotropin, or Growth Hormone (GH), contains a chain of 256 amino acids. In 1970, Li devised a method for synthesizing GH and in doing so set the record for creating the largest synthesized protein molecule.

Demonstration
Locations of the Endocrine Glands

Materials: felt (assorted colors and sizes), 1 m x 0.75 m corrugated cardboard sheet, duct tape, rubber cement, permanent markers. **Procedure:** Construct a felt board as follows. Cover the cardboard with a suitable background color of felt. Wrap the edges of the felt around the back of the cardboard, securing them with duct tape. Using a dark color of felt, cut out a silhouette similar to the one in Figure 41-11 on page 977. You may choose to permanently attach the silhouette to the background using rubber cement. You will obtain the best results by liberally applying the rubber cement to the back of the silhouette along its edges, then working quickly to lay it on the background. Prior to applying the cement, you may want to practice with placement of the silhouette on the background. Cut out models of the endocrine glands from various colors of felt, approximating the scale in Figure 41-11. Use Figure 41-11 and Table 41-2 on pages 984–985 to determine the shapes of the glands. Write each gland's name on the back of the piece with a permanent marker. As you introduce each gland in this section, place the gland on the silhouette, pressing firmly to temporarily attach it. The board may be used later as a method of review, with students putting the glands in their proper places.

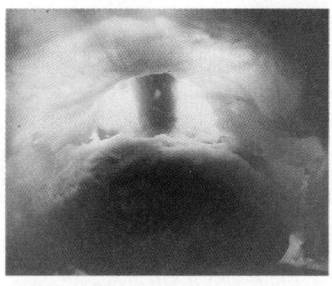

Figure 41-12 Despite its tiny size, the pituitary gland, *above,* consists of two lobes, *right,* which release different hormones.

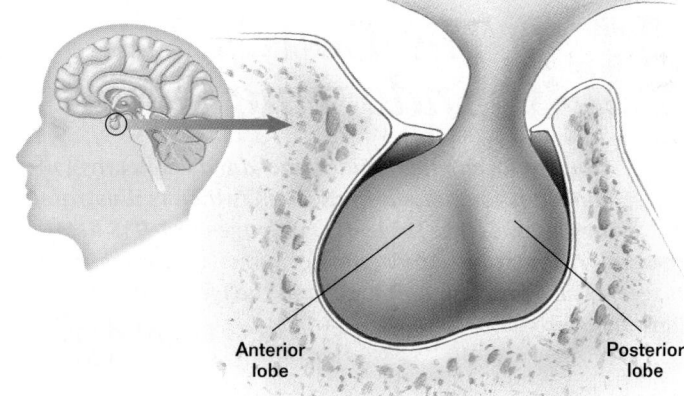

Anterior lobe

Posterior lobe

☐ CAPSULE SUMMARY

The pituitary gland, which is located in the brain, controls many of the body's other endocrine glands. It consists of an anterior lobe, which produces several hormones in response to signals from the hypothalamus, and a posterior lobe, which stores and releases hormones produced by the hypothalamus.

Figure 41-13 The thyroid gland, located in the neck, is wrapped around the windpipe.

As Figure 41-12 illustrates, the pituitary is actually two glands. The front portion of the gland, known as the **anterior lobe,** produces seven major peptide hormones. Each of these hormones is released in response to a particular releasing signal secreted by the hypothalamus. For example, the release of thyroid-stimulating hormone (TSH) by the anterior pituitary is triggered by thyroid-releasing hormone (TRH). The back portion of the gland, known as the **posterior lobe,** stores and releases other peptide hormones, which are produced by the hypothalamus.

Many of the roles of the endocrine glands were first discovered by studies of medical conditions caused by malfunctions of the glands. For example, one key role of the anterior pituitary was discovered in 1909 when the surgical removal of a pituitary tumor cured a South Dakota farmer of **acromegaly,** a growth disorder that causes facial features to thicken. **Gigantism** is another growth disorder that is almost always associated with pituitary tumors. The tallest human being ever recorded—Robert Wadlow, who grew to a height of 8 ft. 11 in. and weighed 475 lb. before he died from infection at age 22—had a pituitary tumor. Pituitary tumors produce giants because such tumor cells produce large amounts of somatotropin, or growth hormone (GH). ☐

The Thyroid Gland Regulates Metabolic Rate

The thyroid gland, which is shaped like a shield, is located just below the Adam's apple in the front of the neck, as seen in Figure 41-13. The name *thyroid* comes from the Greek word *thyros,* which means "shield." The thyroid gland produces several hormones, but the most important is thyroxine.

Thyroxine increases the body's metabolic rate and promotes the normal growth of the brain, bones, and muscles during childhood. Neither a peptide nor a steroid, thyroxine

Figure 41-14 A goiter, *left,* is a swelling in the throat that results from a lack of iodide in the diet. A diet that includes seafood and iodized salt, *right,* provides the iodide necessary for proper thyroid function.

is produced by the addition of iodide to the amino acid thyronine. If iodide salts are lacking in the diet, the thyroid gland becomes greatly enlarged as a result of futile attempts to make more thyroxine and forms a **goiter.** Once common in the United States, goiters, like the one seen in Figure 41-14, are now rare because of the addition of iodide to salt. The underproduction of thyroxine is known as **hypothyroidism.** If it occurs early in childhood, hypothyroidism may cause stunted growth or mental retardation, or both. In adults, hypothyroidism tends to produce a lack of energy, dry skin, and weight gain. Overproduction of thyroid hormones, or **hyperthyroidism,** causes nervousness, sleep disorders, an irregular heart rate, and weight loss.

Specialized cells within the thyroid gland produce another important hormone, **calcitonin** *(kal sih TOH nihn),* which plays a key role in maintaining a proper calcium level in the body. A high level of calcium in the blood stimulates calcitonin production, which in turn stimulates the deposition of calcium in bone tissue and lowers the blood-calcium level. ❑

❑ CAPSULE SUMMARY

The thyroid gland, which is located in the neck, releases thyroxine and calcitonin. Thyroxine regulates growth and metabolism. Calcitonin helps regulate calcium in the body by stimulating the deposition of calcium in bone tissue.

Parathyroid Glands Regulate Blood Calcium

Attached to the thyroid gland, as seen in Figure 41-15, are four **parathyroid glands,** which also help regulate calcium in the body. Small and unobtrusive, they were ignored by researchers until well into this century. The parathyroid glands produce **parathyroid hormone** (PTH). PTH is essential for survival because it helps maintain an adequate supply of calcium in the blood. Recall from Chapter 40 that nerve impulses cause muscles to contract by initiating the release of calcium ions. You cannot live without the muscles that pump your heart, and these muscles cannot function if the blood-calcium level is not kept within narrow limits.

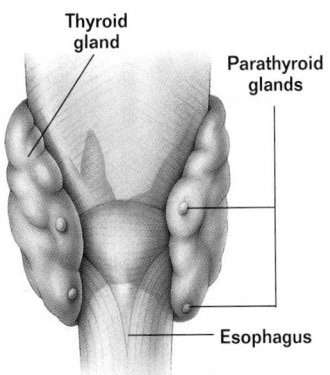

Figure 41-15 The parathyroid glands are located on the back of the thyroid gland.

VISUAL STRATEGY

Figure 41-16

Review the three pathways by which PTH regulates the blood-calcium level. Ask: How do you think the production of PTH is regulated? *(It is probably controlled by a negative-feedback mechanism.)*

Figure 41-16 This graphic organizer shows three ways that parathyroid hormone (PTH) causes the blood-calcium level to rise.

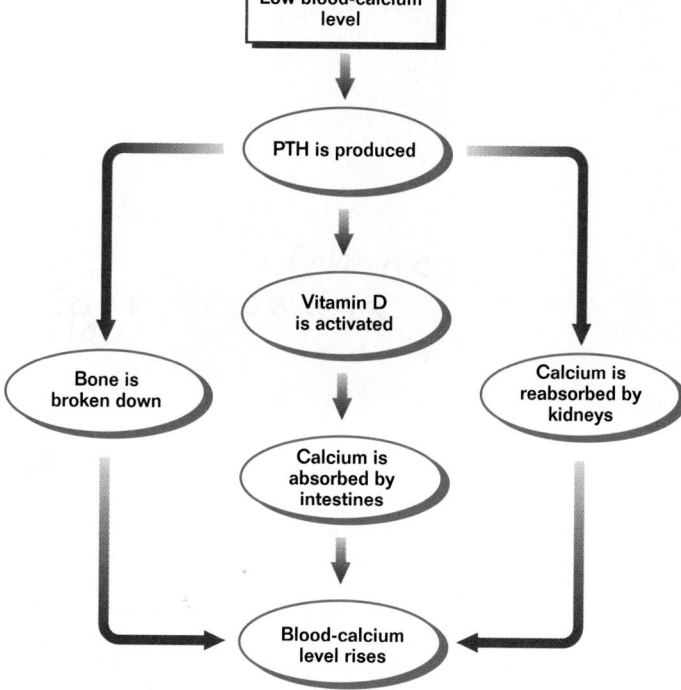

CAPSULE SUMMARY

The parathyroid glands, which are located on the thyroid gland, produce parathyroid hormone (PTH). PTH plays a vital role in regulating the body's blood-calcium supply.

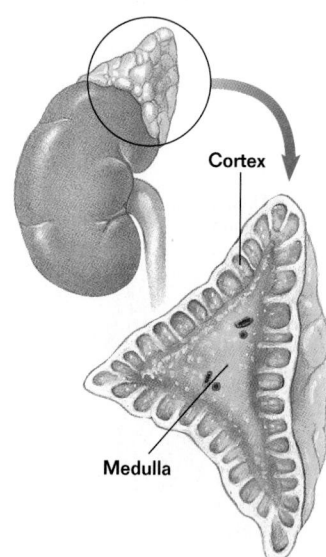

Figure 41-17 The adrenal glands, located on top of each kidney, consist of an inner medulla and an outer cortex.

PTH, which is synthesized in response to a falling level of calcium in the blood, affects the blood-calcium level in three ways, as diagramed in Figure 41-16. First, it ensures that the blood-calcium level never falls too low. When released into the bloodstream, PTH is absorbed by the bones, where it stimulates the osteocytes to dismantle bone tissue and release calcium into the bloodstream. Second, PTH acts on the kidneys, causing them to reabsorb calcium ions from urine. Third, PTH leads to activation of vitamin D, which is necessary for calcium absorption by the intestine.

The Adrenals Are Two Glands in One

Your body has two adrenal glands, one located just above each kidney. Each almond-sized adrenal gland is actually two glands in one, as seen in Figure 41-17. An inner core called the **adrenal medulla** produces the peptide hormones epinephrine (*ehp uh NEF rihn*), also called adrenaline (*uh DRIHN uh lihn*), and norepinephrine. Unlike other endocrine glands, the adrenal medulla is stimulated to release its hormones by nerves that come from the hypothalamus. An outer shell called the **adrenal cortex** produces the steroid hormones cortisol and aldosterone.

The adrenal medulla acts as an emergency warning system in times of stress by releasing **adrenaline** (or epinephrine)

 Did You Know?

Epinephrine (adrenaline) and norepinephrine are closely related chemically. In fact, norepinephrine is a metabolic precursor to epinephrine. These compounds are the same ones used as neurotransmitters by the nervous system.

 Did You Know?

Feedback is often refined by the action of two antagonistic hormones. The pancreas produces glucagon in its beta (B) cells and insulin in its alpha (A) cells. Glucagon has an effect opposite that of insulin. Instead of reducing the amount of glucose in the blood, glucagon increases the blood-glucose level by stimulating the breakdown of glycogen. With a push in either direction, the balance of glucose in the blood can be fine-tuned as needed.

Figure 41-18 The "fight-or-flight" hormones help this stockbroker handle the stress of his high-pressure job. Perspiration, rapid breathing, and an elevated pulse are outward signs that the "fight-or-flight" hormones are at work. Inside, extra blood glucose and oxygen enable his cells to supply the energy needed to keep going.

and **norepinephrine,** the "fight-or-flight" hormones. The effects of these hormones, which prepare the body for action in emergencies, are identical to the effects of the sympathetic nervous system but are longer-lasting. In stressful situations, such as the one in Figure 41-18, the fight-or-flight hormones accelerate heartbeat and increase blood pressure, blood-sugar level, and blood flow to the heart and lungs. These hormones can thus be thought of as extensions of the sympathetic nervous system, which was discussed in Chapter 40.

The adrenal cortex produces hormones that affect your metabolic health and maintain the proper amount of salt in your body. Cortisol (also called hydrocortisone) is produced in response to adrenocorticotropic hormone (ACTH) from the pituitary. It acts on many different cells in the body to maintain nutritional well-being. Cortisol stimulates carbohydrate metabolism and acts to reduce inflammation. Derivatives of this hormone, such as **prednisone** *(PREHD nih sohn)*, are widely used as anti-inflammatory agents.

Aldosterone *(al DAHS tuh rohn)* affects primarily the kidneys by promoting the uptake of sodium and other salts from body fluids, as summarized in Figure 41-19. Recall that sodium ions play crucial roles in nerve conduction and many other bodily functions, such as maintaining blood pressure. Without aldosterone, sodium ions are not retrieved from the fluids removed by the kidneys and are lost in the urine. The resulting loss of salt from the blood causes water to leave the bloodstream and enter cells, and blood pressure falls. Aldosterone affects potassium in the opposite way. If the potassium level rises, aldosterone stimulates the kidneys to secrete potassium ions into the urine. When the aldosterone level is too low, the potassium level in the blood may rise to a dangerous level. Thus, aldosterone, as well as PTH, is essential for survival. ◻

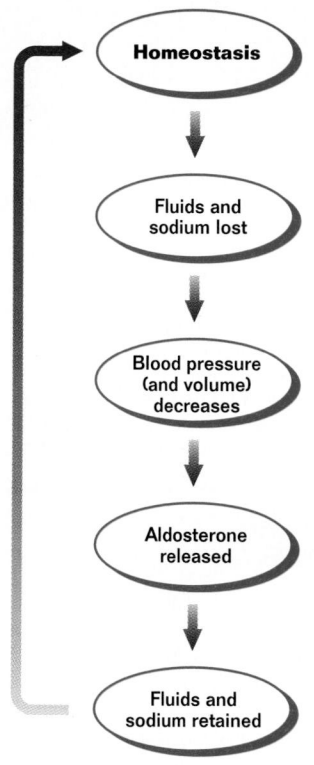

Homeostasis

↓

Fluids and sodium lost

↓

Blood pressure (and volume) decreases

↓

Aldosterone released

↓

Fluids and sodium retained

Figure 41-19 Aldosterone helps maintain the body's fluid and sodium balance by the mechanism shown here.

◻ CAPSULE SUMMARY

The adrenal glands are located on top of each kidney and consist of an outer cortex, which produces steroid hormones, and an inner medulla, which produces the "fight-or-flight" hormones.

SECTION 41-3

👁 VISUAL STRATEGY

Figure 41-18
As students look at this photograph, have them think of similarly stressful situations in their own lives. Ask: What might the outward signs of the fight-or-flight response indicate? *(Students may recognize that these are signs of an increased metabolic rate.)* How does an increased metabolic rate prepare an organism to cope with emergencies? *(Answers will vary but should mention the benefits of greater energy availability due to increased metabolic rate.)*

👁 VISUAL STRATEGY

Figure 41-19
Lead students through this figure. Ask: Why is aldosterone's function so important to healthy functioning of the body? *(Aldosterone functions in maintaining the balance of sodium and potassium, both of which are crucial to nerve transmission.)*

Teaching Tip

What If the body were to secrete too much aldosterone? Hypersecretion of aldosterone is most often caused by tumors of the adrenal glands. Symptoms of this hypersecretion are hypertension and retention of water in the tissues (edema) with increased excretion of potassium in the urine. As potassium loss increases, the sufferer experiences muscle weakness that eventually progress to paralysis. Neurons fail to respond, and the heart begins to beat unevenly, leading to cardiac arrest if the patient's condition isn't stabilized.

❓ Did You Know?

The adrenal cortex in both sexes produces sex hormones—the gonadocorticoids. Most of these hormones are androgens, but some female hormones (estrogen and progesterone) are also produced. Gonadocorticoid production is particularly high in the fetus and during puberty. Production falls rapidly in late puberty and is never again significant in comparison with the production of sex hormones by the gonads. The glucocorticoids may function in maintaining the sex drive in adult women while supplying small amounts of estrogen after menopause.

VISUAL STRATEGY

Figure 41-20
Remind students of the digestive function of the pancreas's exocrine cells. Have students look up the term *islet* in a good dictionary. Ask: Why is the word *islet* a good description for the structures in the pancreas that produce insulin? *(The word* islet *means "little island," an excellent description of the microscopic clusters of endocrine cells —the A and B cells—found interspersed in the "sea" of exocrine gland cells in the pancreas.)*

Application

Health Diabetics are subject to an array of complications, many of which can be life threatening. Type I diabetes may result from an autoimmune disorder. This defect may carry over to other immune system functions—many diabetics' white blood cells are not as effective in fighting disease as those in a nondiabetic. As a result, diabetics, particularly those whose disease is not well controlled, are more susceptible to bacterial and fungal microbes, even those normally benign to the general population. Infection stresses the diabetic, causing a greater demand for insulin. Other complications involve the circulatory and cardiovascular systems. For example, the capillaries of the retina in the eye can weaken and spill blood into the eyeball, sometimes resulting in blindness. Diabetics are also more prone to strokes, heart attacks, and high blood pressure due to their susceptibility to arteriosclerosis and atherosclerosis. Diabetics can also experience kidney disfunctions. Maintaining good control of blood glucose gives diabetics the best chance to achieve a high degree of normalcy in their lives.

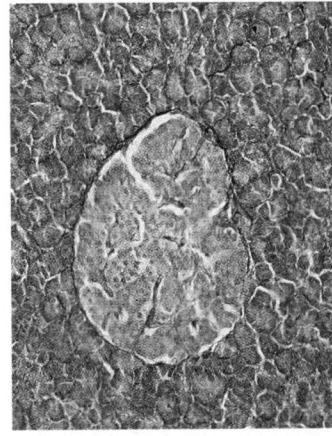

Figure 41-20 Islets of Langerhans are clusters of hormone-secreting cells that are easily recognized among the other cells of the pancreas. Two types of islet cells secrete hormones. The A cells, stained red in this light micrograph, produce glucagon. The B cells, stained purple, produce insulin. Notice that many blood vessels, shown in blue and white, pass through islets of Langerhans.

CAPSULE SUMMARY

The islets of Langerhans, which secrete insulin, are hormone-producing cells found in the pancreas. Either an abnormally low level of insulin in the body or a lack of insulin receptors makes an individual's cells unable to absorb glucose from the blood and causes a condition called diabetes mellitus.

The Pancreas Regulates Blood Sugar

Scattered throughout the pancreas, once thought to be solely an exocrine gland, are clusters of specialized cells that act as endocrine glands. These clusters, which are seen in Figure 41-20 and are called **islets of Langerhans,** produce two peptide hormones that interact to govern the level of glucose in the blood. **Insulin,** a storage hormone, puts away nutrients for leaner times by promoting the accumulation of glycogen in the liver. Glucagon causes liver cells to release glucose that was stored in glycogen. The two hormones thus work together to keep the blood-glucose level within narrow bounds.

Diabetes mellitus is a serious disorder in which the cells of affected individuals are unable to obtain glucose from the blood, causing the blood-glucose level to become very high. Affected individuals may lose weight, suffer brain damage, and ultimately starve to death. About 12 million Americans, and over 100 million people worldwide, have diabetes mellitus, which is the seventh-leading cause of death in the United States and is the leading cause of blindness among adults. It also accounts for one-third of all kidney failures.

There are actually *two* kinds of diabetes mellitus. The warning signs of each are listed in Table 41-1. About 10 percent of affected individuals suffer from **Type I diabetes,** a hereditary autoimmune disease in which the immune system attacks the islets of Langerhans, causing insulin secretion to be abnormally low. Also called juvenile-onset diabetes, Type I diabetes usually develops before age 20. Type I diabetes can be treated with daily injections of insulin. Active research on the possibility of transplanting islets of Langerhans holds the promise of a lasting treatment for Type I diabetes. However, researchers must find a way to prevent the immune system from attacking the pancreas. People with **Type II diabetes** often have an abnormally low number of insulin receptors, while the level of insulin in their blood is often higher than normal. Type II, or adult-onset, diabetes usually develops in people over 40, almost always as a consequence of obesity. In the United States, 90 percent of Type II diabetics are obese. Type II diabetes is usually treated with diet and exercise. Most Type II diabetics do not need daily injections of insulin. ◻

Table 41-1 Warning Signs of Diabetes Mellitus

Type	Symptoms
Type I	Frequent urination, unusual thirst, extreme hunger, unusual weight loss, extreme fatigue, irritability
Type II	Any of the Type I symptoms; frequent infections; blurred vision; cuts or bruises that are slow to heal; tingling or numbness in the hands or feet; recurring skin, gum, or bladder infections

Historical Note

The First Insulin Injection
Insulin was first isolated in 1922 by two doctors working in a Toronto hospital. On January 11, 1922, they injected an extract purified from beef pancreas glands into a 13-year-old boy, a diabetic whose weight had fallen to 65 lb. and who was not expected to survive. Following the injection, the glucose level in the boy's blood fell 25 percent—his cells were taking up glucose. A more potent extract administered later brought the boy's blood-glucose level down to near normal. This was the first instance of successful insulin therapy.

Other Organs and Glands Produce Hormones

In addition to the organs and glands mentioned so far, several other organs and glands produce hormones. Among these are the female and male reproductive organs, the ovaries and the testes. These organs, which also produce gametes, secrete hormones that regulate reproduction. Ovaries secrete **estrogen** and **progesterone,** and testes produce **testosterone.** These hormones not only affect the formation of gametes but also control the development of the secondary sex characteristics, such as breasts in females and coarse facial hair in males.

CONTENT LINK

The reproductive roles of the ovaries and testes and the hormones they produce are described in **Chapter 42.**

Daily Human Body Temperature Variation

Body temperature (°F) vs. Time of day

- Awake period
- Asleep period

100, 99, 98, 97, 96, 95

6 A.M.   6 P.M.   6 A.M.

Figure 41-21 Sleep, *above left,* is an example of a daily biorhythm. A graph, *above,* shows how body temperature fluctuates daily, another type of biorhythm.

The **pineal** (*PIHN ee uhl*) **gland** is a pea-sized gland located in the top of the brain and named for its resemblance to a pine cone. The pineal gland secretes **melatonin,** which is a modified form of the amino acid tryptophan. In hamsters, melatonin regulates reproduction, and in frogs it influences pigmentation. However, the function of the pineal gland in humans is not yet known. Although the pineal gland is not connected directly to the central nervous system, it is connected to the eyes through the sympathetic nervous system. Also, melatonin seems to be released by the human pineal gland as a response to darkness. Therefore, the pineal gland is thought to be involved in establishing daily biorhythms, such as those seen in Figure 41-21. The pineal gland has also been implicated in mood disorders such as winter depression, also called seasonal affective disorder syndrome (SADS), and in a variety of aspects of sexual development.

Other organs that produce hormones and thus act as endocrine glands include the stomach, small intestine, and heart. As discussed in Chapter 38, the stomach and small intestine secrete hormones, such as gastrin, that regulate the release of acids and digestive enzymes and play a key role in digestion. The heart secretes atrial (*AY tree uhl*) natriuretic (*na tree yoo REHT ihk*) factor (ANF). Cells in the blood vessels, kidneys, and adrenal glands have receptors for ANF, which apparently helps regulate blood pressure.

CAPSULE SUMMARY

The pineal gland, which is located in the brain, and several other body organs, such as the ovaries, testes, stomach, and heart, produce hormones and thus act as endocrine glands.

Application

Health Many human processes are affected by the intensity, color mixture, and timing of light. Changes in the regular pattern of any of these variables in a person's life can disrupt normal functions. For example, the relatively rare emotional disorder called seasonal affective disorder syndrome (SADS) is characterized by inappropriate mood swings. SADS seems to be brought on by the reduced amount of daylight in autumn. Sufferers can be helped with phototherapy, which involves exposure to bright light for extended periods each day. Other light-dependent biorhythmic effects include jet lag and the disorientation sometimes experienced by people who work nights. These effects seem to be mediated by changes in the secretion of melatonin. Additionally, the variables mentioned above seem to affect the function of the immune system. Research has shown that ultraviolet light plays a part in the activation of certain kinds of white blood cells.

Teaching Tip
Antagonistic Hormones
Have students use their textbook to identify pairs of hormones that have antagonistic (opposite) effects. You may want to encourage students to do some outside research for this assignment. To summarize their findings, have students prepare a chart similar to the one shown in the *Graphic Organizer* on page 983. Then have students write a paragraph describing the importance of antagonistic hormones in maintaining homeostasis.

Graphic Organizer

Use this graphic organizer with *Teaching Tip:* **Antagonistic Hormones** on page 983.

Hormone pair	Action	Site of action	Site of production
Insulin	Lowers blood sugar	All tissues	Pancreas
Glucagon	Raises blood sugar	Liver, fatty tissue	Pancreas
Continue with others...			

☑ **RESEARCH Update**

Benefits of Androgens in Females

Most people are familiar with the major hormones of "maleness" (androgens) and "femaleness" (estrogens). However, few know that estrogens and androgens exist in both sexes, albeit in very different concentrations. Testosterone, the principal male androgen, ranges in value from 3.0 to 9.0 mg in males and averages about 0.4 mg in females. Testosterone is produced in the adrenal gland, and, in females, it is also produced in the ovaries. Testosterone and other androgens are responsible for the development of secondary sex characteristics in males and, to an extent, in females.

Currently, there is a transatlantic mini-feud over the advantages and disadvantages of administering androgens to females. Why would a physician administer androgens to a female patient? Dr. John Studd, an OB-GYN at Chelsea and Westminster Hospital in London reports very favorable results in patients claiming reduced libido and energy. Many of these patients have had their ovaries removed and therefore have lost a considerable source of androgens. On this side of the Atlantic, Dr. Goeffrey Redmond, president of the Foundation for Developmental Endocrinology, and many of his colleagues believe that an excess of androgens is a greater health problem than a lack of androgens. In females, excess androgens can contribute to baldness, acne, infertility, breast cancer, growth of facial and other body hair, diabetes, high blood pressure, and heart disease. Redmond and other cautious researchers feel that women are better off without them. ☑

Table 41-2 Major Endocrine Glands and Hormones

Gland	Hormone	Target Tissue	Effects
Pituitary gland	**Anterior lobe** Adrenocorticotropic hormone (ACTH)	Adrenal glands	Stimulates the production of steroid hormones
	Follicle-stimulating hormone (FSH)	Ovaries and testes	Regulates the development of male and female gametes; stimulates the production of testosterone (male sex hormone) in males
	Luteinizing hormone (LH)	Ovaries and testes	Stimulates the release of an egg (ovulation) from an ovary; stimulates testosterone production by the testes
	Prolactin	Mammary glands	Stimulates milk production in breasts
	Somatotropin, or growth hormone (GH)	All tissues	Promotes protein synthesis; stimulates growth of muscles and bones
	Thyroid-stimulating hormone (TSH)	Thyroid gland	Stimulates production of thyroxine by the thyroid gland
	Posterior lobe Antidiuretic hormone (ADH)	Kidneys, blood vessels	Stimulates reabsorption of water; constricts blood vessels
	Oxytocin	Mammary glands, uterus	Stimulates uterine contractions and milk secretion
Adrenal glands	**Cortex** Aldosterone	All tissues	Controls salt (sodium and potassium) and water balance
	Cortisol	Kidneys	Stimulates metabolism of carbohydrates, lipids, and proteins; raises blood sugar
	Medulla Epinephrine (adrenaline) and norepinephrine	Skeletal and cardiac muscle, blood vessels	Initiates the response to stress; increases metabolic rate, heart rate, and blood pressure; dilates blood vessels; raises blood sugar
Islets of Langerhans (Pancreas)	Glucagon	Liver, fatty tissues	Stimulates conversion of glycogen to glucose; raises blood sugar
	Insulin	All tissues	Stimulates conversion of glucose to glycogen; lowers blood sugar

Gland	Hormone	Target Tissue	Effects
Parathyroids	Parathyroid hormone	Bone tissue, digestive tract, kidneys	Stimulates breakdown of bone tissue and absorption of calcium by kidneys; raises blood calcium; activates vitamin D
Pineal	Melatonin	Uncertain, possibly ovaries and testes	May regulate biorhythms and moods; may affect the onset of puberty
Thyroid	Calcitonin	Bone tissue	Inhibits loss of calcium from bone; lowers blood calcium
	Thyroxine	All tissues	Raises metabolic rate; necessary for normal growth and development
Ovaries	Estrogen	All tissues, female reproductive structures	Controls development of secondary female sex characteristics and sex organs; initiates preparation of the uterus for pregnancy
	Progesterone	Uterus, breasts	Completes preparation of the uterus for pregnancy; stimulates breast development
Testes	Testosterone	All tissues, male reproductive structures	Controls development of secondary male sex characteristics and sex organs; stimulates sperm formation

Section Review

1. Which endocrine glands are actually two glands in one? Which of these glands produces the greatest number of hormones?

2. What is diabetes mellitus? How is it treated?

3. What are the consequences of an underproduction of thyroxine during childhood? How are adults affected by underproduction of thyroxine?

4. How is parathyroid hormone essential to your health?

Critical Thinking

5. Why is the pituitary gland often referred to as the master gland?

SECTION 41-3

Teaching Tip
Problem Solving

Have students suggest a diagnosis for each of the following problems, based on the information in Table 41-2. A deficiency in which pituitary hormone might cause sterility in both males and females? *(FSH)* An untimely release of which pituitary hormone might cause premature birth? *(oxytocin)* Which hormones have a direct effect on blood sugar? *(cortisol, epinephrine, norepinephrine, glucagon, and insulin)* If a person were to develop a deficiency in cholesterol, which hormones might be affected? *(aldosterone, cortisol, estrogen, progesterone, and testosterone)* Which glands would be implicated if a person began to lose bone mass abnormally? Explain their possible roles in the loss. *(Parathyroid glands might be producing too much PTH, or the thyroid gland might be producing too little calcitonin.)*

Chapter Closure

Have students respond in writing to the following questions: Which systems contain organs that are considered a part of the endocrine system as well as another system? *(the reproductive, nervous, circulatory, and digestive systems)* Which systems are affected by the endocrine system but appear not to be a part of it? *(the integumentary, muscular, and immune systems)* Emphasize that overall health of the body, as maintained by these three systems, does have a bearing on the endocrine system.

Answers to Section Review

1. Both the adrenal glands and the pituitary are two glands in one. The pituitary gland produces the most hormones.

2. Diabetes mellitus is a disorder caused by improper glucose metabolism, which results from a disruption in the mechanism that facilitates the entry of glucose into cells. Type II diabetes, in which the body produces some insulin but is unable to use it effectively, is usually controlled with diet and exercise. Type I diabetes is treated with insulin injections.

3. Hypothyroidism in childhood results in mental retardation and stunted growth (cretinism). In adults, hypothyroidism causes dry skin, lack of energy, depression, and weight gain.

4. Parathyroid hormone helps maintain a proper blood-calcium level.

5. The pituitary gland produces hormones that control the other endocrine glands of the body.

41
CHAPTER REVIEW

CHAPTER REVIEW ANSWERS

Each item in the Chapter Review is correlated by text section in the assignment guide that follows.

ASSIGNMENT GUIDE

Section	Review	Themes Review	Critical Thinking
41-1	1–3, 11, 12	18	21, 22
41-2	4–6, 13		
41-3	7–10, 14–17	18, 19	20–22

Review

Multiple Choice

Code in parentheses indicates section number and objective number.

1. c (41-1.1)
2. c (41-1.3)
3. a (41-1.4)
4. c (41-2.2)
5. b (41- 2.1)
6. d (41-2.3)
7. b (41-3.1)
8. b (41-3.1)
9. d (41-3.1)
10. c (41-3.1)

Completion

11. hormones, nerve impulses (41-1.1)
12. Steroid, peptide (41-1.2)
13. Steroid, peptide (41-2.1, 41-2.2, 41-2.3)
14. posterior, anterior (41-3.1)
15. glucagon, insulin (41-3.1, 41-3.2)
16. II, I (41-3.2)
17. calcium, salt (sodium and potassium) (41-3.3)

---

41 CHAPTER REVIEW ....

acromegaly (978)	glucagon (974)	peptide hormone (969)
adrenal cortex (980)	goiter (979)	pineal gland (983)
adrenal medulla (980)	hormone-receptor complex (975)	pituitary gland (971)
adrenaline (980)	hyperthyroidism (979)	posterior lobe (978)
aldosterone (981)	hypothyroidism (979)	prednisone (981)
anterior lobe (978)	insulin (982)	progesterone (983)
calcitonin (979)	islets of Langerhans (982)	prostaglandin (970)
cortisol (975)	melatonin (983)	second messenger (973)
cyclic AMP (973)	neuroendocrine system (971)	steroid hormone (969)
diabetes mellitus (982)	neuromodulator (970)	target cell (969)
endocrine gland (968)	neuropeptide (970)	testosterone (983)
endorphin (970)	norepinephrine (981)	thyroxine (976)
estrogen (983)	parathyroid gland (979)	Type I diabetes (982)
gigantism (978)	parathyroid hormone (979)	Type II diabetes (982)

Review

Multiple Choice

1. The chemical messengers of the endocrine system are
　a. neurons.　　　c. hormones.
　b. blood cells.　　d. pheromones.

2. The nonendocrine chemical signals secreted by the brain
　a. are never derived from amino acids.
　b. regulate high blood pressure.
　c. include neuropeptides and neurotransmitters.
　d. do not travel through the bloodstream.

3. The production of hormones by glands of the endocrine system is initiated by the
　a. hypothalamus.
　b. nonendocrine glands.
　c. hormone-receptor complex.
　d. medulla of the adrenal glands.

4. Which of the following describes how a steroid hormone produces a response?
　a. The hormone binds to a receptor on the cell membrane.
　b. The hormone first binds to mRNA.
　c. A hormone-receptor complex binds to DNA.
　d. The hormone passes directly into the nucleus.

5. Peptide hormones may use cyclic AMP as a
　a. receptor.
　b. second messenger.
　c. target cell.
　d. coenzyme.

6. The receptor protein for thyroxine is located
　a. in the cytoplasm.
　b. on mRNA.
　c. on the cell membrane.
　d. in the nucleus.

7. Insulin promotes glucose uptake by cells, which leads to
　a. higher blood sugar.
　b. lower blood sugar.
　c. release of additional insulin.
　d. glycogen breakdown.

8. A pituitary tumor in a child could lead to
　a. hyperthyroidism.　c. Type I diabetes.
　b. gigantism.　　　　d. hypothyroidism.

9. What adrenal cortex hormone acts to reduce inflammation?
　a. calcitonin　　c. prostaglandin
　b. aldosterone　d. cortisol

10. Which of the following endocrine glands secretes melatonin and is believed to be involved in establishing biorhythms?
　a. pituitary gland　c. pineal gland
　b. thyroid gland　　d. adrenal gland

Completion

11. Responses caused by _____ are slower but last longer than those caused by _____ .

12. _____ hormones are fat-soluble, and _____ hormones are water-soluble.

13. _____ hormones bind to receptors in the cytoplasm, while _____ hormones remain outside of cells.

14. The _____ pituitary releases hormones that are produced by the hypothalamus. Hormones secreted by the _____ pituitary induce other endocrine glands to secrete particular hormones.

15. Low blood sugar stimulates the release of _____ , while high blood sugar stimulates the release of _____ .

16. A treatment for Type _____ diabetes is proper diet and exercise, while treatment for Type _____ diabetes usually involves insulin injections.

17. Parathyroid hormone regulates _____ level in the blood, and aldosterone regulates _____ level in the blood.

Themes Review

18. **Homeostasis** Explain how your hypothalamus and endocrine glands work together to maintain homeostasis.

19. **Structure and Function** Describe the structure of an adrenal gland, and explain how it acts as two glands in one.

Critical Thinking

20. **Identifying Variables** Suppose that a friend tells you that he or she has recently experienced some of the warning signs of diabetes mellitus. What else could cause symptoms similar to diabetes mellitus. What questions could you ask your friend to determine whether there may be another cause for these symptoms?

21. **Making Inferences** Before iodide was added to table salt, goiters were common among people living in inland regions but were rare among people living in coastal areas. Why do you think this was so?

22. **Making Inferences** The graph below shows the blood glucose levels of three experimental rats, measured over time. At time T_1, two rats received injections of a saline solution plus one hormone, and the control rat was injected with a saline solution only. Which rat (A, B, or C) received insulin? Which rat received glucagon? Which rat was the control?

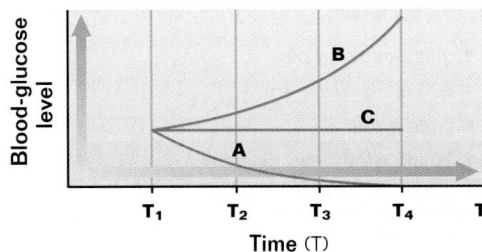

Activities and Projects

23. **Research and Writing** Urine tests that detect steroid use by athletes are being advocated by many high school coaches. Interview several coaches to determine their attitudes toward steroid testing. Write an article for your school newspaper that discusses your findings and explains how steroids affect the body and why doctors oppose their use.

24. **Cooperative Group Project** Using materials such as sturdy cardboard, paint, batteries, flashlight bulbs, wire, and brads, build a model that shows the location of the endocrine glands in the human body. Electrical circuits can be used to "light up" a light at a particular gland when a wire is touched to its name on a key. Display the model in your classroom where other students can use it as a study guide.

Critical Thinking

20. Many other conditions, such as stress, the menstrual cycle, diet, and disease, may produce similar symptoms. Questions should focus on determining whether one of these factors is affecting the student's friend. However, students should be cautioned against self-diagnosis and diagnosing disorders in their friends. Only a doctor can perform the tests necessary to determine whether a person has a disease or disorder. A question that should always be asked in such cases is, "Have you described your symptoms to your doctor?" (41-3.1)

21. People living in coastal regions experienced fewer goiters because they ate sea food, which contains iodide. (41-3.2)

22. Rat A received insulin, Rat B received glucagon, and Rat C was the control. (41-3.2)

Activities and Projects

23. Encourage students to interview both male and female coaches. Doctors oppose the use of steroids because of their dangerous side effects.

24. Different members of each group can take responsibility for collecting materials, drawing or painting the human figure, and hooking up the electric circuit.

Themes Review

18. The pituitary releases hormones in response to signals from the hypothalamus, and stores and releases hormones produced by the hypothalamus. An "inhibiting" hormone from the hypothalamus keeps certain pituitary hormones from reaching an excessive level. The hormones released by the pituitary gland control many of the body's other endocrine glands. (41-1.4, 41-3.1)

19. An adrenal gland has an inner core called the medulla and an outer shell called the cortex. An adrenal gland acts as two glands in one because the two parts produce different types of hormones that also have different functions—the medulla produces the peptide hormones epinephrine and norepinephrine, and the cortex produces the steroid hormones cortisol and aldosterone. (41-3.1)

LABORATORY Investigation Chapter 41

Effects of Thyroxine on Frog Metamorphosis

OBJECTIVE

Determine the effect of the hormone thyroxine on the development of tadpoles.

PROCESS SKILLS

- measuring and comparing anatomical features
- organizing data using tables and graphs

MATERIALS

- glass-marking pen
- 600 mL beakers (6)
- pond water
- 10 mL graduated cylinder
- 0.01% thyroxine solution
- strained spinach
- graph paper marked in 1 mm squares
- petri dish
- small fish net
- 9 tadpoles with hind legs just beginning to show
- 3 pencils in different colors

BACKGROUND

1. What is a hormone?
2. What is metamorphosis?
3. Describe the stages of frog development.
4. What is the importance of thyroxine in human development? What gland secretes thyroxine?
5. Write your own question to explore in your lab report or notebook.

TECHNIQUE

1. Use a glass-marking pen to label three beakers A, B, and C. Add your initials to each.
2. Add 500 mL of pond water to each beaker.
3. Use a graduated cylinder to measure 10 mL of thyroxine solution, and add the solution to beaker A. Add 5 mL of thyroxine solution to beaker B. Add nothing to beaker C.
4. Add approximately 1 cm³ of strained spinach to each beaker.
5. Place a sheet of graph paper, ruled side up, underneath a petri dish.
6. **CAUTION: You will be working with live animals. Be sure to treat them gently and follow directions carefully.** Catch a tadpole with a fish net and place the tadpole in the petri dish. Measure the tadpole's total length, tail length, and body length in millimeters by counting the number of squares it covers on the graph paper. Place the tadpole in beaker A.
7. Repeat step 6 with two more tadpoles. Average the total length, tail length, and body length of the three tadpoles. In the **Records** section, record these averages in a table similar to the table shown on the next page.
8. Repeat step 6 with six more tadpoles, placing three tadpoles in beaker B and three in beaker C.
9. Feed the tadpoles about 1 cm³ of spinach every other day. Be careful to avoid overfeeding. Change the water every four days, adding thyroxine solution to beakers A and B in the original amounts.
10. Measure the tadpoles once a week for three weeks, and average the lengths of the tadpoles in each beaker. Record the average lengths in the table you

---

Preparation

Each lab group will need 9 tadpoles with hind legs just beginning to show, six 600 mL beakers, 3 pencils of different colors, pond water, strained spinach, and graph paper marked in 1 mm squares, 0.01% thryoxine solution, a glass-marking pen, 10 mL graduated cylinder, petri dish, and small fish net. Because developing tadpoles are not always available, order them from a biological supply house well in advance of this investigation. To prepare the 0.01% thyroxine solution, dissolve 10 mg crystalline thyroxine in 5 mL of 1% sodium hydroxide solution. Dilute this solution by adding distilled water to make a total volume of 1 L. Keep the solution refrigerated. If pond water is not available, set out open containers of water the night before the lab session. Baby food is a good source of strained spinach.

Procedural Notes

- This investigation should take one class period to set up and one 15 min. observation period once a week for three weeks.
- This investigation works well when students are divided into groups of three.
- **Safety:** Remind students that tadpoles are living animals and that they should handle the tadpoles with care.

Background Answers

1. Hormones are chemicals that circulate through the bloodstream and affect certain physiological activities. In animals, hormones are secreted by endocrine glands and cells.

2. Metamorphosis is the changing of an organism from an immature stage to a mature stage.

3. Frog eggs (laid in water) hatch into larvae (tadpoles), which first develop hind legs then front legs. As their tails shrink and their legs develop, tadpoles develop lungs and then leave the water to continue growth as a frog.

4. Thyroxine, which is secreted by the thyroid gland, increases the rate of cell metabolism. A thyroxine deficiency can result in hypothyroidism or stunted growth and mental retardation if the deficiency has existed since early childhood. Excessive thyroxine causes hyperthyroidism.

5. How does thyroxine affect the growth and development of tadpoles?

Records

Students should construct and complete a table similar to the one shown on page 989.

Measurement of Tadpole Growth

	Beaker A			Beaker B			Beaker C		
	Avg. total length	Avg. tail length	Avg. body length	Avg. total length	Avg. tail length	Avg. body length	Avg. total length	Avg. tail length	Avg. body length
Initial									
End of week 1									
End of week 2									
End of week 3									
Growth in 1st week									
Growth in 2nd week									
Growth in 3rd week									

constructed in the **Records** section of your report.

11. Calculate the average growth per week for each group of tadpoles. For example, the average growth in total length during the second week is equal to the average total length at the end of week 2 minus the average total length at the end of week 1. Record these values in the appropriate spaces in the table in the **Records** section of your report.

12. Graph your data using different colored pencils for the tadpoles in beaker A, beaker B, and beaker C. In the **Procedure** section of the Vee, briefly summarize the procedure you followed.

 13. Clean up your materials and wash your hands before leaving the lab.

INQUIRY

1. What is the purpose of putting some tadpoles in water without the thyroxine solution?

2. Why are three tadpoles used for each solution, rather than just one?

ANALYSIS

1. What is the effect of thyroxine on tadpole metamorphosis?

2. Which concentration of thyroxine caused the greatest visible change in the tadpoles?

3. How do average body length and tail length change during metamorphosis?

FURTHER INQUIRY

Write a new question that could be explored as a new investigation. The following is an example:

What is the effect of iodine on frog development?

UNIT 9

SCIENCE, TECHNOLOGY, AND SOCIETY

Point of View: Breast Cancer

Objectives

- Develop scientific literacy
- Identify an author's point of view
- Detect bias in a persuasive argument

Background

Each year scientific issues become more important in determining the nature and direction of our society, and yet most citizens have little or no training in evaluating the arguments that surround controversial issues. The purpose of this feature, Point of View, is to provide practice in identifying and evaluating a writer's point of view.

A critical step in this process is detecting bias. Much of the writing on critical issues is persuasive and not objective. A writer will include information that bolsters his or her point of view and suppress evidence that contradicts it.

Instructional Strategies

- As students read this article, remind them that it represents one person's point of view. They must read critically, looking for signs that would indicate biased or slanted writing. In particular, have students look closely for the following:
  1. Does the article present both sides of the issue fairly and objectively?
  2. Is the writer seeking to shape your opinion on the issues?
  3. Does the writer use "loaded" language? Are some words chosen because they have an emotional rather than a logical appeal to the reader?
- Write the words *precautionary principle* on the chalkboard. Ask students what they think it means. Explain that according to this principle, the use and release of

chemicals that may cause harm should be avoided until they have been proved safe. Ask students whether they agree with this principle and whether they think it is a practical principle to try to follow.

- Have students research current methods for creating environmental and health policies for the use of chemicals.

BREAST CANCER

What's Really Behind the Epidemic?

The growth of cancer cells in the body may be related to exposure to certain chemicals, like those routinely applied as pesticides.

BY TRACEY COHEN

The following article presents one person's editorial viewpoint regarding possible conflicts of interest that can influence scientific research priorities. Read this article critically to determine whether the author provides sufficient factual evidence to support her opinion.

About 182,000 women in the United States will be found to have breast cancer this year. Nearly 2 million women have been diagnosed with the disease and perhaps another 1 million have it but do not yet know it. Standard medical treatments have had little impact on the long-range outcome of the disease. The death rate from metastatic breast cancer—disease that has spread to other parts of the body—has remained unchanged for over 40 years. Twenty five percent of women with metastatic breast cancer die within 5 years of their diagnosis and 40 percent die within 10 years. Unfortunately, what has changed is a woman's chance of developing breast cancer at some

time in her life. In 1940, that chance was 1 in 20. Today the National Cancer Institute estimates that by the time a woman is 85 years old, her chance is 1 in 8.

What Causes Breast Cancer?

Scientists have identified several factors that appear to increase a woman's chance of getting breast cancer. Having a close relative like a mother or sister with the disease is one. Another risk factor is increased lifetime exposure to natural or synthetic estrogen. For instance, early puberty, late menopause, having a first child after age 30, or not having children at all increases a woman's exposure to estrogen. A diet high in fats may also put women at risk, but the actual role of dietary fat is still controversial.

Altogether known risk factors account for just 20 to 30 percent of all breast cancers. The majority of women with the disease have none of these. The lack of progress in treatment and the limited number of cases explained by known risk factors is both frightening and frustrating. As a result, some researchers now think that exposure to

pollutants and other chemicals may be responsible for the breast cancer epidemic.

Is Breast Cancer an Environmental Disease?

In the growing body of evidence linking breast cancer to environmental causes, one group of compounds—organochlorines—is highly suspect. Organochlorines include DDT and other pesticides; PCBs, which are used in electrical transformers; and PVC, used to make many kinds of plastic. About 11,000 different organochlorines are used by industries. Hundreds of other organochlorines are formed as byproducts of industrial processes like bleaching paper pulp and disinfecting wastewater. Burning garbage containing plastics and other chlorinated materials also creates organochlorines.

Organochlorines are highly toxic and slow to degrade. Studies have found 177 different organochlorines in samples of fat, blood, semen, and mother's milk taken from people living in the United States and Canada.

Research shows that organochlorines can have a number of harmful effects on the body. Some organochlorines cause genetic mutations that lead to

Then ask them to compare current practice with the precautionary principle. Suppose the precautionary principle were adopted by government agencies. Ask students to predict the effects that using the precautionary principle would have on different aspects of society.

Answers to Analyzing the Issue

1. Answers will vary. Opinions may include the following statements: ties between the chemical companies and the organizations devoted to cancer research are too close; lawmakers may be unduly influenced by lobbyists for the chemical industry.

cancer. Some suppress the immune system, which may also lead to cancer. Others interfere with sex hormones, acting like synthetic estrogen. The International Agency for Research on Cancer reports over 100 organochlorine compounds cause cancer in humans or other animals.

One study supporting the link between breast cancer and environmental pollutants comes from Israel. Before 1976, the breast cancer death rate among Israeli women under age 44 was unusually high. At the same time, high concentrations of three pesticides—DDT, BHC, and lindane—were found in women's breast milk and in commercial milk and dairy products. In 1976, use of the three pesticides was banned. By 1978, DDT levels in breast milk dropped 43 percent, lindane levels dropped 90 percent, and BHC levels dropped 98 percent. Less than a decade later, the breast cancer rate for Israeli women under age 44 had fallen 30 percent. In the time period studied, Israel was the only one of 28 countries observed in which the death rate from breast cancer actually declined.

Pesticides in current use have also been linked to cancer, either in humans or in other animals. Atrazine, used by farmers to kill weeds, has been shown to cause mammary cancers in rodents. It also appears to increase the risk of ovarian cancer in women. Farmers in the United States apply between 70 million and 90 million pounds of atrazine to their fields each year.

Federal law requires that the National Cancer Institute expand its research program on preventing cancers caused by exposures to environmental or workplace carcinogens. But at a congressional budget hearing in October of 1993, NCI Director Samuel Broder testified that the NCI spent only 1% of its almost $2 million budget on environmental cancer studies, and he gave no indications that the NCI

planned to shift its priorities. NCI research will continue to emphasize the treatment and cure of cancer. Of the part of the budget given to researching causes of the disease, the two areas to receive the most emphasis will remain the links to diet and smoking.

Many health specialists have criticized this approach. They want additional funds devoted to research into identifying carcinogens and studying the link between exposure and the incidence of cancer. Some critics have speculated that the ties between the chemical companies and the organizations devoted to cancer research are too close. These critics point to the interlocking relationships among cancer researchers, policy makers, and industry representatives. They question appointments such as that in the 1980s of Armand Hammer, then president of Occidental Petroleum, to a position as chair of the President's Cancer Advisory Board.

The fear often expressed by critics is that lawmakers may be unduly influenced by lobbyists for the chemical industry. Critics also worry that chemical companies have a vested interest in protecting their own industry. Will chemical companies be willing to support research that may identify their products as probable causes of cancer?

Industry spokespeople, on the other hand, claim a role in helping to identify and solve the problem of carcinogens. They argue that the funds donated by the chemical companies for research are essential to finding both the causes and treatment of cancer. The chemical companies view their involvement as stemming from a natural, communal interest—not as a conflict of interest. From this viewpoint, everyone involved is focused on the problem but working from a different perspective.

The issue is complicated, but it is clear that the value of scientific research is its ability to confirm or deny the link between a substance and the incidence of cancer. The problem of how this research will be funded and directed is the issue. To be effective, a researcher needs a degree of independence, and critics of the current policy worry that researchers will be reluctant to "bite the hand that feeds them." As long as chemical companies are directly involved, many health specialists worry that the objectivity of the research may be compromised.

Tracey Cohen is a freelance writer specializing in science and environmental issues.

Analyzing the Issue

1. **Detecting Bias** Do you think the argument presented in the article is convincing? Why or why not? Where does the writer use opinions rather than facts to make a point?

2. **Get the Facts** Go to the library and find at least five books or articles on this subject. Make a list of facts you can use to challenge the writer's argument. Then write a counterargument

supporting a different point of view.

3. **Examine Social Consequences** Find out how many chemicals are made each year. How does the U. S. Environmental Protection Agency determine their safety or risk? Is this method adequate to protect public health and the environment? Explain.

2. Answers will vary. Students' arguments should be supported by the facts they collect. You may wish to give students the list of references used in developing this essay.

3. Hundreds of new chemicals are made each year. Altogether, about 80,000 chemicals are used by various industries. But only 1 or 2 percent of the chemi-

cals have been tested for basic toxicity, and none have been evaluated for widespread hazard. Industry has been successful in insisting that chemicals be proved unsafe only after they are in widespread use, rather than determining their safety beforehand.

SCIENCE, TECHNOLOGY, AND SOCIETY

References

1. Beardsley, T. "A War Not Won." *Scientific American,* Jan. 1994, pp. 130–138.
2. Castleman, M. "Cover Story: Breast Cancer and Environment." *Mother Jones,* May/June 1994, pp. 34–42.
3. Clorfene-Casten, L. "The Environmental Link to Breast Cancer." *Ms.,* May/June 1993, pp. 52–57.
4. Colborn, T., F. S. vom Saal, and A. M. Soto. "Developmental Effects of Endocrine-Disrupting Chemicals in Wildlife and Humans." *Environmental Health Perspectives,* Vol. 101, No. 5, Oct. 1993, pp. 378–384.
5. Epstein, S. S. "Are We Losing the War Against Cancer?" *Congressional Record, Extensions of Remarks,* Sept. 9, 1987, pp. E3449–E3454.
6. Goldberg, K., ed. "House Committee Hits 'Earmarks' for Diseases; Broder Is Lectured on Occupational Exposures." *The Cancer Letter,* Vol. 19, No. 21, May 21, 1993, pp. 1–4.
7. Moss, R. *The Cancer Industry.* Paragon House, New York, 1989.
8. Paulsen, M. "The Cancer Business." *Mother Jones,* May/June 1994, p. 41.
9. Paulsen, M. "The Politics of Cancer: Why the Medical Establishment Blames Victims Instead of Carcinogens." *Utne Reader,* Nov./Dec. 1993, pp. 81–89.
10. Raloff, J. "EcoCancers." *Science News,* Vol. 144, July 3, 1993, pp. 10–13.
11. Thornton, J. *Chlorine, Human Health and the Environment: The Breast Cancer Warning.* Greenpeace, Washington, D.C., 1993.
12. Westin, J. B. and E. Richter. "The Israeli Breast-Cancer Anomaly." *Annals of the New York Academy of Science,* Vol. 609, 1990, pp. 269–279.
13. Wolff, M. S., et al. "Blood Levels of Organochlorine Residues and Risk of Breast Cancer." *Journal of the National Cancer Institute,* Vol. 85, 1993, pp. 648–652.

UNIT 9

REPRODUCTION AND DEVELOPMENT

Block Scheduling Guide

Each block represents about 45–50 minutes of instructional time. The resources cited in a block are coded to a specific program component using the Key on the next page.

Lecture Concepts	Lesson Resources

42-1 The Male Reproductive System pp. 993–996

BLOCK 1

a. The testes produce the male gametes, or sperm, and the male hormone, testosterone. As sperm cells pass through the male reproductive tract, fluids secreted by glands are added to the sperm to make semen.

b. During ejaculation, the penis delivers sperm to the female reproductive tract.

c. Fertilization can be prevented by blocking the path of sperm as they leave the male body.

Lecture Resources
- Opening Question, page 993
- Visual Strategies: Figures 42-1, 42-3
- Overcoming Misconceptions, page 994
- Function of Seminal Fluids, page 995
- Teaching Transparencies: 224, 225, 232
- Holt Biology Videodiscs: Lesson 42-1

Classwork Options
- Effective Reading, page 993
- Human Male Reproductive Structures and Functions, page 996

Assignment Options
- Section Review, page 996
- Chapter Review, questions 1, 3, 11, 18, 21

42-2 The Female Reproductive System pp. 997–1002

BLOCK 2

a. The ovaries produce the female gametes, or ova. At puberty, immature eggs begin to mature one at a time.

b. Fertilization begins in the fallopian tubes. Blocking the path of the sperm to the fallopian tubes prevents fertilization. When a fertilized egg enters the uterus, it implants in the uterine wall.

c. During the ovarian cycle, FSH from the pituitary gland causes an ovum to mature inside a follicle. After ovulation, LH converts a ruptured follicle to a corpus luteum.

d. During the menstrual cycle, estrogen and progesterone produced by the follicle and corpus luteum prepare the uterus for possible pregnancy.

Lecture Resources
- Opening Question, page 997
- Demonstration, page 997
- Accomplishing Fertilization, page 998
- Visual Strategies: Figures 42-7, 42-8, 42-9
- Teaching Transparencies: 224, 226
- Holt Biology Videodiscs: Lesson 42-2

Classwork Options
- Methods of Contraception, page 999
- Hormones and the Female Reproductive Process, page 1002

Assignment Options
- Section Review, page 1002
- Chapter Review, questions 2–5, 12–14, 19, 22

42-3 Human Development pp. 1003–1006

BLOCK 3

a. During the first eight weeks after fertilization, a developing human is called an embryo. After that, the embryo is called a fetus.

b. Human development is most easily disturbed during the first trimester of pregnancy.

c. Growth and neurological development continue after birth, which occurs about nine months after fertilization.

Lecture Resources
- Opening Question, page 1003
- Demonstration: Actual Sizes of a Developing Human
- Visual Strategies: Figures 42-11, 42-12
- Drug Use During Pregnancy, page 1005
- Research Update, page 1006 Teaching Transparencies: 227, 228, 196A

- Holt Biology Videodiscs: Lesson 42-3

Classwork Options
- Laboratory Investigation: Embryonic Development, pages 1012–1013

Assignment Options
- Section Review, page 1006
- Chapter Review, questions 6–8, 15, 20

42-4 Sexually Transmitted Diseases pp. 1007–1009

BLOCK ④

a. Diseases that can be spread by sexual contact are called sexually transmitted diseases (STDs).

b. STDs caused by viruses cannot be cured with antibiotic treatment.

c. STDs caused by bacteria are difficult to detect but can be cured with antibiotic treatment.

Lecture Resources
- ■ Opening Question, page 1007
- ■ Demonstration: Models of Organisms Causing STDs
- ■ Visual Strategies: Figures 42-16, 42-18
- ■ Comparing Viral and Bacterial STDs

Classwork Options
- ■ STDs—A Summary, page 1009
- 🧪 Inquiry Skills Development B34: Embryonic Development

Assignment Options
- ■ Section Review, page 1009
- ■ Chapter Review, questions 9, 10, 16, 17

Review/Enrichment

BLOCK ⑥

- ■ Study Guide: Concept Review and Pretest

Assignment Options
- ■ Activities and Projects, questions 23–25

Assessment Options

BLOCK ⑦

Traditional Assessment
- ■ Chapter 42 Test
- 💿 Test Generator/Assessment Item Listing: Software item bank for preparing customized chapter tests.

Portfolio Assessment
Select student reports for one or more laboratory experiments from this chapter. *The Direct Observation Checklist* on the *BioSources Teaching Resources CD-ROM* should be completed during a laboratory or other cooperative learning experience.

Holt Biology Videodiscs   CONCEPTS OF BIOLOGY

Holt Biology Videodiscs gives you a powerful tool for teaching, review, and assessment. *Concepts of Biology* adds interactive lesson plans and extensive tools for customization using CD-ROM technology.

Use the following topic from *Concepts of Biology* to help you teach this chapter:

- ■ Topic 26: Reproduction and Development

For further information regarding the media on the videodiscs, see *Holt Biology Videodiscs Teacher's Correlation Guide to Biology: Principles and Explorations.*

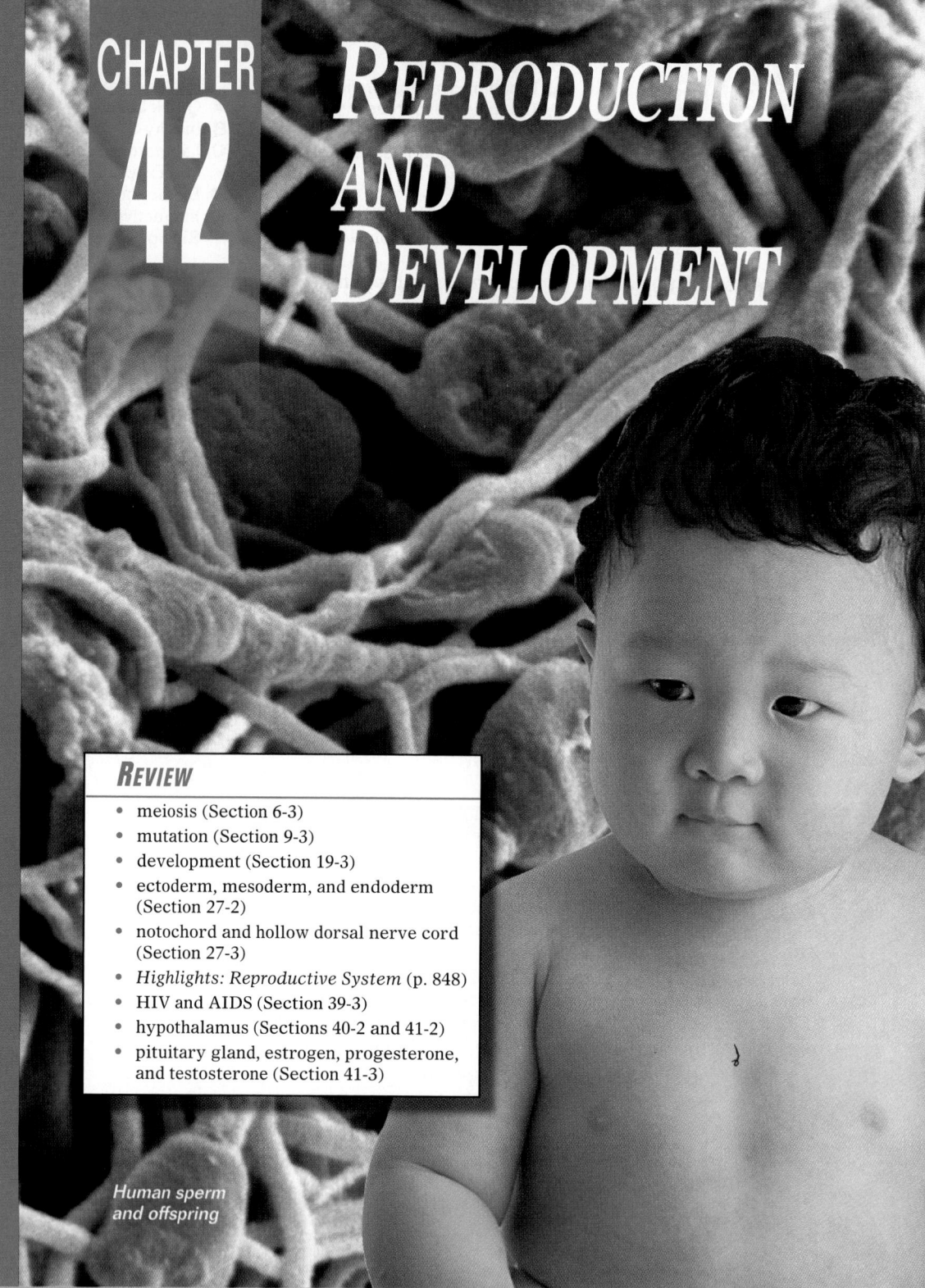

Chapter Themes

Structure and Function

An organism's structures and their ability to function derive from the events of reproduction and development. Figures 42-1, 42-2, 42-3, 42-5, 42-7, and 42-12 illustrate this theme.

Levels of Organization

Beginning with the cell, an organism's levels of organization result from development. For individuals, cell reproduction provides new cells for growth and repair. For populations, reproduction is a means of continuing a species.

Evolution

Development is a type of change over time. The earliest stages of embryonic development are similar among the echinoderms and chordates. These similarities suggest evolutionary relationships.

Tapping Prior Knowledge

- What is reproduction's purpose?
- What are mitosis and meiosis?
- What is development?

Opening Demonstration

Draw a small dot on the board. Ask: How does the size of this dot compare with the size you were when you were conceived? *(It is larger than a fertilized egg.)* Tell students that a fertilized egg is about 0.14 mm (0.005 in.) in diameter. Then show students a doll that is about the size of a newborn baby. Tell students that the average length of a newborn baby is approximately 50 cm (20 in.). Have them calculate a fetus's average rate of growth per month and their own rate of growth per month since birth. Ask: How does the increase in size during prenatal development compare with that of postnatal growth? *(It is much greater.)*

CHAPTER 42
REPRODUCTION AND DEVELOPMENT

REVIEW

- meiosis (Section 6-3)
- mutation (Section 9-3)
- development (Section 19-3)
- ectoderm, mesoderm, and endoderm (Section 27-2)
- notochord and hollow dorsal nerve cord (Section 27-3)
- *Highlights: Reproductive System* (p. 848)
- HIV and AIDS (Section 39-3)
- hypothalamus (Sections 40-2 and 41-2)
- pituitary gland, estrogen, progesterone, and testosterone (Section 41-3)

Human sperm and offspring

Authors' Rationale

The miraculous process of creating another individual is of particular interest to students. A comprehensive understanding of the male and female reproductive systems on the part of both male and female students is also critical. By the time they reach high school, students should already be aware of the workings of the reproductive systems and of the pathologies for which they should be vigilant through the years. Our goal is to present factual information clearly and directly so that students can make informed decisions in the future. Remember, education is the best form of prevention.

42-1 The Male Reproductive System

The male reproductive system produces the male gametes, or sperm, and prepares them for delivery to a female's body so that fertilization, or conception, can occur. Sperm are highly specialized carriers of genetic information. Formed after meiosis, sperm cells are haploid, which means that they have only 23 chromosomes instead of the 46 (diploid number) found in most cells of the human body. As you can see in Figure 42-1, sperm cells consist of a head with very little cytoplasm and a long tail. Digestive enzymes located in the head enable a sperm cell to penetrate an egg. Mitochondria located between the head and tail supply the ATP energy that a sperm cell needs to swim through a female's reproductive system to an egg.

Section Objectives

- Describe the structure of a sperm cell, and explain how it is adapted for its function.
- Discuss the two functions of the testes.
- Trace the path taken by sperm from the testes to the outside of the body.
- Explain how the structure of the penis enables it to deposit sperm inside the female reproductive tract.

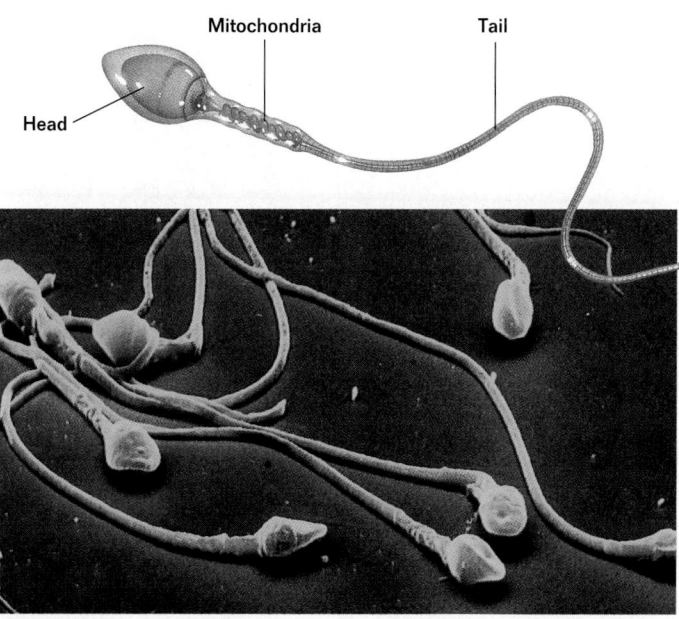

Mitochondria Tail

Head

Figure 42-1 A diagram of a sperm cell, *left,* shows that it is a highly specialized cell that is able to move quickly through a female's reproductive system. The scanning electron micrograph, *bottom left,* shows human sperm at a high magnification.

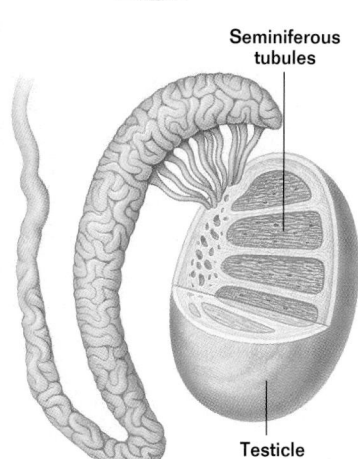

Seminiferous tubules

Testicle

Figure 42-2 A cutaway view of a testicle shows that the testes are filled with tiny coiled tubes, the sperm-producing seminiferous tubules.

Testes Are the Male Reproductive Organs

Two spherical **testes** *(TEHS teez),* or testicles, are the main organs of the male reproductive system. Each testicle contains hundreds of compartments that are packed with large numbers of tightly coiled tubes, as you can see in Figure 42-2. Sperm cells are produced within these tubes, called **seminiferous** *(sehm uh NIHF uhr uhs)* **tubules.** As you learned in

SECTION 42-1

Vocabulary Preview

- testes
- seminiferous tubules
- scrotum
- epididymis
- vas deferens
- seminal vesicles
- prostate gland
- semen
- penis
- ejaculation

Opening Question

Ask students how human males and females differ genetically. Students should recall that females have two X chromosomes and males are XY.

Demonstration

Flagellated Cells

Use an overhead projector or a microscope to display living cells that have flagella, such as *Euglena.* Tell students that, like *Euglena,* human sperm cells also propel themselves with flagella but are much smaller.

VISUAL STRATEGY

Figure 42-1

Tell students that sperm have been called genetic delivery machines. Ask: How does the structure of a sperm cell help it perform this function? *(Sperm cells consist of the head, which is mostly nuclear material, mitochondria, which provide ATP energy for swimming, and a tail, which propels the sperm and its genes toward an awaiting egg.)*

Effective Reading

Studying the Diagrams

Remind students that their understanding of the human reproductive system will be enhanced by studying the diagrams and carefully reading the captions and labels as they read the information in the text.

Did You Know?

Male hormones are called androgens—testosterone and androsterone are examples. Androgens have a generalized anabolic effect; for example, they increase protein synthesis, especially in muscles, and they promote bone growth.

Androgens are derived from cholesterol, as are all steroid hormones. The principal androgen is testosterone. Males synthesize about 6 to 10 mg of testosterone daily; females make about 0.4 mg. The adolescent growth spurt in both males and females probably results from androgens.

Application

 Health The penis is covered by a loose-fitting sheath of skin called the foreskin. Glands under the foreskin secrete a thick, white fluid that may accumulate if the area is not cleansed properly, promoting irritation and infection. An operation called a circumcision removes the foreskin and prevents accumulation of fluids that may cause irritation. Circumcision may also be done for religious reasons or as a puberty rite in some cultures. Once strongly encouraged by physicians in the United States, circumcision is now offered as an optional procedure.

VISUAL STRATEGY

Figure 42-3

Use this diagram to review the major structures and functions of the male reproductive system. Ask students to give examples of how the structure of the reproductive system helps the system accomplish its function. *(Answers will vary. For example, the erectile tissue in the penis enables sperm to be deposited in the female reproductive tract, and the scrotum enables the testes to be kept at a temperature low enough to produce viable sperm.)* Have students write a paragraph describing the life of a sperm cell, from its production in the seminiferous tubules to its exit from the body through the urethra. Encourage students to be creative with their paragraphs.

Chapter 41, the testes also secrete testosterone, the male sex hormone. Testosterone is secreted by cells that are scattered among the connective tissues between the seminiferous tubules. Two pituitary hormones regulate the functioning of the testes. Luteinizing hormone (LH) stimulates the interstitial cells to release testosterone. Follicle stimulating hormone (FSH) stimulates the seminiferous tubules to produce sperm.

The structure of the male reproductive system is illustrated in Figure 42-3. The testes are located outside the body cavity in the **scrotum** *(SKROHT uhm)*, a sac that hangs between the legs. First formed inside the body cavity, the testes descend into the scrotum either before or shortly after birth. Unlike other human body cells, sperm cannot successfully complete their development at the normal human body temperature of 37°C (98°F). In the scrotum, however, the temperature is about three degrees Celsius cooler than in the rest of the body.

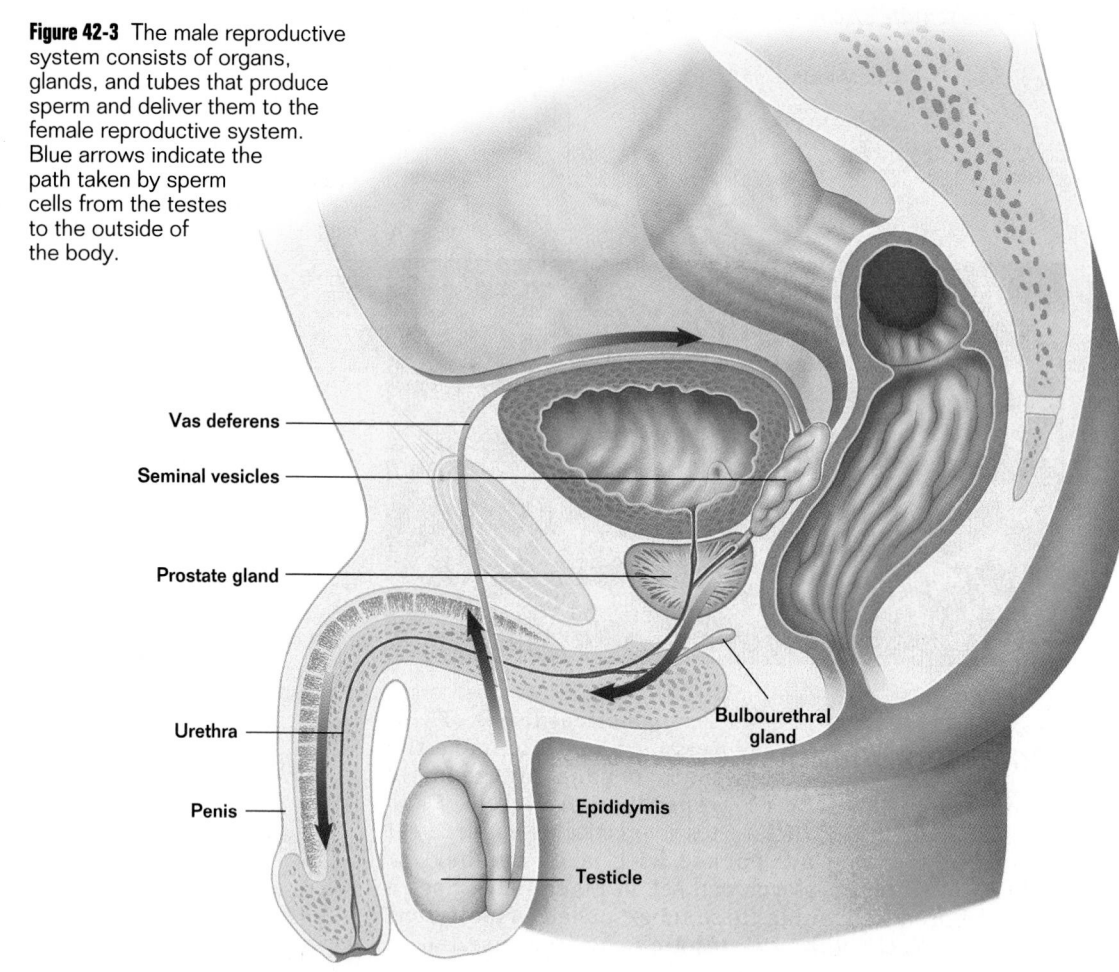

Figure 42-3 The male reproductive system consists of organs, glands, and tubes that produce sperm and deliver them to the female reproductive system. Blue arrows indicate the path taken by sperm cells from the testes to the outside of the body.

Vas deferens

Seminal vesicles

Prostate gland

Urethra

Penis

Bulbourethral gland

Epididymis

Testicle

Overcoming Misconceptions

"Prostate" Versus "Prostrate"

Many students mispronounce the word *prostate*. Emphasize the pronunciation difference between the two words above. A discussion of their respective definitions can help distinguish the two.

 Did You Know?

Boys generally enter puberty about a year later than girls—between the ages of 10 and 14. The ability to ejaculate usually begins between the ages of 11 and 15, but any time between the ages of 8 and 21 is considered normal.

Sperm Are Mixed With Fluids to Make Semen

A typical adult male produces several hundred million sperm each day of his life. After being produced in the seminiferous tubules, the sperm mature as they travel through a series of long tubes. During this journey, indicated by the arrows in Figure 42-3, fluids secreted by several glands are added to the sperm. These fluids nourish the sperm and lubricate the tubes through which they pass on their journey out of the body. Sperm that do not leave the body are broken down, and their materials are absorbed by the body to be recycled.

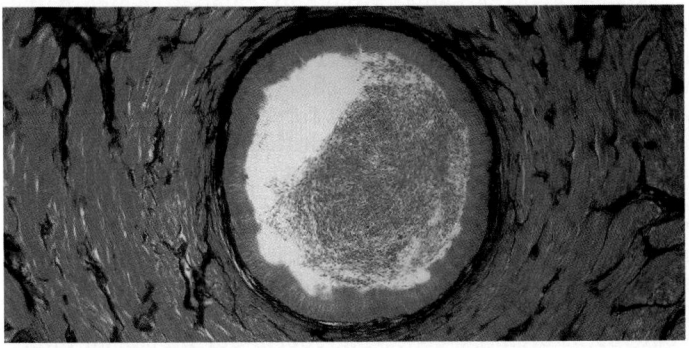

Figure 42-4 This light micrograph shows thousands of sperm cells within a human epididymis.

The full process of sperm production and maturation takes about two and a half months. After the sperm cells are manufactured within a testicle, they begin their long journey out of the body. First, the sperm are delivered to a long coiled tube called the **epididymis** *(ehp uh DIHD ih mihs)*, where they mature and develop the ability to become motile. Sperm do not actually become motile until after they leave the male body. Figure 42-4 shows sperm cells within the epididymis. From the epididymis, the sperm are delivered to another long tube, the **vas deferens** *(vas DEHF uh rehnz)*, where they are stored until they leave the body. Cutting or tying each vas deferens, an operation called a vasectomy *(va SEHK tuh mee)*, prevents sperm from leaving the male body. Sperm leave the body by passing through a tube that leads from the vas deferens to the urethra, where the reproductive and urinary tracts join.

As the sperm pass through the vas deferens and urethra to the outside of the body, the fluid secretions of several glands are added to them. **Seminal vesicles,** which lie between the bladder and the rectum, produce a fluid that is rich in sugars that nourish the sperm cells. The **prostate** *(PRAHS tayt)* **gland,** which is located just below the bladder, secretes an alkaline fluid that will counteract the acids produced by the female reproductive tract. The bulbourethral gland also secretes an alkaline fluid that is added to the mixture. **Semen** is a mixture of these secretions and sperm. ◻

Teaching Tips
Function of Seminal Fluids

Tell students that without the fluids in semen, sperm would not be motile enough to reach an egg. Point out that the fluids contained in semen not only provide a transport medium, but also provide nutrients for the sperm and contain chemicals that activate and protect the sperm. Ask: Why do sperm need to obtain food from seminal fluids? *(Because sperm have an extremely reduced cytoplasm, they cannot carry any stored food.)*

What If the semen were to lack one of its normal components? Students often believe male infertility is caused solely by insufficient sperm production. However, fluids in the semen may also be responsible for infertility. Tell students that insufficient fructose content, hormonal imbalances, pH and prostaglandin concentrations that are too high or too low, and low semen volume all cause infertility. Other causes of infertility include malformations of the sperm and obstructions to the delivery system. Poor circulation in the testes can raise the temperature of the testes, leading to the release of immature sperm in reduced amounts.

 Did You Know?

Hormone-like chemicals called prostaglandins thin the mucus guarding the opening into the uterus and cause uterine contractions that draw the sperm into the opening. Other hormones and enzymes stimulate the motility of sperm, enhancing their chances of reaching an egg. The pH of the semen serves to counteract the acid environment of the vagina.

Teaching Tip

Human Male Reproductive Structures and Functions

Have students prepare a table listing the structures discussed in this section and their functions. The table should include the seminiferous tubules, scrotum, epididymis, vas deferens, seminal vesicles, prostate gland, and penis.

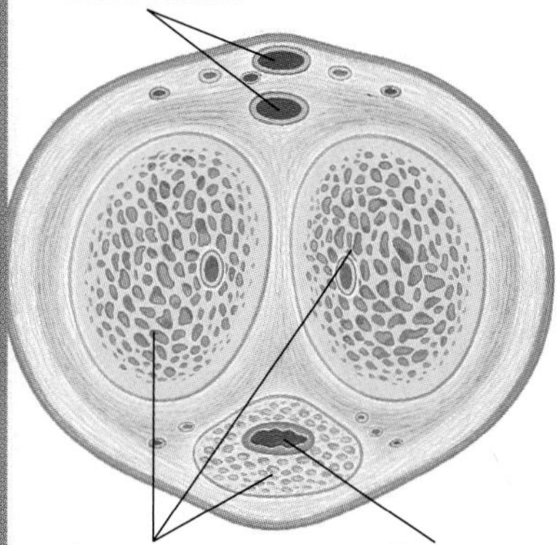

Blood vessels

Spongy tissue   **Urethra**

Figure 42-5 A cross section of a penis shows that it contains three cylinders of spongy tissue that run the length of the penis. When the spaces in the spongy tissue of these cylinders fill with blood, the penis becomes rigid and erect.

CAPSULE SUMMARY

During ejaculation, the penis delivers sperm to the female reproductive tract. Fertilization, which usually requires a high sperm count, can be prevented by abstaining from sexual intercourse or by blocking the path of sperm as they leave the male body.

Sperm Are Delivered by the Penis

The **penis** deposits sperm in the female reproductive tract during sexual intercourse. The penis must harden, or become erect, before it can enter the female reproductive tract. The structure of the penis enables this to happen. As you can see in Figure 42-5, the penis contains two cylinders of spongy tissue that lie side by side. Below and between them is a third cylinder of spongy tissue that surrounds the urethra. Small spaces separate the cells of these spongy tissues. When nerve impulses from the central nervous system cause the small arteries that lead into this tissue to expand, blood collects within these spaces, and the penis becomes rigid and erect.

Sperm exit the penis through **ejaculation,** which is the forceful expulsion of semen. During ejaculation, muscles encircling each vas deferens contract, moving the sperm they contain into the urethra. Eventually, the contractions of muscles at the base of the penis force semen out of the urethra. Only semen passes through the urethra at this time. After they are deposited in a female's body, sperm travel through the female reproductive tract toward the point where fertilization occurs. If, however, sperm are unable to reach an egg, fertilization cannot occur. One way to prevent fertilization is to block the path of the sperm. Covering the penis with a thin rubber sheath called a condom helps prevent fertilization by capturing sperm during sexual intercourse. Abstaining from sexual intercourse is the surest way to prevent fertilization.

About 3.5 mL of semen is expelled during ejaculation. This amount of semen normally contains several hundred million sperm. Still, the odds against any one sperm cell successfully completing the long journey to an egg and fertilizing it are extraordinarily high. Therefore, fertilization requires a high sperm count. Males with less than 20 million sperm per milliliter are generally considered sterile. ☐

Section Review

1. *Describe the structure of a sperm cell.*
2. *What are the two functions of the testes?*
3. *What path do mature sperm follow when they leave the vas deferens?*
4. *How does the structure of a penis enable it to become rigid?*

Critical Thinking

5. *If a male's left vas deferens were to become blocked, would you expect his sperm count to be normal, below normal, or zero? Explain your answer.*

Answers to Section Review

1. A sperm cell consists of a cell membrane, a head containing the nucleus and digestive enzymes, a "neck" containing mitochondria, and a tail.

2. The testes produce testosterone and sperm.

3. After leaving the vas deferens, sperm exit the body through the urethra.

4. The penis has three shafts of spongy tissue that fill with blood and become turgid when stimulated by nerve impulses from the central nervous system.

5. Answers will vary but should be justified. Students will probably suggest that the male's sperm count would be below normal because the sperm produced by only one testicle will become part of the semen.

*I*n addition to producing and maturing the female gametes, or eggs, the female reproductive system plays other important roles in reproduction and development. Fertilization occurs within specialized structures of the female reproductive system. This union of two haploid cells, as seen in Figure 42-6, restores the normal diploid number of 46 chromosomes. Following fertilization, the female reproductive system houses and nourishes a developing embryo (later called a fetus) through the period known as pregnancy.

Section Objectives

- Describe the functions of the female reproductive system.
- Explain how egg production in females differs from sperm production in males.
- Relate the changes that occur during the ovarian cycle to the hypothalamus.
- Relate the changes that occur during the menstrual cycle to the ovarian cycle.

Vocabulary Preview

ovary

puberty

ova (ovum)

fallopian tube

uterus

endometrium

vagina

cervix

implantation

ovarian cycle

follicle

ovulation

corpus luteum

menstrual cycle

menstruation

menopause

Opening Question

Ask students to list advantages of internal fertilization over external fertilization. Students should recall that internal fertilization provides greater protection, better-controlled nutrition, and a more tightly regulated environment.

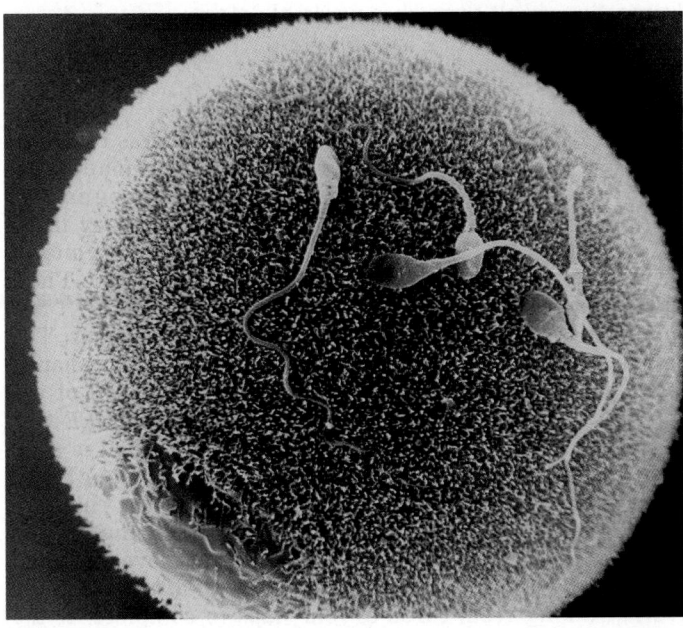

Figure 42-6 This photomicrograph shows five human sperm cells on a human ovum. During fertilization, an ovum and a sperm cell, both haploid (*n*) cells, fuse to form a zygote, which is a diploid (2*n*) cell.

Ovaries Produce Female Gametes

Two egg-shaped **ovaries** are the gamete-producing organs of the female reproductive system. Unlike testes, ovaries are located within the abdominal cavity. And while new sperm cells are constantly formed in males who have reached sexual maturity, females are born with all of the egg cells they will ever make. At birth, a female's ovaries contain some 2 million immature egg cells that have begun the first division of meiosis. Each of these cells has the ability to develop further but does not continue with meiosis. Instead, it waits to receive the proper signal, arrested in prophase of the first meiotic division.

Demonstration

Use models to represent the relative sizes of eggs and sperm, or call students' attention to the photograph in Figure 42-6. Ask: What accounts for the size difference between these cells? *(They have different functions: eggs must have enough food to maintain a zygote until it implants in the uterus, while sperm must efficiently deliver genetic material.)*

 Did You Know?

The difference in the sizes of sperm and eggs is a result of a fundamental difference between the processes of sperm formation (spermatogenesis) and egg formation (oogenesis). During spermatogenesis, equal divisions of the cytoplasm follow meiosis I and meiosis II, resulting in four equal-sized sperm cells. During oogenesis, unequal divisions of the cytoplasm follow meiosis I and meiosis II, resulting in one large egg and three tiny cells that degenerate. In humans, meiosis II occurs just as a sperm penetrates an ovum.

Teaching Tip
Accomplishing Fertilization

An unfertilized ovum can survive for only about 12 to 24 hours after its release from the ovary. Sperm, on the other hand, remain viable in the female reproductive tract for 12 to 48 or even 72 hours. The timing is fairly crucial because fertilization must take place in the upper third of a fallopian tube to be successful. Have students calculate the "window" during which there is the greatest likelihood of fertilization. *(If intercourse occurs no more than 72 hours before ovulation and no later than 24 hours after ovulation, fertilization is likely to take place.)* Caution students that there is no reliable way to determine when ovulation has occurred or will occur.

❑ CAPSULE SUMMARY

The ovaries, which are located in the abdominal cavity, produce the female gametes, or ova. At birth, a female already has all of the immature egg cells she will ever produce. At puberty, immature egg cells begin to mature one at a time.

After a female reaches **puberty,** or reproductive maturity, the first division of meiosis is able to resume. However, only one immature egg cell completes its development each month. The others remain in a developmental holding pattern. When they mature, the egg cells are called **ova** (singular, **ovum**). The Latin word *ovum* means "egg." Only about 400 of the immature egg cells that a woman has at birth will mature. The long period during which immature egg cells exist is one reason that developmental abnormalities occur with increasing frequency in the pregnancies of women over 35 years old. Immature egg cells are exposed to mutation-causing agents throughout a woman's life. After age 35, the odds that a harmful mutation will have occurred become high enough to significantly increase the incidence of abnormalities. There is evidence that aging in men also plays a role in developmental abnormalities, such as Down syndrome, that result from chromosomal defects. ❑

Fertilization Occurs as an Ovum Journeys to the Uterus

When an ovum is released, it begins a journey that may end in pregnancy. The ovum's journey takes it through the organs of the female reproductive system, seen in Figure 42-7. First, cilia sweep the ovum into one of the **fallopian** *(fuh LOH pee uhn)* **tubes,** which lead from the ovaries to the uterus. The **uterus** *(YOO tuh ruhs)* is a hollow, muscular, pear-shaped organ about the size of a small fist. The inner wall of the uterus, which is called the **endometrium** *(ehn doh MEE tree uhm)*, has two layers. The outer layer thickens and is shed monthly, while the inner layer generates another outer layer. Smooth muscles lining the fallopian tubes contract rhythmically. These contractions move the ovum down the tube to the uterus in much the same way that food moves through your intestines. An ovum's journey through a fallopian tube is slow, taking from three to four days to complete. If the ovum is not fertilized within 24 hours, it loses its capacity to develop.

To successfully fertilize an egg, a sperm must make its way far up a fallopian tube, a long journey that only a few complete. Sperm are first deposited within the **vagina** *(vuh JEYE nuh)*, a muscular tube about 7 cm long that leads to the uterus. The entrance to the uterus from the vagina is called the **cervix** *(SUR vihks)*. A soft rubber cap called a diaphragm *(DEYE uh fram)* can be used to cover the cervix and help prevent fertilization by blocking the passage of sperm into the uterus. A diaphragm is more effective when used with a sperm-killing chemical, or spermicide. After swimming through the vagina, the sperm pass through the cervix and uterus and then swim up the fallopian tubes. This journey requires the sperm to

 Did You Know?

Puberty can begin as early as age 9 or 10 and as late as age 15 or 16. The first physical sign of sexual maturity in girls is the development of breasts, usually between ages 9 and 13. In the United States and other Western countries, the mean age of menarche (the start of the first menstrual period) is about 12½.

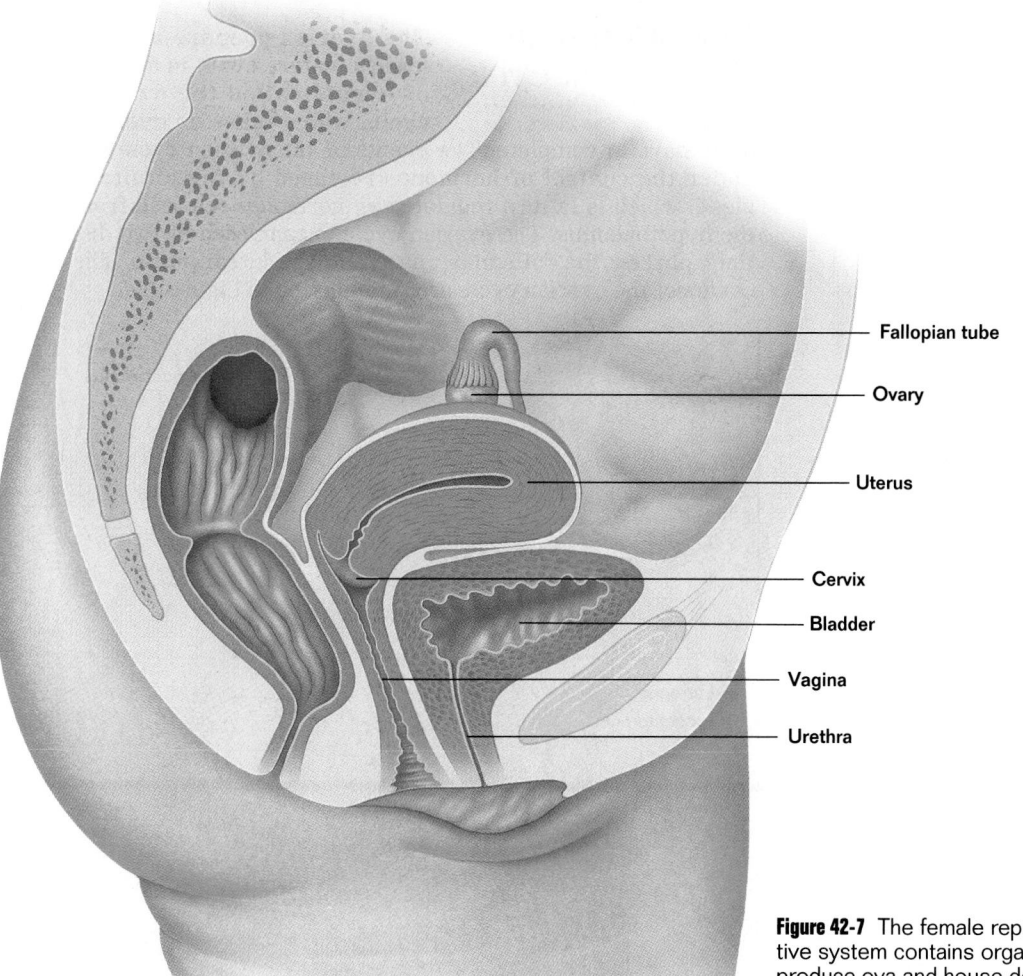

Figure 42-7 labels:
- Fallopian tube
- Ovary
- Uterus
- Cervix
- Bladder
- Vagina
- Urethra

Figure 42-7 The female reproductive system contains organs that produce ova and house developing embryos.

VISUAL STRATEGY

Figure 42-7
Use this diagram to review the major structures and functions of the female reproductive system. Tell students that as a pregnancy progresses, the uterus expands to a size that would fill the part of the abdomen seen in this diagram. Ask: What happens to the mother's internal organs as the fetus grows? *(All of the organs become compressed.)* Point out that compression of the abdominal and thoracic organs can become a problem, especially late in the term. The mother ultimately may experience fatigue, backache, indigestion, circulatory problems, reduced lung capacity, and changes in her gait. Have students write a paragraph that continues the life story of a sperm, begun for **Visual Strategy: Figure 42-3** on page 994, detailing what happens as the sperm makes its way through the female reproductive system toward the possible fertilization of an egg.

Teaching Tip
Methods of Contraception
Divide the class into groups of five students, and ask each member of each group to research one or two methods of contraception to determine the mode of use, method of action, effectiveness, and rate of failure of a variety of contraceptive strategies. Methods of contraception include birth control pills, diaphragms, cervical caps, intrauterine devices (IUDs), contraceptive sponges, spermicides, condoms, the rhythm method, and abstinence. Instruct each group to compile a table of their findings to be shared with the rest of the class. You may want to have each student develop one table summarizing all the contraceptive methods researched.

expend a great deal of energy because they are constantly swimming against the current that carries the ovum to the uterus. Cutting or tying the fallopian tubes, an operation called a tubal ligation *(leye GAY shun)*, prevents fertilization by blocking the sperm's path to the ovum.

When a sperm succeeds in fertilizing an ovum high in the fallopian tube, the fertilized ovum begins to divide and continues on its journey to the uterus. After reaching the uterus, the fertilized ovum (now a hollow ball of cells) burrows into the thick, outer lining of the uterus, an event called **implantation.** Implantation, which marks the beginning of pregnancy, occurs about six days after fertilization. Following implantation, the fertilized ovum starts the long developmental process that eventually leads to the birth of a child. If fertilization does not occur, the outer layer of the endometrium and the unfertilized ovum are shed.

☐ CAPSULE SUMMARY

Fertilization occurs in the fallopian tubes. Blocking the path of sperm to the fallopian tubes prevents fertilization. When a fertilized egg enters the uterus, it implants in the uterine wall.

❓ Did You Know?

Two different mechanisms ensure monospermy, or fertilization by a single sperm. When the cell membrane of a sperm contacts an egg's membrane, sodium channels in the egg membrane open and depolarize the membrane, blocking the fusion of other sperm with the egg membrane. Depolarization also causes calcium ions to enter the egg cytoplasm, activating the egg for cell division. At the same time, bodies called cortical granules spill their contents into the space between the egg membrane and the egg's protective envelope. The contents of the cortical granules bind to water and swell, causing all sperm still in contact with the egg to detach from its membrane.

Figure 42-8
Correlate this figure not only to the accompanying discussion of the ovarian cycle but also to the following discussion of the menstrual cycle on page 1002. Ask: What keeps the uterus and the ovaries synchronized with each other? (*a negative-feedback mechanism*)

Application

🏃 **Sports** The emphasis modern American culture has placed on physical health has led many people to extremes of exercise. Many students think that if exercise is good, lots of exercise must be great. This obsession with fitness leads many students into health problems. While exercise is indeed good for you, it is good only in moderation. In females, an exercise program that is too vigorous can delay menarche (a female's first menstrual flow), disrupt the normal menstrual cycle in adult women, and even cause amenorrhea—a complete cessation of the normal menstrual cycle. These problems, seen extensively in highly trained female athletes, seem to stem from a low body fat content, which impairs the body's ability to produce the estrogens needed for proper hormonal control. The hypothalamus also seems somehow to be blocked. While these effects cease once the athlete assumes a more normal activity regimen, other effects may continue. For example, amenorrhea may cause otherwise healthy, young women to lose bone mass, as do post-menopausal women. In an effort to counteract this loss, sports physicians encourage female athletes to increase their daily intake of calcium to 1.5 g per day, which is about the amount of calcium in a quart of milk.

Eggs Mature in the Ovarian Cycle

The ovaries prepare and release a mature ovum in a series of events called the **ovarian cycle,** which takes an average of 28 days to complete. The events of the ovarian cycle are under the control of hormones released by the pituitary gland, which is in turn regulated by hormones released from the hypothalamus. The ovarian cycle is composed of two distinct phases: the follicular phase and the luteal phase. The events of the ovarian cycle are summarized in Figure 42-8.

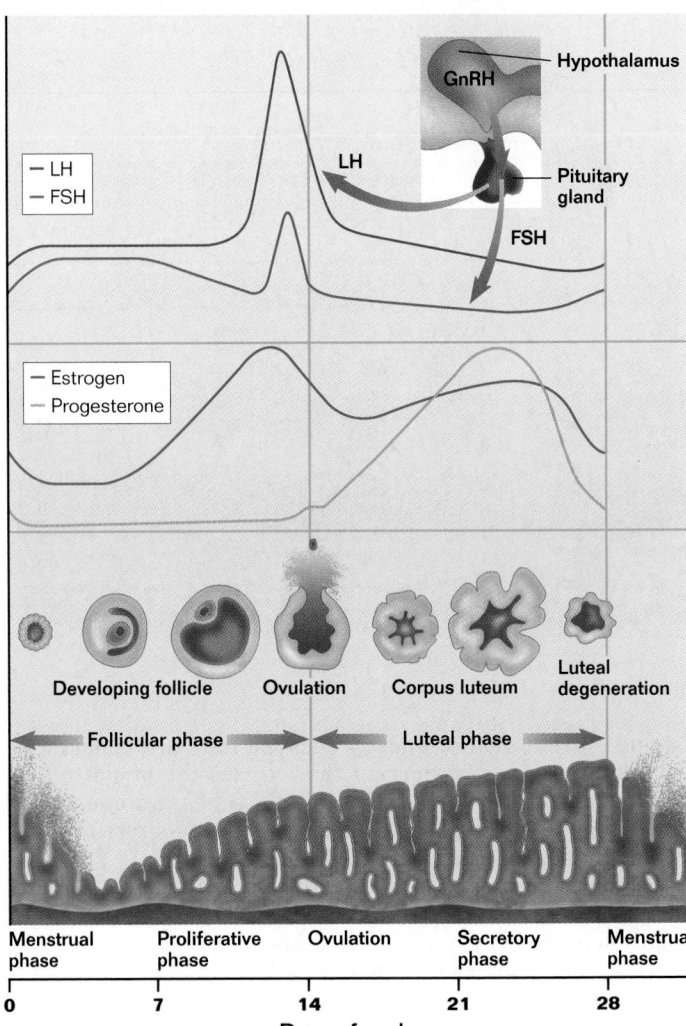

Figure 42-8 This chart summarizes the human ovarian and menstrual cycles. The ovarian cycle, *top*, is regulated by hormones produced by the hypothalamus and pituitary gland. The menstrual cycle, *bottom*, is regulated by the hormones estrogen and progesterone, which are produced by the follicle and corpus luteum.

Triggering the Maturation of an Egg Cell

During the follicular phase of the ovarian cycle, hormones produced by the pituitary gland regulate the completion of an ovum's development. Ova develop within **follicles,** which are clusters of cells that surround the immature egg cells in an ovary. The anterior lobe of the pituitary, after receiving a chemical signal (gonadotropin releasing hormone, or GnRH) from the hypothalamus, starts the cycle by secreting FSH and LH. FSH binds to receptors on the surface of the cells in an immature follicle and triggers the development of the follicle. Together, FSH and LH cause the follicle to begin producing estrogen, which aids in the maturation of the follicle and initiates the thickening of the uterine lining. Normally, only one follicle at a time is able to respond immediately to FSH. The FSH level falls before other immature follicles and egg cells can mature, thus usually allowing only one ovum to mature in every cycle.

The decrease in FSH level is achieved by a negative-feedback mechanism. Neither FSH nor LH carries this signal. Instead, estrogen acts as the messenger. The rising level of estrogen in the bloodstream is detected by the hypothalamus, which responds by commanding the pituitary to cut off the further production of FSH. Thus, the hypothalamus "shuts the door" on further egg development and ordinarily ensures that only one egg cell reaches the final stage of development each month.

A continued rise in the estrogen level initiates the thickening of the uterine lining and signals the end of the follicular phase of the ovarian cycle. The anterior pituitary responds to a high level of estrogen by greatly increasing its secretion of LH. The high level of LH causes the maturation of a follicle and its ovum to be completed and causes the wall of the follicle to burst. When the follicle bursts, the mature ovum within it is released in a process called **ovulation.** Ovulation, seen in Figure 42-9, sends the mature ovum on a journey toward possible fertilization.

Preparing the Body for Fertilization

The luteal phase of the ovarian cycle follows smoothly from the follicular phase. After ovulation, LH causes the ruptured follicle to fill in and become yellowish. The repaired follicle is called a **corpus luteum,** which comes from the Latin words *corpus*, meaning "body," and *luteum*, meaning "yellow." The corpus luteum soon begins to secrete progesterone, which then inhibits FSH and LH (a backup for estrogen in preventing further ovulation). Prescription pills containing synthetic estrogen and progesterone disrupt the ovarian cycle and prevent ovulation. Progesterone is the body's signal to prepare itself for fertilization. If fertilization occurs, the corpus luteum continues to produce progesterone for several weeks. Rising levels of progesterone initiate many physiological changes associated with pregnancy. If, however, fertilization does *not* occur soon after ovulation, production of progesterone slows and eventually ceases, marking the end of the luteal phase. ❏

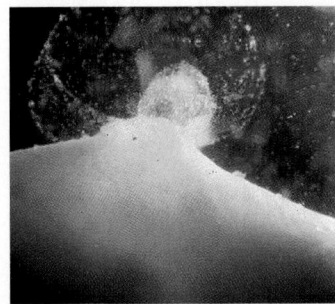
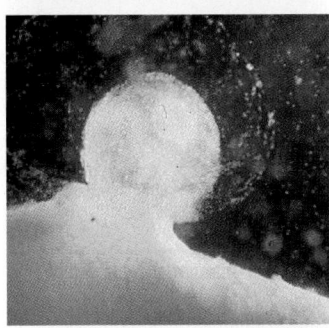
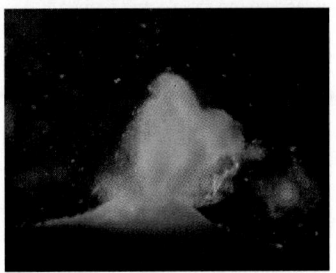

Figure 42-9 In this series of photos showing ovulation, a mature ovum emerges from an ovary, *top* and *center*, and retreats toward a fallopian tube, *bottom*.

❏ CAPSULE SUMMARY

During the ovarian cycle, FSH from the pituitary gland causes an ovum to mature inside a follicle. After ovulation, LH converts a ruptured follicle to a corpus luteum.

SECTION 42-2

◉ VISUAL STRATEGY

Figure 42-9
Point out the ovum (egg) in this series of photographs showing ovulation, the release of an egg from the ovary. Tell students that the material in the center of the bottom photograph is the fluids that are expelled from the ruptured follicle. The egg in this photograph is seen in the lower right-hand corner, retreating toward the fallopian tube.

Application

Health In vitro fertilization is the procedure leading to what is commonly called a test-tube baby. In this expensive procedure, healthy eggs are collected from a woman who has been unable to conceive after alternative therapy. Following a short incubation period, sperm from the intended father are added to the egg in a culture dish. If fertilization takes place, the zygote is incubated in a special solution in a glass dish, allowing it to divide to become a 2- to 8-cell stage. At this time the conceptus (embryo) is transferred to the mother's uterus while the mother receives hormones to promote implantation. Doctors then monitor her hormone levels to determine whether she has become pregnant.

Graphic Organizer

Use this graphic organizer with **Teaching Tip: Hormones and the Female Reproductive Process** on page 1002.

Hormone	Production site	Target organs	Effects
GnRH	Hypothalamus	Pituitary gland	(see text)
FSH	Pituitary gland	Ovary	(see text)
LH	Pituitary gland	Ovary	(see text)
Estrogen	Ovary, adrenal cortex	Uterus	(see text)
Progesterone	Ovary, adrenal cortex	Uterus	(see text)

CHAPTER 42

Teaching Tips
Hormones and the Female Reproductive Process

Have students prepare a table that lists the hormones that regulate the reproductive process in females. The table should include the hormones' production sites and their effects on the menstrual cycle, ovarian cycle, ovaries, and uterus. A sample table is found in the *Graphic Organizer* on page 1001.

What If a woman were to become pregnant?

Pregnancy disrupts the normal homeostasis of the female body. At least four major systems—the digestive, urinary, respiratory, and cardiovascular systems—change in response to a pregnancy. Until a woman can adjust to the elevated progesterone and estrogen levels in her body, she may experience excessive salivation and nausea, which is called morning sickness. Additionally, pressure in her abdomen may cause stomach acid to wash backward into her esophagus, causing chronic heartburn. Total blood volume rises in a pregnant woman to protect against blood loss. Both blood pressure and pulse rate increase, while her heart pumps 20 to 40 percent more blood, depending on the stage of pregnancy. Pressure by the uterus on pelvic blood vessels may cause varicose veins to result from the restricted blood flow back from the legs.

❏ CAPSULE SUMMARY

During the menstrual cycle, which occurs along with the ovarian cycle, estrogen and progesterone produced by the follicle and corpus luteum prepare the uterus for possible pregnancy. If pregnancy does not occur, estrogen and progesterone production stops, and the thickened uterine lining is shed during menstruation.

The Menstrual Cycle Prepares the Uterus for Pregnancy

While changes occur in the ovaries during the ovarian cycle, changes also occur in the uterus. Look back at Figure 42-8 to see how the uterus changes during the ovarian cycle. These changes, which prepare the uterine lining for a possible pregnancy each month, are called the **menstrual cycle.** The term *menstrual* comes from the Latin word *mensis,* meaning "month." Like the ovarian cycle, the menstrual cycle is, on the average, a 28-day cycle. During the period before ovulation, increasing levels of estrogen initiate the thickening of the endometrium. After ovulation, high levels of estrogen and progesterone cause further thickening of the endometrium and maintain this thickened uterine lining. If pregnancy does not occur, however, the levels of estrogen and progesterone decrease. The decreasing levels of these hormones mark the end of the ovarian cycle and cause the thickened outer layer of the endometrium to be shed.

As the outer layer of the endometrium is shed, blood vessels are broken and bleeding results. A mixture of blood and discarded endometrial cells, the menstrual fluid, then leaves the uterus through the vagina. This process, called **menstruation,** usually occurs about 14 days after ovulation. Thus, the end of the menstrual cycle also marks the end of the luteal phase of the ovarian cycle. At the end of the female reproductive cycles, neither estrogen nor progesterone is being produced. In their absence, the pituitary is again stimulated to produce FSH and LH, thus starting another ovarian and menstrual cycle. Between the ages of 45 and 55, menstruation usually ceases. After this event, which is called **menopause,** a woman is no longer able to conceive. ❏

Section Review

1. *What are three functions of the female reproductive system?*
2. *How does egg production in females differ from sperm production in males, and how is this difference significant?*
3. *What roles do the hypothalamus and pituitary gland play in regulating the ovarian cycle?*
4. *What causes the lining of the uterus to thicken and then be shed during the menstrual cycle?*

Critical Thinking
5. *How could the maturation of an egg be blocked during the ovarian cycle?*

Answers to Section Review

1. The female reproductive system produces hormones and eggs, matures eggs, provides a place for fertilization, and nurtures a developing baby.

2. A female is born with all of the eggs she will ever produce, while a male produces gametes continuously after puberty. This difference can be significant because eggs in an older woman have a greater chance of accumulating genetic abnormalities that result in birth defects.

3. The hypothalamus produces GnRH, which stimulates the pituitary to release FSH and LH. These two hormones stimulate the growth, maturation, and release of an egg.

4. Rising levels of estrogen and progesterone cause the uterine lining to thicken. A drop in the levels of these hormones causes the uterine lining to be shed.

5. Increasing levels of estrogen and progesterone in the blood would simulate pregnancy and inhibit the maturation of eggs.

42-3 Human Development

Human development takes an average of nine months from fertilization to birth. The nine months of pregnancy are often divided into three trimesters, or three-month periods. Much of the first trimester is devoted to the formation of a fetus, like the one seen in Figure 42-10. The development of the basic human body plan is essentially completed by the end of the first trimester. The second and third trimesters are devoted mainly to the growth of the fetus. As you study the course of this long developmental period, notice that the most crucial events of development occur very early in pregnancy, many of them before a woman knows that she is pregnant.

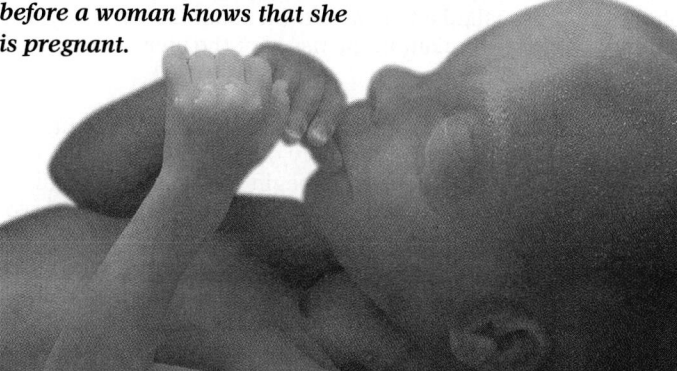

Figure 42-10 At 12 weeks old, this fetus, shown here in the uterus, already looks distinctly human.

Section Objectives

- Describe the primary events of the first month of human embryo development.
- Explain why events that occur early in a pregnancy are more likely to result in birth defects.
- Describe the events that occur after a fetus has formed.

The Embryo Forms During Early Development

Embryonic development in humans is a complex and dynamic process—a symphony of cell movement and change that starts with a single diploid cell. From this single cell, 100 trillion others will arise as the individual develops and grows to adulthood. After fertilization, the developmental journey begins with a series of rapid cell divisions that results in a ball of cells. Some of these cells then migrate to form a structure composed of the three primary tissue layers. The development of the specific tissues of the human body follows.

Week One: Cleavage

The first major event in the development of a human embryo is the rapid division of the zygote into a large number of cells—first two cells, then four, then eight, and so on. During this period of division, called **cleavage,** the overall size of the embryo does not increase; the cells simply become smaller. While cleavage is occurring, the embryo continues its

---

? Did You Know?

In humans, the gestation period (time during which pregnancy occurs) is conventionally given as 280 days. It is counted from the last menstrual period until birth, adding two weeks or so to the actual development time. The embryo stage in human development begins in the third week after fertilization. Prior to this, the developing human is referred to as a *conceptus.* The fetal stage begins in the ninth week. The term *pregnancy* refers to all events from fertilization until a baby is born.

Vocabulary Preview

- trimester
- cleavage
- blastocyst
- amnion
- chorion
- placenta
- gastrulation
- neurulation
- organogenesis
- morphogenesis
- fetal alcohol syndrome, FAS
- fetus

Opening Question

Ask students to list the systems that must develop in a human from the time of conception to birth. Students should include the integumentary, skeletal, muscular, nervous, endocrine, cardiovascular, lymphatic, immune, respiratory, digestive, urinary, and reproductive systems.

Demonstration

Actual Sizes of a Developing Human

Have students place a 0.5 mm dot (about the size of a mechanical pencil mark) at the top of a sheet of paper. Tell them to label the mark "14 days." Then have them draw a line about 1 mm in length below the dot and label it "18 days." Tell them to continue down the sheet in a similar manner, drawing lines 4, 8, 19, 38, and 63 mm in length and labeling them "4 weeks," "6½ weeks," "9 weeks," "11 weeks," and "15 weeks," respectively. Tell students that these are the actual sizes of a developing human through the 15th week.

Figure 42-11

Walk students through this figure, pointing out sites of important stages in early embryonic development such as fertilization, cleavage, and implantation.

Teaching Tip

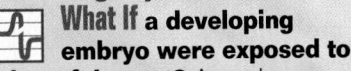

What If a developing embryo were exposed to a harmful agent? An embryo is very sensitive to environmental factors that can cross the placental barrier. Such factors can cause congenital malformations. Factors that cause congenital malformations, which are called teratogens, include drugs such as alcohol, nicotine, anticoagulants, sedatives, medicines for high blood pressure, and some antibiotics, as well as infectious agents such as German measles. For this reason, expectant mothers are cautioned not to take any drugs, even aspirin, without consulting their doctors. Aside from alcohol, the most common teratogen an embryo might encounter is nicotine. Nicotine hinders growth and development by interfering with normal oxygen transport to the embryo.

Figure 42-12

Go over the structure of the placenta, pointing out its maternal and fetal portions. Ask: Why are the fetal arteries shown in blue (denoting deoxygenated blood) and the fetal veins shown in red (denoting oxygenated blood)? *(In the placenta, blood in the fetal arteries has come from the heart and contains carbon dioxide and other wastes, while blood in the veins is returning to the embryo's heart with oxygen from the mother's blood.)*

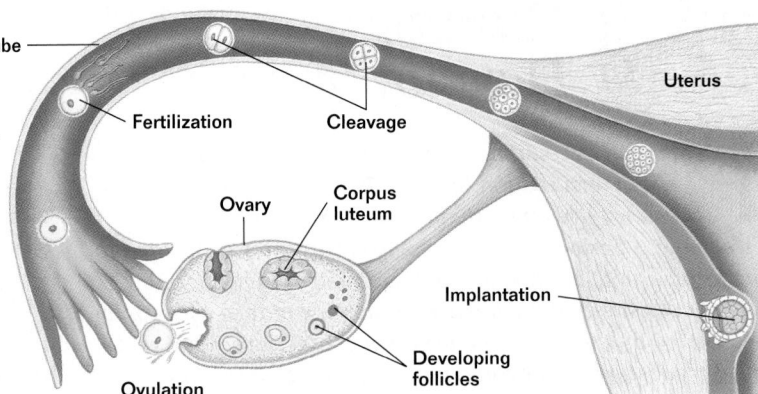

Figure 42-11 Following ovulation, a mature ovum is swept into the nearby fallopian tube. Fertilization occurs at about one-third of the way through the tube. Cleavage begins as the fertilized ovum completes its journey to the uterus, where implantation occurs about six days after fertilization.

Figure 42-12 An embryo (later called a fetus) develops within the fluid-filled amniotic sac and is nourished by its mother through the placenta, *below*. Shaped like a pancake, the placenta forms where chorionic villi (extensions of the chorion containing fetal blood vessels) penetrate the endometrium. The umbilical cord connects the developing embryo (fetus) to the placenta. Food and oxygen pass from the mother's blood into the embryo's blood while the embryo's wastes pass into the mother's blood, but the mother's blood and embryo's blood remain separate.

journey down the fallopian tube, as seen in Figure 42-11. By the time the embryo reaches the uterus, it is a hollow ball of about 100 cells, called a **blastocyst.** On about the sixth day, the embryo begins to implant in the tissue of the uterine lining.

Week Two: Gastrulation

About eight days after fertilization, shortly after implantation, the embryo begins to grow rapidly, and the membranes that will later protect and nourish it begin to form. One of these membranes, the **amnion,** will enclose the developing embryo. Another embryonic membrane, the **chorion,** will interact with uterine tissue to form the **placenta,** which will nourish the growing embryo. To learn more about the placenta, look at Figure 42-12. As the placenta forms, the inner cells of the blastocyst begin a carefully orchestrated migration called **gastrulation.** This migration largely determines the future development of the embryo. By the end of gastrulation, the distribution of cells into the three primary tissue types—endoderm, mesoderm, and ectoderm—has been completed. The ectoderm is destined to form the epidermis and nerve tissue. The mesoderm is destined to form the connective tissue,

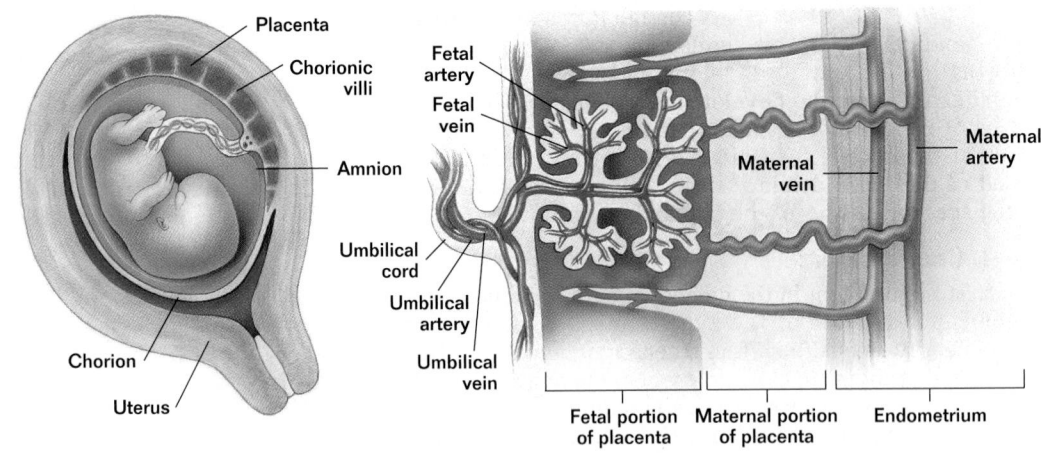

 Did You Know?

As it grows, the placenta secretes hormones that affect the mother. One hormone, **human placental lactogen,** works with estrogen and progesterone to help prepare the breasts for milk production. It also promotes growth of the mother's tissues and causes her cells to use fatty acids instead of glucose as an energy source, thus leaving the glucose in her blood for use by her developing baby. A second placental hormone, **human chorionic thyrotropin,** increases maternal metabolism during pregnancy.

Historical Note

Thalidomide

Thalidomide, a sedative used in Europe in the 1960s as a remedy for morning sickness, caused deformities, such as legs and arms resembling flippers, when taken during the limb-bud differentiation period (days 26 to 56).

muscle, and vascular elements. The endoderm will form the lining of the organs in the gut. One of the characteristic features of all vertebrates—the notochord—also develops from mesoderm during gastrulation.

Week Three: Neurulation

In the third week, the three primary cell types begin to differentiate into the tissues and organs of the body. As in all vertebrates, differentiation begins with the formation of another characteristic feature, the hollow dorsal nerve cord. The process by which the hollow dorsal nerve cord forms from ectoderm is called **neurulation.** While neurulation is occurring, the rest of the basic architecture of the human body is being rapidly determined by changes in the mesoderm and endoderm. By the end of the third week, blood vessels and the gut have begun to develop. At this point, the embryo is about 2 mm (less than 0.1 in.) long.

Week Four: Organogenesis

In the fourth week, the body organs begin to form in a process called **organogenesis.** The developing heart begins a rhythmic beating that stops only at death. At an average of 70 beats per minute, the heart is destined to beat more than 2.5 billion times during a lifetime of about 70 years. The arm and leg buds also begin to form in the fourth week, as you can see in Figure 42-13. The embryo more than doubles in length during this week to about 5 mm. By the end of the fourth week of development, all of the major organs of the human body have begun to form. Although development is now far advanced, many women are not aware that they are pregnant at this stage.

The Second Month: Morphogenesis

During the second month, the final stage in the formation of a human embryo takes place. During **morphogenesis,** the miniature limbs of the embryo assume their adult shapes. Within the body cavity, the major internal organs are evident, including the liver and pancreas. By the end of the second month, the embryo is about 22 mm (less than 1 in.) in length, weighs about 1 g, and is beginning to look distinctly human.

Because so much is happening and the proper course of events can be easily interrupted, early pregnancy is a very crucial time in development. Most miscarriages occur during this period. Alcohol use by pregnant women, especially during early pregnancy, is a leading cause of birth defects, such as the one seen in Figure 42-14. **Fetal alcohol syndrome,** or **FAS,** is a common birth defect (affecting 1 in 750 newborns in the United States), in which a baby is born with a deformed face and often severe mental and physical retardation. While alcohol should be avoided throughout pregnancy, it is particularly important for women to avoid alcohol if they are planning to become pregnant. Organogenesis can also be upset during the first months of pregnancy if the mother contracts rubella (German measles).

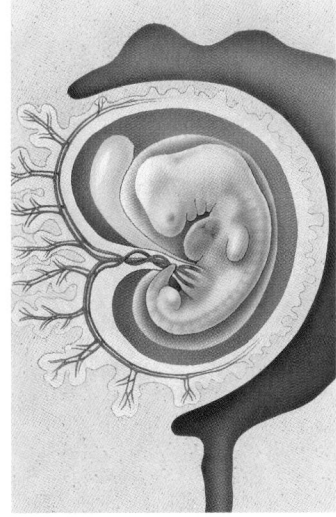

Figure 42-13 A four-week-old human embryo, *above,* has a tail, arm and leg buds, eyes, and a heart that has begun to beat. The placenta and umbilical cord have also formed.

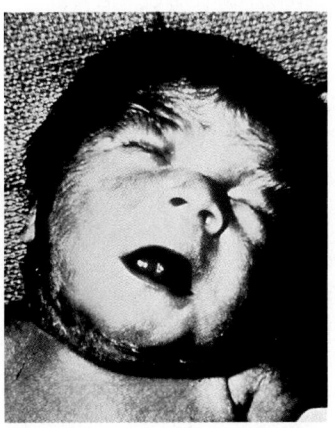

Figure 42-14 This prematurely born infant has several of the signs of fetal alcohol syndrome (FAS), including an abnormally small head and widely spaced eyes.

CONNECTIONS

Chapter 27
Vertebrate Characteristics
Review the characteristics of a vertebrate. Remind students that the notochord is a dorsal, cartilaginous rod found in all vertebrates at sometime during their development. It provides support to the organism. Tell students that the hollow, dorsal nerve cord of a vertebrate eventually becomes enclosed within the vertebral column.

Teaching Tip

Drug Use During Pregnancy
Initiate a class discussion concerning the effects of drugs taken during pregnancy, especially during organogenesis and morphogenesis. Encourage students to speculate about the effects of drugs, including alcohol and tobacco, on the body systems of both a mother and her developing fetus. Ask them to identify ways that the drugs might cause their expected effects. The discussion could easily be based on the knowledge students already have. The important goal of the discussion is not that the students come up with clinical descriptions, but rather that they apply their knowledge of systems to the problem at hand.

Multicultural Perspective

Culture-Specific Disorders

Several genetic disorders appear predominantly in certain cultures. The most well known of these is sickle cell anemia, which is found primarily in people of African descent. Tay-Sachs disease, a disorder of chromosome 15 found most often in Ashkenazi Jews (those of eastern European descent), is a progressive neurological deterioration. Babies with Tay-Sachs disease are often born healthy but begin to lose muscular control at five or six months of age and usually die by the age of four or five years. Other culture-specific genetic disorders include thalassemia, a blood disorder affecting people of Mediterranean heritage, and Gaucher's disease, another disorder found in Ashkenazi Jews.

Fruit Flies and Human Homeobox Genes

Genetic engineering has not turned a man into a fly or vice versa, but scientists have been able to direct fly development with genes borrowed from humans. William McGinnis and Michael Kuziora of Yale University, among many other scientists, conduct studies on the homeobox family of genes. Discovered only recently, these genes direct the development of structures in an organism by turning on certain proteins at specific times. Scientists first discovered homeobox genes in *Drosophila melanogaster* (fruit flies), where they affect the developmental location of body structures. By moving these genes around, scientists found they could cause serious mutations such as legs growing out of the head where antennae should have been.

Once discovered in the fruit fly, scientists started looking elsewhere and soon identified homeobox genes in a number of species, including our own. McGinnis, Kuziora, and other scientists were surprised to find that homeobox genes are extraordinarily similar in many types of organisms (e.g., humans, flies, mice, and roundworms). In fact, these genes are so similar that when researchers put human homeobox genes in fruit flies, the human genes continued to correctly direct fruit fly development. Scientists hope that this research and associated technology may someday be used to prevent, screen for, or correct human birth defects. ☑

The Fetus Grows Rapidly Until Birth

The body plan of a human embryo is essentially complete at eight weeks. From this point on, a developing human is referred to as a **fetus.** By the end of the first trimester, all of the major body organs have differentiated, and the sex of the fetus has been established. During the second and third trimesters, the fetus grows rapidly as its organs are completed and become functional. Table 42-1 describes some of the major events of fetal development.

Table 42-1 Events of Human Fetal Development

Stage	Major Events
First Trimester (0 to 3 months)	Fertilization, cleavage, implantation, gastrulation, neurulation, and organogenesis occur as the embryo becomes a fetus; all major organ systems are formed; the fetus begins to move, but movements cannot be felt
Second Trimester (4 to 6 months)	Skin and hair grow; eyes blink; fetal movements can be felt; arms and legs reach final proportions; heartbeat can be heard
Third Trimester (7 to 9 months)	Substantial increase in size; skin is red and wrinkled; development of the lungs is completed; fingernails and toenails grow; fetus can survive if born during this stage

By the end of the third trimester, a fetus is able to exist outside of its mother's body. Yet the neurological development of the fetus is far from complete. At the end of the third trimester, the fetus is about as large as it can safely be delivered without damage to the mother or child. Physical growth and neurological development continue after birth. ☐

☐ CAPSULE SUMMARY

During the first eight weeks after fertilization, a developing human is called an embryo. After that, the embryo is called a fetus. Human development is most easily disturbed during the first trimester of pregnancy. Growth and neurological development continue after birth, which occurs about nine months after fertilization.

Section Review

1. *What events in human development occur during the first month after fertilization?*
2. *What is the placenta, and what is its function?*
3. *When does an embryo become a fetus?*
4. *At what stage of development can a fetus survive if it is born?*

Critical Thinking

5. *Why can certain drugs so profoundly affect an embryo's development when they are ingested or injected by a woman during early pregnancy?*

Answers to Section Review

1. Cleavage, gastrulation, neurulation, and organogenesis occur during the first month after fertilization.

2. The embryonic chorion and part of the uterine lining make up the placenta, which provides food and oxygen to the fetus and eliminates its wastes.

3. An embryo becomes a fetus at eight weeks.

4. A baby can survive if it is born during the third trimester.

5. Answers will vary but should suggest that drugs may cause mutations or interfere with the normal function of hormones and other control chemicals. Errors in gene replication and cell differentiation may have profound effects in a developing embryo, because cells are rapidly dividing and forming the systems of the body.

42-4 Sexually Transmitted Diseases

Disease-causing agents travel from one host to another in many ways. When certain disease-causing microbes are present in human body fluids, they can be passed from one person to another when these fluids are exchanged during sexual activity. Diseases that are spread by sexual contact are called **sexually transmitted diseases, or STDs**. Both viruses and bacteria can cause STDs.

Section Objectives

■ Identify the causes of AIDS and genital herpes, and describe their symptoms.

■ Identify the causes of syphilis, gonorrhea, and chlamydia, and describe their symptoms.

■ Contrast the treatment and cure rates of viral STDs with those of bacterial STDs.

Viral STDs Cannot Be Cured With Antibiotics

STDs that are caused by viruses are called viral STDs. Because viruses are not affected by antibiotics, viral STDs cannot be treated and cured with antibiotics. Two viral diseases that can be transmitted by sexual activity are AIDS and genital herpes.

Acquired Immune Deficiency Syndrome

Acquired immune deficiency syndrome (AIDS) is a fatal disease caused by the human immunodeficiency virus (HIV). As you learned in Chapter 39, transmission during sexual intercourse is the most common way that people get AIDS. However, HIV can be transmitted in many other ways, and AIDS, therefore, is not exclusively a sexually transmitted disease. HIV destroys the immune system of infected individuals. AIDS patients die from infections and cancers that a healthy immune system normally defeats. The graph in Figure 42-15 shows how AIDS cases among young adults have increased over the last decade. AIDS is now the leading killer of American men between the ages of 25 and 44. In addition to the known AIDS cases and HIV-positive individuals, some 1.5 million more people in the United States are thought to be infected—most unknowingly—with HIV.

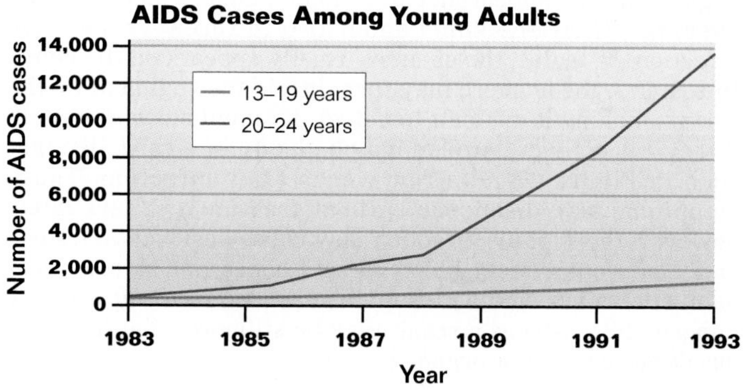

AIDS Cases Among Young Adults

(Graph: Number of AIDS cases vs. Year, 1983–1993; legend: 13–19 years, 20–24 years)

Figure 42-15 As this graph shows, the number of AIDS cases among teenagers and young adults rose sharply from 1983 to 1993. Although the greatest increase in AIDS cases occurred among 20- to 24-year-old individuals, you should keep in mind that most of these individuals contracted HIV during their teens.

SECTION 42-4

Vocabulary Preview

sexually transmitted disease, STD

genital herpes

pelvic inflammatory disease, PID

syphilis

chancre

gonorrhea

chlamydia

Opening Question

Remind students that HIV mutates very quickly. Ask them how this knowledge explains why attempts to make a vaccine for AIDS have failed so far. Students could infer that the rapid mutation rate makes vaccines ineffective because antibodies produced by the body in response to the vaccine would not recognize a mutated HIV.

Demonstration

Models of Organisms Causing STDs

Materials: (per group) 2 small round balloons, 1 very long balloon, 1 foam ball, 1 box of toothpicks, 1 package of gumdrops.

Procedure: *Neisseria gonorrhoeae*, the diplococcus that causes gonorrhea, can be modeled with 2 round balloons joined at their stems. *Treponema pallidum*, the spirochete that causes syphilis, can be modeled by twisting a long balloon around a dowel to form a spiral. HIV can be modeled by studding a foam ball with toothpicks that are tipped with gumdrops. Point out that the gumdrops and toothpicks represent the quickly mutating protein coat that has thwarted development of a vaccine to date. Have students draw and label each of these agents.

Figure 42-16

After students have examined the photograph, remind them that herpes blisters may appear in places where they are not visible, and therefore a person may be infected with genital herpes and be unaware of it. Ask: When genital-herpes blisters heal, is the disease cured? *(No. The virus remains in the body and may cause further outbreaks.)* Tell students that the best way to protect themselves from infection with viral STDs is to abstain from sexual activity.

Teaching Tip

Comparing Viral and Bacterial STDs

Have students compare the location of syphilis blisters with the location of genital-herpes blisters. *(A syphilis rash may appear on any part of the body, while genital-herpes blisters tend to appear on or near the genitals.)* Ask: What is another difference between genital herpes and syphilis? *(Syphilis can be cured with antibiotic treatment, while genital herpes cannot.)* Why should a person seek medical attention if he or she suspects a possible syphilis infection? *(The sooner syphilis is treated, the less serious the damage to the body will be.)* Tell students that by the time destructive lesions appear in the nervous and circulatory systems, serious damage has been done. While treatment is still possible, this damage cannot be reversed.

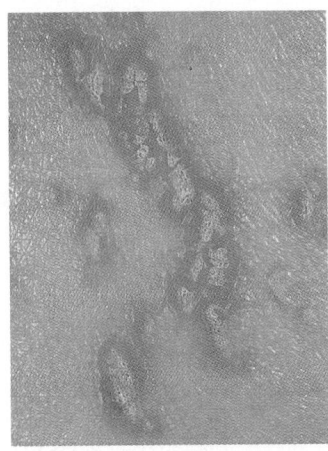

Figure 42-16 Painful blisters such as these are a symptom of genital herpes. In males, genital herpes blisters may appear on the penis, scrotum, or skin near the genitals. In females, genital herpes blisters may develop on the labia (external parts of the genital area), within the vagina, or on the skin near the genital area.

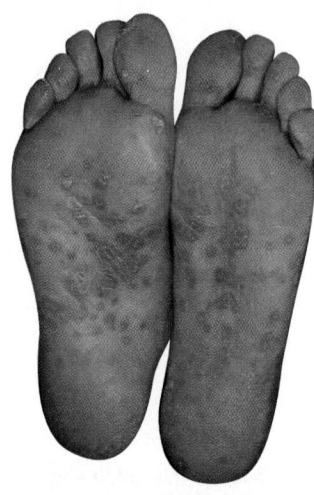

Figure 42-17 A rash such as the one on these feet is a symptom of the second stage of syphilis. A person with syphilis may also develop a fever and swollen lymph glands. Even at this stage, syphilis can easily be cured by treatment with antibiotics.

Genital Herpes

Genital herpes is caused by a herpes simplex virus (HSV). The symptoms of genital herpes include periodic outbreaks of painful blisters in the genital area, as seen in Figure 42-16, and flulike aches and fever. Two types of HSV can cause genital herpes. About 80 percent of genital herpes infections are caused by HSV-2. The rest are caused by HSV-1, which more commonly causes cold sores, or fever blisters, around and in the mouth. Antiviral drugs can temporarily eliminate the blisters, but they cannot eliminate HSV from the body. Although genital herpes is not life threatening, it can have serious consequences. Women with genital herpes have a greater risk of developing cervical cancer. Also, like HIV, HSV can be passed from mother to fetus during pregnancy or birth. A baby infected with HSV may suffer severe damage to its nervous system or even die as a result of the infection.

Bacterial STDs Can Be Difficult to Detect

STDs that are caused by bacteria are called bacterial STDs. Unlike viral STDs, bacterial STDs can be successfully treated and cured with antibiotics. Unfortunately, the early symptoms of most bacterial STDs are very mild and often go undetected. Routine laboratory tests that detect bacterial STDs are available at hospitals and clinics. Early detection and treatment are necessary to prevent serious consequences that can result from infection. For example, untreated bacterial STDs can cause sterility in both men and women. One of the most common causes of infertility (inability to become pregnant) in women is **pelvic inflammatory disease,** or **PID.** PID is a severe inflammation of the uterus, ovaries, fallopian tubes, or pelvic cavity that is usually caused by a bacterial STD. Three major bacterial STDs are syphilis, gonorrhea, and chlamydia.

Syphilis

Syphilis *(SIHF uh lihs)* is a serious bacterial STD that usually begins with the appearance of a small, painless ulcer called a **chancre** *(SHAHNG kuhr)* about two to three weeks after infection. In males, the chancre usually appears on the penis. In females, the chancre may form inside the vagina or on the cervix, making it difficult to detect. If syphilis is not treated, it may cause fever, swollen lymph glands, or a rash, like that seen in Figure 42-17, a few weeks after infection. These symptoms also disappear without treatment. Years later, however, the syphilis infection may cause destructive lesions on the nervous system, blood vessels, bones, and skin. A pregnant woman infected with syphilis can also transmit it to her unborn child, who as a result may be stillborn or suffer serious damage to major organ systems.

❓ Did You Know?

The human herpes simplex virus (HSV) belongs to a group of viruses that includes the Epstein-Barr virus and others. These viruses are all extremely difficult to control. HSV-2, the most common cause of genital herpes, is transmitted via infectious secretions. It has been estimated that one-quarter to one-third of all adult Americans harbor HSV-2 without knowing it.

Historical Note

Gabriel Fallopius

Fallopius, the Italian anatomist who first described the fallopian tubes, developed a condom made of chemically treated linen, not as a means of birth control but as an attempt to prevent the spread of syphilis.

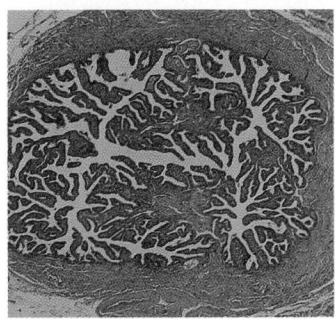 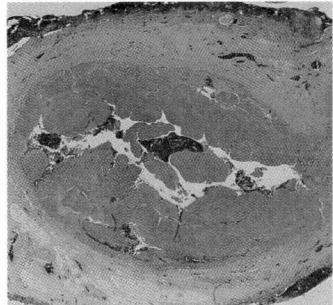

Figure 42-18 A normal fallopian tube, *left,* has a highly folded lining and many spaces through which gametes can pass. In a fallopian tube scarred by PID, *right,* many of these spaces have become filled with tissue, blocking the path of gametes and decreasing the likelihood of fertilization.

Gonorrhea

Gonorrhea *(gahn uh REE uh)* is a bacterial STD that causes painful urination and pus discharge from the penis in males. In females, it sometimes causes a vaginal discharge but more often has no symptoms. In males, untreated gonorrhea can spread to the vas deferens, epididymis, or testes. In females, it can spread to the fallopian tubes and cause scarring that results in infertility. Some strains of gonorrhea are resistant to the more commonly used antibiotics, such as penicillin.

Chlamydia

Chlamydia is the most common bacterial STD in the United States. Between 3 million and 10 million new cases occur each year. The symptoms of chlamydia are similar to those of a mild case of gonorrhea: painful urination in males and vaginal discharge in females. Like gonorrhea, chlamydia often produces no symptoms and often is not diagnosed. Chlamydia, even more so than gonorrhea, can cause scar tissue to form on infected fallopian tubes and close them, leading to infertility. Most cases of PID are the result of gonorrhea or chlamydia infections. Figure 42-18 shows the damage that can be caused by pelvic inflammatory disease.

CAPSULE SUMMARY

Diseases that can be spread by sexual contact are called sexually transmitted diseases (STDs). STDs caused by viruses cannot be cured with antibiotic treatment. STDs caused by bacteria are difficult to detect but can be cured with antibiotic treatment.

Section Review

1. *What are two sexually transmitted diseases caused by viruses?*
2. *Name three symptoms of genital herpes.*
3. *What are three common STDs caused by bacteria? Why is early detection of these diseases important?*
4. *What is one main difference between viral STDs and bacterial STDs?*

Critical Thinking

5. *How can you best protect yourself from contracting an STD?*

CHAPTER REVIEW ANSWERS

Each item in the Chapter Review is correlated by text section in the assignment guide that follows.

ASSIGNMENT GUIDE

Section	Review	Themes Review	Critical Thinking
42-1	1, 3, 11	18	21
42-2	2–5, 12–14	19	22
42-3	6–8, 15	20	
42-4	9, 10, 16, 17		

Review

Multiple Choice

Code in parentheses indicates section number and objective number.

1. c (42-1.3)
2. d (42-2.1)
3. a (42-1.3, 42-2.1)
4. a (42-2.3)
5. d (42-2.3)
6. b (42-3.1)
7. c (42-3.2)
8. b (42-3.3)
9. a (42-4.1)
10. d (42-4.2)

Completion

11. sperm, testosterone (42-1.2)
12. sperm, eggs (42-2.2)
13. ovum (egg) (42-2.3)
14. menstruation (42-2.4)
15. four (42-3.1)
16. immune, virus (42-4.1)
17. chlamydia, gonorrhea (42-4.3)

---

42 CHAPTER REVIEW

Vocabulary

amnion (1004)
blastocyst (1004)
cervix (998)
chancre (1008)
chlamydia (1009)
chorion (1004)
cleavage (1003)
corpus luteum (1001)
ejaculation (996)
endometrium (998)
epididymis (995)
fallopian tube (998)
fetal alcohol syndrome, FAS (1005)
fetus (1006)
follicle (1001)

gastrulation (1004)
genital herpes (1008)
gonorrhea (1009)
implantation (999)
menopause (1002)
menstrual cycle (1002)
menstruation (1002)
morphogenesis (1005)
neurulation (1005)
organogenesis (1005)
ova, ovum (998)
ovarian cycle (1000)
ovary (997)
ovulation (1001)
pelvic inflammatory disease, PID (1008)

penis (996)
placenta (1004)
prostate gland (995)
puberty (998)
scrotum (994)
semen (995)
seminal vesicles (995)
seminiferous tubules (993)
sexually transmitted disease, STD (1007)
syphilis (1008)
testes (993)
trimester (1003)
uterus (998)
vagina (998)
vas deferens (995)

Review

Multiple Choice

1. The correct pathway of sperm is
   a. testes to vas deferens to epididymis.
   b. epididymis to urethra to vas deferens.
   c. testes to epididymis to vas deferens.
   d. urethra to vas deferens to testes.

2. Which of the following is *not* a function of the female reproductive system?
   a. production of gametes
   b. nourishment of the fetus
   c. maturation of eggs
   d. secretion of FSH

3. Fertilization normally takes place in the
   a. fallopian tubes.   c. epididymis.
   b. cervix.   d. vas deferens.

4. The maturation of an ovum is ultimately controlled by GnRH released by the
   a. hypothalamus.   c. fallopian tubes.
   b. corpus luteum.   d. follicles.

5. The follicular phase of the ovarian cycle
   a. occurs when LH levels drop to zero.
   b. starts when fertilization occurs.
   c. causes estrogen levels to drop to zero.
   d. ends when ovulation occurs.

6. An embryo develops endoderm, mesoderm, and ectoderm during
   a. cleavage.   c. neurulation.
   b. gastrulation.   d. organogenesis.

7. Which of the following is *not* true for human development?
   a. Alcohol and drugs taken at any time during pregnancy may damage the embryo or fetus.
   b. Most miscarriages occur during the first trimester of pregnancy.
   c. Birth defects are never the result of drugs and alcohol taken by parents prior to pregnancy.
   d. Normal development may be affected by viral diseases.

8. Morphogenesis occurs during the
   a. first week of development.
   b. second month of development.
   c. third trimester.
   d. fourth week of development.

9. Which of the following sexually transmitted diseases cannot be treated with antibiotics?
   a. genital herpes   c. gonorrhea
   b. syphilis   d. chlamydia

10. A symptom associated with the earliest stage of syphilis is
   a. painful urination.
   b. blisters in the genital area.
   c. fever blisters and cold sores.
   d. a painless chancre.

Completion

11. The testes contain seminiferous tubules, in which the _____ are produced, and cells that secrete the hormone _____ .

12. A male produces _____ throughout his life, but all a female's _____ are produced during fetal development.

13. In the human female, timely reductions in the FSH level will ensure that one _____ will mature in each cycle.

14. The discharge of endometrial cells mixed with blood is called _____ .

15. Organogenesis begins in week _____ following fertilization.

16. AIDS is a disease of the _____ system, apparently caused by a _____ .

17. Two bacterial STDs that cause sterility in females are _____ and _____ .

Themes Review

18. **Structure and Function** Human sperm cells have a tail that whips back and forth and a head that contains digestive enzymes. How do these structures enable a sperm to fertilize an ovum?

19. **Homeostasis** FSH controls final egg development. How is the FSH level regulated by estrogen in the female body? What could happen if the FSH level increases and remains high?

20. **Levels of Organization** During the first four weeks of development, the human embryo undergoes many changes. Describe the changes that occur during each of the first four weeks of development, and identify the names associated with these weekly changes.

Critical Thinking

21. **Making Inferences** A man interested in fathering children has a sperm count of over 60 million for a 3.5 mL sample of semen. As his physician, what would you tell him about the results of the test?

22. **Making Predictions** Data on levels of pituitary and ovarian hormones collected from two women are presented in the table below. Based on the data, which woman could have conceived two or three days after the day the data were collected? Explain.

Hormone	Woman 1	Woman 2
FSH	high	low
LH	high	high
Estrogen	high	moderate
Progesterone	low	high

Activities and Projects

23. **Research and Writing** Research the causes of infertility, and find out about different biotechnological methods that assist conception. Relate your findings in a written report.

24. **Research and Writing** Some scientists do not believe that AIDS is caused by HIV. Look up and read articles by Peter Duesberg and others on this subject. In a written report, evaluate the arguments presented on both sides of this controversial issue.

25. **Multicultural Perspective** Almost every culture in the world celebrates puberty, a child's transition into adolescence. Research the puberty rites of two cultures from different parts of the world. What similarities and differences do they have? Relate your findings in a written report.

Critical Thinking

21. Although 60 million sperm sounds like a high number, the man might be sterile. A normal count for this volume of semen would be at least 70 million sperm. (42-1)

22. Woman 1 is most likely to become pregnant within two or three days of the day the data were collected because the pituitary and ovarian hormone levels indicate that she is about to ovulate. (42-2.3)

Activities and Projects

23. Infertility has several causes, including cervical irregularities, endometriosis, ovulation problems, and low sperm count. In-vitro techniques, microsurgical techniques, and gamete or zygote intrafallopian transfer are methods used to assist conception.

24. Duesberg is an expert on retroviruses like HIV. He is so sure that AIDS is not caused by a virus that he is willing to take injections of HIV to prove his point. His arguments appeared in major newspapers and journals (such as the *Proceedings of the National Academy of Sciences*) in the late 1980s and early 1990s.

25. Answers will vary widely. Students might compare various African and Native American initiation ceremonies, which for boys often require physical endurance and courage, to the Jewish bar mitzvah (for boys) and bat mitzvah (for girls) or the Hispanic celebration of a girl's 15th birthday, which is called fiesta quinceañera.

Themes Review

18. The tail's movement enables the sperm to swim up a fallopian tube to reach an ovum. Digestive enzymes enable the sperm to penetrate an ovum. (42-1.1)

19. A rising level of estrogen causes the hypothalamus to signal the pituitary to stop FSH production. If FSH production is not halted, more than a single egg will achieve maturity, and multiple births of fraternal twins could result. (42-2.3)

20. Cleavage occurs during week one; the cells of the zygote rapidly divide, but the size of the embryo does not increase. Gastrulation occurs in week two; a portion of the cell mass folds inward, forming the three primary tissue types. Neurulation occurs during week three; the three primary tissues begin to develop into specialized tissues and organs, and the nerve cord forms. Organogenesis occurs during week four; the body organs, including the heart, form. (42-3.1)

LABORATORY Investigation Chapter 42

Embryonic Development

Preparation

Obtain prepared slides of the following stages of sea star development: unfertilized egg, zygote, 2-cell stage, 4-cell stage, 8-cell stage, 16-cell stage, 32-cell stage, 64-cell stage, blastula, early gastrula, and young sea star larva. Each lab group will need a set of slides and a compound microscope for viewing the slides. If there are not enough sets of slides for each group to have their own set, you can set up a station for each developmental stage and have lab groups rotate from station to station. Before beginning this investigation, review the stages of human embryonic development that are discussed in this chapter. Also discuss why sea star eggs and embryos are used in this investigation rather than those of mammals. Remind students that the sequence of early embryonic development is only the very beginning of development. In humans, the process from zygote to blastocyst takes about seven days.

OBJECTIVES

- Identify the stages of early animal development.
- Describe the changes that occur during early development.

PROCESS SKILLS

- to observe prepared slides of sea star embryonic development
- to communicate microscopic observations through drawings

MATERIALS

- **prepared slides of sea star development**
- **compound microscope**
- **paper and pencil**

BACKGROUND

1. How does a fertilized egg become a completely developed organism?
2. What are the early stages (first four weeks) in the development of a human embryo?
3. Describe what happens during cleavage.
4. What are the three primary tissue types resulting from gastrulation, and what will they form in the completely developed individual?
5. Write your own question to explore in your lab report or notebook.

TECHNIQUE

1. Most members of the animal kingdom (including both sea stars and humans) begin life as a single cell — the fertilized egg, or zygote. The early stages of development in different species are quite similar. The zygote divides many times during cleavage, the new cells begin to specialize as they become part of specific tissues, and complex structures are formed as the embryo grows into a fully developed organism. Similarities and differences in these early stages of development reflect evolutionary relationships among species.

2. Obtain a set of prepared slides showing sea star eggs at different stages of development. Choose slides labeled unfertilized egg, zygote, 2-cell stage, 4-cell stage, 8-cell stage, 16-cell stage, 32-cell stage, 64-cell stage, blastula, early gastrula, middle gastrula, late gastrula, and young sea star larva. (Note: A blastula is the general term for the embryonic stage resulting from cleavage. A blastocyst in mammals is a modified form of the blastula.) Examine each slide under a low-power compound microscope. For each slide, focus on one good example of the developmental stage listed on the label. Then switch to high power.

3. Draw a diagram of each developmental stage you examine (in chronological order) in the **Records** section of your report. Label each diagram with the name of the stage it represents and at what magnification it was observed. Record your observations as soon as they are made. Do not recopy your diagrams. Draw what you see. Lab drawings do not need to be artistic or elaborate. They should be well organized and include specific detail.

Procedural Notes

- This investigation should take 40 minutes to complete.
- Remind students that their drawings should be informative rather than artistic.
- Have students compare the stages of sea-star development to photos of mammalian stages, if available.

Background Answers

1. A fertilized egg becomes a completely developed organism through cell division and differentiation.

2. The early stages of human embryo development include cleavage (2-cell stage, 4-cell stage, etc.), blastocyst, gastrulation, neurulation, and organogenesis.

3. During cleavage, cell division occurs without growth.

4. The three tissue types are ectoderm, which forms the epidermis and nerve tissue; mesoderm, which forms connective tissue, muscle, and vascular elements; and endoderm, which forms the lining of the gut.

5. What happens to a sea-star egg after it is fertilized?

Records

Students should draw a diagram of each developmental stage they examine.

4. Compare your diagrams to the diagrams of human embryonic stages shown below. In your report , briefly summarize the procedure you followed.

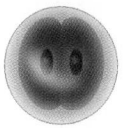
2-cell stage

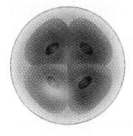
4-cell stage

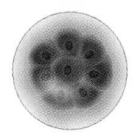
8-cell stage

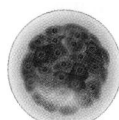
64-cell stage

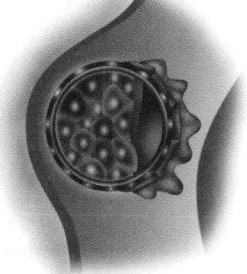
Blastocyst

 5. Clean up your materials and wash your hands before leaving the lab.

INQUIRY

1. Compare the size of the zygote with that of the blastula. At what stage does the embryo become bigger than the zygote?

2. By what stage do all of the cells in the embryo not look exactly like each other?

3. How do cell shape and size change during successive stages of development?

4. Are the cell nuclei the same size, larger, or smaller as the stages progress?

ANALYSIS

1. Why are sea star eggs a good choice for the study of embryonic development?

2. Compare the number of chromosomes in a fertilized sea star egg with the number of chromosomes in one cell of each of the following: 2-cell stage, blastula, gastrula, and adult organism.

3. From your observations of changes in cellular organization, why do you think the blastocoel (the space in the center of the hollow sphere of cells of a blastula) is important during embryonic development?

4. Label the endoderm and ectoderm in your drawing of the late gastrula stage. Describe the developmental fates of these two tissue types.

5. How is the symmetry of a sea star embryo and larva different from the symmetry of an adult sea star? Would you expect to see a similar change in human development?

6. In what ways can sea star embryos be used to study early human development?

7. Describe one way in which the cleavage of echinoderms and mammals is identical.

FURTHER INQUIRY

Write a new question that could be explored as a new investigation. The following is an example:

How is embryonic development in other organisms, such as birds, similar to or different from embryonic development in sea stars?

Inquiry Answers

1. The zygote and the blastula are the same size. By the time the embryo has developed into a gastrula, it is bigger than the zygote.

2. When the embryo reaches the gastrula stage, some cells look different from others.

3. Until the gastrula stage, all cells are roughly spherical. After that, some become flattened, elongated, or irregularly shaped. From the first cleavage until the gastrula stage, the cells get progressively smaller. After that, some cells begin to grow larger.

4. The nuclei remain the same size as the stages progress.

Analysis Answers

1. Sea-star eggs are a better choice for studies of early embryonic development than mammalian eggs because they are larger, more of them are produced, they are easier to culture, and their development is very similar to that of humans.

2. The chromosome number is the same in each cell during embryonic development. In adult organisms, however, all cells have the same number of chromosomes except for eggs and sperm, which are haploid.

3. A blastocoel enables invagination (folding) to occur during gastrulation, when the three primary tissue types form.

4. Check drawings for labels. The endoderm forms the lining of the gut. The ectoderm forms the epidermis and nervous tissue.

5. Both sea-star embryos and larvae exhibit bilateral symmetry. The adult exhibits radial symmetry. No; both human embryos and adults exhibit bilateral symmetry.

6. Sea-star embryos can be used as a model of early human development because they exhibit similar stages of cell division and differentiation.

7. Cleavage in both echinoderms and mammals results in equal-sized cells.

Careers

The following list provides the names and addresses of organizations that can supply information on a broad range of careers that can be pursued with a background in biology.

Agricultural Scientist
American Society of Agronomy
Crop Science Society of America
Soil Science Society of America
677 South Segoe Road
Madison, WI 53711

AIDS Information Specialist
New York Department of Health
Office of Public Health,
 AIDS Institute
Empire State Plaza
Albany, New York 12237-0684

Biochemist
American Society for Biochemistry and Molecular Biology
9650 Rockville Pike
Bethesda, MD 20814

Biological Scientist
American Institute of
 Biological Sciences
730 11th Street NW
Washington, DC 20001

Biomedical Engineer
Biomedical Engineering Society
PO Box 2399
Culver City, CA 90231

Biomedical Researcher
Federation of American
 Societies for
 Experimental Biology
9650 Rockville Pike
Bethesda, MD 20814

Botanist
Botanical Society of America
Department of Genetics
Ohio State University
1735 Neil Avenue
Columbus, OH 43210

Cardiovascular Technologist
Society of Vascular Technology
1101 Connecticut Ave NW,
 Suite 700
Washington, DC 20036-4303

Dentist/Dental Assistant
American Dental Association
211 East Chicago Avenue
Chicago, IL 60611

Diagnostic Medical Sonographer
Society of Diagnostic
 Medical Sonographers
12770 Coit Road, Suite 508
Dallas, TX 75251

Ecologist
Ecological Society of America
Center for Environmental
 Studies
Arizona State University
Tempe, AZ 85287-3211

Emergency Medical Technician/Paramedic
National Association of
 Emergency Medical
 Technicians
9140 Ward Parkway
Kansas City, MO 64114

Entomological Inspector
Entomological Society
 of America
9301 Annapolis Road
Lanham, MD 20706

Environmental Engineer
American Society
 of Sanitary Engineering
PO Box 40362
Bay Village, OH 44140

Epidemiologist
Epidemiology Program Office
MS C08, Centers for Disease
 Control and Prevention
Atlanta, GA 30333

Forester
Society of American Foresters
5400 Grosvenor Lane
Bethesda, MD 20814

Forensics Expert
American Academy of
 Forensic Sciences
218 East Cache La Poudre
Colorado Springs, CO 80901-0669

Genetic Counselor
National Society of Genetic
 Counselors
233 Canterbury Drive
Wallingford, PA 19086

Geneticist
Genetics Society of America
9650 Rockville Pike
Bethesda, MD 20814

Health Inspector
Public Health Service
Department of Health and
 Human Services
200 Independence Avenue SW
Washington, DC 20201

Horticultural Therapist
American Horticultural
 Therapy Association
9200 Wightman Road, Suite 400
Gaithersburg, MD 20879

Landscaper
National Landscape Association
1250 I Street NW, Suite 5000
Washington, DC 20005

Medical Technologist
American Society for
 Medical Technology
2021 L Street NW, Suite 400
Washington, DC 20036

Medical Illustrator
Association of Medical
 Illustrators
1819 Peachtree Street NE,
 Suite 560
Atlanta, GA 30309

Medical Record Administrator
American Health Information
  Management Association
919 N. Michigan Ave., Suite 1400
Chicago, IL 60611-1683

Nurse
American Nurses Association
2420 Pershing Road
Kansas City, MO 64108

Occupational Therapist
American Occupational Therapy
  Association
1383 Piccard Drive
PO Box 1725
Rockville, MD 20849-1725

Oceanographer
Scripps Institute of
  Oceanography
A-033 University of California,
  San Diego
La Jolla, CA 92093

Optometrist
American Optometric
  Association
243 North Lindbergh Boulevard
St. Louis, MO 63141

Parasitologist
American Society of
  Parasitologists
1041 New Hampshire Street
Lawrence, KS 66044

Park Ranger
National Recreation and
  Park Association
2775 South Quincy Street,
  Suite 300
Arlington, VA 22206

Pharmacologist
American Society for
  Pharmacology
9650 Rockville Pike
Bethesda, MD 20814

Physical Therapist
American Physical Therapy
  Association
1111 North Fairfax Street
Alexandria, VA 22314

Physician
American Medical Association
515 North State Street
Chicago, IL 60610

Physiologist
American Physiological Society
9650 Rockville Pike
Bethesda, MD 20814

Pollution-Control Technician
Environmental Protection
  Agency
401 M Street SW
Washington, DC 20005

Public Health Microbiologist
American Society for
  Microbiology
1325 Massachusetts Avenue NW
Washington, DC 20005

Radiation Therapy Technologist
American Society of
  Radiological Technologists
15000 Central Avenue SE
Albuquerque, NM 87123

Respiratory Therapist
American Association for
  Respiratory Care
11030 Ables Lane
Dallas, TX 75229

Science Writer and Editor
National Association of Science
  Writers and Editors
PO Box 294
Greenlawn, NY 11740

Science Teacher
National Science Teachers
  Association
1742 Connecticut Avenue NW
Washington, DC 20009

Soil Scientist
U.S. Department of Agriculture
Soil Conservation Office
PO Box 2890
Washington, DC 20013

Surgical Technologist
Association of Surgical
  Technologists
8307 Shaffer Parkway
Littleton, CO 80127

Toxicologist
Society of Toxicology
1101 14th Street NW, Suite 1100
Washington, DC 20005

Veterinarian
American Veterinary Medical
  Association
930 North Meacham Rd.
Schaumburg, IL 60196

Virologist
American Society for
  Microbiology
1325 Massachusetts Ave. NW
Washington, DC 20005

Wildlife Manager
U.S. Department of Interior
Fish and Wildlife Service
Mail Stop 100 Arlington Square
Washington, DC 20240

Wood Scientist
Society of Wood Science and
  Technology
One Gifford Pinchot Drive
Madison, WI 53705

Zoo Curator
American Association of
  Zoological Parks and
  Aquariums
Oglebay Park, Route 88
Wheeling, WV 26003

Zoologist
American Society of Zoologists
Box 2739
California Lutheran University
Thousand Oaks, CA 91360

Measurement

All measurements in this book are expressed in metric units. Scientists throughout the world use the metric system, and you will always use metric units when you make measurements in the laboratory. The official name of the measurement system is the Système International d'Unités, or International System of Measurements. It is usually referred to simply as SI.

The metric system is a decimal system, that is, all relationships between units of measurement are based on powers of 10. Most units have a prefix that indicates the relationship of that unit to the base unit. For example, a meter equals 100 cm, or 1,000 mm. The lists below show the most commonly used prefixes as well as the main units used for each type of measurement.

Metric Prefixes

Prefix	Symbol	Factor of Base Unit
giga	G	1,000,000,000
mega	M	1,000,000
kilo	k	1,000
hecto	h	100
deka	da	10
deci	d	0.1
centi	c	0.01
milli	m	0.001
micro	μ	0.000001
nano	n	0.000000001

Area

square kilometer (sq.km) = 100 hectares

1 hectare (ha) = 10,000 square meters

1 square meter (sq.m) = 10,000 square centimeters

1 square centimeter (sq.cm) = 100 square millimeters

Mass

1 kilogram (kg) = 1,000 grams

1 gram (g) = derived from kg (base unit of mass)

1 milligram (mg) = 0.001 gram

1 microgram (μg) = 0.000001 gram

Liquid Volume

1 kiloliter (kL) = 1,000 liters

1 liter (L) = base unit of liquid volume

1 milliliter (mL) = 0.001 liter

Note: When measuring liquid volume in a graduated cylinder, be sure to read the measurement at the bottom of the meniscus, or curve.

Length

1 kilometer (km) = 1,000 meters

1 meter (m) = base unit of length

1 centimeter = 0.01 meter

1 millimeter (mm) = 0.001 meter

1 micrometer (μm) = 0.000001 meter

1 nanometer (nm) = 0.000000001 meter

Temperature

In the metric system, temperature is measured on the Celsius (C) scale. On the Celsius scale, 0° is the freezing point of water, and 100° is the boiling point of water. Thus 1°C equals 0.01 of the difference between the freezing point and boiling point of water. You can use the scale shown to convert between the Celsius scale and the Fahrenheit scale, which is commonly used in the United States.

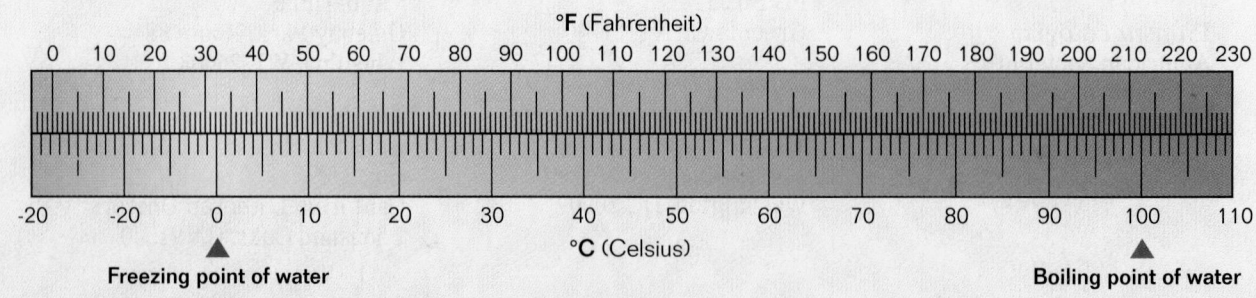

°F (Fahrenheit)
0 10 20 30 40 50 60 70 80 90 100 110 120 130 140 150 160 170 180 190 200 210 220 230

-20 -20 0 10 20 30 40 50 60 70 80 90 100 110
°C (Celsius)

▲ Freezing point of water

▲ Boiling point of water

Safety

You will avoid accidents in the biology laboratory by following directions, handling materials carefully, and taking your work seriously. Before you begin working, be sure that you are familiar with safety procedures and know the location of fire extinguishers and first aid supplies. Be aware of your classmates' safety as well as your own. Never attempt any laboratory procedure without an instructor's direction, and never work in the laboratory by yourself.

Read the text that follows to familiarize yourself with the safety symbols used in your text and the guidelines that you should follow when you see these symbols.

ANIMAL CARE
You are both legally and ethically required to treat animals as humanely as possible.

ANIMAL SAFETY
Wear leather or thick gloves when handling laboratory animals, especially rodents. When working in the field, be aware of poisonous or dangerous animals in the area. Do not touch or approach wild animals.

CAUSTIC SUBSTANCE
Use extreme care when handling caustic substances such as hydrochloric acid. These substances can injure the skin. If your skin comes in contact with a caustic substance, wash thoroughly with water and notify your teacher immediately.

CHEMICAL SAFETY
Use poisonous chemicals with extreme caution. Two harmless chemicals may become poisonous when combined. Never mix chemicals without teacher direction. Never put chemicals in your mouth, and avoid chemical contact with eyes or skin.

CLOTHING PROTECTION
Wear a laboratory apron in the classroom. Confine loose clothing.

ELECTRICAL SAFETY
Watch for loose plugs or worn electric cords. Be sure that cords are not placed where they could cause a fall. Do not use electrical equipment near water or with wet hands.

EXPLOSION DANGER
Many chemicals are explosive when combined, and some will explode when jarred, heated, or exposed to air. When heating chemicals, always point the test tube away from people.

EYE SAFETY
Wear approved safety goggles when working near an open flame or handling chemicals. If any chemical gets into your eyes, flush it out thoroughly with water and notify your instructor immediately.

FIRE SAFETY
Keep combustible materials away from sources of fire. Shield any open flame with an asbestos-protected screen.

GAS PRECAUTION
Do not inhale fumes directly. When instructed to smell a substance, wave fumes toward your nose and inhale gently. Use flammable liquids only in small amounts and in a well-ventilated room or under a fume hood. Always use a fume hood with toxic or flammable fumes. Do not breathe pure gases such as hydrogen, argon, helium, nitrogen, or high concentrations of carbon dioxide.

GLASSWARE SAFETY
Handle glassware carefully. Never attempt to clean up broken glass. Notify your teacher immediately.

HAND SAFETY
Dissect specimens in dissecting pans—never in your hand. Never use chipped or cracked glassware. Wear gloves when working with an open flame or with caustic chemicals.

HEATING SAFETY
Use proper procedures when lighting Bunsen burners. Heat flasks or beakers on a ring stand with a wire gauze between glass and flame. Use only heat-resistant glassware for heating materials or storing hot liquids. Turn off hot plates and open flames when not in use.

HYGIENIC CARE
Always wash your hands after the lab. Keep hands away from your face and mouth. Use sterile technique when transferring bacteria or other microorganisms between cultures or to a microscope slide. Do not open a petri dish to observe or count bacterial colonies.

PLANT SAFETY
Some plants cause ill effects when touched or eaten. Use a good field guide when collecting specimens; never eat any part of an unknown plant.

PROPER WASTE DISPOSAL
Clean up the laboratory after you finish. Follow your teacher's directions for waste disposal, especially for chemicals and microbes. Place broken glass in a special container.

WATER SAFETY
When working near water, always work with a partner or adult. Always wear a life jacket. Do not work near water during stormy weather.

Study Skill: Concept Mapping

To remember information longer and to be able to use it more effectively, you need to move that information into your long-term memory. Concept mapping can help you do this.

Identifying Concepts

Concept mapping helps you understand ideas by showing you their connections to other ideas. It is different from taking notes or making an outline. A concept map not only identifies the major ideas of interest from a chapter or your class notes, but also shows the relationships among the ideas, much as a road map illustrates how highways and other roads link cities. The figure below shows what a concept map looks like.

Suppose you have just finished reading a section of a chapter and wish to make a concept map as a study aid. How do you begin? First you need to identify the concepts in that section. Concepts usually form a picture in your mind. For instance, examine the following words: pool, biking, grass, sky, tree. All of these words are concepts because they create images in your mind. Now read this series of words: the, to, be, with, is, can. Are they concepts? No. They don't form a picture in your mind. They are linking words that connect ideas or concepts. Linking words play an important role in concept mapping because you use them to connect concepts in your map. See if you can identify the concepts and linking words in the figure below.

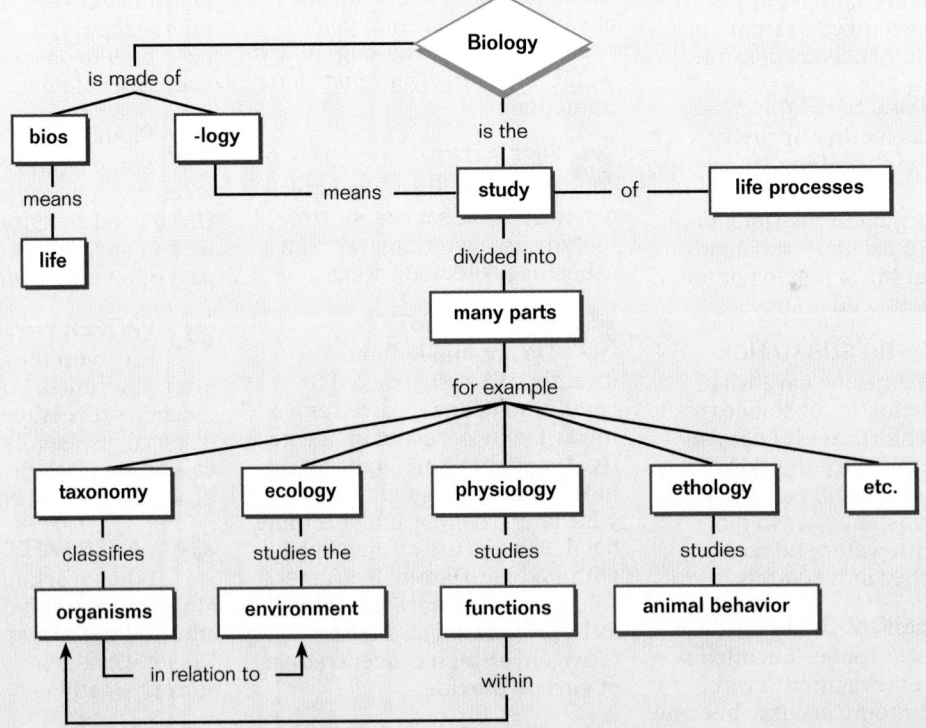

Sorting and Ranking Ideas

Another feature of a concept map is that some concepts or ideas are more general and include other concepts. Such general concepts will be the main ideas in your map. What is the main idea for this list of concepts: Tokyo, Mexico City, Seoul, New York, Bombay? Cities, of course. What is the main idea for this list: car, bus, train, bicycle, truck? Each of these concepts is an example of a vehicle.

Every concept map should have at its top the most general and all-inclusive concept. The smaller, more specific concepts and examples should go below. It might help you to capitalize the first letter of the main concept or idea. You can write the other, more specific, concepts using all lowercase letters. Study the concept map on page 1018 before you try one yourself. Then read the following paragraph and make a list of any important words you think should be learned.

What is life? What is the difference between living and nonliving things? If you were in a wilderness area, it would be easy for you to pick out the living and nonliving things. The animals and plants are the living organisms. Organisms are made up of many substances organized into living systems. The rocks, air, water, and soil you see are nonliving. They contribute substances to the living organisms.

Connecting Ideas

Two general ideas prevail in the preceding paragraph: living and nonliving. You could either make two separate maps, the chief concept of one being "living" and of the other being "nonliving," or you could make one map and use "natural things" as the main idea. Capitalize the first letter and put it at the top of the map. "Natural things" did not actually appear in the paragraph, but frequently the main idea does.

Natural things can be living or nonliving. Therefore, the ideas of "living" and "nonliving" are parts of the main idea and should be placed below it in the map. In the paragraph, you learned that living things are organisms, such as plants and animals, and that they are made up of substances that are organized into living systems. The paragraph also explained that living things are different from but related to nonliving things. So you should make the connection between living and nonliving things near the top of the map. And the rest of the map should explain how living and nonliving things are related. Nonliving things such as air, soil, water, and rocks all contribute substances that are organized into living systems.

Now you have the entire map shown on page 1020. If you had a choice between studying all those sentences or looking at this map, you would probably agree that the map shows the concepts more clearly. This map gives you the main idea more quickly, and it's easier to understand all the ideas because their relationships to other ideas are shown.

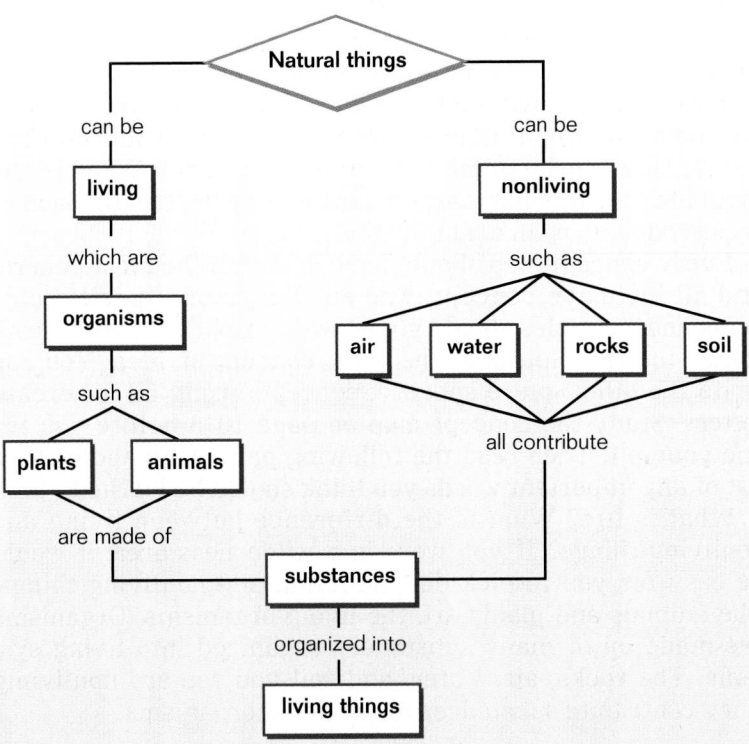

Features of Good Concept Maps

Remember, practice is the key to good concept mapping. You'll get better as you go along. Here are some things to remember that will help.

- A concept map does not have to be symmetrical. It can have more concepts on one side than the other.
- There are no perfectly correct concept maps, only maps that come closer to the meanings you have for those concepts. As the mapmaker, you must make it correct for you.
- Do not put more than three words in a concept box.
- Do not string out more than three boxes in a row or line without branching out.
- Write linkage words connecting every two concepts. Use as few words as it takes to make the connection between the concepts clear.

If the relationships you have made between any two concepts are wrong, your teacher will help you sort out your misconception. Even if you are absolutely correct in your relationships, maps made by your fellow students may be different. These maps could be equally correct, even though they may look nothing like yours. Everyone thinks a little bit differently, and, as a result, other people may see different relationships between certain concepts.

As you practice making concept maps, your teacher will examine your linkage statements more closely. Since these lines represent the relationships between concepts, whatever you write on the line will tell you if you really understand how those two concepts are connected. After you have practiced, your maps should always

- be two dimensional—not just a list of concepts connected by lines.
- show which concepts are more important by their placement on the map and by what concepts branch off them.
- have many branches with no more than three concept boxes in a row and no more than three or four words in a concept box.
- have only concepts in the boxes and only linkage words on the lines.

Evaluating Your Skills

For the first map you make on your own, think about something you know very well. Do you play a team sport or an individual sport? Do you have a hobby? Do you enjoy a particular kind of music? Whatever you choose, this will be the major concept for your next concept map. This will be more fun and easier since you know this topic so well.

Laboratory Skill: Using a Compound Light Microscope

Parts of the Compound Light Microscope

- The **eyepiece** magnifies the image, usually 10×.
- The **low-power objective** magnifies the image even more, such as 4×.
- The **high-power objectives** magnify the image even more, such as 10× and 43×.
- The **nosepiece** holds the objectives and can be turned to change from one magnification to another.
- The **body tube** maintains the correct distance between the eyepiece and the objectives. This is usually about 25 cm (10 in.), the normal distance for reading and viewing objects with the naked eye.
- The **coarse adjustment** moves the body tube up and down in large increments to allow gross positioning and focusing of the objective lens.

- The **fine adjustment** moves the body tube slightly to bring the image into sharp focus.
- The **stage** supports a slide.
- The **stage clips** secure the slide in position for viewing.
- The **diaphragm** (not labeled) controls the amount of light allowed to pass through the object being viewed.
- The **light source** provides light for viewing the image. It can either be a light reflected with a mirror or an incandescent light from a small lamp. NEVER use reflected direct sunlight as a light source.
- The **arm** supports the body tube.
- The **base** supports the microscope.

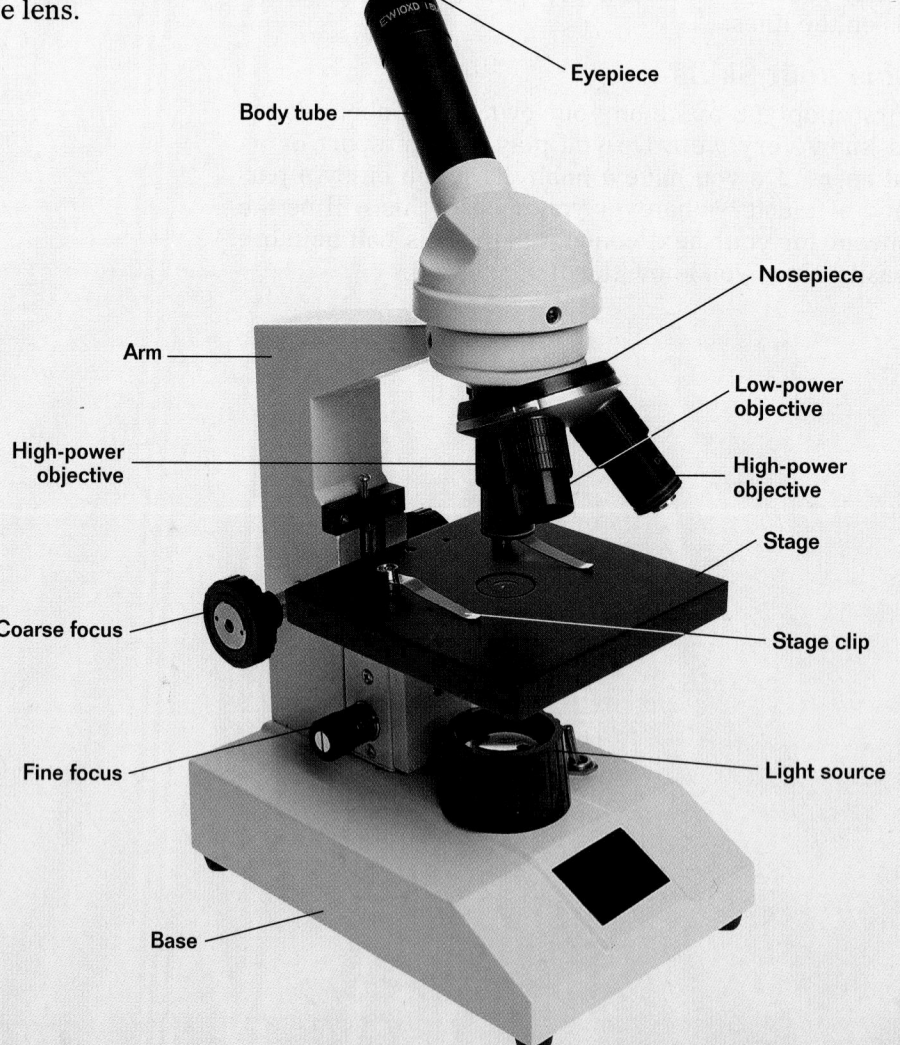

Eyepiece

Body tube

Arm

High-power objective

Coarse focus

Fine focus

Nosepiece

Low-power objective

High-power objective

Stage

Stage clip

Light source

Base

Proper Handling and Use of the Compound Light Microscope

1. Carry the microscope to your lab table using both hands, one beneath the base and the other holding the arm of the microscope. Hold the microscope close to your body.

2. Place the microscope on the lab table, at least 5 cm (2 in.) in from the edge of the table.

3. Check to see what type of light source the microscope has. If the microscope has a lamp, plug it in, making sure that the cord is out of the way. If the microscope has a mirror, adjust it to reflect light through the hole in the stage.

CAUTION: If your microscope has a mirror, do not use direct sunlight as a light source. Direct sunlight can damage your eyes.

4. Adjust the revolving nosepiece so that the low-power objective is in line with the body tube.

5. Place a prepared slide over the hole in the stage, and secure the slide with stage clips.

6. Look through the eyepiece and move the diaphragm to adjust the amount of light coming through the specimen.

7. Now look at the stage from eye level, and slowly turn the coarse adjustment to lower the objective until it almost touches the slide. Do not allow the objective to touch the slide.

8. While looking through the eyepiece, turn the coarse adjustment to raise the objective until the image is in focus. Never focus objectives downward. Use the fine adjustment to achieve a sharply focused image. Keep both eyes open while viewing a slide.

9. Make sure that the image is exactly in the center of your field of vision. Then switch to the high-power objective. Focus the image with the fine adjustment. Never use the coarse adjustment at high power.

10. When you are finished using the microscope, remove the slide. Clean the eyepiece and objectives with lens paper, and return the microscope to its storage area.

Making a Wet Mount

1. Use lens paper to clean a glass slide and coverslip.

2. Place the specimen you wish to observe in the center of the slide.

3. Using a medicine dropper, place one drop of water on the specimen.

4. Position the coverslip so that it is at the edge of the drop of water and at a 45° angle to the slide. Make sure that the water runs along the edge of the coverslip.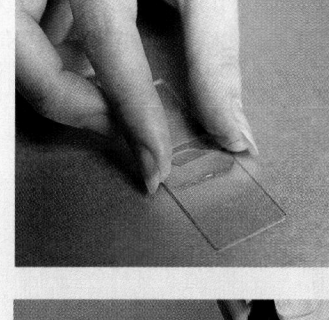

5. Lower the coverslip slowly to avoid trapping air bubbles.

6. If a stain or solution is to be added to a wet mount, place a drop of the staining solution on the microscope slide along one side of the coverslip. Place a small piece of paper towel on the opposite side of the coverslip.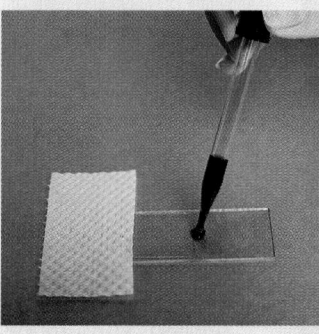

7. As the water evaporates from the slide, add another drop of water by placing the tip of the medicine dropper next to the edge of the coverslip, just as you would if adding stains or solutions to a wet mount. If you have added too much water, remove the excess by using the corner of a paper towel as a blotter. Do not lift the coverslip to add or remove water.

A Six-Kingdom System for the Classification of Organisms

* *

This is a classification of organisms based on the six-kingdom system explained in Chapter 19. Not all groups are shown. Numbers of species are approximate. Also, how the Eubacteria and Archaebacteria should be divided into phyla is controversial. Here, only broad, generally recognized groups are presented; these groups may not correspond to phyla.

KINGDOM EUBACTERIA

Typically unicellular; prokaryotic; without membrane-bound organelles; nutrition mainly by absorption, but some are photosynthetic or chemosynthetic; reproduction usually by fission or budding; about 5,000 species have been recognized, but there are undoubtedly many times that number

Cyanobacteria
Photosynthetic; surrounded by a gooey, deeply pigmented covering; common on land and in the ocean; probably ancestors of chloroplasts in some kinds of protists: *Anabaena, Oscillatoria, Spirulina*

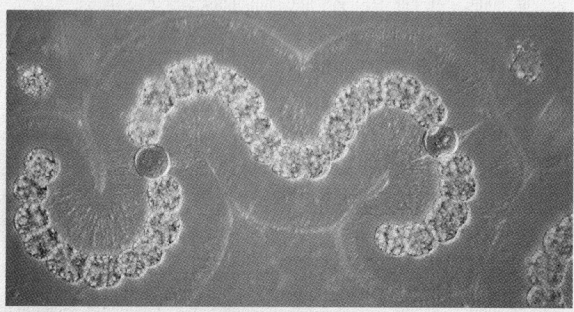

Anabaena

Chemoautotrophs
An ancient group of bacteria that can grow without sunlight or other organisms; derive energy from reduced gases—ammonia (NH_3), methane (CH_4), hydrogen sulfide (H_2S); play critical roles in the Earth's nitrogen cycles: nitrobacteria, sulfur bacteria (*Nitrosomonas, Nitrobacter*)

Enterobacteria
Typically rigid, rod-shaped, heterotrophic bacteria; usually aerobic; have flagella; responsible for many serious diseases of plants and humans, including cholera (*Vibrio cholerae*): *Escherichia coli, Salmonella typhimurium*

Pseudomonads
Straight or curved rods with flagella at one end; very common in soil; many are serious plant pathogens: *Pseudomonas aeruginosa*

Spirochaetes
Long, spiral cells; flagella originating at each end; responsible for several serious diseases, including syphilis (*Treponema pallidum*) and Lyme disease (*Borrelia burgdorferi*)

Actinomycetes
Filamentous bacteria that are often mistaken for fungi; spore-producing; sources of antibiotics including streptomycin, tetracycline, and chloramphenicol; cause dental plaque, leprosy, and tuberculosis: *Mycobacterium tuberculosis*

Rickettsias
Obligate parasites within the cells of vertebrates and arthropods; responsible for Rocky Mountain spotted fever: *Rickettsia rickettsii*

Gliding and budding bacteria
Long, rod-shaped cells that secrete slimy polysaccharides; often aggregate into gliding masses; live mainly in soil: *Myxobacteria*

KINGDOM ARCHAEBACTERIA

Anaerobic and aerobic bacteria adapted to environments with extreme temperatures, acidity, or salt content; prokaryotic; differ from eubacteria in structure of cell membrane and cell wall; RNA polymerase and a ribosomal protein similar to those in eukaryotes, suggesting that archaebacteria are more closely related to eukaryotes than to eubacteria; asexual reproduction only; fewer than 100 named species, divided into three broad groups

Methanogens
Anaerobic methane producers; most species use carbon dioxide as a carbon source; found in the soil, swamps, and the digestive tracts of animals, particularly grazing mammals such as cattle; produce nearly 2 trillion kg (2 billion tons) of methane gas annually

Thermoacidophiles

Inhabit very hot environments that are often very acidic; some species can tolerate temperatures of 110°C (230°F); require sulfur; nearly all are anaerobes: *Sulfolobus*

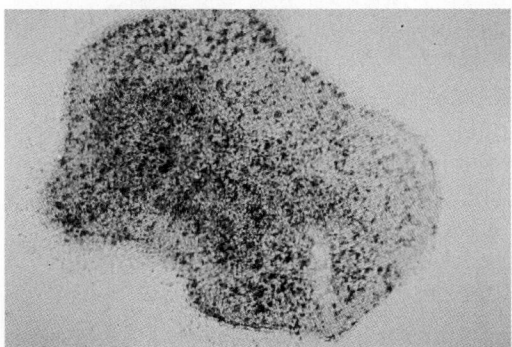

Sulfolobus

Extreme halophiles

Live in environments with very high salt content, including the Dead Sea and the Great Salt Lake; nearly all are aerobic; all are gram-negative

KINGDOM PROTISTA

A catchall kingdom for the eukaryotes that are not plants, fungi, or animals; the most structurally diverse kingdom; unicellular and multicellular representatives; all have a membrane-bound nucleus; nearly all have chromosomes, mitochondria, and internal compartments; many have chloroplasts; most have cell walls; reproduce sexually and asexually; aquatic or parasitic; about 43,000 species

Phylum Rhizopoda

Unicellular and heterotrophic; amorphously shaped cells that move using cytoplasmic extensions called pseudopods: amoebas (*Amoeba, Entamoeba*); about 300 species

Phylum Foraminifera

Unicellular and heterotrophic; marine; have shells of organic material with pores through which many cytoplasmic threads project: forams; about 300 species

Phylum Oomycota

Heterotrophic; unicellular parasites or decomposers; cell walls composed of cellulose, not chitin as in fungi: water molds, white rusts, downy mildews (*Phytophthora*); about 580 species

Phylum Ciliophora

Very complex single cells; heterotrophic; have rows of cilia and two types of cell nuclei: ciliates (*Didinium, Paramecium, Stentor, Vorticella*); about 8,000 species

Phylum Zoomastigina

Mostly unicellular; heterotrophic; all have at least one flagellum: zoomastigotes (*Giardia, Leishmania, Trichonympha, Trypanosoma*); about 3,000 species

Phylum Sporozoa

Unicellular; heterotrophic; nonmotile; spore-forming parasites of animals; have complex life cycles; asexual and sexual reproduction; *Plasmodium* is responsible for malaria, which kills more than 1 million people each year: sporozoans (*Plasmodium, Toxoplasma*); about 3,900 species

Phylum Myxomycota

Heterotrophic; individuals stream along as a multinucleate mass of cytoplasm; when dry or starving, can give rise to spores that start a new individual in a more favorable environment: plasmodial slime molds (*Physarum*); about 500 species

Physarum

Red Algae

Phylum Acrasiomycota
Heterotrophic; amoeba-shaped cells that aggregate into a moving mass called a slug when deprived of food; cells within the slug retain their membranes and do not fuse; a slug produces spores that form new amoebas elsewhere: cellular slime molds *(Dictyostelium)*; about 70 species

Phylum Caryoblastea
Unicellular; lacks mitochondria and chloroplasts; possibly an early stage in the evolution of eukaryotes; lives in low-oxygen conditions in the mud on pond bottoms: *Pelomyxa palustris* is the only species

Phylum Chlorophyta
Unicellular, colonial, and multicellular species; all are photosynthetic; have chlorophylls *a* and *b*; contain chloroplasts very similar to those of plants; most scientists think that plants descended from this group: green algae *(Chlamydomonas, Chlorella, Oedogonium, Spyrogyra, Ulva, Volvox)*; about 7,000 species

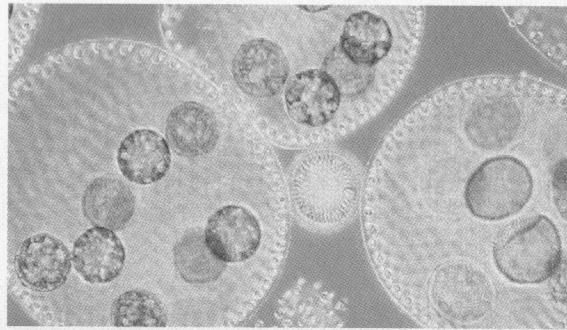

Volvox

Phylum Phaeophyta
Multicellular and photosynthetic; nearly all are marine; contain chlorophylls *a* and *c* and fucoxanthin, the source of their brownish color: brown algae *(Fucus, Laminaria, Postelsia, Sargassum)*; about 1,500 species

Phylum Rhodophyta
Almost all are multicellular; all are photosynthetic; most are marine; contain chlorophyll *a* and phycobilins; chloroplasts probably evolved from symbiotic cyanobacteria: red algae *(Porphyra)*; about 4,000 species

Phylum Bacillariophyta
Unicellular and photosynthetic; secrete a unique shell made of opaline silica that resembles a box with a lid; chloroplasts resemble those of brown algae; contain chlorophylls *a* and *c* and fucoxanthin: diatoms; more than 11,500 species

Phylum Dinoflagellata
Unicellular; heterotrophic and autotrophic species; mostly marine; body enclosed within two cellulose plates; contain chlorophylls *a* and *c* and carotenoids: dinoflagellates *(Gonyaulax, Noctiluca)*; more than 2,100 species

Phylum Euglenophyta
Unicellular; both photosynthetic and heterotrophic species; asexual; most live in fresh water; chloroplasts are similar to those of green algae and are thought to have evolved from the same symbiotic bacteria: euglenoids *(Euglena)*; about 1,000 species

KINGDOM FUNGI

Eukaryotic heterotrophs with nutrition by absorption; all but yeasts are multicellular; nearly all are terrestrial; body is typically composed of filaments (called hyphae) and is multinucleate, with incomplete divisions (called septae) between cells; cell walls made of chitin; reproduction asexual or sexual; about 77,000 species

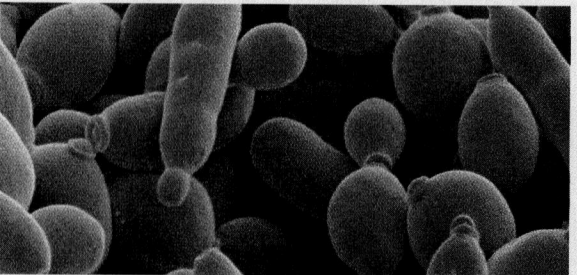

Yeasts

Phylum Ascomycota

Hyphae usually have perforated septae; fusion of hyphae leads to formation of densely interwoven mass that contains characteristic microscopic reproductive structures called asci (singular, ascus); terrestrial, marine, and freshwater species: bread molds, morels, truffles, *Neurospora*, *Saccharomyces*; about 30,000 species

Phylum Basidiomycota

Hyphae usually have incomplete septae; reproduction is typically sexual; fusion of hyphae leads to the formation of densely interwoven reproductive structure (mushroom) with characteristic microscopic, club-shaped structures called basidia (singular, basidium); terrestrial: mushrooms, toadstools, shelf fungi, rusts, smuts; about 16,000 species

A basidiomycete

Phylum Deuteromycota

(also called Fungi Imperfecti); Sexual stages of life cycle not observed; most are probably ascomycetes that have lost the ability to reproduce sexually; includes the molds that produce penicillin *(Penicillium)* and athlete's foot fungus *(Trichophyton)*; about 17,000 species

Phylum Zygomycota

Usually lack septae; fusion of hyphae leads directly to formation of a zygote, which divides by meiosis when it germinates; terrestrial or parasitic: bread molds *(Pilobolus, Rhizopus)*; about 660 species

FUNGAL ASSOCIATIONS

Fungi form symbiotic associations with plants, green algae, and cyanobacteria.

Lichens

Mutualistic relationships between fungi (almost always ascomycetes) and cyanobacteria, green algae, or both; the photosynthetic partners actually live among the hyphae of the fungus; the fungus derives energy from its photosynthetic partners and cannot survive without them; about 15,000 species

Mycorrhizae

Mutualistic relationships between fungi and the roots of plants; 80 percent of all plants have mycorrhizae associated with their roots; the plant provides sugars to the fungi; in return, the fungi serve as accessory roots, greatly increasing the surface area available for the absorption of nutrients; about 5,000 species

KINGDOM PLANTAE

Multicellular, eukaryotic, autotrophic, terrestrial organisms having tissues and organs; cell walls contain cellulose; chlorophylls *a* and *b* present and localized in plastids; all have alternation of generations; about 265,000 species

Phylum Bryophyta

Small and usually found in moist environments; most have simple vascular tissues; gametophyte is the dominant generation; gametophyte lacks roots, stems, and leaves; mosses *(Sphagnum)*; about 10,000 species

Phylum Hepatophyta

Gametophyte is dominant, usually small, and grows close to the ground; gametophyte lacks stomata, vascular tissue, roots, stems, and leaves; sporophyte inconspicuous and parasitic on the gametophyte: liverworts *(Marchantia)*; about 6,000 species

Marchantia

Phylum Anthocerophyta

Gametophyte is dominant generation; usually small and flat; stomata present on sporophyte; lacks vascular tissue, roots, stems, and leaves: hornworts *(Anthoceros):* about 100 species

Phylum Psilotophyta

Vascular system present; seedless; no roots or stems; gametophyte is small and independent; sporophyte is dominant and has small leaves: whisk ferns *(Psilotum);* 21 species

Phylum Lycophyta

Vascular system present; sporophyte dominant and mosslike; has roots, stems, and leaves; gametophyte is small and independent; seedless: club mosses *(Lycopodium, Selaginella);* about 1,000 species

Phylum Sphenophyta

Vascular system present; gametophyte is small and independent; sporophyte dominant; roots present; seedless: horsetails *(Equisetum);* 15 species

Phylum Pterophyta

Vascular system present; gametophyte is small and independent; sporophyte generation dominant; has roots, stems, and leaves; seedless: ferns *(Salvinia);* about 12,000 species

Phylum Cycadophyta

Palmlike gymnosperms; vascular system present; male and female cones produced on different trees; naked seeds; sporophyte dominant; tropical and subtropical: cycads *(Cycas);* about 100 species

Phylum Coniferophyta

Gymnosperms that produce cones; vascular system lacks vessels; leaves usually needles or scales; typically evergreen; sporophyte dominant; ovules exposed at time of pollination; pollen is dispersed by wind: pines, spruces, firs, larches, yews *(Pinus, Taxus);* about 550 species

Lodgepole pine

Phylum Gnetophyta

Specialized gymnosperms; vascular system contains water-conducting vessels; seeds naked; sporophyte dominant; gnetophytes *(Ephedra, Welwitschia);* about 70 species

Phylum Ginkgophyta

Deciduous, gymnosperm tree; vascular system present; has fanlike leaves; sporophyte dominant; produces conelike male reproductive structures and uncovered seeds on different individuals: the ginkgo, *Ginkgo biloba,* is the only species

Phylum Anthophyta

Angiosperms; vascular system present; sporophyte dominant; ovules are fully enclosed by ovary; after fertilization, ovary and seed mature to become fruit; flowers are reproductive structures: flowering plants *(Aster, Prunus, Quercus, Zea);* about 250,000 species

Class Monocotyledones

Embryo has one cotyledon; flower parts in threes; leaf veins parallel; vascular bundles scattered through stem tissue: grasses, sedges, lilies, irises, palms, orchids; around 70,000 species

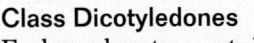

Class Dicotyledones

Embryo has two cotyledons; flower parts in fours or fives; leaves with netlike veins; vascular bundles in orderly arrangement in stems: roses, maples, elms; about 170,000 species

Widow's tears

KINGDOM ANIMALIA

Multicellular, eukaryotic, heterotrophic organisms; nutrition mainly by ingestion; most have specialized tissues, and many have complex organs and organ systems; no cell walls or chloroplasts; sexual reproduction predominates; both aquatic and terrestrial forms; about 1 million species

Phylum Porifera

Asymmetrical; lack tissues and organs; body wall consists of two cell layers, penetrated by numerous pores; internal cavity is lined with unique food-filtering cells called chaonocytes; mostly marine: sponges; about 9,000 species

Purple tube sponges

Phylum Cnidaria

Radially symmetrical and gelatinous; most have distinct tissues; baglike body of two cell layers; marine and freshwater species: hydras, jellyfish, corals, sea anemones (*Hydra, Obelia*); about 10,000 species

Class Hydrozoa

Most have both polyp and medusa stages in life cycle: hydras, Portuguese man-of-war; about 2,700 species

Class Scyphozoa

Exclusively marine; medusa stage dominant: jellyfish; about 200 species

Class Anthozoa

Marine; solitary or colonial; medusa stage absent: sea anemones, corals, sea fans; about 6,200 species

Phylum Ctenophora

Transparent, gelatinous marine animals resembling jellyfish; radially symmetrical: comb jellies; about 100 species

Phylum Platyhelminthes

Body flat and ribbonlike, without true segments; bilaterally symmetrical acoelomates; organs present; three germ layers: flatworms (*Dugesia, Planaria, Schistosoma*); about 20,000 species

Marine flatworm

Class Turbellaria

Mostly free-living aquatic or terrestrial forms: planarians (*Dugesia*)

Class Trematoda

Internal parasites, with mouth at anterior end; often have complex life cycle with alternation of hosts: human blood flukes (*Schistosoma*), human liver fluke (*Chlonorchis sinensis*)

Class Cestoda
Extremely specialized internal parasites; hooked scolex for attaching to host: tapeworms

Phylum Rotifera
Small, wormlike or spherical animals; bilaterally symmetrical; pseudocoelomates; almost all live in fresh water: rotifers; about 1,750 species

Phylum Nematoda
Typically tiny, parasitic, unsegmented worms; body slender and elongated; pseudocoelomates; includes important human parasites such as *Ascaris*, pinworms, hookworms, *Trichinella*, and *Wuchereria*: roundworms; more than 12,000 species

Phylum Loricifera
Minute, bilaterally symmetrical pseudocoelomates; live in spaces between grains of sand; loriciferans *(Nanaloricus mysticus)*; 6 species

Phylum Mollusca
Soft-bodied animals with a true coelom; three-part body plan consisting of foot, visceral mass, and mantle; protostomes; most have a unique rasping tongue called a radula; terrestrial, freshwater, and marine: clams, snails, octopuses, squid, mussels, slugs; about 110,000 species

Class Cephalopoda
Foot modified into tentacles: squids, octopuses, nautilus; over 600 species

Brief squid

Class Bivalvia
Two shells connected by a hinge; no radula; large, wedge-shaped foot: clams, oysters, scallops; about 10,000 species

Class Gastropoda
Visceral mass twisted during development; head, distinct eyes, and tentacles usually present: snails, slugs, whelks; about 80,000 species

Class Polyplacophora
Elongated body and reduced head: chitons; about 600 species

Phylum Rhynchocoela
Bilaterally symmetrical acoelomates; long, typically ribbonlike body: ribbon worms; about 650 species

Phylum Annelida
Serially segmented worms; bilaterally symmetrical; protostomes: segmented worms; about 12,000 species

Class Polychaeta
Fleshy outgrowths called parapodia extend from segments; marine; many bristles: sandworms; about 8,000 species

Fireworm

Class Oligochaeta
Head not well developed; no parapodia; terrestrial and freshwater forms: earthworms; about 3,100 species

Class Hirudinea
Body flattened; no parapodia; usually suckers at both ends; many are external parasites: leeches; about 600 species

Phylum Onychophora
Protostomes; chitinous exoskeletons; wormlike: velvet worms *(Peripatus)*; about 70 species

Phylum Pogonophora
Long, deep-sea worms that live within chitinous tubes on the ocean floor: tube worms, or beard worms; about 100 species

Phylum Arthropoda

Segmented bodies with paired, jointed appendages; bilaterally symmetrical; chitinous exoskeleton; protostomes; aerial, terrestrial, and aquatic forms: arthropods; about 1 million species

Subphylum Chelicerata

Distinguished by absence of antennae and presence of chelicerae; all appendages unbranched; four pairs of walking legs; body has two regions (cephalothorax and abdomen); predominantly terrestrial

Class Merostomata
Cephalothorax covered by protective "shell"; sharp spike on tail: horseshoe crabs (*Limulus*); 5 species

Class Pycnogonida
Small, marine predators or parasites; usually four pairs of legs, but sometimes five or six pairs: sea spiders; about 1,000 species

Class Arachnida
Terrestrial; use book lungs and tracheae for respiration: spiders, scorpions, ticks, mites; about 57,000 species

Subphylum Crustacea

Two pairs of antennae, mandibles, and appendages with two branches; predominantly aquatic

Class Malacostraca
Typically five pairs of legs; two pairs of antennae; most are aquatic: crayfish, lobsters, crabs, shrimp, sow bugs; about 20,000 species

Subphylum Uniramia

Have antennae, mandibles, and unbranched appendages

Class Chilopoda
Body flattened and consisting of 15 to 170 or more segments; one pair of legs attached to each segment: centipedes; about 2,500 species

Class Diplopoda
Elongated body of 15 to 200 segments; two pairs of legs per segment; primarily herbivorous: millipedes; about 10,000 species

Class Insecta

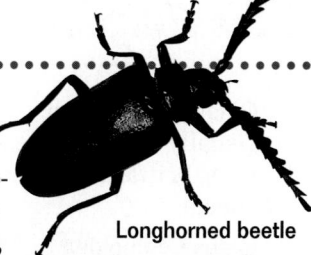

Longhorned beetle

Body has three regions—head, thorax, and abdomen; three pairs of legs, all attached to thorax; usually two pairs of wings: insects; about 750,000 described species, but millions more may exist

Order Thysanura: silverfish
Order Ephemeroptera: mayflies
Order Odonata: dragonflies, damsel flies
Order Orthoptera: grasshoppers, cockroaches, walking sticks, praying mantises, crickets
Order Isoptera: termites
Order Dermaptera: earwigs
Order Mallophaga: chicken lice
Order Anoplura: human body louse
Order Hemiptera: true bugs—water striders, water boatmen, back swimmers, bedbugs, squash bugs, stink bugs, assassin bugs
Order Homoptera: cicadas, aphids, leaf hoppers, scale insects
Order Neuroptera: ant lions, lacewings
Order Coleoptera: beetles—ladybugs, fireflies, boll weevil
Order Lepidoptera: butterflies and moths
Order Diptera: flies, mosquitoes, gnats, midges
Order Siphonaptera: fleas
Order Hymenoptera: bees, ants, wasps, hornets, ichneumon fly

Phylum Echinodermata

Deuterostomes; adults radially symmetrical with five-part body plan; most forms have water vascular system with tube feet for locomotion; marine: echindoderms; about 6,000 species

Class Crinoidea
Mouth faces upward and is surrounded by many arms: sea lilies, feather stars; About 600 species

Class Asteroidea
Body usually with five arms and double rows of tube feet on each arm; mouth directed downward: sea stars; about 1,500 species

Class Ophiuroidea
Usually with five slender, delicate arms or rays: brittle stars, basket star; about 2,000 species

Class Echinoidea
Body spherical, oval, or disk-shaped; arms lacking but five-part body plan still apparent: sea urchins, sand dollars; about 900 species

Class Holothuroidea
Elongated, thickened body with tentacles around the mouth: sea cucumbers; about 1,500 species

Phylum Hemichordata
Wormlike chordates; deuterostomes; body in three regions: acorn worms; about 90 species

Phylum Chordata
Bilaterally symmetrical; deuterostomes; coelom present; have a notochord, a dorsal nerve cord, pharyngeal slits, and a tail at some stage of life; aquatic and terrestrial; about 42,500 species

Subphylum Urochordata
Saclike covering, or tunic, in adults; larvae are free-swimming and have nerve cord and notochord; marine: tunicates; about 1,250 species

Subphylum Cephalochordata
Small and fishlike with a permanent notochord; filter feeders: lancelets (*Branchiostoma*); 21 species

Subphylum Vertebrata
Most of the notochord is replaced by a spinal column composed of vertebrae that protect the dorsal nerve cord; recognizable head containing a brain: vertebrates; about 40,000 species

Class Agnatha
Freshwater or marine eel-like fishes without true jaws, scales, or paired fins; cartilaginous skeleton: lampreys, hagfish; 63 species

Class Chondrichthyes
Fishes with jaws and paired fins; gills present; no swim bladder; cartilaginous skeleton: sharks, rays, skates; about 850 species

Blacktip shark

Class Osteichthyes
Freshwater and marine fishes with gills free and attached to gill arch; jaws and paired fins; bony skeleton; most have swim bladder: bony fishes; about 18,000 species

Class Amphibia
Freshwater or terrestrial; gills present at some stage; skin often slimy and lacking scales; eggs typically laid in water and fertilized externally: amphibians; about 4,200 species

Order Apoda: caecilians
Order Urodela: salamanders and newts
Order Anura: frogs and toads

Class Reptilia
Terrestrial or semiaquatic vertebrates; breathe by lungs at all stages; body covered by scales; most species lay amniotic eggs covered with a protective shell; fertilization internal: reptiles; about 7,000 species

Order Rhynchocephalia: tuataras
Order Chelonia: turtles and tortoises
Order Crocodilia: alligators, crocodiles, gavials, caimans
Order Squamata: lizards and snakes

Gila monster

Class Aves

Body covered with feathers; forelimbs modified into wings; four-chambered heart; endothermic; lay shelled, amniotic eggs: birds; about 9,000 species

Order Gaviiformes: loons

Order Pelecaniformes: pelicans, cormorants, gannets

Order Ciconiiformes: herons, bitterns, egrets, storks, spoonbills, ibises

Order Anseriformes: ducks, geese, swans

Order Falconiformes: hawks, falcons, eagles, kites, vultures

Order Galliformes: pheasants, turkeys, quails, partridges, grouse

Order Gruiformes: cranes, coots, gallinules, rails

Order Charadriiformes: snipes, sandpipers, plovers, gulls, terns, auks, puffins, ibises

Order Procellariiformes: albatrosses, petrels

Order Columbiformes: pigeons and doves

Order Psittaciformes: parrots, parakeets, macaws, cockatoos

Order Cuculiformes: cuckoos, roadrunners

Macaw

Order Strigiformes: owls

Order Caprimulgiformes: goatsuckers, whippoorwills, nighthawks

Order Apodiformes: swifts, hummingbirds

Order Coraciiformes: kingfishers

Order Sphenisciformes: penguins

Order Piciformes: woodpeckers, sapsuckers, flickers, toucans, honeyguides

Order Passeriformes: robins, bluebirds, sparrows, warblers, thrushes

Order Struthioniformes: ostrich

Order Apterygiformes: kiwis

Class Mammalia

Hair on at least part of body; young nourished with milk secreted by mammary glands; endothermic; breathe with lungs; mammals; about 4,000 species

Order Monotremata: duckbill platypus and spiny anteaters

Order Marsupialia: opossums, kangaroos, koalas, wallabies

Order Insectivora: moles and shrews

Order Chiroptera: bats

Order Edentata: armadillos, sloths, anteaters

Order Pholidota: pangolins

Deer mouse

Order Rodentia: squirrels, woodchucks, mice, rats, muskrats, beavers

Order Lagomorpha: rabbits, hares, pikas

Order Carnivora: bears, weasels, mink, otters, skunks, lions, tigers, wolves

Order Cetacea: whales, porpoises, dolphins

Order Sirenia: sea cows, dugongs, manatees

Order Proboscidea: elephants

Order Pinnipedia: seals, sea lions, walruses

Order Perissodactyla: tapirs, rhinoceroses, horses, zebras

Order Artiodactyla: hippopotamuses, camels, llamas, deer, giraffes, cattle, sheep, goats

Order Primates: monkeys, lemurs, gibbons, orangutans, gorillas, chimpanzees, humans

Order Macroscelidea: elephant shrews

Order Scandentia: tree shrews

Order Hyracoidea: hyraxes

Order Dermoptera: flying lemurs

Order Tubulidentata: aardvark

Glossary

Pronunciation Key

Sound	As In	Phonetic Respelling
ahy	bat	(BAT)
ay	face	(FAYS)
ah	lock	(LAHK)
	argue	(AHR gyoo)
ow	out	(OWT)
ch	chapel	(CHAP uhl)
eh	test	(TEHST)
ai	rare	(RAIR)
ee	eat	(EET)
	feet	(FEET)
	ski	(SKEE)
ih	bit	(BIHT)
eye	idea	(eye DEE uh)
y	ripe	(RYP)

Sound	As In	Phonetic Respelling
ihng	going	(GOH ihng)
k	card	(KAHRD)
	kite	(KYT)
ng	anger	(ANG guhr)
oh	over	(OH vuhr)
aw	dog	(DAWG)
	horn	(HAWRN)
oy	foil	(FOYL)
u	pull	(PUL)
oo	pool	(POOL)
s	cell	(SEHL)
	sit	(SIHT)
sh	sheep	(SHEEP)

Sound	As In	Phonetic Respelling
th	that	(THAT)
	thin	(THIHN)
uh	cut	(CUHT)
ur	fern	(FURN)
y	yes	(YEHS)
yoo	globule	(GLAHB yool)
yu	cure	(KYUR)
z	bags	(BAGZ)
zh	treasure	(TREHZH uhr)
uh	medal	(MEHD uhl)
	pencil	(PEHN suhl)
	onion	(UHN yuhn)
uhr	paper	(PAY puhr)

A-B-O system the system used to classify human blood by antigens (881)

acetylcholine neurotransmitter that stimulates skeletal muscles (67)

acetyl-Co A molecule derived from glucose and fatty acid metabolism; compound that enters the Krebs cycle (109)

acid compound that is a proton donor (33)

acid rain precipitation with below-normal pH, often the result of industrial pollution and automobile exhaust (6, 390)

acne a skin condition resulting from clogged oil ducts (869)

acoelomate an animal that lacks a coelom, or body cavity (618)

acromegaly disorder that results from overproduction of growth hormone and causes facial figures to thicken (978)

actin *See* actin filament

actin filament a protein found in a muscle cell that functions in contraction (46)

action potential sudden reversal of voltage across the neuron membrane (941)

activation energy amount of energy required to start a chemical reaction (81)

active site site on an enzyme that attaches to a substrate (83)

active transport movement of a particle through a membrane against a concentration gradient with the use of energy from ATP (62)

adaptation process of becoming adapted to an environment; an anatomical structure, physiological process, or behavioral trait that improves an organism's likelihood of survival and reproduction (249)

adductor muscle attachment between the two valves of a mollusk that causes the shell to open and close (662)

adenine a purine base; a component of nucleotides (169)

adrenal cortex outer shell of the adrenal gland that produces the steroid hormones cortisol and aldosterone (980)

adrenaline peptide hormone released by the adrenal medulla in times of stress; also called epinephrine (980)

adrenal medulla part of the adrenal gland that produces the peptide hormone adrenaline, or epinephrine (980)

adventitious roots roots that grow from aboveground parts of a plant, such as stems and leaves (559)

aerobic term for processes that require oxygen (107)

aggregation a temporary collection of cells that come together for a period of time and then separate (427)

aldosterone hormone that affects the kidneys by promoting the uptake of sodium and other salts from body fluids (981)

allele an alternative form of a gene (146)

allele frequency the relative abundance of an allele of a gene within a population, expressed as a percentage (322)

allergy a major defense mounted by the body's immune system against a harmless substance (930)

allosteric enzyme an enzyme whose shape can be altered by the binding of a signal molecule to its surface (88)

allosteric site the site where the signal molecule binds to an allosteric enzyme's surface (88)

alternation of generations life cycle in which a haploid individual alternates with a diploid individual (524)

alveolus microscopic air sac in the lung where oxygen and carbon dioxide are exchanged (714, 889)

amino acid organic molecule that is the building block of protein (36)

amniocentesis procedure in which a sample of amniotic fluid is withdrawn and tested for genetic abnormalities in a fetus (124)

amnion membrane enclosing the embryo (725, 1004)

amniotic egg watertight, fluid-filled egg in birds, reptiles, and mammals (726)

amoeba a protist that moves using flexible, cytoplasmic extensions (485)

amoebocyte amoeba-like cell that supplies nutrition and removes wastes from sponge body cells (638)

amylase enzyme that breaks down starches into sugars (904)

anabolic steroids synthetic chemicals that resemble testosterone and aid in increasing muscle size and strength (863)

anaerobic term for processes not requiring oxygen (107)

analogous characters similar features of organisms that evolve independently (438)

anaphase stage of cell division in which chromosome copies separate (131)

anapsid type of vertebrate skull that has openings only for the eyes and nostrils (773)

androecium part of a flower that produces male gametes, or pollen grains (536)

anemia condition in which red blood cell production slows down (880)

angiosperm seed plant that reproduces via flowers, which produce seeds within an ovary (531)

annual plant plant that completes its life cycle during one growing season (577)

anterior front end of bilaterally symmetric animal (617)

anterior lobe front portion of the pituitary gland from which seven major hormones are secreted (978)

anther the sac at the tip of the stamen in which pollen grains are formed (536)

antheridium reproductive structures in seedless plants that produce sperm by mitosis (525)

antibiotic substance used as a drug to kill bacteria (469)

antibody defensive protein released by B cells in response to a foreign substance in the body (211, 458, 927)

anticoagulant protein that prevents blood from clotting (211)

anticodon a three-nucleotide sequence on tRNA that recognizes a codon on mRNA (137)

aorta main artery in the body; receives blood from the left ventricle (884)

aortic valve one-way valve that prevents blood in the aorta from flowing back into the left ventricle (884)

apical dominance the inhibition of lateral bud growth by auxin produced in the terminal bud (582)

apical meristem meristem in the tips of stems and roots (573)

appendicular skeleton bones that form the arms and legs (858)

aquifer ground water trapped within porous rock (397)

arachnid any of a large class of arthropods, usually with four pairs of legs, including spiders, scorpions, mites, and ticks (684)

archaebacteria one of the two kingdoms of prokaryotes; represented today by a few groups of bacteria inhabiting extreme environments (274)

archegonium reproductive structure in seedless plants that produces eggs by mitosis (525)

artery vessel that carries blood away from the heart to the body's organs (877)

arteriole a branch of an artery that gives rise to capillaries (877)

arteriosclerosis a cardiovascular disease caused by the formation of hard plaques in the walls of arteries, which leads to decreased elasticity (887)

arthropod members of the phylum Arthropoda, which includes invertebrate animals such as insects, crustaceans, and arachnids (281)

ascocarp compact mass of hyphae constituting the reproductive structure in Ascomycota (506)

ascus sac that forms on the surface of an asocarp in which haploid spores are formed (506)

asexual reproduction reproduction that does not involve a union of gametes (422)

A site a binding site on the ribosome filled by a tRNA molecule carrying an amino acid (187)

aspirin compound originally derived from the white willow tree; used as a pain reliever (601)

asthma a chronic respiratory disorder caused by constricted air passages in the lungs (893)

atherosclerosis a cardiovascular condition caused by a buildup of fat deposits on artery walls (887)

atoll a ring-shaped coral island with a lagoon in the center (646)

atom smallest unit of matter that cannot be broken down by chemical means (29)

ATP-synthetase protein channel in a membrane through which protons are pumped to produce ATP (102)

atrioventricular node a bundle of cardiac cells in the right atrium that helps conduct impulses through the heart (885)

atrium a chamber that receives blood returning to the heart (718)

autoimmune disease disease in which the body loses the ability to distinguish its own cells from foreign cells and as a result, attacks itself (929)

autonomic nervous system the network of motor nerves that regulate cardiac muscle, smooth muscle, and gland activity (944)

autosomes a chromosome that is not directly involved in determining sex (124)

autotroph organism that obtains energy from sunlight or chemicals (78)

auxin a class of plant hormones that stimulates cell elongation, secondary growth, and leaf and fruit development and inhibits lateral bud growth (582)

axial skeleton bones that form the main body axis (858)

axon the elongated extension of a neuron that carries impulses away from the cell body (833, 940)

Glossary

B

bacillus rod-shaped bacterial cell (462)

bacteriophage virus that infects bacteria (167, 457)

balancing selection type of natural selection in which the homozygous genotypes are selected against but the heterozygous genotype is selected for, maintaining both alleles in a population (263)

barrier reef coral reef that serves as a barrier between waves and beaches (646)

basal disk an small area on hydra that secretes a sticky substance, which enables them to adhere to rocks or plants in the water (642)

basal layer a layer of the epidermis that continuously replaces lost cells (867)

base compound that is a proton acceptor (33)

base-pairing rules Chargaff's rules that state the amount of adenine equals the amount of thymine and the amount of guanine equals the amount of cytosine (170)

basidium club-shaped sexual reproductive structure that forms on the gills of basidiomycetes (508)

B cell white blood cell that produces antibodies (925)

biennial plant plant that completes its life cycle within two years (577)

bilaterally symmetric animal configuration with left and right halves that mirror each other (617)

bile salts fat-emulsifying molecules produced by the liver (905)

binary fission form of asexual reproduction that produces identical offspring (126)

binomial nomenclature a system for giving each organism a two-word scientific name that consists of the genus followed by the species (432)

biochemical pathway series of enzyme-catalyzed reactions that perform a specific function (85)

biological species concept group of actually or potentially interbreeding natural populations, reproductively isolated from other such groups (436)

biology science of life (12)

biomass the dry weight of tissue and other organic matter used to determine the amount of energy present in trophic levels (348)

biome major type of terrestrial ecological community such as a grassland and desert (379)

bipedal term used to denote the ability to walk erect on two feet (299)

blastocyst hollow-ball embryo that consists of about 700 cells when it arrives at the uterus (1004)

blood type the antigenic characteristic of the blood of an individual (881)

Bohr effect effect of carbon dioxide on oxygen unloading (890)

brain stem the part of the brain that connects to the spinal cord; contains nerves controlling breathing, swallowing, digestive processes, heartbeat, and blood vessel diameter (946)

bran papery outer coat, or husk, of a grain (592)

bronchus one of the two branches of the trachea that leads to the lungs (889)

bronchiole tiny air passages that connect alveoli to the bronchi (889)

bundle of His network of fibers over which a wave of contraction of the heart is conducted (885)

C

C₃ plant plant that fixes carbon using the Calvin cycle (556)

C₄ plant plant that fixes carbon using an alternative pathway in which the first detectable product is a four-carbon compound (556)

calcitonin hormone produced in the thyroid gland that helps maintain proper calcium levels in the body (979)

Calorie amount of energy needed to raise the temperature of 1 gram of water 1°C. The Calorie used to indicate the energy content of food is a kilocalorie. (900)

Calvin cycle the second major pathway in photosynthesis involving carbon fixation and carbohydrate formation (105)

calyx outermost whorl of a flower; the sepals (536)

cancer a disease characterized by abnormal cell growth (17, 195)

capillaries tiny blood vessels that allow exchange between blood and cells in tissue (877)

capsid a protein sheath that surrounds the nucleic acid core in a virus (456)

capsule a gelatinous outer layer enclosing many bacterial cell walls (463)

carapace shieldlike plate covering the cephalothorax of decapods; dorsal part of a turtle's shell (686, 776)

carbohydrate organic compound composed of carbon, hydrogen, and oxygen; used by living things as an energy source (34, 900)

carbon fixation process by which carbon dioxide is incorporated into organic compounds (105)

carbonic anhydrase enzyme within red blood cells that combines carbon dioxide with water to form carbonic acid (891)

carcinogen cancer-causing substance (195)

carcinoma common type of skin cancer originating in the non-pigment producing cells of the epidermis (868)

carnivore flesh-eating organism (344)

carotenoids yellow and orange plant pigments that aid in photosynthesis (101)

carrying capacity population size that an environment can sustain (318)

cartilage lightweight, strong, and flexible connective tissue (706)

caste role played by individual insect in a colony (693)

catalysis process of increasing a chemical reaction rate through the action of a catalyst (82)

catalyst material that speeds up a chemical reaction without being used itself (82)

cell smallest unit that can perform all the life processes (12)

cell body the part of a neuron that processes incoming information (939)

cell cycle repeating five-phase sequence of eukaryotic cell growth and division (127)

cell membrane bilipid layer that encloses the cytoplasm, essential to the cell's cytoplasm; also called the plasma membrane (26)

cell specialization ability of different cells to perform specific functions in a multicellular organism, such as protection, mobility, and reproduction (429)

cell surface marker membrane protein of a cell that distinguishes it from other cells and foreign matter (57)

cell surface protein protein within a cell's plasma membrane (56)

cell theory principle that states all organisms are made of one or more cells that are derived from other cells (25)

cellular respiration process by which living things obtain energy from the bonds of food molecules (78)

cellulose carbohydrate that is the main structural component of cell walls (34)

central nervous system system composed of the brain and spinal cord (826)

central vacuole membrane-bound cavity in plant cells used for storage (46)

centriole rod-shaped organelle that help move chromosomes during cell division (46)

centromere region joining two chromatids (120)

cephalization process of head development in bilaterally symmetric animals (617)

cephalothorax the mid-body region in arthropods; consists of a head fused with a thorax (680)

cereal grass grown as food for humans and livestock (592)

cerebellum region of the brain that controls coordination and balance (946)

cerebral cortex thin outer layer of the cerebrum (945)

cerebral ganglion the brain contained in the anterior segment of an annelid (625)

cerebrum rounded, wrinkled anterior portion of the brain; center for memory, learning, emotions, and other highly complex functions(945)

cervix portion of the uterus that joins with the vagina (998)

chancre initial sore caused by syphilis (1008)

channel cell surface proteins that loop back and forth through membrane bilayer (57)

character displacement situation in which two potentially competing species differ more where their ranges overlap (368)

chelicera one of the first pair of appendages of arachnids and their relatives (680)

chelicerate arthropods with fangs (680)

chemically gated description of a channel that opens and closes due to the binding of a chemical (67)

chemical reaction process by which the atoms of one or more molecules are rearranged to form molecules of one or more new substances (80)

chemical signal a molecular signal carried by the blood stream (839)

chemiosmosis process by which cells pump protons across a plasma membrane and use the resulting proton gradient to produce ATP (62)

chemosynthesis production of organic compounds using energy contained in inorganic molecules (78)

chitin tough carbohydrate found in many fungi and in the exoskeletons of all arthropods (501)

chlamydia sexually transmitted disease marked by discharge, burning, and pain, caused by bacteria (1009)

chloroflurocarbon any of a group of compounds that contain carbon, chlorine, and florine, often used as coolants, propellants, or foaming agents (392)

chlorophyll the green pigment molecule responsible for trapping light energy in photosynthesis (101)

choanocyte unique flagellated cell surrounded by a collar of microvilli found inside sponges (613)

chorion outer membrane surrounding embryos of birds, reptiles, and mammals that contributes to the development of the placenta (726, 1004)

chorionic villi sampling fetal-testing procedure in which pieces of the chorionic villi in a mother's uterus are removed and examined for genetic abnormalities (124)

chromatid one of a pair of strands of DNA that make up a chromosome during meiosis or mitosis (120)

chromosome cellular structure on which genes are located (44, 119)

chromosome puff material accumulating at a site on a chromosome that is undergoing rapid transcription (192)

chrysalis in insects, protective capsule enclosing the transforming larva (692)

chyme semisolid material in the stomach and duodenum composed of food, HCl, and enzymes (907)

cilia in cells, tightly packed rows of short flagella used for movement (46)

circulatory system network of vessels that carries nutrients and oxygen through the body (621)

cirrhosis degenerative condition of the liver in which cells are replaced with scar tissue and the organ shrinks into a hard mass (960)

cladistics phylogenetic method in which relationships are inferred based on presence of derived characters (438)

cladogram diagram based on patterns of shared, derived traits that shows the evolutionary relationships among groups of organisms (438)

class taxonomic category containing orders with common characteristics (434)

cleavage in development, the rapid, mitotic division of the zygote (1003)

cloning process of growing a large number of genetically identical cells from a single cell (204)

closed circulatory system system in which the blood does not leave the blood vessels and materials pass in and out by diffusing across the walls of the vessels (660)

closed system in thermodynamics, a system that allows no matter or energy to enter or leave (76)

cnidocyte stinging cell used by cnidarians to stun prey (614)

coccus spherical-shaped bacterial cell (462)

cochlea fluid-filled chamber of the inner ear that is involved in the perception of sound (950)

codominance condition in which both alleles for a gene are expressed when present(152)

codon a three-nucleotide sequence in DNA or RNA that encodes an amino acid or signifies a stop signal (186)

coelom fluid-filled body cavity that forms within the mesoderm (621)

coenzyme nonprotein molecule that assists an enzyme in carrying out a reaction (86)

Glossary

coevolution reciprocal evolutionary adjustments between interacting members of an ecosystem (361)

collagen fibrous structural protein present in all multicellular organisms, particularly in connective tissue (35, 832)

collecting duct in the urinary system, the tube into which the renal tubule empties (913)

colon organ that compacts waste for excretion; also called the large intestine (908)

colonial organism collection of cells that are permanently associated but in which little or no integration of cell activities occurs (427)

commensalism ecological interaction in which one species benefits and the other is neither harmed nor helped (363)

common ancestor species from which two or more species diverged (254)

community the many different species that live together in a habitat (338)

compact bone dense, almost solid tissue within a bone (855)

companion cell of plant phloem, cell alongside a sieve tube member that carries out metabolic functions for it (554)

competition ecological interaction between two or more species that use the same scarce resource such as food, light, and water (363)

competitive exclusion local extinction of one species due to competition (369)

complement defensive proteins that circulate in the bloodstream (923)

complementary characteristic of nucleic acids in which the sequence of bases on one strand determines the sequence of bases on the other (171)

complete flower flower that has all four whorls of appendages—sepals, petals, stamens, and pistils (537)

complete protein protein that contains all the essential amino acids needed by the body (595)

compound substance that is made up of more than one kind of atom (30)

cone photoreceptor of the retina of the eye that can detect color (952)

conifer gymnosperm that produces seeds in cones (532)

conjugation temporary union of two protists to exchange nuclear material (482)

consumer organism that must obtain energy to build its molecules by consuming other organisms; heterotroph (97, 343)

continental drift movement of the continents over geologic time (287)

continuous variation a genetic phenomenon in which a trait is controlled by several genes and therefore exhibits a variety of phenotypes (153)

control factor that is held constant throughout an experiment to test a hypothesis (10)

conus arteriosus elongated heart chamber that smooths heart pulsations (718)

convergent evolution process by which unrelated species become similar as they adapt to similar environments (438)

cork cambium lateral meristem of woody plants that produces cork cells of the outer bark (574)

cork cells cells that replace the epidermis of woody stems; part of the bark (558)

cornea transparent outer covering of the eye (950)

corneal layer the outermost layer of the epidermis, consisting mostly of dead cells containing keratin (866)

corolla whorl of flower that consists of the petals (536)

coronary arteries arteries that branch from the aorta and carry oxygenated blood to the heart muscle (885)

corpus luteum structure that forms from the ruptured follicle in the ovary after ovulation and releases hormones (1001)

cortex the outer layer of ground tissue in roots and stems of plants (557)

cortisol steroid compound produced by the adrenal glands in response to stress (975)

cortisone drug used in the treatment of inflammation and allergies, and in birth control pills (602)

cotyledon leaflike structure of a plant embryo in which food is stored (539)

coupled channel form of channel that carries into the cell sodium ions that accumulate outside the cell, as with nerve cells (62)

coupled reaction endergonic reaction that is driven by the splitting of ATP molecules (86)

cranium portion of the skull that encases the brain (858)

crop plant cultivated for use by humans (96)

crossing over the exchange of reciprocal segments of DNA by chromosomes at the beginning of meiosis; source of genetic recombination (133)

cross-pollination transfer of pollen from the male structures of one plant to the female structures of a different plant of the same species (143)

cuticle waxy, watertight outer covering of aboveground parts of a plant (520)

cyanobacteria group of photosynthetic eubacteria (273)

cyclic AMP common second messenger in cells (973)

cystic fibrosis genetic disorder in which excessive amounts of mucus are secreted, blocking intestinal and bronchial ducts and causing difficulty in breathing (18)

cytokinesis division of the cytoplasm (127)

cytokinin plant hormone produced in root tips that stimulate cell division (583)

cytoplasm the material between the cell membrane and the nuclear membrane (26)

cytosine nitrogenous base of the pyrimidine class; component of RNA and DNA (169)

cytoskeleton cytoplasmic network of protein filaments that plays an essential role in cell movement, shape, and division (46)

cytosol liquid portion of the cytoplasm (26)

day-neutral plant plant whose flowering cycle is not affected by day length (585)

deciduous describes trees, shrubs, and woody vines that drop all of their leaves at the end of each growing season (577)

decomposer organism that causes decay (344)

deletion mutation in which a nucleotide or segment of DNA is lost (124, 195)

demography statistical study of populations (316)

dendrite cytoplasmic extension from the body of a neuron (833, 939)

deoxyribonucleic acid. *See* DNA

deoxyribose five-carbon sugar that is a component of DNA nucleotides (169)

depolarization loss of the electrical charge across the membrane of a neuron (939)

depressant drug that decreases the activity of the central nervous system (957)

derived trait unique characteristic of a particular group of organisms (438)

dermal tissue system the outer protective layer of dermal tissues of vascular plants (551)

dermis thick layer of skin beneath the epidermis (867)

detritivore organism that obtains its energy by feeding on dead organisms or wastes (344)

deuterostome an animal whose mouth forms from an opening other than the blastopore (629)

diabetes mellitus serious disorder in which cells are unable to obtain glucose from the blood (982). *See also* Type I diabetes; Type II diabetes

diaphragm in mammals, sheet of muscle at the bottom of the rib cage that aids in respiration (811, 889)

diapsid term used to identify vertebrate skull that has two openings on each side (773)

diastolic pressure the low blood pressure that occurs during relaxation of the heart (886)

diatom photosynthetic unicellular protist of the phylum Bacillariophyta (486)

dicot dicotyledon; angiosperm that has seeds with two cotyledons (539)

differentiation process in which the cells of a multicellular individual become specialized during development (430, 573)

diffusion movement of particles from an area of high concentration to an area of low concentration (58)

digitalis drug used in the treatment of cardiac disorders (602)

dihybrid cross a cross that involves two pairs of contrasting traits (150)

dikaryon a cell that contains a pair of nuclei that stem from different parent cells (504)

dikaryotic in fungi, condition in which two nuclei are present in each cell (504)

dinoflagellate unicellular, photosynthetic protist of the phylum Dinoflagellata (488)

diploid term used to indicate cell containing two homologues of each chromosome (120)

directional selection natural selection that causes the frequency of a particular allele to move in one direction (263, 329)

dispersion the pattern of distribution of organisms in a population (317)

disruptive selection natural selection in which individuals with extreme forms of a trait have an advantage (330)

diurnal term describing animals that are active during the day and sleep at night (296)

divergence accumulation of differences between groups; can lead to the formation of new species (264)

diversity. *See* species diversity

division in taxonomy, an alternative term for phylum (434)

DNA (deoxyribonucleic acid) nucleic acid that stores hereditary information (37)

DNA fingerprint the pattern of bands that result when DNA fragments are separated by gel electrophoresis (208)

DNA polymerase enzyme that catalyzes the replication of DNA (172)

dominant trait trait that is expressed when its allele is homozygous or heterozygous (145)

dormancy condition in which a seed or plant remains inactive for a period of time (585)

dorsal top surface of a bilaterally symmetrical animal (617)

double fertilization process by which two sperm fuse with cells of the megagametophyte, producing both a zygote and an endosperm (539)

double helix spiral-staircase structure characteristic of the DNA molecule (170)

Down syndrome a syndrome of congenital defects, especially mental retardation, resulting from an additional copy of chromosome 21 (122)

duodenum first part of the small intestine (907)

duplication form of mutation in which a chromosome contains an extra copy of a segment of DNA (124)

ecdysis process of shedding and discarding the exoskeleton; also called molting (682)

ecological race population of a species that differs genetically because of adaptations to different living conditions (265)

ecology the study of the interactions of living organisms with one another and with their environment (337)

ecosystem ecological system encompassing a community and all the physical aspects of its habitat (338)

ectoderm in animals, the outer layer of embryonic tissue from which the skin and nervous system develop (614)

ectomycorrhizae mycorrhizae that do not physically penetrate plant roots but wrap around them (513)

ectoparasites parasites that live outside their host (649)

ectothermic referring to an animal whose body temperature is determined by the temperature of the environment (707)

ejaculation the forceful expulsion of semen from the penis (996)

electrical signal a signal carried from one place in the body to another by neurons (839)

electromagnetic spectrum total range of all electromagnetic waves (100)

electron elementary particle with negative electric charge (29)

electron transport chain series of molecules in an inner cell membrane through which high-energy electrons are passed to pump protons across the membrane and generate ATP by chemiosmosis (102)

element substance composed of a single type of atom (29)

emphysema respiratory condition in which the alveoli of the lungs lose their elasticity, making it difficult for them to release air during exhalation (893)

Glossary

emulsification process in which fat globules are broken down into droplets and exposed to enzymes (905)

endergonic reaction reaction that absorbs free energy (81)

endocrine glands ductless glands in the body that release their products directly into the bloodstream (968)

endocytosis process by which extracellular matter is taken up by a cell (64)

endoderm in animals, inner layer of embryonic tissue from which the digestive organs develop (614)

endometrium lining of the inner wall of the uterus (998)

endomycorrhizae mycorrhizae in which the fungal hyphae penetrate the outer cells of the roots (512)

endoparasite parasite that lives inside its host (649)

endoplasmic reticulum (ER) cell membranes in the cytoplasm that transport substances made by the cell (42)

endorphins chemical signals secreted by the brain and nervous tissue and thought to regulate emotions (970)

endoskeleton an internal skeleton (629)

endosperm highly nutritious tissue developed by the seeds of angiosperms (538)

endospore dormant cell enclosed by a tough coating that is highly resistant to environmental stress (463)

endothermic refers to an animal that generates its own body heat (707)

energy capacity for doing work, often described as the ability to make things move or change (75)

enhancer region preceding a eukaryotic gene that must be activated prior to gene expression (193)

enkephalins a class of neurotransmitters secreted by the central nervous system that act as pain relievers (957)

entropy amount of disorder in a system; amount of unavailable energy in a system (77)

envelope outer layer covering the capsid of many kinds of viruses (456)

enzyme protein that catalyzes a chemical reaction (35, 82)

epidermis outermost layer of tissue, consisting of several layers of cells (866)

epididymis long, coiled tube on the surface of the testicle where sperm mature (995)

epiglottis flap of tissue that covers the trachea during swallowing (889)

epithelium protective tissue that covers the body's interior surfaces (829)

equilibrium state in a chemical reaction when the rates of the forward and reverse reactions are equal (58)

erythrocyte oxygen-carrying red blood cell (832, 880)

esophagus tube that connects the mouth to the stomach (906)

essential amino acids amino acids not synthesized by the human body and must be obtained from food (595)

estrogen hormone secreted by the ovaries that regulates maturation of an ovum in reproduction and regulates the development of female characteristics (983)

ethylene plant hormone that causes fruit ripening (583)

eubacteria one of the two kingdoms of prokaryotes (274)

euglenoids members of the phylum Euglenophyta (489)

eukaryotic cell complex cell that has nucleus enclosed by a membrane (26)

eumetazoans animals with both tissues and symmetry (614)

evergreens trees, shrubs, and woody vines that drop only a few leaves at a time throughout the year (577)

evolution change in the genetic makeup of a population or species over time (48)

evolutionary systematics method of constructing phylogenies that involves weighting characters by their presumed evolutionary significance (442)

excretion the process of eliminating wastes (910)

exergonic reaction a reaction that releases free energy (81)

exhalation expulsion of air from the lungs (890)

exocrine glands glands that release their products through ducts (869)

exocytosis releasing materials outside a cell by discharge from waste vacuoles (64)

exon sequence of nucleotides that gets translated and transcribed (174)

exoskeleton hard external covering of some invertebrates (628, 680)

experiment test of a hypothesis in which data are gathered under controlled conditions (9)

exponential growth curve J-shaped curve showing the rapid increase in an exponentially growing population (318)

extensor skeletal muscle that causes a limb to straighten at a joint (862)

external fertilization union of egg and sperm occurring outside the body of either parent (727)

extinct term used to indicate species that have disappeared permanently (251)

extraterrestrial origin origin (of life) elsewhere in the universe than Earth (225)

eyespot organelle containing light-sensitive pigments (481)

F₁ generation the offspring from a cross of two varieties (144)

F₂ generation the offspring from crosses among individuals of the F₁ generation (144)

facilitated diffusion transport of substances through a cell membrane along a concentration gradient with the aid of carrier molecules (61)

fallopian tubes tubelike organs of the female reproductive system that lead from the ovaries to the uterus (998)

family taxonomic category containing genera with similar properties (434)

family tree a diagram that shows the evolutionary relationships among a set of organisms (254)

fat class of organic compound containing carbon, hydrogen, and oxygen and used to construct membranes and to store energy (900)

feedback inhibition negative feedback mechanism in which the end product of a metabolic pathway inhibits an enzyme that catalyzes a reaction previous in the pathway (88)

feedback loop process of surveillance of internal conditions and response used to maintain homeostasis (837)

fermentation anaerobic pathway of cellular respiration that converts pyruvate to either lactic acid or ethyl alcohol and carbon dioxide (111)

fetal alcohol syndrome (FAS) birth defect resulting from alcohol use by the mother during pregnancy (1005)

fetus the developing human from the age of eight weeks until birth (1006)

fibrin clot-forming protein in the blood (881)

fibrous root system root system consisting of roots that are about the same size (559)

filament stalk of a stamen (536)

filtrate in the kidneys, fluid that has passed from the bloodstream into Bowman's capsule (913)

first law of thermodynamics law that states energy cannot be created or destroyed, but only converted from one form to another (76)

flagellum whiplike structure that grows out of a cell and enables it to move (46)

flexor skeletal muscle that causes a limb to bend at a joint (861)

fluke parasitic flatworm of the class Trematoda (649)

follicle cluster of cells that surround the immature egg cells in an ovary (1001)

food chain linear pathway of energy transfer in an ecosystem (97, 344)

food web a network of feeding relationships in an ecosystem (344)

foram marine protist of the phylum Foraminifera (485)

fossil preserved or mineralized remains or traces of an organism that lived long ago (252)

free energy amount of energy available for work, e.g., to drive cell activities (81)

freely movable joint joint that permits the greatest movement of bones (859)

fringing reef coral reef that forms close to the beach (646)

frond long, highly divided leaf of ferns (530)

fruit a mature ovary that contains one or more seeds (540)

fundamental niche the entire range of conditions an organism can tolerate (367)

Fungi Imperfecti group of fungi in which sexual reproduction has not been observed; also known as Deuteromycota (510)

G

gametangium reproductive structure in which gametes form (504)

gamete haploid cell that participates in fertilization by fusing with another haploid cell (120)

gametic meiosis sexual life cycle in which gametes are the only haploid cells; all of the other cells of the individuals in the life cycle are diploid (424)

gametocyte third stage of the Plasmodium life cycle; undergoes a sexual phase in the bloodstreams of infected humans (494)

gametophyte in the life cycle of a plant, the haploid phase that produces gametes (425)

gastrin hormone that regulates the synthesis of HCl by the stomach (907)

gastrulation inward migration of the outer cells of the blastula, resulting in the distribution of cells into three primary tissue layers (1004)

gated channel a channel that opens and closes like a gate in a fence (67)

gel electrophoresis technique that uses an electrical field passed through a gel to separate molecules in a mixture (207)

gemmules food-filled buds produced by freshwater sponges in cold or dry weather (640)

gene section of chromosome that codes for a protein or RNA molecule (119)

gene expression two-stage processing of information from DNA to proteins (183)

gene flow movement of alleles into or out of a population due to the migration of individuals to or from the population (324)

genetic code sequence of nucleotides that specifies the amino acid sequence of a protein (186)

genetic disorder harmful effect, such as sickle cell anemia, produced by mutated genes (154)

genetic drift random change in allele frequency in a population (325)

genetic engineering process of isolating a gene from DNA of one organism and transferring it to the DNA of another organism (203)

genetic polymorphism genetic variation in a population that results from a more than one allele for a gene (327)

genetic recombination rearrangement of genetic material (134)

genetics study of heredity (141)

genital herpes sexually transmitted disease caused by the herpes simplex virus (1008)

genotype genetic constitution of an organism as indicated by its set of alleles (146)

genus taxonomic category containing similar species (431)

germination resumption of growth by a plant embryo (572)

gestation period length of time between fertilization and birth (811)

gibberellin type of plant hormone that regulates growth, especially stem elongation (583)

gigantism condition of excess growth; almost always associated with pituitary tumors (978)

gill of fishes, structure located in the pharynx that is the site of gas exchange (623)

gill filament fingerlike projection from a gill in which gases enter and leave the blood (711)

gizzard a portion of the digestive tube of earthworms where strong muscles grind up the organic material in ingested soil (673)

glial cell *See* supporting cell

global warming increase in global temperatures as a result of increased concentration of greenhouse gases in the atmosphere (394)

glomerulus in the kidney, a cluster of capillaries that receives blood from the renal artery and that serves as a filter (912)

glucagon a peptide hormone produced by the pancreas that causes liver cells to release glucose stored in glycogen (974, 982)

glycogen polymer of glucose used for short-term energy storage (34)

glycolysis biochemical pathway that breaks down glucose into pyruvate (108)

Glossary

glycoprotein protein with carbohydrate molecules attached (456)

goiter enlargement of all or part of the thyroid gland (979)

Golgi apparatus cell organelle of a eukaryotic cell that consists of flattened sacs and collects, packages, and distributes molecules produced by the cell (43)

gonorrhea sexually transmitted disease caused by bacteria that cause inflammation of the mucous membranes in the urinary and reproductive tracts (1009)

gradualism model of evolution in which gradual change over a long period of time leads to macroevolution (258)

grain edible dry fruit of a cereal grass (592)

gram-negative designates a bacterium that does not retain the Gram stain (463)

gram-positive designates a bacterium that retains the Gram stain (463)

granum stack of thylakoids in a chloroplast (99)

gravitropism growth response to gravity (583)

greenhouse effect atmospheric warming resulting from heat trapped by gases such as carbon dioxide (393)

ground tissue system all the tissues of a vascular plant except the dermal tissue and vascular tissue (551)

ground water water found beneath Earth's surface (349)

guanine nitrogen base of the purine class; component of DNA and RNA nucleotides (169)

guard cells pair of specialized cells that border a stoma (521)

gymnosperm seed plant that produces seeds that do not develop within a fruit (531)

gynoecium part of a flower that houses the female gametophytes; the pistils, collectively (537)

H

habitat place where an organism lives (338)

hair follicle specialized dermal structure that produces hair (867)

half-life the period of time that it takes for one-half of a radioisotope to decay (233)

haploid having only one set of chromosomes (120)

Hardy-Weinberg principle principle stating that the frequency of alleles in a population does not change unless evolutionary forces such as selection act on the population (321)

Haversian canal hollow channel surrounded by concentric rings of bone and through which blood vessels and nerve pass (856)

heartwood dark, nonconducting wood in the center of a log (557)

helicase enzyme that unwinds a DNA molecule's double helix before replication (172)

helper T cell white blood cell that activates killer T cells and B cells (925)

hemodialysis treatment that cleanses the blood of patients with kidney failure by machine outside the body (914)

hemoglobin component of red blood cells that binds with and carries oxygen through the body (154)

hemophilia genetic disorder that impairs the blood's ability to clot and can cause excessive bleeding (155)

hepatic vein blood vessel that carries blood collected from the liver toward the heart (909)

herbaceous stem flexible, usually green (nonwoody) stem (557)

herbivore organism that eats only plants or algae (343)

heredity transmission of genetic traits from parent to offspring (14)

hermaphrodite organism that produces both eggs and sperm (641)

heterocyst thick-walled, cyanobacterial cell with enzymes that fix nitrogen gas (N_2) into ammonia (NH_3) (467)

heterotroph organism that cannot make its own food (78)

heterozygous refers to a pair of genes, or an individual, with two different alleles for a trait (146)

histamine chemical released by mast cells in an inflammatory response (930)

histocompatibility antigen marker on a body cell that protects it from attack by the body's immune system (915)

HIV human immunodeficiency virus that causes AIDS (17)

HIV positive condition of an individual whose blood tests detect the presence of antibodies to HIV or of viral proteins (932)

homeostasis maintenance of the internal stability of a cell, organism, or population in its environment (13)

hominid member of the family Hominidae of the order Primates; characterized by opposable thumbs, no tail, longer lower limbs, and erect bipedalism (299)

homologous chromosomes chromosomes that are similar in shape, size, and the genes they carry (120)

homologous structures structures that share a common ancestry (256)

homozygous refers to a pair of genes, or an individual, with two identical alleles for a trait (146)

hormone chemical produced in one part of an organism and then transported to another part of the organism, where it causes a response (57, 582, 840)

hormone-receptor complex complex formed by the binding of a hormone to its receptor protein (975)

Human Genome Project research effort to identify and locate the entire collection of genes in a human cell (209)

hybrid offspring of individuals from two different species (436)

hydrogen bond weak chemical bond in which a slightly positive hydrogen atom in a polar bond of one molecule is attracted to a slightly negative atom (usually oxygen or nitrogen) in a polar bond of another molecule (31)

hyperthyroidism condition resulting from overproduction of thyroxine by the thyroid gland, characterized by nervousness, sleep disorders, irregular heartbeat, and weight loss (979)

hypertonic describes a solution with a higher concentration of solute molecules than the solution across a selectively permeable membrane (60)

hypha slender filament that is part of the body of a multicellular fungus (502)

hypothalamus region of the brain located below the thalamus that produces several hormones and that controls many body activities related to homeostasis (838, 946)

hypothesis proposed explanation (9)

hypothyroidism condition resulting from underproduction of thyroxine by the thyroid gland, characterized in childhood by stunted growth and mental retardation and characterized in adulthood by lack of energy and weight gain (979)

hypotonic describes a solution with a lower concentration of solute molecules than the solution across a selectively permeable membrane (60)

immovable joint tight joint that permits little or no movement of the bones (859)

imperfect flower flower that lacks either a gynoecium or an androecium (537)

implantation burrowing of a blastocycst into the thick outer lining of the uterus (999)

incomplete dominance condition in which a trait is intermediate between two parents (152)

incomplete flower flower that lacks any one of the floral whorls: calyx, corolla, androecium, or gynoecium (537)

incomplete protein food that lack one or more of the essential amino acids (595)

inducer molecule that enables transcription to resume (191)

industrial melanism darkening of populations of organisms over time in response to industrial pollution (261)

inferior vena cava large vein that delivers blood from the lower portion of the body back to the heart (885)

inflammatory response series of events initiated by an injury or local infection that suppress infection and promote healing by removing disease-causing agents and dead cells (923)

in-group in cladograms, closely-related organisms (439)

inhalation part of breathing in which the diaphragm contracts and air moves into the lungs (889)

inhibitory synapse chemical junction where the receptor is a gated potassium channel that is closed at rest (943)

insertion in point mutations, addition of one or more nucleotides to a gene (195)

insertion in the muscular system, end of a muscle attached to a bone that moves when the muscle contracts (861)

insulin peptide hormone produced by the islets of Langerhans in the pancreas that stores excess glucose by promoting the accumulation of glycogen in the liver (982)

integration sum of the interactions among the signals from all the excitatory and inhibitory synapses of a neuron, which tend to cancel or reinforce one another (943)

intercellular coordination adjustment of a cell's activity in response to what other cells are doing (430)

intermediate filament long cytoplasmic protein filament found in the cytoskeletons of many eukaryotic cells (46)

internal fertilization fertilization that occurs within the body of the female parent (728)

internode area between two nodes of a plant stem (557)

interphase period of growth between two mitotic or meiotic divisions of a eukaryotic cell (128)

intron segment of mRNA transcribed from eukaryotic DNA but removed before translation of mRNA into a protein (174)

inversion mutation in which a chromosome fragment rejoins its original chromosome with its nucleotides reversed (124)

ion electrically charged atom or molecule (29)

ionic bond chemical bond joining positive and negative ions (31)

iris ring of tiny muscles that controls the amount of light entering the eye (951)

islets of Langerhans clusters of endocrine cells in the pancreas that secrete the hormone insulin (982)

isolation condition in which two populations of a species are separated so that they cannot interbreed (251)

isopod crustacean with its first thoracic segment fused to the head, seven pairs of similarly-sized walking legs, and a dorsoventrally flattened body (687)

isotonic describes solutions with equal solute concentrations on either side of a selectively permeable membrane (60)

joint junction of two bones (825, 859)

karyotype array of the chromosomes found in an individual's cells arranged in order of size and shape (122)

kidney organ that removes metabolic wastes from the blood and regulates salt and water balance (723)

killer T cell white blood cell that attacks and kills foreign cells and body cells infected by pathogens (925)

kinetic energy energy of an object due to its motion (76)

kinetochore disk of protein on a chromosome's centromere to which microtubules attach during mitosis and meiosis (129)

kingdom taxonomic category that contains phyla with similar characteristics (434)

Krebs cycle cyclic biochemical pathway of cellular respiration that uses pyruvate from glycolysis, releases CO_2, and produces ATP, NADH, and $FADH_2$ (109)

K-**strategist** species characterized by slow maturation, low fertility, slow population growth, and high competitive ability (319)

lac **operon** gene system with an operator gene and three structural genes that control lactose metabolism in *E. coli* (190)

larva independent, immature stage in animal development that emerges from an egg (626)

larynx voice box; structure at the upper end of the trachea containing the vocal cords (888)

lateral bud bud located at a node of a stem that grows into a branch of the stem (557)

lateral line system row of pressure- and vibration-sensing organs running the length of both sides of a fish's body (754)

Glossary

lateral meristem meristem that produces secondary growth in woody plants; vascular cambium or cork cambium (574)

latex milky white sap of certain plants, such as those of the genus *Hevea*, that is used to make natural rubber (603)

law of independent assortment law stating that pairs of genes separate independently of one another in meiosis (147)

law of segregation law stating that pairs of genes separate in meiosis and each gamete receives one gene of a pair (147)

leaflet individual segment of a compound leaf (555)

legume member of the Fabaceae (pea) family of plants; type of fruit produced by members of the pea family (598)

lens part of the eye that focuses light passing through it (951)

lenticle area of loosely-packed cork cells on a woody stem that enables gas exchange (558)

leukemia disease that results in the overproduction of leucocytes (881)

leukocyte white blood cell; the primary cell of the immune system (880)

lichen symbiotic association between a fungus and an alga or cyanobacterium (511)

ligament band of connective tissue that holds together the bones in a joint (825)

light microscope microscope that uses a beam of light passing through one or more lenses (39)

limbic system network of neurons linked to the cerebral cortex that is responsible for many drives and emotions (946)

lipase enzyme that breaks down fat molecules into fatty acids and glycerol (905)

lipid bilayer basic structure of a plasma membrane; composed of two layers of phospholipids (55)

liver large organ that secretes bile and performs a wide variety of functions such as detoxification of poisons and storage and metabolism of carbohydrates, fats, and proteins (909)

lobe-finned fish bony fish with paired fins consisting of long, fleshy, muscular lobes (750)

logistic model model of population growth that assumes finite resource levels limit population growth (318)

long-day plant plant that produces flowers when days become longer than a critical length (585)

loop of Henle U-shaped section of a renal tubule in a kidney (913)

lumber wood from trees that have been cut down and sawed into boards, beams, or planks (600)

lung spongelike respiratory organ of a vertebrate that enables gas exchange between the air and the blood (712)

lung cancer malignant growth of cells of the lungs (893)

lymph fluid found in the intracellular spaces and lymphatic vessels of vertebrates (879)

lymphatic system system of the body that collects and recycles fluids leaked from the cardiovascular system (879)

lymphocyte type of white blood cell that matures in the organs of the lymphatic system (831, 880)

lysosome cell organelle of a eukaryotic cell containing hydrolytic, digestive enzymes (45)

M

macroevolution change that occurs among species over time as new species evolve and old species become extinct (251)

macrophage large white blood cell that engulfs pathogens (831, 880)

magnification enlargement or enlarging of an image (39)

malignant melanoma cancer derived from the pigment-producing melanocytes of the skin (868)

Malpighian tubule slender, fingerlike organ of excretion that opens into the gut of certain arthropods (681)

mandibulate arthropod with jaws (680)

mantle heavy fold of tissue that surrounds the visceral mass of mollusks (623)

mantle cavity space between the mantle and the visceral mass of mollusks (660)

mass extinction episode during which large numbers of species become extinct (278)

medusa free-swimming, bell-shaped, mouth-down body plan of a cnidarian; jellyfish (616)

megagametophyte female gametophyte of seed plants (531)

megakeryocyte giant bone-marrow cell characterized by a large, irregularly lobed nucleus; precursor to platelets (881)

megaspore spore that grows into a megagametophyte (531)

meiosis process in which the nucleus of a cell completes two successive divisions that produce four nuclei, each with a chromosome number that has been reduced by half (132)

melanocyte type of cell in the basal layer of the skin's epidermis that produces the dark pigment melanin (867)

melatonin hormone secreted by the pineal gland (983)

menopause time when a woman stops menstruating, usually between the ages of 45 and 55, and is no longer able to conceive (1002)

menstrual cycle series of hormone-induced changes that prepare the uterine lining for a possible pregnancy each month (1002)

menstruation periodic flow of blood and tissue shed from the outer layer of the endometrium of a woman's uterus that occurs approximately every 28 days (1002)

meristem region (or zone) of actively-dividing undifferentiated plant cells that are capable of developing into specialized plant tissues (528)

merozoite second stage of the life cycle of *Plasmodium* in which the protist divides rapidly and produces millions of cells in the liver of infected humans (494)

mesoderm middle layer of embryonic tissue in animals from which the skeleton and muscles develop (614)

mesophyll ground tissue of a leaf (556)

messenger RNA RNA copy of a gene used as a blueprint for the making of a protein during translation (185)

metabolism sum of all chemical processes occurring in an organism (13, 85)

metamorphosis process of change through which an immature organism passes as it grows to adulthood (692, 728)

metaphase stage of mitosis and meiosis when chromosomes move to and line up at center of a cell (131)

metastasis spread of malignant cells beyond their original site (196)

microevolution change that occurs within a species over time (251)

microgametophyte male gametophyte of seed plants (531)

microsphere tiny, abiotically-produced vesicle formed by a double layer of amino acids (231)

microspore spore that grows into a microgametophyte (531)

microtubule hollow protein fiber in the cytoplasm of a eukaryotic cell involved in cell movement and structure (46)

microvilli cytoplasmic projections that cover the villi and increase the absorptive surface of the small intestine (908)

milk nutrient-rich fluid produced by mammary glands of mammals for the nourishment of their young (729)

mitochondrion cell organelle of a eukaryotic cell that supplies the cell with ATP by performing oxidative respiration (44)

mitosis process in which the nucleus of a cell divides into two nuclei, each with the same number and kind of chromosomes (127)

mitral valve one-way valve between the left atrium and left ventricle of the heart that prevents blood from returning to the left atrium when the left ventricle contracts (884)

model generalized, hypothetical description of a system, such as a population, used to analyze or explain the system (317)

molecule smallest particle of a substance that retains all the properties of the substance and is composed of two or more atoms bonded by the sharing of electrons (30)

mollusk soft-bodied animal of the phylum Mollusca, having an unsegmented body usually surrounded by a mantle and a calcareous shell (622)

monocots monocotyledons; angiosperms with seeds that have a single cotyledon (539)

monohybrid cross cross involving one pair of contrasting traits (149)

morphogenesis process of development in which the parts of the body develop their shape (1005)

motor nerves nerves that transmit commands from the central nervous system to muscles and glands all over the body (944)

mucous membrane membrane covering internal surfaces of the body that secretes mucus and functions in defense (921)

multicellular an organism that consists of more than one cell (13)

multigene family group of genes that has evolved from a single ancestral gene and is characterized by the existence of multiple copies (175)

multiple alleles having more than two alleles (versions of the gene) for a genetic trait (152)

multiple sclerosis disease resulting from the destruction of the myelin sheath covering nerves and characterized by loss of muscular coordination (940)

muscle strain muscle injury resulting from overstretching or tearing of the muscle; also called a "pulled muscle" (863)

mutagen mutation-inducing agent (195)

mutation change in the DNA of a gene or chromosome (124, 154)

mutualism symbiotic association in which both partners benefit (363)

mycelium mass of hyphae forming the body of a fungus (502)

mycorrhiza mutualistic association between a fungus and a plant's roots, in which the fungus absorbs water and nutrients for the plant and the plant supplies food to the fungus (281)

myelin sheath fatty, insulating covering of an axon (940)

myofibril bundle of small fibers within a muscle containing many filaments of the proteins myosin and actin (863)

myosin type of protein filament in muscle tissue that works with actin to enable a muscle to contract (833)

N

NAD⁺ nicotinamide adenine dinucleotide; oxidized form of a coenzyme that acts as an electron carrier in oxidation-reduction reactions of cell metabolism (87)

NADH reduced form of NAD⁺ (87)

NADP⁺ nicotinamide adenine dinucleotide phosphate; oxidized form of a coenzyme that acts as an electron carrier in oxidation-reduction reactions of cell metabolism (103)

NADPH reduced from of NADP⁺ (103)

natural killer cell immune system cell that attack cells infected by microbes (923)

natural selection process by which populations change in response to their environment as individuals better adapted to the environment leave more offspring (249)

negative feedback mechanism used in homeostasis to keep a monitored variable within a certain range using a response that is triggered by fluctuation of the variable (837)

nematocyst barbed harpoon within a cnidocyte of a cnidarian, used to spear prey (614)

nephridium tiny tubular organ of excretion that filters cellular wastes from the coelom of certain invertebrates (661)

nephron tubelike structure in the kidneys that filters wastes from the body and retains useful molecules (723, 912)

nerves bundles of neurons that appear as fine white threads to the naked eye (940)

neuroendocrine system the endocrine and nervous systems considered together (971)

neuromodulator nonendocrine chemical signal that alters a cell's response to a neurotransmitter (970)

neuron cell of nervous tissue that transmits a nerve impulse (832)

neuropeptide chemical signal secreted by the brain and nervous tissue (970)

neurotransmitter chemical messenger in sacs at the end of an axon that carries nerve impulses across a synapse (66, 942)

neurulation developmental process in which the hollow dorsal nerve cord of chordates forms from ectoderm (1005)

neutrophil white blood cell that kills bacteria and itself through the release of chemicals (922)

niche functional role of a species in an ecosystem (365)

nicotine highly addictive stimulant found in the leaves of the tobacco plant (959)

nitrification formation of nitrates by the oxidation of ammonia (467)

Glossary

nitrogen fixation process of combining nitrogen gas with hydrogen to form ammonia (352)

node point on a plant stem from which leaves grow (557)

nondisjunction accident in chromosome separation when one daughter cell receives both chromosomes and the other daughter cell receives none (123)

nonrandom mating mating between individuals of the same genotype (324)

norepinephrine peptide hormone released by the adrenal medulla (981)

normal distribution bell-shaped curve that results when the values of a trait in a population are plotted against their frequency (328)

nuclear envelope double membrane that surrounds the cell nucleus (44)

nuclear pore one of a series of channels that span the nuclear envelope and allow transport of material between the nucleus and cytoplasm (44)

nucleic acid organic molecule that stores information for cell function; DNA or RNA (37)

nucleotide subunit of nucleic acids consisting of a nitrogenous base, a sugar, and a phosphate group (37, 169)

nucleus the organelle that houses the DNA of eukaryotic cells (26)

nutrient mineral element extracted from the soil by plants (580)

nymph juvenile stage of some insects that is a smaller version of the adult (693)

objective lens in a microscope, lens closest to what is being viewed (39)

obligate anaerobe organism that is poisoned by oxygen (418)

observation information that is known from being seen or experienced directly (9)

ocellus single-lens eye of some invertebrates (680)

ocular lens in a microscope, the lens closest to the viewer (39)

ommatidium in a compound eye, a visual unit having a lens and retina (680)

omnivore animal that eats both plants and animals (344)

oncogene gene that, when mutated, can cause a cell to become cancerous (196)

oocyte thick-walled cyst that develops from a zygote during sexual reproduction of sporozoans (492)

oomycete fungus-like protist of the phylum Oomycota (492)

open circulatory system system in which blood leaks out of blood vessels and bathes the body's tissues (660)

open system in thermodynamics, a system that exchanges matter and energy with its surroundings (76)

operator region of DNA that controls RNA polymerase's access to the structural genes (190)

operculum hard plate that covers the gills on each side of the head of bony fishes (755)

operon segment of DNA that controls gene regulation in prokaryotes (190)

opposable thumb the thumb finger that stands out at an angle from the other fingers and can be bent for grasping (297)

order taxonomic category consisting of families with similar characteristics (434)

organ collection of tissues that work together to perform a particular function (426, 618, 829)

organelle subcellular structure that has a special function (26)

organogenesis process of organ formation in the embryo (1005)

organ system group of organs that function together to carry out a major activity of the body (829)

origin the end of the muscle that remains stationary during muscle contraction (861)

osmosis movement of water through a membrane from area of high concentration to an area of low concentration (60)

osmotic pressure increased water pressure that results from osmosis (60)

osteoblast a bone-forming cell (856)

osteocyte living bone cell trapped within spaces in bone (856)

osteoporosis condition of bone loss that produces porous bone (857)

otolith tiny piece of calcium contained in the jellylike fluid in the inner ear that help sense position and movement (950)

out-group in a cladogram, a group of organisms that is only distantly related to the group under consideration (439)

ovarian cycle series of events by which the ovaries prepare and release a mature ovum (1000)

ovary gamete-producing organ of the female reproductive system (997)

ovary of flowering plants, hollow structure at the lower part of the pistil that contains the ovules (537)

oviparous term that describes organisms that produce eggs that hatch outside the mother's body (729)

ovoviviparous term that describes organisms that produce eggs that hatch inside the mother's body (729)

ovulation the release of an ovum from a follicle (1001)

ovule sporophyte structure of a seed plant in which a megaspore forms and develops into a megagametophyte, from which a seed forms (531)

ovum an individual egg cell

oxidation loss of electrons by an atom or molecule (79)

oxidation-reduction reaction chemical reaction in which electrons are passed from one atom or molecule to another (79)

oxidative respiration aerobic chemical reactions that follow glycolysis and that produce large amounts of ATP (108)

ozone molecule containing three oxygen molecules that is present mainly in the upper atmosphere, where it shields the Earth from ultraviolet radiation (280)

P

paleontologist scientist who studies fossils (254)

palisade layer of a leaf, cell layer of the mesophyll consisting of one or more rows of closely packed, columnar ground tissue cells (556)

palps of arachnids, a pair of appendages used to catch and handle prey (684)

pancreas organ in the abdomen that produces digestive enzymes and hormones (907)

Pangaea the single continent of the Triassic period that included all presently known continents (783)

parapodium fleshy appendage of marine annelids (669)

parasitism type of predation in which the predator feeds on but usually does not kill a larger organism (361)

parasympathetic division division of the autonomic system that slows the heartbeat and breathing and promotes digestion and elimination (949)

parathyroid glands small glands attached to the thyroid gland that help regulate calcium in the body (979)

parathyroid hormone (PTH) hormone produced by the parathyroid glands that is essential for maintaining an adequate supply of calcium in the blood (979)

passive transport movement of a substance through a cell membrane without the expenditure of cellular energy (58)

pathogen a disease-causing agent (458)

pectoral girdle portion of the appendicular skeleton that serves as the attachment for the arms (858)

pedigree family history of traits recorded over generations (156)

pellicle in many protozoans, a protein scaffold inside the cell membrane that creates a hard covering (489)

pelvic girdle portion of the appendicular skeleton that serves as an attachment for the legs (858)

pelvic inflammatory disease (PID) severe inflammation of the uterus, ovaries, fallopian tubes, or pelvic cavity usually caused by bacteria (1008)

penis male organ that delivers sperm into female reproductive tract (728, 996)

peptide hormone hormone molecule that consists of chains of amino acids (969)

peptidoglycan the carbohydrate-protein compound that makes up the cell walls of eubacteria (418)

perennial plant plant that lives for more than two years and may produce flowers, fruits, and seeds many times during its life (577)

perfect flower one that has both stamens and pistils (537)

periosteum membrane containing many blood vessels that surrounds individual bones (856)

peripheral nervous system system in humans that includes nerves that bring information to the brain and transmit commands from it (826)

peristaltic contraction contraction of the smooth muscle in the wall of the gut that moves food through the digestive system (906)

peroxisome intracellular, membrane-bound organelle that contains oxidative enzymes and carries out oxidative reactions (45)

petiole slender stalk that supports the blade of a leaf (555)

P generation (parental generation) plants that displayed only one form of a particular trait (144)

phagocytosis process by which cellular or fragmentary organic matter is engulfed by a cell (64)

pharynx upper portion of the throat leading to the esophagus (888, 906)

phenotype observable characteristics of an organism (146)

phenylketonuria (PKU) genetic disorder in which an individual lacks an enzyme that converts the amino acid phenylalanine into the amino acid tyrosine (158)

phloem in plants, soft-walled vascular cells that conduct carbohydrates throughout the plant (527)

pH number indicator of the acidity of a solution, representing the hydrogen ion concentration (6)

phospholipid organic molecule in the plasma membrane of a cell (55)

photon unit of electromagnetic energy (100)

photoperiodism response of a plant to the length of days and nights (584)

photosynthesis process by which organisms use light energy to produce ATP and other organic molecules from inorganic molecules (77)

photosystem cluster of pigments and other associated molecules in a thylakoid membrane that gather light energy and boost an electron to a higher energy state (101)

photosystem I photosystem that boosts electrons to a higher energy state by absorbing light with a wavelength of 700 nm (101)

photosystem II photosystem that boosts electrons to a higher energy state by absorbing light with a wavelength of 680 nm (101)

phototropism growth response to light (583)

pH scale method of relating the hydrogen ion concentration in a solution (33)

phylogeny evolutionary history of a species (438)

phylum a taxonomic category containing classes with similar characteristics (434)

pigment molecule containing atoms that enable it to absorb certain wavelengths of light (101)

pilus short, thick outgrowth of a bacterium that allows it to attach to another bacterium (464)

pineal gland pea-sized gland in the brain that secretes melatonin (983)

pinocytosis uptake of extracellular fluid by a cell using small vesicles derived from the plasma membrane (64)

pistil seed-producing part of a flower, including the ovary, style, and stigma (537)

pith of plant stems, the inner layers of ground tissue (557)

pituitary gland endocrine gland at base of brain whose hormones control endocrine glands elsewhere in the body (971)

placenta organ that nourishes the embryos of placental animals (710, 1004)

plankton biological community drifting freely in the upper waters of the ocean and consisting mostly of microscopic organisms (377)

planula free-swimming larva developed from zygotes produced by cnidarians (644)

plasma noncellular portion of blood (879)

Glossary

plasma membrane thin layer on the surface of all cells, consisting mainly of lipids and proteins (26)

plasmid a circular DNA molecule, usually found in bacteria, that can replicate independently from the main chromosome (204)

plasmodium protist of the phylum Myxomycota; characterized by multinucleate, amoeboid appearance (491)

plastron the bottom, or ventral, portion of a turtle's shell (776)

platelets un-nucleated cell fragments that aid in blood clotting (881)

point mutation mutation in which one or just a few nucleotides in a gene are changed (194)

polar molecule a molecule, such as water, that has positively and negatively charged ends (31)

pollen grain structure, consisting of a few haploid cells surrounded by a thick protective wall, that contains a plant's male gamete (531)

pollen tube structure growing from the pollen grain to an ovule, enabling a sperm to pass directly to an egg (531)

pollination in flowering plants, transportation of pollen grains from a male reproductive structure to a female reproductive structure (531)

polygenic trait characteristic of an organism that is influenced by several genes (328)

polymerase chain reaction (PCR) laboratory technique for making unlimited copies of a gene (208)

polyp in cnidarians, a cylindrical, pipe-shaped animal usually attached to a rock (616)

polypeptide a chain of amino acids (36)

polysaccharide complex carbohydrate composed of three or more monosaccharides (34)

population group of individuals that belong to the same species, live in the same area, and breed with others in the group (248, 315)

population density in a population, the number of individuals in a given area (316)

population size total number of individuals in a population (316)

portal vein vessel that collects blood leaving the small intestine and carries it to the liver (909)

positive feedback in homeostasis, the condition in which a change in a variable causes the body to drive the variable even farther from the initial value (838)

posterior back end of a bilaterally symmetric animal (617)

posterior lobe portion of the pituitary gland that stores and releases peptide hormones produced by the hypothalamus (978)

potential energy the energy an object has because of its position (77)

predation an ecological interaction in which one organism feeds on another (361)

prediction the expected outcome if a hypothesis is accurate (9)

prednisone steroid hormone produced by the adrenal glands; widely used as anti-inflammatory agent (981)

pressure-flow model the model of translocation in which pressure created by water entering a sieve tube by osmosis pushes a sugar solution from a source to a sink (565)

primary growth growth that occurs from the formation of new cells at the tips of plants (573)

primary induction process in the development of specialized tissues in which one of the three primary tissues interacts with another (621)

primary productivity the amount of organic material that the photosynthetic organisms of an ecosystem produce (343)

primary succession succession that occurs in a newly formed habitat that has never before sustained life(340)

primary tissue plant tissue that results from primary growth (573)

primates order of mammals that includes prosimians, monkeys, apes, and humans and is characterized by limb structure, dentition, digital mobility, flat nails, stereoscopic vision, and a well-developed cerebral cortex (295)

principle of competitive exclusion the principle that states that very similar competing species cannot coexist (370)

probability the likelihood that a specific event will occur (148)

producer organism that makes its own food from energy and carbon atoms in its environment; autotroph (97, 343)

product new substance formed as a result of a chemical reaction (80)

progesterone hormone secreted by the ovaries that regulates reproduction (983)

proglottid rectangular body section of tapeworms (650)

prokaryotic cell a cell without a nucleus (26)

promoter specific sequence of DNA that acts as a "start" signal for transcription (184)

prophase stage of mitosis in which chromosomes condense and become visible, the nuclear envelope breaks down, and spindle fibers become visible (131)

prosimian a group of primates that includes lorises, tarsiers, and lemurs (296)

prostaglandins nonendocrine chemical signals that tend to accumulate in areas where tissue is disturbed or injured (970)

prostate gland gland located just below the bladder in males that secretes an alkaline fluid necessary to counteract the acids produced by the female reproductive tract (995)

protease enzyme that attacks individual protein strands and cut them into smaller fragments (905)

protein organic compound formed of one or more chains of polypeptides (35, 900)

protist a member of the kingdom Protista (275)

proton pump mechanism that causes the movement of protons across a plasma membrane resulting in a build up of protons that can be used to do cellular work, such as the production of ATP by chemiosmosis (62)

protostome an animal whose mouth develops from or near the blastopore, an opening in the early embryo (628)

pseudocoelom a body cavity located between the endoderm and the mesoderm (619)

pseudopodium extension of cytoplasm of the amoeba that enables it to move (485)

P site in translation, a binding site on the ribosome held by a tRNA molecule that has just released its amino acid to the end of the polypeptide chain (187)

psoriasis inherited skin disorder in which new epidermal cells are produced about eight times faster than normal (869)

psychoactive drug substance that affects the nervous system (955)

puberty stage of growth in which reproductive maturity is reached (997)

pulmonary valve valve through which the blood enters the pulmonary arteries as the right ventricle contracts (885)

pulmonary vein vessel that carries oxygenated blood from the lungs to the left side of the heart (719, 884)

punctuated equilibrium model of evolution in which short periods of rapid change in species are separated by long periods of little or no change (258)

Punnett square diagram used by biologists to predict the probable outcome of a cross (149)

pupil contractile opening in the center of the iris through which light enters the eye (951)

purine class of organic, nitrogenous molecules in nucleic acids that have a double ring of carbon and nitrogen (169)

pyrimidine class of organic, nitrogenous molecules in nucleic acids that have a single ring of carbon and nitrogen (169)

pyruvate salt of pyruvic acid that is produced by the breakdown of glucose during glycosis (107)

R

radially symmetric arrangement of body parts around a central point (614)

radiant energy energy, such as light, that is transmitted in waves that can travel through a vacuum (100)

radioisotope unstable isotope of an element (233)

radiometric dating dating of objects through the measurement of relative proportions of certain radioisotopes and the products of their radioactive decay (233)

radula rasping tongue-like organ of mollusks used in obtaining food (623)

rain-shadow effect condition of decreased precipitation on the leeward side of a mountain (374)

ray-finned fish bony fish with a swim bladder that can be adjusted to hold differing amounts of air (749)

reactant substance that is the starting material in a chemical reaction (80)

reaction center area of a photosystem where an electron is boosted and leaves a chlorophyll *a* molecule (101)

realized niche the part of its fundamental niche that a species actually occupies (368)

receptor protein special protein on the cell's surface that matches particular molecules in the cell's surroundings, causing the cell to respond in a particular way (57)

recessive trait the trait that is not expressed in F1 generation after crossing (145)

recombinant DNA molecule made from pieces of DNA from different organisms (203)

rectum short, final extension of the large intestine through which feces pass to the anus (909)

red marrow soft tissue that fills spaces in spongy bone and that produces red blood cells in some bones (856)

reducing power ready supply of attachable hydrogen atoms that can be used for reduction (103)

reduction gain of electrons by an atom or molecule (79)

reduction division in meiosis, term used to indicate the reduction of the number of chromosomes (by half) when the cytoplasm divides (134)

reflex sudden, rapid, and involuntary response to a stimulus (948)

renal tubule structure in the kidneys that is connected to Bowman's capsule and that serves as a reabsorption device (913)

replication process of synthesizing a new strand of DNA (172)

replication fork point at which the double helix of DNA separates so that it can be copied (172)

repolarization restoration of the resting potential in a neuron (941)

repressor protein bound to the operator that switches off the lac operon (191)

reproductive isolation prevention of mating between formerly interbreeding groups, or the inability of these groups to produce fertile offspring (266)

resolution in microscopes, the ability to distinguish small, close objects (39)

respiration simultaneous uptake of oxygen and release of carbon dioxide by the lungs (888)

resting potential the difference in voltage across a neuron's membrane before the transmission of a nerve impulse (941)

restriction enzyme bacterial enzyme that cuts DNA at a specific sequence of nucleotides (205)

restriction fragment length polymorphism analysis (RFLP) laboratory technique used to identify base sequences in DNA (208)

reticular formation network of nerves in the brain stem and thalamus connected with other parts of the brain that is essential to consciousness, awareness, and sleep (946)

retina light sensing area at the back of the eye (952)

retrovirus virus that uses reverse transcriptase to transcribe DNA from an RNA template (461)

reverse transcriptase enzyme in a retrovirus that manufactures DNA complementary to the virus's RNA (459)

rheumatoid arthritis painful degeneration of movable joints after attack by cells of the immune system on the joint membranes (859)

Rh factor protein antigen on the surface of red blood cells (882)

rib cage protective structure of bone around the heart and lungs (858)

ribonucleic acid. *See* RNA

ribosomal RNA type of RNA molecule that plays a structural role in ribosomes (185)

ribosome cytoplasmic organelle on which proteins are synthesized (26)

right atrium collection chamber for blood returning from the body to the heart (885)

RNA (ribonucleic acid) a type of nucleic acid that participates in the expression of genes (38, 183)

RNA polymerase enzyme that carries out transcription (184)

Glossary

rod receptor cell of the eye that can detect various shades of gray (952)

root cap protective layer covering the tip of a root (558)

root hair extension of an epidermal cell near the root tip that aids in the absorption of water (558)

root underground portion of a plant (528)

rough ER endoplasmic reticulum with many ribosomes on its surface (42)

r-strategist species characterized by rapid growth, high fertility, short lifespan, and exponential population growth (319)

S

sapwood the lighter wood in a tree trunk, containing xylem cells that conduct water (557)

sarcomere in a muscle, one of the repeating units of myosin and actin filaments bounded by two Z lines (863)

saturated fat fat that contains no C=C bonds (35)

scanning electron microscope (SEM) a microscope that scans the surface of an object with a beam of electrons, enabling the viewer to see three-dimensional images (40)

Schwann cell the supporting cell that wraps around the axon of some neurons (940)

scientific name unique two-word name for a species in taxonomy; the first word is the genus, the second word is the species (432)

scrotum sac outside the body cavity that contains the testicles (994)

secondary compound defensive chemical compound produced by a plant (362)

secondary growth growth that causes a plant to increase its circumference (574)

secondary succession an episode of succession that occurs in areas where there has been previous growth, such as abandoned fields or forest clearings (340)

secondary tissue plant tissue that develops as a result of secondary growth (574)

second law of thermodynamics law that states disorder increases continually in a closed system (77)

second messenger chemical produced in response to the binding of a chemical signal to the outside of a cell; alters the chemical activity within the cell (973)

seed a plant embryo surrounded by a protective coat (534)

seed coat the covering of a seed (534)

seedling a young plant soon after germination (572)

selectively permeable condition in which plasma membrane allows passage of some solutes but not others (61)

self-pollination process by which a plant pollinates itself (143)

semen the mixture of secretions and sperm produced by male reproductive organs (728, 995)

seminal vesicles in males, glands that lie between the bladder and the rectum and that produce fluids rich in sugars to nourish sperm cells (995)

semininferous tubules tightly coiled tubes within the testicle in which sperm form (993)

sensory nerve nerve in the peripheral nervous system that gathers information about the environment and the body's condition (944)

sensory organ organ, such as the eye and ear, composed of sensory receptors (950)

sensory receptor specialized sensory cells that are sensitive to environmental and internal stimuli (950)

sepal a modified leaf that is part of the calyx of a flower (536)

septum in fungi, a wall-like division between cells within a hypha (502)

septum thick wall that divides the atrium or ventricle vertically into right and left halves (720)

setae external bristles of annelids (669)

set point in homeostasis, the normal value of a variable (837)

sex chromosomes chromosomes that differ between males and females (124)

sex-linked trait a trait that is determined by a gene found on the X chromosome (155)

sexual life cycle the production of haploid gametes by meiosis, followed by the union of two gametes in sexual reproduction (424)

sexual reproduction reproduction in which gametes from opposite sexes or mating types unite to form a zygote (132)

sexually transmitted disease (STD) disease that is spread by sexual contact (1007)

shoot aboveground portion of a plant (528)

short-day plant plant that flowers when days become shorter than a critical length (585)

sickle cell anemia condition caused by a mutant allele that produces a defective form of the protein hemoglobin (154)

sieve plate cluster of pores at the end of a sieve tube member (554)

sieve tube in phloem, a continuous strand of sieve tube members stacked end to end (554)

sieve tube member type of conducting cell in the phloem of plants (554)

sink part of a plant to which sugar is delivered (564)

sinoatrial node cluster of cardiac muscle cells that act as the pacemaker for the heart (885)

sinus venosus large collection chamber that sends blood to the atrium of the heart (718)

siphon hollow tube of bivalves used for sucking in and expelling sea water (662)

skin gills small, fingerlike projections that grow between the spines of echinoderms (695)

sliding filament theory explanation of muscle contraction in which myosin and actin filaments overlap and shorten each sarcomere (863)

slightly movable joint a joint that permits limited movement of the bones (859)

smooth ER endoplasmic reticulum with few or no ribosomes (42)

sodium-potassium pump membrane channel through which sodium ions are exchanged with potassium ions, creating an abundance of sodium ions outside the cell wall (62)

solute component of a solution in the lesser amount (58)

solution homogeneous mixture of two or more substances (58)

solvent component of a solution in the greater amount (58)

somatic nervous system motor and sensory nerves that control the skeletal muscles and that are under conscious control (944)

source part of a plant that provides sugar for other parts of the plant (564)

Southern blot technique that uses radioactive labeled RNA or single-stranded DNA as a probe to identify a specific gene (207)

speciation process by which new species are formed (264)

species group of organisms that look alike and are capable of producing fertile offspring in nature (251)

species diversity the number of species in an ecosystem (338)

spicule needle of silica or calcium carbonate in the skeleton of some sponges (640)

spinal cord cable of nerve tissue extending from the base of the brain, through the backbone, to the level just below the ribs (946)

spindle fibers network of hollow protein cables that form between separated centrioles and move chromosomes apart (128)

spine the backbone; a flexible stack of 26 vertebrae (858)

spiracle respiratory opening of certain arthropods that allows for the passage of air into the body (680)

spirillum spiral-shaped bacterial cell (462)

spongin flexible, structural protein fibers in the mesenchyme of some sponges (640)

spongy bone tissue in the human skeleton composed of hardened fibers interspersed with many spaces (855)

spongy layer of a leaf, the lower portion of the mesophyll, consisting of loosely packed, spherical ground tissue cells (556)

spontaneous origin origin (of life) through natural chemical and physical processes (226)

sporangia of fungi, reproductive structure in which spores form sexually (504)

sporangia of protists, reproductive cells that produce haploid zoospores by meiosis (483)

sporangium spore capsule in which haploid spores are produced by meiosis (526)

spore an asexual, resting, reproductive, haploid cell (132)

sporic meiosis a form of meiosis in which certain diploid cells of the sporophyte undergo meiosis to form haploid spores (425)

sporophyte the diploid phase in the life cycle of a plant that produces spores (425)

sporozoite one of three stages of *Plasmodium* that lives in mosquitoes and is injected into humans (494)

sprain torn or overly stretched muscle or connective tissue in a joint (865)

stabilizing selection type of natural selection in which the average form of the trait is favored and becomes more common (329)

stamen male reproductive structure of the flowering plant (536)

starch a storage form of glucose consisting of hundreds of glucose molecules (34)

stem cells specialized cells in bone marrow that constantly produce new red blood cells (880)

sternum the breastbone (858)

steroid a class of lipids that includes cholesterol and some hormones (35)

steroid hormones fat-soluble hormones derived from cholesterol and secreted by the adrenal cortex, testis, ovary, and placenta (969)

stigma swollen, sticky tip of a pistil on which pollen lands and adheres (537)

stimulant substance that excites the central nervous system and speeds up body processes (958)

stomata specialized pores in plant cuticles that enable gas exchange to occur (521)

stroma fluid matrix of the chloroplast (99)

style stalk rising from the ovary of a pistil (537)

subcutaneous tissue layer of fat-rich cells just below the dermis (868)

substitution a type of point mutation in which one nucleotide in a gene is replaced with a different nucleotide (195)

substrate the molecule on which an enzyme acts (82)

succession the regular progression of species replacement that occurs after disturbance or the creation of new habitat (340)

superior vena cava the large vein that returns blood from the upper portion of the body to the heart (885)

supporting cells cells that insulate a neuron and provide it with nutrients (832)

swim bladder the gas- or fat-filled sac of bony fishes that provides buoyancy (706, 749, 754)

swimmeret appendage on the underside of decapods used in swimming and reproduction (686)

symbiosis ecological interaction in which two or more species live together in a close, long-term association (281, 363)

symmetry arrangement of body parts around a point or central axis (612)

sympathetic division division of the autonomic nervous system that dominates in times of stress (949)

synapse tiny gap between two neurons (942)

synapsid term used to identify a vertebrate skull with a single opening on each side (773)

syphilis sexually transmitted disease caused by the bacterium *Treponema pallidum* (1008)

system in science, a collection of related objects that can be studied, such as an organism (76)

systemic circulation loop series of vessels that carries blood to the tissues of the body and back to the heart (883)

systolic pressure highest pressure that occurs during the pumping of the heart (886)

T

tadpole swimming larva of a toad or frog (728)

taproot system root system with a large central root and smaller lateral roots (559)

target cell final destination of hormones, where they produce their effect (969)

taste buds cluster of receptor cells on the tongue that specialize in detecting chemicals (954)

taxonomy science of naming and classifying organisms (431)

Glossary

tegument thick protective covering of endoparasite cells that helps prevent digestion by their host (649)

teleost group of bony fishes with highly mobile fins, thin scales, a swim bladder, and symmetrical tails (749)

telophase stage of mitosis in which a new nuclear envelope forms and spindle fibers disappear (131)

telson tail spine of many decapods (686)

tendinitis inflamed and painful tendons (865)

tendons dense connective tissue attaching muscles to bones (860)

tension-cohesion theory theory that water moves up a plant because of its cohesive properties and because of tension created by the loss of water from the leaves through transpiration (563)

terminal bud bud that grows at the tip of a stem (557)

terminator sequence of bases that tells RNA polymerase to stop transcription (185)

terrestrial term that describes an organism with the ability to live on land (704)

testes main organs of the male reproductive system (993)

testosterone hormone secreted by the testes that regulates the maturation of sperm cells in reproduction and regulates the development of secondary male sex characteristics (983)

thalamus part of the brain that directs incoming sensory impulses to the proper region of the cerebral cortex (945)

theory explanation based on a hypothesis that has been tested many times (11)

theory of endosymbiosis theory that mitochondria and chloroplasts are the descendants of symbiotic eubacteria (419)

thermodynamics study of energy transformations (76)

thigmatropism growth response to touch (584)

thoracic cavity the portion of the coelom containing the lungs and heart (889)

thylakoid internal membrane-bound sac of a chloroplast (99)

thymine nitrogenous base of the pyrimidine class; component of DNA (169)

thyroxine a hormone produced by the thyroid gland that stimulates cell metabolism and promotes growth (976)

tissue group of cells with a common structure and function (426)

tissue culture technique for growing pieces of living tissue in artificial media (576)

torsion twisting of the visceral mass of gastropods during development (664)

trachea tube that carries air from the larynx to the lungs (888)

tracheae in certain arthropods, fine tubes that extend into the interior of the body; used for gas exchange (680)

tracheid a narrow, thick-walled cell of the xylem (553)

transcription stage of gene expression in which the information in DNA is transferred to mRNA (183)

transfer RNA interpreter molecule that translates mRNA sequences into amino acid sequences (185)

transformation the transfer of genetic material from one organism to another; first observed by Griffith (166)

translation stage of gene expression in which the information in mRNA is used to make a protein (183)

translocation form of mutation caused by a chromosome fragment joining a nonhomologous chromosome during cell division (124)

translocation movement of sugar within a plant from a source to a sink (564))

transmission electron microscope (TEM) a microscope that produces a stream of electrons that passes through a specimen and strikes a fluorescent screen (40)

transpiration loss of water from a plant through its stomata (562)

transposon gene that has the ability to move from one chromosomal location to another (176)

tricuspid valve valve between the right atrium and the right ventricle (885)

trimester a three-month period of human pregnancy (1003)

trisomy condition in which a diploid cell has an extra chromosome (122)

trophic level a group of organisms that have the same source of energy; a step in a food chain (343)

tropism growth response in which the direction of growth is determined by the direction from which the stimulus comes (583)

true-breeding displaying only one form of a particular trait in offspring (144)

tuberculosis disease of the respiratory tract caused by the bacterium *Mycobacterium tuberculosis* (469)

tuber a modified underground stem (598)

tumor a mass of cells resulting from the proliferation of a cancerous cell (195)

Type I diabetes hereditary autoimmune disease in which the immune system attacks the islets of Langerhans in the pancreas, causing abnormally low insulin secretion (210, 982)

Type II diabetes disease in which the number of insulin receptors is abnormally low while the level of insulin in the system is often higher than normal (982)

ulcer hole in the lining of the stomach or small intestine caused by excessive acid (907)

unsaturated fat fat that contains C=C bonds (35)

uracil nitrogen-containing base of RNA, complementary to adenine (183)

urea principal nitrogenous waste of mammals and a chief component of urine (911)

ureter tube through which urine produced in the kidneys passes to the bladder (914)

urethra tube through which urine leaves the bladder and exits the body (914)

urinary bladder hollow, muscular sac that stores urine (914)

urine water and metabolic wastes left after filtering process of the kidneys ; expelled from the body (723)

uropod flattened, paddle-like appendage at the end of the abdomen of decapods (686)

uterus hollow, muscular, pear-shaped organ of the female reproductive system in which the embryo and then the fetus develops (998)

V

vaccine substance prepared from killed or weakened pathogens and introduced into a body to produce immunity(165)

vagina muscular tube that leads from the uterus to the outside of the female body (998)

vascular bundle cluster of vascular tissue in a leaf or a herbaceous stem that contains both xylem and phloem (557)

vascular cambium lateral meristem of woody plants that produces secondary vascular tissue (576)

vascular system in plants, tissues that transport water and other materials (522)

vascular tissue group of specialized cells that distribute water and carbohydrates (526)

vascular tissue system of plants, system that conducts water, mineral nutrients, and organic molecules made by photosynthesis (551)

vas deferens tube that connects the epididymis to the urethra and that stores sperm until they leave the body (995)

vector agent used to carry a DNA fragment into a cell (204)

vegetative part any nonreproductive part of a plant (591)

vein large vessel that carries blood toward the heart (877)

veliger second, free-swimming stage of reproduction of marine bivalves (663)

ventral bottom surface of a bilaterally symmetrical animal (617)

ventricle thick-walled heart chamber that pumps blood from the heart (718)

venule small vessel that carries blood from the capillaries to the veins (877)

vertebrate animal with a backbone (283, 632)

vesicle membrane-enclosed sac in a cell's interior (42)

vessel element in plant vascular tissue, xylem cell that conducts water (553)

vessel tube of xylem tissue formed from a series of vessel elements stacked end to end (553)

vestigial structure structure reduced in size and function; considered to be evidence of an organism's evolutionary past (256)

villi fine, fingerlike projections that cover the lining of the small intestine, increasing its absorptive surface (908)

virulent referring to the deadliness of a disease-causing agent (165)

virus a strand of nucleic acid encased in a protein coat that can infect cells and replicate within them (456)

visceral mass central section of a mollusk that contains the body's organs (623)

viviparous term used to describe organisms whose young are born alive from egg cells that develop within the mother's body (729)

voltage gated characteristic of gated channels that are activated by electric signals from nerve cells (68)

W

water vascular system hydraulic system of echinoderms that aids in movement (630)

wheat germ embryo of a wheat grain (593)

whole wheat the endosperm, germ, and bran layer of a grain of wheat (593)

wood pulp wood that is ground, moistened, and used in the manufacture of other products such as paper and rayon (600)

woody stems stiff, usually nongreen stems that contain layers of wood, as opposed to herbaceous stems (557)

X

xylem in plants, hard-walled cells that transport water and dissolved minerals up from the roots (527)

Y

yeast common name given to unicellular ascomycetes (507)

yellow marrow soft tissue filling the cavity of a long bone where fat is stored (856)

Z

Z line location of a protein that crosses a myofibril and anchors one end of each actin filament (863)

zoomastigote unicellular, heterotropic protist that is a member of the phylum Zoomastigina (488)

zoospore haploid cell that results from mitosis of single-cell protists and that develops flagella and cilia and ultimately breaks out of the parent cell (481)

zygosporangium thick-walled sexual structure that characterizes members of the phylum Zygomycota (505)

zygospore diploid zygote that results from the pairing of gametes of opposite mating types (482)

zygote fertilized egg cell (120)

zygotic meiosis life cycle in which the zygote is the only diploid cell and undergoes meiosis immediately after it is formed (424)

Index

Index

Index

Index

Index

Index

Index

Credits